ENCYCLOPEDIA OF FOOD SCIENCE AND TECHNOLOGY

VOLUME 2

Y. H. Hui
Editor-in-Chief

A Wiley-Interscience Publication
John Wiley & Sons, Inc.
New York / Chichester / Brisbane / Toronto / Singapore

In recognition of the importance of preserving what has been
written, it is a policy of John Wiley & Sons, Inc., to have books
of enduring value published in the United States printed on
acid-free paper, and we exert our best efforts to that end.

Copyright © 1992 by John Wiley & Sons, Inc.

All rights reserved. Published simultaneously in Canada.

Reproduction or translation of any part of this work
beyond that permitted by Section 107 or 108 of the
1976 United States Copyright Act without the permission
of the copyright owner is unlawful. Requests for
permission or further information should be addressed to
the Permissions Department, John Wiley & Sons, Inc.

Library of Congress Cataloging in Publication Data:
Encyclopedia of food science and technology / [edited by] Y. H. Hui.
    p.   cm.
  "A Wiley-Interscience publication."
  Includes bibliographical references.

  1. Food industry and trad—Encyclopedias. I. Hui, Y. H. (Yiu H.)
TP368.2.E62 1991
664'.003—dc20   91-22434
                                                    CIP

ISBN 0-471-50541-2 (set)

Printed in the United States of America

10 9 8 7 6 5 4 3 2 1

E

EDIBLE FILMS AND COATINGS

Edible coatings and films fill various needs in foods. The applications range from the use of collagen casings in sausage making, to the use of wax to prevent desiccation of fruits and vegetables, to the use of sugar glazes in bakery products and confections, to the use of internal barriers to prevent moisture migration within foods. The choice of a particular protective coating depends on the specific needs of a given food and storage conditions. Some choices are obvious because of historical use and experience. Other storage or processing issues may require innovative new coating concepts. The literature, especially the patent literature, is filled with suggestions for new edible films and coatings which could be the basis for new processing and preservation technologies. The following summary of current and potentially useful edible barriers is organized according to composition of the coating or film rather than along lines of applications because the work is intended to serve as a reference for film type rather than an applications guide.

PROTEIN FILMS

Collagen Casings

One of the earliest uses of edible films is the use of casings made from animal intestines to restructure comminuted meat. In the modern era, casings made from animal gut have been largely supplanted by casings made from swollen collagen fibers. The function of the sausage casings is to hold the meat batter in a specific form while the meat protein is being heatset. The casings have no real barrier properties; they allow moisture to pass when needed and allow smoke and flavorings to enter the sausage. The fabricated collagen casings offer the advantages of uniformity, processibility, and sanitation over the natural animal intestine casings. The edible reconstituted collagen casings are widely used in modern pork sausages, Polish sausages, and other ethnic sausages. They are popular with meat processors because of their machinability and uniformity, giving consistent weight control. However, edible collagen casings are somewhat tough and tend to split when grilled.

The current extrusion process for fabricating collagen casings was developed in the late 1950s (1). Bovine hide corium, which is more than 90% collagen on a dry basis, is used as the source of collagen. The insoluble collagen in the hide is decalcified and swollen by acidification and then dispersed into a very viscous suspension of collagen fibers through high shear. At this point, a small amount of cellulose and carboxymethylcellulose may be added to strengthen the mechanical properties of the final casing. Limed hides are preferred in this process because they tend to swell more easily in acid due to a reduction in covalent interfiber cross-links and increased free carboxylic acid groups caused by liming. The acidified collagen dispersion, containing about 4% solids, is then formed into the casing shape by extrusion into a coagulation bath of brine or ammonium. A perpendicular shear is applied to the fibers as they exit the extruder to disrupt the laminar alignment of the fibers and produce a more woven fiber structure. The high salt concentration or neutralization by acid causes the collagen fibers to shrink and form a mechanically stable casing. The casings are then washed and a plasticizer such as glycerol or sorbitol is added. The plasticized casing can then be collapsed accordion style so that it fits over the stuffing horn of the sausage extruder.

A collagen casing extrusion process was developed in Germany prior to World War II that employs a higher solids (12%) dispersion of collagen (1). The high viscosity of this dispersion requires a much higher pressure system for extrusion and has had limited acceptance outside Europe.

A new and innovative approach to creating sausage casings is to coextrude the sausage batter and the casing simultaneously (2). The collagen dispersion is prepared in situ and pumped to a special extrusion nozzle. The meat emulsion is pumped through an inner opening in the nozzle and the collagen dispersion is applied to the emerging meat batter through counter rotating concentric cones that serve to orient the collagen fibers in a woven alignment. The continuous encased sausage rope is passed through a brine bath to dehydrate and set the collagen casing. The sausage rope is crimped into links and the sausage is dried and smoked to further set the collagen casing. The coextruded sausage casing is more tender than preformed casings and therefore has better eating qualities, yet it remains more durable toward grilling, with less tendency to split. However, the high initial expense of the coextrusion equipment has limited its acceptance.

Gelatin Films

Gelatin is a partially hydrolyzed form of collagen that is soluble in warm water but retains the properties of thermoreversible gelation and film forming. The words gelatin and films are closely linked because gelatin is the structural component of photographic film emulsions. Gelatin is widely used in the pharmaceutical industry as an edible barrier both in the familiar hard capsule and in soft encapsulated forms. Drugs are also microencapsulated in gelatin matrices and coacervates. Unfortunately, the encapsulating and film forming properties of gelatin that have made it indispensable to the photographic and pharmaceutical industries have not been effectively translated to food industry usage. For photographic films and hard capsules, the gelatin is insolubilized by chemical cross-linking, which is not acceptable in food usage. Food grade gelatin is too easily hydrated and solubilized to form an effective barrier in foods. Because the amine groups on the basic amino acid groups in gelatin react with the aldehydes in flavor and aroma systems, gelatin is unsuitable for encapsulating flavors. Gelatin has outstanding properties as a structural builder in foods and can be used as such as a support in fabricated films or coatings, but other food ingredients must supply the barrier properties in these films.

Zein and Gluten

The structurally similar corn zein and wheat gliadin are soluble in alcohol and insoluble in water, making them good candidates for moisture barriers and coatings of hydrated food systems. Zein is the basis of a commercial product called Cozeen. According to the manufacturer, Cozeen is a suspension of zein and ethyl alcohol in hydrogenated soy and cottonseed oils plasticized with glycerol (3). This suspension is coated onto nuts or candies and dried. As the ethanol evaporates, the corn protein and oil interact to form a hydrophobic barrier and glazelike coating. Coating peanuts in this manner extends their shelf life from 4 weeks to 6 months. The addition of citric acid, BHA, and BHT into the coating system contributes greatly to the shelf-life extension.

Although gluten and zein are insoluble in water, because they are proteins, they hydrate readily and are actually very poor barriers against moisture migration except in very dry systems.

Mussel Protein

The unique structure of the protein component of the attachment beards of marine mussels is the basis of a patent application for a novel medical use of a protein film (4). The solubilized protein is treated with the enzyme catechol oxidase, which oxidizes the many phenolic groups, allowing the protein to polymerize and form a film. Although proposed as a medical treatment for sealing incisions, the concept is chemically similar to the polymerization of edible shellac films and could stimulate the future development of a new class of edible barriers.

CARBOHYDRATE FILMS

Alginate

The use of alginate gels dominates the patent literature on edible films. The alginates are linear copolymers of D-mannuronic acid and L-guluronic acid extracted from seaweed. Di- or trivalent cations are able to form ionic cross-links between carboxylic acid groups on adjacent polymeric chains of alginates and form stable gels. The general procedure for forming an alginate film coating on a food is to first coat the food with a solution of sodium alginate and then treat the coating with a second solution containing calcium to gel and insolubilize the alginate polymeric network. Calcium alginate coatings clearly reduce moisture losses from treated foods, most likely because they contain more water than the coated foodstuff and act as a reservoir of moisture. Early suggestions of oxygen barrier properties of calcium alginate gels have not been confirmed or utilized. The presence of calcium causes these alginate coatings to impart an unpleasant bitter taste to the coated foods.

The majority of the patented uses of alginate films have been in coating meat, poultry, and fish. The treatment of various fish fillets with dextrose alginate solutions followed by dipping in calcium chloride solutions has been described (5). The alginate treated fillets retained moisture and flavor during up to 6 days of refrigerated storage while untreated fillets became dry and dark, and developed an unpleasant odor. Beef and chicken pieces showed similar retention of moisture, tenderness and flavor during storage. The coatings were thought to reduce oxygen availability and reduce bacterial growth, although no supporting measurements were made. A later patent (6) describes a similar sodium alginate–calcium chloride coating for freshly slaughtered beef carcasses that produces multiple benefits including reduced bacterial growth, reduced moisture loss, enhanced carcass chilling, and reduced bone darkening. The inventors proposed that the bacterial growth reduction was a result of the surface bacteria being trapped within the initial high moisture (90%) alginate film. As the alginate lost moisture, down to less than 30%, the bacteria would be contained in this dry state and rendered nonviable. The reduction of bacterial growth was confirmed in inoculation studies.

Research studies at the University of Florida (7,8) confirmed the moisture retaining properties of calcium alginate films but could not confirm the inhibitory effect on bacterial growth. These studies suggested that the alginate coating does not act as a moisture barrier but as a moisture sacrificing agent. steaks that were coated with a calcium alginate film had reduced moisture loss, less off-odor, and better muscle color after 96 h of refrigerated storage than uncoated control steaks. However, the coated steaks lost moisture rapidly after 96 h.

Patent activity covering alginate coatings continues. A Japanese patent (9) describes the treatment of shellfish with alginate gelled by acidification using vinegar. Clam tissue soaked in alginate followed by vinegar and then frozen showed less weight loss and better flavor after 2 weeks frozen storage compared to control clam meat treated with guar gum.

Carrageenan Films

Carrageenans are sulfated polysaccharides extracted from red seaweeds. The highly sulfated ι-carrageenan forms gels in the presence of divalent cations and can form films similar to the calcium alginate films. The less sulfated κ-carrageenan forms thermoreversible gels. Calcium carrageenan coatings for frozen, precooked meat, poultry, and fish have been patented (10). The precooked meat is dipped into a 1% calcium carrageenan solution, picking up about 10% of its weight of carrageenan solution, and then frozen. After 6 months frozen storage, the calcium carrageenan samples scored consistently better in a ten-member panel's hedonic rating than the untreated controls. The carrageenan coating also maintained the quality of the meat, poultry, and fish through a freeze–thaw cycle. No objective measurements of weight loss or flavor and texture changes were reported. As with the alginate films, the carrageenan films probably prevent moisture loss by serving as a reservoir of moisture that is sacrificed.

An imaginative use of carrageenan as an edible protective coating has been demonstrated (11). A pH gradient on a food surface was created using deionized carrageenan as an immobilized polyelectrolyte. The coating contained 1%

carrageenan along with poly(ethylene glycol), water, sorbic acid, and agarose. When this coating was applied to an intermediate moisture cheese analog product, a pH gradient of up to 0.5 pH units was formed by a Donnan equilibrium of counterions. This pH gradient greatly enhances the antimicrobial activity of sorbate without acidifying the food itself. This concept should be the forerunner of future developments in active edible coatings.

Cellulose Derivatives

The modified cellulose ethers, including methyl cellulose, hydroxypropyl cellulose, hydroxypropyl methyl cellulose, and ethyl cellulose, are the basis of many edible film coatings. The methyl celluloses have some hydrophobic character that makes them suitable for components of nonpolar lipid films having moisture barrier properties. The cellulose ethers are good film formers and as such they are often incorporated as the matrix of composite films. They are rarely the active ingredient in the barrier properties of coatings but play a passive, supportive role. Their application will be covered in the section on composite films.

Pectin Films

Pectins are composed primarily of the methyl esters of linear chains of 1,4-α-D-galacturonic acid units. During processing the methyl esters may be hydrolyzed, producing low methoxy pectin. The low methoxy pectins have a divalent cation gelling mechanism similar to that of the alginates and ι-carrageenan, and the use of low methoxy pectins as edible coatings is similar to that of the alginates. The food to be coated is first dipped into a solution of low methoxy pectin and then into a solution of calcium chloride. The thickness of the coating can be controlled by the concentration of the pectin or the viscosity of the pectin solution. These gelled low methoxy pectins have been used as coatings for almonds, candied fruit, and dried fruit (12). The pectin films create a smooth surface for these normally sticky or oily products.

High methoxy pectin forms a gel in the presence of sugars, the basis of many fruit jellies. It has been shown that a pectin gel film could be formed at the boundary of high and low moisture portions of a food if the high moisture phase contained 1 to 3% high methoxy pectin and from 20 to 40% dextrin (13). When the two components of the food are brought together, the high moisture side of the interface loses moisture to the low moisture component and the partial dehydration of the pectin–dextrin mixture causes a gel to form. The interfacial pectin gel retards further moisture migration. These findings should lead to further creative solutions to migration problems through the use of barriers that are created when migration begins.

Starch Films

Starch is composed of two types of glucose polymers: linear amylose and branched amylopectin. Because linear polymers have better film-forming properties, the commercial development of high amylose corn starch in the 1950s led to the investigation of these starches as edible coatings and barriers. Linear polymers like amylose, cellulose, and mannan tend to crystallize, and in their pure form are insoluble in water. The use of amylose in film forming has required high temperatures and pressures or chemical modification of the amylose to form the more soluble hydroxypropyl amylose. In general, candies, fruits, and nuts can be coated with solutions of amylose containing plasticizers such as glycerol or emulsifiers. Of special interest are the reported oxygen barrier properties of amylose films. No measurable oxygen permeability was found in both plasticized (16% glycerol) and unplasticized high amylose films at relative humidities below 100%. At 100% relative humidity, the films showed moderate oxygen permeability (14). Numerous coatings were formulated from hydroxypropylated high amylose starch. Coated almond nut meats did not develop oxidative rancidity. An extruded hydroxypropyl high amylose starch film plasticized with glycerol showed no detectable oxygen transmission in a standard test (15).

Low dextrose equivalent (DE) starch dextrins have well known encapsulating and barrier properties and are used extensively to stabilize spray-dried flavors and aromas. The use of low DE dextrins as coatings for fruits and nuts has been described (16). Apple slices dipped in 40% 15 DE dextrin and dried did not brown, and retained flavor and structure. Apricot halves dipped in 30% 15-DE dextrin and sun dried had superior flavor to controls.

Chitosan Films

Chitin, the exoskeletal glucosamine polymer of insects and shellfish, has always been of interest to the food scientist because of its abundance and uniqueness as a cationic polysaccharide. But chitin has resisted development because of its insolubility in food-related solvents. A modified chitosan has recently been developed that is soluble in water and can be made into a semipermeable film (17,18). N,O-Carboxymethyl chitosan (CM-Chitosan) is produced by reacting chitin with chloroacetic acid. A 0.7 to 2% aqueous solution of the CM-chitosan can be used as a dip or spray to coat fruits such as apples, pears, peaches, or plums. The dried CM-chitosan reduces the influx of oxygen and outgassing of carbon dioxide to retard ripening of picked fruits in storage.

LIPID FILMS

Acetylated Monoglyceride Films

Since their development as a food ingredient in the early 1950s (19), acetylated monoglycerides have been the ingredient of choice in the formation of edible barriers to moisture migration. Acetylated monoglycerides solidify into stable α-form crystals which are waxy and relatively impermeable to moisture migration. The use of acetylated monoglycerides in a water-in-oil emulsion to coat bakery products has been patented (20). It was observed (19) that an unordered network of interlocking ribbonlike crystals formed as the basis of the unusually good moisture barrier properties of the acetylated monoglycerides. Fresh donuts coated with the acetylated monoglyceride water-in-oil emulsion lost only 6% of their weight in 72 h compared to a 19% weight loss by uncoated donuts. Other patents describe the direct application of acetylated monoglycerides to foods to prevent moisture loss: coating of dehydrated apricot pieces by dipping them into molten acetylated monoglyceride so that they may be stored in cake mixes without further loss of moisture (21); and spraying or immersion of fresh cut beef or pork in acetylated monoglyceride and storage at refrigerated temperatures for 25 days without loss of color and with limited weight loss (22). Recent Russian literature substantiates these observations (23).

Wax Coatings

Waxes are widely applied to fresh fruits and vegetables and represent one of the largest uses of edible coatings. Waxes are composed of long chain alcohols and their fatty acid esters. Commercial formulations are proprietary but usually contain paraffin wax, carnauba wax, or beeswax. Waxes prevent moisture losses and improve surface luster. Apples, grapes, pears, prunes, tomatoes, peppers, and cucumbers have a natural wax coating but an added coating can reduce moisture loss by 50%. Waxes, however, do not reduce decay. Excess wax can interfere with oxygen and carbon dioxide gas exchange and cause uneven ripening or decay. The general rule is that when the wax coating reduces moisture losses by more than one-third, gas exchange begins to be affected. Waxes may be applied as emulsions of carnauba or paraffin wax dispersed in water either by spraying or by immersion of the fruit into the emulsion, followed by drying. Citrus fruits are treated with wax dissolved in a solvent such as acetone or ethyl acetate. These wax solutions are sprayed or fogged onto the fruit. Vegetables are often treated with paste waxes of varying melting points and viscosities which are dripped onto brushes which brush the wax onto the vegetables.

A patented wax coating for raisins relies on beeswax plasticized with the ubiquitous acetylated monoglyceride which stabilizes the wax coating against mechanical damaged (24). Another patented coating for fruits and vegetables is composed of wax dispersed in an aqueous hydrophilic polymer system (25). The description implies that the barrier and gas transmission properties of the coating can be controlled by the size of the voids left by evaporation of the aqueous phase. Apparently this exciting possibility has not been developed or implemented.

SHELLAC COATINGS

Shellac is the alcohol soluble exudate of the insect *Laccifer lacca* and is composed primarily of 9,10,16-trihydroxy palmitic acid esters and schelloic acid esters. Edible shellacs are used extensively to coat candies, nuts, and pharmaceuticals. Recent patents (26,27) have extended the utility of shellac coatings by improving their moisture barrier properties. In this process, dry shellac is combined with other ingredients and heated to 130–175°C. The ingredients participate in a further cross-linking of the shellac to improve its barrier properties. The molten reaction mixture is dissolved in a solvent and applied to the food and dried. Using hydroxypropyl cellulose as the reacting ingredient causes a fourfold reduction in water vapor permeability of the shellac film. Using stearic acid or lauric acid can reduce the water vapor permeability by up to 50 times. Chocolate chips coated with these modified shellac films are stable for up to 18 days suspended in yogurt; uncoated chips last only 6 hours. Another recent improvement in the utility of shellac coating for foods addressed the problem of shellac turbidity when applied to foods containing water (28). By adjusting the pH of the ethanol shellac solution to above 7.9 with NaOH, roasted chestnuts can be coated and dried to give a glossy, nonsticky surface. Non-NaOH shellac coatings became sticky and dull.

COMPOSITE FILMS

An obvious step has been to combine the best properties of several film elements into a single film or coating; for example, combining ethyl cellulose, acetylated monoglyceride, and calcium stearate into a extrudable film which has moisture barrier properties and thermal stability (29). An emulsion coating containing methyl cellulose, vegetable oil, beeswax, and glycerol monooleate has been developed (30). The microemulsion technology of high-gloss floor polish has been attempted in films having carnauba wax with palmitic and stearic acids dispersed in a caseinate glycerol or gelatin glycerol carrier (31). The film is set by the addition of an organic acid. These films have good water vapor properties, but are brittle and nontransparent, and have a waxy taste. Composite precast edible barrier films have been developed (32). These films have a hydrophilic polymer layer and a lipid layer (33). Methyl cellulose and hydroxypropyl methyl cellulose are dissolved in water, and ethanol and poly(ethylene glycol) are added. Then stearic and palmitic acids are added. This emulsion is cast as a film 0.25 mm thick and dried. Molten beeswax is spread over the dried film at a thickness of 0.035 mm. The hot wax melts the fatty acid crystals at the surface of the emulsion film and, as the fatty acids and the wax solidify, the fatty acids bind the wax coating to the emulsion film. This composite film was placed between the crust and sauce of a French bread pizza so that the cellulose ether face of the film was oriented toward the bread and the wax face of the film was oriented toward the sauce. This film maintained the moisture gradient between the sauce and bread through an accelerated storage test. When the pizza is heated for consumption, the film melts and is absorbed into the pizza components.

BIBLIOGRAPHY

1. L. L. Hood, "Collagen in Sausage Casings," in A. M. Pearson, T. R. Dutson, and A. J. Bailey, eds., *Advances in Meat Research*, Vol. 4, Van Nostrand Reinhold Co., New York, 1987.
2. R. C. Waldman, "Co-extrusion—High Tech Innovation," *The National Provisioner*, 13–16 (Jan. 12, 1985).
3. "Extending Shelf Life with Edible Films," *Prepared Foods* (Mar. 1987).
4. European Pat. Appl. EP 244,688 (Nov. 11, 1987), C. V. Benedict and P. T. Picciano (to Bio-Polymers Inc.).
5. U. S. Pat. 3,395,024 (July 30, 1968), R. D. Earle.
6. U. S. Pat. 3,991,218 (November 9, 1976), R. D. Earle and D. H. McKee (to Food Research, Inc.).
7. C. R. Lazarus, R. L. West, J. L. Oblinger, and A. Z. Palmer, "Evaluation of a Calcium Alginate Coating and a Protective

Plastic Wrapping for the Control of Lamb Carcass Shrinkage," *Journal of Food Science* **41**, 639–641 (1976).
8. S. K. Williams, J. L. Oblinger, and R. L. West, "Evaluation of a Calcium Alginate Film for Use on Beef Cuts," *Journal of Food Science* **43**, 292–296 (1978).
9. Japanese Pat. JP 62,278,940 (Dec. 3, 1987), S. Kio and co-workers (to Nippon Starch Chemical Co.).
10. U.S. Pat. 4,196,219 (April 1, 1980), C. Shaw and co-workers (to United States of America, United States Army).
11. J. A. Torres, J. O. Bouzas, and M. Karel, "Microbial Stabilization of Intermediate Moisture Food Surfaces II. Control of Surface pH," *Journal of Food Processing and Preservation* **9**, 93–106 (1985).
12. H. A. Swenson, J. C. Miers, T. H. Schultz, and H. S. Owens, "Pectinate and Pectin Coatings II. Application to Nuts and Fruit Products," *Food Technology,* 232–235 (June 1953).
13. U.S. Pat. 4,401,681 (Aug. 30, 1983), L. K. Dahle (to Campbell Soup Co.).
14. A. M. Mark, W. B. Roth, C. L. Mehltretter, and C. E. Rist, "Oxygen Permeability of Amylomaize Starch Films," *Food Technology* 75–77 (Jan. 1966).
15. L. Jokay, G. E. Nelson, and E. L. Powell, "Development of Edible Amylaceous Coatings for Foods," *Food Technology* **21**, 1064–1066 (1967).
16. D. G. Murray and L. R. Luft, "Low-D.E. Corn Starch Hydrolysates," *Food Technology,* 32–40 (Mar. 1973).
17. J. Raloff, "Preserving Fruit with a Chitin Coat," *Science News* **133**, 410 (June 1988).
18. European Pat. Appl. EP 265,561 (May 4, 1988), E. R. Hayes (to Nova Chem Ltd.).
19. R. O. Feuge, "Acetoglycerides—New Fat Products of Potential Value to the Food Industry," *Food Technology,* 314–318 (June 1955).
20. U.S. Pat. 4,293,572 (Oct. 6, 1981), R. Silva and co-workers (to International Telephone and Telegraph Corp.).
21. U.S. Pat. 3,516,836 (June 23, 1970), R. A. Shea (to The Pillsbury Co.).
22. U.S. Pat. 3,851,077 (Nov. 26, 1974), M. Stemmler and H. Stemmler.
23. M. A. Dibirasulaev, E. S. Soloveva, G. P. Vozmitel, and S. A. Dmitriev, "Edible Coatings as an Effective Means of Decreasing the Shrinkage and Preserving the Quality of Meat," *Kholodil'naya Tekhnika,* 28–31 (1988).
24. U.S. Pat. 2,909,435 (Oct. 20, 1959), G. G. Watters and J. E. Brekke (to The United States of America).
25. U.S. Pat. 3,997,674 (Dec. 14, 1976), N. Ukai and co-workers (to Tsukihoshi Kasei Kabushshiki Kaisha).
26. U.S. Pat. 4,661,359 (Apr. 28, 1987), J. Seaborne and D. C. Egberg (to General Mills, Inc.).
27. U.S. Pat. 4,710,228 (Dec. 1, 1987), J. Seaborne and D. C. Egberg (to General Mills, Inc.).
28. Jpn. Pat. JP 63 42,666 (Feb. 23, 1988), M. Asano and co-workers (to Daisho Co., Ltd.).
29. U.S. Pat. 3,471,304 (Oct. 7, 1960), M. M. Hamdy and H. S. White (to Archer-Daniels-Midland Co.).
30. U.S. Pat. 3,483,004 (Dec. 9, 1969), C. D. Bauer and co-workers (to W. R. Grace and Co.).
31. S. Guilbert, "Technology and Application of Edible Protective Films," in M. Mathlouthi, ed., *Food Packaging and Preservation,* Elsevier Applied Science Publishers, London, 1986, pp. 371–394.
32. J. J. Kester and O. R. Fennema, "Edible Films and Coatings: A Review," *Food Technology,* 47–59 (Dec. 1986).
33. International Pat. Appl. WO87/03453 (June 18, 1987), O. R. Fennema and co-workers (to Wisconsin Alumni Research Foundation).

TED R. LINDSTROM
KEISUKE MORIMOTO
CHARLES J. CANTE
General Foods USA
White Plains, New York

EDUCATION. See FOOD CHEMISTRY/BIOCHEMISTRY; FOOD ENGINEERING, AND ENTRIES UNDER FOOD SCIENCE AND TECHNOLOGY; INSTITUTE OF FOOD TECHNOLOGISTS (IFT).

EEL

Eel is a popular food fish in Europe and the Far East, especially in Japan. In some countries, however, eel is not so popular because of its snakelike appearance. In spite of this, the demand for eel has increased considerably for the last two decades.

Annual world production of eels is about 200,000 t. Twenty-five percent of these are wild eels captured mainly in Europe, North America, and Oceania, whereas the remaining 75% are cultured eels.

Japan pioneered the culture of eels more than 150 years ago. Its eel culture technique is now one of the most advanced in the world. Taiwan, which adopted Japanese culture methods, follows closely. In annual eel output, Taiwan, however, surpasses Japan, making it the world's largest supplier of eels (1). Meanwhile, several countries are also developing their own eel culture industry. Fluctuating natural eel stocks and the pollution of the eel's habitat have made world eel production unstable. The culture of eels may be the best way of ensuring the adequate supply of the fish.

BIOLOGY

Species and Characteristics

Freshwater eels are widely distributed throughout the world, mostly in the areas around the Atlantic and Indo-Pacific oceans (Fig. 1) (2). A total of 16 species and six subspecies are recorded, namely, *Anguilla anguilla* (European eel), *A. rostrata* (American eel), *A. japonica* (Japanese eel), *A. marmorata, A. bicolor* (subspecies: *A. bicolor pacifica* and *A. bicolor bicolor*), *A borneensis, A. celebesensis, A, ancestralis, A. obscura, A. australis* (subspecies: *A. australis australis* and *A. australis schmidti*), *A. megastoma, A. mossambica, A. nebulosa* (subspecies *A. nebulosa labiata* and *A. nebulosa nebulosa*), *A. reinhardti, A. interioris,* and *A. dieffenbachi.* However, *A. celebesensis* and *A. ancestralis* are classified as synonyms (3). There are several characteristics that distinguish the world's eel species, including dorsal fin length, vertebral number, color, size, head shape, habitat, and distribution (Table 1).

Dorsal Fin Length. Eels are generally divided into two types according to the proportion of the length from the anterior base of the dorsal fin to the anus and the total

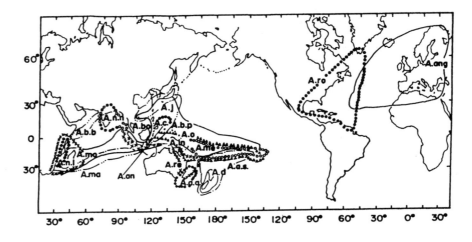

A. ang.	Anguilla anguilla	A. an.	Anguilla ancestralis	A. b.b.	Anguilla b. bicolor
A. ro.	Anguilla rostrata	A. me.	Anguilla megastoma	A. b.p.	Anguilla b. pacifica
A. j.	Anguilla japonica	A. in.	Anguilla interioris	A. o.	Anguilla obscura
A. d.	Anguilla dieffenbachi	A. bo.	Anguilla borneensis	A. a.a.	Anguilla a. australis
A. re.	Anguilla reinhardti	A. n.n.	Anguilla n. nebulosa	A. a.s.	Anguilla a. schmidti
A. ma.	Anguilla marmorata	A. n.l.	Anguilla n. labiata		
A. c.	Anguilla celebesensis	A. mo.	Anguilla mossambica		

Figure 1. Natural distribution of freshwater eels.

body length. The long-fin type has a proportion of 7–17%, while the short-fin type has a proportion of 0–5%. Thirteen of the eel species belong to the long-fin category, three species belong to the short-fin type.

Vertebral Number. Another taxonomic criterion for eel types is the number of vertebrae. Eels have 100–119 vertebrae; the number varies among the different species. The number even varies within each species, which increases with higher latitudes.

Color. Eels can be mottled or plain. There are seven mottled and nine plain species.

Size. There is a considerable difference between female and male eels. The growth of female eels in inland water is much more superior than that of male eels. The females also stay longer in inland water before returning to the sea. Eels can grow at maximum sizes of 27 kg and 200 cm.

Head Shape. Eels are also classified according to the shape of their head. Those with a narrow head and thin, narrow lips are classified as narrow-head type while those with broad head and thick, broad lips are classified as broad-head type (4,5).

Distribution

Of the 16 species of eels, 14 are distributed in the Indo-Pacific Ocean. Some of these eels inhabit the tropical zone, while others abound in the temperate zone. Eels in the Indo-Pacific Ocean are distributed in both the southern and northern hemispheres. The northernmost boundary is at about 45° N in Hokkaido, Japan, while the southernmost limit is at about 50° S in the Auckland Islands, New Zealand. The two other species, the European eel and the American eel inhabit the Atlantic Ocean and abound in the temperate zones of Europe and North America, respectively (Fig. 1).

LIFE HISTORY

The first successful attempt to understand the amazing life cycle and unusual traits of the eel has generally been accredited to two men who devoted most of their lives to studying eels: Danish biologist Johannes Schmidt and his student Vilhelm Ege. The 1904 discovery of a leptocephalus by Schmidt started a 35-year study that helped establish a solid foundation to the ecological study of eels. Nevertheless, many questions concerning the life history of eels still remain unanswered. Although the life history of three species, European eel, American eel and the Japanese eel are well investigated, little is known about the other species.

Eels are catadromous migrants, that is, they spawn at sea and spend most of their lives in inland waters. The spawning ground and migration route vary with each species.

European Eel

The European eel spawns at the Sargasso Sea, 22–30° N, 48–65° W, in the western Atlantic Ocean from early spring through early summer. The prelarvae grow to about 25 mm in June; attenuate; and gradually grow into transparent, leaf-shaped leptocephalus. The leptocephalus drift with the current away from the Sargasso Sea to

Table 1. Major Details of the World's *Anguilla* Species[a]

Dorsal Fin Length	Vertebral Number	Color[b]	Maximum Size of Females kg	cm	Species	Zone[c]		Distribution[d]
Long	110–119	P	6.0	125[e]	A. anguilla	Temp	ATL	Europe, North Africa
Long	103–111	P	6.0	125[e]	A. rostrata	Temp	ATL	United States, Canada, Greenland
Long	112–119	P	6.0	125[e]	A. japonica	Temp	PAC	Japan, Korea, Taiwan, People's Republic of China
Long	109–116	P	20.0	150	A. dieffenbachi	Temp	PAC	New Zealand
Long	104–110	M	18.0	170	A. reinhardti	Trop	PAC	Australia, New Caledonia
Long	100–110	M	27.0	200	A. marmorata[f]	Trop	PAC	Africa, Madagascar, Indonesia, People's Republic of China, Japan, Pacific Islands, Sumatra
Long	101–107	M			A. celebesensis[g]	Trop	PAC	Phillipines
Long	101–106	M			A. ancestralis[g]	Trop	PAC	Celebes
Long	108–116	M	22.0	190	A. megastoma	Trop	PAC	New Caledonia
Long	104–108	M			A. interioris	Trop	PAC	New Guinea, New Caledonia
Long	103–108	P	2.0	90	A. borneensis	Trop	PAC	Brunei
Long		M	10.0	150	A. nebulosa	—	—	—
Long	106–112				A. n. nebulosa	Trop	IND	Sri Lanka, Burma, Sumatra
Long	107–115				A. n. labiata	Trop	IND	South Africa
Long	100–106	P	5.0	125	A. mossambica	Trop	IND	Africa
Short		P	3.0	110	A. bicolor	—	—	—
	106–115	P			A. b. bicolor	Trop	IND	South Africa, Madagascar, Sri Lanka, Burma, Sumatra, Australia
Short	103–111				A. b. pacifica	Trop	PAC	Brunei
Short	101–107	P			A. obscura	Trop	PAC	New Caledonia
Short		P	2.5	95	A. australis			
Short	109–116				A. a. australis	Trop	PAC	Australia
Short	108–115				A. a. schmidti	Temp	PAC	New Caledonia, New Zealand

[a] Refs. 5 and 9.
[b] Color abbreviations are P = plain, M = mottled.
[c] Zone abbreviations are Trop = tropical, Temp = temperate.
[d] Distribution abbreviations are ATL = Atlantic Ocean, IND = Indian Ocean, PAC = Pacific Ocean.
[e] A. anguilla, A. rostrata, and A. japonica are closely related (9).
[f] A. marmorata is the most widely distributed species (9).
[g] Synonymous (3).

the coasts of Europe in about three years. The leptocephalus can attain a body length of 50 mm after a year and about 75 mm at two years old. They metamorphose into slim transparent elvers (Fig. 2) at 2.5 years old. The elvers then head for the coast and enter estuaries.

Both the female and male European eels spend their growing lives in rivers until they grow to 6–12 and 9–19 years old, respectively. When they become sexually mature, the eels migrate downstream and back to the Sargasso Sea to spawn. After spawning, the adults die.

American Eel

The American eel also spawns in the Sargasso Sea. Its elvers only take about 1.5 years to reach the coastal area of North America (6).

Japanese Eel

The spawning ground of the Japanese eel is estimated to be within the eastern part of Taiwan and the southern part of the Ryukyu islands, which may extend to the northeastern part of the Philippines. The spawning season is estimated to be between June and July. At the latter part of their development, the leptocephalus drift with the Kuroshio current from the spawning ground and metamorphose into elvers when they reach the estuaries of Taiwan, the People's Republic of China, Korea, and Japan after about 4–5 months. The migration time is estimated by the daily growth rings in the otolith. Using this

Figure 2. Elvers.

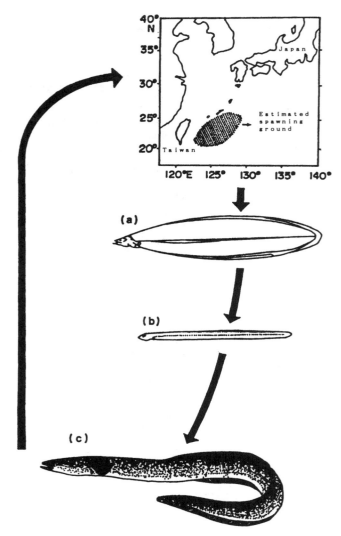

Figure 3. Estimated spawning ground and life cycle of the Japanese eel: (**a**) Leptocephalus, 5.5 cm T.L., drifted by current; (**b**) Elver (ascending) migrate from estuary to river; (**c**) adult (descending) migrate from river to spawning ground (6).

index, the peak upstream migration season for elvers is between December and January. The elvers enter the rivers and live and feed on small fish, shrimp, and aquatic insects for about 5–20 years until they reach adult size. They mature during autumn and winter (September–November). The pectoral fin, dorsal and ventral part of the mature males turn blackish and silvery. While mature males can weigh only about 500 g on maturity, the females can reach 2–3 kg. Sexually mature eels then migrate down the river into the ocean to spawn. Figure 3 illustrates the estimated spawning ground and life cycle of the Japanese eel (7). The Japanese eel is the most popular and most commonly eaten eel, particularly in Japan. Thus the following discussion focuses on the Japanese eel.

ECOLOGY OF ELVERS

The amount of catch for elvers in coastal waters seems to be correlated with temperature. The maximum catch of elvers in the upstream areas has often been during or several days after daily seawater temperature reaches its lowest in winter. The lowest water temperature ever recorded in an eel habitat is 15–16°C.

Maximum catch has also been observed to occur at the same time when salinity has leveled off and the flow of seawater has reached its maximum. The elvers also become most active during this period at night.

The biological rhythm of elver activity was found to follow the lunar cycle in the coastal waters. The peak catches occur only once a month, about the time of the new moon. On the other hand, a semilunar rhythm of the elvers in the rivers was observed, namely, two peak catches occurred in each lunar cycle, one around full moon, the other around new moon. The semilunar rhythm of upstream elvers also coincided with the spring tide (8).

CULTURE TECHNIQUES

Species Cultured

Of the different species of eels in the world, three are valued for its economic importance, namely, the European eel, the American eel, and the Japanese eel. From a marketing point of view, these three species are similar in shape, growth, and behavior.

Elver Collection

The collection of the elvers is the starting point of eel culture. Attempts to propagate eels artificially have not been successful to date. In 1974, the Japanese succeeded in inducing artificial spawning of the eel and hatching the eel eggs. But attempts to rear the elver during the first three weeks of its life have not been as successful. Thus eel fry are still sourced entirely from the wild-caught elvers.

A growing concern within the eel culture industry in recent years is the diminishing number of large seaward running eels, as well as the elvers swimming upstream. This has been attributed mainly to pollution in estuaries, rivers, and streams and the overfishing of elvers. To remedy this, captured elvers are now widely used not only for culturing eels for consumption, but also for restocking in the lower reaches of rivers and streams.

Figure 4. Offshore collection using boat and light to attract elvers.

Elvers are usually captured at nighttime. Light is used to attract them (Fig. 4) and they are caught in scoop nets in shallow water, in a fine-mesh net set across the width of the river, or in trap nets set in the estuaries (9).

Elvers are quite hardy; nevertheless, great care is taken when handling them to prevent mortality due to injury. They are not handled directly. They are placed in boxes lined with wet muslin or hung in mesh cages in the river. These boxes or cages are taken to eel farms within a few hours. The high demand coupled with a supply shortage has resulted in eel's phenomenal price increase.

CULTURE SYSTEMS

Eels can be cultured both extensively and intensively. Most eel farms, however, use the intensive culture system. This culture system is characterized by high stocking densities, stringent water-quality management, additional inputs such as formulated diets to increase fish production, and aggressive disease prevention.

Elver Rearing

There are various sizes and types of tank being used in stocking elvers. The most commonly used, however, is the circular concrete tanks, about 5 m in diameter and 60 cm in depth. These tanks are usually built under greenhouses and heated to above 25°C by thermostatically controlled electric immersion heaters (9). This is to protect the elvers from low temperatures.

The average stocking density used in intensive culture is 1–2 kg of elvers/3.3m². Sorting and restocking is usually done after 20–30 days, when a significant difference in the size of eels is observed. The average density used for the second stocking phase is 1 kg of fish/3.3 m². Frequent sorting and restocking will bring about higher survival rate. These will also help maintain uniform growth and provide better feed conversion rates.

The elvers have to be trained to take formulated feeds. The usual method is to lower baskets containing tubifex worms or minced meat of oyster and clam twice a day, in the early morning and in the late evening under a wooden enclosure lit by a 20–40 W lamp. Elvers do not feed at temperatures below 13°C, thus temperatures must be kept above 13°C. Feeding time is gradually shifted to daytime.

Initially, elvers may not feed. In the long run, however, they learn to feed and eventually, attain rapid growth. After 100–150 days, a size of 100–200 fingerlings/kg is reached.

Feeding Management

Eels are carnivorous fish and thus require high animal protein input. The more animal protein that can be provided, the greater the eels' weight will be. Most eel farmers feed their eels with formulated feeds that are chiefly made of fish meal with added carbohydrates, minerals, vitamins, and other ingredients. Due to the high cost of these feeds, most eel farms supplement these with raw fish such as mackerel, sardine, and anchovy. Trash fish or scraps that are minced using electric grinders are also used widely as feeds. Eels, however, convert formulated feeds more efficiently than they convert raw fish.

Figure 5. Eels feeding on formulated feed paste from a perforated tray lowered into the pond. The tray is lifted out once the eels stop feeding to avoid scraps fouling the water.

Feeding is done once a day, at about 7:00 A.M. during summer and about 10:00 A.M. in winter. The usual feeding rate is 5–15% of the total weight of the eel for minced trash fish and 1.0–3.5% for formulated feeds (10). Nevertheless, factors such as the previous day's feeding condition and water and air temperature as well as water quality are considered when determining the feeding rate. Ideally, feeds should be consumed within 20–30 min.

Precautions must be taken to insure that no unnecessary organic debris are allowed to fall into the ponds. Thus eel farmers do not just broadcast the feeds into the ponds. Instead, a special feeding platform, usually shaded, within the pond is designated for feeding, and the eels learn to swim toward the area each feeding time. Feeding areas are preferably located where the oxygen level is high as this encourages the eels to eat well. Formulated feed is lowered on a perforated tray suspended just under the surface and lifted after feeding is done (Fig. 5). Raw fish, dipped for a few minutes in boiling water to soften the skin, are threaded through the eyes and also lowered and raised once eels have eaten all the flesh. Thus the only organic matter that goes into the pond is that which the eels eat.

Pond and Water Management

Growout ponds are either rectangular or square. In Japan, adult eels are raised in several types of pond that make use of efficient drainage with water temperatures of about 28°C. Until recently, these eel ponds were between 5,000–20,000 m² in area. However, the practice of intensive culture has reduced pond size to about 500–1,000 m², and is expected to be further reduced in the future (9).

In Taiwan, two types of eel pond are used, namely, the hard pond and the soft pond (Fig. 6). The hard pond has a concrete or red brick bottom and dikes and is small (1,000–1,650 m² and water depth of 0.8–1.2 m). The soft pond, on the other hand, has a mud bottom and concrete or brick dikes and is generally large (0.6–1.0 ha and water depth of 2.0–3.0 m). Recently, some farmers have reverted back to using earthen ponds (Fig. 7).

Most eel farms have many ponds of different sizes because eels of different sizes must be sorted to maintain

Figure 6. Soft ponds.

uniform growth. The eels that grow fast are separated from those that grow slow and are transferred to larger ponds.

Pond water quality is also an important aspect of eel culture because it could affect the health and growth of the cultured eels. A well-maintained pond could spell the difference between healthy and fast-growing eels, and poor quality ones.

Pond water color can also be used to gauge water quality. The most favorable color is green because it can show the presence of zooplanktons, phytoplanktons, ideal pH value, high dissolved oxygen concentration, and other factors affecting water quality.

Eel ponds require large amounts of fresh water and are thus generally located in areas with good water supply both quantitatively and qualitatively. Water is kept free from pollution and within pH values of 6.5–8.0. Temperature is kept above 13°C because eels stop feeding below this temperature. Oxygen, which is considered the most critical water characteristic, must be maintained above 1 mL/L because eels cannot sustain life below this level.

To ensure and improve the water quality in eel ponds, various methods may be adopted. These would include allowing more water to flow into the pond to increase the rate of water exchange, installing aerators to increase oxygen supply, and liming to improve water quality. Bottom sediments are also removed at the end of the growing season.

Disease Problems, Prevention, and Treatment

The eel's body surface is covered by thick mucus. Thus the eel generally is highly resistant to diseases, unless its skin is injured or the skin or gill is infected by parasites (11).

Various fungal and bacterial infections can contaminate eels. The most common of these are fungus disease, red disease, branchionephritis, swollen intestine disease, gill disease, and anchor worm disease.

Diseases usually infect eels during those months when changes in seasons take place. These may be caused by a wide variety of factors such as poor pond water quality and inefficient feeding management.

Prevention rather than medication should be the program followed in disease management. Prevention of diseases would include efficient feeding management and proper water management. Moreover, eels are handled with care to avoid skin damage, which is the primary cause of infections. In case of disease outbreak, medicines are added to the food or to the pond water.

Harvesting and Marketing

When eels reach marketable size, they are harvested daily or once every few days, usually at feeding time. The ideal marketable size is 5–6 fish/kg for the Japanese market and more than 6 fish/kg for other markets.

There are several ways of capturing eels in the ponds. The common practice is by placing a net below the feeding platform. Other methods include draining the pond and catching the eels in long net bags or by drawing a seine across the pond.

Harvested eels are sorted into different sizes. Smaller ones are put back to the pond for further growth. The harvested eels are then starved for about two days by holding them in baskets in front of the water inlet of the pond or in midpond, or stacking perforated plastic baskets under showers of trickling water. They are then packed and sent to the market. Starvation removes undigested substances inside the intestine. This minimizes the risk of

Figure 7. Earthern ponds.

Figure 8. Eels readied for the market. Eels are packed in polyethylene bags inflated with oxygen. (Note oxygen tank and corrugated cardboard boxes in background.)

meat contamination with gastroenteric bacteria during cutting and deboning in processing plants.

Eels, when marketed live or quick frozen and glazed, fetch fair prices. When they are transported to the market live, eels are packed in containers with conditions that may vary, depending on the transport time and distance (Fig. 8).

CONSUMPTION AND PROCESSING

Different countries have different preferences and eating habits, and thus have different ways of preparing and eating eels. In Europe, the main eel consumers are the Germans, Dutch, Danes, and Swedes. These consumers commonly prefer smoked eel, which ranks as an expensive and luxurious food. The British also eat eels, but prefer the jellied type.

In the Far East, the main eel-eating country is Japan, where it is a custom to eat roasted eel (*kabayaki*) especially every thirty-first of July, which is the *Ushinohi* (a special eel-eating day for the Japanese). The Japanese believe that the eel is nutritious and can make up for overexhaustion during the summer. Eel is also consumed widely in Taiwan, the People's Republic of China, Korea, and Hong Kong.

Preprocesing

Before eels are processed, they are graded according to size (Fig. 9). One way of preparing them is by putting them in a deep container and sprinkling them with salt. This treatment removes their slime until they die of asphyxiation. Another way is by putting them in fresh water and stunning them with an electric shock (9). Or they may simply be put in cold storage overnight. In this way, eel activities can be slowed down, and thus they can easily be handled for gutting or deboning.

Cutting Methods

There are generally two methods of cutting, namely, *Kanto* style (cut abdominally) and *Kansai* style (cut dorsally). Newly cut eels are usually cleaned of slime by washing them in cold water and scraping. The eels are then gutted. This is done by slitting from the throat through either the belly (*Kanto* style) or up toward the back (*Kansai* style) to the tail, about 2.5 cm beyond the anus. The guts are then emptied and the backbone and head removed. The eel is washed thoroughly to remove traces of slime and blood. Machines specifically designed to eviscerate eels efficiently are also available.

Preparations

Fresh Eels. Fresh eels are cooked in several ways, such as braising and steaming. Steaming, however, is the more popular method. In both Taiwan and Mainland China, eel is either steamed with various vegetables and mushrooms or stewed with Chinese herbs, such as medlar, lovage, and dates. This herb-filled soup is considered as a revitalizing tonic soup for frail or disabled persons.

Smoked Eels. After the eels are cut, they are brined in a saltwater solution, then hot smoked. In this process, the eels are dried and smoked. To allow uniform drying throughout the thickness of the fish, the temperature during smoking is increased gradually. The smoking process may vary from country to country. The finished product is then packed either in boxes or in cans (9).

Jellied Eels. Jellied eels is a traditional British way of preparing eels. First, gutted, cleaned, and cut eels are placed into boiling water and then simmered until the flesh becomes tender. Cooking time depends primarily on size. The cooked pieces marinated with hot liquor are then poured into large bowls containing gelatine dissolved in a small amount of water. The amount of gelatine solution is determined by considering the condition of the eels and their natural capacity to gel. Experience is thus necessary to get the right recipe. Once the mixture has cooled, the pieces in jelly are packed into cartons for immediate fresh consumption. Shelf life can reach two weeks at chilled temperatures. There are various other recipes for preparing jellied eels (9).

Roasted Eels. Roasted eel is a favorite Japanese cuisine. The Japanese prepare roasted eels in several ways, such as, *kabayaki* (roasted eel with seasoning), *shirayaki* (roasted eel without seasoning), *kimoyaki* (roasted eel viscera with seasoning), and *capitalyaki* (roasted eel head with seasoning). Until recently, these preparations were only available in Japanese specialty restaurants. Domestic production, however, has not kept up with the large demand for roasted eels in Japan. Japan thus imports frozen roasted eel from Taiwan, where the roasted eel processing industry has boomed in recent decades, and the People's Republic of China, where the industry is in its initial stages.

Live eels are processed into frozen roasted eel in processing plants. The ideal size of the raw material is between five and six tails per kilogram. Within this size range, the smaller size commands a higher price.

In the cut fillet style, the eel is cut crosswise into three equal sections, with the tail cut lengthwise into two. If the weight is not enough, each half of the tail is stretched together with one of the other sections by using bamboo

Figure 9. Eels being graded according to size.

Figure 10. Frozen roasted eel without seasoning, cut fillet style.

sticks. This step is usually done using a stretching machine. Stretching is done to keep the meat flat during cooking. For the whole fillet style, the fish is not cut into sections after the degutting process; they are, however, also stretched.

After cutting, the fillets are arranged on a conveyor and pass through single-sided or double-sided roasting machines. The fillet may be roasted seasoned or unseasoned (Figs. 10 and 11). In single-sided machines, the inside portion is roasted first, then the eel is turned and the other side is roasted. Liquefied petroleum gas is used as fuel. The appropriate roasting time is about 3–5 min. Roasting indicators are the color of the meat (it should become evenly light scorched), the scorch bubbles (some should appear on the skin; under 3% of the total area), and the central temperature of the meat (it should reach 78°C). The most delicate part in the whole procedure would be in the precooling step because this is where the fillet becomes most susceptible to recontamination. It is thus necessary to control the bacterial drop rate, until the total plate count is under three colony-forming units/min.

Some manufacturers use precooling tunnels and spray cold air into the surface of the fillet; a shorter precooling time is maintained to minimize recontamination. The use of either individual quick freezing or contact freezing equipment is popular. In these types of equipment, the temperature drops to −18°C within 30 min. After the central temperature has reached −30°–35°C, the frozen fillet is then removed from the pan or conveyor, then packed and stored at a temperature of −20°C (12). Comparing the weight of the processed eel (eviscerated but not yet seasoned and roasted) to the total weight of the raw material, the yield for cut and whole fillet may reach as much as 58–60% and 68–70%, respectively.

Sanitary quality control standards for the products include negative amounts of coliforms; total bacterial count of under 3,000/g; and negative presence of residues of contaminated chemicals in the meat, such as malachite green, methylene blue, oxolinic acid, antibiotics, nitrofurans, sulfamides, insecticides, and herbicides.

SKIN PROCESSING

Eel is not only used for food but also for other uses. Eel skin, for example, is processed into leather (Figs. 12 and 13). Although there is no fur on the eel skin, there are scales that are covered by follicles dispersed in the corium (13). The major steps in eel skin processing are liming, decoloring and removal of scales, removal of excess fat, tanning using alum, and glazing. Considering that the fat content of the eel skin is more than 25%, which is even higher than that of cattle, the removal of excess fat and scales are the more important steps in eel skin processing.

FUTURE DEVELOPMENT

The eel is a highly rated food. Compared with other fish, the eel is rich in vitamins A and E, and calcium. The eel is recognized as one of the most nutritious cultured fishes. It is high in energy content, and calorific values, such that it is eaten to create appetite and stamina especially in hot and humid days. Its health benefits make it an ideal fish of suitable value.

Figure 11. Roasted eel dish with seasoning, cut whole fillet style.

Figure 12. Tanned eel skin stained with different kinds of dye.

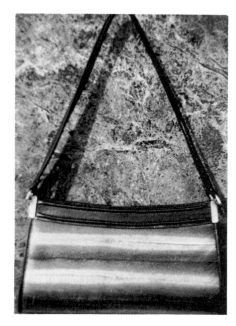

Figure 13. A shoulder bag made from tanned eel skin.

Eel skin, when processed, can be used in leathercrafts. More research, however, must be done in finding ways to efficiently tan its skin to enable it to compete with other leathers.

Interest in the culture of eels has reached a high level in recent years because of its versatility and popularity, and high demand as gourmet food particularly in Japan, where it commands a good price. In eel-eating countries such as Japan, Taiwan, and some parts of Europe (Germany, Denmark, and the UK) enterprising activities concerning eels are in progress. These activities include research into new techniques of eel culture in cold climate, ie, Northern Europe.

Elvers used in eel culture are still sourced from the wild, and are diminishing in numbers. Thus investigations on eel propagation in captivity should be made. Attempts albeit unsuccessful have been made to propagate elvers artificially. The challenge now is for researchers to look for efficient and commercially successful ways to propagate them artificially. Researchers may well benefit from the lessons of past investigations, and studying the natural spawning habits and looking for ways to unlock the mystery of the long life history of the eel. A big challenge, perhaps, but the fish's potentials are so encouraging that the expected hardships will surely be worthwhile.

BIBLIOGRAPHY

1. C. C. Huang, "Taiwan Becomes World Leader in Exports of Eel," *Almanac of Food Industries in Taiwan, R.O.C.*, 56–57 (1990).
2. "Fisheries in Japan, Eels," *Jap. Mar. Prod.* 1977.
3. D. H. J. Castle and G. R. Williamson, "On the Validity of the Freshwater Eel Species, *Anguilla ancestralis* Ege, from Celebes," *Copeia* **2**, 569–570 (1974).
4. L. Bertin, *Eels: A Biological Study*, Cleaver-Hume Press, London, 1956.
5. I. Matsui, *Eel Study: Biology*, Vol. 1, Kosei-Sha Kosei-Kaku, Tokyo, 1972 (in Japanese).
6. F. W. Tesch, *The Eel: Biology and Management of Anguillid Eels*, Chapman and Hall, London, 1977.
7. T. Kafuku and H. Ikenoue, eds., *Modern Methods of Aquaculture in Japan: Developments in Aquaculture and Fisheries*, Vol. 11, Elsevier, Amsterdam, The Netherlands, 1983, pp. 31–43.
8. W. N. Tzeng, "Immigration Timing and Activity Rhythms of the Eel, *Anguilla japonica*, Elvers in the Estuary of Northern Taiwan, with Emphasis on Environmental Influences," *Bull. Jap. Soc. Fish. Oceanogr.* **47–48,** 11–28 (1985).
9. A. Usui, *Eel Culture*, Fishing News (Books) Ltd., Surrey, UK, 1974.
10. T. P. Chen, *Aquaculture Practices in Taiwan*, Fishing News (Books) Ltd., Surrey, UK, 1976, pp. 17–28.
11. S. Egusa, *Epidemic Diseases of Fish*, Kosei-Sha Kosei-Kaku, Tokyo, Japan, 1978 (in Japanese).
12. H. C. Chen, "Frozen Roasted Eel Processing Industry in Taiwan," in J. L. Chuang, B. S. Pun and G. C. Chen, eds., *Fishery Product of Taiwan, JCRR Fisheries Series* **25B,** Joint Commission on Rural Reconstruction, Taiwan, R.O.C., 1977, pp. 20–26.
13. Y. S. Lai and Y. Y. Kuo, "Primary Report on the Eel Skin Tanning," *Bulletin of the Taiwan Fisheries Research Institute* **32,** 439–441 (1980) (in Chinese with English abstract).

I. Chiu Liao
Taiwan Fisheries Research Institute
Keelung, Taiwan

EGGS AND EGG PRODUCTS

DEVELOPMENT OF THE EGG INDUSTRY

Eggs have been a human food source since the beginning of human residence on earth. The first usage of eggs was probably the taking of eggs from nests of wild birds. In primitive societies this continues to be a significant source of high-quality protein. As civilization developed, the birds were domesticated and were kept in pens and shelters. In Western societies the chicken became the primary source of eggs. The growth of the egg industry in selected countries from 1965 to 1985 is shown in Table 1.

Until relatively recent times the keeping of chickens on farms was generally only a sideline to the main business of farm operation. Commercial poultry farms became a primary source of farm income only in the last century. Large poultry farm operations for egg production did not become common until the last half of the twentieth century. The change in size of egg production units in the United States is shown in the data presented in Table 2. In 1984 there were 63 companies with over one million laying hens each. These companies owned 131 million hens or 52.2% of the total layer population of the United States (3).

EGG FORMATION AND STRUCTURE

The egg consists of four distinct parts: shell, shell membranes, albumen or white, and yolk. A schematic drawing

Table 1. Egg Production in Selected Countries[a]

Country	1965	1975	1985
United States	65,692	64,379	68,250
Canada	5,194	5,339	5,855
Mexico	5,000	7,446	18,092
Argentina	2,880	3,480	3,150
Brazil	8,124	6,000	9,000
Venezuela	508	1,721	2,736
France	9,220	13,120	14,910
Federal Republic of Germany	11,930	15,003	13,150
Italy	9,990	11,400	10,900
The Netherlands	4,206	5,320	10,051
Spain	6,320	10,152	10,164
United Kingdom	7,840	13,861	13,117
Poland	6,264	8,013	8,631
USSR	29,000	57,700	77,000
Japan	18,625	29,798	35,700
Australia	2,196	3,384	3,825

[a] Ref. 1 (1967, 1977, 1987). All values in millions of eggs.

Table 2. Changes in Size of Laying Flocks in the United States from 1959 to 1974[a]

	Percentage of all Eggs Sold		
Size of Flock	1959	1964	1974
Less than 400	26.78	10.87	3.64
400 to 1,599	22.58	12.64	1.80
1,600 to 3,199	12.83	7.39	1.14
Over 3,200	37.81	69.10	93.43
Over 20,000	—	30.10	67.74
3,200 to 9,999	—	—	9.59
9,999 to 19,999	—	—	16.10
20,000 to 49,999	—	—	24.02
50,000 to 99,999	—	—	13.14
Over 100,000	—	—	30.58

[a] Ref. 2.

of an egg is shown in Figure 1. The yolk is formed in the ovary. During embryonic development of the female chick tiny ovules appear on the ovary which may later develop into yolks of eggs to be produced by that hen. The number of immature ovules present in the chick at the time of hatching is far in excess of the number of eggs that will be produced by the hen during her lifetime.

Approximately 14 days prior to the laying of an egg, one of the rudimentary ovules in the ovary starts to grow in size. During the next 13 days the yolk develops in its follicular membrane. A suture line forms in the membrane so that the matured yolk can be dropped into the infundibulum of the oviduct without the rupture of any blood vessels in the follicular membrane. A blood spot in an egg is generally caused by incomplete formation of the suture line.

The oviduct consists of five identifiable sections: infundibulum, magnum, isthmus, uterus, and vagina. The infundibulum or funnel collects the yolk or ovum and by peristaltic action moves it on to the magnum region, where the thick albumen is secreted and laid down in layers on the yolk. In the isthmus the shell membranes are formed around the yolk and thick albumen. The membranes, two in number, only loosely fit the enclosed material. Most of the time that the egg is in the oviduct is spent in the uterine section where thin albumen fills the shell membrane sack and the shell is laid down to form the rigid egg protective layer. The shell color is determined by the breed of the hen. Generally, hens with red earlobes produce eggs with brown shells and white eggs producers have white earlobes. The total time spent in the oviduct is usually about 24 hours. Total time for the formation of the egg from the start of rapid growth of the ovule to time of laying varies from 12 to 15 days.

COMPOSITION OF THE EGG

The egg is composed of approximately 10% shell, 30% yolk, and 60% white or albumen. The egg is a very good source of high-quality protein and many minerals and vitamins. The chemical composition of eggs, including the shells, is summarized in Table 3.

The protein of egg white is complete; it contains all of the essential amino acids in well-balanced proportions. The thick white is made up mainly of the proteins ovalbumin, conalbumin, ovomucin, ovoglobulins, ovomucoid, and lysozyme. The structure of the thick white is the result of a complex of ovomucin and lysozyme.

The important yolk proteins are ovovitellin and ovolivetin. The lipid materials in the egg are all in the yolk.

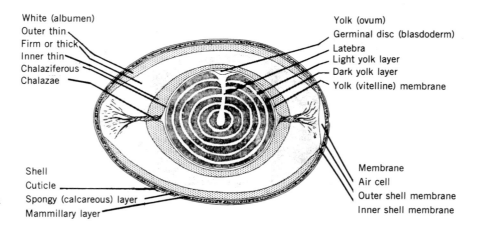

Figure 1. A schematic of the parts of an egg (4).

Table 3. Chemical Composition of the Egg (percentage)[a]

	Water	Protein	Fat	Ash
Whole egg	65.5	11.8	11.6	11.7
White	88.0	11.0	0.2	0.8
Yolk	48.0	17.5	32.5	2.0

	Calcium Carbonate	Magnesium Carbonate	Calcium Phosphate	Organic Matter
Shell	94.0	1.0	1.0	4.0

[a] Ref. 4.

The fatty acid composition can easily be modified by changing the fatty acid makeup in the feed. In most commercial eggs where the hens are fed on a corn and soybean meal diet the fatty acids are about one-third saturated and two-thirds mono- and polyunsaturated. Yolk color is controlled by the level of pigments, mostly xanthophyll, in the feed of the hens.

PRODUCTION PRACTICES

As shown in Table 2, the majority of eggs at this time are produced on large egg production farms. With this change, the level of mechanization used in the production of eggs has increased. Over 86% of all laying hens in the United States are kept in laying cages (5). It is also estimated that about 70% of all hens in the world are kept in cages. The percentages may be decreasing in some countries early in the 1990s as a result of the efforts of animal welfare groups. Legislation has been introduced in many countries and passed in a few that would eliminate the use of cages for the housing of laying hens.

Some elements of the medical profession have emphasized the role of dietary cholesterol in human cardiovascular problems. The negative effect on egg consumption that this emphasis on cholesterol has had has led to a number of research studies of attempts to reduce the cholesterol content of eggs. Genetically it has been possible to achieve only slight reductions. Selection of hens for small yolk size has met with some success. A review on altering cholesterol by feeding (b) concluded that dietary modifications resulted in only minor changes in the cholesterol content of egg yolk. Another report (7) states that including 1.5 or 3.0% of menhaden oil in the laying hen's ration results in a temporary reduction of about 50% in the cholesterol concentration. The fatty acid composition of the eggs was also modified in that the eicosopentenoic acid and docosahexanoic acid (omega-3 fatty acid) content of the yolks was significantly increased when the fish oil was included in the hen's ration.

Other nutrients in the egg, with the exception of zinc and choline, may also be modified in concentration (8).

SHELL EGG PROCESSING

In a modern egg production unit the hens are housed in cages with sloping floors so that eggs will roll from the cage onto an egg gathering belt. The eggs are transported by conveyor belts to a separate unit of the facility where they are washed, inspected for defects by passing them over an intense light source, sized, and packed into cartons for retail or onto trays for bulk packaging. Throughout the operation individual eggs are not touched by any workers. The cartons or trays are packed into master containers by hand.

Some countries ban the washing of eggs to be sold at retail. In others, including the United States and Canada, almost all eggs are washed in a detergent solution at a temperature of about 45°C (113°F) and then sanitized with an approved sanitizing agent, frequently sodium hypochloride.

After the eggs are washed and dried they are conveyed to the candling area of the processing line. Here there is a strong light source under the eggs so that any defects such as cracked shells, dirty or stained shells, or inclusions in the egg might be detected and the defective eggs removed from the conveyor. Candling of eggs is used to segregate eggs into the several grades that are found in the marketplace. The retail grades are AA, A, and B. Quality factors used in grading include the shell (cleanliness and soundness), air cell (size), white (clarity and firmness), yolk (visibility of the outline and freedom from defects), and freedom from visible defects. All grades of eggs are required to have a sound not cracked, shell. Other quality factors are judgment calls, with smallness of air the cell, freedom from stains on the shell, firmness of the white, and dimness of the yolk outlining being considered most desirable. For eggs that do not meet the grade standards two additional groupings are provided: dirty and checked or cracked. Details on the grading standards have been published (9).

There are six size classifications of eggs in the United States. Table 4 gives minimum weight requirements for each of the sizes. On the processing line the sizes are segregated by passing the eggs over a series of scales set so that the jumbo eggs are removed from the line first and the pee wees are removed last.

After the eggs have been mechanically placed in car-

Table 4. Weight Classes of Shell Eggs in the United States

Size Name	Minimum Ounces per Dozen	Minimum Average Ounces per Egg	Minimum Average Grams per Egg
Jumbo	30	2.5	70.0
Extra large	27	2.25	63.8
Large	24	2.0	56.7
Medium	21	1.75	49.6
Small	18	1.5	42.5
Peewee	15	1.25	35.4

tons or on trays many processors spray the tops of the eggs with a food grade mineral oil to preserve the interior quality of the egg during movement through market channels. The oil partially seals the pores of the shell to reduce loss of gases from the egg and to reduce the rate of pH change of the albumen. As a further aid in maintaining quality, eggs are handled at refrigerated temperatures throughout the market channels and consumers are advised to keep eggs in the refrigerator after they purchase them. With good handling throughout market channels and in the home, eggs will remain high quality for several weeks and in a usable condition for several months. The factors that have the greatest effect on quality loss in eggs are time, temperature, humidity, and handling.

CONVERSION OF SHELL EGGS TO LIQUID PRODUCT

An ever increasing number of eggs are being broken at commercial egg breaking plants. The early growth of the egg products industry has been reviewed (10). In the first half of the twentieth century most of the egg breaking plants were located in the North Central and Western Plains states. Since 1960 many of the new breaking plants have been built in the Southeastern states. The yield of liquid products from a case (30 dozen) of large eggs is shown in Table 5. Data in this table also include the yields of solids that might be obtained if the egg products were dehydrated by either spray drying, pan drying, or freeze drying.

The growth of the egg products industry is summarized in Table 6. The shift from dried to liquid since 1970 was likely the result of mandatory pasteurization of all egg products since 1971 (12). The shift will likely accelerate during the next few years, with liquid gaining a percentage of the frozen egg business. This shift will be the result of ultrapasteurization and aseptic packaging of liquid eggs that will give more than 4 weeks usable shelf life for the liquid product (13). The savings in costs of freezing and in time saved by not having to thaw the products before using in food formulations are evident.

The pasteurization of liquid eggs was practiced for a number of years prior the passage of the law making it mandatory for all egg products in the United States. The times and temperatures required for pasteurization in several countries are listed in Table 7. All of these values are for liquid whole egg. Details of pasteurization have been discussed (15). The *Salmonella entriditis* food poisoning incidents in England, Europe, and the United States during 1987 and 1988 will likely further the usage of liquid, pasteurized egg products in place of shell eggs, especially in hospitals and nursing homes.

In the processing of egg white it is necessary to remove the reducing sugars before the liquid can be dried. In early work on dehydration of egg white the process included a natural fermentation by whatever organisms were present in the egg white. A patent was issued in 1931 covering the use of lactic acid bacteria for egg white fermentation. The next development was the use of a yeast, *Saccharomyces apiculatus*, and then the use of an enzyme, glucose oxidase. Each of these methods is still in use in some parts of the world.

The natural egg products of white and yolk are also supplied to the food processing industries in a number of blends and modifications with the addition of sodium chloride or sugar. The chemical composition of some of the egg products is shown in Table 8. The difference in composition between pure yolk and commercial yolk should be emphasized. The difference is due to the inclusion of a significant amount of albumen with the yolk in the commercial product.

NUTRITIONAL AND FUNCTIONAL PROPERTIES OF EGGS

Complete nutritional data on eggs and egg products prepared in several ways have been summarized (16). An abbreviated listing of nutrients in whole egg, yolk, and white is given in Table 9. The value listed for cholesterol is questionable. During the latter half of 1988 and early 1989 the value has been reevaluated using a capillary column gas chromatographic method. The earlier value, 274 g, was obtained using a colorimetric method. A report indicated that a capillary column gas chromatographic method resulted in values about 27% less than the colorimetric method (18). Preliminary data from the 1988–1989 USDA reevaluation indicate that the new value for cholesterol will be about 210 mg per large egg (19).

The various uses for eggs in food products have been reviewed (20). The functional properties include whipping, emulsifying, coagulation, flavor, color, and nutrition. The foam produced during whipping should give large volume and be stable. The time required to get maximum foam volume will vary for eggs of different qualities. Eggs with a high percentage of thick albumen re-

Table 6. Processed Egg Product Production in the United States (in millions of pounds)[a]

Year	Liquid	Frozen	Dried
1985	579	324	97
1984	528	322	76
1983	494	305	87
1982	460	339	80
1980	422	331	81
1977	394	348	75
1974	301	367	73
1970	110	357	281
1965	44	368	216
1960	44	362	176

[a] Ref. 1.

Table 5. Yields of Liquid Egg and Egg Solids per 30-Dozen Case of Large Size Eggs[a]

	Liquid		Solids	
	Kilograms	Pounds	Kilograms	Pounds
White	13.8	30.4	1.67	3.68
Yolk	6.1	13.5	3.12	6.87
Whole	19.9	43.9	4.79	10.55

[a] Ref. 11.

Table 7. Minimum Pasteurization Requirements for Whole Egg Products

Country	Temperature °C	°F	Time, min.	Reference
United Kingdom	64.4	148	2.5	14
Poland	66.1–67.8	151–154	3	15
China (PRC)	63.3	146	2.5	15
Australia	62.5	144.5	2.5	15
Denmark	65–69	149–156.5	1.5–3	15
United States	60	140	3.5	12

quired a longer time for whipping but the foam formed will be more stable. Cakes made using very low quality eggs generally do not have the volume of similar cakes made with higher quality eggs. When egg white is dried there is generally a loss in whipping properties. To overcome the decrease in foam volume, whipping aids are frequently added to the dried egg white.

PRODUCTION OF EGG-RICH CONVENIENCE FOODS

Reference 21 outlines procedures for the manufacture of a number of value-added egg products, some of which will be described here.

Hard Cooked Eggs

The preparation of hard cooked eggs has been a function of kitchens for many years. In about 1970 the production of hard cooked eggs for sale to salad bars, restaurants, and commercial food caterers was begun. The original products were cooked in hot water, partially cooled, peeled, and stored in a solution of citric acid and sodium benzoate. The concentration of the citric acid was initially 2% but has been lowered over the last 20 years to from 0.5 to 0.8%. The lower levels of acid produced eggs with softer cooked whites, much closer to freshly cooked eggs.

Cooking of hard cooked eggs is now being done either in steam or hot water. Peeling of eggs is primarily a hand operation aided by equipment to crack the shells. With improvements in packaging technology, some hard cooked eggs are now marketed as dry packed product.

Deviled Eggs

One use of hard cooked eggs is the production of deviled eggs. A hard cooked egg is cut in half through the midline either longitudinally or horizontally, and the yolk is removed. The yolk is mixed with a salad dressing and other ingredients to produce a filling for the depression left in the white. A patent was issued (22) for mechanization of the entire deviled egg production system.

Scotch Eggs

This product was developed years ago but received very little publicity until relatively recently. A hard cooked egg is wrapped in a thin layer of sausage meat. The sau-

Table 8. Approximate Composition of Selected Egg Products (per 100 g)[a]

| Product | Calories | Composition, g | | | | |
		Water	Protein	Lipid	CHO	Ash
Frozen Eggs						
Whole egg	158	74.57	12.14	11.15	1.20	0.94
White	49	88.07	10.14	Trace	1.23	0.56
Yolk, pure[b]	377	48.2	16.1	34.1	—	1.69
Yolk, commercial[c]	323	55.04	14.52	28.65	0.36	1.43
Yolk, sugared	323	50.82	12.92	25.50	9.49	1.27
Dehydrated Eggs						
Whole egg	594	4.14	45.83	41.81	4.77	3.45
Whole egg, stabilized	615	1.87	48.17	43.95	2.38	3.63
White, stabilized, flakes	351	14.62	76.92	0.04	4.17	4.25
White, stabilized, powder	376	8.54	82.40	0.04	4.47	4.55
Yolk	687	4.65	30.52	61.28	0.39	3.16

[a] Ref. 16.
[b] Ref. 17.
[c] Contains approximately 17% white.

Table 9. Distribution of Nutrients in Large Chicken Eggs (per egg, edible portion)[a,b]

	Whole Egg	Yolk[c]	White
Weight per egg, g	50	17	33
Water, g	37.28	8.29	29.06
Calories	79	63	16
Protein, g	6.07	2.79	3.35
Lipid, g	5.58	5.60	Trace
Carbohydrate, g	0.60	0.04	0.41
Fiber, g	0	0	0
Ash, g	0.47	0.29	0.18
Minerals			
Calcium, mg	28	26	4
Iron, mg	1.04	0.95	0.01
Magnesium, mg	6	3	3
Phosphorus, mg	90	86	4
Potassium, mg	65	15	45
Sodium, mg	69	8	50
Zinc, mg	0.72	0.58	0.01
Vitamins			
Ascorbic acid, mg	0	0	0
Thiamin, mg	0.044	0.043	0.002
Riboflavin, mg	0.150	0.074	0.094
Niacin, mg	0.031	0.012	0.029
Pantothenic acid, mg	0.864	0.753	0.080
Vitamin B_6, mg	0.060	0.053	0.001
Folacin, μg	32	26	5
Vitamin B_{12}, μg	0.773	0.647	0.021
Vitamin A, R.E.	78	94	0
Lipids			
Saturated fatty acids, g	1.67	1.68	0
14:0	0.02	0.02	0
16:0	1.23	1.24	0
18:0	0.43	0.43	0
Monounsaturated FA, g	2.23	2.24	0
16:1	0.19	0.19	0
18:1	2.04	2.05	0
Polyunsaturated FA, g	0.72	0.73	0
18:2	0.62	0.62	0
18:3	0.02	0.02	0
20:4	0.05	0.05	0
Cholesterol, mg	213	213	0
Amino Acids			
Tryptophan, g	0.097	0.041	0.051
Threonine, g	0.298	0.151	0.149
Isoleucine, g	0.380	0.160	0.204
Leucine, g	0.533	0.237	0.291
Methionine, g	0.196	0.171	0.130
Cystine, g	0.145	0.050	0.083
Phenylalanine, g	0.343	0.121	0.210
Tyrosine, g	0.253	0.120	0.134
Valine, g	0.437	0.170	0.251
Arginine, g	0.388	0.193	0.195
Histidine, g	0.147	0.067	0.076
Alanine, g	0.354	0.140	0.215
Aspartic acid, g	0.602	0.233	0.296
Glutamic acid, g	0.773	0.341	0.467
Glycine, g	0.202	0.084	0.125
Proline, g	0.241	0.116	0.126
Serine, g	0.461	0.231	0.247

[a] Ref. 16.
[b] Shell is 12% of weight of egg.
[c] Fresh yolk includes a small proportion of white.

sage is then cooked in a deep fat fryer. In producing this item the sausage used should be very low in fat to get better adhesion of the sausage to the surface of the hard cooked egg. Scotch eggs are a snack food item frequently sold from refrigerated vending machines and are also used for hor-d'ouerves by cutting the finished product into quarters or halves.

Diced Egg Products

Diced eggs were first marketed as a means of utilizing hard cooked eggs that were not perfectly smooth. With the popularity of hard cooked diced eggs on salad bars the demand was greater than the quantity available from the hard cooked eggs. At this point innovative producers separated yolks and whites as liquids and steam cooked trays of the two liquids. The cooked yolk and white were then diced. The product thus prepared gave yolk cubes that held together because of the albumen included in liquid separation. The diced egg products have been marketed as frozen material in most instances. However, where market channels are short, nonfrozen, controlled atmosphere packaging is also being used.

Scrambled Eggs

Scrambled eggs are prepared as fully cooked, freeze-dried products for the camper; as fully cooked, frozen items for the microwavable meal; and as frozen prepared liquid mixes, packaged in cook-in-bag film. The basic formula for scrambled eggs consists of about 70% whole liquid egg and 30% milk with seasonings of salt and pepper. The most significant form of scrambled eggs is the frozen prepared mix packaged in film. This product is used extensively in hospital and nursing home feeding because the cooked product remains hot for an extended period in the film bag.

Omelets

The major difference between omelets and scrambled eggs is that in a true omelet no milk is included in the formula. Water is used instead of the milk. Omelets are produced for distribution as frozen product, either fully cooked or as a premix. The formulation of omelet mixes frequently includes the diced ham, bacon bits, mushrooms, onions, green or red peppers, and a number of other vegetables.

Crepes, Pancakes, and Waffles

Each of these products has a formula including flour and eggs. The crepe mixture is usually richer in eggs than the other two. Each of these products is fully cooked and frozen for distribution. There are also dry mixes for preparation of the products in the home.

Snack Foods from Eggs

A number of egg-rich snack foods have been proposed but, thus far, none have become significant users of eggs. A breaded, fried egg white ring that looks much like an onion ring was patented (23) and a procedure for producing an egg jerky flavored to taste like dried meat jerky was suggested (24). A cookie formulation was prepared so that each cookie contained one egg equivalent. Another snack food that is quite common is yogurt. It is possible to make a yogurt-type product using egg albumen as a partial replacement for the milk. In a sensory comparison of normal yogurt with the egg-substituted product the latter was judged superior for mouthfeel and smoothness.

A number of egg-rich drinks have been produced. The drink sold in greatest quantity is eggnog. This is produced as dried product as well as the refrigerated liquid drink. A number of fruit juices have been mixed with eggs to produce drinks of orange juice and egg, apple juice and egg, and cranberry juice and egg. When preparing any of the drinks in quantity it is recommended that pasteurized liquid eggs be used. The prepared drinks should be kept at temperatures below 5°C (41°F).

MODIFYING THE COMPOSITION OF EGGS BY PROCESSING

The composition of egg products can be modified rather easily by utilizing yolk, white, or whole egg in various amounts. Much interest has been shown in reducing the amount of cholesterol in eggs. The cholesterol can be removed from the yolk by using supercritical extraction techniques with carbon dioxide as the extracting solvent. Much of the research has been with dried egg yolk but it is possible to remove approximately 90% of the cholesterol from the liquid yolk by supercritical extraction techniques.

EGGS AS A SOURCE OF PHARMACEUTICAL PRODUCTS

Lysozyme makes up 3.5% of the egg white. A cation-exchange system consisting of a macroporous, weak acid resin was used to recover most of the lysozyme (25). The system can be designed as a continuous operation. It was found that the residual lysozyme-free egg white possessed superior whipping, gelling, and emulsifying properties compared to native egg albumen. The lysozyme-free albumen has been approved for food product usage by federal agencies in the United States, providing the labeling is appropriate.

Uses for the extracted lysozyme may be in pharmaceutical products or as bactericidal agents in the preservation of food products.

BIBLIOGRAPHY

1. USDA, *Agricultural Statistics*, U.S. Government Printing Office, Washington, D.C., 1961–1987.
2. W. J. Stadelman and O. J. Cotterill, *Egg Science and Technology*, 3rd ed., AVI Publishing Co., Inc., Westport, Conn., 1986.
3. "63 Companies Own 52.2 Percent of Nation's Layers," *Poultry Tribune* **90**(12), 40–41 (1984).
4. USDA, *Egg Grading Manual; USDA, AMS, Agriculture Handbook 75*. U.S. Government Printing Office, Washington, D.C., 1983.
5. M. O. North, *Commercial Chicken Production Manual*, 2nd ed., AVI Publishing Co., Inc., Westport, Conn., 1981.
6. E. C. Naber, "The Effect of Nutrition on the Composition of Eggs," *Poultry Science* **58**, 518–528 (1979).

7. R. L. Adams, D. E. Pratt, J. H. Lin, and W. J. Stadelman, "Introduction of Omega-3 Polyunsaturated Fatty Acids into Eggs," *Abstracts of Papers of Southern Poultry Science Society,* Atlanta, Ga., 1989.
8. W. J. Stadelman and D. E. Pratt, "Factors Influencing Composition of the Hen's Egg," *World's Poultry Science Journal* **45,** 247–266 (1989).
9. USDA, "Regulations Governing the Grading of Shell Eggs and United States Standards, Grades and Weight Classes for Shell Eggs," *USDA, AMS, Poultry Division, 7 CFR, Part 56* (1987).
10. J. W. Koudele and E. C. Heinsohn, "The Egg Products Industry of the United States. I. Historical Highlights, 1900–1959," *Kansas Agricultural Experiment Station Bulletin* **423** (1960).
11. O. J. Cotterill and G. S. Geiger, "Egg Product Yield Trends from Shell Eggs," *Poultry Science* **56,** 1027–1031 (1977).
12. USDA, "Regulations Governing the Inspection of Eggs and Egg Products," *USDA, AMS, Poultry Division, 7 CFR, Part 59* (1984).
13. H. R. Ball, Jr., P. M. Foegeding, K. R. Swartzel, and M. H. Hamid-Samimi, "Functions and Shelflife of Ultrapasteurized, Aseptically Packaged Whole Egg," *Poultry Science* **64** (Suppl. 1), 63 (1985).
14. "The Liquid Egg (pasteurization) regulations 1963," *Statutory Instruments, No. 1503,* H.M. Stationery Office, London, 1963.
15. F. E. Cunningham, "Egg Product Pasteurization," in W. J. Stadelman and O. J. Cotterill, eds., *Egg Science and Technology,* 1st ed., AVI Publishing Co., Inc., Westport, Conn., 1973.
16. USDA, "Composition of Foods. Dairy and Egg Products—Raw, Processed, Prepared," *Agricultural Handbook 8.1. USDA, ARS.* U.S. Government Printing Office, Washington, D.C., 1976; revised, 1989.
17. O. J. Cotterill and J. L. Glauert "Nutrient Values for Shell, Liquid/Frozen and Dehydrated Eggs Derived by Linear Regression Analysis and Conversion Factors," *Poultry Science* **58,** 131–134 (1979).
18. R. S. Beyer and L. S. Jensen, "Cholesterol Content of Eggs as Determined by High-Performance Liquid Chromatography (HPLC)," *Abstracts of Papers of Southern Poultry Science Society,* Atlanta, Ga., 1989.
19. Egg Nutrition Center, "Preliminary Cholesterol Content Data Announced," *News from AEB* (Oct. 17, 1988).
20. R. E. Baldwin, "Functional Properties in Foods," in W. J. Stadelman and O. J. Cotterill, eds. *Egg Science and Technology,* 3rd ed., AVI Publishing Co., Inc., Westport, Conn., 1986.
21. W. J. Stadelman, V. M. Olson, G. A. Shemwell, and S. Pasch, *Egg and Poultry-Meat Processing,* Ellis Horwood, Ltd., Chichester, England, 1988.
22. U.S. Pat. 4,426,400 (Jan. 17, 1984), J. L. Newlin and W. J. Stadelman, (to Purdue University).
23. U.S. Pat. 4,421,770 (Dec. 20, 1983), J. M. Wicker and F. E. Cunningham (to Kansas State University).
24. U.S. Pat. 4,537,788 (Aug. 28, 1985), V. Proctor and F. E. Cunningham (to Kansas State University).
25. E. Li-Chan, S. Nakai, J. Sim, D. B. Bragg, and K. V. Lo, "Lysozyme Separation from Egg White by Cation Exchange Column Chromatography," *Journal of Food Science* **53,** 425–427, 431 (1986).

WILLIAM J. STADELMAN
Purdue University
West Lafayette, Indiana

ELASTINS AND MEAT LIGAMENTS

Elastin and collagen are the principal components of connective tissue. They form a network that is responsible for the transmission of tension and the structure of muscle and other tissues. Elastin forms a lesser proportion of connective tissue than collagen and is not soluble during heating. The ratio of collagen to elastin in connective tissue is dependent on the tissue and its location. This ratio affects the tissue's mechanical properties and its physiological functions (1). Elastin is present in muscle only in small amounts, less than 3% of the total connective tissue. *Musculus semitendinosus,* however, contains more elastin, up to 37% of the total connective tissue. Elastin normally forms fibers and lamellae, is abundant in elastic ligaments, and elastic blood vessels, and is found to a small extent in the skin, lungs, and other organs.

Elastic tissue is often referred to as yellow connective tissue because of its color. It contains the elastin fibers with filamentous, refractive, and fluorescent with a blue–white appearance under ultraviolet light. Because of the large number of nonpolar amino acids in the structure and their hydrophobic nature, the elastin fibers stain poorly with acid or basic dyes but do stain selectively with phenolic dyes such as orcein.

MORPHOLOGY

The basic morphological component of elastin is filament made up of a linear sequence of globular structures with a probable maximum diameter of 9–10 nm. Dehydrated elastin may be 4–6 nm in diameter (2). The freeze-etching technique reveals elastin's structure as a regular, three-dimensional network of filaments. The filaments cross their neighbors every 8–12 globular subunits, giving rise to a network with a large, less-dense central core (3). The filaments of elastin can be disordered and swollen when immersed in 25% glycerol for 8–10 days. The swollen elastin shows a much higher affinity for stains.

CHEMISTRY AND BIOCHEMISTRY

Elastin can be determined from the amount of desmosine or isodesmosine found in the tissue. Its amino acid composition is essentially identical as extracted from various tissues and are not affected by the age of the animal (4). However, compositions differ from animal to animal. Elastin contains 1–2% hydroxyproline. Almost 95% of its amino acids are nonpolar (Table 1). The sulfur amino acids, tryptophan, and tyrosine are present in small amounts or nonexistent. In contrast to collagen, elastin does not contain hydroxylysine. Elastin has about one-third of its amino acids as glycine, but their arrangement in the elastin molecule is unknown. Desmosine and isodesmosine, unique cyclic polymers, are made up of four lysine residues, which connect two to four elastin molecules. This reaction requires the oxidative deamination of three lysine ε-amino groups which give an intermediate followed by condensation with the ε-amino group of a fourth lysyl residue, and is catalyzed by a copper-containing enzyme, lysine oxidase. Therefore, the synthesis of

Table 1. Amino Acid Compositions of Elastins from Selected Tissues of Various Meat Animals

Amino Acid	Residues per 1000 residues			
	Avian, Insol[a]	Bovine, Semitendinosus[b]	Sheep, Vascular Tissue[c]	Yellowfin Tuna, Dark Muscle[d]
Asp	2	7.4	1.8	66.0
Hyp	22	12.4	0.0	6.0
Thr	3	8.2	9.2	51.0
Ser	5	8.3	9.0	53.0
Glu	12	17.1	20.4	134.6
Pro	128	110.9	105.3	65.1
Gly	352	325.9	241.1	128.5
Ala	176	223.1	288.7	91.7
Val	175	145.8	111.7	55.9
Cys/2	1	—	—	17.6
Met	—	—	—	22.3
Ile	19	26.5	17.5	38.2
Leu	47	61.4	57.5	82.9
Tyr	12	8.4	13.4	31.7
Phe	23	29.8	34.5	29.9
Hyl	22	12.4	—	0.0
Lys	4	5.3	30.7	62.2
His	1	0.6	13.6	14.8
Arg	5	5.7	6.8	48.0
isodes	3	2.1[a]	29.3[b]	0.5[a]
Des	3	1.2[a]	7.2[b]	0.3[a]

[a] Ref. 4.
[b] Ref. 5.
[c] Ref. 6.
[d] Ref. 7.

elastin can be blocked by inhibition of this enzyme through dietary copper depletion, or by β-aminopropionitrile, the active lathyrogen of the sweet pea. The synthesis is age dependent; the amount of desmosine and isodesmosine increase and the lysine content decreases with advancing age. The elasticity of elastin is due to the exposure of nonpolar sequences of its molecules to water during the stretching, followed by spontaneous refolding of the molecule on removal of the applied force.

Elastin can be isolated as the residue remaining after the extraction of tissue with 0.1 N NaOH at 98°C for 45 min (4). Elastin is heat stable to a temperature of 140–150°C and is insoluble in a wide range of hydrogen bond breaking solvents at temperatures up to 100°C. It must be processed with harsh enzymatic or chemical treatments in order to solubilize it. This includes KOH/ETOH solubilized κ-2-elastin, oxalic acid solubilized α-elastin, elastase solubilized elastin, and various salt-soluble forms of elastin (9).

Using immunological techniques, two serologically distinct elastin antibody fractions in antiserum prepared against bovine ligamentum nuchae elastin were identified (9). One is a species-specific fraction that bound only to bovine elastin. The other population of antibodies has affinity for elastin from porcine aorta, bovine ligamentum nuchae and aorta, rabbit and hamster lung, and human aorta. The elastin radioimmunoassay can be modified to detect bovine or other elastins, depending on the choice of radiolabeled antigen, in meat products. This assay also permits the study and quantification of elastin synthesis and degradation in a wide range of meat-animal species without having to raise specific antibodies to each one.

The emission spectrum of elastin is similar to that of Type I collagen with peak excitation near 370 nm. Macroscopic ultraviolet fluorometry was used to measure the gristle contents, elastin, and collagen Type I of beef (10).

Fluorescence emissions were measured with a monochromator and a photomultiplier tube. Intact tendons and elastic ligaments had a strong fluorescence emission peak around 440–450 nm and only weak fluorescence around 510 nm. The 510:450 nm ratio was correlated to the amount of gristle in comminuted meats.

INDUSTRIAL APPLICATIONS

The visible gristle in meat is from a number of sources, including tendons (particularly from myotendon junctions), ligaments (such as the ligamentum nuchae in rib roasts), perimysium (particularly in the pennate extensors and flexors of the lower limbs), fasciae (such as the lumbodorsal fascia of the longissimus muscle), and intramuscular vessels (particularly arteries). All these sources consist mainly of Type I collagen or elastin. Elastin-contained gristle retains its tensile strength after typical cooking procedures, such as roasting and broiling (11,12). Fresh meat cuts containing gristle are generally perceived as lower quality products, because of the unpleasant sensation of gristle. However, the elastin concentration is not consistently related to variations in the tenderness of muscles, at least in the bovine species (13), nor is the total concentration of connective tissue components (collagen and elastin). Tenderness is actually more related to soluble collagen and the overall fiber arrangement in the meat.

Aging is the industrial practice of storing meat carcasses above the freezing point for a certain period of time to improve the tenderness. The changes are the result of the physical breakage of muscle and collagen protein fibers by rigor and enzymatic reactions. However, aging has no apparent effect on the structure of elastin tissue and its physical properties.

Tissues with high elastin content, such as elastic ligaments, blood vessels, and lung, are removed in the slaughtering operation. They are processed as offals with other organs and by-products in the rendering industry as a protein source for animal feed. In this application, they may be rendered at high or low temperature, depending on the planned use of the fat. Fat is separated from the solids by centrifugation. The solids are dried and batched with other protein sources for optimum amino acid content and distribution, and then used in animal feed. The make-up of the feed is determined by a linear program based computer program which takes account of the amino acids, fat, carbohydrate, vitamin, and mineral content required in feed formulations.

In the ready-to-eat processed meat industry, elastin tissues and meat ligaments retain their tensile strength and are not gelatinized under most heat-processing conditions. For this reason, elastin tissues in meat are removed, with other gristle, by mechanical means, by hand or mechanical gristle removers. Grinders equipped with gristle and bone removers are widely used in the processed meat industry for coarse-ground, nonemulsified products. In this case, the meat is put through a grinder and the bone chips and gristle are separated from the ground meat. For large-muscle products, such as ham, the meat must be hand trimmed. The separated gristle with elastin and collagen tissue is discarded or chopped fine and used in emulsified products. It is used as a filler rather than functional proteins for economic reasons.

Meat cuts with a high gristle, collagen, and elastin content are not used in large-muscle products. They are used most often in emulsified products to minimize the toughness problem of the gristle. However, elastin and collagen tissues from gristle do not bind water or function as emulsifiers in meat emulsion; therefore, emulsion breakdown occurs more often in these products. To solve these problems, a computer-based, least-cost formulations program, which is widely used in the processed meat industry, has incorporated the connective tissue in individual meat cuts as a quality constraint in the computer simulation of product formulations. But no effort has been made to separate the elastin or ligament contribution as a constraint in the computer program, because of the low content of elastin in most meat cuts.

Unlike collagen, elastin does not swell at mild acid conditions and has no film-forming properties. It is not used in gelatin production because of its poor water solubility and gelation properties. It cannot be substituted for collagen in regenerated collagen casing, or collagen film and tissue. Elastin also has little use in the processed meat industry because it functions poorly as a binder and emulsifier. Because it is found in only small amounts in meats, because it is uneconomical to separate it from other meat tissues, and because it is poor nutritionally, elastin's technological development for use in foods and other industries has been hindered.

BIBLIOGRAPHY

1. R. J. Bagshaw, E. Weit, and R. H. Cox, "Aortic Connective Tissue Content in White Leghorn Females," *Poultry Science*, **65**, 403 (1986).
2. C. Fornieri, I. P. Ronchetti, A. C. Edman, and M. Sjostrom, "Contribution of Cryotechniques to the Study of Elastin Ultrastructure," *Journal of Microscopy* (Oxford) **126**, 97 (1982).
3. M. Morocutti, M. Raspanti, P. Govoni, A. Kadar, and A. Ruggeri, "Ultrastructural Aspects of Freeze-Fractured and Etched Elastin," *Connective Tissue Research* **18**, 55 (1988).
4. H. R. Cross, G. C. Smith, and Z. L. Carpenter, "Quantitative Isolation and Partial Characterization of Elastin in Bovine Muscle Tissue," *Journal of Agricultural and Food Chemistry* **21**, 716 (1973).
5. J. A. Foster "Elastin Structure and Biosynthesis: An Overview," in, *Structural and Contractile Proteins. Part A. Extracellular Matrix*, L. W. Cunningham and D. W. Frederiksen, eds. Academic Press, New York, 1982, pp. 559–570.
6. J. R. Bendall, "The Elastin Content of Various Muscles of Beef Animals," *Journal of the Science of Food and Agriculture* **18**, 553 (1967).
7. S. Kanoh, T. Suzuki, K. Maeyama, T. Takewa, S. Watabe, and K. Hashimoto, "Comparative Studies on Ordinary and Dark Muscles of Tuna Fish," *Bulletin of the Japanese Society of Scientific Fisheries (NIHON SUISAN GAEFFAI-SHI)* **52**(10), 1807 (1986).
8. J. N. Manning, P. F. Davis, N. S. Greenhill, A. J. Sigley, "Salt Soluble Cross-linked Elastin: Formation and Composition of Fibers," *Connective Tissue Research* **13**(4), 313 (1985).
9. R. P. Mecham and G. Lange, "Measurement by Radioimmunoassay of Soluble Elastins from Different Animal Species," *Connective Tissue Research* **7**, 247 (1980).
10. H. J. Swatland, "Fiber-Optic Reflectance and Autofluorescence of Bovine Elastin and Differences between Intramuscular and Extramuscular Tendon," *Journal of Animal Science* **65**, 158 (1987).
11. A. M. Pearson and F. W. Tauber, *Processed Meats*, AVI Publishing Co., Inc., Westport, Conn. 1984.
12. P. P. Purslow, "The Physical Basis of Meat Texture: Observations on the Fracture Behavior of Cooked Bovine M. Semidendinosus," *Meat Science* **12**, 39 (1985).
13. H. R. Cross, Z. L. Carpenter, and G. C. Smith, "Effects of Intramuscular Collagen and Elastin on Bovine Muscle Tenderness," *Food Science* **38**, 998 (1973).

Rudy R. Lin
Swift-Eckrich, Inc.
Downers Grove, Illinois

EMULSIFIERS, STABILIZERS, AND THICKENERS

The food additives that will be discussed in this article fall into three functional categories. Although the definition of each function will be clarified in the discussion, the definitions given in 21 *Code of Federal Regulations* are

§170.3(o)(8). Emulsifiers and emulsifier salts: Substances which modify surface tension in the component phase of an emulsion to establish a uniform dispersion or emulsion.

§170.3(o)(28). Stabilizers and thickeners: Substances used to produce viscous solutions or dispersions, to impart body, improve consistency, or stabilize emulsions, including suspending and bodying agents, setting agents, jellying agents, and bulking agents, etc.

§170.3(o)(29). Surface-active agents: Substances used to modify surface properties of liquid food components for a variety of effects, other than emulsifiers, but including solubilizing agents, dispersants, detergents, wetting agents, rehydration enhancers, whipping agents, foaming agents, and defoaming agents, etc.

Generally emulsifiers and surface-active agents are relatively small molecules (molecular weight less than 1,000 Da), while stabilizers and thickeners are polymers such as gums and proteins. There are exceptions to this; calcium stearate is listed as a GRAS (generally recognized as safe) stabilizer and thickener while gum ghatti is approved for use as a GRAS emulsifier.

FUNCTIONS OF EMULSIFIERS, STABILIZERS, AND THICKENERS

As shown in Figure 1 surfactants have a lipophilic (fat-loving) portion and a hydrophilic (water-loving) portion; for this reason they are sometimes called amphiphilic (both-loving) compounds. The lipophilic part of food surfactants is usually a long chain fatty acid obtained from a food grade fat or oil. The hydrophilic portion is either nonionic (glycerol is shown in Fig. 1), anionic (negatively charged, such as lactate), or amphoteric, carrying both positive and negative charges (serine is shown in Fig. 1). Cationic (positively charged) surfactants are usually bactericidal and somewhat toxic; they are not used as food additives. Examples of the three types indicated are a monoglyceride (nonionic), stearoyl lactylate (anionic), and lecithin (amphoteric). The nonionic surfactants are relatively insensitive to pH and salt concentration in the aqueous phase, while the functionality of the ionic types may be markedly influenced by pH and ionic strength.

Emulsifiers

The formation of an oil-in-water emulsion is outlined in Figure 2. If 100 mL of pure vegetable oil plus 500 mL of water are mixed vigorously to obtain an emulsion in which the average diameter of the oil globules is 1 μm, 600 m^2 (slightly more than 6,400 ft^2) of oil–water interface is generated. Using a purified vegetable oil the interfacial tension versus water is about 3×10^{-6} J/cm^2. To form the emulsion, 18 J of the energy input is converted into interfacial energy. The addition of 1% glycerol monostearate (GMS) to the oil phase will lower γ to about 3×10^{-7} J/cm^2 so that the excess interfacial energy is only 1.8 J. This excess interfacial energy is the driving force behind coalescence of the oil globules and also the force resisting subdivision of oil droplets during mixing (each division increases the amount of interface present in the

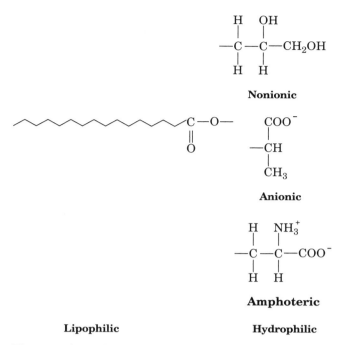

Figure 1. Generalized structure of amphiphilic molecules. The lipophilic chain represents a fatty acid, palmitic acid. The hydrophobic groups are glyceryl, lactyl, and serine.

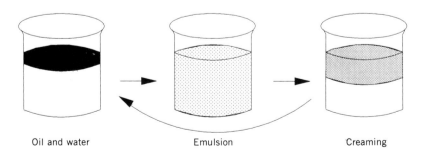

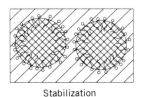

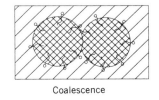

Figure 2. Stages of emulsion formation and breakdown. In the upper figures the dark area represents oil, the clear area water. In the lower figures crosshatched circles represent oil and the diagonally-shaded areas water.

system). In a series of emulsification experiments in which the amount of mixing energy is constant and γ is changed by adding emulsifier, it is found that the average oil droplet diameter parallels γ, ie, as more emulsifier is added γ decreases and so does the average droplet size. The total excess interfacial energy is roughly constant.

Stabilizers

The phenomenon just described is the promotion of emulsion formation; this is not the same as stabilization of emulsions. The difference is seen by reference to Figure 2. After the emulsion is formed, if it is allowed to stand, the oil droplets will rise to the top or cream (assuming the volume ratio of oil to water is low enough for flotation to occur). The rate of creaming is inversely related to droplet diameter and to the viscosity of the aqueous phase; large droplets rise faster than small droplets and rise faster in water than in a viscous gum solution. Emulsifiers that promote formation of smaller diameter drops and additives that increase viscosity give emulsions in which the rate of separation is slower, so in that sense only the emulsion is stabilized. When two oil droplets make contact (facilitated by the creaming process) they immediately coalesce into one larger droplet, reducing total surface area and total excess interfacial energy. In Figure 2 the scheme labeled coalescence indicates this, where the emulsifier used may be efficient at reducing γ even at a low interfacial concentration but does not prevent the oil droplets from touching and coalescing during creaming.

A true emulsion stabilizer prevents the droplet contact that leads to coalescence (Fig. 2). In essence, the thin layer of water between the oil droplets is stabilized by various mechanisms. If the surfactant is anionic, then the surfaces of both oil droplets carry a negative charge and they are mutually repelled by electrostatic effects. This sort of stabilization is sensitive to ionic strength, and a high salt concentration will suppress the electrostatic repulsion, promote contact and coalescence, and lead to rapid emulsion breakdown.

A second kind of stabilization is shown by surfactants in which the hydrophilic portion is quite large, eg, the polyoxyethylene chain of the Tweens or ethoxylated monoglyceride. In this case the chain is anchored at the surface of the oil droplet by the lipophilic tail, but it is strongly hydrated and generates a layer of bound water around the droplet, preventing contact and coalescence. This functionality is relatively insensitive to salt concentration.

A third kind of stabilization is due to simple steric hindrance of contact. The alpha-tending emulsifiers such as propylene glycol monostearate (PGMS) form an actual solid layer at the oil–water interface (1,2). This film physically prevents the contents of oil droplets from coalescing even though their surfaces may be touching. Gums such as gum arabic and gum ghatti stabilize oil-in-water emulsions by a similar mechanism, forming a film of adsorbed polymer around the oil droplet, while some water-soluble proteins perform the same function in mixtures of ground meat and fat for making sausage.

It should be noted that emulsion stabilization is not directly related to the ability to lower interfacial tension.

For example, 1% GMS in the oil phase lowers γ to 3×10^{-7} J/cm^2 but has little effect on coalescence rate, whereas 12% PGMS yields a γ of 9×10^{-7} J/cm^2 but gives an emulsion with excellent long-term stability.

Foam Stabilizers

The generation and stabilization of a foam is subject to the same thermodynamic energy considerations as emulsions; lowering the interfacial (surface) tension favors foam formation. Small surfactant molecules dissolved in the aqueous phase will promote foaming, and the stability of the foam is dependent on the maintenance of a film of water between air bubbles. The lipophilic portion of the surfactant enters the gas phase (rather than an oil phase) but in all other respects the situation is analogous to that where both phases are liquid (ie, emulsions).

Proteins are amphiphilic molecules in that many amino acids (eg, leucine, isoleucine, and valine) contain hydrophobic (lipophilic) side chains while others (glutamic, aspartic, lysine, and arginine) have ionic, hence hydrophilic, side chains. Normally, proteins such as egg albumin in solution are folded in such a way that the hydrophobic side chains are buried in the interior of the molecule in a nonpolar environment, while the hydrophilic side chains are on the surface of the molecule and interact with the polar aqueous environment. Introduction of air bubbles into the solution presents a new possibility for the lowest energy state of the protein, namely for it to unfold with the hydrophobic side chains entering the air phase and the hydrophilic chains remaining in the water phase. The portion of the proteins located in the aqueous phase hold water, preventing it from draining away from this region, hence stabilizing the foam. Whipping aids enhance the ability of the protein to unfold at the air–water interface; the energetics of protein unfolding become more favorable and the ease of foam formation increases. Solutes that increase the viscosity of the water phase, eg, gums, slow the rate of draining and thus increase foam stability.

Thickeners

Thickeners may function in four different ways: (1) by stabilizing an emulsion with a high volume percentage of internal phase, (2) by increasing the viscosity of the external phase, (3) by forming an elastic network in the external phase, and (4) by removing a portion of the external phase. An example of the first kind is mayonnaise, where egg yolk lipoprotein stabilizes an oil-in-water emulsion in which the internal phase (oil) represents more than 70% of the total volume. As the particle size of the oil droplets decreases, the relative amount of water that is immobilized around the surface increases, and the amount of mobile external phase (free water, in a nonthermodynamic sense) decreases, contributing to the high viscosity and body of the final product.

Certain gums form aqueous solutions that have a high viscosity (Table 1). When such a solution is used as the water component of a food formula, the viscosity of the final product reflects the viscosity of the solution. In a simple salad dressing formulation, increased viscosity in

Table 1. Properties of Some Gums

Gum	1% Solution Viscosity, Pa·s	21 CFR
Low-Viscosity Gums		
Arabic	0.002–0.005	184.1330
Ghatti	0.004–0.010	184.1333
Larch	0.002–0.010	172.610
Medium-Viscosity Gums		
Sodium alginate	0.025–0.800	184.724
Tragacanth	0.200–0.500	184.1351
Xanthan	0.800–1.400	172.695
High-Viscosity Gums		
Guar	2.0–3.5	184.1339
Karaya	2.5–3.5	184.1349
Locust bean	3.0–3.5	184.1343
Cellulose Gums		
Sodium carboxymethylcellulose	0.050–5.000	182.1745
Hydroxypropyl methylcellulose	0.020–50.000	172.874
Methylcellulose	0.010–2.000	182.1480
Gel Formers		
Agar	Gel	184.1115
Calcium alginate	Gel	184.1187
Carrageenan	Gel	172.620
Furcelleran	Gel	172.655
Pectin	Gel	184.1588
Gellan	Gel	—

the aqueous phase slows down the flotation rate of the oil droplets formed during shaking and (in this sense) stabilizes the emulsion. Commercial batters (ie, cake, cake donut and fish breading) require a certain viscosity for optimum functionality; gums may be used to impart and control batter viscosity.

Gums and many modified starches in aqueous solution form gels under the proper conditions (Table 1). Gelation is usually dependent on some additional factor (temperature change and presence of divalent cations), and these properties are used during production to generate the desired degree of elasticity in the final food product. The elastic gel network prevents separation of dispersed oil and solids (eg, fruit) as well as giving a pleasing texture to the food.

In most food products the external phase is water or an aqueous solution and the internal phase is a mixture of solid material or emulsified lipid. The viscosity of the product depends in part on the volume ratio of internal to external phases. Insoluble fibrous materials can physically adsorb several times their weight in water, so the addition of a few percent of a material such as cellulose (either alpha or microcrystalline) or hemicellulose (cereal bran) will markedly reduce the amount of water, raising the internal to external phases ratio and hence the viscosity. As an example, a sauce might contain 20% solids (80% water) for an internal to external ratio of 20:80, or 0.25. Adding 5% powdered cellulose will convert about 30% of the water to the solid phase via adsorption, raising the internal to external ratio to 55:50, or 1.1, producing a marked difference in the viscosity of the sauce.

TYPES OF EMULSIFIER

Emulsifiers may be divided into four classes as mentioned earlier: (1) nonionic, (2) anionic, (3) amphoteric, and (4) cationic. The actual item of commerce is seldom exactly like the organic chemical structures, discussed but is usually a mixture of similar compounds derived from natural raw materials. As a simple comparison, dextrose is one single, relatively pure chemical entity, described by the formula $C_6H_{12}O_6$. On the other hand, GMS is made from a hydrogenated natural fat or oil and the saturated fatty acid composition may well be something like 1% C_{12}, 2% C_{14}, 30% C_{16}, 65% C_{18}, and 2% C_{20}. In addition, the monoglyceride will be approximately 92% 1-monoglyceride and 8% 2-monoglyceride, which represents the chemical equilibrium.

Nonionic Emulsifiers

Monoglycerides and Derivatives. The manufacture of monoglycerides and derivatives used by the food industry were estimated to be over 200 million lb in 1981 (3–5). The use of monoglycerides in food products first began in the 1930s when superglycerinated shortening became commercially available. Glycerine was added to ordinary shortening along with a small amount of alkaline catalyst, the mixture was heated causing some interesterification of triglyceride with the glycerine and the catalyst was removed by neutralization and washing with water. The resulting emulsified shortening contained ca 3% monoglyceride and was widely used for making cakes, particularly with high sugar levels. Subsequent use of monoglyceride to retard staling (crumb firming) in bread used plastic monoglyceride made by altering the ratio of glycerine to fat to achieve a final concentration of 50–60% monoglyceride, with most of the remainder being diglyceride. Later developments in monoglyceride technology include (1) distilled monoglyceride, consisting of a minimum of 90% monoglyceride; (2) hydrated monoglyceride, which contains roughly 25% monoglyceride, 3% sodium stearoly lactylate (SSL), and 72% water and is a lamellar mesophase for better water dispersibility; and (3) powdered distilled monoglyceride in which the composition of the original feedstock fat is balanced between saturated and unsaturated fatty acids so that the resulting powder hydrates fairly rapidly during mixing in an aqueous system such as bread dough.

The monoglyceride structure shown in Figure 3 is for 1-monostearin, also called α-monostearin. If the fatty acid is esterified at the middle hydroxyl the compound is 2-monostearin, or β-monostearin. In technical specifications manufacturers usually give the monoglyceride content of their product as percent of α-monoglyceride. The routine analytical method for monoglyceride (6) detects only the 1-isomer; quantitation of the 2-isomer is much more tedious. The total monoglyceride content of a product is about 10% higher than the reported α-monoglyceride content. In a practical sense, however, when the

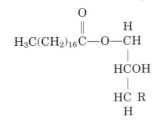

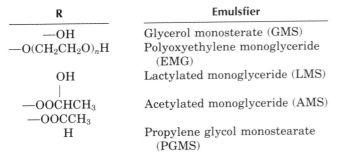

Figure 3. Nonionic emulsifiers based upon glycerol esters and derivates.

functionality and cost effectiveness of various products are being compared, the α-monoglyceride content is a useful number because for all products it equals about 92% of the total monoglyceride present.

The fatty acid composition of monoglyceride reflects the makeup of the triglyceride fat from which it is made. Commercial GMS may contain as little as 65% stearate, if made from fully hydrogenated lard, or as much as 87% stearate, if made from fully hydrogenated soybean oil. The other principal saturated fatty acid will be palmitic, and because hydrogenation is practically never carried out to the extent that all unsaturation is removed (iodine value of zero), a few percent of unsaturated (oleic and elaidic) acid is also usually present. Iodine values for powdered distilled monoglycerides are in the range of 19–36, and for plastic monoglycerides a typical range is 65–75. The unsaturated fatty acids are a mixture of oleic and linoleic and the trans isomers of these acids.

The product of the manufacture of ethoxylated monoglyceride (EMG) is somewhat random in structure. Monoglyceride is treated with ethylene oxide gas under pressure in the presence of alkaline catalyst and at elevated temperatures. Ethylene oxide polymerizes via a series of ether linkages and also forms ether bonds with the free hydroxyl groups on monoglyceride. The average chain length comprises about 20 units ($n = 20$ in Fig. 3). Chains may be attached to hydroxyls at both the two and three positions of the monoglyceride, although many more chains will be located at the α (three) position than at the β (two) position because of the difference in their chemical reactivities. The exact distribution of polymer chain lengths and distribution between α and β positions are functions of reaction conditions, eg, catalyst type and concentration, gas pressure, temperature, agitation, and length of reaction time.

The second group of monoglyceride derivatives, the α-tending emulsifiers (lactylated monoglyceride, acety-

lated monoglyceride, and PGMS), find their main use in cake production. These emulsifiers are dissolved in the shortening phase of the cake formulation and contribute to the emulsification of the shortening in the water phase as well as promoting incorporation of air into the fat phase. The particular property of these emulsifiers that makes them valuable in liquid shortening cakes is that they form a solid film at the oil–water interface, not only stabilizing the emulsion but also keeping the lipid phase from preventing air incorporation (protein-stabilized foam formation) during cake batter mixing.

Sorbitan Derivatives. When the sugar alcohol sorbitol is heated with stearic acid in the presence of a catalyst two reactions occur: sorbitol cyclizes to form the five-membered sorbitan ring and the remaining primary hydroxyl group is esterified by the acid. The resulting sorbitan monostearate (Fig. 4) is oil soluble, has a rather low hydrophilic–lipophilic balance (HLB) value, and is the only one of the many sorbitan esters presently approved for food use in the United States. Other sorbitan esters of importance are the monooleate and the tristearate. Any of the three esters may be reacted with ethylene oxide to give polyoxyethylene derivatives, as indicated in Figure 4, which are water soluble and have a relatively high HLB. The monostearate derivative is known as Polysorbate 60, the tristearate is Polysorbate 65, and the monooleate is Polysorbate 80. The remarks made in connection with EMG regarding the length and location of the polyoxyethylene chains apply to these compounds. The average number of oxyethylene monomers is 20 ($n = 20$), and in the case of the monoesters, chains may be located on more than one hydroxyl group of the sorbitan ring (triester has only one hydroxyl group available for derivatization).

Sorbitan monostearate is used in applications where fat is the continuous phase, ie, in fat-based confectionery coatings, in bakery icings, and in coffee creamers. The polyoxyethylene derivatives have found wider usage. Polysorbate 60 has been used in fluid oil cake shortening systems (4,5,7) generally in combination with GMS and PGMS. The various sorbitan derivatives are often used in combination to obtain a specific desired HLB; the regulations regarding the permitted levels of each emulsifier in the final product are rather complicated; details can be found in 21 *CFR*.

Esters of Polyhydric Compounds. Polyglycerol esters (Fig. 5) have a variety of applications as emulsifiers in the food industry (3,8). The polyglycerol portion is synthesized by heating glycerol in the presence of an alkaline catalyst; ether linkages are formed between the primary hydroxyls of glycerol. In Figure 5 n may take any value, but for food emulsifiers the most common ones are $n = 1$ (triglycerol), $n = 4$ (hexaglycerol), $n = 6$ (octaglycerol), and $n = 8$ (decaglycerol) (in all cases n is an average value for the molecules present in the commercial preparation). The polyglycerol backbone is then esterified to varying extents, either by direct reaction with a fatty acid or by interesterification with a triglyceride fat. Again, the number of acid groups esterified to a polyglycerol molecule varies around

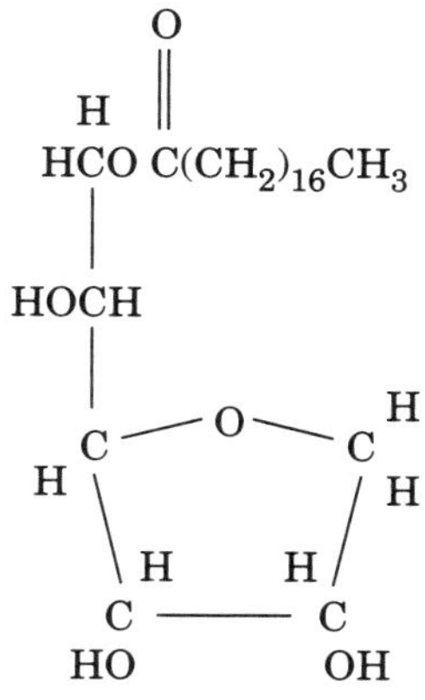

sorbitan monostearate

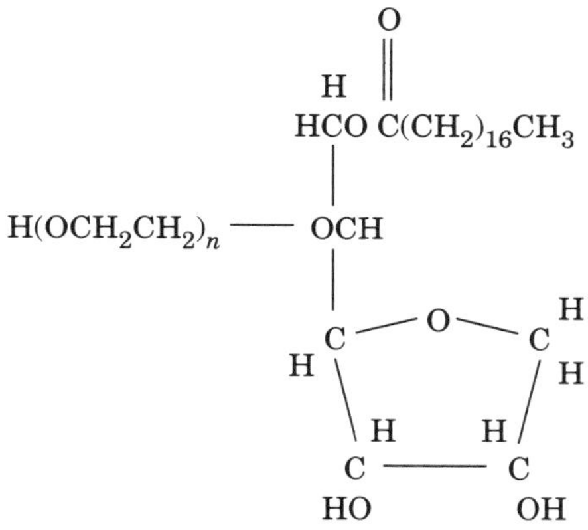

polyoxyethylene (20) sorbitan monostearate (polysorbate 60)

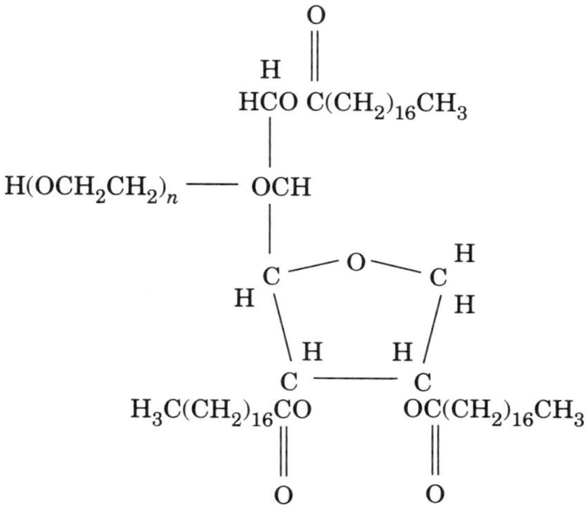

polyoxyethylene (20) sorbitan tristearate (polysorbate 65)

Figure 4. Nonionic emulsifiers based upon sorbitan esters.

some central value, so an octaglycerol octaoleate really should be understood as (approximately octa)glycerol (approximately octa)oleate ester. By good control of feedstocks and reaction conditions manufacturers do manage to keep the properties of their various products relatively constant from batch to batch.

The HLB of these esters depends on the length of the polyglycerol chain (the number of hydrophilic hydroxyl groups present) and the degree of esterification. As examples, decaglycerol monostearate has an HLB of 14.5, while triglycerol tristearate has an HLB of 3.6. Intermediate species have intermediate HLB values, and any desired value may be obtained by appropriate blending. The wide range of possible compositions and HLB values make these materials versatile emulsifiers for food applications.

Sucrose has eight free hydroxyl groups that are potential sites for esterification to fatty acids. Compounds containing six or more fatty acids per sucrose molecule have been proposed for use as noncaloric fat substitutes under the name Olestra; this material acts like a triglyceride fat and has no surfactant properties. Compounds containing one to three fatty acid esters (Fig. 5) act as emulsifiers and are approved for food use in that capacity (9). They are manufactured by the following steps. First, an emulsion is made of fatty acid methyl ester in a concentrated aqueous sucrose solution. The water is then removed under vacuum at elevated temperature. An alkaline catalyst is added and the temperature of the dispersion is slowly raised to 150°C under vacuum, distilling off methanol formed on transesterification. Finally, the reaction mixture is cooled and purified. The degree of esterification is controlled by the reaction conditions, especially the sucrose to methyl ester ratio, and the final product is a mixture of esters. The HLB value of a particular product is lower (more lipophilic) as the degree of esterification increases, as would be expected.

polyglycerol monostearate

sucrose diester

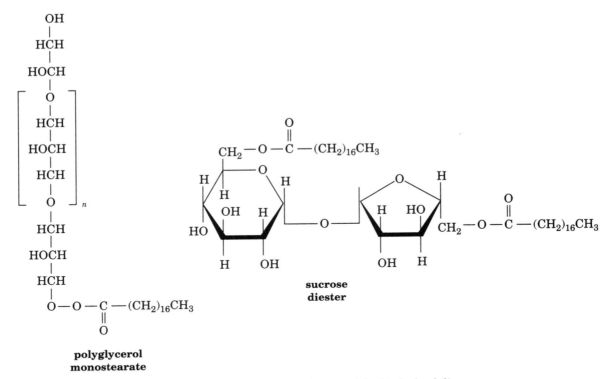

Figure 5. Nonionic emulsifiers based upon polyhydric hydrophilic groups.

Anionic Emulsifiers

Monoglyceride Derivatives. A large number of anionic derivatives of monoglyceride have been synthesized and tested but the two shown in Figure 6 are the most widely used. Others listed in Table 2 include stearyl monoglyceridyl citrate, succistearin (the succinate monoester of propylene glycol monofatty acid ester), monoglyceride citrate, and monoglyceride phosphate. Succinyl monoglyceride (SMG) is manufactured by reacting succinic anhydride with monoglyceride, forming the succinate monoester. Diacetyltartrate ester of monoglyceride (DATEM) is made by mixing monoglyceride, tartaric acid, and acetic anhydride in a defined ratio and reacting under closely controlled conditions. The product is a mixture of the possible esters, with the one shown in Figure 6 forming the largest proportion. The most usual application of these compounds is as dough conditioners, ie, surfactants added to bread dough, which increase the strength and elasticity of gluten in the dough. Bread and other yeast-leavened baked foods made with the addition of dough conditioners typically have larger volume and finer internal texture than products made without such conditioners.

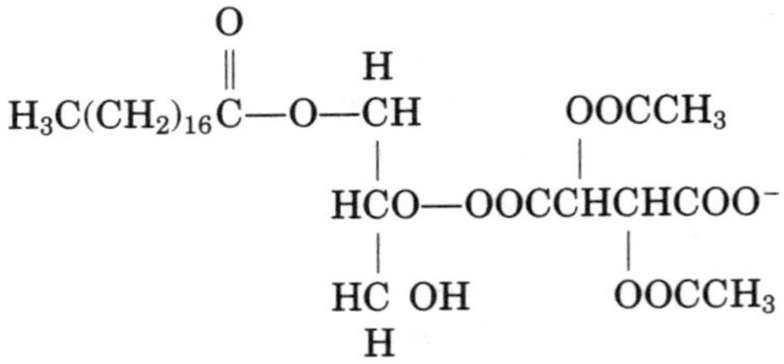

Other Anionic Surfactants. In addition to SMG and DATEM, some other anionic surfactants that have been tried as dough conditioners are shown in Figure 7. Presently SSL is the one most widely used in the United States; sodium stearyl fumarate (10) has been offered but did not find acceptance, and sodium lauryl sulfate is used only as a whipping agent with egg whites.

Lactic acid, having both a carboxylic acid and a hydroxyl function on the same molecule, readily forms an ester with itself. When stearic acid is heated with polylactic acid under the proper reaction conditions and then neutralized with sodium hydroxide a product having the structure shown in Figure 7 is obtained. The monomer lactylic acid shown represents the predominant product; the dilactylic dimer is also present, as well as lactylic trimers and tetramers (11). As with all compounds based on commercial stearic acid derived from fats some per-

succinyl monoglyceride (SMG)

diacetyltartrate ester of monoglyceride (DATEM)

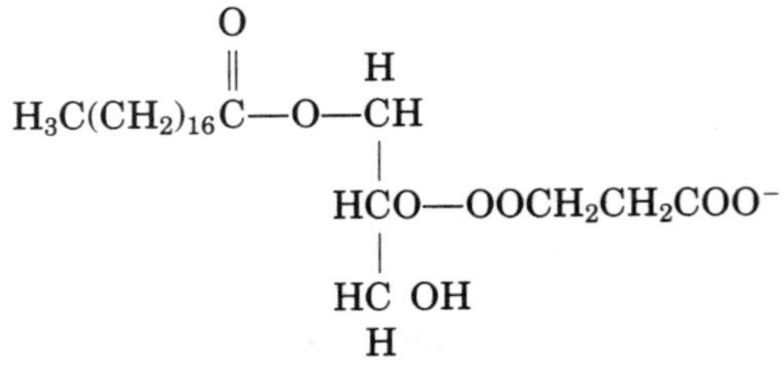

Figure 6. Anionic derivatives of monoglycerides.

Table 2. Emulsifier Uses

Emulsifier	21 CFR	Emulsifiers													Conditioners			Stabilizers					Wetting Agent					Whipping Agent	
		General Use	Cakes, Bakery Mixes	Coffee Creamer	Frozen Desserts	Ice Cream	Whipped Vegetable Oil Toppings	Bakery Fillings	Bakery Icing	Bakery Toppings	Pastry	Pan-Release Agent	Dressing	Cocoa Fat Beverage	Bread Doughs	Dehydrated Potatoes	Processed Cereal	Emulsion Shortening	Antioxidant in Oil	Cheese Substitute	Compound Coatings	Sauces, Gravies	Gelatin Dessert Mix	Dry Beverage Mix	Fruit Juice Drink	Cocoa Powder	Pudding Mix	Egg White	Gelatin
Stearyl monoglyceridyl citrate	172.755	X																											
Succistearin	172.765		X																										
Dioctyl sodium sulfosuccinate	172.810							X	X	X				X												X			
Hydroxylated lecithin	172.814	X																											
Sodium lauryl sulfate	172.822									X	X												X	X	X			X	X
Sodium stearyl fumarate	172.826														X		X												
Acetylated monoglyceride	172.828	X	X																										
Succinylated monoglyceride	172.830														X			X											
Monoglyceride citrate	172.832																		X										
Ethoxylated monoglyceride	172.834		X	X			X	X	X			X			X														
Polysorbate 60	172.836		X	X	X		X	X	X						X			X			X		X	X			X		
Polysorbate 65	172.838		X	X	X	X	X		X				X																
Polysorbate 80	172.840					X	X	X	X							X	X	X					X						
Sorbitan monostearate	172.842		X	X			X	X																					
Calcium stearoyl lactylate	172.844						X	X							X	X					X							X	
Sodium stearoyl lactylate	172.846		X		X		X	X	X		X				X	X				X		X					X		
Fatty acid lactylate	172.848		X					X	X		X									X									
Lactylated propylene glycol monoester	172.850	X																											
Lactylated monoglyceride	172.852	X																											
Polyglycerol esters	172.854	X																											
Propylene glycol monoester	172.856	X																											
Sucrose fatty acid ester	172.859		X		X																								
Glycerol monostearate	182.1324	X																											
Triethyl citrate	182.1911																											X	
Diacetyl tartrate ester of monoglyceride	182.4101	X																											
Monoglycerides	182.4505	X																											
Monoglyceride phosphate	182.4521	X																											
Lecithin	184.1400	X																											

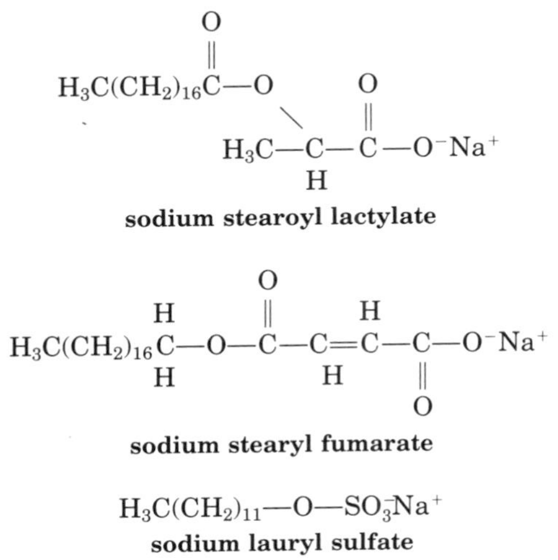

688 EMULSIFIERS, STABILIZERS, AND THICKENERS

$$H_3C(CH_2)_{16}C(=O)-O-CH(CH_3)-C(=O)-O^-Na^+$$

sodium stearoyl lactylate

$$H_3C(CH_2)_{16}CH_2-O-C(=O)-CH=CH-C(=O)-O^-Na^+$$

sodium stearyl fumarate

$$H_3C(CH_2)_{11}-O-SO_3^-Na^+$$

sodium lauryl sulfate

Figure 7. Sodium salts of three anionic emulsifiers.

centage of the fatty acid is palmitic, with small portions of myristic and arachidic acids also found. Although SSL is readily water soluble, the calcium salt is practically insoluble. In this respect it mimics a soap; for example, sodium stearate is water soluble but calcium stearate is oil soluble. Either form may be used, depending on the details of the intended application. The free (unneutralized) form is also approved for certain uses.

Stearyl fumarate is a half ester of fumaric acid with stearyl alcohol (octadecanol). Although it might be expected to have dough-strengthening properties similar to SSL, this was not found to be so in practice and the product was not a commercial success.

The third structure shown in Figure 7 is sodium dodecyl sulfate (SDS), a sulfate ester of the C_{12} alcohol dodecanol. Commercially lauryl alcohol is produced by reduction of coconut oil. The alcohol portion of sodium lauryl sulfate is a mixture of chain lengths; the approximate composition is 8% C_8, 7% C_{10}, 48% C_{12}, 20% C_{14}, 10% C_{16}, and small amounts of longer chains. The most common use of sodium lauryl sulfate is as a whipping aid. It is added to liquid egg whites at a maximum concentration of 0.0125%, or to egg white solids at a level of 0.1%. It promotes the unfolding of egg albumin at the air–water interface and foam stabilization as discussed earlier. (Triethyl citrate, not strictly an emulsifier, is also used as a whipping aid.)

Amphoteric Emulsifiers

Lecithin. During the purification of crude oil extracted from soybean, safflower, or corn germ a phosphatide-rich gum is obtained that is treated and purified to give the various commercial lecithin products available to the food processor today (12–14). The dark crude material is bleached to give a more acceptable light brown color. Treatment with up to 1.5% hydrogen peroxide gives the product known as single-bleached lecithin, and further addition of up to 0.5% benzoyl peroxide yields double-bleached lecithin (14). Reaction with even higher levels of hydrogen peroxide plus lactic acid hydroxylates unsaturated fatty acid side chains at the double bond (yielding, eg, dihydroxystearic acid from oleic acid) giving hydroxylated lecithin, which is more dispersible in cold water than the other types and is more effective as an emulsifier for oil-in-water emulsions (15). Normally some vegetable oil is added to the lecithin to reduce the viscosity to 150 Pa·s;

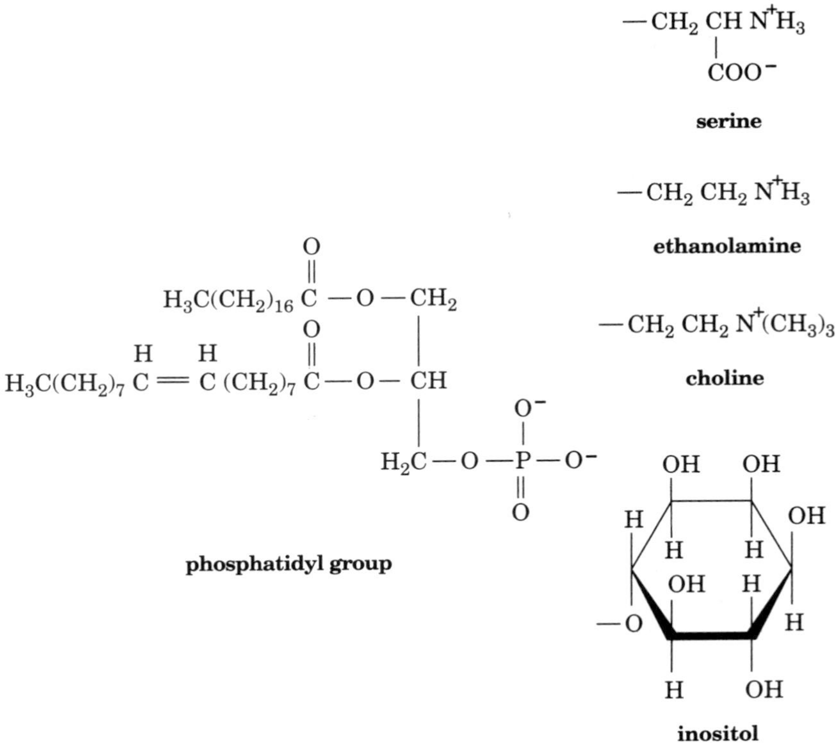

Figure 8. Surface active components of commercial lecithin. Phosphatidylserine is not usually found in vegetable-derived lecithin. Phosphatidyl-ethanolamine and -choline are amphoteric emulsifiers, while phosphatidylinositol is anionic.

this is called standardized fluid lecithin and contains roughly one-third oil and two-thirds phosphatides.

Figure 8 shows the structure of the main surface active components of lecithin. The phosphatidyl group is a phosphate ester of diglyceride. The fatty acid composition of the diglyceride is similar to that of the basic oil (16) so a number of different fatty acids are found, not just the stearic and oleic acids depicted. There is little phosphatidylserine present in soybean lecithin, and the other three derivatives are found in approximately equal amounts. Phosphatidyl ethanolamine (PE) and phosphatidyl choline (PC) are amphoteric surfactants, whereas phosphatidyl inositol (PI) is anionic. The HLB values of the three species are varied: PC having a high, PE an intermediate, and PI a low value. The HLB of the natural blend is around 9 to 10, and emulsifier mixtures having this value will tend to form either, oil-in-water or water-in-oil emulsions, although neither one is highly stable. On the other hand, intermediate HLB emulsifiers are excellent wetting agents and this is a principal application for lecithin.

The emulsifying properties of lecithin can be improved by ethanol fractionation (15). PC is soluble in ethanol, PI is rather insoluble, and PE is partially soluble. Adding lecithin to ethanol gives a soluble and an insoluble fraction. The phosphatide compositions of the two are (1) ethanol soluble; 60% PC, 30% PE, 2% PI, and (2) ethanol insoluble; 4% PC, 29% PE, 55% PI (17). The soluble fraction is effective in promoting and stabilizing oil-in-water emulsions, whereas the insoluble portion promotes and stabilizes water-in-oil emulsions. Only one company (Lucas Meyer, FRG) is using this process to produce industrial food-grade emulsifiers.

Cationic Emulsifiers

Quaternary Ammonium Surfactants. Cationic emulsifiers are not used as food additives. Some of them, such as the quaternary ammonium compound cetyltrimethylammonium bromide (CTAB) (Fig. 9), have bactericidal properties and are used for sanitation of food-processing equipment. The structure is shown to indicate the chemical nature of this group of surfactants and to complete the description of classes of emulsifiers and surfactants.

USES OF EMULSIFIERS

The regulatory status of these food additives falls into one of three groups. The first group, Section 172 additives, includes those additives that have been examined and judged to be allowable for certain uses at concentrations not to exceed a set level. As an example, stearyl monoglyceridyl citrate (21 CFR 172.755) is allowed for use to stabilize shortenings containing emulsifiers. To use this additive as a wetting agent in a dry gelatin dessert mix,

$$CH_3(CH_2)_{15}-N^+-(CH_3)_3Br^-$$
cetyltrimethylammonium bromide (CTAB)

Figure 9. A cationic emulsifier, a bromide salt of a quarternary amine compound.

the manufacturer would have to submit a petition to the Food and Drug Administration (FDA) seeking clearance for this application and could not sell it for this purpose until such clearance was given. Several of the additives listed in Section 172 are cleared for general use in foods, and therefore may be used in any food where they are functional, at a level not to exceed that amount that produces the desired functional effects.

Sections 182 and 184 of 21 CFR deal with additives that are generally recognized as safe (GRAS) and thus can be used in any food. In some cases the regulation allows use at any level that is consistent with good manufacturing practices, while for other additives a tolerance or maximum use level is established. Section 182 (Substances Generally Recognized as Safe) lists additives that are recognized as safe based on a history of use without reported safety problems. If an additive undergoes a specific detailed evaluation and its safety is affirmed, it is then moved to Section 184 (Direct Food Substances Affirmed as Generally Recognized as Safe).

Table 2 lists the emulsifiers approved by the FDA for food use, along with the principal uses for which they are approved (as of April 1, 1988). Not all uses are listed in the table, because many approvals are for unique applications (eg, Polysorbate 80 is approved as a wetting agent in scald water for poultry defeathering). The use categories reflect the main food areas in which emulsifiers are used.

Many emulsifiers have functions that are not directly related to emulsification. For example, monoglycerides are widely used to retard firming (staling) of bread and similar baked goods. In this use the monoglyceride forms a complex with gelatinized starch in the bread crumb decreasing the rate at which recrystallizes. Lecithin is added to melted chocolate intended for coating confectionery goods. The lecithin decreases the viscosity of the chocolate, improving coverage of the confectionery piece at lower use levels of the chocolate. It is not clear that the mechanism of this functionality involves changes in interfacial free energy, ie, classic emulsifier action. Some emulsifiers (succistearin and polyglycerol esters) are used in oils to inhibit or modify fat crystal formation when the oils are cooled. The emulsifier deposits on the surface of the fat microcrystals (nuclei) and interferes with normal crystal growth.

CONCLUSION

Amphiphilic compounds (surfactants and emulsifiers) promote the formation of emulsions by lowering the interfacial free energy (surface tension γ) between the two phases, thus facilitating the subdivision of the internal phase into droplets of smaller diameter. They may be nonionic (monoglyceride), anionic (sodium lauryl sulfate), amphoteric (phosphatidyl choline), or cationic (cetyltrimethylammonium bromide); cationic surfactants are not used as food additives.

Emulsifiers stabilize emulsions by preventing the contact between the droplets that would lead to coalescence. This may occur by three mechanisms: (1) establishing electrostatic charges on the droplet surfaces, (2) generating a layer of bound water (in oil-in-water emulsions) around the surface of the droplet, and (3) forming a solid film at the interface.

Materials that stabilize foams usually perform both functions. They lower the surface tension between the liquid and air phases and also stabilize the thin liquid film between the occluded air bubbles. For food purposes, effective foaming agents are usually proteins (egg white and gelatin) but certain whipping agents can enhance this functionality.

Compounds that act as thickeners or bodying agents function in one of two ways. They may increase the viscosity of the aqueous phase or they may form a gel network within that phase. Viscosity builders are usually gums or modified starches, while gelling agents can be gums, modified starches, or proteins. Some solid materials, eg, microcrystalline cellulose, increase the viscosity in a formulated product by adsorbing water and thus increasing the ratio of internal phase (solids, oil) to external phase (mobile water).

BIBLIOGRAPHY

1. J. C. Wootton, N. B. Howard, J. B. Martin, D. E. McOsker, and J. Holme, "The Role of Emulsifiers in the Incorporation of Air into Layer Cake Batter Systems," *Cereal Chemistry* **44**, 333–343 (1967).
2. C. E. Stauffer, "The Interfacial Properties of Some Propylene Glycol Monoesters," *Journal of Colloid and Interface Science* **27**, 625–633 (1968).
3. H. Birnbaum, "The Monoglycerides: Manufacture, Concentration, Derivatives and Applications," *Bakers Digest* **55**, 6–18 (1981).
4. W. H. Knightly, "Surfactants in Baked Foods: Current Practice and Future Trends," *Cereal Foods World* **33**, 405–412 (1988).
5. D. T. Rusch, "Emulsifiers: Uses in Cereal and Bakery Foods," *Cereal Foods World* **26**, 111–115 (1981).
6. American Association of Cereal Chemists, *Approved Methods of the American Association of Cereal Chemists,* 8th ed., AACC, St. Paul, Minn., 1983, Method 58–45.
7. D. I. Hartnett, "Cake Shortenings," *Journal of the American Oil Chemists' Society* **54**, 557–560 (1977).
8. V. K. Babayan, "Polyglycerol Esters: Unique Additives for the Bakery Industry," *Cereal Foods World* **27**, 510–512 (1982).
9. C. E. Walker, "Food Applications of Sucrose Esters," *Cereal Foods World* **29**, 286–289 (1984).
10. B. A. Brachfeld, J. J. Geminder, and C. P. Hetzel, "Sodium Stearyl Fumarate: A New Dough Improver," *Bakers Digest* **40**, 53–58, 86 (1966).
11. J. B. Lauridsen, "Food Surfactants, Their Structure and Polymorphism," paper presented at the AOCS Short Course on Physical Chemistry of Fats and Oils, May 11–14, Hawaii, 1986.
12. W. J. Wolf and D. J. Sessa, "Lecithin," in M. S. Peterson and A. H. Johnson, eds., *Encyclopedia of Food Technology and Food Science,* AVI Publishing Co., Inc., Westport, Conn., 1978.
13. C. R. Scholfield, "Composition of Soybean Lecithin," *Journal of the American Oil Chemists' Society* **58**, 889–892 (1981).
14. B. F. Szuhaj, "Lecithin Production and Utilization," *Journal of the American Oil Chemists' Society* **60**, 258A–261A (1983).
15. W. Van Nieuwenhuyzen, "Lecithin Production and Properties," *Journal of the American Oil Chemists' Society* **53**, 425–427 (1976).
16. J. Stanley, "Production and Utilization of Lecithin," in K. S. Markley, ed., *Soybeans and Soybean Products,* Interscience Publishers, New York, 1951.
17. O. L. Brekke, "Oil Degumming and Soybean Lecithin," in D. R. Erickson, E. H. Pryde, O. L. Brekke, T. L. Mounts, and R. A. Falb, eds., *Handbook of Soy Oil Processing and Utilization,* American Soybean Association, St. Louis, Mo., 1980.

CLYDE E. STAUFFER
Cincinnati, Ohio

EMULSIFIER TECHNOLOGY IN FOODS

Food emulsion preparation has been practiced in the culinary arts for centuries. Food emulsions prepared for use in the home or restaurant a century ago were relatively crude by today's standards. Such problems as long-time storage, freeze–thaw stability, rapid dispersion, and wetting of powders or maximum rate and desirable degree of aeration of a foam did not concern the chef of 100 years ago. A sauce needed an emulsion stability measured in terms of hours, not months, as would be the case with a commercial preparation sold today.

The most commonly used emulsifying agent used in the culinary arts was egg. Egg yolks, of course, are rich in phospholipids (lecithin) and lipoprotein, naturally occurring emulsifiers. Mayonnaise is the prime example of an egg-stabilized food emulsion.

A food emulsion may be described as a multiple-phase system containing normally immiscible components such as oil and water, and usually other components, including carbohydrates and protein. There are two primary food emulsion types—oil in water and water in oil. WIth an oil-in-water type emulsion, such as milk or ice cream, the fat or oil is known as the disperse, discontinuous, or internal phase. The surrounding aqueous serum phase is known as the continuous or external phase. With butter or margarine, the reverse is true. The aqueous phase dispersed in the fat is the discontinuous or internal phase, whereas the fat is the continuous external phase.

The disperse phase may also be a gas. Dual emulsions are common, wherein an aqueous phase and/or a gas phase may be included within a fat globule, which itself is emulsified into an aqueous phase. A cake batter is a good example of this type of emulsion.

Generally, the emulsion viscosity will be a function of the external phase and is usually low when the external phase is aqueous, as in the case of milk, and high when the external phase is a fat or oil, as would be the case with margarine. However, viscosity will also be a function of the density of packing of the internal phase, affecting its ability to flow. As the internal phase becomes more closely packed, the mobility of the emulsion is greatly reduced. Mayonnaise is an example of an oil-in-water emulsion that is quite viscous, exhibiting little or no flow properties, leading the casual observer to believe it to be a water-in-oil emulsion. The internal oil phase of mayonnaise occupies greater than 70% of its volume. Theoretically, the internal phase can occupy only 74% of the total emulsion volume, if of uniform spherical shape. However, if the internal phase is capable of distortion so that the particles are not spherical, more internal phase may be compressed into the emulsion. The globule shape and size and the presence of a surface electrical charge can contribute greatly to emulsion viscosity.

If the internal phase ratio is low, colloidal stabilizers such as starch, alginate, cellulose, and sugar may be added to the external aqueous phase to increase emulsion viscosity. Reduction of internal-phase particle size, as with the homogenization of milk, will increase viscosity slightly because of the dramatic increase in the internal phase surface area. As an example, a fat globule of unhomogenized milk averaging 8 μm diameter is converted to 512 globules of 1 μm diameter after homogenization with an eight fold increase in surface area.

Most commercial food emulsion contain fat or oil globules with diameters of 0.1–2.5 μm. Most dairy-type emulsions produced commercially today, such as frozen desserts, possess fat globules averaging less than 0.5 μm diameter because of the availability of very efficient homogenizers.

The size of the fat or oil globule in an oil-in-water type emulsion will affect the appearance of the emulsion. In fact, the size of the fat globule in a simple oil-in-water emulsion can be estimated from the appearance of the emulsion, as given in Table 1.

Oil globules tend to coalesce after mixing in an aqueous phase, becoming larger and eventually rising to form a layer on the surface. The rise of the oil phase through the aqueous phase is in accordance with Stokes' law:

$$V = 2\,g\,r^2\,\frac{(d_1 - d_2)}{9\eta}$$

Where V = the velocity of the sphere, r the radius of the sphere, η (eta) the viscosity of the liquid phase, d_2 the density of the liquid phase, d_1 the density of the sphere, and g the force of gravity. Obviously, the larger the oil globules as a result of coalescence, the more rapid the rate of separation will be.

Oil tends to coalesce as the finely divided globules have provided increased surface area with concurrent increase in free surface energy. The system, of course, will tend to that condition with the lowest free energy. It is the function of the surfactant to reduce free energy (surface tension) and to encapsulate the oil by adsorption on the globule surface. The surfactant may also impart an electrical charge to the globule surface if ionic, as would be the case with lecithin. The presence of an electrical charge on the globule further aids in inhibiting coalescence through repellency.

Because of modern food processing techniques, longtime shelf life demands, and the rigors of long-distance distribution, it has been found desirable to add emulsifiers (surface-active agents) to food emulsions to contribute to their stability. Food emulsifiers are esters of fatty acids derived from fats and oils. They may be produced by direct esterification of the fatty acid with a polyhydric alcohol such as glycerine or more likely by glycerolysis of a fat or oil.

Emulsifiers are characterized by the fact that they possess both a lipophilic and a hydrophilic moiety. As a result of this structure, they will orient themselves at oil–water or gas–water interfaces, thereby reducing surface energy.

The alcohol moiety of the molecule provides the water-loving hydroxyl groups; typically propylene glycol, glycerol, sorbitol, and sucrose are used for purposes of esterification with the fatty acid. The functional properties of the surfactant can be further modified by reaction with ethylene oxide to make the ester more hydrophilic or by further esterification with an organic acid such as acetic, citric, or lactic.

The most commonly used surfactants in the production of food emulsions are the glyceryl monoesters such as glyceryl monostearate. As seen in Figure 1, it possesses both the fat-loving hydrocarbon chain, stearic acid, and the two remaining water-loving hydroxyls of the glycerol.

When the surfactant is oriented at the oil–water interface, it acts to reduce interfacial tension, as seen in Figure 2. The reduction in interfacial tension is in large part a function of the hydrophilic character of the emulsifier; the more hydrophilic the emulsifier is, the greater will be the reduction in interfacial tension.

The primary function of an emulsifier in forming a stable emulsion is, therefore, threefold:

1. Reduction of interfacial tension.
2. Formation of a cohesive film encapsulating the fat or oil globule.
3. Imparting an electrical charge to the globule surface.

Although these three characteristics of an emulsifier facilitate the formation of a stable emulsion, surface-active agents have other valuable physical and chemical attributes useful in food technology. The additional functions of surfactants are improving the texture and shelf life of starch-containing products by complex formation

Table 1. Estimation of the Globule Size[a]

Globule Size, μm	Appearance
> 1	Milky white
0.1–1.0	Blue white
0.05–0.1	Gray, semitransparent
< 0.05	Transparent

[a] Ref. 1.

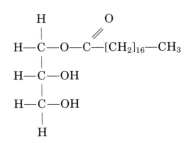

Figure 1. Glyceryl monostearate.

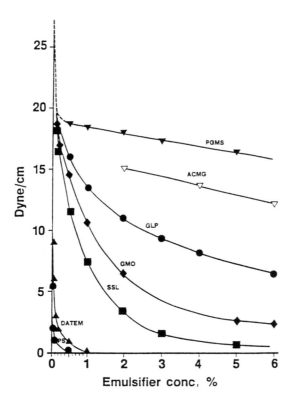

Figure 2. Interfacial tension between soybean oil (SBO) and water at 50°C. Measured by the Du Nouy ring technique. Following emulsifiers are tested: distilled propylene glycol monostearate (PGMS), acetylated monoglycerides, acetylation degree: 0.7 (ACMG), lactylated monoglycerides (GLP), distilled monoglycerides from sunflower oil (GMO), sodium stearoly-2-lactylate (SSL), diacetyl tartaric acid ester of monoglycerides (DATEM), and polysorbate 60 (PS) (2).

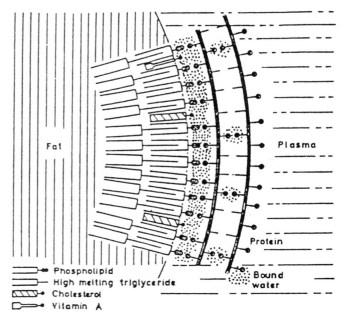

Figure 3. Fat globule surface (3).

with starch components; modification of rheological properties of wheat doughs by interaction with gluten proteins, and improving the consistency and texture of fat-based products by controlling polymorphism and crystal structure of fats (2).

The reduction of surface tension improves the ease with which oil is mixed with water. Surface tension is reduced, under ideal conditions, by a factor of 20–25 with the use of a surfactant requiring less energy input, as in stirring to break the oil phase into fine particles for dispersion throughout the aqueous phase. However, when one realizes that surface energy is increased by a factor of 10^6 when the oil phase is finely emulsified, it is apparent that the reduction of surface tension alone is insufficient to obtain a stable emulsion.

The formation of a rigid film around the oil globule by absorption of a surfactant onto the globule surface makes a major contribution to emulsion stability (Fig. 3). The so-called α-tending surfactants such as propylene glycol monostearate and lactated monoglycerides are particularly effective encapsulating agents. Such surfactants are nonpolymorphic and exist in the crystalline state only in the highly oil-soluble α-form.

When the surfactant is absorbed onto the globule surface, the fat-loving hydrocarbon chain is oriented into the oil with the water-loving moiety, eg, the OH group, at the surface interface. The fat-globule surface then takes on a dual change in character. The surface is now relatively more hydrophilic, with a vastly reduced tendency to coalesce with adjacent oil globules. The tendency to coalescence is further reduced when the fatty acids of the lipophilic moiety are high melting, eg, stearic or palmitic acids, because the surface of the fat globule takes on a melting point approaching that of the encapsulating surfactant.

If the surfactant imparts an electrical charge to the surface of the oil globule, it further aids in contributing to emulsion stability because the likely charged globules tend to repel each other, further reducing the opportunity for coalescence. In addition, repellency aids in showing the separation of the phases (migration of the internal phase). The presence of a surface electrical charge improves the stability of an emulsion when frozen because the emulsified globules are less likely to coalesce, owing to compression and rupture of the surface film when concentrated into the less unfrozen aqueous phase.

The surfactants added to improve food emulsions are generally nonionic and do not impart an electrical charge to the globule surface. Lecithin is an exception. Lecithin, with its component phosphatides, although amphoteric, carries with it a predominantly negative electrical charge; it is relatively anionic.

The relative propensity of a surfactant for oil or water is known as its hydrophile–lipophile balance or HLB for short. The HLB value may be expressed as the weight percent of the hydrophilic moiety of the surfactant molecule divided by 5 to reduce the magnitude of the numbers with which we work. For example, if 50% of the molecular weight is hydrophilic, the HLB value would be 50 ÷ 5 or 10. The HLB values of some surfactants used in food emulsification are given in Table 2. The surfactants with lower

Table 2. Calculated or Determined HLB Number of Some Surfactants Used in Foods

Chemical Name	HLB Number
Glycerol monooleate	2.8
Propylene glycol monostearate	3.4
Glycerol monostearate	2.8
Lecithin	4.2
Sorbitan monostearate	4.7
Glycerol monostearate, self-emulsifying	5.5
Polyoxyethylene (20) sorbitan tristearate	10.5
Polyoxyethylene (20) sorbitan monostearate	14.9
Polyoxyethylene (20) sorbitan monooleate	15.0
Polyoxyethylene (40) stearate	16.9
Sodium lauryl sulfate	~40

HLB values tend to induce the formation of water-in-oil emulsions, such as margarine, those with higher values (>10) tend to induce the formation of oil-in-water emulsions. Some relationships between HLB value and the application of the surfactant to food emulsion are given in Table 3. It should be remembered that food emulsions usually contain a significant quantity of naturally occurring surfactants such as phospholipids; the added emulsifiers are merely supplementing the activity of the naturally occurring ones. HLB may be estimated from the solubility of the surfactant, as seen in Table 4.

The first known significant commercial use of surfactants in foods was the use of lecithin to control the viscosity of molten chocolate in the early 1930s (4). Lecithin was found to form a monolayer over the nonfat constituents of chocolate (cocoa, sugar, milk) with the lipophilic moiety extending into the cocoa butter vehicle, presenting less frictional drag to the flow of the molten fat (5). Less-expensive cocoa butter was required to obtain any given viscosity reduction of molten chocolate.

The incorporation of sorbitan monostearate in chocolate inhibited the formation of bloom on chocolate (6). Bloom is the result of the migration to the surface of chocolate of unstable polymorphs of cocoa butter. As the polymorphs of fat migrate, they leave behind the cocoa fibers, which impart color and ultimately resolidify on the surface of chocolate as light-colored blotches. Sorbitan monostearate forms a monolayer of chocolate nonfat solids, impeding the capillary migration of the unstable cocoa fat polymorph to the surface, thus, inhibiting bloom (7).

Around 1933, the so-called superglycerinated shortenings were introduced in the United States for use in cake batters (8,9). These contained a significant amount of

Table 3. Relationship between HLB and Surfactant End Use

HLB Number	Application
4–6	Emulsifiers for w/o systems
7–9	Wetting agents
8–18	Emulsifiers for o/w systems
13–15	Detergents
15–18	Solubilizers

Table 4. Estimation of HLB by Water Solubility

Action in Water	HLB Range
Not dispersible	1–4
Poor dispersibility	3–8
Milky dispersion after vigorous agitation	6–8
Stable milky dispersion	8–10
Translucent to clear dispersion	10–13
Clear	13+

monoglyceride, eg, 3%, formed in situ by alcoholysis of the fat with glycerine during refining.

Shortenings with added surface-active monoglycerides were found to impart greater structural stability to the cakes, allowing for the inclusion of higher ratios of sugar to flour. Cakes with improved texture, volume, and symmetry, as well as keeping quality, resulted from the use of such shortening.

In 1968, the following definition for cake was proposed (10): Cake is a protein foam stabilized with gelatinized wheat starch and containing fat, emulsifiers, mineral salts, and flour and aerated principally by gases evolved by chemical reaction in situ.

It is well-known that oil is an antifoam that tends to destroy the foam structure of cake by weakening the protein film. Plastic shortenings are composed principally of an oil fraction—usually 70–75%, blended with solid fats for solidity at room temperature. One would expect the shortening to act as an antifoam, and indeed it does, unless properly encapsulated with an emulsifier. The α-tending surfactants such as propylene glycol monostearate and glyceryl lactopalmatate are exceptionally effective emulsifiers, especially when liquid oil is used as the shortening, because of their ability to form an α crystalline membrane around the oil globules, preventing the oil from coming into direct contact with the protein lamellae (2). Monoglycerides are also very effective emulsifiers when added to the batter in the α crystalline dispersible hydrate form. The emulsifiers may be added either directly to the batter in hydrate form or dispersed in the shortening.

Hydrophilic surfactants are added to cake batters to improve the dispersibility of the shortening in the aqueous phase of the batter and to promote aeration of the batter. Naturally occurring surfactants such as the phospholipids from egg and other ingredients also aid in emulsification of the batter.

As early as 1925, it was confirmed that the addition of eggs improved the whipping ability of ice cream mixes (11). It was assumed that the albumin fraction of egg was the aerating agent. In 1928, it was demonstrated that egg albumin had no beneficial effect on the aerating quality of ice cream mix and that improved whipping must be a function of the yolk (12). The emulsification properties of egg yolk were shown to be due to the phospholipid and lipoprotein content of the yolk (13,14). The use of egg yolk in ice cream mixes was demonstrated to increase the negative electrical charge on the milk fat globule (15).

In 1936, a patent was granted on the use of glyceryl monostearate as an emulsifier and whipping aid for ice

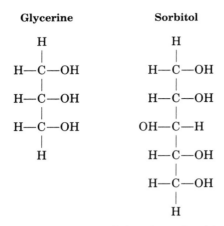

Figure 4. Structures of glycerine and sorbital.

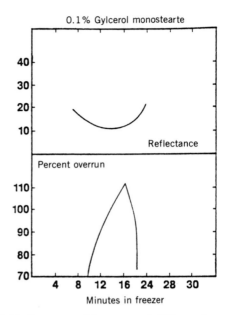

Figure 6. Performance of ice cream stabilizer, glycerol monostearate.

cream (16). It was found that the monoester was as effective at 0.1–0.2% as egg yolk was at 0.5%. It was also found that the use of the monoester provided an ice cream that was drier and stiffer on freezing and thus was extruded and packaged with greater facility.

With the development of a shortage of glycerine during World War II, a substitute for glyceryl monostearate was sought. It was found that fatty acid partial esters of sorbitol, a polyhydric alcohol could be substituted. Not only were the sorbitol esters equivalent, they were found to be slightly superior. Sorbitol differs from glycerine in that it contains six hydroxyls as opposed to the three in glycerine and is, therefore, somewhat more hydrophilic (Fig. 4).

In a search for more hydrophilic surfactants, it was found that ethoxylation further improved hydrophilic character and functionality (Fig. 5). Ethoxylated surfactants were found to be on the order of twice as effective as ice cream emulsifiers as were the lipophilic surfactants (Fig. 6,7). Most hydrophilic ethoxylated surfactants approved for use in foods contain an average of 20 moles of ethylene oxide.

In the production of a whippable emulsion, such as ice cream or whipped toppings, it is desirable to obtain good volume and stiffness. Contrary to published hypothesis, it was demonstrated that a partial breakdown of the emulsion during whipping to produce agglomerated fat globules provides maximum product stiffness (17,18). It was found that the more hydrophilic ethoxylated surfactants based on saturated fatty acids such as stearic provided maximum whippability, whereas those based on unsaturated fatty acids such as oleic provided maximum dryness and stiffness to the aerated product. A turbidimetric procedure was used to prove that surfactants actually accelerated the breakdown of the ice cream mix to provide agglomerated fat globules (Fig. 8) (17).

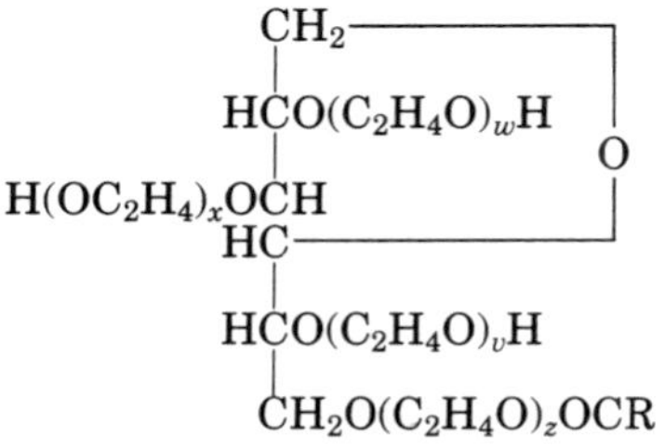

Combining a lipophilic saturated surfactant such as glyceryl monostearate with a hydrophilic unsaturated surfactant such as ethoxylated sorbitan monooleate resulted in a synergistic effect in frozen aerated desserts whereby the combination produced improved whippability and stiffness (19).

Lipophilic, α-tending surfactants such as glyceryl lacto-palmatate and propylene glycol monostearate are especially effective in promoting agglomeration of fat

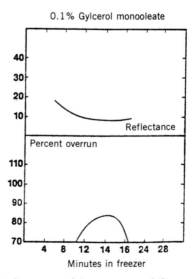

Figure 7. Performance of ice cream stabilizer, glycerol monooleate.

$$CH_2-$$
$$HCO(C_2H_4O)_wH$$
$$H(OC_2H_4)_xOCH \quad O$$
$$HC-$$
$$HCO(C_2H_4O)_yH$$
$$CH_2O(C_2H_4O)_zOCR$$

Figure 5. Polysorbate 60 ethoxylated sorbitan monostearate.

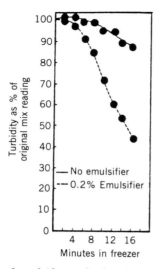

Figure 8. Effect of emulsifier on fat destabiliation in the freezer.

globules (2). The α-tending surfactants are absorbed onto the globule surface, interfering with strong protein bonding and making the protective protein film more easily swept off the surface during whipping, resulting in increased agglomeration of globules.

Margarine is probably the simulated dairy product most widely consumed. It is a water-in-oil emulsion usually containing 80% fat and 20% aqueous phase. 0.5–1.0% of a lipophilic emulsifier system are used to stabilize the emulsion, preventing leakage of the aqueous phase (weeping) and reducing spattering when the margarine is used for frying. Smaller water droplets do not explode into steam with as much force when exposed to frying temperatures.

Another category of simulated dairy products with a large market are milk replacers for young animals. Such emulsions are usually spray-dried in such a manner as to retain the emulsion potential intact, and the emulsion reforms instantly on rehydration.

The largest consumption of surfactants within the food industry is by the baking industry. Aside from the use of surfactants for cake-batter emulsification previously discussed, surfactants are used in years-raised products such as bread and rolls for the purpose of reducing the rate of staling and as dough strengtheners (conditioners).

As long ago as 1852, the French chemist Boussingault (20) proved that bread did not stale by reason of moisture loss, simply by hermetically sealing it in cans, where the crumb firmed as rapidly as that of unsealed loaves. It is now generally conceded that bread crumb firms primarily owing to the retrogradation of the gelatinized wheat starch. The linear amylose fraction (Fig. 9) retrogrades almost immediately after the baked product cools, whereas the branched amylopectin fraction (Fig. 10). retrogrades slowly over a period of days.

The ability of certain surfactants to retard crumb firming is due to the fact that they are able to form complexes with amylose and amylopectin. It is generally accepted that the starch fractions will form a helix around the hydrocarbon chain of surfactants possessing the proper steric configuration (Fig. 11). The inside of the helix is lipophilic owing to the C–H groups, and the exterior is hydrophilic owing to the presence of OH groups. Straight hydrocarbon chains such as stearic acid can accommodate the core of the helix with its diameter of 4.5–6.0 Å. On the other hand, unsaturated fatty acids of the cis configuration cannot enter the helix as they are not straight.

The complexing ability of a number of surfactants have been evaluated (21); some of these are given in Table 5. A good correlation has been established between amylose complexing ability and the effectiveness of a surfactant in retarding crumb firming (22). Apparently, the starch components are not as prone to retrograde, owing to interference with hydrogen bonding when in the helical clathrate form. For complex formation to take place, the surfactant must possess not only the proper steric configuration but must be dispersible on a molecular basis in the aqueous phase of the dough or batter.

Surfactants such as monoglycerides are used because of their starch-complexing ability in other foods containing starch such as pasta, instant potato products, and starch-based desserts. By complexing with the starch, surfactants improve the texture, and cohesive strength and prevent stickiness in cooked pasta; they also, for the same reason, prevent lumpiness and stickiness in rehydrated instant potatoes and starch-based desserts, which would normally occur with the presence of free amylose.

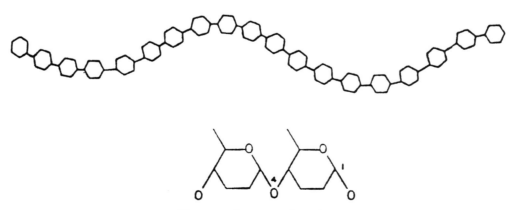

Figure 9. Structure of amylose.

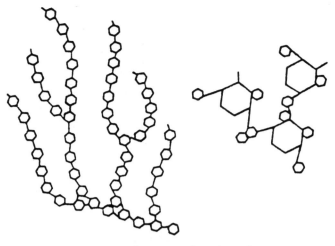

Figure 10. Structure of amylopectin.

Certain surfactants that are either anionic or ethoxylated have the ability to react with proteins such as gluten, thereby improving the cohesive strength and film-forming ability of the protein. They are known as dough conditioners or dough strengtheners.

Because of the improved cohesive strength of the gluten, the dough has improved tolerance to mixing and increased tolerance to the addition of nonwheat proteins. These changes result in improved loaf volume and symmetry, better crumb texture and cell structure, and greater resistance to staling. The best known of these dough conditioners are the sodium and calcium salts of stearoyl lactylic acid and the ethoxylated glyceryl and sorbitan monostearates.

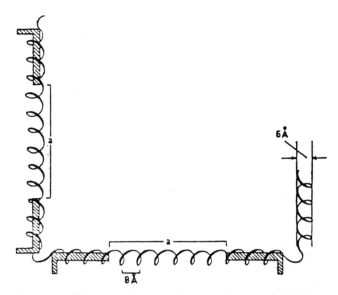

Figure 11. Schematic diagram of a glyceryl monopalmitate/amylose complex a, free space.

Surfactants are used to stabilize salad-dressing emulsions and to emulsify flavor oils. They are used in chewing gum and certain other confections such as caramel to prevent wrapper sticking and sticking to false dentures. Surfactants find use as extrusion aids in snack foods and meat analogues. Surfactants based on unsaturated fatty acids such as glyceryl monostearate are used as antifoams in sugar processing and confectionery manufacture. The market for surfactants in foods is now about 300,000,000 lb annually.

Table 5. Complex Formation between Amylose and Food Emulsifiers[a]

Materials	Amylose Complexing Index
Distilled Monoglycerides, 90–92% 1 Monoester	
Lard, hydrogenated	92
Lard, unhydrogenated, 45% monoolein	35
Soya bean oil, hydrogenated, 85% monostearin	87
Soya bean oil, unhydrogenated, 55% monolinolein	28
Mono-diglycerides, 45% monoester	28
Organic Acid Esters of Monoglycerides	
Lactic acid esters	22
Succinic acid esters	63
Diacetylated tartaric acid esters	49
Distilled propylene glycol monostearate, 90%	15
Sucrose monostearate (commercial sample)	26
Polyoxylethylene-sorbitan-(20)-monostearate (commercial sample)	32
Stearoyl-2-lactylate	79
No stearoyl-2-lactylate	72
Ca stearoyl-2-lactylate	65
Na stearyl-fumarate	67

[a] Ref. 21.

BIBLIOGRAPHY

1. N. J. Lynch and W. C. Griffin, "Food Emulsion," in K. J. Lissant, ed., *Emulsions and Emulsion Technology,* Marcel Dekker, New York, 1974, p. 249.
2. N. Krog, *Journal of the American Oil Chemists' Society* **54**(3), 124 (1977).
3. N. King, Commonev, Agr. Bureau, Farnham Roy, Bucksminster, UK, 1955.
4. Ger. Pat. 530187 (1930), Hansa Mahle.
5. W. Duck, *Pennsylvania Manufacturing Conference Progress Report* 10, November 1955.
6. N. Easton and co-workers, *Journal of Food Technology* **6**(11) (1952).
7. J. W. DuRoss and W. H. Knightly, *Proceedings of the 19th Annual Production Conference* Pennsylvania Manufacturing Conference Association (Sect. 15), 1965.
8. U.S. Pat. 2,132,393–398 (1938), N. S. Coeth and co-workers (to Procter and Gamble Co.).
9. U.S. Pat. 2,132,046 (1938), A. K. Epstein and B. R. Harris (to Procter and Gamble Co.).
10. W. H. Knightly, "Surface Active Lipids in Foods," *Society of Chemical Industry* **32**, 131–157 (1968).
11. H. F. DePew and S. W. Dyer, *Agriculture and Milk Plant Monthly* **14**(8), 87 (1925).
12. O. A. Pettec, *Ice Cream Manufacturing* **2**, 56 (1929).
13. C. C. Walts and C. D. Dahle, *Journal of Agricultural Research* **47**, 967–977 (1933).
14. C. D. Dahle and D. V. Josephson, *Ice Cream Review* **20**(6), 60 (1937).
15. G. C. North, Ph.D. Thesis, University of Wisconsin, 1931.
16. U.S. Pat. 2,065,398 (December 22, 1936), W. D. Roth et al.
17. P. G. Keeney, *Ice Cream Field* **72**(1), 20 (1958).
18. W. H. Knightly, *Ice Cream Trade Journal* **55**(6), 24 (1959).
19. U.S. Pat. 3,124,464 (March 10, 1964), W. H. Knightly and G. P. Lensach.
20. J. B. Boussingault, *Annals Chemistry and Physics* **36**, 470 (1852).
21. N. Krog, *Die Staerke* **23**(6), 206, (1971). 936.
22. N. Krog and B. N. Jensen, *Journal of Food Technology* **5**, 77 (1970).

William H. Knightly
Emulsion Technology, Inc.
Wilmington, Delaware

ENCAPSULATION TECHNIQUES

DEFINITION AND TECHNOLOGY DEVELOPMENT

Encapsulate is most commonly defined in its verb form as the act of enclosing or when used as a noun, it is synonymous with capsule. Microencapsulate applies to enclosure, or encasement in microcapsules.

Terms that are often used interchangeably with encapsulate are coated, or protected, eg, coated leavening, protected salt, and coated acids. All three terms are used when referring to practically any core material that is encased or enclosed in an outer shell. However, there are subtle distinctions that often dictate which term is more properly used. Very small capsules, perhaps 100 μ and less, are practically always called capsules or microcapsules. The adjectives coated or protected usually imply a cheaper, more crudely manufactured capsule. Coated or protected particles are generally larger than 100 μ.

In the late 1800s a pharmacist named Upjohn patented a process still referred to as pan coating (1). Relatively large solid particles were tumbled in a cylindrically shaped mixer as liquid coating material was sprayed on. Typically a sugar in water solution was the liquid used and the process involved alternately spraying and evaporating water, until through deposition of sugar, the rounded polished capsules achieved the desired size, shape, and shell thickness. The process was originally applied to drugs and pharmaceuticals for the purpose of taste masking—sugar coated pills—and is still being used extensively. Its application has expanded into several well-known brands of candies and confections.

Most texts establish pioneer work in microencapsulation as having been done in the 1930s by the National Cash Register Company (Dayton, Ohio) (2). Using a chemical process often termed coacervation, National Cash Register developed a means of occluding a colorless dye in a cross-linked gelatin shell. The particles were deposited (less than 20 μ in a cross section) in a thin layer on the underside of a sheet of paper in contact with a second sheet impregnated with a colorless reagent. Pressure exerted on the laminate by the point of a pen or pencil fractured the microencapsulate releasing the reagent-sensitive dye, producing a colored image. The process, which became commercial in 1954, produced the world's first carbonless duplicating paper. In 1981 production of carbonless copy business forms exceeded 500,000 tons per year (3,4).

The National Cash Register invention and the new coacervation process created interest during the mid-1950s in encapsulation in a wide variety of fields from agriculture to cosmetics and toiletries, from electronic parts to food processing (5). The coacervation process itself was examined and, for some applications, was found technically suitable and economical. For many others, of the myriad potential uses that were conceived, coacervation had technical limitations and was much too costly. However, the interest in coacervation spawned development of new encapsulation technologies; some of which have been found entirely suitable for food ingredient applications.

SPRAY DRYING

Spray drying is probably the oldest encapsulation technique and the most often used for preparation of dry, stable, food additives, particularly flavors. Spray-dried flavors have been available since the mid-1930s (6). The process is economical, flexible, and adaptable to commonly used, readily available processing equipment. Generally it is accomplished in three stages. A flavor, commonly an oil, although water-soluble flavors are also fixed using this technique, is mixed with an edible, food-grade polymeric material, such as gelatin, vegetable gum, modified starch, dextrin, or nongelling proteins often in the ratio of one part flavor to four parts fixatives (5). An emulsifier is added and the mixture is homogenized to produce an oil-in-water emulsion with a small micelle structure (7). The emulsion is atomized using one of several different methods, introducing an aerosol into a column of heated air in a drying chamber. The small droplets develop a spherical shape with the flavor oil gravitating to the center, or core and the aqueous–hydrocolloidal phase forming the outside coating. Rapid evaporation of water from the coating's surface maintains a temperature below 100°C (212°F) even in air columns of much higher temperatures. Fortunately the dwell time (exposure to heat in the chamber) is relatively short, usually only a few seconds (6).

Water-soluble flavors (or other food ingredients) are also fixed or entrapped in edible hydrocolloids but the particles do not have a clearly defined core and coating. Rather the spray-dried particle tends to be homogeneous.

Spray-dried particles are usually produced in particle sizes less than 100 μ, an optimum range for easy water solubility, or dispersability. Where necessary spray dried particles may readily be agglomerated or granulated for dry mixes where greater uniformity of particle size is desirable.

Before the introduction of spray drying, dry flavors were produced by plating out liquid flavors on dry substrates, typically sucrose, dextrose, starches, salt, or gelatin, depending principally on the composition of the mix being flavored. Although this crude method provided good flavor rendition and was obviously inexpensive (really nothing more than mixing), it provided no protection against loss of volatile components and oxygen-sensitive fractions were readily degraded (5).

With the advent of spray-dried flavors, oxygen degradation was significantly reduced as was loss of volatiles through evaporation. The colloidal, polymeric coatings provided at least some measurable protection. Some commercial products began to be referred to as locked-in flavors or protected. Even encapsulated began to be used to describe some spray-dried flavors in the early 1950s.

While spray drying offered a step forward in obtaining stability, the process created a problem in obtaining good flavor rendition (from wet to dry). Air, hot enough to instantly vaporize water, is going to drive off at least some low-boiling molecules, and large volumes of hot air will cause some instant oxidation particularly with sensitive terpene-rich citrus oils. Some answers to the problems associated with spray drying flavors came from the flavor chemist who began tailoring liquid flavors using less oxygen-sensitive molecules and fortifying liquids with excess volatiles to allow for rounding off during drying.

Spray drying remains the most common method for producing dry flavors. Some highly sophisticated flavor labs believe emphatically that they can accomplish anything with spray dehydration that can be done with more recently developed encapsulation techniques. However, the inadequacies of the process are generally recognized and it is the desire for superior dry flavors that have spurred the search for new encapsulation techniques and spread these concepts into many other food ingredient applications.

SOLVENT DEHYDRATION

This encapsulation method involves the preparation of a flavor emulsion or solution with the same polymeric hydrocolloidal coatings used in making spray-dried flavors. The emulsions are homogenized and atomized directly into polar solvents such as isopropenol, ethanol, glycerin, or polyglycols, which act to dehydrate the aerosol particles. Resulting microcapsules are recovered by filtration and vacuum dried at low temperatures (8). The technique offers the benefit of tighter protected flavor capsules, somewhat higher payloads than spray drying, and excellent flavor rendition. On the down side processing equipment is expensive and specialized and processing costs are significantly higher than with spray drying.

ENCAPSULATION BY EXTRUSION

Encapsulation of flavors by extrusion incorporated the basic techniques of solvent dehydration with a novel method for providing excellent protection to the core materials. The process involves the preparation of a low-moisture (5–10%) melt (100°–130°C) of low DE malto dextrin, sugar, and modified edible starch. The flavor to be encapsulated is combined along with an emulsifier and optionally an antioxidant, and mixed with a melt under extreme agitation. The molten emulsion is extruded through holes about 1/64 in. into a cold, isopropanol bath. The melt solidifies and is broken into small rods by agitation. The flavor particles are recovered by filtration, or centrifugation, often mixed with a flow agent and packaged for sale. The literature discusses many variations in the process. Some are designed to increase the payload (typically 8–10% flavor); often flavor capsules are given additional solvent washes to remove traces of surface oils further improving the exceptional stability this process affords to deterioration by oxidation. As such it is particularly useful with citrus-flavored oils (9).

Development of today's encapsulation by extrusion solvent dehydration began more than 30 years ago. It really comes from a process developed in 1956 involving adding flavor oils to a molten solution of sucrose and dextrose, cooling the solution to a hard slab, and subsequently grinding the rock-candy-like product to the desired particle size (10,11). From the development of pulverized rock-candy flavor to the current state-of-the-art flavor encapsulation by extrusion, significant processing and product improvements have been made by a number of inventors covered by at least six important patents. At one point there were three companies in the United States selling what was advertised as extruded flavor capsules. The volume, however, although significant never reached the size that the quality of these encapsulates would seem to merit. High process cost (perhaps as much as six times the cost of spray drying) is the major factor. Improvements in spray-drying technology, particularly improvements in formulating flavors designed for spray dehydration have narrowed the quality gap. When a dry flavor is required spray drying is still the technology of choice unless the need for the superior quality of extruded capsules is great enough to warrant the added costs.

AIR SUSPENSION COATING

Perhaps more commonly termed fluid-bed processing, this technology probably accounts for the second largest commercial production of encapsulates for the food industry—second only to spray dehydration. Solid particles to be spray coated are suspended in a column of moving air of a controlled temperature. The coating material, which may be molten or dissolved in an evaporable solvent, can be a wide range of water-soluble or insoluble polymeric material, including starches, emulsifiers, dextrins, protein derivatives, and lipids atomized through spray nozzles into the air chamber and deposited on the particles of suspended core material. The movement from the bottom of the chamber through the aerosol to the top of the chamber is random allowing for a rather uniform coating of core material. The constant flow of air allows the molten lipid coating material impinging on the particle to cool and harden, or in the case of a solubilized coating, the solvent to evaporate. The partially coated capsule falls to the bottom of the chamber and the cycle is continued until the parameters for complete or near complete encapsulation have been achieved. The finished product is removed from the chamber, cooled, or perhaps put through a final drying procedure prior to packaging (2,4,12).

Fluid-bed encapsulation has achieved outstanding success as applied to food ingredients, probably because of its flexibility in being suitable for many different core materials. For a wide variety of coatings, it is cost-effective for many ingredient applications, although it is not inexpensive. Credit for the development process in the 1950s is generally given to Wurster, who, while a professor of pharmacy at the University of Wisconsin, used the technique for coating pharmaceutical tablets (13). The Wurster process has become popular, aided perhaps by the commercial availability of suitable solid particle fluidizing equipment. There have been many modifications, mostly centering around different methods to dispense coatings to achieve improved encapsulation or special product characteristics. In principle, however, they all adhere to the basic concept of Wurster, spraying aerosol droplets to impinge on and coat a solid particle (12).

CENTRIFUGAL EXTRUSION

The Southwest Research Institute (SWRI) is responsible for developing several generations of a unique encapsulating concept extending back into the 1960s. Two tubes are arranged concentrically to receive a flow of core material through the center tube and coating material through the outer orifice (the liquids are pumpled). The tubes converge in a common orifice (nozzle) through which is extruded a liquid column within a liquid column. The SWRI equipment actually includes multiple nozzles mounted on a rotating shaft so that they spin around a verticle axis of the shaft. Centrifugal force acts to break off the liquid shaft into small particles as they emerge from the nozzles. The process relies on surface tension to form the spherical core and for the outer liquid to form a continuous coating around the core.

Particles are collected in either solids (starches) or liquid solvents, which tend to cushion the impact and protect the particles and may also perform additional functions. Starches absorb excess moisture and coatings. Gelatin coatings may be hardened or solubility reduced through emersion in solvents containing suitable cross-linking agents (2,14).

COACERVATION

Coacervation is a colloidal chemical phenomenon first used as a microencapsulation technique in the early 1940s by Green of the National Cash Register Corp. It is considered by many to be the first true microencapsulation process.

Simple coacervation involves dispersing a colloid, eg, gelatin, in an appropriate solvent, eg, water. A core material, eg, a hydrophobic citrus oil, is dispersed in the mixture with agitation. By one or more method, such as the addition of sodium sulfate and lowering the temperature, the solubility of gelatin in water is reduced, creating a two-phase system. The colloid-rich phase appears as an amorphous cloud in the colloid-poor, more aqueous phase. If left to stand, the minute droplets would coalesce forming a separate liquid layer. However, under proper conditions (and most literature refers to coacervation as an art as well as a science), the coalescence of the polymeric colloid occurs around the suspended core particles of citrus oils, creating small, still unstable microcapsules. Final steps in the process include adding a suitable cross-linking, or hardening, agent such as glutaraldehyde; adjusting the pH; and subsequent collecting, washing, and drying of the now-stable citrus oil encapsulate.

Coacervation as described is also referred to as aqueous-phase separation, or oil-in-water encapsulation. Complex coacervation is possible only at pH values below the isoelectric point of gelatin. It is at these pH values that gelatin becomes positively charged, but gum arabic continues to be negatively charged. A typical complex coacervation process begins with the suspension, or emulsification, of core material in either gelatin or gum arabic solution. Then the gelatin or gum arabic solution (whichever was not used to suspend the core material) is added to the system while mixing continuously. The pH is adjusted to 3.8–4.3, the system is cooled to 5°C (41°F) and the gelled coacervate capsule walls are insolubilized with glutaraldehyde, or another hardening agent. Microcapsules are collected, washed, and dried.

It is possible to microencapsulate hydrophilic core material in oil-soluble coatings. A polar core is dispersed in an organic, nonpolar solvent at an elevated temperature. The coating material is then dissolved in the solvent. Encapsulation is achieved by lowering the temperature and allowing the polymeric coating material to emerge as a separate coacervate phase microencapsulating the core particles. The coating gradually solidifies and remains insoluble in cold solvent. This process is called either water-in-oil microencapsulation, or organic-phase separation. It is used typically in the pharmaceutical industry for encapsulation with ethylcellulose. However, ethylcellulose does not have general approval for use in the food industry.

Coacervation is an efficient, but expensive, microencapsulation technique. When small particle sizes are required it is probably the only process that can produce submicron-size particles. With typical payloads in the range of 85–95%, it might be expected that the process would be economical to allow for its application in the food industry for myriad ingredients; such is not the case. Aside from a few specialized flavor applications there are no current uses for encapsulated food ingredients by the coacervation process; costs are much too high (2–4).

INCLUSION COMPLEXATION

B-Cyclodextrin is a cyclic glucose polymer consisting of seven glucopyranose units linked in a 1->4 position. The molecule is a doughnut-shaped structure with a hollow center, which enables it to form complexes with many flavors, colors, and vitamins. The B-cyclodextrin molecule forms inclusion complexes in the presence of water with compounds that can fit dimensionally within its cavity. Water molecules within the nonpolar center of the B-cyclodextrin molecules are readily substituted by less-polar guest molecules. The resulting complex precipitates from the solution and is recovered by filtration and a subsequent drying step. Core material ranges from 6 to 15% (w/w) (15). The odorless, crystalline complexes will tenaciously bind the guest molecules up to temperatures of 200°C (392°F) without decomposition. The same complexes easily release the bound molecule in the temperature and moisture conditions of the mouth (16).

Inclusion complexes are stable to oxygen, light, and radiation, and even onion and garlic oil complexes are almost odorless. Cyclodextrins—while approved for food applications in Eastern Europe, where they were developed, and in Japan—are not approved for use in foods in Western Europe, or in the United States (17,18).

SPRAY CHILLING

Material to be encapsulated, usually dry solids of a relatively small particle size, are suspended in molten coating material, most commonly fractionated and hydrogenated vegetable oils. The mixture is atomized through spray nozzles into a chamber of tempered air (usually refrigerated) where the droplets solidify into small spherical particles. Spray chilling is applied principally to solid food additives including vitamins and minerals, and also to spray-dried flavors to retard volatilization during thermal processing. Spray-chilled products are commonly used in dry mixes. The process is relatively inexpensive. Spray-chilled capsules provide a controlled release of core material. For systems in which a more positive barrier or shell is required, other encapsulation techniques may be preferable (3,19,20).

BASIC CRITERIA FOR ENCAPSULATING FOOD INGREDIENTS

Economics

When microencapsulation was in its infancy in the early 1970s and food technologists were searching for scraps of information about the concept from any source they could find, there was no lack of apparently good ideas for developing new encapsulated ingredients. Few were so foolish to ignore the cost of encapsulating, but most inventive minds operated on the basis that the real problem was to develop the technology, ie, to make the encapsulate and adapt it to the system. It was generally known that once a prototype is made, the reduction in costs is just a matter of time.

Since 1970 the development of encapsulation technology has been startling; however, the cost of encapsulating has not decreased, but has kept pace with inflation. Some relatively crude, lower—cost methods of encapsulation have been developed, but these have been limited in application where a poor encapsulate was good enough.

Most FDA-approved food ingredients have been encapsulated on a laboratory or small pilot-plant basis, but the real world exists outside the laboratory and there commercial viability of an encapsulate revolves around economics in its broadest sense.

Encapsulated ingredient projects begin with a need or a problem. Sometimes the problem is a matter of poor shelf life. More often a project begins with the concept of an encapsulate that would allow the development of an entirely new product that cannot be available without it. Whatever the need for encapsulation, the wisest first step is not to analyze the various methods of encapsulation that might work, but to look at the project, really a processed food system, in its entirety. Assume that the encapsulate is available and technically satisfactory in all respects. Estimate the cost of the encapsulate, ie, ingredient, coating, and both minimum and maximum processing costs.

Then make a judgment: Under the best technical and economic conditions should this project be taken to commercialization? The answer may be no, and for a variety of reasons having nothing to do with encapsulation. The maze of approvals and evaluations that are prerequisites for taking new processes or products to commercialization can be formidable. But assume that in the best-case scenario the answer is yes. It is then just as important to take a look at the worst case, ie, a product that is more costly than desirable and technical requirements that are not quite fulfilled. A definite yes answer in this case makes the project much more attractive, but a no answer, although perhaps disappointing, at least more clearly defines what is really necessary. Projects having long odds may be abandoned, avoiding an investment and a high-risk gamble.

The evaluation of projects prior to embarking on costly research and development is obviously necessary for all new ideas regardless of whether encapsulation is required. However, there is particular wisdom in carefully evaluating encapsulation projects. All too often problems are solved technically with an encapsulate that never becomes commercial for reasons that may have been obvious from the outset. This is particularly true of systems where low-cost food ingredients have been encapsulated. When a formulator must work with a $0.75/lb item that in its raw form costs only $0.10/lb, it is amazing how many other solutions to the problem can be developed in place of encapsulation.

Most encapsulation companies charge for development and preparation of encapsulated prototypes; not to make a profit or even to recover expenses, but primarily to provide that extra insurance that a project has been well thought out and will have some reasonable chance of becoming commercial. Once technical objectives are met, commitment to any financial investment requires consideration and approvals at a relatively high managerial level.

ENCAPSULATED INGREDIENTS—COMMERCIAL TYPES AND APPLICATIONS

The encapsulated food ingredients that are commonly discussed in the literature may represent just the tip of the iceberg. There are more than 20 companies advertising services in encapsulation, granulation, and agglomeration (21). Much of their work is done on a custom basis, and volumes and applications are unknown. There is also a significant amount of encapsulation being done by companies for captive use on a proprietary basis. However, there has developed since the late 1960s a major market for encapsulated ingredients, in addition to the dry flavor business, that by that time had already been established.

Flavors

In 1971 it was estimated that more than 15,000,000 lb of dry flavors were produced domestically (5). This figure has at least tripled. The primary reason for flavor encapsulation is still to provide a dry form for liquid materials, but as technology has advanced, encapsulated flavors have been developed to provide outstanding flavor rendition, excellent protection from degradation through oxidation, and reduced or controlled volatility to the extent that significant economies can be achieved by reducing high temperature losses through evaporation of expensive flavoring ingredients. Technology has been applied to flavors as diverse as citrus oils, menthol, oleoresins of spices and herbs, cinnamaldehyde, peppermint, natural and artificial cheese and seafood flavors, and many other (2,5).

Acidulants

Encapsulated acidulants were first offered commercially in the early 1970s. Initially dry, granular acids including fumaric, malic, citric, tartaric, and adipic were encapsulated to improve stability of acid-sensitive ingredients in dry mixes. By encapsulating acids with coatings such as hydrogenated vegetable oils, or malto dextrin, it was possible to reduce contact between acids and starches in puddings and pie fillings, thereby reducing potential prehydrolysis and significantly extending shelf life. An attendant benefit with encapsulated citric acid was that it reduced hygroscopicity. Core acids were released with addition of water or milk to a mix or, in the case of products employing hydrogenated vegetable oil coatings, on heating (2,5,22).

Encapsulated acids have been used to prevent undesirable discoloration of acid-sensitive food colors and dyes, degradation of aspartame, control of dusting, as means of regulating the gelling effect of acidulants on acid-sensitive pectins in fruit-based products, and alteration of bulk density. The most recent, and by now the largest application, for encapsulated acidulants is in the meat industry where the use of encapsulated citric acid, lactic acid, and glucono-*delta*-lactone, a gluconic acid precursor, provide a means of direct acidification of processed meat. In the manufacture of fermented sausage the industry has long relied on naturally occurring microorganisms or on *Lactobacillus* starter cultures whose activity on carbohydrates produces lactic acid, thereby lowering the pH to a level adequate to ensure protection against pathogenic microorganisms. The fermentation process can take as long as four to five days, and although it has been used for hundreds of years, it is not without problems, primarily the length of fermentation time and the variability from batch to batch. The potential for direct acidification was obvious, but not easily accomplished. Contact of ground meat with acid causes rapid separation of fat from protein, discoloration, and a change in texture from a soft uniform emulsion to hard, brittle, crumbly particles that are no longer usable. Use of encapsulated acidulants having hydrogenated vegetable oil coatings with appropriate melting points offer a mechanism for releasing acidulants into meat slowly and uniformly throughout the mass as temperature is elevated during cooking. The fermentation step is eliminated. The concept has been expanded to meat systems where acidification is possible to replace high-temperature retorting with short-term pasteurization (23–26). The desirability of using lactic acid as the acidulant by some technologists has extended the creativity of the encapsulators to include liquid lactic acid, which is first absorbed on solid calcium lactate particles prior to tight encapsulation in hydrogenated vegetable oil.

Dough-Conditioning Agents

Encapsulation technology has been applied to conditioners for protection and to control activity until the optimum point in the baking cycle is reached. Ascorbic acid, which is used as an oxidizing agent in strengthening and conditioning bread doughs, is unstable in the presence of water and degrades rapidly. Encapsulated ascorbic acid can be added to typical prefermenting broths with minor loss of potency, through the fermentation stage. During baking the hydrogenated vegetable oil coating melts, releasing the ascorbic acid for its conditioning–strengthening effect at the stage where it is most effective (27).

Leavening

Shelf life of chemically leavened doughs is restricted by reaction prior to proofing or baking. Dry, packaged mixes lose leavening activity particularly in high-moisture systems. The same is true of chemically leavened frozen doughs. Refrigerated and microwaveable frozen doughs offer unique leavening problems in that activity needs to be controlled, not simply prevented. Refrigerated doughs require some leavening initially when the dough is formed, but ideally no more reaction through the packaging stage. Once in the closed containers and held under suitable proofing conditions, rapid leavening is desirable to purge the seal and provide maximum can pressure. Microwaveable frozen products require initial protection from reaction, with rapid, intense leavening during the very brief time that a rise may occur within the microwaving process.

Encapsulated sodium bicarbonate with precise release characteristics in combination with selected leavening acids offers a commercial means of providing protection, and optimizing activity in a wide variety of baking systems. Encapsulation has also been applied to leavening acids, particularly glucono-*delta*-lactone for similar applications.

Protection of Yeast-Leavened Baked Goods

Sensitivity of yeast cells to a variety of ingredients used in baked products has created a highly specialized need for encapsulates. Even at low flavor levels acidulants, eg, citric, lactic, and acetic acids, will reduce the viability of yeast and may further affect the cellular structure of dough by reducing the total effect of yeast leavening. Preservatives such as sorbic acid and calcium proprionate, and even salt may have a similar killing effect on yeast as on the molds they are intended to inhibit. Citric, lactic, and acidic acids encapsulated in hydrogenated vegetable oil (HVO) are being used on a commercial basis to prevent the effect on yeast of low pH, as is HVO-coated sorbic acid. Other encapsulated additives that affect yeast negatively in the raw form have been produced and evaluated on a developmental basis with promising results. Potentially dried yeast cells themselves may one day be commercially available in an encapsulated form. Work has been done along this line, but apparently with limited success. The coating may be extremely useful in maintaining the viability of the cells, but encapsulation itself can have an inhibiting effect on yeast, and certainly the benefits of encapsulation would have to be major to warrant the added cost.

Vitamins and Minerals

A variety of dry mixes are fortified with vitamins and minerals including cereals, infant formulas, pet foods, and beverages. Encapsulation with both lipids and water soluble coatings are used to reduce off-flavors, permit controlled release in the digestive tract, improve flow properties, facilitate tableting, and enhance stability, particularly of vitamins, to extremes of moisture and temperature. Iron is encapsulated as are other minerals to control bulk density, reduce stratification in combination with lighter mix components, and to retard catalytic oxidative rancidity.

One use for an encapsulated vitamin provides such a major product improvement with an accompanying economic advantage that it merits special comment. The aquaculture industry has grown enormously in the last decade. The key to maintaining the health of pond-grown fish is in proper feeding. A number of nutrient supplements are added to fish rations, but one stands out prominently in any discussion of the applications for microencapsulation technology. Fish must ingest vitamin C in an adequate dosage for survival. Fish along with humans, apes, guinea pigs, and the fruit bat lack the enzymes required to synthesize ascorbic acid from protein consumed in their diet. Using raw ascorbic acid for feed fortification requires 5–10 times the level recommended in the normal diet of fish because oxygenated water is used in all three of the basic processes for producing fish feed, which rapidly reduces the potency of ascorbic acid. This reduction is accelerated with temperature and pressure elevation, which is required in all three processing techniques.

Encapsulation of vitamin C in a HVO coating prevents contact of the oxygen-sensitive ascorbic acid with water. The half-life of raw ascorbic acid in fish feed is measured in days. The half-life of the digestible, bioavailable, encapsulated version of vitamin C is measured in months (28–30).

Sodium Chloride

There has been a significant, although not major market, for coated salt as a food ingredient since the late 1960s, primarily in the meat industry where it is used in ground pork sausage where contact with raw salt will affect texture and turns the bright pink color to a dull bluish gray. Fat-coated salt, even when crudely applied, extends shelf life and the improvement is worth the additional cost at least to many manufacturers. As technology has improved so has the quality of HVO-encapsulated salt. Although still restricted from becoming a principal ingredient in the food industry because of the economics involved, applications have been extended to baked products to protect yeast and to retard the stiffening effect induced by raw salt through water absorption. Use in comminuted meat products has also increased to control changes in viscosity and texture produced by water absorption. Maintenance of a free-flowing viscosity of ground beef allows the cost of encapsulation to be offset by savings provided by increased throughput.

FUTURE DEVELOPMENT

A new and better technique to microencapsulate food ingredients may be developed in the next 10 years, and perhaps the new process will be more economical than the current state of the art. However, if the available technology stays the same, the market for encapsulates in the food industry will certainly keep growing at an accelerating rate. The benefits are significant.

The industry is becoming accustomed to what encapsulation can do and as the confidence level grows, so do the opportunities. What is perhaps equally important is knowing what the new technology can do and knowing what it can't do. Fortunately the food industry no longer looks at encapsulation as some far out technology that might be considered if all else fails. The industry has matured to the extent that products are tailored to fit specific applications. It is interesting to observe that with one common core ingredient—citric acid—one of the major ingredient producers offers 11 different grades, the other producer offers 8. Each one differs in variables that are fundamental to encapsulation, eg, payload, particle size, coatings, water solubility, or fat coatings. Other important differences include the coatings, melting points, origins, and costs. Each product is designed to be useful in a specific food ingredient system.

While commercialization of encapsulation technology covers many diverse fields within the broad term Food Industry, the majority of products and the greatest volume is focused on two segments, ie, meat processing and baking. It is interesting to observe that in both, encapsulation has been treated not as a tool to be tried and abandoned if it doesn't perform adequately the first time, but rather as a part of the answer to a food-processing problem, albeit an important part. Most initial attempts to solve problems with encapsulates are only partially successful. It is in the fine tuning, not only of the encapsulates to fit a system but, where possible, of the system's design to fit what can be achieved through encapsulation. The meat and baking industries have been doing this. Other industries such as the dairy and microwaveable entrée industry would seem to be candidates for a similar developmental approach.

A significant part of encapsulation technology is in the selection or development of optimum edible Food and Drug Administration approved coatings to enclose and protect and still release under the desired conditions. There is room for new coating materials, particularly ones that will have higher melting points or will remain intact under higher temperatures than what is currently available. Edible coatings having release points of ca 74°C (165°F) would be particularly interesting.

BIBLIOGRAPHY

1. R. J. Versic, "Flavor Encapsulation, An Overview," *ACS Symposium Flavor Encapsulation, Series 370,* ACS, Washington, D.C., 1988, Chap. 1, p. 5.
2. J. D. Dziezak, "Microencapsulation and Encapsulated Ingredients," *Food Technology* **42**(4), 137 (Apr. 1988).
3. R. J. Versic, "Coacervation for Flavor Encapsulation," in Ref. 1, Chap. 14, pp. 126–131.
4. R. E. Sparks, "Microencapsulation," in M. Grayson, ed., Kirk-Othmer, *Encyclopedia of Chemical Technology,* Vol. 15, 3rd ed., John Wiley & Sons, Inc., New York, 1981.
5. R. E. Graves, "Uses for Microencapsulation in Food Additives," *Cereal Science Today* **17**(4), 107, 1972.
6. G. A. Reineccius, "Spray-Drying of Food Flavors," in Ref 1, Chap. 7.
7. J. Brenner, "The Essence of Spray Dried Flavors: The State of the Art," *Perfumer and Flavorist* **8**, 40 (Apr.–May 1983).
8. R. Zilberboium, I. J. Kopelman, and Y. Talmon, "Microencapsulation by a Dehydrating Liquid: Retention of Paprika Oleoresin and Aromatic Esters," *Food Science* **1**(5), 1301 (1986).
9. Sara J. Risch, "Encapsulation of Flavors by Extrusion", in Ref. 1, pp. 103–109.
10. U.S. Pat. 2,809,895 (Oct. 15, 1957), H. E. Swisher (to Sunkist Growers, Inc.).
11. T. H. Schultz, K. P. Dimick, B. Makower, "Incorporation of Natural Fruit Flavors into Fruit Juice Powder," *Food Technology* **10**, 57–60 (1956).
12. David M. Jones, "Controlling Particle Size and Release Properties," in Ref. 1, pp. 158–176.
13. U.S. Pat. 2,648,609 (Aug. 11, 1953), D. E. Wurster (to Wisconsin Alumni Research Foundation).
14. R. E. Lyle, D. J. Mangold, and W. W. Harlow, "The Making of Microcapsules," *Technology Today,* **3**(3), 13–16 (Sept. 1904).
15. H. B. Heath and G. A. Reineccius, "Flavor Production," in *Flavor Chemistry and Technology,* AVI Publishing Co., Inc., Westport, Conn., 1986.
16. J. S. Pagington, "Molecular Encapsulation with B-Cyclodextrin Food Flavor Ingredients," *Process Packaging* **7**(9) 50 (Sept. 1985).
17. G. A. Reineccius and S. J. Risch, "Encapsulation of Artificial Flavors with B-Cyclodextrin Food Flavor Ingredients," *Process Packaging* **8**(9), 1 (Aug.–Sept. 1986).
18. L. Szente and J. Szejti, "Stabilization of Flavors by Cyclodextrins," in Ref. 1, Chap. 16, pp. 148–157.
19. J. A. Bakan, and J. S. Anderson, "Microencapsulation," in L. Lachman, H. A. Lieberman, and J. L. Kanig, eds., *The Theory and Practice of Industrial Pharmacy,* Lea & Febiger, Philadelphia, 1970, p. 384.
20. D. Blenford, "Fully Protected Food Flavor Ingredients," *Process Packaging* **8**(8), 43 (July 1986).
21. H. Weiss, president, Balchem Corp., unpublished data.
22. L. E. Werner, "Encapsulated Food Acids," *Cereal Food World* **25**(3), 102 (1980).
23. R. E. Graves, "Sausage Fermentation: New Ways to Control Acidulations of Meat," *National Provisioner* **198**(21), 32–34, 49 (May 1988).
24. J. Bacus, "Fermenting Meat, Part I," *Meat Processing* **24**(2), 26–31.
25. J. Bacus, "Fermenting Meat, Part II," *Meat Processing* **24**(3), 32–35 (Mar. 1986).
26. G. J. Jedlicka, "Chemical Acidulation of Semi-Dry Sausage." *paper presented at American Meat Institute Convention,* New Orleans, 1984.
27. U.S. Pat. 3,959,496 (May 25, 1976) S. Jackel and R. V. Diachuk (to Baker Research and Development Co.).
28. C. Andres, "Encapsulated Ingredients," *Food Processing* **38**(12), 44 (Nov. 1977).
29. L. A. Gorton, "Encapsulation Stabilizes Vitamin C in Cupcakes," *Baking Industry* **148**(1805), 20–21 (Jan. 1981).
30. C. J. Pacifico, "A Novel Approach to Vitamin C Nutrition for Fish Rations," *Aquaculture Today,* **2**(4), 20–22 (1989).

Bob Graves
Herb Weiss
Balchem Corporation
Slate Hill, New York

ENERGY

Definition

Energy has been referred to as "an abstraction which cannot be measured directly but only in its transformation from one of its forms of manifestation to another" (1). The chemical energy in food is converted through metabolism to electrical, mechanical, and altered chemical energy with approximately 40% efficiency. The remainder of the available energy is lost as heat. Although it is difficult to measure each of these components per se, the total potential energy of food is fairly easy to quantitate as heat released during complete oxidation. Likewise, the energy used by an organism can be measured by proportion to the amount of heat generated.

Therefore, energy in biological systems is most commonly expressed as a unit of heat, the calorie. One calorie is the amount of heat required to raise 1 g of water from 14.5 to 15.5°C. This is too small a unit to be generally useful to systems larger than mosquitoes; hence, the term calorie in common usage refers to kilocalories (heat required to raise 1 kg of water by 1°C), which may be denoted as kcal, Cal, or large calories.

An alternative measure is the kilojoule, which is a unit of mechanical energy, or work (1 kJ = 1 kg moved over a distance of 1 m with the force of 1 N). This is a logical choice for expressing the requirement for physical activity, which depends on weight carried over distance. However, these numbers are generally calculated from kcal equivalents (1 kcal = 4.18 kJ), which have in turn been calculated from oxygen consumption data. Most American publications have resisted a switch to kJ, although it is widely used in European literature. A review of the kcal/kJ debate may be found in Ref. (1).

Energy measurements of most interest to humans are (1) calories in foods that can become available for tissues for storage and metabolism; (2) energy amounts used for life support, work, and play; and (3) the balance between energy intake from food and usage by the body.

MEASUREMENT OF FOOD CALORIES

Calorie contents of foods are classically measured in a bomb calorimeter, which is a well-insulated vessel in which the dried sample can be completely burned in a pure oxygen atmosphere. The resultant heat is quantitated from the rise in temperature of the water-jacketed container after equilibrium is reached. Heat of combustion values are approximately 4.15 kcal/g for pure carbohydrate, 9.45 kcal/g for fat, 6.9 kcal/g for ethanol, and 5.65 kcal/g for protein. To be translated into available energy, these numbers are adjusted for coefficients of digestibility, as some types of proteins and carbohydrate are resistant to human digestive processes. Protein values are adjusted for the fact that combustion releases heat from amine bonding (1.25 kcal/g) that is not biologically available. The resulting numbers are called physiological fuel values, which are the well-known figures of 4.0 kcal/g of carbohydrate or protein, and 7.0 and 9.0 kcal/g for ethanol and fat, respectively (2). Substances such as cellulose and other dietary fibers also yield heat in a bomb calorimeter, but are not energy substrates for humans. These must be measured and subtracted from the kcal total for biological accuracy.

In practice, most kcal information has been determined by the Atwater analyses (2), which are used as a standard for nutrition labeling of food products. The noncaloric weight of the sample is determined: Total moisture is calculated by simple subtraction of dry from wet weight, and the mineral weight by ashing of the sample. Nitrogen content is determined, generally by Kjeldahl analysis, and multiplied by conversion factors. The average food protein contains 16% nitrogen, so the factor most often used is 6.25. These factors vary according to the amino acid composition of the food, being lower for those proteins with high proportions of amino acids with more than one nitrogen, such as arginine and lysine. Other nitrogenous compounds such as urea, nucleic acids, and polyamines will be measured as protein using this procedure (sometimes called crude protein determination), but the differences are regarded as negligible, at least for calorie determination.

Fat content is most commonly determined by ether extraction, which quantitates all fat-soluble substances. For some foods, acid hydrolysis or saponification methods may give more precise values. Carbohydrates are estimated by subtraction of moisture, mineral, fat, and protein from the total sample weight. Fiber content may be subtracted as a nondigestible component, but this is not required by current food-labeling regulations (3). The calculated carbohydrate fraction, then, tends to be the least accurate of the reported components.

Total energy value in food is determined by multiplication using both the factors for coefficients of digestibility and the heats of combustion of the extracted components (2). Atwater compared energy contents determined by bomb calorimetry with calculated values for 276 food samples and seldom found greater than 3–5% discrepancy. It is also permissible under FDA food labeling regulations to multiply carbohydrate, protein, and fat content by 4, 4, and 9 kcal/g, respectively, unless this abbreviated calculation yields a number more than 20% greater than Atwater values (3).

Components not technically classified as fat, carbohydrate, protein, or ethanol also have calories, but generally contribute a negligible portion of total energy content. These include lactic, acetic, and citric acids, and other intermediates of energy metabolism. Probably only in the case of lemons (citrate) and vinegar (acetate) do these contribute major proportions of available calories, about 15 kcal/100 g. In any case, these organic acids are measured as carbohydrate using the subtraction method of determination.

Although fiber components of the diet, by definition, are not digested by human enzymes, and so are not absorbed, they may be partially degraded by the intestinal flora. Microorganisms in the colon produce short-chain fatty acids from soluble fibers such as pectin and hemicellulose. These volatile fatty acids include propionate, acetate, and butyrate and have been shown to pass from the lumen into circulation, where they are used for energy production in the liver and other tissues. This process may contribute from 4 to 6% of total energy, depending on intake and source of soluble fiber (4).

CALORIE DENSITY

The number of calories in a given amount of food depends, of course, on composition. Foods high in water or dietary fiber (which tends to attract and hold water) will generally be low in calories per unit volume compared to foods that are concentrated in fat or carbohydrate. Thus, calorie density may range from more than 200 kcal/oz for pure fats, such as oil or butter; 160 kcal/oz for chocolate or nuts, concentrated mixtures of fat and carbohydrate; 100 kcal/oz for nearly pure carbohydrates, such as sugar or crackers; and approximately 50 kcal/oz for lean meats, which contain mostly water with 20% protein and < 10% fat. Fruits and vegetables may have less than 5 kcal/oz, owing to very high water content, and almost no fat.

ENERGY IN STORAGE TISSUES

After food energy has been digested and absorbed, it must be taken up by the tissues for metabolism. Dietary components not immediately used for energy production are converted into storage forms. The major energy reserve in animals is in the form of triglyceride, or neutral fat, which is stored in large lipid droplets in adipose tissue. The amount of energy stored in a pound of adipose tissue is approximately 3500 kcal. The typical adult man has 10–20% of his body weight as triglyceride; for women, the normal range is 18–28% of body weight. This means, in theory, that the average adult has at least a 30-day reserve of kcal as fat.

The body maintains a much more limited kcal supply of carbohydrate as glycogen, also known as animal starch, stored mainly in liver and muscle tissue. Tissue proteins can also be degraded and used as calories under some circumstances. Because of the hydrophyllic nature of carbohydrates and proteins, two to three times their weights in water are associated with stored glycogen or protein tissue. When a pound of lean, nonadipose tissue is used for energy, only about 700–800 kcal are provided, much less than the equivalent weight of adipose tissue. This is because pure carbohydrate and protein each yield about 4 kcal/g (vs 9 kcal/g for fat), and the associated water weight is liberated, and usually excreted, when glycogen or lean tissue protein is metabolized.

Most people have only a 300–400 g reserve of liver glycogen, which is the major supply of blood sugar (glucose) between meals. If a low carbohydrate diet, or severely limited calories, continues for more than a day (by which time glycogen is depleted), the body can make this blood sugar from protein metabolism, but not from fat. Many tissues require energy specifically as glucose, so lean tissue protein is metabolized when the glycogen supply is depleted. Thus, limited carbohydrate diets and fasting tend to produce striking initial weight losses due to lean tissue and water loss.

ENERGY METABOLISM

Physiological fuel value is proportional to the amount of energy yielded when substrates are degraded via cellular enzyme pathways. Adenosine triphosphate (ATP) is the major biochemical energy form and is common to all living systems. ATP is an unstable compound, so its production is closely linked to its usage:

Energy-yielding reactions
$$ADP + PO_4 \rightleftharpoons ATP$$
Energy-requiring reactions

There is very little storage of ATP per se, but it may be rapidly produced, eg, by transferring a phosphate moiety from a substrate to the diphosphate form (ADP) (Fig. 1). However, the majority of cellular ATP is produced in the mitochondrial citric acid cycle (or Krebs cycle) by electron transfer in the citric acid cycle to form the B vitamin-containing cofactors NADH and $FADH_2$. These shuttle energy potential to the respiratory chain, or electron transport pathway. The oxidation–reduction reactions of this pathway are coupled to ATP production, yielding 3 ATP per NADH. In the course of these pathways, carbon molecules of the original substrates are removed sequentially, and the terminal electron acceptor, oxygen, is converted to water. Thus, these complex reactions may be abbreviated as:

$$\text{Glucose } (C_6H_{12}O_6) \xrightarrow{6 O_2} 6\ CO_2 + 6\ H_2O + 36\ ATP$$

$$\text{Palmitate (16 carbon fatty acid)} \xrightarrow{23 O_2}$$
$$16\ CO_2 + 23\ H_2O + 131\ ATP$$

ENERGY USAGE

It is possible, though difficult and expensive, to measure human energy expenditure with direct calorimetry, which quantitates the heat generated during life processes. More often, indirect calorimetry, or the measurement of metabolic gas exchange, is used, and expressed as the respiratory quotient (RQ):

$$RQ = \frac{CO_2 \text{ produced}}{O_2 \text{ consumed}}$$

Because of the differences of reduction potentials between glucose and fatty acids, relatively more CO_2 is generated per O_2 used from carbohydrate than from fat. When pure carbohydrate is completely oxidized, an RQ of 1.0 (6 CO_2/6 O_2) results, versus 0.7 (16 CO_2/23 O_2) for the oxidation of pure fat. When protein is used for energy, RQ is approximately 0.8. However, protein is generally assumed to be a minor (about 10%), though constant, portion of calorie usage and is frequently ignored rather than using the more complicated nonprotein RQ calculations (5).

When RQ is known, energy expenditure can be calculated from oxygen consumption: When glucose is used for fuel, 5.01 kcal are used per liter of O_2 consumed, versus 4.69 kcal/l for fat. By measuring gas exchange in various levels of resting and exercise, calorie requirements have been estimated.

ENERGY REQUIREMENTS

Basal

Calculations of energy usage may be subdivided into categories of survival, activity, and adaptive expenditure (6). Basal metabolic rate (BMR) has traditionally been used to express energy required to maintain life functions and tissues, exclusive of movement, stress, or effects of food, and is measured by indirect calorimetry while reclining and relaxed in early morning before breakfast. More recently, this technique has been thought to measure a transient state between sleeping and waking and has

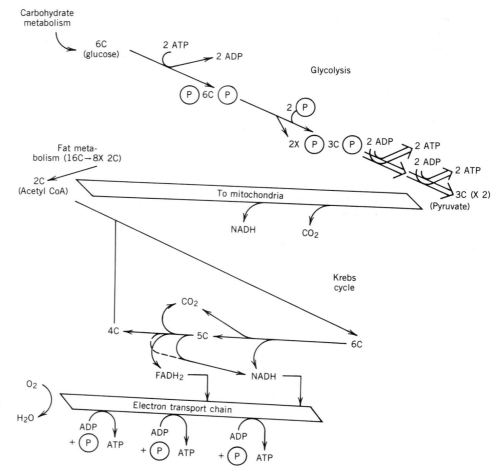

Figure 1. Energy metabolism pathways. The above scheme represents a much-abbreviated version of enzyme pathways involved in ATP production for use in energy-requiring reactions. Ⓟ represents inorganic phosphate ions; C denotes the number of carbons in energy-producing substrates, which ultimately are generated as carbon dioxide (CO_2) in the course of metabolism.

been replaced by resting energy expenditure (REE, or resting metabolic rate, RMR), which may be measured 3–4 hours after a meal with the subject in a relaxed state and is thought to be approximately 10% higher than BMR. In fact, these terms are often used interchangeably in the literature.

BMR level depends on metabolic activity, which is influenced by hormonal status, particularly thyroid hormone. Protein synthesis and maintenance are very expensive in terms of ATP expenditure. Therefore, the greater the proportion of lean body tissue to be maintained, the higher the relative calorie requirement. Lean tissue mass varies in humans with age and sex and accounts for much of the reduced calorie requirement in women and elderly people relative to young men of the same height and weight. Adipose, or fat tissue, requires much less energy for maintenance, and further influences BMR by increasing subcutaneous insulation and reducing heat loss.

BMR may be estimated by many formulas:

1. Standard rough estimate:
   Weight (kg) × h × 1.0 kcal/kg h (for men)
   Body weight (kg) × h × 0.9 kcal/kg h (for women)
2. Harris-Benedict equation for daily expenditure:
   Men: 66.5 + [13.5 × wt (kg)] + [5.0 × ht (cm)] − [6.75 × age]
   Women: 655.1 + [9.56 × wt (kg)] + [1.85 × ht (cm)] − [4.68 × age]

These last calculations are regression equations derived from BMR measurements among a large population and account for sex differences as well as age-related changes in lean body mass. Prediction of an individual BMR from these equations may involve an error in the range of 15%.

Activity

ATP is expended for both skeletal muscle contraction and relaxation. This component, called thermal effect of exercise, or TEE, is in addition to basal needs. It includes incidental activity, such as fidgeting and shivering, as well as intentional exercise. This is the most highly individual and variable portion of energy expenditure.

The energy cost of an individual activity may be only marginally higher than BMR requirement for armchair-type occupations. For strenuous exertion, such as fast running or swimming, expenditure can be seven times BMR. Calculations of kcal required per kg body weight per hour of participation have been obtained from the classic work of Rose (7). More vigorous activities using larger muscle groups have the highest requirements, and greater speeds of walking or running increase the requirement when calculated as a function of time. However, total energy expended by a given person to run a mile is not much greater than that of walking the same distance at the same incline. Therefore, kcal usage can be calculated per distance covered, generally in the range of 70–100 kcal/mile, depending on body weight.

Expenditure for maintenance plus exercise can be calculated by multiplying a BMR estimate by 1.3, for mostly sitting throughout the day, or by 1.5–1.7 for light to moderate activity (8). In practice, the sum total of even exceptionally heavy, long-term activity results in a factor of no more than 2.0–2.4 times BMR; a higher level of sustained activity is exhausting.

Adaptation

The most studied example of adaptive expenditure is dietary thermogenesis, or thermal effects of food (TEF). This an overhead required to process the nutrients that is generally assumed to be about 10% of the consumed calories is in a diet of average composition. TEF consists of an obligatory portion, ie, the energy required to absorb, transport, and store calories until needed for metabolic energy. TEF is greatest for protein, because of the relative cost of protein metabolism, and may exceed 20% of protein calorie intake. It has been calculated that only about 3% of available calories are needed to store dietary fat as triglyceride, whereas 7% of carbohydrate calories are used for glycogen synthesis. If carbohydrate intake is greatly in excess of glycogen storage capacity, approximately 23% of these calories must be used to convert dietary carbohydrate to storage fat. However, on a moderate fat, mixed diet of excess calories, it is unlikely that appreciable fat synthesis occurs. That is, dietary fat may be stored fairly directly in adipose tissue, whereas dietary carbohydrate is made into glycogen, then preferentially oxidized (9).

There is also a facultative, nonobligatory component of TEF, which is less well understood (10). In small mammals, a specialized cell-type called brown adipose tissue expends calories as heat in response to cold exposure and to overfeeding. In some humans, overeating increases thermal energy losses and may be an adaptive mechanism to prevent excessive weight gain. This phenomenon is referred to as luxusconsumption in the older literature, an early observation of elevated TEF during overfeeding. Its effects are seen mostly with carbohydrate and protein, rather than fat, in the diet. The possible mechanisms of facultative thermogenesis include futile cycling, eg, higher than necessary levels of fat, protein, and glycogen breakdown and resynthesis.

The components of energy expenditure may be interactive. Chronically increased physical activity increases the requirement for basal expenditure and may also increase TEF (11). Adaptive energy expenditure may be increased by the effects of smoking and caffeine. Various hormones, including insulin, epinephrine, thyroxine, and sex hormones, have been reported to affect the facultative component of TEF (6).

Estimation of Total Energy Expenditure

Total expenditure in adults may be calculated from its components as:

BMR estimate × activity factor
$$+ \text{TEF (10\% of calorie intake)}$$

Physical expenditure level may, alternatively, be accounted for by adding individual activities.

During growth, pregnancy, and lactation, additional allowances must be made for increased protein synthesis and higher metabolic rates. Recommendations for these conditions may be found in Ref. 8.

Individual Variation

A classic observation by Widdowson is that in any group of 20 or more people of the same sex and age, there will be one whose calorie intake is twice that of another (12). Although the equations shown above work well for an average population, any given person's requirement may be substantially under- or overestimated. Probably 80% of individual variation in basal expenditure can be explained by lean body mass differences (5). However, it is difficult to calculate more than a rough estimate of lean body mass from simple measurements such as height and weight, as in the Harris-Benedict equation.

Although caloric intake recommendations are sometimes made, as in the Recommended Dietary Allowances (8), these are intended as estimates of energy allowances in nutrition planning for groups of people, rather than as suggestions for the individual. In practice, each person must find his/her requirement by trial and error, ie, the calorie level that prevents both weight loss and weight gain.

ENERGY BALANCE: WEIGHT GAIN AND LOSS

In most adults, body weight remains remarkably constant over time. That is, weight as muscle, fat, and organs are maintained at the same level, even though the fat, protein, and glycogen components are constantly used for energy and replaced. This means that dietary calorie intake must match energy requirement fairly closely. If intake exceeds expenditure, excess calories are stored as body fat, which comprises most of weight gain in adults. To lose excess fat, intake must be less than output, which forces the body to metabolize banked triglyceride to make up the calorie deficit. This may be achieved through less calories, more exercise, or both.

Although this is easily said, it is extremely difficult to do. It is estimated that although Americans spend at least $10 billion a year attempting to lose weight, less than 5% of dieters maintain significant weight losses for more than 3 years. The problem of weight control is a complex one that will be addressed here only briefly.

Much of popular dieting relies on quick weight loss programs, which are usually very low in either carbohydrate (< 50 g/day) or calories (< 500 kcal/day), or both. This necessitates glycogen depletion, lean tissue loss, and water weight loss, as stated above. When a normal diet is restored, so are these essential lean tissues. Quick loss is generally followed by a quick gain.

Programs recommended by the nutrition establishment are moderately low in calories (not fewer than 800–1000) and in fat (< 30% of total calories) with adequate carbohydrate (> 100 g/day) and all other nutrients. Many people find such regimens difficult, or boring, so many food products have been marketed as diet aids.

Product labeling for these items may use the following regulated terminology:

Diet or dietetic: No more than 40 kcal/serving, or at least ⅓ fewer calories than a regular product.

Low calorie: No more than 40 kcal/serving or 0.4 kcal/g.

Reduced-calorie: ⅓ less calories than a regular product (the comparison must be shown on the label).

Lean: Meat with less than 10% fat; extra lean means less than 5% fat.

Sugar free, or sugarless: Contains no sucrose (table sugar), but other sugars may be present.

These last terms may be misleading to the average consumer. A meat that is 10% fat by weight may still have 40–50% of its calories from fat and may not be especially low in calories. Sugar-free products may be high in calories from corn syrup, glucose, or any sugars other than sucrose. Products making nutrition or health claims must have full nutrient labeling, so further information should be available on the package to make choices beyond the implied low-calorie claim.

BIBLIOGRAPHY

1. M. Kleiber, "Joules vs. Calories in Nutrition," *Journal of Nutrition* **102**, 309–312 (1972).
2. A. L. Merrill and B. K. Watt, "Energy Value of Foods," *USDA Handbook* 74, 1955.
3. "Food Labeling," *Code of Federal Regulations,* Title 21, Pt. 101, 1990 edition.
4. M. I. McBurney, L. U. Thompson, and D. J. A. Jenkins, "Colonic Fermentation of Some Breads and Its Implication for Energy Availability in Man," *Nutrition Research* **7**, 1229–1241 (1987).
5. E. Jequier, K. Acheson, and Y. Schutz, "Assessment of Energy Expenditure and Fuel Utilization in Man," *Annual Review of Nutrition* **7**, 187–208 (1987).
6. E. A. H. Sims and E. Danforth, Jr., "Expenditure and Storage of Energy in Man," *Journal Clinical Investigation* **79**, 1019–1025 (1987).
7. C. M. Taylor and G. McLeod, *Rose's Laboratory Handbook for Dietetics,* 5th ed. Macmillan, New York, 1949.
8. National Research Council, "Recommended Dietary Allowances," 10th Ed., Food and Nutrition Board/National Research Council, National Academy Press, Washington D.C., 1989.
9. J. P. Flatt, "Dietary Fat, Carbohydrate, and Weight Maintenance: Effect of Exercise," *American Journal of Clinical Nutrition* **45** (suppl), 296–306 (1987).
10. K. J. Acheson, E. Ravussin, J. Wahren, and co-workers, "Thermic Effect of Glucose in Man. Obligatory and Facultative Thermogenesis," *Journal of Clinical Investigation* **74**, 1572–1580 (1984).
11. E. T. Poehlman, C. L. Melby, and S. F. Badylak, "Resting Metabolic Rate and Postprandial Thermogenesis in Highly Trained and Untrained Athletes," *American Journal of Clinical Nutrition* **47**, 793–798 (1988).
12. E. M. Widdowson, "Nutritional Individuality," *Proceedings of the Nutrition Society* **21**, 121–128 (1962).

MARY N. ROSHOLT
JOHN J. CUNNINGHAM
University of Massachusetts
Amherst, Massachusetts

ENERGY MANAGEMENT. See ENERGY USAGE AND FOOD PROCESSING PLANTS; ENERGY USAGE IN THE CANNING INDUSTRY; FOODSERVICE SYSTEMS.

ENERGY USAGE IN FOOD PROCESSING PLANTS

There is a continuing concern about the energy demand by our food system. It is estimated that 17% of U.S. energy consumption is attributed to the food system (1). This figure includes energy used for production through processing, distribution, out-of-home preparation, and in-home preparation. The food industry requires energy for a variety of equipment such as gas fired ovens; dryers; steam boilers; electrical motors; refrigeration units; and heating, ventilation, and air-conditioning systems.

Natural gas is the predominant source of energy used by the U.S. food industry. In the last few decades about 50% of the gross energy used in food processing was from natural gas; 15%, from fuel oil, 13%, from electricity; and about 22%, from propane, butane, other petroleum products, coal, and some renewable energy sources (2). Within the food industry, the principal types of energy use include direct fuel use, steam, and electricity. Nearly 50% of energy use is in the form of direct fuel use. Almost 30% of energy is used to process steam and 10% to heat water. Almost 67% of electrical energy consumption is for generating mechanical power to operate conveyors, pumps, compressors, and other machinery. Refrigeration equipment consumes about 17% of electricity; lights consume about 10%; and heating, ventilation, and air-conditioning use approximately 4% of electricity.

Between 1973 and 1986, the cost of energy escalated dramatically. This trend promoted the installation of heat recovery equipment to conserve energy (3). It also focused the attention of scientists and engineers on the economic feasibility of alternate energy sources in various food-processing situations like cheese processing (4), food dehydration (5), and water heating (6). Energy consumption patterns of various food industries are briefly discussed.

MILK PROCESSING

Considerable variations in energy requirements for milk processing have been reported for different plants. Fuel requirements for producing pasteurized bottled milk ranges from 0.25 to 2.65 MJ/L of milk (7). The energy used for processing cheese during the regular plant operation is approximately 2.13 kJ/kg. Excluding drying and evaporation, energy required to produce 1 kg of cheese is in the range of 3.37 to 17.53 MJ (4). Some of this variation is due to a lack of concern to energy conservation, but there are also operational factors that may explain variations among plants.

Considering energy allocation to individual unit operations separately in a typical processing of pasteurized bottled milk, most of the energy is used in pasteurization step (approximately 23% of electricity and 35% of fuel). In the production of pasteurized bottled milk, bottle washing consumes approximately 58% of the fuel energy and 6% of the electricity used in the process.

In butter production pasteurization consumes approximately 30% of the fuel energy. The churning process and cold storage of butter consume approximately 50% of the electricity required for butter processing.

In yogurt production energy for heating and cooling per kilogram of product averages 1,146 kJ/kg (8). The heating of the base yogurt from 10 to 87.8°C uses 80% of the energy, and the electrical equipment used in product handling and packaging consumes only 6%. The manufacturing process of sour cream is similar to that for yogurt, except the amount of heating is considerably less because the highest product temperature is 22.2°C for sour cream compared to 87.8°C for yogurt. In packaging, the shrink wrap machine is not used for holding the sour cream cartons. The thermal energy ratio is 273 kJ/kg of sour cream and packaging uses less electrical energy primarily because of the lack of the shrink wrap operation. Product handling and packaging uses 49 kJ/kg of sour cream compared to 78 kJ/kg of yogurt. Total energy, thermal plus electrical, is 287 kJ/kg for sour cream compared to 1,224 kJ/kg for yogurt. The above values of energy to manufacture sour cream and yogurt exclude the following: the energy required to pasteurize the raw milk, the energy required to pump the base material into the processing vat, and the energy required in temporary cold storage prior to shipping. Yogurt in the processing vat is cooled from 87.8 to 42.2°C partially with well water. Measurements of the energy taken from the yogurt by the well water showed an average of 212 kJ/L of yogurt in 4.16 m^3 batch. The cooling could be considered free, except for the cost of the water and pumping energy (8).

Spray drying is the common means of converting fluid feedstocks into solids in the form of powders, granules or agglomerates. It is widely used for dehydration in the manufacture of a wide variety of food, pharmaceutical, and chemical products. Among the foods that have been successfully spray dried are milk, whey, cheese, coffee whitener, eggs, soups, baby foods, and fruits. Energy costs constitute a significant fraction of the operating costs. A number of methods have been suggested to decrease energy related costs during the spray-drying process by using insulation to decrease heat losses to the environment and heat recovery from exhaust gases (8). Typically, in the manufacture of spray-dried milk powder, air heating for the spray-drying operation requires between 50 and 80% of total fuel consumption and approximately 30–35% of total electricity consumption in the whole process. The variation in the primary fuel input results from heat recovery steps.

Cleaning in place (CIP) in milk processing plants is one of the operations that takes significant input of energy. In the manufacturing of Cheddar cheese, CIP consumes approximately 31% of the total fuel energy and 18% of the total electricity required for the whole process. In other processing operations such as milk pasteurization, creamery butter, and acid casein processing, the fuel consumption in CIP operation does not exceed 3% of total fuel used in the whole process, and the use of electricity is up to approximately 1% of total electric energy used for the whole process. In spray-dried milk powder production, CIP operation costs approximately 5–8% of the total fuel consumption and 4–5% of the total electricity used in milk powder processing.

LIQUID FOOD CONCENTRATION

Evaporation of liquid foods is most commonly accomplished using multieffect evaporators. The word *effect* indicates vapor flow in the evaporator. The total evaporation costs depend on the steam consumption, the specific area of the evaporator, and the specific product losses (9). The energy efficiency of an evaporator is usually expressed by thermal or steam economy, which is defined as a ratio of quantity of water evaporated to quantity of steam consumed (10). The steam economy of an evaporator increases with the increase in number of evaporator effect, because subsequent effects use vapor from the previous effect as the heating medium. As an example, a single-effect evaporator takes 1.5 kilogram of steam to evaporate 1 kg of water, whereas a double-effect evaporator only consumes 0.75 kg of steam to evaporate the same amount of water (11).

In conventional multieffect evaporators, those without mechanical or thermal recompression, the steam evaporated from the first effect is used as the heating medium for the next effect and so on. A mechanical vapor recompression evaporator is very much like any other evaporator except that steam is recycled. A compressor is installed, which takes the vapor off an effect and pressurizes it to a higher pressure for use as the heating medium. The major energy expense in such a system is electricity. The operating economy of a mechanical vapor recompression unit is equivalent to an 8–20-effect evaporator. All vaporization is carried out at the same temperature. Steam from the boiler and water are only needed for startup, which accounts for the tremendous savings.

The dairy industry carries out the evaporation of whole milk, skim milk, buttermilk, and whey. The steam economy of two-stage falling film plate evaporator with mechanical vapor recompression, when used to concentrate buttermilk and skim milk to 25% solids, was reported to be 6, which was claimed by the evaporator manufacture (12) to be equivalent to an 8–30-effect conventional evaporator, depending on the cost of fuel. The estimated fuel savings were expected to be enough to amortize the rotary compressor in less than one year (12). A 7-effect mechanical vapor recompression evaporator can remove 17.5 kg of water from the feed for each kilogram of steam supplied to the turbine, which gives the steam economy equal to 17.5 (13) while concentrating whey, skim milk, or whole milk.

Frozen concentrated orange juice (FCOJ) is consumed widely in the United States, European Common Market countries, Japan, Venezuela, and other countries. São Paulo, Brazil, and Florida are the two leading producers of FCOJ. Temperature accelerated short-time evaporator (TASTE) evaporators are used extensively for concentrating orange juice (14).

Energy consumption in a four-effect, seven-stage (the word *stage* indicates the flow of orange juice in the evaporator) TASTE evaporator with a capacity of evaporation of 18,140 kg H$_2$O/h ranged between 840 and 1,000 kJ/kg H$_2$O evaporated (15). By introducing one effect and one stage after the fourth stage, the magnitude of energy consumption was reduced to 570–640 kJ/kg of water evaporated. In most orange juice evaporators, the flow of orange juice and the steam pressure are controlled manually to obtain concentrated juice of the desired degree Brix (percent sucrose equals degree Brix). By means of automatic control, the energy consumption in a six-effect, eight-stage evaporator with a capacity to evaporate 9,070 kg H_2O/h was reduced by 6.7%.

The data on energy consumption during the 1978–1981 processing season in a concentrated orange juice plant in Brazil were presented for a plant with a capacity to produce 70,000 t/yr of 65°Brix of frozen concentrated orange juice and 85,000 t/yr of citrus pulp pellets made from orange residues. The plant had on-site storage facilities for 17,600 t/yr of frozen orange juice at $-8°C$. At a separate location away from the plant, 13,650 t/yr of concentrated orange juice were stored at $-25°C$. For the plant as a whole, electricity was accounted for only 10% of the total energy consumed in the plant. Most of the steam generated was used to concentrate orange juice in evaporators; a small portion (16%) was used in the pelletizing operation of citrus pulp (16).

Evaporators are also used to produce tomato products with varying degrees of concentration. The primary products are classified as purées (solids content from 11% to 22%) and paste (solids content from 28% to 45%). A number of energy audits on tomato product evaporators have been performed (17). Daily average performance data are given for single-, double-, and triple-effect evaporators. Average daily steam economies for the single-effect evaporator averaged 0.84 compared to a theoretical average of 0.95. Two double-effect evaporators had average daily steam economies from 0.79 to 2.03. The average daily steam economy measured was 1.45 while the theoretical average was 1.91. As expected, the triple-effect evaporators showed the best steam economies ranging from 1.66 to 3.06. The lower ranges of steam economy were caused by frequent fouling of the heat exchangers by the burning product. The recent evaporators installed in tomato processing plants have four effects. In the multieffect evaporators the energy cost is directly proportional to the temperature difference across each effect. If the temperature difference is cut in half, twice as many effects can be put in, cutting the energy consumption approximately in half. It will also approximately double the equipment cost (18).

Beets are approximately 15% sugar of which about 88% is extracted giving 132 kg sugar/t of beets. The various sugar beet processors in California provided data that showed that for each ton of beets, 3.19 GJ of primary energy is consumed directly in processing (19). Dried pulp, a by-product of sugar beet processing, accounted for large amounts of energy used. Pulp drying requires 0.73 GJ/ton of sugar beets. This value is nearly 23% of the total direct energy consumed during processing of sugar beets. The evaporators used in the sugar concentrating process consumed nearly 44% of the total direct energy required.

Energy requirements were monitored during the production of a beet colorant on a pilot scale (20). Colorant production involved centrifugation, concentration of ferment, vacuum concentration, and spray drying. The first operation was juice extraction. The beets were blanched, abrasion peeled, and washed. They were then dried and immediately comminuted. The product was pumped to a centrifugal separator for separation. The clarified juice was sent to a plate heat and exchanger pasteurized at 88°C for 2 min, then cooled to 2°C. Extraction of the juice from 1 t of beets used 69.3 ± 0.2 kWh of electricity and 226 ± 13 kg of steam. It took 95 ± 5 min and yielded 550 ± 45 kg of juice at 7% solids. Then the juice was vacuum concentrated to 14% total solids. This used 343 ± 14 kg of steam and 5.1 ± 0.2 kWh of electricity.

The second operation in the production of beet colorant was the fermentation of the clarified beet juice. The temperature was maintained at $30 \pm 2°C$ during the continuous fermentation. The juice extracted from 1 t of beets used 10.2 ± 0.2 kWh of electricity. The fermentation removed about 80% of the solids (20).

In the third operation, the fermented juice was centrifically separated. The clarified juice was vacuum concentrated in a single-effect falling film evaporator to 30% soluble solids, cooled to 16°C, and spray dried. The yeast cell slurry was also spray dried. The vacuum evaporation used 396 ± 14 kg of steam and 4.6 ± 0.2 kWh of electricity to concentrate juice from 1 t of beets. Spray drying the juice took 29.6 ± 1.5 m^3 of natural gas and 2.8 ± 0.2 kWh of electricity per ton of beets. Spray drying of yeast cells took 26.6 ± 1.5 m^3 of natural gas and 2.5 ± 0.2 kWh of electricity per ton of beets.

FOOD BLANCHING

Blanching is an important unit operation in the pretreatment of fruits and vegetables. It is used prior to freezing, canning, and dehydration. The blanching process is a relatively fast process. Blanching time is a function of piece size, heating medium, and temperature as well as material packing in the blancher. The most common heating media are steam and water. Design of a blancher has a significant influence on the energy used. The open steam blanchers are approximately 10% less efficient than the sealed units (21). Steam losses from the unsealed entrance and exit account for almost 80% of the energy input to the blancher. The steam blancher with end seals is similar in design to the conventional open steam blancher. Water sprays are positioned inside the ends of the blancher to liquify steam that might escape. They also serve to help cool the product at the exit. In such a design the steam losses are up to approximately 51% of the thermal energy input to the blancher. The thermal energy required to blanch spinach using a conventional steam blancher is 2.12 MJ/kg of product (21). Spinach processed in a blancher with hydrostatic seals requires only 0.95 MJ/kg of product (21). In a commercial blancher, at a spinach processing plant, energy used is 6.5 MJ of natural gas and fuel oil and 0.072 MJ of electricity per 1 kg of raw spinach (22).

Energy consumption data for three commercial water blanchers (tubular, screw conveyor, and water tank) indicate that type of product and type of blancher dictate the energy usage (21,23). For lima beans processed in a tubular water blancher, thermal energy usage is 0.543 MJ/kg of processed beans and thermal efficiency is 44.5%, the highest of all three water blanchers tested (23).

The screw conveyor water blancher processed cauliflower cuts. Thermal energy consumption averages 0.91 MJ/kg of product and its thermal efficiency is 31.2%. Energy accounting data on a commercial blancher of a water tank type at the spinach processing plant indicate that 31% of the energy input goes to heating the product and 69% is lost through incomplete condensation of steam, hot water discharge to a drain, and heat losses by convection and radiation (22).

POULTRY PROCESSING

Poultry processing consists of several operations: live bird holding, hanging, slaughtering, scalding, defeathering, eviscerating and chilling, grading, cutup, and packaging. All processes require some energy either as fossil fuel or electricity. Electricity is used for conveyors, refrigeration, lighting, air-conditioning, pumps, and the mechanical drives. Fossil fuels are used for space heating, production of steam or hot water, and feather singeing. The electrical demand for three selected poultry processing plants indicates that the refrigeration and other mechanical drives account for more than 80% of the total calculated electrical use. Lighting requires 6.8% and fans 7.8% of the total electricity. The rest of the total electrical energy is used for heating and air-conditioning of the plants (24). Calculating the electrical use by major area in those three processing plants, the average electrical use in the eviscerating and chilling areas together is 6.7% of the total. This percentage does not include electricity used for ice making and water chilling. Packing and shipping uses 1.6% of the total electricity, which is about 3.8 kWh/1,000 head (total electricity consumption is 237.5 kWh/1,000 head); but this does not include refrigeration for storage areas. Offal and waste handling equipment and waste treatment use 12.7% of the total electricity, and shops and services account for 11.9% of the total electricity use (24).

In another study (25), the electrical energy use per 1,000 broiler equivalents processed was 212 kWh. The cost of ice making, used for a water chiller, and in transportation accounted for approximately 45% of the total electricity consumption. It was noted that 7.2% of the electricity was used in clean-up operations.

FISH PROCESSING

The fresh and frozen packaged fish industry consumes 0.6–0.8% of the energy required by the food-processing industry (26). The fish plant manufactures fillets; dressed, gilled, and puffed fish; and ground fish scrap, which is sold as animal feed. Electrical energy consumption makes up 80% of the plant's total energy requirements (26). Natural gas is used for space heating purposes. Electrical energy is primarily used to operate the refrigeration equipment for the chilling and freezing of whole fish and processed fish products. In addition, it is used to a lesser extent to power motor drives and air-conditioning eqquipment, to generate hot water for cleaning purposes and to supply lighting. The average energy used ranges between 3.93 and 4.89 MJ/kg of fish product (26).

Ammonia reciprocating compressors are the largest single electricity consumers in the ice production plant. Electrical energy is the main source of energy in that production, representing approximately 85% of its energy requirements. The reported annual energy use in the ice industry was between 581 and 697 kJ/kg in the mid-1970s. This ratio has decreased in recent years due to more energy-efficient equipment and a shift to higher production rates of fragmentary ice and is reported as 380–420 kJ/kg of ice (26).

CANNING

The energy use for canning of tomato products was obtained from the daily amount of tomato received by the plant and the total daily energy consumption (2). Tomato processing consists of several operations. At the receiving station the tomatoes are removed from the gondolas with the use of water. The water containing field dirt of the tomatoes is pumped to a mud-settling tank. The tomatoes are conveyed in a hydraulic flume for additional washing and initial inspection. Water is recirculated in the hydraulic flume. The majority of the electrical energy at the receiving station was consumed by the pump conveying water to the mud-settling tank. The energy use of the receiving operation was 0.32 Wh/kg of raw tomatoes. Using data from six days of processing a thermal energy intensity value of 1,251.4 kJ/kg of tomatoes was calculated. This value represents processing of tomatoes into three products: tomato juice, canned peeled tomatoes, and tomato paste. The electrical energy intensity value is 0.025 kWh/kg of raw tomatoes.

Production of tomato juice involves crushing tomatoes, heating them rapidly to inactivate enzymes, filling juice into cans, and retorting the cans. Most operations consume electricity. Hot-break heaters and retorts use steam. Combining the thermal energy and electrical energy data to a similar base unit of kilojoules, the hot-break heaters and retorts account for almost 97% of the energy consumed in the processing line and the energy intensity of tomato juice production was 1,086.2 kJ/kg of raw tomatoes (2).

Processing of canned, peeled tomatoes requires lye-bath peelers to facilitate peel removal and retorts to sterilize the canned product and conveying equipment. Lye-bath peelers and retorts consume steam. These two operations account for almost 99% of the total energy consumed (2). All other equipment operates with electricity. The energy intensity is 1,300.2 kJ/kg of raw tomatoes.

In tomato paste production, several operations such as sorting, pulping, and finishing are similar to tomato juice production. Steam is used in vertical heat exchangers to preheat tomato juice. A large quantity of steam is used in the evaporator. The evaporator and the heat exchanger used in preconcentration account for the majority of steam consumption. The energy intensity of tomato paste production in the cannery industry is 1,307.2 kJ/kg of raw tomatoes (2).

As a consequence of the seasonal vegetable or fruit production, most of the energy consumption is also seasonal. Energy is also consumed during the nonprocessing season, which adds to the energy cost of processed fruits and vegetables. A study of four vegetable canneries located in Western, New York (27) producing apple sauce, canned beans and carrots, indicate that, on the average, blanching and sterilization consume 9% (theoretical estimate); lighting and electric motors, 21%; and movement of vehicles, 4.1%. The rest of the energy is lost in different ways. A significant portion, 22.6%, is lost in steam generation; 21% is lost to the environment after steam generation; and 22.3% is spent on other activities (losses to start up and operation of process equipment prior to actual processing and inefficiencies of process equipment).

Peach canning involves receiving peaches in bins, dumping fruit into a water tank and then elevating it out with a conveyor. The peaches are then graded for size, rinsed with water and pitted. After pit removal, the peach halves are oriented cup down on a belt conveyor and conveyed to a lye-bath peeler. After exposure to caustic solution for a predetermined time, the fruit are again rinsed. The fruit are allowed to orientate cup up; are reinspected, are size-graded, and filled into cans along with syrup. The canned fruit are then heated in retorts to achieve the desired sterilization (2). Within various operations, transport of fruit and waste consume the most electricity. It was determined that 23% of the total electrical energy consumption occurs in dry conveying whereas 38% is consumed in pumping water to convey fruit and waste products (2). Retorting and lye-bath peeling are the most energy-consumptive unit operations, accounting for 89% and 10%, respectively, of total energy consumption. The energy used by the three lye-bath peelers is 135.7 kJ/kg of canned peaches.

There are four energy-intensive operations in citrus packing: degreening–precooling, wash–wax–drying, holding, and storage. Energy analysis of four citrus packing plants in California showed that electricity and natural gas are two major types of energy source (28). Orange and lemon packing differ primarily in the duration of storage required before shipment. Oranges are shipped as soon as they are harvested and packed in a plant. Storage is used mainly to chill and hold the fruit until shipment. This holding time does not exceed a few days. Lemons might be stored from two weeks to six months. Another important distinction between an orange and a lemon packing plant is in the degreening operation. Lemons degreen naturally in storage, oranges are put into a chamber with air containing ethylene at 18–21°C (28).

In the unload–dump unit, field bins of fruit are conveyed from a receiving area to the dump station where the fruit drop into a water tank. Fruit are conveyed in a water flume to the wash–wax–dry operation, which consists of four steps: manual sorting and removal of rotten fruit, washing and rinsing, waxing and surface moisture removal by heated air, and manual sorting of fruit for juice processing. The above sequence is followed when water-based waxes are used. In cases when solvent-based waxes are used, the surface moisture is removed both prior to and after waxing.

The total energy intensities for orange and lemon packing vary between 0.64 and 1.03 MJ/kg (28). Electrical energy intensities are about the same in citrus packing plants (0.55–0.82 MJ/kg). Natural gas intensities, however, vary from 0.07 to 0.3 MJ/kg. Another study on consumption of energy in lemon, orange and grapefruit packing plants indicates that total energy intensities for plants have a range of 362.6–702.4 kJ/kg (29). Electrical power intensities vary from 163.3 to 583.3 kJ/kg. Natural gas use intensities for citrus packing vary from 53.8 to 159.8 kJ/kg. The possible reasons for the differences in energy used for packing are different harvest times, the refrigerated holding costs, and storage time.

BAKING INDUSTRY

The largest consumers of energy in a bakery are the ovens, heating boilers, steam generators, and refrigerators. The differences among the energy balances of the bakeries can be large. They are influenced, for example, by the size of the bakery, production structure, amount of production, location, and apparatus.

Based on energy measurement in two bakeries in Sweden, energy consumption of 13.96 MJ/kg of bread for a bakery with a capacity of 250,000 kg of bread per year and 4.88 MJ/kg bread for another bakery with a capacity of 3,500,000 kg of bread per year was reported (30). An investigation conducted in the United States (31) reported an energy consumption for the baking industry of 7.26 MJ/kg bread baked. This is based on measurements on a bakery with a capacity of 35,000 kg of bread per day. In another study, an energy consumption of three bakeries on an average was found to be 6.99 MJ/kg bread baked (32).

The difference in the energy consumption figures may be due to several factors including the size of the bakery. A small baker uses more energy per unit production than a large one. If the bakery has many different products, this will also cause an increase in the energy use. The transportation costs are also important. It was shown (32) that 13–21% of the total energy use in the bakeries was for the delivery vans. Thus the distribution range causes differences in energy use between different bakeries.

A German oven manufacturer reported that the energy use of their oven (a tunnel oven with convective heat transfer) was as low as 0.6 MJ/kg baked bread. This low figure is probably explained by an optimal use of the oven and the convective heat transfer. Much less energy consumption was mostly due to less ventilation.

Considering the energy distribution in bread making, the heating of pans and lids used 26% of the energy. The rest of the energy use lost either through ventilation (31%), exhaust gases (13%), or radiation and convection losses from the walls (30%) (33). These figures show that there are some major energy-conserving opportunities in the baking industry. These steps include minimization of the ventilation of the oven, use of materials with lower heat capacities in the pans and lids and use of heat exchangers to recover heat from the hot exhaust gases.

The British-style crumpet is a small, round bakery product made of unsweetened batter and cooked on a griddle. Traditionally, the product is marketed in the frozen form. In 1982, a modified-atmosphere technique was introduced to package crumpets (34). The aim of modified-atmosphere packaging was to reduce energy consumption without adversely affecting quality during storage and marketing of the product. The production of the crumpets involves mixing, pumping, baking, cooling, packaging, freezing, and storage. Electricity for mixing and pumping of batter and power for running associated motors of the griddle, refrigeration units, and packaging machines together total 92.1 kWh for every 1,000 kg of crumpets produced (35). Thermal energy requirements for process heating (baking) and cooling of 1,000 kg crumpets are 1,050 and 160 MJ, respectively. The results (35) indicate that the freezing process requires 440 MJ, an additional 29% energy to produce and cool the crumpets.

A major advantage of modified-atmosphere packaging over frozen crumpets is that the former requires no special conditions for storage, whereas, for the latter, refrigeration is necessary. Storage of 1,000 kg of crumpets uses 7.4 kWh of electricity and 642.8 MJ of refrigeration for 30 days in storage. In terms of total thermal equivalents, frozen storage adds approximately 43% more to the consumption of process energy for crumpets above that of the modified-atmosphere packaging system.

BIBLIOGRAPHY

1. *Energy Consumption in the Food System*, report **XIV, 13392-007-001,** Federal Energy Administration, Washington, D.C., 1975.
2. R. P. Singh, "Energetics of an Industrial Food System," in R. P. Singh, ed., *Energy in Food Processing,* Vol. 1, Elsevier Science Publishing Co., Inc., New York, 1986.
3. D. P. Donhowe, C. H. Amundson, and C. G. Hill, Jr., "Performance of Heat Recovery System for a Spray Dryer," *Journal of Food Process Engineering* **12,** 13–32 (1989).
4. R. K. Singh, D. B. Lund, and F. H. Buelow, "Applications of Solar Energy in Food Processing I: Cheese Processing," *Transactions of the American Society of Agricultural Engineers* **26,** 1562–1569 (1983).
5. R. K. Singh, D. B. Lund, and F. H. Buelow, "Application of Solar Energy in Food Processing II: Food Dehydration," *Transactions of the American Society of Agricultural Engineers* **26,** 1569–1574 (1983).
6. R. K. Singh, D. B. Lund, and F. H. Buelow, "Application of Solar Energy in Food Processing IV: Effect of Collector Type and Hot Water Storage Volume on Economic Feasibility," *Transactions of the American Society of Agricultural Engineers* **26,** 1580–1583 (1983).
7. B. Elsy, "Survey of Energy and Water Usage in Liquid Milk Processing," *Milk Industry* **82**(10), 18–23 (1980).
8. G. H. Brusewitz and R. P. Singh, "Energy Accounting and Conservation in the Manufacture of Yogurt and Sour Cream," *Transactions of the American Society of Agricultural Engineers* **24,** 533–536 (1981).
9. S. Bouman, D. W. Brinkman, P. de Jong, and R. Waalewijn, "Multistage Evaporation in the Dairy Industry: Energy Savings, Product Losses and Cleaning," in S. Bruin, ed., *Preconcentration and Drying of Food Materials,* Elsevier Science Publishing Co., Inc., New York, 1988, pp. 51–60.
10. A. H. Zaida, S. C. Sarma, P. D. Grover, and D. R. Heldman, "Milk Concentration by Direct Contact Heat Exchange," *Journal of Food Process Engineering,* **9,** 63–79 (1986).
11. D. P. Lubelski, S. L. Clark, and M. R. Okos, "Process Modifications to Reduce Energy Usage," in *Energy Management and Membrane Technology in Food and Dairy Processing, Proceedings from the Special Food Engineering Symposium Held in Conjunction with Food and Dairy Expo 83,* American Society of Agricultural Engineers, St. Joseph, Mich., 1983, pp. 7–13.
12. "MVR Evaporator Provides $60,000 Annual Fuel Cost Savings," *Food Process* **40**(1), 92 (1979).
13. P. Standford, "A Milestone in Evaporation Systems," *Dairy Rec.* **84**(7), 89–90, 92, 94, 96, 98 (1983).
14. C. S. Chen, "Citrus Evaporator Technology," *Transactions of the American Society of Agricultural Engineers* **25,** 1457–1463 (1982).
15. C. S. Chen, R. D. Carter, and B. S. Buslig, "Energy Requirements for the TASTE Citrus Juice Evaporator," in *Energy Use Features,* Vol. 4, Pergamon Press, New York, 1979, p. 1841.
16. J. Filho, A. Vitali, C. P. Viegas, and M. A. Rao, "Energy Consumption in a Concentrated Orange Juice Plant," *J. Food Process Eng.* **7,** 77–89 (1984).
17. T. R. Rumsey, T. T. Conant, T. Fortis, E. P. Scott, L. D. Pederson, and W. W. Rose, "Energy Use in Tomato Paste Evaporation," *Journal of Food Process Engineering* **7,** 111–121 (1984).
18. R. W. Cook, "Multiple Effect Evaporation," in R. F. Matthews, ed., *Energy Conservation and Its Relation to Materials Handling in the Food Industry,* Proceedings of the 15th Annual Short Course for the Food Industry, 1975, pp. 78–82.
19. P. K. Avlani, R. P. Singh, and W. J. Chancellor, "Energy Consumption in Sugar Beet Production and Processing in California," *Transactions of the American Society of Agricultural Engineers* **23,** 783–787, 782 (1980).
20. J. E. Block, C. H. Amundson, and J. H. Von Elbe, "Energy Requirements of Beet Colorant Production," *Journal of Food Process Engineering* **5,** 67–75 (1981).
21. E. P. Scott, P. A. Carroad, T. R. Rumsey, J. Horn, J. Buhlert, and W. W. Rose, "Energy Consumption in Steam Blanchers," *Journal of Food Process Engineering* **5,** 77–88 (1981).
22. M. S. Chhinnan, R. P. Singh, L. D. Pedersen, P. A. Carroad, W. W. Rose, and N. L. Jacob, "Analysis of Energy Utilization in Spinach Processing," *Transactions of the American Society of Agricultural Engineers* **23,** 503–507 (1980).
23. T. R. Rumsey, E. P. Scott, and P. A. Carroad, "Energy Consumption in Water Blanching," *Journal of Food Science* **47,** 295–298 (1981).
24. W. K. Whitehead and W. L. Shupe, "Energy Requirements for Processing Poultry," *Transactions of the American Society of Agricultural Engineers* **22,** 889–893 (1979).
25. L. E. Carr, "Identifying Broiler Processing Plant High Electrical Use Area," *Transactions of the American Society of Agricultural Engineers* **24,** 1054–1057 (1981).
26. L. G. Enriquez, G. J. Flick, and W. H. Mashburn, "An Energy Use Analysis of a Fresh and Frozen Fish Processing Company," *Journal of Food Process Engineering* **8,** 213–230 (1986).
27. W. Vergara, M. A. Rao and W. K. Jordan, "Analysis of Direct Energy Usage in Vegetable Canneries," *Transactions of the American Society of Agricultural Engineers* **21,** 1246–1249 (1978).
28. M. Naughton, R. P. Singh, P. Hardt, and T. R. Rumsey, "Energy Use in Citrus Packing Plants," *Transactions of the American Society of Agricultural Engineers* **22,** 188–192 (1979).
29. L. P. Mayou and R. P. Singh, "Energy Use Profiles in Citrus Packing Plants in California," *Transactions of the American Society of Agricultural Engineers* **23,** 234–236, 241 (1980).
30. C. Trägardh and co-workers, "Energy Relations in Some Swedish Food Industries," in P. Linko and co-workers, eds., *Food Process Engineering,* Elsevier Applied Science Publishers, Ltd., Barking, UK, 1980.

31. L. A. Johnson and W. J. Hoover, "Energy Use in Baking Bread," *Bakers Digest* **51**, 58 (1977).
32. G. A. Beech, "Energy Use in Bread Baking," *Journal of the Science of Food and Agriculture* **31**, 289 (1980).
33. A. Christensen, and R. P. Singh, "Energy Consumption in the Baking Industry," in B. M. McKenna, ed., *Engineering and Food,* Vol. 2, Elsevier Applied Science Publishers, Ltd., Barking, UK, 1983, pp. 965–973.
34. B. Ooraikul, "Gas-packaging for a Bakery Product," *Canadian Institute of Food Science Technology Journal,* **15**, 313 (1982).
35. N. Y. Aboagye, B. Ooraikul, R. Lawrence, and E. D. Jackson, "Energy Costs in Modified Atmosphere Packaging and Freezing Processes as Applied to a Baked Product," in M. Le Maguer and P. Jelen, eds., *Food Engineering and Process Applications,* Vol. 2, Elsevier Applied Science Publishers, Ltd., Barking, UK, 1986, pp. 417–425.

S. Cenkowski
D. S. Jayas
University of Manitoba
Winnipeg, Manitoba
Canada

ENERGY USE IN THE CANNING INDUSTRY

DEFINITION OF ENERGY

Energy is the capacity to do work, or, alternatively, anything which is capable of doing work is said to be a source of energy or to possess energy. The major forms of energy are mechanical, potential (by virtue of height), kinetic (by virtue of motion), heat, light, sound, electrical, and chemical. Although energy may be converted into other forms, it cannot be destroyed—a fact which is contained in the thermodynamic principle of energy conservation. This is the basis on which energy balances or audits are carried out on process operations in order to assess the amount of energy utilized and converted.

In the canning industry the energy contained in primary fuels, eg, coal, natural gas, and petroleum, is used to convert water to steam at the required operating pressure. Steam and electricity are the primary forms of energy used in the factory and these, in turn, are used for heating and activating machinery. For a complete energy balance it is necessary to consider water, effluent, and compressed air as secondary energy sources.

UNITS OF ENERGY

In order to quantify the use of fuel and energy it is necessary to be able to express their magnitude in comparable units. This is done by knowing the interrelationships between mechanical, heat, and electrical forms of energy.

Mechanical Units

The basic dynamical unit in the CGS system is the erg or dyne-centimeter, which is defined as the work done when a force of 1 dyne moves through a distance of 1 cm. It is an extremely small unit and, for practical purposes, a unit of 10^7 ergs is used. This is known as the joule (J) and has been adopted by the Système International des Unités (SI). The commonly used multiples are the kilojoule, kJ = 10^3 J; the megajoule, MJ = 10^6 J; the gigajoule: GJ = 10^9 J; and the terajoule, TJ = 10^{12} J. The joule is also known as the newton-meter (Nm), defined in an analogous way to the erg. The common units of energy are the foot poundal, 1 ft-poundal = 0.042 J; the foot pound force, 1 ft-lbf = 1.36 J; and the horsepower hour, 1 hp-h = 2.68 MJ.

Thermal Units

The basic unit is the international calorie (cal), which is defined as the amount of heat required to raise a mass of air-free water of 1 g through 1°C. The SI unit is the joule and 1 cal = 4.1868 J.

The common unit is the British thermal unit (Btu) and this is the amount of heat required to raise 1 lb of water through 1°F: 1 Btu = 1.055 kJ. A frequently encountered unit is the therm, which is equivalent to 10^6 Btu or 105.5 MJ.

For energy audits the most useful unit is specific energy, which is the energy per unit mass of specified material (raw or processed). This is expressed as joules per kilogram, which is equivalent to 0.43×10^{-3} Btu/lb. Some other useful conversions are

1 kcal/kg = 4.18 kJ = 1.8 Btu/lb
1 Btu/ton = 1.16×10^6 J/kg
1 Btu/lb = 2.326 kJ/kg

Electrical Units

The absolute electrical unit of power is the watt (W), which is 1 J/s, ie, the power that creates 1 J of energy in 1 s. The watt is equivalent to 1.34×10^{-3} hp and is equal to 3.42 Btu/h. Electrical energy is usually measured in kilowatt hours (kWh). The SI electrical unit of power is watt second (Ws): 1 kWh = 3.6×10^6 Ws = 3.6 MJ.

ENERGY MEASUREMENT

Every factory possesses information about the consumption of fuel and electricity from financial statements of cost, which are monthly or quarterly. However, in order to manage the use of energy, including fuel, electricity, steam, compressed air, and vacuum, it is necessary to make use of special metering devices (1). Ideally, these should be placed in positions which would record the energy consumed in different parts of the factory; however, this ideal must be limited by the cost of metering and meter reading. In the most comprehensive management schemes, all the data logging and subsequent production of management information can be carried out using various types of computing systems. Consequently, meters that have a digital output which may be used for this purpose should be selected. A further important point concerning the choice of meters is the scale range, which should be selected to suit potential energy reduction, since this is the object of energy management.

In order to maintain the accuracy of meter readings, the instruments should be calibrated on a regularly scheduled basis. Failure to insure that this is carried out can lead to erroneous information.

Electricity Metering

The basic and most widely used device is the watt-hour meter, which measures and registers the integral with respect to time of the power in the circuit to which it is connected. In effect, these devices are electrical motors, the torque developed being proportional to the power. Since the speed of the rotor is proportional to the torque, this makes each revolution of the rotor a measure of the watt hours consumed. In the standard induction type, which uses a disc for a rotor, the integration is achieved using a counter. For three-phase power measurement, multistator devices are used. The most widely used instrument is the traditional induction disc type with analogue dial indicators. These devices are accurate to about ±2% and are adequate for energy management purposes. It is also possible to obtain this type of meter with a pulsed digital output; however, it should be noted that it cannot be retrofitted. Electronic meters are also available which have an accuracy of ±1% and incorporate additional measurements, eg, kWAh (kilowatt amp hour) and power factors. The power factor is a measure of the efficiency with which the electricity supplied to the factory is converted into useful energy. The nearer the power factor is to unity the better; most factories achieve values of between 0.95 and 0.97. In situations where most of the electrical power is used with induction motors, it is necessary to install a bank of capacitors to reduce the power loss in the incoming supply and increase the power factor. Most tariffs penalize consumers with poor power factors, hence the need for careful vigilance. In addition to charging for the total units used, most industrial tariffs also take account of the maximum rate at which electricity is used. This is known as the maximum demand and is measured by recording the amount of electricity used in successive 30-min periods. The maximum demand is agreed on with the supplier for the determination of charges before use, and any infringement carries penalties of additional charges.

Gas and Air Metering

Gas meters measure the volume of gas passing through them, using either a diaphragm expansion device or a rotating vane. Since the volume of a gas is affected by both temperature and pressure, it is essential to correct the supply for factory conditions, using a correction factor related to the standard conditions of supply, ie, a pressure of 10^5 N/m² (1 bar) and 288.6 K (15.6°C). It is also important to know the gross calorific value of the fuel in order to calculate the equivalent energy consumption. The value is given to the consumer by the producer in units of either therm/100 ft³ or MJ/m³ and related to the standard conditions of temperature and pressure. The energy can then be obtained by multiplying by the volume flow rate.

Oil Metering

Positive and semipositive inferential displacement meters are often used for metering the oil feed to steam-raising boilers. In the first type the moving elements, ie, rotary pistons, are sealed so that fluid cannot pass without displacing them, while in the second type the moving membranes have a small but definite clearance which permits the passage of a small portion of the fluid. Alternatively, differential pressure flow meters may be used which incorporate restrictions such as orifices or nozzles. In the case where the burners operate on a circulating oil supply, it is necessary to use two meters: one on the supply and the other on the return from the feed. The flow rates should be corrected to 288 K (15·C) and the calorific value at the same temperature obtained from the suppliers.

Water Metering

For the purposes of factory audits it is useful to include the water usage. The types of meters described in the previous section may be used or, alternatively, a variable area or aperture type with a recording device. Effluent flow is usually measured in a standard V-notch or weir-type device.

Steam Metering

In the canning industry the measurement of steam flow is especially important, first to determine the efficiency of the boilers and second to determine the use for specific operations and individual pieces of equipment.

There are basically five types of flow meter suitable for measuring steam flow rates (2).

1. Orifice-plate meters, which relate the pressure drop across the orifice to the flow rate. These are useful when pressure fluctuations are very small; however, compensation devices may be incorporated in the design. The accuracy of this type depends on the installation being correct, with the orifice having a precision edge. The accuracy falls off significantly at low flow rates.

2. Vortex meters make use of the development of vortices when a fluid flows over an irregularly shaped object, eg, a prism. The frequency of vortex shedding is proportional to the velocity of flow. This type of meter must be installed in long lengths of pipe in order to achieve accurate results. Generally, the meters operate best at high flow rates.

3. Turbine meters work on the principle of converting flow motion into rotary motion, the rate of rotation being proportional to the flow.

4. Variable area flow meters consist of a vertical tapered tube, usually transparent, and a small solid object of circular cross-section which adopts a position in the tube proportional in height to the flow velocity. Metallic tubes may be used with magnetic positioners to locate the position of the float.

5. Pitot tube flow meters consist of a tube which is placed in the flow stream and connected to a manometer, so that the differential pressure developed by the flow can be measured in relation to the static pressure. Pressure compensation for steam can be achieved automatically with this type of flow meter.

An important point with steam flow meters which involve the use of differential pressure measurement using a manometer is that vacuum breakers are required when the factory is closed down to prevent condensate from sucking back into the measuring system.

Degree Days

In order to quantify the effects of weather conditions on space-heating requirements within offices and the factory environment, the degree day is used. This is defined as the daily difference in temperature in degrees Celsius between the base temperature of 15.5°C and the 24-h mean outside temperature. The colder the weather, the higher the value. When the temperature reaches 15.5°C, the value is zero, since internal heating should not be required. A maximum–minimum thermometer is used to determine degree days, the average of the two extremes being used and the daily figures added up to produce a cumulative value for the week or month, as appropriate.

Specific Energy

It is convenient to relate the factory production rate to the energy used. In the canning industry, production is usually measured in cases of two dozen cans; however, for the purposes of energy audits, it is necessary to have a more precise measure. Experience has shown, especially in factories producing a variety of canned goods, that the total weight of finished product that is being handled is the best measure. The unit of specific energy is kilojoules per kilogram (kJ/kg).

Since most factories use several sizes of cans, it is necessary to be able to convert cases of cans into a common unit size. The most convenient unit size is the one which is produced in the largest numbers and all other sizes are related to this size. For example, if a factory produces mostly UT cans (average weight 440 g), the equivalent for A1 cans (average weight 283.5 g) would be 283/440 or 0.64; ie, the weight of a case of A1 cans would be 0.64 times the weight of a case of UT cans of the same product.

MANAGING ENERGY CONSUMPTION

In order to make savings in energy use to increase the profitability of a factory, it is essential to have a strategic management technique. One of the most widely used techniques is known as monitoring and targeting (1). This basically involves the measurement of energy and production, and determining energy saving measures in the factory in order to target reductions in energy. Ideally, the work force should be involved and energy savings dictated from the factory floor by instilling a keen sense of awareness of prevention of waste.

One of the most useful methods of following the trend in energy usage is known as the scattergram technique. This involves plotting energy consumption against production (weight of product) or, for space heating, degree days, on a convenient time basis, eg, by shift or daily. A typical scattergram is shown in Figure 1. The points should be numbered to identify the time; data for different shifts, especially day and night shifts, should be dealt with separately so that correct comparisons are made. After plotting the series of points it is necessary to establish the slope of the best-fit line and the intercept. The former gives the specific energy of production and the latter the base load (for no production). Anomalous points, especially above the line, should be examined and appropriate staff consulted to establish possible causes for poor energy performance. Careful examination of the base load should also be made in order to reduce the zero-production energy requirement. After monitoring production it is possible to target for a reduction in energy usage by plotting a new line about 5% below the previous line. To achieve the new target it is necessary to introduce energy-saving measures, which may be simply good housekeeping practice, eg, not leaving any service running unnecessarily, or improving lagging, steam trap performance, lighting efficiency, etc (see below). The investment necessary for these measures should be planned against the pay-back period,

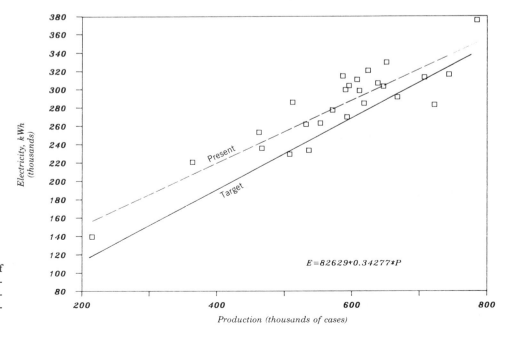

Figure 1. Typical scattergram of electricity consumption vs production of canned foods in cases showing the present achievement and future target.

$E = 82629 + 0.34277*P$

since large projects may not be economical in terms of the financial position of the company.

For other graphical methods and computer data handling, refer to more detailed sources of information (1).

The preceding discussion assumes that adequate submetering has been installed in the factory. This is not always possible and, when this is the case, the method of multiple linear regression analysis may be used (3). From the factory meters it is possible to establish the total individual energy usage and the production corresponding to the same period. A convenient model for this type of analysis is

$$E_T = E_0 + E_1 x_1 + E_2 x_2 + \cdots + E_n x_n$$

where E_T = total energy used per period of time, eg, MJ/day
E_0 = base load, MJ/day
x_n = production rate of product, kg/day
E_n = energy used per unit production of product, n

From the equation it is possible to determine values for the various energy terms: E_0, E_1, E_2, etc.

The technique has wide general applicability and may also be used in factories where there is submonitoring and several different product lines are being operated at the same time. In this case, E_T refers to the particular area being monitored.

FACTORY OPERATIONS AND ENERGY MANAGEMENT

It is convenient to divide the factory into energy accountable areas for the purposes of energy management (1). This may involve considerable alteration to the service supplies, and the cost of this must be taken into account before deciding on the areas and operations to be monitored. A useful method of division of the factory site is (a) preparation, (b) processing–filling, seaming, and retorting, (c) labeling, packaging, and storage, (d) offices and administration, and (e) boiler house. The operations carried out in these areas are as follows:

(a) The preparation area will contain equipment for reception and handling of raw materials, eg, grading, washing, peeling, cutting, homogenization, comminution (meat), blanching, inspection, soaking, extrusion (pasta), and forming; preparation of soups, sauces, and brines; as well as conveying and elevating. Raw materials storage, especially chilled storage, should also be included.

(b) The processing area starts with the prepared materials and should include can conveying, filling, seaming, retorting, and cooling filled containers.

(c) The packaging area involves the labeling, packaging, shrink wrapping, and storage of cans.

(d) The offices, administration, and workshops should be considered together in relation to space heating, ventilation, and small electrical loads.

(e) In the boiler house the efficiency of conversion of fuel to steam is extremely important. The results of a recent survey of UK canning factories (4) show that enormous savings can be made by attention to small details in the operation of the boiler house. Good management of this resource can result in considerable financial rewards and should be made a priority by all factory managers.

ENERGY SAVING MEASURES

These fall into several categories (1) and vary from no cost to high capital investment.

Good Housekeeping Measures

These are essential in a well-managed factory and are the responsibility of every employee. They include simple items such as switching off power when not required, turning off water when not being used, and reporting all leaks to management. Employees should be made aware of the cost of these problems and the financial savings which will be made as a result of prevention.

Good Manufacturing Practice

The correct operation of the process plant in relation to achieving full loading is necessary to reduce energy waste. This also applies to buildings such as storage areas.

Plant and Building Modification

The cost of making modifications and carrying out maintenance is a necessary and essential part of good factory practice. Typical examples include sealing leaks of steam, air, water, and oil; correcting the operation of steam traps; improving lagging of steam, hot water, and refrigeration services; improving the insulation of buildings, and installation of special types of doors.

New Technology

Much of the machinery currently being used in canneries was designed without considering energy use, noise, or environmental aspects. It is therefore often necessary to invest in new equipment which will minimize these problems. The capital cost is often high and the pay-back period considerable, but it may be the best solution to reducing energy usage. Some examples are installation of high-efficiency burners on boilers, high-efficiency motors and lighting, motor controllers for power saving in fixed and variable speed applications, electricity demand controllers, and electronic energy management systems.

Storage Systems

Storage systems are useful in minimizing energy usage by either recirculation and recovery techniques or use of special storage systems. Examples of the former are heat recovery from condensate and preheating using recirculation, and of the latter, electric batteries, steam accumulators, masonry thermal storage, and hot water recovery and storage.

Examples

Table 1 gives some specific examples of energy saving measures, together with a guide to pay-back times, which depend on the prevailing economic situation and costs of energy.

Table 1. Some Examples of Energy Saving Measures

Energy Source		Energy Saving Measures	Pay-Back Time in Years
1. Boiler house	1.1	Boiler scheduling and sequencing by installation of automatic control	1–2
	1.2	Boiler replacement	4–6
	1.3	Alternative fuels and renegotiation of fuel prices	1
	1.4	Boiler isolation to prevent circulation of steam through standing (nonfiring) boilers by fitting nonreturn valves	1
	1.5	Burner improvements	3–5
	1.6	Flue gas monitoring and control	1
	1.7	Increased air temperature	3–4
	1.8	Fuel oil treatment—chemical closing	1
	1.9	Combustion chamber cleaning to increase effective heat transfer	1
	1.10	Fuel oil handling system maintenance	1
	1.11	Feed water preheating and condensate return	2–3
	1.12	Flue gas heat recovery	2–3
	1.13	Automatic blowdown and heat recovery	1
	1.14	Boiler water treatment—total dissolved solids monitoring	3–4
	1.15	Fan and pump speed control	2–3
2. Steam distribution system	2.1	Main header pipework configuration to ensure balanced distribution and workload	1–2
	2.2	Steam accumulation	3–4
	2.3	Steam distribution system: removal of dead pipework and isolation of pipework not in use	1
	2.4	Steam-trap and air-release valve maintenance	1
	2.5	Steam leak elimination	1
	2.6	Insulation of steam pipes	1
	2.7	Steam condensate recovery	1
	2.8	Steam/water cleaning systems: use of alternatives, eg, waste hot water	1
	2.9	Load scheduling in relation to production planning	
3. Compressed air and vacuum supplies	3.1	Intake air relocation to be as cool as possible	1–2
	3.2	Correct trapping and filtering	1–2
	3.3	Heat recovery from compressor	3–4
	3.4	Reduction in air pressure to working pressure required by machinery	1
	3.5	Plant scheduling and automatic loading	1–2
	3.6	Leak correction	1
	3.7	Air drying using waste heat from compressor	1–2
	3.8	Air/vacuum system design	3–4
4. Electricity supplies	4.1	Tariff examination and avoidance of exceeding maximum demand	
	4.2	Power fraction correction	1
	4.3	Motor drives: use of high-efficiency motors and variable-speed drive controllers	2
	4.4	Lighting: high-efficiency lamps	2–3
	4.5	Electronic energy management systems	2–3

ENERGY USE IN CANNING OPERATIONS

For a typical fruit and vegetable canning factory (4) producing 3.6 million cans per week, the overall energy required is approximately 19 pounds of steam per case of 24 UT cans or 0.8 kg/kg product. For a meat canning operation, the figure is slightly higher, ie, 1 kg/kg product, due to the longer processing times required. The corresponding figures for electricity are 0.1 kWh/kg product for a fruit and vegetable cannery and 0.22 kWh/kg product for a meat canning factory. The higher figure for meat canning is due to the use of comminuting machinery and the requirement for chilled storage.

These figures may be broken down further for energy use in different areas of the factory. The preparation area for a fruit and vegetable cannery, which will include steam peeling, syrup and brine preparation, and blanching, requires approximately 0.3 kg steam/kg product and 0.018 kWh electricity/kg product, whereas the processing requires 0.5 kg steam/kg product and 4.8×10^{-4} kWh electricity/kg product. The labeling and packaging area requires 0.035 kWh/kg product only.

For the general preparation operations, the following figures (kg steam/kg product) are a useful guide: pea blanching, 0.12; vegetable blanching (green and broad beans and root vegetables), 0.28; liquor preparation including brines and syrups, 0.15; steam peeling—batch, 0.3 and continuous, 0.43.

These figures are taken from a data base compiled from actual factory operations (4). It is important to note that to achieve low energy use it is necessary to operate the lines at the highest possible loading and to run the canning plant at maximum production rate. This is clearly observed when night production is carried out on a reduced number of lines.

Several other factory investigations have been carried out in detail (5) for tomato products (6), peaches (7), and spinach (8). The overall use of energy for a tomato processing plant handling 1,200–1,350 tons of raw tomatoes per day amounted to 1,163 kJ and 0.028 kWh per kilogram of raw tomatoes. The thermal energy required for different types of products was also measured: 1,086 kJ/kg for tomato juice production; 1,300 kJ/kg for peeled tomatoes; and 1,307 kJ/kg for tomato paste. In relation to individual operations, the lye peeling required 160 kJ/kg and the retorting about 800 kJ/kg for the whole peeled-tomato canning operation, whereas the tomato paste production required 1,000 kJ/kg for the evaporator.

For a spinach canning plant the main uses of steam were: blanching, 1.0 MJ/kg; exhausting, 1.5 mJ/kg; and retorting, 0.5 MJ/kg. The electricity use was also reported for each operation; some figures for the higher consumption of electricity are washing, 4×10^{-3} kWh/kg; blanching, 4×10^{-4} kWh/kg; and retorting, 4×10^{-4} kWh/kg. For a peach canning plant the consumption of steam was 1.6 MJ/kg raw material for retorting and 0.2 MJ/kg for peeling, and electricity usage was 4.75×10^{-4} kWh/kg for retorting and 6×10^{-4} kWh/kg for peeling.

A similar exercise (9) on a meat cannery demonstrated that, for a conventional static retort, the energy required for 4-oz can (530 kJ/kg) and 15.5-oz (568 kJ/kg) can showed that for a fourfold increase in weight, only a 7% increase in energy was required.

In relation to different types of sterilizers, several workers (4,10–13) have published information which is summarized in Table 2.

Considerable energy saving may be achieved (up to 10%) by the use of hot water processing compared with atmospheric steam processing for sterilizing canned peaches (14). However, there is relatively little work reported in the area of heat transfer media.

In relation to energy saving by optimization of processing times and temperatures, energy requirements for batch steam sterilization of pea purée in 307 × 409 cans have been studied (15). Six different times, ranging from 196 to 50 min, and temperatures from 104 to 132°C were used. It was shown that there is a minimum energy requirement of 0.3 kg steam/kg product for a process of 64 min at 121.8°C.

Relatively little work has been reported on the energy requirements for differing types of packages, although it is generally appreciated that pouch and semirigid types, because of their geometry, require less energy than the corresponding volume metallic containers. A comparison of the processing of pouches versus cylindrical metallic containers (303 × 406), which included details of the equipment, labor, and containers economics, has been made (16). This work showed that the retort pouch system was more energy efficient than the metallic container for the processing of peaches in syrup. No doubt for steam sterilization of low-acid foods, greater energy savings may be achieved.

A detailed study (17) of three water and four steam blanchers has shown that the most energy-intensive type was the open-ended steam blancher (2.12 MJ/kg), whereas the conventional screw-type water blancher required 0.91 MJ/kg.

COMPARISON OF ENERGY USED IN CANNING WITH OTHER FOOD PROCESSING OPERATIONS

In order to make comparisons, it is necessary to ensure that all aspects of the processing are included, eg, transport of raw material, processing, packaging, distribution, and storage. Table 3 gives some of the comparisons avail-

Table 2. Relative Energy Usage for Various Types of Sterilizer

System	Casimir (10)	Savage (11)	Jowitt and Thorne (12)	Ferrua and Col (13)	Holdsworth (4)
Batch retort	100	100	100	100	100
Crateless retort		50			
Hydrostatic cooker	21.3	25	56		56–96
Fluidized-bed retort			38		
Microwave sterilizer	1230				
Continuous rotary sterilizer				46	45
Continuous atmospheric sterilizer				64	
Flame sterilizer (batch)	266				
(continuous)	90			88	
Hydrolock cooker					54

Table 3. Some Energy Usage Comparisons

Process	Product	Energy, MJ/kg	Ratio	Reference
Canning	Peas	14.9	1	18
Freezing	Peas (stored 1 mo)	10.0	0.67	
Drying	Peas	10.9	0.73	
Canning	Peas	48.3	1	19
Freezing	Peas (stored 8 mo)	43.6	0.9	
Chilling	Peas	44.7	0.92	
Canning	Peas	27.1	1	20
Freezing	Peas (stored 5.5 mo)	21.6	0.79	
Canning	Corn	40.3	1	21
Freezing	Corn (stored 6+ mo)	38.2	0.95	
Chilling	Corn	49.9	1.24	
Canning	Corn	25.8	1	22
Freezing	Corn (stored 8 mo)	29.5	1.14	
Canning	Snap beans	47.4	1	21
Freezing	Snap beans (stored 6 mo)	44.2	0.93	
Chilling	Snap beans	34.7	0.73	
Canning (home)	Root vegetables	3.56	1	23
Drying (home)	Root vegetables	6.44	1.8	
Freezing (home)	Root vegetables (stored 1 mo)	3.79	1.1	
Freezing (home)	Root vegetables (stored 3 mo)	6.93	1.9	
Freezing (home)	Root vegetables (stored 6 mo)	11.80	3.3	
Freezing (home)	Root vegetables (stored 12 mo)	21.40	6.0	

able in the literature; these are indicative rather than quantitative because of the differing assumptions which various workers made.

In general, canning is the most energy intensive process, although this depends on the product. With frozen foods, the energy required is very dependent on the length of time of frozen storage.

BIBLIOGRAPHY

1. S. D. Holdsworth, C. Dennis, B. Mallalieu, and B. Powell, *Energy Monitoring and Targeting Manual for the Canning Industry* (Technical Manual No. 23), Campden Food and Drink Research Association, Chipping Campden, Gloucestershire, UK, 1988.
2. A. T. J. Hayward, *Flowmeters*, Macmillan, London, 1985.
3. A. C. Cleland, M. D. Earle, and I. F. Boag, "Application of Multilinear Regression to Analysis of Data from Factory Energy Surveys," *Journal of Food Technology* 16, 481–492 (1981).
4. S. D. Holdsworth, *Data Base of Energy Usage in the UK Canning Industry*, Campden Food and Drink Research Association, Chipping Campden, Gloucestershire, UK, 1989.
5. R. P. Singh, "Energy Accounting in Food Canning Industry," in B. A. Stout, ed., *Energy in World Agriculture*, Vol. 1 of Energy in Food Processing, Elsevier, Amsterdam, 1986, Chapt. 10, pp. 174–190.
6. R. P. Singh, P. A. Carroad, M. S. Chhinnan, W. W. Rose, and N. L. Jacob, "Energy Accounting in Canning Tomato Products," *Journal of Food Science* 45, 735–739 (1980).
7. R. P. Singh, P. A. Carroad, M. S. Chhinnan, and W. W. Rose, "Energy Use Quantification in the Canning of Clingstone Peaches," *Journal of Food Science* 45, 723–725 (1980).
8. M. S. Chhinnan, R. P. Singh, L. D. Pedersen, P. A. Carroad, W. W. Rose, and N. L. Jacob, "Analysis of Energy Utilization in Spinach Processing," *Transactions of the American Society of Agricultural Engineers* 23, 503–507 (1978).
9. K. T. H. Farrer, "Some Aspects of Energy Costs in Food Processing," *Food Technology in Australia* 7–13 (Jan. 1977).
10. D. J. Casimir, *Economics of Flame Sterilisation* (Food Research Report No. 54), CSIRO Division of Food Research, Australia, 1971.
11. D. Savage, "Processing Sterilised Products," in *Profitability of Food Processing* (Symposium Proceedings, Series No. 84), Institution of Chemical Engineers, London, 1984, pp. 307–310.
12. R. Jowitt and S. N. Thorne, "Evaluates Variables in Fluidised-Bed Retorting," *Food Engineering* 43(11), 60–62 (1971).
13. J. P. Ferrua and M. H. Col, "Energy Consumption Rates for Sterilisation Equipment," *Canner/Packer* 44–46 (Jan. 1975).
14. M. S. Chhinnan and R. P. Singh, "Energy Conservation in a Continuous Atmospheric Retort," *Lebensmittel-Wissenschaft und Technologie* 14, 122–126 (1981).
15. J. A. Barreiro, C. R. Perez, and C. Guariguata, "Optimisation of Energy Consumption during the Heat Processing of Food," *Journal of Food Engineering* 3, 27–37 (1984).
16. J. R. Williams, J. F. Steffe, and J. R. Black, "Economic Comparison of Canning and Retort Pouch Systems," *Journal of Food Science* 47, 284–290 (1981).
17. W. W. Rose, L. D. Pedersen, P. A. Carroad, and E. P. Scott, *Energy Conservation in the Food Processing Industry, Phase II: Measurement of Energy Used in Blanching*, National Food Processors' Association Research Foundation, Berkeley, California, 1981.
18. P. E. Doe, "Energy Wastage in Food Processing," *Food Technology in Australia*, 158–160 (April 1977).

19. M. A. Rao, "Energy Consumption for Refrigerated, Canned and Frozen Peas," *Journal of Food Process Engineering* **1**, 149–165 (1977).
20. G. Löndahl, "Energy Consumption in Freezing Compared with Food Canning," *Frozen Foods* **32**(10), 36, 38, 42 (1979).
21. M. A. Rao, "Energy Consumption for Refrigerated, Canned and Frozen Snap Beans and Corn," *Journal of Food Process Engineering* **3**, 61–76 (1980).
22. Y. S. Henig and H. M. Schoen, "Frozen Versus Canned Corn," *Food Engineering International* 46–47 (Sept. 1976).
23. F. Drew and K. S. Rhee, "Energy Use, Cost and Product Quality in Preserving Vegetables at Home by Canning, Freezing, and Dehydration," *Journal of Food Science* **45**, 1561–1565 (1980).

General References

G. C. Dryden, ed., *The Efficient Use of Energy*, IPC Science and Technology Press, London, 1975.

Energy Managers' Workbook, Energy Publications, Cambridge, 1982.

P. M. Goodall. *The Efficient Use of Steam*, IPC Science and Technology Press, London, 1980.

D. N. Lapedes, *Encyclopedia of Energy*. McGraw Hill Book Company, Inc., New York, 1976.

O. Lyle, *The Efficient Use of Steam*, HM Stationery Office, London, reprinted 1975.

G. Payne, *The Energy Manager's Handbook*, IPC Science and Technology Press, London, 1977.

J. P. Quale, ed., *Kempe's Engineers Year Book*, Morgan Grampian Book Publishing, London, 1989.

A. Turtell, *Energy Consumption and Cost of Food Processing: A Bibliography*, Technical Note No. 163, Campden Food and Drink Research Association, Chipping Campden, Gloucestershire, UK, 1984.

S. Donald Holdsworth
Campden Food & Drink
Research Association
Gloucestershire, United
Kingdom

ENOLOGY. See Food fermentation; Wine

ENTERAL FEEDING SYSTEMS

Oral ingestion and digestion of an adequate amount of regular foods and beverages still constitutes the best way to ensure the proper nutrient intake of a patient (1). However when a patient is unable or unwilling to ingest a sufficient quantity or variety of such foods, supplemental or complete feedings may be initiated, either enterally (into the gastrointestinal tract) or parenterally (intravenous infusion). Parenteral nutrition will not be addressed in this chapter (2).

Enteral feeding has been defined as the provision of liquid formula diets administered orally or by tube into the stomach or upper gastrointestinal tract (3). Prior to the commercial availability of proprietary enteral feeding products, blenderized foods were used. Such homemade or institutionally prepared feedings are unsatisfactory because of their uncertain nutritional adequacy, homogeneity, and microbiological quality. In addition, these solutions often are too viscous to flow readily through feeding tubes. Because of these shortcomings, most medically supervised enteral feeding now makes use of commercially available defined formula diets. Proprietary defined formulas began to appear in the 1960s and 1970s, after a long period of feeding commonly available foods in liquified form. Interestingly, the first commercially available enteral feeds were elemental diets (formulas requiring very little digestion prior to absorption). These products were tested by NASA scientists who were attempting to develop a residue-free diet for use by astronauts (4,5). Although NASA rejected the product for use in space because of its poor palatability, elemental diets have been used for patients with impaired digestion. For a more complete history of enteral nutrition, see reference 6.

CLINICAL NEEDS AND COMPONENTS OF ENTERAL NUTRITION SYSTEMS

Basic Issues: Elemental Vs Polymeric, Caloric Concentration, Osmolality, and Fiber Content

Most of the enteral products currently on the market are so-called polymeric diets because they are composed of the same intact whole nutrients (proteins, carbohydrates, and lipids) that would occur in a normal diet (Table 1). Because they are designed to replace regular food, they are also referred to as meal-replacement formulas. Individuals fed polymeric formulas must have at least some digestive capability because polymeric formulas require digestion prior to their absorption and metabolism. Such formulas are nutritionally complete, meaning they contain most if not all the nutrients needed to prevent nutritional deficiencies, and defined, meaning that they are formulated with isolated nutrient sources and their composition is precise, according to rigorous specifications (7).

Although elemental formulas were extensively used in the past, polymeric diets now provide nutritional support in many medical conditions that do not entail a serious compromise of the gut's digestive capabilities (8). Elemental diets are desirable for patients with alimentary tract fistulas, cancer patients receiving radiation therapy, nonspecific maldigestion and malabsorption, short-bowel syndrome, inflammatory bowel disease (including Crohn disease and ulcerative colitis), and pancreatic insufficiency (9). Although a number of elemental diets exist, not all are therapeutically equivalent. Some elemental formulas contain oligomers while others contain only monomers. Several researchers suggest that nitrogen is absorbed more rapidly and uniformly in both the healthy and the diseased gut in the form of peptides than from equimolar free amino acid preparations (10–21). Glucose polymers present advantages over glucose because of their low osmolality, rapid hydrolysis, and subsequent rapid intestinal absorption (22). Medium chain triglycerides (MCTs) containing 8 or 10 carbon fatty acids are used because MCTs are well absorbed and transported directly into the portal blood, bypassing the complex steps required for digestion and absorption of long-chain triglycerides (23).

Table 1. Common Components of Enteral Solutions[a]

Carbohydrates	Proteins	Lipids
Corn syrup	Casein	Corn oil
Corn syrup solids	Soy protein	Soybean oil
Hydrolyzed corn starch	Egg white solids	Safflower oil
Sucrose	Skim milk	Canola oil
Malto dextrin	Whey	Fractionated coconut oil
Fructose	Puréed beef	Milk fat
Cereal solids		Beef fat
Fruit and vegetable purées		

[a] Not all of these components are highly purified and some may contribute more than just one macronutrient, for example, puréed beef is a source of fat as well as protein.

Caloric requirements or medical condition dictates the caloric concentration of the formula to be used. For example, a relatively stable adult could be provided with an adequate daily intake of protein, essential fatty acids, minerals, electrolytes, vitamins, and water in about 2,000 mL of a 1 cal/mL product. Contrast this with an elderly patient with congestive heart failure who has a much lower daily calorie requirement and needs fluid restriction, but still must have a full complement of protein, vitamins, and minerals. In this case a concentrated formula would be desirable.

These issues—caloric need and tolerance of formula concentration—must be considered in view of other needs of the patient. For example, one patient may need a low-residue diet prior to colonoscopic examination, while another patient starting on tube feeding postoperatively may tolerate an isotonic formula better than a hypertonic formula, and a chronically tube-fed patient may need a fiber-supplemented formula to maintain normal bowel function.

The osmolality of a formula is directly proportional to the number of particles in the solution. The main determinants of osmolality are the forms of the protein and carbohydrate and the levels of sodium, potassium, and chloride. Formula hyperosmolality has been associated with decreased tolerance, ie, vomiting, nausea, and diarrhea (23). The proposed mechanisms are delayed gastric emptying and fluid shifts the small bowel. Such fluid adjustments occur normally as the result of digestion, but may not be compensated for as readily in a patient with compromised gut function. Formula intolerance can be dealt with by slowing the feeding rate or using a half-calorie formula. Dehydration, another potential complication of tube feeding, can occur because of severe diarrhea, because the concentrating ability of the kidneys is exceeded, or as a result of any other cause of excessive water loss. Dehydration can be corrected by carefully adjusting water intake (23).

The addition of fiber to enteral products represents another step toward satisfying the complete nutritional needs of tube-fed patients. Most enteral products currently on the market are low residue, meaning that all of the components of the product are highly digestible and well absorbed. The addition of fiber to enteral products was done principally to maintain normal bowel function in chronically tube-fed patients. Dietary fiber increases stool bulk as a result of the water-holding capacity of fiber, the osmotic effect of short chain fatty acids, or alteration of colonic bacteria (microflora may account for up to 75% of wet fecal output) (24–28).

Dietary fiber also moderates transit time, that is the time taken by undigested particle to move through the gastrointestinal tract (24,25,28,29). An increase in dietary fiber has been shown to hasten transit time in individuals with slow transit and to prolong it in subjects with normally rapid transit (24,29). Because of this moderating effect of dietary fiber on bowel function, the addition of fiber to the diet has successfully alleviated both diarrhea and constipation in hospitalized or institutionalized patients (30–33).

The physiological effects of fiber are numerous and diverse and vary considerably according to the chemical and physical characteristics of the source of fiber consumed. To achieve the desired physiological effect in the clinical setting (for example, an increase in fecal wet weight) without any of the negative effects (for example, loss of minerals or vitamins) requires a considerable amount of careful testing. At present the only type of fiber that is added to medical nutritional products in amounts sufficient to have a documented effect on bowel function is a polysaccharide extracted from the cotyledon of soybeans. As shown in Table 2, soy polysaccharide is a fairly insoluble fiber source. An insoluble source is necessary because the addition of soluble fibers increases the viscosity of the formula resulting in difficulty in flowing through tubes.

Disease or Condition-Specific Nutritionals

In addition to the basic decisions regarding nutrient structure, calorie content, fiber content, and osmolality, certain clinical conditions may call for special formulations.

Pulmonary Disease. Patients with chronic obstructive pulmonary disease (COPD) develop various degrees of respiratory muscle fatigue, hypoventilation, carbon dioxide retention (hypercapnia), and oxygen depletion (hypoxemia). In the majority of patients who experience difficulty in breathing, the obstruction of airflow is the result of chronic bronchitis or emphysema (37). Typically, these patients experience marked weight loss, which may result both from decreased caloric intake (38,39) and from increased energy expenditure, presumably because of increased work of breathing (38). Weight loss and other parameters of inadequate nutrition are associated with poor outcome from pulmonary disease, such as respiratory failure or lack of success in weaning from the ventilator (40–

Table 2. Typical Fiber Content of Soy Polysaccharide by Various Analytical Methods[a]

Method	Component	g/100 g
AOAC Method	Total dietary fiber (soluble dietary fiber, 5%; insoluble dietary fiber, 95%)	75
Southgate method	Total dietary fiber (lignin, 12%; cellulose, 3%; soluble noncellulose, 5%; insoluble noncellulose, 80%)	75
Englyst method	Total dietary fiber (soluble dietary fiber, 27%; insoluble dietary fiber, 73%/cellulose, 15%; noncellulose, 85%)	74

[a] Refs. 34–36.

43). Because the oxidation of fat results in less carbon dioxide production than the oxidation of a equal caloric load of carbohydrate, formulas high in fat and low in carbohydrate make sense for pulmonary patients. Such formulas can benefit both the ambulatory patient with chronic obstructive lung disease by allowing increased activity (44) as well as the ventilator-dependent patient whose consumption of a high fat formula may aid in weaning from the ventilator (45). Because respirator-dependent patients are prone to develop hypophosphatemia, they may also benefit from phosphate supplementation as part of their nutritional regimen (46,47).

Diabetes. Currently, most diabetes health care specialists use the dietary recommendations proposed by the American Diabetes Association. These guidelines specify that protein should supply 12–20% of total energy intake with carbohydrate providing 50–60% and fat 20–30% (48). The rationale for this recommendation is the desirability of lowering saturated fat and cholesterol intake to reduce cardiovascular risk (49). Most enteral formulas now available comply with these recommendations. Nonetheless, tube feeding or dietary supplementation with standard medical nutritional products can compromise the metabolic control of patients with hyperglycemia. This occurs because rapid gastric emptying of enteral formulas and efficient absorption of nutrients supplied by these products produce a rapid elevation in blood glucose. Standard enteral formulas empty from the stomach at least twice as fast as an isocaloric solid-food meal (50) and produce a peak blood glucose response equivalent to that seen when an equivalent solution of pure glucose was fed (51). While rapid gastric emptying of formulas and rapid absorption of nutrients may be advantageous for patients with normal pancreatic function and insulin sensitivity, they may complicate metabolic control of hyperglycemic patients. Recent research has shown that in normal solid food diets, the substitution of polyunsaturated or monounsaturated fat for carbohydrate can lower mean plasma glucose and triglyceride levels in type II diabetics without a deleterious effect on serum cholesterol levels (52,53). When high- and low-fat enteral formulas were compared in type II diabetics, the high-fat formulas resulted in lower mean increments in the glucose absorption curve and in the insulin response than the low-fat–high-carbohydrate formulas (54). Therefore a low-carbohydrate, high-monounsaturated fat formula seems a reasonable choice for patients with abnormal glucose tolerance.

Chronically Tube-Fed Patients. Elderly patients may require long-term tube feeding as a result of stroke, Alzheimer's disease, upper gastrointestinal surgery, or other chronic debilitating illnesses. Younger patients with neurologic impairments or developmental disabilities may also require long-term support. Typical daily energy intakes of nursing home patients are 1,400 kcal for women and 1,700 kcal for men (55). A study in developmentally disabled institutionalized individuals reported an average energy need of approximately 1,000 kcal/day (56). This suggests that these patients need adequate amounts of protein, vitamins, and minerals supplied in a relatively low calorie base. Long-term tube-fed and parenterally fed, patients often have deficiencies that are infrequent in the general population and in patients that have been fed enterally for only a short time. For example, deficiencies of chromium (57–59), selenium (60,61), and molybdenum (62,63) as well as carnitine (64,65) and taurine (66) have been reported in this population. The supplementation of trace elements and conditionally essential nutrients, as carnitine and taurine have been called, can help normalize serum levels of these nutrients in long-term tube-fed patients (67).

Renal Disease. Renal disease typically results in progressive loss of kidney function until dialysis or kidney transplantation is required. Therefore, the enteral feeding of patients with kidney disease can be divided into two broad categories: predialysis and dialysis patients. One of many features of chronic renal failure is a gradual increase in toxic nitrogenous wastes, the by-products of protein and amino acid metabolism (68). Protein restriction has been advocated to control the rate of production and accumulation of these wastes (68–70) and possibly limit disease progression (68). For patients with partial loss of kidney function, a moderate restriction of protein to 0.55–0.60 g/kg body weight/day is recommended (68). If the loss of kidney function is severe, a very low protein diet of 0.28 g/kg/day has been recommended, supplemented with an essential amino acid mixture or the equivalent keto or hydroxy analogues of essential amino acids (68). In addition to restricting protein, it is necessary to ensure adequate energy intake (a minimum of 35 kcal/kg/day is recommended) (71), as well as phosphorus (68,69,72) and vitamins A and D (68,73–75), while restricting water and electrolytes (76). Supplementation with water-soluble vitamins (68,73), calcium (68,72), and trace elements (68,77) is also recommended. The recommendations for moderate protein restriction can be achieved with an energetically concentrated enteral product containing approximately 6% of calories as protein. Once the disease has progressed to end stage and no kidney function remains, transplantation or dialysis are necessary to insure the survival of the patient. Dialysis, which can be done by filtration of the blood or lavage of the peritoneal cavity, results in losses of various nutrients that need to be replaced in the diet. For example, dialysis patients need to liberalize protein intake to 1.2–1.5 g/kg/day (ca 14–15% of calories) (68).

Cancer. Cachexia or wasting as a result of altered metabolism or decreased intake is a hallmark of cancer. The presence of a malignant tumor can cause a number of physiologic changes that can have an impact on nutritional status. Some tumors cause an increase in basal metabolic rate (78), and changes in glucose (79) and protein (80) metabolism have been reported. Fluid and electrolyte imbalances (81), malabsorption (81), and anorexia (82) are other reported problems. There also can be serious nutritional consequences of cancer therapy: surgery, chemotherapy, or radiation therapy can cause nausea and vomiting, which can sometimes be treated with antiemetic drugs (81,83,85). Creative approaches are needed to maintain food intake, such as offering small high-calorie meals to anorexics, or puréed foods to patients with painful swallowing (86,87). If oral intake is not satisfactory, tube feeding may be necessary; patients with obstructions of the nasopharynx or the esophagus can be fed in different sites as described under tube feeding (Fig. 1).

The maintenance of adequate dietary intake in cancer patients does have benefits. Oral supplementation and tube feeding can help prevent weight loss during therapy (91–92), which is strongly related to the positive outcome of therapy. Aggressive tube feeding during chemotherapy can improve hematopoeitic status (93), which may be related to more active chemotactic and phagocytic function (94). A peptide elemental diet provides protection against radiation damage to the intestinal mucosa (95).

Subjects with Special Nutritional Requirements

Inborn Errors of Metabolism. A number of inborn errors of amino acid metabolism can be managed by means of correct early diagnosis of the problem and judicious nutritional intervention. Consider, for example, the following diseases (the offending amino acid is in parentheses):

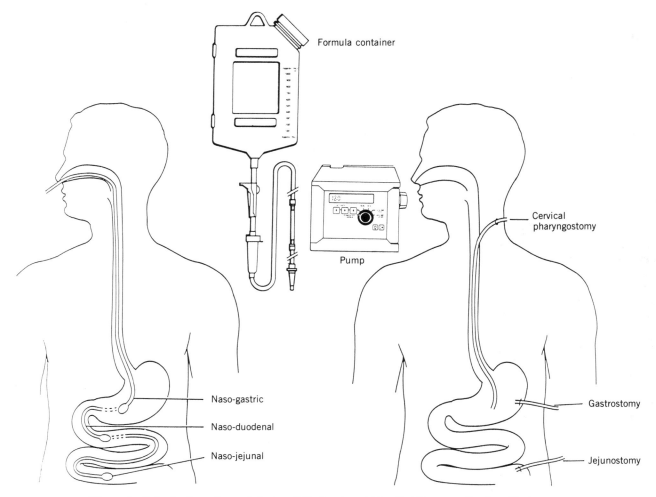

Figure 1. Tube feeding sites and devices (88). Courtesy of Applied Therapeutics, Inc., Vancouver, Washington.

phenylketonuria (phenylalanine), maple syrup urine disease (the branched-chain amino acids leucine, isoleucine, and valine), propionic and methylmalonic acidemias (methionine, threonine, valine, and isoleucine), homocystinuria (methionine), tyrosinemia type I (phenylalanine, tyrosine, and methionine), and alkaptonuria (phenylalanine and tyrosine). The effects of these disorders can be controlled in infants and young children by feeding formulas or medical foods that are low in the offending amino acid. The balance of the nutrients can be adjusted to suit the age of the patient (infant or young child).

Classic phenylketonuria (PKU), the most prevalent of these disorders, is caused by a defect in the enzymatic conversion of phenylalanine to tyrosine leading to toxic blood concentrations of phenylalanine and its metabolites. If dietary restriction of phenylalanine is begun within one to two weeks after birth and continued through life, physical and mental development may be normal. Because the fetus is very sensitive to high blood phenylalanine concentrations, PKU women who wish to become pregnant are advised to remain on a low-phenylalanine diet. In addition, recent evidence suggests that high blood levels of phenylalanine may affect neural function in adults. Special PKU formulas are available for these adults (96).

Obesity. According to the National Center for Health Statistics, as many as one-quarter of adult Americans are overweight (defined as 20% or more above desirable weight) and about one-third of these are severely overweight (40% above ideal body weight) (97,98). Large-scale epidemiological studies have revealed a dramatic increase in the relative risk of morbidity and mortality from chronic diseases in obese subjects (99–101).

The preferred method of reduction depends on how much weight needs to be lost. Because moderately overweight subjects are not usually considered to be at significant health risk, rapid loss of weight is not imperative. In these individuals, the recommended approach to weight loss is a balanced reduced-calorie diet, combined with nutrition education, behavior modification, and physical activity.

Obese subjects, however, have more than just aesthetics to consider and may benefit from a more aggressive weight loss program. Very low calorie diets, providing less than 800 kcal per day, are now considered a medically appropriate treatment for moderately and morbidly obese patients (102–105). These diets should provide 1.2–1.5 g of protein per kilogram of ideal body weight to help preserve lean body mass, less than 50 g of carbohydrate to promote moderate ketosis, and 100% of the USRDA for vitamins and minerals (105–110). Fluid intake should be liberal, with at least one to two quarts per day. Patients on these diets require medical supervision and monitoring throughout weight loss and during adaptation to a solid food diet. There are distinct benefits to a beverage-based very low calorie diet as a dietary treatment, because there are no food decisions or distractions. Studies have confirmed that the use of such a diet in obese subjects results in an average weight loss of 44 lb in 12 weeks (104,111). For successful, long-term maintenance of weight loss, it is essential to pair the diet with a comprehensive treatment program including nutrition education, behavior modification, and increased physical activity (103,104,112).

Athletes. Caloric requirements vary considerably among athletes: a survey of 27 college football players showed that their daily calorie intakes ranged from 19.7 to 62.1 kcal/kg of body weight (113). This makes it difficult to prescribe a standard diet for athletes as a group, even within a fairly homogeneous group participating in only one sport. When daily calorie intake is marginal, muscle carbohydrate stores (glycogen) become depleted, resulting in decreased performance and endurance. Controlled studies have suggested that to prevent glycogen depletion, athletes need to consume 60% of total daily calories as carbohydrates (114,115). This means that there must be an effort to substitute carbohydrate for fat, while protein intake is generally sufficient (15% of dietary calories from protein is more than adequate and typical of a standard American diet). In addition to consuming adequate amounts of carbohydrate, recent research suggests that the timing of carbohydrate ingestion can affect the performance and recovery of athletes. The essential characteristics of a preevent meal for endurance athletes are described in Table 3.

During prolonged activities, such as soccer or distance running, water is the most important replacement factor. However, when an appropriate carbohydrate-containing replacement drink is used instead of water during exercise, energy status and endurance are improved (117). Following the completion of strenuous exercise, muscle glycogen is normally resynthesized to normal levels within approximately 24 h provided sufficient dietary carbohydrate is consumed (118,119). However, athletes on heavy

Table 3. Essential Characteristics of the Preevent Meal for Endurance Athletes[a]

High carbohydrate content	Supports blood glucose during competition; should consume 60–65% of total calories as carbohydrate
Low-fat and low-protein content	Fat slows gastric emptying, protein can aggravate dehydration
Low-salt content	High salt levels can cause greater water losses
Minimal bulk	Bulk foods increase gastrointestinal residue, which can lead to discomfort or diarrhea
Adequate fluid	16–32 oz of water or juice will help insure adequate hydration

[a] Ref. 116.

training schedules (ie, those who compete or train several times in a 24-h period) often have difficulty recovering completely in 24 h. These athletes may benefit from consuming carbohydrate immediately after exercise. If carbohydrate is provided immediately, glycogen synthesis is almost twice as fast than if carbohydrate is consumed 2 h after exercise (120). To fulfill the requirements described above, a number of specially formulated drinks are now available. Runners and triathletes are familiar with the numerous fluid-replacement and energy drinks provided at weekend races. Athletes, and in particular student athletes who combine very high calorie needs with hectic schedules, often benefit from high carbohydrate and liquid meal replacement formulas inserted at strategic points during their daily routine.

Modules. There are a number of single nutrient modules currently available on the market, such as sources of protein, carbohydrate, fat, fiber, vitamins, and minerals. Although the use of a module to modify a liquid requires special handling, it allows for custom blending of unique formulations (121).

DELIVERY SYSTEMS

Nasoenteric tube feeding dates back as far as the fifteenth century; modern use began in the early 1950s with the development of fine polyethylene tubes and feeding pumps (122). In the 1960s, the development of total parenteral nutrition (TPN), which was viewed as more sophisticated and problem free, contributed to reduced use of enteral tube feeding in the 1970s and early 1980s. However, tube feeding is now resurging because of a number of factors. Enteral nutrition is less expensive than TPN (123) and results in fewer serious complications (124). Research shows that enteral feeding maintains gut mass, integrity, and function better than parenteral nutrition (7). The use of small-bore tubes made of soft, biocompatible materials has led to increased patient tolerance of nasally placed tubes.

Most feeding tubes are now made of polyurethane or silicone, which does not disintegrate or become brittle over time and, therefore, does not need frequent replacement. Feeding tubes are most easily inserted into the stomach by way of the nasopharynx. If the patient is at high risk for aspiration, the tube can be passed further down into the duodenum or jejunum, although it is sometimes difficult to get the tube through the pylorus (122). If the transnasal route is not possible or long-term feeding is desired (more than six weeks), tubes can be inserted directly into the esophagus or into the stomach or duodenum through the abdominal wall (tube placements are depicted in Fig. 1). Cervical pharyngostomies or, more rarely, esophagostomies are performed when the access site has been obtained during surgery for head and neck cancer or surgical repair of the maxillofacial area (125). Feeding gastrostomies can be placed surgically or percutaneously (126); the advantage of percutaneous endoscope placement is that it can be done as an outpatient procedure by a gastroenterologist with local anesthesia. Feeding jejunostomies are usually performed in conjunction with other intestinal surgery. Because paralytic ileus associated with abdominal surgery resolves more quickly in the jejunum than in the upper gastrointestinal tract, feeding via jejunostomy facilitates earlier postoperative enteral nutrition (126).

The diameter of feeding tubes is described in French (F) units (F unit = 0.33 mm). Large diameter tubes ($\geq$ 16 F) tend to cause pressure necrosis in the nose, as well as inflammation of throat and esophagus. They may compromise the function of the esophageal sphincter, increasing the potential for pulmonary aspiration. Large tubes may also interfere with swallowing, discouraging oral consumption of liquid supplements. Most adults tolerate nasal tube ranging between 8 and 12 F. High viscosity formulas or those supplemented with fiber require at least a 10 F tube unless a feeding pump is being used. Extremely small caliber tubes ($<$ 6 F) require very low viscosity formulas such as elemental formulas.

Once the tube is in place, formula is usually administered at a slow rate and increased by increments depending on the condition of the patient, until the patient's full needs are met. Formula can be filled into 500 mL or 1,000 mL administration containers (containers can also be purchased prefilled). The formula flows from the container through a feeding set, by gravity or by a pump (a small portable pump is shown), into the feeding tube (88). Formula can be delivered by a continuous or intermittent administration schedule. A continuous schedule is preferred when tube feeding is initiated, when patients have not been fed for several days, when patients are critically ill, and when duodenal or jejunal feeding sites are used (122,124). Intermittent feedings, given five to eight times a day, can simulate a meal pattern. Ambulatory tube-fed patients frequently use an intermittent feeding schedule because it permits more freedom of movement than does continuous feeding. Intermittent feeding may also be preferable in patients with diabetes because it simplifies insulin dosing. Most often, delivery is controlled by means of an enteral feeding pump, thus insuring an even and controlled flow of formula, but it is also possible to infuse by gravity or with the use of a syringe.

PROCESSING TECHNOLOGY

Because enteral formulas are often the sole source of nutrition, insuring the exact nutrient content of the formula is of paramount importance to the manufacturer, in contrast to other food manufacturers where taste or texture are primary concerns. The production of enteral products is governed by rigid specifications for the ingredients and a thorough quality-control (QC) program.

Ingredient Testing

The ingredient QC includes the verification of the following properties of each ingredient: the functional attributes, for example, the viscosity of casein in solution; microbiological properties, such as the presence of spores that could cause spoilage of the product or that are pathogenic; nutrient content, such as the potency of a vitamin mix; and organoleptic parameters. A high-quality source of water is necessary for the formulation of both liquid and spray-dried products.

Processing

Enteral formulas are provided either as ready to feed liquids or as powders requiring reconstitution to a liquid product. The exact processing parameters employed to produce enteral formulas are proprietary, but are described below in general terms.

Liquids

In the high-volume, rapid-turnover environment of hospitals, a shelf life of at least one year is generally preferred for ready-to-feed liquid nutritional products in bottles, metal, or plastic cans, and prefilled large volume ready-to-hang containers. Liquid batches are produced by adding part or all of the protein and fiber, fat-soluble vitamins, and lecithin to the oil(s) used in the product. The oils are often heated to the liquification point to facilitate the mixing of other ingredients. The remaining portion of the protein and fiber is added to water, and the carbohydrate and minerals are dissolved in a separate volume of hot water. These three mixes are then combined to form a concentrate of the final product, which is adjusted for pH and subjected to a combination of the following processing steps, which are similar to those used in dairy processing: heating, homogenization, pasteurization, UHTST (ultra high temperature short time sterilization), deaeration, and clarification. Following testing for major nutrients, the heat-sensitive components such as water-soluble vitamins and flavors are added, and the product is diluted to its final volume. Containers are then filled and sterilized in commercial retorts, either by batch or continuous process. The entire procedure is summarized schematically in Figure 2.

Aseptically filled containers similar to the type used for fruit juices can also be used for enteral formulas, but are not typical because of the shorter shelf life caused by the deterioration of the packaging and product. Following the testing of finished product for final verification of nutrient content and commercial sterility, the batch is released for distribution.

Powders

Powdered products are sometimes the preferred form for certain products or for specific settings. For a given volume of final product, powders are less bulky and less heavy than liquid products and, therefore, are easier to deliver and store. They have a shelf life of up to five years, as compared to the one year shelf life that is typical of liquids. Certain products are available only as powders because of components that cannot undergo processing (eg, some protein sources coagulate if they are heated) or are not stable in liquids (eg, the amino acids of elemental diets are unstable in an aqueous environment).

Powders are manufactured by two methods: spray drying or dry blending, with several possible variations. Spray-dried products may be manufactured from finished liquid products. Dry blending is considerably less expen-

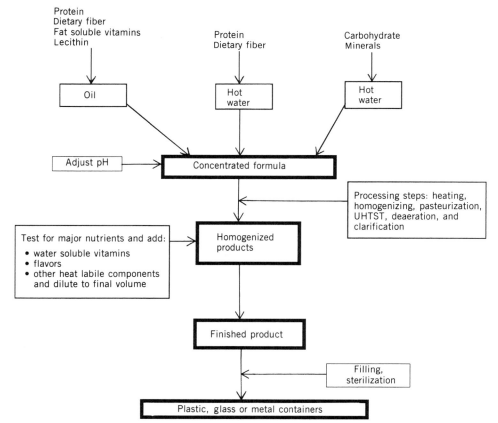

Figure 2. Formulation and processing of a liquid enteral formula.

sive than spray drying, but the products may suffer from nonhomogeneity, and can stratify during shipping and handling. Both types of product can be agglomerated to enhance mixability (agglomeration is a process by which particle size is increased).

REGULATORY ACTIVITIES

Enteral diets that are formulated simply as adult meal replacements are considered to be foods by the Food and Drug Administration (FDA) and are not regulated any differently than standard foods. *Webster's* dictionary defines food as "material consisting essentially of protein, carbohydrate, and fat used in the body of an organism to sustain growth, repair, and vital processes, and to furnish energy; also: such food together with supplementary substances (as minerals, vitamins, and condiments)." Like regular foods, enteral diets are formulated with items that are generally recognized as safe (GRAS) or items approved as food additives by the FDA.

The FDA has a role in approving or advising food manufacturers regarding the health messages for conventional foods. There are two opposing views about the potential effects of regulating such messages: if health claims are not allowed, then the public may not be informed of the potential benefit of consuming certain foods. On the other hand, allowing health messages on foods opens the door for false or exaggerated claims to be made. Recently, the policy of the FDA has been changing toward the allowance of certain health messages. Regulations or food labeling are detailed in two sections of the Code of Federal Regulation (127).

The development of more specialized enteral formulas for specific diseases has placed these formulas under increasingly intense scrutiny by the FDA. In 1988, the orphan drug amendment created a new category of food called medical food which was defined as follows: "The term medical food means a food which is formulated to be consumed or administered enterally under the supervision of a physician and which is intended for the specific dietary management of a disease or condition for which distinctive nutritional requirements, based on recognized scientific principles, are established by medical evaluation" (128). Implicit in this definition is the notion that disease conditions can change an individual's nutritional requirements or food intake leading to a deficiency or overabundance of a nutrient. Although the amendment did issue the above definition of a medical food, under the current federal regulations that implement the Food, Drug and Cosmetic Act, there is no procedural framework for the evaluation of medical food products. This contrasts with the detailed regulations for the development and approval of new drugs and food additives as well as the regulations for the formulation and manufacture of infant formulas specified in the Infant Formula Act of 1980 (129).

BIBLIOGRAPHY

1. Y. H. Hui, "Hospital Diets and Enteral and Parenteral Nutrition," in *Human Nutrition and Diet Therapy*, Wadsworth Health Sciences Division, Monterey, Calif., 1983.
2. J. L. Rombeau and M. D. Caldwell, eds., *Parenteral Nutrition*, W. B. Saunders & Co., Philadelphia, Pa., 1986.
3. American Society for Parenteral and Enteral Nutritional Board, "Guidelines for Use of Enteral Nutrition in the Adult Patient," *JPEN* 11, 435–439 (1987).
4. M. Winitz and co-workers, "Evaluation of Chemical Diets As Nutrition for Man-in-Space," *Nature (London)* 205, 741 (Feb. 1965).
5. W. R. Thompson and co-workers, "Use of the Space Diet in the Management of a Patient with Extreme Short Bowel Syndrome," *Annals of Surgery* 170, 449 (Apr. 1969).
6. H. T. Randall, "The History of Enteral Nutrition," in J. L. Rombeau and M. D. Caldwell, eds., *Enteral and Tube Feeding*, W. B. Saunders & Co., Philadelphia, Pa., 1984.
7. A. F. Roche, ed., *The Gastrointestinal Response to Injury, Starvation, and Enteral Nutrition Report of the Eighth Ross Conference on Medical Research*, Ross Laboratories, Columbus, Ohio, 1988.
8. D. C. Heimburger and R. L. Weinsier, "Guidelines for Evaluating and Categorizing Enteral Feeding Formulas According to Therapeutic Equivalence," *JPEN* 9(1), 61–67 (1985).
9. R. Shapiro, "Historical Development and Limitations of 'Elemental' Diets" in M. E. Shils, ed., *Defined-Formula Diets for Medical Purposes*, American Medical Association, Chicago, Ill., 1977, 1–5.
10. I. L. Craft, D. Geddes, C. W. Hyde, I. J. Wise, and D. M. Matthews, "Absorption and Malabsorption of Glycine and Glycine Peptides in Man," *Gut* 9, 425–437 (1968).
11. S. A. Adibi, M. R. Fogel, and R. M. Agrawal, "Comparison of Free Amino Acid and Dipeptide Absorption in the Jejunum of Sprue Patients. *Gastroenterology* 67, 586–591 (1974).
12. D. M. Matthews and S. A. Adibi, "Peptide Absorption," *Gastroenterology* 71, 151–161 (1976).
13. S. A. Adibi, "Intestinal Transport of Dipeptides in Man: Relative Importance of Hydrolysis and Intact Absorption," *Journal of Clinical Investigation* 50, 2266–2275 (1971).
14. D. B. A. Silk, "Peptide Absorption in Man," *Gut* 67, 586–591 (1974).
15. F. A. Klipstein and J. J. Corcino, "Malabsorption of Essential Amino Acids in Tropical Sprue," *Gastroenterology* 68, 239–244 (1975).
16. S. A. Adibi, "Intestinal Absorption of Amino Acids and Peptides," *Viewpt Dig Dis* 10(4) (1978).
17. M. H. Sleisenger and Y. S. Kim, "Protein Digestion and Absorption," *New England Journal of Medicine* 300, 659–663 (1979).
18. A. M. Asatoor, B. Cheng, K. D. G. Edwards, A. F. Lant, D. M. Matthews, M. D. Milne, F. Navab, and A. J. Richards, "Intestinal Absorption of Two Dipeptides in Hartnup Disease," *Gut* 11, 380–387 (1970).
19. F. Navab and A. M. Asatoor, "Studies on Intestinal Absorption of Amino Acids and a Dipeptide in a Case of Hartnup Disease," *Gut* 11, 373–379 (1970).
20. A. M. Asatoor, B. D. W. Harrison, M. D. Milne, and D. I. Prosser, "Intestinal Absorption of an Arginine-Containing Peptide in Cystinuria," *Gut* 13, 95–98 (1972).
21. M. R. Fogel, M. M. Ravitch, and S. A. Adibi, "Absorptive and Digestive Function of the Jejunum after Jejunoileal Bypass for Treatment of Human Obesity," *Gastroenterology* 71, 729–733 (1976).
22. F. Daum, M. I. Cohen, H. McNamara, and L. Finberg, "Intestinal Osmolality and Carbohydrate Absorption in Rats Treated with Polymerized Glucose," *Pediatric Research* 12, 24–26 (1978).

23. M. M. McBurney and L. S. Young, "Formulas," in Ref. 6.
24. The Royal College of Physicians of London, Medical Aspects of Dietary Fibre, Pitman Medical Ltd., Kent, UK, 1980.
25. J. H. Cummings, "Consequences of the Metabolism of Fiber in the Human Large Intestine," in G. V. Vahouny and D. Kritchevsky, eds., *Dietary Fiber in Health and Disease*, Plenum Press, New York, 1982, pp. 9–22.
26. J. H. Cummings and A. M. Stephen, "The Role of Dietary Fibre in the Human Colon," *Canadian Medical Association Journal* **123**, 1109–1114.
27. D. Burkitt and N. Painter, "Gastrointestinal Transit Times, Stool Weights and Consistency; Intraluminal Pressures," in D. P. Burkitt and H. C. Trowell, eds., *Refined Carbohydrate Foods and Disease; Some Implications of Dietary Fibre*, Academic Press, London, 1975.
28. J. L. Kelsay, A Review of Research on Effects of Fiber Intake on Man, *American Journal of Clinical Nutrition* **31**, 142–159 (1978).
29. G. A. Spiller and R. J. Amen, "Dietary Fiber in Human Nutrition," *Critical Review of Food Science and Nutrition*, 39–70 (Nov. 1975).
30. C. Hull, R. S. Greco, and D. L. Brooks, "Alleviation of Constipation in the Elderly Dietary Fiber Supplementation." *Journal of the American Geriatric Society* **28**, 410–414 (1980).
31. M. Iseminger and P. Hardy, "Bran works!" *Geriatric Nursing*, 402–404 (Nov.–Dec. 1982).
32. H. A. Frank and L. C. Green, "Successful Use of a Bulk Laxative to Control the Diarrhea of Tube Feeding," *Scandinavian Journal of Plastic Reconstructive Surgery* **13**, 193–194 (1979).
33. J. R. Smalley, W. J. Klish, M. A. Campbell, and M. R. Brown, "Use of Psyllium in the Management of Chronic Nonspecific Diarrhea of Childhood," *Journal of Pediatric Gastroenterology and Nutrition* **1**, 361–363 (1982).
34. Association of Official Analytical Chemists, "Total Dietary Fiber in Foods," *Journal of the Association of Official Analytical Chemists* **68**, 339 (1985).
35. D. A. T. Southgate, "The Measurement of Unavailable Carbohydrates: Structural Polysaccharides," in *Determination of Food Carbohydrates*, Elsevier Applied Science Publishers Ltd. Barking, UK, 1976, pp. 61–74.
36. H. N. Englyst and G. J. Hudson, "Colorimetric Method for Routine Measurement of Dietary Fiber as Non-Starch Polysaccharides. A Comparison with Gas-Liquid Chromatography," *Food Chemistry* **24**, 63–76 (1987).
37. R. S. Mitchell and T. L. Petty, "Chronic Obstructive Pulmonat Disease (COPD)," in R. S. Mitchell and T. L. Petty, eds., *Synopsis of Clinical Pulmonary Disease*, 3rd ed, C. V. Mosby Co., St. Louis, 1982.
38. S. E. Brown and R. W. Light, "What Is Now Known About Protein-Energy Depletion: When COPD Patients Are Malnourished." *Journal of Respiratory Diseases*, 36–50 (May 1983).
39. R. J. Browning and A. M. Olsen, "The Functional Gastrointestinal Disorders of Pulmonary Emphysema." *Mayo Clinic Proceedings*, 537–543 (1961).
40. A. G. Driver and M. Lebrun, "Iatrogenic Malnutrition in Patients Receiving Ventilatory Support," *Journal of the American Medical Association* **244**, 2195–2196 (1980).
41. A. G. Driver, M. T. McAlevy, and J. L. Smith, "Nutritional Assessment of Patients with Chronic Obstructive Pulmonary Disease and Acute Respiratory Failure," *Chest* **82**, 568–571 (1982).
42. L. Larca and D. M. Greenbaum, "Effectiveness of Intensive Nutritional Regimens in Patents Who Fail to Wean from Mechanical Ventilation," *Critical Care Medicine* **10**, 297–300 (1982).
43. H. R. Bassili and M. Deitel, "Effect of Nutritional Support on Weaning Patients Off Mechanical Ventilators." *JPEN* **5**, 161–163 (1981).
44. J. D. Frankfort, C. E. Fischer, D. W. Stansbury, D. L. McArthur, and co-workers, "Comparison of Ensure-HN Plus and Pulmocare on maximum exercise performance in patients with chronic airflow obstruction (CAO). *American Review of Respiratory Diseases* **135**, A361 (1987).
45. N. All-Saady, C. Blackmore, and E. D. Bennett, "Nutrition and Monitoring of the Critically Ill Patient. High Fat, Low Carbohydrate Enteral Feeding Reduces PaCO2 and the Period of Ventilation if Ventilated Patients," *Chest* **94**(Suppl. 1), 49S (1988).
46. M. Deitel, V. P. Williams, and T. W. Rice, "Nutrition and the Patient Requiring Mechanical Ventilatory Support," *Journal of the American College of Nutrition* **2**, 25–32 (1983).
47. J. H. Newman, J. A. Neff, and P. Ziporin, "Acute Respiratory Failure Associated with Hypophosphatemia," *New England Journal of Medicine* **296**, 1101–1103 (1977).
48. "Principles of Nutrition and Dietary Recommendations of Individuals with Diabetes Mellitus," *Diabetes Care* **199**, 520–523 (1979).
49. U.S. Department of Health and Human Services, *The Surgeon General's Report on Nutrition and Health. Summary and Recommendations*, U.S. Government Printing Office, Washington, D.C., 1988.
50. C. Steven, D. L. Costill, and B. Maxwell, "Impact of Carbohydrate (CHO) Source and Osmolality on Gastric Emptying Rates of Liquid Nutritionals, Abstracted," *JPEN* **3**, 32 (1979).
51. K. A. Cashmere, D. L. Costill, S. Caaland, and A. L. Hecker, "Serum Endocrine Glucose Response Elicited from Ingestion of Enteral Feedings, Abstracted," *Federal Proceedings, Federation of American Societies for Experimental Biology* **40**, 440 (1981).
52. A. M. Coulston, C. B. Hollenbeck, A. L. M. Swislocki, and co-workers, "Deleterious Metabolic Effects of High-Carbohydrate, Sucrose-Containing Diets in Patients with Non-Insulin-Dependent Diabetes Mellitus," *American Journal of Medicine* **82**, 213–220 (1987).
53. A. Garg, A. Bonanome, S. M. Grundy, and co-workers, "Comparison of a High-Carbohydrate Diet with a High-Monosaturated-Fat Diet in Patients with Non-Insulin Dependent Diabetes Mellitus," *New England Journal of Medicine* **319**, 829–834 (1988).
54. B. L. Thomas, D. C. Laine, and F. C. Goetz, "Glucose and Insulin Response in Diabetic Subjects: Acute Effect of Carbohydrate Level and the Addition of Soy Polysaccharide in Defined-Formula Diets," *American Journal of Clinical Nutrition* **48**, 1048–1052 (1988).
55. C. C. Folds, "Practical Aspects of Nutritional Management of the Elderly," *Clinical Nutrition* **2**, 15–18 (1983).
56. M. Fischer, W. Adkins, L. Hall, and co-workers, "The Effects of Dietary Fibre in a Liquid Diet on Bowel Function of Mentally Retarded Individuals," *Journal of Mental Deficiency Research* **29**, 373–381 (1985).
57. K. N. Jeejeebhoy, R. C. Chu, E. B. Marliss, and co-workers, "Chromium Deficiency Glucose Intolerance, and Neuropathy Reversed by Chromium Supplementation, in a Patient

Receiving Long-Term Total Parenteral Nutrition," *American Journal of Clinical Nutrition* **30**, 531–538 (1977).

58. H. Freund, S. Atamian, and J. E. Fischer, "Chromium Deficiency during Total Parenteral Nutrition," *Journal of the American Medical Association* **241**, 496–498 (1979).

59. R. O. Brown, S. Forloines-Lynn, R. E. Cross, and W. D. Heizer, "Chromium Deficiency after Long-Term Total Parenteral Nutrition," *Dig Dis Sci* **31**, 661–664 (1986).

60. H. J. Cohen, M. E. Chovaniec, D. Mistretta, and S. S. Baker, "Selenium Repletion and Glutathione Periodase: Differential Effects on Plasma and Red Blood Cell Enzyme Activity," *American Journal of Clinical Nutrition* **41**, 735–747 (1985).

61. S. Jacobson and L.-O. Plantin, "Concentration of Selenium in Plasma and Erythrocytes during Total Parenteral Nutrition in Crohn's Disease," *Gut* **26**, 50–54 (1985).

62. N. N. Abumrad, A. J. Scneider, D. Steel, and L. S. Rogers, "Amino Acid Intolerance during Prolonged Total Parenteral Nutrition Reversed by Molybdate Therapy," *American Journal of Clinical Nutrition* **34**, 2551–2559 (1981).

63. G. D. Phillips and V. P. Garnys, "Parenteral Administration of Trace Elements to Critically Ill Patients," *Anaesthesia, Resuscitation and Intensive Care* **9**, 221–225 (1981).

64. A. G. Feller, D. Rudman, P. R. Erve, and co-workers, "Subnormal Concentrations of Serum Selenium and Plasma Carnitine in Chronically Tube-Fed Patients," *American Journal of Clinical Nutrition* **45**, 476–483 (1987).

65. R. K. Chawla, C. J. Berry, M. H. Kutner, and D. Rudman, "Plasma Concentrations of Transsulfuration Pathway Products during Nasoenteral and Intravenous Hyperalimentation of Malnourished Patients," *American Journal of Clinical Nutrition* **4**, 577–584 (1985).

66. H. S. Geggel, M. E. Aent, J. R. Heckenlively, and co-workers, "Nutritional Requirement for Taurine in Patients Receiving Long-Term Parenteral Nutrition," *New England Journal of Medicine* **312**, 142–146 (1985).

67. M. H. Fischer, W. N. Adkins, P. Scaman, and J. A. Maillett, "Improved Selenium, Carnitine and Taurine Status in an Enterally Fed Population," *JPEN* **4**, 270–274 (1990).

68. J. Kopple, in M. E. Shils and V. R. Young, eds., *Modern Nutrition in Health and Disease,* Lea & Febiger, Philadelphia, Pa., pp. 1232–1250.

69. F. J. Zeeman, in, *Clinical Nutrition and Dietetics,* Collamore Press, D. C. Heath and Co., Lexington, Mass., 1983, pp. 221–243.

70. J. D. Kopple, "Abnormal Amino Acid and Protein Metabolism in Uremia," *Kidney International* **14**, 340–348 (1978).

71. J. D. Kopple, F. J. Monteon, and J. K. Shaib, "Effect of Energy Intake on Nitrogen Metabolism in Nondialyzed Patients with Chronic Renal Failure," *Kidney International* **29**, 734–742 (1986).

72. L. S. Ibels, A. C. Alfrey, W. E. Huffer, P. Craswell, and R. Weil III, "Calcification in End-Stage Kidneys," *American Journal of Medicine* **71**, 33–37 (1981).

73. Committee on Dietary Allowance of the Food and Nutritional Board, *Recommended Dietary Allowances,* 9th ed., National Academy of Sciences, Washington, D.C., 1980.

74. F. R. Smith and D. W. S. Goodman, "The Effects of Diseases of the Liver, Thyroid, and Kidneys on the Transport of Vitamin A in Human Plasma," *Journal of Clinical Investigation* **50**, 2426–2436 (1971).

75. H. Yatzidis, P. Digenis, and P. Fountas, "Hypervitaminosis A Accompanying Advanced Chronic Renal Failure," *British Medical Journal* **9**, 352–353 (1975).

76. C. M. Pemberton, K. E. Moxness, M. J. German, J. K. Nelson, and C. F. Gastineau, *Mayo Clinic Diet Manual,* B. C. Decker, Inc., Philadelphia, Pa., pp. 216–220.

77. H. H. Sandstead, "Trace Elements in Uremia and Hemodialysis," *American Journal of Clinical Nutrition* **33**, 1501–1508 (1980).

78. J. Warnold, K. Lundolm, and T. Scherstein, "Energy Balance and Body Composition in Cancer Patients," *Cancer Research* **38**, 1801–1807 (1978).

79. C. P. Holroyde, T. Gabuzda, R. D. Putman, and co-workers, "Altered Glucose Metabolism in Metastatic Carcinoma," *Cancer Research* **35**, 3710–3714 (1975).

80. M. F. Brennan and M. E. Burt, "Nitrogen Metabolism in Cancer Patients," *Cancer Treatment Report* **65**(Suppl. 5), 67–68 (1981).

81. M. E. Shils, "Nutritional Problems Induced by Cancer," *Medical Clinics of North America* **63**, 1009–1025 (1979).

82. A. Theologides, "Anorexia in Cancer: Another Speculation on Its Pathogenesis," *Nutrition and Cancer* **2**, 133–135 (1981).

83. J. G. Harris, "Nausea, Vomiting and Cancer Treatment," *Cancer* **28**, 194–201 (1978).

84. J. Medoff, "A Double-Blind Evaluation of the Anti-Emetic Efficacy of Benzquinamide, Prochlorperazine, and Trimethobenzamide in Office Practice," *Current Theoretical Research* **12**, 706–710 (1970).

85. R. A. Harrington, C. W. Hamilton, R. N. Brogden, and co-workers, "Metoclopramide: An Updated Review of Its Pharmacological Properties and Clinical Use," *Drugs* **25**, 451–494 (1983).

86. U.S. Department of Health, Education, and Welfare, *Eating Hints: Recipes and Tips for Better Nutrition during Cancer Treatment,* National Institutes of Health, Rockville, Md., 1980.

87. E. H. Rosenbaum, C. A. Stitt, H. Drasin, and I. R. Rosenbaum, *Nutrition for the Cancer Patient.* Bull Publishing, Palo Alto, Calif., 1980.

88. L. Y. Young and M. A. Koda-Kimble, eds., *Applied Therapeutics: The Clinical Use of Drugs,* 4th ed., Applied Therapeutics, Inc., Vancouver, Wash., 1988.

89. R. A. Hooley, T. Wheeler, H. Levin, and E. Steiger, "Effect of Long-Term Radiation Therapy On Protein-Calorie Status and Cell-Mediated Immunity, Abstract," *Federation Proceedings,* **40**, 948 (1981).

90. K. A. Rickard, A. Kirksey, R. L. Baehner, and co-workers, "Effectiveness of Enteral and Parenteral Nutrition in the Nutritional Management of Children with Wilms Tumors," *American Journal of Clinical Nutrition* **33**, 2622–2629 (1980).

91. B. C. Walike and J. W. Walike, "Relative Lactose Intolerance. A Clinical Study of Tube-Fed Patients," *Journal of the American Medical Association* **238**, 948–951 (1977).

92. H. L. Greene, G. L. Helink, C. C. Folk, and co-workers, "Nasogastric Tube Feeding at Home: A Method for Adjunctive Nutritional Support of Malnourished Patients," *American Journal of Clinical Nutrition* **34**, 1131–1138 (1981).

93. D. A. Lipschitz and C. O. Mitchell, "Enteral Hyperalimentation and Hemoatopoietic Toxicity Caused by Chemotherapy of Small Cell Lung Cancer, Abstract," *JPEN* **4**, 593 (1980).

94. L. H. Brubaker, D. F. Scott, G. K. Best, and co-workers, "Serum Opsonic Defect in Malnourished Cancer Patients and Improvement Following Nutritional Therapy, Abstract," *Clinical Research* **27**, 774A (1979).

95. A. H. McArdle, E. G. Reid, M. P. Laplante, and C. R. Freeman, "Prophylaxis against Radiation Injury," *Archives of Surgery* **121**, 879–885 (1986).

96. C. R. Scriver, A. L. Beaudet, W. S. Sly, and D. Valle, eds., *The Metabolic Basis of Inherited Disease*, 6th ed., McGraw-Hill Inc., New York, 1989.
97. T. B. Van Itallie, "Implications of Overweight and Obesity in the United States." *Annals of Internal Medicine* **103**, 983–988 (1985).
98. M. F. Najjar and M. Rowland, "Anthropometric Reference Data and Prevalence of Overweight, United States, 1976–80." *Vital and Health Statistics,* Series 11, No. 238, DHHS Publication No. (PHS) 87–1688. National Center for Health Statistics, Public Health Service, 1987.
99. E. A. Lew and L. Garfinkel, "Variations in Mortality by Weight among 750,000 Men and Women," *Journal of Chronic Diseases* **32**, 563–576 (1979).
100. H. B. Hubert, M. Feinleib, P. M. McNamara, and P. Castelli, "Obesity as an Independent Risk Factor for Cardiovascular Disease: A 26-Year Follow-Up of Participants in the Framingham Heart Study," *Circulation* **67**, 968–977 (1983).
101. National Institutes of Health, "Consensus Development Panel on the Health Implications of Obesity," *Annals of Internal Medicine* **103**, 1073–1077 (1985).
102. A. Palgi, J. Read, I. Greenberg, and co-workers, "Multidisciplinary Treatment of Obesity with Protein-Sparing Modified Fast: Results in 668 Outpatients," *American Journal of Public Health* **75**, 1190–1194 (1985).
103. P. G. Lindner and G. L. Blackburn, "Multidisciplinary Approach to Obesity Utilizing Fasting Modified by Protein Sparing Therapy," *Obesity/Bariatric Medicine* **5**, 198–216 (1976).
104. T. A. Wadden, A. J. Stunkard, and K. D. Brownell, "Very-Low-Calorie Diets—Their Efficacy, Safety, and Future." *Annals of Internal Medicine* **99**, 675–684 (1983).
105. G. Sikand, A. Kondo, J. P. Foreyt, and co-workers, "Two-Year Follow-Up of Patients Treated with a Very-Low-Calorie Diet and Exercising Training," *Journal of the American Dietetic Association* **88**, 487–488 (1988).
106. B. R. Bistrian, "Clinical Use of a Protein-Sparing Modified Fast," *Journal of the American Medical Association* **240**, 2290–2302 (1978).
107. J. P. Flatt and G. L. Blackburn, "The Metabolic Fuel Regulatory System—Implications for Protein-Sparing Therapies during Caloric Deprivation and Disease," *American Journal of Clinical Nutrition* **27**, 175–187 (1974).
108. G. L. Blackburn, S. D. Phinney, and L. I. Moldawer, "Mechanisms of Nitrogen Sparing with Severe Calorie-Restricted Diets," *International Journal of Obesity* **5**, 215–216 (1981).
109. L. J. Hoffer, F. R. Bristrian, V. R. Young, and co-workers, "Metabolic Effects of Very Low Calorie Weight Reduction Diets," *Journal of Clinical Investigation* **73**, 750–758 (1984).
110. S. D. Phinney, B. M. LaGrange, M. O'Connel, and E. Danforth, Jr., "Effects of Aerobic Exercise on Energy Expenditure and Nitrogen Balance during Very Low Calorie Dieting," *Metabolism* **37**, 758–765.
111. T. A. Wadden, A. J. Stunkard, K. D. Brownell, and S. C. Day, "Treatment of Obesity by Behavior Therapy and a Very Low Calorie Diet: A Pilot Investigation," *Journal of Consulting and Clinical Psychology* **52**, 692–694 (1984).
112. M. A. Kirschner, "An Eight-Year Experience with a Very-Low-Calorie Formula Diet for Control of Major Obesity," *International Journal of Obesity* **12**, 69–80 (1988).
113. A. C. Grandjean, L. M. Hursh, W. C. Majure, and co-workers, "Nutrition Knowledge and Practices of College Athletes," *Med Sci Sports Exerc* **13**(2), 82 (1981).
114. J. Bergstrom, L. Hermansen, E. Hultman, and co-workers, "Diet, Muscle Glycogen and Physical Performance," *Acta Physiologica Scandinavica* **71**(2), 140–150 (1967).
115. A. L. Hecker, "Nutrition and Physical Performance," in R. H. Strauss, ed., *Drugs, and Performance in Sports,* W. B. Saunders & Co., Philadelphia, Pa., 1987.
116. D. L. Costill and J. M. Miller, "Nutrition for Endurance Sport: Carbohydrate and Fluid Balance," *International Journal of Sports Medicine* **1**, 4 (1980).
117. K. B. Wheeler, "Sports Nutrition for the Primary Care Physician: The Importance of Carbohydrate," *The Physician and Sports Medicine* **17**(5) (1989).
118. E. F. Coyle, A. R. Coggan, M. K. Hemmert, and co-workers, "Muscle Glycogen Utilization during Prolonged Strenuous Exercise When Fed Carbohydrate," *Journal of Applied Physiology* **61**(1), 165–172 (1986).
119. D. L. Costill, W. M. Sherman, W. J. Fink, and co-workers, "The Role of Dietary Carbohydrates in Muscle Glycogen Resynthesis after Strenuous Running," *American Journal of Clinical Nutrition* **34**(9) 1831–1836 (1981).
120. J. L. Ivy, A. L. Katz, C. L. Cutler, and co-workers, "Muscle Glycogen Synthesis after Exercise: Effect of Time of Carbohydrate Ingestion," *Journal of Applied Physiology* **64**(4), 1480–1485 (1988).
121. H. Y. Hui, *Handbook of Enteral and Parenteral Feedings.* John Wiley & Sons, Inc., New York, 1988.
122. J. L. Rombeau and D. O. Jacobs, "Nasoenteric Tube Feeding," in Ref. 6.
123. McArdle and co-workers, "A Rationale for Enteral Feeding as the Preferable Route for Hyperalimentation," *Surgery* **90**, 613–623 (1981).
124. T. Jones, "Enteral Feeding: Techniques of Administration," *Gut* **27**, 47–49 (1986).
125. C. Page, R. Andrassay, and J. Sandler, "Techniques in Delivery of Liquid Diet: Short-Term and Long-Term," in M. Deitel, ed., *Nutrition in Clinical Surgery.* Williams & Wilkins, Baltimore, 1985, pp. 60–87.
126. J. Rombeau, K. Barot, D. Low, and P. Twomey, "Feeding by Tube Enterostomy," in Ref. 6, pp. 275–291.
127. *Regulations for food labeling: conventional foods,* 21 CFR Part 101 *Foods for Special Dietary Use,* 21 CFR Part 105.
128. *Definition of Medical Food,* Orphan Drug Admendments of 1988, 21 USC 360 e(b)(3).
129. Infant Formula Act of 1980 Public Law 96–359.

Stephen R. Behr
Sheila M. Campbell
Elizabeth M. Besozzi
Ross Laboratories
Columbus, Ohio

ENTROPY

Entropy is a term that has come into common use in everyday vocabulary. Gleick (1) noted: "The Second Law [the Entropy Law] is one piece of technical bad news from science that has established itself firmly in the nonscientific culture." For example, Rifkin (2) wrote: "If we continue to ignore the truth of the Entropy Law . . . we shall do so at the risk of our own extinction." Weaver (3) commented:

> The word entropy has begun to be one of those vogue-words that suddenly become popular for no traceable reason. People talk about the entropy of everything—their possessions, their life-styles, their bank accounts, their love affairs. Entropy is both misused and overused. . . . It might be a good idea to develop a clear definition of entropy while it is still possible.

This article is an attempt to give such a definition and show the application of the entropy concept to food science and technology.

Entropy (S) is defined and discussed in nearly all physical chemistry and thermodynamics and even in modern general chemistry textbooks (4–9). Some of these are listed in the Bibliography. Entropy is an extensive property; that is, the entropy of a system is equal to the sum of the entropies of its parts. It is a measure of the thermal energy of a closed system that is not available for conversion into mechanical work. Entropy may be regarded as a measure of randomness of a system; the probability of a given state of a system is its entropy and it can be calculated by the equation

$$S = k \ln W \qquad (1)$$

where k is the Boltzmann constant and W is the total number of configurations that are compatible with a given macroscopic state. The entropy of a system must be positive and can be zero only for a hypothetical perfect crystal at absolute zero, where $W = 1$.

The total thermal energy in a system is the enthalpy (H); the energy available to do work [the Gibbs free energy (G)] at an absolute temperature T is

$$G = H - TS \qquad (2)$$

Equation 1 is often written in differential form as

$$dG = dH - T\,dS - S\,dT \qquad (3)$$

and it is often more useful (and practical) to discuss the change in entropy in a process. Guggenheim (7) has summarized the possible changes in entropy as: "... the entropy of an insulated closed system increases in any natural change, remains constant in any reversible change and is a maximum at equilibrium."

As a fundamental thermodynamic parameter, entropy has relevance to almost every real process. The role of entropy in two processes of importance in the food area, emulsification and dehydration, will be summarized.

In earlier remarks on the emulsification process (10) it was noted that a positive entropy change (the entropy of mixing) accompanies dispersion of oil in water, but most emulsification processes are not spontaneous because water-to-water hydrogen bonds and lipid-to-lipid hydrophobic bonds must be broken when one phase is dispersed in another. For a process to proceed, the Gibbs free energy change for that process (eq. 3) must be negative; therefore, for emulsification to procede, enough work must be done on the system (enough energy must be put into the system) to compensate for the loss in energy by the system when the hydrogen and hydrophobic bonds are broken. Conversely, the second law predicts that, once formed, most emulsions will be unstable, since when an emulsion breaks up, the loss of entropy by the system will be more than balanced by the gain in energy accompanying reformation of the bonds.

The thermodynamic properties of foods in dehydration have been discussed at length (11,12). "The entropy of sorption goes through a minimum around the monomolecular layer coverage, because sorbed water is increasingly localized (and the degree of randomness is lowered) as the first layer is covered." Then, with increasing moisture content, the entropy of the water increases. The familiar hysteresis phenomenon which is evident on desorption can be explained by consideration of entropy changes: "Phenomena that lead to a lower entropy of water while desorption occurs include the entrapment of water in microcapillaries and matrix collapse."

BIBLIOGRAPHY

1. J. Gleick, *Chaos: Making a New Science*, Viking Penguin, Inc., New York, 1987.
2. J. Rifkin, *Entropy: A New World View*, The Viking Press, New York, 1980.
3. J. H. Weaver, "Entropy," in *The World of Physics: A Small Library of the Literature of Physics from Antiquity to the Present. Vol. 1: The Aristotelian Cosmos and the Newtonian System,* Simon and Schuster, New York, 1987.
4. G. M. Barrow, *Physical Chemistry*, 3rd ed., McGraw-Hill Book Co., New York, 1973.
5. H. A. Bent, *The Second Law: An Introduction to Classical and Statistical Thermodynamics,* Oxford University Press, New York, 1965.
6. J. D. Fast, *Entropy: The Significance of the Concept of Entropy and Its Applications in Science and Technology,* Macmillan and Co., London, 1970.
7. E. A. Guggenheim, *Thermodynamics: An Advanced Treatment for Chemists and Physicists,* 5th ed., North-Holland Publishing Co., Amsterdam, 1967.
8. M. A. Lauffer, *Entropy-driven Processes in Biology: Polymerization of Tobacco Mosaic Virus and Similar Reactions,* Springer-Verlag, New York, 1975.
9. G. C. Pimentel and R. D. Spratley, *Understanding Chemistry,* Holden-Day, Inc., Oakland, Calif., 1971.
10. D. N. Holcomb, in M. S. Peterson and A. H. Johnson, eds., *Encyclopedia of Food Science,* AVI Publishing Co., Inc., Westport, Conn., 1978.
11. S. S. H. Rizvi, in M. A. Rao and S. S. H. Rizvi, eds., *Engineering Properties of Foods,* Marcel Dekker, Inc., New York, 1986.
12. J. G. Kapsalis, in L. B. Rockland and L. R. Beuchat, eds., *Water Activity: Theory and Applications to Food,* Marcel Dekker, Inc., New York, 1987.

DAVID N. HOLCOMB
Kraft Technology
Glenview, Illinois

ENVIRONMENT. See HISTORY OF FOODS.
ENVIRONMENTAL POLLUTION. See WASTE MANAGEMENT AND FOOD PROCESSING.

ENZYME ASSAYS FOR FOOD SCIENTISTS

REASONS FOR ASSAYING ENZYMES

In designing enzyme assays it is important to determine the level of sophistication required of the data obtained. One may be led astray by answers which are too simplistic; conversely, excessive time and effort may be expended in obtaining highly precise data to meet "quick-and-dirty" needs. Understanding the basis of various kinds of assays will help in choosing the proper level of assay sophistication to avoid these two kinds of errors.

Characterization of Enzymes for Applications

In industrial food applications the usual requirement of an enzyme is that it produce the desired functionality for the minimum cost. This often implies that offerings by alternative suppliers are assayed to find the best enzyme source for the process in hand. The characterizing factors are (1) rate (activity per gram of enzyme), (2) pH optimum, (3) temperature optimum, and (4) stability under conditions of use.

Rate. By rate, the food processor usually means the amount of modification obtained during the time allowed for enzyme action in the process. This may be different from the initial rate of conversion of substrate to product as defined by an enzymologist. An assay which measures the latter rate may be misleading if the modification occurs during extended incubation in the process.

pH, Temperature. The pH and temperature optimum curves published by suppliers usually confound the true influence of these factors on enzyme catalytic properties with the effect on enzyme stability. Enzyme denaturation is influenced by numerous factors, is usually irreversible, and occurs with first-order kinetics. The optimum curves are constructed using assays which measure the amount of substrate modification over a period of time, and represent a summation of true rate effects plus denaturation during that time. They should be used with extreme caution.

Stability. The presence of substrate stabilizes enzyme against denaturation; thus the real optimum of interest is the stability of enzyme under the conditions of use: time, pH, temperature, substrate concentration, inorganic ions, organic solvents, etc. An assay used to screen enzymes for a particular application should mimic as nearly as possible the actual use conditions to give a reliable estimate of cost/benefit.

Characterization of Enzymes in Raw Materials

Specificity. Often enzymes with similar activities (ie, proteases) are found from different sources and assays are used to characterize them in terms of units per gram. If the different enzymes have varying specificity requirements, the results may be misleading. For example, if two proteases have specificities corresponding to elastase and trypsin, an assay based on azocollagen substrate will give a much higher value for the former enzyme, while one based on casein will favor the latter. If the protein to be modified is something quite different, eg, wheat gluten, neither assay will give a reliable comparison. Substrate specificity may be of major importance, as with proteases, or a negligible factor, as with lipoxygenase. It is best to assume that it is important until the contrary is established.

If an assay is used to monitor production of enzyme from biofermentation, specificity may be overlooked. Here the ideal is the quickest possible assay consistent with reasonable accuracy, and the assumption is that enzyme specificity is constant. A quick assay based on azocollagen may suffice even if the protease being made is similar to trypsin in its specificity.

Characterization of Enzyme Rate Parameters

For a study made for the purpose of establishing basic enzymatic parameters (catalytic rate constant, affinity for substrate, inhibitor binding, etc) the assay must provide appropriate velocity estimates. In general this means the catalytic rate at time zero, ie, when the enzyme and substrate are first combined. Assays involving incubation for a fixed length of time present some difficulties for this purpose. Progress curves, in which the concentration of product is measured at intervals during the incubation, are valid and are not used as often as they might be to determine enzymatic parameters.

THEORETICAL ASPECTS OF ENZYME ASSAYS

Properly judging the nature of the desired assay requires a certain amount of theoretical understanding. This does not have to be in great depth; the requisite fundamentals are easily grasped and applied. The effort will be repaid by improved enzyme assays for routine work and process design (1).

Assay Characteristics

A sound assay, regardless of the level of sophistication required, will have (1) linear dependence on enzyme concentration, (2) adequate consideration of pH and temperature effects, (3) appropriate accuracy, (4) adequate sensitivity, (5) and speed and ease of performance. Unfortunately, many assays emphasize the last factor at the expense of the other four.

Enzyme Linearity. Enzyme linearity is paramount. A curved plot of assay response versus amount of enzyme used indicates that some chemical, physical, and/or enzymological factors have been overlooked. In fixed-time assays the formation of product is often not strictly linear with time because of substrate depletion or product inhibition and this nonlinearity is more pronounced at higher enzyme concentrations. Occasionally the chemical reaction used to quantitate the amount of product formed is not stoichiometric, leading to a nonlinear plot of, eg, spectrophotometric absorbance versus enzyme amount. Numerous nonlinear assays found in the literature indicate an incomplete understanding of the system being used for the assay.

Temperature, pH. Elevated temperatures and extremes of pH contribute to enzyme denaturation during the assay. Temperature optimum curves are always due to this phenomenon, and activity decreases at high or low pH also may be due to enzyme instability. The pH also may effect enzyme catalytic activity, influencing the ionization

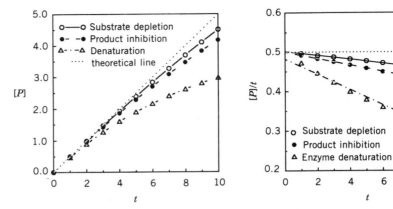

Figure 1. Method of obtaining initial rate from a curved plot of product versus time.

state of the active site and possibly the substrate (2–4). A sound assay will take these factors into account.

Accuracy, Sensitivity. The required levels of accuracy (more properly, precision) and sensitivity should be carefully considered. Measurements of α-amylase activity may have a coefficient of variation of 2% or of 10%, depending on whether the assay is a replicated colorimetric one using a modified substrate (5) or a viscometric one using gelatinized ground grain (6). For standardizing a purified amylase for use in bread production the former would be appropriate, while for finding bins of wheat which have been subjected to sprouting the latter method is quite adequate. Likewise, assays at almost any level of sensitivity may be constructed. Using a fluorescent substrate, picomolar concentrations of trypsin may be assayed (7), while in monitoring the production of microbial protease a simple protein-based assay (8), although some five orders of magnitude less sensitive, will do the job.

Convenience. Speed and ease of performance should be the last factors considered. These are important in many industrial contexts, but a fast, easy, inadequate assay will only result in the quick generation of much useless data. After linearity, precision, enzyme stability, and sensitivity are established, then steps may be taken to increase output.

Initial Rates. Most assays measure the rate of formation of product from substrate, ie, $d[P]/dt$. This measurement may be of the initial rate, $d[P]/dt$ at the initiation of the reaction, or of the amount of product formed during a fixed time of incubation of substrate with enzyme. Fixed-time assays are convenient, in that a large number of samples may be run simultaneously. Unfortunately, they are also more prone to complications leading to the nonlinearity mentioned above.

Accurate initial rate measurements are not contaminated by effects such as substrate depletion, product inhibition, or enzyme denaturation. However, they are not always easy to obtain. A method of finding the initial rate is the following (9). Set up the assay system, take samples at various times t, and measure product concentration $[P]$ at each time. Usually there is a slight downward curve of the plot of $[P]$ versus t (Fig. 1a) which makes the determination of the tangent ($d[P]/dt$) at zero time difficult. A straight line through the plot of $[P]/t$ versus t gives an intercept which is very close to the true initial rate (Fig. 1b). Even in the worst case shown, involving rather rapid enzyme denaturation, the error is only 3.5%.

Determining Rate Parameters

Michaelis-Menten Parameters. In designing assays it is useful to know the maximum rate obtainable with a given amount of enzyme, V_{max}, and the concentration of substrate which gives half that rate, K_M. These are the fundamental parameters in the Michaelis-Menten (M-M) rate equation: $v = V_{max}[S]/(K_M + [S])$, where $[S]$ is substrate concentration and v is the actual rate $d[P]/dt$. The usual procedure is to measure v at several concentrations $[S]$ and then calculate V_{max} and K_M, either by applying a computer program (10) such as HYPER (Fig. 2a) or by fitting a straight line to one of the linear transforms of the M-M equation. The usual transform is the double reciprocal or Lineweaver-Burk plot: $1/v = 1/V_{max} + (K_M/V_{max})(1/[S])$, where $1/v$ is plotted versus $1/[S]$ (Fig. 2b). From statistical considerations, this is the least desirable transform to use (12).

A better plot is the Hanes plot of $[S]/v$ versus $[S]$ (Fig. 2c): $[S]/v = K_M/V_{max} + (1/V_{max})[S]$. The points are spaced along the x axis at the same intervals as in the M-M plot, rather than being crowded together near the y axis. The larger experimental errors inherent in the smaller values of v (at low $[S]$) have less influence on the linear least squares regression line. The Hanes plot should be used in treating v, $[S]$ data graphically; the use of the unsatisfactory Lineweaver-Burk plot should be discontinued.

The M-M equation is a differential equation, ie, v equals $d[P]/dt$ at the instantaneous value of $[S]$, usually taken as the initial substrate concentration. If v is only available as $[P]/t$ from a fixed-time assay, then the value taken for $[S]$ for the above calculations should be the average of the substrate concentration at the beginning and end of the incubation period, $([S]_o + [S]_t)/2$. This approximation gives estimates of V_{max} and K_M which are much closer to the true values than if the initial value of $[S]$ is used (13).

Inhibitors

Two types of enzyme inhibitors are of interest to food scientists: (1) low affinity inhibitors and (2) high affinity

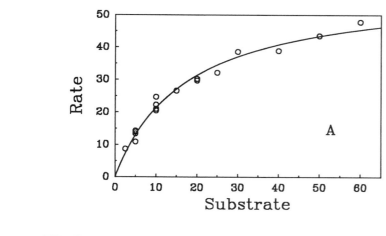
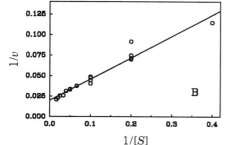
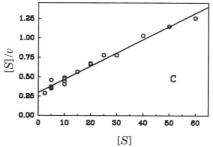

Figure 2. Comparison of Hanes and Lineweaver–Burk plots for determining V_{max} and K_M (data taken from (11)). V_{max}, K_M, and Std. Dev. are, respectively; HYPER—58.55, 17.44, and 1.65; Hanes—58.96, 17.56, and 1.66; Lineweaver–Burk—49.60, 12.85, and 2.54.

inhibitors. The former are effective in the millimolar to micromolar concentration range and readily dissociate from the enzyme; an example is inorganic phosphate inhibiting phytase. The latter are effective in the nanomolar to picomolar concentration range and are bound tightly to the enzyme; an example is the trypsin inhibitor from soybean. For low affinity inhibitors the dissociation constant K_i is a useful parameter to know; for high affinity inhibitors the amount present is usually of more concern.

Inhibition Model. A general equilibrium model for inhibition is shown in Figure 3 (14). If the parameter α is very large, so that the species EIS does not exist, the inhibition is termed *competitive* (inhibitor competes with S for the enzyme). The effect, shown in Figure 4a, is that V_{max} is unchanged, and K_M increases. If $\alpha = 1$ and $\beta = 0$, EIS is formed but does not proceed to product P; this is *noncompetitive* inhibition. As shown in Figure 4a, V_{max} decreases and K_M is unchanged. If α is greater than 1 but not extremely large, mixed inhibition occurs. These types are diagnosed by comparing the Hanes plots in the absence and presence of inhibitor (Fig. 4b).

Inhibitor Constant. If the inhibition is competitive, K_i is determined from the ratio of K_{app} (the apparent value of K_M in the presence of inhibitor of concentration $[I]$) to K_M (no inhibitor present): $K_{app}/K_M = 1 + [I]/K_i$. If inhibition is noncompetitive, the ratio of true maximum velocity to apparent maximum velocity in the presence of inhibitor gives $V_{max}/V_{app} = 1 + [I]/K_i$. Determining K_i, α, and β in the cases of partial and mixed inhibition is too complex to be discussed here.

High Affinity Inhibitors. Measuring the concentration of high affinity inhibitors, eg, soy trypsin inhibitor, is relatively straightforward (15). Trypsin is mixed with aliquots of soy meal extract and after a brief incubation (for formation of the inhibitor–enzyme complex) the amount of uninhibited enzyme remaining is measured by a simple assay. The rate of reaction is plotted versus the size of the extract aliquot; the straight line through the data obtained at lower levels of inhibition intersects the x axis at a point where the amount of enzyme equals the amount of inhibitor present (Fig. 5) (16). In the example shown 1.2 mL of soy meal extract contained a molar amount of trypsin inhibitor equal to the number of moles of trypsin used in each assay tube. If the absolute amount of trypsin were established by a titration assay, the amount of soy trypsin

$$E + S \xrightleftharpoons{K_M} ES \xrightleftharpoons{k} E + P$$
$$+ \qquad\qquad +$$
$$I \qquad\qquad I$$
$$\updownarrow K_i \qquad\qquad \updownarrow \alpha K_i$$
$$EI + S \xrightleftharpoons{\alpha K_M} EIS \xrightleftharpoons{\beta k} EI + P$$

Figure 3. General equilibrium model for reversible enzyme inhibition.

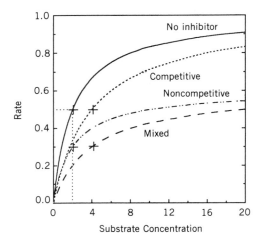
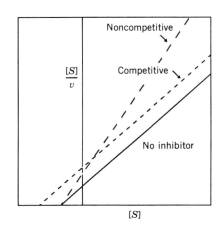

Figure 4. Hyperbolic and Hanes plots showing different modes of inhibition.

inhibitor could be expressed in absolute molar units, rather than arbitrary trypsin inhibitor units. Conversely, if the absolute concentration of high affinity inhibitor were known, this method would serve to measure the molar amount of enzyme present.

Endogenous Inhibitors. Endogenous inhibitors may be present in crude extracts of materials containing the enzyme being assayed, resulting in a marked nonlinearity in the assay (Fig. 6a). The uninhibited rate may be found as follows (17). Let $[e]$ be the amount of enzyme in the largest aliquot of extract used in making the plot of Figure 6. X is the fraction of that aliquot used for each of the other points. A plot is made of X/v_i versus X (Fig. 6b). From the y-axis intercept calculate the uninhibited rate, ie, 1/intercept equals the true rate due to enzyme concentration $[e]$. This method is useful for comparing enzyme amounts in different sources and during enzyme purification until the inhibitor has been removed.

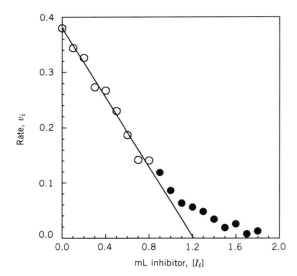

Figure 5. Plot of enzyme rates in presence of high-affinity soy trypsin inhibitor (data taken from (16)).

SPECIFIC ENZYME ASSAYS

Several assays for enzymes most commonly measured by food scientists will be described. These are only examples of the wide range of ingenious methods which have been reported for following the rate of product formation by food-related enzymes. In most cases factors such as pH, temperature, activating ions, time of reaction, and detection methods may be adjusted to fit the specific needs of the project in hand. Thus, these should be considered as starting points for designing assays to meet particular requirements, not only as ways to measure the enzyme activity under investigation.

Proteases and Peptidases

Substrates. Protein substrates for proteases are most often hemoglobin or casein and must be completely soluble in buffer. Casein for protease assays is designated "nach Hammarsten," while hemoglobin "for protease assay" is usually of good quality. Casein precipitates below pH 6, so it is used at neutral to alkaline pHs. Hemoglobin must be denatured before use, either by treatment with acid (if the assay is at acidic pH) (18) or urea (neutral to alkaline pH assay) (19). Gelatin, sometimes used in viscometric assays, is quite heterogeneous so lot-to-lot reproducibility is a concern.

Proteins may be derivatized to fit assay needs. Diazotized protein allows measurement of solubilized peptide with a visible-range colorimeter (20). If the amino groups freed during hydrolysis are quantitated using, eg, TNBS (trinitrobenzenesulfonic acid), the ε-amino groups of lysine give a high blank value which may be removed by making the succinyl (21) or N,N-dimethyl (22) derivative of the protein. The latter is preferable for trypsinlike proteases.

Small synthetic molecules are also useful for assaying proteases. These give a change in spectrophotometric absorbance as they are hydrolyzed, so a continuous assay with its advantages is possible. Table 1 lists a number of small molecule substrates. Most of these are applicable to serine and/or sulfhydryl proteases, with two exceptions; FAGLA is a substrate for metalloprotease (neutral prote-

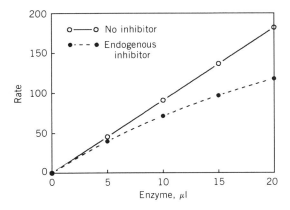
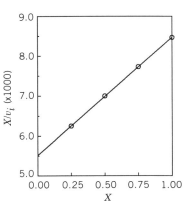

Figure 6. Finding uninhibited rate in presence of endogenous inhibitor.

ase), and Z-Gly-Phe is a carboxypeptidase substrate. Acidic protease may be assayed using a chromogenic peptide. Aminopeptidase is usually assayed using an amino acid derivative such as L-leucine-β-naphthylamide.

Protease Assays. Most assays with protein substrates involve incubation for a set length of time, stopping the reaction with TCA (trichloroacetic acid), and measuring the amount of soluble peptide. Buffered TCA (0.11 M TCA, 18 g/L; 0.22 M sodium acetate, 18 g/L; and 0.33 M acetic acid, 19.8 g/L) gives superior enzyme linearity as compared to a simple aqueous TCA solution (8). Quantitation may be direct (absorbance of filtrate at 275 nm) or colorimetric (Folin-Lowry (30), bicinchoninic acid (31), or TNBS (32)). An example of a casein-based assay is the following (8). To 5 mL casein (12 mg/mL in 0.03 M phosphate buffer, pH 7.5) add 1 mL enzyme solution and after 10 min reaction add 5 mL of buffered TCA. After 30 min the mixture is filtered and absorbance at 275 nm is measured. For Folin-Lowry quantitation, to 1 mL of filtrate add 1 mL alkaline buffer (1 M Na$_2$CO$_3$, 0.25 M NaOH), 0.4 mL copper reagent (0.1% CuSO$_4$·5H$_2$O, 0.2% NaK tartrate), mix, and allow to stand 10 min. Then add 0.75 mL diluted phenol reagent (Folin-Ciocalteau reagent diluted with 3 volumes H$_2$O), mix, wait 10 min, and measure absorbance at 700 nm versus an appropriate reagent blank. A plot of log absorbance versus log protein is linear over the range 3 to 400 μg of protein (eg, bovine serum albumin standard) (33).

An assay using TNBS to measure freed amino groups (32) uses N,N-dimethyl casein (22) (1 mg/mL) in buffer as substrate. To 1 mL in a reaction tube add 0.1 mL enzyme solution, incubate for the desired time, and heat the tube briefly in boiling water to stop the reaction. Add 1 mL 0.4 M phosphate buffer, pH 8.2 (containing 0.25% sodium dodecyl sulfate), 2 mL freshly made TNBS (1 mg/mL in water), and incubate in the dark 1 h at 60°C. Add 4 mL 0.1 N HCl, cool to room temperature, and measure absorbance at 340 nm. The reagent blank contains only buffers and TNBS; a complete assay with zero incubation time gives the correction for free amino groups in substrate and enzyme.

The absorbance change during the reaction of synthetic substrates with enzyme is recorded on a stripchart recorder, giving a continuous trace from which the initial velocity or the first-order reaction rate is obtained. Two assays will be described as examples.

1. TAME is a substrate for trypsin (25). The reaction solution is 1.04 mM TAME (0.394 mg/mL) in 0.04 M Tris buffer, pH 8.1, 10 mM in CaCl$_2$. To 2.9 mL add 0.1 mL enzyme solution and record the increase in absorbance at 244 nm.

2. FAGLA is a useful substrate for metalloproteases such as thermolysin (28). To 100 mL 0.05 M phosphate buffer, pH 7.2, add 76.9 mg FAGLA to give a 2.5 mM solution. This is well below the K_M of thermolysin for this substrate, so the reaction is kinetically first order. In the

Table 1. Absorbance Changes for Some Synthetic Substrates

Substrate	Acronym	λ, nm	Δε, mM	Reference
N-Acetyl-L-tyrosine ethyl ester	ATEE	237	+0.23	23
		275	−0.17	23
N-Benzoyl-L-arginine ethyl ester	BAEE	253	+1.15	23
N-Benzoyl-L-arginine-p-nitroanilide	BAPA	410	+8.8	24
N-Benzoyl-L-tyrosine ethyl ester	BTEE	254	+1.03	25
		256	+0.96	26
N-Tosyl-L-arginine methyl ester·HCL	TAME	244	+0.80	25
		247	+0.41	27
3-(2-Furylacryloyl)-glycyl-L-leucinamide	FAGLA	345	−0.32	28
N-Carbobenzoxy-glycyl-L-phenylalanine	Z-Gly-Phe	224	−1.0	29
N-Carbobenzoxy-L-tyrosine p-nitrophenyl ester	Z-Tyr-pNP	400	+18.8	26

cuvette place 1.5 mL substrate, 1.4 mL buffer, 0.1 mL enzyme, and measure the decrease in absorbance at 345 nm from 0.96 (zero time) toward an infinite-time value of 0.56. Plot log(absorbance − 0.56) versus time; the slope divided by 2.303 gives the first-order rate for the reaction, which is directly proportional to the concentration of enzyme.

Peptidase Assays. Carboxypeptidase is assayed by incubation with an N-acylated dipeptide, followed by measurement of the amino group freed (34). Substrate solution is 1 mM dipeptide (eg, 35.6 mg Z-Phe-Gly per 100 mL) in 0.025 M phosphate buffer, pH 7.2. To 200 µL add 50 µL enzyme, incubate for 1 h, stop by heating in boiling water, and measure free amino groups by reaction with TNBS as described above. Carboxypeptidase may also be assayed continuously using Z-Gly-Phe in 0.05 M Tris buffer, pH 7.5, following the decrease in absorbance at 224 nm (29). Aminopeptidase assays usually employ the β-naphthylamide of an amino acid, and after incubation the freed β-naphthylamine is quantitated by diazotization (35). Substrate stock solution is 2 mM amino acid naphthylamide in 0.01 N HCl. Buffer is 0.025 M phosphate, pH 7.2. Color reagent is the diazonium salt Fast Garnet GBC (1 mg/mL) in 1M acetate buffer, pH 4.2, containing 10% (v/v) Tween 20. To the assay tube add 1.6 mL buffer, 0.2 mL substrate, 0.2 mL enzyme, and incubate 3 h. Then add 1 mL color reagent, allow 5 min for color development, and read absorbance at 525 nm.

Amylases, Glycosidases, and Cellulase

Substrates. Starch, the substrate for amylases, consists of a linear polymer, amylose, and a highly branched polymer, amylopectin. The ratio of these two components varies in starches obtained from different plant sources. They may be separated by treating a solution of gelatinized starch with thymol (36). In certain assays for α-amylase the use of one or the other fraction is preferable. If the assay involves colorimetric quantitation of freed reducing groups, the blank due to reducing ends of the substrate chain may be removed by reduction with $NaBH_4$ (5).

Soluble cellulose substrates for cellulase are the carboxymethyl, hydroxypropyl, or hydroxyethyl derivatives (37). These are available as commercial products, but should be characterized before use to ensure comparability between lots. Insoluble complexes of dyes with starch (38, 39) or cellulose (40) are also available; enzyme activity is assayed by measuring the amount of dye which is solubilized during an incubation period. Small synthetic substrates, usually the p-nitrophenylglycosides, are available for some glycosidases (41).

Amylase Assays. For α-amylase (5), use 1 mL of a 1% solution of reduced starch in 0.02 M acetate buffer, pH 4.7, containing 1 mM $CaCl_2$. Add 1 mL enzyme solution, incubate for the time desired, then add 4 mL each of Reagent A and Reagent B of the neocuproine system (42). Heat in boiling water 12 min, make up to 25 mL, and read absorbance at 450 nm. Reagent A: in 600 mL water dissolve 40 g anhydrous Na_2CO_3, 10 g glycine, and 0.45 g $CuSO_4 \cdot 5H_2O$. Make up to 1 L. Reagent B: dissolve 0.12 g neocuproine (2,9-dimethyl-1,10-phenanthroline·HCl) in 100 mL water. Store in a brown bottle.

Amylopectin forms a complex with I_2 which absorbs at 570 nm. As it is hydrolyzed by α-amylase the color decreases, the basis for a fixed end point assay frequently used in cereals-related industries (43). A continuous assay based on this phenomenon is the following (44). Suspend 1 g soluble starch in 10 mL water, then slowly add this to 50 mL boiling water. Boil gently for 2 min, cool, and make up to 100 mL volume. To 2.5 mL buffered iodine solution (32 mg I_2, 330 mg KI, and 23.4 mg NaCl in 100 mL of 0.01 M phosphate buffer, pH 7.0) add 10 µL enzyme followed by 10 µL starch. Record the decrease in absorbance at 570 nm. This assay was developed for pancreatic α-amylase; for cereal α-amylases the buffer system would be 0.02 M acetate, pH 4.7, 1 mM in $CaCl_2$.

β-Amylase hydrolyzes soluble starch to yield maltose, which is quantitated by colorimetric reaction with DNS (dinitrosalicylic acid) reagent (45). A 1% soluble starch solution is prepared as described above, except that 10 mL 0.16 M acetate buffer, pH 4.8, is added before adjusting to 100 mL total volume. Mix 0.5 mL substrate with 0.5 mL enzyme solution, incubate at 25°C 3 min, then add 1 mL DNS reagent (46). Heat in boiling water 5 min, cool, add 10 mL water, and read absorbance at 540 nm. DNS reagent: in 500 mL water dissolve 10 g NaOH, 10 g 3,5-dinitrosalicylic acid, 2 g phenol, 0.5 g Na_2SO_3, and 200 g NaK tartrate, then make up to 1 L volume.

Cellulase Assays. Cellulase activity may be monitored using the DNS reagent (46). Dissolve 11.4 g CMC (carboxymethylcellulose) in 700 mL water with stirring, then add 7 g citric acid monohydrate, 19.6 g sodium citrate dihydrate, 0.1 g merthiolate, and 0.1 g glucose and make up to 1 L. To 1 mL substrate add 1 mL enzyme solution, incubate 20 min at 50°C, add 3 mL DNS reagent, heat 15 min in boiling water, cool, and read absorbance at 540 nm. Cellulase activity may also be assayed by following the decrease in viscosity of a solution of CMC during incubation with enzyme (47,48). This requires a great deal of care and preparation, and will not be discussed here.

Glycosidase. As an example of assays using nitrophenyl glycosides, that for β-galactosidase is given (49). In the reaction tube combine 50 µL 0.1 M citrate buffer, pH 4.3, 50 µL 0.4% bovine serum albumin in water, 50 µL 20 mM pNP-β-galactoside in water, and 50 µL enzyme solution. After 30 min at 37°C add 2 mL 0.25 M glycine, pH 10, and read the absorbance at 410 nm.

Ester Hydrolases

Lipase. Lipase acts on esters at the interface between water and oil (50); the substrate is a liquid vegetable oil (olive oil, soy oil) purified by percolation over a column of activated alumina or silica gel. It is emulsified with the help of vegetable gum or bile salt. The release of fatty acid is monitored by addition of base to keep the pH constant, or by extraction and colorimetric measurement of the copper soap (51). For the latter assay, make substrate emulsion by adding 2 mL vegetable oil solution (1% in absolute EtOH) to 100 mL 0.025 M Tris buffer, pH 8.8, containing 0.6% sodium deoxycholate. To 5 mL substrate add 0.1 mL enzyme, incubate 5 min, add 1 mL 1 N HCl and 10 mL isooctane, shake, and allow phase separation. To 5 mL of the upper layer add 1 mL copper reagent (5% cupric acetate in water, pH adjusted to 6.1 with pyridine), shake vigorously for 90 s, and allow the phases to separate. Read the absorbance at 715 nm of the upper phase.

Esterase. The best assay for esterase activity is the pH-stat method. The substrate in water or salt solution is adjusted to the desired pH, enzyme is added, and base is continually added to keep the pH constant. The plot of base consumption versus time gives the rate of the reaction. For a spectrophotometric assay the system is lightly buffered at the desired pH, an indicator dye is included, and the change in absorption is measured. An assay for pectin methylesterase (52) uses 0.5% citrus pectin in water adjusted to pH 7.5, an indicator/buffer system of 0.01% bromthymol blue in 3 mM phosphate buffer, pH 7.5, and enzyme in the same buffer. To 2 mL of pectin solution add 0.15 mL indicator/buffer and 0.83 mL water. Record absorbance at 620 nm briefly to establish a baseline, then add 20 μL enzyme and record the rate of absorbance change due to esteratic action. The system (without enzyme) is titrated with galacturonic acid to establish the correspondence between absorbance change and acid concentration.

Phosphatase. Phosphatase activity is monitored by colorimetric quantitation of P_i (inorganic phosphate) formed (53). An example is the assay for acid phosphatase using G-6-P (glucose-6-phosphate) (54). Substrate is 0.2 M maleate buffer, pH 6.7, containing 60 mM G-6-P, 8 mM KF, and 8 mM EDTA. To 1 mL add 1 mL enzyme extract, incubate 20 min at 37°C, and add 1 mL cold 10% TCA to stop. To 1 mL add 2 mL P_i reagent, hold 20 min at 45°C, and read absorbance at 820 nm. Reagent A: 10% (w/v) ascorbic acid in water. Reagent B: 4.2 g ammonium molybdate tetrahydrate in 1 L 1 N H_2SO_4. P_i reagent: 1 part A plus 6 parts B, made fresh daily, and kept in an ice bath until used.

Oxidases

Polyphenoloxidase. Polyphenoloxidase enzymes (eg, tyrosinase) catalyze two reactions: (1) oxygenation of a phenol to o-diphenol (cresolase) and (2) oxidation of a diphenol to o-quinone (catecholase). Substrates commonly used for assays are p-cresol (4-methylphenol), L-tyrosine (4-hydroxy phenylalanine), and p-coumaric acid (4-hydroxycinnamic acid). The corresponding diphenols are 4-methylcatechol, L-dopa, and caffeic acid. Rates are monitored by absorbance changes at specific wavelengths as oxygenation or oxidation occurs. An assay system which differentiates the two kinds of activities is the following (55). For cresolase activity, mix 1.5 mL 0.05 M acetate, pH 4.8, 0.4 mL 1 mM p-cresol, 0.1 mL 1 mM 4-methylcatechol, and 10 μL enzyme. Record increase in absorbance at 291 nm. The change is due only to oxygenation; oxidation of diphenol to quinone produces no absorbance change at 291 nm. For catecholase activity, mix 1.6 mL buffer, 0.4 mL 1 mM 4-methylcatechol, and 10 μL enzyme, recording the decrease in absorbance at 280 nm.

Lipoxygenase. Native lipoxygenase requires activation by hydroperoxydecadienoic acid, the product of its reaction with linoleic acid and oxygen; there is a lag period in the reaction if enzyme is simply mixed with a solution of substrate (56). The delay is removed by including a small amount of product in the stock substrate solution (168 mg linoleic acid and 14 mg hydroperoxylinoleic acid in 100 mL EtOH) (57). For assay, to 2.9 mL 0.2 M borate buffer, pH 9.0, add 50 μL substrate stock, followed by 50 μL enzyme solution. Mix and record the increase in absorbance at 235 nm.

Ascorbic Acid Oxidase (AAO). A simple spectrophotometric assay for AAO is based on the loss of absorbance at 265 nm when ascorbate is oxidized to dehydroascorbate (58,59). To 2.8 mL buffer (0.025 M citrate, 0.05 M phosphate, pH 5.6) add 100 μL substrate solution (0.05% ascorbate, 1% Na_2EDTA, neutralized), 50 μL 1% bovine serum albumin solution, and finally 50 μL enzyme solution. Record the rate of disappearance of absorbance at 265 nm.

Catalase, Peroxidase. For peroxidase, the reaction mixture is 0.01 M phosphate, pH 7.0, containing 25 mg/L H_2O_2 and 2.5 g/L pyrogallol (60). After adding enzyme record the increase at 430 nm due to the formation of purpurogallin. The reaction of catalase with H_2O_2 is a first-order reaction (60). Hydrogen peroxide has a broad absorption band in the far ultraviolet; molar absorbance is 120 at 200 nm and 30 at 250 nm. To 2 mL assay solution (0.05 M phosphate, pH 7.0, containing 2 g/L H_2O_2) add 1 mL diluted catalase solution and record the ultraviolet absorbance. Plot log absorbance versus time; the slope divided by 2.303 equals the first-order rate constant, which is directly proportional to the amount of catalase in the assay.

Other Enzymes

Lysozyme. Hydrolysis of bacterial cell walls by lysozyme causes the cells to lyse; a turbid suspension of cells will slowly clear. The turbidimetric assay (61) uses a suspension of *Micrococcus lysodeikticus* (0.5 g/L) in 0.06 M phosphate, pH 6.2. To 1.5 mL add 0.5 mL 0.3 M NaCl and 1 mL enzyme, place the cuvette in a colorimeter, and record % transmittance (not absorbance) at 540 nm. A plot of %T versus time is linear from about 10 to 40%, and the rate %T/min is proportional to enzyme in the range of 0 to 10 μg.

BIBLIOGRAPHY

1. C. E. Stauffer, *Enzyme Assays for Food Scientists*, Van Nostrand Reinhold, New York, 1989.
2. K. F. Tipton and H. B. F. Dixon, "Effects of pH on Enzymes," *Methods in Enzymology* **63**, 183–233 (1979).

3. W. W. Cleland, "The Use of pH Studies to Determine Chemical Mechanism of Enzyme-Catalyzed Reaction," *Methods in Enzymology* **87**, 390–404 (1982).
4. R. L. Van Etten, "Human Prostatic Acid Phosphatase: A Histidine Phosphatase," *Annals of the New York Academy of Sciences* **390**, 27–51 (1982).
5. D. H. Strumeyer, "A Modified Starch for Use in Amylase Assays," *Analytical Biochemistry* **19**, 61–71 (1967).
6. H. Perten, "Application of the Falling Number Method for Evaluating Alpha-Amylase Activity," *Cereal Chemistry* **41**, 127–140 (1964).
7. M. Monsigny, C. Kieda, and T. Maillet, "Assay for Proteolytic Activity Using a New Fluorogenic Substrate (Peptidyl-3-amino-9-ethylcarbazole)," *EMBO Journal* **1**, 303–306 (1982).
8. B. Hagihara, H. Matsubara, M. Nakai, and K. Okunuki, "Crystalline Bacterial Proteinase. I. Preparation of Crystalline Proteinase of *Bacillus subtilis*," *Journal of Biochemistry* **45**, 185–194 (1958).
9. Ref. 1, p. 93.
10. W. W. Cleland, "Statistical Analysis of Enzyme Kinetic Data," *Methods in Enzymology* **63**, 103–137 (1979).
11. J.-M. Caillat and R. Drapron, "La Lipase du Blé. Caracteristiques de Son Action en Milieux Aqueux et Peu Hydrate," *Annales de Technologie Agricole* **23**, 273–286 (1974).
12. G. L. Atkins and I. A. Nimmo, "Current Trends in the Estimation of Michaelis-Menten Parameters," *Analytical Biochemistry* **104**, 1–9 (1980).
13. H.-J. Lee and I. B. Wilson, "Enzymic Parameters: Measurement of V and K_m," *Biochimica et Biophysica Acta* **242**, 519–522 (1971).
14. Ref. 1, p. 38.
15. M. L. Kakade, J. J. Rackis, J. E. McGhee, and G. Puski, "Determination of Trypsin Inhibitor Activity of Soy Products: A Collaborative Analysis of an Improved Procedure," *Cereal Chemistry* **51**, 376–382 (1974).
16. G. E. Hamerstrand, L. T. Black, and J. D. Glover, "Trypsin Inhibitors in Soy Products: Modification of the Standard Analytical Procedure," *Cereal Chemistry* **58**, 42–45 (1981).
17. Ref. 1, p. 109.
18. American Association of Cereal Chemists (AACC), *Approved Methods of the American Association of Cereal Chemists*, 8th ed., American Association of Cereal Chemists, St. Paul, Minn., 1983, Method.
19. M. L. Anson, "The Estimation of Pepsin, Trypsin, Papain and Cathepsin with Hemoglobin," *Journal of General Physiology* **22**, 79–89 (1938).
20. R. M. Tomarelli, J. Charney, and M. L. Harding, "The Use of Azoalbumin as a Substrate in the Colorimetric Determination of Peptic and Tryptic Activity," *Journal of Laboratory and Clinical Medicine* **34**, 428–433 (1949).
21. C. Schwabe, "A Fluorescent Assay for Proteolytic Activity," *Analytical Biochemistry* **53**, 484–490 (1973).
22. Y. Lin, G. E. Means, and R. E. Feeney, "The Action of Proteolytic Enzymes on N,N-Dimethyl Proteins," *Journal of Biological Chemistry* **244**, 789–793 (1969).
23. G. W. Schwert and Y. Takenaka, "A Spectrophotometric Determination of Trypsin and Chymotrypsin," *Biochimica et Biophysica Acta* **16**, 570–575 (1955).
24. B. F. Erlanger, N. Kowosky, and W. Cohen, "The Preparation and Properties of Two New Chromogenic Substrates of Trypsin," *Archives of Biochemistry and Biophysics* **95**, 271–278 (1961).
25. B. C. W. Hummel, "A Modified Spectrophotometric Determination of Chymotrypsin, Trypsin and Thrombin," *Canadian Journal of Biochemistry and Physiology* **37**, 1393–1399 (1959).
26. K. A. Walsh and P. E. Wilcox, "Serine Proteases," *Methods in Enzymology* **19**, 41–63 (1970).
27. K. A. Walsh, "Trypsinogens and Trypsins of Various Species," *Methods in Enzymology* **19**, 41–63 (1970).
28. J. Feder, "A Spectrophotometric Assay for Neutral Protease," *Biochemical and Biophysical Research Communications* **32**, 326–332 (1968).
29. P. H. Petra, "Bovine Procarboxypeptidase and Carboxypeptidase A," *Methods in Enzymology* **19**, 460–503 (1970).
30. O. H. Lowry, N. J. Rosebrough, A. L. Farr, and R. J. Randall, "Protein Measurement with the Folin Phenol Reagent," *Journal of Biological Chemistry* **193**, 265–275 (1951).
31. P. K. Smith, R. I. Krohn, G. T. Hermanson, A. K. Mallia, F. H. Gartner, M. D. Provenzano, E. K. Fujimoto, N. M. Goeke, B. J. Olson, and D. C. Klenk, "Measurement of Protein Using Bicinchoninic Acid," *Analytical Biochemistry* **150**, 76–85 (1985).
32. J. Adlers-Nissen, "Determination of the Degree of Hydrolysis of Food Protein Hydrolysates by Trinitrobenzenesulfonic Acid," *Journal of Agricultural and Food Chemistry* **27**, 1256–1262 (1979).
33. C. E. Stauffer, "A Linear Standard Curve for the Folin-Lowry Determination of Protein," *Analytical Biochemistry* **69**, 646–648 (1975).
34. J. E. Kruger and K. Preston, "The Distribution of Carboxypeptidases in Anatomical Tissues of Developing and Germinating Wheat Kernels," *Cereal Chemistry* **54**, 167–174 (1977).
35. J. E. Kruger and K. R. Preston, "Changes in Aminopeptidases of Wheat Kernels during Growth and Maturation," *Cereal Chemistry* **55**, 360–372 (1978).
36. H. V. Street and J. R. Close, "Determination of Amylase Activity in Biological Fluids," *Clinica Chimica Acta* **1**, 256–268 (1956).
37. J. J. Child, D. E. Eveleigh, and A. S. Sieben, "Determination of Cellulase Activity Using Hydroxyethylcellulose as Substrate," *Canadian Journal of Biochemistry* **51**, 39–43 (1973).
38. M. Ceska, E. Hultman, and B. Ingleman, "The Determination of Alpha-amylase," *Experientia* **25**, 555–556 (1969).
39. B. Klein, J. A. Foreman, and R. L. Searcy, "The Synthesis and Utilization of Cibachron Blue-Amylose: A New Chromogenic Substrate for Determination of Amylase Activity," *Analytical Biochemistry* **31**, 412–425 (1969).
40. T. K. Ng and J. G. Zeikus, "A Continuous Spectrophotometric Assay for the Determination of Cellulase Solubilizing Activity," *Analytical Biochemistry* **103**, 42–50 (1980).
41. S. Matsubara, T. Ikenaka, and S. Akabori, "Studies on Taka-amylase A. VI. On the α-Maltosidase Activity of Taka-amylase A," *Journal of Biochemistry* **46**, 425–431 (1959).
42. S. Dygert, L. H. Li, D. Florida, and J. A. Thoma, "Determination of Reducing Sugar with Improved Precision," *Analytical Biochemistry* **13**, 367–374 (1965).
43. Ref. 18, Method 22-01.
44. L. B. Marshall and G. D. Christian, "A Rapid Spectrophotometric Method for Iodimetric α-Amylase Assay," *Analytica Chimica Acta* **100**, 223–228 (1978).
45. P. Bernfeld, "Amylases, α and β," *Methods in Enzymology* **1**, 149–158 (1955).
46. G. L. Miller, R. Blum, W. E. Glennon and A. L. Benton, "Measurement of Carboxymethylcellulase Activity," *Analytical Biochemistry* **1**, 127–132 (1960).

47. M. A. Hulme, "Viscometric Determination of Carboxymethylcellulase in Standard International Units," *Archives of Biochemistry and Biophysics* **147**, 49–54 (1971).

48. K. Manning, "Improved Viscometric Assay for Cellulase," *Journal of Biochemical and Biophysical Methods* **5**, 189–202 (1981).

49. J. J. Distler and G. W. Jourdian, "The Purification and Properties of β-Galactosidase from Bovine Testes," *Journal of Biological Chemistry* **248**, 6772–6780 (1973).

50. L. Sarda and P. Desnuelle, "Action de la Lipase Pancreatique sur les Esters en Emulsion," *Biochimica et Biophysica Acta* **30**, 513–521 (1958).

51. D. Y. Kwon and J. S. Rhee, "A Simple and Rapid Colorimetric Method for Determination of Free Fatty Acids for Lipase Assay," *Journal of the American Oil Chemists Society* **63**, 89–92 (1986).

52. A. E. Hagerman and P. J. Austin, "Continuous Spectrophotometric Assay for Plant Pectin Methylesterase," *Journal of Agricultural and Food Chemistry* **34**, 440–444 (1986).

53. P. S. J. Chen, T. Y. Toribara, and H. Warner, "Microdetermination of Phosphorus," *Analytical Chemistry* **28**, 1756–1758 (1956).

54. A. Rossi, M. S. Palma, F. A. Leone, and M. A. Brigliador, "Properties of Acid Phosphatase from Scutella of Germinating Maize Seeds," *Phytochemistry* **20**, 1823–1826 (1981).

55. M. H. Keyes and F. E. Semersky, "A Quantitative Method for the Determination of the Activities of Mushroom Tyrosinase," *Archives of Biochemistry and Biophysics* **148**, 256–261 (1972).

56. B. Axelrod, T. M. Cheesbrough, and S. Laakso, "Lipoxygenase from Soybeans," *Methods of Enzymology* **71**, 441–451 (1981).

57. M. J. Gibian and R. A. Galaway, "Steady-State Kinetics of Lipoxygenase Oxygenation of Unsaturated Fatty Acids," *Biochemistry* **15**, 4209–4214 (1976).

58. E. Racker, "Spectrophotometric Measurements of the Metabolic Formation and Degradation of Thiol Esters and Enediol Compounds," *Biochimica et Biophysica Acta* **9**, 577–578 (1952).

59. T. Tono and S. Fujita, "Determination of Ascorbic Acid by Spectrophotometric Method Based on Difference Spectra. II. Spectrophotometric Determination Based on Difference Spectra of L-Ascorbic Acid in Plant and Animal Foods," *Agricultural Biological Chemistry* **46**, 2953–2959 (1982).

60. B. Chance and A. C. Maehly, "Assay of Catalases and Peroxidases," *Methods in Enzymology* **2**, 764–775 (1955).

61. R. M. J. Parry, R. C. Chandan, and K. H. Shahani, "A Rapid and Sensitive Assay of Muramidase," *Proceedings of the Society for Experimental Biology and Medicine* **119**, 384–386 (1965).

CLYDE E. STAUFFER
Consultant
Cincinnati, Ohio

ENZYMES. See CORN AND CORN PRODUCTS; FLAVOR CHEMISTRY; FOOD SPOILAGE; FOOD UTILIZATION; KINETICS; MICROBIOLOGY; SYRUPS; YEASTS; and entries under GENETIC ENGINEERING.

ENZYMES AS FOOD ADDITIVES

This article describes those enzymes that can be legally used in food processing.

LEGALLY APPROVED ENZYMES FOR FOOD PROCESSING

The following enzymes have been approved by the U.S. Food and Drug Administration as legal food additives to be used in food processing.

Amyloglucosidase Derived from *Rhizopus niveus*

Amyloglucosidase enzyme product, consisting of enzyme derived from *Rhizopus niveus*, and diatomaceous silica as a carrier, may be safely used in food in accordance with the following conditions.

1. *Rhizopus niveus* is classified as follows:
   Class. Phycomycetes
   Order. Mucorales
   Family. Mucoraceae
   Genus. *Rhizopus*
   Species. *niveus*
2. The strain of *Rhizopus niveus* is nonpathogenic and nontoxic in man or other animals.
3. The enzyme is produced by a process that completely removes the organism *Rhizopus niveus* from the amyloglucosidase.
4. The additive is used for degrading gelatinized starch into constituent sugars, in the production of distilled spirits and vinegar.
5. The additive is intended for use at a level not to exceed 0.1% by weight of the gelatinized starch.

Carbohydrase and Cellulase Derived from *Aspergillus niger*

Carbohydrase and cellulase enzyme preparation derived from *Aspergillus niger* may be safely used in food in accordance with the following prescribed conditions.

1. *Aspergillus niger* is classified as follows:
   Class. Deuteromycetes
   Order. Moniliales
   Family. Nomiliaceae
   Genus. *Aspergillus*
   Species. *niger*
2. The strain of *Aspergillus niger* is nonpathogenic and nontoxic in man or other animals.
3. The additive is produced by a process that completely removes the organism *Aspergillus niger* from the carbohydrase and cellulase enzyme product.
4. The additive is intended for use as follows:
   a. For removal of visceral mass (bellies) in clam processing.
   b. As an aid in the removal of the shell from the edible tissue in shrimp processing.
5. The additive is used in an amount not in excess of the minimum required to produce its intended effect.

Carbohydrase Derived from *Rhizopus oryzae*

Carbohydrase from *Rhizopus oryzae* may be safely used in the production of dextrose from starch in accordance with the following prescribed conditions.

1. *Rhizopus oryzae* is classified as follows:
   Class. Phycomycetes
   Order. Mucorales
   Family. Mucoraceae
   Genus. *Rhizopus*
   Species. *oryzae*
2. The strain of *Rhizopus oryzae* is nonpathogenic and nontoxic.
3. The carbohydrase is produced under controlled conditions to maintain nonpathogenicity and nontoxicity, including the absence of aflatoxin.
4. The carbohydrase is produced by a process that completely removes the organism *Rhizopus oryzae* from the carbohydrase product.
5. The carbohydrase is maintained under refrigeration from production to use and is labeled to include the necessity of refrigerated storage.

Catalse Derived from *Micrococcus lysodeikticus*

Bacterial catalase derived from *Micrococcus lysodeikticus* by a pure culture fermentation process may be safely used in destroying and removing hydrogen peroxide used in the manufacture of cheese, in accordance with the following conditions.

1. The organism *Micrococcus lysodeikticus* from which the bacterial catalase is to be derived is demonstrated to be nontoxic and nonpathogenic.
2. The organism *Micrococcus lysodeikticus* is removed from the bacterial catalase prior to use of the bacterial catalase.
3. The bacterial catalase is used in an amount not in excess of the minimum required to produce its intended effect.

Esterase Lipase Derived from *Mucor miehei*

Esterase lipase enzyme, consisting of enzyme derived from *Mucor miehei* var. *Cooney et Emerson* by a pure culture fermentation process, with maltodextrin or sweet whey as a carrier, may be safely used in food in accordance with the following conditions.

1. *Mucor miehei* var. *Cooney et Emerson* is classified as follows:
   Class. Phycomycetes
   Subclass. Zygomycetes
   Order. Mucorales
   Family. Mucoraceae
   Genus. *Mucor*
   Species. *miehei*
   Variety. *Cooney et Emerson*.
2. The strain of *Mucor miehei* var. *Cooney at Emerson* is nonpathogenic and nontoxic in man or other animals.
3. The enzyme is produced by a process that completely removes the organism *Mucor miehei* var. *Cooney et Emerson* from the esterase lipase.
4. The enzyme is used as a flavor enhancer.

5. The enzyme is used at levels not to exceed current good manufacturing practice in the following food categories: cheeses, fat and oils, and milk products. Use of this food ingredient is limited to nonstandardized foods and those foods for which the relevant standards of identity permit such use.
6. The enzyme is used in the minimum amount required to produce its limited technical effect.

α-Galactosidase Derived from *Mortierella vinaceae* var. *raffinoseutilizer*

The food additive α-galactosidase and parent mycelial microorganism *Mortierella vinaceae* var. *raffinoseutilizer* may be safely used in food in accordance with the following conditions.

1. The food additive is the enzyme α-galactosidase and the mycelia of the microorganism *Mortierella vinaceae* var. *raffinoseutilizer*, which produces the enzyme.
2. The nonpathogenic microorganism matches American Type Culture Collection (ATCC) No. 20034, and is classified as follows:
   Class. Phycomycetes
   Order. Mucorales
   Family. Mortierellaceae
   Genus. *Mortierella*
   Species. *vinaceae*
   Variety. *raffinoseutilizer*
3. The additive is used or intended for use in the production of sugar (sucrose) from sugar beets by addition as mycelial pellets to the molasses to increase the yield of sucrose, followed by removal of the spent mycelial pellets by filtration.
4. The enzyme removal is such that there are no enzyme or mycelial residues remaining in the finished sucrose.

Milk-Clotting Enzymes, Microbial

Milk-clotting enzyme produced by a pure-culture fermentation process may be safely used in the production of cheese in accordance with the following prescribed conditions.

1. Milk-clotting enzyme is derived from one of the following organisms by a pure-culture fermentation process:

   Endothia parasitica
   Class. Ascomycetes
   Order. Sphaeriales
   Family. Diaporthacesae
   Genus. *Endothia*
   Species. *parasitica*

   Bacillus cereus
   Class. Schizomycetes
   Order. Eubacteriales
   Family. Bacillaceae

Genus. *Bacillus*
Species. *cereus* (Frandland and Frandland)

Mucor pusillus Lindt
Class. Phycomycetes
Subclass. Zygomycetes
Order. Mucorales
Family. Mucoraceae
Genus. *Mucor*
Species. *pusillus*
Variety. *Lindt*

Mucor miebei Cooney et Emerson
Class. Phycomycetes
Subclass. Zygomycetes
Order. Mucorales
Family. Mucoraceae
Genus. *Mucor*
Species. *miehei*
Variety. *Cooney et Emerson*

2. The strains of organism identified above are non-pathogenic and nontoxic in man or other animals.
3. The additive is produced by a process that completely removes the generating organism from the milk-clotting enzyme product.
4. The additive is used in an amount not in excess of the minimum required to produce its intended effect in the production of those cheeses for which it is permitted by standards of identity established.

Y. H. HUI
Editor-in-Chief

ENZYMOLOGY

Enzymes can be defined as proteins with very specific powers of catalysis. While all enzymes are proteins, some require an additional small molecule, called a coenzyme or cofactor (cofactors can be metal ions or small organic molecules), to function. The role of enzymes in cells means that growth, maturation, storage, processing, and consumption–digestion of food all depend on various enzyme activities. In plants, enzymes are responsible for changes associated with ripening, including alterations in color, flavor, and texture. Enzymatic changes continue after harvesting of plants and also after the death of animals, eg, in the conversion of muscle to meat. These changes affect subsequent food quality. In many cases the action of endogenous enzymes is arrested or controlled by processing and storage.

However, enzymes are also widely used to provide desirable changes in food quality attributes. Enzymes have significant advantages over chemical catalysts, of which the most important are specificity and the ability to work at moderate temperatures. As a consequence, side reactions are minimized and undesirable changes caused by harsh conditions are averted. The value of enzymes used in food processing is already several hundred million dollars per year and will likely expand as new applications are discovered or improved enzymes become available.

KINETICS

The activity of an enzyme is determined by many factors, including enzyme, substrate, and cofactor concentrations; ionic strength; pH; and temperature. For conversion of substrate (S) to product (P) by an enzyme (E), the reaction scheme can be simply represented as:

$$E + S \underset{}{\overset{k_s}{\rightleftharpoons}} ES \overset{k_{cat}}{\longrightarrow} E + P$$

The reaction velocity (V) is then given by the Michaelis-Menten equation

$$V = \frac{k_{cat}[E][S]}{K_m + [S]}$$

where K_m is the substrate concentration at which V equals one-half the maximum velocity (V_{max}). Integration of this equation with respect to time gives

$$V_{max} = K_m \ln\left(\frac{[S]}{[S_t]}\right) + ([S] - [S_t])$$

where $[S]$ and $[S_t]$ are the substrate concentrations at zero time and time t, respectively. This equation is particularly useful in industrial situations where a reaction is allowed to proceed to near completion or equilibrium.

APPLICATIONS IN FOOD PROCESSING

α-Amylase

α-Amylase (E.C. 3.2.1.1) catalyzes random hydrolysis of α-1–4 linkages in amylose and amylopectin to form straight- and branched-chain dextrins, oligosaccharides, and monosaccharides. Bacteria α-amylase is used to partially hydrolyze or thin gelatinized starch, often prior to further degradation to glucose syrups. The enzyme from *Bacillus* spp. is very thermostable and is used at 85–105°C. The enzyme can also be used to hydrolyze starch in sugar cane juice and in the brewing mash.

Fungal α-amylases, derived from *Aspergillus niger* and *A. oryzae* are much less heat stable. They produce large amounts of maltose and maltotriose and some glucose. Their principal applications are production of maltose syrups used in jam and confectionery; as brewing aids to improve fermentability of the mash and in removal of starch haze in beer; supplementation of endogenous α-amylase in bread flour to enhance the rate of fermentation by yeast and reduce dough viscosity, thereby improving loaf volume. Fungal α-amylase has largely replaced the malt α-amylase previously used in these applications.

β-Amylase

β-Amylase (EC 3.2.1.2) splits off β-maltose from the nonreducing ends of starch molecules. In the case of amylopectin, this produces β-limit dextrins, since the enzyme cannot bypass the β-1–6 branch points. Complete hydrolysis of liquified starch produces about 80% maltose and 20% dextrins. Unlike fungal α-amylase, β-amylase does not produce maltotriose. The enzyme is produced from cereal or microbial sources and is used for production of maltose syrup from starch, and in both brewing and baking to produce maltose for fermentation by yeast to CO_2 and to alcohol.

Amyloglucosidase

Amyloglucosidase (glucoamylase) (EC 3.2.1.3) is an exo-amylase which catalyzes the stepwise hydrolysis of α-1–4 linkages in starch, thereby releasing glucose molecules from the nonreducing end. It is used mainly to produce glucose syrups from liquified starch previously treated with α-amylase. Since the α-1–6 linkages in amylopectin are also slowly hydrolyzed, the final conversion to glucose reaches 95–97% w/w, with the remainder being mostly maltose and higher saccharides. The glucose produced is usually used as a syrup, crystallized out, or converted to fructose by glucose isomerase. The enzyme can also be used to hydrolyze residual oligosaccharides in high-fructose corn syrup and in the analysis of the starch content of foods. Glucoamylase is produced commercially from *Aspergillus* or *Rhizopus* spp. It is typically used to process liquified starch at 60°C. Immobilization of the enzyme has been widely studied to provide the benefits of a continuous process. However, the soluble enzyme is relatively cheap, so it continues to be used mostly in that form.

Catalase

Catalase (EC 1.11.1.6) specifically catalyzes the decomposition of hydrogen peroxide to water and oxygen. In the dairy industry, a low concentration of H_2O_2 (up to 0.05% w/w) is used to cold-pasteurize milk destined for cheese making and to preserve milk and whey in some instances where refrigeration is not practical. The H_2O_2 is then destroyed by catalase. Commercial sources include beef liver and *Aspergillus niger*. The enzyme is frequently used in conjunction with glucose oxidase to remove H_2O_2 produced by that enzyme.

Cellulase

Cellulase (EC 3.2.1.4) is the name given to a complex of several enzymes which, acting together, hydrolyze cellulose into β-dextrins and glucose. The native structure of cellulose is a major impediment to enzyme action, so pretreatment by milling or with alkali or steam is often necessary to make the substrate accessible. Currently, the most effective enzyme preparation is obtained from the fungus *Trichoderma reesie*. It contains (1) endo-cellulases, which randomly split internal β-1–4 linkages to produce dextrins and cellobiose, (2) exo-cellulases, which act from the nonreducing end of the polymer to produce cellobiose, and (3) cellobiase (or β-glucosidase), which converts cellobiose to glucose. The latter is especially important since it relieves end-product inhibition of the other enzymes by cellobiose. Cellulase is used primarily to turn cellulosic wastes into glucose which can be fermented to ethanol. It is also used on a small scale to degrade cellulose in foodstuffs.

β-Glucanase

β-Glucanase (EC 3.2.1.6) hydrolyzes β-1–3 or β-1–4 bonds in β-D-glucans. The products are oligosaccharides and glucose. The enzyme is used primarily in brewing where it is added, along with other enzymes, to malted barley at the mash stage. The β-glucanase degrades and solubilizes barley gums (β-glucan polymers) which would otherwise increase the viscosity of the wort, including those contributed by dying yeast cells. These residual glucans can contribute to haze problems in the final product. Commercial sources of β-glucanase include *Bacillus subtilis* and *Aspergillus niger*.

Glucose Isomerase

Glucose isomerase (EC 5.1.3.5) catalyzes conversion of glucose to fructose, thereby increasing sweetness and value. The enzyme is actually a xylose isomerase which also acts on glucose and requires magnesium as a cofactor. The reaction is readily reversible and at equilibrium, a fructose/glucose ratio of 52/48 is achieved. However, in practice it is uneconomical to take the conversion beyond 42% fructose, when the syrup is as sweet as glucose on a solids basis. The product (high-fructose corn syrup) is produced at a rate of several million tons per year and has replaced sucrose and glucose syrups for many products, especially in the United States.

The enzyme is produced commercially from several microbial sources including *Bacillus*, *Actinoplanes*, and *Streptomyces* spp. and is readily immobilized by a variety of methods (1).

Glucose Oxidase

Glucose oxidase (EC 1.1.3.4) catalyzes the reaction

$$\text{glucose} + O_2 \rightarrow \text{gluconic acid} + H_2O_2$$

It is often used in conjunction with catalase, which breaks down H_2O_2 and thereby spares the glucose oxidase from denaturation as well as providing oxygen for use by glucose oxidase. Commercial preparations are derived from *Aspergillus niger* or *Penicillium* spp. Its main uses are to remove glucose from foods, and as an antioxidant to prevent changes in color and flavor, particularly during food storage. It is used for removal of glucose from egg whites and whole eggs. Significant browning and off-flavor development due to Maillard reactions occur if eggs are not desugared prior to drying. Other applications include removal of either dissolved or head-space oxygen from citrus drinks, canned soft drinks, beer, and wine, thereby preventing oxidative deterioration; use as an antioxidant in mayonnaise; and production of gluconic acid. It is often the method of choice for glucose assay since it is highly specific and sensitive.

Invertase

Invertase (β-fructofuranosidase) (EC 3.2.1.6) hydrolyzes sucrose into an equimolar mixture of glucose and fructose. While the same result can be achieved by acid hydrolysis, the syrup produced by enzyme action is more pure and free from discoloration. Invert sugar syrup has several advantages over sucrose syrup: it is slightly sweeter, it does not crystallize at higher concentrations, and the sweetness intensity is stable in acidic foods. Until the advent of high-fructose corn syrup, it was used extensively to replace sucrose in jams and confectionery.

Other applications of invertase include the production of liquid or soft centers in chocolate-coated sucrose candies, recovery of scrap candy, artifical honey, and a humectant to hold moisture in foods. Invertase is produced commercially from yeasts, usually *Saccharomyces* or *Candida* spp., and is relatively cheap to produce. Consequently, it is used as a soluble enzyme and there is little commercial impetus for its application in an immobilized form.

Lactase

Lactase (β-galactosidase) (EC 3.2.1.23) hydrolyzes lactose into glucose and galactose. Small quantities of oligosaccharides containing galactose may also be formed as byproducts. Compared to lactose, the main products are, in combination, three to four times more soluble, about twice as sweet, easier to ferment, and directly absorbed from the intestine. Applications of lactase take advantage of these changes.

Hydrolysis of lactose in milk makes the milk more digestible for those who are lactose intolerant (especially infants) due to low levels of intestinal β-galactosidase. It also prevents crystallization of lactose in concentrated or frozen milk products, such as ice cream. Cheese whey is a major disposal problem, mostly because of its high lactose content. After separating the valuable whey protein, the lactose can be hydrolyzed by soluble or immobilized lactase. The product is then concentrated to give a stable, sweet syrup which may be used in a variety of foods or fermented to ethanol (2).

Lactases produced commercially from yeasts such as *Kluyveromyces marxianus* have a neutral pH-optimum and are used in milk processing. Lactases from molds such as *Aspergillus niger* or *A. oryzae* have an acid pH-optimum and are more suitable for whey processing. Lactases are readily immobilized on a variety of supports for industrial use and for analysis of lactose in dairy products.

Lipases

Lipases (EC 3.1.1.3) hydrolyze ester linkages of triglycerides to give free fatty acids, diglycerides, monoglycerides, and eventually glycerol. The enzyme is widely distributed in food tissues and belongs to the general class of esterases. Lipase action has traditionally been regarded as a problem in food science since the fatty acids produced may be unpalatable or susceptible to oxidation. This is a particular problem in stored cereals such as wheat and rice and in milk, where the release of short-chain fatty acids from milk fat by endogenous lipase leads directly to off-flavors. However, it is recognized that lipase action is also reponsible for some of the desirable flavor in matured cheeses. Consequently, impure preparations (containing both lipase and esterase activities) are produced from molds such as *Mucor, Rhizopus,* and *Aspergillus* spp. for use in accelerated cheese ripening, often in conjunction with proteases. Concentrated cheese flavors produced in this fashion can be used in a variety of snack products. Lipase can also be used in directed transesterification of fats to improve functional properties and uses.

Pectinases

Pectinases are a group of enzymes which act on various pectic substances (pectins) in higher plants. There are three major types of pectinase: polygalacturonase (PG), pectin lyase (transeliminase) (PL), and pectin esterase (PE). Polygalacturonase (EC 3.2.1.15) splits glycosidic bonds within (endo-) or at the end of (exo-) the pectin molecule. Endo-PG action leads to a large decrease in viscosity of pectin solutions. Pectin lyase (EC 4.2.2.10) also splits endo-glycosidic bonds but the transelimination reaction yields a C(4)–C(5) double bond on the nonreducing end (3). Pectin esterase (EC 3.1.1.11) cleaves methanol from carboxyl groups yielding low-methoxy pectin and polygalacturonic acid.

Since pectin is a major structural element in and between plant cell walls, alteration of the size or esterification of pectins can change the texture of fruits and vegetables during ripening and storage. Endogenous pectinases may soften texture during ripening. Microbial pectinases are often responsible for postharvest rotting and decay. However, fungal pectinases, mostly from *Aspergillus* spp. are widely used as processing aids, mainly for extraction and clarification of fruit juices.

Phenolase

Phenolase catalyzes two types of reaction: hydroxylation of phenols (such as tyrosine) to an orthodiphenol (cresolase activity) and oxidation of diphenols to orthoquinones (polyphenol oxidase or catecholase activity). The enzyme occurs widely in fruits and vegetables, where it is separated from phenolic substrates in intact tissue. However, on cutting or other injury and exposure to oxygen, phenolase activity results in rapid browning due to polymerization of the quinones to give brown pigments (melanins) (4).

Enzymatic browning is a major problem in handling and storage of fresh produce. However, in foods such as tea, coffee, and cocoa, and in dried fruits such as raisins and dates, it improves color and flavor. Methods for control of enzymatic browning include inactivation of the enzyme by heat, sulfites, or proteases; use of acidulants to lower the pH and inhibit the enzyme; use of chelators to remove the copper which is essential for activity; and exclusion (or removal) of oxygen by appropriate packaging or glucose oxidase activity.

Proteases

Proteases hydrolyze peptide bonds in proteins and polypeptides to produce smaller peptides and amino acids. Many proteases also have esterase activity. These enzymes vary widely in their substrate specificity and optimum pH range. Enzymatic hydrolysis of protein is employed in a number of industries to create changes in product taste, texture, and appearance, as well as in waste recovery. Plant proteases such as papain and ficin have broad substrate specificity and good thermal stability.

They are used to tenderize meat, to chill-proof beer, and also to recover protein hydrolyzate from scrap fish and bones. Animal proteases are generally more specific than the plant proteases. Trypsin has been used to inhibit the development of an oxidized flavor in stored milk and to solubilize heat-denatured whey protein (5). Pepsin is used primarily as an extender for calf rennet. Fungal proteases from *Aspergillus niger* and *A. oryzae* are used to modify gluten in bread flour to reduce dough viscosity and improve color, texture, and loaf volume. Bacterial proteases from *Bacillus* spp. are widely used in biscuit, cookie, and cracker dough because hydrolysis of gluten yields very elastic doughs which can be spread thinly without rupture. Other uses of microbial proteases include modification of soy and other food proteins to give improved functionality, decolorization of red blood cells to facilitate use of waste blood plasma protein from animal slaughter, and production and modification of gelatin from collagen. Recently there has been interest in using proteases in reverse to synthesize peptides. An example of this is the use of the enzyme thermolysin to produce the dipeptide sweetener aspartame (6).

Pullulanase

Pullulanase (EC 3.2.1.41) and isoamylase (3.2.1.68) act specifically on the α-1–6 bonds of amylopectin in liquified starch to produce maltose and maltotriose. These enzymes improve the yield of mono- and disaccharides from starch when used in conjunction with other amylases. Pullulanase is produced commercially from *Klebsiella pneumonia* and is used in brewing to remove limit dextrins and allow the production of specialty beers such as high alcohol or low calorie.

Rennet

Rennet is obtained from the fourth stomach of unweaned calves and contains several enzymes including pepsin and chymosin (EC 3.44.3). Chymosin is an acid protease used in cheesemaking to coagulate the casein in milk. It catalyzes very specific and limited proteolysis of k-casein, thereby destabilizing the casein micelle and causing subsequent formation of curd. Calf rennet is expensive and is increasingly being replaced by microbial proteases with a high ratio of milk-clotting activity to general proteolytic activity. Suitable proteases have been obtained from *Endothia parasitica*, *Mucor meihei*, and *M. pusillus*. Microbial proteases are now used in about one-third of all cheese production worldwide. Their main disadvantage is that they tend to be heat stable and cause proteolysis in whey products. However, a second generation of modified microbial rennets with lower stability is now available. The gene for chymosin has now been successfully cloned and expressed in several microorganisms (7). Products from this source may well replace both traditional and microbial rennets in the near future.

IMMOBILIZED ENZYMES IN FOODS

Immobilization of enzymes permits repeated or continuous use and prevents the enzyme from entering the product. This leads to savings in costs. Enzymes can be immobilized after isolation or the cells containing the enzyme can be immobilized. Both forms are employed in food processing. Current examples include glucose isomerase (for production of fructose), aminoacylase (for production of L-amino acids from a racemic mixture), α-galactosidase (for removal of raffinose in sugar beet syrup), β-galactosidase (for lactose hydrolysis), and amyloglucosidase (for hydrolysis of starch). The glucose isomerase process is by far the most significant in terms of throughput at present.

BIBLIOGRAPHY

1. S. A. Barker and G. S. Petch, "Enzymatic Processes for High-Fructose Corn Syrup" in A. Laskin, ed., *Enzyme and Immobilized Cells in Biotechnology* Benjamin/Cummings, Menlo Park, Calif., 1985.
2. J. Barry, "A New Source for Alcohol," *Food Manufacture* **58**(11), 63–67 (1983).
3. W. Pilnik and F. M. Rombouts, "Pectic Enzymes," in G. G. Birch, N. Blakeborough, and K. J. Parker, eds., *Enzymes and Food Processing,* Applied Science Publishers, London, 1981.
4. A. B. Learner and B. Fitzgerald, "Biochemistry of Melanin Formation," *Physiological Reviews* **30**, 91–126 (1950).
5. J. C. Monti and R. Jost, "Solubilization of Cheese Whey Protein by Trypsin and a Process to Recover the Active Enzyme from the Digest," *Biotechnology and Bioengineering* **20**, 1173–1185 (1981).
6. K. Nakanishi and R. Matsuno, "Enzymatic Synthesis of Aspartame," in R. D. King and P. S. J. Cheetham, eds., *Food Biotechnology 2,* Elsevier Applied Science, London, 1988.
7. D. Jackson, "Cost Reduction in Food Processing Using Biotechnology," in S. K. Harlander and T. B. Lubuza, eds., *Biotechnology in Food Processing,* Noyes Press, Park Ridge, N.J., 1986.

General References

Cheetham, P. S. J., "The Application of Enzymes in Industry," in A. Wiseman, ed., *Handbook of Enzyme Biotechnology,* 2nd ed., Ellis Horwood, Chichester, U.K., 1985.

Godfrey, T., and J. Reichelt, eds., *Industrial Enzymology,* Nature Press, New York, 1983.

Schwimmer, S., *Source Book of Food Enzymology,* AVI Publishing Co., Inc., Westport, Conn., 1981.

RAYMOND R. MAHONEY
University of Massachusetts
Amherst, Massachusetts

ESSENTIAL OILS: CITRUS AND MINT

Essential oils are flavor and fragrance materials that, by nature, consist primarily of a complicated mixture of terpenes and sesquiterpenes. These materials are obtained from approximately one-third of the known plant families, most often by physical means. The physical means most often used to obtain the essential oil of a plant are inclusive of extraction, vacuum distillation, steam distillation, and, most recently, supercritical gas and solvent extraction. The most widely used method is steam distillation, which greatly takes advantage of the volatility associated with a vast number of essential oils.

Figure 1. Mint oil producing regions.

Once a crude essential oil is obtained, it is often further processed to become a monoterpeneless, sesquiterpeneless product. This is usually done through classic vacuum and steam distillation procedures to concentrate the body or heart of an oil. Owing to the large number of essential oils that are commercially produced (including oils from the families Rutacea, Labiataa, Liliaceae, Pinaceae, Malvaceae, Betulaceae, and others), it would be impossible to cover all essential oils and their composition. The scope of this article is briefly to cover the characteristics of some mint oils from the family Labiatae and a citrus oil from the family Rutaceae—specifically lime oil.

MINT ESSENTIAL OILS

Mint essential oils are primarily obtained by steam distillation of the fresh overground parts of a mature mint plant. American peppermint oil is obtained from *Mentha piperita L. Mentha spicata L.* produces spearmint oil Native type and *Mentha gentilis L., Cardiaca ger.* produces spearmint oil Scotch type. All these oils are obtained in crude form by doing a large steam distillation of the plants after they have been cut and allowed to dry for several days. Both peppermint and spearmint production is concentrated primarily in six states: Washington, Oregon, Idaho, Wisconsin, Indiana, and Michigan (Fig. 1).

Often these states are simply classed into two segments: Midwestern and Far Western. The mint oils produced in each of these regions are often similar in composition, but possess very distinct odor and taste characteristics. These differences are caused by regional variances in temperature, photoperiod, and light intensity (1). All these factors, and many others, effect the biosynthetic interconversions that take place within the plant.

Peppermint Oil

In peppermint oil one of the most important of these interconversion deals with the production of menthone and menthol; whose total composition comprises from 50 to 70% of a peppermint oil. The in vivo reaction pathway that gives rise to menthone and menthol are shown in Figure 2. This reaction pathway was first proposed in 1922 (2) and then later in 1958 (3). Most recently, it has been demonstrated that (+)-*cis*-isopulegone may be an intermediate to the formation of pulegone (4). So it can readily be seen that reduction of the oxygenated monoterpene *d*-pulegone would give rise to *l*-menthone and on further reduction *l*-menthol. These interdependencies make control of the composition of some essential oils an art as much as it is a science. The composition of a typical oil from *Mentha piperita L.* is shown in Table 1. Note that only the major constituents are shown. Further information on composition may be found in Ref. 5.

Spearmint Oil

Spearmint oil is an equally important essential oil for use primarily as a flavoring agent in dentrifices and confectionary products. The primary constituent in spearmint oil is *l*-carvone, with typical levels being 60–80% of an oil. The second largest constituent is *l*-limonene, which typically accounts for 9–20% of the whole oil. A further breakdown of a typical chemical profile for spearmint oil (Scotch type) is shown in Table 2. In addition to their composition,

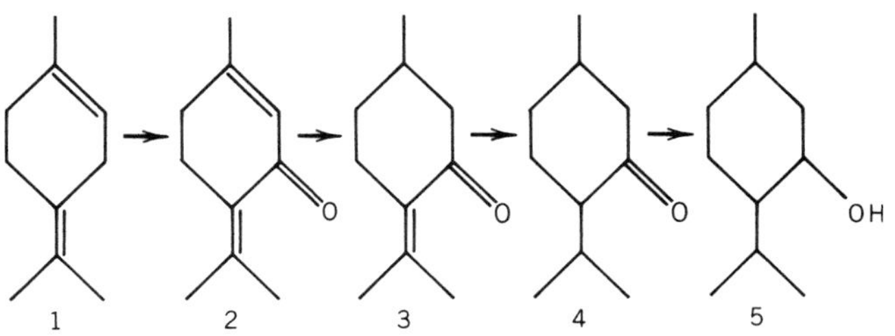

(1) Terpinolene
(2) Piperitenone
(3) *d*-Pulegone
(4) l-Menthone
(5) l-Menthol

Figure 2. In vivo peppermint reaction pathway.

Table 1. Peppermint Oil Composition

Compound	Percent
Dimethylsulfide	0.02
2 Methyl propanal	0.03
3 Methyl propanal	<0.01
2 Methyl butanal	<0.01
3 Methyl butanal	0.15
2 Ethyl furan	0.03
trans-2,5-Diethyl THF	0.02
α-Pinene	0.66
Sabinene	0.42
Myrcene	0.18
α-Terpinene	0.34
Limonene	1.33
1,8 Cineole	4.80
trans-Ocimene	0.03
cis-Ocimene	0.31
G-Terpinene	0.56
trans-2-Hexenal	0.07
para-Cymene	0.10
Terpeniolene	0.16
Hexanol	0.13
3 Octyl acetate	0.03
cis-3-Hexenol	0.01
3-Octanol	0.21
trans-2-Hexenol	0.02
Sabinene hydrate	0.80
Menthone	20.48
Menthofuran	1.67
D-Isomenthone	2.77
B-Bourbonol	0.37
Neomenthylacetate	0.21
Linalool	0.26
cis-Sabinene hydrate	0.07
Menthyl acetate	5.02
Isopulegol	0.07
Neoisomenthyl acetate	0.26
Neomenthol	3.34
β-Caryophyllene	2.13
Terpinene-4-ol	0.98
Neoisoisopulegol	0.03
Neoisomenthol	0.78
Menthol	43.18
Pulegone	0.77
trans-β Farenscene	0.29
Isomenthol	0.19
Humelene	0.03
α-Terpineol	0.16
Germacrene-D	2.29
Piperitone	0.96
Viridiflorol	0.26
Eugenol	0.02
Thymol	0.04

Table 2. Spearmint Oil Composition (Scotch)

Compound	Percent
Dimethylsulfide	<0.01
2-Methyl propanal	0.01
3-Methyl propanal	<0.01
2-Ethyl furan	0.02
2-Methyl butanal	<0.01
3-Methyl butanal	0.08
α-Pinene	0.73
trans-2,5-Diethyl THF	0.11
β-pinene	0.72
Sabinene	0.46
Myrcene	0.90
α-terpinene	0.04
Limonene	17.94
1,8 Cineole	1.45
trans-Ocimene	0.05
cis-Ocimene	0.05
γ-terpinene	0.08
trans-2-Hexenal	0.03
para-Cymene	0.03
Terpeniolene	0.04
Hexenol	0.01
3-Octanal acetate	0.13
cis-3-hexenol	0.02
3-Octanol	1.78
trans-2-Hexenol	0.05
Sabinene hydrate	0.16
Menthone	1.35
D-Isomenthone	0.18
β-Bourbonol	0.87
Linalool	0.31
cis-Sabinene hydrate	0.02
β-Caryophyllene	0.62
Terpinene-4-ol	0.16
Dihydrocarvone	0.95
cis-dihydrocarvylacetate	0.62
D-Terpeniol	0.11
trans-β Farnescene	0.29
trans-Dihydrocarvylacetate	0.09
Humelene	0.11
α-Terpeniol	0.24
γ Mureol	0.02
Germacrene D	0.36
Neodihydrocarveol	0.20
1-Carvone	64.82
cis-Carvyl acetate	0.08
Cubenene	0.09
trans-Carveol	0.28
cis-Carveol	0.23
C-Jasmon	0.15
Viridiflorol	0.01
Eugenol	0.03
Thymol	0.02

typical physical constant ranges for both peppermint and spearmint are given in Table 3.

The current regulatory status of both peppermint oil and spearmint oil is that the Food and Drug Administration evaluation was GRAS, the Flavor Extract Manufacturers Association Evaluation was GRAS, FDA registry number 182.20.

Table 3. Peppermint and Spearmint Physical Ranges

	Peppermint	Spearmint
Specific gravity 25°/25°C	0.896–0.905	0.922–0.934
Optical rotation (100 mm tube)	−22 to −30	≤−48
Refractive index (20°C)	1.459–1.465	1.485–1.494
Solubility		
1 volume in 3 volumes 70% EtOH	Yes	Yes

Table 4. Physical Characteristics of Lime Oils

	Distilled	Centrifuged	Essence
Optical rotation 20°	34–47	35–41	43–46
Specific gravity 20°	0.855–0.863	0.872–0.883	0.853–0.855
Refractive index	1.4745–1.4770	1.4820–1.4885	1.4750–1.4770
Residue	0.2–2.2%	10–19%	0%

CITRUS PRODUCTS

Unlike peppermint and spearmint, there are several different types of products that may be obtained from citrus products, specifically the essential oil of lime. The primary classes of lime oil are distilled oils obtained via steam distillation, cold-pressed oils obtained via centrifugation, and essence oils. Each of these essential oils of lime possesses unique character and varies in color, aroma, physical characteristics, and chemical characteristics. The distilled and essence oils are slightly yellowish; the centrifuged oils possess pigment from light green to dark green.

The odor of each type of oil varies by significant degrees. Distilled oil has a sharp, fresh, terpenic odor; the centrifuged oils tend to have an intensely fresh, rich, and sweet odor profile. The odor of essence oil is fresh but less terpenic than distilled oils and less sharp than centrifuged oils.

The different values for physical characteristics of the oils are given in Table 4. Lime oil is estimated to have approximately 350 constituents (6). These constituents are usually classed into three divisions: hydrocarbon compounds, oxygenated compounds, and nonvolatile residue. Typical compositions for each of the oils are shown in Table 5. Note that only the major constituents are shown.

For the purpose of quality control, analysis results are very important tools in monitoring essential oils. The most common analyses performed on essential oils are optical rotation, refractive index, specific gravity, evapora-

Table 5. Lime Oil Composition

	Distilled	Centrifuge	Essence
Acetaldehyde	0.18	—[a]	—
α-Thujene	0.02	0.40	0.34
α-Pinene	1.32	2.19	2.41
Camphene	0.50	0.09	0.14
Sabinene	—	3.38	0.71
β-Pinene	2.23	20.93	21.17
Myrcene	1.26	1.22	1.27
α-Phellandrene	0.36	0.04	0.08
1,4 Cineole	1.87	<0.01	0.03
α-Terpinene	2.26	0.19	0.46
para-Cymene	2.14	0.18	1.03
Limonene	48.53	45.08	52.52
γ-Terpinene	11.69	10.46	8.46
α-para-Dimethylstyrene	0.24	—	—
Terpinene	7.63	0.49	0.71
Linalool	0.16	0.15	0.36
α-Fenchol	0.75	—	—
Terpinene-1-ol	0.81	—	—
β-Terpineneol	0.62	—	—
Borneol	0.49	—	—
Terpinene-4-ol	0.85	0.05	1.68
α-Terpineneol	6.35	0.19	1.33
γ-Terpineol	0.92	<0.01	0.02
Neral	0.02	1.64	1.08
Geraniol		0.09	0.05
Geranial	0.04	2.75	1.8
Neryl acetate	0.12	0.41	0.12
Geranyl acetate	0.08	0.23	0.12
Dodecanal	0.09	0.16	0.04
β-Caryophyllene	0.52	1.05	0.30
α-Bergmatene	0.73	1.22	0.43
trans,trans-α-Farnesene	1.08	1.73	0.37
β-Bisabolene	1.27	2.04	0.56

[a] — = <0.01.

Table 6. Gas Chromotography Conditions

Instrument	Varian Vista 6000
Detector	FID
Column	Fused silica, capillary, 60-m long 0.2-mm I.D., 0.25-mm film phase SE-30
Sample size	0.2 ml split 150:1

tive residue, and solubility in ethanol. Today's scientist will also find gas chromatography (GC) results an integral tool when evaluating any essential oil. The compositional results presented herein were obtained using GC with the conditions shown in Table 6.

CONCLUSION

Using these and other tools, the flavorist can be assured of the compositional quality of an oil, but it should be stressed that with essential oil usage sensory evaluation is as important as composition. It is at this point that odor and taste evaluation truly separate the art from the science. The best way for an individual to learn to assess the sensory qualities of an essential oil is to practice with someone who is an established expert in the area of interest.

As one can see, there is extreme complexity in essential oils. It is this complexity and balance that truly makes essential oils unique, and for this reason these natural materials will never truly be replaced by synthetic materials in the world of flavors and fragrances.

BIBLIOGRAPHY

1. M. J. Murray, *Peppermint Oil Quality Differences and the Reasons for Them,* 10th International Congress of Essential Oils, 1986.
2. R. E. Kremers, *Journal of Biol Chem* **50**, 31 (1922).
3. R. H. Reitsema, *Journal of American Pharm. Association* (Science Edition) **47** 267 (1958).
4. R. Croteau and K. V. Venkatachalam, *Archives of Biochemistry and Biophysics,* **249**(2), 306–315 (1986).
5. B. Lawrence, *Perfumer and Flavorist,* vol. 5, Allured Publishing Corp. 1980, pp. 8–12.
6. G. Goretti and A. Liberti, *Sui Constituenti degli Olii Essenziali diAgrumi.* ATTl 10. Convegno Nazionale sugli Olii Essenziali Reggio Calabria, Italy, 1972

JEFFREY S. SPENCER
A. M. Todd Co.
Kalamazoo, Michigan

EVAPORATED MILK

HISTORY

Evaporated milk, like other processed canned foods, originated with the experiments of the French scientist Nicholas Appert. Appert, whose work on food preservation began in 1795, was the first to evaporate milk by boiling in an open container and then preserve it by heating the product in a sealed container. Fifty years later another French scientist, Louis Pasteur, laid the scientific foundation for heat preservation through demonstrations that food spoilage could be caused by bacteria and other microorganisms.

Patents dealing with the preservation of milk after evaporation in a vacuum were granted to Gail Borden by the United States and England in 1856. Although these patents applied to concentrating milk without the addition of sugar, Borden's first commercial process was for the manufacture of sweetened condensed milk. The original vacuum equipment developed by Borden is now on display at the Smithsonian Institution.

Gail Borden produced sweetened condensed milk at Wassaic, New York in 1861. By 1865 new plants were opened in Brewster, New York and Elgin, Illinois. In 1866 Charles A. Page and his three brothers built Europe's first commercial sweetened condensed milk plant in Switzerland. Later they expanded their operation by building plants in England and in the United States.

In 1884 a U.S. patent was issued for "an apparatus for preserving milk" and in 1885 the world's first commercial evaporated milk plant was opened in a converted wool factory in Highland, Illinois, where "evaporated cream" was manufactured and sold.

The first advertisement for evaporated milk appeared in 1893, calling the product "a perfect instant food." In 1894 the first recipe booklet for evaporated milk was distributed, and by 1895 the product was popular both in western mining areas of the United States, where fresh milk was scarce, and in the South, where there was little refrigeration. Consumers quickly recognized the value of evaporated milk as a safe, wholesome, and convenient food, as well as a nutritional ingredient for cooking that had the added benefit of storage stability.

THE PRODUCT

Evaporated milk is a canned whole milk concentrate to which a specified quantity of vitamin D has been added and to which vitamin A may be added. It conforms to the Food and Drug Administration (FDA) Standard of Identity (21 CRF 131.130), having a minimum of 7.5% milkfat, 25.0% total milk solids, and 25 IU vitamin D per fluid ounce. Related evaporated milk products are evaporated skimmed milk, evaporated lowfat milk, evaporated filled milk, and evaporated goat milk. Evaporated skimmed milk conforms to the FDA Standard of Identity (21 CFR 131.132) and contains not less than 20.0% total milk solids, not more than 0.5% milkfat, with added vitamins of 25 IU vitamin D and 125 IU vitamin A per fluid ounce. Standards of Identity have not been established for the other evaporated milk products. Their typical compositions are the following:

Evaporated lowfat milk	2.0% milkfat, 18% nonfat milk solids, vitamins A and D added
Evaporated filled milk	6.0% vegetable fat, 17.5% nonfat milk solids, vitamins A and D added

Table 1. Nutritional Content Per One-Half Cup Evaporated Milk

	Original	Skimmed	Lowfat
Calories	170	100	110
Protein, g	8	9	8
Carbohydrates, g	12	14	12
Fat, g	10	Less than 1	3
Percentage of U.S. Recommended Daily Allowance (US RDA)			
Protein	20	20	20
Vitamin A	4	10	10
Vitamin C	a	a	a
Thiamine	2	2	2
Riboflavin	20	20	20
Niacin	a	a	a
Calcium	30	30	30
Iron	a	a	a
Vitamin D	25	25	25
Phosphorus	25	25	25

[a] Contains less than 2% of the US RDA of these nutrients.

Evaporated goat milk   not less than 7.0% milkfat and 15.0% nonfat milk solids, vitamin D added

Table 1 shows the nutritional content of evaporated, evaporated skimmed, and evaporated lowfat milks.

PRODUCT PROCESSING

A typical processing scheme for evaporated milk begins with high quality, fresh whole milk, which then is standardized to produce the exact composition desired in the final product (addition or removal of cream or skimmed milk). The product then is heated, concentrated under reduced pressure in an evaporator, homogenized, and cooled. Vitamins (A and/or D) then are added and the final composition verified. Final standardization is accomplished, when necessary. After cans are filled and sealed, they are sterilized in a 3-phase continuous system consisting of preheater, retort, and cooler; cooled; labeled; and packed for shipment. In the United States, evaporated milk is packed in 5-,12-,20-, and 97-fluid ounce lead-free cans.

INDUSTRY PRODUCTION

Table 2 reflects U.S. production of evaporated milk and related products during the period 1975–1988. The data are based on manufacturers' production reported to the American Dairy Products Institute. Prior to 1983 data were not collected on the production of related evaporated milk products.

BIBLIOGRAPHY

General References

Food and Drug Administration, Department of Health and Human Services, *1988 Code of Federal Regulations* 21: 131.130 Evaporated Milk; 131.132 Evaporated Skimmed Milk, 1988.

C. W. Hall and T. I. Hedrick, *Drying of Milk and Milk Products,* 2nd ed., AVI Publishing Co., Inc., Westport, Conn., 1971.

E. H. Parfitt, "The Development of the Evaporated Milk Industry in the United States," *Journal of Dairy Science* **39**(6), 838 (1956).

R. Seltzer, *The Dairy Industry in America,* Magazines for Industry, New York, 1976.

J. C. FLAKE
WARREN S. CLARK, JR.
American Dairy Products Institute
Chicago, Illinois

Table 2. Evaporated Milk Production[a]

Year	Evaporated Milk	Evaporated Milk and Related Products[b]
1975	20,796,000	
1976	20,362,000	
1977	18,347,000	
1978	17,340,000	
1979	17,110,000	
1980	15,500,000	
1981	15,840,000	
1982	15,050,000	
1983	14,300,000	16,612,000
1984	13,700,000	15,892,000
1985[c]	14,360,000	16,537,000
1986	12,996,000	14,881,000
1987	12,966,000	14,759,000
1988	13,357,000	15,164,000

[a] Figures reported in cases; 48 tall equivalent.
[b] Includes evaporated milk plus evaporated skimmed milk, evaporated lowfat milk, and evaporated filled milk. Data not available before 1983.
[c] Product weight/case since 1985: 40 lb.

EVAPORATION

The primary objectives of evaporation, as a unit operation in food processing, are to reduce the volume of the product by some significant amount with minimum loss of nutrient components, and to preconcentrate liquid foods such as fruit juice, milk, and coffee before the product enters a dehydration process, thus saving energy in subsequent operations and reducing handling (transport, storage, and distribution) costs. Evaporation is accomplished by boiling liquid foods until the desired concentration is obtained, as result of the difference in volatility between the solvent and solutes. In the food industry, the solvent to withdraw is almost always water (1). Evaporation increases the solids content of a food and hence preserves it by a reduction in water activity; however, the flavor and color of a food may be changed during the process. The technical simplicity of evaporation gives it an obvious advantage compared to other methods such as reverse osmosis and freeze concentration. The removal of unpleasant volatile substances, for instance, during evaporation of milk, may also be an advantage. Evaporation of heat-sensitive materials is often accomplished under relatively high vacuum (low pressures), lower temperatures, and short residence time.

The first commercial use of evaporating equipment was made by Borden in 1856 (2). Since then, equipment for evaporation has evolved from open kettles through rotating steam coils to the modern natural or forced circulation evaporators. A wide range of different types of evaporator are available and evaporation can take place under different processing conditions. Commercial evaporation capacities range from 100 kg/h for pilot plants to over 200,000 kg/h for industrial installations (for instance, FMC, San Jose, Calif.; APV, Crawley, UK; Alfa-Laval, Lund, Sweden, and Dedert, Olympia Fields, Ill.).

Evaporation involves a sequence of processes. Computers can be used in the design of individual components and the overall system; however, to compromise between vastly opposed factors, the choice of the optimum design solution must be based on acquired experience and sound judgment. The intensive energy use of the evaporation process poses a continuing challenge in minimizing energy costs while maintaining product quality, especially at times of escalating energy costs. Multiple-effect evaporators and vapor recompression are measures that can be employed to improve the energy efficiency of the evaporation plant.

THEORY

The complete design of evaporator systems requires engineering calculations that involve combined heat and mass balances, and is essentially an application of the first law of thermodynamics. In an evaporator, the steam condenses at constant pressure, and the condensate leaves at this pressure and corresponding equilibrium temperature. The heat from the condensing steam boils water from a liquid that is being concentrated. The water vapor leaving the evaporator is at a lower temperature and pressure than the original steam supply, and can give up its latent heat in a double-effect evaporator to evaporate a lower boiling temperature liquid. This necessary lower boiling temperature can be maintained by operating a second effect of evaporation at a lower pressure than the first, usually under a vacuum. The same principles can be extended to more than two effects in industrial applications.

Steady-state mathematical models with equations that represent heat and mass transfer may be used for design purposes. Steam requirement, heat transfer surface area, and system capacity (production rate) can be determined for a given degree of concentration and processing time. A double-effect evaporator, with feed and steam or vapor streams moving in counterflow arrangement (Fig. 1) will be used here to illustrate the theory.

For the first effect, mass balance that relates solids entering and leaving the evaporator gives

$$\dot{m}_2 X_2 = \dot{m}_1 X_1 \tag{1}$$

and the rate of heat transfer to the evaporator from the steam is

$$\dot{m}_s \lambda_s = U_1 A_1 (T_s - T_1) \tag{2}$$

(The terms in equations are defined in the notation list at the end of the article). Assuming that there are negligible heat losses from the evaporator, the heat balance states that the amount of heat given up by the condensing steam equals the amount of heat used to raise the feed temperature to boiling point and then to boil off the vapor. In other words, the input energy of condensing vapor and feed equals the output energy of vapor and product, thus

$$\dot{m}_2 h_2 + \dot{m}_s \lambda_s = (\dot{m}_2 - \dot{m}_1) H_1 + \dot{m}_1 h_1 \tag{3}$$

Similarly, for the second effect, mass and energy balances are represented by the equations

$$\dot{m}_F X_F = \dot{m}_2 X_2 \tag{4}$$

$$(\dot{m}_2 - \dot{m}_1) \lambda_1 = U_2 A_2 (T_1 - T_2) \tag{5}$$

$$\dot{m}_F h_F + (\dot{m}_2 - \dot{m}_1) \lambda_1 = (\dot{m}_F - \dot{m}_2) H_2 + \dot{m}_2 h_2 \tag{6}$$

The overall heat-transfer coefficient U in equation 2 is defined by the sum of three thermal resistance terms, as represented by

$$\frac{1}{UA_m} = \frac{1}{h_s A_s} + \frac{1}{k A_m} + \frac{1}{h_p A_p} \tag{7}$$

These resistances include resistance in the boundary layer (heat-transfer film) on the heating medium side, resistance to heat conduction in the material making up the heat-transfer surface, and resistance in the heat-transfer film on the product side.

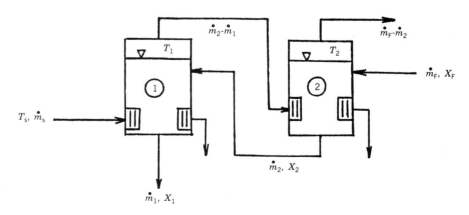

Figure 1. Schematic of double-effect evaporator, backward feed.

In most evaporators, the heating medium will be steam or some other condensing vapor. The resistance to heat transfer on the heating medium side of the heat-transfer surface is normally created by a condensation film, and an expression for condensation heat-transfer coefficient as presented in one study (3) was

$$h_c = 0.94 \left(\frac{k_f^3 \rho_f^2 g \lambda}{L \mu_f (T_s - T_w)} \right)^{1/4} \quad (8)$$

for vertical tubes and

$$h_c = 0.725 \left(\frac{k_f^3 \rho_f^2 g \lambda}{D \mu_f (T_s - T_w)} \right)^{1/4} \quad (9)$$

for horizontal tubes. Both expressions have accounted for heat transfer involved in phase change. The heating-surface material is usually stainless steel, and it is relatively easy to compute the corresponding resistance knowing the thickness and the thermal conductivity of the material. The most complex resistance to heat transfer is due to the resistance film on the product side.

The heat transfer from a heated surface to a boiling liquid is by convection, but the heat flux varies with the temperature difference between the surface and the bulk liquid, as illustrated in Figure 2. When the temperature difference is very small (less than 5°C), the heat transfer is by natural convection in a regime called pool boiling. By increasing the temperature difference, vapor bubbles tend to form at selected locations on the surface, and the regime is called nucleate boiling, which continues until the temperature difference is near 55°C. Heat-transfer coefficient from surface to boiling liquid is increased. Beyond 55°C, a film of vapor bubbles forms and collects near the heat-transfer surface, resulting in a reduced heat flux. At temperature differences greater than 55°C, the film becomes stable; however, radiation heat-transfer mode begins to contribute to the heat flux and leads to greater heat-transfer coefficients. Either pool or nucleate boiling will probably describe the majority of physical situations that may exist in liquid food evaporation (4).

Heat transfer to boiling liquids is normally expressed in terms of dimensionless groups in the same manner as other types of heat transfer (5). Experimental studies with evaporation systems have resulted in a number of empirical expressions for convective heat-transfer coefficients to the product For the case of natural circulation evaporators, the following correlation was developed (6):

$$h = 0.0086 \frac{k_f}{D} \left(\frac{u_m D \rho_f}{\mu_f} \right)^{0.8} \left(\frac{c_f \mu_f}{k_f} \right)^{0.6} \left(\frac{\sigma_f}{\sigma} \right)^{0.33} \quad (10)$$

For the case of forced convection systems, the following expression was suggested (7):

$$h_b = 3.5 h \left(\frac{1}{Z} \right)^{0.5} \quad \text{for} \quad 0.25 < \frac{1}{Z} < 70$$

$$\frac{1}{Z} = \left(\frac{y}{1-y} \right)^{0.9} \left(\frac{\rho_f}{\rho_v} \right)^{0.5} \left(\frac{\mu_v}{\mu_f} \right)^{0.1} \quad (11)$$

where h_b and h are for convective heat transfer with and without boiling, respectively.

In falling-film evaporation three different flow regimes normally exist, depending on the Reynolds number (Re) of the film: laminar, wavy laminar, and turbulent. The truly laminar regime takes place for Re < 20–30, the wavy laminar regime starts at Re = 30–50, and at 1,000–3,000 a fully turbulent flow is established. The Reynolds number for a film is calculated according to the following equation (8):

$$\text{Re} = \frac{4G}{\mu} \quad (12)$$

where G is the mass flow rate per unit width of the wall.

For the laminar region, the local heat-transfer coefficient may be written as (8):

$$h = a \phi \left(\frac{G}{\mu} \right)^{-0.33}$$

$$\phi = \left(\frac{k_f^3 \rho_f^3 g}{\mu^2} \right)^{0.33} \quad (13)$$

Table 1 gives formulas for local heat-transfer coefficients according to different investigators.

Experimental product-side heat-transfer coefficients h_p were determined (9) for thin-film wiped-surface evaporators, which are extensively used for concentrating high-viscosity or heat-sensitive products. Product degradation is minimized by vigorous agitation and continuous removal of liquid films on the heat-transfer wall. The correlation obtained for the transition-flow regime (Re = 100–1,000) was

$$\text{Nu} = 0.0483 \, \text{Re}^{0.586} \text{Pr}^{1.05} \text{Fr}^{0.118} \left(\frac{\mu}{\mu_w} \right)^{-2.93} \quad (14)$$

which is valid for liquid feed rate of 58–87 kg/h and agitation speed of 150–500 rpm. The correlation obtained for

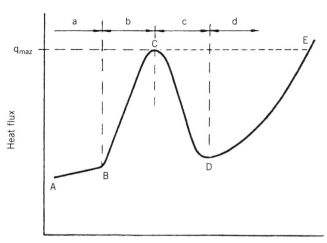

Figure 2. Typical boiling curve illustrating various regimes of boiling near heated surfaces. a, natural convection; b, nucleate boiling; c, transition boiling; d, film boiling; Reprinted with permission from Ref. 4.

Table 1. Local Heat-Transfer Coefficients[a]

Flow Regime	A	ϕ_1	Exponent for Re	Notes
Laminar (Re < 20–30)	0.69	ϕ	−0.33	Nusselt theory
Wavy laminar (Re < 1,000–3,000)	0.61	ϕ	−0.223	
Turbulent (Re > 1,000–3,000)	0.067	$\phi \cdot \mathrm{Pr}$	0.2	Colburn theory based on pipe friction
	0.030	$\phi \cdot \mathrm{Pr}$	0.33	
	0.19	$\phi \cdot \mathrm{Pr}$	0.067	
	0.0078	$\phi \cdot \mathrm{Pr}$	0.4	Pr ≈ 5
	0.0122	$\phi \cdot \mathrm{Pr}$	0.33	Pr ≈ 5

[a] Ref. 8.

the turbulent-flow regime (Re > 1,000) was

$$\mathrm{Nu} = 4.137\, \mathrm{Re}^{0.263} \mathrm{Pr}^{0.325} \mathrm{Fr}^{0.032} \left(\frac{\mu}{\mu_w}\right)^{-0.753} \quad (15)$$

which is valid for liquid feed rate of 87–133 kg/h and agitation speed of 500–1,300 rpm. The authors found that unstable operation of the evaporator occurred when the speed of the wiper blades was below 150 rpm, and rapid increase in h_p was observed for rpm range 250–350. Under similar conditions of evaporation, heat-transfer coefficients were lower for high-viscosity liquids.

A detailed expression for the overall heat-transfer coefficient in multiple-effect falling-film evaporators was worked out (10). This coefficient U was given as a function of the enthalpy and temperature of the liquid and vapor phases, local heat-transfer coefficients, and mass fractions of solute in the inlet and outlet liquid streams. The fundamental mathematical model of multiple-effect evaporators was further applied (11) along with an accurate estimate of U and fouling factor R_d to design a five-effect evaporation unit capable of treating the input citrus juice flow rates indicated in (Figure 3). Results of the calculations are shown in Table 2 for orange and lemon juices that were concentrated from 11.7 to 65°Brix, and from 9.3 to 40°Brix, respectively.

According to one study (12), factors influencing the rate of heat transfer can be summarized as

1. Temperature difference between the steam and the boiling liquid.
2. Fouling of evaporator surfaces.
3. Boundary film thickness.

Factors influencing the economics of evaporation include

1. Loss of concentrate due to foaming and entrainment.
2. Energy expenditure.

The total evaporation cost depends on the steam consumption and the specific surface area of the evaporator. Capital costs increase with the size of the evaporator whereas energy costs decrease with it. If the fouling of the heating surface is expressed as specific product losses and

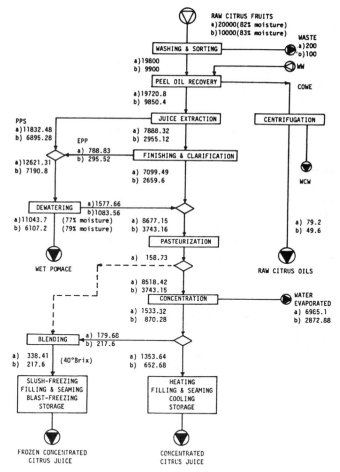

Figure 3. Flow diagram and material balance flow sheet for a concentrated citrus juice production plant with working capacity of 20,000 kg of raw oranges per hour and 10,000 kg of raw lemons per hour. The data reported here are expressed as kgh⁻¹: symbol (a) refers to orange, while symbol (b) to lemons. Stream identification symbols: COWE, citrus oils and water emulsion; EPP excess pulp and peel; PPS, peel, pulp, and seeds; WCW, water and citrus waste; WW, washing water. Reprinted with permission from Ref. 11.

Table 2. Material and Energy Balances for the Evaporation Unit of the Concentrated Citrus Process[a]

Paramater		1st Effect	2nd Effect	3rd Effect	4th Effect	5th Effect	Unit
Internal pressure	a[b]	73.5	63.8	54.2	44.0	6.0	kPa
	b	57.0	39.2	25.0	14.2	6.0	
Juice temperature	a	91.7	88.1	84.2	79.6	41.2	°C
	b	85.9	76.9	66.9	55.5	40.8	
Boiling point rise	a	0.44	0.55	0.77	1.3	4.9	°C
	b	1.21	1.45	1.84	2.57	4.5	
Clean heat-transfer coefficient	a	3,764	3,416	2,971	2,259	130	$Wm^{-2}\,°C^{-1}$
	b	4,061	3,470	2,902	2,188	799	
Input liquid flow rate	a	8,518	7,140	5,765	4,392	3,014	kgh^{-1}
	b	3,743	3,233	2,694	2,123	1,517	
Input liquid concentration	a	11.7	14.0	17.3	22.7	33.1	°Brix
	b	9.3	10.8	12.9	16.4	23.0	
Output liquid flow rate	a	7,140	5,765	4,392	3,014	1,533	kgh^{-1}
	b	3,233	2,694	2,123	1,517	870	
Output liquid concentration	a	14.0	17.3	22.7	33.1	65.0	°Brix
	b	10.8	12.9	16.4	23.0	40.0	
Recirculation ratio	a	0	0	0	0	6	
	b						
Thermal loss	a	0.03	0.03	0.03	0.03	0.03	
	b						
Heat-transfer surface	a	193.3	193.3	193.3	193.3	193.3	m^2
	b						
Fouling factor ($\times 10^3$)	a	0.43	0.43	0.43	0.43	0.43	$m^2\,°CW^{-1}$
	b	4.5	4.5	4.5	4.5	4.5	
Live steam consumption (95°C)	a			1,484			kgh^{-1}
	b			625			
Cooling water consumption (21–33°C)	a			72.3			m^3h^{-1}
	b			29.7			

[a] Adapted with permission from Ref. 11.
[b] Orange juice = a; lemon juice = b.

is reduced from 1.0 to 0.2 kg thin milk/m², for modern evaporators with a specific surface area of 0.12–0.14 m²(kg/h), the production costs will decrease by about 40% for whole milk evaporation and 30% for skim milk evaporation (13). Increased fouling is caused by a nonuniform distribution of liquid to individual heating tubes in falling-film evaporators such that some groups of tubes have more deposit than the others.

Steam economy, defined as the mass of water evaporated per unit mass of steam utilized, is often used as a measure of evaporator performance. A similar definition could be used if another condensing vapor (ammonia, Freon, or diphenyl) is employed as the heating medium.

Steam requirement for double-effect is computed directly from solving the heat and mass balances presented earlier. Equations 1–6 can be used to compute the heat-transfer surface area after the overall heat-transfer coefficients U_1 and U_2 have been evaluated for both effects. The usual procedures for computing steam requirement involve assumptions, such as equal heat flux, equal heat-transfer area, or equal temperature gradient in each effect of the system.

For plants with increasing number of effects resulting in decreased potential temperature difference in each effect, the boiling point rise is of importance when calculating the temperature program of an evaporation plant. For well-defined solutions such as water, the elevation of boiling points is proportional to the molar concentration of the solute, thus

$$\Delta = \frac{R T_{A0}^2 X^1}{\lambda} \qquad (16)$$

Food liquids are normally more complicated and the boiling point rise has to be determined experimentally. For milk, the following values have been given (8):

Concentration, % DM   16   27.5   39   49   62   69   73
Temperature, °C   0.5   1   1.5   2   3   4   5

The Duhring plot for sucrose for various conditions are presented in Figure 4.

EQUIPMENT

Evaporators are basically heat exchangers, that transfer heat from steam or other heating medium to the food. Vapors produced must be separated. The selection of an evaporator should include the following factors (12):

1. Production rate (kilograms of water removed per hour).

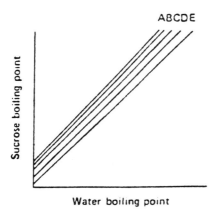

Figure 4. Duhring plot for sucrose for various sucrose concentrations per 100 g of water. A, 1000 g; B, 800 g; C, 600 g; D, 400 g; E, 200 g. Reprinted with permission from Ref. 12.

2. Degree of concentration required (percentage dry solids in the product).
3. Heat sensitivity of product in relation to the residence time and temperature of evaporation.
4. The requirement for volatile recovery facilities.
5. Ease of cleaning.
6. Reliability and simplicity of operation.
7. Size of evaporator.
8. Capital and operating costs.
9. Product quality.

Evaporators are classified as natural circulation evaporators and forced circulation evaporators.

Natural Circulation Evaporators

These may be open or closed pan evaporators, short-tube evaporators (also known as calandria vacuum pans in the food industry) (Fig. 5), or long-tube evaporators. Natural circulation evaporators operate by the thermo-syphon principle. The density difference between the boiling liquor and the circulation leg produces the driving force for liquid circulation. Typical applications include beet sugar, low to moderately viscous liquors, and nonsalting materials.

Pan evaporators are heated directly by gas or electrical resistance wires or indirectly by steam. They have relatively low rates of heat transfer, low energy efficiencies, and cause damage to heat-sensitive foods. However, they have low capital costs and are easy to construct and maintain. They have found wide applications in the preparation of sauces, gravies, and jam or other preserves (12).

Short-tube evaporators consist of a vessel or shell that contains a bundle of tubes. The feed solution is heated by steam condensing on the outside of the tubes. Liquor rises through the tubes, boils, and recirculates by flowing back down through a wide central bore. The vertical arrangement of tubes promotes natural convection currents and, therefore, higher heat transfer rates. The flow velocity is typically 0.3–1 m/s (15). These evaporators have low capital costs and are flexible, although generally unsuited to

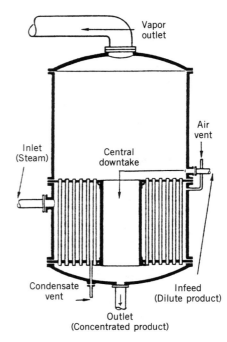

Figure 5. Calandria evaporator. Reprinted with permission from Ref. 14.

high-viscosity solutions. They are widely used for concentrating syrups, salt, and fruit juices.

The long-tube evaporators consist of a vertical bundle of tubes, each up to 50 mm in diameter and 5–15 m high. The length-to-diameter ratio is 70:130 (15). Liquid is preheated almost to boiling point (evaporation temperature) before entering the evaporator. For low-viscosity (less than $0.1 \text{ N} \cdot \text{s/m}^2$) foods such as milk, the thin film of liquor is forced up the evaporator tubes and this arrangement is known as the climbing-film evaporator (Fig. 6). A

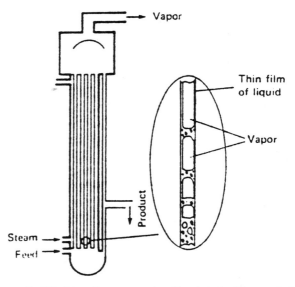

Figure 6. Climbing-film evaporator. Reprinted with permission from Ref. 12.

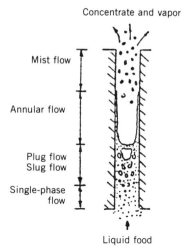
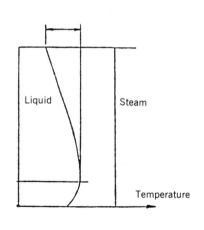

Figure 7. Flow regimes and temperature distribution in a climbing-film evaporator. Reprinted with permission from Ref. 16.

great deal of foaming does not matter, but they are susceptible to caking (15). As steam is admitted to the chest of the evaporator, the liquid reaches the boiling point and evaporation starts; a lot of small vapor bubbles appears in the liquid and starts the two-phase flow. The bubbles expand and push the mixture of liquid and vapor upward, further heat is applied resulting in more vapor formed (plug flow). Farther upward the remaining liquid is maintained as a film on the tube wall with vapor flowing as a core (annular flow). In the final part of the tube, the film is bursting and the liquid appears as droplets in a flow of vapor (mist flow). Pressure drop increases as more vapor is formed, and hence the evaporation temperature drops. This temperature drop must be as small as possible because it leads to higher energy consumption. The flow regime and temperature profile are depicted in Figure 7.

For more viscous foods, or those that are very heat sensitive, the feed is introduced at the top of the tube bundle, and the concentrate and the vapor are leaving the tubes at their bottom. Normally the liquid is superheated when entering the tube, thus the liquid is already at its boiling point and no part of the heating surface is used for preheating. This type of equipment is called the falling-film evaporator (Fig. 8). Only a small amount of product is in the tube, compared with the column of liquid in the climbing-film evaporator. Hydrostatic head loss found in climbing-film evaporators is absent in falling-film evaporators. Thus little temperature drop is seen during the passage through the tube (14). The gravitational force enhances the forces due to expansion of the steam, to produce high flow velocities of up to 200 m/s at the end of the tubes (12). Falling-film evaporators were developed to increase the rate of removal of water and decrease the danger of localized overheating.

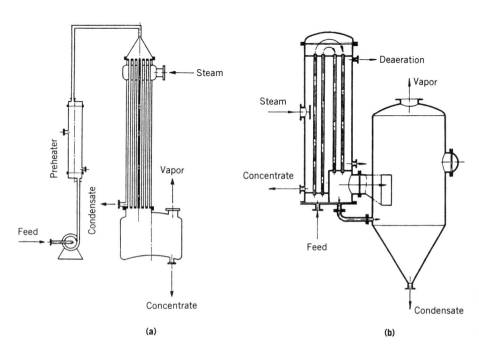

Figure 8. (a) Falling-film evaporator. (b) Falling-film evaporator, with falling and climbing sections. Reprinted with permission from Ref. 15.

A falling-film evaporator consists of the following major components: vertical heating tube bundle with surrounding heating jacket, calandria upper section with product distribution device, and calandria base section with separator. The liquid to be concentrated is fed into the upper part of the calandria where it is distributed evenly on the tube sheet, from which it flows down the interior walls of the vertical tubes as a thin boiling film. The most important factor in the trouble-free operation of this evaporator is the liquid coverage of the heating tubes, that is, the liquid flow rate required to cover each heating tube. Without enough liquid, the liquid film will break near the bottom of the tube. Dry spots develop on the tube, the heating surface becomes coated with a deposit, and eventually the tube may get clogged with burned-on product. The minimum liquid coverage depends on the type of product, required concentration, and operating time. If the coverage falls short of a minimum value, fewer but longer heating tubes may be used, thus giving more liquid per tube while maintaining the same total heating surface area. The next choice is to subdivide the calandria to create more passes on the product side; recirculation of product within a calandria also improves coverage, however, it increases the residence time. Tube length design shall be a compromise between adequate liquid coverage and pressure drop. Short tubes with larger diameters have smaller pressure loss, whereas small diameter tubes lead to high velocity of the two-phase flow, causing high pressure loss and boiling point elevation. To this effect, it has been recommended that tubes between 8 and 12 m in length and with a diameter between 30 and 50 mm be used (16).

Product distribution in the upper part of the calandria can be done in two ways: dynamic distribution or static distribution. Dynamic distribution may be done by a full-cone nozzle in which the kinetic energy of the liquid is used to atomize the fluid into droplets and distribute them on the upper tube sheet. This distribution device, although simple and economic, does not guarantee even distribution. Static distribution devices (Fig. 9) are generally preferred. Even distribution is achieved by having uniform liquid static pressure over all openings in the base of a predistribution bowl, giving uniform wetting of all tubes. The system adapts readily to varying feed conditions by varying the liquid level. The falling-film evaporator is widely used for sugars and syrups, yeast extracts, fruit juices, dairy products and starch processing.

It has been pointed out that both types of long-tube evaporator are characterized by short residence times, high-heat transfer coefficients, and efficient energy use (steam economy = 2.5 to 3.3) in multiple-effect systems (12).

Forced Circulation Evaporators

Forced circulation evaporators have a pump or scraper assembly to move the liquor, usually in thin layers, to enhance heat-transfer rates and shorten residence times. The forced circulation design is generally used in applications where a solid is precipitated during evaporation or a scaling constituent is present.

Recirculated falling-film configurations are used when insufficient feed liquor is available to use the heat-transfer surface with single-pass operation (natural circulation method). A portion of the product liquor is combined with the feed stream and is pumped to the upper liquor chamber. Product retention time is greater than for the single-pass evaporator but is still relatively short as the operating volume is small. These evaporators find applications in moderately heat-sensitive foods.

The plate type or AVP evaporator, developed in the UK in the 1950s, is suitable for viscous liquids (0.3–0.4 N·s/m²), because the food is pumped through the plate stack, consisting of a series of climbing-film and falling-film sections. The heating medium is steam. They are compact and are advantageous in having a small amount of product in the system at any one time. They have the additional advantage of ease of maintenance and inspection.

An example of a high-temperature, short-time (HTST or flash) evaporator is one with rotating heating surface and is shown in Figure 10. It is a mechanical thin-film evaporator. Feed flows to the underside of a stack of rotating hollow cones from a central shaft. The feed liquor is

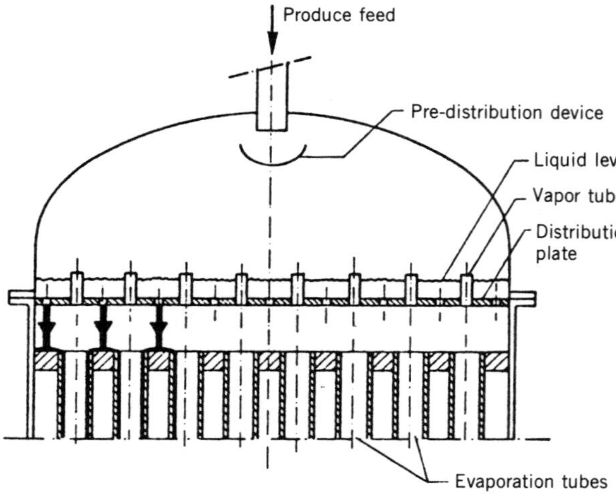

Figure 9. Liquid distribution in a falling-film evaporator. Reprinted with permission from Ref. 8.

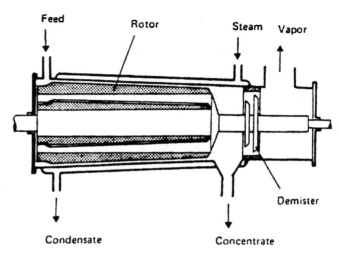

Figure 10. Mechanical thin-film evaporator. Reprinted with permission from Ref. 15.

kept in rapid motion, and the centrifugal force (750–3,000 N) causes the formation of a very thin film (about 0.1 mm thick) in which the liquid remains for only a very short time of 0.6 to 1.6 s (17). Steam condenses on each cone, and heat is conducted rapidly through the thin metal to evaporate the liquor. High rates of heat transfer are obtained as steam condensate is flung from the rotating cones as soon as they are formed. A high degree of flexibility can be achieved by changing the number of cones. This evaporator is used for coffee and tea extracts, meat extract, fruit juices and enzymes for use in food processing.

A schematic cross section of an agitated-film (also called scraped-surface or wiped-surface) evaporator is shown in Figure 11. The product film near the heat-transfer surface is continuously agitated by a rotor to increase the heat-transfer rate on the product side. The evaporator may also consist of a steam jacket surrounding a high-speed rotor, fitted with short blades along its length (Fig. 12). The liquid enters the wide end and is forced through the precise space between the rotor and the wall (heated surface) toward the outlet for the concentrate. The blades agitate the thin film of liquid vigorously, thus promoting high heat-transfer rates and preventing the product from burning onto the hot surface. If the viscosity increases greatly during evaporation, it is better to feed the liquid at the narrow end. This type of equipment is highly suited to viscous (up to 20 N·s/m²), heat-sensitive foods, or to those that are liable to foam or foul evaporator surfaces

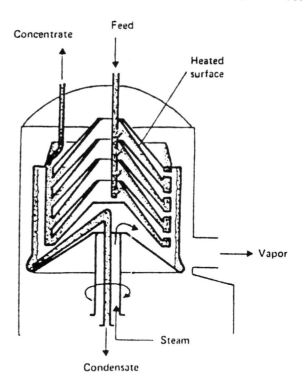

Figure 12. Centri-term evaporator. Reprinted with permission from Ref. 18.

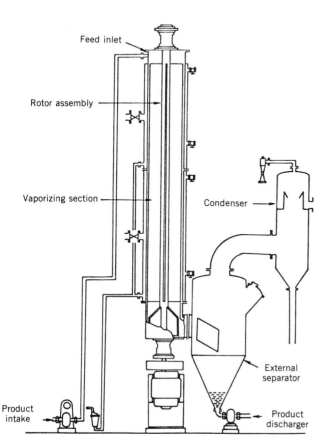

Figure 11. Schematic cross-section of agitated film evaporator. Reprinted with permission from Ref. 4.

(fruit pulps and juices, tomato paste, meat extracts, honey, cocoa mass, coffee, and dairy products). Only single effects are possible, thus giving low steam economy. It is more intended for finishing highly viscous products after preconcentration in other equipment (12).

Energy Efficiency

The energy efficiency of evaporation systems may be improved by the following approaches (19): multiple-effect, vapor recompression, or refrigerant cycle evaporators.

Multiple-Effect Evaporating System. For the evaporation of temperature-sensitive materials it is important to keep the temperature as low as possible. This is achieved by working at a lower pressure. Practically all the evaporators mentioned above can be used at pressures below atmospheric.

To make use of the large amount of heat in the vapor driven off the liquid food, the water vapor can be used as the heating medium in a following evaporator. This second evaporator must then operate at a lower pressure than the first. This is called double-effect evaporation. Repetition of this step produces multiple-effect evaporation. The effects then have progressively lower pressures to maintain the temperature difference between the feed and the heating medium. A multiple-effect evaporating system can be designed in several ways in regard to the direction of flow of the solution. Figure 13 illustrates, with a triple-effect evaporator, the arrangements of forward feed, backward feed, parallel feed, and mixed-feed system.

The liquid in the forward feed system enters the highest temperature evaporator first and exits as a concen-

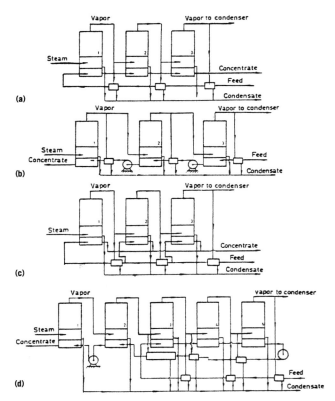

Figure 13. Multiple-effect evaporation. (**a**) Forward feed; (**b**) backward feed; (**c**) parallel feed; (**d**) mixed feed. Reprinted with permission from Ref. 15.

trated liquid from the third evaporator that operates at the lowest temperature and pressure. The liquid is usually preheated by steam or vapor from one of the effects before it enters the first evaporator. The liquid leaving effect 1 has a higher temperature than the liquid in effect 2, and in crossing the throttle valve to effect 2, a small amount of the liquid water will flash to water vapor and the temperature of the liquid will drop to the operating temperature of effect 2.

In the backward feed system, the liquid flows in opposite direction of the forward feed evaporator. This system allows the concentrated liquid to exit at the highest temperature, which is advantageous for products having a high viscosity in the concentrated form. Because the liquid leaving effect 3 has a lower temperature than the liquid in effect 2, a pump is needed to provide additional energy to increase its temperature to the saturation temperature corresponding to effect 2. A similar situation exists between effect 2 and effect 1.

The parallel feed evaporator essentially acts like three separate evaporators, each operating at a different pressure. It has the advantage of simple operation. However, it is the most expensive arrangement, because extraction pumps are required for each effect. Mixed-feed evaporators are useful for viscous food, such as distillery by-products (20). They have the simplicity of forward feed and economy of backward feed. The merits and shortcomings of each arrangement are described in Table 3.

In each evaporator of a multiple-effect installation it is necessary for the temperature of the heating medium to be higher than the boiling temperature; the number of stages is, therefore, limited. Savings in energy consumption decreases with every effect added. Ideally, 1 kg of steam vaporizes 1 kg of water in all effects, so that the steam economy would be 1, 2, 3, and 4, respectively, for one to four effects. In practice, the steam consumption is somewhat higher because of heat losses. Actual steam economy for evaporating milk is 0.8, 1.7, 2.5, and 3.3 (2). The steam economy for evaporating tomato paste was calculated (22); the values were 0.95, 1.91, and 2.60 compared to measured values of 0.84, 1.45, and 2.20 (on average), respectively, for one, two, and three effects.

Many HTST evaporators, under the trade name TASTE (thermally accelerated short time evaporator) have been installed in Florida and elsewhere (23). The initial feed is preheated to about 90°C in a conventional heat exchanger and flashed through a series of evaporators with progressively higher vacuums (15). A typical unit has seven stages plus a vacuum flash cooler arranged so there are four effects. The word *effect* indicates vapor flow in the evaporator, while the word *stage* indicates the flow of product (24). Temperature in the various stages range from 41 to 96°C. These units are small for their

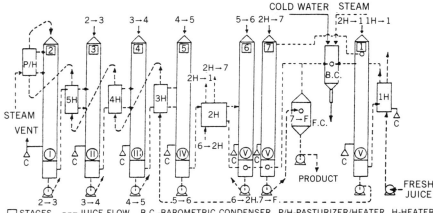

Figure 14. Schematic of a 5-effect, 7-stage TASTE Evaporator. Reprinted with permission from Ref. 24.

Table 3. Advantages and Limitations of Various Methods of Multiple-Effect Evaporation[a]

Arrangement of Effects	Advantages	Limitations
Forward feed	Least expensive, simple to operate, no feed pumps required between effects, lower temperatures with subsequent effects and, therefore, less risk of heat damage to more viscous product	Reduced heat-transfer rate as the feed becomes more viscous, rate of evaporation falls with each effect, best quality steam used on initial feed, which is easiest to evaporate; feed must be introduced at boiling point to prevent loss of economy (if steam supplies sensible heat, less vapor is available for subsequent effects)
Reverse feed	No feed pump initially, best-quality steam used on the most difficult material to concentrate, better economy and heat-transfer rate as effects are not subject to variation in feed temperature and feed meets hotter surfaces as it becomes more concentrated thus partly offsetting increase in viscosity	Interstage pumps necessary, higher risk of heat damage to viscous products as liquor moves more slowly over hotter surfaces, risk of fouling
Mixed-feed	Simplicity of forward feed and economy of backward feed, useful for very viscous foods	More complex and expensive
Parallel	For crystal production, allows greater control over crystallization and prevents the need to pump crystal slurries	More complex and expensive of the arrangements, extraction pumps required for each effect.

[a] Adapted with permission from Ref. 21.

capacity and lack the vapor–liquid separation chambers of the low-temperature evaporators. There is more tendency for hesperidin to build up on the heat exchange surface of the TASTE evaporator in comparison with low temperature units, but because of the low retention time, it can be emptied rapidly. It can be cleaned in place, without dismantling, in minutes.

The steam economy of two tubular TASTE evaporators with four effects and seven stages (Fig. 14) and with nominal evaporation capacity of 27,211 kg (60,000 lb) and 45,352 kg (100,000 lb) of water per hour is 0.85 N; N being the number of effects of the evaporator (24). More stages are incorporated in the last effect because of reduction in product volume as it is concentrated. Similarly, it was found that the steam economy for two plate evaporators (four effects, five stages) (Fig. 15) with a capacity of 13,605 kg (30,000 lb) per hour was 0.82 N.

Vapor Recompression. Thermal recompression with a steam-jet compressor (25) is illustrated in Figure 16. A fraction of the vapor produced is combined with steam entering the heat exchanger in the low-pressure section of

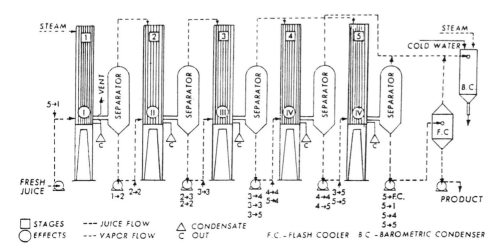

Figure 15. Schematic of a 4-effect, 5-stage plate evaporator. Reprinted with permission from Ref. 24.

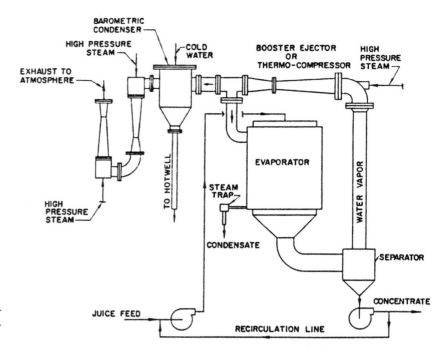

Figure 16. System of concentration using thermal recompression with a steam-jet compressor. Reprinted with permission from Ref. 25.

the Venturi. The remaining vapor is used in subsequent effects. This system is sometimes advantageous when the temperature of steam supplied by the boiler is too high for the foods. By regulating the size of the Venturi the pressure can be controlled and hence condensing temperature of the steam. Part of the kinetic energy associated with the high-velocity flow in the Venturi is used to compress the recirculated vapor. Vapor recompression will decrease steam consumption and increase the steam economy. When steam costs are moderate and electricity costs comparatively high, thermal compression is recommended. This simple method is relatively inexpensive, has a long operating life, and requires little maintenance. However, its characteristic operating curve has only one optimum operating point, thus restricting flexibility between design and actual operating conditions (16).

In a mechanical vapor recompression system (Fig. 17) all of the vapor produced is compressed to a slightly higher pressure by a mechanically driven compressor. The compressed vapor then enters the heat exchanger and on condensation provides heat to drive the evaporation process. This method can virtually eliminate steam requirement, as only a small amount of make-up steam is needed. Yet both the high capital and energy costs to operate the compressor need to be considered; it is about 10 times more expensive than the thermal type. Also, when evaporation pressures become too low, the volume of vapor to be handled increases dramatically, and mechanical compression is not economical. An evaporator operating at 5.0 kPa (33°C) would produce almost four times as much vapor volume as an evaporator operating at 20 kPa (60°C) on the same production rate basis (19). Nevertheless, mechanical vapor recompressors have more or less the same limited flexibility as thermal recompressors. This applies particularly to the high-speed radial turbo compressor, which reaches its surge limit if the actual vapor flow is too small, thus rendering the system unstable (16). Mechanical recompressors are complicated in design, and need careful maintenance and supervision of mechanical vibration.

Refrigerant Cycle Evaporators. Figure 18 shows a schematic system of juice concentration using a technique that combines a vapor compression refrigeration cycle with the evaporation system (25). Condensing water vapor boils liquid refrigerant to form a refrigerant vapor, which is then compressed to a higher temperature level and acts as a heat source to boil the liquid being concentrated (19). The evaporator, for simplicity, is shown as a single-effect

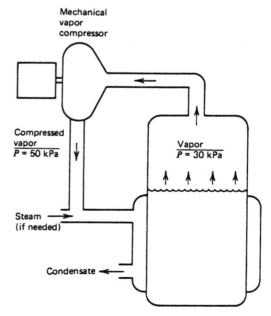

Figure 17. A mechanical vapor recompression system. Reprinted with permission from Ref. 19.

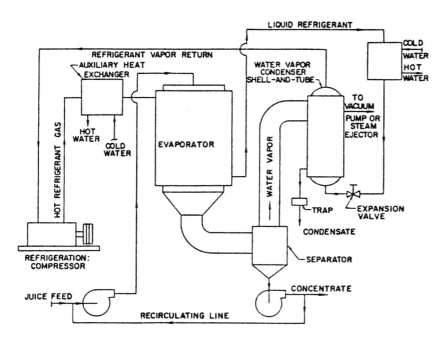

Figure 18. System of concentration using refrigerant cycle evaporators. Reprinted with permission from Ref. 25.

design with means for recirculating the juice. Variations may employ multiple effects. The pressure in the system can be maintained by a relatively small steam ejector or a mechanical vacuum pump. Its advantage over a mechanical vapor recompression system is that the volume of vapor being compressed is much reduced at a given evaporator temperature. However, the refrigerant must be compressed through a wider temperature range because an additional heat exchanger is required.

Auxilliary Equipment

Preheaters. After feeding the product into the evaporation plant via a float balance tank and feed pump, the first process step is to preheat the product. In general, liquid food is stored at between 5 and 20°C for quality preservation; therefore, it must be preheated to at least the evaporation temperature of the first effect, most commonly between 40 and 110°C. Three types of preheater are used: plate, spiral tube, and straight tube heat exchangers (16). If there are solid particles in the product, a plate heat exchanger should not be used, because the particles may block the narrow product passages. The same problem may also occur with protein-containing products, due to the formation of deposits if the preheating temperature exceeds the denaturalization temperature. Secondary flows developed in the spiral tube leads to higher heat-transfer coefficients than straight tube, and a reduced tendency to foul. The disadvantages are that inspection and replacement of the tube is very difficult. Straight tube preheaters are especially useful as intermediate preheaters following first step preheating with plate heat exchanger, where product of high concentration is required, and where equipment inspection is essential.

Separators. Separators are used to separate the mixture of vapor and concentrated product produced in the calandria. A separator should have high separation efficiency, small pressure loss, and short residence time. Because a fine mist of concentrate is produced during vigorous boiling and carried over in the vapor, separators should be designed to minimize entrainment. Large pressure loss is to be avoided lest extra heating surface must be installed in the next effect. The many different separators in use can be classified according to three operating principles: gravity separator, baffle separator, and centrifugal separator.

Condensers. Surplus vapor produced in evaporation can be condensed and reused where possible. Condenser cooling water is normally provided from a cooling tower or the city main; in certain cases, it may be extracted from rivers, lakes, or the sea. The tasks of a condenser are to maintain the heat balance of an evaporation plant, to stabilize the temperature in each effect, and to produce and maintain vacuum.

EFFECT ON FOODS

Nutritional changes take place during processing, the extent varying with the type of food, the process, the plant in use, and the degree of control exercised. Many losses are inevitable, particularly if the process involves heating. Part of the water-soluble nutrients such as the B vitamins, together with lesser and less-important amounts of mineral salts, protein, and even carbohydrate will be precipitated out. Vitamin C is oxidized in air and accelerated by heat, whereas vitamins A and D and niacin are unaffected (20). A comparison of nutrient losses in milk preserved by evaporation and UHT sterilization is shown in Table 4.

Aroma compounds that are more volatile than water will be lost during evaporation. With some products such as fruit juices, the retention of taste and aroma is important, yet in other foods such as cocoa and milk, the loss of

Table 4. Vitamin Losses in Concentrated and UHT Sterilized Milk[a]

Product	Loss (%)				
	Thiamin	Vitamin B_6	Vitamin B_{12}	Folic Acid	Ascorbic Acid
Evaporated milk	20	40	80	25	60
Sweetened condensed milk	10	<10	30	25	25
UHT sterilized milk	<10	<10	<10	<10	<25

[a] Adapted with permission from Ref. 26.

unpleasant volatiles improves the product quality. The color of foods darkens, partly due to an increase in the solids content and partly because the reduction in water activity promotes chemical changes (for example, Maillard browning) (12). As these changes are time and temperature dependent, short residence times and low boiling points produce concentrates with higher qualities. For instance, the Centri-therm mechanical thin-film evaporator produces a concentrate that, when diluted, has sensory and nutritional qualities that are virtually unchanged from those of the feed material (12).

BIBLIOGRAPHY

1. H. A. Leniger, "Concentration by Evaporation," in T. Hoyem and O. Kvale, eds., *Physical, Chemical and Biological Changes in Food Caused by Thermal Processing,* Elsevier Applied Science Publishers, Ltd., Barking, UK, 1980, pp. 54–76.
2. A. W. Farrall, *Food Engineering Systems. Vol. 1, Operations,* AVI Publishing Co., Inc., Westport, Conn., 1976.
3. F. Kreith and W. Z. Black, *Basic Heat Transfer,* Harper & Row, 1980.
4. D. R. Heldman and R. P. Singh. *Food Process Engineering,* 2nd ed., AVI Publishing Co., Inc., Westport, Conn., 1980.
5. *Handbook of Fundamentals,* American Society of Heating, Refrigeration and Air Conditioning Engineers, 1989.
6. E. L. Piret and H. S. Isbin, "Natural Circulation Evaporators," *Chemical Engineering Progress* **50**(60), 305–310 (1954).
7. J. M. Coulson and J. F. Richardson, *Chemical Engineering,* Vol. 2, 3rd ed., Pergamon Press, Elmsford, N.Y., 1976.
8. B. Hallstroem, "Heat Exchange," in *Evaporation, Membrane Filtration and Spray Drying in Milk Powder and Cheese Production, North European Dairy Journal,* 1985, pp. 37–41.
9. K. Stankiewicz and M. A. Rao, "Heat Transfer in Thin-Film Wiped-Surface Evaporation of Model Liquid Foods," *J. Food Process Eng.* **10**, 113–131 (1988).
10. S. Angeletti and M. Moresi, "Modeling of Multiple-Effects Falling-Film Evaporators. *Journal of Food Technology* **18**, 539–563 (1983).
11. M. Moresi, "Economic Study of Concentrated Citrus Juice Production," in B. M. McKenna, ed., *Engineering and Food,* Vol. 2, *Processing Applications,* 1984, pp. 975–991.
12. P. Fellows, *Food Processing Technology—Principles and Practice,* VCH Publishers, New York, 1988.
13. S Bouman, D. W. Brinkman, P. de Jong, and R. Waalewijn, "Multistage Evaporation in the Dairy Industry: Energy Savings, Product Losses and Cleaning," in S. Brun, ed., *Preconcentration and Drying of Food Materials,* 1988, pp. 51–60.
14. M. A. Joslyn and J. L. Heid, *Food Processing Operations—Their Management, Machines, Materials, and Methods,* Vol. 3, AVI Publishing Co., Inc., Westport, Conn., 1964.
15. H. A. Leniger and W. A. Beverloo, *Food Process Engineering,* Reidel Publishing, Boston, 1975.
16. G. Hahn, "Evaporator Design," in D. MacCarthy, ed., *Concentration and Drying of Foods,* 1987, pp. 113–132.
17. P. P. Lewicki and R. Kowalczyk, in P. Linko, Y. Malkki, J. Olkku and J. Larinkari, eds., *Food Processing Systems,* Elsevier Applied Science Publishers, Barking, UK, 1980, pp. 501–505.
18. C. H. Mannheim and N. Passy, "Aroma Retention and Recovery During Concentration of Liquid Foods," Proceedings of the Third Nordic Aroma Symposium, Hemeelinna, 1972.
19. J. C. Batty and S. L. Folkman, *Food Engineering Fundamentals,* John Wiley and Sons, Inc., New York, 1983.
20. A. E. Bender, "Food Processing and Nutritional Values," in S. M. Herschdoerfer, ed., *Quality Control in the Food Industry,* Academic Press, Inc., Orlando, Fla., 1984.
21. J. G. Brennan, J. R. Butters, N. D. Cowell and A. E. V. Lilly, *Food Engineering Operations,* Applied Science, London, 1976.
22. T. R. Rumsey, T. T. Conant, T. Fortis, E. P. Scott, L. D. Pedersen and W. W. Rose, "Energy Use in Tomato Paste Evaporation," *J. Food Process Eng.* **7**, 111–121 (1984).
23. B. S. Luh and J. G. Woodroof, *Commercial Vegetable Processing,* 2nd. ed. van Nostrand Reinhold, N.Y, 1988.
24. J. G. Filho, A. A. Vitali and F. C. P. Viegas, Energy Consumption in a Concentrated Orange Juice Plant. *J. Food Process Eng.* **7**, 77–89, 1984.
25. *HVAC Systems and Applications,* Amer. Soc. Heating, Refrigeration and Air Conditioning Engineers, 1987.
26. J. W. G. Porter and S. Y. Thompson. "Effects of Processing on the Nutritive Value of Milk," Vol. 1 *Proc. Fourth International Conference on Food Science and Technology,* Madrid, 1976.

GLOSSARY

A	Area, m^2
D	Diameter, mm
DM	Dry matter content, %
H	Enthalpy, J/kg
G	Mass flow rate per unit width of wall, kg/m · s
R	Gas constant = 8,314 N · m/g · mol °K
T	Temperature, °C
U	Overall heat transfer coefficient, W/m^2 °C
X	Mass fraction of solute in the liquid

X^1	Molar concentration
Fr	Froude number
Nu	Nusselt number
Pr	Prandtl number
Re	Reynolds number
a	Coefficient used in eq 13
c	Specific heat, J/kg °C
g	Gravitational constant, m/s^2
h	Enthalpy of liquid, J/kg
h_c	Convective heat-transfer coefficient, W/m^2 °C
k	Thermal conductivity, W/m °C
$\dot{m}$	Mass flow rate for evaporator, kg/s
u	Liquid–vapor velocity, m/s
x	Thickness, mm
y	Vapor quality
Δ	Boiling point elevation, °C
λ	Latent heat of vaporization, J/kg
μ	Viscosity, kg/m·s
ρ	Density, kg/m^3
σ	Surface tension, N/m

Subscripts

1	First effect in a double-effect evaporator
2	Second effect in a double-effect evaporator
A	Solvent component of product (water)
F	Feed
b	Boiling
c	Concentration
f	Fluid, or liquid state
m	Log-mean value
p	Product
s	Steam (or other heating medium)
v	Vapor
w	Wall condition
0	Pure state

A. K. Lau
The University of British Columbia
Vancouver, British Columbia
Canada

EVAPORATORS: TECHNOLOGY AND ENGINEERING

TYPES OF EVAPORATORS

In the evaporation process, concentration of a product is accomplished by boiling out a solvent, generally water, so that the end product may be recovered at the optimum solids content consistent with desired product quality and operating economics. It is a unit operation that is used extensively in processing foods, chemicals, pharmaceuticals, fruit juices, dairy products, paper and pulp, and both malt and grain beverages. It also is a unit operation, which, with the possible exception of distillation, is the most energy-intensive.

While the design criteria for evaporators are the same regardless of the industry involved, the question always arises as to whether evaporation is being carried out in the equipment best suited to the duty and whether the equipment is arranged for the most efficient and economical use. As a result, many types of evaporators and many variations of processing techniques have been developed to take into account different product characteristics and operating parameters.

The more common types of evaporators include the following:

1. Batch pan
2. Natural circulation
3. Rising film tubular
4. Falling film tubular
5. Rising–falling film tubular
6. Forced circulation
7. Wiped film
8. Plate equivalents of tubular evaporators

Batch Pan

Next to natural solar evaporation, the batch pan as shown in Figure 1 is one of the oldest methods of concentration. It is somewhat outdated in today's technology but still is used in a few limited applications such as the concentration of jams and jellies where whole fruit is present and in processing some pharmaceutical products. Up until the early 1960s, it also enjoyed wide use in the concentration of corn syrups.

With a batch pan evaporator, product residence time normally is many hours. Therefore, it is essential to boil at low temperatures and high vacuum when a heat-sensitive or thermodegradable product is involved. The batch pan is either jacketed or has internal coils or heaters. Heat-transfer areas normally are quite small as a result of vessel shapes, and heat-transfer coefficients tend to be low under natural convection conditions. Heat transfer is improved by agitation within the vessel. Low surface areas

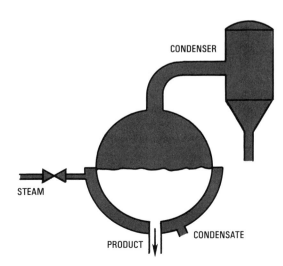

Figure 1. Jacketed batch pan.

together with low HTCs (heat-transfer coefficients) generally limit the evaporation capacity of such a system. In many cases, large temperature differences cannot be used for fear of rapid fouling of the heat-transfer surface. Relatively low evaporation capacities, therefore, limit its use.

Tubular Evaporators

Natural Circulation. Evaporation by natural circulation is achieved through the use of a short tube bundle within the batch pan or by having an external shell and tube heater outside the main vessel as illustrated by Figure 2. The external heater has the advantage that its size is not dependent on the size or shape of the vessel itself. As a result, larger evaporation capacities may be obtained. The most common application for this type of unit is as a reboiler at the base of a distillation column.

Rising Film Tubular. The first of the "modern" types of evaporators is the rising film unit, although its use by industry goes back to the early 1900s. The rising film principle was developed commercially by using a vertical tube with steam condensing on its outside surface (Fig. 3). Liquid inside the tube is brought to a boil, the vapor generated forming a core in the center of the tube. As the fluid moves up the tube, more vapor is formed, resulting in a higher central core velocity that forces the remaining liquid to the tube wall. Higher vapor velocities, in turn, result in thinner and more rapidly moving liquid film. This provides higher HTCs and shorter product residence time.

The development of this principle was a giant step forward in the evaporation field, particularly from the viewpoint of product quality. Its further advantage of high HTCs resulted in reduced heat-transfer area requirements and consequently, in a lower initial capital investment.

Falling Film Tubular. Following development of the rising film principle, it took almost a further half century for a falling film evaporation technique to be perfected (Fig. 4). The main problem was to design an adequate system for the even distribution of liquid to each of the tubes. Distribution in its forerunner, the rising film evaporator, was easy as the bottom bonnet of the calandria always

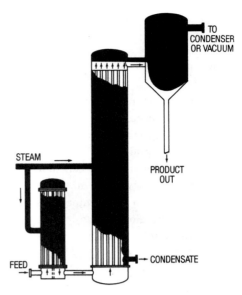

Figure 3. Rising film tubular.

was pumped full of liquid, thus allowing equal flow to each tube.

While all manufacturers have their own techniques, falling film distribution generally is based on the use of a perforated plate positioned above the top tube plate of the calandria. Spreading of liquid to each tube sometimes is further enhanced by generating flash vapor at this point. The falling film evaporator does have the advantage that the film is "going with gravity" instead of against it. This results in a thinner, faster moving film and gives rise to even shorter product contact time and a further improvement in the value of HTC.

The rising film unit normally needs a driving force or temperature difference across the heating surface of at least 25°F to establish a well-developed film, whereas the

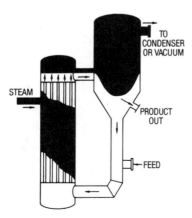

Figure 2. Natural circulation.

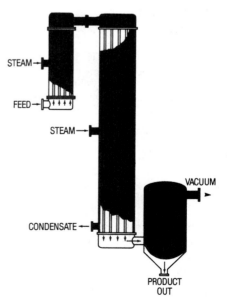

Figure 4. Falling film tubular.

falling film evaporator does not have a driving force limitation. This permits a greater number of evaporator effects to be used within the same overall operating limits, in other words, if steam is available at 2.5 psig corresponding to 220°F and the last effect boiling temperature is 120°F, the total available ΔT is equal to 100°F. This would limit a rising film evaporator to four effects, each with a ΔT of 25°F. It would be feasible, meanwhile, to have as many as ten or more effects using the falling film technique.

Rising–Falling Film Tubular. As illustrated by Figure 5, the rising–falling film evaporator has the advantages of the ease of liquid distribution of the rising film unit coupled with lower head room requirements. The tube bundle is approximately half the height of either a rising or falling film evaporator and the vapor–liquid separator is positioned at the bottom of the calandria.

Forced Circulation. The forced circulation evaporator (Fig. 6) was developed for processing liquors that are susceptible to scaling or crystallizing. Liquid is circulated at a high rate through the heat exchanger; boiling is prevented within the unit by virtue of a hydrostatic head maintained above the top tube plate. As the liquid enters the separator, where the absolute pressure is slightly less than in the tube bundle, the liquid flashes to form a vapor.

The main applications for a forced circulation evaporator are in the concentration of inversely soluble materials, which results in the deposition of solids. In all cases, the temperature rise across the tube bundle is kept as low as possible, generally in the region of 3–5°F. This results in a recirculation ratio as high as 200–330 lb of liquor per pound of water evaporated. These high recirculation rates result in high liquor velocities through the tubes, which helps to minimize the buildup of deposits or crystals along the heating surface. Forced circulation evaporators normally are more expensive than film evaporators because

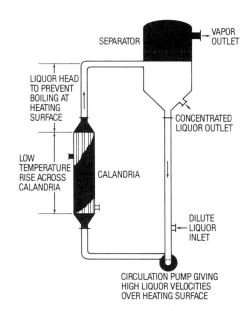

Figure 6. Forced circulation.

of the need for large-bore circulating pipework and large recirculating pumps generally of the axial flow type. Operating costs of such as unit also are considerably higher.

Wiped Film. The wiped or agitated thin film evaporator depicted in Figure 7 has limited applications and is confined mainly to the concentration of very viscous materials and the stripping of solvents down to very low levels. Feed is introduced at the top of the evaporator and is spread by wiper blades on to the vertical cylindrical sur-

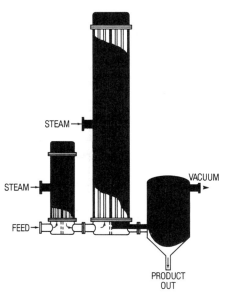

Figure 5. Rising–falling film tubular.

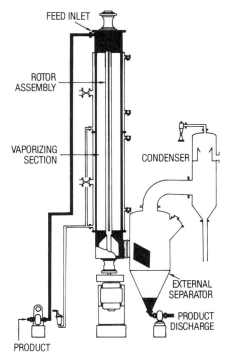

Figure 7. Wiped film.

face inside the unit. Evaporation of the solvent takes place as the thin film moves down the evaporator wall. The heating medium normally is high-pressure steam or oil. A high-temperature heating medium generally is necessary in order to obtain a reasonable evaporation rate since the heat-transfer surface available is relatively small as a direct result of its cylindrical configuration.

The wiped film evaporator is very satisfactory for its limited applications. However, in addition to its small surface area, it also has the disadvantage of requiring moving parts such as the wiper blades, which, together with the bearings of the rotating shaft, need periodic maintenance. Capital costs in terms of dollars per pound of solvent evaporated also are very high.

Plate Evaporators

The plate equivalent of the tubular evaporator is available in four configurations: rising–falling film, falling film, Parvap, and Parflash plate evaporators. All have been developed by APV to handle the concentration of products that have varying characteristics.

Plate-type evaporators now have been sold commercially for nearly 30 years, and during that time over 1600 units manufactured by APV have been installed for concentrating hundreds of different products.

Rising/Falling Film Plate. The original plate type evaporator, the rising–falling plate system basically is a flexible, multiduty unit engineered for medium production runs of heat-sensitive products under sanitary conditions and at the lowest possible capital investment.

Operating on a single-pass rising and falling film principle, as shown in the exploded view of Figure 8, the evaporator consists of a series of gasketed plate processing units within a compact frame. As a thin layer of feed liquor passes over the rising and falling film section in each evaporation unit, it is vaporized on contact with adjacent steam heated plates and is discharged with its vapor to a vapor–liquid separator. All evaporation is accomplished in a matter of seconds within the plate pack. The product then is extracted and the vapor passed to a condenser, the next evaporator effect or a mechanical compressor. Product quality is given maximum protection against thermal degradation by means of high heat-transfer rates, low liquid holding volume, and minimum exposure to high temperatures.

The plate evaporator has a number of advantages over its tubular counterpart. Since it is designed to be erected on a single level with minimum headroom requirements, in many cases it will fit within an existing building with overhead restrictions as low as 12 or 13 ft. It also can be arranged as single or multiple effects without extensive building modifications or structural steel supports and can handle expanded duties by the addition of more plate units. Systems are available for evaporation rates to 35,000 lb/h with efficient in-place cleaning of all stainless, steel product contact surfaces.

Falling Film Plate. The success of the rising–falling film plate evaporator prompted the development of the falling film plate system. This design has several advantages over its predecessor, including even lower residence times and higher evaporation capabilities. While the rising–falling film plate unit is restricted to a maximum of 30,000–35,000 lb/h of water removal, the falling film plate evaporator with its larger vapor ports can accommodate 59,000–60,000 lb/h of evaporation. One of this unique design features as illustrated in Figure 9 is that each side of the plate may be used independently of the other, thus permitting two pass operation within the same frame for even shorter residence time and improved product quality.

Paravap. Especially designed to concentrate foaming liquids or those with high solids content or non-Newtonian viscosity characteristics, the APV Paravap replaces the wiped film evaporator in many cases. It successfully concentrates corn syrups and soap to better than 97–98% total solids or strips hexane and other solvents from vege-

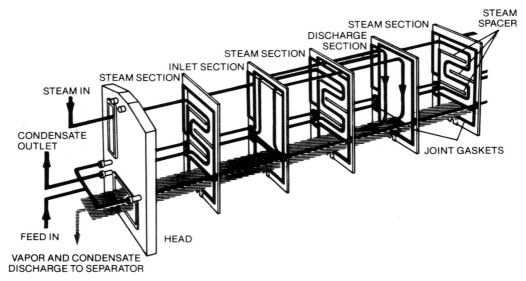

Figure 8. Rising–falling film plate evaporator.

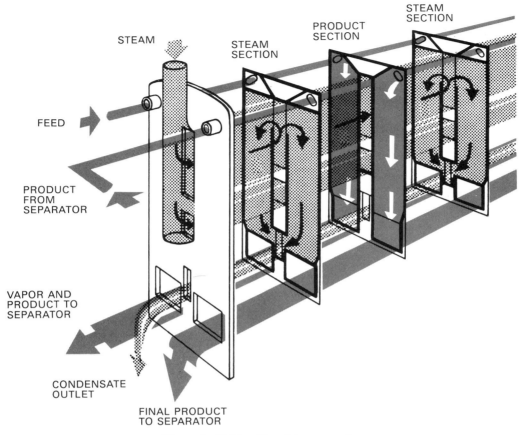

Figure 9. Falling film plate evaporator.

table oils and similar products. Figure 10 shows this type of unit in a simple schematic.

Note that the key element in the system is a plate heat exchanger (Fig. 11). While specifically designed for liquid–liquid applications, it has been found that if a fluid is allowed to vaporize within the plate pack, the small plate gap and corrugated plate pattern create high vapor velocities (Fig. 12). This causes atomization of the liquid within the high-velocity vapor stream, resulting in a greater liquid surface area for mass transfer and enabling low residual solvent concentrations to be realized. Although the final product may be extremely viscous after separation from the vapor, the apparent viscosity within the plate pack is very low since only droplets are being transported in the vapor.

Paraflash. Similar to its tubular counterpart, the APV Paraflash (Fig. 13) is a suppressed boiling forced-circulation evaporator used mainly for concentrating products subject to fouling that is too excessive for a film evaporator. While this system uses a plate heat exchanger, vaporization is not allowed to take place within the exchanger itself. Boiling is suppressed either by a liquid static head above the heat exchanger or by the use of an orifice piece in the discharge line. From a heat-transfer standpoint, the forced-circulation Paraflash is far more efficient than a tubular unit but suffers from a smaller equivalent diameter when large crystal sizes are present.

SELECTION OF EVAPORATORS

In choosing the evaporator best suited to the duty on hand, the following factors should be weighted.

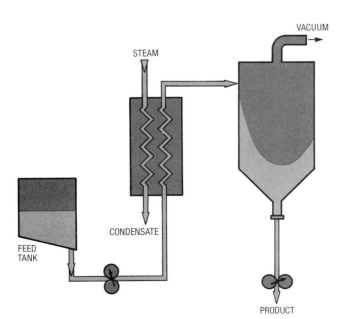

Figure 10. Paravap evaporator.

Figure 11. Plate heat exchanger.

Heat-Transfer Coefficients

The evaporator with the highest heat-transfer coefficients (HTCs) generally is the falling film type. Higher HTCs lead to smaller surface areas, resulting in minimum capital cost. Wherever possible, therefore, a film evaporator should be considered first.

While the biggest single disadvantage of the film evaporator is its susceptibility to scale buildup on the heat-transfer surface, its lower cost offsets this inconvenience on mildly scaling duties, particularly when high evaporative capacities are involved. A typical example of this is spin bath liquors. This solution, which results from the production of rayon and cellulose, consists mainly of sulfuric acid, sodium sulfate and water plus various impurities such as zinc sulfate. Scale builds up over a period of about 1 week and the evaporator then is boiled out.

In food related industries where the nature of most products is heat sensitive, it generally is desirable or necessary to clean out an evaporator at least once per day for sanitary reasons. In the chemical field, however, a minimum run of a week normally is required, and, in many cases, an evaporator will run for months.

Capacities

For evaporation capacities of up to about 60,000 lb/h of water removal, the possibility of using either a plate or tubular system should be considered. The capital invest-

Figure 12. Corrugated plate pattern.

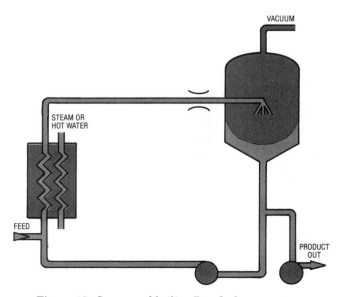

Figure 13. Suppressed boiling Paraflash evaporator.

Figure 14. Single-effect R56 Paraflash concentrates brewers yeast from 13

ment is relatively close, but the lower cost of installing a plate evaporator together with its compactness, expendability, and reduced building requirements generally makes it extremely attractive. Above 60,000 lb/h evaporation, the choice lies among various types of tubular evaporators. Available configurations are multieffect or mechanical vapor recompression.

Materials of Construction

In many cases, the choice of evaporator is determined by the necessary materials of construction (Fig. 15). An example would be a sulfuric acid evaporator, where, for product concentrations of up to 50%, the heating surface normally would be graphite. This immediately indicates use of a tubular evaporator as opposed to plate unit. With satisfactory temperature and concentration conditions, Incoloy and Hastelloy could be employed using either a tube or plate but would be much more expensive than graphite. Separators and piping would be either a fiber-filled phenolic resin or rubber-lined carbon steel. Sulfuric acid between 50 and 72% concentration again would normally use graphite as the heating surface providing temperatures are kept below 230–240°F, but piping and vessels would be lead-lined steel, tantiron, or cast lead. For construction reasons, a graphite tube bundle has to arranged as a rising film or forced-circulation evaporator with the vapor–liquid separator positioned at the top.

Multiple Evaporators

More than one evaporator may be used to concentrate a particular product. For example, there may be a main or bulk evaporator followed by a finishing unit. The main evaporator would be used to remove the majority of the water or solvent to a point where the liquid concentration gives rise to restrictions. Among these are boiling-point elevation, which reduces the overall temperature difference of a multieffect or MVR evaporator, product viscosity resulting in low HTCs, or saturated liquor resulting in crystallization. A forced-circulation crystallizing evaporator very often is the last product stage of a multieffect evaporator.

Consider the example of concentrating a solution from 4 to 60% total solids, a 15:1 concentration ratio. At the 40% total solids level, the viscosity is too high for a film evaporator and the product has to be processed in a Paravap or wiped film unit. However, in concentrating from 4 to 40% or a 10:1 ratio, 90% of the feed has been evaporated in the initial evaporator. The finishing unit now has to concentrate only from 40 to 60% or 1.5:1 ratio. This means that 96.43% of the total evaporation has been achieved in a multieffect or mechanical vapor recompression bulk evaporator at high thermal economy and probably while obtaining good HTCs. Only 3.57% of the duty is confined to a small single-effect evaporator, which handles the most difficult part of the concentration.

Product		Materials of Construction
Most dairy products		304/316 stainless steel
Most food products		304/316
Most fruit juices		304/316
Most pharmaceuticals		304/316
if NaCL in any of above		Monel or titanium
Sulfuric acid	<50%	FFPR/graphite
		RLCS/graphite
	<72%	Lead lined/tantiron or cast lead
Hydrochloric acid		FFPR/graphite/RLCS
Phosphoric acid	(dilute/pure)	316
	(with impurities)	FFPR/graphite/RLCS
Caustic soda	<40%	Stress relieved C.S.
Caustic soda	(with NaCl)	Monel/nickel/NiLCS
Ammonium nitrate		304/304L
Ammonium sulfate		316

FFPR—fiber-filled phenolic resin; RLCS—rubber-lined carbon steel; NiLCS—nickelHined carbon steel.

Figure 15. Guide to materials of construction.

Solution properties:

Specific gravity	1.35
Specific heat	0.62
Thermal conductivity	0.26 Btu units
Viscosity	100 cps

	Tubular Unit	Plate Unit
Heat duty, Btu/hr	5×10^6	5×10^6
Element	1¼ in. OD, 16 SWG, 25-ft tube	APV R10 plate
Total surface, ft^2	2620	2622
Fluid temperature in, °F	160	160
Fluid temperature out, °F	161.08	170
Service temperature, °F	190	190
Recirculation rate, gpm	10,975	1610
Total ΔP, psi	16.9	10.8
BHP absorbed	107.7	10.1

SWG = standard wire gauge; OD = outer diameter; gpm = gallons per minute.

Figure 16. Comparison between forced circulation tubular and plate evaporators.

Crystallization

When the requirement is for a crystallizing evaporator, there never is any doubt that the stage where crystallization takes place has to be a forced-circulation evaporator and, in most cases, a tubular unit. However, when the problem is not one of crystallization but a general scaling of the heat-transfer surface instead, the Paraflash is much more efficient. As shown in Fig. 16, the circulation rate through the plate is considerably less than through a tube and results in much lower horsepower requirements.

If crystals are formed and are too large for a plate evaporator to handle, the designer has to determine whether the tubular forced circulation unit should be a single- or multiple-pass configuration. Normally, forced-circulation tubulars are designed with tube velocities of 6–8 ft/s or higher to minimize fouling. In the Figure 17 comparison between single- and three-pass configurations, three-pass units are shown to have lower circulation rates directly resulting in higher temperature rises, lower log mean temperature differences, and, therefore, greater surface area. Even though the circulation rate is less, the pressure drop on the three-pass versus one-pass is more because of greater overall tube length. This results in larger absorbed horsepower. It generally is concluded, therefore, that a single-pass arrangement is less expensive from both the capital and operating cost standpoints.

HTCs, Forced Circulation, and Film

Estimation of HTCs from physical properties is easier and more reliable for forced-circulation units than for film evaporators. As a general rule, HTCs used for the latter are determined by experience or test work.

EVAPORATION CONFIGURATION AND ENERGY CONSERVATION

Engineers can help cut energy consumption even when evaporation requirements seem to become greater every year. This can be done by

1. Installing a greater number of steam effects to an evaporator.
2. Making use whenever possible of steam thermocompression or mechanical vapor recompression.
3. Making sure that the feed to an evaporator is preheated to the boiling point in the most efficient manner.

DUTY: TRIPLE EFFECT FORCED CIRCULATION EVAPORATOR
FEED: 102,000 LB/HR, CORNSTEEP 6.14% → 51%

	1st EFFECT 1 PASS	1st EFFECT 3 PASS	2nd EFFECT 1 PASS	2nd EFFECT 3 PASS	3rd EFFECT 1 PASS	3rd EFFECT 3 PASS
CIRCULATION RATE GPM	12,275	6,190	12,275	6,190	14,500	4,800
HEAT TRANSFER AREA FT	4,434	5,541	4,434	5,541	4,434	4,656
ABSORBED BHP	50	91	66	113	156	192

	ONE PASS	THREE PASS
EQUIPMENT	$491,280	$540,960
ABSORBED BHP	272	396

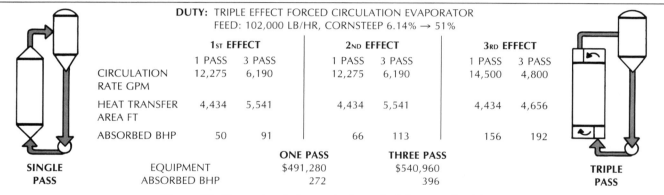

Figure 17. Comparison between single- and triple-pass calandria.

4. Using low-grade heat from the evaporator whenever possible.
5. Insulating equipment to keep losses to a minimum.

Some of these points are illustrated by the Figure 18 schematic of a three-effect evaporator with steam thermocompression that uses evaporator condensate and intereffect vapor for feed preheating. Note that even this vapor has been used efficiently in prior effects and that condenser water also may be used if the feed temperature is low. Figure 19, meanwhile, outlines a four-effect evaporator with thermocompression and the use of a slightly different technique with a spray heating loop.

While a greater number of effects increases initial capital investment, savings in operating costs will justify the expenditure if evaporation capacities are large enough and the operating period long enough. A thermocompressor generally adds the equivalent of one or more effects at relatively little capital cost, although the surface area in the effects between thermocompressor suction and discharge must be increased. The HTC of these effects generally is very good so the increase in surface area is minimized. Care should be taken, however, to design with a low ΔT across the thermocompressor to ensure high entrainment ratios of vapor to steam.

It is also important when using a steam thermocompressor not to have a high-boiling-point elevation in those effects. This will cut down the (ΔT) available for heat transfer. Thermocompressors are somewhat inflexible and do not operate well outside their design conditions. They should not be used when the product may foul the evaporator surface and cause the design pressures and pressure boost across the thermocompressor to rise. Under these conditions, the amount of vapor entrained is reduced and a fall-off in evaporator capacity results.

Effective preheating of the feed also can reduce energy consumption considerably. The capital cost of preheaters is fairly small with the plate heat exchangers being particularly effective owing to their capability for realizing a very close temperature approach to the heating medium, whether it is condenser water, waste evaporator condensate, or intereffect vapor.

While all of these features reduce energy consumption, still better methods are available. In less than a decade and probably within the next 5 yr, the majority of new large-capacity evaporators installed, particularly for bulk duties, will be of the mechanical vapor recompression (MVR) type. The mechanical vapor compressor can be driven by electricity from coal, hydropower, or nuclear power even when all the gas and oil has been exhausted or is too expensive to use.

Briefly examining the thermodynamics involved with MVR, the Figure 20 schematic shows an evaporator with a liquid boiling point of 212°F (atmospheric pressure). All of the water vapor that is boiled off passes to a compressor. In order to keep the energy input to the system as low as possible, the pressure boost across the compressor is limited. In the majority of cases, this pressure boost will correspond to a saturated temperature rise in the region of 15°F or less.

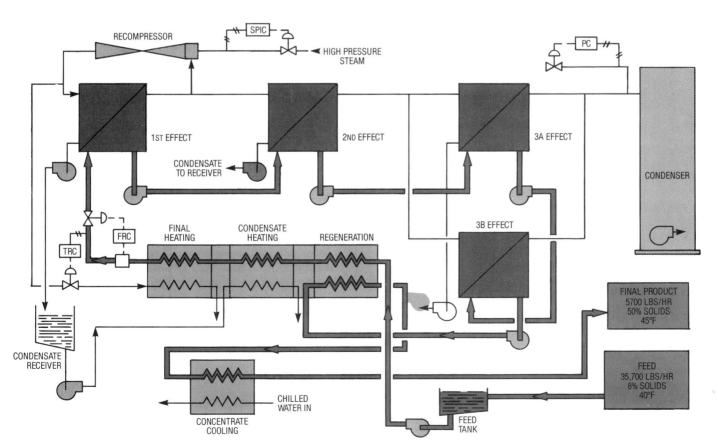

Figure 18. Three-effect steam jet recompression.

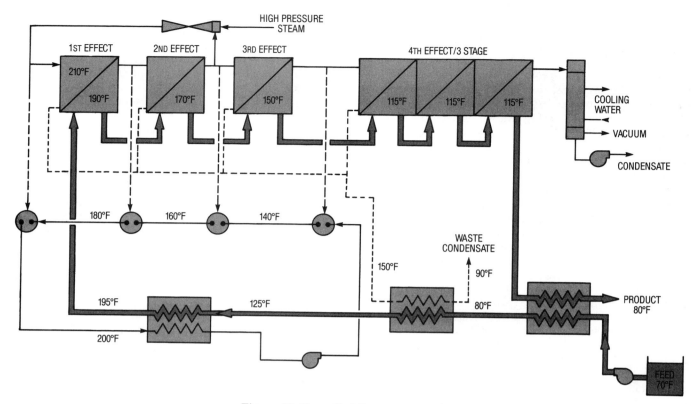

Figure 19. Four-effect thermocompression.

In this example, there is a pressure boost of 4.5 psi across the compressor. Assuming that there is a pressure low of 0.5 psi in the system, the effective pressure on the steam side of the evaporator is 18.7 psia. This compressed water vapor condenses and gives up its latent heat, thus vaporizing more water from the liquid that is being concentrated. The latent heat, thus vaporizing more water from the liquid that is being concentrated. The latent heat of vaporization of water at atmospheric pressure is 970 Btu/h. Note that it requires a theoretical energy input of only 18 Btu/lb to raise the water vapor from 14.7 to 19.2 psia. The theoretical steam economy, therefore, is 970/18 = 54. When compressor efficiency is taken into account, this figure is brought down to between 32 and 35, which is another way of saying that the MVR system is equivalent to a 32–35-effect steam evaporator. Related to energy costs of 4 cents per kilowatt-hour for electricity for the compressor drive and $6.00 per 1000 lb of steam for a

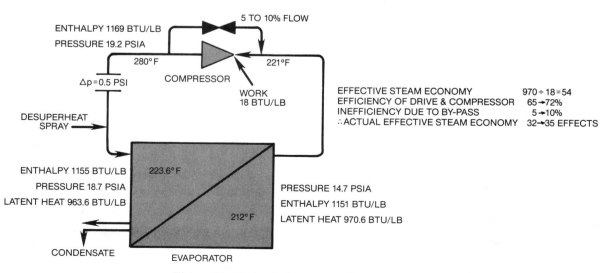

Figure 20. Mechanical recompression evaporator.

conventional steam evaporator, the MVR system then becomes the economic equivalent of just under a 19-effect evaporator.

The MVR has another definite advantage over steam. The condensate is available at high temperature and is ideal for evaporator feed preheating, particularly if the condensate rate is as high as 90% of the feed rate, ie, a 10:1 concentration ratio within the evaporator. There are many such evaporators in operation where the sole energy input to the system is through the compressor with steam requirements limited to approximately 15 min during startup.

While Figure 20 illustrates an evaporator system that uses a compressor, it will be shown that the use of a turbo fan in place of the centrifugal compressor not only yields lower boost but also provides more surface area and better operating economics.

RESIDENCE TIME IN FILM EVAPORATION

Since many pharmaceutical, food and dairy products are extremely heat-sensitive, optimum quality is obtained when processing times and temperatures are kept as low as possible during concentration of the products. The most critical portion of the process occurs during the brief time that the product is in contact with a heat-transfer surface that is hotter than the product itself. To protect against possible thermal degradation, the time–temperature relationship therefore must be considered in selecting the type and operating principle of the evaporator to be used.

For this heat-sensitive type of application, film evaporators have been found to be ideal for two reasons: (1) the product forms a thin film only on the heat-transfer surface rather than occupying the entire volume, and thus residence time within the heat exchanger is greatly reduced, and (2) a film evaporator can operate with as low as a 6°F steam-to-product temperature difference. With both the product and heating surfaces close to the same temperature, localized hot spots are minimized.

As described previously, there are rising film and falling film evaporators as well as combination rising–falling film designs. Both tubular and plate configurations are available.

Rising Film Evaporators

In a rising film design, liquid feed enters the bottom of the heat exchanger, and when evaporation begins, vapor bubbles are formed. As the product continues up either the tubular or plate channels and the evaporation process continues, vapor occupies an increasing amount of the channel. Eventually, the entire center of the channel is filled with vapor while the liquid forms a film on the heat-transfer surface.

The effect of gravity on rising film evaporator is twofold. It acts to keep the liquid from rising in the channel. Further, the weight of the liquid and vapor in the channel pressurizes the fluid at the bottom and with the increased pressure comes an increase in the boiling point. A rising film evaporator therefore requires a larger minimum ΔT than does a falling film unit.

The majority of the liquid residence time occurs in the lower portion of the channel before there is sufficient vapor to form a film. If the liquid is not preheated above the boiling point, there will be no vapor and since a liquid pool will fill the entire area, the residence time will increase.

Falling Film Evaporators

As liquid enters the top of a falling film evaporator, a liquid film formed by gravity flows down the heat-transfer surface. During evaporation, vapor fills the center of the channel and as the momentum of the vapor accelerates the downward movement, the film becomes thinner. Since the vapor is working with gravity, a falling film evaporator produces thinner films and shorter residence times than does a rising film evaporator for any given set of conditions.

Tubular and Plate Film Evaporators

When compared to tubular designs, plate evaporators offer improved residence time since they carry less volume within the heat exchanger. In addition, the height of a plate evaporator is less than that of a tubular system.

ENGINEERING CONVERSION

Table 1 provides data for engineering conversions applicable to the use of an evaporation in food processing.

Estimating Residence Time

It is difficult to estimate the residence time in film evaporators, especially rising film units. Correlations, however, are available to estimate the volume of the channel occupied by liquid. Table 2 describes a number of applicable equations. Equation 1 is recommended for vacuum systems.

For falling film evaporators, the film thickness without vapor shearing can be calculated by equation 2.

Since the film is thin, this can be converted to liquid volume fraction in a tubular evaporator by equation 3.

For a falling film plate evaporator, equation 4 is used. As liquid travels down the plate and evaporation starts, vapors will accelerate the liquid. To account for this action, the rising film correlation is used when the film thickness falls below that of a falling film evaporator. In practice, the film thickness may be less than estimated by either method because gravity and vapor momentum will act on the fluid at the same time.

Once the volume fraction is known, the liquid residence time is calculated by equation 5. In order to account for changing liquid and vapor rates, the volume fraction is calculated at several intervals along the channel length. Evaporation is assumed to be constant along with channel length except for flash due to high feed temperature.

Table 3 shows a comparison of contact times for typical four-effect evaporators handling 80 GPM of feed. The tubular designs are based on 2-in-diameter by 30 ft. Designs using different tube lengths, incidentally, do not change the values for a rising film tubular system.

The given values represent total contact time on the evaporator surface, which is the most crucial part of the processing time. Total residence time would include con-

Table 1. Engineering Conversions

To convert from	to	Multiply by
Heat Capacity		
Calories/(gram)(°C)	Calories/(Gram)(mole)(°C)	Molecular weight
	Btu/(pound)(°F)	1.0
Density		
Gram/milliliter	Pounds/gallon	8.33
	Pounds/Ft^3	62.42
Thermal Conductivity		
Kilocalorie/(hr)(m)(°C)	Btu/(h)(ft)(°F)	0.6719
W/(m)(°C)		0.5778
Viscosity		
Centistokes	Centipoise	Specific Gravity
Dynamic viscosity pound-mass/(ft)(s)	Centipoise	1488.2
Kinematic viscosity cm^2/s	Centistokes	100
Pressure		
Kilopascal	psi	0.14504
Bar		14.504
Inches Hg absolute	psia	0.4912
Atmosphere		14.696
Torr		0.01908
mmHg		0.01908
Enthalpy		
Calorie/gram	Btu/pound-mass	1.8
Work/Energy		
(Kilowatt)(hr)	Btu	3412.1
(Horsepower)(hr)		2544.4
Calorie		0.003968
Heat-Transfer Coefficient		
Kilocalorie/(h)(m^2)(°C)	Btu/(h)(ft^2)(°F)	0.2048
W/(cm^2)(°C)		1761.1

Table 2. Estimating Residence Time

(1) $R_L = 1 - \dfrac{1}{1 + \left(\dfrac{1-y}{y}\right)\left(\dfrac{2\,\rho_v}{\rho_L}\right)^{1/2}}$

(2) $m = \left[\dfrac{3\Gamma\mu}{g\rho_L(\rho_L - \rho_v)}\right]^{1/3}$

(3) $R_L = 4m/d$

(4) $R_L = 2m/z$

(5) $t = AL/R_L q_L$

[a] Symbols: A = cross-sectional area, ft^2; d = tube diameter, ft; g = 32.17 ft/s^2; L = tube length, ft; m = film thickness, ft; R_L = liquid volume fraction (dimensionless); q_L = liquid rate, ft^3/s; t = time, s; z = plate gap, ft; Γ = liquid wetting rate, lb/s · ft; ρ_L = liquid density, lb/ft^3; ρ_v = vapor density; lb/ft^3; μ = liquid viscosity, lb/ft s; y = local weight fraction of vapor (dimensionless).
[b] Refs. 1, 2.

tact in the preheater and separator as well as additional residence within interconnecting piping.

While there are no experimental data available to verify these numbers, experience with falling film plate and tubular evaporators shows that the values are reasonable. It has been noted that equation 2 predicts film thicknesses that are too high as the product viscosity rises so, in actuality, four-effect falling film residence times probably are somewhat shorter than charted.

Summary

Film evaporators offer the dual advantages of low residence time and low temperature difference that help assure a high product quality when concentrating heat-sensitive products. In comparing the different types of film evaporators that are available, falling film designs provide the lowest possible ΔT and the falling film plate evaporator provides the shortest residence time.

Table 3. Residence Time Comparison

Contact time	Rising Film Tubular	Rising Film Plate[a]	Falling Film Tubular	Falling Film Plate
1st effect	88	47[b]	23	16[b]
2nd effect	62	20	22	13
3rd effect	118	30[b]	15[b]	9
4th effect	236[b]	78[b]	123[b]	62[c]
Total contact time	504	175	183	100

[a] Plate gap of 0.3 in. assumed.
[b] Two stages.
[c] Three stages.

PLATE-TYPE EVAPORATORS

To effectively concentrate an increasing variety of products that differ by industry in such characteristics as physical properties, stability, or precipitation of solid matter, equipment manufacturers have engineered a full range of evaporation systems. Included among these are a number of plate-type evaporators.

Plate evaporators initially were developed and introduced by APV in 1957 to provide an alternative to the tubular systems that had been in use for half a century. The differences and advantages were many. The plate evaporator, for example, offers full accessibility to the heat-transfer surfaces. It also provides flexible capacity merely by adding more plate units, shorter product residence time (resulting in a superior quality concentrate), a more compact design with low headroom requirements and low installation cost.

These APV plate evaporation systems are available in four arrangements—rising–falling film, falling film, Paravap, and Paraflash—and may be sized for use in new produce development or for production at pilot plant or full-scale operating levels.

Rising–Falling Film Plate

The principle of operation for the rising–falling film plate evaporator (RFFPE) involves the use of a number of plate packs or units, each consisting of two steam plates and two product plates. As shown in Figure 21, these are hung in a frame resembling that of a plate heat exchanger. The first product passage is a rising pass and the second, a falling pass. The steam plates, meanwhile, are arranged alternately between each product passage.

The product to be evaporated is fed through two parallel feed ports and is equally distributed to each of the rising film annuli. Normally, the feed liquor is introduced at a temperature slightly higher than the evaporation temperature in the plate annuli, and the ensuing flash distributes the feed liquor across the width of the plate. Rising film boiling occurs as heat is transferred from the adjacent steam passage with the vapors that are produced helping to generate a thin, rapidly moving turbulent liquid film.

During operation, the vapor and partially concentrated liquid mixture rises to the top of the first product pass and transfers through a "slot" above one of the adjacent steam passages. The mixture enters the falling film annulus where gravity further assists the film movement and completes the evaporation process. The rapid movement of the thin film is the key to producing low residence time within the evaporator as well as superior heat-transfer coefficients. At the base of the falling film annulus, a rectangular duct connects all of the plate units and transfers the evaporated liquor and generated vapor into a separating device. Steam and condensate ports connect to all the steam annuli (Fig. 22).

The plate evaporator is designed to operate at pressures extending from 10 psig to full vacuum with the use of any number of effects. However, the maximum pressure differential normally experienced between adjacent annuli during single effect operation is 15 psig. This, and the fact that the pressure differential always is from the steam side to the product side, considerably reduces design requirements for supporting the plates. The operating pressures are equivalent to a water vapor saturation temperature range of 245°F downward and thus are compatible with the use of nitrile or butyl rubber gaskets for sealing the plate pack.

Most rising–falling film plate evaporators are used for duties in the food, juice, and dairy industries where the low residence time and 100–200°F operating range temperatures are essential for the production of quality concentrate. An increasing number of plate evaporators, however, are being operated successfully in both pharma-

Figure 21. Rising–falling film plate evaporator in final stages of fabrication.

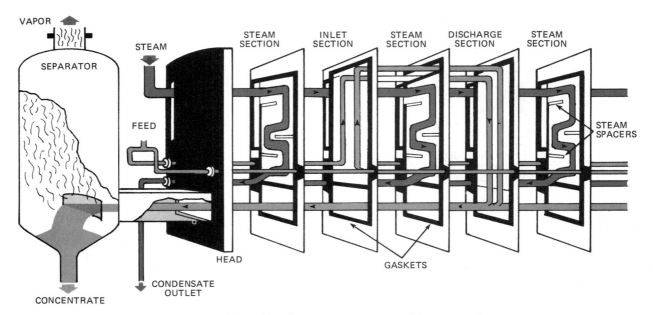

Figure 22. Rising–falling film plate evaporator arranged for one complete pass.

ceutical and chemical plants on such products as antibiotics and inorganic acids. These evaporators are available as multi-effect and/or multi-stage systems to allow relatively high concentration ratios to be carried out in a single pass, nonrecirculatory flow.

The rising–falling film plate evaporator should be given consideration for various applications.

- That require operating temperatures between 80 and 210°F
- That have a capacity range of 1000–35,000 lb/h water removal
- That have a need for future capacity increase since evaporator capabilities can be extended by adding plate units or by the addition of extra effects
- That require the evaporator to be installed in an area that has limited headroom
- Where product quality demands a low time–temperature relationship
- Where suspended solid level is low and feed can be passed through 50 mesh screen

A "junior" version of the evaporator is available for pilot plant and test work and for low-capacity production. If necessary, this can be in multieffect–multistage arrangements such as the system illustrated in Figure 23.

Falling Film Plate

Incorporating all the advantages of the original rising–falling film plate evaporator system with the added benefits of shorter residence time and larger evaporation capabilities, the falling film plate evaporator has gained wide acceptance for the concentration of heat sensitive products. With its larger vapor ports, evaporation capacities typically are up to 50,000–60,000 lb/h.

The falling film plate evaporator consists of gasketed plate units (each with a product and a steam plate) compressed within a frame that is ducted to a separator. The number of plate units used is determined by the duty to be handled.

As shown in Figure 24, one important innovation in this type of evaporator is the patented feed distribution system. Feed liquor first is introduced through an orifice (1) into a chamber (2) above the product plate where mild flashing occurs. This vapor/liquid mixture then passes through a single product transfer hole (3) into a flash chamber (4), which extends across the top of the adjacent steam plate. More flash vapor results as pressure is further reduced and the mixture passes in both directions into the falling film plate annulus through a row of small

Figure 23. Three-effect, four-stage, "junior."

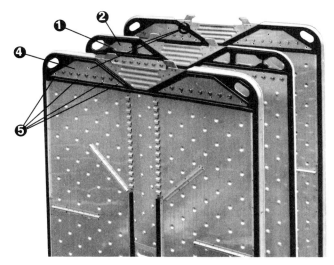

Figure 24. Typical product and steam plate unit for falling film plate evaporator.

distribution holes (5). These ensure an even film flow down the product plate surface where evaporation occurs. A unique feature is the ability to operate the system either in parallel or in series, giving a two-stage capability to each frame. This is particularly advantageous if product recirculation is not desirable.

In the two-stage method of operation, feed enters the left side of the evaporator and passes down the left half of the product plate, where it is heated by steam coming from the steam sections. After the partially concentrated product is discharged in the separator, it is pumped to the right side of the product plate where concentration is completed. The final concentrate is extracted while vapor is discharged to a subsequent evaporator effect or to a condenser.

Paravap

The operating principle of the Paravap represents an application of the thin film, turbulent path evaporation process and in many cases, replaces the wiped or agitated film evaporator. Since it has no mechanical moving parts within the heat transfer area, costly maintenance repairs common to wiped film systems are eliminated. It is especially designed to concentrate liquids with high solids contents or non-Newtonian viscosity characteristics (Fig. 25).

With feed liquor and the heating medium of steam, hot water or hot oil directed into alternate passages within a plate heat exchanger, boiling begins as the liquid contacts the heated plates. The small plate gap and high-velocity flow pattern (Fig. 26) atomizes the feed and provides a greater liquid surface area for mass transfer. Since the vapor carries the feed in the form of minute particles, the apparent viscosity within the heat exchanger plate pack is very low. The final product, however, may be extremely viscous after separation from the vapor.

Advantages include low residence time to minimize thermal degradation, low rates of fouling, and economical operation.

Typical applications: concentrating soap to final product consistency; concentrating apple puree from 25 to 40°

Figure 25. Single-effect R86 Paravap.

brix, grape puree to 50° brix, cherry puree to 60° brix, fruit juices to 95% solids, and sugar or corn syrup solutions to over 97%; solvent stripping duties such as hexane from oil (See also Figs. 25 and 27.)

Paraflash

Operating under a suppressed boiling, forced circulation principle, the APV Paraflash is used to concentrate products subject to fouling too excessive for film evaporators. Unlike the Paravap, vaporization does not occur within the heat-exchanger plate pack. Instead, liquor flashes as it enters a separator, crystallization takes place, and a suspended slurry results. Suppressed boiling combined with high liquid velocities deters scaling on the heat-transfer surface, minimizes cleaning downtime, and promotes longer production runs.

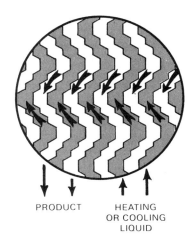

Figure 26. Small plate gap and high-velocity flow pattern.

Figure 27. Single-effect, forced circulation Paraflash.

The Paraflash can be used in single or multiple effects for such products as grape juice (tartrate crystals), coffee, wheat starch, distillery effluent, and brewer's yeast (suspended solids). (See Fig. 27.)

MECHANICAL VAPOR RECOMPRESSION EVAPORATORS

In recent years, MVR technology has been introduced and widely accepted as an effective approach to powering medium to large-capacity evaporators for both chemical and sanitary applications. While experience has shown that the higher capital cost of this equipment relative to that of steam-driven evaporation systems has been offset by significant energy savings, advances in MVR technology now have reduced energy requirements even further.

Definition

Simply stated, with mechanical recompression the water vapor boiled off in the evaporator is passed to an electrically powered compressor and is compressed through 1–3 psi. This raises the temperature of the vapor, which then is used as the heating medium. The difference in enthalpy between the vapors on the heating and process sides is comparatively small, with a resultant reduction in energy consumption and, depending on regional steam and power costs, an operating cost equivalent to at least an 8–30-effect evaporator. Theoretical thermal efficiency may exceed that of a 60-effect steam evaporator.

Typically, Figure 28a shows a single-effect steam evaporator operating at atmospheric pressure (14.7 psia) with vapor being produced at 212°F and sent to a condenser. The heat source for this system is steam at 17.2 psia, condensing at 220°F.

In Figure 28b, the same evaporator is shown operating with mechanical vapor compression. In this case, the vapor at 212°F, 14.7 psia, is sent to a compressor, where its pressure is raised to 17.2 psia. It then is charged to the steam side of the evaporator as the heating medium, condensing at 220°F. Savings are realized in two areas: no steam is required for evaporation, although nominal amounts are required for startup and compressor seals, and since the vapor from the evaporator is not sent to a condenser, cooling water requirements are dramatically reduced.

It will be shown later that excess vapor is produced gas a result of the inefficiencies in the compression cycle. This excess vapor is used to compensate for vent and radiation losses and at times, as an assist to preheating.

Development

To a large extent, the design of MVR evaporators has been dictated by the capabilities of the compressors available at the time. The primary limitation has been the compression ratio (absolute discharge pressure divided by absolute suction pressure) that can be achieved since this determines the temperature difference available for the evaporator. For example, if an evaporator is run at 212°F and atmospheric pressure and a compression ratio of 1.4 is used, the steam pressure at the discharge of the compressor is 1.4×14.7 psia or 20.58 psia. Since water vapor at 20.58 psia condenses at 229.5°F, the allowable temperature difference ΔT for the evaporator is 17.5°F before losses are considered. Table 4 shows the total temperature differences for a number of conditions.

Centrifugal Compressors

The relatively high volume of vapor encountered in evaporators requires the use of centrifugal compressors in most cases. Initially, centrifugal compressors on steam were limited to a compression ratio of 1.4, which, in turn, limited the available temperature difference to between 13 and 17.5°F depending on boiling temperature. Only film evaporators could use MVR since there was not sufficient temperature difference to consider forced-circulation designs. Furthermore, products having significant boiling point elevation were excluded for this technique.

Two trends developed in recent years, however, have increased MVR capabilities and applications.

First improved design allows centrifugal compressors to run at higher speeds, increasing compression ratios approximately to 2.0 to 1. Consequently, a (ΔT of 27–36°F now can be obtained and thus, MVR can be used with forced-circulation systems and for the evaporation of products with boiling-point elevations.

Table 4. Saturated ΔT at Various Compression Ratios, °F

Compression Ratio	Boiling Temperatures		
	130°F	170°F	212°F
1.2	6.9	8.0	9.3
1.4	12.9	15.0	17.5
1.6	18.2	21.2	24.7
1.8	23.0	26.8	31.2
2.0	27.3	31.9	37.2

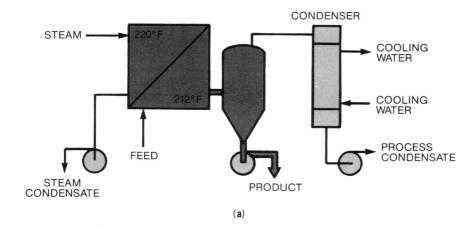

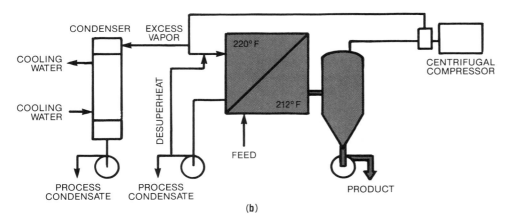

Figure 28. (a) Single-effect evaporator. (b) Single-effect MVR evaporator.

Secondly, advances in evaporator design allow the operation of film evaporators with a lower (ΔT), thereby minimizing energy requirements. For some products, film evaporators can be calculated with a (ΔT (excluding losses) in the range of 6°F.

By increasing the ΔT available and decreasing the ΔT required, it also became possible to design multiple—effect MVR evaporators. Figure 29 illustrates a double-effect MVR system with a forced-circulation finisher operating across all effects. With this arrangement, the flow to the compressor is reduced and the finisher, operating at the higher concentration, takes advantage of the full ΔT available.

Fans

The next development in MVR design was an apparent reversal of prior advances.

As the required ΔT across a film evaporator decreased, it became possible to provide that ΔT by using a fan. This produces a compression ratio on the order of 1.2 to 1, providing approximately 7 or 8°F ΔT (before losses) for evapo-

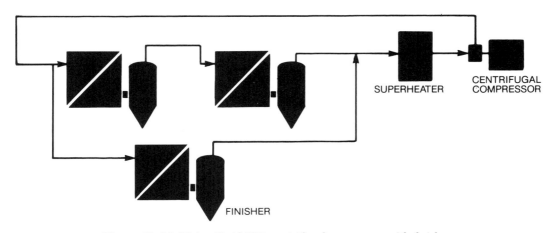

Figure 29. Multiple-effect MVR, centrifugal compressor with finisher.

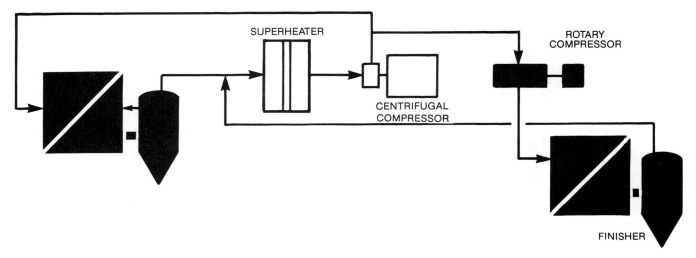

Figure 30. Single-effect MVR evaporator with centrifugal compressor and rotary-blower-driven finisher.

ration. The MVR system previously discussed and shown in Figure 30 is typical of fan use. It should be noted that the fan MVR technique can be used only on single-effect film evaporators handling product without significant boiling point elevation. Where the heat-transfer coefficient of the product is low, it sometimes is better to make use of the higher ΔT available from a centrifugal compressor.

Where fans can be used, however, the horsepower requirement usually is less than in centrifugal compressor designs. Furthermore, fans do not require that the inlet vapor be superheated. Instead, this is done either by a separate heat exchanger using steam or by recycling some of the compressor discharge vapor to the compressor suction.

It is, incidentally, a common practice to use a fan evaporator to concentrate a product up to a point where a large ΔT is required and then to switch to a small steam or MVR finishing evaporator for final concentration.

Other Designs

Rotary blowers are low-capacity positive compressors that occasionally are used with small MVR evaporators of up to approximately 15,000-lb/h capacity. This generally is a finishing-type evaporator.

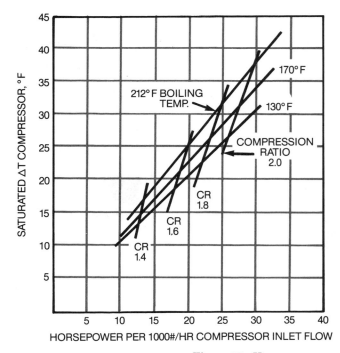

COMPRESSORS							
Boiling Temperature 130°F			Boiling Temperature 170°F			Boiling Temperature 212°F		
CR	ΔT	HP	CR	ΔT	HP	CR	ΔT	HP
1.3	10.00	9.39	1.3	11.65	10.01	1.3	13.53	10.67
1.4	12.91	12.18	1.4	15.03	12.99	1.4	17.47	13.85
1.6	18.21	17.38	1.6	21.23	18.54	1.6	24.69	19.76
1.8	22.98	22.15	1.8	26.81	23.63	1.8	31.20	25.18
2.0	27.31	26.57	2.0	31.89	28.34	2.0	37.16	30.20
2.2	31.30	30.69	2.2	36.57	32.74	2.2	41.31	34.88

Figure 31. Horsepower versus ΔT for centrifugal compressors.

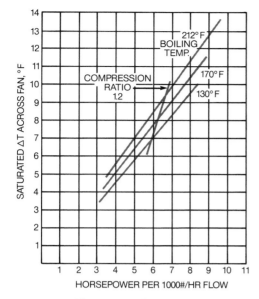

Fans								
Boiling Temperature 130°F			Boiling Temperature 170°F			Boiling Temperature 212°F		
CR	ΔT	HP	CR	ΔT	HP	CR	ΔT	HP
1.1	3.59	2.99	1.1	4.17	3.19	1.1	4.85	3.41
1.2	6.91	5.79	1.2	8.04	6.18	1.2	9.34	6.60
1.3	10.00	8.43	1.3	11.65	9.00	1.3	13.53	9.60

Figure 32. Horsepower versus ΔT for fans. *Note:* Vapor assumed to be superheated 10°F by external steam.

In some cases, two compressors will be arranged in series as is shown in Figure 30. Here, the bulk of the evaporation is done in a single effect MVR system having a centrifugal compressor. Some of the compressor discharge vapor then is further compressed in a blower in order to operate a finishing evaporator.

Estimating Compressor Power Requirements

To calculate estimated power requirements for an MVR compressor power and ΔT values for both centrifugal compressors and fans have been plotted in Figures 31 and 32, respectively. These values compare reasonably well with installed MVR systems.

Table 5, meanwhile, compares the power requirements for different MVR designs. Note that it is possible to calculate an equivalent steam economy by converting horsepower to Btu/h and multiplying by 2545. This value then is divided by the latent heat to arrive at the equivalent steam used.

Thermodynamics

To fully comprehend the potential of MVR evaporation, it is important that the thermodynamics of this technique be examined. This analysis involves Table 6 as well as the steam and MVR evaporator schematics shown in Figures 28a and 29b.

The first two columns of Table 6 give the properties of saturated water vapor 14.7 psia, 212°F and at 17.2 psia, 220°F. These values can be used to estimate the energy requirements for an evaporator operating at these conditions.

Looking first at the steam evaporator shown in Figure 29a, water at 17.2 psia, 220°F has an enthalpy (heat content) of 188.2 Btu/lb while steam at the same conditions

Table 5. Comparison of Typical MVR Designs (Approximate)[a,b]

	Single Effect Fan	Double Effect Centrifugal	Triple Effect Centrifugal
Compression ratio	1.2	1.6	2.0
Total ΔT available, °F	6.9	18.2	27.3
Vapor to compressor, lb/h	60,000	30,000	20,000
Horsepower/1000 lb/h vapor flow	5.79	17.38	26.57
Total Horsepower	374.4	521.4	531.4
Equivalent Btu	884,133	1,326,963	1,352,413
Equivalent steam, lb/h	870	1315	1347
Equivalent steam economy	69	45.6	44.5
Average ΔT per effect before losses, °F	6.9	9.1	9.1

[a] Boiling temperature 130°F; evaporation rate 60,000 lb/h.
[b] *Note:* In this example, the fan horsepower is lower than either of the centrifugal designs, but the lower ΔT required the greater the surface area.

Table 6. Properties of Water Vapor

Pressure, psia	14.7	17.2	17.2
State	Saturated	Saturated	Saturated
Temperature, °F	212	220	243
H—enthalpy vapor, Btu/lb	1150.5	1153.4	1164.6
Latent Heat, Btu/lb	970.3	965.2	—
H—enthalpy liquid, Btu lb	180.2	188.2	—
S—entropy Btu/lb	1.7568	1.7442	1.7596

has an enthalpy of 1153.4 Btu/lb. The difference between the vapor and liquid is the latent heat or 965.2 Btu/lb. In other words, in order to produce one pound of steam at 220°F, 17.2 psia, from water at 220°F, 17.2 psia, the addition of 965.2 Btu/lb of energy is required.

In the case of the MVR evaporator, however, the vapor at 17.2 psia, 220°F (enthalpy 1153.4 Btu/lb) is produced from vapor at 212°F, 14.7 psia (enthalpy 1150.5 Btu/lb). Theoretically, only 2.9 Btu/lb of energy must be added.

The compressor, a fan in this case, operates such that the entropy of the discharge vapor be at least as high as the entropy of the inlet vapor. Since inefficiencies in the compression cycle will result in an exit entropy above that of the inlet entropy, the temperature of the existing vapor and energy input to the compressor must be increased.

Figure 33 is an enthalpy–entropy diagram for water vapor. The vapor at 14.7 psia, 212°F, is shown at point A. During compressions the entropy remains constant (ideally) or increases (actually). With typical efficiencies for this duty, the discharge temperature may be expected to rise to 243°F where the enthalpy is 1164.6 Btu/lb. The energy input from the compressor is 14.1 Btu/lb.

In order to cool the vapor to its condensing temperature of 220°F, 11.2 Btu/lb of heat is removed. This can be done by introducing condensate at 220°F from the steam side of the evaporator. The 11.2 Btu/lb of heat removed from the vapor is absorbed into the condensate, some of which will vaporize and give off what is know as "excess vapor." For every pound of vapor cooled, 11.2 Btu of heat is absorbed in the condensate, which requires 965.2 Btu/lb to boil. Therefore, for each point of vapor leaving the compressor, 11.2/965.2 or 0.0116 lb of excess vapor is available.

This excess vapor is used in several ways. Since there is a slight difference in latent heat between the steam and the vapor, slightly more steam is required than the vapor generated. Other excess vapor is used to cover losses due to radiation and venting, is made available in some instances for preheating, or is sent to a condenser. It is significant to note that the condenser on an MVR evaporator is responsible only for vent and excess vapors. This results in a much lower cooling requirements than is necessary for steam evaporators.

It is possible to calculate an equivalent steam economy for an MVR system. In this example, for every pound of water evaporated, 970.3 Btu is absorbed. The compressor supplies 14.1 Btu but with motor and gear losses, probably requires 14.5 Btu of energy. The equivalent economy (970.3/14.5) is 67 to 1. Since one horsepower is equivalent to 2545 Btu/h, the compressor in the example requires 14.5/2545 or 0.0057 hp/lb · h of evaporation.

It should be noted that pressure losses through the evaporator that must be absorbed by the compressor have

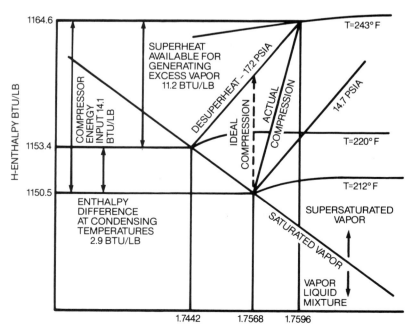

Figure 33. Enthalpy/entropy diagram.

not been considered in this example. These losses would be taken into account by either higher compressor horsepower or lower ΔT over the heat-transfer surface.

BIBLIOGRAPHY

Adapted from:

P. Worrall, "Selection and Design of Evaporators," *CPI Digest* 4(4), 2–10 (Feb. 1977); 5(4), 2–9 (March 1979).

P. Worrall, "Residence Time in Film Evaporation," *CPI Digest* 8(3), 17–19 (Dec. 1986).

P. Worrall, "Plate Type Evaporators," *CPI Digest* 8(4), 12–15 (July 1987).

P. Worrall, "Tubular Evaporators and 3A Sanitary Standards," *APV Evaporator Handbook,* Nov. 1977, pp. 24–26.

P. Worrall, "Advances in MVR Evaporator Technology," *APV Evaporator Handbook,* Nov. 1989, pp. 24–26.

Copyrighted APV Crepaco, Inc. Used with permission

1. HTR1 report, May 1978, p. 7.
2. *Perry's Chemical Engineer's Handbook*, 5th ed., pp. 5–57.

APV CREPACO, INC.
Lake Mills, Wisconsin

EVOLUTION. See HISTORY OF FOODS.
EXOTIC SPECIES. See AQUACULTURE.

EXPERT SYSTEM: CHEESE DEFECTS

DEVELOPMENT OF AN EXPERT SYSTEM

Expert systems are computer programs that use knowledge to solve problems normally solved by a human expert. The body of facts and rules used by expert systems to solve problems is known as the knowledge base. An expert system can use its knowledge base to solve problems that are less structured than those solved with conventional programs. Expert systems are tools used by humans to examine large data bases of knowledge and to arrive at a decision.

With a rule-based expert system, knowledge is not represented by subroutines or other program codes. The control element is separate from the knowledge base. The program acts as an interpreter. This allows the same control structure to be used for more than one knowledge base. Rules can be added or deleted without directly affecting other rules. This degree of flexibility is not available using traditional programming techniques.

Equipment

Early work in artificial intelligence and expert systems required the use of large and expensive hardware. The necessity of dedicating machines to artificial intelligence applications limited their widespread application. As memory capacity of smaller computers increased, expert system tools became available to a wider audience. Large expert system programs such as a medical diagnostic adviser required a mainframe computer for their operation. In order for the many programs designed for personal computers to be used, knowledge domains must be kept small.

Software

Although any programming language can be used for writing expert system programs, LISt Processor (LISP) is the most common. LISP was designed for describing and programming symbol manipulation tasks. Another common symbolic program is Programming in Logic (Prolog), which is also well suited to solving problems that involve objects and relations between logics.

Symbol manipulation programs use symbolic expressions to work with data. If–then rules are common features of expert systems. An example of symbol manipulation in a rule used to identify natural pigments is as follows:

(RULE IDENTIFY
  (IF (PIGMENT IS ORANGE)
    (PIGMENT IS WATER SOLUBLE)
    (PIGMENT IS HEAT LABILE)
    (PIGMENT IS pH SENSITIVE))
  (THEN (PIGMENT IS ANTHOCYANIN)))

In using this rule the program finds the conditions following the if statement, searches for them on a list of assertions, and adds the conclusion to the list.

Expert systems are usually divided into components known as the knowledge base, the inference engine, and the user interface. The inference engine uses the knowledge base to solve the problem. The user interface communicates between the user and the system. The inference engine and the user interface together are known as the shell. Expert system shells have been developed that greatly facilitate development of the knowledge base.

The problem to be solved influences the decision of which tool to use. If a problem requires manipulation of formulas or mathematical calculations, a tool that could easily handle them would be preferred. Specific considerations include the ability to manipulate formulas or mathematical calculations, data base access, and confidence factors.

Identification of the Problem

Expert system tools are not suitable for all problems. A problem needs to have a potential solution. If a trained and experienced specialist in the particular discipline is unable to solve the problem, it is likely that the expert system will also fail. A common recommendation is to determine if the problem can be solved within five minutes to two hours. A problem that can be solved in less than five minutes is too trivial to justify the time and expense of developing the program. Problems that take over two hours to solve are too complex unless they can be segmented into smaller programs.

Assembly of Expertise

A large part of assembling expertise or knowledge is finding order in what appears to be a disorderly decision-making process. Knowledge engineers must determine how experts use their knowledge in solving a problem. The scope of the decision-making process and the outcomes of decisions must be identified. An expert may use several resources, including both facts and assumptions. Confidence factors for various relationships can then be established.

EXAMPLE OF AN EXPERT SYSTEM

The expert system tool selected for this example is LEVEL5 (Information Builders, Inc., New York). LEVEL5 uses a language known as Production Rule Language (PRL). The knowledge base is organized using PRL. Once completed, the knowledge base is compiled before it is run. The compiler translates the programming language into machine language. This has the advantage of requiring less memory with shorter run times.

A number of reserved words and symbols are used in the organization and classification of the content of a PRL knowledge base. The reserved words are recognized by the PRL compiler only when they are capitalized. The PRL program elements must occur in their proper sequence as follows:

1. The title and title display text reference
   TITLE . . . DISPLAY
2. Declaration of facts shared by chained knowledge bases
   SHARED SIMPLEFACE   SHARED NUMERIC
   SHARED ATTRIBUTE   SHARED STRING
3. Data base declarations
   OPEN . . . AS . . . FOR . . . CALLED
4. Fact type declarations
   SIMPLEFACE
   ATTRIBUTE
   NUMERIC
   STRING
5. Parameter initialization statements
   INIT
   REINIT
   FORGET
6. Control element selectors
   FILE   GOAL SELECT
   THRESHOLD   MULTI
   CONFIDENCE   SUPPRESS
   EXHAUSTIVE   UNKNOWN FAIL/
                   CONTINUE/OFF
7. The goals of the knowledge base
   Goal outline
8. The rules of the knowledge base
   RULE . . . IF . . . THEN . . . ELSE
9. Information display text
   DISPLAY
   EXPAND
   TEXT
10. The ending statement
    END

Each group of knowledge base elements must be in the exact location in the PRL text file. However, the order of statements within any single group of knowledge base elements is not important. This feature facilitates modification of the knowledge base. Not all of the groups or reserved words are required in a knowledge base. Those that are most suitable can be selected. The larger portion of the knowledge base is composed of rule statements referred to in statement 8.

Sample Program

A short example will demonstrate how rules are structured. This example will demonstrate one approach to developing a knowledge base using this particular expert system tool. Numerous other useful tools are also available commercially. This example will trace Cheddar cheese defects to their source. Common defects are listed on a grader's score sheet. As a defect is identified, the production and quality personnel need information on the most likely source of the defect. Because many interrelating factors can be involved, a system that considers these factors, along with their relative importance, would be useful. The knowledge base will then be expanded to explain how other features are used. An important requirement is that the conclusions of rules must be identical to the goals.

The following knowledge base contains two rules.

```
!          Short Example Knowledge Base
!          Level Five Research
TITLE CHEESE DEFECT ANALYSIS
!          Set the degree of confidence required for
!          reaching a conclusion to 60
THRESHOLD = 60
!          Define the goals of the knowledge base
1.         The defect is caused by high moisture
2.         The defect is caused by low moisture
!          Specify the rules that support these goals
RULE       High moisture criteria
IF         Weak
OR         Acid
THEN       The defect is caused by high moisture CF 80
RULE       Low moisture criteria
IF         Crumbly
OR         Flat
THEN       The defect is caused by low moisture CF 75
END
```

The exclamation points are editorial operators, or comment lines. The first item that needs to be present is the title. The capitalized word title is a reserved word. Because the title is limited to 60 characters, the display function must be used for additional statements. This same title will be the heading of screen displays and reports.

Threshold. The threshold statement specifies the lowest level of confidence necessary to reach a conclusion or goal. When a user specifies a confidence for a supporting fact in a rule, LEVEL5 compares it with the threshold for the knowledge base. User confidence in each supporting fact must be greater than or equal to the threshold value for the individual rule to continue to be evaluated.

Confidence. With the confidence statement the knowledge engineer can determine whether a user will be asked to give a confidence value for the answer given. This allows for uncertain responses. If confidence is not specified, LEVEL5 will assign a confidence of 100 to a true response and 0 to a false response. Prompts for confidence levels can be established for the entire knowledge base, or for individual facts. The confidence level shown for a fact from the LEVEL5 conclusion display or from the reports system is the product of the user's confidence level times the rule conclusion confidence level.

Goals. When developing an expert system there are two different ways of reasoning. The first approach consists of gathering information and drawing conclusions. It is also referred to as a data-driven method. The second approach begins with a goal or hypothesis. A search is then made for information that supports or contradicts the goal. These methods are referred to as forward chaining and backward chaining, respectively. LEVEL 5 mainly uses a backward-chaining procedure.

A PRL knowledge base must have at least one goal. The goal statement is the major conclusion of the knowledge base. If a hypothesis is being tested, it can be the proof of the hypothesis. All goals are preceded by a goal outline number. A knowledge base may consist of a single goal or multiple goals. Goals can be expanded to include subgoals. The example is continued adding two additional rules:

1. The defect is caused by high moisture.
   1.1 The source of high moisture is the cooking procedure.
   1.2 The source of the high moisture is the Starter quality.
2. The defect is caused by low moisture.
   2.1 The source of low moisture is the cheddaring procedure.
   2.2 The sources of low moisture are the mill characteristics.
3. The defect is caused by poor milk quality.
   3.1 Poor milk quality is caused by improper early acid development.
   3.2 Poor milk quality is caused by inherent off-flavors.
      3.2.1 Off-flavor is caused by poor farm practice.
      3.2.2 Off-flavor is caused by hauling and storage practice.
4. The defect is caused by poor starter culture quality
   4.1 Poor starter quality is caused by slow acid development.
   4.2 Poor starter quality is caused by undesirable fermentation.

Production Rules. Knowledge is represented in the knowledge base by rules. The rules create a cause and effect organization in the knowledge base that attempts to duplicate human reasoning. Rules must have three parts: the rule name, a supporting condition, and a conclusion. The title of the first rule in the example is high-moisture criteria. The supporting conditions of the rule are weak and acid. The conclusion of the rule is the defect is caused by high moisture, and the confidence level for the rule is 80.

For an expert system to be able to prove a goal, that goal must be a conclusion of at least one rule in the knowledge base. For a given goal a subset of rules will have useful knowledge in validating that rule. The job of the expert system is to validate all of the causes specified in the rule. In the process of examining causes as they relate to a goal, the expert system constructs a path consisting of one or more rules. A path is a condition when one rule references another rule. The expert system endeavors to choose the path that provides the greatest accuracy. The measure of accuracy is specified by the confidence levels.

In a knowledge base with a large number of goals and subgoals, the user may have to respond to a large number of questions that may be unrelated to the particular problem at hand. To avoid this, a goal-selection process is possible. The goal-selection option presents the user with the lowest level goals. The user can then select one of the goals as the best path to pursue. The subordinate goals within that goal are then presented to the user. When the user can no longer decide which goal to select, the displayed goals will then be examined by the inference engine. When needed information is required, the user will be questioned.

Expanded Program

The first line of the knowledge base must be the title line. To allow additional text, the DISPLAY function has been used.

> TITLE Cheese Defect Analysis DISPLAY
> CHEESE DEFECT ANALYSIS
> CHEDDAR CHEESE
> This program is designed to receive
> ADSA score card criticisms and by
> conversing with the user trace the
> criticisms to their source.

Elements 2 through 5 are not needed in the sample knowledge base. From heading number 6 the following control elements are selected:

MULTI ALL
THRESHOLD = 60
CONFIDENCE ON

Information in the knowledge base can be represented by various data types. Simple factual statements, numeric data and mathematical expressions, attribute–value associations, and string or character information can all be contained within the LEVEL5 knowledge base. A simple fact is any phrase or statement that can be either true or false. With numeric data comes the capability to assign, compute, and compare values within the knowledge base. The third data type associates a value with a particular attribute. This data type is useful because the attributes can be associated and evaluated with values in other rules in the knowledge base. This increases the efficiency of the knowledge base by reducing the number of required questions. A string fact is any combination of ASCII characters up to 80 characters in length. String data can be passed to external programs or stored in a data base.

Once a particular attribute–value relationship has been verified, LEVEL5 will not pursue other possible relationships that attribute may have. To permit associations with other values, the reserved word MULTI is used. Selected attribute–value facts can be designated individually, or collectively using the ALL after MULTI.

Because a threshold level of 60 is specified, the supporting fact with the lowest user stated confidence will have to be greater than that amount. If it is greater than the threshold, the user confidence will be multiplied by the rule confidence specified for that particular rule. This will be stored by LEVEL 5.

Whether or not the user will be asked to specify a confidence level for his or her response is determined by the CONFIDENCE ON statement. Confidence prompting can be specified either on a global basis as is shown here or on an individual fact basis by using CONFIDENCE ON and CONFIDENCE OFF.

The sample program will be expanded to 10 goals and 10 subgoals listed as follows:

1. High moisture established.
    1.1 High moisture source established.
2. Low moisture established.
    2.1 Low moisture source established.
3. High acid established.
    3.1 High acid source established.
4. Low acid established.
    4.1 Low acid source established.
5. Milk quality established.
    5.1 Milk quality source established.
6. Sanitation established.
    6.1 Sanitation source established.
7. Culture quality established.
    7.1 Culture quality source established.
8. Acid development established.
    8.1 Acid development source established.
9. Press problem established.
    9.1 Press problem source established.
10. Altered surface established.
    10.1 Altered surface source established.

These goals divide the defect sources into general categories. This facilitates development of rules that are related by that particular source. The strategy is to first determine which well-known Cheddar cheese criticisms are associated with which category. For the first goal this is accomplished with the first rule as shown below:

```
RULE    High moisture criteria
IF      Defect IS Open
OR      Defect IS Weak
OR      Defect IS Acid
OR      Defect IS Bitter
OR      Defect IS Unclean
OR      Defect IS Whey
THEN    High moisture established
!
RULE    For determining high moisture source
IF      HM source IS Cook
OR      HM source IS Starter
OR      HM source IS Cut
OR      HM source IS Draw
OR      HM source IS Cheddaring
OR      HM source IS Mill
THEN    High moisture source established
AND     DISPLAY HM sou
AND     DISPLAY continue
!
RULE    Cooking
IF      Cooking IS Too rapid
OR      Cooking IS Too low temperature
OR      Cooking IS Too high temperature
THEN    HM source IS Cook CF 85
!
RULE    Starters
IF      Starter amount IS Too low amount
OR      Starter activity IS Too low activity
THEN    HM source IS Starter CF 85
```

This is an example of the attribute–value fact type. The attribute defect is associated with the various values using the reserved word *is*. If one or more of these high-moisture criterion values is selected, the conclusion is reached and the goal is achieved. The inference engine then moves to goal 1.1 and begins searching for information to achieve that goal. Because the conclusion of the second rule is the same as goal 1.1, that rule is evaluated. The second rule begins looking at specific areas or ingredients in the cheese-manufacturing process, such as the cooking procedure, the cutting procedure, or the addition and quality of the starter. All of the subsequent rules, up until the low moisture heading, list practices by which a high moisture defect could occur. The rule for cooking contains the variables too rapid, too low temperature, and too high temperature. Based on the users response, these variables could become associated with the attribute cooking. If one or more of these variables is chosen by the user, the conclusion HM source is cook is reached. This fact will allow the second rule to reach its conclusion high moisture source established. The reserved word DISPLAY HM sou(rce) follows the conclusion. The text display with this title will then be displayed. In this case the text display contains bar graphs corresponding to the defect sources that have been established. The values of the bars correspond to the confidence figures previously described.

DISPLAY HM sou

High moisture may be the source of your problem because of:

Low amount of starter or activity

BAR HM source IS Starter
Cut too firm or too large
BAR HM source IS Cut
Cook temperature too high or too low
BAR HM source IS Cook
Poor agitation or low acid at draw
BAR HM source is Draw
Blocks too high, too large, not turned enough, or low temperature
BAR HM source IS Cheddaring
Mill acid low
BAR HM source IS Mill

Evaluation

To demonstrate how the program functions, a typical session will be represented as follows. Following the title statement the user is presented with a list of defects (Fig. 1). Any number or combination of defects can be selected. For this example, the grader is examining a sample from a particular vat of cheese that has a weak or soft body and tastes acid or sour. The defects weak and acid are selected. After each selection the question "What is your confidence in this answer?" is shown on the screen, along with a horizontal bar across the bottom of the screen. The bar length adjusts according to the confidence selected. The more predominant defect is the weak body so a confidence of 80 is chosen. The acidic defect is slightly less objectionable than the weak body so it is given a confidence level of 70. A useful function is the EXPL key at the bottom of the screen. Supplemental information that was previously stored in the knowledged base under EXPAND can be displayed here. The information revealed for these two defects are:

EXPAND   Defect IS Weak
A weak soft bodied cheese may be noted by the small amount of pressure needed to break the structure or mash the cheese.

EXPAND   Defect IS Acid
An acid or sour flavor is a quick sensation that disappears leaving the mouth free of any off-flavor.

The next screen requests information about the coagulated milk at cutting (Fig. 2). The decision is made that

Cheese Defect Analysis
Select the defect(s) in appearance, texture, and flavor that you observed.

Open	Curdy	Fermented
Weak	Flat	Feed
Bitter	Acid cut	Rancid
Unclean	Mealy	Weed
Whey	Short	Seamy
Corky	Mottled	
Crumbly	Gassy	

After making your selections press F4 for DONE

Figure 1. Screen showing list of observable defects.

Cheese Defect Analysis
Select the characteristics of the curd at cutting.

Too firm
Too large
Too small
Too soft

After making your selections press F4 for DONE

Figure 2. Screen showing defects at curd cutting.

Cheese Defect Analysis
Select the characteristics that describe the cooking procedure.

Too rapid
Too low temperature
Too high temperature

After making your selections press F4 for DONE

Figure 3. Screen showing list of cooking errors.

none of these factors is a problem, and the next screen is displayed.

Three variables in the cooking process are displayed on the next screen (Fig. 3). Because the user is certain that no errors were made in the cooking procedure, no selection is made. The next screen addresses the amount of starter added (Fig. 4). This possibility is dismissed and the user goes to the next screen.

The next screen also deals with the starter culture but this time with the activity of the culture (Fig. 5). This is defined as the culture's ability to produce lactic acid during the cheese-making process. The user decides that because the culture was several days old, its activity may

Cheese Defect Analysis
Select the phrase that describes the amount of starter culture.

Too high amount
Too low amount

After making your selections press F4 for DONE

Figure 4. Screen showing list of errors in addition of starter culture.

Cheese Defect Analysis
Select the phrase that describes the activity of the starter culture.

Too low activity
Too high activity

What is your confidence in this answer?
                                                    65

After making your selections press F4 for DONE

Figure 5. Screen showing list of errors in starter culture activity.

Cheese Defect Analysis
Select the characteristics of the curd at milling.

Too low acid
Too high acid
Excess stir

What is your confidence in this answer?
95

After making your selections press F4 for DONE

Figure 6. Screen showing list of errors during milling.

have declined. Too low activity is selected and the confidence is set at 65, expressing more uncertainty in this decision than in the decisions made for the first screen. When questioned about the process at the time the curd is milled into small pieces, the user refers to the manufacturing record for information on the acidity at milling (Fig. 6). Alternatively, this information could have been stored in a data base and obtained by the program directly. In this case the user is very confident and selects "too low acid" with a confidence of 95.

The next screen lists several variables at the time of drawing or draining the whey from the cooked curd (Fig. 7). The acid was found to be only slightly slow in developing and a confidence level of 60 is selected. The user decides not to select any of the cheddaring factors (Fig. 8).

At this time the system has sufficient information to make a recommendation (Fig. 9). The general conclusion is that high moisture may be the main source of the weak and acid defects. More specific problem areas of starter, draw acid, and mill acid are then given. The most likely problem area is the acid level at the time of milling. The recommendation can now be made to increase the milling acid. One possibility is to lengthen the cheddaring time to develop more acid. Based on the other conclusions, the problem had already begun by the draw step. Because low starter activity is also listed as a potential source, improving the starter activity may serve to increase the mill acid.

A useful feature at this point is the generation of reports explaining how the inference engine was operating

Cheese Defect Analysis
Select the characteristics of the curd at draw.

Too low acid
Too high acid
Poor agitation
Excess agitation
Too firm
Too soft

What is your confidence in this answer?
60

After making your selections press F4 for DONE

Figure 7. Screen showing list of errors at draw.

Cheese Defect Analysis
Select the procedure during cheddaring.

Blocks piled too high
Blocks not piled high enough
Blocks too large
Blocks too small
Not turning blocks often enough
Too low temperature
Too high temperature

After making your selections press F4 for DONE

Figure 8. Screen showing list of errors during cheddaring.

as it evaluated the knowledge base and user responses. A line of reasoning report is shown as follows:

The following goal was pursued:
  High moisture established.
The following Attribute–Value fact was obtained:
  Defect
  is(are) Weak = True
The following Attribute–Value fact was obtained:
  Defect
  is(are) Acid = True
Rule: High moisture criteria fired.
As a result the following conclusion was reached:
  High moisture established CF = 80.
The following goal was pursued:
  High moisture source established.
The following Attribute–Value fact was obtained:
  Starter activity
  is(are) Too low activity = True

Cheese Defect Analysis
High moisture may be the source of your problem because of:

Low amount of starter or activity
55

Cut too firm or too large
0

Cook temperature too high or too low
0

Poor agitation or low acid at draw
45

Blocks too high, too large, not turned enough, or low temperature at cheddar
0

Mill acid low
76

Figure 9. Screen showing system's analysis of causes of cheese defects.

Rule: Starter fired.
As a result the following conclusion was reached:
  HM source
    is(are) Starter = False
The following Attribute–Value fact was obtained:
  Mill characteristics
    is(are) Too low acid = True
Rule: Milling fired.
As a result the following conclusion was reached:
  HM Source
    is(are) Mill = True
The following Attribute–Value fact was obtained:
  Draw characteristics
    is(are) Too low acid = True
Rule: Drawing fired.
As a result the following conclusion was reached:
  HM source
    is(are) Draw = False
Rule: For determining high moisture source fired.
As a result the following conclusion was reached:
  High moisture source established CF = 76

This report shows the order in which the rules are evaluated. The first goal to be reached is high moisture established. The subgoal, high moisture source established, is then pursued. The conclusion high moisture (HM) source is false because the product of the user confidence of 65 percent or 0.65 and the rule confidence of 0.85 is 0.55. This value, although expressed in the final display, is below the threshold value of 0.60 or 60 percent. The conclusion HM source is(are) Draw is also false for the same reason. Obtaining the fact that mill characteristics is(are) Too low acid = true leads to the conclusion that HM source is(are) Mill. This conclusion is acceptable because the product of the rule and user confidences is greater than 60 percent. The conclusion high moisture source established is finally reached. This satisfies the subgoal 1.1.

Numerous approaches to a problem of this type are possible. Goal statements can be consolidated into more general categories or expanded into more specific goals and subgoals. Rules can be structured in various ways resulting in more specific recommendations. Laboratory analytical results may be useful and can be accessed directly by the system. The structure of the input required from the user can be altered and made more useful to an untrained cheese grader. A major advantage of using an expert system program for problem solving is the independent nature of the rules. Modifications can more easily be made compared to a conventional program. Because the knowledge base is separate from the control mechanism, numerous knowledge bases in diverse areas can be used with the same system.

ROBERT OLSEN
Schreiber Foods, Inc.
Green Bay, Wisconsin

EXTRACTION

Food processing includes a wide variety of unit operations that are designed in a sequence to change a raw material to a finished food product. This can range from simple washing and packaging to extensive mass and heat transfer procedures that are necessary to make a final product fit and wholesome for human consumption. The separation or removal of components from a raw food material is involved in almost every processing procedure. For example, the simple washing of a vegetable is necessary to ensure that contamination from soil, fertilizers, living organisms, pesticides, etc are removed or at least reduced to an acceptable level, making the vegetable safe to eat. Flour is put through a screening process to separate the correct particle size and recycle the particles requiring additional milling. A much more complex procedure for removing or separating a component is found in the process for producing vegetable oils. Corn oil is treated with a solvent to extract the oil from the ground, pulverized corn kernel.

These procedures of removing or separating a component during food processing have been divided into numerous categories to clarify their classification. This greatly simplifies the scientific and engineering studies of various unit operations and the subsequent engineering design, manufacture, and integration of the machinery into an overall food process. Hence the unit operations of separation, distillation, evaporation, dehydration, and filtration all involve the extraction of a component from a liquid, gas, or solid by physical or chemical means. However, each category listed has such different applications of scientific and engineering principles that it is too unwieldy to place them all under a combined unit operation. For this reason, the unit operation of extraction has been considered to be the removal or separation of a component in a food system by contact with another material or phase that has greater affinity for the component being removed. Extraction processes can be carried out on a batch basis or as continuous steady-state operation. As is the case with most unit operations, the continuous operation is the easiest to analyze and design. This is certainly reflected in the commercial operating cost and efficiency of extraction processes used in the food industry (1,2).

The basis for success of extraction processes is the difference in affinity for one component or material over another. For example, water has little affinity for vegetable oil but the oil is completely soluble in an organic solvent. Hence, when an oil-containing food raw material, such as soybeans or corn, is placed in contact with the solvent, oil will be absorbed into the solvent and removed from the food. However, when the food is placed in contact with water, the water will not absorb any oil. Oil will only be released if the food is in hot water, such as during cooking, when the structure of the food is changed so that oil is released. In this case the oil is not extracted by water but the water acts as a vehicle for washing out the oil and floating it to the surface.

There are two primary unit operations to consider under extraction. The first is a solid–liquid extraction in which a soluble component is removed from a solid by a liquid. This process is known as leaching. The second involves the removal of a component in a liquid by another liquid and is called liquid–liquid extraction.

There are three special cases of extraction whereby the solvents are in a different state or are changing state during the extraction process. Steam is used for stripping volatile materials from both solids and liquids and thus actually is the solvent in a solid–vapor or liquid–vapor extraction. Some processes, such as the extraction of soybeans with hexane, introduce the solvent as a vapor and remove the extracted oil solute with the condensed solvent. Finally, a new extraction technique has been developed in which the solvent, a gas under normal temperatures and pressures, is converted to a supercritical liquid by greatly increasing the temperature and pressure until critical temperature and pressure are surpassed. Carbon dioxide in this supercritical state is being used as a solvent for a wide variety of solid–liquid and liquid–liquid extraction of food materials.

PRINCIPLES OF EXTRACTION

The extraction rate at which a component (solute) is transferred from the material (solid or liquid) phase being treated to a solvent phase is the controlling factor involved in extraction. This is related to the mass-transfer coefficient ($K_L A$) as determined by conditions of the process and properties of the raw material, solute, and solvent:

1. The properties of the solvent as related to its affinity for the solute.
2. The immiscibility of the raw material with the solvent.
3. The particle size of the solid and thus the exposed surface area and depth of solvent penetration required (for leaching only).
4. Temperature of the liquids.
5. Degree of contact between liquids, which is a function of the agitation or equipment design.

The driving force is the difference between the saturation concentration of the solute in the raw material and the solvent ($C_s - C$). Hence, the rate of extraction (amount per unit time, dN/dt) would be

$$dN/dt = K_L A (C_s - C) \qquad (1)$$

Extraction involves three separate operations: (1) mixing the raw material and the solvent to bring them into intimate contact, (2) separating the two phases following contact, and (3) removing the solute from the solvent so that the solvent can be recycled. Often the economic viability of a process depends on the third step of removing the solute and reclaiming the solvent. In many cases the solvent is relatively expensive (eg, organic solvent) and good separation and recycling is imperative. When the solvent in leaching is water and because the solute is often a waste material, the solvent would not be recycled.

LIQUID–SOLID EXTRACTION (LEACHING)

As a solvent comes in contact with the solid material, the solvent becomes richer in the solute that is being removed from the solid. The rate at which the solvent becomes richer in the solute depends on the factors in equation 1: contact time, mass transfer coefficient for the specific system, area of the solid exposed to the solvent, and concentration driving force. Table 1 shows typical commercial processes in which a food is leached by a solvent to remove a specific solute. Examples of different food-leaching processes are shown in Table 1. Note that the product or products can be the removed solute (eg, leaching sugar beets, vegetable oil, etc), the solid material remaining after extraction (eg, decaffeinated coffee, cleaned vegetables, etc), or both the remaining solid and solute (eg, fish oil and protein concentrate).

Table 1. Commercial Leaching Operations

Raw Material	Solvent	Solute	Product
Sugar beets	Water	Sugar	Refined sugar
Corn, rapeseed, cottonseed, soybeans	Organic solvent, nonpolar (hexane)	Oil	Edible oil (vegetable)
Vanilla beans	Supercritical carbon dioxide	Vanilla	Vanilla extract
Coffee beans	Supercritical carbon dioxide	Caffeine	Decaffeinated coffee
Roasted coffee beans	Water	Insoluble fraction	Soluble coffee constituents
Vegetables	Steam	Debris	Cleaned vegetables
Fish and meat flesh	Nonpolar solvent (dichloroethane)	Oil	Refined fish oil
	Polar solvent (isopropyl alcohol)	Water–oil	Oil, protein concentrate
Fish and meat flesh, enzyme hydrolysis	Water plus enzymes	Proteins	Water-soluble proteins

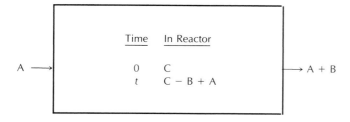

A = Incoming pure solvent
A = Outgoing solvent rich in solute (B)
B = Solute leached from solid (C)
C = Original solid rich in solute (B)
C = Extracted solid depleted of solute with retained solvent (A)
t = Extraction time

Figure 1. Batch solid–liquid extraction (leaching).

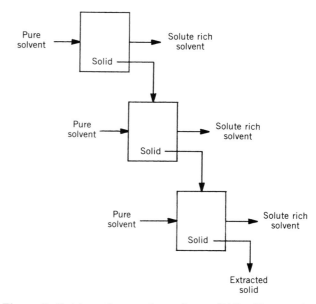

Figure 2. Batch-continuous three phase solid liquid extraction.

A diagram of a semicontinuous solid–liquid extraction system is shown in Figure 1. The pure solvent enters the extractor and emerges rich in solute while the solid that has been placed in the extractor decreases in solute concentration. The rate of reaction decreases during the process as the solid is depleted in solute. Because fresh solvent is passing over the solid, the extraction can continue until essentially all solute is removed from the food. If the process is made continuous by conveying the solid into and out of the reactor, the extraction normally will not be complete because the solid–solvent exposure time is reduced. Throughput can be increased and extraction can be more complete when two or more extractors are used in series. Usually, a series of batch-continuous reactors are connected and fresh solvent is used for each reactor and the solid is moved down the reactor series. Figure 2 shows this multiextraction process in which three reactors are in series. If the remaining solid is one of the end products, it is necessary to remove the residual solvent. Often this secondary process turns out to be another extraction process such as steam stripping. If the solvent is water, the normal process is to remove it by a drying operation.

Leaching is complicated by a process whereby components of a food material are concurrently extracted while they are being made soluble by microbial and chemical action. A good example of this type of process is the batch hydrolysis of minced fish flesh by enzyme action, which changes water-insoluble components into soluble (digested) materials. The resultant water-based protein solution can be dried to produce a high-quality protein product for both animal and human consumption. This technique is a good means for using the tremendous amounts of fish waste from processing, industrial fish, by-catch, and underutilized species (3).

LIQUID–LIQUID EXTRACTION

Liquid–liquid extraction is essentially identical to the leaching process except that the food, solvent, and solute are all in the liquid state. Most commercial processes are normally operated on a continuous basis (Table 2). The liquid food product, rich in the solute to be extracted, enters the reactor countercurrent to the incoming solvent. The outgoing solvent is rich in the solute and the outgoing food is depleted in solute. As shown in Figure 3, the ordinary extraction involves countercurrent flow to maintain the maximum driving force. However, the relationship between the liquids sometimes makes a cocurrent operation more advisable.

The amount of solvent remaining in the food stream is dependent on the partial miscibility of the two liquids. If the retained solvent must be reduced, it is removed by

Table 2. Commercial Liquid–Liquid Extraction Operations

Raw Material	Solvent	Solute	Product
Vegetable oil (cottonseed)	Organic solvent (propane)	Fatty acid (oleic)	Refined vegetable oil
Fish oil	Hot water	Low molecular odor and taste components	Refined fish oil
	Supercritical carbon dioxide	Cholesterol	Low-cholesterol, high omega-3 refined fish oil
Milkfat	Supercritical carbon dioxide	Cholesterol	Low-cholesterol milk

Figure 3. Continuous liquid–liquid extraction.

A = Incoming food product rich in solute (S)
A = Outgoing food depleted of solute (S)
B = Incoming pure solvent
B = Outgoing solvent rich in solute (S)
S = Solute originally in food

steam stripping, evaporation, distillation, adsorption, or chemical means. Likewise, if the solute is to be recovered from the solvent or the solvent is to be recycled, these techniques must be used to the purify the solvent. Distillation is the most common means of solvent recovery.

SPECIAL EXTRACTION OPERATIONS

Processes have been developed for extraction using supercritical fluids as the solvent. Carbon dioxide has been chosen as the supercritical fluid because it is nontoxic, less expensive than many organic solvents, leaves no residue of chemicals, and has a molecular weight and reaction behavior that is desirable. The extraction process using CO_2 is carried out at approximately 35–40°C and several thousand pounds per square inch (psi).

Coffee and tea decaffeination were the first commercial processes carried out with CO_2 in a supercritical state. Other products being extracted by this process include hops and high-quality flavorings such as vanilla, ginger, paprika, rosemary, sage, and celery. Spices from supercritical extraction have been found to have superior antioxidant properties to some of the synthetic products in current use (4).

With the growing concern over cholesterol content of fats and oils, there is considerable interest in removing this fraction by supercritical CO_2. Current research has demonstrated that more than of 95% of the cholesterol can be removed from milkfat and butterfat by operating at pressures up to 34.475×10^6 Pa and 40°C. Many university research groups are examining supercritical extraction, and the pace of research seems to be accelerating as the consuming public indicates its willingness to pay premium prices for the higher quality and healthier foods (5).

Considerable research has been conducted on the extraction of highly unsaturated omega-3 fatty acids (HUFA) from marine oils. These fatty acids contain 20 and 22 carbon atoms in the chain and have been found to be beneficial in preventing or minimizing certain cardiovascular diseases. Most research with omega-3 fatty acids has concentrated on obtaining HUFA from fish oils. However, HUFA is not manufactured by fish but comes from their food. Although marine and freshwater plants produce an oil that is high in omega-3 fatty acids, land plants have no HUFA but produce shorter omega-3 fatty acids and omega-6 fatty acids. Small sea life eat algae; thus the ingested oil works its way up the food chain to the larger fish used for human consumption. An exciting new research program has shown that algae extracted with supercritical CO_2 might be a better source of HUFA concentrates for preparation of high-quality foods. It has been shown that clear golden yellow extracts containing up to 25% HUFA can be obtained by this process (6).

Extraction processes are important factors in realizing added value in food products. Many improvements in flavor, color, and nutritional quality are foreseen through future applications.

BIBLIOGRAPHY

1. D. R. Heldman, *Food Process Engineering,* AVI Publishing Co., Inc., Westport, Conn., 1977, pp. 325–337.
2. G. M. Pigott, G. O. Bucove, and J. G. Ostrander, "Engineering a Plant for Enzymatic Production of Supplemental Fish Protein," *Journal of Food Processing and Pres.,* **2,** 33–54 (1978).
3. D. D. Duxbury, "High-Quality Flavor and Color Extracts Derived by Supercritical Extraction," *Food Processing* **50,** 50–54 (1989).
4. R. M. Sperber, "New Technologies for Cholesterol Reduction," *Food Processing* **50**(11), 154–160 (1989).
5. J. P. Polak, F. Balaban, A. Petlow, and A. J. Phlips, "Supercritical CO_2 Extraction of Lipids from Algae," in K. P. Johnston and J. M. L. Panninger, eds., *Superflow Critical Fluid Science and Technology,* American Chemical Society, Washington, D.C., 1989, Chapter 28.

GEORGE M. PIGOTT
University of Washington
Seattle, Washington

EXTRACTIVE DISTILLATION. See DISTILLATION: TECHNOLOGY AND ENGINEERING.

EXTRUSION AND EXTRUSION COOKING

EXTRUSION

Definition and History of Usage

Extrusion is a process that combines several unit operations, including mixing, kneading, shearing, heating, cooling, shaping, and forming. It involves compressing and working a material to form a semisolid mass under a variety of controlled conditions and then forcing it to pass through a restricted opening such as a shaped hole or slot at a predetermined rate (1). The first known record of its usage was in 1797 when Joseph Bramah used a piston-driven device to manufacture seamless lead pipes (2). After a few years, this innovation was adapted by the food industry for the production of macaroni. The first twin-screw extruder was developed in 1869 by Follows and Bates for sausage manufacture (2). A single-screw extruder, with shallow flights and rotating in a stationary cylindrical barrel, was developed for the Phoenix Gummiwerke A.G. in 1873 (3).

In the mid-1930s, forming extruders were used to mix semolina flour and water and to form pasta products. A few years later, extrusion technology, which combines transport, mixing, and shaping operations, was used to produce the first ready-to-eat breakfast cereal from precooked, oat-flour dough (4). In the late 1930s, Robert Colombo and Carlo Pasquetti developed a corotating and intermeshing twin-screw extruder for mixing cellulose without the use of a solvent (5). In the mid- to late 1940s, the first extrusion cooked, expanded food products, corn snacks, were commercially produced using single-screw extruders (6). Using single-screw extruders for dry expanded pet food, dry expanded ready-to-eat breakfast cereals, and textured vegetable protein was introduced in the 1950s, 1960s, and 1970s, respectively (4,6). In the early 1970s, Creusot-Loire started to develop their twin-screw plastic extruder for food applications. The 1980s have brought rapid commercialization of the production of feeds for aquatic species using extrusion (6).

Food Applications

Many different food products have been produced directly or indirectly by extrusion. Applications of extrusion have been classified into two categories: semifinished products and finished products (7). For semifinished products, extrusion cooking offers economic advantages over traditional processes, such as drum drying, for producing pregelatinized cereal flours, potato starch, and cereal starches. By controlling process conditions to achieve the desired balance of gelatinization and molecular degradation, starches and chemically modified starches with a wide range of cold-water solubility values can be produced (8). Extensive patent literature has evolved on extruded breakfast cereals (9), snack foods (10), and textured food (11–13). Other applications include dry and soft-moist pet foods, precooked and modified starches, flat bread, breadings, croutons, full-fat soy flour, precooked noodles, beverage bases, soup and gravy bases, and confections such as licorice, fruit gums, and chocolate (4,14).

General Description of Equipment

Extrusion equipment can be classified thermodynamically or by pressure development in the extruder (15). According to the former classification, there are (1) autogenous extruders, which convert mechanical energy into heat energy in the flow process; (2) isothermal extruders, where constant temperature is maintained throughout the extruder; and (3) polytropic extruders, which operate between the extreme conditions of (1) and (2). If classified by the manner in which pressure is developed in the extruder, there are (1) direct- or positive-displacement types, which include the ram- or piston-type extruders and the intermeshing counter-rotating twin-screw extruders; and (2) indirect or viscous drag types, which include roller extruders, single-screw extruders, intermeshing corotating twin-screw extruders, and nonintermeshing multiple-screw extruders.

Extruder components have been well described (4,16). They consist of drive, feed assembly, screw, barrel, die-head assembly, cutters, and take-away systems. Figure 1 shows a twin-screw extruder system manufactured by Werner & Pfleiderer with various components. The drive motors are used to rotate the extruder screw in the barrel.

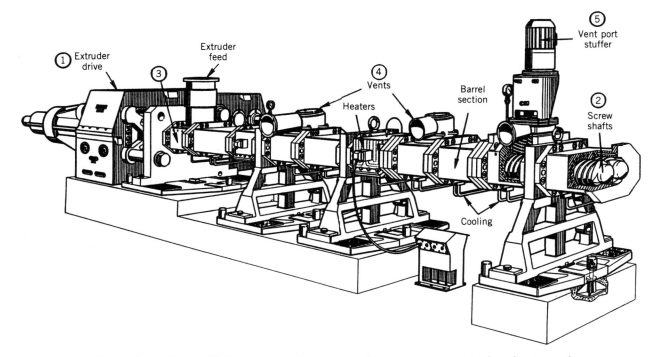

Figure 1. A Werner-Pfleiderer co-rotating, intermeshing twin screw extruder. Courtesy of Werner & Pfleiderer, Stuttgart, FRG.

Either A-C or D-C motors can be used. Depending on the capacity of extruders, the size of the motors can be as small as a few horsepowers for a laboratory extruder or several hundred horsepower for an extruder used for full-scale production of extruded foods. Except for some simple extruders, such as the collet type, most extruder drive motors have a transmission and a speed variation device for controlling the screw rotational speeds. In addition, a thrust bearing is required to support and center the screw and absorb the thrust exerted by the screw, as the screw pushes food forward against a back pressure at the die that can be as high as hundreds of atmospheric pressure.

The feed assembly consists of dry ingredient and liquid ingredient feeders. Preblended dry ingredients are held in hoppers or bins above feeders that can be vibratory, variable-speed auger, or loss-in-weight type. Liquid ingredients are metered through positive displacement pumps, variable orifice, variable head, or water wheels. If a preconditioner is used, dry ingredients are mixed with water, steam, or other ingredients in a closed vessel that can be operated under pressure if needed. The consistent and uniform feeding of dry and liquid ingredients is imperative for the consistent operation of an extruder.

The screw can be one piece or can consist of many individual screw elements assembled in a screw shaft. It is usually divided into feed, transition and metering sections for single-screw extruders. The feed section has deep flights to accept the food ingredients from the feeder. In the transition section, the food ingredients are mixed, heated, and worked into a continuous mass. Thus, the transition section is also called the compression section because the materials are changed from a loose, powdery state to a plasticized dough. This is accomplished by a gradual decrease in the flight depth or in the pitch in the direction of discharge and heating from the barrel and working of the feed material to generate the frictional heat. The metering section has very shallow flights, which increase the shear rate; therefore, the temperature of material increases rapidly in this section.

The twin-screw extruders are classified by the direction of screw rotation—corotating or counter-rotating—and by the depth of screw engagement—fully intermeshing, partially intermeshing, and not intermeshing (16,17). A schematic diagram of fully intermeshing counter- and corotating extruder screws is shown in Figure 2. For fully intermeshing twin-screw extruders, the flight depth is always constant and is equal to the distance of the screw overlap. The screw design consists of single-, twin- and triple-lead types, which refers to the number of flights in parallel along the length of the screw. To improve mixing and increase the conversion of mechanical energy into heat, kneading discs, another type of screw design, are usually used.

The barrel fits tightly around the rotating screw. It is fabricated by bolting or clamping several segments together. An important parameter for the specification of extruder is its L/D ratio, which is the extruder barrel length divided by the extruder barrel bore diameter. The barrels and their liners, in particular, are usually constructed of special hardened alloy to become wear resistant. The interior surface of the barrel is either smooth or grooved. The presence of grooves increases the pumping capability of the extruder.

The die head assembly is located at the end of the extruder barrel. It holds the extruder die plate and sometimes serves as the support for the cutter. The die plate can hold many dies that shape the food product before it emerges from the extruder. The dies are small openings that can be round, annular, slit, or of specially designed shapes, such as alphabets or animals.

EXTRUSION COOKING

Extrusion cooking generally refers to the combination of heating food products in an extruder to create a cooked and shaped product. Raw materials, such as flours, starches, proteins, salt, sugar, and other minor ingredients, are mixed, kneaded, cooked, and worked into a plasticlike dough mass. Heat is applied directly by steam injection or indirectly through a heated barrel or the conversion of mechanical energy. The final process temperature in the cooking extruder can be as high as 200°C, but the residence time is relatively short: 10–60 s (18). Thus, extrusion cooking is also called a high-temperature, short-time (HTST) process.

The first cooking extruder was developed in the late 1940s. This led to a great expansion of the application of extruders in the food field (4). Cooking extruders have many different sizes and shapes and allow a wide range of moisture contents (10–40%), feed ingredients, cooking temperatures, and residence times. In addition, the ability for the extruders to vary the screw, barrel, and die configurations for the cooking and shaping requirements of dif-

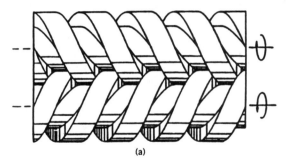
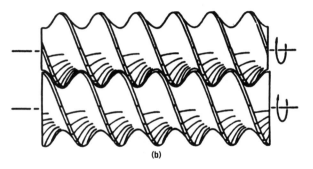

Figure 2. Schematic diagram of fully intermeshing counter- and co-rotating extruder screws: (**a**) counter-rotating screws and (**b**) co-rotating screws.

ferent food products makes extrusion cooking one of the most versatile food processes.

Changes in Food after Extrusion

Because extrusion cooking is a high-temperature, short-time process, it changes the physical and chemical properties of food. Many methods have been developed and applied to characterize extruded products. Expansion of extrudates in both longitudinal and lateral directions can be evaluated by direct physical measurement. Apparent bulk density is determined by weighing the extrudates held in a container with fixed volume. Bulk density or specific volume can be obtained by a sand or rapeseed displacement method. Scanning electron micrographs have been used to examine the extrudate's internal porous structure. Crispness, crunchiness, hardness, and other texture attributes can be determined objectively with Instron or subjectively with human sensory evaluation methods. Color of extrudates can be assessed with a colorimeter. Flavor compounds formed can be determined with gas chromatography-mass spectrometry.

Many analytical tests have been developed to evaluate extruded products. For cereal or starchy products, there are the water absorption index (WAI), the water solubility index (WSI), the amylograph tests, and gelatinization. The WAI is the weight of gel obtained per gram of dry product after dispersing the dry product into water, centrifuging, and decanting the supernatant. The WSI is the amount of dried solid recovered by evaporating the supernatant from the WAI test. The amylograph test determines the viscosities of a slurry of the finely ground, extruded product through a preprogrammed heating and cooling cycle.

Gelatinization of starches is one of the most important changes occurring in cereal foods after extrusion. In the raw or native state, starch exists in the form of granules with many different shapes, ranging from round, oval, to irregular and with sizes between 1 and 100 μm, depending on the sources of starch. The gelatinization process can be briefly described as follows. As water and heat are added during extrusion, the starch granules swell and the amylose in the granules begins to diffuse out. Eventually, the starch granules, consisting now mostly of amylopectin, collapse and become a colloidal gel held in a matrix of amylose. For different types of starch, the temperature range at which gelatinization occurs differs according to the sources of starch. Many methods have been developed to determine the degree of starch gelatinization after extrusion. Some commonly used methods are (1) loss of birefringence (crystallinity) in polarized light, (2) X-ray diffraction patterns, (3) differential scanning calorimetry (DSC), and (4) enzyme susceptibility.

For proteinaceous products, two indices are of interest: the protein dispersibility index (PDI) and the nitrogen solubility index (NSI). The latter measures the percentage of total nitrogen in extrudate that is soluble in water; the former determines the amount of protein in the extrudate that is dispersible in water. Both are conducted under controlled conditions of extraction (4). As the amount of heat treatment during extrusion increases, more protein in extrudate will be denatured, which will result in decreases in both PDI and NSI. Similar to cereal products, texture is also an important quality attribute for extruded proteinaceous product. Both sensory and instrumental methods have been developed. The extrudate can also be examined by scanning electron microscopy to reveal the structural integrity of the protein matrix. Changes in proteins during extrusion can be briefly summarized as follows: (1) Physically, extrusion converts protein bodies into a homogeneous matrix; and (2) chemically, the process recombines storage proteins in some way into structured fibers (19).

Other than starch gelatinization and protein denaturation, many other changes may occur during the extrusion process, such as nonenzymatic browning, enzyme inactivation, and destruction of antinutritional factors and microorganisms.

HTST Extrusion Cookers

HTST extrusion cookers can be classified into two main categories: single-screw and twin-screw cooking extruders. The latter is further divided into corotating and counter-rotating twin-screw extruders. Within the same type of cooking extruders, the mechanical features and operating characteristics may differ substantially from one another because there are many manufacturers supplying each type of cooking extrusion equipment. Following is a partial list of manufacturers who supply HTST extrusion cookers:

Single-screw cooking extruders: Anderson, Bonnot, Buss, Egan, Johnson, Mapimpianti (Mapa), Sprout-Waldron, and Wenger.

Co-rotating twin-screw cooking extruders: APV Baker, Buhler, Clextral, Wenger, and Werner-Pfleiderer.

Counter-rotating twin-screw cooking extruders: Brabender, Cincinnati Milacron, and Textruder.

Methodology for HTST Extrusion Cooking

Before start-up of HTST extrusion cooking the extruder and related equipment must be first assembled. All fasteners should be tight, and all parts are aligned. If there is a torque specification for the retaining bolts, it must be closely followed using a torque wrench. After the barrel and screw(s) are assembled, the screw should be turned by hand to ensure that it moves freely. The die plate with die inserts is then bolted to the end of the extruder barrel. The rotating cutter assembly is installed, and the cutting knives are adjusted to provide a cleanly cut product and an extended life for the blades. Water and/or steam lines are then connected to the extruder barrel as required. Condensate in the steam line must be removed. All feeders and regulators should be checked. If a volumetric feeder is used for dry ingredients, it should be calibrated first for each batch of feed materials. A typical Wenger extrusion cooking system is shown in Figure 3.

To minimize the amount of waste produced during start-up, the cooking extruder is usually brought to operating condition and equilibrium as quickly as possible. This can be accomplished by preheating the barrel and die plate to the desirable operation temperature. The water

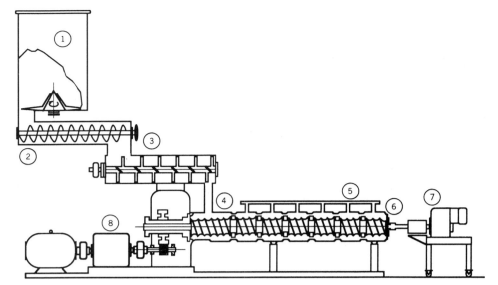

Figure 3. A Wenger extrusion cooking system: 1, bin discharger; 2, feeder; 3, blender/preconditioner; 4, extruder; 5, steam injection ports; 6, die and variable speed cutter; 7, variable speed drive for cutter; and 8, gear drive. Courtesy of Wenger, Sabetha, Kansas.

pump is then started at a rate about 1.5 times the target operating rate. Once water is in the extruder, start the extruder and start the feed stream at a low rate. Bring the screw speed and feed rate up, but reduce the water rate to operating levels gradually. When the barrel and die have reached stable operating temperatures, adjust the water rate, feed rate, and screw speed to obtain the desired product at a moderate torque level (90% or less motor load).

As with start-up, shutdown procedures for HTST cooking extruders must be followed in sequence. Water is gradually increased first to allow uncooked material to purge cooked material from the system. Barrel temperatures are reduced to stop cook reactions. This can prevent burning on after shutdowns. When the percent load meter readings drop to below 25%, slow the screw speed and then shut off the extruder. Also shut off the feeder and water pump. The die plate is then removed by gradually and carefully undoing the bolts as some pressure may still exist behind the die plate.

Applications and Description for Extrusion Cooking of Different Categories of Products

Breakfast Cereals. Four types of breakfast cereals can be produced by the HTST extrusion cooking process: (1) extrusion flaked cereals; (2) extrusion gun-puffed cereals; (3) extrusion shredded cereals; and (4) extrusion expanded cereals (20). Traditional flaked cereals are produced by flaking cooked grits. Extruded flaked cereals differ from them in that the grit for flaking is formed by extruding the mixed ingredients with a feed moisture of 25–35% through die holes and cutting off pellets in the desired size. The pellets are then dried slightly and tempered before flaking. This process allows additional flexibility in the ingredients, such as multigrains or brans, used for the manufacture of flaked breakfast cereals.

Similar to extrusion flaked cereals, the first step of manufacturing extrusion gun-puffed and extrusion shredded cereals is to form pellets with the HTST cooking extruders. The pellets are dried and tempered before gun-puffing or tempered only before shredding. Thus, the pellet moisture content for flaked and gun-puffed cereals is usually lower than that for shredded cereals. After flaking, gun-puffing, or shredding, cereals may be sugar-coated if desired and then dried or toasted in an oven to impart cereals with a desirable toasted color and crisp or crunchy texture.

Extrusion expanded cereals differ from the other three types of extruded cereals in that the cereals are expanded into a porous matrix during the HTST extrusion cooking process. A typical moisture content in feed is about 15–25%, depending on the type of extruder and the extrusion conditions used. The extruded cereals are then sugar-coated if desired and then dried or toasted in an oven as are the other three types of extruded cereals.

Snacks. Many advances have been made over the years since corn collet was commercially developed in the 1940s. For example, extruded snacks now come in many different shapes other than the common collets or curls and in many different flavors other than the basic cheese/oil, and they are produced from potato, wheat, rice flour, or brans, along with starches or modified starches (4).

Collet snacks can be divided into two types, baked or fried, depending on the moisture of the collets (21). Baked corn collets are extruded using a collet extruder at low moisture (< 15%) to produce highly expanded products from degerminated corn grits. After extrusion, the collets are dried (baked) in an oven and enrobed with flavor and oil. Fried collets are extruded from cornmeal, which is moistened to a higher moisture (20–30%). Product expansion takes place during the deep fat frying step instead of during extrusion. The latter partially cooks and forms the product shape.

Other than cornmeal or corn grits, potato- or other grain-based extruded snacks have been developed. Some examples are French-fried potato sticks, flatbread, rice cake, and multigrain snacks. Similar to collet snacks, they can be divided into either baked or fried types. Although a collet extruder can be used in some cases, other snacks, such as crispbread, may require more sophisticated extruders such as HTST single-screw or twin-screw cooking extruders.

Extruded half-products are another type of snack products. Potato and/or other starches and modified starches are extruded to form very dense and thin pellets or collets that are dried at a low temperature for storage stability. The half-products need to be expanded by deep fat frying or baking and then enrobed with flavorings before consumption.

The latest development in extruded snacks involves dual or coextrusion of two different components to form a single piece of snack using a specially designed die. Two extruders may be needed, although occasionally one of the extruders may be substituted by a high-pressure gear pump if the second component is pumpable. This allows the development of many new snacks, eg, dual-textured snack products. The flow rates of the two extruders and the viscosity of the two components must be well-regulated.

Texturized Protein Meat-Like Products. Two types of texturized meat-like products are derived from extrusion processes (4). The first is the meat extender, which can be produced by a single HTST extrusion cooking step using defatted soy flours or soy grits mixed with a variety of additives. The resulting product is highly expanded and shows distinct fiber formation. Once hydrated, the products can be used to extend ground meat or meat products. This product has found extensive applications in pizza toppings, meat sauces, and fabricated food formulations.

The second texturized meat-like product is meat analogues, which can be used to replace meat. The finished product must be dense and have a layered fiber formation similar to that found in real meat. Additional requirements are it should maintain a meat-like character after extensive cooking or retorting and give the appearance and mouthfeel of real meat. The manufacture of meat analogues usually requires multiple extrusion processes or specially cooled dies.

Breadings and Croutons. The manufacture of breadings by the extrusion process has been described (14,22). The starting materials are corn flour, wheat flour, or the second clears of wheat flour. Leavening agents, dry skim milk, salt, emulsifier, and other food additives may be added. Processing conditions during extrusion can be varied to obtained breadings that are functionally similar to traditional breadings. The extrudate is wet milled and sized before drying. Croutons are made in a similar manner. The small cubes are sized at the die of the extruder using a variable-speed knife. Thus, the wet milling stage for the extrudate is not needed.

Pasta. The raw materials used in pasta products is semolina, the purified middlings of durum wheat, or durum flour, water, eggs, and other ingredients (23). Commercially, pasta products are formed by extruder or continuous press with a deep-flighted screw. The screw not only forces the dough through the die, but it also kneads the dough into a homogeneous mass. A very accurate control of semolina flour and water is needed. Traditional pasta products use low-temperature extrusion (< 50°C). If the temperature of the dough exceeds 60°C, the cooking quality of the finished product will be damaged (4,23). Thus, excessive heat generated from mechanical energy input must be removed by circulating cooling water around the extruder barrel.

BIBLIOGRAPHY

1. J. D. Dziezak, "Single- and Twin-Screw Extruders in Food Processing," *Food Technology* **43**(4), 164 (1989).
2. L. P. B. M. Janssen, *Twin Screw Extrusion,* Elsevier Science Publishing, New York, 1978.
3. P. Linko, P. Colonna, and C. Mercier, "High-Temperature, Short-Time Extrusion Cooking," in Y. Pomeranz, ed., *Advances in Cereal Science and Technology,* Vol. 4, AACC, St. Paul, Minn., 1981, pp. 143–235.
4. J. M. Harper, *Extrusion of Foods,* Vol. 1 and 2, CRC Press, Boca Raton, Fla., 1981.
5. F. G. Martelli, *Twin-Screw Extruders,* Van Nostrand Reinhold, New York, 1983.
6. B. W. Hauck and G. R. Huber, "Single Screw vs Twin Screw Extrusion," *Cereal Foods World* **34**(11), 930 (1989).
7. W. Wiedmann and E. Strobel, "Technical and Economic Advantages of Extrusion Cooking," *Technische Mitteilungen Krupp* **2**, 95 (1987).
8. M. G. Fitton, "Extruded Starches—Product Analysis, Structure and Properties," in G. O. Philips, D. J. Wedlock, and P. A. Williams, eds., *Gums and Stabilizers for the Food Industry,* Vol. 3, Elsevier Applied Science, New York, 1985, pp. 213–220.
9. R. Daniels, *Modern Breakfast Cereal Processes,* Noyes Data Corporation, Park Ridge, N.J., 1970.
10. M. H. Gutcho, *Prepared Snack Foods,* Noyes Data Corporation, Park Ridge, N.J., 1973.
11. M. H. Gutcho, *Textured Foods and Allied Products,* Noyes Data Corporation, Park Ridge, N.J., 1973.
12. M. H. Gutcho, *Texture Food Products,* Noyes Data Corporation, Park Ridge, N.J., 1977.
13. M. H. Gutcho, *Textured Protein Foods,* Noyes Data Corporation, Park Ridge, N.J., 1977.
14. O. B. Smith, "Extrusion Cooking of Corn Flours and Starches as Snacks, Breadings, Croutons, Breakfast Cereals, Pastas, Food Thickeners, and Additives," in G. E. Inglett, ed., *Maize: Recent Progress in Chemistry and Technology,* Academic Press, New York, 1982, pp. 193–219.
15. J. L. Rossen and R. C. Miller, "Food Extrusion," *Food Technology* **27**(8), 46 (1973).
16. J. M. Harper, "Food Extruders and Their Applications," in C. Mercier, P. Linko, and J. M. Harper, eds., *Extrusion Cooking,* AACC, St. Paul, Minn., 1989, pp. 1–15.
17. J. Fichtali and F. R. van de Voort, "Fundamental and Practical Aspects of Twin Screw Extrusion," *Cereal Foods World* **34**(11), 921, 1989.
18. E. W. Schuler, "Twin-Screw Extrusion Cooking Systems for Food Processing," *Cereal Foods World* **31**(6), 413, 1986.
19. D. W. Stanley, "Protein Reactions During Extrusion Processing," in C. Mercier, P. Linko, and J. M. Harper, eds., *Extrusion Cooking,* AACC, St. Paul, Minn., 1989, pp. 321–341.
20. R. B. Fast, "Manufacturing Technology of Ready-to-Eat Cereals," in R. B. Fast and E. F. Caldwell, eds., *Breakfast Cereals and How They are Made,* AACC, St. Paul, Minn., 1990, pp. 15–42.

21. S. A. Matz, "Extruding Equipment," in S. A. Matz, *Snack Food Technology,* AVI Publishing Co., Westport, Conn., 1984, pp. 203–230.
22. P. L. Noakes and W. A. Yacu, "Extrusion Cooking of Wheat Flour to Process Breadings," *Cereal Foods World* 33(8), 687 (1988).
23. O. J. Banasik, "Pasta Processing," *Cereal Foods World* 26(4), 166 (1981).

FU-HUNG HSIEH
University of Missouri
Columbia, Missouri

EXTRUSION PROCESSING: TEXTURE AND RHEOLOGY

In recent years extruders have been widely used in the human food and animal feed industry to manufacture products such as textured proteins, snack foods breakfast cereals, pasta, confectioneries, and pet foods. A list of products presently produced using the extrusion process is given in Table 1. The history of food extrusion has been reviewed in Ref. (2).

The two major functions of a food extruder are cooking and forming. In the cooking process, the product is heated through the transfer of heat energy, which is applied by steam injection or heaters and/or by the dissipation of viscous energy through shearing action between the rotating screw and the material. As a result of the cooking process, starches are gelatinized, proteins are denatured, undesirable enzymes are inactivated, and antinutritional substances (eg, trypsin inhibitors in soybeans) are destroyed. Because the temperature reached during the process can be quite high (ca 200°C) and the material takes a relatively short time to travel through the extruder (5 s to a few min), the cooking process is often referred to as high-temperature, short-time (HTST) extrusion. The forming extruder is used to produce special shapes from precooked material. Typically, the operating temperature of forming extruders are lower than that of cooking extruders.

Two major types of extruders are used in the food industry: single-screw and twin-screw extruders. The twin-screw extruders can be further classified according to the type of screws, ie, nonintermeshing or intermeshing and the type of rotation, ie, corotating or counter-rotating. Use of single-screw extruders began in the early years of the industry and they are still used today, though the twin-screw machines are becoming increasingly popular. Twin-screw extruders are comparitively more expensive than single-screw machines of the same production capacity. The flow of material in single-screw extruders depends on the friction between the barrel wall and the material. Hence the type of material that can be used is limited by its properties; proper conditioning of the material is necessary to reduce or eliminate process instability. The corotating, intermeshing twin-screw extruder uses the positive pumping characteristics created by the two intermeshing screws to convey the material forward. This type of mechanism reduces the need to precondition material. Other potential advantages of the corotating twin-screw extruder over the single-screw extruder include (1) narrow residence time distribution, leading to uniform processing, (2) better mixing capabilities, and (3) wide formulations of product, particularly low bulk density and high fat material that cannot be processed using a single-screw machine. The twin-screw extruder provides better control of the process and more uniform product characteristics than does the single-screw extruder. These advantages of the twin-screw machine compensates to some extent for the high initial capital investment required. The advantages and disadvantages of different types of

Table 1. Extrusion Application in the Food and Feed Industry[a]

Human Food		Extrusion Processes	Animal Feed	
Starch-based	Protein-rich	Agrochemicals	Cattle and Pig Feed	Fish Feeds
Pasta (2 types)	Animal protein	Fertilizers	Partial precooked cereals/seeds	Precooked flours (starch and protein) with different bulk densities
Tortilla flours	Sausages	Vegetables/soups	Enzyme-engineered feed	
Bakery products	Scrap meat/fish	Bagasse and other cellulosic material	Complete gelatinized cereals	
Confectioneries	Blood protein			
Candies	Milk protein	Cocoa waste material	Yeast wastes	
Licorices	Germ extrusion	Sucrose-based chemicals	Breaking up of cellulose	
Caramels	Co-extruded meat/starch	Spices	Protein hydrolysates	
Chocolate	Soft cheese	Maltodextrins	Pet foods	
Starches	Vegetable protein	Pulp for paper production		
Partly cooked	Soya/cottonsee/rapeseed	Proteins chemically modified		
Gelatinized	Oilseed meals	Cross-linked starches		
Malt	Health foods	Lactose-based chemicals		
Cereal snacks				
Health foods				
Breakfast cereals				
Flat breads/cookies				
Bran extrusion				
Crumb				
Instant-cooked foods				
Potato products				
Baby foods				

[a] Ref. 1. Courtesy of Elsevier.

extruders have been reviewed in Refs. 2–4. The engineering principles of extruders have also been reviewed (5).

During the HTST process, material typically enters the extruder in a powdered form. The heat, shear, and pressure generated during the process melts the material into a dough-like mass. The material flows from the barrel to a die of a smaller cross section. The pressure at the barrel exit, in the extruder-die setup, can be as high as 3000 psi, depending on the product and process conditions. In the extruder, no water vapor formation occurs at the elevated temperatures attained inside because of high pressures. The material expands as it exits from the die to the atmosphere. This expansion is caused partly by the stresses that are free to relax and partly by the moisture in the material that is flashed-off as vapor, owing to the pressure drop between the die exit and the atmosphere. The expanded product imparts textural and functional characteristics that determine the acceptability of the product. This article reviews the existing literature regarding the textural and rheological properties during extrusion cooking and addresses the relationship between the two.

TEXTURE

The term texture describes a wide range of physical properties of the food product. A product of acceptable texture is usually synonymous with the quality of a product. Table 2 is a glossary of terminology used to describe textural attributes. Textured has been defined as "the attribute of a substance resulting from a combination of physical properties and perceived by senses of touch (including kinaesthesis and mouthfeel), sight and hearing (6)." Texture as defined by the International Organization for Standardization is "all of the rheological and structural (geometric and surface) attributes of a food product perceptible by means of mechanical, tactile and where appropriate, visual and auditory receptors (7)." The following terms have been used to describe product characteristics of extrudates: tender, chewy, soft, tough, brittle, crunchy, crisp, smooth, fine, coarse, puffed, flaky, fibrous, and spongy. One or more of these terms may describe the same behavior.

A variety of properties have been used to quantify or measure the texture and functional properties of extrudates. Some of these are:

1. Bulk density, or density, defined as the mass per unit volume of the extrudate.
2. Breaking strength, defined as the stress required to shear a material. This stress is the ratio of the breaking force to the area of the material. In certain instances it is referred to as the force required to break or shear the sample. The breaking strength is an indirect measure of the stress–strain relationship of the extrudate. It is an important parameter that determines the final quality of the extrudate and can be related to such textural attributes as toughness, chewiness, crunchiness, and brittleness.
3. Viscosity of extrudate paste is measured using amylographs.
4. The extrudate dimension, particularly for snack foods, is another important characteristic. It is often referred to as the expansion ratio or the puff ratio and is defined as the ratio of the area of cross section of the extrudate to the area of cross section of the die. In this article, the swelling of extrudate due to elastic and moisture effects is referred to as extrudate expansion. Extrudate swell is referred to as swelling from all effects other than moisture. Puffing is referred to as the swelling believed to be predominantly from the moisture effect.
5. Water absorption index (WAI) is the weight of sediment formed per unit mass of sample, after the sample is centrifuged and the supernatant is removed. Water absorption capacity (WAC) is the ratio of the weight gained over the original weight after a known mass of powdered extrudate is soaked in water for a fixed period of time at a given temperature. Water solubility index (WSI) is the percentage of the initial sample present in the supernatant obtained from the WAI test.

Table 2. List of Terms and Groups[a]

Structure	Texture	Consistency
Terms Relating to the Behavior of the Material Under Stress or Strain		
Firm	Rubbery	Brittle
Hard	Elastic	Friable
Soft	Plastic	Crumbly
Tough	Sticky	Crunchy
Tender	Adhesive	Crisp
Chewy	Tacky	Thick
Short	Gooey	Thin
Springy	Glutinous	Viscous
	Glutenous	
Terms Relating to the Structure of the Material		
Smooth	Gritty	
Fine	Coarse	
Powdery	Lumpy	
Chalky	Mealy	
Relating to Shape and Arrangement of Structural Elements		
Flaky	Cellular	Glassy
Fibrous	Aerated	Gelatinous
Stringy	Puffed (puffy)	Foamed (foamy)
Pulpy	Crystalline	Spongy
Terms Relating to Mouthfeel Characteristics		
Mouthfeel	Watery	Creamy
Getaway	Juicy	Mushy
Body	Oil	Astringent
Dry	Greasy	Hot
Moist	Waxy	Cold
Wet	Slimy	Cooling

[a] Ref. 6.

The methods used to obtain these properties are not standardized. Because different methods have been used to obtain a property by the same name, procedures used to determine the properties should be reviewed before drawing conclusions.

The textural properties of the extrudate are affected by

the process variables, including temperature, screw speed, screw and barrel dimensions, and product variables (eg, moisture content, composition, and particle size). The properties are evaluated as a function of different parameters, the most common of which includes product moisture, process temperature, and extruder screw speed. Other parameters, some of which may be interrelated, include residence time, die dimensions, screw dimensions, shear rate or shear stress, and product composition. When a number of process conditions are to be evaluated at different levels, the experimental size could be very large. To reduce the experiments to a manageable scale without increasing the experimental error, the response surface methodology (RSM) has been used to develop relationship between dependent and independent variables. According to Meyers (8), "response surface procedures are a collection involving experimental strategy, mathematical methods, and statistical inference which, when combined, enable the experimenter to make an efficient empirical exploration of the system in which he is interested." The RSM helps in finding an approximate function to enable one to predict future response and to determine conditions of optimum response.

In the RSM approach, the entire system is modeled as a black box. The mathematical functions that are developed to correlate the dependent and independent variables are empirical in nature. An advantage of this approach is that a detailed understanding of the changes that occur during the process is not necessary. The disadvantage is that a general relationship is not available, which requires studies for different machines and different formulations of raw material.

Proteins

The basic constituents of all proteins are amino acids. The linear protein chain consists of amino acids linked together by peptide bonds between amino and carboxylic sites. Proteins can have different levels of structure—primary, secondary, and tertiary—that result in a three-dimensional shape. Texturization of proteins can be used to produce two types of products (9). The first are meat analogues that are sold in place of meat. Such products have found acceptance in some countries, such as Japan (10). The second type of textured plant proteins (TPP) is as a meat extender. Several reviews on protein texturization using extruders can be found (9,11–15). During the extrusion process in the presence of water, globular proteins and aluerone granules unravel and align themselves in the presence of the shear field that exists in the extruder (14). Thermal denaturation (loss of native structure) is also expected to occur and is thought to be necessary for texturization (9), although some investigators contend that denaturation is not a necessary step in the texturization process (16).

The mechanism of texturization has been the subject of intense research. It is generally agreed that intermolecular disulfide bonding of proteins is responsible for the structure formation of spun soy fiber (17). It has been reported that the presence of heat and pressure cause the protein molecules in defatted soy meal to disassociate into subunits and become insoluble (18). The mechanism of bond formation has been investigated during extrusion (19). Results indicated that disulfide bonding had no role in extrusion texturization, suggesting that intermolecular amide bonds are instead responsible for texture formation during extrusion processing. The importance of amide bonds in protein texturization was also reported (20).

However, other researchers disagree with these findings. One study reported the formation of intermolecular disulfide bonding when defatted soy meal was extruded at temperatures between 110–150°C and found no evidence of intermolecular peptide bond formation (21). Another study used a single-screw extruder without a die and reported the formation of intermolecular disulfide bond for blends of defatted corn gluten and defatted soy flour extruded at 145°C (22). Others have also refuted the idea that intermolecular peptide bonds are responsible for texturization of extruded soy protein (23). The reason for this disagreement is not clear. The temperatures used by the first investigators (20) were higher than 175°C whereas the latter investigators (21,23) used lower temperatures. It was reported that thermal polymerization of peptide bond formation requires a minimum temperature of 180°C (24). This could be one possible reason for the conflicting results obtained by these researchers.

Most of the work evaluating product quality has used soy proteins. One study extruded defatted soy meal of moisture content 30% (dry basis) at a constant screw speed of 100 rpm with temperatures ranging from 107 to 200°C (25). The product density decreased with temperature (Fig. 1). Shear force and work increased with increasing temperature, but breaking strength increased as temperature increased to 160°C and then decreased with further increase in temperature. Another study used a three-variable, three-level fractional factorial design to measure the effect of extrusion of defatted soy flakes on residual trypsin inhibition activity (TIA), Warner Bratzler shear (WBS), and water absorption capacity (WAC) (26). Lower product moisture decreased residual TIA. Higher temperatures and lower moistures increased

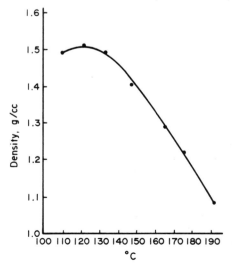

Figure 1. Effect of process temperature on product density. Adapted with permission from Ref. 25.

WAC, but increased feed moisture content decreased WBS.

A third study used both sensory and mechanical means to evaluate the texture of an extruded soybean meal product (27). A five step temperature range of 135–180°C was used in this experiment. A high correlation between the instrumental and sensory parameter was obtained. With increases in temperature, the product became less compact and spongy in appearance and had increased aligned fibers as reflected by cohesiveness. In a subsequent work, these investigators used a four-factor (temperature, screw speed, product moisture, and protein content) three-level RSM to evaluate the textural properties of extruded defatted soybean meal (28). Results from this study indicated that protein level and extrusion temperature were the most important factors affecting WBS values.

The effect of pH (5.5–10.5) of defatted soybean meal and soybean isolate on WBS values (20) indicated that at the extremes of the pH values, the extrudate appeared to have suffered a loss of structural integrity. The maximum shear force and sensory values were obtained at about a pH of 8, as indicated in Figure 2a and b. At pH < 5.5, the extrusion of the product becomes quite difficult, as pH is increased from 5.5 to 8.5 the product becomes tender, less chewy, and rehydrates rapidly (29). These effects have been attributed to the lower solubility of the proteins at acidic pH, which produced a denser product due to the formation of aggregates in the extruder (14).

Independent variables such as moisture content, temperature, and screw speed affect shear strain, stresses, and shear rate. The effect of shear environment on the textural properties of unroasted defatted soy flour was reported (30). Increasing strain enhanced cross-linking in the dough, as indicated by a higher work required to shear the sample. Higher product moisture reduced the density. A high shear rate (or shear stress) in the die achieved by increasing flow rate led to a denser and less absorbant product, caused by the disruption of cross-linking. When low levels of hydrocolloid (sodium alginate or methylcellulose) were added to defatted soy flour, sodium alginate increased chewiness, maximum force, WAC, and bulk density, whereas the addition of the same amount of methylcellulose decreased maximum force and chewiness (31).

A seven-factor, five-level experiment using RSM to optimize the process variables of temperature, screw speed, screw compression ratio, die diameter, and product moisture content during extrusion cooking of defatted soy grits studied the individual and interactive effects of the independent variables (32). Results indicated that high screw compression ratio and maximum temperature decrease between the barrel and the die favored good texture. In an investigation of the effects of product moisture, barrel temperature, and die temperature on the properties of extruded soy isolate and soy flour, it was found that soy isolate required higher pressure to extrude and had higher expansion and a narrower range of textures than did soy flour (33). The peak force measured using a Ottawa texture machine was found to correlate well with protein solubility, as illustrated in Figure 3. The higher barrel and low die temperature resulted in a better texture for both materials.

Although most research has been on soy proteins because of the potential to simulate meat products, some research has evaluated the properties of other high-protein materials after extrusion. These included studies of the effect of extrusion on glandless cottonseed flour and soy meal mixtures (34,35) and on cowpea meal (36) and the effect of process temperature, screw speed, moisture content, and pH on the WAC and nitrogen solubility index

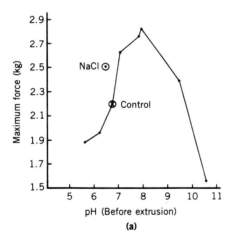

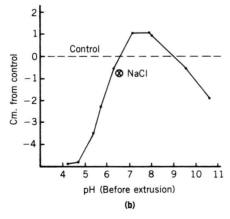

Figure 2. (a) Influence of pH adjustment on Warner-Bratzler shear values of extruded soy protein; (b) influence of pH adjustment on sensory response of extruded soy protein. Adapted with permission from Ref. 20.

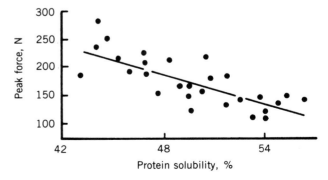

Figure 3. Correlation between OTMS texture and protein solubility for soya isolate. Adapted with permission from Ref. 33.

of cowpea, mung bean, defatted soybean, and air-classified mung bean (37). Protein type and amount were found to affect properties. Results from extrusion of blends of corn and soy proteins have been reported (22,38–40). Increasing the amount of soy protein in the corn–soy mixture led to products with better textural and functional properties.

Starch

Much literature is available on extrusion cooking of starch-based materials. Reference 41 is an excellent review of starch extrusion. Starch consists of α-D-glucose units that are linked to form large macromolecules. Native starch is a mixture of two different glucose polymers—the linear amylose and the branched amylopectin. Under the severe shear environment inside the extruder, the starch molecule loses its granular and crystalline structure (42) and undergoes macromolecular degradation (43–48). The insoluble native starch is partially solubilized after extrusion at room temperature. The magnitude of the transformation is a function of multiple processing parameters and their interactions as well as the type of starches.

The extrudate characteristics of several varieties indicated that wheat starch had the highest expansion ratio at 135 and 225°C (Fig. 4) (49). Increasing product moisture content decreased the expansion ratio (49–54). Increasing temperature increased the expansion ratio up to a point after which it decreased with further increase in temperature (49,55). This decrease can be attributed to the degradation of starch molecules at higher temperatures. Expansion is affected by the dimensions of the die or nozzle (56,57) and product composition. Increasing the length-to-diameter ratio of the nozzle increased the expansion ratio. Increasing the amylose content decreased the expansion ratio (49,53,58). Presence of lipids, proteins, salts, and sugar can also affect the puff ratio (50,54,59–61). Decreasing the lipid content increased the expansion ratio of the product whereas increasing the salt and sugar concentration increased the expansion ratio.

Increasing moisture content increased the breaking strength of extrusion-cooked cornmeal (53). Increasing temperature was found to decrease the breaking strength for nonwaxy corn, but the reverse was true for waxy corn. Increasing the concentration of salt and sugar in the extrudate reduced the breaking strength (54). Results of two studies indicated an inverse relationship between expansion ratio and shear strength (54,55).

Native starches are insoluble in water at room temperature. However, on extrusion these materials are soluble in water and could be added as ingredients in other food mixes to provide an acceptable consistency. The solubility of starches increase with increasing severity of treatment. At lower moisture content, the solubility index is higher (49,62–64). Higher temperature also increases solubility.

A small portion of a considerable body of literature that addresses the effect of process and product variables on properties of extrudates has been cited. Some researchers have used the RSM to locate optimum conditions for the best product texture. Empirical predictive models have been developed to relate product quality to independent parameters. These models or equations suffer from two major drawbacks: (1) they are machine-dependent in that the data obtained from one extruder cannot be extrapolated to other machines, and (2) they do not explain the physical phenomena. Understanding of how the basic physical properties control texture formation is limited.

RHEOLOGY

Although, rheology is defined as the science of deformation and flow of matter, it typically is used to describe the flow of non-Newtonian fluids (65). The flow properties of the fluid can affect important extrusion parameters such as velocity and pressure profiles in the extruder, heat transfer between the walls and the fluid, pressure drop in the die, and energy requirements. These parameters determine the product quality, extruder and die design, and production rate.

During the extrusion process the flow patterns are complex (Fig. 5) and involve six major flow regions (66). These regions are:

1. Metering section, where the material is fully plasticized.
2. The entrance region to the die, where a converging flow field is developed generating large stresses.
3. The region within the die, where the disturbance caused by the entry flow gradually disappears.
4. The steady flow region in the die, where a fully developed flow exists.
5. The exit region, where the vapor bubble nucleation and growth and velocity profile rearrangement occur within the die followed by the expansion of the extrudate outside the die.
6. The free stream region, where the extrudate expansion reaches an equilibrium value.

These flow situations of viscoelastic fluids necessitates the measurement of rheological parameters such as steady shear viscosity, primary and secondary normal stress differences, and elongational viscosity.

Two types of flows occur during extrusion cooking. The

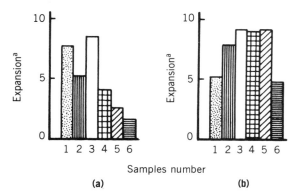

Figure 4. Expansion of products extruded at (a) 135·C and (b) 225·C. Starches are (1) waxy corn; (2) corn; (3) common wheat; (4) rice; (5) amylon 5 (6) amylon 7. Initial moisture content was 22% by weight. Adapted with permission from Ref. 49.

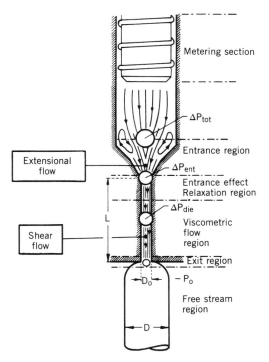

Figure 5. Flow pattern in an extruder barrel and die system. Adapted with permission from Ref. 66.

first is the shear flow that occurs owing to the presence of the walls, as in the screw channel and the die. The second is the extensional flow that is present farther away from the wall where the stream lines converge or diverge, such as at the entrance and exit to the die. The stress (σ) components in shear flow, are given by:

$$\sigma_{12} = \eta_s \dot{\gamma}, \quad \sigma_{23} = \sigma_{13} = 0, \quad \sigma_{11} - \sigma_{22} = N_1, \quad \sigma_{22} - \sigma_{33} = N_2 \quad (1)$$

where 1 is the direction in which the fluid flows, 2 is the direction of the velocity gradient, and 3 is the remaining neutral direction. The shear viscosity (η_s) and the normal stress differences (N_1, N_2) are functions of shear rate. For Newtonian fluids the shear viscosity is a constant and under simple shear flow N_1 and $N_2 = 0$.

For extensional flows, extensional viscosity is defined by:

$$\sigma_{11} - \sigma_{22} = \eta_e \varepsilon \quad (2)$$

Extensional viscosity is constant for a newtonian fluid, but it may be a function of extensional rate for non-Newtonian fluids.

The three types of extensional flows are (1) uniaxial, (2) biaxial, and (3) planar. A cylindrical sample when stretched along its axis results in uniaxial extension. Compression of a cylindrical sample along its axis results in biaxial extension. Stretching a rectangular sheet of a sample along one direction, while keeping one of the remaining dimensions constant, results in planar extension (there are two viscosities in planar extension; see Ref. 67 for details). The ratio of the elongational viscosity (η_e) to the shear viscosity (η_s) is defined as the Trouton ratio. For Newtonian fluid, this ratio is a constant and equal to 3 for uniaxial extension, 4 for planar extension, and 6 for biaxial extension. For non-Newtonian fluids, this ratio is generally much higher and may vary with extensional rate.

The measurement of these rheological parameters during extrusion processing poses special problems. The rotational instruments used in the measurements of rheological parameters of liquids foods cannot be used because the shear rates obtained in these rheometers are several orders of magnitude below those encountered during extrusion and it is difficult to reproduce extrusion-like conditions in the rheometer. Thus on-line measurements of rheological parameters are necessary to avoid these problems.

Shear Viscosity

The rheological property that has received the most attention by scientists is shear viscosity. Some of the published viscosity studies are summarized in Table 3. A die or a viscometer (capillary or slit cross section with multiple pressure transducers located along the wall) attached to the extruder is often used to measure the viscosity. Alternatively, a single pressure transducer mounted near the exit of the barrel, along with capillary dies of different lengths but having the same radius and entry geometry, can be used to obtain the shear viscosity (77).

The first investigators to evaluate the viscosity of doughs during extrusion found that doughs exhibited

Table 3. Summary of Viscosity Studies Conducted During Extrusion-like processes

Product	Method[a]	Moisture, % w.b.	Temperature, °C	Reference
Cooked cereal dough	CDV	25–30	67–100	68
Defatted soy flour	CR	32	35–60	69
Pregelatinized corn flour dough	CD	22–35	90–150	70
Soy fluff	CD	25.5–35.5	100–160	71
Soy flour	CR	25–60	25–120	72
Corn grits	SDV	20–32	100–140	73
Potato grits	SDV	18–42	80–140	73
Corn starch	CDV	20–40	100–170	74
Corn grits	CDV	15–30	130–180	75
Corn grits	SDV	25–45	150–180	76

[a] *Abbreviations:* CD, capillary die attached to the extruder; CR, capillary rheometer; CDV, capillary die viscometer (a capillary die with multiple transducers along its length); SDV, slit die viscometer (a slit die with multiple transducers along its length).

shear thinning behavior (68). The viscosity decreased exponentially with temperature and was found to follow Arrhenius kinetics. Moisture was found to act as a plasticizer. Increased moisture content (M) decreased the dough viscosity. An empirical model of the form below was proposed:

$$\eta_s(\dot{\gamma}, T, M) = \eta^* \dot{\gamma}^{n-1} \exp(\Delta E/RT) \exp(KM) \quad (3)$$

This model is one of the simplest and the most popular. The following model was proposed for constant moisture dough (69):

$$\eta_S(T, \dot{\gamma}, t) = \eta^* \dot{\gamma}^{n-1} \exp(\Delta E/RT) \exp$$

$$\int_0^t k \exp(\Delta E_k/RT(t)) \, dt \quad (4)$$

In equation 4 it is assumed that the temperature is a function of time is known. Reactions such as gelatinization and denaturation can lead to network formation and affect viscosity (parameters k and ΔE_k control the reactions). A log polynomial model was used to express viscosity as a function of moisture content, shear rate, and temperature (71). To account for the shear history in the extruder, an equation was suggested of the form (73):

$$\eta_S(\dot{\gamma}, T, M, N) = \eta^* \dot{\gamma}^{n-1} \exp(\Delta E/RT) \exp(KM) N^{-\alpha} \quad (5)$$

The flow curves obtained using a capillary rheometer (9) and slit-die viscometer were significantly different for food doughs (corn grits and potato flour), whereas for low-density polyethylene the flow curves were the same whether obtained using a capillary rheometer or slit-die viscometer. The flow behavior index was found to be affected by screw speed and temperature. This would indicate that shear history is varying, due to different screw speeds. Hence, the Bagley procedure for obtaining the viscosity during extrusion cooking is invalid because the fluid entering the die is rheologically not the same for each condition. A model has been developed that accounts for the thermal and mechanical energy imparted to the product (shear history) before the viscosity is measured (78).

In addition to these viscosity models, two recent models are worth mentioning. They present almost identical expression for starch and protein doughs based on reaction kinetics (79,80). The models are cumbersome because of the large number of constants (10 or more). One notable feature of these models is the inclusion of yield stress. Another study indicated that soy doughs exhibited yield stress (72). The magnitude of yield stress was found to be a function of temperature.

There are several drawbacks to the existing viscosity models. The model of Harper and co-workers (68) assumes that the decrease in viscosity is due to shear thinning only. Most plastic polymers are modeled by a network of macromolecular entanglements. Shear thinning flow is associated with the decrease of the entanglement density under the influence of deformation of the polymer (81). For food doughs the decrease in viscosity is due to the sum of shear thinning, process history (screw configuration,

residence time), and molecular degradation. A complete model that quantifies the effect of parameters other than product moisture and temperature on shear viscosity is still lacking. These effects are discussed in Ref. 82, though no attempt has been made to quantify them. Another drawback of these models are that the fluid is assumed to be inelastic, ie, normal stress effects are neglected. Because doughs are viscoelastic fluids, inelastic models are inadequate. Evidence has been presented that the pressure drop experienced at the entrance of the die is much greater than that predicted by shear viscosity alone, indicating that elasticity is important (79). As will be discussed later, elasticity is important in the manufacture of products at temperatures below 100°C.

Extruded starches are used as thickening agents or as ingredients in instant foods. The viscosity of powdered extrudate in solvent (water) has also received attention. Typically, shear thinning behavior is observed in all cases. Increasing the concentration of wheat starch from 5 to 9% in solution resulted in the shifting of flow behavior from almost Newtonian to shear thinning (83). For plastic polymers it is known that when the molecular weight is greater than a certain critical molecular weight, non-Newtonian flow behavior is observed (81). Extruded cornstarch in solution exhibited a constant viscosity at high and low shear rates (Fig. 6) (84). Increasing barrel temperature decreased the values of the constant viscosities, whereas increasing moisture content was found to increase the values of constant viscosity, ie, depended on the severity of the extrusion environment.

The effect of emulsifiers on dough viscosity has also been studied (70). The presence of sodium stearoyl-2-lactylate (SSL) was found to affect the viscosity, whereas diacetyltartaric acid ester of monoglyceride did not affect the viscosity. Fat did not affect the viscosity when SSL was present, but it was found to increase the viscosity in the absence of additives.

Normal Stress Difference

It is seen from equation 1 that when N_1 and N_2 are not equal to zero the normal stresses are unequal. The presence of unequal normal stresses can create some interesting phenomena (85). The reason for nonzero values of N_1 and N_2 can be attributed to the anisotropy of fluid microstructure in the flow field and is observed in elastic fluids (86). The ratio of primary normal stress difference to the wall shear stress is an indirect measure of the elasticity of the fluid, ie, the higher the ratio, the more elastic is the fluid.

Measurement of normal stress differences on-line has been a subject of intense research among polymer engineers and rheologists. The two methods used to measure primary normal stress difference are the exit pressure method and the hole-pressure method. Exit pressure is the residual stress at the die exit. For inelastic fluid this pressure is zero, but positive for a viscoelastic fluid. Hole pressure is defined as the difference between the pressure measured by a flush-mounted transducer and a transducer located at the bottom of a slot directly opposite to it (Fig. 7). For a Newtonian fluid, under creeping flow conditions (Reynolds number ~ 0), the hole pressure is zero.

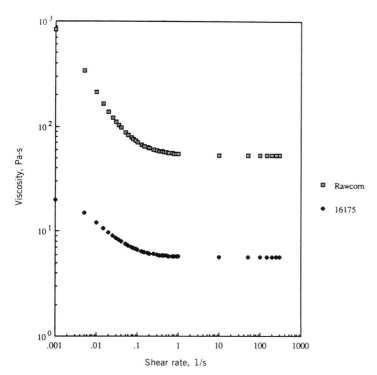

Figure 6. Flow behavior of extruded (16% moisture, 175°C and 150 rpm) and unextruded corn starch at 6% concentration and 60°C (84).

The exit pressure technique has been used to calculate the normal stress differences (87,88). As it is difficult to measure the pressure at the die exit directly, exit pressure is usually obtained by linear extrapolation of the pressure profile from the region of fully developed flow in the die. The expressions for the normal stress differences for a slit die are (89):

$$N_1 = P_{ex}\left[1 + \frac{d\ln P_{ex}}{d\ln \tau_w}\right] \qquad (6)$$

However, this method is controversial. Critical analyses of the method have been done (88–90). The origin of the controversy lies in the fact that for polymers, positive (87,88,91), zero (92), and negative (90,93–97) exit pressures have been obtained. There has been very few published studies using this method on foods. Both positive and negative exit pressures were reported for corn grits, but positive exit pressures only for potato grits (98). Another study examining this method for corn grits showed that products having lower moisture (25–35% db) content gave mostly negative exit pressures, whereas higher moisture values (40 and 45% db) gave positive exit pressures (99). Positive exit pressure was obtained using a capillary die (100). As it is impossible to flush-mount transducers on the walls of the capillary, the presence of hole pressure could lead to erroneous exit pressure values, so the use of a capillary die for this procedure is not recommended.

One assumption in obtaining the expression for the normal stress differences from the exit pressure theory (eq. 6) is that the flow remain fully developed till the die exit. There is evidence that for polymers, the flow does not remain fully developed till the exit (101,102), owing to the presence of exit disturbance. The exit disturbance will lead to a rearrangement of the velocity profile. The velocity rearrangement would lead to an overestimation of the exit pressure (89). Beyond a critical shear stress, the exit disturbance is negligible (88). The value of this stress would depend on the material. For low-density polyethylene this value is 25 kPa. However, there are data in the literature where negative exit pressures were obtained for experiments conducted above the recommended 25 kPa value for wall shear stress. A problem during the extrusion cooking of foods is that at temperatures greater than 100°C the presence of moisture flash at the die exit could cause an additional exit disturbance. Other problems with the exit-pressure method are viscous dissipation in the die and pressure dependence of viscosity of the material. Both these effects could lead to a concave pressure profile and would result in erroneous estimates of exit pressure.

The hole pressure was originally observed as an error in the measurement of normal stress difference using

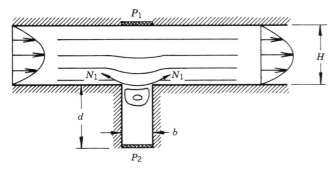

Figure 7. Schematic of the hole-pressure geometry for a transverse slot. The curvature of the streamlines near the mouth of the slot is shown. The fluid elasticity (N_1) results in a tension along the curved streamlines that tends to lift out of the slot.

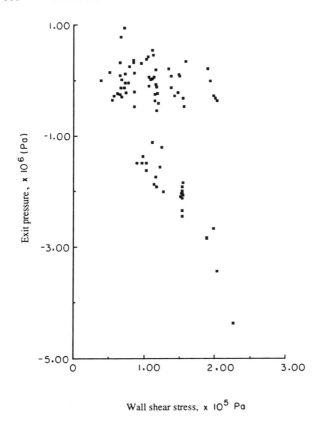

Figure 8. Exit pressure values for corn grits extruded under different conditions (76).

standard geometry rheometers (103). Studies subsequently established the presence of significant hole-pressure errors (104). The interest in the hole pressure arises from the claim that this property can be used for reliable prediction of the elastic properties of viscoelastic fluids. As the fluid passes over a hole or a slot, the stream line for a viscoelastic fluid tends to dip and this leads to the development of normal stresses that tend to lift the fluid (Fig. 8), thus creating a lower pressure at the bottom of the slot or hole (105). This method has been used successfully to measure the normal stress difference for polymer solutions and melts (90,96,106–111). Typically, a slit die with holes or slots of different geometries is used to estimate the normal stress differences. The common hole geometry is a rectangular slot with its length transverse to the direction of the flow. This geometry is known as the transverse slot.

The original equations relating hole pressure and the normal stress difference (112) were later modified by Baird (109) to give the HPB equations. The HPB equations were reinterpreted to give the HPBL equations (90). Depending on the geometry of the hole, the HPBL equations are given by:

$$N_1 = 2 P_{HE} \frac{d \ln (P_{HE})}{d \ln (\tau_w)} \quad \text{(Transverse slot)} \quad (7a)$$

$$N_2 = - P_{HE} \frac{d \ln (P_{HE})}{d \ln (\tau_w)} \quad \text{(Parallel slot)} \quad (7b)$$

$$N_1 - N_2 = 3 P_{HE} \frac{d \ln (P_{HE})}{d \ln (\tau_w)} \quad \text{(Circular hole)} \quad (7c)$$

where P_{HE} signifies the elastic contribution to the hole pressure. Despite violations of some assumptions, reasonably good comparisons of N_1 obtained from standard geometries and slit die were obtained (90,96,111). However, recent numerical studies have shown the HPBL equations to be valid (113,114).

This method is presently under evaluation in the authors' laboratory and has yielded promising results. A potential problem with this method is that for a fluid with yield stress, an error in pressure transmission could occur. For a fluid that has yield stress, higher hole pressure values will be obtained (95). Thus the effect of yield stress on hole-pressure values will have to be accounted for.

Extensional Viscosity

When a fluid flows from a larger to a smaller diameter tube (eg, from the barrel of the extruder to the die), a pressure drop is encountered. For a polymer melt or solution the magnitude of this pressure drop is significantly higher than that obtained for a Newtonian fluid of the same viscosity (87,115). Similar results were reported for corn grits during extrusion cooking (75). The excess pressure drop at the entrance was originally believed to be due to the elastic properties of the fluid. Recent studies for plastic polymers indicate that the flow in the die entry region cannot be explained by melt elasticity alone (116) and that the extensional viscosity is an important parameter that must be considered.

Cogswell (117) was the first to propose a method for obtaining extensional viscosity from the entrance pressure drop method. The entrance pressure drop is calculated by linear extrapolation of the pressure readings along the walls from the fully developed flow region of the capillary or slit to the die entrance. This reading is subtracted from that of the transducer located at the barrel exit or a reservoir. The basis for Cogswell's analysis is that the large entrance pressure drop in the converging flow region is due to the extensional nature of the flow. In addition, the presence of the wall introduces a shearing component. Thus, the entrance pressure drop is assumed to consist of two components: due to shear flow and due to extensional flow. The expression for the shear component of the pressure drop due is given by:

$$\Delta P_{Shear} = \frac{2B}{3n \tan \alpha} \left(\frac{1 + 3n}{4n} \right)^n \dot{\gamma}^n \left(1 - \left(\frac{r_1}{r_0} \right)^{3n} \right) \quad (8)$$

and the extensional component for the pressure drop is:

$$\Delta P_{Ext} = \frac{2A}{3m} \left(\frac{\tan \alpha}{2} \right)^m \dot{\gamma}^m \left(1 - \left(\frac{r_1}{r_0} \right)^{3m} \right) \quad (9)$$

These equations have been derived assuming a power law behavior in both shear and extensional flow. Expressions for sudden contraction in cylindrical and rectangular flow can also be found in Cogswell's paper. Several other expressions have been suggested using differing assumptions (118,119). There is no significant difference from the values of strain rate obtained from these expression for low die angles (<30°). A spherical coordinates system was assumed to arrive at a slightly modified expressions for equations 8 and 9 (120). An alternative analysis for determining the extensional viscosity in planar and uniaxial extension has been proposed from entrance pressure-drop data (121–123).

Studies on extensional viscosity of food doughs are scarce. A stretch-thinning behavior for corn grits using a capillary rheometer has been reported (124). Similar behavior for corn grits has been noted during extrusion cooking (76). Using Gibson's analysis, these authors found that the extensional viscosity decreased with increasing moisture content. The shear component of the entrance pressure was less than 5%. The Trouton ratio for the doughs ranged from 80–100. Not all materials exhibit shear thinning and stretch-thinning behavior. Several polymeric liquids exhibit stretch-thickening behavior while displaying shear thinning behavior in shear flows.

The entrance pressure-drop method for obtaining extensional viscosity has not been universally accepted. The shear viscosity parameters (B and n in eq. 8) are assumed to be known a priori. These parameters can be calculated from the pressure profile and flow-rate data in the die. At large die entrance angles, poor data fit is observed. The assumption used in the calculation of the shear component of the entrance pressure drop can lead to error. However, the magnitude of this error is expected to be small because extensional flow dominates. Also, only average quantities are computed, whereas in reality the stress and the rate of strain will change with position. It should be noted that this method gives only qualitative comparisons and not definitive data. Hence it is important that a comparison be made between the magnitude of the extensional viscosity obtained from entrance pressure calculations with those obtained by other methods. Such a comparison has been made for polymers, and the values obtained were reported to be of the same order of magnitude (125,126).

Dynamic Oscillatory Flow

The dynamic properties of rice dough have been studied using an on-line rheometer (127). Storage and loss moduli increased with increasing frequency and decreasing moisture content. However, these authors reported significant problems with the equipment and recommended several modifications.

TEXTURE AND RHEOLOGY

The relationship between product structure and its texture is well recognized (9). The structure achieved is dependent on the product composition, processing condition, and molecular configuration. Rheological properties are sensitive to changes in product composition, processing condition, and molecular configuration. Thus, rheological properties present themselves as a measure of product structure and hence its texture.

One of the most important characteristics of an extrudate is the expansion ratio. To some extent, extrudate expansion is indirectly related to other textural properties such as breaking strength, bulk density, and WAI. The expansion phenomena can be characterized in part from rheological parameters. Two recent articles discuss the expansion of extrudate from a rheological point of view (100,128).

During HTST (>100°C) extrusion, the moisture that is unable to escape due to the high pressures in the die is free to vaporize at or near the die exit and is assumed to be responsible for the expansion of the extrudate. Expansion ratios as high as 15 have been obtained. Experiments where the extrudates were cooled to below 100°C (as in the forming or half-product process) before exiting the die were reported (100). Extrudate swells (diameter of extrudate to that of the die) of 1.5, 2.27, and 3.00 were obtained for wheat flour, wheat starch, and manioc starch, respectively. When temperatures are below 100°C, no water vapor is released. In such instances the expansion is due to the elasticity of the material. A Newtonian fluid flowing out of a tube would expand at low Reynolds numbers (< 16) but shrink at higher Reynolds numbers (85).

An approach commonly used to explain the extrudate swell is the elastic recoil mechanism. When an elastic fluid is flowing through the tube, it is subjected to extra tension along the stream lines. Once the fluid exits, the walls are no longer present to constrain the flow, which causes the fluid to contract along the stream lines, resulting in radial expansion. The magnitude of this tension is represented by the primary normal stress difference. Expressions for obtaining the extrudate swell have been derived (129). Other factors affecting extrudate swell include surface tension, gravity, inertia, wall slip, temperature effects, flow behavior index, and extensional viscosity. In general, the effect of inertia, surface tension, gravity, and wall slip reduce the extrudate swelling, whereas temperature effects, flow behavior index, elastic recovery, and extensional viscosity increase extrudate swell.

The rheological phenomena involved during extrusion cooking of foods have received some attention. The first attempt to incorporate the effect of elasticity on extrudate expansion qualitatively used pore volume as an indication of elasticity, ie, if the pore volume was small, elastic effects was responsible for the expansion (58). Product morphology was used to estimate dough elasticity (130). These studies, along with that of Alvarez-Martinez and co-workers (131), have speculated that dough elasticity is an important parameter that contributes to the magnitude of the radial expansion, although no quantification has been attempted.

At temperatures above 100°C the extrudate expansion contains contributions from both the elasticity and the moisture flash. A two-stage expansion occurs under these conditions. This phenomenon has been observed and photographed (132). The first stage happens immediately after the extrudate emerges from the die and is due to the elastic effect. A few seconds after the first effect and further downstream, bubble formation due to moisture vapor trying to escape can be observed. For high forces due to moisture effects, the moisture vapor quickly finds its way out. At these conditions the swelling occurs almost instantaneously. Depending on the process conditions, elastic or moisture effects may dominate.

Several investigators (58,131,133) have speculated that radial expansion alone is an insufficient parameter to quantify extrudate expansion. It was proposed that expansion occurs both radially and longitudinally and that a negative correlation exists between the two (58). Longitudinal expansion was defined as the mean length of unit mass of extrudate, or as the ratio of extrudate velocity to that of the velocity of fluid inside the die (58,131). Both these researchers report that longitudinal expansion is inversely proportional to the shear viscosity. However, these authors do not give a reason why shear viscosity should be important in longitudinal expansion.

Consider the flow inside the die, where the melt adheres to the die but flows at a maximum velocity at the center (parabolic profile for a Newtonian fluid). Downstream of the die exit the velocity profile approaches that of a plug flow, with all points having a uniform velocity. This means that the surface layer must be accelerated from rest, and it has been shown that conservation of mass and momentum demands that this acceleration cause an extensional flow in the surface layer while the center is decelerated and compressed (134). This could be a possible reason for the longitudinal expansion. Owing to the extensional nature of the flow, this effects to account for extrudate swelling are necessary (135). Another reason for longitudinal expansion may be the moisture vapor formed in the extrudate. As the vapor tries to escape from the extrudate, it stretches the dough. The magnitude of the expansion due to moisture loss would depend on the extensional viscosity (biaxial extension) of the fluid. This phenomenon is similar to the rise of bread during baking. The vapor pressure of the water will stretch the filament of dough that is cooling in the ambient temperature. If the vapor pressure is sufficient to rupture the extrudate, open structures on the extrudate surface are produced. On cooling, the material dries and solidifies to its final shape.

We must caution against the use of shear viscosity to explain the expansion phenomenon. It has been speculated that lower shear viscosity is favorable for greater expansion (41,136). Data contrary to the findings of these researchers have been presented (132). An increase in dough moisture causes a decrease in shear viscosity, but also causes a reduction in extrudate expansion. Shear viscosity is the least sensitive to molecular weight changes. From a rheological point of view, primary normal stress difference and elongational viscosity are probably far more important. These parameters may be able to better explain some of the conflicting results that exist in the literature and provide an insight as to why lipids, salts, and sugars behave the way they do. Morever, both primary normal stress difference and extensional viscosity are significantly more sensitive to average molecular weight and molecular weight distribution (and hence to the structure of the material) than shear viscosity and could serve as an effective tool in process control.

CONCLUSION

Food texture as practiced in the industry has remained more of an art than a science. The ingredients of dough components, ie, proteins, starches, and lipids interact in a manner that is not fully understood. Evaluating the rheological parameters of the dough during extrusion may help in providing an understanding of the effect of product and process conditions on the texture of the extrudates. Techniques such as the hole pressure and entrance pressure drop for the measurement of elastic and extensional material functions could help provide a better understanding of the relationship between rheology and texture. These parameters could also be useful in developing a complete rheological equation of state (one that will predict both the viscous and elastic properties). Such equations of state would be useful in numerical modeling of the food extrusion process and help in developing a better understanding of the flow patterns in the extruder and aid in process control and product development.

BIBLIOGRAPHY

1. D. J. Van Zuilichem and W. Stolp in C. O'Connor, ed., *Extrusion Technology for the Food Industry*, Elsevier Science Publishing, New York.
2. J. D. Dziezak, "Single and Twin-Screw Extruders in Food Processing," *Food Technology* **43**(4), 164–174 (1989).
3. J. M. Harper, "Food Extruders and Their Applications," in C. Mercier, P. Linko, and J. M Harper, eds., *Extrusion Cooking*, AACC, St. Paul, Minn., 1989, pp. 1–5.
4. B. W. Hauck and G. R. Huber, "Single Screw vs Twin Screw Extrusion," *Cereal Foods World* **34**(11), 934–939 (1989).
5. L. P. B. M. Janssen, "Engineering Aspects of Food Extrusion," in Ref. 3, pp. 17–38.
6. R. Jowitt, "The Terminology of Food Texture," *Journal of Texture Studies* **5**, 351–358 (1974).
7. International Organization for Standardization, "Sensory Analysis Vocabulary, Part IV," Geneva, Switzerland, 1981.
8. R. H. Meyers, *Response Surface Methodology*, Allyn & Bacon, Boston, 1976.
9. J. M. Harper, *Extrusion of Foods*, Vols. 1 and 2, CRC Press, Boca Raton, Fla., 1981.
10. N. Noguchi, "Extrusion Cooking of High-Moisture Protein Foods," in Ref. 3, pp. 343–370.
11. F. F. Huang and C. K. Rha, "Protein Structures and Protein Fibers—A Review," *Polymer Engineering Science* **14**, 81–89 (1974).
12. J. P. Clark, "Texturization by Extrusion," *Journal of Texture Studies* **9**, 109–123 (1978).
13. D. W. Stanley and J. M. deMan, "Structural and Mechanical Properties of Textured Proteins," *Journal of Texture Studies* **9**, 59–76, 1978.
14. J. E. Kinsella, "Texturized Proteins: Fabrication, Flavorings and Nutrition," *CRC Reviews in Food Science and Nutrition* **10**, 147–207 (1978).
15. D. A. Ledward and J. R. Mitchell, "Protein Extrusion—More Questions Than Answers," in J. M. V. Blanshard and J. R. Mitchell, eds., *Food Structure—Its Creation and Evaluation*, Butterworths, Boston, 1988, pp. 219–229.
16. P. R. Sheard and co-workers, "Macromolecular Changes Associated with the Heat Treatment of Soy Isolate," *Journal of Food Technology* **21**, 55–60 (1986).
17. J. J. Kelley and R. Pressey, "Studies with Soybean Protein and Fiber Formation," *Cereal Chemistry* **43**, 195–205, 1966.
18. D. B. Cumming and co-workers, "Fate of Water Soluble Soy Protein During Thermoplastic Extrusion," *Journal of Food Science* **38**, 320–323 (1973).
19. L. D. Burgess and D. W. Stanley, "A Possible Mechanism for Thermal Texturization of Soybean Protein," *Journal of the Institute of Canadian Science and Technology—Alimentation* **9**, 228–231 (1976).
20. R. W. Simonsky and D. W. Stanley, "Texture-Structure Relationships in Textured Soy Protein. V. Influence of pH and Protein Acylation on Extrusion Texturization," *Canadian Institute of Food Science Technology* **15**, 294–301 (1982).

21. D. F. Hager, "Effects of Extrusion upon Soy Concentrate Solubility," *Journal of Agricultural and Food Chemistry* **32**, 293–296 (1984).
22. P. E. Neumann and co-workers, "Uniquely Textured Products Obtained by Co-Extrusion of Corn Gluten Meal and Soy Flour," *Cereal Chemistry* **61**, 439–445 (1984).
23. P. R. Sheard and co-workers, "Role of Carbohydrates in Soya Extrusion," *Journal of Food Technology* **19**, 475–483 (1984).
24. P. Melius, "Thermal Polymerization of Amino Acids Mixtures," *Federation Proceedings, Federation of American Society of Experimental Biologist* **34**, 573, (1975).
25. D. B. Cummings and co-workers, "Texture–Structure Relationships in Texturized Soy Protein. II. Textural Properties and Ultrastructure of an Extruded Soybean Product," *Journal of the Institute of Canadian Science and Technology—Alimentation* **5**, 124–128 (1972).
26. J. M. Aguilera and F. W. Kosikowoski, "Soybean Extruded Product: A Response Surface Analysis," *Journal of Food Science* **41**, 647–651 (1976).
27. T. J. Maurice and co-workers, "Texture–Structure Relationship in Textured Soy Protein. III. Textural Evaluation of Extruded Products," *Canadian Institute of Food Science Technology* **9**, 173–176 (1976).
28. T. J. Maurice and D. W. Stanley, "Texture–Structure Relationships in Texturized Soy Protein, IV. Influence of Process Variables on Extrusion Texturization," *Canadian Institute of Food Science Technology* **11**, 1–6 (1978).
29. O. B. Smith, "Extrusion and Forming: Creating New Foods," *Food Engineering* **47**, 48–50 (August 1975).
30. S. H. Holay and J. M. Harper, "Influence of the Extrusion Shear Environment on Plant Protein Texturization," *Journal of Food Science* **47**, 1869–1874 (1982).
31. G. Boison and co-workers, "Extrusion of Defatted Soy Flour-Hydrocolloid Mixtures," *Journal of Food Technology* **18**, 719–730 (1983).
32. P. J. Frazier and co-workers, "Optimization of Process Variables in Extrusion Texturing of Soya," *Journal of Food Engineering* **2**, 79–103 (1983).
33. P. R. Sheard and co-workers, "Comparison of the Extrusion Cooking of a Soya Isolate and a Soya Flour," *Journal of Food Technolgoy* **20**, 763–771 (1985).
34. M. V. Taranto and co-workers, "Textured Cottonseed and Soy Flours: A Microscopic Analysis," *Journal of Food Science* **43**, 767–771 (1978).
35. M. V. Taranto and co-workers, "Morphological, Ultrastructural and Rheological Evaluation of Soy and Cottonseed Flours Texturized by Extrusion and Nonextrusion Processing," *Journal of Food Science* **43**, 973–979, 984 (1978).
36. M. B. Kennedy and co-workers, "Effects of Feed Moisture and Barrel Temperature on the Rheological Properties of Extruded Cowpea Meal," *Journal of Food Processing and Engineering* **8**, 193–212 (1986).
37. C. B. Pham and P. R. Del Rosario, "Studies on the Development of Texturized Vegetable Products by the Extrusion Process," *Journal of Food Technology* **19**, 535–547 (1984).
38. J. A. Maga and K. Lorenz, "Sensory and Functional Properties of Extruded Corn-Soy Blends," *Lebensmittel Wissonschalf Technologie* **11**, 185–187 (1978).
39. Y. C. Jao and co-workers, "Evaluation of Corn Protein Concentrate: Extrusion Study," *Journal of Food Science* **50**, 1257–1259, 1288 (1985).
40. M. Bhattacharya and co-workers, "Textural Properties of Plant Protein Blends," *Journal of Food Science* **51**, 988–993 (1986).
41. P. Colonna and co-workers, "Extrusion Cooking of Starch and Starchy Products," in Ref. 3, pp. 247–319.
42. P. Colonna and co-workers, "Flow, Mixing and Residence Time Distribution of Maize Starch Within a Twin Screw Extruder with a Longitudinally-Split Barrel," *Journal of Cereal Science* **1**, 115–125 (1983).
43. C. Mercier, "Effect of Extrusion Cooking on Potato Starch Using a Twin Screw French Extruder," *Staerke* **29**, 48–52 (1977).
44. P. Colonna and C. Mercier, "Macromolecular Modifications of Manioc Starch Components by Extrusion Cooking With and Without Lipids," *Carbohydrate Polym* **3**, 87–108 (1983).
45. P. Colonna and co-workers, "Extrusion Cooking and Drum Drying of Wheat Starch. I. Physical and Macromolecular Modifications," *Cereal Chemistry* **61**, 538–543 (1984).
46. V. J. Davidson and co-workers, "Degradation of Wheat Starch in a Single Screw Extruder: Characteristics of Extruded Starch Polymers," *Journal of Food Science* **49**, 453–458 (1984).
47. V. J. Davidson and co-workers, "A Model for Mechanical Degradation of Wheat Starch in a Single Screw Extruder," *Journal of Food Science* **49**, 1154–1157 (1984).
48. L. L. Diosady and co-workers, "Degradation of Wheat Starch in a Single Screw Extruder: Mechanico-kinetic Break Down of Cooked Starch," *Journal of Food Science* **50**, 1697–1699, 1706 (1985).
49. C. Mercier and P. Feillet, "Modification of Carbohydrate Components by Extrusion-Cooking of Cereal Products," *Cereal Chemistry* **52**, 283–297 (1975).
50. J. M. Faubion and R. C. Hoseney, "High Temperature Short-Time Extrusion Cooking of Wheat Starch and Flour. I. Effect of Moisture and Flour Type on Extrudate Properties," *Cereal Chemistry* **59**, 529–533 (1982).
51. M. Gomez and J. M. Aguilera, "A Physicochemical Model for Extrusion of Corn Starch," *Journal of Food Science* **49**, 40–43, 63 (1984).
52. J. Owusu-Ansah and co-workers, "Textural and Microstructural Changes in Corn Starch as a Function of Extrusion Variables," *Canadian Institute of Food Science Technology* **17**, 65–70 (1984).
53. M. Bhattacharya and M. A. Hanna, "Textural Properties of Extrusion Cooked Corn Starch," *Lebensmittel Wissenschaft und Technologie* **20**, 195–201 (1987).
54. F. Hsieh and co-workers, "Effects of Salt, Sugar and Screw Speed on Processing and Product Variables of Corn Meal Extruded with a Twin-Screw Extruder," *Journal of Food Science* **55**, 224–227 (1990).
55. R. Chinnaswamy and M. A. Hanna, "Optimum Extrusion-Cooking Conditions for Maximum Expansion of Corn Starch," *Journal of Food Science* **53**, 834–836, 840 (1988).
56. R. Chinnaswamy and M. A. Hanna, "Die-Nozzle Dimension Effects on Expansion of Corn Starch," *Journal of Food Science* **52**, 1746–1747 (1987).
57. A. L. Hayter and co-workers, "The Physical Properties of Extruded Food Foams," *Journal of Materials Science* **21**, 3729–3736 (1986).
58. B. Launay and J. M. Lisch, "Twin-Screw Extrusion Cooking of Starches: Flow Behavior of Starch Pastes, Expansion and Mechanical Properties of Extrudates," *Journal of Food Engineering* **2**, 259–280 (1983).
59. D. Paton and W. A. Spratt, "Component Interactions in the Extrusion Cooking Process. I. Processing of Chlorinated and Untreated Soft Wheat Flour," *Cereal Chemistry* **55**, 973–980 (1978).
60. J. M. Faubion and R. C. Hoseney, "High Temperature Short-Time Extrusion Cooking of Wheat Starch and Flour.

II. Effect of Protein and Lipid on Extrudate Properties," *Cereal Chemistry* **59,** 533–537 (1982).

61. M. Bhattacharya and M. A. Hanna, "Effect of Lipids on the Properties of Extruded Products," *Journal of Food Science* **53,** 1230–1231 (1988).

62. R. A. Anderson and co-workers, "Gelatinization of Corn Grits by Roll and Extrusion Cooking," *Cereal Science Today* **14,** 4–7, 11–12 (1969).

63. R. A. Anderson and co-workers, "Roll and Extrusion Cooking of Grain Sorghum Grits," *Cereal Science Today* **14,** 372–376 (1969).

64. H. F. Conway, "Extrusion Cooking of Cereals and Soybeans," *Food Product Development* **5,** 14–17, 27–29 (1971).

65. S. Middleman, "Advances in Polymer Science and Engineering: Application to Food Rheology," in C. K. Rha, ed., *Theory, Determination and Control of Physical Properties of Food Materials,* Riedel Publishing, Dordrecht, Holland, 1975, pp. 39–53.

66. J. L. Leblanc, "Recent Progress in Understanding Rubber Processing Through a Rheological Approach," *Progress and Trends in Rheology* **2,** 32–43 (1988).

67. C. J. S. Petrie, "Some Asymptotic Results for Planar Extension," *Journal of Non-Newtonian Fluid Mechanics* **34,** 37–62 (1990).

68. J. M. Harper and co-workers, "Viscosity Model for Cooked Cereal Doughs," *American Institute of Chemical Engineering Symposium Series,* No. 108, 67, 40–43 (1971).

69. C. H. Remsen and J. P. Clark, "A Viscosity Model for a Cooking Dough," *Journals of Food Process Engineering* **2,** 39–63 (1977).

70. N. W. Cervone and J. M. Harper, "Viscosity of an Intermediate Moisture Dough," *Journal of Food Processing Engineering* **2,** 83–95 (1978).

71. Y. C. Jao and co-workers, "Engineering Analysis of Soy Dough Rheology in Extrusion," *Journal of Food Process Engineering* **2,** 97–112 (1978).

72. L. A. Luxenburg and co-workers, "Background Studies in the Modeling of Extrusion Cooking Processes for Soy Doughs," *Biotechnology Progress* **1,** 33–38 (1985).

73. A. Senouci and A. C. Smith, "An Experimental Study of Food Melt Rheology. I. Shear Viscosity Using a Slit Die Viscometer and a Capillary Rheometer," *Rheologica Acta* **27,** 546–554 (1988).

74. L. S. Lai and co-workers, "On Line Rheological Properties of Amylose and Amylopectin Based Starch. The Role of Viscosity on Extrudate Expansion," *Proceedings of the Xth International Congress on Rheology,* **2,** 55–57, (1988).

75. M. Padmanabhan and M. Bhattacharya, "Analysis of Pressure Drop in Extruder Dies," *Journal of Food Science* **54,** 709–713 (1989).

76. M. Bhattacharya and M. Padmanabhan, "Elongational Viscosity During Extrusion Cooking Using a Converging Flow Analysis," Paper 243, Presented at the Institute of Food Technology Annual Meeting, Anaheim, Calif., 1990.

77. E. B. Bagley, "End Corrections in the Capillary Flow of Polyethylene," *Journal of Applied Physics* **28,** 624–627 (1957).

78. B. Vergnes and J. P. Villemaire, "Rheological Behavior of Low Moisture Molten Maize Starch," *Rheologica Acta* **26,** 570–576 (1987).

79. R. G. Morgan and co-workers, "A Generalized Viscosity Model for Extrusion of Protein Doughs," *Journal of Food Process Engineering* **11,** 55–78 (1989).

80. K. L. Mackey and co-workers, "Rheological Modeling of Potato Flours During Extrusion Cooking," *Journal of Food Process Engineering* **12,** 1–11 (1989).

81. G. V. Vinogradov and A. Y. Malkin, *Rheology of Polymers,* Mir Publishers, Moscow, 1980.

82. B. van Lengerich, "Influence of Extrusion Processing on In-Line Rheological Behavior, Structure, and Function of Wheat Starch," in Hamed Faridi and Jon M. Faubion, eds., *Dough Rheology and Baked Product Texture,* Van Nostrand Reinhold, New York, 1989, pp. 421–471.

83. J. L. Doublier and co-workers, "Extrusion Cooking and Drum-Drying of Wheat Starch. II. Rheological Characterization of Wheat Starch," *Cereal Chemistry* **63,** 240–246 (1986).

84. B. Launay and T. Kone, "Twin-Screw Extrusion-Cooking of Corn Starch: Flow Behavior of Starch Pastes," in P. Zeuthen and co-workers, eds., *Thermal Processing and Quality of Foods,* Elsevier Applied Science, London, 1984, pp. 54–61.

85. R. B. Bird and co-workers, *Dynamics of Polymeric Liquids,* Vol. 1, John Wiley & Sons, New York, 1987.

86. H. A. Barnes and co-workers, *An Introduction to Rheology,* Elsevier Science Publishing, Amsterdam, The Netherlands, 1989.

87. C. D. Han, *Rheology in Polymer Processing,* Academic Press, New York, 1976.

88. C. D. Han, "Slit Rheometry," in A. A. Collyer and D. W. Clegg, eds., *Rheological Measurement,* Elsevier Applied Science, New York, 1988, pp. 25–48.

89. D. V. Boger and M. M. Denn, "Capillary and Slit Methods of Normal Stress Measurements," *Journal of Non-Newtonian Fluid Mechanics* **6,** 163–185 (1980).

90. A. S. Lodge and L. de Vargas, "Positive Hold Pressures and Negative Exit Pressures Generated by Molten Polyethylene Flowing Through a Slit Die," *Rheologica Acta* **22,** 151–170 (1983).

91. C. Rauwendaal and F. Fernandez, "Experimental Study and Analysis of a Slit Die Viscometer," *Polymer Engineering Science* **25,** 765–771 (1985).

92. R. Eswaran and co-workers, "A Slit Viscometer for Polymer Melts," *Rheologica Acta* **3,** 83–91 (1963).

93. J. L. Leblanc, "New Slit Die Rheometer. Some Results with a Butadiene-Styrene Block Copolymer," *Polymer* **17,** 235–240, (1976).

94. L. Choplin and P. J. Carreau, "Excess Pressure Losses in Slit," *Journal of Non-Newtonian Fluid Mechanics* **9,** 119–146 (1981).

95. H. M. Laun, "Polymer Melt Rheology with a Slit Die," *Rheologica Acta* **22,** 171–185 (1983).

96. D. G. Baird and co-workers, "Comparison of the Hole Pressure and Exit Pressure Methods for Measuring Polymer Melt Normal Stresses," *Polymer Engineering Science* **26,** 225–232 (1986).

97. N. Y. Tuna and B. A. Finlayson, "Exit Pressure Experiments for Low Density Polyethylene Melts," *Journal of Rheology* **32,** 285–308 (1988).

98. A. Senouci and A. C. Smith, "An Experimental Study of Food Melt Rheology. II. End Pressure Effects," *Rheologica Acta* **27,** 649–655 (1988).

99. M. Bhattacharya and M. Padmanabhan, "On-line Rheological Measurements of Food Dough During Extrusion Cooking," in *Proceedings of the Second International Conference in Extrusion and Rheology of Foods.* Rutgers University, New Brunswick, N.J., 1990.

100. R. C. E. Guy and A. W. Horne, "Extrusion and Co-Extrusion of Cereals," in Ref. 15, pp. 331–349.

101. B. A. Whipple and C. T. Hill, "Velocity Distributions in Die Swell," *American Institute of Chemical Engineering Journal* **24**, 664–678 (1978).
102. M. Gottleib and R. B. Bird, "Exit Effects in Non-Newtonian Liquids. An Experimental Study." *Ind. Eng. Chem. Fund.* **18**, 357–368 (1979).
103. J. M. Broadbent and co-workers, "Possible Systematic Errors in the Measurement of Normal Stress Differences in Polymer Solution in Steady Shear Flow," *Nature* **217**, 55–57 (1968).
104. A. Kaye and co-workers, "Determination of Normal Stress Differences in Steady Shear Flow. II. Flow Birefringence, Viscosity and Normal Stress Data for Polyisobutene Liquid," *Rheologica Acta* **7**, 368–379 (1968).
105. R. I. Tanner and A. C. Pipkin, "Intrinsic Errors in Pressure-Hole Measurement," *Transactions of the Society of Rheology* **13**, 471–484 (1969).
106. A. S. Lodge, "Low-Shear Rate Rheometry and Polymer Quality Control," *Chemical Engineering Communications* **32**, 1–60 (1985).
107. A. S. Lodge, "A New Method of Measuring Multigrade Oil Shear Elasticity and Viscosity at High Shear Rates," *SAE Technical Paper,* Series 872043 (1987).
108. A. S. Lodge and co-workers, "Measurement of the First Normal-Stress Difference at High Shear Rates for a Polyisobutylene/Decalin Solution D2," *Rheologica Acta,* **26**, 516–521 (1987).
109. D. G. Baird, "A Possible Method for Determining Normal Stress Differences from Hole Pressure Error Data," *Transactions of the Society of Rheology* **19**, 147–151 (1975).
110. D. G. Baird, "Fluid Elasticity from Hole Pressure Error Data," *Journal of Applied Polymer Science* **20**, 3155–3173 (1976).
111. R. D. Pike and D. G. Baird, "Evaluation of the Highashitani and Pritchard Analysis of the Hole Pressure Using Flow Birefringence," *Journal of Non-Newtonian Fluid Mechanics,* **16**, 211–231 (1984).
112. K. Higashitani and W. G. Pritchard, "A Kinematic Calculation of Intrinsic Errors in Pressure Measurements Made with Hole," *Transactions of the Society of Rheology* **16**, 687–696 (1972).
113. R. I. Tanner, "Pressure-Hole Errors—An Alternative Approach," *Journal of Non-Newtonian Fluid Mechanics* **28**, 309–318 (1988).
114. M. Yao and D. S. Malkus, "Error Cancellation in HPBL Derivation of Elastic Hole-Pressure Error," *Center of the Mathematical Science,* Technical Summary Report 90–18, University of Wisconsin, Madison (1989).
115. W. Philippoff and F. H. Gaskins, "The Capillary Experiment in Rheology," *Transactions of the Society of Rheology* **2**, 263–284 (1958).
116. S. A. White and D. G. Baird, "The Importance of Extensional Flow Properties on Planar Entry Flow Patterns of Polymer Melt," *Journal of Non-Newtonian Fluid Mechanics* **20**, 93–101 (1986).
117. F. N. Cogswell, "Converging Flow of Polymer Melts in Extrusion Dies," *Polymer Engineering and Science* **12**, 64–73 (1972).
118. D. R. Oliver, "The Prediction of Angle of Convergence for the Flow of Viscoelastic Liquids into Orifices," *The Chemical Engineering Journal* **6**, 265–271 (1973).
119. A. B. Metzner and A. P. Metzner, "Stress Levels in Rapid Extensional Flows of Polymeric Fluids," *Rheologica Acta* **9**, 174–181 (1970).
120. A. G. Gibson "Converging Flow Analysis," in Ref. 88, pp. 49–82.
121. D. M. Binding, "An Approximate Analysis for Contraction and Converging Flows," *Journal of Non-Newtonian Fluid Mechanics* **27**, 173–189, (1988).
122. D. M. Binding and K. Walters, "On the Use of Flow Through a Contraction in Estimating the Extensional Viscosity of Mobile Polymer Solution," *Journal of Non-Newtonian Fluid Mechanics* **30**, 233–250 (1988).
123. D. M. Binding and D. M. Jones, "On the Interpretation of Data from Converging Flow Rheometers," *Rheologica Acta* **28**, 215–222 (1989).
124. A. Senouci and co-workers, "Extensional Rheology in Food Processing," *Progress and Trends in Rheology* **2**, 434–437 (1988).
125. R. N. Shroff and co-workers, "Extensional Flow of Polymer Melts," *Transactions of the Society Rheology* **21**, 429–446 (1977).
126. H. M. Laun and H. Schuch, "Transient Elongational Viscosities and Drawability of Polymer Melts," *Journal of Rheology* **33**, 119–175 (1989).
127. J. F. Steffe and R. G. Morgan, "On-Line Measurements of Dynamic Rheological Properties During Food Extrusion," *Journal of Food Process Engineering* **10**, 21–26 (1987).
128. M. Padmanabhan and M. Bhattacharya, "Extrudate Expansion During Extrusion Cooking of Foods," *Cereal Foods World* **34**, 945–949 (1989).
129. R. I. Tanner, "A Theory of Die Swell," *Journal of Applied Polymer Science* A-2, **8**, 2067–2078 (1970).
130. R. C. Miller, "Low Moisture Extrusion: Effect of Cooking Moisture on Product Characteristics," *Journal of Food Science* **50**, 249–253 (1985).
131. L. Alvarez-Martinez and co-workers, "A General Model for Expansion of Extruded Products," *Journal of Food Science* **53**, 609–615 (1988).
132. R. Chinnaswamy and M. A. Hanna, "Relationship Between Viscosity and Expansion Properties of Variously Extrusion-Cooked Corn Grain Components," *Food Hydrocolloids* **3**, 423–434 (1990).
133. K. B. Park, "Elucidation of the Extrusion Puffing Process," Ph.D. Thesis, Department of Food Science, University of Illinois, Urbana, Il., 1976.
134. S. Richardson, "The Die Swell Phenomenon," *Rheologica Acta* **9**, 193–199 (1970).
135. R. J. Tanner, "Recoverable Elastic Strain and Swelling Ratio," in Ref. 88, pp. 93–117.
136. B. Vergnes and co-workers, "Interrelationships Between Thermomechanical Treatment and Macromolecular Degradation of Maize Starch in a Novel Rheometer with Preshearing," *Journal of Cereal Science* **5**, 189–202 (1987).

NOMENCLATURE

A, B	Constants in equations 8 and 9
b	Slot width
d	Slot height
ΔE	Activation energy of flow
ΔE_k	Activation energy for cooking reaction
H	Slit height
k	Specific reaction velocity constant
K	Coefficient for moisture in equations 3 and 5.
M	Moisture content
m	Power law index in extensional flow

N_1	Primary normal stress difference	t	Time
N_2	Secondary normal stress difference	α	Half-cone angle of the die
n	Power law index in shear flow	$\dot{\varepsilon}$	Elongational strain rate
P_{ex}	Pressure drop in the exit	$\dot{\gamma}$	Shear rate
ΔP_{Shear}	Pressure drop due to viscous effects	η_s	Shear viscosity
ΔP_{Ext}	Pressure drop due to elongational flow	η_e	Elongational viscosity
ΔP_{tot}	Total pressure drop	η^*	Reference viscosity
ΔP_{ent}	Entrance pressure drop	σ_{ij}	Total stress
ΔP_{die}	Pressure drop across the die	τ_{ij}	Extra stress components
P_{HE}	Elastic hole pressure		
Q	Volume flow rate		
r_o	Radius of the barrel		
r_1	Radius at the die		
T	Temperature		

Mrinal Bhattacharva,
Mahesh Padmanabhan
University of Minnesota
St. Paul, Minnesota

F

FADDISM

Fads, by definition, move quickly, and food fads prove no exception to this principle. The U.S. Food and Drug Administration (FDA) estimates place the number of U.S. consumers who are attracted by food fads at nearly 18 million, and it is estimated that they annually spend $2 billion on nutritional supplements, diet books, special food products, and products such as weight-reducing devices.

Promoters of such products rely on emotional appeals, exaggerated claims, and such powers of persuasion as massive advertising campaigns to promote their products. Despite claims for such products that run counter to nutritional facts, the resultant fads are successful for various reasons.

Nutritionists must cope not only with food fads but also with basic misinformation regarding food. Virtually every population center has at least one health food store, and most large chain stores have at least a health food section. The existence of such businesses attests to both the health consciousness of Americans and their affluence, for health foods are some of the most expensive food items and supplements available.

The FDA tried to bring supplements under its regulatory authority. Those efforts were frustrated in 1977 by a health food lobby that successfully promoted legislation to deny the FDA that authority. The lobbying efforts were fueled by the realization that over three-fourths of U.S. households use dietary supplements of some kind. If our use of supplements were based on our need, three-fourths of us would be labeled as malnourished on the basis of our supplement use.

What the FDA found in regard to food quackery was that as soon as they closed down one operation, the promoters were at work on another product. Food fads tend to follow a predictable pattern. One characteristic is that they give almost magical properties to the product. Usually the products are claimed to cure a condition for which no cure exists—or one for which there are numerous sufferers. Cancer and arthritis are common examples of such a condition.

Some nutritionists have pointed out that food fads often conflict from one culture to another. For example, a food may be considered a cancer cure in one culture, a poison in another culture, and a source of sexual potency in a third culture. A current example of conflict within a single culture surrounds laetrile. Promoters herald it as a cancer cure, while nutritionists point out that it is a potentially lethal cyanide compound.

IDENTIFYING A FAD PRODUCT

Labeling on fad products often tends to be misleading. A common example would be a label that lists numerous ingredients and their benefits for a healthy existence but fails to advise that the product provides only a fraction of the recommended dietary allowances of the ingredients.

Fads can also use a negative approach. For example, the idea that enrichment of a food represents a chemical risk has caused some consumers to avoid such products as enriched white bread, despite the fact that enrichment of such products normally involves the addition of basic nutrients. What the fad promoters hope to do in such cases is to scare you away from a popular product so that you will substitute theirs.

A current fad of far-reaching effect argues that all processed foods involve health risks. This fad stresses the use of organically grown or naturally grown products. Efforts to promote such naturally grown products go so far as to claim that food processing causes disease.

The organically grown trend promotes such food products as yogurt, honey, wheat germ, alfalfa sprouts, sunflower seeds, and herbal tea. Promoters argue that their products are free of the harmful effects of chemical fertilizers and pesticides. These products typically cost two to three times as much as their counterparts in grocery stores.

A sideline of health food promoters involves special food preparation equipment. Usually the claim is that the equipment processes food in such a way that little nutrient value is lost. The promoters use large gatherings such as fairs or department store sale days to take advantage of impulse buying.

Additives have given rise to a food fad of avoidance. Since some food additives have been found to have harmful side effects, suspicion surrounds all additives. This is, of course, an unfortunate attitude, for many additives prevent potentially harmful bacterial action. Others make food attractive and palatable. While additives must be constantly evaluated for both short-term and long-term effects, they are regulated ingredients and generally serve very positive purposes.

You may initially ask what the harm can be in people's spending their money for harmless products that fall far short of their claims. One concern is that some people—the elderly, for example—will be bilked out of money that they cannot afford to spend by promoters who exploit their natural fear of illness and dependence. Many promoters direct their appeals at the elderly, knowing that that segment of the population has such fears. A second concern is that individuals relying on worthless products for a cure are wasting valuable time and money that should be spent on known and effective treatments.

TECHNIQUES USED BY QUACKS

Food quacks appeal to the emotions of the consumers, but they do not stop there. Quacks will attempt to shroud their promotions in an air of medical or scientific knowledge. They use titles for themselves, and create their own professional organizations. These are all legal moves, but they are made to mislead consumers. Such efforts tend to be successful because most consumers are not able to distinguish between the legitimate professional groups and those that are strictly profit-motivated.

A favorite technique of food quacks is to draw on scientific fact and enlarge on its significance. For example, quacks like to follow the findings of animal studies. If they found, for example, that a standard additive produced cataracts in laboratory mice after massive doses, quacks would rush to the market with a product that did not contain the additive. Their advertisements would argue that their competitors use cataract-causing additives whereas they do not.

Certain characteristics seem to be shared by most get-rich-quick food promoters. They generally claim to have information not available to the medical community, such as a secret formula, often from some primitive culture, and they claim that their product can almost magically cure diseases or maladies that have proved medically difficult to resolve. We have previously mentioned cancer and arthritis as typical examples. Arteriosclerosis (hardening of the arteries) also tends to be a popular disease for magical cures. Quacks will usually claim that the medical profession tries to silence them because they are cutting into the physician's livelihood. Quacks also talk about soil depletion, food processing dangers, and contaminants. They argue that theirs is the only safe product. They also tend to try to enlarge their potential market by claiming that we are all undernourished, usually owing to such foods as refined sugars, pasteurized milk, or canned foods.

Sales techniques involve almost any approach possible. Food quacks use large crowds to take advantage of impulse buying. They sell door to door with evangelical enthusiasm, generally offering the consumer an opportunity to safeguard the family's health. Introductory offers in popular publications are a common approach. Public lecture and radio and television advertising are becoming increasingly popular as merchandising techniques of food quacks. Books have become especially useful in launching food fads, especially those by authors who hold medical degrees. The authors often appear on talk shows and will frequently feature testimonials by popular figures such as movie stars and athletes.

Many governmental agencies and community agencies devote significant effort to combatting food quackery. The FDA seeks to prevent mislabeling but must necessarily concentrate its efforts on dangerous products. The U.S. Postal Service can prosecute firms that use the mails to promote and sell worthless products by means of false advertising. The American Medical Association has initiated efforts to counter quackery and faddism. It issues statements on products and maintains its own Bureau of Investigation to respond to inquiries about products and fad promoters. The U.S. Federal Trade Commission (FTC) can prosecute for false advertising. Such local organizations as the Better Business Bureau serve educational roles in their communities, as well as corrective roles.

Professional organizations such as the American Public Health Association and the American Dietetic Association have quite active programs for educating the general public regarding nutrition. Nutrition education offers what may be the best hope for combatting quackery and food fads. A constant flow of nutrition information from reliable sources can stem the success of those who prey on ignorance. For those who use half-truths and slightly twisted scientific facts, the challenge is greater. The public must not only learn nutritional facts but also become sufficiently informed to identify quackery themselves. The establishment of community nutrition centers also represents an effective source of nutrition education. Another highly successful approach is the use of nutrition columns in newspapers. Unfortunately, quackery and faddism are also able to utilize the press for their own nutritional claims.

One thing that remains certain is that the battle against nutrition misinformation will continue. New products will regularly surface in the marketplace, and few claims will be made for them. The growing national interest in nutrition should prove helpful in the long run, although that interest also generates consumers for faddism. In the final assessment, profit will remain a significant motivator for faddism and quackery. Nutrition education can make the most long-lasting inroads into such practices.

RELIABLE INFORMATION

With the maze of advertising surrounding food fads, and with cases of downright food quackery included in that advertising, where can the consumer turn for nutrition information? Fortunately, there are a growing number of good sources, some of which have come into being to try to offset the misinformation that is available. Probably the least utilized sources of nutritional information are the professional sources. These include university nutrition departments and related disciplines, medical research institutions, health professionals, government information offices, and the food industry itself. These sources simply have not attracted the attention of consumers, who tend to rely on popular information sources. In defense of consumer choices, though, there are some reasons why professional sources are not popular.

Research Results

Because of strict regulations regarding the use of human subjects, many research studies must rely on animals for experimental investigations. Also, research studies tend to focus on very technical methods. Results are not easily translated into practice directions for consumers. Consumers may have little faith that the nutritional results obtained with rats will necessarily hold true for them; moreover, such research results usually appear in professional journals. Consumers do not typically read such publications and additionally find their technical language difficult to understand.

Government Sources

Government information publications have improved to the point where they are more informative for the typical reader, but most consumers do not know how to obtain these publications. Further, governmental publications still lack the flair that characterizes popular sources.

Recent budget cuts have added a price tag to most of those previously free items. This further alienates interested consumers.

Health Professionals

Physicians are in an idea position to give advice regarding nutritional matters, but many are not trained in nutrition. Further, consumers are reluctant to take on the costs that would likely result from seeking information from their physicians.

Food Industry

Consumers tend to mistrust food industry publications, believing them to be biased. Consumers assume that the food industry is motivated strictly by profit and that such publications are designed to stimulate sales.

Nonprofessional Sources

For the foregoing reasons, professional sources do not prove nearly as popular as newspapers, magazines, television, radio, and word of mouth. Popular sources are easily found, entertaining, and understandable. Although advertisements are rarely untruthful, they certainly do not provide balanced viewpoints. Thus, the consumer's task becomes one of evaluating the information obtained.

EVALUATING NUTRITION INFORMATION

As a consumer, you can take several steps in evaluating food and nutrition claims. The first of these involves determining the promoter's intent. The second step involves determining whether the claim flows logically. The final step focuses on the research on which the claim is based.

Determine whether the promoter is biased. Among the clues to watch for are seductive advertising, unrealistic claims, emotional appeals, and attempts to intimidate you. What kind of publication does the advertisement appear in? Is it credible?

Be attuned to certain natural laws. For example, is it likely that a promoter has discovered a cure for a disease when that cure has eluded researchers for years? Similarly, certain changes take time. If a promoter promises sudden and dramatic change, be suspicious. Watch for the use of words like "may," "probably," or "seems to" and phrases such as "it appears" and "it is thought." These are clues that the advertisement is protecting itself.

More complex techniques involve the use of "buzz words" such as "power," "energy," "new," and "improved." Such words cast a product in favorable light, but look beyond them to determine whether this is justification for the use of such favorable terms.

Has the promoter cited research to support such claims as "new" or "improved"? Has an authority been cited? what are the authority's credentials? Are any data provided? Are the facts recent? It is also important to determine whether the research was conducted with humans or animals. The nature of the research design also proves helpful to the consumer. Researchers who conduct experiments in which human subjects are actually tested to determine the effect of certain treatments typically use one of three research designs: single-blind studies, double-blind studies, and anecdotal records.

Single-blind studies utilize two groups in the experimental study, an experimental group and a control group. If, for example, a supplement is being tested to determine side effects, the experimental group will receive the actual product being tested, while the control group will be receiving an identically appearing product that will have no effect. This "pretend product" is called a placebo.

Single-blind studies have one potential drawback in that those conducting the study know who constitutes the experimental group. There is, therefore, a tendency to see in the experimental group the results hoped for in the study.

In a double-blind study, a third party selects who receives the product being tested and who receives the placebo. Those conducting the research do not know which is which and can therefore be more objective about their findings.

When a researcher draws conclusions on the basis of testimonials from individuals, those brief accounts are called anecdotal records. They are of limited benefit, since no one can be sure of the cause-and-effect relationships. For example, assume that you have frequent headaches and are told that prune juice cures them. You drink prune juice, and the headaches disappear. Did prune cause the change? Could there have been other intervening variables? Could you have been influenced by the "placebo effect"—the tendency to feel better because you expect to feel better?

For the foregoing reasons, anecdotal records are of limited value. However, anecdotes often catch the attention of researchers and point to areas in need for formal research. As a consumer, you should be aware that a personal testimony, even one from a popular individual, can be of limited value and does not necessarily reflect a product's worth.

Newspapers and magazines frequently refer to another type of study that gets significant attention. For example, a newspaper article carries the following statement; "A study done by a certain individual or group shows that Americans who drink more than five cups of coffee daily are likely to have stomach ulcers." Usually, the article refers to a type of study that the public has very little knowledge of, an epidemiological or population study.

In an epidemiological study, the researchers scientifically assemble two groups of patients; for example, 500 in each. All patients in one group have stomach ulcers, with one in the other group. But carefully questioning the patients, the researchers discover that nearly 70% of those who have ulcers drink more than five cups of coffee daily, while in the other group only 40% do so. By evaluating just that much information scientifically—that is, statistically—the researchers come to the conclusion about coffee and ulcers. Note that there is no direct proof of a cause-and-effect relationship. Rather, a scientific "observation" has been made, one that can however, influence consumer behavior and decisions.

BIBLIOGRAPHY

Adapted from: Y. H. Hui, *Principles and Issues in Nutrition,* Copyrighted 1985, Jones and Bartlett Publishers, used with permission.

General References

J. V. Anderson and M. R. Van Nierop, eds., *Basic Nutrition Facts: A Nutrition Reference,* Michigan State University and the Michigan Department of Public Health, East Lansing, Mich., 1989.

F. R. Copper, "Health Claims of Foods—Reflections on the Food/Drug Distinction and on the Law of Misbranding," *American Journal of Clinical Nutrition,* **44,** 560 (1986).

J. D. Gussow and P. R. Thomas, *The Nutrition Debate: Sorting out Some Answers.* Bull Publishing Co., Palo Alto, Calif., 1986.

A. E. Harper, "Scientific Substantiation of Health Claims: How Much is Enough," *Nutrition Today* **24**(2), 17 (1989).

A. E. Harper, 1988. Nutrition: From Myth and Magic to Science," *Nutrition Today* **23**(1), 8 (1988).

V. Herbert, 1987. "Health Claims in Food Labeling and Advertising," *Nutrition Today* **22,** 25 (1987).

M. E. Shils and V. R. Young, eds., *Modern Nutrition in Health and Disease,* Lea & Febiger, Philadelphia, 1988.

P. Szilard, *Food and Nutrition Information Guide,* Libraries Unlimited, Littleton, Colo., 1987.

Y. H. Hui
Editor-in-chief

FANS AND EVAPORATORS. See EVAPORATORS: TECHNOLOGY AND ENGINEERING; EVAPORATION

FAST FREEZING. See FREEZING SYSTEMS FOR THE FOOD INDUSTRY

FATS AND OILS: CHEMISTRY, PHYSICS, AND APPLICATIONS

Fats and oils are the commercially important group of substances classified as lipids. Lipids are compounds usually associated with solubility in nonpolar solvents. They are mostly esters of long-chain fatty acids and alcohols and closely related derivatives. The most important aspect of lipids is the central position of the fatty acids. The scheme of Figure 1 illustrates this (1). The basic components of the lipids (also known as derived lipids) are listed in the central column. In the left-hand column are the phospholipids, most of which contain phosphoric acid groups. The right-hand column includes the compounds most important from a quantitative standpoint in fats and oils. These are mostly esters of fatty acids and glycerol. Fats are solid at room temperature, oils are liquid.

Food fats can be divided into visible and invisible fats (2). More than half of all the fats consumed are in the latter category, that is, those contained in dairy products (excluding butter), eggs, meat, poultry, fish, fruits, vegetables, and grain products. The visible fats include lard, butter, shortening, frying fats and oils, margarines, and salad oils.

Fats and oils may differ considerably in composition, depending on their origin. Fatty acid as well as glyceride composition will influence many of the properties of fats and oils. These properties may be further modified by appropriate processing methods. The fats and oils can be classified broadly into the following groups: animal depot fats, ruminant milk fats, vegetable oils and fats, and marine oils. The processing of fats and oils serves several purposes: to separate the oils or fats from other parts of the raw materials, to remove impurities and undesirable components, to improve stability, to change physical properties, and to provide desirable functional properties.

COMPONENT FATTY ACIDS

Even-numbered straight-chain saturated and unsaturated fatty acids make up the largest proportion of fatty acids in fats and oils. Minor amounts of odd-carbon-number acids, branched-chain acids, and hydroxy acids may also be present. Processed fats, especially hydrogenated fats, may contain a variety of geometric and positional isomers. The division of fatty acids into saturated and unsaturated groups is important because it generally reflects on the melting properties of the fat of which they are a part. There are some exceptions to this rule. Short-chain saturated fatty acids such as those present in milk fats and lauric fats have very low melting points; unsaturated trans isomers have much higher melting points than the cis isomers and are therefore comparable to saturated fatty acids in their effect on melting characteristics.

Some of the important saturated fatty acids are listed in Table 1 with their systematic and common names. The unsaturated fatty acids are listed in Table 2. The naturally occurring unsaturated fatty acids are almost exclusively in the cis form. Trans acids are abundant in ruminant milk fats and in hydrogenated fats.

The depot fats of higher land animals consist mainly of palmitic, oleic, and stearic acids and are high in saturation. The fats of birds are somewhat more complex. The fatty acid compositions of the major food fats of this group are listed in Table 3. The kind of feed consumed by the animals greatly influences the fatty acid composition of the depot fats. For example, the high linolenic acid content of the horse fat in Table 3 is the result of pasture feeding. Animal depot fats are generally low in polyunsaturated fatty acids. The iodine value of beef fat is about 50 and of lard about 60.

Ruminant milk fat has an extremely complex fatty acid composition. The following fatty acids are present in cow's milk fat (3): even and odd saturated acids from 2:0 to 28:0; even and odd monoenoic acids from 10:1 to 26:1 with the exception of 11:1, and-including positional and geometric isomers; even unsaturated fatty acids from 14:2 to 26:2 with some conjugated trans isomers; monobranched fatty acids 9:0 and 11:0 to 25:0; some iso and some anteiso (iso acids have a methyl branch on the penultimate carbon, anteiso on the next to penultimate carbon); multibranched acids from 16:0 to 28:0, both odd and even with three to five methyl branches; and, finally, a number of keto, hydroxy, and cyclic acids.

Marine oils also contain a large number of different fatty acids. As many as 50 to 60 fatty acids have been reported (4). The 14 major ones consist of few saturated fatty acids (14:0, 16:0, and 18:0) and a larger number of

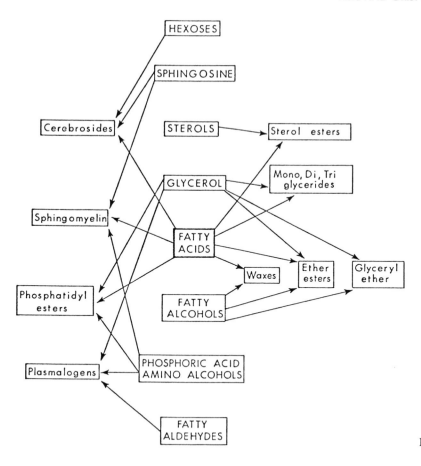

Figure 1. Interrelationship of the lipids.

unsaturated fatty acids with 16 to 22 carbon atoms and up to 6 double bonds. This provides the possibility for many positional isomers. It is customary to number carbon atoms of fatty acids starting from the carboxyl end; however, for biological activity it is interesting to number from the methyl carbon, which is done by using the symbol ω. Three different types of fatty acids can be distinguished, the oleic type with one double bond removed nine carbons from the methyl end ($18:1\omega3$). Three different types of unsaturated fatty acids can be distinguished: the oleic type with one double bond removed nine carbon atoms from their methyl end ($18:1\omega9$ or oleic acid type); the linoleic type with two double bonds removed six carbon atoms from the methyl end ($18:2\omega6$ or linoleic acid type); and the linoleic type with three double bonds removed three carbon atoms from the methyl end ($18:3\omega3$ or lino-

Table 1. Saturated Even- and Odd-Carbon-Numbered Fatty Acids

Systematic Name	Common Name	Formula	Shorthand Description
n-Butanoic	Butyric	$CH_3 \cdot (CH_2)_2 \cdot COOH$	4:0
n-Hexanoic	Caproic	$CH_3 \cdot (CH_2)_4 \cdot COOH$	6:0
n-Octanoic	Caprylic	$CH_3 \cdot (CH_2)_6 \cdot COOH$	8:0
n-Decanoic	Capric	$CH_3 \cdot (CH_2)_8 \cdot COOH$	10:0
n-Dodecanoic	Lauric	$CH_3 \cdot (CH_2)_{10} \cdot COOH$	12:0
n-Tetradecanoic	Myristic	$CH_3 \cdot (CH_2)_{12} \cdot COOH$	14:0
n-Hexadecanoic	Palmitic	$CH_3 \cdot (CH_2)_{14} \cdot COOH$	16:0
n-Octadecanoic	Stearic	$CH_3 \cdot (CH_2)_{16} \cdot COOH$	18:0
n-Eicosanoic	Arachidic	$CH_3 \cdot (CH_2)_{18} \cdot COOH$	20:0
n-Docosanoic	Behenic	$CH_3 \cdot (CH_2)_{20} \cdot COOH$	22:0
n-Pentanoic	Valeric	$CH_3 \cdot (CH_2)_3 \cdot COOH$	5:0
n-Heptanoic	Enanthic	$CH_3 \cdot (CH_2)_5 \cdot COOH$	7:0
n-Nonanoic	Pelargonic	$CH_3 \cdot (CH_2)_7 \cdot COOH$	9:0
n-Undecanoic		$CH_3 \cdot (CH_2)_9 \cdot COOH$	11:0
n-Tridecanoic		$CH_3 \cdot (CH_2)_{11} \cdot COOH$	13:0
n-Pentadecanoic		$CH_3 \cdot (CH_2)_{13} \cdot COOH$	15:0
n-Heptadecanoic	Margaric	$CH_3 \cdot (CH_2)_{15} \cdot COOH$	17:0

Table 2. Unsaturated Fatty Acids

Systematic Name	Common Name	Formula	Shorthand Description
Dec-9-enoic		$CH_2 \cdot CH \cdot (CH_2)_7 \cdot COOH$	10:1
Dodec-9-enoic		$CH_3 \cdot CH_2 \cdot CH=CH \cdot (CH_2)_7 \cdot COOH$	12:1
Tetradec-9-enoic	Myristoleic	$CH_3 \cdot (CH_2)_3 \cdot CH=CH \cdot (CH_2)_7 \cdot COOH$	14:1
Hexadec-9-enoic	Palmitoleic	$CH_3 \cdot (CH_2)_5 \cdot CH=CH \cdot (CH_2)_7 \cdot COOH$	16:1
Octadec-6-enoic	Petroselinic	$CH_3 \cdot (CH_2)_{10} \cdot CH=CH \cdot (CH_2)_4 \cdot COOH$	18:1
Octadec-9-enoic	Oleic	$CH_3 \cdot (CH_2)_7 \cdot CH=CH \cdot (CH_2)_7 \cdot COOH$	18:1
Octadec-11-enoic	Vaccenic	$CH_3 \cdot (CH_2)_5 \cdot CH=CH \cdot (CH_2)_9 \cdot COOH$	18:1
Octadeca-9:12-dienoic	Linoleic	$CH_3 \cdot (CH_2)_4 \cdot (CH:CH \cdot CH_2)_2 \cdot (CH_2)_6 \cdot COOH$	18:2ω6
Octadeca-9:12:15-trienoic	Linolenic	$CH_3 \cdot CH_2 \cdot (CH:CH \cdot CH_2)_3 \cdot (CH_2)_6 \cdot COOH$	18:3ω3
Octadeca-6:9:12-trienoic	Gamma linolenic	$CH_3 \cdot (CH_2)_4 \cdot (CH:CH \cdot CH_2)_3 \cdot (CH_2)_3 \cdot COOH$	18:3ω6
Octadeca-9:11:13-trienoic	Elaeostearic	$CH_3 \cdot (CH_2)_3 \cdot (CH:CH)_3 \cdot (CH_2)_7 \cdot COOH$	20:3
Eicos-9-enoic	Gadoleic	$CH_3 \cdot (CH_2)_9 \cdot CH:CH \cdot (CH_2)_7 \cdot COOH$	20:1
Eicosa-5:8:11:14-tetraenoic	Arachidonic	$CH_3 \cdot (CH_2)_4 \cdot (CH:CH \cdot CH_2)_4 \cdot (CH_2)_2 \cdot COOH$	20:4ω6
Eicosa-11:14:17-pentaenoic acid	EPA	$CH_3 \cdot CH_2 \cdot (CH:CH \cdot CH_2)_5 \cdot (CH_2)_2 \cdot COOH$	20:5ω3
Docosa-13-enoic	Erucic	$CH_3 \cdot (CH_2)_7 \cdot CH:CH \cdot (CH_2)_{11} \cdot COOH$	22:1
Docosa-13:16:19-hexaenoic acid	DHA	$CH_3 \cdot CH_2(CH:CH \cdot CH_2)_6 \cdot (CH_2) \cdot COOH$	22:6ω3

leic acid type). The end structure is usually retained even if additional double bonds are introduced or if additional carbon atoms are added. Thus, linoleic acid (18:2ω6) may be changed into arachidonic acid (20:4ω6) while retaining the ω6 structure which confers essential fatty acid character to the molecule. The latter two types are now often referred to as n-6 and n-3 fatty acids. The high content of polyenoic fatty acids makes fish oils highly susceptible to autoxidation. The component fatty acids of some marine and freshwater fish oils are listed in Table 4 (5).

Considerable interest has developed recently in the health effect of certain n-3 fatty acids, especially eicosapentaenoic acid (EPA) (20:5ω3) and docosahexaenoic acid (DHA) (22:6ω3). These fatty acids can be produced slowly from linolenic acid by herbivore animals, but not by humans. EPA and DHA occur in major amounts in fish from cold deep waters, such as cod, mackerel, tuna, swordfish, sardines, and herring (6,7). Arachidonic acid is the precursor in the human system of prostanoids and leukotrienes.

The vegetable oils and fats can be divided into three groups on the basis of their fatty acid composition. The first group comprises oils containing mainly 16- and 18-carbon fatty acids and includes most of the seed oils—cottonseed oil, peanut oil, sunflower oil, corn oil, sesame oil—as well as palm oil. The second group comprises seed oils containing erucic acid (docos-13-enoic), and includes rapeseed and mustard oil. The third group is that of the vegetable fats, comprising coconut oil and palm kernel oil, which are highly saturated and also known as lauric fats, as well as cocoa butter. The component fatty acids of some of the common vegetable oils and fats are listed in Table 5. palmitic is the most common saturated fatty acid. Oils containing high levels of linolenic acid are susceptible to rapid oxidative deterioration.

The Crucifera seed oils, including rapeseed and mustard oil, are characterized by high levels of erucic acid (docos-13-enoic) and smaller amounts of eicos-11-enoic acid. Plant breeders have succeeded in replacing virtually all of these fatty acids by oleic acid, resulting in what is now known as canola oil (8).

Cocoa butter is unusual in that it contains only three major fatty acids, palmitic, stearic, and oleic, in approximately equal proportions.

COMPONENT GLYCERIDES

When a fat or oil is characterized by the determination of its component fatty acids, there still remains the question of how these acids are distributed among and within the glycerides. The stereospecific numbering (sn) of glycerol for a triacid glyceride is as follows:

$$CH_2O_2CR^1 \qquad (1)$$
$$R^2CO_2CH \qquad (2)$$
$$CH_2O_2CR^3 \qquad (3)$$

Table 3. Component Fatty Acids of Animal Depot Fats

Animal	Fatty Acids, % of Total									
	12:0	14:0	16:0	18:0	20:0	16:1	18:1	18:2	18:3	20:1
Cow		6.3	27.4	14.1			49.6	2.5		
Pig		1.8	21.8	8.9	0.8	4.2	53.4	6.6	0.8	0.8
Sheep		4.6	24.6	30.5			36.0	4.3		
Goat	3.5	2.1	25.5	28.1	2.4		38.4			
Horse	0.4	4.5	25.9	4.7	0.2	6.8	33.7	5.2	16.3	2.3
Chicken	1.9	2.5	36.0	2.4		8.2	48.2	0.8		
Turkey	0.1	0.8	20.0	6.4	1.3	6.2	38.4	23.7	1.6	

Table 4. Major Component Fatty Acids of Some Marine and Freshwater Fish Oils[a,b]

	14:0	16:0	18:0	16:1	18:1	20:1	22:1	18:2ω6	18:3ω3	18:4ω3	20:4ω6	20:5ω3	22:5ω3	22:6ω3
Herring	6.4	12.7	0.9	8.8	12.7	14.1	20.8	1.1	0.6	1.7	0.3	8.4	0.8	4.9
Turbot	6.5	12.0	0.9	15.4	17.4	18.6	17.8	0.5	0.2	0.8	0.1	3.0	0.6	1.9
Sablefish	6.7	11.1	1.9	6.6	29.0	18.1	14.8	0.7	0.2	0.3	0.3	1.4	0.5	1.0
Cod	1.4	19.6	3.8	3.5	13.8	3.0	1.0	0.7	0.1	0.4	2.5	17.0	1.3	29.8
Sole	4.3	16.5	2.4	14.4	12.2	3.9	Tr	0.3	2.0	1.6	4.0	11.9	10.6	7.0
Halibut	0.8	9.6	9.0	2.5	12.3	4.0	5.0	Tr	Tr		1.4	13.0	2.5	37.6
Carp	3.1	16.8	4.3	17.1	28.3	3.9		13.2	2.3		2.5	3.2		
Trout	2.7	20.9	8.3	3.9	18.4			7.3	1.6	3.2	1.7	5.8	Tr	7.0
Catfish	1.0	15.2	3.9	2.9	29.7	0.9		10.0	0.5	0.4	0.8	0.2	0.2	0.6

[a] Ref. 5.
[b] Values are percentage of total; Tr = trace.

The molecule is shown in the Fisher projection with the secondary hydroxyl pointing to the left. The location of fatty acids in the various positions on the glycerol molecule can be determined by stereospecific analysis (9). Several theories of glyceride composition have been proposed, such as even distribution, random distribution, and restricted random distribution. The distribution of fatty acids in the glycerides is of utmost importance for the physical properties of a fat. This is illustrated by pig fat and cocoa butter, which have similar fatty acid composition. In pig fat most of the unsaturation is located in the 1- and 3-positions, in cocoa butter it is in the 2-position (Table 6).

PHOSPHOLIPIDS

All fats and oils and fat-containing foods contain a number of phospholipids. The lowest amounts are present in animal fats such as lard and beef tallow. In some crude vegetable oils, such as cottonseed, corn, and soybean oils, phospholipids may be present at levels of 2 to 3%. Phospholipids are surface active, because they contain a lipophilic and a hydrophilic portion. Since they can be easily hydrated, they can be removed from fats and oils during the refining process. The structure of the most important phospholipids is given in Figure 2. The phospholipids removed from soybean oil are used as emulsifiers in foods. Soybean phospholipids, also known as soy lecithin, contain about 35% lecithin and 65% cephalin. The acyl groups in phospholipids are usually more unsaturated than those of the triglycerides in which they are present. Saturated fatty acids are found mostly in position 1 and unsaturated fatty acids in position 2.

UNSAPONIFIABLE COMPONENTS

The unsaponifiable portion of fats consists of sterols, terpenic alcohols, squalene, and hydrocarbons. In most fats the major unsaponifiable component is sterols. Animal fats contain cholesterol, and in some cases, minor amounts of the other sterols, such as lanosterol. Plant fats and oils contain phytosterols, usually at least three and sometimes four (10). The predominant phytosterol is β-sitosterol; others are campesterol and stigmasterol. The sterols are solids with high melting points; part of the sterols in natural fats are present as esters of fatty acids, part in free form. Cholesterol makes up 99% of the sterols of fish. The sterol content of some fats and oils is given in Table 7.

PROCESSING

In the commercial production of fats and oils, processing is used to separate, purify, and modify the oils and fats to make them suitable for the various functions they fulfill in the food system. A large portion of the seed oils produced in the temperate regions of the world are used in the form of solid fats: margarine, shortening, and frying and baking fats. Hydrogenation is most often used to change oils into fats.

Table 5. Component Fatty Acids of Some Vegetable Oils[a,b]

Oil	Fatty Acid									Total C_{18}
	14:0	16:0	18:0	20:0	22:0	16:1	18:1	18:2	18:3	
Cottonseed	1	29	4	Tr		2	24	40		68
Peanut	Tr	6	5	2	3	Tr	61	22		88
Sunflower		4	3				34	59		96
Corn		13	4	Tr	Tr		29	54		87
Sesame		10	5				40	45		90
Olive	Tr	14	2	Tr		2	64	16		82
Palm	1	48	4				38	9		51
Soybean	Tr	11	4	Tr	Tr		25	51	9	89
Safflower	Tr	8	3	Tr			13	75	1	92

[a] Ref. 10
[b] Values are weight%, Tr = trace.

Table 6. Positional Distribution of Fatty Acids in Pig Fat and Cocoa Butter

Fat	Position	Fatty Acid, mole %					
		14:0	16:0	16:1	18:0	18:1	18:2
Pig fat	1	0.9	9.5	2.4	29.5	51.3	6.4
	2	4.1	72.3	4.8	2.1	13.4	3.3
	3	0	0.4	1.5	7.4	72.7	18.2
Cocoa butter	1		34.0	0.6	50.4	12.3	1.3
	2		1.7	0.2	2.1	87.4	8.6
	3		36.5	0.3	52.8	8.6	0.4

The separation of oils and fats from animal tissues is done by rendering, either dry rendering or steam rendering. The separation of oils from oilseeds usually involves a pretreatment, crushing or flaking of the seeds, followed by pressing (11). This is usually followed by solvent extraction to remove the remainder of the oil and yield a residue with less than 1% residual oil (12).

The crude oils obtained by rendering, pressing, and/or extraction are purified by a series of operations designed to remove impurities that may detract from the quality of the oil. Removal of phospholipids is achieved by degumming (13). The crude oils are treated with steam, which hydrates the phospholipids and makes them settle out. Degumming can also be achieved by using solutions of phosphoric or organic acids. The soybean "gums" are purified and used as food emulsifiers, known as soy lecithin.

Free fatty acids in crude oils are removed by alkali refining (13). Solutions of caustic soda are used to reduce the level of free fatty acid to 0.01 to 0.03%. Care is required to prevent saponification of neutral oil. Removal of free fatty acids can also be achieved by physical refining. This involves treatment of the oils under vacuum with steam. The advantage of physical refining is that the process is similar to deodorization and these processes can be combined.

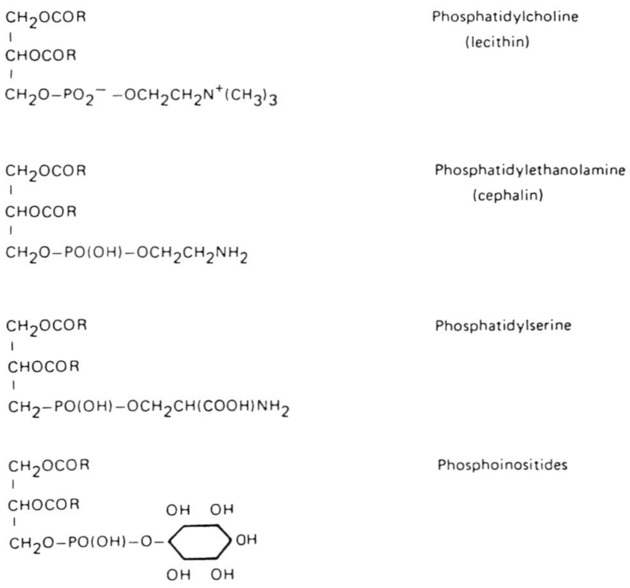

Bleaching is used to remove colored impurities, such as carotenoids and chlorophyll. In the bleaching process the oils are treated with bleaching earth or activated carbon. The yellow-red color of most vegetable oils, mostly carotenoids, is easily removed by bleaching earth. The green and brown pigments are more difficult to remove.

After refining and bleaching, vegetable oils are further processed into margarines, shortenings, and frying and baking fats. Two-thirds of all liquid oils produced in North America are used in the form of fats. Hydrogenation is used to change oils into fats, and involves the reaction of gaseous hydrogen, liquid oil, and solid catalyst under pressure and at high temperature (14). The catalyst used for edible oil hydrogenation is invariably of the activated nickel metal type (15). The hydrogenation reaction can be represented by the following scheme, in which the reacting species are the olefinic substrate (S), the metal catalyst (M), and hydrogen:

$$S + M \rightleftarrows [S-M] \quad (1)$$

$$M + H_2 \rightleftarrows [M-H_2] \quad (2)$$

$$[S-M-H_2] \rightarrow SH_2 + M \quad (3)$$

The intermediates 1, 2, and 3 are organometallic species and are labile and short-lived and cannot usually be isolated. In heterogeneous catalysis the metal surface performs the catalytic function. In theory, the finer the particle size, the more active the catalyst will be. In practice, however, particle size has to be balanced against filterability, since removal of the catalyst at the end of the process should not be too difficult.

```
CH2OCOR                                         Phosphatidylcholine
|                                                  (lecithin)
CHOCOR
|
CH2O-PO2- -OCH2CH2N+(CH3)3


CH2OCOR                                         Phosphatidylethanolamine
|                                                  (cephalin)
CHOCOR
|
CH2O-PO(OH)-OCH2CH2NH2


CH2OCOR                                         Phosphatidylserine
|
CHOCOR
|
CH2-PO(OH)-OCH2CH(COOH)NH2


CH2OCOR                                         Phosphoinositides
|
CHOCOR
|
CH2O-PO(OH)-O-[inositol ring with OH groups]
```

Figure 2. Structure of the major phospholipids.

Table 7. Sterol Content of Fats and Oils

Fat	Sterol, %
Lard	0.12
Beef tallow	0.08
Milk fat	0.3
Herring	0.2–0.6
Cottonseed	1.4
Soybean	0.7
Corn	1.0
Rapeseed	0.4
Coconut	0.08
Cocoa butter	0.2

When hydrogen is added to double bonds in natural fats and the reaction is not carried to completion, a complex mixture of reaction products results. Hydrogenation may be selective or nonselective. Selectivity means that hydrogen is added first to the most unsaturated fatty acids. Selectivity is increased by increasing hydrogenation temperature and decreased by increasing pressure and agitation. Selectively hydrogenated oil is more resistant to oxidation because of the preferential hydrogenation of the linolenic acid.

Another important factor in hydrogenation is the formation of positional and geometric isomers. Formation of trans isomers is rapid and extensive. The isomerization can be understood by the reversible character of chemisorption. When the olefinic bond reacts, two carbon–metal bonds are formed as an intermediate stage. The intermediate may react with an atom of adsorbed hydrogen to yield the half-hydrogenated compound, which remains attached by only one bond. Additional reaction with hydrogen would result in formation of a saturated compound. There is also the possibility that the half-hydrogenated olefin may again attach itself to the catalyst surface at a carbon on either side of the existing bond, with simultaneous loss of hydrogen. Upon desorption of this species a positional or geometric isomer may result. The proportion of trans isomers is high because this is the more stable configuration. Double bond migration occurs in both directions, but probably more extensively in the direction of the terminal methyl group. The hydrogenation of oleate can be represented as follows:

The change from oleate to isooleate involves no change in unsaturation but does result in a considerably higher melting point. This is why the hardening effect of hydrogenation is only partly the result of saturating double bonds; trans-isomer formation has a major effect on hardness.

For example, olive oil with an iodine value of 80 is liquid at room temperature. When soybean oil is hydrogenated to the same iodine value it is a fat with the consistency of lard.

Hydrogenation of linoleate first produces some conjugated dienes, followed by the formation of positional and geometric isomers of oleic acid, and finally stearate:

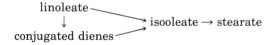

Hydrogenation of linolenate is more complex and is greatly dependent on reaction conditions. The possible reactions can be summarized as follows:

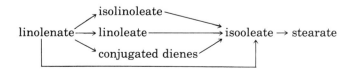

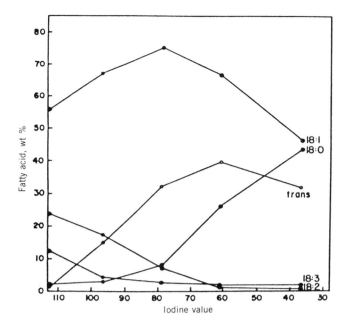

Figure 3. Change in fatty acid composition during hydrogenation of canola oil.

In the nonselective hydrogenation of seed oils, polyunsaturated fatty acids are rapidly isomerized or reduced and trans isomers increase to high levels (Fig. 3).

Interesterification is a process whereby fatty acid radicals can be made to move from one hydroxyl of a glycerol moiety to another one, either within the same glyceride or to another glyceride. The reaction pattern has been described (16):

$$RCOOR^2 + R^1COOR^3 \rightarrow RCOOR^3 + R^1COOR^2$$

The reaction is used in industry to modify the crystallization behavior and the physical properties of fats. The catalysts are usually alkaline and consist of sodium methoxide or alloys of sodium and potassium. At temperatures above the melting points of the reactants, several raw materials may be interesterified together so that new products are produced. If the reaction is carried out below the melting point, so that only the liquid fraction reacts, the process is called directed interesterification. Industrially, lard has been interesterified to improve its properties. Lard has a narrow plastic range, creams poorly, and gives poor cake volume. After interesterification these properties are greatly improved. Acetoglycerides can be prepared by interesterification of natural fats with glyceryl triacetate. The resultant products are waxy, translucent materials that can be used as edible coatings and plasticizers.

Interesterification may provide an alternative to hydrogenation for the production of margarine and shortening fats. Hydrogenation has the disadvantage of forming trans isomers and of losing essential fatty acids. Interesterification of liquid oils with highly saturated fats, obtained by complete hydrogenation or by fractionation, may result in nutritionally more desirable fats.

Ester interchange of fats with a large excess of glycerol, at high temperature, under vacuum, and in the presence of a catalyst, results in an equilibrium mixture of mono-, di-, and triglycerides. After removal of excess glycerol, the mixture is called technical monoglyceride. Technical monoglycerides are used as emulsifying agents in foods. Molecular distillation yields products with well over 90% 1-monoglycerides; these are also widely used in foods.

PHYSICAL PROPERTIES

Fats and oils are long-chain compounds that have particular physical properties of importance in processing and final use. Properties of surface activity, viscosity, solubility, and melting behavior are important in formulating emulsions and fat products such as shortenings and margarines.

The most important aspect of the physical properties of fats and oils is related to the solid–liquid and liquid–solid phase changes; in other words, melting and solidification. In fact, fats can be defined as partially solidified oils. When oils solidify they form crystals, usually of a size in the range of 1 to 10 μm. These fat crystals form a three-dimensional network that lends solid properties to the fat. The nature of this crystal network determines the rheological or textural properties of the product (17). The amount and size of the crystals in a fat determine its physical properties by influencing the density of junction points and, therefore, the strength of the crystal network. The crystals are held together by weak van der Waals forces and these can be disrupted by working or kneading. Upon resting, many of these bonds may be reestablished and the product is said to be thixotropic. Not all bonds are reformed after working, leading to what is known as work softening. The crystals in fats can be observed by polarized light microscopy as shown in Figure 4 (18).

The proportion of solids in a fat is of major importance in determining the rheological properties of a product. Fats may retain their solid character with solid fat content as low as 10%. Desirable spreadability occurs in a narrow range of solid fat content, roughly 15–35%. This is called the plastic range of fats.

Figure 4. Fat crystals in a fat as seen in the polarizing microscope.

Solidification of liquid oil results in a volume contraction and a positive (exothermic) heat effect. Melting of a fat results in volume expansion and a negative (exothermic) heat effect. Traditional methods of determination of solid fat depend on specific volume measurement at a given temperature (dilatometry). The heat effect is used in differential scanning calorimetry to determine melting and solidification properties. Modern methods of solid fat determination are based on nuclear magnetic resonance. Protons in solid fat behave differently in a magnetic field after being excited by radio-frequency energy than protons in liquid fat. This enables rapid and accurate estimation of solid fat content. The dilatometric method is empirical and the results are expressed as solid fat index (SFI).

In addition to crystal content and crystal size and shape, an important factor in the solidification of fats is polymorphism (the existence of different crystal modifications). Polymorphism results from different patterns of molecular packing in fat crystals. Triglycerides may occur in three main forms named α, β', and β, in order of increasing stability. After crystals of a lower melting form have been produced, change into a higher melting form may take place. The change is monotropic, that is, it always proceeds in the sequence from lower to higher stability. When a fat is crystallized in an unstable form and heated at a temperature slightly above its melting point, it may resolidify in a more stable form (18). The melting point of tristearin in the α form is 54.7°C; in the β' form, 64.0°C; and in the β form, 73.3°C. The cross-sectional structures of long-chain compounds in the different polymorphic forms is shown in Figure 5. In the α form the chain axes are randomly oriented and the crystals are of the hexagonal type. In the β' form alternate rows are oriented in opposite directions and the crystal type is orthorhombic. In the β form the rows are oriented in the same direction and the crystal type is triclinic. The polymorphic forms are also distinguished by differences in the angle of tilt of the unit cell in the crystal. X-ray diffraction analysis enables measurement of the short spacings and long spacings of the unit cell. Principal short spacings of the polymorphic forms are: α, 4.15 Å; β', 4.2 and 3.8 Å; 4.6, 3.85, and 3.7 Å (19). Fat crystals of different polymorphs have different crystal habit. That is, they appear to have different shapes when seen in the microscope: α crystals are fragile, transparent platelets of about 5-μm size; β' crystals are tiny needles about 1 μm long; β crystals are large and coarse, averaging 25 to 50 μm in size (20). These crystals can now be made visible by scanning electron microscopy (21).

For the production of margarines and most shortenings, the fat should be present in the β' form. Some fats

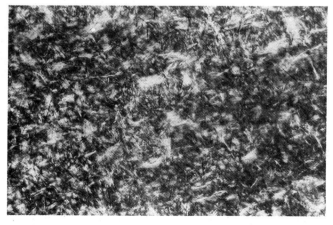

Figure 5. Cross-sectional structures of long-chain compounds. (18). Courtesy of *Journal of the American Oil Chemists' Society*.

have a tendency to convert slowly from the β' to the β form and this results in a coarsening of the crystal structure, making the product unacceptable. Fats which contain a very high level of fatty acids with identical chain lengths are especially prone to this effect. Canola and sunflower oils are notorious in this respect and have well over 90% of C_{18} fatty acids (Table 5).

The melting point of a fat is basically determined by the melting points of its constituent fatty acids. When we speak of the melting point of a fat we mean the end of the melting range, since only simple substances have a sharp melting point. The end of the melting range of a fat is dependent on the method used. Chain length and unsaturation of fatty acids determine their melting point. In addition, the configuration around the double bond is important, as is the arrangement of fatty acids at the different positions of the glycerides. Trans-unsaturated fatty acids have much higher melting points than their cis counterparts, making them more comparable with saturated fatty acids. For example, oleic acid (cis) has a melting point of 13°C and elaidic acid (trans) has a melting point of 44°C.

AUTOXIDATION AND ANTIOXIDANTS

The unsaturated bonds present in all oils and fats represent active centers, which may react with oxygen. This reaction leads to the formation of primary, secondary, and tertiary oxidation products which may result in the fat or fat-containing food becoming unfit for consumption. Oxidation of fats and oils may occur by a free-radical chain reaction known as autoxidation. Another process of oxidative deterioration involves the presence of a sensitizer and exposure to light. This is known as photosensitized oxidation (22,23).

The process of autoxidation and the resulting deterioration in flavor of fats and fatty foods are often described by the term rancidity. Lundberg (24) distinguishes several types of rancidity. When fats are exposed to oxygen, common oxidative rancidity will result in sweet but undesirable odors and flavors which will progressively become more intense and unpleasant. Flavor reversion is the term used for the objectionable flavors that develop in oils containing linolenic acid, especially soybean oil. This type of oxidation is produced with considerably less oxygen than common oxidation.

Among the many factors which affect the rate of oxidation are the following: amount of oxygen present, degree of unsaturation of the oil, presence of pro- and antioxidants, presence of heme-containing molecules and lipoxidase, light exposure, nature of packaging material, and storage temperature.

The autoxidation reaction can be divided into the following three parts: initiation, propagation, and termination. During the initiation, hydrogen is abstracted from an olefinic compound to yield a free radical:

$$RH \rightarrow R^{\cdot} + H^{\cdot}$$

The removal of hydrogen takes place at the carbon atom next to the double bond. The dissociation energy of hydrogen in various olefinic compounds has been listed by Ohloff (25). The value for an isolated double bond is 103 kcal/mol; for two double bonds separated by a methylene group it is only 65 kcal/mol. Once a free radical is formed, it will combine with oxygen to form a peroxy free radical, which can in turn abstract hydrogen from another unsaturated molecule to yield a peroxide and a new free radical, thus starting the propagation reaction. This may be repeated up to several thousand times and has the nature of a chain reaction:

$$R^{\cdot} + O_2 \rightarrow RO_2^{\cdot}$$
$$RO_2^{\cdot} + RH \rightarrow ROOH + R^{\cdot}$$

The propagation reaction can be followed by termination if the free radicals react with themselves to yield nonactive products:

$$R^{\cdot} + R^{\cdot} \rightarrow R\text{---}R$$
$$R^{\cdot} + RO_2^{\cdot} \rightarrow RO_2R$$
$$nRO_2^{\cdot} \rightarrow (RO_2)_n$$

The hydroperoxides formed in the propagation part of the reaction are the primary oxidation products. The hydroperoxide mechanism of autoxidation has been described (26) and reviewed (22). The primary oxidation products are unstable and decompose into secondary oxidation products, mostly carbonyls. The peroxides are not importance in flavor deterioration, which is caused wholly by secondary oxidation products. In the initial stages of the reaction there is a slow increase in the amount of hydroperoxides formed; this is the induction period. At the end of the induction period there is a sudden and rapid increase in peroxide content. The induction period is measured in accelerated tests to determine the storage stability of a fat or oil.

The rate of oxidation depends greatly on the degree of unsaturation. In the series of 18-carbon-atom fatty acids 18:0, 18:1, 18:2, 18:3, the relative rates of oxidation have been reported to be in the ratio of 1:100:1200:2500. The reaction of unsaturated compounds proceeds by the abstraction of hydrogen from a carbon α to the double bond, resulting in a free radical stabilized by resonance. These free radicals are then transformed into a number of isomeric hydroperoxides of the general structure

$$\text{---CH}_2\text{---CH---CH=CH---}$$
$$|$$
$$\text{OOH}$$

In addition to the changes in double bond position, there are isomerizations from cis to trans and 90% of the peroxides formed may be in the trans configuration (24).

The decomposition of hydroperoxides has been described (27). It involves decomposition to alkoxy and hydroxy aldehydes, which are in great part responsible for the oxidized flavor of fats. They are powerful flavor compounds with very low flavor thresholds. For example, 2,4-decadienal has a flavor threshold of less than one part per billion.

Trace metals, especially copper, and to a lesser extent iron, will catalyze fat oxidation; metal deactivators, such as citric acid, can be used to counteract the effect of metals.

Antioxidants may be present naturally or added to fats and oils. Many foods contain natural antioxidants; the tocopherols are the most important of these. They are present in greater amounts in vegetable oils than in animal fats, which may explain the greater stability of the former. Antioxidants react with free radicals, thereby terminating the chain reaction. The synthetic antioxidants are phenolic compounds and limited in number. The four most widely used are BHA (butylated hydroxy anisole), BHT (butylated hydroxy toluene), PG (propyl gallate), and TBHQ (tertiary butyl hydroquinone). BHT is volatile with steam, which makes it less suitable for use in frying oils and baked products. Nonvolatile antioxidants are said to have carry-through properties for these applications.

EMULSIONS AND EMULSIFIERS

An emulsion is a heterogeneous system of two immiscible liquids, one of which is intimately dispersed in the other. The droplets of the dispersed phase generally have a diameter of over 0.1 μm. The emulsions are stabilized by the presence of a third component, the emulsifier, a surface-active agent which is partly soluble in both phases (28). Food emulsions usually contain water and oil as the immiscible phases, giving rise to emulsions of the oil-in-water (O/W) type or water-in-oil (W/O) type. The action of emulsifiers can be enhanced by the presence of stabilizers. Emulsifiers are surface-active compounds that have the ability to reduce the interfacial tension between liquid-liquid and air-liquid interfaces. This ability is the result of an emulsifier's molecular structure: the molecules contain two distinct sections, one having a polar or hydrophilic character, the other having nonpolar or hydrophobic properties. Most surface-active agents reduce the surface tension from about 50 dynes/cm to less than 10 dynes/cm when used in concentrations below 0.2%.

The relative sizes of the hydrophilic and hydrophobic sections of an emulsifier molecule mostly determine its behavior in emulsification. To select the proper emulsifier for a given application, the so-called HLB system was developed (HLB = hydrophile-lipophile balance). It is a numerical expression for the relative simultaneous attraction of an emulsifier for water and for oil. The HLB of an emulsifier is an indication of how it will behave but not how efficient it is. Emulsifiers with low HLB tend to form W/O emulsions, those with intermediate HLB form O/W emulsions, and those with high HLB are solubilizing agents. The HLB value can be either calculated or determined experimentally (29). The scale goes from 0 to 20, at least in theory, since at each end of the scale the compounds would have little emulsifying activity. The HLB value of some commercial nonionic emulsifiers is given in Table 8 (30).

Foods contain many natural emulsifiers, of which phospholipids are the most common.

Emulsions are stabilized by a variety of compounds, mostly macromolecules such as proteins, starches, and gums.

FUTURE DEVELOPMENTS

Fats and oils are important in the diet for supplying energy, but in addition serve to improve palatability by providing a smooth mouthfeel and act as a carrier for fat-soluble flavors and vitamins. Recent developments have indicated the possible changes that are likely to occur in the fats and oils supply in the future. It has been customary to think that natural oils and fats have a relatively constant composition, exemplified in fatty acid composition and a number of chemical and physical constants.

Table 8. HLB Values of Some Commercial Nonionic Emulsifiers

Trade Name	Chemical Designation	HLB
Span 85	Sorbitan trioleate	1.8
Span 65	Sorbitan tristearate	2.1
Atmos 150	Mono- and diglycerides from the glycerolysis of edible fats	3.2
Atmul 500	Mono- and diglycerides from the glycerolysis of edible fats	3.5
Atmul 84	Glycerol monostearate	3.8
Span 80	Sorbitan monooleate	4.3
Span 60	Sorbitan monostearate	4.7
Span 40	Sorbitan monopalmitate	6.7
Span 20	Sorbitan monolaurate	8.6
Tween 61	Poly(oxyethylene sorbitan monostearate)	9.6
Tween 81	Poly(oxyethylene sorbitan monooleate)	10.0
Tween 85	Poly(oxyethylene sorbitan trioleate)	11.0
Arlacel 165	Glycerol monostearate (acid stable, self-emulsifying)	11.0
Myrj 45	Poly(oxyethylene monostearate)	11.1
Atlas G-2127	Poly(oxyethylene monolaurate)	12.8
Myrj 49	Poly(oxyethylene monostearate)	15.0
Myrj 51	Poly(oxyethylene monostearate)	16.0

Ref. 30.

The recent development of canola oil from rapeseed oil by plant breeding is an indication that the composition of oils can be changed drastically. This will be even more likely as biotechnology provides added tools to the plant breeder. High oleic acid sunflower oil and low linolenic soybean oil are already among the possible "new" oils.

Another area where changes are likely to occur is in processing. Although the vegetable oils produced in temperate regions are highly unsaturated, two-thirds of these oils are used in the hydrogenated form for the production of margarine, shortening, and frying fat. By interesterification of naturally occurring solid fat, such as beef fat or palm oil, with these unsaturated oils, a nutritionally superior end product could be obtained. Much research interest is being generated in the use of microbial lipases for interesterification (31).

Because fats are the most concentrated source of energy, yielding 9 kcal/g, there has been interest in developing products which perform the functional properties of fats but supply few or no calories. Several of these fat substitutes are at various stages of development and testing (32). Olestra is a sucrose polyester, a mixture of hexa-, hepta-, and octaesters of sucrose with long-chain fatty acids. It behaves as a fat in food preparation but is not digested or absorbed in the body. Another fat substitute, Simplesse, is produced by a special process called microparticulation from milk or egg white proteins. It can be used as a substitute for fats in certain foods, but cannot be used for frying or baking. Other fat substitutes are made from carbohydrates and supply less than half the calories of fat. Some of these fat substitutes face years of regulatory investigation to establish safety in use.

BIBLIOGRAPHY

1. J. M. de Man, *Principles of Food Chemistry*, 2nd ed., Van Nostrand Reinhold, New York, 1989.
2. L. M. Smith, "Introduction to the Symposium on Milk Lipids," *Journal of the American Oil Chemists' Society* **50**, 175–177 (1973).
3. R. G. Jensen, "Composition of Bovine Milk Lipids," *Journal of the American Oil Chemists' Society* **50**, 186–192 (1973).
4. R. G. Ackman, "The Analysis of Fatty Acids and Related Materials by Gas-Liquid Chromatography," in R. T. Holman, ed., *Progress in the Chemistry of Fats and Other Lipids*, Vol. 12, Pergamon Press, Oxford, 1972.
5. R. G. Ackman, "Marine Lipids and Fatty Acids in Human Nutrition," *FAO Technical Conference on Fishery Products, Japan*, 1973.
6. R. G. Ackman, "The Year of the Fish Oils," *Chemical Industries* March, 139–145 (1988).
7. A. P. Simopoulos, "ω-3 Fatty Acids in Growth and Development and in Health and Disease," *Nutrition Today*, 10–19, (March/April 1988).
8. J. K. Daun, "Composition and Use of Canola Seed, Oil, and Meal," *Cereal Foods World* **29**, 291–293 (1984).
9. A. Kuksis, "Newer Developments in Determination of Structure of Glycerides and Phosphoglycerides," in R. T. Holman, ed., *Progress in the Chemistry of Fats and Other Lipids*, Vol. 12., Pergamon Press, Oxford, 1972.
10. E. Fedeli and G. Jacini, "Lipid Composition of Vegetable Oils," *Advances in Lipid Research* **9**, 335–382 (1971).
11. D. K. Bredeson, "Mechanical Oil Extraction," *Journal of the American Oil Chemists' Society* **60**, 211–213 (1983).
12. P. L. Christensen, "Solvent Extraction: Recent Developments," *Journal of the American Oil Chemists' Society* **60**, 214–215 (1983).
13. G. Haraldsson, "Degumming, Dewaxing and Refining," *Journal of the American Oil Chemists' Society* **60**, 251–256 (1983).
14. R. Larsson, "Hydrogenation Theory: Some Aspects," *Journal of the American Oil Chemists' Society* **60**, 275–281 (1983).
15. M. J. Beckman, "Hydrogenation Practice," *Journal of the American Oil Chemists' Society* **60**, 282–290 (1983).
16. M. W. Formo, "Ester Reactions of Fatty Materials," *Journal of the American Oil Chemists Society* **31**, 548–549 (1954).
17. J. M. de Man and A. M. Beers, "Fat Crystal Networks: Structure and Rheological Properties," *Journal of Texture Studies* **18**, 303–318 (1987).
18. E. S. Lutton, "Technical Lipid Structures," *Journal of the American Oil Chemists' Society* **49**, 1–9 (1972).
19. J. M. de Man, "Microscopy in the Study of Fats and Emulsions," *Food Microstructure* **1**, 209–222 (1982).
20. C. W. Hoerr, "Morphology of Fats, Oils and Shortenings," *Journal of the American Oil Chemists' Society* **37**, 539–546 (1960).
21. A. N. Mostafa, A. K. Smith, and J. M. de Man, "Crystal Structure of Hydrogenated Canola Oil," *Journal of the American Oil Chemists' Society* **62**, 760–762 (1985).
22. E. N. Frankel, "Lipid Oxidation: Mechanisms, Products and Biological Significance," *Journal of the American Oil Chemists' Society* **61**, 1908–1918 (1984).
23. A. Sattar and J. M. de Man, "Photo-oxidation of Milk and Milk Products: A Review," *Critical Reviews in Food Science and Nutrition* **7**, 13–38 (1975).
24. W. O. Lundberg, *Autoxidation and Antioxidants,* John Wiley & Sons, Inc., New York, 1961.
25. G. Ohloff, "Fats are Precursors," in J. Solms, ed., *Functional Properties of Fats in Foods*, Forster Publishing Ltd., Zürich, 1973.
26. E. H. Farmer, "Peroxidation in Relation to Olefinic Structure," *Transactions of the Faraday Society* **42**, 228–236 (1946).
27. M. Keeney, "Secondary Degradation Products," in H. W. Schultz, E. A. Day and R. O. Sinnhuber, eds., *Lipids and Their Oxidation*, AVI Publishing Co., Inc., Westport, Conn., 1962.
28. P. Becher, "Emulsions—Theory and Practice," Van Nostrand Reinhold Co., New York, 1965.
29. S. Friberg, *Food Emulsions*, Marcel Dekker Inc., New York, 1976.
30. W. C. Griffin, "Emulsions," in *Kirk-Othmer Encyclopedia of Chemical Technology*, 2nd ed., Vol. 8, 1965, pp. 117–154.
31. A. R. Macrae, "Lipase Catalyzed Interesterification of Oils and Fats," *Journal of the American Oil Chemists' Society* **60**, 291–294 (1983).
32. J. D. Dziezak, "Fats, Oils and Fat Substitutes," *Food Technology* **43**(7), 66–74 (1989).

General References

Hilditch, T. P., and P. N. Williams, *The Chemical Constitution of Natural Fats*, 4th ed., John Wiley & Sons Inc., New York, 1964.

Kuksis, A., ed., *Handbook of Lipid Research*, Vol. 1. *Fatty Acids and Glycerides*, Plenum Press, New York, 1988.

Mehlenbacher, V. C., *The Analysis of Fats and Oils*, The Garrard Press, Champaign, Ill., 1960.

Patterson, H. B. W., *Hydrogenation of Fats and Oils*, Applied Science Publishers, London, 1983.

Ratledge, C., P. Dawson, and J. Rattray, eds., *Biotechnology for the Oils and Fats Industry*, American Oil Chemists' Society, Champaign, Ill., 1984.

Small, D. M., *Handbook of Lipid Research, Vol. 4. The Physical Chemistry of Lipids*, Plenum Press, New York, 1986.

Swern, D., ed., *Bailey's Industrial Oil and Fat Products*, 4th ed., 3 Volumes, John Wiley & Sons Inc., New York, 1979.

J. M. DE MAN
University of Guelph
Guelph, Ontario, Canada

FATS AND OILS: FLAVORS

Fats and oils that are essential components of shortenings, mayonnaises, and cooking oils contribute unique flavors to foods. Fats and oils are chemically referred to as lipids and occur in food mainly as triglycerides. These fats and oils may have a positive or negative effect on the flavor of foods depending on the chemical reactions taking place during processing and storage. Fats and oils are susceptible to oxidation, which produces undesirable volatile compounds and causes detrimental flavor effects to foods. The oxidation reaction of fats and oils is influenced by light, trace metals, antioxidants, temperature, and fatty acid compositions of fats and oils. Even though thermal oxidation in deep-fat frying may provide good flavor, autoxidation, which produces volatile flavor compounds, is mainly responsible for the deterioration of flavor quality of oils during storage. These volatile flavor compounds include esters, aldehydes, alcohols, ketones, lactones, and hydrocarbons. Unsaturated aldehydes and ketones, which have the lowest sensory threshold, are principally responsible for the undesirable oxidized flavor of food. The separation and identification of these volatile flavor compounds by gas chromatography and mass spectroscopy have increased the knowledge of flavor chemistry of fats and oils.

The isolation, separation, and identification of volatile compounds are difficult due to the complexity of volatile compounds of oils, the variations of concentrations and volatilities of flavor compounds, interaction between the volatile compounds, and reaction between volatile compounds and food components such as proteins and carbohydrates during isolation (1). The mechanisms responsible for the production of volatile flavor compounds during oxidation must be thoroughly studied and understood to improve the flavor quality of fats and oils.

MECHANISMS OF AUTOXIDATION

Autoxidation is a free-radical chain reaction that includes initiation, propagation, and termination steps that are shown below:

$$RH \rightarrow R\cdot + H\cdot$$

initiation

$$R\cdot + O_2 \rightarrow ROO\cdot$$

$$ROO\cdot + RH \rightarrow ROOH + R\cdot$$

propagation

$$ROO\cdot + R\cdot \rightarrow ROOR$$
$$R\cdot + R\cdot \rightarrow R-R$$

termination

Unsaturated fatty acids undergo oxidation by the above free-radical chain reaction to form hydroperoxides. The hydroperoxides are decomposed to form off-flavor compounds, which are called secondary oxidative products (2,3). The reaction of singlet-state fatty acids with triplet-state oxygen is thermodynamically difficult, but a free radical of fatty acids can react with the triplet oxygen molecule. Heat, metal catalysts, ultraviolet, and visible light can accelerate the formation of free radicals of unsaturated fats in the initiation of autoxidation. A hydrogen atom is removed from the methylene group alpha to the double bond of the fatty acid to form a free radical, which has a short lifetime due to its tendency to achieve a stable electron pair. The alkyl radical reacts with oxygen to form a peroxy radical. This peroxy radical reacts with a hydrogen that was removed from another unsaturated fatty acid to form hydroperoxides. The formation of hydroperoxides is usually accompanied by a shift in the position of double bonds due to resonance stabilization of the fatty acid free radical forming isomeric hydroperoxides. A mixture of four isomeric hydroperoxides that have —OOH groups at the 8, 9, 10, and 11 carbon is produced in the autoxidation of oleic acid. Hydroperoxides that are tasteless, colorless and odorless are relatively unstable and readily decompose to produce compounds such as hydrocarbons, alcohols, esters, acids, aldehydes, and ketones having different molecular weights and flavor thresholds. The principal aldehyde compounds formed from the decomposition of the hydroperoxides at the 8, 9, 10, and 11 carbon atom position are 2-undecenal, 2-decenal, nonanal, and octanal, respectively. Oxidized linoleic acid produces hydroperoxides at either the 9 or 13 position and a new conjugated double-bond system is formed at either the 10 or 11 position (4). The carbonyl compounds are mainly responsible for the oxidative flavor of fats and oils.

MECHANISMS OF PHOTOSENSITIZED OXIDATION

Although free-radical autoxidation is the primary mechanism for the formation of volatile flavor compounds, photosensitized oxidation initiated by sensitizers such as chlorophyll in the presence of light should be included in the oxidation mechanisms that contribute to the formation of off-flavor volatile compounds of fats and oils (2). The reaction of fatty acids with oxygen to produce hydroperoxides requires a change in the electron spin state because the unsaturated fatty acid and hydroperoxide are in singlet states while oxygen is in a triplet state. The reaction between singlet-state fatty acids and triplet-state oxygen to form singlet-state hydroperoxides can only occur if enough energy is present to overcome the spin barrier. The presence of electrophilic singlet-state oxygen can directly react with the electron-rich, singlet-state unsaturated fatty acids to form singlet-state hydroperoxides. Naturally occurring pigments, such as chlorophyll in vegetable oils, can absorb ultraviolet and visible light and act as photosensitizers (sens). The sensitizer transfers energy to triplet-state oxygen to form singlet-state oxygen, which reacts with unsaturated fatty acids to form singlet-state hydroperoxides as shown below:

$$sens \xrightarrow{h\nu} {}^1sens^* \rightarrow {}^3sens^*$$

$$^3sens^* + {}^3O_2 \rightarrow {}^1O_2 + {}^1sens$$

$$^1O_2 + RH \rightarrow ROOH$$

Singlet-state oxygen reacts directly with singlet-state double bonds of unsaturated fatty acids by an "ene" reaction to form hydroperoxides as shown below:

Oleate $\xrightarrow{{}^1O_2^*}$

The singlet-state oxygen molecule is joined onto one end of a double bond simultaneously as it abstracts an allylic proton and a new double bond is formed between the allylic position and the other end of the original C=C bond (5). A mixture of conjugated and nonconjugated hydroperoxides is formed by the reaction of singlet oxygen with linoleic acid. Flavor stability of vegetable oils can be decreased by singlet oxygen, which is formed from triplet-state oxygen in the presence of light and chlorophyll. This singlet oxygen oxidation can be minimized by quenchers such as carotene and tocopherol, which deactivates singlet-state oxygen to the ground triplet-state oxygen.

THERMAL OXIDATION OF FATS AND OILS

Although autoxidation is the principle mechanism for the development of off-flavors in fats and oils, the desirable flavor of food develops during deep-fat frying (6). Thermal oxidative decomposition of fats and oils occurs as foods are continuously heated in oil at about 185°C in the presence of air. This mechanism is virtually the same as autoxidation where a hydrogen is abstracted from fatty acids to form a fatty acid free radical, molecular oxygen reacts with the fatty acid free radical to form the peroxy radical and then hydroperoxide. The hydroperoxide is decomposed to form volatile flavor compounds. Flavor compounds formed at deep-fat frying temperatures are different from those produced at room temperature where each particular pathway of flavor compound formation has its own activation energy. When reaction rate (ln K) versus temperature ($1/T$) is plotted for specific flavor compounds, one compound would be produced at a higher rate at deep-fat frying temperature, but when the temperature is decreased to room temperature another flavor compound would be produced at a higher rate. The formation of flavor compounds is dependent on the processing temperature (7). Different fats and oils produce different volatile compounds during deep-fat frying due to their difference in fatty acid composition. Corn and soybean oils produce a large amount of decadienals and unsaturated aldehydes whereas coconut oil produces more saturated aldehydes, methyl ketones, and gamma and delta lactones during deep-fat frying (8). Coconut oil, which contains 90–94% saturated fatty acids has extremely high thermal oxidative stability. It has been reported that the gamma lactones with unsaturation at the two or three position are of particular significance in the deep-fat-fried flavor (9).

FACTORS AFFECTING OXIDATION

The rate of autoxidation is affected by many factors such as the degree of unsaturation of fatty acids, metals, and antioxidants in vegetable oils. For example, linoleic acid, which contains two double bonds, would be oxidized at a faster rate than oleic acid, which contains only one double bond. The rate of autoxidation of oleic to linoleic to linolenic acids has been reported as 1:40–50:100 on the basis of oxygen uptake (10). The bond strength of linoleic acid (52 kcal/mol) is much lower than oleic acid (77 kcal/M) and linoleic acid has a higher reactivity with oxygen than oleic acid. Metals such as copper and iron that exist in two different valence states play an important catalystic role in the oxidation of fats and oils. Metals differ in their ability to catalyze the autoxidation of fats and oils depending on the concentration, reaction temperature, and the polarity of the reaction medium. The metals can be chelated by phospholipids and free fatty acids in vegetable oils and animal fats and the addition of chelating agents can improve the oxidative flavor stability of oils (11). Natural and synthetic compounds that reduce the oxidation of fats and oils are called antioxidants. These compounds minimize oxidation of fats and oils by becoming oxidized themselves or by donating hydrogen to fatty acid free-radicals to terminate free-radical chemical chain reactions. Antioxidants such as butylated hydroxyanisole, butylated hydroxytoluene, tocopherols, and propyl gallate are used to minimize oxidation of fats and oils. The presence of light and oxygen as well as high storage temperature should be controlled to minimize oxidative deterioration of fats and oils.

FLAVOR PROPERTIES OF FATS AND OILS

The types of volatile compounds produced from the oxidation of fats and oils are influenced by the composition of the hydroperoxides and the extent of oxidative cleavage of double bonds contained in fatty acids (12–14). Due to many possible reaction pathways, a variety of volatile compounds such as hydrocarbons, alcohols, furans, aldehydes, ketones, and acids are formed during oxidation. Most of the compounds responsible for oxidized flavors, however, are the aliphatic carbonyl compounds (15). The flavor characteristics and threshold values of aldehydes vary as shown in Table 1.

Table 1. Flavor Characteristics and Threshold Values of Aldehydes

Aldehyde	Description	Threshold Value in Oil (ppm) Odor	Taste
3:0	Sharp–irritating	3.60	1.00
5:0	Sharp–bitter	0.24	0.15
6:0	Green	0.32	0.08
7:0	Oily–putty	3.20	0.05
8:0	Fatty–soapy	0.32	0.04
9:0	Tallowy–soapy	13.5	0.20
10:0	Orange peels	6.70	0.70
5:1	Sharp–paint–green	2.30	1.00
6:1	Green	10.00	0.60
7:1	Putty–fatty	14.00	0.20
8:1	Woodbugs–fatty	7.00	0.15
9:1	Tallowy–cucumber	3.50	0.04
10:1	Tallowy–orange	33.80	0.15
7:2	Frying odor	4.00	0.04
9:2	Fatty–oily	2.50	0.46
10:2	Deep–fried	2.15	0.10

CONTRIBUTION OF SPECIFIC OILS TO FLAVOR

Soybean oil is the most abundant vegetable oil in the United States and contains approximately 7% linolenic acid. Linolenic acid is highly susceptible to oxidation and is mainly responsible for the oxidation of the oil to produce off-flavor compounds. Soybean oil also contains photosensitizers such as chlorophyll that produce singlet oxygen in the presence of light. Singlet oxygen directly reacts with double bonds of fatty acids and oxidation is 1,500 times the rate of autoxidation (16,17). Analysis of headspace has shown that pentane and hexanal are the principal volatile compounds formed from oxidized soybean oil. The flavor characteristics of oxidized soybean oil have been described as grassy or fishy, but a single volatile flavor compound that gives unique grassy or fishy flavor of oxidized soybean oil has not been identified (18). However, trans-hex-2-enal and nona-2,6-dienal have been reported to give general grassylike flavor and deca-trans-2,cis-4,trans-7-trienol and oct-1-en-3-one for a fishylike flavor (9).

Vegetable oils such as corn, cottonseed, peanut, coconut, safflower, and olive oil as well as animal fats such as lard and beef tallow provide unique flavor characteristics to foods. The oil processing, which involves refining, bleaching, and deodorization, removes impurities that contribute to the development of undesirable flavors. Olive oil, which contain 50–83% oleic acid, is popular in the United States as a result of the nutritional implication of monounsaturated fats. The flavor compounds of olive oil have been separated and identified by gas chromatography–mass spectrometry. Hexanal, trans-2-hexenal, 1-hexanol, and 3-methylbutan-1-ol, which are formed by autoxidation, are the principal volatile compounds of olive oil. The flavor of olive oil can be influenced by many factors such as climatic and soil conditions, the maturation process of the olive, and storage conditions (19). Although many volatile compounds have been identified, specific chemical compounds that are responsible for olive oil's flavor have not been determined. Safflower oil, which has 78% linoleic acid, contains the highest content of polyunsaturated fatty acid of all the commercial oils. The high content of linoleic acid increases the susceptibility of the oil to oxidation resulting in the formation of off-flavor compounds (20). Cottonseed oil, which contains 18% oleic acid and 53% linoleic acid, is mainly used in restaurants for frying potato chips, seafoods, and snack foods. The bland flavor of cottonseed oil reportedly does not mask the flavor of the product nor does it revert during deep-fat frying (21).

DETERMINATION OF FLAVOR COMPOUNDS

The isolation, separation, and identification of volatile flavor compounds are important to determine the chemical mechanisms of flavor compounds formation from fats and oils during storage, to identify chemical compounds that are responsible for specific desirable and undesirable flavor, and to develop optimum processing and storage methods that can minimize the undesirable oxidation of fats and oils. The analysis of flavor compounds of fats and oils is difficult because the compounds are volatile, present in low concentrations, and must be isolated with minimum alteration. Food flavors are extremely complex mixtures of volatile compounds that act individually or in combination to produce unique flavors. Any modification of the chemical composition of these compounds could drastically change the flavor characteristics of food. Threshold values are significantly different among flavor compounds, thus the presence of a small quantity of a compound with a low threshold value could have a more significant influence on the flavor than the presence of a large quantity of a particular compound with a high threshold value. Hydrocarbons, for example, have a threshold value of 90–2,150 ppm; alkanals, 0.04–1.0 ppm; and vinyl ketones, 0.00002–0.007 ppm (2). This wide range of threshold values of flavor compounds illustrates how one flavor compound could have a significant influence on the overall flavor regardless of the concentration. Several volatile compounds have been identified to be responsible for specific flavor in oxidized fats as shown in Table 2 (9).

SENSORY EVALUATION

Sensory evaluation is considered the most important test to determine the flavor quality and stability of oils and is used to evaluate the effects of processing conditions, storage time and periods, and packaging environments on the flavor of fats and oils. The most common sensory evalua-

Table 2. Specific Flavor in Oxidized Fats and Cause

Flavor	Cause
Cardboard	Nona-trans-2,trans-6-dienal
Oily	Aldehydes
Painty	Pent-2-enal, aldehydes
Fishy	deca-trans-2,cis-4,trans-7-trienol, oct-1-en-3-one
Grassy	Trans-hex-2-enal, nona-2,6-dienal
Deep-fried	deca-trans-2,trans-4-dienal

Table 3. Sensory Evaluation of Oil

Flavor Score	Description of Flavor
10	Completely bland
9 (good)	Trace of flavor, but not recognizable
8	Nutty, sweet, bacony, buttery
7 (fair)	Beany, hydrogenated
6	Raw, oxidized, musty, weedy, burnt, grassy
5 (poor)	Reverted, rubbery, butter
4	Rancid, painty
3 (very poor)	Fishy, buggy
2	Intensive flavor and objectionable
1 (repulsive)	

tion method of fats and oils is a hedonic scale of 1 to 10; 1 indicates repulsive and 10 completely bland as shown in Table 3 (22).

Sensory evaluation is expensive, subjective, difficult to reproduce, time-consuming, and not easily available. Chemical and physical methods such as peroxide value, thiobarbituric acid, conjugated diene determination, fluorescence, Schaal Oven test, and active oxygen method have been developed to evaluate the flavor quality and stability of fats and oils. The identification of volatile compounds that are responsible for specific flavors such as fishy or grassy has been made possible by correlating sensory evaluations with instrumental analysis such as gas chromatography.

FLAVOR EVALUATION BY GAS CHROMATOGRAPHY

With recent advancement of the isolation techniques of flavor compounds, gas chromatography, and statistical analyses, several evaluations of flavor qualities of oils by gas chromatographic methods have been published (22–24). These papers have reported excellent correlation coefficients of better than 0.9 between sensory scores and predicted sensory scores by gas chromatography of corn oil, soybean oil, sunflower oil, and hydrogenated soybean oil. Gas chromatography has many advantages such as analysis time, convenience, cost, and reproducibility compared to sensory evaluation. A preliminary gas chromatographic analysis of flavor compounds of mayonnaise has been done to predict sensory quality of mayonnaise during storage (25). Various flavor compounds have been identified in mayonnaise including allyl isothiocyanate, acetic acid, ethyl acetate, pentane, and isomers of 2,4-decadienals. As storage time increased, levels of pentane and 2,4-decadienals increased, whereas other volatile concentrations stayed the same. The flavor compound allyl isothiocyanate, the major flavor constituent of mustard, and acetic acid and ethyl acetate, the flavor compounds in vinegar, did not change during the six-month storage period. There was a good correlation coefficient, better than 0.9, between actual sensory scores of mayonnaise and predicted sensory scores by gas chromatography. Therefore, the shelf life of mayonnaise can be predicted by analyzing specific compounds associated with off-flavors such as pentane and 2,4-decadienals during storage periods.

Gas chromatography can also be used to study the effects of storage conditions and processing methods on the oxidation of fats and oils. Soybean and corn oils were exposed to different periods of light to produce a wide range of flavor qualities (26). As the period of light storage was increased, the amount of 2,4-decadienals increased and the sensory scores decreased. The correlation of sensory evaluation and instrumental analysis using gas chromatography was greater than 0.95.

BIBLIOGRAPHY

1. M. Supran, *Lipids as a Source of Flavor*, American Chemical Society, Washington, D.C., 1978.
2. E. N. Frankel, "Chemistry of Autoxidation: Mechanism, Products and Flavor Significance," in David B. Min and Thomas H. Smouse, eds., *Flavor Chemistry of Fats and Oils*, American Oil Chemists' Society, 1985.
3. H. W. -S. Chan, "The mechanism of autoxidation," in H. W. -S. Chan, ed., *Autoxidation of Unsaturated Lipids*, Academic Press, Inc., Orlando, Fla., 1987.
4. Chan and Levett, "Autoxidation of Methyl Linoleate. Separation and Analysis of Isomeric Mixtures of Methyl Linoleate Hydroperoxides and Methyl Hydroxylinoleates," *Lipids*, **12**, 99–104 (1977).
5. H. W. -S. Chan and D. T. Coxon, "Lipid Hydroperoxides," in Ref. 3.
6. W. W. Nawar, "Chemistry of Thermal Oxidation of Lipids," in Ref. 2.
7. S. S. Chang, R. J. Peterson, and C. -T. Ho, "Chemistry of Deep Fried Flavor," in M. K. Supran, ed., *Lipids as a Source of Flavor*, ACS Symp. Ser. **75**, American Chemical Society, Washington, D.C., 1978, p. 18.
8. M. W. Formo, E. Jungermann, F. A. Norris, and N. O. V. Sonntag, "Composition and Characteristics of Individual Fats and Oils," in D. Swern, ed., *Bailey's Industrial Oil and Fat Products*, Vol. 1, John Wiley & Sons, Inc., New York, 1979, p. 289.
9. W. Grosch, "Lipid Degradation Products and Flavour," in I. D. Morton and A. J. Macleod, eds., *Food Flavours, Part A, Introduction*, Elsevier, Amsterdam, The Netherlands, 1982, pp. 325–398.
10. R. T. Holman and O. C. Elmer, "The Rates of Oxidation of Unsaturated Fatty Acids and Esters," *Journal of the American Oil Chemists' Society* **24**, 127 (1947).
11. J. Pokorny, "Major Factors Affecting the Autoxidation of Lipids," Ref. 3.
12. E. Selke, W. K. Rohwedder, and H. J. Dutton, "Volatile Components from Triolein Heated in Air," *Journal of the American Oil Chemists' Society*, **54**, 62–67 (1977).

13. E. N. Frankel, J. Nowakowska, and C. D. Evans, "Formation of Methyl Azelaaldehydrate on Autoxidation of Lipids," *Journal of the American Oil Chemists' Society* **38**, 161–162 (1961).
14. A. C. Noble, and W. W. Nawar, "Identification of Decomposition Products from Autoxidation of Methyl 4,7,10,13,16,19-Docosahexaenoate," *Journal of the American Oil Chemists' Society* **52**, 92–95 (1975).
15. D. A. Forss, "Odor and Flavor Compounds from Lipids," *Prog. Chem. Fats and Other Lipids* **13**, 181 (1972).
16. E. N. Frankel, "Soybean Oil Flavor Stability," in D. R. Erickson, E. H. Pryde, O. L. Brekke, T. L. Mounts, and R. A. Falb, eds., *Handbook of Soy Oil Processing and Utilization*, American Soybean Association, St. Louis, Mo., and American Oil Chemists' Society, Champaign, Ill., 1980.
17. H. R. Rawls, and P. J. Van Santen, "A Possible Role for Singlet Oxygen in the Initiation of Fatty Acid Autoxidation," *Journal of the American Oil Chemists' Society* **47**, 121 (1970).
18. T. Mounts, "Processing of Soybean Oil for Food Uses," *Cereal Foods World* **34**(3), 268 (1989).
19. G. Montedoro, M. Bertuccioli, and G. Anichini, "Flavor of Foods and Beverages," in G. Charalambous and G. Inglett, eds., Academic Press, Orlando, Fla., 1972.
20. T. J. Weiss, *Food Oils and Their Use*, AVI Publishing Co., Inc., Westport, Conn., 1970.
21. J. D. Dziezak, "Fats, Oils, and Fat Substitutes," *Food Technology* **7**, 65–74 (1989).
22. D. B. Min, "Analyses of Flavor Qualities of Vegetable Oils by Gas Chromatography," *Journal of the American Oil Chemists' Society* **60**, 544 (1983).
23. H. P. Dupuy, S. P. Fore, and L. A. Goldblatt, "Direct Gas Chromatographic Examination of Volatiles in Salad Oils and Shortenings," *Journal of the American Oil Chemists' Society* **54**, 340 (1973).
24. J. L. Williams and T. H. Applegate, "Correlation of the Flavor Scores of Vegetable Oils with Volatile Profile Data," *Journal of the American Oil Chemists' Society* **54**, 461 (1977).
25. D. B. Min and D. B. Tickner, "Preliminary Gas Chromatographic Analysis of Flavor Compounds on Mayonnaise," *Journal of the American Oil Chemists' Society* **59**(5) (1982).
26. D. B. Min, "Correlation of Sensory Evaluation and Instrumental Gas Chromatographic Analysis of Edible Oils," *Journal of Food Science* **46**(5), 1453–1456 (1981).

David B. Min
Dondeena G. Bradley
Ohio State University
Columbus, Ohio

FATS AND OILS: PROPERTIES, PROCESSING TECHNOLOGY, AND COMMERCIAL SHORTENINGS

Fats and oils are key functional ingredients in a large variety of food products. Common product applications are baked goods, snacks, icings, confections, fried foods, and imitation dairy products. Fats and oils are also key parts of the human diet providing concentrated sources of energy and essential nutrients. This section will briefly cover the chemistry, processing, physical characteristics, formulation, nutrition, and consumption of fats and oils in food applications.

CHEMISTRY AND STRUCTURE

Fats and oils are predominately mixtures of triglycerides. Triglycerides are the triesters of fatty acids and glycerol formed as a condensation product:

$$3\ RC(=O)-OH + HO-CH_2-CH(OH)-CH_2-OH \rightarrow RC(=O)O-CH_2-CH(OC(=O)R)-CH_2-OC(=O)R + 3\ H_2O$$

Fatty Acid   Glycerol   Triglyceride

Triglycerides may range from solid to liquid consistency at room temperature depending on the fatty acid esters they contain. Solid triglycerides are commonly called fats, and liquid triglycerides, oils. If all three fatty acids that comprise the triglyceride are the same, it is called a single triglyceride; conversely if the fatty acids differ, the triglyceride is mixed.

Triglycerides typically comprise about 95% of fat or oil. The other components include free fatty acids, monoglycerides, diglycerides, phosphatides, sterols, vitamins, natural antioxidants, color-causing pigments, minerals, flavor–odor compounds, and other substances. Free fatty acids are the uncombined or unesterified fatty acids present in a fat or oil. Crude oils may contain several percent fatty acids, which result from hydrolysis. The formation of free fatty acids via hydrolysis leaves partial glycerides behind. Diglycerides result from the loss of one fatty acid from the triglyceride, which may occur at the end or middle position:

$$\text{1,3-Diglyceride} \qquad \text{1,2-Diglyceride}$$

Loss of two fatty acids from the triglyceride results in monoglycerides:

$$\text{1-Monoglyceride} \qquad \text{2-Monoglyceride}$$

Monoglycerides and diglycerides are frequently used in foods as emulsifiers. They are often found in functional cake and icing shortenings. They are produced commercially by the alcoholysis reaction of triglycerides or fatty acids with glycerol or by esterification of glycerol with fatty acids (1).

Phosphatides are found in fat and oil from plant and animal sources. These materials consist of a polyhydric alcohol (usually glycerol) that is esterified with fatty acids, phosphoric acid, and a nitrogen-containing compound. The most common phosphatide is a family of compounds found in soybean oil referred to as lecithin, which includes chemical lecithin (phosphatidyl choline), cephalin (phosphatidyl ethanolamine), and lipositols (phosphatydyl inositol) (2). Sphingomyelins are a group of phosphatides that contain no glycerol (1). Lecithin is typically used as a food additive for its emulsification properties. Phosphatides are essentially removed in the refining process.

Colored materials occur naturally in fats and oils and generally are a result of two classes of pigments. Carotenoids are the source of the yellow–red color found in fats. These are the same family of compounds that give carrots, pumpkins, and apple skins their characteristic colors. Chlorophylls are the source of green color in fats and oils. These are the family of compounds that give leaves their green color and are part of the process of photosynthesis. Level and type of pigment may vary greatly with the oil source. Palm oil, for example, has high levels of carotenoids giving crude palm its characteristic orange color, while crude rapeseed, or canola, oil has high levels of chlorophyll, resulting in a dark green color. Fortunately, processing reduces these color pigments to acceptable levels.

Vegetable oils also contain tocopherols that serve as natural antioxidants, retarding oxidation, and as a source of the essential nutrient vitamin E. Careful processing will remove only a small portion of these materials. Other antioxidants may be added to processed fats to preserve freshness (3,4). In contrast, meat fats contain negligible levels of tocopherols.

Sterols are common minor components of all natural fats and oils. Sterols are crystalline, neutral, unsaponifiable, high-melting point alcohols. Cholesterol, the infamous sterol associated with clogged arteries, is the principal sterol found in animal fats. Vegetable oils only contain trace quantities of cholesterol. Vegetable oils do contain other sterols, such as sitosterol and stigmasterol. The sterols will vary in type and quantity with the vegetable oil source (1).

Fats and oils may contain a number of other minor constituents such as waxes, gums, hydrocarbons, and fatty alcohols. These materials are generally of little importance and are typically handled during processing. Some salad oils, like corn oil or sunflower oil, may require special processing to remove waxes that would otherwise cloud the product when refrigerated (5).

Fatty Acids

The key to understanding the chemical and physical characteristics of a fat is understanding the component fatty acids that comprise the triglyceride. These can vary in chain length (number of carbon atoms), type of carbon–carbon bonding (saturated or unsaturated with hydrogen), and stereochemistry both within the fatty acid and its position in the triglyceride. The most common edible fatty acids are aliphatic, straight chain, saturated, and unsaturated, containing an even number of carbon atoms and ending in a carboxyl group:

$$H_3C-(CH_2)_n-\overset{\overset{O}{\|}}{C}OH$$

Saturated, aliphatic carbon chain   **Carboxyl group**

Small amounts of odd-numbered, straight-chain acids, branched chains, and cyclical are also present in edible vegetable oils (5).

The length of the chain can vary in edible fatty acids from as few as 4 carbons to as many as 24. Most fats predominately contain fatty acids that are 16 or 18 carbons long, although there are some exceptions like coconut oil, palm kernel oil, and butterfat, which contain high percentages of shorter (8–14 carbons) chain fatty acids.

The fatty acids in fats are classified by structure. Saturated fatty acids contain only single carbon–carbon bonds in the chain with hydrogen filling the remaining bonds. Unsaturated fatty acids contain one or more carbon–carbon double bonds. Unsaturated fatty acids may either be monounsaturated (containing only one double bond) or polyunsaturated (containing two or more double bonds):

$$-\underset{H}{\overset{H}{C}}-\underset{H}{\overset{H}{C}}-\underset{H}{\overset{H}{C}}-\underset{H}{\overset{H}{C}}-$$

Saturated

$$-\underset{H}{\overset{H}{C}}=\underset{}{\overset{H}{C}}-\underset{H}{\overset{H}{C}}-\underset{H}{\overset{H}{C}}$$

Monounsaturated

$$-\underset{}{\overset{H}{C}}=\underset{}{\overset{H}{C}}-\underset{H}{\overset{H}{C}}-\underset{}{\overset{H}{C}}=\underset{}{\overset{H}{C}}-$$

Polyunsaturated

Saturated fats are chemically the least reactive. The more polyunsaturated, the more chemically reactive a fatty acid will be. Saturated fatty acids have higher melting points than the corresponding fatty acid of the same chain length.

Fatty acids with double bonds may differ in geometric form. Orientation of the hydrogen atoms around a double bond determines whether the fatty acid is cis or trans. In the cis form, both hydrogens are bonded in the same direction from the carbon. In the trans form, the hydrogens are oriented in opposite directions:

$$-\underset{\underset{H}{|}}{\overset{\overset{H}{|}}{C}}=\underset{\underset{}{}}{\overset{\overset{H}{|}}{C}}-$$

Cis

$$-\underset{\underset{H}{|}}{\overset{\overset{H}{|}}{C}}=\underset{\underset{H}{|}}{\overset{}{C}}-$$

Trans

The trans form of a fatty acid is more stable and has a higher melting point.

The double bond may also vary by position on the chain, although most natural fatty acids have double bonds that commonly occur at fixed positions. When two fatty acids differ only in configuration around a double bond, they are called geometric isomers.

There are two accepted systems of nomenclature to identify the position of double bonds in the chain. The Geneva system numbers the carbons in the chain consecutively starting from the carbon in the carboxyl group, which is number 1. The carbon with the lower number is used to identify the location of the double bond, ie, the carbon closest to the carboxyl group.

Biochemists typically identify the position of the double bond by the omega or "n minus" classification. This system refers to the position of the carbon in the double bond closest to the methyl end of the chain; the carbon of the end methyl group is number 1. Linoleic acid, a fatty acid 18 carbons long with two double bonds may be identified as 9,12-octadecenoic acid by the Geneva system, noting the 9 and 12 carbon atoms as the position of the double bonds from the carboxyl group. This same fatty acid would be an omega-6 or n-6 fatty acid by the alternate system, which identifies the position of the first double bond from the methyl end of the fatty acid. The second double bond would be an omega-9 ($n-9$). When two fatty acids differ only in the position of the double bond, they are called positional isomers. Fatty acids can be both positional and geometric isomers. All of these structural variations can affect the chemical and physical properties of a fatty acid and the resulting triglyceride. The effects of chain length, unsaturation, and geometric isomerism on the melting point are shown in Table 1.

Table 1. Fatty Acid Melting Points

Common Name	Number of Carbon Atoms	Number of Double Bonds	Melting Point, °F
Caproic	6	—	26
Caprylic	8	—	62
Capric	10	—	89
Lauric	12	—	112
Myristic	14	—	130
Myristoleic	14	1	65
Palmitic	16	—	145
Margaric	17	—	142
Stearic	18	—	157
Oleic	18	1	61
Elaidic	18	1 (trans)	111
Linoleic	18	2	20
Linolenic	18	3	9
Arachidic	20	—	168
Behenic	22	—	176
Erucic	22	1	92

Polyunsaturates may also be described by the relative position of the double bonds. Two or more double bonds normally alternate with a single bond or a nonconjugated position. When the two double bonds are adjacent, they are called conjugates.

$$-\overset{\overset{H}{|}}{C}=\overset{\overset{H}{|}}{C}-\overset{\overset{H}{|}}{\underset{\underset{H}{|}}{C}}-\overset{\overset{H}{|}}{C}=\overset{\overset{H}{|}}{C}-$$

Nonconjugated
(cis–cis)

$$-\overset{\overset{H}{|}}{C}=\overset{\overset{H}{|}}{C}-\overset{\overset{H}{|}}{C}=\overset{}{\underset{\underset{H}{|}}{C}}-$$

Conjugated
(cis–trans)

Conjugates are generally more reactive due to the proximity of the two double bonds. Other positional isomers are obviously possible. The double bond locations in unhydrogenated fats and oils usually occur in characteristic positions. Processing, particularly hydrogenation, may alter these locations forming conjugates and other positional isomers.

The physical properties of a triglyceride will be affected by the individual fatty acids it contains and the location of these fatty acids on the glycerol backbone. Some typical source oil fatty acid compositions are shown in Table 2. A triglyceride composed of a single fatty acid type will have uniform distinct physical properties such as a sharp melting point. Fats and oils are mixtures of triglycerides. These mixtures result in less distinct, broader physical properties, eg, a melting point range. The molecular configuration gives rise to characteristic physical properties typical of a given source. Some sources such as cocoa butter have unique properties due to their specific triglyceride structures. Monoglycerides and diglycerides have higher melting points than the corresponding triglycerides, as shown in Table 3.

Solid fats can also orient themselves in different crystal forms or polymorphs depending on how the fat crystals orient themselves in the solid state (6,7). Fats containing a diverse mixture of triglycerides are more likely to remain in lower melting crystal forms. The more singular triglycerides will readily transform successively from lower melting to higher melting forms. The extent and rate of transformation are a function of the triglyceride structure and mixture, temperature of crystallization, and the duration of crystallization at a given temperature. Some fats like cocoa butter may have several high-melting polymorphic forms. Control of these forms is critical to functionality in some fat applications. Solid, plastic shortenings and bakery shortenings require a beta prime (lower melting) crystal form to achieve the plasticity and structure required for those applications. Pourable, opaque liquid shortenings require the beta (higher melting) form, which allows the crystals to flow. Partially hydrogenated soybean oil is typically used in this type of

Table 2. Approximate Fatty Acid Composition, Percent of Total Fatty Acids

Composition	Canola Oil	Coconut Oil	Corn Oil	Cottonseed Oil	Lard	Palm Oil	Peanut Oil	Soybean Oil	Tallow
C6:0		<1							
C8:0		8							
C10:0		6							
C12:0		47							
C14:0		19		<1	2	1			3
C14:1									1
C16:0	4	9	11	22	26	44	11	11	24
C16:1				<1	3				4
C17:0									2
C17:1									<1
C18:0	2	3	2	3	14	5	2	4	19
C18:1	61	6	27	19	44	40	47	24	43
C18:2	21	2	59	54	10	10	32	54	3
C18:3	9		1	1				7	<1
C20:0	1						1		
C20:1	1				<1		2		
C22:0							3		
C22:1	<1								
C24:0							2		

application, whereas partially hydrogenated cottonseed or palm oils are used to obtain stable, beta prime crystal formation. Mixtures of fats with different polymorphic tendencies can be directed to a particular polymorphic form by control of composition and crystallization conditions. Beta prime crystal formation of partially hydrogenated soybean oil in bakery shortenings and margarines is promoted by inclusion of small amounts of beta prime tending, partially hydrogenated palm oil (8–12).

Processing

Animals or oilseeds are the sources of edible fat in the United States. Animal fats are obtained by separating the fat from the tissues in a process called rendering. Heat is applied to separate the fat from the tissue protein. Vegetable fats are obtained by crushing the oil source and extracting the fats by direct compression and filtration. Modern processing utilizes solvent (typically hexane) extraction to improve efficiency and yield. The solvent is recycled after recovery and reused. Oils and fats obtained from this process are called crude because they contain small amounts of nontriglyceride components at this stage. These may or may not be desirable. Crude fats and oils are processed to remove the undesired material and modify the physical properties to obtain useful, functional shortening and oil products.

Two types of refining process are commonly in use today: chemical or physical (steam) refining. Both employ similar steps, but with different objectives. Most vegetable oils and meat fats are chemically refined, but for some vegetable oils, notably coconut and palm, physical refining is more economical. The choice is generally determined by the convenience and economics of the process.

Chemicals, or caustic refining, is intended to remove most of the free fatty acids present in the crude oils. Other impurities such as phospholipid, mucilagenous, and proteinaceous materials are also removed. The process consists of treating the heated crude with an alkali (usually caustic) solution, which converts the free fatty acids into water-soluble soaps. The soaps are removed through centrifugation, which also separates the hydrated impurities. Residual soap and impurities are removed by water washing and centrifugation of the soapy wash water. Residual water is removed by vacuum drying. Caustic strength, temperature, mixing, contact time, and removal are all critical process parameters.

Table 3. Fatty Acid Ester Melting Points, °F[a]

Fatty Acid	Monoglyceride	Diglyceride[b]	Triglyceride
C12:0 Lauric	145	136	116
C14:0 Myristic	159	152	135
C16:0 Palmitic	171	169	146
C18:0 Stearic	179	175	164
C18:1 Oleic	95	71	42
C18:1 Elaidic[c]	137	131	108
C18:2 Linoleic	54	27	8
C18:3 Linolenic	60	10	−12

[a] Ref. 1.
[b] 1,3-Diglyceride.
[c] Trans isomer of oleic acid.

Bleaching is usually the next process step, and as the name implies, its function is to remove color from the oil. The primary source of colored material in oils are the carotenes (red pigments) and chlorophylls (green pigments). These are usually removed by physical absorption on an acid-activated clay or bleaching earth. Other absorbent materials such as activated carbon may also be employed in specific situations. Critical variables are absorbent level, contact time, temperature, and removal of the spent absorbent. Typically, hydrogenation follows bleaching. Hydrogenation physically adds hydrogen to the unsaturated fatty acids. Hydrogenation is a heterogeneous reaction in which gaseous hydrogen is reacted with the oil in the presence of a solid catalyst, usually nickel. The reaction is exothermic. Hydrogenation has two fundamental functions. First, it raises the melting point and solids content of the oil, thus altering its physical properties. Second, it improves oxidative and thermal stability of the oil by reducing the number of labile, polyunsaturated reaction sites of the fatty acids in the triglyceride.

Hydrogenation is an important step in differentiating products for their various applications. By hydrogenating and blending various oils, many different products are obtained from lightly hydrogenated, liquid fry shortening and soft margarine shortening to more heavily hydrogenated, solid bakery shortenings and coating fats. Control of the reaction conditions and stopping the reaction at the proper point are key to obtaining the desired product. Critical parameters are reaction time, temperature, pressure, catalyst type and amount, and gas purity. Vessel design to provide effective mass and heat transfer is also important.

After hydrogenation and blending, the shortening is typically deodorized. Deodorization is a high-temperature, vacuum, steam distillation that removes the small quantities of undesirable flavor and odor compounds. Other volatile compounds such as residual fatty acids and naturally occurring antioxidants will also be removed. Contact time and process conditions are controlled to adequately remove the undesirable components and minimize loss of natural antioxidants. The high temperatures involved (greater than 450°F) usually results in lighter color due to transformation of some of the red and yellow color bodies into nonchromophores. Deodorization can occur directly after bleaching if hydrogenation is unnecessary. Critical process parameters are temperature, vacuum, steam rate, and exposure time. Again, vessel design to provide effective and efficient heat and mass transfer is important.

Plastic and some fluid shortenings are votated, or crystallized, after deodorization. The fat is held just above its melting point and fed under pressure to a scraped-surface heat exchanger where it is chilled. Fat crystal nuclei are generated there and then allowed to slowly crystallize in a mildly agitated vessel, commonly called a B unit, for a controlled time. The pressure is returned to atmospheric just before putting the product into its container. Liquid products can be tempered in-process, whereas solids are typically stored for 24–48 h at their tempering temperature. The shortening appearance, consistency, and crystal structure are controlled in this processing step, and it is critical for good, functional bakery shortenings where crystal structure is important.

Other Processes

There are other processing steps that are specialized. Fractionation is employed to remove solids from the oil at certain temperatures. The most common fractionation process is winterization. During winterization, the oil is cooled or chilled. Any resulting solid material is removed by filtration. The resulting product, a salad oil, is clear and will resist clouding at refrigerator temperatures. This process was practiced by exposing oil to cool winter temperatures for crystallization prior to the availability of commercial refrigeration systems. The oil is usually partially hydrogenated before winterization to improve stability.

Fractionation may be used to separate a mixture of triglycerides into two or more different melting fractions via crystallization. This is typical in the production of hard butters (cocoa butter substitutes) for coating applications and specialty shortenings from palm, palm kernel, and coconut oils (12).

Winterization and fractionation may also employ a solvent to improve the separation. Hexane or acetone are typically used, and the solvent is removed by distillation (stripping) and deodorization. This is commonly referred to as solvent winterization or solvent fractionation (13).

Transesterification, a type of interesterification, is a process that rearranges or redistributes the individual fatty acids on the glycerol backbone of the triglyceride. The process involves the use of a catalyst at low to moderate temperatures. Conditions may be selected to distribute the fatty acids more randomly, thus modifying the physical properties of the shortening. Unlike hydrogenation, this process does not create a significant level of isomers from the fatty acids. Hydrogenation may be combined before or after the process to allow further modification of the shortening.

If crystallization is employed during the process, the least-soluble, higher melting triglycerides can be removed without destroying the catalyst and the mixture will again equilibrate. This process, used to obtain relatively pure glycerides, is called directed interestification (14–16). If excess glycerol is present, monoglycerides and diglycerides are formed. The relative amounts of each at equilibrium are a function of the ratio of fatty acids available to glycerol (17). This process is referred to as glycerolysis.

PROCESSING AIDS AND ADDITIVES

There are a number of food-approved materials that may be added at low levels to improve processing or to protect the oil quality during processing (Table 4). These materials are removed or reduced to minimal levels in the final product. Other additives may also be used to provide a functional effect in the final product. These uses must comply with FDA regulations covering addition levels, methods of addition, permissible residual levels, and labeling.

FATS AND OILS FUNCTIONALITY

Fats and oils provide a number of functional characteristics in a variety of food products. Shortening products can

Table 4. Additives Used in Fats and Oils

Additive	Effect

Antioxidants
  Tocopherols
  Butylated hydroxyanisole (BHA)
  Butylated hydroxytoluene (BHT)
  Tert-butylhydroquinone (TBHQ)
  Propyl gallate
	Retard oxidation

Colors Beta-carotene	Provides yellow color to finished product
Flavors Various acids, aldehydes, and ketones	Provide flavor and odor to the finished product
Phosphoric acid Citric acid	Metal chelators, process hydrating agents, neutralizing agents
Dimethylpolysiloxane (methyl silicone)	Antifoaming agent, high-temperature antioxidant
Polyglycerol esters Oxystearin	Crystal inhibitors, crystal modification

Emulsifiers
  Monoglycerides diglycerides
  Lecithin
  Polyglycerol esters
  Calcium/sodium steroyl lactylate
  Lactylated monoglycerides and diglycerides
Sorbitan esters
Propylene glycol esters
	Emulsification, aeration, dough conditioning, wetting, antispattering, antisticking, separation, and viscosity control

provide lubricity, structure, a heat-transfer medium, a moisture barrier, and aeration. Fat systems can also act as a carrier for flavor, color, and vitamins. Oils themselves provide vital nutrition.

In most applications fat is multifunctional. Frying and cooking fats provide lubricity, a heat-transfer medium, and a flavor medium in some applications (deep-frying, snack frying, and griddling). Fats provide lubricity and aeration in cakes. In icings, shortening not only provides lubricity and aeration, but also structure. The same is true for cream fillers and whipped toppings. In baked goods like pie crust, Danish, and puff pastry, lubricity and structure are provided. Fats provide a moisture barrier as well as lubricity and structure when used in confectionary coatings.

SALAD AND COOKING OILS

Salad and cooking oils are usually prepared from refined, bleached, and deodorized vegetable oils. The oil may be lightly hydrogenated and winterized for added stability. The oil used may be selected for flavor (eg, peanut and olive oils), nutrition (eg, corn and canola oils), and stability. Antioxidants and crystal inhibitors may be added to salad oil (1,8,19). The inhibitors ensure clarity at refrigerated temperatures, which is critical to dressing applications. If the oil is intended for repeated heating and reuse, it may be lightly hydrogenated and winterized or blended with other oils to provide added stability. Dimethylpolysiloxane may also be added for frying applications. Salad oils are typically light in color, resist clouding, and are bland in flavor. Salad dressing must contain a minimum of 30% oil by federal regulation (20). Mayonnaise is an emulsified, semisolid dressing and must contain at least 65% vegetable oil and dried whole eggs or egg yolks (21). These dressings may also contain salt, sugar, vinegar, spices, seasonings, lemon juice, coloring, and other ingredients. Reduced-calorie dressings must contain at least one-third fewer calories than the conventional product. This is usually accomplished by reducing the fat and carbohydrate content and increasing the water content. Gums and thickeners may be added to provide emulsion stability in these products.

FRYING SHORTENINGS

Frying fats must maintain stability under relatively abusive conditions. Frying involves exposure to high temperatures (330–400°F), moisture from the fried foods, extractable materials from the fried food, and oxygen from the atmosphere. Stability is also important to the packaged fried food, especially for items like snack foods. Also important are the textural and flavor effects the shortening imparts to the fried foods. These are a function of the oil source and the physical and chemical characteristics of the shortening.

Early shortenings were derived from animal sources. At the turn of the century, hydrogenation permitted the use of vegetable oils for applications requiring solid, highly stable products. Frying shortenings are now available as meat fats, meat fat–vegetable blends, solid vegetable, and liquid vegetable shortenings. Meat fats, the traditional source, provide good stability, special flavor effects, and are generally economical to use. Poor nutritional at-

Table 5. Typical Fry Shortening Analyses

Analysis	Solid: Meat Fat–Vegetable Blend	Solid: Vegetable, Partially Hydrogenated Soybean	Liquid: Vegetable, Partially Hydrogenated Soybean
Iodine value	52	69	101
Mettler melting point, °F	117	103	95
Solid fat index,			
at 50°F	40	50	5
at 70°F	27	36	4
at 80°F	23	30	3
at 92°F	17	14	2
at 104°F	12	2	2
Lovibond color, red, maximum	1.5	1.5	1.5
Free fatty acid, wt. %, maximum	0.05	0.05	0.05
Peroxide value, meq/kg, maximum	1.0	1.0	1.0
Fatty acid composition, %			
C16	24	11	11
C18	18	11	7
C18:1	41	70	46
C18:2	8	2	28
C18:3	1	trace	3

tributes (cholesterol and saturated fat) have had a negative influence on usage. Blends of meat fats and vegetable oils are used typically to optimize flavor, stability, and textural attributes.

Solid vegetable frying shortenings provide excellent fry stability, particularly for abusive conditions, and flexibility in eating quality through hydrogenation and blending. Nutrition is improved with no cholesterol and about half the saturates of meat fats. Solids are generally high in monounsaturated fat and low in polyunsaturate content.

Pourable, liquid shortenings offer the convenience of easy handling with good fry life. There is some compromise on oxidative stability, but fry stability is quite acceptable for most applications. Food appearance and eating quality will change with a liquid fry shortening due to the change in physical properties. Liquid vegetable frying fats generally offer lower saturated fat content than solid frying fats as well as no cholesterol. Liquids offer some flexibility in the level of monounsaturated and polyunsaturated fat depending on the stability required. Table 5 shows the analytical characteristics of the various fry shortenings.

Fry life is a function of the chemical and physical characteristics of the fry shortening. These are dependent on the source oil and its processing. Fry life is also a function of myriad operational factors such as the type and amount of food fried, the size of the fryers, the time and temperature of frying, frequency of filtration, and cleaning procedures.

GRIDDLE AND PAN FRYING SHORTENINGS

Butter and meat fats were originally used to transfer heat, prevent sticking, and add flavor in these applications. Today, butter and margarines may still be used; however, both can scorch, due to the presence of milk solids; darken the grill or pan with use; and have high levels of waste, due to the water and other nonshortening components. Butter is also relatively expensive.

Current specialty griddle and pan frying shortenings are usually colored and butter-flavored. Vegetable-based products may be semisolid or liquid and contain lecithin, as an antisticking agent; antioxidants; and dimethylpolysiloxane. Coconut oil may be added for abusive high-temperature applications (22). The solid products usually contain from 10 to 50% coconut oil. More recent products may contain salt to enhance flavor, specialized butter-flavor systems, and small amounts of butter itself (23,24). These products are also used for other applications, including gravies, sauces, soups, and basting liquids. Furthermore, they are used as a wash on baked goods and as a lubricant for bun-toasting machines.

BAKERY SHORTENINGS

There are a variety of general and specialized shortenings available for baking applications (1,25). The principal products are bread, rolls, biscuits, cakes, icings, cookies, pastry, doughnuts, and pies. All-purpose shortening is intended to function adequately in a broad variety of applications from baking to frying. It is formulated to have stability, a wide plastic range for proper mixing, and some structure applications like cakes. In cookies, it must provide the proper solids content for cohesion and spread, yet still be a tenderizer. The shortening must have controlled plasticity to sufficiently work into the blend during mixing.

Cake-and-icing shortening is designed to meet the dual needs of cake and icing applications. The shortening must entrap and disperse air in the oil-in-water emulsion that is a cake batter. The fineness and amount of the air bubbles determine the finished cake height and cell structure. The shortening also lubricates the batter components and

Table 6. Typical Shortening Analyses

Analysis	All-Purpose (Soy Base)	Cake and Icing (Soy Base)
Iodine value	74	74
Mettler melting point, °F	120	120
Solid fat index,		
at 50°F	24	24
at 70°F	20	20
at 80°F	19	19
at 92°F	16	16
at 104°F	10	10
Lovibond color, red, maximum	1.0	2.0
Free fatty acid, wt. %, maximum	0.05	0.15
Peroxide value, meq/kg, maximum	1.0	1.0
Alpha monoglyceride, wt. %, maximum	—	3.5

tenderizes the cake. Icings are water-in-oil emulsions with air incorporated during mixing. The shortening must provide the structural properties of the icing (body and firmness) as well as its creaming or aerating ability. As air bubbles become finer and more uniformly distributed, the emulsion becomes more structured. Monoglyceride and diglyceride emulsifiers are typically added to cake and icing shortening to improve the emulsifying properties. Table 6 shows the typical analysis for all-purpose and cake-and-icing shortenings. The primary difference is the presence of the monoglyceride emulsifier and the higher free fatty acid level associated with it.

The development of emulsifier technology has allowed the formulation of specialized cake or icing shortenings and the development of prepared cake mixes. There are a number of different specialized cake shortenings available as either a plastic solid or a liquid. Their formulation is a function of their application: continuous versus batch mixed, rich or lean formulation, liquid content, and sugar content. The shortening base is formulated and processed to obtain optimum emulsifier performance. Emulsifiers generally permit a richer, moister cake due to their capability to incorporate high levels of moisture in the batter (Table 7). Cake-and-icing shortening makes a rich cake, but with relatively low water holding ability due to the low emulsification level. Such cakes are intended for typical retail applications. The specialty plastic or liquid shortenings have increased emulsification, which allows for richer and moister cakes. The specialized mix shortenings contain even higher levels of emulsification and are used for consumer cake mixes, leaner snack-type cake applications, or cake products where long shelf life is required (26–28).

Similarly, emulsifier technology has spawned a variety of specialized icing shortenings. The emulsifiers allow incorporation of high levels of moisture and air into the icing. The shortening must provide the structure to allow the icing to hold this moisture and air as well as retain its shape. Processing of the shortening, particularly formulation and votation, is critical to provide this structure. Formulation and votation (including tempering) are key to obtain the proper crystal polymorphic form and size distribution for superior icing body and structure. Generally, cooler votation conditions improve an icing's ability to aerate as measured by specific gravity. The icing's resistance to separation and slump is also improved. Processing conditions to achieve a shortening's optimum icing performance will vary by product, so that an entirely different set of conditions may be necessary for each individual formulation.

Different levels and types of emulsification are required for formulations with increased moisture and aeration requirements (Table 8). Flat or thick fudge icings require little, if any, aeration. However, butter cream and some decorator icings require some emulsification ability usually accomplished with monoglucerides and diglycerides or low levels of polysorbates. Specialty icing products are formulated to produce fluffier and higher moisture-type icings. Polyglycerol esters or combinations of monoglyceride and diglyceride, polysorbate emulsifier

Table 7. Typical Cake Shortenings

Shortening	Emulsifiers	Moisture Incorporation	Fat Content
Cake and icing	Monoglyceride and diglyceride	Low	Moderate
Specialty cake			
Plastic solid	Propylene glyco monesters and monoglycerides	Moderate	Moderate
Liquid	Lactylated monoglycerides and monoglycerides	Moderate	Moderate
Specialized mix shortenings	Higher levels of propylene glycol monoesters and monoglycerides	High	High

Table 8. Typical Icing Shortenings

Shortening Type	Emulsifier	Icing Application
All-purpose	None	Flat, fudge icings
Cake and icing	Monoglyceride and diglycerides	Butter cream and decorator icings
Specialty icings	Monoglycerides, diglycerides, and polyglycerol esters	Butter cream, fluffy decorator icings, and cream fillers

systems are used. As the level of emulsification increases, the structure provided by the shortening becomes more critical to icing performance.

Breads and Rolls

Shortening is added at relatively low levels to breads for lubricity, softening, and shelf life. Structure is primarily obtained from starch and protein complexes. Emulsifiers are employed as dough conditioners to complex the proteins or starches and improve shelf life (Table 9). High melting, hard fat may also be added to the shortening system to help provide some structure. The emulsifier may be added as part of the shortening or hydrated first by heating in water until a liquid crystaline phase (mesophase) is formed. This mesophase is maintained by cooling the mixture. This process makes the emulsifier more available for complexing.

Biscuit shortenings are formulated to blend as small, discrete chunks into the cold dough, yet still provide sufficient solids for a flaky texture. This requires a uniform plasticity over a wide temperature range. Dry biscuit mixes usually employ an all-purpose type shortening combined with higher melting point shortening flakes to provide a similar functional effect in the finished biscuits.

Pie Crust Shortening

Pie shortenings are also mixed loosely into the cold dough to maintain separate fat layers. This layering produces the desired flaky crust. Overmixing, or mixing too warm, incorporates the shortening intimately, which reduces flakiness, increases toughness, and increases shrinkage. The shortening system contains a loosely bound oil fraction to provide lubricity and tenderness to the crust. Shortening solids content must be high enough to be plastic and provide layering, but must melt rapidly to avoid a waxy mouth feel when eaten.

Lard is the classic choice for pie shortening due to its solids content, plasticity at cool temperatures, and crystal structure. Vegetable shortenings have been formulated to provide similar functional properties with an improved nutritional (cholesterol and saturates) profile (Table 10).

Table 9. Bread Emulsifiers and Dough Conditioners

Monoglycerides and diglycerides	Sodium stearoyl lactylate
Distilled monoglycerides	Calcium stearoyl lactylate
Ethoxylated monoglycerides	Lactylated stearate
Succinylated monoglycerides	Diacetyl tartaric acid esters

Puff Pastry and Danish Shortening

Puff pastry and Danish shortenings are referred to as roll-in products because they are formed from a laminated dough sheet, consisting of alternating layers of dough and shortening, that is traditionally formed by rolling the coated dough sheet and folding it to form layers.

Danish has a higher fat content than puff pastry, and softer-consistency products are used to achieve lubricity. Long refrigerated retarding (dough relaxing) steps are used between folds to minimize toughness and maximize lubicity while maintaining the structural integrity of the fat film. Puff pastry shortening contains higher solids for flakiness due to the lower fat content used in the puff pastry (Table 11). This shortening must be carefully formulated and processed so it remains workable, not stiff, to properly apply to the dough sheet. Furthermore, it must not contain too much solids at higher temperatures to avoid waxiness and maintain lubricity. Dough and shortening temperature are critical. Puff pastry has larger expansion because of the number of layers and the water vapor sealed by the shortening between the pastry sheet layers.

Butter was originally used for these applications. Puff pastry and Danish shortenings are formulated to have similar composition (fat, moisture, and salt) and functionality and may be butter flavored as well. The moisture in the shortening contributes to the flakiness and puffing action during baking.

OTHER APPLICATIONS

There are a number of other specialized shortening applications. Confectionary fats are primarily intended to be

Table 10. Typical Pie Shortening Analysis

Analysis	Lard	Vegetable Pie Shortening
Iodine value	65	80
Solid fat index, %		
at 50°F	27	23
at 70°F	20	20
at 80°F	14	17
at 92°F	5	13
at 104°F	2	4
Wiley melting point, °F	110	120
Saturates, %	42	26
Cholesterol, mg/100 g	95	—

Table 11. Typical Danish and Puff Pastry Shortening Analysis

Flavor	Danish, Buttery	Puff Pastry, Buttery
Moisture, wt. %	17	17
Salt, wt. %	2.7	2.7
Iodine value	75	75
Mettler melting point, °F	112	120
Solid fat index, %		
at 50°F	28	34
at 70°F	21	30
at 104°F	9	15

substitutes for cocoa butter in confections and coatings. These consist of lauric-based substitutes (coconut and palm kernel oils), which may be hydrogenated, fractionated, or interesterified to better match the solids profile and melting point of cocoa butter. Domestic substitutes may be hydrogenated or fractionated as well (12,29–32). Domestic fats generally do not have the same crystal type as cocoa butter, and they do not require tempering. Eating quality, snap, and mouth melt are key performance factors. Appearance (primarily gloss) is also important. In addition, coatings must provide a moisture barrier. These are typically high-solids, low melting point products and may be referred to as hard butters.

There is a category of domestic shortenings used as tropical fat replacements in imitation dairy products like sour cream, dips, nondairy creamers, and whipped toppings. Texture and stability are key to successful shortening formulation for these products. Particularly important are solids content and melting point, which affect texture. Physical and flavor stability must also be maintained.

Peanut butter stabilizers are partially hydrogenated, high-melting fats, or emulsifiers, which are added to fluid peanut butter to prevent oil separation. Excessive levels of stabilizer addition can reduce spreadability, cause cracking and pullaway in the package, and cause poor eating quality, ie, reduced mouth release. Sufficient stabilizer will raise the solids content of the product fat system by 1 to 2% at 104°F. Stabilizers are also used to control the consistency of the finished peanut butter.

BUTTER AND MARGARINES

Butter contains at least 80 wt % butterfat (33), which serves as the matrix sponge for the aqueous phase. The aqueous phase consists of water, casein, milk solids, and minerals. The solids comprise about 1–2 wt % of the butter, and salt may typically be added at concentrations from 1.5 to 3 wt %. Margarines are butter replacements made from shortening or shortening and oil blends with other ingredients to match butter's aqueous phase. Coloring, flavoring, and vitamins A and D may also be added. Margarine must contain 80% fat by federal regulation (34). Margarines are available in solid form, whipped, and as pourable liquids. They are also available in reduced-calorie, diet, or imitation spreads, which have a lower fat content. Both animal and vegetable fats may be used for margarine, but the latter is far more common. Vegetable source margarines are generally lower in saturated fat (about 40% lower than animal fat blends and 60% lower than butterfat) and much higher in polyunsaturated fat content.

U.S. FAT CONSUMPTION

The U.S. Department of Agriculture provides data on the availability of food fat for consumption per capita in the United States. These data measure the amount of fat available and do not consider the quantities wasted or discarded. Fats are separated into visible and invisible categories. The visible category consists of easily identifiable fat like butter, margarine, shortening, and salad oils. The invisible category is ingested as part of other foods and is not easily recognized as fat.

Total fat available in the diet has increased 10.5 lb per capita from 1970 to 1985 (Table 12). Visible fats usage accounts for this increase. Butter and lard availability has decreased significantly in the visible fat category while most of the increase has come from shortening and salad and cooking oils. This is to some degree a reflection of the increase in fast-food and convenience foods, meals not made from scratch at home. Dietary and health concerns are expected to have a negative impact on per capita consumption of visible fat. Dairy fats did not change much, but the availability of egg and meat, poultry, and fish fat decreased during the same period. The invisible category experienced a slight overall decrease.

There was a change in the type of visible fats available from 1970 to 1985. Less animal fats were used, while there was an increase in the use of vegetable fats (Table 13). The amount of fat actually consumed in the U.S. diet is currently about 38% of total calories. The trend for visible fats has been to use lower levels of saturated fat and much higher levels of polyunsaturated fat. However, the high level of saturated fat in the invisible part of the diet, primarily meat and dairy fats, has mediated this change and kept the relative overall level of saturated fat in the diet high.

NUTRITION

Fats are an important part of the human diet along with proteins and carbohydrates. Fats are a concentrated source of energy (9 kcal/g vs 4 kcal/g for proteins and carbohydrates), a source of essential fatty acids, and a source of fat-soluble vitamins (36).

Certain fatty acids cannot be synthesized by the body and are essential for human health. These acids, linoleic and linolenic, must be supplied by the diet. About 2% of linoleic acid is necessary in the human diet to prevent a deficiency, and 3% is considered to be a more adequate intake. The minimum requirement for linolenic acid has been estimated at a little over 0.5% of total calories (36,37).

Estimates of total fat in the U.S. diet are usually around 40% of calories, and will vary depending on the source of the data. Reduction in fat intake is an important part of any weight loss–control diet because fat is such a dense source of calories. Most recent dietary health con-

Table 12. Edible Fat Availability Per Capita[a]

Source	1970	1975	1980	1985
Visible fats				
Butter, fat content	4.3	3.8	3.7	4.0
Lard	4.6	2.9	2.6	1.8
Tallow, direct use	—	—	1.1	1.9
Margarine	8.9	9.0	9.2	8.6
Shortening, baking and frying fats	17.3	17.0	18.2	22.9
Salad and cooking oils	15.4	17.8	21.2	23.5
Other	2.3	2.0	1.5	1.6
Total visible fats	*52.8*	*52.5*	*57.4*	*64.4*
Invisible fats				
Dairy products, excluding butter	15.4	15.1	14.7	15.7
Eggs	3.8	3.4	3.4	3.2
Meat, poultry, and fish	48.0	42.6	47.5	46.1
Fruits and vegetables	0.9	1.0	1.0	1.2
Legumes, nuts, and soy	3.6	3.7	3.2	3.8
Grains	1.6	1.7	1.7	1.8
Miscellaneous	1.8	1.5	1.5	2.2
Total invisible fat	*75.1*	*69.0*	*73.0*	*74.0*
Total fats and oils	*127.9*	*121.5*	*130.4*	*138.4*

[a] Ref. 35.

cerns center on cardiovascular disease, including heart attack and stroke. Diet is just one, and not necessarily the major, risk factor identified from epidemiological studies of cardiovascular disease. Other factors include smoking, high blood pressure (hypertension), obesity, elevated serum cholesterol, male sex, sedentary life style, diabetes, excessive stress, and a positive family history of cardiovascular disease. No direct cause-and-effect relationship has been established for these risk factors, only an increased statistical risk has been shown.

Most of the recent studies involving dietary fat focus on the specific type of fatty acid (saturated, monounsaturated and polyunsaturated) and its relationship to serum cholesterol and lipoproteins (cholesterol carriers). There have been studies that have shown a relationship between carbohydrates, proteins, fiber, and trace minerals and cardiovascular disease (38).

Vegetable fats do not directly contain a significant amount of cholesterol. Animal and dairy fats do contain cholesterol and thus may directly affect serum cholesterol. Much recent work has been done relating specific fatty acid types with their effect on the lipoprotein carriers and, indirectly, serum cholesterol (39–42). Two general classes of lipoprotein carriers are largely involved. Low-density lipoproteins (LDL) contain the largest portion of total cholesterol and are responsible for carrying cholesterol in the blood. High levels of LDLs are associated with increased risk of cardiovascular disease, and they can be thought of as the bad cholesterol. High-density lipoproteins (HDL) transport cholesterol from the blood, and elevated levels of HDL are associated with decreased risk of coronary heart disease. Thus HDLs can be viewed as the good cholesterol (43).

Increased levels of saturated fat in the diet have been shown to increase total cholesterol and LDL cholesterol, thus increasing the risk of heart disease (39). Much negative publicity has been associated with tropical fats (coconut, palm kernel, and palm oil) due to their high saturated fat content. Recent studies show palm oil may be an exception, despite its high saturated fat content (42,44–

Table 13. Fats and Oils Used in Foods, mm Pounds[a]

	Year			
Source	1970–1971	1975–1976	1980–1981	1985–1986
Vegetable oils				
Soybean	5,780	7,431	8,610	10,004
Corn	403	520	624	837
Cottonseed	805	578	555	627
Palm	160	737	291	364
Coconut	220	506	338	333
Peanut	181	229	119	137
Sunflower			79	86
Meat fats				
Tallow	508	595	740	1,021
Lard	1,612	155	415	355
Total fats	*9,669*	*10,898*	*11,908*	*13,972*

[a] Ref. 51.

46). Similarly, meat and dairy fats suffer because their relatively high saturated fat content as well as their cholesterol content.

Diets high in monounsaturated fat and low in saturated fats, the so-called Mediterranean diet, have recently been shown to decrease total and LDL cholesterol while not affecting HDL cholesterol (39,41). Studies suggest this type of diet may be equal to, or better than, a low-fat diet for lowering blood cholesterol. Olive oil and canola oil are common sources of monounsaturates. Diets high in polyunsaturated fats beneficially decrease total and LDL cholesterol, but also decrease HDL cholesterol (39,41). Soybean and corn oil are common polyunsaturated fat sources.

Some animal studies have suggested a relationship between diets high in polyunsaturates and certain types of cancer, particularly breast cancer. Other animal studies indicate a high caloric diet due to increased fat may be related to increased cancer incidence, particularly colon and breast cancer (47,48). Any direct relationship between total caloric intake, fat unsaturation, and cancer incidence has not been proved.

Specific fatty acid isomers have also been studied. Trans isomers, which result from the hydrogenation process, are present at significant levels, 10–35%, in most processed shortenings. Most are present as monounsaturated, trans fatty acids. Concerns about the safety of trans fatty acids in the U.S. diet are not supported by reliable epidemiological and animal data (49,50).

Diets high in omega-3 polyunsaturates, like fish oils, have shown a reduced incidence of coronary heart disease. However, other data show an increased risk of stroke and bleeding tendency with this type of diet (40). More studies are needed to determine the long-term effects of diets supplemented with fish oil. Some vegetable oils (soybean and canola) are sources of omega-3 fatty acids in the diet without the cholesterol and saturated fatty acids associated with fish oil.

Most of the current work and interest centers on tropical fats, specifically palm oil; monounsaturated fatty acids, and omega-3 fatty acids. There is still much controversy concerning nutritional and health effects of fat in our diets. Even the dietary recommendations may differ. Given the current state of knowledge, prudence would dictate moderation in the diet.

BIBLIOGRAPHY

1. D. Swern, ed., *Bailey's Industrial Oil and Fat Products,* Vol. 1, 4th ed., John Wiley & Sons, New York, 1979, pp. 47, 651.
2. D. R. Erickson and co-workers, *Handbook of Soy Processing and Utilization,* American Soybean Association, Mo., and the American Oil Chemists' Society, Ill., 1980, pp. 77–79.
3. *Oils & Fats International,* Vol 2, (1986), p. 31.
4. D. F. Buck, "Antioxidant Applications," *Manufacturing Confectioner* **64**(6), 45–49 (1985).
5. T. H. Applewhite, ed., *Bailey's Industrial Oil and Fat Products,* Vol. 3, John Wiley & Sons, Inc., New York, 1985, pp. 9–10.
6. C. W. Hoerr, *Journal of the American Oil Chemists' Society* **37**(10), 539–546 (1960).
7. E. S. Lutton, *Journal of the American Oil Chemists' Society* **49**(1) (1972).
8. U.S. Pat. 2,530,596 (1950), N. W. Zills and W. H. Schmidt (to Lever Bros. Co.).
9. U.S. Pat. 2,625,478 (1953), K. F. Mattil and F. A. Norris (to Swift & Co.).
10. U.S. Pat. 2,521,242 (1950), P. J. Mitchell, Jr. (to Procter & Gamble Co.).
11. Ref. 5, pp. 91–93.
12. U.S. Pat. 4,127,597 (1978), J. C. Craig, M. F. Kozempel, and S. Elias (to U.S. Secretary of Agriculture).
13. Ref. 5, pp. 18–20.
14. U.S. Pat. 3,232,961 (1966), W. Stein, H. Rutzen, and E. Sussner (to Henkel & Cie).
15. U.S. Pat. 2,875,066 (1959), G. W. Holman and L. H. Going (to Procter & Gamble Co.).
16. U.S. Pat. 2,875,067 (1959), G. W. Holman and L. H. Going (to Procter & Gamble Co.).
17. M. Naudet, *Rev. Fr. Corps Gras* **17**(2), 97–104 (1970).
18. E. R. Sherwin, *Journal of the American Oil Chemists' Society* **53** 430–436 (1976).
19. *U.S. Code of Federal regulations,* Title 21, Part 172, Sections 172.818 and 172.854, 1987.
20. Ref. 19, Part 169, Section 169.150.
21. Ref. 19, Part 169, Section 169.140.
22. U.S. Pat. 4,359,482 (1982), T. G. Crosby (to Procter and Gamble Co.).
23. U.S. Pat. 4,385,076 (1983), T. G. Crosby (to Procter and Gamble Co.).
24. U.S. Pat. 4,384,008 (1983), D. H. Millisor (to Procter and Gamble Co.).
25. E. J. Pyler, *Baking Science and Technology,* Vols. 1 and 2, 3rd ed., Siebel Publishing Co., Chicago, Ill., 1982.
26. U.S. Pat. 4,310,556 (1982), J. L. Suggs, D. F. Buck, H. K. Hobbs (to Eastman Kodak Co.).
27. U.S. Pat. 4,310,557 (1982), J. L. Suggs, D. F. Buck, H. K. Hobbs (to Eastman Kodak Co.).
28. U.S. Pat. 4,242,366 (1980), J. E. Morgan, A. J. Del Vecchio, B. L. Brooking, and D. M. Laverty (to Pillsbury Co.).
29. U.S. Pat. 4,201,718 (1980), J. T. Marsch.
30. U.S. Pat. 4,205,095 (1980), M. Pike, I. G. Barr, and F. Tirtaux.
31. U.S. Pat. 4,268,534 (1981), T. Kawada and Y. Tanaka (to Kao Soap Co., Japan).
32. U.S. Pat. 4,108,879 (1978), S. Minowa, Y. Tokoshima, N. Yasuda, and T. Tanaka (to Ashai Denko Kogyo KK, Japan).
33. Ref. 19, Chapter 9, Section 321a (Butter Act).
34. Ref. 19, Part 166, Section 166.110.
35. Economic Research Service, U.S. Department of Agriculture, Washington, D.C.
36. J. E. Kinsella, *Food Technology* **42**(20), 124–145 (1988).
37. Food and Nutrition Board, National Research Council, *Recommended Daily Allowances,* 9th ed., National Academy of Sciences, Washington, D.C., 1980.
38. E. H. Ahrens, Jr., and W. E. Connor, *American Journal of Clinical Nutrition* **32**(Suppl.), 2621–2748 (1982).
39. S. M. Grundy, *Journal of the American Medical Association* **256**(20), 2849–2858 (1986).
40. A. Leaf and P. C. Weber, *New England Journal of Medicine* **318**(9), 549–557 (1988).
41. F. H. Mattson and S. M. Grundy, *Journal of Lipid Research* **26** 194–202 (1985).
42. A. Bonahome and S. M. Grundy, *New England Journal of Medicine* **318**(19), 1244–1248 (1988).

43. B. Dolan and J. M. Nash, *Time*, 62–66 (Dec. 12, 1988).
44. J. T. Anderson, F. Grande, and A. Keys, *American Journal of Clinical Nutrition* **29**, 1184–1189 (1976).
45. *Nutritional Reviews* **45**, 205–207 (1987).
46. M. Sugano, *Lipids* **40**, 48–51 (1987).
47. C. Ip, D. F. Birt, A. E. Rodgers, and C. Mettlin, eds., *Dietary Fat and Cancer,* Alan R. Liss, Inc., New York, 1986.
48. W. C. Willett and co-workers, *New England Journal of Medicine* **316**, 22–28 (1987).
49. J. E. Hunter, C. Ip, and E. J. Hollenbach, *Nutrition and Cancer* **7**, 199–209 (1985).
50. F. R. Senti, ed., *Health Aspects of Dietary Trans Fatty Acids,* Federation of the American Society for Experimental Biology, Bethesda, Md., 1985.
51. *Oil Crops*, Economic Research Service, USDA, July 1989.

General References

References 1 and 2 are good general references.

W. H. Meyer and co-workers, *Food, Fats, and Oils*, 6th ed., Institute of Shortening and Edible Oils, Inc., Washington, D.C., 1988.

K. S. Markley, *Fatty Acids,* 2nd ed., John Wiley-Interscience, New York, 1960.

T. P. Hilden and P. N. Williams, *The Chemical Constitution of Natural Fats*, 4th ed., John Wiley & Sons, Inc., New York, 1964.

E. W. Eckey, *Vegetable Fats and Oils*, Reinhold Publishing, New York, 1954.

E. Bernadini, *Vegetable Oils and Fat Processing,* Vol. II, B. E. Oil, Rome, 1983.

H. W. Lawson, *Standards for Fats & Oils,* AVI Publishing Co., Inc., Westport, Conn., 1985.

T. J. Weiss, *Food Oils and Their Uses,* 2nd ed., AVI Publishing Co., Inc., Westport, Conn., 1983.

THOMAS G. CROSBY
Bunge Foods Corporation
Bradley, Illinois

FERMENTATION. See ACIDULANTS; BEER; CHOCOLATE AND COCOA; DAIRY FLAVORS; DISTILLED BEVERAGE SPIRITS; FOOD FERMENTATION; FOOD PRESERVATION; PLANT CELL AND TISSUE CULTURE; FOOD RELATED PRODUCTS; RICE; SOYFOODS, FERMENTED; TEA; VEGETABLES: PICKLING AND FERMENTING; YEASTS; and entries under GENETIC ENGINEERING.

FIBER, DIETARY

DEFINITION

While there is no single accepted definition of dietary fiber, there is general consensus that it is that portion of plant material consumed in the diet that is resistant to breakdown (digestion) by human enzymes secreted into the gastrointestinal tract (small intestine). Therefore, dietary fiber is not a single entity, but several. Different materials from different sources, all of which may be classified as dietary fiber, may differ both in their chemical and physical structures and in their physiological, metabolic, and nutritional effects. In the natural state they are accompanied by other substances. Nutritionists divide dietary fiber into two rough categories: soluble fiber and insoluble fiber. Few of the components of dietary fiber are actually physically fibrous in nature.

COMPOSITION

The composition and properties of dietary fiber varies with the source and what is included depends on the method of analysis. As an example of the complexity of fiber composition, a major source of fiber for addition to foods is cereal bran, the outer part of a grain kernel which is isolated through mechanical grinding and sieving. These products of commercial grain processing are composed of several tissue layers, including the pericarp, seed coat, nucellar epidermis, and usually at least parts of the aleurone and subaleurone layers. Table 1 lists the materials that might be analyzed as fiber by one or more methods and designates each as soluble or insoluble fiber, although the definitions of these categories, based on hot-water solubility, are inexact.

ANALYSIS

A number of different methods for determination and analysis of dietary fiber have been developed. The major impediment to general progress in assessing dietary fiber has been the unavailability of a single analytical method that is compatible with the accepted definition of dietary fiber and is accurate, reproducible, rapid, and simple. Multiple methods have evolved because dietary fiber encompasses such a diverse group of materials, there has been a strong desire for a rapid, reliable method to determine the total dietary fiber fraction of foods for labeling purposes, and researchers have sought methods that will give quantitative information on specific individual components. The most widely recognized and commonly used methods for dietary fiber analysis are the following: Association of Official Analytical Chemists (AOAC), American

Table 1. Possible Components of Dietary Fiber

	Class	
	Insoluble	Soluble
Polysaccharides		
Cellulose	X	
Hemicelluloses A[a]	X	
Hemicelluloses B[b]	X	X[b]
Pectic substances[c]	X	X
Beta-glucans[d]		X
Resistant starch[e]	X	
Gums		X
Lignin	X	
Other components		
Phytate		
Phenolic compounds		

[a] Primarily linear xylans, often acidic xylans.
[b] Primarily branch-on-branch arabinoxylans. With xylans, they are components of the primary and secondary cell walls of annual plants. Generally, they are only partially extractable with water but are soluble once extracted, eg, with alkaline solutions from delignified tissue.
[c] Rhamnogalacturonans. Pectins are water soluble, although they generally require acidic solutions for extraction. Protopectins are insoluble.
[d] Glucans with mixed β-D-(1→3) and β-D-(1→4) linkages in various proportions.
[e] Granular or retrograded starch that is resistant to human digestive enzymes.
[f] All food gums with the exception of food starches, modified food starches, and gelatin, all of which are digested by human gastrointestinal enzymes.

Table 2. Approximate Dietary Fiber Content of Various Foodstuffs[a]

Product	Dietary Fiber, Percent Fresh Weight
Bran	44
Whole wheat bread	8.5
Nuts	8.5
Peas and beans (cooked)	6.2
Sweet corn (cooked)	4.7
Carrots, parsnips, and turnips (raw and cooked)	3.6
White bread	2.7
Leafy vegetables (fresh and cooked)	2.6
Fruits (fresh)	1.7
Tomato (fresh)	1.4
Sweet pepper (cooked)	0.9

[a] Ref. 14.

Association of Cereal Chemists, (AACC), Mongeau–Brassard, Theander, Englyst, and modifications thereof.

The AOAC method (1,2) is the official method in a number of countries, including the United States. In this method, the sample is first homogenized, dried, ground, and defatted. Then protein and starch are removed via enzymic digestion. The dried residue is weighed and corrected for ash and protein content. The intention is to give a single value for total dietary fiber (TDF) content. The method has been modified to give separate measurements of soluble and insoluble fiber (3,4).

Another method based on extraction and gravimetric analysis, the official method of the AACC (5), is a modified neutral detergent fiber (NDF) method. The Mongeau–Brassard method (6) gives separate determinations of NDF and soluble fiber, the sum of which gives TDF. The methods of Theander (7) give soluble and insoluble fiber separately and soluble plus insoluble fiber together. The Englyst method (8) determines nonstarch polysaccharides. Comparisons of these methods have been published (3,9).

Determination of the individual components of fiber is more difficult. Methods for isolation and separation of individual polysaccharides based on differential solubilities and their identification and quantitative analysis have been reviewed (10,11).

SOURCES OF DIETARY FIBER

Because dietary fiber arises from the walls of plant cells, it is found in all natural foods obtained from plants, viz, whole grains; leafy vegetables; root vegetables; tomatoes, corn, peppers, etc; legumes; fruits; and nuts. The dietary fiber content of foods depends on the analytical method used. Tables that compare the TDF content of various foods have appeared (12,13). Tables 2 and 3 present some general information on fiber content and composition.

Many foods and beverages are fortified with dietary fiber. Most commonly, fiber is added to cereals and bakery products. Some commercial sources of dietary fiber and their composition are given in Table 4. Not listed in Table 4 are blends such as wheat bran plus corn bran, wheat bran plus date fiber, wheat bran plus fig fiber, wheat bran plus pea flour, and apple fiber plus corn and rice brans; whole dried fruits, such as apples, apricots, cranberries, dates, figs, prunes, peaches, and raisins; whole nuts; or some lesser used fiber sources.

PHYSIOLOGICAL EFFECTS OF FIBER

At this time, many of the claimed benefits of dietary fiber have not been conclusively demonstrated. Although the data are inconclusive and being debated, the interest in foods with a high content of fiber and in fiber-enriched products arises from evidence that dietary fiber reduces the risk of coronary heart disease, reduces the risk of colon cancer and diverticulitis, and improves glucose tolerance and insulin response (18). However, different kinds of fiber effect different physiological responses, and certain fiber sources are more effective than others in bringing about each of these effects. The physiological effects associated with increased ingestion of dietary fiber are complex, and the exact mechanism(s) responsible for them are unknown and under current discussion and research.

The primary effect of dietary fiber is on improving bowel function. In general, dietary fiber holds water and thickens aqueous systems, swells and provides bulk volume, and is an adsorbent. It provides a matrix and a surface for bacterial growth and thereby increases bacterial mass. Dietary fiber produces absorbable metabolites, has cation-exchange properties in some cases, and has beneficial physiological effects. There are considerable differ-

Table 3. Composition of Dietary Fiber[a]

Fiber Source	Percent Dry Weight		
	Noncellulose Polysaccharides	Cellulose	Lignin
Cereals (8 samples)			
Average	75.7	17.4	6.7
Range	71–82	12–22	trace–15
Raw vegetables (11 samples)			
Average	65.6	31.5	2.98
Range	52–76	23–42	trace–13
Fruits (10 samples)			
Average	62.9	19.7	17.4
Range	46–78	9–33	1–38

[a] Refs. 15, 16.

Table 4. Sources and Reported Compositions of Commercial Dietary Fiber Products[a]

Source	TDF, Approximate Percent	Soluble Fiber, Approximate Percent of TDF
Brans		
Barley	45–70	3–20
Corn	50–95	0–5
Oat	22–90	1–50
Rice	35–75	4–40
Wheat	40–66	2–5
Cellulose	96–98	0–4
Fruits		
Apple fiber	44–60	7–31
Citrus fiber	50–70	30
Date fiber	52	
Fig powder	12–64	
Prune fiber	57	4
Gums, water-soluble	80–100	94–100
Legumes		
Lupine flour	60–83	8
Pea flours	45–87	51–55
Peanut flours	55	3
Soy flours	50–75	0–59
Sugar beet fiber	65–80	12–18
Tomato fiber	55	3

[a] Ref. 17.

ences between the effectiveness of soluble and insoluble fiber (18,19) on bowel function, and effectiveness is dependent on the specific composition of the fiber in most cases.

TDF in foods provides a texture that requires more chewing and slows the rate of ingestion. It may cause earlier satiation and longer satiety, thus reducing caloric intake. TDF increases the rate of intestinal flow (decreases transit time, which is the time it takes a meal to pass in the feces after ingestion). TDF increases both fecal mass and the number of bacteria in it.

Insoluble dietary fiber (IDF) acts as a bulking agent that holds water and decreases transit time. It is primarily responsible for the increase in stool weight and water content, although soluble dietary fiber (SDF) also contributes. Increasing the bulk and water content produces softer stools. It also dilutes the contents, ie, reduces the concentration of waste products. This may well be a factor in the ability of dietary fiber to reduce the risk of colon and rectal cancer.

SDF, because of its thickening effect, slows gastric emptying and changes intestinal propulsion. SDF delays the absorption of nutrients, especially glucose, and lowers postprandial blood glucose and insulin concentrations, probably because of its ability to increase the viscosity of the intestinal contents. In the same way, ie, by reducing mixing diffusion, SDF reduces the amount of bile acids reabsorbed; this is probably partially responsible for its effect of lowering blood cholesterol levels.

Bacteria that are responsible for the anaerobic fermentation that takes place in the large intestine (colon) degrade 20–80% of the IDF and almost all the SDF. This fermentation produces more fecal bacteria, increases fecal weight, and results in an increase in the contents of water and acetic, propionic, and butyric acids, the so-called short-chain fatty acids (SCFA). There is evidence that these SCFA, which are absorbed and provide calories, suppress cholesterol biosynthesis, another likely contributing factor to the lowering of blood cholesterol concentrations. There may be other mechanisms operating. It is clear, however, that SDF is more effective than IDF in lowering blood cholesterol concentrations. In addition, barley bran, in particular, contains tocopherols, tocotrienols, and a special triglyceride, all of which suppress cholesterol biosynthesis.

Some adverse physiological effects of high-fiber diets are possible. These include excess gas formation and poor absorption of certain drugs and lipid-soluble vitamins. Although reduced absorption of minerals has been suggested, the majority of data do not support an adverse effect on mineral nutrition.

USE IN FOODS

SDF in the form of food gums is incorporated in small amounts in a wide variety of food products. A number of liquid and moist foods, bakery products, breakfast cereals, and snack foods are enriched with fiber because of the ability of all polysaccharides to bind water and the extra ability of insoluble fiber to hold water in interstitial spaces by capillary action. Also, because of its water-holding capacity, fiber is used to prepare reduced-caloric products. The presence of fiber aids the process of extrusion. Specific products to which fiber may be added in substantial amounts are breads, muffins, hot and cold cereals, extruded snacks, fruit rolls, granola bars, dietetic items, fiber drinks, pasta, and processed meats. Oatmeal has a high natural fiber content.

BIBLIOGRAPHY

1. L. Prosky, N. G. Asp, I. Furda, J. W. DeVries, T. F. Schweizer, and B. F. Harland, *Journal of the Association of Official Analytical Chemists* **68**, 677 (1985).

2. L. Prosky, N. G. Asp, T. F. Schweizer, J. W. DeVries, and I. Furda, *Journal of the Association of Official Analytical Chemists* **71**, 1017 (1988).
3. F. Meuser, P. Sukow, and W. Kulikowsky, *Zeitschrift fuer Lebensmittel-Untersuchung und-Forschung* **181**, 101 (1985).
4. I. Furda, in R. P. Millane, J. N. BeMiller, and R. Chandrasekaran, eds., *Frontiers in Carbohydrate Research*, Vol. 1, Elsevier Applied Science, Publishers, Ltd., Berking, UK, 1989
5. *Approved Methods of the AACC*, American Association of Cereal Chemists, St. Paul, Minn., 1983.
6. R. Mongeau and R. Brassard, *Journal of Food Science* **51**, 1333 (1986).
7. O. Theander and E. A. Westerlund, *Journal of Agriculture and Food Chemistry* **34**, 330 (1986).
8. H. N. Englyst, H. S. Wiggins, and J. H. Cummings, *Analyst* **107**, 307 (1982).
9. J. M. Hall, *Cereal Foods World* **34**, 526 (1989).
10. A. Olson, G. M. Gray, and M.-C. Chiu, *Food Technology* **41**(2), 71 (1987).
11. J. W. Baird, in R. L. Whistler and J. N. BeMiller, eds., *Industrial Gums*, 3rd ed., Academic Press, Inc., Orlando, Fla., 1991.
12. R. H. Matthews and P. R. Pherrson, *Dietary Fiber Content of Selected Foods*, HNIS/PT-106, Human Nutrition Information Service, USDA, Washington, D.C., 1988.
13. *Food Technology* **43**(10), 133 (1989).
14. D. A. T. Southgate, B. Bailey, E. Collingson, and A. F. Walker, *Journal of Human Nutrition* **30**, 303 (1976).
15. D. A. T. Southgate, in G. W. Vahouny and D. Kritchevsky, eds., *Dietary Fiber in Health and Disease*, Plenum Press, New York, 1982.
16. B. O. Schneeman, *Food Technology* **40**, 104 (1986).
17. *Cereal Foods World* **32**, 555 (1987).
18. D. T. Gordon, *Cereal Foods World* **34**, 517 (1989).
19. B. O. Schneeman, *Food Technology* **41**, 81 (1987).

J. N. BeMiller
Purdue University
West Lafayette, Indiana

FILTH AND EXTRANEOUS MATTER IN FOOD

INTRODUCTION

Filth will be defined and the laws and regulations pertaining to it will be described as they have developed over the decades in the United States. Filth in food is derived from food pests, especially certain beetles, moths, and rats. The distribution of these pests around the world coincides almost exactly with the distribution of the human species. A distinction will be made between field pests and storage pests, because the latter are more likely to have significance for human health. Because both field and storage pests are so successful in their efforts to reach human food, various administrative procedures have been devised to cope with their presence in or around food. The most creative of these, the food defect action levels, will be described at length, along with some less-appealing alternative approaches.

It will be seen that the current standards of food purity in the United States have resulted from a remarkable series of interactions among scientists, politicians, lawyers, economists, farmers, public health administrators, and activists of every description. A few important court cases will be discussed to show how the law has been administered by the Food and Drug Administration and how it has been interpreted by the courts.

The goal of every food manufacturer is to produce a safe and clean food from a safe and clean plant. How this objective can be achieved in spite of continuous assaults from food pests will be discussed in detail below. The role of food processing in making or keeping food safe to eat will be noted, along with hazardous situations that should be avoided.

The practice of food protection is neither a static science nor a reflex reaction. It requires creative thinking and contingency planning to meet future problems. Although food protection as practiced in the United States is the envy of the world, there is still much to be learned, and some of these potential problem areas are brought into the discussion.

The points of view concerning food purity expressed here arise largely from the context of a federal regulatory agency in a technologically advanced nation. In third world countries human concerns focus more on food per se than on food purity. As the United Nations urges, the problem of filth in food is dealt with here under the umbrella slogan, Think Globally, Act Locally.

THE FEDERAL FOOD LAWS

Around the turn of the century, the efforts of a great champion of food purity, Wiley, began to pay off. The U.S. Congress passed the Pure Food Law in June of 1906 (1–3). The revisions that strengthened the new law of 1938 (4) came into effect only after a tortuous trip through Congress (5). Since then, the food law has been amended and expanded on several occasions. Now the Federal Food, Drug, and Cosmetic Act (FD&C Act) and related legislation cover a broad array of consumer products (6).

Wiley eventually became disenchanted with the way the food law was being administered (7). He left government in 1912, and 17 years later published a book (8) describing the alleged foibles of certain federal administrators, not the least of whom was President Theodore Roosevelt. Wiley was not the last critic of the food law and its administrators (critics have been numerous and vocal through the decades). But the fact remains that the pure food law in all of its manifestations from 1906 until today stands as a major achievement in the protection of consumer health (9). As great as that achievement was, however, it should be noted that protection of the public's health was not the only objective. The courts have routinely understood that it was the intention of Congress to keep filth (harmful or not) out of food. In other words, Congress wanted to ensure both the hygienic and the aesthetic qualities of food (10,11). Nevertheless, the Food and Drug Administration (FDA) is committed to giving highest priority in its enforcement of the act to public health protection and public health promotion (12).

Two Sections on Filth

The FD&C Act (6) contains two paragraphs that are especially pertinent to this article. The first, 402(a)(3), states that a food shall be deemed to be adulterated if it consists in whole or in part of any filthy, putrid, or decomposed substance or if it is otherwise unfit for food. The second, 402(a)(4), states that a food shall be deemed to be adulterated if it has been prepared, packed, or held under insanitary conditions whereby it may have been contaminated with filth, or whereby it may have been rendered injurious to health.

Extraneous Matter

The courts have always defined filth in its ordinary sense (ie, the dictionary definition) rather than giving the term any specialized, technical meaning (13). The same can be said for extraneous matter, that is, things that do not belong in food, such as filth or any foreign matter in a product as a result of objectional conditions or practices in production, storage, or distribution of food. Included within the meaning of extraneous matter (or materials) are filth (any objectionable matter contributed by animal contamination such as rodent, insect, or bird matter); decomposed material (decayed tissues due to parasitic or nonparasitic causes); and miscellaneous matter such as sand, soil, glass, rust, or other foreign substances (bacteria excluded) (14).

The more common kinds of filth encountered during food inspections and analyses are whole insects and mites and fragments thereof, rodent and other mammalian hairs, feces, feathers and feather barbules, and urine and uric acid (15–19). Other kinds of filth and extraneous matter include molds and other fungi, plant seeds and other foreign plant materials, parasitic worms (helminths), predatory and parasitic arthropods, glass fragments, wood chips, paint chips, metal chips and objects, grease and other lubricants, bits of plastic, fibers (wool, cotton, and synthetic), sand, gravel, and soil (20–25).

Hard Objects in Food

Most hard objects in food (eg, pebbles, sand, glass, metal fragments, ball bearings, and pits) are not considered filth. However, they are illegal if they cause a gritty mouth-feel (26), if consumers would be likely to reject the food as unfit, or if the particles are large enough or sharp enough to be hazardous to chew or swallow. The food defect action levels (DALs) permit very low levels of hard objects in the following specific foods: cloves, dates (pitted), olives (pitted), peanut butter, prunes (pitted), raisins, strawberries, chocolate liquor, and cocoa powder press cake. The sanitary standard for all of these foods is based on the FD&C Act (6), and the occurrence of hard objects in all foods is treated on a case-by-case basis when there is no applicable DAL (27).

Standardized Procedures

Sample collection by FDA investigators provides the basic materials for analysis in agency laboratories (28). Both the procedures for sample collection and the protection of sample integrity have been standardized (29). Analytical procedures in the laboratory are also standardized (30). These methods have been published (14,31) and have appeared in the *Journal of the Association of Official Analytical Chemists*. Several trade associations also issue methods manuals (32–35).

Light Filth Separation by Flotation

Since the 1930s the principal method for separating light filth from food has been by flotation (36). It happens that rat hairs, feather barbules, and insect cuticle are oleophilic, whereas most food materials are hydrophilic. Thus when oil (usually mineral oil) is mixed with a test portion of food in water and then permitted to stand without agitation, the filth elements float into the oil layer and the food materials settle into the water layer. When the oil fraction is passed through a filter paper, the filth elements remain on the paper where they can be identified and counted (37,38). This basic procedure has been quantified and modified to suit many different food products (14,39). The analyst can use several helpful references to identify many of the filth elements recovered from the test sample (40–43).

Chemical Methods for Filth

Reliable chemical methods have been developed to identify the characteristic nitrogenous excretory products of animals (14,44). There is now a reliable test to identify mammalian feces, especially small bits and pieces of rodent pellets, in cornmeal and other grain products and in ground black pepper (45–47). Some preliminary work has been done on the identification of salivary amylase from the gnawed edges of packaging damaged by rodents (48). Gas chromatography has been used to distinguish the cuticular hydrocarbons of German and Asian roaches (the two species are morphological twins) (49,50). A similar technique might be adapted to identify fragments of storage insects and their parasitoids in foods. Good progress has been made on an immunoassay test for drosophilid eggs in fruit juices (51). Several research projects are under way to develop other chemical methods to isolate and identify various kinds of filth in food.

Enforcement of the Act

Most enforcement actions based on the FD&C Act refer to the two paragraphs quoted above [402(a)(3) and (4)] or to comparable sections elsewhere in the act (52–54). Of course, there have been numerous challenges to interpretations of 402(a)(3) and (4) and to other filth sections of the act. In a precedent-setting case (1947) involving butter oil, butter was seized because the cream from which it had been made was grossly contaminated with insect and rodent filth. The claimant wished to recoup his losses by melting down the butter, straining it, and selling the resultant liquid as butter oil. FDA objected to this on the grounds that the butter oil still contained soluble filth from the original adulterants. The court upheld this objection (55), thus making soluble (invisible) filth just as objectionable as particulate filth.

Another important principle emerged from this case, namely, that any food found to be filthy within the meaning of "filth" in the act should be condemned (55). This principle, however, does not always preclude the possibility of bringing the product into compliance by reconditioning (56,57,58).

Particulate filth has been the focus of numerous court cases. One landmark case involved some tomato paste that contained bits and pieces of larval corn earworms. The claimant asserted that the insect fragments could not be considered filth, and the consumer would not so perceive them, because they were too small to see. The court disagreed and the product was kept off the market (59). Since then, a defense of "too small to see" has had little chance of success in the courts.

One litigant claimed that the phrase "may have been" in 402(a)(4) was too vague to be constitutional. The court rejected this contention (60), and the law remains intact as written. This decision upheld the intent of the law to prohibit the presence of filth around food as well as the presence of filth in food.

Since the first enactment of the pure food law in 1906, about 92,000 cases (injunctions, seizures, and prosecutions) have been filed as a result of enforcement actions. This universe of cases is so vast that some lawyers limit their practices to litigation revolving around the FD&C Act. Most cases begin and end in federal district courts (61), but some find their way to courts of appeals, and a few eventually reach the Supreme Court of the United States.

Regulatory Activities

Enforcement actions fall roughly into two categories: actions against products and actions against persons (62). Many alleged violations never even get close to a court hearing for several good reasons. Because the investigatory (inspectional) resources of the FDA are minuscule compared to the vastness of the food industry, FDA now relies on and encourages voluntary compliance by industry (63–66). Many industry leaders take voluntary compliance very seriously. When this positive approach is combined with strict adherence to good manufacturing practice (GMP) statements (52,54,56,57,67–70), FDA investigators often find little to criticize during inspections.

Agency inspections per se are usually beneficial to the facility being inspected, even when defects are found (71). Industry officials often readily and voluntarily correct any defects noted, thus obviating the need for enforcement actions. When there appears to be some reluctance on the part of industry officials to acknowledge defects in their operations, or when the defects are serious, FDA can take a variety of actions to stimulate compliance. When relatively minor problems of insanitation have been noted by an FDA investigator, the agency may generate a Notice of Adverse Findings Letter that underscores the investigator's observations.

Even though recalls are not mentioned in the act and in spite of the fact that recalls relating to food products are technically voluntary on the part of industry (62), the FDA can still be instrumental in prompting a recall when circumstances warrant it (11,61,72,73). To maintain their good reputations, many firms are willing to act promptly to protect both themselves and the public. Recalls are graded according to the severity of the problem (13,74).

Enforcement actions can escalate to more serious levels. The Regulatory Letter is a formal notification that the FDA is prepared to take legal action if the violations cited are not corrected immediately. Sometimes specific lots of a product or even all of the products in a warehouse can be seized to keep them from the marketplace. State regulatory personnel sometimes collaborate with the FDA to keep violative goods off the market. A court injunction can be issued to prevent individuals from engaging in violative practices (61,75,76).

When all else fails, the FDA may cite the firm and then initiate criminal prosecution against individuals who are accused of violating the FD&C Act (61). The several enforcement techniques pertaining to filth in food have been summarized (13) and discussed in detail (11,53,56,57,62,67).

All of the cases mentioned above represent actions against products. Actions against persons are less frequent but always serious. The guilty can expect to pay fines or spend time in prison or both (13,77). One important aspect of actions against persons has been reviewed twice by the Supreme Court. That court has answered the question about whether or not high corporate officials can be held responsible for violative acts committed by subordinate persons to whom they have delegated responsibility. The Court said yes in 1943 when it heard the *Dotterweich* case (11,56,57,78,79), a proceeding that involved adulterated drugs. The second case to be noted involved storage of food.

Park-Acme **Case**

The *Park-Acme* case (11,57,78–80) began in 1970 when FDA investigators found rodents infesting a warehouse operated by Acme Markets in Philadelphia. Similar conditions were observed in the company's Baltimore warehouse when it was inspected in January 1972. A follow-up inspection in March indicated that rodents were still present. The corporation and its president were charged with violation of 402(a)(4). The corporation pled no contest but the president pled not guilty. He was found guilty, but the conviction was overturned on appeal. The FDA took the case to the Supreme Court, which determined that the president was guilty as charged. The Supreme Court's decision vitiated the excuse of delegation and reaffirmed that top executive officers bear complete responsibility for violations that occur in their organizations (79). As a result, at least some high-level corporate officials have developed a more intense personal interest in the design and effectiveness of pest prevention and control operations undertaken in their firms.

De Minimus **Filth**

It has been noted that a defense of "too small to see" has not been well received in the courts (59). But what about a defense of "too little to matter" (*de minimus* filth)? This defense has been used in relation to both 402(a)(3) and 402(a)(4), but more to the former than the latter. The act seems to preclude both defenses when it boldly declares that a food "shall be deemed to be adulterated if it consists . . . of any filthy . . . substance" [(a)(3)], or if it has been "prepared, packed, or held under insanitary conditions whereby it may have become contaminated with filth" [(a)(4)].

The courts have never held to the letter of the law. They have, in fact, on occasion, been remarkably lenient in their interpretation of evidence. For example, evidence gathered in 1972 by FDA investigators showed that a certain food warehouse was infested by rodents. Numerous active rodent burrows were seen around the warehouse; gnawed holes embellished eight bags of flour; and two dead rodents were found near the stored bags. Two of the bags were collected and examined. Both contained rodent excreta pellets. The district court judge, finding nothing compelling in this evidence, acquitted the defendants of all charges including those based on 402(a)(4). Not surprisingly, the FDA appealed, but to no avail. The appellate judge sided with his lower court colleague. But more than that, he toyed with the idea that Congress had intended to imply a range of tolerance in the 402(a)(4) clause (57). Fortunately, that idea died with the case, and other courts have been less forgiving of solid (a)(4) evidence.

The idea of *de minimus* filth has gained a lot more ground as the courts have considered (a)(3) evidence. In one case, *United States v. Gerber Products Co.* (56), a district court jury heard evidence that canned peaches contained moldy peach pulp, worm fragments, and insect fragments. The presiding judge was quite clear in instructing the jury that they had to decide that the filth was either significant or insignificant. The jury chose the latter. Apparently the jury was swayed by that part of the judge's instructions in which he explained "that it was not the intention of Congress to include as a criminal offense the presence of filthy, putrid or decomposed matter in such infinitesimal and inconsequential quantities as even the highest degree of care could not eliminate."

This judge's characterization of *de minimus* filth, along with similar explanations offered from the bench in numerous other cases, eg, *United States v. 1500 cases* (56), has provided the judicial basis for what has come to be known as the food defect action levels (DALs) (10,81). (Note that DALs are not mentioned in the FD&C Act.) Even though exporters and domestic food manufacturers aim at a high level of food purity, it is still impossible, save for an unconscionable waste of raw food materials, to produce certain food products that are completely free of filth. To bridge the gap between this practical reality and the absolute ban on filthy food as stated in the act [402(a)(3)], the FDA has established DALs for approximately 90 categories of foods (26,67). The DALs refer mainly to field infestations and other unavoidable contamination. Infestations of stored foods and related insanitary conditions get very little help from the DALs.

The foregoing cursory commentary on two small paragraphs of the act should in no way be considered exhaustive. Its purpose was merely to highlight what the act says about filth in food in sections 402(a)(3) and (4). Other sections of the act also refer to filth in food. Laws and regulations enforced by other agencies of the federal government deal with filth in food. All of the states and many local jurisdictions have laws, ordinances, and regulations pertaining to filth in food (67,82–87). Similarly, some of the countries that export food to the United States have their own laws and regulations that apply to exported food. The Food and Drug Administration routinely cooperates and collaborates with foreign countries, other federal agencies, the several states and territories, and often with local governments to achieve the overall goal of protecting the nation's food supply (12,88–90). Many nations in the world community accept the provisions of *Codex Alimentarius*, an international system of food standards and regulations sponsored by the United Nations (91,92).

Cooperative agreements between nations are often formalized as Memoranda of Understanding (MOUs). MOUs covering a variety of foods (eg, powdered milk and shellfish) and food containers (ceramicware) have been concluded between about 20 countries and the United States (29); others are under development. These MOUs help exporting nations gain entry to the United States for their products. MOUs place the burden of sample collection and analysis on the foreign nation, thus relieving the FDA of this responsibility and its associated staffing requirements. MOUs often prevent detention of imported products (because the products are manufactured to FDA standards before they leave foreign shores) or at least assist in getting products off an automatic detention list. MOUs provide unmistakable evidence that the FDA is not involved in setting up trade barriers against imported products.

The point of all of this is to emphasize that the administration and enforcement of laws and regulations aimed at protecting food from adulteration by filth are complex, far-reaching, and not easily summarized. Moreover, even the comparatively simple Food and Drugs Act of 1906 was itself the progeny of earlier traditions and laws that are not mentioned at all in this article and are only rarely treated elsewhere (10,93–95).

THE FOOD DEFECT ACTION LEVELS

The procedure for setting DALs has evolved over the decades and continues to evolve. In earlier days it was an informal method based largely on the experience of the person charged with the task of setting the DAL. In the 1970s and 1980s, computer simulation techniques were tried (81). Now a statistical approach is being used. Samples are collected nationwide and analyzed by standard methods (14). The defects data are arranged in a frequency distribution from lowest to highest. The DAL selection process begins at the upper 99% confidence limit of the 95th percentile of this frequency distribution. Nonmathematical factors, such as the effect of unusual weather on a given crop, may influence the setting in one direction or another. Generally, it turns out that the DAL allows about 97–98% of a product to be accepted in the marketplace.

The food DALs are administrative guidelines that are set on the basis of no hazard to health (26). Any products that might be harmful to consumers are acted against on the basis of their hazard to health, whether or not they exceed the action levels. Insanitary manufacturing practices will result in regulatory action whether or not the product is above or below the pertinent defect action level. DALs are resorted to because it never has been possible to grow, harvest and process crops that are totally free of natural defects.

One alternative to establishing DALs in some foods would be to insist on increased use of chemicals to control insects, rodents and other natural contaminants. That alternative is not satisfactory because of the very real danger of exposing consumers to potential hazards from residues of these chemicals. It seems preferable to expose consumers to harmless natural and unavoidable defects, even though such defects may be aesthetically unpleasant. However, in the day-to-day experience of food consumption, only the rare consumer, as the claimant in the tomato paste case (59) was eager to point out, would be aware of finely comminuted, natural and unavoidable defects.

The fact that the FDA has an established action level does not mean that a manufacturer need only stay below that level. The DALs do not represent an average of the defects that occur in any of the food categories. The averages are actually much lower. The levels represent the limit at or above which the FDA will take legal action against the product to remove it from the market. DALs on the list are periodically reviewed and lowered as technology improves (occasionally, a DAL may be raised). Many food industry managers routinely set in-house action levels for their own products that are more stringent than the FDA guidelines.

It is the position of the FDA that compliance with the DALs does not excuse violation of the requirements in section 402(a)(4) of the FD&C Act that food shall not be prepared, packed, or held under insanitary conditions or the requirement that food manufacturers, distributors, and holders must observe good manufacturing practice (52). Evidence indicating that such a violation exists causes the food to be adulterated within the meaning of the act, even though the amounts of natural or unavoidable defects are lower than the currently established DALs. The mixing (blending) of food containing any amount of defective food at or above the current DAL with another lot of the same or another food is usually not permitted and may render the final food unlawful regardless of the defect level of the finished food (11,26,57,96).

Filth, Health, Hunger, and Pesticides

As food moves along the train of events from farm to consumer, it becomes increasingly more valuable per unit of measure. Losses caused by pest infestation become correspondingly more expensive. Both pest contamination and pesticide residues become less tolerable as food gets closer to the consumer. The food industry is forced by law and by public concern to operate within rather narrow limits; insecticide residue action levels (97,98) must not be exceeded, yet enough pesticides (or efficacious nonpesticidal procedures) must be used to keep filth loads (26) below unacceptable levels.

The FDA's cautious, middle-of-the-road approach to filth levels in food might be thought of as the centerpost of a merry-go-round. Around it swirl many divergent and often conflicting ideas about the significance of filth in food. Although there is some merit in many, perhaps all, of these ideas, there are flaws in them as well. Because many of the pertinent facts remain unknown, it is probably best to remain deliberately circumspect (but not stubbornly intransigent) regarding suggestions for change in current policy and philosophy. At the centerpost of the merry-go-round of ideas is the assertion that DALs are "set on the basis of no hazard to health" (26). In other words, the small quantities of filth that the agency allows to remain in food will cause no harm to the consumer. There is no experimental evidence (how could it be acquired?) to support this assumption, but practical experience (in which all consumers have shared) seems to corroborate the correctness of this approach. If new evidence should show that some harm is caused by a filth load below the action level, that level would be promptly adjusted to protect the public. The agency is quite capable of reacting promptly to a threat to the public's health (99).

If a little filth (that is, a level at or below the DAL) is harmless, then what about a little more or even a lot of filth? There is little direct evidence that small amounts of certain adulterants of animal origin in food can cause disease. However, the circumstantial evidence (ie, the well known filthy habits of certain foods pests) should cause us to be cautious until the effects of such contaminants are better understood (100).

Three questions regarding filth in food are somewhat related. What is the significance of harmless filth that at worst is merely aesthetically unpleasant to the consumer? What is the significance of pesticide use in food production? Can the use of pesticides be justified as a way of lowering filth loads to levels below the action thresholds (DALs) when the previous higher levels would have been harmless and no more than aesthetically unpleasant for the consumer? These questions cannot simply be ignored. They are related to major global issues that impact the world's ecosystems and touch the lives of millions of hungry human beings.

Filth and Hunger

"It is estimated that by the turn of the century, 64 countries—29 of them in Africa—will not be able to feed their populations using present technology" (101). Many find it hard to believe that some people around the world have already reached the state where anything that can be eaten must be eaten. People living under such circumstances obviously accept much higher levels of pest damage and pest infestation in their foods than would be considered tolerable in the United States (102).

> Where safety is not a factor, at some point considerations of aesthetics must give way to considerations of human need. Our traditional perceptions of "filth" and "unfit for food" may be expected to change as the world's food supply shrinks in relation to the world's population. Aesthetic considerations increase costs and reduce available food supplies, but they are so ingrained in us they cannot be ignored. Indeed, we may ultimately face many other changes in our notions about "filth" and "fitness" in our food supply if we wish to have sufficient food in future generations. . . . [As] we insist upon "purer" and "cleaner" food, rejecting that which does not meet our standards, we thereby reduce the available food supply and increase its cost (103).

As noted above, it is usually illegal to knowingly mix one lot of food having a violative action level with another having a lower filth load to produce a blend that is below the DAL. "[Where] the level has been exceeded through no fault of manufacturer or processor (and assuming that no health hazard exists), it may become more difficult in the future to justify condemnation of food which violates a defect action level when such food might be blended with other food to produce a level of contaminant below the action figure" (103).

Defects Listed on the Label

If the filth in a given food were in fact harmless, even though aesthetically unpleasant, its significance would be simply economic. Even if such food could be legally placed in commerce, virtually no one would buy it, at least so long as other alternatives were available (10,57). A manufacturer could list the nature and quantity of filth on the label and produce various grades of foods based on the amount of filth present. The lower grades (higher filth loads) would presumably be cheaper to buy. But there is the possibility that some institutional providers might be tempted to buy the cheaper grades and serve them to unknowing consumers (57). The label on the package would state what kinds and amounts of pest damage and filth would be found in the container and what risks, if any, such damage or filth might represent for human health. It would be left to consumers to make their own personal compromise among the aesthetic, hygienic, and economic considerations. In this still-hypothetical situation, any monitoring function of a regulatory agency would consist of determining the accuracy of the label statement (102).

In spite of any merit of the labeling approach, the "American public is not yet prepared to face each new day with label statements of the maggots, mold, rodent pellets, rat hairs, and insects contained in their fruit juice, cereal, bread, jam, and coffee" (10). For those who have ample supplies of food from which to choose, the only significance of harmless filth in food is economic. They can take it or leave it. For the rest of the world, harmless filth is more readily accepted (for many, there is no alternative).

Some observers suggest that those who can choose the food they eat make a poor choice indeed when they select food low in filth that could have been produced only by intensive use of pesticides (104). Among U.S. consumers, fear of pesticide residuals is widespread (105,106). Pesticides in small amounts are perceived by the public to be far more bioreactive than any amount of filth allowed in foods by the DALs.

FOOD PRODUCTION: SHORT CYCLE

The production of food for human and animal consumption is a complex phenomenon in much of the developed world. Basically, it begins with the soil or the sea, that is to say, terrestrial and aquatic substrates are exploited in the production or collection of food. Generally, there are two kinds of production–processing cycles: short and long. In the short cycle, comparatively little is done to the food before it reaches the consumer. The consumer plucks a tomato from a vine in the backyard garden, washes it, and sets it on the table. A wilderness enthusiast pulls a trout from a stream, cleans it, and fries it for supper. Many frozen and fresh products (seafood, fruits, nuts, vegetables) come to consumers by this short cycle. The major complicating factors here are three: the biological quality of the substrate, the history of pesticide use on the commodity and its environs (ie, the chemical quality of the substrate), and the conditions of storage and transportation.

Biological Quality

The biological quality of the substrate depends on whether parasites or other pathogens or spoilage organisms are present in the soil or water or in the harvested product. For vegetables grown in direct contact with the soil (eg, carrots and potatoes), this is not a matter of great consequence because the food is normally washed before use. The microbiological quality of the fresh seafood is directly dependent on the microbiological quality of the substrate. Most consumers can see the larger parasites in seafoods, especially in fresh fish, but are quite unable to assess microbiological quality and must rely on advisories from public health authorities.

Pesticide Residues

Pesticide use is far more complicated. In aquatic environments, the presence of pesticides is more likely to be inadvertent than deliberate (an exception might be pest control in cranberries). The runoff from agricultural or forest lands to which pesticides have been applied carries pesticides into lakes and oceans. Pesticides or their chemical derivatives that are absorbed or ingested by aquatic organisms are passed along through the food chain to animals that may be used as human food. Again, because most consumers cannot determine whether or not pesticide residues are present and in what amounts (107), they must rely on advisories from public health authorities.

Generally, when pesticides have been applied to crops on agricultural lands, the shorter the processing cycle, the greater the likelihood that pesticide residues might be carried over into the finished product. Undesirable residues are not routinely carried over into finished products. Surveys indicate that the occurrence of pesticide residues in foods produced in the United States is not a significant problem (108–110); nevertheless, it is still a significant concern for many consumers (105,106). To help alleviate this concern, Congress recently passed the Pesticide Monitoring Improvements Act (21 U.S.C. 1401–1403).

Fresh or frozen foods such as green beans and spinach or fresh foods such as head lettuce, cabbage, celery, and asparagus undergo only sorting, trimming, and washing before being packaged (if packaging is required) for sale. In the fields, these products are exposed to both pests and pesticides. Some of these pesticides are applied to the surface of the plants where they remain and eventually degrade. Others are absorbed by plant tissues and physically translocated to all parts of the plants. Some of the surface-only pesticides can be washed off; the systemic pesticides cannot. In both cases, the Environmental Protection Agency (EPA) requires a preharvest interval, that is, a minimum number of days between the day of last pesticide application and the day of harvest. The preharvest interval allows time for pesticide degradation. The consumer, of course, is at the mercy of the producer's integrity in strictly adhering to the preharvest interval. Residue surveys of fresh produce show that the preharvest interval is generally respected (111).

Pests that affect short-cycle processing also affect long-cycle processing. These field pests, mostly insects and mites, attack crops where they grow. In small numbers, at least, they tend to stick to the plants on which they feed, thus sometimes surviving the wash step of processing. Surface-only pesticides applied to crops whose edible portions are buried in the soil (eg, carrots, potatoes, and peanuts) or covered by anatomical sheathing (such as sweet corn, lima beans, peanuts, almonds, peas, and beans) are discarded with the inedible portions and are, therefore, less likely to appear as residues in food.

What has been said about surface-only pesticides may also be said about surface-only pests. Pests situated on the surface of edible leaves or fruits (eg, spinach, strawberries, and leaf lettuce) are often washed away during the product's limited processing. Surface-only pests are even more likely to be removed when the edible portions of vegetables and fruits are surrounded by hulls, husks, and peels that are removed or otherwise discarded before processing or at the time of consumption (eg, sweet corn, watermelon, oranges, and lima beans).

Where there are pesticides, there are also people who, for whatever reason, are willing to misuse them (112,113). Reported incidents of misuse are small in number, perhaps because EPA in its administration of the Federal Insecticide, Fungicide, and Rodenticide Act (FIFRA) imposes tight constraints on pesticide use on food and in food facilities (112,114). Moreover, the National Pest Control Association, to which most reputable pest control operators (PCOs) belong, issues Good Practice statements for a great variety of pest prevention and control situations (115). Finally, most food plant managers and PCOs are committed to the philosophy of voluntary compliance (116), and this applies no less to FIFRA than to the FD&C Act.

Insects on the DAL List

The food defect action levels (26) are generally nonspecific about the kinds of arthropods allowed in certain foods; however, the DALs mention (but only by common name) the following species or groups: asparagus beetles and thrips on asparagus; thrips in canned multer; aphids, thrips, and mites on frozen broccoli; aphids and thrips in frozen Brussels sprouts; aphids (probably the hop aphid) in hops; drosophilid eggs in golden bleached raisins; aphids, thrips, and mites in canned and frozen spinach; caterpillars (probably corn earworm, fall armyworm, and southern beet webworm) in canned and frozen spinach; scale insects on salt-cured olives; and mites in mushrooms. There is also a DAL for damage caused by the olive fruit fly to imported green olives, imported black olives and salad olives. Aphids, mites, scale insects, and thrips are excluded from the DAL for apple butter. Thrips, aphids, and mites are excluded from the DAL for canned or frozen berries (thrips in canned multer excepted) (Table 1).

While the subject of arthropods and the DAL list is at hand, it will be convenient to finish the review of the list now by noting the remaining arthropods mentioned in the DALs, all of which in one way or another penetrate the product. In addition to the surface-only insects and mites, the following systemic arthropods eat their way into the plant tissues: maggots (cherry fruit fly, black cherry fruit fly, and western cherry fruit fly) in fresh, canned, or frozen cherries (maggots are excluded from DALs for brined or maraschino cherries); drosophilid eggs and drosophilid maggots in canned citrus juices and in canned tomatoes, tomato juice, tomato paste, and tomato purée; corn earworms and corn borers (probably European corn borer) in canned sweet corn; maggots in mushrooms; cowpea curculio larvae in canned peas, cowpeas, and black-eyed peas; leafminers (probably serpentine leafminer and spinach leafminer) in canned and frozen spinach; and parasitic copepods in redfish (ocean perch) (Table 1).

Storage and Transportation

Food storage and transport may be simple or complex but in any case their management has profound effects on the quality of food when it reaches the consumer. In short-cycle processing, the people involved in harvesting, processing (eg, cleaning, sorting, packaging), shipping, warehousing, and retailing must work within narrow limits of time and temperature. Time–temperature abuse can diminish product quality. On the other hand, strict adherence to good manufacturing practices for specific foods means that the product reaches the consumer in at least an acceptable (marketable) condition.

While it is true that the freezing step (if there is one) in a short-cycle process may kill certain parasites and other microorganisms in food, that is not the major objective of freezing. The major objective is to preserve food, but freezing also results simultaneously in the inactivation or at the least a marked slowing down of biological functions of microorganisms. This means that proliferation and metabolism of pathogenic and spoilage microorganisms are slowed to the level of insignificance. Short-cycle products are designed to have a short shelf life, whether fresh or frozen.

FOOD PRODUCTION: LONG CYCLE

Long-cycle processing is affected by the same three factors as short-cycle processing: biological quality of the substance, pesticide use, and conditions of transportation and storage. Although the human factor affects short-cycle processing, it is much more important in long-cycle processing. The food-handling aspect of short-cycle processing is relatively limited. In long-cycle processing, it is more significant and can become the scene of many potential misadventures if not carefully managed.

Biocidal Step

Long-cycle food processing usually includes a distinct biocidal step. This usually means commercial sterilization, but it could also consist of exposure to an extreme pH, pasteurization, oxygen deprivation, drying, freeze-drying, or complete sterilization by heat or irradiation. Some typical long-cycle products might include a box of

Table 1. Common and Scientific Names and Higher Classification of Selected Organisms[a]

Common Name	Scientific Name	Family	Order
Almond	*Prunus dulcis*	Rosaceae	Rosales
Almond moth	*Cadra cautella*	Pyralidae	Lepidoptera
Ants		Formicidae	Hymenoptera
Aphids		Aphididae	Homoptera
Apple, common	*Malus domestica*	Rosaceae	Rosales
Apple maggot	*Rhagoletis pomonella*	Tephritidae	Diptera
Asian roach	*Blattella asahinai*	Blattellidae	Dictyoptera
Asparagus	*Asparagus officinalis*	Liliaceae	Liliales
Asparagus beetle	*Crioceris asparagi*	Chrysomelidae	Coleoptera
Australian spider beetle	*Ptinus ocellus*	Ptinidae	Coleoptera
Black cherry fruit fly	*Rhagoletis fausta*	Tephritidae	Diptera
Black-eyed pea	*Vigna unguiculata*	Fabaceae	Leguminales
Broccoli	*Brassica oleracea botrytis*	Brassicaceae	Cruciales
Brussels sprouts	*Brassica oleracea gemmifera*	Brassicaceae	Cruciales
Cabbage	*Brassica oleracea capitata*	Brassicaceae	Cruciales
Carrot	*Daucus carota sativus*	Apiaceae	Umbellales
Celery	*Apium graveolens dulce*	Apiaceae	Umbellales
Cheese mite	*Tyrolichus casei*	Acaridae	Acari
Cherry fruit fly (cherry maggot)	*Rhagoletis cingulata*	Tephritidae	Diptera
Cherry, sweet	*Prunus avium*	Rosaceae	Rosales
Cloudberry	*Rubus chamaemorus*	Rosaceae	Rosales
Clove	*Syzygium aromaticum*	Myrtaceae	Myrtales
Cocoa bean	*Theobroma cacao cacao*	Sterculiaceae	Malvales
Codling moth	*Cydia pomonella*	Tortricidae	Lepidoptera
Confused flour beetle	*Tribolium confusum*	Tenebrionidae	Coleoptera
Copepod, parasitic	*Sphyrion lumpi*	Sphyriidae	Siphonostomatoida
Corn earworm	*Helicoverpa zea*	Noctuidae	Lepidoptera
Cowpea curculio	*Chalcodermus aeneus*	Curculionidae	Coleoptera
Cowpea	*Vigna unguiculata*	Fabaceae	Leguminales
Cranberry	*Vaccinium macrocarpon*	Ericaceae	Ericales
Cumin	*Cuminum cyminum*	Apiaceae	Apiales
Date	*Phoenix dactylifera*	Arecaceae	Arecales
Differential grasshopper	*Melanoplus differentialis*	Acrididae	Orthoptera
Driedfruit beetle	*Carpophilus hemipterus*	Nitidulidae	Coleoptera
Drosophilid fruit fly	*Drosophila melanogaster*	Drosophilidae	Diptera
European corn borer	*Ostrinia nubilalis*	Pyralidae	Lepidoptera
Fall armyworm	*Spodoptera frugiperda*	Noctuidae	Lepidoptera
German roach	*Blattella germanica*	Blattellidae	Dictyoptera
Grackle, common	*Quiscalus quiscula*	Icteridae	Passeriformes
Green bean	*Phaseolus vulgaris*	Fabaceae	Leguminales
Greenbug	*Schizaphis graminum*	Aphididae	Homoptera
Hispid cotton rat	*Sigmodon hispidus*	Cricetidae	Rodentia
Hop	*Humulus lupulus*	Cannabaceae	Urticales
Hop aphid	*Phorodon humuli*	Aphididae	Homoptera
House fly	*Musca domestica*	Muscidae	Diptera
House mouse	*Mus musculus*	Muridae	Rodentia
Indianmeal moth	*Plodia interpunctella*	Pyralidae	Lepidoptera
Larder beetle	*Dermestes lardarius*	Dermestidae	Coleoptera
Lettuce	*Lactuca sativa*	Asteraceae	Asterales
Lima bean	*Phaseolus lunatus*	Fabaceae	Leguminales
Little brown bat	*Myotis lucifugus*	Vespertilionidae	Chiroptera
Market fly	*Musca sorbens*	Muscidae	Diptera
Mediterranean flour moth	*Anagasta kuehniella*	Pyralidae	Lepidoptera
Mediterranean fruit fly	*Ceratitis capitata*	Tephritidae	Diptera
Mexican bean beetle	*Epilachna varivestis*	Coccinellidae	Coleoptera
Mites		(Many families)	Acari
Multer (Danish for cloudberry)			
Norway rat	*Rattus norvegicus*	Muridae	Rodentia
Ocean perch	*Sebastes marinus*	Scorpaenidae	Perciformes
Olives	*Olea europaea*	Oleaceae	Loganiales
Onion maggot	*Delia antiqua*	Anthomyiidae	Diptera
Onion thrips	*Thrips tabaci*	Thripidae	Thysanoptera
Orange, sweet	*Citrus sinensis*	Rutaceae	Rutales
Oyster, eastern	*Crassostrea virginica*	Ostreidae	Ostreoida

Table 1. (*continued*)

Common Name	Scientific Name	Family	Order
Pea, garden	*Pisum sativum sativum*	Fabaceae	Leguminales
Peanut	*Arachis hypogea*	Fabaceae	Leguminales
Pepper maggot	*Zonosemata electa*	Tephritidae	Diptera
Pharaoh ant	*Monomorium pharaonis*	Formicidae	Hymenoptera
Pigeon	*Columba livia*	Columbidae	Columbiformes
Plum curculio	*Conotrachelus nenuphar*	Curculionidae	Coleoptera
Potato	*Solanum tuberosum*	Solanaceae	Solanales
Prune plum	*Prunus domestica domestica*	Rosaceae	Rosales
Raisins	*Vitis vinifera*	Vitaceae	Rhamnales
Redfish	*Sebastes marinus*	Scorpaenidae	Perciformes
Red flour beetle	*Tribolium castaneum*	Tenebrionidae	Coleoptera
Rice weevil	*Sitophilus oryzae*	Curculionidae	Coleoptera
San Jose scale	*Quadrispidiotus perniciosus*	Diaspididae	Homoptera
Sawtoothed grain beetle	*Oryzaephilus surinamensis*	Cucujidae	Coleoptera
Scale insects		Coccoidea (superfamily)	Homoptera
Serpentine leafminer	*Liriomyza brassicae*	Agromyzidae	Diptera
Shrimp, northern white	*Penaeus setiferus*	Penaeidae	Decapoda
Silverfish, common	*Lepisma saccharina*	Lepismatidae	Thysanura
Southern beet webworm	*Herpetogramma bipunctalis*	Pyralidae	Lepidoptera
Spinach	*Spinacia oleracea*	Chenopodiaceae	Chenopodiales
Spinach leafminer	*Pegomya hyoscyami*	Anthomyiidae	Diptera
Spring cankerworm	*Palearita vernata*	Geometridae	Lepidoptera
Spotted asparagus beetle	*Criocerus duodecimpunctata*	Chrysomelidae	Coleoptera
Strawberry, garden	*Fragaria × ananassa*	Rosaceae	Rosales
Sweet corn	*Zea mays mays*	Poaceae	Graminales
Tarnished plant bug	*Lygus lineolaris*	Miridae	Hemiptera
Thrips		(Several families)	Thysanoptera
Tomato	*Lycopersicon lycopersicum*	Solanaceae	Solanales
Trout, rainbow	*Salmo gairdneri*	Salmonidae	Salmoniformes
Vegetable leafminer	*Liriomyza sativa*	Agromyzidae	Diptera
Warehouse beetle	*Trogoderma variabile*	Dermestidae	Coleoptera
Watermelon	*Citrullus lanatus lanatus*	Cucurbitaceae	Cucurbitales
Western cherry fruit fly	*Rhagoletis indifferens*	Tephritidae	Diptera
Wheat, bread	*Triticum aestivum*	Poaceae	Graminales
Wheat curl mite	*Eriophyes tulipae*	Eriophyidae	Acari
Wheat stem sawfly	*Cephus cinctus*	Cephidae	Hymenoptera

[a] Refs. 117–118.

breakfast cereal, a bag of white flour, a can of tuna, a can of condensed soup, and a microwavable frozen dinner. Long-cycle products, which often have a long shelf life, are often the result of complex manufacturing procedures, all of which must be carefully controlled, especially those aspects dependent on human intervention.

Biological Quality

Foods harvested from a compromised biological environment are undersirable whether destined for short-cycle or long-cycle processing. But if long-cycle processing has been carried out with honest reference to GMPs, the foods are not dangerous, at least insofar as pathogenic or spoilage microorganisms are concerned. Commercial sterilization (heating under pressure) kills most pathogens and most spoilage microorganisms (other biocidal techniques have to be tried and tested on a case-by-case basis to determine their effectiveness). Thus whether because of commercial sterilization or some other effective biocidal procedure, the resulting food remains safe to eat for the duration of its shelf life, providing package integrity is perfectly maintained (a potential exception to this will be noted later). If it is a frozen product, the appropriate storage temperatures must also be maintained.

Frozen and refrigerated products pose an additional element of risk simply because their quality and wholesomeness depend directly on the strict maintenance of the product within specific temperature ranges (119–121). The quality of long-cycle, nonfrozen food is largely (but not entirely) independent of temperature (eg, freezing, obviously, does nothing for the quality of canned or bottled foods, and thermophilic spoilage bacteria may proliferate when commercially sterilized canned foods are stored at high temperatures).

Pesticide Residues

What has already been said about preharvest (field applied) or runoff pesticides in connection with short-cycle processing also applies to long-cycle processing. Most processing steps (eg, washing, polishing, dilution, heating, and aging) usually result in continuing pesticide degradation and reduction of pesticide residues. The relatively longer time involved in long-cycle processing is also conducive to progressive pesticide degradation. Postharvest application of pesticides (122) takes place in two forms: (1) direct application to raw food materials to protect them from pest attack or at least to suppress pest populations and the damage associated with them (123–125). This would include such techniques as spraying malathion directly on wheat as it is being augured into a bin or fumigating lettuce before shipping it to Japan. (2) Application of pesticides to food-storage, food-processing, and food-handling facilities (126). This would include such things as spraying empty bins of a granary, placing rodent bait boxes in a warehouse, fumigating a flour mill, or spraying a warehouse.

Fumigation

Fumigation is a useful pest control technique in the food industry (127–129). Bulk grain transported in ships may be fumigated during voyages from one port to another (130). Railcars, too, may be sealed and fumigated in transit (129). Fumigation is especially useful for stopping infestations in small quantities of foods or feeds (eg, sawtoothed grain beetles in a pallet load of bagged dog food in a warehouse), but, of course, an entire warehouse may also be effectively fumigated. Because there is no residual pesticidal effect from fumigation, products become susceptible to reinfestation as soon as the fumigant gas clears (131). Any insects, mites, or rodents that happened to be killed within the bulk of the food or feed remain in place after the fumigation has been completed and may be considered filth under section 403(a)(3) of the FD&C Act.

Storage and Transportation

Foods placed in storage or transport are at risk of attack by food pests. Storage sites vary widely in location, design, and construction (132,133). Pest prevention and control techniques obviously must be tailored to accommodate all the demands of climate, geography, and architecture. Everything that applies to storage facilities also applies to modes of transport. Railcars, airplanes, semitrailers, and freighters are simply mobile storage facilities, all capable of moving pests from one point to another (134), a phenomenon that is monitored by the Animal and Plant Health Inspection Service (APHIS) of the United States Department of Agriculture (USDA) and by comparable agencies of many countries around the world (84,135,136).

The ideal situation in regard to food pests and storage facilities is that these places should be pest-free. This ideal should be the objective of every storage–transport manager and, ultimately, of every food handler. Any infestation at any point between the harvester and the consumer is undesirable (and often preventable) and constitutes a potential violation of the FD&C Act or its state-level counterparts.

FIELD PESTS AND STORAGE PESTS

Each link in the chain of food production (field stage, on-farm storage, commercial storage, manufacture, distribution, warehousing, transportation, retailing, and consumer's pantry) has a more or less characteristic set of pest species. As mentioned earlier, pests can be placed two diverse groups, field pests and storage pests, and several members of each group have been named.

Field Pests

Included in the field pest category are vertebrates and arthropods that feed on trees and crops in the field: rodents, eg, hispid cotton rat (137); birds, eg, grackle (138); mite, eg, wheat curl mite; beetles, eg, Mexican bean beetle; moths, eg, spring cankerworm; flies eg, Mediterranean fruit fly; sawflies, eg, wheat stem sawfly; aphids, eg, greenbug; scales, eg, San Jose scale; bugs, eg, tarnished plant bug; thrips, eg, onion thrips; and grasshoppers, eg, differential grasshopper (Table 1). The field pests comprise so vast an assemblage of species that it is difficult to name just a few. This diverse group of species is responsible for major damage to field crops and orchards. The suppression of these pest populations is the focus of a large part of the pesticide and pest control industries.

Apple Butter

As we have seen from our review of arthropods named in the DALs, field pests tend to be either excluded from the count (apple butter and druplet berries) or included in the count but at rather high levels (eg, hops). Two situations are implied here and both bear some relation to the relatively high counts or to the exclusion of some arthropods from the counts. In an ideal world, all apple butter would be manufactured from apples of the same quality as the beautiful specimens displayed at the produce counter of the local supermarket.

In the real world, it seems fairly certain that at least some of the apples used to make apple butter would never make it to the supermarket shelf. These apples might be afflicted with an array of defects that could include rotten spots (from mold), maggots, and scale insects. While the apples were stored in a bin at the orchard, it is possible that a multitude of thrips, aphids, and mites either were attracted to the apples or chanced to fall or crawl on them and became entrapped in the syrupy exudates. Doubtless a similar fate could have befallen numerous flies, beetles, and wasps. It is, therefore, no surprise that on analysis the resulting product sometimes yields high counts of mold and arthropod fragments, even without including mites, aphids, thrips, and scale insects. Furthermore, the comminution of ingredients during the manufacture of apple butter breaks up whole insects and mites and the larger fragments thereof, making recognition (and, therefore, exclusion) of aphids, mites, thrips, and scales more difficult, as well as producing a higher overall fragment count. Of course, all of these hypothetical defects (and probably others, too) were taken into consideration when the DAL for apple butter (and all other comminuted products for which DALs have been set) was developed.

Hops

There are many insect pests of hops but the principal one is the hop aphid (139). Hop aphids attack all succulent parts of the vine, including the catkins (cones and hops) that are harvested for use in brewing beer. One of the criteria used to valuate hops is the number of aphids present per unit measure (140). The DAL (more than 2,500 aphids/10 g) (26) reflects the severity of aphid infestations during one especially bad season. Pesticides are often used to suppress aphid populations, but the grower must respect the preharvest interval to avoid residue problems. In traditional brewing practice, the hops are boiled with the wort and then separated from the resultant liquid. Thus neither the hops nor the aphids carried along with them ever reach the final product. Perhaps that is another (but unstated) reason for the generous DAL. The hops contribute flavor to the beer in the form of essential oils and other complex chemicals (139). Of course, the aphids also contribute some of their own chemicals to the wort and perhaps also eventually to the finished product.

Filth Flies

The term filth flies is familiar to most people. The house fly immediately comes to mind, along with its gustatory counterparts (eg, market fly) in other parts of the world. There are so many kinds of flies that frequent filth and they are so abundant that the term fly is almost synonymous with filth. Other kinds of insects have similar habits to some degree, eg, certain cockroaches, ants, and beetles that visit excrement, wounds, garbage, sewers, drains, carrion, and rotting plant materials. In regard to the subject of this article, there is no quarrel with any of these filth insects (or with certain rodents and birds in this context), save for those that also visit foods or places where foods are stored or handled or where food packaging materials are manufactured and stored.

CATEGORIES OF FOOD PESTS

After a food crop is harvested and entered into some sort of processing, the food material encounters two general classes of pests: those associated mainly with the food itself (eg, rice weevil, cheese mite, almond moth, larder beetle, and Indianmeal moth) and those associated with and often to some degree dependent on buildings, that is, an artificial environment created by people around the food (eg, German cockroach, common silverfish, house fly, Australian spider beetle, house mouse, Norway rat, pigeon, and little brown bat) (Table 1).

In the case of roosting birds and bats and in those instances where dogs and cats are present (without FDA approval!) around food, the contact between pest and food is largely accidental, as when hairs, feathers, and excrement chance to fall on unprotected food or food-contact surfaces. Other pests (eg, Asian roach, pharaoh ant, and Norway rat) are mainly opportunistic in regard to food. Their populations fluctuate with the availability of food, but they also require harborage and water. The third (and most important) category of pest–food relationships is essentially obligatory. The food material provides the pests (eg, confused flour beetle, warehouse beetle, granary weevil, and Mediterranean flour moth) with all the necessities of life (141).

Sanitary Importance of Storage Pests

Pest organisms can also be categorized as those that have sanitary significance and those that do not, although the two groups may overlap in some cases. House flies and German roaches have sanitary significance, but the hop aphid does not; its repertoire of habits and habitats does not include any sources of pathogenic microorganisms. But what about the insects and mites in the obligatory group (storage pests or pantry pests) mentioned above? These arthropods make their homes in various raw or processed foods and are loathe to leave them. They have sanitary significance because the filth-frequenting accidental and opportunistic pests can bring pathogenic and spoilage microorganisms to food infested with the obligatory pests, which in turn inadvertently pick up microorganisms and disseminate them whenever they chance to travel within the food material (142–144).

Some foods such as spices may be exposed to both field and storage pests. When such materials are analyzed early on in their journey from the producing regions of the world to collection points for export to the United States, their filth load consists mainly of field pest fragments. During intervals of storage, some of the typical storage pests may move in and begin to proliferate. These storage populations may continue to expand during transport by ship. Thus analysis on arrival yields evidence of both field and storage pests. All of the fragments are counted and measured against the applicable DAL (the DAL makes no distinction between fragments of field pests and fragments of storage pests). This level has probably already been set rather high to accommodate the contributions of fragments from infestations by field pests (no sanitary significance). But the DAL appears to have been set only modestly high because of the probable exposure of the product to storage pests (ie, pests of sanitary significance).

As noted above, there is a balance between pesticide use (residue tolerance levels) (97,98) and filth loads (DALs) (26). Critics of this balance suggest that pesticide residues at any level are far more dangerous to human health than any number of rat hairs and moth mandibles. This line of thought leads to the suggestion that pesticide residue tolerances should be tightened and DALs should be relaxed, especially in regard to field pests (ie, no sanitary significance) (104). This might result in lower pesticide residue levels, higher rates of blemished (pest-damaged) fruits and vegetables, and perhaps higher filth loads in foods reaching the consumer.

Others have gone further, proposing that agricultural producers should simply abandon pesticides or at least restrict their use to very small quantities (low input) (145,146). Under the total ban on pesticide use, losses to pests would rise from 37% to 46% of the total harvest according to one estimate (147). The last figure seems to be proffered rather facilely, as if it didn't amount to much, but should the notion of giving up nearly half the world's harvest to pests even be entertained? There must be a better way.

INTEGRATED PEST MANAGEMENT

Integrated pest management (IPM) is the better way. IPM may be applied to crops in the field; to stored, raw food grains; to transportation systems; and to food facilities of all kinds, from manufacturing through warehousing to retailing (145,148). IPM merges with, and becomes indistinguishable from, those aspects of the hazard analysis and critical control points (HACCP) system that deal with pest prevention and pest control (100,149). Pesticides may play a greater or lesser role in IPM, depending on local circumstances. The goal of IPM is not to eliminate pesticides (although there may be a distinct trend in that direction), but to suppress or exclude pest populations. This goal requires a comprehensive approach that involves an array of techniques, often including application of pesticides (128,150,151) and often emphasizing pest-exclusion practices. It is entirely probable that the nonpesticidal aspects may become so effective that pesticide applications can be eliminated or used only occasionally. In this scenario, the pests are not allowed to run rampant as the level of pesticide use follows a carefully managed descent.

Use of Beneficial Insects and Mites

One of the controversial aspects of IPM is the proposed use of parasitic and predatory insects and mites to suppress populations of certain insects and mites of the obligatory kind in stored, raw food commodities, mainly cereal grains and in-shell peanuts (107,152–156). Because pesticides can kill the parasites and predators (often called beneficials), pesticides cannot be used in this system. Thus the problem of residues from pesticides applied to stored commodities is avoided (it would be possible, of course, to breed pesticide-resistant parasites and predators).

The presence of myriad parasites and predators naturally inflates the fragment count of samples taken from the raw commodity (the DALs do not distinguish between fragments of beneficials and fragments of storage insects). Sifting, washing, and other cleaning steps of processing remove many of the added beneficials (they are all external with respect to the kernels), as well as the pest insects and mites that are external feeders (some are internal feeders and, therefore, much harder to remove from whole grains). The proponents of this biological control technique argue that the higher fragment counts are far preferable to pesticide residues.

Discussions about how to resolve the beneficials question continue among regulatory officials and scientists at the EPA, USDA, and FDA. Although it seems likely that beneficials will be granted status as pesticides and then exempted from most tolerances or action levels as they apply to bulk-stored cereal grains and legume seeds, the matter remains unresolved as of this writing.

Phytoalexins

Another technique of IPM involves the development of food plants resistant to attack by pest insects and mites (157). In one scenario, the plants achieve resistance by manufacturing chemicals that kill or repel pests, something akin to the action of systemic pesticides (158). It is important to find out if these chemicals are also poisonous to humans. Little attention has been given to this question (159,160), but it is likely that one chemical, a phytoalexin called phaseollin (phaseolin), produced by green beans, remains stable through commercial sterilization (as in canning) and commercial freezing. Nothing is known about the toxicity (if any) of phaseollin to humans.

Pheromones

Great progress has been made in developing effective pest-monitoring systems based on insect and mite pheromones (161–163). The wide-scale use of these pheromone-based monitoring systems in the food industry was greatly hampered by industry fears that the collected insects and mites would be considered to be (a)(4) evidence by FDA. Now this impasse seems to be largely of only historical interest (164). The food industry is free to use this surveillance tool without threat of regulatory interference, and the public will benefit from this because these pheromone traps are more effective than other surveillance techniques. Thus infestations can be discovered earlier, and if pesticides are needed they can be used in smaller quantities because the infestation is in an early phase. The transition of pheromone traps as surveillance tools to the use of pheromones as control agents is under way (165,166). This technique shows great promise as an effective control measure that requires no pesticides.

Potential Problems with Pesticides

There are many serious questions to be resolved concerning the use of pesticides. Opinions range from one extreme to the other (167,168). If pesticides must be abandoned at some point in the future, then present objections to the use of parasitic and predatory insects and mites and to various other nonpesticidal management techniques as well as concerns about minuscule residues of pesticides in food may turn out to be inconsequential in retrospect.

In addition to the persuasive arguments of a strong and growing antipesticide movement (169), progressive contamination of groundwater by pesticides, growing resistance of pests to pesticides, detrimental effects of pesticides on nontarget organisms, and the enormous costs of bring new pesticides to market may collectively tip the balance against the use of pesticides in food production (145,170–172). When and if that happens, it may be necessary to rethink many of our attitudes and regulations concerning filth in food.

Insects as Food

Perhaps the big picture can be made a little clearer if insects per se are thought of as food (102,173). All insects, including cockroaches, Indianmeal moth larvae, and house fly maggots, contain nutrients that could be useful in human nutrition. FDA has no objection to the importation or interstate distribution of insects to be marketed for human consumption (174). The FDA does have a compelling interest in the sanitary production of insects as food, but the agency does nothing to prohibit or in any way discourage the sale of insects as food. The distinction between insects of sanitary significance and those without sanitary significance fades away in this context. Any kind of insect (eg, chocolate-covered ants and chocolate-covered cockroaches) may be offered for sale so long as it is properly labeled and the appropriate GMPs were followed. The insects are considered to be food, not filth, as the act defines those terms (6).

Entomophagy has a long history that does not need to be repeated here (173), but the idea that people actually eat insects intentionally might temper objections to the deliberate mixture of insects, perhaps in a form comparable to fish meal, with other food in the same commercial package. From there, it is only a short step to accepting the idea of an unintentional mixture of field pests (only insects and mites and none with any sanitary significance) and food. This is precisely what is allowed by the DALs for certain foods (26).

AESTHETICS AND HEALTH

This section reconsiders the labeling approach to filth in food. It has been seen that insects per se may be marketed as human food and that certain insects and mites, while not welcomed in food, are, to a small extent, allowed to be there, especially if they fall into the field pest category and lack sanitary significance. Pests (or their fragments) with sanitary significance are decidedly unwelcome both in and around food.

As background, the words *aesthetic* and *hygienic* must be reconsidered. It was the intent of Congress in 1938 that all food covered under the act should be both aesthetically and hygienically acceptable. Subsequent amendments have not changed that concept. It is much easier to define what is hygienic than what is aesthetic in this context. Most people would agree that feces should be barred from food on at least aesthetic grounds. Yet whole fish (including intestinal contents) are used to make fish meal, a wholesome and aesthetically pleasing food by most standards. Sometimes we eat the vein (hindgut) in the tail of shrimp. In both of these examples, the gut contents (feces) have been sterilized by heat. This is not the case with raw oysters, an undisputed gustatory delight in the minds of many.

Feather Barbules and Rat Hairs

All that aside, bird droppings and rat pellets are universally revulsive, no matter if sterile or disguised by food. But what about feather barbules and rat hairs? It can easily be imagined that a bird could fly over food or food-contact surfaces (eg, conveyors and packaging materials) and lose a feather. The feather happens to fall on a conveyor, gets taken up in a manufacturing process, and breaks up into fragments (mainly barbules). An occasional can of finished product yields a feather barbule or two on analysis. Does this have sanitary significance? No, because the feather in its disintegrated state went through a sterilization step. Does it have aesthetic significance? Perhaps if a whole feather were recovered from a can of food, it might. But many barbules are so small that they would go unnoticed by most consumers.

Rat and mouse hairs are also found on occasion in food.

How do they get there? Again, it is hoped that it wouldn't happen, but it is possible that a rat could run along a rafter in a wheat granary or a midnight mouse could check out a conveyor in a bakery. The rodent sheds an occasional hair as it moves about. A hair turns up in a loaf of bread. Does this have sanitary significance? Maybe. Aesthetic significance? No (only an analyst can tell). Is this a probable scenario? Not likely.

Adulteration of Packaging

Concern over the recovery of feather barbules and of rat and mouse hairs from food is not based on the random shedding of a feather or hair. In regard to birds, many people know from personal experience that birds intermittently expel feces as they fly about and as they roost. So where the feature has fallen, it is very likely that feces have also fallen. The feces may follow the same path as the feather and disappear into the corpus of the food being processed. Bird droppings often have an aqueous phase that may be absorbed by permeable packaging. If pathogens are present (a likely event (175,176)), they may be carried into the enclosed food materials. Under this and similar circumstances, there is no doubt about sanitary significance.

If droppings can be seen on the package, aesthetic significance becomes obvious. But packaging may be cleaned up in some cases to the point that all visual evidence of bird droppings has been removed. Food in permeable packaging might be repackaged, but that would not eliminate food contaminated by the liquid phase of bird feces. Food packaged in impermeable packaging (eg, glass bottles and metal cans) is usually exempt from seizure because of superficial adulteration by bird and rodent droppings. However, when the adulteration is blatantly excessive, even bottled and canned foods may be seized.

Any food packaging material that crosses state lines falls within the purview of the FD&C Act. Packaging is a subject unto itself (177) but at least the resistance (if any) of packaging to insect attack is of interest in this article. Much of the important research on this subject has been reviewed (178), and a method for determining direction of penetration has been standardized (179).

Feces in Food

Rats and mice are much more predictable than birds. We can be certain that where rodents are, so also will be urine and feces. These commensal rodents leave a trail of urine as they travel about. Exposed food, food-contact surfaces, and permeable packaging are all at risk from urine contamination. In some instances, pathogens are shed in rodent urine (180,181). Commensal rats and mice characteristically groom themselves by licking their fur. Many of the dislodged hairs thus encountered are swallowed. These hairs eventually become emeshed in the rodent's feces. These pellets of excrement are dropped frequently and intermittently as the rodent moves about and feeds. Whole pellets or at least recognizable fragments are likely to be found only in raw or minimally processed foods. Only popcorn and wheat have DALs for rodent excreta pellets (26).

These products may be sifted, sorted, washed, or otherwise cleaned to separate the product from unwanted foreign matter, including fecal pellets of rodents. Most of us do not eat raw wheat kernels, and popcorn is exploded by high heat before it is eaten. Thus the low levels of rodent fecal pellets allowed in these two commodities, although clearly undesirable, are not a matter of great significance from either the hygienic or the aesthetic point of view.

Capsicum pods, cocoa beans, condimental seeds and several spices have DALs for mammalian excreta (26). This reflects cultural practices in the producing regions (largely outside of the United States). The product (whatever it happens to be) is collected and laid out unprotected on the ground in rural villages. Here the food material is exposed to fecal adulteration by domestic animals and by an assemblage of peridomestic species of wildlife. Evidence of such contamination is occasionally found in the finished products.

Rodent hairs are noted in the DALs for apple butter, capsicum, paprika, chocolate, chocolate liquor, cocoa powder press cake, cornmeal, curry powder, macaroni, noodle products, peanut butter, popcorn, and wheat flour (26). What do these rodent hairs represent? They are largely the recognizable remains of fecal pellets that disintegrated during food processing.

Note that there is both a rodent pellet and a rodent hair DAL for popcorn (26). Because this product is minimally processed, the pellets may reach the packaged finished product intact or only partially disintegrated. On the other hand, there is no pellet DAL for wheat flour. The pellets often present in raw wheat will either be cleaned out of the product or disintegrated to such a degree that only the hairs remain recognizable. Consumers can't see the hairs under normal circumstances, so the aesthetic factor is minimal. What, then, is the sanitary significance of rodent hairs in food? It should be noted here that a bag of wheat flour containing rodent pellets would be treated as a violation of section 402(a)(3) of the act (6).

Contamination before and after the Biocidal Step

The question of sanitary significance must be answered within the context of food processing. Included in both the question and the context are all of the animals associated with filth in or around food. It was explained earlier that a major moment in food processing is the biocidal step. Feces of all kinds (from cats, cows, rats, bats, birds, mice, roaches, and donkeys) come out of the biocidal step sterile. Thus the compelling question is, Did the contamination occur before or after the crucial biocidal step?

If the contamination occurred before the biocidal step, then any misgivings about the resultant food are aesthetic only (with rare exceptions, pathogenic and spoilage microorganisms are killed during the biocidal step). When contamination occurs after the biocidal step, a more risky scenario emerges (182). The main hazard comes from animals that carry pathogenic and spoilage microorganisms to food and food-contact surfaces (158,183–185). Roaches, ants, flies, rats, mice, and (to a lesser extent) birds are especially suspect here. All are capable of carrying microbial pathogens (Table 2) and spoilage organisms, either on or in their bodies, and all have habits conducive to the acquisition and spread of harmful microbes.

Potential Disease Agents in Food

Another set of potential health threats appears to be independent of the biocidal step in food processing. Phaseollin, a phytoalexin and potential toxicant, seems unaffected by the biocidal step. Some animals that contaminate food produce toxins. Quinones, the best known group, elaborated by confused flour beetles and red flour beetles, appear to be toxic under special circumstances in laboratory

Table 2. Selected Examples of Isolations of Microbial Pathogens from Natural Populations of Food Pests

Pathogen	Host or Vector	Reference
Bacillus cereus	Ants	186,187
Campylobacter jejuni	Flies	188
Clostridium perfringens	Ants	186
	Roaches	189
Entamoeba histolytica	Flies	190
	Roaches	191
	Rats	192
Escherichia coli	Ants	186,187,193
	Flies	194
	Roaches	195
Hantavirus	Rats	196
Salmonella	Ants	186,197
	Flies	198–200
	Roaches	197,201–204
	Rats	197,205–208
Shigella	Ants	187
	Flies	199,209
Staphylococcus aureus	Ants	186,187
	Flies	210
	Roaches	195,204
Streptococcus pyogenes	Ants	186

systems (211,212), but nothing is known of their toxicity to people.

Clinical tests have demonstrated that some people have antibodies to several kinds of insects (all pantry pests) commonly found in food (213). The antibody response is presumed to have been triggered by the inadvertent ingestion of these insects with food contaminated after the biocidal step. The effect commercial sterilization would have on these allergens from pantry pests is unknown, but there is a cockroach allergen that remains potent after heating for 1 h at 100°C (214). Dermestid beetle larvae of the genus *Trogoderma* are densely clothed with specialized hairs called hastisetae. In one case a four-month-old infant was diagnosed as having ulcerative colitis after the child had consumed baby cereal infested with larval *Trogoderma*. The damage to the patient's intestinal mucosa was tentatively attributed to the hastisetae (215).

The biocidal step (eg, heating, acidification, and irradiation) might be ineffective in eliminating hazards from phytoalexins, quinones, allergens, and hastisetae. In the case of postprocessing contamination, phytoalexins would never be a factor, but there is some element of danger to the consumer from toxins, allergens, and hastisetae. Without expending exorbitant amounts of effort and money, a food-facility manager, especially at the retail level, can do nothing to protect consumers from toxins, allergens, and hastisetae except to inspect all incoming items diligently and to purchase only from reliable suppliers.

Pest-Free Premises

The situation is entirely different in regard to infesting pests such as rats, mice, roaches, ants, beetles, moths, mites, and birds. A sanitary food facility and an infested food facility are mutually exclusive concepts. An active infestation compromises an otherwise adequate sanitation program. No matter how closely standard cleaning and sanitizing protocols are followed, pests in a food facility will inevitably deposit spoilage microorganisms and, occasionally, pathogens on food and food-contact surfaces. Therefore, the objective of a food manager is clear: Maintain pest-free premises (216).

Index of Laxity in Sanitation Management

The question about the sanitary significance of bird and rodent feces in commercially sterilized food was answered negatively and perhaps cavalierly. Yet there may be a sanitary implication. A rule of thumb in sanitation audits is that food pests (roaches, flies, ants, rats, and birds) are indicators of insanitation. Whenever evidence of pests is found in a food facility, a question arises immediately in the mind of the inspector: "If these people can't get rid of the rats and the bugs, then how are they coping with the microbiological hazards that are far more dangerous and far more difficult to detect and attenuate?" In other words, if managers are insensitive to the obvious bugs, maybe they are also inattentive to those procedures that eliminate microbial contaminants (217) already in food and that prevent the addition of others from the people who handle food.

If everything goes well in a food establishment, the outcome is a safe product. Yet we know that errors happen. Preprocessing errors are usually attenuated by the biocidal step. Postprocessing errors increase the probability of a hazardous product. The presence of pests in food or in a food facility further increases that level of probability. The pests are amplifying and disseminating agents; they move pathogens and spoilage microorganisms from places where they may be fairly innocuous to places where their chances of becoming dangerous are increased. If the new conditions are conducive to microbial proliferation, then a hazardous situation may develop.

Labeling Reconsidered

The labeling approach and the cosmetic use of pesticides are concepts that need further discussion. It has already been noted that the U.S. consumer remains unready for an ingredient label that lists maggots, mold, rodent pellets, rat hairs, and insects (10). An attorney might envision this as a realistic scenario, but for an entomologist the scene is painted with too broad a brush.

This hypothetical label obviously refers to preprocessing contaminants. We have already seen that rodent hairs and rodent pellets often represent one and the same thing. No one, either in this country or anywhere else in the world, should be asked or expected to eat diluted and disguised rodent feces. There is no such thing as a rodent population that cannot be dramatically reduced if management is willing to invest the money and effort to make it happen. It is, therefore, unconscionable and inexcusable for a manger to allow rodents (and birds) free access to a food facility.

Rodent Feces

Too often it is assumed that consumers are unwilling to sacrifice in one area to achieve what they deem to be a higher goal in another. Thus if a few consumers want to buy popcorn that has never been sprayed with insecticides or stored in the presence of rodenticides, and they are also willing to pick out the fecal pellets that are very likely to be there, according to the statement on the hypothetical label (10), then it is hard to find a reason to deny them this opportunity (the economic constraints on such a scheme would exclude such a product from the market, in any case).

Mold

Mold has about as many champions as rat feces, but all things considered, it is not quite so repulsive. About half of the food categories listed in the DALs have action levels for mold, dry rot, or mildew (26). This speaks to the ubiquitous nature of molds and to the frequency and severity of the damage they do to a wide range of foods (218), from field crops all the way to the kitchen refrigerator and pantry. In field situations and in bulk storage, some molds may produce dangerous mycotoxins in peanuts, cereal grains, and other foods (159,219,220). Storage systems that can compensate for the vagaries of weather are needed to prevent the production of mycotoxins. Otherwise, mold prevention focuses on three areas: use of only high-quality (minimal rot) fruits (eg, tomatoes and peaches) (defects determined by the Howard mold count (21)), production machinery sanitation (221), and proper storage of food commodities (222). Thus, although some aspects of mycotoxin production seem to be beyond the scope of human intervention, selection of high-quality ingredients, machinery sanitation, and proper storage are all subject to human control.

Fungicides can be used effectively for mold control (223), especially in field crops, but there are associated risks as well as benefits. Improved yields are desirable, of course, but misuse is a constant potential threat. It may be concluded that molds, although very widespread, can be kept at rather low levels in food if strict attention is paid to GMPs. Again, if consumers preferred fruits grown without fungicides and were willing to purchase and eat the rot-blemished product, no harm would result. No one would suggest that foods contaminated by dangerous levels of mycotoxins should be eaten by either people or animals; such foods are written off as a complete loss.

Maggots

The term maggots means fly larvae. If maggots are seen at all during the course of daily life, they are usually in uncollected garbage, carrion, or excrement. There they play important roles in the process of converting these unpleasant materials into soil. The carrion feeders will infest fresh meats if given the chance, but, of course, they should never be given the opportunity. There is no reason or excuse for any of the carrion or excrement maggots to be in human food, either deliberately or accidentally.

It would be more probable that the maggots listed on the hypothetical label (10) would belong to the kinds of flies that infest fresh fruits and vegetables. Because the damage they cause is so great, so obvious, and so revulsive to many consumers, growers often rely heavily on insecticides to suppress fly populations. Many kinds of flies are involved, but a few of the more typical ones are the vegetable leafminer, onion maggot, apple maggot, and pepper maggot (Table 1) (some other common species are included in the DAL list (26)).

Because it seems unlikely that the term maggot was used in its technical sense on the hypothetical label (10), what was probably meant was larviform immature insects in general. If so, then the label might include representatives of the beetles (grubs) (eg, plum curculio and dried-fruit beetle) and moths (caterpillars) (eg, codling moth) (Table 1), in addition to the larvae of flies.

Food Production with and without Pesticides

Under the present scheme of things, and taking into account all of the nonpesticidal techniques available, consumers are able to purchase unblemished fruits (including tree nuts and peanuts) and vegetables only because pesticides are used to control pests on these crops (158). Without pesticides, some food crops would never reach the market in commercial quantities (224). Many others would suffer either reduced yields or higher levels of pest damage and infestation. The degree of yield reduction or the level of pest damage and infestation could be moderated if suitable (ie, available, affordable, and efficacious) nonpesticidal pest management techniques were applied. Some of these nonchemical IPM techniques would be highly labor intensive (eg, hand-picking pests), resulting in a dramatic increase in price of the specific commodity and a general decrease in quality for most foods so produced. The numbers of maggots listed on the hypothetical label would reflect the intensity with which nonpesticidal techniques had been applied. Lower numbers would necessarily mean higher prices to cover the labor involved in production.

Those who abhor pesticides might readily accept all the consequences of abandoning pesticides or at least of having the opportunity to choose between foods produced with or without pesticides. But it must be recognized that knowledge of how to produce foods without pesticides or with low input of chemicals is extremely limited. However, the outlook for nonpesticidal agriculture (alternative agriculture (145,225)) may not be as gloomy as the foregoing discussion might imply.

Insects

The final term on the hypothetical label is insects (10). Obviously, it is important to know what kinds of insects. The word *insects* on a food label would have such a range of connotations that it would be meaningless if used without qualifying terms. The modifying terms should at least give some idea about the nature of the insects (field pests or storage pests). The issues covered in the section on maggots apply equally to the more general term insects.

Expansion of DALs

The labeling approach may not be necessary. An alternative would be to keep the DAL system but modify and expand it to accommodate methods of food production, ie, one DAL might apply to foods produced without pesticides, while a different DAL would be used for foods produced with pesticides. It might also be necessary to set more specific DALs, associating a given level with a specific pest. The DAL for hop aphids is an example of a specific level already in the official list (26).

The present system is working well under present circumstances. It seems likely that no major change will be required in the short term, but at least some contingencies should be kept in mind to meet a range of potential challenges in the future.

It is important to look much more carefully at the safety of pest-blemished foods and pest-infested foods. Do the phytoalexins generated by plants in defense against pest attacks have any harmful effects on human health (159)? Is there any reason to be concerned about potential allergic responses caused by higher levels of insect parts? How do these seemingly relatively minor threats compare to the dangers (if any) of pesticide residues in food or to the effect of pesticides on the environment and on nontarget organisms, including humans? If it can be shown that the higher levels of pest damage and pest infestation are harmless, then, to answer the question posed earlier, the cosmetic use of pesticides cannot be justified. The associated risks, as understood today, are just too great. On the other hand, can not using pesticides be justified when the resulting reduced yields will be inadequate to fill stomachs that are already hungry? But even while recognizing the urgent nutritional needs of millions of hungry people, can the use of the kinds of pesticides that pollute underground waters to the point that the health of many other millions of people might be harmed be justified?

BIBLIOGRAPHY

1. R. C. Litman and D. S. Litman, "Protection of the American Consumer: The Congressional Battle for the Enactment of the First Federal Food and Drug Law in the United States," *Food Drug Cosmetic Law J.* **37**, 310–329 (1982).
2. R. A. Merrill and P. B. Hutt, *Statutory Supplement to Food and Drug Law—Cases and Materials*, The Foundation Press, Inc., Mineola, N.Y., 1980.
3. J. H. Young, *Pure Food—Securing the Federal Food and Drugs Act of 1906*. Princeton University Press, Princeton, N.J., 1989.
4. J. E. Hoffman, "FDA's Administrative Procedures," in *Seventy-fifth Anniversary Commemorative Volume of Food and Drug Law*, Food and Drug Law Institute, Washington, D.C., 1984.
5. W. F. Janssen, "Crawford," *J. Assoc. Food Drug Off.* **54**, 6–12, 59–67 (1990).
6. Food and Drug Administration, *Federal Food, Drug, and Cosmetic Act, as Amended, and Related Laws*, HHS Publication No. (FDA)89-1051, U.S. Department of Health and Human Services, Rockville, Md., 1989.
7. S. White, "Harvey Wiley Remembered," *J. Assoc. Food Drug Off.* **51**, 214–217 (1987).
8. H. W. Wiley, *The History of a Crime Against the Food Law*, Arno Press, New York, 1976.
9. W. F. Janssen, "Golden Anniversary of the FD&C Act: Consumers 'Never Had it so Good,'" *J. Assoc. Food Drug Off.* **52**, 59–60 (1988).
10. P. B. Hutt, "The Basis and Purpose of Government Regulation of Adulteration and Misbranding of Food," *Food Drug Cosmetic Law J.* **33**, 505–540 (1978).
11. J. T. O'Reilly, *Food and Drug Administration* (Regulatory Manual Series), Shephard's, Inc., Colorado Springs, Colo., 1979.
12. J. P. Hile, "New Theories of Enforcement," *Food Drug Cosmetic Law Journal* **41**, 424–428 (1986).
13. P. M. Brickey, Jr., "The Food and Drug Administration and the Regulation of Food Sanitation," in J. R. Gorham, ed., *Ecology and Management of Food-industry Pests*, Association of Official Analytical Chemists, Arlington, Va., 1991.
14. K. Helrich, ed., *Official Methods of Analysis of the Association of Official Analytical Chemists*, 15th ed., Association of Official Analytical Chemists, Arlington, Va., 1990.
15. J. L. Boese, "Mites," in J. R. Gorham, ed., *Principles of Food Analysis for Filth, Decomposition, and Foreign Matter*, FDA Technical Bulletin **1**, 2nd ed., Association of Official Analytical Chemists, Arlington, Va., 1981.
16. W. V. Eisenberg, "Sources of Food Contaminants," in Ref. 15.
17. J. E. Kvenberg, "Insects," in Ref. 15.
18. J. J. Thrasher, "Detection of Metabolic Products," in Ref. 15.
19. A. W. Vazquez, "Hairs," in Ref. 15.
20. J. W. Bier, "Protozoa and Helminths," in Ref. 15.
21. S. M. Cichowicz, "Analytical Mycology," in Ref. 15.
22. P. B. Mislivec, "Fungi," in Ref. 15.
23. A. E. Schulze, "Analytical Plant Histology," in Ref. 15.
24. E. J. Peel and L. G. Clark, "Foreign Matter in Food," *Food Technol. Austral.* **32**, 18–21 (1980).
25. A. W. Vazquez, "Miscellaneous Filth and Extraneous Matter," in J. R. Gorham, ed., *Training Manual for Analytical Entomology in the Food Industry*, FDA Technical Bulletin **2**, Association of Official Analytical Chemists, Arlington, Va., 1978.
26. Industry Activities Section, *The Food Defect Action Levels*, Food and Drug Administration, Washington, D.C., 1989.
27. "Metal Fragments in Potato Chips Bring Large Fine," *J. Assoc. Food Drug Off.* **53**, 46 (1989).
28. D. F. O'Keefe, Jr., "Legal Issues in Food Establishment Inspections," *Food Drug Cosmetic Law J.* **33**, 121–134 (1978).
29. *Inspection Operations Manual*, Food and Drug Administrator, Rockville, Md., 1985–1991.
30. W. F. Janssen, "For 100 Years, They've Tested the Testers," *FDA Consumer* **18**, 12–15 (Oct. 1984).
31. Center for Food Safety and Applied Nutrition, *Macroanalytical Procedures Manual*, FDA Technical Bulletin No. **5**, Association of Official Analytical Chemists, Arlington, Va., 1984.
32. Approved Methods Committee, *Approved Methods of the American Association of Cereal Chemists*, 2 vols., 8th ed., American Association of Cereal Chemists, Inc., St. Paul, Minn., 1983.
33. E. Kneen, ed., *Methods of Analysis of the American Society of Brewing Chemists*, 7th ed. rev., American Society of Brewing Chemists, St. Paul, Minn., 1981.
34. *Manual of Microscopic Analysis of Feeding Stuffs*, 2nd ed., American Association of Feed Microscopists, Sacramento, Calif., 1987.
35. *Official Analytical Methods of the American Spice Trade Association*, 3rd ed., American Spice Trade Association, Englewood Cliffs, N.J., 1985.
36. K. L. Harris, "An Annotated Bibliography of Methods for the Examination of Foods for Filth," *J. Assoc. Off. Agric. Chem.* **29**, 420–439 (1946).
37. R. G. Dent, "Extraction Methods," in Ref. 25.
38. R. G. Dent, "Elements of Filth Detection," in Ref. 15.
39. R. G. Dent and J. R. Gorham, "Collaborative Study of the Extraction of Light Filth from Canned Crabmeat," *Journal of the Association of Official Analytical Chemists* **59**, 825–826 (1976).
40. J. R. Gorham, ed., *Insect and Mite Pests in Food: An Illustrated Key*, Agriculture Handbook **655**, U.S. Department of Agriculture, Washington, D.C., 1991.
41. O. L. Kurtz and K. L. Harris, *Micro-analytical Entomology for Food Sanitation Control*, Association of Official Agricultural Chemists, Washington, D.C., 1962.
42. D. McClymont-Peace, *Key for Identification of Mandibles of Stored-food Insects*, Association of Official Analytical Chemists, Arlington, Va., 1985.
43. J. P. Sutherland, A. H. Varnam, and M. G. Evans, *A Colour Atlas of Food Quality Control*, Wolfe Publishing Ltd., London, 1986.
44. A. W. Vazquez, "Vertebrate Pests: Birds, Bats, Rodents," in Ref. 25.
45. H. R. Gerber, "Colorimetric Determination of Alkaline Phosphatase as Indicator of Mammalian Feces in Corn Meal: Collaborative Study," *Journal of the Association of Official Analytical Chemists* **69**, 496–498 (1986).
46. H. R. Gerber, "Chemical Test for Mammalian Feces in Grain Products: Collaborative Study," *Journal of the Association of Official Analytical Chemists* **72**, 766–769 (1989).

47. H. R. Gerber, "Detection of Mammalian Feces in Ground Black Pepper," *Abstracts of the 103rd AOAC Annual International Meeting and Exposition,* St. Louis, Mo., Sept. 25–28, 1989, Abst. No. 252.
48. R. L. Heitzman and J. L. Boese, "Confirmation of Rodent Gnawing on Food Packaging by a Salivary Amylase Test," in Ref. 47, Abst. No. 251.
49. D. A. Carlson, "Hydrocarbons for Identification and Phenetic Comparisons: Cockroaches, Honey Bees and Tsetse Flies," *Fla. Entomol.* **71,** 333–345 (1988).
50. D. A. Carlson and R. J. Brenner, "Hydrocarbon-based Discrimination of Three North American *Blattella* Cockroach Species (Orthoptera: Blattellidae) Using Gas Chromatography," *Annals of the Entomological Society of America* **81,** 711–723 (1988).
51. P. E. Kauffman and D. B. Shah, "Enzyme Immunoassay for Detection of *Drosophila melanogaster* Antigens in the Juice of Various Foods," *Journal of the Association of Official Analytical Chemists* **71,** 636–642 (1988).
52. E. Corwin, "Preventing Food Adulteration," *FDA Consumer* **10,** 10–15 (Nov. 1976).
53. D. D. Horner, "The FDA and Quality Assurance," *Pest Control* **49,** 40–42 (Oct. 1981).
54. C. Lecos, "Making 'Clean is Keen' a Warehouse Motto," *FDA Consumer* **14,** 9–11 (Nov. 1980).
55. V. A. Kleinfeld and C. W. Dunn, *Federal Food, Drug, and Cosmetic Act, Judicial and Administrative Record 1938–1949,* Commerce Clearing House, Inc., New York, 1953.
56. T. W. Christopher and W. W. Goodrich, *Cases and Materials on Food and Drug Law,* 2nd ed., Commerce Clearing House, Inc., New York, 1973.
57. R. A. Merrill and P. B. Hutt, *Food and Drug Law—Cases and Materials,* The Foundation Press, Inc., Mineola, N.Y., 1980.
58. Office of Enforcement, "Reconditioning of Foods Adulterated under 402(a)(4)," *Compliance Policy Guide* **7153.04,** Food and Drug Administration, Washington, D.C., 1989.
59. V. A. Kleinfeld and C. W. Dunn, *Federal Food, Drug, and Cosmetic Act, Judicial and Administrative Record 1949–1950,* Commerce Clearing House, Inc., New York, 1951.
60. V. A. Kleinfeld and C. W. Dunn, *Federal Food, Drug, and Cosmetic Act, Judicial and Administrative Record 1951–1952,* Commerce Clearing House, Inc., New York, 1953.
61. J. R. Phelps, "Actions in the Courts," *Food Drug Cosmetic Law J.* **35,** 502–510 (1980).
62. E. M. Pfeifer, "Enforcement," in *Seventy-fifth Anniversary Commemorative Volume of Food and Drug Law,* Food and Drug Law Institute, Washington, D.C., 1984.
63. M. Brannon, "Organizing and Reorganizing FDA," in Ref. 4.
64. Industry Activities Section, *FDA, the Food Industry, & Quality Assurance,* Publication No. (FDA)**79-2114,** Food and Drug Administration, Washington, D.C., 1979.
65. Industry Activities Section, *Do Your Own Establishment Inspection, a Guide to Self Inspection for the Smaller Food Processor and Warehouse,* Publication No. (FDA)**82-2163,** Food and Drug Administration, Washington, D.C., 1984.
66. Industry Activities Section, *Industry Assistance Programs for the Food Industry,* Publication No. (FDA)**84-2140,** Food and Drug Administration, Washington, D.C., 1984.
67. Y. H. Hui, *United States Food Laws, Regulations, and Standards,* 2 vols., 2nd ed., John Wiley & Sons, New York, 1986.
68. T. Quinn, "Food GMP's," *Food Drug Cosmetic Law J.* **35,** 215–220 (1980).
69. F. B. Jacobson, "Current Good Manufacturing Practices," *Candy Industry* **151,** 65–90 (1986).
70. M. R. Johnston and R. C. Lin, "FDA Good Manufacturing Practice Regulations," *Journal of Food Quality* **3,** 109–118 (1980).
71. D. R. Martin, "Inspectional Activities of the Food and Drug Administration," in Ref. 13.
72. M. H. Bozeman, "Recalls—On Making the Best of a Bad Thing," *Food Drug Cosmetic Law J.* **33,** 342–359 (1978).
73. J. Bressler, "What FDA Expects During a Recall," *Food Product Development* **14,** 64–65 (Apr. 1980).
74. C. Lecos, "Determining When a Food Poses a Hazard," *FDA Consumer* **17,** 25–28 (June 1983).
75. M. R. Taylor, "Seizures and Injunctions: Their Role in FDA's Enforcement Programs," *Food Drug Cosmetic Law J.* **33,** 596–606 (1978).
76. D. F. Weeda, "FDA Seizure and Injunction Actions: Judicial Means of Protecting the Public Health," *Food Drug Cosmetic Law J.* **35,** 112–121 (1980).
77. "Rodent and Insect Infestation Earns Candy Company President Fine, Community Service," *J. Assoc. Food Drug Off.* **54,** 50 (1990).
78. Y. H. Hui, *United States Food Laws, Regulations, and Standards,* John Wiley & Sons, New York, 1979.
79. J. S. Kahan, "Criminal Liability under the Federal Food, Drug, and Cosmetic Act—The Large Corporation Perspective," *Food Drug Cosmetic Law J.* **36,** 314–331 (1981).
80. J. A. Levitt, "FDA Inspections and Criminal Responsibility," *Food Drug Cosmetic Law J.* **36,** 469–477 (1981).
81. J. M. Taylor, "Establishment and Use of Defect Action Levels," in *Association of Food Industries 82,* Association of Food Industries, Matawan, N.J., 1982.
82. R. L. Frank and D. R. Johnson, "The USDA's Compliance and Enforcement Programs," *Food Drug Cosmetic Law J.* **44,** 205–230 (1989).
83. D. L. Houston, "Action Through Networking: The USDA Perspective," *J. Assoc. Food Drug Off.* **51,** 184–195 (1987).
84. M. Kenney, D. Orr, and M. J. Shannon, "Regulatory and Inspectional Functions in the U.S. Department of Agriculture," in Ref. 13.
85. N. E. Kirschbaum, "Role of State Government in the Regulation of Food and Drugs," *Food Drug Cosmetic Law J.* **38,** 199–204 (1983).
86. V. Modeland, "America's Food Safety Team: A Look at the Lineup," in *Safety First: Protecting America's Food Supply* (an *FDA Consumer* special report), Food and Drug Administration, Rockville, Md., 1988.
87. V. O. Wodicka, "Role of State & Local Food & Drug Programs," *Assoc. Food Drug Off. Quart. Bull.* **48,** 227–232 (1984).
88. "Something to Hide," *FDA Consumer* **17,** 29 (Mar. 1983).
89. J. M. Taylor, "Federal-State Coordination: The Ultimate in Networking," *J. Assoc. Food Drug Off.* **51,** 206–213 (1987).
90. L. Vinci, "The Importance and Need for Continued Federal-State-Local Communication and Cooperation," *J. Assoc. Food Drug Off.* **51,** 214–217 (1987).
91. *Introducing Codex Alimentarius,* Food and Agriculture Organization of the United Nations, Rome, 1988.
92. Codex Alimentarius Commission, *Codex Alimentarius,* Vol. 1, *Explanatory Notes on the Joint FAO/WHO Food Standards Programme and on the Codex Alimentarius Commission,* Food and Agriculture Organization and World Health Organization, Rome.
93. P. B. Hutt and P. B. Hutt II, "A History of Government Regulation of Adulteration and Misbranding of Food," *Food Drug Cosmetic Law J.* **37,** 310–329 (1982).

94. W. F. Janssen, "America's First Food and Drug Laws," *FDA Consumer* **9**, 12–19 (June 1975).
95. W. F. Janssen, "The Constitution and the Consumer: Discovering the Connections," *J. Assoc. Food Drug Off.* **51**, 218–225 (1987).
96. P. B. Hutt, "FDA Good Manufacturing Practice Regulations for Food," *Cereal Foods World* **26**, 186–187 (1981).
97. R. G. Chesemore, "Action Levels for Residues of Certain Pesticides in Food and Feed," *Federal Register* **55**, 14359–14363 (Apr. 17, 1990).
98. Industry Activities Section, *Action Levels for Poisonous or Deleterious Substances in Human Food and Animal Feed*, Food and Drug Administration, Washington, D.C., 1987.
99. B. Grigg and V. Modeland, "The Cyanide Scare—A Tale of Two Grapes," *FDA Consumer* **23**, 7–11 (July–Aug. 1989).
100. J. R. Gorham, "HACCP and Filth in Food," *Journal of Food Protection* **52**, 674–677 (1989).
101. "Feeding Their Populations—Daunting Challenges Face Tropical Countries," *International Pest Control* **31**, 134 (1989).
102. J. R. Gorham, "Insects as Food," *Bull. Soc. Vector Ecol.* **3**, 11–16 (1976).
103. S. H. McNamara, "Some Legal Aspects of Providing New Foods for Hungry Populations," *Food Product Development* **9**, 54, 57, 59–60 (Mar. 1975).
104. D. Pimentel, E. E. Terhune, W. Dritschilo, D. Gallahan, N. Kinner, D. Nafus, R. Peterson, N. Zareh, J. Misiti, and O. Haber-Schaim, "Pesticides, Insects in Foods, and Cosmetic Standards," *BioScience* **27**, 178–185 (1977).
105. C. Lecos, "Pesticides and Food: Public Worry No. 1," *FDA Consumer* **18**, 12–15 (July–Aug. 1984).
106. J. H. Steele, "Pesticides and Food Safety: Perception vs Reality," *Healthy Animals, Safe Foods, Healthy Man*, in *Proceedings, World Association of Veterinary Food Hygienists 10th (Jubilee) International Symposium* (Stockholm, July 2–7, 1989), pp. 63–76, 1990.
107. F. R. Shank and K. L. Carson, "The Risk of Normal Eating—Risk Communication in Risk Management," *J. Assoc. Food Drug. Off.* **54**, 30–38 (Apr. 1990).
108. Division of Contaminants Chemistry, "Food and Drug Administration Pesticide Program, Residues in Foods—1988," *Journal of the Association of Official Analytical Chemists* **72**, 133A–152A (1989).
109. D. Farley, "Setting Safe Limits on Pesticide Residues," in Ref. 86.
110. Jones, P., *Pesticides and Food Safety*, American Council on Science and Health, New York, 1989.
111. D. K. Small, "Risk Management: A Retail Perspective," *J. Assoc. Food Drug Off.* **54**, 39–44 (Apr. 1990).
112. C. W. Carnevale, "Pesticide Residues in Foods—Results, Risks, and the Law," *Assoc. Food Drug Off. Quart. Bull.* **51**, 105–109 (1987).
113. H. Hopkins, "Rodenticide Twice Misused, Exterminator Gets Jail Term," *FDA Consumer* **17**, 27 (Mar. 1983).
114. J. G. Cummings and R. B. Perfetti, "Regulatory and Residue Issues in the Environmental Protection Agency," in Ref. 13.
115. National Pest Control Association, *Good Practice Statements*, National Pest Control Association, Dunn Loring, Va., 1978.
116. S. S. Balling, "Managing Pesticides for a Safe Food Supply: An Industry Program," *J. Assoc. Food Drug Off.* **54**, 19–29 (Apr. 1990).
117. M. B. Stoetzel, ed., *Common Names of Insects and Related Organisms*, Entomological Society of America, Lanham, Md., 1989.
118. E. E. Terrell, S. R. Hill, J. H. Wiersema, W. R. Rice, *A Checklist of Names for 3,000 Vascular Plants of Economic Importance*, Agriculture Handbook **505**, U.S. Department of Agriculture, Washington, D.C., 1986.
119. J. L. Kornacki and D. A. Gabis, "Microorganisms and Refrigeration Temperatures," *Dairy Food Environ. Sanit.* **10**, 192–195 (1990).
120. L. Moberg, "Good Manufacturing Practices for Refrigerated Foods," *Journal of Food Protection* **52**, 363–367 (1989).
121. National Food Processors Association, "Factors to Be Considered in Establishing Good Manufacturing Practices for the Production of Refrigerated Foods," *Dairy Food Sanit.* **8**, 288–291 (1988).
122. Anon., "Pesticides Bound to Grain May Be Harmful," *New Scientist* **123**, 36 (Aug. 5, 1989).
123. D. P. Gigat and P. Zvoutete, "The Evaluation of Different Insecticides for the Protection of Maize against Some Stored Product Pests," *International Pest Control* **32**, 10–13 (1990).
124. P. K. Harein, "Chemical Control of Insect Pests in Bulk-stored Grains," in Ref. 13.
125. J. T. Snelson, *Grain Protectants*, Australian Centre for International Agricultural Research, Canberra, 1987.
126. F. J. Baur, "Chemical Methods to Control Insect Pests of Processed Foods," in Ref. 13.
127. Anon., "Fumigation Method Leaves No Residue," *World Grain* **7**, 23 (Apr. 1989).
128. M. Voight and L. Benzing, "Phosphine Fumigation—No Detectable Residues," *International Pest Control* **32**, 35 (Mar.–Apr. 1990).
129. V. E. Walter, "Fumigation in the Food Industry," in Ref. 13.
130. R. Davis and R. H. Barrett, "In-transit Shipboard Fumigation of Grain: Research to Regulation," *Cereal Foods World* **31**, 227–229 (1986).
131. J. A. McFarlane, "Factors Affecting Insect Pest Management on Wheat Grain Stored in Tropical Uplands," *Trop. Sci.* **29**, 51–73 (1989).
132. J. Chesky, "New and Old Grain Storing Techniques," *International Pest Control* **27**, 73–74 (1985).
133. J. A. McFarlane, "Postharvest Research in Developing Countries: Issues and Prospects from a Storage Viewpoint," *Postharvest News and Information* **1**, 15–18 (1990).
134. J. R. Gorham and M. Ouye, "Specific Needs of Action Programs: Stored Products," in *Abstract Volume, XVII International Congress of Entomology*, Hamburg, FRG, Aug. 20–26, 1984, Abst. No. S1.2.16, p. 45.
135. R. V. Dowell and R. Gill, "Exotic Invertebrates and Their Effects on California," *Pan-Pacific Entomologist* **65**, 132–145 (1989).
136. H. I. Rainwater and C. A. Smith, "Quarantines—First Line of Defense," in *Protecting Our Food, The Yearbook of Agriculture 1966*, U.S. Department of Agriculture, Washington, D.C., 1966.
137. D. W. Hawthorne, "Cotton Rats," in R. M. Timm, ed., *Prevention and Control of Wildlife Damage*, Great Plains Agricultural Council and University of Nebraska, Lincoln, 1983.
138. R. A. Dolbeer, "Blackbirds," in Ref. 137.
139. J. S. Hough, D. E. Briggs, R. Stevens and T. W. Young, *Malting and Brewing Science*, Vol. 2, *Hopped Wort and Beer*, 2nd ed., Chapman and Hall, London, 1982.
140. "Aphids in Hops," in Ref. 33.
141. J. R. Gorham, "Filth in Food: Implications for Health," *Journal of Milk Food Technology* **38**, 409–418 (1975).
142. E. De Las Casas, "Stored Product Insects and Microorganisms in Grain Ecosystems," *University of Minnesota Agri-

cultural *Experimental Station Technical Bulletin* **310**, 31–34 (1977).

143. F. V. Dunkel, "The Relationship of Insects to the Deterioration of Stored Grain by Fungi," *International Journal of Food Microbiology* **7**, 227–244 (1988).

144. N. Pande and B. S. Mehrotra, "Rice Weevil (*Sitophilus oryzae* Linn.): Vector for Toxigenic Fungi," *Nat. Acad. Sci. Lett.* (India) **11**, 3–4 (1988).

145. Committee on the Role of Alternative Farming Methods in Modern Production Agriculture, Board on Agriculture, National Research Council, *Alternative Agriculture,* National Academy Press, Washington, D.C., 1989.

146. J. McDermott, "Some Heartland Farmers Just Say No to Chemicals," *Smithsonian* **21**, 114–127 (Apr. 1990).

147. "Pesticide Risk Reduction without Food Supply Loss Argued," *Food Chem. News* **32**, 24 (Mar. 19, 1990).

148. E. G. Thompson, "The Integrated Pest Management Approach to Food Protection," *Cereal Foods World* **29**, 149–151 (1984).

149. J. R. Eilers, "New Foods Provide New Food Safety Challenges," *Food Processing* **51**, 104–108 (June 1990).

150. J. A. Gibson, "Review of Rodent and Insect Damage to Stored Products and Non-pesticidal Methods of Control," in D. R. Houghton, R. N. Smith and H. O. W. Eggins, eds., *Biodeterioration 7,* Elsevier Applied Science Publishers, Ltd., Barking, UK, 1988.

151. D. F. Jones and E. G. Thompson, "Integrated Pest Management for the Food Industry," in Ref. 13.

152. "Food Additive Petition Needed for Adding Beneficial Insects to Food," *Food Chemical News* **30**, 7–9 (July 25, 1988).

153. Anon., "Predaceous Insect Issue Weighed by Government Task Force: CBS," *Food Chemical News* **32**, 10–11 (Apr. 2, 1990).

154. "Lake Says FDA Does Not Regulate Field Use of Parasitic Insects," *Food Chemical News* **30**, 43–44 (Jan. 16, 1989).

155. J. H. Brower and M. A. Mullen, "Effects of *Xylocoris flavipes* (Hemiptera: Anthocoridae) Releases on Moth Populations in Experimental Peanut Storages," *J. Entomol. Sci.* **25**, 268–276 (1990).

156. W. A. Bruce, "Mites as Biological Control Agents of Stored Product Pests," in M. A. Hoy, G. L. Cunningham and L. Knutson, eds., *Biological Control of Pests by Mites,* Proceedings of a Conference, University of California, Berkeley, Apr. 5–7, 1982, 1983, pp. 74–78.

157. D. P. Singh, *Breeding for Resistance to Diseases and Insect Pests,* Springer-Verlag, New York, 1986.

158. W. L. Hollis, "Agrichemical Residues in Perspective to Agriculture and Food Risk/Hazards," *J. Assoc. Food Drug Off.* **47**, 110–119 (1983).

159. R. C. Beier, "Natural Pesticides and Bioactive Components in Foods," *Rev. Environ. Contam. Toxicol.* **113**, 47–137 (1990).

160. J. G. Surak, "Phytoalexins and Human Health—a Review," *Proceedings of the Florida State Horticultural Society* **91**, 256–258 (1978).

161. W. Burkholder, "Pheromone Trap Good Practices Statement," *Bull. Assoc. Oper. Millers,* 5433–5434 (Apr. 1989).

162. W. E. Burkholder and D. L. Faustini, "Biological Methods of Survey and Control," in Ref. 13.

163. D. B. Pinniger, "Damage Prevention through Improved Detection of Insect Pests in Stores," in Ref. 150.

164. Gustafson, Inc., "FDA Says Insects Found in Traps Don't Indicate Unsanitary Conditions," *The Grain Probe* **3**, 1 (Apr. 1990).

165. F. Fleurat-Lessard, "Utilisation d'un attractif de synthese pour la surveillance et le piégeage des pyrales Phycitinae dans les locaux de stockage et de conditionnement de denrées alimentaires végétales," *Agronomie* **6**, 567–573 (1986).

166. D. McDonald, "Progress and Prospects in Insect Control," *International Pest Control* **31**, 145–151 (1989).

167. "Senate Agriculture Committee Approves 'Circle of Poison' Legislation," *Food Chemical News* **32**, 63–64 (June 11, 1990).

168. "Pests, Not Pesticides, Pose Greater Food Safety Risk, IFT Says," *Food Chemical News* **32**, 64–65 (June 11, 1990).

169. W. Schulz, "Food for Thought" (editorial), *National Gardening* **13**, 6 (July 1990).

170. Environmental Studies Board, National Research Council, *Pest Control: An Assessment of Present and Alternative Technologies,* Vol. 1, *Contemporary Pest Control Practices and Prospects: The Report of the Executive Committee,* National Academy of Sciences, Washington, D.C., 1975.

171. D. F. Emery, "Water Quality: Problems with an Essential Resource," *Cereal Foods World* **34**, 483–486 (1989).

172. M. T. Wan, "Levels of Selected Pesticides in Farm Ditches Leading to Rivers in the Lower Mainland of British Columbia, *Journal of Environmental Science and Health* **B24**, 183–203 (1989).

173. G. R. DeFoliart, "The Human Use of Insects as Foods and as Animal Feed," *Bull. Entomol. Soc. Amer.* **35**, 22–35 (Spring 1989).

174. P. M. Brickey, Jr. and J. R. Gorham, "Preliminary Comments on Federal Regulations Pertaining to Insects as Food," *Food Insects Newsletter* **2**, 1, 7 (Mar. 1989).

175. Z. Mehr, "Health Hazards from Bird Droppings," *Pest Management Bulletin* **11**, 12–13 (Mar. 1990).

176. H. G. Scott, "Pigeons—Public Health Importance and Control," *Pest Control* **29**, 9–20, 60–61 (Sept. 1961).

177. J. Newton, "Insects and Packaging—A Review," *International Biodeterioration* **24**, 175–187 (1988).

178. H. A. Highland, "Protecting Packages against Insects," in Ref. 13.

179. P. M. Brickey, Jr., J. S. Gecan, and A. Rothschild, "Method for Determining Direction of Insect Boring Through Food Packaging Materials," *Journal of the Association of Official Analytical Chemists* **56**, 640–642 (1973).

180. J. H. Steele, "Bolivian Hemorrhagic Fever (BHF)," in G. W. Beran, ed., *CRC Handbook Series in Zoonoses,* Section B: Viral Zoonoses, Vol. 2, CRC Press, Boca Raton, Fla., 1981.

181. W. G. Winkler and V. J. Lewis, "Lymphocytic Choriomeningitis," in Ref. 180.

182. J. R. Gorham, "Food Pests as Disease Vectors," in Ref. 13.

183. J. R. Gorham, "Foodborne Filth in Human Disease," *Journal of Food Protection* **52**, 674–677 (1989).

184. N. G. Gratz, "Rodents and Human Disease: A Global Appreciation," in I. Prakash, ed., *Rodent Pest Management,* CRC Press, Boca Raton, Fla., 1988.

185. M. J. Klowden and B. Greenberg, "Effects of Antibiotics on the Survival of Salmonella in the American Cockroach," *J. Hyg.* **79**, 339–345 (1977).

186. S. Beatson, "Pharaoh's Ants as Pathogen Vectors in Hospitals," *Lancet* **1**, 425–427 (1972).

187. J. Ipinza-Regla, G. Figueroa, and I. Moreno, "*Iridomyrmex humilis* (Formicidae) y su papel como posible vector de contaminación microbiana en industrias de alimentos," *Folia Entomol. Mex.* **62**, 111–124 (1984).

188. O. Rosef and G. Kapperud, "House Flies (*Musca domestica*) as Possible Vectors of *Campylobacter fetus* Subsp. *jejuni*," *Applied and Environmental Microbiology* **45**, 381–383 (1983).

189. P. B. Cornwell and M. F. Mendes, "Disease Organisms Carried by Oriental Cockroaches in Relation to Acceptable Standards of Hygiene," *International Pest Control* **23**, 72–74 (1981).

190. P. A. Buxton, "The Importance of the House Fly as a Carrier of *E. histolytica*," *British Medical Journal* **1**, 142–144 (1920).

191. R. Bonfante, E. C. Faust, and L. E. Giraldo, "Parasitologic Surveys in Cali, Departamento de Valle, Colombia. IX. Endoparasites of Rodents and Cockroaches in Ward Siloe, Cali, Colombia," *Journal of Parasitology* **47**, 843–846 (1961).

192. J. Andrews, "*Endamoeba histolytica* and Other Protozoa in Wild Rats Caught in Baltimore," *Journal of Parasitology* **20**, 334 (1934).

193. W. Eichler, "Syanthrope Aspekte sur Ökologie der Pharaoameise," *Dtsch. Entomol. Z.* **20**, 425–432 (1973).

194. P. Echeverria, B. A. Harrison, C. Tirapat, and A. McFarland, "Flies as a Source of Enteric Pathogens in a Rural Village in Thailand," *Applied and Environmental Microbiology* **46**, 32–36 (1983).

195. A. Le Guyader, C. Rivault, and J. Chaperon, "Microbial Organisms Carried by Brown-banded Cockroaches in Relation to Their Spatial Distribution in a Hospital," *Epidemiology and Infection* **102**, 485–492 (1989).

196. J. E. Childs, G. E. Glass, G. W. Korch, and J. W. LeDuc, "The Ecology and Epizootiology of Hantaviral Infections in Small Mammal Communities of Baltimore: A Review and Synthesis," *B. Soc. Vector Ecol.* **13**, 113–122 (1988).

197. S. P. Singh, M. S. Sethi, and V. D. Sharma, "The Occurrence of Salmonellae in Rodents, Shrews, Cockroaches and Ants," *Internat. J. Zoonoses* **7**, 58–61 (1980).

198. S. P. Bidawid, J. F. B. Edeson, J. Ibrahim, and R. M. Matossian, "The Role of Non-biting Flies in the Transmission of Enteric Pathogens (*Salmonella* species and *Shigella* species) in Beirut, Lebanon," *Annals of Tropical Medicine and Parasitology* **72**, 117–121 (1978).

199. R. Bolanos, "Frecuencia de *Salmonella* y *Shigella* en moscas domésticas colectadas en la ciudad de San José," *Revista de Biologia Tropical* **7**, 207–210 (1959).

200. E. Hormaeche, C. A. Peluffo, and P. L. Aleppo, "Investigaciones sobre la existencia de bacterias de los géneros *Salmonella* y *Shigella* en las moscas," *An. Inst. Hig. Montevideo* **4**, 75–79 (1950).

201. J. P. Mackey, "Salmonellosis in Dar es Salaam," *East African Medical Journal* **32**, 1–6 (1955).

202. J. I. Okafor, "Bacterial and Fungal Pathogens from the Intestinal Tract of Cockroaches," *Journal of Communicable Diseases* **13**, 128–131 (1981).

203. T. A. Olson and M. E. Rueger, "Experimental Transmission of *Salmonella oranienburg* Through Cockroaches," *Public Health Reports* **65**, 531–540 (1950).

204. M. E. Rueger and T. A. Olsen, "Cockroaches (Blattaria) as Vectors of Food Poisoning and Food Infection Organisms," *Journal of Medical Entomology* **6**, 185–189 (1969).

205. H. Fukumi, "Salmonelloses in Japan," in E. van Oye, ed., *The World Problem of Salmonellosis*, Dr. W. Junk, The Hague, 1964.

206. C. B. Gerichter and I. Sechter, "Animal Sources of Salmonella in Israel," *Israel Journal of Medical Sciences* **6**, 413 (1970).

207. P. A. M. Guinee, E. H. Kampelmacher, A. van Keulen, and A. J. Ophof, "Incidence of *Salmonella* in Brown Rats Caught in and near Slaughterhouses, Farms and Mink Farms," *Zentralblatt für Veterinärmedizin* **B10**, 181–185 (1963).

208. J. Olarte and G. Varela, "Epidemiología de la Salmonelosis en México," in Ref. 205.

209. C. S. Richards, W. B. Jackson, R. M. DeCapito, and P. P. Maier, "Studies on Rates of Recovery of *Shigella* from Domestic Flies and from Humans in Southwestern United States," *American Journal of Tropical Medicine and Hygiene* **10**, 44–48 (1961).

210. O. A. Akinboade, J. O. Hassan, and A. Adejinmi, "Public Health Importance of Market Meat Exposed to Refuse Flies," *Internat. J. Zoonoses* **11**, 111–114 (1984).

211. R. K. Ladisch, S. K. Ladisch, and P. M. Lowe, "Quinoid Secretions in Grain and Flour Beetles," *Nature* **215**, 939–941 (Aug. 26, 1967).

212. R. A. Wirtz, "Food Pests as Disease Agents," in Ref. 13.

213. H. S. Bernton and H. Brown, "Insects as Potential Sources of Ingestant Allergens," *Annals of Allergy* **25**, 381–387 (1967).

214. H. S. Bernton and H. Brown, "Insect Allergy: The Allergenic Potentials of the Cockroach," *Southern Medical Journal* **62**, 1207–1210 (1969).

215. G. T. Okumura, "A Report of Canthariasis and Allergy Caused by *Trogoderma*," *Calif. Vector Views* **14**, 19–22 (1967).

216. Dennis Thayer Associates, "Starting Your Rodent Elimination Program—Good Advice for Food Facilities," *Dairy Food Environ. Sanit.* **10**, 356–357 (1990).

217. M. V. Norcross, "Food Safety—Meeting the Challenges of the '90s," *J. Assoc. Food Drug Off.* **54**, 45–50 (1990).

218. A. L. Snowdon, "A Review of the Nature and Causes of Postharvest Deterioration in Fruits and Vegetables, with Especial Reference to Those in International Trade," in Ref. 150.

219. R. V. Bhat, "Moulds that Can Kill," *World Health*, 20–22 (Mar. 1987).

220. M. Moorman, "Mycotoxins and Food Safety," *Dairy Food Environ. Sanit.* **10**, 207–210 (1990).

221. J. H. Emrick, "Machinery Mold: Indicator of Insanitation in Food Plants," *FDA By-Lines* **7**, 266–277 (1977).

222. J. Lacey, "Grain Storage: The Management of Ecological Change," in Ref. 150.

223. *Fungicides and our Food Supply*, National Agricultural Chemicals Association, Washington, D.C., n.d.

224. *Tough Questions and Frank Answers, A Candid Approach to Food Safety*, National Agricultural Chemicals Association, Washington, D.C., 1989.

225. M. D. Lowe, "Organic Farming Gets Some Respect," *World Watch* **3**, 39–40 (May–June 1990).

General References

"Watchdog!" *World Health*, 23 (Mar. 1987).

R. T. Arbogast, G. L. LeCato, and R. Van Byrd, "External Morphology of Some Eggs of Stored-product Moths (Lepidoptera: Pyralidae, Gelechiidae, Tineidae)," *Internat. J. Insect Morphol. Embryol.* **9**, 165–177 (1980).

E. Auld, "Risk Communication and Food Safety," *Dairy Food Environ. Sanit.* **10**, 352–355 (1990).

H. L. Avallone, "GMP Philosophy of Pharmaceutical Manufacturers," *Assoc. Food Drug Off. Quart. Bull.* **51**, 116–120 (1987).

E. M. Basile, "The Case Law on Inspections," *Food Drug Cosmetic Law J.* **34**, 20–31 (1979).

L. M. Baukin, "GMP Requirements for Building and Facilities and for Equipment," *Food Drug Cosmetic Law J.* **34**, 442–449 (1979).

C. A. Benschoter, "Methyl Bromide Fumigation and Cold Storage as Treatments for California Stone Fruits and Pears Infested with the Caribbean Fruit Fly (Diptera: Tephritidae)," *Journal of Economic Entomology* **81**, 1665–1667 (1988).

D. Bergeron, R. J. Bushway, F. L. Roberts, I. Kornfield, J. Okedi, and A. A. Bushway, "The Nutrient Composition of an Insect Flour Sample from Lake Victoria, Uganda," *J. Food Composit. Anal.* **1**, 373–377 (1988).

F. L. Bryan, "Hazard Analysis Critical Control Point (HACCP) Concept," *Dairy Food Environ. Sanit.* **10**, 416–418 (1990).

A. C. Celeste, "The Inevitable FDA Inspection," *Food Drug Cosmetic Law J.* **34**, 32–39 (1979).

M. Q. Chaudhry, H. Ahmed, and M. Anwar, "Development of an Airtight Polyethylene Enclosure for Integrated Pest Management of Grains, Stored at Farm Level in Pakistan," *Trop. Sci.* **29**, 177–187 (1989).

A. Ciegler, "Mycotoxins, a Limited Review," *J. Assoc. Food Drug Off.* **47**, 75–80 (1983).

L. E. DeBell and D. L. Chesney, "FDA Inspections Process," *Food Drug Cosmetic Law J.* **37**, 244–249 (1982).

G. R. DeFoliart, "Are Processed Insect Food Products Still Commercially Available in the United States?" *Food Insects Newsletter* **1**, 1, 6 (Nov. 1988).

G. M. Doherty, "The Inspections Process," *Food Drug Cosmetic Law J.* **35**, 555–567 (1980).

C. Ely, Jr., "Regulation of Pesticide Residues in Food: Addressing the Critical Issues," *Food Drug Cosmetic Law J.* **40**, 494–498 (1985).

Expert Panel on Food Safety and Nutrition, *Quality of Fruits and Vegetables,* Institute of Food Technologists, Chicago, 1990.

M. M. Feinberg, "FDA Investigator: 'From His Office to Yours'," *Food Drug Cosmetic Law J.* **36**, 486–492 (1981).

J. R. Fleder, "Administrative Inspections by the Food and Drug Administration: Role of the Department of Justice," *Food Drug Cosmetic Law J.* **44**, 297–314 (1989).

J. R. Gorham, "A Rational Look at Insects as Food," *FDA By-Lines* **6**, 231–241 (1976).

J. R. Gorham, "The Significance for Human Health of Insects in Food," *Ann. Rev. Entomol.* **24**, 209–224 (1979).

W. Grigg, "The Making of a Milestone in Consumer Protection," *FDA Consumer* **22**, 30–32 (Oct. 1988); Ibid, 30–32 (Nov. 1988); Ibid, 28–30 (Dec. 1988–Jan. 1989).

F. A. Hegele, "Integrated Pest Management—A Quality Assurance Tool?" *Cereal Foods World* **34**, 296 (1989).

C. O. Jackson, *Food and Drug Legislation in the New Deal,* Princeton University Press, Princeton, N.J., 1970.

W. Janssen, "How the Law Changed in 1938," *FDA Consumer* **22**, 30–31 (Dec. 1988–Jan. 1989).

W. Janssen, "The Squad that Ate Poison," *FDA Consumer* **15**, 6–11 (Dec. 1981–Jan. 1982).

W. F. Janssen, "The Story of the Laws Behind the Labels," *FDA Consumer* **15**, 32–45 (June 1981).

R. J. Kelsey, *Packaging in Today's Society,* 3rd ed., Technomic Publishing Company, Inc., Lancaster, Pa., 1989.

K. T. Khalaf, "Micromorphology of Beetle Elytra, Using Simple Replicas," *Fla. Entomol.* **63**, 307–340 (1980).

R. D. Kiernan, "Factory Inspection: Responding to an Issued FDA Form 483," *Food Drug Cosmetic Law J.* **43**, 699–708 (1988).

B. Kobbe, "Mold Toxins: Hazard to Animal and Human Health," *Calif. Agric.* **33**, 18–19 (Nov.–Dec. 1979).

G. LeCato, and B. R. Flatherty, "Description of Eggs of Selected Species of Stored-product Insects (Coleoptera and Lepidoptera)," *J. Kan. Entomol. Soc.* **47**, 308–317 (1974).

C. H. Lushbough, "A Food Company's Approach to the Inspectional Process," *Food Drug Cosmetic Law J.* **35**, 436–450 (1980).

S. H. McNamara, "The FDA Inspection: What You Need to Know to Protect Your Company," *Food Drug Cosmetic Law J.* **36**, 245–257 (1981).

R. D. Middlekauff, "Pesticide Residues in Food: Legal and Scientific Issues," *Food Drug Cosmetic Law J.* **42**, 251–264 (1987).

S. A. Miller, "The Saga of Chicken Little and Rambo," *J. Assoc. Food Drug Off.* **51**, 196–205 (1987).

J. Moore, "Insect Vacuums Hit the Market," *Ag Consultant* **46**, 18 (June 1990).

C. L. Morin and M. A. Yeager, "Responding to FDA's Insistence on Taking Photographs During an Administrative Inspection," *Food Drug Cosmetic Law J.* **42**, 485–499 (1987).

T. R. Mulvaney, "Good Manufacturing Practice Regulations, Guidelines and Voluntary Programs," paper presented at the Conference for Food Protection, Ann Arbor, Mich., Aug. 17–20, 1986.

A. S. Neely IV, "FDA Inspectional Authority—Is There an Outer Limit? *Food Drug Cosmetic Law J.* **33**, 710–725 (1978).

J. H. Nelson, "Where are *Listeria* Likely to be Found in Dairy Plants?" *Dairy Food Environ. Sanit.* **10**, 344–345 (1990).

J. H. Nicholas, "Problems in the Control of Pesticide Residue on Imported Foods," *Food Drug Cosmetic Law J.* **36**, 573–595 (1981).

D. B. Norton, "The Constitutionality of Warrantless Inspections by the Food and Drug Administration," *Food Drug Cosmetic Law J.* **35**, 25–43 (1980).

Office of Regional Operations, *Analyst Operations Manual,* Food and Drug Administration, Rockville, Md., 1984.

G. E. Peck, "Historical Perspective" [on Good Manufacturing Practices], *Food Drug Cosmetic Law J.* **34**, 450–456 (1979).

J. F. Schaefer, "Inspection Policy," *Food Drug Cosmetic Law J.* **36**, 493–500 (1981).

J. L. Sharp, M. T. Ouye, W. Hart, S. Ingle, G. Hallman, W. Gould, and V. Chew, "Immersion of Florida Mangos in Hot Water as a Quarantine Treatment for Caribbean Fruit Fly (Diptera: Tephritidae)," *Journal of Economic Entomology* **82**, 186–188 (1989).

P. Sharpe, "Buddy Maedgen Is . . . Bugged!!" *Texas Agriculture,* 118–121, 172–177 (Mar. 1989).

K. A. Silver, "The Food and Drug Investigator and the Fourth Amendment," *J. Assoc. Food Drug Off.* **51**, 226–232 (1987).

O. P. Snyder, "Food Safety 2000—Applying HACCP for Food Safety Assurance in the 21st Century," *Dairy Food Environ. Sanit.* **10**, 197–204 (1990).

R. M. Spiller, "How to Handle an FDA Inspection," *Food Drug Cosmetic Law J.* **33**, 101–108 (1978).

R. S. Street, "Is Vacuum Pest Control for Real?" *Agrichem. Age* **34**, 22–23, 26 (Feb. 1990).

J. W. Swanson, "How to Handle the FDA Inspection—The Investigator's View," *Food Drug Cosmetic Law J.* **33**, 109–115 (1978).

J. RICHARD GORHAM
Food and Drug Administration
Washington, D.C.

FILTRATION. See AQUACULTURE ENGINEERING AND CONSTRUCTION; BEER; MEMBRANE FILTRATION SYSTEMS.

FISH AND SHELLFISH MICROBIOLOGY

Humans consume over 1,000 species of fish and shellfish that grow in diverse habitats and geographic regions all over the world (1). These fish and shellfish carry a variety of microorganisms from both aquatic and terrestrial sources. The high levels of moisture, rich nutrients, including free amino acids, other extractable nitrogenous compounds, digestible proteins, and psychrophiles, render seafood easily perishable, often spoiling in a short period of time even under refrigeration. In addition to spoilage microorganisms, seafood may contain various potential pathogens that can threaten the public health.

It is often difficult to maintain the quality of seafood products because there is a considerable distance between consumers and the harvesting areas, which provides opportunities for microbial growth and recontamination. To process fish and shellfish into stable products, low temperature, heat, curing, fermentation, and irradiation can be applied. This article covers the quantitative and qualitative aspects of microorganisms found in fish and shellfish and the factors affecting seafood quality. Organisms involved during the seafood processing are also described and discussed. Emphases are placed on spoilage bacteria, which cause the degradation of products and organisms that present risks to the public health.

MICROORGANISMS IN FINFISH

In healthy fish, muscle tissue or flesh is generally considered sterile. However, the fish surface and certain organs contain various levels of microorganisms: skin, 10^2–10^7/cm^2; intestinal fluid, 10^3–10^8/mL; and gill tissue, 10^3–10^6/mL (2,3). Cold marine water fish mainly carry psychrophilic gram-negative bacteria including *Moraxella*, *Acinetobacter*, *Pseudomonas*, *Flavobacterium*, and *Vibrio* (4,5). Both *Moraxella* and *Acinetobacter* were designated as *Achromobacter* in the past (6–8). The levels of these bacteria vary somewhat depending on season and food ingested (2,5,9). Fish intestines normally contain *Vibrio*, *Moraxella*, *Acinetobacter*, *Pseudomonas*, and *Aeromonas*, in addition to a small number of anaerobic bacteria, including *Clostridium* and *Bacillus*. Warm-water fish carry large numbers of gram-positive, mesophilic bacteria such as *Corynebacterium*, *Bacillus*, *Micrococcus*, and sometimes *Enterobacteriaceae* or even *Salmonella* (4,5).

The flora of fish depends on intrinsic factors (season, fish ground, and species) and extrinsic ones (fishing method, fish handling on board, storage condition, sampling technique, medium, and incubation temperature). Trawled fish usually carry bacterial loads 10–100 times greater than those of lined fish, because fish are dragged for a long time along the sea bottom (4). The physiological condition of fish prior to death has an effect on postharvest quality. When tuna, the fastest swimming fish, are captured in a highly stressful state, the buildup of lactic acid combined with elevated muscle temperature degrade the muscle quality, although the tuna is still acceptable for canning. Salmon harvested by gill netting die after an exhausting struggle, resulting in a shorter period of rigor mortis and deterioration during icing (10). The fish should be handled as soon as possible after being landed on the vessel. Careful handling of fish with gaff hook or forks and avoiding severe physical damage are crucial. Any breaks in the skin and flesh quickly introduce spoilage bacteria, which deteriorate fish quality. Fish should be carefully cleaned, not exposed to sunlight or to the drying effects of wind, and cooled down to the temperature of melting ice (0°C) as quickly as possible (11,12).

Gutted and well-cleaned fish contain fewer bacteria than whole fish. Bleeding and gutting should be done as soon as fish arrive on deck (13). However, bleeding and gutting are not helpful in all fish harvesting operations; blue fish and dogfish are not benefited by this operation (13) if in-plant processing occurs shortly after harvest.

The method of stowing fish on the vessel can have different impacts on fish quality (14,15). Boxing, commonly used in Asia, Norway, and Iceland, provides good fish quality and quicker unloading at dockside, although it results in more labor needed in handling fish on the deck and less storage capacity. Bulking, commonly used in North America, allows for quick operation and maximum storage capacity; however, the pressure often causes poorer quality and decreases shelf life (11,16).

During the transport of fish from fish ground to fish pier, bacteria grow in the fish pen or hold board at various rates depending on fish handling and storage temperature. In the bulking stow in cold weather, a slime accumulates on the bottom of the fish hold. This fish hold slime is constituted mainly of various bacteria with a level ranging from 10^9 to 10^{10}/g (17). The bacterial flora includes *Moraxella*, *Acinetobacter*, *Flavobacter*, pseudomonads, and heavily mucoid corynebacteria, which are the major organisms responsible for the slimy deposits and are critical problems in boat sanitation (17,18). Thorough sanitation of the fish hold after unloading the catch is necessary to insure high quality of raw fish on future trips.

FISH SPOILAGE AND CHANGES IN BACTERIAL FLORA OF FISH DURING COLD STORAGE

The fish's regulatory mechanisms, which prevent invasion of the tissues by bacteria, cease to function after death. Bacteria then invade the fish body through the skin, enter the body cavity and belly walls via intestines, and penetrate the gill tissue and kidney by way of the vascular system. The low molecular substances and soluble proteins yielded from fish body during autolysis after rigor mortis provide rich nutrients for bacterial growth.

Various proteases and other hydrolytic enzymes secreted by psychrophilic and psychrotrophic organisms can act on the fish muscle even at low temperatures (17,19). The factors that influence microbial contamination and growth include fish species and size, method of catch, onboard handling, fishing vessel sanitation, processing, and storage condition (12,20,21). Fish are subject to rapid microbial spoilage if fish handling and storage are inadequate. It is estimated that about 10% of the total world catch is lost due to bacterial spoilage (22). Various microorganisms involved in spoilage are listed in Table 1 in descending order of spoilage activity. Some organism cause spoilage in different degrees depending on the total

Table 1. Microorganisms Associated with Spoilage of Fresh Seafood

Spoilage Activity	Microorganism
High	*Pseudomonas* (*Alteromonas*) *putrefaciens*, *Pseudomonas* (*Alteromonas*) *fluorescens*, other fluorescent pseudomonads, and other pseudomonads
Moderate	*Moraxella*, *Acinetobacter*, and *Alcaligenes*,
Low of active only in specific conditions	*Aerobacter*, *Lactobacillus*, *Flavobacterium*, *Micrococcus*, *Bacillus*, and *Staphylococcus*

microbial flora, fish quality, handling and packaging methods, and storage temperature.

Refrigerated fresh haddock fillets contain about 10^5/g of initial bacteria, predominated by *Moraxella–Acinetobacter* and *Corynebacterium*. After storage at 1°C for 14 days the bacterial number reaches 2.1×10^8/g and seafood enters the spoilage stage. *Pseudomonas* (*Alteromonas*) *putrefaciens* and fluorescent pseudomonads are organisms responsible for the spoilage of haddock at refrigerated temperatures (23). These spoilers account for only about 1% of the total count at the beginning but increase to at least 30% at the stage of spoilage. In other words, whenever *Pseudomonas putrefaciens* and fluorescent pseudomonads reach 30% of the total bacterial count, fish spoilage will result regardless of total bacterial level. When cod is stored at 20°C *Alteromonas* and *Vibrionaceae* will cause spoilage in one day (24).

As *Pseudomonas putrefaciens*, fluorescent pseudomonads, and other potential spoilers increase rapidly in initial spoilage stage; they produce vast amounts of proteolytic and other hydrolytic enzymes (5,25). Various macromolecules of fish body are degraded. Proteins are decomposed by proteases to peptides and amino acids and then further broken down to indole, amines, acids, sulfide compounds, and ammonia (26). Lipases break down lipids to form fatty acids, glycerol, and other products. Nucleotides are decomposed into nitrogenous compounds. Many enzymatic tests can determine microbial spoilage activity in fish, including hydrogen sulfide, gelatin hydrolysis, DNase, RNase, amylase, lipase, and trimethylamine oxide reductase tests and inoculation test on fish juice or fillets (20).

MICROORGANISMS IN SHELLFISH

Shellfish is composed of crustaceans (shrimp, crabs, lobster, crawfish, etc) and mollusks (bivalves, squids, snails, etc). Shellfish normally contains more moisture, greater amount of free amino acids, and more extractable nitrogenous compounds than finfish. These biochemical characteristics facilitate bacterial growth and deteriorative reactions resulting in the rapid spoilage of shellfish (27). Many shellfish grow in estuarine, coastal waters, and aquacultural ponds near residential areas and are hence susceptible to contamination by potential pathogenic organisms. Deterioration of shellfish quality results from enzymatic action from both the tissue and the contaminating organisms. Microorganisms that spoil shellfish are similar to those responsible for finfish spoilage. However, such organisms as *Moraxella–Acinetobacter* and *Lactobacillus* are more active in shellfish than in finfish.

Crustaceans

Shrimp. Among shellfish, shrimp ranks first in value and second in quantity next to crabs (28). Immediately after shrimp death, the tissue enzymes phenolases become active oxidizing tyrosine to bluish black zones or spots at the edges of the shell segments. The dark color is produced by melanin pigments that form on the internal shell surfaces on the underlying shrimp meat. At the same time a variety of bacteria start to proliferate and the growth can be accelerated if the storage temperature is not kept low enough. Removing the heads can reduce 75% of the bacteria. Gulf of Mexico shrimp contain mainly *Moraxella–Acinetobacter*, *Bacillus*, *Micrococcus*, and *Pseudomonas* (29). Most of these bacteria produce hydrolytic enzymes: 62% proteolytic, 35% lipolytic, 18% TMA-O reductive, and 12% indole positive (29–31). Shrimp unloaded from the trawlers have an average bacterial load of 6.0×10^5/g and market shrimp, 3.2×10^6/g. Bacterial counts used for indicating shrimp quality are 1.3×10^6/g, acceptable; 4.5×10^6/g, good; 1.1×10^7/g, fair; and 1.9×10^7/g, poor (32). During iced storage for 16 days, *Moraxella–Acinetobacter* increase from 27 to 82% of the total bacterial count while *Flavobacterium* decrease from 18 to 1.5%; *Micrococcus*, from 34 to 0%; and *Pseudomonas*, from 19 to 17% (33).

There are two putrefactive types of spoilage in shrimp. One is the production of indole, presumably from tryptophan by bacterial action before icing when exposed at a temperature favorable for bacterial growth. After commencement, the decomposition proceeds fairly rapidly even under ice. Indole is heat resistant and is a reliable spoilage indicator of raw material prior to processing. The other type of ammoniacal decomposition is slow and is characterized by an odor of free ammonia (34–36). The reaction is attributed to both microbial and tissue enzymatic activities depending on storage temperature and bacterial composition.

Crabs. The dominant crabs harvested in the United States are blue crab, king crab, Dungeness crab, and tanner–snow crabs. The bacterial flora of freshly caught crabs reflect that of the growing water, season, and geographic location. The hemolymph of healthy blue crabs from Chincoteague Bay, Va., is about one-fifth sterile according to 290 freshly caught crabs tested (37). The organisms found in the hemolymph of blue crabs are *Acinetobacter*, *Aeromonas*, *Pseudomonas*, *Flavobacterium*, *Vibrio*, *Bacillus*, coliforms, and *Clostridium* (38).

Greater numbers of bacterial species are found in Dungeness crabs from Kodiak Island and the Columbia River, waters close to human habitation, whereas the least number of species are found in the tanner crab from the Bering Sea, an area far from human habitation (39). The highest levels of bacteria occur in the gills, 10^3 to 10^7/

g, as compared to 1×10^1 to 4×10^2/g in muscle tissue. Gills of Dungeness crabs from the Bering Sea contain *Moraxella, Acinetobacter, Alcaligenes, Micrococcus,* and *Staphylococcus,* whereas the muscle carries *Micrococcus* and *Staphylococcus* (39)

Crawfish. Spoilage of crawfish is caused by the potential spoilers similar to those detected in other crustaceans. In a total of 280 isolates found from spoiled crawfish, 22.1% were shown to be rapid spoilers; 16.4%, low spoilers; and 61.5%, nonspoilers (40). In the group of rapid spoilers over half were pseudomonads and less than half were *Moraxella–Acinetobacter*. Slow spoilers include *Pseudomonas, Moraxella–Acinetobacter, Alcaligenes, Flavobacterium, Aerobacter, Lactobacillus, Micrococcus,* and *Staphylococcus*. Organisms considered as nonrapid spoilers are *Aerobacter, Bacillus, Flavobacterium, Micrococcus, Sarcina,* and *Staphylococcus* (40,41). It is clear that organisms belonging to the same genus have different activities in spoilage.

Bivalves

Bivalves are mollusks, including oysters, clams, and mussels. They are soft-bodied animals that are enclosed by two rigid, bilaterally symmetrical shells. Bivalves are filter feeders and pass a large volume of water through their gills to obtain oxygen and food. Particulate matter, including microorganisms, from the water is trapped in mucus on the gills, then conveyed to the mouth, and finally to the digestive system. Bivalves, particularly oysters, ingest many microorganisms that can survive the digestive process and accumulate in the animals (42–44). The concentration of microorganisms in bivalves can be tens to hundreds of times as high as that in their growing water (42,45). Consumers and public health regulatory agencies are concerned about the pathogenic organisms found in bivalves that are affected by sewage pollution.

Fecal coliforms are generally used as indicators for bivalve quality and for domestic pollution in shellfish-growing waters (46,47). Following harvest, two microbiological guidelines are applied to determine the acceptability of shellfish meats. Bivalves at wholesale market level should have a 35°C standard plate count (SPC) of < 500,000/g and a most probable numbers (MPN) fecal coliforms of ≤ 230/100 g (46,47).

The microflora of bivalves at harvest is composed of both organisms that are symbiotic with the bivalves and organisms that are filtered from the waer and ingested as food. These microorganisms vary qualitatively and quantitatively, depending on the nutrient level, salinity, temperature, and water quality. The commensal microflora include *Cristispira pectineus,* which colonizes the crystalline style of oysters, and spirochetes such as *Saprospira,* found in the crystalline style, stomach, and intestine of eastern oysters. These commensal organisms are difficult to culture and have no pathological significance to humans (48). Bivalves at harvest normally carry a total plate count of 10^3 to 10^5/g. Soft-shell clams harvested from different growing areas in the Chesapeake Bay contain a geometric mean of SPC, 2.0×10^4–7.2×10^4/g; total coliforms, 1.5×10^3–6.3×10^3/100 g; fecal coliforms, 29–62/100 g; and *E. coli,* 14–27/100 g (49).

The common microflora of bivalves at harvest consists primarily of gram-negative rods including *Pseudomonas* and *Vibrio* species (50,51). The *Flavobacterium–Cytophaga* group occasionally exists in a certain level in oysters. Other organisms in oysters include *Acinetobacter, Corynebacterium, Moraxella, Alcaligenes, Micrococcus,* and *Bacillus*. The microflora of Gulf of Mexico's oysters is dominated by *Vibrio, Aeromonas, Moraxella,* and *Pseudomonas* (52). Low levels of yeasts such as *Rhodotorula rubra, Trichosporon, Candida,* and *Torulopsis* are also frequently encountered in eastern oysters. Several potential pathogenic strains of *Vibrionaceae* are naturally occurring in nonpolluted estuarine waters and may be encountered in bivalves. Coliforms, fecal coliforms, and *E. coli* are the most common contaminating organisms in bivalves at harvest (45,46,49).

Reducing a high microbiological load in bivalves can be accomplished by placing the animals in clean water that is free of undesirable microorganisms and under conditions in which the bivalves will actively feed. This process is called relaying and usually requires about 15 days to reach the satisfactory microbiological quality (53,54). The relaying process is often applied by transferring bivalves harvested from moderately polluted water into approved shellfish-growing water until the animals clean themselves. A second approach is called depuration in which the shellfish are maintained in tanks of clean water with controlled salinity and temperature (55). The water is often recycled through a biofilter to control water quality. The bivalves in this condition can reach the satisfactory microbiological quality within two to three days (53,55). Nevertheless, removal of viruses often does not correlate with fecal coliform elimination even if fecal indicators have a similar reduction rate as do enteric bacterial pathogens (56,57). Ultraviolet irradiation can facilitate the reduction of contaminating bacterial flora in oysters (56). Because hepatitis A may not be eliminated as readily as other enteroviruses during dupuration, hepatitis may still occur by consuming raw depurated shellfish (58).

MICROBIOLOGY AND QUALITY

Fish Products

Fish carries a variety of organisms and should be handled adequately and processed as soon as possible. Refrigeration, freezing, canning, pasteurization, salting, drying, fermentation, curing, and a combination of these methods are commonly used to process seafood into relatively stable, marketable products. Other methods such as irradiation and modified atmospheric preservation (59) have been extensively studied and proved to have potential application.

Refrigerated and Frozen Fish. Refrigeration at 5°C ceases the growth of the mesophiles and as the temperature is further lowered, psychrophiles are eliminated. During refrigeration storage a gradual killing off of the microorganisms occurs. Gram-negative asporogenous pseudomonads are cold sensitive, whereas gram-positives such as micrococci, lactobacilli, and streptococci are more resistant (60,61).

Cooling rate affects the survival of microorganisms. The maximum survival of *E. coli* is at a cooling rate of about 6°C/min, and the minimum at about 100°C/min (62). A similar minimum survival rate has been found for *Streptococcus faecalis*, *Salmonella typhimurium*, *Klebsiella aerogenes*, *Pseudomonas aeruginosa*, and *Azotobacter chroococcum*, but these organisms have an optimum survival rate varying from 7°C/min for *A. chroococcum* to 11°C/min for *P. aeruginosa* (63).

Frozen fish should be stored at or below −20°C and preferably at −30°C. During freezing the number of cells that are inactive ranges from 50 to 90% of the initial bacterial population. Gram-positive bacteria are more resistant to freeze injury than gram-negative ones. Some pathogenic organisms can also have a certain survivability at freezing temperatures (65). Spores are the most resistant microbial entity to freeze damage. Poliovirus inoculated into oysters showed a gradual decline in plaque-forming units during frozen storage at −36°C (64).

Reasons for the cryoinjury of cells are thermal shock, concentration of extracellular solutes, toxic action of concentrated intracellular solutes, dehydration, internal ice formation, and attainment of a minimum cell volume (62). Although cryoinjury results in the cell death, survival of microorganisms occurs and is greater in a supercooled environment than in a frozen one. Some *V. parahaemolyticus* cells, inoculated into oysters, sole fillets, and crabmeat, can persist at −15 or −30°C with a greater survival at −30°C although there is a sharp reduction in viability during freezing (66). Some pathogens such as *Listeria* can survive in freezing temperatures (60).

Cryoprotective agents are substances that can protect bacterial cells during freezing and thawing. Glycerol, dimethyl sulfoxide, egg white, carbohydrates, peptides, serum albumin, meat extract, milk, glutamic acid, malic acid, diethylene glycol, dextran, Tween 80, glucose, polyethylene glycol, and erythritol have cryoprotective functions probably due to reduction of damage to the cell wall and membrane (60).

Canned Fish. Most canned fish are fully processed products such as canned tuna, salmon, sardines, mackerel, fish balls, and other fish. These canned fish are commercially sterile with 12D process to destroy all pathogenic and other organisms, allowing a satisfactory shelf life at room temperature. However, some problems may arise due to the presence of heat-resistant spore formers in the underprocessed products or can leakage from improper seam closure and cross-contamination through cooling water. In oil pack, the oil may protect bacterial spores against heat resulting in nonsterile canned products (67). Flat sour spoilage may occur due to thermophiles, such as *B. stearothermophilus*, which survive processing and multiply during slow cooling and storage at high temperatures. Swollen cans are occasionally encountered when clostria such as *C. sporogenes* survive inadequate processing.

Dried, Salted, and Smoked Fish. Dehydration is an old process that reduces the water activity (A_w) of the fish products below that required for the growth of microorganisms. The process involves drying with or without other preservatives to form dried, salted, or smoked fish.

Dried salted fish, fish bits (fried shredded fish), katsuobushi (dried and smoked skipjack stick), dried shark fins, and dried mullet roe are popular in Asia (68). Smoked salmon, herring, dogfish, and other fish are common fishery products in Europe. Growth of halophilic bacteria or molds may occur, resulting in the spoilage of salted fish. *Aspergillus* and *Penicillium* are major species associated with the color deterioration of salted round herring (68). Halophilic bacteria such as *Halobacterium* and *Halococcus* commonly present in solar salts are most troublesome during the salting and drying process. These bacteria cause red and pink discoloration and induce softness in salted fish.

Surimi-Based Products. Surimi is a mechanically deboned, water-washed frozen fish paste containing cryoprotectants. This high-protein, gel-forming material can be chopped and then mixed with salt, starch, and flavor compounds. The surimi mix is colored, textured, and cooked in two stages to set the gel to process into seafood analogues such as imitation crab, shrimp, lobster, or scallop. Freshly processed crab leg and flaked crab leg analogues contain only 10^2–10^3/g bacteria. During storage bacteria grow and SPC reaches 2×10^8/g at 10°C in 25 days and 10^4–10^6/g at 5°C in six weeks (69). SPC in flaked crab leg increases rapidly to 10^9/g after two and four weeks at 5 and 0°C, respectively. Spoilage and quality deterioration of crab analogues are indicated by number of bacteria (10^7/g), visible slime, odor, and appearance. Slime formation, softened texture, sour odor, and discoloration are consequences of spoilage. *Bacillus* is predominant initially but *Pseudomonas* gradually grows and finally outnumbers other genera at two weeks of storage at 0–5°C. *Bacillus*, which is possibly derived from the ingredient starch, is the major organism throughout the six-day storage at 15°C (69). The spoilage of other fish cake products such as kamaboko can be attributed to *Streptococcus*, *Leuconostoc*, and *Micrococcus* (70,71).

Shellfish Products

After being harvested, shellfish should be kept refrigerated or at low temperature and processed as soon as possible. Shrimp should be beheaded, peeled, or left unshelled and frozen.

Crabmeat. Blue crabs rank first in U.S. crab landings, and their major products are fresh and pasteurized meat. Other crabs caught are king, Dungeness, and tanner-snow crabs from which frozen section, claws and meat, and canned meat are commonly made (72,73).

To process blue crabmeat, live crabs are steam cooked at 121°C for 10 min. Cooked crabs are refrigerated overnight and the meat is removed by hand or machine. The meat is packed in plastic cups for fresh crabmeat or sealed in tin cans to process for pasteurized crabmeat (74). Three kinds of meat are available: lump meat taken from the back fins, claw meat extracted from claws, and regular meat collected from main body (72). In good plant sanitation, fresh crabmeat usually has a geometric mean SPC of 1.5×10^4–4.5×10^4/g, which increases to 1.4×10^5–3.2×10^5/g under poor plant sanitation (75). Cooked crabs should be stored in refrigeration ($< 2°C$), separated from the live crabs. Refrigerated cooked crabs before picking usually contain bacteria of $< 10^4$/g while cooked sponge crabs (gravid females carrying an egg mass) taken from the picking table contain bacterial levels as high as 10^6/g of whole crabs (76,77). Cooked sponge crabs have consistently been found to harbor greater numbers of bacteria than crabs without a sponge (77). Similarly, cooked green crabs, blue crabs that have recently molted and contain a higher level of water than fat crabs, carry higher levels of bacteria than normal crabs into the picking room. They contain higher moisture, which encourages bacterial growth during overnight refrigerated storage (76).

The commercial machine Quik Pik, which mechanically removes the body meat of blue crabs was started on a trial basis in 1978 and now operates successfully in Maryland under proper procedures (76,78). One quick Pik machine can pick 150 lb meat per hour, a rate equal to the work capacity of 30 hand pickers. The cooked crabs are placed in a round, rotating slotted cage to remove legs, fins, and claws, which are dropped through the slots. The crabs then pass through the debacking machine and cleaning device. The cores are loaded on racks for steam heating and then placed in the quik Pik shaker, which vigorously vibrates at 70 oscillations/s for 4 s. All the meat from cores will fall on the collecting belt for further inspection for broken shells and packaging (76).

Machine picking requires constant attention to cleanliness and sanitation to produce a meat product with a satisfactory bacterial quality. The machine should be disassembled and thoroughly cleaned and sanitized at the end of the day. Liquid on the machine during operation creates an aerosol that can greatly contaminate the meat (79). The meat conveyor belt needs continual washing with a tap water spray and sanitation in a chlorine (> 200 ppm) bath. A comparison of the bacterial levels indicates that both the SPC and coagulase positive *S. aureus* for machine-picked meat are lower than for the hand product. However, a higher *E. coli* count is found in machine-picked meat than in hand-picked meat. This is not surprising because machine picking is processed under wet and warm environment whereas hand picking is operated in cool and dry conditions.

The normal shelf life of fresh crabmeat is 7–10 days and may last up to 14 days if meat with a low initial bacterial count (80) is stored under optimum refrigeration temperature. To extend the shelf life of crabmeat, the pasteurization process of holding crabmeat for 1 min at 171°–210°F, depending on the desired shelf life from 1 to 12 months, was patented (81). A process of 185°F for 1 min in the center of a 1-lb can (401 × 301) was found to sufficiently reduce an inoculated 10^8/100 g of *C. botulinum* type E spores to < 6/100 g and to keep the meat nontoxic for 6 months at 40°F (82,83). A recommendation has been made to increase the time at 185°F to 3 min to provide for 12D cook based on the thermal death time studies of type E *C. botulinum* (84–86). Table 2 shows the thermal resistance characteristics of *C. botulinum* type E. For a complete process, an F-value of 31 based on an F 16/185 value and a cooling meat temperature of 55°F within 180 min of heat process following storage at 35°F have been recommended (87).

If storage temperature of pasteurized crabmeat rises above 38°F, surviving bacteria may grow and bacterial spores may germinate leading to spoilage. Any leakage due to can defects can introduce psychrotrophic bacteria from cooling water and result in spoilage during refrigerated storage (89). Poor meat quality with a high level of initial bacteria will increase the chance of survival of potential spoilage bacteria. The spoiled pasteurized crabmeat usually contain both aerobic and anaerobic bacteria ranging from 4.0×10^2 to 5.0×10^8/g. Spoiled crabmeat prepared from machine picking has more anaerobic organisms than hand-picked crabmeat (79).

Bivalves. Heat shock has been attempted to facilitate oyster shucking by quickly passing the stocks through a steam tunnel to slightly open the shells. Although this process increases the shucking rate, disadvantages are yield reduction and high risk of bacterial contamination (90). Canning bivalves serves as a long-term preservation method. When recommended processing times and temperatures are followed, few microbiological problems are encountered (88). However, loss of texture and economical infeasibility are problems with this processing.

Pasteurization of oysters at 72–74°C for 8 min in a flexible pouch results in a relatively stable product with organoleptic properties similar to raw oysters. The products stored at 0.5°C for three months have a low level of

Table 2. Decimal Reduction Time (D_{10}) for Heating *C. botulinum* Type E Spores

Strain	Heating Medium	Temperature Range (°C)	D_{10}-Value Minimum at 82.2°C	Z-Value (°C)	Reference
Beluga	Blue crab meat	73.9–85.0	0.75	6.5	85
Beluga	Crabmeat	73.9–85.0	0.84	6.5	85
Sarotoga	Crabmeat	80.0–82.2	1.90	6.3	87
Alaska	Whitefish, Chubs	73.9–85.0	2.21	7.6	88
Sarotoga	Sardines in tomato sauce	76.7–82.2	2.9	6.3	87
Sarotoga	Tuna in oil	72.2–82.2	6.6	6.1	87
Strain 202	Crabmeat	76.7–85.0	1.16	6.38	85

both aerobic and anaerobic bacteria (91). Immediately after pasteurization, all the surviving bacteria in the oysters are *Bacillus* sp. At storage of five months *Bacillus* continues to dominate the aerobic bacteria while *Clostridium, Corynebacterium, Listeria, Peptostreptococcus,* and *Staphylococus* constitute the facultatively anaerobic bacteria (92). Combining safe preservatives with the pasteurization can provide a safe and longer shelf life of oysters.

SEAFOOD IRRADIATION

Irradiation is a process extending seafood shelf life by exposing seafood to a certain level of ionizing radiation. Studies on the utilization of ionization radiation in food processing began in the early 1940s, and an extensive program implementing this process was launched in the early 1950s, in the United States (93,94). Ionization radiations are a group of corpuscular and electromagnetic radiations of extremely short wavelengths that can cause ejection of electrons from atoms or molecules. Only radiation with strong penetration, including gamma rays, x-rays, and electron beams, is useful in food preservation. Gamma rays from radioactive isotopes such as Co-60 have been the most acceptable radiation source due to its availability, price, and properties. One rad, a unit of irradiation dosage, is equal to 100 ergs of radiation energy absorbed per gram of substrate. A megarad (Mrad) is a million rads, and a kilorad (krad) is a thousand rads.

There are several advantages in using ionization radiation for seafood preservation. The extension of shelf life of seafood by using low-dosage irradiation allows the expansion of markets farther inland where fresh seafood is otherwise unavailable. The long shelf life of irradiated seafood can adjust the seasonal production providing a year-round supply and unfluctuating prices. Irradiation of high-quality catches produces top-quality seafood thus reducing the need to discard or process deteriorating fish to fish meal (94,95). Ionization radiation directly affects living cells by causing breaks in DNA and indirectly affects other cell components. The presence of water increases the degree of the DNA damage. Complex cells are more sensitive to irradiation than simple ones. The death or injury of microorganisms is due to the direct hit on DNA and the indirect effects from ionization and diffusion of free radicals and peroxides produced around the cells.

Food irradiation can be classified into three classes depending on the level of microorganisms to be destroyed (60). Radappertization or Radiation sterilization is the use of high dose with 12D usually more than 2 Mrad for complete destruction of all or practically all of the organisms. Radurization or radiation pasteurization is using low dose of radiation to destroy a sufficient number of organisms to enhance the shelf life of foods. This process usually uses 100–1,000 krad to destroy 90–99% of the organisms. Radicidation is a low-level irradiation treatment that kills nonspore-forming pathogens to reduce or eliminate the food-borne illness problem. This kind of irradiation treatment normally uses 400–600 krad dose for destruction of salmonellae from poultry and red meat or feed but not for spores of *C. botulisum* or *C. perfringens*.

In seafood irradiation, radiation resistance is expressed as the decimal reduction dose (D_{10}), the dose (krad) required for a 90% or 1-log reduction in bacterial count. The radiation resistance of microorganisms varies, depending on strains and species and the D_{10}-values of various organisms are shown in Table 3. The resistance in decending order is yeast, *Micrococcus, Moraxella–Acinetobacter, Flavobacterium,* and *Pseudomonas* (60). The gram-positive bacteria are more resistant than gram-negative ones. Coccoid forms are more radioresistant than rod-shaped cells. Bacterial spores exhibit the most resistance to radiation with the exception of some gram-positive cocci. Compared to bacterial spores, viruses are more or equally resistant to irradiation. Although the amount of radiation needed to inactivate microbial toxins is similar to that for bacterial spores, the radiation level for inactivation of botulinum toxin is much higher than that needed to kill the bacterial cells (98) (Table 3).

In general, D_{10}-values for bacterial vegetable cells (excluding cocci), bacterial spores, *M. radiodurans,* viruses, and bacterial toxins are 20–100, 150–450, 200–600, 400–800, and 200–2,000 krad, respectively (99–104).

Low-dose irradiation treatment of seafoods usually does not induce detectable changes in flavor, texture, and appearance with the exception of a slight loss of flavor in a few products (105,106). The level of irradiation is one factor in determining the type of microorganisms remaining that cause spoilage of irradiated seafood. The spoilage microflora of irradiated fish after a low dose of ≤ 100–150 krad treatment consists mainly of pseudomonads. At dose levels higher than 100 krad, *Maraxella–Acinetobacter* represented the dominant group of organisms in fish spoilage. Irradiation of packaged fish favors the growth of lactic acid bacteria that will be the major spoilers. In general, nonirradiated fish spoils at an aerobic plate count (APC) of 10^6/g whereas low-dose irradiated fish spoils at about 10^8/g (94).

Under the optimum dose of 100–300 krad, fish and shellfish can be maintained three to seven weeks at refrigeration without altering the fresh-product characteristics (Table 4). Compared with the regular 10- to 14-day shelf life of the unirradiated products, low-dose irradiated products have a shelf life two to three times longer (94,109). To insure efficiency and successful processing, the seafood should initially be of high quality and irradiated products should be maintained as near 0°C as possible without freezing.

The safety of irradiated seafood must be evaluated based on the absence of microorganisms and microbial toxins harmful to man, the nutritional contribution of the product, and the absence of any significant amount of toxic compounds formed in the irradiated products (101). Low-dose irradiated seafood, particularly lean fish species, have been found to pose no potential health hazards (94,110). Despite the approval of over 35 countries irradiation in foods and the USDA clearance of low-dose application for poultry irradiation, seafood irradiation has not been permitted. This process can become a successful technique for seafood preservation once it is approved in the United States.

Table 3. Decimal Reduction Doses (D_{10}-Values) for Irradiation of Some Microorganisms and Toxins

Organism or Toxin	Suspending Medium	D_{10} (krad)	Reference
Vibrio parahaemolyticus	Crabmeat	5–12	96
Proteus vulgaris	Oysters	20	95
	Crabmeat	10	95
Aeromonas hydrophila	Blue fish	14–22	96
Escherichia coli	Soft-shell clams	39–42	96
	Mussel	41–48	96
	Oysters	35	95
	Crabmeat	14	95
Salmonella typhimurium	Soft-shell clams	60	96
	Mussel	58–63	96
S. paratyphi A	Oysters	75	95
	Crabmeat	50	95
	Shrimp	85	95
S. paratyphi B	Shrimp	61	97
Shigella flexneri	Mussel	24–34	96
Shigella sonnei	Oysters	27	95
Staphylococcus aureus	Oysters	150	95
	Crabmeat	80	95
	Shrimp	190	95
Adenovirus, echovirus	Eagles medium	410–490	60
Toxins, botulinum type E			
Washed cells	Buffer	1,700–2,100	60
Purified	Buffer	40	60
Staphylococcal			
Enterotoxin B purified	Buffer	2,700	60

SEAFOOD MICROORGANISMS OF PUBLIC HEALTH SIGNIFICANCE

Seafood, particularly shellfish, may contain a variety of pathogenic microorganisms that impose a threat to the consumers' health. These potential pathogens include both indigenous organisms and contaminating organisms. Pathogens may contaminate the seafood after harvest or during processing. Some indigenous pathogens found are Vibrio, Clostridium botulinum type E, Aeromonas, and poisonous phytoplankton such as dinoflagellates (111,112). Extraneous pathogens include Salmonella, Shigella, Listeria, Campylobacter, Staphylococcus aureus, E. coli, Bacillus cereus, Hepatitis A, and Norwalk virus (64,112).

The consumption of bivalves is of greatest concern to the public due to several factors. Bivalves, especially oysters, are frequently consumed raw. The whole animal is consumed rather than just the muscle tissue, as in the case of other raw crustaceans and raw fish. Bivalves are filter feeders that can trap microorganisms from the growing water into their bodies. Because the animals grow in

Table 4. Shelf Life at 0°C of Selected Seafood After Optimal Dose Irradiation Treatment[a]

Seafood	Shelf Life (weeks)	Optimal Dose (Mrad)	Reference
Haddock fillets	3–5	0.15–0.25	107
Cod fillets	4–7	0.10	108
Ocean perch fillets	4	0.15–0.25	107
Mackerel fillets	4–5	0.25	107
English sole fillets	4–5	0.20–0.30	107
Smoked chub	6	0.10	107
Petrale sole fillets	2–3	0.20	107
Shrimp	4	0.15	107
Lobster meat (cooked)	4	0.15	107
King crab meat (cooked)	4–6	0.20	107
Dungeness crab meat (cooked)	3–6	0.20	107
Oyster meat	3–4	0.20	107

[a] Fish or shellfish were packed aerobically in hermetically sealed cans or oxygen-impermeable plastic bags.

near-shore estuarine waters that may encounter sewage pollution, bivalves may become a vector for disease transmission. Based on past records of bivalve shellfish borne disease between 1900 and 1986, there were 12,376 cases of food-borne illnesses in the United States excluding cases of paralytic shellfish poisoning (112). This has been estimated to be only 5–10% of the actual number of cases. These documented cases include 43% gastroenteritis, 26% typhoid, 11% infectious hepatitis, 11% Norwalk virus, and 2% Vibrio. The specific etiologic agents have been identified as *Salmonella typhi, Vibrio cholera* 01, *V. cholera* non-01, *V. vulnificus, V. parahaemolyticus, V. fluvialis, V. mimicus, V. hollisae, V. furnissii, E. coli, Salmonella* sp., *Shigella* sp. *Bacillus cereus, Staphylococcus* sp., *Campylobacter* sp., *Aeromonas hydrophila, Pleisomonas shigelloides*, hepatitis A, Norwalk virus, and other viral agents (112–115).

Food poisoning outbreaks reportedly have often been due to the consumption of seafood. The most common agent involved is *V. parahaemolyticus* (114,116). This species was first isolated in 1950 in Japan from seafood (117,118). Since then it has been implicated in more than 1,000 outbreaks per year in Japan and accounts for over 50% of that country's bacterial food poisoning. *V. parahaemolyticus* strains require a low concentration of salt for growth, but some cultures isolated from fresh water have been reported (119–124). Raw seafood is the major vehicle for the organism in Japan; cooked seafoods that have been recontaminated are the source of implication in the United States (117). Kanagawa reaction is commonly used to differentiate between virulent and avirulent isolates by testing whether a strain can produce a heat-stable hemolysin to lysis a blood agar containing 7% NaCl and mannitol. Cultures isolated from stools of patients are always Kanagawa positive (99%). On the contrary only about 1% of isolates from waters or seafoods is Kanagawa positive (125,126). The mechanism for this discrepancy is still unknown. One of the hypotheses is that there may be a transformation of Kanagawa-positive strains on passage through the intestines. There is a possible competitive advantage for Kanagawa positive strains to proliferate more readily in the intestines. The current method of isolating and identifying Kanagawa-positive strains in seafoods and waters may also leave some undetected (116,126).

The infective dose of *V. parahaemolyticus* for humans ranges from 10^5 to 10^7 viable cells, and a decrease in stomach acidity may lower the infective dose. The incubation period for symptoms is 5–90 h and the duration of the illness is normally 2–10 days. The frequencies of symptoms are diarrhea, 98%; abdominal cramps, 82%; nausea, 71%; vomiting, 52%; headache and fever, 27%; and chills, 24% (116,126). The organism has three biologically active hemolysins, substances that can lyse the animal's blood cells: a heat-stable peptide with 45,000 mol. wt., a heat-labile hemolysin, and a phospholipase (127,128). *V. parahaemolyticus* are heat sensitive and most food-poisoning cases result from cross-contamination or poor sanitation (129–131).

V. cholerae causes a gastrointestinal illness in humans called cholera (132). This species is usually divided into two groups, serotype 01 and non-01; both are found in aquatic environment. The 01 serotype contains two biotypes: classical and El Tor. The classical biotype prevailed worldwide until the 1960s and the El Tor biotype presently predominates, including in the United States (111,133,134). Cholera in the United States is relatively rare but occasional outbreaks occur in southern states such as Texas, Louisiana, and Florida due to consumption of raw or partially cooked molluscs, cooked crabs, and other shellfish (133,134). The infective dose is estimated to be about 10^6 cells. Taking antiacids or medication to lower gastric acidity will lower the infective dose (115). The incubation time varies from 6 h to 5 days. Severe symptoms include profuse watery diarrhea, dehydration, and death in the absence of prompt treatment. In the beginning, the stool is brown with fecal material and quickly acquires the classic rice-water appearance. Enormous amounts of fluid leave the body resulting in dehydration and difficulties in circulation The stool is high in potassium and bicarbonate (111,133). Besides severe diarrhea, victims suffer from thirst, leg cramps, weakness, hoarse speech, and rapid pulse (132,133). In emergency treatment, prompt replacement of fluid and electrolyte losses by intravenous injection is often used. After an initial recovery, oral intake of glucose and electrolytes can improve the condition. In general, *V. cholerae* is sensitive to heat and cold, but it can survive at low temperatures for a certain time (135).

Non-01 cholera is a group of nonagglutinable (NAG) cholerae, which exist naturally in estuarine and coastal waters and also in rivers and brackish waters (128). Only about 5% of this group from seafood and patients isolates in the United States produce cholera toxin. However, the nontoxigenic strains cause gastrointestinal illness with principal symptoms of abdominal cramps, fever, bloody stools, nausea, and vomiting (133). Some human isolates of this group in the United States are from extraintestinal sources including wound infection, ear infection, and primary and secondary septicemia (133). Eating raw oysters has been found to be the major cause of most non-01 *V. cholerae* infection cases in the United States. Other seafood, egg and asparagus salad, or potatoes can occasionally be a vehicle for these organisms (134).

V. vulnificus is widely distributed in the environment and has been found in most U.S. estuarine and coastal waters (136–138). This water and seafood organism is found most frequently at water temperatures $> 20°C$ and low salinity of 0.5–1.6%. Environmental isolates are phenotypically identical to clinical isolates. Some strains show bioluminescence and may also be pathogenic (138). This organism can cause illness and infection through the consumption of contaminated raw or undercooked seafood, particularly mollusks such as oysters and clams. The incubation period for this illness is 16–48 h after ingestion. Symptoms include weakness, chills, fever, hypotension, and fatigue with occasional vomiting and diarrhea. Infection occurs, progresses rapidly, and may cause death in 40–60% of patients (128). Patients may have greater risk of infection if they have skin cuts or suffer chronic liver disease, gastric disease, or hemochromatosis (111,128).

Other potential *Vibrio* pathogens such as *V. hollisae, V. mimicus* and *V. furnissi* have been implicated in seafood-borne illnesses (111). Vehicles for these organisms are shellfish including oysters, clams, shrimp, and crawfish. Common symptoms are diarrhea, nausea, vomiting, and abdominal cramps. Both toxigenic and nontoxigenic strains of *V. mimicus* have been isolated but food-poisoning cases have mostly occurred with nontoxogenic ones. Some strains of these organisms have been newly isolated in the last few years, and more information will be available after further studies (111).

Clostridium botulinum type E is another intrinsic pathogen in seafood causing botulism type E intoxication. Based on the serological classification of the neurotoxin, *C. botulinum* is composed of eight types: A, B, C_1, C_2, D, E, F, and G (139). These types are divided into four groups according to proteolytic activity (140,141). Groups I and II are of particular importance for causing botulism in humans. Group I includes type A and proteolytic strains of types B and F. Group II contains all type E and nonproteolytic strains of types B and F. Type E is distributed in sediments of marshes, lakes, and coastal ocean waters and is common in the intestines of fish and shellfish. The organism does not proliferate in living fish but it may multiply in bottom deposits and in aquatic vegetation when the growth of algae reduces the oxygen level of the water to conditions suitable for *C. botulinum*. Type E can grow and produce toxin at 4°C and is heat sensitive (142,143). It is inhibited by water activity of < 0.975 (5% NaCl) and pH < 5.3 (144). Most outbreaks of botulism associated with fishery products have been with semipreserved products. Smoked, salted, fermented, and canned products are eaten without further cooking and can involve the risk of botulism due to inadequate processing (89,143). Satisfactory heat processing is critical in assuming quality safety of pasteurized crabmeat described above.

Plesiomonas shigelloides is found in fresh surface water and possibly in seawater and is more often isolated in the summer months. This organism has been implicated in human gastroenteritis for 40 yr (145). Seafood that may carry this organism are cuttlefish, raw oysters, salt mackerel, and undercooked oysters. Incubation time for symptoms is one to two days after ingestion of the food. Symptoms include diarrhea, abdominal pain, nausea, chills, fever, vomiting, and headache (145).

Enteric viruses are those animal viruses that are excreted in feces and discharged into domestic sewage. These viruses are more resistant than enteric bacteria to sewage treatment process and various environmental stresses. Several human enteric viruses may be implicated in the viral disease by seafood consumption, such as hepatitis, fever, diarrhea, paralysis, meningitis, and myocarditis. Shellfish may play a significant role as vectors in the transmissions of viral diseases. A study of cockles-borne disease indicated that about one-quarter of the hepatitis A cases in the southeast UK could be caused by shellfish consumption (146,147). In Frankfurt, Germany, about one-fifth of infectious hepatitis was associated with the consumption of contaminated oysters (148). Molluskan shellfish is the major seafood involved in the outbreak of hepatitis A incidence (58) and also attributes to 12.5% of non-A, non-B hepatitis in the United States (114). Outbreaks of Norwalk virus gastroenteritis by shellfish ingestion must be documented in Australia and the United States (97).

According to the Centers of Disease Control annual documentation, nearly 90% of all reported outbreaks of seafood-borne illness are associated with consumption of molluskan shellfish and a very few species of fish (148,149). In geographic distribution, Hawaii has the most cases accounting for 35% of all these seafood-borne illnesses. About half of all these seafood-borne illness cases are found in four states and territories: Hawaii, Puerto Rico, Virgin Islands, and Guam. Another third of these seafood-related illnesses occur in the continental states of New York, California, Washington, Connecticut, and Florida (149).

Chemical intoxication resulting from the harmful metabolites of microorganisms are often implicated in seafood. An allergic food poisoning after eating deteriorated scombroid fish is caused by formation of histamine by microbial enzyme histidine decarboxylase when fish have been exposed to environments favorable to bacterial growth. The organisms involved in histamine poisoning include *Pseudomonas putrefaciens*, *Aeromonas hydrophila*, *Proteus vulgaris*, *Clostridium perfringens*, *Enterobacter aerogenes*, and *Vibrio alginolyticus* (150).

MICROBIOLOGICAL CRITERIA AND INSPECTION

Microbiological criteria are an important but controversial issue in establishing a regulatory food-control program, particularly for seafoods in the United States. The purposes of microbiological criteria are to insure food safety, adhere to good manufacturing practice, and provide established measure for food inspection and marketing. According to the degree of compliance, microbiological criteria include three categories: standard, guideline, and specification (151). A standard is part of a law or ordinance and is a mandatory criterion. A guideline is a criterion for assessing microbiological quality during processing, distribution, and marketing of foods. A specification is used in purchase agreements between buyers and vendors (152,153).

As more countries are recognizing the need to assess the safety and quality of foods due to food-poisoning outbreaks and international trade disputes, several agencies have formed organizations to formulate microbiological criteria. The International Committee on Microbiological Specifications for Foods (ICMSF) was established by the International Association of Microbiological Society (IAMS) in 1962 for this and related purposes (154). The committee assists in establishing microbiological standards implementing method of examination, promoting safe movement of foods in international trade, and mediating disputes caused by disparate criteria. The committee also publishes books of its reports and recommendations (153–155). Many of the ICMSF recommendations were incorporated into the work of Code Alimentarius Commission to implement the Joint FAO/WHO Food Standard Program of the United Nations World Health Organization. The goal of the Joint Food Standard Program is to protect the health of consumers, insure fair practices in the food trade, guide the preparation of draft standards and codes of practice, and promote the incorporation of all food standards by different governments. The program has also published recommended international codes of practice for many common seafood products such as shrimps or prawns and frozen fish (156,157).

To ensure consumer safety, inspection systems are enforced by the regulatory agencies in most of the countries. In the United States, the inspection program is carried out by different levels: local, state, and federal government. The federal regulatory seafood control systems include the Food and Drug Administration (FDA), U.S. Department of Agriculture (USDA), U.S. Department of Commerce (USDC), U.S. Department of Defense (USDD), and the Environmental Protection Agency (EPA). The FDA is in charge of imported seafood inspection, a mandatory surveillance system. The National Marine Fisheries Service of USDC operates the voluntary Fishery Product Inspection Program mainly for domestic seafood industry on a fee-for-service basis. The USDA and other federal agencies are responsible for some inspection systems related to their domain of authority. Much of the domestic seafood products are inspected by the states and municipalities in their respective territories. In recent years, the public concern in seafood safety and confusion among various inspection systems have led to the proposal of a national seafood surveillance system. At the time of this writing the USDA will likely be chosen to head this mandatory seafood inspection program.

Microbiological criteria have been controversial and confusing because of the variation in agency objectives, legislative authorities, unsatisfactory definitions, questions in sampling procedures, methodology, quality requirements, and uninsured consumer protection. In recent years a hazard analysis critical control point (HACCP) approach has been regarded as an effective, practical, and economic program in microbiological monitoring and safety of the products (152,158). The industry should develop the actual details of a HACCP program and insure the implementation of the complete process. In conducting hazard analysis of the products each step of the processing operation should be defined to specify the hazards associated with each step. The preventative measures at each processing step will be identified for elimination of the hazards. The critical control points are thus defined, bringing the hazards under control. The theory behind the HACCP concept, if properly implemented, is to reduce the number of governmental inspections, and overcome many of the weaknesses inherent in traditional inspection schemes (158). When plant personnel have been well trained in HACCP concepts, the quality and safety of products can be enhanced with efficient cost inspection.

Seafood microbiology is a difficult subject because of the many obstacles in this area. A broad spectrum of microorganisms is involved, including indigenous microbial flora organisms that contaminate the catch after it is landed and those that multiply during the process from the water to the dining table. The ambiguous taxonomy of many seafood microorganisms is confusing and impedes the study of these organisms. The problems with growth conditions, such as inadequate growth temperature and media composition, often result in disputable results of seafood sample analysis. The large number of nonculturable viable cells in marine bacteria greatly limits the study of the role of these organisms in the natural aquatic ecosystem and the relation to fish and shellfish.

The fast growth of psychrophilic and psychrotrophic bacteria cause fish and shellfish to spoil rapidly even in refrigeration. In addition to spoilers the potential pathogenic microorganisms present in seafood can proliferate expeditiously and threaten consumer safety. Prompt seafood handling and adequate processing are critical in the control of both spoilage and pathogenic organisms to provide safe and stable products. Good quality control and inspection, particularly the HACCP program, can insure a high quality of seafood products.

Application of modern basic microbiology, biochemistry, and biotechnology can advance fish and shellfish microbiology. DNA probe, immunoassay, monoclonal antibody, and other rapid-detection methods provide accurate and expeditious procedures for determination of seafood pathogens. The study of bacterial cell envelope, a new field that has arisen in the last two decades, can provide information about the functions and physiochemical properties of cell envelope components, which are important in understanding bacterial survival and interaction with other organisms in the aquatic ecosystem. This field of knowledge will also be helpful in establishing new technology to control bacterial growth during processing. Study of nonculturable viable or dormant cells may lead to a new field of study: bacterial survival in the natural environment and its impact on seafood. This can provide another avenue in exploring the key factors behind many nondetectable food pathogens. The progress in seafood microbiology will greatly benefit the seafood industry, providing safe, better quality seafoods.

BIBLIOGRAPHY

1. R. E. Martin, W. H. Doyle, and J. R. Brooker, "Toward an Improved Seafood Nomenclature System," *Mar. Fish. Rev.* **45**(7–9), 1–20 (1983).
2. D. C. Georgala, "The Bacterial Flora of the Skin of North Sea Cod," *Journal of General Microbiology* **18**, 84–91 (1958).
3. J. Liston, "Qualitative Variations in the Bacterial Flora of Flatfish," *Journal of General Microbiology* **15**, 305–314 (1956).
4. J. M. Shewan, "Some Bacteriological Aspects of Handling, Processing and Distribution of Fish," *J Roy. Sanit. Inst.* **59**, 394–421 (1949).
5. J. M. Shewan, "The Microbiology of Sea-Water Fish," in G. Borgstrom, ed., *Fish as Food*, Vol. 1, Academic Press, Inc. Orlando, Fla., 1961.
6. P. Baumann, M. Doudoroff, and R. Y. Stanier, "A Study of the *Moraxella* Group. I. Genus *Moraxella* and the *Neisseria catarrhalis* Group," *Journal of Bacteriology* **95**, 58–73 (1968).
7. P. Baumann, M. Doudoroff, and R. Y. Stanier, "A Study of the *Moraxella* Group. II. Oxidative-Negative Species (Genus *Acinetobacter*)," *Journal of Bacteriology* **95**, 1520–1541 (1968).
8. G. L. Gilardi, "Morphological and Biochemical Differentiation of *Achromobacter* and *Moraxella* (DeBord's Tribe Mineae)," *Applied Microbiology* **16**, 33–38 (1968).
9. J. Liston, "A Quantitative and Qualitative Study of the Bacterial Flora of Skate and Lemon Sole Trawled in the North Sea," Ph.D. dissertation, Aberdeen University, Aberdeen, UK, 1955.
10. B. R. Botta, B. E. Squires, and J. Johnson, "Effect of Bleed-

ing/Gutting Procedures on the Sensory Quality of Fresh Raw Atlantic Cod (*Gadus morhua*)," *Canadian Institute of Food Technology Journal* **19**, 186 (1986).

11. R. M. Brian and D. R. Ward, "Microbiology of Finfish and Finfish Processing," in D. R. Ward and C. R. Hackney, eds., *Microbiology of Marine Food Products*, Van Nostrand Reinhold, Co., Inc., New York, in press.

12. D. R. Ward and N. J. Baj, "Factors Affecting Microbiological Quality of Seafoods," *Food Technology* **42**, 85–89 (1988).

13. W. K. Rodman, "On Board Fish Handling Systems for Offshore Wetfish Trawlers, Work Smarter Not Harder," in W. T. Otwell, compiler, *Proceedings of the First Joint Conference of the Atlantic Fisheries Technology Society*, Florida Sea Grant, University of Florida, 1987.

14. J. J. Connell, *Control of Fish Quality*, Fishing News Books Ltd., Surrey, UK, 1975.

15. R. Nichelson II, *Seafood Quality Control—Boats and Fish Houses: Processing Plant*, Marine Advisory Bulletins, Texas A & M University, Texas, 1972.

16. J. A. Dassow, "Handling Fresh Fish," in M. E. Stansoy, ed., *Industrial Fishery Technology*, Robert Krieger, New York, 1976.

17. T. Chai, "Studies on the Bacterial Flora of Fish Pen Slime," M.S. Thesis, University of Massachusetts, Amherst, 1970.

18. T. Chai, "Usefulness of Electrophoretic Pattern of Cell Envelope Protein as a Taxonomic Tool for Fish Hold Slime *Moraxella* Species," *Applied Environmental Microbiology* **42**, 351–356 (1981).

19. V. Venugopal, "Extracellular Proteases of Contaminant Bacteria in Fish Spoilage: A Review," *J. Food Prot.* **53**, 341–350 (1990).

20. T. Chai, *Fishery Bacteriology*, Joint Commission Rural Reconstruction, Taipei, Taiwan, 1979.

21. H. C. Chen and T. Chai, "Microflora of Drainage from Ice in Fishing Vessel Fish Holds," *Applied Environmental Microbiology* **43**, 1360–1365 (1982).

22. D. G. James, "The Prospects for Fish for the Undernourished Food and Nutrition," *FAO* **12**, 20–27 (1986).

23. T. Chai and co-workers, "Detection and Incidence of Specific Species of Spoilage Bacteria in Fish. II. Relative Incidence of *Pseudomonas putrefaciens* and Fluorescent Pseudomonads on Haddock Fillets," *Applied Microbiology* **16**, 1738–1741 (1968).

24. E. M. Ravesi, J. J. Licciardello, and L. D. Racicot, "Ozone Treatments of Fresh Atlantic Cod, *Gadus morhua*," *Mar. Fish. Ref.* **49**, 37–42 (1987).

25. J. M. Shewan, G. Hobbs, and W. Hodgkiss, "A Determination Scheme for the Identification of Certain Genera of Gram-Negative Bacteria, with Special Reference to the *Pseudomonadeceae*," *Journal of Applied Bacteriology* **23**, 379–390 (1960).

26. J. Liston, "Microbiology in Fishery Science," in J. J. Connel, ed., *Advances in Fish Science and Technology*, Fishing News Books Ltd., Surrey, UK, 1980.

27. E. A. Fieger and A. F. Novak, "Microbiology of Shellfish Deterioration," in Ref. 5.

28. National Marine Fisheries Service, *Fisheries of the United States 1988*, NMFS, U.S. Department of Commerce, Washington, D.C., 1989.

29. O. B. Williams and H. B. Rees, Jr., "The Bacteriology of Gulf Coast Shrimp. III. The Intestinal Flora," *Texas J. Sci.* **4**, 55–58 (1952).

30. O. B. Williams, "Microbiological Examination of Shrimp," *J. Milk and Food Technol.* **12**, 109–110 (1949).

31. O. B. Williams and co-workers, "The Bacteriology of Gulf Coast Shrimp. II. Qualitative Observations on the External Flora," *Texas J. Sci.* **4**, 53–54 (1952).

32. M. Green, "Bacteriology of Shrimp. II. Quantitative Studies of Freshly Caught and Iced Shrimp," *Food Res.* **14**, 372–383 (1949).

33. L. L. Campbell, Jr., and O. B. Williams, "The Bacteriology of Gulf Coast Shrimp. IV. Bacteriological, Chemical, and Organoleptic Changes with Ice Storage," *Food Technology* **6**, 125–126 (1952).

34. B. F. Cobb and co-workers, "Effect of Ice Storage upon the Free Amino Acid Contents of Tails of White Shrimp (*Panaeus satiferus*)," *Journal of Agriculture and Food Chemistry* **22**, 1052–1056 (1974).

35. C. Vanderzant, E. Roz, and R. Nickelson, "Microbial Flora of Gulf of Mexico and Pond Shrimp," *J. Milk Food Technol.* **33**, 346–350 (Aug. 1970).

36. G. Finne, "Enzymatic Ammonia Production in Penaied Shrimp Held on Ice," in R. E. Martin and co-eds., *Chemistry & Biochemistry of Marine Food Products*, AVI Publishing Co., Inc., Westport, Conn., 1982.

37. H. S. Tubiash, R. K. Sizemore, and R. R. Colwell, "Bacterial Flora of the Hemolymph of the Blue Crab, *Callinectes sapidus*: Most Probable Numbers," *Applied Microbiology* **29**, 388–392 (1975).

38. R. K. Sizemore and co-workers, "Bacterial Flora of the Hemolymph of the Blue Crab, *Callinectes sapidus*: Numerical taxonomy," *Applied Environmental Microbiology* **29**, 393–399 (1975).

39. M. A. Faghri and co-workers, "Bacteria Associated with Crabs from Cold Waters with Emphasis on the Occurrence of Potential Human Pathogens," *Applied Environmental Microbiology* **47**, 1054–1061 (1984).

40. N. A. Cox and R. T. Lovell, "Identification and Characterization of the Microflora and Spoilage Bacteria in Freshwater Crayfish, *Procambarus clarkii*. *Journal of Food Science* **38**, 679–681 (1973).

41. R. J. Miget, "Microbiology of Crustacean Processing: Shrimp, Crawfish, Prawns," in Ref. 11.

42. D. W. Cook, "Microbiology of Bivalve Molluskan Shellfish," in Ref. 11.

43. D. W. Cook, "Fate of Euteric Bacteria in Estuarine Sediments and Oysters Feces," *J. Miss. Acad. Sci.* **29**, 71–76 (1984).

44. A. R. Murchelano and J. L. Bishop, "Bacteriological Study of Laboratory-Reared Juvenile American Oysters (*Crassostrea virginica*)," *J. Muertebr. Pathol.* **14**, 321–327 (1975).

45. T. Chai and co-workers, "Comparison of Microbiological Quality between Soft-Shell Clams and Growing Waters," *Journal of Applied Environmental Microbiology*, submitted.

46. Food and Drug Administration, *National Shellfish Sanitation Program Manual of Operation. Part II. Sanitation of the Harvesting and Processing of Shellfish*, FDA U.S. Department of Health and Human Services, Washington, D.C., 1989.

47. J. D. Clem, *Status of Recommended National Shellfish Sanitation Program Bacteriological Criteria for Shucked Oysters at the Wholesale Market Level*, FDA, Washington, D.C., (1983).

48. V. T. Dimitroff, "Spirochaetes in Baltimore Market Oysters," *Journal of Bacteriology* **12**, 135–177 (1926).

49. T. Chai and co-workers, "Microbiological Studies of Chesapeake Bay Soft-Shell Clams (*Mya arenaria*)," *J. Food Prot.* **53**, 1052–1057 (1990).

50. R. R. Colwell and J. Liston, "Microbiology Shellfish: Bacteriological Study of the Natural Flora of Pacific Oysters

(*Crassostrea gigas*)," *Applied Microbiology* **8**, 104–109 (Feb. 1960).

51. T. E. Lovelace and co-workers, "Quantitative and Qualitative Commensal Bacterial Flora of *Crassostrea virginica* in Chesapeake Bay," *Proc. Nat. Shellfish Assoc.* **58**, 82–87 (1968).

52. C. Vanderzant and co-workers, "Microbial Flora and Level of *Vibrio parahaemolyticus* of Oysters (*Crassostrea virginica*), Water and Sediment from Galveston Bay," *J. Milk Food Tech.* **36**, 447–452 (Sept. 1973).

53. G. P. Richards, "Microbial Purification of Shellfish: A Review of Depuration and Relaying," *J. Food Prot.* **51**, 218–251 (Mar. 1988).

54. D. W. Cook and R. D. Ellender, "Relaying to Decrease the Concentration of Oyster-Associated Pathogens," *J. Food Prot.* **49**, 196–202 (Mar. 1986).

55. G. H. Fleet, "Oyster Depuration—A Review," *Food Tech. Aust.* **30**, 444–454 (1978).

56. G. J. Vasconcelos and J. S. Lee, "Microbial Flora of Pacific Oysters (*Crassostrea gigas*) Subjected to Ultraviolet-Irradiated Seawater," *Applied Microbiology* **23**, 11–16 (1972).

57. J. G. Metcalf and co-workers, "Bioaccumulation and Depuration of Enteroviruses by the Soft-Shelled Clam, *Mya arenaria*," *Applied Environmental Microbiology* **38**, 275–282 (1979).

58. G. P. Richards, "Outbreaks of Shellfish-Associated Interic Virus Illness in the United States: Requisite for Development of Viral Guidelines," *J. Food Prot.* **48**, 815 (1985).

59. K. L. Parkin and W. D. Brown, "Preservation of Seafood with Modified Atmospheres," in Ref. 36.

60. G. J. Banwart, *Basic Food Microbiology,* AVI Publishing Co., Inc., Westport, Conn., 1979.

61. K. Schrøder and co-workers, "Psychrotrophic *Lactobactllus plantarum* from Fish and Its Ability to Produce Antibiotic Substances," in J. J. Connell, ed., *Advances in Fish Science and Technology,* Fishing News Books Ltd., Surrey, UK, 1980.

62. P. H. Calcott and R. A. Macleod, "Survival of *Escherichia coli* from Freeze-Thaw Damage: A Theoretical and Practical Study," *Canadian Journal of Microbiology* **20**, 671–681 (1974).

63. P. H. Calcott and co-workers, "The Effect of Cooling and Warm Rates on the Survival of a Variety of Bacteria," *Canadian Journal of Microbiology* **22**, 106–109 (1976).

64. S. W. Weagant and co-workers, "The Incidence of *Listeria* species in Frozen Seafood Products," *J. Food Prot.* **51**, 655–657 (1988).

65. R. Digirolamo and co-workers, "Survival of Virus in Chilled, Frozen, and Processed Oysters," *Applied Microbiology* **20**, 58–63 (1975).

66. H. C. Johnson and J. Liston, "Sensitivity of *Vibrio parahaemolyticus* to Cold in Oysters, Fillets and Crabmeat," *Journal of Food Science* **38**, 437–441 (1973).

67. N. Neufeld, "Influence of Bacteriological Standards on the Quality of Inspected Fisheries Products," in R. Kreuzer, ed., *Fish Inspection and Quality Control,* Fishing News Ltd., London, 1971.

68. B. S. Pan, "Low Moisture Fishery Products," in J-L. Chuang, B. S. Pan, and G-C. Chen, eds., *Fishery Products of Taiwan,* JCRR Fisheries Series **25B**, Taipei, Taiwan, 1977.

69. I. H. Yoon, J. R. Matches, and B. Rasco, "Microbiological and Chemical Changes of Surimi-Based Imitation Crab during Storage," *Journal of Food Science* **53**, 1343–1346 (1988).

70. E. L. Elliot, "Microbiological Quality of Alaska Pollock Serimi," in D. E. Kramer and J. Liston, eds., *Developments in Food Sciences: Seafood Quality Determination,* Elsevier, Amsterdam, The Netherlands, 1987.

71. J. R. Matches and co-workers, "Microbiology of Surimi-Based Products," in Ref. 70.

72. R. R. Cockey and T. Chai, "Microbiology of Crustacea Processing: Crabs," in Ref. 11.

73. J. S. Lee and D. K. Pfeifer, "Microbiological Characteristics of Dungeness Crabs (*Cancer magister*)," *Applied Microbiology* **30**, 72–78 (1975).

74. J. W. Duersch, M. W. Paparella, and R. R. Cockey, *Processing Recommendation for Pasteurization Meat from the Blue Crabs,* Marine Products Laboratory, University of Maryland, Crisfield, Md., 1981.

75. F. A. Phillips and J. T. Peeler, "Bacteriological Survey of the Blue Crab Industry," *Applied Microbiology* **24**, 958–966 (1972).

76. M. W. Paparella, *Information Tips 78-4,* Marine Products Laboratory, University of Maryland, Crisfield, Md., 1978.

77. J. Pace, R. R. Cockey, and T. Chai, "Sources of Spoilage of Pasteurized Crabmeat," *Proceedings of the Interstate Seafood Seminars,* 314–323 (Oct. 1989).

78. R. R. Cockey and T. Chai, "Quik-Pik Machine Processing of Crabmeat and Quality Control," *Proceedings of the Interstate Seafood Seminars 1990,* in press.

79. R. R. Cockey, "Bacteriological Assessment of Machine Picked Meat of the Blue Crabs," *J. Food Prot.* **43**, 172–174 (1980).

80. M. A. Benarde and R. A. Littleford, "Antibiotic Treatment of Crab and Oyster Meats," *Applied Microbiology* **5**, 368–372 (1957).

81. U.S. Pat. 2,546,428 (1951), G. C. Byrd.

82. R. R. Cockey and M. C. Tatro, "Survival Studies with Spores of *Clostridium botulinum* Type E in Pasteurized Meat of the Blue Crabs *Callinectes sapidus*," *Applied Microbiology* **27**, 629–633 (1974).

83. Maryland State Department of Health, *Regulations Governing Crabmeat,* Maryland State Department of Health, Baltimore, Sept. 3, 1957.

84. D. A. Kautter and co-workers, "Incidence of *Cl. botulinum* in Crabmeat from the Blue Crabs," *Applied Microbiology* **28**, 722 (Oct. 1974).

85. R. K. Lynt and co-workers, "Thermal Death Time of *Clostridium botulinum* Type E in Meat of the Blue Crabs," *Journal of Food Science* **42**, 1022–1025 (1977).

86. R. K. Lynt, D. Kautter, and H. Solomon, "Differences and Similarities among Proteolytic and Nonproteolytic Strains of *Clostridium botulinum* A, B, E, and F: A Review," *J. Food Prot.* **45**, 466–474 (Apr. 1982).

87. D. R. Ward and co-workers, *Thermal Processing Pasteurization Manual,* Department of Food Science and Technology, VPI State University, Blacksburg, Va., 1982.

88. E. Tanikawa and S. Poka, "Heat Processing of Shellfish," in G. Borgstrom, ed., *Fish as Food,* Vol. IV, Academic Press, Inc., Orlando, Fla., 1965.

89. M. W. Eklund, "Significance of *Clostridium botulinum* in Fishery Products Preserved Short of Sterilization," *Food Technology* **36**, 107–115 (Dec. 1982).

90. F. Huand and C. E. Hebard, *Proceedings of Engineering and Economics of the Oyster Steaming Shucking Process,* VPI & SU, Hampton, Va., 1980.

91. T. Chai, J. Pace, and T. Cossaboom, "Extension of Shelf-Life of Oysters by Pasteurization in Flexible Pouches," *Journal of Food Science* **49**, 331–333 (Feb. 1984).

92. J. Pace, C. Y. Wu, and T. Chai, "Bacterial Flora in Pasteurized Oysters after Refrigerated Storage," *Journal of Food Science* **53**, 325–327 (Feb. 1988).
93. G. G. Giddings, "Radiation Processing of Fishery Products," *Food Technology* **38**, 61–6, 94–97 (1984).
94. J. J. Licciardello and L. J. Ronsivalli, "Irradiation of Seafoods," in Ref. 36.
95. J. D. Quinn and co-workers, "The Inactivation of Infection and Intoxication Microorganisms by Irradiation in Seafood," in FAO/IAEA, eds., *Microbiological Problems in Food Preservation by Irradiation*, IAEA, Vienna, 1967.
96. J. J. Licciardello, D. L. Dentremont, and R. C. Lundstrom, "Radio-Resistance of Some Bacterial Pathogens in Soft-Shell Clams (*Mya arenaria*) and mussels (*Mytilus edulis*)," *J. Food Prot.* **52**, 407–411 (1989).
97. C. W. Hung and B. S. Pan, "Effect of Gamma Irradiation Dosage on Frozen Shrimp," *Proceedings of the Atlantic Fisheries Technology Conference and International Symposium of Fisheries Technology*, Boston, Mass., 1985.
98. G. Hobbs, "*Clostridium botulinum* in Irradiated Fish," *Fd. Irrad. Inf.* **7**, 39–54 (1977).
99. A. Hobbs, "Toxin production by *Clostridium botulinum* Type E in Fish," in Ref. 97.
100. T. Miura and co-workers, "Radiosensitivity of Type E Botulinum Toxin and Its Protection by Proteins, Nucleic Acids and Some Related Substances," in Ref. 97.
101. *Wholesomeness of Irradiated Food*, Joint FAO/IAEA/WHO Committee Report, WHO Technological Report Series **604-FAO**, Food and Nutrition Series **6**, World Health Organization, Geneva, 1977.
102. R. B. Read, Jr., and J. G. Bradshaw, "γ-Irradiation of Staphylococcal Enterotoxin B," *Applied Microbiology* **15**, 603–605 (1967).
103. R. Sullivan and co-workers, "Inactivation of Thirty Viruses by Gamma Radiation," *Applied Microbiology* **22**, 61–65 (1971).
104. T. Miura and co-workers, "Radiosensitivity of Type E Botulinum Toxin and Its Protection by Proteins, Nucleic Acids, and Some Related Substances," in M. Herzgog, ed., *Proceedings of the First United States–Japan Conference on Toxic Microorganisms*, USDI, Washington, D.C., 1970.
105. E. F. Reben and co-workers, "Biological Evaluation of Protein Quality of Radiation-Pasteurized Haddock, Flounder and Crab," *Journal of Food Science* **33**, 335–337 (1968).
106. R. O. Brook and co-workers, "Preservation of Fresh Unfrozen Fishery Products by Low-Level Radiation. V. The Effects of Radiation Pasteurization on Amino Acids and Vitamins in Haddock Fillets," *Food Technology* **20**, 99–102 (1966).
107. J. W. Slavin and co-workers, "The Quality and Wholesomeness of Radiation-Pasteurized Marine Products with Particular Reference to Fish Fillets," *Iso. Radiat. Technol.* **3**, 365–381 (1966).
108. L. J. Ronsivalli and co-workers, *Study of Irradiated Pasteurized Fishery Products, Maximum Shelf Life Study. B. Radiation Chemistry*, U.S. Atomic Commission Contract No. AT(49-112)-1889, Bureau of Commercial Fisheries Technological Laboratories, Gloucester, Mass., 1970.
109. B. L. Middlebrooks and co-workers, "Effects on Storage Time and Temperature on the Microflora and Amine Development in Spanish Mackerel (*Scomberomorus maculatus*)," *Journal of Food Science* **53**, 1024–1029 (1988).
110. J. H. Skala, E. L. McGown, and P. P. Waring, "Wholesomeness of Irradiated Foods," *J. Food Prot.* **50**, 150–160 (Feb. 1987).
111. C. R. Hackney and A. Dicharry, "Seafood-Borne Bacterial Pathogens of Marine Origin," *Food Technology* **42**, 104–109 (Mar. 1988).
112. S. R. Rippey and T. L. Verber, *Shellfish Borne Disease Outbreaks*. FDA, Shellfish Sanitation Branch, Northeast Technical Services Unit, Davisville, R.I., 1986.
113. J. M. Janda and co-workers, "Current Perspectives on the Epidemiology and Pathogenesis of Clinically Significant *Vibrio* spp.," *Clinical Microbiology Reviews* **1**, 245–267 (1988).
114. M. J. Alter and co-workers, "Sporatic Non-A, Non-B Hepatitis: Frequency and Epidemiology in an Urban U.S. Population," *Journal of Infectious Diseases* **145**, 886–893 (1982).
115. P. Blake, "Vibrio on the Half Shells; What the Walrus and the Carpenter Didn't Know," *Annals of Internal Medicine* **99**, 558–559 (1987).
116. L. R. Beuchat, "*Vibrio parahaemolyticus*: Public Health Significance," *Food Technology* **36**, 80–92 (1982).
117. T. Fujino, "International Symposium on *Vibrio parahaemolyticus*" Saikon Publishing Co., Tokyo, Japan, 1974.
118. J. Sakurai, A. Matsuzaki, and T. Miwatani, "Purification and Characterization of Thermostable Direct Hemolysin of *V. parahaemolyticus*," *Infection and Immunity* **8**, 775–780 (1973).
119. L. R. Beuchat, "Interacting Effects of pH, Temperature, and Salt Concentration on Growth and Survival of *Vibrio parahaemolyticus*," *Applied Microbiology* **25**, 844–846 (May 1973).
120. D. Covert and M. Woodburn, "Relationship of Temperature and Sodium Chloride Concentration to the Survival of *Vibrio parahaemolyticus* in Broth and Fish Homogenate," *Applied Microbiology* **23**, 321–325 (Feb. 1972).
121. D. Golmintz, R. C. Simpson, and D. L. Dubrow, "Effect of Temperature on *Vibrio parahaemolyticus* in Artificially Contaminated Seafood," *Developments in Industrial Microbiology* **15**, 288 (1974).
122. C. Vanderzant and R. Nickelson, "Survival of *Vibrio parahaemolyticus* in Shrimp Tissue under Various Environmental Conditions," *Applied Microbiology* **23**, 34–37 (Jan. 1972).
123. K. Venkateswaran and co-workers, "Characterization of Toxigenic *Vibrio* Isolated from the Freshwater Environment of Hiroshima, Japan," *Applied Environmental Microbiology* **55**, 2613–2618 (Oct. 1989).
124. R. R. Colwell and J. Kaper, "*Vibrio cholerae, V. parahaemolyticus,* and Other Vibrios: Occurrence and Distribution in Chesapeake Bay," *Science* **198**, 394–396 (1978).
125. G. Spite, D. Brown, and R. Twedt, "Isolation of One Enteropathogenic, Kanagawa-Positive Strain of *V. parahaemolyticus* from Seafood Implicated in Acute Gastroenteritic," *Applied Environmental Microbiology* **35**, 1226–1227 (1978).
126. R. M. Twedt and D. F. Brown, "*V. parahaemolyticus*: Infection or Toxicosis?" *J. Milk Food Technol.* **36**, 129–134 (1973).
127. T. Honda and co-workers, "Identification of Lethal Toxin with the Thermostable Direct Hemolysin Produced by *V. parahaemolyticus*, and Some Physiochemical Properties of the Purified Toxin," *Infection and Immunity* **13**, 133–139 (1976).
128. J. Oliver, "Vibrio: On Increasingly Troublesome Genus," *Diagn. Med.* **8**, 43–49 (1985).
129. R. Delmore and P. Crisley. "Thermal Resistance of *Vibrio parahaemolyticus* in Clam Homogenate," *Journal of Food Science* **41**, 899–902 (1979).
130. L. R. Beuchat and R. E. Worthington, "Relationships Be-

tween Heat Resistance and Phospholipid Fatty Acid Composition of *Vibrio parahaemolyticus*," *Applied Environmental Microbiology* **31**, 389–394 (Mar. 1976).

131. R. P. Delmore, Jr., and F. D. Crisley, "Thermal Resistance of *Vibrio parahaemolyticus* in Clam Homogenate," *J. Food Prot.* **42**, 131–134 (1979).

132. R. Sakazaki, "Vibrio Infections," in H. Rieman and F. Bryon, eds., *Foodborne Infection and Intoxication*, Academic Press, Inc., Orlando, Fla., 1979.

133. J. Morris and R. Black, "Cholera and Other Vibrioses in the United States," *New England Journal of Medicine* **312**, 343–350 (1985).

134. Centers for Disease Control, "Cholera in Louisiana—Update," *Morbidity Mortality Weekly Report* **35**, 687 (1986).

135. L. Shultz and co-workers, "Determination of the Thermal Death Time of *Vibrio cholerae* in Blue Crabs (*Callinectes sapidus*)," *J. Food Prot.* **49**, 4–6 (Jan. 1984).

136. J. Oliver, R. Warner, and D. Cleland, "Distribution of *Vibrio vulnificus* and Other Lactose Fermenting Vibrios in the Marine Environment," *Applied Environmental Microbiology* **45**, 985–987 (1983).

137. R. Tilton and R. Ryan, "Clinical and Ecological Characteristics of *Vibrio vulnificus* in the Northeastern United States," *Diagnostic Microbiology Infectious Disease* **6**, 109–117 (1987).

138. J. Oliver and co-workers, "Bioluminescence in a Strain of the Human Pathogenic Bacterium *Vibrio vulnificus*," *Applied Environmental Microbiology* **52**, 1209–1221 (1986).

139. G. Sakagrechi, "Botulism," in Ref. 132.

140. L. Smith, *Botulism: The Organism, Its Toxins, the Disease*. Charles C. Thomas, Springfield, Ill., 1977.

141. J. Simunovic, J. Oblinger, and J. Adams, "Potential for Growth of Nonproteolytic Types of *Clostridium botulinum* in Pasteurized Meat Products. A Review." *J. Food Prot.* **48**, 265–276 (Mar. 1985).

142. NCA, *Thermal Destruction of Type E Clostridium botulinum*, NCA Research Foundation, Washington, D.C., 1973.

143. G. Hobbs, "*Clostridium botulinum* and Its Importance in Fishery Products," *Adv. Food Res.* **22**, 135–156 (1976).

144. A. Emodi and R. Lechowich, "Low Temperature Growth of Type E. *Clostridium botulinum* Spores. I. Effect of Sodium Chloride and pH," *Journal of Food Science* **34**, 78–81 (1969).

145. M. Miller and S. Koburger, "*Plesiomonas shigelloides*: An Opportunistic Food and Waterborne Pathogen," *J. Food Prot.* **48**, 449–457 (May 1985).

146. C. P. Gerba, "Viral Disease Transmission by Seafoods," *Food Technology* **42**, 99–103 (1988).

147. M. C. O'Mahony and co-workers, "Epidemic Hepatitis A from Cockles," *Lancet* **1**, 518–520 (1983).

148. W. Stille and co-workers, "Oyster-Transmitted Hepatitis," *Deutsche Medizinische Wochenschrift* **97**, 145 (1972).

149. Centers for Disease Control, *Annual Summary of Foodborne Diseases*, CDC, U.S. Department of Health and Human Services, Atlanta, Ga., 1979–1986.

150. S. L. Taylor, "Marine Toxins of Microbial Origin," *Food Technology* **42**, 94–98 (1988).

151. MCCFP, National Research Council, *An Evaluation of the Role of Microbiological Criteria for Foods and Food Ingredients*, National Academic Press, Washington, D.C., 1985.

152. E. S. Garrett III, "Microbiological Standards, Guidelines, and Specifications and Inspection of Seafood Products," *Food Technology* **42**, 90–93 (1988).

153. International Committee on Microbiological Specifications for Foods, *Microorganisms in Foods. 2. Sampling for Microbiological Analysis: Principles and Specific Applications*, ICMFS, University of Toronto Press, Toronto, 1978.

154. International Committee on Microbiological Specifications for Foods, *Microorganisms in Foods. 1. Their Significance and Methods of Enumeration*, ICMFS, University of Toronto Press, Toronto, 1978.

155. International Committee on Microbiological Specifications for Foods, *Microorganisms in Foods. 2. Sampling for Microbiological Analysis: Principles and Specific Applications*, ICMFS, University of Toronto Press, Toronto, 1986.

156. Codex Alimentarius Commission, *Recommended International Code of Practice for Shrimps or Prawns*, Joint FAO/WHO, Rome, Italy, 1980.

157. Codex Alimentarius Commission, *Recommended International Code of Practice for Frozen Fish*, Joint FAO/WHO, Rome, Italy, 1980.

158. National Academy of Sciences, *An Evaluation of the Role of Microbiological Criteria for Foods and Food Ingredients*, National Academy Press, Washington, D.C., 1985.

Tuu-jyi Chai
University of Maryland
Cambridge, Maryland

FISH AND SHELLFISH PRODUCTS

THE SEAFOOD CHAIN

It is known that humans have used rather advanced techniques for obtaining and processing seafood throughout recorded history. A main source of protein for the ancient Egyptians was fish from the Nile, Mediterranean, and pond culture. Fish was consumed fresh and salted for preservation by the Greeks. Dried fish became a major source of animal protein in Europe when the Roman church banned meat consumption on Fridays and during Lent.

It is known that American Indians used fish as an important part of their diets 10,000 years ago. The early European settlers coming to the New World brought processing and preservation practices such as drying, salting, pickling, and cooling. During colonial times salted dried cod was exported back to England. A major part of the diet of American northwest Indians and Eskimos was salmon that was sun dried and smoked over a wood fire.

It was not until the 19th century in the Great Lakes region that fish were frozen by using the combination of salt and ice to lower temperatures below freezing. The growth of commercial freezing processes over the last 100 years has made freezing the leading means of preserving seafood for human consumption. As will be seen, the early fresh fish handling and freezing technology was not developed with good knowledge of the biological and physical factors that must be considered. Hence, the fishy odor of fish along with certain rancid and spoilage off-tastes were considered normal in fresh, frozen, and dried products.

Today it is realized that quality control of fish, as much or more than any other food, must be practiced from harvest to table. The seafood chain (Fig. 1) begins with the boat builder or hatchery designer and ends with the consumer. All must practice good techniques of sanitation, temperature control, and packaging protection. As the fish story unfolds, it will be seen that any break in this

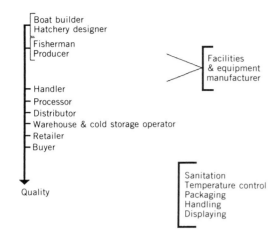

Figure 1. The seafood chain from the harvest to the table.

Table 1. Annual Per Capita Consumption of Live Weight Fish and Shellfish for Human Food, Selected Countries, 1982–1984[a]

Region and Country	Estimated Live Weight Equivalent	
	Kilograms	Pounds
North America		
Canada	21.4	47.2
United States	17.4	36.4
Caribbean		
Barbados	33.6	74.1
Dominican Republic	6.8	15.0
Grenada	19.4	42.8
Haiti	3.7	8.2
Jamaica	17.0	37.5
Latin America		
Argentina	5.7	12.6
Bolivia	1.6	3.5
Brazil	6.2	13.7
Chile	18.7	41.2
Ecuador	14.2	13.1
Mexico	9.9	21.8
Panama	14.4	31.7
Europe		
Denmark	22.0	48.5
France	24.8	54.7
Iceland	88.4	194.9
Norway	46.0	101.4
USSR	27.3	60.2
Near East		
Egypt	5.5	12.1
Israel	14.2	31.3
Lebanon	0.5	1.1
Oman	35.5	78.3
Far East		
China	4.9	10.8
India	3.1	6.8
Japan	74.5	164.2
Philippines	35.7	78.7
Africa		
Congo (Brazaville)	33.4	73.6
Ethiopia	0.1	0.2
Kenya	4.7	10.4
Republic of South Africa	9.6	21.2
Oceania		
Australia	16.0	35.3
New Zealand	12.2	26.9

[a] Ref. 2.

chain can be disastrous to the quality of seafood. If a processing plant, market, restaurant, or home smells fishy, then there has been a break in the chain. Good fish looks good, has a neutral fresh odor, and has firm flesh. Good fish is a healthful, highly nutritious protein food that has an unlimited market.

U.S. FISHERY

There are more than 2,000 species of finfish within the U.S. coastal marine waters, of which about ca 2% are consumed as food fish (1). In fact, the U.S. consumer is familiar with only a few of the some 200 species of finfish and 40 species of shellfish that are consumed throughout the world. This is not surprising when one compares the geographical annual per capita consumption of fish and shellfish as shown in Table 1 (2). Based on live weight of the landed animal, North America consumed 85.6 lb while consumption in other areas of the world ranged from 5 to 15 times that amount. There was little change in the amount of fish consumed by the average U.S. citizen from the early 1900s to 1970s (Table 2). Over the last 15 years, although still considerably below that of the rest of the world, the U.S. per capita consumption has risen about 30%. Most of this increase has been during the last 5 yr. With the increasing emphasis on eating lighter and leaner foods, this U.S. consumption is continuing to rise.

Fishery Landings

The 1988 commercial domestic landings, by species, of marine fish and shellfish in the United States is shown in Table 3 (2). Near-shore landings (0–3 mi offshore) totaled 1,762,059 t (value $2,026,147,000), offshore landings (3–200 mi offshore) totaled 2,953,034 t (value $1,671,279,000), and international water catches totaled 259,735 t (value $313,642,000), for a grand total of 4,974,828 t (value $4,010,068,000). Almost 60% of the total fish and shellfish landings by the United States are caught 3–200 mi offshore. Much of the harvest in these waters is in the Fisheries Conservation Zone, which extends fishing jurisdiction beyond that normally considered international waters.

There is a large variation in the value of different species based on the consumer acceptance and the economics of harvest. Two relatively low-cost species, menhaden and pollock, accounted for 48.3% of the total fish and shellfish landed but had a value amounting to only 8.2% of the total. Menhaden is reduced to fish meal for animal consumption, whereas pollock is the basis for the rapidly growing surimi and pollock fillet industry. Conversely, high-value salmon accounted for only 5.5% of the catch but represented 22.7% of the total value, Shellfish also accounted for a relatively small volume of the total landings in the United States (less than 12%) but represented 37.5% of the total value.

It is important to include the production of fish and

FISH AND SHELLFISH PRODUCTS

Table 2. U.S. Annual Per Capita Consumption of Commercial Fish and Shellfish, 1953–1985[a]

Year	Civilian Resident Population, July 1[b]	Fresh and Frozen[a]	Per Capita Canned[d]	Consumption Cured[e]	Total
1909[f]	90.5	4.3	2.7	4.0	*11.0*
1910	92.4	4.5	2.8	3.9	*11.2*
1915	100.5	5.8	2.4	3.0	*11.2*
1920	106.5	6.3	3.2	2.3	*11.8*
1925	115.8	6.3	3.2	1.6	*11.1*
1930	122.9	5.8	3.4	1.0	*10.2*
1935	127.1	5.1	4.7	0.7	*10.5*
1940	132.1	5.7	4.6	0.7	*11.0*
1945	128.1	6.6	2.6	0.7	*9.9*
1950	150.8	6.3	4.9	0.6	*11.8*
1955	163.0	5.9	3.9	0.7	*10.5*
1960	178.1	5.7	4.0	0.6	*10.3*
1965	191.6	6.0	4.3	0.5	*10.8*
1970	201.9	6.9	4.5	0.4	*11.8*
1975	213.8	7.5	4.3	0.4	*12.2*
1980[g]	225.6	8.0	4.5	0.3	*12.8*
1981[g]	227.7	7.8	4.8	0.3	*12.9*
1982[g]	229.9	7.7	4.3	0.3	*12.3*
1983[g]	232.0	8.0	4.8	0.3	*13.1*
1984[g]	234.4	8.3	5.0	0.3	*13.6*
1985[g]	237.0	9.0	5.2	0.3	*14.5*
1986[g]	239.4	9.0	5.4	0.3	*14.7*
1987[g]	241.5	10.0	5.1	0.3	*15.4*
1988[g]	243.9	9.6	5.1	0.3	*15.0*

[a] Ref. 2.
[b] Resident population for 1909–1929 and civilian resident population for 1930 to date.
[c] Fresh and frozen fish consumption from 1910 to 1928 is estimated. Beginning in 1973, data include consumption of artificially cultivated catfish.
[d] Canned fish consumption for 1910–1920 is estimated. Beginning in 1921, it is based on production reports, packer stocks, and foreign trade statistics for individual years.
[e] Cured fish consumption for 1910–1928 is estimated.
[f] Data for 1909 estimate based on the 1908 census and foreign trade.
[g] Domestic landing data used in calculating these data are preliminary.

Table 3. Commercial Landings of Fish and Shellfish by U.S. Fishing Craft, 1988

Species	Total[b] Metric Tons	Thousand Dollars
Fish		
Alewives		
Atlantic and Gulf	2,561	626
Great Lakes	4,856	191
Anchovies	5,636	2,615
Bluefish	7,644	3,012
Bonito	4,178	1,827
Butterfish	2,468	3,407
Cod		
Atlantic	34,506	42,941
Pacific	232,727	68,858
Croaker	4,810	4,596
Cusk	1,072	1,021
Flounders		
Atlantic and Gulf:		
Blackback	8,211	21,533
Fluke	16,334	44,345
Yellowtail	5,041	13,187
Other	11,616	27,022
Pacific	392,612	87,818
Total flounders	433,814	193,905

Table 3. (*continued*)

Species	Total[b] Metric Tons	Thousand Dollars
Groupers	5,547	21,703
Haddock	2,916	7,030
Hake		
Pacific (whiting)	142,919	15,820
Red	1,739	618
White	5,446	3,639
Halibut	37,017	72,718
Herring, sea		
Atlantic	41,003	5,229
Pacific	59,617	57,431
Jack mackerel	10,227	1,685
Lingcod	2,965	2,385
Mackerel		
Atlantic	12,377	2,722
King	1,954	5,043
Pacific	45,514	7,498
Spanish	1,922	1,479
Menhaden		
Atlantic	360,462	35,162
Gulf	638,721	73,259
Total menhaden	*999,183*	*108,421*
Mullets	14,806	11,218
Ocean perch		
Atlantic	1,066	1,467
Pacific	5,406	2,546
Pollock		
Atlantic	14,992	11,071
Alaska	1,396,833	211,354
Rockfish	58,360	39,368
Sablefish	48,817	91,793
Salmon, Pacific		
Chinook or king	20,716	117,551
Chum or keta	66,436	134,689
Pink	80,053	127,297
Red or sockeye	86,199	437,630
Silver or coho	21,539	93,506
Total salmon	*274,943*	*910,673*
Scup or porgy	6,513	9,572
Sea bass		
Black	2,188	5,144
White	49	218
Sea trout		
Gray	9,314	7,948
Spotted	1,403	3,169
White	168	229
Sharks		
Dogfish	4,568	975
Other	6,622	8,454
Snapper		
Red	1,884	9,496
Other	3,083	11,319
Striped bass	185	517
Swordfish	5,814	42,703
Tilefish	2,066	7,222
Tuna		
Albacore	8,643	16,278
Bigeye	2,472	15,149
Bluefin	1,699	17,305
Skipjack	136,077	149,052
Yellowfin	127,140	182,202
Unclassified	354	546
Total tuna	*276,385*	*380,532*
Whiting	16,134	8,621
Other marine finfishes	119,697	74,953
Other freshwater finfishes	13,271	18,534
Total fish	*4,389,185*	*2,505,515*

Table 3. (continued)

Species	Total[b]	
	Metric Tons	Thousand Dollars
Shellfish		
Clams		
Hard	5,616	67,818
Ocean quahog	21,006	14,921
Soft	3,091	18,717
Surf	28,824	29,183
Other	1,225	4,142
Total clams	*58,757*	*134,781*
Crabs		
Blue, hard	99,184	84,357
Dungeness	21,518	54,771
King	9,513	84,153
Snow (tanner)	66,372	137,052
Other	10,082	23,227
Total crabs	*206,669*	*383,560*
Lobsters		
American	22,064	145,236
Spiny	3,250	23,030
Oysters	14,466	78,498
Scallops		
Bay	258	3,414
Calico	5,383	12,462
Sea	13,860	128,243
Shrimp		
New England	3,078	7,497
South Atlantic	10,997	51,667
Gulf	102,416	414,469
Pacific	33,590	32,401
Total shrimp	*150,081*	*506,034*
Squid		
Atlantic	21,200	16,220
Pacific	36,481	7,689
Other shellfish	52,174	65,386
Total shellfish	*585,643*	*1,504,553*
Grand total, 1988	*4,974,828*	*4,010,068*

[a] Ref. 2.
[b] Note: Landings are in round (live) weight for all items, except clams, oysters, and scallops, which are reported in weight of meat.

shellfish from the rapidly growing aquaculture industry when considering total fish and shellfish availability. Although long established in other parts of the world (particularly Asia) aquaculture is a relatively new and rapidly growing industry in the United States. As shown in Table 4, aquaculture-raised fish accounted for 281,800 t (620 million lb) in 1986 (3).

For many years trout was the leading aquaculture finfish grown in the United States. However, catfish became the dominant species in the 1970s and salmon production passed trout in 1985. Catfish farming in the United States has increased from 1,004 t (22.1 million lb) in 1977 to 134,090 t (295 million lb) in 1988 (4). This represents about 53% of the total aquaculture fish and shellfish and 5% of the total combined wild and aquaculture fish and shellfish.

Imports and Exports

Tables 5 and 6 show the imports and exports of fishery products to and from the United States (2). In 1988 1,350,000 t (2,971 million lb) of edible fishery products were imported while only 36% of that amount, 483,550 t (1,064 million lb), were imported.

Harvesting Gear

Static Gear. Although early fishing depended on spear fishing and later was the basis for sophisticated harpoon systems, modern commercial fishing methods use either static gear or moving gear when a vessel is used for towing or dragging. There are three basic types of static gear used in the commercial fisheries. The hook-and-line technique is familiar to all sports fishermen. The fish is attracted to the hook by an edible bait or by an attractive device such as a feather. As the fish tries to take the food from the bait or attacks the lure, it becomes attached to the barbed hook and cannot shake the hook from its mouth. A technique called long-lining is the major commercial use of hook-and-line fishing. A series of baited hooks are suspended from a buoyed horizontal line. This use of multiple hooks greatly increases the chances of

Table 4. U.S. Private Aquaculture Production and Value[a,b]

Species	1980	1983	1984	1985	1986
Baitfish	22,046	22,046	23,598	24,807	25,247
	(44,000)	(44,000)	(47,045)	(51,280)	(51,522)
Catfish	76,849	220,000	239,800	271,357	326,979
	(53,572)	(132,000)	(191,840)	(189,194)	(228,886)
Clams	561	1,689	1,689	1,588	2,506
	(2,295)	(9,500)	(4,178)	(4,717)	(8,307)
Crayfish	23,917	60,000	59,400	64,999	97,500
	(12,951)	(30,000)	(27,700)	(32,500)	(48,750)
Freshwater prawns	300	275	317	267	178
	(1,200)	(1,500)	(1,698)	(1,540)	(893)
Mussels	—	775	917	928	1,206
	—	(1,500)	(1,584)	(1,248)	(1,725)
Oysters	23,755	23,300	24,549	22,473	24,090
	(37,085)	(31,500)	(38,970)	(39,997)	(42,797)
Pacific salmon	7,616	20,600	45,086	84,305	74,398
	(3,400)	(6,800)	(17,252)	(25,439)	(32,751)
Shrimp	—	255	528	440	1,354
	—	(874)	(1,566)	(1,687)	(3,408)
Trout	48,141	48,400	49,940	50,600	51,000
	(37,474)	(50,000)	(54,435)	(55,154)	(55,590)
Other species	—	7,000	9,900	14,000	15,500
	—	(7,000)	(9,900)	(20,000)	(21,700)

[a] Ref. 3.
[b] Production is in 1,000 lb, and value, in parentheses, is in $1,000.

catching fish and makes the fishing a viable commercial venture.

There are two basic types of trap used to catch fish and shellfish. One is an enclosure placed in the pathway of a moving or migrating animal while the other is a container or pot in which bait is placed. An animal crawls or swims into the tunnel entrance and then cannot escape because it is unable to find its way out of the narrow opening. Traps usually have a one-way wire door that can be easily pushed aside to enter but prevents regression.

Static nets, called gill nets, are a form of static gear closely related to traps. Nets are suspended from a few fathoms up to 50 fathoms into the water from a buoyed line. They are placed in the water where fish are known to be passing. The moving fish hits and penetrates the net. As it realizes that it has hit an obstruction, it tries to pull back. This causes its gills to become entangled in the webbing where the fish is securely held until the fishermen pull the net to the surface. Figures 2–7 show the types of static gear described above (5).

Towed or Dragged Gear. It is often necessary to sweep large areas of the ocean to catch sufficient volumes of fish or shellfish. This is done by trawling, in which large nets are towed through the water, or dredging, in which gear is towed along the bottom.

A trawl net is constructed like a large net bag that has a restricted end called the cod end. As a fishing vessel pulls the funnellike nets, fish are swept into the nets and collected in the bottom or cod end. The vessel must travel at sufficient speed to insure that the fish being collected into the net cannot swim out of the opening. The goal of a trawl operation, lasting from less than an hour to several hours is to fill the cod end. A typical trawl net and a dredge are shown in Figures 8–10 (5).

Encircling Gear. Encircling gear is used for catching large amounts of fish that are schooled or densely concentrated. One end of the net, called a seine net is pulled from the fishing vessel and completes the operation by encircling the school of fish and bringing the end back to the vessel. Most seines (Fig. 11) have a line on the bottom that can be pursed to close the net and prevent fish from escaping. Typical fish that can be exploited by this method of harvest include anchovy, pilchard, sardine, salmon, herring, and tuna.

Fishing Vessels

The subject of fishing vessel design, construction, and operation is a most complex subject and will only be men-

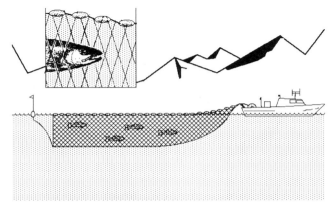

Figure 2. Static fishing gear. Gillnetting—a method of fishing in which fish swim into a suspended net and become entangled by their gills in the webbing. The net can be placed at various depths, depending on the fishery and locale. Courtesy of *Seafood Leader*.

Table 5. U.S. Fishery Products Imports by Principal Items, 1987 and 1988[a,b]

	1987		1988	
Item	Thousand Pounds	Thousand Dollars	Thousand Pounds	Thousand Dollars
Edible Fishery Products				
Fresh and frozen				
Whole or eviscerated				
Cod, cusk, haddock, and flounder	105,158	65,198	85,139	52,717
Halibut	9,295	23,138	11,952	25,432
Salmon	41,902	113,008	50,144	155,173
Tuna				
Albacore	201,988	171,988	195,991	202,967
Other[c]	370,517	158,771	354,156	169,331
Other	214,628	183,022	222,696	203,399
Fillets and steaks				
Flounder	73,003	148,734	58,534	119,996
Groundfish	315,418	570,065	253,187	431,126
Other	232,564	396,324	205,988	358,766
Blocks and slabs	403,577	539,358	303,237	382,482
Shrimp	461,173	1,676,844	489,740	1,725,971
Crabmeat	12,571	67,427	10,821	59,639
Lobster				
American (includes fresh-cooked meat)	38,974	178,069	39,732	183,482
Spiny	41,949	397,854	37,806	363,195
Scallops (meats)	39,934	162,273	32,039	115,706
Analogue products with shellfish	30,539	51,197	24,516	41,569
Other fish and shellfish	98,996	169,855	92,128	163,444
Canned				
Herring, not in oil	5,617	8,726	6,541	10,264
Sardines				
In oil	27,352	35,106	22,813	30,824
Not in oil	37,670	25,470	30,546	23,154
Tuna				
In oil	329	869	318	744
Not in oil	211,356	206,051	244,186	297,922
Balls, cakes, and puddings				
Analogue products without shellfish	4,737	6,786	2,574	4,361
Other	8,797	13,130	5,650	8,376
Abalone	2,790	19,867	2,434	22,487
Clams	13,974	15,288	11,268	13,993
Crabmeat	7,967	20,626	7,720	19,622
Lobsters				
American	637	4,184	594	5,342
Spiny	136	748	52	280
Oysters	32,668	36,144	27,524	39,817
Shrimp	17,132	33,380	14,138	28,730
Other fish and shellfish	57,579	89,499	52,350	80,952
Cured				
Pickled or salted				
Cod, haddock, hake, etc	31,893	60,542	31,361	52,665
Herring	22,213	9,594	20,333	8,618
Other	9,991	22,792	11,784	27,718
Other fish and shellfish	16,108	29,306	10,966	29,119
Total edible fishery products	3,201,132	5,711,233	2,970,958	5,459,383
Inedible Fishery Products				
Meal and scrap	393,730	52,508	265,310	49,567
Fish oils	30,509	18,930	27,688	9,666
Other	—	3,035,026	—	3,353,379
Total inedible fishery products	—	3,106,464	—	3,412,612
Grand total	—	8,817,697	—	8,871,995

[a] Ref. 2.
[b] Note: Data include imports into the United States and Puerto Rico and include landings of tuna by foreign vessels at American Samoa. Statistics on imports are the weight of individual products as exported, ie, fillets, steaks, whole, headed, etc.
[c] Includes loins and disks.

Table 6. U.S. Domestic Fishery Products Exports, by Principal, 1987 and 1988[a,b]

Item	1987 Thousand Pounds	1987 Thousand Dollars	1988 Thousand Pounds	1988 Thousand Dollars
Edible Fishery Products				
Fresh and frozen				
Whole or eviscerated				
Cod, cusk, haddock, and flounder	105,158	65,198	85,139	52,717
Halibut	9,295	23,138	11,952	25,432
Salmon	41,902	113,008	50,144	155,173
Tuna				
Albacore	201,988	171,988	195,991	202,967
Other[c]	370,517	158,771	354,156	169,331
Other	214,628	183,022	222,696	203,399
Fillets and steaks				
Flounder	73,003	148,734	58,534	119,996
Groundfish	315,418	570,065	253,187	431,126
Other	232,564	396,324	205,988	358,766
Blocks and slabs	403,577	539,358	303,237	382,482
Shrimp	461,173	1,676,844	489,740	1,725,971
Crabmeat	12,571	67,427	10,821	59,639
Lobster				
American (includes fresh-cooked meat)	38,974	178,069	39,732	183,482
Spiny	41,949	397,854	37,806	363,195
Scallops (meats)	39,934	162,273	32,039	115,706
Analogue products with shellfish	30,539	51,197	24,516	41,569
Other fish and shellfish	98,996	169,855	92,128	163,444
Canned				
Herring, not in oil	5,617	8,726	6,541	10,264
Sardines				
In oil	27,352	35,106	22,813	30,824
Not in oil	37,670	25,470	30,546	23,154
Tuna				
In oil	329	869	318	744
Not in oil	211,356	206,051	244,186	297,922
Balls, cakes, and puddings				
Analogue products without shellfish	4,737	6,786	2,574	4,361
Other	8,797	13,130	5,650	8,376
Abalone	2,790	19,867	2,434	22,487
Clams	13,974	15,288	11,268	13,993
Crabmeat	7,967	20,626	7,720	19,622
Lobsters				
American	637	4,184	594	5,342
Spiny	136	748	52	280
Oysters	32,668	36,144	27,524	39,817
Shrimp	17,132	33,380	14,138	28,730
Other fish and shellfish	57,579	89,499	52,350	80,952
Cured				
Pickled or salted				
Cod, haddock, hake, etc	31,893	60,542	31,361	52,665
Herring	22,213	9,594	20,333	8,618
Other	9,991	22,792	11,784	27,718
Other fish and shellfish	16,108	29,306	10,966	29,119
Total edible fishery products	*3,201,132*	*5,711,233*	*2,970,958*	*5,459,383*
Inedible Fishery Products				
Meal and scrap	393,730	52,508	265,310	49,567
Fish oils	30,509	18,930	27,688	9,666
Other	—	3,035,026	—	3,353,379
Total inedible fishery products	—	*3,106,464*	—	*3,412,612*
Grand total	—	*8,817,697*	—	*8,871,995*

[a] Ref. 2.

[b] Note: Data include imports into the United States and Puerto Rico and include landings of tuna by foreign vessels at American Samoa. Statistics on imports are the weight of individual products as exported, ie, fillets, steaks, whole, headed, etc.

[c] Includes loins and disks.

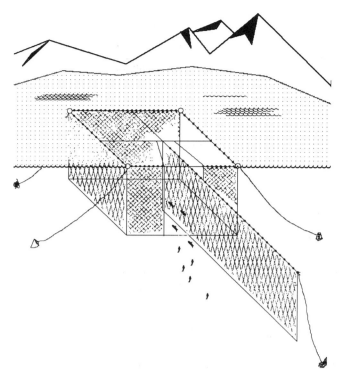

Figure 3. Static fishing gear. Cod traps—set on the ocean floor close to shore, these open-trapped box nets contain a door facing the shore, where seashore feeding cod are deterred by a net fence that directs them to the trap; once inside, they tend to swim in circles. Courtesy of *Seafood Leader*.

tioned here. Fishing vessels range from small outboard motor boats operated by artisan fishermen to large ocean-going vessels, which require a sophisticated crew for efficient operation. The smaller vessels operate close to shore and return with the catch each day.

The length and sophistication of modern near-shore fishing vessels depends on the fishery, the fishing location and distance from shore, the distance that must be covered going to and from the fishing grounds, and the value of the catch. It is common to find vessels 20–50+ ft in length operating in the near-shore fishery and catcher–processor vessels of several hundred feet in length operating offshore on the high seas. The crews number from one to five on near-shore vessels and up to several hundred (including process crew) on the large catcher–processors. Likewise, the cost of fishing vessels varies tremendously, from about $200,000 for a small near-shore vessel to approximately $30,000,000 for the large offshore catcher–processors.

MAINTAINING QUALITY OF THE CATCH

Quality is a term that has many definitions, depending on the background and interests of those queried. To some, the measuring of the product degradation by biological factors such as microorganisms or enzymes is the only means of determining quality. To others the aesthetic values that make a product look good are just as important in defining quality. In reality the biological factors determining safety and nutritional values and the physical fac-

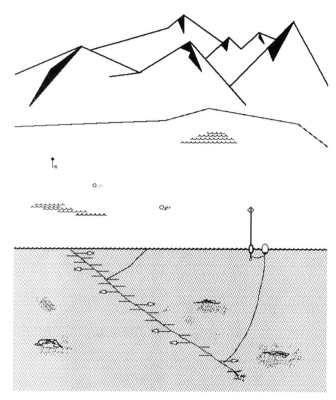

Figure 4. Static fishing gear. Longlining—a method of fishing involving one main line to which a series of shorter lines (gangions) with baited hooks are attached. Used at various depths; ie, surface longlining for pelagic species such as tuna and swordfish, bottom longlining for demersal species as halibut and cod. Courtesy of *Seafood Leader*.

Figure 5. Static fishing gear. Crab traps—framed with iron rods and covered with polyethylene rope webbing, crab trays are usually fished on single lines. Size varies according to fishery; ie, small traps are used for blue crab, large traps are used for king crab. Courtesy of *Seafood Leader*.

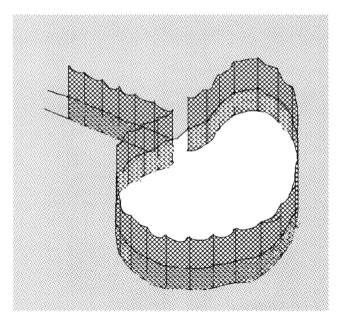

Figure 6. Static fishing gear. Weir fishing—rigid poles are driven into the mud bottom to form a heart-shaped configuration; a straight line of poles, leading from the shoreline to the weir, acts as a barrier directing the fish into the weir. Using skiffs, fishermen first seine the catch, then use a brailer or dip-net to collect them. Courtesy of *Seafood Leader*.

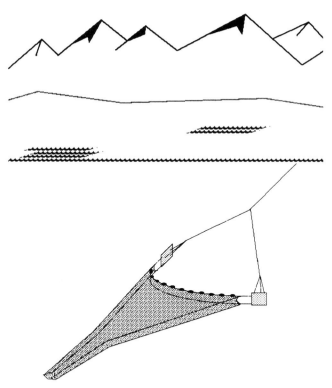

Figure 8. Towed or dragged fishing gear. Otter trawling—a method of fishing in which a large wedge-shaped net is dragged along the ocean bottom; an otter door is attached to each side of the net to hold the net open and keep it horizontal. Fish collect in the cod end (the back) of the net. Courtesy of *Seafood Leader*.

Figure 7. Static fishing gear. Inshore lobster fishing—wooden or wire traps with cotton or nylon twine are set on the ocean floor at various depths, either individually or in strings on a line, baited with either fresh or salted fish. Courtesy of *Seafood Leader*.

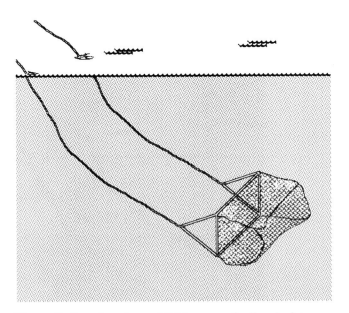

Figure 9. Towed or dragged fishing gear. Scallop dredging—a method of fishing which involves raking a metal frame with teeth and a chainmesh bag across the ocean bottom. Courtesy of *Seafood Leader*.

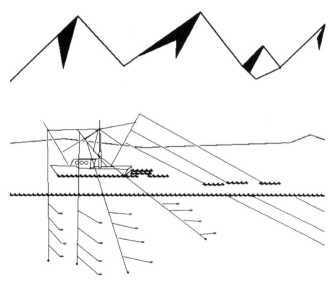

Figure 10. Towed or dragged fishing gear. Trolling—a method of fishing in which several fishing lines with numerous lures are dragged slowing through the water.

tors used in a grading system (ie, size and weight uniformity, color, and blemished surfaces) are integrated to mean quality to most people.

The maintenance of quality or the fresh nature of landed seafood depends on many operations from the catching, landing, and shipboard handling to the transporting, storing, processing, and distributing. A biological specimen can only decrease in quality as it travels through the various steps of a commercial venture. Enzymes and microorganisms cause spoilage and degradation that are irreversible. Physical damage not only affects the appearance of a product, but such damage as skin ruptures allow microorganisms to invade the tissue and cause earlier deterioration. Therefore, all participants in

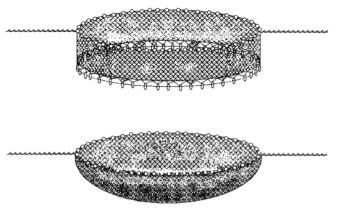

Figure 11. Encircling gear. Purse seining—a method of fishing involving a long, deep net that stands like a fence in the water, supported at the surface by floats and held down by lead lines at the bottom. A person in a skiff takes one end of the net around the school of fish and joins it at the other end, and the vessel hauls in the wire purse line strung through the bottom of the net, forming a purse under the fish. Courtesy of *Seafood Leader*.

the commercial seafood chain are important for maintaining high-quality products.

The Impact of Fishing Methods

Fishing gear is usually designed to give maximum efficiency in catching fish. The total cost of catching a given amount of fish includes vessel cost and operation, fishing gear cost and maintenance, manpower required, and other costs involving machinery and equipment. All factors are combined to give the cost of a given harvest known as the catch per unit effort (CUE). However, the best CUE does not insure good quality in the fish landed, and a processor is somewhat at the mercy of the fishing operation for initial quality of raw materials from the sea.

The type of gear used in fishing has a definite bearing on the quality. When a fish dies after or during vigorous exercise, such as struggling on a hook and line, metabolic activity including that of protein and lipid degrading enzymes, adversely affects the subsequent spoilage rate of the dead animal. If a fish has been feeding and has food in its stomach, the increased metabolic activity greatly accelerates the loss of quality in the slaughtered animal. The so-called soft belly of a fish is caused by this enzyme activity after the fish is dead. Because the amount of time that a hooked fish struggles is directly related to the subsequent spoilage rate, troll-caught fish are often of better quality than longline-caught fish that are allowed to die thrashing in the water.

The quality of gill net-caught fish is extremely variable depending on the length of time that the net is in the water. Hence, there can be a tremendous difference in the quality of fish taken from a given set, the last caught often being of higher quality. This factor is certainly realized by the wholesale buyers in that troll-caught fish consistently command a higher price than gill net-caught fish.

Certain visual or aesthetic factors also affect the market price of fish even though there may not be any real difference in the quality of the flesh. Marks from the web of a gill net are often caused when the fish struggles back and forth to release its entrapped gills. Unless the web causes cuts that allow bacteria to enter the flesh, these marks normally do not adversely affect the biological quality of the flesh. Hence, the word *quality* has different definitions, depending on whether it refers to the biological state of the edible portion or the visual appearance of the fish.

There are several types of fishing gear that cause abrasions and punctures during the catching and subsequent handling. Trawling, dredging, spearing or harpooning, and gillnetting all cause different degrees of damage. Trawling, which accounts for ca 40% of the world's fish catch, exerts extreme pressure on the fish as the cod end becomes full. In addition to scale abrasion, ruptures in the skin and internal portions release gut bacteria and decrease shelf life of the subsequent product.

The best quality is maintained in fish that are caught by trapping. As long as the trapping device is emptied on a reasonable cycle, the fish remains alive and is quickly killed prior to sale or processing. Some species, such as crabs, must be kept alive prior to butchering or there will be a blue color in the meat. This is due to the blood chemistry of crabs; they have a copper complex instead of the heme, or iron, complex found in most animals. If the crab is not butchered live and the blood removed prior to processing, the copper will oxidize, giving a blue color to the meat. Although the aesthetic value of the crab is impaired, the eating quality is not affected. However, as in the case of abraided fish skin, the consumer is not willing to consider blue crabmeat as anything but a low-grade, poor-quality product.

Aquaculture is somewhat akin to catching wild fish in trapping devices. The fish or shellfish are raised in an enclosed area and then removed when ready for market. As in the case of trap-caught fish, farmed fish are live when harvested.

Shipboard Handling of the Catch

The proper handling of fish during harvesting and on shipboard can minimize the adverse effects of gear. Of utmost importance when fish are first landed is that they are segregated and placed in a sanitary chilled environment. Minimizing the bacterial and enzymatic activity by fast reduction of temperature in freshly landed finfish and shellfish is probably the most important step in the entire chain of events that takes a fish from the water to the table.

Figure 12 shows the extreme variation in storage life of fresh and frozen commercial fish prepared for the market. This curve has been compiled from many published sources that give the shelf life of fish as related to the handling methods (6). It has been shown that landed high-quality fish that is chilled rapidly and carefully handled, processed, and packaged can be acceptable for up to three weeks after being caught. On the other end of the scale, a fish that has undergone poor handling on shipboard (eg, 70% of as caught quality) and subsequent marginal handling during the processing and marketing stages is inedible after four days.

Fish that has been properly handled, processed, packaged, and frozen can be held up to one year without significant deterioration. However, there are many factors that must be considered in discussing shelf life of a fresh or frozen product. The species, the oil content, the catching technique, and the state of the fish when harvested are all uncontrollable factors that have a major bearing on the shelf life of a seafood product.

PROCESSING SEAFOOD

Inspecting As-Received Seafoods

Seafoods received in the processing area of a vessel or in a shore-based plant vary tremendously in the state and form. This is the situation when batches of product from different sources or catching vessels are mixed in the received lot. A sensory inspection must be made to insure that the raw material passes the criteria specified by the buyer or processor.

A sensory evaluation utilizes touch, odor, and sight to determine the acceptability of a given lot of seafood (7). An on-site inspection should concentrate on microbial contamination, enzymatic degradation, and other chemical and physical factors that reduce the marketability of the seafood. A faint fresh, nonfishy odor; firm and elastic flesh; bright and full translucent eyes; bright pink gills; and bright and moist skin surface with no heavy deposits of mucus or slime are all properties of a good fresh fish.

In addition to microbial and enzymatic degradation that can take place in improperly handled seafood, the oil in fatty fish that have not been chilled rapidly or adequately protected from the environment are subject to oxidative rancidity. This is both an aesthetic and nutritional

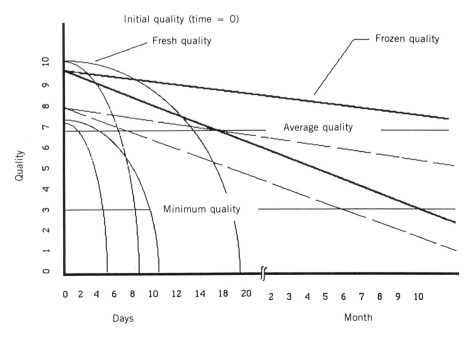

Figure 12. Shelf life of fishery products (12).

problem. Oxidation from the air (and sometimes autooxidation within the seafood) not only affects the odor and taste acceptability of the product but also destroys the omega-3 ($n-3$) fatty acids in the oil that are so important for human nutrition.

Preprocessing As-Received Seafoods

Preprocessing begins on shipboard where the seafood is, at the minimum, segregated and chilled. Many fish destined for the fresh market are butchered and washed. The minimal butchering operation consists of removing the visceral portion and often the gills. The ultimate in shipboard butchering is heading and gutting, which consists of removing the viscera, head, and often the tail and fins. Shrimp are normally iced but sometimes the head is removed on shipboard. Crab are either delivered live directly to the shore plant or are kept alive in seawater tanks until delivery. In all shipboard preprocessing operations the most important procedure is to lower the temperature of the catch as soon as possible after it is taken from the water.

Preprocessing operations not carried out on shipboard are completed in the processing plant. After a second visual inspection, fish are butchered. This primarily consists of eviscerating but also can include scaling, trimming, and further cleaning when necessary for a specific processing operation.

Finfish are portioned for processing or direct marketing by filleting, steaking, or dressing a whole fish or section for roasting or broiling. Depending on the size and sophistication of the processing plant, these operations are carried out either by hand or machine. Crustacea (eg, crayfish, lobster, and crab) and mollusks (eg, clams, oysters, and mussels) are handled and processed quite differently from finfish.

Shrimp are iced on shipboard and unloaded by basket and conveyor into the plant receiving area. In the plant they are segregated and graded as to size by machine, passed over a visual inspection table for removing substandard specimens, cooked, and then headed and peeled (normally by machine). Large prawns are sometimes headed and handled individually on shipboard or in the plant. Those destined for market in an unpealed condition are then frozen. Prawns with head off and not cooked are called headed green prawns in the trade. Shrimp, regardless of whether they are marketed cooked and peeled, green fresh or frozen, or in other forms are sold by the count to designate size. For example 21–25 count shrimp means that there are 21–25 shrimp per lb.

Crab is cooked prior to processing or marketing. It is important that the crab are alive when butchered just before being cooked whole. This is due to the high copper content in the crab blood, which oxidizes to a blue color if allowed to remain in the meat of a dead crab. Crab, depending on the species, are sold in the shell whole (eg, blue and dungeness), segregated with shell on into legs and body portions (eg, king and snow), or as leg and body meat. Meat is removed or shaken from the cracked shell portions by hand.

Bivalves must be alive when purchased and subsequently cooked in a plant, restaurant, or home. A healthy live oyster, clam, or mussel will have a tightly closed shell. Any gapers, eg, those with an open shell, must be discarded. This requirement for handling only live molluscs is important because they normally are not iced after being harvested by digging, picking, dredging, or tonging, but are delivered directly to the receiving station or plant.

Problems involving toxins (eg, paralytic shellfish poisoning) or communicable diseases being transmitted to the consumer by mollusks have become increasingly more prevalent as the coastal waters become more polluted (8). These animals are static so they are particularly vulnerable to any fluctuating environmental pollution problem that may exist. Furthermore, the conditions causing the meat to be inedible cannot be detected by simple in-plant inspection. This has resulted in an intricate system involving surveillance of mollusk-growing areas by federal and local government agencies who are responsible for closing harvest areas when there is a potential problem.

Total Utilization

There is a growing emphasis on improving the total utilization of seafood raw material (8,9). For many years, only the most desirable portion of the fish, often accounting for 20 or 30% (eg, fillets) of the fish, was used. The remainder was considered waste or a raw material for preparing cheap animal feeds. Environmental and economic considerations dictate that this gross misuse of base raw material must be stopped. Hence, the modern attitude is that there is no such thing as fish waste. The portions remaining after the initial edible portion is removed should be considered secondary raw materials that can often equal or exceed the amount of the primary edible portion. The initial preprocessing operation must consider the ultimate total utilization destination of all portions of the raw material. Some of the products that can be prepared or manufactured from secondary raw materials are outlined in Table 7.

The developing operations and markets that use minced flesh for human consumption are beginning to have an impact on the economics of operating and the market for seafood products. Minced fish is used in engineered and formulated foods, the fastest growing segment of the food industry. Sources of minced flesh include

1. Frames (remaining skeleton) from a fillet operation.
2. Industrial fish presently being used for meal and oil.
3. Small fish currently being discarded or converted into fish meal that cannot be economically filleted or otherwise processed.
4. Freshwater fish presently being underutilized.
5. Low-fat fish that do not have good keeping properties due to rapid enzyme action (eg, pollock).

Minced flesh can be used in many products that require a protein base or binder. Such items include sausage (much lower saturated fat and calories than those made from pork), wieners, other cased meats, extruded meat products, nutrition-controlled foods (eg, low fat, high protein, and low sugar or carbohydrate), and other foods requiring ingredients with highly functional properties.

Table 7. Total Utilization of Secondary Raw Materials from the Sea[a]

Raw Material	Product Edible	Product Industrial
Finfish[b]	Steaks Roasts Fillets Minced flesh, surimi, formed foods, and extracted protein Specialty foods Refined oil Roe and milt Food additives	Meal, oil, pet foods, pharmaceutical raw materials, and leather (skins)
Industrial fish[b]	Minced flesh, surimi, formed foods, and extracted protein Specialty foods Refined oil Roe and milt Food additives	Meal, oil, pet foods, pharmaceutical raw materials, and leather (skins)
Shellfish	Whole (as caught) Portions Whole meat Minced meat Specialty products Food additives	Crustacea shell, extracted protein, chitin products, and calcium salts Bivalve shell land fill Meal and oil
Seaweed	Food Food additives	Pharmaceutical raw materials Extracted products for industry

[a] Ref. 8.
[b] Process residues are an important source of recyled high-protein ingredients for on-site preparation of aquaculture-raised fish and shellfish feed.

In the past, fish oils have been by-products from the production of fish meal. This low-grade oil has been used for industrial purposes or for making margarine. The large amount made into margarine has been manufactured in foreign countries because fish oil is not allowed for this purpose in the United States. The recent interest in fish oils for health has created a challenge for researchers to develop satisfactory refining techniques and subsequent edible products from the refined oils.

Some 30 yr ago it was recognized that fish oil has beneficial fatty acids that are active in preventing or minimizing the effects of certain cardiac problems and other diseases. However, it was not until the early 1980s that highly publicized work demonstrated that diets of Greenland Eskimos, high in marine fish and mammals, were associated with greatly reduced numbers of deaths due to ischemic heart disease when compared to populations in more developed countries consuming low fish and high animal fat diets (10). Fish oil, through the consumption of more fish and formulated foods containing fish oil are now considered a valuable contribution to a more healthful diet (11–13).

Heat Processing

Methods of Heating. Seafood, like any other food is cooked or heated to make it taste good by changing the texture and bringing out flavors and odors. If sufficient heat is added to pasteurize the food, microoganisms that can cause public health diseases and illnesses are destroyed or inactivated. Enzymes are also inactivated by heat so that protected cooked food stored under refrigeration is subjected to a minimum of degradation by hydrolysis.

The ultimate in heat processing is sterilization whereby the product is heated for a sufficient time at a given temperature. If the product is hermetically sealed so that there is no postprocessing contamination, it will have an indefinite shelf life free from degradation by microorganisms and enzymes.

Seafood is cooked and pasteurized by standard radiant energy baking ovens, infrared heating ovens, and microwave ovens. The heating in any oven is due to a combination of conduction, convection, and radiation. Food in a standard oven receives conduction heating from the pan or oven rack, convection heating from the air that is heated by the walls of the oven, and radiant heating from the exposed heating units. Infrared ovens do not have open heating elements because there is little convection heating from the air being heated by conduction.

Microwave heating is a specialized form of dielectric radiant heating that has certain advantages over other dielectric methods. Because the molecular polarizations are reversed many millions of times per second, the friction of molecules contacting each other cause heat and subsequent rapid, uniform heating of a food. Microwave heating has risen rapidly over the past decade. There are now microwave ovens in about 80% of U.S. households. Restaurants and institutional kitchens are rapidly increasing the use of this fast way of heating and cooking

Table 8. Effect of Processing on Omega-3 (n −3) Fatty Acids[a]

Species	Product	Lipid, %	Fatty Acid in Oil	
			EPA, %	DHA, %
Cod	Fresh	0.29	18.67	27.98
	Batter and breaded	0.47	12.53	12.25
	Batter and breaded, deep-fried in liquid vegetable oil	5.53	0.03	0.09
	Batter and breaded, deep-fried in solid vegetable fat	9.14	0.21	0.65
	Batter and breaded, deep-fried in beef shortening	7.07	0.66	1.51
Salmon, Sockeye	Fresh	—	6.25	9.73
	Canned	—	5.0	5.35
	Canned (added salmon oil)	—	5.76	8.61

[a] Ref. 9.

foods. The microwave oven with additional convection heating is becoming popular in that it combines the advantages of fast, uniform heating with the advantages of a conventional convection oven.

The growing popularity of microwave ovens has greatly increased the demand for microwavable foods. The major challenge to the food scientist developing microwavable foods is lack of radiant heat that causes the surface of a food to brown. Hence, cooked nondeep-fried potatoes, chicken, white fish, etc remain white on the surface and do not take on the normally expected desirable browning on the surface. Microwavable batter and breadings have been developed that are the color of deep-fried products. Hence, the future of microwaved foods will indeed be insured if low-fat, consumer-acceptable microwavable batter and breaded foods can be developed that take the place of deep-fried products. Fish fillets and formed patties from minced fish base stand to benefit as much from this development as any other food product. Fish sandwiches and batter and breaded fish have been growing in popularity, especially in fast-food restaurants. However, as shown in Table 8, the desirable omega-3 fatty acids suffer and the fat calories are tremendously increased when a fish is deep fried (9).

Commercial sterilization is carried out in steam retorts, which can process at temperatures above that of boiling water, normally at or above 117°C (242°F). The time and temperature required for sterilization must be sufficient to kill *Clostridium botulinum* spores that, when viable, can grow in an anaerobic (nonoxygen) atmosphere and produce lethal toxins. These spores must be held for 32 min at 110°C (230°F) to insure total destruction. Many low pH (high acidity) products such as certain fruits, vinegar-packed foods, and highly acid formulated foods prevent *C. botulinum* spores from growing. Many of these foods do not have to be sterilized at the temperatures required for high pH (low acidity). However, near neutral or high pH vacuum-packed products, such as seafood, are particularly vulnerable to anaerobic spore growth and sterilization must be insured.

To ensure a margin of safety, the sterilization requirement for canned fish is that the geometric center of a can or pouch must be held for 32 min at 116°C (240°F) (14). Each food and each different geometric form of container requires a different total processing time to accomplish sterilization. These processing times, determined by thermal death time laboratory studies, are mandatory for each processing company that is canning hermetically sealed food. Each batch of canned food must be coded and retort processing records kept to prove that the product was sterilized.

It should be emphasized that anaerobic conditions often prevail in a canned food even though an incomplete or no vacuum is drawn on the can prior to sealing. This is a result of the subsequent oxidation of components in a food by the remaining oxygen in the air.

Another precaution that must be practiced by processors is the venting of a steam retort prior to beginning the official retorting time. This is to prevent air pockets from insulating some of the cans so that there is nonuniform temperature in the retort. Of course, hydrostatic retorts that heat containers in a column of water kept above normal atmospheric boiling temperature by hydrostatic water pressure do not have a problem with entrapped air.

Effects of Heating. Fresh or frozen seafood being cooked for a meal should be heated for the minimum time required to improve the texture and the taste for the consumer. The normally dangerous microorganisms, those known as public health disease organisms are destroyed at relatively low temperatures. As shown in Table 9, the most heat resistant of the group, thermophiles, have an optimum growth at 50–60°C (122–150°F). Hence, a seafood is normally safe to eat if the geometric center has been raised above 150°F when it is actually pasteurized.

Overcooking causes heat degradation of nutrients, oxidation of vitamins and oils, and leaching of water-soluble

Table 9. Optimum Growth Temperature Range for Bacterial Groups[a]

Bacterial Group	Optimum Temperature Range
Psychrotrophs	14–20°C (58–68°F)
Mesophiles	30–37°C (86–98°F)
Facultative thermophiles	38–46°C (100–115°F)
Thermophiles	50–66°C (122–150°F)

[a] Ref. 14.

minerals and proteins. The retention of B vitamins, zinc, and iron is particularly important for populations that consume fish as the major source of meat in their diet. In addition, overcooking causes too much water to be released and the drying effect causes flesh to become tough, thus nulifying the desired effect of texture improvement.

Refrigeration and Freezing Technology

As has already been stressed, the most important factor in handling fresh fish is to lower the temperature to just above freezing as soon as it is removed from the water. Because the condition of the harvested fish and the subsequent handling determines the shelf life of a fresh fish, it is not possible to state exact times that a seafood can be held in ice or under refrigeration and be considered a high-quality food. This is shown in the wide range of shelf life that has been published in the literature (Fig. 12). In general, fish with a high oil content or enzyme activity have greatly reduced shelf life and are often of marginal quality after a few days.

High-quality seafood, when frozen properly soon after being removed from the water, is often far superior to fresh fish available on the market. This is due to the fact that all microbial action is stopped and enzyme action is significantly reduced in frozen fish. However, the initial quality of the fish being frozen, the rate of freezing, the temperature at which the frozen seafood is held, and the uniformity of the freezing temperature are all important to maintaining a high-quality product. It is surprising how many people involved in the seafood chain do not understand some of these basic factors in insuring the high quality of a seafood. Thus the constant challenge of those involved in seafood technology is the continual education and reeducation of everyone involved in the commercial seafood chain.

Freezing Seafood. During the freezing of seafood, structural changes take place in the cells and cell walls as well as in components that are between the cells. Many of these adverse changes are caused by water crystals that expand and rupture the cell walls. This allows liquid within the cells to leak out when the flesh is thawed. Hence, free liquid, called drip, exudes from seafood when it is thawed. This loss of free water reduces the water content of the seafood causing economic loss to the seller and greatly reduces the fresh qualities of the thawed product.

As seafood is cooled above and below approximately 28°F in a constant-temperature environment there is a near linear relationship between the temperature decrease and the time. However, as the water in the flesh begins to freeze, there is a long period of time during which the temperature remains almost constant. This period is a critical range for freezing and is caused by heat (heat of fusion) being removed from the fish to freeze water rather than to lower the temperature of the flesh. This relationship is shown in Figure 13.

The longer a product remains in the critical zone, the larger the ice crystals formed in the cells will be. When water is frozen rapidly, small crystals form and do not have time to increase in size due to the nucleation characteristics of a water molecule. Experience has shown that if seafood passes through the critical zone in ca 1.5 h, there will be little damage to cell structure as a result of large crystals formed during freezing. This is demonstrated by microphotographs of cells frozen under different conditions. Figure 14 shows cells (magnified 640 ×) of a rainbow trout that was rapidly frozen and passed through the critical zone in less than 1 h (15). Note the well-defined intact cell structure is similar to that of high-quality fresh flesh. Figure 15 is a similar microphotograph of cells from the other half of the same rainbow trout, which took approximately 5 h to pass through the critical zone. In this case extreme damage to the cell structure caused by large ice crystals can be noted. This is a dramatic demonstration of why fast freezing of seafood is essential to maintaining the quality of frozen products.

There is wide variation in commercial freezing facilities available to the seafood industry, and it is most important that commercial operators carefully study their options when purchasing and installing new equipment.

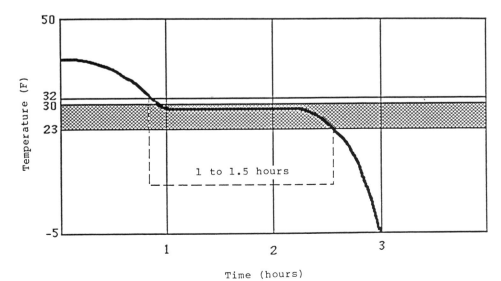

Figure 13. Idealized freezing curve for fish muscle.

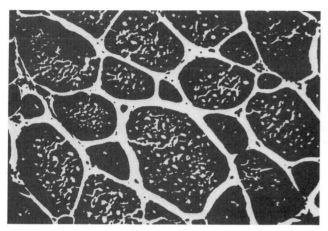

Figure 14. Cell structure of rainbow trout frozen rapidly (× 640) (15).

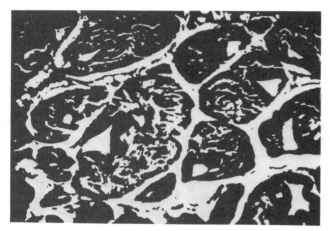

Figure 15. Cell structure of rainbow trout frozen slowly (× 640) (15).

The most efficient methods of freezing involve both conduction and convection freezing in an immersion system. Liquid refrigerants such as the freons have been used for this purpose but have proven too costly. Immersion freezing using refrigerated brine is fast and efficient but a considerable amount of salt is absorbed into the flesh during freezing and can be a problem to product marketing. Refrigerated seawater is used extensively, especially in Alaska for rapidly chilling and holding the fresh catch on shipboard.

Plate freezers rapidly freeze, by conduction, products that are in contact with the plates. Plate freezers are used to their best advantage when the plates can be brought together to contact both sides of a package. This is the case for rectangular packaged seafood portions or prepared formulations. A disadvantage is that irregular items, particularly large fish, can only contact the plate on one side and the exposed side must be frozen by convection. Another disadvantage is using plate freezers for whole fish, because the plate only contacts one side of the large irregular shaped item and causes one side of the finished product to be flat while the other side is the nice rounded shape of the live fish. However, unless the fish are suspended by hooks, this same condition prevails in any freezer where the fish are placed on shelves during freezing.

Convection blast freezing, where cold air is circulated over the products, is less efficient than conduction freezing but can be used for larger volumes. A disadvantage to blast freezing is that the rapidly moving dry air can dehydrate a product that is exposed for any length of time. This can be eliminated by proper packaging protection, but it also greatly decreases the heat transfer rate and lengthens the time of freezing and the subsequent time at which the product is in the critical range.

The problems with blast freezing can often be minimized or eliminated when a combined blast–plate-freezing system is used. This can be demonstrated by comparing Figures 16 and 17 in which chum salmon weighing 4 lb were frozen in two commercial units (16). Figure 16 is the freezing curve for a conventional blast freezer in

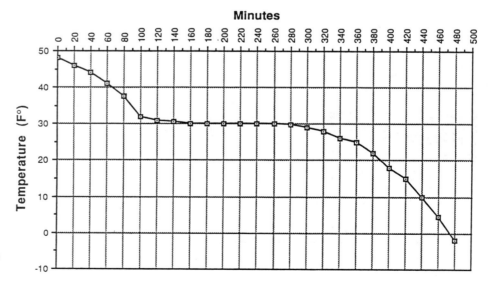

Figure 16. Freezing curve for blast freezing of chum salmon (4 lb).

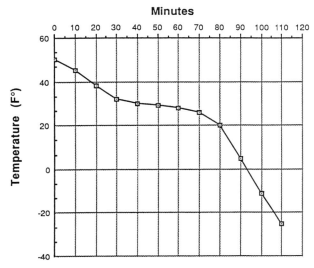

Figure 17. Freezing curve for blast-plate freezing of chum salmon (4 lb).

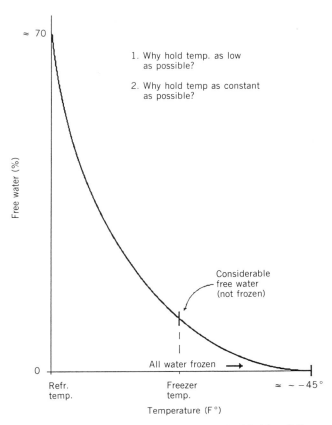

Figure 18. Remaining free water in frozen food held at different temperatures.

which whole (headed and gutted) fish were placed on racks and frozen. Note that the time in the critical range was about 4 h and the total time to reduce the temperature to 0°F was about 8 h. The freezing curve shown in Figure 17 was that of a 4-lb chum salmon in a combined blast and plate freezer. The time in the critical range for this system was about 40 min, and the total freezing time to reach 0°F was slightly over 1.5 h. The fish frozen in the combined blast–plate freezer were far superior to those frozen in the blast freezer alone. Furthermore, the reduction in weight due to drip loss gave a significant economic advantage to the more rapidly frozen fish.

Holding Seafood in Cold Storage. Good freezing practices involving fast freezing and minimal dehydration of fresh seafood insures that high-quality products are delivered to the cold storage for holding. Equally important to the continuing maintenance of high quality is the environment under which the products are kept during the cold storage period. Unless frozen seafood is kept at extremely low temperatures, there is a certain amount of free water (not frozen) remaining in the product. This is a result of the antifreeze effect of soluble salts in the cell, which become more concentrated as water freezes. Depending on the specific food, the physical and chemical conditions of the food, and the composition (including water content), the point at which all of the water is frozen is in the range of $-45°F$. Because few, if any commercial cold storages are held at that low temperature, there is a certain amount of free water remaining in all frozen food (9). This is depicted in Figure 18, indicating that the normal commercial cold storage temperatures are well above $-45°F$. In fact, different commercial cold storage warehouses in the United States range in temperature from slightly above 0°F to a low that seldom is below $-20°F$.

Not only is the average commercial cold storage facility maintained at a temperature at which several percent of the water remain unfrozen but there is normally a significant fluctuation in the base cold storage temperature. Thus when the temperature fluctuates above and below the average, some of the water in the product is continually frozen, thawed, and refrozen. The effect of this fluctuation is to greatly increase the effect of enzyme action that essentially digests the protein in the same manner that it is digested in the gut of an animal. This is emphasized by the fact that continual thawing and refreezing is not limited to the original free water and causes the enzyme action to spread throughout the flesh. This is why the meat in seafood held for long periods can be extremely soft when thawed. This effect is even more noticeable in many home freezer units where temperature fluctuations are greater than those found in commercial facilities. Hence, holding a product at a higher but constant cold storage temperature can result in a better product than when it is held at a fluctuating lower temperature. This concept is depicted in Figure 19.

It can be important to know how much fish a given size cold storage unit will hold. The true density of a seafood ranges from 70 to 80 lb/ft^2. Because frozen fish blocks are essentially composed of solid fish flesh, they have about the same density as the natural flesh. At the other end of the spectrum, individually frozen and loosely packed fish range from 30 to 50 lb/ft^2. A well-run cold storage, allowing for the average distribution of product forms and allowing for air spaces and movement within the room, usually holds about 20–30 lb/ft^2.

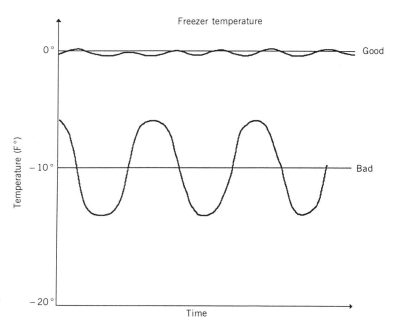

Figure 19. Time-temperature relationship of food during cold storage holding.

Refreezing Seafood. Often it is necessary for a processing plant to receive frozen fish (either whole, butchered, or partially processed portions) for subsequent final processing or reprocessing. This occurs when fish are frozen at sea or other remote areas where total processing is not practical due to limited facilities or economic considerations. A typical example is when fish blocks are shipped to a plant for thawing, trimming, battering and breading, and refreezing. A similar situation occurs when frozen fish or preprocessed raw materials are held in cold storage until they are processed in response to a specific market demand. For example, if fresh salmon are cut into steaks and frozen, there is no opportunity to sell the fish in the fillet form if the demand or price of fillets takes a sudden increase.

It is often said that seafood cannot be thawed and refrozen. This is based on the too-often encountered situation whereby the fresh fish has been abused, the products have not been frozen rapidly, or have been held at high or fluctuating cold storage conditions. However, if the fish have been properly frozen and held in cold storage as described above, there will be little cell damage and the thawed product will have minimum water loss and reduction in quality from that of the original fresh raw material. Thus it can be thawed, reprocessed, and refrozen without significantly altering the quality of the finished product from that of the fresh fish.

The increase of high-seas processing vessels encouraged by the 200-mi limit (Fishery Conservation Zone) has greatly accelerated the final processing of frozen products prepared at sea. The future will probably bring an even greater acceleration of this trend as such products as batter and breaded (nondeep-fried) portions, whole fish, and products engineered and formulated from minced flesh or chunk portions are being developed for microwave cooking. These are inexpensive low-fat products with good sensory properties especially acceptable to the increasingly large number of nutrition-conscious consumers.

Commercial Refrigeration Systems

There is a distinct difference between selecting facilities for freezing or for cold storage for a commercial seafood operation.

Freezing Facilities. The freezing facility is related to the products that will be processed, the throughput of product, and the space availability in the processing area. The differences between various freezing techniques lie in the control of the type of heat transfer between the product and the refrigerant. The types of freezer as related to the products being frozen are as follows.

Natural Convention Freezing. This facility is a room or chamber in which the product is frozen by natural convection. There are minimum problems with dehydration but the freezing rate is slow. Only products not affected by slow freezing should be frozen in this type of freezer. These would include formulated products or very small items where the water loss or cell damage through slow freezing is not a problem.

Combined Conduction and Natural Convection. The addition of freezer plates on which the product is placed greatly accelerates the freezing rate and maintains the advantage of minimum hydration during freezing.

Blast Convection Freezing. Blast freezing is a popular method of freezing irregularly shaped seafood items such as whole or partially dressed fish. A major disadvantage of this facility is that considerable loss of water through hydration can cause an unsightly surface condition known as freezer burn. This hydration, if allowed to continue, can remove a considerable amount of water from the flesh, making the fish inedible. Such is the case when a product is frozen and then allowed to remain (actually to be stored) in the freezer for some length of time before being removed.

When unpackaged products are destined for storage in a cold room with fast moving air, they are glazed to protect against dehydration. This consists of immersing the frozen product in cold water and then allowing the water film to freeze on the surface. The layer of ice protects the fish, because the air removes water from the glaze rather than from the product.

Combined Conduction and Blast Convection Freezing. This facility normally consists of refrigerated plates in a forced-air cold room. This freezer has the advantage of fast conduction freezing assisted by convection air on portions of the seafood that are not in contact with the plate surface. The tremendous increase in freezing rate by this combination of freezing techniques as compared to that of a blast freezer were previously discussed and compared in Figures 16 and 17. Normally the freezing rate is high enough to eliminate the problem of surface dehydration.

Conduction Freezing. When small rectangular products or packages with two parallel surfaces are placed between two refrigerated plates in a cold room or chamber, fast freezing takes place with minimum harm to the products. These types of facility have mechanisms that allow the plates to be vertically adjusted. This allows the plates to open, or separate, during loading and then they come in contact with the product for freezing.

Immersion and Cryogenic Freezing. Immersion in a cold liquid that will not affect the safety of the product (eg, brine, liquid nitrogen, or freon) is the fastest method of freezing. Freezing takes place by both convection from the circulating fluid and conduction from being in direct contact with the liquid. Immersion freezing in extremely cold refrigerants such as nitrogen or freon is called cryogenic freezing. Brine freezing utilizes saturated salt solutions that freeze rapidly due to the combined convection and conduction but takes place at about 0°F.

Cryogenic freezing has two limiting factors for extensive use in freezing seafood products. The first is that the freezing is so rapid that it causes extreme internal tension due to the freezing of the fiberous materials at a differential rate. This causes the flesh to rupture. It can be minimized by allowing the product to temper at room temperature before further handling, packaging, and placing in cold storage. The splitting problem can be fairly well overcome for small items such as fillets and steaks but large fish cannot be satisfactorily frozen by this means. Brine, on the other hand does not have the splitting problem and is a major means of freezing whole tuna fish on high-seas catcher vessels.

The second problem with cryogenic freezing is that it is uneconomical to operate unless the freezer is used continuously, because a considerable amount of heat from the cryogen is lost each time a processing unit is shut down and restarted. Most seafood-processing plants do not operate on an extended basis so that the cost per pound for intermittent freezing limits cryogenic freezing to a few fish processing plants.

Modern cryogenic freezing no longer uses liquid immersion true freezing. It is more economical to spray the liquid refrigerant on the food as it moves along on a conveyor belt. This uses less refrigerant and still gives the same advantages as immersion. Furthermore, when extremely cold liquids are used for freezing, especially in immersion freezing, vapor bubbles are formed that insulate the liquid from the product and prevent the rapid heat transfer expected from such a large temperature difference between the product and the liquid.

Cold Storage Facilities. Cold storage facilities must meet the requirements for long-term storage of commercial seafood products ranging from large whole glazed fish to cases of packaged retail products. Of particular importance in purchasing or contracting for such facilities in a plant or choosing a public cold storage for use include the following:

1. A design with proper insulation and construction materials that will insure a minimum of heat loss through the walls, ceiling, and floor.
2. Properly designed protection to minimize heat loss through doors and other openings while the product is being taken into or out of the cold storage room.
3. Refrigeration machinery that will have sufficient capacity to hold the cold storage at the desired temperature.
4. Sufficient refrigeration capacity to insure that the temperature in the cold storage rooms does not fluctuate to the extent of adversely affecting the products being stored. This is probably the major problem encountered with commercial cold storage facilities. So often a low bid is awarded to a contractor who is the lowest due to cutting back on the amount of refrigeration machinery and thus the ability to maintain constant storage temperature during variable and seasonal outside weather conditions.

Cured and Dried Seafood Products

Control of water activity, defined as the ratio of the vapor pressure of water in a product at any given temperature divided by the vapor pressure of pure water at the same temperature, not only applies to smoking and drying but also to salting, pickling, and product formulation. Product stability, the growth of microorganisms, and chemical reactions occurring during processing and storage are all dependent on the water activity, that is, the ability of water to move and interact with other ingredients in the food. The importance of water activity (a_w) in foods is shown in Table 10. The many facets of water activity and its relation to the preservation, safety, and shelf life of foods has been summarized by a group of internationally recognized experts (17).

Dehydration. Drying or dehydration is a means of controlling water activity by reducing the water content of a product. Many dried products, such as cereal grains, legumes, and many nuts and fruits, are dried by nature in the field prior to harvesting. In the past humans dried seafood products in the sun, long before they were aware that they were controlling water activity. In fact, today many developing countries located in tropical parts of the

Table 10. Importance of Water Activity in Foods[a]

a_w	Phenomena	Food Examples
1.0		Water-rich foods ($a_w = 0.90–1.0$): foods with 40%
0.95		sucrose or 7% NaCl, cooked sausages, bread crumbs, and kippered fish
0.90	General lower limit for bacterial growth	Foods with 55% sucrose or 12% NaCl, dry ham, medium-age cheese, and hard-smoked fish
0.85	Lower limit for growth of most yeast	Intermediate-moisture foods ($a_w = 0.55–0.90$): foods with 65% sucrose or 15% NaCl, salami, old cheese, and salt fish
0.80	Lower limit for activity of most enzymes	Flour, rice (15–17% water), fruitcake, and sweetened condensed milk
0.75	Lower limit for halophilic bacteria	Foods with 26% NaCl (satd), marzipan (15–17% water), and jams
0.70	Lower limit for growth of most xerophilic (dry loving) molds	
0.65	Maximum velocity of Maillard reactions	Rolled oats (10% water)
0.60	Lower limit for growth osmphilic or xerophilic yeasts and molds	Dried fruits (15–20% water), toffees, and caramels (8% water)
0.55	DNA becomes disordered (lower limit for life to continue)	Dried foods (a_w 0–0.55)
0.50		Noodles (12% water), spices (10% water), and fish protein concentrate (10% water)
0.40	Minimum oxidation velocity	Whole egg powder (5% water)
0.30		Crackers and crusts (3–5% water)
0.25	Maximum heat resistance of bacterial spores	
0.20		Whole mild powder (2–3% water), dried vegetables (5% water), and cornflakes (5% water)
0.00	Maximum oxidation velocity	

[a] Ref. 17.

world use the sun as a major means of drying seafood and other food products.

The principal cured fishery products produced in the United States are shown in Table 11 (2). It should be noted that these statistics do not differentiate between smoked, salted, dried, and pickled products due to the fact that all of the processes are based on controlling or reducing water content. It is often difficult to distinguish between process classifications. For example, salted and salted-dried fish have a different final moisture content but the mechanism of removing the water and stabilizing the product are the same, control of water activity.

The most efficient means of drying a seafood product is through dehydration by forced-air drying, vacuum drying, or vacuum freeze-drying. In each case the drying mechanism is a combination of adding heat to increase the temperature and vapor-driving forces between the product and the environment. The drying time is divided into two distinct periods: constant drying rate and falling drying rate. During the constant rate period, all of the heat added to the product is used to evaporate water from the surface and near surface of the product. In this case, there is free water in contact with the environment and drying occurs similarly to that of an open container of water. During the falling rate period, part of the heat energy is imparted to the product to cause water to migrate to the surface. Therefore, the product is heated during this period of the drying cycle.

Excellent highly nutritious dried formulated fish-base products can be prepared from the minced flesh of seafood. Shaped into forms such as patties and air dried, these items have a long and stable shelf life (18).

Curing. Whereas dehydration removes sufficient water to inhibit growth of microorganisms, curing consists of adding sufficient chemicals (eg, sodium chloride, sugars, and acetic acid) to prevent degradation of a product by microorganisms. Although sufficient water is not removed to accomplish this objective, the water activity is reduced to the point where growth is prevented.

Curing methods currently practiced include dry salting, where split fish is covered with salt and the brine liquor is allowed to escape, and pickling where products are immersed in a strong brine, or pickle, allowing salt to penetrate the product and water to be exuded into the brine solution. Low-fat white fish such as cod are dry salted by the heavy (hard) cure and fatty fish are cured in airtight barrels by the Gaspe (light) cure. Figure 20 shows the process for the hard cure; the last step is air drying, which results in a salted-dried product with long-term, room temperature shelf life and a water activity of between 0.75 and 0.85. The Gaspe or light-cured product remains edible only a few days in the wet-stack stage ($a_w = 0.85–0.90$) at room temperatures and must be pressed and mechanically dried for longer-term storage.

Smoking. The age-old practice of smoking has changed drastically over the last few decades. The process as originally practiced by Eskimos and American Indians to preserve fish for the winter months was essentially a drying

Table 11. U.S. Production of Principal Cured Fishery Products[a]

Item	Thousand Pounds	Thousand Dollars
Salted and Pickled		
Cod	2,633	2,509
Halibut	45	73
Herring		
Lake	460	185
Sea	12,190	18,102
Mullet	133	79
Sablefish	1,032	1,404
Salmon	11,386	32,024
Total	27,819	54,376
Smoked and Kippered		
Carp	153	134
Chubs	2,151	3,971
Cod	210	588
Eels	184	669
Halibut	281	630
Herring		
Lake	38	52
Sea	963	1,876
King mackerel	13	52
Lake trout	84	167
Marlin	9	43
Mullet	137	212
Paddlefish	257	469
Pollock, Pacific	900	875
Sablefish	1,880	4,157
Salmon	13,071	63,532
Sturgeon	377	1,338
Trout, unclassified	242	406
Tuna	48	108
Tunalike fish, bonita, yellowtail	4	3
Whitefish	2,910	5,526
Whiting	1,028	485
Unclassified fish and crustaceans	293	746
Total	25,233	86,039
Dried		
Cod	579	971
Shrimp	368	2,064
Total	947	3,035

[a] Ref. 2.

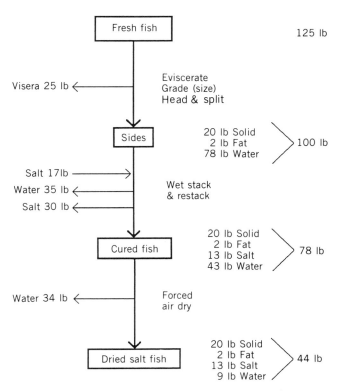

Figure 20. Process for hard curing and drying of fish.

process whereby heat from a fire was used to reduce the moisture content sufficiently for extended storage. The smoke flavor was somewhat incidental to the process. In fact in some areas the fish were dried by the sun and the natural air currents and the smoke was used to prevent flies and insects from consuming and contaminating the product. This dried smoked fish, which takes days to cure, is known as hard smoked.

Today, most smoked fish is smoked for the flavor and there is relatively little loss of water in the process. The change in processing is a reaction to the consumer, who prefers a soft, moist texture rather than the tough texture of a dried product, and to the processor, who cannot afford to tie up large processing areas for longer-term smoking. Furthermore, the minimizing of moisture loss greatly improves the economics of processing and marketing smoked products. Hence, although smoked fish are considered to be processed in the same manner as fish in which water activity is altered, in reality the modern product is a partially or wholly cooked fish that has smoke added as a condiment (19).

Most of the smoked fish prepared in the United States has little shelf life stability beyond that of a fresh fish. The commercial process of smoking involves splitting and cleaning the fish, salting or brining (soaking in a brine solution) to firm the texture of the meat, draining to remove excess moisture, and smoking. Smoking is carried out as a cold smoke, the temperature of the smoke does not rise above 85°F, or as a hot smoke, the smoke is hot enough (eg, 250°F) to raise the center temperature of the fish to above 140°F. It is is also common to smoke the fish with colder smoke and then to raise the smoke and air temperature during the terminal part of the smoking to pasteurize the fish.

Today, much of the smoking is carried out in commercially constructed smoking facilities, or kilns, that have smoke generators (using sawdust), controlled temperature forced air, and humidity control of the air. There are as many specific smoking procedures as there are processors in the business. Each processor has a favorite method, including the use of certain additives in the brine, controlled-temperature drying, smoking, cooking, and cooling. Modern kiln smoking has allowed precise control of the variables in smoking and has removed much of the artisan approach that was so prevalent when all smoking was carried out using open wood fires.

Specialty hard-smoked fish, known as jerky, is prepared by smoking and drying fish to a hard, chewy consistency. This product is popular in the bar and tavern trade and for hikers who wish to carry a meat product that will not spoil during a several-day outing.

Irradiation

The United States is out of step with the rest of the world when considering the use of irradiation for preserving food. During the early 1980s, the World Health Organization gave ionizing radiation its blessing after an extensive review of many years of scientific work and investigation (20). Since that time, many countries in the world have been using irradiation on a wide variety of foods (9). Although there has been steady progress in the United States toward approval of the process, there is still a widespread disagreement between scientific, government, and consumer groups regarding the pros and cons of using this form of energy for processing foods (21).

Two factors probably have equally contributed to this battle, namely the association in the minds of people between ionizing radiation and nuclear warfare, and the DeLaney Clause of the 1958 Food Additive Amendment to the Food, Drug and Cosmetic Law. The amendment, best known for the prohibition of a food additive that is shown to be carcinogenic in a test animal at any dose level, states that irradiation is a food additive and not a food-processing technique. It is interesting to note that the U.S. Department of Agriculture has issued an extensive list of publications that show the safety and wholesomeness of irradiated foods (22).

Some confusion exists about the units used for measuring radiation doses that a product receives. This is because the original unit of measurement, the rad, was changed to kilograys (kGy); 100,000 rad (100 krad or 1 Mrad) is equivalent to 1 kGy (100,000 Gy). As shown in Table 12, the doses for processing range from 0.75–2.5 kGy for pasteurization to 30–40 kGy for sterilization. Note that specific process terminology has been suggested for the various operations. Pasteurization is radurization, sanitization is radicidation, deinfestation is the process for destroying eggs and larvae, and sterilization is radappertization. Two radioisotopes, Co-60 and Ces-137, and electron-accelerator–generated electron beams meet the requirements for producing sufficient energy and intensity to penetrate food and accomplish the four basic processes.

The radurization of fresh seafood has an excellent potential for extending shelf life. Being in the low-dose range, this process should not meet the resistance that is found for the publicized high doses that accomplish sterilization, but result in some off-flavor (23). The extra week of shelf life that can be given to a fresh fish by radurization would be of economic benefit to the entire industry. The extra shelf life would eliminate product loss at the retail level and would extend the market range for delivering fresh fish.

There is every reason to believe that irradiation will eventually become an accepted and extensively used processing technique for seafood, especially seafood for the fresh market, where all parties will benefit. Perhaps during the next decade the radurization of seafoods will be commonplace. After all, 20 yr ago there was the same consternation over the use of microwave ovens, and today they are in 80% of U.S. homes.

Packaging

Proper packaging of products is necessary to protect food from adverse losses and changes in weight, texture, flavor, nutritional components, and protection against contamination and physical damage. With the growing demand for high-quality seafood, the major emphasis is on packaging for fresh and frozen products. Basic types of packaging, in order of increasing ability to protect fresh or frozen food, include paper, coated paper, fiber, foil, films, laminates, and combinations.

One of the primary considerations in selecting packaging for fresh and frozen foods is the gas and water permeability of the material, usually plastic films, plastic containers, and treated paper products. There is always a balance between the cost of the package and the degree of protection given to the seafood. A very low cost uncoated cellophane has a moisture permeability that is so high that frozen and even cold fresh products are rapidly desiccated during holding. A film such as poly(vinyl chloride) (PVC) costs more but gives moisture-barrier protection that improves the overall economics of the processing and marketing chain.

The same economic analysis must be made for gas permeability, because oxygen entering the package tends to oxidize lipids and give off-flavors while some of the desirable volatile taste and odor components may be lost from the product. The fact that some films give good vapor-barrier protection but allow moisture to pass while others act in a reverse manner has encouraged the development of laminated films for food packaging. Laminates for seafood packaging often consist of polyethylene (which is cheap, is heat sealable, is usable over a wide range of temperatures, and is a good moisture barrier) and another film that has a low permeability for gas (eg, CO_2, O_2, and N_2). Examples of laminates used in frozen seafood products include polyethylene laminated with low-gas-permeability poly(vinylidene chloride) (PVDC) or saran), polyester, cellulose, aluminum, or nylon.

Packaging not only protects the food but is also an

Table 12. Dose Ranges for Food Irradiation Processes[a]

Process	Dose Range	
Radurization (pasteurization)	75–250 krad	0.75–2.5 kGy
Radicidation (sanitization)	250–1,000 krad	2.5–10 kGy
Destroy eggs/larvae (deinfestation)	below 100 krad	below 1 kGy
Radappertization (sterilization)	3–4 Mrad	30–40 kGy

[a] Ref. 23.

important factor in the sales appeal of the product. Packaging that shrinks film to give better evacuation of air from the package, overwraps for multiple packages and trays, overwraps to give rigidity to a package, allows see-through to the product, uses vacuum to minimize oxidation of the product, and uses gas flushing to expel further air are all popular methods of protecting processed seafoods. Packaging materials have become quite sophisticated and the accompanying use technology has been developed to meet the requirements for economic applications, long-term storage of products (particularly frozen), and enhanced consumer appeal.

Over the last decade, the market for fresh food products has become increasingly popular. A system combining closely controlled combinations of low temperature, high humidity, and proper ventilation has allowed extensive increase of the shelf life of perishable fresh foods. This system, hypobarics, was awarded the Institute for Food Technology's Technology Industrial Achievement Award in 1979 (24). Although hypobarics is a container system rather than a package, it can be considered a package in that it is used to contain a fresh product during the plant storage and transport time required to move the product to market. Originally developed for fruit, the system was extended to all fresh food items including seafood. Although the system has not been used extensively on seafood, it has been very successful with numerous products and should certainly be mentioned as a potential use for extending the shelf life of fresh seafood.

Another important area involving packaging technology is modified-atmosphere packaging (MAP), which has been shown to increase the shelf life of many perishable foods and formulated products by reported values of from 50 to 400% (25). However, there has been considerable concern by both regulatory authorities and researchers over maintaining the safety of MAP foods as they travel through the channels of commerce. The system, consisting of modifying the gaseous environment with a package, results in various combinations of carbon dioxide, nitrogen, and oxygen. Carbon dioxide is the important factor, replacing much of the oxygen and nitrogen in the air (eg, 75% CO_2, 15% N_2, and 10% O_2). Of concern is maintenance of the proper refrigeration temperatures and the gas ratios for a given product so that microbial growth does not present an undue safety hazard.

A considerable amount of research has been carried out in the MAP of seafood products (26). A significant amount of effort is being expended on developing the processor-distributor combination to improve quality and value of fresh seafoods.

SEAFOOD SAFETY

The discussion of MAP is a good place to end the packaging discussion because it leads to the subject of general safety of foods as presented to the consumer. Often packaging is a major factor involved in these concerns, because the environment created by the combination of processing and packaging often creates potential for adverse microbial growth. There are two programs, one specifically for shellfish and one for refrigerated foods, that are important in reducing the microbiological risks associated with harvesting, handling, processing, and distributing to the consumer.

The National Shellfish Sanitation Program

The National Shellfish Sanitation Program (NSSP) is designed to prevent harvesting of shellfish in polluted waters containing pathogenic organisms or other contaminants (27). The program, begun in 1925 following illnesses culminating in typhoid fever outbreaks caused by contaminated sewage, is administered cooperatively among the federal government, states, and industry through the interstate Shellfish Sanitation Conference. The Food and Drug Administration is responsible for appraising each state's shellfish program to insure that they are complying with the specified requirements. NSSP gives each state the responsibility of defining or classifying its waters from which shellfish are harvested. The waters are tested and analyzed on a continual basis and classified as follows:

1. *Approved.* These waters may be harvested for the direct marketing of shellfish at all times.
2. *Conditionally Approved.* These waters do not meet the criteria for the approved waters at all times but may be harvested when criteria are met.
3. *Restricted.* Shellfish may be harvested from restricted waters if subjected to a suitable purification process.
4. *Prohibited.* Harvest for human consumption cannot occur at any time.

The term harvest limited is used to refer to conditionally approved, restricted, or prohibited waters. A closure area is an area in which some restriction on harvest has been placed (eg, a harvest-limited area).

Hazard Analysis and Critical Control Point Principles

As the popularity and volume of refrigerated foods have been significantly increasing over the last decade, there has been an accompanying increase in the risks of food-borne disease. The overall safety record of a wide variety traditional refrigerated fresh fruits and vegetables; raw meats, poultry, and seafood; pasteurized milk; cured, ready-to-eat meats and seafood; high-moisture cheeses, yogurt, and pickles; and perishable delicatessan products has been good. However, the complexity of food-distribution systems presents risk factors that are increasing. This is especially the case of a new generation of engineered and formulated foods including frozen or restaurant dinners; frozen or delicatessan entrées; dry, frozen, or canned pasta; fresh, refrigerated salads; canned, frozen, or dry gravies; canned or dry soups; and frozen, canned, or refrigerated cooked meats and seafood (28).

In response to this increasing hazard, the National Academy of Science has studied the problem and recommended a system, the hazard analysis and critical control point (HACCP) system, which "provides a more specific and critical approach to the control of microbiological hazard in foods than that provided by traditional inspection and quality control approaches" (29). The system, similiar to the critical control point system used by engineers in planning design, construction, and startup of a building or operation, include the following principles:

1. Describing the product and how the consumer will use it.
2. Preparing a flow diagram for intended manufacturing and distribution of the product.
3. Conducting risk analysis for ingredients, product, and packaging; reducing the risks by making changes to the design; and incorporating these changes into the processing and packaging schemes.
4. Selecting critical control points (CCP) and designating their location on the flow diagram and describing CCP, establishing monitoring.
5. Implementing HACCP in routine activities.

The assurance of food safety is a complex subject that impacts all phases of the commercial food chain, ending with the consumer. HACCP undoubtedly will have a strong influence on the future of many food operations. This impact will especially affect the seafood industry with its myriad safety and sanitation problems, many of which are specific to products from the marine and freshwater environments.

FUTURE DEVELOPMENT

Fish is health food is not a marketing gimmick but truly states the merits of seafood. The demand for this highly nutritious protein food, much avowed for its healthful omega-3 fatty acids, will continue to grow in demand. This pressure on a limited resource will encourage better biological management of the natural wild stocks (including international cooperation), faster expansion of aquaculture operations, and the total utilization of the raw material.

Much of the so-called industrial fish and the waste from present processing operations, currently being reduced to cheap animal foods, can and will be upgraded to human foods. Headed and gutted industrial fish can be deboned to give a highly acceptable minced flesh for engineered and formulated foods. A fish frame from which fillets have been removed contains as much meat as the fillets. This meat can be removed by deboning, and the minced flesh, nutritionally equal to the fillet or to any other minced flesh, is a tremendous source of base protein materials for formulated foods. Modern processing machinery, growing knowledge about the technology of formulated foods, and the demand for high-quality prepared foods support the trend to use previously underutilized seafood and seafood portions for this market.

As has been discussed, the technology of handling, packaging, and transporting fresh seafoods is rapidly developing and the future will see more fresh seafood on the market. In conclusion there is an exciting future for seafood in the United States. The new generation of refrigerated and frozen seafood and formulated products, as well as better quality traditional items, will continue to encourage increased consumption of both wild and farmed fish and shellfish.

BIBLIOGRAPHY

1. S. N. Jhavari, P. S. Karakoltsidis, J. Montecalvo, Jr., and S. M. Constantinides, *Journal of Food Science* **49**, 110 (1984).
2. National Marine Fisheries Service, *Current Fishery Statistics No. 8800,* United States Department of Commerce, Washington, D.C., 1988.
3. U.S. Department of Agriculture, "Aquaculture Situation and Outlook Report," *AQUA 1,* 31 (Oct. 1988).
4. "Farm-Raised Catfish," *Aquaculture Magazine* **18**(4), 81 (1989).
5. L. A. Nielsen and D. L. Johnson, *Fisheries,* American Fisheries Society, Bethesda, Md., 1983.
6. K. S. Hildebrand, unpublished data, 1984.
7. G. M. Pigott, "Total Utilization of Raw Materials from the Sea," *Proceedings of the Conference on Formed Foods,* Brigham Young University, Brighton, Utah, Apr. 1–2, 1985.
8. J. J. Sullivan, M. G. Simon, and W. T. Iwaoka, "Comparison of HPLC and Mouse Bioassay Methods for Determining PSP Toxins in Shellfish," *Journal of Food Science* **48**(4), 1321 (1983).
9. G. M. Pigott and B. W. Tucker, *Seafood: The Effect of Technology on Nutrition,* Marcel Dekker, Inc., New York, 1990.
10. H. O. Bang and J. Dyerberg, "Plasma Lipid and Lipoprotein in Greenlandic West Coast Eskimos," *Acta Medica Scandinavica* **192**, 85–94 (1972).
11. G. M. Pigott and B. W. Tucker, "Science Opens New Horizons for Marine Lipids in Human Nutrition," *Food Reviews International* **3**(1,2), 105–138 (1987).
12. W. E. M. Lands, "Fish and Human Health: A Story Unfolding," *World Aquaculture* **20**(1), 59–62 (1989).
13. B. W. Tucker, "Sterols in Seafood: A Review," *World Aquaculture* **20**(1), 69–72 (1989).
14. *Canned Foods,* The Food Processors Institute, Washington, D.C., 1980.
15. R. A. Bello and G. M. Pigott, "Ultrastructural Study of Skeletal Fish Muscle after Freezing at Different Rates, *Journal of Food Science* **47**(5), 1389–1394 (1982.)
16. G. M. Pigott, insert in Ref. 9.
17. L. B. Rockland and L. R. Beuchet, eds., *Water Activity: Theory and Applications to Food,* Marcel Dekker, Inc., New York, 1987.
18. R. A. Bello and G. M. Pigott, "Dried Fish Patties: Storage Stability and Economic Considerations," *Journal of Food Science* **4**, 247–260 (1980).
19. G. M. Pigott, "Smoking Fish—Special Considerations," *Proceedings of the Smoked Fish Conference Symposium,* University of Alaska and University of Washington Sea Grant, Seattle, Wash., Apr. 27–29, 1981.
20. Codex Alimentarius Commission, *Microbiological Safety of Irradiated Foods,* FAO/World Health Organization Joint Office, Rome, Italy, 1983.
21. G. Giddings, Irradiation: Progress or Peril?" *Prepared Foods,* **158**(9), 62–67 (1989).
22. U.S. Department of Agriculture, *Safety and Wholesomeness of Irradiated Foods,* The National Agricultural Library, Beltsville, Md., 1986.

23. G. M. Pigott, "Radurization of Aquaculture Fish: A Value-Added Processing Technology of the Future," *The Proceedings of Aquaculture International Congress and Exposition,* Vancouver, B.C., Sept. 6–9, 1988.
24. N. H. Mermelstein, "Hypobaric Transport and Storage of Fresh Meats and Produce Earns 1979 IFT Food Technology Industrial Achievement Award," *Food Technology* **33**, 32–40 (1979).
25. J. H. Hotchkiss, "Experimental Approaches to Determining the Safety of Food Packaged in Modified Atmospheres," *Food Technology* **42**(9), 55–64 (1988).
26. M. F. Layrisse and J. P. Matches, "Microbiological and Clinical Changes in Spotted Shrimp Stored under Modified Atmosphere," *Journal of Food Protection* **47**(6), 453–457 (1984).
27. D. L. Leonard, M. A. Broutman, and K. E. Harkness, *The Quality of Shellfish Growing Waters on the East Coast of the United States,* Ocean Assessments Division, NOAA, Rockville, Md., 1989.
28. D. A. Corlett, Jr., "Refrigerated Foods and Use of Hazard Analysis and Critical Control Point Principles," *Food Technology* **43**(2), 91–94 (1989).
29. Food Protection Committee, Subcommittee on Microbiological Criteria, *An Evaluation of the Role of Microbiological Criteria for Foods and Food Ingredients,* National Academy of Sciences, National Research Council, National Academy Press, Washington, D.C., 1985.

GEORGE M. PIGOTT
University of Washington
Seattle, Washington

FISH CAKE (KAMABOKO)

Kamaboko is a typical fish-based food with the form of homogeneous protein gel. When salt is added to fish meat and the mixture is fully kneaded, meat paste is obtained. Then the paste is shaped and heated by steaming, frying, or grilling to make protein-gel. The production of kamaboko is a legacy of the past, which has provided manufactures with much experience in Japan.

TYPES OF KAMABOKO

The origin of kamaboko is far back in the history of Japan. It is even mentioned in some of Japanese documents of about 1,500 years ago. In the course of time, many types of kamaboko have been developed. They differ in respect to heating methods, shapes, and ingredients. Table 1 shows the types of kamaboko (1,2).

Itatsuki-Kamaboko

Itatsuki-kamaboko (Fig. 1) is usually called just kamaboko because it is representative of kamaboko. Seasoned fish-meat paste is piled on a thin wooden board made from pine or Japanese cedar and is fixed to the board by the natural adhesiveness of the meat paste; it is then steamed. Some of the products are broiled instead of steamed. The kamaboko is cut into pieces after the wooden board is removed and eaten with soy sauce. Table 2 gives example of the type of fish and other ingredients used in itatsuki kamaboko.

Table 1. Various Types of Kamaboko

Type	Characteristic
Steamed kamaboko	Heating method
Steamed and broiled kamaboko	
Broiled kamaboko	
Boiled kamaboko	
Fried kamaboko	
Meat paste piled on a thin wooden board (itatsuki-kamaboko)	Shapes
Tubular-shaped (chikuwa)	
Ball, bar-square-shaped (fried kamaboko)	
Leaf-shaped, noodle-shaped	
Starch	Ingredients
Vegetable (green pea, yam, burdock).	
Seaweeds (laminaria, porphyra)	
Others (egg yolk, egg white, squid, boiled quail egg, cheese, ham)	

Fried Kamaboko

Fried kamaboko is made from fish-meat paste mixed with various ingredients, shaped and fried in soybean oil or rapeseed oil. The material for fried kamaboko can be of lessor quality than that of itatsuki-kamaboko. Various kinds of fried kamaboko are made by mixing eggs, squid, or vegetables such as chopped carrots, burdock, green peas, or pepper in different shapes.

Chikuwa

Chikuwa (Fig. 2) is fish-meat paste made by forming a roll of paste around a metal, bamboo, or wooden spit and roasting. Chikuwa of a good quality is white inside and golden brown on the surface. Table 3 gives example of types of fish and other ingredients. Chikuwa is cut into pieces after the spit is removed and eaten with soy sauce.

New Products

Imitation crabmeat, imitation scallop adductor, and imitation shrimp tail meat are made from kamaboko also (3).

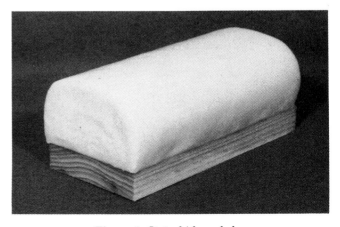

Figure 1. Itatsuki-kamaboko.

Table 2. Type of Fish and Other Ingredients Used in Itatsuki-Kamaboko

Type of Fish	Amount, kg	Ingredient	Amount, kg
White croaker	5	Salt	0.35–0.4
Yellow croaker	5	Potato starch	0–0.5
		Sugar	1.0–0.5
		Sodium glutamate	0.1–0.15
		Sweet sake	0.4–0.5
		Egg white	Optional

The imitation crabmeat is made as follows: frozen surimi (washed mince) is defrosted and kneaded with salt, crab flavor, and seasoning to make paste. This paste is formed into a thin, square kamaboko and cut into threads, or the paste is extruded to a beltlike shape kamaboko, then the kamaboko is cut into thin strips and bound in the thickness of crab-leg muscle. Finally, the product is colored red by natural pigment to make it look like crabmeat. The imitation scallop adductor is made by a similar method. The flavor and seasonings are added to fish-meat paste, and the mixture is formed into a beltlike kamaboko as in the making of imitation crabmeat. The kamaboko is then cut into thin threads, and the threads are bound in the thickness and cut to the length of the size of scallop adductors. In some cases, the products are coated with flour and made ready for frying.

PROCESS AND MECHANISMS OF PRODUCTION

Production process of kamaboko consists of three steps, (1) water washing of fish minced meat, (2) grinding with salt to make meat paste, (3) shaping and heating of the meat paste.

A distinguishing characteristic of kamaboko is its elastic texture called ashi, created by a thermal gelation of fish muscle protein. Main components in forming ashi are myofibrillar proteins, such as myosin and actomyosin, which are present in fish muscle. Myosin plays the most important role in thermal gelation and is essential for the ashi of kamaboko gel (4). Fish myofibrillar proteins account for 40–70% of total muscle protein; therefore their properties largely affect the quality of kamaboko. Prevention of denaturation of these proteins is of great importance, as is also the elimination of the components that may obstruct the forming of ashi. The water-washing of fish minced meat not only rinses off blood and bad fishy odor, but also extracts water-soluble protein, sarcoplasmic protein, which interferes with ashi-forming (5). Sarcoplasmic protein accounts for 30–40% of total muscle protein of fish.

Grinding the meat with sodium chloride is a necessary process for the forming of elastic kamaboko gel. Salt is not used as seasoning, but increases the ionic strength of the meat to solubilize the myofibrillar proteins in the muscle. The minimum concentration of salt necessary for extracting myofibrillar protein from the muscle at around pH 7.0 is about 2.0% of the weight of the meat (6). The ground fish meat with salt is left for a period of time or is heated until it reaches a certain temperature at which myofibrillar protein sol makes a network structure and turns into a gel (7). This phenomena is called setting or suwari and the gel is called suwari gel. The suwari gel is formed during the temperature rise, after elastic kamaboko gel is formed.

Water-Washing Process

A meat separator is used for collecting fish meat from fillet or dressed fish. The collected meat is mixed with cold water, and the mixture is left until the meat settles, then the supernatant is removed. This process is repeated several times. When kamaboko is to be made from small pelagic fish, such as sardine or mackerel, which have no strong ashi, rinsing of the meats in an alkaline solution (0.1–0.5% sodium bicarbonate solution) is useful to help the forming of strong ashi (8).

Grinding and Shaping

The connective tissue, membranes, or small bones in the water-washed meat are removed by a strainer. The meat is ground with sodium chloride, 2.0–3.0%, by a grinder or a silent cutter. A variety of kamaboko can be made by adding ingredients such as starch, sugar, sodium glutamate, and sweet sake after grinding. The plastic and adhesive characteristics of the ground meat facilitate shaping of the various forms required. At factories, shaping processes are done by a shaping machine.

Heating

Shaped meat paste is heated by various heating methods, as mentioned before. Setting is formed during the temperature rise after elastic kamaboko gel is formed. The temperature of the center of kamaboko must be higher than 75°C during heating. However, jelly strength of kamaboko

Figure 2. Chikuwa.

Table 3. Type of Fish and Other Ingredients Used in Chikuwa

Type of Fish	Amount, kg	Ingredient	Amount, kg
Croaker	8	Salt	0.32
Lizard fish	1	Wheat starch	0.7
Sharp-toothed eel	1	Sweet sake	0.5
		Sodium glutamate	0.15
		Egg white	Optional

gel decreases at temperatures higher than 110°C, because of a decomposition of myosin heavy chains (9). After heating, kamaboko is cooled off at room temperature by electric fans or storing in a cold room.

RAW MATERIAL

Any species of fish may serve as material for kamaboko, but the strength of ashi varies according to the fish species and their freshness (10). The main fish species used as good material for commercial kamaboko in Japan are croaker (*Nibea mitsukrii*), lizard fish (*Saurida undosquanis*), sharp-toothed eel (*Muraenesox cinereus*), cutlass fish (*Trichiurus lepturus*), horse mackerel (*Trachurus japonicus*), and flounders. Fatty, small pelagic fish such as sardine and mackerel can be used for materials if they are fresh (11). The pH of meats of these fish species decreases rapidly to about 5.8 after catch, and the speed of denaturation of myofibrillar protein at pH 5.8 becomes four to nine times faster than at pH 7.6 (12). Kamaboko gel-forming abilities of sardine and mackerel are greatly influenced by their freshness. The color of kamaboko derived from these fish species is a bit darker than from white-fleshed fish (13). Lately, frozen surimi from Alaska pollack (*Theragra chalcogramma*) and other species are mainly used.

ADDITIVES AND CHEMICAL COMPONENTS

Amounts and types of additives to be used in kamaboko, such as preservatives and colorings, are for example, the amount of ascorbic acid allowed as a preservative in Japan must not exceed 0.2% of the product. Most kamaboko sold on the market is 5–20% added starch to strengthen its ashi. Potato starch, wheat starch, sweet potato starch, and corn starch are commonly used for this purpose. Vegetable protein is used as an additive to fortify the strength of ashi and to increase the volume. Table 4 shows the chemical compositions of various kinds of kamaboko in the Japanese market. According to the table, except for fried kamaboko, kamabokos have less energy than animal meats. Recently, kamaboko has become not only a traditional food in Japan but also has been discovered as a low calorie and nutritive protein-rich food in many other countries.

BIBLIOGRAPHY

1. T. Suzuki, *Fish and Krill Protein: Processing Technology*, Applied Science Publishers, London, 1981, pp. 62–114.
2. T. Suzuki, "Processing of Fish, Shellfish and Seaweeds," *International Journal of Agriculture Fisheries Technology* 1(1) 45–49 (1989).
3. G. M. Pigott, "Surimi 'The High Tech' Raw Materials from Fish Flesh." *Foods Reviews International* 2(1), 213–246 (1986).
4. N. Nishioka, R. Machida, and Y. Shimizu, "Kamaboko-forming Ability of Dolphinfish Myosin," *Nippon Suisan Gakkaishi (Bulletin of the Japan Society of Sci. Fish.)* 49, 1233–1238 (1983).
5. M. Okada, "Effect of Washing of the Jelly Forming Ability of Fish Meat," *Nippon Suisan Gakkai-shi*, 30, 255 (1964).
6. Y. Shimizu, in Japanese Society Sci. Fish, ed., *White Meat*

Table 4. Chemical Components of Various Kinds of Kamaboko, in 100 g of Kamaboko

	Steamed Kamaboko	Boiled Kamaboko	Fried Kamaboko	Chikuwa
Calories (Kcal)	98	106	149	126
Moisture (g)	74.4	72.0	66.2	69.1
Protein	12.0	16.2	12.3	12.2
Lipids	0.9	0.8	4.5	2.1
Carbohydrate	9.7	7.4	13.9	13.5
Ash	3.0	3.6	3.1	3.1
Calcium (mg)	25	25	60	15
Sodium	1,000	1,200	1,000	1,000
Phosphorus	60	60	70	110
Iron	1.0	1.0	1.5	2.0
Vitamin A (retinol) (μg)	0	0	0	0
B_1 (mg)	0	0	0.05	0.05
B_2	0.01	0.01	0.10	0.08
Niacin	0.5	1.5	0.5	0.7
Vitamin C	0	0	0	0
Sodium chloride (g)	2.5	2.9	2.4	2.4

Fish and Red Meat Fish, Koseisha-Koseikaku, Tokyo, 1976, pp. 106–118.

7. E. Niwa and T. Nakayama, "Theoretical Approaches on the Strengthening of Fish Flesh Gel by Setting," *Nippon Suisan Gakkai-shi* **50,** 1945–1948 (1984).
8. Y. Shimizu, Jap. Pat. Appl. No 40-21224 (1965).
9. H. Kokuryou and N. Seki, "Degradation of Fish Myofibrillar Proteins by Heating at High Temperatures," *Nippon Suisan Gakkai-shi,* **46,** 493–498 (1980).
10. Y. Shimizu, "Kamaboko-forming Ability of Fish Meat," in Y. Shimizu, ed., *Science and Technology of Fish Paste Products,* Koseisha-Koseikaku, Tokyo, 1984, pp. 9–24.
11. T. Suzuki and S. Watabe, "New Processing Technology of Small Pelagic Fish Protein," *Food Review International* **2,** 271–307 (1986–87).
12. K. Arai and R. Takashi, "Studies on Muscular Proteins of Fish—XI. Effect of Freezing on Denaturation of Actomyosin ATPase from Carp Muscle," *Nippon Suisan Gakkai-shi,* **39,** 533–541 (1973).
13. T. Akahane, "Uses of Food Protein Materials from Fatty Species," in *Fatty Fish Utilization: Upgrading from Feed to Food, Proceedings of the National Technical Conference,* Raleigh, N.C., 1988, pp. 265–274.

TANEKO SUZUKI
Nihon University
Kanagawa, Japan

FISHES: ANATOMY AND PHYSIOLOGY

The diversity of forms found among the species of fishes is matched only by the diversity of aquatic environments on the earth. Aside from the shared characteristics of possessing a backbone, being cold blooded, and having gills and fins, the variation in anatomy, physiology, and ecology of all the fish species is extensive indeed. In size, they range from a 1-cm goby to the 15-m whale shark. They are found in waters from nearly 4,000 m above sea level to 7,000 m below sea level as well as in water temperatures from below freezing to 44°C. They are also found in environments that are nearly distilled to those that are very salty. Fish inhabit surface waters in full sunlight as well as very deep waters in complete darkness where pressures are extremely high. While most are strict water breathers, some have adapted to breath air where oxygen in water might be lacking. It may not be surprising, therefore, that there may be more than 25,000 species of fishes, accounting for more than half of all vertebrate species in the animal kingdom. Fishes also play an important role in human life by providing food and recreation. The traditional commercial fishery and now, to an increasing degree, aquaculture provide a high-quality protein source for human consumption. The sport fishery and hobby aquarium industry are both very important sectors of the economy. Feed for agricultural animals and fish oil supplements to human diets are examples of secondary uses of fish products for human benefit.

The purpose of this article is to outline some of the fundamental aspects of the anatomy and physiology of the bony fishes. The reference list at the end of the article serves as a guide to more detailed literature on various subjects covered here. The anatomy and physiology of eight major systems in fish are briefly described. These are the digestive, renal, cardiovascular, respiratory, nervous, endocrine, reproductive, and immune systems. The information is basic and is presented here with the assumption that the reader has some knowledge about vertebrate anatomy and physiology but very little or no knowledge about fish. The intention is to present some basic information and stimulate interest in this important food animal.

GROSS ANATOMY AND ORIENTATION

The anatomy of the salmonid is emphasized here because they are one of the most widely studied genera in terms of physiology and aquaculture science. Figure 1 presents the general orientation and terminology that will be used throughout this article. It also points out the major external features of the animal.

Figure 2 illustrates the general features of the internal organs of the salmonid. The muscle mass is, by far, the largest component of the body. The muscle mass, relative to the body, is larger in fish than in other terrestrial or aerial vertebrates. The dense medium of water requires a large force for locomotion compared to air. That same medium, however, provides substantial structural support such that the fish can maintain neutral bouyancy with a

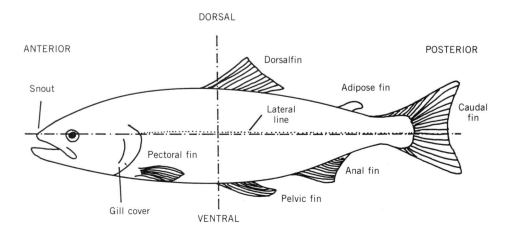

Figure 1. External anatomy.

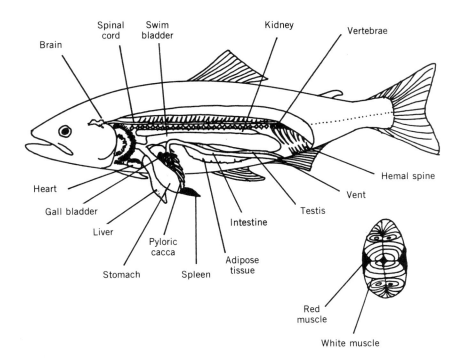

Figure 2. Internal auditing and cross section of trunk.

swim bladder or body lipids. It is also this great mass of muscle that makes the fish uniquely desirable as a food animal. There are two types of muscle in the fish: red and white. In most species of fish less than 10% of the total muscle mass is made up of red muscle. The color-oriented terminology is based on the visual appearance of two types of muscle. The red muscle appears darker because there is up to a 10-fold greater blood capillary to muscle fiber ratio in the red muscle compared to the white muscle. Red muscle is used for sustained swimming, fueled by aerobic glycolysis and lipolysis, while the white muscle is used for burst swimming, fueled by anaerobic glycolysis. The approximate weight distribution, in percent, of the major body parts of a trout are listed below.

Liver	1.2
Spleen	0.1
Intestine	4.7
Heart	1.2
Swim bladder	0.2
Kidney	0.9
Muscle	55.9
Skin	8.7
Axial skeleton	13.5
Gills	2.8
Head	11.8

DIGESTIVE SYSTEM

The anatomy and function of the digestive system in fish is basically the same as in other vertebrates. It is made up of a combination of mechanical and chemical processes. Food is ingested, broken down, and either absorbed into the blood or remains in the gut and is eventually expelled as feces. Differences arise between fish species in feeding habits, and the environment may influence the presence, position, shape, and size of a particular organ. Fish may be divided into three catagories according to the food they eat. Herbivores are specialized and eat detritus and plants. Omnivores eat both plants and animal material, and carnivores feed on larger invertebrates and other fish.

Digestive Organs

The oral cavity is made up of lips, mouth, teeth, and tongue. Fleshy lips are common to herbivores, designed for grazing and straining. Carnivore lips are generally unmodified and thin. Bottom feeders often have barbels, sensory appendages that help guide them when feeding and grazing.

Mouths are large or tubular. Large mouths are usually associated with predatory fish, such as a barracuda, whereas tubular mouths such as that of a seahorse, are better adapted for plankton eaters due to the enhanced ability to suck water and food. Mouth position can be variable depending on the type of feed normally consumed. Bottom feeders have mouths on the ventral surface, surface feeders have a dorsal mouth and bony fish (teleosts) have a terminal mouth.

Teeth are thought to have originally arisen from scales covering the lips. Fish generally have small teeth for gripping and capturing prey. There are four possible types of teeth present in fish, also depending on feeding habits: (1) jaw teeth, (2) palate teeth, (3) tongue teeth, and (4) throat teeth. There is a strong correlation between dentition, feeding habits, and food eaten. Jaw teeth are more developed in carnivores, but they are poorly developed or absent in herbivores, which usually have well-developed throat teeth attached to the gill arches.

The tongue of a fish is very rudimentary. It has no salivary glands but does bear sensory taste buds. Any lubricant fluid in the mouth originates from mucous cells scattered throughout the inner lining, not from the tongue. Plankton eaters also have well-developed gill rakers. This is a sieve-type apparatus attached to the inner edges of the gill arches and serves to filter out phytoplankton, crustaceans, and, in some cases, even diatoms.

A short, broad, and muscular esophagus connects the mouth and elastic stomach. It functions in the transport and taste of food. The musculature of the esophagus tends to be more developed in freshwater fish than in saltwater fish, as it plays a role in minimizing water intake while ingesting food.

The stomach is muscular and motile. The inner surface is lined with gastric glands that secrete digestive enzymes, necessary for the breakdown of food. The size of the stomach is related to the size of food particles and the interval between meals, ie, large food size and infrequent feeding are usually associated with large stomachs, and smaller food particles ingested more constantly are associated with fish with small or no stomachs. The pyloric valve, a ring of muscle, controls the passage of food between the stomach and intestine.

The intestine is a relatively simple tube that has digestive glands and an abundant supply of blood vessels. There are two valves, one at each end. The pyloric valve is at the anterior end and the ilioceacal valve is at the posteior end. Here is where much of the absorption of nutrients takes place. Digestive products are picked up in solution. Intestine length varies with feeding habits: carnivores have short straight intestines, herbivores have long and coiled intestines. Gut length seems more correlated to the amount of indigestible material in the food rather than the nature of the food. A large surface area is realized by extensive infoldings and ridges. The intestine leads to the rectum where vascularization and secretory cells are sparse, but where more mucous cells are present. The anal opening is called a vent and is also the terminal location for the urinary and reproductive ducts. While the common collection of these ducts suggest a corollary with the avian cloaca, it is not the case. The three ducts in fish all empty to the external environment and not to a common internal chamber. Absorption of plant material is probably poor; probably about 20% or less. Hatchery diets for salmonids may be absorbed up to 80%.

Food Movement

Movement of food through the digestive system is aided by peristaltic waves of muscle contraction. Cilia also line the tract and are especially useful in stressful situations when peristalsis may become reduced or even stop. This suggests that peristalsis is under some nervous control. Mucous membranes in the mouth, esophagus, and stomach provide lubrication to aid the passage of the food along the digestive tract.

Associated Organs

The pyloric caeca is an important appendage that opens into the intestine. It has the same epithelium as the intestine and its purpose is to increase the effective surface area. It may have both an absorptive and digestive function. A two or more lobed liver, which serves several metabolic and energetic functions, also produces bile, which aids in digestion. The gallbladder, which opens into the intestine, stores excess bile, which is secreted when required. The pancreas, which is large and distinct in the elasmobranch but diffuse in teleosts, secretes insulin and glucagon in response to nutritional intake, but also secretes digestive enzymes into the intestine. In teleosts, small beads of pancreatic tissue are scattered in the mesentaries near the digestive tract and are supplied with artery, vein, nerve, and pancreatic ducts. Little is known about the pancreatic secretions and most knowledge is based on histological evidence.

Digestive Enzymes

While goblet cells secrete mucus, gland cells throughout the digestive system contribute digestive enzymes necessary for the breakdown of food. The stomach is lined with secretory cells (gastric glands) that secrete hydrochloric acid (HCl) and pepsinogen. These chemicals break down protein molecules into amino acids and are, therefore, especially important to carnivorous fish. The stomach is very acidic (pH 2–4). Pepsin activity is dependent on pH and temperature, whereas HCl secretion is dependent on temperature and meal size. Optimum pepsin activity occurs at pH values of 2 and 4. Stomach distension stimulates gastric secretions.

Both the pyloric caeca and intestinal mucosa are sources of lipase, which breaks down fats into fatty acids and glycerine. Absorption of fat may occur in the anterior of the intestine. Intestinal secretions include proteases such as trypsin, lipase, and carbohydrase. These enzymes work best at neutral to alkaline pH. Bicarbonate is secreted (possibly from the pancreas) to raise the pH of the food coming from the stomach. Bile salts, secreted from the gallbladder may also play a role in adjusting the digestive juices to the proper pH to facilitate the digestive process. Bile salts are detergentlike substances formed from the decomposition of cholesterol and other steroids that function in fat emulsification aiding in the digestion of lipids. Fat soluble vitamins (A, D, E, K) are digested by bile. Digestive proteases are also secreted from the pancreatic tissue.

Most of the food absorption takes place in the intestine and pyloric caeca with the undigested material passing on to the rectum. Factors such as temperature, fat content, and presence of indigestible material affect the passage and digestion and absorption of food.

Innervation of Digestive Organs

Innervation of the digestive organs is not well understood, although it is thought that there is sympathetic innervation from paired ganglia, lateral to the spinal cord to the stomach, intestine, and rectum. This adrenergic system has an inhibitory effect. Parasympathetic innervation is through three cranial nerves to various parts of the digestive tract. Glossopharyngeal (IX) and facial (III) nerves innervate the mouth and esophagus areas. The vagus (X) innervates all visceral portions. These cranial nerves are cholinergic and have a stimulatory effect. Intrinsic nerves form a network inside the tissue of the digestive system. These nerves do not originate in the brain or the spine.

They may assist in peristaltic movement. Intestinal and pancreatic secretions appear to be also under both nervous and hormonal control.

Feeding Types

Predators have well-developed grasping and holding teeth, ie, sharks, pike, and gar, and well-defined stomachs with strong acid secretions and short intestines (relative to herbivores). They depend on their senses of vision or the lateral line sensory organ to detect their prey. Grazers browse on plants and organisms and sometimes on each other. The trout is an example of this, and in crowded ponds they often nip at the fins of the other fish. Strainers usually have well-developed gill rakers that strain out the plankton and crustaceans in the water that flows through the mouth and gills. Suckers suck in their food. They have mouths on the ventral surface of their heads but adaptive lips. An example of this type of feeder is the sturgeon. Parasites live off the body fluids of other fish. An examples of this type of feeder is the lamprey.

Stimuli for Feeding

Internal motivation is driven by factors such as season and temperature, diurnal light, light intensity, and the time and nature of the last feeding. Fish have been known to feed better at dusk and early morning. Due to patterns of growth, fish also feed in spring and through the summer much more voraciously than in the winter. The hypothalamus is also thought to be involved especially where hunger and satiation triggers are concerned. Stimuli for feeding also comes through the senses, such as smell, taste, sight, and the lateral line system.

RENAL SYSTEM

The kidneys are important in filtering undesirable materials out of the blood as well as serving a role in water and ion balance. Active transport, together with epithelial permeability are adjusted in order that ions and other materials vital to the body can be conserved and control over body water content can be exercized.

Basic Anatomy and Physiology

The kidney in salmonidae is made up of two parts running along the anterior–posterior axis between the body cavity and vertebral column. Anterior and posterior portions, also known as head and trunk, respectively, have slightly different functions but there is no visual distinction between them. The anterior kidney is associated with interrenal tissue and chromaffin cells, which are involved in blood cell and hormone production; the posterior kidney is associated with filtration and urine production.

Kidney tissue is made up of individual units called nephrons (Fig. 3). Each nephron is made up of a renal corpuscle, consisting of the glomerulus and Bowman's capsule and a kidney tubule. The glomerulus is a network of afferent and efferent arterioles whose blood supply comes from the dorsal aorta or a venous supply called the renal portal system. The Bowman's capsule encapsulates

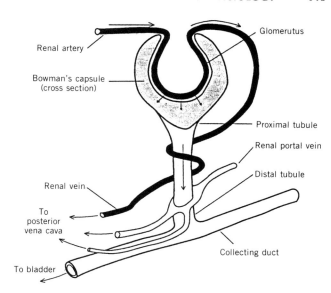

Figure 3. Kidney nephron.

the glomerulus and nonselectively filters solutes and molecules less than 70,000 mol. wt. from the blood. The filtrate is carried through the renal tubule, which is made up of the proximal, intermediate, and distal segments. It is in the tubule that reabsorption of selected electrolytes, minerals, amino acids, glucose, and other plasma organics takes place via a capillary network back into the blood. The remaining fluid containing unreabsorbed constituents flows to the collecting tubule and gathers in the urinary bladder, if present, where reabsorption of more ions may take place before being excreted out the urinary duct.

The process of nonselective filtration and subsequent selective reabsorption is an efficient process, because the body reabsorbs only what it needs. It is an adaptable system, because needs may change over time and through changing circumstances. Although efficient, this system requires energy as many ions are reabsorbed by active transport. Control of kidney function is accomplished by a variety of hormonal mechanisms and includes tissues of the thyroid, kidney, gonads, hypothalamus, and possibly the pituitary. In mammals, it is well established that a renin–angiotensin system and antidiuretic (ADH) hormone regulate renal function for ionic and water regulation. While the exact role of such a system in fishes is not well established, there is evidence that similar mechanisms are found in fishes. In this model, renin is secreted under nervous and hormonal control from a juxtaglomelular apparatus, located near the glomerulus, in response to decreased Na^+ concentration or blood pressure as well as under stressful conditions. Renin is then converted to angiotensin, which increases blood pressure through the constriction of blood vessels and causes aldosterone to be release from the adrenal cortex (of mammals). Aldosterone stimulates ionic uptake from the distal tubule of the nephron. ADH, from the pituitary, controls the amount of water leaving the nephron by controlling the permeability of the epithelium in the collecting duct. An increase in blood osmolarity causes an increase in ADH, which increases the permeability of the epithelium, resulting in the increased absorption of water out of the filtrate, back into the blood.

Freshwater fishes are hyperosmotic regulators. That means that the concentration of ions and other solutes is greater in the blood than in their surrounding water, which in most cases is very dilute. The fish will thus absorb water osmotically from their environment through all permeable epithelia such as the gills, skin, and gut. The regulatory problem is one of getting rid of the excess water and the kidney plays that important role. A large amount of urine is produced, which is dilute and contains creatine, uric acid, and some ions. The volume of urine produced must balance the quantity of water entering the body. Sodium (Na^+) and chloride (Cl^-) ions passively diffuse out of the body across permeable epithelia and are actively taken up, to a large extent, across the gill epithelium.

Electrolyte reabsorption out of the urine takes place across the renal tubule. Na^+ is actively extracted and it appears that Cl^- passively follows. Calcium (Ca^{2+}), magnesium (Mg^{2+}), and other divalent ions must also be reabsorbed, because they are normally absent in the urine of freshwater fish. The reabsorption of these ions is usually accomplished without the osmotic absorption of water. The distal segment, collecting duct and urinary bladder appears to be relatively impermeable to water. Macromolecule reabsorption, including glucose, amino acids, and other plasma organic constituents takes place in the first segment of the proximal tubule. Salts are reabsorbed in the distal segment, and any remaining in the filtrate may be reabsorbed from the urinary bladder. Only a small proportion of total organic nitrogen is excreted via the kidneys, although this appears to be an important excretory pathway for minor nitrogenous products, such as creatine and uric acid. Major nitrogenous products such as ammonia and urea are excreted via the gills.

The kidney in marine fishes plays a crucial role in hypoosmotic osmoregulation. Converse to the freshwater fish, the blood concentrations of ions and solutes in fish blood relative to the concentrated ionic environment of the marine environment is rather dilute. The osmotic problem created by such a gradient is one of dessication, which the fish counteracts by actively taking in water from the environment by drinking and reducing water loss at the kidneys. Between 60 and 80% of ingested water is absorbed through the gut, along with the monovalent ions Na^+, Cl^-, and potassium (K^+). Those excess salts are actively excreted. The gills play a major role in that function, although the kidney excretes Mg^{2+}, sulfate, and other divalent ions. Although less than 20% of the ingested divalent ions are actually absorbed most are passively eliminated via the intestines, that which is absorbed is handled by the kidney tubules.

The kidney nephron of marine fishes often lack the distal segment of the tubule and has less glomeruli. In some cases, such as the goosefish, glomeruli are absent. Because glomeruli are suited to the excretion of large volumes of water the importance of the glomerulus in marine teleosts is greatly reduced. Urine from a marine kidney thus has high osmolality and is low in volume. Much of the water entering the glomerulus reabsorbed to help prevent dehydration, a case opposite to the freshwater kidney. Nitrogenous excretion is similar to the pathway in freshwater fish, with ammonia and urea excreted through the gills and minor nitrogenous products such as creatine and uric acid excreted by the kidney.

Anadromous and catadromous fish must be capable of adjusting their osmotic balance to survive the changes in salinities that they experience in their life cycle. They have glomerular kidneys that can adjust to changing urine volumes, and possess gills and oral membranes capable of both uptake and secretion of certain ions against the prevailing diffusion gradients. The kidneys in those animals are capable of adjusting urine volume and composition on demand. When those fish, for example, are transfered or voluntarily move from fresh water to salt water, the urine composition gradually changes after a few days. Urine flow decreases and osmolality increases as divalent ions are excreted. Sodium and chloride content in the urine decreases as the chloride cells (specialized salt excreting cells on the gills) take over this function. Glomelular filtration rate may temporarily slow down and tubular reabsorption increases.

CARDIOVASCULAR SYSTEM

The cardiovascular system moves the blood around the body. It is basically composed of the heart (cardio-) and network of blood vessels (-vascular) throughout the body. The purpose of blood circulation is the transport of materials between certain locations within the body. Gases such as oxygen are taken up from the water and move from gills to the tissues and others such as carbon dioxide move from tissues back to the gills for excretion. End products of metabolism such as lactate, originate in the muscles and are transported to the liver for breakdown. Glucose made in the liver must be moved to the tissues where it will be used. Materials that are foreign to the body are excreted by gills, released in the urine, or engulfed and destroyed by specialized cells in the blood. Amino acids and other nutrients from digested materials must move from gut to the tissues. Blood cells that are produced in the anterior kidney and spleen must be distributed throughout the body. Because of the vast range of environments in which fish are found, the diversity in the form and function of the cardiovascular systems are as wide ranging. The basic conformity among salmonid fishes are emphasized here.

Heart

The teleost heart has four chambers: the sinus venosus, the atrium, the ventricle, and the bulbus (Fig. 4). The venous blood from the body is received in the sinus venosus. This is a rather flat bag anterior to the transverse septum separating the heart and head compartments from the visceral cavity. The atrium is a thin-walled bag that feeds the blood to the muscular ventricle. The ventricle is the powerhouse that contracts and propels the blood throughout the body. The bulbus is rather muscular as well and helps to dampen pressure waves as the blood pulses out of the ventricle. There are one-way valves that prevent the blood from flowing backward with each con-

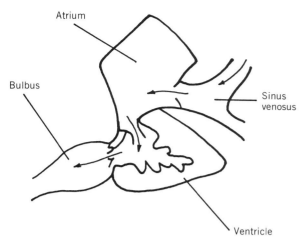

Figure 4. Cross section of a trout heart (1).

traction. Cardiac output in teleosts ranges from 5 to 100 mL/kg per minute with a general mean about 15–30 mL/kg per minute.

Blood flow is accomplished primarily by the heart, which is a pump for the system. The basic route is from the heart to the gills, where gas and ionic exchange with the environment occurs, and then to the rest of the body (Fig. 5). The blood is then pumped back to the heart through a venous network where the cycle begins over again. About 4–6% of body weight is a reasonable approximation of the blood volume. Blood pressures in teleosts are in the range of 30–70 mm Hg. Because the resistance to blood flow increases as the diameter of the blood vessels decreases, blood flow declines from the point where it leaves the heart. For example, the blood vessel network in the gills reduces the pressure found in the ventral aorta by about 40–50%. That resistance may increase with stresses such as low oxygen levels in the water, whereas other conditions such as exercise or elevated levels of adrenaline dilate the blood vessels and decrease the resistance to blood flow.

The Secondary Circulation

The primary circulation in fish comprises the blood vessels that contain red and white blood cells. There is a secondary circulation, sometimes known as the lymphatic or venolymphatic system, that is characterized by low blood cell content and by low pressure. There are direct connections between the primary and secondary circulation through small vessels, often at right angles to the blood vessel of the primary circulation. This physical configuration causes plasma to be skimmed out of the primary circulation and into the secondary circulation. The secondary circulation feeds this fluid, low in cells as a result of the skimming, and fluid from tissue beds back to the primary circulation via the heart. Salmonids have major ducts just under each lateral line and along the dorsal midline. These may serve to collect fluid from the secondary circulation and return it to the heart. Fin and trunk movements may play an important role in pumping venous fluids back to the heart. There is a tail heart in some fishes that aids the movement of this secondary fluid back to the heart.

RESPIRATORY SYSTEM

The respiratory system is extremely varied among the 25,000 species of fishes. This, once again, reflects the wide range of habitats to which fish have managed to adapt. The primary function of the respiratory system is to transport gases between the environment and the tissues of the body that consume oxygen as well as excrete ammonia and carbon dioxide. The vital components in this system are the water flow over the gills outside the fish and the circulatory system inside the body. The consumption of oxygen (O_2) as well as the excretion of carbon dioxide (CO_2) and ammonia (NH_3 and NH_4^+) take place across the gill.

Anatomy

The fish gill has four gill arches on each side of the midline and two rows of primary or gill filaments per arch (Fig. 6). Arranged perpendicularly to the filament are rows of secondary lamellae on both side of each filament. The plates of secondary lamellae form narrow channels through which the water flows. This space is approximately 0.02–0.05 mm wide, 0.2–1.6 mm long, and 0.1–0.5 mm high. This width is particularly important in that one-half of that width is the minimum distance for gases and dissolved materials such as ions to diffuse between water and blood. The surface of the secondary lamellae is the primary surface across which gases, ions, and other dissolved materials pass between water and blood.

Water Flow over the Gills

The water flow over the gills is directed in an anterior to posterior direction. More importantly, it flows in the opposite direction to the flow of blood through the secondary lamellae in the gills. This countercurrent flow maintains the maximum gradients between blood and water for the gases throughout the transit of water through the gills. The mouth or buccal pump is driven by skeletal muscles that control the mouth of the floor and opercular covers. Water, therefore, flows from mouth, over gills, and out the operculum. When the floor of the mouth is lowered, nega-

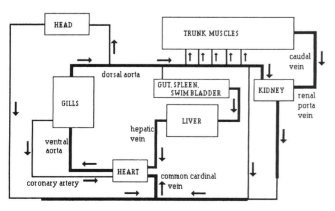

Figure 5. General flow of blood in fish.

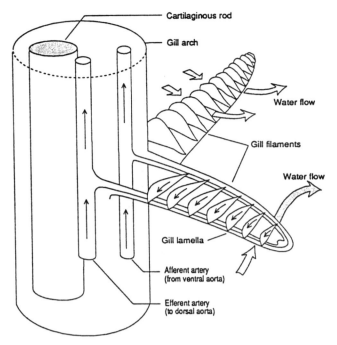

Figure 6. Fish gill with patterns of blood and water flow (2).

tive pressure is created in the buccal cavity and water is sucked into the mouth. As the mouth is closed and the floor of the buccal cavity raised, the water is forced posteriorly, across the gills, through the opercular cavity. This cycle is repeated continuously and the result is a unidirectional flow of water through the mouth and gills.

While this rhythmic ventilation is the general rule, some fish are ram ventilators, that is, they ventilate the gills by keeping their mouths open and swimming forward through the water. Salmonids also do this at moderate to high swimming velocities. There are also other variations on how fish move water across the respiratory surface of the gill for respiration. Some of these variations such as ram ventilation may reduce the energetic costs of ventilation.

Blood

Blood is a suspension of various cells in a solution of proteins and electrolytes, which make up the plasma. It is the vehicle for transporting materials from different locations in the body, as discussed above. The volume of blood in teleosts ranges between 3 and 6% of wet body weight. Fish do not have bone marrow and the origin of the blood cells is the anterior kidney (anterior 30%) and the spleen. These are termed hematopoietic tissues. The production of blood cells is stimulated by conditions such as bleeding and inhibited by starvation. There is an erythropoesis stimulating factor, found in the plasma of fish that stimulates blood cell production. Very little is found in plasma of resting fish.

White blood cells or leukocytes, are involved in the clotting of blood and play an important part in the immune system. The red blood cell, or erythrocyte is important in the transport of oxygen in the body. The packed red cell volume, or hematocrit, is expressed in percent and varies between 20 and 35%. The erythrocyte contains the respiratory pigment, hemoglobin (Hb), which binds oxygen and enables the oxygen content of the blood to be higher than that of simple dissolved oxygen. The concentration of Hb in blood is approximately 7–12 g % (equal to grams/100 mL). The total oxygen content is about 9–14 vol %.

Blood Proteins

The proteins in the plasma serve to maintain the blood osmolarity. It is the source of colloid osmotic pressure. Some other functions are to buffer pH changes in the blood and to transport vitamins, hormones, and inorganic ions, that attach to these molecules in blood, to tissues. Another important function of special plasma proteins are the antibody proteins that play an important role in the specific immune system of the body. In carp, for example 4.15% of the blood weight are protein. Of that weight, 2.82 g is albumin, 0.79 g is globulin, and 0.23 g is fibrinogen.

Response to Environmental Change

Fish respond to environmental changes in different ways. Some fish are known as oxygen regulators because they maintain O_2 consumption at lower O_2 levels by increasing ventilation. Oxygen conformers, on the other hand, adjust O_2 consumption according to ambient O_2 levels. Salmonids are regulators. The following are some effects of environmental changes on the respiratory system and the response of fish.

Temperature increase presents a three-pronged problem for respiration. It increases the O_2 demand with increased metabolism, lowers the solubility of O_2 in water, and lowers the affinity of O_2 to the haemoglobin molecule. Both heart rate and ventilation increase in response to an increase in temperature. The resistance of the peripheral circulation also decreases.

Hypoxia, or lowered environmental O_2, causes an increase in the frequency and amplitude of ventilation in most fish. While there is a decrease in heart rate, there is an increase in the volume of each stroke of the heart (stroke volume), thus maintaining the output of blood from the heart. This results in a large increase in the ratio of water pumped over the gills to the blood perfused through the gills (ventilation-to-perfusion ratio). Oxygen uptake is, therefore, maintained in the face of lowered water O_2 concentration.

Exercise increases cardiac output due to an increase in stroke volume. Ventilation increases and there is a fourfold to eightfold increase in O_2 uptake. The ventilation-to-perfusion ratio remains relatively constant. Acidic conditions generally increase ventilation.

NERVOUS SYSTEM

The nervous system, acting with the endocrine system, is the way a body coordinates its bodily functions. There are sensory input from external as well as internal sources, and nervous output that sends coordinated commands to various parts of the body as nervous impulses and hormones, which can be stimulatory or inhibitory. Finally, there is integration between those two components at various levels including simple automatic reflexes; integration of vital processes, such as breathing; and higher levels, which include complex learning such as in the homing mechanisms of the salmon returning to its freshwater spawning site. The fish nervous system is not completely understood and it is complicated by the many diverse adaptations present. Generalization is difficult but the basics of the fish nervous system are presented here with an emphasis on the bony fish.

Central Nervous System

The central nervous system constitutes the brain and the spinal cord (Fig. 7). The configuration of the fish brain resembles that of other aquatic vertebrates. All the typical nerves and paired lobes are present. While this basic form is consistent among fish species, the relative sizes of the lobes reflect the adaptive response of the particular species to its physical and biological environment. The optic lobes, therefore, in fish with large eyes that live in dimly lit environments are larger than in fish with small eyes. The basic anatomy and function of the brain of a bony fish (including teleosts) is described here.

The Prosencephalon is made up of the telencephalon and diencephalon. The telencephalon is known as the forebrain and contains the paired olfactory lobes. The forebrain is predominantly involved with reception and conduction of smell impulses. The size of the lobes indicates the importance the role of smell has in the fish species. Elasmobranchs (sharks) and bony fishes have quite pronounced olfactory lobes. The forebrain also appears to play a role in fish behavior. The diencephalon is sometimes known as the between brain. It lies in between the olfactory lobes of the telencephalon and contains the pineal organ. There are two possible functions of the pineal organ. It has been suggested that the pineal body is either a photosensory structure or plays a secretory role. The pineal gland may react to the chemical composition of the cerebrospinal fluid or brain tissue by external or internal (endocrine) secretion. Another important component of the diencephalon is the hypothalamus, which is an important area of interaction between the nervous and endocrine systems. Nervous inputs into this area from the brain result in the production and release of many regulatory hormones of the body (more details in the endocrine section). The diencephalon appears to be an important center for incoming and outgoing messages relating to internal homeostasis.

The optic lobes lie in the Mesencephalon and are the central control for vision The optic tectum, which is the nervous center for the optic lobes, is also found in the mesencephalon. The tectum is made up of neurons and the fibers from the optic lobe end here. The tectum correlates visual impressions with muscle reaction such as darting for prey or avoiding objects in the water. The choroid plexus, which serves to nourish the central nervous system, is found in the midbrain. It is filled with cerebrospinal fluid and is richly supplied with blood vessels.

The cerebellum has a role in swimming equilibrium, muscle tone maintenance, and fish orientation. The lateral line and sense organs are associated with the cerebellum. The division between the medulla oblongata and the spinal cord is not distinct. All sensory nerves except smell (I) and sight (II) lead to this center. This acts as a relay center between the spinal cord and the higher brain areas. It also controls certain somatic and visceral functions including respiration, osmoregulation, and swimming equilibrium.

As in other vertebrates the gray matter makes up the central region of the spinal cord and consists of nerve cells. White matter, which surrounds this core, contains the myelinated nerve fibers. The spinal cord receives sensory, afferent (to the brain) fibers by way of the dorsal roots of the spinal nerves and gives off the efferent (from the brain) motor fibers to the ventral roots of the spinal nerves.

Periheral Nervous System

The peripheral nervous system provides a means of communication from the environment where stimuli are received by sensory organs, to the central nervous system and from the central nervous system to the proper effector organs in the body, muscles, or glands. The spinal nerves are paired and carry sensory and motor fibers from and to the fish body. Ganglia act as relay centers along the spinal nerves. Fish have 10 cranial nerves (higher vertebrates have 12). They are similar to spinal nerves except they do not have a ventral or dorsal root and they emerge from the skull. Some cranial nerves are strictly sensory (afferent) and some are strictly motor (efferent).

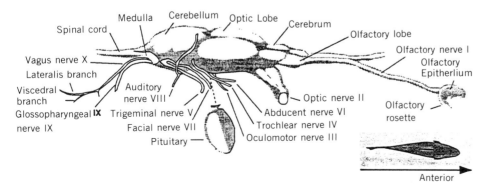

Figure 7. Brain (3).

The autonomic nervous system is part of the peripheral nervous system that innervates smooth muscle, heart muscle, and glands. Within this system is the sympathetic nervous system, made up mostly of spinal nerves, and the parasymphathetic nervous system, made up of cranial nerves. The autonomic nervous system controls many of the vegetative (slow, routine, and automatic) systems of the body. As in mammals, the innumerable voluntary functions of the body are controlled in various centers of the brain.

Coordination within the nervous system involves matching the input stimuli to appropriate actions. Those actions may involve the regulation of heart rate, for example, with various feedback of important information from the cardiovascular system. It involves the coordination of gross behavior such as schooling, where inputs from visual and acoustical sensors are coordinated with gross muscular action involved in swimming, which in itself involves a balancing act involving the labyrinths, fin positions, and trunk movement.

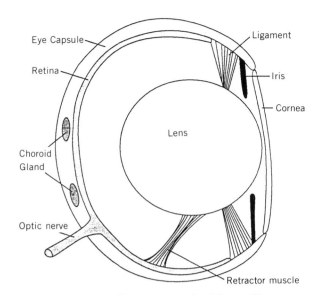

Figure 8. Cross section of a fish eye (2).

Sensory Organs

Fish have an acute sense of smell and rely on the olfactory organ to find food, avoid predators, and to help guide migratory fish to their spawning grounds. Some fish utilize pheromones, chemical substances that can be smelled, to signal alarm to the rest of the school. Catfish, which have an excellent sense of smell, use the specific odor of the slime of individual fish to differentiate fish in a school hierarchy. As the water passes through the nostril over a transverse septum, olfactory epithelium connected to the olfactory nerve signals the smell response to the olfactory lobes in the brain.

Taste buds are the gustatory organs and can be found in the epithelium of the mouth, lips, esophagus and snout. Some fish, mainly those that are not sight feeders, may have taste buds at other places on the body. Free nerve endings from the cranial nerves appear in sensitive regions, suggesting that touch works with taste to stimulate the feeding response. External taste receptors respond to monovalent ions and, to a lesser degree, divalent ions. Internal receptors respond to amino acids, sugars, strong salts, and acids. Both taste and smell are based on organic chemoreception. It also appears that tactile, chemical, and temperature sensations are closely integrated in fish skin.

Sight is extremely important to salmonids and other fish that depend on this sense to detect movements and color changes that signal prey, predator, and mate. The eyes of most bony fishes have a spherical lens, the most powerful shape for a single lens, that in some cases can be changed in convexity and in distance from the retina, the visual layer of the eye (Fig. 8). This is how the fish focus, because very little focusing is done by the cornea. There is little difference in the refractive index, the source of light diffraction, between the cornea and water and, therefore, little focusing is necessary. The cornea does play a role in protecting the rest of the eye from physical pollutants and harm and may help prevent dehydration. The iris controls the amount of light getting through to the retina. Most fish (eg, elasmobranchs) have a fixed iris that contains guanine and melanin. Sight feeders, such as the salmonids, have a well-developed lens muscle and focus by actually moving the lens. The eyeball is ellipsoid and lends a great depth of field, that is the sharp simultaneous focus of objects that are distant and near. This is due to a wide-angle configuration of the components: the flattened eyeball, the round lens, and the protrusion of the lens out of the iris. Light travels through the lens and is focused on the retina. Retinal layers such as the optic nerve fibers, ganglion cells, bipolar cells, and photoreceptor cells, (the cones and rods) all have a role in converting the light to sight images in the brain. The rod cells are sensitive to low light levels but not to color. The cone cells react to strong light and are color sensitive. Photochemical reactions in these cells are transmitted as electrical impulses to the optic nerve and eventually to the optic lobe in the brain. The sensitivity of the retina to particular colors of light depends on the pigments found in the rods and cones in the retina.

The small pineal organ serves a visual as well as a secretory role in fish. It is a small fingerlike organ is located just under the cranium on the dorsal midline of the brain. It serves possible functions in the detection of day length. While it is not vital to life, it may serve an important role in the adjustment of body color to the environment as well as in the synchronization of physiological changes associated with smoltification, in anadromous salmonids, with day length.

The inner ear functions in both equilibrium and hearing. It is found on either side of the medulla oblongata of the brain. The inner ear is made up of two sections, the pars superior and the pars inferior. The pars superior has three semicircular canals filled with fluid and has an ampullae and utriculus. The ampullae contains the receptor tissue, which has sensory cells with long sensory hairs similar to taste buds, and the lateral line. Displacement of the body causes the sensory cells to react and the proper righting mechanism to take place. The utriculus of the pars superior contains an otolith; a calcarious stone. This stone lies on the sensory hairs and responds to the force of gravity. This system works with the retina for maintenance of balance. The pars inferior is the structure of actual sound reception and it is composed of the two vesicles, sacculus and lagena. These vesicles, containing otoliths, are innervated by the auditory nerve. The otoliths will vibrate at a different resonance then the rest of the body when presented with sound waves. This triggers the sensory hairs to send a nervous impulse up the cranial nerve.

The lateral line is a set of sense organs found only in fishes and the aquatic stages of amphibians. The lateral line is innervated by cranial nerves and is closely associated with the auditory centres. The brain relays secondary impulses from the lateral line to the hypothalamus, thalamus, the optic tectum, and the cerebellum. The lateral line helps in the detection of prey, predators or mates by sensing hydrodynamic displacement. The lateral line receptors respond to pressure waves and low frequency vibrations and other localized disturbances.

ENDOCRINE SYSTEM

The endocrine system coordinates physiological processes with each other as well as with the external environment by way of chemical messengers called hormones. The hormones are released in response to changes such as light, temperature and stress. They ellicit and coordinate the appropriate responses from parts of the body, whether it be by a generalized stress response, a fright reaction, maturation of the gonads, smoltification, migration or spawning. Hormones thus play important roles in many physiological processes such as reproduction, osmoregulation, mineral metabolism, and growth. Endocrine glands are ductless tissues that release hormones into the blood.

The effects of hormones may be slow but are persistant in contrast to the rapid and transient action of nerves. It is important to realize the close relationship between the endocrine system and the nervous system. These two systems work together to create a communication network with each system dependent on the other for complete integration. Neurohormones are released by nerve cells, not endocrine glands, but have similar actions. They are carried by the blood to a target organ or to an endocrine gland, which can, in turn, release its own hormone. Neurohormones act as a link between the endocrine and nervous system.

Pituitary Gland

The pituitary gland lies close to the hypothalamus in the brain. It is made up of the infundibulum, the down growth of tissue from the brain, and the hypophysis, the upgrowth of tissue from the roof of the mouth. The hypophysis is differentiated into the adenohypophysis and the infundibular tissue. The adenohypophysis is the site of synthesis, storage, and release into the blood of several different peptide hormones. The infundibular, or neural, tissue forms the neurohypophysis, which is connected to the brain by the infundibular stalk. The neurohypophysis is not a discrete tissue in teleosts and appears to be a storage-release center for the materials that are synthesized in the hypothalamus and then transported to the neurohypophysis along neurosecretory axons.

Growth hormone increases appetite, improves food conversion, increases protein synthesis, decreases nitrogen loss, stimulates fat mobilization and oxidation, and stimulates insulin synthesis and release. Prolactin is necessary for electrolyte and water balance in fresh water. It reduces water absorption and reduces sodium movement through the gill, kidney and urinary bladder. It may also cause the dispersal of xanthophore pigment in the skin of some fish. It is involved in lipid metabolism fat storage and reduces thyroxine levels of the serum. Thyroid stimulating hormone (TSH) regulates thyroid function and is controlled by negative feedback from the levels of circulating thyroid hormones. Adrenocorticotropic hormone regulates the release of adrenocorticoids (steroids) from the interrenal tissue of the kidney. Melanophore stimulating hormone has a small effect on external body pigmentation in teleost. Fish also have sympathetic innervation of melanophores. Gonadotropic hormone, luteinizing hormone, and, to a minor extent, follicle stimulating hormone, have crucial roles to play in gonad development and spawning. Oxytocin is involved in reproduction: egg formation, mating, and spawning. Vasopressin affects diuresis and, therefore, assists in osmoregulation.

It is clear that the pituitary gland is the major gland in the endocrine system, exerting a large influence on processes of the body such as growth, osmoregulation, and reproduction. Target organs of the pituitary hormones are the thyroid gland, interrenal tissue, testis, and ovary. The synthesis of stimuli from the environment and subsequent hormonal release demonstrates the link between the central nervous system and the endocrine system of the body. That link is the hypothalamus–pituitary axis. The hypothalamus receives external stimuli and communicates changes to the adenohypophysis of the pituitary, in close proximity, where the release of the appropriate hormone is elicited.

Thyroid Gland

Thyroid tissue is generally scattered throughout the pharyngeal area, in the head kidney, and in some cases around the eye. The hormones secreted by the thyroid are thyroxine (T4) and triiodothyronine (T3). Thyroid action is linked closely to the pituitary, which secretes TSH. Secretion of TSH is under negative feedback control depending on the levels of circulating thyroid hormones. There also appears to be a strong relationship between thyroid activity and the lunar cycle. Peak thyroxine levels have been shown at the period of the new moon, which is also when peaks in the outward migration of coho salmon have been observed. The surge of thyroxine influences growth, migratory restlessness, and seawater preference.

Ultimo Branchial Gland

This gland is equivalent to the parathyroid gland in terrestrial mammals. It is generally located between the abdominal cavity and sinus venosus. Calcitonin is produced by this gland, which regulates calcium by controlling its conservation and mobilization.

Pancreas

The pancreas of teleost is usually diffuse and scattered around the intestine and spleen, and often extends into the liver. Insulin and glucagon are the two hormones associated with this gland. The action of the pancreas appears to be modified by different nutritional states, particularly in response to changes in protein metabolism. Insulin affects amino acid metabolism and incorporates amino acids into skeletal muscle and may not be as important to glucose homeostasis in teleosts as it is in mammals. It decreases the rate of gluconeogenesis. It may have some role in saltwater survival. Glucagon simulates gluconeogenesis from muscle amino acids.

Renal Tissue and Interrenal Tissue

A number of hormones and tissues are involved in kidney function. Those include renin, angiotensin, and antidiuretic hormone. Their functions are covered in the section on the renal system. Interrenal tissue is located in the anterior kidney tissue, or the head kidney. The primary hormones released are corticosteroids. Under control of the pituitary through the action of adrenocorticotropic hormone (ACTH), the corticosteroids are released in response to appropriate stimuli. Chromaffin cells, the functional equivalents of medullary tissue of the adrenals in mammals, secrete the catecholamines, adrenaline, and noradrenaline.

Corticosteroids such as cortisol are important hormones produced in the interrenal glands. Interrenal tissue response increases with stressful conditions, such as temperature change, exposure to toxic elements, and spawning. This elicits a subsequent increase in cortisol levels. Cortisol is important in regulation of water and electrolyte balance by affecting ATPase activity and sodium flux across the gills. It is known as the saltwater adapting hormone, because it increases renal sodium retention, reduces glomerular filtration rate, and increases permeability of urinary bladder. It is important in smoltification and subsequent saltwater acclimation. Cortisol is also involved in metabolism, by promoting gluconeogenesis and related carbohydrate and protein metabolism. It also plays a role in preventing exhaustion by mobilizing energy reserves. Corticosteroids influence oocyte maturation and seem to be influenced by pituitary gonadotropins.

Catecholamines, such as adrenaline and noradrenaline, respond immediately to stress. Catecholamines cause hyperglycemia (high blood glucose), with a reduction in liver and muscle glycogen in some cases. They also cause systemic vasoconstriction, vasodilation in the gills, and increase the force and rate of the heartbeat.

REPRODUCTIVE SYSTEM

The reproductive systems and strategies in fishes are extremely varied. Examples of the salmonid family are emphasized here. Salmonids reproduce sexually. Pacific salmon spawn only once in their life cycle, whereas the trouts and Atlantic salmon spawn repeatedly during their life span. Unlike other fish, lipid levels have little to do with initiating spawning activity or gamete development in the salmon. Lipid levels are, however, correlated to the distance that animals have to migrate up freshwater streams to reach their spawning grounds. Lipids as well as body proteins are consumed in this migration. The primary environmental factor involved in sexual maturation is day length. The changes in day length are probably more important than absolute day length in the processes that affect reproduction. Temperature also plays an important role in gonadal development in some fishes.

Both the testes and ovaries are located along the midline of the fish, ventral to the swim bladder, extending from the anterior to the posterior end, terminating at the vent. Eggs are commonly released into the peritoneal cavity and reach the outside through a funnel and ovarian duct. Sperm almost always stay inside ducts until they are released.

Egg Size and Fecundity

In general, fish that lay smaller eggs lay many of them. Cod, for instance, lay millions of eggs that incur a high mortality rate. Smaller eggs also tend to hatch in a relatively short time and the hatched young require microscopic food almost immediately because the stored nutrients in the way of yolk is limited or nonexistent. Conversely, fish that lay larger eggs lay fewer eggs that take longer to hatch and have higher survival rates. The young also can survive on the yolk for days or weeks and can ingest large food particles when feeding begins.

There is a correlation between the length of parental care and egg size. That is, the larger and fewer the eggs, the greater the parental care of the young. Herring spawn millions of eggs in kelp beds and abandon them. Salmon lay fewer, larger eggs and bury them in the gravel but still die and do not tend to their young. In the other extreme, some fish produce only a few eggs that are incubated in the body and live young are born that are more or less ready to fend for themselves. The yolk material is solid in salmonids. Their eggs sink in water because they are very dense. The yolk material influences the density and the development time. There is only about 30% water in the yolk material of salmon, whereas the herring egg has oil droplets in the yolk, thus allowing the egg to float. Oviparous fish are egg layers such as salmon, viviparous fish are placental live-bearers as mammals, and ovoviviparous fish are live-bearers but are passive parents.

Hormones

Gametogenesis, the formation of gametes, is dependent on pituitary hormones called gonadotropins. The gondadotropins are leutenizing hormone (LH) and follicle stimulating hormone (FSH). Blocking gonadotropins reduces gametogenesis. These names are taken from mammalian sources. Although both, or analogues of both, may be present in fish, it is LH that predominates. This hormone appears to regulate the maturation and release of the egg.

The follicles probably produce the estrogens, estrone, and estradiol under hormonal control from the pituitary gonadotropin. These hormones control the development of secondary sex characteristics in the female fish. They also serve an important function in the production of vitellogenin in the liver. Vitellogenin is then taken up from the blood and incorporated into the yolk proteins. Progesterone, which maintains the uterine development during pregnancy in mammals, may be absent in fish. The corticosteroids testosterone and androstenodione may control the secondary sex characteristics in the male fish.

Gonads and Gamete Formation

The early development of gametes is similar to other vertebrates. The processes of spermatogenesis, (sperm development) and oogenesis (egg development) proceed in the following manner. In spermatogenesis a diploid spermatogonium, through meiosis, gives rise to haploid spermatocytes, which then become spermatids. A process called spermiogenesis is the final step of differentiation of sperm from spermatids. As Figure 9 shows, the sertoli cells serve as nutritive cells. During this process, the cysts swell and finally rupture, releasing the sperm to the lumen, which is continuous with the sperm duct. The interstitium between the cysts contain interstitial cells, fibroblasts, and blood vessels. In oogenesis, cells destined to become eggs proliferate, enlarge, and each becomes surrounded with follicular, or nutritive cells. The granulosa layer of the follicle cells is thought to give rise to the yolk. The proteins in the yolk are derived from vitellogenin as discussed above. The development of all eggs in the salmonid occurs synchronously, that is all at once.

Ovulation

Ovulation, or the release of all the eggs from the follicles, is triggered by the completion of yolk deposition. Once the ova leaves the follicle, they lose the supply of nutrients provided by the follicular cells. They depend on the ovarian fluid for nutrients and oxygen. The origin, quantity, and turnover rate of the ovarian fluid are unknown, although it could be a dilute ultrafiltrate of plasma.

Spawning

Spawning follows soon after ovulation. The range of spawning behavior as well as the spawning itself in fish is probably wider than that in all land vertebrates put together. This event is probably controlled by a combination of hormones and environmental factors. There may be a special spawning hormone acting directly on the central nervous system. Mammalian reproduction involves both the pituitary gonadotropins, LH and FSH, affecting the gonads and the ovary, which respond by producing estrogen and progesterone, whereas the LH hormone activity in fish predominates and FSH activity is absent or weak in some species.

Fertilization

Fertilization is also varied among species. Generally, one sperm enters the egg via a micropyle, which closes after entry (Fig. 10). This prevents multiple fertilization. Thereafter, the two nuclei unite. The chorion, the tough outer shell, or protective coating, of the egg swells by the egg imbibing water. This water hardening provides the egg with a protective shell permeable to dissolved substances, gases, and water. Water hardening produces a perivitelline space between the cell membrane and the external environment. Eggs and sperm survive only a short time if fertilization does not take place.

Development

Developmental time depends on environmental conditions. The important factors are temperature, salinity, oxygen, and light. All conditions have varying effects, depending on species. The following outline shows the major steps in the development of the fish, from embryonic to adult stages (Fig. 11). It is similar in pattern to the general pattern in vertebrate development.

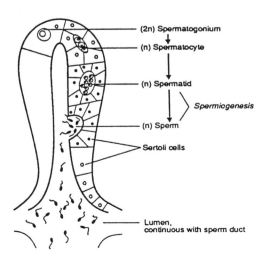

Figure 9. Spermatogenic cyst with several spermatogonial cells.

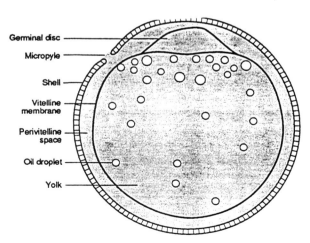

Figure 10. Cross section of a salmonid egg.

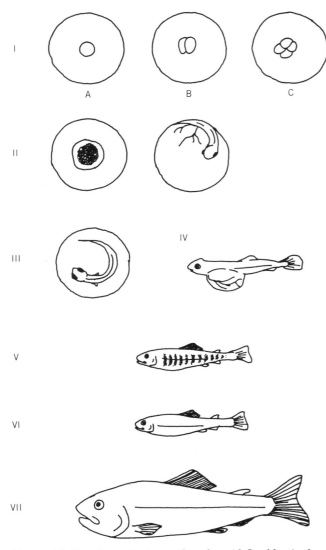

Figure 11. Development stages of a salmonid. I = blastic cleavage; cell division is shown in A, B and C. Cell division continues. II = gastrulation, III = organogenesis, IV = hatching, V = parr, VI = smolt, VII = juvenile to adult.

I. Blastodisc cleavage (2 cells)
II. Gastrulation
    Specialization of cells
    Body tissues develop and grow over the yolk
    Eyed stage
    Basic body form is evident
III. Organogenesis
    Internal organs form
    Fins appear
    Circulatory system develops connection between yolk and internal vasculature
IV. Hatching
    Hatching glands develop around head of embryo and secrete chorionase, an enzyme that dissolves the tough chorion, or egg shell, for the embryo to escape.
V. Parr
    General body form resembles that of the adult
    Dark vertical bars along the lateral body wall
VI. Smolt
    Parr marks disappear
    Body coloration is silver
    Body shape is slightly more elongate than smolt
VII. Juvenile to adult
    General body shape and color is like the smolt
    Size increases as it grows and matures at sea

IMMUNE SYSTEM

The immune system comprises the defense mechanisms of the body that prevent or control disease-producing organisms from entering the body and causing disease. The primary problem of the immune system is to recognize self from nonself. Self refers to all normal physical and chemical components of the body and nonself is everything else. The latter may include bacteria, parasites, viruses, and tumors.

Before discussing immune system in detail, a few key terms should be defined. Immunocompetence is the ability of a fish to mount a specific immune response against a foreign body or agent. The first stages of immunocompetence in salmonid fishes start at 1–2.5 g body weight. It is fully developed at 4 g. Immunological tolerance refers to a condition where the host does not recognize an antigen as being foreign and, therefore, tolerates its presence. Vaccination should not be conducted before immunocompetence is established, or immunotolerance may result. An antibody is a protein manufactured by white blood cells (WBC) that are designed to bind to an antigen for which it was made. An antigen is any substance, usually foreign to the body that is capable of causing the activation of an immune response.

The immune system is made up of two kinds, or lines, of defenses. The first is the non-specific defense and the second is the specific defense. The nonspecific defense mechanisms are also divided into two catagories of those involved in preventing nonself substances from entering the body and those that deal with substances that manage to enter the body. Nonspecific defense mechanisms are mounted against any foreign substance, whereas the specific defense mechanisms are designed to attack very specific antigens. Usually the term immune system refers to this last category.

Nonspecific Defense Mechanisms

Physical and chemical barriers to pathogen entry include the mucus on the skin and the skin itself. Mucus is continuously produced from goblet cells and it provides both a physical and chemical barrier to pathogen entry. It sloughs away pathogens and contains antibacterial proteins such as lysozymes, agglutinins, and lysins. Both the epithelium and the scales that make up the skin provide an effective physical barrier to pathogens. Some pathogens, however, do enter the body. Once the pathogens are inside, the following components of the nonspecific im-

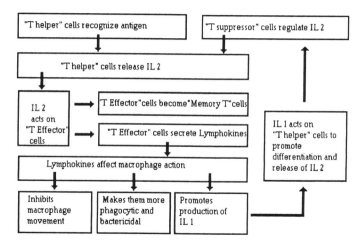

Figure 12. 1. Recognition of antigen by T helper (Th) cell. 2. Th cell releases a soluble messenger cell called interleukin (IL2). 3. Interleukin 2 acts on T effector cells. 4(a). T effector cells secrete lymphokines. 4(b). T effector cells begin differentiating to memory T cells. 5. Lymphokines mediated macrophage action, which results in antigen or antigen–host cell destruction. 6(a). Lymphokine promotes macrophage production of another interleukin (IL1). 6(b). Migration inhibition factor, a lymphokine, inhibits macrophage movement away from the infection locus where antigen was recognized. 6(c). Another lymphokine activates macrophages and makes them more rapidly phagocytic and more strongly bactericidal. 7. IL1 acts on Th cells to promote their differentiation and the release of more IL2. This action is regulated by T suppressor cells.

mune system work to limit growth and multiplication of the pathogen. Serum transferrin is a protein that keeps free iron concentrations low. Growing and replicating bacteria need iron. C-reactive protein is a nonspecific agglutinating protein that coats the surface of the antigen and allows phagocytosis to occur more readily. Interferon is a soluble substance that acts against viruses. It prevents viral DNA replication in an infested host and cell. Cellular nonspecfic defenses are performed by certain WBC such as neutrophils, macrophages and natural killer cells.

Specific Defense Mechanisms

There are two types of specific defense mechanism. Humoral immunity defense involves the production of specific immunoglobulins (Ig), which are globulin proteins that have antibody activity. The cellular immunity involves the stimulation of certain white blood cells to destroy specific antigens. The organs involved include the anterior kidney, thymus and spleen. The head kidney and the spleen have antibody-producing cells as well as phagocytes and macrophages. The spleen produces lymphocytes. Fishes lack bone marrow, and the above tissues act, in place, to supply the necessary elements for a functional immune system. Fish immunoglobins are tetramer molecules as opposed to the pentamers in humans and there is only one class of immunoglobins in fish compared to the two to five classes in mammals. These proteins are found in lymph, blood, mucus, and interstitial fluids.

Humoral Immunity. In this type of specific immune response, antigens are selectively removed from the circulation by the use of antibody-producing white blood cells, which are specific to particular antigen. There are two types of cell involved in humoral immunity. Some of the cells produce the antibody and some act as memory cells. Once exposed to a particular antigen, the memory cells can rapidly grow and multiply when they encounter that antigen again.

Cellular Immunity. In this type of specific immunity, the antigen causes certain groups of primed white blood cells to act specifically against the invading antigen. These WBCs can ingest or destroy the antigen as well as proliferate chemical messengers signaling the presence of nonself material. It seems that immersion-vaccination results in eliciting a particularly strong cellular immunity. This type of immunity is also refered to as cell mediated immunity The following are various cell types involved in cell mediated immunity.

Neutrophils are the first line of defense after antigen invasion. These cells are phagocytic in humans and some fish such as catfish, but they are not in salmonids. They may release a toxic substance in the vicinity of the antigen. They are abundant in the WBC picture. Macrophages have a phagocytic function and also play a significant role as a secretory immune cell. They are more like the human neutrophil in their phagocytic and bactericidal action. Special processing macrophages process and present the antigen to the lymphocytes. The antigen is taken up by these macrophages and processed internally. The antigen is then brought to the surface for lymphocytes to destroy. T-cell lymphocytes are antigen-specific lymphocytes that send chemical messengers that enhance phagocytic and bactericidal activity of macrophages. T-cells do not engulf pathogens, although they do have receptors for antigens. They also enhance macrophage activity. The activities of T cells are restricted by the fact that the antigen must be first processed by a macrophage and presented to this lymphocyte. They are unable to respond directly to free antigen. Their response to an antigen presented to them is to differentiate into effector T-cells and memory T-cells. The coordinated action of the various components of the specific immune system is demonstrated in Figure 12.

BIBLIOGRAPHY

1. D. J. Randall, *American Zoologist* **8**, 179–189 (1968).
2. G. A. Wedemeyer, F. P. Meyer, and L. Smith, *Diseases of Fishes, Book 5: Environmental Stress and Fish Diseases*, TFH Publications, Neptune, N.J., 1976.
3. L. Smith and G. Bell, *A Practical Guide to the Anatomy and Physiology of Pacific Salmon,* Misc. special publication of Environment Canada, Fisheries and Marine Service, Ottawa, 1975.

General References

Reference 3 is a good general reference.

P. J. Bentley, *Comparative Vertebrate Endocrinology*, Cambridge University Press, London, 1976.

P. J. Bentley, *Endocrines and Osmoregulation*, Springer-Verlag, New York, 1971.

C. E. Bond, *Biology of Fishes*, W. B. Saunders Co., Toronto, 1979.

C. P. Hickman, Jr., and B. F. Trump, *Excretion, Ionic Regulation, and Metabolism,* in W. S. Hoar and D. J. Randall, eds., *Fish Physiology*, Vol. 1, Academic Press, Inc., Orlando, Fla., 1969.

W. S. Hoar and D. J. Randall, eds., *Fish Physiology, Vol. IV: The Nervous System, Circulation and Respiration,* Academic Press, Orlando, Fla., 1970.

W. S. Hoar and D. J. Randall, eds., *Fish Physiology, Vol. II: The Endocrine System,* Academic Press, Orlando, Fla., 1969.

K. F. Lagler, J. E. Bardach, R. R. Miller, and D. R. May Passino, *Ichthyology*, John Wiley & Sons, Inc., New York, 1977.

A. J. Matty, *Fish Endocrinology*, Croom Helm, London, 1985.

P. Moyle and J. J. Cech, Jr., *Fishes: An Introduction to Ichthyology*, Prentice-Hall, Press, Englewood Cliffs, N.J., 1982.

C. Pincher, *A Study of Fish*, Duell, Sloan and Pearce, New York, 1948.

G. H. Satchell, *Circulation in Fishes*, Cambridge Press, 1971.

L. Smith, *Introduction to Fish Physiology*, F. F. H. Publications, 1982.

GEORGE KATSUSHI IWAMA
University of British Columbia
Vancouver, B.C., Canada

FISHES: SPECIES OF ECONOMIC IMPORTANCE

This article consists of brief descriptions of fish species of importance to either the commercial or sport fisheries in North America. The description of the each species consists of information about their physical appearance, notes about how they are caught and what food value they hold, their distribution, and their reproductive biology. The common and scientific names used in this article comply with published information (1).

In some cases, limiting the selection of species to those found in North America and of commercial or sport importance restricted the representatives of a particular genus or family. In most cases, however, the omitted species of any genus share most of the major characteristics of those species represented in this review. The amount of information given here about a particular species is largely a function of the body of published knowledge about that fish. This is the reason for any uneven treatment among species. The salmonlike fishes, for example, are probably the most intensively studied family of fishes in the world. The body of knowledge is consequently much larger than a group of fishes such as the ocean perches. Measured units such as body size, weight, fecundity, and age at sexual maturity are estimates in many cases, and are based on measurements on a few individuals. They should, therefore, be considered as rough estimations of the average value. The reference list at the end of the article contains reviews articles and books relevant to the species covered in this article.

ANCHOVY

The anchovies belong to the family Engraulidae. They belong to the same order as the herrings, order Clupeiformes. The anchovies are a schooling fish that are almost exclusively marine, except for some populations that migrate into or become landlocked in freshwater systems. The three species found in North American waters are the northern anchovy, the striped anchovy, and the silver anchovy. The northern anchovy is found off the Pacific Coast and the latter two are Atlantic species that are relatively rare in North America. The anchovies are typically found in temperate to tropical waters of Pacific, Atlantic, and Indian Oceans. They are a small and silver fish that usually measure less than 15 cm, although they have been reported as large as 50 cm. They have characteristically silvery sides with greenish blue to gray backs and a bright strip along the midlateral flank. The anchovies have no lateral line.

The northern anchovy (*Engraulis mordax mordax*) is presently of commercial importance in California. Their distribution extends from Cape San Lucas in Baja California to the Queen Charlotte Islands off the British Columbia coast. The preferred temperature range seems to be between 14.5 and 18.5°C. Although there was a sufficient abundance of northern anchovy off the British Columbia coast in the 1940s, stocks have declined to a point that a viable fishery cannot be sustained there. Populations off the California coast have been increasing in the last 20 yr, where they are captured in schools with nets for the fresh, cured, or canned markets. They have also been caught for the bait and fish meal markets.

Evidence from California, shows that the mature anchovy may spawn several times each year. Spawning takes place at temperatures between about 13 and 17.5°C near the surface and at night. Several thousand eggs are broadcast and fertilized in the open waters. The eggs are ellipsoid and measure about 1.4 mm along the long axis and slightly less than 1 mm in diameter. As the embryo develops, it hangs upside down during development. At those temperatures, the eggs hatch in about two to four days and the alevins are 2.5–3 mm long. The warm waters also enhance the utilization of yolk, taking only about three days for the yolk sac to be absorbed. The young resemble the adults by the time they are about 25 mm long. They become sexually mature when they are about 13 cm. While about half of the two- to three-year-old fish will be mature, all at four years of age are sexually mature. They may live to be seven years old. Like other coastal fishes, the anchovies move to deeper waters during the winter and migrate toward the inshore during the summer. They also spend the daylight hours near the darker bottom and rise toward the surface at night.

The striped anchovy (*Anchoa hepsetus*) is particularly abundant from Chesapeake Bay to the West Indies and Uruguay. Only strays from those areas have ended up in the northern United States and Nova Scotia, Canada. The silver anchovy (*Engraulis eurystole*) is normally found in the Atlantic Ocean from Woods Hole, Mass. to Beaufort, N.C.

STRIPED BASS

The name bass is usually used in reference to the largemouth bass (*Micropterus salmonides*), which has the reputation for being the most popular game fish in North America. The striped bass (*Morone saxatilis*), while also being an extremely popular game fish, is classified under a different family (Fig. 1). Like groups of other fishes, the basses include a number of species that are quite different from each other. The striped bass is not hermaproditic as the sea bass is and it has only two spurs on the opercular bone instead of three. It, therefore, belongs to the Percichthyidae family instead of the Serranidae to which the sea bass belong.

There is no predictable way in which this animal can be caught because of its sporadic feeding habits. The fisherman may be lucky with a particular lure or bait one day but be frustrated without a bite on the identical tackle and bait the next day. The animal is caught with a wide variety of sport fishing tackle. The most common tackle is the spinning rod with a lure that resembles the shad. The striped bass can also be caught on the fly rod with streamers or casting with bait. While it may be caught in open lakes and reservoirs, the most popular locations are at the outfalls of hydroelectric power plants or at the foot of dams. The elusive fish attracts the avid fisherman of inland waters. The present record is a 26.9 kg striper taken from the Colorado River in 1977. In addition to its large size, the popularity of the striped bass as a sport fish is due to the fact that it feeds actively in warm as well as cold temperatures. It also actively feeds in daylight as well as in the dark. This popularity, however, has added to the plight of dwindling populations of this species in Atlantic coastal waters. The effects of pollution and mechanical destruction of their free-floating eggs in dams in combination with the increasing catch by sport fishing has imposed great pressures on populations of this fish.

Although the striped bass is found on both coasts of North America, its native habitat is the Atlantic Coast. The populations in California and Oregon are decedents of a handful of animals that were transplanted from the East Coast into the San Francisco Bay area between 1879 and 1881. The construction of dams, which trapped migrating populations of striped bass, demonstrated that these animals can adapt to a wholly freshwater existence. They can thrive without the saltwater phase of their life cycle. This, along with the development of successful rearing technology in hatcheries, led to the transplantation of populations to inland states throughout North America and to the establishment of successful fisheries in 17 inland states.

The striped bass is omnivorous. The juvenile fish feeds on planktonic organisms, such as copepods, and on insect larvae. After the animals are between 7.6 and 12.7 cm in length, they begin to capture and feed on smaller fishes. The adult will feed on a wider range of organisms including fish, worms, shrimps, and shellfish. The striped bass is an anadromamous fish. That means that, like the costal salmon species, it spends its adult life in salt water and migrates into freshwater streams to spawn. Unlike the salmon that migrate great distances in salt water, however, the bass's saltwater residence is confined to coastal waters.

The female produces about a million eggs, 1 mm in diameter, for every 4.5 kg of body weight. The females are always larger than the males and also live longer. The females can live up to 40 yr, reaching sizes of over 4.5 kg. All animals that are greater than 13.6 kg are almost exclusively females. With the increasing day length and warming temperatures of spring, the migration into freshwater streams occurs. Spawning generally takes place between 4-yr-old females and 2-yr-old males. The spawning takes place in open waters of the stream. The female broadcasts her eggs, which are fertilized by several males. Due to the oil content of the eggs, they are buoyant and float freely in the current of the river and later in the estuary. After a rather short incubation period of two or three days, the alevins hatch with yolk sacs. The alevins obtain nourishment from the yolk for the first four to six days, after which they resemble a small adult. Very much like the salmon young, the 5.1- to 7.6-cm young show vertical parr marks, which are eventually replaced by the characteristic horizontal stripes when they are about 6 in. in length. The juvenile then spends its growing and maturing period in the coastal marine environment until sexual maturity.

COD FISHES

The cods belong to the family Gadidae, which has three subfamilies, the Lotinae, the Phycinae, and the Gadinae.

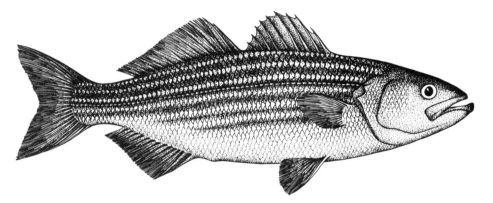

Figure 1. Striped Bass (*Morone saxatilis*). Copyright 1990 by B. Guild-Gillespie.

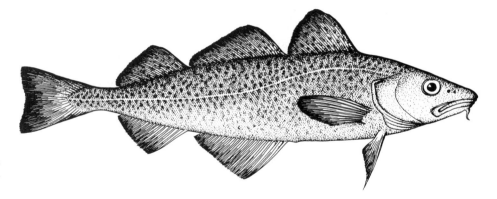

Figure 2. Atlantic cod (*Gadus morhua*). Copyright 1990 by B. Guild-Gillespie.

As Figure 2 shows, the cods have elongate bodies with large heads and mouths. They have well-formed caudal fins and may have two or three dorsal fins as well as one or two anal fins. They are normally dark and colored brown to gray. The cods produce free-floating eggs that do not contain oil globules. There are approximately 55 species of cods belonging to 21 genera. They are primarily marine and inhabit the cooler waters, near the bottom of northern seas. Most migrate inshore during the summer and move to deeper waters in the fall and winter. There are exceptions such as the burbot (*Lota lota*), which is a freshwater fish, and landlocked populations of the Atlantic tomcod (*Microgadus tomcod*). Furthermore, there are species that reside in southern seas. The 7 species discussed here were selected on the basis of their commercial importance as food fishes.

Atlantic Cod

The Atlantic cod (*Gadus morhua*) is one of the most important food fishes of the world. It is Canada's most important commercial species in terms of net value. Atlantic cod is caught commercially in a many ways, including trawl gear, seine, and gill nets, hand lines and jiggers. In 1984, the 463,100 tons of Atlantic cod landed in Canada was valued at $168.6 million. The Atlantic cod is also caught for sport with hook and line. The Atlantic cod cheek is a delicacy, served in the finest restaurants. By-products of the cod include fish meal, cod liver oil, and glue. Between these two extremes, the Atlantic cod is present in almost every food niche. It is sold fresh, frozen, smoked, salted, and canned. Fish and chips and fish sticks are prime market products for the cod.

The Atlantic cod is distributed throughout the cooler northern waters of the Atlantic Ocean. On the eastern shore, the Atlantic cod is found from Iceland, south to the Baltic Sea and to the Bay of Biscay. On the western side of the Atlantic, it is found in the North Atlantic from Greenland and southern Baffin Island, to Cape Hatteras, N.C. The majority of the populations are found on the continental shelf.

The Atlantic cod can be found spawning during most months of the year, from February to December, throughout their extensive range. The northern populations spawn earlier. As an example, spawning in the Grand Bank, which is one of the major fishing grounds for the Atlantic cod, begins in April, peaks in May, and ends in June. Spawning takes place in the open ocean at varying depths. While some may spawn at depths less than 110 m, other Atlantic cod populations will spawn deeper than 182 m. Both eggs and milt are broadcast into the water. Fecundity is high and increases with age. Although a fish 51 cm in length may produce 200,000 eggs, a female 140 cm may produce 12 million eggs. Like the haddock, the eggs are about 1.5 mm in diameter, spherical, transparent, and buoyant. They rise and float freely under the surface during incubation. As for all fish eggs, incubation time depends on temperature. Due to the colder waters that characterize their environment, the eggs take from 40 to 60 days to hatch (-1.5 to $1°C$). The newly hatched young are about 3–6 mm long and remain pelagic until a length of about 25–30 mm. They then descend to the bottom to feed and mature to adult sizes. The average adult caught by the commercial fishery is about 5–6 yr old, 50–60 cm long, and about 1–2 kg. They can live to be over 20 yr old and the largest recorded animal was 95.9 kg.

Pacific Cod

The Pacific cod (*Gadus macrocephalus*) is known as the common or gray cod. Like its counterpart on the Atlantic Coast, it is the most important trawl-caught bottom fish in western Canada. This trawl catch increased to a peak of about 9.1 million kg in 1966. It decreased after that to about 2.3 million kg in 1970. The Pacific cod is sold fresh and frozen. The flesh is filleted, and products such as fish sticks or fillet blocks are made for the domestic and export markets.

The Pacific cod is similar in appearance to the Atlantic cod. In general, it may be more slender and may have a longer barbel under the lower jaw that equals or exceeds the diameter of the eye. The Pacific cod is found on both sides of the Pacific Ocean. Its distribution extends from Santa Monica, Calif. north to Alaska and around into the Okhotsk and Japan seas. It is also found off Korea and in the Yellow Sea to Port Arthur.

The Pacific cod may be found in shallow waters in the spring but migrates into deep water in the fall. They spawn in the winter months. Spawning females range from 40 to 60 cm in length and are two to three years old when they spawn the first time. A 60-cm female may produce about 1.2 million eggs, which measure about 1 mm in diameter. The eggs and milt are broadcast into open waters and the fertilized eggs may float near the surface, depending on the salinity. In the cold northern waters, the eggs may take four weeks to hatch at 2°C. The incubation time is reduced to about 8 days at about 10°C. At 5°C, the yolk sac is absorbed in about 10 days. The young will grow to about 20–50 cm in one year and to about 45–75 cm in about two years.

Haddock

The haddock (*Melanogrammus aeglefinus*) is a commercially valuable food fish of the cod family. Haddock are caught commercially with the otter trawl. They are sold as a fresh, frozen, smoked, and canned product. The major stocks are presently recovering from the overexploitation of the 1960s.

Like the other cods, the haddock is rather dark in color. The color fades from the purplish gray on the dorsal surface, laterally to a light pink below the lateral line, and eventually to a white on the ventral surface. There is a characteristic black blotch on the lateral side posterior to the head and just dorsal to the pectoral fins.

The haddock is found on both sides of the Atlantic. On the eastern side of the North Atlantic, it is distributed from Iceland south to the English Channel. It is found in the White, Kara, Norwegian, and North seas. On the western side of the Atlantic, the haddock is found from southwestern Greenland, south along the east coast of Canada and the United States to about Cape Cod. They migrate between the shallower (30–37 m) warm waters of the banks during the summer and the deeper (55–125 m) waters in the winter.

The haddock spawns in waters up to 90 m in depth, primarily on the Grand, Emerald, Browns, and Georges banks. They spawn between January and May, depending on location. The spawning adults are between 3 and 5 yr old. While many features of the reproductive process in the haddock are characteristic of the cods, observations of aquarium specimens suggest that the reproductive behavior of the male fish is complex, involving color changes, sound production, and courtship displays. The spawning female measures between 40 and 60 cm long and may produce between 230,000 and 1.77 million eggs, respectively. The eggs resemble that of the Atlantic cod. Like the other cods, the eggs and milt are broadcast into the open waters. After fertilization, the eggs float to the surface and float freely until they hatch. In the cooler waters of the North Atlantic, incubation may take between one (10°C) and four (2°C) weeks. The newly hatched alevins are about 4 mm long and grow to young larvae of about 25 mm before they resemble the adult in body form. They start to descend to deeper waters between 40 and 50 mm. Although the average 5-yr-old haddock may measure about 50 cm in length, the fish can live to be over 15 yr in age and an individual 112 cm long and 16.8 kg has been recorded.

Pollock

The pollock (*Pollachius virens*) is also known by other names such as Boston bluefish, coalfish, and green cod. The flesh, which is of a darker color than the cods and haddock, has a richer flavor than its relatives. The adult weighs about 2–5 kg and is fished commercially as an important food fish. The pollock is caught commercially in a number of ways, including trawling, longline, hand line, weirs, and traps. It is sold in fresh, frozen, and smoked forms. It is also fished for sport in the shallower waters with artificial lures.

Like the haddock, the pollock are distributed throughout the North Atlantic. On the east side, they are found from Iceland, south to the Bay of Biscay. On the west side of the Atlantic, the pollock are found from southwestern Greenland to Cape Hatteras, N.C.

At maturity, the male and female are about 3 yr old and are about 50 cm long; although records show that they can live up to 14 yr, weight 70 kg, and measure over 100 cm. The pollock shares much of the reproductive features of other cod fishes. They produce about 225,000 eggs each; the eggs are about 1 mm in diameter, spherical, bouyant, and remain pelagic until hatching. Eggs and milt are released into the open waters at a depth of about 100–200 m. Unlike other cod fishes, the pollock spends more time swimming throughout the water column as opposed to being a primarily bottom fish. Its preferred depth is about 110–180 m. Like the cod fishes, however, the pollock migrates toward shore and shallow waters in the summer and offshore to deeper waters in the winter.

Walleye Pollock

The walleye pollock (*Theragra chalcogramma*) has come to be one of the most important commercial fisheries in the world. It has been described as the most productive single-species fishery in the world. While it once had a limited market for human consumption due to its soft flesh, it is fished intensively today for the roe, frozen fish block and fillets, and processed fish flesh markets. Surimi, a processed minced fish flesh product, is also one of the prime products from the flesh of walleye pollock. These fishery products that derive from the walleye pollock are taken by separate fisheries, although trawling and seining are the common methods of catch. The surimi fishery consists of large factory trawlers or large factory ships with fleets of smaller catcher vessels; walleye pollock of all sizes are taken from the bottom and midwater depths. The freezer fishery consists of large factory trawlers and smaller independent fisherman that harvest walleye pollock off the bottom, along with other species such as turbot and the Pacific cod. The spawning adults are caught for the roe fishery from the midwaters.

The walleye pollock has a typical codlike appearance with three, well-separated dorsal fins and very small or no barbels projecting from the lower jaw. It usually has an olive green to brown color on its back with irregular blotches or mottled appearance. It is lighter on its ventral surface and has silvery sides.

The walleye pollock is distributed from central California, north through the Bering Sea, around to Asian waters to the southern reaches of the Sea of Japan. It has a wide distribution vertically in the water column. It has been found from the surface to depths below 380 m.

Walleye pollock begin to mature sexually at about 2–3 yr of age, when they measure about 25 cm. By the time they are 5–6 yr old and 45 cm, more than 90% are mature.

The fecundity of a 50-cm female is between about 200,000 and 220,000 eggs. Estimates of fecundity range from about 100,000 to 1 million eggs for females between 30 and 70 cm in length, and 225 to 2,000 g, respectively. While spawning occurs during one season of the year, the female will spawn a number of times within that season. The eggs develop and mature in batches over time to allow multiple spawnings. Spawnings takes place mostly between March and April. The eggs measure about 1.4 mm in diameter and float freely in the water, once they are spawned. They are found mostly in the upper 20 m of the water column. Hatching may take place between 14 (5°C) and 24 (2°C) days. The hatched young measure about 4 mm in length and resemble tadpoles, with characteristic markings. They are born with yolk sacs, which are absorbed by the time they attain about 7 mm in length. At 22 mm and about 50 days after hatching, they have all the adult fin rays and are considered juveniles. Adults can live to 15 yr, although most are between 1 and 7 yr of age.

Hake

The red hake (*Urophycis chuss*) and the white hake (*U. tenuis*) are often considered as good food fishes. The white hake grows to a larger size and is distributed over a broader range than the red hake. Characteristics of the white hake are presented here. Similar in size to other cod fishes, the hake is distinguished by the smaller head, large eyes, long dorsal and anal fins and the numerous barbels on the ventral side of the chin. They have a reddish brown color on their back, which fades to lighter shades on the side and white on the belly. The lateral line is pale. While some hake fisheries exist, this species is mostly taken incidental to other fisheries.

The hake live on mud bottoms, 200–1,000 m in depth. They are distributed on the continental slopes in the North Atlantic from Iceland south to North Carolina. Some have strayed as far south as Florida in deep water. Like the other cod fishes, they move to deeper waters in the fall and winter.

The average size of an adult at maturity is between 60 and 70 cm, although specimens over 100 cm have been recorded. Spawning takes place at different times depending on the location, although most takes place in the winter and early spring. Fecundity is high relative to other cods. A 70-cm-long female may produce about 4 million eggs, and one 90 cm long may produce 15 million eggs. The eggs, therefore, are smaller than those of the other cod fishes and measure about 0.75 mm in diameter. The eggs, which are transparent and bouyant, float about the sea until the young alevins, about 2 mm long, hatch. The young grow up to about 80 cm as pelagic fish, after which they migrate to deeper waters to feed and grow to maturity.

Whiting

The name whiting can refer to any one of three species of fish. It was a name originally given to the walleye pollock (*Theragra chalcogramma*) (1). The Pacific tomcod (*Microgadus proximus*) is also commonly called the whiting. While the Pacific tomcod is highly regarded as a tasty food fish, it is not abundant enough for a commercial fishery. The blue whiting (*Micromesistius poutassou*) on the other hand is fished in Europe for food. The blue whiting is rather rare in North America.

The blue whiting is normally found in the northeast Atlantic, off the coast of Greenland and off southern Iceland. It may also be found off Europe between Norway, south to western Mediterranean. Records of the blue whiting in North American range from Sable Island, south to Woods Hole, Mass. The blue whiting is a deep-water fish, caught by otter trawls at depths beyond 183 m off the continental slope during the spring or early summer. The scarce data on the reproductive features of this fish suggest that they resemble, verly closely, reproduction in other species of the Gadidae. It has been suggested that spawning occurs between mid March to mid May (2). Spawning probably occurs in deep waters, deeper than 1,000 m, and the eggs are about 1–1.3 mm in diameter. The few specimens inspected from the northwest Atlantic suggest that the blue whiting is smaller than other Gadids, measuring less than 40 cm.

FLATFISHES

As the name implies, the flatfishes include the fishes that are flat in the dorsoventral axis (Fig. 3). They have both eyes on one side of the body and are often found lying on the bottom of the ocean covered in layer of sand or mud. All the flatfishes are classified in the order Heterosomata and include the soles of the family Soleidae and the flounders in the family Pleuronectidae. This latter family in-

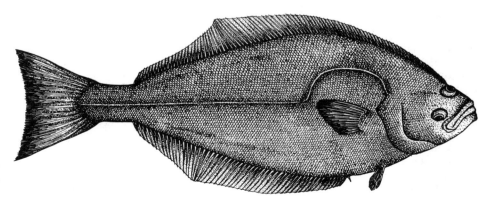

Figure 3. Halibut (*Hippoglossus hippoglossus*). Copyright 1990 by B. Guild-Gillespie.

cludes the commercially important species of the true halibuts in the genus *Hippoglossus*, the turbots in the genus *Reinhardtius*, and the right-eyed flounders in the genus *Platichthys*. The commercially important species tend to be right-sided; that is the side that is visible to the observer is the right side of the fish. The left eye migrates, along with appropriate skeletal, muscular and neural rearrangements, so that both eyes end up on the right, uppermost side of the animal. Although these fishes are well known as benthic animals, fish in the juvenile stages are also pelagic. The reading of rings in the otoliths, or ear stones is sometimes used to age flatfishes. Nine commercially important species are discussed here.

Halibut

Although it seems reasonable at this time to regard both the Atlantic and Pacific halibuts as belonging to the same species *Hippoglossus hippoglossus*, the Pacific halibut is considered in current literature to be a separate species *Hippoglossus stenolepis* (Schmidt) (3,4). There were declining levels of global commercial catches during the 1960s and 1970s. The large increases in bottom trawling in the 1960s and 1970s were significant contributions to the decline. These global trends were even more exaggerated in some local fisheries such as the Atlantic landing in New England. The average annual landing was 4.83 million kg (eviscerated and heads off) in 1879. In 1930–1939 it was 802,858 kg and by the period 1970–1975, it had fallen to 75,296 kg. Regulation and stock management have curtailed this trend and Pacific stocks, for instance, seem to be recovering.

The chief way in which commercially caught halibut is preserved for storage and transport to market is freezing. The animal is eviscerated and has the head off before the intact trunk is frozen. While earlier methods included salting and chilling with ice, current methods of quick freezing and glazing or vacuum packaging in plastic provides a more reliable and consistent method.

Atlantic Halibut. The Atlantic halibut (*Hippoglossus hippoglossus*) is a highly prized food fish. The Atlantic halibut is commercially taken by otter trawls and longlines, with smaller fish being taken by trawls and the larger fish being caught on longlines. Both sizes are taken by fisheries specifically directed at their catch or as incidental catches when trawling or long-lining for cod, haddock, or other ground fishes. The flesh is marketed fresh or frozen. The high price it commands enables fishermen to take relatively small quantities. In spite of this fact, there is evidence suggesting that certain populations are being overfished. The average length of the mature male and female halibut was 84 cm and 98 cm, respectively, in the period 1959 to 1969. Those lengths dropped to 66 cm and 70 cm, respectively, in the years 1970 to 1979. This reduction in average size at maturity is characteristic of populations subjected to heavy harvesting of the larger individuals.

The Atlantic halibut has a typical appearance of a flatfish. It has a flattened body, lying on its left side with its right side showing. The lower side is characteristically white and the right uppermost side is colored various shades of brown. The lateral line is prominent and curved over the pectoral fins. It is a very large fish, with record weights of over 100 kg. The larger fishes are between 30 and 35 yr old. It is the largest of the flatfishes. The females, in particular, have the highest growth rates as well as growth potential among the flatfishes. Females, therefore, become larger than males.

The Atlantic halibut is found on the east and west shores of the north Atlantic Ocean. The distribution on the east shore spans the Bay of Biscay to the Barents Sea. On the west side, the Atlantic halibut is found off the west coast of Greenland, south along the Canadian coast to the coast of Virginia. It prefers cooler waters between 3 and 9°C, at depths of 200–300 m.

Knowledge about reproduction in the halibut is relatively new. Atlantic halibut spawn between February and March around the distribution in Canada. They spawn in deep waters ranging from about 180 m on the western side and about 1,000 m or deeper in the northeastern Atlantic. The mature female, which weighs about 90 kg, may produce over 2 million eggs. The eggs are about 3 mm in diameter, spherical, and neutrally buoyant. They float during development at depths greater than 54 m but the hatched young sink as they develop. These free-floating eggs are most abundant at temperatures between 4.5 and 7°C and salinities between 33.8 and 35.0 ppt. Incubation takes about 16 days at 6°C. Little else is known about the details of the early life stages of the Atlantic halibut.

Pacific Halibut. The Pacific halibut (*Hippoglossus stenolepis*) has had a long commercial history on the Pacific coast of North America. The transcontinental railways across North America facilitated the Pacific halibut catches to be shipped across to eastern destinations and contributed to the near demise of the Pacific populations of this species. At the turn of the century, the total Canadian and U.S. catches was more than 4.54 million kg. Additional technological advancements in fishing vessels and gear increased harvests to a point that by 1930, increases in fishing efficiency no longer yielded greater catches. Through regulation under the International Fisheries Commission, the Pacific halibut stocks have rebuilt over time. The Pacific halibut is commercially caught mainly on hook and line or on other gear that does not involve nets of any kind. Numerous baited hooks are set on lines and rest on the bottom. Those sets can be made by dories from larger mother ships. Longlines of baited hooks can be set and retrieved by the mother ship.

The Pacific halibut is distributed throughout both western and eastern coasts of the Pacific. On the western side, it is found from southern California, north along the western coast to the Bering Sea. It is found off Anadyr, Kuril Islands, and Kamchatka, down to the northern areas of Japan. Like other flatfishes, the Pacific halibut is benthic (living on the bottom) at depths of about 1,100 m.

Reproduction and details of the early life stages in the Pacific halibut are better understood than the Atlantic halibut. Spawning takes place in the winter between November and January between about 275 to 412 m. It is not entirely clear whether females spawn every year or whether they spawn every other year. Both may be possible. Pacific halibut are known to migrate as far as 1,600 km to spawn. The mature female is about between 8 and 16 yr old and about 100–150 cm long. Males are considerably smaller. The female may produce between 600,000 and 1.5 million spherical eggs of about 3–3.5 cm in diameter, which are neutrally bouyant between 100 and 200 m. The egg surface has a honeycombed appearance due to the presence of many small holes. The eggs as well as the newly hatched young are pelagic for four to five months after spawning. Like the Atlantic halibut, the newly hatched alevins start to sink to deeper waters and are found below about 200 m. The alevins are transparent, except for the eyes, and about 8–15 mm long with very large yolk sacs. The young are still symmetrical at this point. When they reach a size of about 18 mm, the left eye starts to migrate to the right side and the yolk sac is barely evident. The young appear as small adults by the time they are about 30 mm. By the time the young are about three to five months old, they rise in the water column to about 100 m. Currents carry the young inshore and they become benthic fish at about six to seven months in age. As they mature, they gradually move to deeper waters. By the time they are about 5–7 yr old, they are established at depths of about 100 m and available to the commercial fishery.

Greenland Halibut. The Greenland halibut (*Reinhardtius hippoglossoides*) is a moderately flat fish that has a rich-tasting flesh. It is also known commonly as the Greenland turbot in the marketplace. Greenland halibut are caught commercially by longlines, gill nets, and otter trawls. It is only moderately flat because both sides of the fish are equally muscled. While it is primarily sold as frozen fillets, it is also sold as a smoked product. It is a moderate-size fish that grows to about 25 kg and 120 cm in Atlantic waters; the Pacific specimens are somewhat smaller. The underside is usually white in a young fish but gray in older fish. The visible side is dark brown to black. It is distinguished by its prominent teeth along a protruding lower jaw.

The Greenland halibut is distributed in cold waters in both the Atlantic and Pacific Oceans. In the northeast Atlantic, it is found from around Iceland and the Greenland Sea to the Arctic Ocean, and both Barents and Norwegian Seas. From there, it is found southward to the Faroe-Shetland Ridge. In the northwest Atlantic, it is distributed from western Greenland, south to Georges Bank. In the Pacific, it is found from northern California to Japan through the Bering and Okhotsk Seas, Sakhalin, and Kamchatka.

The limited knowledge about the details of spawning in this species is, no doubt, due in part to the fact that it spawns in the winter in the far north and at depths of about 650–1,000 m, at temperatures of about 0–3°C, making observation difficult. Spawning takes place in the winter or early spring, depending on location. The Greenland halibut can live to an age between 15 and 20 yr old. Until they are about 5–7 yr old, the males and the females grow at about the same rate. After that, the females grow larger and at a faster rate. All fish over 90 cm are females. The mature female may produce between 30,000 and 300,000 eggs, depending on size. The eggs are larger than the halibut eggs, measuring about 4–4.5 mm in diameter. The eggs are transparent and the newly hatched young rise to depths of about 30 m below the surface, where they live as pelagic fish until they are about 70 mm long. At that stage, they move to deeper waters but do not become benthic like the halibuts; although they are associated close to the bottom.

Yellowtail Flounder. The yellowtail flounder (*Limanda ferruginea*) is valued as a tasty product that is often sold fresh but that may also be sold frozen. It is either fished directly with otter trawls or as an incidental catch on longlines in the American plaice fishery. While it is fished commercially, it is usually just the largest of the population that gets caught because of its small mouth. While it has the lateral line that curves over the pectoral fins, like the halibuts, the yellowtail flounder has a small mouth and the body size is much smaller. An average adult caught in the commercial fishery is about 30 cm long. This fish gets its name from the yellow markings anterior to the tail and at the bases of dorsal and anal fins. It has a white underside and the color of the visible right side may range from a reddish brown to an olive green.

The yellowtail flounder has a relatively limited distribution, compared to other flatfishes. It is found only in the western North Atlantic Ocean, from southern Labrador down to the Chesapeake Bay. The time of spawning depends on location, but mainly occurs in the spring to summer months. The female produces a large number of small eggs. Females that measure 40–45 cm may produce between 1 and 2 million eggs that are slightly less than 1 mm in diameter. The eggs and milt are deposited on, or near the bottom. Fertilization occurs there and the buoyant eggs then float to near the surface and drift during incubation. The incubation time is about five days at about 10°C. The newly hatched alevins are about 2–3.5 mm long and grow to about 11.5–16 mm before they metamorphose from the anatomically symmetrical young to a smaller version of the adult form. The yellowtail flounder may live to 12 yr of age, and individuals have been recorded up to about 60 cm. The typical fish caught commercially is between 4 and 10 yr old, depending on the location of the catch. The females grow at significantly faster rates than the males.

American Plaice. The American plaice (*Hippoglossoides platessoides*) is a white-fleshed fish that is highly enjoyed as a food fish. It is a very important commercial fish that constitutes about one-half of the commercially caught flatfish species in Canada. Most of the northwest Atlantic fishery is Canadian. American plaice is commercially caught with otter trawls, gill nets, and with the Danish seine. Most of the commercial fishing takes place along the Labrador Shelf, Grand Bank, along the southern Gulf of St. Lawrence, and on the banks of the Scotian Shelf. The flesh is normally sold as frozen fillets. It is colored white on the underside and reddish to grayish brown on the visible right side. The maximum body size rarely reaches over 60 cm.

The American plaice is found on both sides of the North Atlantic Ocean. It is distributed from Iceland, south to the

English Channel, on the eastern side and from Baffin Island, south to Rhode Island on the western side. It is normally found between about 75 and 275 m of water, preferring temperatures around freezing. It is benthic and prefers the fine sand or mud bottom. It has a seasonal migratory pattern, which brings it to deeper waters in the winter and to shallower waters in the spring.

Spawning take place between early April and June, depending on location. Spawning takes place at depths of about 90–180 m, at temperatures between 0 and 2.5°C. The mature female produces many small eggs. A 40-cm-long female, 8 yr old, may produce between 250,000 and 300,000 eggs, while a 70-cm-long female has been recorded to produce 1.5 million eggs. Males mature earlier at 4–5 yr old. The eggs measure about 1.5–2.8 mm in diameter. They are bouyant, without oil droplets and float near the surface during incubation. They, therefore, drift widely. At surface temperatures of about 5°C, the young alevins of about 4–6 mm, hatch in about 11–14 days. They transform from the symmetrical young to a small fish resembling the adult form at about 18–34 mm in length. They can live to be 25 yr old, and the record size is 81.2 cm long and 6.3 kg dressed.

Petrale Sole. The petrale sole (*Eopsetta jordani*) is a large flatfish that is prized as an excellent food fish, especially along the west coast of North America. The livers of this fish are a rich source of vitamin A. It is distributed from Bering Sea, south to Baja California. Like other flatfishes, it migrates seasonally and is found at depths of around 70–130 m throughout most of the year, except during the winter, when it moves to deeper waters, around 300–460 m. The petrale sole is white on the underside and olive brown on the visible right side. It has a large mouth and lacks free spines at the origin of the dorsal fin.

The oldest female petrale sole recorded was 25 yr old, whereas the eldest male was 19 yr old. The average spawning female is about 44 cm and the male is about 38 cm. The female produces a large number of relatively large eggs, the number depending on size. Records show a 42-cm female produced 400,000 eggs and a 57-cm female produced 1.2 million eggs. The eggs measure about 1.3 mm in diameter. Spawning takes place between late winter and early spring. Spawning takes place at depths of about 350 m and the bouyant eggs float to shallower depths where they are carried by prevailing currents during incubation. The eggs hatch into 3 mm alevins in about 8–9 days at 7°C and the yolk sac is absorbed within a further 10 days. Samples that measured 22 mm in length showed metamorphosis complete. Other evidence suggests that they settle to their adult benthic existence by the time they are 1–2 yr old.

Rock Sole. The rock sole (*Lepidopsetta bilineata*) is the most commonly used food fish of the smaller flatfishes. There have been several subspecies identified in the Pacific Ocean. At 15 yr of age, the females may reach a length of 60 cm, and the males, 50 cm. It is recognized by a canal formed along the lateral line that arches, typically, over the pectoral fin. It is a right-sided flatfish. Unlike other flatfishes, it tends to inhabit shallower waters. It is commercially taken at depths up to about 200 m, although it is scarce from about 100 m and deeper. Like other flatfishes, however, it does migrate to even shallower waters during the summer. Its distribution extends to both sides of the Pacific Ocean from southern California, up along the North American west coast to the Bering and Okhotsk seas and around to Korea and the Sea of Japan.

Observation off the North American west coast show that spawnings takes place between February and April. The fecundity of females measuring 35 cm and 46 cm in length were estimated at 400,000 and 1.3 million eggs, respectively. The eggs are pigmented yellowish orange and are adhesive. They measure about 1 mm in diameter. Incubation is a function of temperature and may take between 6 and 25 days, corresponding to a temperature range of about 8 to 3°C, respectively. Hatched alevins are about 5 mm long and yolk absorption takes about 10–14 days, influenced again on ambient temperature. Fully metamorphosed young have been observed at about 20 mm in length. The oldest recorded female was 25 yr old, and the eldest male on record was 15 yr old.

Dover Sole. The Dover sole (*Microstomus pacificus*) is highly prized for its quality flesh and excellent keeping qualities in the frozen state. It was originally dismissed as a viable commercial species because of its softness and sliminess. It may be uniformly brown on the visible right side and the underside may range in color from a light to a dark gray. It is characterized by having a lateral line canal that is almost straight as well as the excessive production of slime. While the body is also extraordinarily flaccid, it is known as a very hardy fish. The Dover sole is distributed from northern Baja California, up to the Bering Sea, found mainly on soft substrates. The average body size is about 70 cm.

Spawning in California takes place from November to February. Mature females are approximately 45 cm and males mature at about 40 cm. The female produces a wide range of eggs depending on size. Samples from Oregon showed a 42.5-cm female produced 52,000 eggs, while another 57.5-cm female had 266,000 eggs. The eggs are large, measure 2–2.6 mm in diameter, and have a wrinkled surface. After hatching, the young remain bilaterally symmetrical for several months. Flatfishes usually metamorphose and settle to a near benthic existence at about 20 mm. Dover sole specimens up to 100 mm in length have been observed in a pelagic life stage.

English Sole. The English sole (*Parophrys vetulus*) is characterized by its pointed head and, like the Dover sole, a lateral line canal that is almost straight. It is a moderate-size flatfish that grows to about 60 cm as a female and up to 50 cm as a male. It is colored a uniform brown on the uppermost right side and pale yellow to white on the underside. It has had a long history of being a commercially important species in North America. Its particular iodine flavor, found in some inshore populations, has identified its place in the marketplace. It is typically fished commercially at depths shallower than about 130 m, although its distribution extends to about 300 m.

Spawning off British Columbia occurs between January and March. The range in fecundity is extreme. Records show that a 30-cm female produced 150,000 eggs, and one measuring 44 cm produced 1.9 million eggs. The size of the maturing male and female are rather similar. Mature females measure about 30 cm, whereas the males measure about 26 cm. The eggs are small and measure slightly less than 1 mm in diameter. The eggs float due to the presence of oil droplets of various sizes in the yolk of the egg, start to sink just before hatching. The surface of the egg is covered with small wrinkles and pores. Observations in California show that the incubation time is about 90 days and the new alevins are about 2.8 mm long. Due to the oil droplets, the alevins hang upside down until the yolk is absorbed, in about 10 days, at which time they can swim. They remain pelagic for 6–10 weeks, after which they metamorphose to their adult form and seek the bottom. Young English sole are found in shallow waters but as they mature, the larger fish move to deeper waters. Like the other flatfishes, it tends to inhabit deeper waters in the winter and seek the shallower waters closer to shore in the spring. Another rather peculiar characteristic of this flatfish is that it can migrate long distances. There are records of English sole traveling over 1,000 km, between Vancouver Island and California.

HERRINGS AND SARDINES

The herrings are commercially important in that they support a variety of food fish markets (Fig. 4). Until the late 1960s, the herring was reduced and utilized heavily as a source of oil and fish meal. The herring has and continues to be used for pet food. Herring as food for humans has had a long history. Today, they are sold fresh, frozen, smoked (kippers), and pickled. The smaller Atlantic herring, which are canned, is well known around the world as a sardine. The larger Atlantic herring are canned as kipper snacks and fillets. The roe from herring has recently found a market in Japan as a delicacy item known as kazunoko, as their local herring populations have declined. The herrings belong to the family Clupeidae. Both Atlantic and Pacific herrings are subspecies of the species *Clupea harengus*. The two fish look similar, characterized by a silvery body that is highly compressed laterally, large cycloid scales and large eyes. The pearl essence from the scales was in high demand during the 1940s for use in high-quality paints for aircrafts.

The Pacific sardine (*Sardinops sagax*) is commonly referred to as the pilchard. It is not the fish that North Americans commonly refer to as the sardine. While the pilchard has some commercial value for its oil and for fish meal, very little is canned.

Pacific Herring

The Pacific herring is fished today for its roe and for reduction purposes. Both herring and herring spawn have been fished by the North American natives since 800 BC. The commercial herring fishery that began in the late 1800s was based on the salted herring market. This was replaced by a fishery based on the reduction industry in which the herring carcasses were reduced for oil and meal for commercial feeds for poultry and fish culture. The relatively new market of herring roe for the Japanese market started in the early 1970s. Today, the herring roe goes to Japan, and the carcasses are reduced for oils and meal.

The Pacific herring (*Clupea harengus pallasi*) is distributed throughout the coastal regions of both eastern and western shores of the Pacific Ocean. On the eastern shore, it is found from northern Baja California up to the Beaufort Sea. On the western side, it is found from Korea, north to the Arctic Ocean. The adults are found at depths of 100–150 m. Juveniles are found between 150 and 200 m of water.

The Pacific herring spawn in the late winter through to April, with peak spawning occurring in March. This exclusive spring spawning distinguishes the Pacific herring from their Atlantic counterpart. Most are able to spawn by the time they are about three years of age. The spawning process is very dramatic as the fish broadcast eggs and sperm near shore. The water in which they spawn turns white with the milt from the males, covering as much as 257 km of shore line and at depths from the surface to about 10 m. The spawning area is usually in the intertidal zone, in sheltered bays or on open sand beaches, but not on exposed coastal areas. The texture of the substrate seems to be an important factor in determining the exact location of spawning.

The female produces between 9,000 and 38,000 eggs depending on her size, which can range from about 20 to about 30 cm. The relative fecundity is about 200 eggs per gram body weight per year. In extreme cases, fish can grow to 50 cm and produce over 100,000 eggs. The eggs are about 0.9 mm in diameter before fertilization but expands to about 1.2–1.5 mm in diameter after fertilization

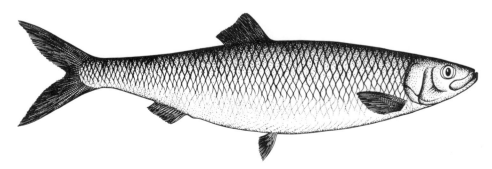

Figure 4. Herring (*Clupea harengus harengus*). Copyright 1990 by B. Guild-Gillespie.

and the absorption of water. They are also very sticky once they are exposed to water. The eggs commonly adhere to aquatic plants such as eelgrasses or rockweed. The incubation time is about 10 days and the newly hatched alevins measure about 7.5 mm in length. The yolk is absorbed within the following two weeks and the fish then begin to feed on planktonic organisms. At that stage, the young do not resemble the adults. They are white, thin, and have large eyes. In about two months following yolk absorption, the young start to resemble the adult form and begin to form schools. By the late summer, the young are about 2.5–4.0 cm long and move to deeper waters in the fall. The Pacific herring can live to be 10 yrs old. During growth and maturation, they may be about 15 mm by the end of their second year, about 20 mm at the end of their fourth year, and about 23 cm at the end of the eighth year of life. They return to shallow waters close to shore as they mature sexually and prepare to spawn. They also tend to have a diurnal migration as rise to the surface in the evening and swim to deeper waters at dawn.

Atlantic Herring

The Atlantic herring (*Cuplea harengus harengus*) looks like its Pacific counterpart with iridescent blue or bluish green back and sides and a silver belly. Like the Pacific herring, its abundance has been declining steadily over time. The Canadian landings, for example, were 528,000 tons in 1968; 250,000 tons in 1975; 177,000 tons in 1980; and 147,000 tons in 1982. The advent of the highly efficient purse seine net has contributed to the harvest pressures on populations of herring. Gill nets, trap nets, and weirs have traditionally been used to catch herring. It is distributed on both sides of the North Atlantic Ocean. On the west side, it is found from Greenland south along the east coast of North America to Cape Hatteras. On the east side of the Atlantic, its distribution extends from Iceland, south to Europe between the White Sea to the Strait of Gibraltar.

Atlantic herring mature to spawn at three to five years of age. There seems to be several discrete stocks that spawn at different times of the year; there are spring, summer, and fall spawning populations. There may be stocks spawning every month between April and November throughout its distribution. Spring spawning occurs in shallower inshore areas, whereas summer and fall spawning occurs in deeper offshore waters. Like the Pacific herring, eggs and milt are broadcast into the open water where fertilization occurs. The eggs then sinks to adhere to bottom plants such as Irish moss and several algal species at depths of 1–4 m. The fecundity can range an order of magnitude from 23,000 to 261,000 eggs, depending on body size and age. The fecundity increases with body size and age up to a certain age, after which egg numbers decline with further aging. Fecundity is also a function of when spawning occurs. Spring spawners produce up to about half the egg numbers of summer and fall spawner. The eggs, however, have a larger yolk mass. This may reflect a strategy to survive the colder months of spring when food supply might be less abundant than in the summer and fall months. The small eggs, which measure from 1 to 1.5 mm in diameter, take from 10 to 30 days to hatch, depending on the ambient water temperature. The hatched alevins are about 4 mm long. They are light sensitive and avoid bright light, seeking deeper waters during the day. Unlike the Pacific herring that stay relatively close to their spawning grounds, Atlantic herring migrate offshore extensively.

MENHADEN

The Atlantic menhaden (*Brevoortia tyrannus*) is a bony, oily fish that is of commercial importance for oil, fish meal, and fertilizer products. It belongs to the family Cludeidae. It is a major protein component in commercial feed fed to cultured fishes. Livestock and poultry feed may also contain menhaden meal. The production of paints, soaps, and certain lubricants may use the menhaden oil. It is not consumed by humans because it has a lot of bones and its oily nature gives off an unpleasant odor when cooked. It is a particularly important fishery from Massachusetts to the Carolinas. It resembles the herring in appearance except for several features. The menhaden is deeper in the body, being more elliptical from a lateral view. It has a large head and lacks teeth. It also has a distinctive black spot posterior to the gill covers with more spots of irregular shapes and sizes along the ventral halves of the flanks. It has silvery sides with a back that can have a blue, brown, or green hue to it. Menhaden are harvested almost exclusively by the purse seine net. Purse seine fishing depends almost completely on the sighting of schools of menhaden at the surface, from aircraft or large vessels. The schooling behavior is an outstanding characteristic of this species, from larval to adult stages.

The menhaden is distinct in its feeding habits in that it is one of few fish that can feed on planktonic organisms. While it is rather common for juvenile fishes to feed on plankton, most fishes use feed higher up the food chain. The menhaden has very fine gill rakers that filter out phytoplankton such as diatoms as well as planktonic crustaceans.

The menhaden is euryhaline. The Atlantic menhaden is mainly an ocean fish that schools off coastal waters of the Atlantic, although it has been reported in fresh water as well. It is distributed from the Gulf of St. Lawrence to southern Florida. Freshwater populations have been reported in the St. John River, New Brunswick, Canada.

The Atlantic menhaden can live up to 12 yr, although specimens over seven yr old are rare. While a few mature at 1 yr of age, 80% are sexually mature at 2 yr of age, and all are sexually mature by the time they are 3 yr old. The body size at 3 yr is approximately 25 cm. Reproduction of the menhaden is similar to the herring. They spawn at sea or in large bays, throughout their distribution. They may also be able to spawn in the St. John River, where they are found year-round. The spawning period for Maine to Massachusetts, its northern distribution, occurs between May and October. Populations south of that area spawn during the other half of the year. The male and female broadcast eggs and sperm into the water where fertilization occurs. The fertilized eggs measure about 1.3–1.9 cm in diameter.

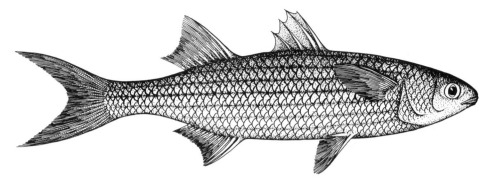

Figure 5. Striped mullet (*Mugil cephalus*). Copyright 1990 by B. Guild-Gillespie.

Fecundity of females measuring 20–35 cm were 38,000–631,000. The eggs are spherical and bouyant, the latter being due to the presence of an oil globule in the yolk. The young are approximately 2.4–4.5 mm in length at hatching.

MULLET

Distributed mainly in the Atlantic Ocean, Black Seas, and the Mediterranean, the order Mullus includes several species of mullet that frequently move up to the Norwegian coast (Fig. 5). Two species, the red mullet (*Mullus barbatus*) and the striped mullet (*Mullus surmuletus*) generally spawn off the coast in the summer and reach sexual maturity in two to three years. They prefer sandy or muddy bottoms. The oil bubbles in the eggs permit them to float in the water. With the arrival of fall season, the juveniles seek greater depths. Preferring sandy or muddy ground, the red mullet has a steep forehead whereas the striped mullet, usually found above sandy ground, has a less-steep forehead. Although teeth are not present in the upper jaw, they are located on the vomer and the gums. Mullets had been an edible delight since Roman times. Although they are small, they commanded a high price and were commonly brought in the dining hall alive.

Nowadays, the commercially important mullet species is another striped mullet (*Mugil cephalus*). It is large and measures 90 cm and weighs 7 kg. This ash gray fish has a dark blue shimmer with 9–10 light longitudinal stripes on the sides of the body. The striped mullet resides in warm seas including the Mediterranean and frequents river mouths or lagoons. Young striped mullets are nourished in salt water or brackish water ponds until they have grown to acceptable size for the consumer market.

The suborder Mugiloidei, family Mugilidae includes several species of mullet that inhabit coastal waters and can acclimatize to brackish water, fresh water, or salt water. They prefer soft ground with a rich plant source located in tidal zones. The generic name *Mugil* (sucker) derives from their feeding habit. They prey on detritus and tiny organisms on the floor. Otherwise, these mullet feed on mussels, snails, planktons, and little organisms that frequent algal populations.

OCEAN PERCH, ROCKFISH, AND REDFISH

The fishes, commonly known as ocean perches, rockfishes, and redfishes belong to the family Scorpaenidae, the scorpion fishes (Fig. 6). In general, these are highly valued food fishes that resemble the freshwater bass in appearance. They are found in all tropical and temperate marine environments around the world. There are 60 genera with over 300 species in this family. The representatives of the popular food fishes in North America belong to the genus *Sebastes*. Fishes in this genus are ovoviviparous, that is they bear live young with some passive nuturing from the mother. The commercially important representatives comprise two species in the Pacific and three species in Atlantic waters.

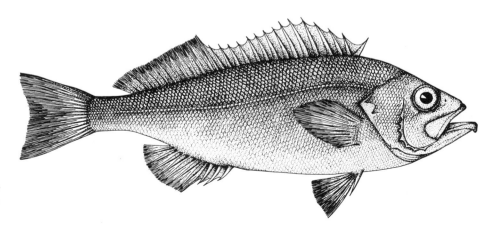

Figure 6. Golden redfish (*Sebastes marinus*). Copyright 1990 by B. Guild-Gillespie.

Pacific Ocean Perch

The Pacific ocean perch (*Sebastes alutus*) constitutes the major species of the Pacific *Sabastes* that is caught commercially for food. The fish may be caught by trawl or hand lines offshore and is mainly sold as fresh and frozen fillets. It can be recognized by a prominent knob, or overhanging tissue, off of the lower lip. It is colored a bright red over the entire body, including fins. The ventral side is usually lighter red. There are olive brown patches below the dorsal fin and on the dorsal side of the caudal peduncle. The Pacific ocean perch is found from southern California north to the Bering Sea. A close relative, the *Sebastes paucispinosus*, is found in the western shores of the Pacific, mainly around northern Honshu in Japan. It is usually found at depths below 130 m and fished most often between about 160 and 300 m.

The mature Pacific ocean perch can be between 30 and 50 cm in length. As for most species, the male is generally smaller, lives longer, and grows more slowly than the female. Spawning occurs in the winter months, and the young are born between January and February. Fecundity can vary an order of magnitude. Fish about 32 cm long and 7 yr old produce about 30,000 eggs, and a female 44 cm long and 20 yr old produces about 300,000 eggs. Unlike the bottom-dwelling adults, the younger are pelagic. Ocean perch have low growth rates. Because of this, the young stay pelagic until the second or third year of life before heading to the bottom to grow to maturity. Adults may live to be 30 yr old.

Yelloweye Rockfish

The yelloweye rockfish (*Sebastes ruberrimus*) are named for their brilliant yellow eyes. They have a characteristic orange-yellow color to their body with a pink tint along the back and sides. The fins are also pink with black margins. The bellies are white with black dots. They are found around reefs from Baja California up the west coast of North America to the Gulf of Alaska. These fish are mostly caught commercially by setlines with live or dead herring bait. They contribute significantly to the white-fleshed fillet market in fresh or frozen states. The yelloweye rockfish can grow to be about twice the size of the ocean perch. They can grow to be about 90 cm. They are normally found at depths between about 50–550 m. The fecundity of a 9 kg female can exceed 2.5 million young. The young off the Washington coast are born in June.

Atlantic Redfishes

There are several species that belong to *Sebastes* in the Atlantic. Four occur in the North Atlantic Ocean, three in the western North Atlantic and one (*Sebastes viviparous*) in the eastern North Atlantic. The three species of Atlantic redfishes found off the North American coast are the Acadian redfish (*Sebastes fasciatus*), the golden redfish (*Sebastes marinus*), and the deepwater redfish (*Sebastes mentella*). They are all commercially valuable as food fishes, although specific fisheries for each species do not exist. They are, rather, fished as a group and marketed as fresh and frozen fillets under the name of ocean perch. The deepwater redfish probably makes up the majority of the catch off the Newfoundland–Labrador coast, while the Acadian redfish is the main species in the commercial catch of redfish off the Georges Bank–Gulf of Maine region. These fishes are caught with otter trawls.

The Acadian redfish has an orange-red body with red fins, with the pelvic and anal fins having a particularly deep red color. There are green-black blotches below the dorsal fin and on the posterior part of the gill covers. There may be green iridescent flecks on the body above the lateral line. The deepwater redfish is bright red all over. The golden redfish can be colored either orange-yellow or gold-yellow.

The distribution of the Atlantic redfishes extends from Iceland, south to about Virginia on the U.S. coast. The golden redfish is more common in the northern range of this distribution but less common in North American waters compared to the Acadian redfish or the deepwater redfish. The golden redfish is common off Greenland, Iceland, Norway, and in the southern Barents Sea. The deepwater redfish has the broadest distribution of the three redfishes and occurs throughout the North Atlantic range of this genus except the North Sea and the Gulf of Maine. It is also found farther offshore than the other two species. The Acadian redfish may be considered the North American redfish because it is most common along the Canadian continental shelf and southward, particularly along the Georges Bank and in the Bay of Fundy and Gulf of Maine waters. The Acadian redfish occupies the shallowest waters of the three species, found most often at depths between about 130 and 370 m. The golden redfish is found from about 300 to about 370 m. The deepwater redfish is found in waters about 350–700 m.

About half of the populations of the Acadian and deepwater redfishes mature at about 19 cm and 30 cm for males and females, respectively. The female golden redfish matures at about 41 cm. There is evidence suggesting the young hatch inside the female golden redfish in April and are born between April and May. The Acadian and deepwater redfishes hatch their young between March and June and release their young between March and June. A study in the Gulf of Maine showed that the Acadian redfish about 30 cm retains about 50,000 fertilized eggs and releases about 15,000 to 20,000 living young. The young of the Acadian redfish grow to be about 8 cm in the first year and grow an average of about 2.5 cm each year up to about 10 yr of age, declining in rate after that. The growth rates in males and females seem to be about the first 20 yr of life.

The Atlantic redfishes, like the Pacific *Sebastes*, grow very slowly. Sections of the otoliths have been used to age these fishes. That data show that the golden and deepwater redfishes can live to be 48 yr old. This depends on the population. The populations of Acadian redfish in the Gulf of Maine may live up to 20 yr and grow to a maximum size of about 46 cm and 1.4 kg.

SALMON

The salmon family Salmonidae comprises numerous landlocked and anadromous species (Fig. 7). They are found in most waters of the northern hemisphere and dominate the

Figure 7. Rainbow trout (*Oncorhynchus mykiss*). Copyright 1990 by B. Guild-Gillespie.

northern waters of North America, Europe, and Asia. This family is made of of three subfamilies, the Salmoninae (salmon, trout, char), the Coregoninae (whitefish), and the Thymallinae (grayling). A common feature among all species is the adipose fin located on the dorsal surface between the dorsal and caudal fins. While representatives of the Coregoninae and Thymallinae belong to this family, the term salmon is usually applied to the salmon and trout of the Salmoninae. This article deals with only those species that are commonly referred to as salmon, which includes species in the genera *Oncorhynchus* and *Salmo*. Even this restriction does not eliminate the confusion between common names and the scientific classification. The Atlantic salmon, for example, is strictly classified as a trout in the genus *Salmo*. The rainbow trout, on the other hand, has just recently been transferred from the genus *Salmo* to *Oncorhynchus*, but it continues to be called a trout. The species covered here include the Pacific salmon species, the rainbow trout, and the Atlantic salmon.

The anadromous species spawn in fresh water where the eggs incubate and where the young spend varying lengths of time. It is common for adults to return to the streambed where they were hatched. During the sexual maturation process, the typically silver fish begins to darken and takes on various colors including black, brown, orange, and red. There is usually a dramatic transformation of the shape of the head and in some instances, the whole trunk. The head of the male usually elongates and there is a pronounced development of teeth. The chum salmon, for instance, develops a large hump back during this process. Many species do not feed once they enter fresh water and begin their migration to their spawning grounds. The spawning process culminates a long journey, sometimes covering thousands of miles, and normally ends in death for the Pacific salmon. The trouts, on the other hand, usually recuperate and may return to spawn a number of times. The Pacific salmon, Atlantic salmon, and steelhead trout are anadromous. The landlocked salmonids discussed in this article include the rainbow trout and the kokanee salmon. While these are the natural life cycles, some species can be grown entirely in fresh water. For example, rainbow trout can be grown quite successfully in seawater netpens and the coho and pink salmon can spend their entire life cycle in fresh water.

Spawning behavior typically involves a high degree of territoriality and aggression between individuals as mates selection takes place. The female will usually create a depression, called a redd, in the gravel bottom by beating the rocks with the side of her tail. After eggs and milt are deposited simultaneously for fertilization to take place, the female will cover the eggs with gravel by moving upstream from the redd and making another redd, causing the displaced gravel to cover the first redd that contains the eggs. The fertilized eggs then incubate in fresh water, which percolates through the gravel, supplying oxygen and carrying away metabolic products such as ammonia and carbon dioxide. The developing embryos as well as the hatched alevins live off yolk carried in a yolk sac while they are still under the gravel. After the yolk is consumed, their mouths open and they begin to emerge from the gravel and feed on organisms in the open waters of the stream. The young of anadromous fishes spend varying lengths of time in fresh water. This fresh water residence ends when the animal undergoes the process of smoltification. This is the total of behavioral, morphological, and physiological changes that occur in the juvenile fish to prepare it for life in salt water. After the process of smoltification, the young migrate to sea to spend the rest of their life cycle growing to sexual maturity.

Pacific Salmon

Chinook Salmon. The chinook salmon (*Oncorhynchus tshawytscha*) is one of the most prized fishes on the west coast of North America. It is also one of the most important commercial species. The chinook salmon is known as the tyee or king salmon when the body size reaches greater than 13.6 kg. This fish has the largest body size of the Pacific salmons. While 13.6 kg is more common as the maximum size, the world's record is 57.3 kg. It is this characteristic combined with its high market value that has made it a popular choice for aquaculture in the northwestern states and on the west coast of Canada. There are two varieties of chinook salmon, the red chinook and the white chinook, which are named according to their flesh color. The red flesh commands the higher price. This segregation is rather unique among the salmon. Chinook salmon are caught commercially with trolling gear, purse seine or gill nets, long lines, and fish wheels. The majority are caught by trollers. Chinook salmon are sold as fresh, frozen, and canned products, with the canned products coming from the net fishery. They are caught for sport with a variety of lures including artificial lures, bait, and flys. Dead herring serve as a desirable bait, trolled deep in the water column.

Chinook salmon adults migrate extensively and are primarily found in the Pacific Ocean from the Ventura River in southern California, north to Point Hope, Alaska.

Limited evidence suggests, for instance, that the Canadian shinook migrate north to the Gulf of Alaska but remain within 160 km of that area until they return south to spawn. This species is less common in the Arctic Ocean and in the Bering and Okhotsk seas and the Sea of Japan. The largest numbers come from the largest rivers such as the Fraser River in British Columbia. Substantial numbers of chinook also come from the Yukon River. While there have been numerous attempts to transplant this species to many locations throughout North America and around the world, it seems that only success was on South Island, New Zealand, where a self-supporting population was established.

The average age of an adult returning to spawn varies from four to seven years, depending on the location. Three- to five-year-old returning adults are more common in southern streams, whereas five- and six-year-old adults are more common in northern streams. It is also quite common for river systems to have more than one stock of chinook salmon returning to it, where there may be spring, fall, and winter runs of returning adults.

The time of spawning depends on the location and the distance the adults must swim upriver to reach the spawning grounds. In the Fraser River, for example, the chinooks spawn between July and November and the adults may travel up to 965 km from the mouth of the river to their spawning grounds. The chinooks that return to the Yukon River spawn between July and August and may travel up to 1930 km to reach their spawning grounds. Other stocks that travel only a few miles, if that, in British Columbia spawn in September and in October. Chinook salmon tend to spawn in deeper waters and in larger gravel than other Pacific salmon species. Each female may carry from 4,000 to 13,000 eggs. The eggs are large for fish eggs and measure about 7 mm in diameter. The eggs hatch the following spring and the alevins spend two to three weeks in the gravel with their yolk sacs. The emergent young remain in the freshwater rivers for varying lengths of time, depending primarily on the water temperature. The young will smolt and migrate to sea after about three months in warmer southern areas such as in southern British Columbia, while chinook spend at least one year in fresh water in northern areas. Chinook juveniles may spend two years in fresh water in the Yukon River. The maturing adults may spend two to three years at sea before returning to fresh water to spawn. Spawning females are, therefore, about four to five years old.

Coho Salmon. The coho salmon (*Oncorhynchus kistuch*) is also an important commercial and sport fish in North America. It is the mainstay of the saltwater sport fishery for salmon. Most of the coho are commercially caught in August by trolling with plugs, spoons, or feathered jigs, although purse seines and gill nets are also used. The trolled catch is sold as the fresh or freshly frozen product, while the netted fish are either canned or smoked. Sport fish are caught in the late summer to early fall months of July to October. The gear used by sport anglers is similar to the commercial trolling gear except it is smaller in size. Coho are also caught on the fly with bucktail flies and with bait such as frozen or pickled herring. Because coho remain silver and continue to feed while traveling upstream to spawn, they are pursued by the sport angler in many freshwater rivers and streams.

The adult distribution extends from southern California to the Gulf of Alaska. The juveniles are found in the fresh waters from Monterey Bay, Calif., to Point Hope, Alaska. Coho salmon are also found in the Anadyr River in the USSR and south from that point to Hokkaido, Japan. This species has been transported to other parts of the world such as Argentina and Chile, with Chile reporting some success in the establishment of naturally reproducing stocks. Coho have also been introduced into the Great Lakes by both Canadians and U.S. citizens. While there are reports of natural reproduction, the maintenance of those stocks are heavily dependent on the annual planting of cultured stocks.

The full-grown adult can weigh from 1.8 to 5.5 kg. It enters fresh waters to spawn from late September to October and may travel up to about 240 km upriver to spawn. Asian populations exhibit more of the separation of summer and fall, or fall to winter runs than those populations returning to North American river systems. Spawning takes place in October or November. The eggs are rather large, measuring about 6–7 mm in diameter. The range in fecundity is about 2,000–3,000 eggs. After the eggs have hatched, the young spends between one and two years growing in fresh water. This again, depends on the location. Coho juveniles in British Columbia spend about one year in fresh water while those in the Yukon spend two years growing in fresh water. After their freshwater residence, the young smolts migrate to sea. The saltwater residence usually lasts about 18 months, although the phenomenon of precocious sexual maturation, or jacking, is predominant in this species. While normal adults will return between three and four years of age, jacks return to spawn at only two years of age. Coho salmon normally stay within about 40 km of the coast.

Chum Salmon. While the chum salmon (*Oncorhynchus keta*), or dog salmon, is a less desirable species for the commercial or sport fisheries, it has maintained an important position in the diets of the native people of North America who capture the fish with nets and fish wheels. It is a preferred species for smoking because of its low fat content. Its pale to white flesh has the lowest fat content of the Pacific salmons. This species has been used the least for transplantations to other geographic areas, perhaps due to its relatively low commercial and sport value. Most of the chum salmon are harvested while they are migrating toward their home streams to spawn. Commercial catches are harvested mainly between August and October with purse seines and gill nets. Some are also caught by trolling gear. The trolled catch is sold fresh. Some are smoked or dry salted for preservation. The bulk of the commercial catch, however, is canned.

The marine adults are found in Pacific and Arctic oceans, in the Sea of Japan, and in the Okhotsk and Bering seas. They probably have the widest distribution among the Pacific salmons. The major spawning areas in North America lie between Puget Sound in Washington State to Kotzebue, Alaska. The geographic range for their spawning covers the west coast of North America from

Oregon to the Mackenzie River. The time frame for adults arriving at the spawning grounds ranges from July, in northern British Columbia to September or even as late as January for streams further south. The location of the spawning grounds, reflecting the degree to which the chum salmon swim into fresh water to spawn, varies considerably. Most arrive at the mouths of rivers in an advanced state of sexual development and spawn relatively close to the ocean. In general, few chum salmon migrate more than 160 km from the mouth of a river to spawn. The Yukon River, however, is an exception; the population ascends over 1,930 km to spawn.

The mature adult is usually between two and four years old. While the adult chum salmon can weigh from 3.6 to 5.5 kg, fish up to 13.7 kg have been reported. The fecundity ranges from about 2,000 to 3,000 eggs and each egg measures about 5–6 mm in diameter. The hatched young spend a period in the gravel feeding off the yolk until the spring when they emerge and migrate directly to sea. Depending on the distance to the ocean, the young may take from a single night up to several months until they reach sea water. The migratory behavior of the young chum are similar to the pink salmon and the young of both species may be found making the seaward journey together. They usually travel by night and hide in the gravel during the day. They are also not dependent on schooling to make the trip, although they do school once they have reached the estuary. Once in the ocean, the young chum will usually spend a few months, until mid to late summer, in the coastal areas before migrating further out to sea. Chum salmon from North America have been captured as far as 4,180 km off the west coast in the North Pacific. Most chum salmon spend two or three years in the ocean before returning to their home streams to spawn. They begin to appear at the mouths of rivers around May or June of their final year at sea.

Pink Salmon. The pink salmon (*Oncorhynchus gorbuscha*) is the most abundant of the Pacific salmon species. It gets its name from the pink color of its flesh. Like the chum salmon, this species is not regarded as being highly desirable compared to the deep red flesh of the sockeye salmon. The size range of the adult is about 1.4–2.3 kg, although animals as large as 6.4 kg have been recorded. A unique characteristic of this species is its fixed, two-year life cycle. There is no overlap between the reproducing stock of one year to the next. This allows two separate stocks to utilize any stream for reproduction. Where there are odd-year stocks as well as even-year stocks using the same river, one often dominates. The Fraser River pink salmon run in British Columbia, for example, is primarily an odd-year stock, while the run in the Queen Charlotte Islands is an even-year stock.

Pink salmon is caught commercially with purse seine and gill nets as well as with trolling gear, the latter contributing only a minor part of the total catch. Fish caught by trolling is sold fresh, but 95% of the commercial catch is canned. Pink salmon is also fished for sport by trolling artificial lures.

Pink salmon is found in the Pacific and Arctic oceans, the Bering and Okhotsk seas, and the Sea of Japan. Its distribution in North America extends from the Sacramento River in California, north to the delta of the Mackenzie River, being most abundant around the central area of this range. It can be found from the surface to depths of about 36.6 m. Transplantations have established self-sustaining populations of pink salmon in northern Europe and in fresh water in Lake Superior. There are transplanted populations along the Atlantic Coast of North America that seem to be self-sustaining.

Pink salmon may spawn in a wide range of locations from tidal areas of certain rivers up to 483 km upstream from the mouths of large rivers. The pink salmon are also known as humpbacks because the sexually mature males develop a large hump behind the head. The snout elongates dramatically and large teeth emerge from the jaws. The adults will appear at the mouths of rivers from June to September to begin their migration to their spawning beds. Pink salmon spawn from mid-July to late October and the eggs hatch from late December to late February. Like the chum salmon, the young emerge from the gravel after the yolk supply is consumed and move directly to sea. The emergent young, which measure about 3.8 cm in fork length, may travel 16 km in a single night to reach the ocean. Their appearance is distinctive for salmon in that they lack the vertical parr marks on their sides. They are colored blue-green on their back and have silver sides. If the journey takes more than a night, the young will hide during the day in the gravel and emerge again at night to be carried by the current to the sea.

Sockeye Salmon. The anadromous sockeye was the first Pacific salmon species to be fished commercially. The sockeye has long been the mainstay of the Pacific salmon fishery. Its deep red flesh and high oil and protein contents have always brought it the highest price among the Pacific salmon species on the market. The sockeye is mainly caught with purse seine and gill nets, although trolling is also used. The product is mainly canned, although it may be sold fresh when the fishery is open. Native people use traditional methods of nets, weirs, and gaffs to harvest the fish. Although the sockeye salmon are not an important sport fish, effective artificial lures have been developed that have served both sport and commercial trollers at the mouth of major sockeye rivers such as the Fraser in British Columbia.

The sockeye salmon (*Oncorhynchus nerka*) is represented by an anadromous form and a landlocked form known as the kokanee. Specific recognition as subspecies has been given to each. The anadromous form i *O. nerka nerka* Walbaum and the kokanee is designated *O. nerka kennerlyi* Suckley. The anatomy of the two subspecies is very similar except for the total body size. The average weight of an adult anadromous sockeye is about 2.7–3.2 kg, whereas the adult kokanee weighs less than 455 g. Adult sockeye have been recorded to be up to 6.8 kg.

The distribution of this species extends from the Kalamath River in California to Point Hope in Alaska. They are also found in Asia from northern Hokkaido, Japan, to the Anadyr River in the USSR. The distribution of the kokanee certainly follows that of the anadromous sockeye. It occurs in lakes where the anadromous salmon must have had access at one time. Both kokanee and most anadromous sockeye return to the spawning grounds as four-

or five-year-old fish. Both species spawn in the fall. The actual time depends on location. Kokanee, for example, spawn in September and October in Kootenay Lake in British Columbia but in November and December in Boulter Lake, Ontario. The major spawning grounds for this species are the watersheds draining into the Fraser, Skeena, and Nass rivers in British Columbia. Anadromous sockeye spawn between July and December, again depending on location.

Sockeye become sexually mature from three to eight years in age. While a one-year freshwater residence and a two- to three-year saltwater residence is normal, precocious males may return to spawn after only one year at sea. Fecundity is highly variable in this species. It ranges from 370 to 1760. While the kokanee egg is naturally small, the egg of the anadromous sockeye is large, measuring about 4.5–5 mm in diameter. After hatching and consuming all the yolk, the young sockeye will spend at least one year in fresh water, although some will spend two or even three years in fresh water before migrating to sea. This variability is, to some extent, responsible for the range in size of adult sockeye. These years in fresh water are spent in nursery lakes. The fry feed on planktonic organisms. Anadromous sockeye smolts will migrate to sea in the spring of their second and fifth year of life. At sea, the maturing sockeye migrate north and northwest, spreading out to the distribution outlined above.

Rainbow Trout. The rainbow trout, steelhead trout, and kamloops trout all belong to the same species, *Oncorhynchus mykiss* formerly *Salmo gairdneri*. This species represents one of the most highly prized game fishes in North America. The names rainbow and kamloops trout refer to the nonanadromous populations of this species, while the steelhead is normally used to refer to the large, anadromous variety of this species. The kamloops trout usually refers to the larger variety of the nonanadromous trout. All varieties of this species are highly prized sport fishes. The flesh is red to pink in fish that have been feeding on planktonic organisms and it tends to be more pale in fish from larger lakes where they feed on fish more than plankton. They take the bait or lure aggressively and fight the line, jumping out of the water many times in the process. Most of the steelhead are caught in fresh water in coastal areas. The commercial catch is usually incidental to the salmon fishery and are almost exclusively canned. The commercial gear involved in this fishery, therefore, consists of gill nets.

Although the native range of this species is the eastern Pacific Ocean and for the freshwater forms, west of the Rocky Mountains from Mexico to Alaska, all forms of the rainbow trout can be found throughout North America in suitable habitats. It has also been introduced successfully in New Zealand, Australia, Tasmania, South America, Africa, Japan, southern Asia, Europe, and Hawaii. It is a very plastic species that adapts to different environments readily and may show variations in its physical and behavioral characteristics throughout its range. It is probably the most studied fish in terms of fish physiology and anatomy.

The nonanadromous rainbow and kamloops trout are naturally spring spawners. They move from the lakes to inlet or outlet streams from about mid-April to June and spawn in streams with beds of fine gravel. This is the timing for North America. Because their distribution spans the globe, the actual month of spawning will depend on when spring occurs at a particular location. The fecundity is about 1,400–2,700 eggs, but this is extremely variable; the range extends from 200 to 12,750. The eggs are 3–5 mm in diameter. The eggs usually hatch in about four to seven weeks and the yolk is consumed in about three to seven days. The young emerge from the gravel between mid June and mid August and may reside in the stream until fall of that year or they may spend as long as three years in the stream. The young of the rainbow and kamloops trout will then migrate upstream or downstream to the lake to feed and mature until they are ready to spawn.

The anadromous steelhead trout populations may spawn in the spring, fall or winter. An established population for a given stream, however, will maintain a consistent pattern from year to year. The fecundity and egg size are similar to the rainbow and kamloops trouts. Most of the young will spend two to three years in fresh water before migrating to sea as smolts in the spring, where they may spend an equal amount of time maturing in the ocean. Many will survive the spawning and return for a second or third spawning.

Atlantic Salmon

The Atlantic salmon (*Salmo salar*) has attracted both commercial and sport fisherman like no other species throughout history; so much so that populations of Atlantic salmon disappeared from the Thames in 1833 and from Lake Ontario in 1890 due to overexploitation. The Atlantic salmon are caught commercially with net and troll gear. They are caught in the sport fishery with live bait as well as on the fly and artificial lures. The intensive interest on this species has, unfortunately, eliminated entire populations of this species from certain watersheds over time. The commercial fishery for the Atlantic salmon off Greenland places a particularly heavy pressure on the Atlantic salmon population at large, because those waters represent the major feeding grounds for all the stocks in the Atlantic Ocean.

The native distribution of the Atlantic salmon is the basin of the North Atlantic Ocean, from the Arctic Circle to Portugal and from northern Quebec south to the Connecticut River. While fertilized eggs and live fish transplantations to the Pacific waters have occurred since 1905, there is no record of the establishment of self-sustaining populations. Like several of the Pacific salmons, there are established landlocked populations of this species in North America. The landlocked Atlantic salmon are sometimes called the ouananiche. There are self-sustaining populations in Sebago Lake, Maine; Lake Ontario; and Lake St. John, Quebec.

The Atlantic salmon is a typical anadromous fish. They spend 1–3 yr in fresh water as juveniles and migrate to sea as smolts where they may spend 1–2 yr growing and maturing. While mortalities in spawning are high, significant numbers of spawners return to spawn two or three times. The normal spawning time is from October to late November. The fish in the southern regions tend to spawn later. The female will produce about 1,540 eggs per kg body weight. The size of returning adults ranges from 2.7 to 6.8 kg. The returning adult spawner will normally have spent 2 yr at sea. The Atlantic salmon generally have longer life spans than the Pacific salmon. Ages up to 11 yr have been reported. While the eggs may hatch in April of the following year, the young alevins will stay in the gravel and consume their yolk until May or June, at which time they emerge into the water column. The period of freshwater residence is variable, depending on the location. The younger parr in North America will spend 2–3 yr in fresh water until they smolt, at about 15.2 cm in length. Populations in Greenland, however, have been recorded to spend as long as 4–8 yr in fresh water, attaining about the same body size.

Unlike the slow growth in fresh water, the Atlantic salmon undergo rapid growth at sea. An exception to the average size of returning adults, stated above, are the precociously matured fish that return after one year at sea. These are often males and are in the 1.4–2.7 kg range. There are many reports of individuals exceeding the normal size range. While the average Atlantic salmon caught commercially is about 4.6 kg, record sizes of 35.9 and 37.7 kg have been reported. While the average landlocked Atlantic salmon weighs about 0.9–1.8 kg, individual fish weighing 16.1 and 20.3 kg have been caught in North American lakes.

SMELT

The smelts are small, slender fishes that belong to the family Osmeridae. This group of fish probably got its name from one characteristic, their smell. The smell, which may be described as that of freshly cut cucumbers, is particularly noticeable when large numbers are caught during spawning runs. The six genera and 10 species are found in the northern hemisphere in the Atlantic, Arctic, and Pacific oceans. They are circumpolar in distribution. Some species in this family spend their entire life cycle in fresh water and others in sea water. There are species, furthermore, that are anadromous. The four species found in North America are the pond smelt (*Hypomesus olidus*), the longfin smelt (*Spirinchus thaleichthys*), the Eulachon (*Thaleichthys pacificus*), and the rainbow smelt (*Osmerus mordax*). The latter two species are discussed here because the existence of both sport and commercial fisheries for them. The former two species play a minor part in any commercial fishery and in normal consumption.

Rainbow Smelt

As Figure 8 shows, the rainbow smelt is a slender fish that may reach lengths of 17.8–20.3 cm, although 35.6-cm fish have been recorded. As the name implies, it is a colorful silver fish with a pale green back and iridescent sides. The rainbow smelt has supported a commercial fishery for over 100 yr. It has always been abundant throughout its distribution. Most fish are caught in the spawning migration by trawl nets. The Atlantic rainbow smelt is also a popular sport fish. It is caught on hook and line through the ice as well as by dip netting or seine netting during the spawning run. It is an anadromous fish but like other anadromous fishes, it can successfully spend its entire life cycle in fresh water. The landlocked forms tend to have a darker color than the marine counterparts. As with other anadromous fishes, they lose the silver appearance and darken when they begin to spawn.

There are two subspecies of the *Osmerus mordax*: the Arctic rainbow smelt (*O. mordax mordax*) and the Atlantic rainbow smelt (*O. mordax dentex*). The Arctic rainbow smelt includes those populations that are in the Pacific Ocean and are distributed from Vancouver Island, north around the state of Alaska, and into the Arctic Ocean to Cape Bathurst. The distribution of the Atlantic smelt was originally restricted to the marine waters from New Jersey to Labrador. The Atlantic smelt today is represented by both anadromous forms and landlocked forms in the Great Lakes.

The anadromous and landlocked forms behave in a similar way in spawning. The spawning adults ascend streams to lay and fertilize their eggs. They begin their ascent in the spring when the ice cover disappears, usually in the months between March and May. Spawners can be up to six years old, but are commonly between two and three years old. Spawning, which usually occurs at night, involves two or more males to each female. While spawning normally takes place in streams, fish will spawn on the shoreline if the currents are too strong to permit migration. Eggs and milt are released into the moving current and the eggs quickly stick within seconds to anything they come in contact with. The fertilized eggs, which measure about 1 mm in diameter, eventually end up resembling balloons on stems as the sticky outer coat washes off except at the point of attachment. Most of the spawners, like many other anadromous fishes, die after spawning. Although reports of fecundity vary, about 494–530 eggs per gram of body weight are produced. The eggs incubate for about two to four weeks in water tempera-

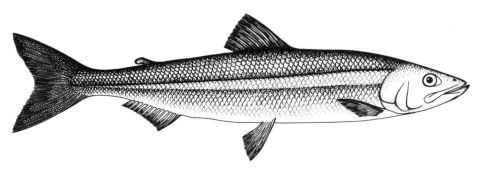

Figure 8. Rainbow smelt (*Osmerus mordax*). Copyright 1990 by B. Guild-Gillespie.

tures between 6 and 10°C. The hatched young, which are about 5 mm long are quickly swept down to the lake or estuary where they continue to feed and grow to a length of about 5 cm. They spend the remaining period up to spawning in the middle of the water column of lakes or in coastal waters close to shore.

Eulachon

Thaleichthys pacificus, the eulachon, means oily fish of the Pacific. Aside from the characteristic odor the oiliness of this fish is its outstanding feature. The oil content of the body is so high that the dried carcass can be lit like a candle. Another common name for this fish is candlefish, because in the days before the luxuries of candles and inexpensive oils, small strips of rags were inserted into the mouths of dried eulachons and lit for light. The rich oils lend a flavorful taste to this fish, which probably accounts for its high value as a food fish. The eulachon resembles the rainbow smelt in appearance, although it is darker in coloration and lacks the iridescence in the body walls. It is typically brown to blue-black on the dorsal surface. The coloration on the ventral surface is white and the sides are a light color, intermediate between the two extremes.

The eulachon is caught by the native Indian community as well as by the commercial fishery. The volume caught by the native fishery has exceeded, and probably still exceeds, the commercial catch. The commercial fishery utilizes drift gill nets and occurs principally in the Fraser River as the eulachon migrates to spawn. The commercial catch is used for human consumption but mostly for feed at fur farms. The native catch has always been based on the oil from the eulachon. It is used for cooking and for curing. Traditionally, the grease trails got their name because they were used to carry the extracted eulachon oil from the fishing grounds to the points of trade.

The eulachon is found only on the west coast of North America, from the Kalamath River in California up to the panhandle of Alaska and to the Pribilof Islands. Like the rainbow smelt, the eulachon is an anadromous fish that migrates from the sea up freshwater streams, soon after the ice cover breaks, to spawn. This occurs from mid March to mid May. The spawning females, which measure about 15.2–17.8 cm in length, produce an average of 25,000 eggs and shed them in the free current along with the milt from the male partners. They are at least three years old when they spawn. Unlike the small rainbow smelt egg, the eulachon produces a relatively large egg of 8–10 mm in diameter, which is irregular in shape. The eulachon eggs have two outer layers. The outer, adhesive, layer breaks on spawning and everts to anchor the egg to coarse particles such as gravel. The outer and inner layers remain attached at one point. While most of the adults die after spawning, some survive to spawn a second time. Like the rainbow smelt, the eggs take about two to three weeks to hatch and the just-hatched young resemble young herring. They measure about 5 mm in length and are slender and nearly transparent. The small young are swept to sea with the current. They move to deeper waters farther from the shore as they mature. Although they normally measure 10.2–17.8 cm in length, individuals up to 30.5 cm have been recorded.

TUNA AND SWORDFISH

Tuna

The true tuna belong to the tribe Thunnini, which belongs to the family Scombridae of the suborder Scombroidei, order Perciformes (Fig. 9). The bullet, frigate, bluefin, albacore, yellowfin, skipjack, blackfin longtail, and the slender tuna are all true tunas. The other tribes in this family include the Sardini (bonitos), the Scomberomorini (seerfish), and the Scombrini (mackerel). The smallest tunas (bullet and frigate) usually weigh less than 3 kg while the largest (bluefin) tunas can exceed 700 kg. The large bluefin tunas can live to 38 yr. This family of fishes possesses remarkable limits for growth. The billfishes, such as the black marlin, grow even larger than the tuna. Commercial fisherman have reported taking black marlin in excess of 3,000 kg, no doubt among the largest fishes in all the seas.

The tunas have been fished for food for thousands of years. The skipjack, yellowfin, and bigeye tunas dominate the commercial catches. Most of the tuna harvests occur in the Pacific Ocean. About 50% of the world catch of tuna is caught by Japan and the United States. The balance is caught by almost every nation in the areas where the tuna are found. They are caught with almost every kind of gear including the pole and line method as well as giant purse seines and long-lining. About 40% of the tuna are caught by the pole and line method where large schools of fish are first baited close to the boat with live anchovies or sardines that are thrown overboard and then caught with barbless lures on single poles. This process is called chumming. The balance of the commercial catch are harvested

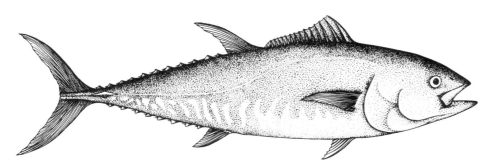

Figure 9. Tuna (*Thunnus thunnus*). Copyright 1990 by B. Guild-Gillespie.

by purse seines (30%) and by long lines (30%). The purse seiner encircles the school of fish up to about a kilometer in diameter while letting out a special net that can be gathered at the bottom, like a purse or bag. As the boat completes the circle, the purse is gathered up to the boat along with the fish. This technique had caused high mortalities in dolphins (also called porpoises by fisherman) that often congregate above schools of yellowfin tuna. Many of the dolphins die because they become entangled in the net and are prevented from surfacing to breath. After this came to public attention in the mid-1960s, purse seine nets and seining methods have been modified to prevent this entanglement and to allow the dolphins to escape from the net. The long-line vessel will let out a line up to 130 km in length with as many as 2,000 baited hooks dangling from this main line, supported by floats. In addition to the commercial catch, the tuna represent one of the most desirable sport fish in the world. It is well known for attracting avid sport fishermen who spend thousands of hours and dollars in pursuit of this challenging and exciting catch.

Tuna is a very expensive food item. It is at least as expensive as beef, relative to protein content. Over 90% of the world catch is consumed by the United States, Japan, France, Spain, Italy, and West Germany. A typical tuna fish sandwich is probably made of albacore, yellowfin, skipjack, or bigeye tuna. Certain species, such as the skipjack, are sold as a dried product in Japan. The flesh of the bluefin and bigeye tunas are valued as raw sashimi in Japan. Record prices for tuna come from Japan where the flesh of such large tuna, with a high oil content, has sold for as much as $26,000 U.S. a ton.

The tuna and the billfishes represent some of the most interesting fishes in the world in that they deviate from the norm. The tuna are found in every temperate and tropical ocean around the world. They are most commonly found in the Atlantic, Pacific, and the Indian oceans. Tuna and billfishes are not only among the largest fishes in the oceans, but are also the fastest swimmers in the world. At high speeds, their body forms represent an extremely high degree of streamlining an hydrodynamic efficiency. At those speeds, their fins retract into grooves and their eyes become flush with the surface of the body. Even the slowest pace for the tuna represents the top speed any human could hope to achieve. Relatives of the tuna, the wahoo and sailfish have been recorded at swimming speeds over 75 and 100 km/h. Not only are these animals most remarkable for their speed but they are also known for the incredible distances they travel. Tagging studies, which tend to yield conservative estimates, have shown individual bluefin tuna and black marlin to travel over 10,000 km, at a rate of over 30 km/day. While most of the distance records are held by the larger bluefin tuna, the skipjack tuna has been recorded to travel over 9,500 km and the yellowfin tuna, over 5,000 km. There are numerous records of individually tagged members of this group of remarkable fish crossing every major ocean in the world.

As stated above, the general body shape of the tuna is one designed for hydrodynamic efficiency. The body is rather round in cross section and laterally oval. The tail is lunate (shaped like a crescent moon), which provides a high degree of forward thrust, with minimum drag. Like many other fishes, the upper body is dark and the ventral side is light in color to aid in camouflage. Body markings change with age and may include vertical bars or markings of different shapes. These markings may serve to identify sex, age, species, and particular stages in the reproductive cycle.

One need for the efficiency in swimming is their mode of respiration. Tuna are ram ventilators. That is, they must swim with their mouths open to pass sufficient water over their gills. One result of this swimming activity is the high energy requirements of this fish. A tuna can consume up to one-quarter of its body weight in food each day. Such a high metabolic rate naturally produces a great amount of heat. The tuna are rather unique among fishes in that four genera of the family regulate body temperature above that of the ambient water. The bluefin tuna, for example, maintains a visceral temperature 10–15°C above ambient water temperature. Maintaining a higher body temperature presumably allows quicker digestion as well as a more rapid mobilization of stored energy reserves. It may be more accurate to state that those tunas that regulate body temperature have specialized mechanisms for dissipating the heat from metabolism. Special circulatory structures enable this heat to be transferred to the external surface.

As stated above, the tuna is highly active fish. The need for high rates of oxygen uptake by the fish to feed this high metabolic rate is met by a larger gill surface area and higher hemoglobin content than most other fishes. The constant swimming motion in combination with pectoral fins that act like hydrofoils maintains the fish at a certain position in the water column. This greatly reduces the need for the swim bladder, which functions in other fishes as a buoyancy organ. The swim bladder is greatly reduced or absent in species of tuna. Unlike most other fishes, there is a greater proportion of red muscle to white muscle mass in the body. White muscle is used for short bursts of anaerobic activity while the red muscle is used for routine swimming in fishes. This makeup provides the machinery with which the tuna can accomplish the incredible journeys over long distances.

Spawning takes place in various regions, depending on the species. For example, the Atlantic mackerel spawns in the northwest Atlantic and the large bluefins spawn in the Straits of Florida and the Gulf of Mexico. Both eggs and sperm are broadcast into the water, near the surface. The eggs of the Atlantic mackerel are found mainly in the top 10 m of water. The female tuna lays about 100,000 eggs per kg body weight; there might be about 10 million eggs from a 100-kg animal. This is not unreasonable because the bluefin tuna can reach a maximum size of about 700 kg. These tiny eggs of about 1 mm float about in the upper layers of the ocean, held bouyant by an oil droplet inside the egg. While the various tuna species are found in waters of a wide range of temperatures, most eggs take only a few days to hatch. As in many other fishes, only about two of those individuals survive to maturity.

Swordfish

The swordfishes belong to the family Xiphidae (Fig. 10). The Xiphidae are related to other giant mackerellike

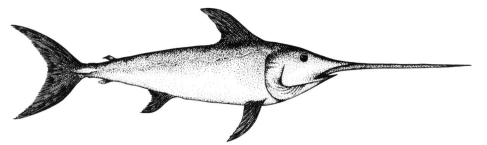

Figure 10. Swordfish (*Xiphias gladius*). Copyright 1990 by B. Guild-Gillespie.

fishes such as the tunas (Scombridae) and the spearfishes (Istiophoridae), in that they all belong to the suborder Scombroidei. The single representative of this species (*Xiphias gladius*) is known as the common swordfish or broadbill. It is highly prized as a game fish and as a food item. Like the tunas, there is evidence that the swordfish has been fished for thousands of years. Its dominant physical feature is the large sword protruding from the upper jaw. While this feature was thought to serve a predatory function, it may function more in the enhancement of swimming speed by reducing the resistance of water to the body, much like the pointed head of a rocket. There are reports, however, of many whales with broken bills of swordfish impaled in their bodies, some through their vertebrae. This fish may be referred to by sport anglers as the broadbill swordfish. The teeth and the spines of the first dorsal fin are lost with age. Its large body is scaleless.

Although an individual weighing 537 kg was caught in 1953 off the Chilean coast, most other record sizes are closer to half that weight. This is relatively small compared to members of the other families in the Scombroidei. The billfishes also represent a minor part of the commercial catch, relative to the tunas. The average annual catch for the swordfish between 1982 and 1985 was estimated at 49,000 t, about 1.6% of the total catch for the Scombroidei.

Unlike the tunas and other billfishes that avoid deep waters, the swordfish moves widely in the water column. When it dives to great depths and returns to the surface, it is often found in semiconsciousness at the surface until it regains equilibrium. It is in this stupor that fishermen have used the harpoon to catch this prize. While this was the popular method of catch until the early 1960s, the majority of today's commercial catch of swordfish is by long line. This billfish also contributes to the specialized market of big game fishing around the world where spending $1,000/day in the pursuit of a sailfish or billfish is not uncommon. The local communities where big-game fishing ports are located may depend heavily on the economy of that sport.

The distribution of the swordfish extends throughout the temperate, subtropical, and tropical waters in all oceans of the world. They are found in coastal as well as oceanic areas. Like the related tunas and spearfishes, the swordfish are highly migratory and have body forms that are very streamlined. The basic physiological and reproductive features of the swordfish are similar to those of the tuna. Spawning is thought to occur throughout the year in the Caribbean Sea, Gulf of Mexico, and off the Florida coast. Although the swordfish egg is somewhat larger than the tuna egg, and the fecundity is slightly lower, the major features of the reproductive system and process are the same as those for the tuna. The eggs are approximately 1.7 mm in diameter and contain a large oil globule. Young hatch out at about 4 mm in length and may grow rapidly, up to 2 mm per day.

BIBLIOGRAPHY

1. *List of Common and Scientific Names from North America and Canada,* 4th ed., American Fisheries Society, 1980.
2. W. A. Clemens and G. V. Wilby, *Fishes of the Pacific Coast of Canada,* 2nd ed., Fish. Res. Board Can. Bulletin **68,** 1961.
3. D. Miller, "The Blue Whiting, *Micromesistius poutassou,* in the Western Atlantic, with Notes on Its Biology," *Copeia* **2,** 301–305 (1966).
4. F. H. Bell, *The Pacific Halibut: The Resource and the Fishery,* Alaska Northwest Publishing Co., Anchorage, 1981.
5. W. B. Scott and M. G. Scott, "Atlantic Fishes of Canada," *Can. Bull. Fish. Aquat. Sci.* **219,** 1–731 (1988).

General References

References 1–4 are good general references.

General

D. L. Alverson, A. T. Pruter, and L. L. Ronholt, *A Study of Demersal Fishes and Fisheries of the Northeastern Pacific Ocean,* H. R. MacMillan Lecture Series in Fisheries, Inst. Fish. University British Columbia, 1964.

G. Godson, *Fishes of the Pacific Coast,* Stanford University Press, Stanford, Calif., 1988.

J. L. Hart, *Pacific Fishes of Canada.* Bull. Fish. Res. Board Can. No. **180,** 1973.

J. G. Hunter, S. T. Leach, D. E. McAllister, and M. B. Steigerwald, "A Distributional Atlas of Records of the Marine Fishes of Arctic Canada in the National Museums of Canada and Arctic Biological Station," *Syllogeus* **52** 1–35 (1984).

H. Kasahara, *Fisheries Resource of the North Pacific Ocean. Part 1.* H. R. MacMillan Lectures in Fisheries, Institute of Fisheries, the University of British Columbia, Vancouver, 1960.

P. A. Larkin and W. E. Ricker, "Canada's Pacific Marine Fisheries. Past Performance and Future Prospects," in *Inventory of the Natural Resources of British Columbia,* 1964, 194–268.

Y. Okada, *Fishes of Japan,* Maruzen Co. Ltd., Tokyo, 1955.

F. S. Russell, *The Eggs and Planktonic Stages of British Marine Fishes,* Academic Press, Inc., Orlando, Fla., 1976.

D. J. Scarratt, ed., *Canadian Atlantic Offshore Fishery Atlas,* Can. Spec. Publ. Fish. Aquat. Sci. **47,** 1982.

W. B. Scott and E. J. Crossman, *Freshwater Fishes of Canada,* Fish. Res. Board Can. Bulletin **184,** 1973.

A. Wheeler, *The Fishes of the British Isles and North-west Europe,* Michigan State University Press, East Lansing, 1969.

N. J. Wilimovsky, "List of the Fishes of Alaska," Stanford Ichthyology Bulletin 4(5), 279–294 (1954).

Specific to Species

S. F. Hildebrand, "Family Engraulidae," in *Fishes of Western North Atlantic, Mem. Sears Found. Mar. Res.,* Vol. 3, Yale University, New Haven, Conn., 1963, pp. 151–249.

J. L. McHugh and J. E. Fitch, "An Annotated List of the Cluepoid Fishes of the Pacific Coast from Alaska to Cape San Lucas, Baja California, *Calif Fish Game* 37(4), 491–495 (1951).

R. H. Boyle, *Bass,* W. W. Norton & Co., New York, 1980.

R. H. Boyle, *The Hudson River, A Natural and Unnatural History,* New York, 1979.

J. N. Cole, *Striper,* Boston, 1978.

N. Karas, *The Complete Book of the Striped Bass,* New York, 1974.

E. C. Raney, E. F. Tresselt, E. H. Hollis, V. D. Vladykov, and D. H. Wallace, *The Striped Bass.* Bulletin of the Bingham Oceanographic Collection 14, Peabody Museum of Natural History, Yale University Press, New Haven, Conn., 1952.

M. P. Fahay, "Guide to the Early Stages of Marine Fishes Occurring in the Western North Atlantic Ocean, Cape Hatteras to the Southern Scotian Shelf; *J. Northw. Atl. Fish. Sci.* 4, 1–423 (1983).

L. S. Incze, C. M. Lynde, S. Kim, and R. Strickland, "Walleye Pollock, *Theragra chalcogramma,* in the Eastern Bering Sea," in N. J. Wilimovsky, L. S. Incze, and S. J. Westerheim, eds., *Species Synopsis: Life Histories of Selected Fish and Shellfish of the Northeast Pacific and Bering Sea,* Washington Sea Grant Program and Fisheries Research Institute, University of Washington, Seattle, 1988, pp. 55–69.

H. A. Innis, *The Cod Fisheries. The History of an International Economy,* Yale University Press, New Haven, Conn., 1940.

H. A. Innis, *The Cod Fisheries,* University of Toronto Press, Toronto, 1954, pp. 1–10.

A. C. Jensen *The Cod,* Fitzhenry and Whiteside, Toronto, 1972.

S. Kim and D. R. Gunderson, "Walleye Pollock, *Theragra chalcogramma,* in the Gulf of Alaska," in H. J. Wilimovsky, L. S. Incze, and S. J. Westrheim, eds., *Species Synopsis: Life Histories of Selected Fish and Shellfish of the Northeast Pacific and Bearing Sea,* Washington Sea Grant Program and Fisheries Research Institute, University of Washington, Seattle, 1988, pp. 70–82.

D. H. Steele, "Pollock (*Pollachius virens* (L.)) in the Bay of Fundy," *J. Fish. Res. Board Can.* 20, 1267–1314 (1963).

C. R. Forrester, *Life History Information Some Groundfish Species,* Fish. Res. Board Can. Tech. Rep. 105, 1969.

F. B. Hagerman, *The Biology of the Dover Sole, Microstomus pacificus (Lockinton),* Calif. Div. Fish. Gam Fish. Bull. 78, 1949.

International Pacific Halibut Commission, *The Pacific Halibut: Biology, Fishery, and Management,* IPHC Technical Report No. 22, 1987.

K. S. Parker, "Pacific Halibut, *Hippoglossus stenolepsis,* in the Gulf of Alaska," in N. J. Wilmovsky, L. S. Incze, and S. J. Westrheim, eds., *Species Synopsis: Life Histories of Selected Fish and Shellfish of the Northeast Pacific and Bering Sea,* Report of the Washington Sea Grant Program and Fisheries Research Institute University of Washington, Seattle, 1988, pp. 94–111.

J. H. S. Blaxter, "The Herring: A Successful Species," *Can. J. Fish. Aquat. Sci.* 42(Suppl. 1), 21–30 (1985).

J. H. S. Blaxter and F. G. T. Holliday, "The Behavior and Physiology of Herring and Other Clupeoids," in F. S. Russell, ed. *Adv. Mar. Biol.,* Vol. 1, Academic Press, Orlando, Fla., 1963, pp. 261–393.

J. H. S. Blaxter and J. R. Hunter, "The Biology of the Clupeoid Fishes," in J. H. S. Blaxter, F. S. Russell, and M. Yonge, eds., *Adv. Mar. Biol.* Vol. 20, Academic Press, Orlando, Fla., 1982, pp. 1–223.

D. J. Grosse and D. E. Hay, "Pacific Herring *Clupea harengus pallasi,* in the Northeast Pacific and Bering Sea," in N. J. Wilimovsky, L. S. Incze, and S. J. Westrheim, eds., *Species Synopsis: Life Histories of Selected Fish and Shellfish of the northeast Pacific and Bering Sea. Washington Sea Grant Program and Fisheries Research Institute, University of Washington Press,* Seattle, 1988, pp. 34–54.

S. Morita, "History of the Herring Fishery and Review of Artificial Propagation Techniques for Herring in Japan," *Can. J. Fish. Aquat. Sci.* 42(Suppl 1), 22–229 (1985).

J. W. Reintjes, *Synopsis of Biological Data on the Atlantic Menhaden, Brevortia Tyrannus,* FAO Species Synop. 42, Cir. 320, Washington, D.C. 1969.

W. Templeman, *Redfish Distribution in the North Atlantic,* Bull. Fish. Res. Board Can. 120, 1959.

B. A. Branson, "Sockeye Salmon," *Oceans* 9(2), 25–29 (1976).

R. E. Foerster, *The Sockeye Salmon, Oncorhynchus nerka,* Fish. Res. Bd. Can. Bull. No. 162, 1968.

W. S. Hoar, *The Chum and Pink Salmon Fisheries of British Columbia, 1917–1947,* Fish. Res. Bd. Can. Bull. No. 90, 1951.

F. Neave, "The Origin and Speciation of *Oncorhynchus,*" *Trans. Roy. Soc. Can.* 52(3), 25–39 (1958).

S. D. Sedgwick, *The Salmon Handbook,* Andre Deutsch Ltd, 1982.

L. Shapovalov and A. C. Taft, *The Life Histories of the Steelhead Rainbow Trout (Salmo gairdneri gairdneri) and Silver Salmon (Oncorhynchus kistuch),* Calif. Fish Bull. No. 98, 1985.

N. J. Wilimovsky, ed., *Symposium on Pink Salmon,* Institute of Fisheries, University of British Columbia, Vancouver, 1962.

M. Fish, "A Review of the Fishes of the Genus *Osmerus* of the California Coast," Proc. U.S. Nat. Mus. 46(2027), 29–297 (1913).

J. L. Hart, *Pacific Fishes of Canada,* Fish. Res. Bd. Can. Bull. No. 180, 1973.

J. L. Hart and J. L. McHugh, *The Smelts (Osmeridae) of British Columbia,* Fish. Res. Board Can. Bull. No. 64, 1944.

M. J. A. Butler, "Plight of the Bluefin Tuna," *National Geographic* 169, 220–239 (1982).

J. Joseph, W. Klawe and P. Murphy, *Tuna and Billfish-Fish Without a Country,* Inter-American Tropical Tuna Commission, Scripps Institution of Oceanography, LaJolla, Calif. 1988.

G. D. Sharp, and A. E. Dizon, eds., *The Physiology Ecology of Tunas,* Academic Press, Inc., Orlando, Fla., 1978.

GEORGE KATSUSHI IWAMA
University of British Columbia
Vancouver, B.C., Canada

FISH, MINCED

DEFINITION AND DEVELOPMENT

Fish flesh is often inefficiently used. For example, the remaining rack with bones after filleting, fillet trimmings, and the by-catch of underutilized fish (eg, those fish that are bony, unpleasant in appearance or too small) are often wasted. To increase the utilization of fish flesh from these sources, mechanical deboners can be used to recover the flesh in the form of a fish hamburgerlike meat or fish mince. Starting with a whole fish, the possible sources of fish mince from the traditional fisheries can be determined. The guts (entrails) of the fish (16% of the total weight) are removed after the fish are caught at sea (1,2). With V-cut fillets (ie, a fillet with the small pinbones removed) 40–45% boneless fillet, 3% skin, and pinbone trim of 3–5% of the landed weight are obtained. In addition, 53% of the landed weight are found in the remaining rack with the attached head. From the trim and frames 7–15% minced fish based on the landed weight are recovered.

AVAILABLE RESOURCES

Types of Fish and Distribution

Many varieties of fish (2–4) are used as raw material for minced fish productions. For example, some of the species used for minced fish (and surimi) production have been or are Alaska pollack (*Theragra chalcogramma*), croaker (*Nibea mitsukurii*), blue whiting (*Micromesistius poutassou*), haddock (*Melanogrammus aeglefinus*), cod (*Gadus morhua*) and other hakes. In Japan, work has been done on some of the fattier species such as sardines. The minced fish recovery of the drum-type flesh separators either from headed and gutted or guillotined Alaska pollack is 25–35% of the whole fish.

Uses

Minced fish can be used as a convenient, cheap, and nutritional food ingredient. Various species of minced fish can be used in many traditional and new fish products or as a replacement for red meat hamburger in many meat products such as (1) battered-breaded products, (2) traditional dishes, (3) ethnic foods, and (4) gourmet fish dishes (5–7).

PROCESSING

Methods, Equipment, and Problems

The freshness of the fish affects the gel-forming ability of minced fish. Basically for surimi-type products, the fish should be processed right after rigor mortis (8,9). Therefore, the raw fish material should be processed with a minimum time delay. Fish deteriorates easily so that the temperature of fish should be maintained as close as possible to 0°C throughout all of the processing procedures. In general, the operating procedures (10) for producing minced fish are

1. Contaminated and damaged fish should be removed before processing.
2. Deheaded and descaled fish might be required to give a higher quality product at the expense of yield loss; larger fish should usually have their backbone partially removed.
3. Fish viscera are not acceptable in minced fish; fish should be properly washed after degutting and prior to any further handling.
4. The presence of any remaining black membrane, lining the stomach cavity of fish, in minced fish products is usually not desirable; the black membrane can be removed with sponges, brushes, or cloth.
5. The product to be deboned is conveyed to the meat-bone separator.

The mechanical deboner is a critical control point in determining the quality of minced fish. Most bone-separation techniques work by forcing the fish flesh, maintained as close as possible to ice temperature, through a perforated drum. The higher the pressure on the deboner, the higher the yield of minced fish (10). However, the elevated pressure can destroy the fish gelling capacity, decrease the water binding capacity, and accelerate lipid rancidity (11). Therefore, a lower pressure produces a higher quality minced fish.

Several mechanical deboning machines are used with fish in the United States. These can be distinguished in terms of operating principles (1,12–14) into essentially four types of fish bone separator: (1) belt–drum, (2) auger–screen, (3) concentric cylinder microgrooved, and (4) hydraulic ram.

The Belt–Drum Deboners. Companies building the belt and drum equipment include Baader (Baader North America Co., Woburn, Mass.), Bibun (Bibun Machine Construction Co., Japan), and Yanagiya (Japan). The fish is fed into a hopper and transferred by a strong, solid rubber conveyor belt to between the belt and the perforated drum. The belt provides an adjustable amount of pressure against the perforated drum, which can be controlled by an eccentric roller. The fish is squeezed through the 3–5 mm holes (in some applications holes as small as 1 mm or as large as 10 mm may be used) into the interior of the drum. Inside the drum, the product may be sprayed with a solution containing such compounds as polyphosphate and the deboned minced fish may be moved by a screw auger to a gathering receptacle, while the impurities such as bones, scales, and skins remain outside the drum and are scraped off of the drum. This waste may be used to produce a fish broth, for hydrolysis, or other potential fishery by-products; however, it is often discarded.

The Drum–Auger Deboners. Beehive (Beehive Machinery Inc., Sandy, Utah) and Yieldmaster (The Kartridge Pak Co., Davenport, Iowa) are examples of companies that make these machines. The dressed fish is carried by a tapered auger through a hopper feeder. The screw feed flow creates the pressure to force the minced fish into the perforated cylinder system that generally has 0.5–1.5 mm holes. To regulate the speed of removal of the waste materials, the pressure inside the deboning head is regulated by a ring valve. The faster the waste is dismissed, the lower the deboning pressure. The deboning pressure affects the quality of the final product, particularly the calcium content and the temperature of the deboned minced fish. When the deboning stress increases, more fish bone is crushed and the higher the bone calcium incorporated into the minced fish. Also, the temperature of minced fish rises due to more friction. If the pressure is high enough, the product can be cooked.

The Two Concentric Cylinders Microgrooved Deboners.

Paoli (Paoli Manufacturing Co., Rockford, Ill.) is an example of the microgrooved concentric cylinder deboner. The yield of the Paoli deboners is 74–91% from the headless, eviscerated fish and 40–60% from the fish frame after filleting.

Fish is pumped directly from a grinder into an in-feed hopper. Then an auger pushes the fish through a drilled perforated cylinder. Two concentric cylinders, the exterior rotating cylinder and the interior microgrooved cylinder, are used in the Paoli separator systems. When the ground fish is fed through the hopper into the deboning drum, a pressure plate constrains the ground fish against the microgrooved drum. The microgrooves prevent the fish fibers from smashing. While an auger in the interior drum carries the minced fish to a collector, the bone residues are expelled in the opposite direction to the open end by the rotation of the drum.

The Hydraulic Ram Deboners.

This type of deboner, eg, Protecon's deboner (Gainesville, Ga), uses a hydraulic ram to squeeze the fish flesh past a screen. All drum–auger deboners incur heat generation in minced fish. However, the hydraulic ram deboners can reduce the heat accumulated in the mince because it is a single-pass system.

After separating, the minced fish need not necessarily be washed if the minced fish is to be used for products such as cakes, fishburgers, sausages, etc. However, washing should be done (1 : 3 to 1 : 7 fish to water) if excessive blood, inorganic salts, deteriorative enzymes and water-soluble materials need to be removed. The washing step helps to increase the shelf life of the minced fish and improve color by removing any remaining blood. The washed mince flesh should be partially dewatered to achieve the desired moisture content. (Without cryoprotectants, such as those used for surimi, ie, sorbitol, sugar, polyphosphate, and salt; it is not clear how well the washed mince will keep when frozen with respect to functional properties.)

The minced fish has storage problems (15,16), particularly if it is derived from gadoid fish such as cod and Alaska pollack. The texture of gadoids deteriorates more rapidly than those of other fish because of some unique chemical reactions. This texture deteriorates faster at $-14°C$ than $-20°C$ and $-40°C$. The spongy, cottony texture of gadoids during frozen storage is due to the enzymatic breakdown of trimethylamine oxide (TMAO) to dimethylamine (DMA) and formaldehyde (FA). The FA may cross-link certain muscle proteins to cause a change in texture that is usually not desirable. The material will lose most of its moisture on the first bite and afterward will feel dry and cottony.

PRODUCT DEVELOPMENT

Factors Affecting the Products

In developing minced fish products, several factors must be taken into account such as fish species, shelf life, economic value, functionality, nutritional benefits, and markets. In general, different fish species can be used for many minced fish products; freshwater fish may be limited by seasons but not saltwater fish. A product's shelf life may be a problem because of product discoloration (lipid oxidation), poor frozen storage (not sufficiently cold temperatures), texture changes (spongy and cottony texture) and bacterial growth.

Functional ingredients (eg, water, oil, salt, sugar, starch, and gum) provide minced fish with the necessary palatability, nutritional value, safety, and shelf life. Minced fish shows a different texture from regular fish flesh because of alteration in some functional properties such as hydration, gelation, and emulsification (10).

Minced fish products have more economic value than fillets because underutilized and industrial fish or food-grade fish waste can be used (7). For the minced fish market, the product can be developed to meet consumers' preferences, eating habits, cultural practices, nutritional value, and price. Some disadvantages, however, may influence the minced fish products market, for instance, these products are in direct competition with traditional ground meats (beef, poultry, pork, and turkey). Other concerns are a possible fall off in sales of breaded seafoods, and a lack of familiarity with the functional properties of fish protein (17). Hence, to expand the minced fish product market, it is necessary to make an effort to: (1) improve product flavor, color, texture, and shelf life using various functional ingredients so as to be able to compete with other meat products; (2) develop health foods; (3) develop ethnic foods; (4) educate consumers about minced fish products from a nutritional point of view; and (5) design versatile and convenient recipes for using minced fish.

TYPES OF PRODUCTS

Frozen Minced Fish Block

Headed and gutted fish, fillets, V-cuts, and frames are all used as raw materials in producing commercial minced fish blocks. To prepare the final frozen minced block, salt, sugar, sodium tripolyphosphate, and water may be blended with the fish flesh (18). Then the mixture is packed into cartons and frozen. This minced frozen block can be used to prepare fish fingers, sticks, and portions. From a nutritional point of view, these products do not present fish in the best light because of the higher fat content since these products are usually breaded-battered or deep fried. However, some commercial frozen minced fish is sold directly to consumers with a recipe on the package; consumers can follow the recipe to make various minced fish dishes.

Traditional Products

Traditional minced fish products such as fish sausages, cakes, patties, balls, loaves, and burgers etc are well established in many countries, made according to local cultural preferences. The ingredients (eg, salt, sugar, starch, polyphosphate, soy protein, potassium sorbate, spices and gums) used for making various products provide functionality to stabilize the minced fish, increase its shelf life, and give desirable flavors and textures (19–21).

In the United States, most minced fish product development has been directed at developing different meatlike products in place of red meat to provide higher protein and less fat in the American diet. These products (including hot dogs, hamburgers, patties, frankfurters, meatballs, meat loaf, and sausages) are able to be served in place of traditional equivalents. For example, crumbled fish meat can be used for tacos, chili, or spaghetti; fish sausages are used for biscuits, pizza, or lasagna (7). Other similar products such as fish–soy balls, loaves, cakes, and patties made of minced fish, soybean curd, and other ingredients are able to provide inexpensive and high-quality proteins (22,23).

Surimi and kamaboko have been very popular and well developed in Japan for many years. Surimi is a semiprocessed mince material that preserves minced fish for later additional processing. It is prepared mainly from Alaska pollack (24). The process involves heading, gutting, bone separation, washing, dewatering, and straining (25). The addition of sugar and polyphosphate prevent denaturation of surimi during frozen storage. Sorbitol is added to retard the browning reaction (26). The gelation characteristic of surimi results from its high concentration of myofibrillar protein (actomyosin) after washing, which produces an elastic and chewy texture (17). However, some water-soluble proteins, vitamins, and minerals are lost after washing minced fish flesh (27).

Kamaboko, one of many surimi-based products, is prepared from surimi by grinding, blending, shaping, and heat sterilizing (15). In addition, surimi can be used for making versatile products such as fish sausages, burgers, hams, imitation crab legs and shrimp products, and salads (24,26,28).

Fish Snacks

Minced fish after specific processing can be made into versatile fish snack items. One of the fish snack products is a smooth-textured, spread-type product characterized by high acceptability, low cost, relatively long refrigerated storage life, and versatility of application (29,30). The basic spread product is a heat processed emulsion of minced fish muscle (nutritional base, emulsifier, and texture) and vegetable oil (texture, lubricity, and carrier for surfactants). Other ingredients include starch (freeze–thaw syneresis and texture), water (spreadability and texture), corn syrup solids (texture and flavor), salt (solubilize protein, texture, and flavor), tripolyphosphate (water-holding capacity), sorbate (mold inhibitor), and various flavor and color spices (flavor and color). Various flavor spreads either in a canned or refrigerated product can be served for appetizers, crackers, chips, and sandwiches etc (30).

Fish–seafood chips and crispies were developed at Cornell University for a market test (1). The ingredients included at least 66% minced fish (either underutilized species or filleting waste), chopped scallop mantels (another by-product) or clams (optional), vegetable oil, soy protein, cracker meal, onion, and seasoning mix (salt, garlic, black pepper, celery seed, and MSG). It is a battered-breaded product that can be prepared for consumption in several ways such as (1) frozen raw for retail sale and then fried by the consumer just before serving or (2) frozen after a short fry by the processor and heated in the oven by the consumer. White sucker (*Catostomus commersoni*), known as mullet in Canada and parts of the United States, was used for this product (31). Fish puffs, similar to fish chips and crispies, are another fish snack product. Fish bits, which combine soy and fish products, are produced using an extruder (32). They are similar to bacon bits and are able to be used for salads.

Gourmet Fish Dishes

Minced fish can also be used for many gourmet fish dishes such as fish Newburg, a crepe filling, gefilte fish, a fish salad, and a gelatin–mayonnaise molded fish salad, which are easily made at home by following recipes (6,33–35). Either saltwater or freshwater fish can be used for these products that taste like meat in the meat-replacement products while tasting like fish in the fish products.

Fish–Seafood Soup

Until recently, the broth usually used for Newburgs and chowders in the United States was chicken broth. However, a good fish broth can be prepared using deboning fish waste and fish heads (6,7) and can replace chicken broth in some recipes. Now commercial fish broth stock is also available for the institutional market.

Different flavors of seafood chowders, eg, Manhattan seafood chowder (MSC) and New England seafood chowder (NESC), have been developed and sold in a few supermarkets as part of Cornell's test marketing (7,26). Generally, around 25–28% minced fish (24% for MSC and 27% for NESC) or scallop mantels are used in the seafood chowders with various other ingredients (34,35).

Canned Minced Fish

Canned minced fish is another way to expand the minced fish market. For example, minced mullet was developed as a lower priced alternative to canned grated tuna fish for a retail market test (36,37). The procedure for canning minced mullet was (1) thaw fish if it is frozen; (2) cook fish with constant stirring until completely cooked; (3) drain off the cookout liquid; (4) pack in cans, adding 1% salt and 5% vegetable broth (37).

Fermented Products

Traditionally, many fermented dairy products are made from milk; fish can also be fermented for the production of cheese (fish cheese) (10,38). The fish being fermented can be obtained from white or fatty raw materials and can be specifically based on mince. Usually, the fish are gutted, minced, and pasteurized to reduce the intrinsic microflora, coagulate the protein and inhibit degradative enzymes. Whey or saccharose are added as an energy source; simultaneously, a starter culture such as *Lactobacillus* is added to carry out the fermentation. The minced fish is then incubated at a proper temperature (eg, 30°C) for one day. During the fermentation, the pH falls to around 4.5. The fish curd is pressed to remove water, brined and then vacuum packed. The final product is like an unripened cheese. Due to the low pH (4.5), the fish cheese has a shelf life of several months at ambient temperatures.

Other Minced Fish Products

A lot of versatile products can still be developed from minced fish. There is no limitation to choose specifically beef, pork, or poultry for preparing dishes; minced fish can also be used. Therefore, for example, minced fish can be prepared for making a fish pie, stuffings, fish rolls, fish noodle casserole, stir-fried fish with bamboo shoots, or seafood quiche (18,34,35,39).

BIOLOGICAL FACTORS AFFECTING MINCED FISH PROPERTIES

The properties of minced fish are affected by the raw materials, ie, the fish themselves. Although any species of fish can be used as a raw material for minced fish, the properties are very dependent on the species, eg, gadoid fish, fatty fish, or a bony fish. The portions of fish used for mincing, eg, fillets with or without skin, or frame also affect the properties. The properties of minced fish are also affected by the mincing process, which causes tissue disruption, mixing of other organs, blood and skin, incorporation of air, and other contamination.

TMAO

The most widely used fish species for minced fish are the gadoid fishes: cod, hake, haddock, pollack, whiting, cusk, etc. Like most marine fish, these species contain TMAO. TMAO is degraded by a NAD-dependent TMAO reductase to trimethylamine (TMA) (40); the latter is characterized as one of the "fishy" odors. TMAO reductase is found in the dark lateral muscles of several fish species (41) and in the cells of many species of bacteria within the family Enterobacteriaceae and other gram-negative facultative anaerobic rods (42). Therefore, TMA formation is a function of the number of bacteria present on the fish and is reflected in the change of sensory quality of ice-stored fish, but this mechanism is not significant during the frozen storage of minced fish.

In frozen minced fish, the focus is on the degradation of TMAO to DMA and FA by the enzyme TMAO demethylase, because FA is believed to cause a cottoniness or sponginess of the fish flesh through cross-linking of muscle proteins during frozen storage (43). Free water is loosely retained by the cross-linked fish tissue causing the product to behave like a sponge; the water can be easily expelled and then reabsorbed. When eating the fish, the moisture is all released on the first bite and the remaining dry sponge then has a cottony texture. TMAO demethylase is mainly found in the kidney (44). Other sources of the enzyme are the blood, the pyloric caeca, and possibly the skin and dark muscle (10). Therefore, incorporation of other fish organs, blood, and skin into a muscle mince may accelerate toughness. One solution is to wash out the TMAO and the enzyme(s). However, this process needs large amounts of water and removes about 25% of the total protein, some of the fish fat, and most of the water-soluble vitamins and minerals. It was reported that the exclusion of air enhances DMA production (45,46). The TMAO demethylase is resistant to freezing but demethylation does not occur at temperatures below −30°C (47).

Therefore, the best way to prevent this reaction from occurring would be to store the minced fish at sufficiently cold temperatures. More information about TMAO, TMAO demethylase, and protein changes related to TMAO degradation can be obtained (16,47–53).

Proteolytic Enzymes

Enzymatic proteolysis of minced fish can cause a significant change in minced fish texture. Fish contain proteolytic enzymes and the fish proteins are highly susceptible to proteolysis. The proteases come mostly from the pyloric cacea and the intestines (54). These proteases are dispersed during mincing. Even when the minced fish are contaminated by a low level of visceral materials, extensive proteolysis can occur. Other proteases such as cathepsins of muscle tissue (55) and microbial enzymes (56) are also involved in minced fish texture and quality changes.

Moisture

Changes in the moisture phase during freezing or frozen storage that lead to changes in proteins in fish may be associated with the action of FA derived from TMAO. The ruptured muscle tissue may also be susceptible to the freeze denaturation of fish proteins. Freezing, especially slow freezing, causes the formation of larger intercellular and intracellular ice crystals, which accelerate dehydration of cells, rupture of membranes, and disorder of the ultrastructure of the cells and tissues. The size and location of ice crystals in frozen fish tissue are influenced by several factors such as the physiological status of the raw and minced fish, the mincing methods, the freezing rate, the storage temperature, the storage time, and temperature fluctuations (57).

It has been reported that there was much more denaturation in muscle frozen postrigor than prerigor (58). A fast freezing rate is recommended so that water inside the cell does not migrate into the extracellular spacing. Temperatures below −30°C for the storage of minced fish are usually recommended. Fluctuations in the storage temperature, even at a quite low temperature, accelerates the growth of the ice crystals.

The most widely used method to protect fish muscle proteins from denaturation during frozen storage is the addition of cryoprotectants. Studies about cryoprotectants have been reviewed (10). Among them, polyphosphates have been extensively used and commercial mixtures are also available for minced fish. The addition of starch has been shown to improve the texture of minced fish during frozen storage (59). Some sugars, sugar alcohols, and polysaccharides are used as synergists of phosphates.

Lipids

Problems related to lipids are the main concern in fatty fish, such as mackerel, herring, sardine, and menhaden, because of their high level of fat and the high unsaturation of the fat. The mincing process allows air to be incorporated into the muscle tissue and enhances lipid oxidation. During mincing, hematin compounds (hemoglobin, myoglobin, and cytochromes), which are known to be effective catalysts of lipid oxidation, are also distributed throughout the muscle. Metal compounds, such as iron, can be released from the fish during mincing. Rancid odors or flavors of oxidized oil are characterized as musty, turnipy, fishy, painty, and cold storage flavor.

Lipid oxidation can be slowed or prevented in minced fish by removal of dark muscles (in which more hematin compounds are found) and fats beneath the skin by deep skinning, low temperature storage, vacuum packaging, modified or controlled atmosphere packaging, glazing, and application of antioxidants. These methods are generally combined. Among them, the incorporation of antioxidants is most common because it is simple, economic, easy to apply, and should be effective when the antioxidant is properly chosen. Among the approved synthetic antioxidants, TBHQ alone or in combination with other antioxidants or synergists has been found to have the highest effectiveness for retarding oxidative rancidity in fish (60-62). Ascorbic acid and erythorbic acid have been shown to be strong antioxidants in fish and minced fish (61-64), but they can be prooxidants in some fish oil products. Tocopherols are natural antioxidants in fish; however, the addition of more tocopherols may accelerate oxidation. Excess tocopherols may act as a prooxidant rather than as an antioxidant (65). The most effective method of retarding lipid oxidation of minced fish is believed to be vacuum packaging (64).

Another aspect of maintaining fish lipid quality is preventing hydrolysis. Lipid hydrolysis in fish occurs by the action of lipases. Triacyl lipase (1 or 3 position specific), phospholipase A2 (lecithinase A or phospholipase A specific), and phospholipase B (lysophospholipase or lecithinase B specific) are the enzymes most involved in lipid hydrolysis in fish (66-68). Lipases originating from microorganisms also take part in the hydrolysis. These enzymes can be distributed throughout the fish muscle by mincing to enhance the hydrolytic activity. The hydrolytic products such as free fatty acids, lysophospholipids, glycerophosphocholine, phosphocholine, choline, and phosphoric acid may not influence fish flavors noticeably by themselves. There has been a controversy on the effect of the hydrolysis products on the rate of oxidative rancidity in fish lipid: do they have an enhancing effect on oxidation or a retarding effect? No correlations have been demonstrated between the development of off-flavors and free fatty acid production (69).

A correlation has been proposed for protein denaturation and the buildup of free fatty acids. The detrimental effect of free fatty acids on the textural quality of frozen-stored fish has been repeatedly reported, but generally not well explained (70,71). Free fatty acids are postulated to react primarily with the myofibrillar proteins. Their binding to sarcoplasmic proteins has not been excluded, but clearly the reaction of free fatty acids does not insolubilize the soluble proteins (72). On the other hand, it has been hypothesized that the presence of moderate levels of lipids (3-10%) may protect the fish proteins, ie, increase their resistance to denaturation, during frozen storage (73).

Methods to reduce free fatty acid formation have not been well developed because lipid hydrolysis is mainly caused by enzymes. Inhibition or inactivation of the enzymes by such treatments as heat, very low temperature frozen storage (the enzymes are believed to be active at frozen storage temperatures), and addition of inhibitors may be possible. However, most of the active inhibitors (mercuric chloride, p-chloromercuribenzoate, phenylmethylsulfonylfluoride, sodium fluoride, sodium hypochlorite, etc) are toxic (66,68).

Microorganisms

The dominant spoilage microorganisms in fish are generally *Pseudomonas* and *Alteromonas*. *Achromobacter, Flavobacterium, Moraxella, Acinetobacter,* and less frequently *Vibrio* are also involved in fish spoilage (74). The microorganisms involved in minced fish are basically similar to those found with intact fish. However, some concerns associated with microorganisms should be taken into account in minced fish. Poor handling of raw fish or the extended holding of the frames or wastes from filleting prior to deboning increases the microbial problems. The contamination of viscera, gills, and skin portions during mincing is accompanied by increased microbial contamination. Hygienic conditions of the mincing equipment in contact with the fish should be optimized to avoid additional microbial contamination. With properly controlled conditions of temperature and prevention of cross-contamination, the mincing process itself does not give too large an increase in microbial counts (75). However, if the raw materials destined for mince are spoiled or semispoiled, the mincing process leads to notable increases in microbial counts by the dispersion of high levels of contamination throughout the muscle tissue (76). It has been suggested that minces processed under even the most controlled conditions are remarkably vulnerable to microbial contamination after mincing (74).

Color

Mincing causes darker colors in minced fish than in the raw materials from which it came due to contamination with skin pigments, black belly membrane, blood, heads, and guts (77). Browning reactions may also occur during frozen storage. Bacterial and nonenzymic (Maillard) browning reactions are involved in nonfrozen fish (77). Oxidation of fats and blood in fish results in discoloration. Carefully controlled mincing using unadulterated fish flesh would be the most important factor in obtaining a lighter and more uniform minced fish. Washing is a commonly used method to remove excessive dark color; however, this process causes the loss of a number of nutrients as mentioned earlier. Alkaline or acid washing is also recommended (78). Hydrogen peroxide is also reported to be a color stabilizer or a bleach (79).

However, darkening should not always be considered a problem. Darker mince products result from blood and dark muscle incorporation; these may be high in iron and omega-3 fatty acids. Also darker mince can be used for products where there is an expectation of a darker color, eg, red meat analogues.

NUTRITIONAL QUALITY OF MINCED FISH

The nutritional characteristics of minced fish depend mostly on two factors: one being the mincing action itself and the other the kind of starting material used for the production of the mince. During mincing, the tissue is disrupted and becomes more susceptible to deteriorative changes than the intact flesh. These changes will also affect the nutritive quality of the minced product.

There are different kinds of fish material that can be used for the production of mince like fillets, whole fish, headed and gutted fish, frames (carcasses left after filleting) and headless frames. The more suitable material for the production of mince for human consumption are fillets, headed and gutted fish, and headless frames. One example of how the starting material would affect the characteristics of the final product is the production of mince from headless frames. This kind of mince will have a different mineral profile than its counterpart made from fillets. For example, iron and calcium levels will be higher in frame minces than in fillet minces (80,81).

One of the earliest studies carried out (82) to determine yield and nutritional characteristics of minced fish, used the whole carcass waste, the machine-separated flesh and the remaining bone–skin fractions from six different fish species: English sole (*Parophrys vetulus*), petrale sole (*Eopsetta jordani*), orange rockfish (*Sebastodes pinniger*), yellowtail rockfish (*Sebastodes flavidus*), true Pacific cod (*Gadus macrocephalus*) and ling cod (*Ophiodon elongatus*).

The ranges in percent of flesh, protein, fat, and ash recovery from whole carcass waste were 49.4–60.2%, 43.7–55.3%, 42.7–66.0% and 13.3–21.6%, respectively. Levels of phosphorus, calcium, strontium, manganese, boron, and chromium were much lower in the mechanically recovered flesh than those found in the whole carcass waste. It was suggested that the calcium content in the minced flesh recovered from carcass waste was mainly affected by differences in the size and bone structure among different species (82). On the other hand, levels of potassium, sodium, and iron were higher in the minced flesh fractions than in the respective whole carcass waste.

The protein efficiency ratio (PER) values for the mechanically separated flesh were significantly higher than the values for whole carcass waste (82). The quality of the protein in the minced flesh fractions obtained from all six species was significantly superior to the casein reference, except for Pacific cod, which was the same as casein.

It was suggested that this improvement in PER was mainly due to the removal of collagenous material found in the skin and bones (83). Previous work (84) also reported that mechanical deboning of poultry tended to remove collagen and skin while incorporating fat into the muscle tissue.

Mechanically separated fish flesh from Atlantic croaker (*Micropogon undulatus*) was found to have significantly higher quantities of sarcoplasmic and nonprotein nitrogen than hand-processed fillets (85). No significant differences for protein, ash, and pH were found between minced flesh and hand-processed fillets. However, fillets had a significantly higher moisture and slightly lower fat content than minced flesh. These differences were attributed to the exposure of fillets to water during rinsing in commercial processing and to a higher concentration of fat associated with skin and bone marrow.

These results (85) were in agreement with an earlier study (86) in which it was found that the proximate composition of fillets and minced flesh from Atlantic croaker was almost the same. It was also determined that there was no statistical difference between the PER and the net protein utilization (NPU) values for mince and fillets (86).

On the other hand, it was pointed out that the bone content was not increased by mechanical deboning as indicated by the ash level (85). This was probably due to the fact that headed and gutted fish were used. When frames are used, more bone is likely to appear in the mince.

Minced fillets from armado (*Nicholsina usta*), an underutilized species found in the Gulf of Mexico, were used to produce salted fish cakes (87). Fish cakes, freshly made and stored for 18 months at an ambient tropical temperature, were evaluated for proximate chemical composition, protein quality (PER and NPU), and bacterial counts. The cakes had an average composition of 30% protein, 60% salt, and 10% moisture. Freshly made cakes were superior, while the 18-month-old cakes were slightly inferior to the reference casein for both PER and NPU. Total plate and halophilic counts, which were initially low, decreased to almost zero after 18 months of storage. A method to produce dehydrated, salted fish–soy cake using mince from flounder frames was also developed (23). The final product had a composition of 49.8% crude protein, 14.4% moisture, 21.1% salt, and 5.3% crude fat.

Hemoglobin, myoglobin, total heme pigments, and nonheme iron concentrations have been measured in phosphate buffer extracts of headed and gutted mince and fillets from mullet (*Mugil cephalus*) (88). It was found that mechanical deboning increased the hemoglobin and nonheme iron content of the minced flesh. The ratios of hemoglobin to myoglobin varied from 0.54 to 0.61 and 1.19 to 1.21 for fillets and mince, respectively. These results agreed with other work (89), which found that mechanical deboning of poultry increased the levels of hemoglobin but did not affect myoglobin concentration. It was concluded that the increase in hemoglobin in mechanically deboned chicken was due to the presence of bone marrow (89).

Various samples of mechanically processed products (MPP), bone residues, and beef semimembranosus muscle (BSM) were analyzed (90). MPP had higher ash and calcium values than intact beef tissue because of the presence of bone powder. Calcium content varied from 0.46 to 1.15% in MPP compared to 0.01% for BSM.

It was also reported that iron content (mg/100 g) ranged from 3.8 to 8 for MPP and 3.2 for BSM. The amino acid content showed an increase in the amount of proline, glycine, and hydroxyproline and a lower content of sulfur-containing amino acids (90), suggesting that the major protein in the bone residue is collagen. Collagen would come from the bone, tendon, and other connective tissue removed by the mechanical deboner. However, the amount of collagen would vary with yield and the type of equipment used. The bone residue obtained from mechanically deboned fish frames was suggested to presumably have collagen as a major protein (83).

A study on dried fish patties made from a mixture of several comminuted fish species (25% of each): rockfish, Pacific cod, ling cod, and Pacific herring (*Cuplea harengus pallasi*), with structured vegetable protein, modified tapioca starch, sodium chloride, or sodium sorbate, and antioxidants was conducted (91). The final mixture had 89.5% minced fish. The recovered flesh ranged from 40.5 to 63.8%. The dried fish patties had a proximate composition of 55.5% protein, 19% fat, 15.4% carbohydrates, and 5.5% moisture. The protein quality of the dried fish patties as measured by PER was superior to the casein reference, with a 2.9 value. Total bacterial counts were around 10²–10³ cfu/g during a four-month storage period at 25°C. Low counts of yeast and molds were also observed (91).

The chemical composition and frozen storage (−18°C) stability of the filleted and minced forms of flesh from spot (*Leiostomus xanthurus*) have been studied (92). The results indicated that maximum nutritional and frozen storage stability values were obtained from spot harvested from October to February when protein content is at its maximum, fat content at a minimum, and fatty acids are relatively stable. It was also pointed out, however, that a sensory panel preferred filleted spot over minced spot for all months of harvest and storage primarily due to the dark color of the minced flesh.

A study on the nutritional properties of fillets and minced flesh from Alaska pollack, spot, silver sea trout (*Cynoscion nothus*), and Atlantic croaker showed that mechanical deboning of dressed carcasses produced minced flesh that had protein of equal nutritional quality to that of fillets by PER, net protein ratio (NPR), amino acid chemical score, and protein digestibility (93). When high pressure mechanical deboning was used, significantly lower PER and NPR values for croaker minced flesh obtained from frames were observed. The minces thus obtained had lower protein and higher ash and oil content as compared to fillets and mince obtained from headed and gutted fish.

It can be suggested that the higher ash content resulted from more bone powder coming out of the frames, increasing the calcium and lowering the PER and NPR values. In addition, experiments showed that lower PER values were associated with high calcium content (90). Moreover, the amount of lean meat present in the material before deboning might affect PER values. The higher the amount of lean meat, the better the PER value obtained (90).

The composition, nutritive value, and sensory attributes of fish sticks produced from textured soy protein (TSP) and minced flesh obtained from Alaska pollack, Atlantic cod, spot, and Atlantic croaker were also investigated (94). A ratio of 3.3:1 minced fish to TSP resulted in fish sticks with good sensory and nutritive characteristics. Lysine and methionine were approximately 20% lower in the soy-supplemented fish sticks than in the sticks made from pure minced fish. All of the samples except for the Alaska pollack and the Alaska pollack with TSP had significantly better PER values than casein. The pollack samples had the same PER values as casein. It is, however, interesting to point out that pollack supplemented sticks had a higher oil and a lower protein content than the rest of the samples.

The chemical composition and frozen storage stability of weakfish (*Cynoscion regalis*) as fillets, minced flesh, and washed minced flesh were studied (95). The results showed that mincing increased rancidity during storage but that it was minimized by washing. Washing the minced fish improved its storage stability, color, flavor, odor, and overall appearance. Washing, on the other hand, resulted in a 3–4% loss of solids (protein, fat, ash). It was also reported that the ash content was lowest for washed minced samples due to leaching of minerals, washing out of scales and small bone particles, and the increase of moisture content. It was concluded that washing of the minced flesh did not appreciably affect its nutritional value while improving flavor, odor, and color. No studies, however, to determine PER and mineral profile as affected by the washing treatment were carried out.

Washing minces to remove lipids, blood, and other prooxidants was also reported to improve color and flavor stability of minced flesh (63,96). On the other hand, it was pointed out that there was a loss of soluble protein from this treatment and that it was believed that adverse texture changes during frozen storage might occur (63,95).

The nutritional characteristics of fillets and unwashed and washed mince flesh obtained from a mixture of rockfish species (*Sebastedes melanops, S. flavidus, S. rubriviactus, S. pinniger*, and *S. proriger*) were investigated (97). It was observed that washing minced flesh resulted in a 37% loss of solids; a 10-fold increase over that found by another study (95). The greatest reductions were found in the ash (80%) and lipid (65%) in the washed flesh. Washing, however, did not affect amino acid composition and PER values were higher than the casein reference. The mineral composition of unwashed minced flesh was similar to that of the fillets (97). Washing, on the other hand, greatly changed the mineral composition of the flesh. Phosphorous, potassium, and sodium levels were reduced, while iron, copper, zinc, and chromium levels were increased in the washed minced flesh.

Two explanations for this change in the mineral profile were offered (83): (1) that the metal ions may have been attached to protein molecules that were not soluble in water or (2) the water used contained high amounts of the minerals in question so that they became an integral part of the minced flesh (83). A third explanation, especially for iron and copper, might be cross-contamination by the equipment used to wash and collect the mince.

Minced fish flesh as well as the intact fillet from three species of the Gadidae family were evaluated for their cholesterol and calcium content (80). Minces obtained from headless cod frames were high in cholesterol (82.9–87.8 mg/100 g). Whether or not the frames had kidney tissue present did not make any significant difference. Skinless cod fillets had a cholesterol content of 48.1 mg/100 g, suggesting that the cholesterol in the headless cod frames might come from the bone marrow. Mince obtained from whiting commercially scaled and headed and gutted (H&G) with kidney tissue had a significantly higher amount of cholesterol (68.1 mg/100 g) than commercially scaled skinless fillets or hand-scaled skinless fillets, cleaned H&G, and commercially H&G with kidney tissue (42.2–57.0 mg/100 g) (80). Red hake (*Urophycis chuss*) showed no significant differences in cholesterol content (ranging from 35.1 to 49.0 mg/100 g) for any treatment. Based on these results, the minces were rated as low cholesterol (red hake), low to medium cholesterol (whiting) and medium to high cholesterol (cod frames) (80). Furthermore, because mechanical deboning is more drastic than hand deboning, more cholesterol-containing material could be extracted into the mince (98).

The ash content was also significantly higher for cod

frames (1.92%) with kidney tissue than for frames without kidney tissue (0.88%) (80). This suggests that removal of the kidney tissue can take away significant amounts of minerals, especially iron and copper. No differences in ash content between treatments with the red hake or whiting were found. On the other hand, calcium content was higher in cleaned frames (38.4 mg/100 g) than in kidney-containing frames (28.6 mg/100 g). Both kinds of frames, however, had more calcium than skinless fillets (4.5 mg/100 g). High levels of calcium were reported for commercially scaled minced whiting (about 110 mg/100 g) compared with filleted flesh (29.5 mg/100 g) (80). Hand scaling, on the other hand, reduced calcium levels for minced flesh close to those of skinless fillets (about 31 mg/100 g). Red hake showed no difference in calcium content with any treatment.

Vitamin A and C were present in marrow and although these values were higher than those reported for muscle they were nutritionally insignificant (99). Marrow often constitutes 10–30% of mechanically separated meat (MSM) and in the United States only 2% MSM is allowed in processed meat (99).

One study (100), which followed earlier work (101), used minced fish to produce fish cakes, using lightly salted minced Atlantic cod. It was found that the addition of 13% salt to cod mince showed several advantages over the previously employed 25% salt treatment (102); especially the gelatinous–fibrous texture that had shape stability after drying and cooking. The dried product had a final composition of 30% moisture, 22% salt, and 48% protein. After rehydration and cooking, the product had a composition of 72% moisture, 26% protein, and 2% salt.

From a nutritional point of view minced fish obtained from headed and gutted fish or from headless frames has the potential to provide good-quality protein and an array of essential minerals, especially iron and calcium. The only drawback to using frames could be the cholesterol content (80 mg/100). This can probably be lowered by reducing the percentage yield because higher pressures could cause cholesterol to be extracted into the mince from the marrow. The higher pressures might also reduce the protein quality.

Minced fish has proven its nutritional significance when used as a material for the manufacture of different products that are economically feasible to market. This is the ultimate utilization goal for minced fish.

BIBLIOGRAPHY

1. J. M. Regenstein, "Minced Fish, A Critical Examination of the Cornell Experience," *Seafood American* 2, 1–9 (1981).
2. J. M. Regenstein and C. E. Regenstein, unpublished data.
3. "Pacific Pollock (*Theragra chalcogramma*): Resources, Fisheries, Products and Markets," NRC Contract No. 81-ABC-00289, 1981, pp. 40–131.
4. J. Nielesen and T. Borresen, "Minced Fish in the Nordic Countries," in R. E. Martin and R. L. Collette, eds., *Proceedings of the International Symposium on Engineered Seafood Including Surimi.* Seattle, Wash., 1985, pp. 639–654.
5. T. Suzuki, "Fish and Krill Protein: Processing Technology," Elsevier Applied Science Publishers, Ltd., Barking, UK, 1981.
6. J. M. Regenstein, "Minced Fish," in J. J. Connell, ed., *Advances in Fish Science and Technology,* Fishing News Books, Farnham, UK, 1979, pp. 192–199.
7. J. M. Regenstein and C. E. Regenstein, "Using Frozen Minced Fish," in Ref. 4, pp. 53–73.
8. S. C. Sonu, "Surimi," NOAA Technical Memorandum NMFS, 1986.
9. "Reference Manual to Codes of Practice for Fish and Fishery Products," FAO Fisheries Circular No. C750, 1982, pp. 94–206.
10. G. J. Grantham, "Minced Fish Technology: A Review," FAO Fisheries Technical Paper No. 216, Rome, 1981, pp. 9–38.
11. X. Hua, H. O. Hultin, S. D. Kelleher, W. W. Nawar, and Y. Young, "Changes in N-3 Fatty-Acids Curing Fish Processing," *Abstract of Papers of the American Chemical Society* **196,** 121 (1988).
12. R. E. Martin, "Third National Technical Seminar on Mechanical Recovery & Utilization of Fish Flesh," National Fisheries Institute, Washington, DC, 1980.
13. "8 Firms Offer Deboning Machines; How They Work," *The National Provisoner* 72(14), 12–50 (1975).
14. R. M. Ryan, "Ryan/Bibun Machinery for Surimi and Surimi-Based Food: The Best of East and West," in *Seventy-Seventh Annual Convention of the National Food Processors Association,* Surimi-Based Foods and Seafood Analogs, Washington, D.C., 1984, pp. 41–71.
15. Y. L. Hsieh and J. M. Regenstein, "Texture Changes of Frozen Stored Cod and Ocean Perch Minces," *Journal of Food Science* **54,** 824–826 (1989).
16. A. D. Samson, J. M. Regenstein, and W. M. Laird, "Measuring Textural Changes in Frozen Minced Cod Flesh," *Journal of Food Biochemistry* **9,** 147–159 (1985).
17. J. Vondruska, "Market Trends and Outlook for Surimi-Based Foods," in *International Symposium on Engineered Seafoods,* Seattle, Wash., 1985, pp. 1–28.
18. L. Long, S. L. Komarik, and D. Tressler, "Section 9 Fish Products," in *Food Products Formulary Vol. 1: Meats, Poultry, Fish, Shellfish,* 2nd ed., AVI Publishing Co., Inc. Westport, Conn., 1982, pp. 294–328.
19. E. Brotsky and E. William, "Use of Polyphosphates in Minced Fish," in Ref. 4, pp. 299–305.
20. R. Clark, "Hydrocolloid Applications in Fabricated Minced Fish Products," in Ref. 4, pp. 284–298.
21. D. B. Westerly, C. D. Decker, and S. K. Holt, "Gelling Proteins," in Ref. 4, pp. 324–347.
22. K. H. Moledina, *Effects of Some Treatments on Quality of Frozen Stored Mechanically Deboned Flounder Meat (MDFM) and Use of MDFM for Preparation of Dehydrated Salted Fish-Soy Cake,* Master's thesis, Cornell University, Ithaca, N.Y., 1975.
23. K. H. Moledina, J. M. Regenstein, R. C. Baker, and K. H. Steinkraus, "A Process for the Preparation of Dehydrated Salted Fish-Soy Cakes," *Journal of Food Science* **42,** 765–767 (1977).
24. P. Sumpeno, "Surimi-Prospects in Developing Countries," *INFOFISH International* **5,** 29–32 (1989).
25. D. Miyauchi, G. Kudo, and M. Patashnik, "A Semi-Processed Wet Fish Protein," *Mar. Fish Rev.* **35**(12), 7–9 (1973).
26. F. W. Wheaton and T. B. Lawson, "By-products," in *Processing Aquatic Food Products,* Wiley-Interscience, New York, 1985, pp. 425–465.
27. J. M. Regenstein and T. C. Lanier, "Surimi: Boon or Boondoggle?" *Seafood Leader* **6,** 152–162 (1986).

28. M. Yamamoto and I. Miyake, "Kamaboko," J. R. Brooker and R. E. Martin, compilers, *Third National Technical Seminar on the Mechanical Recovery and Utilization of Fish Flesh, Abstracts,* Raleigh, N.C., 1980.

29. M. Patashnik, K. Kudo, and D. Miyauchi, "Smooth, White Spread from Separated Fish Flesh Forms a Base for Flavored Dips, Snack Items," *Food Product Development* 7(6), 82–91 (1973).

30. R. C. Baker and C. E. Bruce, "Fish Spreads," *Development of Products from Minced Fish,* Booklet 15, New York Sea Grant Institutes, Albany, N.Y., 1985.

31. R. C. Baker, J. M. Regenstein, and J. M. Darfler, "Seafood Crispies," *Development of Products from Minced Fish,* Booklet 2, New York Sea Grant Institute, Albany, N.Y., 1976.

32. S. K. Williams and J. B. Bacus, "Engineered Seafood Products," in Ref. 4, pp. 349–356.

33. J. M. Regenstein, "Minced Pollock: the Economical, Versatile, and Nutritious Meat," in C. W. Mecklenburg, ed., *Alaska Pollock: Is It a Red Herring?* Alaska Fisheries Development Foundation, Anchorage, Alaska, 1981, pp. 173–187.

34. R. C. Baker and J. M. Darfler, "Tasty Dishes from Minced Fish," *Development of Products from Minced Fish,* Booklet 5, New York Sea Grant Institute, Albany, N.Y., 1979.

35. J. M. Regenstein and C. E. Regenstein, "Choose Your Title: Kosher Minced Fish Cooking; Fish Cooking with a Food Processor; International Fish Recipes," New York Sea Grant Institute, Albany, N.Y., 1982.

36. D. C. Goodrich and D. B. Whitaker, "Retail Market Tests of Canned Minced Fish," A. E. Res. 80–5, Department of Agricultural Economics, Cornell University, Ithaca, N.Y., 1980.

37. R. C. Baker, J. M. Darfler, and E. J. Mulnix, "Canned Minced Fish," *Development of Products from Minced Fish,* Booklet 8, New York Sea Grant Institute, Albany, N.Y., 1980.

38. L. Herborg and S. Johansen, "Fish Cheese: the Preservation of Minced Fish by Fermentation," in *Proceedings of the Conference on the Handling, Processing and Marketing of Tropical Fish, Food Science and Technology Abstracts* 10, 2R48 (1978).

39. R. C. Baker and D. S. Kline, "Seafood Quiche," *Development of Products from Underutilized Species of Fish,* Booklet 12, New York Sea Grant Institute, Albany, N.Y., 1983.

40. T. Unemoto, M. Hayashi, K. Miyaki, and M. Hayashi, "Intracellular Localization and Properties of Trimethylamine-N-Oxide Reductase in *Vibrio parahaemolyticus,*" *Biochimica et Biophysical Acta* 110, 319–328 (1965).

41. T. Tokunaga, "Trimethylamine Oxide and Its Decomposition in the Bloody Muscle of Fish II. Production of DMA and TMA During Storage," *Bull. Japan. Soc. Sci. Fish.* 36, 510–515 (1970).

42. Z. Sikorski and S. Kostuch, "Trimethylamine-N-Oxide Demethylase: Its Occurrence, Properties and Role in Technological Changes in Frozen Fish," *Food Chemistry* 9, 213–222 (1982).

43. K. Amano, K. Amada, and M. Bito, "Detection of Formaldehyde in Gadoid Fish," *Bull. Japan. Soc. Sci. Fish.* 29, 695–701 (1963).

44. K. Tomioka, J. Ogushi, and K. Endo, "Studies on Dimethylamine in Foods—II. Enzymatic Formation of Dimethylamine from Trimethylamine Oxide," *Bull. Japan. Soc. Sci. Fish.* 40, 1021–1026 (1974).

45. R. C. Lundstrom, F. F. Correia, and K. A. Welheim, "Dimethylamine in Fresh Red Hake (*Urophycis chuss*): the Effect of Packaging Material Oxygen Permeability and Cellular Damage," *Journal Food Biochemistry* 6, 229–241 (1981).

46. P. Reece, "The Role of Oxygen in the Production of Formaldehyde in Frozen Minced Cod Muscle," *Journal of the Science of Food and Agriculture* 34, 1108–1112 (1983).

47. K. Harada, "Studies on Enzymes Catalyzing the Formation of Formaldehyde and Dimethylamine in Fish and Shellfish," *J. Shim. Univ. Fish.* 23(3), 157–163 (1975).

48. K. Yamada, K. Harada, and K. Amano, "Biological Formation of Formaldehyde and Dimethylamine in Fish and Shellfish. VIII. Requirement of a Cofactor in the Enzyme System," *Bull. Japan. Soc. Sci. Fish.* 35, 227–231 (1969).

49. K. Harada and K. Yamada, "Some Properties of a Formaldehyde and Dimethylamine-Forming Enzyme Obtained from *Barbatia virescens,*" *Suisan Daigakko Kenkyu Hokoku* 19, 95–103 (1971).

50. D. J. B. da Ponte, J. P. Roozen, and W. Pilnik, "Effects of Additions on the Stability of Frozen Stored Minced Fillets of Whiting: I. Various Anionic Hydrocolloids," *Journal of Food Quality* 8, 51–68 (1985).

51. Y. J. Owusu-Ansah and H. O. Hultin, "Effect of In-Situ Formaldehyde Production on Solubility and Cross-Linking of Proteins of Minced Red Hake Muscle During Frozen Storage," *Journal of Food Biochemistry* 11, 17–39 (1987).

52. H. Rehbein, "Relevance of Trimethylamine Oxide Demethylase Activity and Haemoglobin Content to Formaldehyde Production and Texture Deterioration in Frozen Stored Minced Fish Muscle," *Journal of the Science of Food and Agriculture* 43, 261–276 (1988).

53. K. Ragnarsson and J. M. Regenstein, "Changes in Electrophoretic Patterns of Gadoid and Non-Gadoid Fish Muscle During Frozen Storage," *Journal of Food Science* 54, 819–823 (1989).

54. V. A. Krosing and K. A. Kask, "Proteolytic Activity of Baltic Spat Enzymes," *Tallinna Poleuteh. Inst. Toim.* (A) 331, 9–13 (1973).

55. C. S. Cheng, D. D. Hamann, and N. B. Webb, "Effect of Thermal Processing on Minced Fish Gel Texture," *Journal of Food Science* 44, 1080–1086 (1979).

56. P. Norberg and B. von Hofsten, "Chromatography of a Halophilic Enzyme on Hydroxyapatite in 3.4 M Sodium Chloride," *Biochemical et Biophysical Acta* 220, 132–133 (1970).

57. R. M. Love, "Ice Formation in Muscle," in J. Hawthorn and E. J. Rolfe, eds., *Low Temperature Biology of Foodstuffs,* Pergamon Press, Oxford, UK, 1968.

58. Y. Fukuda, Z. Tarakita, and K. Arai, "Effect of Freshness of Chub Mackerel on the Freeze-Denaturation of Myofibrillar Protein," *Bull. Japan. Soc. Sci. Fish.* 50, 845–852 (1984).

59. X. Xu and J. M. Regenstein, "Preventing Textural Changes in Frozen Gadoid Minces with Corn Starch and Sodium Bicarbonate," in R. E. Martin and R. L. Collette, eds., *Proceedings of the International Symposium on Engineering Seafood Including Surimi,* National Fisheries Institute, Washington, D.C., 1985, pp. 74–83.

60. C. W. Sweet, "Activity of Antioxidants in Fresh Fish," *Journal of Food Science* 38, 1260–1261 (1973).

61. J. C. Deng, R. F. Mathews, and C. M. Watts, "Effect of Chemical and Physical Treatments on Rancidity Development of Frozen Mullet Fillets," *Journal of Food Science* 42, 344–347 (1977).

62. J. J. Licciardello, E. M. Ravesi, and M. G. Allsup, "Stabilization of the Flavor of Frozen Minced Whiting: I. Effect of Various Antioxidants," *Mar. Fish Rev.* 44(8), 15–21 (1982).

63. D. G. Iredale and R. K. York, "Effects of Chemical Additives on Extending the Shelf Life of Frozen Minced Sucker Flesh," *J. Fish. Res. Bd. Can.* **34**, 420–425 (1977).
64. K. T. Hwang and J. M. Regenstein, "Protection of Menhaden Mince Lipids from Rancidity During Frozen Storage," *Journal of Food Science* **54**, 1120–1124 (1989).
65. R. M. Parkhust, W. A. Skinner, and P. A. Strum, "The Effect of Various Concentrations of Tocopherols and Tocopherol Mixtures on the Oxidative Stability of a Sample of Lard." *JAOCS* **45**, 641–642 (1968).
66. E. Bilinski and Y. C. Lau, "Lipolytic Activity Toward Long-Chain Triglycerides in Lateral Line Muscle of Rainbow Trout (*Salmo gairdneri*)," *J. Fish. Res. Bd. Can.* **26**, 1857–1866 (1969).
67. M. A. Audley, K. J. Shetty, and J. E. Kinsella, "Isolation and Properties of Phospholipase A from Pollock Muscle," *Journal of Food Science* **43**, 1771–1775 (1978).
68. M. Yurkowski and H. Brockerhoff, "Lysolecithinase of Cod Muscle," *J. Fish. Res. Bd. Can.* **22**, 643–652 (1965).
69. R. Hardy, "Fish Lipids. Part 2.," in J. J. Connell, ed., *Advances in Fish Science and Technology,* Fishing News Books, Farnham, Surrey, UK, 1980.
70. M. L. Anderson and E. M. Ravesi, "Reaction of Free Fatty Acids with Protein in Cod Muscle Frozen and Stored at −29°C after Aging in Ice," *J. Fish. Res. Bd. Can.* **26**, 2727–2736 (1969).
71. F. J. King, M. L. Anderson, and M. A. Steinberg, "Reaction of Cod Actomyosin with Linoleic and Linolenic Acids," *Journal of Food Science* **27**, 363–366 (1962).
72. W. J. Dyer, M. L. Morton, D. I. Fraser, and E. G. Bligh, "Storage of Frozen Rosefish Fillets," *J. Fish. Res. Bd. Can.* **13**, 569–579 (1956).
73. S. Y. K. Shenouda, "Theories of Protein Denaturation During Frozen Storage of Fish Flesh," *Advances in Food Research* **26**, 275–311 (1980).
74. J. Liston, "Microbiology in Fishery Science," in Ref. 69.
75. M. Raccach and R. Baker, "Microbial Properties of Mechanically Deboned fish flesh," *Journal of Food Science* **43**, 1675–1677 (1978).
76. R. Nickelson II, G. Finne, M. O. Hanna, and C. Vanderzant, "Minced Fish Flesh From Non-Traditional Gulf of Mexico Finfish Species: Bacteriology," *Journal of Food Science* **45**, 1321–1326 (1980).
77. C. A. Jauregui and R. C. Baker, "Discoloration Problems in Mechanically Deboned Fish," *Journal of Food Science* **45**, 1068–1069 (1980).
78. S. K. Gill, J. R. Dingle, B. Smith-Lau, and D. W. Stanley, "Improved Utilization of Fish Protein—Quality Enhancement of Mechanically Deboned Fish," *J. Can. Inst. Food Sci. Technol.* **12**, 200–202 (1979).
79. K. W. Young, S. L. Neuman, A. S. MacGill, and R. Hardy, "The Use of Dilute Solutions of Hydrogen Peroxide to Whiten Fish Flesh," in Ref. 69.
80. J. Kryznowek, D. Peton, and K. Wiggin, "Proximate Composition, Cholesterol, and Calcium Content in Mechanically Separated Fish Flesh from Three Species of the Gadidae Family," *Journal of Food Science* **49**, 1182–1185 (1984).
81. J. Gomez-Basauri and J. M. Regenstein, unpublished data.
82. D. L. Crawford, D. K. Law, and J. K. Babbitt, "Nutritional Characteristics of Marine Food Fish Carcass Waste and Machine-Separated Flesh," *Journal of Agricultural and Food Chemistry* **20**, 1048–1051 (1972).
83. V. D. Sidwell, "Nutritional Quality of Minced Fish Products," in Ref. 4, pp. 477–490.
84. L. D. Satterlee, G. W. Frowning, and D. M. Janky, "Influence of Skin Content on Composition of Mechanically Deboned Poultry Meat," *Journal of Food Science* **36**, 979 (1971).
85. N. B. Webb, E. R. Hardy, G. G. Giddings, and A. J. Howell, "Influence of Mechanical Separation upon Proximate Composition, Functional Properties and Textural Characteristics of Frozen Atlantic Croaker Muscle Tissue," *Journal of Food Science* **41**, 1277–1281 (1976).
86. A. A. Choy, W. W. Meinke, and K. F. Mattil, "Comparative Nutritive Value of Fillets and Minced Fish from Some Species of Fish," paper presented at the 35th Annual Meeting of the Institute of Food Technologists. Chicago, Ill., 1975.
87. F. R. Del Valle and J. T. R. Nickerson, "A Quick Salting Process for Fish 1. Evolution of the Process," *Food Technology* **22**(8), 104 (1968).
88. D. A. Silberstein and D. A. Lillard, "Factors Affecting the Autoxidation of Lipids in Mechanically Deboned Fish," *Journal of Food Science* **43**, 764–766 (1978).
89. G. W. Froning and F. Johnson, "Improving the Quality of Mechanically Deboned Fowl Meat by Centrifugation," *Journal of Food Science* **38**, 279–281 (1973).
90. R. A. Field, Y. Chang, and W. G. Kruggel, "Protein Quality of Mechanically Processed (Species) Product and Bone Residue. *Journal of Food Science* **44**, 690–692, 695 (1979).
91. R. A. Bello and G. M. Pigott, "Dried Fish Patties: Storage Stability and Economic Considerations," *J. Food Proc. & Pres.* **4**, 247–260 (1980).
92. M. E. Waters, "Chemical Composition and Frozen Storage Stability of Spot, *Leiostomus xanthurus*," *Mar. Fish. Rev.* **44**(11), 14–21 (1982).
93. W. W. Meinke, G. Finne, R. Nickelson II, and R. Martin, "Nutritive Value of Fillets and Minced Flesh from Alaska Pollock and Some Underutilized Finfish Species from the Gulf of Mexico," *Journal of Agricultural and Food Chemistry* **30**, 477–480 (1982).
94. W. W. Meinke, G. Finne, R. Nickelson, and R. Martin, "Composition, Nutritive Value, and Sensory Attributes of Fish Sticks Prepared from Minced Fish Flesh Fortified with Textured Soy Proteins," *Mar. Fish. Rev.* **45**(7-8-9), 34–37 (1983).
95. M. E. Waters, "Chemical Composition and Frozen Storage Stability of Weakfish, *Cynoscion regalis*," *Mar. Fish. Rev.* **45**(7-8-9), 27–33 (1983).
96. E. Miyauchi, D. Patashnik, and G. Kudo, "Frozen Storage Keeping Quality of Minced Black Rockfish (*Sebastes Spp.*) Improved by Cold-Water Washing and Use of Fish Binder," *Journal of Food Science* **40**, 592 (1975).
97. G. A. Adu, J. K. Babbitt, and D. L. Crawford, "Effect of Washing on the Nutritional and Quality Characteristics of Dried Minced Rockfish Flesh," *Journal of Food Science* **48**, 1053–1055, 1060 (1983).
98. J. A. Nettleton, in *Seafood Nutrition: Facts, Issues and Marketing of Nutrition in Fish and Shellfish,* Osprey Books. Huntington, N.Y., 1985, pp. 109–110.
99. U.S. Department of Agriculture, "Mechanically Separated Meat. Standards and Labelling Requirements," *Federal Register* **47**, 28214, 1982.
100. E. G. Bligh and R. Duclos-Rendell, "Chemical and Physical Characteristics of Lightly Salted Minced Cod (*Gadus Morhua*)," *Journal of Food Science* **51**, 76–78 (1986).
101. F. R. Del Valle, H. Bourges, R. Haas, and H. Gaona, "Proximate Analysis, Protein Quality and Microbial Counts of Quick-Salted, Freshly Made and Stored Fish Cakes," *Journal of Food Science* **41**, 975–976 (1976).

102. M. B. Wojtowiez, M. G. Fierheller, R. Legendre, and W. L. Regier, "A Technique for Salting Lean Minced Fish," *Fisheries and Marine Service Technical Report* No. **731,** Fisheries and Environment Canada, Ottawa, Ont., 1977.

Yu-Tsyr Li Hsieh
Yin Liang Hsieh
Keum Taek Hwang
Juan Gomez-Basauri
Joe M. Regenstein
Cornell University
Ithaca, New York

FLAVOR CHEMISTRY

OVERVIEW OF FLAVOR COMPOUNDS

Flavors

Flavors possess a variety of chemical groups and structures. They can be heterocyclic, carbocyclic, terpenoid, aromatic, etc. The overall flavor of foods is due to carbohydrates, lipids, and proteins; however, specific flavors can be elicited by numerous other classes of compounds, such as alcohols, aldehydes, ketones, and various heterocyclic compounds (pyrazines, pyrroles, pyridines, etc). Flavor components in food range in number from 50 to 250 compounds in fresh products such as fruits and vegetables and to more than double this number in foods subjected to heat or enzyme treatment; for example, more than 700 compounds have been reported in roast coffee aroma (1). The investigation of around 200 different food products has led to the identification of nearly 5,000 compounds (2,3), the vast majority of them was by gas chromatography/mass spectrometry (GC/MS) analysis.

Flavor sensation may be due to a single compound (4) or to a group of compounds. Single compounds are called flavor notes; examples are 4-hydroxy-3-methoxybenzaldehyde (vanillin) (Fig. 1), the character note for vanilla flavor, and 3-phenyl-2-propenal (cinnamaldehyde) (Fig. 2) for cinnamon flavor. On the other hand, a group of compounds representing a particular flavor is called the flavor profile of that food product (5). Although a large number of flavor compounds are usually identifiable in a flavor profile, in general only a small number have a significant effect on the overall flavor. These compounds are often called character impact compounds. An example is 2-isobutylthiazole (Fig. 3), the character impact compound found in tomato. The relative concentration of each component in a flavor profile is crucial for its imitation. Flavors can be biosynthesized naturally in foods, such as in apples, bananas, and other fruits and vegetable, or they can be produced from precursors during processing or thermal treatment such as baking, roasting, or frying. Furthermore, flavors can be generated by enzymatic modifications such as in cheese or by microbial fermentation as in butter.

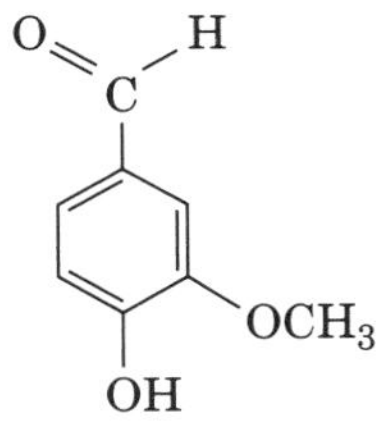

The origin of natural flavors and their precursor compounds is found in animal and plant tissue, so flavor research interfaces with other disciplines in its effort to extract and identify these compounds (6,7).

Flavor Enhancers

Monosodium glutamate (MSG), 5'-ribonucleotides such as 5'-inosine monophosphate (5'-IMP), and 2-methyl-3-hydroxy-4-(4H)-pyrone (maltol) (Fig. 4) are called flavor enhancers. The actual mechanism of flavor enhancement is not known. These substances contribute a delicious or umami taste to foods when used at levels in excess of their detection limits and enhance flavors of food at levels below their detection thresholds. Their effects are prominent and desirable in the flavors of vegetables, dairy products, meats, poultry, and fish.

Essential Oils

Essential oils are also known as volatile oils, or essences. When exposed to air they evaporate at room temperature. They are usually complex mixtures of a wide variety of organic compounds (such as hydrocarbons, terpenes, alcohols, ketones, phenols, aldehydes, and esters). Essential oils are obtained by steam distillation or solvent extraction from many odorous plant sources such as clove, cinnamon, orange, lemon, jasmine, and rose. The organic constituents of essential oils are synthesized by the plant during its normal growth.

Resins

Resins are natural products that either can be obtained directly from the plant as exudates or are prepared by alcohol extraction of plants that contain resinous materials. Naturally occurring resins are solids or semisolids at room temperature. They are soluble in alcohol or basic solutions but insoluble in water. They are usually noncrystalline and soften or melt on heating. Chemically, they are complex oxidation products of terpenes. They rarely occur in nature without being mixed with gums and/or volatile oils forming oleoresins.

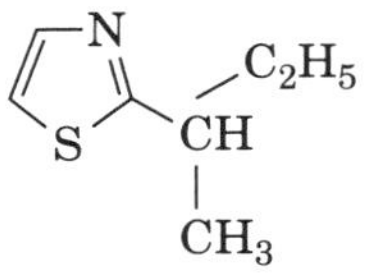

Figure 1. 4-Hydroxy-3-methoxybenzaldehyde (Vanillin).

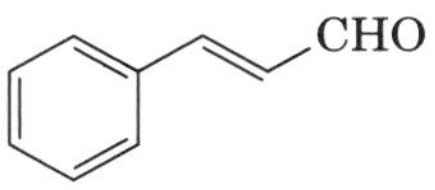

Figure 2. 3-Phenyl-2-propenal (Cinnamaldehyde).

Figure 3. 2-Isobutylthiazole.

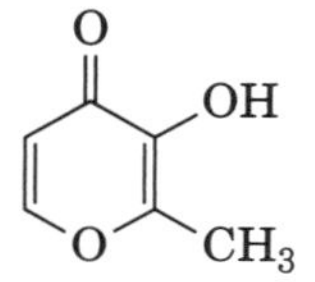

Figure 4. 2-Methyl-3-hydroxy-4(4H)-pyrone (Maltol).

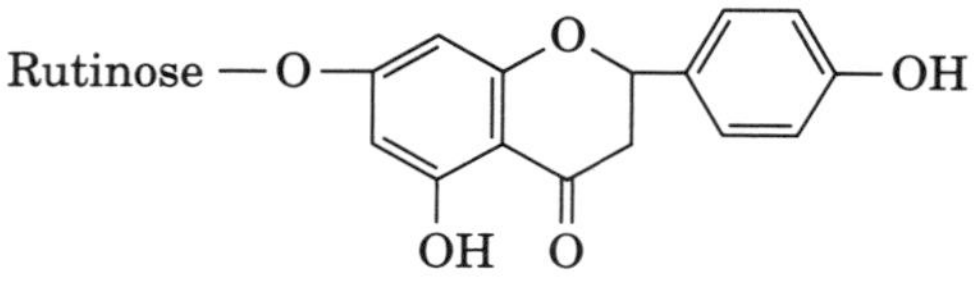

Figure 6. Naringin a bitter tasting glycoside.

SENSORY BASIS OF FLAVOR

Flavor can be defined as the combined perception of taste and smell. It involves receptors in both the oral and nasal cavities. In the oral cavity the taste buds are mainly distributed on the surface of the tongue. Each taste bud consists of a barrel-like structure in which the taste cells are packed. Taste receptors, which are found on the surface of the taste cells, are linked to the brain by way of cranial nerves, which carry the nerve impulses to the brain after the neurotransmitters are released from the taste cells. This process is initiated by the formation of the taste compound–receptor complex. Olfactory cells, on the other hand, are situated in the upper part of the nasal cavity. Their receptors perform a similar function to that of the taste cells through their own nerve fibers, which transmit neural impulses from receptors directly to the olfactory bulb in the brain. There is a common view that there are four primary taste qualities: sweet, sour, bitter, and salty, and seven primary odors: camphoraceous, musky, floral, peppermint, ethereal, pungent, and putrid (8).

Sweet Taste

Sweet taste is produced by several different classes of compounds (Fig. 5), including sugars, polyhydric alcohols, α amino acids, and synthetic sweeteners.

Sour Taste

Sour taste results from the presence of hydrogen ions on the tongue. However, sourness and acidity (pH) are not directly related, but there is some correlation. Two acids having the same pH do not produce the same degree of response.

Salty Taste

Salt taste is stimulated by most soluble salts having low molecular weights.

Bitter Taste

Three major classes of organic compounds encountered in food materials are associated with bitterness: alkaloids, glycosides (Fig. 6), and peptides.

Astringency

A taste-related phenomena perceived as a dry feeling in the mouth along with a coarse puckering of the oral tissue, astringency is due to tannins or polyphenols interacting with the proteins in the saliva to form precipitates or aggregates. Astringency may be a desirable flavor property, such as in tea and red wine.

Pungency

Certain compounds (Fig. 7) found in several spices and vegetables cause characteristic hot, sharp, burning, and stinging sensations that are known collectively as pungency.

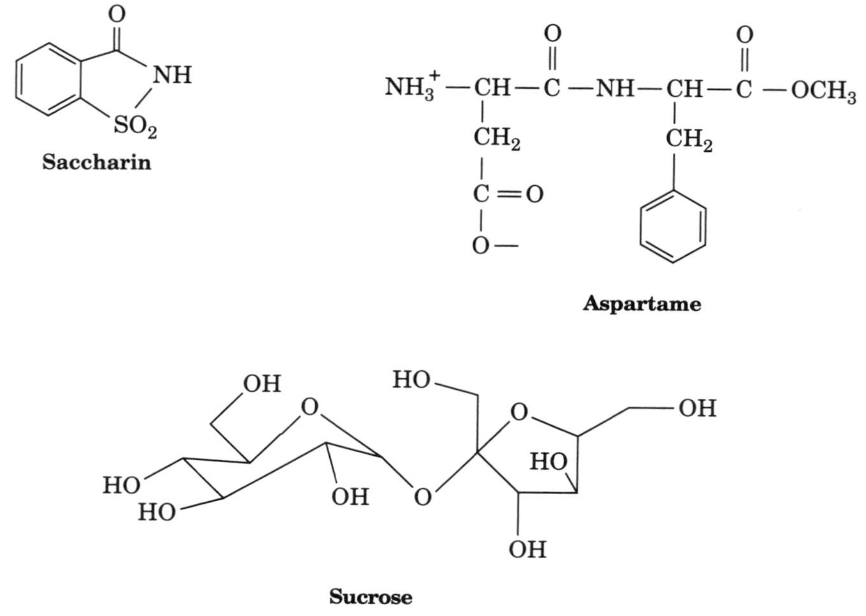

Figure 5. Examples of sweet tasting compounds (**a**) saccharin; (**b**) aspartame; (**c**) sucrose.

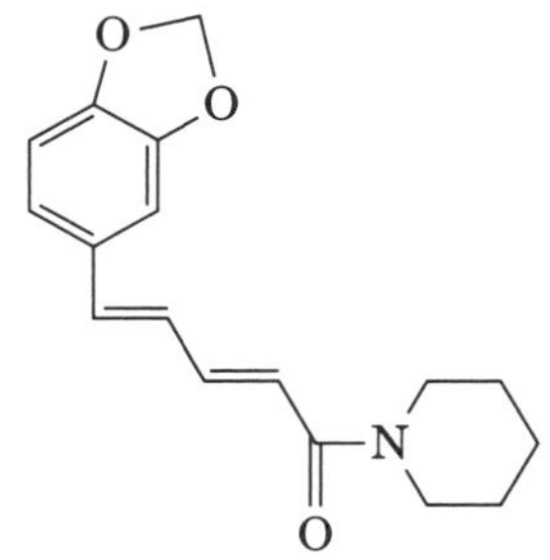

Piperine
Component responsible
for the "hot" sensation of pepper

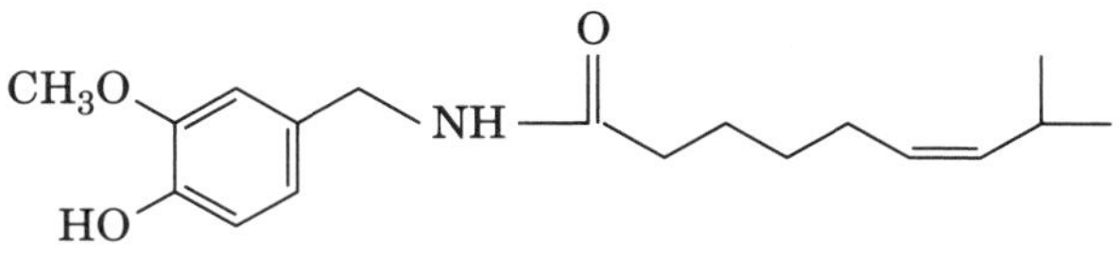

Capsaicin
The pungent principle of
red pepper

Figure 7. Examples of compounds causing pungency (**a**) piperine-component responsible for the hot sensation of pepper; (**b**) the pungent principle of red pepper.

Cooling Effect

Cooling sensations occur when certain chemicals (Fig. 8) contact the nasal or oral tissues and stimulate a specific receptor. This effect is most commonly associated with mintlike flavors, including peppermint, spearmint, and wintergreen.

CLASSIFICATION AND ORIGIN OF FLAVOR COMPOUNDS

Classification of Flavor Compounds Based on Their Mode of Formation: Biogenetic and Thermogenetic

One way of classifying flavor compounds is based on their mode of formation; flavors can be generated either naturally (9) or by heat treatment during food processing (10). Natural flavors are mainly the secondary metabolites of the living tissue, formed during the natural growth cycle of the organism by the action of enzymes, whereas flavors produced by heat treatment are the result of thermal degradation and oxidation of various food ingredients and their complex interactions. The initial or primary precursors of flavors are the polymers found in food such as proteins, lipids, polysaccharides, and DNA, which can undergo either enzymatic or thermal hydrolysis to produce the intermediate precursors, mainly dimers and monomers. These, in turn, can undergo different biotransformation reactions during normal metabolic growth and produce metabolites that have specific flavor qualities that remain in situ when the plant is harvested and produce the perceived flavor effect when consumed. On the other hand, during thermal processing of food products, the intermediate precursors formed by thermal hydrolysis, can undergo complex chemical transformations, and further degradations, to generate cooked flavor (Fig. 9). The type of flavor effect depends on the conditions of processing and the type of the initial precursors found in the particular food product. In terms of composition, thermally generated flavors are richer in heterocyclic compounds compared to enzymatically produced flavors.

Precursors of Flavor Compounds in Food

Meat and Related Products. Raw meat has no particular appealing flavor. However, during various processes of cooking, the initial precursors in meat produce water-soluble intermediate precursors, such as glycopeptides, free nucleotides, peptides, amino acids, amino sugars, free sugars, and fatty acids. The meat flavor, therefore, is the combined effect of the chemicals produced from the thermal degradation of these intermediates and the products formed from the amino–carbonyl interactions, such as between the reducing sugars and amino acids; this reaction is known as the Maillard reaction or nonenzymatic browning as it is responsible also for the brown color produced when foods are heated (10,11). During thermal degradation amino acids and sugars produce complex mixtures of compounds. Sulfur-containing amino acids are specially important for the generation of meat aroma; cysteine, cystine, and methionine produce sulfur-containing small molecules, such as hydrogen sulfide, 3-(methylthio)-propionaldehyde, and 2-mercaptoethylamine, that play an

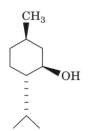

Figure 8. (-)-Menthol causes a cooling sensation.

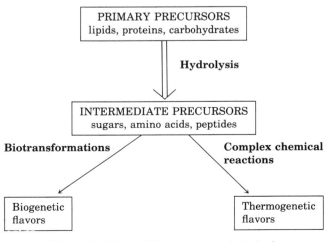

Figure 9. Origin of flavor compounds in food.

2-Methyl-3-(Methylthio)furan

2-Methyl-1,3-dithiolane

1,2,3-Trithiacylohept-5-ene

Thiazole

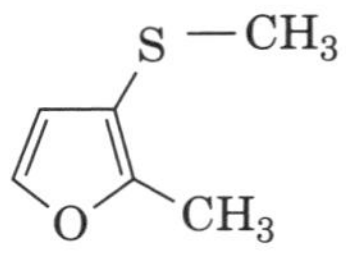

Figure 10. Aroma compounds identified in cooked beef (**a**) 2-Methyl-3-(methylthio)furan; (**b**) 2-Methyl-1,3-dithiolane; (**c**) 1,2,3-trithiacylohept-5-ene; (**d**) Thiazole.

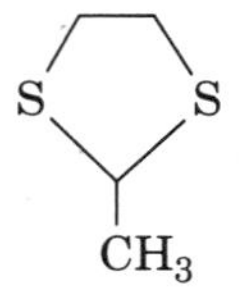

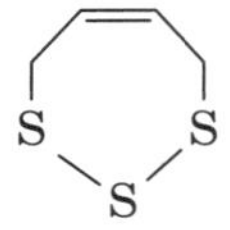

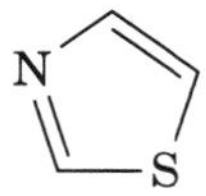

important role in the aroma of meat products (Fig. 10). Thermal degradation of thiamine produces furans, thiophenes, and thiazoles, important heterocyclic compounds with meatlike flavor. Thermal oxidation of unsaturated fatty acids leads to the formation of different aldehydes.

Nonalcoholic Beverages: Tea and Coffee. Tea and coffee are examples of food products in which both modes of flavor formation, thermal and enzymatic, play a role in the generation of flavor. During the initial fermentation of green tea leaves, for example, important intermediate precursors are produced by the action of endogenous enzymes to produce polyphenols, carotenoids, and unsaturated fatty acids, such as lenolenic acid, which under the action of lipoxygenase enzymes can produce important flavor aldehydes such as (*Z*)-3-hexenal and (*E*)-2-hexenal. In the heat-treatment stage tea leaves are dried at 85°C and coffee beans are roasted at > 180°C to produce numerous volatile aroma compounds (Fig. 11) by thermal degradation and Maillard reaction processes. The main intermediate precursors involved during roasting of coffee are sugars, amino acids, fatty acids, peptides, amines, phenolic acids such as 3-(3′,4′-dihydroxyphenyl)propenoic acid (caffeic acid), and 1-methylpyridinium-3-carboxylate (trigonelline). In addition, tea and coffee contain alkaloids such as 1,3,7-trimethylxanthine (caffeine) and 3,7-dimethylxanthine (theobromine) that impart a bitter taste to the beverage.

2-Methylfuran

2-Ethyl-6-Methylpyrazine

Pyridine

2-Pentanone

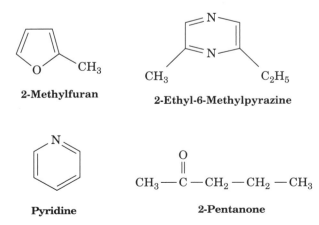

Figure 11. Some volatiles identified in coffee aroma (**a**) 2-Methylfuran; (**b**) 2-Ethyl-6-methylpyrazine; (**c**) Pyridine; (**d**) 2-Pentanone.

Alcoholic Beverages: Wine. There are three main sources of flavor compounds in wine: some are already present in the grape, some are formed during the fermentation process, and some may form during aging by Maillard-type reactions. The type of wood in which the wine is stored can also contribute to the overall flavor of the wine. Most of the 300 or so flavor compounds identified in different wines, such as 2-methyl-1-propanol, 1-hexanol, 2-phenylethanol, ethyl acetate, and ethyl lactate, are formed during fermentation. The natural aroma of some grape varieties results from a mixture of different terpenes such as (*E*)-3,7-dimethyl-2,6-octadien-1-ol (geraniol), 3,7-dimethyl-1,6-octadien-3-ol (linalool), (*Z*)-3,7-dimethyl-2,6-octadien-1-ol (nerol) (Fig. 12). Sugars such as arabinose, xylose, mannose, and rhamnose produce furans during the aging process by reacting with the amino acids such as proline, alanine, and asparagine by the Maillard reaction (12,13).

Fruits and Vegetables: Tomato. Generally, the aroma of all fresh fruits and vegetables are produced by the action of enzymes on the hydrolysis products of proteins, lipids, and carbohydrates. Free fatty acids in tomato and other vegetables are generated by the action of phospholipases. Then, lipoxygenase enzymes convert the unsaturated fatty acids into important flavor aldehydes such as (Z)-3-hexenal, hexanal, and alcohols such as 3-hexenol. Amino acids can also be converted into aldehydes and alcohols by the action of enzymes, for example, leucine is converted into 3-methyl-butanol and 3-methyl-butanol; carotenoids in tomato can be converted into flavor ketones such as 2,6-dimethyl-2,6-undecadien-10-one.

Milk and Dairy Products. More than 200 aroma compounds have been identified in differently processed milk. Some of these compounds have their origin in the raw milk, and some of them are formed during processing due to the degradation of the main precursors found in the raw milk such as fats, carbohydrates (mainly lactose), and proteins. These precursors undergo two types of transformations: (1) chemical, such as oxidations, and (2) biochemical with the participation of the enzymes and the microbial flora contained in the milk or by exogenous bacteria and

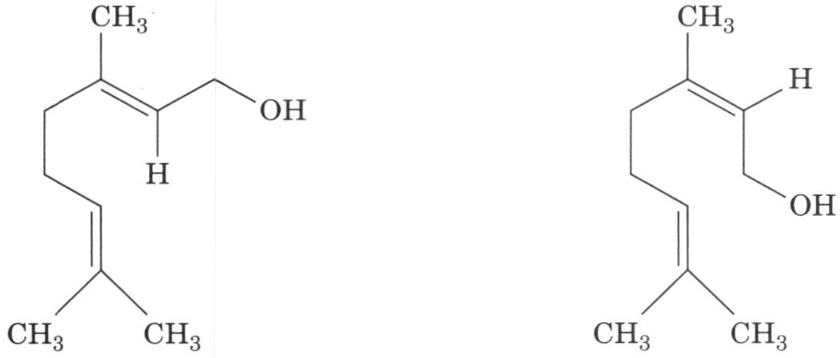

(E)-3,7-Dimethyl-2,6-octadiene-1-ol (Gereniol)

(Z)-3,7-Dimethyl-2,6-octadiene-1-ol (Nerol)

Figure 12. Terpenoids in wine (**a**) (E)-3,7-Dimethyl-2,6-octadiene-1-ol (Gereniol); (**b**) (Z)-3,7-Dimethyl-2,6-octadiene-1-ol (Nerol).

microbial flora, such as during preparation of cheese and yoghurt. The most important precursor responsible for the production of milk aroma is the lipid. Milk triglycerides are hydrolyzed by bacterial lipases to produce free fatty acids, which in turn are oxidized into different carbonyl compounds, among them many methyl ketones responsible for the milk aroma. Other important carbonyl compounds include, oct-1-en-3-one (responsible for the metallic aroma), (Z)-4-heptenal (characteristic odor of cream), and γ and δ lactones formed by the thermal oxidation of fatty acids (14).

OCCURRENCE AND ORGANOLEPTIC PROPERTIES OF HETEROCYCLIC COMPOUNDS IN FOOD

Although heterocyclic compounds are present only in minute amounts in foods, they constitute the most important character impact compounds because of their low threshold values. They are extremely important as compounding ingredients and in the development of new flavors. Almost half the 5,000 flavor compounds identified till now are heterocyclic in nature (15).

Furans

Furans are mainly associated with caramel-like, sweet, fruity, nutty, meaty, and burnt odor impressions. Because of their olfactory properties, many furans are commercially important flavoring chemicals. Furans are formed in food by thermal degradation of carbohydrates and by the Maillard reaction. They are present in nearly all food aromas and essential oils. The most abundant furans are 2- and 2,5-disubstituted, such as 2-methylfuran, which is found in coffee aroma. The 2-substituted furans with aldehyde, ketone, or alcohol functional groups generally have fruity aromas with the mild flavor of caramel when added in small amounts to nonalcoholic beverages and ice creams. 3-Acetyl-2,5-dimethylfuran is used in imitation nut flavors (Fig. 13).

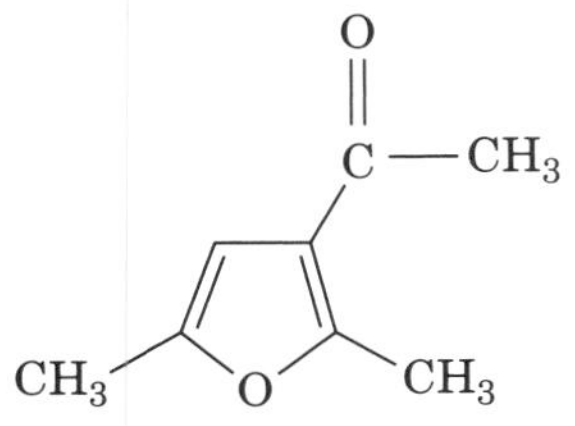

Thiophenes

Thiophenes can significantly contribute to the sensory properties of foods, but they are not as numerous as furans. They have been detected in boiled and canned beef, cooked chicken, asparagus, leeks, roasted peanuts, popcorn, rice, bread, and coffee. Of the 46 sulfur-containing compounds identified in pressure-cooked lean ground beef, 20 were thiophenes. The most frequently found thiophenes are the parent compound itself, 2-alkylated, 2- and 3-acylated derivatives. They are described as being pungent and green-sweet. 2-Acetyl-3-methylthiophene (Fig. 14) is described in a patent as imparting a honey-like flavor to syrup bases, and 3-acetyl-2,5-dimethylthiophene is recommended in another patent for improving the aroma of tobacco and perfumes.

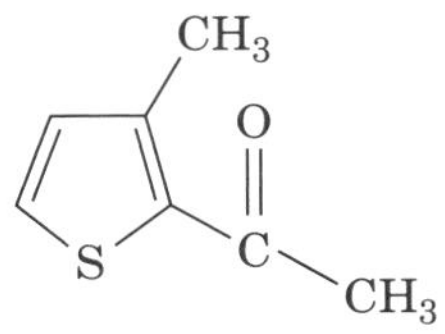

Pyrroles

Pyrroles are among the most widespread heterocyclic compounds found in food flavors such as coffee, roasted peanuts, popcorn, and tobacco smoke. A few pyrroles have also been identified in cooked beef, cooked asparagus, leeks, and bread. They are mainly formed via the Maillard reaction during cooking. N- and 2-substituted pyrroles are more common. They impart a burnt character note to the food product. 1-Pyrroline significantly enhances butter flavor of manufactured margarine (Fig. 15).

Thiazoles

Thiazoles possess extraordinary potent sensory properties that can be described as green, roasted, or nutty. They play an important role in the flavors of meat products, vegetables, passion fruit, roasted products, milk, coffee,

Figure 13. 3-Acetyl-2,5-dimethylfuran.

Figure 14. 2-Acetyl-3-methylthiophene.

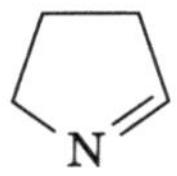

Figure 15. 1-Pyrroline.

rum, and whiskey. Thiazoles can be formed in food products either naturally or through the Maillard reaction. 2-Isobutylthiazole (Fig. 3), the characteristic aroma of fresh tomato, when added to canned tomato can produce a more intense flavor. 2,4,5-Trisubstituted thiazoles usually have roasted and meaty flavor characteristics. 2-Hydroxymethyl-4-methylthiazole is a patented flavor compound to produce woody and burnt flavor notes.

Pyrazines

Pyrazines represent 4% of all aroma compounds used as flavoring agents. They are the most widely distributed heterocyclic compound in food flavors, found in more than 50 food products of vegetable and animal origin whether processed or not. Pyrazines are described as having nutty, roasted, green, and fruity flavors. They principally occur in cooked beef, potato, mushroom, roasted nuts, bread, cheese, coffee, and some alcoholic beverages. 2-Alkyl-3-methoxypyrazines are widespread in vegetables; the characteristic aroma of green pepper is due to 2-i-butyl-3-methoxypyrazine (Fig. 16). 2-Methoxy-3-ethylpyrazine imparts a potato flavor to food products. 2,5-Dimethylpyrazine has the flavor of fried chicken, whereas 2-ethoxy-6-methylpyrazine gives the strong aroma and flavor of fresh pineapple.

MAILLARD REACTION AND FORMATION OF HETEROCYCLIC FLAVOR COMPOUNDS IN FOOD

Maillard Reaction

Flavors can be produced either naturally by the action of enzymes or by thermal processing due to the interaction of different food components or their pyrolytic degradation. The interaction of reducing sugars with amino-containing components and their subsequent reactions is termed nonenzymatic browning or Maillard reaction (16–19). This reaction is considered to be the most important reaction in food chemistry because, in addition to flavors, it is also responsible for the formation of color, antioxidants, carcinogens, etc. It can also reduce the nutritional value of foods by effectively decreasing the concentration of essential amino acids. The Maillard reaction results in the formation of distinctive brown color and aroma of broiled, baked, and roasted food products. Consequently, many unpleasant tasting raw foods can be transformed by Maillard reactions into desirable products via processes such as bread baking, coffee roasting, and chocolate manufacture. The Maillard reaction is initiated by the interaction of the open chain form of the reducing sugars with amino acids resulting in the formation of a Schiff base that exists in equilibrium with glycosylamino acid. The equilibrium constant for this reaction is unfavorable, but the glycosylamino acid slowly undergoes a rearrangement reaction to yield a relatively stable derivative. The type of derivative formed depends on the reducing sugar. Aldoses undergo Amadori rearrangement to produce 1-(amino acid)-1-deoxy-2-ketoses, whereas ketoses undergo Heyns rearrangement to produce 2-(amino acid)-2-deoxy-1-aldose. Both rearrangements are acid catalyzed; the carboxyl group of the amino acids provides the internal acid catalyst. The net result of this reaction is the transformation of an aldose into a ketose and *vice versa via* the formation of N-glycosides (Fig. 17). This initial stage of the Maillard reaction is well documented and understood; however, the subsequent reactions of these rearrangement products that produce flavors and colors are not very well defined. The rearrangement products themselves are colorless, nonvolatile compounds that do not impart to the food products any specific flavor qualities. However, during the thermal processing of foods, they decompose to produce different reactive intermediates that interact further with other food components to produce a wide range of heterocyclic compounds. Among others, these heterocyclic compounds possess potent flavor qualities at very low concentrations that makes them important contributors to the flavor of baked, roasted, and cooked food. Eventually, some of the intermediates formed will polymerize to form brown-colored melanoidins that are characteristic of cooked food products.

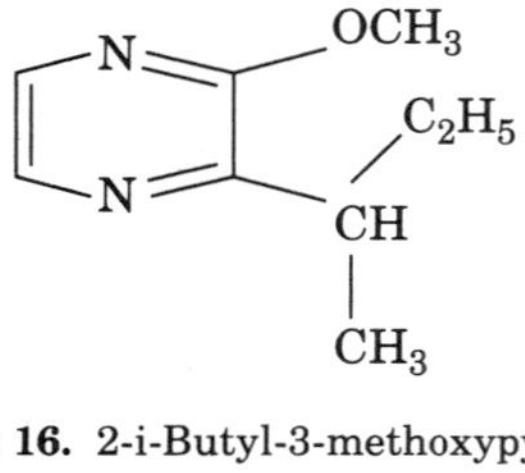

Further Reactions of Amadori and Heyns Rearrangement Products

The mechanism of decomposition of Amadori and Heyns rearrangement products (AHRP) to produce heterocyclic flavor compounds is not well understood owing to the complexity of the food matrix. However, based on the chemistry of carbohydrates, some decomposition pathways have been proposed that involve 1,2- and 2,3-enolizations of the open chain form of the AHRPs followed by β-eliminations to produce 1,2- and 2,3-dicarbonyl compounds. These reactive intermediates can undergo many reactions, including dehydrations, cyclizations to produce furans, retroaldolizations, and further reactions with nitrogen and sulfur nucleophiles to produce N- and S-heterocyclic compounds, such as pyrazines by the Strecker degradation (16).

FLAVOR FORMATION BY ENZYMES AND MICROORGANISMS

In contrast to thermally produced flavors which comprise of a large number of heterocyclic compounds, flavors generated by enzymes in vegetables and fruits consist mainly of aldehydes, ketones, esters, alcohols, terpens, terpenoids and S-containing aliphatic and aromatic volatile com-

Figure 16. 2-i-Butyl-3-methoxypyrazine.

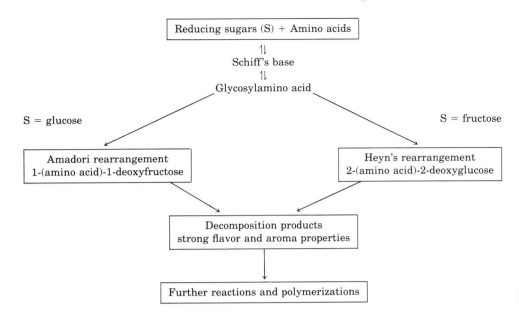

Figure 17. Maillard reaction.

pounds; they can be produced from non-volatile precursors by the action of different enzymes in the intact tissue before harvest or due to the disruption of the cell tissue so that the compartmentalized endogenous enzymes and the substrates interact. Alternatively, flavors can be produced by the action of exogenous enzymes or microorganisms during fermentation (20).

Origins of Enzymatically Produced Flavors in Vegetables

Aldehydes. Aldehydes containing less than five carbon atoms usually are associated with the development of off-flavors. However, unsaturated alkenals containing 5 to 10 carbon atoms impart desirable flavor attributes to fresh vegetables; cis- and trans-2-hexenals are both potent aroma compounds of green leaf character. The melon-like odor of cucumbers are due to cis-2-nonenal (Fig. 18). Aliphatic aldehydes may arise and accumulate in vegetables from precursors such as esters, amino acids, and free fatty acids. Esters can be hydrolyzed by esterases and followed by enzymatic oxidation (NADP-dependent dehydrogenases) to produce aldehydes. Amino acids can be converted into α-keto acids by the action of transaminases, followed by enzymatic decarboxylation to produce flavor aldehydes. Triglycerides in plant tissue can be hydrolyzed by lipases to produce free fatty acids, which in turn can produce long chain aldehydes by the action of lipoxygenases and lyases.

Sulfur Volatiles. Sulfur-containing volatiles are produced mainly by alliaceous (onion, garlic, leek, etc.) and cruciferous (cabbage, mustard, broccoli, horseradish, etc.) plants. Alliaceous vegetables produce disulfides and related volatiles, whereas cruciferous vegetables generate isothiocyanates. The main precursor of disulfides in alliaceous plants are S-substituted L-cysteine sulfoxides, which on the action of allinases produce thiosulfinates (Fig. 19), the principle component responsible for the odor of fresh alliaceous plants. On heating, the thiosulfinates are converted into corresponding disulfides, which are responsible for the cooked flavor. The pungent taste of cruciferous vegetables, on the other hand, are caused by isothiocyanates (Fig. 20). The enzymes responsible for the release of these compounds are thioglucosidases, which act on different thioglucosides found in these vegetables to produce the corresponding isothiocyanates. The enzymes and the substrates in these vegetables are compartmentalized; the tissue of the plant should be disrupted to bring the substrate close to the enzyme for the generation of the flavor compounds.

Origins of Enzymatically Produced Flavors in Fruits

About 2,000 distinct fruit volatiles have been isolated till now. They consist mainly of esters, aromatic aldehydes,

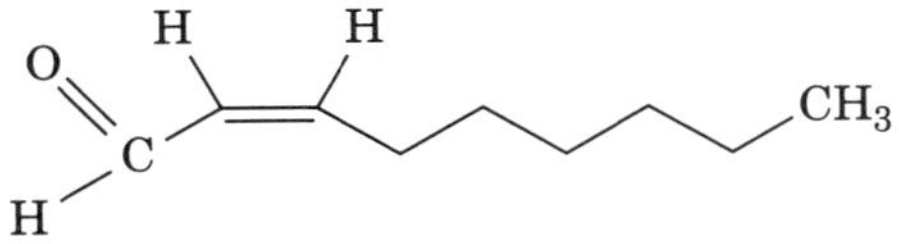

Figure 18. cis-2-Nonenal.

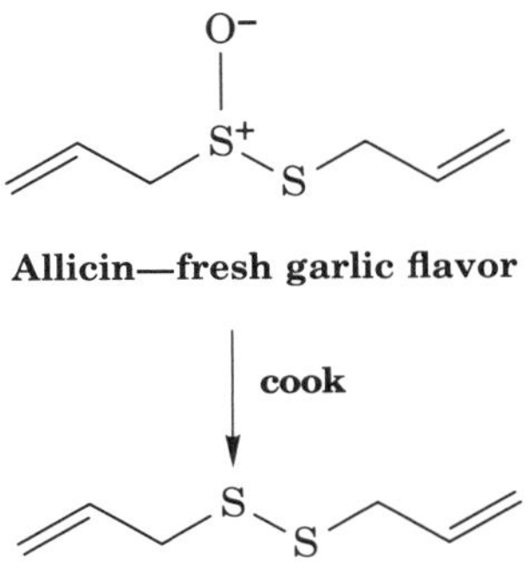

Figure 19. Components responsible for the garlic flavor.

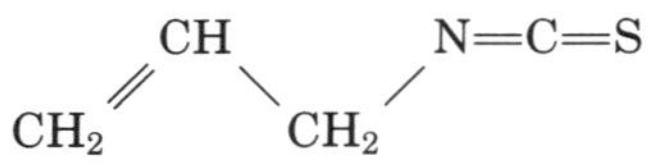

CH₂=CH−CH₂−N=C=S

Figure 20. Allylisothiocyanate, principle hot component in mustard.

lactones, alcohols, terpenoids, and some thioesters. Unlike fresh vegetables, the aroma is more frequently preformed and arises directly from the intact fruit.

Esters and Aromatic Aldehydes. Esters make the most important contribution to what we usually perceive as fruit flavors. They are present in higher levels than any other class of compound and also they are present in a greater variety compared to any other class. Most of the esters contribute to the character impact of the resultant aroma. Certain esters have been associated with the aroma of specific fruits, such as methyl butyrate with apple, isopentyl acetate with banana, and ethyl butyrate with orange (Fig. 21). Esters are biosynthesized by enzymatic esterification of carboxylic acids with alcohols through the action of such enzymes as esterases and acyl CoA-alcohol transacetylase. In general, the most abundant alcohol in the volatile fraction of a fruit is usually present in the most abundant ester. Aromatic aldehydes sometimes constitute the main flavor component of certain fruits. Benzaldehyde (Fig. 22), for example, is the principle flavor compound in bitter almond. It is also present in the volatiles of peaches, apricots, cherries, and plums. They are generally produced from cyanogenic glycosides by the action of β-glucosidases, which hydrolyzes the glycoside into the free sugar and (R)-mandelonitrile. The latter then undergoes an elimination reaction in the presence of hydroxynitrile lyase to produce the benzaldehyde. 4-Hydroxy-3-methoxybenzaldehyde (vanillin), another aromatic aldehyde, is also produced by the same series of enzymes.

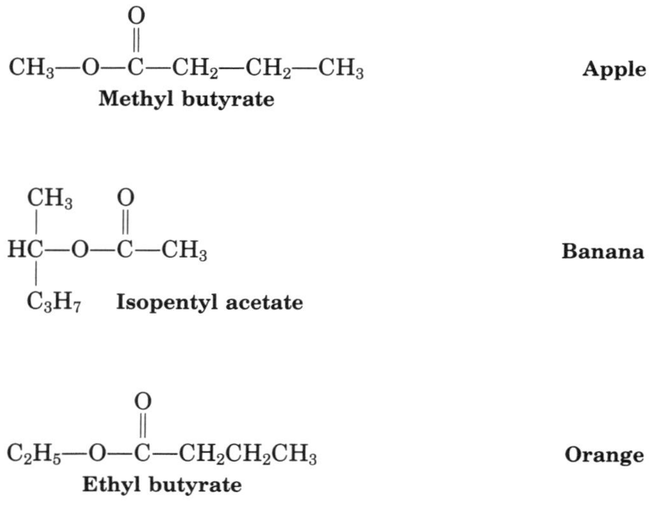

$$CH_3-O-\overset{O}{\underset{\parallel}{C}}-CH_2-CH_2-CH_3 \qquad \text{Apple}$$
Methyl butyrate

$$\underset{C_3H_7}{\overset{CH_3}{\underset{|}{HC}}}-O-\overset{O}{\underset{\parallel}{C}}-CH_3 \qquad \text{Banana}$$
Isopentyl acetate

$$C_2H_5-O-\overset{O}{\underset{\parallel}{C}}-CH_2CH_2CH_3 \qquad \text{Orange}$$
Ethyl butyrate

Figure 21. Esters with specific fruit flavor (**a**) Methylbutyrate (apple); (**b**) Isopentylacetate (banana); (**c**) Ethylbutyrate (orange).

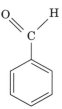

Figure 22. Benzaldehyde.

Origins of Enzymatically Produced Flavors in Dairy Products: Milk and Cheese

In contrast to vegetables and fruits, in which the endogenous enzymes are responsible for the release of aroma compounds, dairy flavors are generated by, in addition to endogenous enzymes, added microbial enzymes that play a predominant role in the formation of flavors. In general, the flavor of dairy products originates from microbial, enzymatic, and chemical transformations (mostly oxidations)—the relative importance of which is not always understood. These transformations give rise to a series of volatile and nonvolatile compounds, some of which have been shown to correlate well with some typical dairy flavor notes and some have little indication of their real contribution. For any dairy aroma, the total number of compounds identified is far less than in those foods subjected to thermal treatment. Thousands of heterocyclic compounds present in such foods are missing in fermented ones. The characteristic dairy food flavors are usually related to the occurrence of relatively few key components.

Milk furnishes the flavor precursors for all the fermented products (cheese, yogurt, etc.) made from it. The milk lactose, in addition to being a nutrient for the growth of the starter microorganisms, is also the principle precursor of 2,3-butadione (diacetyl), an important constituent of cultured products. It imparts the characteristic creamy note to most cheeses and other dairy products. It is found in butter, unripened soft cheeses such as cottage cheese, and also in Swiss cheese. Milk proteins (caseins) are the source of bitter peptides that contribute to the aroma of different cheeses. The sweet flavor of Swiss cheese is due to a complex formed between calcium (or magnesium) and peptides. The burned note of Gruyère cheese is attributed to peptides complexed with 2,5-dimethyl-4-hydroxy-3-[2H]furanone; these peptides and amino acids are produced during the ripening of cheese by the action of proteolytic enzymes on the milk caseines. Some amino acids can be oxidized into corresponding carboxylic acids, such as alanine, and into propionic acid, which is responsible in part for the aroma and taste of Swiss cheese. Although carbohydrates and proteins play an important role as precursors of cheese aroma, fatty acids are the key precursors for both desirable and undesirable aroma. C4–C10 fatty acids impart well-known pungent notes to the cheese whereas 2-methylketones, lactones, and dairy-associated aldehydes such as cis-4-heptenal, others, are important precursors to other flavors. In addition, they act as a medium for the action of enzymes responsible for the formation of aroma and they serve as a repository for all the other aroma compounds. As such, they determine the relative composition of the aroma vapor above the cheese.

ANALYSIS OF FLAVOR COMPOUNDS

The analysis of food flavors is complicated by several factors. The flavor-active compounds are found in very low concentrations, ranging from few 100 ppm for strongly flavored food products to less than 10 ppm for weakly flavored foods. In addition, they represent a wide range of functional groups with widely differing physical and chemical properties. Moreover, some of the components are thermally labile whereas others are highly reactive and may be lost if great care is not exercised during their isolation and concentration as well as in the subsequent analysis. A further complication that puts great demands on the analytic methodology is the fact that the trace components, even when present at levels of ppm or less, may sometimes make a greater contribution to the flavor than components present in vastly greater amounts. The economic importance of such low threshold value flavor compounds lies in the reduced cost of production of highly flavored food products. Despite impressive data available on the chemical composition of food flavors, there is still a lack of complete understanding of what determines the flavor of many foods. This may be due—for some products at least—to the failure of the analytic techniques of the past, including gas chromatography (GC) and mass spectrometry (MS), to register and identify important trace components. However, major advances in both GC and MS technology—such as the introduction of inert thermostable capillary columns; gentle, nonvaporizing on-column injection methods; improved interfaces between GC and MS; sophisticated facilities for spectral acquisition, storage, data manipulation, and library search—make modern GC/MS an analytic tool of outstanding powers, capable of acquisition of useful mass spectral data from the narrowest of GC peaks.

Establishing the chemical composition and the structure of flavor compounds found in food is important for correlating structures to sensory properties and for understanding the mechanisms by which flavors are formed from their precursors. Although analysis of food flavors is similar in many respects to analysis of any other mixtures, there are special problems associated with it, such as the low concentration of flavor components and their extreme complexity, frequently containing several hundred components of different functional groups. In some cases, trace components in an aroma have far greater sensory importance than other components present in larger amounts; hence, sample preparation and concentration is specially important for analysis of flavor compounds.

Sample Preparation

There are two different approaches to preparing samples for flavor analysis. The first approach is total volatile analysis, which attempts to isolate and concentrate from target food products all the chemicals that could possibly contribute to the flavor. This can be achieved by different distillation and extraction methods. The most appropriate method is largely dictated by the nature of the product under study and by the extant to which its flavors are affected adversely by heat. Fresh fruits and other heat-sensitive products, for example, can be distilled at low temperatures under vacuum, whereas, heat-stable products can be steam distilled and simultaneously extracted into an organic solvent by using Likens and Nickerson type apparatus. The second approach is headspace analysis which aims at analysis of volatile compounds found at equilibrium in the vapor phase above the food. The second approach is faster and simpler and requires small quantities of the food product for analysis. In addition the headspace contains volatiles in the same relative concentrations as one actually inhales. The headspace volatiles can be analyzed by injecting a small volume of the volatiles directly into a gas chromatograph or after concentration by cryogenic trapping or by the use of porous polymers such as Porapak and Chromosorb or by activated charcoal (21).

Chemical Methods of Analysis

The identification of the flavor components present in a sample can be established rapidly by GC/MS analysis. However, in complex mixtures, important trace components are often either poorly separated from or totally masked by other components. Consequently, interpretation of their mass spectra may be difficult to perform. To overcome this problem, the sample can be modified before, during, or after GC/MS analysis to simplify its complexity. Modification of the sample before GC/MS analysis can be achieved by preliminary fractionation of its components into polar, nonpolar, and weakly polar fractions or into acidic, basic, and neutral fractions by column chromatography using silica gel. Carbonyl compounds as a group are of special significance as they occur in a wide range of food products of plant and animal origin. As a group they can be isolated by treating the sample with an acidified solution of 2,4-dinitrophenylhydrazine and analyzed as their 2,4-dinitrophenylhydrazone derivatives by GC, high performance liquid chromatography (HPLC), TLC, etc. Modification of the sample injected onto a GC/MS can be performed on-line by means of chemical abstractors deposited on deactivated solid supports contained in a coiled tubing and placed before the analytic column of the GC. The function of the abstractors is to remove components bearing specific functional groups by reacting with them. The absence of a particular peak is an indication of the presence of the functional group when the resulting chromatograms are compared with those obtained in the absence of the abstractor. Alternatively, to achieve greater resolution for incompletely separated peaks, components of each chromatographic peak can be recovered by different trapping techniques and analyzed again using different columns. Trapping techniques may be very simple, such as collection of peaks in cooled glass capillary tubes inserted into the GC column or by use of porous layer open tubular (PLOT) glass capillaries containing a layer of diatomacous earth support permanently fused to the wall of the capillary tube (21).

Gas Chromatography/Mass Spectrometry

Combined gas chromatography/mass spectrometry remains the most powerful technique available to chemist for the separation and identification of complex mixtures, because it generates the maximum amount of structural information for the smallest amount of sample in the shortest time. For more than 20 years it has been the mainstay technique in the analysis of flavor components in food and still is despite recent advances in complementary techniques such as GC/IR (22).

The separation of the complex mixture into its components is achieved on the gas chromatographic column, which can be divided into two types: packed and open tubular (or capillary). The separated components are then introduced through the GC/MS interface into the ionization chamber of the mass spectrometer, where the molecules are bombarded with a beam of energetic electrons (electron impact mode), causing them to ionize and fragment in a way that is characteristic of the molecule. The resulting mixture of ions are then separated on the basis of their mass/charge ratio (m/z), and their relative abundances are recorded. The results are then displayed as a plot of ion abundance versus m/z, which is called the mass spectrum. This represents the characteristic fingerprint of the molecule. The mass spectrum of an unknown compound then can be used to identify the unknown structure by comparison to other known mass spectra. This can be done conveniently by computers.

BIBLIOGRAPHY

1. S. Van Straten and F. De Vrijer, J. C. de Beauveser, eds., *List of Volatile Compounds in Food*. 3rd ed., suppl. 1–8, Central Institute for Nutrition and Food Research, TNO, Zeist, The Netherlands, 1977–1980.
2. S. Van Straten, H. Maarse, J. C. de Beauveser, and C. A. Visscher, eds., *Volatiles in Food*, Qualitative data, 5th ed., suppl. I, Division for Nutrition and Food Research, TNO, Zeist, The Netherlands, 1983.
3. H. Maarse, and C. A. Visscher, eds., *Volatile Compounds in Food*. Qualitative data, 5th edition, supplement I, Division for Nutrition and Food Research, TNO, The Netherlands, 1984.
4. P. Z. Bedoukian, *Perfumary and Flavoring Synthetics*, Allured Publishing Corporation, Wheaton, Ill., 1986.
5. L. B. Sjöström, *The Flavor Profile*, A.D. Little Inc., Cambridge, Mass., 1972.
6. H. B. Heath, *Source Book of Flavors*, AVI Publishing Co., Westport, Conn., 1981.
7. H. B. Heath and G. Reineccius, *Flavor Chemistry and Technology*, AVI Publishing Co., Westport, Conn., 1986.
8. G. G. Birch and M. G. Lindley, eds., *Developments in Food Flavors*, Elsevier Applied Science, London, 1986. p. 1.
9. T. H. Parliment and R. Croteau, eds., *Biogeneration of Aroma*, American Chemical Society (ACS symposium series, 317), Washington, DC, 1986.
10. T. H. Parliment, R. J. McGorrin, and C. T. Ho, eds., *Thermal Generation of Aromas*, American Chemical Society (ACS symposium series 409), Washington, DC, 1989.
11. W. G. Moody, Beef Flavor—A Review, *Food Technology* 37(5), 227–232, 1983.
12. M. A. Amerine and C. S. Ough, *Wine and Must Analysis*, John Wiley & Sons, Inc., New York, 1980.
13. J. R. Piggott and A. Paterson, eds., *Distilled Beverage Flavour*, VCH, Ellis Horwood Ltd., Chichester, UK, 1989.
14. Ref. 8., p. 151.
15. G. Vernin, ed., *Chemistry of Heterocyclic Compounds in Flavours and Aromas*, Ellis Horwood Ltd., Chichester, UK, 1982.
16. C. Eriksson, ed., *Maillard Reactions in Food: Chemical, Physiological, and Technological Aspects*, Progress in Food and Nutrition Science, Vol. 5, Pergamon Press, Oxford, UK, 1981.
17. G. R. Waller and M. S. Feather, eds., *The Maillard Reaction in Foods and Nutrition*, American Chemical Society (ACS symposium series 215), Washington, DC, 1983.
18. M. Fujimaki, M. Namiki, and H. Koto, eds., *Amino Carbonyl Reactions in Food and Biological Systems*. Developments in Food Science, No. 13, Elsevier, 1986.
19. P. A. Finot, H. U. Aeschbacher, R. F. Hurrell, and R. Liardon, eds., *The Maillard Reaction in Food Processing, Human Nutrition and Physiology*, Birkhäuser Verlag, Basel, FRG, 1990.
20. Ref. 8, p. 89.
21. I. D. Morton and A. J. Macleod, eds., *Food Flavours*, Part A, Elsevier Scientific Publication Co., New York, 1982.
22. J. Gilbert, ed., *Applications of Mass Spectrometry in Food Science*, Elsevier Applied Science, London, 1987.

Varoujan A. Yaylayan
McGill University
Quebec, Canada

FLAVORS. See Acidulants; Beer; Chocolate and cocoa; Confections; Dairy flavors; Essential oils; Citrus and mint; Poultry flavors; Seafoods, flavors and quality; Taste and odor; Vanilla extract; and entries under Sensory science.

FOAMS AND SILICONE IN FOOD PROCESSING

Potato chips, jams and jellies, beers and wines, fruits and vegetables. What do all these foods have in common? The manufacturing processes for all these products commonly experience unwanted foaming.

FOAMS

Foam can be defined as a gas dispersed in a liquid at a ratio such that the mixture's bulk density approaches that of a gas rather than a liquid (1). Foaming can occur for several reasons, stemming from both mechanical and chemical origins. Examples of mechanical contributors of foam formation are agitation near a gas–liquid interface, excessive vacuums or pressure gradients, or the free fall of liquids (2). The addition of certain chemicals that act as surface-active materials, or surfactants, may also produce foaming within a process. The existence of starch in a food process is an example of this. Foam formation can cause many problems within a process, including reducing the capacity of open vats, increasing the need for head space in closed systems, slowing the time needed to drain liquids or dried foods, interfering with process instruments, or it can increase housekeeping costs by introducing the need for maintenance personnel to deal with safety hazards such as slippery floors and the overflowing of process tanks (2).

Two basic actions can be taken to stop the formation of

foam in a food process. The first involves mechanical means of eliminating the foam, such as heating, centrifuging, spraying, or ultrasonic vibration (3). The other option is to introduce chemical antifoams or defoamers to the process. This article will focus on the chemical method of foam control. Defoamers are materials that specialize in destroying existing foam immediately. Antifoams are materials that are designed as long-lasting means of preventing the formation of foam. Examples of foam control products range from liquid products such as hydrocarbons (kerosene, vegetable oils, organic phosphates, and acetylenic glycols), polyethers, fluorocarbons, and silicones, to solid products such as fatty amides, hydrocarbon waxes, solid fatty acids, and esters (3). The remainder of this article will focus on silicone-based antifoam products.

SILICONE ANTIFOAMS

Two theories exist to describe how antifoams work. Both are based on the fact that pure liquids do not foam. The presence of some stabilizing material is necessary in the surface layer of foam films to allow them to remain stable. These theories are also based on the premise that antifoaming agents either replace or modify these stabilizing materials. The first theory states that the antifoam disperses in the form of fine drops into the liquid film between the bubbles and spreads as a thick duplex film, ie, a film that is thick enough to have two definite surfaces. The tension created by the spreading duplex film leads to the rupture of the original liquid film (bubble wall), destroying the foam.

The second theory is similar to the first and states that the antifoam produces a less cohesive, mixed monolayer on the surface of the liquid film. Because this monolayer is of less coherence than the original film stabilizing monolayer, it will cause the destabilization of the bubble wall (3). In either case, it is important to note that antifoams must be surface active, that they must be able to spread to the entire foaming surface of the process, and that they must remain insoluble to continue to be active.

Since the discovery of polydimethylsiloxane in 1943 by scientists of what is now Dow Corning Corp., there have been many applications found for this unique substance. Applied as antifoams since shortly after their discovery, these silicone-based products are able to provide benefits that most organic defoamers cannot (4). Due to the inert nature of the polydimethylsiloxane molecule, silicone-based antifoams will not react with most process media. This inert nature also allows the antifoam to remain effective longer than reactive organic material. Silicones exhibit a low surface tension (< 20 dynes/cm) (5), allowing them to spread quickly and evenly over the foaming surface, an important requirement for an antifoam already mentioned.

Both of these attributes contribute to the most important benefit of using silicone antifoams: their efficiency. Silicone-based antifoams are effective usually in amounts less than 100 parts active silicone per million parts foaming media in industrial applications and in as little as 5–10 ppm in food-grade systems.

Silicone antifoams are available as three basic types of product. First, the polydimethylsiloxane, or silicone oil, can be used as an effective antifoam for many nonaqueous-type processes. Second, the silicone oil can be compounded with hydrophobic silica to form a product that can be useful in both aqueous and nonaqueous systems. It has been theorized that the addition of this silica gives an added benefit of physically disrupting the foam interlayer. The third and most effective product for aqueous systems is an emulsion of the silicone–silica compound. An emulsion is the most effective carrier of a compound in a water environment, allowing for ease in dilution and dispersion.

Silicones can be used in many processes. There are silicone-based antifoams that are permissible for use in direct contact with food, as per the Food and Drug Administration's regulation #173.340, which allows for up to 10 parts of active silicone per million parts of food in the final product. Certain silicones also meet the U.S. Department of Agriculture's specifications for federally inspected foods. Many are also kosher approved. Silicone antifoams usually will not impart any taste or smell to a product. Because these antifoams are essentially inert and are used in such low amounts, they rarely affect process reactions.

BASIC GUIDELINES

There are several basic guidelines to follow when using silicone antifoams to be the most effective as possible. Silicone antifoams should be prediluted and mixed in with a small amount of the foaming medium prior to introduction into the system. This allows the antifoam to disperse evenly and can prevent shocking of an emulsion, which could cause the emulsion particles to coalesce, or stick together. The antifoam should be dispensed in such a manner as to allow it to disperse evenly and reach the surface of the foaming site. This could involve spraying the antifoam onto the surface, or providing slight agitation when adding the antifoam to allow it to reach the surface. The antifoam should be added prior to the point where foaming occurs. This allows the product to act as it was designed; that is, as an antifoam rather than a defoamer.

When using silicone emulsified antifoams, there are additional precautions that should be taken that involve the amount of shear that the antifoam is subjected to. Because this oil-based antifoam is emulsified using surfactants and thickeners, the product can be destroyed by disturbing the emulsion particles, causing coagulation or separation. High levels of mechanical shear can do this through turbulent agitation, violent shaking, or pumping through gear type, centrifugal pumps. Care should be taken to provide gentle agitation when prediluting, or when adding the antifoam to the foaming process.

FUTURE DEVELOPMENT

Significant advancements continue to be made in the development of silicone-based antifoams. Recent developments have focused on areas involving organic–silicone blends. Many newly commercialized products use this new technology and may impart even more efficiency to current antifoam product lines. Work is also being conducted on varying the types of delivery systems for silicone antifoams. Instead of water-based emulsions, solid media are being investigated as delivery systems for the antifoam compound to the foaming interface. This could provide benefits to separation processes using ultrafiltration. Typically, emulsifiers used in antifoam emulsions have shown a tendency to clog and even damage many types of crossflow ultrafiltration membranes.

As the food industry continues to expand with quicker, more convenient processed foods, so too will the opportunities grow for foam control. Currently, an area experiencing growth is the microwavable foods segment. This is also an area where foaming has been a problem. Many manufacturers desire microwaved food to resemble food that has been prepared by conventional means. Although the formation of foam on food or drinks that have been microwaved does not affect the taste of these products, it is not aesthetically pleasing. The use of silicone antifoams can alleviate these types of problem.

BIBLIOGRAPHY

1. A. P. Kouloheris, "Foam: Friend and Foe," *Chemical Engineering* (1987).
2. J. B. McGee, "Selecting Chemical Defoamers and Antifoams," *Chemical Engineering* (1989).
3. M. J. Owen, "Antifoaming Agents," in *Encyclopedia of Polymer Science and Engineering*, Vol. 2, 2nd ed., John Wiley & Sons, Inc., New York.
4. C. C. Currie, "What Silicones Can Do for the Food Processor," *The Canner* (1952).
5. Internal Dow Corning Surface Tension Studies, TIS #1952–I0030–119.

CHRIS COMBS
Dow Corning Corporation
Midland, Michigan

FOOD ADDITIVES

Food additives have been the topic of vast attention over the past decade by large numbers of concerned groups. Once the principal domain of scientific endeavors, food additives are now *de rigueur,* a lively conversational subject for the public at large. No longer an exclusive issue within the United States, concerns over the safety of food additives have been voiced on a global scale by the scientific community, the food and chemical industries, regulatory agencies, legislative branches of governments, activists groups, and, ultimately, the consuming public. In the United States, occasional congressional intervention threatens the authority of the Food and Drug Administration (FDA), which is empowered to safeguard public health.

The topic of food additives can be discussed from various perspectives including: risk and benefits issues; food additives of major public concerns; United States laws and regulations; and the technical uses of food additives in food processing. The risk and benefits issues are discussed under those entries discussing food toxicology and risk analysis. Specific food additives of major concerns are discussed under individual entry (for those that are included). This article discusses the technical uses of food additives in food processing with a brief discussion of the basic laws and regulations currently in effect in the United States. Furthermore, this article assumes the following position. The addition of chemicals to foods is necessary in the production, processing, and marketing of many foods. When properly used, they are not harmful and contribute greatly to the abundance, variety, stability, flavor, and appearance of the food supply. The problem is to ensure that such chemicals are safe for continuous use, that they serve a useful technological purpose, and that they are not employed to mask with fraudulent intent a defective inferior product.

DEFINITIONS

Defining food additives is not a straightforward task since there are two important factors that can affect description. In the technical sector, the term "food additives" has been used to broadly describe any chemical substance used to augment or modify the characteristics of a food product. To the food scientists, colorants, preservatives (including antimicrobial agents, antioxidants, and sequestrants), emulsifiers, thickeners, acidulants, enzymes, flavors, and nonnutritive sweeteners are traditionally included in the technical definition. Basic foodstuffs with nutritional value are generally excluded from the technical definition although some view vitamins, minerals, amino acids, and protein fractions as additives when added to foods to upgrade nutritional value, ie, fortification. For the most part, the food science community views food additives as useful chemicals of natural or synthetic origins employed in small amounts—a few parts per million (ppm) to perhaps as much as 0.5–1.0%—although there are exceptions (eg, acidulants such as citric acid can be used at a few percent in some products). Present working suggests that food additives and food ingredients are synonymous. The latter generally is considered to be the broader term, covering both additive and nonadditive materials in a food product.

However, the legal definition in this country is fairly narrow. In 1958, a Food Additives Amendment was added to the original Federal Food, Drug, and Cosmetic Act passed by Congress in 1938. This amendment contains four important aspects:

1. It defines what a food additive is.
2. It authorizes the FDA to license the use of food additives.
3. It permits the proposed use of a food additive if it can be proven safe (by the food processor) when present at a certain level in the food.
4. If a food additive is a proven carcinogen in animals or humans, its addition to food is not permitted. This provision is popularly known as the Delaney Amendment.

Legally, a food additive is defined as:

Any substance the intended use of which results or may reasonably be expected to result, directly or indirectly, in its becoming a component or otherwise affecting the characteristics of any food (including any substance intended for use in producing, manufacturing, packing, processing, preparing, treating, packaging, transporting, or holding food; and including any source of radiation intended for any such use), if such substance is not generally recognized, among experts qualified by scientific training and experience to evaluate its safety, as having been adequately shown through scientific procedures ... to be safe under the conditions of its intended use.

According to the Federal Food, Drug, and Cosmetic Act, if a chemical substance is legally a food additive then its use will be restricted. The party interested in proposing a use for the substance is required to submit a petition to the FDA, requesting the issuance of a regulation that prescribes various requirements such as the exact chemical structure of the substance, the food to which it can be added, the amount permitted, and other criteria. However, the legal definition for a food additive makes two exceptions:

1. Certain substances are specifically excluded from the legal definition of a food additive, for example, pesticide and related substances, color additives, and new animal drugs. They are regulated by other laws and regulations.
2. Certain chemical substances, although technically food additives, are exempt from following the restrictive regulations in their usage. These are listed as follows:
   a. A chemical substance will not be legally a food additive if it is generally recognized as safe (GRAS) under the conditions of intended use, by scientists qualified by experience and training to evaluate food safety. The regulations define the necessary expertise as "sufficient training and experience in biology, medicine, pharmacology, physiology, toxicology, veterinary medicine, or other appropriate science to recognize and evaluate the behavior and effects of chemical substances in the diet of man and animals." Substances generally recognized as safe are commonly called GRAS, as shown by the acronym. The law also specifically indicates that the general recognition of safety of such a substance may rely on experience based on common use in food prior to January 1, 1958.
   b. A chemical substance will not be legally a food additive if it is used in accordance with sanctions granted prior to the passage of the Food Additives Amendment. In actual practice, these "prior sanctioned" chemicals may be divided into three categories.
      i. Those used in food products that have been standardized (eg, peanut butter, ice cream).
      ii. Those approved by the United States Department of Agriculture for use under the Meat Inspection Act and the Poultry Products Inspection Act.
      iii. Those approved specifically by the FDA before the passage of the Food Additives Amendment. The approval was usually granted by a written statement, such as letters, memoranda, and other evidence through direct communications.

A legal food additive can be an intentional additive, in which case it is added purposefully to food to achieve certain specific technical functions. It can also be an incidental additive, in which case it is not intended to be in food but has become a part of the food through some aspects of food production, processing, storage, or packaging. The latter occurrence has no function in the finished product. Thus, the first type is sometimes called direct and the second, indirect, food additives. Also, the irradiation of food with radioactive rays is considered to add a food additive since it is usually used to destroy bacteria and thus achieve a technical function.

CATEGORIES OF FOOD PRODUCTS AND EFFECTS OF FOOD ADDITIVES

The FDA has established more than 40 categories of food products to group specific, related foods together for the purpose of establishing tolerances or limitations for the use of direct human food ingredients. For example, three such categories are as follows:

1. Beverages and beverage bases, nonalcoholic, including only special or spiced teas, soft drinks, coffee substitutes, and fruit- and vegetable-flavored gelatin.
2. Fish products, including all prepared main dishes, salads, appetizers, frozen multicourse meals, and spreads containing fish, shellfish, and other aquatic animals, but not fresh fish.
3. Processed fruits and fruit juices, including all commercially processed fruits, citrus, berries, and mixtures; salads, juices, and juice punches, concentrates, dilutions, "ades," and drink substitutes therefrom.

The FDA uses the following terms adopted from the National Academy of Sciences/National Research Council national survey of food industries to describe the physical or technical effects for which direct human food ingredients may be added to foods:

1. Anticaking and free-flow agents
2. Antimicrobial agents
3. Antioxidants
4. Colors and coloring adjuncts
5. Curing and pickling agents
6. Dough strengtheners
7. Drying agents
8. Emulsifiers and emulsifier salts
9. Enzymes
10. Firming agents
11. Flavor enhancers
12. Flavoring agents and adjuvants
13. Flour-treating agents
14. Formulation aids
15. Fumigants

16. Hemectants
17. Leavening agents
18. Lubricants and release agents
19. Nonnutritive sweeteners
20. Nutrient supplements
21. Nutritive sweeteners
22. Oxidizing and reducing agents
23. pH control agents
24. Processing aids
25. Propellants, aerating agents, and gases
26. Sequestrants
27. Solvents and vehicles
28. Stabilizers and thickeners
29. Surface-active agents
30. Surface-finishing agents
31. Synergists
32. Texturizers

LEGAL APPROVAL

When a food manufacturer wants to introduce a new chemical as a direct food additive or a new use for an old one, the law requires him or her to file a petition with the FDA and request the issuance of a regulation sanctioning the proposed use of the substance. Some of the information needed in the petition includes the following:

1. Chemical identity and composition of the substance and its physical, chemical, and biological properties.
2. Amount specified for use and purposes for which it is proposed, together with all directions, recommendations, and suggestions regarding the proposed use.
3. Data establishing that the food additive will have the intended physical or other technical effect or that it may reasonably be expected to become a component of the food.
4. Practicable methods to determine the amount of the food additive in the raw, processed, and/or finished food.
5. Full reports of investigations made with respect to the safety of the food additive. The reports ordinarily should include detailed data derived from appropriate animal and other biological experiments in which the methods used and the results obtained are clearly set forth. The petition should not omit without explanation any reports of investigations that would bias an evaluation of the safety of the food additive.
6. Proposed tolerances for the food additive, if tolerances are required, so as to ensure its safety.
7. The environmental impact analysis report assessing the environmental impact of the manufacturing process and the ultimate use or consumption of the food additive.

How will the FDA decide about the safety of a food additive in the petition? Some basic considerations are as follows:

1. The probable consumption of the additive and of any substance formed in or on food because of the use of the additive.
2. The cumulative effect of such additive in the diet of man or animals, taking into account any chemically or pharmacologically related substance or substances in such diet.
3. Safety factors which in the opinion of experts qualified by scientific training and experience to evaluate the safety of food additives are generally recognized as appropriate for the use of animal experimentation data.

The FDA recommends two important guidelines on safety to any petitioner for a food additive. First, the petitioner should follow the principles and procedures for establishing the safety of food additives stated in current publications of the National Academy of Science/National Research Council, although this is not binding. Second, a food additive for human use will not be granted a tolerance that will exceed 1/100th of the maximum amount demonstrated to be without harm to experimental animals. The FDA regards this as a safety factor.

It is expected that, after an adequate and unbiased review of the petition, the FDA will not issue a regulation for the proposed use of the food additive if the data confirm one or more of the following:

1. The safety of the proposed use is not established.
2. Deception of the consumer may result from the intended use.
3. The intended technical or physical effect will not be accompanied by the proposed use.
4. It may possibly not be safe for a person to accumulate the substance taken in from more than one food source.
5. The substance can cause cancer in animals.

How will the FDA determine the eligibility for the classification of a substance as GRAS? The FDA has issued the following guidelines:

1. If a chemical substance satisfies the following conditions, it "will ordinarily be regarded as GRAS without specific inclusion" in any official list:
   a. A food ingredient of natural biological origin that has been widely consumed for its nutrient properties in the United States prior to January 1, 1958 without known detrimental effects and for which no known safety hazard exists.
   b. Same as (a), except that the substance is one that is "subject only to conventional process as practiced prior to January 1, 1958."
2. The status of the following food ingredients has to be reviewed and affirmed as GRAS. In case such a substance is not affirmed, the FDA will either determine it to be a legal food additive or subject it to the prior-sanction clause.
   a. Any substance of natural biological origin that has been widely consumed for its nutrient properties in the United States prior to January 1, 1958 without known detrimental effect. No health hazard is known about this substance and

it has been modified by processes first introduced into commercial use after January 1, 1958, which may reasonably be expected to significantly alter the composition of the substance.
  b. Any substance of natural biological origin that has been widely consumed for its nutrient properties in the United States prior to January 1, 1958 without known detrimental effect. No health hazard is known about this substance and it has had significant alteration of composition by breeding or selection after January 1, 1958, where the change may be reasonably expected to alter the nutritive value or the concentration of toxic constituents.
  c. Distillates, isolates, extracts, concentrates, of extracts and reaction products of GRAS substances.
  d. Substances not of a natural biological origin, including those showing evidence of being identical to a GRAS counterpart of natural biological origin.
  e. Substances of natural biological origin intended for consumption for other than their nutrient properties.
3. The GRAS classification will not be assigned if a substance has no history of food use or if its safe use requires setting a limitation.

What is the process of affirmation of a GRAS substance status? Such a process may be initiated by the FDA or any other interested party. The latter is required to file a petition with the FDA. In either case, all available information will be published in the *Federal Register* and a specific period of time allotted for receiving public comments. The FDA will study the whole file including the comments and make a decision. Some of the information required in the petition is as follows:

1. A description of the substance, such as name or chemical structure.
2. Use of the substance, such as date when use began and foods in which used.
3. Methods for detecting the substance in food, with details such as their sensitivity and reproducibility.
4. Information to establish the safety and functionality of the substance in food, such as published scientific literature and other supporting data.

The affirmation process is less complex than a food additive petition since it does not require the animal feeding studies that are mandatory for a food additive petition.

Any group of scientists, whether government employed or not, may render their opinion or judgment about the general recognition of safety on a chemical substance. Of course, any private and independent determination of such a status may be jeopardized if the FDA and/or their scientists disagree, in which case the substance will not generally be regarded as safe. However, a private group, the Flavor and Extract Manufacturer's Association, has published long lists of substances used in food flavors that have been determined as GRAS by a panel of independent scientists. Their opinions are published in the journal *Food Technology*. The FDA has expressed both explicit and tacit approval of the decisions of this expert panel.

We will first study those food additives permitted for direct addition to food for human consumption. Regulations prescribing conditions under which food additive substances may be safely used always specify that normal good manufacturing practices be exercised. The latter are defined to include the following:

1. The quantity of the substance used does not exceed the amount reasonably required to accomplish its intended physical, nutritive, or other technical effect in food.
2. The substance is of appropriate food grade and is prepared and handled as a food ingredient.
3. The existence of any regulation prescribing the safe conditions of use for a nutrient substance does not constitute a finding that it is useful or required as a supplement to the diet of humans.
4. The substance must also comply with all other provisions of the Federal Food, Drug, and Cosmetic Act.

Accordingly, the list of food additives for direct addition to food is classified into eight types which include:

1. Food preservatives, eg, sodium nitrite, BHT.
2. Coatings, films, and related substances, eg, polyacrylamide.
3. Special dietary and nutritional additives, eg, vitamins.
4. Anticaking agents, eg, calcium silicate, silicon dioxide.
5. Flavoring agents and related substances, eg, disodium inosinate.
6. Gums, chewing gum bases, and related substances, eg, furcelleran.
7. Other specific usage additives, eg, calcium lignosulfonate.
8. Multipurpose additives, eg, aspartame, glycine.

Those secondary direct food additives permitted in food for human consumption are divided into four different types.

1. Polymer substances for food treatment, eg, acrylate/acrylamide resins and polyvinylpolypyrrolidone.
2. Enzyme preparations and microorganisms, eg, amylglucosidase (derived from *Ryizopus niveus*) and milk-clotting enzyme.
3. Solvents, lubricants, release agents, and related substances, eg, acetone, hexane, hydrogenated sperm oil, and trichloroethylene.
4. Specific usage additives, eg, boiler water additives, defoaming agents, dichlorodifluoromethane, and sodium methyl sulfate.

The next group is indirect food additives. Again, the usage of such substances must follow, in addition to other requirements, good manufacturing practices that include the following. First, the quantity of any such substance that may be added to food as a result of use in articles that contact food:

1. Is not permitted to exceed any prescribed limitations
2. Is not permitted to exceed, where no limits are specified, the amount that results from use of the substance in a portion not more than reasonably required to accomplish the intended physical or technical effect in the food-contact article and
3. Is not intended to accomplish any physical or technical effect in the food itself, except as permitted by regulation

Second, any substance used as a component of articles that contact food should be of a purity suitable for its intended use. Third, the substance must not violate other provisions of the Federal Food, Drug, and Cosmetic Act. Different types of indirect food additives with examples are as follows:

1. As components of adhesives, such as calcium ethyl acetoacetate 1,4-butanediol modified with adipic acid.
2. As components of coatings, such as acrylate ester copolymer coatings and poly(vinyl fluoride) resins.
3. As components of paper and paperboard, such as slimicides, sodium nitrate/urea complex, and alkyl ketene dimers.
4. As basic components of single- and repeated use food contact surfaces, such as cellophane, ethylene-acrylic acid copolymers, isobutylene polymers, nylon resins, and poly(vinyl alcohol) film.
5. As components of articles intended for repeated use, such as ultrafiltration membranes and textiles and textile fibers.
6. Controlling growth of microorganisms, such as sanitizing solutions.
7. Antioxidants and stabilizers, such as octyltin stabilizers in vinyl chloride plastics.
8. Certain adjuvants and production aids, such as animal glue, hydrogenated castor oil, synthetic fatty alcohols, and petrolatum.

As mentioned earlier, radiation is legally considered a food additive. Regulations prescribing irradiation in the production, processing, and handling of food are briefly discussed below.

Some sources of radiation for the inspection of foods before and after packaging and for controlling food processing may be safely used under certain conditions, eg, if the radiation source is from X-ray tubes producing X-radiation from operation of the tube source at energy levels of 300 kV peak or lower. The label of the sources should specifically identify the source of information and the maximum energy of radiation emitted by X-ray tube sources. The gamma radiation for the treatment of certain foods may be safely used under certain conditions. The radiation source consists of sealed units containing the isotope cobalt-60 or cesium-137. The technique is permitted for use on wheat, wheat flour from unirradiated wheat, and white potatoes. To ensure safe use, the label should bear a statement "Treated with ionizing radiation" or "Treated with gamma radiation" on retail packages. When gamma radiation is used in the treatment of prepackaged foods, the packaging materials that may be subject to a dose of radiation, not to exceed 1 Mrad (megarad), include nitrocellulose-coated or vinylidene chloride copolymer-coated cellophane, glassine paper, and acrylonitrile copolymers. All these materials must also comply with indirect food additive regulations. There are other examples.

There are food additives permitted in food on an interim basis or in contact with food pending additional study. Substances having a history of use in food for human consumption or in food-contact surfaces may at any time have their safety or functionality brought into question by new information that in itself is not conclusive. An interim food additive regulation for the use of any such substance is promulgated if the following criteria are satisfied. First, new information raises a substantial question about the safety or functionality of the substance. Second, there is a reasonable certainty that the substance is not harmful. Third, no harm to the public health will result from the continued use of the substance for a limited period of time while the question raised is being resolved by further study. No interim additive regulation will be promulgated if the following conditions prevail. The new information is conclusive with respect to the question raised. There is a reasonable likelihood that the substance is harmful. Continued use of the substance will result in harm to the public health.

An interim food additive petition may be initiated by the FDA or by an outsider. The FDA uses the answers to the following questions to determine the interim status.

1. Is the interim food additive regulation justified?
2. What are the types of study necessary and appropriate to resolve questions raised about a substance?
3. Do the interim results indicate the reasonable likelihood that a health hazard exists?
4. Do the data available at the conclusion of those studies justify a food additive regulation?

Some examples of interim food additives are acrylonitrile copolymers, mannitol, brominated vegetable oil, and saccharin and its salts.

It should be noted that even for the list of prior-sanctioned food ingredients there are regulations and requirements that control their usage. The list divides the substances into the following classes:

1. Certain substances employed in the manufacture of food packaging materials.
2. Antioxidants, eg, gum guaiac, propyl gallate.
3. Antimycotics, such as sorbic acid, sodium benzoate.
4. Driers, eg, cobalt linoleate, iron caprylate.
5. Drying oils as components of finished resins, eg, linseed oil.
6. Plasticizers, eg, diphenyl-2-ethylhexyl phosphate.
7. Release agents, eg, oleamide, stearamide, and linoleamide.

8. Stabilizers, eg, ammonium citrate, calcium phosphate.
9. Substances used in the manufacture of paper and paperboard products used in food packaging, eg, borax, titanium dioxide.
10. Acrylonitrile copolymers and resins.

Substances generally recognized as safe are classified into the following:

1. Spices and other natural seasonings and flavorings such as paprika, mace, capers, and grains of paradise.
2. Essential oils, oleoresins (solvent free), and natural extractives (including distillates), such as cloverleaf, hops, and immortelle.
3. Natural substances used in conjunction with spices and other natural seasonings and flavorings, such as brown algae.
4. Natural extractives (solvent free) used in conjunction with spices, seasonings, and flavorings, such as peach kernel and quince seed.
5. Certain other spices, seasonings, essential oils, oleoresins, and natural extracts, such as white and green cognac oil and musk.
6. Synthetic flavoring substances and adjuvants, such as acetoin, 1-malic acid, and diacetyl(2,3-butandeione).
7. Substances migrating from cotton and cotton fabrics used in dry food packing, such as corn dextrin, sorbose, urea, and zinc chloride.
8. Substances migrating to food from paper and paperboard products, such as aluminum oleate, reduced iron, and sulfamic acid.
9. Adjuvants for pesticide chemicals.
10. Multipurpose GRAS food substances, such as tartaric acid, caffeine, pepsin, rennet, triacetin, and beeswax.
11. Anticaking agents, such as calcium silicate and sodium aluminosilicate.
12. Chemical preservatives, such as ascorbic acid and sodium bisulfite.
13. Emulsifying agents, such as cholic acid and ox bile extract.
14. Nutrients and/or dietary supplements, such as carotene, inositol, and lysine.
15. Sequestrants, such as calcium phytate, tetrasodium pyrophosphate, and stearyl citrate.
16. Stabilizers such as agar/agar and sodium alginate.

Some examples of direct food substances affirmed as GRAS are benzoic acid, methylparaben, and oil of rue. Some examples of indirect food substances affirmed as GRAS are acacia (gum arabic) and pulp.

SUBSTANCES PROHIBITED FROM USE

There are certain substances prohibited from use in human food. The FDA has determined that they present a potential risk to the public health or have not been shown by adequate scientific data to be safe for use in human food. Use of any of these substances in violation of regulations causes the food involved to be adulterated and thus a violative product. For easy reference purposes, the list includes only some of those substances prohibited from use in human food and is not a complete list. The substances prohibited from direct addition or use as human food include calamus and its derivatives, cobaltous salt and its derivatives, coumarin, cyclamate and its derivatives, diethylpyrocarbonate, dulcin, monochloroacetic acid, nordihydroguairetic acid, P-4000, safrole, and thiourea. The substances prohibited from indirect addition to human food through food-contact surfaces include flectol H, mercaptoimidazoline, and 2-mercaptoimidazoline, 4, 4'-methylenebis (2-chloroanaline).

TECHNICAL DESCRIPTIONS

The remaining part of this article gives a brief description for some of food additives. Detailed information can be obtained from the references at the end of this article.

Anticaking Agents

As the name implies, these substances are added to finely powdered or crystalline food products to prevent their caking and, hence, to maintain their free-flowing properties. Such chemicals as calcium aluminum silicate, calcium phosphate tribasic, magnesium carbonate, magnesium silicate, magnesium stearate, and sodium aluminum silicate at a level of 1.0% have been suggested as effective anticaking agents in free-running salt and in dry powdered mixes. Colloidal silicon dioxide is also used at a level of 1.0% as an anticaking agent in salt. The decahydrate of sodium ferrocyanide at the extremely low level of 5 ppm, this chemical causes the formation of star-shaped crystals of sodium chloride rather than the usual cubical ones, the star-shaped crystals known the trade as dendritic salt possess nonsegregating properties, hence, they are less likely to cake.

Antifoaming Agents

A troublesome frothing or foaming occurs during certain processing steps in the manufacture of many food items. Fortunately, foaming may be largely suppressed or completely eliminated by the use of small quantities, generally about 10 ppm, of dimethylpolysiloxane as an antifoaming agent. This compound has been suggested for use in the preparation of certain meat products, bakery products, confections, dairy products, fats and oils, fruit juice products, jams and jellies, molasses, soups, starches, syrups, wines, and pickles. The mono- and diglycerides have also been suggested as antifoaming agents.

Antioxidants

See the specific entry on antioxidants in this encyclopedia.

Bleaching, Maturing, and Dough-Conditioning Agents

Because some chemicals serve as both bleaching and maturing agents, and other are referred to as dough-conditioning agents or bread improvers, it is perhaps desirable to consider all of them under one heading.

Bleaching agents are used in the production of certain cheeses, processed fruits, and meat products and to neutralize color that may be present naturally in fats and oils. However, in view of the fact that they are of special importance to the flour-milling and baking industry, these chemicals are discussed here only as they are used in these foods. Freshly milled wheat flour has a yellowish color, caused by the presence of small quantities of carotenoid and other pigments. This flour also lacks the quality necessary to make an elastic, stable dough. When such flour is stored and allowed to age for several months, it gradually becomes whiter because of oxidation, and matures, making it acceptable for baking. For many years, the natural bleaching and aging was the only means millers and bakers had of producing the desired material. These natural processes were slow an and costly and did not always yield a consistently satisfactory product. Then it was discovered that certain oxidizing agent incorporated into the flour in small amounts brought about rapid improvements in its color and bread-making properties.

Colorants

Colorants, both synthetic and derived from natural sources, are used in many foods to enhance their appearance. Although current estimates indicate that 80% of the food colorants are synthetic, those of natural origin are gaining popularity principally due to adverse industry and consumer reactions to the safety issues surrounding the synthetic colorants. Because of their high tinctorial strength, synthetic colorants of the azo and triphenylmethane class are used at levels of few ppm to ca 300–500 ppm, whereas naturally derived colorants require much higher levels, often a few percent as in the case of caramel color.

Emulsifying, Gelling, Stabilizing, and Thickening Agents

This group of food additives contains a large number of compounds that influence the texture, homogeneity, or stability of the foods in which they are used. Because in many cases the distinction between an emulsifier, a gelling agent, a stabilizer, or a thickener is not too easily defined, it is desirable to consider them together.

The texture of ice cream and other frozen desserts is dependent in part on the size of the ice crystals in the product. Smooth texture results from rapid freezing with the formation of numerous small crystals; a coarse texture results when the crystals form slowly and are large. The growth of large crystals is minimized by the use of stabilizers that increase viscosity. The viscosity of chocolate milk is increased by the use of stabilizers to prevent sedimentation of the dense cocoa particles.

The use of certain emulsifiers in baked goods reportedly results in improved texture and eating quality, prolonged palatability and other keeping qualities, improved flavor, and improved appearance and uniformity.

Process cheese is a blend of selected cheeses comminuted and melted into a plastic mass that firms on cooling. Unless the butterfat is emulsified, it may separate as an oil during the heating process and prevent the blending of the components. Emulsification is aided by the presence of certain soluble salts that are not true emulsifiers but cause emulsification indirectly through their effect on the proteins. These salts also impart smooth texture, improved slicing properties, and freedom from fat leakage in the finished cheese.

Gelling agents are used in jams, jellies, and marmalades to provide a product with satisfactory consistency. Soft drinks may contain thickeners to provide a product with body or consistency. Brewed beverages often require a stabilizer to regulate their foaming qualities. Emulsifiers improve the keeping qualities and homogeneity of candies and confections. Chocolate candy tends to bloom or change in surface color if exposed to temperature changes because the cocoa butter tends to separate from the chocolate; an demulsifier helps to prevent this condition by keeping the fat in a more stable emulsion within the chocolate. Some of the compounds commonly used as emulsifiers are discussed below.

The polysaccharides that are used widely in foods may be classified according to origin: (1) plant and tree exudates, eg, acacia gum (gum arabic), karaya gum, tragacanth gum; (2) marine plant extracts (seaweed), eg, agar-agar, algin, and its derivatives, carrageenan (Irish moss extractive), furcellaran, oat gum; (4) fruit and vegetable extract, eg, pectin; and (5) fermentation products, eg, xanthum gum.

These gums and pectin are used in amounts of 0.1–1.0% with some specialty foods containing up to 5% or higher. They are added to such foods as cream cheese, chocolate drinks, French dressing, pickles and relishes, salad dressing, ice cream, sherbet, certain meat products, jams and jellies, pie fillings, fountain syrups, confectionery, dessert gels, and pudding mixes.

Cellulose derivatives such as methyl cellulose, carboxymethyl cellulose, (hydroxypropyl) methyl cellulose, and sodium carboxymethyl cellulose are increasingly being used. In such foods as dressings, chocolate drinks, process cheese, ice cream, sherbet, the amount generally added is ca 1%.

Various mineral emulsifiers suggested for process cheese include calcium citrate, tricalcium phosphate, potassium citrate, dipotassium hydrogen phosphate, sodium hydrogen pyrophosphate, sodium gluconate, sodium hexametaphosphate, mono-, di-, and trisodium phosphates, sodium potassium tartrate, tetrasodium pyrophosphate, and sodium tartrate.

The mono- and diglycerides of fatty acids, as well as acetylated monoglycerides, acetylated tartaric acid ester of mono- and diglycerides, and lactylated mono- and diglycerides, are used in foods as emulsifiers and stabilizers. The mono- and diglycerides themselves are more widely used than the derivatives, being added to such foods as bread, shortening, process cheese, ice cream, chocolate milk, and confections in amounts ranging up to 20%.

The so-called synthetic emulsifying agents include such products as polyoxy-ethylene sorbitan monooleate, polyoxyethylene sorbitan monostearate, polyoxyethylene stearate, and sorbitan monostearate, which are added in amounts of 1% or less to such foods as ice cream, cakes, confectionery icings, pie fillings, whipped vegetable oil toppings, chocolate drinks, and beverage bases.

Gelatin is widely employed in the food industry because of its elastic consistency, because it has the property of holding air and water, or because it can inhibit the crystallization or graining of sugar. It is used in certain meat products, process cheese, frozen desserts, confectionery, etc. The amounts required for good manufacturing practice may be only 0.5% for certain foods but for others as much as 10% may be required.

For proper gel formation of certain food products, a source of calcium ions is required. Calcium salts used for this purpose include calcium citrate, calcium gluconate, calcium glycerophosphate, calcium hypophosphite, di- and tricalcium phosphates, and calcium sulfate.

Other emulsifiers, stabilizers, thickening agents, or gelling agents are cholic acid, deoxycholic acid, and glycocholic acid, (or bile extract); as well as sodium lauryl sulfate and triethyl citrate, used in the processing of egg whites; lecithin, used in bread, margarine, process cheese, ice cream, etc; stearyl monoglyceridyl citrate, used in shortening.

Food Enzymes

Enzymes play a significant role in a very wide range of food processing technology. Cheese making, bread making, brewing, glucose production and the confection industry, fruit juice and soft drink manufacture, meat tenderizing, egg white preparation, and the manufacture of cereal products are examples for food processes utilizing enzymes.

Proteases. Proteolytic enzymes, such as ficin, papain, bromelain, and fungal proteases are used for tenderizing meat. One such technique involves the injection of papain into cattle just prior to slaughter; this practice is reported to provide uniform distribution of the enzyme to all parts of the animal body. Pancreatin, a mixture of two proteinases as well as other enzymes, is used in the production of gelatin and protein hydrolyzates, and rennin is used in the manufacture of certain cheeses.

Pepsin, papain, and other enzymes are used in the brewing industry to stabilize and chill proof beer. Proteolytic enzymes aid in the separation of residual meat from the bones of animal carcasses; they render cereal proteins soluble in the production of instant cereals, and they help to wet milk and similar solids for quicker reconstitution. Fungal proteases improve bread grain, texture, and compressibility of bread crumbs. They also improve color and increase loaf volume.

Carbohydrases. Invertase is used in the confection industry to make syrupy centers for chocolate and for the production of invert sugar, which is increasing in usefulness in the food industry. Amylases break down gelatinized starches in the production of corn syrup and dextrose sugars. They can also convert acid-modified starches.

Glucose Oxidases. Glucose oxidases are used to stabilize egg solids (by desugaring yolks, whites, and whole eggs); to remove oxygen from carbonated and still beverages, and food dressings; and to prevent undesirable flavor changes.

Catalases. Catalases destroy residual hydrogen peroxide employed in the cold sterilization of milk for the manufacture of Cheddar, Colby, granular, Swiss, and washed-curd cheese.

Lipases. Lipases have been used for the treatment of butterfat to produce flavored substances that are used to flavor bakery products and confections. The latest use of enzymes by the food industry is in the conversion of corn syrup to high fructose syrups. Enzymes are insolubilized by binding onto various carriers. Following hydrolysis, the corn syrup is contacted with the bound enzyme that converts glucose into fructose. The process is unique in that the enzyme used in the process does not become part of the finished product.

Firming Agents

Fruits and vegetables contain pectin components that are relatively insoluble and that form a firm gel around the fibrous tissues of the fruit and prevent its collapse. Addition of calcium salts causes the formation of calcium pectate gel, which supports the tissues and affords protection against softening during processing. The calcium salt is sometimes added to the canned vegetable in the form of a combination tablet containing both salt and calcium chloride.

Canned vegetables, canned apples, frozen apples, and canned tomatoes are sometimes treated during processing with calcium chloride, calcium citrate, mono-calcium dihydrogen phosphate, or calcium sulfate to prevent them from becoming soft and disintegrating. One suggested level of use of these calcium salts is 0.026%, calculated as calcium in the final food product. In canned potatoes, calcium chloride and calcium citrate at a level of 0.5% (calculated as calcium) are reportedly used.

Aluminum sulfate, ammonium aluminum sulfate, potassium aluminum sulfate, and sodium aluminum sulfate are reported to be of value as firming agents in pickles and relishes. A more recently introduced firming agent is aluminum sulfate for canned crabmeat, lobster, salmon, shrimp, and tuna. Calcium chloride is reported to act as a canning agent in cheddar and cottage cheese.

Flavoring Agents

Because of the importance of the flavor of a food product, food manufacturers use a variety of flavoring substances. As a result, flavoring agents, both natural and synthetic, make up one of the largest groups of food additives.

Flavor Enhancers

Although monosodium glutamate (MSG) has been recognized and used in the food industry as a flavor-enhancing product for many years, two 5'-ribonucleotides—disodium inosinate and disodium guanylate—were introduced by Japanese scientists in 1964. Like MSG, these compounds enhance the natural flavoring substances of a food without contributing a characteristic flavor of their own. With MSG, they are used principally in meat and vegetable processed foods at levels of from 0.01 to 0.03%. Maltol and ethyl maltol are used to enhance the flavor of fruits and fruital food compositions.

Glazing and Polishing Agents

Glazes and polishes are used on coated confections to give luster to the otherwise dull coating. Chemicals that are used for this purpose include acetylated monoglycerides, beeswax, carnauba wax, gum arabic, magnesium silicate, mineral oil, petrolatum, shellac, spermaceti wax, and zein, generally at levels of ca 0.4%, with the exception of mineral oil and petrolatum which are used at 0.15%, and Zein at 1.0%.

Humectants

These are hygroscopic materials that prevent loss of moisture when incorporated into foodstuffs. The principal humectants include glycerol, mono- and diglycerides of fatty acids, propylene, glycol, and sorbitol. Pectins also act as humectants. Shredded cocoanut, marshmallows, and certain confections are examples of foods in which humectants are employed.

Nonnutritive Sweeteners

For nearly a decade this particular category of food additives has been charged with controversy and emotion owning to the implied safety risks associated with their use by a large segment of the population. Originally intended for those who had to restrict the intake of carbohydrates, the use of nonnutritive sweeteners by weight-conscious individuals has grown to unprecedented levels.

Nutrients

Where a group of people or a population is deficient in a specific essential nutrient, its addition to the diet can be justified, eg, potassium iodide and thiamine (vitamin B_1).

The enrichment of flour and bread products is now a widely accepted procedure. In the United States, the standard for enriched flour reads in part as follows: (1) it contains in each kilogram 6.4 mg (2.9 mg/lb) of thiamine, 4.0 mg (1.8 mg/lb) of riboflavin, 53 mg (24 mg/lb) of niacin, and not less than 28.7 mg and not more than 36.4 (13.0–16.5 mg/lb) of iron; (2) it may contain added calcium in such quantity that the total calcium content is 2.1 2 g/kb (960 mg/lb). Fortunately, the technological means whereby these materials may be uniformly added to foods are available to the food processor.

pH Adjusting Agents

Another large group of chemical additives that are widely used in foods might be considered under the broad heading of pH adjusting agents. Other terms that describe these chemicals include acidulants, acids, alkalies, buffers, and neutralizers. These chemicals are used in the baking industry as chemical leavening agents, in soft drinks to provide tartness, in certain dairy products to adjust the acidity, in cheese spreads for emulsification, in confectionery products as flavoring, to control the degree of inversion of sugars, and to control the texture in the processing of chocolates, and in jams, jellies, and marmalades to provide proper gel formation.

Preservatives

These chemicals may be divided into two main groups for discussion purposes: antioxidants and antimicrobials. It should be clearly understood, however, that although preservatives serve a very useful purpose in foods, they cannot take the place of cleanliness in food processing. The use of preservatives will not improve the quality of inferior material, nor will preservatives, once spoilage has set in, enable a processor to make a wholesome product out of an unwholesome one.

Antimicrobial Agents

Benzoic Acid. Benzoic acid and its sodium salt are among the bacteriostatic or germicidal agents most widely used in foods. They are included on the permitted lists of at least 30 countries throughout the world for a great variety of foods. Calcium and potassium salts are on the permitted list of certain European countries. Actually, it is the benzoic acid that is the effective agent, since sodium benzoate must be used in an acid medium to be effective. The acidity of the medium is significant; eg, a decrease in pH from 7 to 3.5 may increase five- to tenfold the antiseptic and bactericidal action. Benzoic acid is more effective against yeasts than against molds. Benzoic acid or sodium benzoate is used in several countries in such foods as soft drinks, including fruit drinks, margarine, certain fish products, roe, fruit juices, jams, marmalade, pickles, and certain vegetable products. Benzoic acid is generally used at a level of about 1000 ppm but as little as 100 ppm appears to be effective in certain foods.

p-Hydroxybenzoic Acid Esters. Butyl, ethyl, methyl, and propyl p-hydroxy-benzoates and the sodium derivatives of the methyl and propyl esters (p-NaOC6H4COOR) have been suggested as antimicrobials in foods; the methyl ester is more effective against molds, whereas the propyl ester is more active against yeasts. These esters have been suggested for use as an alternative preserving agent for benzoic acid in the same foods and at approximately the same levels.

Propionates. Although the baking process destroys the spores of molds and most species of bacteria present in flour and other ingredients, bakery products are constantly exposed to spores present in the air and on baking equipment. Despite sanitary precautions, molds frequently develop during the summer when the temperature and humidity favor their growth. Flour contains the spores of the bacterium, *Bacillus mesentericus,* which are not killed by the baking temperature. Under summer storage conditions, these bacteria become active and produce a condition called rope, which renders bread inedible. The calcium and sodium salts of propionic acid are widely used as mold and rope inhibitors in bread and bakery products at levels of ca 2000 ppm. The propionates have also found use in the prevention of mold in certain cheeses and on certain fruit and vegetable products.

Sulfur Dioxide and Sulfites. These compounds have been employed as food preservatives since the time of the early Egyptians and Romans. Sulfur dioxide is unique in being the most effective inhibitor of the deterioration of dried fruits and fruit juices. It is used widely in the fermentation industry to prevent spoilage by microorganisms and as a selective inhibitor of undesirable organisms. It is also used as an antioxidant and as an inhibitor of enzyme-catalyzed oxidative discoloration and of nonenzymic browning during the preparation, storage, and distribution of many food products. Refer to Y. H. Hui, *United States Food Laws, Regulations, and Standards,* 1986 for a review of the chemistry and technology of the preservation of fruit and vegetable products with sulfur dioxide and sulfites.

It is desirable that the initial sulfur dioxide levels in dried, cut fruits be 200–2500 ppm to provide satisfactory storage life and to maintain a high quality during prolonged storage. During storage the sulfur dioxide content decreases, due in part to the volatility of the gas and in part to oxidation of the sulfite to the inert sulfate. When cut fruits are prepared for consumption by cooking, the level of SO_2 is still further reduced. Levels of 1000 ppm or less of sulfur dioxide are used on dried vegetables.

In the production of wine, sulfur dioxide serves not only as an antimicrobial agent but also as an antioxidant and as a specific remedy for casse brune, a brown discoloration. Residual sulfur dioxide is present in certain wines up to a level of 350 ppm.

In addition to sulfur dioxide, the following sulfur compounds are reported to be used as antimicrobial agents: potassium bisulfite, potassium metabiosulfite, sodium bidulfite, sodium metabisulfite, sodium sulfite, and sodium dithionite.

Sorbic Acid. Sorbic acid and its calcium, potassium, and sodium salts are selective growth inhibitors for certain molds, yeasts, and bacteria that cause spoilage in food products. Sorbic acid or its salts are reported to be effective antimicrobial agents in such foods as cheese products, pickles, certain fish products, carbonated beverages, margarine, and certain fruit and vegetable products, including wines. The levels of use are in the range of 100–1000 ppm.

Nitrates and Nitrites of Potassium and Sodium. These salts are often present in the brines used for curing certain meat products, and appear to act as a color fixative as well as to retard microbial growth (KNO_2, $NaNO_2$, $NaNO_3$, and $NaNO_3$). Nitric oxide formed from nitrite combines with the pigment, myoglobin, of meat to form nitric oxide myoglobin, which when heated, is converted into a more stable compound that gives to cured meats the desired pink color. As a result of the cancer risks associated with the formation of nitrosamines, nitrate and nitrite levels are being reduced and alternative preservatives are being sought.

Diethyl Pyrocarbonate. Diethyl pyrocarbonate, $C_2H_5OC(O)\text{-}OC(O)OC_2H_5$ has been suggested as an antimicrobial agent. It is reported to be effective against the common spoilage organisms of various beverages at levels of 10–100 ppm. It is different from other preservatives, however, in that it reacts rapidly with water, thereby decomposing into ethyl alcohol and carbon dioxide. Many of the applications of this chemical have been in wines and, to a lesser extent, in beers, fruit juices, and carbonated beverages. However, there is evidence to show that diethyl pyrocarbonate does not hydrolyze completely to CO_2 and ethynol. Intermediate alkylating agents are also formed and this presents a serious safety risk. Consequently, it is not permitted for use in foods in the United States and many other nations.

Miscellaneous Antimicrobial Agents. Mention should be made of sodium diacetate as a mold and rope inhibitor in baked goods; diphenyl as a fungistatic that migrates from wrapping material to inhibit growth of mold-causing decay of citrus fruits; hexamethylenetetramine as a preservative for certain fish products; and *o*-phenylphenol and sodium *o*-phenylphenoxide in the postharvest treatment of fruits and vegetables to protect against microbial damage.

Antioxidants

A serious type of spoilage in foods containing oils or fats is the development of rancidity. Although rancidity may arise because of oxidative, hydrolytic, or ketonic degradation, it is generally due to oxidation through the formation of peroxides at the double bonds of fat molecules with subsequent breakdown of these peroxides to form aldehydes, ketones, and acids of lower molecular weight. The substances used to retard oxidation in foods may be classified into two groups: antioxidants and synergists.

Among the more common oil-soluble antioxidants are butylated hydroxytoluene (BHT), butylated hyroxyanisole (BHA), tert-butylhydroquinone (TBHQ), and propyl gallate (PG); in Europe, other gallate esters are also permitted (eg, isopropyl, octyl, and dodecyl). TBHQ is the most recently approved antioxidant for use in foods in the United States, and nordihydroguaiaretic acid (NGA) was deleted some years ago. Other antioxidants used by the food industry include; ascorbic acid (water soluble) and its oil-soluble esters (eg, ascorbyl palmitate), the tocopherols, gum guaiac, and lecithin.

For the most part, BHT, BHA, and TBHQ are employed alone, as blends with one another, and in conjunction with synergists such as citric acid or other metal chelating agents. Use levels are generally set at 100–200 ppm of total antioxidant based on the fat or oil content of foods. However, certain foods have specific exemptions permitting higher use levels.

The safety of BHT is under investigation and it is currently on interim status in the United States, pending further information. However, industrial use of BHT has been replaced largely by BHA and TBHQ. The gallates, although quite effective, are prone to discoloring food substrates; they also may impart off-flavor. BHT, BHA, and TBHQ have low temperature volatility (150–170°C) and are readily lost from thermal processes generating steam.

Synergists used in conjunction with antioxidants are of two types: metal-chelating agents and chain terminators. In the first instance, metals that act as pro-oxidant catalysts (eg, copper, iron, etc) are effectively neutralized using citric acid, phosphoric acid, EDTA, etc. The most popular metal chelator used with oil-soluble antioxidants is citric acid. Since this is water soluble, it is incorporated into fats and oils via special antioxidant formulations (liquid concentrates) containing glycerol monooleate and propylene glycol as cosolvents. These solvents also afford the opportunity to prepare blends of various antioxidants. Less popular is the use of sulfur-containing chain terminators as synergists, eg, thiodipropionic acid and dilauryl thiodipropionate.

Release Agents

Release agents are used to prevent confectionery and, to a lesser extent, baked goods from sticking to the container in which they are heated. They are also used to prevent pieces of confection from adhering to each other. Included are acetylated monoglycerides, calcium stearate, magnesium carbonate, magnesium silicate, magnesium stearate, mannitol, mineral oil, mono- and diglycerides, sorbitol, and stearic acid.

Sequestering Agents

Ethylenediaminetetraacetic acid (EDTA) is a sequestering, or metal-complexing, agent that has taken its place among the other sequestering (stabilizing) agents as a full-fledged tool of the food industry. This compound, as the disodium or calcium disodium salt, promotes color retention in dried bananas, beans, chick peas, canned clams, pecan pie filling, frozen potatoes, and canned shrimp; it improves flavor retention in canned carbonated beverages, dressings, mayonnaise, margarine, and sauces; it retards struvite formation in canned crabmeat and shrimp; and protects against rancidity in dressings, mayonnaise, sauces, and sandwich spreads. It is used at levels of 33–800 ppm. Other chemicals that may be included in this category are citric acid, stearyl citrate, and calcium monoisopropyl citrate, used in margarine; sodium hexametaphosphate, phosphoric acid, and various calcium, potassium, and sodium phosphates.

Starch-Modifying Agents

The food processor who wishes to use starch as an ingredient of a food product is no longer restricted to the starch as it was obtained from the plant source. A large variety of modified starches are available, each altered in a way that makes it suitable for a specific purpose.

In many food-manufacturing procedures, starches are required that retain their thickening qualities under a variety of conditions. If heat sterilization of a food product is employed, a starch with high temperature stability is required. In foods such as fruit-pie fillings and salad dressings, a thickener is needed that will not be adversely affected by the fruit acids or vinegar present. There are cases in which the thickened food product receives mechanical agitation sufficient to decrease the thickening properties of ordinary starches. Homogenized gravies and sauces, eg, require a thickener that will not break down under severe mechanical agitation; the so-called cross-linked starches provide a stable thickener required in foods of this type. The chemicals, eg, vinyl acetate, acetic anhydride, and succinic anhydride, used to modify the starches, or the modified starches themselves, may be considered to be food additives. However, these thickeners are either completely removed from the starches, or remain only as traces after processing.

Water-Correcting Agents

Water used in the brewing industry is often corrected to a uniform mineral salt content that corresponds to water known to give the most satisfactory final product. A wide variety of salts are used for this purpose, including mono- and diammonium phosphates, calcium chloride, calcium hydroxide, calcium oxide, calcium dihydrogen phosphate, calcium sulfate, magnesium sulfate, potassium aluminum sulfate, potassium chloride, potassium sulfate, sodium bisulfate, and mono-, di-, and trisodium phosphates. Some of the chemicals in addition to standardizing the salt content, also control the acidity, thus providing uniform conditions for yeast fermentation.

BIBLIOGRAPHY

This article was adapted from T. Furia, "Food Additives," M. Grayson, ed., Kirk-Othmer, *Encyclopedia of Chemical Technology*, Vol. 11, 3rd ed., John Wiley & Sons, Inc., New York, 1979; and Y. H. Hui, *United States Food Laws, Regulations, and Standards*, Vols. 1, 2, Wiley-Interscience, New York, 1986. Note all specific citations have been removed. A user must refer to the original text to obtain specific references.

General References

A. T. Branen, and co-workers, *Food Additives*, Marcel Dekker, New York, 1990.

J. M. Concon, *Food Toxicology, Contaminants, and Additives*, Vols. 1 and 2, Marcel Dekker, New York, 1987.

J. N. Hathcock, ed., 1987. *Nutritional Toxicology*, Vol. 2 Academic Press, Orlando, Fla., 1987.

R. J. Lewis, Sr., *Food Additives Handbook*, Van Nostrand Reinhold, New York, 1989.

F. R. Senti, "Food Additives and Contaminants," in *Modern Nutrition in Health and Disease*, 7th ed., M. E. Shils and V. R. Young, eds., Lea & Febiger, Philadelphia, Pa., 1988.

Y. H. HUI
EDITOR-IN-CHIEF

FOOD ALLERGY

DEFINITIONS

Allergy refers to an excess sensitivity to substances or conditions such as food; hair; cloth; biological, chemical, or mechanical agents; emotional excitement; extremes of temperature; and so on. The hypersensitivity and abnormal reactions associated with allergies produce various symptoms in affected people. The substance that triggers an allergic reaction is called an allergen or antigen, and it may enter the body through ingestion, injection, respiration, or physical contact.

In food allergies, the offending substance is usually, although not always, a protein. After ingestion, it is absorbed into the circulatory system, where it encounters the body's immunological system. If this is the first exposure to the antigen, there are no overt clinical signs. Instead, the presence of an allergen causes the body to form antibodies, which are made up of four classes of immunoglobulins (Ig): IgA, IgE, IgG, and IgM. The organs, tissues, and blood of all healthy people contain antibodies that either circulate or remain attached to the cells where they are formed. When the body encounters the antigen a second time, the specific antibody will complex with it. Because the resulting complexes may or may not elicit clinical manifestations, merely identifying a specific immunoglobin in the circulatory system will not indicate whether a person is allergic to a specific food antigen.

The human intestine is coated with the antibody IgA, which can protect a person from developing a food allergy. However, infants under seven months old have a lower amount of intestinal IgA. The mucosa thus permits incompletely digested protein molecules to enter. These can then enter the circulation and cause antibodies to form.

Children can also develop a food allergy called the delayed allergic reaction or hyperactivity. The classic sign of this is the tension–fatigue syndrome. Children with the syndrome have a dull face, pallor, infraorbital circles, and nasal stuffiness. A delayed food allergy symptom is more difficult to diagnose than an immediate one.

Although food allergy is not age specific, it is more prevalent during childhood. Because a reaction to food can impose stress and interfere with nutrient ingestion, absorption, and digestion, the growth and development of children with food allergies can be delayed. Half of the adult patients with food allergy claim that they had a childhood allergy as well. Apparently, a childhood food allergy rarely disappears completely in an adult. If a newborn baby develops hypersensitivity in the first five to eight days of life, the pregnant mother was probably eating a large quantity of potentially offending foods, such as milk, eggs, chocolate, or wheat. The child becomes sensitized in the womb, and the allergic tendency may either continue into adult life or gradually decrease.

In clinical medicine, it is extremely important to differentiate food allergy from food intolerance. The former relates to the immunosystem of the body, while the latter is the direct result of maldigestion and malabsorption due to a lack of intestinal enzyme(s) or an indirect intestinal reaction because of psychological maladjustment.

MANAGEMENT

About 2–8% of the U.S. population have some form of food allergy. The clinical management of food allergy is controversial and has many problems. For instance, a food allergy is influenced by the amount of allergen consumed, whether the allergen is cooked or raw, and the cumulative effects from successive ingestions of the allergen. A person with a food allergy also tends to be allergic to one or more of the following: pollen, mold, wool, cosmetics dust, and other inhalable items. Because these substances are so common, they are difficult to avoid. Other difficulties in allergy management are as follows: (1) if a person is allergic to food, even a very small amount can produce a reaction; (2) some patients allergic to an item at one time are not at another; (3) some patients react to an allergen only when they are tired, frustrated, or emotionally upset; and (4) although protein is suspected to be the substance most likely to allergy, people can be allergic to almost any food chemical.

In managing patients with food allergy, there are two basic objectives. First, the offending substance must be identified. Patients should then be placed on a monitored antiallergic diet to ensure adequate nutrient intake, especially young patients whose growth and development may be adversely affected by the allergy. The clinical reactions of patients allergic to a food vary from relatively mild ones such as skin rash, itchy eyes, or headache to more severe ones such as abdominal cramps, diarrhea, vomiting, and loss of appetite. Other symptoms include cough, asthma, bronchitis, purpura, urticaria, dermatitis, and various problems affecting the digestive tract (vomiting, colic ulceration of colon, etc). In children, undernutrition and arrested development may occur.

Milk Allergy

Many individuals of all ages develop an allergy as well as an intolerance to milk and milk products. The reaction may occur when a person is sick (eg, with infection, alcoholism, surgery, or trauma). Thus dietitians and nurses should always check to see whether a patient can tolerate milk. Someone allergic to milk must also avoid many foods that contain milk products. Ingesting regular homogenized fresh milk can damage the digestive mucosa of some susceptible individuals, especially children. The damaged cells bleed continuously but only minute amounts of blood are lost. The result is occult blood loss in the stool and iron-deficiency anemia. Professionals do not agree about whether this phenomenon is an allergic reaction. In rare cases, penicillin used in cows to prevent or control mastitis may leave a residue in milk. Consequently, some individuals who are allergic to the pencillin may have an allergic reaction to the inoculated cow's milk.

Breast milk is much preferred over cow's milk for feeding a baby in a family whose members have allergies. Cow's milk contains the protein B-lactoglobulin, which may trigger an allergic reaction, but breast milk does not. If an infant has symptoms of milk allergy, special formulas with soy or another protein source as a base can be safely substituted.

However, breast-feeding does have one problem when it is used to prevent an infant from having an allergic reaction to cow's milk. If the child is also allergic to substances such as cheese, crab, or chocolate, the mother can in effect feed them to her child via breast milk if she ingests them herself. Therefore, the breast-fed child may show allergic reactions.

Food allergies are difficult to test for and subsequently to diagnose and confirm. Furthermore, patients with an allergic reaction to one food may in reality be allergic to many others that contain a common ingredient. When an infant is allergic to a formula, it is usually assumed that the protein is responsible. In reality, it could be the vegetable oil base.

When patients complain about food allergies they should first provide the nurse, dietition, or nutritionist with a detailed food history. A careful study of a patient's food preference may identify the culprit. Patient reactions to food color and additives (which are found in many processed foods), laxatives, and salicylate-related chemicals should also be noted. Allergens are likely to be found in some of these items. Also, for a defined period, the patient should keep a complete record of food eaten.

Currently, three types of test are available for diagnosing food allergy. None of them is guaranteed to identify an offending substance, although they can provide some information about the patient's reaction toward different foods. These three procedures are skin testing, fasting, and elimination diets.

Skin testing for food allergy involves exposing the skin to the suspected offending chemical (from a food). If the patient is allergic, local swelling and inflammation develop within a few minutes to a day after application. However, some clinicians use this test for preliminary diagnosis while others consider it useless and unreliable. The fasting test is more reliable. After a period of fasting under medical supervision, the patient is given one food at a time to identify the offending substance. Cost and risk of nutrient deficiencies result in limited application of this test.

One method of diagnosing food allergy that has been somewhat successful is the use of progressive elimination diets. The patient is given a basal diet that is expected to produce no or minimal symptoms. If no reaction occurs, the patient is maintained on the diet for two to four weeks. Then other foods are added one at a time. Each new food is used for at least 5–10 days before another one is added. If the patient reacts to a newly added food, an offending allergen is positively identified. If the elimination diet is short term, there should be no concern about its nutritional adequacy, although nutrient intakes should be monitored. However, nutrient supplements are especially important for children.

Consultation with a nurse, dietitian, or nutritionist is important, particularly when such a diet is first implemented. At least one family member should also be involved. If the diet is to succeed, the diet instructions and implementation must be completely understood. Although a progressive elimination diet does not always yield positive results, total compliance is mandatory to maximize its effectiveness.

COMMON OFFENDERS

Although food allergy rarely constitutes a serious, life-threatening concern, it results in chronic illness for many sufferers. This problem can be significantly eliminated if attention is paid to the most common allergens and the manifestations of allergic reactions.

Cow's Milk

The allergen in cow's milk is probably the most common. A susceptible person may be allergic to whole, skim, evaporated, or dried milk, as well as to milk-containing products such as ice cream, cheese, custard, cream and creamed foods, and yogurt. If exquisite milk allergy exists, even butter and bread can create a reaction. Symptoms can include either or both constipation and diarrhea, abdominal pain, nasal and bronchial congestion, asthma, headache, foul breath, sweating, fatigue, and tension.

Kola Nut Products

Chocolate (cocoa) and cola are products obtained from the kola nut. An allergy to one almost always means an allergy to the other as well. Symptoms most commonly include headache, asthma, gastrointestinal allergy, nasal allergy, and eczema.

Corn

Because corn syrup is widely used commercially, corn allergy can result from a wide variety of foods. Candy, chewing gum, prepared meats, cookies, rolls, doughnuts, some breads, canned fruits, jams, jellies, some fruit juices, ice cream, and sweetened cereals all utilize corn syrup. In addition, whole corn, cornstarch, corn flour, corn oil, and cornmeal can cause allergic reactions to such foods as cereals, tortillas, tamales, enchiladas, soups, beer, whiskey, fish sticks, and pancake or waffle mixes. Symptoms can be bizarre, ranging from allergic tension to allergic fatigue. Headache can take the form of migraine.

Eggs

Those with severe allergy to eggs can react to even their odor. Egg allergy can also cause reaction to vaccines, because they are often grown on chicken embryos. Allergic reactions are generally to such foods as eggs themselves, baked goods, candies, mayonnaise, creamy dressings, meat loaf, breaded foods, and noodles. Symptoms can be widely varied, as with milk. Egg allergy often results in urticaria (hives), although, like chocolate, larger amounts are usually necessary to produce that symptom. Other symptoms include headache, gastrointestinal allergy, eczema, and asthma.

Peas (Legumes)

The larger family of plants that are collectively known as peas include peanuts, soybeans, beans, and peas. Peanuts tend to be the greatest offender, and dried beans and peas cause more difficulties than fresh ones. Products that can cause selected allergy reaction are honey (made from the offending plants) and licorice, a legume. Soybean allergy presents a problem similar to corn owing to its widespread use in the form of soybean concentrate or soybean oil. Legume allergies can be quite severe, even resulting in shock. They commonly cause headache and can be especially troublesome for asthma patients, urticaria patients, and agioderma sufferers (blood vessel swelling and spasm).

Citrus Fruits

Oranges, lemons, limes, grapefruit, and tangerines can cause eczema and hives and often cause asthma. They commonly cause canker sores (aphthous stomatitis). Although citrus fruit allergy does not cause allergy to artificial orange and lemon–lime drinks, if patients are allergic to citric acid in the fruits then they will also react to tart artificial drinks and may also react to pineapple.

Tomatoes

This fruit, commonly called a vegetable, can cause hives, eczema, and canker sores. It also causes asthma. In addition to its natural form, it can be encountered in soups, pizza, catsup, salads, meat loaf, and tomato paste or tomato juice.

Wheat and Other Grains

Wheat, rice, barley, oats, millet, and rye are known allergens, with wheat the commonest of the group. Wheat occurs in many dietary products. All common baked goods, cream sauce, macaroni, noodles, pie crust, cereals, chili, and breaded foods contain wheat. Reaction to wheat and its related grains can be severe. Asthma and gastrointestinal disturbances are the most common reactions.

Spices

Of various spices that can cause allergic reaction, cinnamon is generally the most potent. It occurs in catsup, chewing gum, candy, cookies, cakes, rolls, prepared meats, and pies. Bay leaf allergy generally occurs as well, because this spice is related to cinnamon. Pumpkin pie reactions are also common owing to the high cinnamon content in them. Other spices most frequently mentioned as allergens are black pepper, white pepper, oregano, the mints, paprika, and cumin.

Artificial Food Colors

Although various artificial food colors have been implicated in such problems as hyperactive syndrome in children, as allergies the two most common offenders are amaranth (red dye) and tartrazine (yellow dye). Amaranth is most often encountered, but reactions to tartrazine tend to be more severe. Food colors occur in carbonated beverages, some breakfast drinks, bubble gum, flavored ice foods, gelatin desserts, and such medications as antibiotic syrups.

Other Food Allergens

Any food is capable of producing an allergic reaction. However, those offenders often mentioned after the top 10 are pork and beef, onion and garlic, white potatoes, fish, coffee, shrimp, bananas, and walnuts and pecans. Vegetables, other than those already mentioned, rarely cause allergic reactions. Fruits that usually are safe include cranberries, blueberries, figs, cherries, apricots, and plums. Chicken, turkey, lamb, and rabbit prove the safest meats. Tea, olives, sugar, and tapioca are also relatively safe foods, although some herbal teas can cause unique difficulties.

BIBLIOGRAPHY

This article has been adapted from Y. H. Hui, *Human Nutrition and Diet Therapy,* Jones and Bartlett Publishers, Inc. Monterey, Calif., 1983. Used with permission.

S. T. Bahna and D. C. Heiner, *Allergies to Milk,* Grune & Stratton, New York, 1980.

J. C. Breneman, ed., *Handbook of Food Allergies,* Marcel Dekker Inc., New York, 1987.

J. Brostoff and S. J. Challacombe, eds., *Food Allergy and Intolerance,* Saunders, Philadelphia, 1987.

S. N. Butkus and co-workers, "Food Allergies: Immunological Reactions to Food," *Journal of the American Dietetic Association* **85**(5), 601 (1986).

A. J. Cant, "Food Allergy in Childhood," *Human Nutrition: Applied Nutrition* **39A**, 277 (1985).

R. K. Chandra, "Food Allergy," in M. E. Shils and V. R. Young, eds., *Modern Nutrition in Health and Disease,* 7th ed, Lea & Febiger, Philadelphia, 1988.

R. K. Chandra, ed., *Food Allergy* Nutrition Research Foundation, St. John's N.F., Canada, 1987.

F. M. Dong, *All About Food Allergy,* Stickley, Philadelphia, 1984.

C. D. May, "Food Allergy: Perspective, Principles, Practical Management," *Nutrition Today* **15**, 28 (1980).

R. M. Sly, *Pediatric Allergy,* Medical Examination Publishing, New York, 1981.

L. I. Thomas, *Caring and Cooking for the Allergic Child,* Sterling Publishing, New York, 1980.

Y. H. HUI
EDITOR-IN-CHIEF

FOOD ANALYSIS

The basic purpose of an analysis is to determine the mass (weight) of a component in a sample. The numerical result of the analysis is expressed as a weight percentage or in other units that are equivalent to the mass/mass ratio. The mass (weight) of a component in a food sample is calculated from a determination of a parameter whose magnitude is a function of the mass of the specific component in the sample.

Some properties are basically mass dependent. Absorption of light or other forms of radiant energy is a function of the number of molecules, atoms, or ions in the absorbing species. Certain properties, such as specific gravity and refractive index, are not mass dependent. They can, however, be used indirectly for mass determination. Thus one can determine the concentration of ethanol in aqueous solutions by a density determination. Refractive index is used routinely to determine soluble solids (mainly sugars) in syrups and jams.

Some mass-dependent properties may be characteristic of several or even of a single component and may be used for selective and specific assays. Examples are light absorption, polarization, or radioactivity. Some properties have both a magnitude and specificity parameter (nuclear magnetic resonance and infrared spectroscopy). Such properties are of great analytical value because they provide selective determinations of a relatively large number of substances.

INFORMATION SOURCES

Keeping informed of current developments is important to all professional workers, particularly those in a rapidly expanding area such as food analysis.

Periodicals are the principal means for the exchange of scientific information. There are two basic types of scientific journals: (1) primary sources, which contain reports of original research and include detailed descriptions of the experimental procedures and the data derived from them, and (2) secondary sources consisting of abstracts and reviews, which provide a condensed source of information.

For general background information, several encyclopedias devoted to chemistry and technology are available. These include the technical and detailed *Kirk-Othmer Encyclopedia of Chemical Technology* (Wiley-Interscience) and *Ullmanns Enzyklopädie der technischen Chemie* (Urban and Schwarzenberg). Ullmann's work is available in English (Verlag Chemie).

An excellent single-volume source of information is *The Merck Index*. This compilation, revised every several years, contains formulas, preparations, and properties of over 10,000 chemicals.

In the identification and determination of food composition, the analyst often has to determine the physical properties of the substance. Relevant reference data can be obtained from various standard tables, handbooks, and some of the newer encyclopedias and dictionaries. Information on physical properties is often published by manufacturers of chemicals. In addition to industrial laboratories, the U.S. National Bureau of Standards publishes authoritative compilations of physicochemical data.

Reviews

Critical reviews of the current knowledge in a particular field are comprehensive surveys of current knowledge on a specific subject. One of the best known is the *Advances in Chemistry Series* published by the American Chemical Society. Another monograph series is published in *Annals of the New York Academy of Sciences*. Both series are based largely on symposia organized by scientific societies. Since 1959, the Chemical Abstracts Service of the American Chemical Society has published annually a *Bibliography of Chemical Reviews*.

Since 1904, the Chemical Society (England) has published authoritative summaries of the previous year's important papers in *Annual Reports on the Progress of Chemistry*. A parallel series, *Reports on the Progress of Applied Chemistry*, is published by the Society of the Chemical Industry. Each year, the April issue of *Analytical Chemistry* is devoted to review papers; various aspects of food analysis are covered every second year.

Theses

Since 1938, theses from United States universities have been processed for microfilming by University Microfilms of Ann Arbor, Michigan. Abstracts of up to 600 words of such theses are published in *Dissertation Abstracts*. An annual list of United States masters theses in the pure and applied sciences has been published since 1955–1956 by the Thermophysical Properties Research Center, Purdue University, Lafayette, Indiana.

Symposia

Programs of conferences and symposia are published in several journals. A comprehensive list is published in *Science*; more selective and limited lists are published in *Food Technology, & Chemical & Engineering News*.

Abstracts of papers presented at scientific meetings generally are published for members in the form of books by the sponsoring scientific society (eg, American Chemical Society or Federation of Biological Sciences) or are included in periodicals (eg, *Food Technology* for Institute of Food Technologists, *Cereal Foods World,* for American Association of Cereal Chemists, and *Journal of the American Oil Chemists' Society,* for American Oil Chemists Society.

Trade Publications

Some trade publications are the best sources for properties and applications of specialized equipment and chemicals. Several manufacturers periodically issue bibliographies and abstracts of technical/scientific articles in a specific area; some prepare detailed handbooks giving specifications, properties, and details of analytical procedures. Industrial manuals are indispensable in installing, using, and servicing equipment. Several scientific journals periodically prepare lists of major commercial supply houses. A "Comprehensive Guide to Scientific Instruments" is published annually in *Science*.

Abstracts and Bibliographic Lists

Chemical Abstracts publishes over 200,000 abstracts a year, selected from about 10,000 journals in over 50 languages. The section on biochemistry covers various aspects of biology and chemistry of materials of plant, animal, and microbial origin. Included are informative abstracts of original scientific papers, patents, and some reviews, as well as lists of theses, monographs, books, reviews, and proceedings.

Some of the recent monographs, reviews, and textbooks are listed in the bibliography (1–6).

Another most comprehensive and useful source is *Food Science and Technology Abstracts*.

Standard Methods

Food analysts use primarily methods approved by various associations such as the Association of Official Analytical Chemists, the American Association of Cereal Chemists, the American Oil Chemists' Society, the American Public Health Association, and the American Society of Brewing Chemists (and their national or international counterparts) (7–9). Most of the methods recommended by the United States and other organizations have been developed after years of collaborative testing and are considered reliable and official.

Modern Information Retrieval Systems

During the past few decades, computerized systems for storing and retrieving scientific information have been developed. An awareness of the systems available and how to use them properly can significantly reduce the time and effort involved in a literature search of a particular field.

Several data bases of particular interest to food scientists and analysts are available. They include *AGRICOLA* (USDA, National Agricultural Library, Beltsville, Md.), *BIOSIS PREVIEWS* (Biosciences Information Services, Philadelphia, Pa.), *CA Search* (Chemical Abstracts Service, Columbus, Ohio), *CAB Abstracts* (Commonwealth Agricultural Bureau, Slough, U.K.), *CRIS* (USDA, Washington, D.C.), *Dissertation Abstracts Online* (University Microfilms International, Ann Arbor, Mich.), and *Food*

Science and Technology Abstracts (International Food Information Service, Reading, UK).

SAMPLING

Valid analyses require adequate methods of food sampling and preserving. The best sample will be as close as possible in all its intrinsic properties to the bulk of the sample from which it was taken. The major steps in sampling are identifying the population from which the sample is to be obtained, selecting and obtaining gross samples, and reducing the gross sample to a laboratory-size sample suitable for analysis. Both manual and continuous (mechanical) samplers are available. The care, time, and effort devoted to the preparation of samples for analyses are critical to obtaining the required information. Grinding of dry or moist materials, enzymic and chemical procedures to disintegrate various types of materials, enzyme inactivation (depending on the type of analysis), minimizing lipid changes, and controlling oxidative and microbial attack are all part of a sound sample preparation (10–11).

In reporting analytical results, both the reference basis and the units used to express the results must be considered.

The reliability of an analytical method depends on the following factors:

(a) Specificity—absence of interfering substances that yield a measurement of the same kind as the substance being determined.
(b) Accuracy—the degree to which a mean estimate approaches a true estimate of an analyzed substance.
(c) Precision—the degree to which a determination of a substance yields an analytical true measurement of that substance.
(d) Sensitivity—the ratio between the amount of a substance and the magnitude of instrumental response, or the smallest measurable compositional difference between two samples.

ANALYTICAL METHODS

The basis for many common instrumental techniques of analysis is the interaction of radiation with matter. The instruments are used to measure light in the visible region or ultraviolet and infrared radiation. Numerous sensitive methods are based on the phenomenon of luminescence, systems that can be made to emit light. The two major divisions of analytical importance are fluorescence and phosphorescence.

The heat energy from a flame can raise an electron in some atoms from the ground state to an excited state. Each element has a different set of energy levels and the wavelengths of emitted radiations can be used to identify and measure elements in flame photometry. An atomic absorption instrument contains a source of radiation whose wavelengths match those required to excite the atoms in the flame, thus providing a more efficient means of atom excitation. A disadvantage of atomic absorption spectroscopy is that only one element can be determined at a time. In flame photometry or emission spectroscopy several elements can be determined simultaneously.

The numerous types of x-ray analyses include absorption, diffraction, fluorescence, emission, electron probe, and nondispersive techniques. X-ray fluorescence and nondispersive methods are today the most important analytical techniques.

The basis of potentiometric methods is the use of reference electrodes (calomel, silver–silver chloride) and indicator electrodes (glass; ion-selective electrodes of various types: solid-state, liquid–liquid membrane, enzyme, bimetallic). Potentiometry consists of measuring the potential difference between a reference electrode and an indicator electrode at various intervals during a titration. In coulometric determinations the reagent is electrically generated at the surface of a working electrode immersed in a solution with a counter electrode. In conductivity measurements use is made of two small parallel platinum foil electrodes about 1 cm apart. When a voltage is applied to the two electrodes in a solution containing ions, the anions move toward the anode and the cations toward the cathode. The amount of current passing between the two electrodes is a function of the concentration, temperature, and kind of ions.

The principle of conductivity is used in detectors for various forms of chromatography: ion, high-performance liquid, and gas. Electrophoresis involves movement of charged colloidal particles and macromolecular ions under the influence of an electric field. Zone electrophoresis (on solid supports and in gels) is distinguished from paper electrophoresis (of spots or streaks on a strip). Specialized variations include disk electrophoresis, isotachophoresis (separation of sample components on the basis of differences in net mobility), and isoelectric focusing (separation in a pH gradient influenced by an electric field). Various modifications of voltammetry (polarography) utilize the principle of a dropping mercury electrode in modern electroanalytical assays.

In mass spectroscopy we produce ions by bombarding organic molecules with high-energy electrons. The ions are accelerated so that they can be separated, detected, and measured according to their mass or velocity. Mass spectrometry is useful in the identification and determination of the components in mixtures of organic materials. Modern uses include direct attachment for analyses of components separated by several chromatography techniques.

Nuclear magnetic resonance (nmr) spectra are used widely to fingerprint molecules in solution: wide band nmr found wide application in the determination of moisture, oil, or the solid fat content of agricultural commodities and foods.

Chromatographic methods are one of the greatest achievements of modern analytical chemistry. No other separation technique can match chromatography for simplicity, efficiency, and versatility. Types of chromatography are listed in Table 1.

In these methods, separations depend on gross chemical and/or physical differences among the substances that are to be separated. In gel filtration, separation is based on differences in molecular size and shape. In ion-exchange chromatography, separations are based on dif-

Table 1. Types, Phases, and Principles of Chromatography

Type	Phase Stationary	Phase Mobile	Physical Principal
Column, thin layer, paper	Solid	Liquid	Adsorption partition, ion exchange gel permeation
Column, thin layer, paper	Liquid	Liquid	Partition
Gas–solid	Solid	Gas	Adsorption
Gas–liquid	Liquid	Gas	Partition

ferences in electrical properties. In affinity chromatography (including immobilized enzymes), separations are based on specific interactions between pairs of substances such as a macromolecule and its substrate, cofactor, allosteric effector, or inhibitor. In principle, a ligand is attached covalently to a water-insoluble matrix that has been designed (or is available) specifically to adsorb from a mixture only the component(s) with an affinity for the ligand. All other components pass freely through the adsorbent. The adsorbed component(s) can then be eluted after some change in conditions such as pH or ionic strength.

The original column chromatographic techniques employed glass columns and either gravity flow or slight vacuum to move the mobile phase through the column. To speed up the otherwise flexible procedure, steel columns and high pressures are now used widely in high-pressure (or performance) liquid chromatography (HPLC). Capillary HPLC is a new dimension in sensitivity, reproducibility, and specificity of separation.

Some of the advantages of paper chromatography (ie, for separation of simple carbohydrates, amino acids, and peptides) are speed and simplicity. In two-dimensional or ion-exchange paper chromatography, specificity can be increased. Thin layer chromatography (well suited for separation of lipids) has the advantages of high speed, sensitivity, versatility, and capacity for scale-up separations (including preparative). Gas–liquid chromatography has been (and in many cases still is) the method of choice for separation of naturally volatile compounds or compounds that can be volatilized. Its use in identification of flavor components is particularly noteworthy.

Many analytical methods for foods require extraction as part of a cleanup procedure (to eliminate interfering substances), as a concentration step, or as an aid in identification of a component in a mixture. Batch, continuous, and discontinuous countercurrent techniques are available for liquid–liquid and liquid–solid extractions.

Centrifugation is used for separation of solids from liquids and from immiscible solvents and for resolution of emulsions that are formed during extraction. Ultracentrifugation (centrifugation at high speeds) is useful for concentrating high molecular weight materials and for estimating their molecular weights. Special preparative techniques are density gradient, stabilized moving boundary, zone, and isopycnic gradient ultracentrifugation. Determination of density is made primarily on liquids. It includes pycnometric and buoyancy methods.

Refractive index measurements, especially in combination with density determination, can be used for identification of unknown substances or the structures of known substances. Refractometers of the Abbe type, which measure the critical angle of the sodium D line, are used widely in analyses of lipids, sugar solutions, and mixtures of natural compounds. For very precise measurements, differential refractometry and interferometry are used.

Polarimetry (measurement of plane polarized light) is used in determining optical rotation using a monochromatic source of radiation (sodium vapor lamp and mercury vapor lamp). A modification of a polarimeter is the quartz wedge saccharimeter used extensively by the sugar industry.

Rheology is concerned with stress-strain relations of materials that show behavior intermediate between that of solids and liquids, i.e., practically all foods. Measurements of rheological properties of foods can be based on either the analytical or the integral approach. In the first, the properties of a material are related to such simple systems as Newtonian fluids or Hookean solids. In the integral approach, a simple empirical relation between stress, strain, and time is established. The relation measures, basically, the way in which rheological properties vary under a specific system of applied forces. The instruments used provide a measure of food texture and consistency (12–15).

In recent years, serology, immunochemistry, and immunoelectrophoresis techniques are increasingly used to provide valuable information concerning the identity and nature of chemical components of biological materials. Methods involving labeling of antigens and antibodies include fluorescence, radio, and enzyme immunoassays (EIA). Numerous variants of the two main EIA methods—ELISA (enzyme-linked immunosorbent assay) and EMIT (enzyme-multiplied immunoassay technique)—have found many applications in food analyses. They include identification of food sources; rapid and specific assay of micro amounts of drugs and biological substances in human biological fluids; and detection of pollutants and toxins in fish, of diseases in slaughter animals, and of environmental contaminants.

The major advantages of enzymatic analyses (analyses with the aid of enzymes) lie in their ability to react specifically with individual components of a mixture. This avoids lengthy separations of the components and reduces the time required for an analysis. The required sample size is small. Since mild analysis conditions are employed, labile compounds can be detected and measured.

Because some of the nutritional requirements of microorganisms and experimental animals are similar, it is possible to employ analytical microbiology to determine some substances that are essential constituents of living cells. Microorganisms, as reagents, have been used to determine amino acids, vitamins, nucleic acids, heavy metals, growth factors, nutritional value of proteins, and antibiotics. The basic principle is that in the presence of limiting amounts of certain compounds, the amount of microbial growth is a function of the amounts of those compounds. The microorganisms used for assay are primarily bacteria, but yeasts, fungi, and protozoans also have been used. The assay methods include diffusion in a gel (radius of a growth or inhibition zone), turbidimetric and dilution methods, gravimetric methods, and metabolic response methods.

Although many new, rapid, and relatively selective methods (ie, hplc) have been developed, immunochemical, enzymatic, and microbiological methods continue to be used both for confirmation and for differentiation among various forms that differ in biological activity.

MAJOR FOOD COMPONENTS

The methods described in the previous section are used in the determination of the composition of foods, including the major components (moisture, carbohydrates, proteins, lipids, and minerals). Those components are included in standard tables of composition of foods.

Moisture. Moisture determination is one of the most important and most widely used measurements in the processing and testing of foods: The amount of moisture is inversely related to the amount of dry matter in a food; moisture has a large (and mostly, critical) effect on stability and quality; moisture determination is important in evaluation of materials balance or of processing systems; and moisture content must be known to determine the nutritional value of a food or to express results of analytical determinations on a common moisture basis.

Commonly used procedures for determining moisture involve thermal drying procedures (forced air oven, vacuum oven, infrared drying). While distillation methods have been applied for over 100 years, their use has decreased. They are used only in very special cases. Among chemical methods the Karl-Fischer titration is most widely used and accepted. Several physical methods are available: infrared determination, gas chromatographic and nuclear magnetic methods, and several electrical methods for rapid, routine determination (based on conductivity, resistance, and capacitance).

Carbohydrates. Carbohydrates are the most abundant and widely distributed food components. They include mono-, di-, oligo-, and polysaccharides. The latter can be divided into structural polysaccharides (cellulose, hemicellulose, lignin) and nutrient polysaccharides (starch, glycogen).

Assay methods for mono- and oligo-saccharides include chemical, colorimetric, chromatographic, electrophoretic, optical, and biochemical procedures. Many procedures involve preliminary separation by chromatographic and electrophoretic techniques prior to the classical chemical or optical assays. The use of enzymatic assays is gaining in popularity.

Starch can be extracted with perchloric acid and precipitated with iodine, and the starch liberated from the complex can be determined colorimetrically. Several methods are available to determine starch in extracts of hot, concentrated solutions of calcium chloride. Combinations of acid and enzymatic hydrolysis or hydrolysis by several enzymes followed by specific assays of glucose have been described.

The determination of structural polysaccharides is still fraught with many difficulties because the materials contain many undefined polymers varying widely in size and compositions. In food composition tables, carbohydrates are reported by difference after subtracting from 100% moisture, protein, oil, and ash content; in some cases fiber (crude, total, and dietary) is determined separately (16).

Protein. Methods of determining the total protein content of foods (crude protein) are empirical in nature. Proteins are polymers of about 24 amino acids. Amino acids can be determined, after hydrolysis of proteins, by colorimetric, enzymatic, microbiological, and chromatographic methods. The most powerful methods of amino acid analysis in protein hydrolysates are based on ion exchange and gas chromatography.

The most common procedure for a protein assay depends on determining a specific element or group in the protein, and calculating the protein content by using an experimentally established factor. The elements or groups most commonly used are nitrogen, certain amino acids, or the peptide linkage. It is assumed that the constituent determined is present entirely in the protein fraction and that the empirical factor for conversion is constant. Determination of protein on the basis of nitrogen content can be made by the Kjeldahl method or by the Dumas method. Other assays include the biuret method (formation of a purple complex between copper salts in alkali solutions and compounds containing at least two peptide bonds), the Lowry method (color formed after interaction of proteins with a phenol reagent and copper under alkaline conditions), direct spectrophotometric methods (at 280 nm, 210 nm, or two wavelengths), nephelometric and turbidimetric methods, dye binding techniques, as well as many instrumental assays (nmr, near infrared reflectance spectroscopy, neutron activation, proton activation).

Lipids. Determinations of lipids are based on their sparing solubility in water and their considerable solubility in organic solvents. Successful extraction requires that bonds between lipids and other compounds be broken so that the lipids are free and solubilized. Generally, such solubility is attained when polarities of the lipid and the solvent are similar. Ethyl ether and petroleum ether are two of the common extraction solvents. Combinations of alcohol and ether, a ternary mixture of chloroform–methanol–water, or water-saturated butanol are examples of effective extractants. The extracts are rich, however, in nonlipid components that must be subsequently separated. Direct extraction is often carried out in a Soxhlet or modifications of it for rapid assay. Numerous indirect methods for rapid determination of lipids (near infrared reflectance spectroscopy, nmr, x-ray absorption, specific gravity, dielectric properties) are available. The isolated lipids can be characterized by fractionation (countercurrent distribution, column chromatography, or thin layer chromatography) or numerous physical measurements of fats for identification, for checking purity, and for the control of certain aspects of processing. Rancidity tests have been developed to describe and determine objectionable flavors and odors caused by either hydrolytic or oxidative changes in the fat (17).

Mineral Components. Ash is the inorganic residue from the incineration of organic matter. It is a measure of total mineral components in a food (18). Depending on the nature of the mineral components, various ashing procedures can be employed. For determining the individual mineral components, wet ashing is frequently used. Those methods are especially useful for assay of trace elements. The elements in a digest can be determined by emission spectroscopy, flame photometry, atomic absorption spectroscopy, neutron activation analysis, x-ray spectroscopy, glass electrode, and by many colorimetric, turbidimetric, polarographic, and electron probe microanalyzer methods. Some are applicable to determination of individual elements in situ, in combination with scanning electron microscopy.

OVERALL EVALUATION OF FOODS

In addition to the quantitative aspects of food composition, the food analyst is interested in its nutritional value (either through direct feeding studies or from computations based on composition), presence of toxic components, and in sensory attributes (appearance, color, taste, odor, texture, and overall acceptance) (19). Finally, many raw materials and products are evaluated by practical performance tests (flour in breadmaking, oils and fats in production of margarine and mayonnaise, sugars and jellying substances in production of confections, and many others).

BIBLIOGRAPHY

1. Y. Pomeranz and C. E. Meloan, *Food Analysis, Theory and Practice,* 2nd ed., AVI–Van Nostrand Reinhold Co., New York, 1987.
2. D. W. Gruenwedel and J. R. Whitaker, eds., *Food Analyses, Principles and Techniques,* Vols. I–V, Marcel Dekker, New York, 1984.
3. W. Baltes, P. B. Czedik-Eysenberg, and W. Pfannhauser, eds., *Recent Developments in Food Analysis,* Verlag Chemie, Weinheim, FRG, 1982.
4. R. D. King, ed., *Developments in Food Analysis Techniques,* Elsevier Applied Science Publishers, London, 1984.
5. G. G. Birch and K. J. Parker, eds., *Control of Food Quality and Food Analysis,* Elsevier Applied Science Publishers, London, 1984.
6. W. Diemair, ed., *Handbook of Food Chemistry, Food Analysis,* Springer Verlag, Berlin, 1967.
7. AACC, *Cereal Laboratory Methods,* 8th ed., American Association of Cereal Chemists, St. Paul, Minn., 1983.
8. AOAC, *Official Methods of Analysis of the Association of Official Analytical Chemists,* 13th ed., Association of Official Analytical Chemists, Washington, D.C., 1980.
9. AOCS, *Official and Tentative Methods of The American Oil Chemists' Society,* 3rd ed., American Oil Chemists' Society, Chicago, Ill., 1982.
10. F. M. Garfield, N. Palmer, and G. Schwartzman, eds., *Optimizing Chemical Laboratory Performance through the Application of Quality Assurance Principles,* Association of Official Analytical Chemists, Washington, D.C., 1980.
11. W. Horwitz, "Evaluation of Analytical Methods for Regulation of Foods and Drugs," *Analytical Chemistry,* **54,** 67A, 68A, 70A, 72A, 74A, 76A (1982).
12. M. C. Bourne, *Food Texture and Viscosity; Concept and Measurement,* Academic Press, New York, 1982.
13. J. M. de Man, P. W. Voisey, V. F. Rasper, and D. W. Stanley, eds., *Rheology and Texture in Food Quality,* AVI Publishing Co., Inc., Westport, Conn., 1976.
14. N. W. Mohsenin, *Physical Properties of Plant and Animal Materials,* Gordon & Breach Science Publishers, New York, 1980.
15. M. Peleg and E. B. Bagley, eds. *Physical Properties of Foods,* AVI Publishing Co., Inc., Westport, Conn., 1983.
16. W. P. T. James and O. Theander, eds., *The Analysis of Dietary Fiber in Food,* Marcel Dekker, New York, 1981.
17. J. C. Allen and R. J. Hamilton, eds., *Rancidity in Foods,* Applied Science Publishers, London, 1983.
18. J. Gilbert, ed., *Analysis of Food Contaminants,* Elsevier Applied Science Publishers, London, 1984.
19. M. A. Amerine, R. M. Pangborn, and E. B. Roessler, *Principles of Sensory Evaluation of Foods,* Academic Press, New York, 1965.

Y. Pomeranz
Washington State University
Pullman, Washington

FOODBORNE DISEASES

All through history, human beings no doubt were affected by a great variety of foodborne diseases. No one knows for certain the number of cases of foodborne intoxications and infections occurring annually in the world. In the United States, the estimation is about 24 million cases per year, which means about 1 in 10 U.S. citizens is affected by foodborne disease each year. In countries with poor sanitation, one can only surmise that the number of foodborne disease cases is much higher. The current concern in public health is food safety. Consumers are much more aware of the great potential for large-scale foodborne outbreaks and demand safer food supply. Food microbiologists are charged with the responsibility of studying the occurrence, enumeration, isolation, detection, characterization, prevention, and control of foodborne microorganisms from food, water, and the environment.

DEFINITIONS

Food intoxication is the ingestion of preformed toxic compounds by susceptible persons who later become ill.

Food infection is the ingestion of large numbers of viable, pathogenic microorganisms by susceptible persons who later become ill.

Food poisoning is the ingestion of contaminated food containing either chemical preformed toxins or live microbes by susceptible persons who later become ill.

Foodborne outbreak is the consumption of contaminated food from one source by one or many people who later become ill.

Foodborne disease case is the consumption of contaminated food by one susceptible person who later became ill. An outbreak can have one case or 100,000 cases.

Endemic is the usual cases of illness in a community.

Epidemic is an unusual, large number of cases of illness from a single source in a community.

Pandemic is a disease affecting the entire world.

Epidemiology is the study of diseases in a population using statistical methods. Epidemiologists study patterns of diseases and their causative agent in terms of a population, whereas a physician treats individual patients.

Etiologic agent is the agent that caused a specific disease.

FOOD INTOXICATIONS

Chemical intoxications are usually the result of accidents. People have been poisoned by inorganic compounds such as antimony, arsenic, cyanide, cadmium, lead, selenium, and mercury. The symptoms usually occur rapidly (a few minutes or hours), and reactions are usually violent in cases of ingestion of large doses of the toxic compounds. Immediate medical assistance is essential for the victims in such cases.

BACTERIAL INTOXICATION

Clostridium botulinum Intoxication

The first recorded outbreak of botulism was in 1793 involving sausages (botulus) in Germany. Since that time, many outbreaks all over the world have been reported. For example, in the United States between 1971 and 1985, three outbreaks were recorded with 485 cases and 55 deaths. The organism is a gram-positive, anaerobic, spore-forming rod that can grow at temperatures from 3.5°C to as high as 50°C. Most strains will grow well at 30°C. The spores formed by this organism can reside in soil, water, and the environment and can be transmitted to foods. Foods involved in botulism cases usually are improperly home-canned medium or low acid foods. Information since 1899 indicates that about 70% of the outbreaks can be traced to improperly processed home-canned foods and 9% to commercially processed food, with the other outbreaks from unknown sources. Symptoms develop 18 to 96 h after ingestion of toxic foods. They include vomiting, nausea, fatigue, dizziness, vertigo, headache, dryness of mouth, muscle paralysis, and death by asphyxiation. There are several types of botulin toxins (types A, B, C_1, C_2, D, E, F, and G). These are large molecular weight proteins (about 1 million dalton). The most important ones affecting human beings are toxins A, B, and E. These toxins are among the most toxic materials produced by a biologic system. It was estimated that one pure ounce of toxin can kill 200 million people. Treatment is by administration of monovalent E, bivalent AB, trivalent ABE, or polyvalent ABCDEF antisera. Fortunately, the toxins are heat sensitive. Boiling of the toxin for 10 minutes will destroy it. The toxins can be detected by animal tests using mice as well as immunologic tests using specific antibodies (gel diffusion tests, ELISA, RIA tests, etc). The key to preventing botulism is to know the composition (pH, A_w, oxidation-reduction potential, presence of inhibitory compounds, etc) of the food and to utilize proper time and temperature for processing as well as correct packaging and storage of the processed food. All high-moisture foods processed and then stored under anaerobic conditions should be subject to close scrutiny to avoid the possibility of *C. botulinum* surviving and later germinating and producing the toxins. When in doubt, always boil the suspected food for 10 minutes before discarding it.

Staphylococcus aureus

Staphylococcus aureus is a gram-positive facultative anaerobic coccus occurring in clusters. The organisms is ubiquitous and can be found in human skin, nose, hair, etc. The organisms, when allowed to grow in food, may produce a class of low molecular weight (ca 30,000 daltons) protein toxins called staphylococcal enterotoxins (A, B, C_1, C_2, C_3, D, and E). These toxins, when ingested by a susceptible person, will cause severe nausea, vomiting, abdominal cramps, diarrhea, and prostration about 4 to 6 h after consumption. Recovery is about 24 to 72 h. Victims will not die, but may wish they had, as the reactions are very violent. These toxins are heat stable. Once formed in food, boiling will not destroy the preformed toxins. The organism is not a good competitor compared with other spoilage organisms (eg, *Pseudomonas*); however, in the absence of competitors, such as in salty food or processed foods, the organism can grow and produce the heat-stable toxins. It is therefore essential to prevent *S. aureus* from growing in the food by proper refrigeration of cooked foods or keeping hot food hot. They can produce enough toxin in 4 h at room temperature to cause a problem.

Aspergillus

Aspergillus flavus and *A parasiticus* are molds that can produce the carcinogenic toxins called aflatoxins. In 1960 in England, 100,000 turkeys died of unknown causes, and the disease was called Turkey X disease. After much work, the contaminant was found to have originated from peanut meals from Brazil. The organisms responsible for producing the toxic compounds were identified as *A. flavus*. Later, *A. parasiticus* was also found to be able to produce the toxin. The mold can grow between 7.5°C and 40°C with optimal temperature at 24–28°C. In 1 to 3 days of growth, the organism can produce the toxins. The primary toxins are B_1, B_2, G_1, and G_2. B and G indicate that the toxins fluoresce blue or green under ultraviolet light. When cows consume B_1 and B_2 toxins, they can modify the toxins and excrete the toxins as M_1 and M_2 in milk.

Spores of these molds are ubiquitous; the organisms has been found to grow in rice, sorghum, peanut, corn, wheat, and soybean corps, as well as animal feed. Because the toxins are carcinogenic, they are under strict government scrutiny. Currently the allowed limit is 20 ppb for foods. Although no food-related aflatoxin cases have been reported in the United States, there were cases reported in Southeast Asia when people consumed heavily contaminated food. The toxins can be detected by animal tests using ducklings or chick embryo. Thin layer chromatography and high performance liquid chromatography can also be used to detect these toxins. Recently, monoconal antibodies have been employed to detect these toxins with great rapidity (10 to 30 min) and sensitivity (1 ppb and lower).

Exotoxins Versus Endotoxins

It is necessary to differentiate these toxins before a discussion on foodborne infections. *Exotoxins* are toxins produced by an organism and later released into the environment. Ingestion of these preformed toxins causes foodborne intoxication. These toxins are protein toxins mainly produced by gram-positive organisms. Because they are proteins, they can be neutralized by corresponding antibodies and detected by a variety of immunologic methods. These toxins are relatively heat sensitive (except the staphylococcal enterotoxins as described earlier). These toxins also have distinct pharmacology. Examples of exotoxins are staphylococcal enterotoxins (affecting the intestinal tracts) and botulinum neurotoxins (affecting the nervous system).

Endotoxins are part of the cell wall material of gram-negative cells. Every gram-negative bacterium examined has endotoxins. These are complex molecules containing protein, carbohydrate, and lipid. The protein moiety determines antigenicity, the carbohydrate moiety determines immunologic specificity, and the lipid moiety causes toxicity. Unlike exotoxins, antibodies will not neutralize toxicity because the toxic part is the lipid. All endotoxins have the same action and are released when the gram-negative bacterium lyses. These endotoxins cause fever by acting as *exogenous pyrogens*. The exogenous pyrogen, when absorbed into the bloodstream, causes injury to the leukocytes, which in turn releases an *endogenous pyrogen*. This endogenous pyrogen stimulates the thermoregulatory center of the brain at the hypothalamus and causes fever. Therefore, fever production by a patient is indicative of a foodborne infection case. Endotoxins can be detected by the limulus amebocyte lysate (LAL) test. In the presence of endotoxins, the LAL will form a gel. The reaction takes about 1 h. Currently, hospital materials should be pyrogen-free and LAL is the standard test for pyrogens in the hospital supplies and environment.

BACTERIAL INFECTION

Clostridium perfringens

Clostridium perfringens occupies an interesting position as being both a foodborne infection agent as well as a foodborne intoxication agent. On the one hand, the susceptible person has to ingest large numbers of viable *C. perfringens* before coming down with a food-poisoning case, and on the other hand, the organism produces an enterotoxin to cause the illness. *C. perfringens* is a gram-positive anaerobic spore-forming rod. Its generation time in ideal conditions is as short as 9 min, making it the fastest growing organism known. Spores of the organism distribute widely in nature and can easily contaminate foods. Most of the incidences of *C. perfringens* food poisoning involve meats prepared in large quantities one day and consumed the next day while the food is held at lukewarm temperatures. In such conditions, most vegetative cells of competitors die off while the spores of *C. perfringens* have a chance to germinate and grow into large numbers. The organisms, when ingested by a susceptible person, will start to sporulate owing to the favorable anaerobic conditions in the small intestine. The gene coded for sporulation also controls the release of an enterotoxin that is responsible for the mild diarrhea characteristics of *C. perfringens* food poisoning. Symptoms occur 8–20 h after ingestion of a large number of viable *C. perfringens* and include acute abdominal pain, diarrhea, nausea with rare vomiting. The symptoms are milder than those caused by *Salmonella*. Detection of this organism is by anaerobic cultivation of food using differential anaerobic agar such as tryptose sulfite cycloserine agar. *C. perfringens* forms black colonies in this agar medium. *C. perfringens* causes about one-fifth of all foodborne cases in the United States annually.

Salmonella

Salmonella is the classic example of foodborne infection. *Salmonella enteritidis* was isolated in 1884 and still is an important foodborne organism. The organism is a gram-negative, facultative anaerobic, non-spore-forming rod, motile by peritrichous flagella. It does not ferment lactose and sucrose but ferments dulcitol, mannitol, and glucose. The organism is heat sensitive but can tolerate a variety of chemicals such as brilliant green, sodium lauryl sulfite, selenite, and tetrathionate. These compounds have been used for the selective isolation of this organism from food and water. To confirm the isolate as a *Salmonella*, one must perform serologic tests. Currently, there are more than 2,000 serotypes of *Salmonella*, and each type is potentially pathogenic. *Salmonella* has been found in water, ice, milk, dairy products, shellfish, poultry, and poultry meat products, eggs and egg products, animal feed, pets, etc. Human beings can be carriers of this organism. It has been estimated that 4% of the general public carries this organism. There are actually three types of diseases caused by *Salmonella*: Enteric fever caused by *S. typhosa* (typhoid fever), in which the organism, ingested with food, finds its way into the bloodstream and disseminates to the kidney and is excreted in the stools; septicemia caused by *S. cholera-suis,* in which the organism causes blood poisoning; and gastroenteritis caused by *S. typhimurium* and *S. enteritidis,* a true foodborne infection. In the last case, large numbers of live *Salmonella* are ingested with food and, in 1 to 3 days, liberate the endotoxins that cause localized violent irritation of the mucous membrane with no invasion of the bloodstream and no distribution to other organs. Symptoms of salmonellosis occur 12–24 h after ingestion of food containing 1 to 10 million *Salmonella* per gram and include nausea, vomiting, headache, chills, diarrhea, and fever. The illness lasts for 2–3 days. Most patients recover; however, death can occur in the very old, the very young, and those with compromised immune systems.

Because processed food is not allowed to have *Salmonella* present, it becomes necessary for the food industry to closely monitor the occurrence of this organism in foods. The conventional methods of detection of *Salmonella* includes pre-enrichment, enrichment, selective enrichment, physiologic tests, and, finally, serologic tests. The entire sequence takes about 7 days. Currently, efforts are being made to shorten the detection time by such methods as ELISA test and DNA probes. It is possible now to provide a negative screen in about 48 h. However, once a sample is deemed positive, the conventional procedure is needed to confirm the presence of *Salmonella*. Because *Salmonella* is heat sensitive, proper cooking will destroy the organism. Also, proper chilling, refrigeration, and good sanitation will minimize the problem. *Salmonella* remains one of the most important food pathogens in our food supply.

Shigella

Shigella is a gram-negative, facultative anaerobic non-spore-forming rod quite often confused with *Salmonella* in the bacteriologic diagnostic process. It is nonmotile and hydrogen sulfide negative. The colonies are smaller than *Salmonella*. This organism is very important in waterborne diseases, especially in tropical countries where sanitation conditions are poor. The organism is transmitted by water, food, humans, and animals. The "4 F's" involved in the transmission of *Shigella* are food, finger, feces, and flies. One to four days after ingestion of the organisms, there will be an inflammation of walls of the large intestines and ileum. Invasion of the blood is rare. Bloody stool will occur, owing to superficial ulceration. The cell wall of *Shigella*, when lysed, will release endotoxins. In addition, *S. dysenteriae* produces an exotoxin that is a highly toxic neurotoxin. This toxin can be neutralized by specific antibody. Mortality rate of shigellosis is higher than that of salmonellosis. Prevention of shigellosis can be achieved by sanitation, good hygiene, treatment of water, prevention of contamination, detection of carriers, and isolation of patients from the general public.

Vibrio comma

Vibrio comma was a very important disease-causing organism in the late 19th century and early 20th century. It is still important in places with poor sanitary conditions. The organism can cause pandemic infection. It is a gram-negative, curved rod that looks like a comma under the microscope. No spore is formed. *V. comma* grows well in alkaline medium and is actively motile by a single polar flagellum. The organism is endemic in India and Southeast Asia and is spread by person-to-person contact, water, milk, food, and insects. The organism produces enterotoxins and endotoxins in the intestines and causes severe irritation to the mucous membranes with resultant outflow of fluid and salts and impairs the sodium pump of mammalian cells, thus causing severe diarrhea, dehydration, acidosis, shock, and even death. The mortality rate may be as high as 25–50%. The most effective therapy is replacement of water and electrolytes to correct severe dehydration and salt depletion. *Vibrio cholerae* remains a dreaded communicable disease in many parts of the world, and much education and public health work needs to be done to reduce human suffering from this organism.

Vibrio parahemolyticus

Vibrio parahemolyticus is an organism that has caused many cases of foodborne disease in Japan for many years. Most of the original reports and research works were in Japanese and not readily understandable or available to microbiologists in the West. U.S. scientists started working on the organism around 1969. In 1971, three outbreaks of this organism occurred in the United States. The organism is a gram-negative curved rod and is halophilic, grows best in 3–4% salt medium and can grow in 8% salt also. The growth temperature range is 15–40°C, and pH range is 5–9.6. The organism is sensitive to streptomycin, tetracycline, chloramphenicol, and novobiocin, but resistant to polymyxin and colistin. The Kanagawa positive strains hemolyze human blood. Environmental strains are negative for this test. The organism is distributed in fish and shellfish from seawater as well as from fresh water. Most of the outbreaks are recorded in the summer months when the water is warm in the Northern Hemisphere. Symptoms of the disease occur about 12-h after ingestion of a large number of viable cells (10^5/g) and include abdominal pain, diarrhea, vomiting, mild chills, and headache. The symptoms are similar to those of salmonellosis but more severe. It has been described that salmonellosis affects the abdomen of the patient whereas *Vibrio parahemolyticus* infection affects the stomach of the patient. Detection of the organism is best achieved by good selective medium such as BTB-salt-Teepol agar. Prevention of the occurrence of infection by this organism is by adequate cooking of seafood.

Bacillus cereus

Bacillus cereus and other *Bacillus* species have been implicated in foodborne diseases only in recent years, although these organisms have been suspected as agents of foodborne illness for a long time. These are gram-positive, aerobic, spore-forming rods occurring widely in nature and contaminating foods easily. Because of the general resistance of spores of this organism and the prolific biochemical activity of the vegetative cells, it can be considered one of the most important environmental bacterial contaminants of foods. Two distinct clinical symptoms are caused by this organism. The diarrheal syndrome occurs after 12–24 h of ingestion of large numbers of viable *B. cereus* and includes abdominal pain, watery diarrhea, rectal tenesmus, and nausea without vomiting. The diarrheal syndrome is the result of consuming proteinaceous foods, pudding, sauces, and vegetables. The emetic syndrome causes illness almost exclusively associated with cooked rice and is characterized by a rapid onset (1–5 h) with nausea, vomiting, and malaise. The two syndromes are related to two separate toxins: diarrheal enterotoxins and emetic toxins.

Other *Bacillus* suspected of causing foodborne diseases include *Bacillus licheniformis* and *Bacillus subtilis,* in which large numbers (10^6–10^7 organisms/gram of food) of these organisms are ingested by susceptible persons. Control of *Bacillus* food poisoning is complicated by the ubiquitous nature of this organism. The best measures are to prevent the spore from germinating and to prevent multiplication of vegetative cells in cooked and ready-to-eat foods. Freshly cooked food eaten hot immediately after cooking should not be a problem. However, slow reheating of previously cooked rice products should be treated with caution. Refrigeration of leftover cooked rice products is highly recommended as a preventive measure.

Campylobacter jejuni

Campylobacter jejuni, recognized as an emerging pathogen in the past 10 years, has been reported as the most common bacterial cause of gastrointestinal infection in humans, even surpassing rates of illness caused by *Salmonella* and *Shigella*. *Campylobacter* was originally called *Vibrio fetus*, because it was first recognized as an agent of infertility and abortion in sheep and cattle. The organism is a member of the family *Spirillaceae* because of the physiologic and morphologic similarities to *Spirillum*. The organism is a gram-negative, slender, curved bacteria that is motile by a single polar flagellum. It neither ferments nor oxidizes carbohydrates, is oxidase-positive, reduces nitrates, but will not hydrolyze gelatin or urea and is methyl red and Voges-Proskauer reaction negative. It will grow between 25 and 43°C. The organism is an obligate microaerophile that grows optimally in 5% oxygen. This attribute has been used for isolation of the organism. The incubation time of *C. jejuni* food poisoning ranges from 2 to 5 days; the duration of the sickness may be up to 10 days. The patient will exhibit enteritis, fever, malaise, abdominal pain, and headache. The stools become liquid and foul smelling. Blood, bile, and mucus discharge may occur in serious cases. The organism has a worldwide distribution, with outbreaks related to milk, poultry, eggs, red meat, pork, and water reported. Control and prevention of this organism is by proper food processing techniques (heating, cooling, chemical treatment of foods, etc) since this is a fragile organism. Its prevalence can be attributed to postprocessing contaminations of food. Again, good sanitation and hygiene should reduce the incidence of this organism in our food supplies.

Escherichia coli

Escherichia coli is one of the most common bacteria in our environment. Most people do not think of *E. coli* as a food pathogen; however, recent research and information indicates that some strains of *E. coli* can indeed cause severe foodborne disease. *E. coli* is a gram-negative, facultative anaerobic, non-spore-forming rod that occurs widely in nature as well as in intestines of humans and animals. It is glucose- and lactose-positive, indole- and methyl-red positive but, Voges-Proskauer and citrate-negative. The most useful way to classify the species is by serotyping, using antibodies against O, H, and K antigens of various strains of *E. coli*. Most *E. coli* isolated from the environment are not pathogenic. However, a group of *E. coli* has been defined as EPEC or enteropathogenic *E. coli*—diarrheagenic *E. coli* belonging to serogroups epidemiologically incriminated as pathogens, but pathogenic mechanisms have not been proven to be related either to heat-liable enterotoxins (LT) or heat-stable enterotoxins (ST) or to *Shigella*-like invasiveness. The serotypes included in EPEC are 018ab, 018ac, 026, 044, 055, 086, 0111, 0114, 0119, 0125, 0126, 0127, 0128ab, 0142, and 0158. Another newly recognized pathogenic *E. coli* is the vero cytotoxin-producing *E. coli* 0157:H7. Another class of pathogenic *E. coli* is called enteroinvasive *E. coli* (EIEC), which resembles *Shigella* by producing an invasive, dysenteric form of diarrheal illness in humans. The serogroups associated with EIEC are 028ac, 029, 0124, 0136, 0143, 0144, 0152, 0164, and 0167. The last pathogenic group of *E. coli* is called the enterotoxigenic *E. coli* (ETEC). This group of organisms produces one or both of two well-established enterotoxins: heat-labile enterotoxin (LT) and a nonantigenic, heat-stable enterotoxin (ST). Serogroups include 06, 08, 015, 020, and 027.

Detection of EPEC, EIEC, and ETEC in foods can be accomplished by common procedures used in coliform isolation at 44–45°C (for fecal coliform). In the case of *E. coli* 0157:H7; however, 44–45°C will not allow this pathogenic *E. coli* to grow. A variety of methods are available for the isolation of these pathogenic *E. coli* (see FOOD MICROBIOLOGY).

Prevention and control of pathogenic *E. coli* is best done by education of food-handlers, who should adhere to strict hygienic practices. Fecal and other waste materials from humans and animals should be decontaminated and not allowed in contact with water and food supplies.

Yersinia enterocolitica

Yersinia enterocolitica is a gram-negative, facultative anaerobic, non-spore-forming bacterium, sucrose-positive, rhamnose-negative, indole-positive, motile at 20°C but not at 37°C, and highly virulent to mice. Serotyping is very important in separating this organism from other closely related gram-negative bacteria. Although *Y. enterocolitica* has an optimal growth temperature at around 32–34°C, it is often isolated on enteric agars at 22–25°C. It grows slowly in simple glucose-salts medium but grows much better with supplements such as methionine or cysteine and thiamine. One important aspect of this organism is that it can grow in vacuum-packaged meat under refrigeration because it is a facultative anaerobe and is a psychrotroph. After ingestion of large numbers of this organism, the susceptible person can develop fever, abdominal pain, and diarrhea, with nausea and vomiting occurring less frequently. More serious intestinal disorders include enteritis, terminal ileitis, and mesenteric lymphadenitis. Extraintestinal infections of *Y. enterocolitica* have been reported, including septicemia, arthritis, erythema nodosum, sarcoidosis, skin infection, and eye infection.

Foods suspected of being a source of yersiniosis in the United States include chocolate milk, milk powder, chow mein, tofu, and pasteurized milk. Pork products have also been suspected.

Isolation of this organism typically goes through an enrichment step using nutrient broth or Rappaport broth and then through a plating medium using an enteric agar (SS, XLD, DCL, etc). An excellent agar for this purpose is the CIN agar (cefsulodin-irgasan-novobiocin agar). Control of yersiniosis is proper handling of raw and cooked food of all types, especially pork products, and of water for food processing.

Listeria monocytogenes

Listeria monocytogenes has developed into a very important food pathogen in the past 10 years from the standpoint of economic and public health impact. The organism is a small, short, gram-positive non-spore forming rod. It is motile by a characteristic tumbling motion or slightly rotating fashion. The organism grows on simple laboratory media in the pH range between 5 and 9. On solid agar, the colonies are translucent, dewdrop-like, and bluish when viewed by 45° incident transmitted light (Henry's illumination step). Biochemically, this organism can be confused with such organisms as *Lactobacillus, Brochothrix, Erysipelohrix,* and *Kurthia*. A variety of biochemical tests have been devised to separate *L. monocytogenes* from other *Listeria* species, such as *L. innocua, L. welshimeri,* and *L. murrayi*. Serotyping is also important in the identification of this organism, the most important ones being 1/2a, 1/2b, 1/2c, 3a, 3b, 3c, and 4b. *Listeria* is a psychrotroph capable of growing at temperatures as low as 2.5°C and as high as 44°C. Because dairy products have been implicated in outbreaks of listeriosis, much research has been directed toward cheese and milk products. The organism has been found to survive the processing of cottage cheese, cheddar cheese, and colby cheese. A question of great concern is whether *L. monocytogenes* can survive the current pasteurization temperature of milk (ie, 63°C for 30 min or 72°C for 15 s). Data on this issue are still inconclusive, and research on his topic is still ongoing. It is important to note that, at present the time and temperature regulation for pasteurization of milk has not been affected by the possible heat resistance of *L. monocytogenes*. The disease starts with infection of the intestine—the infective dose is not known at this point. Patients may develop transitory flulike symptoms such as malaise, diarrhea, and mild fever. In severe cases, virulent strains are capable of multiplying in macrophages and later producing septicemia. When this occurs, the bacteria can affect the central nervous system, the heart, the eyes, and may invade the fetus of pregnant women and result in abortion, stillbirth, or neonatal sepsis.

In recent years, several well-documented cases of listeriosis have been reported: Nova Scotia (1981), Massachusetts (1983), and the most well-known one involving Mexican-style soft cheese in southern California (1985).

L. monocytogenes has been isolated in a variety of commodities, including poultry carcasses, meat and chopped beef, dry sausages, milk and milk products, cheese, vegetables, and surface water. Control measures include controlling occurrence of the organism in the raw food materials, transporting vehicles, and food-processing plants (especially in controlling cross contamination of raw and finished products), and in practicing good general sanitation of the entire food-processing environment and regular monitoring of the occurrence of this organism in the food-processing facilities. Because the organism is killed by heat, proper cooking of food will also help reduce risks.

Aeromonas hydrophia

Aeromonas hydrophia has been associated with foodborne infection, although the evidence is not conclusive. The organism is a facultative anaerobic, gram-negative, motile rod. Biochemically, it is similar to *E. coli* and *Klebsiella*. The optimal temperature for growth is 28°C and the maximum is 42°C. Many strains can grow at 5°C, which is a temperature usually considered adequate to prevent growth of foodborne pathogens.

Diseases caused by *A. hydrophila* include gastroenteritis (choleralike illness and dysenterylike illness) and extraintestinal infections such as septicemia and meningitis. This organism has been isolated from fish, shrimp, crabs, scallops, oysters, red meats, poultry, raw milk, vacuum-packaged pork and beef, and even bottled mineral water.

Because the organism is a psychrotroph, cold storage is not an adequate preventive measure. Proper heating of food offers sufficient protection against this organism. Consumption of undercooked food or raw food such as raw shellfish is discouraged.

Plesiomonas shigelloides

Plesiomonas shigelloides has been a suspect in foodborne disease cases. The organism is gram-negative, facultative anaerobic, catalase negative, and fermentative. It is oxidase-positive and can be differentiated from bacteria in the family *Enterobacteriaceae* by this test, since the latter is oxidase-negative. The organism also resembles *Shigella* but can be differentiated from *Shigella* by being motile. It is capable of producing many diseases, ranging from enteritis to meningitis.

Gastroenteritis by *P. shigellosides* is characterized by diarrhea, abdominal pain, nausea, chills, fever, headache, and vomiting after an incubation time of 1–2 days. Symptoms last for a week or longer. All reported food involved with cases of gastroenteritis were from aquatic origin (salted fish, crabs, and oysters). The organism can be isolated from a variety of sources, including humans, birds, fish, reptiles, and crustaceans. The true nature of this organism as a foodborne agent is not fully known because the organism has not been well studied to date.

Miscellaneous Bacterial Foodborne Pathogens

Many other microbes are suspected of being foodborne pathogens. However, they are not currently being labeled as true foodborne pathogens owing to a lack of reports of these organisms, as well as a lack of isolation methods and research on these organisms. Many of these organisms may be identified as foodborne pathogens in the future. Among these organisms are the gram-negative bacteria *Citrobacter, Edwardsiella, Enterobacter, Klebsiella, Hafnia, Kluyvera, Proteus, Providencia, Morganella, Serratia, Vibrios* and *Pseudomonas* and the gram-positive bacteria *Corynebacterium, Streptococcus,* and other species of *Bacillus* and *Clostridium*. Miscellaneous organisms include *Brucella, Mycobacterium* (T, B), *Coxiella burnetii* (Q-fever), and *Leptospirosis, Erysipelas,* and *Tularemia.*

Foodborne Viruses

Foodborne viruses are much less studied by food microbiologists than are bacteria and fungi owing to the difficulty of cultivating these entities, as conventional bacteriologic media will not allow these particles to grow. There are, no doubt, many foodborne outbreaks and cases caused by a variety of viruses, but scientists in many cases are not able to identify the sources of the infection. Viruses that have been incriminated in foodborne diseases include hepatitis A virus (oysters, clams, doughnuts, sandwiches, and salad), Norwalk virus (oysters), polio virus (milk and oysters), ECHO virus (oysters), enteroviruses (oysters), and coxsackievirus (oysters). Much more research needs to be done in the field of food virology to help reduce the incidences of foodborne diseases caused by viruses.

Nonmicrobial Foodborne Disease Agents

Consumption of food containing other living organisms can directly and indirectly cause foodborne diseases as well. Among nonmicrobial foodborne disease agents are scombroid fish (associated with high level of histamine), cestodes (flatworms such as *Taenia saginata, T. solium,* and *Diphyllobothrium latum*), nematodes (hookworm such as *Trichinella spiralis*), trematodes (fluke such as *Clonorchis sinensis*), Protozoa (such as *Toxoplasma gondii*), shellfish (indirectly toxin from the diinoflagellate *Gonyaulax catenella*), ciguatera (from eating fish such as barracudas, groupers, and sea basses that feed on toxic algae), and other poisonous fish (such as puffer fish and moray eel).

SUMMARY

Food safety is everybody's responsibility. Scientists are charged with identifying the agents causing foodborne infections and intoxications and studying the mechanisms of the intoxication and infection as well as working on the isolation, enumeration, characterization, and identification of the causative agents. The food industry uses this basic knowledge and applies it to good manufacturing practices to produce wholesome, nutritious, and safe foods. The consumer must also be educated in the handling of raw and cooked food at the point of purchase, as well as preparation of the food and final consumption. All three parties are responsible for the food safety of all involved.

BIBLIOGRAPHY

General References

J. C. Ayres, O. Mundt, and W. E. Sandine, *Microbiology of Foods,* W. H. Freeman and Co., San Francisco, 1980.

G. J. Banwart, *Basic Food Microbiology,* AVI Publishing Company, Inc., Westport, Conn., 1979.

J. E. L. Corry, D. Roberts, and F. A. Skinner, *Isolation and Identification Methods for Food Poisoning Organisms,* Academic Press, New York, 1982.

M. P. Doyle, *Food Borne Bacterial Pathogens.* Marcel Dekker, Inc., New York, 1989.

Food and Drug Administration, *Bacteriology Analytical Manual,* 6th ed., Association of Official Analytical Chemists, Arlington, Vir., 1984.

D. Y. C. Fung, "Types of Microorganisms," In F. E. Cunningham and N. A. Cox, eds., *Microbiology of Poultry Meat Products,* Academic Press, New York, 1987.

D. Y. C. Fung, "Rapid Methods for Determining the Bacterial Quality of Red Meats," *Journal of Environmental Health* **46**(5):226–228, 1984.

W. F. Harrigan and E. McCance, *Laboratory Methods in Food and Dairy Microbiology,* Academic Press, New York, 1976.

C.-G. Heden and T. Illeni, eds. *Automation in Microbiology and Immunology,* John Wiley & Sons, New York, 1975.

C.-G. Heden and T. Illeni, eds. *New Approaches to the Identification of Microorganisms,* John Wiley & Sons. New York, 1975.

B. C. Hobbs and J. H. B. Christian. *The Microbiological Safety of Food,* Academic Press, New York, 1973.

International Commission on Microbiological Specifications for Foods, *Microorganisms in Foods, vol. 2,* University of Toronto Press, Toronto, Canada, 1974.

J. M. Jay, *Modern Food Microbiology,* 3rd ed. Van Nostrand Reinhold, New York, 1986.

E. H. Marth, ed., *Standard Methods for the Examination of Dairy Products,* American Public Health Association, Washington, D.C., 1978.

B. M. Mitruka, *Methods of Detection and Identification of Bacteria,* CRC Press, Inc., Cleveland, Ohio, 1976.

J. T. Nickerson and A. J. Sinskey, *Microbiology of Foods and Food Processing,* American Elsevier Publishing Co., New York, 1972.

M. D. Pierson and N. J. Stern, *Foodborne Microorganisms and Their Toxins: Developing Methodology,* Marcel Dekker, Inc., New York, 1985.

M. Recheigl Jr., *Handbook of Naturally Occurring Food Toxicants,* CRC Press, Boca Raton, Fla., 1983.

H. Rieman and F. L. Bryan, eds. *Food-Borne Infections and Intoxicants,* Academic Press, New York, 1979.

T. A. Roberts and F. A. Skinner, *Food Microbiology Advances and Prospects,* Academic Press, New York, 1983.

A. N. Sharpe, *Food Microbiology: A Framework for the Future,* Charles C Thomas, Springfield, Ill., 1980.

J. H. Silliker, R. P. Elliott, A. C. Baird-Parker, F. L. Bryan, J. H. B. Christian, D. S. Clark, J. C. Olson, Jr., and T. A. Roberts, *Microbial Ecology of Foods,* vol. 2, Academic Press, New York, 1980.

S. L. Taylor and R. A. Scanlan, *Food Toxicology,* Marcel Dekker, Inc., New York, 1989.

R. C. Tilton, *Rapid Methods and Automation in Microbiology,* American Society of Microbiologists, Washington, D.C., 1982.

C. Vanderzant and D. Splittstoesser, *Compendium of Methods for the Microbiological Examination of Foods.* American Public Health Association, Washington, D.C., 1991.

H. H. Weiser, G. J. Mountney, and W. A. Gould, *Practical Food Microbiology and Technology,* AVI Publishing Company, Inc., Westport, Conn., 1971.

Daniel Y. C. Fung
Kansas State University
Manhattan, Kansas

FOODBORNE MICROORGANISMS: DETECTION AND IDENTIFICATION

The many methods that are used to enumerate microorganisms or to detect their toxic products in foods can be placed in two groups. One group includes those methods that require the organisms to be viable, while the other includes methods that detect cells that may be nonliving, or parts and products of cells, and the two groups are listed in Table 1 along with the reported minimum detectable numbers of cells or products for each. In both groups are methods that are used primarily for the detection and enumeration or microorganisms, whereas some are used primarily to identify microorganisms.

DETECTION METHODS THAT REQUIRE VIABLE CELLS

This group of methods includes some of the oldest in use along with some that were developed only in the last two decades. Because they all depend on dividing cells, their speed is dependant directly upon the growth rate of the organisms of interest. Their chief value is that they may be used to determine the number of viable or colony-forming units (cfu) in a food product, and the number of organisms found depends on the factors of culture media employed, temperature and time of incubation, incubation environment (aerobic or anaerobic), pH, and a_w of culture medium. Each of those listed in Table 1 is further discussed as follows.

Standard/Aerobic Plate Count (SPC, APC)

Along with direct microscopic count, this is one of the oldest methods for enumerating bacteria in foods and food specimens. Official methods for determining cfu's are described in standard references (1,2) and no further details are presented here.

There are two basic ways to conduct an SPC: pour and surface plating of samples or diluents. By the former, diluent is placed in empty petri dishes followed by pouring of cooled molten agar and mixing with diluted sample. With surface plating, diluents are spread over the surface of dried agar media, typically with bent glass rods (hockey sticks). The following factors may be considered in deciding whether to use the pour or surface plate method, assuming the same time and conditions of incubation.

1. Size of colonies: smaller on pour, larger on surface.
2. Spreading of colonies: less pronounced on pour.
3. Enumeration of psychrotrophs: better with surface.
4. Crowding of colonies on plate: more with surface.
5. Use of selective/differential media: better with surface.
6. Colonial features of colonies: better with surface.
7. Microaerophilic organisms: better with pour.
8. Strict aerobes: better with surface.
9. Colony Pigmentation: better observed with surface.
10. Strict anaerobes: better with pour.
11. Subculturing of colonies: better with surface.

Reproducibility: Generally Better With Pour

Although surface plating has some advantages over pour plating, the latter method is much more widely used. The way in which food specimens are homogenized for cfu determinations is an important decision that affects the numbers of organisms found. Brisk shaking by hand or by use of mechanical shakers provides suitable results when comminuted meats or powdered food specimens are used, but reproducibility is not always good. Homogenizing with a Waring blender is a widely used method and this

Table 1. Examples of the Two Broad Categories of Enumeration/Detection Methods and their Minimum Response Cell Numbers

	Minimum Numbers
A. *Viable/respiring Cells Required*	
1. Standard plate count	1
2. Most probable number (MPN)	<1
3. Dye reductions	~10^{5a}
4. Hydrophobic grid membrane filter (HGMF)	<10
5. Microcolony-DEFT	10^3
6. Electrical impedance	10^6–10^7
7. Radiometry	10^4–10^5
8. Catalasemeter	~10^4
9. Microcalorimetry	~10^4
B. *Viable Cells are Not Needed*	
1. Direct microscopic count (DMC)	10^4
2. Direct epifluorescent filter technique (DEFT)	10^3–10^4
3. Fluorescent antibody	10^6–10^7
4. ATP assay	10^5–10^6
5. *Limulus* amoebocyte lysate test (LAL)	~10^2
6. Radioimmuoassay	<10^3
7. ELISA, EIA	10^5–10^6
8. Thermostable nuclease (TNase)	10^7
9. DNA probes	10^6–10^7

[a] Four hours for resazurin reduction.

device allows one to homogenize chunks of food as well as comminuted products. One drawback to using this method is the heat build up that occurs in the container when blending is carried out for more than about 2 min. Other drawbacks to the use of the Waring blender have been discussed elsewhere (3). The Stomacher has emerged as the method of choice for food homogenizations. Developed by Sharpe and Jackson (4), this device has been compared to the Waring blender and other methods and found to give better results overall (5–7).

Spiral Plater. The use of the Spiral plating device is another way to achieve surface plating, and a large number of investigators have shown that the proper use of this device achieves results comparable to those by the more traditional methods (8–11). Among the advantages it offers is the use of only one petri dish to effect the enumeration of cfu's over a wide range, and dilution blanks are not needed. The plating of solid foods can present problems unless care is taken to prevent particles from clogging the Spiral plater stylus.

Dry Film. A more recent surface-plating methodology is the use of dry film such as Petrifilm where an inoculum is spread over the surface of a prescribed area impregnated with culture medium. The colonies that develop are smaller than for the classical surface plate method, but their enumeration is aided by color development from tetrazolium in the film. This method has been shown to give results comparable to those by pour or surface plating in petri dishes (12,13).

Other Methods. Other variations of methods that allow one to enumerate cfu's include roll tube methods that are especially useful for enumerating strict anaerobes, and Rodac or contact plates that may be used to determine cfu's by their direct application to food plant surfaces. A common problem with contact plates employing nonselective media is the spreading of colony growth and also overgrowth by molds, but this can be minimized by using selective culture media for specific groups of organisms (14), or using properly dried contact plates when nonselective media are used.

Most Probable Numbers (MPN)

This is a statistical method that can be used to enumerate all viable cells in a food product or specific indicators, pathogens, or fungi, depending on the media employed. MPN is run either as a 3- or 5-tube method and results are generally higher than by either pour or surface Plating methods. Details of MPN methodology are presented in *Standard Methods* (2). Some of the problems encountered with plate count methods are avoided by the use of MPN. For example, problems presented by spreading colonies as well as those of inaccurate counts are avoided. On the down side, no information is provided on colony features or the types of organisms when nonselective media are employed. As noted above, MPN counts are generally higher than plate counts; and large quantities of glassware are needed (3). MPN is the method of choice for enumerating coliforms or *Escherichia coli* in foods and waters.

Dye Reductions

Although these methods cannot be used to make precise determinations of microbial numbers, they can be used to make estimates of certain organisms within number ranges. The dyes most commonly used are methylene blue, resazurin, and tetrazolium, and detailed procedures for their use are described in *Standard Methods* (2).

Dye reduction methods are based on the fact that respiring microorganisms effect their reduction resulting in color changes. Methylene blue changes from blue to white, resazurin from slate blue to pink or white, and tetrazolium from colorless to its red formazan. In general, the color changes are essentially linear in the $\log_{10}$ 5 to 8 cfu/g range, although not all organisms are equal in their reductive abilities.

Dye reduction methods are normally employed as screens. Resazurin has been shown to be an excellent screen for $\log_{10}$ 5.0 cfu/g for raw milk (15,16), meats (17–19), poultry (20), and frozen shrimps (21). Methylene blue has a long history of use as a screen for raw milk (15) where the time (in hours) it takes for the blue color to be reduced to its colorless form is used as an approximation of overall microbial load. When compared to nitroblue tetrazolium and indophenyl nitrophenyl tetrazolium, resazurin produced faster results (18). Using surface samples from sheep carcasses, resazurin was reduced in 300 min by 18,000 cfu/cm^2, nitroblue in 600 min by 21,000 cfu/cm^2 whereas with indophenyl nitrophenyl tetrazolium 18,000 cfu/cm^2 effected reduction in 660 min (18).

These methods are both simple to use and inexpensive, but as noted above, some groups of organisms are more efficient in reducing the oxidized forms than others. Their use has been described (1) and discussed elsewhere (3).

Hydrophobic Grid Membrane Filter (HGMF)

A hydrophobic grid membrane filter consists of 1600 wax grids on a single membrane filter that restricts growth and colony size to individual grids (22–24). On one HGMF, from 10 to 9×10^4 cells can be enumerated by an MPN procedure, and enumeration can be automated. The method can detect as few as 10 cells/g and results can be achieved in 24 h or so, and it can be used to enumerate all cfu's or specific groups such as indicator organisms (25–27), yeasts and molds (28), or pathogens such as salmonellae (29). This method has been automated (28) and given AOAC approval (25).

A typical application of HGMF consists of filtering 1 ml of a 1:10 homogenized sample through a HGMF membrane and placing the membrane on a suitable medium for incubation overnight to allow colonies to develop. The grids that contain colonies are enumerated and MPN is calculated.

Microcolony Direct Epifluorescent Filter Technique (Microcolony-DEFT)

The DEFT, described shortly is a method for determining viable and nonviable cells, whereas microcolony-DEFT is a variation used to determine viable cells only. Typically, food homogenates are filtered through DEFT membranes and the membranes are then placed on the surface of appropriate media and incubated for 3 to 6 h (3 h for gram-negative and 6 h for gram-positive bacteria). The microcolonies on the filters are stained and enumerated (30). The latter authors showed this method to be satisfactory for enumerating coliforms, pseudomonads, and staphylococci, and that as few as 10^3/g could be detected with results within 8 h.

In another variation of DEFT, Rodrigues and Kroll (31) devised a microcolony epifluorescence microscopy method that combines DEFT with the hydrophobic grid membrane filter (HGMF). By this method, nonenzyme detergent-treated samples are filtered through Nucleopore polycarbonate membranes. The membranes are transferred to the surface of selective agar and incubated for 3 h for gram-negative and 6 h for gram-positive bacteria. The membranes are then stained with acridine orange and the microcolonies are enumerated by epifluorescence microscopy. Results were achieved in less than 6 h without resuscitation for injured organisms, and in about 12 h when a resuscitation step was employed.

Electrical Impedance

Impedance does not lend itself to the enumeration of microbial numbers per se, but it can be used to indicate the presence of respiring cells in the range of about 10^4/g or above. The method is based on the measurement of impedance decrease caused by respiring microorganisms in suitable culture substrates, and it lends itself to the screening of food products for the presence or absence of a certain minimum number of cells (32,33). The lowest number of viable bacterial cells that can elicit an impedance response is in the 10^6–10^7/ml range. If a product contains a lower number of cells/milliliter, say, 10^3/g, the time it takes this number to reach the threshold level is the impedance detection time (IDT).

As a screen, impedance measurement may be used to determine how long it takes (in hours) for IDT where the longer the time the lower the number of cells in the food sample. By use of known numbers of organisms in appropriate substrates, one can determine cutoff times to reflect, eg, 10^5 cells/g. All aerobic bacteria (total numbers) may be detected as well as specific groups of organisms by using the appropriate growth media (34). Results can be obtained within hours, depending on initial numbers, and the method lends itself to automated data collection. This method has been used to detect coliforms in foods (35–37), to estimate the flora of frozen orange juice concentrate (38), raw meats (39), and milk (40).

Radiometry

This method employs the use of ^{14}C-labeled fermentable substrates with the radiolabel on a carbon atom that respiring organisms can release as $^{14}CO_2$. In general, the larger the number of cells the more $^{14}CO_2$ released, and approximate numbers of cells are determined by the amount of radioactive carbon dioxide released. It normally requires 10^4–10^5 coliforms to release enough $^{14}CO_2$ from lactose to be detected.

Although radiometry is not widely used, it has been shown to be applicable to the detection of the microbial flora of foods (41,42) and to the specific detection of salmonellae (43).

Radiometry is used as a screen in much the same way as impedance and dye reduction methods, where the time required to produce a threshold quantity of $^{14}CO_2$ is a function of the number of cells that can release CO_2 from the labeled substrate. The radiolabel and the necessity for radioactivity counters make this method unpopular in the food industry.

Catalasemeter

This disk flotation device was designed by Gagnon et al. (44) and further developed at the University of Quebec (45) to detect the volume of oxygen released when microbially produced catalase acts on hydrogen peroxide. The catalasemeter consists of the preparation of food extracts that contain catalase (formed by catalase-positive organisms) followed by soaking a filter paper disk with the enzyme preparation. The filter paper disk is immediately placed in a tube containing 3% hydrogen peroxide. The disk-containing tube is placed in a photometer and the time in seconds that elapses between the initial fall of the disk to the bottom of the tube and its subsequent rise, due to the buoyancy caused by bubbles of oxygen at its surface, is recorded as flotation time. When large numbers of catalase-positive organisms are in the food preparation, more catalase is present to produce more oxygen from the hydrogen peroxide and subsequently the shorter the flotation time.

The Catalasemeter as described above was employed by Dodds et al. (46) to make rapid assessments of the microbial quality of vacuum-packaged cooked turkey products, and cfu's of 10^4/g or higher could be detected accurately in 300 sec, with longer and less accurate flotation times required with lower numbers. Overall, the test correlated well with cfu's of catalase-positive Enterobacteriaceae with a correlation coefficient of .804.

Microcalorimetry

This method has received only limited study, but it offers potential as a screen for ca 10^4 microorganisms/g in canned foods as well as in ground beef. It is based on the fact that respiring cells emit heat and that accurate measurements of the heat correspond to cell numbers. Either a batch or flow type microcalorimeter is used, and some of these instruments are sensitive to a heat flow of 0.01 cal/h (47). The organisms that grow in canned foods can be detected by observing and measuring the thermogram produced when sensors are placed on the outside of cans (48). About 10^4 cells/g are necessary to give a detectable response (3), and specific groups of organisms, such as lactic acid bacteria (49) and yeasts (50,51) can be detected or characterized under appropriate conditions.

DETECTION OF ORGANISMS AND/OR THEIR PRODUCTS BY METHODS THAT DO NOT REQUIRE VIABLE CELLS

Unlike some of the enumeration methods that can be used to make rather precise determinations of microorganisms in foods, most of those that do not require cell viability can be used only to determine numbers at or above their respective threshold ranges. In addition to the latter, some of these methods lend themselves to the identification of certain groups of organisms, and they are discussed further in this regard in a later section. Their application for enumeration of cells or the detection of cell products is discussed as follows.

Direct Microscopic Count (DMC) Methods

In addition to being the oldest ways to determine the numbers of microorganisms in foods and other specimens, DMCs are the fastest. For food use, the two oldest and most widely used DMC methods are the Breed method for bacteria, and the Howard mold count slide for molds.

The Breed method consists of adding 0.01 ml of a sample to a 1-cm^2 area on a special microscope slide followed by drying, fixing, staining, and viewing with the oil immersion objective of a calibrated compound microscope (1). The method is widely used for raw milk, and it can be used for other food products such as powdered eggs. Results can be obtained in about 5 min. but a minimum of around 10^4 cells/ml are needed. Results obtained by this method are generally always higher than for viable cell methods such as APC since no distinctions can be made between viable and nonviable in cells (3). In spite of its drawbacks, this method is valuable when one wishes to know in the shortest possible time if a given level of microorganisms exists in a food product.

A slide method that may be used to detect and enumerate only viable cells was developed by Betts et al (52). Key to the method is the use of the tetrazolium salt 2-(p-iodophenyl-3-(p-nitrophenyl)5-phenyl tetrazolium chloride (INT). Cells are exposed to filter-sterilized INT for 10 min at 37°C in a water bath followed by filtration on 0.45 μm membranes. Following drying of membranes for 10 min at 50°C, the special membranes are mounted in cotton seed oil and viewed with coverslip in place. The method was found to be workable for pure cultures of bacteria and yeasts. Betts et al. found that the INT method underestimated APC by 1–1.5 log cycles when compared to APC using milk. By use of fluorescence microscopy and Viablue (modified aniline blue fluochrome), viable yeast cells could be differentiated from nonviable cells (53,54).

The Howard mold count is a slide method specifically developed for monitoring tomato products for molds (55). A similar method has been developed for *Geotrichum candidum* in fruits and vegetable products. These methods can be used to assess the prevalence of fungal contamination by observing their mycelia or mycelial fragments, but the values obtained rarely correspond to viable mold counts when plating methods are employed (3).

Direct Epifluorescent Filter Technique (DEFT)

DEFT is a more recent modification of the classical DMC method. By the basic method, a diluted food homogenate is filtered through a 5-μm nylon filter. The filtrate is collected and treated with 2 ml Triton X-100 and 0.5 ml trypsin. After incubation, the treated filtrate is passed through a 0.6-μm Nucleopore polycarbonate membrane and the filter is stained with acridine orange. Following drying, the stained cells are enumerated by epifluorescence microscopy and the number of cells/g calculated by multiplying the average number/field by the microscope factor (56–58). Results can be obtained in 25–30 min, and numbers as low as around 6,000 cfu/g can be obtained from meats and milk products.

DEFT has been used successively not only with milk but to estimate numbers of microorganisms on meat and poultry (58) and on food contact surfaces (59). It has been adapted to enumerate viable gram-negative and all gram-positive bacteria in milk in about 10 min (60). The further adaptation of DEFT to the enumeration of viable cells was previously discussed.

Fluorescent Antibody (FA)

This is a microscope slide method that finds its widest use in food microbiology in the examination of foods for salmonellae (61,62), although the method can be used for any organisms to which an antibody can be made. By this method, an antibody is made to the organism of interest and made fluorescent by coupling it to a fluorescent compound such as fluorescein isocyanate. Following the preparation of slide smears, drying, and fixing, the coupled antibody–fluorescent dye is added. If the organism to which the antibody is made is present, it will react with the cells and evidence for the reaction is ascertained by observing for fluorescence under a microscope with oil immersion objective. The FA technique can be run as a direct or an indirect method, with the latter being preferred by many. By the latter, the homologous antibody is not coupled with fluorescent label but instead, an antibody to this antibody is prepared and coupled. This method obviates the need to prepare antibody for each organism of interest. More recent modifications of FA include the use of monoclonal antibodies, which appear to reduce false positive reactions (3).

This method is rapid but requires the presence of about 10^6–10^7 cells/ml for success in finding positive cells on the slide. It is official by AOAC (63), and it is more satisfactory in the hands of some than others. False positive results may run as high as 8–10% and false negatives may be as high as 1–3%, depending on the experience of the microscopist. The FA technique is more of a screen for the presence of targeted organisms in the 10^6–10^7/ml range than an enumeration method per se. It can be automated (64) and viable salmonellae cells can be enumerated when combined with a microcolony technique (65). Its utility as a method to identify given organisms is obvious.

ATP Assay

This method is based on the fact that all cells contain ATP and that the quantity detected in a specimen is referable to a given number of cells. The presence and amount of ATP are detected by use of the firefly luciferin-luciferase assay with the use of a photometer to measure emitted light (66,67). The problems that must be dealt with are that (1) the quantity of ATP varies for different bacterial cells, with the amount ranging from 0.1 to 4.0 fg/bacterial cell and for yeast cells from 13 to 100 fg/cell and (2) background ATP from food substances must be excluded from the analysis. By one commercially available system, background ATP can be eliminated and ATP assays can be run in an automated manner.

It generally requires 10^5–10^6 cells/g in order to have enough ATP to emit consistent detectable light, but results can be obtained within 1 h depending on the initial number of cells. The method has been shown to be comparable to APC for the aerobic flora of ground beef (68) over the APC range of 10^5–10^6 cells/g and to be applicable to seafoods (69), and yeasts in beverages (70). Sensitivity can be increased for beverages by filtration and concentration of cells.

Limulus Amoebocyte Lysate (LAL) Test

The LAL method is applicable to the determination of lipopolysaccharides (LPS) or endotoxins in foods, and as such the detection of gram-negative bacteria, both viable and nonviable. It employs the lysate from the amoebocytes of the horseshoe crab (*Limulus polyphemous*), which is the most specific substance known for detecting LPS (71). Because all gram-negative bacteria produce LPS while gram-positives do not, LAL is a test for gram-negative bacteria.

The LAL test can be run in several ways, and these have been presented and discussed elsewhere (72). The most commonly used is a tube gelation method where the food specimen is diluted in pyrogen-free water and 0.1 or 0.2 ml amounts are added to similar quantities of the LAL reagent in pyrogen-free tubes (all glassware and reagents must be pyrogen-free for use by this method). Following incubation at 37°C for 1 h, positive tests are indicated by the presence of firm gels. The chromogenic substrate method consists of using synthetic substrates that produce a color change in the presence of LPS and the LAL reagent, and the color change is read spectrophotometrically. The latter method can be run in 30 min, and it lends itself to automation (72,73).

Commercially available LAL reagents can detect 1.0 pg of LPS, and since a typical *Escherichia coli* cell contains about 3.0 fg of LPS, the LAL test should be able to respond to about 300 cells. Studies with pseudomonads from meats have shown the test to be capable of detecting 10^2 cfu/ml (74), and it has been shown to be effective for testing meats (75), raw milk (76), seafoods (77), sugars and some processed meats (46).

Since the total gram-negative flora of a food product can be determined within 1 h by use of LAL, this test has been adapted to the approximation of the total bacterial count using fresh ground beef (72,78). By this procedure, the gram-negative bacteria are represented by LAL values, whereas gram-positives are estimated by multiplying respective gram-negative numbers by predetermined ratio values of gram negatives to gram-positives. Estimated total bacteria by this procedure have been shown to compare favorably to the total count by use of APC methods (74,78).

Since the LAL test can be conducted in 1 h or less, it can be used to make rapid assessments of the microbial load or overall sanitation of a food product. In this application, low LAL values are more meaningful than high values since the latter could represent nonviable cells.

Radioimmunoassay (RIA)

This method does not lend itself to the enumeration of microbial cells per se, but it can be used to detect/quantitate toxic products of microbial cells. Briefly, RIA consists of adding a radioactive label (eg, ^{125}I) to a soluble antigen followed by reaction of the antigen with its homologous antibody bound to a solid such as polystyrene. After washing, the adherence of the radiolabel to the polystyrene is assessed by use of a radioactivity counter.

RIA is one of the most sensitive methods known for detecting staphylococcal enterotoxins (79,80), enterotoxins of *E coli* (81), *Vibrio cholerae* (82), mycotoxins (83,84), and other similar products (3). It is sensitive to 0.1 ng of staphylococcal enterotoxin A, 0.5 ng of aflatoxin B_1, and 20 ppb ochratoxin A. The most serious drawbacks to this method is the need for radioactive compounds and the consequent need for a radioactivity counter.

Enzyme Immunoassay, Enzyme-linked Immunoabsorbent Assay (EIA, ELISA)

In its simplest form, ELISA can be viewed as RIA without a radioactive element. Instead of the latter, an enzyme substrate reaction is used to indicate an homologous antigen—antibody reaction. Because of the absence of radioactivity, ELISA is much more widely used to detect microbial toxins in foods, and it has been used to detect salmonellae cells in food products (85–87). For the latter, 10^4–10^5 cells/ml may be detected but the general minimum is around 10^6/ml. It can detect 0.1 ng/ml of *Clostridium perfringens* enterotoxin, and 10 pg/ml or less of aflatoxin M_1 (88) and other mycotoxins (89–91). ELISA tests to detect botulinal toxins Type A (92), Type E (93), and Type G (94) have been developed as well as tests for staphylococcal enterotoxins (95–97).

Although ELISA methods can be run with either polyclonal or monoclonal antibodies, the latter tend to give better results (98). A commercial method for detecting *Listeria monocytogenes* employing monoclonal antibodies is available (99). A monoclonal method for salmonellae has been shown capable of detecting as few as 10^6 cells/ml (100).

Thermostable Nuclease (TNase)

This test is specifically designed for *Staphylococcus aureus* and it can be run in 1–2 h. All strains of *S. aureus* that produce enterotoxins also produce this heat-stable nuclease, although many TNase-positive strains do not produce enterotoxins (101).

The microslide method consists of a layer of agar containing toluidine blue O and DNA into which holes are made for test samples. After heating to 100°C, food or food homogenates are placed in the holes and the slide is incubated at 37°C for 1–2 h. If TNase is present, the DNA is degraded and this is manifested by a change in color of the dye (102). It generally takes about 5×10^7 S. aureus cells to produce detectable amounts of TNase and this enzyme is generally produced before enterotoxins. Thus, this test can be used to screen foods for the presence of enterotoxins even in the absence of viable S. aureus cells (103,104). Occasional false positive tests may occur from *Streptococcus faecalis* strains that also produce TNase.

DNA Probes

Although DNA probes are most often used to detect specific organisms, their use may be viewed as screens for the organism in question since 10^6–10^7 cells/ml are needed for positive tests. In a typical application, one selects a unique DNA sequence of the organism of interest and tags a single strand with a radioisotope. DNA fragments of unknown organisms are prepared by use of restriction endonucleases. After separating the fragment strands, they are transferred to cellulose nitrate filters and hybridized to the radioactive probe. After gentle washing to remove unreacted probe DNA, the presence of the radiolabel is detected by autoradiography or by use of a radioactivity counter.

A number of investigators have developed and tested DNA probes for salmonella (105,106), listeriae (107,108), and other food-borne pathogens and found the method to be satisfactory for detecting the presence of target organisms in mixed culture. Commercial methods exist for salmonellae and listeriae, but they generally require 44 + h when used on products that contain low numbers of target organisms.

DNA probes are used in colony hybridization methods where micro- or macrocolonies of the target organism are allowed to develop directly on membranes that are incubated on suitable agar plates. Following treatment of the colonies to separate and fix DNA strands, the DNA probe is applied to the membrane and the radiolabel is tested for. Colony hybridizations have been employed with success to detect *Listeria monocytogenes*, enterotoxigenic strains of *E. coli* (109,110), and *Yersinia enterocolitica*.

One of the drawbacks to the use of DNA probens in the food industry is the presence of radioactive elements. One of the most promising nonisotope probes employs biotin and its detection using a streptavidine-alkaline phosphatase conjugate to produce an insoluble color precipitate in the presence of a dye. One such method was developed (111) and shown to be workable for *E. coli* and to detect 2×10^7 cells/g in about 30 h. The nonisotopic detection system required about 3 h for results.

BIBLIOGRAPHY

1. E. H. Marth, ed., *Standard Methods for the Examination of Dairy Products*, 14th Ed., American Public Health Association, Washington, D.C., 1978.
2. M. L. Speck, ed., *Compendium of Methods for the Microbiological Examination of Foods*, 2nd Ed., American Public Health Association, Washington, D.C., 1984.
3. J. M. Jay, *Modern Food Microbiology*, 3rd ed., Van Nostrand Reinhold, New York, 1986.
4. A. N. Sharpe and A. K. Jackson, "Stomaching: A New Concept in Bacteriological Sample Preparation," *Applied Microbiology* **24**, 175–178 (1978).
5. B. S. Emswiler, C. J. Pierson, and A. W. Kotula, "Stomaching vs Blending. A Comparison of Two Techniques for the Homogenization of Meat Samples for Microbiological Analysis," *Food Technology* **31**(10), 40–42 (1977).
6. W. H. Andrews, C. R. Wilson, P. L. Poelma, A. Romero, R. A. Ruce, A. P. Duran, P. D. McClure, and D. E. Gentile, "Usefulness of the Stomacher in a Microbiological Regulatory Laboratory," *Applied Environmental Microbiology* **35**, 89–93 (1978).
7. J. M. Jay and S. Margitic, "Comparison of Homogenizing, Shaking, and Blending on the Recovery of Microorganisms and Endotoxins from Fresh and Frozen Ground beef as Assessed by Plate Counts and the *Limulus* Amoebocyte Lysate Test," *Applied Environmental Microbiology* **38**, 879–884 (1979).
8. J. E. Gilchrist, J. E. Campbell, C. B. Donnelly, J. T. Peeler, and J. M. Delaney, "Spiral Plate Method for Bacterial Determination," *Applied Microbiology* **25**, 244–252 (1973).
9. C. B. Donnelly, J. E. Gilchrist, J. T. Peeler, and J. E. Campbell, "Spiral Plate Count Method for the Examination of Raw and Pasteurized Milk," *Applied Environmental Microbiology* **32**, 21–27 (1976).
10. B. Jarvis, V. H. Lach, and J. M. Wood, "Evaluation of the Spiral Plate Maker for the Enumeration of Micro-organisms in Foods," *Journal of Applied Bacteriology* **43**, 149–157 (1977).
11. Association of Official Analytic Chemists "Spiral Plate Method for Bacterial Count: Official First Action," *Journal of the Association of Official Analytical Chemists* **60**, 493–494 (1977).
12. C. L. Nelson, T. L. Fox, and F. F. Busta, "Evaluation of Dry Medium Film (Petrifilm VRB) for Coliform Enumeration," *Journal of Food Protection* **47**, 520–525 (1984).
13. L. B. Smith, T. L. Fox, and F. F. Busta, "Comparison of a Dry Medium Culture Plate (Petrifilm SM plates) Method to the Aerobic Plate Count Method for Enumeration of Mesophilic Aerobic Colony-forming Units in Fresh Ground Beef," *Journal of Food Protection* **48**, 1044–1045 (1985).
14. M. P. deFigueiredo and J. M. Jay, "Coliforms, Enterococci, and Other Microbial Indicators," in M. P. deFigueiredo and D. F. Splittstoesser, eds., *Food Microbiology: Public Health and Spoilage Aspects*, Avi Publishing, Westport, Conn., 1976.
15. C. K. Johns, "Place of the Methylene Blue and Resazurin Reduction Tests in a Milk Control Program," *American Journal of Public Health* **29**, 239–247 (1939).
16. R. Dabbah, W. A. Moats, S. R. Tatini, and J. C. Olson, Jr., "Evaluation of the Resazurin Reduction One-Hour Test for Grading Milk Intended for Manufacturing Purposes," *Journal of Milk Food Technology* **32**, 44–48 (1969).
17. P. J. Dodsworth and A. G. Kempton, "Rapid Measurement of Meat Quality by Resazurin Reduction. II. Industrial Application," *Canadian Institute of Food Science Technology Journal* **10**, 158–160 (1977).
18. D. N. Rao and V. S. Murthy, "Rapid Dye Reduction Tests for the Determination of Microbiological Quality of Meat," *Journal of Food Technology* **21**, 151–157 (1986).
19. R. A. Holley, S. M. Smith, and A. G. Kempton, "Rapid Measurement of Meat Quality by Resazurin Reduction. I. Factors Affecting Test Validity." *Canadian Institute of Food Science Technology Journal* **10**, 153–157 (1977).

20. H. W. Walker, W. J. Coffin, and J. C. Ayres, "A Resazurin Reduction Test for Determination of Microbiological Quality of Processed Poultry," *Food Technology* **13**, 578–581, (1959).
21. R. Kummerlin, "Technical Note: Resazurin Test for Microbiological Control of Deep-Frozen Shrimps," *Journal of Food Technology* **17**, 513–515 (1982).
22. A. N. Sharpe and G. L. Michaud, "Hydrophobic Grid-Membrane Filters: New Approach to Microbioligcal Enumeration," *Applied Microbiology* **28**, 223–225 (1974).
23. A. N. Sharpe and G. L. Michaud, "Enumeration of High Numbers of Bacteria Using Hydrophobic Grid-Membrane Filters," *Applied Microbiology* **30**, 519–524 (1975).
24. A. N. Sharpe, M. P. Diotte, I. Dudas, S. Malcolm, and P. I. Peterkin, "Colony Counting on Hydrophobic Grid-Membrane Filters" *Canadian Journal of Microbiology* **29**, 797–802 (1983).
25. A. O. A. C., "Enumeration of Coliforms in Selected Foods, Hydrophobic Grid Membrane Filter Method, Official First Action," *Journal of the Association of Official Analytical Chemists* **66**, 547–548 (1983).
26. P. Entis, "Enumeration of Coliforms in Non-Fat Dry Milk and Canned Custard by Hydrophobic Grid Membrane Filter Method: Collaborative Study," *Journal of the Association of Official Analytical Chemists* **66**, 897–904 (1983).
27. M. H. Brodsky, P. Entis, A. N. Sharpe, and G. A. Jarvis, "Enumeration of Indicator Organisms in Foods Using the Automated Hydrophobic Grid Membrane Filter Technique," *Journal of Food Protection* **45**, 292–296 (1982).
28. M. H. Brodsky, P. Entis, M. P. Entis, A. N. Sharpe, and G. A. Jarvis, "Determination of Aerobic Plate and Yeast and Mold Counts in Foods Using an Automated Hydrophobic Grid Membrane Filter Technique," *Journal of Food Protection* **45**, 301–304 (1982).
29. P. Entis, "Rapid Hydrophobic Grid Membrane Filter Method for *Salmonella* Detection in Selected Foods," *Journal of the Association of Official Analytical Chemists* **68**, 555 (1985).
30. U. M. Rodrigues and R. G. Kroll, "Rapid Selective Enumeration of Bacteria in Foods Using a Micro-Colony Epifluorescence Microscopy Technique," *Journal of Applied Bacteriology* **64**, 65–78 (1988).
31. U. M. Rodrigues and R. G. Kroll, "Microcolony Epifluorescence Microscopy for Selective Enumeration of Injured Bacteria in Frozen and Heat-Treated Foods," *Applied Environmental Microbiology* **55**, 778–787 (1989).
32. D. Hardy, S. J. Kraeger, S. W. Dufour, and P. Cady, "Rapid Detection of Microbial Contamination in Frozen Vegetables by Automated Impedance Measurements," *Applied Environmental Microbiology* **34**, 14–17 (1977).
33. P. Cady, D. Hardy, S. Martins, S. W. Dufour, and S. J. Kraeger, "Automated Impedance Measurements for Rapid Screening of Milk Microbial Content," *Journal of Food Protection* **41**, 277–283 (1978).
34. J. M. Wood, V. Lach, and B. Jarvis, "Detection of Food-Associated Microbes Using Electrical Impedance Measurements," *Journal of Applied Bacteriology* **43**, xiv–xv (1977).
35. M. P. Silverman and E. F. Munoz, "Automated Electrical Impedance Technique for Rapid Enumeration of Fecal Coliforms in Effluents From Sewage Treatment Plants," *Applied Environmental Microbiology* **37**, 521–526 (1979).
36. S. B. Martins and M. J. Selby, "Evaluation of a Rapid Method for the Quantitative Estimation of Coliforms in Meat by Impedimetric Procedures," *Applied Environmental Microbiology* **39**, 518–524 (1980).
37. R. Firstenberg-Eden and C. S. Klein, "Evaluation of a Rapid Impedimetric Procedure for the Quantitative Estimation of Coli-Forms" *Journal of Food Science* **48**, 1307–1311 (1983).
38. J. L. Weihe, S. L. Seist, and W. S. Hatcher, Jr., "Estimation of Microbial Populations in Frozen Concentrated Orange Juice Using Automated Impedance Measurements," *Journal of Food Science* **49**, 243–245 (1984).
39. R. Firstenberg-Eden, "Rapid Estimation of the Number of Microorganisms in Raw Meat by Impedance Measurement," *Food Technology* **37**(1), 64–70 (1983).
40. R. Firstenberg-Eden and M. K. Tricarico, "Impedimetric Determination of Total, Mesophilic and Psychrotrophic Counts in Raw Milk," *Journal of Food Science* **48**, 1750–1754 (1983).
41. R. A. Lampi, D. A. Mikelson, D. B. Rowley, J. J. Previte, and R. E. Wells, "Radiometry and Microcalorimetry—Techniques for the Rapid Detection of Foodborne Microorganisms," *Food Technology* **28**(10), 52–55 (1974).
42. D. B. Rowley, J. J. Previte, and H. P. Srinivasa, "A Radiometric Method for Rapid Screening of Cooked Foods for Microbial Acceptability," *Journal of Food Science* **43**, 1720–1722 (1978).
43. B. J. Stewart, M. J. Eyles, and W. G. Murrell, "Rapid Radiometric Method for Detection of *Salmonella* in Foods," *Applied Environmental Microbiology* **40**, 223–230 (1980).
44. M. Gagnon, W. M. Hunting, and B. Esselen, "New Method for Catalase Determination," *Analytical Chemistry* **31**, 144–146 (1959).
45. R. Charbonneau, J. Therrien, and M. Gagnon, "Detection and Measurement of Bacterial Catalase by the Disk-Flotation Method Using the Catalasemeter," *Canadian Journal of Microbiology* **21**, 580–582 (1975).
46. K. L. Dodds, R. A. Holley, and A. G. Kempton, "Evaluation of the Catalase and *Limulus* Amoebocyte Lysate Tests for Rapid Determination of the Microbial Quality of Vacuum-Packed Cooked Turkey," *Canadian Institute of Food Science Techology Journal* **16**, 167–172 (1983).
47. W. W. Forrest, "Microcalorimetry," *Methods in Microbiology* **6B**, 285–318 (1972).
48. L. E. Sacks and E. Menefee, "Thermal Detection of Spoilage in Canned Foods," *Journal of Food Science* **37**, 928–931 (1972).
49. T. Fugita, P. R. Mond, and I. Wadso, "Calorimetric Identification of Several Strains of Lactic Acid Bacteria," *Journal of Dairy Research* **45**, 457–463 (1978).
50. B. F. Perry, A. E. Beezer, and R. J. Miles, "Flow Microcalorimetric Studies of Yeast growth: Fundamental aspects," *Journal of Applied Bacteriology* **47**, 527–537 (1979).
51. B. F. Perry, A. E. Beezer, and R. J. Miles, "Characterization of Commercial Yeast Strains By Flow Microcalorimetry," *Journal of Applied Bacteriology* **54**, 183–189 (1983).
52. R. P. Betts, P. Bankes, and J. G. Board, "Rapid Enumeration of Viable Micro-Organisms by Staining and Direct Microscopy," *Letters in Applied Microbiology* **9**, 199–202 (1989).
53. T. C. Hutcheson, T. McKay, L. Farr, and B. Seddon, "Evaluation of the Strain Viablue for the Rapid Estimation of Viable Yeast Cells," *Letters in Applied Microbiology* **6**, 85–88 (1988).
54. H. A. Koch, R. Bandler, and R. R. Gibson, "Fluorescence Microscopy Procedure for Quantification of Yeasts in Beverages," *Applied Environmental Microbiology* **52**, 599–601 (1986).

55. Association of Official Analytical Chemists, *Official Methods of Analysis,* 15th ed., Vol. 1, AOAC, Arlington, Va., 1990.
56. J. E. Hobbie, R. J. Daley, and S. Jasper, "Use of Nucleopore Filters for Counting Bacteria by Fluorescence Microscopy," *Applied Environmental Microbiology* 33, 1225–1228 (1977).
57. G. L. Pettipher, R. Mansell, C. H. McKinnon, and C. M. Cousins, "Rapid Membrane Filtration-Epifluorescent Microscopy Technique for Direct Enumeration of Bacteria in Raw Milk," *Applied Environmental Microbiology* 39, 423–429 (1980).
58. B. G. Shaw, C. D. Harding, W. H. Hudson, and L. Farr, "Rapid Estimation of Microbial Numbers on Meat and Poultry by Direct Epifluorescent Filter Technique," *Journal of Food Protection* 50, 652–657 (1987).
59. J. T. Holah, R. P. Betts, and R. H. Thorpe, "The Use of Direct Epiflourescent Microscopy (DEM) and the Direct Epifluorescent Filter Technique (DEFT) to Assess Microbial Populations on Food Contact Surfaces," *Journal of Applied Bacteriology* 65, 215–221 (1988).
60. U. M. Rodrigues and R. G. Kroll, "The Direct Epifluorescent Filter Technique (DEFT): Increased Selectivity, Sensitivity and Rapidity," *Journal of Applied Bacteriology* 59, 493–499 (1985).
61. J. M. Goepfert, M. E. Mann, and R. Hicks, "One-day Fluorescent-Antibody Procedure for Detecting Salmonellae in Frozen and Dried Foods," *Applied Microbiology* 20, 977–983 (1970).
62. B. M. Thomason, "Rapid Detection of *Salmonella* Microcolonies by Fluorescent Antibody," *Applied Microbiology* 22, 1064–1069 (1971).
63. L. D. Fantasia, J. P. Schrade, J. F. Yager, and D. Debler, "Fluorescent Antibody Method for the Detection of *Salmonella*: Development, Evaluation, and Collaborative Study," *Journal of the Association of Official Analytical Chemists* 58, 828–844 (1975).
64. T. E. Munson, J. P. Schrade, N. B. Bisciello, Jr., L. D. Fantasia, W. R. Hartung, and J. J. O'Connor, "Evaluation of an Automated Fluorescence Antibody Procedure for Detection of *Salmonella* in Foods and Feeds," *Applied Environmental Microbiology* 31, 514–521 (1976).
65. U. M. Rodrigues and R. G. Kroll, "Rapid Detection of Salmonellas in Raw Meats using a Fluorescent Antibody-Microcolony technique," *Journal of Applied Bacteriology* 68, 213–223 (1990).
66. A. N. Sharpe, M. N. Woodrow, and A. K. Jackson, "Adenosinetriphosphate (ATP) Levels in Foods Contaminated by Bacteria," *Journal of Applied Bacteriology* 33, 758–767 (1970).
67. G. A. Kimmich, J. Randles, and J. S. Brand, "Assay of Picomole Amounts of ATP, ADP, and AMP Using the Luciferase Enzyme System," *Analytical Biochemistry* 69, 187–206 (1975).
68. C. J. Stannard and J. M. Wood, "The Rapid Estimation of Microbial Contamination of Raw Meat by Measurement of Adenosinetriphosphate (ATP)," *Journal of Applied Bacteriology* 55, 429–438 (1983).
69. D. R. Ward, K. A. LaRocco, and D. J. Hopson, "Adenosine Triphosphate Bioluminescent Assay to Enumerate Bacterial Numbers on Fresh Fish," *Journal of Food Protection* 49, 647–650 (1986).
70. K. A. LaRocco, P. Galligan, K. J. Littel, and A. Spurgash, "A Rapid Bioluminescent ATP Method for Determining Yeast Contamination in a Carbonated Beverage," *Food Technology* 39(7), 49–52 (1985).
71. J. D. Sullivan, Jr., F. W. Valois, and S. W. Watson, "Endotoxins: The Limulus Amoebocyte Lystae System, in A. W. Bernheimer, ed., *Mechanisms in Bacterial Toxinology,* Wiley, New York, 1976.
72. J. M. Jay, "The *Limulus* Amoebocyte Lysate (LAL) Test, in M. R. Adams and C. F. A. Hope, eds., *Progress in Industrial Microbiology: Rapid Methods in Food Microbiology,* Elsevier, Amsterdam, 1989.
73. K. Tsuji, P. A. Martin, and D. M. Bussey, "Automation of Chromogenic Substrate *Limulus* Amebocyte Lysate Assay Method for Endotoxin by Robotic System." *Applied Environmental Microbiology* 48, 550–555 (1984).
74. H. J. Fallowfield and J. T. Patterson, "Potential Value of the *Limulus* Lysate Assay for the Measurement of Meat Spoilage," *Journal of Food Technology,* 20, 467–479 (1985).
75. J. M. Jay, S. Margitic, A. L. Shereda, and H. V. Covington, "Determining Endotoxin Content of Ground Beef by the *Limulus* Amoebocyte Lysate Test as a Rapid Indicator of Microbial Quality," *Applied Environmental Microbiology* 38, 885–890 (1979).
76. K. -J. Zaadhof and G. Terplan, "Der *Limulus*-test—ein Verfahren zur Buerteilung der Mikrobiologischen Qualität von Milch und Milchprodukten," *Deutsch Molkerzeit.* 34, 1094–1098 (1981).
77. J. D. Sullivan, Jr., P. E. Ellis, R. G. Lee, W. S. Combs, Jr., and S. W. Watson, "Comparison of the *Limulus* Amoebocyte Lysate Test with Plate Counts and Chemical Analyses for Assessment of the Quality of Lean Fish," *Applied Environmental Microbiology* 45, 720–722 (1983).
78. J. M. Jay, "Rapid Estimation of Microbial Numbers in Fresh Ground Beef By Use of the *Limulus* Test," *Journal of Food Protection* 44, 275–278 (1981).
79. B. A. Miller, R. F. Reiser, and M. S. Bergdoll, "Detection of Staphylococcal Enterotoxins A, B, C, D, and E in Foods By Radioimmunoassay, Using Staphylococcal Cells Containing Protein A as Immunoadsorbent," *Applied Environmental Microbiology* 36, 421–426 (1978).
80. H. Robern, M. Dighton, Y. Yano, and N. Dickie, "Double-Antibody Radioimmunoassay for Staphylococcal Enterotoxin C_2," *Applied Microbiology* 30, 525–529 (1975).
81. R. A. Giannella, K. W. Drake, and M. Luttrell, "Development of a Radioimmunoassay for *Escherichia coli* Heat-Stable Enterotoxin: Comparison With the Suckling Mouse Bioassay," *Infectious Immunology* 33, 186–192 (1981).
82. D. B. Shah, P. E. Kauffman, B. K. Boutin, and C. H. Johnson, "Detection of Heat-Labile-Enterotoxin-Producing Colonies of *Escherichia coli* and *Vibrio cholerae* by Solid-Phase Sandwich Radioimmunoassay," *Journal of Clinical Microbiology* 16, 504–508 (1982).
83. O. Aalund, K. Brunfeldt, B. Hald, P. Krogh, and K. Poulsen, "A Radioimmunoassay for Ochratoxin A: A Preliminary Investigation," *Acta Pathologic Microbiologica Scandinavica Section* 83, 390–392 (1975).
84. J. J. Pestka, V. Li, W. O. Harder, and F. S. Chu, "Comparison of Radioimmunoassay and Enzyme-Linked Immunosorbent Assay for Determining Aflatoxin M_1 in Milk," *Journal of the Association of Official Analytical Chemists* 65, 294–301 (1981).
85. S. A. Minnich, P. A. Hartman, and R. C. Heimsch, "Enzyme Immunoassay for Detection of Salmonellae in Foods," *Applied Environmental Microbiology* 43, 877–883 (1982).
86. B. J. Robison, C. I. Pretzman, and J. A. Mattingly, "Enzyme Immunoassay in Which a Myeloma Protein is Used for Detection of Salmonellae," *Applied Environmental Microbiology* 45, 1816–1821 (1983).

87. R. S. Flowers, K. -H. Chen, B. J. Robison, J. A. Mattingly, D. A. Gabis, and J. H. Silliker, "Comparison of *Salmonella* Bio-EnzaBead Immunoassay Method and Conventional Culture Procedure for Detection of *Salmonella* in Crustaceans," *Journal of Food Protection* **50**, 386–389 (1987).
88. W. J. Hu, N. Woychik, and F. S. Chu. "ELISA of Picogram Quantities of Aflatoxin M_1 in Urine and Milk," *Journal of Food Protection,* **47**, 126–127 (1984).
89. J. J. Pestka, P. K. Gaur, and F. S. Chu, "Quantitation of Aflatoxin B_1 and Aflatoxin B_1 Antibody by an Enzyme-Linked Immunosorbent Microassay," *Applied Environmental Microbiology* **40**, 1027–1031 (1980).
90. J. J. Pestka, B. W. Steinert, and F. S. Chu, "Enzyme-Linked Immunosorbent Assay for Detection of Ochratoxin A," *Applied Environmental Microbiology* **41**, 1472–1474 (1981).
91. J. J. Pestka and F. S. Chu, "Enzyme-Linked Immunosorbent Assay of Mycotoxins Using Nylon Bead and Terasaki Plate Solid Phases," *Journal of Food Protection,* **47**, 305–308 (1984).
92. S. Notermans, J. Dufrenne, and M. van Schothorst, "Enzyme-Linked Immunosorbent Assay for Detection of *Clostridium botulinum* Type A," *Japanese Journal of Medical Science Biology* **31**, 81–85 (1978).
93. S. Notermans, J. Dufrenne, and S. Kozaki, "Enzyme-Linked Immunosorbent Assay for Detection of *Clostridium botulinum* Type E Toxin," *Applied Environmental Microbiology* **37**, 1173–1175 (1979).
94. G. E. Lewis, Jr., S. S. Kulinski, D. W. Reichard, and J. F. Metzger, "Detection of *Clostridium botulinum* Type G Toxin by Enzyme-Linked Immunosorbent Assay," *Applied Environmental Microbiology* **42**, 1018–1022 (1981).
95. G. C. Saunders and M. L. Bartlett, "Double-Antibody Solid-Phase Enzyme Immunoassay for the Detection of Staphylococcal Enterotoxin A," *Applied Environmental Microbiology* **34**, 518–522 (1977).
96. G. Stiffler-Rosenberg and H. Fey, "Simple Assay for Staphylococcal Enterotoxins A, B, and C: Modification of Enzyme-Linked Immunosorbent Assay," *Journal of Clinical Microbiology* **8**, 473–479 (1978).
97. S. Notermans, R. Boot, P. D. Tips, and M. P. DeNooij, "Extraction of Staphylococcal Enterotoxins (SE) from minced meat and subsequent detection of SE with Enzyme-Linked Immunosorbent Assay (ELISA)," *Journal of Food Protection,* **46**, 238–241 (1983).
98. J. M. Farber and J. I. Speirs, "Monoclonal Antibodies Directed Against the Flagellar Antigens of *Listeria* Species and Their Potential in EIA-Based Methods," *Journal of Food Protection* **50**, 479–484 (1987).
99. B. T. Butnam, M. C. Plank, R. J. Durham, and J. M. Mattingly, "Monoclonal Antibodies Which Identify a Genus-Specific *Listeria* Antigen," *Applied Environmental Microbiology* **54**, 1564–1569 (1988).
100. J. A. Mattingly, "An Enzyme Immunoassay for the Detection of All *Salmonella* Using a Combination of Myeloma Protein as a Hybridoma Antibody," *Journal of Immunological Methods* **73**, 147–156 (1984).
101. C. E. Park, H. B. El Derea, and M. K. Rayman, "Evaluation of Staphylococcal Thermonuclease (TNase) Assay as a Means of Screening Foods for Growth of Staphylococci and Possible Enterotoxin Production," *Canadian Journal of Microbiology* **24**, 1135–1139 (1978).
102. A. Koupal and R. H. Deibel, "Rapid Qualitative Method for Detecting Staphylococcal Nuclease in Foods," *Applied Environmental Microbiology* **35**, 1193–1197 (1978).
103. J. F. Kamman and S. R. Tatini, "Optimal Conditions for Assay of Staphylococcal Nuclease," *Journal of Food Science* **42**, 421–424 (1977).
104. B. S. Emswiler-Rose, R. W. Johnston, M. E. Harris, and W. H. Lee, "Rapid Detection of Staphylococcal Thermonuclease on Casings of Naturally Contaminated Fermented Sausages," *Applied Environmental Microbiology* **40**, 13–18 (1980).
105. R. Fitts, M. L. Diamond, C. Hamilton, and M. Neri, "DNA-DNA Hybridization Assay for Detection of *Salmonella* Spp. in Foods," *Applied Environmental Microbiology* **46**, 1146–1151 (1983).
106. R. S. Flowers, M. A. Mozzola, M. S. Curiale, D. A. Gabis, and J. H. Silliker, "Comparative Study of a DNA Hybridization Method and the Conventional Culture Procedure for Detection of *Salmonella* in Foods," *Journal of Food Science* **52**, 781–785 (1987).
107. S. Notermans, t. Chakraborty, M. Leimeister-Wächter, J. Dufrenne, K. J. Heuvelman, H. Maas, W. Jansen, K. Wernars, and P. Guinee, "Specific Gene Probe for Detection of Biotyped and Serotyped *Listeria* Strains," *Applied Environmental Microbiology* **55**, 902–906 (1989).
108. A. R. Datta, B. A. Wentz, D. Shook, and M. W. Trucksess, "Synthetic Oligodeoxyribonucleotide Probes for Detection of *Listeria Monocytogenes*," *Applied Environmental Microbiology* **54**, 2933–2937 (1988).
109. S. L. Moseley, P. Echeverria, J. Seriwatana, C. Tirapat, W. Chaicumpa, T. Sakuldaipeara, and S. Falkow, "Identification of Enterotoxigenic *Escherichia coli* by Colony Hybridization Using Three Enterotoxin Gene Probes," *Journal of Infectious Disease* **145**, 863–869 (1982).
110. W. E. Hill, J. M. Madden, B. A. McCardell, D. B. Shah, J. A. Jagow, W. L. Payne, and B. K. Boutin, "Foodborne Enterotoxigenic *Escherichia coli:* Detection and Enumeration by DNA Colony Hybridization," *Applied Environmental Microbiology* **45**, 1324–1330 (1983).
111. S. Dovey and K. J. Towner, "A Biotinylated DNA Probe to Detect Bacterial Cells in Artificially Contaminated Foodstuffs," *Journal of Applied Bacteriology* **66**, 43–47 (1989).

JAMES M. JAY
Wayne State University
Detroit, Michigan

FOOD CHEMISTRY/BIOCHEMISTRY

Food science and technology can be defined as the application of engineering, microbiology, and chemistry/biochemistry principles to food. The corollary of this definition is that food chemistry/biochemistry is one of three basic disciplines that make up food science and technology. If the definitions of the words themselves are considered (1), biochemistry is defined as "the chemistry of plant and animal life"; chemistry as "a science that deals with the composition, structure, and properties of substances and of the transformations that they undergo"; and food as "material consisting of carbohydrates, fats, proteins, and supplementary substances (as minerals, vitamins) that is taken or absorbed into the body of an organism in order to sustain growth, repair, and all vital processes and to furnish energy for all activity of the organism."

Clearly the main focus of food chemistry/biochemistry is food, but to define all of the facets implied by the terms food chemistry/biochemistry is very difficult. It is difficult, not only because of the breadth of the discipline, but because of the rapid evolution of food chemistry/biochemistry in the twentieth century.

EDUCATION

One indication of the rapid evolution of food chemistry/biochemistry is in the teaching of the discipline. In the early 1970s there were virtually no general food chemistry/biochemistry textbooks. From the mid-1970s on, however, there was an explosive growth in such books (Table 1). Most of the books have chapters or sections on water, carbohydrates, lipids, amino acids and proteins, enzymes, vitamins and minerals, color, flavor, texture, toxic compounds, and additives. Besides the trend to more general textbooks in the food chemistry/biochemistry area, there is a trend to increased complexity and proficiency in these books. For example, books such as D. W. S. Wong's *Mechanism and Theory in Food Chemistry* differ little in appearance from any advanced chemistry or biochemistry textbook.

With the increase in complexity of the discipline, increased specialization is necessary. It is interesting that as early as 1944 the need for specialization in food science and technology was recognized by educators in the field (2). The guidelines for a satisfactory curriculum for food science and technology in 1944 were:

1. That the students first acquire as a foundation certain basic courses applicable to all phases of food technology; such as courses in chemistry, physics, mathematics, microbiology, and biological chemistry.
2. That it is impossible for any person to become proficient in all phases of food technology and therefore it would be better to train students in the basic sciences with perhaps a limited knowledge of their application to food technology, rather than to turn out students with a superficial knowledge of a large number of food industries and operations but lacking in basic training.
3. That the students' specialized training in the application of these basic science principles to the field of food technology be reserved for periods toward the end of their college training.
4. That the student acquire some knowledge of the principles of engineering, either in the field of mechanical engineering or in the field of chemical engineering.
5. That the student acquire training in subject matter auxiliary to the scientific training but very necessary to proper functioning in industry; such as law, business, accounting, statistics, personnel relationships, English, report writing, record keeping, geographic agriculture.
6. That it is probably not desirable for all students to receive the same training. Some might specialize in the sanitary aspects of food technology, others in the biochemical, physical, organic, or the engineering, but all should have such knowledge of the field as a whole as to enable them to understand the literature and language of the whole field.
7. That during summer vacations a student work in food-processing plants or perhaps drop out during his/her college course to gain a year of practical experience in a processing plant.
8. That probably five years of college work will be required (rather than 4 for an undergraduate degree).

From these recommendations it is obvious that the best educational choice for someone interested in food chemistry/biochemistry would be to first obtain extensive training in chemistry or biochemistry. This preliminary training could mean as much as obtaining a degree in one of these disciplines before carrying out further study on the application of chemistry/biochemistry principles to food.

PUBLICATIONS

There is no doubt that parallel with the increase in educational material, there has been an increase in publications in the food chemistry/biochemistry discipline. As in the educational area, there is indication that the discipline is becoming increasingly specialized and complex.

It is instructive to look at trend-setting journals in the food chemistry/biochemistry area. Although it is difficult

Table 1. Some General Food Chemistry/Biochemistry Textbooks

Author or Editor	Title	Publisher	Year
F. A. Lee	*Basic Food Chemistry*	AVI Publishing Co., Westport Conn.	1975
O. R. Fennema (ed.)	*Food Chemistry*, 2nd ed.	Marcel Dekker, Inc. New York	1985 (original version, 1976)
C. Zapsalis and R. A. Beck	*Food Chemistry and Nutritional Biochemistry*	Originally John Wiley & Sons, New York, now Macmillan, London	1985
H. D. Belitz and W. Grosch	*Food Chemistry* (transl. from German by D. Hadziyev)	Springer Verlag, Berlin	1986
J. M. deMan	*Principles of Food Chemistry*, 2nd ed.	AVI Publishing Co., Westport, Conn.	1990 (original version, 1980)
D. W. S. Wong	*Mechanism and Theory in Food Chemistry*	AVI Publishing Co., Westport, Conn.	1989

Table 2. 1988 Food Science and Technology Journals Rated by Impact Factor[a]

Rank	Journal Title	Impact Factor
1	CRC Critical Reviews in Food Science	1.826
2	Journal of Cereal Science	1.484
3	Journal of Dairy Science	1.423
4	Biotechnology Progress	1.283
5	Journal of Food Protection	1.224
6	Journal of Dairy Research	1.182
7	Journal of Agricultural and Food Chemistry	1.165
8	Food Technology	1.004
9	Netherlands Milk and Dairy Journal	0.962
10	Food and Chemical Toxicology	0.887
11	Journal of the American Oil Chemists Society	0.880
12	Food Additives and Contaminants	0.857
13	International Journal of Food Microbiology	0.827
14	Zeitshrift fur Lebensmitteluntersuchung und-Forshung	0.825
15	Journal of Food Science	0.781
16	Journal of Micronutrient Analysis	0.756
17	Journal of Fermentation Technology	0.737
18	Cereal Chemistry	0.720
19	Journal of the Science of Food and Agriculture	0.705
20	Archiv Fur Lebensmittel Hygiene	0.671

[a] Ref. 3.

to define a journal as solely dealing with food chemistry/biochemistry, leading journals in food science and technology have been rated by the Institute for Scientific Information on the basis of impact factor (top 20 journals are given in Table 2) (3). Impact factor is defined as the number of yearly citations of journal articles divided by the number of yearly articles. Therefore, an impact factor of 1.00 means that if a journal contained 50 papers in a given year, there would also be 50 citations of former journal papers for the same year.

Review journals such as *CRC Critical Reviews in Food Science* or *Food Technology* will always have somewhat higher impact factors because reviews are generally larger, well-referenced papers and preferentially cited over many smaller papers in a field. Also, some journals are not considered primarily food technology journals and may not appear on the list for this reason. Examples of this latter case are the *Journal of the Association of Official Analytical Chemists* (1.083 impact factor) and *Agricultural and Biological Chemistry* (0.913 impact factor). However, impact factors do give a measure of the importance of various journals independent of the number of papers that appear in a journal (ie, large journals that have more total citations are rated on an equivalent scale to small journals because of the per article nature of impact factors).

From Table 2 then, it is striking that the major journals in the food science and technology area are generally commodity journals with dairy technology represented by 3 journals in the top 10. In fact, in North America many of the present food science departments were originally dairy departments. This points out that research in food chemistry/biochemistry often has a commodity bias.

The highest rated general food science and technology journal, which deals primarily with chemistry on the list (Table 2) is the *Journal of Agricultural and Food Chemistry*. This journal is also rated with the highest impact factor among agricultural journals (3). A further indication of the increased specialization is the recent practice by this journal to group papers into topics. The largest topic group is food chemistry/biochemistry. Other topics include analytical methods, biotechnology, environmental chemistry, chemical changes during processing/storage, chemistry of crop and animal protection, composition of foods/feeds, flavors and aromas, and nutrition/toxicology. Although separated from food chemistry/biochemistry by name, many of these other topics would also be considered to involve at least some food chemistry/biochemistry.

So which topics are covered by the food chemistry/biochemistry discipline? There may be considerable controversy, but in looking at food chemistry/biochemistry textbooks and journals, topics include those given in Table 3. Of course the inclusion of a topic in this table does not imply that food chemistry/biochemistry alone deals this topic, but that there is at least some involvement of food chemistry/biochemistry in the topic. Also, Table 3 implies no weighting. For example, postharvest physiology of plants and postmortem changes in animals are huge topic areas involving numerous changes within even one food system. These two topics also highlight another point about food chemistry/biochemistry. Other disciplines such as biochemistry, botany, plant science, animal science, and zoology deal mainly in living, growing, reproducing systems; while food chemistry/biochemistry deals mainly in postmortem or postharvest senescence changes in animals and plants, as well as the chemistry and biochemistry of disrupted tissue.

Another factor to note in the topics given in Table 3 is the inclusion of food analysis. No one can properly study food chemical and biochemical reactions without at least a working knowledge of analytic methods. In fact, most scientists in food chemistry/biochemistry find that they must become familiar with a wide variety of techniques borrowed from other disciplines such as chemistry, physics, biochemistry, and immunology. In addition to the complication of learning a variety of unrelated techniques in

Table 3. Food Chemistry/Biochemistry Topics

Water

Physical properties, pH
Hydrogen bonding, solubility of food components
Water activity

Carbohydrates

Basic structures and naming conventions
Biosynthesis and biodegradation, metabolism
Chemistry of reducing and nonreducing sugars
Nonenzymatic browning
Natural homo- and heteropolysaccharides, sources, biosynthesis, and uses
Chemical modification of natural polysaccharides and their uses

Lipids

Basic structures and naming
Biosynthesis and biodegradation, metabolism
Lipid reactions, hydrolytic and oxidative rancidity
Crystal structure and physical properties
Refining, hydrogenation, and interesterification reactions

Amino Acids and Proteins

Basic structure and naming
Biosynthesis and biodegradation, metabolism
Protein structure
Denaturation conditions
Protein chemical changes in processing

Enzymes

Catalytic action, stereochemical specificity
Food enzymes, action, kinetics, stability, uses, and sources
Enzymatic browning and rancidity
Enzyme immobilization
Postharvest physiology of plants
Postmortem changes in animals

Vitamins and Minerals

Chemistry and interaction of vitamins and minerals with food components under processing conditions
Biosynthesis and biodegradation, metabolism

Color

Natural food pigments and their chemistry
Artificial colors and their metabolism
Color measurement

Flavor

Biosynthesis of flavor components
Sweetening and other flavor theories
Physiology of taste and smell

Texture

Food colloids
Gel formation
Emulsifiers and foaming agents
Viscosity

Toxic Compounds

Natural microbial, plant, and animal toxins
Synthetic food toxins
Radionucleotides and toxic metals
Residues
Metabolism and excretion of toxic compounds

Table (*continued*)

Food Additives

Sequestrants and their chemistry
Artificial and natural sweeteners
Antioxidants
Acidulants
Bases, leavening agents
Antibiotic and probiotic agents
Gases and Propellants
Other food additives

Food Analysis

Analysis of any of the above food components using:
    Wet chemical and physical methods,
    Chromatography techniques including high performance liquid chromatography (HPLC), gas chromatography (GC)
    Spectroscopy techniques, visible (absorption and reflectance), ultraviolet, fluorescent, infrared, atomic absorption and emission, mass spectrometry (MS), nuclear magnetic resonance (NMR)
    Electroanalytical techniques, selective electrodes, coulometry, and voltammetry, electrophoresis
    Enzymatic methods, enzymatic electrodes, enzyme immunoassays (EIA)
    Thermal methods, differential scanning calorimetry, cryoscopy
    Rheology methods

food analysis, there is the complication of working with very complex food matrices. This diversity of analytic techniques and food matrices has also led to increasing specialization of food chemistry/biochemistry in the analytic area.

Finally what does the future hold for food chemistry/biochemistry? The indication is that as food products and processing methods become more diverse and consumer concerns with respect to nutrition, toxicology, and food contaminants become greater, food chemistry/biochemistry can only grow as a discipline.

BIBLIOGRAPHY

1. P. B. Gove, *Webster's Third New International Dictionary* G. and C. Merriam Company, Springfield, Mass., 1967.
2. O. Fennema, "Educational Programs in Food Science: A Continuing Struggle for Legitimacy, Respect, and Recognition," *Food Technology* 43(9), 170–182, (1989).
3. E. Garfield, *SCI Journal Citation Reports,* vol. 19, Institute for Scientific Information Inc., Philadelphia, 1988, p. 369.

Peter Sporns
University of Alberta
Edmonton, Alberta, Canada

FOOD CHEMISTRY: MECHANISM AND THEORY

Food chemistry deals with the chemical identity of food components and chemical reactions governing the changes and performance of individual or interacting food components during handling, processing, and storage. Although the individual constituents may often be readily identified, the interactions of food components are extremely complex.

Food chemistry is a branch of chemistry with its foundation built on chemical principles and reaction mechanisms, and a comprehension of the subject often requires thorough understanding and application of knowledge from various chemistry and chemistry-related disciplines. In this regard, food chemistry is quite similar to biochemistry, except that the former relates chemistry to food systems and the latter to biological systems. For example, a biochemist may be interested in elucidating the molecular structures of the wheat storage proteins gliadin and glutenin; however, for a food chemist the results drawn from these structural studies are related to functional properties, such as changes and effects in dough quality and baking. Another example is found in the Maillard reaction. An organic chemist may investigate details of the reaction pathways. A biochemist is more likely to be interested in the reaction in relation to aging of certain vital proteins, such as lens crystalline, collagen, and elastin. A food chemist is also interested in the Maillard reaction, but more in linking the reaction mechanism to physical and chemical changes in food systems such as flavor development, browning, and nutritional loss.

In this article reaction mechanisms of sufficient importance and their current developments are presented. Only the most recent and relevant references will be included. A number of good background references have been published (1–5).

CARBOHYDRATES

The Maillard Reaction

In 1912 the French chemist Maillard first observed that yellow-brown pigments formed when sugars reacted with amino acids, peptides, and proteins in a heated solution. Food chemists have recognized the practical relevance of this reaction to many chemical and physical changes during processing and storage of food. The first English-language review on the Maillard chemistry in food systems was published in 1951 (6). Since then, numerous reviews on this subject have appeared (7–9). The biological importance of this reaction has been recognized only in the last 20 yr. It is now well established that the reaction is linked to glycosylated hemoglobin (HbA_{1c}) in diabetes, hardened lens crystallins in cataract disease, and a number of other aging proteins (10,11).

The Maillard reaction comprises a series of reactions: (1) formation of glycosylamine via a Schiff base reaction between a reducing sugar and the amino group of an amino acid, (2) Amadori rearrangement in which glycosylamine is converted to ketosamine, and (3) enolization (C1–C2, or C2–C3), followed by cyclodehydration. In general, under acidic conditions the nitrogen is protonated, and 1,2-enolization is assisted by the positively charged nitrogen acting as an electron sink. Alkali and strongly basic amines favor 2,3-enolization (Fig. 1).

The actual reactions are far more complicated than those outlined above, and there are many variations in the pathway. The initial step in the Amadori rearrangement is suggested as N-protonation. However, addition of the proton to the ring oxygen has also been proposed. Most discussions on the Maillard reaction concern the monosubstituted amines, but the ketoseamine formed in the reaction can also react with another molecule of an aldose resulting in disubstitution. Another variation is the Heynes rearrangement in the conversion of D-fructosylamine (a ketosamine) to 2-amino-2-deoxy-D-glucose (an aldosylamine).

In recent years a new pathway that involves sugar fragmentation and free-radical formation prior to the Amadori rearrangement, has been suggested (12) (Fig. 2). The radical has been structurally identified to be N,N'-disubstituted pyrazine cation radical, and it is formed by the dimerization of a two-carbon enaminol product from the cleavage of glycosylamine.

Very recently, direct dehydration of the Amadori compound has been proposed as an alternative to enolization (13–15) (Fig. 3). In this mechanism the Amadori compound undergoes a trans-elimination at C2–C3, followed by a second dehydration at C3–C4, to form a hydroxypyran and finally a pyrylium ion. The highly electrophilic pyrylium ion can undergo various nucleophilic additions, ring opening and recyclization.

Maillard reaction products have diverse structures and are involved in various secondary reactions. A compilation of 450 volatile Maillard reaction products and related

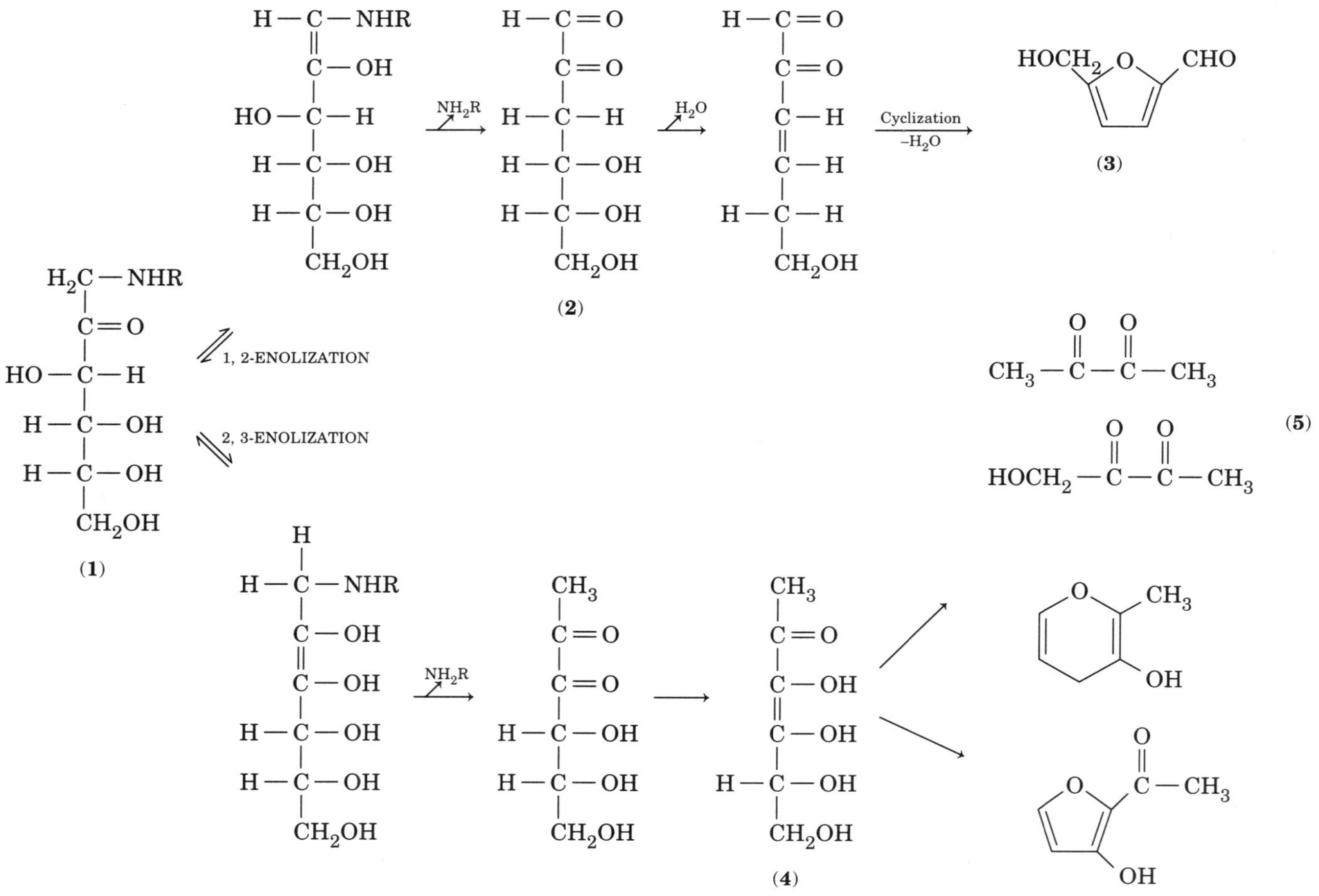

Figure 1. Reaction pathways of Amadori product. (1) Amadro product, (2) deoxyglycosulose, (3) furaldehyde, (4) glycosulos-3-ene, and (5) reductones.

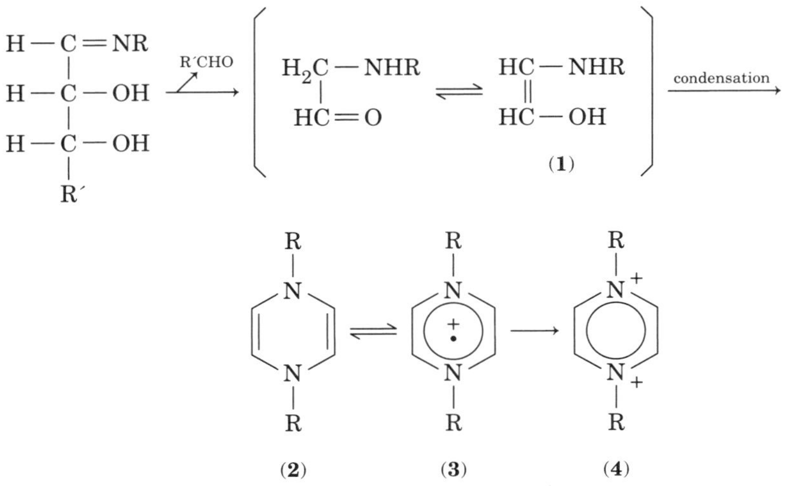

Figure 2. Fragmentation and free-radical formation in the Maillard reaction (12). (1) Enaminol, (2) dialkyldihydropyrazine, (3) dialkylpyrazine cation radical, and (4) dialkylpyrazinium compound.

compounds (16) and reviews on this subject are available (7–9). Dicarbonyl compounds, such as 3-deoxyglycosulose generated by the 1,2-enolization pathway and the glycosulos-3-ene formed via the 2,3-enolization are the key intermediates for subsequent degradative reactions relating to color and flavor production. One of the well-known pathways is the Strecker degradation in which the carbonyl forms a Schiff base with the α-amino group of an amino acid. Enolization, decarboxylation, and hydrolysis yield an aldehyde corresponding to the original amino acid with one less carbon. Aldehydes derived from this degradative pathway constitute many important flavor compounds in food systems. Compounds generated by degradation of the dicarbonyl compounds include pyrrole, pyrazine, oxazoles and derivatives, pyrrolines, pyrrolidines, pyrones, thiazole, and thiazoline.

Sulfite inhibition of nonenzymatic browning also involves the dicarbonyl intermediates reacting with sulfur oxoanion to form stable sulfonate product (17). The sulfur oxoanion may replace the C4 hydroxy group of the deoxyglycosulose or undergo 1,4-addition to the double bond of the α,β-unsaturated glycosulos-3-ene.

Formation of melanoidins is caused by polymerization of unsaturated carbonyl compounds. Condensation be-

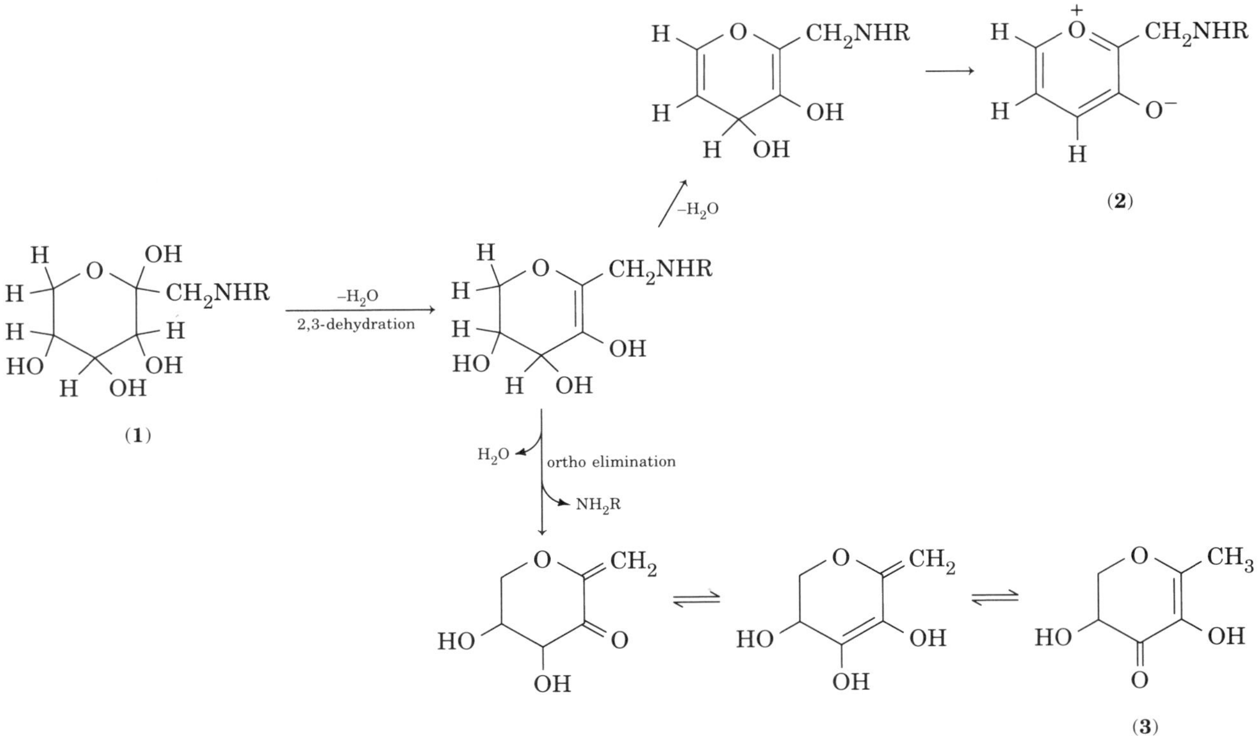

Figure 3. Direct sequential dehydration of Amadori product (13,15). (1) Amadori product, (2) pyrylium ion, and (3) 2,3-dihydro-3,5-dihydroxy-6-methyl-4H-pyran-4-one.

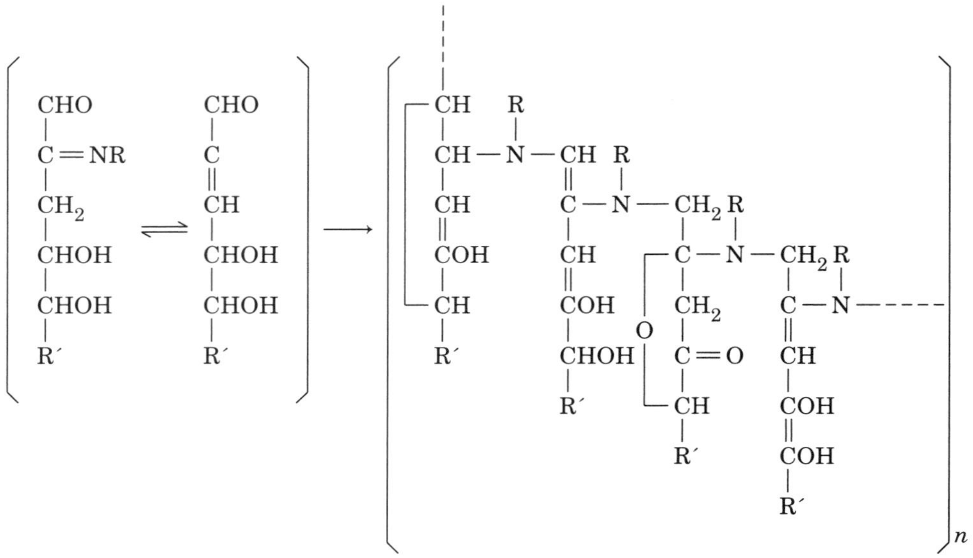

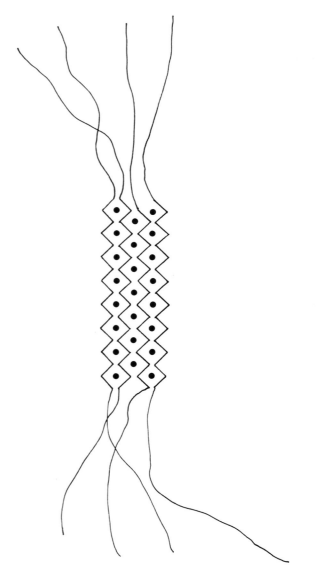

Figure 4. Repeating unit of melanoidin (18). R–NH$_2$ = amine and R' = H or CH$_2$OH.

tween 3-deoxyglycosulose and its enamine, which is a reductone, has been suggested to be the structural unit of the polymer (18,19) (Fig. 4), although polymers of repeating units of furan or pyrrole have also been proposed.

The Maillard reaction offers a perfect example of how basic organic chemistry (studies of the reaction mechanism between sugar and amino acids and the structural identification of possible intermediate compounds) has made it possible to explain some of the most interesting and important interactions in food systems.

Polysaccharide Polymers

The last two decades have witnessed an increasing interest in the relationship between the molecular structure and functional mechanism of food polysaccharides. Growth in knowledge of polymer chemistry has made a dramatic change in the way food scientists interpret the structure and function of food polysaccharides.

One milestone in this area is the elucidation of the gelling mechanism of alginate in the early 1970s (20). The mechanism of gelation has been shown to be an interaction between calcium ions and polyguluronate blocks of alginate polymer. Such interactions provide junctions that cross-link alginate polymers into a three-dimensional network, known as the egg box model (Fig. 5).

The concept of junction zone linking of long chain polymers has since been included in explanations of various kinds of gelation. For example, carrageenan polymers are known to associate by intermolecular double helices. Gelation occurs with the subsequent aggregation of these helices providing cross-linking junctions to build a continuous network (21,22). The number of charged sulfate groups along the polymer contributes to the degree of aggregation and the characteristics of the gel.

Carrageenan forms stable complexes with κ-casein via the interaction between sulfate anions and the highly positively charged region of the protein. The synergistic interaction between κ-carrageenan and locust bean gum has also been interpreted to involve junction cross-linking.

Figure 5. The egg box model (20). Line = alginate polymer and dark circle = calcium ion.

High methoxy pectin gels at low pH and in the presence of a cosolute such as sucrose. At low pH, the carboxyl groups are protonated. As a result, the decrease in electrostatic repulsion and addition of a cosolute lower the water solvation of the polymer. Both factors increase hydrophobic interaction and association of the polymers into cross-linking junction zones (23).

Starch gelation represents a more complex system that has received continuous attention from various disciplines. The structure of crystalline amylose (of B-form starch), as originally elucidated by x-ray diffraction, consists of two parallel strands of right-handed, six-folded helices packed in an antiparallel double-helix (24). Recent work has suggested a parallel packing of left-handed, parallel-stranded double helices for crystalline amylose from both A- and B-form starch (25,26). In solution the conformation of amylose assumes a random coil containing short segments of loose and irregular helical structure (27,28).

Various models have been proposed for the structure of amylopectin. Evidence from recent studies suggests a cluster-type model such that amylopectin is composed of clusters of oriented chains with the branching points collected together toward the reducing end (29).

Gelation occurs when starch granules in suspension are heated above the gelatinization temperature and cooled. Heating causes the granules to swell irreversibly and is accompanied by the solubilization of amylose while most of the amylopectin is retained.

On cooling, the starch gel formed has a composite structure of swollen amylopectin granules distributed in a matrix of amylose gel (30). Retrogradation involves crystallization of both the amylose in the gel phase and the amylopectin in the granules. Amylose molecules associate through hydrogen bonding into an insoluble precipitate. Amylopectin also exhibits interchain association, but the amylopectin molecule has an average chain length of 20–25 glucose units and a degree of polymerization of approximately 10^3. Interchain association between polymers can only extend for 15–20 glucose units before it is interrupted by branching points. Crystallization of the amylopectin fraction is slow and results in a gradual increase in the rigidity of the granules and hence the amylose gel matrix. Reversion of a starch gel to the granular crystalline state is termed retrogradation, a process chiefly responsible for the staleness of bread.

It is important to note that in a gel the intermolecular association of polymers usually involves extensive segments of the polymeric chains held together by hydrogen bonding, electrostatic forces and hydrophobic or ionic interactions to form cross-linking junctions. Depending on the gel, a variety of rheological properties can exist. The entire field of polysaccharide gels has generated a rapid expansion in the scope of basic research and an intensified interest in the relationship between the molecular arrangement and gelling characteristics has occurred.

LIPIDS

Lipid Oxidation

Lipid autoxidation is a free-radical chain reaction involving hydroperoxides. The mechanism has been extensively investigated since the first systematic study by Bateman and Bolland in the 1940s. In 1970 a possible role for singlet oxygen in the initiation of lipid oxidation was proposed (31). In the presence of a suitable sensitizer, lipid oxidation proceeds via the ene reaction in which the dioxygen molecule is added to the olefinic carbon with a subsequent shift in the position of the double bond, and both conjugated and nonconjugated hydroperoxides are formed. In contrast, the classic free-radical mechanism produces only the conjugated hydroperoxides. The distribution of hydroperoxides is different from that found in the free-radical reaction.

In linoleate autoxidation the product is composed predominantly of the 9- and 13-hydroperoxides, because the stability of the conjugated diene system favors the oxygen attack at the end position. Experimentally, however, four hydroperoxides are obtained: two with trans, cis and the other two with trans, trans stereochemistry (13-hydroperoxy-9-*cis*,11-*trans*-octadecadienoic; 13-hydroxyperoxy-9-*trans*,11-*trans*-octadecadienoic; 9-hydroxyperoxy-10-*trans*,12-*cis*-octadecadienoic; and 9-hydroperoxy-10-*trans*,12-*trans*-octadecadienoic). A unified mechanism has been proposed to account for the formation of these products (32) (Fig. 6). In this scheme the peroxy radicals initially formed have the trans, cis configuration and exist in two conformational isomers. The normal pathway involves the abstraction of hydrogen from another linoleate by either isomer to yield trans,cis hydroperoxides. The alternative pathway involves the loss of oxygen from conformers to give either the original pentadienyl radical or a new carbon radical in which the stereochemistry of the partial double bond is inverted. Oxygen addition to this new pentadienyl radical yields a trans, trans-diene peroxy radical, which ultimately gives the trans, trans-conjugated hydroperoxide.

Because autoxidation involves free radicals, radical scavengers should effectively terminate the chain reaction. Most antioxidants are substituted phenolic compounds that act by transferring hydrogen to lipid peroxy radicals. The resulting aryloxy radical of the antioxidant is stable and unreactive in oxidative reactions. The resonance stability of the aryloxy radical depends on the substitution groups. Electron-releasing groups decrease the transition energy for the formation of the aryloxy radical. Bulky substituents stabilize the aryloxy radical but also create steric hindrance making the antioxidant less accessible to the lipid peroxy radical (33).

The Physical Chemistry of Lipids

Another area of great interest to food chemists concerns the physical chemistry of lipids, including polymorphisms, crystal habits, emulsions. Triglycerides exhibit multiple crystalline structures. For example, tristearin melts at 54.7, 64.0, and 73.3°C representing the transition from the less-stable α form to the more-stable β' and β forms. The α form has a hexagonal crystal subcell. The intermediate melting β' form is orthorhombic, and the highest melting polymorph, the β form, has a triclinic subcell. This multiplicity of molecular conformation and packing influences the fluidity, texture, and appearance of the product. Alpha crystals are fragile platelets of 5 μ in

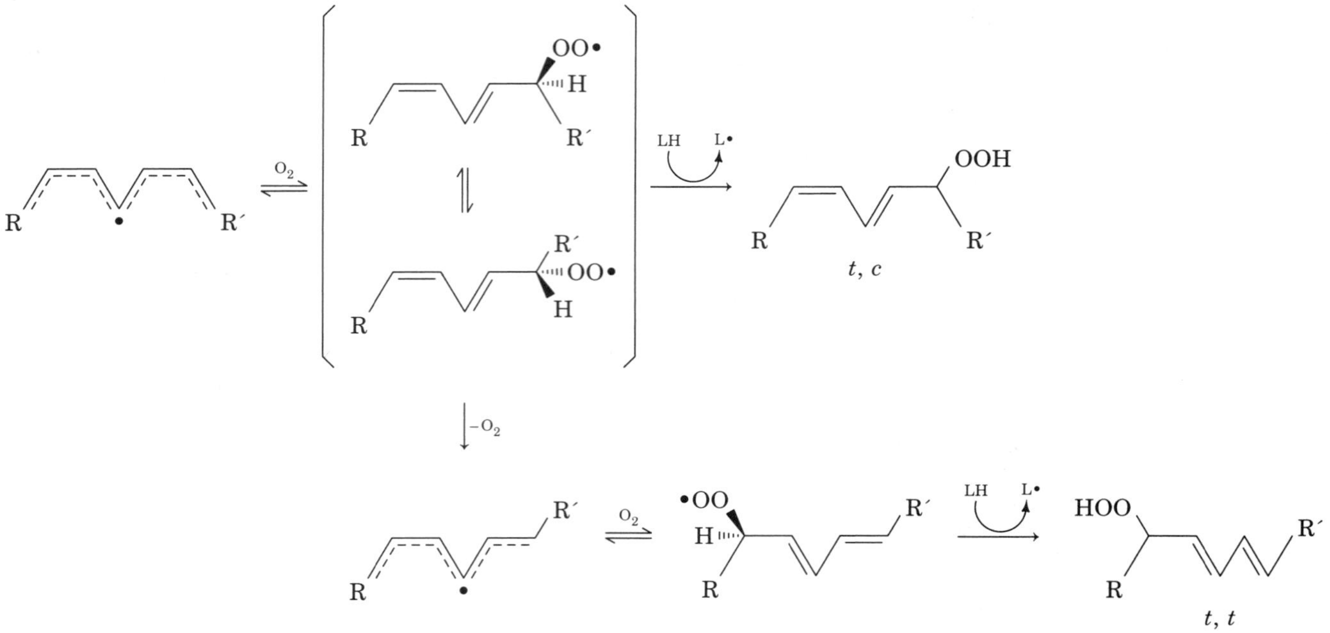

Figure 6. A unified mechanism for lipid autoxidation (32).

size. Beta crystals tend to be large and coarse with a diameter of 25–50 μ. Fats in the β' form appear as tiny needles of approximately 1 μ in length. The β' crystals can incorporate a large amount of air bubbles providing a smooth texture to oil products, such as margarine and shortening. The fatty acid composition as well as the position of a particular fatty acid in the glyceride affect the crystal habit of a fat. Interesterification usually causes a β' to β conversion.

The principal advance in the current concept of emulsion stability came in the last decade with the understanding of the factors controlling attractive–repulsive forces and their interaction with distances between disperse particles. Flocculation and coalescence of oil droplets in an emulsion are dependent on the balance between van der Waals attraction and electrostatic repulsion. The combination of these two gives a net potential energy versus distance curve with a potential minimum at certain intermediate ranges where repulsion may be greater than attraction and a measure of stability exists (34). At long and short ranges the net potential is always attraction.

An emulsifier consists of both hydrophilic and hydrophobic segments in the same molecule. Food-grade emulsifiers are usually partial esters of fatty acids, polyols, and water-soluble organic acids. When an emulsifier is dispersed in water and heated, a liquid crystalline mesophase is formed (35). The mesophase assumes a lamella, hexagonal, or cubic structure depending on the type of emulsifier and temperature. In a ternary system such as an emulsifier in an oil—water system, similar types of mesophase are formed (36). The ordered layer of this liquid crystalline phase stabilizes the oil-in-water emulsion by forming a film at the interface. It decreases the attractive forces between oil droplets and provides a steric barrier against coalescence between the droplets.

PROTEINS

Chemical Reactions in Processing

Because proteins contain many amino acids with reactive side chain groups, it is expected that a variety of reactions may occur during food processing. One of the most extensively studied reactions is alkali degradation. Alkaline treatment is used in the food industry for peeling, solubilization, and texturization of food proteins and manufacture of gelatin, sausage casings, and tortillas.

The most well-studied reaction is β-elimination in which the α-hydrogen of an amino acid residue is abstracted by the hydroxide ion (37) (Fig. 7). In protein-bound cysteine the resulting product is a dehydroalanine, which is an α,β-unsaturated compound. Nucleophilic side groups, such as lysyl ε-amino and cysteinyl sulfur, react with dehydroalanine via Michael addition leading to new cross-inkings in the proteins and loss of certain essential amino acids.

Alkaline treatment also causes hydrolysis of the amide groups in asparagine and glutamine and the guanidino group in arginine. Racemization of amino acids is also detected. Protein-bound amino acids are more susceptible to α-hydrogen abstraction and hence racemization to the D-form (38). Prolonged heating results in isopeptide formation between the ε-amino group of lysine and the carbonyl group of aspartic or glutamic acid, or the amide groups of glutamine or asparagine.

Protein Structure and Functionality

For a food chemist the structure of proteins is quite often viewed in the context of functionality in a food system. For example, the chemistry of muscle fibers and the mechanism of muscle contraction are related to rigor mortis and

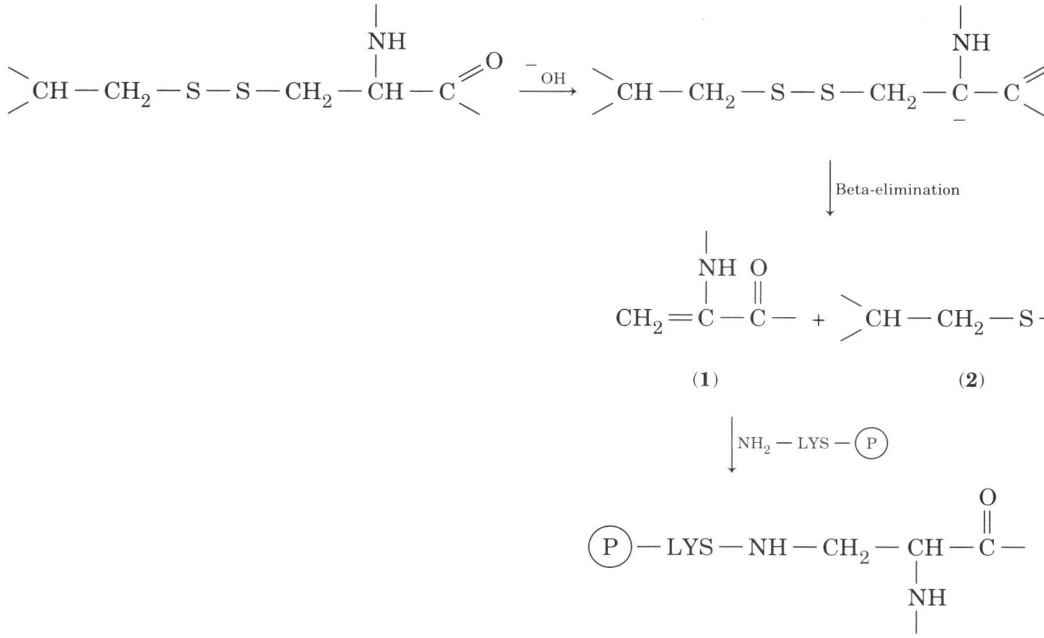

$$\text{>CH-CH}_2\text{-S-S-CH}_2\text{-CH(NH)-C(=O)<} \xrightarrow{^-OH} \text{>CH-CH}_2\text{-S-S-CH}_2\text{-C(NH)-C(=O)<}$$

$$\downarrow \text{Beta-elimination}$$

$$\text{CH}_2\text{=C(NH)-C(=O)-} \quad + \quad \text{>CH-CH}_2\text{-S-S}^-$$
$$(1) \qquad\qquad\qquad (2)$$

$$\downarrow \text{NH}_2-\text{LYS}-\text{(P)}$$

$$\text{(P)-LYS-NH-CH}_2\text{-CH(NH)-C(=O)-}$$

Figure 7. Reaction of β-elimination in alkali degradation of proteins. **(1)** Dehydroalanine, **(2)** persulfide product, and **(3)** lysinoalanine.

postmortem tenderness of meat (39). The onset of rigor mortis follows rapid depletion of ATP and breakdown in the regulatory system that controls the calcium level in muscle. Increasing concentration of calcium in the sarcoplasm induces contraction while lack of ATP in the system stops the dissociation of the actin–myosin complex formed. The muscle loses its natural extensibility, and this postmortem change is known as rigor mortis. Postmortem tenderness, however, is related to proteolysis of the muscle proteins. The acid proteases, such as the cathepsins, have received much attention in this respect. Recently the calcium-dependent proteinase, a calcium activated factor, has been linked to the causes of postmortem tenderization (40). In the last decade there have been significant advances in the understanding of the molecular structure as well as the morphology and mechanism of muscle cytoskeletal proteins (41–43). These new developments in the basic knowledge of muscle proteins will inevitably affect the way their functionality in food systems is interpreted.

Despite the importance of cereal seed proteins, their molecular structure has been little understood until recently. In the last few years the complete amino acid sequence of gliadin (α-type) has been determined (44). The high molecular weight glutenin genes have been sequenced and expressed in *Escherichia coli* (45–47). The low molecular weight glutenin subunits have been mapped by two-dimensional gel electrophoresis and N-terminal sequencing (48). Analysis of the amino acid sequence of the high molecular weight glutenins is especially revealing. The protein molecule contains a large central, repetitive region rich in β-turns, which forms an elastic β-spiral structure. Several cysteine residues are located at the α-helical region near the N- and C-terminal ends. Intermolecular disulfide bonds between these cysteine residues cross-link the glutenin subunits into gluten polymers with the extensive β-spiral regions in between them. Over the years several models have been proposed for the structure of glutenin (49), but none of these adequately explains the characteristic elastic properties of glutenins. Therefore, the recent results on the protein sequence and structure obtained by cloning studies are of immense value in providing a precise molecular basis for the specific role of the high molecular weight glutenins. For the first time characteristic physical properties of glutenin can be related to its secondary structure.

Bovine casein micelles exist in large spherical, colloidal particles of 500–3,000 Å in diameter and 10^7–3×10^{10} in particle weight. A micelle is assembled from submicelles containing 25–30 casein molecules. A number of casein micelle models have been proposed. The current model suggests that submicelles are bound by electrostatic interaction via colloidal calcium phosphate through their ester phosphate groups (50,51). Because κ-casein is almost phosphate free, binding occurs only among the other caseins in the submicelle. Submicelles with a low level of κ-casein are oriented in the interior, and the surface of the micelle is covered entirely with submicelles having a high content of κ-casein.

The cDNA sequences for the four major caseins (α_{s1}, α_{s2}, β, and κ) are known. Suggestions have been made to improve the functionality of casein using genetic engineering techniques (52). These suggestions include (1) alteration of the proportion of κ-casein to enhance the stability of casein micelles, (2) construction of an additional cleavage site in casein for chymosin to alter the rheological effects of proteolysis, (3) phosphorylation of casein to create additional phosphate groups for stabilization, and

(4) deletion of a polar segment from the otherwise nonpolar N-terminus of κ-casein to enhance its amphiphilicity.

A similar strategy has been applied to the whey protein β-lactoglobulin. It has been postulated that the thermal instability of this protein is a result of the unfolding of the polypeptide segment (residues 115–125) containing the free cysteine-121 (cys-121), and the subsequent sulfhydryl–disulfide exchange with the disulfides (residues 65–160 and 106–143) or with κ-casein in a milk system. Hence, deletion or substitution of the cys-121 by site-directed mutagenesis is expected to enhance the thermal stability and the functional properties of β-lactoglobulin.

COLORANTS

There are basically two categories of color compounds: conjugated polyenes and metalloporphyrins. The former includes carotenoids, annatto, anthocyanins, betanain, dyes, and lakes. The effect of conjugation is to lower the $\pi-\pi^*$ transition energy from the highest occupied molecular orbital to the lowest unoccupied molecular orbital. Increased conjugation in the molecule shifts the absorption maximum to a higher wavelength. Substituent groups with lone pairs of electrons tend to increase π conjugation by resonance.

Carotenoids consist of a basic structure of eight repeating isoprene units, which are highly conjugated. In the conjugated system the terminal double bond has the highest electron density and is most susceptible to oxidative attack. Degradation proceeds from the end to the center of the molecule and results in a progressive shortening of the polyene chain (53).

Anthocyanins are flavonoid compounds characterized by the flavylium nucleus. The flavylium form is stabilized by resonance with the positive charge delocalized throughout the entire structure and produces an intensely colored anthocyanin molecule. The structural transformation of anthocyanins in aqueous medium was thoroughly investigated in the early 1980s. Flavylium salts exist in equilibrium in different forms: flavylium cation (AH^+), quinoidal base (A), carbinol pseudobase (B), and chalcone (C). The equilibrium between AH^+ and A involves the transfer of proton from the C5, C7 or C4' hydroxyl groups to a water molecule. In the hydration reaction, the water molecule is preferentially added to the C2 of the pyrylium ring of AH^+, resulting in colorless B. The conversion of B to C is a base-catalyzed tautomerization (Fig. 8). Because most isolated natural anthocyanins when placed in slightly acidic medium (pH 3–6) exist largely in the colorless forms (B and C), these fundamental studies should have practical implications. Evidence indicates that subsitution pattern at various positions of the anthocyanin molecule influences the equilibrium constants of these reactions and hence the distribution of the colored (AH^+ and A) and colorless (B and C) species (54).

The two best-known examples of metalloporphyrins found in food are the myoglobin and chlorophylls. A porphyrin metal complex possesses 19 π-electrons in an 18-atom ring. The main effect of the metal on the transitions is the conjugation of the metal $p\pi$ orbital with the porphyrin π orbital (55). The splitting of the δ orbitals of the metal ion because of the porphyrin results in additional loss of degeneracy from the theoretically predicted octahedral symmetry. In oxymyoglobin heme complex the iron coordination positions are directed to the four porphyrin nitrogens and to histidine F8 and O_2 (or H_2O) in the fifth and the sixth positions, respectively. The roles of the protein globin are to stabilize the steric and electronic configuration of the iron heme, and to facilitate the back-bond-

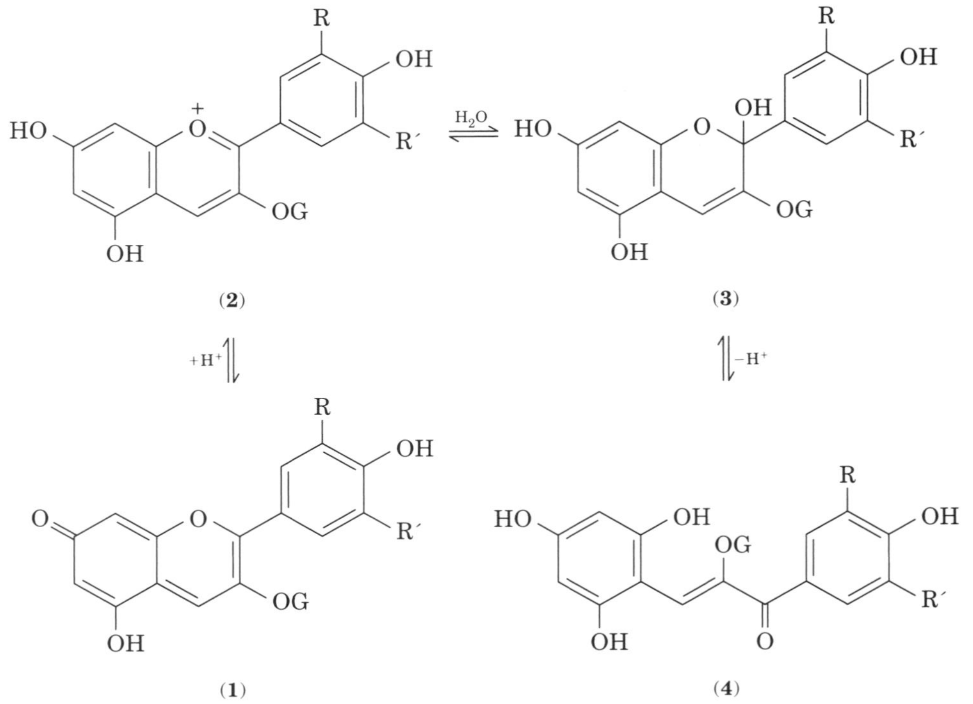

Figure 8. Resonance stabilization of flavylium cation (54). (1) Quinoidal base (blue), (2) flavylium cation (red), (3) carbinol pseudobase, and (4) chalcone.

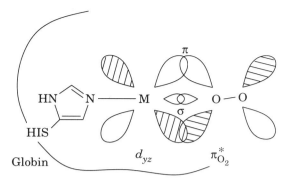

Figure 9. Back bonding of electrons from the transition metal to π^* orbital of oxygen.

ing of electrons from the iron to the π^* orbital of the oxygen (Fig. 9). The hydrophobicity of the heme pocket excludes the binding of ionic ligands such as CN^- and OH^-. The closely packed amino acid side chains restrict the size and orientation of the ligand (56).

Chlorophylls (unlike myoglobin, which contains a transition metal ion) are porphyrins complexed with the alkali earth metal Mg^{2+}. The compounds are hydrophobic because of the long chain C_{20} phytol esterified to the propionic acid side chain at C7. The magnesium atom of chlorophyll can readily be replaced by weak acids or other metals, such as copper, zinc, and iron. The free base obtained after removal of the metal is pheophytin with the transition intensity shifted to a lower wavelength. Hydrolysis of the phytyl chain by alkali or enzyme yields the chlorophyllides, which are water soluble and green colored and have spectral characteristics similar to those of the chlorophylls. Removal of magnesium from the chlorophyllides yields pheophorbides, which have the same spectral properties as those of the pheophytins. These types of conversion have frequently been implicated as causes of color losses in processed green vegetables (57,58). Copper complexes of pheophytin and chlorophyllide are stable to acid and are used as food colors in some European countries (59).

Currently there is a considerable interest in colorants from natural resources. Cape jasmine (*Gardenia jasminoides*) has been investigated for the production of the carotenoid crocin. The flower is also rich in iridoids and flavanoids. Another example is the mold *Monascus purpureus*, which has been used in Asian countries for centuries. The pigments produced are a mixture of red, yellow and purple polyketides (60). Pigments extracted from algae, yeasts, and insects have also been investigated for potential use as natural colorants (61,62). Attempts have also been made to produce anthocyanins and other pigments in plant tissue cultures. Beets (*Beta vulgaris*) have also recently been used as a source for obtaining natural colorants.

FLAVORS

Flavor is the sensation produced by a material and perceived principally by the senses of taste and smell. Over the years scientists have attempted to relate the chemical senses to the molecular structures of various compounds in an effort to develop a coherent theory. In 1967 it was proposed that the glucophore unit, which is responsible for sweetness in compounds, should consist of an AH, B hydrogen bonding system where H is an acidic proton and B is an electronegative atom or center (63). In 1972, the unit was supplemented with a dispersion bonding component designated as the γ component. The distance parameters of the resulting tripartite structure are A,B = 2.6; B,γ = 5.5; and A,γ = 3.5 Å. The orbital distance between the AH proton and B is 3.0 Å. In a three-dimensional picture, the glucophore binds to the receptor site and the sweet taste is initiated by intermolecular hydrogen bonding between the glucophore and a similar AH, B unit on the receptor. The γ component acts to align the molecule to the receptor. The locations of AH, B, and γ components in many sweet compounds are known. For example, in aspartame the protonated α-amino group and the ionized β-carboxyl group of the aspartyl residue represent the AH and B units, respectively (Fig. 10). The phenylanine end of the molecule represents the γ component. Only the L-L isomer of aspartame is sweet, and this type of enantiomeric effect in sweetness is also shown in several simple amino acids. To account for this fact, the receptor AH,B site is believed to consist of a spatial barrier. Hence, the

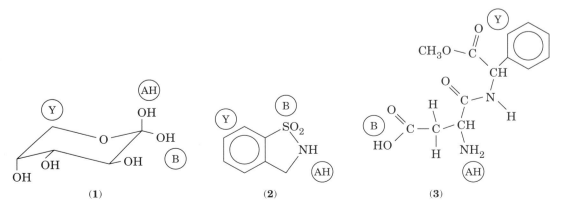

Figure 10. Assignment of glucophore unit in sweet compounds (64). (**1**) Fructose, (**2**) saccharin, and (**3**) aspartame.

L-D isomer of aspartame has the methyl ester group so positioned that the molecule cannot fit into the pocket for interaction (64). A similar concept is also applied to the aminosulfonates, such as saccharin and acesulfame K. Ring substitution experiments on acesulfame K indicate a loss of sweetness when the length of the hydrophobic group on the nitrogen exceeds 0.7 Å, suggesting a bulky substituent would create a steric hindrance (65).

Another group of sweeteners worthy of attention are the sweet proteins, which are approximately 100,000 times sweeter than sucrose on a molar basis and several times sweeter on a weight basis. Thaumatin I, the most-studied sweet protein, consists of a single polypeptide of 207 amino acids (mol wt 22,000), with eight disulfides and no histidine. The largest domain is a flattened β barrel formed by 11 antiparallel β strands, with 6 residues per strand on the average. Attached to this are domains II and III with loops linked and stabilized by disulfide bonds (66). Monellin, the other known sweet protein, consists of subunits I and II of 50 and 44 amino acid residues, respectively. The protein consists of a 5-stranded, antiparallel β sheet (2 β strands from subunit 1 and 3 from subunit II) where each strand contains an average of 10 residues (67). There is only one sulfhydryl group (in subunit II) and, in contrast to thaumatin I, no disulfide bond. Despite the intense sweetness of both proteins, there is little similarity in their three-dimensional structures. However, antibodies raised against thaumatin also cross-react with monellin (68). Five homologous amino acid sequences are found in thaumatin I and monellin; two are topologically similar and one of these is located in a β bend protruding from the surface of the molecule. This tripeptide sequence (-Glu-Tyr-Gly-) has been postulated to be the active site responsible for the sweet taste (69,70). The structure–function relationship of these proteins needs further investigation.

A similar AH,B concept has been employed to explain bitterness with little success. The structural requirements for bitter taste have been proposed, and it has been shown that, in contrast to the AH,B system for sweeteners, bitter compounds need only one polar group (electrophilic or nucleophilic with a negative charge) and a hydrophobic group because the intensity of bitter taste depends on the size and shape of the hydrophobic part of the molecule (71,72). Although the monopolar–hydrophobic concept for bitter taste requires more research, the importance of hydrophobicity to bitterness has been well established. Introduction of an additional hydrophobic group onto a sweet compound often results in modification of the taste from sweet to bitter.

Several theories have been presented over the years to correlate the molecular structure with the perceived odor quality of a chemical compound. The infrared theory has had some success in correlating odor quality with low-energy molecular vibrations (73). The site-fitting theory attempts to link the size, shape, and electronic status of a molecule to a complementary receptor site (74). It has been suggested that although the overall shape and size of an odorous compound is responsible for the sensory perception, the functional group determines the orientation and the affinity of the molecule to the receptor (75). Molecular connectivity terms have been derived to relate the quantitative description of certain aspects of molecular topology to several classes of odor compounds and found significant correlations between the two (76). Another study on the qualitative structure activity relationship has been conducted using computerized pattern recognition techniques. Both chemical composition and the molecular shape seem to be important contributing factors. In addition to the properties mentioned, it is also evident that the chirality and hydrophobicity of the compound are important for odor intensity (77). Recently, theories of olfactory quality coding have been proposed in that the code of odor quality depends on the responses of a population rather than of individual receptor cells to an odorous compound. Such population responses result in spatial and temporal patterns of receptor cell responses as odorant passes through the nasal passages (78).

Current knowledge about olfactory reception suggests that initial interaction between the odorant and the chemosensory membrane involves protein receptors. Binding of an odorant to an odorant receptor activates a G protein–adenylate cyclase cascade, which results in the generation of cAMP. The cAMP causes the opening of an ion channel either directly or by phosphorylation via a cAMP-dependent protein kinase (79). There are many known examples in which enantiomeric forms of chiral compounds have different odors. Experiments on the modification of odor in enantiomeric compounds seem to support the hypothesis that the initial interaction occurs between an odorant molecule and a protein receptor (80). Until now there has been a lack of a clear relationship between chemical structure and odor quality, and it is impossible to predict the odor quality of a compound with a known structure.

NATURAL TOXICANTS

There are a large number of natural toxic chemicals in the daily diet. Toxicants may be chemical constituents of the food itself, contaminants from microbial infestation, or degradation products formed during food storage and processing. Toxic chemicals are commonly present in all plants, where they act as protective agents against microbial attack, insects, and predators (81). Although a generally low level of these compounds together with a varied diet usually eliminates the high risk of intoxication, the presence of many of these natural pesticides in food seems to be receiving very little attention from the general public.

One well known example of natural pesticides is the glycoalkaloids solanine and chaconine, found in potatoes at the level of approximately 10 mg/100 g fresh weight (82). These glycoalkaloids are strong inhibitors of cholinesterase, and cause neurological disorder symptoms. Lethal doses for humans range from 3 to 6 mg/kg body weight, and doses of greater than 2 mg/kg are normally considered toxic.

Another class of toxicants, glucosinolates, occurs predominantly in vegetables, such as cabbage, cauliflower, radish, mustard greens, brussels sprouts, and broccoli. Glucosinolates are hydrolyzed by the enzyme thioglucoside glycohydrolase when plant tissues are crushed. The unstable aglycone produced is converted to isothiocyanate

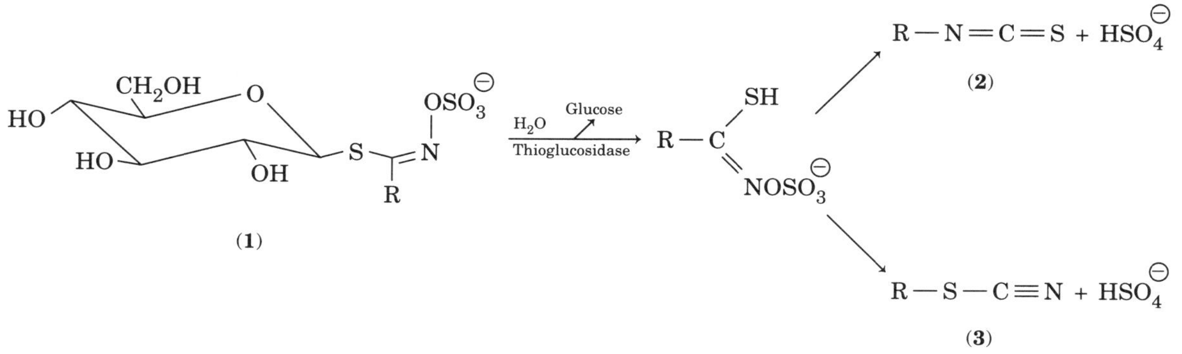

Figure 11. Degradation of glucosinolate (83). **(1)** Glucosinolate, **(2)** isothiocyanate, and **(3)** thiocyanate.

in a process similar to the Lossen rearrangement or thiocyanate (83) (Fig. 11). The latter compound is a goitrogen.

A number of amino acids, peptides, and proteins constitute an interesting class of toxicants. There are more than 250 nonprotein amino acids found in plants, and many are structurally similar to the protein amino acids. For example, β-pyrazol-1-ylalane, pipecolic acid, homoarginie, and α,β-diaminobutyric acid are analogues of histidine, proline, arginine, and lysine, respectively. Toxicity of these nonprotein amino acids is usually caused by competition with structurally similar protein amino acids for proteins, resulting in functionally inactive products, or in interference with protein synthesis.

The notorious toxins produced by *Clostridium botulinum* are neurotoxins that act at cholinergic nerve endings to block the release of acetylcholine. The lethal dose is in the range of 1–5 ng/kg of body weight. The molecular structure of these proteins is known, but details of the mechanism of action must be clarified. *Botulinum* toxins are synthesized as an inactive precursor, which is a single polypeptide with a molecular weight of approximately 150,000 dalton. The protein is cleaved by an endogenous protease into an active molecule consisting of a heavy

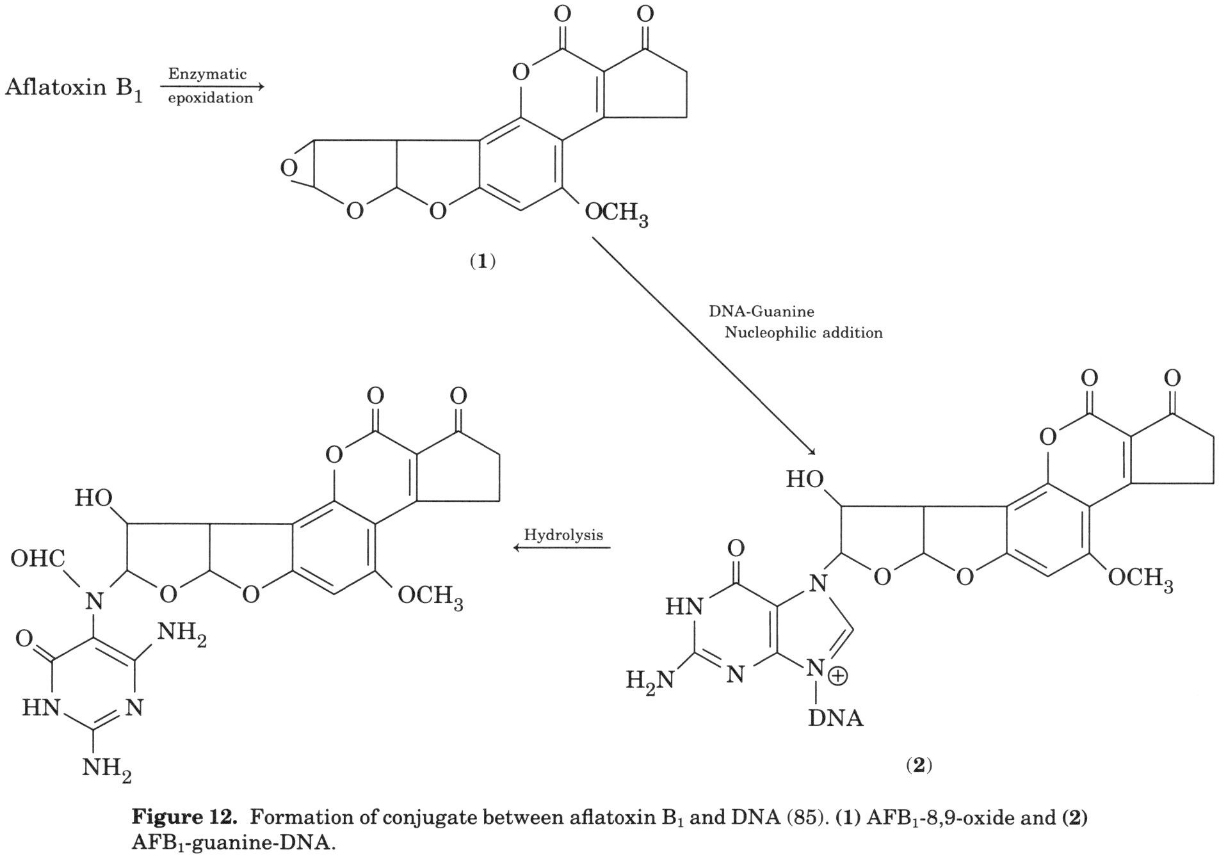

Figure 12. Formation of conjugate between aflatoxin B_1 and DNA (85). **(1)** AFB_1-8,9-oxide and **(2)** AFB_1-guanine-DNA.

chain and a light chain linked by a disulfide bond. It is now believed that the toxin initially binds to gangliosides (sialic acid-containing glycolipids) on the plasma membrane of the cholinergic nerve, before the toxins forms a ternary complex with a toxin-specific protein receptor (84). A membrane translocation occurs via receptor-mediated endocytosis.

Mycotoxins are toxic fungal metabolites that often contaminate peanuts, cereals and dairy products. The most extensively studied mycotoxins are the aflatoxins produced by *Aspergillus* and *Penicillium*. The toxicity of aflatoxin B_1 is caused by its metabolism to the reactive AFB_1-8,9-oxide, which forms covalently linked conjugates with DNA and proteins (85) (Fig. 12) or undergoes hydrolysis to AFB_1-8,9-dihydrodiol. The latter product is a dialdehyde phenolate ion stabilized by resonance and forms a Schiff base with proteins.

Polycyclic aromatic hydrocarbons are known carcinogens found in grilled meat products. These compounds include benzo[a]pyrene, benzo[a]fluoranthene, benzo[a]anthracene, and chrysene. They are metabolized by enzymatic activation to form a dihydrodiolepoxide, which binds DNA. Heterocyclic amines found in high temperature cooking are enzymatically converted to *O*-acyl derivative, which covalently binds DNA, resulting in mispairing in replication and transcription (86).

Many *N*-nitrosamines are found in cured meat and fish, including *N*-nitrosodimethylamine, *N*-nitrosodiethylamine, *N*-nitrosopyrrolidine, and *N*-nitrosopiperidine. Metabolically nitrosamines undergo enzymatic α-hydroxylation followed by *C–N* bond cleavage to yield alkydiazohydroxide, which further breaks down to alkyldiazonium ion. Both alkyldiazohydroxide and alkyldiazonium ion react with DNA to form covalent conjugates (87).

This brief overview clearly indicates the dynamic interactions among various interrelated disciplines. The very fact that interactions among food components are complex reflects in the scope and diversity of the underlying chemistry. It is often inadequate to treat food chemistry in a descriptive manner. Approaches must be taken to analyze the kind of chemical mechanisms involved and to follow the sequence of steps in terms of chemical equations and principles. Food chemistry encompasses reaction mechanisms involving free radicals, transition metals, coordination chemistry, stereochemistry, molecular structures, biopolymers, and colloids in addition to common organic chemical reactions. The process of searching for a fundamental understanding of how food components react and interact is indeed an exciting challenge for food scientists.

BIBLIOGRAPHY

1. D. W. S. Wong, *Mechanism and Theory in Food Chemistry*, Van Nostrand Reinhold, Co., Inc., New York, 1989.
2. H.-D. Belitz and W. Grosch, *Food Chemistry*, 2nd ed., Springer-Verlag, New York, 1987.
3. J. M. deMan, *Principles of Food Chemistry*, 2nd ed., Van Nostrand Reinhold, Co., Inc., New York, 1990.
4. O. R. Fennema, *Food Chemistry*, 2nd ed., Marcel Dekker, Inc., New York, 1985.
5. A. L. Branen, P. M. Davidson, and S. Salminen, *Food Additives*, Marcel Dekker, Inc., New York, 1990.
6. J. P. Danehy and W. W. Pigman, "Reactions Between Sugars and Nitrogenous Compounds and Their Relationship to Certain Food Problems," *Adv. Food Res.* **3**, 241–290 (1951).
7. E. Dworschak, "Nonenzymatic Browning and Its Effect on Protein Nutrition," *CRC Crit Rev. Food Sci. & Nutr.* **13**(1), 1–40 (1980).
8. J. P. Danehy, "Maillard Reactions: Nonenzymatic Browning in Food Systems with Special Reference to the Development of Flavors," *Adv. Food Res.* **30**, 77–138 (1986).
9. M. Namiki, "Chemistry of Maillard Reactions: Recent Studies on the Browning Reaction Mechanism and the Development of Antioxidants and Mutagens," *Adv. Food Res.* **32**, 115–184 (1988).
10. A. Cerami, H. Vlassara, and M. Brownlee, "Glucose and Aging," *Scientific American* **256**(5), 90–96 (1987).
11. A. Cerami, "Aging of Proteins and Nucleic Acids: What Is the Role of Glucose?" *TIBS* **11**, 311–314 (1986).
12. M. Namiki and T. Hayashi, "A New Mechanism of the Maillard Reaction Involving Sugar Fragmentation and Free Radical Formation," in G. R. Waller and M. S. Feather, eds., *The Maillard Reaction in Foods and Nutrition*, American Chemical Society Symposium Series 215, American Chemical Society, Washington, D.C., 1983.
13. V. Yaylayan and P. Sporns, "Novel Mechanisms for the Decomposition of 1-(amino acid)-1-deoxy-D-fructoses (Amadori Compounds): A Mass Spectrometer Approach," *Food Chemistry* **26**, 283–305 (1987).
14. V. Yaylayan and P. Sporns, "Diagnostic Ion Series for the Identification of Amadori Rearrangement Products by MS Techniques Based on Electron-Impact Ionization," *Journal of Agricultural and Food Chemistry* **37**, 978–981 (1989).
15. V. Yaylayan, "In Search of Alternative Mechanisms for the Maillard Reaction," *Trends Food Sci. & Technol.* **1**(1), 20–22 (1990).
16. S. Fors, "Sensory Properties of Volatile Maillard Reaction Products and Related Compounds," in Ref. 12.
17. B. L. Wedzicha and D. J. McWeeney, "Non-Enzymatic Browning Reactions of Ascorbic Acid and Their Inhibition. The Production of 3-deoxy-4-sulphopentosulose in Mixtures of Ascorbic Acid, Glycine and Bisulphite Ion," *Journal of the Science of Food and Agriculture* **35**, 577–587 (1974).
18. H. Kato and H. Tsuchida, "Estimation of Melanoidin Structure by Pyrolysis and Oxidation," *Prog. Food Nutr. Sci.* **5**, 147–156 (1981).
19. H. Kato, S. B. Kim, and F. Hayase, "Estimation of the Partial Chemical Structures of Melanoidins by Oxidative Degradation and ^{13}C CP-MAS NMR," *Dev. Food Sci.* **13**, 215–223 (1986).
20. G. T. Grant, E. R. Morris, D. A. Rees, P. J. C. Smith, and D. Thorn, "Biological Interactions Between Polysaccharides and Divalent Cations: The Egg Box Model," *Federation of European Biochemical Societies Letters* **32**, 195–198 (1973).
21. E. R. Morris, D. A. Rees, and G. R. Robinson, "Cation-Specific Aggregation of Carrageenan Helices; Domain Model of Polymer Gel Structure," *Journal of Molecular Biology* **138**, 349–362 (1980).
22. G. Robinson, E. R. Morris, and D. A. Rees, "Role of Double Helices in Carrageenan Gelation: The Domain Model," *J. Chem. Soc. Chem. Comm.* **1980**(4), 152–153 (1980).
23. D. Oakenfull and A. Scott, "Hydrophobic Interaction in the Gelation of High Methoxyl Pectins," *Journal of Food Science* **49**, 1093–1098 (1984).

24. H.-C. H. Wu and A. Sarko, "The Double-Helical Molecular Structure of Crystalline β-Amylase," *Carbohydrate Research* **61**, 7–25 (1978).
25. A. Imberty and S. Perez, "A Revisit to the Three Dimensional Structure of β-Type Starch," *Biopolymers* **27**, 1205–1221 (1988).
26. A. Imberty, H. Chanzy, and S. Perez, "The Double-Helical Nature of the Crystalline Part of A-Starch," *Journal of Molecular Biology* **201**, 365–378 (1988).
27. R. C. Jordan, D. A. Brant, and A. Cesaro, "A Monte Carlo Study of the Amylose Chain Conformation," *Biopolymers* **17**, 2617–2632 (1978).
28. A. Neszmelyi and J. Hollo, "Some Aspects of the Structure of Starch—A 3-D Molecular Modelling Approach," *Starch/Starke* **41**, 1–3 (1988).
29. D. J. Manners, "Recent Development in Our Understanding of Amylopectin Structure," *Carbohydrate Polymers* **11**, 87–112 (1989).
30. M. Miles, V. J. Morris, P. D. Orford, and S. G. Ring, "The Role of Amylose and Amylopectin in the Gelation and Retrogradation of Starch," *Carbohydrate Research* **135**, 271–281 (1985).
31. H. R. Rawls and P. J. Van Santen, "A Possible Role for Singlet Oxygen in the Initiation of Fatty Acid Autoxidation," *Journal of the American Oil Chemists' Society* **47**, 121–125 (1970).
32. N. A. Porter, "Mechanisms for the Autoxidation of Polyunsaturated Lipids," *Accounts of Chemical Research* **19**, 262–268 (1986).
33. G. Scott, "Antioxidants In Vitro and Vivo," *Chemistry in Britain* **21**(7), 648–653 (1985).
34. S. Friberg, "Emulsion Stability," in S. Friberg, ed., *Food Emulsions*, Marcel Dekker, Inc., New York, 1976.
35. N. Krog, "Theoretical Aspects of Surfactants in Relation to Their Use in Breadmaking," *Cereal Chemistry* **58**, 158–164 (1981).
36. L. Hernquist, "Polar Lipids in Emulsions and Microemulsions," in E. Dickinson, ed., *Food Emulsions and Foams*, Royal Society of Chemistry, London, 1987.
37. J. R. Whitaker and R. E. Feeney, "Chemical and Physical Modification of Proteins by the Hydroxide Ion," *CRC Crit. Rev. Food Sci. Nutr.* **19**, 173–212 (1983).
38. R. Liardon and S. Ledermann, "Racemization Kinetics of Free and Protein-Bound Amino Acids Under Moderate Alkaline Treatment," *Journal of Agricultural and Food Chemistry* **34**, 557–565 (1986).
39. J. C. Acton, G. R. Ziegler, and D. L. Burge, Jr., "Functionality of Muscle Constituents in the Processing of Comminuted Meat Products," *CRC Crit. Ref. Food Sci. & Nutr.* **18**(2), 99–121 (1983).
40. D. E. Goll, Y. Otsuka, P. A. Nagainis, J. D. Shannon, S. K. Sathe, and M. Muguruma, "Role of Muscle Proteinases in Maintenance of Muscle Integrity and Mass," *Journal of Food Biochemistry* **7**, 137–177 (1983).
41. K. Maruyama, "Myofibrillar Cytoskeletal Proteins of Vertebrate Striated Muscle," in L. Ralston, ed., *Developments in Meat Science—3*, Elsevier Applied Science Publishers, Ltd., Barking, UK, 1985.
42. H. Kajiyama, "Shape of the Myosin Head in the Rigor Complex," *Journal of Molecular Biology* **204**, 639–652 (1988).
43. Y.-J. Cho, J. Liu, and S. E. Hitchcock-DeGregori, "The Amino Terminus of Muscle Tropomyosin Is a Major Determinant for Function," *Journal of Biological Chemistry* **265**, 538–545 (1990).
44. D. D. Kasarda, T. W. Okita, J. E. Bernardin, P. A. Baecker, C. C. Nimmo, E. J.-L. Lew, M. D. Dietler, and F. E. Greene, "Nucleic Acid (cDNA) and Amino Acid Sequences of α-Type Gliadins from Wheat (Triticum aestivum)," *Proceedings of the National Academy of Sciences of the United States of America* **81**, 4712–4716 (1984).
45. J. Forde, J.-M. Malpica, N. G. Halford, P. R. Shewry, O. D. Anderson, F. C. Greene, and B. J. Miflin, *Nucl. Acids Res.* **13**, 6817–6832 (1985).
46. R. D. Thompson, D. Bartels and N. P. Harberd, *Nucl. Acids Res.* **13**, 6833–6846 (1985).
47. G. Galili, "Heterologous Expression of a Wheat High Molecular Weight Glutenin Gene in Escherichia Coli," *Proceedings of the National Academy of Sciences of the United States of America* **86**, 7756–7760 (1989).
48. H. P. Tao and D. D. Kasarda, "Two-Dimensional Gel Mapping and N-Terminal Sequencing of LMW-Glutenin Subunits," *Journal of Experimental Botany* **40**, 1015–1020 (1989).
49. B. J. Miflin, J. M. Field, and P. R. Shewry, "Cereal Storage Proteins and Their Effect on Technological Properties," in J. Daussant, J. Mosse, and J. Vaughan, eds., *Seed Proteins*, Academic Press, Inc., Orlando, Fla., 1983.
50. D. G. Schmidt, "Association of Caseins and Casein Micelle Structure," in P. F. Fox, ed., *Developments in Dairy Chemistry*, Elsevier Applied Science Publishers, Ltd., Barking, UK, 1982.
51. T. Aoki, "Incorporation of Individual Casein Constituents into Casein Aggregates Cross-Linked by Colloidal Calcium Phosphate in Artificial Casein Micelles," *Journal of Dairy Research* **56**, 613–618 (1989).
52. Y. Kang and T. Richardson, "Genetic Engineering of Caseins," *Food Technology* **39**(10), 89–94 (1985).
53. C. Marty and C. Berset, "Factors Affecting the Thermal Degradation of All-trans-β-carotene," *Journal of Agricultural and Food Chemistry* **38**, 1063–1067 (1990).
54. R. Brouillard, G. A. Iacobucci, and J. G. Sweeny, "Chemistry of Anthocyanin Pigments. 9. UV-Visible Spectrophotometric Determination of the Acidity Constants of Apigeninidin and Three Related 3-Deoxyflavylium Salts," *Journal of the American Chemical Society* **104**, 7585–7590 (1982).
55. D. J. Livingston and W. D. Brown, "The Chemistry of Myoglobin and Its Reactions," *Food Technology* **35**(5), 244–252 (1981).
56. D. Ladikos and B. L. Wedzicha, "The Chemistry and Stability of the Haem-Protein Complex in Relation to Meat," *Food Chemistry* **29**, 143–155 (1988).
57. H. K. Lichtenthalev, "Chlorophylls and Carotenoids: Pigments of Photosynthetic Biomembranes," *Methods in Enzymology* **148**, 350–382 (1987).
58. S. J. Schwatz and T. V. Lorenzo, "Chlorophylls in Foods," *CRC Rev. Food Sci. & Nutr.* **29**(1), 1–27 (1990).
59. A. H. Humphrey, "Chlorophyll," *Food Chemistry* **5**, 57–67 (1980).
60. A. J. Taylor, "Natural Colors in Food," in J. Walford, ed., *Developments in Food Colors—II*, Elsevier, Applied Science Publishers, Ltd., Barking, UK, 1984.
61. F. J. Francis, "Lesser-Known Food Colorants," *Food Technology* **41**(4), 62–68 (1987).
62. K. Spears, "Developments in Food Colourings: The Natural Alternatives," *TIBTECH* **6**, 283–288 (1988).
63. R. S. Shallenberg and T. E. Acree, "Molecular Theory of Sweet Taste," *Nature* **216**, 480–482 (1967).
64. F. Lelj, T. Tancredi, P. A. Temussi, and C. Toniolo, "Interaction of α-L-aspartyl-L-phenylalanine Methyl Ester with the

Receptor Site of the Sweet Taste Bud," *Journal of the American Chemical Society* **98**, 6669–6675 (1976).
65. F. Pautet and C. Nofre, "Correlation of Chemical Structure and Taste in the Cyclamate Series and the Steric Nature of the Chemoreceptor Site," *Zeitschrift fuer Lebensmittel-Untersuchung und-Furschung* **166**, 167–170 (1978).
66. A. M. deVos, M. Hatada, H. van der Wel, H. Krabbendam, A. F. Peerdeman, and S.-H. Kim, "Three-Dimensional Structure of Thaumatin I, An Intensely Sweet Protein," *Proceedings of the National Academy of Sciences of the United States of America* **82**, 1406–1409 (1985).
67. C. Ogata, M. Hatada, G. Tomlinson, W.-C. Shin, and S.-H. Kim, "Crystal Structure of the Intensely Sweet Protein Monellin," *Nature* **328**, 739–742 (1987).
68. C. A. M. Hough and J. A. Edwardson, "Antibodies to Thaumatin As a Model of the Sweet Taste Receptor," *Nature* **271**, 381–383 (1978).
69. H. van der Wel, "Physiological Action and Structure Characteristics of the Sweet-Tasting Proteins Thaumatin and Monellin," *TIBS* **5**, 122–123 (1980).
70. S. H. Kim, A. deVos, and C. Ogata, "Crystal Structures of Two Intensely Sweet Proteins," *TIBS* **13**, 13–15 (1988).
71. H.-D. Belitz, W. Chen, H. Jugel, W. Stempfl, R. Treleano, and H. Wieser, "Structural Requirements for Sweet and Bitter Taste," in P. Schreier, ed., *Flavour '81, 3rd Weurman Symposium: Proceedings of the International Conference, Munich, April 28–30, 1981,* Walter de Gruyter, Co., Berlin, 1981.
72. H. D. Belitz, W. Chen, H. Jugel, H. Stempfl, R. Treleano, and H. Wieser, "QSAR of Bitter Tasting Compounds," *Chemistry and Industry* **1**, 23–26 (1983).
73. R. H. Wright, "Odour and Molecular Volume," *Chem. Senses* **7**, 211–213 (1982).
74. J. E. Amoore, *Molecular Basis of Odor,* Charles C Thomas, Springfield, Ill. 1970.
75. M. G. J. Beets, "Relationship of Chemical Structure to Odor and Taste," in *Proceedings, Third International Congress on Food Science and Technology,* Institute of Food Technologists, Chicago, Ill., 1971.
76. L. M. Kier, T. D. Paolo, and L. H. Hall, "Structure-Activity Studies on Odor Molecules Using Molecular Connectivity," *Journal of Theoretical Biology* **67**, 585–595 (1977).
77. G. Ohloff, "Chemistry of Odor Stimuli," *Experienta* **42**, 271–279 (1986).
78. J. A. Marunak, "The Sense of Smell," in J. R. Piggott, ed., *Sensory Analysis of Foods,* Elsevier Applied Science Publishers, Ltd., Barking, UK, 1988.
79. D. Lancet, and U. Pace, "The Molecular Basis of Odor Recognition," *TIBS* **12**, 63–66 (1987).
80. W. Pickenhagen, "Enantioselectivity in Odor Perception," in R. Teranishi, R. G. Buttery and F. Shahidi, eds., *Flavor Chemistry, Trends and Developments,* American Chemical Society Symposium Series **388**, American Chemical Society, Washington, D.C., 1989.
81. B. N. Ames, "Chemicals, Cancers, Causalities and Cautions," *Chemtech* **19**(10), 590–598 (1989).
82. S. F. Osman, "Glycoalkaloids in Potatoes," *Food Chemistry* **11**, 235–247 (1983).
83. G. R. Fenwick, R. K. Heaney, and W. J. Mullin, "Glucosinolates and Their Breakdown Products in Food and Food Plants," *CRC Crit. Rev. Food Technol.* **18**, 123–200 (1983).
84. C. Montecucco, "How Do Tetanus and Botulinum Toxins Bind to Neuronal Membranes," *TIBS* **11**, 314–317 (1986).
85. R. G. Croy and G. N. Wogan, "Temporal Patterns of Covalent DNA Adducts in Rat Liver After Single and Multiple Doses of Aflatoxin B_1," *Cancer Research* **41**, 197–203 (1981).
86. Y. Hashimoto, K. Shudo, and T. Okamoto, "Activation of a Mutagen, 3-Amino-methyl-5H-pyrido[4,3-b]indole. Identification of 3-Hydroxyamino-1-methyl-5H-pyrido[4,3-b]indole and Its Reaction With DNA," *Biochemical and Biophysical Research Communications* **96**, 355–362 (1980).
87. M. C. Archer, "Reactive Intermediates from Nitrosamines," in R. Snyder, ed., *Biological Reactive Intermediates—II, Advances in Experimental Medicine and Biology,* Vol. 136B, Plenum Press, New York, 1982.

Dominic W. S. Wong
Attila E. Pavlath
United States Department of Agriculture
Albany, California

FOOD CROPS: NONDESTRUCTIVE QUALITY EVALUATION

Objective methods of measuring quality of food crops were introduced in the 1930s, with some of the methods still in use. These methods as well as many of the current new methods generally make a measurement of a chemical or physical parameter of the food crop that is associated or correlated with quality and does not measure quality directly. Measurement by some of the methods require destruction and sacrifice of the sample, which restricts the use to a random sample for predicting quality of the surrounding population. Nondestructive methods of measurement are preferred over the destructive methods to monitor quality over a time period in research programs or so that quality of each item can be categorized in a commercial grading line. Various methods are available for measuring quality nondestructively, and a few are used commercially while others remain in the developmental stage.

DEFINITIONS

Nondestructive. Nondestructive refers to a method that makes an analysis of the food crop without altering its physical or chemical condition so that the food crop can be stored or be marketable after the analysis.

Quality. Quality is considered the degree of excellence. The degree of excellence is based on a composite of those characteristics that differentiate individual units of a product and have significance in determining the degree of acceptability of that unit to the buyer.

Quality attributes of greatest interest are color, texture, and flavor. Intensity of these attributes is commonly determined by making an objective measurement of the physical or chemical constituents associated with the attribute. For example, measurement of pigment content is used to describe intensity of color, the sugar to acid ratio is used to indicate intensity of flavor, and density is used to describe firmness.

Evaluation. Evaluation refers to assessment of quality based on the unit or value expressed by the measuring instrument.

PRINCIPLES USED FOR NONDESTRUCTIVE OBJECTIVE QUALITY EVALUATION

The fundamental principles used in the nondestructive objective methods of measuring quality are the measurements of light or sound. These principles are applied in the following methods of measurement.

1. Reflectance
2. Transmitted radiation
   a. Transmittance
   b. Interactance/body transmittance
3. Fluorescence
4. Delayed light emission
5. Imaging
6. X ray
7. Acoustic impulse
8. Resonance
9. Ultrasound
10. Deformation/compression

DESCRIPTION AND APPLICATION OF METHODS

Reflectance

Reflectance, known also as diffuse reflectance, is the light reflected from below the skin of an illuminated food crop or a few millimeters below the surface of an illuminated ground sample. An incandescent bulb is used as source of illumination and is positioned on the sample so that the detector responds only to the light reflected from below the skin or surface of the sample. This technology is used in grading lines of packing houses to sort tomatoes, oranges, lemons, and apples according to defined color categories.

Diffuse spectral reflectance in the visible to near-infrared wavelength range is used to detect blight, soft rot, and scab in potato tubers (1). Effectiveness is dependent on cleanliness and uniformity of potato color. Diffuse reflectance in the 640 to 750 nm wavelength range is used to predict maturity of lettuce and oranges (2,3). With both commodities, maturity is based on the chlorophyll content, which has peak absorption at 660 nm. Near-infrared diffuse reflectance spectra is used to predict starch, protein, oil, and moisture contents of grains and oil seeds (4). The second derivative of the spectra is often used to derive the multivariate linear regression equations that are used in predicting the chemical component.

Transmittance

Transmittance is the light that is not absorbed by the illuminated food crop and is transmitted through the sample. Transmittance data on scattering samples are expressed in optical density (OD) $= \log_{10}(E_0/E)$ where E_0 is the incident energy and E is the transmitted energy. The OD includes the energy loss from reflectance and scatter as well as from absorption. Most samples contain more than one compound that absorbs radiation in the spectral region under test. To minimize the effects from all but the desired compound, the difference in OD (Δ OD) or ratio of OD (R) at 2 wavelengths is used to relate spectrophotometric measurements to quality attributes.

Defects are detected by transmittance when light absorption by the defect is different than that of surrounding healthy tissue. Internal defects that can be detected in this method include hollow heart in potato tubers, core breakdown in pears, water core in apples, freeze damage in citrus, and smut content on wheat. Hollow heart is based on Δ OD(800–710 nm), core breakdown on Δ OD(690–740 nm), water core on Δ OD(810–760 nm), and smut on wheat Δ OD(890–930 nm). Surface defects that can be quantified include bruise on apples ($R_{400-450\,nm}$), defect on lemons ($R_{580-650\,nm}$), scab on dried prunes ($R_{1300-1700\,nm}$), and scab on potatoes ($R_{590-750\,nm}$) (5). Among the above, water core is the only defect that is a desirable attribute at harvest because of the juicy and sweet condition, but becomes undesirable during storage owing to breakdown and discoloration of tissue.

Maturity is predicted by transmittance when physical or chemical components that change with maturity can be measured. Δ OD(690–740 nm) readings of pears are closely associated with firmness changes, which are used to identify maturity (6). Maturity of tomato is based on chlorophyll and lycopene content, which can be predicted by Δ OD(710–780 nm) and Δ OD(570–780 nm) values, respectively (7). Effectiveness of prediction depends on uniformity of pigmentation. For example, some tomato fruit have a dark green variegated shoulder that will affect the transmittance values and consequently give misleading values for predicting maturity.

Maturity of apples, blueberries, papaya, and peaches can also be predicted by transmittance based on pigment content (5). The wavelengths selected for the measurements and the effectiveness are dependent on the composition of the fruit. In peaches the density of stone as well as the amount of anthocyanin around the stone must be considered when the measurement is based on the chlorophyll content. With apples, the potential of water core condition affecting chlorophyll readings must be kept in mind.

Interactance and Body Transmittance

Interactance and body transmittance are terms used synonymously to describe a method that was initially termed body reflectance. Transmittance readings by this method are attained at an angle from the direction of illumination. The depth of transmittance for such measurement is only a few millimeters below the skin or surface of a sample, so the application of this interactance/body transmittance technique is limited to commodities where the tissue condition near the skin or surface can be correlated with the quality attribute of interest. This technology was developed because of the limitation or restriction of the transmittance technique. In the in-line transmittance, light had to be transmitted through the commodity and a very sensitive detector was required to measure small changes at 10–12 OD. This was not feasible with many commodities.

The interactance technique is used to predict maturity and composition of food crops (8). Maturity of papaya is predicted by determining the chlorophyll, carotenoid, and soluble solids content. The wavelengths of measurements are 588 and 620 nm, 520 and 643 nm, and 582 and 714 nm with corresponding constants, respectively. Dry matter content of onions is determined by measurements at near-infrared wavelengths. The second derivative of the spectra at wavelengths from 700 to 1000 nm is used to derive the multiple regression equation for predicting dry matter of onions ranging from 5 to 25%. The wavelengths selected for the analysis imply that the absorption band is associated with carbohydrates as the main dry matter in onions.

Quantity of water, protein, oil, and carbohydrates in oil seeds and grains can be predicted from interactance measurements (4). Mathematical handling of data to predict the quantity is similar to that used in the diffuse reflectance method.

Fluorescence

Fluorescence is the light that is emitted by a sample at wavelength different from the illuminated wavelength. Total fluorescence emission is used to categorize tomato genotypes according to susceptibility to injury at chilling temperatures (9). Fluorescence emission peaks characteristic of the photosystem II antennae (686 nm), reaction center (696 nm), and photosystem I antenna (730–740 nm) were used to study chlorophyll organization during the growth and ripening periods of fruit (10).

Delayed Light Emission

Delayed light emission (DLE) is a measurement of light energy that is emitted after a delay by an illuminated sample. It is used to predict color, maturity, and injury caused by chilling temperatures. Tomatoes, persimmons, and apricots were sorted according to color or maturity by measuring the decay of DLE after single illumination (11,12). Optimum duration of dark periods before illumination, illumination periods, and decay periods of DLE differ and need to be defined for each type of fruit. For example, optimum dark periods are 15, 20, and 5 min, optimum illumination periods are 2, 1, and 300 s, and DLE decay periods are 2, 1.5, and 5 s for persimmons, apricots, and tomatoes, respectively. Maximum DLE response occurs near 650 nm, which is near the absorption band of chlorophyll.

Chilling temperatures are deleterious to chlorophyll activities, with changes in activities detectable by DLE. Prediction of the degree of stress caused by chilling temperatures can be made more effectively by repetitive DLE measurements rather than following decay from a single illumination (13). For effective measurements with cucumbers, the sample is kept in the dark for at least 30 min before illumination, then illuminated for 7 ms, kept in the dark 0.5 ms, DLE measured for 7 ms, and kept in dark 0.5 ms. This cycle, taking 15 ms, is then repeated for 15 s. The DLE amplitude increases with time, and the maximum amplitude is lower with cucumbers held at lower chilling temperatures. The maximum amplitude could be used to predict the injury, but a more effective prediction can be made by using the slope of change within the first 1.5 s.

Imaging

Imaging is a new technology that is being examined to measure surface defects or internal defects or injuries. An algorithm of the gray level of pixels is used to determine the extent of defect or physical injury on the surface. Progress in surface imaging has been made in separating peaches according to blush color (14), separating good and defective dried prunes (15), and recognizing bruises from scab, hail damage, bird pecks, and insect stings on apples (16). Instrumentation for imaging is complex, which limits the application of technology.

The nuclear magnetic resonance (NMR) imaging method for radiology is used to detect internal defects of food crops. Cross-sectional images of defects are reconstructed on the basis of uneven distribution of mobile water and its NMR relaxation times, spin-lattice, and spin-spin. These data are able to describe areas and can be used to monitor development of water core in apples (17) and core breakdown in pears (18). This technology will be useful particularly in research to associate physical changes with biochemical and physiologic changes of defects for obtaining information to better understand the mechanism of injury.

X-ray

X-ray technology has been modified to recognize rapidly internal characteristics of commodities on grading lines or on mechanical harvesters. The technology can detect hollow heart of potatoes (19) and other defects that cause changes in material density. Absorption of X rays varies directly with density. X ray is used to determine the size and density of lettuce for selecting mature heads on a mechanical harvester.

Acoustic Impulse

Acoustic impulse is a measure of acoustic energy emitted from a sample that has been impacted by a controlled hammer mechanism. The emitted sound is picked up by a condenser microphone and amplified by a linear amplifier or a sound level meter to be recorded on a data recorder. Indices measured are the natural frequency. The indices correlate significantly with elastic modulus and breaking stress data from the force-deformation curve of the compression test of apples and watermelons (20).

Mechanical Resonance and Vibration Transmissibility

Mechanical resonance and vibration transmissibility are also methods of using sound to analyze the quality of a product (21). In the mechanical resonance method, the sample is vibrated at selected or at a wide range of frequency wavelengths such that the resonance response can be measured. The resonant frequencies are associated with different characteristic modes of vibration. Resonance tests are used to measure firmness/texture of peaches, pears, oranges, apples, and tomatoes. Transmissibility is the ratio of transmitted force to driving force during forced vibration testing. It is strongly influenced by the damping characteristics of the test material. Firm commodities have high transmissibility, whereas soft products have low transmissibility. The measurements have been used to predict anthocyanin in content of raspberries and turgor of strawberry fruit.

FUTURE DEVELOPMENT

Nondestructive methods for evaluating quality of food products are continually improving. Improvement results with development of new technologies that become available. Some of the methods are incorporated into grading lines to sort fruit according to color or other quality attributes. Application of some methods into the grading lines is restricted owing to cost or lack of speed. However, with continual development of new technologies in this area, as noted in the past, cost of some of the methods will become reasonable and allow greater opportunity for research and commercial application.

BIBLIOGRAPHY

1. R. L. Porteous, A. Y. Muir, and R. L. Wastie, "The Identification of Diseases and Defects in Potato Tubers from Measurements of Optical Spectral Reflectance," *Journal of Agricultural Engineering Research* **26**:151–160, 1981.
2. E. J. Brach, C. T. Phan, B. Poushinsky, J. J. Jasmin, and C. B. Aube, "Lettuce Maturity Detection in the Visible (380–720 nm), Far Red (680–750) and Near Infrared (800–1850) Wavelength Band (*Lactuca sativa*)," *Agron. Sci. Prod. Veg. et de L'environ* **2**:685–694, 1982.
3. Y. Chuma, T. Shiga, and K. Morita, "Evaluation of Surface Color of Japanese Persimmon Fruits by Light Reflectance (Mechanized Grading Systems)," *Journal Society Agri. Machinery, Japan* **42**(1):115–120, 1980.
4. K. H. Norris, "Instrumental Techniques for Measuring Quality of Agricultural Crops," p. 471 in M. Lieberman, ed., *Postharvest Physiology and Crop Preservation*, Plenum Press, New York, 1983.
5. S. Gunasekaran, R. Paulsen, and G. C. Shove, "Optical Method for Nondestructive Quality Evaluation of Agricultural and Biological Materials, *Journal of Agricultural Engineering Research*, **32**:209–241, 1985.
6. C. Y. Wang and J. T. Worthington, "A Nondestructive Method for Measuring Ripeness and Detecting Core Breakdown in 'Bartlett' Pears," *Journal of the American Society for Horticultural Science* **104**:629–631.
7. A. E. Watada, K. H. Norris, J. T. Worthington, and D. R. Massie, "Estimation of Chlorophyll and Carotenoid Contents of Whole Tomato by Light Absorbance Technique," *Journal of Food Science* **41**:329–332, 1976.
8. G. S. Birth, G. G. Dull, J. B. Magee, H. T. Chan, and C. G. Cavaletto, "An Optical Method for Estimating Papaya Maturity," *Journal of the American Society for Horticultural Science* **109**:62–66, 1984.
9. T. L. Kamps, T. G. Isleib, R. C. Herner, and K. S. Sink, "Evaluation of Techniques to Measure Chilling Injury in Tomato," *HortScience* **22**(6):1309–1312, 1987.
10. J. Gross and I. Ohad, "*In vivo* Fluorescence Spectroscopy of Chlorophyll in Various Unripe and Ripe Fruit," *Photochemistry and Photobiology* **37**:195–200, 1983.
11. Y. Chuma, K. Nakaji, and W. F. McClure, "Delayed Light Emission as a Means of Automatic Color Sorting of Persimmon Fruit, DLE Fundamental Characteristics of Persimmon Fruits," *Journal of the Faculty of Agriculture*, Kyushu University, Japan **27**(1–2), 13–20, 1982.
12. Y. Chuma, K. Nakaji, and A. Takagawa, "Delayed Light Emission as a Means of Automatic Sorting of Tomatoes," *Journal of Faculty of Agriculture*, Kyushu University, Japan **26**(4), 221–234, 1982.
13. J. A. Abbott and D. R. Massie, "Delayed Light Emission for the Early Detection of Chilling Stress in Cucumber and Bell Pepper," *Journal of American Society for Horticultural Science* **110**, 42–47, 1985.
14. B. K. Miller and M. J. Delwiche, "A Color Vision System for Peach Grading," Paper, American Society of Agricultural Engineering, No. 88-6025, 1988.
15. M. J. Delwiche, S. Tang, and J. F. Thompson, "Prune Defect Detection by Line-scan Imaging," Paper, American Society of Agricultural Engineering, No. 88-3024, 1988.
16. G. Rehkugler and J. A. Throop, "Image Processing Algorithm for Apple Defect Detection," Paper, American Society of Agricultural Engineering, No. 87-3041, 1987.
17. S. Y. Wang, P. C. Wang, and M. Faust, "Nondestructive Detection of Watercore in Apple with Nuclear Magnetic Resonance Imaging," *Scientia Horticulturae* **35**:227–234, 1988.
18. C. Y. Wang and P. C. Wang, "Nondestructive Detection of Core Breakdown in 'Bartlett' Pears with Nuclear Magnetic Resonance Imaging," *HortScience* **24**, 1, 1989.
19. E. E. Finney and K. H. Norris, "X-ray Scans for Detecting Hollow Heart in Potatoes," *American Potato Journal* **55**:95–105, 1978.
20. H. Yamamoto, M. Iwamoto, and S. Haginuma, "Nondestructive Acoustic Impulse Response Method for Measuring Internal Quality of Apples and Watermelons," *Journal of the Japanese Society for Horticultural Science* **50**(2), 247–261, 1981.
21. E. E. Finney and J. A. Abbott, "Methods for Testing the Dynamic Mechanical Response of Solid Foods," *Journal of Food Quality* **2**, 55–74, 1978.

ALLEY E. WAFADA
Horticultural Crops Quality
Laboratory/USDA
Beltsville, Maryland

FOOD CROPS: POSTHARVEST DETERIORATION

Harvest is a major event for any food that is derived from plants. As a plant part is severed from the plant it may lose its source of supply of nutrients and its repository for metabolic waste products. Until the detached plant part undergoes conventional processing, however, it continues to live and respire as well as to senesce and die. Postharvest handling techniques have been developed to slow the senescing processes for the satisfaction and convenience of the consumer.

Food crops can be roughly divided into two categories: agronomic and horticultural. Agronomic (field) crops are primarily the grains and oilseeds that are harvested in a semidry to dry state and tend to be relatively stable to handling and storage as long as they are protected from moisture and insects. Horticultural (garden) crops, which comprise fruits, vegetables, and nuts, tend to be much more perishable requiring sophisticated handling systems to transport them from field to consumer.

Estimates vary widely on how much of a harvested crop is actually consumed (1,2,3). Postharvest handling techniques and means of food processing and preservation are designed to minimize these losses. Edible product can be lost in the field during harvesting and loading operations; during transport to market or processing plant; during short-, intermediate-, or long-term storage; during handling and preparation; or during consumption itself by rejection of the product by the consumer. Losses may be complete, resulting in discarding of the entire item (eg, a rotten apple), or partial, resulting in a paring away of part of that item (eg, removal of outer leaves of lettuce). Losses are frequently more subtle, resulting in loss of acceptability, nutritional value, or economic value of the product. To understand losses of product and changes in quality, it is necessary to understand the handling system from harvest to consumer for fresh product and from harvest to the initiation of processing in processed foods (4,5).

Because fresh products are living, respiring, senescing, dying tissue, losses in quality begin to occur at the point of harvest and continue to the point of consumption. Respiration is defined as "the overall process by which stored organic materials (carbohydrates, proteins, fats) are broken into simple end products with a release of energy" (4), while senescence describes those processes leading to death. Quality is defined as the "composite of those characteristics that differentiate individual units of a product and have significance in determining the degree of acceptability of that unit by the buyer" (6). Shelf life is the length of time the product can be maintained at an acceptable level of quality.

Although most harvested products are at their peak of quality at harvest, a special class of fruits, known as climacteric fruits, will actually continue to ripen after detachment from the plant. These fruits (eg, apples, bananas, pears, and tomatoes) will develop color, flavor, and texture attributes during postharvest handling, although, in some cases, the full flavor potential does not develop when ripened off the plant. The quality characteristics important to the consumer when buying a product (purchase quality or external quality) are not the same as those characteristics important to the consumer when eating the product (consumption quality or internal quality). Other quality characteristics may not be directly discernable to the consumer but the perception of characteristics such as nutritive value and product safety may influence the purchase of a product. Shelf life tends to be more a function of purchase quality than consumption quality.

CAUSES OF LOSSES

Several factors contribute to losses of fresh products during postharvest handling. Mechanical injury might result in damage such as cuts or bruises that create unsightly defects resulting in product loss. Such damage may not be immediately evident. The softening and discoloration associated with bruising takes time to develop. Thus it is sometimes difficult to determine when and where the damage was incurred and how to take corrective action.

Microbes can invade plant products, particularly those items that have been mechanically injured, causing decay that can quickly spread from item to item within bulk containers. The microorganisms may be present in the product in the field or may be introduced during handling operations, particularly any steps that involve contact of the product with water. Insects can be as destructive to harvested products as they are to crops in the field.

As handling systems become more integrated and more firms become dependent on computers, the demand for standardization of packages becomes greater. The trucker wishes to stack the maximum amount of product on the truck to reduce unit transportation cost. Recommended stacking patterns have been designed with standard cartons to permit maximum loads with adequate flow of refrigerated air. The wholesaler and retailer expect a specific number of items in that standard carton to maintain inventory control. Unfortunately nature's products do not always conform to uniform package dimensions. Overfilling of containers can lead to crushed product while underfilling may lead to product shifting in transport and greater chances for impact damage.

As mentioned above, physiological processes continue in harvested products. Respiration, an oxygen-consuming, carbon-dioxide–evolving process, leads to degradation of compounds within the product leading to the accumulation of metabolic waste products. These waste products may be desirable in the development of fruit flavors or undesirable in the development of off-flavors in other products. Fruits and vegetables also transpire (lose water), which can result in wilting or loss of turgor; these are undesirable characteristics in products such as lettuce or broccoli. Physiological disorders can develop during handling and storage due to nutritional deficiencies during growth and development preharvest, response to stress conditions during postharvest handling or an interaction of preharvest and postharvest factors.

Some fruits and vegetables are susceptible to storage at low temperature and will develop a physiological disorder known as chilling injury. Chilling injury is evidenced in different ways in different crops but can result in inhibited or abnormal ripening, surface lesions or pitting, increased susceptibility to decay, browning discoloration, and off-flavor development among other manifestations. Chilling-susceptible products include bananas, beans, cucumbers, grapefruit, melons, and tomatoes (7). Physiological disorders also result from improper mineral nutrition during growth and development; during exposure to low levels of oxygen (O_2), high levels of carbon dioxide (CO_2), or high levels of ethylene (C_2H_4) during storage; or an interaction of these and other factors. The etiology of these disorders is complex and the degree of susceptibility varies widely among species.

PREVENTION OF LOSSES

The goal of the postharvest handler is to minimize the losses of quality of a product between the time it is harvested and the time it is either processed or consumed. The first line of defense is prevention of the initial injury by physical means. The next line of defense is to manipulate the environmental conditions of handling. Chemical compounds are also available to protect the products. Finally, sophisticated packaging techniques are being developed and used that employ one or more of the above lines of defense.

Physical protection of products from injury are employed to minimize mechanical damage. The easiest way to control a problem is usually to prevent that problem from occurring initially. Reduction of mechanical injury can be achieved by decreasing the number and height of falls of an item during handling whether individually or in a container. In processes where falls are unavoidable, cushioning of the impact can minimize damage. Cushioning can be achieved by foam padding, liquid foam, water, or even product that will be discarded. Impact damage of one fruit on another can be reduced by decreasing dumping operations and using spacing bars in conveyor lines to prevent fruit-to-fruit contact. Vibration damage tends to increase with an increase in the size of the bulk handling of products and the distance from field (or orchard) to packing facility. Packing the product in wholesale or retail packages close to the field, reducing or eliminating the use of tractor-drawn vehicles from field to packinghouse, and using paved roads for transport of fresh product will reduce damage due to vibration.

Likewise, microbial decay and insect damage can be reduced by physical protection. Any means of preventing mechanical damage will slow microbial decay and to a lesser extent insect damage. Once endogenous protective barriers such as a rind or peel are penetrated chances for further injury are increased. A greater incidence of pests and disease in the field will be reflected in greater problems during postharvest handling. By leaving these problems in the field, similar problems during handling and storage will be reduced. Physical removal of diseased or infected items during sorting or grading to prevent the spread of pathogens reduces subsequent infections. The relatively short handling periods of fruits and vegetables are such that insect infestation is not usually a problem. With grain products, however, storage times are long and insects present a much more serious threat. Physical barriers represent a part of the prevention of loss by insects.

In addition, screening products from light affects quality. Light enhances chlorophyll breakdown and thus speeds yellowing of green vegetables. Light also enhances chlorophyll synthesis in nongreen vegetables such as potatoes. This greening is a quality defect as it is associated with coincidental synthesis of toxic alkaloids such as solanine. Light enhances β-carotene synthesis in tomatoes, but the effect on color quality is much more significant during the preharvest period than postharvest.

Manipulation of environmental conditions is also an important tool available to postharvest handlers. In general, lowering the temperature while maintaining high relative humidity increases the shelf life of a product by reducing the rates of respiration and transpiration. Compositions of gaseous atmospheres can be either modified or controlled to slow ripening and senescence.

Proper temperature control is probably the most powerful tool in preventing postharvest losses. As the temperature is lowered, rates of respiration and transpiration decrease. The growth of microorganisms is also slowed by lower temperatures. For many products, then, storage at temperatures as close to freezing as possible will extend their shelf life. Freezing is to be avoided at all costs as the inadvertent freezing and thawing of fresh products will cause severe quality damage. Quick cooling after harvest to remove field heat is imperative in products such as strawberries and green vegetables that respire quickly and are highly perishable. Hydrocooling or icing are used for products that can withstand water, but it must be understood that water is an excellent vehicle for transmitting microorganisms. Forced-air cooling is another effective method, while vacuum cooling is used for high-value products with a large surface area such as lettuce. Slower cooling such as room cooling is permissible for products being stored for a longer period of time such as apples, but the final temperature should be as close to the optimum as possible. When calculating refrigeration requirements, it must be remembered that respiring plant material evolves heat, known as the heat of respiration.

Prevention of chilling injury can be achieved by storing susceptible commodities at temperatures above the critical storage temperature, which ranges from 4°C for snap beans to 15°C for bananas. A complete list of optimal storage temperatures is available (8). In commercial practice, however, a compromise temperature between 5 and 10°C is frequently used to store most produce. At this temperature the damage to chilling-susceptible product is assumed to be not too severe while the decrease in shelf life of nonsusceptible product is assumed to be not of economic significance. Ice is usually added to green vegetables to lower the temperature without changing room temperature. An unspoken goal of handlers thus becomes to move the product through the process as rapidly as possible to avoid the expiration of shelf life.

Maintenance of a high relative humidity lowers transpiration of heavily transpiring products. Just as each commodity has an optimal storage temperature, it also has an optimal relative humidity. If the relative humidity is too high, however, microbial growth is enhanced. Many products are relatively resistant to microbial attack as long as no standing water accumulates on the surface. Rapid changes in temperatures of a product can contribute to water accumulation. The ice added to cool products mentioned previously also serves to increase relative humidity. The products in which ice is added tend to be not as susceptible to decay transmission through standing water by virtue of the presence of endogenous waxes.

Food additives are effective agents for protection of plant products, but they are coming under greater scrutiny as consumers become more wary about anything chemical. Many currently used compounds are being reviewed by regulatory agencies, and the prospect for approval of new additives in the near term is not likely. Microbial inhibitors such as benomyl and potassium sorbate help prevent the growth of spoilage microorganisms and thus slow the onset of decay. Fumigants such as ethylene dibromide (EDB) have been used to disinfest products from insects. External waxes are applied to porous fruits such as citrus fruits and cucumbers to slow water loss. These waxes may also enhance appearance by providing gloss, although a segment of the consuming population will avoid any product coated with wax.

Although not a chemical as such, food irradiation induces chemical changes similar to conventional processing and is considered as a food additive by the FDA. Low-dose radiation has been approved in many countries for inhibition of sprouting in potatoes and onions as well as for extension of shelf life of fresh fruits and vegetables. In some crops, like strawberries, irradiation is effective in extending shelf life, but in others damage is incurred at doses required to increase shelf life. Irradiation does possess a great potential for disinfestation of products from insects without the use of chemical fumigants if the consumer perceives irradiated product as safer than fumigated product.

Shelf life of fresh products may also be controlled by changing the composition of the atmospheric gases. A lowering of the oxygen composition and an increase of carbon dioxide can lead to an inhibition of respiration and other metabolic processes. In some crops such as pears and apples, long-term storage is enhanced by maintaining altered gaseous compositions. Other crops including lettuce and most root crops are highly susceptible to CO_2 and thus not as promising for atmospheric manipulation. Controlled-atmosphere storage usually occurs in large rooms where the gaseous composition is monitored and changed to maintain the composition of the environment within limits. In modified-atmosphere storage the initial gaseous composition is established but this composition changes due to respiration of the product. Because the mechanism of physiological response to atmosphere modification is not well understood, selection of optimal gaseous composition for a particular crop is largely based on empirical investigation.

Recent advances in film technology have given new life to the use of packages during postharvest handling and storage. Packaging protects the product by confining it as a barrier to pests, by permitting atmospheric modification and by providing instructions for optimal handling. The type of container employed and its function may change as the product travels through the handling system. As a general rule the less a product is transferred from one container to another the fewer are the opportunities for damage.

Plastics are being widely used in the packaging of fresh fruits and vegetables. They may be employed primarily for containerization and prevention of moisture loss around stacked pallets of cartons or they may be used to modify atmospheres within boxes, retail packages or around individual produce items. Barrier films have different gas-transmission properties to permit or exclude specific gases such as water vapor, carbon dioxide, and oxygen, depending on the requirements of the particular crop. These films permit in-package atmosphere modification, which extends the advantages of this technology to the supermarket shelves. Determination of the best initial composition of gases has been limited by the variation of individual items in response to differing atmospheres. When accumulation of CO_2 or C_2H_4 could be detrimental to product quality, small packets of additives can be placed in the package to absorb the deleterious gases. These packets, although they may contain toxic chemicals, offer the advantage of not coming in contact with the food item.

A logical extension of the modified-atmosphere package is the shrinking of a plastic film tightly around the individual produce item. Although generally considered a type of modified-atmosphere storage, individual plastic films are really more analogous to externally applied waxes, which change the diffusion properties of the item with the external atmosphere. These films thus slow transpiration while modifying the internal gaseous composition. Shelf life of some products such as lemons is dramatically extended while other products develop off-odors and -flavors. Edible films such as sucrose polyesters are currently being studied to determine the potential benefits of their unique barrier properties.

BIBLIOGRAPHY

1. M. C. Bourne, *Post Harvest Food Losses—The Neglected Dimension in Increasing the World Food Supply,* Cornell International Agriculture Mimeograph 53, Ithaca, N.Y., 1977.
2. National Academy of Sciences, *Postharvest Food Losses in Developing Countries,* National Academy of Sciences, Washington, D.C., 1978.
3. E. B. Pantastico and O. K. Bautista, "Postharvest Handling of Tropical Vegetable Crops," *HortScience* **11**, 122–124 (1976).
4. A. A. Kader and co-workers, *Postharvest Technology of Horticultural Crops,* University of California, Davis, 1985.
5. S. E. Prussia and co-workers, "A Systems Approach for Interdisciplinary Postharvest Research on Horticulture Crops," Georgia Experimental Station Research Report **514,** 1986.
6. A. Kramer and B. A. Twigg, *Quality Control for the Food Industry,* 3rd ed., AVI Publishing Co., Inc., Westport, Conn., 1970.
7. R. L. Jackman and co-workers, "Chilling Injury, a Review of Quality Aspects," *Journal of Food Quality* **11**, 253–278 (1988).
8. R. E. Hardenburg, A. E. Watada, and C. Y. Wang, *The Commercial Storage of Fruits, Vegetables and Florist and Nursery Stocks,* rev. ed., USDA Agriculture Handbook No. **66,** 1986.

General References

B. H. Ashby, *Protecting Perishable Foods During Transport by Motortruck,* rev. ed., USDA Agriculture Handbook No. **105,** 1970.

Council for Agricultural Science and Technology, *Ionizing Energy in Food Processing and Pest Control: II. Applications, Report R115,* Council for Agricultural Science and Technology, Ames, Iowa, 1989.

A. A. Kader, D. Zagory, and E. L. Kerbel, "Modified Atmosphere Packaging of Fruits and Vegetables," *Critical Reviews in Food Science and Nutrition* **28,** 1–30 (1989).

T. P. Labuza and W. M. Breene, "Applications of 'Active Packaging' for Improvement of Shelf-Life and Nutritional Quality of Fresh and Extended Shelf-Life Foods," *J. Food Proc. Pres.* **13,** 1–70 (1989).

B. M. Lund, "The Effect of Bacteria on Post-Harvest Quality of Vegetables and Fruits, with Particular Reference to Spoilage," in M. E. Rhodes-Roberts and F. A. Skinner, eds., *Bacteria and Plants,* Academic Press, Inc., Orlando, Fla., 1982.

M. O'Brien, B. F. Cargill, and R. B. Fridley, *Principles and Practices for Harvesting and Handling Fruits and Nuts,* AVI Publishing Co., Inc., Westport, Conn., 1983.

H. E. Pattee, *Evaluation of Quality of Fruits and Vegetables,* AVI Publishing Co., Inc., Westport, Conn., 1985.

K. Peleg, *Produce Handling, Packaging and Distribution,* AVI Publishing Co., Inc., Westport, Conn., 1985.

D. Schoorl and J. E. Holt, "Fresh Fruit and Vegetable Distribution—Management of Quality," *Sci. Hort* **17**, 1–8 (1982).

R. L. Shewfelt, "Quality of Minimally Processed Fruits and Vegetables," *Journal of Food Quality* **10**, 143–156 (1987).

A. E. Watada and co-workers. "Terminology for the Description of Developmental Stages of Horticultural Crops," *HortScience* **19**, 20–21 (1984).

J. Weichmann, *Postharvest Physiology of Vegetables,* Marcel Dekker, Inc., New York, 1987.

ROBERT L. SHEWFELT
University of Georgia
Griffin, Georgia

FOOD CROPS: SENSORY EVALUATION

One goal of food science is to maintain and improve the quality of the food supply. Before the effectiveness of a new process or the role of a new ingredient can be evaluated, a means of assessing the quality of a food product must be available. Quality has been defined as "the composite of those characteristics that differentiate individual units of a product, and have significance in determining the degree of acceptability of that unit by the buyer" (1). Thus quality is related directly to consumer acceptance, which is a difficult entity to quantify. Usually consumer acceptance is estimated by sensory evaluation, which is defined as "a scientific discipline used to evoke, measure, analyze, and interpret reactions to those characteristics of foods and materials as they are perceived by the senses of sight, smell, taste, touch and hearing" (2).

Although sensory panels are a good source of information, they are costly to train and maintain for use in long-term research studies or in routine quality analysis. Thus, where possible, objective measures using instruments or standard chemical analyses are sought in an attempt to estimate sensory characteristics and, in turn, estimate consumer acceptability.

Sensory evaluation is a useful tool in estimating the quality of a food product. The field has evolved into a reasonably exact science that can provide meaningful results if guidelines for sample presentation, questionnaires, and statistical analysis are followed. It can also result in reams of meaningless computer printouts if it is not treated as a serious part of a research investigation. Before conducting any sensory evaluation the specific objectives of the test must be clearly defined and the test must be carefully designed to meet those objectives. The simple objective of many experiments is to determine consumer acceptance of a series of products. This objective is seldom met in most sensory tests. Tests specifically designed to meet this objective usually sacrifice valuable information about specific attributes responsible for acceptability.

SENSORY ATTRIBUTES OF FOOD CROPS

Color, flavor, and texture are the three main classifications of sensory attributes of food products. Changes in these attributes can be affected by the genetic makeup of the plant species and cultivar, preharvest cultural factors, postharvest handling and storage conditions, and the type and extent of processing and preparation. The perception of sensory attributes and their relative importance in consumer acceptance varies widely from crop to crop. A consumer will readily accept a yellow apple, for example, but may not accept a yellow tomato. Likewise a crisp apple is desirable but not a crisp peach.

Color and other appearance factors such as gloss, size, shape, and absence of defects influence the consumer's perception of a product. Appearance of a product can play a principal role in the consumer's decision to purchase it and can influence perception by other senses. For example, it has been shown that beverages at the same level of sugar were perceived as being sweeter with an increase in dark coloration (3). Color is primarily a function of pigmentation in the plant tissue and the interaction of the pigments and other cellular constituents. Water-soluble pigments include the anthocyanins (reds, blues, and purples) and betalains (reds and yellows). Fat-soluble pigments, located primarily in plastids, include the chlorophylls (green) and carotenoids (yellows, oranges, and reds). Browning in plant products is primarily attributed to polyphenoloxidase activity although polymerization of anthocyanins and degradation of chlorophyll can also lead to browning.

Flavor is composed of two components: taste and odor. Taste is classified into four basic sensations: sweet, sour, salty, and bitter. Sweetness is attributed to the sugars present while sourness is related to the concentration of organic acids. Perception of sweetness or sourness in fruits is usually related to the balance between sugar and acid, although in tomatoes total flavor impact is dependent on total sugar and acid concentration (4). Saltiness is not usually an attribute of a fresh product, although tomatoes harvested from plants subjected to salt-stress during growth and development are higher in sodium and have enhanced flavor than those grown conventionally (5). Bitterness in plant products is attributed to several compounds including naringin and limonin in citrus products.

Odors are created by volatile compounds that are perceived by the nose. In many fruits and vegetables that have a distinctive flavor, a character-impact compound predominates such as benzaldehyde in cherries. The full flavor of a fresh product and its subtle variations are the result of the combined contributions of many compounds. Total flavor perception is thus a function of the composition of taste and odor compounds and their interaction.

Texture can be perceived by the consumer in two ways: touch and mouth feel. As mentioned above, certain fruits and vegetables are desirable when firm and crisp while others are not acceptable unless they are soft. In most cases the texture is related to the maturity and flavor development of the product. Firmness is one external cue the consumer uses to estimate internal quality. Firmness is a function of the complex carbohydrates in the cell wall structure. Fruits soften as the soluble pectins are broken down by enzymes such as polygalacturonase and pectinmethylesterase. When a product is eaten mouth-feel characteristics such as hardness, chewiness, and juiciness contribute to the perception of that product. These characteristics are a function of the cell wall structure and moisture content. Associated with the texture may also be the sound characteristics that enhance the appeal of the product such as the crunch of crisp celery or carrots.

PANEL EVALUATION OF SENSORY ATTRIBUTES

Panels used in sensory tests range from one to two experts up to hundreds or thousands of consumers. The objective of the test dictates the type of panel, number of judges, and training of judges required. Detailed guidelines are available to those desiring to conduct sensory tests (2,6).

Most food companies and food science departments have special facilities for sensory evaluation as shown in Figure 1. Partitioned booths are designed to prevent individual panelists from communicating with each other. Booths must have a means for receiving the samples from the food preparation area without disturbing other panelists, adequate space for product evaluation, a means of scoring the samples such as a paper ballot or computer keyboard, a slight positive air pressure, and adequate and proper lighting. Colored lights are sometimes used to screen out differences in color that might bias the judges. Not all sensory tests are conducted in such sensory laboratories. Highly trained panels that develop flavor or texture profiles may evaluate products around a table reaching a consensus on each product. Open interaction between judges is also an important aspect of consumer focus groups.

Attention must also be given to the means of food preparation. Food must be prepared in a sanitary environment and presented to the panelists in the form normally encountered by the consumer. Fresh fruits and vegetables should be evaluated for visual quality characteristics in the form they would appear in the grocery store if purchase quality is to be evaluated. Consumption quality should be evaluated in raw or cooked form as the crop is normally consumed.

Sensory tests may be divided into those that compare products and those that evaluate specific attributes of products. Product comparisons may involve the detection of preferences, differences, or both. Preference tests are designed to determine the product preferred by the greatest number of panelists in a series of samples. Judges may be asked to rank the samples in order of preference or to evaluate each sample on a fixed Hedonic scale (eg, like extremely to dislike extremely). To be able to draw inferences on preference to the general population, representative panels of at least 50 and preferably more than 100 judges are required. These judges should be untrained consumers representing a wide range of demographic characteristics. Difference tests usually require smaller panels and are designed to tell if differences exist between two or more samples. These panelists are more highly trained, and sensory scientists have developed tests such as triangle test to compensate for guessing. Sometimes investigators are interested in knowing which sample is preferred by a panel when they differ. Once again, if preference is desired the same rules apply as described for preference panels. Sensory scientists are not in agreement as to whether preference and difference tests can be conducted simultaneously.

If only difference or preference is determined for a product, the investigator has no information on why a sample is different or preferred. To determine the effect of individual attribute scaling, magnitude estimation or profiling techniques are used (7). One type of scaling technique known as quantitative descriptive analysis (QDA) asks panelists to mark a point on an unstructured line that corresponds to the intensity of that particular attribute (eg, 0 = not sweet, 75 = moderately sweet, and 150 = very sweet) (8). Magnitude estimation asks the panelists to estimate how much more of an attribute one sample has than another (eg, if sample A is three times as sweet as sample B but only half as sweet as sample C the panelists could assign them scores of A = 30, B = 10, and C = 60). In these panels some training of judges is required so that panelists know and are in agreement on basic terminology and scaling methodology. More highly trained panels are also used for these tests or for more in-depth product profiling such as flavor or texture profiles.

To determine the extent and significance of sensory evaluation results, statistical analysis must be employed. Straightforward statistical methods are available for standard sensory tests (9), although all statistical analyses need to be incorporated into the design of the original experiment and panelist questionnaires. Sophisticated multivariate techniques can efficiently reduce the useful

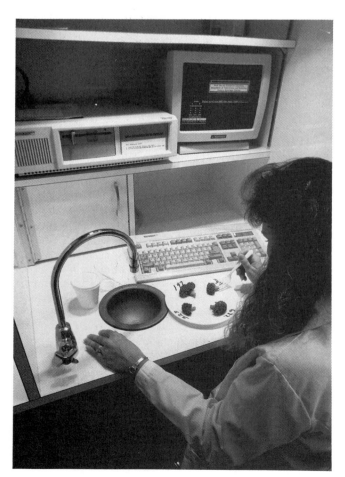

Figure 1. A typical sensory evaluation booth equipped for data entry by computer.

information available from sensory panel data to meaningful terms (10).

OBJECTIVE MEASURES OF SENSORY ATTRIBUTES

Sensory panels place severe constraints on quality evaluation in research, product development, or quality assurance settings. The number of samples a judge can evaluate before fatigue affects judgment is limited. Time spent by an employee judge in sensory panel training and evaluation is usually time spent away from a primary job. Panels generally must be convened only at certain times of day to avoid the interference of meals and breaks. Such constraints provide incentives to find instrumental or wet chemistry methods that don't rely on human subjective judgment. Thus there is a concerted effort to find objective measures that can be used to predict human sensory perception.

Food color is measured by an instrument called a colorimeter, which represents the color of the product as a point in a three-dimensional space. In the most frequently used color scale (L, a, and b or its close relative L^*, a^*, and b^*) each measurement provides a reading on the lightness (L), green–red (a), and blue–yellow (b) axes. Lightness or darkness is signified by L. A negative a corresponds to a sample with more green character than red, while a positive a represents a sample with more red than green character. Likewise a negative b indicates the predominance of blue character and a positive b, yellow character. Mathematical conversions can provide functions that more closely relate these measurements to sensory perception. Hue, or the color name we use to describe a product (red, orange, etc), is described by the trigonometric function $\tan^{-1} b/a$ known as the hue angle. A hue angle of 0° represents red, while 90° represents yellow. Chroma, which is related to brightness, is described by the geometric function $\sqrt{a^2 + b^2}$ or the hypotenuse of the right triangle in the a,b plane formed when the point is connected with either the a or b axis. The colorimeter is a useful instrument in measuring the color of samples that are reasonably uniform in color. Unfortunately most food crops have a wide variation in color within a sample. This difficulty is frequently overcome by taking an average of several measurements to provide an average color. Good correlations between hue and lightness with color perception can be obtained if differences tend to be pronounced. When color changes are subtle in nonuniform samples, the colorimeter is not as useful. Pigment analysis is tedious and tends to give poor correlations with sensory panel data.

Several instruments have been developed to measure textural properties of food products. Before texture is measured, it is necessary first to determine the attributes important to the consumer. If firmness is important, then puncture or deformation tests are used. Shear measurement gives an indication of turgor pressure and of the first bite. An instrumental counterpart to the texture profile panel has been developed that has been used in the assessment of food products. Puncture, deformation, and shear methods provide a reasonable assessment of firmness-to-the-touch measures and general textural properties but are not as effective in predicting mouth feel. Deformation measurements, when applicable, offer the advantage that they are nondestructive thus allowing repetitive measurements on the same sample (11). Most instrumental textural measurements are measures of force or energy and should be reported as newtons (N) and joules (J), respectively (12).

Sweetness is usually measured by determining the percent soluble solids, or degrees Brix, by refractometer, which approximates the percent sugar present. Sourness is related to the total acidity, which is determined by titration with a base and expressed in terms of the predominant organic acid present. Because sugar and acid are both frequently present in fresh fruits the perception of sweetness or sourness usually depends on which class of compounds predominates. Thus the Brix-to-acid ratio (BAR) is usually used to estimate sweetness or sourness. Volatile aroma compounds are usually measured by gas chromatography. With technical advances in the field, rapid separation of volatiles can be achieved, and the compounds are identified with the help of a mass spectrometer. Identification of the primary flavor impact compound(s) and which other volatiles contribute to aroma is more difficult. An odor-threshold technique has been developed that has been a useful tool (13), but it ignores possible interactions between compounds.

SENSORY ATTRIBUTES AND ACCEPTANCE

The ultimate objective for most tests that measure sensory attributes of food crops is an estimation of consumer acceptance. Few studies have been performed to directly determine the specific factors important in consumer acceptance of these crops. Unfortunately, the emphasis of many tests is placed on the attributes that are most easily measured by panel or instruments with the implicit assumption that if it can be measured it must be important.

Meaningful sensory evaluation of food crops starts with identification of the attributes of that crop with greatest relevance to consumer acceptance, both during purchase and consumption. Next, appropriate sensory tests should be designed to measure these attributes. Once sensory methods have been developed to discriminate between differing levels of quality, instrumental and chemical methods can be pursued to provide objective estimates of these characteristics permitting greater flexibility and larger scale studies. It should also be remembered that quality or acceptance of a product is rarely a composite of a given set of attributes but frequently a function of the acceptability of those attributes important to that consumer. An item that fails to be acceptable for purchase will not be consumed, and, likewise an item unacceptable during consumption will affect future purchases of similar product.

BIBLIOGRAPHY

1. A. Kramer and B. A. Twigg, *Quality Control for the Food Industry,* 3rd ed., AVI Publishing Co., Inc., Westport, Conn., 1970.
2. IFT, "Sensory Evaluation Guide for Testing Food and Beverage Products," *Food Technology* **35**(11), 50–59 (1981).

3. F. M. Clydesdale, "The Influence of Color on Sensory Perception and Food Choices," in J. Walford, ed., *Developments in Food Colours,* Vol. 2, Elsevier Applied Science Publishers, Ltd., Barking, UK, 1984.
4. R. A. Jones and S. J. Scott, "Improvement of Tomato Flavor by Genetically Increasing Sugar and Amino Contents," *Euphytica* **32,** 845–855 (1983).
5. Y. Mizrahi, "Effect of Salinity on Tomato Fruit Ripening," *Plant Physiology* **69,** 966–970 (1982).
6. E. Larmond, *Laboratory Methods for Sensory Evaluation of Food,* Food Research International, Publication **1637,** Canada Department of Agriculture, Ottawa, Ont., 1977.
7. H. R. Moskowitz, *Product Testing and Sensory Evaluation of Foods*: *Marketing and R&D Approaches,* Food and Nutrition Press, Trumbull, Conn., 1983.
8. H. Stone and co-workers, "Sensory Evaluation by Quantitative Descriptive Analysis," *Food Technology* **28**(11), 24–34 (1974).
9. M. O'Mahony, *Sensory Evaluation of Food—Statistical Methods and Procedures,* Marcel Dekker, Inc., New York, 1986.
10. A. V. A. Resurreccion, "Applications of Multivariate Methods in Food Quality Evaluation," *Food Technology* **42**(11), 128–136 (1988).
11. M. C. Bourne, "Use of the Penetrometer for Deformation Testing of Foods," *Journal of Food Science* **38,** 720–721 (1973).
12. M. C. Bourne, *Food Texture and Viscosity,* Academic Press, Inc., Orlando, Fla., 1982.
13. D. G. Guadagni, R. G. Buttery, and J. Harris, "Odour Intensities of Hop Oil Components," *Journal of the Science of Food and Agriculture* **17,** 142–144.

General References

F. M. Clydesdale, "Colorimetry—Methodology and Applications," *Critical Reviews in Food Science and Nutrition* **10,** 243–301 (1978).

O. Fennema, *Food Chemistry,* 2nd ed., Marcel Dekker, Inc., New York, 1985.

J. Gross, *Pigments in Fruits,* Academic Press, Inc., Orlando Fla., 1986.

R. Hunter, *The Measurement of Appearance,* Wiley-Interscience, New York, 1987.

H. R. Moskowitz, *New Directions for Product Testing and Sensory Analysis of Foods,* Food and Nutrition Press, Trumbull, Conn., 1985.

R. L. Shewfelt, "Flavor and Color of Fruits as Affected by Processing," in J. G. Woodroof and B. S. Luh, eds., *Commercial Fruit Processing,* AVI Publishing Co., Inc., Westport, Conn., 1986.

R. Teranishi and F. Shahidi, eds., *Flavor Chemistry, Trends and Developments,* American Chemical Society, Washington, D.C., 1989.

M. Meilgaard, G. V. Civille, and B. T. Carr, *Sensory Evaluation Techniques,* 2 vols., CRC Press, Inc., Boca Raton, Fla., 1987.

J. L. Sidel and H. Stone, *Sensory Evaluation Practices,* Academic Press, Inc., Orlando, Fla., 1985.

Z. M. Vickers and M. C. Bourne, "Crispness in Foods. A review," *Journal of Food Science* **41,** 1153–1157 (1976).

H. Martens and H. Russwarm, *Food Research and Data Analysis,* Elsevier Applied Science Publishers, Ltd., Barking, UK, 1983.

ROBERT L. SHEWFELT
University of Georgia
Griffin, Georgia

FOOD CROPS: STORAGE

This article describes various factors to be considered during the storage of fresh fruits and vegetables and some supplemental methods used with cold storage. Space limitations preclude detailed prescriptions of storage for specific crops. This information is available in other publications (1–4).

Tissues of fruits and vegetables after harvest are still alive and continue to respire, metabolize, and change. Successful storage of fruits and vegetables extends the marketing period of these commodities and minimizes economic losses because most of the crop is utilized. Good storage conditions will maintain crop quality, reduce deterioration, render excellent salable commodities, and avoid causing injury to tissues by chilling, freezing or ammonia fumes.

FACTORS AFFECTING STORAGE LIFE

The most important factor affecting storage life is the quality of the product at harvest. Other factors to be considered during storage are temperature, humidity, and sanitation. Optimal storage conditions for fresh food crops will change from time to time, as changes in technology and popularity of different cultivars occur.

Product Quality

The quality of produce at harvest greatly affects the length of storage. The commodity should be harvested at the recommended stage of maturity (which differs with commodity) for maximal storage and should be free of physical damage, as skin breaks and bruises can lead to increased susceptibility to water loss and to bacterial and fungal attack. The extent of incipient infections by pathogens should be determined prior to storage, because such infections can cause rot and decay during storage.

Other factors to be considered in the determination of storage life and conditions are cultivar, environmental conditions during growth, cultural practices, handling practices before storage, and transit conditions. Allowances for improper handling, unfavorable growing conditions, and poor or long-distance transit conditions should be made in determining optimal storage conditions and predicting storage life. Furthermore, different parts of plants are used as food, including roots, stems, fruits, flowers, and leaves, and these parts have different storage requirements.

Respiration

Plant tissues respire, using oxygen in the conversion of carbohydrates to carbon dioxide, water, and energy. The energy is primarily given off as heat. Respiration rates and the amount of heat released increase as storage temperature increases, about two to three times for every 10°C (18°F) rise in temperature. The storage life of a commodity is inversely proportional to the rate of respiration. Therefore, as temperature increases, respiration rate increases, and storage life decreases. Respiration rates differ with plant species, cultivar, morphological tissue, and growing conditions. Rates gradually decrease after harvest for many fruits and vegetables, but in certain crops respiration rates gradually increase as they ripen, peak to a climacteric, and then decrease (5).

Temperature

Temperature affects the rates of produce metabolism and respiration, spore germination and pathogen growth, softening, moisture loss, and undesirable growth, such as sprouting. For every 10°C increase in temperature, the rate of deterioration or undesirable growth increases two to three times. Refrigerated storage is therefore recommended for many perishable commodities to slow the changes listed above (6). For optimal storage, storage room temperatures should be held constant, with variations of less than 1–2°C. The recommended storage temperature for many crops is 0°C, while temperatures below 0°C will cause freezing damage. Crops sensitive to chilling should be stored at 12.5°C; temperatures below 12.5°C but above 0°C will injure these commodities. Storage temperatures higher than optimal will shorten the expected shelf life of the produce. Temperature fluctuations can cause condensation on the product, leading to an increase in pathogen growth and subsequent decay.

Precooling

The first step in good temperature management of perishable commodities is precooling after harvest and before shipment, storage, or processing (7–10). Prompt precooling can reduce commodity respiration, water loss, metabolism, and pathogen growth. This is done using special equipment and rooms, when rapid cooling is essential and economically beneficial. Precooling can be accomplished in refrigerated storage rooms equipped with proper control of airflow.

Precooling methods quickly transfer heat from the produce to a cooling medium, such as water, ice, or air. The rate of transfer depends on the availability of the cooling medium to the commodity, the temperature difference between the medium and the commodity, the type of medium, and the flow rate of the medium. The commercial methods used include hydrocooling, vacuum cooling, air cooling, and icing. The methods used depend on the type of produce, the availability of equipment, facilities, containers, cost, and the proximity to market. See Table 1 for a list of methods of cooling for different commodities.

Table 1. Methods of Cooling for Different Commodities

Cooling Method	Commodities Cooled
Hydrocooling	Asparagus, cantaloupes, celery, cucumbers, green peas, peaches, peppers, radishes
Vacuum cooling	Asparagus, artichokes, broccoli, Brussels sprouts, cabbage, cauliflower, celery, endive, escarole, head lettuce, parsley, spinach, sweet corn
Air cooling	Cauliflower, cucumbers, grapes, melons, peppers, strawberries, tomatoes
Icing	Broccoli, Brussels sprouts, cantaloupes, carrots, green onions, kale, radishes

Humidity

Maintenance of proper humidity around the commodity is necessary during precooling and storage. Fruits and vegetables transpire and lose water during precooling, storage, and transit. They will wilt or shrivel if humidity is too low. Humidity is generally expressed as percent relative humidity (% RH), the ratio of water vapor pressure in the air to saturation vapor pressure at a certain temperature (11). As temperature increases, the capacity of air to hold water increases. For most vegetables, a relative humidity of 98–100% are recommended to prevent wilting. Water may be sprinkled on the floor of a storage room or misted in the air if it is necessary to increase the relative humidity around a commodity. Other practices to insure adequate humidity include using good room insulation, preventing air leaks, and providing sufficient cooling surface to keep the difference between refrigerant surface temperature and commodity temperature as small as possible.

Sanitation

Sanitary conditions must be maintained in storage facilities to prevent the growth of contaminating organisms. Under high humidity conditions, molds grow on the surfaces of packages, walls, and ceilings, and although they may not directly decay produce, they may produce unwanted substances that accelerate decay, give off-flavors to the commodity, or create favorable environments for decay organisms. Periodic and thorough cleaning of storage surfaces and good air circulation help prevent the growth of surface molds. Whitewash can be sprayed on ceilings and walls when storage rooms are empty and before a new load comes in. Surfaces may also be scrubbed with sodium hypochlorite or trisodium phosphate, with subsequent rinsing and airing before new commodities are loaded. Fungicidal paint or fumigation with a mixture of 85% carbon dioxide and 15% ethylene can also be used. If odors have developed in a room, air purification can be accomplished by absorbing the odors with 6–14 mesh-activated coconut-shell carbon or washing the air with water (12).

INJURIES

Certain injuries to produce can occur during refrigerated storage under suboptimal conditions and include those due to chilling, freezing, or ammonia. Chilling injury usually occurs to commodities of tropical origin at temperatures below 10–13°C (50–55.4°F), although certain temperate commodities are also susceptible (13–15). Table 2 lists some commodities susceptible to chilling injury and their symptoms. Often symptoms cannot be detected at low temperatures and only become visible several days after the commodity has been removed to warmer temperatures. The extent of damage depends on length of exposure and chilling temperature. It may occur in a commodity exposed a short amount of time to a temperature considerably below the danger zone, but not if it is exposed longer to a higher temperature in the danger zone. Also, chilling injury may be cumulative, and chilling during transit or in the field before harvest may add to the effects

Table 2. Commodities Susceptible to Chilling Injury and Their Symptoms[a]

Commodity	Lowest Safe Temperature (°C)	Symptoms
Avocados	4.5–13.0	Grayish brown flesh discoloration
Bananas	11.5–13.0	Dull color when ripe
Cantaloupes	2–5	Pitting, surface decay
Cucumbers	7	Pitting, decay
Eggplants	7	Surface scald, *Alternaria* rot, black seeds
Lemons	11–13	Pitting, red blotch
Peppers	7	Sheet pitting, *Alternaria* rot on pods and calyxes, seed blackening
Potatoes	10	Browning, sweetening
Tomatoes, ripe	7–10	Water soaking, softening, decay

[a] Ref. 1.

of storage chilling. Chilling injury has been reduced by treatments such as intermittent warming, temperature preconditioning, controlled atmosphere or hypobaric storage, waxing, and film packaging.

Freezing injury occurs when ice crystals form in produce tissues, and is manifested by a mushy, limp, and water-soaked appearance. The extent of freezing injury depends on exposure time and temperature and varies with different commodities.

Ammonia injury occurs when ammonia escaping from direct-expansion refrigeration units come in contact with the commodity. Injury is evident as brown to greenish black discoloration of the outer tissues and, in severe cases, as discoloration and softening of inner tissues (16). Daily odor checks or installation of ammonia detection systems help prevent ammonia injury to produce. Fumes can be removed by aeration and washing the air of the storage room with water.

SUPPLEMENTAL METHODS TO COLD STORAGE

Although refrigeration is the most effective method for retarding decay of food crops, supplements can be used to control the growth of pathogens, slow respiration rates, control physiological disorders and sprouting, reduce moisture loss, or retard ripening and senescence. Supplemental methods in use include chemical treatments, controlled- and modified-atmosphere storage, waxes, irradiation, and protective packaging.

Chemical treatments are used if they are nontoxic to humans and the commodity, cost-effective, effective for their intended purpose, and approved by federal regulatory agencies, for example, the use of growth regulators to control sprouting of potatoes and onions during storage. Calcium chloride is used to control bitter pit and maintain firmness in apples. Fungicides and bactericides, such as chlorine and sodium o-phenylphenate, are often added to wash water, soak tanks, and hydrocoolers to reduce growth of pathogens prior to and during refrigerated storage (17).

In controlled-atmosphere (CA) storage, commodities are held in atmospheres of specified proportions of oxygen, carbon dioxide, or nitrogen, different from ambient air. This is accomplished by adding or scrubbing oxygen or carbon dioxide in airtight containers or rooms (18–21). In modified-atmosphere storage, storage atmospheres are different from that of ambient air, but are not controlled precisely.

Waxing is used to reduce moisture loss and improve the appearance of the commodity. Fungicides can also be mixed with the wax to retard decay. The effectiveness of the wax depends on the thickness of the coat and the uniformity of application.

Gamma irradiation has been used to disinfect papayas, mangoes, and grapefruit. Commercial use is limited because of the cost and size of necessary equipment and consumer acceptance. Dosages of 1.5–2 kilogray can control decay and sprouting in certain commodities (22), but can also cause discoloration, pitting, softening, abnormal ripening, and flavor loss (23).

Protective packaging safeguards produce against physical damage during packing, such as that acquired from dropping of containers, overfilling of containers, or movement of the commodity within containers (24). Protective packaging must be sturdy and moisture tolerant. Packaging materials include plastic foam, waxed fiberboard, film box liners, wooden pallet bins, nailed wooden boxes, wirebound veneer crates, and perforated polyethylene or polypropylene bags.

FUTURE DEVELOPMENTS

Technologies are continually being researched and developed to improve the storage life of food crops. Future developments should result in more economical and energy-effective cooling methods, more efficient transportation and distribution systems, improved shipping containers, and increased usage of controlled- and modified-atmosphere storage.

BIBLIOGRAPHY

1. R. E. Hardenburg, A. E. Watada, and C. Y. Wang, *The Commercial Storage of Fruits, Vegetables, and Florist and Nursery Crops*, USDA Agricultural Handbook, **66**, rev., U.S. Government Printing Office, Washington, D.C., 1986.
2. A. A. Kader and co-workers, *Postharvest Technology of Horticultural Crops*, Cooperative Extension Publication **3311**, University of California, Berkeley, 1985.
3. A. L. Ryall and W. J. Lipton, *Handling, Transportation and Storage of Fruits and Vegetables*, Vol. 1, *Vegetables and Melons*, 2nd ed., AVI Publishing Co., Inc., Westport, Conn., 1979.
4. A. L. Ryall and W. T. Pentzer, *Handling, Transportation and Storage of Fruits and Vegetables*, Vol. 2., *Fruits and Tree Nuts*, 2nd ed., AVI Publishing Co., Inc., Westport, Conn., 1982.
5. J. B. Biale and R. E. Young, "Respiration and Ripening in Fruits—Retrospect and Prospect," in J. Friend and M. J. C. Rhodes, eds., *Recent Advances in the Biochemistry of Fruit and Vegetables*, Academic Press, Inc., Orlando, Fla., 1981.

6. J. A. Bartsch and G. D. Blanpied, *Refrigerated Storage for Horticultural Crops*, Agric. Eng. Ext. Bull. **448**, Cornell University, Ithaca, N.Y., 1984.
7. F. E. Henry, A. H. Bennett, and R. H. Segall, "Hydroaircooling: A New Concept for Precooling Pallet Loads of Vegetables," *ASHRAE Trans.* **82**(part 2), 541–547 (1976).
8. F. M. R. Isenberg, R. F. Kasmire, and J. Parson, *Vacuum Cooling Vegetables,* Cornell University Cooperative Extension Information Bulletin **186**, 1982.
9. F. G. Mitchell, R. Guillou, and R. A. Parson, *Commercial Cooling of Fruits and Vegetables,* California Agricultural Experimental Station Extension Service Manual **43**, 1972.
10. I. J. Pflug and co-workers, "Precooling of Fruits and Vegetables," *ASHRAE Symp.* SF-4-70, 1970.
11. J. J. Gaffney, "Humidity: Basic Principles and Measurement Techniques," *HortScience* **13**, 551–555 (1978).
12. H. Hansen and J. Kuprianoff, "Some Experiences with Air-Washing in Cold and Gas Stores of Pears and Apples," *Internat. Inst. Refrig. Bul. Sup.* **1**, 337–340 (1961).
13. H. M. Couey, "Chilling Injury of Crops of Tropical and Subtropical Origin," *HortScience* **17**, 162–165 (1982).
14. J. M. Lyons, "Chilling Injury in Plants," *Annual Review of Plant Physiology* **24**, 445–466 (1973).
15. L. L. Morris, "Chilling Injury of Horticultural Crops: An Overview," *HortScience* **17**, 161–162 (1982).
16. E. Brennan, I. Leone, and R. H. Daines, "Ammonia Injury to Apples and Peaches in Storage," *Plant Dis. Rptr.* **46**, 792–795 (1962).
17. J. W. Eckert, "Postharvest Diseases of Fresh Fruits and Vegetables—Etiology and Control," in N. F. Haard and D. K. Salunkhe, eds., *Symposium: Postharvest Biology and Handling of Fruits and Vegetables,* AVI Publishing Co., Inc., Westport, Conn., 1975.
18. P. E. Brecht, "Use of Controlled Atmospheres to Retard Deterioration of Produce," *Food Technology* **34**, 45–50 (1980).
19. F. M. R. Isenberg, "Controlled-Atmosphere Storage of Vegetables," *Hortic. Rev.* **1**, 337–394 (1979).
20. D. G. Richardson and M. Meheriuk, eds., *Controlled Atmospheres for Storage and Transport of Perishable Agricultural Commodities, Proceedings of the Third National CA Research Conference, Beaverton, Oreg., July 1981.*
21. R. M. Smock, "Controlled atmosphere storage of fruits," *Hortic Rev.* **1**, 301–336 (1979).
22. J. H. Moy, "Potential of Gamma Irradiation of Fruits: A Review," *Journal of Food Technology* **12**, 449–457 (1977).
23. R. A. Dennison and E. M. Ahmed, "Irradiation Treatment of Fruits and Vegetables," in Ref. 17.
24. F. G. Mitchell, *Packaging Horticultural Crops*, Univ. Calif. Coop. Ext. Perishables Handling **52**, 1983, pp. 2–4.

CINDY B. S. TONG
ALLEY E. WATADA
USDA/ARS
Beltsville, Maryland

FOOD CROPS: VARIETAL DIFFERENCES, MATURATION, RIPENING, AND SENESCENCE

With the advent of nutritional labeling, most consumers tend to think that products have uniform composition. Scientists more familiar with the derivation of nutritional labels realize that they represent a conservative estimate of composition in a formulated food product due to the inherent variation in composition of ingredients and other factors. Composition of nutrients and other components is more variable in raw agricultural crops than in formulated foods due to differences in cultivar, maturity at harvest, and the stage of ripening and senescence (1). Terms critical to an understanding of this subject are defined in Table 1.

VARIETAL DIFFERENCES

Humans eat a wide variety of plant products thanks to the large number of plant species cultivated for food. Most individual fruits, vegetables, nuts, oilseeds, and grains represent separate species. A notable exception is *Brassica oleracea*, a species comprising the crops of broccoli, Brussels sprouts, cabbage, cauliflower, collards, kale, and kohlrabi. Although not as diverse as *B. oleracea*, most species have variations in cultivars that are apparent in differences in growth response, pest and disease resistance, stress response, nutritional quality, and shipping stability.

The concept of varietal differences is based on Mendel's classic study of genetics in peas and is the basis of modern efforts in crop improvement. Varietal differences are evident in the growth patterns of the crop in the field such as the height of the plant, its branching characteristics, color and morphology of flower or fruit, and the yield potential for edible product at harvest. Some cultivars offer greater resistance to pests and disease at harvest. Resistance can be achieved as a result of removal of chemical compounds that attract insects or synthesis of compounds that are toxic to the disease microorganisms or pest. Unfortunately, the same compounds that are toxic to the pest may adversely affect quality, as in bird-resistant sorghum (5), or may even be toxic to humans (6).

Cultivars adapt in different ways to changes in growing locations and environmental conditions. Some cultivars will thrive in one location and suffer in others. Temperature, cultural practices, and soil conditions such as fertility, compaction, and moisture content tend to affect cultivars differently. Certain cultivars within a species may show greater resistance to water, salt, and low-temperature or high-temperature stresses. The response to stress by plants is often reflected by changes in the edible portion of the tissue (eg, tomato plants resistant to low-temperature stress tend to produce fruit resistant to low temperatures during storage). Varietal differences have also been noted in the nutritional composition of crops at harvest and the ability of a crop to maintain marketable quality during handling, shipping, and storage.

Crops may be improved by selecting for desired traits from existing cultivars or wild strains. The most-celebrated modern example of crop improvement was the development of high-yield rice cultivars by Borlaug resulting in the Green Revolution. In addition to conventional breeding techniques, biotechnology offers the powerful tools of cell culture and recombinant DNA in the development of better crops. Cell culture techniques permit rapid

Table 1. Definition of Terms

Cell and tissue culture	Culture of isolated plant cells or detached fragments of plant tissue on a nutrient medium under aseptic conditions (2)
Cultivar (cultivated variety)	Group of plants within a particular cultivated species that is distinguished by a character or group of characters and that maintains its identity when propagated either asexually or sexually (3)
Genetic engineering	In plants, the transfer of DNA from a donor plant species to a recipient species by means of a bacterial plasmid, virus, or other vector (2)
Horticultural maturity	The stage of development when a plant or plant part possesses two prerequisites for utilization by consumers for a particular purpose (4)
Maturation	The stage of development leading to the attainment of physiological and horticultural maturity (4)
Physiological maturity	The stage of development when a plant or plant part will continue ontogeny even if detached (4)
Ripening	The composite of the processes that occur from the latter stages of growth and development through the early stages of senescence and results in characteristic aesthetic quality and food quality as evidenced by changes in composition, color, texture, and other sensory attributes (4)
Senescence	Those processes that follow physiological maturity or horticultural maturity and lead to death of tissue (4)
Species	Comprises a group of plants (within a genus) that often exhibit many more morphological similarities than do members of the genus (3)

screening of potential strains if a reliable index for the desirable trait(s) is available. Genetic engineering involving the use of recombinant DNA or plasmid transfer permits the direct manipulation of genetic material within the plant, promising the development of "designer plants" with the characteristics desired. Recent research with tomatoes has lead to the development of a fruit with modified levels of polygalacturonase (an enzyme involved in softening) activity (7). This study illustrates both the potential and limitations of current biotechnological approaches as the genetic modification did not affect tomato softening indicating that the physiology of softening is not sufficiently understood. It is likely that as rapid advances are made in biotechnological approaches, the limiting factor in quality improvement of crops will be the understanding of quality and the physiological basis of quality changes.

Many examples demonstrate the importance of varietal differences in the marketplace. Differences in color and sweetness in fresh apple (*Malus domestica*) cultivars help satisfy regional and individual tastes. Other cultivars are preferred for specific processing or home-cooking applications. The length of the peach season from a particular growing area is due to a succession of early-, middle-, and late-maturing cultivars, and different cultivars are adapted to different climatic regions. Clingstone cultivars provide a firmer peach for canning with a better appearance although many prefer the flavor of freestone peaches for the fresh market. The cultivar of choice for a grape grower depends on the intended processing application: fresh, juice, jelly, raisins, or wine. Supersweet cultivars of corn were developed to increase sugar content or slow the hydrolysis of sucrose during postharvest handling and storage. Processing cultivars of tomatoes must have a high solids content, commercial cultivars that are shipped across country must be able to withstand the rigors of handling and storage, while cultivars grown in backyard gardens must have high flavor impact. Hard cultivars of wheat (*Triticum aestivum*) are higher in protein and used in breads while soft cultivars are used in other applications such as biscuits, cookies, and cakes.

Mechanization of harvesting operations has been achieved for several commodities. Certain species are more amenable to mechanical harvest than others, and processed products are more likely to have been adapted to mechanical harvesting than fresh products. Within a species, certain cultivars are more adaptable to mechanical harvest than others. The characteristics important in ease of mechanical harvesting include the uniformity of size and maturity at harvest, decreased susceptibility to bruising and other damage, and the adaptability to fully mechanized production and handling systems. The most successful mechanization efforts have come from research that has integrated the breeding, quality, and engineering aspects of the production and handling systems.

MATURATION

Harvest is a critical point for the quality of any product derived from plants. The optimum harvest time varies with the period of growth and development of a plant from crop to crop. For example celery (stems), cabbage (leaves), and broccoli (inflorescences) are harvested during vegetative growth while cucumber and beans are harvested as partially developed fruits. Maturity at harvest or the stage of development of the crop at harvest can be viewed from either a biological or a market perspective. Full maturity from the biological perspective is known as physiological maturity, whereas the preferred stage of development for eating enjoyment is known as horticultural maturity (4).

Many crops present a narrow window for optimal harvest. Maturity of the crop at harvest can affect the yield, stability during handling and storage, susceptibility to bruising or other damage, nutrient composition and the purchase and consumption quality of the product. All of these factors should be considered in determining the best time to harvest. With most vegetables this decision requires a compromise between harvesting early to obtain higher quality or waiting to achieve higher yields (eg, toughening of snap beans) with the economic considerations predominating. Because little or no flavor development occurs after harvest of nonclimacteric fruit, it is important to harvest these crops as close to the peak of flavor development as possible. Climacteric fruits such as apples and tomatoes, which continue to ripen after harvest, present a more complex situation. If they are harvested too early they may not develop full flavor potential while ripening off the plant. If they are harvested too late their physical condition and appearance will not be acceptable for purchase. The marketplace offers greater incentives to the grower who harvests too early than to the one who harvests too late. In crops where this window between too early and too late is very narrow, the consumer may not have the opportunity to enjoy top-quality produce.

RIPENING

The desirable attributes associated with high-quality fruit develop during ripening. Nonclimacteric fruits develop these attributes while still attached to the plant and are characterized by low levels of carbon dioxide (CO_2) and ethylene (C_2H_4) evolution. Little further ripening occurs in nonclimacteric fruit after harvest.

Climacteric fruits are characterized by a rapid respiratory burst of greatly increased CO_2 and C_2H_4 at the initiation of ripening. The climacteric rise, triggered by C_2H_4, can occur on or off the plant. The fruit will generate its own C_2H_4 from aminocyclopropane carboxylic acid (ACC) by ethylene-forming enzyme (EFE). ACC is generated from S-adenosyl methionine (SAM) by ACC synthase (8). If the fruit is harvested in a physiologically mature state prior to the increased generation of ACC and C_2H_4 and is kept in an environment free of C_2H_4, it can be maintained in a mature, unripe state during transportation and storage. Then ripening can be artificially triggered by using an external source of C_2H_4, eg, the catalytic conversion of ethanol to C_2H_4 under controlled conditions in ripening rooms. This process permits the shipment of green bananas from the tropics to geographically distant markets where ripening is triggered close to the consumer, long-distance truck shipment of green tomatoes for ripening close to the distribution point, and controlled-atmosphere storage of apples prior to ripening to permit year-round consumption.

Desirable changes associated with ripening include changes in appearance, flavor, texture, and nutritive value. Color changes are the result of the degradation of chlorophyll (green), which unmasks xanthophylls and carotenes (yellow and orange), and the biosynthesis of anthocyanins (red, blue, or purple) and the carotenoid lycopene (red). The development of sweetness in fruits is the result of a disappearance of sour (acids) and bitter compounds and the accumulation of sweet (sugars) compounds. Flavor is a combination of taste (sweet, sour, and bitter) and aroma that results from the accumulation of volatile compounds during ripening producing the characteristic flavor we associate with a specific fruit. Fruits that soften during ripening undergo a degradation of cell wall compounds, primarily the pectins, by a series of hydrolase enzymes (9). Vitamins (particularly A and C) accumulate during ripening, thereby enhancing the nutritional quality of the product. Although there is still controversy in the scientific literature, there is evidence that some climacteric fruits ripened off the plant do not develop the full flavor or nutritional potential as when ripened on the plant.

SENESCENCE

Plants senesce by predictable, controlled patterns that lead to death. Senescence is distinct, although not always distinguishable, from aging (longevity) (10). Plant organs also go through senescence, which is frequently enhanced by detachment. Ripening, as described above and defined in Table 1, is a special case of the early phase of senescence. In somewhat oversimplified terms ripening tends to lead to improvement of the quality attributes of edible fruit tissue while other senescence processes tend to lead to quality deterioration of edible plant tissue. The objectives of postharvest handling and storage is to control ripening and slow senescence thus providing the consumer with a product of optimum quality at a reasonable price.

The most obvious signs of senescence in vegetables are loss of color and tissue softening. Unlike similar changes noted in ripening fruits that are considered to be improvements in quality, yellowing and softening of green vegetable tissue is considered detrimental. Yellowing, as in ripening, involves the unmasking of yellow pigments associated with the loss of chlorophyll as chloroplasts are converted to chromoplasts. Likewise, softening is associated with breakdown of cell wall components such as pectin and cellulose.

The cellular physiology of senescence is complex and just beginning to be understood. Although it was previously believed that senescence resulted from the release of organic acids and hydrolases within the cell, it is now generally recognized that senescence is controlled, at least in part, by genetics. Senescence affects gene expression, protein and nucleic acid degradation, chemical composition and physical properties of membranes, and the structure and function of plant organelles (10). Although predominantly a degradative process, it is misleading to view senescence solely in the context of degradation. Metabolic processes function as a series of delicate balances between biosynthetic and degradative reactions. During growth and development as well as ripening, accumulation of a compound is the net result of biosynthesis exceeding degradation. Likewise, during senescence net loss of a compound may be achieved by increased degradation, decreased biosynthesis or a change in the balance of the two processes.

Ultrastructure of senescing plant tissue reveals changes in chloroplasts such as swelling of thylakoids to be some of the earliest changes. Mitochondria and vacuoles are affected later in the process. Changes in the nucleus are generally observed late in senescence before disruption of the plasma membrane leading to death (10). Of particular interest in the study of senescence are the increase in protease enzymes and also loss of membrane integrity. Membranes serve as internal barriers within the cell regulating the flow of ions and metabolites. During senescence and aging these membranes lose their integrity leading to increased permeability. A current theory suggests that membrane lipids are degraded by hydrolytic and peroxidative mechanisms resulting in increased permeability and decreased function of membrane-bound proteins.

Not all degradation of plant tissue is attributable to senescence. Mechanical damage can cause physical rupture of membranes leading to mixing of enzymes and substrates resulting in browning and discoloration we observe as a bruise. Inadequate nutrition of the plant during growth and development can lead to deficiencies within a detached plant organ resulting in a physiological disorder. Low temperatures and controlled atmospheres slow respiration and senescence of many plant organs, but certain species may be susceptible to physiological disorders resulting from exposure to low temperatures, low O_2, high CO_2, or elevated C_2H_4. In addition, excess water loss can lead to loss of turgor pressure resulting in undesirable textural properties.

BIBLIOGRAPHY

1. R. L. Shewfelt, "Sources of Variation in the Nutrient Content of Agricultural Commodities from the Farm to the Consumer," *Journal of Food Quality* **13**, (1990).
2. J. M. Poehlman, *Breeding Field Crops*, 3rd ed., AVI Publishing Co., Inc., Westport, Conn., 1986.
3. J. Janick, *Horticultural Science*, 4th ed., W. H. Freeman, New York, 1986.
4. A. E. Watada and co-workers, "Terminology for the Description of Developmental Stages of Horticultural Crops," *HortScience* **19**, 20–21 (1984).
5. D. K. Salunkhe, S. S. Kadam, and J. K. Chavan, "Nutritional Quality of Proteins in Grain Sorghum," *Qual. Plant.* **27**, 187–205 (1977).
6. B. N. Ames and L. S. Gold, "Pesticides, Risk and Applesauce," *Science* **244**, 755–757 (1989).
7. J. J. Giovannoni and co-workers, "Expression of a Chimeric Polygalacturonase Gene in Transgenic *rin* (Ripening Inhibitor) Tomato Fruit Results in Polyuronide Degradation But Not Fruit Softening," *Plant Cell* **1**, 53–63 (1989).
8. S. F. Yang, "Regulation of Ethylene Biosynthesis," *HortScience* **15**, 238–243 (1980).
9. D. J. Huber, "The Role of Cell Wall Hydrolases in Fruit Softening," *Hort Rev.* **5**, 169–219 (1983).
10. L. D. Nooden, "The Phenomena of Senescence and Aging," in L. D. Nooden and A. C. Leopold, eds., *Senescence and Aging in Plants*, Academic Press, Orlando, Fla., 1988.

General References

J. Gross, *Pigments in Fruits*, Academic Press, Orlando, Fla., 1986.
R. C. Hoseney, *Principles of Cereal Science and Technology*, American Association of Cereal Chemists, Inc. St. Paul, Minn., 1986.
J. J. Jen, *Quality Factors of Fruits and Vegetables*, American Chemical Society, Washington, D.C., 1989.
A. A. Kader and co-workers, *Postharvest Technology of Horticultural Crops*, University of California, Davis, 1985.
L. D. Nooden and A. C. Leopold, eds., *Senescence and Aging in Plants*, Academic Press, Inc., Orlando, Fla., 1988.
M. O'Brien, B. F. Cargill, and R. B. Fridley, eds., *Principles and Practices for Harvesting & Handling Fruits & Vegetables*, AVI Publishing Co., Inc., Westport, Conn., 1983.
H. E. Pattee, ed., *Evaluation of Quality of Fruits and Vegetables*, AVI Publishing Co., Inc., Westport, Conn., 1986.
L. C. Peirce, *Vegetables—Characteristics, Production, and Marketing*, John Wiley & Sons., Inc., New York, 1987.
J. Weichmann, *Postharvest Physiology of Vegetables*, Marcel Dekker, Inc., New York, 1987.
J. G. Woodroof and B. S. Luh, eds., *Commercial Fruit Processing*, 2nd ed., AVI Publishing Co., Inc., Westport, Conn., 1986.

ROBERT L. SHEWFELT
University of Georgia
Griffin, Georgia

FOOD ENGINEERING

The largest business in the world is the supplying of food for the ever increasing population. This includes growing and harvesting, transporting, handling, storing, processing and preserving, packaging, distribution, and marketing. Over the last 50 years the various phases of the food business have grown from small family-type enterprises to gigantic, increasingly sophisticated, and integrated food-supply systems. The need for this change was dictated by the increasing concentrations of people in large urban areas where the very livelihood of people now is wholly dependent on large amounts of food being made available. Because volume production of foodstuffs is usually remote from these large consumer areas, efficient mass production and transportation of food supplies is required. The vertically integrated food industry that has grown from these demands, probably more than any other human activity, requires the support of diversified, well-rounded teams of scientists, engineers, economists, and marketing specialists. Food engineering, a relatively new field virtually nonexistent 20 years ago, was defined as a discipline in the 1950s by a number of engineering-educated food scientists who realized that few engineers were adequately trained for the growing complexity of the world food industry.

Food is composed of a large variety of complex chemical materials; it requires the basic chemical-engineering–defined unit operations and processes for all steps in the above-outlined sequence from the raw material to the table. Engineers in most other industries are basically applied physical scientists. Functional engineers in the food industry must also be extremely knowledgeable in the biological sciences as applied to the food industry including sanitation, spoilage, public health, and environmental control.

The modern world can no longer enjoy the small, isolated, and nonintegrated food-production system of the traditional Mom and Pop operations, each involved in a single phase of the food industry. The logistic requirements and the complexity of feeding a world in which many countries are unable to produce sufficient food for their own populations have created a demand for a more scientific- and engineering-based approach. Today the food engineer must have the ability to integrate all phases of food production into a smoothly functioning industry.

PROFESSIONAL RESPONSIBILITIES OF A FOOD ENGINEER

A principal responsibility of the food engineering profession, whether it be in industry or academia, is to develop and improve the relatively few basic steps involved in the processing of food and to apply this knowledge to commercial operations. This includes maintaining or creating the aesthetics or sales appeal (including acceptable sensory properties) of a product and insuring the safety of the food for human or animal consumption through good manufacturing processes (GMP) as prescribed by the U.S. Food and Drug Administration.

Food engineers are involved in a wide spectrum of activities, including not only operations but engineering feasibility and pre-engineering studies for updating existing facilities and planning new operations. Other responsibilities include designing process machinery and equipment layouts and integration, developing and improving unit operations and processes, ensuring food sanitation and safety, and maintaining and upgrading facilities. The planning operations also incorporate logistics of supplying, storing, and transporting raw materials or ingredients and finished products.

A growing responsibility in food engineering is the knowledge and application of computer process control techniques to many manufacturing phases of food production. The feedback of information in closed-loop systems is particularly important because the biological state of food is the primary factor in controlling the quality of the final product.

EDUCATIONAL REQUIREMENTS OF A FOOD ENGINEER

The basic principles governing the engineering aspects of food manufacturing and processing are the same as those required by all engineering disciplines. Normally engineers trained in other branches of the profession do not have the biological background necessary to deal with food handling and processing. This includes food spoilage, plant and food sanitation, by-product utilization, prevention of nutritional degradation during processing, design of processes to prevent environmental impact by processing wastes, and other areas involving the biological properties of food. The complex and often competing chemical, biochemical, and microbiological reactions and their effects on the physical properties that control many food processing operations must be considered in relationship to food handling, processing, packaging, storing, and transporting.

Some universities are attempting to teach food engineering in a four-year course culminating in a Bachelor of Science degree. However, it is most difficult to cover the required and extremely varied subject matter and produce a well-rounded food engineer in four years. The best background consists of a basic degree in chemical, mechanical, or civil engineering followed by graduate work in food science and food engineering. However, because food engineering covers several specialties, a trained food scientist can acquire the engineering knowledge by taking a judicious selection of engineering courses slanted toward the specific interest of the student such as material of construction, machine design, structures, unit operations and processes, environment, and computer courses in process control. It should be noted that these subjects generally include formal course work in chemical, civil, and mechanical engineering.

The applied science of food engineering requires knowledge of basic microbiology, chemistry, and the nutritional value of foods. Special courses in food engineering are necessary to combine these principles in the unit operations and processes required for all phases of food production, processing, and storage.

RESEARCH AND DEVELOPMENT AREAS IN FOOD ENGINEERING

In the United States there has been relatively little financial support given to food engineering programs for fundamental investigation of the complex composition and physical properties that must be considered in the design, construction, and operation of food processing operations. European countries have taken a world leadership position in the design, manufacturing, and marketing of food-processing equipment as the support given to process engineers parallels the development of innovative new technologies (eg, aseptic processing and controlled atmosphere packaging).

In general, engineering is the applied aspect of basic science. The basic laws of physics, chemistry, and biology as expressed in the language of mathematics show the relations between many of the elements composing a system whether it be a small piece of machinery or the planet on which humans live. However, due to many shortcomings of the precise knowledge of these relationships, balanced mathematical expressions are not determinable without the use of experimentally designed constants and secondary relationships that balance the equations and make them useful for application to practical situations. Hence, the application of basic engineering principles implies the use of experimentally determined information combined with the basic scientific laws. These relationships can then be practically applied to design a successful plant, to design and build machinery and equipment for economical operations, and to complete records and information of production performance so that future operations can be improved or developed.

The problems of using the engineer's tools in the food industry are all too traditional. In the past, food-processing and food-handling machinery and equipment were often based on history and tradition rather than efficient functioning, easy maintenance, and sanitary requirements. Modern day food production requires the reliance on basic engineering data, often either not available or in need of sophistication, to study and commercialize processing techniques. Some of the important areas of study for the food engineer include

1. The changes in weight and composition that occur in food during processing.
2. The pack densities, specific gravities, and compressibilities that are brought about in loading bins, storage tanks, and vats.
3. The safe loading depths for fresh or raw products during transportation and storage.
4. The flow properties for various foods in different forms and process states.
5. The properties of viscous material such as purées, pulps, slurries, etc.
6. The specific heat, heat capacity, and heat conductivity of foods in all states of preservation, especially when simultaneous mass and heat transfer occur.
7. The effect of particle size on various processes, which has a bearing on many processes such as spray drying, grinding, filtering surries and miscella, and disintegration operations.

Both mechanical and chemical unit operations and processes are necessary in food production. Mechanical operations include washing, sizing, sorting, grinding, mixing, polishing, grading, packaging, and materials handling. Processes for preserving (and often to improve texture and flavor) include a combination of chemical and physical unit operations and processes. These include (1) adding heat for cooking, pasteurizing, or sterilizing; (2) removing heat for cooling (refrigerating), freezing, or storing in a frozen state; (3) adding chemicals to control water activity, aid in fermentation or other processes (eg, enzyme hydrolysis), and improve sensory characteristics and acceptance; (4) removing water to control water activity; and (5) irradiating to pasteurize or sterilize. Most food processing uses a combination of these steps. For example, drying is normally carried out by adding heat to vaporize the water, whether it be under vacuum or atmospheric pressure. Of course, the primary purpose of these operations is to control or destruct microorganisms and their metabolic products and to reduce certain detrimental chemical and physical reactions. However, many undesirable chemical and physical reactions within foods, such as oxidative rancidity, vitamin deterioration, and certain texture changes, are often caused or accelerated by the processing techniques.

FOOD ENGINEERING AS RELATED TO OTHER DISCIPLINES AND PROFESSIONAL SOCIETIES

The field of food science and technology is composed of three related and overlapping disciplines that have similar basic background science requirements. However, today the four-year university backgrounds of a food scientist and a food technologist are identical. All are well versed in the basic sciences including general and organic chemistry, physics, mathematics through calculus, and microbiology as well the food science courses that stress these basics as applied to food. The separation comes after graduation and the start of a career. A food scientist concentrates on basic chemistry and microbiology of foods whereas the food technologist is more involved in the practical aspects of handling and processing foods. Many of the more basically oriented researchers specialize in the biological relationships of foods and, therefore, must have particularly strong backgrounds in biology, biochemistry, and microbiology.

The third partner on the team is the food engineer who is basically a food technologist with an engineering background. The engineering discipline became a necessary addition to insure that engineers involved in the food industry had the food science and technology background required for an active participation in the increasing sophistication of the world food industry.

The close relationship between food scientists, food technologists, and food engineers is shown by their professional societies. The Institute of Food Science and Technology (IFT) was founded in 1939. Several decades later a food engineering division was formed in the IFT. Many of the founders were food technologists who, through experience or further training, had gained a background in applied engineering principles as related to food processing. Today the food engineer has found more direct relationships with other engineers and are active in the Association of Agriculture Engineers, the American Society of Chemical Engineers, and several other societies. The food engineer is now established as a strong and indispensable member of the team.

GEORGE M. PIGOTT
University of Washington
Seattle, Washington

FOOD FERMENTATION

DEFINITION

Food fermentation is the study of microbial activity, usually anaerobic, on suitable substrates under controlled or uncontrolled conditions resulting in the production of desirable foods or beverages that are characteristically more stable, palatable, and nutritious than the raw substrate.

HISTORY

The history of food fermentation paralleled the development of microbiology and food microbiology. Traditionally many foods were prepared by fermentation but the reasons behind the success or failure of the processes were not known. After the work of Pasteur and others, who demonstrated that a specific microorganism (eg, yeast) acting on a suitable substrate (grape juice) will produce a desirable product (wine), the science of food fermentation began. Now many food fermentation principles and practices are well established and food companies can predictably produce consistently good-quality fermented products. With the advances in genetic engineering, old processes are being improved and new ones are being discovered. Also, many indigenous fermented foods (such as some Oriental foods and African tribal foods) and their processes are not well known and are areas for future investigation.

Although the principles of fermentation in many foods are understood in a laboratory setting, the scaling-up of these processes to commercially successful operations is complicated. A detailed account of all aspects of anaerobic fermentation, including methodology of anaerobic cultivation; mutation and genetic engineering of anaerobic bacteria; industrially important strains and pathways, biochemistry, kinetics, and transport in anaerobic fermentation; bioenergetics of anaerobic processes; data collection and analysis; mixed culture interactions; and design and application of anaerobic systems has been published (1).

MICROORGANISMS USED IN FERMENTED FOODS

The number of microbial species used in fermented foods is surprisingly small. Of the thousands of species of microorganisms in nature, only the following genera are well utilized by food fermentation industry (2).

Bacteria. Acetobacter, Streptococcus, Leuconostoc, Pediococcus, Lactobacillus, and Propionibacterium.

Yeast. Saccharomyces (especially *S. cerevisiae* and *S. carlsbergensis*), Candida, and Torula.

Mold. Aspergillus, Penicillium, Rhizopus, and Mucor.

TYPES OF FERMENTATION PROCESSES

Single-Culture Fermentation

The key to success of single-culture fermentation is to provide the culture with a sterile substrate and environment with no contamination during the fermentation process. Examples include wine making, beer making, bread making, single-culture fermented dairy products, and vinegar production.

Mixed Pure Culture Fermentation

Some products need a mixture of known cultures. The mixed pure cultures can be a controlled mixture of bacteria or bacteria with a combination of yeast and mold. Yogurt making and cheese making are good examples of this type of mixed fermentation.

Mixed Natural Culture Fermentation

In many parts of the world, especially in developing countries, the flora of indigenous fermented foods are mixed natural cultures. The interactions of those microbes are exceedingly complex and the success of such fermentations depends on following traditional processes and not on scientific principles because many of the responsible cultures have not been isolated and studied. The principal fermentation reactions in foods are listed in Table 1. This list does not include reactions in industrial fermentation of solvents, acids, etc.

Fermented Liquid and Semisolid Dairy Products

Cultured Buttermilk. Pasteurized skim milk inoculated with 0.5% mixture of *Streptococcus* (*Lactococcus*) *lactis, S. (L.) cremoris, Leuconostoc cremoris*, or *L. dextranicum*, incubated at 22°C for 14–18 h results in a product with pH 4.5 acid curds with pleasant aroma and flavor.

Table 1. The Principal Fermentation Reactions in Foods

Lactic Acid Fermentation

Homofermentative: $C_6H_{12}O_6$ (glucose) to 2 $CH_3CHOHCOOH$ (lactic acid)

Heterofermentative: $C_6H_{12}O_6$ (glucose) to $CH_3CHOHCOOH$ (lactic acid) + CO_2 + C_2H_5OH (ethyl alcohol)

Propionic Acid Fermentation

3 $C_6H_{12}O_6$ (glucose) to 6 $CH_3CHOHCOOH$ (lactic acid)
3 $CH_3CHOHCOOH$ (lactic acid) to 2 $CH_3CH_2CO_2H$ (propionic acid) + CH_3COOH (acetic acid) + CO_2 + H_2O

Citric Acid Fermentation

$CH_2COOHHOCCOOHCH_2COOH$ (citric acid) to 2 $CH_3COCOOH$ (pyruvic acid) to $CH_3COCHOHCH_3$ (acetylmethylcarbinol) + 2 CO_2
Acetylmethylcabinol can be oxidized to $CH_3COCOCH_3$ (diacetyl) or reduced to $CH_3CHOHCHOHCH_3$ (2,3,buytlene glycol)

Alcoholic Fermentation

$C_6H_{12}O_6$ (glucose) to 2 C_2H_5OH (ethyl alcohol) + 2 CO_2

Butyric Acid Fermentation

$C_6H_{12}O_6$ (glucose) to CH_3COOH (acetic acid) + $CH_3CH_2CH_2COOH$ (butyric acid) + CH_3CH_2OH (ethyl alcohol) + $CH_3(CH_2)_2CH_2OH$ (butyl alcohol) + CH_3COCH_3 (acetone) + CO_2 + H_2

Gassy Fermentation

2 $C_6H_{12}O_6$ (glucose) + H_2O to 2 $CH_3CHOHCOOH$ (lactic acid) + CH_3COOH (acetic acid) + C_2H_5OH (ethyl alcohol) + 2 CO_2 + 2H_2

Acetic Acid Formation (Oxidative)

C_2H_5OH (ethyl alcohol) + H_2O O_2 to CH_3COOH (acetic acid)

Sour Cream. Milk with 19% milkfat and 0.2% citric acid inoculated with 1% *Streptococcus* (*Lactococcus*) *lactis* and *S.* (*L.*) *cremoris* incubated at 22°C for 14–16 h results in a product with pH 4.5 that is, creamy and sour and good for baked potatoes and other foods.

Acidophilus Milk. Skim milk inoculated with 5% *Lactobacillus acidophilus* incubated at 38°C for 18–24 h results in an acidic (1% acid) sour milk product.

Sweet Acidophilus Milk. Skim milk inoculated with a large percentage of *Lactobacillus acidophilus* results in 5×10^6 *L. acidophilus* per mL of *live* culture in milk. The product *is not fermented* and must be kept cold (4°C) until sold. This product has no acid taste but is claimed to relieve gastrointestinal problems.

Bulgaris Milk. Whole milk inoculated with 2% *Lactobacillus bulgaricus* incubated at 38°C for 10–12 h results in a sour (2–4% acid) product, claimed to relieve gastrointestinal problems.

Yogurt. Milk with added 2–4% nonfat dry milk powder inoculated with 5% combination of *Lactobacillus bulgaricus* plus *Streptococcus thermophilus* (1:1 ratio) incubated at 45°C for 3–6 h or 32°C for 12–14 h results in a smooth, viscus gel with a delicate walnutlike flavor. The product must be chilled immediately. A variety of fruits, berries, and flavors can be added to the product. This is the fastest-growing product in the dairy industry.

Kefir. Goat, sheep, or cow's milk inoculated with kefir grain (a mixture of *Saccharomyces kefir, Torula, Lactobacillus caucasicus, Leuconostoc,* etc) incubated at 22°C for 12–18 h results in a product with 1% alcohol and 0.8% acid. The product is strained through cheesecloth. The milk foams and fizzles like beer.

Koumiss. Mare's milk (which has 5 times more vitamin C than cow's milk) inoculated with *Lactobacillus bulgaricus* plus *Torula* yeast incubated at 22°C for 12–18 h results in a drink with 1–2.5% alcohol and 0.7–1.8% acid.

THERAPEUTIC VALUE OF CULTURED MILK

In 1907 Metchnikoff wrote the book *Prolongation of Life* and claimed that drinking sour milk regularly gave people in Southeast Europe long life and good health, because premature old age was a result of absorption of toxins produced by bacteria in the large intestines. Large volumes of sour milk with live cultures changed the balance of microflora in the intestines so that toxic compounds were not produced and people lived longer. At a minimum, drinking live cultures of *Lactobacillus acidophilus* or *L. bulgaricus* is not harmful. For people taking antibiotics after major surgery, which tend to sterilize the intestines, these live organisms might very well be beneficial. People with gastrointestinal problems may find sour milk helpful in suppressing diarrhea. This is still a controversal area.

FERMENTED SOLID DAIRY PRODUCTS—CHEESE

Cheese is made by the coagulation or precipitation of milk protein by acid produced by starter cultures (*Streptococcus* (*Lactoccus*) *lactis* and *S.* (*L.*) *cremoris*) with the aid of rennin (from the stomach of calves) added to the milk. A food-grade dye, annatto, is often added to give the cheese the familiar yellow color. When milk proteins coagulate, the resultant liquid portion is called whey. From 10 lb of milk only 1 lb of cheese is made and 9 lb are discarded as whey. After the curd is formed, they can be poured into perforated molds lined with cheesecloth and the whey is drained. For harder cheese the curd is cooked, stirred, cut, and pressed. The more whey expressed from the protein mass the harder the resultant cheese will be. Salt is often added to the curd to give flavor as well as to prevent undesirable microorganisms from growing. Cheese is then ripened for several months or even years before consumption.

Classification of Cheese

There are literally hundreds of cheese varieties being made in the world. Most of the names are from the town or city in which the cheese originated. From a texture point of view cheese can be classified into the following:

Soft Unriped or Riped Cheeses. Cottage cheese is an example of soft unripened cheese. It can be made by acid produced by the starter cultures along with rennin or by direct set using food-grade acid such as meso-lactide and D-glucono-*delta*-lactone.

Soft Ripened Cheeses. Limburger cheese is an example of soft-ripened cheese. The soft cheese is placed on wooden shelves and a surface bacterium (*Brevibacterium linens*) grows on the surface of the cheese and produces a brownish red surface growth. Protein is broken down into ammonia and gives the strong flavor to this cheese. Camembert cheese is another soft cheese but it is ripened by a surface mold, *Penicillium camemberti,* with a mixture of bacteria. This famous French cheese is called the queen of cheese.

Semisoft Ripened Cheeses. Several mold-ripened cheeses are classified under this category. They are all ripened by *Penicillium roqueforti,* a mold that grows throughout the cheese mass. They are also called blue-veined cheese. Roquefort cheese is called the king of cheese. Other cheeses in this group are Stilton, Gorgonzola, and bleu.

Hard Cheeses. These cheeses are well ripened and hard. Among the most important cheeses are Cheddar, a bacterial-ripened cheese and Swiss cheeses, which are cheeses with eyes. The eyes are caused by CO_2 produced by *Propionibacterium shermanii* during the ripening stage.

Cheese Varieties and Descriptions

Principal cheese varieties and descriptions are as follows (3).

American. A term used to identify the group that includes Cheddar, Colby, etc, popularized in the United States.

Bleu. The French name for cheese similar to Roquefort but either not made in Roquefort, France, or not made from ewe's milk.

Blue. Roquefort-type cheese made in the United States and Canada.

Brick. Of American origin, made from whole milk, with a mild but pungent and sweet flavor.

Cheddar. Most important cheese, accounts for 75% of cheese made in the United States. The name came from the town Cheddar in the UK where it was first made. In addition, cheddaring is the name of a step in the manufacturing process (piling and repiling of curd) and the common shape. Most cheddars are ripened for 60 days and some for a year or more (sharp and very sharp cheddar).

Colby. Similar to Cheddar, has a softer body and more open texture.

Cottage. A soft unripened cheese made from skim milk. Flavoring materials such as peppers, olives, pimientos or garlic may be added. When more than 4% of fat is added it is called creamed cottage cheese.

Cream. A soft, rich, unripened cheese made of cream or a mixture of cream and milk.

Edam. Made from whole milk. Has a mild flavor and firm body. Usually shaped like a flattened ball and covered with red coloring or red paraffin.

Gouda. Similar to Edam except that it contains more fat and is usually packaged like Edam.

Limberger. A soft, surface-ripened cheese with a characteristic strong flavor and aroma.

Parmesan. A very hard cheese that will keep almost indefinitely. It is used as grated cheese on salads, soups, and with pasta.

Process Cheese. Made by grinding and mixing together by heating and stirring one or more cheese of the same or different varieties, together with an emulsifying agent into a homogenous mass. At least one-third of all cheese made in the United States, except soft, unripened cheese, is process cheese.

Process Cheese Food. Is made in the same way as process cheese, except that certain dairy products (cream, milk, skim milk, cheese whey, or whey albumin) may be added. At least 51% of the weight of the finished product must be cheese.

Process Cheese Spread. Made in the same way as process cheese food except that it contains more moisture and less fat and must be spreadable at a temperature of 21.1°C (70°F). Fruits, vegetables, or meats may be added.

Roquefort. A cheese made only in the Roquefort area of France from ewe's milk. Characterized by sharp, peppery flavor and blue-green veins throughout the cheese. Ripened principally by blue mold in the interior.

Swiss. A hard cheese with an elastic body, nutlike flavor. Best known by the holes or eyes that develop as the cheese ripens.

ALCOHOLIC BEVERAGES

In 1989 the per capita consumption of alcohol beverages in the U.S. was 2.4 gal of wine, 24 gal of beer, and 1.6 gal of spirits. Alcoholic beverages accounted for about $20 billion of value of food-processing shipments in 1985. Alcoholic beverages can be classified as fermented, not distilled and distilled.

Fermented, Not-Distilled Beverages

Wine. By definition wine is a fermented product from fruits. By far the most important wine is produced from grapes, although it can be produced from apples, pears, berries, and other fruits. *Vitis vinifera* and *V. labrusca* are the most important species. Grape growing is a science itself (viticulture) and wine making is called enology.

To obtain a good wine, the enologist must consider the species and varieties used, the climate and soil conditions that dictate the vintage of a good wine, and the time to pick the grapes to make the wine. Europe produces about 80% of the world's wine with North and South America producing about 14%; Africa, 4%; Asia, 1%; and Oceania, 1%. Whereas in Europe people consume about 30 gal of wine a year, the U.S. population consumes 2–3 gal/y. Grapes are first picked in autumn, when the amount of sugar in the grape stabilizes, then the grapes are stemmed and crushed to separate and remove the leaves and stems. The resultant materials, juice, skin, pulp, etc, is called *must* (L. *mustum*, "new wine"), which is the source of the wine.

Next SO_2 is added at 50–100 ppm to condition the must for its ability to inhibit undesirable organisms and its antifungal and antioxidative properties. The correct amount of sugar is added to define the final concentration of alcohol in the wine. The conditioned must will then be inoculated with a wine yeast (usually *Saccharomyces cerevisiae* var. *ellipsoideus* Montrachet strain) at 1% inoculum level. For red wine the must can be fermented on the skin, which allows extraction of the red anthocyanin by alcohol to give the red color to the wine. Alternatively the must can be hot pressed (the must is heated to 62.7°C and the juice is pressed out while hot), which also extracts the red color. To make white wine, white grapes can be used or, if red grapes are used, a cold press will yield a white juice. Pressing of the wine is an important step. In a traditional wine press, when the must is poured into the press some juice flows through the system without pressing. This Free-run juice yields the best wine. The harder the grapes are pressed, the poorer the quality of the wine produced. At this point it is important to check the acidity of the must. The liquid should be about pH 3.6 and 0.7–0.9 g acid/100 mL of juice. If adjustment is needed, tartaric acid is used to acidify the must and water is used to dilute the juice. This is the amelioration step. The juice is now ready for secondary fermentation.

Secondary fermentation involves putting the fermenting must into a closed container (cooperage) and applying an airlock to prevent oxygen from entering the system. Because heat is produced (exothermic reaction) the fermenting must is kept at 60–70°F for red wine and 55–65°F for white wine.

Fermentation will continue for two to six weeks depending on the amount of dryness (reduction of sugar) desired in the finished wine. Before the completion of fermentation, several racking steps may be done. Racking is a process in which the wine is separated from lees (dead yeast cells and insoluble materials at the bottom of the cooperage). The first racking is done after fermentation to desired dryness, (an appropriate amount of sugar is fermented to alcohol), followed by racking after dropping the insoluble tartaric in cold temperature (30–35°F), and finally, between clarification and bottling. Clarification can be achieved by the addition of fining agents such as gelatin, casein, bentonite, and polyclar AT. Enzymes may also be added to degrade pectin. The new wine is put into a wooden cooperage for aging. The biochemistry involved in aging is complex but at least one year of aging is necessary for a good wine to be produced. The exact time for bottling of wine is determined by the wine master of the particular winery. Wine can be pasteurized (55.5–65.5°C for a few seconds) but more often it is bottled without heat treatment, because heat tends to destroy flavor compounds. Wine continues to improve in the bottle. Ten years of aging is optimum.

Classification of Wines

There are hundreds of varieties of wines produced around the world and no classification can cover all of them. The following is a functional classification of wines.

Dry or Sweet Wines. In dry wine all fermentable sugar is converted to alcohol. Typically the wine has about 8–12% alcohol. Examples are Chablis, Riesling, Burgundy, and Chianti. Sweet wine has some unfermented sugar. The alcohol is about 13–15%. Examples are Tokay and Sauterne.

Fortified or Unfortified. In fortified wine more alcohol is added to the product. Some examples are Sherry, Port, Muscatel, and Champagne. Unfortified wine is a wine in which all the alcohol resulted from fermentation.

Sparkling or Still. Sparkling wine is bottled before fermentation has ceased and liberates effervescence after the bottle is opened. Champagne made in France is the true example of fermentation in the bottle. Injection of CO_2 into a wine to achieve effervescence is called crackling. Still wines are those in which fermentation ceased before bottling and no CO_2 exists.

Red or White. Fermentation of the skin of red grape will achieve red wine such as Claret, Burgundy, Port, and Chianti. Reisling is a good example of a white wine.

Generic Wine. Wines blended from several varieties of grapes are called generic wines. Examples are Burgundy, Chianti, Chablis, and Claret.

Variety Wine. Wines from predominating (at least 75%) one type of grape variety are variety wines. Examples are Cabernet, Pinot Noir, Zinfandel, and Chardonnay.

Vintage Wine. Vintage wine must have at least 95% of the juice coming from the year claimed.

Prevention of Wine Spoilage

Oxygen is the worst enemy of wine because it promotes the growth of *Acetobacter,* which oxidizes alcohol to acetic acid. Oxygen also exidizes wine color from white to heavy amber or from rich purple to tawny brown and affects the flavor of the wine. Dirty equipment and environment contributes to spoilage of wine directly by introducing microorganisms or indirectly due to residues of unwanted chemicals. Undesirable microorganisms during the fermentation process or in aging and storage can also contaminate wine.

Beer

Although beer is more complicated to produce, it was developed earlier in history than wine (some 6,000 years ago). By definition beer is an alcoholic drink derived from grains. The most common source of beer is barley with rice and corn used as adjuncts. The per capita consumption in Europe is about 32 gal of beer a year whereas in the United States the per capita consumption is about 25 gal. Beer is made by malting barley to obtain amylase then by mashing, which is a process in which more grain (adjunct) and water are added to the malt. The mixture is heated slowly to first achieve proteolysis (35°C for 1 h) and then starch hydrolysis (67–68°C for 20–30 min), finally, the enzymes are inactivated (mashing off at 75–80°C). The main purpose is to allow alpha- and beta-amylase to degrade amylose and amylopectin to yield glucose and maltose for fermentation by yeast, because yeast cannot degrade starch directly. The liquid (which can be considered as nutrient broth for yeast) is separated from the husks of the grains by lautering and sparging. The resultant clear fluid is called wort. Before addition of yeast for fermentation the wort is boiled and hops (*Humulus lupuus*) is added to give the bitter flavor of beer. After cooling to 8.8°C, yeast is added (pitching). *Saccharomyces carlsbergensis* is the yeast of choice for lager beer making. Fermentation continues for 3–4 days at which time a lot of foam is developed in the fermentation vessel (*Kräusen* formation). After 5 days the *Kräusen* collapses and after 10 to 12 days fermentation ceases. At this stage the beer is called green beer. Yeast cells can be recovered from animal feed and CO_2 is collected and later injected into finished beer for foam formation. The green beer is aged for about three months before bottled for sale. Beer can be sold in unpasteurized form (draft beer) and has a short shelf life. Most of the beer on the market is pasteurized at 57.2–60°C for 15–20 min in bottles and cans. Some beers are filtered and not pasteurized. Beer made in the above manner is called lager beer and the alcohol content is about 3.5%. Using other yeast (*S. cerevisiae*) ale can be made that is about 6% alcohol. European beers are heavier because they use only barley, which has more protein than the procedure used in the United States where barley is mixed with rice and corn adjuncts.

Light Beer and Dry Beer. By using enzymes such as glucoamylase, all of the starch in the wort is converted to fermentable sugar thus no sugar is left in the beer after fermentation. The beer becomes lighter in calories than the regular beer, which has residual starch. Dry beer started in Japan. Japanese beer has a sweeter taste than Western beer and it is not blended. By a process of superattenuated fermentation, prolonged fermentation occurs to get rid of all the sugars. In dry beer making no enzyme is added to reduce the carbohydrate.

Sake Making

Sake, a pale white drink, is closely related to Sinto religion and is often called rice wine. Technically it is a beer because it is made of rice. The rice is first inoculated with spores of *Aspergillus oryzae* (Koji) which is the source of amlyase to degrade starch. The Koji is then added to more rice along with yeast to achieve fermentation. The resultant liquid has a 14–20% alcohol content with no gas.

Beverage From Distilled Fermented Liquids

Because natural fermentation can achieve only about 12–15% alcohol and in extreme cases like sake, 20%, drinks of higher alcoholic content are achieved by concentrating alcohol through the process of distillation.

Whiskeys. Whiskey includes rye, bourbon, Scotch, Irish, Canadian Pisco, and vodka, which are distilled from fermented grains. Rums (Cuba, Puerto Rico, Philippines, Jamaica, Barbados, Martinique, and New England) are distilled from sugarcane and molasses. Tequila is distilled from agave. Brandies are distilled from a variety of fermented fruit juices such as grape (Cognac, Armagnac, and brandy), apple (calvados and apple jack), cherry (kirsch and cherry brandy) and plum (slivovitz and Micabelle). In addition, distillation with flavored compounds provide such drinks as gins, liqueurs, absinthe, aquavit, and bitters.

Fermented Products From Vegetables

Vegetables have low buffering capacity. After harvest vegetables have a heterogeneous population including *Pseudomonas, Bacillus, Chromobacterium,* and a variety of enteric organisms; lactic acid bacteria (the organisms responsible for fermentation) exists in relatively small numbers. However, once the material is placed under anaerobic conditions with the addition of salt, lactic acid bacteria quickly predominates and initiates a favorable fermentation process. Such is the case in sauerkraut making. Shredded cabbage has 3–6% sugar (glucose, fructose, and sucrose). The first group of organisms growing are *Enterobacter cloacae, Erwinia herbicola,* and *Leuconostoc mesenteroides* (a heterofermentative lactic). As the lactic acid increases the enterics dies off and *Leuconostoc mementeroides* predominates producing favorable compounds for the development of *Lactobacillus plantarum* (a homofermentative lactic). This organism continues to produce lactic acid to a level of 1.5–2.0% and completes the fermentation process. Vegetable fermentation represents a form of natural fermentation because, regardless of the type of starter culture used, *Lactobacillus plantarum* always ends up as the principal organism in a properly treated fermentation vet.

The fermentation of cucumbers in the making of dill pickles and salt stock pickles is another example of vegetable fermentation. The microbial succession of cucumber fermentation closely follows the pattern achieved in sauerkraut fermentation. Almost any kind of vegetable can be fermented. In the Orient, mixed vegetable fermentation is popular.

Fermented Products From Meats

Since ancient times many types of fermented meat product have been developed. However, scientific understanding of fermented sausage resulted only after the advancement of meat microbiology in the last 100 years. Currently most meat is consumed unfermented. However, many varieties of meat products are preserved in the form of sausages (such as frankfurters, bologna, Vienna sausage, loaves and luncheons meats, bratwurst, brockwurst, and braunschweiger). Only 6% of U.S. sausages are fermented. They are either fermented sausages (pepperoni and salami) or semidry fermented sausages (cervelat and Thruinger). The basic ingredients for fermented sausage are ground meat (one type or a mixture of a few types), salt (2.5–3.5%), sugar (1%, for fermentation), spices, nitrites, ascorbate and erythorbate, and lactic acid bacteria. The traditional process of applying cultures is to let chance contaminants start the fermentation. This process is highly unreliable because ground meat contains many types of microorganism other than lactic acid bacteria. The second process is by back-slopping, when the batter from a previous successful batch is added to a fresh sausage batch to initiate the fermentation. This method is more reliable than the traditional method but is still subject to unwanted variations. The third and most reliable process is by the addition of starter cultures. Lactic acid bacteria such as *Pediococcus* and *Lactobacillus* have been successfully used to make fermented sausages. These bacteria in the presence of sugar and the anaerobic environment provided by the casing of the sausage quickly produce lactic acid and also develop the tangy flavor typical of the taste of fermented sausage. Fermented sausages can also be smoked to add distinctive flavors.

Fermented Cereal Foods—Bread

According to many people of the world, bread is the staff of life. Bread is only one type of fermented cereal product. Other products include rolls, Danish pastry, crackers, doughnuts, pretzels, etc. The value of bread and bakery products was about $21 billion in 1985 in the United States, the majority of which was bread. The definition of a bread is a product of moistened, kneaded, fermented, and baked meal of flour (mainly from wheat) with appropriate added ingredients. Yeast (*Saccharomyes cerevisiae*) is the organism to facilitate the fermentation process to make bread. Yeast can be supplied to the baking industry in a compressed, bulk, or active dried yeast form. Bread is made by many different processes but the most popular method is the sponge-dough method. After all the ingredients are mixed for the sponge part (65% of total flour, water, yeast, and yeast nutrients) the sponge is kept at 25°C for 4.5 h for the first fermentation. The volume increases to four to five times and the temperature increases to 30°C due to the exothermic reaction of fermentation. The rest of the ingredients (flour, water, sweetner, fat, dairy product, crumb softener, rope and mold inhibitor, dough improver, and enrichment) are then added to make the final dough. The dough is mixed at 72 rpm to produce a smooth cohesive dough that has a glossary sheen. The mixed dough is allowed to rest for 20–30 min then the dough is divided and rounded and put into molds. In the proof box (35–43 °C at 80–90% humidity for 60 min) the dough expands to the desired volume due to the fermentation of sugar with the development of CO_2, which is trapped by the elastic gluten of the dough. Baking is the last step in bread making. During baking the heat of the process expands the trapped gas in the dough matrix and causes the dough fabric to ovenspring. Enzymes are active until the dough reaches 75°C, when the gluten matrix coagulates and dough structure is set into the form of the bread. When the bread surface reaches 130–140°C, sugar and soluble protein react chemically to give the crust color and textures. The center of the bread does not exceed 100°C. After cooling, slicing, wrapping, and distributing, the bread reaches consumers. In many places in Europe bread is baked a couple of times a day and consumers eat bread fresh from the oven. In the United States, breads in supermarkets are usually half a day to a day old before consumers purchase the products. One of the problems of the science of bread making is staling. After years of research the exact cause of staling is still unresolved.

Production of Vinegar

Vinegar is literally a result of souring (*aigre*) of wine (*vin*). The origin of vinegar no doubt followed the production of wine, because a bad batch of wine will result in some form of vinegar. Vinegar has been used as a flavoring agent, food preservation agent, and even as medicine. Both Eastern and Western cultures have records of vinegar in ancient history. Although vinegar production is always treated as a part of food fermentation it is, in reality, an oxidative process. In the presence of molecular oxygen and *Acetobacter aceti,* or related species, alcohol is oxidized to acetic acid. Historically vinegar was made by the let-alone process or the field process; poor quality wine was allowed to sit in the open air for oxidation to occur. Currently the methods to make vinegar are the trickling process and the submerge culture generator. In the trickling process, wine is trickled into a Frings-type generator where the liquid comes in contact with wood shavings thereby exposing it to large amounts of oxygen. The alcohol is then oxidized by the *Acetobacter* in the system. At the bottom of the system, the collected liquid contains about 12% vinegar. This process takes about two weeks. The submerged culture system takes about 35 h to achieve the same concentration of vinegar. In this closed system, air is pumped into the generator containing wine and the organisms in the mixture actively oxidize alcohol to vinegar. Vinegar can be pasteurized to render it more stable in storage.

Mold-Modified Foods

In many parts of the world the indigenous fermented products are modified by molds along with yeast and bacteria. These products are usually the result of uncontrolled, naturally mixed culture fermentation. Much research needs to be done on these foods to ascertain the microbes involved in the process, the safety of the products, and the economical feasibility of large-scale production in modern industrial plants. Some of the more important foods in this category are listed below.

Soy Sauce. A product of the Orient from fermentation of whole or defatted soybean and soybean products along with roasted wheat using a combination of *Aspergillus oryzae, Pediococcus halophilus, Lactobacillus delbrueckii, Torulopsis versatilis,* and *Saccharomyce rouxii* resulting in a dark reddish liquid, salty taste, and an important flavoring agent.

Miso. This is a product of the Orient from fermentation of whole soybeans mixed with rice or barley using *Aspergillus oryzae, Saccharomyces rouxii, Torulopsis etchellsii,* and *Pediococcus halophilus* to produce a dark reddish, smooth paste with a strong flavor and salty taste.

Hamanatto. A product of the Orient from fermentation of whole soybeans and wheat flour by *Aspergillus, Streptococcus,* and *Pediococcus*. These black soft beans have a salty flavor and are used as a condiment.

Sufu. A product of China by fermentation of soybean curd (tofu) using *Actinomucor elegans* and *Mucor dispersus* resulting in salty and strongly flavored cream cheese–type cubes.

Tempeh. A product of Indonesia and vicinity using whole soybeans inoculated with *Rhizopus oligosporus* resulting in a soft bean cake bound by mycelia of the mold. It has a clean taste and can be fried, cooked, or eaten as is.

Natto. A product of Japan by fermentation of whole soybeans with *Bacillus subtilis* var. *natto,* results in beans covered with viscous, sticky polymers produced by the bacteria. Strong ammonia odor, eaten with or without further cooking.

Other mold-modified foods include *bongkrek* (coconut cake, Indonesia), *ontjom* (peanut cake, Indonesia), *lao-chao* (glutinous rice, China and Indonesia), *ang-kak* (Red rice, China and the Philippines), *idli* (rice, India), *doza* (rice, India), *trahana* (parboiled wheat, meat, and yogurt, Turkey), *injera*, (teff, Ethiopia), *kishk* (wheat and milk, Egypt and Middle East), *gari* (cassava roots, West African and Nigeria), *ogi* (maize, Benin and Nigeria), *mahewu* (maize, South Africa), *pozol* (maize, Mexico), and many others.

Other food products are also fermented such as coffee, cacao, vanilla, tea, citron, ginger, mead, fish sauces (*nu'óc mǎn*), poi, olives, *pidan* (hundred-year-old egg), etc. The field of food fermentation can even include such processes as production of single-cell protin, lactic acid, citric acid, glutamic acid, lysine, antibiotics, vitamins, lipids, ascorbic acid, etc. Many areas of food fermentation overlap with industrial fermentation and biochemical engineering.

In conclusion, food fermentation is an important field of study. Although much has been learned about some well-known fermentation processes such as beer, wine, cheese, and bread, much needs to be studied concerning mold-modified foods, indigenous foods, and improvement of existing processes. With the rapid advancement of genetic engineering, some dramatic changes in the field of food fermentation should occur in the near future.

BIBLIOGRAPHY

1. L. E. Erickson and D. Y. C. Fung, ed. *Handbook on Anaerobic Fermentations,* Marcel Dekker, Inc., New York, 1988.
2. S. E. Gailliland, *Bacterial Starter Cultures for Food,* CRC Press, Boca Raton, Fla., 1985.
3. G. Sauder, *Cheese Varieties and Descriptions,* USDA Agricultural Handbook No. **54,** Washington, D.C., 1978.

General References

M. A. Amerine, H. W. Berg, R. E. Kunkee, C. S. Ough, V. L. Singleton, and A. D. Webb, *Technology of Wine Making,* 4th ed., AVI Publishing Co., Inc., Westport, Conn., 1980.

J. C. Ayres, O. Mundt, and W. E. Sandine, *Microbiology of Foods,* W. H. Freeman & Co., New York, 1980.

G. J. Banwart, *Basic Food Microbiology,* AVI Publishing Co., Inc., Westport, Conn., 1979.

L. R. Beuchat, *Food and Beverage Mycology,* AVI Publishing Co., Inc., Westport, Conn., 1978.

D. R. Buege and R. C. Cassens, *Manufacturing Summer Sausage,* Extension. No. **A 3058,** University of Wisconsin, Madison, 1980.

W. Crueger and A. Crueger, *Biotechnology: A Textbook on Industrial Microbiology.* Sinauer Asso. Inc., Sunderland, Mass., 1984.

D. E. Emmons and S. L. Tucky, "Cottage Cheese and Other Cultured Milk Products," *Pfizer Cheeses Monograph* **3,** (1967).

J. M. Jay, *Modern Food Microbiology,* 3rd ed., Van Nostrand Reinhold Co., Inc., New York, 1986.

F. Kosikowski, *Cheese and Fermented Milk Foods,* Edwards Brothers Inc. Ann Arbor, Mich., 1977.

P. Z. Margalith, *Flavor Microbiology,* Charles C Thomas, Springfield, Ill., 1981.

N. F. Olson, "Ripened Semisoft Cheese," *Pfizer Cheese Monograph* **4,** (1969).

C. S. Pederson, *Microbiology of Food Fermentation,* 2nd ed., AVI Publishing Co., Inc., Westport, Conn., 1979.

H. J. Peppler and D. Perlman, *Microbial Technology,* Vol. 11, Academic Press, Inc., Orlando Fla., 1979.

G. W. Reinbold, "Italian Cheese Varieties," *Pfizer Cheese Monograph* **1,** (1963).

G. W. Reinbold, "Swiss Cheese Varieties," *Pfizer Cheese Monograph* **5,** (1972).

G. W. Reinbold and H. L. Wilson, "American Cheese Varieties," *Pfizer Cheese Monograph* **2,** (1965).

R. K. Robinson, *Dairy Microbiology,* Vols. 1 and 2, Elsevier Applied Science Publishers, Ltd., Barking, UK, 1976.

J. H. Silliker, R. P. Elliott, A. C. Baird-Parker, F. L. Bryan, J. H. B. Christian, D. S. Clark, J. C. Olson, Jr., and T. A. Roberts, *Microbial Ecology of Foods,* Vols. 1 and 2, Academic Press, Inc., Orlando, Fla., 1980.

P. F. Stanbury and A. Whitaker, *Principles of Fermentation Technology,* Pergamon Press, Elsmford, N.Y., 1984.

K. H. Steinkraus, *Handbook of Indigenous Fermented Foods,* Marcel Dekker, Inc., N.Y., 1983.

R. P. Vine, *Commercial Winemaking: Processing and Controls,* AVI Publishing Co., Inc., Westport, Conn., 1981.

D. I. C. Wang, C. L. Cooney, A. L. Demain, P. Dunnill, A. E. Humprey, and M. D. Lilly, *Fermentation and Enzyme Technology,* John Wiley & Sons, Inc., New York, 1979.

H. H. Weiser, G. J. Mountney, and W. A. Gould, *Practical Food Microbiology and Technology,* AVI Publishing Co., Inc., Westport, Conn., 1971.

G. H. Wilster, *Practical Cheese Making,* 12th ed., Oregon State University Book Store, Corvallis, 1974.

DANIEL Y. C. FUNG
Kansas State University
Manhattan, Kansas

FOOD FREEZING

Freezing is one of the most common methods for preserving foods and has the potential to maintain the quality of foods close to their fresh status. Freezing, thermal processing, and dehydration are the three methods most commonly used for long-term preservation; in most instances, freezing is regarded as superior to the other two methods (1). Properly conducted freezing can retain the flavor, color, texture, and nutritive value of food by reducing the activity of microorganisms and enzyme systems.

Refrigeration was being used back in ancient times long before its commercial application. One of the earliest reports on refrigeration was in the "She King," Chinese poetry of 1100 BC (2). At first, natural ice obtained from mountain snows, frozen rivers, and ponds in the winter was used for refrigeration. Farmers and fishermen have preserved their meat, game, and fish by using nature's refrigeration. The application of cold preservation was later extended to eggs, vegetables, and fruits.

REFRIGERATION HISTORY

The early history of the development of refrigeration has been reviewed and summarized (1–3). The important events in the development of the U.S. frozen food industry have also been tabulated (1). Some of the major events are presented in Table 1.

The utilization of cool conditions for food storage is older than history. Studies on refrigeration and freezing effects on foods and insects were reported as early as 1683 by Boyle. However, fish was one of the first foods to benefit from refrigeration. In 1838, the captain of a Gloucester, Mass., fishing smack used ice to preserve a catch of halibut until it could reach port and be marketed (4). Natural ice harvested in the winter was used for refrigeration at first until the development of thermodynamics in the 1800s; this development provided the fundamentals of mechanical systems for ice making.

The first major step in the development of mechanical refrigeration was forced by a need rather than for food preservation. John Gorrie, who set up a hospital in Apalachicola, Fl., used ice to help cure malaria. In 1844, after much experimentation, he developed and put into use his air/compression/expansion device that employed the Joule–Thompson effect to cool the air in a cylinder (2). Modern commercial ice-making plants went online about 1876. Mechanical warehouse refrigeration first became of practical importance around 1890, however, large-scale construction of refrigerated warehouses did not begin until about 1915. The early refrigeration systems were too large to be put into individual homes. In the late 1930s and early 1940s, the development of small-scale compressors and manufacturing of relatively cheap refrigeration units made home refrigeration possible. The need for a better and safe refrigerant also led to the development of Freon by the Du Pont Company.

Commercial consumer-pack frozen foods started in 1923 when Clarence Birdseye entered the field and created a frozen food business. In 1930, the list of commercial frozen foods consisted mainly of fish, small fruits, poultry, meats, eggs, and a few vegetables. The major development for frozen food quality improvement is attributed to M. Joslyn and W. V. Cruess of the University of California, Davis, for introducing a blanching treatment before freezing to destroy enzymes. During the past 50 years, frozen foods have made tremendous gains due to the research efforts of frozen food processors, packaging manufacturers, and scientists at universities and federal government agencies.

FUNDAMENTALS OF FREEZING

Freezing is a process that reduces the temperature of a product to 0°C or lower. For long-term preservation, food freezing usually reduces the temperature of food to −18°C

Table 1. Important Events in the Development of the United States Frozen Food Industry[a]

Year	Event
500 BC–1800 AD	Cooling achieved by use of snow, natural ice, air in cold climates, evaporative cooling of water, and radiative cooling.
1820	By this time, natural ice had come into general use as an article of commerce and was used on a large scale for food preservation.
1851	Jacob Fussell of Baltimore, Md., first sold ice cream on a significant commercial scale in the United States.
1861	Thomas S. Mort established what is believed to be the world's first cold-storage plant in Darling Harbor, Australia.
1864	Ferdinand Carre patented an ammonia compression machine in France.
1870	Ammonia compression machines were brought to a level of practicality almost simultaneously by Dr. Carl Linde in Germany and David Boyle in the United States.
ca 1880	Ammonia compression machines and insulated rooms began to be used in the United States.
ca 1916	Scientific work on the methodology of freezing foods began in earnest.
1923	The Quick Freezing industry had its origin with the founding of a freezing company by Clarence Birdseye.
1929	M. A. Joslyn and W. V. Cruess reported on the need to blanch vegetables prior to freezing.
	The Birdseye organization was bought by Postum Co. (now General Foods Corp.) marking the real beginning of marketing of frozen foods through retail stores in the United States.
1930	By this time, mechanical refrigeration had assumed an almost indispensable part of the food distribution system in the United States.
1949	Mechanically refrigerated railroad cars came into use about this time.
Early 1960s	Fluidized-bed freezers and individually quick-frozen (IQF) foods began to assume a position of importance in the United States.
1962	Liquid nitrogen food freezers were first used commercially.
1968	Freezant-12 freezers were first used commercially.

[a] Ref. 1.

or lower. The primary component of food is water. When the temperature of a food is reduced, it will first pass its freezing point and become supercooled. No ice is formed at this stage. Then the formation of the ice nucleus releases heat to bring the supercooled temperature back to the freezing point. Once ice formation has begun, the ice crystals propagate in parallel with heat removal. The rate of heat removal will determine the size and location of ice crystals.

At a high freezing rate, the propagation rate of the initial ice crystals is slower than the rate of heat removal, which will then cause the development of supercooling. Accompanying this supercooling is an increased frequency of nucleation. The result is an increase in the number of ice crystals formed (5). For a slower freezing process, the propagation of ice can keep pace with heat removal resulting in formation of fewer nuclei and fewer, larger ice crystals. Fast freezing also does not allow sufficient time for water translocation from within cells, so ice forms within the cells. However, slow freezing generally causes ice crystals to form at extracellular locations. Generally, fast freezing forms intracellular ice crystals and causes less damage than extracellular ice crystals.

Frozen food quality is not totally controlled by the initial freezing process. During frozen storage, the process of maturation begins. Small ice crystals gradually reduce their size, whereas large ice crystals grow in size. The number of ice crystals decreases and their average size increases with time (5). The advantages of high quality achieved by a fast freezing rate will gradually be reduced during storage.

Another problem particularly associated with frozen vegetables is the development of tough texture and hay-like flavor and aroma due to enzymatic activity during storage. Blanching treatment by boiling water or steam immediately before freezing can inactivate enzymes and retain the high quality of frozen vegetables.

REFRIGERATION PRINCIPLES

There are a number of ways to reduce the temperature of foods. The two most common methods are the vapor compression and absorption refrigeration cycles. The vapor compression refrigeration system utilizes mechanical energy and is most commonly used in warehouse operations. The transfer of heat in an absorption refrigeration system is often achieved by using a steam engine.

Thermodynamics

To understand the operation of a refrigeration system, it is necessary to understand the principles of thermodynamics. Thermodynamics is the study of energy, its transformations, and its relation to states of matter. The unit of measurement representing the amount of heat is the calorie. A calorie is the amount of heat required to raise the temperature of 1 g of water 1°C.

Refrigeration involves removal of heat from an object or area. Heat always flows from a warm to a cold body and follows the first and second law of thermodynamics. The first law of thermodynamics is often called the law of the conservation of energy in which energy may be neither created nor destroyed. This law can be stated in many different ways. In one way, although energy assumes many forms, the total amount of energy remains constant, and when energy disappears in one form it appears simultaneously in other form(s) (6). This implies that different forms of energy are interconvertible, and a definite numerical ratio exists for each conversion. Heat is a form of energy that transfers from one system to another by temperature difference. The second law of thermodynamics states that no system can receive heat at a given temperature and reject it at a higher temperature without receiving work from the surroundings. A refrigeration system transfers heat from a low temperature region to a high temperature region. This transfer can only be achieved when work is done as stated in the second law of the thermodynamics.

The characteristic thermodynamic properties such as internal energy and enthalpy are not directly measurable. However, at states of equilibrium these properties are functions of measurable parameters like pressure, temperature, and volume. An ideal gas is a hypothetical gas in which no molecular forces exist and also satisfies the relationship of $PV = RT$. For an ideal gas, the internal energy is independent of pressure (P) and volume (V) but is a function of temperature (T). The concept of an ideal gas is extended to the development of simple equations that are also applicable to actual gases. Several thermodynamic relationships important to understanding a refrigeration system are described as follows:

1. The constant-temperature (isothermal) process. In this process temperature remains constant and pressure (P) is inversely related to volume (V).
2. The constant-pressure (isobaric) process. In this process pressure remains constant and temperature is related to volume.
3. The adiabatic process. In such a process no heat transfer takes place between the system and its surroundings.

Thermodynamic Diagrams

A thermodynamic diagram is one in which the temperature, pressure, volume, enthalpy, and entropy relationships of a substance are represented by a single chart. The most commonly used diagrams are temperature-entropy and pressure-enthalpy.

Calculations of heat and work in a mechanical refrigeration system require knowledge of the thermodynamic properties of the refrigerant. One useful way to present this information is the pressure-enthalpy diagram. The diagram is composed of information pertaining to eight constant properties. A skeleton pressure-enthalpy diagram is presented in Figure 1. The lines for saturated liquid (**g**) and saturated vapor (**h**) and the critical point are marked. Between the saturated liquid and saturated vapor lines is the two-phase region where liquid and vapor phases coexist. Along any vertical line such as (**a**) is the constant enthalpy line where refrigerant has the same amount of heat. The horizontal line (**b**) is the constant pressure line. Line (**c**) is the constant temperature line. In

1044 FOOD FREEZING

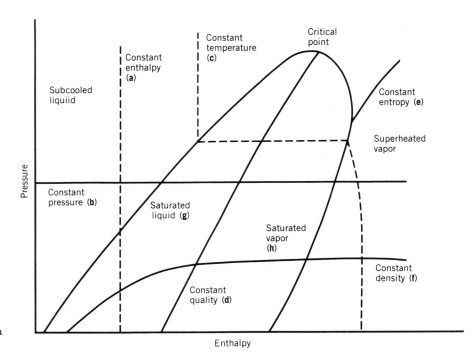

Figure 1. Diagrammatic representation of a pressure–enthalpy diagram.

the two-phase region there is a constant quality line (**d**). On this line the refrigerant has the same liquid to vapor ratio. Fifty percent ($x = 0.5$) quality means that the refrigerant is 50% liquid and 50% vapor. Another important property is the constant entropy line (**e**), which has been used for compression calculation. Constant density information (**f**) is also available on the diagram. Diagrams similar to the pressure-enthalpy diagram presented in Figure 1 are available for most of the refrigerants and are very useful in selecting the refrigerant for a particular system.

Compression Refrigeration Cycles

The Carnot cycle was developed by Sadi Carnot in 1824 to describe a device that produces work from heat in a cyclical process. A refrigeration cycle is a reversed Carnot cycle that uses mechanical work to remove heat from a system in a cyclical process. The vapor compression systems absorb heat as latent heat of vaporization of a refrigerant. The vaporized refrigerant can be recycled by compression and condensation and returns to a liquid state. This process can be illustrated in temperature-entropy or pressure-enthalpy diagrams (Fig. 2). The refrigerant leaves the receiver (point 4) as a high-pressure, medium-temperature, saturated liquid and enters the expansion valve where it expands adiabatically to point 1. At this stage the liquid refrigerant expands to evaporator pressure, and some of the refrigerant flashes and cools the remainder of the liquid refrigerant to evaporator temperature. Then the low-pressure, low-temperature refrigerant enters the evaporator where it absorbs heat at constant pressure to the saturated vapor state (point 2). The low-pressure, low-temperature, saturated vapor is then compressed adiabatically to point 3. The high-temperature, high-pressure, superheated vapor then enters the condenser where it is first desuperheated and then condensed at constant pres-

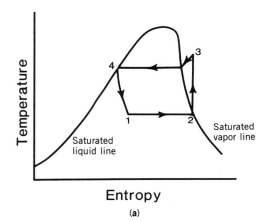

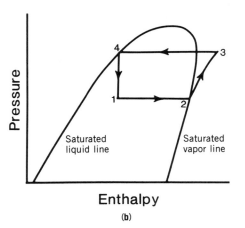

Figure 2. Thermodynamic diagrams of a basic vapor compression cycle: (**a**) temperature–entropy system; (**b**) pressure–enthalpy diagram: 1 to 2 is an isothermal expansion, 2 to 3 is an adiabatic compression, 3 to 4 is an isothermal compression, 4 to 1 is an adiabatic expansion.

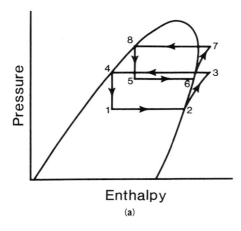

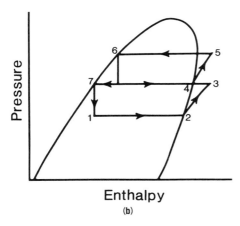

Figure 3. Pressure–enthalpy diagram of multistage systems: (a) cascade system: 1 to 4 is a low pressure–enthalpy cycle, 5 to 8 is a high pressure–enthalpy cycle; (b) two-stage compressor system: 2 to 3 is a first stage compression, 4 to 5 is a second stage compression.

sure. At the end of the condenser is the completion of the theoretical single-stage cycle.

Two other refrigeration systems can be used in order to achieve an extremely low temperature. One way is by connecting two systems in series (cascade system) allowing the evaporator of the high pressure-enthalpy cycle to remove heat from the condenser of the low pressure-enthalpy cycle (Fig. 3a). Cascade systems normally use different types of refrigerants in each of the systems.

Another way to achieve super-low temperature is to use two compressors connected in series to pump the very low vapor pressure refrigerant to the condensing pressure (Fig. 3b). Compressing the refrigerant in two stages results in a more efficient compression operation (lower compression ratio at each stage).

Absorption Refrigeration Cycles

An absorption refrigeration system uses heat energy instead of mechanical compression to complete the refrigeration cycle. The condenser, receiver, and evaporator are similar to those used in the vapor compression system except that the compressor is replaced by a heater and a generator. The refrigerant vapor from the evaporator is absorbed by an absorbent (liquid carrier). The refrigerant–absorbent solution is then pressurized in the solution pump to the condenser pressure. Heat provided from a heater then distills the refrigerant from the refrigerant–absorbent solution at the elevated pressure. The high-pressure refrigerant vapor then passes to the condenser as for the vapor compression system. The most common combination is ammonia and water for refrigerant and absorbent, respectively.

Thermoelectric Cooling

Thermoelectric cooling uses the same principle as thermocouples, where two dissimilar metal wires are soldered together at the ends. If one of the wires is cut and the cut ends connected to a source of direct current, the resulting current flow causes heat to be absorbed at one junction and released at the other end. Using this principle, the wire couple can then be used as a refrigerating device by placing the heat-absorbing end in the space to be cooled and using air or water to remove heat released at the other end (7).

Direct Contact with Refrigerant

In the refrigeration cycles, the refrigerant travels in a closed system. A cooling effect is achieved through some kind of heat exchange surface. Cooling efficiency can be improved if refrigerant can directly contact the material to be cooled. Direct contact with ice or cold water is a means of temperature reduction. However, a product can only be cooled down to 0°C using ice.

Secondary Refrigerants. For this type of system, heat is transferred to a secondary coolant and then removed by another mechanical refrigeration system. This secondary fluid has its freezing point below the temperature to which it is required to freeze the product. Secondary refrigerants can be brines (calcium chloride or sodium chloride solutions), glycols, or the CFC, dichlorodifluromethane. Propylene glycol or brines can only be used in contact with prepackaged products to avoid product contamination. Liquid food freezing systems (LFF) using dichlorodifluromethane (permitted for use in direct contact with food providing it is food grade) were introduced in the 1960s and have been used successfully on corn on the cob, shrimp, green beans, and peas. This method freezes faster even than cryogens and represents the ultimate in speed of freezing. However, CFC has the potential to cause ozone depletion. All LFF systems are in the process of being replaced due to the phaseout of all CFCs by the end of twentieth century.

Cryogenic Refrigeration. Cryogenic means very cold, and some of the most commonly used cryogenic refrigerants and their properties are listed in Table 2. Due to their low temperature and high heat absorption during vaporization, the cryogenics produce rapid freezing in the materials to be frozen. Carbon dioxide, in liquid or solid form or liquid nitrogen are the cryogens used in food freezing. They can be used either for crust freezing or complete freezing. Crust freezing, by immersing in the cryogen for a few seconds, is employed for chilled poultry before distri-

Table 2. Thermal Properties of Some Cryogenic Refrigerants

Property	Cryogenic Refrigerant	
	Liquid Nitrogen	Liquid Carbon Dioxide
Boiling point, °C	−195.4	−78.6
Specific heat of vapor, kJ/kg·K	1.03	0.837
Heat of vaporization, kJ/kg	199	573
Heat removal to −18°C, kJ/kg	384	623

bution and for delicate fruits (like raspberries) so that they can withstand the turbulence inside a blast freezer. Care must be taken in complete freezing using a cryogen as many products will become fragile at the very low temperatures and shatter as the central portion expands on freezing. Liquid nitrogen freezing tunnels use rest periods between the intervals of spraying with the cryogen to allow the food to freeze evenly. Although these total loss refrigerants are expensive in operating costs, they have the advantages of low capital cost and maintenance and requiring comparatively unskilled operators.

REFRIGERATION SYSTEMS

Mechanical System

The major components of a vapor compression system are compressor, condenser, evaporator, and controls. Any standard type of compressor can be used for the refrigeration compressor; the most common type is the reciprocating compressor. Rotary and centrifugal compressors have also been used.

The condenser and evaporator are simply heat exchangers. The condenser usually serves to reject heat from the system. Condensers can be classified according to the cooling medium as (1) water cooled, (2) air cooled and (3) evaporative (air and water cooled). A water-cooled condenser uses water as the cooling medium and has three configurations of heat exchange arrangement: shell and tube, shell and coil, and tube in tube. An air-cooled condenser has the high-pressure, high-temperature refrigerant circulating through the condenser coil and uses air moved by free convection, wind, or fans to remove heat from the refrigerant. An evaporative condenser is a combination of a water-cooled and air-cooled condenser that rejects heat through the evaporation of water into an air stream traveling across a condenser coil.

Absorption System

Absorption machines are heat-operated refrigeration machines. The size of an absorption machine chosen is based on the flow rate and the inlet and exit temperatures of the liquid to be chilled.

Capital investment and maintenance of the compressor are significant costs associated with a vapor compression system. For absorption refrigeration the liquid pumping operation is less expensive than compressing a vapor. Absorption systems operate quietly, have few moving parts, and can be operated in areas where electricity is unavailable. However, absorption refrigeration requires more equipment for both vapor absorption and refrigerant regeneration steps. For the ammonia–water system, a distillation column is needed for removal of water vapor from ammonia. Heat removal at the absorption step and heating at the regeneration step add to the operation cost. As a result, the absorption refrigeration system is economically suitable on an industrial scale when the supply of waste heat is free.

Thermoelectric Cooling

The development of semiconductor materials in recent years has provided materials that give a greater thermal effect than ordinary thermocouple wires. A thermoelectric cooling system has the advantage of no moving parts and is ideally suited for use aboard aircraft or space vehicles (7). However, the application is limited due to excessive system cost and low efficiency.

Direct Contact with Refrigerant

Secondary Refrigerants. For the direct expansion refrigeration systems, the refrigerant is circulated to various evaporators located in the cold rooms. This arrangement means long pipes under pressure that are subject to leaks. Refrigerant piping is expensive to install and maintain. Long refrigerant suction lines also account for considerable pressure drop, which decreases machine capacity and increases refrigeration operating costs (8). Advantages of a brine system are confinement of all the refrigerant to a single engine room and simple control.

The disadvantages of a brine refrigeration system are that the refrigerant temperature must be lower than the brine temperature; this means a heavier load to the mechanical system than would be necessary for an equivalent direct expansion system. Another disadvantage is the requirement of a double heat transfer, which lowers the heat transfer efficiency and thus results in additional power cost to circulate brine solution.

Cryogenic Refrigeration. A cryogenic refrigeration system has the advantage of easy installation, low maintenance cost during off-production season than the mechanical system, and high end product quality due to its high freezing rate. However, the very low temperature of cryogenic refrigerants may cause food to crack due to an instantaneous drop in the surface temperature and prevention of further volume expansion. The very high temperature difference between food and cryogenic medium produces a film of vapor surrounding the food and prevents further effective heat transfer.

Refrigerator Capacity

To evaluate the capacity and performance of a refrigeration system, some terms need to be defined.

Refrigeration Ton. The standard unit of refrigerating capacity is known as a ton of refrigeration. One standard ton is derived on the basis of the removal of the latent heat of fusion from 1000 kg of water at 0°C to produce 1000 kg of ice at the same temperature in 24 h. The latent heat of ice fusion is equal to 79.68 kcal per kg. Therefore, the standard ton of refrigeration is equal to 79,680 kcal per 24 h or 3,320 kcal per h.

Enthalpy, kcal/kg. Enthalpy is a measurement to quantify the total amount of heat in a product. It is a relative value and generally the enthalpy at 0°C is defined as 0.

Refrigerating Effect, kcal/kg. If the operating condition of a refrigeration system is known, then the net refrigerating effect can be calculated from the pressure-enthalpy diagram as

$$RE = h_g - h_f \quad (1)$$

where RE is the refrigerating effect (kcal/kg), h_g is the enthalpy of vapor refrigerant leaving the evaporator, and h_f is the enthalpy of refrigerant entering the evaporator.

Quantity of the Refrigerant Circulated per Ton of Capacity, kg/min ton. This can be determined by the ratio of one standard ton to the refrigerating effect.

$$\text{Mass flow (kg/min ton)} = 55.33/RE \quad (2)$$

Heat of Compression, kcal/kg. This is the refrigerant enthalpy difference between discharge and inlet of the compressor.

Work of Compression, kcal/min ton or hp/ton. This is the product of heat of compression and the refrigerant mass flow. It can be converted to horsepower per refrigeration ton by dividing the work of compression by 10.7.

Compression Ratio. The compression ratio of a refrigerant compressor is the ratio of initial volume to the final volume of refrigerant at discharge. As volume is inversely correlated with pressure, compression ratio can also be calculated from the final pressure over the initial pressure.

Condenser Heat Load, kcal/kg. This can be determined by subtracting the enthalpy of the saturated liquid leaving the condenser from the enthalpy of the superheated vapor entering the condenser.

Coefficient of Performance. This is the ratio of output to input. The system output can be obtained from the refrigerating effect, and the system input can be obtained from the heat of compression. However, the heat input by the compressor is less than the electrical energy put into the motor due to friction losses in the compressor. The overall coefficient of performance usually is only 60% of the theoretical.

Refrigeration System Consideration

Requirements for each individual refrigeration system may vary. Factors involved in designing a refrigeration system include: (1) length of the operation season, (2) refrigeration load and the ability to handle wide load fluctuations in very short time intervals, (3) frost control for continuous operation, (4) choice of appropriate cooling medium, (5) temperature level to which process fluid must be cooled, (6) energy source for driving the refrigeration unit, (7) system efficiency and maintainability, (8) space availability, and (9) consideration of operating pressures including the choice of refrigerants, type of condensers, and staging, ie single, compound, or cascade.

One of the major considerations for system selection is the compression ratio. A compressor operating at a high compression ratio has a hot discharge gas stream that may cause the oil and refrigerant to break down. Operation at high compression ratios also harms the compressor and its components. Because of these factors, manufacturers usually limit the compression ratio of multicylinder machines to approximately 9, vertical single-acting slow speed machines to 13, and rotary screw compressors to 19 (9). When the compression ratio is too high, multistage systems are selected.

The approximate temperature ranges for the most commonly used refrigerants are available in most of the refrigeration handbooks. Selection of an appropriate refrigerant should consider the operation temperature levels. Sometimes, more than one refrigerant may be usable at a given evaporating temperature; however, one refrigerant may be more advantageous to use than another because of its low operating cost. Flammability of a refrigerant should also be taken into consideration when designing a refrigeration system.

The most commonly used compressor for a small single-stage or multistage system is the reciprocating compressor. Many factors must be considered in selecting the most economical compressor to assure a satisfactory system. These factors include system size and capacity requirements, location, equipment noise, part-load and full-load operation, time, and cost associated with servicing the unit and breakdown (9).

A condenser can be selected from among air-cooled, water-cooled, and evaporative types. However, the condenser must be sized for the maximum amount of expected heat rejection.

Several basic types of evaporators are available for refrigeration systems. A direct-expansion evaporator is commonly used because of its ease of returning oil and small refrigerant charge requirement.

FOOD FREEZING

Food freezing is the process of reducing the temperature of a product to below 0°C. The temperature reduction process can be divided into three stages: cooling, freezing, and

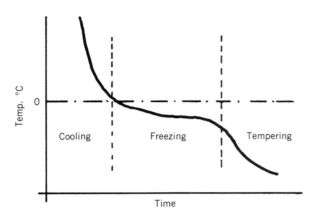

Figure 4. A typical freezing curve.

tempering (Fig. 4). The cooling stage reflects the reduction of product temperature from the initial temperature to the first freezing point, but there is no change of phase. The heat removed in this stage is referred to as sensible heat. The freezing stage represents where most of the water crystallization occurs, and heat removed is called latent heat. Latent heat is the heat required to be extracted from food in order to change the status of water to ice. As in the case of water, 80 cal of heat need to be removed to change the state of 0°C water to ice. The tempering stage further reduces a product's sensible heat until the product reaches its final temperature.

Refrigeration Requirement for Freezing

Proper selection or design of the refrigeration system is based on knowledge of the amount of heat necessary to be removed by the refrigeration system. The total heat load for the refrigeration system can be considered as the sum of four components represented as follows:

$$H = H_s + H_u + H_l + H_f \qquad (3)$$

where H is the total enthalpy change required to reduce the product temperature from its initial temperature above freezing to a final storage temperature below freezing, H_s is the sensible heat that must be removed to reduce the product temperature to the initial freezing point, H_u is the sensible heat that must be removed to reduce the temperature of the unfrozen portion of the product below its initial freezing point, H_l is latent heat removal, and H_f is the sensible heat that must be removed to further reduce the temperature of the frozen portion of the product. In order to estimate the refrigeration requirement, thermal property information of foods is important for calculating the refrigeration system load and the rate of freezing foods. Some of these important properties are discussed as follows:

Water Content. Water content is not a thermal property of food; however, it significantly influences the total enthalpy change and other thermal properties. Food material is a complex system. Not all the water in food is freezable, and this characteristic will influence the calculation of latent heat removal. The amount of unfreezable water (m_i) at any temperature during freezing (T_i) can be predicted (10) based on the following relationship:

$$\frac{L \times M_A}{R}\left(\frac{1}{T_o} - \frac{1}{T_i}\right) = \ln\left(\frac{m_i/M_A}{m_i/M_A + m_b/M_B}\right) \qquad (4)$$

where the mass fraction of the unfrozen water (m_i) can be calculated from latent heat of fusion (L), gas constant (R), freezing point of pure water (T_o), product freezing temperature (T_i), molecular weight of water (M_A), mass fraction of product solute in solution (m_b), and effective molecular weight of product solute (M_B). Latent heat removal (H_l) can then be calculated from:

$$H_l = L(m_a - m_i) \qquad (5)$$

where m_a is the mass fraction of water in food.

Freezing Point. The freezing point of pure water is 0°C; however, the freezing point of foods varies with the composition but generally ranges between −1 and −5°C. As a food product is frozen, ice crystals form, resulting in increased solute concentration in the unfrozen phase and depression of the freezing point. There is no single freezing point for foods as compared with pure water due to the continuous concentration process in the unfrozen phase during freezing. For the purpose of estimating the refrigeration requirement, the initial freezing point can be used for calculating sensible heat removal.

Specific Heat. Specific heat is the ratio of the heat required to raise the temperature of unit mass of the product one degree to the heat required to raise the temperature of unit mass of water one degree. Information on specific heat can be found in the literature (11). One of the simplest methods to measure specific heat is the method of mixtures. A test sample with known mass and temperature is dropped into a calorimeter of known specific heat that contains water at a known temperature and weight. The specific heat of test sample can then be calculated based on heat balance. Detailed information on measuring and calculating specific heat can be found in the literature (12). The specific heat of a particular product is a function of temperature but for practical applications can be assumed to depend only on either frozen or unfrozen phases. Sensible heat removal above the initial freezing point (H_s) can be calculated as follows:

$$H_s = Cp_l(T_i - T_f) \qquad (6)$$

where Cp_l is the specific heat of food above freezing, T_i is the initial product temperature, and T_f is the initial freezing point. Sensible heat removal for the unfrozen portion (H_u) and the frozen portion (H_f) of the product to below its initial freezing point can be estimated as follows:

$$H_u + H_f = Cp_s(T_f - T_t) \qquad (7)$$

where Cp_s is the specific heat of food below freezing and T_t is the final product temperature.

The amount of heat to be removed per unit mass of product can then be calculated by the summation of equations (5), (6), and (7). The total amount of heat removed in the refrigeration system per unit time is the mass flow rate times the heat removal per unit mass.

Freezing Time Prediction

Successful design of a refrigeration system to produce high quality frozen food depends upon the ability to predict the freezing time of a product. Freezing time information can be used to (1) determine heat loading for refrigeration requirements of a freezer, (2) obtain a satisfactory design of a process for production of frozen foods, and (3) control the quality of foodstuffs that is affected by freezing rate (13). The terms "freezing rate" and "freezing time" are related; generally the higher the rate the shorter the time.

Two definitions of freezing time were suggested by the International Institute of Refrigeration (IIR) in 1971 (14). One definition, nominal freezing time, is the time elapsed between the surface reaching 0°C and the thermal center reaching 10°C colder than the temperature of initial ice formation at the thermal center. The second definition is effective freezing time. This is the time required to lower the temperature of a product from its initial temperature to a given temperature at its thermal center. Effective freezing time is related to the total time the food is placed in the freezer. Thus, effective freezing time is related to the physical freezer capacity. Most of the prediction methods available in the literature are to predict effective freezing time.

Freezing time prediction in foods is very complex due to changes occurring within the product and the corresponding influence on product properties. Other reasons for the complexity include: food composition may exhibit anisotropy, there is no clear endpoint to the removal of latent heat, thermal properties of foods are temperature dependent, and the convective heat transfer coefficient is difficult to measure. Methods for predicting freezing time can be divided into two groups, analytical and numerical, and have been reviewed in the literature (15–17). Analytical methods are derived on theoretical grounds with some restrictive assumptions but have the advantages of simplicity, ease of understanding, minimal thermal data requirements, and wider availability.

Probably the least complex method to predict freezing time was originally derived by Plank (18) as follows:

$$t_f = \frac{\rho L}{(T_f - T_a)} \left(P \frac{a}{h} + R \frac{a^2}{k} \right) \tag{8}$$

where t_f is the freezing time, L is the latent heat of water fraction, ρ is the density, T_f is the initial freezing temperature, T_a is the freezing medium temperature, a is the characteristic dimension, h is the surface heat transfer coefficient, k is the thermal conductivity, P and R are $\frac{1}{2}$ and $\frac{1}{8}$, respectively, for infinite slab, $\frac{1}{4}$ and $\frac{1}{16}$ for infinite cylinder, and $\frac{1}{6}$ and $\frac{1}{24}$ for sphere.

Two approaches are available in the literature for multidimensional regular and irregular geometry. One approach is to find a global multiplying factor (EHTD, equivalent heat transfer dimensionality) to apply to the freezing time for one of the basic shapes (19). The other approach is to find the mean conducting path to account for internal resistance effect and irregular geometries (20).

Numerical methods are mathematical abstractions of the complex heat transfer problem that are solved by using a high-speed computer. Numerical methods can predict the freezing time more accurately than analytical methods and will also provide the time-temperature history information of the whole process. However, numerical methods are more complex than analytical methods and require more product and process knowledge and substantial effort to implement.

Refrigeration Load During Storage

The heat load in a refrigeration system can be classified into four categories: wall heat losses, air exchange or door loss, product load, and incidental loss.

Wall Heat Losses. Wall heat losses vary with the type and thickness of insulation used, construction, outside wall area, temperature difference between refrigerated space and ambient air, and outside wind speed. It is a combination of convection and conduction heat transfer through a wall.

Air Exchange or Door Loss. Infiltration of outside hot air with inside cold air is the single largest refrigeration load item. The amount of heat exchange estimation is based on experimental measurement of outside and inside air temperatures, placement of the information collected on psychrometric charts, and calculation of the total heat exchange.

Product Load. Product heat load includes the heat of respiration and the removal of sensible heat to reduce the product temperature to that of storage temperature.

Incidental Loss. These kinds of losses include heat added to the storage room from workers, lights and electric motors. For estimation purposes, the incidental losses usually add into the refrigeration system some additional 10 to 15% of the other three previously discussed loads.

BIBLIOGRAPHY

1. O. Fennema, "The U.S. Frozen Food Industry: 1776–1976," *Food Technology* 30(6), 56–68 (1976).
2. T. P. Labuza and A. E. Sloan, "Forces of Change: From Osiris to Open Dating," *Food Technol.* 35(7), 34–43 (1981).
3. R. V. Enochian and W. R. Woolrich, "The Rise of Frozen Foods," in *Fundamentals of Food Freezing*, N. W. Desrosier and D. K. Tressler, eds., AVI Publishing Co., Inc., Westport, Conn., 1977.
4. H. T. Cook, "Refrigerated Storage," in C. W. Hall, A. W. Farrall, and A. L. Rippen, eds., *Encyclopedia of Food Engineering*, AVI Publishing Co., Inc., Westport, Conn., 1986.
5. D. S. Reid, "Fundamental Physicochemical Aspects of Freezing," *Food Technology* 37(4), 110–115 (1983).
6. J. M. Smith and H. C. Van Ness, *Introduction to Chemical Engineering Thermodynamics*, 3rd ed., McGraw-Hill Inc., New York, 1975.

7. E. L. Watson and J. C. Harper, *Elements of Food Engineering*, 2nd ed., AVI Publishing Co., Inc., Westport, Conn., 1988.
8. E. R. Hallowell, *Cold and Freezer Storage Manual*, 2nd ed., AVI Publishing Co., Inc., Westport, Conn., 1980.
9. ASHRAE Handbook. "Engineered Refrigeration Systems (Industrial and Commercial)," in *Refrigeration Systems and Applications,* American Society of Heating, Refrigeration and Air Conditioning Engineers, Inc., Atlanta, Ga., 1986.
10. D. R. Heldman, "Predicting the Relationship Between Unfrozen Water Fraction and Temperature during Food Freezing Using Freezing Point Depression," *Transactions of the American Society of Agricultural Engineers* **17**(1), 63–66 (1974).
11. S. L. Polley, O. P. Snyder, and P. Kotnour, "A Compilation of Thermal Properties of Foods," *Food Technology* **34**(11), 76–94 (1980).
12. N. N. Mohsenin, *Thermal Properties of Foods and Agricultural Materials.* Gordon and Breach, Inc., New York, 1980.
13. E. Kinder and J. Lamb, "The Prediction of Freezing Times of Foodstuffs," *Meat Research Institute Symposium,* Bristol, UK, **3**, 17.1–17.6 (1974).
14. "Recommendation for the Processing and Handling of Frozen Foods," International Institute of Refrigeration, Paris, 14–16 (1972).
15. Y.-C. Hung, "Prediction of Cooling and Freezing Times," *Food Technology* **44**(5), 137 (1990).
16. "Cooling and Freezing Times of Foods," in *ASHRAE Handbook Fundamentals,* American Society of Heating, Refrigeration and Air Conditioning Engineers, Inc., Atlanta, Ga., 1989.
17. A. C. Cleland, *Food Refrigeration Process—Analysis, Design and Simulation,* Elsevier Applied Science Co., London, 1990.
18. R. Z. Plank, 1913, cited by A. J. Ede, "The Calculation of the Rate of Freezing and Thawing of Foodstuffs, *Modern Refrigeration* **52**(3), 52–55 (1949).
19. D. J. Cleland, A. C. Cleland, and R. L. Earle, "Prediction of Freezing and Thawing Times for Multi-dimensional Shapes by Simple Formulae. Part 2: Irregular Shapes, *International Journal of Refrigeration* **10**, 234 (1987).
20. Q. T. Pham, "Simplified Equation for Predicting the Freezing Time of Foodstuffs," *Journal of Food Technology* **21**, 209 (1986).

General References

R. Thevenot, *History of Refrigeration,* International Institute of Refrigeration, Paris, 1979.

D. R. Heldman and P. R. Singh, *Food Process Engineering,* 2nd ed. AVI Publishing Co., Inc., Westport, Conn., 1981.

A. D. Althouse, C. H. Turnquist, and A. F. Bracciano, *Modern Refrigeration and Airconditioning,* The Goodheart-Willcox Co., Inc. South Holland, Il., 1988.

M. K. Karel, O. R. Fennema, and D. B. Lund, *Principles of Food Science,* Part II. *Physical Principles of Food Preservation,* Marcel Dekker, Inc., New York, 1975.

W. B. Gosney, *Principles of Refrigeration,* Cambridge University Press, Cambridge, UK, 1982.

M. R. Okas, *Physical and Chemical Properties of Food,* ASAE Publication No. QO986, St. Joseph, Mich., 1987.

Refrigeration Systems and Applications, American Society of Heating, Refrigeration and Air Conditioning Engineers, Inc., Atlanta, Ga., 1986.

Equipment, American Society of Heating, Refrigeration and Air Conditioning Engineers, Inc., Atlanta, Ga., 1988.

Fundamentals, American Society of Heating, Refrigeration and Air Conditioning Engineers, Inc., Atlanta, Ga., 1989.

YEN-CON HUNG
University of Georgia
Griffin, Georgia

FOOD INDUSTRY ECONOMIC DEVELOPMENT

The food system in the United States is an intricate web of activities involved in the production, processing, distribution, and marketing of foodstuffs to the consumer. The food industry, in turn, is the aggregate of enterprises (firms, cooperatives, institutions, agencies, etc) that perform these activities. So effective are these enterprises that, while the variety of foodstuffs available to today's U.S. consumers is unparalleled in human history (the average U.S. chain supermarket offered over 18,000 items in 1989), the average American household in 1987 only dedicated 12.1 percent of its disposable income to the purchase of food (1).

Most discussions of the food industry and its developmental trends are organized in a sequential fashion beginning with the production industries and proceeding through to the point in the system at which products are made available to consumers. Such an organization obviously parallels the flow of goods through the system. It also, however, indirectly casts consumers as passive participants who purchase whatever this system makes available. Contemporary authorities now recognize that in the United States consumers are, in fact, actively involved in the development of the food system (2).

Most discussions of the system (3) describe two primary channels by which consumers receive food: grocery (including supermarket and other retail establishments) and food service (restaurants, institutions, etc). The two grocery channels have been further subdivided into branded and unbranded channels (4). With the recent growth in supermarket-prepared take-out foods, home delivery by restaurants, and branded produce and meat products, these traditional classifications increasingly appear moot. Consumers today purchase food in various stages of preparation from myriad retailing establishments and, by making purchase decisions, send information back through the system, which affects the development of the enterprises involved. It is this point that is often given little recognition by economists interested in evaluating the development of the food industry.

Nonetheless, it is useful for consideration of current industry performance and trends to classify the industry in accordance with the traditional treatment of economists. These classifications generally describe five principal sectors of the food industry: production, processing (or manufacturing), transportation, marketing (wholesaling and retailing), and food service. Individual firms may be involved in more than one sector, depending on the degree of vertical integration, and may also be involved in other industries, depending on the degree of horizontal integration. Proper consideration of the food industry must note the significant contribution made by the input industries,

Table 1. Standard Industrial Classification (SIC) Codes of Principal Sectors of the Food Industry[a]

Group	Definition
01	Agricultural production—crops
02	Agricultural production—livestock and animal specialties
07	Agricultural services
20	Manufacturing—food and kindred products
26	Paper and allied products
SIC 2656	Sanitary food containers, except folding
SIC 2657	Folding paperboard boxes, including sanitary
35	Manufacturing—industrial and commercial machinery and computer equipment
SIC 3523	Farm machinery and equipment
SIC 3556	Food products machinery
51	Wholesale trade—nondurable goods
SIC 514	Groceries and related products
SIC 515	Farm product raw materials
SIC 518	Beer, wine, and distilled alcoholic beverages
SIC 5191	Farm supplies
54	Retail trade—food stores
58	Retail trade—eating and drinking places
59	Miscellaneous Retail
SIC 5921	Liquor stores

[a] Ref. 5.

such as manufacturers of agricultural chemicals and food additives; financial institutions providing capital; and service enterprises, such as marketing and advertising agencies. Table 1 lists the industries involved in these sectors using the current Standard Industrial Classification (SIC) system.

HISTORY

In the overall history of humans, the beginnings of the food industry coincide with the dawn of civilization. Anthropologists generally agree that the development of agriculture, or controlled production of foodstuffs, is at the very foundation of civilization. From this point, the development of ever-increasing sophistication in storage, processing, distribution, and marketing has progressively allowed humans to be less constrained by the acquisition of food and permitted the development of other social, cultural, and technological dimensions of civilization (6).

At the time of its colonization and founding, the United States was primarily an agrarian economy. During the nineteenth and early twentieth centuries, the United States followed the rest of the Western world in the adaptation of modern processing technologies, the first of which was canning of perishable foods as initially developed in France by Appert. A rapid proliferation of companies engaged in processing and marketing of foods ensued. It was during this period that many of today's food companies had their inception, including Kraft, Borden, Heinz, Kellogg, Post, and Busch. Vertical and horizontal integration and the quest for improved operating and marketing efficiency and effectiveness led to gradual consolidation of the industry participants during most of the twentieth century. The era of consolidation has had two high points. The first occurred in the first third of the twentieth century, when the diverse corporate food giants were created.

The second high point is contemporary and coincides with the financially driven wave of mergers and leveraged buyouts affecting the entire business community. There are detailed reports on the history of the food industry in the United States by a number of authors (4), (7), (8). Students of the industry may consult such references to gain a more thorough understanding of the significant events that have shaped the development of this industry.

CURRENT INDUSTRY SITUATION

Total U.S. consumer expenditures for food were approximately $461 billion in 1985, representing slightly less than 20% of total consumer expenditures of $2.6 trillion. Food expenditures, in fact, are the largest of the consumer expenditure categories as reported by the U.S. Bureau of Economic Analysis (9). However, as shown in Table 2, food expenditures as a percent of disposable income is trending downward.

Total employment in the food system, including farming, stood at 16.4 million in 1987 (10). Employment in the food system in 1980 represented 12.9% of the total U.S. civilian work force (4). With the addition of employment in input industries, it has been estimated that more than 18% of the total U.S. work force is involved in the food system. Trends indicate a decline of the work force in the production of food, a relatively stable work force in food processing, and increasing employment in wholesaling and retailing, which should continue in the future (Fig. 1).

Table 3 lists the number of establishments, by sector in the food industry for recent years. In most sectors, the trend in the number of establishments parallels the employment trend. One characteristic of the food system is the relatively small number of firms in processing contrasted with the high numbers of businesses in production and marketing. While not all goods pass through the pro-

Table 2. Food Expenditures As a Share of Disposable Personal Income, 1967–1987[a]

Year	Disposal Personal Income Billions of Dollars	Food at Home		Food Away from Home		Total	
		Billions of Dollars	Percent	Billions of Dollars	Percent	Billions of Dollars	Percent
1967	562.1	60.3	10.7	19.8	3.5	80.1	14.2
1972	839.6	84.4	10.1	30.1	3.6	114.5	13.7
1977	1,379.3	131.6	9.5	56.1	4.1	187.7	13.6
1982	2,261.4	197.7	8.7	98.7	4.4	296.4	13.1
1987	3,209.7	244.9	7.6	143.8	4.5	388.7	12.1

[a] Ref. 1.

Table 3. Number of Food-System Business Establishments, 1972–1987[a]

Year	Number of Farms	Number of Manufacturing Establishments	Number of Wholesale Establishments	Number of Retail Establishments	Number of Food-Service Establishments	Total
1972	2,869,710	28,103	60,363	173,084	287,250	3,418,510
1977	2,455,830	26,656	59,473	171,592	308,614	3,022,165
1982	2,400,550	22,130	58,766	176,219	319,873	2,977,538
1987	2,176,110	20,201	42,075	190,706	391,303	2,820,395

[a] Refs. 11–14.

cessing sector (some fresh commodities do not), the vast majority do. The processing sector becomes a pinch point in that a relatively few number of firms control the flow of goods and have, until recently, exercised significantly greater influence on system performance and development than firms at the opposite ends of the system. Recent advances in information gathering and analysis at point of purchase are, however, increasingly shifting this balance of power to retailers (2,15).

Economists generally favor the use of the concept of value added to gauge industry performance and development. Value added represents the difference between what is paid to other industries for goods and services and the value of outputs. The sum of value added by the entire economy represents the gross national product. Table 4 summarizes estimates of the value added by sectors of the food industry for 1982 and 1987. Also included is a similar estimate for the year 1947 (4). Despite some disagreement between these sources in estimating value added, it is apparent that the value added by the food industry represents a large part of the U.S. gross national product. One source estimates that the food industry and its input industries represented approximately 14.5% of the gross national product in 1982 (4). Equally significant is the shift between relative value added by the principal sectors from 1947 to 1982. The relative reduction in economic significance of the production sector is a trend that will likely continue into the twenty-first century, while growth in the significance of the processing, marketing, and food service sectors will also continue.

Table 5 lists value added as a percentage of sales for certain product categories in the processing sector in recent years. A wide range in value added exists with respect to the products produced. In 1987, for example, the value added by processing averaged about 37% of the total value of shipments, but ranged from about 10% for soybean oil processing to nearly 75% for cereal breakfast foods. Similarly, as shown in Table 6, considerable differences exist among states with respect to the average value added to their agricultural commodities. Inasmuch as economic development is ultimately the quest for enhancement of the ability to add value, a better understanding of the reasons for current product and geographical differences should offer important clues as to how to positively

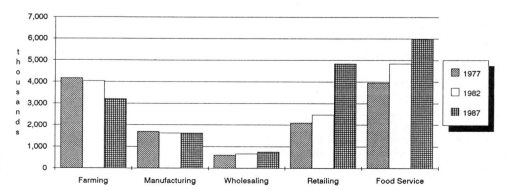

Figure 1. Food system employment, by sector, 1977–1987. From Ref. 9.

Table 4. Value Added by Industry Sector, Selected Years (in Billions of Dollars)

Industry Sector	1947[a]	1972[a]	1982[a]	1982[b]	1987[b]
Farming	19.5	22.0	71.1	48.1	51.0
Processing and manufacturing	11.3	30.6	74.1	65.0	85.7
Retailing and wholesaling	30.0	37.9	169.8	87.0	108.3
Transportation	15.3	6.5	30.1	16.9	20.5
Eating and drinking places	6.0	18.5	42.6	46.8	63.5
Other supporting sectors[c]		49.9		121.6	165.3
Total sector value added	82.1	165.4	387.7	385.4	494.3
U.S. gross national product	252.0	1,212.8	3,166.0	3,166.0	4,526.7
Total sector divided by GNP	0.326	0.136	0.122	0.122	0.109

[a] Ref. 4.
[b] Ref. 16.
[c] Includes auxiliary activities such as packaging.

influence economic development in the food industry. One important statistic in this regard is relatively low expenditure by food firms for research and development (Table 7) as compared with other industries (10), which does not speak well of the food industry's capacity for internal development. Similarly, other indexes of development potential or tendencies, such as capital expenditures and growth in productivity suggest that the food industry has not positioned itself as well for economic development as have other industries (9).

Table 5. Value Added Compared to Value of Shipments for Selected SICs, 1982 and 1987[a]

	Value Added by Manufacture (Millions of Dollars)	Value of Shipments (Millions of Dollars)	Value Added as a percent of Shipments
SIC 20—industry totals and average			
1987	121,997.3	329,784.3	37.0
1982	88,391.9	280,502.2	31.5
SIC 2011—meat packing			
1987	5,155.6	44,939.0	11.5
1982	5,824.6	44,853.6	13.0
SIC 2021—butter			
1987	155.5	1,420.2	10.9
1982	135.6	1,686.8	8.0
SIC 2024—ice cream and frozen desserts			
1987	1,262.3	3,914.6	32.2
1982	910.4	2,855.1	31.9
SIC 2038—frozen specialties			
1987	2,770.7	5,544.9	50.0
1982	2,066.2	5,033.9	41.0
SIC 2043—cereal breakfast foods			
1987	4,904.5	6,565.7	74.7
1982	2,622.8	4,131.9	63.5
SIC 2047—dog and cat food			
1987	2,737.9	5,056.7	54.1
1982	2,281.4	4,402.2	51.8
SIC 2048—animal feed			
1987	2,504.3	11,454.4	21.9
1982	2,188.4	11,298.1	19.4
SIC 2075—soybean oil milling			
1987	1,024.0	9,238.2	11.1
1982	678.2	8,603.6	7.9
SIC 2079—shortening and table oils			
1987	1,295.4	4,270.3	30.3
1982	1,261.7	4,905.6	25.7
SIC 2087—flavoring extract and syrups			
1987	3,282.6	4,634.5	70.8
1982	2,669.8	4,236.8	63.0
SIC 2091—canned and cured fish and seafood			
1987	289.3	797.8	36.3
1982	613.4	1,849.1	33.2

[a] Ref. 11.

Table 6. Value Added from Manufacture and Cash Receipts from Farming, by State, 1977, 1982, and 1986[a]

State	1977			1982			1986		
	Cash Receipts from Farming, in Millions of Dollars	Value Added from Manufacturing in Millions of Dollars	Value Added Divided by Cash Receipts	Cash Receipts from Farming, in Millions of Dollars	Value Added from Manufacturing in Millions of Dollars	Value Added Divided by Cash Receipts	Cash Receipts from Farming, in Millions of Dollars	Value Added from Manufacturing in Millions of Dollars	Value Added Divided by Cash Receipts
Alabama	1,496	549	0.367	2,272	845	0.372	2,009	1,122	0.558
Alaska	11	199	18.091	16	185	11.563	29	158	5.448
Arizona	1,198	194	0.162	1,657	349	0.211	1,495	659	0.441
Arkansas	2,469	620	0.251	3,417	1,096	0.321	3,022	1,329	0.440
California	9,370	6,578	0.702	14,321	10,964	0.766	14,049	13,462	0.958
Colorado	2,060	749	0.364	3,004	1,065	0.355	3,109	1,740	0.560
Connecticut	234	409	1.748	309	546	1.767	372	821	2.207
Delaware	261	282	1.080	405	477	1.178	520	590	1.135
Florida	2,626	1,646	0.627	4,250	2,915	0.686	4,688	3,651	0.779
Georgia	2,194	1,362	0.621	3,210	2,457	0.765	3,206	3,160	0.986
Hawaii	325	351	1.080	484	399	0.824	575	574	0.998
Idaho	1,170	339	0.290	2,102	657	0.313	1,925	755	0.392
Illinois	5,792	4,726	0.816	7,434	6,810	0.916	6,880	8,357	1.215
Indiana	3,239	1,572	0.485	4,586	2,064	0.450	4,110	2,967	0.722
Iowa	7,065	1,833	0.259	10,343	3,131	0.303	9,106	3,219	0.354
Kansas	3,849	681	0.177	5,809	1,202	0.207	5,425	1,653	0.305
Kentucky	1,806	939	0.520	2,903	1,347	0.464	2,389	2,499	1.046
Louisiana	1,257	778	0.619	1,844	1,105	0.599	1,372	2,642	1.926
Maine	418	22	0.053	408	331	0.811	365	570	1.562
Maryland	657	1,024	1.559	1,055	1,337	1.267	1,186	2,651	2.235
Massachusetts	214	786	3.673	341	1,161	3.405	423	1,617	3.823
Michigan	1,824	2,114	1.159	2,863	2,974	1.039	2,664	3,830	1.438
Minnesota	4,323	1,466	0.339	6,672	2,392	0.359	6,074	2,774	0.457
Mississippi	1,714	444	0.259	2,431	766	0.315	1,785	859	0.481
Missouri	2,870	1,529	0.533	3,673	2,501	0.681	3,516	3,208	0.912
Montana	957	104	0.109	1,633	142	0.087	1,213	154	0.127
Nebraska	3,980	830	0.209	7,087	1,557	0.220	6,928	1,943	0.280
Nevada	147	66	0.449	236	127	0.538	232	167	0.720
New Hampshire	79	108	1.367	104	188	1.808	109	270	2.477
New Jersey	351	1,996	5.687	516	3,266	6.329	580	3,907	6.736
New Mexico	791	104	0.131	960	133	0.139	1,010	156	0.154
New York	1,725	3,311	1.919	2,588	4,703	1.817	2,533	5,360	2.116
North Carolina	2,622	978	0.373	4,112	1,930	0.469	3,782	3,088	0.816
North Dakota	1,539	147	0.096	2,710	228	0.084	2,299	210	0.091
Ohio	2,794	2,802	1.003	3,674	4,178	1.137	3,610	5,874	1.627
Oklahoma	1,926	375	0.195	3,131	588	0.188	2,622	724	0.276
Oregon	1,034	668	0.646	1,775	1,128	0.635	1,784	1,203	0.674
Pennsylvania	1,903	3,205	1.684	2,991	4,950	1.655	3,165	6,430	2.032
Rhode Island	26	86	3.308	33	118	3.576	75	147	1.960
South Carolina	784	307	0.392	1,156	535	0.463	894	825	0.923
South Dakota	1,610	226	0.140	2,587	423	0.164	2,463	591	0.240
Tennessee	1,370	1,280	0.934	2,113	2,161	1.023	1,924	3,076	1.599
Texas	6,910	3,041	0.440	9,680	5,329	0.551	8,444	6,798	0.805
Utah	363	198	0.545	542	389	0.718	570	599	1.051
Vermont	267	60	0.225	408	93	0.228	398	177	0.445
Virginia	1,004	1,204	1.199	1,680	2,059	1.226	1,613	2,607	1.616
Washington	1,708	1,040	0.609	3,023	1,499	0.496	2,793	1,918	0.687
West Virginia	148	148	1.000	225	225	1.000	227	212	0.934
Wisconsin	3,152	2,304	0.731	5,247	3,324	0.634	5,057	4,251	0.841
Wyoming	452	32	0.071	530	65	0.123	566	NA	NA
Total	96,084	55,812	0.581	144,550	88,414	0.612	135,185	115,554	0.855

[a] Refs. 17 and 18.

Table 7. Research and Development Expenditures in Food Manufacturing Versus All Industry, Selected Years in Millions of Dollars[a]

Category	1972	1975	1977	1980	1982	1984
Food manufacturing	259	335	415	620	762	834
All industries total	19,552	24,187	29,825	44,505	57,996	71,137
Food manufacturing as a percent of All Industries Total	1.32	1.39	1.39	1.39	1.31	1.17

[a] Refs. 10, 19, and 20.

Production Sector

Much of the U.S. food policy is directed at influencing the activity of the production sector. Agricultural price-support programs, for example, seek to provide assurance of production of essential commodities by guaranteeing producers minimum prices. The philosophical argument for much of this policy rests on the premise that a dependable and affordable food supply is essential for social and economic stability. As production agriculture becomes progressively less significant in the economy, in terms of both number of farms (Table 3) and relative value added (Table 4), and as pressures mount to constrain the growing U.S. budget deficit, continuation of many of the food-policy programs is increasingly threatened.

Within the production sector, economic development activity has historically focused on improving production efficiency rather than on enhancing the value of the outputs. Most of the technical support for economic development in the production sector has not been provided by the agricultural establishment itself but rather by publicly supported research and technology transfer through land-grant universities and government research laboratories. Public expenditures for research in support of food system development have traditionally been directed almost exclusively to the production sector (17).

In recent years, some commodity groups representing large numbers of producers have collectively begun to support research and development activities. Many such groups currently support collective market development programs as well, principally through generic advertising and promotion. Increasingly these development activities are aimed at influencing the food system after the farm gate, unlike the publicly supported research aimed primarily at improving production sector efficiency.

Processing Sector

The processing sector has been the subject of extensive research. One report highlights several salient features concerning the present condition of this sector and the outlook for the near future (10). The most significant trend identified is the decline in numbers of plants, projected to decrease to between 13,000 and 14,000 by the year 2000. Employment in this sector is also declining, but not as rapidly, because plants that remain in operation tend to be larger. Productivity in this sector is high, however, with the value of shipments per employee ($200,000) ranking second highest of the principal manufacturing categories.

Consolidation, as measured by both number of companies and number of establishments, is a trend likely to continue in the processing sector. The degree of concentration in this sector is, however, more significant than these numbers alone indicate. It has been reported that the 50 largest processors held 75% of the assets in the sector in 1987 (10). Similarly, spending on research, development, and advertising and capital expenditures are concentrated in a small number of firms. This leaves the vast majority of firms in this sector at a serious disadvantage in capacity for development.

Despite this concentration, the processing sector is the only sector in the food industry with significant internal research and development activity. Recent trends, however, indicate a relative reduction in research expenditures by food processors (21). Moreover, research expenditures may be shifting away from new products and technology toward improved quality control and resolution of consumer concerns for safety (22).

A number of states have identified food processing for special emphasis in industry attraction and retention programs, as well as for other economic development attention. It remains to be seen whether these targeted efforts will be effective. A report on the food-processing sector projects that above average growth in food processing will occur in the U.S. manufacturing heartland and the mountain states, with below-average growth in the South, if present trends continue (10). These growth projections are based on anticipated increases in shipments, not necessarily number of jobs or plants, however. Because most state-level economic development efforts focus heavily on employment statistics, even states with above average growth in shipments may be disappointed with results of their programs.

Marketing Sector

Total grocery store sales have risen steadily in recent years, even when adjusted for inflation (23) (Fig. 2). The rate of increase, however, has been on a downward trend for most of the 1980s. In fact, the consumer price index for food rose faster during this period than did grocery sales, which explains why retailer's profits as a percent of sales declined. Despite a slight recovery from this situation in 1988, significant pressures are forecast on retailers to achieve real economic growth. This situation appears to exist in part due to the continuing decline in the percent of disposable income consumers spend on food at home (Table 2). Retailers are likely to emphasize strategies aimed at winning back consumers from food service establishments and to increase consumer spending by emphasizing high-value products, including perishable foods (produce and refrigerated products) and prepared foods for take-out or to eat in the store.

The retail grocery industry is moving to fewer but larger establishments. Between 1979 and 1988, conventional grocery establishments saw their share of retail grocery sales drop from 83.2 to 50.1%, while hyperstores, warehouse stores, and other types of superstores increased correspondingly. As previously stated, employment is increasing in retailing (Fig. 1). These trends are likely to continue in the future. Interestingly, as the grocery industry moves to larger establishments, significant growth in convenience stores, at the other end of the size spectrum, has also occurred and is forecast for the rest of the twentieth century. In 1988, there were approximately 55,000 such establishments. Grocery sales by convenience stores represented only 7.8% of total grocery sales in 1988, however (23).

The number of wholesaling establishments in grocery and other food products, including alcoholic beverages, has declined dramatically recent years. Employment in wholesaling, however, is on an upward trend, increasing from 624,000 in 1977 to 825,000 in 1982 (9).

Management in both wholesaling and retailing ex-

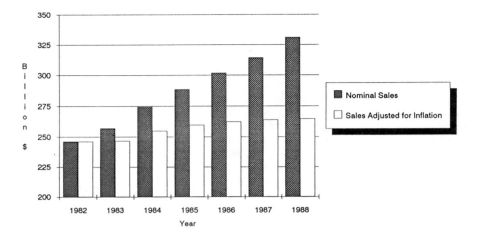

Figure 2. Food store sales, 1982–1988.

pressed concern, in a recent poll (23) that labor availability and costs will likely be the principal problem facing this sector in the 1990s. This sector, as is the case with retail trade in general, traditionally offers lower wages and salaries to its workers than other industry groups, depending on a steady input of entry-level employees to sustain growth and concomitantly tolerating employee flux. A 1987 study by Cornell University, for example, indicated an annual turnover rate among part-time supermarket employees, (who comprise more than 60% of the work force) to be greater than 100% (24). Demographic changes resulting in a decline in the number of entry-level employees and the simultaneous economic expansion in the manufacturing industries in the 1980s have indeed put pressure on retailers to be able to sustain an affordable labor supply. The food-marketing sector is likely to emphasize worker retention and productivity increases through training and development, as well as automation, to sustain real growth in the 1990s.

Food-Service Sector

The food-service sector is of special interest to those concerned with food system economic development for three principal reasons: this sector has shown the most dramatic increases in value added and employment in recent years compared to the other sectors, the food-service sector uses a substantial portion of the outputs of the processing and marketing sectors, and consumer expenditures for food away from home as a percentage of disposable income have increased (Table 2).

This sector has been divided into three categories: commercial feeding, institutional feeding, and military feeding (25). Together these categories generated approximately $227 billion in sales in 1989. The commercial category, with sales of $202 billion, is the largest of the categories. Institutional feeding and military food service generated approximately $24 billion and $1 billion, respectively.

Within the commercial category, the majority of sales are represented by two types of establishment: restaurants ($75 billion) and limited-menu eating places ($64 billion). With increasing mobility and pressures on meal preparation time as more women have entered the workplace, fast-food restaurants, which comprise the bulk of the limited-menu establishments, have captured an increasing percentage of sales in the commercial category. Similarly, the number of fast-food establishments has increased dramatically in recent decades; by 1986 the number of such establishments virtually equaled conventional restaurants (11).

Sales for fast-food restaurants are projected to grow at a faster pace into the early 1990s than sales for full-service restaurants, 3.9% compounded real growth rate versus 2.8%. Interestingly, however, food-service sales in other retail establishments (grocery stores, supermarket, and convenience stores) are projected to grow at an 8.7% rate over this same time period (25). This latter statistic points to what many in the food-service sector see as the most serious threat to their economic growth in the 1990s: increased competition for consumer food expenditures, including food away from home, by enterprises traditionally associated with the retailing sector. In fact, as businesses in the retailing and food-service sectors compete for the growing product categories, such as take-out and prepared foods, the traditional broad classifications of the food system begin to break down.

Other issues likely to confront the food-service sector in the 1990s are real estate availability and cost and, similar to the marketing sector, declining availability of labor from the sector's traditional labor pool (26,27). Real estate is of special concern to the fast-food category, where continued expansion is dependent on proliferation of establishments. As with the retailing sector, the food-service sector depends on entry-level, often part-time, workers at a low wage rate for a large percentage of its work force. Also similar to the retailing sector, a large employee flux has been characteristic of the industry, as high as an annual 300% turnover in the fast-food category, for example (26). Improved employee recruitment, retention, and productivity will be at the heart economic growth of the food-service sector in the closing years of the twentieth century.

THE FOOD INDUSTRY IN CHANGE

General

The U.S. food industry is in a period of change. While many researchers report on the significance of changes in eating patterns resulting from social and cultural upheaval, others debate how, where, and what U.S. citizens will eat as they enter the twenty-first century and as their society changes dramatically from traditional patterns (28). Even those who express a conservative outlook, however, agree with the more radical forecasters that the industry itself not only will, but must, change as a result of powerful external forces. To predict the characteristics of the food industry in the twenty-first century, it is important to use more than extrapolations of economic trends and to consider the potential impact that, at least, the most significant of these forces might have.

Certain areas of change consistently appear in discussions on the future food industry. These are globalization, mergers and acquisitions, new technologies, and the changing U.S. population. Forecasting is always uncertain, and it is difficult to project quantitatively the impact these areas of change might have on the food industry. It is impossible to predict the yet-to-be discovered information or technology, for example, that might alter the course of events in a monumental fashion, yet it must be expected that such discoveries can occur. Each individual firm's success, if not survival, and the development of the industry will be shaped by the industry's resilience and willingness to accept and adapt to changes, indeed, to take advantage of them.

Globalization

At the end of World War II, the U.S. gross national product represented more than 50% of the combined gross national products of all world economies; in the late 1980s, the U.S. share had fallen to 22% (29), pointing to the increasing international economic parity that has developed in the late twentieth century. The economic power the United States once projected on the rest of the world has given way to an emerging world economy of which the United States is only a part. Increasingly, events outside the U.S. national borders influence and affect domestic companies. This will probably continue to be true in the U.S. food industry.

The effects of globalization on the U.S. food system have been and will continue to be manifold. Discussions of these effects tend to focus on two areas: international trade and foreign ownership of U.S. food companies. Government statistics point out the U.S. mix of export products in recent years has been roughly 30% processed foods and 70% raw commodities, whereas U.S. imports have been essentially 50% of each (9). Such data are often used as arguments to suggest a weakness in the U.S. approach to international markets for food products. Specifically, it has been argued that a significant contribution to the U.S. gross national product is lost by not adding value to commodities prior to shipment (30). It should also be noted, however, that the United States does in fact enjoy a trade surplus in food products (more than $5.5 billion in 1985, for example), mainly due to commodity exports. Some analysts have expressed concern that the United States may be entering a period of growing trade deficits in processed foods, partially as a result of the tradition by U.S. processors to locate facilities offshore rather than export to foreign markets from U.S. facilities (31).

Foreign ownership of U.S. food companies has expanded dramatically. In 1987, for example, total foreign investment in U.S. companies was over $16 billion, nearly 60% of which came from The Netherlands and the UK (Table 8). The impact of this trend will likely express itself with increasing significance in the coming years as the new owners implement business development approaches and strategies consistent with their home countries.

Technological Change

Much of the early development of the U.S. food system was catalyzed by the development and introduction of new technologies, especially in production and processing. A list of the most significant food science innovations from 1939 to 1989 has been prepared and includes (1) aseptic processing and packaging, (2) minimum safe canning processes for vegetables, (3) the microwave oven, (4) frozen concentrated citrus juices, (5) controlled atmosphere packaging for fruits and vegetables, (6) freeze drying, (7) frozen meals, (8) the concept of water activity, (9) food fortification, and (10) ultrahigh temperature processing of milk and other products (32). Entire businesses have been developed on the basis of such technologies, as well as other significant technologies that did not make the list.

In the latter half of the twentieth century, food companies, in particular processors, have concentrated most of their research efforts on the development and proliferation of new products, rather than development of technology. New product introductions have increased steadily in recent years, from 5,617 new products in 1985 to 9,192 in 1989 (Table 9). Some predict this trend to continue, with over 20,000 new product introductions per year forecasted for the early twenty-first century (34). Most of these new products do not represent a great technological change. In

Table 8. Foreign Direct Investment in U.S. Food Products Industry, 1980 and 1987[a]

Country of Origin	1980	1987	1980	1987
	Millions of Dollars		Percent of Total	
Total foreign investment	4,896	16,004	—	—
Canada	—	430	—	2.7
The Netherlands	225	5,492	4.6	34.3
UK	1,098	4,156	22.4	26.0
Other Europe	765	4,242	15.6	26.5
Japan	98	164	2.0	1.0
Austrailia, New Zealand, and South Africa	4	1,342	0.1	8.4
Latin America	101	24	2.1	0.1
Other countries	2,605	584	53.2	3.6

[a] From Ref. 19.

Table 9. New Food Product Introductions, 1985–1989[a]

Food Category	1989	1988	1987	1986	1985
Baby foods	53	55	10	38	14
Bakery foods	1,155	968	931	681	553
Baking ingredients	233	212	157	137	142
Beverages	913	936	832	697	625
Breakfast cereals	118	97	92	62	56
Condiments	1,355	1,310	1,367	1,179	904
Candy, gum, and snacks	1,701	1,608	1,145	811	1,146
Dairy products	1,348	854	1,132	852	671
Desserts	69	39	56	101	62
Entrées	694	613	691	441	409
Fruits and vegetables	214	262	185	194	195
Pet Foods	126	100	82	80	103
Processed meats, poultry, and fish	509	548	581	401	383
Side dishes	489	402	435	292	187
Soups	215	179	170	141	167
Total Food	*9,192*	*8,183*	*7,866*	*6,107*	*5,617*

[a] Ref. 33.

fact, by most measures, the food industry has not positioned itself for significant internal technology development. Research and development investments are low compared to other industries, public investment in food research for postproduction industries is low, and food-industry employment of engineers and scientists is considerably lower than other industries (10). Still, productivity in the food industry has grown at an annual rate of 3% since 1963, largely due to borrowing technologies from other industries.

It is impossible to forecast technological breakthroughs that might occur. Yet, a number of emerging technologies do appear poised to have significant impact on the food industry, including biotechnology, separation techniques, new processing technologies, and information processing–automation.

Biotechnology is, in fact, a group of technologies, and its effects will likely be extremely diverse (35,36). Production agriculture, for example, will probably see the development of crops with unique characteristics, including disease and pest resistance and improved yield and quality. Animal agriculture will see the introduction of growth regulators and perhaps genetic manipulation of the animals themselves. Via gene splicing, it will be possible to produce food ingredients and enzymes or, perhaps, major food components by use of microorganisms. New diagnostic and analytical procedures will be introduced to ensure the safety and quality of foodstuffs more adequately. These are but a few of the likely outcomes of current biotechnology research that could impact the food industry.

Separation techniques allow basic commodities to be fragmented efficiently into components for restructuring or incorporation into other foods or for removal of undesirable compounds. Corn wet milling, introduced in the United States in the mid-twentieth century, is an example. The emerging separation technologies include passive membrane filtration (reverse osmosis, ultrafiltration, and microfiltration) and supercritical fluid extraction. These technologies have already achieved some commercialization in the food industry.

New processing technologies should provide more efficient means of converting commodities into high-quality foods. Emerging technologies in this category include microwave and ohmic heating, as well as irradiation. The future of irradiation in the United States is jeopardized by consumer resistance due to safety concerns, which appear to be unwarranted on the basis of experience in other countries (37).

Information processing and automation often go unnoticed in technological forecasts. Yet, information processing, in particular at the retail level and including the development of direct product profitability (DPP) strategies, has effectively shifted the balance of power in the food system from processors to retailers. New information processing hardware and software, coupled to online sensing and control devices may encourage increased vertical integration in the industry, or at the very least, strategic alliances, by linking retail activity with distribution, processing, and, perhaps, production. At the processing level, computer integrated manufacturing (CIM), which integrates order entry, scheduling, operations, inventory control, and other operations management activities, has already been implemented in many facilities. Further development of this area could have profound effects on food industry organization and performance (38).

It is important to recognize that technological change itself will be shaped by the marketplace. Technologies will be developed and successfully introduced only when an economic advantage is afforded or a need in the marketplace is met and when consumers are willing to accept them.

Changes in the U.S. Population

Much has been written about the changing U.S. population, especially regarding age distribution (Fig. 3) and its effect on U.S. business and industry. The food industry is no exception. Broadly, the effects are twofold, affecting consumer demand for foodstuffs and in changing the makeup of the work force.

Consumer demand, in fact, will be the principal source of growth and development in the food industry in the

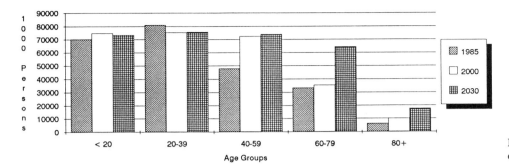

Figure 3. Projected age distribution of the U.S. population, 1985–2000.

future (10). Economists do not forecast significant changes in total caloric intake, but do anticipate consumer demand to be affected by increases and demographic changes in the U.S. population, shifts in income distribution, and changes in consumer preferences. Increases in total consumer demand can be fairly accurately forecasted based on Census Bureau projections of the U.S. populations (39). Demographic, income, and preference shifts, however, quantitatively and qualitatively affect demand. These projections have been made (Table 10), but generally they are based on the premise that consumers of certain age and income groups in the future will exhibit the same preference as persons of similar age and income groups in today's population. This premise is not universally accepted by marketers. In fact, changes in preference are almost certain to occur. Some preference changes will arise from continued immigration and assimilation of new cultural values. Others will arise from changes in household patterns. Most, however, will arise from poorly understood social phenomena. These changes in preference may influence future food industry development more than the others. Thus the ability to quickly react to changing preferences will be significant in industry development.

The changing work force has been noted, especially with respect to declining numbers of potential workers in the age groups normally associated with entry-level jobs. Another shift has been and will be the increasing assimilation of women, minorities, and immigrants into the work force (Fig. 4). The changing work-force composition will place special demands on human resource management and development in the coming years, especially in the food industry, where a high percentage of entry-level workers are employed.

Mergers and Acquisitions

The period from the late 1970s through the 1980s in U.S. business will likely be remembered as the period of intensive activity in mergers and acquisitions. As previously stated, this was not the first such period for the food industry. However, unlike mergers of the early twentieth century, which largely enhanced operating efficiency, the latter-day consolidations have generally been motivated by the desire for short-term financial gain by a small number of investors. It is difficult to point to a precise event or series of events that initiated the latter trend, but activity appears to have suddenly increased in the mid 1970s (10). Analysts often cite the hostile takeover of the ESB Co. of Philadelphia, overseen by large Wall Street investment firms, as a turning point in business, in which hostile takeovers became respectable (41). The increasing use of high-yield (junk) bonds provided increased financial liquidity to support these types of takeovers, leading to a characterizing form of acquisition of this period known as the leveraged buyout (LBO). It has been said that this new financial liquidity, in fact, led to asset liquidity, in which the assets of companies were bought and sold almost as readily as their products.

Figure 5 summarizes recent food industry merger activity in four sectors. It is noteworthy that merger activity in food manufacturing has been much more prevalent than in the other sectors. Not evident in this presentation, however, is the fact that certain recent mergers have been

Table 10. Percent Changes in Food Expenditures, 1980–2000[a]

Category	Change Due to Age Distribution	Change Due to Income	Net Change—All Effects
Food, general	1.7	21.7	22.7
Vegetables	4.5	13.5	18.7
Fruits	4.1	13.4	18.1
Beef	3.0	13.4	16.1
Pork	5.7	8.2	14.2
Fats and oils	4.2	9.8	13.6
Poultry	5.5	4.4	10.8
Cereals	3.1	8.4	10.4
Sugars	2.1	8.2	9.8
Dairy	2.2	7.7	9.0

[a] Ref. 40.

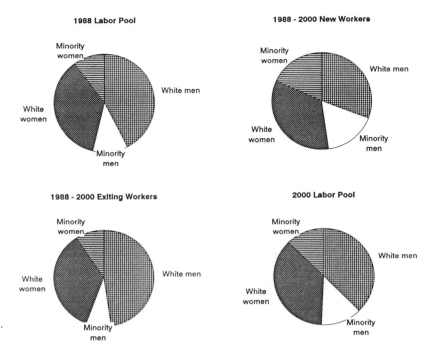

Figure 4. The changing labor pool, 1988–2000. From Ref. 41.

on a much larger scale than in the past. Of concern to economic developers is the effect this new wave of restructuring has had and will have on the food industry. Most recent commentary on this subject by economists has suggested a negative impact on rates of growth in productivity and a significant reduction in research and development (10,42). In addition, some concern has been expressed for the trauma such activity causes to the employees of affected firms. Professional organizations and associations, such as the Institute of Food Technologists, have called the U.S. government to increase their oversight of this activity.

It is uncertain whether this trend will continue, but one study suggested that food companies are particularly susceptible to leveraged buyouts for four reasons: (1) food stocks have been chronically undervalued, (2) the food industry is more predictable and less cyclical than most industries, (3) food companies generally generate high cash flows, and (4) purchase of brands, often available through food company buyouts, is particularly attractive (42). Furthermore, the long-term decline in the dollar against most foreign currencies makes the assets of U.S. food companies especially attractive to foreign companies looking to increase penetration into U.S. markets.

Economic Development Programs

In recent years, economic development has received increasing direct support from the public sector. As communities, states, and the nation seek to expand their economic bases, programs and direct support to the private sector are being offered by public agencies and institutions.

In the food system, the public has traditionally supported most research for production agriculture, as previously mentioned. More recently, public sector interest in supporting value added research and development has increased. A number of states have funded such activities and, in addition, created organizations or centers at universities to focus such activities. There is not a broad consensus as to the worthiness or probable effectiveness of such programs (43). In parallel with these centers for re-

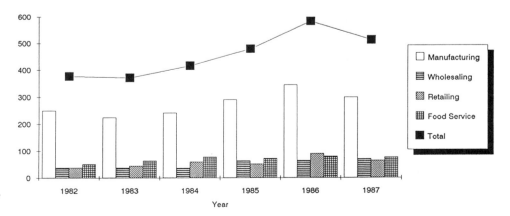

Figure 5. Food industry mergers, 1982–1987. From Ref. 16.

search and development has been the creation of outreach or extension units to provide customized business and technical assistance to food companies. Often these latter units serve as the focus for value added research as well as the extension mission.

Publicly supported economic development also includes broader activities such as development of capital, tax incentive programs for expansion and relocation, industry attraction efforts, publicly supported advertising and promotion for agricultural products, and grants to communities for training, construction and economic analyses. Many of these programs are in part politically inspired, and focus on job creation and new business starts (or business relocated). Some concern has been expressed for many of these programs operating at the community and state level as being zero-sum for the nation as a whole (44), because they may not result in incremental contributions to the gross national product.

Many economists, in fact, believe that real economic development is only achieved through increases to the gross national product or, as it is often stated, wealth creation. Economic development in the U.S. food industry in this sense will be achieved through improved productivity, which in turn, will arise from improvements in efficiency and effectiveness. Success will be measured by continued increases in value added by the various sectors.

BIBLIOGRAPHY

1. J. J. Putnam, *Food Consumption, Prices and Expenditures, 1966–87*, U.S. Department of Agriculture, Washington, D.C. 1989.
2. W. Bishop, "The Retail Profit Wedge," *Marketing Communications* 12(10), 64–70 (1987).
3. T. Pierson, unpublished data.
4. J. M. Connor, R. T. Rogers, B. W. Marion, and W. F. Mueller, *The Food Manufacturing Industries*, Lexington Books, Lexington, Mass., 1985.
5. National Technical Information Service, *Standard Industrial Classification Manual 1987*, Office of Management and Budget, Springfield, Va., 1987.
6. C. A. Reed, *Origins of Agriculture*, Mouton Publishers, The Hague, 1977.
7. S. A. Goldblith, "Fifty Years of Progress in Food Science and Technology: From Art Based on Experience to Technology Based on Science," *Food Technology* 43(9), 88–102 (1989).
8. E. C. Haupe, Jr., and M. Wittenberg, *Development of the Food Industry*, McGraw-Hill, Inc., New York, 1964.
9. U.S. Department of Commerce, *Statistical Abstract of the United States*, U.S. Government Printing Office, Washington, D.C., 1987.
10. Research Committee, Institute of Food Technologists, "The Growth and Impact of the Food Processing Industry: A Summary Report," *Food Technology* 42(5), 95–110 (1988).
11. U.S. Department of Commerce, Bureau of the Census, *1987 Census of Manufactures*, U.S. Government Printing Office, Washington, D.C., 1989.
12. U.S. Department of Commerce, Bureau of the Census, *1982 Census of Wholesale Trade*, U.S. Government Printing Office, Washington, D.C., 1985.
13. U.S. Department of Commerce, Bureau of the Census, *1982 Census of Retail Trade*, U.S. Government Printing Office, Washington, D.C., 1985.
14. U.S. Department of Commerce, Bureau of the Census, *1987 Census of the Wholesale Trade*, U.S. Government Printing Office, Washington, D.C., 1989.
15. P. Donegan, "DPP: Still Going Strong," *Progressive Grocer* 67(12), 39–45 (1988).
16. U.S. Department of Agriculture, *Food Marketing Review, 1988*, Commodity Economics Division, Economic Research Service, Agricultural Economics Report No. 614.
17. J. N. Ferris, "A Description of Michigan Agriculture and the Food Industry," Michigan State University Agricultural Economics Staff Paper No. 87-36, (1987).
18. U.S. Department of Agriculture, *Agricultural Statistics 1988*, U.S. Government Printing Office, Washington, D.C., 1988.
19. U.S. Department of Commerce, *Statistical Abstract of the United States*, U.S. Government Printing Office, Washington, D.C., 1989.
20. National Science Foundation, *Research and Development in Industry, 1981*, U.S. Government Printing Office, Washington, D.C., 1983.
21. R. J. Swientek, "Top 100© R & D Trends," *Food Processing* 49(8), 30–34 (1988).
22. D. D. Duxbury, "R & D Directions for the 1990s," *Food Processing* 49(8), 19–26 (1988).
23. "Fifty-Sixth Annual Report of the Grocery Industry," *Progressive Grocer* 68(4), 4–50 (1989).
24. M. Duff, "If You Notice Part-Timers Leaving, It's Too Late," *Supermarket Business* 44(5), 90–140 (1989).
25. "Industry Summary," *Restaurant Business* 87(14), 72–126 (1988).
26. D. Nichols, "The Fast Food Dilemma," *Incentive Marketing* 8(3), 46–49 (1985).
27. "Trends in Full-Service Restaurants," *Restaurant Management* 2(1), 38–40 (1988).
28. N. Byal, "Guess What's Coming Home for Dinner," paper presented at the Seventh Annual Conference on New Products, Gorman Publishing, Inc., Ft. Lauderdale, Fla., (Oct. 1989).
29. National Research Council, *Investing in Research*, National Academy Press, Washington, D.C., 1989.
30. G. Mathia and R. Elleson, "Trading High Value Products," *National Food Review* 38, 14–21 (1987).
31. R. G. McGovern, "Worldwide Consumer Trends and the Competitive Position of the U.S. Food Industry," *Food Technology* 42(9), 119–126 (1988).
32. Institute of Food Technologists, "Top Ten Food Science Innovations, 1939–1989," *Food Technology* 43(9), 308 (1989).
33. *Gorman's New Product News* 25(12) 4 (1990).
34. M. Friedman and L. Dornblaser, "Slower Rate in 1988," *Grocery Marketing* 55(1), 48 (1989).
35. *Biotechnology and the U.S. Food Industry*, Technomic Publishing Company, Lancaster, Pa., 1989.
36. S. Harlander, "Food Biotechnology: Yesterday, Today and Tomorrow," *Food Technology* 43(9), 196–206 (1989).
37. C. M. Bruhn and H. G. Shultz, "Consumer Awareness and Outlook for Acceptance of Food Irradiation," *Food Technology* 43(7), 93–94 (1988).
38. K. Kolbe, "How CIM Is Restructuring the American Food Industry," *Food Processing* 50(1), 49–58 (1989).
39. D. J. Bogue, *The Population of the United States*, The Free Press, New York, 1985.
40. The Food Institute, *Demographic Directions for Food Marketing*, American Institute of Food Distribution, Fairlawn, N.J., 1987.

41. J. Solomon, "Firms Grapple with Language Barriers," *Wall Street Journal,* Nov. 7, 1989, p. B1.
42. J. D. Dziezek, "Food Industry Mergers: A Perspective of Our Work Force," *Food Technology* **43**(6), 68–76 (1989).
43. G. W. Williams, ed., *Value-Added Research Investments: Boon or Boondoggle?* Texas Agricultural Market Research and Development Center, Special Series Report No. **88-1-89**, College Station, Tex., 1989.
44. J. N. Ferris, "A Proposal for the Food Industry Institute at Michigan State University," The Food Industry Institute, East Lansing, Mich., 1985.

WILLIAM C. HAINES
M. RATLIFF
Michigan State University
East Lansing, Michigan

FOOD LABELS. See FILTH AND EXTRANEOUS MATTER IN FOOD; FOOD LAWS AND REGULATIONS.

FOOD LAWS AND REGULATIONS PART I: AGRICULTURAL PRODUCTS

The topic of food laws and regulations is always a difficult one to discuss in a technical encyclopedia of this nature because space limitation prohibits details and because the question of which country to discuss is always a problem. Although most publications present data on international recommendations, their usefulness is limited for the simple reason that each country is interested in enforcing its own laws and regulations. This article concentrates on the laws and regulations applicable to the United States so that at least two groups of professionals will benefit from the information. Food processors in the United States and in those countries that export foods and beverages to this country. With few exceptions, laws and regulations in the United States also apply to imported items.

In the United States two federal agencies that promulgate mandatory regulations directly affecting the processing of foods and beverages include the U.S. Department of Agriculture (USDA) and the U.S. Food and Drug Administration (USFDA). This article will discuss those laws and regulations governing agriculture. Part II of this article will discuss processed foods and beverages. Note that all laws are described in the *United States Codes* (*USC*) and the implementing regulations are described in the *Code of Federal Regulations* (CFR).

MAJOR LAWS

Most laws and regulations affecting agricultural products are enforced by the USDA. One major legislation administrated by the USDA is the Agricultural Marketing Act of 1946. The act authorizes the USDA to perform many functions. One is to help the private marketing systems move food and other farm products from producer to consumer quickly, efficiently, and fairly. From this perspective, the USDA is responsible for:

1. Establishing standards for grades indicating quality of food and farm products and providing grading services for most of these commodities.
2. Administering marketing agreements and orders for fruits, vegetables, related specialty crops, and milk.
3. Providing market reporting services on grain, livestock, meat, poultry, eggs, fruits, vegetables, honey, peanuts, ornamental crops, dairy products, cotton, and tobacco.
4. Promoting fair trade practices in marketing agriculture products.
5. Purchasing food for USDA food-assistance programs.
6. Overseeing research and promotion programs conducted by industry for several commodities.
7. Enhancing food safety.
8. Conducting research to find better ways of marketing farm products at lowest possible cost.

Due to space limitations, this article will be concerned with two aspects of the Agricultural Marketing Act: food standards and grades, and food safety.

Standardization and Grading

The USDA develops grade standards for meat; cattle; swine; wool; poultry; eggs; dairy products; fresh, frozen, canned, and dried fruits and vegetables; cotton; tobacco; spirits of turpentine; and rosin. These standards and USDA's grading services provide buyers and sellers with an impartial appraisal of the quality of what is being sold and help farmers receive fair prices for their products. Grading services for food, livestock, cotton, and naval stores are available on request and for a fee paid by the users. USDA carries out many of the commodity grading services in cooperation with individual state departments of agriculture.

In relation to grading and standardization, the Agricultural Marketing Act has the following objectives:

1. Authorizing grading, inspection, and certification of all agricultural products; services are permissive or voluntary.
2. Making certifications *prima facie* evidence in court.
3. Prohibiting misrepresentation with respect to such services.

Enhancing Food Safety

The USDA protects the safety of agricultural products in two ways:

1. Using the Agricultural Marketing Act, the USDA provides voluntary services for the inspection, standardization, and grading of agricultural food products: fruits, vegetables, dairy products, poultry products, and meat products; voluntary services mean that the USDA will perform certain functions for a user for a prescheduled fee.
2. Using a number of other laws, the USDA enforces mandatory inspections for the safety of meat and poultry products, eggs and egg products, and grains.

The individual groups of agricultural products and the voluntary services and mandatory regulations governing them will be discussed in the following sections.

FRESH FRUITS AND VEGETABLES

One major responsibility of the USDA is to insure the quality of fruits, vegetables, and related products sold in the United States. It standardizes, grades, and inspects fruits and vegetables. These services are voluntary, the users pay the fees, and the Agricultural Marketing Act of 1946 provides the basic statutory authority. Some aspects of the voluntary grading and inspection of fresh fruits, vegetables, and related products are discussed in the following paragraphs.

Regulations

Official inspection services are offered on a fee-for-service basis. In terminal markets and at shipping points, the inspector examines a commodity and issues a certificate that identifies its condition or quality and indicates its compliance with federal specifications or grades or those designed by the seller or buyer. These certificates are legally acceptable as *prima facie* evidence in all federal and most state courts.

The USDA has cooperative agreements with various state agencies to render federal–state inspection services. Under these agreements, thousands of federally licensed officials perform inspection work at points of origin. The USDA establishes basic inspection policies and procedures and furnishes overall direction.

Inspection service for fresh fruits and vegetables means the following:

1. Determines and certifies or identifies as to the grade, quality, and condition of fresh fruits or vegetables and related products, including the condition of the container.
2. Performs related services such as reporting the temperatures of loads or lots of fresh products.
3. Observes conditions under which a product is being packed, observes plant sanitation as a prerequisite to inspection of the packed product either on a continuous or periodic basis, or checkloads the inspected product in connection with the marketing of the product.
4. Issues inspection certificates or reports related to the above functions and responsibilities.

Products will be inspected at appropriate points: shipping points, destination markets, and other destination points. Inspection is provided at shipping points in all areas covered by cooperative agreements entered into on behalf of the USDA with cooperating federal–state inspection agencies providing for this inspection work pursuant to authority contained in a federal statute. Inspection service will also be provided to areas approved by USDA headquarters. Inspection service is available at destination or central markets in which an inspection office is located. In regard to other destination points, inspection is made at any point conveniently reached from a terminal market in which an inspection office is located to the extent inspection personnel are available.

Continuous inspection service is made available to packers at shipping points and in terminal markets. The service was originally designed to aid packers who pack in consumer-size containers, but at present it may be used by packers who pack in containers of any size. During the full-time packing conducted at the packing plant, one or more inspectors are assigned to observe plant conditions and preparation and packing of the product. The inspector makes frequent quality checks on the commodity from the packing lines and examines samples of the packed products to determine whether they comply with the U.S. grade or specifications for which the commodity is being packed. The fee charged for any type of service follows a prearranged schedule. Both the U.S. grade standards and the inspection services are permissive and their use voluntary. Inspection and certification service for quality and condition shall be based on the appropriate standards promulgated by USDA, applicable standards prescribed by the laws of the state where the particular product was produced, specifications of any governmental agency, written buyer and seller contract specifications, or any written specification by an applicant that is approved by the USDA.

The USDA has developed detailed requirements for plants operating under continuous inspection on a contract basis. The premises shall be free from conditions objectionable to packing operations, including, but not limited to, litter, waste, and refuse within the immediate vicinity of the plant buildings; excessively dusty roads, yards, or parking lots; and poorly drained areas. The packing plant buildings shall be properly constructed and maintained in a sanitary condition. Each packing plant shall be equipped with adequate sanitary facilities and accommodations. All equipment used for receiving, washing, grading, packaging, or storing shall be of such design, material, and construction that it may be kept clean. The survey also includes a study of the operations and operating procedures.

Grade Standards

The USDA has established U.S. standards for the following: basic trading grades, grades of products for processing, and grades of nuts and special products; consumer standards for certain products; and specifications for damaged products.

PROCESSED FRUITS AND VEGETABLES

The USDA has established numerous grade standards for different categories of processed fruits and vegetables. The USDA provides voluntary services for the inspection and grading of processed fruits, vegetables, and related products. There are three basic types of inspection service: lot inspection, continuous inspection, and pack certification. Lot inspection means the inspection and grading of specific lots of processed fruits and vegetables that may be located in plant warehouses, commercial storage, railway cars, trucks, or any other conveyance, storage facility, or other location consistent with industry practices. Generally, under lot inspection the inspector does not have knowledge of conditions and practices under which the product is packed, grading is limited to examination of the finished processed product only.

Continuous inspection is inspection and grading services conducted in an approved plant where one or more inspectors are present at all times the plant is in operation. The inspectors make in-process checks on the preparation, processing, packing, and warehousing of all products under contract and insure compliance with sanitary requirements.

Pack certification is the conduct of inspection and grading services in an approved plant whereby one or more inspectors may make inspections of the preparation and processing of products but are not required to be present at all times the plant is in operation. Inspection and grading can be done under two types of contract. Under a designated lot contract, inspectors will grade and certify only those lots designated by the applicant. Under a quality assurance contract, inspectors will use information available from the applicant's quality-control records to certify lots, as requested, and will grade lots at random as often as necessary to verify the reliability of the applicant's quality-control system.

Prior to a plant being approved, or the inauguration of in-plant inspection services, the USDA will make a survey and inspection of the plant where such inspection services are to be performed to determine whether the plant and methods of operation are suitable and adequate for the performance of service in accordance with the terms of contract and the following requirements.

The plant survey and inspection will be based on the regulations issued by the U.S. Food and Drug Administration in manufacture, processing, packing, or holding. When the plant meets the requirements for the survey, inspection services may be inaugurated at a time mutually satisfactory to the plant management and USDA. When the plant fails the requirements of the survey, contract services shall be withheld until corrective action is completed to the satisfaction of the USDA.

MEAT QUALITY

Services

The USDA provides voluntary grading and certification services for meat. It also issues established standards for different livestocks, their carcasses and products. Grading service under the regulations consists of the determination and certification and other identification, requested by application, of the class, grade, or other quality of livestock under applicable standards issued by the USDA. Three kinds of grading service are specifically indicated in the regulations: meat acceptance service, carcass data services, and beef carcass evaluation service.

Large-volume meat buyers such as the military, prisons, hospitals, restaurants, schools, hotels, steamship lines, and airlines depend heavily on the acceptance service. Generally, the buyer specifies what is wanted, awards the contract by bid, and mandates that the products be certified by federal graders. The fee for the service is usually paid by the supplier who, in turn, adds the cost to the price that the buyer pays. The cost usually adds up to about a fraction of a cent per pound. For a small price, the buyer is guaranteed the quality of the meat.

The meat acceptance service of the division is based on USDA-approved Institutional Meat Purchase Specifications (IMPS). These specifications are the end products of extensive testing and development conducted by the division in conjunction with various segments of the meat industry. At present there are specifications available for a number of meat products. They have been prepared to make their use as easy as possible; each item can be referred to simply by number. Also, each item listed in these specifications is described in careful detail. When a purchasing agent orders any item, for instance, cubed steaks, there is no doubt on the part of the supplier or the federal meat grader as to exactly what is wanted. The armed forces have incorporated such specifications directly into their subsistence procurement reference sources. The major IMPS are listed below:

1. Institutional Meat Purchase Specifications—General Requirements.
2. Institutional Meat Purchase Specifications for Fresh Beef.
3. Institutional Meat Purchase Specifications for Fresh Lamb and Mutton.
4. Institutional Meat Purchase Specifications for Fresh Veal and Calf.
5. Institutional Meat Purchase Specifications for Fresh Pork.
6. Institutional Meat Purchase Specifications for Cured, Cured, and Smoked and Fully Cooked Pork Products.
7. Institutional Meat Purchase Specifications for Cured, Dried, and Smoked Beef Products.
8. Institutional Meat Purchase Specifications for Edible By-Products.
9. Institutional Meat Purchase Specifications for Sausage Products.
10. Institutional Meat Purchase Specifications for Portion-Cut Meat Products.

The carcass data service provides cattle feeders and producers with carcass data that determine the quality and yield grade factors of the carcasses and their by-products. This is especially important to persons who may not own the carcass, but who were financially interested in the live animal at some point during its development. The service works in the following way.

Producers and feeders purchase specially designed ear tags used for official identification from an authorized source. Each animal whose carcass data is needed will be assigned one of the bright orange, shield-shaped, and serially numbered ear tags. In the slaughterhouse, the federal meat inspector removes the tag from the ear, attaches it to the carcass, and then informs the official meat grader. After the usual chilling process, the tagged carcass will be evaluated by the grader with respect to quality and yield grade factors. The information will be recorded on a special form that is then sent to a carcass data center in the headquarters of the USDA. The center processes the data and returns them to the cooperator–cattle producer and feeder organizations, state departments of agriculture, and so on, from which the tags were initially purchased. The cooperator then forwards the data to the producer or feeder who originally purchased the ear tag. A fee is charged for each data from processed and returned. Some of the important value-determining carcass characteristics include conformation; maturity; marbling; quality grade; packer's warm carcass weight; adjusted fat thickness; rib eye area; kidney, pelvic, and heart fat; yield grade; and evaluation data.

A third grading service is the USDA's beef carcass evaluation service. This program attempts to certify detailed carcass information from specific slaughter cattle. The service is used extensively by breed associations, beef cattle producers, agricultural experimental stations, and others. The carcass evaluation service is based on positive identification of the live animal and its carcass. It is suitable, therefore, for use in sire evaluation and other performance-testing programs.

Other information or characteristics of the carcass can also be obtained if they are requested. For a nominal fee, an estimate of rib eye area and yield grade may also be made if such information is needed. When less-detailed information is requested, such as only carcass grade and weight, it may be furnished on a regular grading certificate.

Grade Misrepresentation

Misuse and misrepresentation of USDA grade names in a serious offense. It is the policy of the USDA to:

1. Eliminate and prevent misuse and misrepresentation of USDA grade names at all marketing levels between meat packer and the consumer.
2. Conduct frequent reviews of grade labeling and advertising at retail outlets.
3. Review all reported violations promptly.

According to the law:

> Whoever knowingly shall falsely make, issue, alter, forge, or counterfeit any official certificate, memorandum, mark, or other identification, or device for making such mark or identification, with respect to inspection, class, grade, quality, size, quantity, or condition, issued or authorized under this section or knowingly cause or procure, or aid, assist in, or be a party to, such false making, issuing, altering, forging, or counterfeiting, or whoever knowingly shall possess, without promptly notifying the Secretary of Agriculture or his representative, utter, publish, or use as true, or cause to be uttered, published, or used as true, any such falsely made, altered, forged, or counterfeited official certificate, memorandum, mark, identification, or device, or whoever knowingly represents that an agricultural product has been officially inspected or graded (by an authorized inspector or grader) under the authority of this section when such commodity has in fact not been so graded or inspected shall be fined not more than $1,000 or imprisoned not more than one year, or both.

Accordingly, the USDA meat graders will conduct periodic reviews of all retail outlets selling meat in their areas. They will also carry out the policy regarding misuse and misrepresentation of USDA grade names. Finally a preliminary review of all violations will be conducted and appropriate action, as outlined below, will be taken.

There are four major types of violation involving misuse and misrepresentation of USDA grade names: bribery, unauthorized application of roller brands or stamps, counterfeit devices, and false advertising or misbranding.

POULTRY QUALITY AND EGG GRADING

The USDA provides both voluntary and mandatory services and derives its authority from two federal statutes. The Agricultural Marketing Act authorizes the USDA to perform voluntary inspection and grading services for poultry, eggs, and rabbits. The fee for these services is paid by the user. The Egg Products Inspection Act authorizes the mandatory inspection of egg products and surveillance for shell eggs. The U.S. government assumes the cost of these programs. The grading service provided by the USDA includes

1. Determination of the class, quality, quantity, or condition of any product by examining each unit or representative unit drawn by a grader.
2. Issuance of a grading certificate with respect to the examination.
3. Identification of the graded product.
4. Determination that, with reference to an official plant, the product has been processed, handled, and packaged in accordance with regulations.
5. Any regrading or appeal grading of a previously graded product.

Poultry means any kind of domesticated bird, including, but not limited to, chickens, turkeys, ducks, geese, pigeons, and guineas. A rabbit means any domesticated rabbit, whether dead or live. Ready-to-cook poultry is any dressed poultry from which the protruding pinfeathers, vestigial feathers (hair or down as the case may be), head, shanks, crop, oil gland, trachea, esophagus, entrails, reproductive organs, and lungs have been removed; with or without the giblets; and that is ready to cook without need of further processing. It also includes any cut-up or disjointed portion of such poultry or any edible part. A ready-to-cook rabbit is any rabbit that has been slaughtered for human consumption, from which the head, blood, skin, feet, and inedible viscera have been removed and that needs no further processing before cooking.

Voluntary Inspection of Egg Products

The voluntary inspection and grading of egg products is designed for any service required but not covered by the mandatory service. For example, the regulations provide for the following kinds of services:

1. Inspection of the processing of products containing eggs in official plants.
2. Sampling and laboratory analyses of products.
3. Quantity and condition inspection of products.
4. Laboratory analysis of samples (with or without added ingredients) of products submitted to the laboratory by the applicant.

The regulations also present the official U.S. standards for palatability scores for dried whole eggs. It describes in detail the preparation of samples for the palatability test and palatability scores for dried whole eggs. The USDA also administers the regulations for grading of shell eggs and U.S. standards, grades, and weight classes for shell eggs.

Egg Grading Services

The egg grading program is voluntary. If an egg packing plant wants its eggs to carry the U.S. grade shield such as Grade AA, A, or B, it may employ the program. The official grade shield certifies that the eggs have been graded for quality and size under federal–state supervision.

The grading service for shell eggs requires an official grader to determine the class, quality, quantity, or condition of a large lot of eggs by examining specific and representative samples and to issue an official grading certificate. Sometimes the grading service involves continuous supervision, in an official plant, of the handling or packaging of any product. The grading service is based on the U.S. standards, grades, and weight classes for shell eggs. Also, with proper authorization, grading service may be rendered with respect to products bought and sold on the basis of institutional contract specifications or those of the applicant.

An example of egg grading is as follows. In order for cartoned eggs to be grade labeled, several requirements must be met. They are listed as follows:

1. The eggs must be graded by an identified under the supervision of a licensed grader; the eggs may be graded by an authorized company employee but then must be check graded by the supervising grader.
2. Eggs to be packed in consumer packages and marked U.S. Grade AA or A must be packed from eggs of current production.
3. Eggs to be packed in consumer packages and identified as U.S. Grade AA, Fresh Fancy, or U.S. Grade A under the quality-control program must be packed from eggs produced during a specified testing period.
4. Establishments that pack eggs with official marks in consumer packages must meet the facility and operating and requirements set forth in the USDA regulations.

Mandatory Egg Inspection

The Egg Products Inspection Act contains the following features:

1. Mandatory inspection of egg products, which include whole eggs, whites, yolks, and various blends in liquid, frozen, and dried forms.
2. Thorough and continuous inspection of the product processing plants and the entire operation from the selection of a shell egg for breaking to the final products; a final check is applied to the product after pasteurization, cooling, and packaging.
3. Mandatory application to eggs and egg products in interstate, foreign, and intrastate commerce.
4. Regulation of the distribution of certain qualities of shell eggs in commerce; use of nationwide standards, grades, and weight classes for eggs in commerce; restricted shell eggs are defined to include dirties, checks (eggs with cracked shells but not leaking), incubator rejectors, inedibles, leakers, and eggs unsuitable for human consumption; in addition, they are regulated with reference to their selective disposal and distribution. For example, the act provides authority to destroy all restricted eggs except checks and dirty eggs that may be shipped to officially inspected plants to be properly segregated and processed.
5. Prohibition of individual states from requiring state origin labeling of shell eggs except for noncontiguous areas (Hawaii, Alaska, Puerto Rico, and the Virgin Islands) of the United States; individual states may require the name, address, and licence number of the person processing and packaging eggs.
6. Mandatory application to those that break only a few cases of eggs each week, as well as to highly mechanized operations producing many thousand pounds of egg products a day. Registration is required of all individuals who handle (grade and pack) shell eggs for retail stores, households, restaurants, institutions, or food manufacturers; producer–packers with more than 3,000 hens, who pack eggs that ultimately reach the consumers; and all hatcheries.
7. Cooperation between the federal government and state agencies to enforce various provisions of the act, with the state to be reimbursed for services performed.
8. Equal treatment of imported egg and egg products and domestic products.
9. Provision of requirements for exemptions, record keeping, detentions, and penalties for violations.

Regulations

The regulations for implementing the Egg Products Inspection Act contain the following basic outline:

Classifications of shell eggs used in the processing of egg products
Egg cleaning operations
Breaking room facilities
Breaking room operations
Liquid egg cooling
Freezing facilities
Freezing operations
Defrosting facilities
Defrosting operations
Spray process drying facilities
Spray process drying operations
Spray process powder, definitions and requirements
Albumen flake process drying facilities
Albumen flake process drying operations

Drying, blending, packaging, and heat treatment rooms and facilities

Dried egg storage

Washing and sanitizing room or area facilities

Cleaning and sanitizing requirements

Health and hygiene of personnel

Pasteurization of liquid eggs

Heat treatment of dried whites

Laboratory tests and analyses

Inspection and disposition of restricted eggs

Prohibition of disposition of restricted eggs

Disposition of restricted eggs

Inspection of egg handlers

Identification of restricted eggs or egg products not intended for human consumption

Identification of restricted eggs

Nest run or current receipt eggs

Identification of inedible unwholesome, or adulterated egg products

According to the plant requirements, the plant should be free from objectionable odors, dust, and smoke-laden air. Rooms should be kept free from refuse, rubbish, waste materials, odors, and insects, rodents and from any conditions that may constitute a source of odors or engenders insects and rodents. There should be an efficient drainage and plumbing system for the plant and premises. Drains and gutters should be properly installed with approved traps and vents. The water supply (both hot and cold) should be ample, clean and potable, with adequate pressure and facilities for its distribution throughout the plant or portion utilized for egg processing and handling operations and protected against contamination and pollution.

All equipment and utensils used should be of appropriate design, material, and construction. These conditions permit the examination, segregation, and processing of eggs and egg products to be efficient, sanitary, and satisfactory. They allow easy access to all parts for thorough cleaning and sanitizing. Whenever practicable, all equipment should be made of metal or other impervious material that will not affect the product by chemical action or physical contact.

Any operation involving processing, storing, and handling of shell eggs, ingredients, and egg products should be sanitary and conducted as rapidly as practicable. Pasteurization, heat treatment, stabilization and other process should follow approved requirements. All operations should prevent any deterioration of the egg products. All loss and inedible eggs or egg products are placed in a container clearly labeled inedible and containing a sufficient amount of approved denaturant or decharacterant, such as FD&C brown, blue, black, or green colors; meat and fish by-products; grain and milling by-products; or any other approved substance that will accomplish the purposes. Shell eggs should be crushed and the substance should be dispersed through the product in amounts sufficient to give the product a distinctive appearance or color.

DAIRY PRODUCTS

The USDA offers programs that cover voluntary services relating to dairy products. The authority to carry out the service is provided by some provisions of the Agricultural Marketing Act of 1946. The USDA has issued the following types of regulations:

1. Regulations governing the inspection and grading services of manufactured or processed dairy products.
2. General specifications for dairy plants approved for USDA inspection and grading services.
3. U.S. standards for grade of various dairy products.

Each category of information is briefly discussed in the following sections.

Inspection and Grading Services

The USDA offers four major programs: plant surveys, inspection and grading, laboratory service, and the resident grading and quality-control service. An interested party must make a request and be qualified for any voluntary services; fees are based on a scale. To carry out its inspection and grading services, the USDA depends heavily on the grade standards they have been established for various dairy products.

The plant survey is the basic program. A plant must meet the requirements of prescribed specifications. This will qualify the plant for other services such as grading, inspection, sampling, testing, and certification of various dairy products.

The inspection and grading of dairy products are used extensively by dairy manufacturers, buyers, wholesalers, processors, restaurant owners, and others. This service makes extensive use of the USDA's grade standards. The USDA also extends its acceptance service of grading and inspection to volume dairy product buyers such as schools, restaurants, steamship and airline companies, Veterans Administration, Department of Defense, and state and local government agencies (eg, hospitals, nursing homes, and prisons). This acceptance service is similar to that for fruits and vegetables, livestock, and poultry products. The division examines and certifies that a supplier's delivery has complied with contract specifications. Official stamps, certificates, and so on are also required.

The laboratory service monitors the class, quality, conditions, and keeping properties of all dairy products. Specialists in the laboratory establishment make chemical and bacteriological determinations of the products. A number of such laboratories are maintained in different parts of the country. They also perform tests on dairy and nondairy products for industry and other federal agencies on a fee-for-service basis. The dairy technicians have the training and equipment to check almost any kind of food product.

The resident grading and quality-control service aims at serving the approved dairy plants. It provides quality checks on incoming raw material, in-processing controls, and sanitation. Grading and certification of the finished product are provided by an inspector stationed at the plant on a full-time basis. The program thus combines plant survey, inspection, grading, and laboratory services. To obtain the service, the plant must be approved and possess an approved laboratory.

The USDA has issued detailed specifications for each plant that manufactures, processes, and packages any of the following categories of dairy products: nonfat dry milk, instant nonfat dry milk, dry whole milk, dry buttermilk, butter and related products, cheese, cottage cheese, frozen desserts, pasteurized process cheese and related products, whey, whey products, and lactose.

Grades Standards

The USDA has promulgated grade standards for butter, Cheddar cheese, Colby cheese, dry buttermilk, dry whole milk, dry whey, edible dry casein (acid), ice cream, instant nonfat dry milk, Monterey Jack cheese, roller process nonfat dry milk, spray process nonfat dry milk, scorched particle standards for dry milk, sediment standards for milk and milk products, Swiss cheese, and bulk American cheese for manufacturing.

MEAT AND POULTRY INSPECTION

The Meat Poultry Inspection Program (MPIP) of the USDA inspects the slaughtering of certain domestic livestock and poultry and the processing of meat and poultry products. It has jurisdiction from the time that livestock and poultry are received at the slaughtering establishments until the finished products are distributed in commerce to consumers or for other purposes. This prevents the sale and distribution of adulterated or misbranded meat and poultry products, and it applies to all products in interstate or foreign commerce. The program reviews foreign inspection systems and packing plants that export meat and poultry to this country and reinspects imported products at ports of entry.

This program derives its authority from four major federal statutes: the Federal Meat Inspection Act, the Poultry Products Inspection Act, the Agricultural Marketing Act, and the Human Slaughter Act. The MPIP has also established criteria and procedures for designating establishments with operations within the jurisdiction of a state that would clearly endanger the public health, the disposition of products, and the application of regulations. Such an establishment will be placed under federal inspection. For example, the endangering of public health applies if any meat or meat food product prepared at the establishment is adulterated in any of the following respects:

1. It bears or contains a pesticide chemical, food additive, or color additive that is legally unsafe or was intentionally subjected to radiation in an illegal manner.
2. It contains an added poisonous substance that may be harmful to health.
3. It consists of any filthy, putrid, or decomposed substance; was prepared from a carcass affected with a disease transmissible to humans and its condemnation would be required; or is a ready-to-eat pork product that has not been treated to destroy trichinae.
4. It has been prepared under insanitary conditions.
5. It is the product of an animal that died other than by slaughter.
6. Its container is composed of harmful substances.

There are seven basic requirements that must be adequate in all federal plants, and state plants must also meet them if they are to be judged on an equal basis. Federal inspectors are required to indicate and certify their findings with respect to these requirements. The seven requirements are

Antemortem and postmortem inspection
Reinspection (processing)
Sanitation
Potable water
Sewage and waste disposal control
Pest control
Condemned and inedible material control

Corrective action must be taken immediately for any of the deficiencies identified. Some additional undesirable operations are

1. Use of nonpotable water in edible product departments of the establishment.
2. Improper sanitation that results in (a) bacterial growth and development in or on the product, (b) foreign matter entering the product, or (c) failure to control vermin and insects.
3. Use of unsound meat or poultry in processing food products.

Poultry Products

The regulations for poultry are fairly similar to those for meat. The USDA has issued definitions and standards of identity or composition for

Poultry meat content standards for certain poultry products
Canned bones poultry and baby or geritric food
Poultry dinners (frozen) and pies
Poultry rolls
[Kind] burgers and [kind] patties
[Kind] à la Kiev
[Kind] steak or fillet
[Kind] baked or [Kind] roasted
[Kind] barbecued
[Kind] barbecued prepared with moist heat
Breaded products
Other poultry dishes and specialty items
Maximum percent of skin in certain poultry products

Ready-to-cook poultry products to which solutions are added

Standards for kinds and classes, and for cuts of raw poultry

Definition and standard for turkey ham

Certain requirements governing operating procedures

All operations and procedures involving the processing and other handling or storing of any poultry product should be clean and result in sanitary processing, proper inspection, and the production of unadulterated poultry and poultry products. Any room compartment, or other place in an official establishment where any poultry product is processed should not house or handle any adulterating materials.

When thawing frozen ready-to-cook poultry in water, absorption of moisture should be avoided. Some approved thawing methods are as follows:

1. Use continuous running tap water below 70°F. Complete thawing is necessary to permit thorough examination before any further processing.

2. Placing frozen ready-to-cook poultry into cooking kettles without prior thawing is permitted only when a representative sample of the entire lot has been thawed and found to be sound and unadulterated. Thawing may be accomplished in cookers where the water is heated to enable the cooking process to begin immediately following completion of thawing. Any thawing technique should not result in a weight larger than the frozen weight. When whole carcasses or parts are thawed for repackaging as parts, it is not acceptable to recool the parts in slush ice. However, they may be held in tanks of crushed ice with the drains open, pending further processing or packaging.

3. The poultry may be thawed in recirculated water at less than 50°F for an optimal period.

Cut-up poultry should be processed from chilled carcasses, and the parts should not be rechilled in ice, water, or both. They may be held temporarily in crushed ice pending further processing and packaging. The ice must be in drained containers.

Any offal resulting from evisceration of the poultry should be removed from the premise as often as possible. Poultry products should be packaged in containers that are clean and with no potential adulterants and can withstand normal distribution without breaking or changing shape. Barrels or other containers used for packaging poultry products should be lined with paper or other material that remains intact when moistened.

When certain poultry products are in an official establishment or being transported between establishments, they should be placed under protective coverings. The coverings should be capable of excluding any foreign substances such as dust and dirt and should take into consideration the means of transporting the products.

Temperatures and procedures necessary to chill and freeze ready-to-cook poultry are important. They must insure prompt removal of animal heat, preserve the condition and wholesomeness of the poultry, and insure that the products are not adulterated. A description of the chilling and freezing procedures used at the official establishment is required to be filed with the inspector in charge. Some selected aspects are presented in the following paragraphs.

All poultry that is slaughtered and eviscerated should be chilled immediately after processing so that the internal temperature is reduced to 40°F or less unless such product is to be frozen or cooked immediately at the establishment. Eviscerated poultry to be shipped from the establishment in packaged form should be maintained at 40°F or less, except that during further processing and packaging operations, the internal temperature may rise to a maximum of 55°F. However, immediately after packaging, the poultry should be placed under refrigeration at a temperature that will promptly lower the internal temperature of the product to 40°F or less, or the poultry should be placed in a freezer. Poultry held at the plant in packaged form in excess of 24 h should be held in a room at a temperature of 36°F or less.

Poultry carcasses and their major parts should be chilled to 40°F or lower within the times specified. For example, 4 h for a carcass less than 4 lb, 6 h for 4–8 lb carcass and 8 h for a carcass over 8 lb.

Only ice produced from potable water may be used for ice and water chilling. The ice should be labeled and stored in a sanitary manner. If block type, the ice should be washed by spraying all surfaces with clean water before crushing. The temperature of the chilling media in the warmest part of any poultry chilling system should not exceed 65°F or any specified temperature in the regulations.

To facilitate continuous processing operations, poultry carcasses and major parts may be held overnight in chilling tanks containing water-saturated ice, refrigerated water, or other approved cooling media that will maintain al poultry in the tanks at 40°F or lower. Practices (such as reicing, recirculation of the chilling medium, holding the product in refrigerated rooms, or using increased amounts of ice) shall be employed that will result in all of the poultry in the chilling tanks being maintained at 40°F or lower throughout the holding period. Giblets should be chilled to 40°F or lower within 2 h after they are removed from the inedible viscera. Any acceptable methods of chilling the poultry carcass may be followed in cooling giblets.

The term chicken meat, unless modified by an appropriate adjective, is construed to mean deboned white and dark meat, whereas the term chicken may include other edible parts such as skin and fat not in excess of their natural proportion, in addition to the chicken meat. If the term chicken meat is listed and the product also contains skin, giblets, or fat, it is necessary to list each such ingredient.

Special Services Related to Meat and Other Products

The following special services related to meat and other products, are available.

1. *Identification Service.* A meat or other product that is federally inspected and passed domestically or on im-

portation is officially marked for identification. Such products may be divided into smaller or combined into larger units and still maintain its legal identity if the handling has been supervised by MPIP personnel and the resulting products marked and identified officially.

2. *Certification Service.* At the request of a purchaser, supplier, exporter, or others, inspectors may make certification regarding livestock products intended for human consumption (including casing) to be exported, as meeting conditions or standards that are not or are in addition to those ordinarily imposed by federal or state regulations and laws.

3. *Food Inspection Service.* An inspection and certification service for wholesomeness related to the manufacturing of a food article may be furnished on application. Applicable regulations of MPIP apply to the preparation, labeling, and certification of the food article prepared under such service.

4. *Reindeer Inspection Service.* The service inspects and certifies the wholesomeness relating to the slaughter of reindeer. All applicable regulations of the MPIP apply to such slaughter and the preparation, labeling, and certification of the reindeer meat and products prepared under such service.

Humane Slaughter Act

Another important federal statute administered jointly by the MPIP and other offices within the USDA is the Humane Slaughter Act of 1958. This act establishes the use of humane methods of slaughter as a policy of the United States and requires that packers who wish to sell meat and meat products to agencies of the federal government must use USDA-approved methods of slaughtering. Section 2 of the Act contains the following provision:

> No method of slaughtering or handling in connection with slaughtering shall be deemed to comply with the public policy of the United States unless it is humane. Either of the following two methods of slaughtering and handling are hereby found to be humane:
>
> (a) In the case of cattle, calves, horses, mules, sheep, swine, and other livestock, all animals are rendered insensible to pain by a single blow or gunshot or an electrical, chemical, or other means that is rapid and effective, before being shackled, hoisted, thrown, cast, or cut; or
>
> (b) By slaughtering in accordance with the ritual requirements of the Jewish faith or any other religious faith that prescribes a method of slaughter whereby the animal suffers loss of consciousness by anemia of the brain caused by the simultaneous and instantaneous severance of the carotid arteries with a sharp instrument.

The USDA has promulgated regulations to implement the act. For example, the federal government provides instruction on the slaughtering of animals with chemical (carbon dioxide), mechanical (captive bolt or gunshot), and electrical (stunning with electric current) means.

Meat and Poultry Packing Plants: Construction and Layout

The federal meat and poultry inspection regulations contain sections dealing with sanitation and facilities, eg, packing plants construction and layout. These regulations form the basis for all decisions relating to such subjects. Like most regulations, however, they require interpretation and judgment in their application. Thus the USDA has made recommendations and suggestions regarding the application for approval of packing plants.

The recommendations are the result of past experiences of government and industry in designing, building, altering, and maintaining meat and poultry packing plants for operation under federal inspection. For the most part, they are applicable to both large and small volume operators. Proposals that are submitted for approval are evaluated on the basis of those practices necessary to process meat and poultry in a sanitary manner and to provide for inspection needs.

The scope and nature of the operations are taken into account in deciding the facilities and equipment necessary for each individual project. Some variation is allowable in existing construction if meat and poultry can otherwise be processed in a sanitary manner. It is strongly recommended that these guidelines be followed in all new construction.

The MPIP has issued a list of specifications or notations covering meat and poultry inspection requirements. This list is for the convenience of architects and engineers. Specifications accompanying drawings submitted for approval are usually selected from the list. The guidelines are designed for use by industry managers, engineers, architects, and federal inspectors, both for construction of new plants and for modification of existing plants. Changes in federal inspection regulations or requirements obviously supersede these guidelines. In evaluation an application for approval of a meat packing plant, the MPIP will give considerations to, among others, the following;

- Description of plans and specifications
- Location of establishment
- Water, supply, plant drainage, and sewage disposal system
- Plant lighting, ventilation, refrigeration, and equipment
- Hand-washing facilities, sterilizers, drinking fountains, and connections for clean-up hoses
- Facilities for processing edible product
- Design, equipment, and operation of meat and poultry slaughtering departments and related areas
- Required cattle slaughtering facilities
- Required cattle slaughtering facilities
- Required sheep, goat, and calf slaughtering facilities
- Required hog slaughtering facilities
- Required horse slaughtering facilities
- Required poultry slaughtering facilities
- Welfare facilities for plant employees
- Inspector's office and welfare facilities

The MPIP has provided other suggestions including

- Suggested notes on specifications to accompany drawings

Summary of suggested principal distances

Sample specifications for poultry plant drawings

Plans and diagrams

Locator key, to be included on drawings that do not show an entire floor

Suggested format for showing room finishes and details

Operations key, to be included on drawings to show location of operations

Cattle carcass on bleeding rail

Dressed cattle carcass sides

700-lb steer carcass and relation to various foot platforms

700-lb steer carcass and relation to flight-top inspection table

Cattle viscera inspection table and boot sterilizing facilities

Pipe safety fence for cattle dry landing area

Dressed sheep carcass and relation to moving-top viscera inspection table

Dressed sheep and lamb carcasses

Dressed calf carcass and relation to moving-top viscera inspection table

Dressed calf carcasses

Hog carcass and relation to various foot platforms

Hog carcass and relation to moving-top inspection table, 300 or more hourly slaughtering rate

Inspection facilities for stationary table layout, hogs, sheep, and calves

Inspection facilities for small moving-top table layout, hogs, sheep, and calves

Inspection facilities for up to 240 hogs, calves, or sheep per hour

Inspection facilities for large layout, hogs

Hog casing stripping facilities

Suggested small on-the-rail layout for cattle and hogs

Suggested small plant layout

Suggested on-the-rail layout with inspection facilities for 10 cattle per hour

Suggested on-the-rail layout with inspection facilities for 20 cattle per hour

Suggested layout with inspection facilities for 60 cattle per hour

Suggested layout with inspection facilities for 120 cattle per hour

Suggested slaughtering department with inspection facilities for all species

Suggested layout with inspection facilities for 125 hogs per hour

Details of cattle head conveyor

Typical electric stunning facilities for hogs

Suggested layout with inspection facilities for 660 hogs per hour

Suggested layout with inspection facilities for hot skinning 125 calves per hour

Cattle tripe washing facilities

Cattle viscera separating table

Cattle head flushing compartment

Final inspector's desk over moving viscera inspection table

Suggested two-inspector layout for young chicken slaughter

Modified traditional layout for chicken slaughter

Accepted Meat and Poultry Equipment

The MPIP reviews plant equipment, including materials used in the construction and provisions for its sanitary maintenance in proposed use. Basically, the concern is sanitary design, construction, installation, and maintenance. Equipment acceptable in sanitary design and construction may not be accepted unless its operation and output meet certain requirements. Such approval is needed to insure that the right equipment is based at the beginning instead of correcting any problems that result later. This is a constructive and preventive approach with advantages for the processor, equipment merchant, consumer, and inspection service.

The evaluation and approval apply to used as well as new equipment. Any equipment already installed, capable of producing acceptable products, and kept clean do not need approval. It must, however, be acceptable to the inspector in charge at the point of use, who may also exempt other equipment, including simple hand tools, equipment used for preparing packaging materials, equipment for handling or transporting packaged goods, equipment used in inedible departments, cleaning systems, utensil and equipment cleaning machinery, and many others.

Guidelines for Obtaining Authorization of Compounds to Be Used in Meat and Poultry Plants

The MPIP has issued guidelines for obtaining authorization of compounds to be used in meat and poultry plants. They assist manufacturers or distributors in applying for authorization to use certain chemicals or other materials in establishments operating under the USDA–MPIP program. On receipt of applications, consideration will be given to the suitability of preparations and their safety for use as directed.

Two federal laws require the maintenance of safe and sanitary conditions in federally inspected meat and poultry plants. These two laws are the Federal Meat Inspection Act as amended by the Wholesome Meat Act of 1967 and the Poultry Products Inspection Act as amended by the Wholesome Poultry Products Act of 1968. These acts are enforced by the MPIP.

The inspection program calls for authorization of the use of substances and compounds in the plants, because misuse of such material may result in adulteration or unwholesomeness of meat and poultry being processed. The scope of the compound evaluation program is national and international in significance. All chemicals produced anywhere within the United States for marketing to federally inspected meat and poultry plants must be evaluated by the USDA. In addition, chemicals produced outside of the United States for marketing to U.S. plants or to plants exporting meat or poultry products to the United States may require such evaluation. Although the MPIP deals mainly with firms that supply chemicals to federally inspected meat and poultry plants, its primary responsibility is to the federal inspectors in those plants. In that respect, the MPIP's primary consideration is to provide inspectors with continual assurance that chemicals used in federally inspected plants are authorized for use and that their proper use will not result in the adulteration or contamination of food products. For example, compounds that must be evaluated include

1. Maintenance and cleaning chemicals, sanitizers, pesticides, and certain other substances.
2. Packaging materials, marking and branding inks, coatings applied to equipment or structural members a prior to installation in the plant, and all materials such as metal alloys, plastics, belting, hose, etc proposed for use in association with processing facilities and equipment

Other chemicals proposed for use in meat and poultry plants are shell egg cleaning, defoaming, destaining, and sanitizing compounds.

Compounds that do not require evaluation are

Denaturants
Outdoor pest control substances
Compounds used in offices and other similar nonprocessing areas
Compounds used in cafeterias or other retail food-service areas
Compounds used in heating system
Compounds used in the treatment of certain inedible materials
Compounds used in holding pens
Compounds that are not applicable for evaluation
Compounds used in sewage or waste water systems outside the plant
Compounds used in secondary cooling loops
Compounds used on the exterior of buildings or immediate surrounding areas
Compounds used for cleaning or maintaining of the exterior of vehicles

The MPIP has issued guidelines regarding compounds and their acceptable uses, including

Cleaning compounds
Compounds for laundry use
Compounds used in inedible product processing areas and nonprocessing areas
Sanitizing compounds
Compounds for employee hand care
Pesticides
Potable water treatment compounds
Cooling and retort water treatment compounds
Boiler treatment compounds
Compounds for steamlines or primary cooling water loops
Poultry scald media
Hog scald media
Tripe processing compounds
Fruit and vegetable washing compounds
Lubricants
Sewer and drain treatment compounds
Absorbant–antislip compounds
Paints or other resinous or polymeric coatings
Solvents

The following compounds are not permitted as ingredients in meat and poultry products: hazardous substances, potentially harmful compounds, and odorous compounds.

FEDERAL GRAIN INSPECTION SERVICE

The Federal Grain Inspection Service (FGIS) is a major unit within the USDA. This service administers the voluntary and mandatory requirements of the U.S. Grain Standards Act and the voluntary requirements of the Agricultural Marketing Act of 1946.

Grain is an essential source of the world's total supply of human food and animal feed and is merchandised in interstate and foreign commerce. It is the policy of the U.S. government, for the promotion and protection of such commerce in the interests of producers, merchandisers, warehousemen, processors, and consumers of grain, and the general welfare of the U.S. population, to maintain the quality of grains. Thus, the responsibility of the FGIS is to provide for the establishment of official U.S. standards for grain, to promote the uniform application by official inspection personnel, to provide for an official inspection system for grain, and to regulate the weighing and the certification of the weight of grain shipped in interstate or foreign commerce in the manner described in this article with the objectives that grain may be marketed in an orderly and timely manner and that grain trading may be facilitated.

Among other provisions, the U.S. Grain Standards Act:

1. Provides for a national official inspection system for grain.
2. Establishes official U.S. standards of quality, quantity, and condition for corn, wheat, rye, oats, barley, flaxseed, grain sorghum, soybeans, mixed grain, and other grains if the circumstances warrant; standards may be revoked if necessary, and the act requires uniform application of standards by inspection personnel.
3. Provides for mandatory inspection, weighing, and certification for grains to be exported; all grains to be transferred out of or into an export elevator at an export port location are required to be officially weighed, and must have official or other approved standards. Official inspection and weighing may be waived under certain circumstances.
4. Provides for the mandatory certification of the grade and weight of a lot of grain that has been

officially inspected and weighed while being transferred into or out of a grain elevator, warehouse, or other storage or handling facility.

5. Establishes standards for accurate weighing and weight certification procedures and controls, including safeguards over equipment calibration and maintenance for grain shipped in interstate or foreign commerce.
6. Authorizes permissive or voluntary inspection of any grain engaged in domestic commerce in accordance with established standards or other approved criteria; an interested party must apply for the service.
7. Makes official certificates for inspection *prima facie* evidence in a court of law.
8. Permits inspection by official samples, submitted samples, or otherwise in commerce within the United States and Canada.
9. Permits the use of modern mechanical and electronic grain-sampling equipment.
10. Authorizes agreements with the Canadian government with respect to the official inspection and weighing in Canadian ports of U.S. export grain transhipped through Canadian ports.
11. Requires the operator of a grain handling facility to comply with a number of prerequisite requirements before official weighing is rendered by the service.
12. Requires the testing of all equipment used in the sampling, grading, inspection, and weighing of grain located at all grain elevators, warehouses, or other storage or handling facilities at which official inspection or weighing services are provided by the service.
13. Provides for licensing of grain inspectors for three-year renewable periods.
14. Authorizes the refusal, renewal, suspension, or revocation of licenses.
15. Authorizes the refusal of inspection and weighing services and the institution of civil penalties.
16. Requires the keeping of and access to records for all transactions involving grain, such as inspection, weighing, purchases, sales, transportation, storage, handling, treating, cleaning, drying, and blending.
17. Requires the registration of all persons engaged in the business of buying, handling, weighing, or transporting grain for sale in foreign commerce; there are many exemptions to this requirement, for example, no registration is needed for any grain producer who only incidentally or occasionally sells or transports grain that has been purchased.
18. Prohibits any state or other political subdivision to require the inspection or description in accordance with any standards of kind, class, quality, condition, or other characteristics of grain as a condition of shipment, or sale, of such grain in interstate or foreign commerce, or require any license for, or impose any other restrictions upon, the performance of any official inspection or weighing function . . . by official inspection personnel.

The FGIS has the following responsibilities and functions:

1. Administer and direct the program to provide federal grain inspection and weighing at export locations, state and private grain inspection and weighing at inland locations, and emergency federal grain inspection and weighing at inland locations.
2. Provide for federal and state certification of weights at export points and state and private certification of weights at inland points.
3. Provide for registration of grain firms engaged in buying or selling of grain for foreign commerce and handling, weighing, and transportation of grain foreign commerce.
4. Monitor the quality of U.S. grain shipped to foreign ports.
5. Develop and revise U.S. standards for grain and assigned Agricultural Marketing Act of 1946 commodities.
6. Provide for a compliance program involving regulatory activities: designation and delegation, licensing, registration, and auditing of records.
7. Administer and direct an inspection and grading program of rice, beans, peas, lentils, and other assigned processed products under the Agricultural Marketing Act of 1946.

In relation to grain standards, the USDA has been assigned the following functions:

1. Compile and evaluate data on the need for new standards or the adequacy of existing standards; prepare impact studies on proposed amendments or new standards.
2. Maintain FGIS library of historical material on standards, research, reports, etc.
3. Interpret existing standards, research reports, etc.
4. Review proposed inspection procedures for conformance to standards.
5. Serve as primary contact for cooperative research agreements with other units.
6. Develop new and objective tests and methods for determining grain quality.
7. Provide reference standards for FGIS methods and develop new reference standards as required.
8. Develop criteria and recommend specifications for mechanical or electronic instrumentation to improve reliability of grain inspection.

In relation to field management, the USDA has the following responsibilities:

1. Direct and coordinate inspection and weighing programs, including study, development, implementation, and review of procedures and program activities.
2. Provide technical direction and coordination of field offices, delegated and designated states, and official agencies.

3. Develop standards for the technical training of official personnel and conduct training for field offices and official agency personnel regarding new procedures, programs, etc.
4. Develop procedures for licensing delegated and designated personnel.
5. Develop and implement an inspection equipment program including specifications, testing procurement, calibration, maintenance, and repair of inspection and weighing equipment.
6. Perform laboratory and testing analyses on processed or nongraded commodities.
7. Supervise laboratories under contract with the federal government to perform analysis work.
8. Conduct testing of protein and moisture instruments.
9. Perform appeal inspections and certification of results.
10. Develop, implement, and maintain procedures to insure uniformity and accuracy of inspection results among field offices and designated agencies.

Mandatory Regulations

The FGIS has promulgated general regulations to enforce the Federal Grain Inspection Act. The purpose of this system is to promote the uniform and accurate application of the official grain standards and to provide inspection services that may be required by the act or desired by the grain industry. According to the regulations and the act, official inspection is interpreted to include the following:

1. Performed by official inspection personnel.
2. May be original inspection, reinspection, or appeal inspection.
3. Determination and certification of any or all of the following:
   a. The kind, class, quality, or condition of grain in accordance with established standards.
   b. The condition of vessels and other carriers or receptacles for the transportation of grain insofar as it may affect the quality or condition of such grains.
   c. On request of any interested party applying for inspection, the quantity of sacks of grain or other factors related to grain under other approved criteria.
4. The technique of determination includes sampling, inspecting, examining, testing, grading, and other methods.

Official weighing is interpreted as follows:

1. Performed by official inspection personnel.
2. Determination and certification of the quantity of a lot of grain under established standards and based on the following:
   a. Actual performance of weighing or its physical supervision, including the physical inspection and testing for accuracy of the weights and scales and the physical inspection of premises at which the weighing is performed.
   b. Monitoring of the discharge of grain into the elevator or conveyance.

The regulations specifically emphasize that certain kinds of inspection service are not authorized and cannot be performed. For example, any agricultural commodity or processed grain product not covered by the official grain standards cannot be inspected. Also, a grain cannot be inspected on the basis of an unofficial standard.

There are at least 10 types of official inspection service. Some examples include: official sample (lot inspection), type sample (lot inspection), warehouseman's sample (lot inspection), submitted sample inspection, quality information inspection, check weighing, sampling, stowage examination, and other miscellaneous services. An explanation for some of these services follows.

Under the official sampling (lot inspection) the official inspector samples an identified lot of grain, inspects the grain in the sample, and issues an official certificate about the examination. Under the warehouseman's sample (lot inspection) a licensed employee of a grain elevator or warehouse samples an identified lot of grain and submits the sample and a report to an official inspection agency. The official inspector then inspects the grain in the sample, grades the sample, files a report, and issues a certificate. Under the quality information inspection, the inspector makes frequent and periodic examinations of grain being loaded aboard or discharged from a ship, for official grade or factors. The inspector files a report and issues a certificate about the result.

One of the major tasks is to obtain an official sample of grain. With some exceptions, such as exporting grain, official samples of grain may be obtained by licensed employees of official inspection agencies, authorized employees of the service, licensed employees of grain elevators or warehouses, and other individuals who are licensed to sample grain under a contract with the service.

Some details of the method and order of inspection services are described as follows. First, all sampling, testing, grading, and related inspection services are required to be performed by official inspection personnel or other designated agents. Second, such inspection will be based on a list of items that must be considered in the order described: careful inspection of grain, prescribed tests and examination, review of previous inspection if any, and other pertinent and necessary information. Third, the sampling must be proportional or random and determinations must be based on a careful and accurate examination, testing, or analysis. The source of the sample must be clearly detailed. There are other details such as record of documents, order of service, and conflict of interest. They are all carefully described in the regulations.

The FGIS has also issued regulations governing official performance requirements for grain inspection equipment and the official performance and procedural requirements for grain weighing equipment and related grain handling systems. The FGIS has promulgated mandatory official grade standards for a number of grains.

BIBLIOGRAPHY

General References

Y. H. Hui, *United States Food Laws, Regulations, and Standards*, 2 vols., 2nd ed, Wiley-Interscience, New York, 1986.

Y. H. Hui, *United States Regulations for Processed Fruits and Vegetables*, Wiley-Interscience, New York, 1988.

Y. H. HUI
EDITOR-IN-CHIEF

FOOD LAWS AND REGULATIONS, PART II: FOODS AND BEVERAGES

The mission of the U.S. Food and Drug Administration (FDA) is to enforce laws enacted by the U.S. Congress and regulations promulgated by the agency to protect the consumer's health, safety, and pocketbook in relation to all forms of processed foods and beverages. Before the laws and regulations enforced by this agency are discussed, it is important to realize that all information pertains to foods and beverages produced in or imported into this country. Two legal abbreviations used in this article USC for the *United States Codes* and CFR for *Code of Federal Regulations*. USC is the legal document that describes the laws and statutes of the United States. CFR is the legal document that describes those regulations promulgated in relation to those laws stated in the USC.

This article is a summary of the basic information. The information discussed in this article cannot and should not be used as *prima facie* evidence in a court of law. The laws enforced by the FDA are discussed first; they include

1. The Federal Food, Drug, and Cosmetic Act (21 *USC* 301–392) and the Fair Packaging Labeling Act (15 *USC* 1451–1461), which apply to foods and drugs for humans or animals, cosmetics, and medical devices.
2. Sections of the Public Health Service Act, relating to electronic products that emit radiation, such as x rays, lasers, microwave ovens, and television sets (42 *USC* 263b–263n).

The Federal Food, Drug, and Cosmetic Act is the basic food and drug law of the United States. With numerous amendments it is the most extensive law of its kind in the world. Many of the states in the United States have laws similar to the federal law and some have provisions to add automatically any new federal requirements.

The law is intended to assure the consumer that foods are pure and wholesome, safe to eat, and produced under sanitary conditions; that drugs and devices are safe and effective for their intended uses; that cosmetics are safe and made from appropriate ingredients; and that all labeling and packaging is truthful, informative, and not deceptive. Another law, the Fair Packaging and Labeling Act, affects the contents and placement of information required on the package.

PROHIBITED ACTS (VIOLATIONS)

The Federal Food, Drug, and Cosmetic Act prohibits distribution in, or importation to, the United States of articles that are adulterated or misbranded. The term adulterated includes products that are defective, unsafe, filthy, or produced under unsanitary conditions (Sections 402, 501, 601). Misbranded includes statements, designs, or pictures in labeling that are false or misleading, and failure to provide required information in labeling (Sections 403, 502, 602). Detailed definitions of adulteration and misbranding are in the law itself, and hundreds of court decisions have interpreted them. The law also prohibits the distribution of any article required to be approved by FDA if such approval has not been given, the refusal to provide required reports, and the refusal to allow inspection of regulated facilities pursuant to Section 704.

The Prescription Drug Marketing Act, which became law in 1988, amended the Federal Food, Drug, and Cosmetic Act to prohibit (1) reimportation of U.S. produced prescription drugs by persons other than the manufacturer (except when authorized by the secretary for emergency medical care); (2) sale, purchase, or trade of drug samples; (3) sale, purchase, or trade of drugs by hospitals, health care entities, or charitable organizations with certain exceptions; (4) distribution of drug samples, except as provided in the act; and (5) the distribution of prescription drugs in interstate commerce without a state wholesaler's license.

PREMARKET TESTING AND APPROVALS

The Federal Food, Drug, and Cosmetic Act and the Public Health Service Act require manufacturers of certain consumer products to establish, before marketing, that such products meet the safety and effectiveness requirements of the law and are properly labeled. Substances added to food must be generally recognized as safe, prior sanctioned, or approved by specific FDA regulations based on scientific data. Samples of color additives must be tested and certified by FDA laboratories. Residues of pesticide chemicals in food commodities must not exceed safe tolerances set by the Environmental Protection Agency (EPA) and enforced by the FDA. Such premarketing clearances are based on scientific data provided by manufacturers, subject to review and acceptance by government scientists for scope and adequacy. Submission of false data to secure approvals is a criminal violation of the laws that prohibit giving false information the government.

FDA regulations prescribe the type and extent of premarket testing that must be conducted, depending on the legal requirements applicable to the particular product and on the technology available to fulfill those requirements. Testing may include physical and chemical studies, nonclinical laboratory studies, animal tests, and clinical trials on humans.

The importance of the toxicological and other data derived from such investigation demands that they be conducted according to scientifically sound protocols and procedures. FDA regulations (21 CFR 58) prescribing good laboratory practices for nonclinical research should be complied with. Inquiries should be addressed to the FDA.

IMPORTS

Although the legal requirements that must be met are the same for imported and domestic products, the enforcement procedures are necessarily different. Imported products regulated by the FDA are subject to inspection at the time of entry through U.S. Customs. Shipments found not to comply with the laws and regulations are subject to detention. They must be brought into compliance, destroyed, or reexported.

At the discretion of the FDA, an importer may be permitted to bring an illegal importation into compliance with the law before final decision is made as to whether it may be admitted. Any sorting, reprocessing, or relabeling must be supervised by an FDA investigator at the expense of the importer. Both foreign shippers and importers in the United States should realize that conditional release of an illegal importation to bring it into compliance is not a right but a privilege. Abuse of the privilege, such as repeated shipments of the same illegal article, may result in denial of the privilege in the case of subsequent importations.

The FDA has provided the following suggestions to foreign exporters and U.S. importers to expedite entries. Packaged foods, drugs, and cosmetics must be fully labeled. Unlabeled or partially labeled goods must necessarily be detained because of lack of the mandatory label information. This causes delays that might not otherwise occur. Products known to be illegal should not be imported on the theory that they can be brought into compliance with the law on arrival. Conditional releases for that purpose are discretionary under the Federal Food, Drug, and Cosmetic Act, and if the privilege is granted, delays and added expense to importers inevitably occur before final release or denial of release. Obviously objectionable conditions responsible for detentions, such as infestation of foods with insects and other vermin, should be corrected at the source.

U.S. Customs entries must be made promptly on arrival of shipments at the port of entry. The FDA cannot ordinarily act on an importation until entry is actually filed and the Custom Service has notified the agency's local office. A list of these offices can be obtained from the FDA or U.S. Customs. Entry documents should be flagged for the attention of the FDA representative at the appropriate office.

Foreign shippers conducting business transactions with the importing trade of the United States, as well as the importing purchasers, should bear in mind that there is a definite obligation to comply fully with U.S. laws affecting these commodities. Sustained effort should be made to bring about fundamental correction of any existing conditions that adversely affect purity, strength, or quality of such articles.

Such corrective measures might well be undertaken by individual producers and associations of producers, and possibly by foreign government agencies in the countries of production. By such procedures, eradication of causes of deterioration and contamination in a number of commodities has already been successfully accomplished in various parts of the world, with a resulting great reduction in number of adverse actions by the FDA. Individuals, associations, and various foreign government agencies have, in a number of instances, instituted systems of sampling and examination of specific lots intended for shipment to the United States for the purpose of predetermining whether they meet the requirements of the U.S. laws.

EXPORTS

Many U.S. producers are interested in exporting some of their products. If the item is intended for export only, meets the specifications of the foreign purchaser, is not in conflict with the laws of the country to which it is to be shipped, and is properly labeled, it is exempt from the adulteration and misbranding provisions of the act (Section 801(d)).

INTERSTATE SHIPMENTS

Within the United States compliance with the Federal Food, Drug, and Cosmetic Act is secured through periodic inspections of facilities and products, analysis of samples, educational activities, and legal proceedings. When violations are discovered, there are many regulatory procedures available. Adulterated or misbranded products may be voluntarily destroyed or recalled from the market by the shipper, or may be seized by U.S. marshals on orders obtained by the FDA from federal district courts. Persons or firms responsible for violations may be prosecuted in the federal courts and if found guilty may be fined or imprisoned. Continued violations may be prohibited by federal court injunctions. Violation of an injunction is punishable as contempt of court. Any or all types of regulatory procedure may be employed, depending on the circumstances.

When FDA investigators (consumer safety officers) observe unsanitary conditions or practices that may result in violation, they leave a written report of their observations with management. Manufacturers, by correcting these conditions or practices promptly, may bring their operations into compliance. Investigators may also make suggestions regarding other types of compliance problems, but they are not experts in all of the technical fields regulated by the agency. FDA investigators will also report any voluntary corrective action they witness during an inspection, or which management may bring to their attention.

Product Recalls

Product recalls have become a major means of consumer protection under the law. The FDA prefers when possible to promote compliance by means other than going into court. Recall of a violative product from the market by the manufacturer is generally the fastest and most effective way to protect the public. A recall may be voluntarily initiated by the manufacturer or shipper of the product or at the request of FDA. The first step, when a product must be recalled, is for the manufacturer or distributor to get in touch with the nearest FDA field office. Accurate and complete production and shipping records are vital to the success of a product recall. Products should be code labeled to show date and place of manufacture.

It is recommended that all manufacturers develop plans that can be put into effect if a recall emergency arises. Guidelines on FDA recall procedure and industry responsibilities are provided by the *Code of Federal Regulations* (21 *CFR* 7). Special provisions on recalls of infant formulas are in the Public Health Service Act and the Federal Food, Drug, and Cosmetic Act. While cooperation in a recall may make court proceedings unnecessary to remove the product from the market, it does not relieve a person or firm from liability for violations.

Seizure

Seizure is a civil court action against goods to remove them from the channels of commerce. After seizure, the goods may not be altered, used, or moved, except by permission of the court. The owner or claimant of the seized merchandise is usually given about 30 days by the court to decide on a course of action. The accused may do nothing, in which case the goods will be disposed of by the court. If the decisions to contest the government's charges by filing a claim and answering the charges and the case will be requesting permission of the court to bring the goods into compliance with the law. In the latter option, the owner of the goods is required to provide a bone (money deposit), to insure that the orders of the court will be carried out, and must pay for FDA supervision of any compliance procedure.

Regulations

Regulations issued by the FDA are an important part of the food and drug law. Especially important are such regulations as the current Good Manufacturing Practice Regulations, which set requirements for sanitation, inspection of materials and finished products, and other quality controls. FDA Food Standards set specifications for many foods. Such regulations help both consumers and industry by indicating what must be done to insure acceptable products. All FDA regulations are updated and republished annually in *Code of Federal Regulations*, Title 21.

REGISTRATION REQUIREMENTS

The Federal Food, Drug, and Cosmetic Act and FDA regulations require the registration of establishments and products in two major food categories: low-acid canned foods and acidified foods. All commercial processors of heat-processed low-acid canned foods and acidified foods are required to register the establishments and file processing information for all such products with the FDA on appropriate forms. Registration and process filing is required for both U.S. establishments and those in other countries that export such foods to the United States (21 *CFR* 108.25 and 108.35).

Registration forms and process filing forms and instructions are obtainable from the FDA. New establishments must register within 10 days from the date the plant begins processing. Processes must be filed no later than 60 days after registration and before packing a new product. Modifications and changes in previously filed processes are made through an amended submission. Foreign firms must register and file processing information before shipping any low-acid canned food and acidified food to the United States. Failure to comply may result in legal proceedings against U.S. firms or their products or the detention of shipments from foreign firms.

Full text of the low-acid canned food and acidified food regulations is in the *Code of Federal Regulations*, Title 21, parts 108, 113, and 114. These regulations can be obtained from the FDA. The regulations provide a comprehensive guide to the requirements for the processing of low-acid canned foods and acidified foods.

COLOR ADDITIVES

The Federal Food, Drug, and Cosmetic act provides that foods are adulterated if they contain color additives that have not been proved safe to the satisfaction of the FDA for the particular use. A color additive is a dye, pigment, or other substance, whether synthetic or derived from a vegetable, animal, mineral, or other source that imparts a color when added or applied to a food (201(f)). Regulations (21 *CFR* 73, 74, and 81) list the approved color additives and the conditions under which they may be safely used, including the amounts that may be used when limitations are necessary. Testing and certification by the FDA of each batch of color is required before that batch can be used, unless the color additive is specifically exempted by regulation. The manufacturer who wants to use color additives in foods should check the regulations to ascertain which colors have been listed for various uses. Before using a color read the label, which is required to contain sufficient information to insure safe use.

Manufacturers of certifiable colors may address requests for certification of batches of such colors to the FDA. Certification is not limited to colors made by U.S. manufacturers. Requests will not be received from foreign manufacturers if signed by both such manufacturers and their agents residing in the United States. Certification of a color by an official agency of a foreign country cannot, under the provisions of the law, be accepted as a substitute for certification by the FDA. Copies of regulations governing the listing, certification, and use of colors in foods shipped in interstate commerce or offered for entry into the United States are available from the FDA.

SCIENTIFIC FUNCTIONS: METHODS OF ANALYSIS

Modern scientific methods are required to enforce laws such as the Federal Food, Drug, and Cosmetic Act. Insuring the wholesomeness of foods would be impractical without reliable methods of laboratory analyses to determine whether products are up to standard. Food scientists in both government and industry must know the normal composition of products to distinguish them from those that are defective. They investigate the toxicity of ingredients, study the causes of food poisoning, test the potency of vitamins and thousands of drugs. Their investigation also cover the adequacy of controls over processing, packaging, and storage practices. Such research requires experts in many fields, but especially chemistry, microbiology, microanalysis, pharmacology, and both human and veterinary medicine. Any action taken by the FDA must be based on scientific facts that can be supported in court. The principal authority relied on for laboratory methods is the *Official Method of Analysis* of the Association of Official Analytical Chemists (AAOAC). This compendium of tested analytical methods, published since 1895, is the leading internationally recognized guide to analytical procedures for law enforcement.

PRINCIPAL REQUIREMENTS OF FOOD LAW

Section 201(f) of the of the Federal Food, Drug, and Cosmetic Act defines food as follows:

> The term "food" means: (1) articles used for food or drink for man or other animals; (2) chewing gum; and (3) articles used for components of any such article.

Below is a synopsis of the principal requirements of the act relating to foods, in nonlegal language. The numbers in parentheses are the pertinent sections of the statute itself or sections in the *Code of Federal Regulations*.

Health Safeguards

1. A food is illegal (adulterated) if it bears or contains an added poisonous or deleterious (harmful) substance that may render it injurious to health (402(a)(1)).
2. A food is illegal if it bears or contains a naturally occurring poisonous or deleterious substance that ordinarily renders it injurious to health (402(a)(1)).
3. Food additives (201(s)) must be determined to be safe by the FDA before they may be used in a food or become a part of a food as a result of processing, packaging, transporting, or holding the food (409).
4. Raw agricultural products are illegal if they contain residues of pesticides not authorized by the EPA or that are in excess of tolerances established by regulations of the EPA (408).
5. A food is illegal if it is prepared, packed, or held under unsanitary conditions whereby it may have been rendered injurious to health (402(a)(4)).
6. Food containers must be free from any poisonous or deleterious substance that may cause the contents to be injurious to health (402(a)(6)); some packaging materials, for example, plastic or vinyl containers, may be considered food additives and subject to regulations (409).
7. Only those colors found safe by the FDA may be added to food (706). A food is illegal if it bears or contains an unsafe color(s) (402(c)); unless exempt by regulation, colors for use in food must be from batches tested and certified by the FDA (706(c)).
8. A food is illegal if any part of it is filthy, putrid, or decomposed (402(a)(3)).
9. A food is illegal if it is prepared, packed, or held under unsanitary conditions whereby it may have become contaminated with filth (402(a)(4)).
10. A food is illegal if it is the product of a disease animal or one that has died by any means other than slaughter (402(a)(5)).

Economic Safeguards

Damage or inferiority in food must not be concealed in any manner (402(b)(3)). An example would be artificial coloring or flavoring added to a food to make it appear a better value than it is, as in the case of yellow coloring used to make a food appear to contain more eggs than it actually contains. Food labels or labeling (circulars, etc) must not be false or misleading in any particular (403(a)). Labeling is misleading not only if it contains false or misleading statements but also if it fails to reveal material facts (201(n)). A food must not be sold under the name of another food (403(b)). For example, canned bonito cannot be labeled tuna fish.'

A substance recognized as being a valuable constituent of a food must not be omitted or abstracted in whole or in part, nor may any substance be substituted for the food in whole or in part (402(b)(1) and (2)). For example, an article may not be labeled milk or whole milk if part of the butterfat has been skimmed.

Food containers must not be so made, formed, or filled as to be misleading (403(d)). A closed package cannot be filled to less than its capacity. A food for which a standard of fill or container has been prescribed (402) must comply with the fill requirements, and if the fill falls below that which is specified its label must bear a statement that it falls below such standard (403(b)(2)).

Required Label Statements

The law states that required label information must be conspicuously displayed and in terms that the ordinary consumer is likely to read and understand under ordinary conditions of purchase and use (403(f)). Details concerning type sizes, location, etc of required label information are contained in the FDA regulations (21 *CFR* 101), which cover the requirements of both the Federal Food, Drug, and Cosmetic Act and the Fair Packaging and Labeling Act. Food labeling requirements of the regulations are summarized as follows. If the label of a food bears representations in a foreign language, the label must bear all of the required statements in the foreign language, as well as in English. (Note that the Tariff Act of 1930 requires all imported articles to be marked with the English name of the country of origin.) If the food is packaged, the following statements must appear on the label in the English language.

1. The name, street address, city, state, and zip code of either the manufacturer, packer, or distributor.
2. An accurate statement of the net amount of food in the package.

3. The common or usual name of a food.
4. The ingredients in a food

Location. The name, street address, city, state, and zip code of either the manufacturer, packer, or distributor. The street address may be omitted by a firm listed in a current city or telephone directory. A firm whose address is outside the United States may omit the zip code. If the food is not manufactured by the person or company whose name appears on the label, the name must be qualified with an expression such as "Manufactured for" or "Distributed by."

Net Amount. An accurate statement of the net amount of food in the package. The required units of measure are the avoirdupois pound and the U.S. gallon but metric system measurements may also be used, if desired, in addition to the required declaration in English units. The quantity of contents declaration must appear on the principal display panel of the label in lines generally parallel to the base of the package when displayed for sale. If the area of the principal display panel of the package is larger than 5 in.^{2}, the quantity of contents must appear within the lower 30% of the label. The declaration must be in a type size based on the area of the principal display panel of the package (as listed in 21 *CFR* 101.105) and must be separated from the other information.

The net weight on packages containing between 1 lb (avoirdupois) and 4 lb must be declared first in total avoirdupois ounces followed by a second statement in parentheses in terms of pounds and ounces, or pounds and common or decimal fractions of the pound. Two examples are "Net Weight 24 ounces ($1\frac{1}{2}$ pounds)" and "Net Weight 24 ounces (1.5 pound)." The contents of packages containing less than 1 lb must be expressed as total ounces.

Drained weight rather than net weight is required on some products packed in a liquid that is not consumed as food, such as olives in brine. Net volume of liquid products in packages containing 1 pint or more and less than 1 U.S. gal must be declared first in total fluid ounces followed by a second statement in parentheses in terms of quarts, pints, and fluid ounces or fractions of the pint or quart. For example, "40 fluid ounces (1.25 quarts)" and "40 fluid ounces ($1\frac{1}{4}$ quarts)." Volume of packages containing less than 1 pint must be declared in fluid ounces.

Packages 4 lb or larger or 1 gal or larger need not have their contents expressed in terms of total ounces; however, for such packages the contents must be stated in the largest unit of weight or measure, with any remainder in ounces or common or decimal fractions of the pound. In the case of gallons, the remainder should be given in quarts, pints, and fluid ounces or decimal fractions of the gallon. If the label of any food package also represents the contents in terms of the number of servings, the size of each serving must be indicated.

Name. The common or usual name of a food must appear on the principal display panel, in bold type and in lines generally parallel to the base of the package as it is displayed. The form of the product must also be included, such as sliced, whole, or chopped, unless shown by a picture or unless the product is visible through the container. If there is a standard for the food, the complete name designated in the standard must be used, limitations must be labeled as such (403(e) and 21 *CFR* 101.3).

Ingredients. The ingredients in a food must be listed by their common names in order of their predominance by weight unless the food is standardized, in which case the label must include only those ingredients that the standard makes optional. Most ingredients in standardized foods are optional and, therefore, must be listed on the label. The word *ingredients* does not refer to the chemical composition, but means the individual food components of a mixed food. If a certain ingredient is the characterizing one in a food (eg, shrimp in shrimp cocktail) the percent of that ingredient may be required as part of the name of the food.

Food additives and colors are required to be listed, as ingredients, but the law exempts butter, cheese, and ice cream from having to show the use of color. Spices, flavors, and colors may be listed as such, without naming the specific materials, but any artificial colors or flavors must be identified as such, and certain coal-tar colors must be named specifically (403(i) and 403(k)).

FOODS FOR SPECIAL DIETARY USES

Section 403(j) of the Food, Drug, and Cosmetic Act classes a food as misbranded

> if it purports to be or is represented for special dietary uses, unless its label bears such information concerning its vitamin, mineral, and other dietary properties as the Secretary determines to be, and by regulations prescribes as, necessary in order fully to inform purchasers as to its value for such uses.

Section 411(c)(3) of the act defines special dietary use as a particular use for which a food purports or is represented to be used, including but not limited to the following:

1. Supplying a special dietary need that exists by reason of a physical, physiological, pathological, or other condition, including but not limited to the conditions of disease, convalescence, pregnancy, lactation, infancy, allergic hypersensitivity to food, underweight, or the need to control the intake of sodium.
2. Supply a vitamin, mineral, or other ingredient for use by humans to supplement the diet by increasing the total dietary intake.
3. Supplying a special dietary need by reason of being a food for use as the sole item of the diet.

Regulations under this section of the act (21 *CFR* 105) prescribe appropriate information and statements that must be given on the labels of foods in this class.

Importers and foreign shippers should consult the detailed regulations under sections 403(j) and 411 and 412 before importing foods represented by labeling or otherwise as special dietary foods. When special dietary foods are labeled with claims of disease prevention, treatment, mitigation, cure, or diagnosis, they must comply with the drug provisions of the act.

Infant Formulas

The Infant Formula Act of 1980 and subsequent amendments in 1986 (Section 412 of the Federal Food, Drug, and Cosmetic Act) establish nutrient requirements for infant formulas as defined by Section 201(aa) of the Federal Food, Drug, and Cosmetic Act, and provide the FDA with authority to establish Good Manufacturing Practices (GMPs) and requirements for nutrient quantity, nutrient quality control, record keeping, and reporting and recall of infant formulas that pose a potential hazard to health. The act also extends the FDA's factory inspection authority to permit access to complaint files and other manufacturers' records, quality-control records, and test results necessary to determine compliance with the act.

The act specifies that an infant formula is adulterated (1) if it fails to provide nutrients as required; (2) if it fails to meet the nutrient quality factors required by regulation; (3) if the processing is not in compliance with the appropriate GMP and quality-control procedures or record retention requirements as prescribed by regulation; or (4) if it otherwise fails to comply with Section 402 of the Food, Drug, and Cosmetic Act.

The act also requires manufacturers of infant formulas to notify the FDA 90 days before any charitable or commercial distribution of any new infant formula or any infant formula that has had a major change to its formulation or processing. Under the authority of the act, the FDA has promulgated regulations that specify infant formula nutrient quality-control procedures (21 *CFR* 106), the labeling of infant formula, the terms and conditions under which certain infant formula may be exempt from some of the act's requirements, and nutrient specifications for infant formula and infant formula recall regulations (21 *CFR* 107). The FDA is also developing regulations on record retention and Good Manufacturing Practice procedures.

NUTRITION LABELING

Whenever a vitamin, mineral, or protein is added to a food product (as in fortification or enrichment) or when the label, labeling, or advertising claim some nutritional value (such as rich in vitamin C), the label must have a panel of information about the nutritional quality of that food. The kind of information that must be provided is described in the *Code of Federal Regulations* (21 *CFR* 101.9). As a minimum, the required labeling calls for listing the size of a serving, the number of servings in a container, the amount of calories, protein, carbohydrate, and fat per serving, and the percentages per serving of the U.S. Recommended Daily Allowances (USRDA) of protein and seven vitamins and minerals. Other nutrients recognized as essential in the human diet may be listed if they contribute at least 2% of the USRDA.

For the benefit of persons who are on a fat-modified diet, the information may include declarations of cholesterol content and the content of saturated and polyunsaturated fatty acids (21 *CFR* 101.25). The USRDAs are derived from dietary standards established by the National Academy of Sciences to serve as goals for good nutrition.

Sodium labeling regulations (21 *CFR* 101.13) provide for information on food labels to enable consumers to avoid excessive salt and sodium in their diets. Studies have shown that high blood pressure, a leading cause of heart attacks and stroke, may be associated with excessive sodium intake in susceptible individuals. The rules amend FDA's food labeling regulations to require that nutrition labeling include the amount of sodium in milligrams per serving. Sodium information may also be voluntarily provided in the absence of nutrition labeling. Quantitative standards and label terminology for sodium declarations are established as follows.

Sodium-free products contain less than 5 mg per serving.

Very low sodium products contain 35 mg or less per serving.

Low-sodium products contain 140 mg or less per serving.

Reduced-sodium products are processed to reduce the usual level of sodium by 75%.

Unsalted products are processed without salt when the food normally is processed with salt.

SANITATION REQUIREMENTS

One of the basic purposes of the Food, Drug, and Cosmetic Act is protection of the public from products that may be deleterious, that are unclean or decomposed, or that have been exposed to unsanitary conditions that may contaminate the product with filth or may render it injurious to health. Sanitation provisions of the Food, Drug, and Cosmetic Act go further than to prohibit trade in products that are carriers of disease. The law also requires that foods be produced in sanitary facilities. It prohibits distribution of foods that may contain repulsive or offensive matter considered as filth regardless of whether such objectionable substances can be detected in the laboratory. Filth includes contaminants such as rat, mouse, and other animal hairs and excreta, whole insects, insect parts and excreta, parasitic worms, pollution from the excrement of humans and animals, as well as other extraneous materials that, because of their repulsiveness would not knowingly be eaten or used. The presence of such filth renders foods adulterated, whether or not harm to health can be shown.

The law thus requires that food be protected from contamination at all stages of production. Such protection includes extermination and exclusion of rodents, inspection and sorting of raw materials to eliminate the insect-infested and decomposed portions, quick handling and proper storage to prevent insect development or contamination, the use of clean equipment, the control of possible sources of sewage pollution, and supervision of personnel who prepare foods so that acts of misconduct may not defile the products they handle.

Foods that are free from contamination when they are shipped sometimes become contaminated en route and must be detailed or seized. This emphasizes the importance of insisting on proper storage conditions in vessels, railroad cars, or other conveyances. While the shipper may be blameless, the law requires action against illegal merchandise no matter where it may have become illegal. All shippers should pack their products so as to protect them against spoilage or contamination en route and should urge carriers to protect the merchandise by maintaining sanitary conditions and segregating food from other cargo that might contaminate it. For example, vessels transporting foods may also carry ore concentrates and poisonous insecticides. Improper cargo handling or disasters at sea have resulted in shipments becoming seriously contaminated, with detentions required.

Where import shipments become contaminated after customs entry and landing (for example, in truck accidents, fires, and barge sinkings), legal actions are not taken under the import provisions of the law, but by seizure proceedings in a federal district court, as with domestic interstate shipments (304).

Fumigation of commodities already infested with insects will not result in a legal product, because dead insects and evidence of past insect activity are objectionable. Fumigation may be employed where necessary, to prevent infestation, but care is required to prevent buildup of nonpermitted chemical residues from fumigation.

Current Good Manufacturing Practice Regulations

To explain what is needed to maintain sanitary conditions in food establishments, the FDA published a set of current Good Manufacturing Practice Regulations. These tell what kinds of buildings, facilities, equipment, and maintenance are needed and the errors to avoid to insure sanitation. They also deal with such matters as building design and construction, light, ventilation, toilet and washing facilities, cleaning of equipment, materials handling, and vermin control. Food firms that do not have copies of these regulations are urged to request them by writing to the FDA.

Many food materials are intended for further processing and manufacture into finished foods. Such processing in no way relieves the raw materials from the requirements of cleanliness and freedom from deleterious impurities.

Tolerances for Filth

Many inquiries are received by the Food and Drug Administering as to permitted variations from absolute cleanliness or soundness in foods. The act does not explicitly provide for "tolerances" for filth or decomposition in foods. It states that a food is adulterated if it consists in whole or in part of a filthy, putrid, or decomposed substance. This does not mean that a food must be condemned because of the presence of foreign matter in amounts below the irreducible minimum after all possible precautions have been taken. The FDA recognizes that it is not possible to grow, harvest, and process crops that are totally free of natural defects. The alternative (to increase the use of chemicals to control insects, rodents, and other sources of contamination) is not acceptable because of the potential health hazards from chemical residues. To resolve the problem the FDA has published a list of defect action levels, stating the amounts of contamination that will subject the food to enforcement action. Copies may be obtained by request to the FDA.

The food defect actions levels are established at levels that pose no hazard to health and may be changed from time to time. Any products that might be harmful to consumers are subject to regulatory action, whether or not they exceed the defect levels. In addition, manufacturing processes that do not conform to the FDA's current Good Manufacturing Practice Regulations will result in regulatory action by the FDA whether or not the product exceeds the defect action level. The levels are not averages, which are actually much lower, and FDA continues to lower the action levels as industry performance improves. The mixing of food to dilute contamination is prohibited and renders the product illegal regardless of any defect level in the final product.

FOOD ADDITIVES

Food additives are substances that directly or indirectly become components of food or that may otherwise affect the characteristics of the food. The term specifically includes any substance intended for use in producing, manufacturing, packing, processing, preparing, treating, transporting, or holding the food, and any source of radiation intended for any such use (409).

Excluded from the definition of a food additive are

1. Substances generally recognized as safe by qualified experts.
2. Substances used in accordance with a previous approval (prior sanction) under either the Federal Food, Drug, and Cosmetic Act, the Poultry Products Inspection Act (21 *USC* 451), or the Meat Inspection Act.
3. Pesticide chemicals in or on raw agricultural products.
4. A color additive.

Manufacturers or importers not certain whether the chemicals or other ingredients used in their foods are subject to the safety clearance requirements of the Food Additives Amendment may seek an opinion from the FDA. If premarket approval is required, this may mean that studies, including animal feeding tests, must be carried out in accordance with recognized scientific procedures, and the result submitted to the FDA for evaluation. General principles for evaluating the safety of food additives are published in 21 *CFR* 170.20. Detailed instructions for preparing a food additive petition are in 21 *CFR* 171. The Delaney clause in the law (409(c)(3)) provides that no food additive may be found safe it is produces cancer when ingested by man or animals, or if it is shown by other appropriate tests to be a cancer-producing agent; however, such an ingredient may be used in animal feeds if it causes no harm to the animal and if there are no residues of the ingredient in the meat or other edible products reaching the consumer. This later provision is primarily applicable to veterinary drugs added to animal feed.

If the FDA concludes from the evidence submitted to it that the additive will be safe, a regulation permitting its use will be issued. This regulation may specify the amount of the substance that may be present in or on the foods, the foods in which it is permitted, the manner of use, and any special labeling required. A substance cleared under the Food Addictive Regulations is still subject to all the general requirements of the Food, Drug, and Cosmetic Act.

Artificially sweetened products are required to be labeled as Special Dietary Foods (21 *CFR* 105.66). The foods permitted to contain such sweeteners, and the amounts, are specified in the Food Additive Regulations (21 *CFR* 180.37).

HOUSEWARES

Manufacturers of food-contact articles for use in the home or in food-service establishments should make sure that nothing from the articles imparts flavor, color, odor, toxicity, or other undesirable characteristics to food, thereby rendering the food adulterated. Although food packaging materials are subject to regulation as good additives, ordinary housewares are not. Such housewares include dishes, flatware, beverage glasses, mugs, cooking utensils, cutlery, and electrical appliances. This means that manufacturers are not required to submit data to the FDA showing that the materials used are safe. They also are not required to preclear their housewares with FDA. But housewares are not exempt from the general safety provisions of the Federal Food, Drug, and Cosmetic Act. Regulatory actions have been taken against cookware and ceramic dinnerware containing lead or cadmium.

FOOD STANDARDS

Food standards are a necessity to both consumers and the food industry. They maintain the nutritional values and the general quality of a large part of the national food supply. Without standards, different foods would have the same names and the same foods different names, both situations confusing and misleading consumers and creating unfair competition.

Section 401 of the Federal Food, Drug, and Cosmetic Act requires that whenever such action will promote honesty and fair dealing in the interest of consumers, regulations shall be promulgated fixing and establishing for any food, under its common or usual name so far as practicable, a reasonable definition and standard of identity, a reasonable standard of quality, and a reasonable standard of fill of container. However, no definition and standard of identity or standard of quality may be established for fresh or dried fruits, fresh or dried vegetables, or butter, except that definitions and standards of identity may be established for avocados, cantaloupes, citrus fruits, and melons.

Standards of Identity

Standards of identity define a given food product, specify its name, determine the ingredients that must be used or may be used, and regulate which ingredients must be declared on the label. Standards of quality are minimum standards only and establish specifications for quality requirements. Fill-of-container standards define how full the container must be and how this is measured. FDA standards are based on the assumption that the food is properly prepared from clean, sound materials. Standards do not usually relate to such factors as deleterious impurities, filth, and decomposition. There are exceptions. For example, the standards for whole egg and yolk products and for egg white products require these products to be pasteurized or otherwise treated to destroy all viable *Salmonella* microorganisms. Some standards for foods set nutritional requirements such as those for enriched bread, or nonfat dry milk with added vitamins A and D, etc. A food that is represented or purports to be a food for which a standard of identity has been promulgated must comply with the specifications of the standard in every respect.

Imitation Foods

A food that imitates another food is misbranded unless its label bears in type of uniform size and prominence the word *imitation* and immediately thereafter the name of the food imitated (403(c)). Under the law a food is an imitation if it is a substitute for, and resembles, another food but is nutritionally inferior to that food. If the food is not nutritionally inferior to the food it imitates, it is permitted to be labeled descriptively according to the regulations in 21 *CFR* 101.3(e) and need not bear the term imitation.

Standards of Quality

Standards of quality established under the Food, Drug, and Cosmetic Act must not be confused with standards for grades that are published by the U.S. Department of Agriculture for meat and other agricultural products and the U.S. Department of the Interior for fishery products. A standard of quality under the Food, Drug, and Cosmetic Act is a minimum standard only. If a food for which a standard of quality or fill of container has been promulgated falls below such standard, it must bear in a prescribed size and style of type label statements showing it to be substandard in quality or in fill of container, for example, the label may contain the statement "below standard in quality, good food—not high grade."

The U.S. Department of Agriculture grades are usually designated Grade A, or Fancy; Grade B, or Choice; and Grade C, or Standard. The U.S. Department of the Interior grades for fishery products are usually designated Grade A, Grade B, and Grade C. These grade designations are not required by the Food, Drug, and Cosmetic Act to be stated on the labels, but if they are stated, the product must comply with the specifications for the declared grade.

Fill-of-Container Standards

Fill-of-container standards established for certain foods designate the quantity in terms of the solid or liquid components or both. The existing fill-of-container standards for canned fruits and vegetables may be grouped as follows:

1. Those that require the maximum practicable quantity of the solid food that can be sealed in the container and processed by heat without crushing or breaking such component (limited to canned peaches, pears, apricots, and cherries).
2. Those requiring a minimum quantity of the solid food in the container after processing; the quantity is commonly expressed either as a minimum drained weight for a given container size or as a percentage of the water capacity of the container.
3. Those requiring that the food, including both solids and liquid packing medium, shall occupy not less than 90% of the total capacity of the container.
4. Those requiring both a minimum drained weight and the 90% minimum fill.
5. Those requiring a minimum volume of the solid component irrespective of the quantity of liquid (for example, canned green peas and canned field peas).

FDA food standards govern both labeling and composition and should be consulted for detailed specifications. The standards are published in the animal editions of the *CFR* (Title 21, Parts 103–169).

International Food Standards (the *Codex Alimentarius*) have been developed by committees of the World Health Organization and the Foreign Agriculture Organization of the United Nations, of which the United States is a member. The U.S. government may adopt Codex standards in whole or in part, in accordance with the procedures outlined in the preceding paragraph. The following sections provide pertinent information on requirements for specific groups of foods.

Canned Foods

Low-Acid Canned Foods and Acidified Foods Regulations. Special regulations apply to the manufacture of heat processed low-acid canned foods and acidified foods (21 *CFR* 108, 113, and 114). The purpose of these regulations is to insure safety from harmful bacteria or their toxins, especially the deadly *Clostridium botulinum*. This can only be accomplished by adequate processing, controls, and appropriate processing methods, such as cooking the food at the proper temperatures for sufficient times, adequately acidifying the food, or controlling water activity.

Low-acid canned foods are heat-processed foods other than alcoholic beverages, that have an acidity greater than pH 4.6 and a water activity (a_w) greater than 0.85 and that are packaged in hermetically sealed containers. Water activity is a measure of the water available for microbial growth. A hermetically sealed container is any package, regardless of its composition (ie, metal, glass, plastic, polyethylene-lined cardboard, etc), that is capable of maintaining the commercial sterility of its contents after processing. Acidified foods are low-acid foods to which acids or acid foods are added to reduce the pH to 4.6 or below (increase the acidity) and that have a water activity greater than 0.85. Pimientos, artichokes, some puddings, and some sauces are examples of acidified foods. Questions about product status can be referred to the FDA.

The regulations, first adopted in 1973 and revised in 1979, are based on proposals made by the canning industry following FDA investigations of deaths and illness from botulism associated with lax practices in low-acid canned food and acidified food processing. All commercial processors of low-acid canned foods and acidified foods are required to register their establishments and file processing information for all products with the FDA, using appropriate forms. Registration and process filing is required for both U.S. establishments and those in other countries that export such foods to the United States (21 *CFR* 108.25 and 108.35). Registration and processing forms and information are obtainable on request from the FDA. In addition to the registration and process filing requirements, processors must comply with other mandatory provisions of 21 *CFR* 108 as well as Parts 113 and 114. The mandatory provisions are always preceded by the word *shall*.

Canned Fruits and Fruit Juices. Standard of identity, quality, and fill of container have been promulgated for a number of canned fruits and fruit juices. The specific standards should be consulted by anyone intending to ship canned fruit to the United States. Labels on canned fruits and fruit juices that meet the minimum quality standards, and canned fruits and fruit juices for which no standards of quality have been promulgated, need not make reference to quality, but if they do, the product must correspond to the usual understanding of the labeled grade. Particular care must be taken to use the terms Fancy or Grade A only on those products meeting the specifications established for such grades by the U.S. Department of Agriculture.

Fill-of-Container Standards. Fill-of-container standards have been promulgated for a number of canned fruits and fruit juices, but in packing any other canned fruit the container must be well filled with the fruit and with only enough parking medium added to fill the interstices; otherwise the container may be deceptive and prohibited by the act. In judging the finished product, due allowance is made for natural shrinkage in processing.

The standards of fill for canned peaches, pears, apricots, and cherries require that the fill of the solid food component be the maximum practicable quantity that can be sealed in the container and processed by heat without crushing or breaking such component. Standards for canned fruit cocktail, grapefruit, and plums specify minimum drained weights for the solid food component, expressed as a percentage of the water capacity of the container. These drained weight requirements are as follows: fruit cocktail, 65%; grapefruit, 50%; whole plums, 50%; and plum halves, 55%. An additional 90% fill requirement based on the total capacity of the container for the solid food and the liquid packing medium have been established for plums. This 90% fill requirement also applies to applesauce in metal containers (85% fill for applesauce in glass), crushed pineapple, and pineapple juice.

Minimum drained weight requirements for canned pineapple provide that crushed pineapple may be labeled as heavy pack or solid pack. The standard of quality requires a minimum drained weight for crushed pineapple of 63% of the net weight. All fruit used for canning or juice should be mature and sound, that is, free from insect infestation, moldiness, or other forms of decomposition.

Canned Vegetables. Canned vegetable products must be prepared from sound, wholesome raw materials free from decomposition. Definitions and standards of identity for a wide variety of canned vegetables provide that "the food is processed by heat, in an appropriate manner, before or after being sealed in a container so as to prevent spoilage." The importance of adequate heat processing of canned vegetables, particularly the nonacid types, is emphasized by the special regulations for low-acid canned foods. Cans of food that have become swollen or otherwise abnormal are adulterated and should be destroyed.

Standards of Quality. Standards of quality have been promulgated for many vegetables. These are minimum standards only and establish specifications for quality factors such as tenderness, color, and freedom from defects. If the food does not meet these standards, it must be labeled in bold type "Below Standard in Quality" followed by the statement "Good Food—Not High Grade," or a statement showing in what respect the product fails to meet the standard, such as excessively broken, or excessive peel (21 *CFR* 130.14). The standards of quality for canned vegetables supplement the identity standards. In other words, the identity standards define what the product is and set floors and ceilings on important ingredients, while the quality standards provide for special labeling for any product that is not up to the usual expectations of the consumer but is nevertheless a wholesome food.

Fill of Container. If a fill-of-container standard has not been promulgated for a canned vegetable, the container must nevertheless be well filled with the vegetable, with only enough packing medium to fill the interstices. In the case of canned tomatoes, no added water is needed or permitted.

The standards of fill for canned tomatoes and canned corn require that the solid food and the liquid packing medium occupy not less than 90% of the total capacity of the container. The standard for canned corn also specifies a minimum drained weight of 61% of the water capacity of the container for the corn ingredient, while the standard of quality for canned tomatoes specifies a minimum drained weight of 50% of the water capacity of the container for the solid tomato ingredient. Canned mushrooms must meet minimum drained weight requirements, stated in ounces, for various can sizes; and canned green peas and field peas must comply with volumetric fill requirements that state that if the peas and liquid packing medium are poured from the container and returned thereto, the leveled peas (irrespective of the liquid) completely fill the container.

Tomato Products. Shippers of tomato products (canned tomatoes, juice, paste, purée, and catsup) should consult the standard of identity for these items (21 *CFR* 155.190-4). Attention is called to the salt-free tomato solids requirements for purée and paste, and to the fact that neither artificial color nor preservatives are permitted in any of these products. Tomato juice is unconcentrated; tomato purée must contain not less than 8% and tomato paste not less than 24% salt-free tomato solids.

These tomato products are occasionally contaminated with rot because of failure to remove decayed tomatoes from the raw material entering the cannery. Flies and worms are also filth contaminants of tomato products. The preparation of a clean tomato product requires proper washing, sorting, and trimming of the tomatoes and frequent cleaning of the cannery equipment, such as tables, utensils, vats, and pipelines.

In judging whether tomato products have been properly prepared to eliminate rot and decay, the FDA uses the Howard mold count test, refuses admission to import shipments, and takes action against domestic shipments if mold filaments are present in excess of amounts stated in the food defect action levels. Methods of testing tomato products are given in the official methods of the AAOAC.

Dried Fruits and Vegetables

What has been said about preventing contamination of food applies particularly to perishable products such as dried fruits and vegetables that are subject to attack by insects or animals, or to deterioration resulting in moldiness or other forms of decomposition.

Dried Figs and Dates. Dried figs, both domestic and foreign, are subject to insect infestation during their growth and when stored under unsuitable conditions. Figs may also become moldy if not properly stored and handled. The industry has made substantial progress in eliminating conditions that result in contamination or spoilage of figs, but it is still necessary to refuse entry to some dried figs and fig paste because of insect or rodent contamination, mold, sourness, or fermentation.

Dried dates are refused entry if insect infested, if they contain other forms of filth, or are moldy or decomposed. The presence of unpitted dates, or dates containing broken pieces of pits, in shipments of pitted dates is also a cause of refusing shipments.

Dried Mushrooms. Only edible species of dried mushrooms may be offered for import. The most common bar to entry, however, is insect infestation, usually by flies or maggots. Dried wild mushrooms should be handled by people who know how to sort out insect-infested mushrooms and those not clearly identifiable as edible species. If insect infestation is so heavy in a particular growing area that it is impractical to sort out the insect infested ones, then the mushrooms from that area should not be offered for entry. The mushrooms should be protected during drying and storage to prevent their contamination with insects, rodent and bird filth, or other objectionable material. Canned mushrooms should be essentially free of insect infestation. Since the canned product is prepared from domesticated varieties grown under enclosure, the careful producer can readily prevent access by insects.

Fresh Fruits

Apples and other fruits bearing excessive residues from insecticide sprays or dusts are adulterated under the federal law. Pineapples showing or likely to show the internal condition known as brown heart, or black heart, should not be offered for entry into the United States. Blueberries and huckleberries sometimes contain small larvae that render them unfit for food. Fruit from infested areas should be avoided. Fresh blueberries should be held and transported under conditions that will prevent mold or other types of spoilage.

Pesticidal Residues on Raw Agricultural Commodities

Raw agricultural commodity means any food in its raw or natural state, including all unprocessed fruits, vegetables, nuts, and grains. Foods that have been washed, colored, waxed, or otherwise treated in their unpeeled natural form are considered to be unprocessed. Products of this kind containing pesticide residues are in violation of the Federal Food, Drug, and Cosmetic Act unless: (1) the pesticide chemical has been exempted from the requirement of a residue tolerance or (2) a tolerance has been established for the particular pesticide on the specific food and the residue does not exceed the tolerance (408).

Processed foods that contain any residue of a pesticide that is not exempted or for which no tolerance has been established are adulterated under Section 402(a)(2)(C) of the act. If a tolerance has been established, a pesticide residue in the processed food does not adulterate the ready-to-eat food if the residue does not exceed the tolerance established for the raw agricultural commodity. The applicable regulations are in 21 *CFR* 180 and 193.

Tolerances for pesticidal residues on many raw agricultural commodities have been established under Section 408 of the law. Tolerances are established, revoked or changed, as the facts warrant such action, by the Environmental Protection Agency. Firms considering offering for entry into the United States foods that may contain pesticidal residues should write to the FDA for current information concerning the enforcement of tolerances for residues on raw agricultural products.

Fruit Jams (Preserves), Jellies, Butters, and Marmalades

The standards of identity for jams and jellies require that these products be prepared by mixing not less than 45 parts by weight of certain specified fruits (or fruit juice in the case of jelly), and 47 parts by weight of other designated fruits, to each 55 parts by weight of sugar or other optional nutritive carbohydrate sweetening ingredient. Only sufficient pectin may be added to jams and jellies to compensate for a deficiency, if any, of the natural pectin content of the particular fruit. The standards also require that for both jams (preserves) and jellies, the finished product must be concentrated to not less than 65% soluble solids.

Standards of identity have also been established for artificially sweetened jams and jellies, and for these products the fruit ingredient must be not less than 55% by weight of the finished food product. The standard of identity for fruit butters defines fruit butters as the smooth, semisolid foods made from not less than five parts by weight of fruit ingredient to each two parts by weight of sweetening ingredient. As is the case with jams and jellies, only sufficient pectin may be added to compensate for a deficiency, if any, of the natural pectin content of the particular fruit. The fruit butter standard requires that the finished product must be concentrated to not less than 43% soluble solids.

There is no formal standard of identity for citrus marmalade. However, the FDA expects a product labeled sweet orange marmalade to be prepared by mixing at least 30 lb of fruit (peel and juice) to each 70 lb sweetening ingredients. Sour or bitter (Seville) orange marmalade, lemon marmalade, and lime marmalade would be prepared by mixing at least 25 lb of fruit (peel and juice) to each 75 lb of sweetening ingredient. The amount of peel should not be in excess of the amounts normally associated with fruit. The product should be concentrated to not less than 65% soluble solids. Such products would not be regarded as misbranded.

Jams, jellies, and similar fruit products should, of course, be prepared only from sound fruit. Decayed or decomposed fruits and insect-contaminated fruits should be sorted out and discarded.

Confectionery (Candy)

The utmost in cleanliness and sanitation should govern the manufacture of confectionery, because both the raw materials and finished products tend to attract rodents and insects. Stored materials should be inspected frequently and carefully. Equipment should be washed thoroughly and kept free from accumulation of materials that would attract vermin.

The presence of nonpermitted colors and misbranding by the incorrect listing of ingredients also constitute grounds for detentions or seizure. Fruit-type confectionery containing any artificial fruit flavor must be labeled as artificially flavored in the manner specified by the regulations (21 *CFR* 101.22(i)(2)).

Nonnutritive objects, with the exception of such objects that are of practical functional value to the confectionery product and do not render the product injurious or hazardous to health, and alcohol, other than that from flavoring extracts not in excess of 0.5%, are specifically prohibited in confectionery. Manufacturers who use colors should make sure these colors are authorized for use in the United States, and, if they require certification, that they are from a batch certified as suitable for use in foods by the FDA. Information on color additives and color certification is discussed earlier.

Cocoa Products

Standards of identity have been promulgated under the Federal Food, Drug, and Cosmetic Act for approximately 40 cacoa products. If a cocoa product is labeled as one of the standardized products listed in the regulations, the shipper or buyer should take care to insure that it conforms to the standard for that product (21 *CFR* 163).

Excess shell is frequently the cause of detention of various standardized cacoa products. Cocoa beans offered for import sometimes show damage that bars them from entry. Beans with mold permeating the bean or showing clouds of spores when cracked are objectionable, as are wormy, insect-infested beans that may show live or dead insects, webbing, or insect excreta. Moldy and insect-infested beans should be removed from the lot before shipment. Care should be taken to store beans before and during shipment so that mold and insect contamination are prevented. Some shipments have been encountered that contained animal excreta, evidence of storage or handling under unsanitary conditions.

Beverages and Beverage Materials

Alcoholic Beverages. Beer, wines, liquors, and other alcoholic beverages are specifically subject to the Federal Alcohol Administration Act, which is enforced by the Bureau of Alcohol, Tobacco and Firearms of the U.S. Treasury Department. Accordingly, questions of labeling and of composition should be taken up with the bureau rather than with the FDA. This is not the case for cooking wines or for certain other alcoholic beverages, such as diluted wine beverages and cider beverages having less than 7% alcohol by volume, which are solely within the jurisdiction of FDA.

Questions dealing with the presence of deleterious substances in alcoholic beverages (such as excess fusel oil and excess aldehydes in whisky, methyl alcohol in brandy, glass splinters from defective bottles, toxic ingredients, such as arsenic, lead, or fluorine, resulting from the spraying of the fruit used in wine manufacture, and residues of toxic clarifying substances) are also dealt with by the FDA, as are questions relating to sanitation and filth. The importation of absinth and any other liquors or liqueurs that contain an excess of *Artemisia absinthium* is prohibited.

Nonalcoholic Beverages. Products sold as fruit juices may be sweetened if the label states the presence of the added sugar or other wholesome, nutritive sweetening agent (saccharin is not a nutritive sweetening agent), but they should not contain added water. Fruit juices should be manufactured only from clean, sound fruit in clean equipment. Proper preparation involves thorough washing, sorting to remove wormy or spoiled fruit, and trimming. Flies should not come in contact with the fruit or equipment.

Fruit juice beverages containing added water are subject to all the requirements for nonstandard foods, particularly the use of a common or usual name (21 *CFR* 102.5). This must include a declaration of the percent of any characterizing ingredient that has a material bearing on the price of consumer acceptance of the product. Diluted orange juice beverages are specifically required to have the percentage of juice expressed in increments of 5% (21 *CFR* 102.32).

Noncarbonated beverages containing no fruit or vegetable juice must be named as required by the regulation (102.5(a)), and if the label in any way suggests that fruit or vegetable juice is present there must be an added statement such as "contains no ——— juice," with the space filled in with the name of the fruit or vegetable indicated by the flavor, color, or labeling of the product.

Nonalcoholic carbonated beverages are subject to the standard for soda water (21 *CFR* 165.175). Caffeine from kola nut extract or other natural caffeine-containing extracts may be used in any soda water but not to exceed 0.02% by weight of the finished food. All optional ingredients are required to be declared on the label.

Bottled Waters. The FDA has established standards of quality for bottled drinking water (21 *CFR* 103.35). All bottled waters other than mineral waters must comply with these standards. It is anticipated that mineral waters, by their very nature, would exceed physical and chemical limits prescribed in the Bottled Drinking Water Standard and are, therefore, exempt from the standard. In no case, however, will FDA permit the presence of any substance in bottled mineral water at any level deemed toxic. FDA has also established current Good Manufacturing Practice Regulations for processing and bottling waters (21 *CFR* 129).

The listing on the labels of mineral waters of minerals that are present in insignificant amounts is misleading. Such a listing of ingredients may convey an impression of nutritional or therapeutic benefit not possessed by water. If nutritional or health claims are made in labeling, then the requirements of the nutritional labeling regulations apply. All bottled waters offered for import should be

1. Obtained from sources free from pollution.
2. Bottled or otherwise prepared for the market under sanitary conditions, with special reference to the condition of the bottles or other containers and of their stoppers.
3. Free from microorganisms of the coliform group.
4. Of good sanitary quality when judged by the result of a bacteriological examination to determine the numbers of bacteria growing on gelatin at 20°C and on agar at 37°C.
5. Of good sanitary quality when judged by the results of a sanitary chemical analysis.

Some of the types of misbranding found on mineral water labels are

1. Objectionable therapeutic claims.
2. False and misleading statements other than therapeutic claims.
3. Label statement of analysis inaccurate or otherwise misleading.
4. Undeclared addition of salts, carbon dioxide, etc.
5. Labeling of artificially prepared waters as mineral water.

Tea. Tea (*Thea sinensis*) is not only subject to the Federal Food, Drug, and Cosmetic Act but also to the Tea Importation Act. Under the latter law, tea offered for entry must meet the standards of purity, quality, and fitness for consumption prescribed under the act.

Beverages brewed from the leaves of other plants cannot properly be labeled as tea but should be labeled for what they are. For example, mate (or matte) or the beverage prepared from it should not be labeled unqualifiedly as tea or South American tea, because such names might mislead consumers in the United States who may not be familiar with the fact that it is an entirely different plant and beverage from the tea commonly sold in the United States. Like tea or coffee, mate may be mildly stimulating because of its caffeine content, but it should not be represented as capable of producing health, energy, endurance, or other physiological effects.

Coffee. Green coffee imported into the United States should be held at all times under sanitary conditions to prevent contamination by insects, rats, and mice. The following conditions class coffee as objectionable and subject to refusal of entry: coffee berries that are black, part black, moldy, water damaged, insect eaten or infested, brown (sour), Quakers (immature berries), shriveled, and sailors. Coffee berries that contain foreign material, such as pods, sticks, stones, trash, and sweepings of spilled coffee in ships' holds and on docks are also refused entry. Contamination by ores and other poisonous materials in the cargo also has resulted in detentions.

Fishery Products

Fish and shellfish are imported in the fresh, frozen, pickled, canned, salted, dried, or smoked conditions. The raw materials are by nature extremely perishable and must be handled rapidly under adequate refrigeration if decomposition is to be avoided. The conditions of handling are also likely to result in contamination unless particular safeguards are imposed.

Sometimes preservatives have been used to prevent or retard spoilage in fish. Any preservatives used must be one accepted by FDA as safe and must be declared on the label. Chemical contamination of lakes, rivers, and the oceans has been found to concentrate in certain species of fish. Excessive residues of pesticides, mercury, and other heavy metals are prohibited.

Names for Seafoods. To prevent substitution of one kind of seafood for another, and consequent deception of the consumer, it is essential that labels bear names that accurately identify the products designated. Words such as *fish*, *shellfish*, and *mollusc* are not sufficient; the name of the specific seafood must be used. Many fish, crustaceans, and molluscs have well-established common, or usual, names throughout the United States (for example, pollock, cod, shrimp, and oyster). These may not be replaced with other names, even though the other names may be used in some areas or countries. Neither may they be replaced with coined names, even though the coined names may be considered more attractive or to have greater sales appeal.

A more difficult problem is deciding what constitutes the proper designation for a seafood that has not previously been marketed in the United States and thus that has not acquired an established common or usual name in this country. In selecting an appropriate name for such a product, full consideration must be given to its proper biological classification and to avoiding a designation that duplicates or may be confused with the common or usual name in this country of some other species.

Labels on frozen fishery products for which no standards of quality have been promulgated need not make reference to quality, but if they do, the product must correspond to the usual understanding of the labeled grade. Particular care must be taken not to use the term Grade A on products that do not meet the established understanding of these terms in the United States.

Canned Fish. Canned fish generally are low-acid canned foods and packers are, therefore, subject to those registration requirements. Failure to declare the presence of added salt or the kinds of oil used as the packing medium in canned fish has resulted in the detention of fish products. If permitted artificial colors or chemical preservatives are used, their presence must be conspicuously declared in the labeling. Artificial coloring is not permitted if it conceals damage or inferiority or if it makes the product appear better or of greater value than it is.

The packing of canned fish and fish products with excessive amounts of packing medium such as anchovies in oil, the container should be as full as possible of fish with the minimum amount of oil. The fact that the oil may be equal in value or even more expensive than the fish does not affect this principle. Canned lobster paste and similar products have been encountered that were deceptively packaged because of excessive headspace, ie, excessive space between the lid of the can and the surface of the food in the can.

Canned Pacific Salmon. Canned Pacific salmon is required to comply with standards of identity and fill of container. The standards establish the species and names required on labels and the permitted styles of pack (21 *CFR* 161.170).

Anchovies. Products represented as anchovies should consist of fish of the family *Engraulidae*. Other small fish, such as small herring and herringlike fish that may superficially resemble anchovies cannot properly be labeled as anchovies. The product should be prepared from sound raw material and the salting or curing should be conducted in such a manner that spoilage does not occur.

Sardines. The term sardines is permitted in the labeling of the canned products prepared from small-size clupeoid fish. The sea herring (*Clupea harengus*), the European pilchards (*Sardina pilchardus* or *Clupea pilchardus*), and the brisling or sprat (*Clupea sprattus*) are commonly packed in small-size cans and labeled as sardines. The terms brisling sardines and sild sardines are permissible in the labeling of canned small brisling and herring, respectively. Large-size herring cannot be labeled sardines. These canned products must be free from all forms of decomposition such as feedy, belly-blown fish and must be adequately processed to prevent active spoilage by microorganisms. Fish are called feedy fish when their stomachs are filled with feed at the time the fish are taken from the water. Such fish deteriorate rapidly until the viscera and thin belly wall disintegrate producing a characteristic ragged appearance called belly-blown.

Tuna. A standard of identity defines the species of fish that may be canned under the name tuna (21 *CFR* 161.190(a)). There is also a standard for fill of container of canned tuna (21 *CFR* 161.190(c)). The standards provide for various styles of pack, including solid pack, chunk or chunk style, flakes, and grated tuna. Provision is also made for various types of packing medium, certain specified seasonings and flavorings, color designations, and methods for determining fill of containers.

The standard of fill of container for canned tuna specifies minimum values for weights of the pressed cake of canned tuna depending on the form of the tuna ingredient and the can size.

The canned fish *Sarda chilenis*, commonly known as the bonito or bonita may not be labeled as tuna because it is not a true tuna but must be labeled as bonito or bonita. The fish *Seriola dorsalis*, commonly known as yellowtail, must be labeled as yellowtail and may not be designated as tuna.

Fresh and Frozen Fish Fillets. These products are highly perishable and require extraordinary care if decomposition is to be avoided. The manufacturer must exercise extreme care in the selection of raw materials to remove any unfit, decomposed material and then to maintain the product in a sound, wholesome condition.

Caviar and Fish Roe. The name caviar unqualified may be applied only to the eggs of the sturgeon prepared by a special process. Fish roe prepared from the eggs of other varieties of fish, prepared by the special process for caviar, must be labeled to show the name of the fish from which they are prepared, for example, whitefish caviar. If the product contains an artificial color, it must be an approved color and its presence must be stated on the label conspicuously. No artificial color should be used that makes the product appear to be better or of greater value than it is. If a chemical preservative is employed, it must be one approved by FDA and it must be declared on the label.

Shellfish. Certification. Because raw shellfish (clams, mussels, and oysters) may transmit intestinal disease such as typhoid fever or be carriers of natural or chemical toxins, it is most important that they be obtained from unpolluted waters and produced, handled, and distributed in a sanitary manner. Shellfish must comply with the general requirements of the Federal Food, Drug, and Cosmetic Act and also with requirements of state health agencies cooperating in the National Shellfish Sanitation Program (NSSP) administered by FDA and in the Interstate Shellfish Sanitation Conference (ISSC).

Shellfish harvesting is prohibited in areas contaminated by sewage or industrial wastes. To enforce this prohibition, these areas are patrolled and warning signs are posted by state health or fishery control agencies. State inspectors also check shellfish harvesting boats and shucking plants before issuing approval certificates, which are equivalent to a state license, to operate. Plants having approved certificates are required to place their certification number on each container or package of shellfish shipped. The number indicates the shipper is under state inspection and meets the requirements of the NSSP. It also serves the important purpose of identifying and tracing shipments found to be contaminated. Shippers are also required to keep records showing the origin and disposition of all shellfish handled and to make these records available to the control authorities.

Certain shellfish, particularly mussels and clams, may contain a naturally occurring marine toxin derived from plankton organisms on which the shellfish feed. Such toxic shellfish may cause illness or even death. State shellfish control agencies also maintain surveillance of shellfish growing areas for marine biotoxins. The areas where these biotoxins periodically occur are from Maine to Massachusetts, Florida, and the west coast including Alaska.

Standards of identity have been set for raw oysters, Pacific oysters, and canned oysters. The standards define the sizes of oysters and prescribe the methods of washing and draining to prevent adulteration with excess water. A fill-of-container standard for canned oysters fixes the drained weight at not less than 59% of the water capacity of the container (21 *CFR* 161.130–161.145).

Imports. Imported fresh and fresh frozen oysters, clams, and mussels are certified under the auspices of the National Shellfish Sanitation Program through bilateral agreements with the country of origin. Canada, Japan, the Republic of Korea, Iceland, Mexico, England, Australia, and New Zealand now have such agreements.

Rock Lobster, Spiny Lobster, and Sea Crayfish. The sea crayfish (*Palinurus vulgaris*) is frequently imported into the United States in the form of frozen tails, frozen cooked meat, or canned meat. By long usage, the terms rock lobster and spiny lobster have been established as common or usual names for these products. No objection has been offered to either of these terms, providing the modifying words *rock* or *spiny* are used in direct connection with the word *lobster* in type of equal size and prominence.

In examination of imports, decomposition has sometimes been detected in all three forms of the product. In the canned product, decomposition resulted from the packing of decomposed raw material and also from active bacterial spoilage. In the frozen cooked products, detentions have been necessary because of the presence of microorganisms indicative of pollution with human or animal filth as well as of decomposition.

Shrimp. Standards set minimum requirements for canned wet and dry pack shrimp and frozen raw breaded shrimp (21 *CFR* 161.175). Canned shrimp must comply with the regulations for low-acid canned foods discussed earlier in this article. There is also a standard of identity for frozen raw lightly breaded shrimp (21 *CFR* 161.176).

Meat and Poultry

Meat and Meat Food Products. Meat or meat products derived from cattle, sheep, swine, goats, and horses are subject to the provisions of the wholesome Meat Act enforced by the Food Safety and Inspection Service of the USDA as certain provision of the Food, Drug, and Cosmetic Act. Foreign meat products must originate in countries that have approved meat inspection programs. Each shipment must be properly certified by the foreign country and on inspection by the federal meat inspectors found to be completely sound, wholesome, and fit for human food before entry into the United States. Requests for inspection should be made to the USDA. The imported product after admission into the United States becomes a domestic article subject not only to the Federal Meat Inspection Act but also to the Federal Food, Drug, and Cosmetic Act to the extent the provisions of the Meat Inspection Act do not apply. Wild game, however, is subject to the requirements of the Federal Food, Drug, and Cosmetic Act and its regulations.

Poultry and Poultry Products. Poultry and poultry products offered for importation are subject to the Wholesome Poultry Act also enforced by the Food Safety and Inspection Service to which inquiries concerning such products should be addressed. The term poultry means any live or slaughtered domesticated bird (chickens, turkeys, ducks, geese, or guineas). The term poultry product means any poultry that has been slaughtered for human food, from which the blood, feathers, feet, head, and viscera have been removed in accordance with the rules and regulations promulgated by the secretary of agriculture; any edible part of poultry; and, unless exempted by the secretary, any human food product consisting of any edible part of poultry, separately or in combination with other ingredients.

Poultry and poultry products are also subject to the Federal Food, Drug, and Cosmetic Act to the extent to which the provisions of the Poultry Products Inspection Act do not apply. Soups generally are under the jurisdiction of the USDA. However, those containing small amounts of cooked meat, poultry, or broth as flavoring ingredients, are subject to regulation by FDA.

Gelatin. Gelatin is not held to be a meat food product amendable to the Federal Imported Meat Act. The provisions of the Food, Drug, and Cosmetic Act require it to be prepared from clean, sound, wholesome, raw materials, handled under sanitary conditions. Gelatin should not contain glue, or other gelatinlike material with a low gelatinous characteristic, nor have a disagreeable odor or taste. It should be so prepared and marketed as to be free from spoilage, filth, or putridity. It should be free from such preservatives, metals, and salts of metals as may render its use injurious to health.

Dairy Products.

Milk, cream, and dairy products made from them offer ideal conditions for the growth and survival of microorganisms. It is essential that milk be obtained from animals free from diseases and that all materials used in manufactured dairy products be handled under sanitary conditions at all times to prevent contamination with disease-producing and spoilage organisms. While pasteurization offers a considerable safeguard against disease transmission, contamination by workers suffering from disease, by rodents or by unclean equipment or the addition of unpasteurized milk matter, spoilage organisms, such as undesirable bacteria, yeasts, and molds capable of causing decomposition of the raw materials or of the finished products, may be contributed by unsanitary handling. Products prepared from decomposed raw materials, undergoing active spoilage, or contaminated with disease-producing organisms are adulterated and subject to legal action.

Milk Safety. A safe and wholesome supply of fresh milk and cream for the U.S. consumer is the objective of the Federal–State Milk Sanitation Program administered by FDA through Interstate Milk Shippers (IMS) Agreements. In this program the producers of Grade A pasteurized milk are required to pass inspections and be rated by cooperating state agencies. The ratings appear in the IMS List published by the FDA and revised quarterly. The list is used by state authorities and milk buyers to insure the safety of milk shipped from other states.

Uniformity and adequacy of state milk production regulations is obtained through a model Pasteurized Milk Ordinance (PMO) developed and revised under FDA leadership. The PMO is the basis for the milk sanitation laws and regulations of 49 states, the District of Columbia, and more than 2,000 local communities.

Mandatory pasteurization for all milk and milk products in final package form intended for direct human consumption after distribution in interstate commerce is a federal requirement. The regulation (21 *CFR* 1240.61) defines pasteurization as heating and holding every particle of milk or milk product in properly designed and operated equipment at times and temperatures specified in the regulation. Exemptions are provided for acceptable alternative methods, such as aging of certain cheeses.

Cream and Milk for Manufacturing. Cream and milk used for the manufacture of dairy products should be clean and free from decomposition and from residues of pesticides and drugs. It is important to keep the cream for as short a period as possible, in a clean place, and to use clean utensils in handling and processing. Dairymen are advised to use pesticides on cows and in dairy barns only as directed and to avoid feeds that carry pesticidal residues. When treating cows with drugs for mastitis or other diseases, the milk should not be sold until after the recommended period following the end of dosage.

Nonfat Dry Milk. Nonfat dry milk is defined in a standard provided by Act of Congress on July 2, 1956, as follows:

> Nonfat dry milk is the product resulting from the removal of fat and water from milk and contains the lactose, milk proteins, and milk minerals in the same relative proportions as in the fresh milk from which it is made. It contains not over 5 per centum by weight of moisture. The fat content is not over $1\frac{1}{2}$ per centum by weight unless otherwise indicated. The term 'milk,' when used herein, means sweet milk of cows.

In addition, a standard of identity has been established for nonfat dry milk fortified with vitamins A and D.

Import Milk Act. In addition to being subject to the requirements of the Food, Drug, and Cosmetic Act, milk and cream (including sweetened condensed milk) offered for import into the continental United States are subject to another federal act, the Import Milk Act, enforced by the FDA. Such products may be imported only under permit after certain sanitary and other prerequisites have been fulfilled.

Butter. Butter is defined in the standard provided by Act of Congress of March 4, 1923, as follows:

> "Butter" shall be understood to mean the food product usually known as butter, and which is made exclusively from milk or cream or both, with or without common salt, and with or without additional coloring matter, and containing not less than 80 percent by weight of milk fat, all tolerances having been allowed for.

Butter is examined for evidence of the use of dirty cream or milk and for mold, which indicates the use of decomposed cream. Chemical additives and artificial flavor are not permitted.

Cheese. Cheese may be contaminated with insect and rodent filth during the handling of the milk, during the manufacture of the cheese, or during storage. This must be guarded against. Particular care should be taken not to use milk that may be contaminated with pesticides from forage fed to the animals or carelessly used in barns. Large quantities of cheese have been detained because of such contamination.

Standards of identity have been promulgated for most natural cheeses, process cheeses, cheese foods, and cheese spreads (21 *CFR* 133). Where a standard has been adopted for a particular variety of cheese, all cheeses belonging to the variety must comply with the standard and be labeled with the name prescribed in the standard.

Most of the standards prescribe limits for moisture and for fat. A few natural cheeses are required to be made from pasteurized milk. Most, however, may be made from either raw milk or pasteurized milk. When made from raw milk they are required to be aged for 60 days or longer. The 60 days aging is for the purpose of insuring the safety of the cheese. Requirements for longer aging are made for cheeses that need long aging to develop the characteristics of the variety. Shippers of cheese and foods made of cheese should consult the FDA standards before making shipments to the United States.

Evaporated Milk and Sweetened Condensed Milk. Under the established standard of identity for evaporated milk (21 *CFR* 131.130) this product must contain not less than 7.5% milk fat and not less than 25.5% total milk solids. Sweetened condensed milk, for which a standard has also been established (21 *CFR* 131.120), must contain not less than 28% total milk solids, and not less than 8.5% milk fat. Sweetened condensed milk is also subject to the Import Milk Act, but evaporated milk (which is sterilized by heat) is not.

Nuts and Nut Products

Nuts are adulterated if they are infested or insect damaged, moldy, rancid, or dirty. Empty or worthless unshelled nuts should be removed by careful hand sorting or by machinery. Care should be taken to eliminate infested, dirty, moldy, or rancid nuts from the shipment. Conditions that may cause nuts to be refused admission are described below.

Insect Infestation. Nuts are infested if they contain live or dead insect, whether larvae, pupae, or adults, or if they show definite evidence of insect feeding or cutting, or if insect excreta pellets are present.

Dirt. Nut meats may become dirty because of lack of cleanliness in cracking, sorting, and packaging.

Mold. Nut meats occasionally are moldy in the shell and bear fruiting mold or mold hyphae.

Rancidity. Nuts in this class have an abnormal flavor characterized by rancidity; rancid nuts are frequently soft and have a yellow, dark, or oily appearance.

Extraneous Material. Stems, shells, stones, and excreta should not be present.

Defect action levels have been established for tree nuts. Deliberate mixing of good and bad lots to result in defects under these levels is prohibited even though the percentage of defects in the mixed lots is less than the defect action level.

Aflatoxins. The aflatoxins are a group of chemically related substances produced naturally by the growth of certain common molds. The aflatoxins, especially aflatoxin B1, are highly toxic, primarily causing acute liver damage in exposed animals. Aflatoxin B1 also exhibits highly potent cancer-producing properties in certain species of experimental animals. Studies of certain population groups reveal that the consumption of aflatoxin-containing foods is associated with liver cancer in humans. The presence of excess aflatoxin levels in nuts and other products is a significant public health problem and is a basis for seizing or refusing imports of products containing it.

Bitter Almonds. Because of their toxicity, bitter almonds may not be marketed in the United States for unrestricted use. Shipments of sweet almonds may not contain more than 5% bitter almonds. Almond paste and pastes made from other kernels should contain less than 25 ppm of hydrocyanic acid (HCN) naturally occurring in the kernels.

Nuts and nut meats must be prepared and stored under sanitary conditions to prevent contamination by insects, rodents, or other animals. Nuts imported for pressing of food oil must be just as clean and sound as nuts intended to be eaten as such or to be used in manufactured foods.

Standards for Nut Products. Mixed tree nuts, shelled nuts, and peanut butter are subject to FDA standards (21 *CFR* 164). The standards establish such factors as the proportions of various kinds of nuts and the label designations for mixed nuts, the fill of container for shelled nuts, and the ingredients and labeling for peanut butter. All packers and shippers of nut products should be aware of the requirements of these standards.

Edible Oils

Olive oil is the edible oil expressed from the sound, mature fruit of the olive tree. Refined or extracted oils are not entitled to the unqualified name olive oil. Other vegetable oils should be labeled by their common or usual name, such as cottonseed, sunflower, peanut, and sesame. Mixtures of edible oils should be labeled to show all the oils present and the names should be listed on the labels in the descending order of predominance in the product. The terms vegetable oil or shortening are not permitted to be used in food labeling without disclosing the source of each oil or fat used in the product (31 *CFR* 101.4(b)(14)). Pictures, designs, or statements on the labeling must not be misleading as to the kind or amount of oils present or as to their origin.

Cod liver oil is a drug as well as a food because it is recognized in the *United States Pharmacopeia* (*USP*). Its value as a food, whether intended for human or animal use, depends mainly on its vitamin D content. Articles offered for entry as cod liver oil must comply with the identity standard prescribed by the *USP* and conform to the other specifications set forth in that official compendium.

Margarine, Mayonnaise and Salad Dressings

Standards of identity have been established for margarine, mayonnaise, French dressing, and salad dressing. The margarine standard is in 21 *CFR* 166 and the standards for mayonnaise, French dressing, and salad dressing are in part 169. There are also many nonstandard dressings, subject to the general food labeling requirements. Mineral oil is not a permitted ingredient in any of these foods (21 *CFR* 169).

Olives

Pitted and stuffed olives containing more than an unavoidable minimum of pits or pit fragments are regarded as in violation of the Food, Drug, and Cosmetic Act. While it may not be possible to completely eliminate this problem, dealers have been put on notice that they must take steps to reduce it to the extent that is reasonable and feasible.

Spices, Spice Seeds, and Herbs

This group includes food materials that particularly need protection from various animal and insect pests. They may also become moldy or otherwise decomposed unless properly prepared and stored. The U.S. food law requires emphasis on the principle of clean food, not cleaned food. One of the most serious consequences of failure to protect herbs and spices is contaminations with excreta from rats, mice, birds, chickens, or other animals. Emphasis should be placed on harvesting, storing, handling, packing, and shipping under conditions that will prevent contamination.

The same basic principle of prevention applies in the case of insects. Gauze netting spread over foods drying in the open may be necessary to keep insects out of storage or packing places. Careful cleaning and fumigation of premises and equipment before a new crop is put into a storage space may save it from contamination by insects surviving from the previous crop. The use of infested secondhand bags is another common source of trouble.

While insecticides and fumigants have their function (for example, in preparing a storage space for spices or other foods), a product that is already infested is not made acceptable for food by fumigation. Insects are objectionable in food even though they may have been killed. Most insecticides and fumigants are poisonous, and if they contaminate food, the food becomes adulterated and subject to legal action. While most fumigants are volatile, they may nevertheless result in contamination of the food or adversely affect nutritive values. Legal tolerances for such residues have been issued (21 *CFR* 193).

In some cases herbs or spices may be used for drug purposes and they then become subject to the drug provisions of the Food, Drug, and Cosmetic Act. Those spices or spice oils that are in the *United States Phamacopeia* or the *National Formulary* are subject to the standards set forth in these compendia when used for drug purposes.

Many herbs once thought to have medicinal value continue to be marked for various purposes. If no therapeutic claims are made or implied in the labeling or other promotional material, such products are regarded as foods and subject only to the food provisions of the law. For example, the herb ginseng is permitted to be sold as a tea.

Herbs are not necessarily harmless, contrary to common belief. Many such plants are toxic and may be extremely dangerous. The FDA believes it important to prevent the marketing and use of herbs for medicinal purposes if they have not been determined to be safe and effective for their intended uses, as required by law.

Spices and herbs must be the genuine products indicated by their common names on the labels. If obtained from or mixed with material from other plants, they are both adulterated and misbranded. The identity of herbs and spices is established by their botanical names. For example, the herb labeled as sage is *Salvia officinalis* L. As a guide to identity of food spice products, the FDA uses the following definitions.

Spices are aromatic vegetable substances used for the seasoning of food. They are true to name, and from them no portion of any volatile oil or other flavoring principle has been removed. Onions, garlic, and celery are regarded as foods, not spices, even if dried.

Allspice. The dried nearly ripe fruit of *Pimentia officinalis* Lindl.

Anise. The dried fruit of *Pimpinella anisum* L.

Bay Leaves. The dried leaves of *Laurus nobilis* L.

Caraway Seed. The dried fruit of *Carum carvi* L.

Cardamom. The dried, nearly ripe fruit of Elettaria cardomomum Maton.

Cinnamon. The dried bark of cultivated varieties of *Cinnamomum zeylanicum* Nees or of *C. cassia* (L.) Blume, from which the outer layers may or may not have been removed.

Ceylon Cinnamon. The dried inner bark of cultivated varieties of *Cinnamomum zeylanicum* Nees.

Saigon Cinnamon, Cassia. The dried bark of cultivated varieties of *Cinnamomum cassia* (L.) Blume.

Cloves. The dried flower buds of *Caryophyllus aromaticus* L.

Coriander. The dried fruit of *Coriandrum sativum* L.

Cumin Seed. The dried fruit of *Cuminum cyminum* L.

Ginger. The washed and dried, or decorticated and dried, rhizome of *Zingiber officinale* Roscoe.

Mace. The dried arillus of *Myristica fragrans* Houtt.

Macassar Mace, Papua Mace. The dried arillus of *Myristica argentea* Warb.

Marjoram, Leaf Marjoram. The dried leaves, with or without a small proportion of the flowering tops, of *Majorana hortensis* Moench.

Nutmeg. The dried seed of *Myristica fragrans* Houtt, deprived of its testa, with or without a thin coating of lime (CaO).

Macassar Nutmeg, Papua Nutmeg, Male Nutmeg, Long Nutmeg. The dried seed of *Myristica argentea* Warb, deprived of its testa.

Paprika. The dried, ripe fruit of *Capsicum annuum* L.

Black Pepper. The dried, immature berry of *Piper nigrum* L.

White Pepper. The dried mature berry of *Piper nigrum* L. from which the outer coating or the inner coatings have been removed.

Saffron. The dried stigma of *Crocus sativus* L.

Sage. The dried leaf of *Salvia officinalis* L.

Tarragon. The dried leaves and flowering tops of *Artemisia dracunculus* L.

Thyme. The dried leaves and flowering tops of *Thymus vulgaris* L.

RETAIL FOOD PROTECTION

The FDA relies mainly on state and local authorities to insure the safety of food passing through retail channels such as restaurants, cafeterias, supermarkets, snack bars, and vending machines. Given the size and diversity of the retail food industry, FDA directs its efforts to promote effective state and local regulation. It assists and supports the development of model food protection laws and uniform standards, by providing technical assistance, exchanging information, conducting training programs for regulatory personnel, and, on request, evaluating state programs.

Preventing the mishandling of food after it reaches the retailer is the primary concern of this joint federal–state effort. Food-borne disease has numerous causes, but results most often from improper holding and storage temperatures, inadequate cooking, poor hygienic practices by food handlers, and lack of cleanliness in facilities. Inquiries should be addressed to the FDA.

INTERSTATE TRAVEL SANITATION

Sanitation requirements protecting passengers and crew on interstate carriers are contained in the regulations for Control of Communicable Diseases (21 *CFR* Part 1240) and Interstate Conveyance Sanitation (21 *CFR* Part 1250). These regulations contain specific requirements for equipment and operations for handling food, water, and waste both on conveyances (aircraft, buses, railroads, and vessels) and those located elsewhere, ie, support facilities such as caterers and commissaries, watering points and waste servicing areas. The regulations also specify requirements for reviewing plans and inspecting construction of equipment, conveyances, and support facilities. The carriers are required to use only equipment and support facilities which have been approved by the Agency.

BIBLIOGRAPHY

General References

Requirements of Laws and Regulations Enforced by the U.S. Food and Drug Administration, DHHS Publication No. (FDA) 89-1115, U.S. Food and Drug Administration, Washington, D.C., 1989.

Y. H. Hui, *United States Food Laws, Regulations, and Standards*, 2 vols., 2nd ed., Wiley-Interscience, New York, 1986.

Y. H. Hui, *United States Regulations for Processed Fruits and Vegetables*, Wiley-Interscience, New York, 1988.

Y. H. Hui
EDITOR-IN-CHIEF

FOOD MARKETING

Most of the efforts and resources in food science and technology are devoted to making food production faster, more efficient, of higher quality, and, of course, less expensive. Food marketing is concerned with a number of other important activities such as the identification of consumer needs; the design of need-satisfying food products and services; the pricing, distribution, and communications (including advertising and promotion) of food and allied products. Most people think of food marketing as synonymous with advertising, and while this is the most obvious activity, it is by no means the most important. Marketing is more than just a set of techniques to entice people to buy a product, it is a philosophic approach to doing business.

MARKETING PHILOSOPHY

The marketing philosophy has been described by numerous authors, in literally hundreds of textbooks. Some of the more classic definitions are "marketing is the performance of activities that seek to accomplish an organization's objectives by anticipating customer or client needs and directing a flow of need-satisfying goods and services from producer to customer or client" (1) and "marketing is a social and managerial process by which individuals and groups obtain what they need and want through creating and exchanging products and value with others" (2). In 1985, the American Marketing Association approved this definition: "Marketing is the process of planning and executing the conception, pricing, promotion, and distributing of ideas, goods, and services to create exchanges that satisfy individual and organizational objectives." (3). A simpler definition that captures the essence of marketing is as follows: "Marketing is the anticipation, understanding, and satisfaction of customer needs while realizing the organization's objectives." The key to all of these definitions is that marketing involves activities that begin with assessing what the consumer needs and wants and translates that into food products that can be sold for a profit. Food marketers believe that the most profitable way to sell products is to offer the consumer the products that they value the most. Food marketers understand that what they make in the factories may not be the same thing consumers are buying at the checkout line. Although not involved in the food industry, Charles Revson, of Revlon Cosmetics, summed up this concept as follows: "In the factories we make cosmetics, at the counters we sell hope." McDonald's makes hamburgers but its customers are buying quick and convenient appetite satisfaction. Food marketers must understand and focus on what consumers want to buy and not simply what food marketers make. But producing the right product is not enough. Food marketers are also concerned with moving that product to the point of consumption (distribution) and making sure that intermediate (wholesalers and retailers) as well as final (consumers) buyers know all the differential benefits derived from product usage. This differential advantage represents the primary basis for persuading intermediate and final buyers to purchase and use a particular food product. The tools available for persuasion include advertising, personal selling, consumer, and trade promotions.

Production vs Marketing vs Sales

After World War II, consumers had considerable spending power, pent-up demand, and too few goods to purchase. Firms found instant success by producing for the mass market. During the production era, food marketers focused on low-cost products to satisfy excess demand. An example is TV dinners. The earliest versions were frequently bland and cheap. Companies that made these products focused on cost reduction and just getting the product to market. However, as demand decreased and competition increased, firms shifted to a sales orientation. While the company made only minor changes to the TV dinners, they increased sales promotions and gave price deals to increase sales.

One point of confusion is the difference between sales and marketing. In the 1960s when marketing came into vogue, many companies promoted their sales managers to marketing managers. Marketing to many was just modern-day sales. Nothing could be further from the truth; marketing is a way of doing business. Marketing focuses on the customer and profits, and it sets the stage so that when the sales force is asked to sell, the odds are in their favor. Marketing ensures that the product the sales force is selling is one that the customer wants, in the type of package, in the proper stores, and in the right sizes, etc. The sales force is responsible for getting the volume. Marketing makes sure that the volume is profitable.

As the TV dinner people found that increased selling activity increased sales, it also meant lower margins and profits. Food marketers began to ask the question, "What do consumers really want from frozen prepared foods?" It was not the traditional TV dinner that they had been pushing on the public in the previous years. Food marketers, not the production department, defined what should be done to make the product more attractive (and, therefore, more profitable). Today's frozen entrées and dinners are much better, and many consumers say a delicious, product choice. By taking a marketing approach, the company wins with higher profits and the consumer wins with better products.

THE MARKETING MANAGEMENT PROCESS

The marketing management process consists of analyzing market opportunities, researching and selecting target markets, developing marketing strategies, planning marketing tactics, and implementing and controlling the marketing effort.

The steps in the marketing management process are as follows:

1. Identify market opportunities.
2. Select target market(s) and position the product in the target market.
3. Develop a strategy for each target market.
4. Plan the tactics to accomplish the strategy.
5. Provide a procedure to implement the strategy.
6. Put in place a system for monitoring performance.
7. Evaluate and modify as needed.

MARKETING STRATEGY

Marketing strategy is how food marketers get from where they are to where they want to be. Many people confuse marketing strategy with simply beating the competition. However, marketing strategy is serving customers' real needs. Competitive realities are what possible strategies are tested against. Marketing strategy involves manipulating the marketing mix in the context of environmental variables to satisfy the needs of the target market. The target market is a fairly homogeneous (similar) group of customers to whom a company wishes to appeal. The marketing mix represents the controllable variables the company manipulates to satisfy the target market. The marketing mix is made up of four basic variables generally known as the four *P*'s product, place, promotion, and price.

MARKETING ENVIRONMENT

The marketing environment consists of forces that are not controllable but that influence both the consumer and the food marketer. Although the list of such environmental, uncontrollable or external forces is endless, the most common factors confronting the marketer are as follows.

1. *Competition.* Competitive analysis has been in vogue for the last five years. Understanding competitive strengths and weaknesses is a major input to marketing strategy today. A number of books have focused on just this topic (4,5).

2. *Demographics (Population Descriptors).* Changes in demographics has also had a major impact on food marketing. Baby boomers and the aging of America are just two examples of demographic trends that will lead to new food marketing techniques. *American Demographics* is one journal that provides an excellent overview of changes in American demographics.

3. *Politics.* No area of business is unaffected by politics. The implication to the food industry is through the USDA, FDA, FTC, FCC, and more recently the DEP.

4. *The Economy.* Recession, depression, inflation, a booming economy, unemployment, etc all affect the ability of food marketers to carry out their mission. Even more important to the food business is Wall Street, which frequently determines corporate policies based on current

stock prices. Companies attempting to raise stock prices by cutting costs and raising profits often find competitors taking market share from them.

5. *The Physical Environment.* This usually refers to available resources. Probably the most important variables are oil prices and availability of farmland.

6. *Technology.* Little needs to be said about technology. It has influenced every aspect of food marketing. At retail, computerized shelf space management, talking shopping carts, and in-store videos will change the nature of the shopping experience. Distribution is already taking advantage of satellite dishes on trucks to monitor the exact location and contents of the trucks. Warehouses are using scanning technology to position and control product movement. Food manufacturers have similarly benefited from technology. The *Journal of the International Food Technologists (IFT)* addresses technology in food processing.

7. *Social–Cultural Factors.* Of all factors changing food marketing, the changing American family is number one. Women leaving home to join the work force has made convenience a priority. More meals are being purchased from alternative outlets than ever before. No segment of the food industry has been unaffected by these changes (6).

Food marketers employ the controllable variables, the four *P*'s, to address the uncontrollable forces that influence the success of a firm's marketing strategy.

MARKETING MIX (THE FOUR *P*'S)

Product

In the area of food marketing, a product can be described as the need-satisfying offering of a firm. Because consumers are buying satisfaction, companies must be concerned with product quality. From a marketing perspective, quality means the ability of a product to satisfy a customer's needs or requirements. The product area of the four *P*'s is concerned with developing the right product for a specific target market. A product can either be a physical good, a service, or a blend of both. Products in the food industry are usually limited to consumer or industrial products. Consumer products are those products meant for the final consumer. Industrial products are products meant for use in producing other products.

Product Development

Product development involves offering new or improved products for a present market. In the food industry, competition is strong and dynamic, making it essential for companies to keep developing new product. New product planning is not optional, it must be done just to survive in today's changing markets. A new product is one that is new in any way for the company concerned. According to the Federal Trade Commission (FTC), a product can only be called new for six months and only if it is entirely new or changed in a functionally significant or substantial respect. The cost of new product introductions can be high. Experts estimate that consumer-product companies spend at least $20 million to introduce a new brand and 70–80% of these new brands are flops (7).

Place (Distribution)

The place component of the four *P*'s can be thought of as making products available in the right quantities and locations when the customers want them. A product is not much good to a consumer if it is not available when and where it is wanted. Place requires the selection and use of marketing specialists to provide target consumers with the product. Most consumer goods in our economy are distributed through multiple institutions, or middlemen, which are commonly referred to as marketing channels, or channels of distribution. A channel of distribution is any series of firms or individuals who participate in the flow of goods and services from producer to final user or consumer. Marketing intermediaries fit into one of the following categories.

1. *Merchant wholesalers* take title to (own) the goods they sell and sell primarily to other resellers (retailers), industrial, and commercial customers rather than to individual consumers.

2. *Agent middlemen,* such as manufacturer's representatives and brokers, also sell to other resellers and industrial or commercial customers, but they do not take title to the goods they sell. They usually specialize in the selling function and represent client manufacturers on a commission basis.

3. *Retailers* sell goods and services directly to final consumers for their personal, nonbusiness use.

4. *Facilitating agencies,* such as advertising agencies, marketing research firms, collection agencies, and railroads, specialize in one or more marketing functions on a fee-for-service basis to help their clients perform those functions more effectively and efficiently (8).

Choosing the correct channel of distribution is critical in getting products to the target market's place. Because in the food industry physical goods are almost always involved, place requires physical distribution (PD) decisions. Physical distribution is the transporting and storing of goods to match target customers' needs with a firm's marketing mix (9). From the customer point of view, the concern is not how the product was stored or moved, but rather what is the customer service level, how rapidly and dependably a firm can deliver what the customer wants. It is important for food marketers to understand the customer's point of view. Physical distribution is usually the invisible part of marketing and only gets the customer's attention if something goes wrong.

Promotion

Promotion is concerned with telling the target market about the right product. A promotion is a direct inducement that offers an extra value or incentive for the product to the sales force, distributors, or the consumer with the primary objective of creating an immediate sale (10). Promotions are aimed at both consumers and middlemen for the purpose of keeping goods moving smoothly through the pipeline. This is accomplished by using a combination of push and pull strategies. A push strategy provides incentives to the middlemen to buy and resell the product, therefore pushing the goods on to the next stage of the pipeline to the ultimate consumer. A pull strategy provides incentives to the consumer to pull the product out of the end of the pipeline at the retailer level. Food marketers should use a combination of both strategies so that the product will flow easily from the manufacturer to the consumer. The ultimate goal of promotions is to induce behavior. Sales promotions refer to promotional activities other than advertising, publicity, and personal selling that stimulate interest, trial or purchase by final customers or others in the channel.

Consumer Promotion. Examples of consumer promotions are coupons, sweepstakes, contests, product samples, refunds, rebates, tie-ins, premiums, bonus packs, trade-ins, and exhibitions. These promotions are directed at consumers who purchase products at the retail level and are designed to provide them with an inducement to purchase the marketer's brand. Consumer promotions are part of a promotional pull strategy and work along with advertising to encourage consumers to create a demand for a particular brand.

Trade Promotion. Trade promotions in the food industry are sometimes known as promotions to the retailer. This term may be misleading as these promotions are targeted to distributors, wholesalers, and retailers. Trade promotions are critical because product cannot be sold to consumers if it is not first sold to the middlemen. Because the average supermarket stocks more than 10,000 items, promotions need creativity to break through the clutter. Trade promotions include activities such as promotional allowances, dealer incentives, point-of-purchase displays, sales contests and sweepstakes, and trade shows. Trade promotions are designed to motivate distributors and retailers to carry a product and make an extra effort to promote it to their customers (push strategy).

One of the most commonly used trade promotions in the food industry is the promotional allowance. These are payments made by a manufacturer to resellers (often off-invoice) for merchandising its products or running in-store promotional programs such as reduced shelf prices, special displays, or in-store advertising. Another common but controversial promotion is the slotting allowance (sometimes known as a stocking allowance or street money). Slotting allowances are the fees that are often demanded by the retailers to gain admission into their stores. The position taken by the retailer is that with increased competition, a proliferation of new products, and small profit margins, they are required to ask for these fees, claiming that they will be used to promote the products, redesign shelves, and reprogram computers. Most manufacturers feel differently about how slotting fees are used. One food industry source estimated that 70% of all slotting fees go directly to the retailers' bottom line (11). Many food manufacturers view slotting allowances as little more than corporate bribery.

Traditionally, food companies have spent more of their promotional dollars on trade promotions than on con-

Table 1. Annual Survey of Promotional Practices[a]

	Consumer Promotions	Trade Promotions
1987	26.1%	39.9%
1988	26.3%	39.6%
1989	27.1%	39.4%

[a] Ref. 12.

sumer promotions, in an effort to win shelf space in the midst of a rising tide of new products. Recently, however, expenditures on promotions to consumers have been growing while expenditures on promotions to the trade have been declining slowly. Of the three major areas of a company's promotional spending, consumer promotions, trade promotions, and media spending, the three-year trend shown in Table 1 has appeared (12).

As an example, Kellogg Co., the world's leading cereal manufacturer, is one of the companies that intends to tilt its marketing strategy away from the trade and back toward the consumer. This comes on the heels of sluggish domestic sales. In addition to Kellogg Co., Kraft General Foods has decided to step up advertising for its Maxwell House coffee, a brand that stopped advertising in 1987 and lost its number one position in the coffee business (13) (Table 2).

Advertising

Simply stated, advertising tells the target market that the right product is available, at the right price, and at the right place and time. Advertising is defined as any paid form of nonpersonal presentation of ideas, goods, or services by an identified sponsor. It includes the use of such media as magazines, newspapers, radio and television, signs, and direct mail (14). The most common vehicles of advertising used by food companies are network and spot television, radio, and magazines. According to one source

the food industry is expected to be the biggest ad spender in 1990 with a budget of $8.4 billion, according to a study by Schonfeld & Associates of Evanston, Ill. The 107 publicly owned restaurant and fast-food chains are forecast to spend 8% more this year, to more than $1.5 billion. Supermarket chains will raise spending by 5.6% to $1.8 billion, with Kroger Co. the biggest at $225 million. Ad spending for health and beauty aids is expected to rise by 11% to more than $11.2 billion with Bristol-Myers Co. leading the way at $1.1 billion (15).

Table 2. Biggest Ad Spenders in the Prepared Food Category[a,b]

Rank	Company
1	Kellogg Company
2	General Mills Inc.
3	Philip Morris Companies Inc.
4	Quaker Oats Company
5	Campbell Soup Company

[a] For a one-year time period ending in June 1989.
[b] Total advertising expenditures for prepared foods for this time period was $1.2 billion. (16)

Personal Selling

Personal selling is defined as direct communication between sellers and potential customers. Personal selling is usually face-to-face, but communication can take place over the telephone. The strength of personal selling in food marketing is its flexibility, as it provides immediate feedback that helps salespeople adapt to the current situation.

In total there are more than 11 million consumer salespersons in the United States, with another 9 million salespersons in the industrial marketing area. How important are salespersons? One study, for example, found that executives of industrial firms rated the sales function as 5 times more important than advertising in their marketing mixes; for consumer durables marketers, sales was rated 1.8 times as important as advertising, while for consumer nondurables, advertising and person selling were rated as about equally important (17).

The basic steps of personal selling are as follows; prospecting, planning sales presentations, making sales presentations, and following up after the sale. Prospecting involves following all the leads in the target market to identify potential customers. Once the prospect has been located, it is necessary to make a sales presentation that is the salesperson's effort to make a sale.

Price

The fourth P a food marketer must be concerned with is price. When a company sets a price, they must take into consideration the kind of competition in the target market and the cost of the whole marketing mix. Other considerations to be made in pricing decisions are customer reaction to possible prices, current practices as to markups, discounts, and other terms of sale, and finally, legal restrictions on pricing. Price is an important consideration in food marketing because if a customer will not accept the price, all of the planning decisions and efforts could be wasted.

MARKETING RESEARCH

Finally, food marketing usually encompasses marketing research. Marketing researchers represent the information-gathering arm of the firm and helps to determine exactly what products consumers want as well as how to best distribute, advertise, and promote those products. Marketing research is also used extensively to measure performance. Do consumers really understand the benefits of the product, how much is being sold and to whom, etc? Simply stated, marketing research is intended to provide timely and relevant information to improve marketing decision making.

In attempting to meet the needs of consumers, food marketers need to keep up with all of the changes taking place in their markets. Most food marketers rely on marketing research to help them make decisions about the marketing plan. The American Marketing Association adopted this definition of marketing research in 1987:

> Marketing Research is the function which links the consumer, customer, and public to the marketer through information—information used to identify and define marketing opportunities and problems; generate, refine, and evaluate marketing actions; monitor marketing performance; and improve understanding of marketing as a process.

Marketing Research specifies the information required to address these issues; designs the method for collection information; manages and implements the data collection process; analyzes the results; and communicates the findings and their implications (19).

Food marketers must use marketing research to help them make good decisions. One way is to use the scientific method—a decision-making approach that focuses on being objective in testing ideas before accepting them. This way a marketer does not assume his or her intuition is correct without evidence to support it. The marketing research process is a five-step application of the scientific method. It includes

1. Defining the problem.
2. Analyzing the situation.
3. Getting problem-specific data.
4. Interpreting the data.
5. Solving the problem.

CONCLUSION

In order for the food industry to maintain its preeminence, all the parts must not only work individually, but work together. While the engineers and food technologists are developing new products, food marketers must ensure that the research and development is market directed. Making food less expensive to produce is only valuable if people want and will buy the less-expensive product. Food marketing is the discipline that brings the consumer into the corporate planning process. It ensures that the products desired will not only be available, but that consumers will know about the benefits of the product and where to get it. Food marketing is no more important than any other food business function, but it is essential in today's food business environment.

BIBLIOGRAPHY

1. E. J. McCarthy, *Basic Marketing,* Richard D. Irwin, Inc., Homewood, Ill., 1990, p. 8.
2. P. Kotler, *Principles of Marketing,* Prentice Hall, Inc., Englewood, N.J., 1989, p. 5.
3. *Marketing News,* 1 (Mar. 1, 1985).
4. A. Ries and J. Trout, *Marketing Warfare,* McGraw-Hill Inc., New York, 1983.
5. M. Porter, *Competition in Global Industries,* Harvard Business School Press, Boston, 1986.
6. J. Stanton, "2001: A Food Odyssey," *HEIB,* 1990
7. Ref. 1, p. 261.
8. H. Boyd, *Marketing Management,* Richard D. Irwin, Inc., Homewood, Ill., 1990, p. 22.
9. Ref. 1, p. 344.
10. L. Haugh, "Defining and Redefining," *Advertising Age,* M-44 (Feb. 14, 1983).

11. J. Dagnoli and L. Freeman, "Marketers Seek Slotting-Fee Truce," *Advertising Age*, 12 (Feb. 22, 1988).
12. Donnelly Marketing, *12th Annual Survey of Promotional Practices*, Stamford, Conn., 1990, p. 3.
13. R. Gibson, "Kellogg Shifts Strategy to Pull Consumers In," *The Wall Street Journal*, Jan. 22, 1990, p. 131.
14. Ref. 1, p. 366.
15. *Ad Week's Marketing Week*, July 3, 1989, p. 6.
16. B. Bagot, "1990 Industry Outlooks," *Marketing and Media Decisions*, 33 (Jan. 1990).
17. W. Wilkie, *Consumer Behavior*, John Wiley & Sons, Inc., New York, 1990, p. 492.
18. D. Lehmann, *Market Research and Analysis*, Richard D. Irwin, Inc., Homewood, Ill., 1989, p. 3.

John L. Stanton
Richard J. George
Carol A. Gallagher
Saint Joseph's University
Philadelphia, Pennsylvania

FOOD MICROBIOLOGY

HISTORY

Food microbiology is the study of all aspects of microbial actions on food and food products, both directly and indirectly, related to the welfare of mankind. Topics included in food microbiology are history of food microbiology, number and kinds of microbes in foods, intrinsic and extrinsic parameters of foods, methodologies, food spoilage, food preservation, and food-borne pathogens.

The precise time at which humans started to realize the role of microbes in food products cannot be determined. However, it is safe to assert that humans noticed the results of microbial action such as spoilage of food and food poisoning early in the history of food-gathering and food-producing periods. Humans experienced and noticed the changes occurring in foods without knowing the reasons behind such activities until the development of the science of microbiology. The real beginning of food microbiology coincided with the development of microbiology, especially Pasteur's work on food fermentation processes. Early studies in food microbiology centered on dairy bacteriology. It was only about 20 years ago that the field of food microbiology became a recognized field. Today it has a great impact on issues such as food safety, genetic engineering of new food products, methodologies in applied microbiology, food technology, and preservation technologies.

NUMBERS AND KINDS OF MICROBES IN FOODS

The number and kinds of microbes in foods depend greatly on the food products and the conditions under which they are stored or processed. The microbial profiles of raw food products and processed products are completely different and pose different sets of problems in terms of food spoilage potential and food safety issues. In general, raw food products tend to have a heterogeneous population whereas processed food usually contains those organisms that can survive the processes and subsequent storage conditions. Microorganisms are very small, and when they occur in foods they may be in the millions per gram, milliliter, or square centimeter. A convenient guide for categorizing microbial loads on meat surfaces is as follows: microbial load on meat surface is considered low when the count is log 0–2 colony-forming units (CFU)/cm^2; intermediate, log 3–4 CFU/cm^2; high, log 5–6 CFU/cm^2, and very high, log 7 CFU/cm^2 (1). Using this guide, samples with log 0–4 CFU/cm^2 would be considered as acceptable; samples with log 5–6 CFU/cm^2 would be considered questionable; and samples with log 7 CFU/cm^2 and above would be considered unacceptable from a food spoilage standpoint. Generally, log 7 CFU/g, ml, or cm^2 is considered the index of spoilage, since food will exhibit odor and/or slime at counts above log 7.5 CFU/g, ml, or cm^2. This guide does not include identification of potential pathogens in foods such as *Salmonella*, *Staphylococcus aureus* or *Clostridium perfringens*.

The number of microorganisms in food is usually monitored by putting a known volume of liquid, weight of solid, or surface area sample into a sterile liquid diluent (usually a buffered liquid) and then appropriately diluting the sample such that when the sample is placed in a petri dish the number of organisms in the petri dish is in the range between 25 to 250. The agar medium used to grow commonly occurring microorganisms is called the standard plate count agar. A count is always converted to per ml, g, or cm^2 for ease of comparison of data from laboratory to laboratory. The result only provides a count of all organisms that can grow in this agar under the time and incubation temperature (usually 48 h at 32°C). A differential count can also be made. By using different agars specifically designed for specific organisms, one can make a specific count such as a coliform count by using a violet red bile agar. This agar contains inhibitory compounds such that other organisms cannot grow, and when coliform organisms grow they will exhibit a characteristic color, shape, and size in the agar. By developing a variety of differential agars, scientists can monitor the occurrence of different type of organisms in food products. Thus one can make a *Staphylococcus aureus* count, a lactobacillus count, a *Clostridium perfringens* count, etc. These types of differential counts give the food microbiologist a clearer picture of the specific population of microbes in certain food products.

To pinpoint the exact organism in a food product, a food microbiologist must take a colony and purify it, then perform a variety of morphologic, biochemical, and physiologic tests to ascertain the genus and species of the microorganism in question. The identification scheme would include morphology (rod, sphere, spiral, fruiting bodies); gram reaction (gram-positive or gram-negative); biochemical tests (carbohydrate fermentation, enzyme production); pigment production (yellow, red, blue, black, gray, green, etc); Nutritional requirements (organic, inorganic, complex, simple); temperature requirements (psychrophiles, 0-10C; mesophiles, 10-45C; thermophiles 45-75C; and psychrotrophs, 0-30C) chrophiles, 0-10C; mesophile, 10-45C; thermophiles; pH requirements; fermentation products (acid, alcohols, etc); antibiotic sensitivity; gas requirement (aerobic, anaerobic, facultative); pathogenicity; serology (serotyping with specific antibodies).

After these tests are made, the food microbiologist can determine the exact nature of the organism in terms of genus and species. In many instances the number of specific pathogens in food is quite small. For example, there may be only 10 *Salmonella* among hundreds and thousands of other organisms in a food product, yet the food microbiologist must be able to detect them. To achieve this, food microbiologists have developed many elaborate pre-enrichment, enrichment, and selective enrichment procedures so that target organisms (eg, *Salmonella*) will grow while other organisms are suppressed. For each target organism a separate scheme is developed. Consult the General References at the end of this article for detail information on methodologies in food microbiology.

The kind of microorganisms in food depend on the food and the conditions under which they were made, processed, or stored. Microorganisms in food can be classified as bacteria, yeasts, molds, and viruses.

Bacteria. Bacteria are unicellular organisms ranging in size from 0.1 to 2.0 μm and occurring in rod, spherical, or curved shapes. They divide asexually by binary fission. Although no sexual stage occurs, bacteria can exchange genetic information from one to another.

Yeasts. Yeasts are unicellular fungi occurring singularly with round or oval shape. Asexual reproduction is by budding, and sexual reproduction is by sexual spores.

Molds. Molds are multicellular filamentous fungi, highly structured and organized in morphology. Reproduction is by sexual and asexual stages. They usually grow so profusely that humans can see them on foods (fizzy masses) without the need of magnification.

Viruses. Viruses are submicroscopic entities that cannot reproduce without a living host. They occur with a protein coat enclosing a coil of DNA or RNA material. They can infect bacteria, plants, and animals. The shapes and sizes vary greatly among different groups of viruses.

Bacteria of interests in food microbiology can be divided into the following groupings:

Lactic Acid Bacteria

These are gram-positive, non-spore-forming bacteria producing lactic acid as the major or sole product of fermentation. As a group they are important in food spoilage because they cause souring and discoloration. However, they are also very important in pickling, cheese making, fermented dairy products, and silage technologies. The major genera of lactic acid bacteria include *Streptococcus Pediococcus, Leuconostoc,* and *Lactobacillus.*

Streptococcus. *Streptococcus* spp. such as *S. lactis, S. cremoris, S. thermophilus,* and *S. diacetilactis* are important starter cultures in the dairy industry. Some *Streptococcus* are pathogenic, such as *S. pyogenes* and *S. faecalis.*

Pediococcus. *Pediococcus* produces large quantities of lactic acid and is very important as a starter culture in the curing of meat.

Leuconostoc. *Leuconostoc* produces gas as well as slime in the presence of sugar. Although they are important spoilage organisms, they are also important producers of flavor compounds in dairy products.

Lactobacillus. *Lactobacillus* is a heterogeneous group of organisms consisting of slender, gram-positive rods. One group is homofermentative (producing large quantity of lactic acid) and another group is heterofermentative (producing acid and gas). *Lactobacillus* spp. are important in dairy, meat, and silage fermentation but are undesirable as spoilage organisms because of the production of large quantities of lactic acid in a variety of food products during storage.

Aerobic, Gram-Positive Catalase-Positive Cocci

These bacteria are gram-positive cocci in pairs, short chains, or clusters. They produce catalase (a very active enzyme) and form acid from carbohydrate. Some of them are quite heat resistant and salt tolerant and produce a variety of colors in food and culture media. Among them are important food pathogens and spoilage organisms.

Micrococcus. *Micrococcus* spp. are aerobic, gram-positive cocci occurring widely in nature. Many species can grow under refrigeration and on inadequately sanitized equipment.

Staphylococcus. *Staphylococcus* is an important genus of gram-positive cocci. This is a facultative anaerobic organism that occurs widely in nature and on human skin. They produce a variety of extracellular enzymes and metabolites. The most important metabolite they produce is a group of heat-stable toxins called enterotoxins, which are the agents of staphylococcal intoxications. The organism is salt tolerant and can spoil a variety of foods besides being an important food pathogen.

Spore-Forming Bacteria

The two important spore forming bacteria in food microbiology are *Bacillus* and *Clostridium.*

Bacillus. *Bacillus* is a gram-positive, aerobic spore-forming bacteria. It occurs widely in nature and in soil. It forms large spores. Bacterial spores are highly resistant to all forms of food processing techniques that usually kill vegetative cells. *Bacillus* possesses a wide range of physiologic activities, including fermentation of sugar, peptonization of protein, hydrolysis of starch, and rennin coagulation of milk. *Bacillus anthracis* is an important animal pathogen. Many *Bacillus* spp. are environmental contaminants and enter the food chain through air, water, and surface contacts. Some important *Bacillus* spp. are *B. subtilis, B. cereus,* and *B. stearothermophilus.*

Clostridium. *Clostridium* species are gram-positive, anaerobic spore formers. Some are highly anaerobic and can be killed in the presence of molecular oxygen whereas others are aerotolerant. Important species in food spoilage include *C. butyricum, C. putrefaciens,* and *C. sporogenes.* From the standpoint of food safety, *C. botulinum* is the most important because it produces a group of highly toxic protein toxins called botulin. These toxins are responsible for the often fatal disease called botulism. The canning industry has spent millions of dollars designing time and temperature treatments for canned goods specifically aimed at killing the spores of this organism. Fortunately, the toxin is heat sensitive. Boiling the toxin for 10 min will render it inactive.

Another species of *Clostridium* important in food microbiology is *C. perfringens.* Although this organism is less toxic than *C. botulinum, C. perfringens* accounts for about one-fifth of all food poisoning cases in the United States annually. It produces a toxin that causes a mild diarrhea in humans.

Gram-Positive Irregularly Shaped Bacteria

This group of bacteria has irregular shapes and is gram-positive. Among them is *Propionibacterium,* which is a small, anaerobic, pleomorphic rod (cells with irregular shapes). *P. shermanii* is the organism responsible in forming eyes in Swiss-type cheeses. They also impart desirable flavor in cheese fermentation. *Corynebacterium* is a pleomorphic rod, arranged in Chinese letter morphology when observed under the microscope. Most of the species are environmental contaminants. *C. diphtheriae* is the agent responsible for diphtheria in humans. *Microbacterium* is a small, aerobic, heat-resistant rod. It can withstand 80°C for 10 min and is important in the spoilage of vacuum-packaged meat products.

Gram-Negative Polarly Flagellated Bacteria

From the standpoint of food spoilage, *Pseudomonas* is considered one of the most important organisms. This gram-negative organism is a prolific metabolizer of organic compounds. The organism can grow in refrigerated temperatures (psychrotrophic) and form slime, fluorescent compounds, and pigments in cold-stored foods. It is responsible for the spoilage of chicken, meat, fish, vegetables, and all kinds of foods kept in cold storage. Because it is aerobic it usually is not responsible for the spoilage of canned or vacuum-packaged foods.

Acetobacter is also a member of this group of gram-negative bacteria. This organism oxidizes alcohol to acetic acid. In vinegar making it is desirable; however, in wine making it is the most important organism causing souring. *Photobacterium* can cause phosphorescence of meat and fish when incubated in suitable conditions. *Halobacterium* can grow in salt concentrations as high as 30%. It can produce pigments and spoil salty fish.

Gram-Negative Short Rods

This heterogeneous group of organisms consists of gram-negative small rods, and when motile, possesses peritrichous (all over the cell) flagella. They are facultative anaerobes that are found in water, soil, human and animal environments, and in the food chain. Some of the organisms are exceedingly important in food microbiology. Many of them are food pathogens and food spoilage organisms. The most important family in this group is the *Enterobacteriaceae.* According to the newest classification there are 14 genera in this family: *Escherichia, Shigella Salmonella, Citrobacter, Klebsiella, Enterobacter, Erwinia, Serratia, Hafnia, Edwardsiella, Proteus, Providencia, Morganella,* and, *Yersinia.*

Escherichia. *Escherichia* is a true fecal coliform; the type species is the well-known *E. coli.* Although some species can cause food-borne disease, the most important role of this organism is as an indicator of fecal contamination. Its presence indicates the potential presence of other more pathogenic enteric organisms and is highly undesirable. This is one of the most monitored organisms in food and water microbiology.

Salmonella. *Salmonella* is an organism of great concern to the food industry. It is ubiquitous in the animal population and especially in poultry flocks. Some raw poultry products harbor *Salmonella,* thus all poultry products should be well cooked before consumption. The organism, when consumed in large number (1 million), can cause a disease called salmonellosis, which includes vomiting, nausea, diarrhea, chills, and fever. Mortality can occur in the very old, the very young, or the immunocompromised. There are more than 2,000 serotypes of *Salmonella* reported. Many new rapid methods are now being developed to detect this organism. The conventional method takes about 5 to 7 days.

All members of the *Enterobacteriaceae* are potentially pathogenic. A more thorough discussion of bacterial pathogens is recorded in a different section of this volume. It is important to note that occurrence of members of the *Enterobacteriaceae* in food is generally undesirable. Proper food-handling will prevent or retard their growth and keep food safer for the consumers.

Yeasts and Molds

In general, yeasts and molds are considered spoilage organisms in food microbiology. When conditions are not favorable for the growth of bacteria, yeasts and molds will take over and spoil the food items. For example, in citrus foods the pH is too low for bacterial growth, so yeasts and molds are more active in spoiling those foods. Another example is dry goods. Bread is spoiled by mold more easily than by bacteria because molds can grow with much less water than can bacteria.

The only real food-poisoning concern of molds is the possible production of mycotoxins and aflatoxins by some molds such as *Aspergillus flavus* and *A. paraciticus.* Some of these compounds are carcinogenic and thus are of concern to regulatory agencies. In the area of food fermentation, yeasts and molds are of great importance because a large variety of fermented foods are produced by direct or indirect activities of yeasts and molds (see FOOD FERMENTATION).

INTRINSIC AND EXTRINSIC PARAMETERS OF FOODS

All foods possess a set of conditions called intrinsic parameters. These parameters can be influenced by another set of conditions called extrinsic parameters. Together, these two groups of parameters have great influence on the number and kinds of microorganisms occurring in and on a food and their physiologic activities. Intrinsic parameters of food include pH, moisture, oxidation–reduction potential (presence or absence of oxygen), nutrient content, occurrence of antimicrobial constituents, and biologic structures.

All microorganisms have a minimum, maximum, and optimal pH tolerance, a moisture requirement, an oxygen tension requirement, and a nutrient requirement. By knowing these parameters, one can predict the presence and growth potential of specific microorganisms in certain types of foods. A pH of 4.5 is considered the demarcation line between acidic foods (less than pH 4.5) and basic foods (more than pH 4.5). Yeast and molds can grow down to pH 1 whereas bacteria cannot grow below pH 3. Thus acidic foods such as citrus fruits and carbonated soft drinks will be spoiled more by yeasts and molds than by bacteria. On the other hand, in a more basic food (above pH 4.5) bacteria will outgrow yeasts and molds owing to their higher metabolic rates in a favorable growth environment. Moisture content is another important parameter. This is usually expressed as water activity (A_w). Most moist foods are in the range of 0.95–1.00 A_w. When the A_w drops to 0.9, most spoilage bacteria reach their minimum level. Most spoilage yeasts have their minimum at 0.88, and molds have theirs at 0.80. Thus in dry food products yeasts and molds grow much better than bacteria and in moist food bacteria will outgrow yeasts and molds. The role of oxygen tension in and around food also has a great impact on the type of organisms growing there. Bacteria can be aerobic, anaerobic, or facultative anaerobic, so they can grow in a variety of oxygen levels (although different types will grow in different oxygen tension environments). Yeast can grow both aerobically and anaerobically. Most molds, however, cannot grow anaerobically. In a properly vacuum-packaged food, for example, one should not find mold growing. The amount of oxygen measured in terms of oxidation–reduction potential also dictates the types of bacteria that can grow in the food. Disrupting the oxygen tension of a food (eg, grinding a piece of meat to make ground beef from a steak) makes it easier for aerobic organisms to spoil the food.

Nutrient content (water; source of energy for metabolism; source of nitrogen, vitamins, and growth factors; and minerals) of different foods will support different types of microbes. In general, a food nutritious for human consumption is also a good source of nutrients for microbes. Some foods have natural antimicrobial compounds, such as eugenol in cloves, allicin in garlic, and lysozyme in egg, that can suppress the growth of some microbes. Biologic structures of some foods are also important for the prevention of microbial invasion. An example is the skin of an apple. When the apple is bruised, microbes can easily enter the fruit and spoil it.

Extrinsic parameters of food also play an important role in the activities of microbes. Temperature of storage greatly influences the growth of different classes of microbes. The amount of moisture in the environment (relative humidity) also influences the absorption of moisture or the dehydration of the food during storage and thus also influences the growth of different organisms. Varying the gaseous environment in storage will also change the types and growth rates of different organisms during storage of the food items. And last, the length of time of food storage also influences the spoilage potential by microbes in the food.

Thus intrinsic and extrinsic parameters of food are of great concern to food microbiologists. Skillful manipulation of these parameters by food microbiologists will result in more stable, nutritious, fresher, and safer foods for the consumer.

FOOD PRESERVATION TECHNIQUES

Food preservation techniques can be grouped under drying, low-temperature freeze-drying, high temperature, radiation, and chemical treatments. The technologies of these processes are recorded in other articles of this encyclopedia. In this article only their effects on microorganisms are discussed.

Drying is the most widely used method of food preservation in the world. Meat, fish, cereals, fruits, and vegetables are dried and preserved for a long time. Controlled dehydration and sun drying of foods removes water from foods so that microorganisms cannot grow. This process does not sterilize the food, so when water is reintroduced (rehydration), microbial growth may resume and may result in spoilage of food. Spores of bacteria and mold are known to survive long periods of time in dried foods.

Low-temperature preservation of food is based on retardation of enzymatic activity of microbes. As temperature decreases, the enzymatic activity of microbes also decrease and eventually stops at around freezing temperature and below. Freezing will kill approximately 10% of the initial microbial population, but the remaining population can survive for a long time. Refrigeration temperature (0 to 10°C) will retard growth of most organisms; however, a group of organisms called psychrotrophs will grow slowly under refrigeration temperature and may eventually spoil the cold-stored food. Psychrotrophic bacteria (*Pseudomonas, Micrococcus,* etc), yeasts (*Candida, Debaryomyces,* etc), and molds (*Penicillium, Mucor,* etc) can spoil foods under prolonged storage. Although most psychrotrophs are nonpathogenic, some pathogens such as *C. botulinum* and *Listeria monocytogenes* have been found to grow at or around 4 to 6°C. Freeze-drying or lyophilization of foods depends on the unique property of water—the triple point of water. Food is first frozen and then a vacuum is applied to sublime the water out of the food mass. Freeze-dried foods need not be refrigerated because microorganisms cannot grow without water. However, rehydrated food will have the same spoilage potential as the original food. Freeze-drying does not kill microorganisms effectively. In fact, freeze-drying is the best method to preserve bacterial cultures.

High-temperature preservation depends on heat coagulation of proteins and enzymes thus killing microorganisms. Pasteurization, a form of high-temperature preservation, usually refers to treatment of food at 63°C for 30 min or 72°C for 15 s. These time and temperature combinations are designed to kill most vegetative cells in milk, especially *Mycobacterium tuberculosis* and *Coxiella burnetti*. Thermoduric organisms are those that can survive pasteurization and later grow and spoil the pasteurized food. To reach sterilization temperature, food must be cooked under pressure in sealed containers. This is the practice used in commercial canning. The purpose is to achieve time and temperature combinations such that the most heat-resistant spores of *C. botulinum* are destroyed in the specific food item being canned.

Radiation treatment of food is of two types—ionizing radiation and nonionizing radiation. Ionizing radiation such as α, β, γ, and x-ray kill microorganisms by breaking chemical bonds of essential macromolecules such as DNA (target theory) or by the ionization of water, which results in the formation of highly reactive free radicals such as HO^- or H_2O^- capable of splitting chemical bonds in the microorganisms. Ionizing radiation destroys microorganisms without generation of heat, thus it is called cold sterilization. Interest in radiation preservation of food have been fluctuating in the past 30 years. Recently there seems to be a resurging interest on an international scale in the use of radiation for food preservation.

Nonionizing radiation includes ultraviolet (UV) treatment and microwave treatment. UV has poor penetration and thus is used only for surface decontamination. Microwave treatment of food has gained great popularity. It is estimated that 75% of the homes in the United States have a microwave oven. The waves at 2,540 MHz when applied to foods bound back and forth and create vibration of asymmetric, dielectric molecules (such as water), which generates heat. It is this heat that cooks food as well as kills microorganisms. There may exist some as yet unexplained mechanism of microwave destruction of microorganisms besides purely the effect of heat (2). Microwave cooking will continue to be an important method for homes and institutions, and it is an effective means of destroying microorganisms in foods as long as enough microwave exposure time is given to the food.

Chemicals are used to kill organisms (bactericidal) or to prevent them from growing (bacteriostatic). Many chemicals are used to treat equipment and the environment such as sanitizers, disinfectants, strong acids and bases, and halogens. In terms of food science, the subject of food additives is of great interest. Food additives are compounds added to foods for purposes such as improvement or modification of flavor, texture, rheology, color, pH change, water-holding capacity, and emulsification. Some of these compounds (lactic acid, acetic acid, propionic acid, sorbate, etc) are used as preservatives because they can kill microorganisms in foods. Compounds approved for use in foods are controlled by the Food and Drug Administration and are listed on the generally recognized as safe (GRAS)list. An up-date of the list was made in 1984 (3).

There are many recent developments in methodologies pertaining to food microbiology. Rapid methods and automation in microbiology have been the subject of many national and international symposia. The four major developments are miniaturization of conventional procedures and development of diagnostic kits, such as Fung's mini systems, API, Enterotube, Spectrum 10, IDS, Minitek, MicroID, and Biolog; modification of viable cell count procedures, such as 3M Petrifilm, Redigel, Isogrid, Spiral System, and DEFT test; development of alternative approaches for the estimation of microbial populations, such as the use of adenosine triphosphate, electrical impedance or conductance, microcalorimetry, or radiometry to indirectly measure biomass (bacteria, yeasts, and molds) in food products; and identification of microbes by novel and sophisticated instruments and procedures, such as Vitek system, DNA Probe (Genetrak), ELISA (Orgnon Teknika, Tecra), motility enrichment (BioControl), protein profiles, and fatty acid analysis.

The field of food microbiology is very important for food science and technology. It is one of the central disciplines in food science. Food microbiologists are called on to solve microbiologic problems related to other branches of food science and technology. The future of food microbiology is very bright indeed.

BIBLIOGRAPHY

1. D. Y. C. Fung, C. L. Kastner, M. C. Hunt, M. E. Dikeman, and D. H. Kropf, "Mesophilic and psychrotrophic bacterial populations on hot-boned and conventionally processed beef," *Journal of Food Protection* **43,** 547–550, 1980.
2. D. Y. C. Fung and F. E. Cunningham, "Effects of Microwave Cooking on Microorganisms in Foods," *Journal of Food Protection* **43,** 641–650, 1980.
3. B. L. Oser and B. K. Bernard, "13. GRAS Substances," *Food Technology* **38**(10), 66, 1984.

General References

Associations of Official Analytical Chemists, *Official Methods of Analysis,* AOAC Arlington, Vir., 1988.

J. C. Ayres, O. Mundt, and W. E. Sandine. *Microbiology of Foods,* W. H. Freeman and Co., San Francisco, 1980.

G. J. Banwart, *Basic Food Microbiology.* AVI Publishing Company, Inc., Westport, Conn., 1979.

J. E. L. Corry, D. Roberts, and F. A. Skinner, *Isolation and Identification Methods for Food Poisoning Organisms,* Academic Press, New York, 1982.

Food and Drug Administration, *Bacteriology Analytical Manual,* 6th ed., Association of Official Analytical Chemists, Arlington, Vir., 1984.

D. Y. C. Fung, "Microbiology of Batter and Breading," in D. R. Suderman and F E. Cunningham, eds., *Batter and Breading Technology,* AVI Publishing Co., Westport, Conn., 1983, pp. 106–119.

D. Y. C. Fung, "Rapid Methods for Determining the Bacterial Quality of Red Meats." *Journal of Environmental Health,* **46**(5), 226–228, 1984.

D. Y. C. Fung, "Types of Microorganisms," in F. E. Cunningham and N. A. Cox, eds., *Microbiology of Poultry Meat Products,* Academic Press, New York, 1987.

D. Y. C. Fung and C. L. Kastner, "Microwave Cooking and Meat Microbiology." *American Meat Scientists Association Reciprocal Meat Conference Proceedings* **35,** 81–85, 1983.

W. F. Harrigan and M. E. McCance. *Laboratory Methods in Food and Dairy Microbiology,* Academic Press, New York, 1976.

P. A. Hartman, *Miniaturized Microbiological Methods,* Academic Press, New York, 1968.

C. G. Heden and T. Illeni, eds., *Automation in Microbiology and Immunology,* John Wiley & Sons, Inc., New York, 1975.

C. G. Heden and T. Illeni, eds., *New Approaches to the Identification of Microorganisms,* John Wiley & Sons, Inc., New York, 1975.

B. C. Hobbs and J. H. B. Christian, *The Microbiological Safety of Food,* Academic Press, New York, 1973.

International Commission on Microbiological Specifications for Foods, *Microorganisms in Foods,* vol. 2, University of Toronto Press, Toronto, Canada, 1974.

J. M. Jay, *Modern Food Microbiology,* 3rd ed., Van Nostrand Reinhold, New York, 1986.

E. H. Marth, ed., *Standard Methods for the Examination of Dairy Products,* American Public Health Association, Washington, D.C., 1978.

B. M. Mitruka, *Methods of Detection and Identification of Bacteria,* CRC Press, Inc., Cleveland, Ohio, 1976.

J. T. Nickerson and A. J. Sinskey, *Microbiology of Foods and Food Processing,* American Elsevier Publishing Co., New York, 1972.

H. Rieman and F. L. Bryan, eds., *Food-Borne Infections and Intoxications,* Academic Press, New York, 1979.

T. A. Roberts and F. A. Skinner, *Food Microbiology Advances and Prospects,* Academic Press, New York, 1983.

A. N. Sharpe, *Food Microbiology A Framework for the Future,* Charles C Thomas, Springfield, Ill., 1980.

J. H. Silliker, R. P. Elliott, A. C. Baird-Parker, F. L. Bryan, J. H. B. Christian, D. S. Clark, J. C. Olson, Jr., and T. A. Roberts, *Microbial Ecology of Foods,* vol. 1, Academic Press, New York, 1980.

J. H. Silliker, R. P. Elliott, A. C. Baird-Parker, F. L. Bryan, J. H. B. Christian, D. S. Clark, J. C. Olson, Jr., and T. A. Roberts, *Microbial Ecology of Foods,* vol. 2, Academic Press, New York, 1980.

R. C. Tilton, *Rapid Methods and Automation in Microbiology,* American Society of Microbiology, Washington, D.C., 1982.

C. Vanderzant and D. Splittstoesser, *Compendium of Methods for the Microbiological Examination of Foods,* American Public Health Association, Washington, D.C., 1990.

H. H. Weiser, G. J. Mountney, and W. A. Gould, *Practical Food Microbiology and Technology,* AVI Publishing Company, Inc., Westport, Conn., 1971.

Daniel Y. C. Fung,
Kansas State University
Manhattan, Kansas

FOOD MICROBIOLOGY AND RAPID METHODS

THE NEED FOR RAPID RESULTS IN FOOD MICROBIOLOGY

Bacterial food-borne infections are a major problem throughout the world. In England and Wales alone the annual number of formal notifications has increased steadily over the last decade, from 8,000 cases of food poisoning in 1977 to 40,000 in 1988 (1). It has been estimated that less than 10% of actual outbreak cases are reported and estimates of 10–20 million food poisoning cases a year have been made for the United States (2). Conventional microbiological methods, which are labor intensive and time consuming, are no longer adequate for the detection of pathogens and new diagnostic methods are finding increasing application in routine diagnosis.

Food testing is done to ensure that it is microbiologically safe and wholesome. Tests may also be done to check that raw materials and finished products meet specifications or comply with regulations. Specifications may relate to numbers of specific microorganisms or to the total number of microorganisms contained in a sample. The acceptability of a product is assessed after results have been obtained and compared to specifications. Microbiological tests rely on the growth of microorganisms to a detectable level. This can take one or two days although some selective tests may take more than a week. During this time, large volumes of food, to which the tests relate, will have been produced and stored. Short-shelf-life products may have been distributed, bought, and consumed before results are obtained. Some microbiological testing is therefore retrospective and results are obtained too late to be effective. Products with a longer shelf life can be stored until results are obtained but storage is costly and distribution of products is delayed. There is therefore a need for rapid methods to detect, enumerate, and identify specific microorganisms which may affect the quality of the product or the health of the consumer (3).

THE ADVANTAGES AND DISADVANTAGES OF USING CONVENTIONAL METHODS

All microbiological methods have a detection limit, that is, the lowest number of microbial cells that can be detected by a particular method. Some conventional tests can detect one *Salmonella* cell contained within 25 g of product which may also contain many millions of non-*Salmonella* bacterial cells (4). Sharpe (5) calculated the mass fraction of one *Salmonella* cell in 25 g of meat to be 2×10^{-14}, all other parts being physically, chemically, and biochemically very similar to the *Salmonella* cell. Conventional tests achieve this sensitivity using only simple equipment such as pipettes and Petri dishes. They rely on the rapid multiplication of bacterial cells in liquid media and the formation of colonies on solid media that can be seen with the naked eye. These amplification steps take time, as illustrated by the conventional test for the detection of low numbers of *Salmonella* in a food sample. Pre-enrichment and selective enrichment stages are needed to increase the numbers of *Salmonella* cells present and inhibit other bacterial growth (6,7). All these enrichment stages usually take about two or three days. Bacterial cells present in a sample after enrichment are detected using selective and differential agar plates. Each bacterial cell capable of growth on the selective agar multiplies, forming millions of cells in a colony. It takes about one day before the colony is visible to the naked eye. Judgment is then needed to decide if the colony morphology on the selective medium indicates the presence of bacteria that could possibly be *Salmonella*. The laboratory technician can report the absence of *Salmonella* in a 25-g sample no earlier than three days after sampling. The presence of bacteria that may be *Salmonella* can be detected three to five days after sampling. It can be another two days before the identity of the bacteria are confirmed biochemically and serologically giving a total test time of seven days for one sample. Conventional test methods also have the disadvantage of being labor intensive, limiting the number of samples that can be processed at one time.

Researchers have tried to reduce the time needed for conventional tests by shortening or combining the enrichment procedures (8,9) but have had only limited success. When *Salmonella* are present in foods, they are usually only present in small numbers and cells are often injured by processes such as heating, freezing, refrigeration or drying (10). The food may also have a low pH, low water activity, or may contain natural or added inhibitory substances. Consequently, food samples tested by conventional methods must undergo pre-enrichment to allow the recovery and growth of injured cells.

Selective agar media used for the presumptive identification of *Salmonella* contain indicators that respond to lactose and/or sucrose utilization and/or production of hydrogen sulphide (11). They also contain inhibitors such as sodium desoxycholate or brilliant green to inhibit the growth of some other bacteria. Elevated incubation temperatures also inhibit some non-Salmonellae. Selective media used for the presumptive identification of other pathogens are based on similar selective and elective methods. These methods only indicate the presence of bacteria that are capable of growth under selective conditions; their identity must be confirmed.

Many rapid methods have been developed to replace the selective agars and/or confirmatory tests. Some rapid tests are rapid screening procedures for negative samples, allowing the produce to be distributed one or two days earlier than conventional testing can allow. Presumptive pathogens from positive samples still need to be identified to confirm the presence or absence of the test organism. Rapid test kits are available that identify bacteria isolated from foods. These kits can be in the form of miniaturized, ready-to-use biochemical tests that rely on the growth of bacteria and usually need one or two days incubation. Other tests use antigen/antibody reactions and may give results in 1 or 2 min, allowing the confirmation of presumptive pathogens one or more days earlier than conventional tests.

Some test kits do not rely on the growth of bacteria for detection or identification; instead, they use antibody or nucleic acid hybridization techniques. As with conventional methods, they have a detection limit and require sample enrichment to give a detectable number of target cells in a sample. Results for enriched samples are obtained within the working day and products showing negative for the test organism can be identified two days after sampling. Presumptive Salmonellae from samples giving positive results still need confirmation by serology and biochemical tests. These methods not only give results earlier than conventional methods but they also save time in media preparation. Many samples can be processed at one time and some tests can be automated, making them less labor intensive and allowing larger numbers of samples to be tested.

Rapid methods can save time when testing for low numbers of pathogens in food, but they still need enrichments to raise the number of pathogens to a detectable level. If the detection level of test kits could be improved to detect the low numbers of pathogens in food, enrichment would be unnecessary.

The main disadvantages of conventional test methods are:

1. Media preparation time
2. Total test time
3. Labor-intensive procedures
4. Low sample throughput
5. High cost
6. Subjectivity

Rapid methods have overcome most, if not all, of these problems but there are still advantages of conventional test methods:

1. Low technology
2. Low capital outlay
3. Sensitivity
4. Well-documented methods
5. Acceptability
6. Accuracy

Low Technology

All conventional food microbiology test methods rely on the growth of microorganisms in media producing a visible end point, usually the growth of a colony on solid agar that may be accompanied by a color change in the medium. Many rapid test kits rely on equally simple techniques using nothing more sophisticated than a pipette and having a visible end point. Other rapid tests may need more skillful interpretation of results, including the use of instruments for their analysis.

Low Capital Outlay

Conventional test methods require incubators, balances, waterbaths, and other nonspecialized apparatus. Many rapid methods require no additional apparatus to that found in a laboratory equipped for conventional testing. Some rapid methods require microtiter plate washers and readers, spectrophotometers, and other specialized instruments not normally found in a food microbiology laboratory.

Sensitivity

Conventional test methods can be very sensitive, detecting as few as one specific bacterial cell in a 25-g product. Rapid test kits are at best as sensitive as conventional methods when used in conjunction with enrichment techniques.

Well-Documented Methods

Conventional test methods are based on cultural techniques developed in the nineteenth century (12). Methods are constantly revised to improve recovery of target organisms from specific products, as the numerous articles in relevant journals testify (13–19). Rapid methods do not have a long history of development and application. There may not be specific applications for a new test kit with certain foods and a food microbiology laboratory may have neither the time nor the expertise to develop protocols for its product.

Acceptability

Food microbiologists test each type of food for a specific microorganism by reference to a manual of methods. This manual will contain standard methods that will most probably have been derived from official guidelines drawn up by the British Standards Institution, the U.S. Food and Drug Administration (7), or other government bodies. These methods have been found to be effective for the detection of microorganisms in foods and most are based on conventional techniques. The use of rapid methods, such as the identification of *Salmonella* (20) and enterotoxic *Escherichia coli* (21) by DNA colony hybridization, and the detection of *Salmonella* (22) and the enterotoxins of *E. coli* and *Clostridium perfringens* by enzyme immunoassay (23) are now included in the *Bacteriological Analytical Manual* (7). The Food and Drug Administration (FDA) will view these rapid procedures as viable alternatives to conventional tests provided that their effectiveness has been demonstrated (24).

Accuracy

Conventional methods for the enumeration of total or specific microorganisms in food rely on the plate count. Food samples or their dilutions are added to solid media in Petri dishes and incubated for a specified time at a specified temperature. The agar plates are then examined and visible colonies counted to calculate the number of bacteria in the original sample. Raw materials and finished produce are accepted or rejected on the basis of these counts, with reference to specifications for each food type and test done. The precision and accuracy of the plate count technique are low and are affected by many factors, both intrinsic to the procedure and implicit in the nature of the sample under examination (25).

The results from plate counts only give the numbers of bacterial cells present in that particular sample that were able to grow on the chosen medium at the chosen temperature of incubation and were able to multiply and form visible colonies within the chosen period of incubation. The distribution of bacterial cells throughout a product, range of microorganisms present and their condition can affect results. The product may have been frozen, refrigerated, dried, or heated and may have a low pH, low water activity or contain inhibitory substances (10). These factors could affect the enumeration of bacteria both by conventional methods and by rapid methods that rely on the growth of bacteria for their detection.

Conventional methods for the detection, enumeration, or identification of specific microorganisms have other limitations apart from those already described for plate counts. Media for the detection of specific bacteria contain ingredients that inhibit the growth of other bacteria present in the sample. These selective agents are often also inhibitory to some extent to the target cells that may already be injured. Selective media also contain indicators to the biochemistry of the target organisms. These rely on visible changes to the media as ingredients are altered by the cell's metabolism. The presence of mixed cultures on selective agar can give misleading results and the biochemistry of isolates from food samples may be atypical. Both conventional and rapid cultural tests may give false-positive and false-negative results. False-positive results can often be eliminated by confirmatory tests for the identification of the isolates. From a human safety point of view, these are less serious than false-negative results (24). False negatives can be reduced when testing for *Salmonella* by the use of two or more selective broths and agars.

RAPID METHODS FOR THE DETECTION OF MICROORGANISMS IN FOOD

Increased microbiological surveillance of foods by the food industry and by government regulating authorities and the introduction of Hazard Analysis and Critical Control Point (HACCP) system (1) has led to increased testing of food for pathogens such as *Salmonella* and *Listeria monocytogenes*. The resulting increased workload on the microbiology laboratory has emphasized the disadvantages of conventional cultural techniques. The need to process many samples at once and to obtain results for perishable products more quickly has led the search for alternative methods.

The use of rapid test methods can increase sample throughput by:

1. Reduction in media preparation time
2. Reduction in sample incubation time
3. Processing of many samples at one time
4. Use of detection methods that do not rely on growth
5. Use of rapid methods for the identification of food isolates

Rapid Test Kits Based on Cultural Methods

Some rapid test kits based on conventional cultural techniques are available that contain dehydrated media, reducing preparation time. Some tests also have shorter total incubation times than conventional methods. The Oxoid *Salmonella* Rapid Test is a ready to use unit for the detection of *Salmonella* that needs little media preparation (26,27). A *Salmonella* elective medium and 1 ml of an 18-h pre-enriched sample are added to the unit and incubated for 24 h. During this incubation period, motile *Salmonella* migrate through two tubes containing semisolid media to selective media in the top of the tubes, giving a color change. Tubes producing this positive reaction are tested further with a latex test. Positive results from this test indicate the presence of *Salmonella*, but this must be confirmed by traditional biochemical and serological tests. Negative results are obtained two days after sampling, allowing the release of products earlier than those tested by conventional methods. Conventional cultural and serological confirmation of presumptive *Salmonella* cannot be done directly from the rapid test. The indicator media must be subcultured into selective enrichment media for confirmatory tests. This could delay the release of some products if samples positive by the Oxoid Rapid Test are proved negative by confirmatory tests.

Another rapid test kit based on cultural techniques is the Colilert test (Palintest Ltd., Gateshead, England) for the detection of coliforms and *E. coli* in water. The conventional test methods for *E. coli* can take between four and six days (7,28); the Colilert test indicates the presence of both coliforms and *E. coli* in under 24 h (29). The test uses a conventional selective liquid culture medium with the addition of two indicator substrates: ONPG (ortho-nitrophenyl-β-D-galactopyranoide) and MUG (4-methylumbelliferyl-β-D-glucoronide). Coliforms cleave the colorless ONPG, releasing ONP, which is yellow when unbound. *E. coli* cleaves MUG, releasing methylumbelliferone that fluoresces when free. Coliform-positive samples appear yellow, whereas *E. coli*-positive samples fluoresce blue under ultraviolet (UV) (366 nm) light. Care must be taken with the interpretation of results as not all *E. coli* are detected using MUG. Most enterohemorrhagic *E. coli* are β-glucoronidase negative (30) including the serotype 0157:H7, which has recently emerged as a significant food-borne pathogen (31).

Enzyme Immunoassays

Immunoassays detect specific microorganisms or their metabolites using antibody techniques. Some early kits used radioisotopes as a detection method (radio immunoassays), but food laboratories were reluctant to use these kits. A nonradioactive alternative was developed and enzyme immunoassays (EIA) are now used that give a colored end point. These are acceptable to the food industry and allow either visual or spectrophotometric reading of results.

Most commercially available EIAs use an antibody sandwich technique for the capture and detection of target bacteria or their toxins. Large molecules, such as microbial proteins, contain several epitopes or binding sites and two antibodies that do not compete for the same epitope can be used to capture and label the target molecule. A simplified diagram (Fig. 1) shows how the EIA sandwich technique works. The test kit is supplied with the capture antibody already attached to a solid surface; in this case the surface is the base and sides of a microtiter plate well (some earlier kits used magnetic beads, BioEnzabead Screen Kit, Organon Teknika Ltd., Cambridge, UK (32)). A suitably enriched sample is added to the well and the target antigen (such as a specific *Salmonella* protein) attaches to the capture antibody. The well is washed, leaving only the target attached to the capture antibody. A second specific antibody is added to the well and this binds to the target. This detection antibody is conjugated to an enzyme. Excess antibody:enzyme conjugate is washed away leaving an antibody:antigen:antibody sandwich. With some tests, the sample and conjugate are added to the well at the same time. A colorless substrate is added that is converted to a colored product in the presence of the enzyme and is detected either visually or by using a microtiter plate reader. The assay takes between 2 and 2½

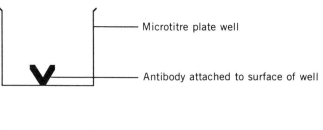
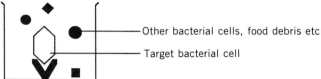
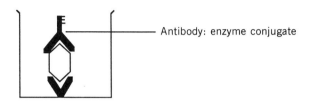
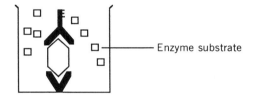
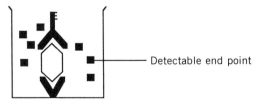

Figure 1. Enzyme immunoassay using the antibody sandwich technique.

h and results indicate the presence or absence of the target bacterial cells. Results indicating the presence of the target cells are presumptive and need to be confirmed biochemically and by serology. EIA tests for *Listeria* and *Salmonella* in foods are done on suitably enriched samples, allowing produce giving negative results to be released two days after sampling, compared to four or five days for conventional tests.

Current cultural methods for *Listeria* and *Salmonella* are costly and time consuming. *Listeria monocytogenes* has only recently been recognized as a serious food-borne pathogen to humans (33–35) and methods of isolation and enumeration are constantly being reviewed (36,37). As with *Salmonella*, there is a requirement at the present time for the absence of *Listeria* in a 25-g food sample, necessitating lengthy cultural processes to increase the number of target cells to a detectable level. Competing flora and interference from food components are problems that need to be overcome in both cultural and rapid test methods, and careful use of EIAs is essential to prevent false results (37).

EIAs available for the detection of pathogens such as *Listeria* and *Salmonella* include the TECRA *Salmonella* or *Listeria* Visual Immunoassay diagnostic kit (Bioenterprises Pty. Ltd., Roseville, NSW, Australia) and the *Listeria* TEK and *Salmonella* TEK Enzyme Linked Immuno-Sorbent Assay (ELISA) (Organon Teknika Ltd., Cambridge, UK). These assays are done using removable well strips in a microtiter plate format that is particularly useful for laboratories doing only a few tests each day. Results from the TECRA assays can be read visually so expensive equipment is not needed. This means that results from TECRA assays are subjective whereas the Organon Teknika ELISA test results are analyzed objectively using a plate reader. A microtiter plate washer and incubator are both optional but aid the throughput of large numbers of samples.

Salmonella EIAs have been given First Action AOAC Approval for the detection of *Salmonella* in foods (38).

Fluorescent Immunoassays

Fluorescent Immunoassays (FIAs) are immunoassays that have a fluorescent, rather than a colorimetric, end point. Samples are pre-enriched and selectively enriched as for EIAs and added to antibody coated wells. The procedure is the same as for sandwich EIAs except that enzyme activity is detected using a fluorescent substrate. The wells are washed either manually or by using an automated wash instrument, and positive samples are detected using a microplate fluorimeter.

Immunodiffusion

The *Salmonella* 1–2 Test (Biocontrol Systems Inc. Bothwell, Wash.) is an AOAC-approved method for the identification of *Salmonella* that uses Immunoband technology. The test consists of a plastic disposable container with a main motility chamber containing a semisolid motility medium and a side chamber containing a selective enrichment broth (Fig. 2). A pre-enriched food sample is inoculated into the enrichment broth and *Salmonella* polyclonal flagellar antibody is inoculated into the motility medium. The antibody diffuses through the motility medium and motile *Salmonella* cells migrate from the enrichment medium and through the motility medium. The antibody binds to the *Salmonella* cells and immobilises them, forming a visible zone or band of precipitation, indicating the presence of *Salmonella* (39). Time taken from sampling to obtaining results is less than 72 h and the test can detect as few as one or two *Salmonella* organisms in a 25-g product. No specialized equipment is needed and the band of precipitation remains visible for up to three days.

Tests on food (40–42) gave a high false negative rate with some foods but not others. This highlights the necessity of ensuring that the test method has been evaluated for the food under analysis. Tests that rely on motility for the detection of *Salmonella* preclude the detection of nonmotile strains such as *S. pullorum* and this will result in some false-negative results (41).

Rapid Test Kits Based on Nucleic Acid Hybridization Techniques

A method for the rapid detection of *Salmonella* in foods using nucleic acid hybridization was described by Fitts and co-workers in 1983 (43). *Salmonella*-specific DNA probes were labeled with a radioactive isotope and positive samples were detected using a scintillation counter or beta detector (44). DNA hybridization was recognized as an AOAC-approved method (20) but was not widely used in the food industry because of the radioactive labeling. Recently, DNA probes with a colorimetric end point have been developed (45) for the detection of *Salmonella, E. coli*, and *Listeria* in foods and these kits are now commercially available (GENE-TRAK Systems, Framingham, Mass.). A colorimetric DNA probe to *Campylobacter jejuni* is available from Du Pont (E.I. du Pont de Nemours & Co. (Inc.), Boston, Mass.).

Analysis of samples using this method requires $2\frac{1}{2}$ to 3 h after conventional pre-enrichment and selective enrichment (and postenrichment for *Salmonella* testing), so results can be obtained two days after sampling. Suitably enriched samples are added to a test tube and the bacterial cells contained in the sample are chemically lysed. Specific DNA probes are added that are targeted against regions of ribosomal RNA unique to the organism to be detected (Fig. 3). Early assays targeted chromosomal DNA present in low numbers (between one and four) in each cell. Assays now target rRNA that are present in high copy numbers (5,000–20,000) in each cell. If the target organism is present in the sample, the capture and reporter probes hybridize with the target rRNA. A dipstick (coated with polydeoxythymidylic acid (dT)) is placed in the test tube and binds to the capture probe (which is tailed with deoxyadenosine monophosphate (dA)). After incubation, the dipstick is washed to remove unbound nucleic acids and cell debris. A conjugate (antifluorescein antibody linked to horseradish peroxidase) is added that binds to the fluorescein label on the reporter probe. Excess conjugate is washed away and the bound complex is detected by the addition of an enzyme substrate and chromogen.

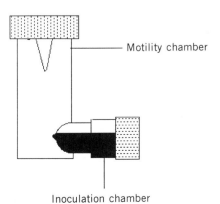

Figure 2. Immunodiffusion using the *salmonella* 1–2 test.

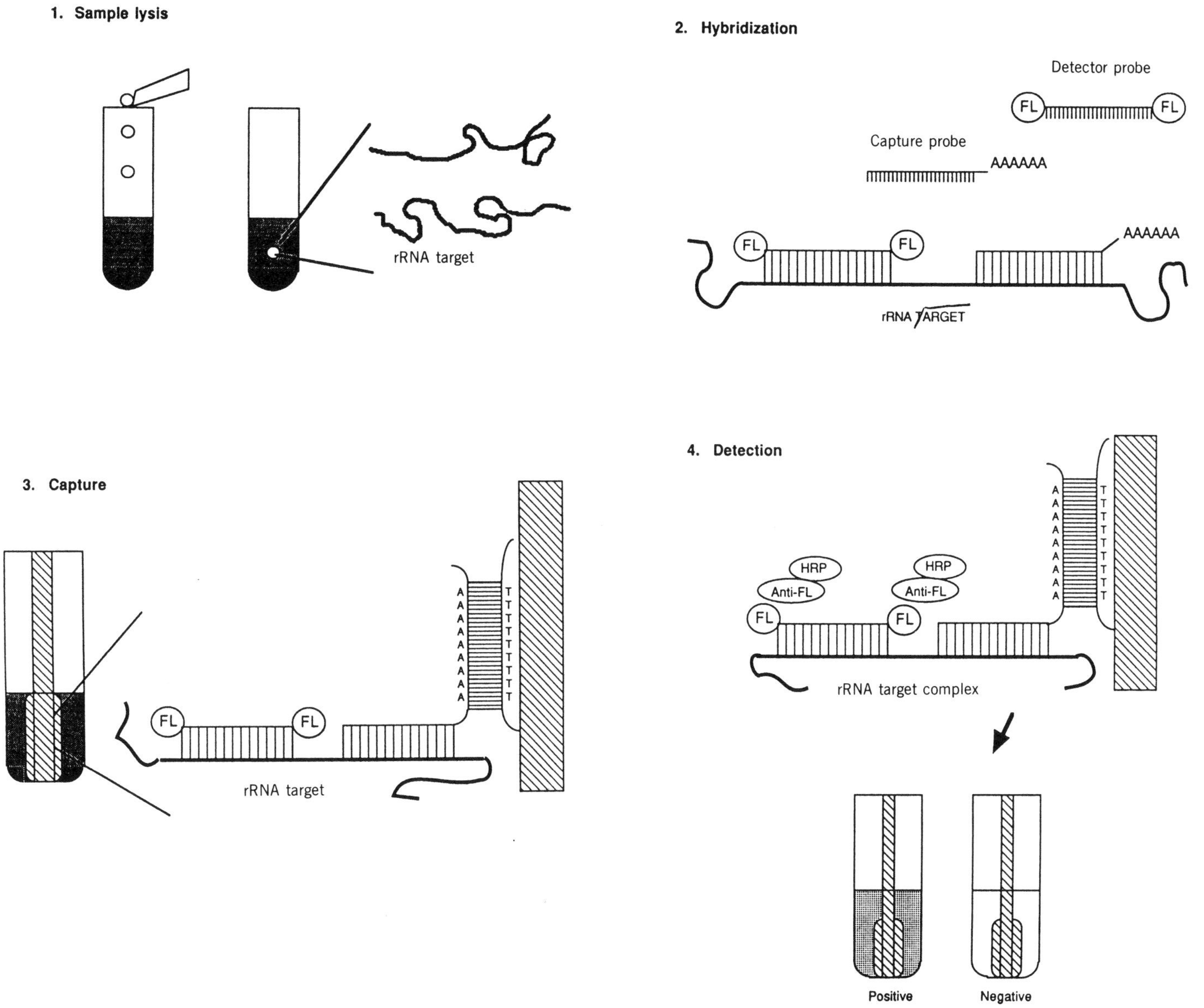

Figure 3. Nucleic acid hybridization assay using the GENE-TRAK test kit.

If the target is present in the sample, a color develops that is measured photometrically. Samples giving positive results should be confirmed using biochemical and serological methods of identification.

The DNA hybridization method has been shown to be at least as effective as conventional methods. The presence of presumptive target organisms is determined two days after sampling and product proving negative for the test organisms can be released earlier than when tested by conventional methods.

Nucleic acid probes with a chemiluminescent end point are available for the identification of Campylobacter (AccuProbe Campylobacter Culture Confirmation Test, GenProbe Inc., San Diego, Calif.). The single stranded DNA probe is targeted to specific ribosomal RNA sequences unique to *C. jejuni, C. coli,* and *C. laridis,* which are the three Campylobacter species most frequently isolated as human enteric pathogens. However, the test does not differentiate between these three species. Detection is based on a chemiluminescent acridinium ester and the light produced is measured using a luminometer.

RAPID METHODS FOR THE IDENTIFICATION AND CONFIRMATION OF MICROORGANISMS

Results from conventional and rapid methods for the detection of specific microorganisms in food samples give only presumptive identification of isolates that need to be confirmed by further testing. Conventional methods of identification include colony and cell morphology and biochemical, enzymological and serological tests. Rapid test kits have been developed for the identification of bacterial isolates that are based on conventional methods. These kits save time in media preparation and many samples can often be tested at one time. Rapid tests have also been developed that are based on immunological techniques and these can often give results in minutes.

Rapid identification kits allow the early release of negative samples, especially when used in conjunction with rapid detection methods. As with conventional methods of identification, rapid tests should only be done using pure cultures; contamination will give inaccurate results. Tests for specific groups of bacteria should only be used for that bacterial group.

Rapid Tests Based on Cultural Methods

These kits are miniaturized, ready-to-use tests with dehydrated substrates and indicators contained in disposable wells or tubes. A range of individual tests are contained in one unit which confirm the identity of a specific bacterial group such as Enterobacteriaceae, *Salmonella* or *Staphylococcus aureus*. Many of these tests were originally used by clinical laboratories but have been evaluated for the food laboratory (46,47). They save time in media preparation, are easy to use, and give reliable results. Because these tests are based on cultural techniques, results are not always obtained more rapidly than with conventional methods.

Commercially available tests include the API identification systems (API-bio Mérieux (UK) Ltd., Basingstoke, UK) for Enterobacteriaceae (API 20E), nonenteric gram-negative bacilli (API 2ONE), and for other groups of microorganisms. Most tests give results after 18–24 h incubation. Other tests available based on cultural methods include the Minitek system (BBL Beckton Dickinson UK Ltd., Oxford) for the detection of Enterobacteriaceae and the MICRO-ID Listeria system (Organon Teknika Ltd.) for the detection of *Listeria monocytogenes* and other species. With these kits, identification is made from color reactions of paper disks impregnated with dehydrated substrates and indicators. Flow Laboratories Ltd. (Rickmansworth, UK) produce a Titertek Enterobac (TTE) system for the identification of Enterobacteriaceae. This is a cultural method in microtiter plate format and results are obtained after 18–24 h. Flow laboratories and API-bio Mérieux produce semiautomated versions of their identification systems. Vitek Systems Inc. (Missouri) produce a fully automated Auto Microbic System for the identification of *Listeria monocytogenes*, Enterobacteriaceae, and other groups of microorganisms. This system uses plastic cards with wells containing substrates that are inoculated automatically and scanned hourly. Identifications can be confirmed within 24 h. Roche Diagnostica (Roche Products Ltd., Welwyn Garden, UK) produce the Enterotube 11 system for the rapid identification of Enterobacteriaceae within 18–24 h. Wells containing culture media are inoculated simultaneously from a single colony and results read within 24 h. There are many more commercially available identification kits and Cox and others (47) gives examples of those that have been used with food isolates.

Rapid Test Kits Based on Enzyme Activity

These kits are based on the presence of preformed bacterial enzymes that can be detected within a few hours by biochemical tests (48). The MICRO-ID enteric identification system (Organon Teknika Ltd.) consists of impregnated filter papers and can identify Enterobacteriaceae within 4 h. The system has been evaluated with food isolates and gave good agreement with conventional tests. Rapid kits that indicate enzyme activity and give results in a few hours are also available from API. These include a 2-h test to differentiate *Salmonella, Shigella,* and *Yersinia enterocolitica* from other food isolates (API Z) and a 2-h test for the detection of *E. coli* (RAPIDEC coli).

Rapid Tests Based on Latex Agglutination Techniques

Latex agglutination kits are simple, easy-to-use tests that confirm the identity of bacterial isolates within a few minutes. They are used in conjunction with, or instead of, conventional cultural methods of identification that normally take up to 24 h. Latex particles are sensitized with antibodies specific to particular antigens on the cell wall of the target organism or group of organisms.

A drop of pure culture is applied to test areas (usually on a black card) and mixed with a sensitized latex solution. Agglutination is detected by the formation of visible granular clumps, indicating the presence of the target organism. Negative controls contain nonsensitized latex particles for the detection of nonspecific agglutination reactions. These identification tests are only for use with pure isolates. They have a sensitivity of about 10^7 cells/ml and results are not quantitative.

Several latex test kits are available for the detection of *Salmonella* from foods, including the Oxoid Salmonella Latex Test (Oxoid Ltd.) and the Micro Screen *Salmonella* Latex Slide Agglutination Test (Mercia Diagnostics Ltd., Guildford, UK). Other latex test kits use colored latex particles to identify the *Salmonella* serogroup (Wellcolex Colour *Salmonella* Test, Wellcome Diagnostics, Dartford, UK; Spectate Salmonella Test, May and Baker Diagnostics Ltd., Glasgow, UK). The Spectate Test has been evaluated for the detection of *Salmonella* from foods (49) and the rapid test compared well with conventional biochemical and serological tests. Latex agglutination tests are also available for *Campylobacter* (Microscreen, Mercia Diagnostics Ltd.), *S. aureus* (Staph Rapid Test, Roche Products Ltd., Staphaurex, Wellcome Diagnostics), *Shigella* (Wellcolex Colour Shigella Test, Wellcome Diagnostics) *Yersinia enterocolitica,* (PROGEN Biotechnik GmbH, Heidelberg), and *E. coli* 0157, (Oxoid Ltd.). These are just a few examples of the many latex agglutination test kits available.

RAPID TEST KITS FOR THE DETECTION OF BACTERIAL TOXINS IN FOOD

The detection of preformed enterotoxins in foods is important for the quality appraisal of foods during manufacture (50). Staphylococcal food poisoning is one of the most common types of food-borne illness and results from the ingestion of enterotoxin produced during growth of enterotoxigenic strains of *S. aureus* in food. Staphylococcal enterotoxin is heat stable, withstanding heat that destroys living *S. aureus* cells. It may therefore be necessary to test heated foods for the presence of enterotoxin even though no living *S. aureus* cells are detected (51).

Traditional methods for the detection of staphylococcal enterotoxins in foods, such as the immunodiffusion technique (52), are labor intensive, time consuming, and often require considerable technical expertise and facilities (53). Developments in immunoassays have resulted in simpler, more sensitive assays for toxin detection and enzyme immunoassay (EIA) kits are now available for the

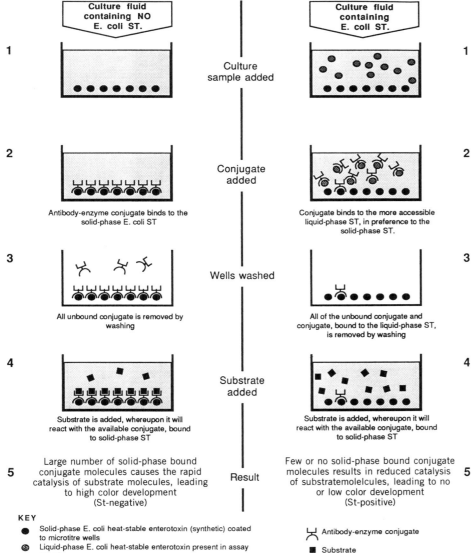

Figure 4. Competitive enzyme immunoassay.

detection of staphylococcal enterotoxins (TECRA SET visual immunoassay, Bioenterprises Pty. Ltd.; SET EIA, Labor Dr. W. Bommeli, Bern, Switzerland), (*E. coli* heat stable enterotoxin (*E. coli* ST EIA Oxoid Ltd.), *Clostridium perfringens* enterotoxin (54), *Bacillus cereus* diarrhoeal toxin, and others.

Enzyme immunoassays for toxin detection work in much the same way as those for the detection of microorganisms do. The TECRA SET ELISA is a visual immunoassay using an antibody sandwich technique in microtiter plate format allowing up to 46 tests to be done at one time. Tests are done manually and results obtained within 4 h. The test kits have been evaluated for use with foods. The SET-EIA kit uses color-coded polystyrene beads coated with antibodies specific for staphylococcal enterotoxin A, B, C, or D. Food samples are incubated overnight with the antibody-coated beads, then washed to remove unbound material. Beads are incubated with an antibody enzyme conjugate, washed, and then a colorless substrate is added that gives a colored end point. The intensity of the color is proportional to the enterotoxin concentration and is measured photometrically. The detection limit of the test is dependent on the extraction method used for the food sample and lies between 0.1 and 10 ng toxin/ml. The SET-EIA kit differs from the TECRA kit in that the former utilizes separate antibodies for the detection of each enterotoxin whereas the latter employs polyvalent antisera for the simultaneous detection of all toxin types.

The ST-EIA test for *E. coli* heat-stable toxin uses a competitive EIA technique instead of the traditional infant mouse assay (53). The test is done in microtiter plate wells precoated with *E. coli* heat-stable toxin (Fig. 4). A sample is added to the wells, followed by an antibody enzyme conjugate. If no toxin is present in the sample, the conjugate binds only to the toxin coating the wells. If toxin is present in the sample, it will compete with the bound toxin for the complex. After washing the sample to remove either unbound conjugate or that bound to toxin, a substrate is added. If no toxin is present in the sample, a color change is seen. If toxin is present in the sample little or no color is seen. Results are analyzed using a microtiter plate reader.

Other kits available for toxin detection are those based on Reversed Passive Latex Agglutination (RPLA) available from Oxoid Ltd., Basingstoke, UK, and include those for the detection of *S. aureus* entertoxin A, B, C, and D (SET-RPLA) and *Bacillus cereus* enterotoxin (BCET-RPLA) in food. These tests are done in microtiter plate wells and use antibody-coated latex particles. Dilutions of food samples are added to wells and mixed with a latex suspension. If the toxin is present, agglutination occurs as the latex and toxin form a lattice structure that appears as a granular layer across the bottom of the well (Fig. 5). In negative wells, the latex particles settle as a tight pink button. These results are obtained after 24-h incubation, although a recent modification allows results to be obtained within the working day (53). The sensitivity of the test depends on the dilution of the food sample but is reported as being 0.25 ng/ml. This is not as sensitive as EIAs (0.1 ng/ml), but the RPLA test is simpler and requires less equipment than EIAs. EIAs and RPLA assays

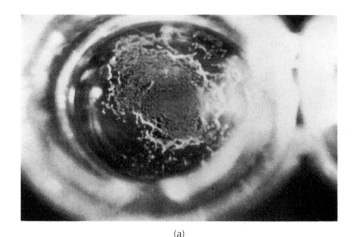
(a)

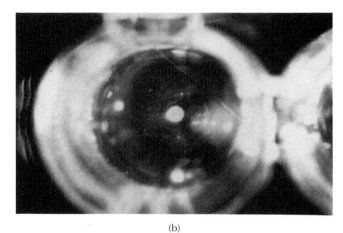
(b)

Figure 5. *Staphylococcus aureus* toxin detection using the RPLA (a) positive and (b) negative results.

have both been evaluated for the detection of toxins in foods and results have been compared to those from conventional tests (51,54–56).

BIBLIOGRAPHY

1. "The Microbiological Safety of Food Part 1," (The Richmond Report). Report of the Committee on the Microbiological Safety of Food, Her Majesty's Stationery Office, London, 1990.

2. G. J. Banwart, *Basic Food Microbiology*, 2nd Ed. Van Nostrand Reinhold, New York, 1989.

3. B. Jarvis and M. C. Easter, "Approaches to the Detection of Pathogens in Industrial Quality Testing," in A. Balows, R. C. Tilton, and A. Turano, eds., *Rapid Methods and Automation in Microbiology and Immunology*, Brixia Academic Press, Brescia, 1989, pp. 309–319.

4. A. C. Baird-Parker, "The Present and Future Role of Rapid Microbiological Methods in Assuring Food Safety," in A. Balows, R. C. Tilton, and A. Turano, eds., *Rapid Methods and Automation in Microbiology and Immunology*, Brixia Academic Press, Brescia, 1989, pp. 276–281.

5. A. N. Sharpe, *Food Microbiology: A Framework for the Future*, Charles C Thomas, Springfield, Ill., 1980.

6. British Standard 4285. "Microbiological Examination for Dairy Purposes Part 3, Methods for Detection and/or Enumeration of Specific Groups of Microorganisms Section 3.9, Detection of *Salmonella*," 1987.
7. Food and Drug Administration, *Bacteriological Analytical Manual*, 6th Ed., *Association of the Official Analytical Chemists*, Arlington, Va., 1984.
8. H. Rappold, R. F. Bolderdijk, and J. M. De Smedt, "Rapid Cultural Method to Detect *Salmonella* in Foods," *Journal of Food Protection* 47(1), 46–48 (1984).
9. W. H. Sveum and P. A. Hartman, "Time-Release Capsule Method for the Detection of Salmonellae in Foods and Feeds," *Applied Environmental Microbiology* 33(3), 630–634 (1977).
10. G. F. Ibrahim and G. H. Fleet, "Detection of Salmonellae Using Accelerated Methods," *International Journal of Food and Microbiology* 2, 259–272 (1985).
11. *The Oxoid Manual*, 5th Ed., Oxoid Ltd., Basingstoke, UK, 1982.
12. R. Y. Stainer, E. A. Adelberg, and J. L. Ingraham, *General Microbiology*, 4th Ed., Macmillan Press, London, 1978.
13. J. R. Hall, "Enhanced Recovery of *Salmonella montivideo* from Onion Powder by the Addition of Potassium Sulfite to Lactose Broth, *Journal of the Association of Official Analytical Chemists* 52(5), 940–942 (1969).
14. W. Edel and E. H. Kampelmacher, "Comparative Studies on the Isolation of "Sublethally Injured" Salmonellae in Nine European Laboratories," *Bulletin of the World Health Organisation* 48, 167–174 (1973).
15. W. H. Andrews, R. W. Clyde, P. L. Poelma, A. Romero, F. D. McClure, and D. E. Gentile, "Enumeration of Coliforms and *Salmonella* in Food Prepared by Blending and Stomaching," *Journal of the Association of Official Analytical Chemists* 61(6), 1324–1327 (1978).
16. P. Vassiliadis, "The Rappaport-Vassiliadis (R.V.) Enrichment Medium for the Isolation of Salmonellas: An Overview. *Journal of Applied Bacteriology* 54, 69–76 (1983).
17. J. Y. D'Aoust, "Salmonella Detection in Foods: Present Status and Research Needs for the Future," *Journal of Food Protection* 47(1) 78–81 (1984).
18. J. Y. D'Aoust, "Effective Enrichment—Plating Conditions for Detection of Salmonella in Foods," *Journal of Food Protection* 47(8), 588–590 (1984).
19. H. J. Beckers, F. M. Van Leusden, and R. Peters, "Comparison of Muller-Kauffmann's Tetrathionate Broth and Modified Rappaport's Medium for Isolation of Salmonella," *Journal of Food Safety* 8, 1–9 (1986).
20. "Rapid Identification of *Salmonella* by DNA Hybridization," in *FDA Bacteriological Analytical Manual*, 6th Ed., Association of Official Analytical Chemists, Arlington, Va. (1987).
21. W. E. Hill, "Detection of Pathogenic Bacteria by DNA Colony Hybridization," in *FDA Bacteriological Analytical Manual*, 6th Ed., Association of Official Analytical Chemists, Arlington, Va., 1987.
22. "Rapid Identification of *Salmonella* in Foods by Enzyme Immunoassay," in *FDA Bacteriological Analytical Manual*, 6th Ed., Association of Official Analytical Chemists, Arlington, Va., 1987.
23. D. B. Shah, J. C. Wimsatt, and P. E. Kauffman, "Enzyme Linked Immunosorbent Assay (ELISA)," in *FDA Bacteriological Analytical Manual*, 6th Ed., Association of Official Analytical Chemists, Arlington, Va., 1984.
24. W. H. Andrews, "FDA's Views on Rapid Methods in Food Microbiology," in Rapid Microbiological Methods. Twenty-First Annual Symposium, Special Report No. 60, D. L. Downing and Y. D. Hang, eds. New York State Agricultural and Experimental Station, Cornell University, Ithaca, New York, 1987, pp. 24–27.
25. B. Jarvis and M. C. Easter, "Rapid Methods in the Assessment of Microbiological Quality: Experiences and Needs," *Journal of Applied Bacteriology*, Symposium Supplement 115S–126S (1987).
26. R. Holbrook, J. M. Anderson, A. C. Baird-Parker, L. M. Dobbs, D. Sawhney, S. H. Stuchbury, and D. Swaine, "Rapid Detection of Salmonella in Foods—A Convenient Two Day Procedure," *Letters in Applied Microbiology* 8(4), 139–142 (1989).
27. R. Holbrook, J. M. Anderson, A. C. Baird-Parker, and S. H. Stuchbury, "Comparative Evaluation of the Oxoid Salmonella Rapid Test with Three Other Rapid Salmonella Methods," *Letters in Applied Microbiology* 9(5), 161–164 (1989).
28. British Standard 4283, Microbiological Examination for Dairy Purposes Part 3, Methods for Detection and/or Enumeration of Specific Groups of Microorganisms, Section 3.8 Enumeration of Presumptive *Escherichia coli*, 1988.
29. S. Stratman, "Rapid Specific Environmental Coliform Monitoring," *American Laboratory*, 60–64 (July 1988).
30. H. Hofstra and J. H. J. Huis in't Veld, "Methods for the Detection and Isolation of *Escherichia coli* Including Pathogenic Strains," *Journal of Applied Bacteriology*, Symposium Supplement, 197S–212S (1988).
31. R. A. Szabo, E. C. D. Todd, and A. Jean, "Method to Isolate *Escherichia coli* 0157:H7 from Food," *Journal of Food Protection* 49(10) 768–772 (1986).
32. R. S. Flowers, K. Eckner, D. A. Gabis, B. J. Robinson, J. A. Mattingley, and J. H. Silliker, "Enzyme Immunoassay for Detection of *Salmonella* in Foods: Collaborative Study," *Journal of the Association of Official Analytical Chemists* 69(5), 786–798 (1986).
33. P. K. Cassiday and R. E. Brackett, "Methods and Media to Isolate and Enumerate *Listeria monocytogenes*: A review," *Journal of Food Protection* 52(3), 207–214 (1989).
34. S. M. James, S. L. Fannin, B. A. Agee, B. Hall, E. Parker, J. Vogt, G. Run, J. Williams, and L. Lieb, "Listeriosis Outbreak Associated with Mexican Style Cheese—California," *Morbidity and Mortality Weekly Report* 34, 357–359 (1985).
35. W. F. Schlech, P. M. Lavigne, R. A. Bartolussi, A. C. Allen, E. V. Haldane, A. J. Wort, A. W. Hightower, S. E. Johnson, S. H. King, E. S. Nicholls and C. V. Broome, "Epidemic Listeriosis—Evidence for Transmission by Food," *New England Journal of Medicine* 308, 203–206 (1983).
36. B. Swaminathan, P. S. Hayes, V. A. Przybyszewski, and B. D. Plikaytis, "Evaluation of Enrichment and Plating Media for Isolating *Listeria monocytogenes*," *Journal of the Association of Official Analytical Chemists* 71(3), 663–668 (1988).
37. J. E. Heisick, F. M. Harrell, E. H. Peterson, S. McLaughlin, D. E. Wagner, I. V. Wesley, and J. Bryner, "Comparison of Four Procedures to Detect *Listeria* spp. in Foods," *Journal of Food Protection* 52(3), 154–157 (1989).
38. R. S. Flowers, M. J. Klatt, S. L. Keelan, B. Swaminathan, W. D. Gehle, and H. E. Chandonnet, "Fluorescent Enzyme Immunoassay for Rapid Screening of *Salmonella* in Foods: Collaborative Study. *Journal of the Association of Analytical Chemists* 72(2), 318–325 (1989).
39. W. H. Andrews, "Recent Developments in the Microbiological Analysis of Foods," *Food Laboratory Newsletter No. 13*, pp. 34–38 (1988).

40. G. Allen, F. B. Satchell, W. H. Andrews, and V. R. Bruce, "Abbreviated Selective Enrichment, Post Enrichment and a Rapid Immunodiffusion Method for Recovery of *Salmonella* from Instant Non Fat Dry Milk," *Journal of Food Protection* **52**(5), 350–355 (1989).
41. J. S. Dickson, "Enumeration of Salmonellae by Most-Probable-Number Using the *Salmonella* 1–2 Test," *Journal of Food Protection* **52**(6), 388–391 (1989).
42. J. Y. D'Aoust and A. M. Sewell, "Reliability of the Immunodiffusion 1–2 Test™ System for Detection of *Salmonella* in Foods," *Journal of Food Protection* **51**(11), 853–856 (1988).
43. R. Fitts, M. Diamond, C. Hamilton, and M. Neri, "DNA-DNA Hybridization Assay for Detection of *Salmonella* spp. in Foods," *Applied Environmental Microbiology* **46**, 1146 (1983).
44. R. S. Flowers, M. J. Klatt, M. A. Mozola, M. S. Curiale, D. A. Gabis, and J. H. Silliker, "DNA Hybridization Assay for Detection of *Salmonella* in Foods: Collaborative Study," *Journal of the Association Official Analytical Chemists* **70**(3), 521–529 (1987).
45. M. Mozola, D. Halbert, S. Chan, H. Y. Hsu, A. Johnson, W. King, S. Wilson, R. P. Betts, P. Bankes, and J. G. Banks, "Detection of Foodborne Bacterial Pathogens Using a Colorimetric DNA Hybridization Method." *The Society for Applied Bacteriology technical series No 28*, 203–216 (1991).
46. N. A. Cox, "Evaluation of Diagnostic Kits—API, Enterotube, Micro-ID, Minitek, AMS," in *Rapid Microbiological Methods*. Twenty-First Annual Symposium, Special Report No. 60, D. L. Downing and Y. D. Hang, eds. New York State Agricultural and Experimental Station, Cornell University, Ithaca, New York, pp. 13–16 (1987).
47. N. A. Cox, D. J. C. Fung, J. S. Bailey, P. A. Hartman, and P. C. Vasavada, "Miniaturized Kits, Immunoassays and DNA Hybridization for Recognition and Identification of Foodborne Bacteria," *Dairy and Food Sanitation* **7**(12), 628–631 (1987).
48. B. J. Robinson, "Use of an Enzyme Immunoassay and a Rapid Biochemical Identification System for Detection of *Salmonella* in Low Moisture Foods, in *Rapid Microbiological Methods*. Twenty-First Annual Symposium, Special Report No. 60, D. L. Downing and Y. D. Hang, eds. New York State Agricultural and Experimental Station, Cornell University, Ithaca, New York, pp. 17–19 (1987).
49. C. Clark, A. A. G. Candlish, and W. Steell, "Detection of *Salmonella* in Foods Using a Novel Coloured Latex Test," *Food and Agricultural Immunology* **1**, 3–9 (1989).
50. S. A. Rose, P. Bankes, and M. F. Stringer, "Detection of Staphylococcal Enterotoxins in Dairy Products by the Reversal Passive Latex Agglutination (SET-RPLA) Kit," *International Journal of Food Microbiology* **8**, 65–72 (1989).
51. S. Ewald, "Evaluation of Enzyme-linked Immunosorbent Assay (ELISA) for Detection of Staphylococcal Enterotoxin in Foods," *International Journal of Food Microbiology* **6**, 141–153 (1988).
52. E. P. Casman, R. W. Bennet, A. E. Dorsey, and J. E. Stone, "The Microslide Gel Double Diffusion Test for the Detection of Assay of Staphylococcal Enterotoxin," *Health Laboratory Science* **6**, 185 (1969).
53. P. Bankes and S. A. Rose, "Rapid Detection of Staphylococcal Enterotoxins in Foods with a Modification of the Reversed Passive Latex Agglutination Assay," *Journal of Applied Bacteriology* **67**, 395–399 (1989).
54. P. R. Berry, A. A. Wieneke, J. C. Rodhouse, and R. J. Gilbert, "Use of Commercial Kits for the Detection of *Clostridium perfringens* and *Staphylococcus aureus* Enterotoxins," J. M. Grange, A. Fox, and N. L. Morgan, eds., *Immunological Techniques in Microbiology,* The Society for Applied Bacteriology Technical Series No. 24, 1987.
55. C. E. Park and R. Szabo, "Evaluation of the Reversed Passive Latex Agglutination (RPLA) Test Kits for Detection of Staphylococcal Enterotoxins A, B, C, and D in foods," *Canadian Journal of Microbiology* **32**, 723–727 (1986).
56. A. A. Wieneke and R. J. Gilbert, "Comparison of Four Methods for the Detection of Staphylococcal Enterotoxin in Foods from Outbreaks of Food Poisoning," *International Journal of Food Microbiology,* **4**, 135–143 (1987).

PAMELA BANKES
Campden Food & Drink
Research Association
Gloucestershire, United
Kingdom

FOOD MICROSTRUCTURE

Microscopy has been used for many years in controlling the adulteration or contamination of food products (1,2). Such applications as identification of insect fragments or other extraneous matter remain of great importance; however, such techniques have been adequately covered elsewhere (1,3–5) and will not be considered further in this article, which will be primarily concerned with the application of light and electron microscopy and image analysis for elucidating and understanding the microstructure of foods.

LIGHT MICROSCOPY

Fundamentals of light microscopy have been covered in many previous books and reviews (1,6,7). While microscope design evolved over centuries, microscopes similar in principle to today's were available at least as early as the 1600s. The use of the light microscope for studies of food materials has proven invaluable to food scientists. The differential staining of whole samples and sectioned material remains an important diagnostic tool for initial investigations of most food systems. Within the last couple of decades, an important advancement was made with the development of the confocal scanning light microscope (8,9). Application of the confocal scanning light microscope to foods has been demonstrated (10).

ELECTRON MICROSCOPY

Electron microscopes are generally of the transmission or scanning type, or can be a combination of the two. Electrons are emitted from some type of filament; tungsten filaments are the most common, although lanthanum hexaboride filaments are also frequently used. The electrons are focused into a small-diameter beam and directed onto the specimen. In transmission electron microscopy (tem), a thin section is used so that the beam of electrons is transmitted through the specimen and forms an image on a viewing screen, photographic plate or other detector under the specimen. In scanning electron microscopy (sem), the beam is scanned across a small area of the surface of the specimen and causes electrons of varying energy, x-rays, and other radiation to be emitted by the sample. Usually low-energy secondary electrons are collected from each point of the surface and processed through a photo-multiplier–amplifier system to yield signals that are used to form an image of the surface of the sample.

Figure 1 shows the ultrastructure of a typical commercial mayonnaise and demonstrates an example of electron microscopic research with foods. Food samples are easily damaged in preparing them for electron microscopy, and efforts have been made to develop or adapt preparation techniques that do not induce artifacts in the specimen's microstructure. The mayonnaise sample was prepared for transmission electron microscopy using an encapsulation method (11). In this procedure, the specimen is encapsulated in an agar tube prior to processing for microscopy. The fixatives (gluteraldehyde and osmium tetraoxide) diffuse through the agar and fix the sample with minimal dilution or other changes in its microstructure.

Mayonnaise is required by its standards of identity to contain at least 65% oil and most commercial mayonnaises contain about 80% oil emulsified in water. The discontinuous phase in a emulsion generally exists as spherical droplets, but at such high levels of oil most of the droplets are angular and almost all of the oil particles are in contact with other oil particles. The continuous phase (egg and salt in water) is only about 20% of the product, but it provides the long-term stability typical of mayonnaise. A stable boundary is needed to prevent droplets from coalescing, and in mayonnaise it is the egg protein that aids as an emulsifier and prevents such coalescence. Electron microscopic data may be used to develop and test hypotheses concerning the mechanisms by which egg and other emulsifiers stabilize food emulsions.

IMAGE ANALYSIS

Micrographs such as that shown in Figure 1 provide a qualitative understanding of the structure of food products and the forces that maintain such structure. It is possible to obtain more quantitative information from micrographs by various processes called image analysis. Image analysis techniques have facilitated correlation of food microstructure with other microanalytical data. These techniques cannot provide information that is not already in the micrograph, but because of the tedium of the thousands of measurements that must be made to extract quantitative information from micrographs, generally such extensive analysis has not been done. For example, particle size distributions can be obtained from micrographs of powders or other particulate material by measuring the sizes of hundreds of particles in a micrograph field. Such measurements, when done manually, can take a full day for the analysis of one micrograph, but can be done in seconds with modern image analysis equipment. Computer-based image analysis systems having sophisticated statistical analysis software are commercially available and detailed analyses of micrographs are now more routinely performed on food products.

CONCLUSIONS

It should be possible to draw correlations between microstructural and rheological properties of foods with increasing availability and application of microscopic and rheological techniques and instrumentation in the food science and technology area (12,13). Then, given the microstructure of a cheese, for example, it will be possible to predict the consumer's perception of the product's texture. This increased research capability will be invaluable in food product development and quality control. Increasing availability (and affordability) of low-temperature attachments for light and electron microscopes will increase the application of cryotechniques, which have already been demonstrated with a number of foods (5,14). The use of cryotechniques (techniques performed under supercooled conditions) will doubtless be extremely important in future studies of food microstructure. Cryofixation offers a means by which samples may be fixed rapidly using physical cryo methods (immersed in liquid nitrogen, liquid helium, propane, etc) rather than chemical methods (glutaraldehyde and osmium tetroxide). Chemical fixation imparts various artifacts sometimes found absent when cryo methods are substituted. The absence of most artifacts using cryopreservation enables a more accurate interpretation of data pertaining to the sample. There exist several cryo-associated techniques that follow cryofixation including cryomicrotomy (thick sectioning), cryo-ultramicrotomy (thin sectioning), cryo-light microscopy,

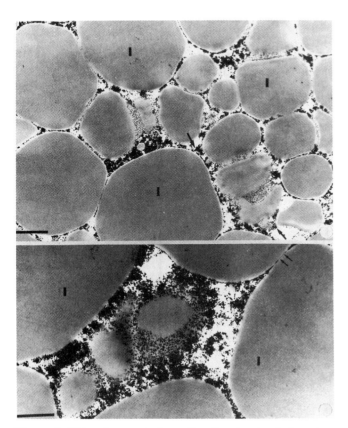

Figure 1. This micrograph shows the ultrastructural characteristics of mayonnaise. The bar represents 1 μ.

cryo-transmission electron microscopy, cryo-scanning electron microscopy, and freeze fracture. The use of filipin binding to localize cholesterol in the milk fat membrane was demonstrated using freeze fracture (15), and further application of such cryotechniques will lead to elucidation of other food products.

Applications of confocal scanning light microscopy to food materials will increase. This new and powerful technique requires minimal sample processing, gives better resolution than ordinary light microscopy and allows optical sectioning of specimens so that information about their interior structure can be obtained. The availability and affordability of confocal scanning light microscopes will increase, and methods now being developed to use this instrument on biological specimens will be adapted to food specimens, just as other biological microscopy methods are now finding application to foods.

BIBLIOGRAPHY

1. C. Sterling, "The Light Microscope in Food Analysis," in D. W. Gruenwedel and J. R. Whitaker, eds., *Food Analysis: Principles and Techniques*, Vol. 2: *Physicochemical Techniques*, Marcel Dekker, Inc., New York, 1984.
2. K. L. Harris, ed., *Microscopic-Analytical Methods in Food and Drug Control*, FDA Technical Bulletin No. 1, U.S. Dept. of Health, Education and Welfare, Food and Drug Administration, Washington, D.C., 1960.
3. J. R. Gorham, ed., *Training Manual for Analytical Entomology in the Food Industry*, FDA Technical Bulletin No. 2, U.S. Dept. of Health, Education and Welfare, Food and Drug Administration, Washington, D.C., 1977.
4. D. N. Holcomb and M. Kalab, eds., *Studies of Food Microstructure*, SEM, Inc., Chicago, 1981.
5. M. Kalab, S. H. Cohen, E. A. Davis, D. N. Holcomb, and W. J. Wolf, eds., *Food Microstructure*, Vols. 1–7, Scanning Microscopy International, Chicago, 1982–1988.
6. M. Spencer, *Fundamentals of Light Microscopy*, Cambridge University Press, Cambridge, UK, 1982.
7. D. Simpson and W. G. Simpson, *An Introduction to Applications of Light Microscopy in Analysis*, Royal Society of Chemistry, London, 1988.
8. J. Pawley, ed., *The Handbook of Biological Confocal Microscopy*, IMR Press, Madison, Wis., 1989.
9. G. S. Kino and T. R. Corle, "Confocal Scanning Optical Microscopy," *Physics Today*, 55 (Sept. 1989).
10. I. Heertje, P. van der Vlist, J. C. J. Blonk, H. A. C. M. Hendrickx, and C. J. Brakenhoff, "Confocal Scanning Laser Microscopy in Food Research: Some Observations," *Food Microstructure* 6, 115 (1987).
11. R. K. Salyaev, "A Method of Fixation and Embedding of Liquid and Fragile Materials in Agar Microcapsulae," *Proceedings of the Fourth European Regional Conference on the Electron Microscope* II, 37 (1968).
12. D. W. Stanley and M. A. Tung, "Microstructure of Food and Its Relation to Texture," in J. M. deMan, P. W. Voisey, V. F. Rasper, and D. W. Stanley, eds., *Rheology and Texture in Food Quality*, AVI Publishing Co., Inc., Westport, Conn., 1976.
13. D. W. Stanley, "Food Texture and Microstructure," in H. R. Moskowitz, ed., *Food Texture: Instrumental and Sensory Measurement*, Marcel Dekker, Inc., New York, 1987.
14. K. G. Berger and G. W. White, "Ice Cream," in J. G. Vaughan, ed., *Food Microscopy*, Academic Press, Orlando, Fla., 1979.
15. R. W. Martin, Jr., "Electron Microscopic Localization of Cholesterol in Bovine Milk Fat Globules," *Food Microstructure* 8, 3 (1989).

DAVID N. HOLCOMB
ROGER S. UNGER
ROBERT W. MARTIN, JR.
Kraft Technology Center
Glenview, Illinois

FOOD PACKAGING. See FILTH AND EXTRANEOUS MATTER IN FOOD; PACKAGING, MODIFIED ATMOSPHERE PACKAGING.

FOOD PESTS. See FILTH AND EXTRANEOUS MATTERS IN FOOD; FOOD LAWS AND REGULATIONS.

FOOD PLANT DESIGN AND CONSTRUCTION

An approach proven successful in the design and subsequent construction of a new or renovated production facility includes three basic steps: a preliminary study referred to as the feasibility study, the engineering and architectural design phase, and the subsequent project/construction management phase (See Table 1.)

PHASE 1—FEASIBILITY STUDY

The objective of the feasibility study phase is to establish a facility model on which a financial analysis for a given project may be based. The study is a logical development of a project whose summary is a complete financial analysis of the facility in projected production. The feasibility study imposes a discipline on the production, engineering, and financial personnel involved in the project. They are focused away from generalities and must begin to decide on a specific criterion of design. This criterion becomes the basis for the engineering detail. The production and operating criterion, together with the engineering detail, forms the basis for a draft of a concept. The draft is then costed for the capital required. A projection of operating costs for the facility is developed, which is part of the financial analysis.

Develop Process Parameters

Present and Historical Sales Data. The sales history in the case of an existing facility, along with projected growth rates, is used to establish required production rates. New ventures require a sales projection. The seasonality of sales, especially in the food industry, is very important as it establishes the minimum and maximum production requirements. Once the minimum and maximum production is determined, a processing system can be designed for optimum utilization of equipment year round. Minimum inventory level consistent with efficient customer service must be established and interfaced with production levels to establish optimum production rates consistent with minimum inventory costs.

Table 1. The Three Phases of the Design and Construction of a Food Plant

Phase I—Feasibility Study

A. Develop process parameters
   1. Percent sales data
   2. Projected sales data
   3. Raw ingredient utilization and procurement
   4. Operations and production
   5. Packaging–filling
   6. Packaging material utilization and procurement
   7. Materials handling methods
   8. Finished product storage–distribution
B. Develop utility parameters
   1. Water utilization
   2. Gas or oil utilization
   3. Refrigeration utilization
   4. Electrical utilization
   5. Sewage–wastewater
C. Develop site parameters
   1. Existing lot
   2. Ideal lot
   3. Survey
   4. Soil borings
   5. Projected plant configuration
   6. Owner's building style
D. Cost estimate

Phase II—Engineering–Architectural Design

A. Process specifications–engineering
   1. Process and instrumentation flow diagram
   2. Cleaning–CIP instrumentation flow diagram
   3. Process equipment arrangements
   4. Process piping details
   5. Process equipment details
   6. Process installation definition
B. Utility Specifications–Engineering
   1. Mechanical flow diagrams
   2. Mechanical equipment arrangements
   3. Mechanical equipment details
   4. Mechanical piping details
   5. Mechanical installation definition
   6. Electrical single-line diagrams
   7. Electrical equipment arrangements
   8. Electrical equipment details
   9. Electrical equipment definition
   10. Electrical equipment installation
C. Site–facility specifications and engineering
   1. Final site layout
   2. Final architectural layout
   3. Final architectural details
   4. Preliminary application for zoning and building permits
D. Cost Estimate
   Projected process equipment–facility cost
E. Financial analysis
   Develop pro forma financial analysis of the operation for 5 yr

Phase III—Project–Construction Management

A. Bid
   1. Site/work
   2. Facility
   3. Process equipment
   4. Mechanical equipment
   5. Electrical equipment
   6. Process–mechanical–electrical

Table 1. *(continued)*

B. Site management
   1. Scheduling of site activities
   2. Verification of equipment
   3. Verification of installation
   4. Coordination of contractors–owner
   5. Certified drawings
   6. Billing
   7. Startup
   8. Turnover to client

Projected Sales Data. The plant to be constructed should satisfy the projected sales requirement for some future date. This forecast is based on client supplied growth rate projections for each product category.

The design basis year is usually designated as the last year in a period defined as the sum of design time plus construction period plus 5 yr.

The opportunity to interview the various plant managers and supervisors should not be passed up, as quite often, valuable information that may impact the new facility's design can be ascertained from these individuals. Their comments often have meaning to a plant designer that might not be significant to a person whose specialty is production.

The use of the projected sales as a design basis should coincide with corporate objectives and marketing strategies.

Expansions beyond the year used as the design basis is provided for by making provision in the plant design for growth by expanding the facility and/or adding production capacity.

Raw Ingredient Utilization and Procurement. The usage levels and purchasing quantities of raw ingredients must be evaluated to determine liquid and dry ingredient storage and receiving requirements. Additional information, such as ordering procedures and minimum inventory levels of finished goods and raw materials, allows for the correct allocation of warehouse space as well as the design of the liquid storage systems.

Operations and Production. Information regarding production rates, operating days (number and length), and shift arrangements are incorporated in a prototype production schedule. This information allows for the development of a process flow schematic and the line production rates. The equipment sizing is based on an evaluation of minimum and maximum production rates. The equipment selected must economically handle the production range.

Methods used in existing facilities are often established to overcome a process shortcoming. It is important to prevent these improvised methods, which are seldom the most efficient solution, from becoming part of the design of a new facility. Often, these procedures are discovered when asking production personnel "Why is it done this way?" Prior to the construction of the facility, each of these methods should be evaluated for potential improvement.

Packaging and Filling. The type and rate of commercially available filling equipment must be evaluated for both the present and projected sales data. The number of packaging–filling lines for both the minimum and maximum production periods will be defined.

The level of automation at the packaging operation should be discussed at length with the owner and the production personnel.

Automation. The level of plant automation must be established during this, the preliminary design, phase. The obvious parameters must be established. What are the existing systems on which the client relies for management informations systems? How should production control systems relate to the established management reporting system? What is the ordering system now in use? Should the ordering systems be changed? How are the order filling systems and the billing systems related?

Production equipment lists can be developed from this information. The capital cost of the selected equipment can be established. All utilities required by the production equipment can be established and operating costs evaluated for inclusion in the financial analysis.

Packaging Material Utilization and Procurement. The usage levels and purchasing methods for packaging must be evaluated to determine storage and receiving requirements. Unique storage requirements such as temperature or environment control must be defined.

The packaging materials storage capacity will reflect such factors as proximity to supplier, long-term purchasing agreements, and volume discount purchases.

Materials Handling Methods. Materials handling methods for both raw and finished product must be established. Finished product may be handled effectively either manually or automatically depending on the availability of equipment, equipment and installation costs, operating efficiencies, and savings in manpower.

Finished Product Storage–Distribution. Distribution systems and shipping schedules impact product storage, order picking, and loadout requirements. It is important to evaluate during this project definition stage what, if any, revisions in the product storage, order picking, and loadout procedures are indicated. The revised production and warehousing methods required to meet the increased sales often render the current method of order picking and loadout obsolete.

Develop Utility Parameters

Water Utilization. The projected water utilization can be determined from the preliminary process flow design and parameters. Water usage as an ingredient can be ascertained from the product formulas. Water usage for personnel, plant and equipment cleaning, and equipment services can be estimated.

The projected water usage will be a factor in the site selection as well as a factor in the cost analysis of the projected facility.

Gas or Oil Utilization. Again, the basis of preliminary process flow design and parameters, the projected energy utilization of the facility can be determined. Whether oil or gas is selected depends on local cost and availability considerations.

Refrigeration. The refrigeration capacity that will be required for the new facility can be estimated from the finished product and raw ingredient cooling rates, finished product and raw ingredient storage requirements (freezer or cooler), and other facility requirements such as air-conditioning. The type of system will depend on the projected load and local requirements.

Electrical Usage. Electrical utilization estimates will be based on connected motor horsepower and general facility requirements. An appropriate diversity factor will be used when establishing the appropriate facility electrical supply requirements. The local power company is usually contacted at this stage to establish any limitations on power usage.

Sewage–Wastewater. A prime consideration in the location and design of a processing facility is the disposal of the sewage–wastewater. The sewage loading can be estimated from the processing parameters either as a percentage of loss or by published industry standards. As a small processing facility of certain types can generate sewage loading equivalent to that of a small city, the projected sewage load is critical to the determination of site location. A significant capital cost and operating cost can be encountered if a waste treatment facility is required on site. Municipal treatment facility charges must be included in the financial analysis.

Develop Site Parameters

Existing Lot. The evaluation of a process facility is based on either the expansion of an existing facility or the development of a specific lot. When a specific site is known, the zoning, building, and utility requirements can be determined from local authorities.

Factors such as setbacks, maximum facility heights, material or design requirements, and the availability of utilities will be investigated for potential limitations to the new facility.

The orientation of the facility on the site must be evaluated as it impacts truck and automobile traffic on the lot and their access from adjacent roadways.

Ideal Lot. If the projected facility design is based on an unknown location, an ideal lot size and configuration will be developed.

With the lot configuration, process parameters, utility requirements, truck and automobile access requirements, and sewage requirements, a property search to locate a suitable lot can be initiated. Proximity to operating staff, raw ingredients, packaging materials, and finished product distribution requirements must also be taken into account and considered.

Survey. On the basis of a selected site, a survey and topographical study of the property will be obtained. The study will indicate the lot outline, location of utilities and sewers, and general site topography. The preliminary building design and cost estimate will be developed from these data.

Soil Borings. After the site–facility plan is developed, soil borings will be made at selected points. Borings are generally conducted in areas of high truck traffic, building footings, and heavy equipment locations.

The borings will provide data on the subsurface conditions that allow the architect–structural designer to develop the facility design.

Projected Plant Configuration. A preliminary plan illustrating the production facility on the existing lot is developed. When presented, this preliminary plan provides the visualization of the concept as developed. This then becomes the basis for future refinement of the production methods, as well as all factors of design, which has been based on interpretation of the engineering analysis of the data assembled above. The amended plan becomes the basis for the capital and operating financial analysis. This information can then be used by the client for presentation to lending institutions and local municipal agencies.

Owner's Building Style. The final design of a process facility can vary greatly to suit the owner's needs. The owner may consider the facility as part of an overall marketing plan and as a result desire a showplace facility. This type of facility will often include visitor areas and aisles for plant tours.

The owner may consider the facility as a functional operation only and thus may minimize the investment.

Cost Estimate

An estimate of the capital cost of process and process related equipment and services can be determined from this preliminary design. In order to project an operating cost associated with the facility, an estimate must be produced of the various services required, such as electrical, water, refrigeration, steam, air, sprinkler, sewage, and utilities.

The feasibility study summarizes all the assumptions, projected costs, and operating procedures, as well as the accepted parameters on which final design will be based. The feasibility study becomes a written record of what has been developed as a cooperative effort of the client and the engineering group.

A schedule of capital required to complete the project is compiled. The data are required for a financial analysis of the project.

The financial aspects of the study includes an analysis of projected capital requirement during construction and any additional capital that would be required during the first 5-yr period, to meet the projected production requirements. Pro forma operating statements, staffing requirements, inventory costs, and other scheduled operating costs are the basis of a pro forma 5-yr operating statement, as well as pro forma profit–loss statements for 5 yr.

This analysis is usually the basis for the decision as to whether to execute the project.

The finished document will serve as the basis for the development of a complete set of bid plans and specifications. It is a reference document for future production personnel as well as a basis for future plant expansions and changes.

The feasibility study defines the parameters required in an environmental impact statement.

This preliminary phase of the project is most important as it develops a document that contains the agreed-on plant criterion. These criteria—the data on which the client may base a decision as to whether the project is financially justified—is then communicated to the various designers. With this, the preliminary phase of the project is completed and the initiation of the second phase, the design phase, is then practical.

PHASE II—ENGINEERING–ARCHITECTURAL DESIGN

The preliminary drawings, which were developed during the feasibility study, are used to locate ground borings to help predict subgrade conditions generally and specifically where heavy loads will be located on the property. Once the results of these tests are complete, the normal architectural and engineering phase of the project can begin.

The new facility should be designed, from start to finish, for and around the process. The concept of building the largest structure possible on a given lot is impractical in terms of cost, equipment layout, and efficient materials flow. The facility should be designed to accommodate the projected growth in production and future expansions. Two very important items that must be incorporated into the design are flexibility and future expansion. Although the plant is being designed for a production rate projected 5–7 yr in the future, one must consider that new products will be developed and existing products discontinued. The layout of the processing area should be accomplished in such a fashion that any changes in projected production can be made with minimal cost or downtime. The design of the facility should take into account expansions that will occur at a period beyond the design year.

The concept that the building houses and facilitates production impacts the architectural design of the facility. The structural steel should be designed to support equipment and pipes located in the preliminary design. Floor drains should be placed where heavy flow is encountered. Footings and foundations should be provided to facilitate the planned expansion.

Equipment layout is of equal importance. Access for maintenance is an obvious concern, but the changes to accommodate future growth are often neglected. The least expensive time to plan for the future is during the initial design phase.

Designing the facility for the process allows for a very efficient equipment layout and reduces the overall cost of the facility. Openings for piping and conveyors must be coordinated with the architectural designer. When all process and mechanical piping are considered in the planning stages, it provides the architect with sufficient information to identify the areas of concentrated loads and therefore diminish the extra costs in either additional changes or the "over design" of all areas of the plant to accommodate loadings in specific areas. The detailed process layout allows the designer to accurately assess the needs for various room finishes and process area heating, ventilating, and air-conditioning (HVAC) requirements. The importance of thorough planning, coordination, and communication between the building designers and the process engineers becomes evident.

Once the design of the facility and process is complete, a detailed set of drawings and specifications should be available for the construction phase of the project.

Process Specifications/Engineering

Process and Instrumentation Flow Diagram. The process flow and instrumentation flow diagram will indicate all process equipment type, size, and operating rate. Instrumentation required for process control and alarm will be identified.

Each device will be identified by a unique item number. This number will be used through project completion for shipping and receiving, installation, records, and project cost control.

Cleaning–CIP Instrumentation Flow Diagram. To complement the process and instrumentation flow diagram, the CIP or cleaning requirements will be identified. The cleaning requirements will define the process–cleaning interfaces such as swing connections and routing plates.

Process Equipment Arrangements. The process equipment arrangement drawings locate all major process equipment. The equipment arrangement, along with equipment detail drawings and the specification, defines which contractor will install the device.

Process Piping Detail Drawings. Typical piping details, isometrics, pipe support methods, and piping routes will be identified. These drawings are used for material takeoff.

Process Equipment Details. Each major piece of process equipment will be defined by an equipment specification. This description is based on performance and should include process rates, special material handling or processing techniques, special design features, and regulatory agency compliance requirements.

Process Installation Details. The installation responsibility for each piece of processing equipment is defined. This definition will include responsibility for supply of the equipment, receiving and storage responsibility, and installation responsibility.

Utility Specifications–Engineering

Mechanical Flow Diagrams. The mechanical flow diagrams will indicate all mechanical equipment type, size, and operating rates. Instrumentation for mechanical control and alarm will be identified.

As the mechanical system is a support for the process system, the relationship of the process to the mechanical system will be defined.

Each mechanical device is also assigned a unique item number for project control purposes.

Mechanical Equipment Arrangements. The mechanical equipment arrangement drawings locate all mechanical equipment. The mechanical equipment arrangement, along with the equipment detail drawings, allows the appropriate subcontractor to install the device.

Mechanical Equipment Details. Each major piece of mechanical equipment will be defined by an equipment specification. This description is based on performance and should include process rates, design features, and regulatory agency compliance requirements.

Mechanical Piping Details. Typical mechanical piping details, pipe support methods, insulation, and mechanical piping routes will be identified. These details are used for material takeoff.

Mechanical Installation Details. The installation responsibility for each piece of mechanical equipment is defined. This definition will include responsibility for supply of the equipment, receiving and storage responsibility, and installation responsibility.

Electrical Single-Line Diagrams. Electrical distribution for all process, mechanical, and facility equipment drives are indicated on the single-line diagram. The total connected electrical load can be distributed to various use points in the most cost-effective manner.

Electrical Equipment Arrangements. The electrical equipment, switchgear, distribution, motor control centers, low-voltage panels, and lighting panels will be located.

The relationship of the switchgear to incoming power and transformer will be defined.

Electrical Equipment Details. Typical electrical details and conduit arrangements will be identified.

These details are used for material takeoff.

Electrical Equipment Definition. Each major electrical device—transformer, switchgear, motor control center, mechanical control panel, process control panels, and lighting panels—will be defined for equipment purchase and installation.

Electrical Equipment Installation. The installation responsibility for each piece of electrical equipment is defined. This definition will include responsibility for supply of the equipment, receiving and storage responsibility, and installation responsibility.

Site Facility Specifications and Engineering

Final Site Layout. The site layout indicates plant location; site access routes by truck, employees, and visitors; areas for future expansion; etc.

Final Architectural Layout. The final plant layout will indicate room sizes, internal plant traffic flow, employee service, and office facilities.

Final Architectural Details. The final architectural details will defined materials of construction, wall finishes, and special process-related requirements. The definition of materials and subcontractor responsibility will serve as the contract bid basis.

Preliminary Application for Zoning and Building Permits. The site layout, architectural details, and layouts will be used to obtain construction and building permits. These will normally be reviewed by a local government agency for compliance with site zoning and building requirements.

In a "fast-track" project, the building permit procedure will begin prior to design completion. Footings, foundations, and primary structural components can proceed while final engineering is completed.

Cost Estimate

At the completion of the detailed process, mechanical and facility design phase, the cost estimate prepared during the feasibility study will be reevaluated. Significant variance from the original estimate may impact the viability of the project. The impact of the variance can be analyzed prior to further commitment of capital.

PHASE III—PROJECT–CONSTRUCTION MANAGEMENT

The final phase involves the project–construction management. It is the project–construction managers' responsibility to supervise and coordinate the entire installation. This includes the handling of all paperwork and documentation, supervision of installation, and preparation of submittals to government regulatory authorities, which may be necessary to gain required approvals in order to place the facility into operation. Generally, the construction management of an installation includes the tasks and requirements described in the following paragraphs.

Bid

Sitework. Using the site layout, architectural drawings and specifications, the construction manager will obtain bids from qualified subcontractors. The qualification list is generally based on local, financially stable contractors.

Facility. Using the architectural drawings and specifications, the construction manager will obtain bids from qualified subcontractors. Depending on the scope of the project, the facility work may be handled on an overall general contract basis or as an individual subcontract basis.

Process Equipment. During the feasibility study phase of the project, the long-lead process equipment needs were defined. On specialty processing equipment, delivery times can reach one year.

These long-lead items will be purchased according to their delivery time frame and requirement at the site. Ideally, the process equipment will arrive on site, unload, and be placed in or near the intended location.

Mechanical Equipment. Again, items will be purchased according to their needs at the site. As some mechanical equipment may require special rigging because of size or installation location, it is generally required on site prior to the process equipment.

Deliveries of large pieces of equipment installed in the same areas should be coordinated to minimize rigging time and cost.

Electrical Equipment. Electrical equipment can be purchased as individual pieces or an overall contract. As the availability of a single local contractor who can handle high-voltage power, facility and power wiring, and process control and instrumentation wiring is unlikely, these are generally handled as separate subcontracts.

Process–Mechanical–Electrical Installation. Using the installation detail drawings and specifications, the construction manager will obtain bids for the installation; the owner usually receives the bids directly.

Site Management

Scheduling of Site Activities. The construction manager will administer all subcontracts and supervise the installation to ensure that the work is being performed in compliance with the contract specifications, drawings, and schedule.

The construction manager will coordinate all contractors on the project and conduct project meetings with project reports designed to obtain results as well as maintain records, will prioritize ordering sequence for items with longest lead times, and will place orders with suppliers to insure that the equipment is supplied in accordance with the specifications.

Verification of Equipment. The construction manager will provide expediting services. This is to ensure that all necessary equipment and materials arrive on site in a timely fashion avoiding delays in the schedule.

The construction manager will verify each piece of equipment received for compliance with specification and purchase order.

Verification of Installation. The construction manager will coordinate with governmental agencies to ensure adherence to applicable local codes and obtain required permits and will secure waivers of liens from contractors and material vendors.

Coordination of Contractors and Owner. The construction manager will maintain all accounting records that relate the budgeted costs to actual costs throughout the project, will coordinate any necessary change orders to ensure that they have been priced and approved by the client prior to being executed, will provide procurement services, and will consult with the client on preferred bidders and prequalify potential vendors of equipment.

Certified Drawings. All major equipment will require certified drawings. These drawings will be checked by the construction management team to ensure compliance with the specifications.

Billing. All material supplier and contractor invoices will be compiled into a single billing. Every cost item will be substantiated by appropriate documentation such as invoices and receipts. The contract manager will ensure that all vendors are paid on a timely basis to derive maximum benefits from cash discounts, and so on.

Startup. The construction manager will coordinate the startup services of the equipment vendors, subsystem suppliers, and installation contractors.

Turnover to Client. Following satisfaction of the contractual obligation, a complete operating facility, the construction manager will transfer as-built process, facility, mechanical, and electrical drawings and all records received during the project, such as operating and maintenance manuals and spare parts lists.

BIBLIOGRAPHY

Adapted from: A. Cukurs, "Food Plant Design and Construction," APV Crepaco, Inc., 1989.
Copyrighted APV Crepaco, Inc. Used with permission.

APV CREPACO, INC.
Lake Hills, Wisconsin

FOOD POISONING. See FOODBORNE DISEASES

FOOD PRESERVATION

Matching the supply of food as produced by the agricultural sector with the demand for food by consumers in both time and space necessitates the use of a variety of preservation techniques. Food is being produced in larger quantities by fewer people in rural areas often very distant from the urban consumer. The production of agricultural commodities follows cyclical patterns of supply, dictated by such things as cropping times and yearly fluctuations. The requirements on the food supply by the consumer include safety, quality, adequate shelf life, variety, and convenience. Thus, the demand for food tends to be more constant than its production, food needs to be transported from the production sector to the consuming markets, food needs to be transformed from its raw state as produced by the agricultural sector into a vast array of consumer goods, and throughout this process quality must be maintained, safety must be guaranteed, and the economics must be favorable, minimizing losses and waste. This matching of supply and demand in both space and time both defines and introduces the field of food preservation.

Food in its many forms is subject to very rapid deterioration beginning soon after harvest. The factors contributing to this process include biological deterioration and postharvest loss from bacteria, yeasts, molds, insects, and rodents, and chemical breakdown of food components catalyzed by enzymes, light, or oxygen. Preservation techniques attempt to control these deterioration processes through the destruction of microorganisms present in the food, through the manipulation of factors essential for the continued growth of microorganisms, or through the control of factors responsible for chemical deteriorations. It is difficult to define the time frame necessary to consider a food preserved. Milk can be preserved through pasteurization and refrigeration to attain a shelf life of 10–14 days. Milk can also be preserved through spray drying to attain a shelf life of a year or longer. Thus a food might be considered preserved when it is effectively moved from production to consumption, maintaining safety and quality, minimizing losses, and being delivered in a form that is convenient and acceptable to the consumer.

Although new preservation processes have been developed and traditional processes have been modified, food preservation has been practiced for millenia. Since the beginning of mankind, the need to gather food has existed. Processes such as sun-drying, salting, smoking, or food fermentations are considered to be early forms of food preservation. The technology of food preservation increased dramatically with the discovery of microorganisms late in the eighteenth century and the appreciation of their role in food deterioration that followed. See Ref. 1 for a complete history of food preservation.

CAUSES AND MANIFESTATIONS OF FOOD DETERIORATION

Microorganisms

Microorganisms are ubiquitous living organisms that need nutrients, moisture, appropriate oxygen conditions, and favorable pH ranges to grow. Microorganisms can contaminate our food supply from the point of production or harvest until the time of consumption. Many microbial food contaminants are native to the soil or animal environment from which the food is derived, and many are added unintentionally through handling practices. The components of most foodstuffs serve as ideal nutrients for the growth of microorganisms. This growth leads to potentially harmful populations of microorganisms; potentially harmful build-up of microbial metabolites, eg, toxins, enzymes, polysaccharides, or pigments; and a deterioration in food quality resulting from the metabolism of food constituents, eg, amino acid or fatty acid release from proteins or lipids. Microorganisms important to food deterioration include bacteria, yeasts, and molds.

It is important to distinguish between pathogenic and nonpathogenic species. Pathogens are organisms that cause disease. A number of human pathogens can be transmitted through food. They include among many others *Salmonella spp.*, *Staphylococcus aureus* and *Clostridium botulinum*. Mycotoxins are fungal metabolites, some of which are potentially toxic or carcinogenic to humans, eg, aflatoxin from *Aspergillus flavus*. It is essential that all pathogenic species be controlled in preserved foods to eliminate food-borne disease. Most microorganisms, however, are not pathogenic but can cause food spoilage if allowed to grow, primarily as a result of off-flavors and odors from metabolic by-products. For further reading on food microbiology, see Refs. 2–5.

Enzymes

Enzymes are naturally present in most raw food commodities. A number of enzymes catalyze reactions that are detrimental to product quality. These include off-flavor production by lipoxygenases, lipases, and proteases; textural changes due to pectic enzymes and cellulases; color changes due to polyphenol oxidase, chlorophyllase, and peroxidase; and nutritional changes due to ascorbic acid oxidase or thiaminase. A mild heating process such as blanching or pasteurization is usually sufficient to destroy any harmful enzymes that may be present in a food.

Pests

Insect and rodent infestations and losses can occur readily in open fields during growth and production and in bulk stores of food exposed or partially exposed to the environment. This is especially prevalent in developing countries where storage conditions are less than adequate. Insects can attack grains, reducing their nutritive value and imparting a sour taste to flour, peas, beans, meat, fish, cheese, etc. Insects damage far more food than they consume owing to the deposition of larvae in webbings, rendering much of the food store as waste. Rodents are also a serious concern in food storage, also damaging far more than they consume through adulteration with droppings, filth, and hair, and potential transmission of disease.

Temperature

Uncontrolled heat and cold can cause detrimental reactions to occur in foods. Increased but moderate temperatures, 30–40°C, accelerate the rates of reactions, such as oxidation, that lead to the production of off-flavors and odors, color changes, and nutrient loss, and significantly decrease the generation times of microorganisms, leading to enhanced populations. Decreased temperatures, in the 0–15°C range, can lead to detrimental physiological reactions in certain susceptible products. These include chill injury in tomatoes or bananas, which result in color and texture changes and increase in microbial susceptibility; cold shortening in meats, resulting in tough carcasses after slaughter; chill sweetening in potatoes, resulting in increased glucose contents and enhanced browning during frying; and staling of bread, which occurs much more readily at these temperatures.

Moisture

Detrimental reactions in foods are related more to the water activity of the food than its moisture content. Water activity is defined as the equilibrium relative humidity of the food, or a measure of free water in the food, that water which is able to participate in reactions. Both rates of microbial growth and chemical reactions are influenced by water activity. Thus the gain or loss of moisture from a food due to its conditions of storage, particularly relative humidity, may influence the water activity and thus the microbial and chemical stability of the product. Additionally, changes in the relative humidity of storage can also affect structure and texture of the food. High relative humidities of storage can favor splitting or cracking of the skins of some fruits, whereas low relative humidities can favor wilting and shrivelling of fruits and vegetables, leading to a loss of weight and thus economic value.

Oxygen

The oxygen requirements for both the storage of foods and for microorganisms varies considerably. Bacteria have very strict oxygen requirements and are classified accordingly as aerobic, anaerobic, or facultative organisms with wider tolerances. Oxygen is also an initiator of chemical reactions that are detrimental to various food constituents, including lipids, vitamins, pigments, and some amino acids. However, reduced oxygen contents can impair the physiological function of food tissues, such as fruits, vegetables, or meat. Hence control of oxygen contents through modified atmospheres and protective packaging can help to reduce food deterioration.

Light

Exposure to light can be detrimental to some food products. Light can promote oxidative rancidity of lipids; oxidation of milk leading to breakdown of proteins and formation of unpleasant volatiles; changes in various pigments, eg, myoglobin in red meats; breakdown of vitamins such as vitamin A, riboflavin, and ascorbic acid leading to nutrient loss; or the development of potentially toxic light-induced glycoalkyloids such as solanine in potatoes that have turned green owing to increased chlorophyll content (from light exposure).

Time

Time itself plays a very great role in food deterioration. While some preservation methods such as canning are intended to produce food for years of storage, other preservation methods such as pasteurization and refrigeration produce shelf lives of days to weeks. Such processes aim to slow but not eliminate deteriorations. Recontamination becomes possible unless protective packaging is employed or after package integrity is broken. Food deteriorations described above will substantially reduce the shelf life of a product, but virtually all foods will become unconsumable over time.

PRESERVATION PRINCIPLES AND APPLICATIONS

Preservation techniques put into practice the control mechanisms for reducing food deterioration. Often, various methods are combined to ensure safety and preservative action while maintaining maximal quality and stability. For example, fluid milk is preserved through a combination of pasteurization and subsequent refrigeration, aseptic thermal processing is usually combined with aseptic packaging, and freezing is usually accompanied by a blanching pretreatment to reduce enzymatic activity. For more detailed general information of preservation techniques, see Refs. 6–13.

Physical Preservation

Thermal Methods. A number of preservation processes use heat to extend the shelf life of foods. High temperature preservation methods performed commercially are controlled processes that include canning, aseptic processing, pasteurization, and blanching. Microorganisms differ in their heat resistance and are classified as psychrotrophs, mesophiles, or thermophiles according to their tolerance to heat. However, all microorganisms can be destroyed by the application of heat, and each organism has a certain time/temperature relationship associated with it to ensure a given reduction in its population. The most intense heating preservation process would render the food sterile. Sterilization refers to the complete destruction of all microorganisms. Complete sterility is difficult to achieve and often leads to a reduction in the quality attributes of the food, since most food components such as proteins or vitamins are also heat sensitive. Commercial sterility has been defined as the destruction of all pathogens and spoilage organisms in a food. Canning, thermal retorting in aluminum cans or flexible pouches, and aseptic processing, thermal processing before packaging followed by aseptic packaging techniques, target commercial sterility as their goal. Provided the food is maintained in the commercially sterile state after processing, a shelf life of 2 years or more can be achieved. For more information on canning, see Refs. 14–16. A review of aseptic processing techniques can be found in Ref. 17.

Pasteurization is a low order, time- and temperature-dependant heating process that is designed to destroy all pathogens present in the food, to reduce the bacterial load in the case of milk and eggs, to reduce the yeast and mold count in the case of beer, wine, and fruit juices, and to extend the shelf life. Pasteurization can be performed before packaging, as in milk, or subsequent to packaging, as in beer. Blanching is another low-order heat process used primarily as a pretreatment step in freezing to destroy enzymes and in canning to wilt and cleanse tissue and expel tissue gas.

Low temperature methods include both refrigeration and freezing. Storage below 15°C but above freezing retards growth of microorganisms, retards metabolic activities of animal tissues postslaughter and plant tissues postharvest, retards deteriorative chemical reactions such as oxidation and enzyme-catalyzed reactions, and retards moisture loss. Unlike heating preservation, cold preservation does not destroy microorganisms, only retards their growth. The foods are still perishable and organisms will grow more rapidly once conditions become favorable. The shelf life can be extended from less than 1 week for highly perishable products such as raw milk or ripe tomatoes to more than 6 months for more durable products such as onions or smoked meats. Freezing preservation is achieved through both the low temperatures, which inhibit microbial growth and rates of reactions, and the reduction in water content as a result of ice crystallization. Shelf life of frozen foods can range from 3 months to a year or longer, but is still limited by enzymatic activity, oxidation, and dehydration. Commercial food freezing facilities use several methods, including air freezing through sharp (natural convection), blast, or fluidized bed (forced convection) techniques; indirect contact systems such as plate freezers or scraped surface freezers; or direct contact freezing systems using low-temperature liquids or cryogens. Detailed information on food freezing can be found in Refs. 18–20.

Reduction of Water Content. Microorganisms need favorable moisture conditions within the food, water activity greater than 0.6 to 0.8, for their growth. It is therefore possible to manipulate the water activity of the food to inhibit microbial action. Concentration of liquid food products through thermal evaporation, freeze-concentration, osmotic dehydration, and membrane processes achieves lowered water contents and, provided that water activity has been lowered sufficiently, offers preservation action. Thermal concentration removes water in the form of vapor from the liquid. It is usually performed in multiple-effect falling-film vacuum evaporators designed for thermal efficiency and product quality. Freeze concentration removes water from liquid foods in the form of ice and is particularly suited to foods such as fruit juices with depressed freezing points due to high sugar concentrations within the food. Osmotic dehydration has been used for fruit slices in sugar solutions whereby water will migrate from the fruit slice into the sugar solution owing to the high osmotic pressure of the solution. Membrane processes remove water from foods in the form of liquid water, due to the presence of a semipermeable membrane and the imposition of a pressure gradient. Solvent and low molecular weight solute, depending on the membrane pore size, pass through the membrane in the permeate stream while the higher molecular weight solutes are concentrated in the retentate stream.

The nearly complete removal of water through dehydration by solar, cabinet, tunnel, drum, or spray drying methods also offers a form of preservation by reduction of available water for microbial growth. Dehydration occurs under controlled conditions that cause minimal changes in the food properties. The food can then be consumed dried, as in some dried fruits or meats, but is more likely to be rehydrated before consumption, as in dried milk and eggs, instant potato flakes, or instant coffee. The reconstituted product should resemble as closely as possible the quality of the original food. In addition to preservation, the drying of foods decreases the weight and bulk of the original food and adds a measure of convenience to the product. Sun drying has been practiced for centuries and is still employed for the dehydration of grains, seaweeds, raisins, and other foods, particularly in developing countries. Most commercial fruit and vegetable operations employ continuous tunnel or belt dehydration systems that use heated air as the drying medium. The majority of liquid foods, such as skim milk, cake and soup mixes, flavors, purees, juices or instant coffee, are dried in spray dryers that atomize the usually preconcentrated liquid product into tiny droplets that dry rapidly in the surrounding heated environment. Freeze-drying removes water from a frozen food through sublimation under vacuum and is particularly suited to thermally sensitive products, such as instant coffee or convenience-type prepared entrees. See Refs. 21–23 for more information on food dehydration and Refs. 24 and 25 for detailed information on freeze-drying.

Oxygen Control. Because of the strict oxygen requirements for bacterial growth and the participation of oxygen in a number of chemical reactions, oxygen control can act as a means of food preservation. Controlled and modified atmosphere storage of foods are techniques to maintain gaseous atmospheres with strictly controlled oxygen contents. The controlled or modified atmosphere can be maintained in warehouses for bulk foods, eg, apples, often before further processing, or can be maintained at the microatmospheric level within a food package. Food packaging also offers protective barriers to food against the action of contaminating microorganisms, pests, moisture, oxygen, and light. The packaging necessary to maintain preservation is usually chosen to accompany the particular process. Examples include multilaminate flexible packaging for aseptically processed foods or rigid aluminum cans for retorting. Information on food packaging can be found in Refs. 26–30.

Radiation. The use of nuclear energy in the form of γ radiations, short wavelengths emitted by unstable isotopes of cobalt 60, or cesium 137 to inactivate microorganisms has been a developing technology since 1945. The main goal of irradiation is to extend the shelf life of foods where heat or chemical means are unfeasible owing to the nature or geographic location of the food. Major potential applications of this process include spices, owing to the heat-sensitive volatile flavor components; insect disinfestation of grains and fruit; extended shelf life of fruits, vegetables, fish, shellfish, and meat products; sterilized diets for military, space, and medical uses; and animal feeds and moist pet foods. The irradiation occurs in an enclosed chamber in which the product can be exposed to an even distribution of the penetrating γ rays for the necessary time to accomplish microorganism inactivation. The safety of irradiated foods has been extensively studied and proved (31–34).

Chemical Preservation

Intermediate Moisture Foods. An intermediate moisture food (IMF) is one that can be eaten as is, without rehydration, and yet is shelf stable without refrigeration or thermal processing. Whereas most foods have water activities in the range 0.9–1.0, IM foods rely on water activities in the range 0.65–0.85, below that required for the growth of the most tolerant organisms, for their preservation effects. Included in this category of foods are jams and jellies, fruit cakes, pepperoni, sweetened condensed milk, marshmallows, soft cookies, and many others. Sugar and other humectants, water-absorbing compounds such as sorbitol, glycerol, starches, or gelatin, can be used to formulate these foods. Although the technology can produce a range of products with acceptable texture, many of which have been in existence since historical times, the flavor profile created by the various humectants has been the major limiting factor in new IMF product development. Refer to Refs. 35 and 36 for further information on IMF.

pH Control. Acids can be used to lower the pH of foods to below the tolerable range for microorganisms. Acid can also enhance the lethality of heating processes. *Clostridium botulinum*, the organism of concern in commercial canning processes, will not grow at less than pH 4.5, and thus it is not necessary to thermally process high acid foods (pH < 4.5) under the same rigid time temperature standards as is the case with low acid foods (pH > 4.5). The addition of acid to such foods as soft drinks and the production of acid in some food fermentations are effective controls of microbial growth. However, pH control is normally associated with some other means of preservation as well, as the palatability of many foods and chemical stability of their constituents (eg, proteins) also decreases at low pH levels.

Chemical Addition. Salt can be added to foods for its contribution to the preservation of the food, eg, butter, fish, or cured meat products such as bacon. The action of salt results from the osmotic pressure created in the aqueous environment surrounding the microbial cell, in an analogous manner to the addition of sugar in intermediate moisture foods or the use of sugar syrups for osmotic dehydration of fruits. Plasmolysis, the partial dehydration of the cell, results and the viability of the microorganism is thus destroyed. Smoke is also a type of chemical preservative that has been used since historical times to preserve foods, especially meat and fish products. The action of smoke results from the formation of small amounts of preservative chemicals and the internal temperatures and dehydration of tissues associated with the hot-smoking process. Cold-smoking at temperatures less than 30°C relies solely on the formation of bactericidal chemicals and is usually associated with other means of preservation such as salting, refrigeration, or packaging. Chemical preservatives, such as benzoic acid or sodium benzoate, sorbic acid or potassium sorbate, sodium nitrite or nitrate, and sulfur dioxide, are permitted at low levels in some foods as preservative agents against microbial growth. The use of chemical preservatives and other food additives are closely regulated by governmental agencies (37,38).

Biological Preservation

Fermentations. Unlike the processes described above, food fermentations have as their goal an increase in the numbers of microorganisms present in a food. The traditional foods of many countries rely on fermentation processes, and fermentation is an historical but important means of food preservation throughout the world. Fermented foods are preserved through the action of a particular organism, unique to each given commodity, on a particular substrate within the food product, primarily carbohydrates but also proteins and lipids. The conditions of fermentation favor the growth of the desirable organism, which is often added in the form of a pure culture, and cause the competitive disappearance of undesirable spoilage or pathogenic organisms. The metabolic by-products of the fermentation change conditions such as pH or oxygen content within the food, which also act to inhibit the undesirable organisms. They include lactic and other acids, ethanol and other alcohols, gases such as CO_2, and a variety of other compounds at low levels that are responsible for the unique flavor characteristics of the particular product. Examples of food fermentations include the production of alcohols by yeasts in wine, cider and beer, and the production of lactic acid by bacteria in fermented milks, sour cream, yogurt, fermented meats, pickles, sauerkraut, and vinegar (39–41).

BIBLIOGRAPHY

1. S. Thorne, *The History of Food Preservation,* Barnes & Noble Books, Totowa, N.J., 1986.
2. J. C. Ayres, J. O. Mundt, and W. E. Sandine, *Microbiology of Foods,* W. H. Freeman and Company, San Francisco, 1980.
3. G. J. Banwart, *Basic Food Microbiology,* AVI Publishing Co., Westport, Conn., 1979.
4. W. C. Frazier and D. C. Westhoff, *Food Microbiology,* 4th ed., McGraw-Hill, New York, 1988.
5. J. M. Jay, *Modern Food Microbiology,* 3rd ed., Van Nostrand Reinhold, New York, 1986.
6. N. W. Desrosier and J. N. Desrosier, *The Technology of Food Preservation,* 4th ed., AVI Publishing Co., Westport, Conn., 1977.
7. P. Fellows, *Food Processing Technology: Principles and Practice,* VCH Publishers, New York, 1988.
8. M. L. Fields, *Laboratory Manual in Food Preservation,* AVI Publishing Co., Westport, Conn., 1977.
9. P. Jelen, *Introduction to Food Processing,* Reston Publishing Co., Reston, Va., 1985.
10. M. Karel, O. R. Fennema, and D. B. Lund, *Physical Principles of Food Preservation,* Marcel Dekker, New York, 1975.
11. E. Karmas and R. S. Harris, eds., *Nutritional Evaluation of Food Processing,* 3rd ed., Van Nostrand Reinhold, New York, 1988.
12. N. N. Potter, *Food Science,* 4th ed., AVI Publishing Co., Westport, Conn., 1986.
13. S. Thorne, *Developments in Food Preservation,* Elsevier Applied Science Publishers, New York, 1982.
14. A. C. Hersom and E. D. Hulland, *Canned Foods: Thermal Processing and Microbiology,* 7th ed., Chemical Publishing Co., New York, 1980.
15. J. M. Jackson and B. M. Shinn, *Fundamentals of Food Canning Technology,* AVI Publishing Co., Westport, Conn., 1979.
16. A. Lopez, *A Complete Course in Canning,* 11th ed., Canning Trade, Inc., Baltimore, Md, 1981.
17. A. C. Hersom, "Aseptic Processing and Packaging of Food," *Food Review International* 1(2), 215, 1985.
18. N. W. Desrosier and D. K. Tressler, *Fundamentals of Food Freezing,* AVI Publishing Co., Westport, Conn., 1977.
19. O. R. Fennema, W. D. Powrie, and E. H. Marth, *Low Temperature Preservation of Foods and Living Matter,* Marcel Dekker, New York, 1973.
20. D. K. Tressler, W. B. van Arsdel, and M. J. Copley, *The Freezing Preservation of Foods,* 4th ed., AVI Publishing Co., Westport, Conn., 1968.
21. D. MacCarthy, ed., *Concentration and Drying of Foods,* Elsevier Applied Science Publishers, New York, 1986.
22. A. S. Mujumdar, *Handbook of Industrial Drying,* Marcel Dekker, New York, 1987.
23. W. B. van Arsdel, M. J. Copley, and A. I. Morgan, *Food Dehydration,* 2nd ed., AVI Publishing Co., Westport, Conn., 1973.
24. S. H. Goldblith, L. R. Rey, and W. W. Rothmayr, eds., *Freeze Drying and Advanced Food Technology,* Academic Press, New York, 1975.
25. J. D. Mellor, *Fundamentals of Freeze Drying,* Academic Press, New York, 1978.
26. J. F. Hanlon, *Handbook of Package Engineering,* 2nd ed., McGraw-Hill, New York, 1984.
27. M. Mathlouthi, ed., *Food Packaging and Preservation: Theory and Practice,* Elsevier Applied Science Publishers, New York, 1986.
28. F. A. Paine and H. Y. Paine, *Handbook of Food Packaging,* L. Hill Co., Glasgow, Scotland, 1983.
29. S. J. Palling, *Developments in Food Packaging,* Applied Science, London, 1980.
30. S. Sacharow and R. C. Griffin, *Principles of Food Packaging,* 2nd ed., AVI Publishing Co., Westport, Conn., 1980.
31. Council for Agricultural Science and Technology, *Ionizing Energy in Food Processing and Pest Control. I. Wholesomeness of Food Treated with Ionizing Energy,* Task Force Report 109, 1986, *II. Applications,* Task Force Report 115, 1989, Council for Agricultural Science and Technology (CAST), Ames, Iowa.
32. P. S. Elias and A. J. Cohen, eds., *Recent Advances in Food Irradiation,* Elsevier Biomedical Press, Amsterdam, 1983.
33. J. Farkas, *Irradiation of Dry Food Ingredients,* CRC Press, Boca Raton, Fla., 1988.
34. E. S. Josephson and M. S. Peterson, eds., *Preservation of Food by Ionizing Radiation,* CRC Press, Boca Raton, Fla., 1982.
35. R. Davies, G. G. Birch, and K. J. Parker, eds., *Intermediate Moisture Foods,* Applied Science Publishers, London, 1976.
36. J. A. Troller and J. H. B. Christian, *Water Activity and Food,* Academic Press, New York, 1978.
37. E. Lueck, *Antimicrobial Food Additives: Characteristics, Uses, Effects,* Springer-Verlag, New York, 1980.
38. R. H. Tilbury, ed., *Developments in Food Preservatives,* Applied Science Publishers, London, 1980.
39. C. S. Pederson, *Microbiology of Food Fermentations,* 2nd ed., AVI Publishing Co., Westport, Conn., 1979.
40. K. H. Steinkraus, ed., *Handbook of Indigenous Fermented Foods,* Marcel Dekker, New York, 1983.
41. B. J. B. Wood, ed., *Microbiology of Fermented Foods,* Elsevier Applied Science Publishers, New York, 1985.

H. Douglas Goff
University of Guelph
Guelph, Ontario

FOOD PROCESSING

The food industry is a specialized chemical-processing industry. Individual chemical nutrients are dispersed among thousands of additional compounds arranged in highly individualistic forms characteristic of the foods consumed by various geographic, ethnic, and political population groups. These materials must be protected against damage resulting from exposure to light, heat, cold, microbes, oxygen, animal pests, bruising, and chemical contamination. Protection begins with raw materials often available only a few days each year at locations far from population centers. A product varies according to weather, local labor supplies, and political condition.

Table 1. Separation Unit Operations of the Food Industry Arranged by Mode of Separation and Phases Being Separated

Phases to Be Separated	Mode of Separation		
	Physical	Chemical	Mechanical
Gas–gas			
Gas–liquid	Condensing and blanching		Deaeration
Gas–solid	Blanching		Controlled atmosphere, deaeration, compression, and densifying
Liquid–liquid	Distillation and membrane filtration	Solvent extraction	Centrifugation
Liquid–solid	Foaming, coagulation, and drying	Ion exchange, solvent extraction, and flocculation	11 Unit operations[a]
Solid–solid	Dialysis and freeze concentration sublimation		34 Unit operations[b]

[a] Mechanical separations include churning, centrifugation, clarification, draining, expelling, filtration, flotation, pressing racking, rendering, and skimming.
[b] Mechanical separations include abrading, boning, cutting, coring, crushing, dividing, defeathering, eviscerating, flaking, finishing, filleting, grinding, harvesting, husking, hulling, inspection, milling, peeling, picking, pitting, pulverizing, slicing, sieving, sorting, shredding, sizing, sifting, shelling, scarifying, sampling, shucking, stemming, vining, and winnowing.

The technologies used to manufacture the hundreds of individual foods for a varied diet are too numerous to describe in detail. They cover the principles involved in food-process engineering, from harvesting to retail distribution, and those leading to a stable, nutritious, and safe food supply. In the last decade, genetic engineering introduced the exciting field of food biotechnology. This discussion of food processing will consider traditional food engineering and processing.

Conventional food processing, regardless of type of food, can be divided into three classes; separation, assembly, and preservation. These can occur at harvest, at the food-processing plant, or even at the point of retail sales. Increasing emphasis is being placed on field processing as exemplified by the widespread use of mechanical harvesters fitted with cleaning, sorting, color measurement, and size-grading systems. The known food-processing operations are grouped in Tables 1, 2, and 3.

Table 2. Assembly Unit Operations of the Food Industry Arranged by Mode of Assembly and Phases Being Assembled

Phases to Be Separated	Mode of Assembly		
	Physical	Chemical	Mechanical
Gas–gas			
Liquid–gas			Aerating, foaming, carbonating, and whipping
Solid–gas	Baking, puffing, and agglomeration	Humidifying, oxidizing, and proofing	Extruding and aerating
Liquid–liquid		Neutralization and acidification	Emulsifying, homogenization, dispensing, mixing and pumping
Liquid–solid	Crystallization	Soaking, malting, rehydration, and acidification	Dispersing, dissolving, immersion, mixing, and pumping
Solid–solid	Braising and roasting	Aging	7 Unit operations[a]
Gas–liquid–solid		Sprouting	Weighing and blending

[a] Mechanical assembly operations for solid–solid mixtures, include coating, enrobing, filling, forming, molding, pelleting, and stuffing.

Table 3. Preservation Unit Operations of the Food Industry Arranged by Mode of Preservation and Spoilage Vectors

Spoilage Vector	Mode of Preservation				
	Physical			Chemical	Mechanical
	Heat	Cold	Drying		
Microbes	Frying, boiling, pasteurizing, and retorting	Cooling and freezing	Dehydration, desiccation, evaporation, and lyophilization	Acidification, brining, fermentation, irradiation, isomerization, pickling, and smoking	Cleaning, centrifugation, and washing
Enzymes	Blanching, boiling, and scalding	Hydrocooling		Acidification	
Chemical	Exhausting and deodorizing	Chilling		Hydrogenation, esterification, and smoking	Cleaning, degassing, and washing
Economic pests	Pasteurization	Freezing	Desiccation	Fumigating	Cleaning, compressing, and washing
Mechanical damage		Freezing			Aspiration, cleaning, centrifugation, sorting, and washing

Foods can be categorized as living tissue, or raw foods, and nonliving tissue. Living tissue foods include fresh fruits, vegetables, meats, and grains. Processors for preserving living tissue are considered separately from those used for the preservation of nonliving tissue, eg, canned, frozen, and dried foods.

FOOD COMPOSITION AND STANDARDS

Chemical Composition

Virtually all foods are derived from living tissues, although individual nutrients and additives, including lipids, carbohydrates, amino acids, and vitamins, can be synthesized. A nutritionally adequate dietary regimen of pure nutrients would be neither economically nor aesthetically useful except for medical applications such as intravenous feeding. Foods consist of hundreds of compounds because they are derived from the life processes of living tissues. The components of food-composition studies are the nutrients that are essential to sustain life processes. Information on specific compounds in foods can be obtained from the literature, as can tables listing the nutrient composition of commodities, refined food components (eg, sucrose and gelatin), and processed as well as standardized formulated foods (eg, bread and margarine). Tables of nutrient composition are available for most foods found worldwide and contain quantitative data on moisture and caloric value (kJ or kcal × 4.184) and on lipid, protein, carbohydrate, nonnutritive fiber, ash, mineral, and vitamin contents. Tables of food composition are useful only as a first approximation to the actual nutritive value and gross chemical composition of specific foods. Foods derived from living tissue are of extremely variable. composition. The concentration of any nutrient is affected by horticultural, genetic, harvesting, handling, storage, and distribution factors. Furthermore, chemical analytical procedures may not correlate with the biological activity of available nutrients in the food. Tabular data should be used only as a first approximation of the composition and nutritive values of foods.

Functional and Conformational Data

Foods are selected not only on the basis of nutritional content but for their functional and conformational attributes, which include toughness, tenderness, and fiber content; style of cut; color and surface appearance; odor and flavor; microbial content; defects and extraneous matter; adaptability to freezing or heat treatment; genetic and varietal factors; geographical or regional production area; date of production (vintage); portion of plant or animal used; method of preservation; key ingredients; emulsifying capacity; water-binding capacity; and foaming capacity.

Food Standards and Food Grades

Food standards are legal regulations specifying the wholesomeness; labeling requirements; and, for certain foods, the composition, nutritive value, packaging, starting materials, methods of production, storage, and distribution. All foods entering interstate commerce in the United States are covered by food and drug regulations that specify levels of wholesomeness and labeling declarations and prohibit the use of certain chemicals and manufacturing practices. In addition, food and drug regulations specify the composition of a number of foods, thereby giving them a standard of identity. Foods manufactured to meet the standard of identity need not carry a list of ingredients. Manufactured foods meeting the performance characteristics of the standardized food but not the compositional requirements must be labeled as imitation. Food and drug regulations pertinent to food manufacturing-practice, nutritive requirements, standards of identity, labeling, and other regulations are available in Title 21 of the *Code of Federal Regulations*. In the United States, regulations are enforced by the Food and Drug Administration using periodic plant inspection and product sampling at all levels of production and distribution.

The regulation of the meat and poultry processing industry is under the jurisdiction of the Department of Agriculture, which has complete authority for the review, approval, and regulation of all aspects of the industry in interstate and, under certain circumstances, intrastate commerce. Detailed regulations exist for equipment, plant layout, construction, processing operations, packaging, labeling, product formulation, storage, quality assurance, and distribution practices. Regulations are enforced by resident inspectors, and those covering meat and poultry processing inspection procedures are published.

Grades and Grading

The Department of Agriculture also provides grading services for fruits, vegetables, meats, poultry, dairy products, and grains. Services may be obtained on a contractual basis for inclusive grading or for grading of selected lots of product. Grades are not mandatory as are identity standards. Grades provide a rational basis for the evaluation of the worth of processed foods for trading, financial, or contractual purposes where the buyer or seller cannot sample or inspect the products themselves. All grades and grading procedures, fees, labeling and identification procedures, and sampling plans) (including those for canned, frozen, and dried fruits and vegetables) are specified. Because changes in regulations take place frequently, the regional or the Washington, D.C., office of the FDA or the USDA should be consulted for the latest regulations.

PRESERVATION OF LIVING TISSUE

The genetically controlled taste, color, nutritive value, cell structure, and conformation of fresh foods are usually best at harvest. Protection of these qualities is achieved by retarding or inhibiting the detrimental action of microbes and enzymes, and other chemical degradation, while maintaining the integrity of the membranes, enzyme systems, and gross structure of the food. Specifically, the following must be accomplished

1. The growth rate of microbes normally associated with the product at harvest or during handling and transport must be inhibited or reduced.
2. Production of deteriorative enzymes or enzyme systems responsible for softening, color loss, or flavor changes must be inhibited or their rate of action must be reduced. Enzymes may associated with the product or with incidental microbial contamination.
3. Loss of moisture and deterioration by chemical contamination must be prevented. Volatile organic compounds generated through respiratory action or present in storage often must be removed.
4. Higher life forms, insects, mites, etc, must be inactivated or eliminated.
5. Bruising and various forms of mechanical damage must be prevented.

Processing operations required for the distribution of living tissue depends on the following factors: inherent storage potential of the tissue (this can be improved through genetic engineering, eg, by improving skin strength, resistance to microbes, or reducing respiration rate), desired storage life, degree of handling and shipping required, and intended use of the tissue (eg, further processing, food service, or retail consumption).

The intensity of processing operations is limited by the tissue itself in terms of temperature range, water activity, respiratory gas composition, mechanical stresses, and concentrations of chemicals. Temperature has the greatest effect on shelf life, because the increased activation energies (and hence the reaction rates) for microbial growth and enzyme activity are two to five times greater than those of most deteriorative chemical reactions. Except for certain fruits and vegetables, storage should be to 0°C without ice formation (Table 4) When shipping distances are great, expensive and highly perishable commodities are routinely shipped by air freight.

PRESERVATION OF NONLIVING TISSUE

Most processed foods, eg, canned, frozen, and dried products, are marketed as nonliving tissue or manufactured foods. Separation, assembly, and preservation operations determine the final quality of the product. Nonliving tissue products must be defined in terms of composition, absence of defects, nutritional value, and fill of container, rather than genetic quality

Typical processing temperatures, pressures, and pH values are shown in Table 5. Because most preservation operations allow little or no residual microbial or enzyme activity, deterioration during storage is chemical. The relatively low activation energies of deteriorative chemical reactions compared to enzyme reaction makes practical the storage of sterile foods at 25°C for months without appreciable loss of original nutrients and quality factors. Often foods are manufactured with a specified economic shelf life. Expiration dates are used to insure adequate quality while minimizing inventory costs. The effect of various processing operations on the nutritive content of foods has been reported. Measurable changes in the vitamin content of foods usually indicate significant changes in flavor, color, and structure.

HEAT TREATMENT

Populations of all life forms show characteristic death rates when exposed to elevated temperatures, pressures, and concentrations of certain chemicals. Enzymes show similar inactivation kinetics. The benefits of heat preservation, eg, the high inactivation rates of microbes and deteriorative enzymes, considerably outweigh the drawbacks, eg, heat-induced losses of desirable food nutrients losses of desirable food nutrients, structures, colors, and flavors.

Table 4. Storage Conditions and Approximate Useful Storage Life for Selected Living-Tissue Foods

Food	Temperature, °C	Relative Humidity, %	Atmosphere Composition, vol %	Useful Storage, d
Apples	0	90–95	air	100–250
	0		1–5% CO_2, 2–3% O_2	>180
Bananas	14	90–95	air	7–10
Beans, snap	6	95	air	7–10
Beets (topped)	0	>95	air	90–150
Broccoli	0	>95	air	10–14
	0	>95	5% CO_2, 3.5% O_2	40
Cabbage	0	>95	air	30–180
Carrots (topped)	0	>95	air	30–270
Celery	0	95	air	30–60
Cherries	0	>90	air	15
Corn, sweet	0	95	air	4–8
Cranberries	3	90–95	air	60–120
Cucumbers	10	90–95	air	10–14
Eggs	0	85–92	air	150–180
Grapefruit	10–15	85–90	air	40–60
Grapes	0	80–95	air	20–180
Lettuce	0	>95	air	15–20
Meat	0	95	air	3–15
Melons	7–10	90–95	air	5–15
Onions	0	60–75	air	200
Oranges	0	85–90	air	50–80
Oysters (shucked)	0	100 (wet ice)	—	7–10
Papaya	10	85–90	air	5–20
Peaches	0	90	air	15–30
Pears	0	90–95	air	60–180
Peas	0	90–95	air	5–15
Peppers green	10	92–95	air	15
Potatoes	4	90	air	150–250
Poultry	0	>95	air	15
Pumpkin	10–12	70–75	air	90
Radishes	0	>90	air	60–120
Raspberries	0	>90	air	2
Shrimp	0	ice	air	10–12
Spinach	0	>90	air	10
Strawberries	0	>90	air	5
Sweet potato	12–16	85–90	air	90–200
Tomatoes, green	12–16	90	air	15–20
Turnips	0	>90	air	100–150

The heat treatment needed to inactivate microbes or enzymes can be calculated from the energy for activation and the known rate of inactivation at a given temperature, usually 121°C. These values depend on the food system and its pH, water activity, and chemical profile. Typical activation energies and rates at 121°C for heat-resistant microbes are 209–335 kJ/mol (50–80 kcal/mol) and 0.1–10/min (one log 10 cycle reduction in 0.1–10 min). Commercial heat preservation operations assume a starting concentration of *Clostridium botulinum* spores (mixed varieties) of 10–12/g and an inactivation rate of 5–10/min at 121°C.

Preservation heating times at any temperature can be determined by integrating the lethal temperature–time effects at the slowest heating point in the package. Thus if the geometry of the container and the thermophysical properties of the food are known, it is possible to calculate the heating time necessary to insure a safe level of microbes or residual enzyme activity in the food and the residual concentration of desired nutrients. In general, higher life forms (eg, insect eggs, mites, etc) are killed by even mild heat treatments. Computer programs are available for determining safe heat-preservation operations when given information about the product, including its initial temperature, container characteristics, fill, and heat-exchange system (pure steam, steam–air, water, etc).

The hydrogen ion is extremely toxic to most microbes. Foods having a pH below 4.5 can be sterilized by heating to 100°C with a limited holding period.

Aseptic preservation involves performing separate heat preservation operations on the food and the containers prior to assembly. Liquid products can be heated and cooled rapidly under optimum conditions of heat transfer

Table 5. Typical Temperatures, Pressures, and pH Values in Food-Processing Operations

Temperature	Example of Use	Limits of Application
Heat-Transfer Media		
Cryogenic freezing		
Liquid nitrogen, −196°C	Rapid freezing to minimize cell damage, moisture loss	Cost, stress cracking during freezing
Solid carbon dioxide, −78.5°C	Rapid freezing	Cost
Air, plate, aqueous base freezants, −40 to −5°C	Commercial freezing; parasite and insect destruction, commercial freezing, storage; freeze drying	Foods not suitable for freezing
Refrigerated storage, 0–10°C	Commercial refrigerated storage	Microbial growth possible
Room temperature storage, air, 20–40°C	Canned and dry food storage	Suitable only for preserved and packaged foods
Water, air, atmospheric steam, 50–100°C	Pasteurization, milk, eggs, blanching, sterilization of acid foods (pH 4.5); air drying (product temperature); "cooking"	Excessive times will cause poor color, flavor, and structure
Steam, 110–130°C	Thermal sterilization of nonacid foods; destruction of antinutritive factors	Rapid thermal degradation of nutrients, pigments, structure, flavors
Oil, steam, infrared radiation, 180°C	Frying, roasting, baking, generation of browning reaction products	Short duration surface treatments
Various infrared radiation sources, 180°C	Surface heating for flash drying, peeling	Charring, pyrolysis without precision control
Pressure[a]		
<0.5 kPa	Freeze drying	Cost
0.5–4 kPa	Hypobaric storage, deaeration, vacuum concentration, vacuum cooling	Cost
101 kPa	Most processing operations	
0.1–1 MPa	Steam sterilization, over pressure for glass and pouch packs; carbonated and aerosol packaged foods	Cost
>1 MPa	Extruders, hydraulic pressing, homogenization, potential microbial and enzyme inactivation at 1 GPa (1,000 atm) and higher	Cost, specialized products
Hydrogen-Ion Concentration		
pH 1	Acid hydrolysis	Neutralize to pH range of 7
pH 2.5	Lemons, limes, vinegar, organic acids	Taste
pH 3–4.5	Fruits, acidified foods	Noncompatability of foods; protein denaturation
pH 7–3.5	Normal pH of most foods	
pH 8–9	Solubilization of certain proteins for extraction, alkali process cocoa	Neutralize to normal pH for use
pH 12	Limed corn	Color, flavor, nutrient loss
33% sodium hydroxide	Peeling of fruits and vegetables	Surface treatment only, neutralize to normal pH

[a] To convert kPa to mm Hg, multiply by 7.5; to convert MPa to atm, divide by 0.101.

in specialized heat exchangers. The sterile product is packaged under sterile filling conditions into sterile unit retail packages, drums, or bulk-storage tanks.

FREEZING

The rate of loss of color, flavor, structure, and nutrients in foods is a function of temperature; thus lower storage temperatures prolong the useful life of foods. However, below 0°C, the free water in food forms ice crystals as a function of moisture content, solute composition, and storage temperature. Supercooling is a transient phenomenon.

Ice formation is both beneficial and detrimental. Benefits include strengthening of structures and removal of free moisture, which reduces water activity. Benefits, however, are often far outweighed by the deleterious effects of ice crystal formation, the partial dehydration of the tissue surrounding the ice crystal, and the freeze concentration of potential reactants. Ice crystals disrupt cell structures mechanically, and the increased concentration of cell electrolytes can result in the chemical denaturation of proteins.

The technology of food freezing emphasizes as short a passage time through the temperature zone of maximum ice crystal formation as possible. The formation of as small an ice crystal as possible minimizes the mechanical disruption of cells and possibly reduces the effects of solute concentration damage. Rapid freezing can only be accomplished by large temperature differences and high heat transfer coefficients.

Many processed foods and certain animal products tolerate freezing and thawing because their structures can accommodate ice crystallization, movement of water, and the related changes in solute concentrations. Starches can be modified to form gels that accept several freezing and thawing cycles without breakdown. By contrast, most fruits and vegetables lose significant structural quality on freezing because their rigid cell structures fail to accommodate to ice crystal formation. However, it is not possible to store foods at temperatures low enough to insure complete conversion of all water to ice; as a result, commercial frozen food storage temperatures represent an economic balance between storage costs (time, energy, and capital investment) and projected shelf life.

The freezing process disrupts tissue structures allows cell contents to become mixed so that undesirable enzyme reactions can take place at significant rates even at storage of $-18°C$. These reactions can generate off-flavors, reduce nutrient concentrations, and cause major changes in the structure and appearance of foods. The amount of free liquid or drip found after a freeze–thaw cycle is a good indication of the structural damage.

Heat treatment (blanching) prior to freezing eliminates enzyme-mediated changes in color, flavor, and structure. Most deteriorative enzymes are inactivated by exposure to a temperature of $100°C$ for 1–5 min. The enzymatic oxidative deterioration of frozen fruits can be inhibited with sulfur dioxide, sucrose, and combinations of citric acid, sodium chloride, and ascorbic acid (preceded by vacuum removal of oxygen if heat is not used).

Most frozen foods have a useful storage life of one year at $-18°C$ however, foods high in fat, eg, sausage products, may become rancid in two weeks. Frozen storage can result in moisture loss from the food through a freeze-drying process, because the heat-transfer surfaces used to maintain storage temperatures are at a lower temperature than the storage area. For this reason, frozen foods must be protected against drying by a moisture barrier. In addition, foods subject to oxidative deterioration must be protected from air.

Freezing-preservation equipment can be classified by method of heat-transfer. Usually air is the heat-exchange medium for freezing foods. Foods are loaded on a belt or vibrating conveyor and passed through air flowing upward at up to 5 m/s at temperatures as low as $-40°C$. The air is recycled through coils and fans located next to the conveyor and returned through the conveyor. Because the air has a partial water vapor pressure lower than the food, freeze-drying can occur, and some of the water in the product is removed and deposited as ice on the heat-exchange coils. As a freezing medium, air has other drawbacks. The low gas–solid heat-transfer coefficient and the heat capacity of air require either low temperatures or high velocities to obtain needed high heat fluxes. Low air temperatures increase refrigeration costs and high air velocities generate additional fan heat loads. For these and other reasons, freezing by conduction or by liquid heat-transfer methods are less costly in terms of capital investment and energy.

Liquid heat-transfer media that are used for direct-immersion freezing include food-grade dichlorodifluromethane, nitrous oxide, and water solutions of various edible salts, sugars, alcohols, acid, and esters. Liquid heat-transfer agents offer a high–heat-transfer coefficient and reduced pumping costs, eliminate product desiccation, and allow operation at high low side equipment temperatures. Drawbacks include possible changes in food flavor and costs of processing. Although dichlorodifluromethane offers major operating advantages as compared to other heat-transfer media, cost and environmental concerns have reduced its potential usefulness as an ideal direct-immersion freezant. There is a need for a direct-immersion liquid freezant that is safe, low cost, thermodynamically efficient, and compatible with foods with respect to flavor, color, and odor.

Conduction freezing between chilled plates is a cost-effective method of heat removal provided the product can be assembled in a geometry compatible with the plate surfaces. Packages having semi-infinite slab geometry are loaded between stacks of platens through which refrigerant is circulated. Good heat transfer is maintained by maintaining a pressure on the stack of platens.

Other freezing methods use direct immersion in liquid nitrogen, exposure to solidified carbon dioxide, and immersions of packaged products in liquid freezants, eg, sodium or calcium chloride brines, methanol, or propylene glycol solutions.

The quality of frozen food is related to storage temperature; however, because constant storage temperatures are not always feasible, the shelf life is often determined by the highest temperature and total length of time that food is exposed to that temperature before use. Maintaining a $-18°C$ storage environment from time of freezing until use continues to be a major technical problem facing the frozen-food industry.

REDUCTION OF WATER ACTIVITY (DEHYDRATION)

Microbes require a specific minimum level of water activity a/w, defined as the relative humidity (measured in equilibrium with the food) for growth and reproduction at a given temperature and substrate composition. Foods possess characteristic equilibrium relationships between water activity and moisture content at given temperatures. Preservation against microbial spoilage by dehydration requires a moisture content equal to a water activity below 0.65. This contrasts with food concentration where water is removed for reasons other than for effective reduction in water activity (Fig. 1).

Living tissue, when dried to a water activity below

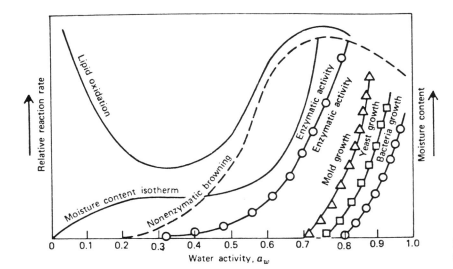

Figure 1. Relative rate of food deteriorative factors as a function of water activity a_w.

0.97, suffers irreversible disruption of metabolic processes. Deteriorative chemical reactions, enzyme-catalyzed or not, are generally a function of water activity and reactant concentration. Thus nonenzymatic browning (Maillard reaction), oxidation, and internal rearrangements (eg, staling and protein cross-linking) can increase in rate as water activity is reduced. Many reactions show a minimum rate in the range of $a_w = 0.4$. The role of water activity and product stability has been reported.

Prior to dehydration, foods are usually heat treated to inactivate enzyme systems. Those foods susceptible to rapid nonenzymatic browning resulting from high concentrations of reducing sugars are treated with sulfur dioxide, and products subject to oxidative rancidity can be treated with antioxidants and packaged to prevent exposure to oxygen. Low temperature storage (5°C) reduces chemical deterioration. Blends of dehydrated products must be assembled from ingredients having the same water activity.

High-quality dried foods can be obtained only if the drying system is designed to match the rate of release of moisture from the food at a predetermined maximum wet-bulb temperature. Typically, continuous-belt dryers are staged to provide three or more zones into which the product is repiled into progressively deeper beds. Each zone operates at a dry- and wet-bulb temperature, a throughflow (upflow and downflow through a bed of materials) air velocity, and a bed depth that optimize product quality, energy use, and production rate.

Liquids and pastes are commonly dried in spray, drum, or freeze dryers. Particulate foods can be dried in continuous conveyor systems, fluidized beds, freeze dryers, or batch tunnel systems. Some fruits, salted animal products, nuts, berries, and many field crops are sun-dried.

MECHANICAL PRESERVATION METHODS

Mechanical food preservation methods are characterized by separation operations, which tend to affect food quality factors less than other methods, as exemplified by the production of sterile draft beer by filtration.

Packaging is the primary mechanical preservation method; barrier materials protect preserved foods from external spoilage factors during storage and distribution. Other mechanical preservation methods include sieving, filtration, air classification, washing, sorting, ultrafiltration, centrifugation, extraction, stripping, flotation, etc. Size grading, defect sorting, and washing are routinely used to reduce or eliminate microbes, enzymes, chemicals, insects, and other matter. Virtually all food materials require repetition of these operations between harvest and preservation.

Washing equipment is designed to accommodate the type of food being washed. Leafy vegetable washers, for example, use a series of rotating paddles that transfer the material along the water surface in a tank so that sand can settle to the bottom of the tank. Fruit washers have reels or belts fitted with high-pressure water sprays the mechanically loosen dirt and debris. Recently, a new type of washer was developed for tomatoes, tree fruit, and root vegetables. The unit has spinning soft-rubber disks that are mounted in rows to give an intensive wiping action at the surface of the food as it rolls over the disk array. Disk washers reduce water consumption in tomato washing from 1,500 L/t (360 gal/short ton) to as little as 20 L/t (5 gal/short ton). Other food washing equipment use combinations of sprays, brushes, and mechanical agitators. For example, froth flotation removes foreign particulate organic matter from cut corn and peas. Approved wetting agents, eg, sodium lauryl sulfate, are often incorporated in wash water to improve the efficiency of the washing operation. The proper chlorination of wash and transport water also is extremely important in maintaining low levels of microbes. Break-point chlorination insures adequate but not excessive residual free chlorine (because >5 ppm of free chlorine can spoil flavors of some foods). This method of chlorination is implemented by adding sufficient chlorine to oxidize organic materials in the water and leave a residual concentration to insure antimicrobial activity. Water used in the final washing of food materials processed under FDA jurisdiction must be suitable for drinking.

Liquid foods free of suspended solids may be sterilized

by filtering spoilage microbes. The filtered product, eg, wine, beer, and certain fruit juices, is packaged under aseptic conditions in a sterilized container. Centrifugation can be used to reduce microbial contamination and to remove extraneous organic matter in liquid foods such as milk.

Ultrahigh pressure causes the death of microbes and inactivates enzymes at a rate proportional to concentration and temperature. Significant rates have been achieved experimentally at temperatures only slightly above 25°C; however, no commercial applications have been reported.

CHEMICAL PRESERVATION METHODS

Addition of chemicals usually occurs in conjunction with other preservation methods because they enhance the effectiveness of heat, refrigeration, drying, or packaging. Examples include the addition of antioxidants to fried foods to reduce the need for more expensive oxygen-impermeable flexible packaging and the use of sulfur dioxide to retard the development of brown color in frozen or dried apples.

The extreme toxicity of the hydrogen ion toward food-spoilage microbes and its tolerance by humans has made it a preferred food preservative. Studies have shown that various organic acids exert a strong inhibitory effect on the growth of microbes beyond the activity of the hydrogen ion itself. Lactic, propionic, acetic, sorbic, and benzoic acids; their sodium or potassium salts; and certain derivatives of benzoic acid find extensive use as yeast, mold, and bacterial inhibitors in bread, beverages, and sauces. Combinations of hydrogen ion, organic acids, mild heat treatment, and exclusion of oxygen have allowed the simple and safe preservation of foods not tolerant of conventional heat sterilization above 100°C, freezing, or dehydration.

Traditional food-preservation methods were based largely on drying in conjunction with preservation with locally available chemical preservatives. These included salt, organic acids generated by a lactic or acetic acid fermentation (as in pickling), wood smoke, and ethanol derived from the fermentation of substances containing sugar. Many traditionally preserved products are available in the marketplace, although modern preservation methods predominate because they are convenient.

Chemical preservatives and chemical preservation methods used in conjunction with other preservation methods for meat and poultry are controlled by USDA regulations. The FDA has jurisdiction over the use of chemical preservatives in other foods. Additives are reviewed continually for heat safety.

The FDA has reviewed the use of ionizing radiation as a food preservation method since the 1960s. Ionizing radiation is considered to be a chemical preservation method and must be cleared for use not only on a product-by-product basis but also on a dosage basis. In the United States, a dose of up to 100 Gy (10,000 rad) may be used to inhibit potato sprouting, a dose of as much as 500 Gy may be used to destroy insect life in grain products. There have been feeding studies to obtain clearance for beef sterilization using doses up to 6×10^4 Gy.

TOXICOLOGICAL IMPLICATIONS OF FOOD DETERIORATION AND SPOILAGE

A primary objective food processing is to insure a safe food supply. Certain harvested foods can contain naturally occurring poisonous substances, compounds with pharmacological effects, or compounds that can interfere with the utilization of nutrients present in other foods when they are consumed together.

Foods just harvested may contain or become contaminated with spore or vegetative forms of microbes, parasites, or the wastes of higher life forms. Careless posthavest handling, storage, processing, and distribution operations provide additional opportunities for microbial growth or contamination. Microbial growth can result in the release of toxins, the consumer has no way of identifying contaminated foods and, consequentially, food-poisoning outbreaks occur. Fortunately, some toxins are heat sensitive and are inactivated during the cooking process. However, only good manufacturing, handling, and distribution practices insure the prevention of microbial growth. These practices must be carried out from the time of harvest to the time of consumption.

A third source of potential health hazards in food supplies are toxic compounds formed during the chemical degradation of processed and stored foods. Oxidative breakdown products of unsaturated fatty acids have been implicated as potential carcinogens, eg, malonaldehyde.

In addition to the above, foods must be protected from contamination by pesticides and other chemicals used for crop protection; from exposure to hazardous chemicals and extraneous materials, eg, asbestos in air and water supplies; and radioactive isotopes. Many toxic compounds can be found as soil contaminants or as contaminants in lakes, oceans, or ground water supplies and can be concentrated in the edible tissues of plants and animals. The most effective method to insure the absence of these hazards is chemical analysis of representative samples of the raw material prior to purchase. Analyses can lead to prevention of contamination above prescribed levels by heavy metals (mercury and lead), DDT, polychlorinated biphenyls (PCBs), S-90, growth-promoting estrogens in meat, and other contaminants.

FOOD-PROCESSING FACILITIES

Factors important in the selection of sites for food-processing operations include availability of raw materials, abundant potable water supply, low-cost waste disposal facilities, adequate low-cost energy supply, adequate seasonal and nonseasonal labor supply, ease of access to rail and truck transportation, proximity to consumers, and adequate storage areas. The relative importance of each of these factors depends on the length of the processing season, the ease of storage and handling of raw materials, the perishability of the finished products, and the complexity of the processing operation.

Highly perishable salad vegetables can be trimmed, inspected, bagged, and vacuum cooled using mobile equipment in or at the edge of the growing area. Processing of frozen french fries requires raw potato storage capacities of up to 10 months; thus plants tend to be located near principal potato-growing areas, eg, the northwestern United States. Continued developments in mechanical harvesting and mobile harvester separation operations for fruits and vegetables will cause further processing plant shifts toward concentrated growing areas where irrigation, climate, and level terrain make mobile processing feasible.

As processing technology improves, an increasing number of plants are relocating away from eastern U.S. population centers toward sources of raw materials. This shift is particularly noticeable in the meat industry and has occurred concurrently with the development of vacuum-packed cut beef, which provides more than a 50% reduction in storage and shipment space over conventional quarter or sides.

Certain food products are highly perishable, have marginal value or have only regional acceptance. Dairy products, bread, soft drinks, beer, glass-packed items, and certain cured-meat products are usually manufactured from more stable raw materials or are locally produced starting materials and are distributed daily. With such distribution, production schedules are determined by local weather forecasts, by day of the week and holidays, and by seasonal factors so that returns of stale or overage products can be minimized. Considerable progress has been made in the analysis of perishable-food distribution practices. A universal product code applicable to all processed foods for retail sale will further improve distribution efficiencies by allowing automatic retail checkout and continuous inventory review.

EQUIPMENT STANDARDS

Standards for food-processing equipment are specified by the USDA in the dairy, meat, poultry, and egg processing industries. Equipment and processing methods used in the manufacturing of other foods are also covered. Where standards exist, eg, 3-A Sanitary Standards, they have taken the form of specifications for the design of food-processing equipment and for certain operational practices. Requirements state that surfaces must not harbor food materials or prevent the easy removal of these materials during the cleaning process; there should be no openings where food can accumulate and allow microbial growth, particularly openings not easily cleaned during routine equipment cleaning operations; equipment must be capable of visual inspection throughout the food contact area and must be easily disassembled (if necessary) for cleaning; all materials of construction that contact the food must be approved for use, or be inert to migration into the food and *vice versa*; interior equipment surfaces should be self-draining; and exterior surfaces should be easily cleaned and free of openings where material may accumulate.

The FDA does not approve specific pieces of food-processing equipment but provides approval of a process or equipment after reviewing pertinent information. Meat and poultry processing operations must have specific Food Safety and Quality Service (USDA) approved equipment and processes. The USDA approves specific pieces of equipment meeting their design, operating, and sanitation requirements.

FOOD PACKAGING

Glass, woven, natural fiber, and ceramic containers have been used to store food since antiquity. However, the invention of time-plated steel containers (ca 1800) provided the food industry with its first functional and disposable packaging system. Advances in metal container manufacturing continue to maintain the status of the cylindrical can as the primary package for heat-processed, shelf-stable foods and beverages. Notable developments include substitution of electrolytic time plating for hot dipping, new organic linings to greatly reduce the need for time, reduced steel thickness, and improved steel composition for better strength and corrosion resistance. Developments in coating technology have resulted in the availability of two-piece drawn aluminum and steel containers and of numerous easy-open systems. New glass containers have increased strength and have continued to reduce package weight and cost.

Rigid glass or metal containers have a major processing advantage: they can be filled and sealed at speeds in excess of 1,000/min. Thus the relatively higher container cost can be recovered to some degree by higher production rates. Their rigidity (ie, sturdiness) facilitates stacking and handling during distribution and storage.

Paper, plastics, and aluminum have provided the food industry with container material that can be tailored to specific barrier needs. Ease of forming, sealing, opening, and decoration; strength; and low weight have made them preferred materials for refrigerated, dry, and frozen products.

Resistance to thermal treatment and desired barrier and mechanical qualities are obtained by laminating or coextruding appropriate combinations of paper, aluminum foil, and plastic. Lamination and coextrusion seal transparent packages against moisture and gas permeation, and strengthen the packaging.

Pouches formed by heat-sealing paper–foil–plastic laminates on three sides, filling, and then heat-sealing the fourth side have found extensive use with moisture- and oxygen-sensitive dry foods as well as with high-moisture thermally processed foods. Pouches capable of withstanding saturated steam or steam–air mixtures at 121°C for several hours are finding greater use as substitutes for metal and glass containers. The advantage of pouches for heat sterilization is their slab rather than cylinder shape. Their thinner cross section allows a shorter heat-processing time than cylindrical containers of the same capacity. Mechanical advances in forming, filling, sealing, and handling pouches will result in continued substitution of laminates for rigid metal and glass containers.

Plastics and plastic-coated papers are widely used for beverages including carbonated drinks. Several aseptic systems are marketed for bulk packing of liquids in multiliter plastic bags.

Semirigid containers are also gaining increased acceptance, particularly as replacements for the #10 can (2.72 kg or 96 oz). Initial developments have produced a lightweight drawn-metal container measuring 24 × 30 × 6 cm with a double-seamed top. The thickness of this container (6 cm) is such that during heat sterilization heat travels only 3 cm by conduction vs about 8 cm in the conventional #10 can. Shorter heat treatments allow the preservation of a greater variety of speciality products. The half-steam-table size eliminates cleaning of pans, because the opened unit serves as a serving tray. Other semirigid systems have been developed to replace the steel can for single-portion service. These containers generally use a heat-sealed or glued paper–foil–plastic laminated closure.

Aerosol containers have played a specialized role in the food industry as dispensers for such foods as whipped toppings (using nitrous oxide as a propellant), cake frosting, and barbecue sauces. Special filling techniques and product formulations are required because the aerosol package cannot be heat sterilized once it has been sealed.

Packaging forms an integral part of most food processing. For this reason, studies of storage conditions over a range of temperatures, relative humidities, and handling conditions must be made to determine the suitability of a package. Packaging in contact with foods must be tested for migration of packaging components under actual conditions of use, and pickup of off-flavor must be evaluated on a product-by-product basis. Because of potential migration of packaging materials into foods, current food and drug regulations should be consulted prior to marketing foods packaged in nonstandard materials.

This article was adapted from M. Grayson, ed., *Kirk-Othmer, Encyclopedia of Chemical Technology,* Vol. 11, 3rd ed., John Wiley & Sons, Inc., New York, 1979. Note: All specific citations have been removed. A user must refer to the original text to obtain specific references.

General References

P. Jelen, *Introduction to Food Processing,* Prentice-Hall Press, Englewood Cliffs, N.J., 1989.

M. Le Maguer and P. Jelen, eds., *Food Engineering and Process Applications, Vol. II: Unit Operations,* Elsevier Science Publishing Co., Inc., New York, 1986.

M. Le Maguer and P. Jelen, *Food Engineering and Process Applications, Vol. 1: Transport Phenomena,* Elsevier Science Publishing Co., Inc., New York, 1986.

D. MacCarthy, ed., *Concentration and Drying of Foods,* Elsevier Science Publishing Co., Inc., New York, 1986.

M. Mathlouthi, ed., *Food Packaging and Preservation,* Elsevier Science Publishing Co., Inc., New York, 1986.

B. M. McKenna, ed., *Engineering and Food, Vol. I: Engineering Sciences in the Food Industry,* Elsevier Science Publishing Co., Inc., New York, 1984.

B. M. McKenna, ed., *Engineering and Food, Vol. II: Processing Applications,* Elsevier Science Publishing Co., Inc., New York, 1984.

N. N. Porter, *Food Science,* Van Nostrand Reinhold Co., Inc., New York, 1986.

R. P. Singh and D. R. Heldman, *Introduction to Food Engineering,* Academic Press, Inc., Orlando, Fla.

E. L. Watson, *Elements of Food Engineering,* Van Nostrand Reinhold Co., Inc., New York, 1988.

Y. H. HUI
EDITOR-IN-CHIEF

FOOD PROCESSING: STANDARD INDUSTRIAL CLASSIFICATION

When we are in a grocery store, we are faced with an incredible selection of processed foods. For food manufacturers, this illustrates only one point: the amount of research and development that precedes their marketing. For regulatory agencies such as the Food and Drug Administration and the Department of Agriculture, this confirms the amount of work needed to assure the safety and wholesomeness of these products. However, for some other federal agencies, this poses an entirely different headache. For example, the Department of Labor (DOL) and the Environmental Protection Agency (EPA) must classify and give a name to each establishment that manufactures each product in order to enforce their legal mandates. To them it is a huge logistic nightmare. For example, the DOL must have access to a record of the number of employees suffering injuries in each type or category of food processing plants. To fill this need, the executive branch of the government has developed the Standard Industrial Classification (SIC) for food processing plants. The *SIC Manual* is one of the most useful federal publications available to both government agencies and the industries. One example of its usefulness is in assisting the allocation of resources. For example, a government agency can plan its budget accordingly by using the SIC to determine the number of food processing plants it has to inspect annually. Another application is the use of SIC to determine the employment status of the food processing profession. Such classification applies only to food processing establishments in the United States and this article discusses the SIC in details.

STANDARD INDUSTRIAL CLASSIFICATION MANUAL

The SIC Manual is issued by the Office of Management and Budget (OMB) of the Executive Office of the President of the United States. The manual divides industries in the United States into major groups. One major group 20 covers food and kindred products. This group includes establishments manufacturing or processing foods and beverages for human consumption, and certain related products, such as manufactured ice, chewing gum, vegetable and animal fats and oils, and prepared feeds for animals and fowls. Products described as dietetic are classified in the same manner as nondietetic products, eg, as candy, canned fruits, and cookies. Chemical sweeteners are classified in Major Group 28.

Major Group 20 is subdivided according to the following criteria (Table 1):

Table 1. Standard Industrial Classification (SIC) and Subdivisions for Major Group 20: Food and Kindred Products

Industrial Group No.: Industrial Group Name
    Industrial No.: Industrial Name

201: Meat products
    2011: Meat-packing plants
    2013: Sausages and other prepared meat products
    2015: Poultry slaughtering and processing

Table 1. *(continued)*

Industrial Group No.: Industrial Group Name
Industrial No.: Industrial Name

202: Dairy products
  2021: Creamery butter
  2022: Natural, processed, and imitation cheese
  2023: Dry, condensed, and evaporated dairy products
  2024: Ice cream and frozen desserts
  2026: Fluid milk
203: Canned, frozen, and preserved fruits, vegetables, and food specialties
  2032: Canned specialties
  2033: Canned fruits, vegetables, preserves, jams, and jellies
  2034: Dried and dehydrated fruits, vegetables, and soup mixes
  2035: Pickled fruits and vegetables, vegetable sauces and seasonings, and salad dressings
  2037: Frozen fruits, fruit juices, and vegetables
  2038: Frozen specialties, not elsewhere classified
204: Grain mill products
  2041: Flour and other grain mill products
  2043: Cereal breakfast foods
  2044: Rice milling
  2045: Prepared flour mixes and doughs
  2046: Wet corn milling
  2047: Dog and cat food
  2048: Prepared feeds and feed ingredients for animals and fowls, except dogs and cats
205: Bakery products
  2051: Bread and other bakery products, except cookies and crackers
  2052: Cookies and crackers
  2053: Frozen bakery products, except bread
206: Sugar and confectionery products
  2061: Cane sugar, except refining
  2062: Cane sugar refining
  2063: Beet sugar
  2064: Candy and other confectionery products
  2066: Chocolate and cocoa products
  2067: Chewing gum
  2068: Salted and roasted nuts and seeds
207: Fats and oils
  2074: Cottonseed oil mills
  2075: Soybean oil mills
  2076: Vegetable oil mills, except corn, cottonseed, and soybean
  2077: Animal and marine fats and oils
  2079: Shortening, table oils, margarine, and other edible fats and oils, not elsewhere classified
208: Beverages
  2082: Malt beverages
  2083: Malt
  2084: Wines, brandy, and brandy spirits
  2085: Distilled and blended liquors
  2086: Bottled and canned soft drinks and carbonated waters
  2087: Flavoring extracts and flavoring syrups, not elsewhere classified
209: Miscellaneous food preparations and kindred products
  2091: Canned and cured fish and seafoods
  2092: Prepared fresh or frozen fish and seafoods
  2095: Roasted coffee
  2096: Potato chips, corn chips, and similar snacks
  2097: Manufactured ice
  2098: Macaroni, spaghetti, vermicelli, and noodles
  2099: Food preparations, not elsewhere classified

The following discusses each industry under food and kindred products. The abbreviations used are: * = The item has been made in the same establishment as the basic material; ** = The item has been made from purchased materials or materials transferred from another establishment; IGN = Industry Group No; IN = Industry No.

MEAT PRODUCTS (IGN 201)

Meat Packing Plants (IG 2011)

Establishments primarily engaged in the slaughtering, for their own account or on a contract basis for the trade, of cattle, hogs, sheep, lambs, and calves for meat to be sold or to be used on the same premises in canning, cooking, curing, and freezing, and in making sausage, lard, and other products. Also included in this industry are establishments primarily engaged in slaughtering horses for human consumption. Establishments primarily engaged in slaughtering, dressing, and packing poultry, rabbits, and other small game are classified in Industry 2015; and those primarily engaged in slaughtering and processing animals not for human consumption are classified in Industry 2048. Establishments primarily engaged in manufacturing sausages and meat specialties from purchased meats are classified in Industry 2013; and establishments primarily engaged in canning meat for baby food are classified in Industry 2032.

1. Bacon, slab and sliced*
2. Beef*
3. Blood meal
4. Boxed beef*
5. Canned meats, except baby foods and animals feeds*
6. Corned beef*
7. Cured meats*
8. Dried meats*
9. Frankfurters, except poultry*
10. Hams, except poultry*
11. Hides and skins, cured or uncured
12. Horse meat for human consumption*
13. Lamb*
14. Lard*
15. Luncheon meat, except poultry*
16. Meat extracts*
17. Meat packing plants
18. Meat*
19. Mutton*
20. Pork*
21. Sausages*
22. Slaughtering plants: except animals not for human consumption
23. Variety meats edible organs*
24. Veal*

Sausages and Other Prepared Meat Products (IN = 2013)

Establishments primarily engaged in manufacturing sausages, cured meats, smoked meats, canned meats, frozen meats, and other prepared meats and meat specialties, from purchased carcasses and other materials. Prepared meat plants operated by packing houses as separate establishments are also included in this industry. Establishments primarily engaged in canning or otherwise processing poultry, rabbits, and other small game are classified in Industry 2015. Establishments primarily engaged in canning meat for baby food are classified in Industry 2032. Establishments primarily engaged in the cutting up and resale of purchased fresh carcasses, for the trade, (including boxed beef) are classified in Wholesale Trade, Industry 5147.

1. Bacon, slab and sliced**
2. Beef**
3. Bologna**
4. Calf's-foot jelly
5. Canned meats, except baby foods and animal feeds**
6. Corned beef**
7. Corned meats**
8. Cured meats: brined, dried, and salted**
9. Dried meats**
10. Frankfurters, except poultry**
11. Hams, except poultry**
12. Headcheese**
13. Lard**
14. Luncheon meat, except poultry**
15. Meat extracts**
16. Meat products: cooked, cured, frozen, smoked, and spiced**
17. Pastrami**
18. Pigs' feet, cooked and pickled**
19. Pork: pickled, cured, salted, or smoked**
20. Potted meats**
21. Puddings, meat**
22. Sandwich spreads, meat**
23. Sausage casings, collage
24. Sausages**
25. Scrapple**
26. Smoked meats**
27. Spreads, sandwich: meat**
28. Stew, beef and lamb**
29. Tripe**
30. Vienna sausage**

Poultry Slaughtering and Processing (IN 2015)

Establishments primarily engaged in slaughtering, dressing, packing, freezing, and canning poultry, rabbits, and other small game, or in manufacturing products from such meats, for their own account or on a contract basis for the trade. This industry also includes the drying, freezing, and breaking of eggs. Establishments primarily engaged in cleaning, oil treating, packing, and grading of eggs are classified in Wholesale Trade, Industry 5144; and those engaged in the cutting up and resale of purchased fresh carcasses are classified in Wholesale and Retail Trade.

1. Chickens, processed: fresh, frozen, canned, or cooked
2. Chickens: slaughtering and dressing
3. Ducks, processed: fresh, frozen, canned, or cooked
4. Ducks: slaughtering and dressing
5. Egg albumen
6. Egg substitutes made from eggs
7. Eggs: canned, dehydrated, desiccated, frozen, and processed
8. Eggs: drying, freezing, and breaking
9. Frankfurters, poultry
10. Game, small: fresh, frozen, canned, or cooked
11. Game, small: slaughtering and dressing
12. Geese, processed: fresh, frozen, canned, or cooked
13. Geese: slaughtering and dressing
14. Ham, poultry
15. Luncheon meat, poultry
16. Poultry, processed: fresh, frozen, canned, or cooked
17. Poultry: slaughtering and dressing
18. Rabbits, processed: fresh, frozen, canned, or cooked
19. Rabbits, slaughtering and dressing
20. Turkeys, processed: fresh, frozen, canned, or cooked
21. Turkeys: slaughtering and dressing

DAIRY PRODUCTS (IGN 202)

This industry group includes establishments primarily engaged in:

1. Manufacturing creamery butter; natural, processed, and imitation cheese; dry, condensed, and evaporated milk; ice cream and frozen dairy desserts; and special dairy products, such as yogurt and malted milk; and
2. Processing (pasteurizing, homogenizing, vitaminizing, bottling) fluid milk and cream for wholesale or retail distribution.

Independently operated milk-receiving stations primarily engaged in the assembly and reshipment of bulk milk for use in manufacturing or processing plants are classified in Industry 5143.

Creamery Butter (IN 2021)

Establishments primarily engaged in manufacturing creamery butter.

1. Anhydrous butterfat
2. Butter oil

3. Butter powder
4. Butter, creamery and whey
5. Butterfat, anhydrous

Natural, Processed, and Imitation Cheese (IN 2022)

Establishments primarily engaged in manufacturing natural cheese (except cottage cheese), processed cheese, cheese foods, cheese spreads, and cheese analogs (imitations and substitutes). These establishments also produce byproducts, such as raw liquid whey. Establishments primarily engaged in manufacturing cottage cheese are classified in Industry 2026, and those manufacturing cheese-based salad dressings are classified in Industry 2035.

1. Cheese analogs
2. Cheese products, imitation or substitutes
3. Cheese spreads, pastes, and cheeselike preparations
4. Cheese, except cottage cheese
5. Cheese, imitation or substitutes
6. Cheese, processed
7. Dips, cheese based
8. Processed cheese
9. Sandwich spreads, cheese
10. Whey, raw: liquid

Dry, Condensed, and Evaporated Dairy Products (IN 2023)

Establishments primarily engaged in manufacturing dry, condensed, and evaporated dairy products. Included in this industry are establishments primarily engaged in manufacturing mixes for the preparation of frozen ice cream and ice milk and dairy- and nondairy-based cream substitutes and dietary supplements.

1. Baby formula: fresh, processed, and bottled
2. Buttermilk: concentrated, condensed, dried, evaporated, and powdered
3. Casein, dry and wet
4. Cream substitutes
5. Cream: dried, powdered, and canned
6. Dietary supplements, dairy and nondairy based
7. Dry milk products: whole milk, nonfat milk, buttermilk, whey, and cream
8. Eggnog, canned: nonalcoholic
9. Ice cream mix, unfrozen: liquid or dry
10. Ice milk mix, unfrozen: liquid or dry
11. Lactose, edible
12. Malted milk
13. Milk, whole: canned
14. Milk: concentrated, condensed, dried, evaporated, and powdered
15. Milkshake mix
16. Skim milk: concentrated, dried, and powdered
17. Sugar of mix
18. Whey: concentrated, condensed, dried, evaporated, and powdered
19. Whipped topping, dry mix
20. Yogurt mix

Ice Cream and Frozen Desserts (IN 2024)

Establishments primarily engaged in manufacturing ice cream and other frozen desserts. Establishments primarily engaged in manufacturing frozen bakery products, such as cakes and pies, are classified in Industry 2053.

1. Custard, frozen
2. Desserts, frozen: except bakery
3. Fruit pops, frozen
4. Ice cream: eg, bulk, packaged, molded, on sticks
5. Ice milk: eg, bulk, packaged, molded, on sticks
6. Ices and sherbets
7. Juice pops, frozen
8. Millorine
9. Parfait
10. Pops, dessert: frozen-flavored ice, fruit, pudding, and gelatin
11. Pudding pops, frozen
12. Sherbets and ices
13. Spumoni
14. Tofu frozen desserts
15. Yogurt, frozen

Fluid Milk

Establishments primarily engaged in processing (eg, pasteurizing, homogenizing, vitaminizing, bottling) fluid milk and cream, and related products, including cottage cheese, yogurt (except frozen), and other fermented milk. Establishments primarily engaged in manufacturing dry mix whipped toppings are classified in Industry 2023; those producing frozen whipped toppings are classified in Industry 2038; and those producing frozen yogurt are classified in Industry 2024.

1. Buttermilk, cultured
2. Chocolate milk
3. Cottage cheese, including pot, bakers', and farmers' cheese
4. Cream, aerated
5. Cream, bottled
6. Cream, sour
7. Dips, sour cream based
8. Eggnog, fresh: nonalcoholic
9. Flavored milk drinks
10. Half and half
11. Milk processing (pasteurizing, homogenizing, vitaminizing, bottling)
12. Milk production, except farm
13. Milk, acidophilus
14. Milk, bottled

15. Milk, flavored
16. Milk, reconstituted
17. Milk, ultrahigh temperature
18. Sour cream
19. Whipped cream
20. Whipped topping, except frozen or dry mix
21. Yogurt, except frozen

CANNED, FROZEN, AND PRESERVED FRUITS, VEGETABLES, AND FOOD SPECIALTIES (IGN 203)

The canned products of this industry group are distinguished by their processing rather than by the container. The products may be shipped in bulk or in individual cans, bottles, retort pouch packages, or other containers.

Canned Specialties (IG 2032)

Establishments primarily engaged in canning specialty products, such as baby foods, nationality speciality foods, and soups, except seafood. Establishments primarily engaged in canning seafoods are classified in Industry 2091.

1. Baby foods (including meats), canned
2. Bean sprouts, canned
3. Beans, baked: with or without meat—canned
4. Broth, except seafood: canned
5. Chicken broth and soup, canned
6. Chili con carne, canned
7. Chinese foods, canned
8. Chop suey, canned
9. Chow mein, canned
10. Enchiladas, canned
11. Food specialties, canned
12. Italian foods, canned
13. Macaroni, canned
14. Mexican foods, canned
15. Mincemeat, canned
16. Nationality specialty foods, canned
17. Native foods, canned
18. Pasta, canned
19. Puddings, except meat: canned
20. Ravioli, canned
21. Soups, except seafood: canned
22. Ravioli, canned
23. Soups, except seafood: canned
24. Spaghetti, canned
25. Spanish foods, canned
26. Tamales, canned
27. Tortillas, canned

Canned Fruits, Vegetables, Preserves, Jams, and Jellies (IN 2033)

Establishments primarily engaged in canning fruits, vegetables, and fruit and vegetable juices; and in manufacturing catsup and similar tomato sauces, or natural and imitation preserves, jams, and jellies. Establishments primarily engaged in canning seafoods are classified in Industry 2091; and those manufacturing canned specialties, such as baby foods and soups, except seafood, are classified in Industry 2032.

1. Artichokes in olive oil, canned
2. Barbecue sauce
3. Catsup
4. Cherries, maraschino
5. Chili sauce, tomato
6. Fruit butters
7. Fruit pie mixes
8. fruits, canned
9. Hominy, canned
10. Jams, including imitation
11. Jellies, edible: including imitation
12. Juice, fruit: concentrated-hot pack
13. Juices, fresh: fruit or vegetable
14. Juices, fruit and vegetable: canned or fresh
15. Ketchup
16. Marmalade
17. Mushrooms, canned
18. Nectars, fruit
19. Olives, including stuffed: canned
20. Pastes, fruit and vegetable
21. Preserves, including imitation
22. Purees, fruit and vegetable
23. Sauces, tomato based
24. Sauerkraut, canned
25. Seasonings (prepared sauces), tomato
26. Spaghetti sauce
27. Tomato juice and cocktails, canned
28. Tomato paste
29. Tomato sauce
30. Vegetable pie mixes
31. Vegetables, canned

Dried and Dehydrated Fruits, Vegetables, and Soup Mixes (IN 2034)

Establishments primarily engaged in sun drying or artificially dehydrating fruits and vegetables, or in manufacturing packaged soup mixes from dehydrated ingredients. Establishments primarily engaged in the grading and marketing of farm dried fruits, such as prunes and raisins, are classified in Wholesale Trade, Industry 5149.

1. Dates, dried
2. Dehydrated fruits, vegetables, and soups
3. Fruit flour, meal, and powders
4. Fruits, sulphured
5. Olives, dried
6. Potato flakes, granules, and other dehydrated potato products

7. Prunes, dried
8. Raisins
9. Soup mixes
10. Soup powders
11. Vegetable flour, meal, and powders
12. Vegetables, sulphured

Pickled Fruits and Vegetables, Vegetable Sauces and Seasonings, and Salad Dressings (IN 2035)

Establishments primarily engaged in pickling and brining fruits and vegetables and in manufacturing salad dressings, vegetable relishes, sauces, and seasonings. Establishments primarily engaged in manufacturing catsup and similar tomato sauces are classified in Industry 2033, and those packing purchased pickles and olives are classified in Wholesale or Retail Trade. Establishments primarily engaged in manufacturing dry salad dressing and dry sauce mixes are classified in Industry 2099.

1. Blue cheese dressing
2. Brining of fruits and vegetables
3. Cherries, brined
4. French dressing
5. Fruits, pickled and brined
6. Horseradish, prepared
7. Mayonnaise
8. Mustard, prepared (wet)
9. Olives, brined: bulk
10. Onions, pickled
11. Pickles and pickle salting
12. Relishes, fruit and vegetable
13. Russian dressing
14. Salad dressings, except dry mixes
15. Sandwich spreads, salad dressing base
16. Sauces, meat (seasoning): except tomato and dry
17. Sauces, seafood: except tomato and dry
18. Sauerkraut, bulk
19. Seasonings (prepared sauces), vegetable: except tomato and dry
20. Soy sauce
21. Thousand Island dressing
22. Vegetable sauces, except tomato
23. Vegetables, pickled and brined
24. Vinegar pickles and relishes
25. Worcestershire sauce

Frozen Fruits, Fruit Juices, and Vegetables (IN 2037)

Establishments primarily engaged in freezing fruits, fruit juices, and vegetables. These establishments also produce important byproducts such as fresh or dried citrus pulp.

1. Concentrates, frozen fruit juice
2. Dried citrus pulp
3. Frozen fruits, fruit juices, and vegetables
4. Fruit juices, frozen
5. Fruits, quick frozen and cold pack (frozen)
6. Vegetables, quick frozen and cold pack (frozen)

Frozen Specialties Not Elsewhere Classified (IN 2038)

Establishments primarily engaged in manufacturing frozen food specialties, not elsewhere classified, such as frozen dinners and frozen pizza. The manufacture of some important frozen foods and specialties is classified elsewhere. For example, establishments primarily engaged in manufacturing frozen dairy specialties are classified in Industry Group 202, those manufacturing frozen bakery products are classified in Industry Group 205, those manufacturing frozen fruits and vegetables are classified in Industry group 205, those manufacturing frozen fruits and vegetables are classified in Industry 2037, and those manufacturing frozen fish and seafood specialties are classified in Industry 2092.

1. Dinners, frozen: packaged
2. French toast, frozen
3. Frozen dinners, packaged
4. Meals, frozen
5. Native foods, frozen
6. Pizza, frozen
7. Soups, frozen: except seafood
8. Spaghetti and meatballs, frozen
9. Waffles, frozen
10. Whipped topping, frozen

GRAIN MILL PRODUCTS (IGN 204)

Flour and Other Grain Mill Products (IN 2041)

Establishments primarily engaged in milling flour or meal from grain, except rice. The products of flour mills may be sold plain or in the form of prepared mixes or doughs for specific purposes. Establishments primarily engaged in manufacturing prepared flour mixes or doughs from purchased ingredients are classified in Industry 2045, and those milling rice are classified in Industry 2044.

1. Bran and middlings, except rice
2. Bread and bread-type roll mixes*
3. Buckwheat flour
4. Cake flour*
5. Cereals, cracked grain*
6. Corn grits and lakes for brewers' use
7. Dough, biscuit*
8. Doughs, refrigerated or frozen*
9. Durum flour
10. Farina, except breakfast food*
11. Flour mills, cereals: except rice
12. Flour mixes*
13. Flour: buckwheat, corn, graham, rye, and wheat
14. Frozen doughs*
15. Graham flour

16. Granular wheat flour
17. Grits and flakes, corn: for brewers' use
18. Hominy grits, except breakfast food
19. Meal, corn
20. Milling of grains, dry, except rice
21. Mixes, flour: eg, pancake, cake, biscuit, doughnut*
22. Pancake batter, refrigerated or frozen*
23. Pizza mixes and prepared dough*
24. Semolina (flour)
25. Wheat germ
26. Wheat mill feed

Cereal Breakfast Foods (IN 2043)

Establishments primarily engaged in manufacturing cereal breakfast foods and related preparations, except breakfast bars. Establishments primarily engaged in manufacturing granola bars and other types of breakfast bars are classified in Industry 2064.

1. Breakfast foods, cereal
2. Coffee substitutes made from grain
3. Corn flakes
4. Corn hulled (cereal breakfast food)
5. Farina, cereal breakfast food
6. Granola, except bars and clusters
7. Hominy grits prepared as cereal breakfast food
8. Infants' foods, cereal type
9. Oatmeal (cereal breakfast food)
10. Oats, rolled (cereal breakfast food)
11. Rice breakfast foods
12. Wheat flakes

Rice Milling (IN 2044)

Establishments primarily engaged in cleaning and polishing rice, and in manufacturing rice flour or meal. Other important products of this industry include brown rice, milled rice (including polished rice), rice polish, and rice bran.

1. Flour, rice
2. Milling of rice
3. Polishing of rice
4. Rice bran, flour, and meal
5. Rice cleaning and polishing
6. Rice polish
7. Rice, brewers'
8. Rice, brown
9. Rice vitamin and mineral enriched

Prepared Flour Mixes and Doughs (IN 2045)

Establishments primarily engaged in preparing flour mixes or doughs from purchased flour. Establishments primarily engaged in milling flour from grain and producing mixes or doughs are classified in Industry 2091.

1. Biscuit mixes and doughs**
2. Bread and bread-type roll mixes**
3. Cake flour**
4. Cake mixes**
5. Dough, biscuit**
6. Doughnut mixes**
7. Doughs, refrigerated or frozen**
8. Flour: blended or self-rising**
9. Frozen doughs**
10. Gingerbread mixes**
11. Mixes, flour: eg, pancake, cake, biscuit, doughnut**
12. Pancake batter, refrigerated or frozen**
13. Pancake mixes**
14. Pizza mixes and doughs**

Wet Corn Milling (IN 2046)

Establishments primarily engaged in milling corn or sorghum grain (milo) by the wet process, and producing starch, syrup, oil, sugar, and byproducts, such as gluten feed and meal. Also included in this industry are establishments primarily engaged in manufacturing starch from other vegetable sources (eg, potatoes, wheat). Establishments primarily engaged in manufacturing table syrups from corn syrup and other ingredients, and those manufacturing starch-based dessert powders, are classified in Industry 2099.

1. Corn oil cake and meal
2. Corn starch
3. Corn syrup (including dried), unmixed
4. Dextrine
5. Dextrose
6. Feed, gluten
7. Fructose
8. Glucose
9. High fructose syrup
10. Hydrol
11. Meal, gluten
12. Oil, corn: crude and refined
13. Potato starch
14. Rice starch
15. Starch, instant
16. Starch, liquid
17. Starches, edible and industrial
18. Steep water concentrate
19. Sugar, corn
20. Tapioca
21. Wheat gluten
22. Wheat starch

Dog and Cat Food (IN 2047)

Establishments primarily engaged in manufacturing dog and cat food from cereal, meat, and other ingredients. These preparations may be canned, frozen, or dry. Establishments primarily engaged in manufacturing feed for animals other than dogs and cats are classified in Industry 2048.

Prepared Feeds and Feed Ingredients for Animals and Fowls, Except Dogs and Cats (IN 2048)

Establishments primarily engaged in manufacturing prepared feeds and feed ingredients and adjuncts for animals and fowls, except dogs and cats. Included in this industry are poultry and livestock feed and feed ingredients, such as alfalfa meal, feed supplements, and feed concentrates and feed premixes. Also included are establishments primarily engaged in slaughtering animals for animal feed. Establishments primarily engaged in slaughtering animals for human consumption are classified in Industry Group 201. Establishments primarily engaged in manufacturing dog and cat foods are classified in Industry 2047.

1. Alfalfa, cubed
2. Alfalfa, prepared as feed for animals
3. Animal feeds, prepared: except dogs and cats
4. Bird food, prepared
5. Buttermilk emulsion for animal food
6. Chicken feeds, prepared
7. Citrus seed meal
8. Earthworm food and bedding
9. Feed concentrates
10. Feed premixes
11. Feed supplements
12. Feeds, prepared (including mineral): for animals and fowls—except dogs and cats
13. Feeds, specialty: mice, guinea pigs, minks, etc
14. Fish food
15. Hay, cubed
16. Horsemeat, except for human consumption
17. Kelp mean and pellets
18. Livestock feeds, supplements, and concentrates
19. Meal, bone: prepared as feed for animals and fowls
20. Mineral feed supplements
21. Oats: crimped, pulverized, and rolled: except breakfast food
22. Oyster shells, ground: used as feed for animals and fowls
23. Pet food, except dog and cat: canned, frozen, and dry
24. Poultry feeds, supplements, and concentrates
25. Shell crushing for feed
26. Slaughtering of animals, except for human consumption
27. Stock feeds, dry

BAKERY PRODUCTS (IGN 205)

Bread and Other Bakery Products, Except Cookies and Crackers (IN 2051)

Establishments primarily engaged in manufacturing fresh or frozen bread and bread-type rolls and fresh cakes, pies, pastries and other similar perishable bakery products. Establishments primarily engaged in producing dry bakery products, such as biscuits, crackers, and cookies, are classified in Industry 2052. Establishments primarily engaged in manufacturing frozen bakery products, except bread and bread-type rolls, are classified in Industry 2053. Establishments producing bakery products primarily for direct sale on the premises to household consumers are classified in Retail Trade, Industry 5461.

1. Bagels
2. Bakery products, fresh: bread, cakes, doughnuts, and pastries
3. Bakery products, partially cooked: except frozen
4. Biscuits, baked: baking powder and raised
5. Bread, brown: Boston and other—canned
6. Bread, including frozen
7. Buns, bread-type (eg, hamburger, hot dog), including frozen
8. Buns, sweet, except frozen
9. Cakes, bakery, except frozen
10. Charlotte Russe (bakery product), except frozen
11. Croissants, except frozen
12. Crullers, except frozen
13. Doughnuts, except frozen
14. Frozen bread and bread-type rolls
15. Knishes, except frozen
16. Pastries, except frozen: eg, Danish, French
17. Pies, bakery, except frozen
18. Rolls, bread-type, including frozen
19. Rolls, sweet, except frozen
20. Sponge goods, bakery, except frozen
21. Sweet yeast goods, except frozen

Cookies and Crackers (IN 2052)

Establishments primarily engaged in manufacturing fresh cookies, crackers, pretzels, and similar dry bakery products. Establishments primarily engaged in producing other fresh bakery products are classified in Industry 2051.

1. Bakery products, dry: eg, biscuits, crackers, pretzels
2. Biscuits, baked: dry, except baking powder and raised
3. Communion wafers
4. Cones, ice cream
5. Cookies
6. Cracker meal and crumbs
7. Crackers: eg, graham, soda
8. Matzoths
9. Pretzels
10. Rusk
11. Saltines
12. Zwieback

Frozen Bakery Products, Except Bread (IN 2053)

Establishments primarily engaged in manufacturing frozen bakery products, except bread and bread-type rolls. Establishments primarily engaged in manufacturing frozen bread and bread-type rolls are classified in Industry 2051.

1. Bakery products, frozen: except bread and bread-type rolls
2. Cakes, frozen: pound, layer, and cheese
3. Croissants, frozen
4. Doughnuts, frozen
5. Pies, bakery, frozen
6. Sweet yeast goods, frozen

SUGAR AND CONFECTIONERY PRODUCTS (IGN 206)

Cane Sugar, except Refining (IN 2061)

Establishments primarily engaged in manufacturing raw sugar, syrup, or finished (granulated or clarified) cane sugar from sugar cane. Establishments primarily engaged in refining sugar from purchased raw cane sugar or sugar syrup are classified in Industry 2062.

1. Cane sugar, made from sugarcane
2. Molasses, blackstrap: made from sugarcane
3. Molasses, made from sugarcane
4. Sugar, granulated: made from sugarcane
5. Sugar, invert: made from sugarcane
6. Sugar, powdered: made from sugarcane
7. Sugar, raw: made from sugarcane
8. Syrup, cane: made from sugarcane

Cane Sugar Refining (IN 2062)

Establishments primarily engaged in refining purchased raw cane sugar and sugar syrup.

1. Molasses, blackstrap: made from purchased raw cane sugar or sugar syrup
2. Refineries, cane sugar
3. Sugar, granulated: made from purchased raw cane sugar or sugar syrup
4. Sugar, invert: made from purchased raw cane sugar or sugar syrup
5. Sugar, powdered: made from purchased raw cane sugar or sugar syrup
6. Sugar, refined: made from purchased raw can sugar or sugar syrup
7. Syrup, made from purchased raw can sugar or sugar syrup

Beet Sugar (IN 2063)

Establishments primarily engaged in manufacturing sugar from sugar beets.

1. Beet pulp, dried
2. Beet sugar, made from sugar beets
3. Molasses beet pulp
4. Molasses, made from sugar beets
5. Sugar, granulated: made from sugar beets
6. Sugar, invert: made from sugar beets
7. Sugar, liquid: made from sugar beets
8. Sugar, powdered: made from sugar beets
9. Syrup, made from sugar beets

Candy and Other Confectionery Products (IN 2064)

Establishments primarily engaged in manufacturing candy, including chocolate candy, other confections, and related products. Establishments primarily engaged in manufacturing solid chocolate bars from cacao beans are classified in Industry 2066, those manufacturing chewing gum are classified in Industry 2067, and those primarily engaged in roasting and salting nuts are classified in Industry 2068. Establishments primarily engaged in manufacturing confectionery for direct sale on the premises to household consumers are classified in Retail Trade, Industry 5441.

1. Bars, candy: including chocolate-covered bars
2. Breakfast bars
3. Cake ornaments, confectionery
4. Candy, except solid chocolate
5. Chewing candy, except chewing gum
6. Chocolate bars, from purchased cocoa or chocolate
7. Chocolate candy, except solid chocolate
8. Confectionery
9. Cough drops, except pharmaceutical preparations
10. Dates: chocolate covered, sugared, and stuffed
11. Fruit peel products: candied, glazed glace, and crystallized
12. Fruits: candied, glazed, and crystallized
13. Fudge (candy)
14. Granola bars and clusters
15. Halvah (candy)
16. Licorice candy
17. Lozenges, candy: nonmedicated
18. Marshmallows
19. Marzipan (candy)
20. Nuts, candy covered
21. Nuts, glace
22. Popcorn balls and candy-covered popcorn products

Chocolate and Cocoa Products (IN 2066)

Establishments primarily engaged in shelling, roasting, and grinding cacao beans for the purpose of making chocolate liquor, from which cocoa powder and cocoa butter are derived, and in the further manufacturing of solid chocolate bars, chocolate coatings, and other chocolate and cocoa products. Also included is the manufacture of similar products, except candy, from purchased chocolate or cocoa. Establishments primarily engaged in manufacturing candy from purchased cocoa products are classified in Industry 2064.

1. Baking chocolate
2. Bars, candy: solid chocolate

3. Cacao bean products: chocolate, cocoa butter, and cocoa
4. Cacao beans: shelling, roasting, and grinding for making chocolate liquor
5. Candy, solid chocolate
6. Chocolate bars, solid: from cacao beans
7. Chocolate coatings and syrups
8. Chocolate liquor
9. Chocolate syrup
10. Chocolate, instant
11. Chocolate, sweetened or unsweetened
12. Chocolate butter
13. Chocolate mix, instant
14. Chocolate powdered: mixed with other substances

Chewing Gum (IN 2067)

Establishments primarily engaged in manufacturing salted, roasted, dried, cooked, or canned nuts or in processing grains or seeds in a similar manner for snack purposes. Establishments primarily engaged in manufacturing confectionery-coated nuts are classified in Industry 2064, and those manufacturing peanut butter are classified in Industry 2099.

1. Nuts, dehydrated or dried
2. Nuts: salted, roasted, cooked, or canned
3. Seeds: salted, roasted, cooked, or canned

FATS AND OILS (IGN 207)

Cottonseed Oil Mills (IN 2074)

Establishments primarily engaged in manufacturing cottonseed oil, cake, meal, and linters, or in processing purchased cottonseed oil other than into edible cooking oils. Establishments primarily engaged in refining cottonseed oil into edible cooking oils are classified in Industry 2079.

1. Cottonseed oil, cake, and meal: made in cottonseed oil mills
2. Cottonseed oil, deodorized
3. Lecithin, cottonseed

Soybean Oil Mills (IN 2075)

Establishments primarily engaged in manufacturing soybean oil, cake, and meal, and soybean protein isolates and concentrates, or in processing purchased soybean oil other than into edible cooking oils. Establishments primarily engaged in refining soybean oil into edible cooking oils are classified in Industry 2079.

1. Lecithin, soybean
2. Soybean flour and grits
3. Soybean oil, cake, and meal
4. Soybean oil, deodorized
5. Soybean protein concentrates
6. Soybean protein isolates

Vegetable Oil Mills, Except Corn, Cottonseed, and Soybean (IN 2076)

Establishments primarily engaged in manufacturing vegetable oils, cake and meal, except corn, cottonseed, and soybean, or in processing similar purchased oils other than into edible cooking oils. Establishments primarily engaged in manufacturing corn oil and its byproducts are classified in Industry 2046, those that are refining vegetable oils into edible cooking oils are classified in Industry 2079, and those refining these oils for medicinal purposes are classified in Industry 2833.

1. Castor oil and pomace
2. Coconut oil
3. Linseed oil, cake, and meal
4. Oils, vegetable: except corn, cottonseed, and soybean
5. Oiticica oil
6. Palm kernel oil
7. Peanut oil, cake, and meal
8. Safflower oil
9. Sunflower seed oil
10. Tallow, vegetable
11. Tung oil
12. Walnut oil

Animal and Marine Fats and Oils (IN 2077)

Establishments primarily engaged in manufacturing animal oils, including fish oil and other marine animal oils, and fish and animal meal; and those rendering inedible stearin, grease, and tallow from animal fat, bones, and meat scraps. Establishments primarily engaged in manufacturing lard and edible tallow and stearin are classified in Industry Group 201; those refining marine animal oils for medicinal purposes are classified in Industry 2833; and those manufacturing fatty acids are classified in Industry 2899.

1. Feather meal
2. Fish liver oils, crude
3. Fish meal
4. Fish oil and fish oil meal
5. Grease rendering, inedible
6. Meal, meat and bone: not prepared as feed
7. Meat and bone meal and tankage
8. Oils, animal
9. Oils, fish and marine animal: eg, herring, menhaden, whale (refined), sardine
10. Rendering plants, inedible grease and tallow
11. Stearin, animal: inedible
12. Tallow rendering, inedible

Shortening, Table Oils, Margarine, and Other Edible Fats and Oils, Not Elsewhere Classified (IN 2079)

Establishments primarily engaged in manufacturing shortening, table oils, margarine, and other edible fats and oils, not elsewhere classified. Establishments primarily engaged in producing corn oil are classified in Industry 2046.

1. Baking and frying fats (shortening)
2. Cottonseed cooking and salad oil
3. Margarine oil, except corn
4. Margarine, including imitation
5. Margarine-butter blend
6. Nut margarine
7. Oil, hydrogenated: edible
8. Oil, partially hydrogenated: edible
9. Oil, vegetable winter stearin
10. Olive oil
11. Peanut cooking and salad oil
12. Shortenings, compound and vegetable
13. Soybean cooking and salad oil
14. Vegetable cooking and salad oils, except corn oil: refined

BEVERAGES (IGN 208)

Malt Beverages (IN 2082)

Establishments primarily engaged in manufacturing malt beverages. Establishments primarily engaged in bottling purchased malt beverages are classified in Industry 5181.

1. Ale
2. Beer (alcoholic beverage)
3. Breweries
4. Brewers' grain
5. Liquors, malt
6. Malt extract, liquors, and syrups
7. Near beer
8. Porter (alcoholic beverage)
9. Stout (alcoholic beverage)

Malt (IN 2083)

Establishments primarily engaged in manufacturing malt or malt byproducts from barley or other grains.

1. Malt byproducts
2. Malt: barley, rye, wheat, and corn
3. Malthouses
4. Sprouts, made in malthouses

Wines, Brandy, and Brandy Spirits (IN 2084)

Establishments primarily engaged in manufacturing wines, brandy, and brandy spirits. This industry also includes bonded wine cellars that are engaged in blending wines. Establishments primarily bottling purchased wines, brandy, and brandy spirits but that do not manufacture wines and brandy, are classified in Wholesale Trade, Industry 5182.

1. Brandy
2. Brandy spirits
3. Wine cellars, bonded: engaged in blending wines
4. Wine coolers (beverages)
5. Wines

Distilled and Blended Liquors (IN 2085)

Establishments primarily engaged in manufacturing alcoholic liquors by distillation, and in manufacturing cordials and alcoholic cocktails by blending processes or by mixing liquors and other ingredients. Establishments primarily engaged in manufacturing industrial alcohol are classified in Industry 2869, and those only bottling purchased liquors are classified in Wholesale Trade, Industry 5182.

1. Applejack
2. Cocktails, alcoholic
3. Cordials, alcoholic
4. Distillers' dried grains and solubles
5. Eggnog, alcoholic
6. Ethyl alcohol for medicinal and beverage purposes
7. Gin (alcoholic beverage)
8. Grain alcohol for medicinal and beverage purposes
9. Liquors: distilled and blended—except brandy
10. Rum
11. Spirits, neutral, except fruit—for beverage purposes
12. Vodka
13. Whiskey: bourbon, rye, scotch type, and corn

Bottles and Canned Soft Drinks and Carbonated Waters (IN 2086)

Establishments primarily engaged in manufacturing soft drinks and carbonated waters. Establishments primarily engaged in manufacturing fruit and vegetable juices are classified in Industry Group 203; those manufacturing fruit syrups for flavoring are classified in Industry 2087; and those manufacturing nonalcoholic cider are classified in Industry 2099. Establishments primarily engaged in bottling natural spring waters are classified in Wholesale Trade, Industry 5149.

1. Beer, birch and root: bottled or canned
2. Carbonated beverages, nonalcoholic: bottled or canned
3. Drinks, fruit: bottled, canned, or fresh
4. Ginger ale, bottled or canned
5. Iced tea, bottled or canned
6. Lemonade: bottled, canned, or fresh
7. Mineral water, carbonated: bottled or canned
8. Soft drinks, bottled or canned
9. Tea, iced: bottled or canned
10. Water, pasteurized: bottled or canned

Flavoring Extracts and Flavoring Syrups, Not Elsewhere Classified (IN 2087)

Establishments primarily engaged in manufacturing flavoring extracts, syrups, powders, and related products, not elsewhere classified, for soda fountain use or for the manufacture of soft drinks, and colors for bakers' and confectioners' use. Establishments primarily engaged in manufacturing chocolate syrup are classified in Industry 2066.

1. Beverage bases
2. Bitters (flavoring concentrates)
3. Burnt sugar (food color)
4. Cocktail mixes, nonalcoholic
5. Coffee flavorings and syrups
6. Colors for bakers' and confectioners' use, except synthetic
7. Cordials, nonalcoholic
8. Drink powders and concentrates
9. Flavoring concentrates
10. Flavoring extracts, pastes, powders, and syrups
11. Food colorings, except synthetic
12. Food glace, for glazing foods
13. Fruit juices, concentrated: for fountain use
14. Fruits, crushed: for soda fountain use

MISCELLANEOUS FOOD PREPARATIONS AND KINDRED PRODUCTS (IGN 209)

Canned and Cured Fish and Seafoods (IN 2091)

Establishments primarily engaged in cooking and canning fish, shrimp, oysters, clams, crabs, and other seafoods, including soups; and those engaged in smoking, salting, drying, or otherwise curing fish and other seafoods for the trade. Establishments primarily engaged in shucking and packing fresh oysters in nonsealed containers, or in freezing or preparing fresh fish, are classified in Industry 2092.

1. Canned fish, crustacea, and mollusks
2. Caviar, canned
3. Chowders, fish and seafood: canned
4. Clam bouillon, broth, chowder, juice: bottled or canned
5. Codfish: smoked, salted, dried, and pickled
6. Crab meat, canned and cured
7. Finnan haddie (smoked haddock)
8. Fish and seafood cakes: canned
9. Fish egg bait, canned
10. Fish, canned and cured
11. Fish: cured, dried, pickled, salted, and smoked
12. Herring: smoked, salted, dried, and pickled
13. Oysters, canned and cured
14. Salmon: smoked, salted, dried, canned, and pickled
15. Sardines, canned
16. Seafood products, canned and cured
17. Shellfish, canned and cured
18. Shrimp, canned and cured
19. Soups, fish and seafood: canned
20. Stews, fish and seafood: canned
21. Tuna fish, canned

Prepared Fresh or Frozen Fish and Seafoods (IN 2092)

Establishments primarily engaged in preparing fresh and raw or cooked frozen fish and other seafoods and seafood preparations, such as soups, stews, chowders, fishcakes, crabcakes, and shrimpcakes. Prepared fresh fish are eviscerated or processed by removal of heads, fins, or scales. This industry also includes establishments primarily engaged in the shucking and packing of fresh oysters in nonsealed containers.

1. Chowders, fish and seafood: frozen
2. Crabcakes, frozen
3. Crabmeat picking
4. Crabmeat, fresh: packed in nonsealed containers
5. Fish and seafood cakes, frozen
6. Fish fillets
7. Fish sticks
8. Fish: fresh and frozen, prepared
9. Oysters, fresh: shucking and packing nonsealed containers
10. Seafoods, fresh and frozen
11. Shellfish, fresh and frozen
12. Shellfish, fresh: shucked, picked, or packed
13. Shrimp, fresh and frozen
14. Soups, fish and seafood: frozen
15. Stews, fish and seafood: frozen

Roasted Coffee (IN 2095)

Establishments primarily engaged in roasting coffee, and in manufacturing coffee concentrates and extracts in powdered, liquid, or frozen form, including freeze dried. Coffee roasting by wholesale grocers is classified in Wholesale Trade, Industry 5149.

1. Coffee extracts
2. Coffee roasting, except by wholesale grocers
3. Coffee, ground: mixed with grain or chicory
4. Coffee, instant and freeze dried

Potato Chips, Corn Chips, and Similar Snacks (IN 2096)

Establishments primarily engaged in manufacturing potato chips, corn chips, and similar snacks. Establishments primarily engaged in manufacturing pretzels and crackers are classified in Industry 2052; those manufacturing candy covered popcorn are classified in Industry 2064; those manufacturing salted, roasted, cooked or canned nuts and seeds are classified in Industry 2068; and those manufacturing packaged unpopped popcorn are classified in Industry 2099.

1. Cheese curls and puffs
2. Corn chips and related corn snacks
3. Popcorn, popped: except candy covered
4. Pork rinds
5. Potato chips and related corn snacks
6. Potato sticks

Manufactured Ice (IN 2097)

Establishments primarily engaged in manufacturing ice for sale. Establishments primarily engaged in manufacturing dry ice are classified in Industry 2813.

1. Block ice
2. Ice cubes
3. Ice plants, operated by public utilities
4. Ice, manufactured or artificial: except dry ice

Macaroni, Spaghetti, Vermicelli, and Noodles (IN 2098)

Establishments primarily engaged in manufacturing dry macaroni, spaghetti, vermicelli, and noodles. Establishments primarily engaged in manufacturing canned macaroni and spaghetti are classified in Industry 2032, and those manufacturing fried noodles, such as Chinese noodles, are classified in Industry 2099.

Macaroni and products, dry: eg, alphabets, rings, seashells

1. Noodles: egg, plain, and water
2. Spaghetti, dry
3. Vermicelli

Food Preparations, not Elsewhere Classified (IN 2099)

Establishments primarily engaged in manufacturing prepared foods and miscellaneous food specialties, not elsewhere classified, such as baking powder, yeast, and other leavening compounds; peanut butter; packaged tea, including instant; ground spices; and vinegar and cider. Also included in this industry are establishments primarily engaged in manufacturing dry preparations, except flour mixes, consisting of pasta, rice, potatoes, textured vegetable protein, and similar products that are packaged with other ingredients to be prepared and cooked by the consumer. Establishments primarily engaged in manufacturing flour mixes are classified in Industry Group 204.

1. Almond pastes
2. Baking powder
3. Bouillon cubes
4. Box lunches for sale off premises
5. Bread crumbs, not made in bakeries
6. Butter, renovated and processed
7. Cake frosting mixes, dry
8. Chicory root, dried
9. Chili pepper or powder
10. Chinese noodles
11. Cider, nonalcoholic
12. Coconut, desiccated and shredded
13. Cole slaw in bulk
14. Cracker sandwiches made from purchased crackers
15. Desserts, ready to mix
16. Dips, except cheese and sour cream based
17. Emulsifiers, food
18. Fillings, cake or pie: except fruits, vegetables, and meat
19. Frosting, prepared
20. Gelatin dessert preparations
21. Gravy mixes, dry
22. Honey, strained and bottled
23. Jelly, corncob (gelatin)
24. Leavening compounds, prepared
25. Marshmallow creme
26. Meat seasonings, except sauces
27. Molasses, mixed or blended**
28. Noodles, fried (eg, Chinese)
29. Noodles, uncooked: packaged with other ingredients
30. Pancake syrup, blended and mixed
31. Pasta, uncooked: packaged with other ingredients
32. Peanut butter
33. Pectin
34. Pepper
35. Pizza, refrigerated: not frozen
36. Popcorn, packaged: except popped
37. Potatoes, dried: packaged with other ingredients
38. Potatoes, peeled for the trade
39. Rice, uncooked: packaged with other ingredients
40. Salad dressing mixes, dry
41. Salads, fresh or refrigerated
42. Sandwiches, assembled and packaged: for wholesale market
43. Sauce mixes, dry
44. Sorghum, including custom refining
45. Spices, including grinding
46. Sugar grinding
47. Sugar, industrial maple: made in plants producing maple syrup
48. Sugar, powdered**
49. Syrups, sweetening: honey, maple syrup, sorghum
50. Tea blending
51. Tofu, except frozen desserts
52. Tortillas, fresh or refrigerated
53. Vegetables peeled for the trade
54. Vinegar
55. Yeast

BIBLIOGRAPHY

"Standard Industrial Classification Manual," Office of Management and Budget, Executive Office of the President. U.S. Government Printing Office, Washington, D.C., 1987.

Y. H. HUI
EDITOR-IN-CHIEF

FOOD PROCESSING: TECHNOLOGY, ENGINEERING, AND MANAGEMENT

The technology of food is intertwined with the business of selling food. Although this is not unique to food technology, having insight into the food business will provide some insight into the motivation it provides for the development of new technologies and the maintenance of old ones. Decision making in the food industry requires knowledge of both the business of selling food and the technology of food. The right information about the technology and the business, coupled with the organized approaches to decision making, can yield important benefits to those responsible for making decisions in the industry.

Some generalizations about the food business are that it is high volume, low margin, multiple product, transportation intensive, and end-user marketing intensive. Because of the need for food to be ubiquitous, the business requires multiple distribution points and complicated distribution networks. The nature of food as material, and as a perception of the consumer supports and constrains the food business. These materials provide the means to deliver safe, palatable, and profitable product to the purchaser. This same technology constrains the business by specifying what can and cannot be done with food materials while maintaining a viable product. The reality is that the technology and the business of selling food are inseparable.

Because profit per sale is low, food businesses must make many sales to gain a reasonable profit. This emphasis on volume requires food technology to provide efficiency in production and speed in production. This makes the production facility challenge one of turning salable food stuffs out by the ton. In the food industry, great value is placed on technological innovation that enables automated handling of materials and scale-up of processes and procedures from small to large volumes. In such a high volume business, small improvements in efficiency of the technology or distribution or the decision making can have important consequences for the profit (or even survival) of an enterprise.

Most food companies see opportunities for growth in the introduction of new products. This has led to varieties of products from single companies and much product differentiation within categories of products. The technology to develop, manufacture, and distribute these new product entries must be invented and applied in order for such growth to happen. This perception of how growth is attained, and the perception that growth is important, provide important spurs for technological development.

The fact that food must be ubiquitous leads to situations in which the business and the technology cannot be separated. Although lengthening shelf life is a regular goal and concern of the food technologist, the distribution system must be designed around what the technology and the consumer will bear. Some fresh food stuffs are flown to market, while other products are shipped by train or barge. The network of warehouses, transportation corridors, vehicles, and storage facilities all must accommodate to the realities of the food as material and to the requirement that consumer's expectations are matched in food products once delivered. The business accommodates to the realities of the product, whereas the technologist tries to change the product to achieve new advantages for the business. Because food must be everywhere there are people, the marketing of food must be ubiquitous—ultimately, food products are consumed teaspoon by teaspoon and each individual consumer must be sold.

The other technological reality that affects the food business is the need for uniform output products when the inputs are variable. The biological systems that generate the inputs to food products require modification in order to yield uniform outputs. The modification is usually called processing. Besides turning food out by the ton, food technology must also provide a means to control the performance of the inputs so that the consumer's perception of the product is the same container to container and bite by bite.

The nature of the business of food has important consequences for the technologist, and the technology of food has important consequences for the business person. Decisions, whether they are viewed as business oriented or technology oriented, should not be seen in isolation from either arena.

THE OPTIMIZING DECISION MAKER

Those responsible for the policy and assets of a food business are optimizers. These people try to achieve the best results in the situation they face—the business and technological situation. If they are conscientious managers this means they will attempt to use resources efficiently. The usual procedure when optimizing is to array the alternatives, eliminate the infeasible ones from consideration, and then measure the remaining alternatives until a best one is determined (1). Sometimes this is an informal procedure, with little information gathering and minor mathematics, and sometimes it is very formal with structured information requirements and complex mathematics. The science associated with food materials often structures information requirements, the technology provides numbers and determines what is feasible, and the business objectives provide the measuring stick to choose between alternatives. This kind of decision making links the business to the technology in important and useful ways. Not only is the approach structured and systematic, but also for many kinds of decisions there are formal (mathematical) methods that can assist the decision maker.

Decisions that affect the efficiency of a food industry enterprise are made at many levels in the business. Sometimes they are one-time occurrences (should we buy that warehouse?) and sometimes they are recurring (how long should we process that product and at what temperature?). Because decisions become routine, they are sometimes dismissed as trivial or unimportant. This is not always the case, as routine decisions about food formulations, vehicle scheduling, processing parameters, production scheduling, and the like have important impacts on costs, profits, and the safety of the food product. If a recurring decision is not optimal, the result is regular losses to the business. In many cases, formal models can be built for these decision situations and formal methods can be applied to give the decision maker good information to act on. Although the optimization opportunity itself is constrained by the quality of the information available to the user (ie, is the information (model) accurate? is the information set complete with respect to the decision?), it still will be the best the food manager can do in the situation he or she faces.

An important piece of optimizing is the objective measure (2). This device allows the user to choose between alternatives—providing a measure of which one is better. There is sometimes a perceived conflict between the objectives of the food technologist and the food industry manager. If the technologist insists on some kind of quality objective—best product, highest quality product—this comes in conflict with profit objectives of the business. The inputs required to achieve such objectives are often so expensive that the product price becomes more than the market will bear. The contribution of food technology in this business context must be to specify what is feasible. Otherwise, the decision will be driven by two (or more) objectives and will never be optimal.

EXAMPLES

Allocation of Milk Resources in Cheese Making. In a step in the manufacture of cheddar cheese, a variety of milk resources are blended in a cheese vat with other ingredients (3). Making the blend is called standardizing and the resulting blend is called cheese milk. The blend is allowed to coagulate before other processing steps are taken. The milk resources used are valuable; the cheesemaker pays to acquire them. Furthermore, the cheese yield and cheese quality are affected by which combination and proportions of resources are used in the blend. In the United States, the composition of the resulting product (which is determined by the composition of the cheese milk) is also constrained by government regulations and definitions. A recurring decision that the cheesemaker faces is determining the constituent amounts for each of the inputs to the cheese milk. The decision is constrained in many ways, but it is made regularly with the ordinary business objective of making a profit.

Technological investigation has determined a cheese yield formula (4) that allows the cheesemaker to predict the cheese yield if enough detail is known about the casein, fat, salt retention, and water retention of this step in the process. This technological model requires that the casein-to-fat ratio of the cheese milk be restricted to a small range. If this is done, the formula can be used to predict pounds of output for a given set of inputs. The potential inputs to the cheese milk include milk, condensed skim milk, nonfat dry milk, whey protein concentrate, cream, frozen cream, water, and other ingredients. Almost all the inputs affect the casein-to-fat ratio and thus the cheese yield. Because each of these resources will have different prices at different times, the cheesemaker can optimize the cheese milk—make the optimal choice of resources for the current set of price and availability circumstances.

This optimization opportunity may be modeled as a linear program (5). The variables are the inputs to the cheese milk, the constraints may be constructed out of these variables to restrict the casein-to-fat ratio and to reflect government regulations, company policies, resource availability, and other restrictions on the cheese product. An objective measure may be constructed from the variables and corresponding costs coupled with the sales value of the resulting product. The advantage of casting this as a linear program is that the resulting model can then be solved by computer to quickly determine the optimal constituents of the cheese milk. The result will simultaneously reflect the technological, legal, and economic realities facing the cheesemaker. Solutions generated this way have been shown to have a profound impact on the profitability of the cheesemaker. Without such a model and the computer support, cheesemakers can make good guesses, but the cannot be assured of regularly choosing optimal inputs to the cheese vat.

In this example we have a high volume product (the vats may hold 20,000 lb of cheese milk) and variable inputs (the fat and solids content of incoming milk varies from farm to farm and animal to animal) and the recurring decision of how to standardize the cheese milk to achieve uniform outputs. Because of its recurring nature, this decision has important economic consequences for cheesemaker. Investigation of cheese technology provided the model (the cheese yield formula) that enables the application of optimization methods to assist with the decision of how to standardize the cheese milk.

Formulating a Cheese Topping. When a new product is fielded, there is the opportunity to establish its formulation. This decision affects the economics associated with the product and the consumer's perception of the product. The challenge is to devise a formulation that is technologically feasible and commercially viable. Because there are many alternatives to any formulation some strategy has to be adopted to determine the desired formulation. An optimization strategy gives a good way to weed through alternatives and come up with an attractive formulation.

Cheese toppings may be manufactured by making processed cheese in such a way that the substance has many potential consistencies and textures. Such material may be mixed with other ingredients such as onions, chives, bacon bits, red pepper, and many more alternatives. The resulting product may be used by consumers as a topping for vegetables, potatoes, and salads. Properly formulated, it may be used hot or cold in a variety of ways. Details about the formulation of the plain cheese topping may be found in Hanrez-LaGrange (6). For this illustration we will consider the problem of determining what proportions of the noncheese ingredients added to the plain cheese toppings will make the best product. Preliminary studies have guided the decision makers to the point of determining what combination of red pepper pieces and nacho flavor will make the best cheese topping.

With a new product in the laboratory or pilot stage, determination of an objective measure for optimization is difficult. This measure will be used to compare alternative formulations, but it should indicate the performance of the product when it is ultimately marketed. Because this amounts to a forecast, the measure is beset with intrinsic variability problems of all forecasts. A procedure is to ask a sample of consumers to taste and then rate the product. Presuming this sample represents the intended market, the formulator may use the rating as a guide for adjusting the formulation. The connection between the formulator and the consumer panelist is a ballot requiring the panelist's responses. The questions used to solicit these responses can be used to fashion an objective measure. Although there are many ways to phrase such questions, care should be taken that the responses somehow relate to the potential market for the product. One possibility is ask the consumer to rate the product as acceptable or not acceptable. If the sample of tasters represents the target market for the product, the set of purchasers will be a subset of the set of acceptors. The bigger the set of acceptors, the larger potential market. Actual sales will depend on many factors not controllable by manipulating the formulation, such as advertising, distribution, price, and the competition. This acceptor set size measure can be taken for any potential formulation and becomes a way to compare formulations in order to discard the less desirable ones. Other measures are possible, but for this example, the acceptor set size will be used as the objective measure.

The strategy, then, is to ask a sample of consumers whether various formulations of this product with varying amounts of red peppers and nacho flavoring are acceptable. The objective is to find a formulation that will maximize the acceptor set size when the product is marketed. The acceptor set size is presumed to be a function of the formulation. Because the intention is that the product is to be sold in large volume, this function may be presumed to be continuous and differentiable over a reasonable range of values for input amounts of red pepper and nacho flavor. For such a function the mathematical construct called the gradient exists and may be used as a guide in determining a better (bigger acceptor set size) formulation. The gradient of a function is a vector of partial derivatives of the function with respect to the variables. It has the property that it always points in the direction in which the function is increasing most rapidly. The advantage here is that we may estimate the partial derivatives of this function without knowing an explicit formula for the acceptor set size function. Thus, the gradient may be estimated and improvements to the acceptor set size function may be made by determining the gradient, taking a step in that direction, and retesting to determine how to adjust the formulation further. In symbols:

$A = f(x,y)$ is the acceptor set size function
$\text{grad}(A) = (f_x, f_y)$

where

x = the amount of nacho flavor
y = the amount of red pepper

If (x_0, y_0) is the initial formulation, then

$$(x_1, y_1) = (x_0, y_0) + k \times \text{grad}(A)|(x_0, y_0),$$

where k = the step size in the gradient direction. This process of estimating the gradient and determining the next test formulation ends when there is no more improvement possible (the gradient is at or near (0,0)) or when the noise in the data overpowers the information in the estimate of the gradient.

Partial derivatives are measures of rates of change in the direction parallel to the axes of the chosen variable. To estimate the value of the partial derivative at a given formulation requires that functional estimates be generated at higher and lower values of the chosen variable while the other variables are held constant. In this example, the initial formulation was $x_0 = 3\%$ and $y_0 = 13.75\%$. To obtain estimate of the partial derivative required functional estimates for 2%, 3%, and 4% nacho flavor while holding the amount of red pepper in the mix at a constant 13.75%. To get the other partial derivatives required holding nacho flavor at a constant 3% and getting functional estimates of the acceptor set size with red pepper at 8.75%, 13.75%, and 18.75%. The results of the initial test are shown in Figure 1.

The gradient vector for this test was estimated as:

$$\text{grad}(A)|(x_0, y_0) = (0.50, -0.38)$$

which implies that a better formulation is possible with more nacho flavor and less red pepper. Under various assumptions about the step size the next step in the optimization process was to test with the formulation set at:

$$(x_1, y_1) = (4.35\%, 12.67\%)$$

The functional estimates for the acceptor set size are shown in Figure 2.

Further adjustments in this formulation are possible providing that big enough samples of consumers can be used to generate the accuracy necessary to make the results useful. This approach can be used for more than two variables, but the number of formulations for which data must be gathered also grows.

This sketch of a gradient search method leaves out many details of how the estimates are generated and used to determine the next formulation. More information may be found in Hanrez-LaGrange and Hanrez-LaGrange and Norback (7). This approach has some similarity to response surface methods (RSM). The main difference is that in RSM, a formula for the objective measure must be generated, and this estimation of the objective measure then may be searched for a maximizing value. Because this function is maximized over the entire range of feasible formulations, it usually requires more expensive data collection than direct estimates of the gradient for a spe-

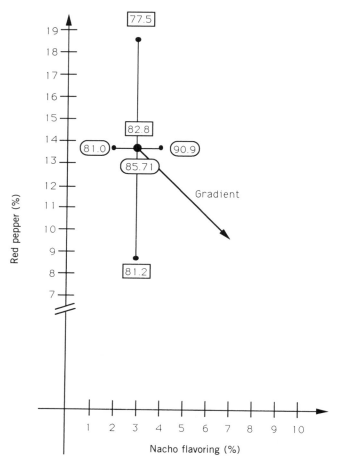

Figure 1. First test with consumer panelists. Numbers enclosed 81.2 indicate acceptor set sizer for the corresponding formulation. Resulting gradient is 0.5, −0.38).

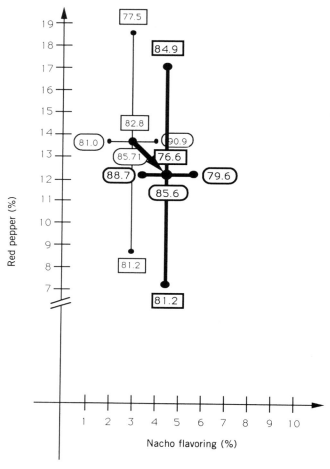

Figure 2. Second consumer test. Note the formulation is a step in the gradient direction from the first test.

cific formulation and is of little value beyond its use in finding a maximum.

The weaknesses in this method are in the determination of the step size at each iteration, in the noise intrinsic to sampling from a population of consumers, and the fact that it may be suboptimal if the objective measure is multimodal. The latter difficulty can be avoided by careful prescreening of the product and its potential formulations followed by restriction of the feasible formulations so that only one maximum may occur over the range of ingredient values being considered. In every test case, the procedure has yielded good solutions to formulation problems in two or three iterations. The gradient search procedure may be used for any functional objective measure, not just for the acceptor set size measure.

Making and adjusting formulations will have important economic consequences in high volume situations. In this case, the food technology constrained the formulation, whereas the business objective of making the acceptor set size largest adjusted the formulation among the feasible alternatives. The situation could also be constrained by the cost of the inputs to the product, by company policy, or by government regulation. Decision support methods (gradient search) were applied to assist in determining the right formulation for the product. Such methods could be applied again if the product were to be reformulated for a substitute ingredient. Small variations in a formulation can have important consequences to the consumer's response to the product and to its profitability. With infinitely many formulations to choose from, these methods provide an organized means of choosing a good one.

Managing Material Flow and Batching. Production scheduling, production quality control and inventory management in the food industry lead to a focus on the flow of materials through the production facility. Many important decisions depend on accurate and timely information regarding the amount and quality of materials that are undergoing some step in the manufacturing sequence. This information is used by the manager to determine product costs, procurement, or harvest requirements and to project output product amounts and to assist in setting prices. Product safety, good product qualities, and product profitability require that the manufacturing sequence conform to the requirements of the food as material and that the processing facilities be managed efficiently. Because of the low margin per sale, the throughput for such facilities must be large so that small improvements in efficiency will have important consequences for profitability. In many food production facilities, the manufacturing sequence is further complicated because the technology or the economics require batch processing. Although the ideal is seen to be "continuous processing," the technology may not be available as in the manufacture of cheddar cheese, the blending of processed cheese, or the processing of low-acid foods. In other cases, the continuous technology is too expensive or otherwise infeasible to be adopted as in the retort processing of low-acid foods. This leads to complications in achieving target productions, especially in cases where batches of intermediate materials are used in multiple products. The decision making surrounding procurement, production scheduling, and quality control is strongly connected to the technology of the food being processed. This technology once again determines the manufacturing sequence and constrains the amount of throughput.

The flow of materials in canning requires that incoming vegetables be washed, sorted by grades, placed in a container-which is partially filled with brine—then retort processed before being cooled, labeled, and placed in cases. Although not all of these steps are used in every manufacturing sequence, brining and retort processing are both batch processing steps. The flow of materials provides a structure to organize information regarding this manufacturing sequence. Such information can be fed forward to control subsequent steps in the processing. For example, the size of peas (as well as the variety peas and the formulation of the brining solution) will affect the time and temperature settings for the subsequent retort processing. Amounts of materials and corresponding costs may be kept in this same flow of materials structure as well as information required by government regulation or company policy.

Using flow of materials structure provides great organizational advantage. Although it is possible to do the arithmetic and track the resources by hand for a few products, with many products that use many intermediate products the mass of detail that accumulates is soon overwhelming. Using the flow of materials as a structure provides more than organizational convenience, however. The development of the so called Gozinto matrices can be done from the flow of materials structure (8,9) and their connection to materials requirements planning (MRP) means that accurate projections of materials use, costs, and inventory amounts can be conveniently made. Although the time phasing of inventory and production is not as important a problem in the food industry as it is in fabrication industries, tracking the of the use of intermediate products, product costing, and batching are.

A Gozinto matrix is created through a procedure that organizes production information into a lower triangular, invertible matrix. The rows and columns of this matrix correspond to products and inputs to products, organized so that no row corresponding to an input to a product occurs above the row corresponding to the product in the matrix. This means that the bottom rows in the matrix correspond to ingredients purchased or harvested from outside the production facility, whereas the top rows correspond to the output products of the facility. The rows in between correspond to intermediate products which may find there way into many output products. More details may be found in Mize and co-workers (9).

Inverting this matrix and doing some elementary matrix algebra provides a means of anticipating and tracking inventories and production. For specified target production, the precise input needs of each ingredient or intermediate product may be determined. Inventories would be kept a minimum if we could procure or manufacture exactly these quantities of input materials. But getting exactly the quantity of intermediate products required is complicated by batch processing and by the use of batch output in multiple products. The amount of brine needed for a target number of cases of canned vegetables is rarely an exact number of batches of brine. More likely, the target will require 12.37 batches of brine, or some other in-between number. Is it better in these cases to make 12 batches or 13 batches or to make partial batches (ie, .37 batches)?

If we make more batches than the target, we will have excess quantity of batch output, which must be stored or discarded, or used in excess production of final product, which then must be stored. If we make less than the target production, we may not meet demand (causing stock outs) and opportunity costs associated with lost sales. If we make partial batches, we face the same labor and fixed costs associated with full batches but get less output. Furthermore, partial batches often imply important technological changes—especially in the food industry—since batch inputs may not scale linearly. Salt and spices are good examples of inputs for which it is not good enough to simply cut the amounts in half when the batch is half the size. If this technological information is not available, product quality will vary when the batch is rescaled.

In these circumstances, what is the best the production manager can do? If the costs described above are understood and can be estimated, this situation can be rendered as an optimization model. The decision is to determine how many batches of input are best for target production amounts. The difficult part is the construction of an objective measure, which takes into account the costs of the different alternatives as well as potential revenues from the products simultaneously. Measuring profits for a production run will require determining the revenue from each of the products made and subtracting ingredient costs, batching costs, the costs from overproduction and underproduction.

Maximize $f(x,b)$

$$f(x,b) = \sum_{i=1}^{n} p_i x_i - \left[\sum_{i=1}^{n} e_i x_i^m + \sum_{j=1}^{n} c_j B_j + \sum_{i=1}^{n} z_i(x_i - a_i) t_{1i} \right.$$
$$\left. + \sum_{i=1}^{n} (z_i - 1)(x_i - a_i) t_{2i} + \sum_{j=1}^{m} (B_j^m - b_j) t_{3j} \right] \quad (1)$$

$$= \sum_{i=1}^{n} [p_i x_i - e_i x_i - z_i(x_i - a_i) t_{1i}$$
$$- (z_i - 1)(x_i - a_i) t_{2i}]$$
$$- \sum_{j=1}^{m} [c_j B_j - (B_j - b_j) t_{3j}] \quad (2)$$

or

$$f(x,b) = (p - e)^t x - c^t B - (z \cdot t_1)^t(x - a)$$
$$- [(z - 1) \cdot t_2]^t(x - a) - t_3^t (B - b) \quad (3)$$

subject to

$$b_j = \sum_{i=1}^{n} x_i r_{ij} \quad (b^t = x^t R) \quad (4)$$

$$b < q \quad (5)$$

$$\alpha a \leq x \leq \beta a \quad (6)$$

$$[b_j] = B_j \quad (7)$$

$$0 \leq B_j - b_j \leq d_j \quad (8)$$

$$z_i = 1 \quad \text{if } x_i - a_i \geq 0, \quad z_i = 0, \text{ otherwise.} \quad (9)$$

$$\Sigma z_i(x_i - a_i)v_i \leq w \quad (10)$$

$$i = 1, 2, \ldots, n, \quad j = 1, 2, \ldots, m \quad (11)$$

where

p_i = selling price per unit of product i, $p = (p_1 p_2 \ldots p_n)^t$
e_1 = ingredient cost price per unit of product i, $e = (e_1 e_2 \ldots e_n)^t$
x_i = number of units of product i, $x = (x_1 x_2 \ldots x_n)^t$
c_j = labor and utility costs per unit (single occurrence) of batch type j
r_{ij} = per unit requirements of product i for batch type j

$$R = \begin{bmatrix} r_{11} & r_{11} & \cdots & r_{1m} \\ r_{21} & r_{22} & \cdots & r_{2m} \\ \cdot & \cdot & & \cdot \\ \cdot & \cdot & & \cdot \\ \cdot & \cdot & & \cdot \\ r_{n1} & r_{n2} & \cdots & r_{nm} \end{bmatrix} = [r_{ij}]_{n \times m}$$

m = number of batch types
n = number of units of batch type j required for the production of products ($b_j = \sum_{i=1}^{n} r_{ij} x_i$), $b = (b_1 b_2 \ldots b_m)^t$
B_j = the nearest integer not less than b_j (ie, whole batch corresponding to partial batch b_j), $B = (B_1 B_2 \ldots B_n)^t$
q_j = number of units of batch type j constrained by the ingredient availability or production capacities $q = (q_1 q_2 \ldots q_m)^t$
a_1 = number of units of product i in a production target $a = (a_1 a_2 \ldots a_n)^t$
d_j = a minimum limit on the partial batch size of batch type j. The limit is set up high enough to keep economics of batching scale and product quality attributes.
t_{1i} = a penalty for overproduction per unit of product i
$t_1 = (t_{1i} t_{12} \ldots t_{1n})^t$
t_{21} = a penalty for underproduction per unit of product i
$t_2 = (t_{21} t_{22} \ldots t_{2n})^t$
t_{3j} = a penalty for partial batches per unit of batch type j
$t_3 = (t_{31} t_{32} \ldots t_{3m})^t$
α = lower limit ratio of acceptable production
β = upper limit ratio of acceptable production
v_1 = space requirement per unit of product i
w = maximum space allowed for inventories

The goal of the model is to find not only a product mix, but also a batch mix, which maximize the penaltied profits. Thus, the objective function is denoted by $f(x,b)$, where x and b represent a product mix and a batch mix, respectively. In the objective function and constraints, z_1 is used to integer variables as needed and other symbols except x_i, b_j, and B_j represent constants.

Although other models are possible, this example shows how such a model might be constructed. Such models may require that certain values be integers but still may be solved by computer methods. The result will be optimal choices regarding the number of batches produced which reflect the costs and revenues that the production facility faces, rather than the intuition of the manager.

Extending the Shelf Life of Fresh Seafood. An ongoing effort by the purveyors of fresh seafood is the opening of inland markets. This can be done by speedier transportation, but this turns out to be costly and sometimes hard to arrange if these products are to enter ordinary marketing channels. This business objective can also be pursued by improving the holding technologies for fresh seafood products (2). With such technologies the product can come through shipping and handling and still be desirable to the consumer. Perhaps the most common of these technologies is refrigeration, but this too has limits, which still leaves large markets untapped.

Two emerging technologies are: modified atmosphere packaging and the use of a sorbate dip on the product. With modified atmosphere packaging, the product is enclosed by a barrier material and the atmosphere over the product is replaced by one without oxygen. The permeability of the barrier material to oxygen is a controllable variable—the packaging material must be chosen (and paid for) by the seafood seller. The sorbate dip technology requires that the product be coated with a sorbate solution of variable concentration. Because these two technologies can be used simultaneously, a reasonable question is what combination of both would be most beneficial to the seafood seller?

This was investigated by gradient search methods. Consumer response data were taken by asking consumers to taste and then to rate the sample tasted as acceptable or unacceptable. (Other data were taken as well.) The purpose was to determine the consumer response to applications of combinations of these technologies by estimating the acceptor set size over time. The system was constrained by the limits on the sorbate concentration that seemed acceptable and by available barrier materials. The acceptor set size function was assumed to be differentiable and the data taken were used to estimate the gradient of this function. The gradient of any differentiable function has the property that it points in the direction of the steepest ascent of the function. Thus, after each completed test, new information was available that indicated what adjustment in the technologies could be made to make the combined application better. This approach is an example of a gradient search method and is a decision-making tool that is broadly applied in optimization, often as part of a computer algorithm. The novelty here is that data collection must be done before each iteration of the algorithm making the whole procedure considerably slower than most algorithms.

A starting permeability of the barrier film and concentration of sorbate is chosen. Tests are run for higher and lower permeabilities at this sorbate concentration and for higher and lower sorbate concentrations at this permeability. The resulting data allows the estimation of the gradient vector at the starting point. If the gradient vector is at or near zero, little improvement is possible, otherwise a step in the direction of the gradient will improve the performance of the combined technologies. New values for permeability and concentration are chosen and the test is repeated.

The application of this procedure characterized consumer response through time to different combinations of technologies. Important questions not addressed include especially the economic ones—is the extra expense of implementing and using the technology worth the resulting extension in shelf life? This would take a more complicated model and perhaps different computational methods.

CONCLUSION

The examples given here give some ideas how food technology may be used to shape the structure of decision making models. When there are many alternatives to choose from, the search for a best one is often profitable. The business of food technology and the food technologist in this decision making realm is to constrain the set of alternatives. The objectives of the organization can then drive the decision making process to the choice of an appropriate alternative. This optimization philosophy helps keep all participants in food industry decision making pulling in the same direction, even though interests and expertise of the individuals may be vastly different.

What is seldom investigated is this connection between the technology of food and the decision making done by food industry managers. If this connection is managed properly, important consequences for the efficiency and profitability of the business can be realized. The procurement of materials, the production, marketing and distribution of food products all depend on the nature of food as material and the consumer's perception of that material as each spoonful is consumed. Approaching this from an optimization perspective—arraying the alternatives, measuring the alternatives, and choosing a desirable one—is a useful and profitable way to solve decision-making problems.

BIBLIOGRAPHY

1. N. M. Gordon and J. P. Norback, "Choosing Objective Measures when Using Sensory Methods for Optimization and Product Positioning," *Food Technology* 39(11), 96, 98–101 (1985).
2. W. F. Sharp, J. P. Norback, and D. A. Stuiber, "Using a New Measure to Define Shelf Life of Fresh Whitefish," *Journal of Food Science* 51(4), 936–939, 959 (1986).
3. G. L. Kerrigan and J. P. Norback, "Linear Programming in the Allocation of Milk Resources for Cheese Making," *Journal of Dairy Science* 69(8), 1432–1440 (1986).
4. L. L. Van Slyke and W. V. Price, *Cheese*, Ridgeview Publishing Co., Independence, Ohio (1979).
5. J. P. Norback and S. R. Evans, "Optimization and Food Formulation," *Food Technology* 37(4), 73–76, 78, 80 (1983).
6. V. Hanrez-LaGrange, *An Optimization Procedure with Sensory Analysis Data in the Development of Process Cheese Toppings*, Master's Thesis, University of Wisconsin, Madison, Wisconsin (1987).
7. V. Hanrez-LaGrange and J. P. Norback, "Product Optimization and Acceptor Set Size," *Journal of Sensory Studies*, 2, 119–136, (1987).
8. A. Vazsonyi, *Scientific Programming in Business and Industry*, Wiley, New York, pp. 258–267 (1958).
9. J. H. Mize, C. R. White, and G. H. Brooks, *Operations Planning and Control*, Prentice-Hall, Englewood Cliffs, New Jersey (1971).
10. S. Dano, *Linear Programming in Industry*, 4th ed., Springer-Verlag/Wien, New York.

JOHN P. NORBACK
University of Wisconsin
Madison, Wisconsin

FOOD SAFETY. See CANNING: REGULATORY AND SAFETY CONSIDERATIONS; FILTH AND EXTRANEOUS MATTER IN FOOD; FOODBORNE DISEASES; FOODBORNE MICROORGANISMS; FOOD TOXICITY; GRAINS AND PROTECTANTS; IMMUNOLOGICAL METHODOLOGY; INDICATOR ORGANISMS: DETECTION AND ENUMERATION IN FOODS; LOW-ACID AND ACIDIFIED FOODS; TOXICOLOGY AND RISK ASSESSMENT.

FOOD SCIENCE AND TECHNOLOGY: DEFINITION AND DEVELOPMENT

Various individuals and agencies have defined food science and technology in different ways. All definitions basically resolve to specifying that food technology is the application of the principles and facts of science, engineering, and mathematics to the processing, preservation, storage, and utilization of foods. Food science, on the other hand, deals chiefly with the acquiring of new knowledge to elucidate the course of reactions or changes occurring in foods whether natural or induced by handling procedures. In theory, there need not be any benefit intended to be derived from the new knowledge, merely the acquiring of it, but in practice there usually is some hoped for benefit such as greater stability of the food, a gain in acceptability or nutritional value, or a reduction in the likelihood that spoilage will occur.

Food science and technology arose as separate disciplines through the coalescing of those portions of other sciences that dealt with foods. Traditionally, food chemists, food microbiologists, and other kinds of scientists studied or taught foods from their particular points of view, but food processing transcends any one of the traditional disciplines. There are still food chemists and food microbiologists; they generally analyze foods. But when food processing is involved, knowledge of only one branch of science is not enough. There is a need for scientists and engineers whose knowledge of the various fields germane to some aspect of food preservation enables them to couple principles from these traditional fields with principles of food processing to effect the most suitable solution to a food processing problem. To illustrate, if an individual trained in one of the traditional branches of engineering were required to devise a new procedure for the handling of food, he or she would normally opt for the most efficient engineering procedure. The most efficient engineering procedure might, however, be one that favored the growth of spoilage organisms. A food engineer would recognize that possibility and select the next best option, one less efficient from an engineering point of view but one least likely to permit spoilage bacteria to flourish. In effect, a food scientist or a food technologist is a generalist in science and engineering rather than being an adherent of only one branch of science.

Food is a chemical substance, but it is more than that. Its own enzymes cause it to undergo change as illustrated by fruits or vegetables first ripening, then ultimately decaying. Enzymic action may arise when a second substance or agent acts on the food. An illustration of that is the manifold changes microorganisms often induce; the changes may be wanted or unwanted. When wanted, special knowledge and skill is needed to cause the microorganisms to ferment or otherwise change the original substance into the food sought to be produced. When unwanted, special strictures in handling of the food are required to avert microbiological degradation, or spoilage. Because foods themselves are intrinsically complex and subject to various forms of change (biological, chemical, nutritional, physical, and sensory), it was quite logical that a science arose that was designed to pull together the various kinds of knowledge that impinge in some way on food and food processing.

The coalescing of the interests of those who conduct research on foods and food processing came somewhat late in science. Not until 1939 when the Institute of Food Technologists was founded did the term food technologist come into general use. The two terms, food science and technology, started to be regularly coupled somewhat later, chiefly when institutions used the joint term to designate the department that taught the two disciplines, disciplines so interlinked they function as one and are generally taught that way.

Not only was food science and technology rather late in being recognized as a separate field, it has differed from many other sciences in the scope of its outlook. As most sciences have developed, they have looked inward toward greater and greater specialization. Notwithstanding that, most practitioners within a field must keep abreast of developments elsewhere in the great area to which their science belongs. A microbiologist, for example, generally attempts to stay knowledgeable of developments elsewhere in the biological and biochemical fields. The scientist thus looks outward to other areas of science. Food scientists and technologists must look outward with vigor. They cannot confine themselves to one major area such as the biological, biochemical, or physical sciences or engineering. Because food and food processing involves all the principal aspects of science and engineering, the food scientist or technologist must stay abreast of developments in several rather diverse areas to practice the profession most effectively. The profession is thus one of unusual challenge. Because so many sciences impinge on food processes and because food is processed in so many forms today, there is an even greater need for food science and technology than there was when it was first constituted a new and separate field. Food scientists and technologists are the individuals most fit by training to be able to recognize, study, use, and interpret the myriad interrelated processes that occur whenever food is preserved, stored, or utilized.

JOHN J. POWERS
University of Georgia
Athens, Georgia

FOOD SCIENCE AND TECHNOLOGY: THE PROFESSION

Food scientists and technologists are employed in academia, by government and industry, and as private consultants. As such, their efforts have a profound effect on the quality, safety, and economy of foods available to consumers. It has been reported that in 1985 U.S. food and beverage expenditures as a proportion of household disposable income were approximately 18% (1). In many countries consumers must spend considerably more of their disposable income for food. Other countries that are in a range approaching that of the United States are also developed countries that have good research and development capabilities in food science and technology. Since the end of World War II, the value-added part of the food industry has increased substantially, and in 1980 it surpassed agriculture's contribution. Developments in food processing are largely a result of the efforts of food scientists and technologists, because they are the ones who originate new products and processing techniques. That the value-added portion of the food industry should have surpassed agriculture itself is significant, because U.S. agriculture is known worldwide for its efficiency.

EARLY FOOD SCIENCE RESEARCH

Aspects of that which would be known today as food science and technology have existed for decades. Appert's development in 1810 of the process of canning (2,3) was an epochal event. The process wasn't called canning then and Appert did not really know the principle on which his process worked, but canning has had a major impact on food preservation ever since its development. It was the first of the purposely invented processes. Other methods developed earlier, such as drying and fermentation, go back to antiquity and were a result of the evolution of procedures over centuries rather than purposeful application of the scientific method. Pasteur's study on the spoilage of wine and his description of how to avoid such spoilage in 1864 (4) persists not only because of the scientific importance of his findings but also because the term pasteurization is so much a part of our vocabulary. There were other early studies on food spoilage. In 1897 Prescott and Underwood published a seminal study of the spoilage of canned clams and lobsters (5). There was also a 1898 paper of theirs that dealt with the spoilage of canned corn. Russell carried out what is now believed to be the first scientific work on spoilage of canned foods. His work dealt with canned peas; it was published in 1895, but little noticed for some time because it was buried in the *Twelfth Annual Report of the Wisconsin Agricultural Experiment Station*. A study leading to a truly scientific breakthrough was that of the National Canners Association (now called the National Food Processors Association) and other cooperating agencies (6). Pasteur's research and that of Prescott and Underwood and of Russell all contributed substantially to knowledge concerning the spoilage of canned foods, but there was not a full understanding of the heat resistance of spoilage organisms and their spores and especially of the process of imparting the proper, but minimum, amount of heat needed to destroy the spores in a bulk material, such as the contents of a can, until ca 1920 (6). Thermal death rates were determined, how much heat is put into a can and extracted from it with different combinations of time and temperature was determined, and a procedure to calculate just how long and at what temperature the cans needed to be heated to provide the minimum amount of heat needed to destroy the spores was devised. These principles contributed to fundamental knowledge and are the basis of today's profession.

THE FIRST OF THE TEACHING DEPARTMENTS

Just a little bit earlier than investigations such as that of the National Canners Association in the early 1920s, universities began to teach phases of food preservation, which would be called food science today. Dairy processing and meats processing were taught during the 19th century, but the subject matter tended to be taught from the point of view of that commodity alone. In the 1910s, things changed. At four different institutions, predecessor departments to food science and technology arose. Like dairy manufacturers their first departmental names were descriptive of the commodity area from which they came, but the pioneer teachers of food preservation recognized that the principles and facts that applied to their products had application to other commodities; thus they tended to broaden their approach by teaching food preservation across commodity and discipline lines. At the University of Massachusetts, the departmental designation was horticultural manufacturers; at the University of California, Fruit Products. The same identification of the originating field applied at Oregon State University and at the Massachusetts Institute of Technology. Among those who instituted formal teaching of what is now called food science were Walter Chenoweth at Massachusetts, William V. Cruess at California, Ernest Wiegand at Oregon State University, and Samuel C. Prescott at MIT. The early teachers found that they could not truly teach food preservation unless they looked at the food and processes from the points of view of several disciplines such as chemistry, microbiology, and engineering. Not until the 1910s did these pioneers begin to think in terms of teaching food preservation, not from the point of view of just the commodity, but from the points of view of basic disciplines that impinged in some manner on food preservation, eg, bacteriology in Prescott's case (5). That kind of thinking led to their departments becoming the first of the present food science and technology departments. In defense of the dairy departments and other commodity-oriented departments of that time, it must be pointed out that in the last century there was less of what would be called further processing than today, when one type of commodity is incorporated into another commodity to produce still a third form of food; consequently there was less reason to look across commodity lines.

Before discussing the evolution of the term food science and technology and the roles food science and technology departments play today in providing the industry with the kinds of trained men and women it needs, another early pioneer in the teaching of food science and technology principles should be mentioned. Fellers was one of the most influential of the early teachers and researchers (7). From the early 1920s on, he evinced both in his teaching and research the concept that science and technology as they relate to food are one, that there are particular facts associated with particular foods, but the same principles and methods apply whether the food be apples, confections, or fish; tortilla, veal, or zucchini. Incidentally, Prescott, Cruess, and Fellers were all early presidents of the Institute of Food Technologists.

FOOD SCIENCE AND TECHNOLOGY: AS A TITLE

The founding of the Institute of Food Technologists in 1939 (8) led to general use of the term food technologist. A few individuals had used the term earlier; the author of a book published in 1937, called himself a food technologist on the title page of that book (2). By 1945, the original four departments that had taught the subject under different titles had changed their departmental names to food science, food science and technology, or to some variant such as food science and nutrition. Several other food science departments were organized in the 15 years after World War II. A few of the departments started *ex nihilo*; most had their origins in some commodity department or represented amalgamations where phases of food science taught in various departments were transferred to form a new department.

RECOMMENDED CURRICULUM

The founding of the Institute of Food Technologists (IFT) was not only the chief force bringing into use the term food science and technology but was also the force that ultimately pulled together food industry needs and teaching curricula. In 1958 the institute sponsored a conference between industrial leaders and university personnel to devise a curriculum that would answer industry's needs for personnel well trained in food science (9). There a model curriculum was devised; in essence it was a consensus between the things industry would like graduates to have versus the very practical matter that there is a limit to the number of courses a curriculum can have and still permit students to graduate in four years. In 1962 and 1965 additional conferences were held; they resulted in the IFT adopting its minimum recommended curriculum (10). It was revised in 1977 (11). The continuing struggle to teach all the things industry would like graduates to have, which is especially difficult because the food industry encompasses so many kinds of industrial operations and the budgetary restraints and other limits placed on departments attempting to provide the soundest education possible to students, has been discussed (12). The Education Committee of the IFT constantly moniters the curriculum. Science, engineering, and industrial processes do not stand still. The recommended curriculum is frequently reexamined to insure that it is still pertinent to the kinds of duties food scientists and technologists will probably have once on the job.

The intent of the curriculum recommended by the IFT (10,11) is that students prior to taking food science courses should have acquired a fairly broad background in the sciences undergirding food science. Students are thus expected to have taken courses in inorganic, organic, and quantitative chemistry and in biochemistry; have at least a general course in biology; an introductory course in microbiology; a year of physics; mathematics through partial derivatives and simple differential equations; a course in statistics; a year of English, oral and written; and not less than 15 semester hours in fields such as history, government, economics, literature, sociology, philosophy, psychology, or the fine arts. The necessity for the science and mathematics prerequisites is obvious. If students are to secure the maximum benefit from instruction in food science, they must have a sufficient and sound background in the basic science(s) underlying each of the different types of course in food science to follow so that specialized courses can be taught at the appropriate level of comprehensiveness and depth. The courses in the humanities and the other nonscience areas are to insure that the student receives a general education, not one solely in science. The minimum of instruction in food science must consist of food chemistry, food analysis, food microbiology, a year of food processing, and two (preferably three) courses in food engineering. For each of the prerequisites and the food science courses there are specifications as to the subject matter the course must include. The institute also specifies other requirements such as the food science and technology unit should preferably be administratively independent and have an adequate budget. The faculty members should be trained in the fields of food science that they teach and they should be encouraged to participate in professional associations and to engage in research. Teaching loads should not be so high as to interfere with adequate preparation for teaching. There are other requirements or suggested conditions, all intended to enable the faculty member to provide a sound level of instruction and to keep abreast of developments in the field. Not only has the Education Committee of the IFT been involved but other individuals and groups have examined various aspects of the educational process (13–15) in an effort to foresee changes occurring, or needed, to keep up with the needs of industry, institutions, students, and the long-term well-being of the United States.

There are approximately 60 institutions in the United States that claim to teach food science and technology. Of that number about 50 meet the minimum curriculum recommended by the IFT (16). The curriculum recommended is just that; it is not accreditation. Accreditation does exist for some sciences and certain other fields, but the IFT has not sought to foster that kind of program conformity, and in general universities themselves are not in favor of it. Accreditation often ties the hands of universities. Sometimes a university must supply monies to an accredited department to enable it to keep its accreditation; yet there may be more pressing needs for funds elsewhere in the university. The IFT has a recommended minimum curriculum chiefly to insure that individuals graduating from a food science program bring to the job the kind of qualifications employers have a right to expect. That is a matter of self-interest on the part of a professional society.

IFT SCHOLARSHIP/FELLOWSHIP PROGRAM

A second purpose of the IFT's minimum curriculum is to provide authorization for a student to receive an IFT scholarship or fellowship. The IFT puts teeth in its recommended curriculum by awarding scholarships only to students enrolled in a department complying with the recommended curriculum. The teeth have a little more bite to them than that. The student must not only be enrolled in a qualifying department but the student must be enrolled in a curriculum meeting the minimum standard, not in some other curriculum available in the department. Some food science and technology departments have more than one program of study. The Education Committee of the IFT moniters compliance with the minimum curriculum yearly. If a university fails to meet the minimum, its students lose their eligibility to receive IFT scholarships. Although the Canadian Institute for Food Science and Technology has a scholarship program of its own, students at Canadian institutions that have a curriculum that conforms to the IFT's recommended curriculum are also eligible to compete for the IFT scholarships.

For some years, female applicants have had a far greater success rate at receiving scholarships than male candidates. At least 85% of the scholarships and fellowships presently go to women. The program was initiated in 1954 (8); today about 110 individuals receive awards. It has been reported that there were 3,474 food science majors enrolled for the bachelor's degree in 1986–1987 (15). During that year, 903 graduated with the bachelor's degree, 260 with the master of science degree, and 132 earned Ph.D.s. In 1986–1987, the total number of graduate students in food science was 1,676.

FOOD SCIENCE TEACHING ABROAD

Quite naturally the United States is not the only country that has institutions teaching food science and technology. Nearly all the developed countries do, and many of the developing countries have at least one institution teaching the subject. In 1980 the International Union of Food Science and Technology (IUFoST) published a list of the professional organizations and universities teaching professional courses in food science in various countries (17). In the industrially advanced countries, such as Japan and Germany, the list is quite extensive. The list for the developing countries is more restrictive, but in the aggregate there are many institutions in the world teaching at least some elements of food science and technology. Unfortunately, the list is not complete. The University of Pretoria in South Africa, for example, is not listed. Because of omissions and programs established since 1980, IUFoST is in the process of updating its list.

POSITIONS FILLED BY FOOD SCIENTISTS AND TECHNOLOGISTS

The types of job available to food scientists and technologists cover the gamut of responsibilities from examination of raw products to final sales. Many firms have quality-control operations wherein raw products are subjected to various types of analysis (chemical, physical, nutritional, and sensory) to be sure the raw products and other ingredients conform to the company's specifications or governmental regulations. The same types of analysis are performed on the finished product. A key to repeat sales is uniformity. A purchaser wants the product to be the same from one time of purchase to the next. One of the purposes of quality-assurance programs is to insure that any difference in products from successive batches or days of production is so minor as not to be perceptible by the consumer. That objective cannot always be met, but quality-assurance programs are designed to permit release to the market of only those products the company considers to be as uniform in composition and characteristics as is reasonably attainable.

For many years, jobs in production were generally closed to women. Today that is not so. Within the last 15 years, the opportunities for women to fill production positions has increased dramatically. The same is true as to technical service and sales, the latter whether technical or not. Packaging firms, chemical companies, flavor houses, and other kinds of companies selling to food firms often employ technically trained individuals to provide advice to purchasers of the supplier's products to be sure the product supplied is being used correctly, to do troubleshooting if there are difficulties in use, or just as a service to place the selling firm in a better competitive position. Those who occupy technical service positions naturally play a key role in the welfare of a company because their attitude and the way they perform their jobs are among the most frequent sources of interaction between the buyer and the seller.

Food scientists and technologists are also employed by trade organizations. Sometimes the positions they fill may be much the same as if they were employed by an individual firm. Some trade organizations analyze the products of their members at intervals to provide guidance to the individual firm, to gather information to defend their members from any government regulation that the members believe unnecessary or undesirable, to permit claims to be made about some attribute of the product, and for other purposes. In many instances food technologists are employed by consulting firms to conduct analyses of various types or to provide advice on some technical matter.

Food scientists and technologists are frequently employed by research and development (R&D) departments. Openings in research and development departments are not confined to those holding a doctorate degree, but R&D openings are quite often filled by Ph.D.s. An area where there is increasing opportunity for employment is with the fast-food firms, because they are constantly trying to bring new products or new forms of a product onto the market to gain a competitive advantage. Quite often those positions are filled by those holding a bachelor's or master's degree.

Food technologists find employment in editorial positions. The Institute of Food Technologists, which publishes Food Technology, employs editors not only to edit manuscripts submitted by others but also sometimes to gather information and to write about some new process or goings-on in the trade to update its members on the latest happenings, be they packaging, governmental operations, or other topics of importance or interest to the members. The same kinds of editorial functions are carried on by many of the trade magazines.

Reference has been made to greater opportunities in production now available to women. The same thing is happening in sensory analysis except for a different reason. For many years most sensory positions were filled by women and they still are. Today's female graduates are often highly trained in statistics and other scientific fields, such as psychology, critical to carrying on sensory analysis in its most advanced forms; that was not true 30 years ago. The change today is that women are in more demand because they are more adequately trained to fill positions requiring the exercise of independent judgment whereas when they were taught in a cookbook fashion about sensory analysis, their chief role was to perform a task under the direction of someone else. Sensory analysis today carries more responsibility. Companies are much more aware that sensory analysis is among the most important analyses any firm performs. If a food is not acceptable, a purchaser may turn against that company and not make future purchases. Unless it is acceptable upon prolonged use, the product will not last on the market. Today's sensory technologists are almost invariably individuals who know how to set up trials and to evaluate acceptability in all its ramifications. The most significant change is that now they have the responsibility to anticipate questions that will later be asked, make provisions to answer those questions insofar as possible, and then be able to interpret the findings in an objective and comprehensive manner.

SIZE, COMPLEXITY, AND IMPORTANCE OF FOOD PROCESSING

The number of scientifically and technologically trained individuals needed in any field is governed partially by the size of the industry and partially by the complexity of its operations. Because food is so common and many households engage in food preservation in some slight way at least, the complexity of food-processing operations is often not fully realized. Heat sterilization as in canning or pasteurization seems like a simple operation, but all the variables that go into deciding whether a given process is adequate to destroy food-poisoning organisms and other kinds of spoilage organisms, while at the same time not putting in any more heat than is required to avoid damaging the food unnecessarily, has required the best of scientific effort and thought to accomplish. A seemingly simple process like making jelly so that it gels properly every time is a process still not fully understood because the properties of pectin are still not known. Early in the 20th century women canned foods and made jellies, but the processes were really an art. Even in large-scale food preservation, art still exists in the compounding of modern-day foods, such as fast foods, but science is even more important, because a food firm must think of safety and have the food invariably turn out right. Because there are so many issues of safety, wholesomeness, nutritive value, and acceptability, and because processes that seem to be simple on the surface are in fact highly intricate ones many food scientists and technologists are needed. In 1987–1988, under the auspices of the IFT, a study of the scope and size of the food industry was conducted (1). The food industry is a powerhouse; it powers many phases of the economy, provides jobs, and massively contributes to the Gross National Product. The only two manufacturing industries in the United States that increase the value-added portion of the economy more are the nonelectrical and the electrical–electronic machinery industries. They barely exceed the food industry in the value-added contribution made to the economy; in 1985 the value-added side of the food industry contributed $104 billion. If the food industry is also given credit for the share of the transportation industry it generates, the food portion of the wholesale–retail trade, printing, packaging, and a host of other industries, its effect on the economy of the United States is even more impressive. Food scientists and technologists find job opportunities not only in food processing itself but in associated industries such as packaging and industrial firms producing ingredients used in foods.

BASIC CAUSES OF COMPLEXITY

Even if the food processing industry did not exist, foodstuffs obviously would have to be considered a complex substance indeed just because of the wide range of food materials that exist. Because they are products of nature, all foodstuffs go through a process of maturation then decay. They deteriorate chemically and biologically. When food processing is added as a variable, complexity is increased exponentially. In spite of the complexity of food production and preservation operations, the U.S. food industry turns out a remarkably safe and inexpensive product. Today U.S. consumers spend 17–18 cents out of each dollar of their disposable income on food, well below that of many other economies; in 1947, they spent 32 cents. U.S. agriculture can take some of the credit for that; much of the credit should go to the efforts of food scientists and technologists.

There are, of course, those who maintain that the food supply is not safe because of the use of food additives. There is no question that ingredients that are now considered unsafe have been used in the past, and quite possibly some of today's ingredients will be similarly judged in the future. That comes about in part because analytical methods are becoming increasingly sensitive. Substances that could not have been detected in the parts per billion or trillion range in years past now often are detectable. If the substance is one that in any amount induces cancer, it must be forbidden under the Delaney Clause of the 1958 Food Additives Amendment to the Food, Drug and Cosmetic Act of 1938; yet in the parts per billion range it may well be without effect. Many of nature's ingredients would likewise be unsafe if they were to be judged by the same standard applied to added ingredients (18–20). The purpose here is not to argue that U.S. foods are safe or unsafe, merely to indicate the complexity involved in making that judgment. "Only the Dose Makes the Poison," which the Swiss physician Paracelsus pronounced in the 16th century, is an aphorism as true today as it was in his time. Many substances indigenous to food are toxic (18). The dilemma that would be faced if naturally occurring toxic substances in foods had to meet the same hurdles as do added substances has been illustrated (20). An example is the effect it would have on the dinner table. The foods in a typical menu were listed and then subtracted, one by one, according to the types of natural toxicant they contained. When finished, the plates were bare save for crackers, rolls, sour cream, and patty shells. Today, the crackers might well be frowned on if they were salted and the sour cream for its cholesterol content.

Although food scientists and technologists are not toxicologists, their advice is often needed to make a toxicological decision because they are the ones who understand the effect different processing operations are likely to have on added substances. Just as is true for the retention of nutrients, they are the ones who know about the probable effect of different operations such as washing, blanching, or thermal processing, for example, on the stability or ease of removal of a substance. The roles food scientists and technologists play are not only challenging personally but are vital to keeping the U.S. food supply abundant, safe, wholesome, and consumptive of a minor part of the income of most U.S. families. Food scientists and technologists occupy a central position in determining this happy state of affairs. They are the ones who devise new processes, improve the effectiveness, and reduce the costs of present procedures. Many of the procedures leading to greater convenience for the consumer are costly. They often are instituted by marketing personnel. The efforts of food scientists and technologists offset in part what would be even higher costs if their input were lacking. In fact, were the efforts of food scientists and technologists not involved, the convenience features would generally be unattainable. For good or bad, depending on the merit one assigns to cost and convenience, food scientists and technologists are key players in such developments.

PROBLEMS IN THE INDUSTRY

Not all industrial forces are stimulatory of job opportunities for food scientists and technology. Leveraged buyouts (LBOs) have caused dislocations and graduating Ph.D.s have been hit pretty hard. Many of the LBOs have resulted in 200–500 research personnel being fired all at one time. When newly released doctorates with industrial experience are seeking to relocate, new graduates with no industrial experience find themselves at a disadvantage. The problem will partially correct itself with time. Some companies that disbanded certain groups, such as sensory departments, thinking the work could be done cheaper at the headquarters of the new owner soon found otherwise and the former operations had to be restored. There is no doubt, however, that as companies become bigger and bigger the trend will be to hire fewer employees than if there were a greater number of companies. Fortunately for those who are well educated, as food scientists and technologists must be, the demand becomes greater the more sophisticated industrial operations become.

A problem of great significance to the United States is that leveraged buyouts and some other industrial changes result in fewer dollars going into research (14). The money is needed to pay for the greater interest charges ensuing from that type of consolidation. Because U.S. industry is not putting into research the seedbed effort required for future gains in productivity, the United States is losing some of its former competitive advantage. That problem is especially critical on the value-added side of the food industry. The U.S. government has never allocated to value-added research the same kind of resources it has provided to production agriculture. While agriculture increased its value-added portion by 3.6 times from 1947 to 1982, food processing increased 6.6 times (1). The food industry made the gain it did largely from its own efforts; agriculture for years has had substantial research support from the U.S. Department of Agriculture. There are opportunities for food scientists and technologists in the research arms and some other functions of the federal government, but the kinds and numbers are not commensurate with the importance of food processing to the economic welfare of the United States. A consequence of research lagging in the United States is that many of our processed foods are imports rather than the domestically produced items they once were. The United States has become an importer of food-processing machinery instead of the exporter it once was. The production of food-processing machinery has little impact on the job market for food scientists and technologists (somewhat more so for food engineers), but loss of any kind of industrial endeavor is symptomatic of a general industrial weakness.

Notwithstanding some dislocations and undesirable trends, which like the complexity of the industry itself have scarcely been touched on, the food industry is still a major employer. As the industry becomes more and more complex in its scientific features and in the use of computer operations, any diminutions in jobs caused by industrial forces outside the control of food scientists and technologists should be offset by a more critical need for their services.

FUTURE FOOD PRODUCT DEVELOPMENTS

The American food supply has changed dramatically during the last 20 years. Convenience foods are far more abundant and come in a wider range of forms. Concern about the nutritional aspects of food are more prevalent today than they were. Part of that comes about because the Food and Drug Administration relaxed its prior stance on health-claim labeling. Because of the excesses of some companies, retightening of regulations dealing with health claims is under way. Eventually there will be nutrient delivery packages, customized for particular situations as a consequence of the understanding of interactions within the food, not just at absorption sites or within the body (21). Some of the technological advances will come by way of biotechnology (22). The task will be partially that of the nutritionist, but food technologists and engineers will be required, because they are the ones who understand the intricacies of evenly distributing small amounts of added substances, such as for enrichment purposes, throughout a bulk material.

In 1984 the Institute of Food Technologists conducted a special workshop on research needs (23). Among the recommendations for needed research was one specifying that a basic understanding must be acquired concerning the molecular and structural properties of food and how they affect the conversion, processing, distribution, storage, and acceptance of foods. Similarly, it was concluded that a basic understanding of the scientific basis of molecular and cellular mechanisms for inhibition and control of enzymatic, microbial, and entomological activities in foods must be secured. The subcommittee on research needs listed a whole series of processing techniques that must be investigated more fully—such as dewatering processes, utilization of supercritical extraction procedures, and separation of valuable constituents from complex mixtures—if food is to meet challenges of the future. At least 16 areas needing research were specified.

In engineering the same sort of examination was undertaken. A basic understanding of chemical, physical, and transport properties of food ingredients, products, processes, and packages must be acquired. So too is there a need for real-time sensors to monitor food operations. Biosensors will come into being (23). Development of robot technology should have high priority, especially where present operations are carried on in an environment possibly hostile to the health of workers. Biotechnology is, of course, already at work in the food industry to originate new strains of plant materials, eg, tomatoes, or to yield higher amounts of milk (24), the last not without question in the minds of many in the dairy industry.

Diet and health were referred to above. Among recommendations made by the subcommittee dealing with these subjects was one that a greater understanding be acquired of relations between nutrients and sensory acceptance, and moods and behavior. Food safety requires evaluation, both as to methods of testing and of establishing suitable rules to judge safety (19). The environment is coming increasingly to the fore. It has taken a long time for some of the concerns of Carson to be widely recognized, and the causes are not always as she envisioned them, but her book *Silent Spring* did set concerns to scintilating (25). The food industry is sometimes one of the contributors to a decline in the wholesomeness of the environment as when waste water is not properly managed or as when water is used more copiously than needed. The food industry must look more critically at its operations and conduct technological research for its own welfare in terms of cost, the availability of clean water, and air that does not contaminate a food product when used in pneumatic or drying operations.

The IFT report (23) is replete with issues critical to the future well-being of the food industry, the training of its scientists and technologists, and the interaction of the food industry with society. Any young man or woman having an interest in sciences and mathematics should consider a career in food science. There is scarcely a science that is not brought into play in some manner in the production, processing, storage, or utilization of food. For the betterment of those in industrial countries, basic and applied science are needed to foster still more advanced technologies. At the village level in developing countries, appropriate science and technology is needed so that these countries can begin, through their own efforts, to take better care of their citizens. The IFT report makes clear that research and all forms of scientific endeavor must go deeper and deeper into the unknown if significant advances are to be made in nutrition, health, efficiencies of processing, and at the myriad other junctures where science, technology, and the food supply meet.

BIBLIOGRAPHY

1. J. M. Connor, *Food Processing: An Industrial Powerhouse in Transition,* Lexington Books, D. C. Heath and Company, Lexington, Mass., 1988.
2. A. W. Bitting, *Appetizing or the Art of Canning: Its History and Development,* The Trade Pressroom, San Francisco, 1937.
3. S. A. Goldblith, "The Science and Technology of Thermal Processing, Part 1," *Food Technology* **25,** 1256–1262 (1971).
4. M. Frobisher, Jr., *Fundamentals of Bacteriology,* W. B. Saunders Co., Philadelphia, 1937, pp. 443–448.
5. S. A. Goldblith, "The Science and Technology of Thermal Processing, Part 2," *Food Technology* **26,** 64–69 (1972).
6. W. D. Bigelow, G. S. Bohart, A. C. Richardson, and C. O. Ball, *Heat Penetration in Processing Canned Foods,* Bulletin **16-L,** National Canners Association, Berkeley, Calif., 1920.
7. R. L. Hall, "Pioneers in Food Science and Technology: 'Giants in the Earth,'" *Food Technology* **43,** 186–195 (1989).
8. N. H. Mermelstein, "History of the Institute of Food Technologists: The First 50 Years," *Food Technology* **43,** 14–18, 35–52 (1989).
9. R. M. Schaffner, "What Training Should a Four-Year Food Technology Student Receive?" *Food Technology* **12**(9), 7–14 (1958).
10. "Report of IFT Council," *Food Technology* **20,** 1567–1569 (1966).
11. "IFT Undergraduate Curriculum Minimum Standards—1977 Revision," *Food Technology* **44,** 32–40 (1990).
12. O. Fennema, "Education Programs in Food Science: A Continuing Struggle for Legitimacy, Respect, and Recognition," *Food Technology* **43,** 170–182 (1989).
13. T. P. Labuza, "Mission 2000: IFT's Future," *Food Technology* **43,** 68–84 (1989).
14. P. Labuza and D. R. Lineback, "The University-Industrial Relationship in Food Science and Technology," *Food Technology* **41**(12), 74–91 (1987).
15. T. P. Labuza and D. P. Lineback, "The University-Industrial Relationship in Food Science and Technology," paper delivered at Food and Drug Law Institute "Food Update," Wesley Chapel, Fla., 1987.
16. Institute of Food Technologists, unpublished data.
17. International Union of Food Science & Technology, *Directory of Courses and Professional Organizations in Food Science and Technology,* 2nd ed., rev., CSIRO Division of Food Research, North Ryde, Australia, 1980.
18. National Academy of Sciences, *Toxicants Occurring Naturally in Foods,* Washington, D.C., 1973.
19. R. L. Hall and S. L. Taylor, "Food Toxicology and Safety Evaluations: Changing Perspectives and a Challenge for the Future," *Food Technology* **43,** 270–279 (1989).
20. R. L. Hall, "Toxic Substances Naturally Present in Food," *Food Product Development* **4,** 66–71 (1970).
21. F. M. Clydesdale, "Present and Future of Food Science and Technology in Industrialized Countries," *Food Technology* **43**(9), 134–144, 146 (1989).
22. S. Harlander, "Food Biotechnology: Yesterday, Today, and Tomorrow," *Food Technology* **43**(9), 196–206 (1989).
23. "Word Worthy, Chemical Sensors Play Growing Role in Analytical Sciences," *Chemical and Engineering News* **68,** 24–26 (Jan. 29, 1990).
24. I. Hart, J. Bines, S. James, and S. Morant. "The Effect of Injecting or Infusing Low Doses of Bovine Growth Hormone on Milk Yield, Milk Composition, and the Quantity of Hormone in the Milk Serum of Cows," *Animal Prod.* **40,** 243 (1985).
25. R. Carson, *Silent Spring,* Houghton Mifflin Co., Boston, 1962.

JOHN J. POWERS
University of Georgia
Athens, Georgia

FOODSERVICE SYSTEMS

OVERVIEW

The foodservice industry is an active, dynamic system energized by its external environment. Without this energy, the system fails. Foodservice, large or small, can be approached as a system of interrelated activities that make up the total system. As in any system, the total system is influenced by the basic operating activities.

Using the systems approach, inputs in the foodservice system would be the operational, physical, and human resources required to produce the operation objectives, a food(s) product. The activity, procurement, and production of food is the finished product; the quality of food and service are the output of the system. An internal control would be the menu. The federal and state laws and regulations governing the operation would be external controls. The foodservice system continually receives feedback—energy—from the internal and external environment. This feedback is provided by the customer comments, plate waste, cost, and frequency of selection of a food item that determines the effectiveness of the system. This external feedback is the energy that keeps the system going.

Any foodservice operation that supplies prepared food to a customer, such as institutions, fast-food operations, and restaurants, can be classified as a system. The development of new food technologies in the 1960's in food production, packaging, transportation; changes in life styles such as two-income families and increased mobility of the population, and the development of computers made possible the growth of modern foodservice. This large and complex foodservice industry accounts for more than 37% of consumer expenditures for food. It is the largest retail employer in the nation, with more than 8 million employees. In the past, the largest number of foodservice employees were women and teenagers. Today a growing number of senior citizens are being employed and are replacing the teenage employee segment.

The foodservice industry can be divided into three groups: commercial, institutional, and military feeding. The commercial feeding is the largest segment of the foodservice industry and includes those operations that are open to the public and supply meal service on a regular basis for a profit. The institutional feeding group includes all institutional organizations—business, education and government—that operate their own foodservice. Foodservice is an auxiliary service that complements the organization mission. Some of these establishments operate for profit, but profit is not the primary goal for the foodservice operation. Military feeding includes the foodservice to the troops, noncommissioned officers and officers clubs, and military exchanges foodservice.

Foodservice systems are designed for efficient purchasing, storage, preparation, delivery of product, service, and, in some cases, transportation. Recipes are standardized for each type of operation. Methods of preparation are controlled using assembly-line techniques in some facilities. Elaborate training systems are established, and computerized costing, purchasing, and controls are used. Larger operations may use sophisticated, computerized pressure cookers, fryers, grills, cooking vats and ovens. Microwave ovens are used to reheat or reconstitute many food items in many commercial and institutional operations.

Restaurants can prepare cooked-to-order food or use prepared food already portioned and shaped, eg, ready-cut potatoes for frying or meat and desserts that are preportioned and ready to prepare and/or serve. The larger restaurants and fast-food chains order food from a central location where whole meals are prepared, packaged in a manner so that they remain fresh. Some large chains order direct from the manufacturer. The pre-prepared food/meals are distributed to other establishments, sometimes at a great distance from the preparation facility. Preservation methods include fast freezing, precooking, and dehydrating.

Many successful foodservice operations have identified a particular segment of the population and focused on specific consumer desires. The take-out phenomena of foodservice grew from recognizing that needs of a particular consumer population were not being met. The ability to have complete meals delivered to the home has included the gourmet take-out, as well as pizza and other ethnic foods. The larger franchise chain operations use marketing research and advertising strategies to identify and increase customer sales.

Commercial foodservice includes eating and drinking operations, food and lodging facilities, food contractors, and various other foodservice operations for profit. For example, foodservice in drugs stores, grocery stores, convenient stores, (C-stores), recreational and sports centers.

Full-service commercial operations include fine dining establishments, family restaurants, coffee shops, cafeterias, specialty restaurants, and membership clubs. The chain operations provide more theme/specialty foodservice. The fast-food foodservice includes specialties in sandwich/hamburgers, seafood, and ethnic foods. This very competitive market uses marketing promotions and currently expanded menus to increase sales. In the 1980s more emphasis was placed on the take-out element of the fast-food operations. Some fast-food operations may be totally self-owned with many units, others are franchised operations. The franchised operations, owned by local operators, present a challenge to the parent company. Foodservice in hotel restaurants focuses on the business traveler as the primary customer. A trend in the all-suite hotels is to emphasize their foodservice operation. The hotel foodservice operations include snack bars, coffee shops, beverage bars, fine restaurants, catering, and room service.

Foodservice in department stores, supermarkets, and discount houses and other retail outlets make up another portion of the commercial market. C-stores foodservice have been the fastest growing segment of the commercial foodservice. Presently, 83% of all C-stores offer foodservice. The C-stores provide a one-stop service for the fast life-style of their targeted population. Department stores have been quick to expand to the take-out foodservice market to increase revenue. Many department stores provide gourmet take-out food items for their customers.

The foodservice operation in health-care institutions can be operated by in-house management or by a food contractor. Increasing costs, competition, and change of population have provided many challenges in the health-care industry. Because of the increasing number of older patients who require complicated dietary regimens and modified diets, operating expenses of hospitals, convalescent homes, and long-term care facilities have continued to rise. Foodservice operations in the health-care industry have had to become more efficient and productive. Feeding facilities for employees, the public, and catering are viewed as potential revenue sources.

Changes in the public's eating behaviors have also influenced changes in college and university feeding. College costs, fewer college-age persons, and more community colleges have had an impact on the foodservice operation. Nontraditional board plans, à la carte programs, cash operations, and a variety of menus are innovations that have developed. A variety of foodservice operations are available to students today. Foodservice may be obtained in residence hall dining rooms that provide meals for students living in the residence halls or off-campus, student union buildings, recreational and meeting facilities, bookstores, vending machines, and fast-service units with limited menus. Special dining halls for faculty and guests and catering activities are other types of foodservices that may be provided.

Centralization of various functions has also been a trend among college and university foodservice operations. For example, food purchasing, storage, prepreparation, and some production may be centralized. The preprepared item or raw material may then be transported to a satellite unit. Standardization and computerized operations have made the college and university foodservice operations more efficient and cost effective.

School lunch in the United States has a long history. The rejection of young men for military service in World War II because of poor nutritional status led the impetus to the development of the National School Lunch Program. In the beginning, this program only provided loans for communities to pay labor costs of school lunch. Since the passage of the National School Lunch Act in 1946, more than 23.5 million children have received a school lunch, about 50% of whom have received free or reduced price meals. The National School Lunch Act had a dual purpose, "to safeguard the health and well-being of the nation's children and to encourage the domestic consumption of nutritious agricultural products." Federal regulations control the meal pattern and provide for joint administration of the program through a federal-state-local school district relationship.

Legislation enacted since 1946 has had an impact on the size and direction of school lunch. Legislation has made the Child Nutrition Act, which provides a breakfast program, available to all schools. About 85% of the children participating receive a free breakfast. The program was targeted for low-income children in inner city schools. Another part of the federal Child Nutrition Program is the Summer Foodservice and Special Milk Program.

The Child and Adult Care Food Program is another federal program that supplies federal funds for meals for children and adults in day-care centers. This program is also controlled by federal regulations that specify nutritional requirements and administrative guidelines.

Utilization of surplus farm commodities was one of the initial purposes of the National School Lunch Program. The average amount spent on each student per year in 1988 was $92. Schools receiving commodities have a unique managerial challenge in planning for use and storage of these commodities.

School foodservice programs vary in type of organization and management from the one unit, one manager operation to a centrally managed unit in a large city school system. Many large school systems have centralized food preparation centers that distribute food in bulk or preplated, hot or refrigerated, for service to students in schools with satellite service centers. Since early 1970, program regulations have changed, allowing schools to negotiate with food contractors to provide meals. The number of foodservice contractors in school foodservice has increased in the 1980s.

Commercial and industrial organizations also offer employee feeding programs. The increase in white collar workers and gasoline prices have increased foodservice sales in commercial and office buildings. Industrial plants have seen a decline in foodservice for blue collar and manufacturing employees.

Food contractors use marketing strategies, eg, on-site bakeries, espresso coffee centers, take-home food items, and games, that offer free drinks or meals as rewards, to encourage sales. A variety of settings are used, from elegant executive dining rooms to vending operations.

In the transportation segment of foodservice, food contractors provide a major portion of in-transit airline feeding. Menus and food items are adapted to specifications of the various airline carriers. Sales in the transportation feeding for passenger and/or cargo liners and railroads have declined. The major thrust in transportation foodservice has been efficient productivity.

MENU SYSTEM

The foodservice operation begins and ends with the menu, a detailed list of foods to be served. A total list of food items offered by the establishment, is provided by a menu. The menu is also a guide for the kitchen and wait people and a printed list for the customers. The menu determines the foods to be purchased, equipment needed for production, personnel required for production and service as well as the work schedules of the employees. The menu is the marketing tool of the establishment; it influences the design of the kitchen and implements the organizational goals of the establishment.

Creative menu planning uses originality, imagination, and knowledge. Menu planning must be appealing, interesting, and kept within the budget. Advances in food technology make it possible to select foods from many forms—frozen or chilled, partially or completely prepared, preportioned or bulk, or prepared from raw materials.

The menu must be consistent with the goals of the organization and meet the food preferences of the identified customer. If the goal is to provide a nutritious meal for a specific customer, then meeting the nutritional requirement of the customer is important. Schools, day-care centers, hospitals, and other government institutions provide nutritious meals for their clientele.

Many commercial foodservice operations are providing healthy menus for their clientele/guest. A National Restaurant Association report found that restaurant menus have become more healthy. To accommodate the changing food habits and life-styles of their guests, foodservice operators are planning a more flexible meal schedule as well as healthy menus, eg, planning for early dinners, with lower prices, to attract the senior citizen population. Managers planning menus for any foodservice operation must be aware of the food preferences, food habits and cultural make-up of their customers. Monitoring current trends in eating behaviors is also necessary.

The amount of money available for food production must be known before planning the menu. This amount should be based on anticipated revenues from the sale of food and beverages. Food and beverage sales will generate money to be budgeted for the foodservice operation. This income must cover purchase of raw food, labor, and other operating expenses. In a profit operation it will also allow for appropriate profit. In a nonprofit foodservice operation, a raw food cost allowance per person, per meal or per day may be determined. Balancing high- and low-cost items may be important to provide menu variety and budget stability.

The facilities and equipment available should be considered when planning menus. Kind and size of ovens, kettles, and steamers, as well as the refrigerator and freezer space must be considered. The small equipment, such as baking pans, steam-table inserts, and small utensils used in production must also be considered.

The food preparation skill of the employees and the time the employee is available is an important factor when planning menus. Consideration must be given to the amount of hand preparation required for each food item, balancing high labor-intense food items with preprepared food items. Employees should be scheduled so that their skills can be used to the best advantage. The workload should be balanced each day and each week. Other factors to consider when planning menus include variety of foods that would be acceptable to the customer; variety of foods in terms of texture, flavor, methods of preparation; striving for contrast and maximum color presentation; and variety of fresh foods, when in season.

Presentation of the menu is as diverse as the number of foodservice operations: from a menu board, verbal presentation of food items available by the wait person, computer printout, or a very elegant printed menu on quality paper. Menus may be changed daily or cycle menus may be used.

Cycle menus are a series of planned menus used for a definite period of time, then repeated. Food preparation can become more efficient by using cycle menus. This allows for preparation procedures to be standardized. The cycle menu allows the employees a chance to become more efficient through repeated use of recipes. Most facilities repeat cycle menus every 2 to 8 weeks. The use of cycle menus can provide an aid in menu evaluation and improvement. Forecasting is more accurate because a cycle menu has items that appear in the same grouping each time they are produced. Cycle menus are used in institutional and military feeding and can be used in commercial feeding. Feedback and flexibility must be incorporated when using cyclical menus. Any type of menu should be monitored to respond to changing desires of the customer, season of year, and holidays.

Other types of menus that can be used in foodservice programs include table d'hôte prix fixe menu, a complete meal for one price; à la carte menu, food items priced separately; limited menu, offers a small number of selections, under 10 items; du jour menu, planned every day. A specialty of the house may be offered daily.

To determine effectiveness of the menu, many foodservice operations prepare a method to evaluate menus. This may be a record of food purchased, customer preference surveys, plate waste checks, or verbal feedback from customers. Production employees can contribute to the evaluation of menus. Providing a routine employee menu evaluation allows the employees some ownership/involvement in the production of the menus.

PURCHASING, RECEIVING, AND STORAGE SYSTEMS

To maintain an effective and cost-efficient foodservice operation the purchasing and inventory system must be carefully planned. Purchasing becomes more predictable, which leads to control. Selecting the establishment's suppliers on an objective basis and developing specifications and formal buying procedures contribute to efficient management.

The menu governs the requirements in purchasing, receiving, storing, delivering, preparing, holding, serving, and cleaning. Although the menu is the initiator, the establishment's image, types of service, and storage space will be considered when making purchasing decisions. Financial resources will be a factor in purchasing decisions. The purchasing manager will need to consider availability of food, number of suppliers in the area, skill of production employees, and the quality and value of food items.

Menus should be planned to combine food items that can be delivered at the same time to reduce cost. Careful use of competitive buying practices will help control food cost and upgrade the quality of the menu. Purchase specifications should be used to ensure the purchase of quality foods at competitive prices. Food items should be inspected on delivery to assure specifications are met. Whenever possible, foods that are federally graded and/or inspected should be purchased. Perishability of food and amount of storage space should be considered.

Records of the approximate cost per serving of each menu item in order to determine the cost of each item should be mandatory. Careful selection of a supplier is important to the purchasing manager. Vendors should be chosen who are service-oriented and financially sound. The supplier's warehouse should be evaluated on size, inventory, sanitation, personnel, and delivery equipment. Frozen food should be delivered in appropriate refrigerated trucks. Fresh produce should be delivered under appropriate conditions.

Clear specifications provide effective communication tools between the foodservice operation and the food distributor. The operation's specification is the contract between the two organizations, therefore, the specifications should be precise and detailed when appropriate. The formal, or competitive, purchasing method is used most frequently when purchasing large amounts. This method is mandatory for state and government operations. A written bid system is primary to a formal purchasing method. Purchase specifications are sent to several suppliers. In case of state or federal institutions, the formal bid opening will be publicly posted. The formal purchasing method provides control over foodservice purchasing. Informal purchasing methods can be used when purchasing small amounts.

Designing effective food receiving procedures provides control and predictability. Receiving procedures that involve daily reports, a tagging system, and invoice stamps are appropriate management tools. Providing a checking system to evaluate the accuracy of the foodservice receiving system is important.

Adequate storage and procedures for storage can reduce spoilage and product theft. Storage areas, including dry, refrigerated, and frozen, should be checked, and procedures for checking should be developed for control. A key and lock system is normally used.

Two types of inventory systems can be used, physical and perpetual, for inventory/product control. Physical inventory systems are essential; a perpetual inventory system can be used on expensive food items. A product requisition form provides efficient control of inventory.

PRODUCTION SYSTEM

Basic preparation standards and procedures are necessary in any foodservice system. A standard product can be developed consistently, and waste can be avoided. Pre-prepared foods can be used and will reduce the need for skilled labor, as well as reduce energy requirements.

Standardized recipes, or recipes adjusted for the foodservice operation, allow the production manager to predict quality, quantity, and portion cost of the finished product. The use of standardized recipes simplifies purchasing also. Employees must understand the importance of following standard recipes.

The controlling of portion sizes is necessary to maintain cost per portion. Decreasing the time between product preparation and serving prevents loss of nutrients and quality of food item. This is not always possible. Methods and schedules need to be coordinated in order to use employees time efficiently and to produce quality food products. It is essential to be aware of time/temperature control during the holding period. Good sanitation procedures are necessary during all food preparation.

The menu is the input and control, where production planning begins. The production system transforms the raw product into the salable output of the foodservice system. Forecasting is the prediction of the food needs for the meal, day, or defined period based on established trends. Forecasting is a procedure that can limit food waste and maintain cost controls. Past sales trends can be used along with present and future developments.

Other variables that need to be considered besides past sales are weather, time of year, and current economics. The object of forecasting is to minimize waste. Accurate forecasting will prevent leftovers and running out of food. Today, precosting can be done with the aid of a computerized system. All foodservice systems use standard calculations to analyze product cost and to determine the cost per portion for serving a food item.

The amount of food to prepare and the scheduling of preparation and work assignments are included in a work production schedule. The production schedule is vital to the efficient foodservice system. Special calculations can be used to determine the amounts of foods to prepare. Careful planning uses equipment, space, products, and employees more effectively. The production manager must be familiar with the steps and time each procedure takes to produce a quality food item. Until the final cooking, many steps should be planned early, even the day before, and coordinated with other production and service needs. Again, sanitation procedures should be used at all times during the production and holding procedures. Time schedules should allow time for heat to penetrate to the center of large pans. Final cooking should be planned to minimize the time before the serving. Batch cooking, especially of vegetables, should be used.

Production meetings with appropriate staff and employees should be regularly scheduled. Open communication is necessary for an efficient and effective kitchen. A written production schedule is necessary for a productive kitchen staff.

To maintain food quality, the purchase of quality food is the first step. Using standardized recipes, supervision of food production and critical tasting are other factors in producing quality food items. The food needs to be safe, nutritious, and pleasing to the targeted customer for the foodservice operation to be successful.

THE SERVICE SYSTEM

The waitperson, or server, communicates the establishment's attitude toward the customer. Some foodservice operations use the term guest to emphasize the value placed on their clientele. The server spends more time with the customer and may be the first and last person with whom the customer has contact. Good servers are aware of the customer's discomfort and/or dissatisfaction. Management should be notified of any customer complaint.

A full-service restaurant practices one of the four basic service styles, or a combination of several. Family-style service uses serving dishes placed on the table, and the customers serve themselves. This is also known as, English service. Smorgasbords and self-serve salad bars are forms of the family-style service. Plate service, or American service, is generally used by most restaurants. All the food is put on plates in the kitchen and served in the dining room. Bread and butter may be served on a bread and butter plate. French service, or table-side service, is another style. This service is characterized by the service of food from a rolling, heated cart by a chef de rang (waiter) and commis de rang (assistant). The waiter and assistant prepare finishing touches to the food on the cart at the table; the food is served to the customer from the cart. During platter service, or Russian service, the food is fully prepared and cut into portions in the kitchen. The waiter serves the food on platters directly to the customer. Both French and Russian service allows for showmanship and drama at the customer's table. More equipment, utensils, and skilled waiters are needed for this type of service. Portion control may also be a problem.

HUMAN RESOURCE SYSTEM

Foodservice operations have had a negative production record in the past, but the trend is to reverse this image. The foodservice industry is learning to maximize its best resource, the people who are employed in the industry. To increase productivity and decrease employee turnover should be the objective of a good foodservice manager. Effective recruitment, selection, orientation, and training will enhance these objectives.

A plan for recruiting and selecting new employees communicates the knowledge and skill needed for employment. Employee orientation and training programs, developed by the fast-food chains, have emphasized the need for training programs in all foodservice operations. Managers need to understand the principles of learning to maximize training efforts.

Developing standards of performance to evaluate employee and staff can increase the employee's involvement and motivation. Management by objective can be used for staff personnel and is an effective personal motivator. Management by walking around (MBWA) can be an effective way to catch an employee doing something right. MBWA adds to more positive employee performance evaluations. Using the MBWA technique will allow the manager to correct inappropriate behavior or work procedures on the spot. An employee manual with the foodservice operation's policies and procedures is necessary for good employee-employer relationship. A positive approach, humor, and a good self-image can be valuable assets for a foodservice manager. Providing a positive work environment and paying attention to the employee and the employee's contributions to the organization's goals can be the most effective motivator in the industry.

EQUIPMENT AND ENERGY SYSTEM

A knowledge of equipment and energy is necessary to provide efficient transformation of the raw food to quality food products. Equipment must be properly operated and maintained to preserve the integrity of the equipment and its value. Training of personnel in the care and use of the equipment is important for preserving the equipment. When purchasing a piece of equipment, consideration should be given to the initial cost, installation, maintenance, personnel using the equipment, and energy cost. Calculations can be used to determine these answers as well as the equipment cost over time.

In energy management, a planned and organized program is necessary for an efficient organization. An equipment maintenance checklist can provide a starting point. Many equipment manufacturers and utility companies have tips for energy conservation.

QUALITY ASSURANCE/CONTROL SYSTEM

Quality assurance (QA) programs allow foodservice operations to be proactive rather than reactive. Satisfaction of the customer is the goal of the operation. Quality assurance programs ensure the attainment of that goal. The use of the hazard analysis critical control point (HACCP) concept has been recommended as a preventive approach to quality control in foodservice systems. Quality audits are an important component of a QA program and provide feedback. These audits allow management to control the output of the system.

BIBLIOGRAPHY

General References

"Market Trends," *Restaurants and Institutions*. **99**:32, 1989.

L. J. Minor and R. F. Cichy, *Foodservice Systems Management*, AVI Publishing Co., Westport, Conn., 1984.

M. C. Spears and A. G. Vaden, *Foodservice Organizations, A Managerial and Systems Approach*. Macmillan, New York, 1985.

B. B. West and L. Wood. *Foodservice in Institutions,* 6th ed., Macmillan, New York, 1988.

DOROTHY POND-SMITH
Washington State University
Pullman, Washington

FOOD SPOILAGE

Spoilage is a major cause of food loss. Much of the food that is produced fails to reach the consumer. Moreover, food often becomes either unpalatable or unsafe to eat after it is purchased. Thus, spoilage is a problem for both the food industry and the consumer (1).

Although most people think they know what spoilage is, the term spoilage often conveys different images to different people. In the broadest sense, spoilage refers to any change in a food that makes the food unacceptable for consumption. Usually these changes are real and involve obvious defects in sensory characteristics such as color, odor, or flavor. In other cases, the changes may be equally real but more difficult to detect. The presence of harmful bacteria or the loss of nutritional quality falls into this category. Finally, food is often considered spoiled because it is perceived to be of unacceptable quality. This type of spoilage is often difficult to describe and quantify. Consequently, changes that are considered spoilage to one person may not be considered so by another. Often, consumers even disagree on whether or not a food is spoiled when changes are obvious. An example would be the souring of milk into yogurt. Some consumers consider yogurt nothing more than spoiled milk while others consider it a treat. Thus, one should always provide the rationale for categorizing a food as spoiled.

CAUSES OF FOOD SPOILAGE

The reasons foods spoil are many and can be both intrinsic and extrinsic. That is, the cause of the spoilage can come from external sources or from the food itself. In general, there are three major causes of food spoilage. These are microbial growth, biochemical or physiologic deterioration, and physical damage. The relative importance of each cause differs for particular foods. In many foods, however, more than one cause contributes to the spoilage of the product. Foods that are highly susceptible to spoilage are termed perishable.

Microbiological Spoilage

Microorganisms are a major cause of food spoilage. Microbes are a normal part of our environment and can be found almost everywhere on earth, including in foods. The types of microorganisms that are responsible for spoiling foods include bacteria, molds and yeasts. However, not all microorganisms present on food will cause spoilage. Most microorganisms present on foods are essentially hitchhikers and are usually harmless. However, some normally harmless microbes can cause spoilage if conditions are right. These organisms, known as opportunists, are normally not a problem in foods. Unlike the opportunists, a few kinds of microorganisms are consistently responsible for causing the bulk of food spoilage. These true spoilage organisms have the ability to actively degrade the food. It is this process of degradation that causes the changes that we associate with spoilage. Finally, microorganisms are able to cause disease in humans or animals. These organisms, known as pathogens, are among the least common organisms involved in food spoilage. Nevertheless, the serious consequences of their presence in foods makes them of most concern to food processors and consumers.

Microorganisms are not engaged in a sinister plot to spoil our foods. They are simply trying to survive like any other creature. Like other living things, microorganisms require nutrients, water, and minerals to survive and grow. Microorganisms degrade foods to obtain these requirements. Foods that are especially moist and nutritious are most favorable for the growth of microorganisms and are the ones most susceptible to spoilage. However, many other factors will also influence the growth and survival of microorganisms in foods. These factors include acidity of the food, the presence of natural or added antimicrobial chemicals, and the amount of usable water present.

Different types of microorganisms are often best adapted to specific growth habitats (2). In general, bacteria are adapted to fast growth in nonacidic (pH 5.0–8.0) and moist conditions. Some bacteria grow at refrigeration (psychrotrophs) or hot temperatures (thermophiles), but most (mesophiles) grow best at temperatures from about 15 to 40°C. In contrast, molds grow more slowly but are better adapted to growth in relatively acidic (pH 2.5) or dry conditions. Yeasts are similar to molds, but many also tolerate conditions containing high concentrations of sugar. However, exceptions exist for all three groups.

The type of microbial spoilage that one finds varies with the food in question. Because most nuts and grains are usually dried before storage, they are quite resistant to microbial spoilage. However, microorganisms will cause spoilage if these foods are improperly dried or not kept dry. Even improperly dried nuts and grains are usually dry enough to prohibit the growth of bacteria. Thus, these products are most often spoiled by molds. Moldy nuts and grains usually, but not always, have a fuzzy appearance due to the presence of the mold's mycelium (fuzz). Whether obvious or not, the presence of molds on nuts and grains is a serious concern because some molds can produce dangerous toxins (mycotoxins).

Fresh fruits and vegetables are very moist and thus highly susceptible to microbial spoilage. However, these two foods differ significantly in several ways. Fruits are normally quite acidic and often contain relatively high concentrations of sugars. In contrast, vegetables usually possess a neutral pH and contain less sugar. Consequently, the type of organisms responsible for spoiling fruits and vegetables are also quite different. The acidity of fruits makes them quite resistant to bacterial growth, but these conditions support growth of molds. Mold spoilage in fruits usually appears as a darkened or rotted area accompanied by obvious mycelium. Rotting occurs because molds produce enzymes, such as pectinase, that digest the pectin and cause softening of tissue. Yeasts often cause secondary spoilage once mold spoilage or other damage has occurred and allowed them access to sugar-laden inner tissues.

Both bacteria and molds are important causes of vegetable spoilage. However, bacteria can grow faster than molds and are therefore a more likely cause of spoilage. The symptoms of mold spoilage in vegetables are similar to that in fruits. Bacteria usually produce a type of spoilage in vegetables known as wet or soft rots. Soft rots initially begin as small sunken areas on the vegetables. Eventually, the sunken area becomes soft and wet and spreads to encompass the entire vegetable. As with molds, these symptoms are caused by pectinolytic enzymes produced by spoilage bacteria. Some bacteria are especially proficient at producing these enzymes and are common sources of spoilage. These genera include *Erwinia*, *Pseudomonas*, *Xanthomonas*, and *Clostridium*. Of these, *Pseudomonas* is the most common in refrigerated products (3).

Meats, poultry, seafoods, and dairy products are very perishable because they contain virtually all the nutrients and moisture microorganisms require. These foods are usually spoiled by psychrotrophic bacteria, although molds do cause some spoilage of dried and fermented meats and fermented dairy products. *Pseudomonas* species and related bacteria cause most of the spoilage in meats, poultry, and fish. These bacteria are capable of producing many degradative enzymes, especially those that degrade proteins (proteases) and lipids (lipases). Consequently, symptoms of spoilage in these foods are usually similar. Spoiled meats, poultry, and fish usually develop putrid odors as they spoil and often become slimy or discolored. Putrid odors develop when proteolytic enzymes degrade proteins, and rancidity is caused when lipids are broken down by lipolytic enzymes. The slimy appearance often results from bacteria producing slime on the surface of the food.

Spoiled dairy products may also develop off-flavors or odors. The specific defect often differs with the dairy product. Spoiled pasteurized fluid milks usually taste and smell bitter, sour, rancid, and, occasionally, putrid. In general, *Pseudomonas* and related bacteria cause these defects in fluid milk. As with meats, spoilage bacteria in milk produce proteolytic and lipolytic enzymes that attack proteins and lipids, respectively. Various amino acids and peptides resulting from proteolysis give the bitter or putrid flavors to spoiled milk. Souring is caused when lactic acid bacteria ferment milk sugar (lactose) into lactic acid. In addition to flavor changes, the souring also causes the milk to clot into what is often called clabbered or curdled milk. On occasion, milk can also become slimy, a condition known as ropy milk.

Fermented dairy products, such as cheeses, can be spoiled by both bacteria and fungi. Moreover, gas-producing bacteria can cause holes, gas pockets, cracking, and other defects to develop in cheeses. Fungi, especially molds, are a common problem in cheeses, cottage cheese, and yogurt. Usually, these fungi cause the food to become fuzzy and unappetizing. However, some molds are very proteolytic or lipolytic and can cause major flavor or textural defects. In addition, the potential for mycotoxin production also exists. Yeasts can also cause some dairy foods to become discolored, frothy, and off-flavored.

Most people are unaware that food-poisoning bacteria can grow to dangerous levels in foods without affecting flavor, odor, or color. Usually, it is the nonpathogenic spoilage bacteria present that cause changes in sensory qualities. Because consumers instinctively reject eating foods with abnormal odors or flavors, these changes act as a valuable warning that the food may not be safe to eat. However, cooking can remove this safety factor by killing the spoilage, as well as the pathogenic, bacteria in the food. If pathogenic bacteria then recontaminate the cooked food, they can grow and make the food unsafe to eat.

Although pathogenic microorganisms cause the majority of food poisonings, the growth of nonpathogenic spoilage bacteria can also cause illness. A notable example of this situation is an illness known as scombroid food poisoning (4). This disease is a form of histamine poisoning. It is especially a problem in fish, such as tuna or mackerel, that have undergone bacterial spoilage. The breakdown of proteins resulting from bacterial growth causes the accumulation of histamine in fish tissues. Consumers who eat fish that have been spoiled in this way normally show symptoms typical of strong allergic reactions.

Biochemical/Physiologic Spoilage

All foods, in their most basic forms, are simply composites of various naturally occurring chemicals. Because they are chemical mixtures, foods are subject to chemical reactions and biochemical processes. Many of these reactions and processes lead to what consumers view as deterioration in quality.

Microbiological spoilage notwithstanding, some foods are naturally more perishable than others. Fresh fruits and vegetables are notable in this regard because they continue to live, respire, and change after harvest. However, these changes are not always bad. Some fruits, such as tomatoes and bananas, continue to ripen after harvest. Fruits that behave in this way are often termed climacteric. The ripening process in climacteric fruits usually improves the sensory characteristics of these products immediately after harvest. At some point, however, the fruit becomes overripe and sensory quality deteriorates. Non-climacteric fruits and vegetables, such as citrus fruits or beans, do not ripen or improve after harvest. Any changes that occur are usually degradative. This degradative metabolic activity, known as senescence, results in changes that many consumers often find objectionable and associate with spoilage.

Many factors affect the respiration, ripening, and senescence of fruits and vegetables (5). Of these, ethylene is thought to exert a major influence, particularly with climacteric fruits. This compound acts as a plant hormone and is involved in the initiation and acceleration of ripening. Ethylene gas is widely used to promote ripening of certain fruits, most notably bananas. However, fruits and vegetables also produce trace amounts of ethylene during senescence and as a result of cellular disruption. Ethylene can promote premature spoilage in sensitive products exposed to excessive amounts of the gas. In addition, physiologic spoilage will also occur in ethylene-sensitive products that are stored together with fruits producing large amounts of the compound.

Several techniques are used to minimize spoilage in fresh produce. One technique being increasingly used is modified atmosphere storage. This technique involves storing foods in atmospheres containing enhanced carbon dioxide and reduced oxygen concentrations. Storing fresh produce in this way enhances shelf life by slowing the rate of respiration and retarding the ripening process. The mechanism for the enhanced shelf life is related to the general equation for respiration:

$$\text{Glucose} + \text{oxygen} \rightarrow \text{carbon dioxide} + \text{water}$$

According to this formula, reducing oxygen or increasing carbon dioxide should theoretically slow down respiration and ripening (6).

The optimum concentrations of gases required for extension of shelf life varies with different products (7). For most fresh produce items, at least 1–3% oxygen is required to maintain a minimum level of respiration. Using lower oxygen concentrations causes respiration to cease and the tissues to become anaerobic. This environment often leads to the development of off-flavors (5). The usual concentration of carbon dioxide necessary to extend shelf life is about 5–20%. However, the exact amount of carbon dioxide used is critical to maintaining product quality. In fact, using an elevated carbon dioxide concentration is not always desirable. An excessive concentration of carbon dioxide causes uneven ripening (tomatoes), off-flavors (broccoli and cauliflower), surface pitting (asparagus), or discoloration (celery).

Refrigeration is another common technique used to extend the shelf life of fresh produce. In addition to inhibiting microorganisms, chilling works by slowing down respiration in the product. In general, the lower the temperature at which a product is stored, the slower it will respire and spoil. However, the minimum temperature at which a given fresh produce item is stored varies with the product. Refrigeration will actually increase respiration in some fruits and vegetables (5). In addition, some fruits and vegetables are damaged by low but nonfreezing temperatures. This damage, often termed chilling injury, can become irreversible and often only appears when the affected product warms to nonrefrigeration temperatures. The simplest way to avoid chill injury is to store susceptible products above the minimum temperature at which damage occurs. However, many products are damaged by unexpected cold weather or defective refrigeration equipment. Thus, agricultural scientists are also developing chill tolerant cultivars of normally chill-sensitive products.

A variety of crops are susceptible to chilling injury, but fruits and vegetables grown in warm climates are among those most commonly cited. Symptoms of chilling injury likewise vary with the product and degree of damage. Typical symptoms can include disruption or arrest of normal ripening, development of off-flavors, or darkening or pitting of tissues. For example, tomatoes held at temperatures below 13°C fail to develop normal red color and typical fresh tomato flavor. The exact reasons why these defects occur is as yet unknown. However, membrane damage is suspected as the most likely mechanism for chill injury (5).

Spoilage can also be attributed to a deficiency or excess of water in foods. Dehydration causes many defects related to water. This condition results in unsightly wilting and also makes fresh produce more susceptible to decay by microorganisms. Dehydration can also be a major concern to food wholesalers and retailers because many foods are sold by weight. Losses in moisture that might be almost imperceptible to consumers can nevertheless represent significant economic losses to those selling the product. A common form of dehydration that is often not recognized as such is freezer burn. This defect often occurs when foods are frozen without proper wrapping or packaging. The actual damage occurs when ice sublimates out of the tissue and causes surface dehydration. Freezer burn usually appears as discolored and toughened areas on the exterior of affected foods.

An excess of water in or on a food can also lead to spoilage. Water that accumulates on the surface of foods serves both as a growth and dispersal medium for microorganisms. This problem is especially troublesome with fruits and vegetables because they expel water vapor (transpire) when they respire. This vapor sometime accumulates on the surface of the foods, especially during refrigerated storage.

Various techniques are used to prevent an excess or deficiency of water from spoiling foods. Storing foods in the presence of high relative humidities minimizes dehydration of foods. However, employing excessively high relative humidities encourages mold growth. For most foods, a relative humidity of about 90% prevents dehydration but also minimizes microbial growth. Various sealants, barrier films, and other new packaging techniques are also used to maintain optimum moisture in foods.

Enzymes

Enzymes are proteins that act as highly efficient catalysts in biochemical reactions. Enzymes are produced by virtually every living organism and are therefore also naturally present in foods. The specific functions of enzymes vary widely. Some are primarily involved in biosynthesis, whereas others act primarily in degradative processes. Both types of enzymes are functioning within the cell, but spoilage problems develop in foods when the degradative enzymes begin to predominate (8).

Enzymes can affect almost every quality attribute of foods. Many flavor defects that exist in foods are caused by enzymatic activity. High-fat foods often become rancid and are especially sensitive to oxidative enzymes. Enzymes known as lipases are responsible for splitting lipid molecules. The result of this lipolysis is the production of biochemical compounds associated with rancid flavor. The rancidity that develops in butter after improper processing of cream is one example of the effects of lipase.

Defects in texture can also be attributed to enzymes. Pectinases cause softening of many types of plant foods. Similarly, native proteolytic enzymes can cause meats to take on a mushy texture if they are aged too long.

Colors of foods can also be affected by enzymes. An obvious example of color changes resulting from enzymes is the darkening or bruising of cut fruits and vegetables. This darkening results when enzymes known as polyphenoloxidases react with oxygen and phenolic compounds in the fruit to form dark pigments. Conversely, enzymes can also cause breakdowns of colors in foods. Chlorophyllases and anthocyanases can degrade the green and reddish colors, respectively, in vegetables and fruits (9). Finally, important nutrients can also be degraded by enzymes. For example, thiaminases naturally present in fish catalyze the degradation of thiamin (vitamin B_1) and oranges contain ascorbic acid oxidase, a cause of ascorbate (vitamin C) degradation in citrus juices.

Because enzymes are important factors in spoilage, food processors often take steps to avoid or minimize degradative enzymatic reactions. In most foods, enzymes are compartmentalized within cells and cellular organelles (8). Any procedure that causes foods to be cut or crushed will allow enzymes to escape their compartments and possibly cause spoilage. Thus, an important factor in preventing spoilage by enzymes is to avoid tissue damage. However, cutting, slicing, and other handling procedures are integral parts of food processing. Therefore, additional ways to minimize enzymolysis must also be used. Various food additives, such as acids, antioxidants, or chelators stop or retard the actions of some enzymes, For example, sodium bisulfite or citric acid are often used to prevent enzymatic browning on potatoes and salad vegetables.

Thermal processing, or blanching, is another technique often used to prevent enzymatic reactions. For example, vegetables destined for frozen storage are usually blanched with steam or hot water at 80–100°C for several minutes before freezing. Eliminating blanching can allow enzymes to cause detrimental changes in flavor, color, texture, and nutritional value during storage. However, the minimum amount of heat necessary varies with different enzymes. Some bacterial proteases, for instance, can survive normal pasteurization processes used with milk and cause off-flavors during storage.

The presence of oxygen can cause changes in foods that are often perceived as spoilage. Many of these changes are catalyzed by oxidative enzymes, but others are caused simply by the presence of oxygen. The color of fresh meats, for example, undergoes several changes (9). Myoglobin, the primary meat pigment, changes from a purplish color to bright red on exposure to oxygen. The oxidized form of myoglobin, oxymyoglobin, continues to become oxygenated and converts to brownish metmyoglobin on further exposure to air. Consumers often reject this brownish colored meat because they perceive it to be old or of poor quality. Oxygen can also affect the flavor of meats in a negative way. A defect commonly known as warmed-over flavor occurs in meats that are cooked, refrigerated, and then later reheated before consumption. This defect is particularly a problem in industries, such as airlines, which rely on this type of food preparation. Warmed-over flavors result when lipids react with oxygen and other meat components to bring about the rapid development of rancidity (10). Other high-fat foods, such as potato chips or breakfast cereals, are also subject to oxidative rancidity and associated off-flavors. Consequently, such foods often must rely on antioxidants or special oxygen impermeable packaging to protect them from oxidation.

Light can cause similar reactions as those mentioned for oxygen. Like oxygen, light primarily reacts with food lipids and causes them to become oxidized and rancid. A classic example of light-induced flavor changes is the development of cardboard flavor in milk. This defect, also known as sunshine flavor, was common in milks that were delivered to homes in clear glass bottles. The name for this defect arose because milks developed an off-flavor after being set outside homes in the morning sun. Today, a similar problem occurs when milk packaged in plastic bottles is stored in refrigerated cabinets illuminated by florescent lights. Light also causes reactions that degrade vitamins. Riboflavin (vitamin B_2) is one casualty of light degradation. Other vitamins, such as ascorbic acid, are similarly lost to oxidation. Various types of food packages that block all or certain wavelengths of light are often used by food processors to prevent light-induced oxidation.

Physical or Mechanical Spoilage

Some people do not consider physical damage a type of spoilage. However, damaged food products are often less likely to be purchased or consumed and therefore fit the definition for spoiled foods. There are many ways that foods can be damaged, but most can be categorized into two broad categories. The first category is mechanical damage. This occurs when inert objects strike, abrade, or puncture the food. The equipment used to harvest or process foods often causes damage. In addition, severe weather can also directly or indirectly damage some foods. Hail is a common and well-known means by which weather can damage foods. In addition, strong winds can blow sand and debris against the product or allow products to rub against one other, causing abrasions or bruises. Freeze damage is a type of spoilage that can be considered physical and can be related to either weather or improper storage conditions. Freezing causes damage when ice crystals form within tissues and then rupture or puncture the cells of plants or meats. Foods subjected to freeze damage usually become mushy or discolored without the toughening associated with freezer burn. Freezer burn, discussed previously, is also considered by some to be a form of physical damage.

Living creatures can also damage foods. The most notable creatures in this regard are insects. Many insects feed on plant products during growing and harvesting and can cause extensive loss. In addition, some insects, such as grain beetles or various weevils, also damage foods during storage. However, plant products are not the only type of foods plagued by insects. Certain cured meats, such as country-style hams, are often damaged by insects that lay their eggs in the meat. The larvae arising from the eggs burrow and feed on the meat, causing a corky appearance. Both small and large animals are also responsible for damaging foods. Mice and rats not only cause damage when they chew but further make the food unusable when they urinate and defecate on foods. Larger animals, such as deer, can damage fruits or vegetables with their hoofs when they walk or run through fields.

Physical damage usually has immediate and obvious consequences. However, damage can also have more subtle or delayed effects on the food. Punctures and abrasions are not only cosmetic defects but damage protective skins or peels of fruits and vegetables. This condition allows microorganisms to more easily gain access to inner tissue and increases microbial spoilage later. In addition, physical abuse or insults can cause a hidden or latent damage. In such cases, symptoms of the abuse are not immediately obvious but show up days or weeks later. Although not damage in the strict sense, many foods can also be spoiled by physical adsorption of odors. Those foods that are high in lipids are particularly sensitive. For example, milk can acquire off-flavors from unclean barns, and nuts can become tainted with the flavors of other products with which they are stored. Consequently, packaging or other barriers to migration of flavors must often be used in such products.

An important concept to remember is that the onset of one type of spoilage may increase the chance that others will occur. For example, insects often serve as vectors for specific types of microbial spoilage. For example, the common fruit fly, *Drosophila malanogaster,* often serves to spread the mold that causes *Rhizopus* soft rot in vegetables (11). Thus, spoilage should be thought of in the broadest sense whereby all forms of spoilage are either directly or indirectly related.

Food spoilage has a major impact on all aspects of agriculture, from production through consumption. Spoilage still causes the majority of food loss despite the mammoth amount of research on food processing and preservation done by food scientists. This research has solved some problems and allowed at least the developed countries to have the greatest and safest supply of food ever. Nevertheless, the development of new varieties of foods and food-processing technologies will continue to make food spoilage an ever-present problem.

BIBLIOGRAPHY

1. E. C. D. Todd, "Impact of Spoilage and Foodborne Disease on National and International Economies," *International Journal of Food Microbiology* **4**, 83–100, 1987.
2. G. Hobbs, "Ecology of Food Microorganisms," *Microbial Ecology* **12**, 15–30, 1986.

3. B. M. Lund, "The Effect of Bacteria on Post-harvest Quality of Vegetables," in P. W. Goodenough and R. K. Arkin, eds., *Quality in Stored and Processed Vegetables and Fruit,* Academic Press, New York, 1982, pp. 287–300.
4. S. H. Arnold and W. D. Brown, "Histamine (?) Toxicity from Fish Products," *Advances in Food Research* 24:114–154, 1987.
5. R. S. Rolle and G. W. Chism III, "Physiological Consequences of Minimally Processed Fruits and Vegetables," *Journal of Food Quality* 10, 157–177, 1987.
6. D. Zagory and A. A. Kader, "Modified Atmosphere Packaging of Fresh Produce," *Journal of Food Science* 42(9), 70–77, 1988.
7. P. Brecht. "Use of Controlled Atmospheres to Retard Deterioration of Produce," *Food Technology* 34(3), 45–50, 1980.
8. T. R. Richardson, "Enzymes" in O. R. Fennema, ed., *Principles of Food Science. Part 1, Food Chemistry,* Dekker, New York, 1976, pp. 285–345.
9. F. M. Clydesdale and F. J. Francis. "Pigments," in O. R. Fennema, ed., *Principles of Food Science, Part 1, Food Chemistry,* Dekker, New York, 1976, pp. 385–426.
10. A. Asghar, J. I. Gray, D. J. Buckley, A. M. Pearson, and A. M. Booren, "Perspectives on warmed-over flavor," *Food Technology* 42(6), 102–108, 1988.
11. R. E. Brackett. "Vegetables and Related Products," in L. R. Beuchat, ed, *Food and Beverage Mycology,* 2nd ed, Van Nostrand, New York, 1987, pp. 129–154.

General References

N. W. Desrosier and J. N. Desrosier, *The Technology of Food Preservation,* 4th ed., AVI Publishing Co., Westport, Conn., 1982.

M. P. Doyle, *Foodborne Bacterial Pathogens,* Dekker, New York, 1988.

J. M. Jay, "Food Spoilage," in *Modern Food Microbiology,* 3rd ed., Van Nostrand Reinhold, New York, 1986, pp. 191–255.

R. H. H. Willis, T. H. Lee, D. Graham, W. B. McGlasson, and E. G. Hall, *Postharvest. An Introduction to the Physiology and Handling of Fruit and Vegetables,* AVI Publishing Co., Westport, Conn., 1981.

ROBERT E. BRACKETT
University of Georgia
Griffin, Georgia

FOOD STRUCTURE AND MILK PRODUCTS

Milk is a complex solution of salts, lactose, and proteins in which corpuscular proteins (casein micelles) and fat globules are dispersed. Biochemical and structural aspects of milk synthesis and secretion are discussed in another work (1). Only a small part of the milk produced is consumed in the fluid state, whereas the greater part is processed into various food products (2). Milk products may be divided into those based on protein (yogurt, cheeses) (3) and products based on fat (cream, butter) (4).

Milk proteins have a unique ability to curdle, ie, to form a gel. Curdling is induced by proteolytic enzymes, particularly chymosin (rennet) (5,6), lactic acid (produced by bacterial cultures) (7) and other acids (8,9), and also by heating (10,11). Each milk gel has a solid structure consisting of a protein matrix composed of casein as the main structural component. This matrix may contain additional components such as whey proteins, fat globules, lactic acid bacteria, and/or other ingredients, which are either the integral parts of the gel structure or are interspersed in the matrix depending on the conditions under which the gel had been formed. The ability to immobilize the liquid phase (milk serum, whey) is one of the most important properties of the gel matrix. By modifying this ability, it is possible to manufacture stable milk products with a high water content (yogurt) or products having a low water content (hard cheeses).

Milk fat is part of a great variety of milk products based on protein and is the main structure-forming component of products based on fat such as cream, whipped cream, ice cream, cream cheese, and butter. Differences in the composition and structures of products based on protein and products based on fat are considerable and contribute to the great variety of milk products.

The structure of the milk products determines their other properties such as firmness, spreadability, elasticity, viscosity, and susceptibility to syneresis, which are globally called texture. The understanding of the processes that lead to the development of the structure and the relationships between structure and texture (12) is important to various aspects of research and development as well as manufacture (13).

The structure of milk products is generally studied by microscopy. Optical microscopy may be used in various forms, including traditional methods (2), fluorescence microscopy (14,15), and confocal scanning laser microscopy (cslm) (16) developed recently. Electron microscopy consists of two major types, scanning electron microscopy (sem) and transmission electron microscopy (tem). Individual microscopical techniques are suited to meet particular needs (17–19) (Table 1). Examination of each sample by at least two different techniques and comparison of the results increases the probability that the micrographs obtained portray the original structure correctly (20). Although optical microscopy (unlike electron microscopy) does not provide high resolution, it is useful for the evaluation of the structure of the product in general and is also used to characterize the chemical nature of the individual components by specific staining procedures. Fluorescence microscopy is based on the ability of some structures to fluoresce in ultraviolet light in their native states or following their coupling with fluorescent stains (fluorophores) (15). Fluorescence microscopy provides a higher resolution and sensitivity of detection than does brightfield microscopy.

Unlike optical microscopy, electron microscopy is carried out *in vacuo*. The samples must be treated in such a way that they do not release vapors or gases when inserted into the microscope. This condition may be met by chemical fixation and drying of the sample for conventional SEM (20), or by freezing it at $-150°C$ in Freon 12 or at $-205°C$ in nitrogen slush and examining it by cryo-sem (21). For tem, the sample is embedded in a resin and sectioned, or a freeze-fractured sample is replicated with platinum and carbon (tem) (18,22). Additional sem and

Table 1. Type of Microscopy Used for Figures in This Article

Figure	Subject	Type of Microscopy Used
1a	Casein micelles	Unidirectional metal shadowing, tem
1b	Casein micelles	Thin section, tem
1c	Casein micelle	Rotary metal shadowing, tem
2a	Protein in milk	Freeze-fracturing + metal shadowing, tem
2b	Protein in milk	Freeze-fracturing + metal shadowing, tem
2c	Protein in milk	Freeze-fracturing + metal shadowing, tem
3a	Casein micelle	Rotary metal shadowing, tem
3b	Casein micelles	Negative staining, tem
3c	Casein micelles	Negative staining, tem
4a	Milk foam	Thin section, tem
4b	Milk foam after 1 h	Thin section, tem
5a	Coagulating casein micelles	Rotary shadowing, tem
5b	Yogurt	Thin section, tem
6a	Yogurt (10% total solids)	Conventional sem
6b	Yogurt (20% total solids)	Conventional sem
6c	Milk gel	Conventional sem
7a	Whey–protein in yogurt	Thin section, tem
7b	Mucogenic bacteria in yogurt	Conventional sem
7c	Mucogenic bacteria in cottage cheese	Freeze-fracturing, drying, rotary coating, tem
8a	Kefir grains	Macrophotography
8b	Microorganisms in kefir	Conventional sem
8c	Kefir drink	Conventional sem
9a	UF retentate gel	Thin section, tem
9b	UF retentate gel	Thin section, tem
10a	Homogenized retentate gel	Thin section, tem
10b	Nonhomogenized retentate gel	Thin section, tem
10c	Retentate gel	Lipid fixation, conventional sem
10d	Whole-milk curd	Freeze-fracturing, metal shadowing, replication, tem
11a	Curd granule junction in cheese	Conventional sem
11b	Curd granule junctions in cheese	Microphotography
11c	Cheddar cheese junctions	Microphotography
12a	Young Cheddar cheese	Thin section, tem
12b	Young Cheddar cheese	Conventional sem
13a	Camembert cheese—hyphae	Conventional sem
13b	Penicillium camemberti	Cryo-sem
14	White cheese	Thin section, tem
15a	Paneer cheese fried in oil	Thin section, tem
15b	Fried Paneer cooked in water	Thin section, tem
16	Ricotta cheese	Thin section, tem
17a	Process cheese	Thin section, tem
17b	Process cheese	Conventional sem
18a	Process cheese	Thin section, tem
18b	Process cheese	Thin section, tem
18c	Process cheese	Thin section, tem
19	White cheese in process cheese	Thin section, tem
20	Excessively heated process cheese	Thin section, tem
21a	Whipped cream	Cryo-sem
21b	Whipped cream	Cryo-sem
21c	Whipped cream	Thin section, tem
22	Ice cream	Cryo-sem
23a	Traditional cream cheese	Thin section, tem
23b	Newly formulated cream cheese	Thin section, tem
23c	Grittiness in a cheese spread	Conventional sem
24	Butter	Freeze-fracturing, metal shadowing, replication, tem
25a	Skim milk powder	Conventional sem
25b	Buttermilk powder	Conventional sem
26a	Spray-dried yogurt	Conventional sem
26b	Spray-dried UF milk retentate	Conventional sem
26c	Spray-dried UF milk permeate	Conventional sem
27a	Lactose crystals on milk powder	Conventional sem
27b	Instant skim milk powder	Conventional sem
28a	Gouda cheese stained for fat	Confocal scanning laser microscopy
28b	Gouda cheese stained for protein	Confocal scanning laser microscopy

FOOD STRUCTURE AND MILK PRODUCTS

tem methods exist for special purposes (17,18). While tem provides images of the internal structure, sem provides images of surfaces (23); internal structure may be studied by SEM in samples after they had been fractured.

In 1977, a bibliography on microscopy of milk and milk powders was compiled (24) and in 1979, a review on the structure of milk products was published (2). Recently, the structure of dairy products was dealt with in a broader context of microstructural principles applicable to food processing and engineering (25). Papers on the microstructure of milk products are published in most food science journals and, particularly, in *Food Structure* (26).

MILK

Casein is a major milk protein. It is present in milk in the form of globular particles called micelles (Figs. 1 and 2). Because of their small dimensions (diameter 100–300 nm) (2, 29–43), casein micelles cannot be seen using a light microscope, but TEM shows them as globules apparently composed of submicelles (Fig. 1c). The submicelles, 10–20 nm in diameter (44), are held together by calcium phosphate (2,28). Some authors (28,45) do not accept this view and consider the casein micelle to be a continuous structure of threadlike protein strands. Various structural models have been proposed for the casein micelles (46–51) and in some of them, specific positions have been allocated to the individual components such as α_{s1}-, β-, and κ-caseins. Intact molecules of κ-casein and the presence of calcium phosphate are generally recognized as essential for the integrity of the casein micelles. In bovine milk, cleavage of the polypeptidic chain between Phe_{105} and Met_{106} by proteolytic enzymes results in the disintegration of κ-casein into para-κ-casein (fragment 1–105) and a glycomacropeptide or caseinomacropeptide (fragment 106–169) (52). Consequently, the micelles deprived of their stabilizing factor aggregate and form a gel (5) that undergoes structural changes during aging (53).

Acid-induced gelation was recently studied by Heertje and co-workers (28). Casein micelles at various stages of gelation were rapidly frozen, freeze-fractured, replicated with platinum and carbon, and examined by tem. As the pH of the milk was decreased below 5.9, the micelles started to disintegrate as the result of the solubilization of calcium phosphate. Its release resulted in the dissociation of weakly bound β- and κ-caseins from the micelles whereas their framework (α_s-casein) remained intact (54). At pH 5.2, reassociation of the caseins without contraction took place and on additional lowering of pH to 4.8, β-casein acquired a positive charge and associated with α_s-casein, which was negatively charged. The new particles formed were completely different from the original micelles (Figs. 2a–c).

In fresh milk, the micelles have smooth surfaces (Fig. 1c). In milk heated at $\geq 85°C$ for ≥ 10 min, β-lactoglobulin in the milk serum interacts with κ-casein (present in the micelles) and forms an insoluble complex (55). This complex imparts a ragged topography to casein micelle surfaces (Fig. 3a) (4). Differences in the nature of casein micelles in heated and unheated milk play an important role in curdling of the milk (see the sections on cultured milk

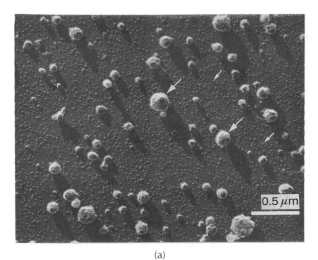

(a)

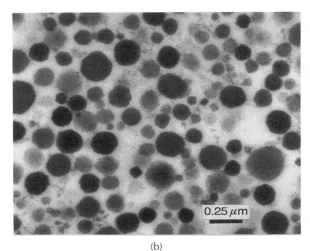

(b)

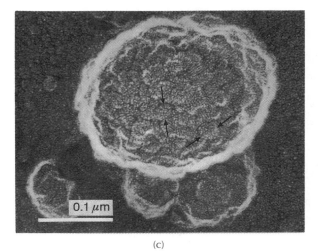

(c)

Figure 1. Casein micelles in milk. (**a**) Metal shadowing gives the casein micelles (large arrows) a three-dimensional appearance. Small arrows: submicellar protein. (**b**) In thin sections the micelles appear as disks. (**c**) Casein micelles in unheated milk have smooth surfaces. Submicellar structure (arrows) is noticeable at high magnification. [Rotary shadowing; reprinted by permission from Scanning Microscopy International (SMI) (27).]

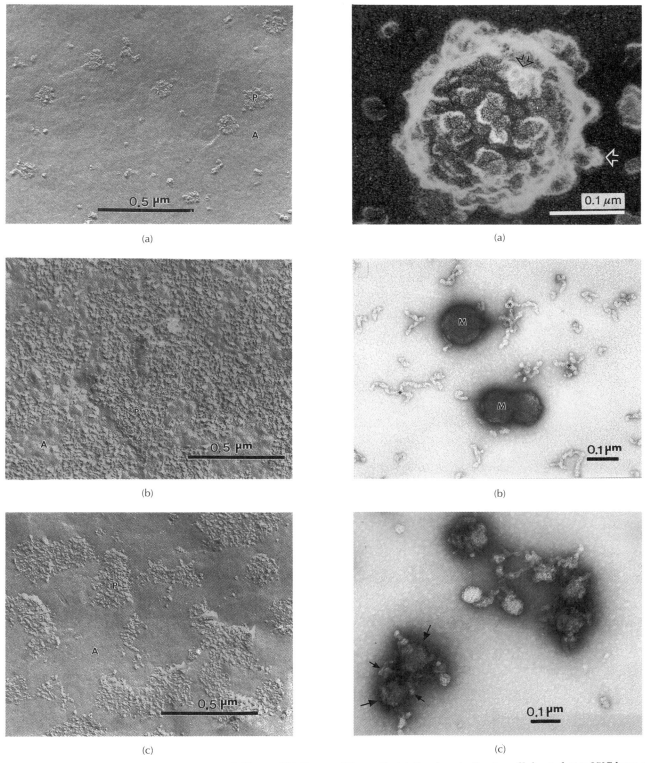

Figure 2. Casein micelle system as a function of pH. (**a**) pH 5.5; (**b**) pH 4.8; (**c**) pH 4.5. A, aqueous phase; P, protein. [Freeze-fracturing and replication; courtesy of I. Heertje (28); reprinted by permission from SMI.]

Figure 3. (**a**) Casein micelles in milk heated at ≥85°C have ragged surfaces because aggregates containing denatured β-lactoglobulin (arrows) become attached to them. [Rotary shadowing; reprinted by permission from SMI (27).] (**b**) and (**c**) Heat-induced interaction between β-lactoglobulin (asterisks or small arrows) and casein micelles (M or large arrows). (**b**) There is no interaction at pH 7. (**c**) Both components interact at pH 6.5. [Negative staining; courtesy of I. Heertje (28); reprinted by permission from SMI.]

products and cheese). The interaction of denatured β-lactoglobulin with casein micelles depends on pH: at pH 7.0, there is no interaction (Fig. 3b) but at pH 6.5, both components interact (Fig. 3c) (28).

Heating of milk to 200°C for 3 min (27) and interacting it with glutaraldehyde at this temperature did not confirm a hypothesis (56) that at high temperature, casein micelles disintegrate into submicellar particles that would reaggregate on cooling into particles larger than the original casein micelles. Some deformations caused by heating, however, were noticed.

Ultra-high-temperature (UHT) sterilized evaporated milk is, in general, susceptible to gelation on storage. Electron microscopy reveals the deformation of micelle surfaces and the formation of tendrils protruding from the micelles before the onset of gelation (10, 57–59).

Milk foams when poured or vigorously agitated. In the foam, casein micelles are attached to an electron-dense membrane, about 5 nm thick, which develops at the liquid–gas interface in the air bubbles (Fig. 4a) (60). In a foam allowed to stand quiescently and to disintegrate, the air bubbles burst and membrane fragments with a single layer of casein micelles attached to them are released into the milk in the form of so-called bubble ghosts. With time, the micelles completely dissociate from the air–serum interface (Fig. 4b) and the membrane residues are found in the milk. Brooker (60) assumes on the basis of high-performance liquid chromatography (hplc) that the interface consists of a mixture of globular whey proteins and some soluble caseins.

Homogenization of milk reduces fat globule dimensions in order to prevent spontaneous separation of cream. The combined surface of the small fat droplets is five- to sixfold higher than the initial total fat particle surface. The original fat globule membranes are fragmented and the newly formed fat surface is rapidly coated with surface-active material consisting of caseins and whey proteins present in the milk (61). The reduced dimensions of the fat globules associated with their heavy coating of protein including submicellar casein (62) lead to some clustering of the fat droplets and affect the buoyancy of the clusters to the point that they may sediment in the ultracentrifuge (63).

CULTURED MILK PRODUCTS

Using bacteria, molds, or yeasts, milk may be transformed into a variety of cultured products such as yogurt, kefir, koumiss, acidophilus milk, cultured buttermilk, and sour cream (64,65). The presence of lactic acid is a characteristic common to all types of cultured milk products. The curd may be very thin in koumiss but is very thick in strained yogurt called labneh (66). Microstructural studies have been most extensive with yogurt, apparently in response to markedly increased production in recent years.

Yogurt

Yogurt is a popular milk product produced by culturing (fermenting) condensed or fortified milk using *Lactobacillus delbrueckii* subspecies *bulgaricus* and *Streptococcus thermophilus* cultures. The product has a high moisture content (82–86%) and is marketed in cups in solid, "set-style," or liquid, "stirred-style," forms. Yogurt is available in plain and fruit-flavored varieties and in combination with canned fruits.

Milk destined for yogurt manufacture has traditionally been concentrated by heating to increase its total solids content in order to make a firm product which is resistant to syneresis, ie, separation of whey from the gel. Although the solids content may be increased by fortifying the milk with milk powder or other ingredients (65,67–69), heating was shown to be of essential importance for the yogurt microstructure (70) with which other properties are closely related.

The κ-casein–β-lactoglobulin complex, which develops in milk heated at ≥85°C (55) and alters casein micelle surfaces, prevents the micelles from clotting in the form of large clusters (71). Instead, they associate in the form of chains (Fig. 5a,b) and these chains become interconnected to form the gel matrix. The matrix is characterized by a uniform distribution of small void spaces (pores) filled

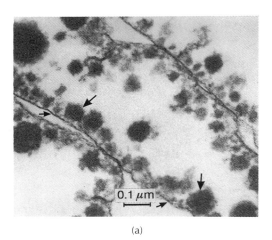

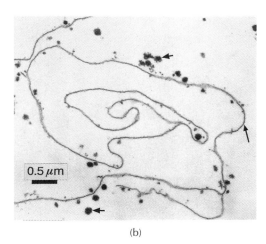

Figure 4. (a) Casein micelles (large arrows) adhere to a gas–liquid interface (smaller arrows) in the air bubbles in milk foams. [Thin section; courtesy of B. E. Brooker (60); reprinted by permission from SMI.] (b) With time (1 h), casein micelles (large arrows) dissociate from the gas–liquid interface (smaller arrows) in milk foams. [Thin section; courtesy of B. E. Brooker (60); reprinted by permission from SMI.]

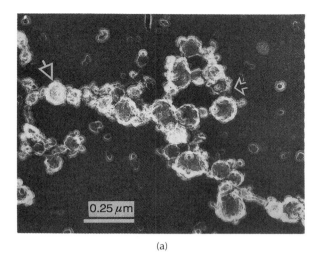

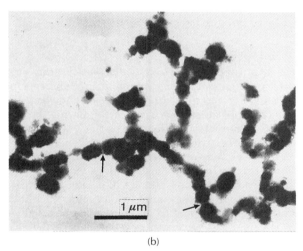

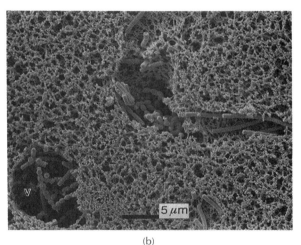

Figure 5. (a) Casein micelles form chains (closed arrow) and the chains form loops (open arrow) in a developing yogurt. [Rotary shadowing; reprinted by permission from SMI (70).] (b) The protein matrix of yogurt is composed of interconnected casein particle chains, in which the casein particles are tightly attached (arrows) to each other. (Thin section.)

with the liquid phase (Figs. 6a,b). In contrast, gels made from unheated milk consist of large clusters of casein particles, have large void spaces, and easily release the whey (Fig. 6c).

The density of the yogurt matrix and its water-binding capacity increase with increased protein content (Figs. 6a,b) (72). Whey proteins in the form of a concentrate may also be used to fortify yogurt milk; they become incorporated in the casein particle chains and separate them from each other (Fig. 7a). (68). Thickening agents such as modified starch, various gums, gelatin, or pectin also increase water binding and some of them affect the microstructure whereas others (gelatin) cause no visible changes (67).

Bacterial colonies that develop in yogurt are usually surrounded by void spaces that are more clearly noticeable in products having a higher total solids content (Fig. 6b). The origin of the void spaces has not yet been determined. It is probable that proteolytic activity of the bacteria or the development of lactic acid play a role.

Figure 6. (a) The density of the protein matrix in yogurt depends on the protein content in the yogurt milk, 3.4% protein, (b) 6.7% protein. V, void spaces; large void spaces are associated with the colonies of lactic acid bacteria. (c) The protein matrix of an acidulated milk gel made from unheated milk consists of casein micelle clusters and large void spaces. (Conventional sem.)

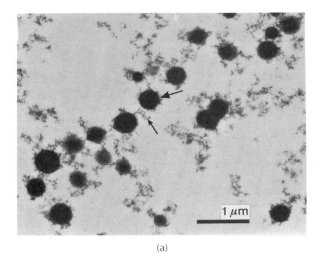

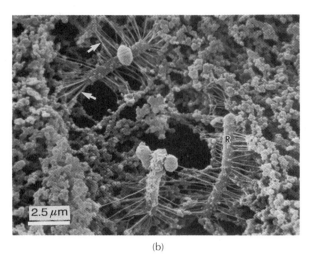

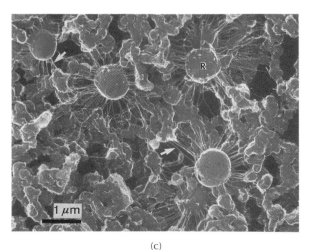

Figure 7. (a) Whey proteins (small arrow) used as an ingredient form bridges between casein particles (large arrow) in yogurt. (Thin section.) (b) Mucogenic lactic acid bacteria (R) appear to be attached by filaments (arrows) to the protein matrix of yogurt (conventional sem) and (c) in cottage cheese (freeze-fracturing, drying, replication, tem).

Ropy (mucogenic, slime-producing) lactic acid bacteria produce exocellular polysaccharides (69,73–75) that are released into the yogurt where they increase the viscosity and improve water retention. The mucus that surrounds the bacteria appears in the micrographs (Figs 7**b**) as filaments connecting the bacteria with the casein matrix; this appearance is the result of sample preparation (dehydration) for electron microscopy, when the mucus shrinks. Ropy microorganisms may be used in the production of yogurt and other cultured milk products (Fig. 7**c**) but are not always included in the bacterial cultures.

Labneh

Labneh is produced by straining yogurt and increasing the solids content above 20%. The pore dimensions are thus reduced and the microstructure resembles that of yogurt made from milk having a higher total solids content. Tamime and co-workers (66) prepared labneh from ultrafiltered milk and showed its structure.

Kefir

Kefir is a cultured product (65), which was initially made from the milk of goat, sheep, or cow using (and reusing) kefir grains (Fig. 8**a**) consisting of bacteria and yeast cells (76–78) (Fig. 8**b**) embedded in a polysaccharide (kefiran) matrix. The characteristic feature of the grains is their ability to sustain the fermentation of milk. The yeast cells (*Saccharomyces kefir* and *Torula kefir*), which constitute 5–10% of the grain microflora, produce carbon dioxide, which causes kefir to foam and fizz like beer. In addition, the yeast cells produce ethanol, which may reach a 1% concentration in the product. Recently, freeze-dried cultures for a single use have been developed (76) to facilitate kefir production. The protein matrix in kefir (Fig. 8**c**) resembles that in yoghurt (Figs. 6,**a,b**,7**a**).

ULTRAFILTRATION MILK RETENTATES

Ultrafiltration (uf) concentrates high-molecular-mass (proteins) and corpuscular components (fat globules) of milk or whey in the form of retentate by removing a portion of the serum phase (permeate) devoid of molecules above a certain (5000–20,000 daltons) molecular mass cutoff. Concentrations of the low-molecular-mass components (lactose and minerals) are evenly divided between the two fractions. Ultrafiltration thus considerably alters the chemical composition of the milk. By addition of water to the retentate during ultrafiltration (diafiltration), the protein content of the retentate may be increased up to 70% on the dry basis. Although uf does not affect the dimensions of casein micelles (79), the altered composition of the retentate may make it behave in a different way from milk during cheese manufacture (65). A comparison of firmness and microstructure of curd obtained by coagulating homogenized and nonhomogenized uf retentates using rennet and microbial proteases showed a good correlation between both parameters (80). Curd firmness was highest with rennet (Fig. 9**a**) and *Mucor miehei* protease, was lower with *M. pusillus* and *Endothia parasitica* proteases, and was lowest with *Bacillus polymixa* protease (Fig.

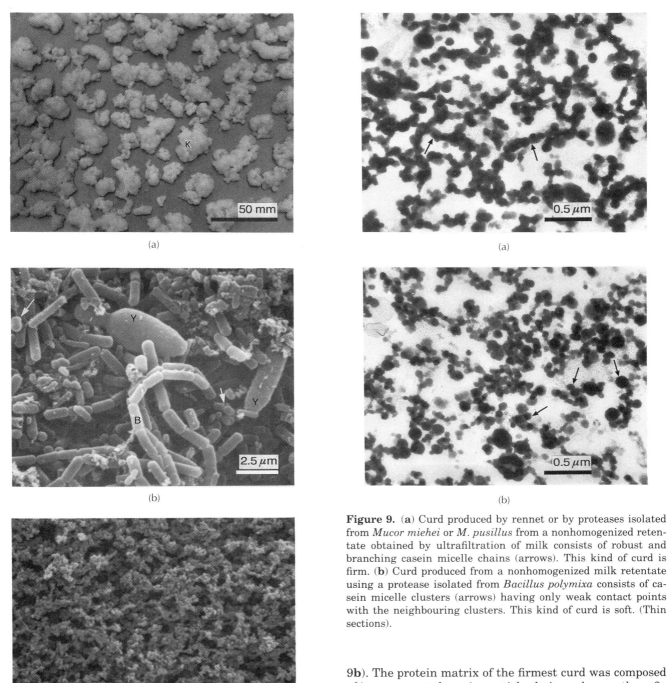

Figure 8. (a) Kefir gains (K). (Macrophotography.) (b) A variety of microorganisms is the active part of kefir grains. B, bacilli; Y, yeast cells; arrows, cocci. (c) The microstructure of kefir resembles that of yogurt. (Conventional sem).

Figure 9. (a) Curd produced by rennet or by proteases isolated from *Mucor miehei* or *M. pusillus* from a nonhomogenized retentate obtained by ultrafiltration of milk consists of robust and branching casein micelle chains (arrows). This kind of curd is firm. (b) Curd produced from a nonhomogenized milk retentate using a protease isolated from *Bacillus polymixa* consists of casein micelle clusters (arrows) having only weak contact points with the neighbouring clusters. This kind of curd is soft. (Thin sections).

9b). The protein matrix of the firmest curd was composed of interconnected casein particle chains, whereas the softest curd consisted of casein particle clusters with a low extent of branching.

Association of protein with minute fat particles in curd made from homogenized retentate (Fig. 10a) resulted in higher firmness compared to large fat globule clusters freely dispersed in curd made from nonhomogenized retentate (Figs. 10b,c) (80).

CHEESE

Cheese is a milk product formed by the coagulation of milk proteins using proteolytic enzymes (chymosin, also called rennet, pepsin (81), or proteases of microbial origin

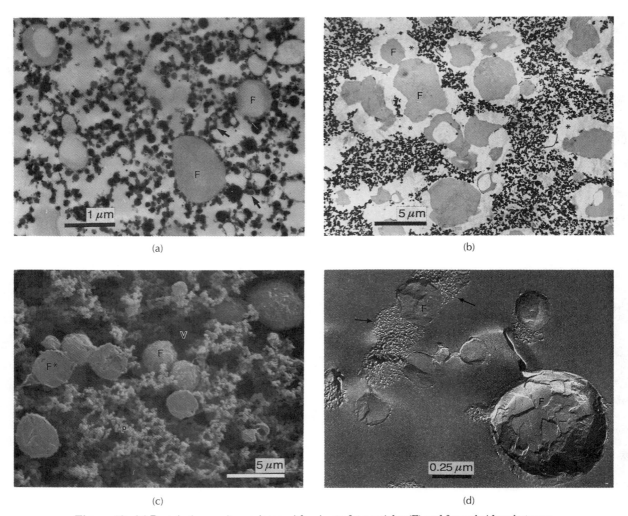

Figure 10. (a) Protein (arrows) associates with minute fat particles (F) and forms bridges between them in curd made from homogenized retentate, thus increasing curd firmness. (b) Fat globules (F) and their clusters are dispersed in the protein matrix (P) of curd made from nonhomogenized retentate and are surrounded with void spaces (asterisks or V); thin section; (c) conventional sem (F*, fractured fat globules). (d) Milk curd consists of casein particle clusters (arrows) and fat globules (F); fat crystals are noticeable in the large fat globule (F). (Freeze-fracturing, replication, tem.)

(80,82)) and lactic acid produced from lactose by bacterial starter cultures. The casein-based curd formed is separated from whey and ripened using various microorganisms. Changes in the heat treatment of the milk, variations in mechanical processing of the curd (eg, cheddaring, stretching), use of bacteria or fungi to ripen the cheese, and varying the ripening conditions result in the production of many cheese varieties (65) that differ from each other in their characteristic features, including structure (3,83).

Curd

Curd develops from casein micelles following the destabilization of κ-casein and solubilization of calcium phosphate by coagulation of the micelles and formation of a three-dimensional network (matrix). Clustering of the casein micelles induced by rennet commences at a considerably higher electric charge than acid curd formation (71). As mentioned in the section on cultured milk products, renneted curd (made from milk not heated above 85°C) consists of casein micelle clusters. Void spaces in the protein matrix are filled with whey that is easily released on heating (cooking). Fat globules are dispersed in the protein matrix (Figs. 10b–d) (80,84). A comparison of curd structures in cottage cheese coagulated with and without microbial rennet has been published (82).

Cutting of the milk coagulum exposes the fat globules along the cut surfaces and results in the formation of surface layers as areas depleted of fat (85). When cooked, the granules form a base of cottage cheese. Contrary to former belief, microscopical examination showed that no hypothetical "skin" develops at the granule surface (3,86,87) in good-quality cottage cheese. Defects may develop when a thick layer of sludge develops at the vat bottom, but the curd-cutting stage is not reached because the pH does not drop below 5.2. Electron-microscopic examination of the sludge revealed that the bacteria were infected by bacteriophage and thus unable to produce enough lactic acid to bring pH down below 5.2; micellar casein produced a structure resembling a precipitate rather than a gel network (88). Sludging does not develop in yogurt manufacture because bacteriophages are inactivated by heating the milk.

Hard Cheeses

Matting leads to the fusion of cooked curd granules and formation of stirred-curd cheese. As the curd particles fuse with each other, their junctions become noticeable (Fig. 11a). Curd cut with wire knives and curd broken by a stirrer produce different curd granule junction patterns; those in the former curd are more uniform (Fig. 11b) (89). In Cheddar cheese, an additional kind of so-called milled curd junctions develops as the result of milling and pressing cheddared curd (Fig. 11c); traditional, mechanized, and automated processes used in the manufacture of Cheddar cheese impart characteristic junction patterns on the products by which they may be distinguished (90). Bacterial colonies in Cheddar cheese are usually associated with milled curd junctions (91). Curd junction patterns in Cheddar cheese made from homogenized low-fat milk are weaker because the area depleted of small fat globules is narrow and the difference in the fat content between the low-fat granule interior and the fat-depleted granule junction is small (92). In Swiss cheese, differences in the junction patterns of vertical and horizontal sections were used as indicators of cheese deformation (93); similar studies were later extended to other Swiss cheeses (94).

Cooking puts the casein particles in the protein matrix into a closer contact. Drainage of whey and pressing and ripening of the curd result in the formation of a compact protein mass of hard cheese in which fat globules and lactic acid bacteria are interspersed (Figs. 12a,b); occasionally, minute calcium salt crystals may also be present (95). Fat globules have a tendency to appear in clusters and their membranes remain intact, particularly in young cheese (Fig. 12a). Cheddaring leads to longitudinally oriented protein (85).

The development of structure in cheeses was studied from various aspects (2,6,83,97–108). Particularly important are studies of the effects of concentrated milk (eg, by ultrafiltration) in cheesemaking (83). In regular milk, the mean free path of casein micelles is 3 micelle diameters (41) but in fourfold concentrated milk the mean free path is less than 1 micelle diameter (83). As the concentration factor in the milk is increased, deficiency of proteases may develop in the cheese during ripening and the structure of the cheese becomes progressively coarser.

Bacterial surface-ripened cheeses such as Limburger, Saint Paulin (84), and Tilsit (109,110) usually have strong flavor, which is imparted to the cheeses by the microorganisms used (*Brevibacterium linens*). These cheeses are best produced small in order to obtain a ripened interior while the surface is not yet overripened, although the size is not as critical as with mold-ripened cheeses (65). The development of structure was studied in Olomouc cheesecakes (99) made from skim milk. Lactic acid initially pro-

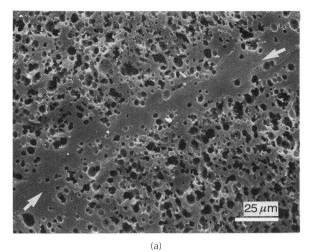

(a)

(b)

(c)

Figure 11. (a) Curd granule junctions (arrows) in cheese are areas depleted of fat. (Conventional sem.) (b) Curd granule junction (arrows) patterns are relatively uniform in stirred-curd cheese made from curd cut manually using wire knives. (Microphotography.) (c) Milled curd junctions (large arrows) develop in Cheddar cheese in addition to the curd granule junctions (small arrows). (Microphotography.)

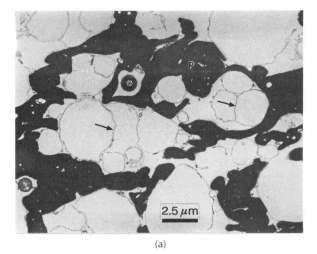

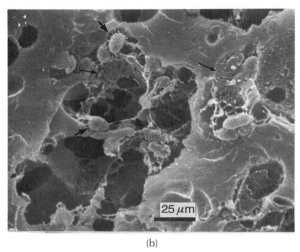

Figure 12. (a) Young hard cheese consists of a compact protein (P) matrix with dispersed fat globules (arrows point to fat globule membranes) in the form of clusters. Bacteria (asterisks) are also present. (Thin section.) (b) Conventional sem shows the protein matrix of hard cheese with residues of fat globule membranes (small arrows) around void spaces originating from the extraction of fat during sample preparation; large arrows point to bacteria.

duced by the starter bacterial culture is oxidized by surface microorganisms of the genera *Torulopsis, Candida,* and *Oospora,* consequently increasing pH of the cheese. As proteolysis proceeds, a golden-orange mucous layer, which is rich in the microorganisms, develops on cake surfaces and the cakes become translucent and acquire a characteristic piquant flavor (111).

In some cheeses, so-called eyes develop as the result of carbon dioxide production by propionic acid bacteria. The formation of the eyes in cheese was reviewed by Kosikowski (44, pp. 261–291). While the eyes are particularly desirable in Emmental, Swiss, and Gruyere cheeses, they would be considered a defect in Cheddar, Camembert, and most other cheeses (65).

Soft Cheeses

Soft cheeses are also called mold-ripened cheeses because fungi such as *Penicillium camemberti, P. roqueforti, P. glaucum,* and *Mucor rasmusen,* rather than bacteria are used to ripen them (65). Of the great variety of mold-ripened cheeses produced, Blue (blue-veined) and Camembert cheeses are probably best known.

In the production of Blue cheese, air passages are bored into a markedly salted curd made from milk which had been inoculated with *P. roqueforti* spores. The mould then grows in these passages and produces the blue venation. Homogenized milk was reported to yield better Blue cheese than nonhomogenized milk because the curd is more porous, thus facilitating the development of the mold mycelium (112).

Camembert cheese is made by dipping firm curd wheels, 2.5 cm thick, into a suspension that contains *P. camemberti* spores and incubation at a high air humidity (65). As the spores germinate, fungal hyphae penetrate the curd (Fig. 13a) and thus contribute to its ripening and softening through proteolysis (109,113). A thick white mat develops on the cheese surface (Fig. 13b) (114). During the ripening of the cheese, calcium phosphate crystallizes and becomes concentrated in a thin layer near the

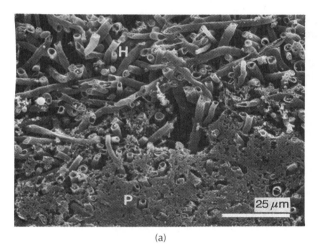

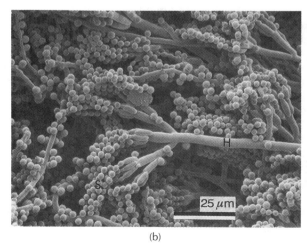

Figure 13. (a) Hyphae (H) of the *Penicillium camemberti* mould penetrate the protein matrix (P) in Camembert cheese. (Conventional sem.) (b) *Penicillium camemberti* hyphae (H) and spores (S) form a thick white mat on the surface of Camembert cheese. (Cryo-sem; courtesy of P. Allan-Wojtas.)

cheese surface. Using microscopy and digital x-ray spectrometry, Brooker (115) explained that pH elevated in the superficial cheese layer by the deamination of amino acids by the microorganisms results in the precipitation of calcium phosphate from the aqueous phase. The precipitation (and recrystallization) causes a calcium and phosphate gradient to develop between the interior and the superficial layer of the cheese. Consequently, more calcium and phosphate diffuses from the interior and precipitates at the surface.

A study of the microstructure of soft Meshanger cheese which liquefies within 3 weeks, showed no visible differences in the protein matrices of ripened hard and soft cheeses except that the process are accelerated. In Meshanger cheese, the matrix becomes compact after 7 days, shows a granular structure after 14 days, and at 21 days the fine granularity apparently consists of casein submicelles (116).

Unripened Cheeses

Curd may also be obtained by coagulating hot milk with an acid, such as citric or hydrochloric acids. This curd is processed into unripened cheeses such as White cheese (117), its Latin American equivalent, Queso blanco, and Indian Paneer cheese (9). Provided that the milk had been heated to ≥85°C for at least 10 min and pH of the cured is approximately 5.5, the curd has a unique *core-and-shell* structure (Fig. 14) (which was initially called core-and-lining structure) (8). Although the exact mechanism of the development of this structure is not known, the presence of β-lactoglobulin and the milk salt system are essential prerequisites (118). The structure is very stable; it becomes compacted in Paneer cheese fried in vegetable oil (Fig. 15a) but is fully restored in Paneer cheese subsequently cooked in water (Fig. 15b) (9). The core-and-shell structure may even be found in process cheese in which White cheese was used as an ingredient (117,119).

The structure of whey protein curd in cheeses such as Mysost or Ricotta (Fig. 10b) is markedly different from casein curd in that it is considerably finer (Fig. 16) irre-

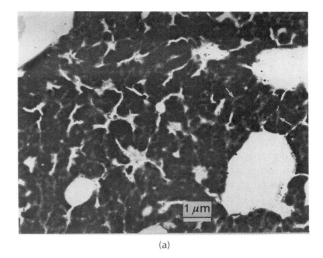

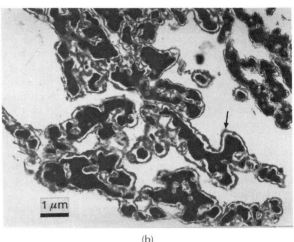

Figure 15. (a) The core-and-shell structure (arrows) of casein particles is very stable and may be found in Indian Paneer cheese even if it had been fried and (b) subsequently cooked in water for 5 min. (Thin sections.)

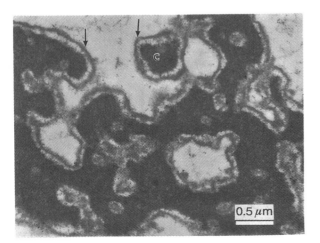

Figure 14. White cheese obtained by coagulating hot milk with an acid to final pH of 5.5 has a characteristic core-and-shell structure (arrows) of the casein particles (C). (Thin section.)

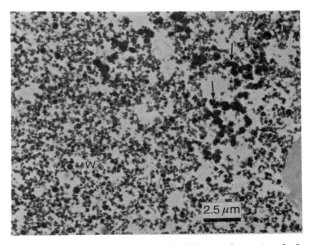

Figure 16. Whey protein (W) curd in Ricotta cheese (made from a mixture of 90% whey protein and 10% milk protein (122) has a fine structure composed of whey proteins. Arrows point to casein particles that originated from the milk. (Thin section.)

spective of whether the whey proteins have been coagulated by heat (120) or acid (121,122).

PROCESS CHEESE

Since its introduction on the food market at the beginning of this century, process cheese has gained popularity over natural cheese because of several advantages such as great diversity of type, shape, flavor, reduced cost of storage and transportation, and extended shelf life (119).

Processing, which consists of heating and stirring in the presence of melting salts (123), markedly alters the structure of natural cheese. The characteristic features of the natural cheese vanish and new features characteristic of processed cheese are developed (29,124,125).

Boháč (126,127) selected natural cheeses for processing on the basis of their interactions with melting salt solutions during heating to 85–90°C recorded by a movie camera attached to a polarizing microscope. Several phenomena were observed: in some cheese samples, diffuse zones consisting of protein and fat developed or the samples disintegrated along the curd granule junctions as the temperature was increased above 70°C. Other cheese samples contracted at 60–70°C but did not melt until the temperature reached 95–98°C. Rapid diffusion of ripe or overripened cheeses into the melting salt solution was commonly observed even before the melting temperature had been reached. The assessment of the suitability of a particular natural cheese for processing was based on the temperature at which the cheese melted and on the amount of fat released from the sample.

Heating and stirring results in the disintegration of the fat globule membranes that cover individual fat globules in natural cheese, aggregation of the fat droplets, and their subsequent emulsification into smaller fat particles (128,129). Dimensions of the fat particles in the finished processed cheese depend on several factors such as the use of the melting salt or the passage of the process cheese through a homogenizer (125). Provided that other parameters such as pH and the water and fat contents are the same, the decrease in the fat particle dimensions is correlated with the increase in firmness of the process cheese. Melting salts are not emulsifiers but they restore the emulsifying capacity of cheese proteins indirectly by sequestering calcium from the calcium caseinate complex (119,124). The emulsifying ability of casein in cheese may be understood by viewing the molecules as having one end nonpolar, ie., lipophilic, and the other end (which contains calcium phosphate) as polar, ie, hydrophilic. Replacing calcium with sodium increases the solubility of casein in water as well as its emulsifying ability. However, the use of salts having a strong affinity for calcium, (eg, tetrasodium pyrophosphate) may result in an emulsion where the total surface of the fine fat particles is so large that there would be not enough protein to cover them all; consequently, the surplus fat would separate from the hard nonmeltable cheese. Salts having a lesser affinity for calcium (sodium hexametaphosphate, disodium phosphate, sodium citrate) produce a softer and more meltable process cheese.

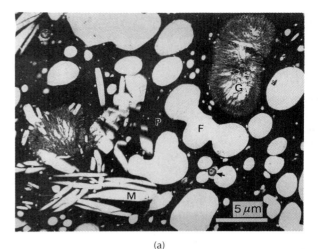

(a)

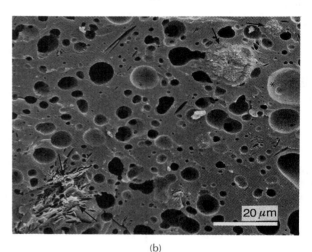

(b)

Figure 17. Characteristic structure of process cheese as visualized by tem (a) and sem (b). In (a), P, protein; G, calcium phosphate crystals (already present in the natural cheese); M, melting salt crystals; F, fat particles, either in the process of emulsification (marked particle) or already emulsified. In (b), small arrows point to calcium phosphate crystals, large arrows point to imprints of melting salt crystals, and asterisks indicate fat globules.

Emulsification is arrested by cooling and solidification of the blend. In some process cheeses, part of the fat particles may be seen to be incompletely emulsified (Figs. 17a,b). Although the original fat globule membranes in natural cheese are destroyed by processing, new protein coating develops on the surface of emulsified fat droplets (Fig. 18a) (117,130).

Electron-microscopic studies of protein matrices in process cheese showed that a soft product (made with a mixture of 1.0% sodium citrate and 1.5% polyphosphate) consisted predominantly of individual protein particles whereas the protein matrix of a hard cheese (made with 2.2% polyphosphate) contained a high proportion of protein in the form of long strands. The authors of this finding (131,132) assumed that the protein strands in the hard-type process cheese contributed to the ability to retain its shape on heating. Heertje and co-workers (130) observed protein strands in pasteurized process cheese having a long creaming time (Fig. 18b) but not in a similar cheese having a short creaming time (Fig. 18c). Protein strands were also found in a soft process cheese made using a combined phosphate–polyphosphate emulsifying agent and direct steam heating (119). Tinyakov (133) reported that process cheese contains microvacuoles, the variation in size and shape of which depends on the type of the cheese being examined. A fibrous structure resulted in cheese made using sodium citrate.

In general, it is not possible to determine, by electron microscopy, which varieties of natural cheese were used to prepare the process cheese blend. The exception is White cheese, made by acidifying hot milk to pH of approximately 5.5 and draining the whey. This cheese has a characteristic and remarkably stable core-and-shell ultrastructure of the casein particles (8,118). The presence of ≥8% of White cheese in process cheese was detected by TEM (Fig. 19) (117).

The structure of process cheese changes under the effects of various treatments. Klostermeyer and Buchheim (125) showed that an increase in the rate of the stirrer from 120 to 1500 rpm during processing markedly reduces the dimensions of the fat particles; as a result, homogenized process cheese appears optically lighter than nonhomogenized process cheese. A homogenizing effect on the protein matrix of process cheese is also noticeable. Excessive heating has a different effect on the microstructure of process cheese. If problems develop in a process cheese plant during the packaging of the product, the process cheese transported through heated pipes is held at a high temperature for some time until the problems are solved. Excessive heating causes the process cheese to harden and lose its meltability. Such cheese is ground and reworked, ie, added in small quantities to new blends of natural cheeses. However, because of its changed structure, this reworked cheese imparts new properties on the process cheese in which it is present. The development of characteristic osmiophilic areas in excessively heated cheese

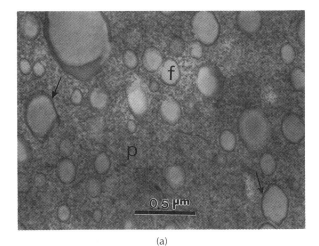

(a)

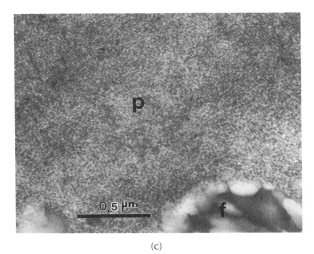

(b)

(c)

Figure 18. (a) Protein coating (arrows) develops on the surfaces of freshly emulsified fat particles (f) in the protein matrix (p) of process cheese. (b) At high magnification, protein strands (arrows) may be visualized by tem in pasteurized process cheese at a long creaming time of 160 min but (c) not in a process cheese at a short creaming time of 8 min (f, fat globules; p, protein matrix). (Thin sections; courtesy of I. Heertje.)

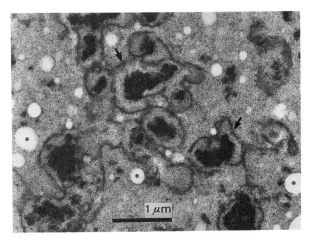

Figure 19. The presence of the core-and-shell structure (arrows) of casein particles in process cheese indicates that White cheese obtained by coagulating hot milk with an acid was used as an ingredient, asterisks, fat. (Thin section.)

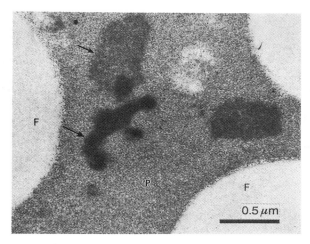

Figure 20. Excessive heating (eg, 80°C for 5 h) causes the development of osmiophilic areas. Small arrow points to an area developed as the result of the presence of trisodium phosphate; larger arrow points to an area developed in the presence of sodium citrate (134). (Thin section).

(Fig. 20) is one indicator of structural changes (134); this indicator may be used to detect the presence of rework of this kind in finished process cheese. It is not yet clear whether the osmiophilic areas originate as a result of physical changes (compaction) or chemical changes (increased affinity to osmium tetroxide) in the cheese proteins. Some of the osmiophilic areas indicate the presence of spray-dried whey powder used as an ingredient in process cheese that did not fully dissolve in the blend (117). Baking of natural cheese or process cheese as an ingredient of pizza resulted in melting of the cheeses and considerable changes in their structure but did not lead to the the development of new osmiophilic areas (135).

In order to make cheese processing more economical, attempts have been made to replace part of the ripened cheese in the blend with a "cheese base" prepared by ultrafiltration and diafiltration of milk to reduce the lactose content and increase the protein content, and acidification of the concentrate to lower the calcium content in the casein micelles. Addition of a proteolytic enzyme to some extent simulated natural ripening in the cheese base and resulted in its open structure. Unless the proportion of this cheese was too high, most experimental process cheeses made from two types of cheese base were acceptable and their structures resembled the traditional process cheese (136).

Microscopical studies have also been carried out to find the causes of structural defects in process cheese such as the development of salt crystals (123,137). Small white crystals identified as calcium citrate may occasionally develop on the surface of process cheese made using sodium citrate as early as 1 week after manufacture (138). This defect may be avoided by elimination of sodium citrate from the emulsifying salts.

Melting salt crystals are seen in micrographs of process cheese as the result of using them in excess or because the crystals did not dissolve completely in the cheese during processing (139). Melting salt crystals usually appear to be different from the crystals of calcium phosphate that are present initially in the natural cheese (119,124). The crystals, which are water-soluble, are washed out from the cheese during preparation for electron microscopy and leave void spaces (imprints) in the cheese protein matrix (Figs. 17a,b).

MILK PRODUCTS BASED ON FAT

Fat is present in milk in the form of fat globules, 0.5–10 μm in diameter. Their properties and structure have been studied and the findings reviewed (2,35,63,140–144).

Fat Globules

Fat globules are encased in fat globule membranes, which are the rearranged and restructured membranes of the secretory cells in the mammary gland (142). The microstructure of the fat globule membranes depends on the length of time elapsed since the fat globule had been secreted from the mammary gland (145). The stability of fat globules in milk depends on the continuity of the secondary membrane, which is more elastic than the initial membrane and readily corrugates (143). The microstructure of fat globules was studied in detail particularly by Buchheim (146–148), who reported that highly organized, paracrystalline domains exhibiting hexagonal symmetry are present at the interface between the triglyceride core and the fat globule envelope. These domains are the imprints of organized proteins, most probably xanthine oxidase (146).

Cream

Cream is a product obtained from milk by centrifugation and consists of all the milk components with a considerably higher fat concentration. Various forms of cream (coffee cream, whipped cream, ice cream, sour cream, cream cheese, etc.) and butter differ in their structure from products based on milk protein by that they consist of a large proportion of fat in the form of oil-in-water emulsions (cream), water-in-oil emulsions (butter), or foams (whipped cream, ice cream). These products lack a rigid protein matrix and, therefore, are usually fluid or, at higher concentrations, spreadable (4). Depending on the fat content, the following types are distinguished: coffee cream (10–18%), whipping cream (24–35%), double cream (40–50%), and plastic cream (80–85%).

When added to hot coffee, homogenized cream forms a homogeneous milky mixture irrespective of the nature and temperature of the coffee. A defect called feathering, ie, formation of curd floating on the coffee surface, was found by electron microscopy (147,149) to be a complex problem, where the effects of pH, presence of stabilizing salts (a mixture of phosphate and citrate), homogenization, and thermal treatment of the cream were interrelated. In cream free of stabilizing salts, casein and denatured whey proteins are adsorbed on the fat globules; the salts decrease the protein adsorption and increase the "free protein" content in the serum (147). Whey proteins adhering to the fat globules following homogenization

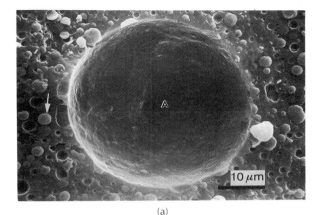

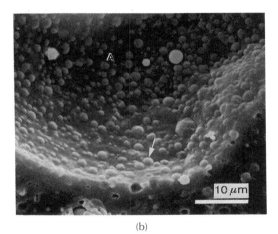

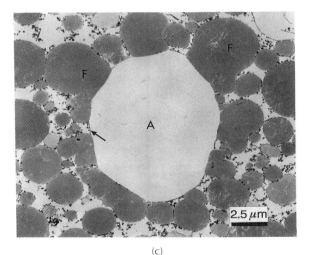

Figure 21. (a) At the early stage of cream whipping (10 s), very few fat globules penetrate the air–water interface of the air cells (A). (b) In well-whipped cream, numerous fat globules (arrow) protrude into the lumen of each air cell (A). [Cryo-sem; courtesy of B. E. Brooker (150); reprinted by permission from SMI.] (c) An air cell in ice cream 20 s after the start of whipping. The air cell (A) surface is stabilized by adsorbed fat globules (F) and by an air–water interface (arrow). The adsorbed fat globules are partially coalesced. [Thin section; courtesy of B. E. Brooker (150); reprinted with permission from SMI.]

were easier to remove by washing before pasteurization than was the adhering casein. After pasteurization and storage, the proteins were more tightly bound to the fat globules (61).

Whipping incorporates a large volume of air into the cream and produces a foam; the oil-in-water emulsion is transformed into a three-phase system. During the initial stages of whipping (150), air bubbles are stabilized mostly by β-casein and to some extent by whey proteins (Fig. 21a). The crucial step is the coalescence of the fat–water interface of each fat globule with the interfacial layer at the air bubble surface, putting the fat in direct contact with the air (Fig. 21b). The loss of primary and/or secondary fat globule membranes during whipping and the adsorption of soluble whey proteins modify the fat globule surfaces. In the finished whipped cream, the surface of each air bubble is stabilized by variable amounts of adsorbed fat and by the original air–water interface of adsorbed proteins (150). The fat globules frequently protrude into the air cells (Fig. 21b) (21,150,151). The stabilization of the air cell surface is particularly evident in Fig. 21c (150). In homogenized cream, considerably longer time is required to reach the point of maximum foam strength than in nonhomogenized cream. Overwhipping leads to the collapse of the foam, disappearance of the air bubbles, and formation of butter grains (21). The nature of the milk fat globule membrane before whipping is one of the factors that affect the susceptibility of whipped, pasteurized, nonhomogenized dairy cream to collapse (152).

Ice Cream

The composition and structure of ice cream (4) are considerably more complex than whipped cream, although cream is the major constituent in both products. A mix consisting of sugar, corn syrup, milk solids, emulsifiers, and additional ingredients is aged before aeration and freezing. Emulsifiers produce a dry and smooth textured ice cream by controlling the adsorption of protein on the fat globule surfaces (153). Aging promotes crystallization of fat as well as hydration of the stabilizers, and adsorption of proteins on fat globule surfaces. Changes in the ice cream emulsion during freezing are numerous. Structure and texture of ice cream have been reviewed in varying detail (154,155). Freezing produces ice crystals, but they are not noticeable in the mouth unless their length exceeds 40–50 μm; in a smooth ice cream, they are shorter than 20 μm (Fig. 22) (156). About 50% of ice cream water remains unfrozen at −5.5°C, ie, at the temperature when ice cream is consumed. For storage and handling, a lower temperature is used whereby the liquid water content is reduced to approximately 20%. Fluctuation of temperature promotes the recrystallization of ice resulting in a coarse structure, this defect may be avoided by the use of cryoprotective agents such as locust bean gum, guar gum, or carboxymethyl cellulose.

Lactose crystals may be present in ice cream, which contains a high proportion of whey solids. Crystals longer than 16 μm impart the sensation of sandiness in ice cream.

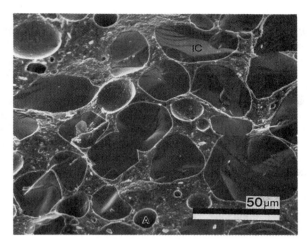

Figure 22. Ice cream is a fat-rich foam (A, air cell) and a coarse dispersion of ice crystals (IC). [Cryo-sem; courtesy of K. B. Caldwell (156); reprinted by permission from SMI.]

The structure of ice cream in which dairy cream was replaced with vegetable oil was also studied (155).

Cream Cheese

In the traditional system of manufacturing, a cream mix is ripened using a lactic bacterial culture. When pH of the curd drops to 4.6, the curd is heated to 52–63°C, drained, and cold- or hot-packed. Cream cheese production has recently been simplified by using a mix in which the total solids content is increased to equal the final desired composition of the product (156). There is no whey produced to be drained. Stirring and homogenization of the mix disrupt the curd and the fine particles produced impart spreadability to the product. Another simplification consists of mixing high-fat cultured cream with acid or renneted curd and homogenizing the mixture (101). Each manufacturing method leads to a distinct microstructure of the product. In the traditional procedure, small fat globules are aggregated in clusters and protein is concentrated at the cluster surfaces (Fig. 23a). Differences in the manufacturing practice are reflected in the dimensions of the fat globules, the fat globule clusters, and the distribution of protein (157,158). In a cream cheese spread produced by the wheyless process, a greater part of the fat is in the form of large globules; protein is uniformly distributed around smaller fat globules (Fig. 23b). New formulation based on acid curd of White cheese shows the characteristic core-and-shell structure of the casein particles (indicating that the curd was obtained by coagulating hot milk with an acid to final pH of 5.5) mixed with high-fat cream (159).

These results show that each of the three systems of manufacture produces a different structure in the cream cheese. Addition of melting salts to cream cheese (160) without heating led to the disintegration of the fat–protein aggregates in the cheese and formation of a very fine homogeneous fat distribution in the casein–whey protein matrix. Electron microscopy was instrumental to a solution of a problem of gritty curd (acid-coagulated as well as renneted) used as the major ingredient in a cream cheese

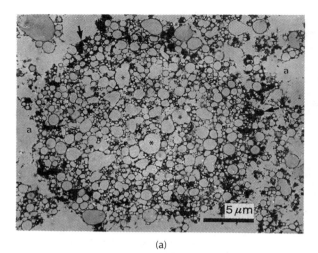

(a)

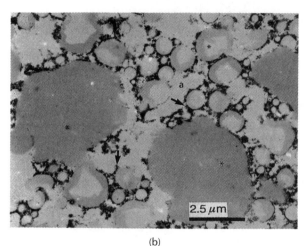

(b)

(c)

Figure 23. Microstructure of cream cheese. (**a**) In traditional cream cheese, clusters of small fat globules (asterisks) are cemented by protein, a large proportion of which is present at the surface (arrows) of the clusters (a, aqueous phase). (**b**) In newly formulated cream cheese, protein (arrow) is evenly distributed among the fat globules (asterisks) (a, aqueous phase). [Thin section; reprinted with permission from SMI (157).] (**c**) Grittiness in a cream cheese spread was caused by compact protein particles (arrow). [Conventional SEM; reprinted with permission from SMI (101).]

spread. The gritty particles developed in the curd as the result of insufficient heating of the milk and resembled cheese particles (Fig. 23c). A more severe heat treatment of the milk (90°C for 10 min) prevented this artifacts from developing (101).

Butter

Churning at 7–18°C of sweet and microbiologically ripened cream disrupts the fat globule membranes and causes the milk fat to coalesce. Butter is obtained by separating the fat-containing butter grains thus formed from aqueous buttermilk. During this process, an oil-in-water emulsion (cream) is converted into a water-in-oil emulsion (butter) (2). Butter is formed by a fat phase, in which droplets (≤10 μm in diameter) of the aqueous phase, intact fat globules, and fat globule membrane fragments are dispersed (Fig. 24). The interglobular phase is a mixture of liquid oil, fat crystal aggregates, and fat globule membrane residues (161). The aqueous phase consists of buttermilk or wash water. Fat globules present in butter have been classified by their appearance in replicas of freeze-fractured butter (148) into four types: (1) fat globules containing a very small amount of crystalline fat in the form of thin monomolecular layers, (2) relatively thin crystalline shells at the globule surface and individual crystals or crystal aggregates in the interior, (3) liquid fat in the interior and thick (≤0.5 μm) crystalline shell consisting of parallel monomolecular layers at globule surfaces, and (4) high concentration of crystalline aggregates in the interior and thick shells at globule surfaces. By subjecting cream to various temperature regimens during ripening, it is possible to control the redistribution of saturated fat crystals inside the fat globules (162). The cold–warm–cold regimen consists of cooling pasteurized cream to 6°C, addition of bacterial culture, incubation for 3 h, increasing the temperature to 20.5°C until pH of 5.3 is reached, and cooling to 14°C; the warm–cold–cold regimen consists of cooling pasteurized cream to 23°C, inoculation, after reaching pH of 5.3 cooling to 6°C for 3 h and then increasing the temperature to 13°C. The cold–warm–cold regimen leads to an increased proportion of the type 4 fat globules which are resistant to churning; the resulting butter is soft; ie, it spreads easily. In contrast, the warm–cold–cold regimen results in large numbers of type 2 fat globules, which break easily on churning and the free fat causes the butter to have a hard consistency. This control of butter consistency is considered to be superior and considerably less expensive than supplementing the cows' diet with unsaturated fats (162,163).

DRIED MILK PRODUCTS

Spray drying has become the prevailing method for the preservation of skim milk (Fig. 25a), whole milk (164), and other fluid products such as whey (165,166), whey protein concentrates (167), buttermilk (Fig. 25b) (168), sodium caseinate (169), cheese (170), and yogurt (Fig. 26a) for future use. Retentates (Fig. 26b) (171) and permeates (Fig. 26c) (172) from the ultrafiltration of milk have

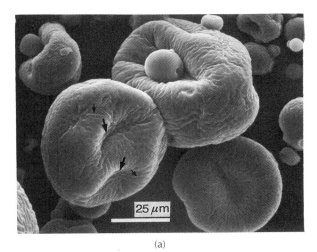

(a)

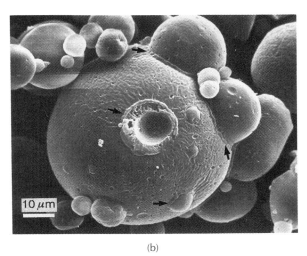

(b)

Figure 24. Butter consists of a fat phase (F) in which fat globules (FG) and occasional droplets of the aqueous phase (W) are dispersed. [Freeze-fracturing, metal-shadowing, replication, tem; courtesy of W. Buchheim (144) and Milchwissenschaft.]

Figure 25. (a) Spray-dried skim milk. Large arrows, deep wrinkles; small arrows, fine wrinkles. (Conventional sem.) (b) Spray-dried buttermilk. Arrows point to rims around embedded smaller powder particles (13). (Conventional sem.)

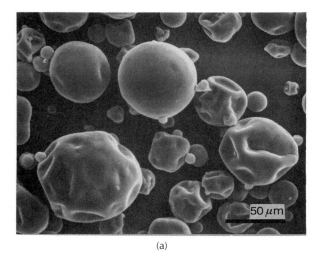

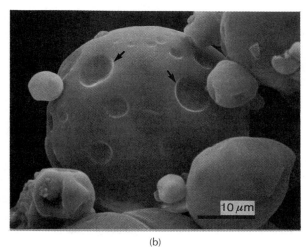

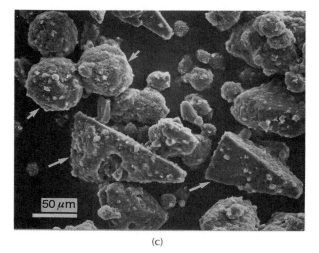

Figure 26. (a) Spray-dried yogurt. (Conventional sem.) (b) Spray-dried retentate obtained from ultrafiltration of milk. Arrows point to dimples in the particle surface. (Conventional sem.) (c) Spray-dried permeate obtained by the ultrafiltration of milk contains lactose crystals (large arrows) and globular aggregates of minute crystals with spray-dried mother liquor (smaller arrows). (Conventional sem.)

recently been added to the list of spray-dried milk products. To reduce the cost of spray drying, a great part of water present in the fluid product is first removed by evaporation, increasing the total solids content to 35–50%. Crystallization of lactose takes place in whey (165) and permeates obtained by ultrafiltration of milk (Fig. 26c) (172).

Spray-dried milk particles are globules several micrometers to several tens of micrometers in diameter consisting of a lactose–salt continuous phase (173) in which casein micelles and fat particles are dispersed. The particles also contain vacuoles with occluded air inside. Vacuoles in spray-dried milk particles originate as the result of the incorporation of air in concentrated milk before and during spray drying. Too much occluded air in the powders is undesirable because it results in low bulk density of the powder. Powder particle dimensions are affected by the concentration and viscosity of the milk, the atomizing system, and the temperature of drying. Fundamental information on the composition and structure of dry milk particles was provided by King (173) and Buma (164) and was extended by other authors (165,166,169,170,174–179). The surfaces of spray-dried milk powders are usually characteristically wrinkled (Fig. 25a) and these features are believed to reflect chemical composition of the powders (164,180). Fine wrinkles have been attributed to the presence of casein micelles since they were not noticeable on smooth particles of spray-dried lactose solutions (181). In a similar way, craterlike structure were attributed to spray-dried buttermilk (Fig. 25b) (180). It is probable, however, that particle interactions during drying also markedly affect some of the features (171). Deep wrinkles have been attributed (180) to high inlet air temperatures in the spray drier and also to large differences between the temperatures of the milk and the inlet air. Small droplets dry and harden more rapidly than large droplets. On collision, a solid particle may become embedded in a large, not fully dried and still soft particle (Fig. 25b). The impact of the collision transforms the material of the larger particle into the rim surrounding the smaller particle. In a similar way, dents found in spray-dried retentates from the ultrafiltration of milk indicate that the collisions took place while the larger particles were no longer sticky. Apparently, the small, hard, dried droplets bounced after the collisions and left their impressions in the large particle surfaces (Fig. 26b).

Casein micelles were found (176) to retain their globular nature in the powders. Fat globules, however, apparently undergo some changes. Fat present in whole-milk powders may be found in four forms (164): (1) surface fat present particularly in surface folds, (2) outer-layer fat that can be reached directly by fat solvents, (3) capillary fat that can be reached by fat solvents through capillary pores and cracks, and (4) dissolution or second-echelon fat that can be reached by solvents through the holes left by dissolved fat globules. These findings made by SEM were confirmed by freeze-fracturing and replication (174,175).

The physical state of lactose markedly affects the properties of milk powders. It has been generally accepted that in spray-dried milk powders, lactose is present partly in an anhydrous, amorphous, glassy state and partly as α-

Table 2. Applications of Microscopical Methods to Dairy Science

Method	Characteristics	Figures[a]
Optical (light) microscopy		11b,c, 12a
	Advantages: Samples examined in hydrated form or embedded; ability to chemically analyze samples by color using specific stains; examination of overall structure; detection and measurement of fat globules, bacteria; examination of powdered products	
	Weak points: Low depth of field limits sample thickness; low magnification (1000×) and resolution (~0.5 μm) (inability to resolve individual casein micelles)	
Polarized light microscopy	Detection of crystals, starch granules and their gelatinization	
Fluorescence microscopy	Chemical analysis, high specificity	
Confocal scanning laser microscopy (cslm)		28a,b
	Advantages: Samples examined in original state or following fixation and staining; focussing at predermined levels below the sample surface (optical sectioning); particularly suitable for emulsions	
	Weak points: Methodology (including image analysis software for the interpretation of results) for this new technique has yet to be developed	
Scanning electron microscopy (sem)		
	Advantages: Great depth of focus, high resolution, relatively easy preparation of samples; freeze-fracturing used to reveal internal structure	
	Weak points: Samples must be dried (for conventional SEM) or frozen (for cryo-sem)	
Conventional sem	Depending on the sample, some components may be extracted during sample preparation. Suitable for all products except fluid milk and cream; viscous samples may be encapsulated; dried samples examined in the original state	6a–c, 7b, 8b,c, 10c, 11a, 12b, 13a, 17b, 23c, 25a,b, 26a–c, 27a,b
Cryo-sem	Suitable for high-fat, whipped, and frozen products; reveals the present aqueous phase	13b, 21a,b, 23a
Transmission electron microscopy (tem)		
	Advantages: Very high magnification (>100,000×) and resolution (several nm) (ability to show the ultrastructure of casein micelles, details of membranes, bacterial structures, bacteriophage)	
	Weak points: Microscopes are expensive, preparation of samples is laborious and costly	
Negative staining	Suitable for suspensions and emulsions	3b,c
Metal shadowing	Suitable for suspensions and emulsions	1a, 3a, 5a, 7b
Thin sectioning	Suitable for all products; most common technique, but some components are better preserved than others depending on the procedure used	1b, 4a,b, 5b, 7a, 9a,b, 10a,b, 12a, 14, 15a,b, 16, 17a, 18a–c, 19, 20, 21c, 23a,b
Freeze fracturing + replication	Provides similar information as cryo-sem at a considerably higher resolution; particularly suitable for high-fat products. Weak points: requires special equipment and is laborious	2a–c, 10d, 24

[a] Figures presented in this article; see Table 1.

lactose hydrate in microcrystalline form. Anhydrous lactose rapidly absorbs water if the powders are exposed to a humid atmosphere and crystallizes in the α-hydrate form. This phenomenon is associated with caking of the powder (Fig. 27a). Controlled crystallization of lactose stabilizes the powders and is commercially used as part of the so-called instantization process, which was designed to improve the dissolution of milk powders. Instantization results in the agglomeration of the powder particles (Fig. 27b) and partial conversion of amorphous lactose into a microcrystalline form. It is based on maintaining a high moisture content in the powder particles during the agglomeration or instantization process and removing this additional moisture by additional drying.

Powders with a high lactose content (66%, of which 95% was in the amorphous state and 5% in the crystalline state), such as spray-dried whey, are highly susceptible to lactose recrystallization when exposed to lower levels of relative humidity (40% at 25°C) (166) than powders with a low lactose content (10.7%) such as spray-dried ultrafiltration milk retentates, which were resistant to lactose recrystallization even when exposed to 85% relative humid-

Figure 27. (a) Exposure of spray-dried milk powder to high humidity results in the crystallization of lactose (arrows) on particle surfaces. (Conventional sem.) (b) Instant milk powders consist of aggregated powder particles resulting from controlled lactose crystallization. (Conventional sem.)

ity for 3 days (171). The susceptibility to recrystallization increases with an increased proportion of amorphous lactose. Ultrafiltration permeate powders with a large proportion of α-hydrate crystals were, therefore, relatively resistant to atmospheric humidity.

CONTEMPORARY AND FUTURE RESEARCH

Microscopical studies in dairy science are currently expanding at a rapid pace. New techniques are either adapted from other disciplines or are developed specifically for milk products.

Considering various changes in our diet and the rapid development of instrumental techniques, the interest in finding correlations between food structure and functional behavior of food components will increase, particularly because some relationships have already been established (12). The new topics to be covered will probably include relationships between food structure and sensory attributes (eg, grittiness, sandiness, smoothness). Microscopy

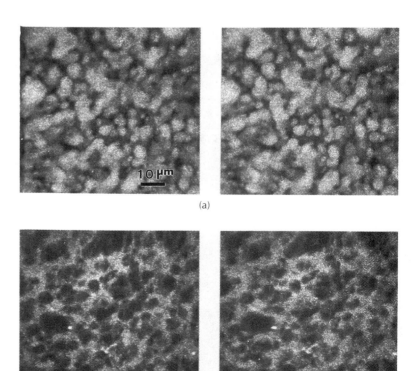

Figure 28. Pairs of stereo micrographs (confocal scanning laser microscopy) of (a) Gouda cheese stained for fat and (b) for protein. [Courtesy of I. Heertje (16); reprinted with permission from SMI.]

and related techniques will be used to improve our understanding of the relationships between origin, processing, and storage of the foods and their ingredients. Quantification of microstructure, ie, conversion of images into numerical sets using image analysis, will be used to express the dimensions, shapes, and distribution of structural elements found in the food under study and to reveal important relationships, eg, between the dimensions of fat globules in process cheese and process cheese firmness, effects of cryoprotective agents on ice crystal dimensions in ice cream and correlations between the ice crystal dimensions and sensory attributes in ice cream, provide quantitative data on casein particle chains and clusters in the curd, etc.

Elemental analysis techniques [such as energy-dispersive spectrometry (edx), electron spectroscopic imaging (esi), electron energy loss spectrometry (eels)] will be used to a greater extent to determine changes in the presence and distribution of various elements (calcium, phosphorus, iron, etc) as the result of processing or spoilage of dairy products as well as nonmilk foods. Immunocytochemical methods using colloidal gold (47) will be particularly suitable for analytical purposes as they are used to locate characteristic proteins and polysaccharides.

Scanning tunneling microscopy makes it possible to reveal submolecular structure. Initially, it was used in semiconductor research but now is making an important impact also in the biological field, including food research (182).

Confocal scanning laser microscopy (Figs 28a,b) (16) will become a routine procedure, particularly for its ease of use and the ability to examine samples at various depths under the surface without need for fixation and to provide, under these conditions, three-dimensional images. Fluorescence microscopy (15) has an excellent potential to be used to a greater extent as the demand for routine structural and compositional analysis is increasing.

Acknowledgments: The author thanks Dr. H. D. Goff, Dr. D. N. Holcomb, Dr. H. W. Modler, and Mrs. Paula Allan-Wojtas for useful suggestions. The author is grateful to the following for permission to reproduce their micrographs: Mrs. Paula Allan-Wojtas (Fig. 13a), Dr. B. E. Brooker (Figs. 4a,b, 21a–c), Dr. W. Buchheim (Fig. 24), Miss K. B. Caldwell (Fig. 23a), and Dr. I. Heertje (Figs. 2a–c, 3b,c, 18a–c, 28a,b). Skillful technical assistance provided by Miss Gisèle Larocque and assistance with photography provided by Mr. Sierk Itz is acknowledged. Electron Microscope Unit, Research Branch, Agriculture Canada in Ottawa provided facilities. Contribution 864 from the Food Research Centre.

BIBLIOGRAPHY

1. S. C. Nickerson and R. M. Akers, "Biochemical and Ultrastructural Aspects of Milk Synthesis and Secretion," *International Journal of Biochemistry* **16**(8), 855–865 (1984).
2. B. E. Brooker, "Milk and Its Products," in J. G. Vaughan, ed., *Food Microscopy*, Academic Press, New York, 1979, pp. 273–311.
3. M. Kaláb, "Microstructure of Dairy Foods. 1. Milk Products Based on Protein," *Journal of Dairy Science* **62**(8), 1352–1364 (1979).
4. M. Kaláb, "Microstructure of Dairy Foods. 2. Milk Products Based on Fat," *Journal of Dairy Science* **68**, 3234–3248 (1985).
5. M. L. Green, "The Formation and Structure of Milk Protein Gels," *Food Chemistry* **6**, 41–49 (1980).
6. M. L. Green, K. R. Langley, R. J. Marshall, B. E. Brooker, A. Willis, and J. F. V. Vincent, "Mechanical Properties of Cheese, Cheese Analogues and Protein Gels in Relation to Composition and Microstructure," *Food Microstructure* **5**(1), 169–180 (1986).
7. F. L. Davies, P. A. Shankar, B. E. Brooker, and D. G. Hobbs, "A Heat-Induced Change in the Ultrastructure of Milk and Its Effect on Gel Formation in Yoghurt," *Journal of Dairy Research* **45**, 53–58 (1978).
8. V. R. Harwalkar and M. Kaláb, "Effect of Acidulants and Temperature on Microstructure, Firmness and Susceptibility to Syneresis in Skim Milk Gels," *Scanning Electron Microscopy* **1981**(III), 503–513 (1981).
9. M. Kaláb, S. K. Gupta, H. K. Desai, and G. R. Patil, "Development of Microstructure in Raw, Fried, and Fried and Cooked Paneer Made from Buffalo, Cow, and Mixed Milks," *Food Microstructure* **7**(1), 83–91 (1988).
10. V. R. Harwalkar, "Age Gelatinization of Sterilized Milks," in P. F. Fox, ed., *Developments in Dairy Chemistry,* Part 1: *Proteins,* Elsevier Applied Science Pubishers, London, 1982, pp. 229–269.
11. M. Kaláb and V. R. Harwalkar, "Milk Gel Structure. II. Relation between Firmness and Ultrastructure of Heat-Induced Skim-Milk Gels Containing 40–60% Total Solids," *Journal of Dairy Research* **41**, 131–135 (1974).
12. D. W. Stanley, "Food Texture and Microstructure," in H. R. Moskowitz, ed., *Food Texture. Instrumental and Sensory Measurement,* Marcel Dekker, New York, 1987, pp. 35–64.
13. M. Kaláb and M. Carić, "Food Microstructure—Evaluation of Interactions of Milk Components in Food Systems," in *Proceedings of 23rd International Dairy Congress,* Montreal, 1990, pp. 1457–1480.
14. P. D. Shimmin, "Observation of Fat Distribution in Cheese by Incident Light Fluorescence Microscopy," *Australian Journal of Dairy Technology* **37**, 33–34 (1982).
15. S. H. Yiu, "A Fluorescence Microscopic Study of Cheese," *Food Microstructure* **4**(1), 99–106 (1985).
16. I. Heertje, P. van der Vlist, J. C. G. Blonk, H. A. C. Hendrickx, and G. J. Brakenhoff, "Confocal Scanning Laser Microscopy in Food Research: Some Observations," *Food Microstructure* **6**(2), 115–120 (1987).
17. M. Kaláb, "Electron Microscopy of Milk Products: A Review of Techniques," *Scanning Electron Microscopy* **1981**(III), 453–472 (1981).
18. M. Kaláb, "Electron Microscopy of Foods," in E. B. Bagley and M. Peleg, eds., *Physical Properties of Foods,* AVI Publishing Co., Westport, Conn., 1983, pp. 43–104.
19. D. G. Schmidt, "Electron Microscopy of Milk and Milk Products: Problems and Possibilities," *Food Microstructure* **1**(2), 151–165 (1982).
20. M. Kaláb, "Artefacts in Conventional Scanning Electron Microscopy of Some Milk Products," *Food Microstructure* **3**(1), 95–111 (1984).
21. D. G. Schmidt and A. C. M. van Hooydonk, "A Scanning Electron Microscopical Investigation of the Whipping of Cream," *Scanning Electron Microscopy* **1980**(III), 653–658 (1980).
22. W. Buchheim, "Aspects of Sample Preparation for Freeze-Fracture/Freeze-Etch Studies of Proteins and Lipids in Food

Systems. A Review," *Food Microstructure* 1(2), 189–208 (1982).
23. M. Kaláb, "Scanning Electron Microscopy of Dairy Products: An Overview," *Scanning Electron Microscopy* 1979(III), 261–272 (1979).
24. G. W. White and A. J. Shenton, "Food Microscopy (An Annotated Bibliography), Part II E: Major Ingredients: Milk and Milk Powder," *Journal of the Association of Publ. Analysts* 15, 33–37 (1977).
25. J. M. Aguilera and D. W. Stanley, "Microstructural Principles of Food Processing & Engineering," Elsevier Applied Science Publishers, London, 1990, pp. 144–235.
26. D. N. Holcomb, "Food Microstructure—Cumulative Index," *Food Structure* 9(2), 155–173 (1990).
27. V. R. Harwalkar, P. Allan-Wojtas, and M. Kaláb, "Effect of Heating to 200°C on Casein Micelles in Milk: A Metal-Shadowing and Negative Staining Electron Microscope Study," *Food Microstructure* 8(2), 217–224 (1989).
28. I. Heertje, J. Visser, and P. Smits, "Structure Formation in Acid Milk Gels," *Food Microstructure* 4(2), 267–277 (1985).
29. G. G. Calapaj, "An Electron Microscope Study on the Ultrastructure of Bovine and Human Casein Micelles in Fresh and Acidified Milk," *Journal of Dairy Research* 35, 1–6 (1968).
30. R. J. Carroll, M. P. Thompson, and P. Melnychyn, "Gelation of Concentrated Skim-Milk: Electron Microscopic Study," *Journal of Dairy Science* 54(9), 1245–1252 (1971).
31. R. J. Carroll, M. P. Thompson, and G. C. Nutting, "Glutaraldehyde Fixation of Casein Micelles for Electron Microscopy," *Journal of Dairy Science* 51(12), 1903–1908 (1968).
32. P. F. Fox and J. Guiney, "Casein Micelle Structure: Susceptibility of Various Casein Systems to Proteolysis," *Journal of Dairy Research* 40, 229–234 (1973).
33. N. W. Freeman and M. E. Mangino, "Effects of Ultra High Temperature Processing on Size and Appearance of Casein Micelles in Bovine Milk," *Journal of Dairy Science* 64, 1772–1780 (1981).
34. C. Holt, "The Size Distribution of Bovine Casein Micelles: A Review," *Food Microstructure* 4(1), 1–10 (1985).
35. E. Knoop, "Electron Microscopic Studies on the Structure of Milk Fat and Protein," *Milchwissenschaft* 27(6), 364–373 (1972).
36. M. E. Mangino and N. W. Freeman, "Statistically Reproducible Evaluation of Size of Casein Micelles in Raw and Processed Milks," *Journal of Dairy Science* 64(10), 2025–2030 (1981).
37. T. C. A. McGann, W. Buchheim, R. D. Kearney, and T. Richardson, "Composition and Ultrastructure of Calcium Phosphate-Citrate Complexes in Bovine Milk Systems," *Biochimica et Biophysica Acta* 760, 415–420 (1983).
38. T. C. A. McGann, W. J. Donnelly, R. D. Kearney, and W. Buchheim, "Composition and Size Distribution of Bovine Casein Micelles," *Biochimica et Biophysica Acta* 630, 261–270 (1980).
39. D. Rose and J. R. Colvin, "Appearance and Size of Micelles from Bovine Milk," *Journal of Dairy Science* 49, 1091–1097 (1966).
40. M. Rüegg and B. Blanc, "Influence of Pasteurization and UHT Processing upon the Size Distribution of Casein Micelles in Milk," *Milchwissenschaft* 33(6), 364–366 (1978).
41. D. G. Schmidt and W. Buchheim, "Particle Size Distribution in Casein Solutions," *Netherlands Milk Dairy Journal* 30, 17–28 (1976).
42. D. G. Schmidt, P. Walstra, and W. Buchheim, "The Size Distribution of Casein Micelles in Cow's Milk," *Netherlands Milk Dairy Journal* 27, 128–142 (1973).
43. E. Uusi-Rauva, J-A. Rautavaara, and M. Antila, "Effects of Various Temperature Treatments on Casein Micelles. An Electron Microscopic Study Using Negative Staining," *Meijeritiet. Aikakauskirja* 31, 15–25 (in German).
44. T. Kimura, S. Taneya, and K. Kanaya, "Observation of Internal Structure of Casein Submicelles by Means of Ion Beam Sputtering," *Milchwissenschaft* 34(9), 521–524 (1979).
45. J. Garnier, "Models of Casein Micelle Structure," *Netherlands Milk Dairy Journal* 27, 240–248 (1973).
46. H. M. Farrell, Jr., "Models for Casein Micelle Formation," *Journal of Dairy Science* 56, 1195–1206 (1973).
47. M. Horisberger and M. Vauthey, "Localization of κ-Casein on Thin Sections of Casein Micelles by the Gold Method," *Histochemistry* 80(9), 9–12 (1984).
48. D. J. McMahon and R. J. Brown, "Composition, Structure, and Integrity of Casein Micelles: A Review," *Journal of Dairy Science* 67(3), 499–512 (1984).
49. T. A. J. Payens, "Association of Caseins and Their Possible Relation to Structure of the Casein Micelle," *Journal of Dairy Science* 49, 1317–1324 (1966).
50. D. G. Schmidt, "Association of Caseins and Casein Micelle Structure," in P. F. Fox, ed., *Developments in Dairy Chemistry*, Part 1, *Proteins*, Elsevier Applied Science Publishers, London, 1982, pp. 61–86.
51. P. Walstra, "The Voluminosity of Bovine Casein Micelles and Some of Its Implications," *Journal of Dairy Research* 46, 317–323 (1979).
52. R. Jenness, "Inter-Species Comparison of Milk Proteins," in P. F. Fox, ed., *Developments in Dairy Chemistry*, Part 1, *Proteins*, Elsevier Applied Science Publishers, London, 1982, pp. 87–114.
53. A. M. Knoop and H. K. Peters, "Structural Changes of Rennet Curds during Aging," *Kieler Milchwirt. Forschungsber.* 27(4), 315–330 (1975) (in German).
54. S. H. C. Lin, S. L. Leong, R. K. Dewan, V. A. Bloomfield, and C. V. Morr, "Effect of Calcium Ion on the Structure of Native Bovine Casein Micelles," *Biochemistry* 11, 1818–1821 (1972).
55. C. A. Zittle, M. P. Thompson, J. H. Custer, and J. Cerbulis, "κ-Casein—β-Lactoglobulin Interaction in Solution when Heated." *Journal of Dairy Science* 45(7), 807–810 (1962).
56. H. J. Hostettler, K. Imhof, and J. Stein, "Studies on the Effect of Heat Treatment and Lyophilisation on the State of Distribution and Physiological Properties of Milk Proteins with Special Consideration to the Heat Treatment Conditions Applied in the Uperisation. 1. Effect on the State of Distribution of Milk Proteins," *Milchwissenschaft* 20(4), 189–198 (1965) (in German).
57. A. T. Andrews, B. E. Brooker, and D. G. Hobbs, "Properties of Aseptically Packed Ultra-Heat-Treated Milk. Electron Microscopic Examination of Changes Occurring during Storage," *Journal of Dairy Research* 44, 283–292 (1977).
58. V. R. Harwalkar and H. J. Vreeman, "Effect of Added Phosphates and Storage on Changes in Ultra-High Temperature Short-Time Sterilized Concentrated Skim Milk. 2. Micelle Structure," *Netherlands Milk Dairy Journal* 32, 204–216 (1978).
59. S. Henstra and D. G. Schmidt, "On the Structure of the Fat-Protein Complex in Homogenized Cow's Milk," *Netherlands Milk Dairy Journal* 24, 45–51 (1970).

60. B. E. Brooker, "Observations on the Air-Serum Interface of Milk Foams," *Food Microstructure* **4**(2), 289–296 (1985).
61. D. F. Darling and D. W. Butcher, "Milk-Fat Globule Membrane in Homogenized Cream," *Journal of Dairy Research* **45**, 197–208 (1978).
62. H. Heintzberger, J. Koops, and D. Westerbeek, "Gelation of Sterilized Canned Evaporated Milk," *Netherlands Milk Dairy Journal* **26**, 31–40 (1972).
63. H. Mulder and P. Walstra, *The Milk Fat Globule. Emulsion Science as Applied to Milk Products and Comparable Foods*, Commonwealth Agric. Bureaux, Farnham Royal, Bucks, England, 1974, 296 pp.
64. H. J. Klupsch, *Acidulated Milk Products—Mixed Milk Drinks and Desserts*, Theodor Mann Publishing House, Gelsenkirchen-Buer, Federal Republic of Germany, 1984, pp. 99–116 (in German).
65. F. Kosikowski, *Cheese and Fermented Milk Foods*, Edward Brothers, Inc., Ann Arbor, Mich., 1977.
66. A. Y. Tamime, M. Kaláb, and G. Davies, "Rheology and Microstructure of Strained Yoghurt (Labneh) Made from Cow's Milk by Three Different Methods," *Food Microstructure* **8**(1), 125–135 (1989).
67. M. Kaláb, D. B. Emmons, and A. G. Sargant, "Milk Gel Structure. IV. Microstructure of Yoghurts in Relation to the Presence of Thickening Agents," *Journal of Dairy Research* **42**, 453–458 (1975).
68. H. W. Modler and M. Kaláb, "Microstructure of Yogurt Stabilized with Milk Proteins," *Journal of Dairy Science* **66**(3), 430–437 (1983).
69. A. Y. Tamime, M. Kaláb, and G. Davies, "Microstructure of Set-Style Yoghurt Manufactured from Cow's Milk Fortified by Various Methods," *Food Microstructure* **3**(1) 83–92 (1984).
70. M. Kaláb, P. Allan-Wojtas, and B. E. Phipps-Tood, "Development of Microstructure in Set-Style Nonfat Yoghurt—A Review," *Food Microstructure* **2**(1), 51–66 (1983).
71. A. M. Knoop and H. K. Peters, "Formation of the Curd Structure in Rennet and Acid Gelation of Milk," *Kieler Milchwirt. Forschungsber.* **27**, 227–248 (1975) (in German).
72. V. R. Harwalkar and M. Kaláb, "Relationship between Microstructure and Susceptibility to Syneresis in Yoghurt Made from Reconstituted Nonfat Dry Milk," *Food Microstructure* **5**(2), 287–294 (1986).
73. B. E. Brooker, "Cytochemical Observations on the Extracellular Carbohydrate Produced by *Streptococcus cremoris*," *Journal of Dairy Research* **43**, 283–290 (1976).
74. B. E. Brooker, "Electron Microscopy of Dextrans Produced by Lactic Acid Bacteria," in R. C. W. Berkeley, G. W. Gooday, and D. C. Ellwood, eds., *Microbial Polysaccharides and Polysaccharidases*, Academic Press, New York, 1979, pp. 85–115.
75. S. M. Schellhaass and H. A. Morris, "Rheological and Scanning Electron Microscopic Examination of Skim Milk Gels Obtained by Fermenting with Ropy and Non-Ropy Strains of Lactic Acid Bacteria," *Food Microstructure* **4**(2), 279–287 (1985).
76. C. L. Duitschaever, N. Kemp, and A. K. Smith, "Microscopic Studies of the Microflora of Kefir Grains and of Kefir Made by Different Methods," *Milchwissenschaft* **43**(8), 479–481 (1988).
77. V. M. Marshall, W. M. Cole, and B. E. Brooker, "Observations on the Structure of Kefir Grains and the Distribution of the Microflora," *Journal of Applied Bacteriology* **57**, 491–497 (1984).
78. I. Molska, J. Kocon, and S. Zmarlicki, "Electron Microscopic Studies on Structure and Microflora of Kefir Grains," *Acta Alimentaria Polonica* **6**(3), 145–154 (1980).
79. W. Buchheim and D. Prokopek, "Electron Microscopic Examination of Ultrafiltration Concentrates of Skim Milk and Cheese Prepared from It. 1. Behaviour of Casein Micelles during Ultrafiltration," *Milchwissenschaft* **31**(8), 462–465 (1976) (in German).
80. D. Dj. Gavarić, M. Carić, and M. Kaláb, "Effects of Protein Concentration in Ultrafiltration Milk Retentates and the Type of Protease Used for Coagulation on the Microstructure of Resulting Gels," *Food Microstructure* **8**(1), 53–66 (1989).
81. M. F. Eino, D. A. Biggs, D. M. Irvine, and D. W. Stanley, "A Comparison of Microstructure of Cheddar Cheese Curd Manufactured with Calf Rennet, Bovine Pepsin, and Porcine Pepsin," *Journal of Dairy Research* **43**, 113–115 (1976).
82. J. R. Bishop, A. B. Bodine, and J. J. Janzen, "Electron Microscopic Comparison of Curd Microstructures of Cottage Cheese Coagulated with and without Microbial Rennet," *Cultured Dairy Products Journal* **18**(3), 14–16 (1983).
83. M. L. Green, A. Turvey, and D. G. Hobbs, "Development of Structure and Texture in Cheddar Cheese," *Journal of Dairy Research* **48**, 343–355 (1981).
84. M. Rousseau, "Changes in the Microstructure of Saint Paulin Cheese during Manufacture Studied by Scanning Electron Microscopy," *Food Microstructure* **7**(1), 105–113 (1988).
85. M. Kaláb, "Milk Gel Structure, VI. Cheese Texture and Microstructure," *Milchwissenschaft* **32**, 449–458 (1977).
86. J. Glaser, P. Carroad, and W. L. Dunkley, "Surface Structure of Cottage Cheese Curd by Electron Microscopy," *Journal of Dairy Science* **62**, 1058–1068 (1979).
87. J. Glaser, P. A. Carroad, and W. L. Dunkley, "Electron Microscopic Studies of Casein Micelles and Curd Microstructure in Cottage Cheese," *Journal of Dairy Science* **63**, 37–48 (1980).
88. B. E. Brooker, "Electron Microscopy of Normal and Defective Cottage Cheese Curd," *Journal of the Society of Dairy Technology* **39**(3), 85–88 (1986).
89. M. Kaláb, R. J. Lowrie, and D. Nichols, "Detection of Curd Granule and Milled Curd Junctions in Cheddar Cheese," *Journal of Dairy Science* **65**, 1117–1121 (1982).
90. R. J. Lowrie, M. Kaláb, and D. Nichols, "Curd Granule and Milled Curd Junction Patterns in Cheddar Cheese Made by Traditional and Mechanized Processes," *Journal of Dairy Science* **65**, 1122–1129 (1982).
91. C. G. Rammell, "The Distribution of Bacteria in New Zealand Cheddar Cheese," *Journal of Dairy Research* **27**, 341–351 (1960).
92. D. B. Emmons, M. Kaláb, E. Larmond, and R. J. Lowrie, "Milk Gel Structure. X. Texture and Microstructure in Cheddar Cheese Made from Whole Milk and from Homogenised Low-Fat Milk," *Journal of Texture Studies* **11**, 15–34 (1980).
93. M. Rüegg, U. Moor, and J. Schnider, "On the Size Distribution and Shape of the Curd Granules in Emmental Cheese," *Schweiz. Milchw. Forschung* **14**(1), 3–7 (1985) (in German).
94. M. Rüegg and U. Moor, "The Size Distribution and Shape of Curd Granules in Traditional Swiss Hard and Semi-Hard Cheeses," *Food Microstructure* **6**(1), 35–46 (1987).
95. B. E. Brooker, D. G. Hobbs, and A. Turvey, "Observations on the Microscopic Crystalline Inclusions in Cheddar Cheese," *Journal of Dairy Research* **42**, 341–348 (1975).

96. V. Botazzi, B. Battistotti, and F. Bianchi, "The Microscopic Crystalline Inclusions in Grana Cheese and Their X-Ray Microanalysis," *Milchwissenschaft* **37**(10), 577–580 (1982).

97. E. Flückiger and P. Schilt, "Formation of Salt Crystals in Swiss Cheese," *Milchwissenschaft* **18**, 437–442 (1963) (in German).

98. D. M. Hall and L. K. Creamer, "A Study of the Sub-Microscopic Structure of Cheddar, Cheshire and Gouda Cheese by Electron Microscopy," *New Zealand Journal of Dairy Science and Technology* **7**, 95–102 (1972).

99. M. Kaláb and V. Palo, "Development of Microstructure in Olomouc Cheese Cakes: Electron Microscopic Study," *Milchwissenschaft* **42**(4), 207–211 (1987).

100. Y. Lee and R. T. Marshall, "Microstructure and Texture of Process Cheese, Milk Curds and Caseinate Curds Containing Native or Boiled Soy Proteins," *Journal of Dairy Science* **64**, 2311–2317 (1981).

101. H. W. Modler, S. H. Yiu, U. K. Bollinger, and M. Kaláb, "Grittiness in a Pasteurized Cheese Spread: A Microscopic Study," *Food Microstructure* **8**(2), 201–210 (1989).

102. M. Nanni and S. Annibaldi, "Observations on the Microstructure of Parmigiano-Reggiano Cheese: I. Curd Granules, *Scienza e Tecnica Lattiero-Casearia* **33**(2), 81–94 (1982) (in Italian).

103. M. M. Omar and W. Buchheim, "Composition and Microstructure of Soft Brine Cheese Made from Instant Whole Milk Powder," *Food Microstructure* **2**(1), 43–50 (1983).

104. M. Rüegg and B. Blanc, "Electron Microscopial Study of Curd and Cheese," *Schweiz. Milchwirt. Forschung* **1**, 1–8 (1972) (in German).

105. M. Rüegg, U. Moor, and B. Blanc, "Changes in the Fine Structure of Ripening Gruyere Cheese. A Scanning Electron Microscope Study," *Milchwissenschaft* **35**(6), 329–335 (1980) (in German).

106. M. Rüegg, R. Sieber, and B. Blanc, "The Structure of Cheese as Observed by a Scanning Electron Microscope," *Schweiz. Milchwirt. Forschung* **3**, 1–5 (1974) (in German).

107. M. V. Taranto, P. J. Wan, S. L. Chen, and K. C. Rhee, "Morphological, Ultrastructural and Rheological Characterization of Cheddar and Mozzarella Cheese," *Scanning Electron Microscopy* **1979**(III), 273–278 (1979).

108. C. J. Washam, T. J. Kerr, and V. J. Hurst, "Microstructure of Various Chemical Compounds Crystallized in Cheddar Cheese," *Journal of Food Protection* **45**(7), 594–596 (1982).

109. A. M. Knoop and W. Buchheim, "Different Development of the Structure in Harz, Tilsit, and Camembert Cheeses during Ripening," *Milchwissenschaft* **35**(8), 482–488 (1980) (in German).

110. A. M. Knoop, D. Prokopek, and K. H. Peters, "Submicroscopical Structural Changes during Ripening of Tilsit Cheese Obtained by Different Manufacturing Procedures," *Kieler Milchwirt. Forschungsber.* **31**(2), 97–113 (1979) (in German).

111. H. Mair-Waldburg, *Handbook on Cheese,* Volkswirtschaftlicher Verlag GmbH, Kempten (Allgäu), Federal Republic of Germany, 1974, pp. 636–638 (in German).

112. K. M. K. Kebary and H. A. Morris, "Porosity, Specific Gravity and Fat Dispersion in Blue Cheeses," *Food Microstructure* **7**(2), 153–160 (1987).

113. A. M. Knoop and K. H. Peters, "Submicroscopical Structural Variations during Ripening of Camembert Cheese," *Milchwissenschaft* **26**(4), 193–198 (1971) (in German).

114. M. Rousseau, "Study of the Surface Flora of Traditional Camembert Cheese by Scanning Electron Microscopy," *Milchwissenschaft* **39**(3), 129–135 (1984).

115. B. E. Brooker, "The Crystallization of Calcium Phosphate at the Surface of Mould-Ripened Cheeses," *Food Microstructure* **6**(1), 25–33 (1987).

116. L. L. de Jong, "Protein Breakdown in Soft Cheese and Its Relation to Consistency. 3. The Micellar Structure of Meshanger," *Netherlands Milk Dairy Journal* **32**, 15–25 (1978).

117. M. Kaláb, H. W. Modler, M. Carić, and S. Milanović, "Structure, Meltability, and Firmness of Process Cheese Containing White Cheese," *Food Structure* **10**(2) (1991), in press.

118. V. R. Harwalkar and M. Kaláb, "The Role of β-Lactoglobulin in the Development of the Core-and-Lining Structure of Casein Particles in Acid-Heat-Induced Milk Gels," *Food Microstructure* **7**(2), 173–179 (1988).

119. M. Carić and M. Kaláb, "Processed Cheese Products," in P. F. Fox, ed., *Cheese: Chemistry, Physics and Microbiology.* Vol. 2. *Major Cheese Groups,* Elsevier Applied Science, London, 1987, pp. 339–383.

120. W. Buchheim and P. Jelen, "Microstructure of Heat-Coagulated Whey Protein Curd," *Milchwissenschaft* **31**(10), 589–592 (1976).

121. P. Jelen and W. Buchheim, "Stability of Whey Protein upon Heating in Acidic Conditions," *Milchwissenschaft* **39**(4), 215–218 (1984).

122. M. Kaláb and H. W. Modler, "Milg Gel Structure. XV. Electron Microscopy of Whey-Protein-Based Cream Cheese Spread," *Milchwissenschaft* **40**(4), 193–196 (1985).

123. A. Meyer, *Process Cheese Manufacture,* 1st ed., Food Trade Press, London, 1973, 330 pp.

124. M. Carić, M. Gantar, and M. Kaláb, "Effects of Emulsifying Agents on the Microstructure and Other Characteristics of Process Cheese—A Review," *Food Microstructure* **4**(2), 297–312 (1985).

125. H. Klostermeyer and W. Buchheim, "Microstructure of Processed Cheese Products," *Kieler Milchwirt. Forschungsber.* **40**(4), 219–231 (1988) (in German).

126. V. Boháč, "Application of Microscopy to Cheese Testing," in L. Forman, ed., *The Collection of Papers from the Dairy Research Institute in Prague 1978–1983,* Technical Information Centre for the Food Industry, Prague, Czechoslovakia, 1984, pp. 203–212 (in Czech).

127. V. Boháč, "Microscopic Methods in Cheese Production," *Mljekarstvo* **36**(4), 114–120 (1986) (in Croatian).

128. A. A. Rayan, "Microstructure and Rheology of Process Cheese," *Diss. Int. B.* **41**(8), 2954 (1980).

129. A. A. Rayan, M. Kaláb, and C. A. Ernstrom, "Microstructure and Rheology of Process Cheese," *Scanning Electron Microscopy* **1980**(III), 635–643 (1980).

130. I. Heertje, M. J. Boskamp, F. van Kleef, and F. H. Gortenmaker, "The Microstructure of Process Cheese," *Netherlands Milk Dairy Journal* **35**, 177–179 (1981).

131. T. Kimura and S. Taneya, "Electron Microscopic Observation of Casein Particles in Cheese," *Journal of Electron Microscopy* **24**(2), 115–117 (1978).

132. S. Taneya, T. Kimura, T. Izutsu, and W. Buchheim, "The Submicroscopic Structure of Processed Cheese with Different Melting Properties," *Milchwissenschaft* **35**, 479–481 (1980).

133. V. G. Tinyakov, "Study of the Submicrostructure of Cheese," *Izv. Vyss. Ucheb. Zaved., Pishch. Tekhnol.* **1964**(5), 62–63(1964) (in Russian).

134. M. Kaláb, J. Yun and S. H. Yiu, "Textural Properties and Microstructure of Process Cheese Food Rework," *Food Microstructure* **6**(2), 181–192 (1987).

135. A. Paquet and M. Kaláb, "Amino Acid Composition and Structure of Cheese Baked as a Pizza Ingredient in Conventional and Microwave Ovens," *Food Microstructure* **7**(1), 93–103 (1988).

136. A. Y. Tamime, M. Kaláb, G. Davies, and M. F. Younis, "Microstructure and Firmness of Processed Cheese Manufactured from Cheddar Cheese and Skim Milk Powder Cheese Base," *Food Structure* **9**(1), 23–37 (1990).

137. H. Klostermeyer, G. Uhlmann, and K. Merkenich, "Formation of Crystals in Process Cheese. II. Identification of a New Citrate," *Milchwissenschaft* **39**(4), 195–197 (1984) (in German).

138. H. A. Morris, P. B. Manning, and R. Jenness, "Calcium Citrate Surface in Process Cheese," *Journal of Dairy Science* **52**, 900 (1962).

139. G. Uhlmann, H. Klostermeyer, and K. Merkenich, "Formation of Crystals in Process Cheese—I. The Phenomenon and Its Sources," *Milchwissenschaft* **38**(10), 582–585 (1983) (in German).

140. W. Buchheim and P. Dejmek, "Milk and Dairy-Type Emulsions," in K. Larsson and S. Friberg, eds., *Food Emulsions*, 2nd ed. (revised and expanded), Marcel Dekker, New York, 1990, pp. 203–246.

141. D. F. Darling, "Recent Advances in the Destabilization of Dairy Emulsions," *Journal of Dairy Research* **49**, 695–712 (1982).

142. A. V. McPherson and B. J. Kitchen, "Reviews of the Progress of Dairy Science: The Bovine Milk Fat Globule Membrane—Its Formation, Composition, Structure and Behaviour in Milk and Dairy Products," *Journal of Dairy Research* **50**, 107–133 (1983).

143. P. Pinto da Silva, A. Peixoto de Menezes, and I. H. Mather, "Structure and Dynamics of the Bovine Milk Fat Globule Membrane Viewed by Freeze Fracture," *Experimental Cell Research* **125**, 127–139 (1980).

144. D. Precht and W. Buchheim, "Electron Microscopic Studies on the Physical Structure of Spreadable Fats. 1. The Microstructure of Fat Globules in Butter," *Milchwissenschaft* **34**(12), 745–749 (1979); "2. The Microstructure of the Interglobular Fat Phase in Butter," ibid. **35**(6), 399–402 (1980); "3. The Aqueous Phase in Butter," ibid. **35**(11), 684–690 (1980) (in German).

145. F. B. P. Wooding, "The Structure of the Milk Fat Globule Membrane," *Journal of Ultrastructure Research* **37**, 388–400 (1971).

146. W. Buchheim, "Membranes of Milk Fat Globules—Ultrastructural, Biochemical, and Technological Aspects," *Kieler Milchwirt. Forschungsber.* **38**(4), 227–246 (1986).

147. W. Buchheim, G. Falk, and A. Hinz, "Ultrastructural Aspects of PhysicoChemical Properties of Ultra-High-Temperature (UHT)-Treated Coffee Cream," *Food Microstructure* **5**(1), 181–192 (1986).

148. W. Buchheim and D. Precht, "Electron Microscopic Study on the Crystallization Processes in Fat Globules during the Ripening of Cream," *Milchwissenschaft* **34**(11), 657–662 (1979) (in German).

149. M. Anderson, B. E. Brooker, T. E. Cawston, and G. C. Cheeseman, "Changes during Storage in Stability and Composition of Ultra-Heat-Treated Aseptically Packed Cream of 18% Fat Content," *Journal of Dairy Research* **44**, 111–123 (1977).

150. B. E. Brooker, M. Anderson, and A. T. Andrews, "The Development of Structure in Whipped Cream," *Food Microstructure* **5**(2), 277–285 (1986).

151. N. Krog, N. M. Barfod, and W. Buchheim, "Protein-Fat-Surfactant Interactions in Whippable Emulsions," in E. Dickinson, ed., *Food Emulsions and Foams*, Royal Society of Chemists, London, 1987, pp. 144–157.

152. M. Anderson, B. E. Brooker, and E. C. Needs, "The Role of Proteins in the Stabilization/Destabilization of Dairy Foams," in E. Dickinson, ed., *Food Emulsions and Foams*, Royal Society of Chemists, London, 1987, pp. 100–109.

153. H. D. Goff, M. Liboff, W. K. Jordan, and J. E. Kinsella, "The Effects of Polysorbate 80 on the Fat Emulsion in Ice Cream Mix: Evidence from Transmission Electron Microscopy Studies," *Food Microstructure* **6**(2), 193–198 (1987).

154. W. S. Arbuckle, *Ice Cream*, 4th ed., AVI Publishing Co., Westport, Conn., 1976, pp. 49–92.

155. K. G. Berger, "Ice Cream," in K. Larsson and S. Fridberg, eds., *Food Emulsions*, 2nd ed. (revised and expanded), Marcel Dekker, New York, 1990, pp. 367–444.

156. K. B. Caldwell and H. D. Goff, "The Use of Low-Temperature Scanning Electron Microscopy to Study the Microstructure of Ice Cream," *Food Structure* **10**(4) (1991).

157. M. Kaláb, A. G. Sargant, and D. A. Froehlich, "Electron Microscopy and Sensory Evaluation of Commercial Cream Cheese," *Scanning Electron Microscopy* **1981**(III), 473–482, 514 (1981).

158. T. Ohashi, S. Nagai, K. Masaoka, S. Haga, K. Yamauchi, and N. F. Olson, "Physical Properties and Microstructure of Cream Cheese," *Nippon Shokuhin Kogyo Gakkaishi* **30**(5), 303–307 (1983).

159. M. Kaláb and H. W. Modler, "Development of Microstructure in a Cream Cheese Based on Queso Blanco Cheese," *Food Microstructure* **4**(1), 89–98 (1985).

160. W. Buchheim and J. Thomasow, "Structural Changes in Cream Cheese Induced by Thermal Processing and Emulsifying Salts," *North European Dairy Journal* **50**(2), 38–44 (1984).

161. A. C. Juriaanse and I. Heertje, "Microstructure of Shortenings, Margarine and Butter—A Review," *Food Microstructure* **7**(2), 181–188 (1988).

162. D. Precht and K. H. Peters, "The Consistency of Butter. I. Electron Microscopic Studies on the Effect of Various Cream Ripening Temperatures on the Frequency of Certain Fat Globule Types in the Cream," *Milchwissenschaft* **36**(1), 616–620 (1981) (in German).

163. D. Precht and K. H. Peters, "The Consistency of Butter. II. Relationships between the Submicroscopic Structures of Fat Globules in Cream and Butter and the Consistency [of Butter] as Dependent on Special Physical Methods of Cream Ripening," *Milchwissenschaft* **36**(11), 673–676 (1981) (in German).

164. T. J. Buma, "Free Fat and Physical Structure of Spray-Dried Whole Milk," Thesis, University of Wageningen, The Netherlands, also published in *Netherlands Milk Dairy Journal* **22**, 22–27 (1968) and **25**, 33–42, 42–52, 53–72, 88–106, 107–122, 123–140, 151–158, 159–174 (1971).

165. Z. Saito, "Lactose Crystallization in Commercial Whey Powders and in Spray-Dried Lactose," *Food Microstructure* **7**(1), 75–81 (1988).

166. M. Saltmarch and T. P. Labuza, "SEM Investigation of the Effect of Lactose Crystallization on the Storage Properties of Spray-Dried Whey," *Scanning Electron Microscopy* **1980**(III), 659–665 (1980).

167. P. Jelen, M. Kaláb, and R. I. W. Greig, "Water-Holding Capacity and Microstructure of Heat-Coagulated Whey Protein Powders," *Milchwissenschaft* **34**(6), 351–356 (1979).

168. M. Kaláb and F. Comer, "Detection of Buttermilk Solids in Meat Binders by Electron Microscopy," *Food Microstructure* **1**(1), 49–54 (1982).

169. T. J. Buma and S. Henstra, "Particle Structure of Spray-Dried Milk Products Observed by a Scanning Electron Microscope," *Netherlands Milk Dairy Journal* **25**, 75–80 (1971).
170. T. J. Kerr, C. J. Washam, A. L. Evans, and W. E. Rigsby, "Structural Characterization of Spray-Dried Dairy Products by Scanning Electron Microscopy," *Dev. Ind. Microbiology* **24**, 475–484 (1983).
171. M. Kaláb, M. Carić, M. Zaher, and V. R. Harwalkar, "Composition and Some Properties of Spray-Dried Retentates Obtained by the Ultrafiltration of Milk," *Food Microstructure* **8**(2), 225–233 (1989).
172. M. Carić, S. Milanović, and M. Kaláb, "Storage Quality of Demineralized Spray-Dried Permeate Powders," *Milchwissenschaft* (in press).
173. N. King, "The Physical Structure of Dried Milk," *Dairy Science Abstracts* **27**, 91–104 (1965).
174. W. Buchheim, "A Comparison of the Microstructure of Dried Milk Products by Freeze-Fracturing Powder Suspensions in Non-Aqueous Media," *Scanning Electron Microscopy* **1981**(III), 493–502 (1981).
175. W. Buchheim, "Electron Microscopic Localization of Solvent-Extractable Fat in Agglomerated Spray-Dried Whole Milk Powder Particles," *Food Microstructure* **1**(2), 233–238 (1982).
176. H. R. Müller, "Electron Microscopic Studies of Milk and Milk Products. 1. Elucidation of the Milk Powder Structure," *Milchwissenschaft* **19**, 345–356 (1964) (in German).
177. K. Roetmann, "Crystalline Lactose and the Structure of Spray-Dried Milk Products as Observed by Scanning Electron Microscopy," *Netherlands Milk Dairy Journal* **33**, 1–11 (1979).
178. Z. Saito, "Particle Structure in Spray-Dried Whole Milk and in Instant Skim Milk Powder as Related to Lactose Crystallization," *Food Microstructure* **4**(2), 333–340 (1985).
179. S. Warburton and S. W. Pixton, "The Moisture Relations of Spray-Dried Skimmed Milk," *Journal of Stored Products Research* **14**, 143–158 (1978).
180. M. Carić and M. Kaláb, "Effects of Drying Techniques on Milk Powders Quality and Microstructure. A Review," *Food Microstructure* **6**(2), 171–180 (1987).
181. T. J. Buma and S. Henstra, "Particle Structure of Spray-Dried Caseinate and Spray-Dried Lactose as Observed by a Scanning Electron Microscope," *Netherlands Milk Dairy Journal* **25**, 278–281 (1971).
182. M. J. Miles, "The Application of STM/AFM to Biological Molecules," *Microscopy and Analysis* 7–9 (July 1990).

MILOSLAV KALÁB
Food Research Centre,
Agriculture Canada
Ottawa, Ontario, Canada

FOOD SURFACE SANITATION

Of paramount importance in food manufacture is the freedom of microbial (spoilage and pathogenic microorganisms) and foreign body contamination in the final product. Such contamination may arise from the constituent raw materials or the processing environment, which includes food contact surfaces, the air, people, and pests. Failure to control these factors may lead to product recalls, loss of sales or profits, adverse publicity, and, if regulatory requirements have been infringed, fines, sanctions, or ultimately site closure or loss of production/export license.

The sanitation of surfaces, when undertaken correctly, is cost effective, easy to manage, and, if diligently applied, can reduce the risk of microbial or foreign body contamination of product. When incidents of such contamination occur, it is often not easy to trace and rectify the exact source and as surface sanitation is relatively cheap, it can provide management with a good tool to reduce this risk. In this context, surface sanitation:

1. Removes microorganisms, or material conducive to microbial growth. This reduces the chance of contamination by pathogens and extends the shelf life of some products.
2. Removes materials that could lead to foreign body contamination or could provide food or shelter for pests.

Surface sanitation is also implemented in food processing to provide a wide range of additional benefits including:

1. Reduces waste and improves the appearance and quality of product by removing product left on lines that may deteriorate and reenter subsequent production runs.
2. Increases process performance in some areas (eg, plate and scrape surface heat exchangers).
3. Extends the life of, and prevents damage to, equipment and services.
4. Provides a safe and clean working environment for employees and thus increases morale and productivity.
5. Presents a favorable image to customers and the public.

In this article, food surfaces are defined as both food contact and environmental surfaces. Environmental surfaces are included as, although they are not in direct contact with the product, contamination may be transferred from them to the product by people, pests, the air, or cleaning procedures. Only hard surfaces are considered eg, equipment, floors, walls, and utensils as other surfaces, eg, protective clothing or skin, would be traditionally dealt with under personal hygiene.

It is further assumed that the surfaces addressed have been designed hygienically. Poor hygienic design will restrict the efficiency of even the most effective cleaning procedure and may vitiate any subsequent disinfection programs. The principles of hygienic design are comprehensively described elsewhere (1,2,3).

FOOD SOILS

Debris will build up on surfaces throughout the production period; this will require subsequent removal by cleaning. This debris may be the result of normal production, spillages, line jams, maintenance, packaging, or general dust and dirt. Such undesirable material, which may include food residues, microorganisms, and foreign matter,

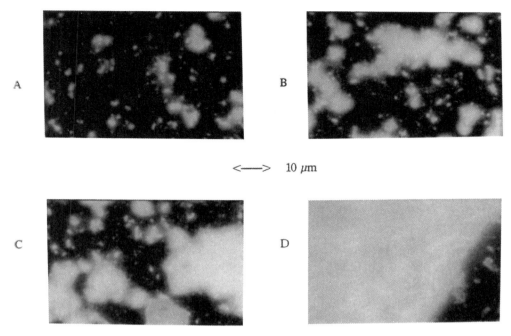

Figure 1. Build up of a bacterial biofilm on a baked bean production line. Photographs were taken using epiflourescent microscopy such that bacteria fluoresced orange while the background remained dark: A, 4 h, B, 8 h; C, 12 h; and D, 16 h.

is referred to as soil. In practical terms, a soil is anything in the wrong place at the wrong time; peas on a conveyor during production are product but after production or on the floor are soil.

Product soils are usually easy to visualize and are characterized primarily by the product type eg, protein, fat or carbohydrate. The process however is also important as a given product may present a variety of cleaning problems, depending on whether it is dry, wet, heat treated, frozen or the length of time it is left prior to cleaning.

Microbial soils cannot generally be observed by the eye but require microscopic examination. Bacterial attachment to surfaces is well documented (4,5), and the influence of bacterial growth on surfaces, termed biofilms, in the food industry has been discussed (6,7). Examination of stainless steel coupons attached to production lines, by epifluorescent microscopy, has been used to assess microbial levels in a range of food soils (8). An example of a heavy biofilm buildup over 16 h is shown in Figure 1 for a baked bean soil. After 12 h or so, this biofilm appeared to be of a brown vegetable appearance, but the photographs show clearly that it is bacterial in nature. A thorough understanding of a soil's characteristics is therefore required to ensure a successful and economic sanitation program.

FUNDAMENTALS

The process of sanitation is intended to remove all undesirable material (food residues, microorganisms, foreign bodies, and cleaning chemicals) from surfaces in an economical manner, to a level at which any residues remaining are of minimal risk to the quality or safety of the product. Sanitation is divided into two broad areas, the cleaning of open and closed surfaces, though the principles are essentially the same. Open surface sanitation refers to all equipment and environmental surfaces that are readily cleaned manually. Closed surface sanitation, generally referred to as clean-in-place or CIP, is undertaken where manual cleaning is difficult, impractical or impossible (e.g., tanks, homogenizers and pipelines).

Sanitation programs are a combination of four major factors:

- Mechanical or kinetic energy
- Chemical energy
- Temperature or thermal energy
- Time

Mechanical or kinetic energy is employed to physically remove soils and may include manual brushing (physical abrasion), pressure jet washing (fluid abrasion), or the circulation of fluid in CIP systems (turbulent flow). Of these methods, physical abrasion is the most efficient in terms of energy transfer (9); for turbulent flow, a mean velocity of 1.5 m/s should be achieved (10).

Chemical energy is fundamental to both the cleaning and disinfection elements of sanitation. In cleaning, chemicals are used to break down soils so that the soils are less tenacious and to suspend them in solution to allow them to be rinsed away. In disinfection, chemicals are used to reduce the viability of microorganisms remaining on surfaces after cleaning.

Temperature or thermal energy is important for several reasons. Cleaning and disinfection chemical effects increase with temperature linearly and approximately double for every 10°C rise. Temperatures above the melting point of fatty or oily soils are used to break down and emulsify these deposits and high temperatures, particularly in CIP systems, have a disinfection effect in their own right.

Time is a factor that is often overlooked in sanitation systems. It is an essential prerequisite for the previously discussed energy forms and generally the longer the time period employed, the more efficient the process. Time can also be used to reduce the degree of energy input required from other sources when precleaning soaking is undertaken.

The combinations of these four factors varies for different cleaning systems such that if one energy source is restricted, this shortfall may be compensated for by utilizing greater inputs from the others. For example, in CIP cleaning, the energy that can be derived from mechanical energy is low but much higher temperatures and chemical concentrations are possible than can be safely used in open surface cleaning. The influence of chemical (detergents), temperature, and mechanical energy (pressure washing) has been described for open surfaces (11–14) and for CIP systems (15).

Soil removal from surfaces has been shown to basically follow first-order reaction kinetics (16,17) such that the decrease in the log of the mass of soil per unit area remaining is linear with respect to cleaning time (Fig. 2). This approximation is only valid in the central portion of the plot; it has been reported (18) that in practice, soil removal is initially faster and ultimately slower than that which a first-order reaction predicts (dotted line in Fig. 2). The reasons for this are unclear although initially, unadhered, gross soil is usually easily removed whereas ultimately, soils held within surface imperfections or otherwise shadowed from cleaning effects would be more difficult to remove.

As routine cleaning operations are therefore not 100% efficient over multiple soiling/cleaning cycles, soil deposits will accumulate on surfaces. During this phase, cleaning will become less efficient and attached microbial numbers will increase. This situation is usually controlled by the application of a periodic clean (19), the object of which is to periodically return the surface bound soil accumulation to an acceptable base level (Fig. 3). This is achieved by increasing cleaning time and/or energy input (eg, higher temperatures, alternative chemicals, or manual scrubbing) and is the basis of many food processors weekend clean down.

SANITATION PROCEDURE

The principal stages involved in a typical sanitation procedure are as follows:

1. Preparation. Dismantle equipment as far as is practicable or necessary and/or remove unwanted utensils/equipment. Protect electric or other sensitive systems and/or screen off other lines or areas to prevent transfer of debris by the sanitation process.

2. Gross soil removal. Where appropriate, remove all loose or gross soil by eg, brushing, shoveling, scraping, or vacuuming. Wherever possible, soil on floors and walls should be picked up rather than washed to drains.

3. Prerinse. Rinse with low pressure cold water to remove loose small debris. Hot water can be used for fatty soils, but too high a temperature may coagulate proteins.

4. Cleaning. Apply cleaning chemicals, temperature, and mechanical energy to remove adhered soils.

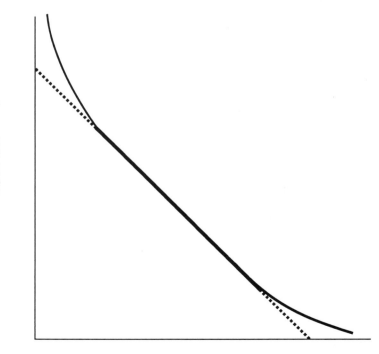

Figure 2. Removal of soil from surfaces with cleaning time.

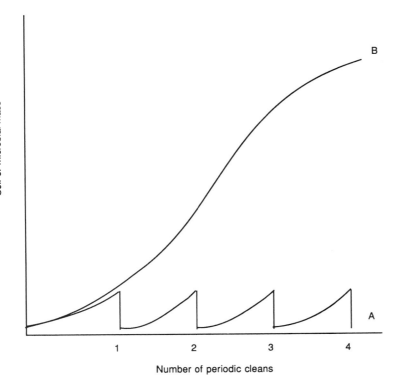

Figure 3. Build up of soil and/or microorganisms A, with periodic cleans and b, without periodic cleans (19).

5. Interrinse. Rinse with low pressure cold water to remove soil detached by cleaning operations and cleaning chemical residues.
6. Disinfection. Apply chemical disinfectants to remove or reduce the viability of remaining microorganisms to as low a level as possible.
7. Postrinse. Rinse with low pressure cold water to remove disinfectant residues if required.
8. Intercycle conditions. Remove excess water and/or do everything necessary to prevent the growth of microorganisms in the period up until the next production process or the next use of the equipment/area.

Although broadly similar for all general sanitation procedures, alternatives may be used. In CIP systems a second cleaning cycle may be added if both an alkaline and acidic phase are incorporated. With only light soiling to be removed, it may be appropriate to combine stages 4–6 by using a detergent-sanitizer, a chemical with both cleaning and antimicrobial properties. This should only be used for light soils, however, as with normal or heavy soils the antimicrobial properties will be quickly lost. Stages 3, 5, and 7 may be omitted in dry cleaning operations where organic solvents replace water in the cleaning and, where appropriate, the disinfection stages. Because of the general nature of this article, specialist dry cleaning is not discussed further.

CHEMICALS

This section gives a brief and general introduction to the chemicals used in sanitation. For further information, readers are directed to the specific articles on detergents and disinfectants in this encyclopedia. Sanitation chemicals, because of the procedure outlined in the previous section, are usually employed as cleaning or disinfection agents. This section is therefore divided into cleaning and disinfection subsections. Some chemicals routinely used, such as quaternary ammonium compounds (QUATS) or iodophores, have both cleaning and biocidal properties, although in this section they are described for their primary function only.

Cleaning

Unfortunately no single cleaning agent is able to perform all the functions necessary to facilitate an efficient cleaning program, so a cleaning solution, or detergent, is blended from some of the following characteristic components: water, surface active agents inorganic alkalis inorganic and organic acid, and sequestering agents.

Water. Water is the basic ingredient of most cleaning systems as it provides the cheapest readily available transport medium for removing soils. It also has dissolving powers to remove ionic and water soluble compounds such as salts and sugars, will help emulsify fats at temperatures above their melting point, and can be used as an abrasive agent when high pressure washing. It is, however, a poor wetting agent and cannot dissolve nonionic compounds.

Surface Active Agents. Surface active agents, wetting agents or surfactants are composed of a long, nonpolar (hydrophobic or lyophilic) chain or tail and a polar (hydrophilic or lyophobic) head as illustrated in Figure 4. Surfactants may be anionic, cationic, or nonionic, depending on their ionic charge in solution. The polar end is able to

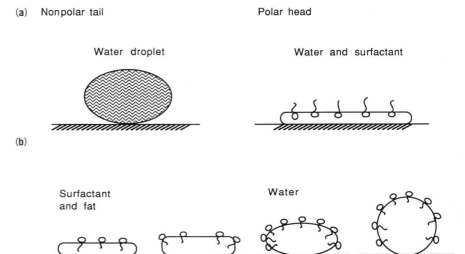

Figure 4. Schematic diagrams showing: (a) structure of surfactant molecule, (b) surfactant improving the wettability of water, and (c) surfactant removing fat from a surface and dispersing it in suspension.

penetrate into water as the ionic charges are greater in magnitude than the weaker hydrogen bonding between the water molecules. The nonpolar end is unable to easily break apart the water molecules' hydrogen bonding but can enter other nonpolar compounds such as fats and oils. This aids cleaning in two ways: it reduces water surface tension and removes and suspends fats.

Surface tension is reduced as the polar head enters the water and breaks down hydrogen bonding. If a surfactant is added to a drop of water on a surface, the surface tension of the water is reduced and the drop collapses to wet the surface (Fig. 4). Increased wettability allows greater water penetration into soils and hence better cleaning action. Fats and oils are removed from surfaces as the polar head of the surfactant molecule dissolves in the water, whereas the hydrophobic end dissolves in the fat. Due to the forces acting on the fat–water interface, the fat particle tends to form a sphere, as this has the lowest surface area for a given volume. In so doing, the fat particle will roll up, become detached from the surface, and remain in suspension (Fig. 4).

Inorganic Alkalies. Alkalies are useful cleaning agents because they are cheap, break down proteins, saponify fats, and may be bactericidal. A typical saponification reaction is shown below in which a nonpolar R group is changed to a soluble form that has good surfactant properties.

$$\begin{array}{l} CH_2COO-R \\ CHCOO-R \\ CH_2COO-R \end{array} + NaOH \longrightarrow \begin{array}{l} CH_2OH \\ CHOH \\ CH_2OH \end{array} + 3R-COO^-Na^+$$

Triglyceride Sodium   Soap
fat   hydroxide Glycerol

Generally, the stronger the alkali the greater the degree of saponification, although this is a compromise, as corrosiveness also increases with alkali strength. Alkaline compounds can also precipitate scum, a reaction of the soap with water hardness ions (Ca^{2+} and Mg^{2+}), and may have poor rinsability. For some applications, alkaline detergents are chlorinated, as at high pH chlorine increases peptizing of proteins and may reduce mineral deposition. Chlorine at high pH, however, is also very corrosive.

Inorganic and Organic Acids. Acids are not generally used in the food industry as they have little dissolving power for fats, oils, and proteins. They are useful, however, in making soluble mineral scales such as hard-water deposits, beer stone, and milk stone. A typical reaction would be as follows:

$$CaCO_3 + 2HCl \longrightarrow CaCl_2 + H_2O + CO_2$$
Insoluble   **Soluble**

As with alkalis, the stronger the acid the more corrosive it is. Strong acids (mineral acids) include hydrochloric, nitric, and phosphoric acid, whereas weaker organic acids typically used include citric, lactic, and acetic acid.

Sequestering Agents. Sequestering agents are employed to sequester or chelate mineral salts by forming soluble complexes with them. Their primary use is in the control of water hardness ions although they are also useful in maintenance of alkaline conditions by buffering, emulsification of oils, and fats and increasing rinsability. Sequestrants can be both organic and inorganic, organic sequestrants being usually based on polyphosphates and inorganic chelating agents commonly being the potassium or sodium salts of ethylene diamine tetraacetic acid (EDTA).

A general-purpose food detergent may contain a strong alkali to saponify fats, weaker alkali builders to aid corrosion resistance, surfactants to improve wetting, dispersion, and rinsability and sequestrants to control hard water ions. Ideally, the detergent should also be safe, nontainting, stable, noncorrosive, biodegradable, and

Table 1. Solubility Characteristics and Cleaning Procedures Recommended for a Range of Soil Types[a]

Food (or Soil)	Solubility Characteristics	Cleaning Procedure Recommended
Sugars, organic acids, salt	Water soluble	Mildly alkaline detergent
High protein foods (meat, poultry, fish)	Water soluble Alkali soluble Slight acid soluble	Chlorinated alkaline detergent
Fatty foods (fat meat, butter, margarine, oils)	Water insoluble Alkali soluble	Mildly alkaline detergent; if ineffective, use strong alkali
Stone-forming foods, mineral scale (milk products, beer, spinach)	Water insoluble Acid soluble Alkali insoluble	Chlorinated cleaner or mildly alkaline cleaner; alternate with acid cleaner on each 5th day
Heat-precipitated water hardness	Water insoluble Alkali insoluble Acid soluble	Acid cleaner
Starch foods, tomatoes, fruits, vegetables	Partly water soluble Alkali soluble	Mildly alkaline detergent

[a] Ref. 22.

cheap. The detergent chosen for a particular application will depend on the soil to be removed; the solubility characteristics and cleaning procedure recommended for a range of food soils is shown in Table 1.

Disinfection

The cleaning portion of the sanitation cycle has been shown to reduce bacterial numbers on surfaces by up to 99.9% or three log orders (20). Work in our own laboratories has shown that with detergent soaks and pressure washing, up to 4–5 log orders may be removed. However, given that bacterial numbers on surfaces could be between 10^7 and 10^{10} organisms/cm (8,20), viable bacteria are likely to be present on surfaces after cleaning. The aim of disinfection procedures is to remove or reduce the viability of these remaining microorganisms.

If possible, temperature is used as a disinfectant as it penetrates into surfaces, is noncorrosive, is nonselective to microbial types, and is easily measured and leaves no residue (16). Whereas high temperatures are often used as disinfectants in CIP systems, their use on open surfaces is usually uneconomic, hazardous, or impossible. In such cases, chemical biocides are employed and, as with cleaning chemicals, no single disinfectant satisfies all the performance requirements. Biocides are rarely mixed, however, so a choice has to be made from a limited number of disinfectant types. Universally used biocides include chlorine releasing components, quaternary ammonium compounds, iodine compounds, amphoterics, and peracetic acid.

Chlorine is the most widespread and cheapest disinfectant used in the food industry and is available in fast-acting (chlorine gas, hypochlorites) or slow-releasing forms (e.g., chloramines, dichlorodimethylhydantoin). Quaternary ammonium compounds (QUATS or QAC's) are based on ammonium salts with substituted hydrogen atoms and a chlorine or bromine anion, whereas iodophores are soluble complexes between elemental iodine (active ingredient) and nonionic surface active agents. Amphoterics are based on the amino acid glycine, often incorporating an imidazole group. Peracetic acid may be used by itself or formulated with hydrogen peroxide. A range of characteristics for examples of these disinfectant types is shown in Table 2. Other disinfectants used to a limited extent include biguanides, formaldehyde, ozone, chlorine dioxide, and bromine; however, biocides successfully used in other industries, eg, phenolics or metal ion-based products, are not used for food applications due to safety or taint problems.

Disinfectant concentration and contact time are important considerations in the reduction of microbial viability. The relationship between death and concentration is not linear but follows a sigmoidal curve dependent on the resistance of organisms within the population. Disinfectants do not, therefore, necessarily kill all microorganisms in a population, and increasing concentration may not enhance this effect. For disinfectants to be effective, they must find, bind to, and transverse microbial cell envelopes before they reach their target site (21). Sufficient contact time is therefore critical to give good results. Amphoterics and QUATS may be left on surfaces for extended periods between production runs without rinsing, as they are FDA approved as indirect food additives at low concentration.

The performance of the cleaning procedure may influence disinfection efficiency. Any soil or cleaning chemical residues remaining may protect microorganisms from disinfectant penetration or may react with the disinfectant and destroy its antimicrobial abilities. Biocides are best used within their specified pH range, although performance can generally be increased by increasing the temperature.

METHODS

Open Surfaces

After gross soil removal, open surfaces can be cleaned and disinfected in their normal position or dismantled and/or transferred to a separate area (cleaned-out-of-place, COP). Sanitation procedures can be undertaken by hand using simple tools eg, brushes or cloths (manual cleaning) or by

Table 2. Characteristics of Some Universal Disinfectants

Property		Chlorine	QUAT	Iodophore	Amphoteric	Peracetic Acid
Microorganism control	Gram positive	++	++	++	++	++
	Gram negative	++	+	++	++	++
	Spores	+	−	+/−	−	++
	Yeast	++	++	++	++	++
Developed microbial resistance		−	+	−	+	−
Inactivation by	Organic matter	++	+	+	+	+
	Water hardness	−	+	−	−	−
Detergency properties		−	++	+	+	−
Residual film formation		−	++	+/−	++	−
Foaming potential		−	++	+	++	−
Problems with	Taints	+/−	−	+/−	−	+/−
	Stability	+/−	−	−	−	+/−
	Corrosion	+	−	+	−	−
	Safety	+	−	+	−	++
	Other chemicals	−	+	+	−	−
Cost		−	++	+	++	+

^a Key: − no effect (or problem).
+ effect.
++ large effect.

using specialized equipment designed to cover larger areas more rapidly.

Manual cleaning is often undertaken because no specialist equipment is required and high levels of mechanical energy input can be used at exactly the right point. Sanitation procedures can also be undertaken quickly (over small surface areas) and in places that alternative techniques could not reach or may damage. Due to operator safety, however, only low levels of temperature and chemical energy can be applied and it is costly in time and labour to cover large surface areas. Only tools specifically designed for food sanitation should be used, eg, plastic-handled brushes with stiff, colored nylon bristles, and they should be regularly cleaned and stored in a disinfectant solution to reduce cross-contamination in use.

Manual cleaning of smaller, dismantled items can be improved using COP techniques. When soak tanks are used, items can be subjected to higher temperatures and chemical concentrations for an extended time period prior to manual cleaning and disinfection. For an automated approach, specialized equipment can be used that also applies the mechanical energy input, eg, tray washers and tunnel dish washers.

For large areas of open surface, sanitation equipment is designed to disperse chemicals and/or provide mechanical energy. Chemicals may be applied as mists, foams, or gels, whereas mechanical energy is provided by water jets or scrubbing actions. The use of high temperature is rarely used for cleaning or disinfections, because of the excessive energy requirements needed to heat soil and/or surfaces to suitable temperatures.

Mist spraying of chemicals onto surfaces is undertaken using small, hand-pumped containers, knapsack sprayers (Fig. 5) or pressure washing systems at low pressure. As for the other chemical application methods, pressure washing systems can be mobile units (Fig. 5), wall mounted serving one or more outlets, or centralized where one unit may supply many outlets via a ring main. Chemicals are added, usually by venturi injectors, either at the pumps of mobile or wall-mounted units, at the outlets of high pressure water ring main systems or via separate low pressure chemical ring mains. Misting is only able to wet vertical smooth surfaces and therefore only small quantities can be applied. Only weak chemicals can be applied, as the technique tends to form aerosols, so for cleaning purposes, misting is only useful for light soiling. The detergent and loosened soil are removed via high or low pressure water rinsing. Once the surfaces have been cleaned, however, misting is a very useful and most widely applied technique for applying disinfectants.

If more tenacious soiling is to be removed, detergents

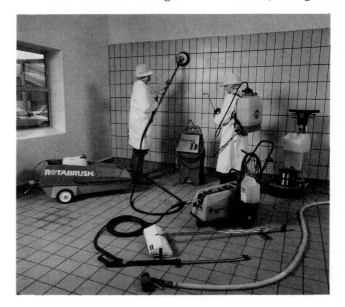

Figure 5. A range of cleaning equipment suitable for use in food hygiene. Clockwise from left: electrically driven scrubber brush, gel applicator, detergent or disinfectant mist sprayer, floor scrubber, mobile high pressure/low volume washer with standard lance connected, shrouded lance to reduce aerosol spread, low pressure/high volume gun.

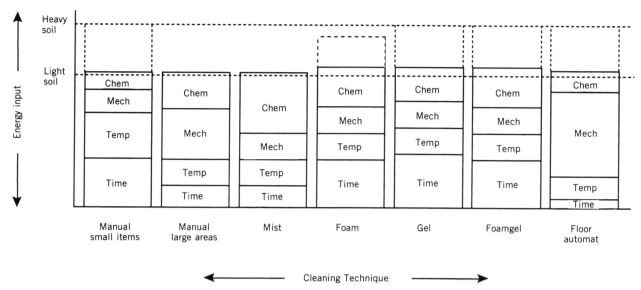

Figure 6. Relative energy inputs for a range of cleaning techniques. The dotted lines above each technique represent their abilities to cope with heavy soiling (9).

may be applied as a foam. Foams are more viscous than mists, are not as prone to aerosol formation, and remain on vertical surfaces for around 5–20 min. This allows the use of more concentrated detergents, longer contact times, and because of the nature of the foam, a more consistent application of chemicals as it is easier to spot areas that have been missed. Foams can be generated and applied by the entrapment of air in high pressure systems or by the addition of compressed air in low pressure equipment, and are rinsed away by high or low pressure water. Although the visible nature of the foam should aid complete removal from surfaces, its foaming capabilities can make rinsing difficult.

The use of thixotropic gels has recently been introduced to further extend contact time over foams and improve rinsability. Gels are typically fluid at high and low concentrations but become thick and gelatinous at a concentration of around 5–10%. They are easily mixed and applied through high and low pressure systems, foaming equipment, or portable electrically pumped units (Fig. 5), remain on surfaces for extended periods (hours) and are easily diluted and rinsed away with high or low pressure water. Gels are usually more expensive than foams however.

The proportions of input from each energy source for a range of open surface cleaning techniques are shown in Figure 6. The figure illustrates the various methods' abilities to satisfactorily remove both light and heavy soiling from surfaces.

High pressure–low volume water systems, in which water is typically pumped at pressures up to 100 bar through a 15° nozzle, are widely used in the food industry. Water jets confer high mechanical energy, can be used on a wide range of equipment and environmental surfaces, are not limited to flat surfaces, will penetrate into surface irregularities, and can mix and apply sanitation chemicals. Soil removal does not necessarily increase linearly with pressure, as illustrated for a bacterial soil in Table 3, and may decrease. This is related to the water droplet size emerging from the nozzle, for which there will be an optimum size and hence impingement, for each soil type, (22). Droplet size reduces with increasing pressure and pressures of around 50 bar are satisfactory for many operations. Care must be taken when using pressure washers as they are able to transfer soil and bacteria from one surface to another over large distances and may damage electrical installations or other delicate equipment.

For the cleaning of floors and other predominantly flat surfaces eg, walls, equipment that produces a mechanical scrubbing action may be used. Such equipment may include water-driven attachments to high pressure systems, electrically operated small diameter (23 cm) brushes, and floor scrubbers (Fig. 5). With these techniques, high mechanical input is combined with chemical energy after which surfaces may be rinsed with low pressure water. Larger and more sophisticated scrubber dryers, or floor automats, have a squeegee and vacuum unit mounted behind the scrubbing brush(es) to suck up the resulting soil and detergent emulsion. These devices provide a rapid and effective clean, and produce a dry surface that is safer and less conducive to microbial growth but can only be used efficiently in food processing areas designed for their

Table 3. The Effect of Water Pressure on Spray Cleaning Efficiency

Pressure (bar)	2	25	50	75	98
Microorganisms remaining/cm² (as measured by DEM)	2.9×10^5	3.1×10^4	1.9×10^4	7.7×10^3	3.3×10^4

use. All mechanical scrubbing units should be cleaned and disinfected after use to avoid cross-contamination.

Enclosed Surfaces, CIP

This section gives a brief and general introduction to CIP and readers are directed to the specific article on CIP in this encyclopedia, for further information. As a technique, CIP is well established, especially in liquid food processing, and is achieved by the sequential circulation of water, detergents, and disinfectants through processing equipment that remains assembled. Mechanical or kinetic energy is provided by turbulent flow in pipes and jet impingement in tanks and vessels. CIP systems are best employed to clean process equipment that has been designed specifically for CIP cleaning and should supply detergents at the correct temperature, concentration, and velocity in the right place for the required contact time.

It is fundamental with enclosed systems that hygienic design and product safety from sanitation chemicals is addressed. Attention must be paid to hygienic design so that areas where product and microorganisms can lodge, out of the flow of cleaning chemicals and disinfectants, are eliminated (3). Special devices should be built into the system, eg, key pieces, flow plates, swing bends, and electrical interlocks to ensure that product contamination by CIP solutions cannot occur, and a single valve must never be relied upon to safeguard product (23).

There are three main types of CIP systems: single use, reuse, and multiuse. Single-use systems are the most basic and are so called because the detergent is only used once and then drained. A simple system is shown in Figure 7. The system can be located close to the equipment to be cleaned, to eliminate long pipe runs, and is low on capital cost. It may be operated automatically or manually and the final rinse may be collected to use as the initial rinse in the following cleaning cycle. Although it is expensive on detergent, a single-use system may be the only system applicable if the product soil to be removed is such that detergent contamination occurs to a degree that detergent recovery is impractical.

Reuse systems are generally more complex and may be used to clean more than one process. They are designed to recover the cleaning fluids for reuse and are best employed where product soil is light or where much of the soil can be removed with the prerinse. Reuse systems are normally run automatically with programmed cleaning cycles for each circuit and there is, therefore, the need for timing, temperature and dosing control equipment, filters, storage vessels, and recording instrumentation. With large installations, the CIP unit can be placed out of the process area, or centralized, which may be advantageous for some applications. This can, however, lead to extensive pipe runs, which require large solution volumes and are prone to heat loss and expansion problems. This can be partially overcome in decentralized systems in which the detergent solution is mixed, heated, and distributed centrally, but rinse waters and other chemicals are heated and/or mixed and supplied close to the process equipment to be cleaned.

Multiuse systems combine the best design features of both single-use and reuse systems and are used where versatility in temperature, cleaning fluid type, concentration, and the ability to recover or drain cleaning fluids are required.

As a cleaning technique, CIP has many advantages. Cleaning schedules are adhered to automatically and use

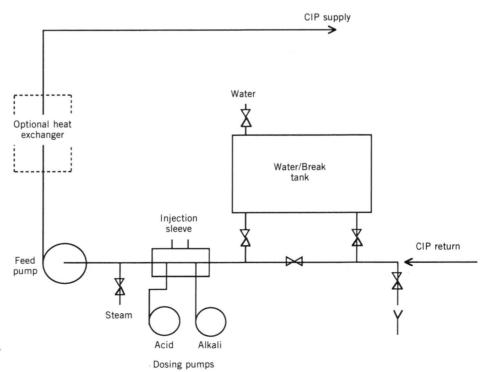

Figure 7. A simple, single use CIP system.

optimum cleaning fluids, concentrations, and temperatures to help achieve consistently satisfactory results. Processes can be cleaned as soon as production finishes to reduce down time and the life of equipment is extended due to reduced dismantling. Manual input can be reduced and operator safety increased through less contact with hazardous materials or work places. The major disadvantages are initial capital cost and the inflexibility of some systems if manufacturing processes are altered.

EVALUATION OF EFFECTIVENESS OF SANITATION SYSTEMS

As with other important aspects of safe and wholesome food production, sanitation programs require regular evaluation of their effectiveness as part of a specified quality assurance system. Traditionally, this has been undertaken on two levels—an immediate assessment of the performance of a sanitation program by sensory evaluation and a historical measurement of microbial surface populations if surfaces are visibly clean. Sensory evaluations are used as a process control to immediately rectify obviously poor sanitation. Bacterial evaluations may be used to optimize sanitation procedures and ensure compliance with microbial standards and in hygiene inspection and troubleshooting exercises. Recently, rapid methods have been developed to assess microbial surface populations in a time relevant to process control and in this context, only techniques that provide estimates in less than 15–20 min are considered.

As with the inherent faults of end product analysis in describing production quality, reliance should not be placed on assessment techniques that can only sample a very small proportion of plant surfaces. Assessment techniques should, rather, be designed to monitor the effectiveness of an integrated HACCP-type (Hazard Analysis and Critical Control Point) approach to sanitation in which such critical control points may include detergent and disinfection concentration, solution temperature, application procedures, and chemical stock rotation.

Sensory evaluation involves visual inspection of surfaces under good lighting, feeling for greasy or encrusted surfaces, and smelling for product and/or offensive odors. If these assessments indicate the presence of product residues, no further analysis is required and microbiological examination will be misleading. If no product residues are detected, microbiological techniques may be applied. All of the microbiological techniques appropriate for food factory use involve the sampling of microorganisms from contaminated surfaces and their culture using standard agar plating methods. Microorganisms may be collected via cotton or alginate swabs (after which they are resuspended by vortex mixing or dissolution), water rinses for larger enclosed areas, or directly using agar contact plates. Traditional techniques have been reviewed (24–26) since which time few improvements have occurred other than with contact methods which are now available commercially as premade sampling units.

Traditional microbiological techniques all require a minimum of 24–48 h to provide results and these only provide a historical view of the efficiency of sanitation programs. Rapid methodology was devised to ascertain sanitation efficiency and for some products, when to clean, in a time relevant to process control. The two most common techniques are epifluorescent microscopy and the ATP technique. With epifluorescent microscopy, microorganisms may be sampled by swabbing and filtering the resuspension media or filtering rinses (direct epifluorescent filter technique, DEFT) or by enumerating the surface of coupons attached to product contact surfaces (direct epifluorescent microscopy, DEM). After swabbing, estimates may be made of soil and/or microorganisms by analysis of total or microbial adenosine triphosphate (ATP) levels. In the majority of practical applications, where an assessment of cleaning is required, analysis for total ATP is preferred on the assumption that any residues, soil or microorganisms, should have been removed. The use of DEM and DEFT has been described for assessing open surface hygiene (8,27), rinses (28), and the use of ATP (29,30).

The accuracy of a range of surface hygiene assessment techniques as compared to the most accurate method, DEM, has been described (31), and is shown in Table 4. The table shows the ability of a number of commercial contact, self-prepared (DIY) contact, traditional swabbing and rapid techniques (ATP and DEFT) to predict a DEM count of 10^4, 10^6 and 10^8 bacteria/cm^2. The results indicate that only above 10^6 bacteria/cm^2 are these methods accurate and they are therefore useful for indicating gross

Table 4. Accuracy in the Estimation of Surface Microbial Populations for Various Hygiene Assessment Methods as Compared to DEM

Method Used	Repeats	Prediction of a DEM count/cm^2			2SD^2
		10^4	10^6	10^8	
Commercial contact A	58	3.42	6.18	8.94	+/− 1.11
DEFT swab	108	2.06	4.46	6.86	+/− 1.22
Commercial contact B	30	3.57	6.01	8.45	+/− 1.77
ATP swab	31	2.36	4.66	6.96	+/− 2.22
DIY contact	30	4.76	6.15	7.55	+/− 2.33
Commercial contact/swab	17	1.50	3.04	4.59	+/− 2.81
Cotton swab	70	1.47	4.41	7.35	+/− 2.88
Alginate swab	70	0.61	3.55	6.49	+/− 4.06
Commercial contact C	46	2.81	5.35	7.89	+/− 5.54

Table 5. Attributes of a Range of Surface Hygiene Assessment Techniques

Attribute Commercial	DEM	DEFT Swab	ATP Swab	TVC Swab	CONTACT DIY	CONTACT
Analysis for	Total count	Viable/total	Viable/soil	Viable count	Viable count	Viable count
Preparation time	5 min	15 min	10 min	1 h	1h	1 min
Test time	5–10 min	15 min	2 min	25 min	1 min	1 min
Surface requirement	Coupon	Any	Any	Any	Flat only	Flat only
Field application	Research Yes	Yes	Yes	Yes	Yes	Yes
Degree of competence	Micro training	Micro training	Basic training	Basic micro	Basic micro	Little/none
Facilities required	U.V. microscope	U.V. microscope	Luminometer (portable)	Basic lab	Basic lab	Incubator
Speed of result	<5 min	<5 min	<1 min	>1–2 days	>1–2 days	>1–2 days
Test cost	50p/$1	50p/$1	£1/$2	50p/$1	50/$1	£1–4/$2–8
Initial cost	£2–5,000 $4–10,000	£3–15,000 $6–30,000	£2.5–10,000 $5–20,000	Low	Low	Very low

contamination only. For cotton swabs, the most widely used method in practice, a result of zero is more likely than a result of 10^4 when assessing surface populations of 10^4 bacteria/cm^2. The rapid methods available are at least as accurate as traditional methods and as shown in Table 5, which compiles a range of attributes for typical assessment methods, the choice between rapid and traditional techniques is a balance of cost and speed of result.

EVALUATION OF THE EFFECTIVENESS OF DISINFECTION

The effectiveness of a disinfectant can only be undertaken by assessing viable microbial levels before and after exposure. Ideally, this should be carried out in the field, but due to problems with reproducibility and scale (assessing only a small area of the total plant), this may take many weeks before satisfactory results are achieved. Disinfectant testing is, therefore, usually undertaken in the laboratory under strictly defined conditions to enable reproducible and rapid results.

At their simplest, disinfectant test methods consist of adding a known concentration of disinfectant to a known inoculum of bacteria for a defined time and temperature period and assessing surviving bacterial numbers. Microorganism survival may be assessed by inoculating a sample of the disinfectant/microorganism mixture into traditional culture medium and enumerating any microcolonies produced or by the use of a disinfectant concentration/microorganism mixture dilution range and recording the disinfectant dilution that just prevents microbial growth. These disinfectant tests, referred to as suspension tests, are simple to undertake but can only be used to screen potential products for disinfectant action or to confirm the biocidal abilities of known disinfectants as a quality control procedure.

Table 6. European Suspension Test Results for a Range of Commonly Used Disinfectants Tested at their Manufacturers Recommended Concentrations (MRC)

Product	MRC %	Organic Load, %	Log Reduction					Result
			Pseudomonas aeruginosa	*Proteus mirabilis*	*Staphylococcus aureus*	*Enterococcus faecium*	*Saccharomyces cerevisiae*	
Peracid	0.1	0.03	7, 6	7, 7	7, 7	6, 5	5, 1, 1	Fail
		1.0	7, 1, 5	7, 6, 7	7, 1, 6	6, 0, 5	5, 0, 0	Fail
Peracid	0.1	0.03	5, 5	7, 7	7, 7	7, 7	>4, 5	Pass
		1.0	5, 5	7, 7	6, 7	7, 7	>4, 5	Pass
HOCL	0.25	0.03	7, 7	7, 7	7, 6, 6	5, 6	4, 5, 2	Fail
		1.0	4, 3	2, 2	2, 2	0, 0	3, 5, 3	Fail
HOCL	0.3	0.03	6, 6	5, 4, 4	6, 5	6, 6	>4, 3	Fail
		1.0	3, 3	0, 3, 2	1, 4	2, 2	0, 0	Fail
QUAT	0.2	0.03	2, 2	3, 5, 4	7, 7	6, 5	5, 3, 4	Fail
		1.0	2, 2	3, 1	7, 6	5, 5	4, 1	Fail
QUAT	1.0	0.03	5, 6	6, 5	5, 7	5, 5	5, 5	Pass
		1.0	5, 7	5, 6	7, 7	4, 6, 6	4, 5, 5	Pass
Biguanide	0.5	0.03	2, 5, 5	2, 5, 5	2, 6, 7	3, 6, 7	5, 4	Pass
		1.0	2, 1	2, 1	2, 2	1, 1	2, 1	Fail
Biguanide	0.5	0.03	6, 6	5, 6	7, 5	7, 7	1, 2	Fail
		1.0	6, 6	4, 3	4, 4	7, 7	1, 1	Fail
Iodophore	0.2	0.03	7, 7	7, 7	7, 6	7, 6	4, 4	Fail
		1.0	1, 1	3, 2	4, 2	1, 1	0, 0	Fail
Iodophore	1.0	0.03	4, 4	7, 7	5, 5	3, 4	2, 3	Fail
		1.0	2, 3	3, 5, 5	5, 5	3, 3	1, 1	Fail
Amphoteric	1.0	0.03	5, 5	4, 7, 7	7, 7	6, 7	4, 5, 5	Pass
		1.0	4, 5, 7	6, 4, 7	7, 7	4, 7, 6	2, 5, 3	Fail
Amphoteric	1.0	0.03	7, 6	7, 7	7, 7	6, 6	5, 5	Pass
		1.0	7, 6	6, 6	7, 7	6, 6	5, 5	Pass

More extensive suspension test methodologies are available that simulate some of the conditions under which disinfectants may have to work. Such conditions may include the presence of soil, different contact times, alternative temperatures, a variety of bacterial types and the degree of water hardness (32). The results of suspension tests have been used to ascertain recommended in-use disinfectant concentrations, but in reality they only highlight conditions under which disinfectants that have failed should not be used.

Through the auspices of national standard organizations, many countries have standard disinfectant testing procedures, including Australia, New Zealand, South Africa, France, Germany, The Netherlands, the UK, and the United States (33). There is little standardization between countries, however, and to address this issue, with the planned harmonization of Europe in 1992, the European Suspension Test (EST) (34) was published with some agreement between European countries. To pass the EST, a disinfectant must reduce the viability of 5 test organisms by 5 log orders in 5 min at both low (0.03% bovine albumin) and high (1.0% bovine albumin) soil levels. Results for two examples each of a range of disinfectant types are shown in Table 6. These results illustrate a number of points including: the relative resistance of microbial species, the variation of disinfectant action within a single microbial species, the influence of organic load and hence the necessity to thoroughly clean prior to disinfection in practice, and the range of biocidal action between and within product types. In its early stages of development, a collaborative trial in 10 European countries found the EST to be sufficiently repeatable and reproducible to be adopted internationally (35).

Microorganisms are not primarily found in the food production environment in suspension. In an attempt to more closely simulate an in-use conditions surface, tests have been developed in which microorganisms are dried onto surfaces prior to disinfection (36–38). Although these tests provide a stronger challenge to the disinfectant (39), there are problems with the tests in that the drying process may alter microbial resistance, it is difficult to remove a consistent proportion of microorganisms for enumeration after disinfection and again, microorganisms are rarely dried onto surfaces in practice. A surface test has recently been described (40) in which disinfectants are tested against bacterial biofilms developed on stainless steel samples and in which the viability of the biofilm was assessed, while surface bound using a Malthus microbiological growth analyser (Malthus Instruments Ltd. (Radiometer), UK). The relative resistance of bacteria when attached to surfaces as compared to in suspension is shown in Table 7. A range of disinfectants was tested under the conditions of the EST at their manufacturers' recommended concentration (MRC) for bacteria in suspension and at the MRC and 10 and 100 times this concentration, when attached to a surface. The results showed that bacteria were 10 to 100 times more resistant to biocides when attached to a surface and that biocides that performed well in suspension did not necessarily perform well on surfaces. Care should therefore be exercised in transferring the results from suspension tests to in-use applications.

Table 7. The Relative Resistance of Bacteria Attached to a Surface (Malthus) and in Suspension (EST)[a]

Product Type	Conc. (%)	EST 0.03	EST 1.0	Malthus 0.03	Malthus 1.0
Peracid	0.1	Pass	Pass	Pass	Fail
	1.0			Pass	Pass
	10.0			Pass	Pass
Iodophore	0.2	Pass	Fail	Fail	Fail
	2.0			Pass	Pass
	20.0			Pass	Pass
Biguanide	0.5	Pass	Fail	Fail	Fail
	5.0			Pass	Fail
	50.0			Pass	Pass
HOCL	0.25	Pass	Pass	Fail	Fail
	2.25			Fail	Fail
	25.0			Pass	Pass
Amphoteric	1.0	Pass	Pass	Fail	Fail
	10.0			Pass	Fail
QUAT	1.0	Pass	Pass	Fail	Fail
	10.0			Fail	Fail

[a] For both the EST and Malthus tests, three organisms were used—*Pseudomonas aeruginosa*, *Proteus mirabilis*, and *Staphylococcus aureus*; a pass criteria of a 5 log reduction in 5 min for all three organisms was used.

FUTURE DEVELOPMENT

Sanitation is likely to feature much more prominently in the eyes of management in the future due to their desire to seek increased standards of hygiene. These have arisen from:

1. The realization of the importance of environmental routes of infection to product for both microbial and foreign body contaminants.
2. Pressure from customers and consumers for higher standards of hygiene.
3. Trends in food production towards short shelf-life products that intrinsically demand higher standards of hygiene.
4. The ability to sell product of a low microbiological load at a premium.
5. Increased plant size, where single mistakes will incur large financial losses.
6. Legislation.

Demand for increased standards of hygiene could lead to further advances in rapid methodology to determine surface hygiene in a time relevant to process control, chemical testing schemes that more closely simulate in-use conditions, improved application techniques that are more efficient and reduce cross-contamination via cleaning methods, the design of equipment and processing areas with more thought to ease of cleanability, and the increased use of automation and/or robotics for both production and sanitation operations.

Other trends may be influenced by environmental and energy restrictions. This will lead to the use of chemicals that leave no undesirable environmental residues, a reduction in water usage, and an associated move towards dry cleaning methods such as ring-main vacuum lines, and as for hygiene reasons, more efficient and less manual sanitation application techniques.

BIBLIOGRAPHY

1. Technical Manual No. 7, *Hygienic Design of Food Processing Equipment,* Campden Food and Drink Research Association, Chipping Campden, UK, 1983.
2. Technical Manual No. 8, *Hygienic Design of Post Process Can Handling Equipment,* Campden Food and Drink Research Association, Chipping Campden, UK, 1985.
3. Technical Manual No. 17, *Hygienic Design of Liquid Handling Equipment for the Food Industry,* Campden Food and Drink Research Association, Chipping Campden, UK, 1987.
4. R. C. W. Berkeley, J. M. Lynch, P. R. Rutter, and B. Vincent, *Microbial Adhesion to Surface,* Ellis Horwood Ltd., Chichester, UK, 1980.
5. K. C. M. Marshall, *Microbial Adhesion and Aggregation,* UK, Springer-Verlag, Berlin, 1984.
6. S. Bouman, D. B. Lund, F. M. Driessen, and D. G. Schmidt, "Growth of Thermoresistant *Streptococci* and Deposition of Milk Constituents on Plates of Heat Exchangers during Long Operating Times, *Journal of Food Protection* **45,** 806–812 (1982).
7. S. J. Lewis and A. Gilmour, "Microflora Associated with the Internal Surfaces of Rubber and Stainless Steel Milk Transfer Pipeline," *Journal of Applied Bacteriology* **62,** 327–333 (1987).
8. J. T. Holah, R. P. Betts and R. H. Thorpe, "The Use of Epifluorescence Microscopy to Determine Surface Hygiene," *International Biodeterioration* **25,** 147–153 (1989).
9. M. T. Offiler, "Open Plant Cleaning: Equipment and Methods," in proceedings of *Hygiene for the 90s* Campden Food and Drink Research Association, Chipping Campden, UK, November 7–8, 1990.
10. D. A. Timperley, "The Effect of Reynolds Number and Mean Velocity of Flow on the Cleaning-in-Place of Pipelines," in Proceedings of *Fundamentals and Applications of Surface Phenomena Associated with Fouling and Cleaning in Food Processing,* Lund University, Lund, Sweden, April 6–9, 1981.
11. D. G. Dunsmore, "Bacteriological Control of Food Equipment Surfaces by Cleaning Systems. 1. Detergent Effects," *Journal of Food Protection* **44,** 15–20 (1981).
12. W. L. Shupe, J. S. Bailey, W. K. Whitehead, and J. E. Thompson, "Cleaning Poultry Fat from Stainless Steel Flat Plates," *Transactions of the American Society of Agricultural Engineers* **25,** 1446–1449 (1982).
13. M. E. Anderson, H. E. Huff, and R. T. Marshall, "Removal of Animal Fat from Food Grade Belting as Affected by Pressure and Temperature of Sprayed Water," *Journal of Food Protection* **44,** 246–248 (1985).
14. N. E. Middlemiss, C. A. Nunes, J. E. Sorensen, and G. Paquette, "Effect of a Water Rinse and a Detergent Wash on Milkfat and Milk Protein Soils," *Journal of Food Protection* **48,** 257–260 (1985).
15. D. A. Timperley and C. N. M. Smeulders, "Cleaning of Dairy HTST Plate Heat Exchangers: Optimisation of the Single-Stage Procedure," *Journal of the Society of Dairy Technology* **41,** 4–7 (1965).
16. W. G. Jennings, "Theory and Practice of Hard-Surface Cleaning," *Advances in Food Research* **14,** 325–459 (1965).
17. H. J. Schlussler, "Zur Kinetik von Reinigungsvorgangen an festen Oberflachen," Symposium uber *Reinigen und Desinfizieren lebensmittel verarbeitender Anlagen,* Karlsruhe, West Germany, 1975.
18. M. Loncin, "Modelling in Cleaning, Disinfection and Rinsing," in proceedings of *Mathematical Modelling in Food Processing,* Lund Institute of Technology, Lund, UK, 7–9 September, 1977.
19. D. G. Dunsmore, A. Twomey, W. G. Whittlestone, and H. W. Morgan, "Design and Performance of Systems for Cleaning Product Contact Surfaces of Food Equipment: A Review," *Journal of Food Protection* **44,** 220–240 (1981).
20. H. Mrozek, "Development Trends with Disinfection in the Food Industry," *Deutsche Molkerei-Zeitung* **12,** 348–352 (1982).
21. R. Klemperer, "Tests for Disinfectants: Principals and Problems," in *Disinfectants: Their Assessment and Industrial Use,* Scientific Symposia Ltd., London, 1982.
22. R. P. Elliott, "Cleaning and Sanitation," in *Principals of Food Processing Sanitation,* A. M. Katsuyama, ed., The Food Processors Institute, USA, 1980.
23. D. A. Timperley, "Cleaning in Place (CIP)," *Journal of the Society of Dairy Technology* **42,** 32–33 (1989).
24. J. T. Patterson, "Microbiological Assessment of Surfaces," *Journal of Food Technology* **6,** 63–72 (1971).
25. J. D. Baldock, "Microbiological Monitoring of the Food Plant: Methods to Assess Bacterial Contamination on Surfaces," *Journal of Milk Food Technology* **37,** 361–368 (1974).
26. S. M. Kulkarni, R. B. Maxy, and R. G. Arnold, "Evaluation of Soil Deposition and Removal Processes: An Interpretive View," *Journal of Dairy Science* **58,** 1922–1936 (1975).
27. J. T. Holah, R. P. Betts, and R. H. Thorpe, "The Use of Direct Epifluorescent Microscopy (DEM) and the Direct Epifluorescent Filter Technique (DEFT) to Assess Microbial Populations on Food Contact Surfaces," *Journal of Applied Bacteriology* **65,** 215–221 (1988).
28. C. H. McKinon and R. Mansell, "Rapid Counting of Bacteria in Rinses of Milking Equipment by a Membrane Filtration Epiflourescent Microscopy Technique," *Journal of Applied Bacteriology* **51,** 363–367 (1981).
29. P. Thompson, "Rapid Hygiene Analysis Using ATP Bioluminescence," *European Food and Drink Review,* Spring Issue, 42–48 (1989).
30. W. J. Simpson, "Instant Assessment of Brewery Hygiene Using ATP Bioluminescence," *Brewers Guardian* **118,** 20–22 (1989).
31. J. T. Holah, "Monitoring the Hygienic Status of Surfaces," In Proceedings of *Hygiene for the 90s,* Campden Food and Drink Research Association, Chipping Campden, UK, 1990.
32. A. D. Russel, "Factors Influencing the Efficiency of Antimicrobial Agents," in *Principles and Practice of Disinfection,* A. D. Russell, W. B. Hugo, and G. A. J. Ayliffe, Blackwell Scientific Publications, London, 1982.
33. *Review of Worldwide Disinfectant Test Methods.* British Association for Chemical Specialties, Sutton, UK, 1989.
34. Test Methods for the Antimicrobial Activity of Disinfectants in Food Hygiene. Council of Europe, Strasbourg, 1987.
35. B. Van Klingeren, A. B. Leussink, and W. Pullen, "A European Collaborative Study on the Repeatability and the Reproducibility of the Standard Suspension Test for the Evaluation of Disinfectants in Food Hygiene," *Report of the National Institute of Public Health and Environmental Hygiene,* No. 35901001, 1981.
36. *Tube Test for the Evaluation of Detergent-Disinfectants for Dairy Equipment,* International Standard FIL-1DF44. Brussels. International Dairy Federation, 1967.

37. "AOAC Use-Dilution Method," *Official Methods of Analysis of the Association of Official Analytical Chemists*, 14th Edition, S. Williams, ed. Association of Official Analytical Chemists, Inc., Arlington, Va., 1984.
38. *AFNOR T72-190 Germ-Carrier Method*. Paris: l'Association Française de Normalisation, 1986.
39. S. F. Bloomfield, "Disinfectant Testing in Relation to the Food Industry," in Proceedings of *Hygiene for the 90s*, Campden Food and Drink Research Association, Chipping, Campden, UK, 1990.
40. J. T. Holah, C. Higgs, S. Robinson, D. Worthington, and H. Spenceley, "A Conductance-based Surface Disinfectant Test for Food Hygiene," *Letters in Applied Bacteriology* 11, 255–259 (1990).

General References

P. R. Hayes, *Food Microbiology and Hygiene*, Elsevier Applied Science Publishers, New York, 1985.

A. M. Katsuyama, *Principals of Food Processing Sanitation*, The Food Processors Institute, Washington, D.C., 1980.

N. G. Marriot, *Principals of Food Sanitation*, AVI Publishing Co., Westport, Conn., 1985.

A. J. D. Romney, *Cleaning in Place*, The Society of Dairy Technology, Huntingdon, Cambridgeshire, U.K., 1990.

J. T. HOLAH
Campden Food and Drink
Research Association
Chipping Campden, United Kingdom

FOOD TOXICOLOGY

Food toxicology is a scientific discipline that studies the nature, sources, and formation of toxic substances in foods, their deleterious effects on consumers, the manifestations and mechanisms of these effects, and the identification of the limits of safety and thus the regulation of these substances. Because the public is very concerned about the quality and safety of foods, it is very important to have knowledge related to the principles of food toxicology and food safety. Because food toxicology is a subdivision of toxicology, it is important to present the general principles of toxicology before efforts are made to discuss food toxicology.

TOXICOLOGY: PRINCIPLES, DEFINITIONS, AND SCOPE

Toxicology is the study of the adverse effects of substances on living organisms. It is a multidisciplinary field of study dealing with the detection, occurrence, properties, effects, and regulations (DOPER) of toxic compounds. It therefore involves an understanding of chemical reactions as well as biological mechanisms of toxic actions.

A toxicological study usually consists of four elements: (1) a chemical agent capable of causing a deleterious response; (2) a biological system with which the chemical agent may interact to produce the deleterious response; (3) a means by which the chemical agent and the biological system are permitted to interact, ie, there must exist a receptor site; and (4) a response that can be used to quantitate the deleterious effect on the biological system.

Two aspects of interaction between substances and living organisms are of importance: the influence of the substances on the living organism and the influences of the organism on the substances. The chemical agent capable of causing a deleterious effect in the user is defined as a poison or toxicant. A toxicant will exert toxicity, which is defined as the capacity to produce toxic injury to cells or tissues, only at appropriate conditions when the biological system is exposed to a certain dose of the toxicant. In effect, Paracelsus (1493–1541) noted, "All substances are poisons; there is none which is not a poison. The right dose differentiates a poison and a remedy." Thus, as a rule, a substance is a toxicant only in toxic doses; virtually any substance, even pure water or sugar, is poisonous when taken in great excess. The capability to detect subtoxic levels of toxicants in biological system, such as in the plasma or urine, is of particular importance because, once known, further exposure can be avoided. In toxicology, exposure is the total amount of toxicants received by the biological system of interest. It can be expressed as the product of concentration and duration. Toxic effects in a biological system are not produced by a chemical agent unless that agent or its biotransformation products reach appropriate sites in the body at a concentration and for a length of time sufficient to produce the toxic manifestation.

Among chemicals there is a wide spectrum of doses needed to produce deleterious effects such as cellular injuries, organ damage, or death. Botulinum toxin is extremely toxic and will produce death in test animals in microgram doses (LD$_{50}$ 0.01 µg/kg). By comparison, ethyl alcohol is harmless; it has a LD$_{50}$ value of 10 g/kg. LD$_{50}$ is the dosage (mg/kg body weight) causing death in 50% of the tested animal population. Thus, one way to classify toxicants is using the toxicity rating or class based on the probable lethal dose of the toxicants for humans (Table 1). Although such classifications are only qualitative, they serve a purpose to the toxicologist who responds to the question, "How toxic is this chemical?"

Toxic agents can be classified in terms of their physical state (gas, dusts, or liquid), their labeling requirements (explosive, flammable, or oxidizer), or their chemistry (aromatic amines or halogenated hydrocarbons). They may also be classified in terms of their target organ (liver, kidney, hematopoietic system, etc), their use (pesticide, solvent, food additive, etc), their source (animal and plant

Table 1. Potential Toxicity Rating Chart

Toxicity Rating or Class	Probably Lethal Oral Dosage for Humans
Practically nontoxic	>15 g/kg
Slightly toxic	5–15 g/kg
Moderately toxic	0.5–5 g/kg
Very toxic	50–500 mg/kg
Extremely toxic	5–50 mg/kg
Supertoxic	<5 mg/kg

Source: Adapted from Ref. 1.

toxicants), and their effects (cancer, mutation, teratogenic, liver injury, etc) (1).

The route or site of exposure also affects the toxicity of a chemical agent to the biological system. Toxic agents may gain access to the body through the gastrointestinal tract (ingestion) and via the lungs (inhalation), skin (topical), or other parenteral (other than the alimentary canal) routes. Toxic agents generally elicit the greatest effect and produce the most rapid response when given by the intravenous route. The next most important route is inhalation, followed by intraperitoneal, subcutaneous, intramuscular, intradermal, oral, and topical routes. The duration of animal exposure to toxicants is usually divided into four categories: acute, subacute, subchronic, and chronic. Acute exposure is defined as exposure to a toxic chemical for fewer than 24 h. Although acute exposure usually refers to a single administration, repeated exposures may be given within a 24-h period for some slightly toxic or practically nontoxic chemicals. Subacute exposure refers to repeated exposure to a toxic chemical for 1 month or less, subchronic exposure to repeated exposure for 1–3 months, and chronic exposure to repeated exposure for more than 3 months.

The biological systems used for toxicity testing can be whole animals, including humans, dogs, and rodents, or they can be tissues or organs in culture, cell cultures, cell-free systems, eukaryotes (such as yeast and *Aspergillus*) and prokaryotes (such as *Salmonella typhimurium* and *Escherichia coli*), and plants. The age, sex, strain, and nutritional and disease status of the animal species all affect the outcome of the toxic effect. The use of enzyme inducers or inhibitors to modulate drug metabolism enzyme systems also affects animal susceptibility and toxicity to environmental chemicals. Whole animals are used to determine LD_{50}, organ target toxicity, and carcinogenicity, as well as teratogenicity of test compounds. Insects are used in conjunction with pesticide development and for the detection of environmental mutagens. Eukaryotes and prokaryotes are now widely used for the determination of mutagenic/carcinogenic (cancer-inducing) potencies of environmental toxicants, drug impurities, and processed reaction products present in foods. Currently, there is a trend in toxicology to use cell cultures as an alternative model system to animals. In this case, the TC_{50} (the toxic concentration that will induce poisonous effects to 50% of the cell population) is used to indicate the potential toxicity of the test compound. Cell-free systems are used to study the biochemical mechanism of toxicity. Plants are useful for routine monitoring of pollutants in the air.

Thus, the occurrence of a toxic response is dependent on the chemical and physical properties of the agent, the duration of exposure, and the susceptibility of the biological system or subject.

The single most important factor that determines the potential harmfulness of a toxicant is the relationship between the concentration of the toxic agent and the effect produced in the biological system. This is referred to as the dose-response relationship. Toxic responses will not occur unless the chemical interacts with the receptor site(s). The degree of the response is related to the concentration of the agent at the reactive site, which in turn is related to the dose administered. In addition, the toxic response should be quantifiable. For example, the degree of inhibition of cholinesterase in blood is used to measure the toxicity of organophosphate insecticide. Serum glutamic-oxaloacetic transaminase (SGOT) and glutamic pyruvic transaminase (SGPT) levels in blood are used to reflect liver damage caused by hepatotoxicants.

A graphic expression of the typical dose-response relationship is shown in Figure 1. A sigmoidal response curve is obtained when the dosage is plotted on a logarithmic scale. The response may be applied to an individual, a system, or a fraction of a population; it ranges from 0 to 100%. The minimally effective dose of any toxicant that evokes a stated all-or-none response is called the threshold dose. It is through the use of this dose-response relationship (including the conversion of the percent response to normal equivalent deviations or the similar Probit transformation) that the toxicologist is able to obtain the LD_{50} of a toxicant if mortality is used as an end point. For detailed information related to theoretical background and determination of the quantal response in toxicity tests, refer to Refs. 2–7.

Animals are exposed to chemicals in the air they breathe or in the water they (fish) are living in. In this case, the concentration of chemical in the air or water that causes death to 50% of the test animals is referred to as LC_{50}. Many solvents exhibit appreciable volatility under conditions of use. Workers are consequently exposed to solvent vapor. Vapor concentrations are expressed as parts of vapor per million parts of contaminated air (ppm) by volume at room temperature and pressure. To help establish a safe working environment for the protection of the health of industrial workers handling organic solvents, the American Conference of Governmental Industrial Hygienists (ACGIH) has established threshold limit values (TLVs). These are defined as airborne concentra-

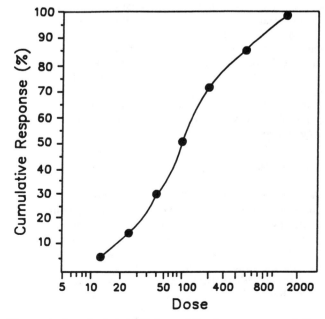

Figure 1. A typical sigmoid form of the dose-response relationship. Dosage is most often expressed as mg/kg and plotted on a log scale.

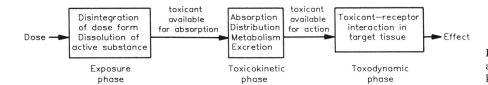

Figure 2. The three phases of toxicant action: the exposure phase, the toxicokinetic phase, and the toxodynamic phase.

tions of substances that represent conditions under which it is believed that nearly all workers may be exposed day after day without adverse effect. TLVs have been derived based on the best available information from industrial experience and studies in animals and human volunteers. Three categories of TLVs have been developed by the ACGIH: (1) time-weighted average (TWA)—a value for a normal 8-h workday and 40-h workweek; (2) short-term exposure limit (STEL)—a value for a short period of time (usually 15 mins); and (3) ceiling (TLV-C)—a value that should not be exceeded even briefly (8).

After administration to a test animal, toxicants usually undergo a series of complex processes, including absorption, distribution, metabolism, and excretion, before they exert their toxic effects (Fig. 2). During the exposure phase, the toxicant may undergo chemical alteration to compounds that may be more or less toxic than the parent compound. For example, hydrolysis of esters can take place in the gastrointestinal tract through the action of intestinal microflora. In this connection, azo compounds can be reduced to the more toxic aromatic amines.

Although the oral route is important for the ingestion of accidental and criminal toxicants, the respiratory system is more important for the uptake of occupational chemicals and air pollutants. Absorption through the skin is also important in occupational poisoning. Only a fraction of the toxic compound absorbed reaches the target tissue and the site of action. The toxicokinetic phase includes all the processes involved in the relationship between the effective dose of a toxicant and the concentration present at the various body fluid compartments and the target tissue. During the toxicokinetic phase, two types of processes play an important role:

1. Distribution processes that involve absorption, distribution to the organs, and excretion. Toxicants are transported and then may bind to protein carriers or tissue components. The principles of pharmacokinetics apply to this distribution process of toxicants.
2. Biotransformation of toxicants. This usually involves the bioactivation of the toxic agent. This metabolic biotransformation is accompanied by changes in chemical properties such as hydrophilicity and lipophilicity, and which in turn affect their distribution in the organisms, the binding to macromolecules (such as proteins and DNA), and excretion.

Metabolism of toxicants mainly occurs in the liver but may also occur in other tissues, such as the lung, kidney, skin, and gonads. Through enzymatic biotransformation processes, the lipophilic compounds are converted to more watersoluble metabolites. Two types of enzymatic reactions are involved in toxicant metabolism: phase I reactions, which involve oxidation, reduction, and hydrolysis; and phase II reactions, which consist of conjugation reactions. Phase I reactions generally convert compounds to derivatives that are more water soluble than the parent molecule. The reactions occur mainly via two oxidative enzyme systems, the cytochrome P-450-system (the mixed-function oxygenase) and the mixed-function amine oxidase. More important than these particular conversions is that these two systems also add or expose functional groups such as -OH, -SH, -NH$_2$, and -COOH, which promote the compound's covalent conjugation with endogenous moieties such as glucuronic acid, sulfate, and amino acids through the actions of phase II reaction enzymes. These conjugated secondary metabolites possess increased water solubility and significant ionization properties at physiologic pH, that, in turn, facilitate their secretion or transfer across hepatic, renal, and intestinal membranes (9).

The toxodynamic phase comprises the action of the toxicant molecules on the specific sites of action, ie, the receptors, and the expression of the observed toxic effect. The target organ on which the toxicant acts and the effector organ in which the effect is induced, or on which the effect is observed, need not be identical. The concentration of the active toxicant metabolite reached in the receptor determines to what degree a biological action will be elicited. The toxic effect observed in the biological system can be the result of interference with the normal function of the enzyme systems, blockade of the oxygen transport by hemoglobin, interference with the general functions of the cell, interference with DNA, RNA and protein synthesis, hypersensitivity reactions, and direct chemical irritation of tissues.

Certain toxicants are carcinogenic, ie, they are capable of inducing cancers in test animals and humans. Many carcinogens undergo enzymatic activation to reactive ultimate carcinogens that are electrophilic and are capable of covalent interaction with cellular macromolecules, including DNA. If the damaged DNA is not repaired, the genome lesions are expressed in replicated cells that later will transform into abnormal cells. These abnormally altered cells, if not removed by the body's homeostatic system, will undergo limited proliferation to form preneoplastic lesions that may then react in the presence of a promotor, resulting in the growth and progression of a neoplasm. In addition to these secondary carcinogens, there are also primary carcinogens that are reactive and do not require metabolic activation. Most industrial intermediates and chemotherapeutic drugs are direct-acting carcinogens; they are electrophilic and can react directly with nucleophilic DNA to form DNA adducts. Examples of direct-acting carcinogens include ethylene imine, dimethyl sulfate, mustard gas, benzyl chloride, and methyl iodide. For detailed discussions on carcinogenesis mechanisms, refer to Refs. 10–12.

FOOD TOXICOLOGY

Food toxicology is a specialized area of the discipline of toxicology that deals with the DOPER of toxic substances in foods. It is the study of the nature, sources, and formation of toxic substances in foods, their deleterious effects on consumers, the manifestations and mechanisms of these effects, and the identification of the limits of the safety and thus the regulation of these substances.

The safety of our food supply has always been the subject of discussion in newspaper and magazine articles, radio and television reports, and scientific and technical literature. The recent controversy surrounding the potential carcinogenic activity of the growth regulator alar (daminozide) residue in apples has triggered the public's resistance to apple consumption, and thus caused millions of dollars loss to the apple industry. The Food and Drug Administration (FDA), the US Department of Agriculture (USDA), the Environmental Protection Agency (EPA), and even Congress were involved, one way or another, to find proper measures to assess the potential toxicity of alar residue in apples, to regulate the use of such a potentially toxic compound in the food, and to help restore the public's confidence in government regulation processes.

Although the public generally considers pesticide residues as the major hazard of concern related to food safety problems in terms of risk to human health, it is actually a less significant hazard than environmental contaminants. Based on the three major risk criteria of severity, incidence, and onset of toxic effects, food safety hazards have been grouped and ranked from the greatest to least risk into seven classes (13): (1) foodborne toxigenic and pathogenic microorganisms, (2) malnutrition, (3) environmental contaminants, (4) naturally occurring toxicants, (5) reaction products, (6) pesticide residues, and (7) food additives. The toxicants present in food can be of biological origin, chemical contamination in nature, or produced during cooking, processing, or radiation. The toxicity of food toxicants is affected by endogenous factors such as the nature of the compounds, the dose, the frequency of exposure, the route of exposure, the presence of other nutrients or drugs, and various environmental factors; as well as endogenous factors such as the physiology and morphology of the gastrointestinal tract, the nature of the gastrointestinal microflora, and the metabolic activity of the body. It is important to point out that oral ingestion is the major route of exposure for food toxicants. The intestinal microflora play important roles in inducing the formation of some toxicants such as nitrosamines and cyanogenic glycosides.

Foodborne Hazards of Microbial Origin

Foodborne disease agents are characterized by their diversity. Some produce their effects through toxic metabolites resulting from the growth of microorganisms in the food before ingestion (eg, staphylococcal food poisoning and botulism). Others produce adverse effects through ingestion of living microorganisms. The rapid replication of microbes and consequent damage to the intestinal tract cause the well-known abdominal discomfort (e.g., *Salmonella*, *Vibrio parahemolyticus,* and *Listeria monocytogenes*). Still others require the ingestion of large numbers of living microorganisms that sporulate in the gastrointestinal tract and release a toxin (eg, *Clostridium perfringens* poisoning). The main source of these hazards may be on the farm, during food processing, or more likely, during food service preparation or preparation at home. The severity of the toxic effects ranges from temporary discomfort to the acute lethality of botulism. Infants, the elderly, and persons on immunosuppressive or chemotherapeutic drugs are thought to be more susceptible to the toxic effects of these microbial agents.

Environmental Contaminants

Toxicants included in this category are the trace elements and organometallic compounds (eg, mercury, lead, and cadmium), as well as a variety of organic compounds (eg, polychlorinated biphenyls [PCBs]). These toxicants tend to have common behavioral characteristics though they differ quite diversely in chemical structure. Environmental toxicants tend to be stable and thus persistent in the environment. They tend to bioaccumulate in the food chain and can be biotransformed with increasing toxicity in humans. Lead, mercury, and PCBs have been shown to cause major toxic effects in infants and young children because of the greater retention of the compounds as well as their greater frequency of exposure to the toxicants.

Naturally Occurring Toxicants

Toxic compounds of this category are products of the metabolic processes of animals, plants, and microorganisms from which the food products and nutrients are derived. Toxicants of animal origin include saxitoxin (paralytic shellfish poisoning), tetrodotoxin (puffer fish poisoning), ciguatoxin (ciguatera poisoning), and histamine (scombroid poisoning). Toxicants of plant origin range from the oxalates in spinach, through cyanogenic glycosides in cassava. The toxins produced by fungus on infected grains and other foods are also considered naturally occurring toxicants. They include aflatoxins, ochratoxin, zearalenone, and the tricothecene toxins. Many of these mycotoxins (toxins produced by fungal cultures) are carcinogenic, mutagenic, and teratogenic. Humans may be exposed to these naturally occurring toxicants either through direct consumption or through secondary exposure from the edible by-products of food-producing animals. Because some of these toxicants are able to cause delayed toxicity, including carcinogenicity, it is important to detect and recognize their presence in food in order to avoid consumption.

Reaction Products

Toxic reaction products are produced after cooking, processing, or charcoal broiling of foods. Carcinogenic nitrosamines are produced in bacon during cooking from the reaction of nitrite with secondary amines. Potent mutagenic and potentially carcinogenic heterocyclic amines such as Try-P-1 (3-amino-1,4-dimethyl-5H-pyrido[4,3-

b]indole), Glu-P-1 (2-amino-6-methyldipyrido[1,2-a:3′2′-d]imidazole), IQ (2-amino-3-methylimidazo[4,5-f]quinoline) and MeIQ(2-amino-3,4-dimethyl-imidazo[4,5-f]quinoline) (14) are found in charred surfaces of broiled fish and meat owing to pyrolysis of amino acids, peptides, and proteins. Benzo(a)pyrene accumulates in beefsteak during charcoal broiling. Chlorinated organic compounds can be produced during water chlorination. The real significance of these toxic compounds to human health and the effects of other nutrients on the toxicity of these compounds remains to be elucidated.

Pesticide Residues

Although public concern regarding pesticides on food has always been the subject of the lay press, there is little evidence of danger from pesticide residues on food, mainly owing to the vigorous regulatory program of the government agencies. However, the recent incidence of intentional misuse of aldicarb on watermelons and alar on apples, and the detection of high levels of pesticide residues in imported vegetables and fruits clearly indicate the need to better enforce current regulations for pesticide use on agricultural products.

Food Additives

The health hazard of food additives is the most controversial issue. If public concern were used as the criterion in ranking food toxicant hazards, food additives would rank highest. A food additive is defined as a substance or mixture of substances, other than a basic foodstuff, that is present in a food as a result of any aspect of production, processing, storage, or packaging. Today, more than 2500 different additives are intentionally added to produce desired effects. They can be divided into six major categories: preservatives, nutritional additives, flavoring agents, coloring agents, texturizing agents, and miscellaneous additives. The majority of direct food additives, predominantly spices and flavors, are generally recognized as safe (GRAS) substances that have been used for hundreds of years and found to present no significant hazard with normal human food uses. However, the use of some food additives such as cyclamates, Red No. 2 and Violet No. 1, is banned owing to their potential carcinogenic activity. In recent years there are more concerns regarding the potential short- and long-term risks of consuming food additives. People are wondering if the use of some of the additives are responsible for hypersensitivity reactions and the use of food dyes for hyperactivity in children, though there is still no experimental evidence to indicate the connection. The final decision on the safety of food additives is under the auspices of government agencies that will act on available information of potential risks based on the results of safety testing, metabolism of the compounds, and benefits to the public.

SAFETY TESTING SYSTEMS

Before the discussion of safety evaluation process and the determination of risk assessment, it is important to discuss briefly the assay systems that can be used for qualitative and quantitative detection of toxicants in foods. Analytical chemical methods, biochemical approaches, and the use of animal systems or the alternatives have all been used to serve the purpose. Thin-layer chromatography, high-performance liquid chromatography, and fluorometry have been applied to isolate and identify aflatoxin metabolites, paralytic shellfish poisoning toxins, histamine, environmental contaminants, and some of the naturally occurring plant toxicants. Serum cholinesterase activity has been used to determine organophosphate poisoning. Mice are used to assay for saxitoxin or tetrodotoxin in mussels or puffer fish gonads. In addition to the determination of acute and subacute toxicities, animal model systems can be used to determine the toxicokinetics of toxicant distribution and excretion, the dietary effect on toxicant metabolism, and the mutagenic, teratogenic, and carcinogenic potentials of food toxicants. Animal models can also be used to study irritation effects on the skin or in the eyes and effects on behavior. Table 2 lists the types of toxicologic tests that are conducted using the animal model system.

Carcinogenicity of food toxicants can be determined using results of epidemiologic studies and lifetime animal testings. Epidemiologic study of cancer incidences in Japanese population, Hawaiian Japanese populations, and the second or third generation of American Japanese population has helped establish the diet—cancer relationship. However, epidemiologic studies have several weaknesses: (1) it takes too long (more than 15 y) to reveal the etiologic agent responsible for cancer induction; (2) it is very costly to conduct; and (3) it can only identify the

Table 2. Types of Toxicologic Tests

Acute Tests (single exposure or dose)
  Determination of median lethal dose (LD_{50})
  Acute physiologic changes (blood pressure, pupil dilation, etc.)

Subacute Tests (continuous exposure or daily doses)
  3-month duration
  Two or more species (one nonrodent)
  Three dose levels (minimum)
  Administration by intended or likely route
  Health evaluation including body weights, complete physical examination, blood chemistry, hematology, urinalysis, and performance tests
  Complete autopsy and histopathology on all animals

Chronic Tests (continuous exposure or daily doses)
  2-y duration (minimum)
  Two species selected for sensitivity from previous tests
  Two dose levels (minimum)
  Administered by likely route of exposure
  Health evaluation including body weights, complete physical examination, blood chemistry, hematology, urinalysis, and performance tests
  Complete autopsy and histopathology on all animals

Special Tests
  Carcinogenicity
  Mutagenicity
  Teratogenicity
  Reproduction (all aspects other than teratogenicity)
  Potentiation
  Skin and eye effects
  Behavioral effect

possible cause of a cancer after it has occurred. Epidemiologic studies cannot determine whether a chemical will harm people until the damage (cancer formation) has been done. Lifetime rodent testings are used by the National Toxicology Program to test the carcinogenic potential of environmental toxicants and food additives. However, such testings require about 3 y plus the use of large numbers of animals and thus tremendous costs (>$1.5 million/compound). It is therefore important to develop and use alternatives that can provide reliable information for determining the carcinogenic potential of food toxicants.

Short-term assay systems and the study on structure–cancer-causing activity relationship are used to explore the carcinogenicity potential of food toxicants. The use of the latter approach still needs a great deal of research to enable us to understand the metabolism of toxicants and their derivatives and carcinogenesis mechanisms including initiation and promotion. The use of structure–activity relationships to elaborate the carcinogenic potential of the toxicants is still at an early stage of development. Recently, the mutagenic/carcinogenic potentials of environmental chemicals and food toxicants have been extensively evaluated using short-term assay systems. Most carcinogens are believed to exert their toxic effect through a damaging action on genetic makeups leading to mutations. The mutagenic/carcinogenic potential of toxicants can be determined from their ability to induce mutations in prokaryotes (*Escherichia coli*, *Salmonella typhimurium*), eukaryotes (*Neurospora crassa*), Chinese hamster ovary cells, and even animals (host-mediated assay to use body fluid or injected microbes); to cause damage in DNA molecules (unscheduled DNA synthesis) and chromosomal structures (chromosomal aberrations and sister-chromatid exchanges) in cultured cells, germ cells or bone marrow; and to initiate cellular transformation of cultured cells. Short-term assay systems have provided valuable results leading to re-evaluation of the carcinogenicity of 2-(2-furyl)-3-(5-nitrofural)acrylamide (AF-2), an antimicrobial additive previously used widely in Japanese foods, and to help resolve the dispute over the carcinogenicity of saccharin in the United States.

SAFETY EVALUATION AND RISK ASSESSMENT

Four different laws—the Food, Drug, and Cosmetic Act (FD&C Act), the Meat Inspection Act, the Poultry Products Inspection Act, and the Federal Insecticide, Fungicide, and Rodenticide Act—are routinely employed by the FDA, USDA, and EPA to govern the safety of the food we eat. The regulatory agencies use results obtained from test animals exposed to high doses of the substance, following the suggested scheme of the safety decision tree (Fig. 3) to develop the safe regulation levels for human use. Three approaches are currently employed by regulatory agencies: the safety factor, the automatic prohibition, and the quantitative risk assessment (QRA) with some form of risk-benefit analysis (15).

A "safe" level can be established by dividing the highest test dose at which no observed adverse effect (NOAE) occurs in chronic animal toxicity studies to a safety factor of 10, 100, or 1,000. The resulting smaller dose then becomes the maximum dose considered acceptable for hu-

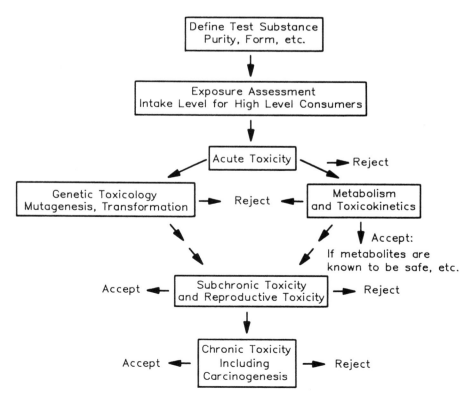

Figure 3. Safety decision tree. Adapted from Ref. 15.

man exposure. A safety factor of 10 is often used when valid, chronic exposure data in humans are available. The factor of 10 takes into consideration human variability in susceptibility. When chronic exposure data in humans are not available for the specific substance of interest, one must extrapolate the safety level from chronic exposure studies in laboratory animals, and the safety factor of 100 is used: 10 for extrapolation from animals to humans, and 10 for individual variation in humans. For chemicals with no good chronic exposure data available, an uncertainty factor of 1,000 is used (15).

The regulation of naturally occurring substances in foods by FDA follows the approach described above. To successfully prosecute an adulteration action against a food product containing a naturally occurring toxicant, the FDA must weigh the relative danger and importance of the food involved and must demonstrate a probability of harm to a significant number of consumers. Because there is no clear distinction between carcinogens (either animal or human) and other types of toxicants under the safety standard contained in section 402(a)(1) of the FD&C Act for naturally occurring food constituents, the FDA is free to ignore naturally occurring carcinogens in foods that do not pose a significant probability of harm to human health.

The automatic prohibition approach is embodied in the Delaney clause of the FD&C Act, which states that "no additive shall be deemed to be safe if it is found to induce cancer when ingested by man or animal, or if it is found, to induce cancer in man or animal . . . " (Food Additive Amendment of 1958, section 409). Food additives are regulated under these vigorous guidelines. Carcinogenic animal drugs are also not allowed in feed if they harm the animal or if the residues are detected in foods by FDA-approved methods of chemical analysis.

Quantitative risk assessment (QRA) has been used to help establish acceptable tolerance levels for added poisonous or deleterious substances, such as pesticide residues or environmental contaminants, which are required in the production of food or cannot be avoided by good manufacturing practices. QRA is a mathematical technique used to estimate from animal test results the probability of risk at the low doses to which people are normally exposed. The regulatory agency weighs the risk estimates against the corresponding implications to the food supply when reducing or eliminating such a substance from foods. This risk-benefit analysis process helps agencies determine what level of risk, if any, is acceptable, and identify the dose that corresponds with the acceptable level of risk. This dose then becomes the maximum amount of a substance permitted in foods. The FDA has authority under section 406 of the FD&C Act to establish tolerances for PCBs in milk and fish and has issued action levels for a number of environmental contaminants in food, including aflatoxin in corn and peanuts, lead in pottery, and mercury in fish (16).

The establishment of the safety standards for different categories of food constituents is thus complex. The controversial issue over whether the Delaney clause is justified in light of modern scientific advances also complicates the regulation of food toxicants. Nevertheless, there is an underlying logic to the federal food safety scheme, and it has proven remarkably durable and effective (16).

FUTURE DEVELOPMENT

Although there still exists a food safety problem, most of the problems are caused by pathogenic and toxigenic microorganisms, fungal toxins, and naturally occurring toxic compounds of edible plants and animals. The incidences of food poisoning due to environmental contaminants, pesticide residues, reaction products, and food additives are rare and preventable with reasonable care. For example, the production of mutagenic reaction products can be mitigated dramatically by altering cooking conditions and practices. It is therefore important to put emphasis on the identification of real food hazards and to learn how to control them. More basic research is needed in order to prevent food safety hazards from occurring. With the progress in technology (eg, food irradiation, biotechnology) and the awareness of nutritional requirements for subpopulations such as the elderly, infants, and cancer patients, new problems related to food safety may emerge. Examples of areas that need vigorous research efforts include studies of (1) the effect of processing on the formation of toxic compounds as well as losses of nutrients in fortified (nutritional supplemented) foods, (2) the interactions of nutrients and food toxicants, (3) safety and wholesomeness of irradiated foods, (4) movement of potential toxicants from packaging material to foods, (5) allergic and sensitivity reactions to food components, and (6) the formation of any new toxicants in genetically engineered foods (17). It is also important to develop rapid assay methods for qualitative and quantitative detection of food toxicants. Polyclonal- or monoclonal-antibody based enzyme-linked immunosorbent assay (ELISA) has been used to detect microbial toxins, mycotoxins, and pesticide and animal drug residues in foods.

The establishment of sound food safety practice thus requires the cooperation of government regulatory agencies, food industries, scientists from research institutes, and consumer protection groups. It is also very important to develop a mechanism to convey the facts about food safety to the public, who in general, has been subjected to a great deal of misinformation and supposition that was based not on fact but simply on the biases of the communicator. It is through the sound education process and information delivery system that the occurrence of food safety problems is minimized.

BIBLIOGRAPHY

1. C. D. Klaassen, "Principles of Toxicology," in C. D. Klaassen, M. O. Amdur, and J. Doull, eds., *Casarett and Doull's Toxicology: The Basic Science of Poisons*, 3rd ed., Macmillan, New York, 1986.
2. C. I. Bliss, "Insecticide Assays," in O. Kempthorne, T. A. Bancroft, J. W. Gowen, and J. L. Lush, eds., *Statistics and Mathematics in Biology*, Hafner, New York, 1964.
3. P. K. Chan, G. P. O'Hara, and A. W. Hayes, "Principles and Methods for Acute and Subacute Toxicity," in A. W. Hayes,

ed., *Principles and Methods of Toxicology*, student ed., Raven Press, New York, 1984.
4. P. B. Dews and J. Berkson, "On the Error of Bioassay with Quantal Response," in Ref. 2.
5. C. W. Dunnett, "Biostatistics in Pharmacological Testing," in A. Burger, ed., *Selected Pharmacological Testing Methods*, Edward Arnold, London, 1968.
6. D. J. Finney, *Probit Analysis*, 3rd ed., Cambridge University Press, Cambridge, UK, 1971.
7. L. C. Miller, "The Quantal Response in Toxicity Tests," in Ref. 2.
8. L. S. Andrews and R. Snyder, "Toxic Effects of Solvents and Vapors," in Ref. 1.
9. I. G. Sipes and A. J. Gandolfi, "Biotransformation of Toxicants," in Ref. 1.
10. G. Williams and J. H. Weisburger, "Chemical Carcinogens," in Ref. 1.
11. W. G. Flamm and R. J. Lorentzen, eds., *Mechanisms and Toxicity of Chemical Carcinogens and Mutagens*, Princeton Scientific Publishing, Princeton, N.J., 1985.
12. C. E. Searle, *Chemical Carcinogens*, ACS monograph 182, American Chemical Society, Washington, D.C., 1984.
13. E. M. Foster, "Is There a Food Safety Crisis?," *Food Technology*, **36**(8), 82–93, 1982.
14. T. Sumura, K. Wakabayashi, M. Nagao, and H. Ohgaki, "Heterocyclic Amines in Cooked Food," in S. Taylor and R. Scanlan, eds., *Food Toxicology: A Perspective on the Relative Risks*, Dekker, New York, 1989.
15. Food Safety Council, *A Proposed Food Safety Evaluation Process*, The Nutrition Foundation, Inc., Washington, D.C., 1982.
16. C. Ely, Jr., "Regulatory Distinctions Between Naturally Occurring and Added Substances in Food," in Ref. 14.
17. E. L. Korwek, "FDA Regulation of Food Ingredients Produced by Biotechnology," *Food Technology*, **40**(10), 70–74, 1986.

CHENG I. WEI
University of Florida
Gainesville, Florida

FOOD UTILIZATION

Most people take the process of eating for granted, knowing that the body will take care of itself. But between eating and the cellular utilization of dietary nutrients, hundreds of thousands of metabolic processes take place. Food must first be digested, or broken down into particles of a size and chemical composition that the body can readily absorb. Absorption takes place mostly in the small intestine, where specialized cells transfer digested nutrients to the blood and lymph vessels. In some cases, special changes are needed so that the nutrients can be transported to the cells where they are to be used or further processed. Within the cells, the nutrients are either stored or metabolized, that is, broken down into simpler components for energy or excretion (catabolism), or used to synthesize new materials for cellular growth, maintenance, or repair (anabolism).

The complicated processes involved are different for each nutrient, although the paths that certain nutrients take intersect at various points. The chemical details of metabolism belong to the disciplines of biochemistry and physiology; standard texts in these subjects can be consulted for detailed information. This article provides a brief overview of what happens in the body to the foods that are eaten, with particular attention to the three major nutrients: carbohydrates, proteins, and fats.

THE ALIMENTARY SYSTEM

The alimentary or digestive system is a long tube that consists of the mouth, esophagus, stomach, small intestine, colon, rectum, and anus. Some important accessory organs connected to the digestive tract are the salivary glands, gallbladder, pancreas, and liver. Along this tract, foods are broken down into smaller units, both physically and chemically, and then absorbed for use by the body. Figure 1 shows the general outline of the entire human digestive system.

The Food Path

The food placed in the mouth is chewed, softened, and swallowed. In the stomach, it is churned and then propelled into the small intestine where it is mixed with the bile from the gallbladder and digestive enzymes from the intestinal walls. The products of this digestion are partly or completely absorbed into either the portal vein or the lacteal system.

In the mouth, chewing (mastication) reduces large food lumps into smaller pieces and mixes them with saliva. This wetting and homogenizing facilitates later digestion. In clinical dietetics, edentulous (toothless) patients or

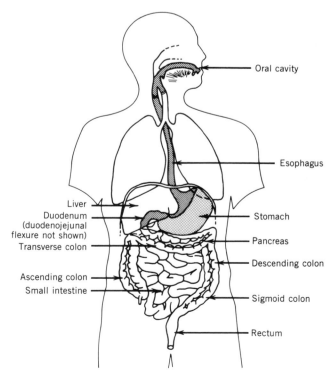

Figure 1. The digestive system of the human body.

those with reduced saliva secretion have trouble eating dry foods and require a soft, moist diet. Saliva facilitates swallowing and movements of the tongue and lips, keeps the mouth moist and clean, serves as a solvent for taste bud stimulants, acts as an oral bugger, provides some antibiotic activity, and inhibits loss of calcium from the teeth by maintaining a neutral pH. Saliva contains ptyalin (salivary amylase, a digestive enzyme) and mucin (a glycoprotein). Mucin lubricates food and ptyalin digests carbohydrates to a small extent. Each day the salivary gland makes about 1,500 mL of saliva (ca 5.5–6 cups).

The bolus of food is propelled forward by rhythmic contractions of the entire intestinal system. These peristaltic waves move the food from the mouth, through the esophagus, and into the stomach. Certain individuals, especially nervous people, tend to swallow air when eating. When part of the air is expelled through the mouth, belching results; the remaining air is expelled as flatus. If too much air is swallowed, there will be abdominal discomfort.

From the mouth, food travels through the esophagus, the stomach cardia, the stomach body, the greater curvature, the pylorus, and the duodenum. These are all parts of the stomach, where food is well mixed. Figure 2 shows the general structure of this organ and the site of specific secretions. The acid, mucus, and pepsin cause partial digestion, and peristalsis mixes up the food. The food is then released gradually through the pylorus into the duodenum.

The gastrointestinal (GI) system, the stomach and intestines, breaks down complex carbohydrates, proteins, and fats into absorbable units, mainly in the small intestine. Vitamins, minerals, fluids, and most nonessential nutrients are also digested and absorbed to varying degrees. Foods are digested by enzymes secreted by different parts of the GI system. Table 1 summarizes the major digestive enzymes and their actions. After the digestive process is complete, nutrients are ready for absorption, which occurs mainly at the small intestine. The absorption of each nutrient is discussed later.

After the nutrients have been absorbed, they enter the circulation in two ways. Most fat-soluble nutrients enter the lacteal or lymphatic system, which eventually joins the systemic blood circulation at the thoracic duct. Other nutrients enter the hepatic portal vein and are received by the liver, which eventually releases them to the bloodstream.

Enzymes and Coenzymes

After digestion and absorption, the nutrients exist as hexoses (mainly glucose and fructose), fatty acids, glycerols, and amino acids and are then metabolized in various fashions. Many of the metabolic processes require the presence of a catalyst, a substance that can facilitate a chemical reaction. Although participating in the process, it may or may not undergo physical, chemical, or other modification itself. Nonetheless, the catalyst usually returns to its original form after the reaction.

In the body, most biological reactions require a special class of catalysts: the protein catalysts or enzymes. Each enzyme catalyzes only one or a small number of reactions. There are many enzymes, each with a specific responsibility. Without enzymes, most biological reactions would proceed at a very slow speed. Coenzymes are accessory substances that facilitate the working of an enzyme, mainly by acting as carriers for products of the reaction. In this case, the enzyme is composed of two parts: a protein (apoenzyme) and a nonprotein (cofactor or coenzyme). Many coenzymes contain vitamins or slightly modified vitamins as the major ingredient. A coenzyme can catalyze many types of reaction. Some coenzymes transfer hydrogens; others transfer groups other than hydrogens. Table 2 describes the characteristics of the former; Table 3, those of the latter. Because most of the metabolic reactions discussed in this article involve coenzymes, a knowledge of the biochemical role of vitamins is important.

CARBOHYDRATES

Starch, cellulose, and their derivatives are the only polysaccharides consumed to any extent by humans. The major simple sugars ingested include the monosaccharides (such as glucose and fructose in honey and fruit juices) and disaccharides (such as maltose in beer, lactose in milk, and sucrose in table sugar). Also ingested are dextrins, sugar alcohol, and trisaccharides and tetrasaccharides, although in very small quantities. The sections that follow discuss briefly the digestion and absorption pathways of these carbohydrates, the involvement of carbohydrates in energy formation and storage, the specific processes by which carbohydrates are broken down or synthesized by the body, and the regulation of glucose levels in the blood.

Digestion and Absorption

Starch is partially hydrolyzed by ptyalin in the mouth (Table 1 and Fig. 3). The short stay in the oral cavity permits only dextrins and small polysaccharide fragments to break off from the starch molecules, and the action of ptyalin is terminated by the acid in the stomach. In the small intestine, all digestible carbohydrates are reduced to monosaccharides, namely, glucose, fructose, galactose, mannose, and pentoses. Currently, it is believed that the final hydrolysis, or digestion, of disaccha-

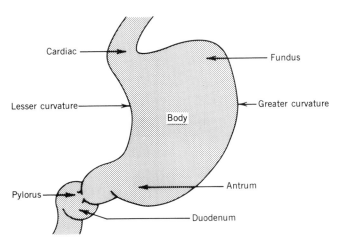

Figure 2. General structure of the stomach.

Table 1. Characteristics of the Enzymatic System of Digestion

Location	Food or Substrate	Products of Digestion	Enzyme(s) Involved in Digestion		
			Name(s)	Source(s)	Active in Acid–Base (pH)
Mouth	Starch	Maltose, dextrins, disaccharides, monosaccharides, branched oligosaccharides	Ptyalin, or salivary amylase	Salivary glands	Slightly acidic (6.7)
Stomach	Protein	Proteoses, peptones, polypeptides, dipeptides, amino acids	Pepsin	Peptic or chief cells of stomach	Acidic (1.6–2.4)
	Milk casein	Milk coagulation	Rennin	Stomach mucosa	Acidic (4.0); requires calcium for activity
	Fat	Triglycerides; some monoglycerides and diglycerides, glycerol, fatty acids	Gastric lipase	Stomach mucosa	Acidic
Small intestine (mainly duodenum and jejunum)[a]	Protein proteoses, peptones, etc	Polypeptides, dipeptides, etc	Trypsin (activated trypsinogen)	Exocrine gland of pancreas	Alkaline (7.9)
	Proteoses, peptones, etc	Polypeptides, dipeptides, etc	Chymotrypsin (activated chymotrypsinogen)	Exocrine gland of pancreas	Alkaline (8.0)
	Polypeptides with free carboxyl groups	Lower peptides, free amino acids	Carboxypeptidase	Exocrine gland of pancreas	
	Fibrous protein	Peptides, amino acids	Elastase	Exocrine gland of pancreas	
	Carbohydrate Starch, dextrins	Maltose, isomaltose, monosaccharides, dextrins	α-amylase (amylopsin)	Exocrine gland of pancrease	Slightly alkaline (7.1)
	Fat Triglycerides	Monoglycerides and diglycerides, glycerol, fatty acids	Lipase (steapsin)	Exocrine gland of pancreas	Alkaline (8.0)
	Cholesterol	Cholesterol esters	Cholesterol esterase	Exocrine gland of pancreas	
	Nucleic acids Ribonucleic acid	Nucleotides Ribonucleotides	Ribonuclease	Exocrine gland of pancreas	
	Deoxyribonucleic acid	Deoxyribonucleotides	Deoxyribonuclease	Exocrine gland of pancreas	
Small intestine (mainly jejunum and ileum)	Protein Polypeptides	Amino acids	Carboxypeptidase, aminopeptidase, dipeptidase	Brush border of the small intestine	
	Carbohydrate Sucrose	Glucose, fructose	Sucrase	Brush border of the small intestine	Acidic–alkaline (5.0–7.0)
	Dextrin (isomaltose)	Glucose	α-dextrinase (isomaltase)	Brush border of the small intestine	
	Maltose	Glucose	Maltase	Brush border of the small intestine	Acidic (5.8–6.2)

Table 1. (continued)

| Location | Food or Substrate | Products of Digestion | Enzyme(s) Involved in Digestion | | Active in Acid–Base (pH) |
			Name(s)	Source(s)	
	Lactose	Glucose, galactose	Lactase	Brush border of the small intestine	Acidic (5.4–6.0)
	Fat				
	Monoglycerides	Glycerol, fatty acids	Lipase (enteric)	Brush border of the small intestine	
	Lecithin	Glycerol, fatty acids	Lecithinase	Brush border of the small intestine	
	Nucleotides	Nucleosides, phosphate	Nucleotidase	Brush border of the small intestine	
	Nucleosides	Purines, pyrimides, pentose	Nucleosidase	Brush border of the small intestine	
	Organic phosphates	Free phosphates	Phosphatase	Brush border of the small intestine	Alkaline (8.6)

^a The food is not grouped together (eg, all fat, all proteins, etc), instead the food is placed in an order that follows the sequence of digestion along the duodenum to jejunum. This attempts to present the digestive enzymes in their expected sequence of action.

rides to monosaccharides occurs in the intestinal mucosal walls. In humans, all nondigestible carbohydrates such as lignin, hemicellulose, and cellulose are passed into the colon, where they are mainly fermented to release gas. By contrast, ruminants (animals such as cattle, sheep, and goats) have the ability to digest fiber.

Most of the monosaccharides are absorbed before the food residue reaches the end of the ileum. Absorption is principally carried out by active transport (requiring energy), although diffusion (passive movement) also occurs. Glucose and galactose enter the blood five times faster than mannose and pentoses, whereas fructose is absorbed about two to three times faster. Glucose entry into the blood may be as high as 129 g/h.

The monosaccharides traverse the intestinal and portal veins to reach the liver, from which they may eventually be released into blood circulation. In the liver, most of the fructose and galactose are converted to glucose, which is the main simple sugar in the blood, although fructose and galactose may be present throughout the bloodstream if a person consumes a large amount of them. An actively nursing mother also has some lactose in the blood, because this disaccharide is manufactured by the active mammary tissues. In the liver, part of the glucose is released to the circulation, part converted to glycogen for storage, part changed to other essential substances required by the body, and part oxidized to energy. Normally, a peak plasma glucose level of 120 to 140 mg/100 mL is reached within 60 min after a mixed meal.

After glucose has reached the bloodstream, some enters cells to give energy, and some is converted to glycogen in tissues such as muscle. Glycogen is found in many organs in the body, but the liver and muscle are the major storage sites. Most of the glucose is used to provide energy through a three-stage process: glycolysis, the citric acid cycle, and the respiratory chain.

Energy Formation and Storage

Everything a body does requires energy. Nature has provided the animal body with a wide spectrum of methods that permit energy either to be released to be stored and released to satisfy its energy need. There are five different energy systems known to operate in animal cells.

Direct Release of Energy (Mainly as Heat)

glucose (or fat or protein) + oxygen = carbon dioxide
+ water + energy to be stored + heat

Energy Stored as Adenosine Triphosphate (ATP). One technique of storing part of the energy released from the

Table 2. Characteristics of Coenzymes That Transfer Hydrogens

Enzyme System	Coenzyme	Vitamin Component	Nonvitamin Component
Dehydrogenase	Flavin adenine dinucleotide (FAD)	Riboflavin (vitamin B_2)	Adenine, ribose, phosphate
Dehydrogenase	Nicotinamide adenine dinucleotide (NAD)	Niacin	Adenine, ribose, phosphate
Part of dehydrogenase	Lipoic acid (thiotic acid)	None	Lipoic acid
Respiratory chain	Coenzyme Q	None	Quinone (vitamin E–related substance)

1220 FOOD UTILIZATION

Table 3. Characteristics of Coenzymes That Transfer Nonhydrogen Groups

Enzyme System	Coenzyme	Vitamin Component	Nonvitamin Component
Transaminases, decarboxylase	Pyridoxal phosphate	Vitamin B_6	None
Part of dehydrogenase	Lipoic acid (lipoamide)	None	Lipolic acid
Dehydrogenase	Coenzyme A	Pantothenic acid	β-mercaptoethylamine, adenine, ribose, phosphate
Cocarboxylase (decarboxylase, transketolase)	Thiamin pyrophosphate	Thiamin	Phosphate
Methyl transferase	5-methyltetrahydrofolate	Folic acid	None
Transmethylase	Coenzyme B_{12}	Vitamin B_{12}	Adenine, ribose
Carboxylase	Carboxyl–biotin complex	Biotin	None

oxidation of foodstuffs is incorporating it into ester bonds between certain organic compounds and phosphoric acid groups. The resulting substances are called high-energy phosphate compounds, the most important of which is probably adenosine triphosphate. This ubiquitous molecule is considered the energy powerhouse of the body. It releases its energy in the following reactions:

$$ATP + H_2O = ADP + P + 7.5 \text{ kcal}$$
$$ADP + H_2O = AMP + P + 7.5 \text{ kcal}$$

where ATP = adenosine diphosphate, P = inorganic phosphate, and AMP = adenosine monophosphate. Theoretically, the conversion of 1 mol of ATP to AMP can produce 15 kcal. However, within the body ATP is normally changed only to ADP when energy is needed. The energy released from this process can be used for such work as organ building, heartbeat, transportation across cell membranes, and muscle contraction. Sometimes these compounds are called active phosphate carriers and dischargers. Figure 4 summarizes the role of ATP in body energy and metabolism.

Energy Stored in Creatine Phosphate, or Phosphocreatine. Creatine phosphate is another energy-rich phosphate compound found in muscle. It can contribute to muscle energy metabolism in two ways:

$$\text{creatine phosphate} + H_2O = \text{creatine} + P + 7.5 \text{ kcal}$$

$$\text{creatine phosphate} + H_2O \xrightarrow{ADP \rightleftarrows ATP} \text{creatine}$$

In the first reaction the energy is released directly. In the second, the energy is transferred to ADP and later released for muscular or other work. Creatine phosphate is sometimes called an active phosphate carrier. Similar substances are 1,3-diphosphoglyceric acid and phosphoenolpyruvic acid.

Energy Stored in Active Acetate. The active acetate is the substance acetyl-CoA, which participates in intermediate metabolism. In terms of energy, formation of 1 mol of acetyl-CoA is equivalent to that of 1 mol of ATP.

Low-Energy Phosphate Compounds. Not all organic phosphates are of the high-energy type. Substances such

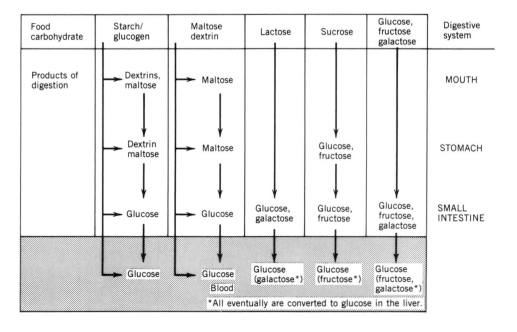

Figure 3. Carbohydrate digestion.

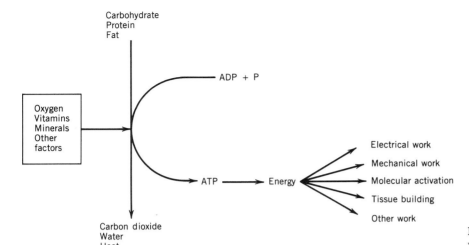

Figure 4. The role of ATP in body energy metabolism.

as glucose-6-phosphate are phosphate compounds carrying a small amount of stored energy, such as 2–3 kcal/mol. If the above energy systems are summarized the oxidation or complete metabolism of glucose will yield the following:

$$C_6H_{12}O_6 \text{ (glucose)} + 6\ O_2 = 6\ CO_2 + 6\ H_2O + \text{heat} + 38\ ATP$$

Cellular Metabolism Processes

Within the cells, a series of complex biochemical processes is needed to degrade glucose (carbohydrate) to release the energy needed by the body. Three biological processes are involved: glycolysis, the citric acid cycle, and the respiratory chain. The glucose involved derives from food and internal production, which occurs mainly in the liver. In the cells of the liver, stored glycogen may be converted to glucose by the process of glycogenolysis; in the process of gluconeogenesis, glucose is synthesized from noncarbohydrate sources. However, as will be seen later, the muscles can also indirectly contribute the energy from cellular metabolism by participating in these processes. These processes are briefly defined in Table 4 and discussed in the following sections.

Glycolysis. The first step of carbohydrate metabolism is glycolysis, in which the six-carbon glucose is converted to a three-carbon substance (pyruvic or lactic acid), as indicated in Figure 5. Figure 6 illustrates the intermediate metabolic steps during the transformation of glucose to pyruvic acid. The interconversion between pyruvic and lactic acid occurs mainly in the muscle and will be discussed later. During the process of glycolysis, the conversion of 1 mol of glucose to 2 mol of pyruvic acid generates four hydrogen atoms and 8 mol of ATP. The hydrogen atoms released are eventually converted to water; the details are discussed below. Before the pyruvic acid can be converted to carbon dioxide, water, and energy, it must be transformed into a highly versatile metabolite, the two-carbon substance acetyl-CoA (Fig. 6). This transformation is irreversible.

Citric Acid Cycle. The citric acid cycle is a series of chemical reactions that metabolizes acetyl-CoA to carbon

Table 4. Definitions of Some Metabolic Terms

Term	Definition
Glycolysis	The breaking down of hexoses (six-carbon sugars), mainly glucose, into three-carbon substances (pyruvic or lactic acid); the process is sometimes termed the Embden–Meyerhof pathway
Glycogenesis	The formation of glycogen from glucose
Glycogenolysis	The breaking down of glycogen into glucose and its metabolites
Gluconeogenesis	The synthesis of glucose (and thus glycogen) from noncarbohydrate sources, such as lactate, glycerol, and amino acids
Citric acid cycle	Also termed the Krebs Cycle or tricarboxylic acid cycle; the process whereby carbohydrate, fat, and protein is completely oxidized to carbon dioxide, water, and energy; accomplished with the assistance of the respiratory chain
Respiratory chain	The transport of hydrogen atoms from biological oxidation for acceptance by oxygen atoms to form water molecules

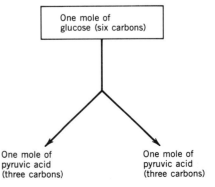

Figure 5. The overall result of glycolysis.

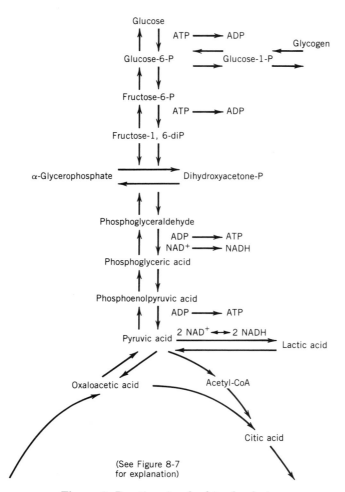

Figure 6. Reactions involved in glycolysis.

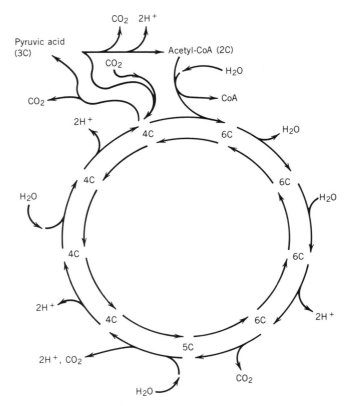

Figure 7. Outline of the citric acid cycle (C refers to a carbon atom).

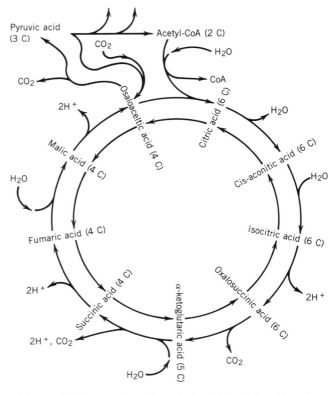

Figure 8. Components and products of the citric acid cycle.

dioxide and hydrogen atoms, as indicated in Figures 7 and 8. In each cycle, two carbon dioxide molecules and four pairs of hydrogen atoms are put through the respiratory chain, together with the electrons, to generate 12 mol of ATP and 4 mol of water from oxygen. This cycle is the major link, or common path, in the transformation of carbohydrate, fat, and protein to carbon dioxide and water. Metabolites of the three nutrients enter the cycle at different strategic points. Because the cycle requires the respiratory chain to complete its work, it will not function in the absence of oxygen (anaerobically).

As indicated above, pyruvic acid may be removed from the glycolysis process by being converted to lactic acid. If so, there must be a source of hydrogen atoms, which are normally obtained from the production of phosphoglyceraldhyde (Fig. 6). In this case, glucose metabolism and energy (ATP) production can continue for a while without oxygen (that is, without going through the citric acid cycle). This anaerobic respiration occurs in muscle where an occasional burst of energy is needed. The lactic acid that accumulates is converted back to pyruvic acid when the oxygen supply is restored, in which case the citric acid cycle is reactivated. The soreness of muscle from heavy work or exercise results from the presence of a large amount of lactic acid.

Respiratory Chain. Biological oxidation, or the respiratory chain, is a very complicated process whereby hydrogen atoms released by substances through oxidation are transported by a number of intermediates until the hydrogen atoms are accepted by oxygen to produce water. The respiratory chain involves both oxidation and reduction. Oxidation is the process whereby a substrate either takes up oxygen or loses hydrogen. The substrate is oxidized, while the source substance that provides the oxygen or accepts hydrogen is the oxidizing agent. Reduction is the reverse process, whereby a substrate loses oxygen or accepts hydrogen. This substrate is reduced and the source substance that gains oxygen or loses hydrogen is the reducing agent.

The respiratory chain involves coenzymes (Tables 2 and 3) and consists of the following series of events:

1. Hydrogen atoms are released from a substrate, a process requiring an energy source.
2. The coenzyme NAD accepts hydrogen (NAD to $NADH_2$).
3. The hydrogen in NAD is accepted by FAD (FAD to $FADH_2$).
4. The hydrogen in $FADH_2$ is accepted by coenzyme Q.
5. The hydrogen in $CoQH_2$ is released as hydrogen by losing an electron, and a chain of cytochromes becomes reduced by accepting the electron.
6. The electron is transferred to molecular oxygen (O_2).
7. The negatively charged oxygen (O_2) reacts with two protons (H^-) to form water.

The entire process of the respiratory chain, including the formation of ATPs, is illustrated in Figure 9. The complete conversion of glucose to carbon dioxide, water, and energy is shown in Figure 10. The amount of energy used or stored is shown in Tables 5 and 6.

Glycogenesis and Glycogenolysis. In plants, carbohydrate is stored as starch, a polysaccharide of glucose. In animals, carbohydrate is stored in the form of glycogen, also a polysaccharide of glucose. The amount of glycogen stored in the body depends on the diet and the physiological status of the animal. In man, glycogen is produced and stored mainly in the liver and muscle. The process, known as glycogenesis, occurs readily when adequate glucose is present.

However, when the glucose concentration in the liver and muscle decreases, their glycogen content must be broken down to provide glucose for energy. This reverse process of glycogenesis is glycogenolysis. Its occurrence in the muscle is slightly different from that in the liver. In the liver, the glycogen can be directly degraded to glucose. However, in the muscle, the enzyme for the final step is missing so that glycogen is degraded only to glucose-6-phosphate, which has to be converted to pyruvic and lactic

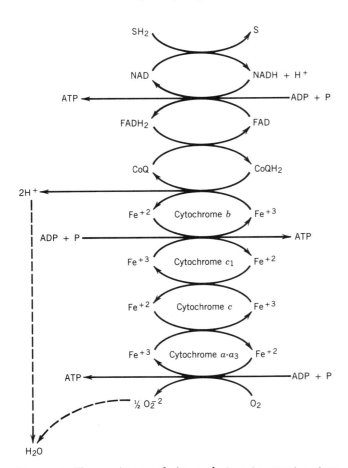

Figure 9. The respiratory chain or electron transport system (S = substrate).

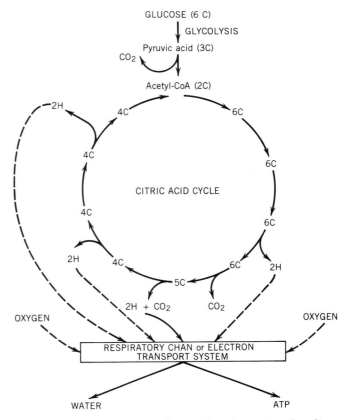

Figure 10. The three stages of converting glucose to carbon dioxide, water, and ATP.

Table 5. Energy Production during the Conversion of 1 Mol of Glucose to Carbon Dioxide and Water in the Presence of Oxygen

Sequence	Product of Specific Reaction	Phosphorylation[a]			ATP	
		SD	SP	OP	Loss	Gain
Glycolysis	Glucose-6-P	x			1	
	Fructose-1,6-diP	x			1	
	Phosphoglyceric acid		x			2
	Phosphoglyceraldehyde			x		6
	Pyruvic acid		x			2
Citric acid cycle	Acetyl-CoA			x		6
	Oxalosuccinic acid			x		6
	Succinyl-CoA			x		6
	Succinic acid		x			2
	Fumarate			x		4
	Oxaloacetate			x		6
Total number of ATP gained during the oxidation of 1 Mol of glucose under aerobic conditions						38

[a] SD = substrate dephosphorylation; SP = substrate phosphorylation; OP = oxidative phosphorylation.

acids instead (Fig. 6). These can be converted to glucose via the citric cycle.

Gluconeogenesis. Gluconeogenesis is the synthesis of glucose (and thus glycogen) from noncarbohydrate sources such as amino acids, fatty acids, glycerol, and lactic and pyruvic acids. Figure 11 shows how muscle glycogen can be converted to glucose in the liver in spite of the muscle's lack of the appropriate enzyme. Figures 11 and 12 show how protein and fat can be converted to glucose. The interrelationship among the three nutrients, carbohydrate, protein, and fat, will be discussed at length in later sections.

Regulation of Blood Glucose

Another important aspect of what happens to carbohydrates in the body is the balancing of blood glucose (or blood sugar) levels. In a normal person, blood glucose fluctuates within narrow limits: between 70 and 100 mg/100 mL of blood. This is achieved by a balance between the supply and removal of blood glucose. If blood glucose drops below the norm, hypoglycemia occurs. In a healthy individual, the blood sugar is restored to normal by the provision of glucose from three sources. Simply eating additional carbohydrates increases the absorption of monosaccharides, and the liver can then release more glucose. Second, the glycogen in liver and muscle may be degraded (glycogenolysis) to form more glucose. Third, protein and fat may be degraded to provide glucose (gluconeogenesis).

If a person's blood glucose rises above the norm, hyperglycemia occurs. If the person is in normal health, the body spontaneously lowers the blood glucose levels in one or more of the following ways: (1) more insulin is released to drive glucose into cells for oxidation, (2) more glycogen is formed (glycogenesis) in the liver and muscle, (3) more glucose is changed to fat (lipogenesis) in fat cells, and (4) more glucose is excreted in the urine (glucosuria).

PROTEINS

Dietary protein exists in three forms. The major portion is conjugated with other substances, a small fraction is associated with fats and carbohydrates, and only a very small part exists as free protein, such as that in egg white. To be used by the cells, all proteins must be broken down into their constituent amino acids. In the sections that follow, the paths of protein digestion, absorption, and metabolism will be traced.

Digestion

Digestion of protein begins in the stomach (Table 1 and Fig. 13) where acid activates the pepsinogen (an enzyme)

Table 6. Energy Production during the Conversion of 1 Mol of Glucose to Carbon Dioxide and Water in the Presence of Oxygen

Sequence	Product of Specific Reaction	Phosphorylation[a]			ATP		Hydrogen Atoms
		SD	SP	OP	Loss	Gain	
Glycolysis	Glucose-6-P	x			1		
	Fructose-1,6-diP	x			1		
	Phosphoglyceraldehyde			x			Released
	Phosphoglyceric acid		x			2	
	Pyruvic acid		x			2	
	Lactic acid						Accepted

[a] SD = substrate dephosphorylation; SP = substrate phosphorylation; OP = oxidative phosphorylation.

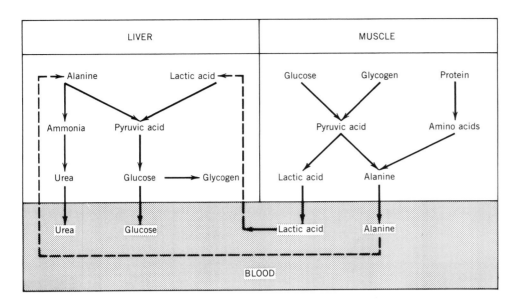

Figure 11. The role of liver and muscle in gluconeogenesis.

to release pepsin (another enzyme). The pepsin cleaves peptide linkages in the protein to produce polypeptides, each with two or more amino acids. When the stomach content reaches the duodenum, the pH is raised to about 6.5 by the presence of alkaline pancreatic juice. In the small intestine, chymotrypsin and trypsin hydrolyze most of the protein molecules to form small polypeptides and dipeptides. These are further digested to form free amino acids by the pancreatic enzyme carboxypeptidase and the intestinal enzymes aminopeptidase and dipeptidase. Those small peptides not split into individual amino acids may gain enter to mucosal cells to be digested later, for body tissues can utilize only amino acids.

Because protein molecules also contain nonpeptide linkages, they must be denatured first (by heat or stomach acid) before the appropriate enzymes can digest them. The denaturation process exposes the protein molecules, providing more surface area for enzymatic action. On the other hand, excess heating or cooking can also reform some other linkages, making digestion more difficult.

Absorption

Normally only amino acids (and certain small peptides) are absorbed. In the small intestine all free amino acids, whether ingested or digested products, are absorbed via the mucosal cells into the hepatic portal vein. The D-amino acids are not absorbed as well as the L forms. The L-amino acids, the biologically active ones, are absorbed by active transport (a process requiring energy). Absorption occurs along the entire small intestine, with the slowest rate along the ileum. The stomach and colon may also absorb some amino acids. About 20–30% of ingested proteins are unabsorbed and excreted in the stools.

Although normally only amino acids are absorbed, it is well known that in small infants some undigested pro-

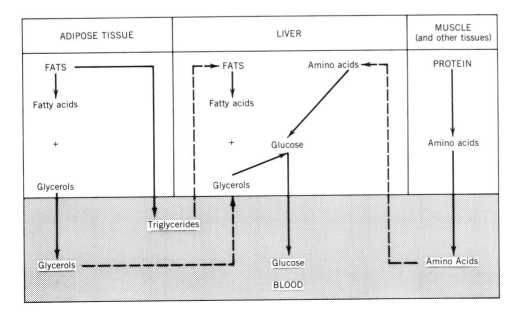

Figure 12. An overview of glucose formation from fat and protein.

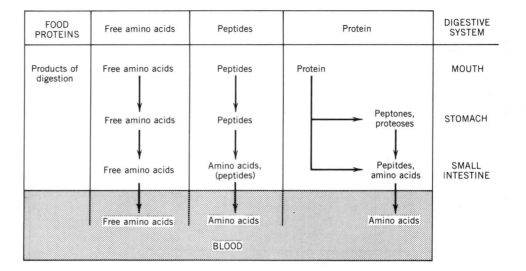

Figure 13. Protein digestion.

teins are also absorbed. The subsequent antigen–antibody interaction causes the child to develop an allergic reaction when ingesting the same protein foods later. This explains the high allergic incidence among infants to foods such as eggs and cereals. If adults show allergy to ingested protein foods, they are still probably capable of absorbing whole protein molecules. For the majority of the population, this ability disappears with age.

Amino acids seem to be utilized best when they are absorbed in accordance with the body need for growth and function. Any excess absorbed will not be stored but excreted in the urine or metabolized to ammonia. Studies have shown that absorption and utilization of proteins are optimized when adults evenly distribute their protein intake throughout the day.

In some cases, erratic patterns of protein absorption may be due to stomach irregularities rather than irregular protein ingestion, for the stomach regulates the emptying of the nutrient into the small intestine. In patients with partially or completely removed stomachs, the rate of protein emptying is so disturbed that much of this valuable nutrient is lost in the fecal waste or degraded by intestinal flora. When undigested amino acids are decarboxylated by intestinal bacteria, the important chemicals such as histamine, tyramine, ammonia, and similar substances are formed, many of which are absorbed. These substances are undesirable in excessive quantities. The ammonia formed by deamination plays a critical role in certain pathological conditions, for the absorbed ammonia can cause body deterioration. The absorption of amino acids is also impaired if the intestinal mucosa is damaged, as occurs in sprue, ulcerative colitis, and resection of a moderate amount of the small intestine. With time, however, there is a functional adaptation of the intestinal mucosa.

Metabolism

The liver may release some absorbed amino acids for body metabolism. Body amino acids may be catabolized or used in body repair, building, and maintenance. The importance of protein to the body is illustrated by its numerous functions and the complexity of its metabolism. If a normal person has a regular diet, three major factors determine the direction of protein metabolism: the quality and quantity of protein consumed, the amount of calories ingested, and the physiological and nutritional status of the body.

Protein metabolism revolves around a body pool of amino acids that are continuously released by protein hydrolysis and resynthesized. The amino acids ingested are the same as those released in the body. Body tissues can utilize only amino acids, not small peptides. About 50–100 g of body protein turns over daily, ranging from the slowest rate in the collagen to the fastest rate in the intestinal mucosa. Any amino acids filtered through the kidneys are reabsorbed, although certain congenital defects in the kidney tubules may interfere with this process. During pregnancy, infancy, childhood, and other conditions of growth, protein synthesis exceeds degradation. Individuals in these categories, therefore, require a large pool of amino acids.

Figure 14 illustrates the general catabolism, or degradation, of protein to amino acids and other metabolites. Figure 15 gives the general outline of protein formation in the body, and Figure 16 provides an overview of protein metabolism. The sections that follow discuss certain aspects of protein metabolism: degradation, protein synthesis, nitrogen balance, and metabolism of creatine and creatinine.

Protein Degradation. Protein is degraded to its individual amino acids in the muscle and other tissues, but the major site of actual destruction (catabolism) of each amino acid is the liver. The amino acids released from all other organs, especially the muscles, reach the blood and are diverted to the liver for degradation (Figure 14). Under normal circumstances, the catabolism of protein is balanced by its formation, although during stresses such as starvation and disease destruction can outstrip synthesis.

The procedure for the oxidation of an amino acid begins with deamination, a process in which the ammonia or

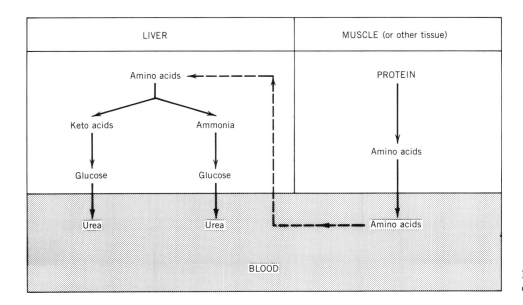

Figure 14. An overview of protein catabolism (degradation).

amino group of the acid is removed so that only the carbon skeleton of the amino acid is left. Two metabolites are formed: the keto acids and ammonia (Fig. 14). The ammonia is changed to urea, which is released to the blood and eventually excreted in the kidney. Normally, we excrete about 20–30 g of urea in the urine each day. Because urea forms exclusively in the liver, advanced liver disorders raise the blood urea nitrogen urea nitrogen level.

The keto acids formed from deamination enter the citric acid cycle to be oxidized. The exact process whereby keto acids and ammonia are formed varies with each individual amino acid, although the goal is the same. Figure 17 shows the points at which each of the acids and their keto acids enter the citric acid cycle or are transformed to pyruvate to form glucose. Within the citric acid cycle, the keto acids may be made to form glucose (gluconeogenesis). The keto acids are also tied to fat metabolism by the interconversion of keto acids and fatty acids. In sum, when protein is degraded amino acids are formed. Some amino acids circulate, some are oxidized, some are converted to glucose, and some are directed to other paths.

Protein Synthesis. Protein synthesis is the process of linking different amino acids by their amino groups (peptide linkages) to form a molecule of protein. This molecule must contain the right number of amino acids in the appropriate pattern and sequence needed for a specific body structure, such as hair, skin, muscle, tears, saliva, enzymes, or hormones. The nucleic acids DNA and RNA control this process, which takes place in cell nucleus cytoplasm. The formation of each specific protein requires a highly specific genetic code, the details of which may be obtained from a standard biochemistry or molecular biology text.

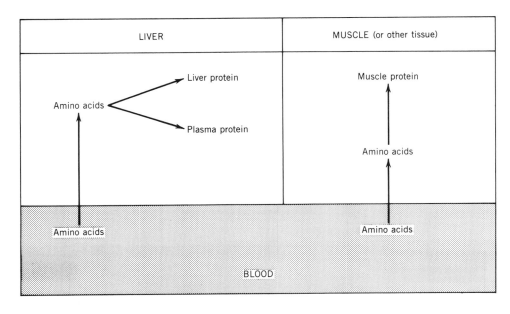

Figure 15. An overview of protein anabolism (synthesis).

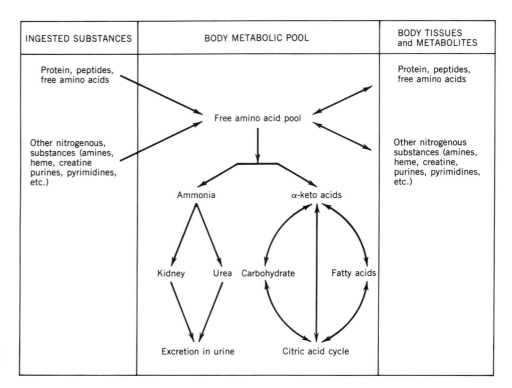

Figure 16. An overview of protein metabolism including nonprotein nitrogenous substances.

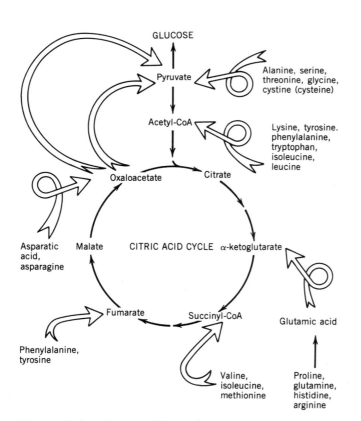

Figure 17. Introduction of the carbon structures of individual amino acids into the citric acid cycle.

Although it occurs in nearly all cells, protein synthesis in some is more frequent and intense than in others. The organs that seem to synthesize the most protein are the liver, muscle, and those cells, tissues, and organs that manufacture or secrete enzymes, hormones, and other protein substances. One exception is the synthesis of plasma proteins by the liver. This occurs because red blood cells have no nuclei and are thus unable to synthesize protein themselves.

For a protein to be synthesized, the appropriate type and number of amino acids must be available. There are 8–10 amino acids whose carbon skeletons cannot be synthesized or manufactured by the body. They are called the essential amino acids and must be supplied in the diet. In addition, there are 10–15 amino acids that the body can manufacture (including their carbon skeletons), which are the nonessential amino acids. If the appropriate carbon skeleton of an amino acid is present, the body can add, subtract, and transfer the amino group until the right amino acid is formed. Protein synthesis is normally preceded by massive deamination, amination, and transamination to obtain the appropriate amino acids.

Nitrogen Balance. Of the three major nutrients protein is the most important in the sense that it makes up lean body mass. Another factor that makes protein important is that body protein turnover is tremendous. For example, the formation of hair, skin, saliva, and sweat all involve large losses of protein that must be constantly replaced. One acceptable technique to determine if there is enough protein in the diet is to measure the ingestion and excretion of nitrogen (nitrogen balance), because all protein has a relatively constant content of this element. Theoretically, if a young adult is in normal health and a proper stage of development, the amount of nitrogen consumed should be equal to the amount excreted. This is nitrogen equilibrium. However, depending on age, physiological condition, dietary intake, and other factors, some people are in positive nitrogen balance and others are in negative nitrogen balance.

Metabolism of Creatine and Creatinine. Body muscles perform work and may occasionally be required to provide a sudden burst of energy. They achieve this by the hydrolysis of two unique high-energy bonds. One is ATP and the other is phosphocreatine (or creatine phosphate), both of which were mentioned above. The body, mainly the muscles, contains about 100–150 g of phosphocreatine and creatine. Although not an amino acid, creatine is a unique nitrogenous chemical derived from three amino acids: arginine, glycine, and methionine. Creatine, which is water soluble, is found in meat and meat products (such as extracts, soups, and gravies). Food contains little or no phosphocreatine, because of its easy degradation to creatine (or creatinine, another metabolite) and phosphoric acid (Fig. 18).

Phosphocreatine is a high-energy compound that provides instant energy when the muscles need it. If it is depleted, ATP is then used. Phosphocreatine seems to help maintain the ATP levels. After exertion and after all stored energy is used, creatine is rephosphorylated to form phosphocreatine.

Although an important body constituent, creatine is usually not excreted as such. Instead, the spontaneous, nonenzymatic dephosphorylation of phosphocreatine produces creatinine, in an irreversible process. Every healthy individual excretes a constant amount of creatinine, which is thus considered to be a normal waste product.

The amount in the urine also reflects the amount of active muscle mass in the host. A woman excretes about 15–22 mg/kg, and a man about 20–26 mg/kg. Because creatinine is normally excreted rapidly, any increase in its level in the blood is a sign of kidney malfunction. Physicians attempting to make sure of the proper collection of urine frequently exploit the constancy of creatinine excretion. The volume of urine collected should reflect a 24-h excretion if the level of creatinine is within the normal range.

Although creatine is normally not excreted in the urine, children and women occasionally do dispose of the chemical in this manner. The excretion rate for women is especially high during and after pregnancy. The urinary level of creatine is also high in patients with diabetes, hyperthyroidism, and fever and those experiencing malnutrition or simple starvation. This reflects the degradation of musculature.

FATS

The digestion, absorption, transportation, and cellular metabolism of fats follow yet a third set of chemical and biological pathways in the body. However, they intersect the pathways of carbohydrates and proteins at several points. The following sections briefly describe the general fate of fats in the body.

Digestion, Absorption, and Transportation

Ingested fat meets its first significant digestive enzyme in the duodenum, where the exocrine gland of the pancreas provides the most important lipase (Table 1 and Fig. 19). The lipases in the saliva, stomach, and small intestine have only a small effect on fat digestion, as shown by the tremendous reduction in fat digestion when the pancreas is disabled. When the exocrine gland of the pancreas is not working properly, undigested and unabsorbed fat causes steatorrhea (bulky, clay-colored, fatty stools). The combined detergent actions of bile salts (from the gallbladder), fatty acids, and glycerides emulsify fat, thus facilitating its digestion by lipase.

In the small intestine, half of the ingested triglycerides are hydrolyzed by lipase to form free fatty acids and glycerols. The rest are changed to monoglycerides and a small amount of diglycerides. The monoglycerides are absorbed into the intestinal mucosa, where they are further hydrolyzed to glycerols and free fatty acids.

Free fatty acids in the intestine are absorbed in two ways. Fatty acids with less than 10–12 carbon atoms pass directly from the intestinal lumen, through mucosal cells, into the portal vein, and to the liver. Here some are released into circulation as free fatty acids, some are converted to triglycerides for deposition, and some are circulated in the blood as glycerides or as fatty acids that reside within the complex of lipoproteins. Fatty acids with more than 10–21 carbon atoms are absorbed into the mucosal cells, where they are regrouped with glycerols to form triglycerides. The triglycerides attach themselves to very low density lipoproteins to form chylomicrons, which enter the systemic circulation via the lymph and thoracic duct. Chylomicrons are fat globules 1 μm in diameter and visible under the microscope.

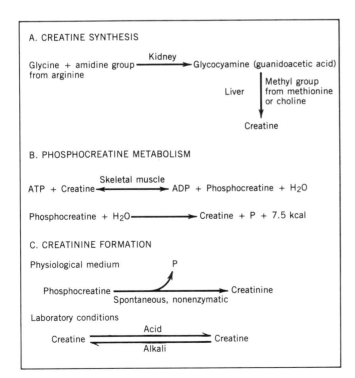

Figure 18. Metabolism of creatine and creatinine.

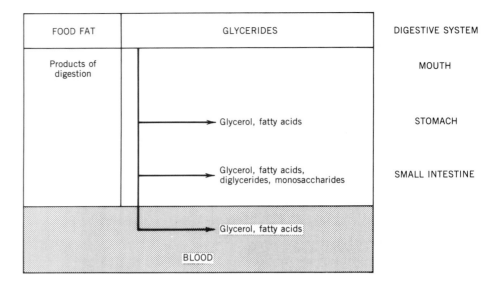

Figure 19. Triglyceride digestion.

Other fatty substances are absorbed to varying degrees. For example, animal sterols are absorbed easier than plant sterols. Pancreatic secretion, fatty acids, and bile salts, which together emulsify and esterify cholesterol, are necessary for cholesterol absorption of cholesterol. It is currently believed that cholesterol is absorbed mainly in the ileum. Like the triglycerides, absorbed cholesterol is incorporated into the chylomicrons, which reach the systemic circulation.

Within two to three hours after the ingestion of food containing short-chain fatty acids, the blood level of chylomicrons remains unchanged, although it may rise sharply if the meal contains long-chain fatty acids. Normally, after a mixed meal, the plasma develops a milky appearance because of the presence of chylomicrons in the blood. This is sometimes known as lipemia. In the presence of the enzyme lipoprotein lipas, these plasma chylomicrons are cleared and their contents diverted to the liver and adipose tissue.

Blood plasma, therefore, contains fat in the following forms: fatty acids, glycerol, glycerides, cholesterol, cholesterol esters, and phospholipids. These forms are bound to the albumin, α-globulin, and β-globulin fractions of the plasma proteins. The resulting lipid–protein complexes have varying densities. The highest densities occur in those with the most protein and least lipid; the lowest densities occur in those with the least protein and most lipid. Consequently, the complexes are classified into high-density, low-density, and very low density lipoproteins. In general, very low density lipoproteins carry mainly triglycerides; low-density lipoproteins carry mainly cholesterol; and α-lipoproteins carry phospholipids, albumin, and free fatty acids. For a normal person, about 95% of ingested fat is absorbed, mainly in the duodenum and jejunum, with some absorption by the ileum. About 5% of fecal waste is fat, which comes from the diet, cell debris, and bacterial synthesis.

Although most of the fats are emptied into the lymphatic system after absorption and eventually reach the systemic circulation, the bile salts separate from the fats and travel through the portal vein into the liver. There they are reincorporated into the bile. Bile salts are thus cycled through the enterohepatic circulation (the liver, gallbladder, intestinal lumen, portal vein, and back to the liver). About 80–90% of bile salts in the intestinal lumen are reabsorbed in this way; the rest are lost in the stool.

Cellular Metabolism

The adult body distributes fats to two main locations: the membranes and other structural parts of cells (commonly called structural fats) and the fat cells (neutral fats), which are mainly white. Infants have some brown fat cells, which can regulate body temperature by producing heat to support the baby's higher metabolic rate. Neutral body fat contains mainly triglycerides, plus small amounts of diglycerides and monoglycerides, which are important metabolic intermediates. Consequently, triglycerides are the main form of stored energy.

Fat Degradation. Stored fat is degraded as needed to provide energy. Fat degradation occurs in two major stages: hydrolysis of glycerides and oxidation of fatty acids. In the adipose tissues, glycerides are hydrolyzed by a lipase to form fatty acids and glycerols. Both of these are released into the circulation for transport to the liver, where further hydrolysis may occur. When triglycerides are hydrolyzed, the released glycerols can be converted to phosphoglyceraldehyde in the liver (Figs. 6 and 20). This compound can in turn be converted to either carbon dioxide and water or glucose.

The process of oxidizing the fatty acids to carbon dioxide, water, and energy is called β-oxidation, or alternate oxidation. It occurs mainly in the mitochondria of liver cells. The carbon chain is broken down by the successive removal of two-carbon fragments from the carboxyl end to form acetic acids. These can combine with CoA to form acetyl-CoA, which can enter the citric acid cycle to be oxidized (Figs. 8, 17, and 20). When the fatty acids are reduced to acetyl-CoA, hydrogen atoms are also released, which can be passed on to the respiratory chain. When fatty acids are completely oxidized, they generate more

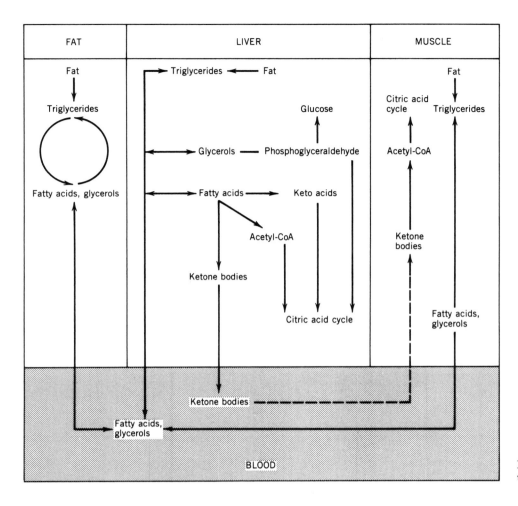

Figure 20. An overview of fat metabolism.

ATP than the molecular equivalent of carbohydrate because less oxygen is present. This explains why fat has a higher caloric value. However, unsaturated fatty acids generate less energy than the molecular equivalent of saturated fatty acids because less hydrogen is present in the former.

Most of the naturally occurring fatty acids are even numbered, and thus their oxidation always produces acetyl-CoA at the end. However, if the fatty acids happen to be odd chained, propionyl-CoA is formed instead. Propionyl-CoA can also enter the citric acid cycle if the coenzyme with vitamin B_{12} is available.

Fat Synthesis. Fat synthesis takes place in two major stages: the formation of fatty acids and the formation of triglycerides. Fatty acids synthesis is achieved in two places: the mitochondria and the cytoplasm. Within the mitochondria, β-oxidation is reversed and two-carbon units are added until the appropriate fatty acids are formed. Outside the micochondria, in the cytoplasm, another form of fatty acid synthesis occurs. Here the starting compound is acetyl-CoA which serves as the end of the fatty acid molecule. The remaining carbons are incorporated as two-carbon units derived from the malonyl group. The incorporation is accompanied by simultaneous recarboxylation. The fatty acids formed are mainly 12–14 carbons long and rarely more than 16. The body can synthesize unsaturated fatty acids from the saturated ones by removing hydrogen, although it is unable to synthesize the essential ones.

In the adipose tissues, fatty acids combine with glycerol to form triglycerides, or neutral fats. This reaction occurs in the mitochondria. Figure 20 summarizes the information on fat synthesis.

As indicated earlier, glycerol can be converted to glucose (gluconeogenesis). However, acetyl-CoA cannot be converted to pyruvic acid (Fig. 10). Although keto acids can enter the citric acid cycle, there is very little net conversion of fat to carbohydrate in the body, with the exception of the small amount of phosphogyceraldehyde formed from glycerol.

Ketone Bodies. During the normal process of β-oxidation of fatty acids, the liver has the appropriate enzyme to remove the CoA from acetoacetyl-CoA to form acetoacetic acid. Acetones and β-hydroxybutyric acids can be formed from acetoacetic acids. The last three compounds are collectively called ketone bodies. The small amount of ketone bodies normally made by the liver is transported by the circulation to the muscle for conversion to acetyl-CoA, which is put through the citric acid cycle (Fig. 20). Acetone is eliminated via urination and respiration. Because under normal circumstances the ketone bodies are metabolized as soon as they are formed, a person rarely excretes more than 1 mg of ketone each day, and blood levels are usually less than 1 mg/100 mL.

However, the ketones can accumulate under certain conditions, and the resulting clinical condition is known as ketosis. The main cause of ketosis is the accumulation of acetyl-CoA because the citric acid cycle in the liver is not operating at its normal or optimal efficiency. The most common cause is a sequence of events called intracellular carbohydrate starvation. First, decreased supply of glucose leads to a reduction in pyruvic acid, acetyl-CoA, and cellular energy supply. Second, for compensation, fatty acid oxidation is increased to provide energy with an accumulation of acetyl-CoA. Third, the oversupply of acetyl-CoA leads to the formation of ketone bodies.

Glucose supply to cells is reduced in people with diabetes mellitus and people who undergo dietary alterations such as high-fat–low-carbohydrate intake or simple starvation. In a diabetic patient, the lack of insulin prevents glucose from entering cells. When a persons's diet is low in calories, high in fat, or low in carbohydrate, a similar metabolic pattern takes place. The inadequate intake of carbohydrate means a low supply of glucose to cells, and ketosis may develop. However, an intravenous introduction of glucose counteracts ketosis, which is why carbohydrate is an antiketogenic agent.

Cholesterol Metabolism. Body metabolism of cholesterol is of special concern to nutritional scientists. Dietary cholesterol comes mainly from animal products such as fats, eggs, and organ meats. Sterols are obtained from plant foods, although they are not absorbed by the human digestive system. Ingested cholesterol is readily absorbed via the lymphatic system after esterification in the intestinal mucosa (with fatty acids). The body can also synthesize cholesterol, mainly in the intestinal mucosa and liver. It is currently believed that the amount of cholesterol synthesized by the body is inversely related to the quantity consumed. However, the problem of regulating serum cholesterol level by reducing dietary intake is a much debated issue.

Synthesis and degradation of cholesterol occur simultaneously and continuously. The body removes cholesterol by conjugating it with taurine or glycine in the liver and excreting it in the bile, although the enterohepatic circulation makes sure that some cholesterol is reabsorbed.

NUCLEIC ACIDS

Practically everything humans eat contains nucleic acids, which occur in cell chromosomes; nucleic acids are responsible for heredity. During digestion, ingested nucleic acids are initially cleaved into nucleotides by pancreatic nucleases (Fig. 21 and Table 1). Next, the small intestine secretes nucleotidase, which hydrolyzes nucleotides to form phosphoric acid and nucleosides. The latter are split by intestinal nucleosidases to form sugars, purine, and pyrimidine, all of which are absorbed by active transport. Undigested large molecules of nucleic acids are excreted in the stool.

Nucleic acids can also be synthesized. In the body, purines, pyrimidines, and sugars are put together to form ribonucleic acid (RNA), deoxyribonucleic acid (DNA), nicotinamide adenine dinucleotide (NAD), and other related substances. However, the liver can also make pyrimidines and purines. Figure 22 shows the origins of atoms in synthesized purine, and Figure 23 those of pyrimidine.

Within a cell, although DNA is stable throughout life,

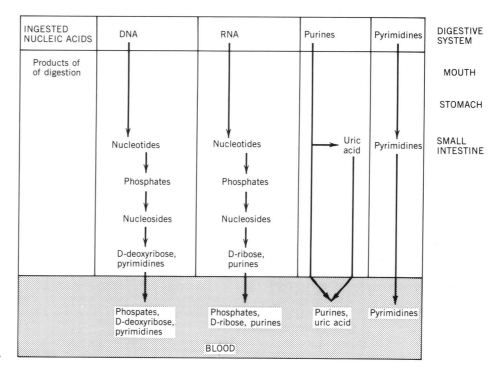

Figure 21. Digestion of nucleic acids.

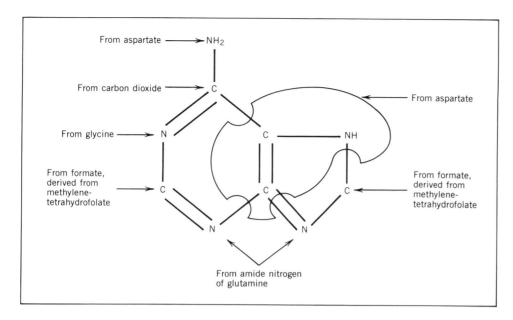

Figure 22. Origins of the atoms in purine.

RNA is in constant equilibrium with a metabolic pool. Purines and pyrimidines may be excreted as such in the urine or be metabolized to uric acid (purines) or carbon dioxide and ammonia (pyrimidines).

Uric acid in the body comes from two sources: synthesis from glycine and degradation of purines. In humans, uric acid is excreted in the urine, although in most other mammals it is converted to allantoin before excretion. The normal blood level of uric acid is 4 mg/100 mL. The kidney reabsorbs much of the filtered uric acid, but the body excretes about 1 g of uric acid in 24 h. A standard reference text should be consulted for additional information on DNA, RNA, and molecular genetics.

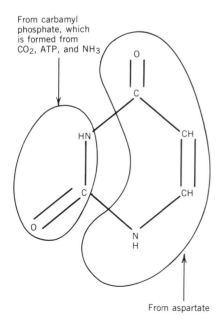

Figure 23. Origins of the atoms in pyrimidine.

WATER, VITAMINS, AND MINERALS

From the stomach to the colon, water passes freely and reversibly between the intestinal lumen and body compartments, although less so in the stomach than elsewhere in the gastrointestinal tract. Water moves in or out of the intestinal lumen to insure osmotic equilibrium on the two sides. In general, the osmolality of the contents of the small intestine resembles that of the plasma. After nutrients in the intestine have been absorbed, the excess water in the lumen is passed out with fecal waste to maintain osmotic equilibrium. Sodium moves freely according to the concentration gradient on the two sides of the mucosal cells. In the colon, sodium moves from body to luman in accordance with the osmotic gradient.

All water-soluble vitamins are absorbed along the small intestine. Except for vitamin B_{12}, a healthy person can absorb these vitamins rapidly. All fat-soluble vitamins require the presence of pancreatic enzyme, bile salts, glycerides, and fatty acids for absorption, as does fat itself. In a healthy person, all essential minerals are absorbed easily by the body, although the extent varies with individual minerals.

BIBLIOGRAPHY

This article has been adapted from Y. H. Hui, *Human Nutrition and Diet Therapy*. Jones and Bartlett, Boston, 1983. Used with permission.

General References

D. A. Bender, *Amino Acid Metabolism*, 2nd ed., John Wiley & Sons, Inc., New York, 1985.

E. Braunwall and co-workers, *Harrison's Principles of Internal Medicine*, 11th ed., McGraw-Hill, Inc., New York, 1987.

M. D. Cashman, "Principles of Digestive Physiology for Clinical Nutrition," *Nutr. Clin. Pract.* 1(5), 241 (1986).

J. J. Cunningham, *Introduction to Nutritional Physiology*, G F. Stickley, Philadelphia, 1983.

G. L. Eastwood. *Core Textbook of Gastroenterology*, Lippincott, Philadelphia, 1984.

S. I. Fox, *Human Physiology*, 2nd ed., W. C. Brown, Dubuque, Iowa, 1987.

E. Goldberger, *A Primer of Water, Electrolyte and Acid-Base Syndromes*, 7th ed., Lea & Febiger, Philadelphia, 1986.

A. Guyton, *Textbook of Medical Physiology*, 3rd ed., Saunders, Philadelphia, 1986.

R. V. Heatley and co-workers, eds., *Clinical Nutrition in Gastroenterology*, Churchill, London, 1986.

Y. H. Hui, *Human Nutrition and Diet Therapy*, Jones and Bartlett Publishers, Boston, 1983.

J. O. Hunter and V. A. Jones, eds., *Food and the Gut*, Bailliere Tindall, London, 1985.

M. C. Linder, *Nutritional Biochemistry in Metabolism*. Elsevier Science Publishing Co., Inc., New York, 1985.

R. K. Murray and co-workers, *Harper's Biochemistry*. Appleton & Lange, Norwalk, Conn., 1988.

R. Pike and M. Brown, *Nutrition: An Integrated Approach*, 3rd ed., John Wiley & Sons, Inc., New York, 1985.

W. K. Stephenson, *Concepts in Biochemistry*, 3rd ed., John Wiley & Sons, Inc., New York, 1988.

L. Stryer, *Biochemistry*, 3rd ed., W. H. Freeman, New York, 1988.

G. V. Vahouny and co-workers, eds., *Dietary Fiber: Basic and Clinical Aspects*, Plenum Press, New York, 1986.

J. West, *Best and Taylor's Physiological Basis of Medical Practice*, 11th ed., Williams & Wilkins, Baltimore, Md., 1985.

P. C. Wilson and H. L. Greene, "The Gastrointestinal Tract: Portal to Nutrient Utilization," in M. E. Shils and V. R. Young, eds., *Modern Nutrition in Health and Disease*, 7th ed., Lea & Febiger, Philadelphia, 1988.

Y. H. HUI
EDITOR-IN-CHIEF

FREEZE DRYING AND FREEZE CONCENTRATION

The advantages of dried foods are described in the article DEHYDRATION. The advantages of concentration include the following:

1. It is used to preconcentrate foods (eg, fruit juice, milk, and coffee) prior to drying, freezing, or sterilization and hence to reduce their weight and volume. This saves energy in subsequent operations and reduces storage, transport, and distribution costs.

2. There is greater convenience for the consumer (eg, fruit drinks for dilution, soups, tomato or garlic pastes) or for the manufacturer (eg, liquid pectin, fruit concentrates for use in ice cream or baked goods).

The heat used to dry foods or concentrate liquids by boiling removes water and therefore preserves the food by the reduction in water activity. However, the heat also causes a loss of sensory and nutritional qualities. In freeze drying and freeze concentration a similar preservative effect is achieved by reduction in water activity without heating the food, and consequently nutritional and sensory qualities are retained more effectively. However, both operations are slower than conventional dehydration or evaporation. Energy costs for refrigeration are high, and, in freeze drying, the production of a partial vacuum is an additional expense. This, together with a relatively high capital investment, results in high production costs for freeze-dried and freeze-concentrated foods.

Freeze drying is the more important operation commercially and is used to dry expensive foods that have delicate aromas or textures (eg, coffee, mushrooms, herbs and spices, fruit juices, meat, seafoods, vegetables, and complete meals for military rations or expeditions). In addition, microbial cultures for use in food processing are freeze dried for long-term storage prior to inoculum generation. Products that are concentrated by freeze concentration include fruit juices, vinegar, and pickle liquors. Freeze concentration is also used to preconcentrate coffee extract prior to freeze drying and to increase the alcohol content of wine.

FREEZE DRYING (LYOPHILIZATION)

The main differences between freeze drying and conventional hot-air drying are shown in Table 1.

Theory

The first stage of freeze drying is to freeze the food in conventional freezing equipment. The type of equipment used depends on the nature of the food. Small pieces of

Table 1. Differences between Freeze Drying and Conventional Hot-Air Drying

Conventional Drying	Freeze Drying
Successful for easily dried foods (vegetables and grains)	Successful for most foods but limited to those that are difficult to dry by other methods
Meat generally unsatisfactory	Successful with cooked and raw meats
Temperature range 37–93°C	Temperatures below freezing point
Atmospheric pressures	Reduced pressures (27–133 Pa)
Evaporation of water from surface of food	Sublimation of water from ice front
Movement of solutes and sometimes case hardening	Minimal solute movement
Stresses in solid foods cause structural damage and shrinkage	Minimal structural changes or shrinkage
Slow, incomplete rehydration	Rapid complete rehydration
Solid or porous dried particles often having a higher density than the original food	Porous dried particles having a lower density than original food
Odor and flavor frequently abnormal	Odor and flavor usually normal
Color frequently darker	Color usually normal
Reduce nutritional value	Nutrients largely retained
Costs generally low	Costs generally high; up to 4 times those of conventional drying

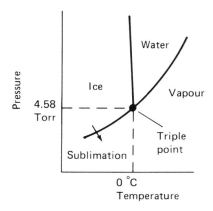

Figure 1. Phase diagram for water showing sublimation of ice.

food are frozen rapidly to produce small ice crystals and to reduce damage to the cell structure of the food. In liquid foods, slow freezing is used to form an ice crystal lattice, which provides channels for the movement of water vapor.

If the water vapor pressure of a food is held below 4.58 torr (610.5 Pa) and the water is frozen, when the food is heated the solid ice sublimes directly to vapour without melting (Fig. 1). Water vapor is continuously removed from food by keeping the pressure in the freeze-dryer cabinet below the vapor pressure at the surface of the ice, removing vapor with a vacuum pump, and condensing it on refrigeration coils. As drying proceeds, a sublimation front moves into the food (Fig. 2). The latent heat of sublimation is either conducted through the food to the sublimation front or produced in the food by microwaves.

Water vapor travels out of the food through channels formed by the sublimed ice and is removed. Foods are dried in two stages; first by sublimation to approximately 15% moisture content (wet-weight basis), and then by

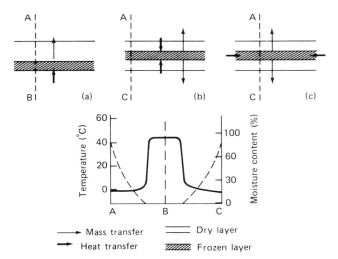

Figure 2. Heat and moisture transfer during freeze drying; (**a**) heat transfer through the frozen layer; (**b**) heat transfer from hot surfaces or radiant heaters through the dry layer; (**c**) heat generated in the ice by microwaves. The graphs show changes in temperature (---) and moisture content (——) along the line $A \to B \to C$ through each sample.

evaporative drying (desorption) of unfrozen water to 2% moisture content (wet-weight basis). Desorption is achieved by raising the temperature in the drier to near-ambient temperature while retaining the low pressure.

In some liquid foods (eg, fruit juices and concentrated coffee extract), the formation of a glassy vitreous state on freezing causes difficulties in vapor transfer. Thus either the liquid is frozen as a foam (vaccum-puff freeze drying), or the juice is dried together with the pulp. Both methods produce channels through the food for the vapor to escape. In a third method, frozen juice is ground to produce granules, which both dry more rapidly and allow better control over the particle size of the dried food.

The rate of drying depends mostly on the resistance of the food to heat transfer and to a lesser extent of the resistances to vapor flow (mass transfer) from the sublimation front.

Rate of Heat Transfer

There are three methods of transferring heat to the sublimation front.

1. Heat transfer through the frozen layer (Fig. 2a). The rate of heat transfer depends on the thickness and thermal conductivity of the ice layer. As drying proceeds, the thickness of the ice is reduced and the rate of heat transfer increases. The surface temperature is limited to avoid melting the ice.

2. Heat transfer through the dried layer (Fig. 2b). The rate of heat transfer to the sublimation front depends on the thickness and area of the food, the thermal conductivity of the dry layer, and the temperature difference between the surface of the food and ice front. At constant cabinet pressure the temperature of the ice front remains constant. The dried layer of food has a very low thermal conductivity, and therefore offers a high resistance to heat flow. As drying proceeds, this layer becomes thicker and the resistance increases. As in other unit operations, a reduction in the size or thickness of food and an increase in the temperature difference increase the rate of heat transfer. However, in freeze drying, the surface temperature is limited to 40–65°C, to avoid denaturation of proteins and other chemical changes that would reduce the quality of the food.

3. Heating by microwaves (Fig. 2c). Heat is generated at the ice front, and the rate of heat transfer is not influenced by the thermal conductivity of ice or dry food or the thickness of the dry layer. However, microwave heating is less easily controlled.

Rate of Mass Transfer

When heat reaches the sublimation front, it raises the temperature and the water vapor pressure of the ice. Vapor then moves through the dried food to a region of low vapor pressure in the drying chamber. As 1 g of ice forms 2 m³ of vapor at 67 Pa, in commercial freeze drying, it is necessary to remove several hundred cubic meters of vapor per second through the pores in the dry food. The factors that control the water vapor pressure gradient are (1) the pressure in the drying chamber; (2) the tempera-

Table 2. Collapse Temperatures for Selected Foods in Freeze Drying[a]

Food	Collapse Temperature, °C
25% coffee extract	−20
22% apple juice	−41.5
16% grape juice	−46

[a] Ref. 1.

ture of the vapour condenser, both of which should be as low as economically possible; and (3) the temperature of ice at the sublimation front, which should be as high as possible, without melting. In practice, the lowest economical chamber pressure is approximately 13 Pa and the lowest condenser temperature is approximately −35°C. Theoretically the temperature of the ice could be raised to just below the freezing point. However, above a certain critical temperature (Table 2) the concentrated solutes in the food are sufficiently mobile to flow under the forces operating within the food structure. When this occurs, there is an instantaneous irreversible collapse of the food structure, which restricts the rate of vapor transfer and effectively ends the drying operation. In practice, there is therefore a maximum ice temperature, a minimum condenser temperature, and a minimum chamber pressure, and these control the rate of mass transfer.

The moisture content falls from the initial high level in the frozen zone to a lower level in the dried layer (Fig. 2), which depends on the water vapor pressure in the cabinet. When heat is transferred through the dry layer, the relationship between the pressure in the cabinet and the pressure at the ice surface is

$$P_i = P_s + \frac{k_d}{b\lambda_s}(\theta_s - \theta_i) \quad (1)$$

where P_i [in pascals (Pa)] is the partial pressure of water at the sublimation front, P_s (Pa) the partial pressure of water at the surface, k_d (W/m·K) the thermal conductivity of the dry layer, b (kg/s·m) the permeability of the dry layer, λ_s (J/kg) the latent heat of sublimation, θ_s (°C) the surface temperature, and θ_i (°C) the temperature at the sublimation front (°C). The factors that control the drying time are related by

$$t_d = \frac{x^2 \rho (M_1 - M_2)\lambda_s}{8 k_d (\theta_s - \theta_i)} \quad (2)$$

where t_d (s) is the drying time, x (m) the thickness of food, ρ (kg/m³) the bulk density of dry food, M_1 the initial moisture content, and M_2 the final moisture content in the dry layer (2).

Sample Problem. In this sample problem food with an initial moisture content of 400% (dry-weight basis) is poured into 1-cm layers in a tray placed in a freeze drier operating at 40 Pa. It is to be dried to 8% moisture (dry-weight basis) at a maximum surface temperature of 55°C. Assuming that the pressure at the ice front remains constant at 78 Pa, calculate the drying time. (*Additional data:* The dried food has a thermal conductivity of 0.03 W/m·K, a density of 470 kg·m³, a permeability of 2.4 × 10^{-8} kg/s·m, and a latent heat of sublimation of 2.95 × 10^3 kJ/kg.)

Solution. From equation 1,

$$78 = 40 + \frac{0.03}{2.4 \times 10^{-8} \times 2.95 \times 10^6}(55 - \theta_1)$$
$$= 40 + 0.42(55 - \theta_1)$$

Therefore

$$\theta_1 = 35.7°C$$

From equation 2,

$$t_d = \frac{(0.005)^2 \, 470 \, (4 - 0.08) \, 2.95 \times 10^6}{8 \times 0.03 \, [55 - (-35.7)]}$$
$$= 6238.5 \text{ s}$$
$$= 1.7 \text{ h}$$

Equipment

Freeze driers consist of a vacuum chamber that contains trays to hold the food during drying and heaters to supply latent heat of sublimation. Refrigeration coils are used to condense the vapor. They are fitted with automatic defrosting devices to keep the maximum area of coils free of ice for vapor condensation. This is necessary because the major part of the energy input is used in refrigeration of the condensers, and the economics of freeze drying is therefore determined by the efficiency of the condenser:

$$\text{Efficiency} = \frac{\text{temperature of sublimation}}{\text{refrigeration temperature in the condenser}}$$

Vacuum pumps remove noncondensible vapors.

Different types of dryer are characterized by the method used to supply heat to the surface of the food. Conduction and radiation types are used commercially. Microwave freeze drying is under development. (Convection heating is not important in the partial vacuum of the freeze drier cabinet.) Both batch and continuous versions are found for each type of dryer:

In general the advantages of batch processing are (1) greater flexibility in being able to change product types or production rates, (2) lower capital costs for equipment, and (3) simpler operation and control. The main disadvantages are (1) higher labor costs, (2) higher operation costs for energy and less efficient use of materials and energy, (3) higher floor space requirement, and (4) lower product uniformity.

Batch processing is used when regular changes in product formulation are required throughout the day or week, small quantities of food are produced, or production is intermittent throughout the year and higher capital costs of continuous equipment cannot be justified.

Conversely, continuous operation has lower flexibility, although developments in automatic control have improved the ease and speed of changeover to different products or production rates. Capital costs are higher than

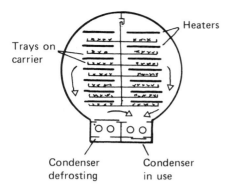

Figure 3. Continuous freeze dryer (3).

batch equipment, but savings in energy, space, and labor permit a rapid payback of the capital cost, provided high production rates can be maintained and plant utilization is therefore improved. Continuous processing is used when there is sufficient demand for a product to permit high production rates for a substantial part of the workday throughout most of the year.

In batch drying, the product is sealed into the drying chamber and the heater temperature is maintained at 100–120°C for initial drying and then gradually reduced over a drying period of 6–8 h. The precise drying conditions are determined for individual foods, but the surface temperature of the food does not exceed 60°C. In continuous freeze drying, trays of food enter and leave the dryer through vacuum locks. A stack of trays, interspersed by heater plates (Fig. 3), is moved on guide rails through heating zones in a long vacuum chamber.

Heater temperatures and product residence times in each zone are preprogrammed for individual foods, and microprocessors are used to monitor and control process time, temperature, and pressure in the chamber and the temperature at the product surface. Details of drying equipment are given in another source (4).

Contact Freeze Dryers

Food is placed onto ribbed trays that rest on heater plates (Fig. 4a). Contact freeze dryers dry more slowly than do other designs because heat is transferred by conduction to one side of the food. There is uneven contact between the frozen food and the heated surface, which reduces the rate of heat transfer. There is also a pressure drop through the food, which results in differences between the drying rates of the top and bottom layers. The vapor velocity is of the order of 3 m/s, and fine particles of product may be carried over in the vapor and lost. However, contact freeze dryers have higher capacity than do other types.

Accelerated Freeze Dryers

In this equipment, food is held between two layers of expanded metal mesh and subjected to a slight pressure on both sides (Fig. 4b). Heat is transferred more rapidly into food by mesh than by solid plates, and vapor escapes more easily from the surface of the food. Both mechanisms cause a reduction in drying times compared with contact methods.

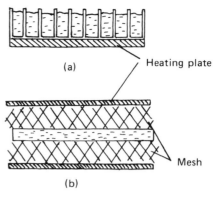

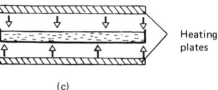

Figure 4. Freeze-drying methods; (a) conduction through ribbed tray; (b) expanded mesh for accelerated freeze drying; (c) radiant heating of flat trays (3).

Radiation Freeze Dryers

Infrared radiation from radiant heaters is used to heat shallow layers of food on flat trays (Fig. 4c). Heating is more uniform than in conduction types because surface irregularities on the food have less effect on the rate of heat transfer. There is no pressure drop through the food, and constant drying conditions are therefore created. Vapor movement is approximately 1 m/s, and there is little risk of product carryover. Close contact between the food and heaters is not necessary and flat trays are used, which are more economical and easier to clean.

Microwave and Dielectric Freeze Dryers

Microwave and dielectric heaters have potential use in freeze drying but are not yet used on a commercial scale. Investigations of microwave freeze drying have been reviewed (5). Microwave drying is difficult to control because water has a higher loss factor than ice and any local melting of the ice causes "runaway" overheating in a chain reaction. However, heating is rapid and does not overheat the surface, which produces minimum heat damage and no surface browning. The heating equipment is small, compact, clean in operation, and suited to automatic control.

A modification of freeze drying is reversible freeze-dried compression. Food is freeze-dried to remove 90% of the moisture and is then compressed into bars using a pressure of 69,000 kPa. The residual moisture keeps the food elastic during compression, and the food is then vacuum-dried. When packaged in inert gas these foods are reported to have a shelf life of 5 yr. They are used in military rations (eg, a 0.3-kg meal consisting of separate bars of pepperoni, stew, granola dessert, and an orange drink). The bars reconstitute rapidly, during which time

Figure 5. Porous structure of freeze-dried food.

the compressed food "groans, rumbles, quivers and eventually assumes its normal shape and size" (6).

Effects on Foods

Freeze-dried foods have a very high retention of sensory and nutritional qualities and a shelf life of longer than 12 months at ambient temperatures when correctly packaged. Volatile aroma compounds are not present in the pure water of ice crystals. They are not therefore entrained in the water vapor produced by sublimation and are trapped in the food matrix. As a result, aroma retention of 80–100% is possible. Theories of volatile retention are discussed in detail elsewhere (7).

The texture of freeze-dried foods is well maintained; there is little shrinking and no case hardening. The open porous structure (Figs. 5, 6) allows rapid and full rehydration, but it is fragile and requires protection from mechanical damage. There are only minor changes to proteins, starches, or other carbohydrates. However, the open porous structure of the food may allow oxygen to enter and cause oxidative deterioration of lipids. Food is therefore packaged in an inert gas. Changes in thiamine and ascorbic acid content during freeze drying are moderate, and there are negligible losses of other vitamins (Table 3). However, losses due to preparation procedures may substantially effect the final nutritional quality of a food.

Commercially, freeze drying is used to produce powders from liquid foods (eg, fruit juices and especially coffee extract) and solid foods, including meat, fruits, and vegetables.

High-quality coffee extracts are preconcentrated by freeze concentration to about 38% (w/w) solids. This retains volatiles better than do other methods and improves the economics of drying by removing more water than in low-concentration extracts. The concentrate is then frozen and the frozen slabs are pregranulated and sieved to produce granules of the required size. The bulk density of the

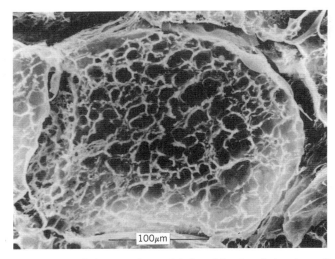

Figure 6. Boiled potato freeze-dried and fractured showing cell filled with starch gel. Porous nature of starch gel is caused by freeze drying, as is the cell wall separation. Courtesy of IFR.

granules can be controlled by foaming the slush-frozen extract with nitrogen gas before final freezing and freeze drying. The color of the product is influenced by the rate and temperature of freezing and the pressure in the freeze-drying chamber.

The heterogeneous nature of meat does not permit drying of the fatty tissue and muscle together. Because fatty tissue contains 10–50% water compared with 70–75% for lean meat, some fatty tissue may melt during the early stages of freeze drying. This causes blocking of some of the porous dry meat by liquid fat, which reduces vapor transfer. Meat should therefore be trimmed of excess fat before being freeze-dried.

Lean meat is prepared for freeze drying by being molded into blocks, frozen, and cut into thin slices 10–15 mm thick. The cut surfaces are at right angles to the direction of the muscle fibers so that the sublimation front moves with the grain during drying and remains parallel to the direction of heat flow. Heat transfer and vapor transfer are therefore assisted. Drying is not so rapid where the heat flows across the grain because the vapor has difficulty in escaping.

Varieties of fruit intended for freeze drying should have high solids content, good color, and full flavor. Apples and pears are peeled and cored, and stone fruits are split and the stone removed and then sliced or diced and

Table 3. Vitamin Losses during Freeze Drying, %[a]

Food	Vitamin C	Vitamin A	Thiamin	Riboflavin	Folic acid	Niacin	Pantothenic Acid
Beans (green)	26–60	0–24	—	0	—	10	—
Peas	8–30	5	0	—	—	0	10
Orange juice	3	3–5	—	—	—	—	—
Beef	—	—	2	0	+[a]	0	13
Pork	—	—	<10	0	—	0	56

[a] Adapted from Ref. 8.
[b] + = apparent increase.

dipped in 0.1% sodium sulfite solution to prevent enzymatic browning prior to freezing. Strawberries, raspberries, and currants are washed and, in some cases (eg, blackcurrants, the skins of which have a low permeability to water vapor) the skins are slit to improve the drying rate. Abrasion of the skins following freezing is used to reduce the thickness of the skin or to wear it away in patches, which also increases the drying rate.

In vegetables of all types there are problems of preparation and storage prior to drying. For example, peas are highly perishable and should be prepared for freezing as soon as possible after harvest, whereas brussels sprouts may be stored for a few hours prior to drying and potatoes are stable for several months.

The main preparative operations for vegetables are preliminary washing to remove adhering earth, peeling, trimming, cutting, final washing of cut material, and blanching. Root vegetables are sliced and diced and leafy vegetables are shredded. Exceptions are brussels sprouts, small carrots, and pickling onions, which are dried whole. Leafy vegetables, spinach, and peas are washed after cutting. Similarly, cut potatoes are washed to remove the starch grains from ruptured cells left by cutting to ensure good final quality.

Cut vegetables are blanched to inactivate enzymes using hot water or steam. Steam blanching gives lower leaching losses than does hot water (~3%, vs. 13%) and it is therefore preferred. Some vegetables, such as onions and leeks, are not blanched before drying since some enzymatic activity is important for the development of full flavour. Air cooling is preferred to water cooling to reduce leaching losses prior to drying.

Freeze drying has generally been recognized as a very expensive preservation method, but still in most cases raw material costs, packaging costs, and distribution costs are the major expenses. In many cases the special advantages that freeze drying gives outweigh the extra production costs incurred. The low weight, high quality, and stability at ambient temperatures make freeze dried foods competitive when the distribution system is not a streamlined one. Camps and expeditions are extreme examples of this type.

In cases where other preservation methods result in large losses of product, freeze drying is also economical. For example, the better quality of stored spices and herbs that are used in small quantities at intervals can make freeze drying the preferred option.

FREEZE CONCENTRATION

Freeze concentration of liquid foods involves the fractional crystallization of water to ice and subsequent removal of the ice. This is achieved by freezing, followed by mechanical separation techniques or washing columns. Freeze concentration comes closest to the ideal of selectively removing water from a food without alteration of other components. In particular, the low temperatures used in the process cause a high retention of volatile aroma compounds. However, the process has high refrigeration costs, high capital costs for equipment required to handle the frozen solids, high operating costs, and low production rates, compared with concentration by boiling. The degree of concentration achieved is higher than in membrane processes, but lower than concentration by boiling. As a result of these limitations, freeze concentration is used only for high-value juices or extracts (9).

Theory

The freezing point of a food is the temperature at which a minute crystal of ice exists in equilibrium with the surrounding water. However, before an ice crystal can form, a nucleus of water molecules must be present. Nucleation therefore preceded ice crystal formation. The production of nuclei depends on the type of food and the rate at which heat is removed. High rates of heat transfer produce large numbers of nuclei. Water molecules migrate to existing nuclei in preference to forming new nuclei. Rapid freezing therefore produces a large number of small ice crystals and vice versa. The rate of ice crystal growth is controlled by the rate of heat transfer for the majority of the freezing period. The rate of mass transfer (of water molecules moving to the growing crystal and of solutes moving away from the crystal) does not control the rate of crystal growth except toward the end of the freezing period when solutes become more concentrated.

In freeze concentration it is desirable for ice crystals to grow as large as is economically possible, to reduce the amount of concentrated liquor entrained with the crystals. This is achieved in a paddle crystallizer by stirring a thick slurry of ice crystals and allowing the large crystals to grow at the expense of smaller ones (10). Details of the effect of solute concentration and supercooling on the rate of nucleation and crystal growth are described by Thijssen (11). Calculations of the degree of solute concentration obtained by a given reduction in the freezing point of a solution are used to produce freezing point curves for different products (Fig. 7).

The efficiency of crystal separation from the concentrated liquor is determined by the degree of clumping of the crystals and amount of liquor entrained. Efficiency of separation is calculated using

$$\eta_{sep} = x_{mix} \frac{x_l - x_i}{x_l - x_j}$$

where η_{sep} (%) is the efficiency of separation, x_{mix} the weight fraction of ice in the frozen mixture before separa-

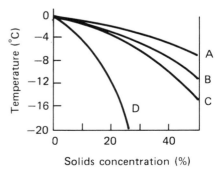

Figure 7. Freezing point curves: curve A, coffee extract; curve B, apple juice; curve C, blackcurrant juice; curve D, wine (12).

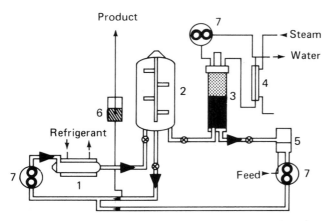

Figure 8. Freeze concentration plant; (1) scraped-surface heat exchanger; (2) mixing vessel; (3) wash column; (4) melting device; (5) storage tank; (6) expansion vessel; (7) pump (12).

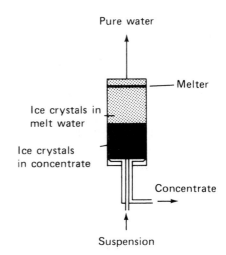

Figure 9. Wash column (11).

tion, x_1 the weight fraction of solids in liquor after freezing, x_i the weight fraction of solids in ice after separation and x_j the weight fraction of juice before freezing.

Separation efficiencies of 50% for centrifuging, 71% for vacuum filtration, 89–95% for filter pressing, and 99.5% for wash columns are reported by Mellor (13).

As the temperature falls, individual solutes reach saturation point and crystallize out. The temperature at which a crystal of an individual solute exists in equilibrium with the unfrozen liquor and ice is its eutectic temperature (eg, for glucose this is $-5°C$; for sucrose, $-14°C$; for sodium chloride, $-21.13°C$; for calcium chloride, $-55°C$). However, it is difficult to identify individual eutectic temperatures in the complex mixture of solutes in foods, and the term "final eutectic temperature" is used. This is the lowest eutectic temperature of the solutes in a food. Maximum ice crystal formation is not possible until this temperature is reached. Commercial foods are not frozen to such low temperatures and unfrozen water is therefore always present.

Equipment

The basic components of a freeze concentration unit (shown in Fig. 8) are (1) a direct freezing system (eg, solid carbon dioxide) or indirect equipment (eg, a scraped surface heat exchanger) to freeze the liquid food, (2) a mixing vessel to allow the ice crystals to grow, and (3) a separator to remove the crystals from the concentrated solution.

Scraped-surface freezers are similar in design to equipment used for evaporation and heat sterilization but are refrigerated with ammonia, brine, or a flurocarbon refrigerant. The rotor scrapes frozen food continuously from the wall of the freezer. Feed liquor is introduced between the rotor and the cold surface. Freezing takes place rapidly as a thin film of liquor is swept through the machine by the rotor blades. The blades keep the film agitated and thus promote high rates of heat transfer.

The mixing vessel typically contains a slowly rotating paddle agitator. Separation is achieved by centrifugation, filtration, filter pressing, or wash columns. Wash columns operate by feeding the ice–concentrate slurry into the bottom of a vertical enclosed cylinder. The majority of the concentrate drains through the crystals and is removed. The ice crystals are melted by a heater at the top of the column, and some of the melt water drains through the bed of ice crystals to remove entrained concentrate (Fig. 9). Detailed descriptions of wash columns are given by Mellor (13).

Concentration takes place in either single-stage or, more commonly, multistage equipment. Multistage concentrators have lower energy consumption and higher production rates. Improvements in techniques for gener-

Table 4. Comparison of Energy Efficiency and Degree of Concentration in Different Methods of Concentration[a]

	Steam Equivalent[b]	Maximum Concentration Possible, %
Ultrafiltration	0.001	28
Reverse osmosis	0.028	30
Freeze concentration	0.090–0.386	40
Evaporation: triple effect without aroma recovery	0.370	80
Triple effect with aroma recovery	0.510	80

[a] Ref. 11.
[b] Cost per kilogram of water removed divided by equivalent cost of steam.

ating large ice crystals and more efficient washing have increased the maximum obtainable concentration to 45% solids (12). The energy consumption and the degree of concentration achieved by freeze concentration in comparison with other methods of concentration are shown in Table 4.

Acknowledgments: Reproduced with permission from Food Processing Technology by P. Fellows published in 1988 by Ellis Horwood Limited, Chichester, Sussex, England. Grateful acknowledgment is made for information supplied by the AFRC Institute of Food Research, Colney Lane, Norwich, UK.

BIBLIOGRAPHY

1. R. J. Bellows and C. J. King, "Freeze Drying of Aqueous Solutions: Maximum Allowable Operating Temperatures," *Cryobiology* **9**, 559 (1972).
2. M. Karel, "Fundamentals of Dehydration Processes," in A. Spicer, ed., *Advances in Preconcentration and Dehydration,* Elsevier Applied Science Publishers, London, 1974, pp. 45–94.
3. J. Rolfgaard, "Freeze Drying: Processing, Costs and Applications," in A. Turner, ed., *Food Technology International Europe,* Sterling, London, 1987, pp. 47–49.
4. J. Lorentzen, "Freeze Drying: The Process, Equipment and Products," in S. Thorne, ed., *Developments in Food Preservation,* Vol. 1. Elsevier Applied Science Publishers, London, 1981, pp. 153–175.
5. U. Rosenberg and W. Bogl, "Microwave Thawing, Drying and Baking in the Food Industry," *Food Technology (USA)* (June) 92–99 (1987).
6. H. G. A. Unger, "Revolution in Freeze Drying," *Food Processing Industry* 20 (April 1982).
7. M. Karel, "Dehydration of Foods," in O. R. Fennema, ed., *Principles of Food Science,* Part 2, *Physical Principles of Food Preservation,* Marcel Dekker, New York, 1975, pp. 359–395.
8. J. M. Flink, "Effect of Processing on Nutritive Value of Food: Freeze-Drying," in M. Rechcigl, ed., *Handbook of the Nutritive Value of Processed Food,* Vol. 1, CRC Press, Boca Raton, Fla., 1982, pp. 45–62.
9. H. A. C. Thijssen, "Freeze Concentration of Liquid Foods, Freeze Concentration of Fruit Juices," *Food Technology* 3(5), (1982).
10. J. G. Muller, "Freeze Concentration of Food Liquids: Theory, Practice and Economics," *Food Technology* **21**, 49–52, 54–56, 58, 60, 61 (1967).
11. H. A. C. Thijssen, "Freeze Concentration," in A. Spicer, ed., *Advances in Preconcentration and Dehydration,* Elsevier Applied Science Publishers, London, 1974.
12. H. G. Kessler, "Energy Aspects of Food Preconcentration," in D. MacCarthy, ed., *Concentration and Drying of Foods,* Elsevier Applied Science Publishers, Barking, Essex, England, 1986, pp. 147–163.
13. J. D. Mellor, *Fundamentals of Freeze Drying,* Academic Press, London, 1978, pp. 257–288.

P. Fellows
Intermediate Technology
Development Group
Rugby, England

FREEZING SYSTEMS FOR THE FOOD INDUSTRY

FUNDAMENTALS

During storage foods are subjected to changes that affect food quality and that sooner or later will lead to severe deterioration and eventually spoilage of the foods. These changes are caused by microorganisms and chemical and physical reactions. Often a combination or interaction of different reactions will cause changes, lowering the quality primarily by changing the sensory properties of the product.

The wide variety of circumstances, including type of food and ingredients, will determine the type of changes that will dominate. In meat and fish products or other foods rich in protein, changes caused by microorganisms will dominate while food rich in fat is more susceptible to chemical and biochemical change. Physical changes occur in all types of food.

The purpose of all food preservation methods is to inhibit or decrease the speed of reaction responsible for the deterioration. All of these reactions are among other factors influenced by temperature. The speed of reaction is decreased at lower temperatures. Cooling and chill storage therefore are perhaps the most important methods of enhancing the storage life of most food products. But even at temperatures near the freezing point some reactions, including growth of many microorganisms, continue at a rate that will limit the preservability to a relatively short period of time.

When the food temperature is $<-10°C$ microbiologic growth will cease. Chemical, biochemical, and physical reactions will still continue at very low temperatures but at a slow pace. Storage life is substantially enhanced as compared to storage at chilled or ambient temperature.

HISTORY OF REFRIGERATION

Refrigeration applied both above and below the freezing point of foods has been used for thousands of years to preserve foods and to increase comfort. Historians estimate that caves were used for food storage about 100,000 yr ago (1).

Temperature inside the caves is naturally low as a result of the evaporation of water, which is often present.

An Egyptian frieze from 3000 BC shows a slave waving a fan in front of a clay pot. The cooling effect from vaporization of water was also utilized in this case. Egyptians also utilized terrestrial radiation toward space at night under a clear sky to produce ice.

Ice was used locally when available. Later it was harvested during wintertime and stored for the summer season. In 1100 BC a Chinese poem mentioned ice houses. Ice and snow were transported over great distances, eg, from the Apennines to Rome, and caravans were transporting ice and snow from Lebanon to the palaces of the Califs of Damascus, Baghdad, and the sultans in Cairo.

In many countries ice was believed to be a gift of the gods to humans. This was confirmed as late as during the mid-nineteenth century, when the American John Goorie in 1844 managed to produce ice with an air compressor but did not dare to publish the invention under his own name. Instead, under a pseudonym he wrote a scientific article describing an ice machine as a future possibility. The New York newspaper, *The Globe*, shortly after published an article titled "Some lunatic in Florida believes his machine can make ice equally good as the All Mighty."

Obviously hunters and gatherers living in a cold climate did use the natural freezing of their food products in order to preserve them over long periods of time.

Exactly when the temperature-decreasing effect—with the addition of certain salts to water—was discovered is not known. There is, however, reason to believe that the method was used in India in the fourth century AD. During the fourteenth and fifteenth century a number of European scientists were working with salt solutions and managed to achieve temperatures as low as -15 to $-20°C$. With those salt mixtures the stage between natural and artificial cooling was passed.

It was not until 1755, however, that the first apparatus for making ice was constructed by William Cullen. Vaporization of water at reduced pressure was utilized.

During the first half of the nineteenth century four events of fundamental importance for the refrigeration industry took place: (1) the systematic work on the liquefaction of gases, (2) the genesis of thermodynamics originated by Nicolas Carnot in 1824 on the invention of the refrigeration machine using compression of a liquefiable gas (often referred to as the Carnot engine), (3) Jacob Perkins's work in mechanical refrigeration in 1834, and (4) the air cycle machine (Goorie, 1844); the latter two inventions remained undeveloped for two decades.

The start of industrial freezing of food is often set at around 1880, even if the first industrial installations were made some 20 years earlier. During 1870–1880 frozen meat was transported from the southern hemisphere to Europe. Initially those endeavors were unsuccessful. In beef cargo from Buenos Aires to Rouen (port on Seine River) in France as much as 25% in weight was lost by sublimation and the quality was unacceptable. The breakthrough came in 1877 when a shipment of frozen meat was brought in from Buenos Aires to Marseilles on the steamer *Paraguay* (2). This shipment was followed by another one from South America to New York in 1879, and in 1880 the steamer *Strathleven* made the journey from Sydney and Melbourne to London.

As compared to the quality achieved today, much was to be wished for. The freezing was carried out very slowly and the storage temperatures were high compared to the storage temperatures used today. In 1915 the German scientist Rudolf Plank showed the importance of rapid freezing in experiments on fish. This knowledge was soon applied for other products.

Although fish, meat, poultry, and berries have long been subjected to preservation by freezing, frozen-state deliveries to consumers did not take place until the development of "quick freezing" in the mid-1920s. The date is often set at the October 14, 1924, the day when Clarence Birdseye received a patent for a plate freezer. This apparatus was revolutionary, as it was now possible to quick freeze packaged foods for the retail outlets. In 1929 the American company Postum/General Foods bought Birdseye's Company, and in the following year the first consumer package of frozen foods were marketed in Springfield, Massachusetts (3).

In 1938 frozen foods were introduced to Europe and the year after, British companies were licensed to produce quick-frozen foods. After World War II the market expanded rapidly throughout all industrialized countries.

FUNDAMENTAL CHANGES DURING FREEZING

The importance of early freezing is related to the need to decrease the rate of the deterioration processes caused by chemical, biochemical, and physical reactions as well as microbiological activity. Chemical and biochemical reactions influence the product quality not during the freezing process, but during subsequent storage. The importance of quick freezing is some times argued from a sensory point of view (4). It must not, however, be forgotten that quick freezing is most important from a technical–economical–operational point of view. With reference to quality, the rate of freezing determines the size of weight loss and in some cases also the microbiologic quality of the product. The drip loss or loss of product juice on thawing is determined by the rate of freezing as well.

The freezing process may be seen as a lowering of the product temperature from its original value to the storage temperature in question. However, from a technical–economical–operational point of view a more strict definition is needed.

Ice Crystallization

The major component of most foods is water, which in the freezing process is transferred from a liquid to a solid state. This transfer obviously result in numerous changes.

Most food products consist of or contain animal and/or vegetable cells forming biological tissues. The water solution of the tissue is contained between the cells—intercellular fluid—and within the cells—intracellular fluid.

When the food product is cooled below 0°C ice begins to form at the initial freezing point. The temperature at which freezing starts depends on the concentration of dissolved substances—salts and other solubles—present. The concentration is higher within than outside the cells. The cell membrane acts as an osmotic barrier and maintains the difference in concentration.

When the product is frozen the first ice crystals are formed outside the cell since the freezing point is higher because of the more diluted fluid here than inside the cell. Once started, the rate of ice crystallization is a function of the speed of heat removal as well as the diffusion of water from within the cell to the intercellular space. If the freezing rate is low, few crystallization centers—nucelli—are formed in the intercellular space. During the freezing process the cell looses water by diffusion through the membrane, and this water crystallizes to ice on the surface on the crystals already formed outside the cell. As in slow freezing, there are few nucleins formed; those existing as crystals grow to a relatively large size.

As the cells loose water, the remaining solution within the cells becomes increasingly concentrated and the cell volume shrinks causing the cell wall to partly or entirely collapse. The large ice crystals formed outside the cell wall occupy a larger volume than does the corresponding amount of water, and therefore they will exert a physical pressure on the cell wall. In some cases this pressure can be sufficient to damage the cell wall and contribute to an increased drip loss on thawing.

By increasing freezing rates a larger number of ice crystallization nuclei are formed, which results in a much smaller size of the final crystals as compared to slow freezing. However, even in the case of high freezing rates most of the crystals are formed outside the cells. Only at extremely high freezing rates not obtainable in commercial freezing of food products, small crystals are formed uniformly throughout the tissue both externally and internally with respect to the cell.

As the food products are cooled down below the initial freezing point, an increasing amount of water is turned into ice and the residual solutions become more concentrated. The relation of water frozen out as ice and the concentration of the remaining solution have an impact on the preservability of a number of food products (5).

The size of the ice crystals has long been regarded as crucial for the quality of the frozen product. It appears, however, from experience as well as a number of investigations that the differences in ice crystal size and distribution have little effect on the sensory properties of the food product when presented to the consumer, provided up-to-date equipment and good commercial practice have been used (6).

The freezing rate is not negligible, however. On the contrary, in good commercial practice the freezing time must be determined for each product in order to safeguard against microbiologic growth, which is most important from a safety point of view and low controlled weight losses, which is important from an economic point of view.

The practical result of different freezing rates—size and locations of ice crystals—can be seen as a difference in drip loss of water or "product juice" when a product is thawed. Loss of juice results in a more or less pronounced loss of texture, flavors, and—in most cases—nutrients. For this reason the drip loss is often used as an indication of the quality loss during freezing and subsequent storage. The relation between speed of freezing and drip loss for two different foods is illustrated in Figure 1.

For strawberries it is rather obvious that the consumer will not readily accept the product with a 20% drip loss if a product with only 8–10% drip loss is available. The latter level of drip loss can be achieved in modern freezers usually employing the fluidization technique.

When comparing slices of beef even very slow freezing rates gave a small drip loss hardly noticeable to the consumer. The small improvement achieved by quick freezing is normally not observed.

Microbiology

With reference to temperature requirements, bacterias are divided into four basic groups according to type of growth: thermophilic, mesophilic, psychrophilic, and psychrotrophic. Of those the two latter groups are of special interest in food spoilage at low temperatures. Psychro-

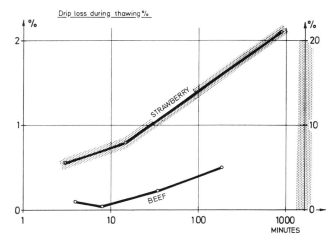

Figure 1. Drip loss during thawing of sliced beef and strawberries, frozen at various rates, to an equalization temperature of $-18°C$ (ref. 6).

philics are often an important cause of spoilage of protein-rich foods such as meat, fish, poultry; psychrophilic bacteria grow well at temperatures above 0°C.

The optimum temperature for growth, however, is much higher. Psychrotrophics are also able to grow close to the freezing point but the optimum temperature is higher than that for psychrophilics. Lowering the temperature will slow down the growth rate of all bacteria, and at temperatures used in commercial storage of frozen foods all microbiologic growth ceases completely. The time to decrease the temperature to below the freezing point is critical. A product temperature of $-10°C$ is normally considered safe with regard to microbiologic growth.

During freezing and frozen storage some bacteria are impaired and even destroyed. Under certain circumstances the death of bacteria will cause a considerable decrease in the total number of viable cells in the frozen products. Since some species are more susceptible to freezing injuries than others there may also be a change in the relation between various species (7).

From a microbiologic point of view the total flow from the production to the consumption must be regarded. During this flow, food products are subjected to various temperatures and other growth-affecting factors. Large variations occur from product to product; meat and meat products have been chosen as examples in the following discussion. The general concept is obviously valid for most food products.

During slaughter and subsequent handling the surface of the meat is infected by microorganisms originating from the animal itself and from the environment.

The number of organisms present is dependent on the hygienic conditions, but the flora consists to a great extent of spoilage organisms that thrive on the meat surface and multiply.

Rapid cooling reduces the rate of growth substantially, but many of the psychrophilic and psychrotropic organisms will grow even at chill temperatures. These organisms depend on free oxygen for their metabolism, and since oxygen is available in the surface layer only, no growth will occur in the interior of the meat.

Freezing of carcass meat therefore normally will not cause any serious problems from a microbiologic point of view. In commercial freezing the freezing rate is fast enough to stop the growth at the surface. As the microorganisms cannot develop under the surface layer, the freezing rate is of less importance microbiologically.

Processing, like cutting and mincing, increases the microbiologic contamination as the surface/volume ratio increases. The freezing rate becomes more critical. A common pack in the industry today is the 30-kg carton normally frozen in a traditional air-blast tunnel. It is then essential that a good air circulation be provided in order not to prolong the freezing time.

Freezing times for cartons subjected to both adequate and inadequate airflow are compared in Figure 2.

Inadequate airflow has been achieved by placing the meat cartons directly on pallets with only a small wooden spacer (30 mm) between each layer instead of being placed on freezing racks or with rigid layer separators with a minimum height of 50 nm.

If spacers used as in the improper airflow system do not cover the total carton area, some cartons may collapse, which will prevent airflow through the different layers.

The freezing time is obviously prolonged. There is also a risk that cartons are removed from the air-blast tunnel before the freezing is completed if blocking of the air channels is not visible from the outside. In the latter case there may be substantial growth of microorganisms during subsequent storage, ie, until a sufficiently low temperature has been reached.

In most cases a freezing time of 24–36 h down to $-10°C$ in the center of wholesale cut meat will not cause any microbiologic problems. If the degree of cutting is increased to smaller cuts, such a long freezing time may become a major problem. Those products should preferably be frozen integrated in the processing line before packaging or in very small packages that allow for a much faster freezing.

As most prepared foods involve a high degree of processing as well as mixing of different ingredients the freezing becomes very important. The general pattern of the growth of microorganisms in the production of prepared foods is illustrated in Figure 3.

During storage of raw material as well as during handling and preparation, microorganism growth will take

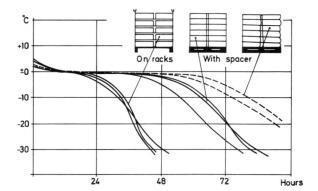

Figure 2. Recorded temperature decrease during the freezing of cartons in an air-blast tunnel (air temperature $-38°C$, front air velocity 1.5 m/s, size of cartons $160 \times 400 \times 600$ mm) (ref. 7).

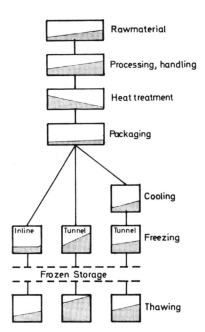

Figure 3. Growth pattern of bacteria during different steps in the processing of precooked frozen foods (ref. 7).

place. If the preparation is followed by heat treatment, the total number of microorganisms will be reduced.

At this point the product could be handled in two different ways: either packaged and frozen in batch-operated equipment or frozen in-line and then packed. If the products are placed on racks and then transported to a freezing tunnel for freezing, there may be a time lapse, resulting in a marked growth of microorganisms. If the product has been heat treated, it will pass through the temperature zone of optimum microbiologic activity.

A chilling operation immediately after the heat treatment—which means that the products are cooled down to below $+10°C$, then packaged and frozen—will reduce the growth of bacteria. However also in this case microbiologic growth can be recorded. If the products are instead frozen in-line immediately after the heat treatment and then packaged, there will be almost no increase in the number of bacteria present.

In-line freezing is, of course, even more essential when prepared foods are processed without heat treatment prior to the freezing process.

A low microbiologic load in the frozen state will directly influence the preservability of the food product after thawing. All measures taken to arrest growth of microorganisms prior to freezing are therefore beneficial.

Compared to batch freezing in a tunnel, the modern inline freezer provides much faster freezing, and even more important, the in-line process itself minimizes the delays in product flow from preparation to freezing and through the temperature zone, which is critical from a microbiologic point of view.

Desiccation

During the freezing process it is unavoidable that a certain evaporation of water from the surface takes place, resulting in both a quality loss and a weight loss. Only if the product is tightly enclosed in a water vapor inpermeable packaging material, evaporation can be completely avoided. If there are small spaces between the product and the packaging material ice is deposited in these.

Freezers that are poorly designed and improperly used may cause a weight loss in the order of 5–7%, while properly designed and used equipment will cause no more than 0.5–1.5% weight loss. As the total freezing cost often is no greater than 3–5% of the product value, it is obvious that dehydration losses are of great importance in any comparison of different freezing methods.

There is a definite correlation between the degree of dehydration loss and the rate of freezing. In Figure 4, the temperature gradient in a product with a thickness of 2 b and in the air surrounding the product is plotted against the partial pressure of water vapor in moist air.

The size of the dehydration is influenced by a number of factors related to the biological materials as well as to physical handling, temperature of the heat-transfer medium, dimensions of the product, etc. The rate of evaporation from the product is determined by the conditions at the surface.

In a simplified consideration the evaporation rate may be regarded as proportional to the vapor pressure Δp_2. The curve corresponds to the "relative" humidity of the product surface. This concept is introduced in order to represent the diffusion resistance that may exists in cell walls, etc.

During infinitely slow freezing the surface temperature approaches the average temperature t_3. Then the evaporation rate increases to the value Δp_3. Hence it is clear that $t_3 - t_2$ should be as great as possible in order to obtain low dehydration losses. Three principal factors contribute to this:

- b large. In thick products the surface temperature will be low during most of the freezing process.
- t_1 low. The lower the temperature of the ambient air, the more curved the temperature gradient in the product will be.
- High heat-transfer rate. A low air temperature has no appreciable effect if the heat-transfer rate is too low. Therefore, it is extremely important that the heat-transfer conditions be favorable. This is more important for thinner products.

Wet products generate water vapor at a rate proportional to the difference between the vapor pressure at the surface of the product and that of the surrounding air. In a product that has a more or less dry surface there is a resistance in the cell walls against diffusion of vapor from the interior of the product to the surface and the air. This results in a reduction of vapor pressure at the surface of the product.

The water vapor pressure decreases rapidly when the temperature is reduced, which means that the dehydration will be less the colder the heat-transfer medium is. The importance of a low temperature during freezing is well illustrated in Figure 5, where the accumulated weight loss during different freezing tests with a temperature varying from −13 to −35°C is plotted against the core temperature of the product. The diagram illustrates both the importance of the latent heat zone and the drastic influence of the air temperature.

Glazing for Protection of Product Quality. In order to improve shelf life by preventing desiccation and oxidative changes, it has become standard practice to glaze certain individually frozen products, eg, shrimp, after freezing. The product quality is greatly improved as the thin ice layer prevents the product from the changes mentioned above. The glazing is carried out by spraying the frozen

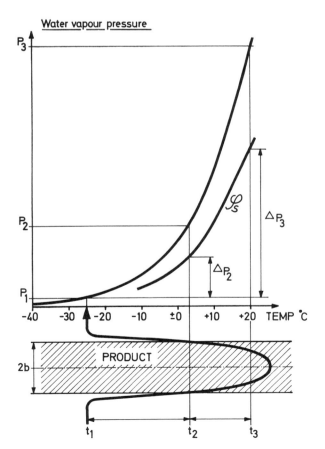

Figure 4. Dehydration mechanism during freezing (ref. 8).

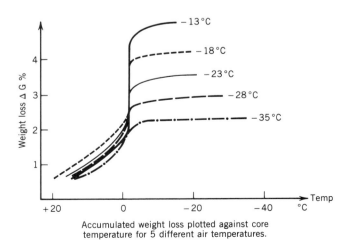

Accumulated weight loss plotted against core temperature for 5 different air temperatures.

Figure 5. Accumulated dehydration during freezing at different air temperatures (ref. 9).

product with cold water, which immediately freezes on the surface. Even if the product leaves the freezer at a low outfeed temperature, the product temperature may increase considerably after the glazing operation. This high-end temperature may cause clumping of the product when packed, and the slow decrease in temperature during subsequent storage and distribution may lead to noticeable texture changes on thawing.

To avoid these problems special equipment has been developed—GLAZoFREEZE—that is designed to lower the product temperature of the glazed product immediately after the glazing operation. The equipment utilizes the fluidization technique. In GLAZoFREEZE the fluidization must be specially gentle to the product in order to avoid damage of the glazed surface.

DEFINITIONS

Definitions of the freezing process, freezing time, freezing rate, and speed of freezing are most useful in comparisons of different systems and equipment. The International Institute of Refrigeration gives the following definitions (10).

Freezing Process. During the freezing process different parts of a product will pass through the various stages at different times. If a particular location in the product is considered, three stages of temperature changes can be defined.

Prefreezing Stage. This is the interval between the time at which a high-temperature product is subjected to a freezing process and the time at which the water starts to crystallize.

Freezing Stage. The period during which the temperature at the considered location is almost constant because the heat that being extracted is causing the majority of water to change phase into ice.

Reduction to Storage Temperature. This is the period during which the temperature is reduced from the temperature at which most of the freezable water has been converted to ice to the intended final temperature. The final temperature can result when the storage temperature is reached in any part of the product, including the thermal center or the equalization temperature. The equalization temperature is the temperature achieved under adiabatic conditions—without heat exchange with the environment.

Freezing Time. The freezing time is defined as the time elapsed from the start of the prefreezing stage until the final temperature is reached. The freezing time depends not only on the initial and final temperature of the product and the quantity of heat removed but also on the dimensions—especially the thickness—and shape of the product as well as on the heat-transfer process and its temperature.

Freezing Rate. For a product or a package the freezing rate (degrees Celsius per hour) is the difference between the initial and the final temperatures divided by the freezing time. In a given point of a product the local freezing rate is equal to the difference between the initial temperature and the desired temperature divided by the time elapsed until the moment at which the latter temperature is achieved in this particular point.

Speed of Ice Front Movement. The freezing rate may be evaluated by the speed of movement of the ice (in centimeters per hour) through a product. This speed is faster near the surface and slower toward the center. As a result, reported freezing rates from different sources are not necessarily comparable.

Practical Freezing Time. Besides the above-mentioned definitions the industry has developed a common definition of freezing time for practical commercial purposes.

From a practical point of view the freezing time is defined as the time required to lower the temperature of a product to an equalization temperature of $-18°C$ under adiabatic conditions. This definition determines the commercial capacity of the freezing equipment.

The time the product is held in the freezer is known as standard freezing time (SFT) or holding time. In Figure 6 ET stands for equalization time. In determining capacity of equipment there will be an equalization to $-18°C$. However, in practical life this temperature equalization takes place during subsequent handling and storage and the equalization temperature will be that of the surrounding environment. From a theoretical point of view the total freezing time (TFT) for a product is SFT + ET.

Important is that contrary to what is commonly believed the product temperature is not brought down to $-18°C$ in the center of the product in the freezer. This temperature or preferably a lower temperature is achieved during subsequent handling, packaging, and storage.

Freezer Capacity. The investment in any freezing equipment is based on requirements to freeze a certain quantity of food per hour. As a principle the following relation is valid for any type of freezer:

$$C = Q/F = Vq/F$$

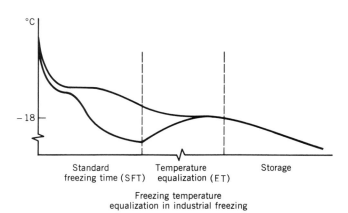

Figure 6. Practical commercial definition of the freezing process.

where C = the capacity of the freezer expressed as tons per hour
 Q = the quantity, in tons, of the product that can be accommodated in the freezer
 F = the holding time, in hours, of the product *in the particular freezer*
 V = the volume, in cubic meters, of the product that can be accommodated in the freezer
 q = the volume weight, in tons per cubic meter, of the product

The holding time refers to the temperature change that is required, ie, usually from incoming temperature to $-18°C$ equalization temperature.

It is important to understand the fundamental importance of the holding time for the capacity of a freezer. Every product has different holding times. The amount of heat to be extracted per kilogram of product usually varies only little within a group of products, eg, vegetables, but the dimensions of each product particle have a drastic influence (see Fig. 7).

As all heat must be extracted from the product through its surface, the relation between surface and weight, the specific surface in square meters per kilogram, is of great interest. Volume and weight are entirely proportional to each other and therefore can substitute each other in this discussion. In Figure 7 can be seen that the freezing time is inversely proportional to the specific surface for particles.

As a practical example a simple rack tunnel with a capacity of one ton per hour of spinach purée in 45-mm-thick packages can be taken. It may have the following capacities:

Spinach, packaged, 45-mm layer	1 ton/h
Peas, unpackaged, 30 mm layer	1.5 ton/h
Peas, packaged, 45 mm layer	0.6 ton/s
Parsley, packaged, 45 mm layer	0.4 ton/h
TV-dinner, packaged, 35 mm layer	0.2 ton/h

In the case of parsley and TV dinners the capacity is reduced primarily because of the low volume weight and secondarily because of the poor heat transfer inside the packages.

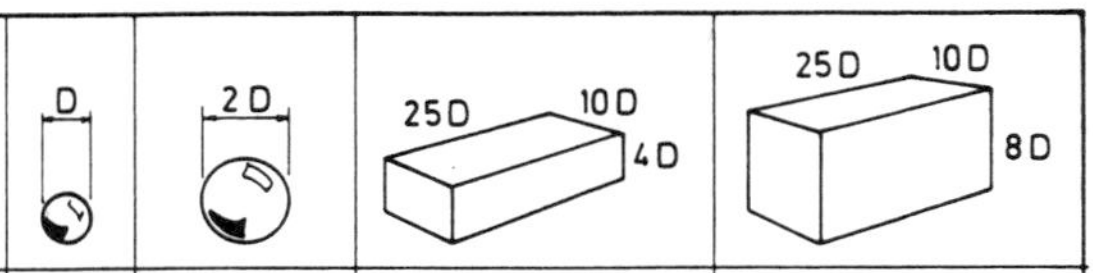

	D	$2D$	$25D \times 10D \times 4D$	$25D \times 10D \times 8D$
$V =$	$\dfrac{\pi \cdot D^3}{6}$	$8 \cdot \dfrac{\pi \cdot D^3}{6}$	$1000 \cdot 0{,}56\,D^3$	$2000 \cdot 0{,}56\,D^3$
$F =$	πD^2	$4\pi D^2$	$780\,D^2$	$1060\,D^2$
$\dfrac{F}{V} =$	$6 \cdot \dfrac{1}{D}$	$3 \cdot \dfrac{1}{D}$	$1{,}40 \cdot \dfrac{1}{D}$	$0{,}94 \cdot \dfrac{1}{D}$
$I =$	1	2	4,3	6,4
$T =$	6 min	12 min	2,5 h	5,5 h

Figure 7. Heat to be extracted during freezing of different package sizes (where V = volume, F = surface, I = inverted specific surface, T = practical holding time, F/V = specific surface).

Design Capacity—Working Capacity. For an in-line operation with short freezing time, (eg, 6–10 min), it is important to distinguish between design capacity and working capacity.

Design capacity refers to temperature reduction in volume per hour, according to specifications, if the product feed is entirely steady and continuous. In reality the load will be fluctuating. Because the temperature is influenced by fluctuations as short as 3–5 min, the product feed will have to be cut back until the peaks do not exceed 100% of the design capacity. This means that the average product flow will be only 80–90% of the design capacity.

Other reductions are required since every minute of every working hour cannot be utilized. When the operation is started in the morning it takes time to build up the product flow to 100%. If there are any production breaks for lunch or shift changes, the product feed will have to be stopped some time in advance.

Finally there may be stoppages in the product flow because of breakdowns anywhere along the processing line. In total, the working capacity may be as low as 70% of design capacity with respect to one and the same product. To obtain maximum production out of a line all links in the chain must meet required capacity.

Frequently a freezer may be used for several different products. For such products the above working capacity will have to be multiplied with the capacity factor, which relates the capacity of that particular product to the capacity of the product stated in the design specification. For products with very low capacity factor, the operating conditions usually differ so much that a separate working capacity must be established.

Precooling. Precooling is defined as the cooling of the product before it enters the freezer. The process of precooling has a positive effect on the energy required to freeze a product and generally reduces the frost load on the coils as well. Precooling can be done by blowing ambient air over the product, blowing ambient air that has been evaporatively cooled by water sprays, immersion in cold water (refrigerated or nonrefrigerated), or blowing refrigerated air over the product.

Sanitation during precooling processes is an important consideration because the conveyor system could be at a temperature for all or part of the cycle that can allow bacterial buildup.

Precooling reduces the load on the low-temperature refrigeration system, which reduces the power required by the low-temperature system. If a refrigerated precooler is used, the precooling heat is removed by a refrigeration system that consumes less power per unit of heat removed from the product, and a net power savings is achieved.

FREEZING EQUIPMENT

Today's freezing equipment can be divided into two main groups: integrated in the processing line and operating in batches.

According to the heat-transfer method, there are basically three main types of equipment:

- Air-blast freezers, which use air for heat transfer. Because air is the most common freezing media, this method of heat transfer has probably the largest range of designs.
- Contact freezers. Heat transfers occurs through conduction. A refrigerated surface is placed in direct contact with the product or package to carry away the heat. Alternatively, the product is immersed in a cold liquid—brine.
- Cryogenic freezers. These freezers use liquefiable gases, nitrogen or carbon dioxide to produce vapors that precool and freeze the products.

Combinations of these heat-transfer methods can be seen in special designs.

The freezing equipment can also be divided into two main groups with reference to the product: Individually quick-frozen (IQF) and packaged products.

All of these methods are used in the food processing industry; however, the preferred systems are those that can be operated in-line, integrated with the processing and packaging operations.

The freezing equipment must be designed to accommodate the three stages of the freezing process and should ideally optimize the total process.

The following design criteria are of special interest (see below under "Major Considerations in Freezer Design"):

Product quality
Hygiene
Minimum product losses
Reliable and simple operation
Simple maintenance
Economy, freezing cost

In the following sections the freezing equipment will be discussed with the basis on the method of heat transfer.

Air-Blast Freezers

Air is the most common freezing medium, and for that reason a number of designs can be found. Even if a storage room should never be considered as a piece of freezing equipment, it is sometimes used for this purpose. However, freezing in a storage room involves so many disadvantages that it should be used only in exceptional cases. The freezing is so slow that the quality of almost all products will be affected adversely.

Sharp Freezer, Blast Room

Basically, a sharp freezer or blast room is a cold storage room that has been especially constructed and equipped to operate at low temperatures for freezing. Even if this room is equipped with extra refrigeration capacity as well as fans for air circulation, there is normally no controlled airflow over the products, and for that reason freezing is normally slow. This type of equipment is, however, still used sometimes for bulk products such as beef quarters, but not for processed food products.

Tunnel Freezers

In tunnel freezers refrigerated air is circulated over the product, which is placed on trays or special spacers that stand in or pass through the tunnel in racks or trolleys.

The racks or spacers are arranged to provide an air space between each layer of trays. The racks or trolleys can be moved in and out of the freezer manually or by a fork-lift truck pushed through the tunnel by a pushing mechanism or slide-through. Tunnels are also used for freezing hanging meat carcasses carried on a suspension conveyor or in especially designed racks.

Practically all products can be frozen in a tunnel freezer. Whole, diced, and sliced vegetables may be frozen in cartons or unpacked in a 30–40-mm-deep layer on trays. Spinach, broccoli, meat patties, fish fillets, and prepared foods are frozen in packages in this type of equipment. It is, however, important to recognize that both refrigeration capacity and arrangements for air circulation have been designed for a specific product range. This means that if the tunnel freezer is designed for freezing meat carcasses or bulk meat cartons, the tunnel design is not appropriate for handling unpacked products such as vegetables. The result is often a increase in weight loss that influences both the sensory properties of the product and the economics of the freezing operation.

The flexibility of this type of equipment is balanced by high manpower requirements and a considerable weight loss if improperly used. The high manpower is caused by the handling, releasing, cleaning, and transportation of the trays and of the racks.

The tunnel should always be filled with product in such way that uniform airflow is obtained over all products to be frozen (see Fig. 8).

Mechanized Freezing Tunnels

A certain degree of mechanization is achieved when the racks are fitted with casters or wheels. The racks or trolleys are usually moved on rails by a pushing mechanism often powered hydraulically. Mechanized tunnel freezers are known as push-through tunnels, carrier freezers, and sliding-tray freezers. A typical design is shown in Figure 9.

As in any other tunnel freezer, the products are placed on trays that, in turn, are stacked on trolleys. The tunnel contains one or two rows of trolleys that are pushed forward stepwise by each other on rails in line with the product line. When a trolley leaves the freezer, the frozen products are removed from the trays and the trolleys are returned to the loading station.

The design and construction of the freezer is common to most tunnel freezers. The evaporation coils are placed on a steel frame standing on the insulated floor. They have a thin spacing, which varies with the depth of the coils. The spacing is wide at the air inlet and narrow at the outlet. This ensures an even frost buildup on the coils without detriment to the airflow. The coils are furnished with liquid and suction headers and arranged for pump circulation of the refrigerant. In some installations arrangements are also made for gravity feed. Standard defrosting is effected by hot gas or by leaving the doors open and running the fans during the night.

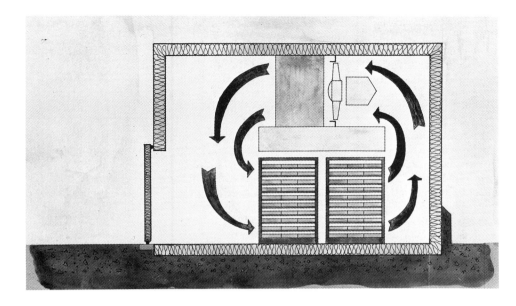

Figure 8. Stationary freezing tunnel.

The fans circulate the air down between the coil and the wall through the coil, through the trolleys and the products, and then the air is deflected up along the wall and back to the fans above the sheeting.

Tunnel freezers are built for capacities varying from a few hundred kilograms to several tons per hour. The freezing time in this type of equipment varies considerably from a few hours when freezing unpackaged vegetables in a thin layer to 48 h freezing meat carcasses.

The sliding-tray freezer is basically a tunnel consisting of one huge rack accommodating many large trays on each tier. At one end of the construction there is an elevating mechanism that lifts the arriving tray to the top tier, where it is pushed in, forcing all the other trays on this tier to advance one step. The tray at the far end is pushed onto an elevator that brings it down one tier where the tray is entered. Consequently, on every odd tier the trays will be advancing, and on every even tier they will be returning. For each tray that enters, all trays will advance one step.

All mechanisms are usually hydraulically powered. Outside the freezer enclosure automatic loading and unloading of the trays may be arranged.

The freezing trays are in the traveling-tray freezer connected to one or two sturdy roller chains at each end. These are arranged to move the trays forward to a set of sprockets elevating the trays one tier while maintaining their horizontal position. This type is constructed as large as those mentioned previously, but requires more space because the minimum pitch between the tiers is decided by the sprockets, which are 200–300 mm in diameter.

The carrier freezer may be regarded as two push-through tunnels, one on top of the other. At the stop section a row of carriers is pushed forward, while it is returned in the lower section. At both ends there are elevating mechanisms. A carrier is similar to a bookcase with shelves. When it is indexed up at the loading end of the freezer, the products on one tier at a time are pushed off the shelf onto a discharge conveyor.

When the carrier is indexed up the next time, this shelf is level with the loading belt from which new products are transferred to the carrier (see Fig. 10).

The carriers may be designed for almost any pitch between the tiers and for any length and width, giving maxi-

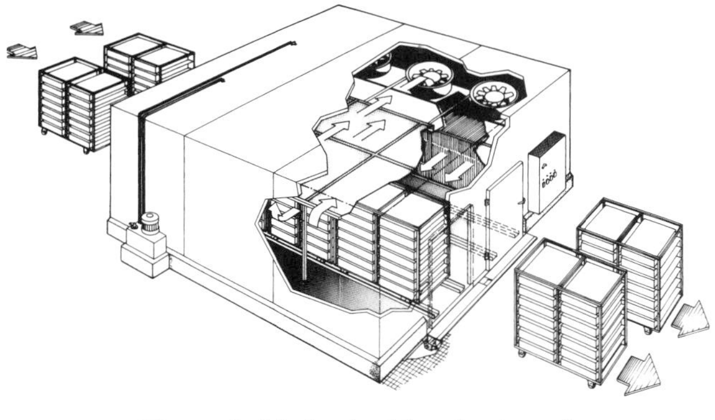

Figure 9. Mechanized freezing tunnel.

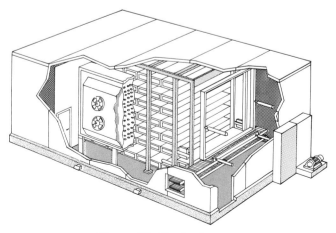

Figure 10. Carrier freezer.

mum compactness. The loading and unloading may be manual or fully automatic.

Another automatic freezing tunnel is the reciprocating spiral freezer, which consists of two parallel sets of rails, one of which is fixed. In between there is a set of movable rails. Initially the products rest on the fixed set. The movable set lifts the products clear of the fixed set, advances one stroke, descends to leave the products resting on the fixed set again, and then returns to the initial position.

Both sets are arranged to form a large spiral, the fixed set supported from an external steel structure and the movable set fitted to a central cylinder. This provides a reciprocating motion around the vertical axis as well as up and down.

The infeed can be arranged very easily for a range of carton sizes provided the utilization of the conveyor area is limited. Small items must be placed on trays that are loaded, unloaded, and transferred from outfeed to infeed separately.

As the total product load is accelerated and decelerated for each stroke, a relatively slow motion is necessitated, making this type of freezer suitable mainly for intermediate and large package sizes.

All freezer designs described above are intended primarily for packaged products. Attempts to freeze fish fillets, meat patties, etc individually on trays invariably have been only moderate successful because of a number of problems, primarily the following:

1. Products stick to the trays. This causes damage and weight losses if products are removed mechanically. An alternative is to heat the trays to release the products. This requires complicated equipment and causes reduced capacity.
2. Hygiene. The trays must be washed after removal of the frozen products if acceptable hygiene is to be obtained.
3. The handling of trays from outfeed to infeed is costly whether it is manual or automatic.

The principal advantages of automatic air-blast freezers in comparison with automatic plate freezers, which are usually the alternative, are the following:

4. Products of much varied thickness may be frozen simultaneously or immediately after another.
5. Products do not need to be square in shape.
6. Higher capacities per unit are possible.

The automatic or mechanised tunnel freezer generally has the same advantages and disadvantages as the classical tunnel except that it is slightly better suited as an in-line freezer.

Labor costs can be reduced and the flexibility is somewhat better as different products can be handled at the same time by different tracks having different dwell times.

Belt Freezers

The first belt freezers basically consisted of a wire mesh belt conveyor in a blast room, which satisfied the need for a continuous product flow. In addition to the disadvantage of poor heat transfer in a blast room, many mechanical problems arose.

Modern belt freezers normally utilize vertical airflow, forcing the air through the product layer, which creates good contact with all product particles. A condition is, however, that the product be evenly distributed over the entire belt area. Where the product layer is thin or nonexisting, there is less resistance to the air, which will concentrate to these areas and bypass the thicker product layer. This phenomenon, called channeling, may result in poorly frozen products and thus must be avoided by careful and even spreading of the product across total belt width under all operating conditions. Single-belt freezers designed for freezing unpacked products can be designed to achieve a fluidized freezing.

In order to decrease the necessary floor space of a single-belt freezer the belts can be stacked above each other as in a multitier belt freezer or a spiral belt freezer. The latter being the most important modern belt freezer equipment. A typical multitier freezer is shown in Figure 11.

A multitier freezer can be used where the factory layout requires a straight-through product flow and where available space is narrow. A freezer consists of three conveyor systems positioned one above the other with fans and coils positioned above the top belt supported by the same steel structure that carries the conveyor system.

Products are fed onto the top belt and transported through the cooling zone into the freezing zone and to the opposite end of the freezer. Here the products are transferred from the top belt to the second via a stainless-steel transfer chute. On a second belt they are conveyed back through the freezing zone to the infeed, where they are transferred to the third belt. The third belt takes the products through the freezing zone and delivers the frozen products to the outfeed, from which the products leave the freezer. This arrangement has the advantage that the product, after being surface frozen on the first belt, may be stacked in a rather deep bed on the lower belts. Thus the total belt area required can be reduced.

By vertically stacking the belt in tiers a minimum of floor space is occupied by the freezer. This is the case in the modern spiral belt freezer (Fig. 12), which maximizes the belt surface area in a given floor space. This is achieved by using a belt that can be bend laterally around a rotating drum. The belt in supported by rails and driven by the friction against the drum. The most advanced and refined spiral freezers operate with a low tension drive system.

The continuous-belt design eliminates product transfer points where product damage can occur within the freezing system. The products are placed on the belt outside the freezer where they can be monitored and will remain in the same spot until leaving the freezer. The flexibility of the belt used allows for more than one infeed and outfeed with one and the same belt, and infeed and outfeed may be arranged in any direction desired to suit the layout of the processing line.

In the most modern version of spiral belt freezers a self-stacking belt—FRIGoBELT—where each tier rests directly on the vertical side links of the tier beneath is used. This construction eliminates the need for rails and runners and allows more tiers of belt to be installed in a given

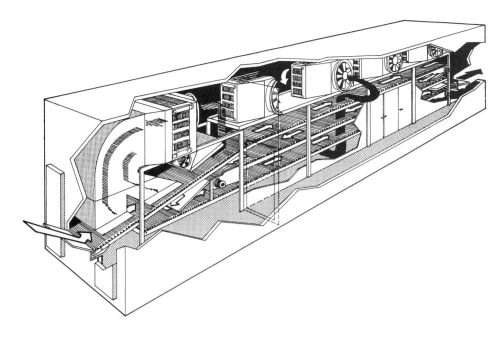

Figure 11. Multitier freezer (TRIo-FREEZE).

space. The whole stack turns as a unit. Products cannot roll and blow out of the closed freezing zone and cannot become stuck within freezer because there are no stationary structural parts to snag them. Equally important is the improvements of the hygienic conditions of the enclosed product zone. The belt in contact with the product is regularly cleaned and dried in an external washing unit outside the freezer.

The side links of the belt will also serve to channel the vertical airflow in the freezing zone. The air is blown down over the upward-moving products in a countercurrent flow, which is a very efficient form of heat transfer.

The spiral belt freezer provides great flexibility with regard to the product range to be handled. Both unpackaged and packaged products are frozen and typical products are meat patties, fish cakes, fish fillets, and bakery products, which all may be frozen raw or prepared.

Fluidized-Bed Freezers

Previously freezing of vegetables took place in a plate freezer or tunnel freezer, and and the result was more or less a block frozen product that was hard to thaw and rather inconvenient in handling. The use of "cluster busters" in order to obtain a more free-flowing product caused considerable mechanical damage. Belt freezers were introduced soon after World War II, but in order to meet the high freezing demands those freezers became rather huge. In the early 1960s fluidized freezing was introduced after years of experiments and tests, and it was possible for the first time to quick-freeze vegetables individually very fast in a commercial application.

Fluidization occurs when particles of fairly uniform shape and size are subjected to an upward airstream. At a certain air velocity the particles will float in the airstream, each one separated from the other but surrounded by air and free to move. In this state the mass of particles can be compared to a fluid.

If a mass is held in a container that is fed in one end and the other end is lower, the mass will move toward the lower end as long as more products are added. By utilizing low-temperature air to achieve the fluidization the products are frozen and simultaneously conveyed by the same air without the aid of a conveyor (see Fig. 13).

The use of the fluidization principle gives a number of advantages in comparison with the use of a belt freezer. The product is always individually quick-frozen (IQF); this applies also to products with a tendency to stick together, eg, French-style green beans, sliced carrots, and sliced cucumber.

The freezer is totally independent of fluctuations in load. If the freezer is partly loaded, the air distribution can be the same as for the full load, ie, with no hazard of channeling. The variability of freezing with products is greatly improved because a deep fluidized bed can accept

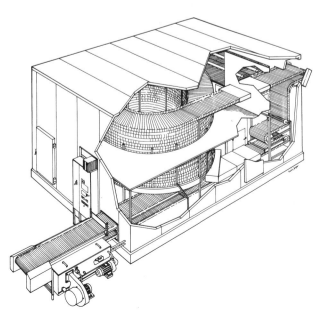

Figure 12. Spiral freezer (GYROCOMPACT).

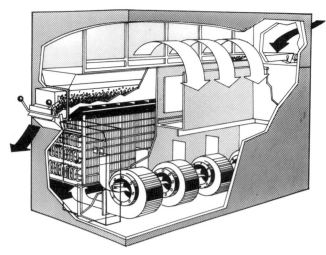

Figure 13. Fluidized-bed freezer (FLoFREEZE).

products with more surface water. Consequently, there is no hazard of belt damage if a dewatering screen is broken down temporarily.

One type of fluidized-bed freezer combines the fluidized bed with a conveyor belt. The freezer operates in two stages: a crust freezing zone and a finishing freezing zone. In the former the product is carried on a fluidized bed that guarantees an efficient heat transfer, a quick crust freezing, and particle separation. The crust frozen product is then conveyed on a belt through the second freezing zone.

The fluidization technique achieves a very efficient air–product contact, which gives heat-transfer rates which are much higher than those for conventional air-blast freezing tunnels or belt freezers. The efficiency of the heat removal can also be seen in the physical dimensions of the equipment, which are generally one-third of the comparable belt freezer.

The fluidized-bed freezer is in-line equipment suitable not only for vegetables, berries, and other fruits but also for processed products such as French fried potatoes, peeled cooked shrimps, diced meat, and meatballs.

Contact Freezers

In a contact freezer the product is either in direct contact with the freezing media—immersed—or indirectly by being in contact with a belt or plate containing the freezing media.

Immersion Freezers

The immersion freezers consists of a tank with a cooled freezing medium, such as salt, sugar, or alcohol solution in water or other nontoxic mixtures of water and solutes. The product is immersed in this brine or sprayed while being conveyed through the tank.

This type of equipment has been quite commonly used for surface freezing of turkeys and other poultry on markets where a light color is demanded. Final freezing is accomplished in a separate blast tunnel or during cold storage. The latter, however, may jeopardize quality because of slow core freezing. It is necessary to protect the product from contact with the brine by using high-quality packaging materials with absolutely tight seal. Brine residues on the packages are washed off with water at the freezer exit.

A sodium chloride brine was earlier sometimes used in direct contact with the product in the fishing industry. For freezing tuna fish, for example, it is still used in some places.

Indirect-Contact Freezers

The most commonly contact freezer is the plate freezer, where the product is pressed between hollow metal plates that are positioned horizontally or vertically, with a refrigerant circulating through them.

Another type of freezer utilizes two belts with the refrigerant circulating outside the belts or alternatively placing the product on a single belt. All these arrangements provide a very good heat transfer, which is reflected in short freezing times, provided the product itself is a good heat conductor.

The advantage of good heat transfer at the surface is gradually reduced with increasing product thickness. For this reason it is often limited to a maximum of 50–60 mm. It is further important that the packages are well filled and if metal trays are used to carry the packs that these are not distorted.

The pressure from the plates or belts maintain throughout the freezing process practically eliminates what is known as "bulging" and the frozen packs will maintain their regular shape within very close tolerances.

Plate Freezers

There are two main types of plate freezer: the horizontal and the vertical plate freezer. Either type can be manual or automatic. The typical manual horizontal plate freezer contains 15–20 plates. The product is placed on metal trays or in other systems metal frames and transported to the freezer where they are manually loaded between the plates.

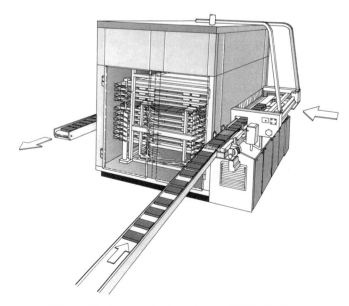

Figure 14. Automatic plate freezer (AUToPLATE).

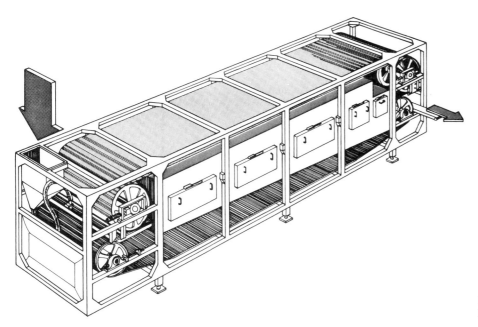

Figure 15. Contact band freezer (PELLo-FREEZE).

In order to obtain automatic operation of the horizontal plate freezer the whole battery of plates is moved up and down in an elevator system. At the loading conveyor level the plates are separated, and a row of packages that have accumulated on a transport conveyor from the processing and packaging are pushed in between the open plates, simultaneously discharging a row of frozen products at the opposite side. This cycle is repeated until all frozen packages at this level has been replaced. Then the space between the two plates are closed, all plates are indexed up, and the next set of plates are opened for loading and unloading. A typical automatic plate freezer is shown in Figure 14.

The vertical plate freezer is used mainly for freezing products in blocks weighing 10–15 kg and has been specially developed for freezing fish at sea. The freezer consists of a number of vertical freezing plates forming partitions in a container with an open top. The product is simply fed in from the top and the blocks after freezing is discharged either to the side, upwards, or down through the bottom. Usually this mechanism is automized. The discharge of the products is enhanced by a short period of gas defrost at the end of the freezing cycle and the use of compressed air or a hydraulic system to force out the product.

Band Freezers

Single-band and double-band freezers are designed to freeze thin product layers. The freezers can either be straight forward bands as shown in Figure 15 or as a drum (Fig. 16).

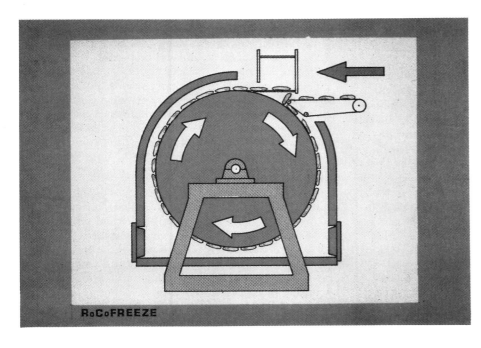

Figure 16. Direct-contact band freezer—drum freezer.

The band freezer illustrated in Figure 16 is designed to freeze and form liquids and semiliquids into individual pellets as an in-line operation.

The product is formed and frozen between two endless stainless-steel bands, of which the top band is flat and the lower band is corrugated with flexible seals on each side. The product is supplied on the corrugated band by a spreading device, after which the flat band is brought in contact with the product, thus totally enclosing it. After the freezing–forming zone the two bands are separated. The liquid is now frozen to a mat and passes through the final forming operation. The product finally enters the outfeed conveyor in an IQF form.

A monopropylene glycol–water brine is used as an intermediate freezing medium. The brine is circulated by pumps from a sump below the freezing pump via a brine cooler to the freezing zone, where it passes over the outer surfaces of the band with high velocity.

Typical products frozen in band freezers are chopped spinach purée, fruits pulps, egg yoke, sauces, and soups.

The drum freezer can be viewed as a more compact band freezer. Distinction is made between both vertical and horizontal drums. This type of freezer is also known as a rotary freezer.

Cryogenic Freezers

Cryogenic freezers differ from all other types of freezers in one fundamental respect: they are not connected to a refrigeration plant. The heat-transfer medium is nitrogen or carbon dioxide liquefied in large industrial installations and shipped to the food freezing factory in low-temperature well-insulated pressure vessels.

The design of cryogenic freezers has improved significantly in recent years. As for all types of freezers also this type can be found as straight-belt, multitier, and spiral belt as well as immersion designs. In principle the same basic equipment can be used for both gases, but with slight modifications. The size and mobility of cryogenic freezers allow for flexibility in design and redesigning a processing plant.

Key attributes of the equipment are high heat-transfer rates, low investment costs, and rapid installation and startup. Especially interesting applications are those for chilling, firming, or crusting products.

A typical belt liquid nitrogen (LIN) freezer is shown in Figure 17. Liquid nitrogen at $-196°C$ is sprayed into the freezer in which the atmosphere is circulated by small fans. The freezant or liquid nitrogen partially evaporates immediately leaving the spray nozzles and on contact with the products. The cold gas is circulated by fans toward the infeed end precooling the products entering the freezer and thereafter extracted by an exhaust fan.

The freezant thus passes in countercurrent to the movement of products on the belt and giving a high heat-transfer efficiency, which is an advantage in terms of quality for some special products.

However, the quick freezing may also result in cracking of the product surface if sufficient precautions are not taken. The freezant consumption is in the range of 1.2–2.0 kg per kilogram of product. The capacity can vary from 150 to 1000 kg/h, and typical products are meat cuts, fish fillets, seafood, fruit, berries, pies, and pastries.

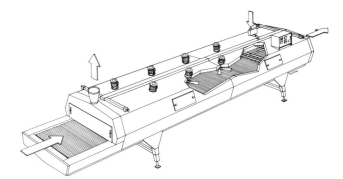

Figure 17. Cryogenic freezer (AGA FREEZE).

Prefreezer

A recent development utilizing liquid nitrogen as freezing medium is an apparatus designed for quick crust freezing of extremely wet, sticky, or sensitive products that can then be easily handled in a spiral belt freezer or a fluidized-bed freezer for completion of the freezing process without deformation or breakage. The freezers also offer a possibility of freezing products that are difficult to freeze in conventional systems (see Fig. 18).

The products are frozen by means of direct immersion in liquid nitrogen, which gives an almost instantaneous freezing of the surface. The product is fed vertically into an IQF tank with a continuous flow of liquid nitrogen.

In this stream the products are gently received and separated at the same time as a very thin layer of the product surface is frozen. From this first step the products drop down on a belt and are fed into a bath of liquid nitrogen that together with a spray completes the crust freezing. The liquid nitrogen is then separated from the product and collected in a sump from where it is recirculated by means of a specially designed pump.

Carbon Dioxide Freezer

Liquid carbon dioxide is normally stored under high pressure. At atmospheric pressure it exists only as a solid or a gas. When the liquid is released to the atmosphere, 50% of the liquid becomes dry-ice snow and 50% vapor both at $-79°C$. Because of these unusual properties, carbon dioxide freezer designs vary widely.

In a LIN freezer the cold gas phase is used to precool the product before it is exposed to the nitrogen spray. As liquid carbon dioxide forms snow that needs time for sublimation, the injection is moved closer to the product infeed as compared to the LIN freezer.

THERMODYNAMIC PROPERTIES

An important consideration in all designs of food processing involving heat exchange is the thermal properties of the food products to be handled in the system. The variable composition and structure of foods influence those properties. During storage the chemical and physical properties, the time, and the temperatures are the most important factors. The composition of the food product varies with species, growth condition, age, feed, harvest, slaughter, catch, handling, and processing, as well as stor-

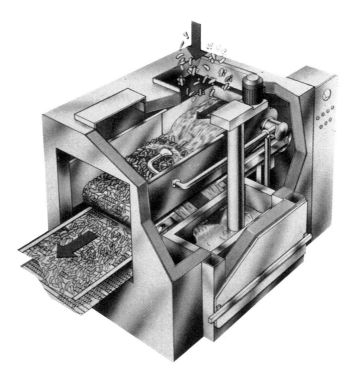

Figure 18. Prefreezer (CRUSToFREEZE).

age conditions. All of these factors influence the thermal properties.

For these reasons it is easily understood that the values of thermal properties are not exact values but most often estimates. The more detailed a description of the food product measured the more accurate the values given. Another important consideration when using experimentally found thermal properties is the difference in the methods of measurement, which may place limitations on the value of the data (4). Computer programs have been developed to estimate the thermal physical properties from the knowledge of the specifications of the products, such as chemical composition, temperature, and density.

Values of specific heat and latent heat of fusion are often calculated directly from the water content of the product. This is the case for the values given in Table 1.

The values of water content in Table 1 are averaged for the product. The water content for fruit and vegetables varies with the stage of development or maturity when harvested and also with type of species, growing conditions, and the storage conditions after harvest. The values given in the table apply to mature product shortly after harvest. For meat the water content values are for the time of slaughter or after the aging period. In reality the water content varies considerably, not only between different animals but also between different muscles from the same animal. For processed products the water content depends on the specific process used.

The freezing points given in Table 1 are based on experiments in which the product has been cooled slowly until freezing occurred. For fruits and vegetables the highest temperature at which the product freeze are given and for other foods average freezing temperature is shown.

With reference to specific heat it should be observed that this is a function of temperature. The value given in Table 1 are from 0°C. In a unfrozen product the specific heat will be slightly lower as temperature rise, and in frozen foods there is a large change in specific heat as temperature decreases. The latter is, of course, related to the changes in composition, in particular the water content. When calculating specific heat of a frozen product it is assumed that the water is frozen to ice and that the specific heat involved is that of ice. This assumption is not totally correct. The freezing of most foods that is transferring from liquid water to ice is a gradual process that occurs over a wide temperature range.

With reference to the latent heat of fusion given in Table 1, these values are also subject to error because they do not consider the chemical composition other than water content. They are, in other words, the product of the heat of fusional water and the water content.

The variations that occur in values given on thermal properties should be taken into account in practical applications. Today a comprehensive work is carried out at several research institutes and universities, eg, at Campden Food and Drink Research Association, UK (12) to compare different data. Besides Table 1, a compilation of thermal properties presented by Polley and co-workers is of interest (13).

The definitions of the thermal properties normally listed are

Water content—the mass of water in the product divided by the total mass expressed in percentage.

Table 1. Thermal and Related Properties of Food and Food Materials[a]

Food or Food Material	Water Content % (mass)[b]	Highest Freezing Point, °C[d]	Specific Heat[c] Above Freezing, kJ/kg·°C	Specific Heat[c] Below Freezing, kJ/kg·°C	Latent Heat of Fusion[e] kJ/kg
Vegetables					
Artichokes, Globe	84	−1.2	3.78	1.90	281
Artichokes, Jerusalem	80	−2.5	3.68	1.85	268
Asparagus	93	−0.6	4.00	2.01	312
Beans, snap	89	−0.7	3.90	1.96	298
Beans, lima	67	−0.6	3.35	1.68	224
Beans, dried	11	—	1.95	0.98	37
Beets, roots	88	−1.1	3.88	1.95	295
Broccoli	90	−0.6	3.93	1.97	302
Brussels sprouts	85	−0.8	3.80	1.91	285
Cabbage, late	92	−0.9	3.98	2.00	308
Carrots, roots	88	−1.4	3.88	1.95	295
Cauliflower	92	−0.8	3.98	2.00	308
Celeriac	88	−0.9	3.88	1.95	295
Celery	94	−0.5	4.03	2.02	315
Collards	87	−0.8	3.85	1.94	291
Corn, sweet	74	−0.6	3.53	1.77	248
Cucumbers	96	−0.5	4.08	2.05	322
Eggplant	93	−0.8	4.00	2.01	312
Endive (escarole)	93	−0.1	4.00	2.01	312
Garlic	61	−0.8	3.20	1.61	204
Ginger, rhizomes	87	—	3.85	1.94	291
Horseradish	75	−1.8	3.55	1.79	251
Kale	87	−0.5	3.85	1.94	291
Kohlrabi	90	−1.0	3.93	1.97	302
Leeks	85	−0.7	3.80	1.91	285
Lettuce	95	−0.2	4.06	2.04	318
Mushrooms	91	−0.9	3.95	1.99	305
Okra	90	−1.8	3.93	1.97	302
Onions, green	89	−0.9	3.90	1.96	298
Onions, dry	88	−0.8	3.88	1.95	295
Parsely	85	−1.1	3.80	1.91	285
Parsnips	79	−0.9	3.65	1.84	265
Peas, green	74	−0.6	3.53	1.77	248
Peas, dried	12	—	1.97	0.99	40
Peppers, dried	12	—	1.97	0.99	40
Peppers, sweet	92	−0.7	3.98	2.00	308
Potatoes, early	81	−0.6	3.70	1.86	271
Potatoes, main crop	78	−0.6	3.63	1.82	261
Potatoes, sweet	69	−1.3	3.40	1.71	231
Pumpkins	91	−0.8	3.95	1.99	305
Radishes	95	−0.7	4.06	2.04	318
Rhubarb	95	−0.9	4.06	2.04	318
Rutabagas	89	−1.1	3.90	1.96	298
Salsify	79	−1.1	3.65	1.84	265
Spinach	93	−0.3	4.00	2.01	312
Squash, summer	94	−0.5	4.03	2.02	315
Squash, winter	85	−0.8	3.80	1.91	285
Tomatoes, mature green	93	−0.6	4.00	2.01	312
Tomatoes, ripe	94	−0.5	4.03	2.02	315
Turnip greens	90	−0.2	3.93	1.97	302
Turnip	92	−1.1	3.98	2.00	308
Watercress	93	−0.3	4.00	2.01	312
Yams	74	—	3.53	1.77	248
Fruits					
Apples, fresh	84	−1.1	3.78	1.90	281
Apples, dried	24	—	2.27	1.14	80
Apricots	85	−1.1	3.80	1.91	285
Avocados	65	−0.3	3.30	1.66	218
Bananas	75	−0.8	3.55	1.79	251

Table 1. (continued)

Food or Food Material	Water Content % (mass)[b]	Highest Freezing Point, °C[d]	Specific Heat[c]		Latent Heat of Fusion[e] kJ/kg
			Above Freezing, kJ/kg·°C	Below Freezing, kJ/kg·°C	
Blackberries	85	−0.8	3.80	1.91	285
Blueberries	82	−1.6	3.73	1.87	275
Cantaloupes	92	−1.2	3.98	2.00	308
Cherries, sour	84	−1.7	3.78	1.90	281
Cherries, sweet	80	−1.8	3.68	1.85	268
Cranberries	87	−0.9	3.85	1.94	291
Currants	85	−1.0	3.80	1.91	285
Dates, cured	20	−15.7	2.17	1.09	67
Figs, fresh	78	−2.4	3.63	1.82	261
Figs, dried	23	—	2.25	1.13	77
Gooseberries	89	−1.1	3.90	1.96	298
Grapefruit	89	−1.1	3.90	1.96	298
Grapes, American	82	−1.6	3.73	1.87	275
Grapes, Vinifera	82	−2.1	3.73	1.87	275
Lemons	89	−1.4	3.90	1.96	298
Limes	86	−1.6	3.83	1.92	288
Mangoes	81	−0.9	3.70	1.86	271
Melons, Casaba	93	−1.1	4.00	2.01	312
Melons, Crenshaw	93	−1.1	4.00	2.01	312
Melons, honeydew	93	−0.9	4.00	2.01	312
Melons, Persian	93	−0.8	4.00	2.01	312
Melons, watermelon	93	−0.4	4.00	2.01	312
Nectarines	82	−0.9	3.73	1.87	275
Olives	75	−1.4	3.55	1.79	251
Oranges	87	−0.8	3.85	1.94	292
Peaches, fresh	89	−0.9	3.90	1.96	298
Peaches, dried	25	—	2.30	1.16	84
Pears	83	−1.6	3.75	1.89	278
Persimmons	78	−2.2	3.63	1.82	261
Pineapples	85	−1.0	3.80	1.91	285
Plums	86	−0.8	3.83	1.92	288
Pomegranates	82	−3.0	3.73	1.87	275
Prunes	28	—	2.37	1.19	94
Quinces	85	−2.0	3.80	1.91	285
Raisins	18	—	2.12	1.07	60
Raspberries	81	−0.6	3.70	1.86	271
Strawberries	90	−0.8	3.93	1.97	302
Tangerines	87	−1.1	3.85	1.94	291
Whole Fish					
Haddock, cod	78	−2.2	3.63	1.82	261
Halibut	75	−2.2	3.55	1.79	251
Herring, kippered	70	−2.2	3.43	1.72	235
Herring, smoked	64	−2.2	3.28	1.65	214
Menhaden	62	−2.2	3.23	1.62	208
Salmon	64	−2.2	3.28	1.65	214
Tuna	70	−2.2	3.43	1.72	235
Fish Fillets or Steaks					
Haddock, cod, perch	80	−2.2	3.68	1.85	268
Hake, whiting	82	−2.2	3.73	1.87	275
Pollock	79	−2.2	3.65	1.84	265
Mackerel	57	−2.2	3.10	1.56	191
Shellfish					
Scallop, meat	80	−2.2	3.68	1.85	268
Shrimp	83	−2.2	3.75	1.89	278
Lobster, American	79	−2.2	3.65	1.84	265
Oysters, clams, meat, and liquor	87	−2.2	3.85	1.94	291
Oyster in shell	80	−2.8	3.68	1.85	268

Table 1. (continued)

Food or Food Material	Water Content % (mass)[b]	Highest Freezing Point, °C[d]	Specific Heat[c] Above Freezing, kJ/kg·°C	Specific Heat[c] Below Freezing, kJ/kg·°C	Latent Heat of Fusion[e] kJ/kg
Beef					
Carcass (60% lean)	49	−1.7	2.90	1.46	164
Carcass (54% lean)	45	−2.2	2.80	1.41	151
Sirloin, retail cut	56	—	3.08	1.55	188
Round, retail cut	67	—	3.35	1.68	224
Dried, chipped	48	—	2.88	1.44	161
Liver	70	−1.7	3.43	1.72	235
Veal, carcass (81% lean)	66	—	3.33	1.67	221
Pork					
Bacon	19	—	2.15	1.08	64
Ham, light cure	57	—	3.10	1.56	191
Ham, country cure	42	—	2.72	1.37	141
Carcass (47% lean)	37	—	2.60	1.31	124
Bellies (33% lean)	30	—	2.42	1.22	101
Backfat (100% fat)	8	—	1.87	0.94	27
Shoulder (67% lean)	49	−2.2	2.90	1.46	164
Ham (74% lean)	56	−1.7	3.08	1.55	188
Sausage, links or bulk	38	—	2.62	1.32	1.27
Sausage, country style, smoked	50	−3.9	2.93	1.47	168
Sausage, frankfurters	56	−1.7	3.08	1.55	188
Sausage, Polish style	54	—	3.03	1.52	181
Lamb					
Composite of cuts (67% lean)	61	−1.9	3.20	1.61	204
Leg (83% lean)	65	—	3.30	1.66	218
Dairy Products					
Butter	16	—	2.07	1.04	54
Cheese, Camembert	52	—	2.98	1.50	174
Cheese, Cheddar	37	−12.9	2.60	1.31	124
Cheese, cottage (uncreamed)	79	−1.2	3.65	1.84	265
Cheese, cream	51	—	2.95	1.48	171
Cheese, Limburger	45	−7.4	2.80	1.41	151
Cheese, Roquefort	40	−16.3	2.67	1.34	134
Cheese, Swiss	39	−10.0	2.65	1.33	131
Cheese, processed American	40	−6.9	2.68	1.34	134
Cream, half-and-half	80	—	3.68	1.85	268
Cream, table	72	−2.2	3.48	1.75	241
Cream, whipping, heavy	57	—	3.10	1.56	191
Ice cream, (10% fat)	63	−5.6	3.25	1.63	211
Milk, canned, condensed, sweetened	27	−15.0	2.35	1.18	90
Milk, evaporated, unsweetened	74	−1.4	3.53	1.77	248
Milk, dried (whole)	2	—	1.72	0.87	7
Milk, dried (nonfat)	3	—	1.75	0.88	10
Milk, fluid (3.7% fat)	87	−0.6	3.85	1.94	291
Milk, fluid (skim)	91	—	3.95	1.99	305
Whey, dried	5	—	1.80	0.90	17
Poultry Products					
Egg, whole (fresh)	74	−0.6	3.53	1.77	247
Eggs, white	88	−0.6	3.88	1.95	295
Eggs, yolks	51	−0.6	2.95	1.48	171
Eggs, yolks (sugared)	51	−3.9	2.95	1.48	171
Eggs, yolks (salted)	50	−17.2	2.93	1.47	168
Eggs, dried (whole)	4	—	1.77	0.89	13
Eggs, dried (white)	9	—	1.90	0.95	30
Chicken	74	−2.8	3.53	1.77	248
Turkey	64	—	3.28	1.65	214
Duck	69	—	3.40	1.71	231

Table 1. (continued)

Food or Food Material	Water Content % (mass)[b]	Highest Freezing Point, °C[d]	Specific Heat[c] Above Freezing, kJ/kg·°C	Specific Heat[c] Below Freezing, kJ/kg·°C	Latent Heat of Fusion[e] kJ/kg
Miscellaneous					
Honey	17	—	2.10	1.68	57
Maple syrup	33	—	2.50	1.26	111
Popcorn, unpopped	10	—	1.92	0.97	34
Yeast, baker's, compressed	71	—	3.45	1.73	238
Candy					
Milk chocolate	1	—	1.70	0.35	3
Peanut brittle	2	—	1.72	0.87	7
Fudge, vanilla	10	—	1.92	0.97	34
Marshmallows	17	—	2.10	1.05	57
Nuts, Shelled					
Peanuts (with skins)	6	—	1.82	0.92	20
Peanuts (with skins, roasted)	2	—	1.72	0.87	7
Pecans	3	—	1.75	0.88	10
Almonds	5	—	1.80	0.90	17
Walnuts, english	4	—	1.78	0.89	13
Filberts	6	—	1.82	0.92	20

[a] Reprinted with permission of the American Society of Heating, Refrigerating and Air-Conditioning Engineers from the 1989 *ASHRAE Handbook—Fundamentals* (11).
[b] Water contents of fruits and vegetables are from Lutz and Hardenburg (1968) except for Jerusalem artichokes; dried beans; and peas, yams, dried apples, figs, peaches, prunes, and raisins; the latter are from Watt and Merrill (1963). Water contents of meats, dairy, and poultry products, miscellaneous candy, and nuts are also from Watt and Merrill; water contents of eggs (yolks, salted) and fish are from ASHRAE (1972, 1974, and 1978).
[c] Freezing points of fruits and vegetables are from Whiteman (1957), and average freezing points of other foods are from ASHRAE (1972, 1974, and 1978).
[d] Specific heat was calculated from Siebel's formulas (1892).
[e] Latent heat of fusion was obtained by multiplying water content expressed in decimal form by 144, the heat of fusion of water in Btu/lb.

Average freezing point—the temperature at which the liquid and solid state of a product are in equilibrium expressed in Celsius.

Latent heat—the quantity of heat necessary to change 1 kilogram of liquid to solid without change of temperature measured as kJ/kg.

Specific heat—the amount of heat needed to rise the temperature of 1 kg of a food product 1° measured as kJ/kg, °C.

Heat of respiration—the quality of heat generated per 24 h measured as kJ/24 h, kg.

The large variations that occur for thermal properties of food products explain why theoretical calculations on freezing time are difficult and should not be used to determine the requirements for a specific freezer. Those requirements should be determined by practical tests.

DETERMINATION OF HOLDING TIME

For calculation of the capacity of any freezer, the necessary holding time is essential. For bulk products such as peas, green beans, French fried potatoes, and fish sticks the capacity of standard equipment is usually specified separately. This is also the case with packaged products that are homogeneous such as spinach purée and fish fillet blocks. For other products holding time must be determined before capacity can be stated.

In the literature many different formulas can be found for accurate determination of holding times. In reality, however, they are of little use, because products differ so much in composition and form that the work to transform the characteristics of a product into mathematical terms is a lot more time-consuming than test freezing of the very product itself. This can usually be done in less time than it takes to analyze the composition of the product.

Of course, such test freezing should be carried out under controlled conditions that do correspond to those that can actually be achieved in production. A suitable pilot freezer for different airflow directions is shown in Figure 19.

MAJOR CONSIDERATIONS IN FREEZER DESIGN

Safety of Personnel

The safety of personnel that operate, clean, and service a freezer should be a main consideration in the freezer design. In too many instances, serious injuries have been the result of unsafe design and operation of freezing machinery.

Mechanical hazards exist in the conveyor drive systems, fans, and other areas. The machinery must be de-

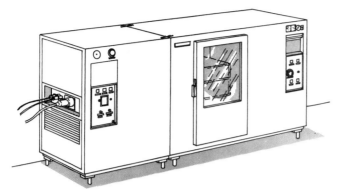

Figure 19. Test freezer.

signed so that all areas of a freezer can be easily cleaned and inspected while ensuring that personnel are suitably protected. All drives and fans must be fully guarded so that no worker individual or worker's clothing can reach any part of the machine and get caught or crushed. Emergency stop switches need to be located throughout the machine so that if someone does get caught in the machinery the machine can be stopped quickly to minimize the injury. Fan guards must be designed sufficiently open to remain unrestricted by frost buildup while still preventing personnel from getting their hands or clothing into the fan.

The extreme cold found in most modern freezers represents a hazard to personnel in the form of hypothermia and frostbite. The high air velocities found in efficient freezers that use air as the product heat-transfer medium greatly increases heat transfer and the wind-chill factor, which will quickly freeze exposed skin and draw heat from the body at a very high rate. Exposure to these low temperatures must be limited to a tolerable period, and the fans must be shut off whenever a person needs to enter such a freeze.

Cryogenic freezers must not be entered during operation; even a very brief exposure to the cryogen can quickly cause frostbite as a result of the extremely high heat transfer between the boiling liquid or sublimating solid and warm flesh.

Cryogenic freezers, which generally employ either liquid nitrogen or carbon dioxide in snow form, must be evacuated and refilled with air prior to entry by personnel. The freezers purge all the air out during operation, which results in an oxygen content insufficient to sustain consciousness or life. Such machines must be locked out during operation to prevent access and risk of asphyxiation.

Noise can reach levels that damage hearing in some freezers. Care must be taken to avoid exposure to noise levels above 95 dB by wearing a suitable form of hearing protection such as earplugs or earmuffs.

Noise can be reduced to safe levels by proper selection of fans. Excessive noise is generally a result of using a fan at a pressure above which it is designed to operate, causing cavitation.

Machinery Protection

The hostile environment in which freezers operate render human surveillance of their operation almost impossible. The machinery therefore must operate in an extremely dependable manner and have a number of detection devices mounted and operational so that if something does go wrong, the damage will be minimized.

The most common problems encountered in freezers are caused by ice accumulation, product jams, and operational errors. Ice accumulation can occur as a result of poor defrost procedure, excessive moisture on the conveyor carrying the product into the freezer, or poor startup and shutdown procedures. Ice can accumulate in locations in the machine where it impedes the safe operation of the freezer by jamming a part of by forcing the conveyor or other parts out of their normal operating position. Product jams can occur because the product to be frozen is improperly loaded onto the conveyor or the conveyor can be blocked by an obstruction in the product path. In either case good operation of the freezer and sound design can alleviate the problem. Operational errors are symptomized by a wide variety of problems. Typical symptoms are frozen conveyors due to improper startup, unbalanced fans, or an ice-plugged coil due to improper defrost.

Excessive pressure in the refrigeration piping and the coils can lead to a failure and subsequent loss of refrigerant and potential safety hazard. During normal operation the refrigeration piping in the freezer is under low pressure, so the risk of a ruptured pipe is small at that time. When a freezer is defrosted, the temperature of the coil is raised in order to melt the frost of the coil. If the refrigeration piping is not properly designed and installed, potentially dangerous pressures can develop in the coil. It is imperative that suitable relief devices be installed in the coil so that if the operator makes an error the pressure of the refrigerant in the coil cannot go above a safe level.

Product Safety

In a typical freezer application the value of the product that passes through the freezer in a period of just a few weeks can exceed the cost of the freezer. It is therefore sound practice to ensure that the product is not damaged or contaminated by the freezer.

Contamination can occur as a result of improper cleaning, debris such as surface coatings, or particles created by wear. Even if the contaminant is not harmful or toxic, its presence can render the product unsalable.

Product contamination can be minimized by sound freezer design in which wear debris cannot be generated in a location where it can get into the product as it is being conveyed through the freezer. Typical sources of contamination are wear debris created between the conveyor belt and the wear strip supporting the belt, dripping of condensed water at entries and discharges from freezers, flaking coatings, and leaking fluids such as hydraulic oil.

Product handling within the freezer can physically damage the product. Symptoms of products that have been damaged within the freezer include clumping together of pieces intended to be separate and free-flowing (IQF), pieces of the product torn from the conveying device as a result of the product being frozen to the conveyor, collision damage as a result of the product itself, and buildup of ice on the product.

Clumping together of product is generally a result of improper handling by the freezer as a result of poor design or improper operation. It is important that relative motion between the pieces to be frozen is maintained while the surface of the product freezes. The relative motion between the particles can be achieved by fluidization, mechanical agitation, or immersion in a boiling liquid that boils at a temperature well below the product freezing point.

In order to avoid damage to a product as it is removed from the conveyor that is transporting it through the freezer, it is essential that the product be solidly frozen and that the product does not adhere to the conveyor excessively.

Excessive adhesion can be avoided by careful selection of the conveyor material or by transporting the product in a manner such that it does not rest on a solid conveyor during the period that the surface of the product is being frozen.

Collision damage is generally a result of excessive loading of product onto the conveyor, which can result in the product colliding with other products inside the freezer or colliding with stationary structures within the freezer. Such collisions frequently result in product jams, which can destroy a considerable amount of product at each occurrence. To avoid such problems even feed equipment should be used.

Ice buildup on products can result in an insightly product. This condition can occur when excessive free moisture is transported into the freezer and is transferred from unfrozen to frozen product in the early stages of freezing. This can be controlled by minimizing the free moisture going into the freezer with the product and by controlling the motion of the product once it is inside the freezer.

Hygiene in the Freezer Environment

Proper hygiene in the freezer environment must be defined by the user. Factors such as the sensitivity of the product to various degrees of contamination must be considered in setting sanitary standards. For example, if a cooked product is frozen that requires no additional cooking prior to consumption, extreme hygienic procedures and very sanitary equipment must be employed because of the high risks to the consumer. Contamination of the product can assume many forms. Typical forms of contamination that must be considered are bacterial, wear debris, pieces of a different product, and foreign debris.

Bacterial contamination is generally the result of bacterially contaminated food coming in contact with the product and attaching itself. This can occur if the conveying system has not been adequately cleaned and sanitized.

Wear debris will be created in all freezing and processing machinery. Whether or not the product is considered contaminated is a function of how obvious the debris is and what the composition is. Stainless steel rubbing on either a plastic such as polyethelyne or directly on another piece of stainless steel will generate a significant amount of dark-gray powder. If this gray powder is deposited on the product such that it is visible macroscopically, the product is considered contaminated.

Contamination can also occur if previously frozen food or any food particle is deposited onto a product that is being frozen. This can be the result of inadequate cleaning

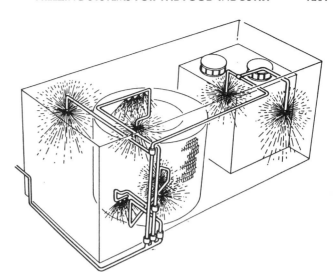

Figure 20. Cleaning in place (CIP).

between changes in product changes on the freezer or production line. A clean design and thorough cleaning can alleviate this problem.

Automatic cleaning of freezing machinery permits cleaning with less labor and reduces the need to access difficult areas and dangerous locations while still providing adequate cleaning. Additionally, automatic cleaning reduces labor and the necessary cleaning time (see Fig. 20).

Cleaning solutions must be selected carefully to adequately clean the machinery without damaging the materials or surface coatings used on the freezer. After cleaning, a sanitizing agent is sometimes employed to sterilize the machine.

Materials for Freezing Machinery

Hygienic standards demand that all food machinery including freezers be constructed of nontoxic materials that permit use of aggressive cleaning agents such as mild caustics without corroding. Typical materials used presently are 300 series stainless steel, galvanized steel, aluminum, and various food-grade plastics.

Special considerations must be taken into account in selection of materials for specific functions in the freezer.

Surfaces that come in contact with the product must be smooth and totally noncorrosive and not adhere to the product, either frozen or thawed. Stainless steel and plastics are generally used for product contact.

Materials through which heat must pass must have a high thermal conductivity. Applications that fall under this category are heat-transfer coils, flat metal belts used on contact freezers, and plates used on horizontal and vertical plate freezers.

High-insulation properties are required in practically all freezers to separate the cold environment from the ambient air as well as to prevent the formation of condensation on the warm external walls. Generally the insulating walls are of a panel construction with either metal or fiber-glass skins and a plastic low thermal conductivity core. The panel is generally bonded together to give suitable mechanical properties and also to prevent the ingress of moisture. If moisture were to enter the panel, deterioration of the panels thermal and mechanical properties could result because of the freeze–thaw cycles encountered in a typical freezer.

Lubricants selected for food freezer applications must be selected on the basis of both their low-temperature properties and their toxicity. Gearboxes in the freezer must be located so that any leakage of lubricant cannot contaminate the product. Greases and oils used in close proximity to the product or product-carrying surface must be edible as insurance against accidental contamination over the life of the freezer is impossible. Lubricants must remain viscous enough over the entire operating range they are to be exposed (even when the freezer is held at extremely low temperatures without product) to permit dependable operation of the machinery. This generally requires the use of synthetic lubricants at temperatures below 7°C. Lubricants should also have the ability to maintain their properties with considerable moisture content as the cleanup and thawing will result in significant water accumulation in the lubricant. Lubricants must generally be changed at short intervals as a result of water contamination.

Coatings used in freezers must be permanent, or contamination of the product can result. Painted coatings requiring meticulous preparation in application are frequently fragile and require routine maintenance. Metallic coatings such as galvanizing, flame spraying, and plating are generally durable but are not as durable as a component made entirely of noncorrosive material. When coating a material with zinc, consideration should be given the thickness of the coating desired as the zinc will be consumed over time depending on its thickness and the environment. Zinc may not be in contact with the food product.

Materials at freezer temperatures generally undergo large changes in their physical properties, such as brittleness and size. The property changes can result in large stress-induced distortions and breakage unless adequately accounted for.

Mechanical Efficiency

The mechanical efficiency of the freezer is a measure of how much work goes into the mechanical devices within the freezer. The mechanical devices typically consists of conveyor drives, fans, and other powered devices. It is important to realize that any energy put into the freezer either by the work performed in the freezer or by removing heat from the product must be removed by either the refrigeration system or the cryogen. Therefore, a significant multiplier must be applied to the cost of adding extra energy to the freezer environment.

The quantity, types, and power consumption of energy-consuming devices in a given freezer vary greatly between freezer types. Fans are generally the largest power-consuming component in Fan-equipped freezers. Fan efficiencies are defined as the ratio of useful fan work to the actual power consumed by the fan motor and can vary from 40 to 70% depending on the selection of the fan. Conveyor drives are generally relatively small power consumers.

Coils

Coils are employed in almost all types of freezers that transfer heat by circulating cold air over the product and serve the function of transferring the heat from the air to the refrigerant. The efficiency of such a freezer is greatly influenced by the design of the coil as the design determines the difference between the air and refrigerant temperatures, which has a large bearing on the initial and operating cost of the refrigeration system. Coil efficiency is a function of materials of construction, configuration of the surface and refrigerant pipes, air velocity, and refrigerant circulation.

In addition to heat-transfer considerations, freezer coils must be designed with sanitation, corrosion, frost buildup, and speed of defrosting in mind.

In order to remain operational as long as possible, coils must be designed to accept frost without becoming plugged up prematurely. This is generally done by careful selection of the spaces between the heat-transfer surfaces or by continuous defrosting.

Continuous defrosting can be accomplished by installing multiple banks of coils and shutting off one bank and defrosting it while the others remain operational, blowing the frost from the coils with compressed air during operation, or rinsing the frost from an operating coil with a mixture of glycol and water. In such cases a glycol concentration is used to remove the water. A continuous (automatic) defrosting system is illustrated in Figure 21.

Fans

Fans for freezers should be selected to provide the highest possible economy over the life of the freezer taking into account dependability, efficiency, and initial costs.

Fans positioned such that the air flows over them before the coil are subject to large frost buildup. Frost buildup on a fan will result in imbalance, which will impose large stresses on the fan wheel and the motor. The fan must be designed to accept the resultant imbalance without failing.

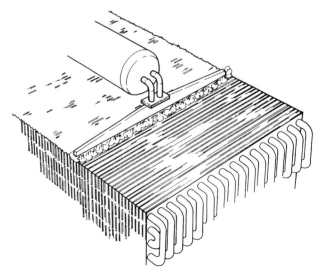

Figure 21. Automatic defrosting (ADF).

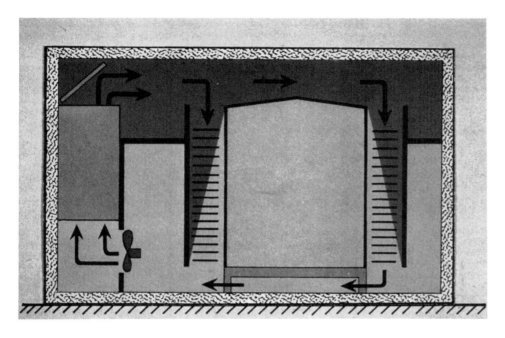

Figure 22. Example of efficient airflow in a freezer.

Fan efficiency is determined by how the fan is selected and applied. Inlet and discharge effects can greatly alter a fan's performance. The air pressure and volume of air that pass through the fan are normally presented graphically by the fan producer. A fan must be selected to give the correct quantity of air over a fully normal operating range, which can vary significantly from the time when a freezer has been recently defrosted and lightly loaded with product to when the freezer is frosted up and heavily loaded with product.

Fan motors must be selected and designed to take the mechanical stresses imposed on them as well as to power the fan wheel over a wide operating range. Fan motor bearings should be specially lubricated to run freely over a full range of operating temperatures. The bearings need to be adequate to tolerate a significant amount unbalance as discussed before.

Electrical Considerations

Dependable and safe electrical installations require special attention in a freezer environment because of the high moisture and large temperature changes that cause water to migrate and condensate on electrical wiring and components. Electrical devices and wiring need to be protected from moisture by ensuring that they always remain above the dewpoint under all operating conditions and by protecting them from water entry during defrosting and cleaning. Electrical panels external to the freezer need to be protected from the cleanup conditions found in most plants by taking special precautions in sealing and ventilation for cooling.

Airflow

In freezers employing air as the heat-transfer medium the design of the airflow has a major influence on the performance of the freezer and on the food product quality. The significant factors to consider in regard to airflow are quantity, evenness of distribution over the product, and power consumed by the fans.

The quantity of the air circulated influences the performance of the coils, the change in air temperatures through the freezer, and heat-transfer rates between the air and the product, all assuming that the areas through which the air flows remain constant. An efficient airflow is illustrated in Figure 22.

The distribution of the air over the product should be carefully controlled to give the designed freezing rate uniformly over all the product passing through the freezer.

The deterioration of the airflow through the freezer with increased coil frosting and heavy product loading needs to be taken into account when designing the air system so that the freezer maintains full performance under the full range of operating conditions.

REFRIGERATION SYSTEMS

The Second Law of Thermodynamics states that heat can be raised from a low to a high temperature level only by expenditure of work. This means that energy inherently flows only from high to low temperatures. From the long list of refrigeration processes, only two are normally of importance for the food industry: (1) a closed mechanical refrigeration system containing a compressor, a condenser, an expansion valve, and an evaporator and (2) an open cryogenic refrigeration system using either liquid nitrogen (N_2), or carbondioxide (CO_2).

Mechanical Refrigeration

The main elements in a mechanical refrigeration system are shown in Figure 23. The closed system is filled with a refrigerant like Freon or ammonia. Refrigerant as a gas is pulled from the evaporator (1) to the compressor (2) driven

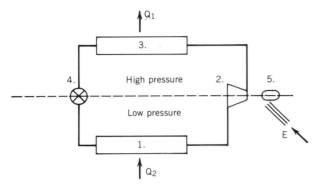

Figure 23. Mechanical refrigeration system: (1) evaporator; (2) compressor; (3) condenser; (4) expansion valve; (5) motor.

by a motor (5). The compressor discharge the gas to the condensing pressure and the gas condensates to liquid in the condenser as heat is removed from the condenser to the surrounding air or to water circulating through the condenser.

From the condenser the liquid goes through the expansion valve. The expansion valve is regulating the flow of refrigerant being evaporated in the evaporator and also keeping the pressure between high and low levels.

Energy picked up in the evaporator (Q_2) plus energy introduced to the compressor shaft (E) equals the energy delivered from the condenser (Q_1):

$$Q_2 + E = Q_1$$

The system described is the same as used in heat pumps. The main difference between the two systems is that in a refrigeration application the heat pickup in the evaporator is the objective, while in a heat pump the main goal is to produce heat in the condenser.

Heat recovery from refrigeration systems will be of greater importance as energy costs rice and environmental questions require an increasing amount of attention. Whenever a demand for heat exists near a refrigeration plant, one form or another of heat recovery from the refrigeration plant should be considered.

One- or Two-Stage Systems

The simple arrangement illustrated in Figure 23 is generally used for all installations where the purpose is to keep the temperature around the freezing point or higher.

For storage temperatures around $-30°C$ and for freezing equipment it is normally economical to use a two-stage system. Such a system contains, in principle, two subsystems, as shown in Figure 23, working together. The condensing pressure for the low-stage compressor is about the same as the evaporation pressure for the high-stage compressor.

The vapor pressure (in megapascals) for refrigerants R12, R22, R134A, and R717 (ammonia, NH_3) as a function of the temperature in Celsius is shown in Table 2. Assuming that ammonia R717 is used, an evaporation temperature of $-30°C$ corresponds to a pressure of 0.119 MPa and a condensing temperature of $+35°C$ corresponds to a pressure of 1.36 MPa. This gives a compression ratio of 11.4, which is uneconomically high. For economical operation the ratio should not be more than 8–10; therefore, in this case a two-stage system should be used.

The intermediate pressure is selected so that the two compressors will work with the same compression ratio. This means in this case that the intermediate pressure should be about 0.44 MPa, corresponding to a temperature of 0°C.

The Refrigerant

Refrigerants are the vital working fluid in a refrigeration system. They absorb heat from one area and dissipate heat into another. The design of the refrigeration equipment is influenced by the properties of the refrigerant selected.

A refrigerant must satisfy many requirements, some of which do not directly relate to its ability to transfer heat. The environmental consequences of a refrigerant that leaks from a system must also be considered. Because of their stability, fully halogenated compounds, called CFCs, persist in the atmosphere for many years and eventually diffuse into the statosphere. According to several scientists the existence of CFCs in the stratosphere has a detrimental effect on the ozone layer. The ozone layer protects the earth from too much ultraviolet radiation. A depletion of ozone might mean a dangerous increase of uv radiation on earth. This fact, coupled with the CFC contribution to the greenhouse effect, has led most nations to urge the industry to minimize the leakage of CFCs into the atmosphere and gradually to phase out their use.

Several of the large chemical companies are at present heavily engaged in finding good substitutes for the CFCs. One such substitute is refrigerant 134a, which is listed in Table 2.

The present available substitutes do not have the same properties as, for example, R12. The users therefore have

Table 2. Saturated Vapor Pressure of Some Refrigerants, MPa[a]

Temperature, °C	Refrigerant[b]			
	R12	R22	R134A	R717
−50	0.0391	0.0645	0.0299	0.0408
−40	0.0642	0.105	0.0516	0.0715
−30	0.100	0.164	0.0847	0.119
−20	0.152	0.245	0.133	0.190
−10	0.219	0.355	0.201	0.290
0	0.309	0.498	0.293	0.428
+10	0.424	0.681	0.415	0.613
+20	0.567	0.910	0.572	0.855
+30	0.745	1.192	0.770	1.164
+40	0.959	1.534	1.016	1.551
+50	1.217	1.943	1.318	2.028

[a] Reprinted with permission of ASHRAE from the 1990 *ASHRAE Handbook—Refrigeration* (14).
[b] R12 = dichlorodifluoromethane, CCl_2F_2, R22 = dichlorodifluoromethane, $CHClF_2$, R134A = tetrafluoroethane, CF_3CH_2F, R717 = ammonia, NH_3.

Figure 24. Package refrigeration unit. Refrigeration capacity = 400 kW at −40°C evaporation temperature and 35°C condensation temperature. Length = 12 m; width = 2.1 m; height = 3 m.

to accept the fact that a plant designed for the use of R12 will have some limitations when operated with, eg, R134a, and in most cases modifications are necessary.

Most likely the use of R717 (ammonia) will be more and more frequent, particularly for large industrial applications.

For commercial installations such as those in department stores most likely we will see a central refrigeration plant using ammonia chilling brine being pump-circulated to the various locations where refrigeration is needed. By using a brine the risk of ammonia gas leaking and coming in contact with the public will be minimized.

Compressors

Piston compressors of various sizes dominated in the refrigeration industry until about 25 years ago, when screw compressors were first used also for refrigeration applications.

Today screw compressors are commonly used for industrial applications, while for small systems such as refrigerators and home freezers the use of hermetic piston compressors is still widespread.

Large compressors are often equipped with a capacity-control mechanism, which makes it possible to maintain constant evaporation temperatures even when the demand varies. It should, however, be noted that the efficiency generally decreases when the compressor works with reduced capacity.

The efficiency of refrigeration systems is influenced by many factors other than the compressor. Only when all details are well designed and tuned together can an optimal result be achieved.

Engine Room as Package Unit or Site-Built

In the early days of the refrigeration industry all engine rooms were site-built. Today, however, it is more common for the refrigeration plant to be delivered as a package unit (see Fig. 24). This offers the benefit of quick deliveries with very little work on-site. If the package unit is designed by an experienced supplier, it also means less risk for functional problems during startup. Finally it offers a greater flexibility than a site-built installation, which tends to be more important as the industry must accommodate changing demands.

Economical Operation Conditions

Two basic facts are important for anyone responsible for the operation of an industrial refrigeration plant.

Out of an existing refrigeration installation more refrigeration capacity is achieved if the evaporation temperature is elevated. This is clearly illustrated in Figure 25. Therefore, it is important to make sure that the evaporator surfaces are kept as clean as possible from ice and that good air circulation is obtained over the whole surface. It is also important that the refrigerant be in good contact with every part of the evaporator.

The condition of the condensor is also of importance for

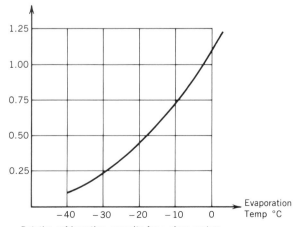

Relative refrigeration capacity for a given system (compressor) and constant condensing temperature

Figure 25. Refrigeration capacity as a function of evaporation temperature.

1266 FREEZING SYSTEMS FOR THE FOOD INDUSTRY

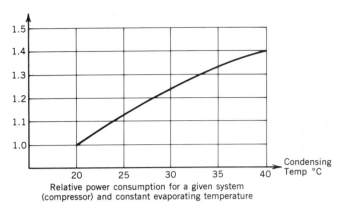

Figure 26. Electric power consumption as function of condensing temperature. Constant condensing temperature.

a good economy. As can be seen from Figure 26 the electrical power consumption goes down with lowering the condensing temperature. For efficient operation the evaporation temperature should be kept high and the condensing temperature low.

With the help of an automatic process control system it is possible to improve the efficiency of a refrigeration installation, including the operation of the freezing equipment.

Figure 27 illustrates a typical example of how a process line is shown on a data screen being part of an automatic process control system.

Cryogenic Refrigeration Systems

The second refrigeration process of importance for the food industry is the cryogenic system. In contrast to the mechanical refrigeration system described above, the cryogenic system is open, which means that the refrigerant is consumed and not recirculated. Liquid nitrogen or liquid or solid carbondioxide (CO_2) are normally used as refrigerants.

In a cryogenic system the refrigeration machinery for production of LIN or liquid CO_2 are large industrial units normally located at a distance remote from the food processing factory. With help of special transport equipment the cryogenics are transported from the place of manufacturing to storage tanks at the food factories.

From an economic viewpoint the principal difference between a mechanical refrigeration system and a cryogenic system is that for the former the investment cost for the food producer is fairly high while the running cost is low. The opposite goes for cryogenic systems. This fact is illustrated in Figure 28.

THE ECONOMICS OF FREEZING

The freezer is usually the largest investment in a processing line. The freezer's impact on operating costs as well as on the final product quality and cost is similarly high. For the investor, it is essential to ensure that the design of the freezing system allows it to be properly integrated into the process and consistently operated at an optimal level.

Reliability

The value of the food products that pass through a freezer in a few weeks' time is often much higher than the investment cost of the freezer itself. This makes reliability a crucial consideration for the food processor.

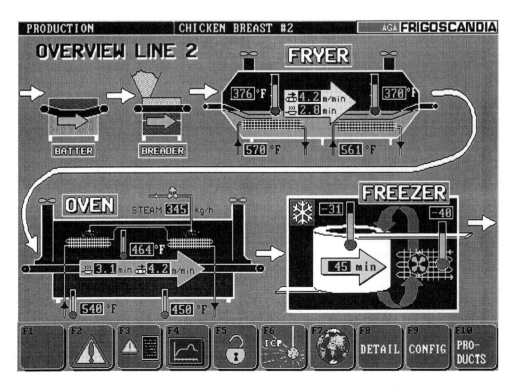

Figure 27. Screen illustration of part of a processing line being controlled automatically.

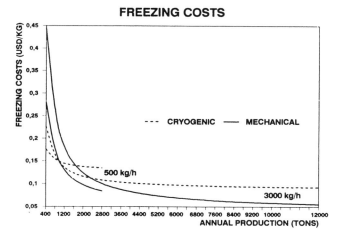

Figure 28. Freezing cost as function of utilization per year for mechanical freezing and cryogenic freezing. Constant evaporating temperature.

Not only is the time the freezer is in operation important, but also the amount of time it is working to full capacity. A breakdown of any part of the line stops the whole line, and product that moves through a line without freezing properly may loose much or most of its market value.

With some systems, product blows out of the freezing zone, lodging in and freezing to other parts of the equipment. This can jam up the line and eventually cause damage.

Dehydration

Dehydration losses will always be present in any freezing system. The evaporation of water vapor from unpacked products during freezing becomes evident as frost builds up on evaporation surfaces. This frost is also caused by excessive infiltration of warm, moist air into the freezer. Still air inside the diffusion-tight carton often creates larger dehydration losses than does the unpackaged products frozen in an IQF freezer. Heat transfer is poor because no air circulation occurs within the package. The result is an evaporation of moisture that can be significant; however, the frost may stay inside the carton.

A poorly designed freezing tunnel may operate with dehydration losses of 3–4%, while a well-designed tunnel can be built to operate with losses of 0.5–1.5%. Liquid nitrogen tunnels normally operate with a dehydration loss of about 0.2–1.25%. This loss occurs when the nitrogen gas is circulated over the product at the infeed end of the freezer.

Freezing Cost Comparison

The cost of a freezing system comes both from the purchase price and the true cost of operation over the lifetime of the system. High operating costs can offset the advantages of a low initial purchase price. Consequently, it is essential to consider all the factors before making an investment, ie, to consider a freezer's "all-in economy."

In making preinvestment analyses, a comparison should be made of all pertinent factors involved, eg, capital, power, operation, cleaning, and maintenance, dehydration, and downtime costs. Figure 29 shows the relative costs when the real-life data from a GYRoCOMPACT were compared with the data from a conventional spiral freezer.

The graphs in this cost study (based on freezing of 1500 kg/h of hamburgers) clearly illustrates that the real cost of a freezer is not merely a question of the initial capital outlay. All the other factors involved in the economics of freezing contribute to the all-in cost picture. Figure 30 shows the effect of an efficient operation over a 5-yr period.

With all the crucial factors taken into account, the somewhat higher purchase price of one freezer was more than offset by the unique module design concept of the same freezer, including a particular belt design indicating major overall cost savings. The extra investment has a payback time of about 1 yr. Thereafter, the continued cost savings means profit for the food processor.

Flexibility and Upgradability

With the competitiveness of today's consumer market, timing is of decisive importance. A system that is easy to modify will make it possible for the processor to respond quickly to the requirements of a new market and to expand or relocate the system easily. A freezing system built as a permanent fixture that cannot be readily expanded, easily removed, or conveniently traded in or resold could cost a great deal more at a later date as needs change.

Maintenance and Service Support

Freezers create an extremely harsh environment for mechanical parts. Optimum performance from even the most efficient freezer requires proper maintenance. Attention must be paid to how complicated and time-consuming such maintenance will be, and what effect this will have on valuable processing time.

Another important factor in the economics of freezing is backup support and a wide range of related services. In judging a freezer supplier's support capability, consideration should be given to the level of service offered in the

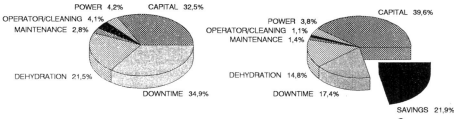

Figure 29. Relative freezing cost.

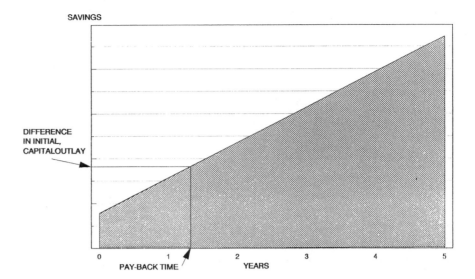

Figure 30. Accumulated annual savings.

planning and installation stages as well as after the deal is complete.

The processor will also need to consider where the supplier's service centers are located, and what commitment of service resources they are prepared to make. This will determine how quickly a service technician can respond in an emergency.

BIBLIOGRAPHY

1. R. Thévenot, *A History of Refrigeration throughout the World,* International Institute of Refrigeration, Paris, 1979.
2. W. R. Woolrich, *The Men Who Created Cold,* Exposition Press, New York, 1967.
3. O. E. Anderson, Jr., *Refrigeration in America,* Princeton University Press, Princeton, N.J., 1953.
4. M. Jul, *The Quality of Frozen Foods,* Academic Press, Orlando, Fla., 1984.
5. F. Lindelöv, "Chemical reactions in Frozen Cured Meat Products. TTT—Examination of Cured and Uncured Pork Bellies," in *Proceedings of Commissions C1 and C2 Meeting* (Karlsruhe, W. Germany, International Institute of Refrigeration, Paris, 1977.
6. S. Åström and G. Löndahl, "Air Blast In-line Freezing versus Ultra Rapid Freezing. A Comparison of Freezing Results with Some Various Vegetables and Prepared Foods," in *Proceedings of Commissions 4 and 5 Meeting* (Budapest), International Institute of Refrigeration, Paris 1969.
7. G. Löndahl and T. Nilsson, "Microbiological Aspects of Freezing of Foods," *International Journal of Refrigeration,* 1(1), 1978.
8. S. Åström, "Freezing Equipment Influence on Weight Losses," in *SOS/70 Proceedings, 3rd International Congress of Food Science and Technology* (Washington, D.C., 1970), Institute of Food Technologists Chicago 1971.
9. G. Lundgren and K. Nilsson, "Vattenförluster vid infrysning av livsmedel" ("Water Loss during Freezing of Foods"), *Livsmedelsteknik,* No. 9, 1969 (Sweden).
10. *Recommendations for the Processing and Handling of Frozen Foods,* International Institute of Refrigeration, Paris, 1986.
11. American Society of Heating, Refrigeration and Air-Conditioning Engineers, *ASHRAE Handbook—Fundamentals,* 1989.
12. S. D. Holdsworth, CFDRA Technical Memorandum No. 452, Campden Food & Drink Research Association, Chipping Camden, UK, 1979.
13. S. L. Polley, O. P. Snyder, and P. Kotnour, "A Compilatision of Thermal Properties of Foods," *Food Technology* (Nov. 1980).
14. American Society of Heating, Refrigeration and Air-Conditioning Engineers, *ASHRAE Handbook—Refrigeration,* 1990.

JIM HEBER, GÖRAN LÖNDAHL,
PER OSKAR PERSSON, and
LEIF RYNNEL
Helsingborg, Sweden

FROZEN FOOD. See entries under REFRIGERATED FOODS.

FRUIT DEHYDRATION

Drying fruits is one of the oldest techniques of food preservation, it was used by Persians, Greeks, and Egyptians since ancient times. Sun drying was the method applied in early times and it still accounts for a significant part of the dried fruit consumed in the world today. However, since the latter part of the nineteenth century mechanically dehydrated fruits are produced in increasingly larger quantities.

Dehydration, or drying, is a preservation process that involves the reversible removal of water from the fruit tissues and thereby extending the storage life because microorganisms and the native enzymes of fruits are deprived of the moisture necessary for their activity. However, it is not a sterilizing process and means must be provided to preserve the low moisture equilibrium and prevent the fruit tissues from regaining moisture until such time as deliberate reconstitution is desired.

Although preservation is usually the principal reason for dehydration, other considerations are often important, for example, significant reductions in the weight and bulk of fruits for economical transport and storage. Recently the development of an increasing variety of convenience food such as instant beverage, breakfast cereal, healthy fruit snack, etc. [often depend on the availability of high-quality dehydrated fruits. Development of new forms of dried fruit and successful preparation of intermediate moisture products has led to products that are stable to storage and pleasant to eat directly.

The following countries are important dried fruit producers: Australia, Argentina, Chile, France, Greece, Iran, Portugal, Spain, South Africa, Turkey, the United States, and Yugoslavia. In the United States more than 90% of the dried fruit output is produced in California. Apples, apricots, dates, figs, peaches, pears, prunes, and raisins represent the significant volume of dried fruits, with raisins (363,000 tons, dry basis, in 1988) accounting for about 60% of the total volume. Dried fruits particularly apples, are also produced in Washington and Oregon.

PRINCIPLES

Dehydration involves simultaneous heat transfer and moisture diffusion (mass transfer). The conversion of liquid (or solid in the case of freeze drying) to vapor demands the supply of latent heat to the product; this can be achieved by a variety of methods: conduction by contact with a heated metal plate, convection from heated air, radiation from an infrared source, or microwave energy. The process may be accelerated by the application of vacuum.

Dehydration Terms

Dried is the term applied to all dried products, regardless of the method of drying. Dehydration refers to the use of mechanical equipment and artificial methods under carefully controlled conditions of temperature, humidity, and airflow. Although the term dehydrated does not refer to any specific moisture content in the finished product, it is usually considered to imply virtually complete water removal to a range of 1–5% moisture. Products with such low water content can be stored in a moistureproof-package at room temperature for periods well in excess of two years with no detectable change in quality.

Evaporation refers to the use of the sun or forced-air driers to evaporate moisture away from fruit to a fairly stable product. The moisture level of evaporated fruit is approximately 15–25%. In general, sun drying will not lower the moisture content of fruit below 15%, therefore, the shelf life of such fruit products does not exceed one year, unless they are held in cold storage. For the extended shelf life of most evaporated fruits sulfur treatment (except for prunes, unbleached raisins, and dates) or the use of a chemical mold inhibitor is necessary.

Vacuum drying is a method of drying in a vacuum chamber under reduced atmospheric pressure to remove water from the fruit at less than the boiling point of water under ambient conditions. Freeze drying is a method of drying in which the fruit is frozen and then dried by sublimation in a vacuum chamber under high vacuum (an absolute pressure of less than 4.6 mm Hg, the vapor pressure of ice and water at 0°C, is necessary) to around 2% moisture.

Intermediate-moisture fruits (IMF) are dried on the basis of water activity level rather than on the percentage of moisture. Restraint of the water molecules to a degree that prohibits spoilage by microorganisms occurs at different moisture contents depending on the amount and nature of the dissolved material present and to some degree on the insoluble components. Food may be classified as an IMF if it has water activity level greater than a common low moisture fruit (0.2) and less than that of most fresh fruit (0.85). In practice most IMF have a water activity level in the range of 0.65–0.85 and contain 15–30% moisture.

Water activity a_w is defined as the ratio of the water vapor pressure in equilibrium with a food to the vapor pressure of water at the same temperature. Bacteria will not grow if a_w is below 0.9, and yeast and molds are inhibited below a_w of 0.7.

Dehydrofreezing is a hybrid process that combines the best features of both drying and freezing. The process consists of drying fruit to about 50% of its original weight then this intermediate material is quick frozen by standard air-blast freezing techniques. The quality of dehydrofrozen fruit is equal to that of frozen products, as the drying process is discontinued at a stage where quality impairment usually does not occur. The advantages of the process are a 50% reduction in storage and freight charges and even greater savings in packaging costs in comparison with traditional frozen products.

DRYING METHODS

Selection of a drying method for a given fruit is determined by quality requirements, raw material characteristics, and economic factors. Several drying methods are commercially used, each better suited for a particular situation.

The selection of drying methods depends on the following factors:

Form of raw material: liquid, paste, slurry, pulp, thick liquid, and large versus small aggregates.

Properties of raw material: sensitive to oxidation; sensitive to temperature damage, thermoplastic residues, etc.

Desired product characteristics: powder, instant solubility, retention of shape, etc.

Value of the finished product: low, medium, or high price.

The following three basic types of drying process are used for fruits:

Sun drying, or solar drying of fruit crops is still practiced for certain fruits, such as prunes, grapes, and dates.

Atmospheric dehydration processes including (1) stationary, or batch processes (using kiln, tower, and cabinet dryers) are used for apples and prunes and (2) continuous processes (eg, tunnel, belt trough, conveyor, and fluidized bed) are used for apples; spray dryers are suitable for making powders from fruit juice concentrates and drum, or roller, dryers are useful for drying fruit juice concentrates, slurries, or pastes.

Subatmospheric dehydration techniques include (1) vacuum dehydration processes, which are useful for the processing of low-moisture fruits with high sugar content such as peaches, pears, and apricots, and (2) freeze drying, which ensures high flavor retention and minimal damage to product structure and nutritional value, and the finished product with an open structure permits fast and nearly complete rehydration.

Sun Drying

Sun drying of fruit crops is still practiced unchanged from ancient times. This method is limited to climates with hot sun and dry atmosphere, and to certain fruits such as prunes, grapes, dates, figs, apricots, and pears. These crops are processed without much technical aid, by simply spreading fruits on the ground, on racks, or roofs, and exposing them to the sun until dry. Because sun-dried products generally have moisture levels no lower than 15–20%, they have a limited shelf life. It is a slow process, unsuitable for producing high-quality products, and the finished products are often contaminated by dust, dirt, and insects.

Solar Drying

In recent years considerable interest has been focused on the use of solar energy because the rapid increase of fuel costs. In this method solar energy is used alone or may be supplemented by an auxiliary energy source. A simple method of accelerating the sundrying rate of fruit is to paint the trays black. A solar trough, application of mirrors to increase solar energy, and indirect solar dryers applying solar collectors are also recommended to improve the efficiency of this process.

Kiln Dryer

The natural draft from rising heated air used for drying fruit is the oldest type of dehydration equipment still in commercial use. Kiln dryers generally have two levels: gas burners on the lower floor provide heat, and the warm air rises through a slotted floor to the upper level. Food materials such as apple slices are spread out on the slotted floor in a layer about 25 cm deep and turned over periodically by fork.

Cabinet Drying

Cabinet driers are arranged for batch operation and are usually held at constant temperature, tough humidity may decrease during the drying process. They consist of an enclosed chamber fitted with a variable heater, fan for air stirring, together with deflectors for airflow adjustment, outlet air louvers, and adjustable inlet air louvers. Cabinet drying is a particularly useful research tool for establishing the drying characteristics of a new product, prior to a large-scale commercial run.

Tunnel Dryer

The equipment is similar to a cabinet dryer, except that it allows a continuous operation along a rectangular tunnel through which move tray-loaded trucks. The tunnel is supplied with a current of heated air that is introduced at one end. A tunnel dryer provides rapid drying without injury to the fruits and permits a uniform drying process; therefore, it is widely used in drying fruit.

Tunnel driers are classified by the direction in which the air traverses the product. In a parallel flow unit, the fresh material encounters the driest and warmest air initially, and leaves the drier at the coldest end; in a countercurrent flow unit, the air direction is opposite to the movement of the product, so the dry product leaving the drier encounters hot dry air. Multistage driers consisting of three, four, or five drier stages are also used. Such systems are flexible and can achieve close to optimum drying conditions for a wide variety of products.

Conveyor Dryers

A conveyor dryer is a continuous processing equipment and consists of an endless belt that carries the material to be dried through a tunnel of warm circulating air. The speed of the conveyor is variable to suit both the product and the heat conditions. Furthermore, process conditions are usually controlled by designing the system in sections, thus allowing different flow rates, humidities, and temperatures to be set in each section, and by rotating the product when it moves from one section of the belt to the next.

This drying method has the advantage of essentially automatic operation, which minimizes labor requirements. The conveyor dryer is best adapted to the large-scale drying of a single commodity for the whole operating season. It is not well suited to operations in which the raw material or the drying conditions are changed frequently, because extensive hours of start-up and shutdowns make it difficult to produce satisfactory products.

Spray Dryers

Spray drying involves the dispersion of liquid or slurry in a stream of heated air. Followed by collection of the dried particles after their separation from the air. The process is widely used to dehydrate fruit juices.

The general construction of a spray dryer incorporates four main features: (1) heater and at least one fan to produce air at the required temperature and velocity, (2) atomizer or jet to produce liquid particles of the required size, (3) chamber in which the liquid droplets are brought into intimate contact with the hot air, and (4) means of removing the product from the airstream. The final product is delivered as a free-flowing powder.

Drum Dryers

In drum drying, which is suitable for a wide range of liquid, slurried and puréed products, a thin layer of product is applied to the surface of a slowly revolving heated drum. In the course of about 300 degrees of one revolution, the moisture in the product is flashed off, and the dried material is scrapped off the drum by a stationary or reciprocating blade at some point on the periphery. The residence time of the product in the drier is on the order of 2 to a few minutes.

Drum drying is an inexpensive method, however its commercial application is limited to less heat sensitive products. Its usefulness for fruit dehydration is quite limited because the high temperature required, usually above 120°C, imparts a cooked flavor and off-color to most fruit products. Also, the high sugar content of most fruit juices makes them difficult to remove from the drum dryers because of their high thermoplasticity.

Vacuum Dryers

Dehydration under vacuum has certain special merits. Drying can be carried out at lower temperatures than with air drying and heat damage is minimized. Furthermore, oxidation during drying is virtually eliminated.

Vacuum-drying systems have the following components: (1) vacuum chamber, (2) heat supply, (3) vacuum producing unit, and (4) device to collect water vapor as it evaporates from the food. The vacuum shelf dryer, the simplest type of vacuum dryer, is used to process a wide range of fruit products, including liquid, pastes, discrete particles, chunks, slices, and wedges. The equipment consists of a vacuum chamber containing a number of shelves arranged to supply heat to the product and to support the trays on which the product is loaded into the chamber. The shelves may be heated electrically or, more often, by circulating heated fluid through them. The vacuum chamber is connected to suitable vacuum-producing equipment, located outside the vacuum chamber, which may be a vacuum pump or a steam ejector.

Another essential part of a vacuum dryer that has a vacuum pump is a condenser, which collects water vapor to prevent it from entering the pump. A steam ejector, which is often used to create vacuum, is an aspirator in which high-velocity steam jetting past an opening draws air and water vapor from the vacuum chamber.

Freeze Dryers

In freeze drying the moisture is removed from the fruit by sublimation, ie, converting ice directly into water vapor. Therefore, no transfer of liquid occurs from the center of the mass to the surface. As drying proceeds, the ice layer gradually recedes toward the center, leaving vacant spaces formerly occupied by ice crystals. The processing involves two basic steps. The raw fruit is first frozen in the conventional manner and then dried to around 2% moisture in a vacuum chamber while still frozen. The most common type of freeze-drying equipment is a batch chamber system similar to a vacuum shelf dryer but with special features to meet the needs of the freeze-drying process.

Freeze dehydration produces the highest quality fruit products obtainable by any drying method. The porous nonshrunken structure of the freeze-dried product facilitates rapid and nearly complete rehydration when water is added. The low processing temperatures and the rapid transition minimize the extent of various degradative reactions such as nonenzymatic browning, and also help to reduce the loss of flavor substances. But because of the requirements of high capital investment, processing costs, and the need for special packaging, this process is suitable only for high-value products.

PREDRYING TREATMENTS

Preparation of raw fruit for drying is similar to that of canning and freezing. It includes washing, sorting for size and maturity, peeling, and cutting into halves wedges slices, cubes, nuggets, etc. Alkali dipping is used for raisins and prunes. Blanching is used for some fruits.

Sulfur is applied to help preserve the color of dried fruits such as apples, pears, peaches, and apricots. Several sulfite salts and SO_2 gas are generally recognized as safe (GRAS) for use in foods by the FDA. Because the use of sulfur compounds is unpopular, alternative treatments to retard enzymatic browning and other oxidative reactions during drying have been investigated. To date none of these alternatives can completely replace sulfur.

POSTDRYING TREATMENTS

Treatments of finished fruits vary with the kind of fruit and intended use of the products. The addition of an anticaking agent, usually calcium stearate at less than 0.1% concentration, is necessary to prevent caking of dehydrated fruits.

Sweating of dehydrated fruits is a treatment to equalize moisture of the batch. It is usually accomplished by keeping the dried fruit in bins or boxes. Bins are also used for secondary drying to reduce moisture levels of particulate fruits from 10–15% to 3–5%, a range at which drying rates are limited by slow diffusion of water. Temperatures of about 40°C and airflow provided by a blower fan, of about 33 m/min suit the nearly dry product. These conditions minimize the risk of heat damage at a stage when fruit products are most susceptible to degradation.

Screening is often required to remove the unwanted size portion of the dried product, which may be used in other products. Sometimes the fines represent a loss. The removal of unwanted size pieces, or fines, is usually accomplished by passing the dry product over a vibrating wire cloth or perforated metal screens and collecting the fractions separately. The acceptable fraction passes onto the final inspection operation.

The dried product is inspected to remove foreign materials, discolored pieces, or other imperfections such as skin, carpel, or stem particles. Manual and visual selection of most dehydrated products is necessary and is carried out by inspectors while the product is moving along on a continuous belt. In addition to inspectors, magnetic devices are usually installed over the belt to remove metal contaminants.

Packaging. The shelf life of a dehydrated fruit product is influenced to a large extent by its packaging, which must protect the dehydrated product against moisture, light, air, dust, microflora, foreign odor, insects, and rodents. To avoid caking, air-conditioned and dehumidified rooms (below 30% RH) are required to package low-moisture fruits with high sugar content.

Freeze-dried fruits must be packed in inert gas to insure storage stability. Nitrogen gas is commonly used to extend the storage stability of oxygen-sensitive products. In inert gas packaging, a head-space oxygen level of 1–2% is targeted.

PROCEDURES FOR SELECTED COMMODITIES

Raisins

Approximately 25% of the U.S. grape production is made into raisins, processed entirely in California. Raisins are the most important dried fruit crop.

Cultivation for raisin grapes is a little different than for fresh market grapes. Thompson Seedless, Muscat, Black Corinth, and Sultan grapes are the principal varieties processed for raisins. The fruit is hand picked in August from the vine in bunches and set on paper to be dried in the sun for two to four weeks. After the raisins have dried to a moisture content of 9–12%, they are loaded into sweat boxes to equalize and then shipped to the processing plant.

After hand harvest the Golden Bleach raisins are dipped in 0.5% hot lye solution for 3–6 s, sulfured for 4 h, and dried in the sun or in forced-air dryers. Soda-dipped raisins are dipped for 30–60 s in 4% solution of soda ash and Na_2CO_3 at 35–38°C and then processed in the same manner as Golden Bleach, but are not sulfured. From this point, the processing is the same as for natural-dried raisins.

Raisins are processed by a series of screening, destemming, and air separations. These processes are repeated until lightweight particles are removed. Small raisins may then utilized in distilled alcohol products or used in cereal or bakery products.

The raisins are next washed and sent through a dewatering operation, which removes excess surface water. They go through another destemming operation, and their moisture content is adjusted to 18% by water sprays. The fruits are sorted by a mechanical sorter and packaged for storage or shipment.

Prunes

Although plums are one of the most widely distributed fruits in the United States, only one type of plum, the French type, is designated a prune. La Petite d'Agen variety brought to the Santa Clara Valley of California by a French nurseryman 1856, known as California French prune, is used for drying and is grown almost entirely in California. Prunes are the second most important dried fruit crop. The California production account for 98% of the U.S. production and 70% of the world's supply.

Most of California's prune production is harvested in late August mostly by machine. The soluble solids content of juice must reach 22% prior to harvest. Immediately after harvesting, the orchard-ripened fruit is taken to the dehydrator yard where it is washed, placed on large wooden trays, and dehydrated to about 18% moisture in forced-draft tunnel dehydrators. The drying process requires 24–36 h, depending on the size and solids content of the fruit. Three pounds of plum will yield 1 lb of prune.

Dried prunes are processed through a series of screening, grading, and washing steps. Grading involves separation according to size, ranging from 23 to 150 prunes per pound. Hand sorting for cull removal follows, after which the prunes are conveyed to a blancher where they are held from 8 to 20 min to inactivate enzymes and preserve color and flavor. Potassium sorbate and fresh water are then sprayed onto the prunes to maintain proper moisture content and to add further preservative. Fruit to be pitted is sent through automatic pitting machines that either squeeze the pit out with mechanical fingers or punch it out. The pitted or unpitted prunes are again hand sorted for rejects, automatically weighed into boxes or sacks, sprayed with potassium sorbate preservative and sealed. Other popular prune products include prune juice (a water leachate of the prune) and prune paste used in baking and confections.

Apples

Apples are either dried immediately after harvesting or after being stored in cold or controlled-atmosphere storage until a convenient processing time occurs. The best apples for drying are Gravenstein, Pippin, and Golden Delicious varieties. Only artificial dryers are used by the commercial apple-drying industry. Kiln, tunnel, or continuous-belt dryers are commonly used to produce dried apples. Dehydrated (low-moisture) apples are processed in forced-air dryers, such as the continuous-belt dryer, using evaporated fruit as raw material. Some apples are vacuum dried for snack or other specialty product applications.

The process involves sizing, coring, and slicing the fruit pieces to 0.95–1.3 cm ($\frac{3}{8}$–$\frac{1}{2}$ in.) in thickness. The fruit is usually dipped into sodium sulfite solution, then dried in a kiln, tunnel, or continuous-belt dryer to approximately 16–25% moisture. The fruit slices during drying are exposed to the fumes of burning sulfur. Drying in the kiln requires 14–18 h at 65–74°C (150–165°F).

If a tunnel dryer is used, the air entering the tunnel is at 74°C (165°F) with a relative humidity of 25%. At the outlet, the air is at about 54°C (130°F) with a relative humidity of 35%.

If the product is processed to a low moisture state (less than 5% moisture) the dried apples are cut to the desired size, usually 0.64 and 0.85 cm ($\frac{1}{4}$ and $\frac{3}{8}$ in.) dice. Frequently the fruit pieces are instantized by compression or perforation and dried to the final moisture content in a continuous-belt air dryer or in vacuum dryers. An estimated 50% of the sulfur dioxide applied in the kiln-drying process is lost during the secondary drying due to volatilization.

Approximately 700 kg of fresh unpeeled applies will yield 100 kg of evaporated fruit. If the product is processed to a low-moisture state, 100 kg of evaporated apples (24% moisture) will yield approximately 77 kg of dehydrated (low-moisture) apples. Thus 100 kg of fresh unpeeled fruit will yield about 10–11 kg of dehydrated apples.

In the past most dried apples were treated with sulfur. But because the recent controversy about the use of sulfur, untreated apples are available in health food stores.

Apricots

Apricots are picked for drying from mid June to early July when they are fully tree ripened. The best varieties for drying are Royal, Blenheim, and Moorparks. California produces more than 90% of the U.S. crop and 10% of the world crop.

Apricots are either sun dried, or dried in tunnel dryers. The fruit is halved the pit is removed, and the fruit is placed cup up on a flat, wooden tray. The filled trays are exposed to sulfur dioxide fumes for about 12 h. After sulfuring the trays are transferred to a field, exposing the fruit to full sun. Apricots are allowed to dry in this manner for one day, then the individual trays are transferred to a shady area and stacked 3–4 ft high. They are allowed to dry in the stack for approximately one week; then they are removed from the trays, placed into boxes or bins, and ultimately delivered to a packing plant.

Freestone Peaches and Nectarines

Freestone peaches and nectarines are harvested and processed similarly to apricots. After sulfuring, the fruits are placed in full sun for two to three days or longer, depending on weather conditions, at which time they are transferred to shady stack storage and dried for several more weeks. Finally, they are removed from the trays, transferred to boxes or bins, and delivered to the packing plant.

The primary peach for drying is the Lovell, a highly colored variety. Faye Elberta is also dried. Sungrade and Le Grande nectarines are preferred for drying, both are freestone fruits.

Dehydrated (low-moisture) apricots and peaches are processed from the sun-dried (evaporated) fruit to a limited extent. Because of the high sugar content and the sensitivity of the yellow pigment to heat, a vacuum process is necessary to dehydrate these fruits to less than 5% moisture. Vacuum shelf dryers are used for the process. The evaporated fruit halves are sliced or diced before loading on the drying trays.

Figs

The principal varieties grown for drying are Calimyrna, Mission, Adriatic, and Kadota. Figs are usually allowed to dry partially on the trees. In some cases, the trees are lightly shaken at intervals. Figs are usually mechanically gathered from the ground and are typically dry enough to be loosely packed in boxes or bins, although sometimes they are further dried on trays in the sun to a moisture content of approximately 17–18%.

After being transported to the plant in sweat boxes or bins, the figs are normally screened, graded for size, and then inspected to remove insect damaged fruits. The screening is necessary to divide the figs into the required finished product styles. After the first sorting and grading operation, the figs to be stored for later processing are packed in boxes and placed in an airtight chamber and fumigated. This operation is repeated several times over a two-week holding period.

The figs to be processed are conveyed through a coldwater reel washer to remove dust and foreign material. They are then directed to a processing unit where they are immersed in hot water for 5–10 min. Soak time depends on size and variety of fruit being processed. The figs at this point have adsorbed some water and, because of increased susceptibility to mold, are sprayed with potassium sorbate. They are conveyed, typically, over a dewatering belt where they may either be put into small plastic tubs to equilibrate or placed into retorts directly. The retorting process includes exposure to live steam for 2 or 3 min, which further softens the figs. The fruit is air cooled and packaged.

Pears

Pears that are to be dried are allowed to ripen on the tree. Summer Bartlett variety is used for drying in California. The fruits are hand picked and transported to cutting sheds where they are cored and halved. Placed cup up on wooden racks, they are stored overnight in sulfur houses where they are exposed to burning sulfur to prevent browning. The pear halves are removed from the sulfur house, dried in the sun for four to eight days, and then transferred to stacked storage for an additional two to three weeks.

Once dried, the fruit is delivered to the packing plant, where it is usually processed to fill orders. The dried fruit from the field may sometimes be stored as long as several years before being repacked. Pears are little known as a dried fruit and most commonly appear in packages of mixed dry fruits.

BIBLIOGRAPHY

General References

H. R. Bolin, "Relation of Moisture to Water Activity in Prunes and Raisins," *Journal of Food Science* **45**(5), 1190–1196 (1980).

H.R. Bolin and D. K. Salunkhe, "Food Dehydration by Solar Energy," *Critical Reviews in Food Science and Nutrition* **13**, 327–354 (1982).

M. Gee, D. F. Farkas, and A. R. Rahman, "Some Aspects for the Development of Intermediate Moisture Foods," *Food Technology*, **31**,(4), 58–84 (1977).

S. A. Goldblith, *Freeze Drying and Advanced Food Technology*, Academic Press, Orlando, Fla., 1975.

J. A. Kitson, "A Continuous Process for Dehydrofreezing Apples, *Journal of Can. Inst. Food Technol.* **3**, 136–138 (1970).

A. L. Moyles, "Drying of Apple Purées," *Journal of Food Science* **46**, 939–942 (1981).

W. Napper, *Fruits and Tree Nuts. Situation and Outlook Yearbook*. USDA Economic Research Service **TFS-250**, Aug. 1989.

L. P. Somogyi and B. S. Luh, "Dehydration of Fruits," in J. G. Woodroof and B. S. Luh, eds., *Commercial Fruit Processing*, 2nd. ed., AVI Publishing Co., Inc. Westport, Conn., 1986.

L. P. Somogyi, "Prunes, a Fiber-Rich Ingredient," *Cereal Foods World* **32**(8), 541–544 (1987).

S. L. Taylor and R. K. Bush, "Sulfites as Food Ingredients," *Food Technology* **40**(6), 47–52 (1986).

W. B. Van Arsdel, M. J. Copley, and A. I. Morgan, Jr., *Food Dehydration,* Vols. 1 and 2, 2nd. ed., AVI Publishing Co., Inc., Westport, Conn., 1973.

Laszlo P. Somogyi
SRI International
Menlo Park, California

FRUIT PRESERVES AND JELLIES

The fruit preserve and jelly categories are certainly a major milestone in food evolution. Their history dates back to ancient times, when a confection or a dessert has been documented as a part of the meal. The use of sugar widened the possibilities for preserving fruits. In fact, during colonial times, a jam sometimes formed while fruit was being boiled. This jam or gel formed when the correct proportions of pectin, sugars, and acids occurred. Jam and jelly production was once considered an art; now it is a science. More is known about the components necessary to produce this kind of gel. This knowledge has led to other applications such as stabilized fruit fillings and sauces, processed fruit juices, canned fruits, frozen desserts and confections.

DEFINITIONS AND STANDARDS

The name preserves covers a broad range of products including jams, butters, marmalades, and conserves, as well as ordinary preserves. Preserves contain the largest fruit pieces, whereas jams contain smaller pieces which are crushed or chopped with added acid. Fruit butters are made of fruit pulp cooked to a smooth consistency. They are pressed through a coarse strainer and are more concentrated than jams. Scorching can be a problem because of their high viscosity. Marmalades have the characteristics of both jellies and preserves. They contain thin citrus peel or fruit pieces and are chiefly made from citrus fruits, alone or in combination with other fruits. Conserves are similar to jams, except that two or more fruits are cooked together and raisins and nuts can be added. Jellies are in a class by themselves. They are clear, sparkling spreads in which fruit juice as the source of flavor, and, in some cases, the thickening agent.

The Federal Standards and Definitions do not differentiate between preserves and jams (1,2). A preserve is minimally 45 parts prepared fruit with 55 parts of sugar and is concentrated to 65% or higher solids, resulting in a semisolid product. Jellies are similar to preserves, with 45 parts of clarified fruit juice and 55 parts of sugar, resulting in a minimum of 65% solids. Both categories can utilize a maximum of 25% corn syrup for sweetness as well as pectin and acid to achieve the gelling texture required. Fruit butters are prepared from mixtures containing not less than 5 parts by weight of fruit to 2 parts of sugar.

GELATION–PECTIN MECHANISM

Typical Gel Formations

Gelation, the formation of the polymer network which gives commercial fruit preserves and jellies their texture, is dependent on four essential ingredients: pectin, sugar, acid, and water, added in the correct proportions. A pectin gel is a system resembling a sponge filled with water. The polymer is in a partially dissolved, partially precipitated state. The chain molecules are locally joined by limited crystallization, forming a three-dimensional network in which water, sugar, and other solutes are held. Some fruits such as tart apples, red and black raspberries, oranges, and cranberries have enough pectin and acid present. Still others, such as ripe apples and plums, contain sufficient pectin, but lack enough acid. Pectin or acid must be added when using most fruits. Sugar is always needed when high-methoxyl pectin is used.

Since fruits vary widely with regard to maturity, climatic conditions, and storage, it can be difficult to ensure the proper composition. Fruit should be picked just before processing to ensure taste and texture. It should be picked as ripe fruit in the early morning to ensure quality. Overripe fruit will have reduced sugar quality and the pectin will suffer molecular breakdown from enzyme activity. If fresh fruit is not available, frozen, cold-pack, or canned fruit can be used for jams and preserves.

The juice of grapes, currants, lemons, sour oranges, and grapefruits contains sufficient pectin and acid for jelly manufacture. Strawberries, rhubarb, and apricots usually contain sufficient acid, but may lack pectin. On the other hand, sweet cherries and quinces may lack acid yet have enough pectin. Commercial pectin, either liquid or powder, can be added as a supplement. The viscosity of a fruit juice is an index of its gelling power.

Pectin is found in the flesh, skins, and seeds of most fruits. It can be extracted when fruit is boiled. Pectin is a complex carbohydrate consisting of polygalacturonic acid chains having a wide variety of molecular weights. The chains contain some carboxyl groups which are partially methylated, forming the ester known as pectin.

Generally, a degree of methylation (DM) of 50% divides commercial pectin into two main groups: high-methoxyl (HM) pectins and low-methoxyl (LM) pectins. The LM-pectin group includes both conventional and amidated versions. The HM-pectins are the predominant choice for the standard jellies and preserves. The LM-pectins can be utilized in low-sugar fruit spreads.

The degree of methylation of pectin has a critical influence on the solution and gelation characteristics of preserves and jellies. The highest DM that can be achieved by extraction of the natural raw material is about 75%. Pectins of DM ranging from 0 to 70% are produced by demethoxylation in the manufacturing process.

The DM of HM-pectins controls their relative speed of gelation; hence the terms slow-set and rapid-set high-methoxyl pectin. If a higher degree of methylation of pectin is used, the higher will be the pH required for a fast set. A fast set is necessary to suspend fruit pieces and prevent fruit flotation or sinking. A slow set is necessary for a clear jelly, so that air bubbles are removed. The pectin's quality is standardized on the basis of its 150° USA-Sag Grade. This method measures the sag of standard 65% sugar–water gels under controlled conditions. The 150° standard means that under controlled conditions, 1 pound of 150-grade pectin will gel 150 pounds of sugar.

Another method for testing pectin gels is the Voland-Stevens LFRA texture analyzer in which the gel's elasticity is exceeded. The Tarr-Baker gelometer was the original jelly strength tester. However, after it was discontinued in 1965, the Voland-Stevens analyzer was declared an acceptable replacement because of its operational similarity, portability, compactness, reproducibility, and ease of use and calibration (3). This flexible instrument can be attached to a printer to measure elasticity and other attributes of a gel. This method is comparable to the SAG method.

Jellies are usually produced at a pH of 3.1 and jams at 3.3. High-methoxyl pectin can gel sugar solutions with a minimum of ca 55% soluble solids within a pH range of approximately 2.0 to 3.4. For any soluble-solids with a value above 55%, there is a pH value at which gelation is optimal for a particular HM-pectin and a pH range within which gelation can be controlled.

Sugars have a general dehydration effect on HM-pectin solubility. At higher solids values, there is less water available to act as a solvent for the pectin. Hence, there is an increased tendency to gel. Because gelation relies on the proper balance of soluble solids and pH in the medium, it is possible to compensate for a reduction in soluble solids by reducing the pH. Any HM-pectin can gel rapidly or slowly and the rate can be controlled by the soluble solids and pH.

An attempt was made to quantify fruit content in jams by combining chemical composition data, particularly the inorganic elements which are stable to processing such as ash, magnesium, and potassium, with the rheological forces such as yield stress and flow index in a regression analysis that could explain 90% of the variability in fruit content (4).

Novel Gel Systems

The typical manufacturing methods for jams and jellies use the four necessary components: fruit, pectin, sugar, and acid. Some combination pectin sources have been devised. In one (5), an emulsifier is added to the sugar surface and blended with the bulk of the sugar. A very fine pectin is then mixed with acid and combined with the emulsifier–sugar complex. The emulsifier acts not only as a glue, causing the fine pectin particles to adhere to the sugar surface, but it also acts as an antifoaming agent and a dispersing agent in the final gel production.

Another convenient product, a one-step pectin gelling composition, has been developed (6). Pectin particles are mixed with moistened coarse sugar particles. Acid may be added in a dry form and will adhere to the sugar or it can be predissolved in water and sprayed onto the mix. Much less pectin can be used in this method as compared to dry blending fine pectin, acid, and sugar. This is due to the use of larger sugar particles and smaller pectin particles, so that the pectin dissolves faster while the concentration of dissolved sugar solids is retarded.

The LM-pectin offers another gel formation method for preserves and jellies. However, the gel formed does not conform to the Federal Standard of Identity. The use of this pectin does result in reduced-sugar jams, jellies, and preserves. The LM-pectin group includes both conventional (acid demethylated) and amidated types. These pectins require calcium ions for gelation, not sugar or acid.

The gelation of LM-pectin is controlled primarily by the reaction of a divalent cation with the acid groups of the pectin chains. LM-pectin can be used at solids levels as low as 10%. The pH range for LM-pectin is 3.0–6.0 because the role of acid is minimized. In order for successful gelation to occur, 50 to 100% of the acid groups must be complexed with calcium. The amidated pectin, which has fewer free acid groups, requires less calcium for gelation and relies on hydrogen bonding between the amide and free acid groups. Amidated pectins form gels that are more rigid than those formed with conventional pectins. Conventional pectins produce a thickening effect and are aptly used for jams and preserves; amidated pectins are used for jellies.

CARBOHYDRATE SWEETENERS

In HM-pectin systems, sugar accounts for more than 50% of the total weight and 80% of the total solids in a jam. It contributes solids; maintains microbiological shelf life; provides sweetness, body, and mouthfeel; contributes to gelation; and adds color and shine to the jam.

Other sugars which can be used are glucose syrup, dextrose, invert sugar syrup, and honey. When other sugars are substituted for sucrose in jam, the effects on HM-pectin gelation are as follows:

- Inversion of sucrose reduces gel strength and lowers the gelling temperature.
- Glucose syrup usually reduces the gel strength. High-dextrose-equivalent (DE) glucose syrups decrease gelling temperature; regular-DE syrups increase it.
- Sugar alcohols such as sorbitol and xylitol are used in dietary products. Sorbitol jams can be made with HM-pectin, soluble solids of 65%, and a pH of 3.0. Xylitol has limited solubility. At the 39% limit of solubility, gelation with HM-pectin can be obtained when the pH is lowered to 2.7.

A study was developed to compare some effects of gelling agents and sweeteners in high- and low-sugar-content carbohydrate gels (7). High-methoxyl pectin, low-methoxyl pectin, carrageenan, and alginate gels were the gelling agents, while sucrose and high-fructose corn syrup (HFCS) were the sweeteners. Soluble solids ranged from 35 to 65% with polydextrose as the bulking agent. The properties compared were bound water, water activity, syneresis, texture, and overall taste. The alginate–HFCS gel (35% soluble solids) closely simulated HM-pectin gels because of the comparable spreadability properties. The water binding property of sucrose exceeded HFCS with most gel systems, except where a combination of LM-pectin and carrageenan was used. The water binding served as an index for predicting syneresis or weeping, spreadability, and shear. However, a synergy of carrageenan with pectin or alginate with pectin resulted in increased bound water compared to individual gums.

Another study to improve nonsugar jam systems used the addition of xylitol or saccharin to veltol to improve the color and taste of apricot jam with minimal changes during storage (8). The use of aspartame in fruit spreads was minimally documented until its stability and effectiveness were measured in 1986 (9). A high-performance liquid chromatographic method was the quality control tool for monitoring levels of aspartame. This study predicted an average half-life of 168 days for aspartame in a fruit spread kept at 25°C, which is in agreement with a prediction of 170 days. Stability at 25°C was found to be pH dependent. Other studies showed that the Maillard reaction was not a factor contributing to the aspartame loss at 25°C. It occurs only at higher temperatures.

Gel strength and gelling temperatures of both amide and conventional LM pectins are influenced by the type of sugar used in the gel. Gels prepared with HFCSs have significantly lower gel strengths at all calcium levels than gels prepared with sucrose. However the use of 42- and 62-DE corn syrups give higher gel strengths than sucrose in LM pectin formulations.

PROCESSING TECHNIQUES

Traditional Process

The traditional process used for preserve and jelly manufacture is the open kettle, batch boiling technique. The boiling process, in addition to removing excess water, also partially inverts the sugar, develops the flavor and texture, and destroys yeast and mold. In jelly manufacture, the fruit is boiled to extract the pectin and destroy the pectin-hydrolyzing enzymes. The juice is then separated by straining or pressing, and the press cake is boiled with more water to obtain more pectin. Pectin deficiency is remedied by the addition of commercial pectin. Added pectin needs to be dispersed with sugar to ensure uniform distribution. Either liquid or dry sugar is added.

A second boiling step is necessary to concentrate the juice to the critical point for gel formation of particular pectin–sugar–acid system being used (10). Extended boiling causes acid volatilization, pectin breakdown, and losses in flavor and color. Vacuum concentration (50–60°C) produces a higher quality jelly than atmospheric pressure boiling (105°C). A refractive index reading indicating soluble solids content is the point at which the concentration stops. A flow diagram for these processes can be found in reference 11.

The pH of the jelly will determine the set temperature of the pectin. The setting temperature of a jelly at pH 3.0 can be lowered approximately 10°C with rapid-set pectin or 20°C with slow-set pectin by decreasing the acidity to pH 3.25 (12).

For preserves and jams, the same procedure is used except that the fruit pulp is not strained. Rapid-set pectin is perferred to suspend the fruit more evenly and to minimize settling out.

Both products are packed hot, at about 85°C, into containers which are then sealed. Hot sterilized jars with hot sterilized lids and caps can also be used. Once filled, the jars are turned over to heat the lids and then returned to the upright position. The hot water bath technique can be used in place of the above procedure. The jars are placed in a water bath and boiled from 5 to 15 minutes, depending on the fruit. This is a better method for deterring mold growth for fruit preserves and jellies. The USDA sanctions the use of paraffin or a 2-piece metal lid and screw-band for sealing jellies, but it highly recommends processing them for extra safety precaution.

The continuous process utilizes a premix for its efficiency. The APV system uses a plate evaporator for jellies. The Alfa-Laval system uses a scraped-surface heat exchanger for preserves because of the fruit pieces involved. The soluble solids content or Brix degree is monitored by either an inline unit or an automatic unit with electrical feedback to control the evaporator (13,14).

The filling temperature for these processes should be 85–95°C. This will ensure proper setting, fruit distribution, and a sterile product. A rotary multiple-piston displacement machine is used for filling. Speeds range from 100 to 600 jars per minute. Jars are washed and preheated before filling. Capping occurs immediately afterward, ensuring a vacuum seal.

The pack will be sterile in most cases if it is filled at not less than 85°C and capped using a steam flow closure. If a steam flow closure is not used, the sterile pack will require the use of a steam-sterilizing unit to cool it. Jars can be cooled continuously using water sprays of about 60°C to avoid thermal shock. Subsequently, 20°C water is used to finish this process. The jar temperature should be above ambient. The vacuum seal is checked by a non-vacuum detector to ensure a hermetic seal. Jars are passed through a visual inspection point to locate and remove jars with unfavorable attributes such as foreign material, floating fruit pieces, and bubble formation. The jars are then passed to labeling machines, packed in cases or trays, and shrink-wrapped.

The order of addition of ingredients is very important. As in vacuum cooking, a slow-set pectin is preferred to limit the chance of preset with the pectin. Most manufacturers use pectin solutions that are easily prepared and dissolve much more effectively than powder. The pectin solution can be added before or after concentration of the batch. Addition of the pectin after concentration results in a faster cooking rate due to the lower viscosity of the batch during concentration. For jelly manufacture, it is best to add the pectin solution before cooking is completed. The addition of 15–25% corn syrup deters crystallization from occuring due to sugar inversion resulting from low-temperature vacuum cooking (15). Low-sugar jams require less cooking than jellies and can use larger quantities of HM-pectin to improve gel quality.

Modern Processing Techniques

A new process to replace canning has been developed in Sweden by Alfastar (16). This multitherm process is said to preserve food for several months without chemicals as well as to achieve a fresher tasting product. The process is rapid, with even heating through the product. The processing temperature is 150°C and can be reached in less than a minute.

The microwave oven is the latest method for processing jams and jellies for the homemaker (17). An oversized container must be used for this process to avoid boiling over. Fruit, sugar, and some butter are mixed and allowed to stand for 30 minutes. The butter will help to deter the frothiness which may develop. The mixture is microwaved until it boils, with frequent stirring. It is then cooked for 10–13 minutes more in the microwave. Jams produced from this mixture keep well in the refrigerator for several months. They can also be canned for greater safety.

The no-cook freezer jams are by far the easiest of these processing techniques. The fruit is mixed with an appropriate amount of sugar. This technique does not rely on pectin as much because pectin is not heated to form the gel bonding that cooking at high temperatures creates. Lemon juice, if any, is added to the pectin and then it is stirred into the sugar–fruit mix. The mixture is placed in sterilized containers and covered with 2-piece metal lids, and kept at room temperature for 24 hours before placing in the freezer. Once opened, it can be stored up to 3 weeks in the refrigerator.

QUALITY PARAMETERS

The overall quality of fruit preserves and jellies has increased because of improvements in processing techniques, increased knowledge of fruit characteristics, and competitive situations. Fruit quality control is most important because it affects flavor, odor and color of the preserves and jellies.

The following criteria are important for manufacturing a quality product:

- Fruit appearance, ripeness, and solids must be optimized.
- Fruit juice must be clarified properly to ensure a clear jelly.
- Appropriate pectin grade, 120 to 200, must be used.
- Sugar assay and appearance must be appropriate.
- Corn syrup buffering capacity, solids, and appearance must meet minimum specifications.

Processing must be monitored in the areas of appearance, flavor, color, viscosity, pH, and solids. Powder pectin use results in a dark, rich color and stiff gel. On the contrary, liquid pectin, which is less concentrated, makes a less stiff product. Typically, high-DM pectin must be conditioned to increase its set time and to optimize the DM. This pectin follows first-order kinetics in its stability.

Jelly quality attributes are similar to those of jam except that jelly is clear and bright, does not contain fruit pieces, holds its shape when unmolded, and cuts easily with a spoon. A stiff jelly is so firm that the mold will retain the mold shape.

Some problems that can occur are the following:

Cloudy Jelly. Unclarified juice, underripe fruit, or pouring so slowly into containers that gelling occurs can result in a cloudy jelly.

Color Changes. Darkening at the top of the jars can be caused by storing them in too warm a place or by an imperfect jar seal.

Color Fading. Fading can occur with red fruits if they are stored in too warm and too bright areas or stored too long. The natural colorants in the fruit are highly susceptible to high temperatures and light. Another possible cause of color fading could be that the processing was not sufficient to either destroy the enzymes which can affect color, or that the processing time elevated the temperature, causing color destruction. Trapped air bubbles can also contribute to the chemical changes caused by oxidation.

Crystal Formation. An excess of sugar can "seed" the jelly when HM-pectin is used. This excess sugar comes from overcooking, too little acid, or from undercooking the recipe. Tartrate crystals can form in grape jelly, if juice is left to stand in the cold for several hours before being used. Moreover, if the glass interior is scratched, seeding can occur.

Floating Fruit. This can result either from undercooking or from not driving off enough water to create the viscous gel necessary to maintain even fruit distribution. Fruit pieces not properly cut or not ripe enough can also lead to floating fruit.

Gummy Jelly. Gummy jelly can result from overcooking and creating invert sugar.

Mold. The appearance of mold can be the result of imperfectly sealed jars and airborne contamination, if the full sugar complement was not used. The water activity created makes a favorable environment for contamination from the jars if they were not properly sterilized or simply underprocessed. A change in the appearance and off-odor or fermented smell will not necessarily occur. However, mold is often seen before taste is affected.

Weeping Jelly. Syneresis in jelly can be overcome by not overcooking, not storing in a warm place, and using the appropriate amount of pectin or acid.

Stiff or Tough Jelly. Overcooking or using too much added pectin delivers a tough jelly.

Jelly Failures. An improper balance has occurred when a gel is not formed. Inaccurate measurement, insufficient cooking, overcooking, or increasing the recipe prevents the pectin from building its network.

In a study (18), reduced boiling time improved both the aroma and flavor of fruit preserves. However, it was noted that the retention of flavor and color can be protected during the shelf life by means of modified packaging and appropriate storage practices, thus eliminating light and oxygen and storing at 15°C.

CURRENT TRENDS WORLDWIDE

The jam, jelly, and preserve market is expected to reach $1.2 billion by 1990, with the upscale market showing the greatest potential for growth. Gourmet fruit spreads, preserves, jams, and jellies, including more imports and exotic flavors, are the new products. According to *Food & Beverage Marketing*, the estimated growth rate for these products is 3% per year. The nutritional and health benefits take the forms of less sugar and more fruit; no sugar is added because high-sugar fruit juice is used instead of 100% fruit. A new, more convenient packaged powdered pectin has sugar added to it to be used as a sugar and pectin mix for preparing jams and jellies at home. A new additive which is awaiting FDA approval is polydextrose. This will be a bulking agent used in lower sugar spreads. Argentina has a line of dietetic jams, the Netherlands markets lower calorie jams and preserves, and Japan produces a new jelly drink.

The packaging revolution has also affected this market in the form of squeezable plastic containers for convenience, trays and containers of layered foil materials which add longer shelf life to products, and contemporary plastic jelly jars and lids featuring dinosaurs.

SUMMARY

Fruit preserves and jellies have become segmented over the last few years and will continue to grow via the gourmet marketplace. They have grown right along with the new processes, continuous operation, aseptic processing, and microwaving. There is still opportunity for growth in refining current gel systems and developing more convenient ones for either commercial or home use.

BIBLIOGRAPHY

1. "United States Standards for Grades of Fruit Preserves (or Jams)," 4th ed., Dept. of Agriculture Food Safety and Quality Service, Washington, D.C., Jan. 4, 1980.
2. "United States Standards for Grades of Fruit Jelly," 2nd ed., Dept. of Agriculture, Food Safety and Quality Service, Washington, D.C., Sept. 3, 1979.
3. S. A. Angalet, "Evaluation of Voland Stevens LFRA Texture Analyzer for Measuring the Strength of Pectin Sugar Jellies," *Journal of Texture Studies* 17, 87–96 (1986).
4. E. Costell, E. Carbonell, L. Duran, "Chemical Composition and Rheological Behavior of Strawberry Jams," *Acta Alimentaria* 16(4), 319–330 (1987).
5. U.S. Pat 4,686,106 (Aug. 11, 1987), R. Ehrlich and R. Cox (to General Foods Corporation).
6. U.S. Pat. 4,800,096 (Jan. 24, 1989), D. DiGiovacchino, R. Carlson, R. Jonas, and S. Marion (to General Foods Corporation).
7. D. L. Gerdes, E. E. Burns, and L. S. Harrow, "Some Effects of Gelling Agents and Sweetners on High and Low Sugar Content Carbohydrate Gels," *Lebensmittel-Wissenschaft und Technologie* 20, 282–285 (1987).
8. M. Ragab, "Characteristics of Apricot Jam Sweetened with Saccharin and Xylitol," *Food Chemistry* 23, 55–64 (1987).
9. M. C. Dever and H. J. T. Beveridge, D. B. Cumming, D. R. MacGregor, "Measurement and Stability of Aspartame in Fruit Spread," *Canadian Institute of Food Science Technology Journal* 19(2), 86–88 (1986).
10. M. Glicksman, "Pectins," *Gum Technology in the Food Industry,* Academic Press, New York, 1969, pp. 159–190.
11. R. W. Broomfield, "Preserves," *Food Industries Manual,* 22nd ed., Avi Publishing Co., Inc., Westport, Conn., 1988, pp. 335–355.
12. R. M. Ehrlich, "Controlling Gel Quality by Choice and Proper Use of Pectin," *Food Product Development* 2(1), 36–42 (1968).
13. "Continuous Processing Systems," Chilton Food Engineering, Chilton Co., Feb. 1984, pp. 90–92.
14. U.S. Pat 4,562,085 (Dec. 31, 1985), F. Ruggiero (to Alfa-Laval, Inc.).
15. D. Tressler, "Jams, Jellies, Marmalades, and Preserves, Candied and Glaced Fruits, Fruit Syrups and Sauces," *Fruit and Nut Products,* Vol. 3, Avi Publishing Co., Inc., Westport, Conn., 1976, pp. 76–98.
16. "Why the Food Industry May Blow Hot and Cold," *Financial Times,* 24, (July 10, 1986).
17. L. Brandt, *Canning, Freezing and Drying,* Lane Publishing Co., Menlo Park, Calif., 1981.
18. D. O'Beirne and S. Egan, "Some Effects of Reduced Boiling Time on the Quality of Fruit Preserves," *Lebensittel-Wissenschaft and Technologie* 20, 241–244, (1987).
19. D. O'Beirne and G. Kelly, "Some Physical and Organoleptic Effects of Adding Skin Milk Powder and Cream to Fruit Preserves," *Lebensmittel-Wissenschaft und Technologie* 18, 47–51 (1985).

MARNIE L. DEGREGORIO
CHARLES J. CANTE
General Foods USA
White Plains, New York

FRYING TECHNOLOGY

Frying is a process of dehydrating food from the surface inward. The process uses triglyceride-based oil (lipid) from an animal or vegetable origin to transfer thermal energy from a heat source to food immersed in the oil. The efficiency of heat transfer is mediated by surfactant chemical species (wetting agents) included or formed in the oil that controls the contact time between hydrophobic oils and aqueous foods.

A dynamic balance occurs between water movement from and oil movement into the frying food at a given temperature (typically 150–190°C) and food immersion time. The dynamic balance and the kinetics of the process are further mediated by the state of degradation of the oil as influenced by its exposure to primarily heat, oxygen, water, and chemicals and particles from the frying food.

ENGINEERING PRINCIPLES

It is possible to calculate many aspects of heat and mass transfer most important to the frying process based on the expected loss of water and the mass of food to be dehydrated. Unfortunately, these calculations do not take into account the constant changes occurring in a degrading heat transfer medium (the oil) and the accumulation of surfactant species in the oil due to both food and process influences.

There are five stages in the life of a frying oil that produce, in sequence, raw, cooked, and overcooked food. The following is a description of how frying proceeds and why surfactant chemicals control the kinetics and dynamics of the frying process. From the perspective of physical chemists and process specialists, the cooking of food in an oil can be reduced to simple engineering principles with parallel simple measurement and control procedures. This is a new paradigm of frying and is different from that of the paradigm of organic chemists and food scientists who initially studied the complexities of frying oil and food chemistry.

The model of understanding frying in terms of physical chemistry and engineering leads to the belief that the foremost way to judge frying and frying oils is by evaluating the physical properties of fried foods. Only the process variables affecting the physical properties of fried foods can be controlled in the engineering sense. Temperature profiles, water loss from, and oil absorption into food are amenable to process control. On rare occasions, the pressure over the frying oil is also controlled.

Elements such as flavor development and typical finished food color are not primarily controlled by the process. Rather they are dependent on the source of the oil, the content and type of surfactants, the type of food fried, and a range of organic reactions, only some of which depend directly on process variables.

Two traditional means of controlling the transfer of thermal energy to frying food are heater temperature control (at designed energy flux) and residence time of the food in the heated oil bath. Overheated oil at the surface of the heaters is reduced to carbon deposits (coke) on the heater surfaces. As the carbon layer builds and becomes an insulating jacket, the heater cycle on times and temperatures are increased to keep the oil at frying temperature. A new oil somewhat resists this process, and heat transfer rates and ratios are essentially a static system. As the oil degrades, however, and products such as thermally formed and food-formed surfactants increase, the dynamics of the heat-transfer system accelerates. An oil makes successively better contact with the food exterior, and thus excessively dehydrates that layer. Thermal energy is expended to convert more and more surface water to steam. This ever-deepening dehydration phenomenon robs energy that otherwise would go to heat and cook the interior of the food.

The time food spends in heated oil can be varied to achieve a particular degree of cooking. Times vary from blanching for 30 s to cooking a chicken for 20 min. Manual placement and removal of basketloads of food controls residence time in smaller fryers. Larger, or continuous, fryers often use conveyor systems to control frying food residence time. Unfortunately, heater designs and residence time variations leave out considerations of the chemical changes that take place in the oil with time. Such changes have been described by analytical and organic chemists in depth, but physicochemical changes that affect heat and mass transfer of the oil with use have been largely, although not completely, ignored. While electromechanical devices for controlling fryer heaters and residence time are well developed (based on the engineering concept that the thermal properties of oils are constant), further analysis of frying systems has ignored available information related to chemical changes taking place in the frying oil.

FRYERS

Fryers are basically oil vats with heaters. The heaters transfer thermal energy into frying oil. The frying oil in turn transfers heat energy to the food's surface. Heat not used up at the surface is then conducted into the interior of frying food. Heaters are turned on and then off as predetermined low and high set points of temperature (energy) in the oil are reached.

The heaters operate at temperatures higher than the maximum set point temperature of the oil. The energy density (flux) of the heater's surface is high to heat the oil quickly to operating temperature regardless of energy losses to the environment. Cycle rates are adjusted by a set point thermocouple to counter overall heat loss at the heater surface, which includes energy dissipated to oil, fryer machinery, and also to further heat-loss in the oil due to food loading.

Engineering of heater configuration is usually based on thermal calculations supposing that frying is really a food dehydration process. Heater design calculations entail determining the specific heat of the dry mass of the food, the percent water, change of state of water from ice to water, and water to steam, as well as the specific heat of the frying oil (figured as a constant representing new oil). For production fryers, heater placement and energy density in the oil vat are designed to compensate for food-loading points (high density) and take-away points (low density) and to compensate for heat sinks such as conveyers, crumb collectors, and filtering systems.

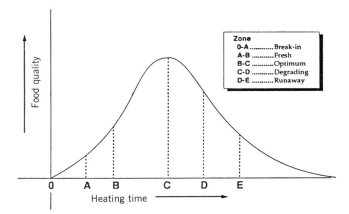

Figure 1. Frying oil quality curve.

FOOD AND OIL QUALITY INTERACTIONS

A broad model describing all the qualities of frying oils and foods prepared in these oils is a continuum of frying oil degradation changes that could be associated with fried food qualities. Five stages of frying oil degradation are identified.

Figure 1 suggests that all frying oils behave similarly in the frying environment, and that their bulk effects change slowly, even though the minor products of degradation may have a strong relationship to consumer acceptance of foods. Furthermore, each oil degradation stage can have different associated analytical values depending on the food being cooked. For instance, the degraded oil stage end points for potato chips, breaded vegetables, and fast food are found at 13%, 17%, and 25% total polar materials (TPM) content, respectively.

Within narrow windows, these TPM chemical index values seem relatively invariant across the food industry and are closely linked with the heating history of the oils. However, specific chemical markers used as process measurement end point specification values such as percent free fatty acids (FFA) and other relatively minor degradation products vary with the mix of foods being fried. The percent FFA end point values for the degraded stage are 0.5%, 2.0%, and 5.0%, respectively, for snack food, processed meat food, and fast-food industry segments. A generalized table for fast foods is shown in Table 1.

FRYING OIL DEGRADATION

The simple hydrolysis of oils (triglycerides) due to heat and moisture at ordinary frying temperatures forms quantities of free fatty acids. The free fatty acids then react with oxygen, each other, food juices, and a variety of intermediate degradation products in the oil. These new chemicals in frying oil are either volatile (and leave) or are soluble (and some are even colloidally suspended) in the triglycerides that comprise the oil. Monoglycerides and diglycerides formed after hydrolytic splitting off of fatty acids are generally more reactive than the original triglycerides. Acrolein, formed by the dehydration reaction of glycerol may be an important toxicant or irritant in the oil and frying vapor, but it is rarely investigated.

Table 1. Values for Fast-Food Degradation by Stages

Parameter	0	Break-In	Fresh	Optimum	Degraded	Runaway
Total polars, %	<4	10	15	20	25	35
FFA, %	0.02	0.5	1.0	3.0	5.0	8.0
Monoglycerides and di-glycerides, %	2	6	7	3	1	0.5
"Soap," ppm	0–7	10	35	65	>150	>200
Polymer, %	0.5	2	5	12	17	25
OFA, %	0.01	0.08	0.2	0.7	1.0	2.0
Metals, ppm	1	5	10	15	25	50

The degradation of frying oil is also accelerated by the soluble chemicals either left in the oil as residues, additives, and artifacts of refining, or are introduced into the oil by frying food. As plant and animal cells (whole food) explode during cooking, the watery (plasma) juices of the cells introduce foreign chemicals into the oil. A similar process occurs when watery systems in formed and fabricated foods are introduced into hot frying oil.

Exposure to heat, light, crumbs, reactive metals, and food juices cause frying oil to react more and more rapidly with its own degradation products, with oxygen from the air, and with food-introduced contaminants. The ongoing combination of chemical changes finally causes a fry chef or line operator to dispose of a batch of oil because poor oil makes poor-quality fried foods.

Frying oils cook poorly when they are degraded by physical (thermal) stress and chemical changes. With use, an oil changes its ability to hold heat and increases its rate of transferring heat. A fresh oil, by comparison, holds heat well and acts like a thermal reservoir. Fresh oil also slowly releases heat to the surface of frying food. These are the most common chemical tests performed by inspectors in food production facilities, restaurants, and laboratories to indicate the approach of the specified discard point of a used frying oil.

Total Polar Materials (Chemical Index)

As an oil breaks down, a group of polar materials includes most of the nontriglycerides. A column chromatography method allows the determination of the percent polar materials in frying oils in the laboratory. The determination may also be made with a quick test on site; the quick test is then correlated to the standard method.

Polymer in Oil (Chemical Marker)

The formation of polymer in frying oils is directly related to the time an oil is exposed to frying temperatures. At elevated temperatures unsaturated fatty acids in triglycerides, and also those fatty acids free in solution combine with oxygen from the air. Some of the oxidized fatty acids cross-link to form high molecular weight polar compounds known as dimers. The dimeric triglycerides and fatty acids again cross-link during further exposure to frying conditions and form polar trimers, tetramers and oligomers. Observation supports the hypothesis that no later than at about 50 h of heating and frying the oil becomes saturated with polymer, and any excess polymer plates out onto the walls of the fryer and onto the fry basket.

Free Fatty Acids

The measurement of free fatty acids as an indicator of oil quality is a widespread practice because it is easy to do and the percentage of free fatty acids does rise with increasing use of a single batch of frying oil. However, the measurement is nearly meaningless because much of the free fatty acids are continually being changed to volatiles and nonacidic, nonvolatile decomposition products. The measured percent of free fatty acids just documents what is left over from other degradation processes. In addition, when fresh oil is used to top up a fryer, the free fatty acids in the original batch are diluted to further confuse the possible relationship between oil quality and oil use.

An accepted reason to measure percent FFA is to be able to predict the approximate resistance to oxidation of the oil on a potato chip distributed in a plastic bag. The free fatty acids are readily oxidized to yield a rancid odor.

Surfactants

Surfactants (wetting agents) in frying oil control the rate and amount of water removed from the surface of cooking food, overall oil pickup, and, by indirect (conductive and perhaps convective) forces, the degree of cooking of the interior. Although fatty acids themselves cannot be surfactants, degradation products of fatty acids can be, as exemplified by the soaps formed by combination of fatty acids with metals leached from food cells and food ingredients. These parts per million of soaps and other surfactants, such as monoglycerides, diglycerides, phospholipids, and oxypolymer fractions, control how (and even if) a food will fry in oils. The efficiency of heat transfer from heater to oil to food, the oil soakage into food, and the foaming tendency of the oil (oxygen incorporation route) are mediated by traces of surfactants.

Color in Oil

The color of an oil is not related to its ability to deliver light-colored, delicious food. The color bodies do not seem to enter into the chemistry and physics of frying. They are inert with respect to the cooking process. The color of an oil is not related to the color of food fried in that oil. A layer of oil on a food's surface is so thin it does not contribute significant color to the food. Removing color from an oil is not equivalent to restoring cooking quality to the oil.

FOOD QUALITY

Very fresh oil does not cook well (food remains relatively raw) until it is broken-in by frying a few portions of food, or by adding a small amount of old oil to the new. Food fried in moderately used oil develops a well-cooked interior, a crisp exterior with a minimum of oil soakage, and typical golden colors and tempting flavors. Food fried in degraded oil does not cook completely in the interior, develops an excessively hard exterior, and suffers from oil soakage, dark coloration, and off-flavor.

Food fried in fresher oil is crisp, tasty, and well received. Food fried in degraded oil is limp and rejected. The quality of fried food from differently aged oil is often adjusted for by varying heater temperature and food residence time. These latter adjustments, however, never fully compensate for changes in the cooking properties of frying oil.

Oil in Prefried Foods

The coatings of battered and breaded foods introduce soluble contaminants into the oil. When prefried foods are refried, the aged oil in the food comingles with the new frying oil and again introduces foreign chemicals. This latter case is especially important to food-service operators and is difficult to guard against. Some prefried-food vendors even tout the fact that their products are heavy with oil and will provide the food service fryer with extra free oil. This free oil is often loaded with soluble contaminant chemicals such as surfactants. Often a food-service operation complains to its frying-oil vendor that a new batch of oil was no good because it failed early in use. Actually, the prefried food was the culprit and the wrong vendor was blamed. An opportunity, therefore, always exists to better control the production and increase the quality of fried foods by identifying the causative agent of variation in oil and its relationship to the process of frying.

NUTRITION AND TOXICOLOGY

The volatile and nonvolatile (possibly toxic) organic decomposition products of frying oils that have been extensively studied and commented on amount to, at most, a few percent of the total mass of heated frying oil. Other, less-studied components can be present in used oils at large percentages (20% or more, for polymer) and have severe operating and economic consequences for fryer operators and consumers. It seems appropriate at this time to study the larger chemical variables and their consequences in frying technology.

REGULATORY

Recently, regulatory agencies and fast-food chains have shown an interest in having a simple means to determine when an oil should be discarded. Even though abused cooking oil consumed with fried food has not been shown to cause tumorigenesis or chronic toxicity in humans, it is the feeling among many regulatory officials and nutritionists that it would be a wise practice to discard frying oils from food production before their total polar materials content has risen to 30% by weight of the oil.

When a frying oil contains about 25% polar materials, numerous inspectors and restaurant operators feel the oil is abused, and that such oil incorporated into food causes the food to be out of quality specification. The total polar material content of oil included in food can then be regarded as a chemical index of fried-food quality. A food may be considered to be adulterated with respect to label declaration when the listed oil is only about 70–75% pure.

FRYING RESEARCH

Frying research has traditionally focused on the volatile and nonvolatile degradation chemicals formed in oils as a consequence of heating the oils in air. Summarizations of important areas of study at an earlier time have been published (1,2). The degradation pathways and products of heated frying oils have been described (3). Important areas of frying oil analysis have been suggested (4). Diverse tools have been used to attempt to find relationships between different breakdown products in frying oils (5). Facets of oil use influencing fried-food quality have been studied (6,7). A simple surfactant formed in frying oils has been found to be related to oil soakage into foods (8). Oil flavor chemistry has been described (9). Early evidence of a toxic material formed in abused frying oils has been presented (10). A summary of the nutrition and toxicology of frying oils has been published (11).

No researcher however, has been able to specifically assign some element or elements in the complex mix of degradation chemicals in oils to the overall physical and sensory quality of food produced in aging frying oils. There is no textbook statement that says if this element is examined in an oil and controlled, the physical and sensory quality of fried food can be controlled. Flavor and taste components, and rates and origins of organic reactions have been intensely studied, but the public as often as not eats poor-quality fried food.

PROCESS OPTIMIZATION

The dynamics of heat and mass transfer while an oil is undergoing chemical changes have not been well studied. Selective absorption of deleterious materials from frying oil solution has been proposed by industry as a way to lengthen the service life of oils. Optimization of the frying process needs to be reduced to principles and rules for computerization in expert systems.

FRYING OIL SUBSTITUTES

To obtain either formulation compatabilies or reduced calories per serving, the food industry has frequently funded research on synthetic oils to be used either as replacers, extenders, or sensory enhancers. The group of discovered oil mimetics copy some sensory properties, but cannot replace the functionality of frying oils in resistance to heat and oxidation. Oil replacers are created to substitute for natural triglyceride oils functionally, but are not bioavailable and so do not contribute to nutrition. The lack of bioavailability is often touted as yielding a low-calorie oil, but the behavior to date has been to remove fat-soluble materials, such as some vitamins, from the gut as the material passes through unchanged. The similarity between synthetics and the nonabsorbable polymer formed in abused oil is striking.

FRYING OIL CONTROVERSIES

Frying research has generally been conducted in ways that illustrate the dynamics and kinetics of oil degradation in nearly fresh oils or nearly abused oils. The middle ground, which is of most concern to industry and commerce, has been scarcely touched. Thus the emphasis has been on the onset of degradation or on thoroughly degraded systems that are not typical of food service or food processing by frying. Problems with processed oils such as flavor and color reversion, hydrogenation flavor, color associated with bitterness, high solids content, trans-fatty acids, unusual oxidation products, and processing residues and additives are rarely discussed. The importance of saturates and cholesterol and the potential negatives of highly polyunsaturated oils, animal-source oils, tropical oils, and heavily hydrogenated oils are not yet resolved.

BIBLIOGRAPHY

1. H. Roth and S. P. Rock, *Bakers Digest* **46**(4), 38 (1972).
2. *Ibid.*, (5), 38 (1972).
3. C. W. Fritsch, *Journal of the American Oil Chemists' Society* **58**, 272 (1981).
4. S. G. Stevenson, M. Vaisey-Genser, and N. A. M. Eskin, *Journal of the American Oil Chemists' Society* **61**, 1102 (1984).
5. P. Wu and W. W. Nawar, *Journal of the American Oil Chemists' Society* **63**, 1363 (1986).
6. L. M. Smith, A. J. Clifford, C. L. Hamblin, and R. K. Creveling, *Journal of the American Oil Chemists' Society* **63**, 1017 (1986).
7. J. Mancini-Filbo, L. M. Smith, R. K. Creveling, and H. F. Al-Shaikh, *Journal of the American Oil Chemists' Society* **63**, 1452 (1986).
8. M. M. Blumenthal and J. R. Stockler, *Journal of the American Oil Chemists' Society* **63**, 687 (1986).
9. S. S. Chang, R. J. Peterson, and C. T. Ho, *Journal of the American Oil Chemists' Society* **55**, 718 (1978).
10. D. Firestone, W. Horowitz, L. Friedman, and G. M. Shue, *Journal of the American Oil Chemists' Society* **38**, 253 (1961).
11. E. G. Perkins and W. J. Visek, eds., *Dietary Fats and Health*, American Oil Chemists' Society, Champaign, Ill., 1983.

MICHAEL M. BLUMENTHAL
Libra Laboratories, Inc.
Piscataway, New Jersey

FUEL. See ENERGY USAGES IN FOOD PROCESSING PLANTS: ENERGY/USAGE IN THE CANNING INDUSTRY.

FUMIGANTS

Fumigants have been used for centuries as a method of controlling stored insect pests. Because some insects deposit their eggs inside the grain kernel, fumigants must possess certain properties so that they exist in a gaseous state to penetrate the kernel to effectively control the infestation. In recent years the number of available fumigant products has diminished and now industry must take greater care in the use of the available products and exercise greater caution so that misuse cannot occur. Improper sealing of an area to be fumigated, poor application techniques coupled with wrong dosages, and inadequate fumigation time always result in failure. One of the principal concerns with poor fumigations is the development of resistant insects.

Newer government regulations require that applicators become more knowledgeable in regard to fumigant chemicals and their application. Safety of the applicator is a prime consideration for any fumigation, both during application of the chemical and the subsequent aeration of the fumigated commodity. The total procedure will only be successful if the space to be fumigated has been properly sealed so that the gas concentration can be maintained for the required time period to insure total control. Improper sealing will result in loss of gas during the fumigation, so it is recommended that the applicator take gas readings during the fumigation to determine gas release and concentration. It is always surprising to hear how much time was taken for a fumigation that failed only to hear that some area was not sealed and that gas readings were not taken. Applications of this nature can only result in the development of resistance because generations of insects become exposed to nonlethal concentrations. The industry still has sufficient means to control stored product pests, but these methods must be used effectively and with proper techniques.

Each fumigation is different and, therefore, the selection of the appropriate fumigant to be used is crucial. It is especially significant that the properties of each chemical is considered for both success and to determine how best to apply the product so that all areas of the space to be fumigated have sufficient gas concentration during the treatment period. Fumigant products vary in mode of action depending on many factors including temperature, relative humidity, sorption, diffusion, and penetration.

A good fumigation will require prefumigation planning. All aspects of the fumigation must be decided so that success will be achieved. Among the factors to be considered are safety of the applicator, proper sealing of the structure, properties of the fumigant to be used, application equipment or techniques, gas readings, placement of warning signs, posting of guards, and notification of emergency personnel in the area in the event of an episode. The proper method of aeration of the treated commodity must also be analyzed and planned so that the release of the fumigant will not result in exposure or risk to anyone. Once all aspects of the fumigation have been established, it then is necessary to choose which fumigant will be used from among those available.

AVAILABLE FUMIGANTS

Metal Phosphides

There are two metal phosphide products that are presently used as a source of hydrogen phosphide: aluminum phosphide and magnesium phosphide.

Aluminum Phosphide. This product is manufactured as a tablet or pellet and is packaged in a formulation in a bag or sachet or as a tablet or pellet prepac. The paraffin-coated, 3-g tablets will produce 1 g of hydrogen phosphide gas when exposed to certain heat and humidity conditions. The pellets are one-fifth the size of the tablets and produce 200 mg of hydrogen phosphide gas. This gas has a density of 1.18 as compared to air and, as a result, has excellent diffusing and penetrating properties. The tableted product usually contains ammonium carbamate, which will break down into ammonia and carbon dioxide when exposed to heat. The gases escape from the tablet or pellet by causing ruptures in the paraffin coating. The ammonia also acts as a warning gas to the applicator. Atmospheric moisture then penetrates the ruptured coating to react with the aluminum phosphide resulting in the release of hydrogen phosphide gas, which is the resultant fumigant. Tablet and pellet prepacs release hydrogen phosphide in an identical manner. These formulations cause a delayed release until appreciable gas concentrations are reached and act as a safeguard to the applicator. Bag or sachet formulations also have a delayed release; atmospheric moisture must penetrate the paper or cloth packaging material before any toxic gas is released. It is important to follow the manufacturer's applicator's manual for use of aluminum phosphide because low temperature and low relative humidity can result in only partial reaction and a minimum of gas release. The result will be low concentration of gas, a good chance at fumigation failure, and possible problems on deactivation of the unreacted metal phosphide. When temperatures are lower or relative humidity is lower, then it is wise to consider the use of a magnesium phosphide product.

Magnesium Phosphide. These products are produced and marketed as either tablets or pellets or as magnesium phosphide impregnated in a plastic matrix. No bag or sachet formulation of this chemical is available. There is a magnesium phosphide pellet prepac product that is used only as a spot fumigant for plant processing equipment. The reactivity of magnesium phosphide with atmospheric moisture is so much faster than aluminum phosphide, it can be used in conditions of high temperatures, low relative humidity or low temperatures and low to medium relative humidity.

PROPERTIES OF METAL PHOSPHIDES

There are several properties of metal phosphide fumigant products that must be considered in any application.

1. Its density is such that any leakages will result in loss of gas.
2. Hydrogen phosphide gas can self-ignite at a concentration of 17,900 ppm if there is an ignition source available. The reaction between a metal phosphide and atmospheric moisture, which yields hydrogen phosphide, is exothermic and, therefore, the piling of tablets during fumigation is to be avoided as the heat generated from this exothermic reaction can be the ignition source.
3. Hydrogen phosphide reacts with certain metals, especially with copper, so care must be taken during fumigation to prevent corrosion of electrical and electronic equipment and instruments.
4. Note that 1 g of hydrogen phosphide gas per 1,000 ft^3 will result in a concentration of 25 ppm. Therefore, if a dose of 20 tablets/1,000 ft^3 is applied, the theoretical maximum concentration would be 500 ppm (20 g × 25 = 500 ppm), which is equivalent to a dose of 0.7 g/m^3.
5. Aluminum or magnesium phosphide tablets and pellets that contain ammonium carbamate also release ammonia and, therefore, can be phytotoxic to living plants if sufficient concentration is present as well as darkening some nut meats when direct contact is made with the product. The darkening of nut meat is usually reversible. It would be better to use the magnesium phosphide impregnated in a plastic matrix (Fumi-Cel/Fumi-Strip) for some of these fumigations because it contains no ammonium salts.
6. Hydrogen phosphide has no effect on germination and, therefore, can safely be used to fumigate various seeds.
7. Hydrogen phosphide has no real CT product and fumigations must be no less than three days for complete control.
8. Do not fumigate if commodity temperature is below 5°C (40°F).
9. A fumigated commodity or space must be aerated until there is 0.3 ppm or less of hydrogen phosphide before reentry is allowed without respiratory protection.
10. Repeated fumigation with hydrogen phosphide will not result in a buildup of residues if the product is properly aerated following fumigation.

Finally, it must also be noted that hydrogen phosphide produced from metal phosphide product has a garlic or carbidelike odor. One word of caution in this regard is to emphasize that pure hydrogen phosphide has no odor, and the odor noted during a fumigation is an impurity and can be sorbed on a commodity, therefore, the lack of odor cannot be considered as a key for the lack of hydrogen phosphide being present. A measuring device must be used to determine concentration.

Metal phosphide products can be applied directly to raw agricultural commodity and animal feeds and have a tolerance of 0.1 ppm for this purpose. No direct contact can be made with these products on processed food so packaged metal phosphide products such as prepacs, bags, sachets, or Fumi-Cel/Fumi-Strip are used for these applications, and the approved tolerance is 0.01 ppm. Care should always be taken when deactivating residual materials as fumigation conditions could have resulted in having some unreacted phosphides in the residuals. The manufacturer's applicator's manual should be consulted prior to initiating this activity.

The half-life of hydrogen phosphide in the atmosphere is short and depends on climatic condition; it is in the range of 5–28 h. Hydrogen phosphide is known to be oxidized to phosphate both in the atmosphere and during fumigation.

Toxicology

Recent studies have shown that hydrogen phosphide is an acute toxic material and has no chronic effects. Effects shown by experimental animals as a result of subchronic exposure were shown to be reversible. This substance also had no teratological or mutagenic action when these studies were completed.

Respiratory Protection

If the application is made inside a structure, then gas readings must be taken and a gas mask with approved canister must be worn if the levels exceed 0.3 ppm. An air pack must be worn when the levels exceed 15 ppm. These same requirements also apply to aeration.

METHYL BROMIDE

Methyl bromide is packaged as a liquid under pressure in either small cans or in various sizes of steel cylinders. It has a density of approximately 3.3 times that of air and, as a result, should be used with fans to prevent stratification, resulting in fumigation failures in the higher portion of the area. To prevent some of this phenomenon from occurring, it is usually recommended that the product be released near top of the site to be fumigated whether it is a silo or mill. If fans are not employed, there is a possibility that the upper portion of the site will not be successfully fumigated. If bagged commodities are to be fumigated under a tarpaulin or in a truck or container, the entry tube should be leakproof and a pan should be placed under the end of the tube so that no liquid methyl bromide is allowed to come into contact with the bagged commodity, which can result in staining.

The properties of methyl bromide that must be considered prior to fumigation are as follows.

1. Its density is such that fans should be used to recirculate the gas for even distribution.
2. A heat exchanger may be necessary to properly volatize the methyl bromide.
3. Methyl bromide is nonflammable but can result in the formulation of hydrobromic acid when open flames are present. Hydrobromic acid can be corrosive.
4. Methyl bromide can be sorbed under certain condition and can react chemically with certain compounds resulting in damage or bad odors. These materials include certain foodstuffs, rubber goods, furs, leather goods, woolens, rayons, various paper products, photographic chemicals, cinder blocks, charcoal or any material that contains active sulfur compounds.
5. Methyl bromide can effect germination of seeds and should be used for this purpose with caution.
6. Methyl bromide does have a true CT product, and fumigation usually can be completed in 12–24 h depending on dosage.
7. Normally fumigation with methyl bromide is not recommended when the commodity temperature is lower than 15.5°C (60°F).
8. Following fumigation, aeration must continue until the level of methyl bromide is below 5 ppm. If the level exceeds this amount, then proper respiratory equipment must be worn.
9. Repeated fumigations with methyl bromide can result in residues exceeding tolerances established by federal agencies for various commodities.

Finally, it should be noted that methyl bromide is odorless and proper measuring instrumentation is required to measure concentration. The halide detector is not sensitive enough to be used to determine levels suitable for reentry into an aerated facility. Aeration may have to be prolonged due to sorption of methyl bromide to commodities or materials in the structure. Fans will help in the aeration process. The half-life of methyl bromide can be 15–18 months depending on climatic conditions.

Toxicology

Methyl bromide is known to be a chronic toxic chemical, that is, the effects due to exposure are cumulative and are not reversible. Major studies such as chronic toxicity, teratology, pharmacokinetics, and mutagenecity are being conducted to fill data gaps for regulatory purposes.

Respiratory Protection

If application is made inside a structure, then an air pack must be worn for any level above 5 ppm. This same restriction applies to aeration.

CONTROLLED ATMOSPHERES

The gases primarily used for controlled-atmosphere fumigations are carbon dioxide, nitrogen, or a mixture of these gases. These gases, in effect, are applied to reduce the oxygen content in a storage.

Carbon Dioxide

The addition of this gas not only reduces the oxygen content but also acts directly on stored pests by acting as a dessicant on the insect body fluids. Carbon dioxide is usually applied from trucks or tanks that contain liquid carbon dioxide. Appropriate vaporizers and regulators are then used to supply gaseous carbon dioxide to the structure to be fumigated. The gas is usually applied at the top of a silo and because it is denser than air the gas will move down through the grain mass. The silo or warehouse to be fumigated must be tightly sealed, and carbon dioxide must be added daily to maintain a 60–70% concentration to be efficient. A system must also be used in the structure that will not allow the pressure inside to build up during application. Usually a vent is placed in the roof or a hatch or portion of polyethylene used to seal roof vents is opened during application.

Nitrogen

Nitrogen can also be used as an inert gas to control oxygen concentrations. A nitrogen generator is required, and normally the oxygen concentration must be reduced to approximately 1%. Fumigation time can vary but usually requires approximately 10 days for good control.

Inert Atmosphere Generators

There are several inert atmosphere generators that result in a mixture of nitrogen and carbon dioxide being applied to the structure to be fumigated.

To use a controlled atmosphere for fumigation, the following factors must be considered.

1. Temperature of the commodity and within the storage area should be 21°C (70°F) or higher to have an effective fumigation. Lower temperature will effect the mortality of the insect present in the structure or commodity.
2. A vent or pressure relief system must be present in the warehouse or silo to prevent structural damage due to pressure build up.
3. Application time can take up to 12 h with daily recharging, which will normally take 3–4 h.
4. Total fumigation time can be from 4 to 10 days depending on temperature and gas tightness of the structure.
5. Even though these gases are considered to be nontoxic, an air pack must be used when entering a facility that has low oxygen content during fumigation.

MIXTURES OF GASES

Research is being conducted by several groups to try to determine if mixtures of gases can result in more favorable fumigation conditions with a shorter duration. Mixtures such as hydrogen phosphide and carbon dioxide, methyl bromide and carbon dioxide, or increased temperature and carbon dioxide are being tested. Early studies show that some mixtures may be promising, but additional studies will have to be performed. Elevated temperature and carbon dioxide combinations have not been successful to date.

BIBLIOGRAPHY

General References

E. J. Bond, *Manual of Fumigation for Insect Control,* Food and Agriculture Organization of the United Nations, Rome, 1984.

Truman's Scientific Guide to Pest Control Operations, Purdue University/Edgell Communications.

JEREMIAH B. SULLIVAN
Degesch America, Inc.
Weyers Cave, Virginia

G

GAME. See Wildlife and game.
GAS CHROMATOGRAPHY. See Antibiotics in food of animal origin; Pesticide residue in foods.
GAS OR OIL UTILIZATION. See Food plant design and construction.

GELATIN

Gelatin type A and type B are obtained by the partial hydrolysis of collagen, the chief protein component in skins, bones, hides, and white connective tissues of the animal body. Type A is produced by acid processing of collagenous raw materials and has an isoelectric point between pH 7 and 9, whereas type B is produced by alkaline or lime processing and has an isoelectric point between pH 4.8 and 5.2. Mixtures of types A and B as well as gelatins produced by modifications of these processes may exhibit isoelectric points outside of the stated ranges (1). Gelatin is a hydrolysis product obtained by hot water extraction and does not exist in nature. Gelatin is used widely in the food industry, which takes advantage of its unique properties such as reversible gel-to-sol transition of aqueous solutions, viscosity of warm aqueous solutions, and capability to act as a protective colloid.

PHYSICAL AND CHEMICAL PROPERTIES

Commercial gelatin produced in the United States ranges from coarse granules to fine powder. In Europe edible gelatin is available in the form of thin sheets. Dry commercial gelatin contains about 9–12% moisture and is an essentially tasteless, odorless, brittle solid with specific gravity between 1.3 and 1.4. The physical and chemical properties of gelatin are measured on aqueous gelatin solutions and are functions of (1) the source of collagen, (2) the method of manufacture, (3) conditions during extraction and concentration, (4) thermal history, (5) pH, and (6) the chemical nature of impurities or additives. Gelatin is classified as a derived protein because it is obtained from collagen by a controlled partial hydrolysis. This makes gelatin a heterogeneous mixture of polypeptides; properties desired are obtained by blending products from selected extracts.

Solubility

In commercial applications gelatin is consumed as a solution. Gelatin is soluble in water, acetic acid, and aqueous solutions of polyhydric alcohols such as glycerol, propylene glycol, sorbitol, and mannitol. The viscosity of aqueous gelatin solutions increases with increasing concentrations and decreasing temperatures.

Dry gelatin absorbs water exothermally. The rate and degree of swelling is a characteristic of the particular gelatin. Swelled gelatin granules dissolve rapidly in water above 35°C. Gelatin freeze dried with water-soluble sugar dissolves rapidly in cold water but then reverts quickly to the stable gel form. A more detailed discussion of gelatin solutions and solvation has been published (2). Many water-soluble organic solvents are compatible with gelatin but interfere with gelling properties (3).

Fast-dissolving gelatins have obvious advantages in the food industry. This has been accomplished by the use of sugar, fine grind, and hot temperature drying of gelatin. In the United States most edible gelatin is sold in the form of fine grind, but a market for thin gelatin sheets exists in Europe.

Stability

Dry gelatin stored in airtight containers at room temperature has a shelf life of many years. However, it decomposes above 100°C. For complete combustion, temperatures above 500°C are required. Aqueous solutions or gels of gelatin are highly susceptible to bacterial growth and breakdown by proteolytic enzymes. Stability is a function of pH and electrolytes and decreases with increasing temperature because of hydrolysis.

Swelling

The swelling property of gelatin is important in its solvation. The swelling of gelatin does not fit a simplified equation for polymers (4) where the swelling depends primarily on the number of cross-links and the interaction constant of polymer and solvent. In gelatin the cross-links are believed to involve multiple interactions between chains that form fibrils; this reduces swelling (5). The stability of those cross-links depends on pH, temperature, time, and electrolytes.

At pHs below the isoelectric point, the choice of anions can control swelling, whereas above the isoelectric point, proper choice of cations reduces swelling. The anions break hydrogen bonds, which probably accounts for increased swelling. The rate of swelling follows approximately a second-order equation (6).

CHEMICAL COMPOSITION AND STRUCTURE

Gelatin is not a single chemical substance. The main constituents of gelatin are large and complex polypeptide molecules of the same amino acid composition as the parent collagen, covering a broad molecular weight distribution range. Most commercial gelatin contains molecular species from 15,000 to above 400,000 (7), with the average between 50,000 and 70,000 mol wt.

Analysis shows the presence of amino acids from 0.2% tyrosine to 30.5% glycine (see Amino acids). The five most common amino acids are glycine, 26.4–30.5%; proline, 14.8–18%; hydroxyproline, 13.3–14.5%; glutamic acid, 11.1–11.7%; and alanine, 8.6–11.3%. The remaining amino acids in decreasing order are arginine, aspartic acid, lysine, serine, leucine, valine, phenylalanine, threonine, isoleucine, hydroxylysine, histidine, methionine, and tyrosine (8). Warm gelatin solutions are more levorotatory than expected on the basis of the amino acid composition, indicating Gly-Pro-Pro and Gly-Pro-Hypro sequences (9). The o-chain form of gelatin behaves in a solution like a random-coil polymer, whereas the gel form may contain as much as 70% helical conformation. The remaining molecules in nonhelical conformation are linking helical regions together in the matrix. Structures have been studied with the aid of an electron microscope (7). The structure of the gel is a combination of fine and coarse interchain networks; the ratio depends on the temperature during the polymer–polymer and polymer–solvent interaction leading to bond formation. The rigidity of the gel is approximately proportional to the square of the gelatin concentration. Presence of crystallites, indicated by x-ray diffraction pattern, are believed to be at the junctions of the polypeptide chains (10). The molecular weight of a single o-chain is about 96,500.

Amphoteric Character

The amphoteric character of gelatin is a result of the functional groups of the amino acids and the terminal amino and carboxyl groups created during hydrolysis. The distribution of these groups and the resulting isoionic point is determined by the manufacturing process. It can be measured conveniently by mixed-bed ion-exchange resin and is expressed in pH units. Titrating the free acidic or basic groups gives an indication of the amphoteric characteristics of gelatin. The amphoteric character and isoionic point of type A gelatin depend on the acid-processing time; a typical isoionic point is between pH 8.0 and 9.0. Type B gelatin exhibits more reproducible amphoteric characteristics, reaching an isoionic point of 5.2 after four weeks of liming, which drops to 4.8 after a prolonged or more vigorous liming process. The isoionic point is reproducible, whereas the isoelectric point depends on the salts present. The isoelectric point can be found by determining a pH value at which gelatin solution exhibits maximum turbidity (11,12).

Gelation

A most useful and unique property of gelatin solution is its capability to form reversible gel–sols. A gelatin solution gels at temperatures below 35°C. This conversion temperature is determined as setting point (sol to gel) or melting point (gel to sol). Commercial gelatins melt between 30 and 35°C, with the setting point being lower by 2–5°C. For melting point determination (13), the end point is reached when colored carbon tetrachloride slides halfway down the gelatin–test tube interface. Several methods have been used to determine the setting point of gelatin but all are time-consuming (14).

MANUFACTURE AND PROCESSING

Technology for gelatin manufacture was developed in the early 1920s. Acid and lime processes have separate facilities and are not interchangeable. In the past, bones and ossein (decalcified bone) have been supplied by India and South America. Today, bones from slaughterhouses and meat-packing plants are important sources. The supply of bones has been greatly increased since the meat-packing industry introduced packaged and fabricated meats, assisted by the growth of fast-food restaurants. Dried and rendered bones yield about 14–18% gelatin, whereas pork skins yield about 18–22%.

Most type A gelatin is made from pork skins, yielding grease as a by-product, which is also marketable. The process includes macerating the skins; washing to remove extraneous matter; swelling for 10–30 h in 1–5% hydrochloric, phosphoric, or sulfuric acid; reduction of acidity by washing; adjustment of pH; and pumping to extraction tanks. There, four to five extractions are made at temperatures increasing from 55–65°C for the first extract to 95–100°C for the last extract. Each extraction lasts about 4–8 h. The grease is then removed and the gelatin solution is filtered and for some applications deionized. Concentration to 30–40% solid is carried out in two stages by continuous vacuum concentration. The viscous solution is chilled, noodled, and dried at 30–60°C on a continuous wire-mesh belt. The dry gelatin is then ground and blended to specification.

Type B gelatin is made mostly from bones, but also from hides and skins. The bones for type B gelatin are crushed and degreased at the rendering facilities, which are usually located at a meat-packing plant. Rendered bone pieces (0.5–2 cm) with less than 1% fat are demineralized by 4–7% hydrochloric acid in 5–14 days. Dibasic calcium phosphate is precipitated and recovered from the spent liquor. The demineralized bones (ossein) are washed and pumped to liming tanks where they are agitated daily for 3–16 weeks in a lime slurry. After washing for 15–30 h to remove the lime, the ossein is acidified to pH 5–7 with hydrochloric, sulfuric, phosphoric, or acetic acid. Then the processing for type A gelatin is followed. Cleanliness is important to avoid contamination by bacteria or proteolytic enzymes.

Bovine hides and skins are substantial sources of raw material for type B gelatin and are supplied in the form of splits, trimmings of dehaired hide, raw hide pieces, or salted hide pieces. Like pork skins, the skins are cut to smaller pieces before being processed. Sometimes the term calfskin gelatin is used to describe hide gelatin. The liming of hides usually takes a little longer than the liming of ossein from bone. In the United States 25–30% of the gelatin derives from hides and skins. In recent years acid-processed hide gelatin has been introduced into the edible gelatin market.

All equipment should be made of stainless steel. The liming tanks, however, can be either concrete or wood. Properly lined iron tanks are often used for the washing and souring operations.

Most gelatin plants operate around the clock. The product is tested in batches and again as blends to confirm individual customer orders.

ECONOMIC ASPECTS

World gelatin production is believed to be about 115,000 t. The United States produces about 31,000 t, followed by

France, the FRG, the UK, and Japan. The United States food industry consumes ca 20,000 t, with an annual growth rate of 0.5%. Of the gelatin produced in the United States, 55% is acid processed (type A).

ANALYTICAL AND TEST METHODS

Gelatin is identified by a positive test for hydroxyproline, turbidity with tannic acid, or a yellow precipitate with acidic potassium dichromate or trinitrophenol. A 5% aqueous solution exhibits reversible gel-to-sol formation between 10 and 60°C.

Elemental analysis of commercial gelatin was reported as carbon, 50.5%; hydrogen, 6.8%; nitrogen, 17%; and oxygen 25.2% (15). A purer sample analyzed for 18.2–18.4% nitrogen (8,16). Standard testing procedures for viscosity, pH, ash, moisture, heavy metals, arsenic, bacteria, and gel strength (with the Bloom Gelometer) have been described (17–19).

USES

Food Products

Gelatin formulations in the food industry use almost exclusively water or aqueous polyhydric alcohols as solvents for candy, marshmallow, or dessert preparations (see FOOD ADDITIVES). In dairy products and frozen foods, gelatin's protective colloidal action prevents crystallization of ice and sugar. In general, a 250-bloom gelatin is added at concentrations ranging from 0.25% in frozen cream pies to 0.5% in ice cream; the use of gelatin in ice cream has greatly diminished in recent years. In sour cream and cottage cheese, gelatin inhibits water separation.

Marshmallows contain as much as 1.5% gelatin to restrain the crystallization of sugar, thereby keeping the marshmallow soft and plastic. Gelatin also increases viscosity and stabilizes the foam in the manufacturing process. Many lozenges, wafers, and candy coatings contain up to 1% gelatin. In these instances, gelatin decreases the dissolution rate. Substantial gelatin use has been developed for gummy bears and in low-calorie spreads such as margarine and butter. In meat products, such as canned hams, various luncheon meats, corned beef, chicken rolls, jellied beef, and other similar products, gelatin in 1–5% concentration helps to retain the natural juices.

The largest use of edible gelatin, however, is in the preparation of gelatin desserts in 1.5–2.5% concentrations. The gelatin is sold either premixed with sugar and flavorings or as unflavored gelatin. Most edible gelatin is type A, but type B is also used. The GRAS status of edible gelatin has been granted by the FDA. Limits have been set on microbial contamination by *Salmonella* and *E. coli* and on heavy metals such as lead, sulfur dioxide, and arsenic (18).

Gelatin can be a source of essential amino acids (except for tryptophan, methionine, and cystine) when used as a diet supplement and therapeutic agent. As such, it has been widely used in muscular disorders, peptic ulcers, and infant feeding and to spur nail growth.

BIBLIOGRAPHY

1. *Proposed Definition for Food Grade Gelatin in 1979*, Gelatin Manufacturers Institute of America, Inc., New York.
2. C. A. Finch and A. Joblin, "Physical Properties of Gelatin," in A. G. Ward and A. Courts, eds., *The Science and Technology of Gelatin*, Academic Press, Inc., Orlando, Fla., 1977, Chapt. 8; A. R. Krogh, "Swelling, Absorptions, and Photographic Uses of Gelatin," Chapt. 14.
3. J. Q. Umberger, *Photogr. Sci. Eng.* **11**, 385 (1967).
4. P. J. Flory and J. J. Behner, *Chemical Physics* **11**, 513 (1943).
5. I. Tompka, J. Bohonek, J. Spuehler, and M. Ribeaud, *J. Photogr. Sci.* **23**, 97 (1975).
6. A. Libicky and D. I. Bermane In R. J. Cox, ed., *Photographic Gelatin*, Academic Press, Inc., Orlando, Fla., 1972, pp. 29–48.
7. A. Courts, *Biochem. J.* **58**, 70 (1954).
8. J. E. Eastoe in G. N Ramachandran, ed., *Treatise on Collagen*, Vol. 1, Academic Press, Inc., Orlando, Fla., 1967, pp. 1–72.
9. J. Josse and W. F. Harrington, *Journal of Molecular Biology (London)* **9**, 269 (1964).
10. T. Fujii, *Bull. Soc. Sci. Photogr.* **16**, 274 (1966).
11. J. E. Eastoe and A. Courts, *Practical Analytical Methods for Connective Tissue Proteins*, Spon, London, 1963, Chapt. 6.
12. A. Veis "The Macromolecular Chemistry of Gelatin" in B. L. Horecker, N. D. Kaplan, and H. E. Scheraga, eds., *Molecular Biology*, Vol. V, Academic Press, Inc., Orlando, Fla., 1964, p.112.
13. *Sampling and Testing Gelatins*, British Standards **6**, 757 (1975).
14. F. W. Wainewright, *GGRA Bull.* **17**(3), 10 (1966).
15. C. R. Smith, *Journal of the American Chemical Society* **43**, 1350 (1921).
16. J. E. Eastoe, *Biochemical Journal* **61**, 589 (1955).
17. W. Horowitz, ed., *Official Methods of Analysis*, AOAC, Washington, D.C., 1986.
18. *The United States Pharmacopeia XX, (USP XX-NFXV)*, The United States Pharmacopeial Convention, Inc., Rockville, Md., 1986.
19. Gelatin Manufactures of America, Inc., *Standard Methods for the Sampling and Testing of Gelatin*, GMIA, Inc., New York, 1986.

General References

A. G. Ward and A. Courts, eds., *The Science and Technology of Gelatin*, Academic Press, Inc., Orlando, Fla., 1977.

Gelatin, Gelatin Manufacturers Institute of America, Inc., New York, 1982.

R. J. Croome and F. G. Clegg, *Photographic Gelatin*, Focal Press, New York, 1965.

A. Veis, *The Macromolecular Structure of Gelatin*, Academic Press, Orlando, Fla., 1964.

T. H. James, *The Theory of the Photographic Process*, 4th ed., Macmillan, New York, 1977.

R. J. Cox, ed., *Photographic Gelatin*, Academic Press, Inc., Orlando, Fla., 1972.

R. J. Cox, ed., *Photographic Gelatin-II*, Academic Press, Inc., Orlando, Fla., 1976.

FELIX VIRO
Kind & Knox
Sioux City, Iowa

GENETIC ENGINEERING PART I: PRINCIPLES AND APPLICATIONS

Genetic engineering is one of the latest additions to the discipline of food science and technology. As we enter the next century, genetic engineering will revolutionize our notions of food, from production to processing. In the words of Douglas McCormick, the editor of *Bio/Technology*, "By most standards, all of this [biotechnology] is still brand new. Yet biotechnology is the product of intellectual earthquakes that change things so completely that it is difficult, after the event, to remember that once the landscape was different. It is easy to forget that the change is recent, and more change is just around the corner" (1).

HISTORY

The discovery of DNA ligase in 1967 followed by those for DNA restriction endonucleases and RNA reverse transcriptase in 1970 ought to be the beginnings of genetic engineering. Also, in 1970 the first record for total synthesis of a gene in vitro was produced. In 1972 the first in vitro recombinant DNA molecule was generated, and a year later the first plasmid vector mediated transformation was demonstrated. In 1975, the methodology for Southern blot was published. The blotting system (Fig. 1) allows for isolation and identification of DNA fragments by gel electrophoresis and their transfer (or blotting) onto a membrane filter for direct hybridization to a single-stranded radioactively labeled or biotin-linked probe. Four years later, the discovery of introns and exons in eukaryotic genes changed our concepts of gene transcripts, and the same year rapid DNA sequencing methods became available. The decade following these outstanding achievements saw a rise in the commercial application of genetic engineering and production of human, plant, and microbial processes or products (2). It is amazing how rapid the growth of this discipline has been and what the future holds.

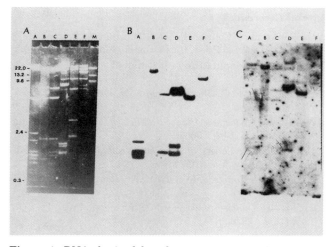

Figure 1. DNA obtained from human genome (A) has been hybridized by either (B) DNA or (C) RNA by Southern blotting method. Photograph courtesy of J. M. MacPherson.

PRINCIPLES OF GENETICS

The hereditary material of most cellular organisms is DNA, although in many bacteriophages and plant or animal viruses it could be either DNA or RNA. DNA is largely the basis for the preservation of the instructions for the organization, structure, and functioning of living cells. Historically, it was realized that the passage of hereditary traits to one's offspring depended on those that were found in the parents. In the early 1860s the Austrian monk Gregor Mendel not only experimentally confirmed this in pea plants but also established the principal rules governing the appearance of specific traits, ie, the parental factors (subsequently these became known as genes) and their assortment into individual progeny. He also made the distinction between the external appearance of such traits (later named phenotypes) and those that are reflections of the composition (later named genotypes). Mendel also advanced the idea of the expression of traits when they were in single copies from those that required two copies (later named dominant and recessive genes). The advent of microscopy and the discovery of dyes that could differentially stain cellular parts permitted the examination of the anatomical features of cells including the nucleus. The hereditary material, housed in distinct and organized structures, were named chromosomes. The number of chromosomes per cell are constant in any species. In most cells, the chromosomes are found in pairs of homologous structures, varying in number from a few to 2–3 dozen. The concept of ploidy was developed to reflect the presence of haploid (N) and diploid (2N) cells, such as those found in bacteria and gametes or sex cells and in nonsomatic or fertilized cells, respectively. The N was defined to be the number of nonhomologous chromosomes found in a cell.

Analysis of Hereditary Material and Its Exchanges

In the 1940–1950s the nature of DNA was elucidated by the use of quantitative genetic experiments, application of radioisotope to analytical biochemistry, and the discovery of the electron microscope. These tools permitted the visualization of DNA and demonstrated its many activities: replication, repair, exchange, and recombination.

The structures of prokaryotic and eukaryotic cells and their genetic apparati have been investigated in great detail (3,4). The best-studied prokaryotic cell is that of the bacterium *Escherichia coli*, which is a cylindrical cell, 1×0.5 μm in size (Fig. 2) and which can grow and double in number every 20 min. The *E. coli* chromosome is in a covalently closed circular (ccc) form about 1 mm long and is made of 4,500,000 base pairs (4,500 kilobase pairs, kb). The chromosomal DNA is found in a folded configuration called the nucleoid or folded chromosome. Often DNA of extra chromosomal origin called plasmids, which are autonomously replicating pieces of DNA about 1/100 of the chromosome size, are found in the cytoplasm. They frequently contain genes for antibiotic resistance, conjugation, and production of proteins nonessential to normal cell functioning. Some bacterial plasmids are nonconjugative, yet others are conjugative and permit their and/or other plasmid's transfer within the species or a broad

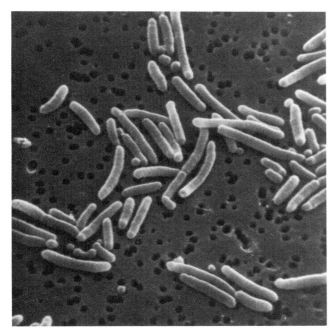

Figure 2. The simple binary fission of *Esherichia coli* cells along with the production of DNA-less minicells (26).

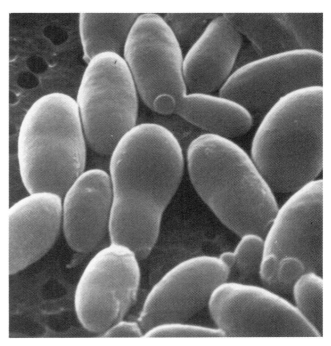

Figure 3. The yeast *Kluyveromyces marxianus* showing budding of a daughter cell and the location of previous bud scars.

range of hosts from different genera. Certain conjugative plasmids can transfer DNA between kingdoms, eg, the Ti plasmids of the bacterium *Agrobacterium tumefaciens* can be transferred to dicotyledonous and monocotylidenous plants or to the yeast *Saccharomyces cerevisiae*. In all cases, transfer of plasmids requires cell to cell contact and the presence of transfer and mobilizing genes.

Cells of the yeasts *S. cerevisiae* and *Kluyveromyces marxianus* (Fig. 3) are quasispherical, 3–5 μm in diameter, and can double hourly. Yeasts may be haploid or diploid, depending on their life-cycle stage. All eukaryotic cell chromosomes contain some basic proteins or histones, which wrap around the DNA to form nucleosomes. *S. cerevisiae* contains 17 linear chromosomes ranging from 150 to 2,500 kbs. Individual chromosomes can be separated using the recently developed PFGE or pulsed-field and other similar gel electrophoresis technique. Eukaryotic cells may also contain membrane-bound intracellular organelles such as mitochondria (Mtc), endoplasmic reticulum (Er), or chloroplasts (Chl). In Mtc and Chl a ccc DNA of 20–90 kb (Fig. 4) is compartmentalized and expressed. mtDNA are believed to have been derived from some prokaryotic cell genome. During cell division, all DNA, whether organellic, plasmid borne, or chromosomal, are divided equally and partitioned between the two daughter cells.

Bacterial cells are able to accept DNA from another parent or donor and undergo in vivo recombination. This was studied initially in the 1950s through the 1970s. These studies used whole cells and relied on natural exchange such as mating, transformation, and recombination occurring in vivo between donor DNA and recipient cells. During the 1960–1980s, after many of the requirements of DNA metabolism were discovered, it became possible to perform in vitro the synthesis, breakage, and joining the DNA from homologous or heterologous origins. These discoveries gave birth to the science and production of in vitro recombinant DNA (rDNA) molecules (5), which when transferred into living cells could express new traits. The terms transformation, transduction, transfection, and conjugation are used to indicate gene transfer into a cell. Transformation and transfection involve transfer of naked donor DNA and a recipient cell, whereas conjugation and transduction require the presence of a donor cell and a bacteriophage, respectively. Both natural and $CaCl_2$ + heat induced transformation of many microorganisms by rDNA are now possible. Natural transformation of certain bacteria eg, *Bacillus subtilis* by linear du-

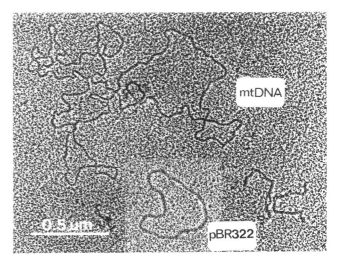

Figure 4. Covalently closed circular DNA molecules from the mitochondrion of a fungus and *E. coli* plasmid pBR322.

plex chromosomal DNA, and induced transformation of others eg, *E. coli* using ccc-plasmid DNA, has become commonplace in genetic engineering. Transformation of many animal and plant cells is also possible, although the underlying mechanisms are quite different from those in bacteria. Transformation of cells can occur by intergeneric fusion of two individual cells from, in the order of their discovery, plant, microbial, and animal origins. Microbial and plant-cell fusion requires the production of protoplasts or cells without external surface layer(s). Protoplasts are osmotically unstable; in hypotonic environments they lyse, but in the presence of stabilizers (sugars, salts) they remain intact. When two protoplasts of different genotypes are brought into contact in the presence of a fusogenic substance (eg, polyethylene glycol 6,000) their membranes fuse, causing cytoplasmic and nuclear mixing events to occur. A transient fusant contains chromosomes from both parents; subsequently, karyogamy and recombination of nuclear material and chromosomes can take place. With animal cells the fusogen could be an animal virus.

Analysis of the DNA as Genetic Code and Its Functions

The flow of genetic information is in general from DNA through transcription to RNA (messenger, transfer, and ribosomal RNA) and through translation of mRNA to proteins (Fig. 5). This was known as the central dogma of molecular biology until the discovery of the enzyme reverse transcriptase, which could synthesize a DNA copy (cDNA) from an RNA molecule. DNA and RNA molecules contain four bases; two purines, adenine (A) and guanine (G); and two pyrimidines, cytosine (C) and thymine (T) in DNA and uracil (U) in RNA. These bases are connected to the sugar deoxyribose (DNA) or ribose (RNA) to form a deoxynucleoside or nucleoside and are phosphorylated to form deoxyribonucleotides (DNA) or ribonucleotides (RNA) respectively (Fig. 6). The double helical or Watson-Crick form of base pairing for A:T and G:C in DNA was the first to be discovered (Fig. 7), but recently, a new Watson-Crick base pairing called κ and π with a hydrogen bonding pattern (6) has increased the genetic alphabet from four to six letters. Many of the natural bases can be modified by addition of organic groups, eg, methylcytosine. Nucleotides are linked through phosphodiester linkage, and each polymer of DNA has a 3' to 5' polarity. The ratio of (G + C)/(A + T) is known by the designation (G + C) content or %(G + C) and reflects taxonomic relatedness and molecular characteristics. The (G + C) content

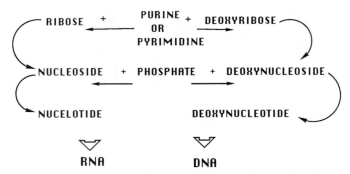

Figure 6. The biosynthesis of nucleosides and nucleotides for their use in DNA or RNA.

of two DNAs determines their separation during buoyant density-gradient centrifugation. Additionally, double-stranded (ds) DNA of higher %(G + C) will have a higher melting temperature (Tm) than that with a higher %(A + T). Through melting dsDNA can be dissociated into two single-stranded (ss) DNAs, and through cooling down they will regain the complimentary ds structure. This process forms the basis for several DNA:DNA or DNA:RNA hybridization techniques. In vitro, every DNA molecule has a topological feature. The open-ended DNA molecule is either in ss or ds form and can be in rod shape. DNA molecules can bend; eg, a small polymer of 242 base pairs (bp) or larger can go from a linear (open-ended) to a circular (ccc) form by the joining of its open ends with ligase. More complex forms of dsDNA are also found. DNA molecules are capable of assuming many forms, including winding or unwinding, depending on their physical and enzymic environments, eg, presence of enzymes called topoisomerases (for twisting) and endonucleases (for nicking and relaxing the twisted DNA). In addition, the sequence of bases within a ssDNA could create structures of

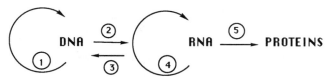

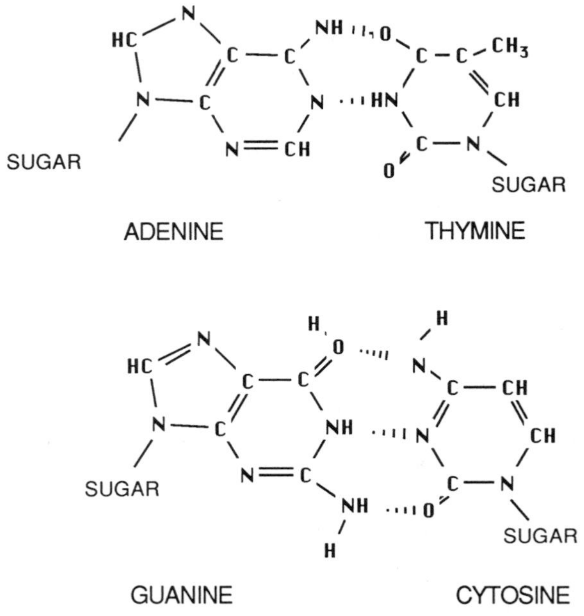

Figure 5. Flow of genetic information as perceived in molecular biology: (**1**) DNA replictes through DNA polymerase system; (**2**) transcription of DNA into RNA occurs via RNA polymerase system; (**3**) RNA is copied into DNA via reverse transcriptase action; (**4**) RNA as an enzyme can act on itself; (**5**) RNA being translated into proteins by the translational system.

Figure 7. The base pairing between A:T and G:C in DNA.

their own. When a sequence of —AAAAAAAGCTTTT-TTT—from a DNA duplex is allowed to separate into two ss polymers, each can generate a hairpin-like structure through base pairing of all A and T residues. Such sequences are thought to be recognition segments within DNA for its interaction with regulatory proteins. RNA structures also have unique organizations such as folding on itself and forming of a hairpin-like structure (eg, tRNA).

All DNA synthesis is enzymatic and utilizes the four deoxyribonucleotides. The pathways for the biosynthesis of deoxyribonucleotide triphosphates and degradation of DNA are well known. The enzyme responsible for DNA biosynthesis is DNA polymerase. This enzyme appears in many forms in prokaryotic (I, II, and III) and eukaryotic (α, β and γ) cells. These enzymes specifically function for DNA synthesis either for repair or replication in a 5′ to 3′ direction. During the synthesis of DNA each deoxynucleotide is joined to the previous one through a phosphodiester bond. DNA polymerization occurs at a maximum speed of 100 and 1000 nucleotides per second for bacterial and animal cells, respectively. The open ends of the DNA at 3′ and 5′ position could be joined by DNA ligase (7).

A gene is a segment of DNA or, as in certain viruses, RNA, made of a stretch of bases that, respectively, code for an RNA or a polypeptide molecule. Synthesis of a polypeptide from the transcript of a gene has been visualized (Fig. 8). It requires translational machinery to join amino acids through their amino- and carboxy- termini. In some viruses and eukaryotic organisms some genes and hence their RNAs are made of translatable and intervening sequences (known as exons and introns, respectively). The genetic code is read as a codon of three bases at a time. There are 64 codons, of which 61 code for amino acids and 3 for termination. There is some degeneracy in certain codons, eg, the amino acid glycine is coded for by the triplets GGA, GGG, GGU, and GGC. Different organisms have a bias in codon usage; ie, they may prefer to use one set of these over the other for the synthesis of polypeptides. Polypeptides in their nascent form are self-assembled into functional structures through the physicochemical properties of the constituent amino acids. Other polypeptides require a special class of proteins called nucleoplasmins, heat-shock proteins, and chaperonins for assembly (8). Although the latter assembling proteins are not components of the final structure they are engaged in the formation of three-dimensionally specific molecules, capable of functioning, catalysis, and other properties. The stability properties of proteins reside in noncovalent interactions that occur within the protein molecule. Mutational replacement of amino acids within certain regions of a protein could affect its conformation and hence function, eg, thermostability and reaction rates (9).

Mutations and Selections

Most heritable variation of phenotypes is explained by mutations in genotypes. Improved mutant organisms have been used in industrial microbiology and food and fermentation technology. The initial attempts at strain improvements for food technology were through selection for spontaneously occurring variants from the original strain. Subsequently, fundamental studies on the occurrence and induction of mutations, enrichment and isolation of mutants, site-specific mutagenesis, and use of genetic recombination have made this process more manageable (10). In vivo mutations in microorganisms can occur spontaneously or with the mediation of mutagens at frequencies of $10^{-(7-9)}$ or $10^{-(4-6)}$ range, respectively. Mutations alter DNA base sequences, by adding, deleting, or substituting base(s) and often affect the structure and hence the function of proteins. Mutagenic agents fall into three classes: physical, chemical, and biological. A special class of plasmids that codes for antibiotic resistance gene(s) and that is capable of insertion into any DNA is known by the name transposon (TN). Insertion of a TN into a gene inactivates the gene, and its removal restores the gene's function. This form of mutagenesis is called transposon mutagenesis and has wide application in genetic engineering (11). The end result of in vivo mutagenesis is alterations in DNA sequence(s), which must be fixed, replicated, and segregated. The rare mutant clone is then screened, enriched, purified, and characterized. One problem with in vivo mutagenesis is the difficulty in targeting specific genes. Certain industrially useful mutants are hard to grow or maintain because of the multiple or deleterious mutations they acquire during random mutagenesis. These problems can be eliminated by using in vitro mutagenesis. Here, a gene sequence is isolated, cloned into a suitable vector, and usually treated with a chemical mutagen, to specifically alter a base. Alternately, an oligonucleotide with a modified sequence is synthesized and then inserted into a gene. Through these approaches it is now possible to generate gene- or site-specific mutations. Recombination between two mutants with different characteristics can generate new organisms for industrial exploitation.

Elements and Regulation of Gene Expression

Gene expression requires template DNA, which is divided into regulatory and structural regions, and RNA transcription to produce the intermediate RNA (mRNA,

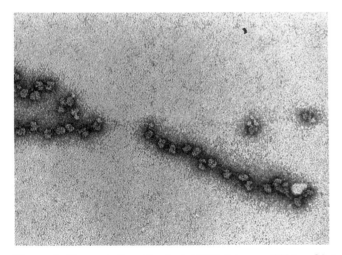

Figure 8. Transcription of bacterial DNA into an mRNA and its translation by polysomes into polypeptide. Photograph courtesy of B. A. Hamakalo.

rRNA, and tRNA). The primary function of mRNA is to serve as a blueprint for translation of a genetic message into a polypeptide (11,12). In addition to the above processes, in 1981–1982 it was found that certain RNA molecules can function as catalysts or a ribozyme for self-splicing (13). This discovery has changed our dogmatic views of catalysis being solely the domain of protein enzymes.

To regulate their growth and metabolism, organisms must determine when and how much cellular constituents are needed. This task is achieved by controlling gene expression (3,11). The following elements regulate and control prokaryotic gene expression (Fig. 9). The operator is a DNA sequence where a repressor protein binds and prevents transcriptional occurrence. Repressor–operator recognition involves a complex code of DNA base-pair sequences. Induction refers to expression of a gene subsequent to the removal of the repressor, usually by the addition of specific inducers. The promoter is located immediately in front of the gene and is the RNA polymerase recognition and transcriptional initiation site. Natural promoter efficiency varies, some are stronger than others. During the transcription of prokaryotic genes, RNA polymerase identifies in the 5' to 3' direction of the sense strand of the promoter before the initiation codon, a recognition sequence of bases, TTGACA or -35 sequence, and a TATAAT (the Pribnow box) sequence at -10 base pairs for its subsequent binding. Mutationally induced base changes in these two sequences can have mild or severe effects on transcription. Through its specificity for DNA transcriptional initiation sites, binding the RNA polymerase generates a localized melting of a small segment of the DNA and the initiation of RNA synthesis, usually starting with GTP or ATP. There is continuous polymerization of ribonucleotides into an RNA transcript until the terminus of the gene is reached. At the terminus, either a sequence of six uridine residues and a hairpin-like structure or a sequence lacking such bases but requiring a termination (rho) factor signal the cessation of transcription. Some structural genes contain a leader and a trailer sequence, respectively, immediately after the promoter and before the terminator regions. At the end of transcription, RNA polymerase is freed to repeat the cycle.

In certain eukaryotic genes and hence their mRNAs (Fig. 10) there are some sequences of about 1,000 bp inserted (introns) between the coding sequences (exons).

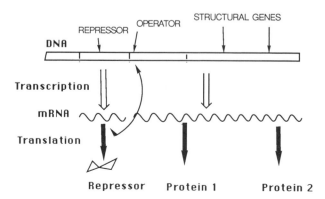

Figure 9. Gene expression in prokaryotic organisms.

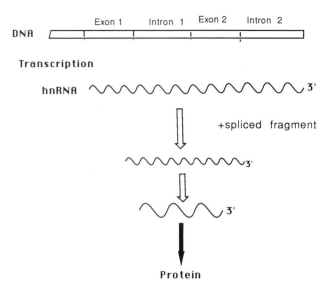

Figure 10. Transcription of an eukaryotic gene containing introns and exons.

Transcription begins at the promoter (5') of a gene and proceeds through the exons and introns (if present) to the 3' terminus, producing a primary mRNA transcript or heterogeneous nuclear RNA. In eukaryotic cells, several processing steps must occur to change the primary mRNA transcript into functional mRNA and allow its translation into proteins. Any introns present are subsequently removed or spliced out. The actual ending of the transcription in some cases could be several hundred base pairs beyond the polyadenylation site of the 3' end of a mRNA. The third step required to change a primary mRNA transcript to a functional mRNA is the addition of a 5' cap.

Eukaryotic promoters are recognized by specific DNA binding proteins of less than 100 amino acid residues in size that bind to DNA and activate or repress gene transcription. Most bacterial RNA polymerases are large multimeric proteins, eg, E. coli enzyme contains two α, one β, and one β' subunits. This holoenzyme has additional subunits, the $\partial 70$, ω, and the σ subunits, which are integral to its functioning. In sporulating bacteria, eg, B. subtilis, σ subunits are responsible for the selection of sporulation promoters. The RNA polymerases of eukaryotic organisms differ from the above both in molecular size and number of enzyme species. As many as 2–4 RNA polymerases can be found in eukaryotic cells, and in several cases their β and β' subunits have conserved sequences. In higher eukaryotes, RNA polymerase I initiates transcription for rRNA genes from nucleolus, and RNA polymerases II and III transcribe mRNA, tRNA, and 5S rRNA from nuclear matrix (14). Functions of RNA can be regulated by antisense RNA, which is an RNA complementary to the sequence in mRNA that through complementary, stable base pairing, prevents its translation (15). Antisense RNA also regulates or even prevents gene expression in prokaryotic, plant, and animal cells.

Whether prokaryotic or eukaryotic, the mRNAs are translated into proteins using ribosomes, charged tRNA molecules, and auxiliary proteins (11). Usually several

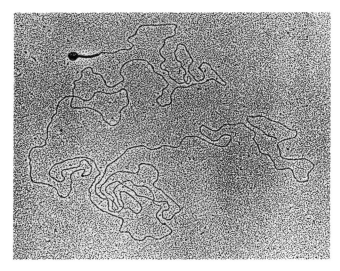

Figure 11. The bacteriophage lambda of *Esherichia coli* ejecting its DNA.

ribosomes (or a polysome) are present on a mRNA. In prokaryotes transcription and translation occur concurrently within the cytoplasm; but in eukaryotic cells translation occurs outside the nucleus. If a nonsense codon is reached either at the end or through mutations within a gene, the translation is halted and a partial polypeptide results. Translational control includes the ribosome binding sequence (Shine-Delgarno sequences) and attenuation, which is prevention of translation of mRNA because of its folding back on itself. After its transcription, the mRNA is degraded. Messenger stability, among other things, depends on the presence of AU rich sequences at the 3′-end, resulting in shorter half-lives than those lacking such sequences. Finally, protein stability and half-life depend on proteases that degrade not only normal but also abnormal (mutant) proteins. In eukaryotic cells protein stability is dependent on the small protein ubiquitin, which when conjugated to proteins leads to their rapid degradation.

Genetic Exchange and Recombination In Vivo

Genetic information can be exchanged in vivo among various organisms (3,4,16). The simplest forms of genetic exchange—transformation, transduction, and conjugation—are those found among bacteria. Only conjugation requires cell to cell contact. Conjugation was initially discovered in the common enteric bacterium *E. coli,* but it occurs in many genera of bacteria and between members of two kingdoms, eg, *Agrobacterium tumefaciens* and many plants species (17), as well as *E. coli* and *S. cerevisiae* (18). In certain bacteria such as Streptococcus, mating depends on the presence of special mating phermones, which are small polypeptides promoting mating aggregation. In fungi, mating occurs when two haploid cells of different mating type fuse to allow genetic exchange to occur. In yeasts this requires special mating control locus (MAT) and a switching loci (HML or HMR), which are silent cassettes that introduce the *a* and *α* mating-cell types through direct transposition of required genetic cassettes. In the absence of sexual cycle, certain fungi use parasexual mating events. In transformation, whether natural or induced, naked DNA or DNA contained in membranous vesicles called transformosomes are transferred into a recipient cell. In generalized and specialized transduction, DNA packaged into a bacteriophage (Fig. 11) is injected into a bacterial cell. The latter two modes of genetic exchange are used in recombinant DNA technology for packaging of rDNA (eg, bacteriophage lambda) or transformation of commercially available competent *E. coli* cells. Once inside a cell, homologous DNA can undergo recombination using recombinational enzymes to generate hybrid DNA molecules. All mechanisms of genetic exchange are currently used in food technology.

GENETIC ENGINEERING TECHNOLOGIES

Instrumentation and Tools

Central to and fundamental in the process of genetic engineering is the isolation, manipulation, insertion into, and expression of DNA in a host cell (19,20). The instrumentation and tools needed for the technical tasks are shown in Table 1.

Organisms

Genetic engineering technologies have been used and/or applied to most simple bacteriophages; human, animal, plant, and insect viruses; bacteria; fungi; yeasts; plants and animals including humans. Microorganisms (19) and plants (21) are the most frequently used organisms as far as food technology, production, and processing are concerned. While the production of microbial cells through fermentation technology has been a long established art,

Table 1. Basic Steps in a Simple Genetic Cloning

Step	Systems Needed
1. Propagation of organisms	Growth chambers, incubators, fermenters, greenhouses
2. DNA extraction and purification	French pressure cell, cell lytic enzymes, spectrophotometer, preparative centrifugation, cesium chloride density ultracentrifugation
3. DNA library preparation	Restriction enzymes
4. DNA size separation	Ultracentrifugation or electrophoresis
5. Vector and probe DNA preparation	Polymerase chain reaction or cDNA technology DNA manipulative enzymes electrophoresis
6. DNA sequencing	Maxam-Gilbert, Sanger's and other sequencing methods
7. Introduction of rDNA into host	Electroporation, biobolistics transformation
8. Identification of transformed cells	Selective media, reporter cells genes, Southern blots, Northern blots, Western blots, enzyme linked antibodies, etc
9. Gene location	OFAGE, CHEF, various blots

that of plant cells in cultures is rather new. In the 1970s several groups studied the ability of plant cells and tissues to grow in liquid nutritive media. Cultures from diverse origins such as (1) organs (roots, flowers, anthers), (2) meristems (shoot, leaf), (3) callus (undifferentiated cell mass), (4) cells (homogenized tissues), and (5) protoplasts were found to grow on defined salts media containing a carbon source, vitamins, plant hormones, and various other nitrogenous substances. Plant cell biotechnology and certainly production of plant products of economic value could not have matured without the advances in plant cell culture systems (21). Construction of transgenic plants and regeneration of the whole plants or selection of somaclonal variants are a few examples of success that can be cited.

Isolation and Preparation of DNA

DNA from most cellular organisms can be isolated through the disruption of cell membranes and/or walls by using lytic enzymes or other physicochemical forces (osmotic pressure, shear forces, and ultrasound). DNA of high (chromosomal) and low molecular weights (from plasmids, mitochondria, and chloroplasts) can be extracted. Because cellular lysates are rich in DNA degrading enzymes, such reactions are performed in the presence of the chelating agent EDTA, or cold temperature. DNA is then purified from proteins and other cellular constituents by extraction with a mixture of phenol-chloroform and is recovered by ethyl alcohol or isopropanol precipitation. To purify further, high-speed dye (ethydium bromide or bisbenzimide) buoyant density ultracentrifugation is performed and, to characterize the plasmid or other DNA agarose, gel electrophoretic techniques are employed (Fig. 12). Here DNAs migrate according to their molecular weight from the negative to the positive electrode, ie, the smaller molecules move further from the origin. If such DNAs are cut by any one of some 600 restriction endonucleases, enzymes that cleave phosphodiester linkages at specific sequences within DNA, the resultant fragments can be separated in an electrophoretic separation gel, according to their sizes (Fig. 13). Location of particular sequences can be determined by the specific enzyme employed and detected by various blotting methods. Often the electrophoretic separation of DNA followed by the bromide staining methodology is used for visualization of the physical size of DNA molecules, although the same can be achieved by electron microscopic analysis of a molecule for its size (Figs. 4 and 11) or heteroduplex analysis. Table 2 lists most commonly used enzymes for cell lysis, DNA extraction, and subsequent rDNA construction. The sequence of a particular DNA molecule can be determined (Fig. 14) by chemical (Maxam-Gilbert and Sanger) or enzymatic (Klenow polymerase, Taq polymerase–polymerase chain reaction or PCR) systems, some of which are now automated and commercially available both for synthesis and sequencing (19–23).

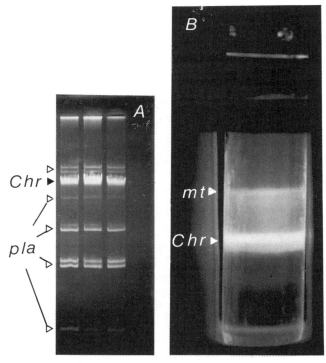

Figure 12. Separation of chromosomal and plasmid DNA from the bacterium *Bacillus thuringiensis* HD-1 by agarose gel electrophoresis (A) and chromosomal and mitochondrial DNA from the fungus *Beauveria bassiana* by cesium chloride bisbenzimide gradient centrifugation (B).

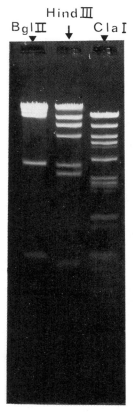

Figure 13. Restriction endonuclease digestion of bacteriophage lambda DNA into fragments. The names of enzymes are indicated on the top of each track.

Table 2. Enzymological Aspects of rDNA Work

Purpose	Enzyme	Function
Cell lysis	Lysozyme, cellulase	Removes cell walls
Proteases	Pronase, protease K	Removes proteins
RNases	RNase H	Elimination of RNA
Nick translate	E. coli DNA polymerase I	Synthesize DNA
	Klenow fragment	
	T4 DNA polymerase	
Copy DNA	Reverse transcriptase	(cDNA) from mRNA
Process DNA	Nuclease Bal31	Exonuclease digestion
	Mung-bean nuclease	
	Restriction endonucleases	Endonuclease cuts
	DNA methylases	Methylated bases
	Phosphatases	Remove 5' phosphate
Ligases	Polynucleotide ligase	Joins ends of DNA or RNA
Polymerase chain reaction	Taq polymerase	Produce oligonucleotides

Cloned Genes in Vitro

Gene cloning is the use of experimental techniques that generate rDNA molecules with a desired gene and its incorporation and expression in a cell. With the use of rDNA technology one can produce a genomic DNA library and hence a fragment containing any gene(s) from any source (19–23). Cloned genes are essential foundations of biotechnology business. To clone a gene (Fig. 15), the genetic information, whether DNA or RNA, is processed by restriction endonucleases or reverse transcriptase, as the case may be. The ends of DNA fragment containing the desired gene may be processed to have complementary ssDNA tails, poly A and poly T-linkers, or other specified sequences on opposing DNA strands. Finally T4-ligase can be used to join flush-ends of two DNA molecules. Several criteria are applicable in deciding on a choice for cloning in vitro, the most important of which is the vector system. Depending on fragment size to be cloned, specific vectors must be employed. Thus, for fragments of DNA with sizes of < 4 kb, < 10 kb, < 23 kb, and < 46 kb the recommended vectors are phage M13, bacterial plasmids (pEMBL, pBR, etc), lambda phage, (Fig. 11) and cosmids (22). Cosmids, which are a combination of a plasmid and

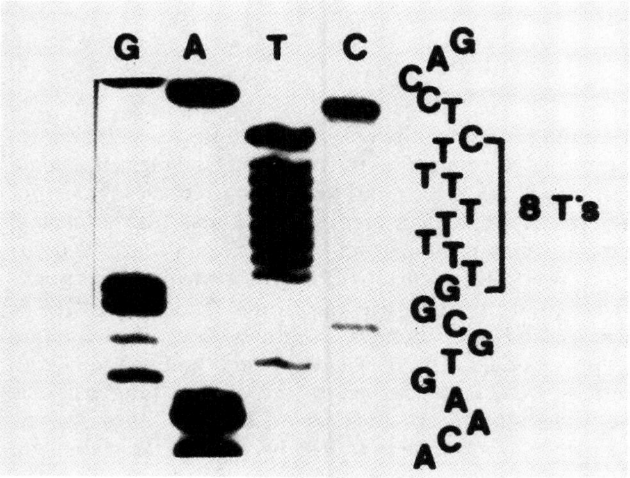

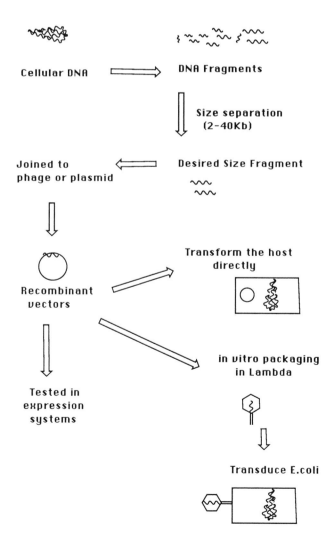

Figure 14. Sample autoradiogram of DNA sequencing reaction pattern by the Sanger method. Photograph courtesy of J. M. MacPherson.

Figure 15. A summary of the basic gene cloning experiment using E. coli as the ultimate host.

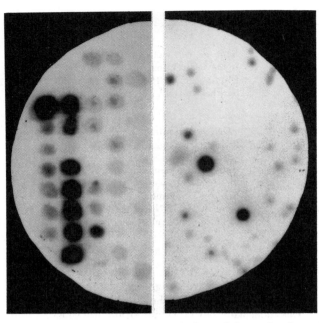

Figure 16. Autoradiogram of two halves of plates showing colony lift (left half) and plaque lift (right half). Photograph courtesy of J. M. MacPherson.

bacteriophage lambda-based vectors, can be made to contain large DNA fragments from the genomic library of an organism. The number of clones needed for fragments of 35 kb from *E. coli* genome would be 340; from *S. cerevisiae*, 6,000; and from tomato plant cell, 60,000. Often the cloning strategies, eg, choice of available restriction sites, reporter genes (to indicate the expression of cloned gene), presence of one or more selectable phenotypes, autonomous replication in two or more hosts (shuttle vectors), nature of the experiment (eg, sequencing, preparation of probes, study of gene regulation), and exploitation and safety consideration (biohazard consideration, placement of nonconjugability functions, vector suicide system) decide the particular details of the system. Finally, in deciding about the choice for the vector, the question of cloned gene copy number (low or high) and ssDNA (used for heteroduplexing and sequence analysis) versus dsDNA phages should be made in advance. In recent years more advanced vectors have been produced, eg, expression vectors and cassette vectors. They contain a promoter, a terminator, and ribosome-binding sites, with the additional feature of a restriction site where a desired structural gene will be cut and inserted. The strong regulatory sequences aid significantly in the expression of the gene in such a tailormade expression vector system. This system offers great advantages for the production of rDNA-derived foreign products from cells. Additional uses of cassette vectors occur in those cases where partial removal of the reading frame from a structural gene has been made. After the start, the chimeric DNA, containing fragments of two different organisms, is inserted. The foreign gene product is recovered as a fused (chimeric) polypeptide containing a short segment of bacterial DNA followed by the foreign polypeptide.

To examine the polypeptide products from rDNA, several in vitro and in vivo, eg, whole bacterial cell, minicell, and maxicell, systems can be used (24,25). Minicells are unique in that they are polar buds from rod-shaped bacteria (Fig. 2) lacking chromosomal DNA and therefore permit the screening of only those proteins that have been derived from the rDNA (27). After a recombinant DNA is generated, it is introduced into the new host through one of many techniques of gene transfer, such as electroporation, particle gun or biolistics, microinjection, microlaser technique, and liposome fusion (19,20,27). Although the choice of a bacterial host for rDNA reflects the special laboratory vs industrial usage, for food technology, the GRAS status is important and strains of *E. coli* and *Bacillus* spp. with multiple gene deletions are available. The transformed cell is then identified by the expression of the recombinant chimera under some selective growth condition or by using a variety of DNA, RNA, and immunological probes for colony or plaque hybridization (Fig. 16) and detection (19,22).

FUTURE DEVELOPMENT

Genomic Banks and Sequences

Genetic engineering tools have deciphered the sequences of thousands of genes and the total sequences of many viruses, plasmids, human mitochondrion, and tobacco chloroplast DNA (28). Many research laboratories have generated such material in libraries, eg, the 1985 EMBL and GeneBank collection of nucleic acid sequences comprises a four-volume set containing thousands of genes (29). The availability of many gene sequences has permitted inspection and identification of gene regulatory element binding sites. One key element is the open reading frame (ORF) and the other is unidentified reading frame (URF). ORFs can locate ribosome attachment sites and, hence through in vitro mutagenesis, base sequence changes within ORFs can be made without much need for formal genetic analysis (ie, mutants). Naturally occurring and engineered genes could be utilized from such gene banks.

Engineered Enzymes

The main source for the enzymes used in general, and the food industry in particular, are those enzymes that come from microbial, plant, and animal sources. Enzymes not only transform food ingredients but also play a crucial role in restoring the lost flavor and aroma that can occur in food processing. Various lipolytic, proteolytic, and multicomponent enzymes are being used for the addition of flavors of natural chemical origin to food. Thus, single-flavor compounds such as fatty acids, ketones, diacetyl, acetaldehyde, lactones, esters, pyrazines, mustard oils, and flavor enhancers such as nucleotides, amino acids, and terpens can all be produced by microbial enzymes for aroma and flavoring of food (30). In addition to the desirable flavors, enzymes such as lipoxygenases contribute to the off-flavoring of corn oil, while exerting a positive effect on the flavor of tomatoes (31). Genetic engineering could control the undesirable elements of many such enzymes without creating a dilemma for the internal functions of a cell. Through the knowledge of the DNA and protein sequence for many enzymes, the genetic analysis for the physical requirements of an enzyme (eg, active, catalytic, and ligand binding sites), and use of computer modeling the production of ideal enzymes should be possible (23,32). Production of a functional enzyme of minimal size or with improved thermotolerance, whether through solid-state peptide synthesis or synthetic oligonucleotides and rDNA technology, should be possible within the near future (33).

Transgenic Animals and Plants

Since the mid-1970s, a number of experiments in genetic engineering have been aimed at the production of transgenic animals and plants. Earlier experiments dealt with the direct injection of nuclei, micronuclei, and DNA from a variety of sources into plant or animal cells. More recent techniques have been developed to transfer DNA constructs into plant or animal cells by such diverse methods as microinjection (plants and animal cells), conjugal transfer (plant cells), electroporation (microbial and plant cells), biolistics or microprojectile, and other less commonly known techniques. Construction of transgenic plants containing viral, bacterial, and other eukaryotic genes presenting desirable traits such as resistance to disease, insects, herbicides, and tolerance to certain environmental factors has been possible for a majority of edible crops and other commercial plants. Several transgenic plants (including tobacco, tomato, corn, canola or rape, and cotton) containing an insecticidal toxin gene of a bacterium have been constructed (34). Through combined knowledge of flowering-plant pigment and hormone biosynthesis, it is now possible to isolate, manipulate, and transfer genes for floral shape, and color, in the petunia (35). The extension of this knowledge toward derivation of food coloring and flavoring substances should be an exciting possibility. Use of antisense RNA in the area of genetic engineering of flowering plants and ripening of climacteric fruit and vegetables (eg, tomatoes) has been achieved (36). Embryo transfer technology (19,23) and production of transgenic animals by microinjection of foreign genes into eggs (transgenesis) to increase animal productivity (meat, milk, wool) has become possible in many animals. Scientists in the United Kingdom and China, after injection of about 1 million copies of rat growth hormone gene into fish eggs, noticed that about 5% of fish hatched were transgenic, showing the presence of rat growth hormone. In sheep, goats, pigs, and cattle, presence of growth hormone produces leaner meat. In Edinburgh, Scotland, researchers have had success in creating transgenic sheep that produced milk with human antihemophilic substance, factor IX, and have obtained sheep β-lactoglobulin from transgenic mice.

Genetic engineering, as described briefly here, is a subject about which courses are given and textbooks are written. What is clear from this article is that which was said at the beginning, these changes are recent and indeed more change is just around the corner.

BIBLIOGRAPHY

1. D. McCormick, "Genome know-how," *Bio/Technology* **8**, 5 (1990).
2. Congress of the United States: Office of Technology Assessment, *Impact of Applied Genetics: Microorganisms, Plants and Animals,* U.S. Government Printing Office, Washington, D.C., 1981.
3. B. Lewin, *Genes,* vol. 4, Oxford University Press, Oxford, UK, 1990.
4. E. A. Birge, *Bacterial and Bacteriophage Genetics: An Introduction.* Springer-Verlag, New York, 1988.
5. S. Cohen, A. Chang, H. Boyer, and R. Helling, "Construction of biologically functional bacterial plasmids in vitro, *Proceedings of the National Academy of Sciences of the United States of America* **70**, 3240–3244 (1973).
6. J. A. Piccirilli, T. Krauch, S. E. Moroney, and S. A. Benner, "Enzymatic incorporation of a new base pair into DNA and RNA extends the genetic alphabet," *Nature* **343**, 33–37 (1990).
7. A. Kornberg, *DNA Replication,* W. H. Freeman, San Francisco, 1980.
8. R. Ellis and S. Hemmingsen, "Molecular chaperonins: proteins essential for the biogenesis of some macromolecular structures," *Trends in Biochemistry* **14**, 339–342 (1989).
9. B. P. Wasserman, "Thermostable enzyme production," *Food Technology* **38**, 78–89 (1984).
10. A. L. Demain and N. A. Solomon, eds., *Manual of Industrial Microbiology and Biotechnology,* American Society for Microbiology, Washington, D.C., 1986.
11. J. D. Watson, J. Tooze, and D. T. Kurtz, *Recombinant DNA: A Short Course.* W. H. Freeman, New York, 1983.
12. J. E. Darnell, Jr., "RNA," *Scientific American* **253**, 68–78 (1985).
13. T. R. Cech, "The chemistry of self-splicing RNA and RNA enzymes," *Science* **234**, 1532–1539 (1987).
14. N. J. Proudfoot, "How RNA polymerase II terminates transcription in higher eukaryotes," *Trends in Biochemistry* **14**, 105–109 (1989).
15. P. J. Green, O. Pines, and M. Inouye, "The role of antisense RNA in gene regulation," *Annual Reviews of Biochemistry* **55**, 569–597 (1986).
16. B. W. Bainbridge, *Genetics of Microbes,* 2nd ed, Blackie & Sons, London, 1987.
17. D. M. Raineri, P. Bottino, M. P. Gordon, and E. W. Nester," *Agrobacterium-* mediated transformation of rice (*Oryza sativa* L.)," *Bio/Technology* **8**, 33–37 (1990).
18. J. A. Heinemann and G. F. Sprague, Jr., "Bacterial conjugative plasmid mobilize DNA transfer between bacteria and yeast," *Nature* **340**, 205–209 (1989).
19. S. L. Berger and A. R. Kimmel, eds., *Guide to Molecular Cloning Techniques,* vol. 152, *Methods in Enzymology.* Academic Press, San Diego, Calif., 1987.
20. T. Manniatis, E. F. Fritsch, and J. Sambrook, *Molecular Cloning: Laboratory Manual.* Cold Spring Harbor Laboratory, Cold Spring Harbor, NY, 1981.
21. R. Teutonico and D. Knorr. "Plant tissue culture: food applications and the potential reduction of nutritional stress factors," *Food Technology* **38**, 120–127 (1984).
22. T. A. Brown, *Gene Clonning. An Introduction,* Van Nostrand Reinhold (UK), Berkshire, UK, 1986.

23. J. M. Walker and E. B. Ginglod, *Molecular Biology and Biotechnology*, The Royal Society of Chemistry Special Publication 54, Burlington House, London, 1985.
24. E. Jay, J. Rommens, and G. Jay, "Synthesis of mammalian proteins in bacteria," in P. N. Cheremisinoff and R. P. Ouellete, eds., *Biotechnology Handbook*, Technomic Publishing, Lancaster, Penn., 1985, pp. 388–400.
25. G. G. Khachatourians and C. M. S. Berezowsky, "Expression of recombinant DNA functional products in *Escherichia coli* anucleate minicells," in M. Moo-Young, J. D. Bullock, C. L. Cooney, and B. R. Glick, eds., *Biotechnology Advances*, Vol. 4, Pergamon Press, Oxford, UK, 1986, pp. 75–93.
26. G. G. Khachatourians, "The use of anucleated minicells in biotechnology: an overview," Ref. 24, pp. 308–318.
27. U. Zimmermann, "Electrofusion of cells: principles and industrial potential," *Trends in Biotechnology* **1**, 149–155 (1983).
28. J. C. Mooeres "Current approaches to DNA sequencing," *Analytical Biochemistry* **163**, 1–8 (1987).
29. EMBL Nucleotide Sequence Data Library and GenBank, *Nucleotide Sequences 1985. A Compilation from the GeneBank and EMBL Data Libraries*," IRL Press, Washington, D.C. 1985.
30. I. L. Gatfield, "Production of flavor and aroma compounds by biotechnology," *Food Technology* **38**, 110–122 (1988).
31. R. J. Moshy, "Impact of biotechnology on food product development," *Food Technology* **35**, 113–118 (1985).
32. G. G. Khachatourians and A. R. McCurdy, "Biotechnology: applications of genetics to food production," in D. Knorr, ed., *Impact of Biotechnology on Food Production and Processing*, Marcell Dekker, New York, 1986, pp. 1–19.
33. W. H. Pitcher "Genetic modification of enzymes used in food processing," *Food Technology* **36**, 62–63, 69 (1986).
34. C. S. Gasser and R. T. Fraley, "Genetically engineering plants for crop improvement," *Science* **244**, 1293–1299 (1989).
35. J. Mol, A. Stuitje, A. Gerats, A. van der Krol, R. Jorgansen, "Saying it with genes: molecular flower breeding," *Trends in Biotechnology* **7**, 148–153 (1989).
36. M. Kramer, R. E. Sheehy, W. R. Hiatt, "Progress towards the genetic engineering of tomato fruit softening," *Trends in Biotechnology* **7**, 190–194 (1989).

George G. Khachatourians
University of Saskatchewan
Saskatoon, Canada

GENETIC ENGINEERING PART II: ENZYME CLONING

The food-processing industry is the oldest and the largest user of biotechnological processes, dating back to pre-Biblical times. Many traditional food processes, such as the production of alcoholic beverages, vinegar, bread, cheese, and preservatives are based on fermentation using biotechnological principles. Biotechnology can be defined as the integrated use of biochemistry, microbiology, and biochemical engineering to achieve the technological application of the capacities of microbial and cultured cells (animal and plant). Fermentation is any process that uses a microorganism to produce a product of commercial value by transformation. The total annual turnover in biobased processes is well over $100 billion; two-thirds of this figure is accounted for by alcoholic beverages. Cheese and fermented milk products represent the second most important category, accounting for approximately 25%, while others make up about 10% (Table 1). Fermentation is also used to produce many food ingredients such as amino acids, flavors, vitamins, and processing aids such as enzymes.

Almost half of increased crop yields and dramatic improvements in the efficiency of livestock production have come about through conventional genetic improvements, such as bulk breeding and selection techniques. In recent decades, these techniques have been extended to microorganisms and have contributed to the development of strong fermentation technologies in the pharmaceutical, food, and beverage industries. Traditionally, the biotechnological process based on classical microbial fermentation has been augmented by simple genetic manipulation using a mutagenic agent such as ionizing radiation, ultraviolet light, or various chemicals (nitrous acid, nitrosoguanidine, etc) to improve microorganisms for food fermentation and to enhance the production of bioingredients. Such exposure to mutagenic agents usually involves subjecting the cells to a mutagenic dose that results in the death of the vast majority of the cells. The survivors of this exposure may then contain some mutants, a very small proportion of which may be superior producers. It is not possible to predetermine the gene that will be affected by the mutagen, and it is difficult to differentiate the few superior producers from the many inferior producers found among the survivors of a mutation treatment.

During the last 10 years or so, new techniques such as recombinant DNA (rDNA), cell fusion (monoclonal antibody), plant cell culture, and bioprocess engineering (bioreactor, immobilization) have been developed. It is, however, the discovery of genetic engineering via rDNA technology that is responsible for the current biotechnology boom. Not only do these techniques offer the prospect of improving existing processes and products, but they are enabling us to develop totally new products and new processes that were not previously possible using standard mutation techniques. Recombinant DNA technology was an outgrowth of basic research on restriction enzymes and enzymes involved in DNA replication (2). In 1973, scientists for the first time successfully transferred DNA (deoxyribonucleic acid: the carriers of genetic information) from one life form into another (3). The new technology has spawned a new industry and prompted a dramatic refocusing of the research directions of established companies. The development of cloning techniques gave rise to a wave of far-reaching speculation on their impact as well as concern over hazards, but commercial interest focused on less controversial targets such as the bacterial production of therapeutic products.

Until recently, only a few rDNA products have gone into production commercially; these include human insulin, human somatotropin (human growth hormone), urokinase, blood-clotting factor (factor VIII), livestock vaccine (colibacillosis), and interferon α_2. However, there is a very large number of potentially useful products, and more than 97 are under development (4). There are 7,000 rDNA patent applications pending, more than twice the

Table 1. The Estimated Value of the World Output of Major Biobased Food Products[a]

Products	Value (million $)	Percentage of Total
1. Alcoholic beverages (beer, wine, spirits, sake, etc)	67×10^3	65
2. Cheese/fermented dairy products	25×10^3	25
3. Enzymes and food ingredients (amino acids, citric acid, sweeteners, vitamins, baker's yeast, etc)	6×10^3	6
4. Fermented foods (pickles, vinegar, soy sauce, miso, etc)	4×10^3	4
Total	$100 billions	

[a] Ref. 1.

number of patents pending at the end of 1984 (5). Many of the highlights of genetic engineering occurred in the area of human health-care products, because of a high added value, but a great proportion also dealt with food production.

Until recently, the production of food through conventional technologies has, in general, been sufficient in developed countries. There is, however, a trend toward a continuous expansion of both the population and the world food markets, indicating a requirement for a different food production strategy. As we move toward the 21st century, scientists will be looking more and more to the new technologies (rDNA) to improve the efficiency and the stability of animal, crop, and food production and to increase profitability. New biotechnology probably has its greatest impact at the top of the food chain in both agronomic and nonagronomic sectors, through the production, modification/improvement, and preservation of raw materials, as well as through the production of food additives or processing aids (Fig. 1).

In livestock and animal cell culture (multiplication of animal cells in in vitro culture), genetic engineering in conjunction with embryo transfers and gene insertions promises great improvement in animal health and productivity. rDNA technology enables the microbiological production of large amounts of biologically active components such as hormones and exceptionally pure and inexpensive vaccines. Several animal health-care products developed via rDNA technology are already on the market and more are under development. However, the application of rDNA technology for improvements of this type often is dependent on a fuller understanding of the biological system being controlled or manipulated. Therefore, near-term progress toward these goals will be slow.

Plant biotechnology is being directed toward meeting three goals: increased crop yield, improved crop quality, and reduced production costs. However, progress is slow in this area of research because plants are genetically and physiologically more complex than single-cell organisms such as bacteria and yeasts. Genetic engineering does not displace conventional plant breeding but simply permits more rapid progress. Examples of genetic engineering that have been useful to the food-processing industry include increasing the solids content of tomato and potato and altering the chemical composition of vegetable oils in domestic oilseed crops (6). Genetically engineered tomatoes remain firm for shipping purposes due to the reduction of polygalacturonase levels, an enzyme that stimulates softening. Also, by increasing the amount and protein contents of seeds, new technology can improve

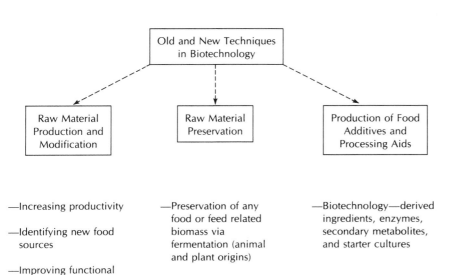

Figure 1. Role and classification of food biotechnology.

their nutritional value. Suspension culture, in which plant cells are submerged in a liquid medium (such as microorganisms), can be useful for the production of important natural substances such as drugs, flavors/fragrances, and chemicals (7). Plant cells cannot compete with bacterial or yeast cell culture with respect to cost of production, but certain unique compounds synthesized exclusively by plant cells may have commercial value.

ENZYME CLONING AND FOOD FERMENTATION

Fermentation is the critical process by which living microorganisms are used to manufacture commercial products. Typically it is performed in a large tank containing a liquid medium with the nutrients and other ingredients needed to propagate the producing microorganisms. The design of a fermentation process is preceded by selection and development (perhaps by genetic engineering) of the microorganism to be used. New products including food and beverage additives such as enzymes, flavors, and colorants have resulted from the more recent scientific advances in fermentation. These products have a relatively high market value as a result of the complex or unique processes used in their manufacture. Several food and beverage products that are produced by fermentation processes are listed (Table 2). These products have traditionally resulted from fermentation produced by a mixture of microorganisms occurring in the environment of the process. There is an increasing tendency to use pure culture strain in these fermentation processes to avoid unwanted extraneous microorganisms that can cause quality defects and spoilage. The contribution of biotechnology to food manufacturing has been to provide improved practices and greater sophistication in the use of microorganisms.

The opportunity for the future is in making further improvements to microbial strains for particular jobs by genetic manipulation and by further developments in fermentation technology. This way we may develop new food products and also improve the yield and quality of existing ones.

The traditional way of increasing fermentation yield has been through mutation. Mutation alters the control function so that synthesis of the product continues at a higher rate. Mutants of *Cornyebacterium glutamicum* have been used extensively for the commercial production of glutamate as well as several amino acids. Under biotin-limiting conditions (ie, the growth of the cells was limited by the availability of biotin) glutamate was excreted from the cells and accumulated in the medium up to a concen-

Table 2. Some Important Fermentation Products that are Produced by Live Microbial Cells[a]

Products	Microorganisms	Attributes
Alcoholic Beverages		
Beer	*Saccharomyces cerevisiae*	Flavor, low calories
	Saccharomyces carlsbergenesis	
Whiskey	*Saccharomyces cerevisiae*	
Wine/brandy	*Saccharomyces ellipsoideus*	
Rum	*Saccharomyces cerevisiae*	
Dairy Products		
Cheese	*Streptococcus lactis,*	Flavor, texture, nutrition
	Sr. cremoris	
Yogurt	*Streptoccocus thermophilus*	Flavor, texture, nutrition
	Lactobacillus bulgaricus	
Acidophilus milk	*Lactobaccilus acidophilus*	Flavor, texture, nutrition
Kefir	*Streptococcus lactis*	Flavor, alcohol
	Lactobacillus cancasicus	
	Leuconostoc sp., yeasts	
Koumiss	*Lactobacillus acidophilus*	Flavor, alcohol
	Lactobacillus bulgaricus	
	Saccharomyces sp.	
Yakult	*Lactobacillus casei*	Flavor, acid
Meat Products		
Sausages	*Lactobacillus plantarum*	Flavor, acid
	Pediococcus cerevisiae	
	Micrococcus sp.	
Vegetable Products		
Sour dough	*Lactobacillus sanfrancisco*	Flavor, nutrition
Soy sauce	*Lactobacillus delbrucckii*	Flavor, nutrition
Sauerkraut/pickles	*Lactobacillus plantarum*	Flavor, storage stability
	Pediococcus cerevisiae	

[a] Ref. 1.

tration of 50 g/dm³. Normally, the level of a particular metabolite is controlled through the inhibition of the activity of key enzymes in the biosynthetic route and the repression of their synthesis by the end product(s) when present in the cell at a concentration sufficient to meet the organism's requirements. Organisms used for commercial production of primary metabolites are rarely modified at only one genetic site. It is often necessary to alter several control sites to achieve overproduction of the desired compound. The task is far easier for strains producing primary metabolites (eg, amino acids, nucleotides, vitamins, alcohol, organic acids) than it is for those producing secondary metabolites (eg, enzymes, antibiotics, flavors). While primary metabolites are essential for life and reproduction of cells, secondary metabolites seem unessential for growth and reproduction. There is evidence that the production of such substances represents a form of waste disposal by the cells.

With the exception of the food industry, only a few commercial fermentation processes use wild strains isolated directly from nature. Mutants that are specifically adapted to the fermentation process are used in the production of antibiotics, enzymes, amino acids, and other substances. Through an extensive strain development program that may require several years, yield increases up to 100 times or more can usually be attained. For example, the yield of products derived from the activity of one or a few genes can be increased simply by raising the gene dose. However, in the secondary metabolites that are frequently the end result of complex, highly regulated biosynthetic processes, a variety of changes in the genome may be necessary to obtain the high-yielding mutants.

The genetic information from two genotypes can be brought together into a new genotype through genetic recombination. This is another effective means for industrial-strain improvement. However, relatively few industrially important organisms exhibit sexual reproduction as such. The attainment of recombinants through the sexual process has been confined to commercial mushrooms and the yeasts used in baking, brewing, and distilling industries. Recently developed experimental methodologies such as protoplast fusion have extended the number of organisms in which two genotypes can be recombined. Protoplasts are cells devoid of their cell walls and may be prepared by subjecting cells to the actions of wall-degrading enzymes in isotonic solutions. Protoplast fusion has been achieved with the filamentous fungi, yeasts, *Bacillus* sp., *Brevibacterium flavum*, and Streptomyces (9). One of the most significant approaches to strain improvement can be anticipated from the use of in vitro recombinant DNA technology. The principle of this technique is that genetic information DNA is isolated from a donor organism and is cleaved by restriction enzymes into individual pieces. The fragments are then ligated with DNA of a suitable vector DNA (plasmid) using another enzyme, DNA ligase, and transferred with this vector into another cell either by transforming protoplasts or by transforming cells that have been made competent through $CaCl_2$ treatment. Currently, there are three possible cloning routes to make a particular product. These may be referred to as the messenger-RNA route, the DNA route, and the protein route (9).

Yeast-Based Fermentation

Yeasts, which are used in baking, beer, wine, distilled beverages, flavor ingredients, and for specialized feeds, are basic products of biotechnology. Yeast biotechnology is closely linked to the production of ethanol and beverages that contain alcohol. The largest single market for yeast is associated with the strains used for baking. The production and sales of major yeast-based industries are over 60 billion dollars (Table 1). New developments continue to be evaluated for specialized yeast applications, such as production of low-alcohol beer, fermentation of wine, and conversion of carbohydrates to ethyl alcohol. The development of genetic engineering for food fermentations has proceeded much more slowly than in the health-care area. The reasons for this are not difficult to find:

1. Expensive R&D may be very difficult to justify against the limited benefits envisioned for process improvement or the added value of a new product.
2. Fermented foods are highly complex products, thus an improvement in processing efficiency is only useful if it can be accomplished without detrimental effects to product quality.
3. Consumer sensitivity to the use of genetically engineered organisms.

However, the ability of the rDNA approach to bring about genetic change in a highly specific and well-controlled way offers the possibility of introducing novel characteristics into brewing strains without damaging their existing desirable capabilities. Moreover, useful cloned genes and protocols for their transfer and expression could be utilized widely with brewing strains and with other industrial yeasts. Specific targets are in:

1. Reducing raw material costs.
2. Increasing usage of cheaper materials.
3. Increasing the efficiency and productivity of fermentation as well as increasing the product quality.

Enhancing the value of surplus yeast and developing new products or new areas of profitable enterprise are of considerable future significance for the brewing industry. Now the industry is employing modern genetic techniques to improve the genetic makeup of yeasts in brewing and wine-making (6). In brewing, yeast cells have been genetically manipulated to improve their ability to ferment dextrins for the production of a low-carbohydrate light beer without adding a commercial enzyme (glucoamylase) from *Aspergillus niger*. The modification of yeasts for wine-making has produced strains with increased alcohol tolerance and strains with sedimentation characteristics that allow for easier separation of yeasts from wine. A few companies are involved in exploring more effective proteases, which break down protein to improve beer clarity and hence improve the marketability of the product. An additional application of biotechnology in brewing involves the genetic engineering of bacteria that manufacture enzymes, notably amylases, that break down starch more efficiently during malting. This decreases the cost of the malting process.

$\mathscr{L}$-amylase alone liquefies the starch, but glucoamylase plays an essential role in complete starch fermentation as the enzyme possesses both $\mathscr{L}$-1, 4 and $\mathscr{L}$-1, 6 hydrolytic activities. The combination of glucoamylase activities derived from *Aspergillus* and *Rhizopus* makes this an enzyme of choice in the commercial saccharification process for the production of high-fructose corn syrup, ethanol, and light beer (10).

A strain of baker's or brewer's yeast, *S. cerevisiae*, engineered to express one of the *Saccharomyces diastaticus* glucoamylases, could shorten or eliminate separate hydrolysis steps in current industrial processes. The yeast enzyme is favored for the construction of amylolytic strains because it is more thermolabile than glucoamylase from *Aspergillus*. Therefore, any changes in product quality associated with residual enzymes could be more readily prevented by mild pasteurization treatments. A possible problem, however, with the yeast enzyme is that unlike the glucoamylases of other fungi, it possesses no debranching activity ($\mathscr{L}$-1, 6) with starch or dextrins.

S. diastaticus is closely allied to *S. cerevisiae* except that the former produces extracellular glucoamylase. *S. diastaticus* carrying any one of the unlinked STA genes (STA 1, 2, 3) produces extracellular glucoamylase isoenzymes I, II, III, respectively (11). By comparison, *S. cerevisiae* produces an intracellular glucoamylase but only during sporulation. The STA 1, 2, and 3 glucoamylase genes from *S. diastaticus* were cloned into *S. cerevisiae* by complementing a STA$^-$ strain to STA$^+$ (11,12). The shotgun strategy employed for cloning of glucoamylase gene from *S. diastaticus* to *S. cerevisiae* consists of (Fig. 2):

1. Preparation of total DNA from *S. diastaticus*, carrying a STA gene (STA 3 gene in this case).
2. Partial digestion of the DNA with the restriction enzyme Sau3AI and separation and isolation of fragments between 5 and 10 kb by sucrose gradient centrifugation.
3. Preparation of plasmid YEp 13, a shuttle vector (a shuttle vector is able to replicate in two different hosts, in this case yeasts and *E. coli*), which carries AP^r (ampicillin resistant) and Tc^r (tetracycline resistant) genes for *E. coli* and also the yeast Leu2 gene; total digestion of this plasmid with Bam HI, followed by treatment with alkaline phosphatase to prevent recirculization of DNA.
4. Ligation of the Sau3AI digested DNA fragments into the BamHI sites of the YEp13 (Sau3AI and BamHI have the same sticky ends).
5. Transformation of *E. coli* RRI with the resulting rDNA molecules to Ap^r.
6. Pooling of the transformed colonies and preparation of plasmid DNA from transformed *E. coli* (Ap^r, Tc^s).
7. Transformation of *S. cerevisiae* LL20 (Leu2$^-$, STA$^-$) with the *S. diastaticus* plasmid bank to leucine prototrophy.
8. Selection of transformants containing the STA gene by screening for clear zones around colonies on minimal starch agar plates.

The glucoamylase gene fragment thus cloned has a restriction map different from the glucoamylase genes from the same organism cloned by others. The genes are localized within a 3.9 kb fragment.

From the large number of colonies produced by transformation, the selection of a few colonies (perhaps only one in several thousand) containing a particular fragment of recombinant DNA is not an easy job. One of the most useful methods to find a particular clone is known as colony hybridization. This method depends on the availability of a radioactively labeled probe, which is a nucleic acid with a sequence complementary to at least part of the desired DNA. With the advent of cheap and rapid methods for the synthesis of a desired long oligonucleotide has come the possibility of producing synthetic probes. The amino acid sequence of the enzyme or protein can be analyzed; a pool of oligonucleotides containing all possible coding sequences derived from the peptide sequence can be constructed; and a genomic library of the clone can be screened with this pool of oligonucleotides as a probe.

Before the widespread use of genetically engineered yeast strains is seen, some work still needs to be done, particularly in the areas of

1. Genetic analysis of industrial strains.
2. Devising construction strategies for stabilizing newly introduced characteristics.
3. Evaluation of new strains for commercial use as well as the formal hurdles of food safety.
4. The secretion mechanism of foreign proteins by yeasts (9).

Further specific processes identified as desirable targets for improvement by recombinant DNA technology for yeast industries are

1. Alcohol-tolerant yeasts for wine-making.
2. Freeze-tolerant yeasts for baking.
3. Sedimentation characteristics that allow for easier separation of yeasts from wine.
4. Improvements in current amylases to break down starch more efficiently during malting.
5. Improvement of current yeasts to ferment dextrins for the production of light beer with less carbohydrate.

At present, the conversion of plant biomass such as starch, cellulose, or hemicellulose depends on the addition of hydrolytic enzymes (eg, $\mathscr{L}$-amylase, glucoamylase, xylose isomerase, cellulases) before fermentation. Therefore, it is of great commercial interest to construct new yeast strains that can convert polysaccharides of plant biomass directly into fermentable sugars. Most recently, a *Saccharomyces cerevisiae* strain was transformed with DNA encoding the genes for $\mathscr{L}$-amylase and glucoamylase from *Schwannimyces occidentalis* (13). Both genes were expressed, and the gene products were secreted. The newly constructed *S. cerevisiae* containing *S. occidentalis* enzymes could be useful for the production of low-carbohydrate beer owing to their high fermentation rate, ethanol

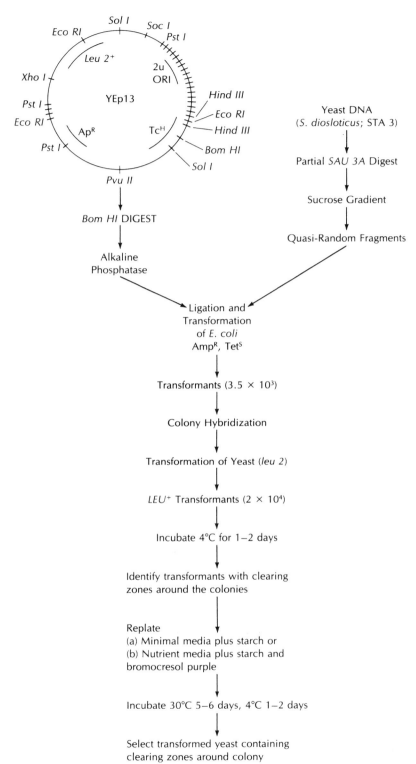

Figure 2. Cloning and expression of glucomylase (STA 3) gene from *S. diastaticus* into *S. cerevisiae*.

tolerance, and the newly acquired ability to degrade starch entirely.

Lactic Acid Bacteria Based Fermentation

Conversion of carbohydrates to lactic acid by lactic acid bacteria is the most important fermentation in the food industry. The characteristic aroma, flavor, and texture of fermented foods (eg, dairy, meat, vegetables) are often due to the growth of lactic acid bacteria. Fermented foods are enjoying increased popularity as convenient, nutritious, stable, natural, and healthy foods (1). The lactic acid bacteria (LAB) are a diverse group of microorganisms that have the common feature of producing mainly lactic acid from their sugar metabolism. Many species of bacteria produce lactic acid, but the ones that produce it in large amounts are considered as the lactics. Today, lactic acid bacteria include the genus Lactobacillus, Lactococcus (Streptococcus), Leuconostoc, and Pediococcus, with the first two types being the most frequently used genera in food products. The lactics are the principal organisms involved in the manufacture of cheese, yogurt, sour cream, cultured butter, sauerkraut, and dry sausages.

Many different products based on lactic cultures are known in different parts of the world (1). They also play a role in wine-making (malolactic fermentation), bread-making (sour dough), and in the production of spirits. More recently, lactic cultures, per se, are being marketed for their health benefits. Certain health benefits such as reduction of lactose intolerance, stimulation of the non-specific immune response, production of antimicrobial substances, antitumor or anticarcinogenic substances, and cholesterol reduction have been reported. Most recently, genetically engineered strains have been produced that can degrade cholesterol. The feeding of nonfermented acidophilus milk to experimental animals containing these bacteria significantly reduced the serum blood cholesterol (14). Thus, the lactic acid bacteria are of great economic importance. Although the per capita consumption of milk has not increased in North America during the last decade, the consumption of cultured milk products has increased 300-fold (1). The current world market for fermented milk products is approximately 25 billion dollars.

The economic importance of this group of microorganisms clearly demands the development of new technology to improve and stabilize the desirable characteristics of these starter cultures. Recent advances in understanding the genetics of cheese starter bacteria may be expected to yield specialized strains designed not only to make cheese but to ripen it quickly. Many desirable new traits and properties have been obtained by classical genetics (1), but totally new genes and gene products cannot be introduced by these techniques. Complicated nutritional requirements and enzyme regulation of lactic acid bacteria often make classical genetics tedious and make strain improvement more difficult.

Conjugal exchange of genetic information between bacteria presents another possible approach to strain improvement. Conjugal transfer of pAMB1 plasmid (erythromycin resistance) from *Streptococcus* into *Lactobacillus casei, L. reuteri, L. acidophilus,* and *L. salivarius* (or a lactase plasmid from *L. casei*) has been reported. Fusion of protoplasts is also possible for the generation of hybrid strains of lactic acid bacteria with new traits and also as a mean of introducing a transposon (movable genetic element) into lactic acid bacteria on a suicide vector. Once established, the transposon can be useful in tagging genes for molecular cloning. Conjugation or transduction techniques can be coupled to circumvent a transformation provided that the desired host can be conjugated to a transformable host. For example *Rhizobium* can be conjugated to *E. coli* and transformed readily.

Transduction systems of streptococci for lactose metabolism, streptomycin, or erythromycin resistance have been reported. However, the value of transduction as a gene transfer method is limited owing to both the small amount of DNA transferred because of the head capacity of the phage and the limited number of relevant phage receptor strains. Transduction can be very useful for making fine structural changes in the genetic material; for example, it can facilitate the introduction of new lactose or proteinase genes into the bacterial chromosome.

The knowledge of transformation systems of lactic acid bacteria has advanced significantly in recent years. Transformation is essential for an in vitro introduction of a vector into a host, which is a prerequisite to a genetic engineering application. However, protoplast transformation is inefficient in starter and lactobacilli strains, probably owing to the poor regeneration frequency of protoplast, the presence of restriction modification systems, or the inability of the selected vector to function. Particular emphasis has been placed on the construction of shuttle vectors that replicate and express their antibiotic resistance markers in lactic bacteria as well as other hosts.

However, most recently electroporation has been demonstrated to be a very efficient method for physically introducing DNA into gram-positive microorganisms, such as lactococcus and lactobacillus. Electroporation uses short electric pulses of a certain field strength, which alter the permeability of membranes in such a way that DNA molecules can enter the cell.

The major metabolic pathways for the catabolism of lactose, casein, and citrate by lactococcus have been investigated. Efficient growth and acid production in milk is dependant on the presence of the plasmid. Expressed genes include P-β-galactosidase, sucrose-6-p-hydrolase, citrate permease, bacteriophage lysin, proteinase, β-galactosidase from *E. coli,* bovine prochymosin, and egg-white lysozyme in lactococcus lactis (15). Also, heterologous expression of the *E. coli* β-galactosidase gene in *Streptococcus lactis* (16) and expression of β-galactosidase gene of *Streptococcus thermophilus* in *E. coli* and *Saccharomyces cerevisiae* (17) have been reported.

A transformation was also tried by introducing the lactose plasmid DNA into lactobacilli, which gave up to 10^4–10^5 transformants/μg DNA. The transformation of both *L. bulgaricus* and *L. helveticus* has recently been reported using a chimeric vector (a piece of foreign DNA is inserted into a vector), PLBC 104 EM (18). Even if a useful transformation system is found for lactobacilli, questions still remain with regard to the stability and expression of foreign DNA sequences in *Lactobacillus* species.

It is clear that genetic studies of lactic bacteria are becoming increasingly popular. The goal of the studies is to manipulate strains for improved performance in existing fermentations and to design new strains. Significant developments require a major ongoing research commitment, but present knowledge suggests that a number of desirable objectives are attainable (Table 3). It is envisaged that manipulation of the genes' encoding enzymes affecting lactose, protein, and citrate metabolism will allow stabilization and amplification of these traits in useful strains. Studies on the cloning of aminopeptidase and C_8-esterase from streptococci into lactobacilli strains have been started, and these may be useful in avoiding flavor (bitter) defects and in accelerated cheese ripening (19).

Chymosin is an aspartyl proteinase (EC 3.4.23.4) found in the fourth stomach of the unweaned calf where it

Table 3. Possible Targets of Genetic Engineering in Modifying Starter Cultures

Improvement of existing catalytic functions and traits
  Development of lactic cultures having specific activities (proteolytic, acidifying, maturing, etc)
  Elimination of culture rotation
  Stabilization of plasmids by insertion into chromosome
  Development of phage-resistant cultures
  Enhancement of slime formation (improved body and texture in fermented milk products)
  Enhancement of tolerance to freeze-drying
Introduction and expression of heterologous genes
  Proteases to replace rennin
  Safe host for production of food additives (eg, enzymes, colors, preservatives, stabilizers, flavor)
  Phage-resistant genes
  Antibiotic (nisin) genes
  Drug-resistant genes
  Cholesterol-degrading enzymes
  Added nutritional value
  Enhanced probiotic effect
New applications
  Natural secondary metabolites production
  New fermented foods and feeds
  Vaccine and pharmaceuticals

cleaves K-casein, resulting in the clotting of milk. This process is also used commercially in the first stage of cheesemaking. By adding a new metabolic trait such as renninlike enzymes, one can eliminate the need for chymosin in cheesemaking.

Two plasmid coded mechanisms of phage resistance have recently been discovered. Other possible application might be the beneficial role of *L. acidophilus* in human and animal health. *L. acidophilus* can provide a more acceptable alternative to the use of antibiotics and various chemicals for the control of certain intestinal diseases and for growth promotion (20).

Lactic acid bacteria (LAB) are easily cultured on a large scale and can ferment inexpensive feedstocks to produce secondary metabolites such as antibiotic nisin (preservative). Because they are already used and have proven to be safe, they would be acceptable hosts for the production of food additives. Furthermore, whey or other food by-products can be upgraded with the genetically engineered LAB for the production of valuable food additives such as enzymes, biomass, organic acids, and vitamins. The possibilities are limited only by our imagination.

New technologies, however, present safety concerns. The FDA intends to apply existing requirements and procedures that are applicable to food additives and GRAS (generally regarded as safe) substances. The antibiotic marker used in most vector constructions cannot be used in engineered food organisms, especially LAB, except in development steps; it will have to be replaced by other metabolic or nisin antibiotic markers. Although the regulations are clear regarding food-grade microorganisms as producers of end products or metabolites, new ground may have to be broken for advocating dissemination of engineered organisms in human foods.

If one can satisfy the most stringent criteria of health and safety, the next decade of research with the lactics should be exciting and the economic returns great.

Enzymes and Food Additives Derived from Other Microorganisms

Fermentation has been applied in the food industries to add nutrition, to improve the flavor of final product, or to perform certain functions in processing operations by using the catalytic properties of enzymes. Enzymes are proteins synthesized by living systems. They are important as catalysts in synthetic and degradative reactions. With the help of a multitude of enzymes, microorganisms are able to carry out a huge number of chemical reaction steps necessary for the growth and maintenance of the cell (metabolism). Many microorganisms synthesize additional, often complex, substances that appear to play no part in the growth process. These are often considered to be secondary metabolites, such as flavors and fragrance substances.

The advantages of employing microbial enzymes for industrial processes derive from:

1. The existence of different types of enzyme activities.
2. The rapidity and stability of production via microbial fermentation as an inexpensive, reproducible, and safe source.
3. An improvement in yield by genetic or protein engineering much more easily than with plant or animal cultures.

Large-scale applications of enzymes are summarized in Table 4. For comprehensive treatment of applications and of genetic engineering, Ref. 21 is a major source.

With the recent surge of interest in biotechnology, considerable emphasis has been placed on advances in genetic engineering as well as in immobilized enzymes in various forms. And now, with the aid of microorganisms changed by recombinant DNA techniques, industry is gearing up to large-scale production of enzymes. The major benefits of rDNA technology can be seen in products of high value such as enzymes and other food additives (eg, flavoring agents, sweeteners, preservatives, vitamins, colors, single-cell protein) as shown in Table 5. In the food industry, the gene encoding calf rennin, sweeteners (thaumatin, aspartame), and many enzymes have been cloned and produced in suitable hosts, and commercial production processes are currently under aggressive development. By 1995, most of the production of enzymes, flavoring agents, sweeteners, and preservatives will likely be based on rDNA technology (21).

For the enzyme and food industries, the application of rDNA technology is becoming a significant component of the corporate commercial production and development process. Although no commercial enzymes or other food additives were on the market by 1989, genetically cloned rennet in *E. coli* (Pfizer Inc.) *Kluyveromyces lactis* (Gist-Brocades), and *Aspergillus* sp. (Genentech) were awaiting FDA approval. After cloning the $\mathscr{L}$-amylase gene from *B. amyloliquefaciens* using plasmid pUB110 into *B. subtilis*, the yield of $\mathscr{L}$-amylase was enhanced about 2,000-fold and about 5 g enzyme protein was produced per liter of culture supernatant (20). Despite the great number of cloned enzymes, reports on a large-scale production of genetically engineered microorganisms are not currently available. One reason for the lack of this kind of information may be

Table 4. Applications of Important Commercial Food Enzymes

Enzymes	Sources	Applications
Starch Processing		
α-Amylase (300)[a]	*Bacillus licheniformis*, *Aspergillus* sp.	Starch liquefaction, alcohol production
β-Amylase (10,000)	Plant (malt)	Maltose production, alcohol production
Glucoamylase (300)	*Aspergillus* sp.	Starch saccharification, brewing, baking
Glucose isomerase (50)	*Bacillus coagulans*	High-fructose corn syrup sweeteners
Invertase	*Saccharomyces cerevisiae*	Invert sugar, sugar confectionary
Pullanase	*Klebsiella* sp.	Debranching of starch, brewing
Dairy Processing		
Rennet (2)	Stomach of calves	Cheese manufacture (milk coagulation)
Microbial rennet (10)	*Mucor miehei*	Cheese manufacture (milk coagulation)
Lipase/esterase	Fungal, bacterial, animal	Cheese ripening, milk fat modification, sausage ripening
Protease/peptidase (10)	*Aspergillus niger*	Cheese ripening
Lactase	*Kluyveromyces*, *Aspergillus*	Lactose hydrolysis
Fruit/Vegetable Processing		
Pectinase (collective term) (15)	*Aspergillus*	Extraction/clarification of fruit juices
Cellulase (collective term)	*Trichoderma*, *Aspergillus*	Fruit and vegetable processing

[a] Numbers mean world production in tons of pure enzyme protein (compiled data).

the high demand for stability of recombinant plasmids under large-scale fermentation conditions.

Basically, overproduction of the food enzymes and additives can be achieved by two different approaches for exogenous products and endogenous products (22). Exogenous products mainly comprise the enzymes, flavors, and food additives. The strategy for cloning and expression of genes encoding exogenous products is relatively simple; it is similar to primary-metabolite-producing strains. Ethanols, amino acids, vitamins, solvents, and nucleotides are examples of primary products obtained from the essential metabolism of microbes. The necessary experimental procedures for cloning a foreign gene in a suitable host microorganism may be complex, but they are fully developed.

The application in vitro of rDNA technology to the improvement of secondary metabolite formation is not as advanced as it is in the primary metabolite field. Examples of secondary metabolites, which are usually not essential for cell metabolism, are antibiotics, flavor substances, and polysaccharides. The main reasons for this lack of progress is the dearth of information on the basic genetics of secondary metabolite production, coupled with difficulties in the study of product production. However, one can overcome this problem by developing methods to secrete the product from the producing microorganism.

Less expensive production of a variety of metabolites already in commercial manufacture will become possible. Genetic engineering offers great promise for industrially important enzymes because enzymes are direct gene products. Potential applications in the field of enzyme production include augmentation of yield by insertion of multiple gene copies or a stronger promoter. The further possibility of expressing specific enzyme genes from a donor organism in a host, with simpler growth requirements, has obvious attractions for the enzyme manufacturer. Although this technology is still developing, there is an impressive list of commercial enzyme genes cloned and expressed in a variety of host microorganisms (21).

For example, rennin (chymosin), an important coagulant for cheese-making, has about a $100 million market per year. Although microbial rennet is already competing successfully with shrinking supplies of natural rennin, they cause off-flavors and adverse effects on shelf life of cheese. The reason is that they have a broader specificity and a higher stability than natural rennet. The gene-encoding calf rennin was introduced in suitable hosts such as *E. coli*, *Saccharomyces*, and *Aspergillus*. However, rennin was either produced at a low yield or made in an inactive or insoluble form (inclusion body), requiring costly renaturation steps (Fig. 3). At least 1 g enzyme/L is required to compete with existing production processes for calf rennin.

Although whey products have found a number of applications in the food industry, the fermentation of lactose in whey (by-product of cheese-making) for the production of baker's biomass is an attractive process not currently being exploited. *S. cerevisiae* cannot utilize lactose directly and thus the lactase gene can be introduced together with the cloned lactose permease gene into *S. cerevisiae*, leading to the ability of the transformants to ferment lactose (17,24).

As another example of rDN technology in food additives, vitamin C (300 tons world market/year) has previously been made by multistep chemical processes coupled with microbial fermentation. By this cofermentation process two unrelated microbial species (*Erwinia herbicola* and *Cornybacterium* sp.) carry out tandem fermentation processes to produce the required steps in vitamin C synthesis. Most recently, Genentech Inc. (25) has successfully cloned and expressed 2-keto-L-gulonic acid (2-KLG) in *Erwinia herbicola*, which resulted in the direct conversion of glucose to 2-KLG, a key intermediate in the synthesis of vitamin C (ascorbic acid).

Protein engineering will enable improvements to be made on industrial enzymes. Enzymes can be tailored genetically through site-directed mutagenesis to function

Table 5. Recombinant DNA Products that are Potentially Useful as Food Ingredients[a]

Products	Uses
Amino Acids/Nucleotides	
Glutamate (320×10^3)[b]	Flavorants
Tryptophane[c] (100)	Food enrichment
Aspartate (1000)	Aspartame
Phenylalanine (100)	Aspartame
Methionine (180×10^3)	Feed additive
Lysine (40×10^3)	Food and feed enrichment
IMP and GMP	Flavorants
Enzymes[d]	
Rennet[e], ℒ-amylase[f]	
Sweeteners	
Aspartame[f]	Low-calorie sweetener
Thaumatin[f]	Low-calorie sweetener
Monellin	Low-calorie sweetener
Fructose	Sweetener
Preservatives	
Nisin[f]	Preservative
Cecropin	Preservative
Lactoperoxidase	Preservative
Organic acids (acetic, lactic acids)	Preservative
Vitamins	
B-2, B-12, D, E, Nicotinic acid, C[e]	Food enrichment
Flavors	
Esters, lactone, menthol	Fruit flavors
Diacetyle	Buttery flavor
Pigments	
Astaxanthine	Red pigment
Monascin	Polyketide pigment
β-carotene	Carotene pigment
Polysaccharides	
Dextrans	Thickening agent
Xantham gum	Thickening agent
Single-Cell Protein	
Yeast	Brewing, baking, wine-making
Lactobacilli, lactococci	Dairy and meat fermentation
Methylophilus methylotrophus[f]	Animal and human food supplement
Algae	Animal and human food supplement

[a] Compiled data.
[b] Numbers mean world production in tons of pure amino acids.
[c] Commercially available rDNA product (Japan).
[d] See Table 4.
[e] FDA approval awaits.
[f] Successfully produced in the laboratory and on a pilot scale.

much more dramatically and efficiently in a given food-processing system (eg, acidic conditions, thermostability). Site-directed mutagenesis, in which point mutations are introduced into a specific position in a gene, allows the biotechnologist to change any amino acid in protein to any other specific amino acid, with the aim of altering the properties of enzymes to suit industrial requirements.

Interest will continue in the microbial conversion of plant waste materials consisting chiefly of cellulose and lignin to chemicals that can be used as fuels or animal feeds. The objective is to clone genes that encode enzymes that could degrade these substances more efficiently than could enzymes found in naturally occurring microbes.

Although it is clear that new developments in genetic

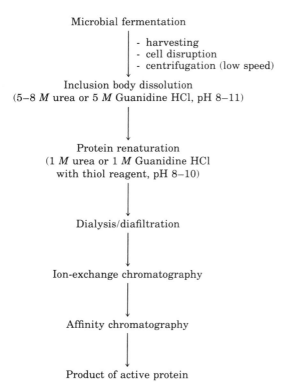

Figure 3. Recovery of inclusion bodies and purification of active protein.

engineering will lead to important new industrial processes, such processes will experience a number of important constraints, once they reach commercialization. Such aspects as economics, difficulties in culturing modified microorganisms on a large scale, and the requirement for any substance produced to be thoroughly purified are the principal issues. Fortunately, there is a vast storehouse of experience in fermentation technologies in the food, chemical, and pharmaceutical industries that can be built on to bring new microbial products to market.

POTENTIAL PROBLEMS

Genetic engineering has reached the stage where it is relatively straightforward to isolate and clone the gene for any well-characterized protein. However, there are some obstacles to the development and introduction of recombinant DNA into the food fermentation industry. The main problems that remain to be solved include (1) expression and plasmid stability, (2) recovery of product, (3) safety, and (4) economics.

Expression and Plasmid Stability

There is a need to obtain high levels of expression of the gene in a foreign host. In several instances, the product has been found to be toxic to the host microorganism. Also, the protein product of the gene might be unstable within the host cell. It may, therefore, be necessary to use a hybrid gene as a way of tagging the protein with a protective polypeptide or to modify it to prevent degradation of foreign proteins. If the gene is plasmid-linked, steps must be taken to ensure that the plasmid is stably maintained in the cell culture either by integration of plasmid DNA at homologous sites in the chromosomal DNA or by including another gene in the plasmid that will apply a selective pressure in favor of plasmid-containing cells.

Recovery of Product

Even when expression of a gene on a stable plasmid has been achieved, the ease of product recovery is a major factor in determining the profitability of a process. The problems associated with recovering active enzymes or proteins from cultured cells were mentioned in the earlier example of the inclusion body. The process of development from gene to market for most of the rDNA products is a difficult and complicated one, as the fermentation yields in scaling-up and purification of the protein may be variable. Consequently, downstream processing and its associated operational limitations and high costs are now the major obstacles in the race to develop and to place commercially viable rDNA products in the market.

Safety

Governmental regulation is still undergoing clarification on the safety of any genetically cloned system, and the process of obtaining regulatory agency approvals remains expensive and time-consuming. However, in recognition of the fact that some rDNA products have already been commercialized without any side effects, the initial concerns over possible health hazards have largely been put to rest and guidelines are being progressively relaxed.

Economics

The food industry tends to be very conservative in adopting new approaches to existing processes; because of modest profit margins, it is difficult to justify the expensive R&D needed. Most of all, there is no guarantee that a product will be profitable because it is made by genetically engineered microorganisms. Much of the investment in genetic engineering is speculative, except where there is no satisfactory alternative source of a product in obvious demand.

None of these problems will prevent progress being made in the application of genetic engineering to the food industry, but these points must be considered in any effort to apply rDNA developments to the food industry. It is also important to consider that before the recent biotechnology boom a principal application of these techniques was to investigate gene structure and to function as research tools in basic science, usually in university laboratories setting. One of the main obstacles of rDNA technology is that the basic biochemistry underlying those features we would like to modify is insufficiently understood in many cases.

SCALE-UP AND DOWNSTREAM PROCESSING

Certain aspects of rDNA fermentation and protein recovery require careful consideration not only during the initial cloning and expression but also before scale-up is considered. Factors to be considered include host strain and promotor selection, vector maintenance, host-vector/gene

Table 6. Type of Data Required from Laboratory and Pilot Plant Experiments for Effective Scale-Up into Production

Oxygen Uptake Data Related to Product Formation
  KL_a (volumetric oxygen transfer coefficient, h^{-1})
  Dissolved oxygen tension (DOT)
  Growth rate
Nutrient Uptake Data Related to Product Formation
  K_s (Substrate specific constant)
  Growth rate
Variability of Environmental Parameter Rate Constants
  pH
  DOT
  Specific rate data
  Temperature
Fluid Characteristics During Course of Fermentation
CO_2 Effects (including back pressure)
Define Geometry for Estimation of Mixing Behavior
Define Pilot Plant Control System
Estimation of Cooling Requirements

interactions, containment requirements, and product recovery (25). The choice of microorganism, fermentation conditions (eg, culture medium, morphology, growth rate) and the ability to select for secretion of the desired protein influence the choice of operations for recovery and purification of the product and affect the yields achieved at various stages during downstream processing.

Scale-up is simply the conversion of growth at a laboratory scale to an industrial process level. It is usually not feasible to take fermentation conditions that have worked in the laboratory and directly apply them on a commercial scale. This is because success in scale-up is evaluated on the basis of maximal yield in terms of minimal operating cost and time. This is especially true for recombinant DNA products in the highly competitive field of biotechnology. Although many parameters have been evaluated for use as scale-up criteria, there is no general formula because of the variation in each fermentation process. The most useful information concerning scale effects has been obtained using a series of geometrically similar fermentors. However, in reality, scale-up is not usually done with geometrically similar fermentors in the laboratory, pilot plant, and industrial settings. Sometimes even more difficult is the problem of scaling down a process from the production plant to carry out experiments on a laboratory scale.

Although there is no intent to give an in-depth review of critical parameters, certain points that have a major significance for fermentation are (1) agitation and aeration, (2) mass and heat transfer, (3) sterilization, and (4) process monitoring and control (25,26). The main data required from laboratory and pilot plant experiments for effective scale-up into production are shown in Table 6. All fermentors/bioreactors involve heterogeneous systems consisting mostly of two phases (gas and liquid). Therefore, there must be a means to transfer heat and mass between these phases in order to create optimal conditions for cell growth and biosynthesis. This must be achieved economically and aseptically.

Not all genetically engineered proteins require massive cultures. The optimized expression systems of bacteria and yeast as well as improved fed-batch culture can produce enormous biomass (more than 100 g dry wt/L) using small equipment (27). Currently, scale-up modeling is done at the 100-L level, and data generated in this fermentor are comfortably scalable to 1,000–2,000 L. At least one company is growing a recombinant strain of *E. coli* on a 30-m^3 scale for production of a pharmaceutically active polypeptide (28,29). The scale has to satisfy market demands, and thus some recombinant proteins will require larger scales than do therapeutic enzymes or monoclonal antibodies, which represent the low end (kilograms per year).

During the development of a new or modified process, much of the excitement lies in coercing the microorganisms to overproduce the desired proteins and there is a tendency to pay less attention to downstream processing. However, for recombinant DNA products, there is now an even greater emphasis on the efficiency of the recovery process. This is probably due to the fact that rDNA products require not only large quantities of proteins but also very high levels of purity to start clinical trials.

Downstream processing comprises a variety of processes used for the recovery and purification of products from microbial or cell cultures. Although most of the current commercially important enzymes are produced extracellularly by microorganisms, nearly all rDNA enzymes are intracellular. A common sequence of purification steps for pilot and large-scale processes thus consists of extraction, nucleic acid removal, precipitation, followed by one or more chromatographic steps to achieve further purification (see Fig. 3). Before the purification steps, cell harvesting is required using centrifugal sedimentation or membrane techniques such as cross-flow filtration or ultrafiltration. Recombinant DNA technology has led to the increasing observation of insoluble inclusion bodies (IBs) that require specific technology for solubilization and renaturation, particularly on a large scale. Many proteins have been purified from inclusion bodies, and some have yielded crystal structures. It now may be possible to purify IBs proteins by direct addition of affinity resins, without denaturation and refolding steps (30).

The modification of genes with signal sequences for secretion has been explored so that their products are secreted. Secretion of rDNA products into the medium is now possible from both gram-negative and gram-positive bacteria as well as from yeast (31,32,33). In principle, the secretion of the desired protein should simplify downstream processing, but proteases are often excreted, which could cause severe product degradation. Protection against this protease is essential for the high expression of some proteins.

Finally, techniques are needed to purify proteins to increased homogeneity to ensure complete removal of undesired proteins. Significant advances have been made on automated purification systems using a variety of techniques (high-resolution spherical matrixes, aqueous two-phase, large-scale affinity, immunoaffinity, ultrafiltration, and ion-exchange). There are now encouraging signs that changes can be made to the fermentation process not only to enhance product titers but also to improve product recovery.

It is important to recognize that the high yield made possible by recombinant DNA technology can affect both fermentation and downstream processing. Therefore, successful process development in this new field requires concerted efforts in genetic manipulation and process development contributed to by scientists and engineers from diversified disciplines.

FUTURE DEVELOPMENTS

This article outlined the present state of the art of genetic engineering in food fermentation and identified some of its applications and limitations, the technical and economic problems, as well as possible areas for future R&D. The selection of review material hinged mainly on the potential importance of lactic acid bacteria and yeasts in industrial fermentations and on the production of bioingredients by other microorganisms.

By application of in vitro rDNA methods, the normal barriers to genetic exchange between unrelated organisms have been broken; thus, the prospects for the rational application of genetics to industrial microbiology are completely changed. The application of rDNA technology may have a major impact on reliable and economic production of novel food additives. A full understanding of the mechanisms of secretion and posttranslational modifications will help in the efficient production of fully active and stable polypeptides. Better insights into the scaling-up of fermentation using genetically cloned strains will become available with increasing production of rDNA products in the market. The in vivo and in vitro engineering of entire biosynthetic pathways will be an order of magnitude more complex than the cloning and expression of a single protein gene.

It must be emphasized that rDNA technology cannot be pursued in isolation; the ideal industrial strain is likely to be developed by a combination of traditional with new ones. It is unlikely that genetic engineering itself will revolutionize the food fermentation industries, but most likely it will provide additional tools to develop better microorganisms for the production of food additives and for the enhancement of food fermentation.

BIBLIOGRAPHY

1. B. H. Lee, "The biotechnological aspects of Lactobacilli in food fermentation," *Trends in Food Biotechnology* (1989).
2. H. O. Smith and K. W. Wilcox, "A restriction enzyme from Hemophilus—influence Part I: Purification and general properties," *Journal of Molecular Biology* **51**, 379–391 (1970).
3. S. N. Cohen and co-workers, "Construction of biologically functional bacterial plasmids in vitro," *Proceedings of the National Academy of Science of the United States of America* **70**, 3240 (1973).
4. C. E. Morris, "Biotechnology blooms," *Food Engineering* 51–58 (July 1989).
5. L. S. Williams and R. J. Crawford, "Biotechnology: an assessment of advances, opportunities and challenges," *SIM News* **39**, 204–208 (1989).
6. S. K. Harlander and T. P. Labuza, *Biotechnology in Food Processing*, Noyes Publications, New Jersey, 1986.
7. D. Knorr, *Food Biotechnology*, Marcel Dekker, New York, 1986.
8. M. D. Trevan and co-workers, *Biotechnology: The Biological Principles*, Taylor & Francis, New York, 1989.
9. C. P. Hollenberg and H. Sahm, *BIOTECH I: Microbial Genetic Engineering and Enzyme Technology*, Gustav Fisher, New York, 1987.
10. I. Russell and co-workers, "The genetic modification of brewer's yeast and other industrial yeast strains," in S. K. Harlander and T. P. Labuza, eds., *Food Biotechnology*, Noyes Publications, New Jersey, 1986.
11. B. H. Lee and co-workers, Cloning and expression of glucoamylase (STA 3) from *S. diastaticus* into *S. cerevisiae*, personal communication, 1987.
12. J. A. Erratt and A. Nasim, "Cloning and expression of a *S. diastaticus* glucoamylase gene in *S. cerevisiae* and *Schizosaccharomyces pombe*," *Journal of Bacteriology* **166**, 484–490 (1986).
13. C. P. Hollenberg, "Improvement of barkers and brewer's yeast by gene technology," *International Conference on Biotechnology and Food*, Feb. 20–24, 1989, Hohenheim University, Stuttgart, Germany.
14. W. M. DeVos, "Gene cloning and expression in lactic streptococci, *FEMS Microbiology Reviews* **46**, 281–295 (1987).
15. M. J. Gasson and co-workers, "Genetics and genetic engineering of lactic acid bacteria," in G. Durand, L. Bobichon, and J. Fleurent, eds., *8th International Biotechnology Symposium*, Vol. 1, Paris, 1988.
16. S. G. Kim and C. A. Batt, "Heterologous expression and stability of the *E. coli* β-galactosidase gene in *S. lactis* by translation fusion," *Food Microbiology* **5**, 59–73 (1988).
17. B. H. Lee and co-workers, "Cloning and expression of β-galactosidase gene of *Streptococcus thermophilus* in *Saccharomyces cerevisiae*," *Biotechnology Letters* **12**, 499–504 (1990).
18. C. A. Batt, "Genetic engineering of *Lactobacillus*," *Food Technology* **40**, 95–98, 112 (1986).
19. B. H. Lee and co-workers, "Molecular cloning of Capyrate (C_8)-esterase gene from *Lactobacillus casei* LLG in *E. coli* and *Lactococcus lactis*," *Third Symposium on Lactic Acid Bacteria*, September 17–21. Wageningen, The Netherlands, 1990.
20. J. B. Luchansky and co-workers, "Genetic transfer systems for delivery of plasmid DNA to *Lactobacillus acidophilus* ADH: conjugation, electroporation and transduction," *Journal of Dairy Science* **72**, 1408–1417 (1989).
21. J. F. Kennedy, *Enzyme Technology*, Vol. 7a in H-J. Rehm and G. Reed, eds., *Biotechnology*, VCH Publishers, 1987.
22. *Biotechnology and the U.S. Food Industry, A Study Prepared by the Office of Planning and Evaluation, and the Center for Food Safety and Applied Nutrition*, U.S., F.D.A., Technomic Publishing, Lancaster, Penn., 1988.
23. Y-L. Lin, "Genetic engineering and process development for production of food processing enzymes and additives," *Food Technology* **40**, 104–112 (1986).
24. K. Sreekrishna and R. C. Dickson, "Construction of strains of *Saccharomyces cerevisiae* that grow on lactose," *Proceedings of the National Academy of Science of the United States of America* **82**, 7909–7913 (1985).
25. K. Stafford, "Metabolic pathway engineering for ascorbic acid production: fermentation optimization," *46th Annual Meeting of SIM & Proceeding*, August 13–18, p. 111, 1989.
26. D. Naveh, "Scale-up of fermentation for recombinant DNA products," *Food Technology* **39**, 102–109 (1985).
27. L. Demain and N. A. Solomon, *Manual of Industrial Microbiology and Biotechnology*, American Society for Microbiology, Washington, D.C., 1986.
28. L. Eppstein and co-workers, "Increased biomass production in a benchtop fermentor," *Biotechnology* **7**, 1178–1181 (1989).
29. J. V. Brunt, "How big is big enough?," *Biotechnology* **6**, 480–485 (1988).

30. N. M. Fish and M. D. Lilly, "The interactions between fermentation and protein recovery," *Biotechnology* **2**, 623–627 (1984).
31. A. Hoess and co-workers, "Recovery of soluble, biologically active recombinant proteins from total bacterial lysates using ion exchange resin," *Biotechnology* **6**, 1214–1217 (1988).
32. M. Ratner, "Protein expression in yeast," *Biotechnology* **7**, 1129–1133 (1989).
33. C. H. Schein, "Production of soluble recombinant proteins in bacteria," *Biotechnology* **7**, 1141–1149 (1989).

BYONG H. LEE
McGill University
Ste Anne de Bellevue, Canada

GENETIC ENGINEERING PART III: FOOD FLAVORS

Alcohol, acetic acid, propionic acid, and lactic acid fermentation most frequently characterize the empirical biotechnologies producing or refining food and food ingredients. Based on accumulated knowledge, a great number of nonvolatile flavor compounds such as acidulants, amino acids, 5′-nucleotides, and carbohydrates are now manufactured on an industrial scale. Although the volatile flavor of many foods is clearly determined by enzymatic activities, not much attention has been paid for a long time to the volatile flavor fraction. During the past years, however, an increasing number of papers and patents on the biogeneration of aromas appeared, indicating a new trend in both academic and industrial flavor research. This recent development has multiple roots:

- Volatile flavors and fragrances command a growing (currently ca $5 billion) market. The increasing demand of the food and beverage industry for natural ingredients suggests that agricultural sources with their inherent instabilities will no longer satisfy the need.
- The legal definitions of the Food and Drug Administration and of the Council of the European Communities classify fermentation products (with few restrictions) as natural compounds.
- The sensory properties of many flavors strongly depend on the spatial fine structure of the molecules. The enantiomeric forms of, eg, linalool, carvone, rose oxide, nootkatone, patchoulol, and of some oxathianes exhibit different, sometimes opposite sensory characteristics. A chiral (bio)catalyst would offer the most adequate synthetic approach.
- Recent advances in cell biology and bioengineering—particularly the ability to alter the biocatalyst's properties genetically—possess great potential for creating new products.
- The physiologically most active constituents of natural flavors (character impact components) are often present in conventional sources in traces and in bound form, thus rendering an economic isolation difficult.

According to the increasing structural complexity of the biocatalyst, this article highlights some recent achievements in the biogeneration of volatiles using enzymes, microbial and fungal cells, and plant cells. Selected patents, partly reflecting the actual commercial status, will be discussed.

ENZYME MIXTURES AND PURIFIED ENZYMES

Microbial enzymes, products of bioprocesses themselves, are integrated parts of modern food technologies. In view of their regiospecifity and enantioselectivity, substrate specifity, and ability to function at statistically cold temperatures, it would seem obvious to use them for the generation of volatile flavors.

Indirect Action on Food Flavor

Most industrially used enzymes belong to main group three (hydrolases). The application of pure or combined lipases, amylases, glucoamylases, cellulases, invertases, proteases, or ribonucleases may affect the overall flavor of food in various ways: (*1*) by direct accumulation of monomer flavor compounds from polymer or conjugated precursors; (*2*) by increasing the pool of precursors for a subsequent thermal flavor generation (reaction flavors); and, (*3*) by liberating previously adsorbed compounds from the hydrolyzed polymer carrier molecules. Phosphatases and glycosidases were applied to liberate volatile flavors directly from their odorless transport or storage forms (Table 1). A rDNA encoding for an anthocyanase/endo-β-glucosidase from *Aspergillus niger* that was active at elevated temperature and in high concentrations of ethanol was described to enhance the flavor of wine (6). Glucose oxidase was useful not only for desugaring, but also for preparing pure rhamnose, an important flavor precursor, from a glucose/rhamnose mixture (7). The addition of enzymes for the removal or prevention of off-flavors, eg, the enzymic debittering of citrus products, may also be interpreted as an indirect action of enzymes on the final sensory quality of the food product.

Classical Fermented Food

Numerous enzymes of various microorganisms must act together in a well-balanced mode to bring about the typi-

Table 1. Volatile Flavors from Plant Sources Liberated by Glycosidases

Source	Products	References
Grape juice	Geraniol, nerol, linalool, linalool oxides	1
Tea leaf	(3Z)-Hexenol, 2-phenylethanol	2
Apple fruit	Alkanols, eugenol, octadiols, norisoprenoids	3,4
Papaya fruit	Benzaldehyde, benzyl isothiocyanate	5

Table 2. Flavors Resulting from Enzymatic Hydrolysis or Reverse Hydrolysis

Substrates	Products	References
Racemic carboxylic acids	Chiral alcohols	12,13
Acid plus alcohol	S-Carboxylic acid esters	14
Acetocarboxylic acid esters	Chiral hydroxy esters	15
Acid plus alcohol/esters	Geranyl butanoate, etc	16
Amino acids	Oligopeptides	17

cal sensory characteristics of the complex substrates of fermented food. As a result, enzyme and substrate mixtures, or purified lipases and proteases and defined substrates, were used with special emphasis on dairy flavors. Proteolytic, lipolytic, and further biopolymer degrading activities were isolated from a yeast culture and combined with the inactivated producer cells to obtain a food flavor (8). A butter flavor was generated by the action of a lipase, isolated from *Chromobacterium*, on milk fat (9). As the claimed gain in flavor intensity is seldomly quantified by the authors, the success of certain bioprocesses is difficult to assess. If, in a food-type substrate, the addition of a single enzyme caused a fortified flavor, the traditional strain should benefit from a transmission of the corresponding gene or from the use of gene dosage effects. Several recent reviews deal with rDNA techniques for the elucidation and improvement of relevant properties of lactic acid bacteria and yeasts (eg, 10,11).

Concerted Enzymatic Production of Flavor

Reverse hydrolysis has developed into one of the most intensively researched fields of enzyme technology. Flavor applications included the use of esterase/lipase preparations from bacterial, fungal, and mammalian cells for the production of chiral aliphatic esters, terpenol esters, and lactones with fruity and flowery odors (Table 2). Similarly, proteases were used for the production of aspartam, a sweetener, and for peptides possessing umami or sodium glutamate-like flavors (17). Vice versa, the selective hydrolysis of racemic amides or esters yielded chiral amines and alcohols.

Since the early 1970s, important industrial applications of hydrolases are the enzymatic resolution of racemic amino acids and the asymmetric hydrolysis of menthyl esters to produce optically pure l-menthol, a bulk compound of flavor chemistry. More than 50 lipases of different origin were characterized with respect to their efficiency and selectivity. Only tertiary alcohols and branched-chain and benzoic acids were usually excluded from the wide range of substrates. Subjects of recent work were the effects of solvent, pH, temperature, surfactants, water content, immobilization of the enzyme, pretreatment of the solid support, mass transfer–reaction interactions, isozymes, and chemical modification of the enzyme. Proofed protocols for ester synthesis can now be obtained from some of the enzyme manufacturers.

Water, essential for the hydration of the enzyme, is detrimental in excess and will lead to incomplete conversion. Various techniques for water limitation were described, such as the use of molecular sieve, distillation, condensation on a cold surface, and reduced pressure.

The step from hydrolytic/reverse hydrolytic reactions toward serial enzymatic reactions is, sooner or later, bound to require cofactor-dependent enzymes (Table 3). With an ADH/NAD$^+$/CH$_3$CHO- system for the production of geranial, 1,500 cycles of regeneration of the cofactor were achieved by coimmobilization of NAD. The bienzymatic method for cofactor regeneration instead of the coupled substrate method was suggested to convert ethanol to acetaldehyde, cinnamic alcohol to cinnamaldehyde, and leucine to 3-methylbutanal and 3-methylbutanol. In the latter case (20), *Streptococcus* enzymes were coencapsulated with substrate, NAD, and enzymes of *Gluconobacter oxidans*, which oxidize ethanol to acetic acid in a milk-fat coat. Cheeses that contained the enzyme capsules exhibited a stronger taste than controls. Such coimmobilisates of biocatalysts of different origin or of biocatalyst and substrate belong to a second generation of immobilized biocatalysts; they seem to be especially promising in the area of flavor production with its often poorly water-soluble substrates and products.

MICROBIAL CELLS

Microbial cells may be regarded as a very natural form of immobilized enzymes that, moreover, provide self-regeneration of catalytic activities and of all cofactors, active transport systems for substrates, and an optimal spatial arrangement of enzyme chains. Empirical biotechnologies using *Schizophyta* and *Protoascomycotina* have been applied since ancient times to improve keeping quality, digestibility, and flavor of foods and beverages.

Table 3. Cofactor Coupled Reactions Generating Volatile Flavor

System Components	Product	References
Geraniol/acetaldehyde/NAD$^+$/ADH	Geranial	18
CinnamylOH/Octanal/NAD$^+$/ADH/Lip-DH/DCIP	Cinnamaldehyde	19
L-Leucine/EtOH/NAD$^+$/enzyme mixture	3-Methylbutanal	20

Table 4. Food Strains and Flavor Formation

Substrate	Catalyst	Flavor	References
Citrate	Streptococcus lactis	Diacetyl, etc	22
Whey/amino acids	Lactobacillus helveticus	Furaneol, etc	23
Milk fat/Methionine	Brevibactacterium linens	Methanethiol	24
Koji	Pediococcus/yeast	Soy sauce	25
Octanoic acid	Penicillium roqueforti	2-Heptanone	26
Citronellol	Botrytis cinerea	Terpenols, etc	27
Standard medium	Saccharomyces cerevisiae	Terpenols	28

Food Strains

Fatty acids and other simple carbonyl compounds were the predominant volatile compounds produced by *Lactobacilli, Pediococci*, or *Streptococci* (21). *Penicillium* strains used in the manufacture of cheese and raw sausages added methylketones and, on mechanical stress, C_8-compounds to the spectrum. Typical yeast volatiles were esters, thio-compounds, and reduced carbonyls. The fermentation of food strains on classical or on chemically defined substrates usually resulted in the production of these common flavor molecules. More interesting compounds were not formed or formed in low amounts (<mg/L) only.

However, some recent applications demonstrate that industry is willing to exploit the capabilities and the operational advantages of microbial cells, despite their genetic limitations. The examples selected (Table 4) reflect some of the actual trends: new substrates and old strains, eg, for the malolactic fermentation of fruit and vegetable juices; better defined, often continuously operated systems for better understanding of biochemical regulation phenomena; cell recycle and immobilized systems for reuse of the biocatalyst (22,24); and the use of mutants (28) or mixed cultures to enhance or supplement the metabolic abilities of conventional strains.

The reported generation of furaneol [2,5-dimethyl-4-hydroxy-3(2H)-furanone, (23)] in aerated culture deserves special attention. This heterocyclic impact compound with an intense pineapple-caramel note (depending on its actual concentration) belongs to the most potent flavor compounds, but is rather unstable in aqueous media. Special precautions will have to be taken for its quantitative recovery from a submerged culture. The role of *Penicillia* in blue-cheese flavor formation is still under investigation. Substrates were pure fatty acid, ripened curd, or lipolyzed milk fat. Experiments to stabilize the active fungal spores in continuous culture by entrapment into alginate gel were successful (26).

Little is known regarding cured meat flavors in mold-fermented sausages. Supplementing catalytic activities probably need to be present to form the complete spectrum of volatiles. This view is supported by some procedures for the formation of cheese flavor, which used an additional protease or a lipase or an additional microorganism. A fermentation of soy sauce from koji with immobilized whole cells of lactic acid bacteria of the genus *Pediococcus* and yeast cells to produce a soy sauce flavor was reported by a Japanese group (25). The Ca alginate entrapped cells were kept in three-column reactors with a working volume of 280 L. Thereby, the ecological situation was better taken into account. Because mixed populations are no longer regarded as the antithesis of good experimental work, new perspectives have opened up. In all cases where the biodegradation of polymer or benzoic structures leads to the formation of flavors, a stimulating effect of the various forms of mutualism or commensalism within microbial communities may be expected.

Nonfood Organisms

The range of accessible volatile flavors was considerably expanded by using nonfood species (Table 5). Proliferating nonfood yeasts substituted purified, immobilized lipase or alcohol dehydrogenase in the production of carboxylic acid esters or aliphatic aldehydes including acetaldehyde, an impact component of citrus and yogurt flavors (29,30). The need to provide all those pathways with energy and substrates that are not directly required for an intended conversion reaction may be compensated by the cell's self-regeneration of active biocatalyst.

Just when cofactor-dependent steps are aimed at, intact cells still appear to be the catalysts of choice. A good example is the conversion of fusel oil constituents to carboxylic acids such as 2-methylbutanoic acid (31). A subsequent conversion of this acid to its ethyl ester will yield one of the most potent fruit-flavor compounds. Strains of *Penicillium, Pseudomonas*, and *Aspergillus niger* substituted the action of plant enzymes in degrading terpenoid structures such as carotenoids, resulting in, eg, tea and tobacco flavor constituents. Genetically altered *Pseudo-*

Table 5. Volatile Flavors from Nonfood Organisms

Substrate	Catalyst	Flavor	Reference
Fusel oil	Hansenula mrakii	Various esters	29
Vegetable and fruit juice	Various yeasts	Fruity odor	30
C_3 to C_{100}-Alcohols	Gluconobacter roseus	Corresponding acids	31
Monoterpenols	Pseudomonas strains	Fragments	32
Standard medium	Thigosaccharomyces/ Candida	4-Ethyl guaiacol	33

Table 6. Volatiles from Cultured *Ascomycetes* and *Basidiomycetes*

Species	Product	References
Morchella	1-Octene-3-ol	34
Trametes, Polyporus	Methyl anthranilate	35
Lentinus lepideus	Methyl cinnamate, sesquiterpenes	36
Poria cocos	Linalool, nerolidol, bisabolol	37
Sporobolomyces odorous	4-Decanolide	38
Tyromyces sambuceus	4-Decanolide, alcohols	39
Monilia fructicola	4-Olides, 5-olides, phenylethanol	40
Trichoderma, Ceratocystis	5-olides	41

monas strains with special plasmids were described for effecting this catabolism (32). Protoplast fusion, another prominent novel technique to recombine genetic information, was used to overcome unstable cocultivations and to produce single compounds such as 4-ethyl guaiacol, a seasoning agent (33). Problems encountered with hybrid cells are the sometimes difficult selection and insufficient genome stability.

Higher Fungi

Eumycetes, taxonomically located between less-organized microorganisms with limited potential for flavor formation and the complex seed plants, appear to be a kind of golden mean for the biogeneration of flavors. They grow well in certain synthetic liquid media and offer a tremendous metabolic diversity; moreover, some producer strains are related to edible species. This should facilitate approval of the products by the authorities.

Not only mushroomlike compounds such as 1-octene-3-ol (34), but also fruity, floral, spicy, and even chocolate-like (due to pyrazines) odors were observed (Table 6). Volatiles of almost every chemical class were synthesized de novo or via bioconversions by submerged cultured mycelia. Depending on strain and composition of the nutrient medium, frequently found volatiles were phenolics, terpenes, and lactones. Previously reported low yields are not an inherent feature of fungal cells, but were due to suboptimal culture conditions. An aerated culture of *Monilia fructicola* grown for 3–5 days at 28°C in a yeast/malt medium produced 200 mg to 1 g/L of a peach aroma basically consisting of lactones (40). The example of *Polyporus durus* demonstrates that these saprophytic organisms are very adaptable to changing culture conditions; thus, they are amenable to a concerted optimization (Table 7) (42). The presence of a re-esterified coconut oil fraction strongly stimulated the formation of volatiles. More than 60 constituents, among them almost 20 lactones, were accumulated in the culture broth. A total concentration of volatiles of ca 800 mg/L was reached. Capillary gas chromatography on a chiral and achiral phases showed a high (> 98% ee) enantiomeric purity of some of the chiral lactones. Similar and pronounced effects of triglycerides were reported for yeasts and for other microorganisms. Generally, the supplementation of the fungal cultures with suitable (natural) precursor substrates, customary in numerous bioprocesses for pharmaceuticals, was an efficient tool to further increase the yields of volatiles. With respect to the vast number of strains, their metabolic flexibility under changing conditions, and the broad bioengineering experience, the potential of fungal cells to produce commercial food flavors can hardly be overestimated.

PLANT CATALYSTS

Most of the volatile flavors processed by the food industry are directly or indirectly based on plant metabolism. Pressing, distillation, extraction, and chromatography are the actual techniques for obtaining raw materials of natural flavors from intact plants.

Crude Enzymes and Homogenates

Crude enzyme preparations from fruits and vegetables were able to partly restore the fresh odor impression in processed food by converting nonvolatile odorless precursors to volatile flavors. Developed in the late 1950s, this so-called flavorese enzyme concept failed for economic and biogenetic reasons. Successors used by-products of food processing as a less expensive source of enzymes. Following up earlier work with apple peel, a Hungarian group developed a preparative scale generation of apple flavor (43). A vigorous formation of the volatile C_6-compounds typical of plant lipoxygenase/hydroperoxide lyase systems occurred in homogenates of some apple cultivars and grass species, and yields of the main volatile compound, (2E)-hexenal, reached some 100 mg/kg/h under optimized conditions (44,45). (2E)-hexenal imparts pleasant green notes to various food flavors. A recently published process obtained (2E)-hexenal and related volatiles by contacting a soybean homogenate with linseed oil and a lipase in the presence of O_2. The homogenate was reacted at 1,000 rpm

Table 7. Stepwise Improvement of 4-Octanolide Production by *Polyporus durus*[a]

Parameter	Productivity, mg/L/day
Standard medium	Traces
N-Source	0.2
Transfer to submerged cultivation	0.6
C-Source, pO_2	3
Addition of precursor, detergent	16
Preparation and size of inoculum	31
Semicontinuous cultivation	46

[a] Ref. 42.

at 40°C for 2 h, and the volatiles were then recovered by twofold distillation (46).

Callus and Suspension Culture

Derived from a wounding-induced tissue (primary callus), a continuously growing sterile culture of plant cells can be established by mechanical separation and further subculturing on a synthetic medium (callus culture). After transfer to a liquid medium, cells of a macroorganism are finally reduced to the single-cell level (suspension culture). According to its origin, the somatic cell is provided with a complete genome and able to develop into a whole plant again. This totipotency was translated into the idea that all genetic functions, including, eg, essential oil formation, would be expressed in such cells. It is clear now that the wish was father to the thought; the very morphology of the cultured cells differs completely from the original state (dedifferentiation). Consequently, most plant-cell cultures do not contain essential oils or other volatile flavors under standard cultivation conditions, even though they have the genetic ability to do so.

Some basic research on why cell cultures do not accumulate the typical essential oils has been carried out. A crucial difficulty may be a low activity of certain key enzymes such as geraniol phosphatase. If single enzymes were rate-limiting, it should be possible to select for higher yields by making use of the somaclonal variation that builds up in plant-cell cultures. Favored by the color of the product, selection procedures were successfully performed for shikonin, a red naphthoquinone. Increased levels of essential oils were reported for mutant plants of *Pelargonium* and *Pogostemon* sp. regenerated from ethyl methane sulfonate treated callus cultures (47); this type of approach is somewhat restricted by the laborious procedures and assays required.

A detailed study on the synthesis of lower terpenes in callus cultures of essential oil-bearing plants showed all species to possess high prenyltransferase and pyrophosphate isomerase activities, and most of the cultures to possess further activities of the mevalonate pathway. A significant accumulation of products, however, was not observed. In other cell cultures monoterpenes were only detectable in the presence of a suspended lipophilic trap simulating natural accumulation sites. Feeding experiments with suspension cultures of *Citrus* cells that did not produce terpenes showed that exogenous terpenes were substrates of rapidly proceeding biotransformations (48). Added limonene, eg, disappeared from the suspension within a few hours. Losses due to volatility or autoxidation were experimentally excluded. The primary transformation product, (E)-1,2-limonene epoxide, and (Z)-2,8-menthadien-1-ol, probably resulting from epoxide hydrolysis and elimination of water in 2,3-position, and two further monoterpene alcohols were the detectable volatile metabolites. An oxidative degradation of the terpene could be concluded from the structure of the volatile and nonvolatile (fatty acids, keto acids, etc) conversion products and from the respiratory rise of the cells on addition of the terpene hydrocarbon. Thus, limonene accumulation was a priori prevented by self-protection mechanisms of the living cell, even if the biosynthesis would have been operating. High terpene epoxidase and epoxide hydratase activities were likewise observed in cell-free extracts from other plant-cell cultures and in numerous microbial cultures, eg, of the genus *Pseudomonas*. These results indicate that either the removal of metabolites from intracellular equilibria or the stability of products in the surrounding medium has been insufficient.

If one would succeed in effectively trapping the intermediate volatile products of exogenous precursors, plant-cell culture could be used in specific bioconversions, especially oxyfunctionalization reactions. Some examples are compiled in Table 8. Of economic relevance could be the regioselective biotransformation of valencene by cells of *Citrus* sp. to yield nootkatone. After 6 hours of incubation, almost 70% of the substrate was transformed with no other detectable volatile by-products. Natural nootkatone with its grapefruit-like odor and bitter taste is a sought-after flavor compound with limited availability.

Despite these achievements, it should not be overlooked that de novo syntheses as well as biotransformations using plant cells will always have to compete not only with field plants, but also with microbial and enzyme technologies. A possible way to increase the yields by reactivating repressed genes of the secondary metabolism would be to redifferentiate the cultured cells. Plant cells are not very adaptable to extreme chemical conditions, yet they can be forced into morphological and biochemical specialization by suitable phytoeffector ratios in the nutrient medium. Although endogenous effectors may interfere with morphogenesis, the reformation of root- and shoot-typical secondary metabolites in specialized tissues developed from cell cultures was demonstrated: Adventitious

Table 8. Biotransformation and Bioconversion of Exogenous Terpenes by Plant Suspension Cultures[a]

Substrate	Species	Product
Limonene	*Citrus limon*	1,2(E)-Limonene epoxide, etc
Farnesol	*C. limon*	2,3-Dihydro-(6E)-farnesol
Valencene	*Citrus*	Nootkatone
Borneol	*Salvia officinalis*	Camphor
1,4-Cineol	*S. officinalis*	8-Hydroxy-1,4-cineol
Nerol	*Arthemisia abs.*	Geraniol, Geranic acid, etc
Menthone	*Mentha piperita*	Menthol, etc

[a] Ref. 48.

Table 9. Some Volatile Constituents of in vitro Cells of *Ruta graveolens*[a]

Compound, µg/kg fruit/wt	Dark	9,000 lux
Root-typical		
Geijerene	480	160
Pregeijerene	1,040	230
Leaf-typical		
2-Undecanone	80	1,660
2-Tridecanone	20	210
n-Nonyl acetate	20	130
2-Undecyl acetate	50	230
n-Octyl 3-methylbutanoate	< 10	110
2-Undecyl 3-methylbutanoate	< 1	120

[a] Ref. 52.

buds originated from cultured *Anthemis* (49) or *Ruta* contained essential oil constituents. Thus, the problem of dedifferentiation was bypassed, but the loss of unlimited growth (part of the definition of any cell culture) and severe bioengineering complications have to be taken into the bargain. The addition of fungal homogenates to growing plant cell cultures was discussed as a means of triggering metabolic cascades. Microbial insult is known to induce antibiotic phytochemical (elicitor) synthesis in plant cells, and, as many flavor compounds interfere with fungal growth, their synthesis may be induced by simulating a fungal attack.

A physical stimulus to secondary metabolism in plant cells that has been neglected for a long time is light. Although not always sufficient, the transition from heterotrophic to phototrophic growth induced the formation of leaf-typical constituents, such as pigments, quinones, or essential oils: Chlorophyllous calli of *Psidium guajava* (50) and various other fruits (51) produced volatile C_6-alcohols and related volatiles. In suspension-cultured cells of *Ruta graveolens* the root-typical irregular terpenes dominated under heterotrophic conditions, while light induced the formation of ketones and numerous aliphatic esters (Table 9) (52). A heterotrophic culture of *Coleonema album* was devoid of any essential oil constituents, whereas the light-grown pendant accumulated numerous mono- and sesquiterpenes (53).

Cell immobilization is a rapidly expanding area of biotechnology. The characteristics of continuous operation, reuse of the biocatalyst, ease of process control, and improved biocatalyst stability should prove especially beneficial for cultured plant cells: Immobilized cells are similar in many respects to the tissues of intact plants. Aggregated cells face heterogeneous microenvironments, release intracellular products, and divide slower. These and other factors may redirect plant secondary metabolism. An impressive demonstration of the advantages of immobilized plant cells in flavor production is a culture of *Capsicum frutescens* that showed increased formation of capsaicin when immobilized on blocks of reticulate polyurethane foam support (54). The use of immobilized plant cells in column or membrane reactors may offer better perspectives for a future application on an industrial scale.

Specialized Plant Cells

The biotechnological ideal of an immobilized plant cell with active flavor metabolism, available in any amount, was created by nature: mature fruits. It has been known for a long time that aged fruit tissues are able to take up exogenous substrates, metabolizing them to flavor compounds. The same holds true for intact fruits and volatile substrates. A procedure was developed to expose fruits during storage to vapors of precursors of volatile flavors. By analogy with controlled atmosphere (CA) storage, the new biotechnological concept was termed PA (for precursor atmosphere) storage. On exposing pear fruits (eg, to a mixture of aliphatic alcohols) the concentrations of numerous volatiles increased. The concentration of the impact component hexyl acetate was fourfold after 8 h of PA storage, whereas the untreated control remained almost constant.

Another example refers to the drying of banana tissue (55). Although many compounds of banana flavor showed good retention in the starchy matrix, the concentrations of the highly volatile carboxylic esters decreased by 50–80% during freeze-drying. When bananas were submitted to PA-storage in 3-methylbutanol vapors before dehydration, the fruits rapidly accumulated volatile impact components. By managing endogenous organized enzymes and their substrates, a dried fruit with an amount of 3-methylbutyl esters comparable to that of the genuine fresh fruit was finally obtained (Table 10). As a result, a preceeding PA storage can compensate for flavor deficiencies caused by one-sided breeding, by improper transport and storage, or by physical losses during the often inevitable thermal-processing operations.

COMMERCIAL BIOFLAVORS

It seems obvious from the references cited and the data presented that novel bioprocesses have become an alternative source of a wide range of high-prized volatile flavors. A couple of companies already advertise 100%

Table 10. Effect of a Preceeding Precursor Atmosphere (PA) Storage on Impact Volatiles of Freeze-Dried Banana Slices[a]

3-Methylbutyl Ester, µg/100 g fresh wt	Fresh Fruit	Freeze-Dried	
		Without	With PA-Storage
Acetate	7,460	2,860	6,990
2-Methylpropanoate	70	27	60
Butanoate	600	250	960

[a] Ref. 55.

natural compounds. If these were obtained by physical treatments of plant (or other natural) sources, or if they in fact originate from a bioreactor, remains open sometimes. True products of concerted bioprocesses may be diacetyl (butter note, from starter cultures), cheese flavors, yeast products (meaty and savory notes), fatty acids and related alcohols and esters (many different odor notes, from lipase technology or intact microorganisms), C_6- and C_8-alcohols and related carbonyls (green and mushroom notes, from plant and fungal homogenates), and some specialities such as lactones from complex biosystems.

The legal definitions (and the average consumer) clearly discriminate artificial flavors against natural ones; this fact has undoubtedly contributed to the recent surge of activity in bioflavor research. The legal discrimination itself, however, must be regarded as artificial, because "no compounds are made on earth other than those permitted by the laws of nature" (56). The existing legal restrictions might be challenged less by the public with its unfounded reservations than by problems in analytically distinguishing natural from synthetic. Sophisticated analytical tools that are available comprise multidimensional chiral gas chromatography (57), high resolution nuclear magnetic resonance-spectroscopy, and isotope ratio mass spectroscopy (58). Despite this selection of analytical options, some volatile compounds are still difficult to differentiate.

Independent from the legal situation, the superiority of biocatalysts to generate complex and chiral molecules is now widely recognized. Volatile flavors will, therefore, have to be listed among the target compounds of future biotechnologies. A second field of application is already well established: the classic food biotechnologies. They will continue to profit by modern developments in gene recombination and delivery, bioreactor and sensor design, on-line control, and related techniques. Pioneering work on lactic acid bacteria demonstrated that some of the metabolic routes to flavors proceed plasmid mediated (10,21). The improvement in fungal and plant strains, by contrast, is still restricted by several fundamental scientific and applied hurdles. The use of eucaryotic or mitochondrial genome fragments and the chromosomal integration of rDNA may characterize lines of future research. Special emphasis should be put on safety testing (59) of food generated by genetically manipulated strains, particularly if these are to remain in the consumed products.

BIBLIOGRAPHY

1. A. P. Aryan, B. Wilson, C. R. Strauss, and P. J. Williams, "The properties of glycosidases of *Vitis vinifera* and a comparison of their β-glucosidase activity with that of exogenous enzymes," *American Journal of Enology and Viticulture* 38, 182–188 (1987).
2. N. Fischer, S. Nitz, and F. Drawert, "Freie und gebundene Aromastoffee in grünem und schwarzen Tee," *Zeitschrift fuer Lebensmittel-Untersuchung und-Forschung* 185, 195–201 (1987).
3. W. Schwab and P. Schreier, "Simultaneous enzyme catalysis extraction (SECE)," *Journal of Agricultural and Food Chemistry* 36, 1238–1242 (1988).
4. R. G. Berger, G. R. Dettweiler, and F. Drawert, "Zum Vorkommen von C_8-Diolen in Äpfeln und Apfelsäften," *Deutsche Lebensmittel Rundschau* 84, 344–347 (1988).
5. W. Schwab, C. Mahr, and P. Schreier, "Enzymic hydrolysis of bound aroma components from Carica papaya fruit," *Journal of Agricultural and Food Chemistry* 37, 1009–1012 (1989).
6. Eur. Pat. Appl. 307,071 (March 15, 1989), Yissum Res. Dev. Co.
7. Eur. Pat. Appl. 317,036 (May 24, 1989), Unilever N.V.
8. Eur. Pat. 191,513 (August 20, 1986), Unilever N.V.
9. Jpn. Pat. 6402,549 (January 6, 1989), T. Hasegawa Co. Ltd.
10. H. Teuber, "The use of genetically manipulated microorganisms in food: opportunities and limitations," *European Congress on Biotechnology* 4, 383–391 (1987).
11. D. R. Berry, "Manipulation of flavour production by yeast: physiologial and genetic approaches," in J. R. Piggott, A. Paterson, eds., *Distilled Beverage Flavour*, Ellis Horwood, Chichester, UK, 1989, pp. 299–307.
12. Jpn. Pat. 3214,198 (September 6, 1988), Kanegafuchi Chem.
13. K. Laumen, D. Breitgoff, and M. P. Schneider, "Enzymic preparation of enantiomerically pure secondary alcohols," *Journal of Chemical Society Chemical Communications* 22, 1459–1461 (1988).
14. Jpn. Pat. 3245,695 (October 12, 1988), Showa Shell Petrol.
15. B. I. Glänzer, K. Faber, H. Griegel, M. Röhr, and W. Wöhrer, "Enantioselective hydrolysis of esters of secondary alcohols using lyophylized baker's yeast," *Enzyme Microbiology Technology* 10, 744–749 (1988).
16. G. Langrand, C. Triantaphylides, and J. Baratti, "Lipase catalyzed formation of flavor esters," *Biotechnology Letters* 10, 549–554 (1988).
17. K. Aso, T. Uemura, and Y. Shiokawa, "Protease catalyzed synthesis of oligo-L-glutamic acid from L-glutamic acid diethyl ester," *Agricultural and Biological Chemistry* 52, 2443–2449 (1988).
18. M. D. Legoy, H. S. Kim, and D. Thomas, "Use of ADH for flavor aldehyde production," *Process Biochemistry* 20, 145–148 (1985).
19. J. S. Deetz and J. D. Rozzell, "Enzyme-catalysed reactions in non-aqueous media," *TIBTECH* 6, 15–19 (1988).
20. S. D. Braun and N. F. Olsen, "Microencapsulation of cell-free extracts to demonstrate the feasibility of heterogenous enzyme systems and cofactor recycling for development of flavor in cheese," *Journal of Dairy Science* 69, 1202–1208 (1986).
21. V. M. Marshall, "Lactic acid bacteria: Starters for flavour," *FEMS Microbiology Revue* 46, 327–336 (1987).
22. P. Schmitt, C. Couvreur, J. F. Cavin, H. Prevost, and C. Divies, "Citrate utilization by free and immobilized *Streptococcus lactis* ssp. diacetylactis in continuous culture," *Applied Microbiology and Biotechnology* 29, 430–436 (1988).
23. J. Kowalewska, H. Zelazowska, A. Babuchowski, E. G. Hammond, B. A. Glatz, and F. Ross, "Isolation of aroma bearing material from *Lactobacillus helveticus* culture and cheese," *Journal of Dairy Science* 68, 2165–2171 (1985).
24. S. C. Kim and N. F. Olsen, "Production of methanethiol in milk fat-coated microcapsules containing *Brevibacterium linens* and methionine," *Journal of Dairy Science* 56, 799–811 (1989).
25. K. Osaki, Y. Okamoto, T. Akao, S. Nagata, and H. Takamatsu, "Fermentation of soy sauce with immobilized whole cells," *Journal of Food Science* 50, 1289–1292 (1985).
26. C. Larroche and J. B. Gros, "Batch and continuous 2-heptanone production by Ca alginate and Eudragit RL entrapped

spores of *Penicillium roquefortii*," *Biotechnology and Bioengineering* **34**, 30–38 (1989).
27. P. Brunerie, I. Benda, G. Hock, and P. Schreier, "Bioconversion of citronellol by *Botrytis cinerea*," *Applied Microbiology and Biotechnology* **27**, 6–10 (1988).
28. Eur. Pat. 313,465 (April 26, 1989), Pernod Ricard.
29. L. Janssens, H. L. De Pooter, L. Demey, E. J. Vandamme, and N. M. Schamp, "Fusel oil as a precursor for the microbiological production of fruity flavours," *Med. Fac. Land-bouww. Rijksuniv. Gent* **54**, 1387–1392 (1989).
30. Eur. Pat 277,062 (August 3, 1988), Pernod Ricard.
31. Eur. Pat. 289, 822 (November 9, 1988), Haarmann & Reimer GmbH.
32. U.S. Pat 4800,158 (January 24, 1989), Microlife-Technics.
33. Jpn. Pat. 1191,679 (August 1, 1989), Kikkoman.
34. Ger. Pat. 3708,932 (September 29, 1988), Hüls.
35. World Pat. 8900,203 (January 12, 1989), BASF.
36. H. P. Hanssen and W. R. Abraham, "Fragrance compounds from fungal liquid cultures. I. Sesquiterpenoids from *Lentinus lepideus*," *European Congress on Biotechnology* **3**, 291–293 (1987).
37. R. G. Berger, S. Hädrich-Meyer, and F. Drawert, "High productivity fermentation of volatile flavours using fungal cultures," in S. C. Bhattacharyya, N. Sen, K. L. Sethi, eds., *11th International Congress on Essential Oils, Fragrances and Flavours*, Vol. 3, pp. 127–133, Oxford & IBH Publ., New Delhi, 1989.
38. Eur. Pat. 258,993 (March 9, 1988), Unilever N.V.
39. G-F. Kapfer, R. G. Berger, and F. Drawert, "Production of 4-decanolide by semi-continuous fermentation of *Tyromyces sambuceus*," *Biotechnology Letters* **11**, 561–566 (1989).
40. Eur. Pat. 283,950 (September 28, 1988), Hoechst AG.
41. Fr. Pat 2,603,048 (February 26, 1988), M. Brizard & Roger Int. S.A.
42. R. G. Berger, K. Neuhäuser, and F. Drawert, "Odorous constituents of *Polyporus durus* (Basidiomycetes)," *Zeitschrift fuer Naturforschung* **41c**, 963–970 (1986).
43. L. Vámos-Vigyázó, N. Kiss-Kutz, and A. Hersiczky, "Preparative scale generation of apple flavour from by-products of processing," in *3rd International Flavour Symposium*, Keksemet, Hungary, 1984, pp. 34–35.
44. R. G. Berger, A. Kler, and F. Drawert, "The C_6-aldehyde forming system in *Agropyron repens*," *Biochimica et Biophysica Acta* **883**, 523–530 (1986).
45. F. Drawert, A. Kler, and R. G. Berger, "Optimierung der Ausbeuten von (2E)-Hexenal bei pflanzlichen Gewebehomogenaten," *Lebensmittel Wissenschaft-Technologie* **19**, 426–431 (1986).
46. U. S. Pat. 4,769,243 (September 6, 1988), Takasago Perfumery Co.
47. Jpn. Pat. 2054,796 (March 10, 1987), Shiseido.
48. R. Godelmann, PhD Thesis, Technical University of Munich, 1985.
49. Jpn. Pat. 1228,413 (September 12, 1989), Tokyo Gas.
50. T. N. Prabha, M. S. Narayanan, and M. V. Patwardhan, "Flavor formation in callus culture of guava (*Psidium guajava*) fruit," *Journal of the Science of Food and Agriculture* **50**, 105–110 (1989).
51. R. G. Berger, A. Kler, and F. Drawert, "C_6-aldehyde formation from linolenic acid in fruit cells cultured in vitro," *Plant Cell Tissue and Organ Culture* **8**, 147–151 (1987).
52. M. Jordan, C. H. Rolfs, W. Barz, R. G. Berger, H. Kollmannsberger, and F. Drawert, "Characterization of the volatile constituents from heterotrophic cell suspension cultures of *Ruta graveolens*," *Zeitschrift fuer Naturforschung* **41c**, 809–812 (1986).
53. R. G. Berger, Z. Akkan, and F. Drawert, "The essential oil of *Coleonema album* and of a photomixotrophic culture derived thereof," *Zeitschrift fur Naturforschurg* **45c**, 63–67 (1990).
54. M. M. Yeoman, "Use of immobilised plant cells for the production of food flavours and colours," in *Biotechnology in the Food Industry*, Online Publ., Pinner, Oxford, 1986, pp. 23–27.
55. R. G. Berger, F. Drawert, and H. Kollmannsberger, "PA-Lagerung zur Kompensation von Aromaverlusten bei der Gefriertrocknung von Bananenscheiben," *Zeitschrift fuer Lebensmittel-Unterschung-Forschung* **183**, 169–171 (1986).
56. R. Teranishi, "New trends and developments in flavor chemistry," in R. Teranishi, R. G. Buttery, and F. Shahidi, eds., *Flavor Chemistry*, ACS Symposium Series 388, American Chemical Society, Washington D.C., 1989, p. 20.
57. S. Nitz, H. Kollmannsberger, and F. Drawert, "Enantiomere γ-Lactone in Passionfrüchten und Passionsfruchtprodukten," *Chemic Mikrobiologic Technologic Lebensmittel* **12**, 105–110 (1989).
58. M. Butzenlechner, A. Rossmann, and H-L. Schmidt, "Assignment of bitter almond oil to natural and synthetic sources by stable isotope ratio analysis," *Journal of Agricultural and Food Chemistry* **37**, 410–412 (1989).
59. D. Anderson and W. F. J. Cuthbertson, "Safety testing of novel food products generated by biotechnology and genetic manipulation," *Biotechnology and Genetic Engineering Revue* **5**, 369–395 (1987).

R. G. BERGER,
Institut für
Lebensmittelchemie
der Universität Hannover
Hannover, Germany

GENETIC ENGINEERING PART IV: FOOD MICROBIOLOGY

One of the newer branches of biotechnology in the food industry is genetic engineering (1,2). It is becoming an important research tool in food microbiology, counterbalancing the era of metabolic engineering where microorganisms were exploited for various end products and enzymes (3).

Although the metabolic approach has produced a number of sophisticated processes, there are limitations to this technology. Development of starter cultures for use in the dairy and meat industries, for example, can be costly and time consuming. Theoretically, application of genetic engineering should be able to produce strains more efficiently and with a greater variety of desirable properties (4). The principle uses of genetic engineering in food microbiology are (1) cloning of genes from an organism to a more desirable one, (2) insertion of a promotor region to amplify natural genes in an organism, and (3) site-directed mutagenesis (5).

Progress toward genetically improving organisms for food use has been slow, partly because of the unavailability of successful cloning procedures and the lack of under-

Table 1. Enzymes for Food Use from Microbial Hosts Targeted for Genetic Engineering[a]

Commodity	Possible Enzymes
Dairy products	Lipases
	Proteases
	Chymosin
	Lactases
Fish hydrolysates	Proteases
Meat tenderization	Proteases
Gelatin production	
Vegetable/fruit processing	Pectinases
	Amylases
	Polygalacturonase
Starch processing	α- and β-amylase
	Glucoamylase
	Glucose isomerase
	Pullanases
	α-D-Glucosidase
	Cyclomaltodextrin-D-glucotransferase

[a] Refs. 6—8.

standing of gene regulation and expression. This is certainly true for microorganisms such as *Lactobacillus* and to a lesser-extent *Bacillus*. Developments in genetic engineering in the pharmaceutical industry (2) are now being applied to target certain food-processing enzymes (Table 1) (6). Another lesson learned from this industry is to widen the options available by using both bacterial and eukaryotic host/vector systems. Although much research has focused on the merits of bacterial systems, recent discoveries have suggested eukaryotic hosts such as yeasts might be preferred (9). These microorganisms are better suited to formal genetics and are able to secrete proteins in an active form. Organisms such as *Saccharomyces cerevisiae* are well established in fermentation technology, do not produce toxins, and therefore are highly competitive with bacterial host/vector systems. Yeasts, however, do not give high rates of expression and also present problems regarding plasmid stability. One of the most promising applications of genetic engineering is site-directed mutagenesis, which makes possible protein engineering. With this technique one or more amino acids in a protein can be modified (eg, at the active site) so as to change its functionality. Potentially, it is possible to modify a number of properties of proteins, including pH or temperature optima and thermostability.

One of the most exciting areas of development is the recent progress in the use of filamentous fungi as hosts for the expression of foreign genes (10). This system has a number of advantages, including natural secretion of a wide range of proteins and the fact that many species are regarded as safe by regulatory authorities. A number of mammalian genes have been expressed in filamentous fungi, such as *Aspergillus nidulans* and *A. awamori*, including calf chymosin and lysozyme (10).

This article briefly reviews applications of genetic engineering and food microbiology. The methods and concepts used in such engineering have been outlined in the articles on Genetic Engineering parts I–III and V. Where possible, review articles have been used so that the reader can pursue some aspects in further depth.

MEATS

The main areas of interest in meat microbiology are starter culture development and improvement of product texture. The purpose of starter culture addition to meat is to generate the necessary levels of acidity. This is usually achieved by the production of lactic acid and to a lesser extent acetic acid. These acids provide characteristic tastes as well as antimicrobial effects (11). Organisms used in meat starters are *Lactobacillus* (usually *L. plantarum*) and *Pediococcus*. Both not only produce lactic acid but also an array of antibacterial agents that inhibit undesirable spoilage or pathogenic organisms. They may also provide other benefits such as lowering histamine and nitrite levels and may extend the shelf life of many meat products. The most formidable task in genetic engineering is the development of new starter cultures for meat fermentation (Table 2). Research in this area has made slow progress compared to the dairy lactics. This has been due in part to lack of genetic transformation systems in *Lactobacillus*. The technique of electroporation has been successful in transforming a number of *Lactobacillus* species, usually with constructed vectors (13). Such techniques now should help considerably in understanding the plasmid biology and genetics of both *Pediococcus* and *Lactobacillus*.

Important to the development of meat starter cultures is the understanding of the genetics and synthesis of bacteriocins. These compounds are proteins that have an antibioticlike activity against strains closely related to the starter strain. Some bacteriocins have been reported to have a wider range of activity. These are not true bacteriocins but more like the antibiotic nisin produced by *Streptococcus lactis*. This compound is active against many gram-positive organisms. See Ref. 14 for a recent review of bacteriocins of the lactic acid bacteria. The most important bacteriocins produced by meat starter organisms are probably lactolin and plantaricin A, both from *L. plantarum*. The synthesis of these compounds does not appear to be plasmid-mediated, and both have a narrow range of activity (14). Recent work on plantaricin A suggests it would be useful in controlling closely related *Lactobacillus* species and might therefore have application in preventing meat spoilage. It would also be useful as a chromosomal marker in genetic studies. Bacteriocins from *Pediococcus* have been more closely studied and have a

Table 2. Possible Target Genes for Recombinant DNA Technology in Lactic Acid Bacteria[a]

Phage resistance
Detoxification of carcinogens (eg, reduction of nitrite)
Catalase
Flavors
Colonizing factors
Nisin and other natural antimicrobial agents
Drug synthesis
Cholesterol metabolism
Cellulase
Collagenase
Proteinases

[a] Refs. 4, 12.

wider range of activity. Genetic studies with *Pediococcus pentosaceus* FBB61 and 63 have linked bacteriocin production to certain plasmids (13.6 Md and 10.5 Md, respectively). The bacteriocin pediocin A, produced by *P. pentosaceus* FBB61, has been shown to have a wide range of antimicrobial activity, including activity against organisms such as *Clostridium botulinum* and *Listeria monocytogenes* (14). Given these properties, the development of pediocin production by meat starter cultures is destined to become a prime focus for genetic engineering strategies.

Some lactobacilli can enzymatically reduce sodium nitrite during the fermentation of meat (15). Reduction of nitrite by meat starter cultures is a desirable property that aims to limit the production of carcinogens such as nitrosamines. This property has not been linked to plasmids but is believed to be chromosomally mediated (unpublished data). The use of conjugative transposons may provide the means to study these genes (16). Recent work has also shown that incorporation of activities such as nitrite reductase and catalase by recombinant DNA techniques would be useful additions to the properties of meat starter cultures (17). Catalase activity would limit spoilage by greening during the ripening of fermented meats.

The development of enzymes modified by protein engineering would be beneficial for tenderization. The current use of plant enzymes by topical application to fresh meats has not been completely successful (8). These authors suggest antemortem injection of a thermophilic protease might be useful. The degree of control of the tenderization process would rely on the dosage and cooking temperatures.

FISH

The production of protein concentrates is an important adjunct to traditional fish-processing operations. In these processes undesirable species of fish not acceptable to the consumer are ground up and treated with proteases (18). The formation of bitter peptides during hydrolysis has, however, been a major concern. It is now claimed that in newer processes using selected proteases from *Bacillus licheniformis* or *B. amyloliquefaciens* bitter flavor formation is avoided (19). The production of proteases by *Bacillus* species has been reviewed (20). Two important proteases are serine proteases or subtilisins and neutral proteases. In fish processing the subtilisins are the most important enzymes in the manufacture of protein hydrolysates. The subtilisin gene from *B. amyloliquefaciens* has been cloned into *B. subtilis* 168 as host. The expression of the gene is triggered in the late logarithmic phase (ie, developmentally regulated) in a fashion similar to the genes for sporulation. The regulation of the protease has also been clarified (20). Applications for subtilisin in fish processing are limited owing to its sensitivity to oxidation (21). The oxidation of a single amino acid residue Met222 has been found largely responsible for inactivation of subtilisin by oxidizing agents. By using techniques in protein engineering (cassette mutagenesis), a series of subtilisin analogues has been generated in which Met222 was replaced by 19 possible amino acid alternatives. It has been determined that nonoxidizable substitutes permit retention of enzymatic activity (22). The Cys222 mutant demonstrated greater specific activity and stability to hydrogen peroxide than the wild type.

Fermented fish products, though not popular in the Western world, make an important contribution to diets of third-world nations. The lactic acid fermentation of fish is seen as a low-cost method of preservation. A carbohydrate source such as rice is used in the fermentation process. Genetic engineering of starter cultures used in fish fermentations may have potential. Although acid production is critical, other factors such as production of hydrogen peroxide and bacteriocins, cyanide utilization by bacteria in cassava products, and presence of starch-splitting enzymes are also important. Some of these attributes could be targets for future genetic improvement in starter cultures (23).

DAIRY

The dairy industry have been a major beneficiary of product improvements brought about by application of genetic engineering techniques (Table 2). Most progress in genetic research has been with the lactic streptococci (12). This group and the genus *Lactobacillus* are the principal microorganisms used for the culture of dairy products. They are used in a range of products, including cheese, yogurt, buttermilk, and sour cream. In cheese manufacture, for example, *Streptococcus* species are used as starter cultures and are added to milk for acid production. Starter cultures are used in conjunction with a clotting enzyme such as chymosin to coagulate the milk proteins. The resulting curds are pressed, salted, and then stored to allow for ripening of the cheese. Starter cultures not only produce acid but also contribute to flavors as a result of their proteolytic activity. They also secrete antimicrobial agents to control potential pathogens in milk (24). These important properties, along with bacteriophage sensitivity, are either mediated by or linked to the presence of plasmids in the starter cultures. Plasmid-linkage has simplified the study of genetics in streptococci, as the relevent genes can be identified more easily. It also means, however, these important traits are unstable because sometimes plasmids may not be transferred during cell division. Bacteriophage attack of starter cultures is the main cause of slow acid production and usually results in failure of the process (25). Plasmid-directed mechanisms for bacteriophage defense in lactic streptococci have been reported (26). Current research has demonstrated that the lactic streptococci have a number of mechanisms to prevent bacteriophage infection. These mechanisms include restriction and modification enzyme systems and interference with phage-adsorption sites. The requirement for bacteriophage-resistant strains has resulted in development of mutant selection techniques. By this approach starter cultures are challenged with a number of bacteriophages and resistant strains are then exposed to additional series of phages. In this way phage-resistant isolates and mutants with resistance to more than 400 virulent phages have been obtained (24). Another possible technique is to isolate and identify the genes responsible for phage resistance and clone them into a plasmid vector that can transform the lactic streptococci. It is suspected that because of the large number of phage resistance mechanisms and phage types several genes would be involved (25).

One of the long-term targets in genetic engineering of dairy cultures is to improve rates of acid production and substrate utilization (27). The ability of lactic streptococci to convert lactose to lactic acid is of great importance to the dairy industry and therefore has been studied extensively. Several of the genes encoding enzymes involved in lactose catabolism have been cloned from plasmid DNA (24). The expression of these genes in several different hosts has led to some understanding of how they are regulated. Improved rates of acid production can only come about by understanding gene expression and regulation in the lactic streptococci. Further genetic engineering strategies applied to the lactic streptococci aim to develop cultures that lead to dairy products with improved textures and flavors. This will include construction of strains that produce unique flavors and extracellular polysaccharides to change the mouthfeel of the product (28). Some strains of lactic streptococci produce proteins that have antimicrobial activity against closely related strains (bacteriocins) or antibioticlike substances such as nisin. These nisinlike molecules have a wider range of activity. Their functions may or may not be mediated by or linked with plasmid DNA. Bacteriocin production by *Streptococcus cremoris*, for example, has been linked to a 37.6 Md plasmid (24). These antimicrobial products would be useful markers in food-grade vectors, as antibiotic resistance markers would not be permitted for food use. Such developments would require further isolation of the bacteriocin/nisin genes and characterization of the gene product. Whether bacteriocins are plasmid-mediated or plasmid-linked (ie, regulated) needs to be clarified.

Application of genetic engineering techniques to the lactobacilli has, to date, been slow. The lactobacilli have not been generously endowed with recognizable plasmid-mediated/linked functions, making the study of genetics difficult. Recent advances in transformation systems, particularly by electroporation, might improve the situation (16). *Lactobacillus* species play an important role in the dairy industry (eg, in the production of yogurt and cheese ripening). There have also been claims of a therapeutic role for lactobacilli in dairy products (29). They may also impart nutritional benefits to the consumer. Genetic engineering of *Lactobacillus* has been discussed by one researcher, who suggested that energy limitation of the cell may preclude extensive genetic manipulation (30). A more progressive viewpoint regarding genetic engineering of lactobacilli suggests that while their metabolism limits biosynthetic capacity, efforts can be made to manipulate key enzymes and metabolic control points (4). It may also be possible to change the regulation of specific genes and consolidate unstable plasmid characteristics into the chromosome. Recent work on transposition in the lactic streptococci may lay the groundwork for the genetic analysis of the chromosome of other lactic acid bacteria, including lactobacilli (16).

During the manufacture of cheese a clotting enzyme, chymosin, is used to coagulate milk proteins. This protease is very much in short supply and has received considerable attention regarding commercial application of recombinant DNA technology (31). Microbial replacements for chymosin present problems related to heat stability, production of lower yields, and flavor defects (25). Chemical modification of these enzymes with oxidizing agents has improved heat stability as well as yields (19). Although the chymosin gene has been cloned into a nonpathogenic strain of *Escherichia coli*, the commercial success of this strain will depend on increasing yields of active enzyme and reducing the costs of purification. Chymosin production has also been reported in *Saccharomyces cerevisiae* and the filamentous fungi *Aspergillus nidulans* and *A. awamori* (21). Good yields have been reported, and the enzyme was secreted properly folded and active.

The enzyme lactase (β-galactosidase) has a wide range of uses in the dairy industry and is used in the reduction of lactose. Strains of *Streptococcus* and *Lactobacillus* produce this enzyme (11). Certain lactobacilli are capable of colonizing the gut (4). It is possible that with genetic manipulation such strains could be made to produce lactase in the gut and assist in the digestion of milk products in lactose-intolerant persons (25).

EGGS

Two aspects arising out of egg processing may have some application for genetic engineering. The first is the removal of cholesterol from egg yolk. Cholesterol-degrading organisms isolated from foods of animal origin have been studied (32). Strains that have proved useful have been identified as species of *Rhodococcus equi*. Strain number 23 was shown to degrade cholesterol via 4-cholesten-3-one into nonsteroid components. The use of such a process has implications for the preparation of low-cholesterol egg-yolk powder. The cholesterol-degrading enzyme system from *Rhodococcus equi* has not been studied further.

Lysozyme is an important enzyme in egg white that provides protection against bacterial infection. Potentially it might have uses as a bactericidal agent in food processing. This enzyme may be immobilized and used to reduce pathogens in heat-sensitive liquids, possibly in combination with mild heat treatment. The stability of lysozyme from hen egg white is dependent on the structure of selected amino acid sequences in the protein. The mechanism of irreversible thermal inactivation involves the deamidation of aspartic acid residues (13). Protein engineering can be applied to replace the heat-sensitive amino acids with those of glutamine and improve the heat resistance of lysozyme without altering the molecular conformation or activity.

FRUITS AND VEGETABLES

The main organisms of note in vegetable fermentations are the lactic acid bacteria. These organisms, indigenous to plant surfaces, are suitably adapted for the fermentation of fruit products such as olives and a whole range of vegetable products (33). The fermentations are brought about by a succession of organisms consisting of both homofermentative and heterofermentative lactic acid bacteria. In cucumber fermentation, *Pediococcus cerevisiae*,

Lactobacillus plantarum, and *L. brevis* all play a role in acidification and flavor development in the final product. As for many fermentations, a greater degree of control can be introduced by use of starter cultures. Such cultures need to have certain traits such as rapid acidification of substrate and competitive growth. For a strain to predominate in the fermentation, bacteriocin production would be an asset. Bacteriocins have been identified in *Pediococcus* and *Lactobacillus* (14). The genetics of these organisms with respect to bacteriocin production has not been sufficiently clarified (see Meats). As yet no bacteriocin or antimicrobial agent produced by *Pediococcus* has been reported to show activity against gram-negative bacteria. This property would be essential in vegetable fermentations where spoilage by such bacteria can be a problem. Likewise, it would be useful to isolate strains of antimicrobial-producing *Pediococci* or *L. plantarum* with activity against yeasts and molds. This approach would improve the storage stability of fermented vegetables and fruits (33). These developments can come about through selection and genetic engineering.

Cellulose enzyme technology has moved at an impressive pace (34). These enzymes can be used to process pulped fruit and vegetables. Many cellulase producers have been identified, especially from the fungus kingdom. The fungus *Trichoderma* produces an array of cellulase complexes, and mutants that hyperproduce these enzymes have been isolated (34). These include three classes of cellulose enzymes from *Trichoderma versei* that hydrolyze crystalline cellulose. Endoglucanase enzymes are used to decrease the extent of polymerization of cellulose or to produce glucose (35). Attempts to clone the enzyme exocellobiohydrolase 1 into yeasts have been described (35). Expression of the endoglucanase gene from *Clostridium thermocellum* in *L. plantarum* has been reported (36). Using a recombinant plasmid pM25, containing the cel E gene from *Clostridium thermocellum*, the transformed *L. plantarum* was able to express 94% of the normal enzyme activity. The pM25 plasmid, however, was not stable in the host *L. plantarum*.

Other enzymes have application to vegetable and fruit processing (37). Some of the more important ones are the pectolytic enzymes. These enzymes consist of polymethyl galacturonases, pectinlyases, and pectin esterases. Commercial pectinases are, however, unable to hydrolyze the rhamnogalacturan backbone completely (19). Recently, a new enzyme system termed SPS-ase (soya polysaccharide degrading complex) produced by *Aspergillus aculeatus* has been found to degrade cell walls. This enzyme has great potential as an aid in the liquefaction of vegetables and fruits (eg, apples). Such products will have better stability and quality. The reduced requirement for thermal treatments normally given in processing will assist in maintaining color, vitamin levels, flavor, and aroma of the product. The SPS-ase enzyme system has great potential and may lend itself to genetic studies. There is little information regarding the cloning of the pectolytic enzymes, possibly because of their complex nature. Recent work described the relationship between the pel genes of the pelADE cluster in *Erwinia chrysanthemi* (38).

Many genetic studies are moving away from processing enzymes and toward the understanding of enzymes involved in fruit ripening. During fruit softening the role of pectin-degrading enzymes is important. A recent review shows that studies have focused on the control of production of the polygalacturase enzyme (PG), ie, the PG gene and its expression (39). Such studies have suggested that PG may play an important role in fruit ripening.

CEREALS

Starch is the nutritional reservoir in plants and cereal crops, including the seeds of wheat, rice, maize, and barley. The chemical composition of this polysaccharide consists of a mixture of linear and branched homopolymers of D-glucose. Starch is widely used in the food industry (eg, bread, fruit juices, and other beverages). Microbial sources of enzymes that use starch are numerous and include *Aspergillus*, *Rhizopus*, *Klebsiella*, and *Bacillus* (11). In the future, screening of other microorganisms for such enzymes should continue to complement genetic engineering. A number of enzymes that hydrolyze starch may be targeted for genetic manipulation; these are indicated in Table 1. The enzyme products are shown in Table 3. The cloning and characterization of some enzymes from filamentous fungi have been described (35). One of the first enzymes to be cloned from a *Bacillus* species was the α-amylase gene (31). A host *B. subtilis* has been shown to secrete α-amylase in low quantities. Thermostable α-amylase from species such as *Bacillus* require Ca^{++} for enzyme activity. The presence of Ca^{++}, however, does have a denaturing effect on the starch-processing enzymes glucoamylase and glucose isomerase (7). Alternative approaches would be to genetically engineer enzymes such as glucoamylase that are not denatured by Ca^{++}. A mutant of *B. stearothermophilus* that retains most of its α-amylase activity in the absence of Ca^{++} has been described (7). The enzyme β-amylase used in the production of high-maltose syrups has been found in *Bacillus* species. This discovery alleviates some of the problems associated with limited supply from plant sources. A mutant strain of *B. cereus* has been found to hyperproduce β-amylase at least 200-fold more than the wild type. The cloning of the plant β-amylase into bacteria or yeast is an alternative approach. The enzyme glucoamylase used in the saccharification process with α-amylase has been actively studied (35). Cloning of the glucoamylase gene from *Aspergillus*

Table 3. Enzymes Used in Hydrolysis of Starch or Starch Products[a]

Enzymes	Sites of Activity or Products
α-Amylase	Dextrins
β-Amylase	Maltose
α-D-Glucosidase	Glucose, glucose syrups
Glucoamylase	Glucose
Iso-amylase	Debranching enzyme
Pullalanase	Debranching enzyme
Glucose isomerase	Glucose = fructose
Cyclomaltodextrin-D-glucotransferase	Cyclodextrins

[a] Ref. 7, 19.

awamori into a vector and transformation of yeasts has met with little success. The use of glucoamylase in the initial starch liquefaction or gelation stage would assist the process by avoiding pH and temperature adjustments. Progress on the development of thermostable enzymes, particularly glucoamylase, has been described (34). This is an ideal enzyme for application of protein-engineering strategies. Similarly, pullulanases, which hydrolyze the $\alpha(1\rightarrow 6)$ linkages, can be used to improve starch-conversion yields. These enzymes also have limited thermal stability (maximum 50°C). One potential source of a thermostable enzyme is *B. stearothermophilus*. Modifications to such an enzyme should be possible through protein engineering, providing the structural basis of its activity, stability, and specificity is known (7). A recent review described novel highly thermostable pullulanases from thermophiles (40). Pullulanase activity has been found in species of *Clostridium, Thermus, Thermobacterioides,* and *Thermoactinomyces*. Some of these enzymes are stable and active above 90°C. The commercial potential of archaebacterial thermostable enzymes has been reviewed (41). Work is now in progress to develop mutants hyperproductive for pullulanases from *C. thermohydrosulfuricum*. Attempts are being made to clone and overproduce pullulanase from this organism using *B. subtilis* as the host cell. Genetic engineering will produce greater quantities of starch-degrading enzymes with improved specificity and stability at lower costs.

MICROBIAL BIOMASS

Biomass recovery has been carried out for some time in the food industry. Agro-industrial wastes such as whey, cellulose wastes, citrus, molasses, gluten, and starch offer ideal opportunities for the development of biomass (11). The low costs of these substrates is attractive for the manufacture of single-cell protein (SCP). Production costs of SCP made from these substrates must compare favorably with those of cheap sources such as soybeans and fishmeal. If SCP can be produced as a by-product of other biotechnologies, then SCP production becomes economical. By-product biomass does present problems when used for SCP production. There are nutritional concerns related to amino acid profiles and protein levels, as well as toxicological considerations regarding nucleic acid content. Some of these problems may be resolved by genetic engineering. There are several possible approaches (42). When SCP is produced as a by-product of a fermentation process, the desired SCP properties could be activated in the spent yeast by means of an external triggering agent during the secondary fermentation. This triggering action could activate genes coding for properties such as cell lysis, nucleic acid hydrolysis, or synthesis of specific enriched proteins or amino acids. The triggering agents could be chemical or physical (eg, heating). The chemical triggering agent would have to be linked to a promotor, eg, lactose (lac) or tryptophane (try). The addition of these nutrients would trigger genes inserted after these control sequences. Generally, strong promotors such as lac or try have been commonly used for overproduction of enzymes. Cheaper agents such as cellulose or gluten could also be used as triggers. Sporulation promotors have been cloned and can be used for expression of foreign genes (6). These promotors could be modified for SCP production as they are activated after the stationary phase of growth (ie, would be activated in the spent microorganism). Physical triggers such as heating do have some advantages. If the temperature were sufficiently high enough, intracellular nucleases could be activated. Because high levels of nucleic acids are undesirable in SCP products, such a system could also be used to produce nucleotides for flavoring agents (42). Microorganisms already established as useful for SCP include some yeasts and selected microfungi.

OTHER IMPORTANT APPLICATIONS

Interestingly, genetic engineering has had a surprising effect in supporting some of the traditional approaches to food microbiology. It has given greater impetus to aspects of taxonomy, species identification and detection, as well as population studies. Some areas once considered out of favor are now flourishing. Ribosomal RNA (16S) in species serves as a useful chronometer for the measurement of phylogenetic relations (43). Computer-assisted sequence comparisons of bacterial 16S rRNA have revealed that highly conserved regions are interspersed by regions of moderate and low homology, even with respect to closely related species. Based on this information, 16S rRNA oligonucleotide probes have been developed to allocate organisms to primary lines of evolutionary descent. This technique has allowed the food microbiologist to examine relationships between such groups as *Listeria, Streptococcus,* and *Brochothrix*. The typing system (ribotyping) (44) using DNA fingerprinting with 16S rRNA gene probes provides the opportunity to investigate the molecular epidemiology of both diverse bacteria and isolates within strains of the same species. Such a tool has tremendous implications in tracking the source of microorganisms in foodborne outbreaks. The use of rRNA as the target molecule that may be present in excess of 10,000 copies per cell can result in a sensitivity of at least 100 times greater than bacterial DNA targets. Current reviews of DNA diagnostic technology have indicated just how far this branch of clinical microbiology has reached (45,46). Further applications of synthetic oligonucleotide probes to foodborne pathogens are still being developed. In general, studies have found good agreement between traditional methods and conventional DNA probes and synthetic oligonucleotide probes (47).

The advent of the polymerase chain reaction (PCR) has the potential for revolutionizing diagnostic food microbiology. This reaction is a primer-mediated enzymatic amplification of a specific genome DNA target (48). Two oligonucleotide primers that flank the region to be amplified are selected. The sample is denatured by heating in the presence of a large molar excess of the two oligonucleotides and the four deoxyribonucleotide phosphates. Cooling and addition of DNA polymerase 1 allows reannealing of the oligonucleotides to occur with the target DNA strands, which is followed by primer extension. The newly synthesized DNA strands become the new template for the PCR primers, and the cycle of denaturation, primer annealing, and extension are repeated 20–30 times, resulting in exponential accumulation of the target sequence. The modified procedure uses a thermostable DNA polymerase, Taq, from *Thermus aquaticus* (48). This enzyme enables the amplification to be performed at higher temperatures, significantly improving specificity, yield, sensitivity, and length of products amplified. The technique has been used to identify enterotoxigenic *E. coli* by amplifying the heat-labile toxin (LT) gene, a single bacterium being detected following 30 cycles of amplification (49). Using PCR amplification, enrichment procedures may be eliminated because assays can be performed on original samples of food containing mixed flora as cell growth is not required for detection of specific bacteria (50).

Acknowledgements: The authors thank Sue Sprowl and Jeanne Hogeterp for typing the manuscript. Thanks are also expressed to Susan Read for the discussions on DNA probes.

BIBLIOGRAPHY

1. S. Harlander, "Food biotechnology: yesterday, today and tomorrow," *Food Technology* **43**, 196–206 (1989).
2. J. Davies, "Genetic engineering: processes and products," *Trends in Biotechnology* **6**, s7–s11, 1988.
3. D. Knorr, "Food biotechnology. Its organization and potential," *Food Technology* **41**, 95–100 (1987).
4. B. M. Chassy, "Prospects for the genetic manipulation of lactobacilli," *FEMS Microbiology Reviews* **46**, 297–312 (1987).
5. S. A. Boffey, "Techniques of genetic engineering," in *Biotechnology. The Biological Principles,* Open University Press, Milton Keynes, 1987, chp. 11.
6. Y. L. Lin, "Genetic engineering and process development for production of food processing enzymes and additives," *Food Technology* **40**, 104–110 (1986).
7. J. F. Kennedy, V. M. Cabalda, and C. A. White, "Enzymic starch utilization and genetic engineering," *Trends in Biotechnology* **6**, 184–189 (1988).
8. D. Cowan, R. Daniel, and H. Morgan, "Thermophilic protease: properties and potential applications," *Trends in Biotechnology* **3**, 68–72 (1985).
9. K. Esser and J. Kämper, "Transformation systems in yeasts: fundamentals and application in biotechnology," *Process Biochem.* **23**, 36–42 (1988).
10. G. Saunders, T. Picknelt, M. Tuite, and M. Ward, "Heterologous gene expression in filamentous fungi," *Trends in Biotechnology* **7**, 283–287 (1989).
11. D. L. Collins-Thompson, J. D. Cunningham, and J. T. Trevors, "Food microbiology and biotechnology: an update," in P. N. Cheremisinoff and R. P. Ouellete, eds., *Biotechnology—Applications and Research,* Technomic Publishing, Lancaster, Penn., 1985.
12. W. E. Sandine, "Looking backwards and forward at the practical applications of genetic research on lactic acid bacteria," *FEMS Microbiology Reviews* **46**, 205–220 (1987).
13. J. B. Luchansky, P. M. Muriana, and T. R. Klaenhammer, "Application of electroporation for transfer of plasmid DNA to *Lactobacillus, Leuconostoc, Listeria, Pediococcus, Bacillus, Staphylococcus, Enterococcus* and *Propionibacterium,*" *Molecular Microbiology* **2**, 637–646 (1988).
14. T. R. Klaenhammer, "Bacteriocins of lactic acid bacteria," *Biochimie* **70**, 337–349 (1988).
15. K. L. Dodds and D. L. Collins-Thompson, "Production of N_2O and CO_2 during the reduction of NO_2^- by *Lactobacillus lactis* TS4," *Applied and Environmental Microbiology* **50**, 1550–1552 (1985).
16. G. F. Fitzgerald and M. L. Gasson, "In vivo gene transfer systems and transposons," *Biochimie* **70**, 489–502 (1988).
17. G. Wolf and W. P. Hammes, "Effect of hematin on the activities of nitrite reductase and catalase in lactobacilli," *Archives of Microbiology* **149**, 220–224 (1988).
18. V. Mohr, "Enzyme technology in the meat and fish industries," *Process Biochem.* **15**(16), 18–21 (1980).
19. J. Alder-Nissen, "Newer uses of microbial enzymes in food production," *Trends in Biotechnology* **5**, 170–174 (1987).
20. W. E. Workman, J. H. McLinden, and D. H. Dean, "Genetic engineering applications to biotechnology in the genus *Bacillus,*" *CRC Critical Reviews in Biotechnology* **3**, 199–228 (1984).
21. R. L. Jackman and R. Y. Yada, "Protein engineering: methodology, applications and status," *Food Biotechnology* **1**, 167–223 (1987).
22. D. A. Estell, T. P. Graycar, and J. A. Wells, "Engineering an enzyme by site-directed mutagenesis to be resistant to chemical oxidation," *Journal of Biological Chemistry* **260**, 6518–6521 (1985).
23. R. D. Cooke, D. R. Twiddy, and P. J. Alan Reilly, "Lactic acid fermentation as a low cost means of food preservation in tropical countries," *FEMS Microbiology Reviews* **46**, 369–379 (1987).
24. G. Venema and J. Kok, "Improving dairy starter cultures," *Trends in Biotechnology* **5**, 144–149 (1987).
25. S. Harlander, "Applications in the dairy industry," *Journal of the American Oil Chemists Society* **65**, 1727–1729 (1988).
26. T. R. Klaenhammer, "Plasmid-directed mechanisms for bacteriophage defense in lactic streptococci," *FEMS Microbiology Reviews* **46**, 313–325 (1987).
27. J. T. Barach, "What's new in genetic engineering of dairy starter cultures and dairy enzymes," *Food Technology* **39**, 73–79, 84 (1985).
28. J. M. Marshall, "Lactic acid bacteria: starter for flavour," *FEMS Microbiology Reviews* **46**, 327–336 (1987).
29. C. F. Fernandes, K. M. Shahani, and M. A. Amer, "Therapeutic role of dietary lactobacilli and lactobacillic fermented dairy products," *FEMS Microbiology Reviews* **46**, 343–356 (1987).
30. C. A. Batt, "Genetic engineering of *Lactobacillus,*" *Food Technology* **40**, 95–112 (1986).
31. W. H. Pitcher, "Genetic modification of enzymes used in food processing," *Food Technology* **40**, 62–69 (1986).
32. K. Watenabe, H. Aihara, and Nakamura, "Degradation of cholesterol in lard by the extracellular and cell-bound enzymes from *Rhodococcus equi* no. 23," *Lebensm Wiss. und Technol* **22**, 98–99 (1989).
33. M. A. Daeschel, R. E. Anderson, and H. P. Fleming, "Microbial ecology of fermenting plant material," *FEMS Microbiology Reviews* **46**, 357–367 (1987).
34. B. P. Wasserman, "Thermostable enzyme production," *Food Technology* **38**, 78–89 (1984).
35. T. J. White, J. H. Meade, S. P. Shoemaker, K. E. Koths, and M. A. Innis. "Enzyme cloning for the food fermentation industry," *Food Technology* **38**, 90–95, 98 (1984).
36. E. M. Bates, H. J. Gilbert, G. P. Hazlewood, J. Huckle, J. I. Laurie, and S. P. Mann, "Expression of a *Clostridium thermocellum* endoglucanase gene in *Lactobacillus plantarum,*" *Applied and Environmental Microbiology* **55**, 2095–2097 (1989).

37. F. M. Rombouts and W. Pilnik, "Enzymes in fruit and vegetable juices technology. Use of enzymes as a processing aid," *Process Biochem* **13**(8), 9–13 (1978).
38. F. van Gijsegem, "Relationship between the pel genes of the pelADE cluster in *Erwinia chrysanthemi* strain B374," *Molecular Microbiology* **3**, 1415–1424 (1989).
39. M. Kramer, R. E. Sheehy, and W. R. Hiatt, "Progress toward the genetic engineering of tomato fruit softening," *Trends in Biotechnology* **7**, 191–194 (1989).
40. B. C. Saha and J. G. Zeikus, "Novel highly thermostable pullulanase from thermophiles," *Trends in Biotechnology* **7**, 234–239 (1989).
41. D. W. Hough and M. J. Danson, "A review: Archaebacteria: ancient organisms with commercial potential," *Letters in Applied Microbiology* **9**, 33–39 (1989).
42. I. Goldberg, "Future prospects of genetically engineering single cell protein," *Trends in Biotechnology* **6**, 32–34 (1988).
43. C. R. Woese, "Bacterial evolution," *Microbiology Reviews* **51**, 221–271 (1987).
44. T. L. Stull, J. J. Li Puma, and T. D. Edlind, "A broad spectrum probe for molecular epidemiology of bacteria: ribosomal RNA," *Journal of Infectious Diseases* **157**, 280–286 (1988).
45. J. J. Pasternak, "Microbial DNA diagnostic technology," *Biotechnology Advances* **6**, 683–695 (1988).
46. J. Walker and G. Dougan, "A review: DNA probes. A new role in diagnostic microbiology," *Journal of Applied Bacteriology* **67**, 229–238 (1989).
47. W. E. Hill, B. A. Wentz, and W. L. Payne, "DNA colony hybridization method using synthetic oligonucleotides to detect enterotoxigenic *Escherichia coli*: collaborative study," *Journal of the Association of Official Analytical Chemists* **69**, 531–535 (1986).
48. R. K. Saiki, D. H. Gelfand, S. Stoffel, S. J. Scharf, R. Higuchi, G. T. Horn, K. B. Mullis, and H. A. Erlich, "Primer-directed enzymatic amplification of DNA with a thermostable DNA polymerase," *Science* **239**, 487–491 (1988).
49. D. M. Olive, "Detection of enterotoxigenic *Escherichia coli* after polymerase chain reaction amplification with a thermostable DNA polymerase," *Journal of Clinical Microbiology* **27**, 261–265 (1989).
50. R. J. Steffan and R. M. Atlas, "DNA amplification to enhance detection of genetically engineered bacteria in environmental samples," *Applied and Environmental Microbiology* **54**, 2185–2191 (1988).

DAVID L. COLLINS-THOMPSON
PETER J. SLADE
University of Guelph
Guelph, Canada

GENETIC ENGINEERING PART V: YEASTS AND ETHANOL PRODUCTION

BACKGROUND INFORMATION

Transformation of renewable plant biomass, including organic wastes, to fuels or chemicals is attractive in light of the continual decline of fossil fuel reserves and the increased dependence on foreign supplies. Cellulose and hemicellulose comprise more than 70% of plant biomass and occur in wastes from agriculture and forestry as well as vegetable and fruit processing and municipal wastes. After enzymatic or acid hydrolysis, plant biomass yields mostly cellobiose, xylose, and glucose, as well as smaller amounts of mannose, galactose, and arabinose (1).

These complex and simple carbohydrates are therefore available for fermentations by bacteria and yeasts. Yeasts are commercially valuable to a variety of industries. These microorganisms, mainly known in the baking and brewing industries, play a role in animal nutrition (single-cell protein) and in production of alcohol fuels, chemical feedstocks, and, more recently, enzymes. Although species of *Saccharomyces* are the most widely studied yeasts, both genetically and biochemically, they do have certain drawbacks, such as the limited secretion of products, low ethanol tolerance, and restricted carbohydrate utilization. Other yeasts, which respond to unique environmental stimuli, may actually be preferred. Previously, organisms with desirable traits have been selected by screening culture collections or by mutagenesis and subsequent isolation of the desired strains. However, mutations are difficult to achieve with commercial polyploid yeast strains. The polyploid nature hinders mutagenesis and selection of dominant characteristics, yet it is important for maintaining the stability of the strain.

With the development of molecular biological techniques, improvement of polyploid and imperfect yeast strains may be obtained by circumventing the problems associated with conventional genetic methods. Fusion of protoplasts within a single species or across the species barrier has been successful. Unfortunately, fusion products are often unstable and may express undesirable parental traits. The best opportunity for successful, stable gene expression may lie with the use of recombinant DNA technology that inserts desirable genes into vectors (plasmids) and then transfers the plasmids to host organisms. Expression of the desired trait can be optimized through the selection of suitable vectors that will increase either intracellular or extracellular levels of products.

IMPROVING ETHANOL PRODUCTION

Modification of yeasts to increase ethanol production for liquid fuels and chemical feedstock industries is an open area for research. Improvement of strains could overcome low tolerance to ethanol and high substrate concentrations, which are characteristics of most commercial yeast strains. If the two characteristics are related, then an osmotolerant strain could possibly be altered to achieve high ethanol and substrate tolerance more easily than a strain that is not tolerant to high osmotic conditions. Characteristics such as flocculence and relative stability to fluctuations in temperature and pH may also be transferable.

Another area for strain improvement, one that could substantially reduce ethanol production costs, is that of increasing the variety of carbohydrates that yeasts will economically ferment. Starch, cellulose, and xylose are abundant carbohydrates found in plant materials. Other fungi, as well as bacteria, have been identified as being capable of breaking down these complex carbohydrates. Combining the genetic material of these various organisms could result in a unique individual capable of coding for the enzymes necessary to attack a number of carbohydrates simultaneously.

In the immediate future the primary goals for increased alcohol production are the development of organisms that have high osmotolerance; degrade cellulose, starch, and xylose; and have high tolerance to ethanol. Such organisms could make ethanol production more competitive.

GENETIC IMPROVEMENT BY PROTOPLAST FUSION

The development of the protoplast fusion technique allows genetic manipulation of strains that normally do not mate. In this procedure, the cells are treated with an enzyme, often zymolyase (Miles Laboratories, Elkhart, Ind.) or gluculase (Endo Laboratories, Garden City, N.Y.), to remove most of the cell wall (Fig. 1). The resulting spheroplast is very fragile and must be maintained in an osmotically suitable buffer, usually 0.8–1.2 M sorbitol, pH 7. After a gentle washing with this buffer the spheroplasts are subjected to a calcium chloride or lithium acetate treatment. The actual fusion of the spheroplasts occurs when polyethylene glycol is added or when the cells are subjected to an electrical field (electrofusion). The fused cells are placed in a suitable medium to undergo cell wall regeneration, then placed on selective media to eliminate the parental cells.

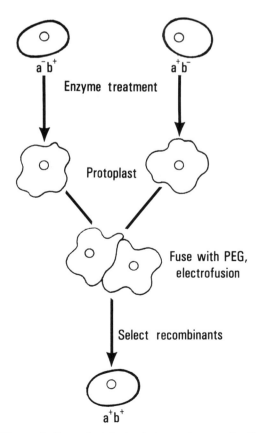

Figure 1. Protoplast fusion between two yeast cells.

A major problem with protoplast fusion is the selection of the fused cells from the parentals. To aid in selecting fused cells, the two parentals must have some unique, easily identifiable trait. This may take the form of an amino acid or vitamin requirement. However, since single gene traits are often difficult to obtain in a polyploid strain, other identifiable characteristics, such as ability to grow on a unique carbohydrate, may be substituted. The choice must be such that the parentals have a low reversion rate of the marker gene, for the successful rate of fusion is often quite low (1/10,000).

In spite of the difficulties of working with industrial yeast strains, success with protoplast fusion has been achieved by using nutritional requirers (2–4) and differential carbohydrates (5–9). The hybrids are often unstable but, once a stable isolate is selected, analysis of the DNA content may be performed. In general, the nuclear contents of a hybrid are often one parental type with a few genes or an extra chromosome of the other parental; therefore, there is no significant increase in the amount of DNA present in the fused product (8). Occasionally, a doubling of the nuclear contents in the fused cell occurs (9).

Efforts to improve ethanol production by fusion have been successful between *S. cerevisiae* and *S. diastaticus*, the latter being a variant of *S. cerevisiae* which produces enzymes that degrade starch. However, most reports thus far describe the success of the fusion, leading to the conclusion that the achievement of better fermentations is apparently limited. However, one *S. cerevisiae*–*S. diastaticus* fusion product did produce more ethanol than either parent, presumably by utilizing both the starch and the glucose (7), but it is unknown what actually caused the greater increase in ethanol.

Increases in ethanol production rates for fused products over parentals have been noted in other strains. A fusion of an osmotolerant species, *Zygosaccharomyces* (*Saccharomyces*) *mellis,* with *S. cerevisiae* produced a better fermentor (as measured by CO_2 evolution) than either parent (3). Brewing strains have also been used to obtain better ethanol producers (10–12).

In many studies involving protoplast fusion, respiratory-deficient (RD) strains are used. Respiratory deficiency is characterized by a loss of mitochondrial function so that oxidative processes no longer occur and the cell must therefore function fermentatively. RD strains can easily be detected by their lack of growth on glycerol, a carbohydrate most yeasts use oxidatively. The genetic lesion for respiratory deficiency can be located within the mitochondrial genome (forming *rho*⁻ or *mit*⁻ mutants) or in the nuclear genome (forming *pet*⁻) mutants. Unless the lesion for respiratory deficiency is clearly defined (which is often not the case), fusion of an RD cell with a respiratory-competent cell may produce a recombinant that is due to either nuclear or mitochondrial complementation, a fact that may be important in the cell's oxidative processes. However, respiratory deficiency has proven to be a suitable marker for interspecific crosses involving *S. cerevisiae* and *Candida utilis* (2,13) and *S. cerevisiae* and *S. diastaticus* (6,7,14). If the end result is an increase in ethanol production, the source of the lesion, at least initially, may not be important.

TRANSFORMATION BY RECOMBINANT DNA TECHNIQUES

Although a good deal of work has been done using protoplast fusion methods, it is obvious that successful production of improved yeast strains has been limited. Transformation of yeast by recombinant DNA techniques is more likely to produce stable isolates carrying a desired characteristic. However, to increase the chance of successful transformation, one needs to know the background of both the donor and recipient cells. Increasing the knowledge of the basic genetic background by knowing the biochemical pathways, cofactor requirements, genetic controls, and genes involved will allow for more successful genetic manipulation.

The basic techniques for yeast transformation are presented in Figure 2. Nuclear DNA of the donor cell is isolated, restricted with an endonuclease (such as by partial digestion with Sau3A to give a range in the sizes of pieces), and mixed with a plasmid that has been cut once by a comparable recognition-site enzyme. The mixture of DNA fragments undergoes ligation (to fuse the cut ends) and the plasmids are then inserted into an *Escherichia coli* host. The cells are placed onto restrictive media (in the example in Fig. 2, ampicillin and tetracycline), and the cells containing recombinant plasmids (ampicillin-resistant, tetracycline-sensitive) are isolated. After amplification of the recombinant plasmids (15), yeast cells are transformed and the desired yeasts are selected.

A suitable plasmid that can transform the host cell is essential to developing a cloning system. A transforming plasmid must contain at least three elements: an origin of replication recognized by the host DNA polymerase; a promoter region recognized by the host RNA polymerase; and a functional gene that serves as a genetic marker. The latter factor can be a nutritional marker, such as the *leu*2 gene in the plasmid YEp13 (16), resistance to copper (17), or resistance to antibiotics such as G418 (18). Even genes coding for an enzyme can be used as a plasmid marker, such as using the lactose permease gene from *Kluyveromyces lactis* to mark transformation in *S. cerevisiae* (19). The origin of plasmid replication may also come from different types of DNA, depending on the host employed. Autonomously replicating sequences (ARS) from *Candida* (20), *Kluyveromyces* (21), and *Schizosaccharomyces* (22), as well as the 2μ origin from *Saccharomyces*, have all been used successfully.

While the majority of research has been done on circular plasmids, linear plasmids are also useful. To increase their meiotic stability, telomeric sequences (23,24) or centromeric regions (CEN sequences) may be added (25). In this way the plasmid acts like a minichromosome and is stable during mitosis and meiosis. Circular plasmids, although they may be present in high numbers, may be lost during cell division.

The second element for a successful plasmid, that of having a suitable promoter region, can be obtained from the host genome or from a related organism that carries an easily identifiable gene. An example of this is the use of the β-galactosidase promoter and gene from *E. coli*. If there is successful transcription and translation of this genetic material in the host, the colonies turn blue on a suitable substrate. A desired gene could then be inserted into a β-galactosidase gene, expression would result from the β-galactosidase promoter region, and the recombinant colony would be scored as white in color.

Despite the problems involved in transforming yeasts with plasmids, limited goals have been reached. Successful transformation resulting in increased alcohol production has not been attained yet, in part because not all the processes involved are understood. Limited areas and specific problems must be attacked first. The utilization of

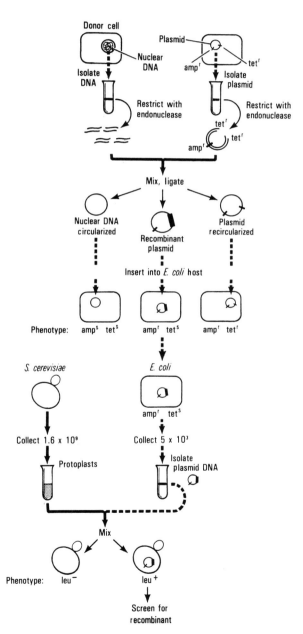

Figure 2. Simplified flow diagram for isolation of yeast cells carrying recombinant plasmids. Donor DNA is mixed with plasmid DNA and inserted into *E. coli*, and recombinant molecules are selected from amp^r tet^s cells. Recombinant plasmids are then inserted into the host yeast protoplasts.

complex carbohydrates by yeasts is one area. The three major complex carbohydrates constituting biomass are cellulose, starch, and xylose. These are potential surplus carbon sources and are renewable resources. The literature on the research for conversion of these complex carbohydrates into useful compounds is extensive, and the reader is referred to reviews on cellulose fermentation (26,27) and starch fermentations (28–30); research on xylose fermentation will be discussed below.

XYLOSE FERMENTATION

The application of protoplast fusion and recombinant DNA techniques to xylose fermentation has been expanding rapidly. Because xylose represents a large fraction of biomass that could be used for production of chemicals, it is one area that is amenable to improvement by these techniques. Until 1980 no yeasts had been identified that could ferment xylose to ethanol. However, since then, screening programs have identified several yeasts and other fungi that produce alcohol from xylose (31–38). Three of these, *Pachysolen tannophilus*, *Pichia stipitis*, and *Candida shehatae*, have been studied in detail (39–41). *Pachysolen tannophilus* is perhaps the species most amenable to genetic analyses, even though it is homothallic. Limited genetic analyses have been done with *P. tannophilus* (42) by adjusting the culture conditions to encourage spore formation and tetrad analysis. Mutants have been induced that show improved ethanol production over the parental (43–45) yet have not overcome all of the problems associated with xylose fermentation. *Candida* is an imperfect yeast and is, as such, more amenable to the newer methods of genetic recombination.

In the first step of the xylose-degradation pathway (Fig. 3), xylose is converted to xylulose. Bacteria generally accomplish this in one step using xylose isomerase; however, yeast commonly use a two-step process producing the intermediate xylitol (1). In yeast, xylose reductase, requiring the cofactor NADPH, reduces xylose to xylitol, which is then converted to xylulose by the NAD-requiring xylitol dehydrogenase (46). This two-step process is very demanding of the supply of NADPH and NAD and can easily upset the cellular NAD(P)H/NAD(P) pool (47). This cofactor specificity for the xylose degrading enzymes in yeasts is a serious limitation to the use of these yeasts for the commercial production of ethanol.

Increasing ethanol yields from xylose may be accomplished by inserting the gene for xylose isomerase into these yeasts. Since, at least in *P. tannophilus*, no significant amount of xylose isomerase activity has been detected (48), it appears that this would be an ideal area for strain improvement. To date, however, success has been limited. Cloning of the isomerase gene has involved procaryotic systems, including *E. coli*, *Bacillus subtilis*, *Streptomyces violaceoniger*, and *Xanthomonas*. The Clarke-Carbon *E. coli* gene bank has been used to isolate recombinant plasmids that would complement xylose-negative mutants of *E. coli* and *Salmonella typhimurium* (49). A number of clones were identified that contained the region of DNA coding for xylose isomerase activity. One such plasmid contained a region including the genes for D-xylose isomerase and D-xylulose kinase. These two genes have also been cloned from *S. violaceoniger* (50) and transformants of *S. violaceoniger* have a higher isomerase and kinase activity than do the wild types, perhaps due to the high copy number of the plasmid. The xylose isomerase gene has also been isolated from the *E. coli* clone bank and inserted into a plasmid that is capable of transforming *Schizosaccharomyces pombe* (51). The transformed yeast cells are capable of expressing the xylose isomerase gene and producing ethanol (52). The xylose isomerase region from the *E. coli* clone bank has also been isolated and the gene subcloned into YEp13, a plasmid which can transform both *E. coli* and *S. cerevisiae* with high efficiency (53). Because of the apparent inability of the bacterial promoter to direct transcription in yeast cells, subsequent work has dealt with inserting a eukaryotic promoter in

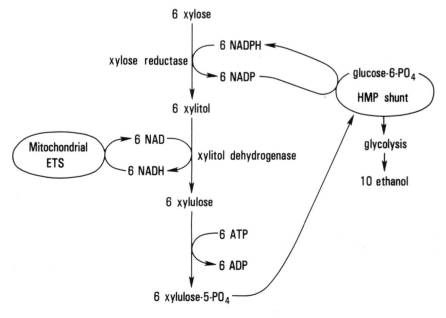

Figure 3. Simplified pathway of xylose degradation in *P. tannophilus*.

front of the bacterial xylose isomerase gene (54). The *E. coli* isomerase gene has been fused to the yeast ADH1 promoter and xylose isomerase mRNA as well as an immunologically cross-reactive xylose isomerase polypeptide in *S. cerevisiae* was detected (55); however, activity of the enzyme remained low. Transposons have also been used to develop clone banks containing the xylose isomerase gene from *E. coli* (56) and *B. subtilis* (57). The *B. subtilis* xylose isomerase and xylulokinase genes have also been cloned into *E. coli* and yeast, and, although immunogenic proteins were made, enzymatic activity remained low (58).

Perhaps the most promising work on producing recombinant organisms for fermenting xylose to ethanol is that of introducing the genes necessary for alcohol production into *E. coli*. Plasmids containing the genes for pyruvate decarboxylase and alcohol dehydrogenase were introduced from *Zymomonas* into *E. coli* and have succeeded in altering the fermentation products of the host to include ethanol (59). Refinements on the plasmids and screening for better hosts produced an *E. coli* proficient in alcohol production from the substrates glucose, lactose, and xylose (60). From an economic standpoint, perhaps this bacterial construction is more suitable for commercial application than trying to improve yeasts.

However, if yeasts are to be improved, several problems associated with the use of bacterial genes in eucaryotic systems must be overcome. One problem is that the promoter of a procaryotic gene may not be recognized by a eucaryotic polymerase. Other factors involved in gene expression are inducer–repressor sequences, enzymatic cofactor requirements, processing enzymes (of RNA and proteins), and transport factors. All in all, obtaining expression of a foreign gene and having it fit properly into a sequence of events of a pathway is a complex process. It may be more advantageous to clone a eucaryotic isomerase gene instead of a procaryotic one into *Saccharomyces*.

If xylose isomerase cloning attempts continue to meet with limited success in producing a *Saccharomyces* capable of converting xylose to ethanol, improvement of strains may be accomplished by cloning the eucaryotic genes for xylose reductase and xylitol dehydrogenase. However, instead of one gene, two or possibly more genes must be transferred. *Pachysolen* appears to have two genes coding for xylose reductase. One gene codes for an enzyme that requires NADPH as cofactor (61), and another codes for an enzyme that binds with NADH (62) or both NADH and NADPH (63,64). *Pichia stipitis*, however, appears to have a single enzyme which has a dual cofactor specificity for either NADPH or NADH (65,66).

The type of xylose reductase (dual or single cofactor specificity) encoded by the yeast cells most likely determines the cells' reactions to the presence of oxygen. *Pichia stipitis*, containing a dual cofactor-specific xylose reductase, can use xylose aerobically or anaerobically. Presumably xylose reductase supplies the NAD necessary for the xylitol dehydrogenase. *Pachysolen*, which produces mainly NADPH-specific xylose reductase, ferments xylose better aerobically when the NAD required by xylitol dehydrogenase is supplied by mitochondrial activity. The limitation of xylose fermentation in *Pachysolen* at the step requiring NAD and xylitol dehydrogenase is suggested by the observation that, even in cells with little detectable xylose reductase activity, xylitol accumulates in the medium (48).

Further support for the theory that NAD is a limiting factor shows that on addition of the hydrogen-accepting compound acetone, the ethanol yield of *Pachysolen* increases by 50–70% over that of control cultures (67). This helps to explain why *Pachysolen* cultures do better under aerobic or semiaerobic conditions, where the oxygen acts as a hydrogen acceptor with regeneration of NAD. Therefore, it might be more advantageous to clone the xylose reductase gene from *Pichia stipitis* into a host yeast, because the xylose reductase enzyme will provide the NAD necessary for the second cloned gene (xylitol dehydrogenase) and will not be a drain on the cellular pool of NAD. However, the recombinant cells (*P. stipitis* genes and *S. cerevisiae* as host) might still require oxygen, for it has been reported that *S. cerevisiae* requires functioning mitochondria to metabolize xylulose to ethanol (68).

In spite of the complexity of xylose metabolism in yeast, protoplast fusion of *C. shehatae* has resulted in an increase in ploidy level and a slightly increased level of ethanol production from xylose (69). Protoplast fusion between *C. shehatae* and *P. stipitis* resulted in approximately 20% stable fusants, with several fusants fermenting xylose more rapidly than either parental (70). However, it may be that *C. shehatae* is not a suitable host for increasing xylose fermentation by the method of introducing high copy numbers of the xylose reductase and xylitol dehydrogenase genes because of high requirements for cellular NAD.

The lack of a suitable transformation system in the three xylose-fermenting yeasts also needs to be defined. Suitable selectable markers are difficult to obtain. Presently used transformation conditions include nutritional mutants with low reversion rates ($<1 \times 10^{-9}$), a clone bank containing *Sau*3A fragments of *Pachysolen* nuclear DNA, and the plasmid YEp13. These conditions may be suitable for transformation by other plasmids, perhaps by one carrying the xylose isomerase gene. *Pachysolen* may then yield ethanol at near theoretical rates.

However, other fermentation parameters may be important in selecting the yeast most suitable for genetic manipulation. A commercially acceptable yeast would need to produce 50 to 60 g/L ethanol within 36 h from xylose with a yield of at least 0.4 g/g (71). When *C. shehatae*, *P. stipitis*, and *P. tannophilus* are compared on their ability to ferment xylose (39,41), *C. shehatae* and *P. stipitis* are the superior strains. However, these strains are less amenable to genetic manipulation (39). Perhaps with the development of selectable plasmids with eucaryotic promoters and genes, the genetic enhancement of any one of the three xylose-fermenting yeasts will produce a high-ethanol-tolerant, rapidly fermenting, low-oxygen-requiring yeast strain.

BIBLIOGRAPHY

This article has been adapted from N. J. Alexander, "Genetic Manipulation of Yeasts for Ethanol Production from Xylose," *Food Technology*, **40**, 99 (1986).

1. T. W. Jeffries, "Utilization of Xylose by Bacteria, Yeasts, and Fungi," *Advances in Biochemical Engineering* **27**, 1 (1983).
2. C. Perez, C. Vallin, and J. Benitez, "Hybridization of *Saccharomyces cerevisiae* with *Candida utilis* through Protoplast Fusion," *Current Genetics* **8**, 575 (1984).
3. R. Legmann and P. Margalith, "Interspecific Protoplast Fusion of *Saccharomyces cerevisiae* and *Saccharomyces mellis*," *European Journal of Applied Microbiology* **18**, 320 (1983).
4. E. Johannsen, L. Halland, and A. Opperman, "Protoplast Fusion within the Genus *Kluyveromyces* van der Walt emend. van der Walt," *Canadian Journal of Microbiology* **30**, 540 (1984).
5. J. F. T. Spencer, C. Bizeau, N. Reynolds, and D. M. Spencer, "The Use of Mitochondrial Mutants in Hybridization of Industrial Yeast Strains," *Current Genetics* **9**, 649 (1985).
6. L. I. Figueroa, M. F. Richard and M. R. van Broock, "Interspecific Protoplast Fusion of the Baker's Yeast *Saccharomyces cerevisiae* and *Saccharomyces diastaticus*," *Biotechnology Letters* **6**, 269 (1984).
7. L. I. Figueroa, M. A. Cabada, and M. R. van Broock, "Alcoholic Fermentation of Starch Containing Media Using Yeast Protoplast Fusion Products," *Biotechnology Letters* **7**, 837 (1985).
8. M. Taya, H. Honda, and T. Kobayashi, "Lactose-Utilizing Hybrid Strain Derived from *Saccharomyces cerevisiae* and *Kluyveromyces lactis* by Protoplast Fusion," *Agricultural and Biological Chemistry* **48**, 2239 (1984).
9. D. P. Groves and S. G. Oliver, "Formation of Intergeneric Hybrids of Yeast by Protoplast Fusion of *Yarrowia* and *Kluyveromyces* Species," *Current Genetics* **8**, 49 (1984).
10. M. Johansson and J. E. Sjostrom, "Ethanol and Glycerol Production under Aerobic Conditions by Wild-Type, Respiratory-Deficient Mutants and a Fusion Product of *Saccharomyces cerevisiae*," *Applied Microbiology and Biotechnology* **20**, 105 (1984).
11. T. Seki, S. Myoga, S. Limtong, S. Uedono, J. Kumnuanta, and H. Taguchi, "Genetic Construction of Yeast Strains for High Ethanol Production," *Biotechnology Letters* **5**, 351 (1983).
12. C. J. Panchal, A. Harbison, I. Russell, and G. G. Stewart, "Ethanol Production by Genetically Modified Strains of *Saccharomyces*," *Biotechnology Letters* **4**, 33 (1982).
13. M. S. Richard and M. R. van Broock, "Protoplast Fusion between a Petite Strain of *Candida utilis* and *Saccharomyces cerevisiae* Respiratory-Competent Cells," *Current Microbiology* **10**, 117 (1984).
14. J. F. T. Spencer, P. Laud, and D. M. Spencer, "The Use of Mitochondrial Mutants in the Isolation of Hybrids Involving Industrial Yeast Strains. II. Use in Isolation of Hybrids Obtained by Protoplast Fusion," *Molecular and General Genetics* **178**, 651 (1980).
15. T. Maniatis, E. F. Fritsch, and J. Sambrook, *Molecular Cloning: A Laboratory Manual*, Cold Spring Harbor Laboratory, Cold Spring Harbor, N.Y., 1982.
16. J. R. Broach, J. N. Strathern, and J. B. Hicks, "Transformation in Yeast: Development of a Hybrid Cloning Vector and Isolation of the *can1* Gene," *Gene* **8**, 121 (1979).
17. B. A. Cantwell, J. C. Hurley, N. Brazil, N. Murphy, and D. J. McConnell, "Transformation of Brewing Yeast Strains—A Practical Application," presented at Symposium on Biochemistry and Molecular Biology of Industrial Yeasts, Kalamazoo, Mich., Oct. 1985.
18. R. Yocum, R. Daves, and S. Hanley, "Novel Vectors for Transforming Industrial Yeast Strains," presented at Symposium on Biochemistry and Molecular Biology of Industrial Yeasts, Kalamazoo, Mich., Oct. 1985.
19. K. Sreekrishna and R. C. Dickson, "Construction of Strains of *Saccharomyces cerevisiae* That Grow on Lactose," *Proceedings of the National Academy of Sciences (USA)* **82**, 7909 (1985).
20. N. W. Y. Ho, H. C. Gao, J. J. Huang, P. E. Stevis, S. F. Chang, and G. T. Tsao, "Development of a Cloning System for *Candida* Species," *Biotechnology and Bioengineering Symposium* **14**, 295 (1984).
21. S. Das and C. P. Hollenberg, "A High-Frequency Transformation System for the Yeast *Kluyveromyces lactis*," *Current Genetics* **6**, 123 (1982).
22. D. Beach and P. Nurse "High-Frequency Transformation of the Fission Yeast *Schizosaccharomyces pombe*," *Nature (London)* **290**, 140 (1981).
23. J. W. Szostak and E. H. Blackburn, "Cloning Yeast Telomeres on Linear Plasmic Vectors," *Cell* **29**, 245 (1982).
24. A. M. Guerrini, F. Ascenzioni, C. Tribioli, and P. Donini, "Transformation of *Saccharomyces cerevisiae* and *Schizosaccharomyces pombe* with Linear Plasmids Containing 2μ Sequences," *The EMBO Journal* **4**, 1569 (1985).
25. L. Clarke and J. Carbon, "Isolation of a Yeast Centromere and Construction of Functional Small Circular Chromosomes," *Nature (London)* **287**, 504 (1980).
26. B. S. Montenecourt and D. E. Eveleigh, "Fungal Carbohydrases: Amylases and Cellulases," in J. W. Bennett and L. L. Lasure, eds., *Gene Manipulations in Fungi*, Academic Press, New York, 1985, p. 491.
27. T. M. Wood, "Properties and Mode of Action of Cellulases," *Biotechnology and Bioengineering Symposium* **5**, 111 (1975).
28. D. French, "Amylases: Enzymatic Mechanisms," in A. Hollaender, ed., *Trends in the Biology of Fermentations for Fuels and Chemicals*, Plenum Press, New York, 1981, p. 151.
29. G. G. Stewart, "The Genetic Manipulation of Industrial Yeast Strains," *Canadian Journal of Microbiology* **27**, 973 (1981).
30. G. G. Stewart, C. J. Panchal, I. Russell, and A. M. Sills, "Advances in Ethanol from Sugars and Starch—A Panoramic Paper," in H. E. Duckworth and E. A. Thompson, eds., *International Symposium on Ethanol from Biomass*," Royal Society of Canada, Ottawa, 1983, p. 4.
31. H. P. Schneider, Y. Wang, Y. K. Chan, and R. Maleszka, "Conversion of D-Xylose into Ethanol by the Yeast *Pachysolen tannophilus*," *Biotechnology Letters* **3**, 89 (1981).
32. P. J. Slininger, R. J. Bothast, J. R. VanCauwenberge, and C. P. Kurtzman, "Conversion of D-Xylose to Ethanol by the Yeast *Pachysolen tannophilus*," *Biotechnology and Bioengineering* **24**, 371 (1982).
33. M. L. Suihko and M. Drazic, "Pentose Fermentation by Yeasts," *Biotechnology Letters* **5**, 107 (1983).
34. A. Toivola, D. Yarrow, E. van den Bosch, J. P. van Dijken, and W. A. Scheffers, "Alcoholic Fermentation of D-Xylose by Yeasts," *Applied and Environmental Microbiology* **47**, 221 (1984).
35. J. C. du Preez and J. P. van der Walt, "Fermentation of D-Xylose to Ethanol by a Strain of *Candida shehatae*," *Biotechnology Letters* **5**, 357 (1983).
36. Y. Morikawa, S. Takasawa, I. Masunaga, and K. Takayama, "Ethanol Productions from D-Xylose and Cellobiose by *Kluyveromyces cellobiovorus*," *Biotechnology and Bioengineering* **27**, 509 (1985).
37. J. N. Nigam, R. S. Ireland, A. Margaritis, and M. A. LaChance, "Isolation and Screening of Yeasts That Ferment D-

Xylose Directly to Ethanol," *Applied and Environmental Microbiology* **50**, 1486 (1985).
38. J. F. Wu, S. M. Lastick and D. M. Updegraff, "Ethanol Production from Sugars Derived from Plant Biomass by a Novel Fungus," *Nature (London)* **321**, 887 (1986).
39. B. A. Prior, S. G. Kilian, and J. C. du Preez, "Fermentation of D-Xylose by the Yeasts *Candida shehatae* and *Pichia stipitis*," *Process Biochemistry* **24**, 21 (1989).
40. K. Skoog and B. Hahn-Hagerdal, "Xylose Fermentation," *Enzyme Microb. Technol.* **10**, 66 (1988).
41. P. J. Slininger, P. L. Bolen, and C. P. Kurtzman, "*Pachysolen tannophilus*: Properties and Process Considerations for Ethanol Production from D-Xylose. *Enzyme Microb. Technol.* **9**, 5 (1987).
42. A. P. James and D. M. Zahab "A Genetic System for *Pachysolen tannophilus*, A Pentose-Fermenting Yeast," *Journal of General Microbiology* **128**, 2297 (1982).
43. T. Clark, N. Wedlock, A. P. James, K. Deverell, and R. J. Thornton, "Strain Improvement of the Xylose-Fermenting Yeast *Pachysolen tannophilus* by Hybridization of Two Mutant Strains," *Biotechnology Letters* **8**, 801 (1986).
44. H. Lee, A. P. James, D. M. Zahab, G. Mahmourides R. Maleszka, H. Schneider, "Mutants of *Pachysolen tannophilus* with Improved Production of Ethanol from D-Xylose," *Applied and Environmental Microbiology* **51**, 1252 (1986).
45. T. W. Jeffries, "Mutants of *Pachysolen tannophilus* Showing Enhanced Rates of Growth and Ethanol Formation from D-Xylose," *Enzyme Microb. Technol.* **6**, 254 (1984).
46. K. L. Smiley and P. L. Bolen, Demonstration of D-Xylose Reductase and D-Xylitol Dehydrogenase in *Pachysolen tannophilus*," *Biotechnology Letters* **4**, 607 (1982).
47. P. M. Bruinenberg, P. H. M. de Bot, J. P. van Dijken, and W. A. Scheffers, "The Role of Redox Balances in the Anaerobic Fermentation of Xylose by Yeasts," *European Journal of Applied Microbiology and Biotechnology* **18**, 287 (1983).
48. N. J. Alexander, "Temperature Sensitivity of the Induction of Xylose Reductase in *Pachysolen tannophilus*," *Biotechnology and Bioengineering* **27**, 1739 (1985).
49. R. Maleszka, P. Y. Wang, and H. Schneider, "A Col E1 Hybrid Plasmid Containing *Escherichia coli* Genes Complementing D-Xylose Negative Mutants of *Escherichia coli* and *Salmonella typhimurium*," *Canadian Journal of Biochemistry* **60**, 144 (1982).
50. T. Marcel, D. Drocourt, and G. Tiraby, "Cloning of the Glucose Isomerase (D-Xylose Isomerase) and Xylulose Kinase Genes of *Streptomyces violaceoniger*," *Molecular and General Genetics* **208**, 121 (1987).
51. P. P. Ueng, K. J. Volpp, J. V. Tucker, C. W. Gong, and L. F. Chen, "Molecular Cloning of the *Escherichia coli* Gene Encoding Xylose Isomerase," *Biotechnology Letters* **7**, 153 (1985).
52. E.-C. Chan, P. P. Ueng, and L. Chen, "D-Xylose Fermentation to Ethanol by *Schizosaccharomyces pombe* Cloned with Xylose Isomerase Gene," *Biotechnology Letters* **8**, 231 (1986).
53. N. W. Y. Ho, S. Rosenfield, P. Stevis, and G. T. Tsao, "Purification and Characterization of the D-Xylose Isomerase Gene from *Escherichia coli*," *Enzyme Microb. Technol.* **5**, 417 (1983).
54. N. W. Y. Ho, personal communication, 1985.
55. A. V. Sarthy, B. L. McConaughy, Z. Lobo, J. A. Sundstrom, C. E. Furlong, and B. D. Hall, "Expression of the *Escherichia coli* Xylose Isomerase Gene in *Saccharomyces cerevisiae*," *Applied and Environmental Microbiology* **53**, 1996 (1987).
56. V. B. Lawlis, M. S. Dennis, E. Y. Chen, D. H. Smith, and D. J. Henner, "Cloning and Sequencing of the Xylose Isomerase and Xylulose Kinase Genes of *Escherichia coli*," *Applied and Environmental Microbiology* **47**, 15 (1984).
57. M. Wilhelm and C. P. Hollenberg, "Selective Cloning of *Bacillus subtilis* Xylose Isomerase and Xylulokinase in *Escherichia coli* Genes by IS5-Mediated Expression," *The EMBO Journal* **3**, 2555 (1984).
58. C. P. Hollenberg, "Construction of Pentose-Fermenting Strains of *Saccharomyces*," European Brewery Convention, *Monograph* 12: 199 (1987).
59. L. O. Ingram, T. Conway, D. P. Clark, G. W. Sewell, and J. F. Preston, "Genetic Engineering of Ethanol Production in *Escherichia coli*," *Applied and Environmental Microbiology* **53**, 2420 (1987).
60. F. Alterthum and L. O. Ingram, "Efficient Ethanol Production from Glucose, Lactose, and Xylose by Recombinant *Escherichia coli*," *Applied and Environmental Microbiology* **5**, 1943 (1989).
61. G. Ditzelmuller, C. P. Kubicek, W. Wohrer, and M. Rohr, "Xylose Metabolism in *Pachysolen tannophilus*: Purification and Properties of Xylose Reductase," *Canadian Journal of Microbiology* **30**, 1330 (1984).
62. G. Ditzelmuller, E. M. Kubicek-Pranz, M. Rohr and C. P. Kubicek, "NADPH-Specific and NADH-Specific Xylose Reduction Is Catalyzed by Two Separate Enzymes in *Pachysolen tannophilus*," *Applied Microbiology and Biotechnology* **22**, 297 (1985).
63. C. Verduyn, J. Frank, J. P. van Dijken, and W. A. Scheffers, "Multiple Forms of Xylose Reductase in *Packysolen tannophilus* CBS4044," *FEMS Microbiology Letters*, **30**, 313 (1985).
64. P. L. Bolen, J. A. Bietz, and R. W. Detroy, "Aldose Reductase in the Yeast *Pachysolen tannophilus*: Purification, Characterization and N-Terminal Sequence," *Biotechnology and Bioengineering Symposium* **15**, 129 (1985).
65. P. M. Bruinenberg, P. H. M. de Bot, J. P. van Dijken, and W. A. Scheffers, "NADH-Linked Aldose Reductase: The Key to Anaerobic Alcoholic Fermentation of Xylose by Yeasts," *Applied Microbiology and Biotechnology* **19**, 256 (1984).
66. C. Verduyn, R. van Kleef, J. Frank, H. Schreuder, J. P. van Dijken, and W. A. Scheffers, "Properties of the NAD(P)H-Dependent Xylose Reductase from the Xylose-Fermenting Yeast *Pichia stipitis*," *Biochemical Journal* **226**, 669 (1985).
67. N. J. Alexander, "Acetone Stimulation of Ethanol Production from D-Xylose by *Pachysolen tannophilus*," *Applied Microbiology and Biotechnology* **25**, 203 (1986).
68. R. Maleszka and H. Schneider, "Involvement of Oxygen and Mitochondrial Function in the Metabolism of D-Xylulose by *Saccharomyces cerevisiae*," *Archives of Biochemistry and Biophysics* **228**, 22 (1984).
69. E. Johannsen, L. Eagle, and G. Bredenhann, "Protoplast Fusion Used for Construction of Presumptive Polyploids of the D-Xylose Fermenting Yeast *Candida shehatae*," *Current Genetics* **9**, 313 (1985).
70. A. S. Gupthar and H. M. Garnett, "Hybridization of *Pichia stipitis* with Its Presumptive Imperfect Partner *Candida shehatae*," *Current Genetics* **12**, 199 (1987).
71. T. W. Jeffries, "Emerging Technology for Fermenting D-Xylose," *Trends in Biotechnology* **3**, 208 (1985).

NANCY J. ALEXANDER
NCAUR/USDA
Peoria, Illinois

GRAINS AND PROTECTANTS

Seven protectants are available for use as a postharvest application to grains in the United States: malathion, Reldan, Actellic, Diacon, Dipel, synergized pyrethroids, and diatomaceous earth (Table 1). Three of these are the most commonly used, but not all can be applied to all types of grain produced.

Malathion, an organophosphate (OP) insecticide, has been labeled for application to grain since 1957. It can be applied to all major grain commodities in addition to some minor crops (1). The label rate of application for stored grains is 10.4 ppm and the established EPA tolerance is 8.0 ppm. In addition to its use as a grain protectant, use patterns in crop production and vector insect control exist.

Acceptable daily intake (ADI) calculations are used by the EPA to regulate nononcogenic compounds. ADI is an estimate of the daily exposure dose that is likely to be consumed without deleterious effect even if continued exposure occurs over a lifetime. Intake of commodities of particular pesticides are represented as percentages of the ADI. Although in future actions by the EPA, ADI will be referred to as reference dose (rfd), it is not used at this writing because no EPA documents have used this terminology on pesticide registrations. The total ADI for all registered uses of an active ingredient should not exceed 100%. For Malathion the calculated ADI is 507% for the general population (1). In addition, in the most susceptible age group (nonnursing infants aged one to six years) the ADI exceeds 1,100%. Malathion, with its broad use patterns, was registered prior to the EPA's use of ADI as a regulatory tool. Many of the uses of Malathion were eliminated during the reregistration process and the application to grain was withdrawn. One factor influencing this decision was extensive documentation of stored grain insect resistance to malathion (2–10).

In 1985, almost 30 years after the introduction of organophosphates as protectants, Reldan (chlorpyrifos-methyl) received a registration for direct application to grain (1). This registration required expenditures in excess of $5 million to satisfy data requirements. Reldan can be applied to all major grains except corn. Corn will be included after some additional studies requested by the EPA are completed. It is important to note the only use pattern for Reldan in the United States is as a protectant for grain (including bin and warehouse application). A similar situation exists on another OP, Actellic (pirimiphos-methyl). Neither of these two compounds exceed the 100% calculated ADI value.

Actellic may be applied to corn and milo (sorghum) only. It was registered in 1986 (11). It is anticipated that use on other crops will be expanded following submission of additional data to the EPA.

Synergized pyrethrums (naturally occurring) as well as diatomaceous earth have been registered as a protectants for many years, but have not been used extensively due to limited availability, high cost, and physically irritating dermal effects to exposed workers. Generally these are used as preventative surface applications.

Diatomaceous earth is a siliceous earth composed of the cell walls of diatoms. This insecticide acts by removing the waterproof layer, or exoskeleton of the insect (12). This causes continuous loss of lipids from the exoskeleton, thereby allowing body fluids to evaporate, which leads to death. Its only insecticidal form is a dust, which is capable of causing pulmonary fibrosis in humans.

Synergized pyrethrums have a high acute toxicity to insects but low toxicity to mammals. These compounds are extracted from chrysanthemum flowers and have an instantaneous knockdown but the insect recovers from the initial toxic effect (13). To counter the recovery of the insect, piperonyl butoxide is added as a synergist to continue the blockage of nerve impulses of the insects. Pyrethrums are limited in supply and expensive compared to synthetic protectants. This explains why there is such limited use of them as a grain protectant.

A biological compound, Dipel (*Bacillus thuringiensis*) was registered for use on grain and other stored commodities in 1977. *Bacillus thuringiensis* is a naturally occurring pathogen isolated from insects and is exempt from tolerance and ADI consideration (1). The current rate of application calculates to 3.17 ppm with respect to traditional pesticide tolerances, which equals 46.02 billion International Units of *Bacillus thuringiensis* per million

Table 1. Protectants, United States, Raw AG Commodities

Product	Crops	Label Rate (ppm)	Tolerance (ppm)	ADI, Percent
Malathion, 1957	Barley, corn, oats, wheat, sorghum rice, rye, sunflowers, and almonds	10.4	8.0	507 1,133 (1–6)
Reldan, 1985	Wheat, milo (sorghum), rice, barley, and oats	6.0	6.0	< 100 < 100 (1–6)
Actellic, 1986	Corn and sorghum (milo)	6.0–8.0	8.0	< 100 < 100 (1–6)
Diacon, 1988	Corn, wheat, grain, sorghum (milo) barley, rice, oats, and peanuts	5.0	5.0	< 100 < 100 (1–6)
Dipel, (b.t.) 1977	Grain, soybean, seed, popcorn, birdseed, herbs, and spices	3.17	Exempt 46.02	N/A BIU
Synergized pyrethroids	Wheat, corn, rye, and sorghum (milo)	3 + 20	3 + 20	< 100
Diatomaceous earth	Grains	N/A	N/A	N/A

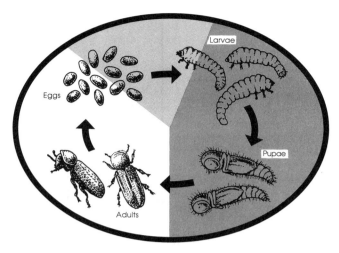

Figure 1. Effects of IGR's on growth stage of stored product insects (shown in shaded areas).

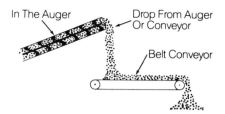

Figure 2. Suggested points at which chemical may be applied.

pounds of grain. The grain surface application is for Lepidoptera pests (Indian meal moth, Angoumois grain moth, and Mediterranean flour moth) because this is the only group of grain pests that current *Bacillus thuringiensis* strains affect. In less than 10 years, resistance had been documented to Lepidoptera pests (14).

Diacon (methoprene), which is an insect-grown regulator (IGR), interrupts the molting process of insects and received registration in 1988 (1). Diacon has a low mammalian toxicity and a tolerance of 5 ppm on all of the major grain crops as well as uses in vector and household insect control. As indicated, Diacon affects insect molting and egg hatch in some species (Fig. 1), which are unique to the immature state of insects. This product does not eliminate existing adults, but has a secondary effect on the F_1 generation by preventing population explosions of most stored grain insects. The rice weevil and granary weevil are not affected by the compound; therefore, it may be necessary to apply it in combination with other protectants to achieve good protection when weevils are present.

In addition to protectants for direct application to grain, some products exist for application to grain storage structures only (15). These include methoxychlor (bin spray only), Resmethrin and piperonyl butoxide (crack and crevice spray), and Vapona (DDVP) (Table 2). In countries other than the United States, additional products are registered for use as grain products and include the following: carbaryl (Sevin), Bromophos, Fenitrothion, Bioresmethrin and PBO, Etremphos, Methromphos, Deltamethrin, and Permethrin (16).

Research has developed pheromones, attractants, repellents, and traps to detect insects in grain (17). This new technology is much more sensitive to low insect population and detects insects when traditional grain trier and sieve monitoring reveal no insects in the grain. These new methods of detection can assist in the proper timing of protectant and fumigant application and can help grain storers to avoid insect damage.

Currently, protectants formulated for direct application to grain are emulsifiable concentrates (EC) or ready to apply as dry material. The ECs are diluted in water or FDA approved mineral or soybean oil. Liquid solutions are applied through gravity-flow systems or pressurized systems on farms. In commercial facilities pressurized pump systems are used to treat from 35.25 to 7,048 m^3/h (1,000 to 200,000 bu/h) (Fig. 2).

The dry formulations are metered into the auger with mechanical applicators or dispersed on the grain surface and cut into the rain with a scoop before unloading from a truck. As the grain is augured to the bin, distribution throughout the grain is adequate for intended protection. Only one application of a protectant is suggested to avoid over tolerance applications of protectants.

The results from previous studies on failures with protectants has been discussed (18). Proper application as close to the entry into the storage bin as possible is important to avoid losses of protectants during handling. Excessive grain moisture and temperature also effect protectant performance. With malathion, "a 10 parts per million treatment on grain at 10% moisture content and 15°C degrades to 6.3 parts per million in one year. The same treatment on 14% moisture grain at 26°C degrades to 0.2 parts per million in one year" (18). A natural degradation of OPs occurs after application, which adds to the food safety factor on raw agricultural commodities. Because all evaluations for protectant registrations are made at the full application rate, this also reduces the probability of any effects of protectants at the end use of the treated commodity.

To illustrate the challenge of maintaining grain quality, a study on granary weevils (Fig. 2) indicates the reproductive potential of 30 adults can multiply to over 11,000 adults in six months (19). In the same report using the rice weevil (Fig. 3), the influence of temperature and

Table 2. Bin, Crack, and Crevice Products, United States

Product	Use	Tolerance (ppm)
Methoxychlor	Bin	2
Resmethrin	Bin, crack, and crevice; space	3
Piperonyl butoxide	Crack and crevice; space	20
DDVP	Space	0.5

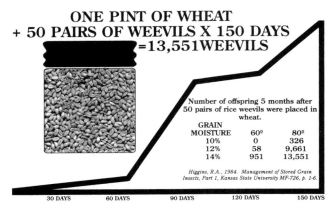

Figure 3. Number of offspring 5 months after 50 pairs of rice weevils were placed in wheat.

moisture is illustrated with low moisture having a negative population effect and high moisture a positive effect. In comparison low and high temperatures have a negative effect on populations of the rice weevil.

Data presented in one report (Table 3) show the population growth potential of one female lesser grain borer is 0.5 million adults in six months (20). In studies conducted at the Gustafson Seed and Grain Technology Laboratory in 1987–1988 (Fig. 4) the percentage of insect-damaged kernels is directly related to the insect population-to-grain ratio. This study showed that with 2 insects per 200 g and 50 insects per 200 g of wheat grain, the percent of insect-damaged kernels (IDK) was 18 and 97%, respectively.

In May 1988 new Federal Grain Inspection Service (FGIS) grading standards went into effect (21). These new standards make it even more critical that grain be in good condition when it is marketed. The more stringent standards mean that in wheat, rye, and triticale, one live weevil and one or more other live insects (OLI) injurious to stored grain in a sample can cause that load of grain to be graded infested (Table 4). In corn, barley, oats, sorghum, soybeans, sunflower seed, or mixed grain, the infested grade is two or more live weevils or one live weevil and five or more OLI injurious to stored grain, or 10 or more OLI injurious to stored grain (Table 4).

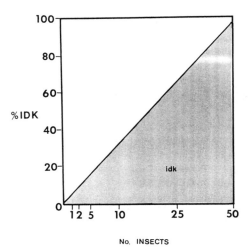

Figure 4. Relationship of insect numbers and % insect damaged kernels (IDK).

An additional FGIS revision established an IDK limit of 32 damaged kernels per 1,000 g of wheat. Prior to the FGIS revisions when IDK was not included as a part of the grade, a load of wheat could have significant damage and still have graded U.S. #1. Under the new standards if a sample (100 g) reveals 32 or more IDK, the load is graded U.S. Sample Grade wheat. The grading agency then notifies the FDA and the load or lot is declared unfit for human consumption, forcing it to be marketed as feed grain or for alcohol production. This market can be from $0.20 to $1.50 per bushel less than food-grade wheat.

With the possibility of so many insects infesting grain, the complex life cycles of insects (Table 5), different adult life spans, and reproductive potentials, it is probable the presence of insects in grain and their damage is so prevalent that it insure grain quality, growers and grain storage agents must move toward a complete grain-management program. An integrated pest management (IPM) program must include

1. High grain standards (recent FGIS changes have increased the awareness of the need for better management).

Table 3. Factors for Stored-Grain Insects[a]

Insect	Total Adult Life	Total Number of Eggs Layed	Life Cycle (Days)
Confused flour beetle	365	450	40
Flat grain beetle	365	240	52
Lesser grain borer	180	500	30
Rice weevil	185	400	30
Granary weevil	240	50	40
Saw-toothed grain beetle	300	280	50

Ref. 20.

Table 4. Insect Infestation Tolerance Comparison[a]

Grain	Insect Standard
Corn, sorghum, soybeans, barley, oats, sunflowers, mixed grain	Two or more live weevils,[b] or one weevil and five or more other live insects,[c] or ten or more other live insects
Wheat, rye	Two live insects[d]

[a] This table illustrates the number of live insects that, if detected in a sample, will result in an infested grade under the FGIS insect-infestation standards.
[b] Weevils are rice, granary, maize, and cowped weevils as well as the lesser grain borer.
[c] Other live insects include the grain and flour beetles, moths, and vetch bruchids.
[d] Live insects are weevils plus other live insects.

Table 5. Stored-Grain Insects—Life Cycles and Characteristics

Insect	Life Cycle, Egg to Egg	Adult Life	Number of Eggs	Penetrate Packages	Fly
Confused flour beetle	40	365	450	Yes	No
Flat grain beetle	45	365	240	Yes (weak)	Yes
Lesser grain borer	30	180	300–500	Yes	Yes
Rice weevil	30	90–185	300–400	Yes (weak)	Yes
Granary weevil	30–50	210–240	30–50	Yes (weak)	No
Saw-toothed grain beetle	20–80	135–300	280	Yes (weak)	No

2. Sanitation and preventative methods.
3. Grain protectant applications.
4. The use of aeration to cool the grain mass.
5. Regular inspection and monitoring of the grain.
6. Control of insects with fumigants when populations become high from immigration or poor prevention.
7. A good marketing system that offers a realistic profit and quality product.

Since the implementation of the FGIS changes, growers and commercial grain stores are becoming more involved in IPM programs and have taken steps to reduce insect populations instead of reacting after large populations are detected and IDK are present. Protectants are a vital part of the total effort to maintain grain quality from harvest through storage and to its end use.

BIBLIOGRAPHY

General References

1. Environmental Protection Agency, "*Bacillus thuringiensis Berliner,* Viable Spores for Residues," *Federal Register* 40 *CFR*, part 180.1011, 1988, p. 456; "Established Tolerances for Residue of the Insecticide Methoprene," *Federal Register* V 53–139, 1988, pp. 27391–27392; "Guidance for the Reregistration of Pesticide Products Containing Malathion as the Active Ingredient," Case-248, p. 315; "Malathion Tolerances for Residues," *Federal Register* 40 *CFR*, part 180.111, pp. 333–335; "Pirimiphos-methyl, Final Rule," *Federal Register,* V.1–51, pp. 28223–28229.
2. P. C. Bansode and W. V. Campbell, "Evaluation of North Carolina Field Strains of the Red Flour Beetle for Resistance to Malathion and Other Organophosphorus Compounds," *Journal of Economic Entomology* 72, 331–333 (1979).
3. R. W. Beeman, W. E. Speirs, and B. A. Schmidt, "Malathion Resistance in Indian Meal Moths Infesting Stored Corn and Wheat in the Northcentral United States," *Journal of Economic Entomology* 75, 950–954 (1982).
4. E. A. Parking, "The Onset of Insecticide Resistance Among Field Populations of Stored-Product Insects," *Pyretholds,* Vol. 1, Academic Press, Inc. Orlando, Fla, 1965, pp. 3–8.
5. I. C. Pasalu, G. K. Girish, and K. Krishnamurthy, "Status of Insecticide Resistance in Insect Pests of Stored Products," *Bull. Grain Technol.* 12, 50–59 (1974).
6. R. D. Speirs, L. M. Redlinger, and H. P. Boles, "Malathion Resistance in the Red Flour Beetle," *Journal of Economic Entomology* 60, 1373–1374 (1967).
7. C. L. Storey, D. B. Sauer, and D. Walker, "Insects and Fungi in Wheat, Corn and Oats Stored on the Farm," *Journal of Economic Entomology* 77, 784–788 (1984).
8. R. G. Strong, D. E. Sbur, and G. J. Partida, "The Toxicity and Residual Effectiveness of Malathion and Diazinon Used for Protection of Stored Wheat," 60, 500–505 (1967).
9. L. E. Vincent and D. L. Lindgren, "Susceptibility of Laboratory and Field-Collected Cultures of the Confused Flour Beetle and Red Flour Beetle to Malathion and Pyrethrins," *Journal of Economic Entomology* 60, 173–174 (1967).
10. J. L. Zettler, "Malathion Resistance in *Tribolium castaneum* Collected from Stored Peanuts," 67, 339–340 (1974); "Insecticide Resistance in Selected Stored-Product Insects Infesting Peanuts in the Southeastern United States," 75, 359–362 (1982).
11. Environmental Protection Agency, "Chlorpyrifos-Methyl Tolerances for Residues," *Federal Register* 40 *CFR*, part 180.419, pp. 426–427.
12. J. L. Zitter, and L. M. Redlinger, "Anthropod Pest Management with Residual Insecticides," in F. J. Baur, ed., *Insect Management for Food Storage and Processing,* American Association of Cereal Chemists, St. Paul, Minn., 1984, pp. 111–130.
13. R. D. Obrien, *Insecticide Actions and Metabolism; Pyrethroids,* Academic Press, Inc., Orlando, Fla., 1967, pp. 168–171.
14. W. H. McGaughey, "Resistance of Storage Pests to *Bacillus thuringiensis,*" paper presented at the Eighteenth International Congress of Entomology, 1988.
15. J. G. Touhey, "Future Development and Availability of Pesticides for Grain: How Pesticides are Registered by EPA," paper presented at USDA-FGIS Grain Insect Interagency Task Force Meeting, July 21, 1988, Washington, D.C.
16. D. R. Wilkin, unpublished data, 1988.
17. W. E. Burkholder, "Stored-Product Insect behavior and Pheromone Studies: Keys to Successful Monitoring and Trapping," in *Proceedings of the Third International Working Conference on Stored-Product Entomology,* 1984, pp. 20–23.
18. P. K. Harein, "Effective Control of Insects in Stored Grain," paper presented at the Fifty-Ninth Annual International Conference of the Grain Elevator and Processing Society, 1988.
19. R. A. Higgins, unpublished data, 1984.
20. F. J. Bauer, *Insect Management for Food Storage and Processing,* 1984, pp. 54–67.

21. U.S. Department of Agriculture, Federal Grain Inspection Service, "Grain Standards: Official U.S. Standards, Handling Practices and Insect Infestation, Final Rules," *Federal Register* V.52 No. 125, 1988, pp. 2424–2442.

J. Terry Pitts
Gustafson, Inc.
Plano, Texas

GRAIN SPIRITS. See Distilled beverage spirits.
GRAS. See Food laws and regulations; Food additives, Food toxicology; Toxicology and risk assessment; and entries under Genetic engineering.

GUMS

Food gums are water-soluble or -dispersible polysaccharides (glycans) and their derivatives (see Carbohydrates) and gelatin (Refs. 1–14). In general, they thicken or gel aqueous systems at low concentration. Although starches and starch derivatives have these properties, they are most often considered in a separate category (see Starch). Polysaccharide food gums can be classified by source (Table 1) or by structure (see Carbohydrates).

The usefulness of gums is based on their physical properties, in particular their capacity to thicken and/or gel aqueous systems and otherwise control water. Because all gums modify the flow of aqueous solutions, dispersions, and suspensions, the choice of which gum to use for a particular application often depends on its secondary characteristics. These secondary characteristics are responsible for their utilization as binders, bodying agents, bulking agents, crystallization inhibitors, clarifying agents, cloud agents, emulsifying agents, emulsion stabilizers, encapsulating agents, film formers, flocculating agents, foam stabilizers, gelling materials, mold release agents, protective colloids, suspending agents, suspension stabilizers, swelling agents, syneresis inhibitors, texturing agents, and whipping agents, in coatings, and for water absorption and binding. Most gums are not true emulsification agents but stabilize emulsions and suspensions by increasing the viscosity and the yield value of the aqueous phase. Some also exhibit protective colloid action.

Food gums are tasteless, odorless, colorless, and nontoxic. All (except the starches, starch derivatives, and gelatin) are essentially noncaloric and are classified as soluble fiber (see Fiber, dietary).

Although we often speak of gum solutions, gums, in general, do not form true solutions because of their large molecular size and intermolecular interactions. Rather, they form molecular dispersions; hence, the term hydrocolloid is often used interchangeably with gum. As a result, the rheology (flow characteristics and gel properties) of gum solutions is a function of particle size, shape, flexibility, ease of deformation and solvation, and the presence and magnitude of charges. Particles may be dispersed molecules and/or aggregated clusters of molecules. In general, gum solutions usually exhibit shear thinning, ie, are pseudoplastic, occasionally thixotropic. The variables that affect their rheology are polymer structure, molecular weight, concentration, shear rate, temperature, pH, and the concentration of salts, other solutes, and sequestrants. Factors that affect dispersion and dissolution are pH, presence of salts, presence of other solutes, gum type, particle size, physical form of particles, shear rate, and method of dispersion (mixing efficiency).

Most gums are available in a wide range of viscosity grades. If viscosity is the goal, a high-viscosity grade at low solids concentration is generally used; if binding or protective colloid action, for example, is the goal, a low-viscosity grade at high solids concentration is generally used.

In general, food gum gels are composed of 99.0–99.5% water and 0.5–1.0% gum. Important characteristics of gels are means of gelation (chemical gelation, thermal gelation), reversibility, texture (brittle, elastic, plastic), rigidity (rigid or firm, soft or mushy), tendency for syneresis, and cutable or spreadable.

Classes of gums that are the principal ones used in foods in the United States are described below. Also described are gellan gum and konjac mannan, only recently introduced in the United States, furcellaran, used primarily in Europe, and hydroxypropylcellulose, which finds only limited use. Not included are the seldom used gum ghatti and gum karaya; tara gum, a galactomannan similar to guar and locust bean gums; fermentation gums used in Japan but not in the United States, namely, curdlan, pullulan, and scleroglucan; nor microcrystalline cellulose, a non-water-soluble polysaccharide and, hence, not a true gum (see Carbohydrates). For β-glucan, see Fiber, dietary.

Table 1. Classification of Polysaccharide Food Gums by Source

Class	Examples
Seed gums	Corn starch, guar gum, locust bean gum
Tuber and root gums	Potato starch, tapioca starch, konjac mannan
Seaweed extracts	Algins, carrageenans, agar
Plant extracts	Pectins
Exudate gums	Gum arabic, gum tragacanth, gum ghatti, gum karaya
Fermentation (microbial) gums	Xanthan, gellan, curdlan, scleroglucan, pullulan
Derived gums	Carboxymethylcelluloses, hydroxypropylmethylcelluloses, methylcelluloses, starch acetates, starch phosphates, hydroxypropylstarches, dextrinized starches

ALGINS

Algins (alginates) (1,2) are extracted from brown algae (*Phaeophyceae*). Algins are salts (generally sodium) or esters (propylene glycol) of alginic acid. Alginic acid is a generic term for polymers of D-mannuronic acid and L-guluronic acid (see Carbohydrates). In alginic acid molecules, at least three different types of polymer segment exist: poly(β-D-mannopyranosyluronic acid) segments, poly(α-L-gulopyranosyluronic acid) segments, and segments with alternating sugar units. The ratios of the constituent monomers and the chain segments vary with the source. Alginates are linear polymers. The degree of polymerization (molecular weight) is controlled and varied in commercial products.

The specific properties of algins depend on the percentage of each type of building block. An important and useful property of sodium, potassium, and ammonium alginates is their ability to form gels on reaction with calcium ions. Different types of gels are formed with alginates from different sources. Alginates with a higher percentage of poly(guluronic acid) segments form the more rigid, more brittle gels that tend to undergo syneresis. Alginates with a higher percentage of poly(mannuronic acid) segments form the more elastic, more deformable gels that have a reduced tendency to undergo syneresis.

Algins are most often used as thickeners and are available in a number of viscosity grades. Algin solutions are pseudoplastic and exhibit shear thinning over a wide range of shear rates. Solutions of propylene glycol alginate are somewhat thixotropic and are much less sensitive to pH and polyvalent cations. The specific properties exhibited by an algin solution depend on the ratio of monomeric units, the concentration and type of cations in solution, the temperature, and the degree of polymerization.

Sodium alginate is used as a coating for frozen fish to prevent moisture loss and freezer burn, in fountain syrups, in dry soup and sauce mixes, and in meringues to prevent syneresis. Calcium alginate/alginic acid gels are found in structured foods, such as fruit pieces, onion rings, pimientos for green olives, shrimp pieces, and jelly-type bakery fillings. Propylene glycol alginate (which can be labeled algin derivative) is used as a beer foam stabilizer; as a stabilizer in cottage cheese, cheesecake, pourable salad dressings, buttered pancake syrups, tartar sauces, sandwich spreads, and chocolate milk drink mixes; as a thickener for low-pH syrups; and in relishes to hold water. Alginic acid is used to prepare soft, thixotropic, nonmelting dessert gels, tomato aspics, and pie fillings.

CARBOXYMETHYLCELLULOSE (CMC)

Carboxymethylcellulose (CMC) (1,2) is the sodium salt of the carboxymethyl ether of cellulose (see CARBOHYDRATES). In an ingredient list, it can be designated cellulose gum, CMC, sodium CMC, sodium carboxymethyl cellulose, or carboxymethyl cellulose.

Modified polysaccharides, such as CMC, have a degree of substitution (DS). There are, on the average, three hydroxyl groups per polysaccharide hexopyranosyl unit. (In the case of cellulose, each β-D-glucopyranosyl unit has three hydroxyl groups; see (CARBOHYDRATES). The average number of hydroxyl groups derivatized (with a carboxymethyl ether group in the case of CMC) per monomeric unit is the DS; the maximum DS of a cellulose derivative is 3.0. The solution characteristics of CMC are primarily determined by the DS, the average chain length or degree of polymerization (DP), and the uniformity of substitution. The most widely used types have a DS of 0.7 or an average of 7 carboxymethyl groups per 10 β-D-glucopyranosyl units. Many viscosity grades of CMC are manufactured. (see Fig. 1).

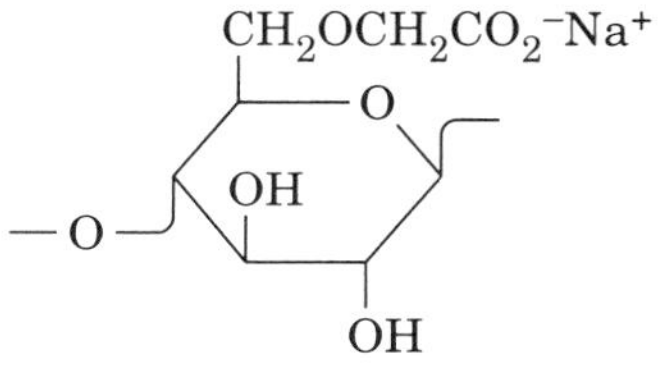

Figure 1. A representative unit of sodium carboxymethylcellulose, ie, a representative β-D-glucopyranosyl unit containing a carboxymethyl ether group.

Carboxymethylcellulose hydrates rapidly and forms clear, stable solutions. Viscosity building is its most important property. Solutions can be either pseudoplastic or thixotropic depending on the type of product used. They are stable over a wide pH range (pH 4–10). As the pH drops below 4, the viscosity increases significantly.

Carboxymethylcellulose helps to solubilize various proteins and to stabilize their solutions. It has a synergistic viscosity building effect when used with casein, soy protein, guar gum, and hydroxypropylcelluloses.

Carboxymethylcellulose is the most extensively used of the food gums. It is used, for example, as a retarder of ice crystal growth in ice creams and related products (its principal use); as a physiologically inert and noncaloric thickening and bodying agent in dietetic foods; in cake and related mixes to hold moisture, enhance organoleptic properties, and extend shelf life; as an extrusion aid; in batters for viscosity control, adhesion, and suspension stabilization; in icings, frostings, and glazes to hold moisture and retard sugar crystallization; as a thickener in pancake syrup; as a suspending agent and suspension stabilizer in hot cocoa and fruit drink mixes; as a water binder in slightly moist pet foods; and in many other areas.

CARRAGEENANS, AGAR, AND FURCELLARAN

Carrageenan is a generic term applied to polysaccharides extracted from a number of closely related species of red seaweeds; agar and furcellaran are also red seaweed extracts and are members of the same larger family (1,2). All polysaccharides in this family are derivatives of linear galactans; all have alternating monosaccharide units and linkages. In agar, the monosaccharide units are D-galactopyranosyl and 3,6-anhydro-L-galactopyranosyl units. In furcellaran and κ- and ι-carrageenans, they are D-galactopyranosyl and 3,6-anhydro-D-galactopyranosyl units. λ-Carrageenan contains only D-galactopyranosyl units. Agar contains little or no sulfate half-ester groups. Other members of the family are sulfated: in increasing order, furcellaran ($\sim$14%), κ-($\sim$25%), ι-($\sim$30%), λ-($\sim$35%) carrageenan. Each polymer type is heterogeneous; ie, none contains an exact repeating unit structure. Because of the general structures, agar is the least soluble of this family of polysaccharides and λ-carrageenan the most soluble. Agar forms the strongest gels. λ-Carrageenan does not gel; it is the only member of this family that is nongelling. (See Fig. 2).

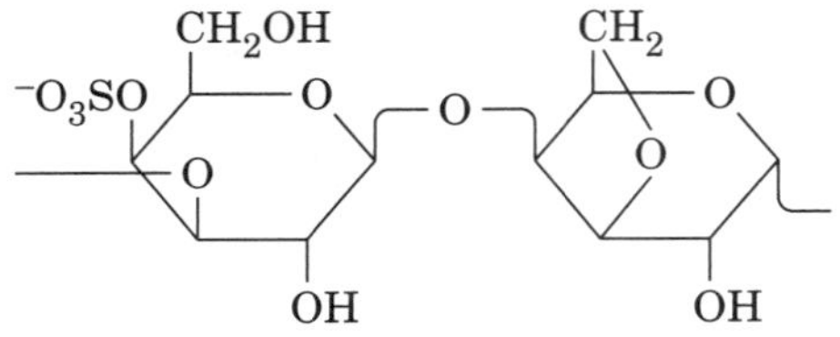

Figure 2. A representative disaccharide unit of κ-carrageenan.

Commercial carrageenans are composed primarily of three types of polymer: κ-, ι-, and λ-carrageenans. The composition and properties of any preparation is dependent on the species collected, growth conditions, and treatment during production. Products are blended and standardized with respect to any of several properties during manufacture.

The primary use of carrageenans is as a gelling or thickening agent in milk-based products. A useful property is their ability to form water and milk gels. Carrageenans are blended to provide products that will form a wide variety of gels—clear gels and turbid gels, rigid gels and elastic gels, tough gels and tender gels, heat-stable gels and thermally reversible gels, and gels that undergo syneresis and gels that do not. Carrageenan gels do not require refrigeration, do not melt at room temperature, and are freeze–thaw-stable. Carrageenans form complexes with proteins.

As pH values decrease below 6, carrageenan solutions become increasingly unstable when heated. The loss of viscosity is due to polymer chain cleavage and, hence, is irreversible.

Only the sodium salts of κ-type carrageenans are soluble in cold water. Carrageenans react with milk proteins. The thickening effect of κ-carrageenan in milk is 5–10 times greater than it is in water; at a concentration of 0.025% in milk, a weak thixotropic gel is formed via the interaction of κ-carrageenan with κ-casein micelles. This property finds use in the preparation of chocolate milk (to keep the cocoa suspended) and ice cream (to prevent whey-off). Other (stabilizing) interactions are important in producing evaporated milk, infant formulas, and whipped cream required to be freeze–thaw-stable.

There is a synergistic effect between κ-carrageenan and locust bean gum; the two gums together produce a much more elastic gel with markedly greater gel strength and less syneresis. Such gels find use in canned pet foods. The κ-carrageenan/locust bean gum mixture is also used to provide body and fruit suspension in yogurt. The synergistic interaction of κ-carrageenan with konjac mannan is even greater.

ι-Type carrageenans are a little more soluble than are the κ-type carrageenans; yet still only the sodium salt form is soluble in cold water, and other salt forms are soluble only at temperatures above 71°C (160°F). Dispersions of ι-carrageenan are thixotropic. ι-Carrageenan forms elastic, syneresis-free, thermally reversible gels that are stable to repeated freeze–thaw cycling. It is used in ready-to-eat milk gels and, when blended with κ-carrageenan, in water-dessert gels (where refrigeration is unavailable), whipped toppings, and cooked flans. Various formulations for both cooked and cold-prepared eggless custards incorporate blends of carrageenans.

All salts of λ-type carrageenans are soluble. λ-Carrageenan is nongelling and is used as an emulsion stabilizer in such products as whipped cream, instant breakfast drinks, milkshakes (κ/λ mixture), and nondairy coffee creamers.

Agar (agar-agar) is the least soluble of this class of polysaccharides; it can be dispersed only at temperatures above 100°C (212°F). When its dispersions are cooled, strong, brittle, turbid gels form. Agar gels remelt when heated, undergo syneresis, and are unstable to freeze–thaw cycles. Agar is used in bakery icings, no-oil salad dressings, and low-fat yogurt.

Furcellaran (Danish agar) is used primarily in Europe. The properties of furcellaran gels lie between those made with carrageenans and those made with agar, most closely resembling those made with κ-carrageenan. Furcellaran forms a weak gel with calcium ions. The gels are somewhat opalescent and thermoreversible. Furcellaran interacts with proteins and will form milk gels. Use of locust bean gum together with furcellaran produces a more elastic gel with higher gel strength and less syneresis. Furcellaran is used primarily in milk puddings and eggless custards and in cake-covering jellies, eg, flan jelly, Tortenguss, apricoture, and nappage.

GELATIN

Gelatin is a protein rather than a polysaccharide. It is produced by acid (type A) or alkali (type B) treatment of collagen obtained from pig skin, cattle hides, and bones. While gelatin contains 18 amino acids, about 95% of the structure is composed of 7. The amino acid sequence of most of the chain is a tripeptide repeating unit of (glycine)—(proline or hydroxyproline)—(alanine, glutamic acid, aspartic acid, or arginine). Gelatin contains no tryptophan.

The important functional properties in food applications of gelatin are gel strength, gelling temperature, setting rate, pH, and isoelectric pH. Typical values for some of these properties are

Type	Gel Strength, g Bloom	pH	Isoelectric pH
A	50–300	3.8–5.5	7.0–9.0
B	50–275	5.0–7.5	4.7–5.0

The presence of ash, sulfur dioxide, and peroxides can affect product appearance, ie, cause gel cloudiness and/or discoloration, but not functional properties.

The gelling temperature of gelatin solutions is related to concentration.

Below pH 5, types A and B gelatins have a net positive charge and will interact with negatively charged polymers such as agar, carrageenans, algins, low-methoxyl (LM) pectins, and gum arabic. With gum arabic, gelatin forms a well-known coacervate.

Although gelatin could be used in a variety of food products, its use is essentially limited to preparation of meltable-in-the-mouth water-dessert gels and aspics, jellied meats, and marshmallows (foam-maker). It can also be used to form emulsions; as an ice cream stabilizer; as a clarifying agent for wines, vinegar, and ciders and other fruit juices; and as a fixing agent for flavors and vitamins. It is not used as a thickener.

GELLAN GUM

Gellan gum is the deacylated form of an extracellular, microbial polysaccharide (gellan) produced by *Pseudomonas elodea*. It is composed of a linear tetrasaccharide repeating unit of D-glucose–D-glucuronic acid–D-glucose–L-rhamnose. Cooling of hot solutions of the commercial, low-acyl gellan (gellan gum) forms firm and brittle gels of texture similar to that of agar and κ-carrageenan gels.

Gellan gum requires cations for gelation; but unlike other ion-sensitive gelling polysaccharides, it gels with a wide variety of cations, including hydrogen ions. Divalent ions such as Ca^{2+} and Mg^{2+} are, however, much more efficient at forming gels than are monovalent ions such as Na^+ and K^+. Like agar gels, gellan gum gels exhibit marked setting and melting temperature hysteresis; and by selection of the ion and its concentration, both temperatures can be controlled. Gellan gum is functional in milk.

Gellan gum should find use in confectioneries, structured foods, pie and bakery fillings, bakery icings, canned frostings, dairy products, and pet foods.

GUAR AND LOCUST BEAN (CAROB) GUMS

Commercial guar gum is the ground endosperm of seeds of the guar plant (5). The guar plant resembles the soy plant, and guar seeds are produced much as soy beans are (1,2). Commercial guar gum usually contains 80–85% guaran, 10–14% moisture, 3–5% protein, 1–2% fiber, 0.5–1.0% ash, and 0.4–1.0% lipid.

Guaran is the purified polysaccharide from guar gum. It is composed of D-galactose and D-mannose and is, therefore, a galactomannan. It has a linear backbone chain of β-D-mannopyranosyl units, approximately every other unit of which (on the average) is substituted with an α-D-galactopyranosyl unit. The mannan chain is rather evenly substituted with D-galactopyranosyl units but still contains some unsubstituted ("smooth") regions (6).

Like guaran and the endosperm polysaccharides of other legumes, locust bean (carob) gum is also a galactomannan. Like guaran, it has a linear mannan backbone. However, in locust bean gum, only approximately one out of four β-D-mannopyranosyl units is substituted with an α-D-galactopyranosyl unit. The locust bean gum molecule contains "smooth" regions that contain no α-D-galactopyranosyl side-chain units and "hairy" regions in which most main-chain units contain single-unit α-D-galactopyranosyl branches (6).

Commercial locust bean gum is the ground endosperm of the seeds of the locust (carob) tree. The tree, which grows primarily in the Mediterranean region, is slow to mature and does not begin to bear until it is about 15 yr old. Hence, the supply of locust bean seeds is limited and is not expected to increase. Locust bean gum usually contains 80–85% galactomannan, 10–13% moisture, ~5% protein, ~2% lipid, ~1% fiber, and ~1% ash.

Another galactomannan related to guar and locust bean gums, tara gum, comes from northern regions of Africa and South America and is used to a slight extent in Europe.

Guar gum forms very high viscosity, pseudoplastic solutions at low concentrations. Because commercial guar gum contains protein, fiber, and lipids, its solutions are cloudy.

Particle size is important in the use of guar gums. Coarse granulations are used for rapid, easy dispersion; fine granulations give rapid hydration. All forms will develop some additional viscosity after their dispersions have been heated.

Guar gum is nonionic. As such, the viscosity of its solutions is not greatly affected by pH. The most rapid hydration occurs at pH 6–9.

Locust bean gum has low cold-water solubility and is generally used when delayed viscosity development is desired. Only when dispersions of locust bean gum are heated [eg, to 85°C (185°F)] and cooled is high viscosity obtained.

The general properties of locust bean gum are similar to those of guar gum. A difference is its synergism with κ-carrageenan, furcellaran, and xanthan, with which it forms gels.

Guar gum has a weak synergistic effect on the viscosity of xanthan solutions and an even weaker effect on carrageenan and carboxymethylcellulose solutions.

Guar gum is used in foods as a thickener and a binder of water. It is often used in combination with other gums, eg, with carrageenan and/or xanthan plus locust bean gum in ice creams, frozen novelties, whipped toppings, sour cream, cottage cheese, and low-fat yogurts and with xanthan in pickle relish and sauces such as pizza and Sloppy Joe sauce. Both guar gum and locust bean (carob) gum are used (usually in combination with xanthan gum) as a stabilizer for ice cream to thicken the mix, prevent heat shock and ice crystal formation, and impart smoothness and chewiness. Other applications of these blends are in the manufacture of canned frostings, potato chip dips, cream cheese, cottage cheese dressings, other processed cheese products, and starch-based gum candies. Guar gum is used in instant oatmeal and dry soup mixes, egg substitutes, dipping batters, stuffed olives, and pet foods. Locust bean gum is almost always used in combination with one or more other gums. In combination with xanthan and starch, it is used to make chewy fruit confections. Locust bean gum–carrageenan blends are used in pet food gels.

GUM ARABIC

Of the several gums that are dried, gummy exudations collected by hand from various trees and shrubs, only gum arabic, also called gum acacia and acacia gum, is still in significant use (1,2).

Gum arabic preparations are mixtures of highly branched, branch-on-branch, acidic protein-polysaccharides that occur as mixed salts. Their specific compositions and structures can vary with species, season, and climate.

Gum arabic is collected by hand from various species of *Acacia* for the most part in the region just south of the Sahara Desert. Dried tears are sorted into lots based on clarity, color, and gross impurities. Representative samples of each lot are analyzed for ash content, water-insoluble impurities, and viscosity. The lots are then cleaned, converted into grains or powders of varying mesh sizes, and blended. The most highly purified grade is produced by spray drying a clarified solution.

Gum arabic is unique among gums because of its high solubility and the low viscosity and Newtonian flow of its solutions. While most other gums form highly viscous solutions at 1–2% concentration, 20% solutions of gum arabic resemble a thin sugar syrup in body and flow properties. Solutions of gum arabic are slightly acidic (pH 4.5–5.0).

Gum arabic is an effective emulsion stabilizer because of its protective colloid action and is often used in the preparation of baker's emulsions of citrus and other essential oils. A major application is in the preparation of dry fixed flavor powders prepared by adding citrus oils and other fruit or imitation flavors to gum arabic solutions and spray drying the resulting emulsions. Some gum arabic is also used in the preparation of confectioneries and lozenges.

GUM TRAGACANTH

Gum tragacanth (1,2), the dried exudate of *Astragalus* species produced from natural or deliberately made wounds in the roots, trunks, and/or branches of the small bushes, comes for the most part from Iran and Turkey, and is relatively expensive. Its use, therefore, is quite limited.

Gum tragacanth contains two polysaccharides. One (60–70%), termed both tragacanthic acid and bassorin, only swells in water, forming a gel. Tragacanthic acid contains D-galactose, D-xylose, L-fucose, and D-galacturonic acid and covalently bound protein; the skeletal structure is that of a highly branched galactan. The minor polysaccharide is a neutral arabinogalactan in which L-arabinose is the predominant monosaccharide; it most probably consists of a core galactan chain to which highly branched arabinan chains are attached.

The most important physical properties of gum tragacanth are its relative acid stability, ability to lower surface and interfacial tensions, and hydration to a gel. For these reasons, the primary use of gum tragacanth is as an emulsifying agent and water controller in low-pH foods.

Gum tragacanth is found in some pourable salad dressings and pickle relishes and in certain bakery products.

HYDROXYPROPYLCELLULOSES

Hydroxypropylcelluloses (1,2) are cellulose ethers prepared by reacting cellulose with propylene oxide. The products are characterized in terms of moles of substitution (MS) rather than DS. MS is used because the reaction of a propylene oxide molecule with cellulose leads to the formation of a new hydroxyl group with which another alkylene oxide molecule can react to form an oligomeric side chain. Therefore, there is no limit to the moles of substituent that can be added to each D-glucopyranosyl unit. MS denotes the average number of moles of alkylene oxide that has reacted per D-glucopyranosyl unit. (See Fig. 3.)

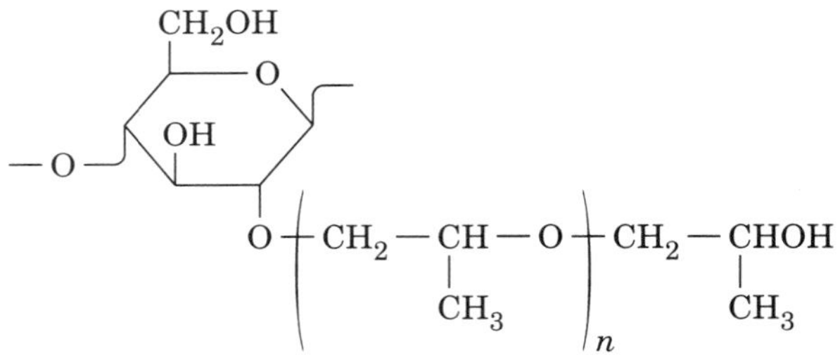

Figure 3. A representative β-D-glucopyranosyl unit of cellulose containing a hydroxypropyl group or a poly(propylene oxide) chain ($n = 0, 1, 2, 3$, etc).

In general, the MS controls the solubility of a hydroxypropylcellulose. Commercially available hydroxypropylcellulose is insoluble in hot water (>40–50°C [>105–120°F]), soluble in cold water, and compatible with several oils. It is produced in a wide range of viscosity grades and forms clear, smooth, uniform solutions. Like other cellulose gums, hydroxypropylcellulose solutions are Newtonian at low shear rates and become pseuodplastic as the shear rate is increased. The shear rate at which solutions change from being Newtonian to being pseudoplastic increases with increasing molecular weight and decreases with increasing concentration. Also, the higher-molecular-weight products are more affected by shear.

Because they are nonionic gums, hydroxypropylcelluloses are unaffected by pH. Flexible, nontacky, heat-sealable packaging films and sheets can be produced from hydroxypropylcelluloses by conventional extrusion techniques.

Hydroxypropylcelluloses, because of their ability to reduce surface and interfacial tension, are used as emulsifying agents, emulsion stabilizers, and whipping aids. Because hydroxypropylcellulose forms a strong, edible film that provides a barrier to oxygen and water vapor, it can be used to coat nuts and confections.

METHYLCELLULOSES AND HYDROXYPROPYLMETHYLCELLULOSES

Methylcelluloses contain methoxyl groups in place of some of the hydroxyl groups along the cellulose molecule; hydroxypropylmethylcelluloses contain, in addition to methyl ether groups, hydroxypropyl ether groups along the cellulose chain (1,2). The properties of methyl- and hydroxypropylmethylcelluloses are primarily a function of the amount of each type of substituent group and the molecular-weight distribution.

Methylcelluloses are made by reacting cellulose with methyl chloride until the DS reaches 1.1–2.2. Hydroxypropylmethylcelluloses are made by using propylene oxide in addition to methyl chloride in the reaction; hydroxypropyl group MS levels in commercial products are 0.02–0.3.

Members of this family of gums are cold-water soluble. Conversion of some of the hydroxyl groups of cellulose molecules into methyl ether groups increases the water solubility of the cellulose molecule and reduces its ability to aggregate, ie, reduces intermolecular interactions. Solubility and solution stability is increased even more when hydroxypropyl groups are added to methylcellulose.

The most interesting property of these nonionic products is thermal gelation. Solutions of members of this family of gums decrease in viscosity when heated, as do solutions of most other polysaccharides. However, unlike solutions of other gums, when a certain temperature is reached (depending on the specific product), the viscosity will increase rapidly and the solution will gel. Gelation can occur at various temperatures from 45 to 90°C (115 to 195°F), depending on the viscosity type, DS/MS, and proportions of methyl and hydroxypropyl substituent groups. The thermal gelation is reversible.

Members of this family are generally referred to simply as methylcellulose. At times they are identified on labels as "carbohydrate gum" (hydroxypropylmethylcellulose) and "modified vegetable gum" (methylcellulose).

There is some synergistic interaction between methylcelluloses and xanthan. Surfactants greatly increase the viscosity of methylcellulose solutions.

Methylcelluloses reduce surface and interfacial tension and can, therefore, be used to stabilize emulsions and make foams.

Methylcelluloses will form high-strength films that are clear, water-soluble, and oil- and grease-resistant and have low oxygen and moisture vapor transmission rates.

Methylcelluloses are used in the preparation of dietetic foods, in baked goods for their gluten-like properties, in dipping batters (where film formation, low oil migration, moisture retention, thermal gelation, and tackiness are important properties), in whipped toppings, in frozen desserts and novelties, in canned fruit juice and fruit drink mixes (as a bodying agent, drying aid, emulsion stabilizer, and/or cloud agent), and as a binder and lubricator in extrusion processes.

KONJAC MANNAN

Konjac mannan is the principal component of konjac flour, the product of commerce. Konjac flour, the powdered tuber (2–3 yr of age) of *Amorphophallus konjac*, is used in several traditional Japanese foods, primarily in noodles and gels.

Konjac mannan is a glucomannan. It is a linear polymer of β-D-mannopyranosyl and β-D-glucopyranosyl (ratio 1.6:1.0) units all linked (1→4). Removal of some of the naturally occurring acetate ester groups with alkali (treatment with lime water) is required for gelation.

Konjac flour hydrates only slowly in room temperature water; so unlike other gums, it disperses easily in water. When fully hydrated [90°C (195°F)], it forms solutions of high viscosity and pseudoplasticity.

Konjac mannan, by itself, when deacetylated, forms elastic gels that are stable in boiling water. In fact, the gel strength increases reversibly on heating. This accounts for its traditional use in the stabilization of noodles that can be cooked in an autoclave. In addition, konjac mannan exhibits a very strong synergism with κ-carrageenan, forming elastic, thermally reversible gels after heating and cooling. It also will form a gel via synergistic interaction with xanthan and interacts synergistically with starches to increase viscosity.

Konjac flour will form heat-stable, flexible, transparent or opaque, high-strength, protective coatings and films.

PECTINS

There is no universally agreed upon and accepted definition of pectin (1,2,7). The definitions presented here are those generally in current commercial use.

Pectic acids are galacturonoglycans [poly(α-D-galactopyranosyluronic acids)] without, or with only a negligible content of, methyl ester groups. Pectic acids may have varying degrees of neutralization. Salts of pectic acids are pectates.

Pectinic acids are galacturonoglycans with various, but greater than negligible, contents of methyl ester groups. Pectinic acids may have varying degrees of neutralization. Salts of pectinic acids are pectinates.

Pectins are mixtures of polysaccharides that originate from plants, contain pectinic acids as major components, are water-soluble, and whose solutions will gel under suitable conditions.

Pectins are subdivided according to their degree of esterification (DE), a designation of the percentage of carboxyl groups esterified with methanol. Pectins with DE >50% are high-methoxyl (HM) pectins; those with DE <50% are low-methoxyl (LM) pectins. In some LM pectins, termed amidated pectins, some carboxyl groups have been converted into carboxamide groups. The degree of amidation (DA) indicates the percentage of carboxyl groups in the amide form.

The principal and key feature of all pectin molecules is a linear chain of α-D-galactopyranosyluronic acid units. In all commercial pectins, some of the carboxyl groups are in the methyl ester form; some of or all the remaining carboxylic acid groups may be in a carboxylate salt form. In amidated pectins, some will be in the amide form. The DE strongly influences the solubility, gel-forming ability, conditions required for gelation, gelling temperature, and gel properties of the preparation.

Pectins are soluble in hot water. The importance of pectin is predominantly the result of its unique ability to form spreadable gels when a hot solution is cooled. HM pectin gels are formed by the addition of sugar to a hot, acidic (pH ~3) solution of pectin in a fruit juice. The pH of greatest stability of pectin is ~4.

LM pectins will gel only in the presence of calcium ions. Increasing the concentration of calcium ions increases the gelling temperature and gel strength.

Any system containing pectin at potential gelling conditions (ie, necessary concentration of an appropriate pectin, pH, concentration of cosolutes, and concentration of divalent cations) must be prepared at a temperature above the gelling temperature. The temperature at which gelation occurs on cooling is the gelling temperature.

The primary use of pectins is in the preparation of spreadable gels (jams, jellies, marmalades, and preserves, including the preserves in fruit yogurt and the filling for chocolates) from fruit juices or whole fruits, with or without sugar. Pectins are also used in preparation of canned fruit juices, cheese spreads, frostings, low-calorie salad dressings, and chewable fruit candies.

XANTHAN

Xanthan (1,2) (also known as xanthan gum) is a widely and extensively used food gum. It is the extracellular (exocellular) polysaccharide produced by *Xanthomonas campestris*. Its characteristics vary with variations in the strain of the organism and the fermentation conditions used.

Xanthan has a linear main chain that has the same structure as does cellulose (see CARBOHYDRATES). In xanthan, every second β-D-glucopyranosyl unit of the cellulosic backbone is substituted with a trisaccharide unit. About half of the trisaccharide side-chain units carry a terminal pyruvic acid cyclic acetal group.

Xanthan solutions are extremely pseudoplastic and have high yield values. These properties make xanthan almost ideal for the stabilization of aqueous dispersions, suspensions, and emulsions.

While other polysaccharide solutions decrease in viscosity when they are heated, xanthan solutions containing a small amount of salt ($\geq 0.1\%$) change little in viscosity over the temperature range 0–95°C (32–205°F).

Although xanthan is anionic, pH has almost no effect on the viscosity of its solutions.

A synergistic viscosity increase results from the interaction of xanthan with κ-type carrageenans, methylcelluloses, locust bean gum, and konjac mannan. The latter two combinations form thermally reversible gels when hot solutions of these polysaccharides are cooled.

Its temperature stability, pH stability, ability to improve gloss (sheen) and retain moisture, and bland flavor; the high yield value, stability to salts, high pseudoplasticity, and stable viscosity in the temperature range 0–95°C (32–205°F) of its solutions; the extraordinary increase in solution viscosity it imparts with increases in concentration; and its synergistic interactions give xanthan wide applications in food products, often in combination with modified food starch. Among products in which it is found are pourable salad dressings; reduced-calorie pourable and spoonable dressings, including reduced-calorie mayonnaise; dry beverage mixes; pizza, taco, tartar, and barbecue sauces; relishes; whipped toppings; marshmallow creme; chocolate and other syrups; butterscotch and fudge toppings; egg substitutes; sauces and gravies, including dry mixes for; snack foods; cake mixes; refrigerated doughs; frozen foods; sandwich spreads; and bakery products.

Blends of xanthan and locust bean gum and/or guar gum are excellent ice cream stabilizers and are often used in combination with a carboxymethylcellulose and carrageenan. Other applications of these blends are in the preparation of canned frostings, potato chip dips, cream cheese, cottage cheese dressings, other processed cheese products, sauces such as pizza and Sloppy Joe sauce, and starch-based gum candies. The synergistic interaction with konjac mannan is even greater.

BIBLIOGRAPHY

1. R. L. Whistler and J. N. BeMiller, eds., *Industrial Gums*, 2nd ed., Academic Press, New York, 1973; 3rd ed., 1991.
2. M. Glicksman, *Food Hydrocolloids*, CRC Press, Boca Raton, Fla., Vol. 1, 1982; Vol. 2, 1983; Vol. 3, 1986.
3. G. O. Phillips, D. J. Wedlock, and P. A. Williams, eds., *Gums and Stabilisers for the Food Industry*, Pergamon Press, Oxford, 1982; Vol. 2, 1984; Elsevier Applied Science Publishers, London, Vol. 3, 1986; IRL Press, Oxford, Vol. 4, 1988.
4. J. R. Mitchell and D. A. Ledward, eds., *Functional Properties of Food Macromolecules*, Elsevier Applied Science Publishers, London, 1986.
5. R. L. Whistler and T. Hymowitz, *Guar: Agronomy, Production, Industrial Use, and Nutrition*, Purdue University Press, West Lafayette, Ind., 1979.
6. B. V. McCleary, A. H. Clark, I. C. M. Dea, and D. A. Rees, *Carbohydrate Research* **139**, 237–260 (1985) and references cited therein.
7. M. L. Fishman and J. Jen, eds., *Chemistry and Function of Pectins*, American Chemical Society Symposium Series, **310** (1986).

J. N. BeMiller
Purdue University
West Lafayette, Indiana

H

HACCP PRINCIPLES FOR FOOD PRODUCTION

On November 1989, the U.S. National Advisory Committee on Microbiological Criteria for Foods [U.S. Department of Agriculture (USDA)] adapted the document *HACCP Principles for Food Production*.

This document defines HACCP as a systematic approach to be used in food production as a means to ensure food safety. Seven basic principles underlie the concept. These principles include an assessment of the inherent risks that may be present from harvest through ultimate consumption. Six hazard characteristics and a ranking schematic are used to identify those points throughout the food production and distribution system whereby control must be exercised in order to reduce or eliminate potential risks. A guide for HACCP plan development and critical control point (CCP) identification are noted. Further, the document points out the additional areas that are to be included in the HACCP plan—the need to establish critical limits that must be met at each CCP, appropriate monitoring procedures, corrective action procedures to be taken if a deviation is encountered, recordkeeping, and verification activities.

This article reproduces a slightly modified version of this document.

PREAMBLE

The National Advisory Committee on Microbiological Criteria for Foods (Committee) endorses the Hazard Analysis and Critical Control Point (HACCP) System as an effective and rational approach to the assurance of food safety. In the application of HACCP, the use of microbiological testing is seldom an effective means of monitoring critical control points (CCP) because of the time required to obtain results. In most instances, monitoring of CCP can best be accomplished through the use of physical and chemical tests and through visual observations. Microbiological criteria do, however, play a role in verifying that the overall HACCP system is working.

The Committee believes that the HACCP principles should be standardized to create uniformity in its work, and in training and applying the HACCP system by industry and regulatory authorities. In accordance with the National Academy of Sciences recommendation, the HACCP system must be developed by each food establishment and tailored to its individual product, processing and distribution conditions.

DEFINITIONS

Continuous Monitoring. Uninterrupted recording of data such as a recording of temperature on a strip chart.

Control Point. Any point in a specific food system where loss of control does not lead to an unacceptable health risk.

Critical Control Point. Any point or procedure in a specific food system where loss of control may result in an unacceptable health risk.

Critical Defect. A defect that may result in hazardous or unsafe conditions for individuals using and depending on the product.

Critical Limit. One or more prescribed tolerances that must be met to ensure that a critical control point effectively controls a microbiologic health hazard.

Deviation. Failure to meet a required critical limit for a CCP.

HACCP Plan. The written document that delineates the formal procedures to be followed in accordance with these general principles.

HACCP System. The result of the implementation of the HACCP principles.

Hazard. Any biological, chemical, or physical property that may cause an unacceptable consumer health risk.

Monitoring. A planned sequence of observations or measurements of critical limits designed to produce an accurate record and intended to ensure that the critical limit maintains product safety.

Risk. An estimate of the likely occurrence of a hazard or danger.

Risk Category. One of six categories prioritizing risk based on food hazards.

Sensitive Ingredient. Any ingredient historically associated with a known microbiologic hazard.

Significant Risk. Posing moderate likelihood of causing an unacceptable health risk.

Spot Check. Supplemental tests performed on a random basis.

Verification. Methods, procedures, and tests used to determine if the HACCP system is in compliance with the HACCP plan.

PURPOSE AND PRINCIPLES

HACCP is a systemic approach to food safety, consisting of the seven following principles:

1. Assess hazards and risks associated with growing, harvesting, raw materials and ingredients, processing, manufacturing, distribution, marketing, preparation and consumption of the food.

2. Determine CCPs required to control the identified hazards.
3. Establish the critical limits that must be met at each identified CCP.
4. Establish procedures to monitor CCP.
5. Establish corrective action to be taken when there is a deviation identified by monitoring a CCP.
6. Establish effective record-keeping systems that document the HACCP plan.
7. Establish procedures for verification that the HACCP system is working correctly.

Principle Number 1

Principle No. 1. Assess hazards associated with growing, harvesting, raw materials and ingredients, processing, manufacturing, distribution, marketing, preparation, and consumption of the food.

Description. Provides for a systematic evaluation of a specific food and its ingredients or components to determine the risk from hazardous microorganisms or their toxins. Hazard analysis is most useful for guiding the safe design of a food product and defining the CCPs that eliminate or control hazardous microorganisms or their toxins at any point during the entire production sequence. The hazard assessment is a two-part process consisting of ranking a food according to six hazard characteristics, followed by the assignment of risk category, which is based on the ranking.

Ranking according to hazard characteristics is based on assessing a food in terms of whether (1) the product contains microbiologically sensitive ingredients, (1) the process does not contain a controlled processing step that effectively destroys harmful microorganisms, (3) there is significant risk of postprocessing contamination with harmful microorganisms or their toxins, (4) there is substantial potential for abusive handling in distribution or in consumer handling or preparation that could render the product harmful when consumed, or (5) there is no terminal heat process after packaging or when cooked in the home. Ranking according to these six characteristics results in the assignment of risk categories based on how many of the characteristics are present.

The risk categories are utilized for recognizing the hazard risk for ingredients and how they must be treated or processed to reduce the risk for the entire food production–distribution sequence.

The hazard assessment procedure is ideally conducted after developing a working description of the product, establishing the types of raw materials and ingredients required for preparation of the product, and preparing a diagram for the food production sequence. The two-part assessment of hazard analysis and assignment of risk categories is conducted according to the procedure described in the following paragraphs.

Hazard Analysis and Assignment of Risk Categories. Rank the food according to hazard characteristics A through F, using a plus (+) to indicate a potential hazard. The number of pulses will determine the risk category. A model diagram outlining this concept is given under Section 4.1.3 in *HACCP Principles for Food Production*. As indicated, if the product falls under hazard class A, it should automatically be considered risk category VI.

Hazard A. A special class that applies to nonsterile products designated and intended for consumption by at-risk populations, eg, infants, the aged, the infirm, or immunocompromised individuals.

Hazard B. The product contains "sensitive ingredients" in terms of microbiological hazards.

Hazard C. The process does not contain a controlled processing step that effectively destroys harmful microorganisms.

Hazard D. The product is subject to recontamination after processing before packaging.

Hazard E. There is substantial potential for abusive handling in distribution or in consumer handling that could render the product harmful when consumed.

Hazard F. There is no terminal heat process after packaging or when cooked in the home.

Note: Hazards can also be stated for chemical or physical hazards, particularly if a food is subject to them.

Assignment of Risk Category (Based on Ranking by Hazard Characteristics)

Category VI. A special category that applies to nonsterile products designated and intended for consumption by at-risk populations, eg, infants, the aged, the infirm, or immunocompromised individuals. All six hazard characteristics must be considered.

Category V. Food products subject to all five general hazard characteristics. Hazard class B, C, D, E, F.

Category IV. Food products subject to four general hazard characteristics.

Category III. Food products subject to three of the general hazard characteristics.

Category II. Food products subject to two of the general hazard characteristics.

Category I. Food products subject to one of the general hazard characteristics.

Category O. Hazard class—No hazard.

Note: Ingredients are treated in the same manner in respect to how they are received at the plant, *before* processing. This permits determination of how to reduce risk in the food system.

It is recommended that a chart be utilized that provides assessment of a food by hazard characteristic and risk category. A format for this chart is given as follows:

Food Ingredient or Product	Hazard Characteristics (A, B, C, D, E, F)	Risk Category (VI, V, IV, III, II, I, 0)
T	A + (special category)[a]	VI
U	Five + (B–F)	V
V	Four + (B–F)	IV
W	Three + (B–F)	III
X	Two + (B–F)	II
Y	One + (B–F)	I
Z	No + s	0

[a] Hazard characteristic A automatically is risk category VI, but any combination of B–F may also be present.

Principle Number 2

Principle No. 2. Determine CCP required to control the identified hazards.

Description. A CCP is defined as any point or procedure in a specific food system where loss of control may result in an unacceptable health risk. CCP must be established where control can be exercised. All hazards identified by the hazard analysis must be controlled at some point(s) in the food production sequence, from harvesting and growing raw materials to the ultimate consumption of the food.

Critical control points are located at any point in a food sequence where hazardous microorganisms need to be destroyed or controlled. For example, a specified heat process, at a given time and temperature to destroy a specified microbiologic pathogen, is a CCP. Likewise, refrigeration required to prevent hazardous organisms from growing, of the adjustment of a food to a pH necessary to prevent toxin formation is a CCP.

Types of CCP may include, but are not limited to, cooking, chilling, sanitizing, formulation control, prevention of cross contamination, employee hygiene, and environmental hygiene.

Critical control points must be carefully developed and documented. In addition, they must be used only for purposes of product safety. They should not be confused with control points that do not control safety. For comparison, a control point is defined as any point in a specific food system where loss of control does not lead to an unacceptable health risk.

Principle Number 3

Principle No. 3. Establish the critical limits that must be met at each identified CCP.

Description. A critical limit is defined as one or more prescribed tolerances that must be met to ensure that a CCP effectively controls a microbiologic health hazard. There may be more than one critical limit for a CCP. If any one of those critical limits is out of control, the CCP will be out of control and a potential hazard can exist. The criteria most frequently utilized for critical limits are temperature, time, humidity, moisture level (Aw), pH, titratable acidity, preservatives, salt concentration, available chlorine, viscosity, and in some cases, sensorial information such as texture, aroma, and visual appearance. Many different types of limit information may be needed for safe control of a CCP.

For example, the cooking of meat patties should be designed to eliminate the most heat-resistant vegetative pathogen that could reasonably be expected to be in the product. The critical limits must be specified for temperature, time, and meat patty thickness. Technical development of these critical limits requires accurate information on the probable maximum numbers of these microorganisms in the meat, use of additional ingredients, and the potential for recontamination.

The relationship between the CCP and its critical limits for the meat patty example is as follows:

Critical Control Point	Critical Limits
Meat patty cooked to destroy the most heat-resistant pathogen, based on lethality tests; the minimum lethal cook will usually be designated "to reach an internal patty temperature of x for time y"	Minimum operating temperature of cooker to achieve microbiological lethality at center of coldest patty; time to achieve lethality (belt speed expressed at rpm); patty thickness; Other possible critical limits: Oven humidity Patty composition Cooker sanitation

This example illustrates that the type and number of critical limits will vary depending on the type of cooking system and equipment used for meat patties.

Principle Number 4

Principle No. 4. Establish procedures to monitor CCP.

Description. Monitoring is the scheduled testing or observation of a CCP and its limits. Monitoring results must be documented. From the monitoring standpoint, failure to control a CCP is a critical defect.

A critical defect is defined as a defect that may result in hazardous or unsafe conditions for individuals using and depending on the product. Because of the potentially serious consequences of a critical defect, monitoring procedures must be extremely effective.

Ideally, monitoring should be at the 100% level. Continuous monitoring is possible with many types of physical and chemical methods. For example, the temperature and time for the scheduled thermal process of low-acid canned foods is recorded continuously on temperature-recording charts. If the temperature falls below the scheduled temperature or the time is insufficient, as recorded on the chart, the retort load is restrained as a process deviation. Likewise, pH measurement may be done continually in fluids or by testing of a batch before processing. There are many ways to monitor CCP limits on a continuous or batch basis and record the data on charts. The high reliability of continuous monitoring is always preferred when feasible. It requires careful calibration of equipment.

When it is not possible to monitor a critical limit on a full-time basis, it is necessary to establish that the monitoring interval will be reliable enough to indicate that the hazard is under control. Statistically designed data collection systems or sampling systems lend themselves to this purpose. However, statistical procedures are most useful for measuring and reducing the variation in food formulations, manufacturing equipment, and measuring devices. Thus, they increase the reliability of the system.

When using statistical process control, it is important to recognize that there is no tolerance for exceeding a critical limit. For example, when a pH of 4.6 or less is required for product safety, no single product unit may have a pH above 4.6. To compensate for variation, the maximum of the product may be targeted at a pH below 4.6. Statistical process control can be applied to understand variation in the system, and assure that no unit exceeds a pH of 4.6. Statistical audits can be based on this concept.

Most monitoring procedures for CCP will need to be done rapidly because they relate to on-line processes and there will not be time for lengthy analytical testing. Microbiological testing is seldom effective for monitoring CCP due to their time-consuming nature. Therefore, physical and chemical measurements are preferred because they may be done rapidly and can indicate microbiologic control of the process.

Physical and chemical measurements that may be utilized by monitoring include:

Temperature
Time
pH
Sanitations at CCP
Specific preventive measures for cross contamination
Specific food handling procedures
Moisture level

Spot checks are useful for supplementing the monitoring of certain CCP and their respective limits. They may be used to check incoming precertified ingredients and assess equipment and environmental sanitation, airborne contamination, cleaning and sanitizing of gloves and any place where follow-up is needed. Spot checks may consist of physical and chemical tests and, where needed, microbiologic tests.

With certain foods, microbiologically sensitive ingredients, or imports, there may be no alternative to microbiologic testing. However, a sampling frequency that is adequate for reliable detection of low levels of pathogens is seldom possible because of the large number of samples needed. For this reason, microbiologic testing has limitations in a HACCP system, but is valuable as a means of establishing and randomly verifying the effectiveness of control at CCP (challenge tests, spot checking, or troubleshooting.)

All records and documents associated with CCP monitoring must be signed by the person doing the monitoring and signed by a responsible official of the company.

Principle Number 5

Principle No. 5. Establish corrective action to be taken when there is a deviation identified by monitoring of a CCP.

Description. Actions taken must eliminate the actual or potential hazard that was created by deviation from the HACCP plan, and ensure safe disposition of the product involved. Because of the variations in CCP for different foods and the diversity of possible deviations, specific corrective actions must be developed for each CCP in the HACCP plan. The actions must demonstrate that the CCP has been brought under control. Deviation procedures must be documented in the HACCP plan and agreed to by the appropriate regulatory agency prior to approval of the plan.

Should a deviation occur, the plant will place the product on hold pending completion of appropriate corrective actions and analyses. In instances where it may be difficult to determine the safety of the product, then the testing and final disposition must be agreed to by the government. In instances not associated with safety, government consultation is not required.

Identification of deviant lots and corrective actions taken to ensure safety of these lots must be noted in the HACCP record and remain on file for a reasonable period after the expiration date or expected shelf life of the product.

Principle Number 6

Principle No. 6. Establish effective record-keeping systems that document the HACCP plan.

Description. The HACCP plan must be on file at the food establishment. Additionally, it should include documentation relating to CCP and any action on critical deviations and disposition of product. These materials are to be made available to government inspectors on request. The HACCP plan clearly designates records that will be available for government inspection. Certain records that deal with the functioning of the HACCP system and proprietary information are not necessarily available to regulatory agencies.

Generally, the types of records utilized in the total HACCP system will include the following:
(*Note:* Only those records pertaining to CCP must be made available to regulatory agencies.)

Ingredients

- Supplier certification documenting compliance with processor's specifications
- Processor audit records verifying supplier compliance
- Storage temperature record for temperature-sensitive ingredients
- Storage time records of limited shelf life ingredients

Records relating to product safety

- Sufficient data and records to establish the efficacy of barriers in maintaining product safety

- Sufficient data and records establishing the safe shelf life of the product
- Documentation of the adequacy of the processing procedures from a knowledgeable process authority

Processing

- Records from all monitored CCPs
- System records verifying the continued adequacy of the processes

Packaging

- Records indicating compliance with specifications of packaging materials
- Records indicating compliance with sealing specifications

Storage and Distribution

- Temperature records
- Records showing no product shipped after shelf life date on temperature-sensitive products

Deviation File. Modification to the HACCP plan file indicating approved revisions and changes in ingredients, formulations, processing, packaging and distribution control, as needed.

Principle Number 7

Principle No. 7. Establish procedures for verification that the HACCP system is working correctly.

Description. Verification consists of methods, procedures, and tests used to determine that the HACCP system is in compliance with the HACCP plan. Both the producer and the regulatory agency have a role in verifying HACCP plan compliance. Verification confirms that all hazards were identified in the HACCP plan when it was developed. Verification measures may include physical, chemical, and sensory methods and testing for conformance with microbiologic criteria when established.

Examples of verification activities include but are not limited to

- Establishment of appropriate verification inspection schedules.
- Review of the HACCP plan.
- Review the CCP records.
- Review deviations and dispositions.
- Visual inspections of operations to observe if CCP are under control.
- Random sample collection and analysis.
- Written record of verification inspections that certifies compliance with the HACCP plan or deviations from the plan and the corrective actions taken.

Verification inspections should be conducted when:

- Routinely, or on an unannounced basis to ensure selected CCP are under control.

- It is determined that intensive coverage of a specific commodity is needed because of new information on food safety issues requiring assurance that the HACCP plan remains effective.
- Foods produced have been implicated as a vehicle of foodborne disease.
- Requested on a consultative basis or established criteria have not been met.

Elements that must be included in verification inspection reports:

- Existence of an approved HACCP plan and designation of person(s) responsible for administering and updating the HACCP plan.
- All records and documents associated with CCP monitoring must be signed by the person monitoring and approved by a responsible official of the firm:
    Direct monitoring data of the CCP while in operation
    Certification that monitoring equipment is properly calibrated and in working order
    Deviation procedures
- Any sample analysis for attributes confirming that CCP are under control to include physical, chemical, microbiologic, or organoleptic methods.

IMPLEMENTATION GUIDE

Describe the food and its intended use.
Develop a flow diagram for the production of the food.
Perform a hazard assessment (Principle 1).
  a. Ingredients prior to any processing step.
  b. End product.
Select CCP (Principle 2).
  a. Enter on flow diagram in numerical order.
  b. List CCP number and description.
Establish critical limits (Principle 3).
Establish monitoring requirements (Principle 4).
Establish corrective action to be taken when there is a deviation identified by monitoring of a CCP (Principle 5).
Establish effective recordkeeping systems that document the HACCP plan (Principle 6).
Establish procedures for industrial and governmental verification that the HACCP system is working properly. Verification measure may include physical, chemical and sensory methods, and when needed, establishment of microbiologic criteria (Principle 7).

Y. H. HUI
EDITOR-IN-CHIEF

HEALTHCARE FOODSERVICE. See FOODSERVICE SYSTEMS.

HEAT

When two bodies are at different temperatures, there is a transfer of heat energy from the body having the higher temperature t_1 to the body having the lower temperature t_2. Hence, the state of energy of the colder body is increased and that of the warmer body decreased. This situation of unsteady-state heat transfer continues with the driving force $t_1 - t_2$, or δt, decreasing until both bodies are at the same temperature and there is no unbalanced state to give a driving force. Such is the case when a hot or warm food is placed in a container of cold water to decrease the temperature of the food.

Steady state heat transfer occurs when the temperature driving force Δt remains constant and the rate of heat transfer between two bodies is constant. An example of this type of steady-state transfer is boiling of water on a hot element of a stove. The element temperature is kept constant by a continuous flow of electrical energy being converted to heat and the boiling water stays constant at the boiling point of water (100°C at standard atmospheric pressure). On the other hand, if the heating element is kept constant and a food is being heated below the boiling point, there is an unsteady-state heat transfer, which results in a temperature rise of the food.

The last two examples also illustrate that there are different types of heat within a substance. Sensible heat is the amount of heat that can be added or removed from a given mass of product between two temperatures of the product (t_1 to t_2) without changing the state of the body (eg, heating or cooling a food by cooking or refrigerating). Latent heat refers to the amount of heat necessary to change a given mass from one state to another (eg, boiling water, freezing food products). Sensible heat results in a temperature rise within a body, whereas latent heat is the heat (at constant temperature) necessary to change the state.

There are numerous systems used in the world to quantify mass and energy, including the English (Imperial) system and several systems using metric units. This can be well demonstrated by the measurement of sensible heat. Each food or material requires a different amount of heat, called the specific heat (heat capacity), to raise or lower a given mass to a given temperature. Through laboratory research, specific heats have been determined and given in different unit systems. A British thermal unit (English system) is defined as the amount of heat required to raise the temperature of 1 lb of water 1°F. A kilocalorie in the metric system is the amount of heat required to raise 1 kg of water 1°C.

Several international organizations have attempted to standardize the unit systems, symbols, and quantities to prevent the confusion that often exists (1,2). The result is the Systeme International d'Unites, or the SI system, in which base metric units have been defined. Table 1 gives the SI system unit definitions along with factors for converting from the Imperial system to SI units.

In the SI system, the basic unit of force is the newton (N), the force that gives a mass of 1 kg an acceleration of 1 m/s. The basic unit of energy is the joule (J), the work done when the point of application of 1 N is displaced by a distance of 1 m. Hence specific heat c is expressed in the three principal systems as:

$$c = 1 \text{ Btu/lb°F} = 1 \text{ cal/g°C} = 4.1865 \text{ J/g K}$$

The mathematical relationship between heat, temperature, and specific heat in a food can be expressed in SI units as:

$$Q = M \int_{t_1}^{t_2} c \, dt$$

where Q = heat (heat gained or lost in KJ), M = mass (kg), c = specific heat (kJ/kg°C), and dt = temperature change (°C).

If the specific heat is at constant pressure it is designated as c_p. If the process is carried out at constant volume (eg, a container of compressed gas), it is designated as c_v. For liquid and solid food, the difference between c_p and c_v is negligible. Because over the range of temperatures for food processes the c_p is essentially constant, heat in a system is normally calculated as:

$$Q = (M)(c_p)(t_1 - t_2) = M c_p \Delta t$$

Thus if 100 kg of apples with a specific heat of 3.6 kJ/kg/K) are cooled from tree temperature of 18°C to 5°C, the amount of heat removed would be

$$Q = (100 \text{ kg})(3.6 \text{ kJ/kg·K})(18-5 \text{ K}) = 4{,}680 \text{ kJ}$$

Table 1. Base Units of SI System (Metric) and Conversion from Imperial to SI units

Measurable Quantity	SI Base Unit	SI Symbol	Imperial Base Unit	Imperial Base Symbol	Conversion Factor (Imperial × Factor = SI)
Length	Meter	m	Foot	ft	0.30480
Mass	Kilogram	kg	Pound mass	lb	0.453592
Time	Second	s	Second	s	1.0
Temperature	Degree Kelvin	K	Degree Rankine	R	0.55556
Electric current	Ampere	A	Ampere	A	1.0
Amount of substance	Mole	mol	Mole	mol	1.0

As a point of reference, 1 Btu equals 1.055 kJ; therefore, this would be equal to 4,436 Btu.

When learning the SI system, which will eventually be the world standard, it is useful to remember some basic approximate conversion factors to visualize the relationships between values. This is especially true for the United States where the English system has become so well entrenched. It is helpful to remember that 1 ft is approximately 0.3 m, 1 Btu is approximately 1 kJ, 1 lbf is approximately 4.5 N, 1 Btu/h is approximately equal to 0.3 W, and 1 Btu/lb is approximately 2.33 kJ/kg. If the equivalent English and SI values are visualized when working a problem in SI units, it will become easier to think in both systems.

PROCESSING FOODS BY ADDING OR REMOVING HEAT

The addition or removal of heat from a food increases shelf life and insures the safety of the product. Heat is added to make a food more acceptable (improving sensory characteristics), to reduce microorganism populations, to lessen enzyme activity (pasteurization or cooking), or to kill microorganism and completely inactivate the enzymes (sterilizing). Heat is also added during dehydration processes to remove water through vaporization and thus lower the water activity a. Heat is removed to slow the growth of microorganisms and the action of enzymes (cooling or refrigerating) or to change the state (freezing), which prevents growth of microorganisms and further reduces enzyme action.

CHANGING THE STATE OF A FOOD PRODUCT

There are three states in which a food can exist, namely solid, liquid, or vapor. Because most natural foods are in some state of equilibrium with water, most of the significant changes of state in food products involve water. Food processes involving a change in state include freezing, heating to the point of evaporating a component (eg, vaporization or dehydration), condensing vapors to liquids (eg, solvent extraction), and subliming water directly from the frozen state to vapor.

Freezing Food

Freezing foods involves removing heat, resulting in the changing of liquid water in the food to ice. This has been demonstrated in the frozen fish industry where sensible heat is removed until the temperature reaches approximately −1°C (30°F). It then takes about 45 min during the so-called critical period to remove the latent heat of fusion from the water. After this period, the temperature drops rapidly as sensible heat is again removed (see FISH AND SHELLFISH PRODUCTS, FIG. 9). It should be noted that foods are not pure substances so do not have precise temperatures at which there is a change in phase. This is because water has dissolved materials that increase or decrease in amount as a food is being processed and some of the water is bound to components of the food rather than being free to flow and act like pure water.

Vaporization Processes

The three vaporization processes involve the change in basic characteristics of a food or material by applying heat or other forms of energy that is converted to heat within the product.

Evaporation. Evaporation is the concentration of a liquid by supplying sufficient energy to vaporize the more volatile component or components. The process is used to concentrate solutions for stability or prior to further processing and to recover solvent from an extraction system.

Drying or Dehydrating. Drying or dehydration involves adding energy, usually heat, to vaporize a liquid, usually water, from a solid food. Foods are dried by supplying sufficient heat energy to vaporize water. This (1) preserves the product by reducing water activity; (2) reduces the cost or difficulty of packaging, handling, storing and shipment; and (3) produces convenience items (eg, instant coffee). In certain cases energy is supplied through other energy forms and then converted to heat in the product. For example, in processing by microwave heating, the energy is supplied to the system in the form of hertzian waves that are absorbed by the food, the resulting friction between vibrating molecules converts the wave energy to heat energy.

There are two distinct periods of drying in which different mechanisms control the rate of drying. During the first period, the constant rate period, the heat transfer predominates; all of the heat added is directly used in evaporating water and the rate of drying is independent of the nature of the food. In this case the moisture movement near the surface is rapid enough to maintain a saturated condition at the surface and the temperature of the food remains constant.

Each food has a critical moisture content at which the moisture can not migrate to the surface by diffusion rapidly enough to utilize all of the heat for evaporation. At this point the second phase, or falling rate period, begins and the food begins to absorb heat and rise in temperature.

A special form of drying involves sublimation of water directly from a frozen product to vapor, that is without passing through the liquid state. This is accomplished by placing a frozen food in a chamber under high vacuum. if the partial pressure of water vapor in the chamber is maintained below that of the ice at 0°C or 4.58 mm Hg, the water does not thaw prior to becoming a vapor.

Distillation. Distillation is the separation of two or more liquids through vaporizing the more volatile component or components. A principal use of this process in the food industry is in the recovery of a product after solvent extraction. In the case of organic solvent extraction of oilseeds, the solvent is recovered for reuse and the oil is retained as a pure food product.

Heat energy is certainly the most important factor involved in the processing of food products. In dealing with the science and engineering aspects of the food industry it is necessary to be constantly aware of the units involved in measuring the amount of energy being added to or removed from a food product. Unit equations (those showing the units as well as the numerical values) insure the consistency of units and dimensions.

BIBLIOGRAPHY

1. R. P. Singh and D. R. Heldman, *Introduction to Food Engineering,* Academic Press, Inc., Orlando, Fla., 1984.
2. H. Wolf, *Heat Transfer,* Harper & Row, New York, 1983.

GEORGE M. PIGOTT
University of Washington
Seattle, Washington

HEAT EXCHANGERS

Previous discussions on heat and the mechanisms for heating or cooling foods have emphasized that virtually every food-processing operation depends on transferring heat. The operations involving heat transfer use a wide variety of heat exchangers for heating or cooling products and for operational aspects of auxiliary equipment. Although there are many types of heat-exchanger equipment used in the food industry, there are relatively few principles that govern the heat transfer and operation of the equipment.

There are two basic classifications of heat exchangers. One is the contact type in which there is direct physical contact between the food product and the heating or cooling medium. The other is the noncontact type in which the heat is transferred through a body that separates the product from the heating or cooling source. Within these two categories there are many proprietary designs and models of heat exchangers depending on the specific requirement for transferring heat to or from a given type of food product.

Capital investment, safety, and economics of processing are important operational factors to consider when purchasing or installing exchanger equipment. Of equal value in determining the type or model of equipment to install is the effect on the physical and chemical properties of the end products. Maintaining or improving nutritional value, aesthetics, safety of the products, and sensory attributes affect the marketability.

Paramount to the successful long-term operation of heat exchangers in meeting the above goals is the efficiency and ease of sanitizing the entire equipment. This is especially important for closed systems that can not be dismantled while cleaning. The operation of cleaning in place (CEP) requires sanitary design that insures complete sanitation when cleaning liquids are circulated through the exchangers. Hence, engineering design is the controlling factor in insuring successful heat-transfer operations involving food processing.

FOOD-PROCESSING OPERATIONS REQUIRING HEAT EXCHANGERS

It is common to think of all heat exchangers as being composed of two adjoining compartments through which fluids are flowing, each cooling or heating the other. This concept is natural because the basic engineering approach to studying heat transfer is to discuss each type of heat exchange operation as a specific process (eg, product freezing, refrigeration systems, batch heating, nonsteady-state operations, etc) with steady-state, liquid–liquid heat exchangers being studied under the subject of heat exchangers. However, when processing food, it is important to think of heat exchangers as being the total range of equipment that is used for cooking, blanching, pasteurizing, sterilizing, cooling, freezing, and cold-storage holding. In addition, there are many processes, such as drying and extraction, in which the transfer of heat is important to accomplish the basic goal of the process. Food products that are heated or cooled during processing, including the heat transfer medium, cover a wide range of vapors, liquids, and solids. Table 1 indicates how machinery and equipment must be used for the transfer of heat in the many different food-processing operations.

In addition to the type of heat transfer taking place

Table 1. The State of a Food Product as Related to Classification of Heat Exchanger

Food Form	Heat Transfer Media	Classification	Example
Vapor	Liquid	Noncontact	Condensing vapors
Liquid	Vapor	Contact	Steam infusion, steam injection
		Noncontact	Heating by steam (condensing), refrigerant cooling
	Liquid	Noncontact	Transfering heat between liquids
	Solid	Contact	Melting ice
		Noncontact	Heating on metal hot plate in a container
Solid	Vapor	Noncontact	Extrusion (heating by steam jacket)
		Contact	Drying, cooking
	Liquid	Contact	Blanching, immersion freezing, deep-fat frying, poaching, steeping
		Noncontact	Extrusion (heating by liquid jacket)
	Solid	Contact	Dry ice cooling, freezing on plates
		Noncontact	Cooking on stove
Any food	None	Noncontact	Radiant heating, irradiation

during processing, the physical and chemical condition of the food before, during, and after heating or cooling must be considered. Products requiring different approaches to heat transfer during the food-processing operation include

1. Dense, hard, solid foods (eg, potatoes) that can stand considerable mechanical abuse during processing but become somewhat fragile during the final phase of heating.
2. Soft, solid foods (eg, tomatoes) that cannot stand any mechanical abuse during or after cooking.
3. Purées or soft items that flow like a liquid but congeal or gel to a solid when heated (eg, surimi being processed into seafood analogues and extruded cereals).
4. Liquids that range from heat stable to extremely heat labile.
5. Foods, particularly those high in protein, that congeal with heat and cause baked-on deposits on the heat-transfer surfaces.

Processing requirements for these types of food range from individual batch cooking to large-scale continuous sterilizing. Many products must also be heated and then cooled to insure maximum retention of nutrients and desired product form.

HEAT-EXCHANGER SYSTEMS AND PRINCIPLES OF OPERATION

Heat is transferred to or from a food in batch and continuous systems. Batch systems involve unsteady-state heat transfer whereby the food being heated or cooled begins at a given temperature and increases or decreases until the desired temperature is reached. The heat-transfer medium can vary in temperature (eg, a hot surface or liquid that changes temperature as it gives or receives heat) or can be at steady state (eg, condensing steam).

Batch steady-state systems involve a series of batch systems that give the overall effect of steady state processing. For example, liquid-filled tanks in series can be stepwise heated by batch heating but can be connected so that the end result of the overall heating is a steady-state emission of constant-temperature liquid continuously flowing at the same rate as the cold liquid entering the first tank.

A true steady-state system is found in flowing liquids or viscous solids whereby mass flow rate, temperature, pressure, and physical properties of the food and the heat transfer medium are constant at any given cross section.

Heating and Cooling Liquids

Batch Heating. Typical batch heating of a liquid food takes place in a steam-jacketed kettle. While the food is being heated, the system is under unsteady-state conditions. After the food reaches the desired processing temperature and is heated at a constant temperature, steady-state conditions are reached. Assuming that the final temperature of the food is to be maintained somewhat below that of the condensing steam, steady-state conditions prevail as the steam flow rate is adjusted to maintain the processing conditions. As is the practice for all commercial heat-exchanger equipment, the outside of the kettle is lagged to minimize heat loss to the surroundings and isolate the environment of the unit operation.

Large kettles (tanks) being used for heating a liquid often have coils in the tank to transfer heat. This greatly improves the heat-transfer efficiency by increasing the heat-transfer surface above that of the outside wall receiving heat from the jacketed heating (or cooling) source. An additional improvement of the heat transfer in tanks can be realized by installing mechanical stirring equipment, which increases heat transfer above that of natural convection.

Tubular Heat Exchangers. The simplest continuous heat exchange occurs when two fluids of different temperature are flowing through concentric pipes or tubes. The flow in steady-state heat exchangers can be either cocurrent or cocurrent (parallel) flow. During countercurrent flow, one stream (liquid or vapor) is introduced at the opposite end of the unit. By controlling the flow rates, it is possible to heat the cold liquid above the outlet temperature of the entering hot stream. Conversely, when the two liquids are introduced at the same point, the stream being heated can never leave at a temperature above that of the stream being cooled. This cocurrent system is normally less efficient than a countercurrent system because the temperature difference driving force can become quite small as the temperatures of the two streams meet. However, there are circumstances whereby cocurrent flow can be used to insure that a heat-sensitive material does not rise above a certain temperature during processing. In the case of using steam to heat a flowing liquid, the food is heated while the condensing steam is maintained at the saturation temperature of the steam. As in the case of a steam kettle cooker, the most common and efficient heating medium is condensing steam. Shell-and-tube heat exchangers are essentially improved tubular heat exchangers where a few to many tubes replace the single concentric inner tube.

Whereas the heat transfer coefficient is constant under these steady-state conditions, the temperature driving force (ΔT) is calculated as the log mean temperature difference

$$\Delta T_M = \frac{(T_S - T_{f_1}) - (T_S - T_{f_2})}{\ln \frac{(T_S - T_{f_1})}{(T_S - T_{f_2})}} \qquad (1)$$

and the steady-state heat transfer is calculated as

$$Q = UA\Delta T_M \qquad (2)$$

where T_S = temperature of condensing steam, T_f = temperature of liquid food being heated, U = overall heat transfer coefficient, and ΔT_M = log mean temperature driving force.

Plate Heat Exchangers. Plate heat exchangers solve one of the principal processing problems encountered with tubular and shell-and-tube heat exchangers, that of sanitation. It is virtually impossible to thoroughly clean and sanitize closed system exchangers when a food liquid or slurry is passed through the larger diameter or the shell side where velocity is low. The basic units of plate exchangers are stainless steel (for sanitation and corrosion resistance) plates that are pressed, machined, or formed to accomplish several special design features. The contour of a stack of the plates is such that, when a formed gasket is placed between each plate, a heat exchanger allowing two liquid streams to flow between plates is formed. In practical operation, the plates are suspended from horizontal rails or pipes that allow them to be brought together and tightly compressed during operation or separated for cleaning and maintenance.

These types of heat exchanger are used extensively in the dairy industry. The exchanger is highly efficient due to the high turbulence and minimum volume flowing between the channels. Capacity can be increased to any flow rate desired (eg, 10,000 kg/h) by increasing the number of plates in the frame. When operated at the proper flow rate, the velocity of liquid through the small channels decreases the tenacious baked-on deposits that are caused when colloidal suspended components (eg, proteins in milk) contact hot surfaces. Also, the plates can be separated for thorough cleaning and sanitizing during the maintenance periods.

The disadvantages of plate heat exchangers is the high initial cost compared to tubular type exchangers and the high cleaning and maintenance cost of taking the exchanger apart for each cleaning period. Also, the maximum velocity is limited by the small cross-sectional area and the pressure limitations of the gasketed plates. The minimum velocity is determined by the varying cross section, which allows dead spots of low velocity and subsequent bake-on of the suspended or dissolved solids.

Direct Steam Heating. Direct contact between steam and a food is the most efficient means of heating by directly transferring the latent heat of vaporization to the food. However, the product must be able to sustain the dilution effect caused by the added water resulting from the condensed steam that remains in the heated product. Furthermore, special consideration must be given to producing steam that is safe for human consumption.

When steam is added to a product, the process is known as steam injection. When the product is sprayed into a chamber of steam, thus adding the product to the steam, the process is called steam infusion.

Falling Film Evaporators. When a film of liquid is allowed to flow down a heated wall, the transfer of heat is extremely rapid. This type of unit is used for rapid heating of a fluid that is being evaporated. However, due to the limitation of flow rate as compared to other types of heat exchanger, this method is not often used solely for heating a product.

Heating and Cooling Solids

Batch Heating. Batch heating of solids is carried out in the type of heat-exchange facilities normally associated with cooking foods. Broiling, roasting, or baking operations are accomplished by a combination of radiant heat, convection, and conduction, the predominant mechanism of heat transfer depending on the specific commercial equipment. Ovens are heated by elements emitting radiation at a wavelength of 5×10^{-6} m at a temperature of 250–400°C. A more effective radiant heating occurs in chambers heated by an electric bulb infrared source. In this case the air is not heated by the source so that little heating of the food is due to convection or conduction heating.

Microwave heating is a specialized form of dielectric radiant heating that has many advantages over other dielectric methods because there is no requirement for critical spacing between the food and the capacitor plates. Heating is accomplished by the friction of excited molecules rubbing against each other as the strong alternating energy reverses the polarization of the molecules in the food many millions of times per second. The advantages of microwave heating include extremely rapid heating of the food, uniform distribution of the heat throughout the entire food mass, high efficiency, and good control of the energy being added to the food.

Many products are cooked and heated by convection heating when immersed in hot vegetable oil (deep frying) or poached in hot or boiling water. Batches of products are often heated by placing a container of the product on a hot surface (eg, hot plate or element of a stove) or in a steam environment. Steam retort canning of foods is a good example of batch heating by steam. Hermetically sealed cans are placed in baskets and placed in a steam chamber that can be closed and pressurized, normally at about 10 psi (117°C or 242°F). Condensing steam transfers energy to the outside of the can by conduction and convection, whereby the heat is conducted through the can by conduction and then to the food. Solid packs with no free liquid are heated in the can by conduction while a pack with free water transfers heat by both conduction and convection. After being held at the required temperature and time to accomplish sterilization, the cans are removed and air cooled. Sensitive products that tend to scorch or decrease in nutritional value during long air cooling periods are often pressure cooled with water prior to being removed from the retort.

Continuous Heating. More efficient production control, increased product throughput, and improved processed food quality can be accomplished when processes are upgraded from batch to continuous. Modern processing of solid foods involves large continuous production lines utilizing continuous baking ovens (eg, bakery products), deep-frying tanks (eg, french-fried potatoes), microwave ovens, radiant heat chambers, and steam chambers.

Batch Cooling or Freezing. Removing heat from foods, with the exception of cooling in tubular type exchangers, requires considerably different types of facilities than heating. This is due to the nature of the recycling refrigerant or the cryogenic liquids used to remove heat from a food. An additional factor is the psychrometric properties of air that are often recycled near the humidity saturation point in refrigeration facilities in which natural or forced convection is involved in the process.

The basic unit involved with refrigeration cooling of foods is a heat exchanger in which a refrigerant is introduced through an expansion valve into the coils that are located in the cooling or freezing chamber. Thus the cooling or freezing heat exchanger has a cold gaseous refrigerant on one side of the coil wall and the food or heat-transfer medium on the other. Air or liquid brines are the normal mediums used for transferring heat by convection from the food to the refrigeration coils. When the food product is in direct contact with the refrigeration coils, heat is transferred directly by conduction.

Blast Cooling and Freezing Facilities. Forced-air convection is used in blast refrigeration to transfer heat from refrigerated coils to the product. A blast freezer or cooler is actually a double heat exchanger. One exchange takes place between the product and direct contact with air flowing past, and the second is the cooling of the recirculating air as it passes over the freezing coils. During the first phase of this cycle, cold air is circulated over the product and through the freezing chamber, which results in an increase in humidity due to the humidity driving force H between the product environment and the colder refrigerant coil. Thus the air removes water from the freezing chamber and deposits it as ice on the refrigeration coils during the second phase when the air is cooled below the saturation point. If the product being frozen is not completely protected from the air, desiccation will occur in the product. Also, if the doors to the chamber are not completely sealed when closed, moist warm air will enter and further complicate the moisture transfer problem. Of course, as the moisture, in the form of ice, builds up on the coil it acts as a heat-transfer barrier and greatly reduces the efficiency of the system. Hence, the advantage of blast refrigeration is the relatively simple facility required while the main disadvantage is desiccation of the food and buildup of ice deposits that must be removed from the refrigeration coils. The ice greatly reduces the efficiency of heat transfer and increases the cost of operation.

Contact Plate Freezers. Many solid food products are frozen by conduction on freezer shelves, called plates, that contain circulating refrigerant. Thus plate freezing involves heat exchange between a solid food and a vapor refrigerant. Efficient plate freezing is limited to foods and food packages that have flat surfaces (eg, rectangular packages of vegetable) because irregular geometries (eg, turkeys) cannot contact the flat freezer plate. The efficiency of freezing suitable packages is further increased by plate freezer systems in which the plates can be adjusted after loading to contact both the top and the bottom of the package.

Immersion Freezing in Brine. Saturated brine solutions have freezing points well below the freezing point of water and can be used efficiently to freeze products, particularly irregularly shaped items such as turkeys and fish. The product is immersed in a cold brine solution that is maintained at the low temperature by freezing coils. As in the case of blast freezing, there is a two-phase heat transfer. The refrigerant takes heat from the brine and the brine removes heat from the food by conduction and convection. This heat exchange is between a solid food and a liquid.

Cryogenic Freezing. Immersion of a solid food in a liquid refrigerant is similar to freezing in a brine except that there is a much higher temperature driving force between the liquid (eg, liquid ammonia or freon) and the product. Thus the freezing is rapid. Batch freezing by this method is not ordinarily carried out commercially because the cost is prohibitive. Due to the extremely fast freezing, a cryogen immersion frozen product must be carefully tempered before further handling and processing or the internal stresses produced will cause cracking of the frozen item.

Continuous Cooling or Freezing. As in the case of heating a food, continuous freezing is much more efficient than batch freezing. Many modern freezing operations involve continuous lines whereby conveyors carry a food through a freezing apparatus. This includes blast freezing and cryogenic freezing. One improvement over immersion cryogenic freezing is the continuous freezing in tunnels in which a liquid cryogenic is flowed over the product. In this operation the liquid expands to a vapor through nozzles directed toward the moving product line. This freezing of a solid by direct contact with a vapor greatly reduces the amount of refrigerant used during processing. However, the cost of the liquid cryogens used (ammonia, freon, carbon dioxide, and nitrogen) are such that only continuous processing lines operating long hours can be justified economically.

There are many different types of heat exchanger and auxiliary equipment available to the food processor. The length of the processing season, the type of food or food product, the value of the raw materials, the cost of utilities at the processing location, environmental factors, and common sense are all factors that must enter into the plans for food-processing operations. Judicial selection of the equipment and facilities for a given food and a given process are necessary to insure that the highest quality product is produced efficiently and economically.

GEORGE M. PIGOTT
University of Washington
Seattle, Washington

HEAT EXCHANGERS: FOULING

THE FOULING FACTOR

In view of its complexity and variability and the need to carry out experimental work on a long-term basis under actual operating conditions, fouling remains a somewhat neglected issue among the technical aspects of heat transfer. Still, the importance of carefully predicting fouling resistance in both tubular and plate heat-exchanger calculations cannot be overstressed. This is well illustrated by examples A and B.

A Typical Water/Water Tubular Design. Clean Overall Coefficient 500 Btu/h·ft²·°F

Fouling Resistance, h·ft²·°F/Btu	Dirty Coefficient, Btu/h·ft²·°F	Extra Surface, Required, %
0.0002	455	10
0.0005	400	25
0.001	333	50
0.002	250	100

B R405 Paraflow, Water/Water Duty, Overall Coefficient 1000 Btu/h·ft²·°F; Single Pass-Pressure Loss 9 psig

Fouling Resistance, h·ft²·°F/Btu	Dirty Coefficient, Btu/h·ft²·°F	Extra Surface, Required, %
0.0002	833	20
0.0005	666	50
0.001	500	100
0.002	333	200

Note that for a typical water–water duty in a plate heat exchanger, it would be necessary to double the size of the unit if a fouling factor of 0.0005 were used on each side of the plate (ie, a total fouling of 0.001).

Although fouling is of great importance, there are relatively little accurate data available and the rather conservative figures quoted in Kern (*Process Heat Transfer*) are used all too frequently. It also may be said that many of the high fouling resistances quoted have been obtained from poorly operated plants. If a clean exchanger, for example, is started and run at the designed inlet water temperature, it will exceed its duty. To overcome this, plant personnel tends to turn down the cooling-water flow rate and thereby reduce turbulence in the exchanger. This encourages fouling and even though the water flow rate eventually is turned up to design, the damage will have been done. It is probable that if the design flow rate had been maintained from the onset, the ultimate fouling resistance would have been lower. A similar effect can happen if the cooling-water inlet temperature falls below the design figure and the flow rate is again turned down.

SIX TYPES OF FOULING

Generally speaking, the types of fouling experienced in most CPI (cleaning in place) operations can be divided into six fairly distinct categories. First is crystallization—the most common type of fouling that occurs in many process streams, particularly cooling-tower water. Frequently superimposed with crystallization is sedimentation, which usually is caused by deposits of particulate matter such as clay, sand, or rust. From chemical reaction and polymerization often comes a buildup of organic products and polymers. The surface temperature and presence of reactants, particularly oxygen, can have a very significant effect. Coking occurs on high-temperature surfaces and is the result of hydrocarbon deposits. Organic material growth usually is superimposed with crystallization and sedimentation and is common to sea water systems. And corrosion of the heat-transfer surface itself produces an added thermal resistance as well as a surface roughness.

In the design of the plate heat exchanger, fouling due to coking is of no significance since the unit cannot be used at such high temperatures. Corrosion also is irrelevant since the metals used in these units are noncorrosive. The other four types of fouling, however, are most important. With certain fluids such as cooling-tower water, fouling can result from a combination of crystallization, sedimentation, and organic material growth.

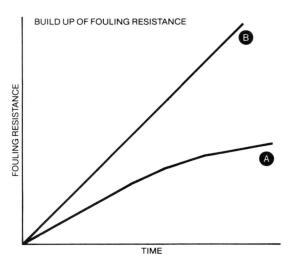

Figure 1. Buildup of fouling resistance.

A FUNCTION OF TIME

From Figure 1, it is apparent that the fouling process is time-dependant with zero fouling initially. The fouling then builds up quite rapidly and in most cases and levels off at a certain time to an asymptotic value as represented by curve A. At this point, the rate of deposition is equal to that of removal. Not all fouling levels off, however, and curve B shows that at a certain time the exchanger would have to be taken off line for cleaning. It should be noted that a Paraflow is a particularly useful exchanger for this type of duty because of the ease of access to the plates and the simplicity of cleaning.

In the case of crystallization and suspended solid fouling, the process usually is of the type A. However, when the fouling is of the crystallization type with a pure compound crystallizing out, the fouling approaches type B and the equipment must be cleaned at frequent intervals. In one particularly severe fouling application, three Series HMB Paraflows are on a 4½ hour cycle and the units are cleaned in place for 1½ h in each cycle.

Biological growth can present a potentially hazardous fouling since it can provide a more sticky surface with which to bond other foulants. In many cases, however, treatment of the fluid can reduce the amount of biological growth. The use of germicides or poisons to kill bacteria can help.

LOWER RESISTANCE

It generally is considered that resistance due to fouling is lower with Paraflow plate heat exchangers than with tubular units. This is the result of five Paraflow advantages:

1. There is a high degree of turbulence, which increases the rate of foulant removal and results in a lower asymptotic value of fouling resistance.
2. The velocity profile across a plate is good. There are no zones of low velocity compared with certain areas on the shell side of tubular exchangers.

3. Corrosion is maintained at an absolute minimum.
4. A smooth heat-transfer surface can be obtained.
5. In certain cooling duties using water to cool organics, the very high water film coefficient maintains a moderately low metal surface temperature, which helps prevent crystallization growth of the inversely soluble compounds in the cooling water.

The most important of these is turbulence. HTRI (Heat Transfer Research Incorporated) has shown that for tubular heat exchangers, fouling is a function of flow velocity and friction factor. Although flow velocities are low with the plate heat exchanger, friction factors are very high and this results in lower fouling resistance. The effect of velocity and turbulence is plotted in Figure 2.

Marriot of Alfa Laval has produced a table showing values of fouling for a number of plate heat-exchanger duties.

Fluid	Fouling Resistance (hr·ft²·°F/Btu)
Water	
Demineralized or distilled	0.00005
Soft	0.00010
Hard	0.00025
Cooling tower (treated)	0.00020
Sea (coastal) or estuary	0.00025
Sea (ocean)	0.00015
River, canal	0.00025
Engine jacket	0.00030
Oils, lubricating	0.00010–0.00025
Oils, vegetable	0.00010–0.00030
Solvents, organic	0.00005–0.00015
Steam	0.00005
Process fluid, general	0.00005–0.00030

These probably represent about one half to one-fifth of the figures used for tubulars as quoted in Kern, but it must be noted that the Kern figures probably are conservative, even for tubular exchangers.

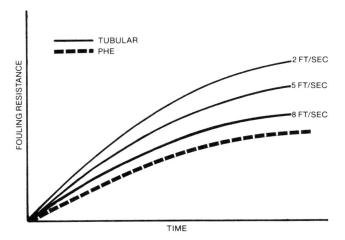

Figure 2. Effect of velocity and turbulence.

APV, meanwhile, has carried out test work that tends to confirm that fouling varies for different plates with the more turbulent type of plate providing the lower fouling resistances. In testing an R405 heating a multicomponent aqueous solution containing inverse solubility salts (ie, salts whose water-solubility decreases with increasing temperature), it was learned that the rate of fouling in the Paraflow was substantially less than that inside the tubes of a tubular exchanger. The tubular unit had to be cleaned every 3 or 4 days while the Paraflow required cleaning about once a month.

Additional tests on cooling water fouling were sponsored by APV at Heat Transfer Research, Inc. (HTRI) with the test fluid being a typical treated cooling tower water. Results of these experiments follow.

OBJECTIVES

In many streams, fouling is an unavoidable by-product of the heat-transfer process. Fouling deposits can assume numerous types such as crystallization, sedimentation, corrosion, and polymerization. Systematic research on fouling is relatively recent and extremely limited (1–3) and almost exclusively concentrated on water—the most common fluid with fouling tendencies. The few research data that exist usually are proprietary or obtained from qualitative observations in plants. However, it is generally recognized that the prime variables affecting fouling buildup are flow velocity, surface temperature, and surface material. In the case of water, the water quality and treatment must also be considered.

The importance of fouling on the design of heat exchangers can be seen from the rate equation

$$\frac{1}{U} = \frac{1}{h_1} + \frac{1}{h_2} + R_f$$

where U = overall heat transfer coefficient, Btu/h·ft²·°F
h_1, h_2 = film coefficients of the two heat-transferring fluids, Btu/h·ft²·°F
R_f = fouling resistance, h·ft²·°F/Btu

It is obvious from equation 1 that the higher the film coefficients, the greater effect the fouling resistance will have on the overall coefficient and therefore on the size of the exchanger.

In tubular heat exchangers, water-side heat-transfer coefficients in the order of magnitude of 1000 Btu/h·ft²·°F are quite common. In plate exchangers, the coefficients are substantially higher, typically around 2000 Btu/h·ft²·°F. Assuming that both types of equipment operate with water–water systems, overall clean coefficients of 500 and 1000 Btu/h·ft²·°F respectively are obtained. Using a typical fouling resistance of 0.001 h·ft²·°F/Btu (equal to a coefficient of 1000 Btu/h·ft²·°F), the inclusion of the fouling will cause the tubular exchanger size to increase by a factor of 4, while for the plate exchanger, the corresponding factor is 7.

This example clearly demonstrates the crucial importance of fouling, especially in plate exchangers. Yet, fouling resistances that are unrealistically high often are specified and invariably have been based on experiences derived from tubular equipment. The common source of water fouling resistances in TEMA which recommends R_f values spanning a tremendous range between 0.0015 and 0.005 h·ft²·°F/Btu.

Flow velocity, as mentioned earlier, is a crucial operating parameter that influences the fouling behavior. For flow inside the tubes, the definition of flow velocity and the velocity profile is straightforward. But in plate exchangers, flow velocity is characterized by constant fluctuations as the fluid passes over the corrugations. It is postulated that this induces turbulence that is superimposed on the flow velocity as a factor that diminishes fouling tendencies. This has been observed qualitatively in practical applications and confirmed by unpublished APV research.

TEST APPARATUS AND CONDITIONS

The plate heat exchanger (PHE) tested was an APV Model 405 using APV Type R40 stainless-steel plates 45 in. high and 18 in. wide. The plate heat-transfer area was 4 ft² with a nominal gap between plates of 0.12 in. with the plate corrugations, the maximum gap was 0.24 in. Seven plates were used, creating three countercurrent passages each of the cooling water and the heating medium. A schematic diagram of the installation is shown in Figure 3.

The PHE was mounted on the HTRI Shellside Fouling Research Unit (SFRU) together with two small stainless-steel shell-and-tube exchangers. The PHE was heated using hot steam condensate, and the fouling was determined from the degradation of the overall heat-transfer coefficient. Simultaneously, tests also were run on an HTRI Portable Fouling Research Unit (PFRU), which uses electrically heated rods with the cooling water flowing in an annulus. The SFRU and PFRU are described (5) elsewhere.

The fouling tests were conducted at a major petrochemical plant in the Houston, Texas area. The test units were installed near the cooling-tower basin on the plant. The cooling-tower operating characteristics are summarized in Table 1. The 140,000-gpm system had two 1600-gpm sidestream filters. Filter backwash accounted for most of the blowdown. The makeup water for the cooling-water system came from several sources and was clarified with alum outside the plant. The water treatment used is as follows:

- Chromate-zinc-based inhibitor for corrosion control (20–25 ppm chromate)
- Organic phosphonate and polymer combination as dispersant (2 ppm organic phosphonate)
- Polyphosphate as anodic passivator (5–6 ppm total inorganic phosphate)
- Chlorine for biological control (continuous feeding of 386 lb/day)
- Biocide for biological control (15 ppm biocide once per week)
- Sulfuric acid for pH control (pH 6–6.5)

TEST PROCEDURE

The velocity for the PHE is defined by

$$V = \frac{w}{\rho A_c}$$

where V = velocity
w = mass flow rate
ρ = fluid density
A_c = cross-sectional flow area

The cross-sectional flow area is based on the compressed gasket spacing. In other words, the flow are is computed as if there were no corrugations but smooth plates instead. The surface temperature is defined as the fluid–deposit interface temperature. Since the PHE was heated by constant temperature steam condensate, the metal temperature remained constant and the surface temperature decreased as fouling built up. In addition, because of the counterflow nature of the PHE, the surface temperature was not constant from inlet to outlet. The surface temperature reported here is the initial surface temperature at the midpoint of the plate.

Conditions of operation were selected so that all the SFRU test exchangers operated roughly at the same nominal flow velocity and surface temperature. During the operation, weekly water samples were taken for chemical analysis. At the end of each test, deposits were photo-

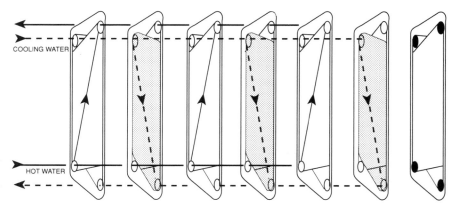

Figure 3. APV plate heat exchanger with three parallel countercurrent parallel passages for both the cooling water and the hot water.

Table 1. Cooling-Water System Characteristics

Tower and Circulation System	Description	
Circulation rate	140,000 gpm	
Temperature difference across tower	23.4°F	
Number of sidestream filters	2	
Corrosion Rate from Carbon Steel Coupon Tests	<1 mil/yr	
Water Description	**Cooling-Tower Water**	**Makeup Water**
Treatment	10 cycle concentration	Unclarified San Jacinto River 1000 gpm
	Blowdown from coolers and sidestream filters	Clarified San Jacinto River 2400 gpm
		Lissie Sand Well 475 gpm
Typical Composition		
Total hardness as $CaCO_3$, ppm	520	48
Calcium as $CaCO_3$, ppm	420	40
Magnesium as $CaCO_3$, ppm	120	8
Methyl Orange Alkalinity as $CaCO_3$, ppm	20	34
Sulfate as SO_4, ppm	1600	34
Chloride as Cl, ppm	800	48
Silica as SiO_2, ppm	150	18
Total inorganic phosphate as PO_4, ppm	10	
Orthophosphate as PO_4, ppm	7	
pH	6	7.5
Specific conductance, micromhos, 18C	5000	300
Chromate as CrO_4, ppm	25	
Chromium as Cr, ppm	0.5	
Soluble zinc as Zn, ppm	2.5	
Total iron as Fe, ppm		0.8
Suspended solids, ppm	100	10

graphed and sampled for chemical analysis. The exchanger plates then were cleaned and the unit reassembled and new test conditions established.

TEST AND RESULT DESCRIPTION

Five test series were performed testing several velocities and surface temperatures (eg, see test 2 in Fig. 4). Typical fouling time histories are shown in Figure 5 along with the operating conditions. Notice that the PHE fouling resistances establish a stable asymptotic value after about 600–900 h of operation. The results of the other tests are shown in Figure 6 only as the values of the asymptotic fouling resistances, as these are the data required for design.

The chemical analysis of the fouling deposit was made after each test and the results are summarized in Table 2. The primary elements found were phosphorus, zinc, and chrome from the water treatment and silicon from suspended solids in the water.

The photograph of the plates of the PHE shown in Figure 4 at the termination of test 2 indicates that fouling occurred only in the upper third of the plates. This is the region near the hot-water inlet and cold-water outlet, region of high surface temperature. Figure 5 shows a surface temperature profile for test 2 conditions of 2.8 ft/s and an inlet bulk temperature of 90°F. From the water chem-

Table 2. Fouling Deposit Analysis, %

	Series 2	Series 3	Series 4	Series 5	Series 6
Loss of ignition	—	25	25	17	21
Phosphorus	19	24	13	12	22
Aluminum	2	2	8	6	4
Silicon	1	4	18	16	13
Calcium	4	8	3	15	9
Chrome	5	13	15	15	9
Iron	2	7	4	5	13
Zinc	2	20	12	13	7
Sodium	—	1	1	—	1
Magnesium	—	—	1	—	1

Figure 4. Fouled plates, test 2.

istry parameters, the saturation temperature for calcium phosphate above which precipitation is expected was calculated to be 163°F (6). The shaded region on Fig. 7 indicates the expected precipitation region and corresponds closely to the fouled region seen in Figure 4.

The surface temperatures during some of the tests were higher than those experienced in normal PHE operation. This is the result of the setup condition criteria that the three exchangers on the SFRU operate at the same midpoint surface temperature. Since the ratio of flow rate to surface area for the shell side of the shell and tube exchangers is twice that for the PHE, the cooling water had a longer residence time in the PHE than in the tubular exchanger. As a result, the lower flow rate to surface area ratio yields a larger temperature rise of the cooling water. The result is the steeper surface temperature profile shown in Figure 7. Although cocurrent operation was investigated as a correction for this problem, the required higher hot-water flow rate was not available. Consequently, the PHE data in some of these tests was slightly penalized because part of the surface is above the critical temperature for tricalcium phosphate precipitation, a condition that normally would not be encountered in industrial operations.

1. The loss of ignition (LOI) is performed by drying the sample to constant weight at 220°F (105°C) and then combusting at 1470°F (800°C). LOI represents bound water and organics present in the deposit. An organic test on samples of test series 5 showed that about $\frac{1}{3}$ of the LOI was organic.

2. Aluminum, silicon, and iron come from the suspended solids. Aluminum comes from operation of the clarifier. The silicon is not from magnesium silicate since little magnesium is in the analyses.

3. Chrome, zinc, and phosphorus are derived from the water treatment.

4. Low values of calcium indicate little or no tricalcium phosphate. Most of the calcium comes from association with zinc and polyphosphates in the treatment.

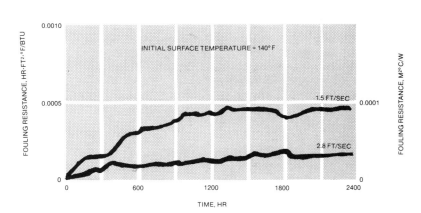

Figure 5. Typical PHE fouling curves.

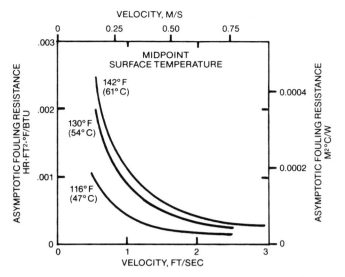

Figure 6. Asymptotic fouling resistance versus velocity with surface temperature as a parameter.

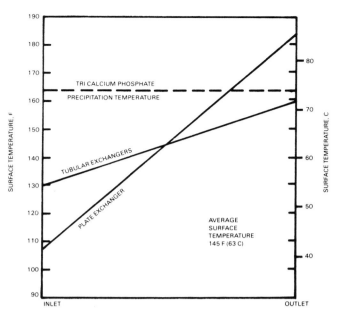

Figure 7. Surface temperature profile with indicated calcium phosphate precipitation.

It is difficult to make meaningful comparisons between the fouling tendencies of different types of heat exchangers. Unless carefully done, the results are misleading. This is particularly true when attempting to compare the fouling experience in the PHE to that of tubeside operation as in the PFRU. The geometries, surface area, and operational characteristics are very different even though the cooling water may be the same. Such comparisons often are attempted using TEMA recommended fouling resistances.

TEMA recommends a fouling resistance of 0.002 h·ft²·°F/Btu for the water system used. However, the maximum fouling resistance measured in the PHE was less than 0.0005 h·ft²·°F/Btu-only 25% of the TEMA recommendation. This confirms the earlier assumption that applying TEMA-recommended fouling resistances for a shell and tube exchangers to the PHE seriously handicaps the performance ratings. Because of the inherently high heat-transfer coefficients, the effects of fouling resistances are more pronounced. Consider a PHE operating at 1.5 ft/s. From Table 3, the fouled overall coefficient using a TEMA fouling resistance of 0.002 h·ft²·°F/Btu is about one half of the overall coefficient using the measured fouling resistance of this investigation. This example illustrates the need for caution in using TEMA fouling recommendations for equipment other than shell-and-tube exchangers.

Figure 8 compares the performance of the PHE with typical tube-side data and with the TEMA-recommended fouling resistance.

A direct comparison of the PHE and the shell-side exchangers can be made through the deposit analyses. A sample analysis is given in Table 4. The amounts of phosphates, calcium, chrome, and zinc are somewhat higher for the PHE. The operation of the PHE at a higher surface temperature than the shell-side units resulted in more crystallization fouling. However, the amount of silicon for the shell-side unit is much higher than for the PHE. The high silicon concentration is due to sedimentation in the baffle-shell corners of the bundle, while the high turbulence promoted by the corrugated plates in the PHE minimizes sedimentation fouling. In heavily sediment-laden waters, the PHE would be especially superior.

CONCLUSIONS

Although the tests were performed on one water system only experience with tubular data indicates that the overall trends generally are valid; ie, plate exchangers should be designed to substantially lower values of fouling than would be used on tubular equipment for the same stream.

Table 3. Effects of Fouling Resistance on PHE Performance

	Overall Heat-Transfer Coefficient		
		Fouling Coefficient	
Operating Velocity (Both Sides)	Clean	TEMA 0.002 h·ft²·°F/Btu (0.00035 m²°C/W)	Present Study 0.0005 h·ft²·°F/Btu (0.00009 m²°C/W)
1.5 ft/s	1078 Btu/h·ft²·°F	341 Btu/h·ft²·°F	700 Btu/h·ft²·°F
0.45 m/s	6121 W/m²·°C	1936 W/m²·°C	3975 W/m²·°C

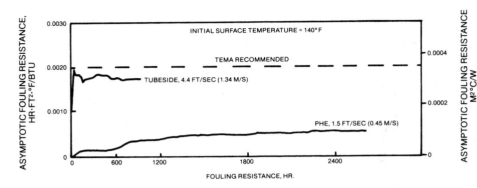

Figure 8. Comparison of tube-side and PHE fouling.

However, the comparison is not always that straightforward, as flow velocity itself is not a valid criterion. The wall shear stress in a PHE operating at 2.8 ft/s is equivalent to that in a tube-side exchanger operating at 8.2 ft/s. On the basis of typical industrial velocities, it was found that the PHE at 1.5 ft/s fouls about one-half as much as a tube-side exchanger operating at 5.9 ft/s. Furthermore, because of the high-heat-transfer coefficients typical to the PHE, surface temperature may differ from those in tubular equipment and should be carefully watched.

The turbulence inducing corrugation pattern prevalent in heat-exchanger plates produces very high local velocities. This results in high friction factors and therefore high shearing. This high shear, in turn, results in less fouling in plate exchangers than in tubular units. For cooling-water duties, Heat Transfer Research Incorporated (HTRI) has shown that the plate exchanger fouls at a much lower rate than either the tube side or the shell side of tubulars. TEMA recommendations for fouling therefore should never be used for plate units since they probably are five times the value found in practice. In particular, the plate exchanger is far less susceptible to fouling with heavy sedimented waters. This is contrary to popular belief. It must be noted, however, that all particular matter must be significantly smaller than the plate gap and generally particles over 0.1 in. in diameter cannot be handled in any plate heat exchanger.

It has been shown that quoting a high fouling resistance can negate a plate heat exchanger design by adding large amounts of surface and thereby overriding the benefits of the high coefficients. Fouling design resistance, therefore, should be chosen with care, keeping in mind that with a Paraflow unit, it always is possible to add or subtract surface to meet exact fouling conditions.

Disclaimer: As a general policy, HTRI as the project operator disclaims responsibility for any calculations or designs resulting from the use of the date presented.

BIBLIOGRAPHY

Adapted from:

J. D. Usher and A. Cooper, "Paraflow Seminar, Part Four," *CPI Digest* **3**(1): 6–9 (1974).

J. D. Usher and A. Cooper, "Cooling Water Fouling in Plate Heat Exchangers," *CPI Digest* **5**(4): 2–9 (1979).

Copyrighted APV Crepaco, Inc. Used with permission.

1. J. Taborek, T. Aoki, R. B. Ritter, J. W. Palen, and J. G. Knudsen, "Fouling—The Major Unresolved Problem in Heat Transfer, Parts I and II," *CEP* **68**, (2,7): (1972).
2. J. W. Suitor, W. J. Marner, and R. B. Ritter, "The History and Status of Research in Fouling of Heat Exchangers in Cooling Water Service," Paper No. 76-CSME/CSChE—19 presented at the 16th National Heat Transfer Conference, St. Louis, Missouri, August 8–11, 1976.
3. R. W. Morse and J. G. Knudsen, "Effect of Alkalinity of the Scaling of Simulated Cooling Tower Water," Paper No. 76-CSME/CSChE—24 presented at the 16th National Heat Transfer Conference, St. Louis, Missouri, August 8–11, 1976.
4. *Standards of Tubular Exchanger Manufacturers Association,* 5th Ed., New York, 1968.
5. P. Fisher, J. W. Suitor, and R. B. Ritter, "Fouling Measurement Techniques and Apparatus," *CEP,* **71**(7): 66 (1975).
6. J. Green and J. A. Holmes, "Calculation of the pH of Saturation of Tricalcium Phosphate," *Journal of the American Water Works Association* **39**(11): (1947).

APV Crepaco, Inc.
Lake Mills, Wisconsin

HEAT EXCHANGERS: PARAFLOW

PRINCIPLES AND APPLICATIONS

The Paraflow is the original plate-type heat exchanger designed by APV to provide maximum efficiency and cost-

Table 4. Comparison of Deposit Analyses from Plate Exchanger and Shell-Side Test Exchanger for Series 5

	Fouling Deposit Analysis, %	
	Plate	Shell-Side
Loss on ignition	17	24
Phosphorus	12	9
Aluminum	6	7
Silicon	16	33
Calcium	15	7
Chrome	15	9
Iron	5	4
Zinc	13	8

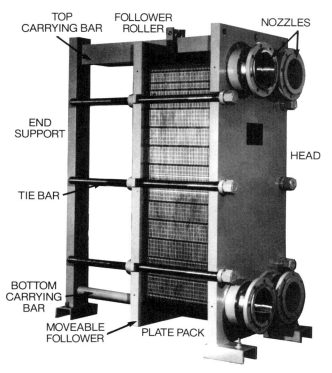

Figure 1. The APV CREPACO Paraflow plate heat exchanger.

Figure 3. Cutaway of Paraflow plate shows turbulence during passage of product and service liquids.

effectiveness in handling thermal duties while minimizing maintenance downtime and floor space requirements.

Frame, Plates, and Gaskets

The Paraflow plate heat exchanger as shown in Figure 1 consists of a stationary head and end support connected by a top carrying bar and bottom guide rail. These form a rigid frame that supports the plates and moveable follower. In most units, plates are securely compressed between the head the follower by means of tie bars on either side of the exchanger. In a few models, central tightening spindles working against a reinforced end support are used for compression. When Paraflows are opened, the follower moves easily along the top bar with the aid of a bearing supported roller to allow full access to each individual plate.

With the exception of some sanitary models that are clad with stainless steel, Paraflow frames are fabricated of carbon steel and are finished in chemical-resistant epoxy paint. Frame ports accept bushings of stainless steel or alternative metals which, with various types of flanged or sanitary connections, form the inlet and outlet nozzles. By using intermediate connector plates as shown in Figure 2, units can be divided into separate sections to accommodate multiple duties within a single frame.

The closely spaced metal heat-transfer plates have troughs or corrugations that induce turbulence to the liquids flowing as a thin stream between the plates (Fig. 3).

The plates have corner ports which in the complete plate pack form a manifold for even fluid distribution to the individual plate passages (Fig. 4).

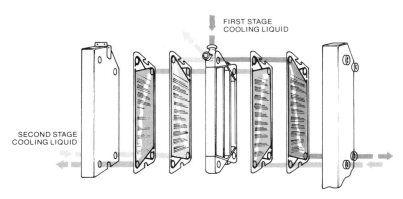

Figure 2. Two-section Paraflow with connector plate.

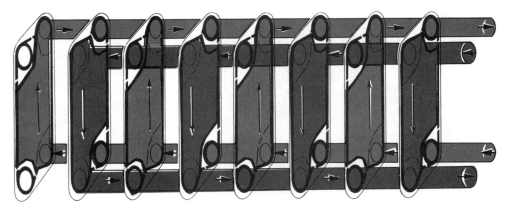

Figure 4. Single-pass countercurrent flow.

The seal between the plates is established by a peripheral gasket that also separates the thruport and flow areas with a double barrier. The interspace is vented to atmosphere to prevent cross-contamination in the rare event of leakage (Fig. 5).

As an exclusive feature, Paraflow heat-exchanger plates have interlocking gaskets in which upstanding lugs and scallops are sited intermittently around the outside edges. These scallops ensure that there are no unsupported portions of the gaskets and, in combination with the patented form of pressed groove, provide mechanical plate-to-plate support for the sealing system. The upstanding lugs (Fig. 6) maintain plate alignment in the Paraflow during pack closure and operation. The groove form provides 100% peripheral support of the gasket, leaving none of the material exposed to the outside. In addition, the gasket–groove design minimizes gasket exposure to the process liquid.

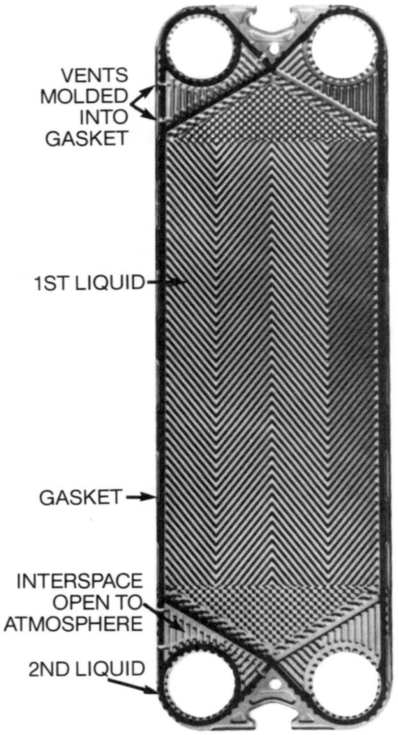

Plate Arrangement

Comparison of Paraflow plate arrangement to the tube and shell-side arrangement in a shell and tube exchanger is charted in Fig. 7. Essentially, the number of passes on the tube side of a tubular unit can be compared with the number of passes on a plate heat exchanger. The number of tubes per pass also can be equated with the number of passages per pass for the Paraflow. However, the comparison with the shell side usually is more difficult since with a Paraflow, the total number of passages available for the flow of one fluid must equal those available for the other fluid to within ±1. The number of cross passes on a shell, however, can be related to the number of plate passes and since the number of passages/pass for a plate is an indication of the flow area, this can be equated to the shell diameter. This is not a perfect comparison but it does show the relative parameters for each exchanger.

Figure 5. Gasket showing separation of throughport and flow areas.

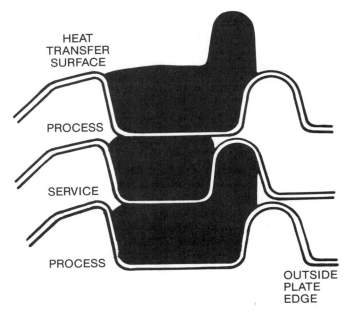

Figure 6. Exclusive interlocking gasket.

	Shell and Tube	Plate Equivalent	
Tube side	No. of passes	No. of passes	Side 1
Shell side	No. of tubes/pass No. of cross passes (No. of baffles +1)	No. of passages/pass No. of passes	Side 2
	Shell diameter	No. of passages/pass	

Figure 7. Pass arrangement comparison: plate versus tubular.

With regard to flow patterns, the Paraflow advantage over shell and tube designs is the ability to have equal passes on each side in full countercurrent flow, thus obtaining maximum utilization of the temperature difference between the two fluids. This feature is particularly important in heat-recovery processes with close temperature approaches and even in cases with temperature crossovers.

Whenever the thermal duty permits, it is desirable to use single-pass, countercurrent flow for an extremely efficient performance. Since the flow is pure counterflow, correction factors required on the LMTD approach unity. Furthermore, with all connections located at the head, the follower is easily moved and plates are more readily accessible.

Plate Construction

Depending on type, some plates employ diagonal flow while others are designed for vertical flow (Fig. 8). Plates are pressed in thicknesses between 0.020 and 0.036 in. (0.5–0.9 mm) and the degree of mechanical loading is im-

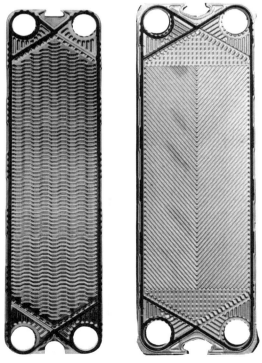

Figure 8. Diagonal and vertical flow patterns.

Figure 9. Corrugations pressed into plates are perpendicular to the liquid flow.

portant. The most severe case occurs when one process liquid is operating at the highest working pressure and the other at zero pressure. The maximum pressure differential is applied across the plate and results in a considerable unbalanced load that tends to close the typical 0.1–0.2-in. gap. It is essential, therefore, that some form of interplate support be provided to maintain the gap and this is done by two different plate forms.

One method is to press pips into a plate with deep washboard corrugations to provide contact points for about every 1–3 in.^{2} of heat-transfer surface (Fig. 9). Another is the chevron plate of relatively shallow corrugations with support maintained by the peak–peak contact (Fig. 10). Alternate plates are arranged so that corrugations cross to provide a contact point for every 0.2–1 in.^{2} of area. The plate then can handle a large differential pressure, and the cross pattern forms a tortuous path that promotes substantial liquid turbulence and thus a very high heat transfer coefficient.

Mixing and Variable-Length Plates

To obtain optimum thermal and pressure drop performance while using a minimum number of heat-exchanger plates, mixing and variable-length plates are available for several APV Paraflow plate heat-exchanger models. These plates are manufactured to the standard widths specified for the particular heat exchanger involved but are offered in different corrugation patterns and plate lengths.

Since each type of plate has its own predictable performance characteristics, it is possible to calculate heat-transfer surface that more precisely matches the required

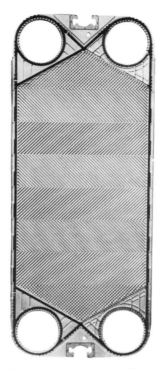

Figure 10. Troughs are formed at opposite angles to the centerline in adjacent plates.

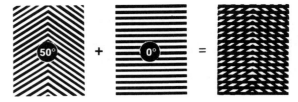

Figure 12. High HTU passage.

thermal duty without oversizing the exchanger. This results in the use of fewer plates and a smaller, less expensive exchanger frame.

To achieve mixing, plates that have been pressed with different corrugation angles are combined within a single heat-exchanger frame. This results in flow passages that differ significantly in their flow characteristics and thus heat-transfer capability from passages created by using plates that have the same corrugation pattern.

For example, a plate pack (Fig. 11) of standard plates that have a typical 50° corrugation angle (to horizontal) develops a fixed level of thermal performance (HTU) per unit length. As plates of 0° angle (Fig. 12) are substituted into the plate pack up to a maximum of 50% of the total number of plates, the thermal performance progressively increases to a level that typically is twice that of a pack containing only 50° angle plates.

Thus, it is possible for a given plate length to fine-tune the Paraflow design in a single or even multiple-pass arrangement exactly to the thermal and pressure drop requirements of the application.

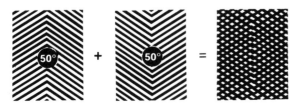

Figure 11. Low HTU passage.

Of more recent development are plates of fixed width with variable lengths that extend the range of heat-transfer performance in terms of HTU. This is proportional to the effective length of the plate and typically, provides a range of 3–1 from the longest to the shortest plate in the series. As shown in Figure 13, mixing also is available in plates of varied lengths and further increases the performance range of the variable-length plate by a factor of approximately 2.

This extreme flexibility of combining mixing and variable-length plates allows more duties to be handled by a single-pass design, maintaining all connections on the stationary head of the exchanger to simplify piping and unit maintenance.

Plate Size and Frame Capacity

Paraflow plates are available with effective heat-transfer area from 0.28 to 50 ft^2, and up to 600 of any one size can be contained in a single standard frame. The largest Paraflow can provide in excess of 30,000 ft^2 of surface area. Flow ports are sized in proportion to the plate area and control the maximum permissible liquid throughput Figure 14. Flow capacity of the individual Paraflow, based on a maximum port velocity of 20-ft/s ranges from 15 gpm in the "junior" to 11,000 gpm in the Model SR235. This velocity is at first sight somewhat high compared to conventional pipework practice. However, the high fluid velocity is very localized in the exchanger and progressively is reduced as distribution into the flow passages occurs from the port manifold. If pipe runs are long, it is not uncommon to see reducers fitted in the piping at the inlet and exit connections of high-throughput machines.

Plate Materials

Paraflow plates may be pressed from 304 or 316 stainless steels, Avesta 254SMO or 254SLX, nickel 200, Hastelloy B-2, C-276 or G-3, Incoloy 825, Inconel 625, Monel 400, titanium or titanium–palladium as required to provide suitable corrosion resistance to the streams being handled.

Gasket Materials

As detailed in Figure 15, various gasket materials are available as standard that have chemical and temperature resistance coupled with excellent sealing properties. These qualities are achieved by specifically compounding and molding the elastomers for long-term performance in the APV Paraflow.

Figure 13. Variable-length plates with mixing options.

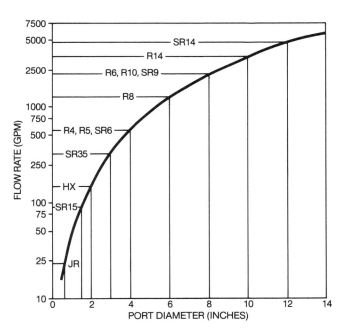

Figure 14. Throughput vs. port diameter at 14 ft/s.

Since the temperatures shown are not absolute, gasket material selection must take into consideration the chemical composition of the streams involved as well as the operating cycles.

THERMAL PERFORMANCE

The Paraflow plate heat exchanger is used most extensively in liquid–liquid duties under turbulent flow conditions. In addition, it is particularly effective for laminar flow heat transfer and is used in condensing, gas cooling, and evaporating applications.

Turbulent Flow

For plate heat transfer in turbulent flow, thermal performance can best be exemplified by a Dittus–Boelter-type equation:

$$\mathrm{Nu} = (C)(\mathrm{Re})^n (\mathrm{Pr})^m \left(\frac{\mu}{\mu_w}\right)^m$$

Gasket Material	Approximate Maximum Operating Temperature	Application
Paracril (medium nitrile)	275°F (135°C)	General aqueous service, aliphatic hydrocarbons
Paratemp (EPDM)	300°F (150°F)	High temperature resistance for a wide range of chemicals and steam
Paradur (fluoroelastomer)	400°F (205°C)	Mineral oils, fuels, vegetable and animal oils
Paraflor (fluoroelastomer)	400°F (205°C)	Steam, sulfuric acids

Figure 15. Paraflow gasket guide.

where Nu = Nusselt number hD_e/k

Re = Reynolds number vD_e/μ

Pr = Prandtl number $C_p\mu/k$

D_e = equivalent diameter (2× average plate gap)

$\left(\dfrac{\mu}{\mu_w}\right)$ = Sieder–Tate correction factor

and reported values of the constant and exponents are

$$C = 0.15-0.40, \quad m = 0.30-0.45$$
$$n = 0.65-0.85, \quad x = 0.05-0.20$$

Typical velocities in plate heat exchangers for waterlike fluids in turbulent flow are 1–3 ft/s (0.3–0.9 m/s), but true velocities in certain regions will be higher by a factor of ≥4 due to the effect of the corrugations. All heat-transfer and pressure drop relationships are, however, based on either a velocity calculated from the average plate gap or on the flow rate per passage.

Figure 16 illustrates the effect of velocity for water at 60°F on heat-transfer coefficients. This graph also plots pressure drop against velocity under the same conditions. The film coefficients are very high and can be obtained for a moderate pressure drop.

One particularly important feature of the Paraflow is that the turbulence induced by the troughs reduces the Reynolds number at which the flow becomes laminar. If the characteristic length dimension in the Reynolds number is taken at twice the average gap between plates, the Re number at which the flow becomes laminar varies from about 100 to 400 according to the type of plate.

To achieve these high coefficients, it is necessary to expend energy. With the plate unit, the friction factors normally encountered are in the range of 10–400 times those inside a tube for the same Reynolds number. However, nominal velocities are low and plate lengths do not exceed 7.5 ft so that the term $(V^2)L/(2g)$ in the pressure drop equation is much smaller than one normally would encounter in tubulars. In addition, single-pass operation will achieve many duties so that the pressure drop is efficiently used and not wasted on losses due to flow direction changes.

The friction factor is correlated with the equations:

$$f = \frac{B}{(\text{Re})^y}, \quad \Delta p = \frac{fL_\rho V^2}{2gd}$$

where y varies from 0.1 to 0.4 according to the plate and B is a constant characteristic of the plate.

If the overall heat-transfer equation $Q = UA\,\Delta T$ is used to calculate the heat duty, it is necessary to know the overall coefficient U, the surface area A, and the mean temperature difference ΔT.

The overall coefficient U can be calculated from

$$\frac{1}{U} = r_{\text{fh}} + r_{\text{fc}} + r_w + r_{\text{dh}} + r_{\text{dc}}$$

The values of r_{fh} and r_{fc} (the film resistances for the hot and cold fluids, respectively) can be calculated from the Dittus–Boelter equations described previously and the wall metal resistance r_w can be calculated from the average metal thickness and thermal conductivity. The fouling resistances of the hot and cold fluids r_{dh} and r_{dc} often are based on experience.

The value taken for A is the developed area after pressing. That is the total area available for heat transfer and, because of the corrugations, will be greater than the projected area of the plate, ie, 1.81 ft² versus 1.45 ft² for an HX plate.

The value of ΔT is calculated from the logarithmic mean temperature difference multiplied by a correction factor. With single pass operation, this factor is about 1 except for plate packs of less than 20 when the end effect has a significant bearing on the calculation. This is due to the fact that the passage at either end of the plate pack only transfers heat from one side and therefore the heat load is reduced.

When the plate unit is arranged for multiple-pass use, a further correction factor must be applied. Even when two passes are countercurrent to two other passes, at least one of them must experience cocurrent flow. This correction factor is shown in Figure 17 against a number of heat transfer units (HTU = temperature rise of the process fluid divided by the mean temperature difference). As indicated, whenever unequal passes are used, the correction factor calls for a considerable increase in area. This is particularly important when unequal flow conditions are handled. If high and low flow rates are to be handled, the necessary velocities must be maintained with the low fluid flow rate by using an increased number of passes. Although the plate unit is most efficient when the flow ratio between two fluids is in the range of 0.7–1.4, other ratios can be handled with unequal passes. This is done, however, at the expense of the LMTD factor.

The issue of how to specify a fouling resistance for a plate heat exchanger is difficult to resolve. Manufacturers generally specify 5% excess HTU for low fouling duties, 10% for moderate fouling, and 15–20% excess for high fouling. The allowed excess surface almost always is suffi-

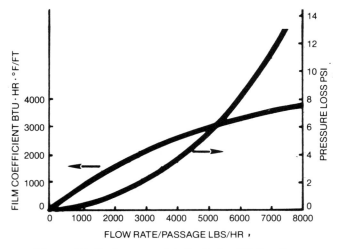

Figure 16. Performance details: Series HX Paraflow.

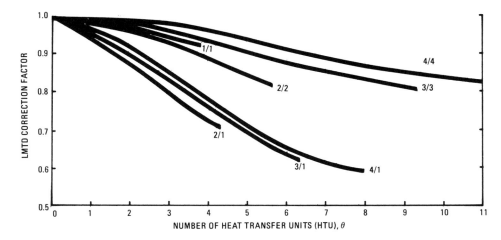

Figure 17. LMTD correction factor.

cient even though in many cases, it represents a low absolute value of fouling in terms of (Btu/h · ft² · °F)⁻¹. If a high fouling resistance is specified, extra plates have to be added, usually in parallel. This results in lower velocities, more extreme temperatures during startup, and the probability of higher fouling rates. Because of the narrow plate gaps and, in particular, because of the small entrance and exit flow areas, fouling in a plate usually causes more problems by the increase in pressure drop and/or the lowering of flow rates than by causing large reductions in heat-transfer performance. Increasing the number of plates does not increase the gap or throat area, and a plate unit sometimes can foul more quickly when oversurfaced.

The customer who does not have considerable experience with both the process and the plate heat exchanger should allow the equipment manufacturer to advise on fouling and the minimum velocity at which the exchanger should operate.

Laminar Flow

The other area suitable for the plate heat exchanger is that of laminar flow heat transfer. It has been pointed out already that the Paraflow can save surface by handling fairly viscous fluids in turbulent flow because the critical Reynolds number is low. Once the viscosity exceeds 20–50 cP, however, most plate heat exchanger designs fall into the viscous flow range. Considering only Newtonian fluids since most chemical duties fall into this category, in laminar ducted flow the flow can be said to be one of three types: (1) fully developed velocity and temperature profiles (ie, the limiting Nusselt case), (2) fully developed velocity profile with developing temperature profile (ie, the thermal entrance region), or (3) the simultaneous development of the velocity and temperature profiles.

The first type is of interest only when considering fluids of low Prandtl number, and this seldom exists with normal plate heat-exchanger applications. The third is relevant only for fluids such as gases that have a Prandtl number of about 1. Therefore, consider type 2.

As a rough guide for plate heat exchangers, the ratio of the hydrodynamic entrance length to the corresponding thermal entrance length is given by

$$\frac{L_{th}}{L_{hyd}} = 1.7 \, Pr$$

Plate the heat transfer for laminar flow follows the Dittus–Boelter equation in this form:

$$Nu = c \left(\frac{Re \cdot Pr \cdot De}{L}\right)^{1/3} \left(\frac{\mu}{\mu_w}\right)^x$$

where L = nominal plate length
c = constant for each plate (usually in the range 1.86–4.50)
x = exponent varying from 0.1 to 0.2 depending on plate type

For pressure loss, the friction factor can be taken as $f = a/Re$ where a is a constant characteristic of the plate.

It can be seen that for heat transfer, the plate heat exchanger is ideal because the value of d is small and the film coefficients are proportional to $d^{-2/3}$. Unfortunately, however, the pressure loss is proportional to d^{-4} and the pressure drop is sacrificed to achieve the heat transfer.

From these correlations, it is possible to calculate the film heat-transfer coefficient and the pressure loss for laminar flow. This coefficient combined with the metal coefficient and the calculated coefficient for the service fluid together with the fouling resistance then are used to produce the overall coefficient. As with turbulent flow, an allowance has to be made to use the LMTD to allow for either end effect correction for small plate packs and/or concurrency caused by having concurrent flow in some passages. This is particularly important for laminar flow since these exchangers usually have more than one pass.

Beyond Liquid–Liquid

Over many years, APV has built up considerable experience in the design and use of Paraflow plate heat exchangers for process applications that fall outside the normal turbulent flow that is common in chemical operations. The Paraflow, for example, can be used in laminar flow duties, for the evaporation of fluids with relatively high viscosities, for cooling various gases, and for condensing applications where pressure drop parameters are not overly restrictive.

Condensing

One of the most important heat-transfer processes if possible, is the condensation of vapors—a duty that often is carried out on the shell side of a tubular exchanger but is entirely feasible in the plate-type unit. Generally speaking, the determining factor is pressure drop.

For those condensing duties where permissible pressure loss is less than one pound per square inch, there is no doubt but that the tubular unit is most efficient. Under such pressure drop conditions, only a portion of the length of a Paraflow plate would be used and substantial surface area would be wasted. However, when less restrictive pressure drops are available the plate heat exchanger becomes an excellent condenser since very high heat-transfer coefficients are obtained and the condensation can be carried out in a single pass across the plate.

The pressure drop of condensing steam in the passages of plate heat exchangers has been investigated experimentally for a series of different Paraflow plates. As indicated in Figure 18, which provides data for a typical unit, the drop obtained is plotted against steam flow rate per passage for a number of inlet steam pressures.

It is interesting to note that for a set steam flow rate and a given duty, the steam pressure drop is higher when the liquid and steam are in countercurrent rather than cocurrent flow. This is due to differences in temperature profile.

Figure 19 shows that for equal duties and flows, the temperature difference for countercurrent flow is lower at the steam inlet than at the outlet with most of the steam condensation taking place in the lower half of the plate. The reverse holds true for cocurrent flow. In this case, most of the steam condenses in the top half of the plate, the mean vapor velocity is lower, and a reduction in pressure drop of 10–40% occurs. This difference in pressure drop becomes lower for duties where the final approach temperature between the steam and process fluid becomes larger.

The pressure drop of condensing steam therefore is a function of steam flow rate, pressure, and temperature difference. Since the steam pressure drop affects the saturation temperature of the steam, the mean temperature difference in turn becomes a function of steam pressure drop. This is particularly important when vacuum steam is being used since small changes in steam pressure can

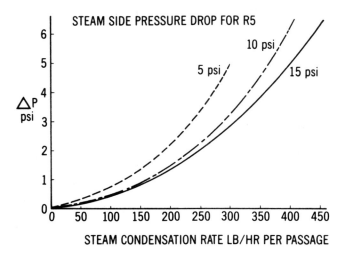

Figure 18. Steam-side pressure drop for R5.

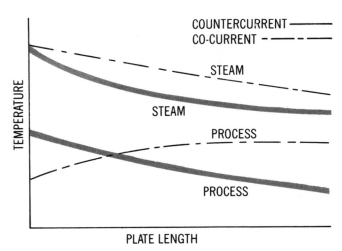

Figure 19. Temperature profile during condensation of steam.

give significant changes in the temperature at which the steam condenses.

By using an APV computer program and a Martinelli–Lockhart-type approach to the problem, it has been possible to correlate the pressure loss to a high degree of accuracy. Figure 20 cites a typical performance of a steam heated Series R4 Paraflow. From this experimental run during which the exchanger was equipped with only a small number of plates, it can be seen that for a 4–5-psi pressure drop and above, the plate is completely used. Below that figure, however, there is insufficient pressure drop available to fully use the entire plate and part of the surface therefore is flooded to reduce the pressure loss. At only one psi allowable pressure drop, only 60% of the plate is used for heat transfer, which is not particularly economic.

This example, however, well illustrates the application of a plate heat exchanger to condensing duties. If sufficient pressure loss is available, then the plate type unit is a good condenser. The overall coefficient of 770 Btu/h · ft² · °F for 4–5-psi pressure loss is much higher than a coefficient of 450–500 Btu/h · ft² · °F, which could be expected in a tubular exchanger for this type of duty. However, the tubular design would, for shell-side condensation, be less dependent on available pressure loss and for a 1-psi drop, a 450–500 Btu overall coefficient still could be obtained. With the plate, the calculated coefficient at this pressure is 746 Btu but the effective coefficient based on total area is only 60% of that figure or 445 Btu/h · ft² · °F.

Gas Cooling

Plate heat exchangers also are used for gas cooling with units in service for cooling moist air, hydrogen, and chlorine. The problems are similar to those of steam heating since the gas velocity changes along the length of the plate due either to condensation or pressure fluctuations. Designs usually are restricted by pressure drop, so machines with low pressure drop plates are recommended. A typical allowable pressure loss would be 0.5 psi with rather low gas velocities giving overall heat transfer coefficients in the region of 50 Btu/h · ft² · °F.

	1	2	3	4
Available pressure loss, psi	1	2	3	4
Total duty, Btu/h	207,000	256,000	320,000	333,000
Fraction of plate flooded	40	30	4	0
Effective overall heat-transfer coefficient, clean	445	520	725	770
Pressure loss, psi	1	2	4	4.5

Figure 20. Steam heating in an R4 Paraflow (water flow rate 16,000 lb/h, inlet water temperature 216°F, Inlet steam temperature 250°F, total number of plates 7).

Evaporation

The plate heat exchanger also can be used for evaporation of highly viscous fluids when as a Paravap the evaporation occurs in the plate or as a Paraflash the liquid flashes after leaving the plate. Applications generally have been restricted to the soap and food industries. The advantage of these units is their ability to concentrate viscous fluids of up to 5000 cP.

CONCLUSION

It has been shown that the plate heat exchanger is a relatively simple machine on which to carry out a thermal design. Unlike the shell side of a tubular exchanger where predicting performance depends on baffle–shell leakage, baffle–tube leakage, and leakage around the bundle, it is not possible to have bypass streams on a Paraflow. The only major problem is that the pressure loss through the ports can cause unequal distribution in the plate pack. This is overcome by limiting the port velocity and by using a port pressure loss correlation in the design to allow for the effect of unequal distribution.

The flow in a plate also is far more uniform than on the shell side. Furthermore, there are no problems in calculation of heat transfer in the window, across the bundle, or allowing for dead spots as is the case with tubular exchangers. As a result, the prediction of performance is simple and very reliable once the initial correlations have been established.

APV Crepaco, Inc.
Lake Mills, Wisconsin

Adapted from: J. D. Usher and A. Cooper, "Paraflow Seminar, Part One," *CPI Digest* 2(2):2–7 (1973). Copyrighted APV Crepaco, Inc. Used with permission.

HEAT EXCHANGERS: PLATE VERSUS TUBULAR

SELECTION

In forming a comparison between plate and tubular heat exchangers, there are a number of guidelines that will generally assist in the selection of the optimum exchanger for any application. In summary, these are:

1. For liquid–liquid duties, the plate heat exchanger usually has a higher overall heat-transfer coefficient and often the required pressure loss will be no higher.
2. The effective mean temperature difference will usually be higher with the plate heat exchanger.
3. Although the tube is the best shape of flow conduit for withstanding pressure, it is entirely the wrong shape for optimum heat-transfer performance since it has the smallest surface area per unit of cross sectional flow area.
4. Because of the restrictions in the flow area of the ports on plate units, it is usually difficult, unless a moderate pressure loss is available, to produce economic designs when it is necessary to handle large quantities of low density fluids such as vapors and gases.
5. A plate heat exchanger will usually occupy considerably less floor space than a tubular for the same duty.
6. From a mechanical viewpoint, the plate passage is not the optimum and gasketed plate units are not made for operating pressures much in excess of 300 psig.
7. For most materials of construction, sheet metal for plates is less expensive per unit area than tube of the same thickness.
8. When materials other than mild steel are required, the plate will usually be more economical than the tube for the application.
9. When mild steel construction is acceptable and when a close temperature approach is not required, the tubular heat exchanger will often be the most economic solution since the plate heat exchanger is rarely made in mild steel.
10. Plate heat exchangers are limited by the necessity that the gasket be elastomeric. Even compressed asbestos fiber gaskets contain about 6% rubber. The maximum operating temperature therefore is usually limited to 500°F.

Heat-Transfer Coefficients

Higher overall heat-transfer coefficients are obtained with the plate heat exchanger compared with a tubular for a similar loss of pressure because the shell side of a tubular exchanger is basically a poor design from a thermal point of view. Considerable pressure drop is used without much benefit in heat transfer because of the turbulence in the separated region at the rear of the tube. Additionally, large areas of tubes even in a well designed tubular unit are partially bypassed by liquid and low heat-transfer areas thus are created.

Bypassing in a plate-type exchanger is less of a problem, and more use is made of the flow separation that occurs over the plate troughs since the reattachment point on the plate gives rise to an area of very high heat transfer.

For most duties, the fluids have to make fewer passes across the plates than would be required through tubes or

	Tubular	Plate
Size	3 shells in series: 20-ft tubes, 1 shell-side pass, 4 tube passes, 2238 tubes, 12,100-ft² area	R10 Paraflow with 313 plates arranged for 3 process passes and 3 service passes, 3630 ft² of area
Overall coefficient	270 Btu/h · ft² · °F	700 Btu/h · ft² · °F
MTD	7.72°F	9.0°F
Pressure loss	10–19 psi	10.8–8.8 psi

Figure 1. Duty: demineralized water/seawater. To cool 864,000 lb/h of water from 107°F to 82.4°F using 773,000 lb/h of 72°F seawater.

in passes across the shell. Since a plate unit can carry out the duty with one pass for both fluids in many cases, the reduction in the number of required passes means less pressure lost due to entrance and exit losses and consequently, more effective use of the pressure.

Mean Temperature Difference

A further advantage of the plate heat exchanger is that the effective mean temperature difference is usually higher than with the tubular unit. Since the tubular is always a mixture of cross and contraflow in multipass arrangements, substantial correction factors have to be applied to the log mean temperature difference. In the plate heat exchanger where both fluids take the same number of passes through the unit the LMTD correction factor is usually in excess of 0.95. As is illustrated in Figure 1, this factor is particularly important when a close or relatively close temperature approach is required.

In practice, it is probable that the sea water flow rate would have been increased to reduce the number of shells in series if a tubular had to be designed for this duty. While this would reduce the cost of the tubular unit, it would result in increased operating costs.

Design Case Studies

Figure 2 covering a number of case studies on plate versus tubular design demonstrates the remarkable heat-transfer performance that can be obtained from Paraflow units. Even for low to moderate available pressure loss, the plate heat exchanger usually will be smaller than a corresponding tubular.

Because of the high heat-transfer rates, the controlling resistance usually is fouling, so an allowance of 20–50% extra surface has been made based on APV experience.

One limitation of the plate heat exchanger is that it is rarely made in mild steel; the most inexpensive material of construction is stainless. Therefore, even when the plate surface requirement is much lower than the tubular on some heat-transfer duties, the tubular will be less expensive when mild steel construction is acceptable. If a close temperature approach is required, however, the plate unit always will cost less and where stainless steel or more exotic materials are required for process reasons, the plate unit usually will cost less.

Physical Size

One important advantage of the plate over a tubular unit is that for a particular duty, the plate heat exchanger will be physically smaller and require far less floor space. This is shown most graphically in Figure 3, where a Series R405 Paraflow is being installed next to tubular units that need twice the amount of space for the identical duty. It is further illustrated in the volumetric comparisons of Figure 4.

If the tightened plate pack of a Paraflow is regarded as a rectangular box, each cubic foot contains 50–100 ft² of heat-transfer area according to the type of plate used. Allowing for metal thickness, the contained liquid is some 80% of this volume. Thus, the total of both heating and cooling media is about 5 gals. Expressed in another way, the liquid hold up per square foot for each stream varies according to plate type from about 0.06 gal down to 0.03 gal.

By comparison, one cubic foot of tubular exchanger of equilateral triangular pitch with a tube pitch–tube diameter ratio of 1.5 has a surface area of 10 ft² for 2-in.-OD tubes or 40 ft² for ½-in.-OD tubes. The average contained liquid is proportionately 0.27 gal/ft² down to 0.07 gal/ft² of heat-exchanger area with no allowance for the headers. If the heat-transfer coefficient ratio between plate and tubular is conservatively taken as 2, the plate exchanger vol-

Case Study	Fluid A, lb/h	Fluid B, lb/h	Duty, °F	Tubular Design Pressure Loss			Plate Design Pressure Loss			
				A	B	ft²	A	B	ft²	type
A	13,380 hydrocarbon	53,383 water	320 A 120 / 120 B 92	0.42	3.05	301	.3	2.2	92	HX
B	864,000 water	77,300 seawater	107 A 82.4 / 99.1 B 71.6	19	19	12,100	10.8	8.8	3,630	R10
C	17,500 solvent	194,000 water	140 A 104 / 95 B 79	2.9	4.5	1,830	3.0	3.7	445	R5
D	148,650 desalter effluent	472,500 salt water	222 A 100 / 106 B 68	10.8	8.0	1,500	4.5	10	380	R4

Figure 2. Case studies of plate heat exchangers and tubular designs. All tubular designs were carried out with the aid of the HTRI program ST3.

Figure 3. At the Glidden Durkee Baltimore plant, a Series R405 Paraflow is being readied to cool 1000 gpm of deionized water. This unit requires only half the space of a tubular exchanger for the same duty.

ume to meet a given duty varies from $\frac{1}{10}$ to $\frac{1}{5}$ of that of the tubular. For a lower tube pitch–tube diameter ratio of 1.25, the comparison becomes $\frac{1}{7}$ to $\frac{1}{4}$. These facts demonstrate why the Paraflow plate heat exchanger is referred to as "compact."

Thermal Limitations

While it would appear offhand that the plate heat exchanger always provides a better performance at usually a lower price than the tubular exchanger, consideration must be given to the thermal as well as mechanical limitations of the plate-type machine. These usually are based on allowable pressure loss.

For single-phase liquid–liquid duties, the plate heat exchanger can be designed for moderately low pressure loss. However, if the pressure loss across any plate passage that has liquid flowing downward is lower than the available liquid static head, the plate will not run full and performance therefore will be reduced. This is termed low plate rate. Use of a plate below the minimum plate rate is inadvisable since it causes a wastage of surface area and results in unreliable operation. It is, however, possible to function below the minimum plate rate in a single-pass arrangement by making sure that the low plate rate is operated with a climbing liquid flow.

These problems are not quite so severe with a tubular exchanger and therefore, operation at a moderately lower available pressure loss is possible.

Conclusion

To summarize, the gasketed plate heat exchanger generally will be the most economical heat exchanger for liquid–liquid duties provided the material of construction is not mild steel and provided the operating temperatures and pressures are below 500°F and 300 psig respectively. For other types of duty such as gas cooling, condensation or boiling, the plate heat exchanger can be a very economical type of unit if the pressure loss allowed is sufficient to utilize the very high heat-transfer performance characteristics of the plate.

RECOVERY OF PROCESS HEAT

Since it is quite clear that never again will energy be as inexpensive as in the past, it therefore is necessary to conserve this natural resource—to recover more of the process heat that currently is dissipated to waterways and atmosphere. Some of this heat can be recovered with the aid of high-efficiency heat exchangers, which can operate economically with a close temperature approach at relatively low pumping powers. One type of unit that is particularly suited for this duty is the APV Paraflow plate heat exchanger. For many applications, this equipment can transfer heat with almost true countercurrent flow coupled with high coefficients to provide efficient and inexpensive heat transfer.

Unfortunately, the plate heat exchanger has been considered by many chemical engineers to be suitable only for hygienic heat-transfer duties. This, of course, is not so. Nowadays, many more plate type units are sold for chemical and industrial use than are sold for hygienic applications. A further mistake is the claim that the plate heat exchanger can be used only for duties when the volumetric flows of the two fluids are similar. Again, this is not true, although it must be stated that the plate heat exchanger is at its most efficient when flows are similar.

Basic Considerations

Since plate and tubular heat exchangers are the most widely used types of heat-transfer equipment, it is well to draw a brief comparison of their respective heat-recovery capabilities for the energy-conscious plant manager.

While the plate heat exchanger does have mechanical limitations with regard to withstanding high operating pressures above 300 psig, it is more efficient thermally than shell and tube units, especially for liquid–liquid du-

	Plate		Tube			
Plate Pitch, in.	Heat-Transfer Area per ft³ of Exchanger, ft²	Liquid Contained per ft² of Heat-Transfer Area, gal	Tube Diameter, in.	Ratio Tube Pitch‡ Tube Diameter	Heat-Transfer Area per ft³ of Exchanger, ft²	Liquid Containment per ft² of Heat-Transfer Area (Average of Both Sides), gal
$\frac{1}{4}$	50	0.06	2	1.5	10	0.32
$\frac{1}{8}$	100	0.03	$\frac{1}{2}$	1.5	40	0.085

Figure 4. Volumetric comparison, plate versus tube.

ties. In many waste heat recovery applications, however, both pressure and temperature generally are moderate and the plate-type unit is an excellent choice since its thermal performance advantage becomes very significant for low-temperature approach duties. Higher overall-heat transfer coefficients are obtained with the plate unit for a similar loss of pressure because the shell side of a tubular basically is a poor design from a thermal point of view. Pressure drop is used without much benefit in heat transfer on the shell side due to the flow reversing direction after each cross pass. In addition, even in a well-designed tubular heat exchanger, large areas of tubes are partially bypassed by liquid and areas of low heat transfer thus are created. Conversely, bypassing of the heat transfer area is far less of a problem in a plate unit. The pressure loss is used more efficiently in producing heat transfer since the fluid flows at low velocity but with high turbulence in thin streams between the plates.

For most duties, the fluids also have to make fewer passes across the plates than would be required either through tubes or in passes across the shell. In many cases, the plate heat exchanger can carry out the duty with one pass for both fluids. Since there are fewer passes, less pressure is lost as a result of entrance and exit losses and the pressure is used more effectively.

A further advantage of the plate heat exchanger is that the effective mean temperature difference usually is higher than with the tubular. Since in multipass arrangements the tubular is always a mixture of cross and contraflow, substantial correction factors have to be applied to the log mean temperature difference. In the plate unit for applications where both fluids take the same number of passes through the exchanger, the LMTD correction factor approaches unity. This is particularly important when a close or even relatively close temperature approach is required.

Thermal Performance Data

Although the plate heat exchanger now is widely used throughout industry, precise thermal performance characteristics are proprietary and thus unavailable. It is possible, however, to size a unit approximately for turbulent flow liquid–liquid duties by use of generalized correlations that apply to a typical plate heat exchanger. The basis of this method is to calculate the heat-exchanger area required for a given duty by assuming that all the available pressure loss is consumed and that any size unit is available to provide this surface area.

For a typical plate heat exchanger, the heat transfer can be predicted in turbulent flow by the following equation

$$\frac{hD_e}{k} = 0.28\left(\frac{GD_e}{\mu}\right)^{0.65}\left(\frac{Cp\mu}{k}\right)^{0.4} \quad (1)$$

$$f = 2.5\left(\frac{GD_e}{\mu}\right)^{-0.3} \quad (2)$$

$$\Delta P = \frac{2fG^2L}{g\rho D_e} \quad (3)$$

The pressure loss can be predicted from equations 2 and 3.

Obviously, equation 1 cannot represent accurately the performance of the many different types of plate heat exchangers that are manufactured. However, plates that have higher or lower heat-transfer performance than given in equation 1 usually will give correspondingly higher or lower friction factors in equation 2. Experience indicates that the relationship for pressure loss and heat transfer is reasonably consistent for well-designed plates. In the Appendix, equations A1–A3 are further developed to show that it is possible for a given duty and allowable pressure loss to predict the required surface area. This technique has been used for a number of years to provide an approximate starting point for design purposes and has given answers to ±20%. For accurate designs, however, it is necessary to consult the manufacturer.

Heat-Recovery Duties

In any heat-recovery application, it always is necessary to consider the savings in the cost of heat against the cost of the heat exchanger and the pumping of fluids. Each case must be treated individually since costs for heat, electricity, pumps, etc will vary from location to location.

One point is obvious. Any increase in heat recovery, and thus heat load, results in a decrease in LMTD and considering a constant heat-transfer coefficient, subsequently in the cost of the exchanger. This effect is tabulated in Figure 5. Because the cost of an exchanger increases considerably for relatively small gains in

Heat Recovered, Btu/h	%	Temperature, °F	12 lb/in.² Pressure Loss					25 lb/in.² Pressure Loss				
			LMTD	Factor	Actual MTD	Area, ft²	Relative Price, $	LMTD	Factor	Actual MTD	Area, ft²	Relative Price, $
3,600,000	60	200 ► 140 160 ◄ 100	40	0.985	39.4	85	5,500	40	0.985	39.4	74	5,400
4,500,000	75	200 ► 125 175 ◄ 100	25	0.975	24.4	197	6,450	25	0.980	24.5	172	6,250
5,100,000	85	200 ► 115 185 ◄ 100	15	0.965	14.4	425	8,400	15	0.965	14.4	362	7,850
5,400,000	90	200 ► 110 190 ◄ 100	10	0.95	9.5	739	11,000	10	0.92	9.2	629	10,100
5,700,000	95	200 ► 105 195 ◄ 100	5					5	0.83	4.16	1580	18,100

Hot liquid 60,000 lb/h of water at 200°F; cold liquid 60,000 lb/h of water at 60°F.

Figure 5. Effect of percentage heat recovery on PHE cost.

Figure 6. Two shrouded Paraflow units provide an optimum return on investment with energy and cost saving regeneration of 88 and 81%.

recovered heat above the 90% level, such applications, even with the plate heat exchanger, must be studied closely to verify economic gain. The economic break-even point is far lower for a tubular exchanger. Situations where it is advantageous to go above 90% recovery usually involve duties where higher heat recovery reduces subsequent heating or cooling of the process stream. High steam or refrigeration costs therefore justify these higher heat recoveries.

As shown in Figure 5, the cost of increasing heat recovery from 85 to 90% at a constant pressure loss of 12 lb/in.^{2} is $2600. From a practical standpoint, going from 90 to 95% requires a significantly higher pressure loss and nearly doubles the exchanger cost. However, even with this 95% heat recovery and assuming steam costs at $6.00/1000#, payback on the plate heat exchanger would take 530 hs.

The plate heat exchanger thus provides a most economic solution for recovering heat (Fig. 6). This degree of heat recovery cannot be achieved economically in a tubular exchanger since the presence of cross flow and multipass on the tube side causes the LMTD correction factors to become very small or, alternately, requires more than one shell in series. This is shown in the Figure 7 comparison.

As detailed, this example illustrates that the plate heat exchanger has considerable thermal and therefore price advantage over the tubular exchanger for a heat recovery of 70%. Since the overall heat-transfer coefficient and the effective mean temperature difference both are much higher for the plate unit, reduced surface area is needed. Furthermore, because of cross-flow temperature difference problems in the tubular, three shells in series were needed to handle the duty within the surface area quoted.

Duty: To heat 1,300,000 lb/h of water from 73 to 97°F using 1,300,000 lb/h of water at 107°F.
Available pressure loss 15 lb/in.^{2} for both streams.

Heat recovery = 70.5%
Number of heat-transfer units (HTU) = 2.4

	Plate	Tubular	
Heat-transfer area	4820	13,100	ft^2
Heat-transfer coefficient (clean)	734	386	Btu/h · ft^2 · °F
Heat-transfer coefficient (dirty)	641	271	Btu/h · ft^2 · °F
Effective mean temperature difference	10.0	8.7	°F
Pressure drop: hot fluid/cold fluid	4.6/4.7	15/15	lb/in^2
Fouling resistance: hot fluid/cold fluid	0.0001/0.0001	0.0005/0.0005	(Btu/h · ft^2 · °F)$^{-1}$
Pass arrangement	1/1	4 tube side 3 shells in series	baffled
Approximate price	$95,000 Stainless steel	$135,000 Mild steel	

This example demonstrates that quite high heat-transfer coefficients can be obtained from a PHE with only a moderate pressure loss.

Figure 7. Heat recovery—comparison between plate and tubular heat exchangers.

Using only two shells would have resulted in a further 40% increase in surface area.

The small size of the plate heat exchanger also results in a saving of space and a lower liquid holdup. For this type of heat recovery duty, a stainless-steel plate heat exchanger almost always will be less expensive than a mild steel tubular unit. Although the tubular exchanger physically will be capable of withstanding higher temperatures and pressures, there is a considerable and for the most part unnecessary penalty to pay for these features both in price and size.

For heat recovery duties in excess of 70%, the plate heat exchanger will become increasingly more economical than the tubular.

Typical Applications

One of the more common uses of regeneration is found in many of the nation's breweries, where it is necessary to cool huge amounts of hot wort before it is dicharged to fermentation tanks. Typical in scope is an operation where 850 barrels per hour of wort (220,000 lbs/h) are cooled from 200 to 50°F by means of 242,000 lbs/h (R = 1.1:1) of water entering at 35°F and being heated to 165°F. As a result; approximately 33,000,000 Btu/h are saved, there is an excellent water balance for use throughout the brewery, and no water is discharged to the sewer.

For a dairy, regeneration usually involves the transfer of heat from pasteurized milk to cooler raw milk entering the system. After initially heating 100,000 lb/hr of milk from 40 to 170°F by high temperature, short time pasteurization, 90% regeneration permits cooling of the milk back down to 40°F with a savings of 11,700,000 Btu/h. Only 10% of the total heat or cooling must be supplied, and no cooling medium such as city water is used and discarded.

Chemically, there are many and varied regenerative applications. For desalination, APV has supplied a number of Paraflows, which are virtually perpetual motion machines. These units achieve 95% regeneration in heating 87,000 lb/h (175 gpm) from 70 to 197°F while cooling a secondary stream flowing at 78,000 lb/h from 214 to 73°F. Savings in the Btu load in this case are 11,050,000 per hour.

While flow rates in hot oil applications are quite low in comparison, temperatures are very high. It is possible to cool 400 gph of vegetable oil from 446 to 226°F while heating 400 GPH of oil from 200 to 400°F with 80% regeneration.

In the production of caustic soda, where very corrosive product streams are encountered, Paraflows with nickel plates are being used to cool 10,000 lb/h of 72% NaOH from 292 to 210°F while heating 14,000 lb/h of 50% NaOH from 120 to 169°F.

Cost and Efficiency

To examine a hypothetical case of Paraflow regeneration from the viewpoint of efficiency and dollar savings, consider the following process duty:

Duty Heat 100 GPM of fluid #1 from 50°F to 200°F while cooling 100 gpm of fluid 2 from 200 to 50°F

Under ordinary conditions, steam required to heat fluid #1 would be in the nature of 7500 lb/h with an equivalent cooling requirement for the second stream of 625 tons of refrigeration. Using 85% regeneration, however, the energy needs are drastically reduced.

At this point in the process, fluid 2 in a conventional system has been cooled to only 90°F by means of 85°F city water and will require supplemental refrigeration for final cooling to 50°F. At the same time, fluid 2 in an APV regenerative system has been cooled to 72.5°F by means of 85% regeneration and must be cooled further to 50°F.

Duty	Cool from 90 to 50°F	Cool from 72.5 to 50°F
Supplemental refrigeration required	$\frac{50{,}000 \times 40}{12{,}000} =$ 166.5 tons	93.5 tons
Assuming average electrical cost of 4¢/kWh	$166.5 \times 3.5 =$ 583.8 kWh	$93.5 \times 3.5 =$ 327.25 kWh
Annual supplemental refrigeration cost	$\frac{4}{100} \times 583.8 \times 24 \times 365 =$ $204,571	$\frac{4}{100} \times 327.25 \times 24 \times 365 =$ $114,669

Energy Needs, Fluid 1

	Conventional	85% Regeneration
	Heat 100 GPM 50 → 200°F	Heat 100 GPM 50 → 177.5°F by cooling fluid 2 from 200 to 72.5°F
Steam required	$\frac{50{,}000 \times 150}{10_3} =$ 7500 lb/h to heat to 200°F	$\frac{50{,}000 \times 22.5}{10^3} =$ 1125 lb/h supplemental heat to raise temperature to 200°F
Assuming average steam cost: $6.00 per 1000#		
Steam cost/h	7.5 × $6.00 = $45.00/h	1.125 × $6.00 = $6.75/h
Annual steam cost	$45.00 × 24 × 365 = $394,200	$6.75 × 24 × 365 = $59,130

Energy Needs, Fluid 2

	Conventional	85% Regeneration
	Cool 100 GPM 200 → 50°F	Cool 100 GPM 200 → 72.5°F by heating fluid 1 from 50 to 177.5°F
Equivalent refrigeration required	$\frac{7,500,000}{12,000} = $ 625 tons to cool to 50°F	.15 × 625 = 93.5 tons supplemental cooling to lower temperature to 50°F
Using available 85°F cooling-tower water	200 gpm to cool 200 → 90°F 200 × 60 = 12,000 GPH	None None
Assuming average water cost of $0.05 per 1000 gal		
Water cost/h	12 × .05 = $.60/h	None
Annual water cost	$.60 × 24 × 365 = $5250	None

Definition

The number of heat-transfer units is defined as the temperature rise of fluid one divided by the mean temperature difference. A further term for the HTU is the temperature ratio (TR), ie,

$$\text{HTU} = \frac{t_1 - t_2}{\text{MTD}}$$

where MTD = mean temperature difference.

Appendix

For heat transfer in a typical PHE with turbulent flow one can write the heat transfer performance in terms of a dimensionless Dittus–Boelter equation

$$\frac{hD_e}{k} = 0.28\left(\frac{GD_e}{\mu}\right)^{0.65}\left(\frac{Cp\mu}{k}\right)^{0.4}\left(\frac{\mu}{\mu_w}\right)^{0.14} \quad (A1)$$

For applications in turbulent flow it is usually sufficiently accurate to omit the Sieder–Tate viscosity ratio and therefore the equation reduces to

$$\frac{hD_e}{k} = 0.28\left(\frac{GD_e}{\mu}\right)^{0.65}\left(\frac{Cp\mu}{k}\right)^{0.4} \quad (A2)$$

The pressure drop can be predicted in a similar exchanger by equations A3 and A4.

$$f = 2.5\left(\frac{GD_e}{\mu}\right)^{-0.3} \quad (A3)$$

$$\Delta P = \frac{2fG^2 L}{g\rho D_e} \quad (A4)$$

To solve these equations for a particular duty it is necessary to know G, L, and D_e, and it is shown below how these can be eliminated to produce a general equation.

For any plate,

$$G = \frac{m}{A_f} \quad (A5)$$

where m is the total mass flow rate and A_f the total flow area.

D_e is defined by APV as four times the flow area in a plate divided by the wetted perimeter. Since the plate gap is small compared with the width, then:

$$D_e = \frac{4 \times A_f}{\text{wetted perimeter}} \quad (A6)$$

but since $A_s = L \times$ wetted perimeter where A_s is the total surface area:

$$L = \frac{A_s}{4A_f} D_e \quad (A7)$$

and it is possible to eliminate L from equation A4.

Cost Comparison[a]

	Original Cost, $	Regenerative System Cost, $	Regenerative Savings, $
Steam	394,200	59,130	335,070 annually
Tower cooling water	5,250	0	5,250 annually
Refrigeration	204,571	114,669	89,902 annually
	604,021	173,799	430,222

[a] Capital cost of a Paraflow plate heat exchanger for the above duty would be approximately $17,000.

Similarly, by substituting G using equation A5 in equations A2, A3, and A4 and then by rearranging the equations to eliminate A_f, it is possible to arrive at equation A8 for the film heat-transfer coefficient.

$$h = \frac{J_2}{J_1^{0.241}} \left(\frac{m}{A_s}\right)^{0.241} D_e^{-0.28} \qquad (A8)$$

where

$$J_1 = \frac{2.5\mu^{0.3}}{2g\rho\Delta P}$$

and

$$J_2 = 0.28\left(\frac{Cp\mu}{k}\right)^{0.4} k\mu^{-0.65}$$

That is, the film heat-transfer coefficient is expressed only in terms of the surface area and equivalent diameter. A computer program solves equation A8 using a constant value of D_e and using an empirical factor Z to account for port pressure loss and other deviations from equations A2–A4. The equation is solved using the HTU approach where

$$\text{HTU} = \frac{(t_1 - t_2)}{\text{MTD}} = \frac{UA_s}{m_1 c_1}$$

That is, the number of heat-transfer units is the temperature rise of fluid 1 divided by the mean temperature difference.

U is defined as

$$\frac{1}{U} = \frac{1}{h_1} + \frac{1}{h_2} + \text{metal resistance}$$

where h_1 and h_2 are calculated from an empirical modification of the constants in equation A8. The powers in equation A8 are not modified.

BIBLIOGRAPHY

Adapted from:

J. D. Usher and A. Cooper, "Paraflow Seminar, Part Two," *CPI Digest* **2**(3): 6–10 (1974).

J. D. Usher and A. Cooper, "Thermal Performance," *CPI Digest* **2**(4): 10–14 (1974).

J. D. Usher and A. Cooper, "Recovery of Process Heat," *CPI Digest* **3**(2): 2–7 (1974).

Copyrighted APV Crepaco, Inc. Used with permission.

APV Crepaco, Inc.
Lake Mills, Wisconsin

HEAT EXCHANGERS: SCRAPED SURFACE

BASIC CONSIDERATIONS

The scraped-surface heat exchanger (SSHE) is a specialized piece of heat-transfer equipment that was patented around 1926 by Clarence Vogt in an effort to develop a more efficient freezer for making ice cream. The design incorporated a scraping action to prevent buildup of frozen ice cream on heat-transfer surfaces. The concept was successful. Thermal efficiency was improved and production capacity greatly increased.

Since that time, the SSHE has become essential for numerous processes in the dairy and food industries. Many of these applications are similar to the ice cream problem in that, as the product is heated or cooled, it fouls heat-transfer surfaces and reduces efficiency. Such applications, for example, include cooling peanut butter or plasticizing shortening and margarine. In each instance, fat crystals that form are like the ice crystals found in ice cream. Without the scraping of the heat exchanger sweeping the solidified fats away from the surface, heat-exchange efficiency would fall off very rapidly. Conversely, in heating applications where a product tends to "burn on," the scraping action removes product from the surface before fouling can occur. Typical heating examples are aseptic cheese sauces and aseptic puddings where processing temperatures approach 300°F (150°C).

The modern SSHE also is capable of processing products containing particulates. Currently, standard units can handle ½-in. (13-mm) particles without excessive breakage while special designs are available to accommodate 1-in. (26-mm) particulates.

While more expensive than other types of heat exchanger, the SSHE is the best and only thermal approach in hundreds of applications where high viscosity, large particulates, crystallization, and burn-on are problems that must be considered.

A scraped-surface heat exchanger (Fig. 1) basically consists of a jacketed cylinder fitted with a rotating shaft on which scraper blades are mounted. Product is pumped through the cylinder while a heating or cooling medium is circulated in the annular space between the cylinder and the jacket. The blades are fixed to pins that allow them to swing freely. No springs are necessary since centrifugal force holds the blades in position against the inside of the cylinder wall as product constantly is swept away from the heat-transfer surface and new product exposed to treatment.

The standard horizontal exchanger generally has from one to three independently functioning jacketed cylinders mounted on a heavy steel base (Fig. 2, 3). A stainless-steel casing covers the cylinders, base and drives to form a completely enclosed system. Vertical units also are available for use when floor space is at a premium (Fig. 4).

Many processes require more than one unit, and in such cases where multiple cylinders are used, product always should be piped in series. Heat transfer will be higher because of the higher flow through each cylinder. Parallel flow arrangements fail because there is no way to ensure equal flow to each cylinder unless individual pumps are used for each circuit. It should be noted that a scraped surface heat exchanger does not do any pumping. A pump is required to move product through the unit.

Dashers

The shaft which carries the scraper blades is called a dasher by some manufacturers and a mutator by others—

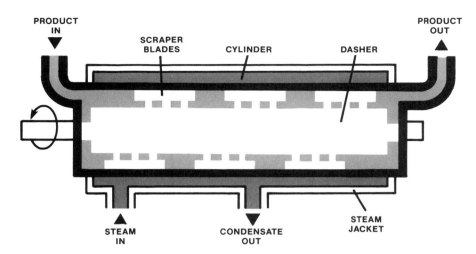

Figure 1. Cutaway of horizontal SSHE.

the "dasher" terminology being derived from early days when the exchanger initially was used as an ice cream freezer.

Dashers are engineered to achieve high heat-transfer coefficients with minimum power consumption and are supported by heavy-duty bearings located outside the product contact zone. Three standard designs (Fig. 5a, b, 6) provide product flow spaces of different sizes to accommodate different product viscosities, dwell time, level of blending, and size of particulates. For margarine or plasticizing applications, dashers with internal water circulation reduce adhesion of product to the dasher surface.

Typical dasher speeds vary from 60 to 420 rpm with standard motors providing a choice of three drive methods—direct-driven hydraulic, belt-driven electric, or direct-driven gearhead.

Scraper Blades

Designed to promote the rapid removal of product from cylinder walls while enhancing product agitation and mixing, scraper blades are available in a selection of materials and configurations. Most common materials are stainless steel and plastics since the blade is the wearing part and as such, must be softer than the cylinder wall or lining.

Blade selection generally is determined by product temperature, pressure, and formulation as well as by the cylinder material and service media being used.

Heat-Exchange Cylinders

To provide optimum performance and economy of operation, heat-exchange cylinders are available in a selection of sizes and materials of construction.

The most common diameter for the scraped SSHE cylinder is 6 in. (152 mm) with lengths established at 48 in. (1220 mm) and 72 in. (1830 mm). There are, however, other sizes available. Some are 4 in. (102 mm) in diameter with lengths of 60 in. (1520 mm) (Fig. 7) and 120 in. (305 mm) while others are of 8 in. (203-mm) diameter and lengths to 84 in. (2130 mm).

Figure 2. Model 2HD-648 SSHE with water flush seals.

Figure 3. A pair of scraped–surface heat exchangers provide an added dimension to an aseptic system by permitting the processing of fluids containing particulates.

Figure 4. Vertical Model VExHD-884 SSHE.

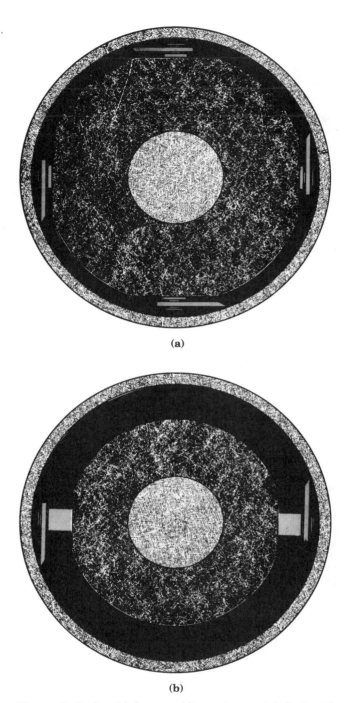

Figure 5. Dasher–blade assemblies cutaway. (a) Series 55; (b) Series 45.

The most common materials of construction are stainless steel, nickel with or without chrome plating, and, more recently, a bimetallic combination.

Since the cost per square foot of heat-transfer surface is higher for a SSHE than other types of heat exchangers, it is essential that cylinder material with the highest feasible heat-transfer coefficient be selected. This selection, however, must be tempered by consideration of the compatibility between cylinder and scraper blade materials and the susceptibility of these materials to acid attack and corrosion.

As charted in Table 1, nickel exhibits the best thermal conductivity while stainless steel is the least conductive. Even so, nickel is not suitable for all applications. It is relatively soft when compared to typical metal scraper blade materials and would wear rapidly if this combination were to be used. Since SSHE always are designed to make the blades rather than the more expensive cylinders the wearing part, plastic blades must be used with nickel

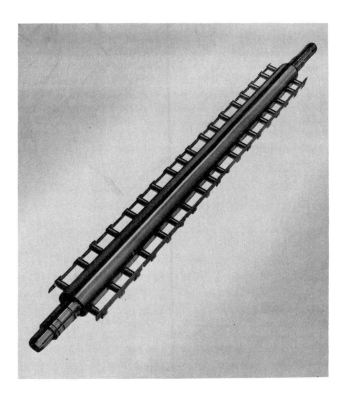

Figure 6. Series 30 dasher for viscous products or particulates.

cylinders if only for reasons of economics and reduced downtime. For the same reasons, only plastic blades are run on stainless-steel cylinders.

To retain the superior heat-transfer characteristics of nickel while benefiting from the extended durability of steel blades, nickel cylinders may be chrome-plated. The scraping surface then becomes the hardest material in use and can accommodate any of the common blade materials. The sole drawback is that the chrome is sensitive to salt and acid attack.

Bimetallic cylinders offer some advantages over those of any other material, whether nickel, chromed nickel, or stainless. Produced from two different materials, the bimetallic cylinder usually is made by centrifugally casting a hard, corrosion-resistant alloy inside a tube that has high tensile strength and thermal conductivity approaching nickel. While the lining is not quite as hard as chrome, it is hard enough to withstand abrasion from both stainless-steel and plastic blades. One disadvantage other than a slight loss in heat transfer is that the inner alloy coating is susceptible to strong acid corrosion.

The selection of a cylinder with a compatible blade ma-

Table 1. Heat Transfer Coefficients of SSHE Cylinder Metals

Material	K, Btu/h · °F · in.
Nickel	52
Chrome nickel	52
Bimetallic	31
Stainless steel	9.4

Figure 7. Four-barrel vertical scraped-surface heat exchanger is designed to cool meat gravies from 190 to 70°F. Unit capacity is 5000 lb/h.

terial is compounded by the products to be processed and the cleaning procedures used. For example, on slush freezing applications or where media temperature is below freezing, plastic blades can be abraded within hours as hard ice is scraped away. And as already noted, despite being the hardest cylinder material in use, chrome plated nickel is subject to attack by common acid CIP solutions and does not stand up well to salty products over extended periods of time. Table 2 shows the difference in corrosion for various materials and acids, and it is worth noting that it is the cleaning regimen rather than the product that causes problems in some cases. A change in CIP procedures may well extend the life or allow the use of nickel or chrome nickel cylinder material. In effect, much thought and experience must go into selecting the best cylinder and blade combinations—a selection best left up to the equipment manufacturer.

Cylinder Jackets–Media

The three common media used in a SSHE are water, steam, and refrigerants such as ammonia and Freon. Each requires a slightly different cylinder jacket design for optimum performance.

Table 2. Corrosion of SSHE Cylinder Metals

Acid	Common Use/Source	Material Corrosion at 140°F (Maximum in./yr)		
		Nickel	Chrome	316 SS
Nitric	Cleaning	0.050	nr[a]	0.002
Citric	Product	0.020	nr	0.002
Acetic	Product	0.020	nr	0.002
Phosphoric	Cleaning	0.050	nr	0.002
Malic	Product	0.020	nr	0.020
NaCl	Product	0.020	nr	0.020

[a] nr = not recommended.

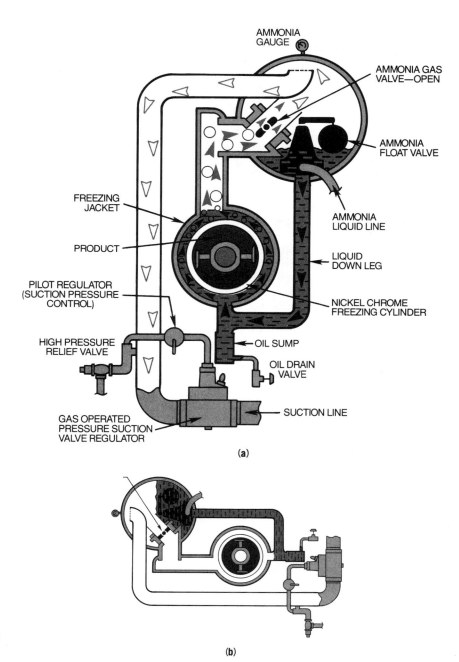

Figure 8. Full flooded SSHE refrigeration system (**a**) shown in operation, (**b**) shown shut down.

The water jacket is designed with a small spacing between the jacket and heat exchange cylinder. This induces high liquid media velocities that cause turbulence, which in turn, improves heat-transfer efficiency. It also minimizes fouling on the media side. Normally, countercurrent flow is recommended between product and media. In many cases, however, because of the high flow rate of media over product and the high temperature differences between media and product, the difference in performance between countercurrent and cocurrent flow is small.

On steam units, media enters the jacket via a header that distributes the steam over the entire length of the cylinder. Condensate runs to the bottom of the jacket, where it is collected and removed by a steam trap. Direction of product flow is immaterial.

When ammonia or Freon is used, the refrigerant is handled differentially by various manufacturers. Most exchanger designs are of the flooded type with media being fed at the bottom of the jacket and boiling occurring within the jacket. Since the key to efficient heat transfer is to maintain a wetted surface on the cylinder, not all of the refrigerant is allowed to evaporate. The combination of gas and liquid is carried over into a surge tank, where the phases are separated and the liquid recycled to the bottom of the jacket. Liquid level is maintained by a valve that allows more liquid to enter the surge tank to replace that which has been evaporated.

One advantage of the flooded arrangement as illustrated in Figure 8 is the incorporation of a quick shutoff valve for instant stop/instant start. Although the valve is open during normal operation, it may be closed when a freeze-up is imminent or the operator want to stop cooling. Since the vapor cannot go up and out, it expands and rapidly pushes the liquid refrigerant down and away from the cylinder. When liquid no longer touches the cylinder and the surface no longer is wetted, all freezing stops. To restart the process, the valve is opened and with liquid quickly filling the jacket area, freezing begins again.

The primary advantage of the SSHE over other types of heat-transfer equipment is its ability to accommodate product viscosities of 500,000 cP and higher and to handle particulates up to 1 inch (26 mm) in size provided that they are not shear-sensitive. Where the product undergoes a major change of state, ie, liquid to gel, or of viscosity as occurs during cooling, the SSHE often is used as a finisher for either the heating or cooling mode in conjunction with other methods of heat transfer.

THEORY AND CALCULATIONS

The operating principle of scraped-surface heat exchange is based on the constant movement of a product away from the heat-exchange surface in order to minimize the formation of films that resist heat transfer.

In considering the scraped-surface heat exchanger, four types of thermal exchange are taken into account:

- *Sensible Heat.* The heat produced by the increase or decrease in temperature of a product (without change of state).
- *Latent Heat.* Heat exchange associated with a physical change in the material being processed.
- *Heat of Reaction.* The heat given off (exothermic) or taken up (endothermic) when two or more chemicals react.
- *Mechanical Heat.* Power is consumed in turning the dasher of a SSHE. Most of this energy is absorbed as heat energy into the product within the heat exchanger.

The ideal form of heat transfer occurs when one product at an elevated temperature is brought into direct contact with another material at a lower temperature. The warmer product gives up its heat without loss of energy and at a rate equivalent to the ability to mix or disperse the two materials.

In practice, however, it is rare that two materials may be brought into direct contact. As a general rule, there is an intervening heat-transfer surface such as a tank or tube wall. This surface presents problems to heat transfer because it resists the passage of heat. It further induces resistances related to hydraulic drag and the buildup of deposits or other films that further retard the passage of heat. Figure 9 illustrates the various resistances that may be encountered.

Since these resistances are in series, they are additive. The adverse effect, however, may be reduced by agitating the fluids on both sides of the wall. Since the SSHE eliminates buildup of deposits on the product side of the tube wall, the resistance to heat transfer is minimized.

The total resistance to heat transfer then is

$$R = R_1 + R_2 + R_3$$

The difference in temperatures between the product and the media is the driving force to push across this resistance. Therefore, the equation for heat transfer is:

$$\text{heat-transfer rate, Btu/h} \cdot \text{ft}^2 = \frac{\text{temperature, °F}}{\text{resistance}}$$

This equation can be developed into a useful form by letting the temperature difference be ΔT and resistance be R. Therefore

$$\text{heat-transfer rate} = \Delta T / R$$

Furthermore, by taking the inverse of resistance, which is conductance, $1/R = U$. The heat-transfer equation now is heat-transfer rate = $U \Delta T$.

Heat-exchange U values in a scraped-surface heat exchanger consist of three parts as shown in this formula:

$$U = \frac{1}{R_1 + R_2 + R_3} = \frac{1}{1/h_p + t/k + 1/h_m}$$

where: —h_p is the heat-transfer coefficient for the product and ranges from 200 to 800 Btu/h · ft² · °F.
—t/k is the heat-transfer coefficient through the cylinder wall (t usually is between $\frac{1}{4}$ and $\frac{1}{8}$ in. and k varies depending on the material.

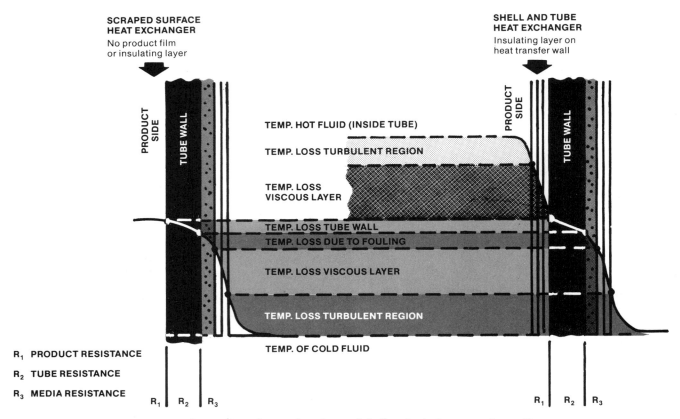

Figure 9. Comparison of scraped surface and shell and tube heat transfer profiles.

—h_m is the heat-transfer coefficient for the media and ranges from 800 (water media) to 2000 (steam or ammonia) Btu/h · ft² · °F.

The overall U value then varies from 150 to 420 Btu/h · ft² · °F. Unfortunately, the h_p is difficult to determine theoretically and, because h_p tends to influence the overall U value much more than h_m and t/k, manufacturers size the heat exchangers using a U that is determined from lab test data or from production runs. Attempts have been made to predetermine the expected U value, but many factors influence the heat-transfer rate:

SSHE geometry An SSHE can be equipped with different diameter dashers. The larger the dasher diameter, the smaller the annular space in which the product travels. A small annular spacing will improve heat transfer.

Viscosity SSHE units normally can handle viscosities of up to 1,000,000 cP. The greater the viscosity, the lower the U value. Products with viscosities that vary greatly at different temperatures; eg, peanut butter and cooking–cooling starch for salad dressings, usually produce lower U values. This is due to mass rotation. It normally is encountered in cooling applications where the low-viscosity warm product short-circuits through the heat-exchange cylinder while the cooler, thicker product rotates with the dasher.

Dasher speed High dasher speeds improve heat transfer but heat generated by high rpm is counter-productive in cooling applications.

Number of scraper blades Dashers typically are equipped with two or four rows of scraper blades. Units with four rows of blades provide higher heat-transfer rates at any given dasher rpm. When supplying two rows of blades, many manufacturers compensate by increasing the dasher speed.

Specific heat The lower the specific heat, the lower the U values.

Thermal conductivity The lower the thermal conductivity, the lower the U value.

Flow rate The higher the product flow rate, the higher the U value. Flow rates below 5 gpm are considered low. See Figure 10.

Applications Heating applications produce higher U values than do cooling applications.

Cylinder materials The cylinder materials, as mentioned previously, can be nickel, stainless steel or bimetallic. Figure 11 charts the differences in overall U values when cylinder thickness or materials are changed. Nickel is the most

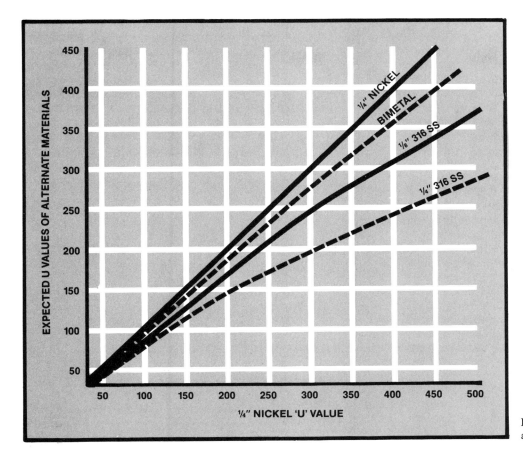

Figure 10. Comparison of nickel and other materials.

efficient and is used as the basis to compare the other materials commonly available.

Once a U value for an application has been determined, a calculation to size the heat exchanger can be made.

A Practical Formula

Deriving the practical scraped-surface heat exchange formula is achieved as follows:

$$\text{heat-transfer rate} = U\,\Delta T$$

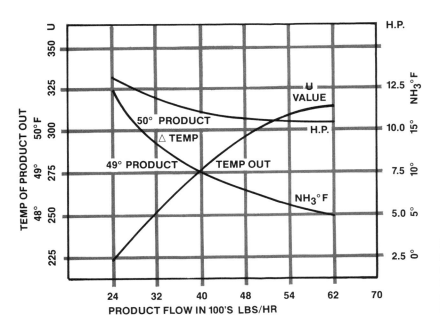

Figure 11. Curves represent a 6-ft^2 SSHE margarine cooler with 4½-in.-diameter dasher. Discharge temperature 49–50°F flow rate increased from 2400 to 6200 lb/h; U value increases with flow rate and eventually levels off. Ammonia temperature is lowered to maintain proper exit temperature.

Product heat load Q (heat load being transferred from the product to the media) will be equal to the heat-transfer rate times the heat transfer area A. Thus

$$Q = UA\Delta T$$

where A = area, ft^2.
The product heat load is equated

$$Q = \text{(product) lb/h} \times \text{temperature change} \times \text{SpHt} \\ + \text{latent heat} + \text{heat of reaction} \\ + \text{mechanical heat}$$

In the majority of applications, heat of reaction does not occur. Therefore, by eliminating this factor and combining equations, the generally accepted formula for calculating scraped-surface heat exchangers is

$$\text{area} = \frac{\text{product flow} \times (T_{\text{in}} - T_{\text{out}}) \times \text{SpHt} + \text{latent heat} + \text{hp} \times 2545 \text{ Btu/h/hp}}{U \times \text{LMTD}}$$

$$\text{LMTD} = \frac{(T_{\text{in}} - T_{\text{media out}}) - (T_{\text{out}} - T_{\text{media in}})}{\ln\left[\frac{(T_{\text{in}} - T_{\text{media out}})}{(T_{\text{out}} - T_{\text{media in}})}\right]}$$

For steam, ammonia, and Freon units,

$$T_{\text{media in}} = T_{\text{media out}}$$

Area	Square feet of surface required
Flow	Pounds per hour of product
T_{in}	Temperature of product in (°F)
T_{out}	Temperature of product out (°F)
$T_{\text{media in}}$	Temperature of media in (°F)
$T_{\text{media out}}$	Temperature of media out (°F)
Latent heat	Heat removed from ice formation or fat crystallization Btu/hr
hp	Dasher drive horsepower. Use only on cooling applications; disregard on heating. Note that horsepower is converted to Btu by the conversion factor 2545 Btu/h/HP
U	U value Btu/h · ft^2 × °F
LMTD	Log mean temperature difference

Sample Calculation

Product	Gravy
Specific heat	0.8
Application	Cool from 200 to 60°F using 5°F ammonia
U value	280
Flow rate	5000 lb/h

Assume 15 hp is required.

$$\text{LMTD} = \frac{(200 - 5) - (60 - 5)}{\ln[(200 - 5)/(60 - 5)]} = \frac{140}{1.266} = 110$$

$$\text{area} = \frac{5000 \text{ lb/h} \times (200 - 60°) \times 0.8 + 15 \text{ hp} \times 2545}{110 \times 280}$$

$$= \frac{598{,}175}{30{,}800} = 19.4 \text{ ft}^2 \text{ required}$$

SELECTION OF COMPONENTS

The scraped-surface heat exchanger is used on a wide range of diverse applications, many requiring different materials of construction and component design. Choices are based on a number of criteria involving both the compatibility of materials with the product to be processed and compatibility between various heat-exchanger components and materials. To better understand the reasoning behind component and material selection, it is necessary to know the equipment options available, the properties of the product, and the process temperatures. The four major areas of concern are cylinder materials, blade materials, seal selection, and dasher sizing.

Selecting Cylinder Materials

Since the per square foot cost of SSHE heat-transfer surface is relatively expensive, it is desirable to select heat-transfer cylinder material that provides the highest heat transfer coefficient. This will maximize the overall thermal performance for any given application. The commonly used materials are stainless steel, nickel with or without chrome plating, and bimetallic combinations. Nickel offers the highest HTC with a coefficient five times that of stainless steel.

To contain internal product pressures and prevent implosion by media pressure, cylinders generally are designed with a quarter inch wall thickness. For lower-pressure applications, however, an ⅛-in. wall occasionally is used to improve the performance of stainless-steel cylinders.

Since nickel has such a high heat-transfer coefficient, it might be expected that this material would have universal application. Unfortunately, however, nickel is susceptible to acid corrosion, so stainless-steel cylinders must be used with high acid type products. Furthermore, nickel is relatively soft and wears rapidly if subjected to constant abrasion by metal blades. Since simple economics dictate that the blades rather than the more expensive cylinders accept the wear, plastic blades must be used with nickel cylinders. Also, because most metal blades are of a stainless alloy and there is not enough difference in hardness between blade and cylinder, plastic blades are used with stainless steel cylinders.

To benefit from both the heat transfer characteristics of nickel and the durability of steel blades, nickel cylinders may be chrome-plated. This provides a scraping surface harder than any of the common blade materials but still cannot be used for all applications since the chrome also is subject to salt and acid attack.

An alternate to nickel, chromed nickel or stainless is the bimetallic cylinder (Fig. 12). With a hard corrosion-resistant alloy centrifugally cast within a tube having high tensile strength and thermal conductivity higher than stainless and approaching that of nickel, the bimetallic cylinder provides an acceptable compromise. While slightly softer than chrome, the inner liner is hard enough for use with any blade material. One limitation on a bimetallic cylinder is that it can be damaged by strong CIP acids, especially nitric acid.

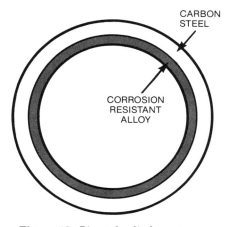

Figure 12. Bimetal cylinder cutaway.

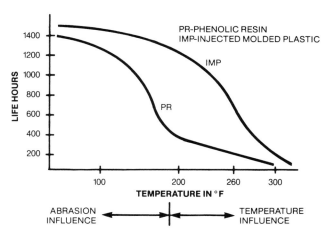

Figure 14. Temperature–wear relationship.

Selecting Blade Materials

While scraper blades are available in plastic or metal (Fig. 13), the determination of which blade will provide optimum service is influenced by the cylinder material, by the media being used, and by product temperature and formulation.

Generally speaking, plastic scraper blades can be used in conjunction with all types of cylinder materials while metal blades are run only against chrome plated surfaces or bimetallic cylinders. Limitations on plastic blades are wear life, temperature, and certain solvents. Metal blades, on the other hand, provide a longer wear life, can be sharpened to extend their use, and are not affected by temperature and solvents.

Plastic Blades

While the most common material for plastic blades is fiber reinforced phenolic resin, new proprietary injected molded plastic materials also are available.

Figure 13. Available scraper blades.

As may be seen from the curves plotted in Fig. 14, injected molded plastic blades can withstand higher temperatures than can those of phenolic resin. Note, however, that there is a break point at about 180°F where temperature becomes the important factor to wear. Below 180°F, abrasion caused by the rubbing of plastic on metal is the prime wear related factor. Since the plastic blade is a wearing part regardless of the application, it must be replaced periodically to preclude breakage and possible product contamination by plastic fragments.

As stated previously, metal blades are not affected by processing temperatures. They are made of any number of tempered stainless alloys such as 17-4 PH or 410 stainless steel, and this hardening improves the wear life of the blades.

This is especially important for slush freezing applications where there is an added abrasive action caused by the scraping of ice from the heat-transfer surface. Since ice impacts the scraping edge as the blade sweeps the surface, it is important that the blade be maintained with a sharp edge (Fig. 15). For ice cream freezing applications, resharpening normally is done on a monthly basis to ensure optimum performance and the blades are replaced every one or two years. In other applications, metal blades may be sharpened every 3–12 months and replaced every 1–3 yr.

Selecting Seals and Materials

Shaft seals tend to be the most troublesome maintenance area on a scraped-surface heat exchanger. SSHE applications differ widely, and no one seal can handle the variety of temperatures, pressures, and product characteristics constantly encountered.

Most manufacturers offer a number of shaft seals, each designed to satisfy the demands of certain applications. The simplest and least expensive seal to maintain is the O-ring seal (Fig. 16a). This type of seal is recommended for use with products having high lubricity, eg, margarine, lard, shortening, ice cream mix, certain meat products, oils, and fats. Normally these products are not processed at temperatures above 140°F. High temperatures above 180°F shorten the effective life of an O-ring by caus-

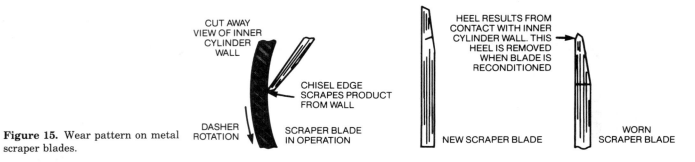

Figure 15. Wear pattern on metal scraper blades.

ing it to lose elasticity. The sealing action of an O-ring seal depends on its ability to be sufficiently flexible to fill the gap between the rotating shafts and the static O-ring groove.

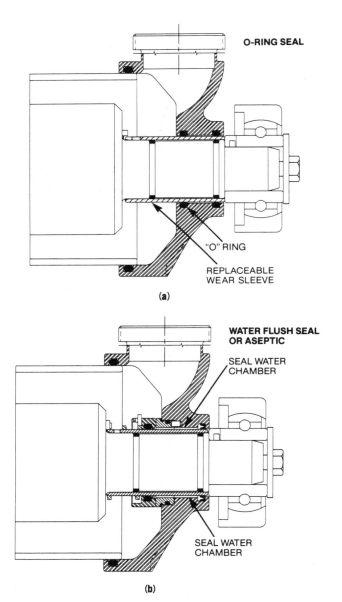

Figure 16. Shaft seals: (a) simple O-ring seal; (b) rotary mechanical seal with water flush or stream aseptic outer chamber.

While the advantage of the O-ring is that it is inexpensive to replace when worn, the disadvantage is that it wears rapidly. Although daily replacement is not uncommon, this is not a problem when frequent equipment inspection is scheduled. The material of the O-ring depends on the fats or oils in the product to be processed. Typical materials are Buna N, Viton, and EPDM.

Rotary face seals are for all other applications (Fig. 16b). These can handle high temperatures and abrasive products, and offer longer wear life than O-rings. Rotary seals can be broken down into two groups—those with and without a seal water chamber.

The seal water chamber allows water or steam to flow around the edges of the seal faces and is recommended for use when handling products that tend to crystallize or dry out in the sealing area. Examples would be liquid sugars, tomato products, candies, fondants, and processed cheese. Water flushing increases the useful life of the seal. Steam is injected into the chamber for aseptic applications since FDA standards require a sterile barrier media on all rotating shaft seals in low acid aseptic processes. The steam, a sterial medium, prevents sterilized products from being contaminated should there be a seal leak. Many manufacturers build an integral water chamber into their equipment with the chamber to be used only as required.

Critical to the life of a seal is the materials used for the seal faces. The typical seal material is hardened stainless steel on carbon. Alternates for the carbon can be phenolic resins, plastics, silica carbide, or ceramics. Other more exotic seal materials to replace the hardened stainless include tungsten carbide and chrome oxide. Each has advantages and disadvantages, such as wear life, and cost. Normally, the manufacturer recommends the seal type and seal materials based on test work and field experience.

Selecting Dashers

Dashers or shafts that carry the scraper blades are available in diameters from $1\frac{1}{2}$ to $6\frac{1}{2}$ in. and are mounted within heat-exchanger cylinders that, in turn, are manufactured in diameters from 4 to 8 in. Since the most widely used cylinder has an ID of 6 in., the following is based on that model.

The best heat-transfer efficiency is provided when the annular space between a dasher and a cylinder is small. For example (Fig. 17), a $5\frac{1}{2}$-in. dasher within a 6-in. cylinder provides a quarter inch annular product flow space, which causes a high axial flow velocity, induces turbu-

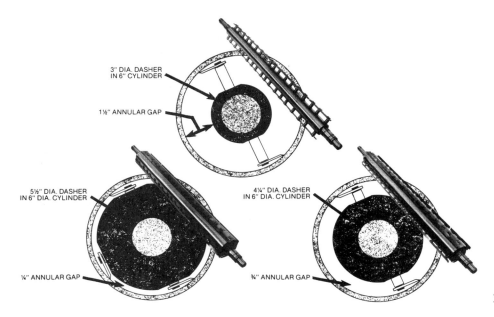

Figure 17. Dasher comparisons.

lence, and offers a short product residence time. This is particularly useful in crystallizing applications such as for processing shortening and margarine since these products essentially are subcooled and allowed to crystallize after leaving the heat exchanger. Dashers for these products are water circulated.

The disadvantages of the 5½-in. dasher are that it cannot handle viscous products because of high product pressure drop and cannot accommodate products containing particulates over a ¼ in. in size. In such cases, the 4¼-in. dasher with its ¾-in. annular product flow space is recommended. The smaller the dasher diameter, the larger the annular space, and the larger the particulates that can be processed. As the dasher diameter decreases, so does the effective U value for a given dasher speed.

Dasher speed is another design consideration. Typical dasher speeds vary from 60 to 420 rpm, and, in general, the higher the dasher speed, the higher the U value (Fig. 18). While this is desirable, increased dasher speed also increases the motor load, which can cause problems in cooling applications. Recalling the heat-transfer formula given previously, heat removed when cooling comes from both the product and the motor load required to turn the dasher. In some applications, the motor load actually can equal the product heat load. Therefore, in some cases a slower dasher speed nets a more beneficial end result.

The graph in Figure 19 compares horsepower and rpm involving various products. All are cooling applications performed on a 6-in. diameter by 72-in.-length unit. Note how small changes in dasher speed on marshmallow change the horsepower requirement significantly while the power requirements vary little while cooling cheese sauces at different speeds. It should be pointed out that horsepower requirements do not necessarily follow viscosity of the product. Rather, the term tenacity would best describe the measurement that would equate to the horsepower—rpm relationship. A good example is corn syrup versus a cooked starch slurry base for salad dressings. At the same viscosity, the corn syrup has much more tenacity in retarding dasher rotation than does the cooked starch.

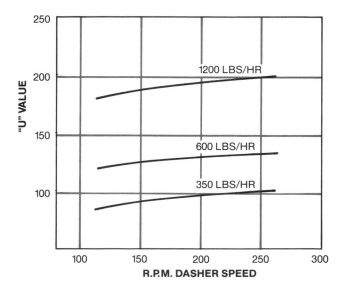

Figure 18. U value versus dasher speed and product flow.

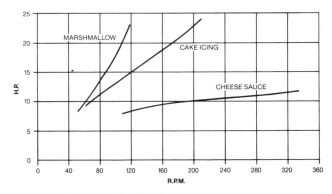

Figure 19. Dasher horsepower versus rpm.

To reduce motor loads while processing tenacious products, slow dasher speeds are required but the dasher cannot be run so slowly that the U value is reduced significantly. The properties of such products normally generate a low U value at any dasher speed, usually in the range of 120–180 Btu/h · ft² °F. At very slow speeds, the U value is further reduced from these already low levels and an economic sizing is not possible. It also should be noted that a certain minimum dasher speed is required to centrifugally "throw" and hold the scraper blades against the heat transfer cylinder. Therefore, there is a definite optimum dasher rpm that maximizes the net capacity of the heat exchanger for tenacious-type products in cooling applications.

Another reason for using low dasher speeds is retention of product identity. When large particulates are being processed with a sauce or gravy, fast-turning dashers tend to cause breakage of the particulates. The dasher speed therefore is not always determined by the desired net heat-transfer efficiency but by limitations of the product.

In both the tenacious and particulate-type products, field measurements and/or lab test data are required to determine optimum dasher speeds.

SYSTEMS APPLICATIONS

As new products emerge from development labs across the country, many processors have determined that the scraped-surface heat exchanger is essential to effective, cost-efficient operations. In some cases, this type of heat exchanger is used for continuous and uniform heating and cooling of highly viscous pastes, fillings that contain up to 80% total solids, or products carrying discrete particles of up to 1 in. in size. In others, the SSHE has been selected for texturizing, gelling, whipping, plasticizing, and crystallizing duties. A few of the more interesting applications are detailed on the following pages.

Mechanical Deboning

In the mechanical deboning of turkey, chicken, and red meats, the heat that is generated in the deboning process must be dissipated to prevent product spoilage. This is done best by means of a scraped-surface heat exchanger.

The mechanical deboning machine (MDM) works on the principle of separating hard material such as bone, gristle, sinew, and cartilage from soft, whole meat. Following prebaking or grinding to reduce large bones to ⅜-in. size, the meat and bones are compressed by an auger inside the deboner and are forced under high pressure into a perforated tube. The soft meat passes through the many small holes, thus separating it from the bones. The meat is emulsionlike in consistency and drops into a hoppered pump. This deboning action requires considerable mechanical working of the meat and, in doing so, increases product temperature by 10–20°F. A scraped-surface heat exchanger (Fig. 20) is used to cool the meat from 55 to 35°F since a temperature of 40°F or less is required to prevent rapid bacterial growth and subsequent spoilage.

When used as a meat emulsion chiller, the SSHE operates on ammonia and is equipped with chrome nickel cylinders and a dasher and blade assembly specifically designed for viscous or particulate products. The typical flow

Figure 20. Three-cylinder SSHE uses ammonia to cool 6000 lb/h of MDM turkey from 50 to 35°F.

rate of product through the chiller ranges from 2500 to 6000 lb/h. Since turkey and red meat are lower in moisture than MDM chicken, they are more viscous. A chiller for these products, in general, operates at two-thirds the capacity of a unit processing chicken and requires a larger dasher drive motor.

Margarine

While a typical system generally consists of a series of storage and blending tanks combined with a crystallizer, high-pressure pumps, and both plate and scraped-surface heat exchangers, the latter is the key to the production of high-quality margarine.

As indicated by the schematic, (Fig. 21) basic oils and fats are drawn from storage and blended with water, salts, emulsifiers, coloring, preservatives, and flavorings for containment in surge tanks at temperatures ranging from 110 to 130°F. If milk ingredients are involved, a Paraflow plate heat exchanger is used for legal pasteurization and cooling before the blending phase. The final mixture then is pumped through an SSHE unit for cooling to approximately 45°F.

While the scraped surface exchanger generally operates on ammonia and is equipped with chromed nickel cylinders and metal blades, the special feature for margarine production is the water-circulated dasher. Since margarine is a crystallizing application, product crystals would tend to stick and collect on the surface of a standard dasher. Such a buildup would increase not only product pressure drop through the exchanger but also the motor load and accelerate overall wear on seals and bearing components. By passing water at about 100°F through the hollow dasher, enough heat is provided to prevent crystal formation while still not contributing significantly to the overall heat load.

Following the cooling cycle, the viscous product is pumped from the scraped-surface chiller through a static crystallizer for table-grade, stick-type margarine or to a

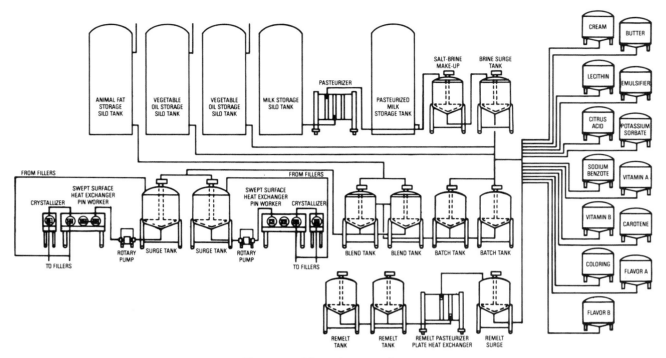

Figure 21. Margarine manufacture.

pinworker crystallizer for tub, soft table, or bakery product. When an acceptable crystal structure has been achieved, the product is formed and packaged and allowed to gain its final viscosity during a quiescent period.

Plasticizing Shortenings

In the manufacture of solid shortening, one primary objective is to make a product that remains plastic and workable over a wide temperature range, specifically between 60 and 90°F. To do this, hydrogenated or hard fats are added to formulations to extend the plastic range. Solid shortening actually is a suspension of high-melting-point fats in liquid fat or oil. Different fats will melt or solidify as is shown in the solid fats index (SFI) charted in Table 3. At room temperatures of 68°F (20°C), the typical shortening is only about 30% solid.

Two methods commonly are used to achieve desired crystal structure, size, and dispersion when producing plasticized shortening. One is for full-scale production (Fig. 22); the other, for limited "in house" applications.

For normal production, an A or scraped-surface heat exchanger is used to cool oil from 140°F (60°C) (no solids)

Figure 22. 10,000 lbs/hr system. Control panel, SSHE with water circulated dashers, pinworker, pumping unit with surge tank.

Table 3. SFI Values for Typical Shortenings

		Melting Point		SFI Value				
				10°C	21.1°C	26.7°C	33.3°C	37.8°C
Shortening	Plastic Range	°C	°F	50°F	70°F	80°F	92°F	100°F
High stability[a]	Narrow	43	109	44	28	22	11	5
All-purpose[a]	Wide	51	124	28	23	22	18	15
14% hardfat in cottonseed oil[b]	Wide	51	124	16	14	14	12	11

[a] Source: Ref. 1.
[b] Source: Ref. 2.

to 60–70°F (16–21°C). This subcooling is done rapidly, and only about 5% of the fats are crystallized. The shortening then enters a B or pinworker unit (Fig. 23), which acts as an agitated holding tube of about 3-min duration. During this time, the remaining crystals form with the latent heat of crystallization causing the shortening to heat up. Typically, if the shortening enters at 65°F (18°C), it will exit at 80°F (27°C). The now plasticized shortening is ready for filling into 50-lb cubes. Most plasticizing systems usually include a shell-and-tube exchanger to precool product from 140°F (60°C) to 120°F (49°C) and reduce the heat load on the SSHE. Nitrogen also is usually added before the SSHE to give the final product a white color.

For bakeries that plasticize shortening for in-house use, a single unit combining the features of A and B machines commonly is used. As shown in Figure 24, this involves an open type dasher shaft. The dasher is hollow and has wide slots at regular intervals. A set of beater bars acts as the pinworker, holding and working the shortening to a high degree in order to produce a soft, readily flowable texture that can be handled in pump lines and tanks downstream.

Marshmallow

Marshmallow is an aerated food product typically formulated from 45% sucrose, 27% corn syrup, 23% water, gelatin, invert sugar, and flavoring. (See Fig. 25.)

As shown in the flow schematic, a production system generally employs multiple tanks, an aerating mixer, and a scraped-surface heat exchanger. A premix tank first is used to blend and heat sugars to 245°F before the solution is transferred to another blending tank for cooling to 155°F. Gelatin and 150°F tempered water, meanwhile, are thoroughly mixed by means of a high shear agitator and blended with the sugar solution. The resulting syrup is charged to a surge tank and pumped as required to a high shear aerating mixer where it is beaten to a weight of between 35 and 48 oz/gal (specific gravity of 0.26–0.36). The 155°F marshmallow mix then flows to an SSHE for chilled water cooling to 90°F.

Marshmallow is a difficult duty for a scraped-surface heat exchanger because of the extremely high viscosity and the low heat-transfer properties of the product. Operating pressures in the mixer and heat-exchanger range

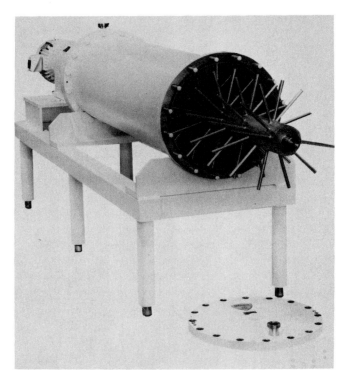

Figure 23. Pinworker with cover removed to expose shaft and pins.

Figure 24. Plasticizer with combination scraped-surface heat exchanger and pinworker.

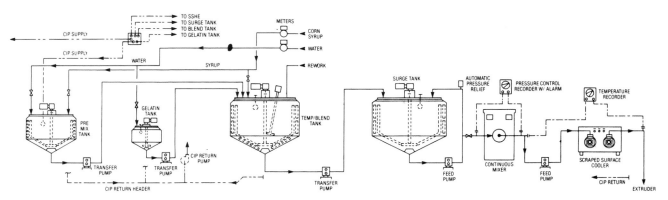

Figure 25. Marshmallow production system.

Figure 26. Scraped-surface heat exchanger cooks and cools ground beef for frozen food entrees. Ground beef is pumped into the rear of the left hand cylinder. Steam is used to head from 45 to 185°F. Cooked product flows out the front, through a holder tube, and into the rear of the right hand cylinder where ammonia at −40°F cools it to below 50°F.

from 150 to 300 psi to handle product viscosity and to force the marshmallow through an extrusion head to a belt where it is shaped and dusted with starch. The low product density results in a low U value and because of viscosity, a high torque is required to turn the dasher. Note in the Figure 19 dasher horsepower—rpm chart how dramatically the dasher hp changes relative speed. Too high an rpm results in excessive motor load that makes cooling more difficult because the motor heat has to be removed by the heat exchanger. Since even the optimum dasher speed results in about one half of the cooling load coming from the motor input, selecting the proper dasher speed for this application is very important. Care also must be given to the selection of shaft seals. Since marshmallow is mostly sugar, seals with abrasive resistant, high hardness qualities should be used and the seals should be water flushed to prevent crystals from building up on the seal faces.

Cooking–Cooling Ground Meat

While ground beef generally is cooked in steam-jacketed kettles for moderate size production runs, the tendency is to install scraped-surface heat exchangers for both cooking and cooling duties when production requirements exceed 10,000 lb/day (Fig. 26). Not only does the SSHE provide faster cooking; it also improves product yield since moisture and fat loss is far less than the 15–20% decline in product volume that occurs with the open-kettle method. (See Fig. 27.)

The typical meat cooking system consists of a single auger feed hopper with a leveling ribbon followed in line by a two-cylinder SSHE, one cylinder with 275°F steam and the other with 0°F ammonia. Both the auger feeder and SSHE are hydraulically driven from a central power source. After meat is ground, spiced, and blended, it is fed to the heat exchanger by means of a single-discharge au-

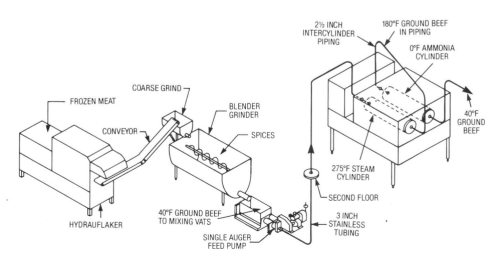

Figure 27. Ground beef cooking and cooling.

ger designed specifically to move viscous, sticky products directly into a rotary pump. The auger maintains a constant stuffing pressure at the rotary pump inlet by varying the operating speed while a leveling ribbon maintains a uniform meat level in the feed hopper. The scraped surface blade and dasher design promote rapid removal of product from the cylinder walls and enhance agitation and mixing during both cooking and cooling phases. Assembly and disassembly for blade inspection and/or replacement is quick and easy. Meat remaining in the auger feed pump, pipelines, or the scraped surface heat exchanger is easily recovered and, once clear of product, the closed system and piping are easily cleaned with a pump circulation loop.

Products with Particulates

When processing sauces, gravies, or juices containing pieces of meat, vegetables, or fruit, one of the key elements is maintaining product identity by reducing or eliminating particulate breakage. This becomes more of a concern if the particulates exceed ¼ in. in size and the carrier is a thin liquid such as broth or soup. Thick sauces, on the other hand, seem to have a cushioning effect with less product damage occurring during passage through the system.

While the dominant type of heat exchanger used for particulate-type products is the scraped-surface unit (Fig. 28), each case still must be considered individually. Many processes require multiple SSHE cylinders that should be piped in series rather than in parallel. This eliminates the need for individual product pumps while ensuring higher heat-transfer rates and an equal flow to each cylinder.

To reduce particle breakage, three design areas must be considered—the gap between the dasher and scraper blade, the sizing of SSHE inlets and outlets, and the dasher operating speed.

The distance between the underside of the blade and the dasher shaft (Fig. 29) should be equal or greater than the largest particle to be processed. This allows passage without particle damage as the blade sweeps by. The disadvantage of a wide annular space between the dasher and the cylinder wall is a longer product retention period

Figure 29. To minimize product damage, clearance between the blade and the dasher shaft should be equal or greater than the size of the particle that is being processed.

and a reduction in exchanger efficiency. The faster the product travels through the cylinder, the higher the U value attained. This results in improved exchanger performance with lower residence time and less product damage.

The sizing of inlets and outlets is directly related to product quality. Using the largest possible ports reduces not only pressure drop through the system but also shearing effect. Furthermore, passage of product is eased and the possibility of contact with inlet walls is minimized.

Finally, there is the matter of dasher speed. As shown by the curve in Figure 30, while higher dasher speed results in a higher U value, it also can cause increased

Figure 28. Three-cylinder SSHE aseptically heats and cools filling with ¼–½-in. pieces of fruit.

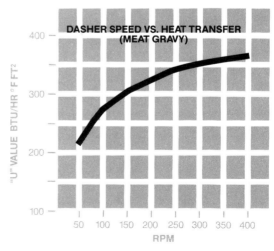

Figure 30. Dasher speed versus heat transfer (meat gravy).

Table 4. Partial List of Products Processed

Product	Heat	Cool	Product	Heat	Cool
Candies, Frostings, and Fillings			*Sauces, Starches, Gravies and Puddings*		
Cake frosting		X	Aseptic cheese sauce	X	X
Cookie filling		X	Aseptic puddings	X	X
Fondant		X	Chili	X	X
Marshmallow		X	Enchilada filling		X
			Gravies and sauces		X
Dairy Products			Salad dressing starch	X	X
Concentrated milk	X		*Slush Freezing for Freeze Concentration*		
Cottage cheese curd		X	Coffee or tea extract		X
Cream cheese		X	Grape juice		X
Ice cream mix	X		Orange juice		X
Melting butter	X		Various aroma extracts		X
Processed cheese	X		Vinegar		X
Quark	X	X	Wine		X
Whey concentrate	X	X			
Whipped butter		X	*Tomato Products*		
Yogurt	X	X	Barbeque sauce	X	X
			Tomato paste	X	X
Fats and Oils			Tomato sauces	X	X
Margarine	X	X	¼-in. diced tomatoes	X	
Shortening					
Plasticizing Bulk		X	*Miscellaneous Products*		
In-house		X	Acetone—deep-cooling		X
Vegetable oil	X	X	Adhesives and glues		X
			Aseptic diced peppers	X	X
Fruit: Concentrates, Juices, and Fillings			Autolyzed yeast	X	
Aseptic applesauce	X	X	Baby foods	X	X
Aseptic bananas	X	X	Candle wax		X
Aseptic fruit fillings	X	X	Hand cream		X
Aseptic fruit purees	X	X	Honey		X
Citrus pulp		X	Ink		X
Jelly and jam preserves	X	X	Lipstick	X	X
Juice concentrate		X	Liquid antacid	X	X
Pumpkin and squash	X	X	Lithium grease		X
Pumpkin pie filling		X	Paints	X	X
			Peanut butter		X
Meat Products			Photographic gelatin		X
Cooking bacon	X		Pickle relish	X	X
Gelatin concentration		X	Poi	X	
Lard		X	Resins	X	X
Low-temperature rendered meat		X	Shampoo		X
Meat pie fillings	X	X	Shaving cream		X
Mechanically deboned			Soup	X	X
Chicken		X	Synthetic polymers	X	X
Red meat		X	Toothpaste		X
Turkey		X	Vanilla wafer batter		X
Sausage and hamburger	X	X	Wallpaper paste	X	X

particulate damage. Therefore, higher dasher rpm and the corresponding high heat-transfer efficiency will apply only to products with small particulates. When products contain delicate particulates of ½–¾ in., a variable-speed drive is used to reduce dasher revolutions to about 120 rpm. While this protects product quality, it also lowers exchanger efficiency. Consequently, striking a balance between performance and quality should be determined by lab runs so that optimum operation at full scale becomes a matter of a proper dasher speed setting.

Peanut Butter

For the production of creamy or chunky peanut butter, a process system will consist of surge tanks, a deaerator, scraped-surface heat exchanger, ingredient feeder with in-line blender, and transfer pumps.

After mixing roasted nuts with a stabilizer, salt, and sugar to the desired formula, the product is ground and discharged at a temperature of about 150–200·F. Deaeration follows to eliminate air pockets, which initiate oil separation. During cooling from an SSHE inlet temperature of 140–190°F to a discharge temperature of 85–95·F, the stabilizer is solidified in finely divided crystalline form and uniformly distributed throughout the mixture. At the proper temperature, the peanut butter becomes a viscous, extrudable mass. Crystal change continues and further solidification occurs after filling. For chunky style, a chunk feeder is located between the SSHE and filler or the transfer pump and deaerator.

Table 4 gives a listing of other products routinely processed using SSHEs.

BIBLIOGRAPHY

Adapted from: SSHE Handbook, SSH-1087. APV Crepaco, Inc. Copyrighted APV Crepaco, Inc. Used with permission.

APV Crepaco, Inc.
Lake Mills, Wisconsin

HEAT SEALING. See Retort pouch.

HEAT TRANSFER

BACKGROUND

Almost all food processes depend on or are affected by heat being added or removed at some stage in the operation. The efficient and effective utilization of heat results in economic savings, minimum adverse affects on nutrient components, higher quality consumer-ready products, and minimum effect on the many environmental factors associated with food processing. This efficient and effective use of heat depends on knowledge and subsequent application of the heat transfer mechanisms involved in heating, cooling, and changing the state of foods and food products. The rate at which heat is transferred depends on the type of product, the condition of the product, and the type of heat transfer by which heat enters or is removed from the product.

Other than maintaining a food under handling and storage conditions that insure the highest quality product, the only option involving heat transfer is the control of processing conditions to give the desired rate of heat transfer. This means considering the three basic mechanisms that control the rate of heat transfer and selecting processing conditions and facilities that optimize the process. Hence a thorough knowledge of heat transfer mechanisms and the relationship between heat transfer rates and other physical and chemical factors is probably the most important consideration involved in the processing of foods. This is further emphasized by the fact that the control of mass and energy balances and fluid flow mechanics, the basics for designing, constructing, and operating the machinery, equipment, and facilities involved in food handling, storing, and processing, is dependent on considerations involving the mechanisms and rate of heat transfer.

Heat is transferred from one body to another by three different mechanisms, namely conduction, convection, and radiation. In fact, most processing facilities utilize two or all three of these means of transferring heat from or to a product. Before considering the mathematical relationships that describe heat transfer and allow the control of the amount and rate of transfer, it is well to have a visual understanding of how each functions.

Conduction heat transfer takes place when two bodies at different temperatures are in contact with each other. Through this direct contact, heat energy is transferred from particle to particle between the solid bodies with no bulk movement of material. When one places a hand on an object having a temperature different from body temperature there is a flow of heat. Hence, when the object is a blackboard in the classroom at room temperature, there is a sensation of the blackboard being cold. Conversely, if one touches a warm element on a stove, there is an immediate sensation of heat being transferred to the hand. The difference in temperatures between the two objects, as well as the characteristics of the conducting body, determines the rate of transfer. In the case of the blackboard, the temperature driving force was probably not more than 20°F (11°C), while the driving force between a warm stove-element and the hand is considerably higher. Hence the sensation of heat will be detected and create considerably more reaction than placing a hand on the blackboard. Other things to consider include the nature of the material to which heat is being transferred. This determines how fast the heat will penetrate as well as the temperature gradient within the material. In general, conduction is highly desired for food being frozen, especially those forms having flat surfaces that can insure maximum contact with plate freezers. In most other cases of freezing irregularly shaped items and heating, cooking, or sterilizing, a combination of conduction and other transfer mechanisms is more efficient and controllable.

Convection heat transfer depends on the bulk movement and mixing of liquids that are initially at different mass temperatures, or on the contact of a solid with a moving liquid stream of a different temperature. These two basic types of convection must be considered together in most heat transfer involved with processing food products. Regardless of whether hot liquids are being mixed in a batch or continuous basis, the temperature of the final combined liquid mixture reaches an equilibrium temperature somewhere between the original temperatures of the two liquids. If a solid is involved, the temperature equilibrium occurs between the bulk of the liquid and the surface of the solid. The velocity of a liquid flowing past a solid affects the rate of transfer between the two materials. Natural or free convection is caused by density gradients (thermal expansion) formed when a liquid is changing in temperature. These can not be controlled and are often a detriment to processing since the rate of transfer is at the mercy of the naturally rising or mixing streams. Forced convection, the pumping or blowing of a liquid or gas over a surface, can be controlled and the heat transfer rates can be predicted. Hence most food processes utilizing convection heat transfer depend on facilities and equipment that use forced convection. This can be demonstrated by considering the wind chill factor, by which a wind causes a person to feel colder than expected at a given outside temperature. For example, in the winter one might bear a weather report in which the outside temperature is given as 0°C and the wind chill factor makes it feel as if the temperature were −15°C. This is because the wind is a forced convection whereas the 0°C temperature is measured under shielded or still conditions.

Many food processing operations include a combination of conduction and convection heat transfer. This is demonstrated in cooling and freezing curves in which blast freezing or convection is the method of heat transfer (see Figure 10 in article "Fish and Shellfish Products") and a plate or conduction heat transfer is combined with blast freezing or convection (see Figure 9 in article "Fish and Shellfish Products"). The critical period during which heat of fusion is being removed between about 30 to 22°F (−2 to −6°C) is 220 min in Figure 10 of "Fish and Shellfish Products". The total time for cooling, freezing, and dropping from 50 to 0°F (10 to −18°C) is 472 min in Figure 10 and 93 min in Figure 9. As has been discussed in "Fish and Shellfish Products," the more rapid freezing as shown in Figure 9 resulted in high quality frozen fish, similar to the fresh product. Conversely, the slower freezing in Figure 10 resulted in considerable cell degradation that greatly reduced the quality of the end product.

The third type of heat transfer used for heat processing of foods is radiation, the transmission of electromagnetic energy through space. Whereas conduction and convection are dependent on a physical medium through which to transfer heat energy, radiation requires no carrier to transfer wave energy from a surface of one body to another. The amount of energy transmitted is dependent on area and the nature of the exposed surface and the temperature of the body. The amount of radiation absorbed or deflected also depends on the nature of the absorbing body and the body temperature. Hence each body receives radiation from every other body, the amount depending on how much the bodies can "see" of each other and the ability of the body to radiate or deflect and absorb radiation (emissivity or absorptivity).

There are several different types of radiation used in the food industry, each used for a specific reason. The radiation heating from a warm or hot body to a food is akin to that of a person standing in the shade or the sun on a hot day. The ambient air temperature may be the same in both positions; however, a person becomes much warmer in the sun where the direct radiation is being added to the body in addition to the convection heating occurring from the surrounding air. The radiating body for heating a food by radiation can be the hot element in an oven or grill (roasting, baking, broiling). In this case the surface receives energy so rapidly that it can not be conducted into the food fast enough to prevent the desired browning (actually scorching and/or Maillard browning reaction) of the surface.

Other types of radiant energy are used in processing food. Some of these are becoming as prevalent as the conventional concept of cooking or processing by exposure to a hot element or heat source. When an alternating current is passed through a conductor, energy is emitted in the form of waves having a specific wavelength and frequency. Microwave heating is a good example of this process, whereby the energy absorbed by a food is converted into heat due to the friction of moving molecules or atoms.

Ionizing radiation from x rays and gamma rays, while not heating a food during a normal exposure time, can destroy microorganisms and thus accomplish the aim of pasteurizing or sterilizing a food.

PROCESSING FOODS BY CONDUCTION HEAT TRANSFER

The relationship between the factors involving heat transfer by conduction can be intuitively realized by considering the flow of heat energy. One would reason that heat flow (q) perpendicular to the direction of flow would be proportional to the area (A) of contact or flow and the temperature difference (dt) along the path of flow. Likewise, the thickness of the food or path through which the heat is flowing (dx) would be inversely proportional to this flow. Although modern mathematics allow one to derive such relationships, the basic conductive heat transfer relationship, known as Fourier's law, was derived empirically and confirmed by experimentation.

In addition to the factors, discussed above, the physical and chemical properties of a material have a significant effect on heat flow and can be represented by an experimentally determined proportionality constant called the thermal conductivity (k). The thermal conductivity is related to the number of free electrons in a material, varying from being high in metals to low in gases. It can be considered the flow of heat per unit time (watts or Btu per hour) through a given area (square meters or square feet) per unit thickness (meters or feet). The mathematical relationships involving conduction heat transfer can be represented by Fourier's law:

$$q = -kA \, dt/dx \qquad (1)$$

The units of each item in SI and English units are shown in Table 1. It is conventional to place a negative sign in

Table 1. Selected Coefficients of Thermal Conductivity

Property	SI Units	English Units
q = heat flow	watts (W)	Btu heat flow Btu/h
k = thermal conductivity	(W)(m)/(K)(m²)	Btu/h of ft²/ft
A = area	m²	ft²
dt = temperature difference	°C	°F
dx = thickness	m	ft

front of the equation to signify a positive heat flow from the higher to the lower temperature.

Although the complete mathematical analysis of heat transfer can become extremely complicated when it comes to considering complex three-dimensional flow and integrations over total volumes, most problems involving conduction in food processing systems can be greatly simplified. In most steady-state cases, considering the relatively short range of temperature changes, the uniformity in area over the distance of heat conduction, and the uniformity in thermal conductivity over these temperature ranges, equation 1 can be simplified to

$$q = -kA(t_1 - t_2)/\Delta x = kA\frac{\Delta t}{\Delta x} \quad (2)$$

Equation 2 is typical of many problems in nature that involve a driving force analogy. In the case of heat transfer, the rate of transferring some discrete quantity of energy q is equal to the driving force DF that makes the movement happen divided by the resistance R to this movement, or

$$q = DF/R \quad (3)$$

A common use of this relationship is in the transfer of electrical energy, where the current flowing (I, amperes) is equal to the driving force (E, volts) divided by the resistance (R, ohms), or

$$I = E/R \quad (4)$$

The resistance in the heat transfer equation (2) equals the thickness (Δx), which is directly proportional to resistance to heat transfer divided by the area and thermal conductivity, both of which decrease the resistance,

$$R = \Delta x/kA \quad (5)$$

The reciprocal of the resistance is the conductance (C), where

$$C = 1/R = kA/\Delta x \quad (6)$$

Since the total heat being transferred through each of a series of materials in contact is the same for each material,

$$q = q_1 = q_2 = q_x = -\frac{k_1 A_1 \Delta t}{x_1} = -\frac{k_2 A_2 \Delta t_2}{x_2} \quad (7)$$
$$= -\frac{k_x A_x \Delta t_x}{x_x}$$

and considering that

$$q = -\frac{\Delta t}{R} = -\frac{\Delta t}{\Delta x/kA} \quad (8)$$

then

$$q = q_1 = q_2 = q_x = -\frac{t_1 - t_2}{x_1/k_1 A_1} = -\frac{t_2 - t_3}{x_2/k_2 A_2} \quad (9)$$
$$= -\frac{t_3 - t_x}{x_x/k_x A_x}$$

Combining to obtain the overall resistance

$$R = (x_1/k_1 A_1) + (x_2/k_2 A_2) + (x_x/k_x A_x) \quad (10)$$

and the overall driving force

$$\Delta t = t_1 - t_x$$

the overall heat transfer equation becomes

$$q = -\frac{\Delta t}{R} = -\frac{t_1 - t_x}{[(x_1/k_1 A_1) + (x_2/k_2 A_2) + (x_x/k_x A_x)]} \quad (11)$$

When, as is often the case, $A_1 = A_2 = A_x$, the equation reduces to

$$q = -\frac{\Delta t}{R} = -\frac{t_1 - t_x}{(1/A)[(x_1/k_1) + (x_2/k_2) + (x_x/k_x)]} \quad (12)$$

Example 1. Using the conductance heat transfer equation, calculate the amount of heat lost through 2 m² (21.528 ft²) of a cold storage wall that is composed of three layers of material. The wall consists of 4-in. (0.1016-m) thick concrete block (x_c) on the outside, 6-in. (0.1524-m) thick polystyrene insulation (x_p), and 2-in. (0.0508-m) thick fir wood (x_f) on the inside. The inside wall of the cold room is $-40°F$ ($-40°C$) and the outside wall is 68°F 20°C). From Table 2, the thermal conductivities are:

$k_c = 0.76$ W-m/m²-K $= 5.27$ (Btu)(in.)/ft²)(h)(°F)
$k_p = 0.038$ W-m/m²-K $= 0.263$ (Btu)(in.)/(ft²)(h)(°F)
$k_f = 0.11$ W-m/m²-K $= 0.76$ (Btu)(in.)/(ft²)(h)(°F)

Using equation 12 where the area is constant (Fig. 1),

$$q = -\frac{t_i - t_o}{(1/A)[(x_c/k_c) + (x_p/k_p) + (x_f/k_f)]}$$

In SI units:

$$q = -\frac{(-40) - (20)}{(1/2)[0.1016/0.76 + 0.1524/0.038 + 0.0508/0.11]}$$

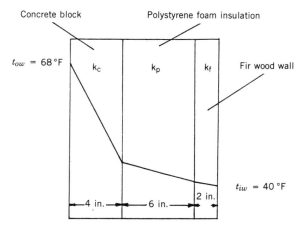

Figure 1. Temperature profile of wall in cold storage room.

Table 2. Thermal Conductivity of Selected Foods and Processing Facility Materials[a]

Product	Temperature, °C	Thermal Conductivity, k[b]	
		$\frac{(W)(m)}{(m^2)(K)}$	$\frac{(Btu)(in.)}{(ft^2)(h)(°F)}$
Apple	2–36	0.393	2.72
Aluminum	0	230	1594
Beef, lean	7	0.476	3.20
perpendicular to fiber	62	0.485	3.36
parallel to fiber	8	0.431	2.99
	61	0.447	3.10
Butter	46	0.197	1.37
Concrete, cinder	23	0.76	5.27
Fish muscle	0–10	0.557	3.86
Ice	0	2.3	15.9
Milk	37	0.530	3.67
Olive oil	15	0.189	1.31
	100	0.163	1.13
Polystyrene	24	0.038	0.263
Potato, raw	1–32	0.554	3.84
White fir	23	0.11	0.76
White pine	15	0.15	1.04

[a] Refs. 1, 2.
[b] 1 (Btu)(in.)/(ft²)(h)(°F) = 0.144228 (W)(m)/(m²)(K).

carrying the unit with the numerical equation

$$q = \frac{°C \text{ or } K}{(1/m)[(m/Wm/m^2K) + (m/Wm/m^2K) + (m/Wm/m^2K)]} = \text{watts}$$

$q = 26.05$ W lost through 2 m² of the wall

Likewise, in English units:

$$q = \frac{(-40) - (68)}{(1/21.528)[4/5.27 + 6/0.263 + 2/0.76]}$$

carrying the unit with the numerical equation

$$q = \frac{°F}{1/ft[\Sigma(in./Btu)(in.)/ft^2(h)(°F)]} = \text{Btu/h}$$

$q = 88.74$ Btu/h lost through 2 m² of the wall

Checking units by converting English to SI:

$q = (88.74 \text{ Btu/h})(0.29307 \text{ W-h/Btu}) = 26$ W

PROCESSING FOODS BY CONVECTION HEAT TRANSFER

Convection heat transfer is concerned with the mixing of fluids or the transfer of heat from a fluid to a surface. Therefore there is no solid-body thickness through which heat must be transferred, as in conduction. In normal food processing operations the concern is usually to transfer heat between a liquid and the surface of a solid or a fluid. Newton's law of cooling, which also applies to heating, defines the driving force as the temperature difference between the surface temperature (t_s) and the bulk temperature (t_f) of the fluid medium:

$$q = -hA(t_s - t_f) \quad (13)$$

In this case the temperature of the surface is higher than the bulk temperature of the liquid and h is the convective heat transfer coefficient (commonly called the film coefficient), the experimentally determined proportionality constant that takes into consideration the flowing characteristics and liquid–solid interface effects. The units are W/m²-K or Btu/(h)(ft²)(°F). Note that there is no dx or thickness factor so that the resistance R corresponding to conduction heat transfer is

$$R = 1/hA \quad (14)$$

Figure 2 depicts the type of temperature profile that occurs when a hot fluid is transferring heat to a solid (or liquid) surface. The equilibrium is between the bulk temperature of the liquid (t_f) and the surface of the solid (t_s). Note that the temperature drops in a nonlinear manner. This is caused by the very thin layer of fluid that is almost stagnant because of the friction of the moving fluid, often referred to as the edge effect. This layer is flowing in the streamlined region and, since there is essentially no mix-

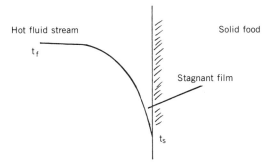

Figure 2. Temperature profile of hot flowing liquid near the surface of solid being heated.

Table 3. Approximate Range of Convection Heat Transfer Coefficients[a]

Medium	Conduction Heat Transfer Film Coefficient, h	
	W/m²-K	Btu/(ft²)(h)(°F)[b]
Gases		
natural convection	2.8–28	0.9–9
forced convection	11–110	3.5–35
Viscous liquids, Forced convection	56–560	17–178
Water, forced convection	560–5600	180–1,800
Boiling water	1,700–28,400	540–9,000
Condensing steam	5,600–1.1 × 10	1,800–3.5 × 10

[a] Refs. 1, 2.
[b] 1 Btu/(ft²)(h)(°F) = 3.15459 W/m²-K.

ing with the main liquid stream, the heat transfer through this layer is actually by conduction but without a thickness (dx) to consider. The magnitude of the edge or fictitious film effect varies widely depending on the solid surface and explains why this coefficient must be experimentally determined for any given condition. Some select convection film coefficients are given in Table 3. The wide range of film coefficient values is due to the surface and geometry over which the fluid is passing and the turbulence (Reynolds number) of the flowing fluid.

Over a period of time, scale deposits build up on the walls of vessels and pipes. The effects of these films or deposits are determined experimentally and expressed in units of the film coefficient h. Therefore films and scale-deposit film coefficients can be handled in the same manner as the thermal conductance of a flowing liquid.

Steady-state heat transfer processes occur in the operations that involve continuous heating or cooling of foods. These include continuous freezing tunnels, deep frying, and pasteurizing and sterilizing liquids. Many processing procedures in which a solid is being heated in a continuous moving system (eg, freezing or cooking a product on a moving belt) can be treated as a total steady-state system by using specific locations on the belt as base points for heat balances. In most of these cases steam, hot water or oil, or a refrigerant is the source for adding or removing heat, so that convection heat transfer is an important factor in these processes.

The only procedures that must be handled as batch, unsteady-state systems are those in which a product is being heated or cooled in place by a steady-state or non-steady-state source (eg, cooking in an oven, retorting canned foods, and freezing a product on a freezer plate). Normally the information required to calculate heat transfer in these situations is collected experimentally and then applied. For example, the thermal death time to insure the sterilization of canned food is experimentally determined for each size of container and product and then applied to the commercial facilities. The actual retorting time is then increased by a sufficient time to insure that there is no error in obtaining sterilization. In the case of sterilizing products, the National Food Processors Association (in coordination with the Food and Drug Administration) provides sterilizing time requirements to the processors. The processors must then record and maintain the records of retort time and temperature for each batch processed to insure compliance with the retorting requirements. These records are available in legal cases involving undercooking.

COMBINED CONVECTION AND CONDUCTION HEAT TRANSFER

Conduction and convection have been discussed separately in order to emphasize the different mechanisms that control the heating and cooling of food products. However, seldom is a food processing operation carried out in which the heat transfer is not a combination of conduction and convection. Although radiation is also present in most of these operations, the effect is usually quite insignificant unless the process is specifically a radiation process.

Figure 3 extends the transfer of heat from a liquid to a solid surface, through the solid, and then to another liquid. This could be the situation occurring in a heat exchanger where a hot liquid is being used to warm another liquid, with the solid being the wall of the heat exchanger. The resistances to heat flow through the hot fluid (R_h), solid pipe wall (R_w), and fluid being heated (R_c) can be summed to give a total resistance

$$R = R_h + R_w + R_c \tag{15}$$

Although only three media are used in the example, there is no limit to the number of resistances that can be summed for a given problem.

An overall heat transfer coefficient U combining conduction and convection heat transfer coefficients now can be defined from equation 15, where, although only three media are used in the example, there is no limit to the summing of as many resistances as are necessary in a given situation

$$1/UA = 1/h_h A_h + x_w/k_w A_w + 1/h_c A_c \tag{16}$$

or

$$UA = \frac{1}{1/h_h A_h + x_w/k_w A_w + 1/h_c A_c} \tag{17}$$

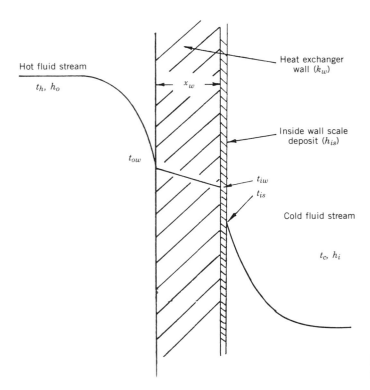

Figure 3. Temperature profile of hot liquid (t_h) transferring heat through heat exchanger wall to another colder liquid (t_c).

Thus the overall heat transfer equation becomes

$$q = -UA\,\Delta t \qquad (18)$$

As noted, U and A are not independent since, if the areas of the heat transfer surfaces vary (eg, a pipe heat exchanger), A must be based on a mean area (A_m). For example, in the process of transferring heat through a pipe that has an inside scale deposit with a determined h_{is}, the overall conductance can be calculated as

$$UA = \frac{1}{1/h_o A_{ow} + x_w/k_w A_m + 1/h_{is} A_{is} + 1/h_i A_{iw}} \qquad (19)$$

where

$$A_{iw} = A_{is} = \pi D L$$
$$A_{ow} = A_{os} = \pi D L$$
$$A_m = (A_{ow} - A_{iw})/\ln(A_{ow}/A_{iw})$$

multiplying the denominator by A_m/A_m gives

$$UA = \frac{1}{(1/A_m)[A_m/h_o A_{ow} + x_w/k_w + A_m/h_{is} A_{io} + A_m/h_i A_{iw}]} \qquad (20)$$

and rearranging gives an expression for UA in a circular cross section where heat is being transferred:

$$UA = \frac{A_m}{A_m/h_o A_{ow} + x_w/k_w + A_m/A_{iw}h_{is} + A_m/A_{iw}h_{iw}} \qquad (21)$$

In this example, the effect of the scale deposit, h_{iw}, is on the inside wall of the pipe. There are commonly five resistances to heat transfer caused by the (1) heat transfer wall and (2) inside (i) and (3) outside (o) of the heat transfer wall, (4) 2 fluids (h_o and h_i), and (5) 2 fouling scales (h_{is} and h_{os}). Often there is a fouling scale on the surface of a transfer surface caused by oxidation or burn-on during operation. Protein foods are particularly bad in that the heat causes denaturation and subsequent adherence to the surface. When gas comes into contact with heat transfer surfaces (eg, a fire tube boiler) heavy scales can be formed. In fact, the maintenance and cleaning of heat exchanger surfaces is a continuous chore, usually accomplished by chemical cleaners. However, in extreme cases (especially in boiler tubes), a periodic rodding out is necessary to maintain the efficiency of the heat transfer.

Boilers are particularly important to the food processing industry since they are the source of steam for many heat transfer systems. The boiler is a heat exchanger that converts a liquid to a vapor. The vapor, after being condensed in a food processing heat exchanger, is normally collected and recycled. Often there are fouling scales or film deposits on both sides of a pipe or plate heat exchanger. This must be recognized in the calculation of the overall heat transfer coefficient.

Example 2. In determining the heat transfer characteristics of a heat exchanger, it is important to have a clear picture of the physical exchanger prior to calculations. For example, in a pipe heat exchanger, it is common to determine the UA for a unit length of pipe. An exchanger can then be sized to meet the heat transfer requirements. Consider a heat exchanger of stainless steel tubes with the cross section shown in Figure 4. Each tube is 1-in. OD

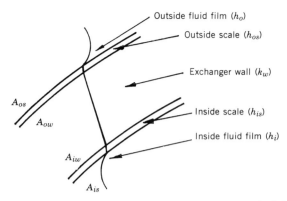

Figure 4. Temperature profile for heat exchanger in which heat transfer area is not constant.

with the following dimensions:

D_i = inside diameter = 0.782 in. (0.0199 m)
D_o = outside diameter = 1.000 in. (0.0254 m)
x_w = wall thickness = 0.109 in. (0.00277)

The heat transfer coefficients are

inside: h_i = 120 Btu/(h)(ft²)(°F) or 681 W/m²-K
inside fouling: h_{is} = 660 Btu/(h)(ft²)(°F) or 3748 W/m²-K
outside fouling: h_{os} = 1000 Btu/(h)(ft²)(°F) or 5678 W/m²-K
outside: h_o = 1500 Btu/(h)(ft²)(°F) or 8517 W/m²-K

and the conductivity of the stainless steel tube is

k = 9.4 (Btu)(ft)/(h)(ft²)(°F) or 16.3 W-m/m²-K

Determine UA for the tubes in this heat exchanger. Basis:

(a) Consider a 1-ft length (L) of tube.
(b) Consider a 1-m length (L) of tube.

Since the area of the tube cross section is expanding perpendicularly to the heat transfer, a mean area (A_m) value must be used in determining UA, and the basic equation becomes

$$UA = \frac{A_m}{(A_m/h_o A_{ow}) + (A_m/h_{os} A_{ow}) + (x_w/k_w) + (A_m/h_{is} A_{iw}) + (A_m/h_i A_{iw})}$$

Calculation of areas in English units:

$A_{ow} = A_{os} = \pi DL = \pi(1)(1)/12 = 0.262$ ft²
$A_{iw} = A_{is} = \pi DL = \pi(0.782)(1)/12 = 0.205$ ft²
$A = \dfrac{A_{ow} - A_{iw}}{\ln(A_{ow}/A_{iw})} = \dfrac{0.262 - 0.205}{\ln(0.262/0.205)}$
  = 0.232 ft²

Calculation of areas in SI units:

$A_{ow} = A_{os} = \pi DL = \pi(0.0254)(1) = 0.0798$ m²
$A_{iw} = A_{is} = \pi DL = \pi(0.0199)(1) = 0.0625$ m²
$A = \dfrac{A_{ow} - A_{iw}}{\ln(A_{ow}/A_{iw})} = \dfrac{0.0798 - 0.0625}{\ln(0.0798/0.0625)}$
  = 0.0708 m²

UA in English units:

$$UA = \frac{A_m}{[\Sigma A/hA + x/k]}$$

$$= \frac{0.232}{\dfrac{0.232}{(1500)(0.262)} + \dfrac{0.232}{(1000)(0.262)} + \dfrac{0.109}{(12)(9.4)} + \dfrac{0.232}{(660)(0.205)} + \dfrac{0.232}{(120)(0.205)}}$$

= 17.1 Btu/(h)(°F)(ft length)

UA in SI units:

$$UA = \frac{0.0708}{\dfrac{0.0708}{(8517)(0.0798)} + \dfrac{0.0708}{(5678)(0.0798)} + \dfrac{0.00277}{(16.3)} + \dfrac{0.0708}{(3748)(0.0625)} + \dfrac{0.0708}{(681)(0.0625)}}$$

= 29.55 W/(K)(m length)

RADIATION HEAT TRANSFER

Since all bodies continuously emit thermal radiation and absorb radiation from other bodies around them, both the received and emitted energy (combining emitted, received, and reflected energies) must be considered when determining the temperature state of a body. This total energy interchange between bodies is significantly affected by the nature of the bodies and the conditions in which these bodies exist. The ability of a body to absorb and emit energy is measured by its emissivity, the fraction absorbed or emitted compared to an ideal perfect radiator or so-called blackbody that has an emissivity (ε) of 1.0. It has been determined that the amount of energy emission from a body is proportional to the fourth power of the absolute temperature. A heat transfer equation can be balanced by the experimentally determined proportionality constant known as the Stefan-Boltzmann constant (σ). Hence, for a blackbody

$$q = \sigma T^4 \qquad (22)$$

where

q = W/m² or Btu/(h)(ft²)
σ = 5.6697 × 10⁻⁸ W/m²-K⁴
  = 0.173 × 10⁻⁸ Btu/(h)(ft²)(°R)⁴
T = K or °R

In reality, there is no such thing as a blackbody that has an emissivity of 1.0, so the practical equation must take into account the experimentally determined emissivity (fraction of 1.0), resulting in the usable equation for radiation

$$q = \sigma \in T^4 \quad (23)$$

Example 3. A person wearing a white shirt and pants ($\varepsilon = 0.9$) at room temperature will radiate energy.

In SI units:

$$q = (5.6697)(10^{-8})(0.9)(W/m^2K^4)(293)^4(K)^4$$
$$= 376 \text{ W/m}^2$$

In English units:

$$q = (0.1713)(10^{-8})(0.9)[\text{Btu}/(h)(ft^2)(°R)^4](528)^4(°R)^4$$
$$= 119.8 \text{ Btu}/(h)(ft^2)$$

Checking units:

$$q = (119.8)(3.1546) = 377 \text{ W/m}^2$$

where

$$1 \text{ Btu}/(h)(ft^2) = 3.1546 \text{ W/m}^2$$

Since there is an interchange between two bodies, the net amount of radiation received from a source is the radiation received minus the radiation returned or emitted. The combination of the heat being absorbed by a food and the amount being emitted gives an equation or the net heat transfer to a food from a radiation source:

$$q = \sigma\{[(\varepsilon_S \alpha_R F_{S\text{-}R})(T_S)^4] - [(\varepsilon_R \alpha_S F_{R\text{-}S})(T_R)^4]\} \quad (24)$$

where σ = Stefan-Boltzmann constant
ε_S = source emissivity
α_R = receiver absorptivity
$F_{S\text{-}R}$ = view factor, source to receiver
T_S = temperature of source
ε_R = receiver emissivity
α_S = source absorptivity
$F_{R\text{-}S}$ = view factor, receiver to source
T_R = temperature of receiver

The amount of energy absorbed from a body depends on how much of the radiating body can be "seen" by the receiver. In most situations involving the processing of food by radiant heat transfer, the view factor is equal to unity since the source is directly below or adjacent to the receiver. Also, the emissivity is essentially the same as the absorptivity in a food processing operation. The emissivities for many surfaces have been determined experimentally and are available in standard data tables. The emissivity of many foods is approximately 0.9, while hot metal (as in the case of oven heating units) has a wide range emissivity values, from as low as 0.1 to 0.6 or higher.

Example 4. Bread is being baked on a stainless steel sheet. The top is cooking nicely but the bottom is not getting done. The oven (ceiling, wall, and floor) temperature is 400°F (204.4°C) and the emissivity and absorptivity are 0.90

(a) The bottom of the stainless steel sheet has emissivity–absorptivity of 0.25. Assuming a view factor of 1.0, what is the net rate of radiant heat transferred to the bottom of the sheet? Assume that the surface of the bread while cooking and vaporizing water is 212°F (100°C) and that the conduction heat transfer to the stainless steel sheet is to a surface at the same temperature.

(b) What is the net rate of heat transfer to the top of the bread, which has emissivity–absorptivity equal to 0.90?

Solution. (a) By substitting the values in equation 22, the net heat that is being transferred to the bottom of the sheet is:

In English units:

$$q = (0.1713)(10)^{-8}[(0.9)(0.25)(1.0)(860)^4 - (0.25)(0.9)(1.0)(672)^4]$$
$$= 132 \text{ Btu}/(h)(ft^2)$$

In SI units:

$$q = (5.6697)(10)^{-8}[(0.9)(0.25)(1.0)(478)^4 - (0.25)(0.9)(1.0)(373)^4]$$
$$= 417 \text{ W/m}^2$$

Checking units:

$$(132) \times (3.1546) = 417 \text{ W/m}^2$$

$$\frac{\text{Btu}}{(\text{hr})(ft^2)} \frac{(W)(\text{hr})(ft^2)}{(m^2)\text{Btu}} = \text{W/m}^2$$

(b) By substituting the values in equation 22, the net heat that is being transferred to the top of the bread is:

In English units:

$$q = (0.1713)(10)^{-8}[(0.9)(0.9)(1.0)(860)^4 - (0.9)(0.9)(1.0)(672)^4]$$
$$= 476 \text{ Btu}/(h)(ft^2)$$

In SI units:

$$q = (5.6697)(10)^{-8}[(0.9)(0.9)(1.0)(478)^4 - (0.9)(0.9)(1.0)(373)^4]$$
$$= 1508 \text{ W/m}^2$$

It is clear that the top portion of the bread is receiving more than 3.5 times the amount of heat being received on the bottom where it is in contact with the stainless steel sheet. This explains why the top, the portion receiving the most heat, is baking at a faster rate than the bottom. It should be noted that often cookies or items being baked in an oven burn on the bottom and are obviously being heated at a faster rate than the top of the item. This is due to the fact that often the metal container is in contact with a hot shelf and is thus receiving heat by conduction.

TRANSFERRING HEAT IN FOODS

Heat transfer was introduced as being an important unit operation in almost every food process. In fact, the transfer of heat to and from a food is certainly the most important factor to consider during processing, whether it occurs in a food processing operation or through natural temperature changes during holding and storage. The type of heat transfer and the rate at which the transfer occurs affect the nutritional value, product safety, and sensory properties of every food that is consumed by humans and animals.

The processes of canning, pasteurizing, and cooking depend on transferring heat into a product to elevate the temperature to a level that will destroy microorganisms. Heat sterilization of a product (eg, pouch packaging and canning) under anaerobic conditions requires closely controlled time and temperature process conditions to ensure that spores resistant to heat (eg, *Clostridium botulinum*) do not survive. Pasteurization temperature and time are just as important to ensure that food-borne diseases are not spread by the food. Cooking a food not only improves sensory acceptance but reduces the microbial content.

Combined heating and cooling operations are commonly carried out during processing or handling of foods. The cooling normally follows the required heating time and temperature period to minimize damage to nutritional and sensory attributes after the heat has accomplished the intended purpose. For example, some fruits and vegetables are blanched prior to freezing to inactivate enzymes that cause degradation and greatly reduced shelf life. Pasteurization to kill microorganisms that might cause public health problems leaves many other microorganisms that will continue to grow. Therefore, pasteurized products (eg, milk) must be held at refrigeration temperatures to ensure a reasonable shelf life. Cooked foods that are not consumed immediately after cooking must be stored in the refrigeration for the same reason.

Many food processes require that heat be added to a food for a purpose other than reducing microorganism contamination or enzyme activity. Vaporization processes such as evaporation, dehydration, and distillation utilize heat energy to vaporize a component, usually water, to change the characteristics of the food. Hence a concentrated juice or a dried cereal product may be subjected to an elevated temperature during the processing.

Often during food processing mechanical energy is converted to heat energy by friction from the particles of food as they rub together. The resulting energy release from such operations as cutting, grinding, screening, and extruding can significantly increase the temperature of a product.

Nature is responsible for adding the heat that is necessary for drying grains in the field to the extent that they can be harvested and stored for long periods of time before being finally processed and consumed. During extremely hot harvest times many fruits and vegetables can be severely damaged during the time it takes to harvest and transport them to a processing plant.

In all cases, whether the heat be intentionally added for safety and to improve texture or unavoidably added as a result of handling or processing conditions, a certain amount of nutritional damage occurs during the heat processing of a food. Depending on the food water content, pH, and oxygen exposure, heat can severely degrade many vitamins, denature proteins, cause free radicals to form, and adversely affect quality attributes such as texture and flavor.

In summary, the transfer of heat directly or indirectly effects the acceptance, quality, and safety of a food. A major responsibility of the food scientist and engineer working with food processes is to optimize the time and temperature conditions of processing and minimize the adverse side affects of heat and other processing conditions to insure that the consumer receives the highest quality food at the minimum cost.

BIBLIOGRAPHY

1. E. L. Watson and J. C. Harper, *Elements of Food Engineering*, 2nd ed., Van Nostrand Reinhold Company, New York, 1988.
2. R. P. Singh and D. R. Heldman, *Introduction to Food Engineering*, Academic Press, New York, 1984.

General Reference

Heat Transfer, H. Wolf, Harper & Row, Publishers, New York, 1983.

GEORGE M. PIGOTT
University of Washington
Seattle, Washington

HERBS. See SPICES AND HERBS: NATURAL EXTRACTION; SPICES AND SEASONINGS; VEGETABLE PRODUCTION
HIDES. See TANNINS.

HISTORY OF FOODS

Writing a history of food is a formidable task that involves identifying, describing, and discussing what people ate where and at what time, where it came from, and how it was obtained. This article attempts to sketch a broad picture of the human quest for food through time and, primarily, will be an archaeological odyssey for reasons discussed below. It treats the history of food as the general history of the human subsistence economy.

Food is an integral part of the human experience, and because the human experience is long, so must be the story of food. The greater part of the human experience occurred long before the domestication of plants and animals. The invention of writing came considerably after the development of agriculture. For some two million years humanity wrote nothing that would aid us in reconstructing its subsistence. Even after writing was invented some 5500 yrs ago, very little was written about food for a long time. Therefore the history of food is for the most part based on archaeological materials.

A history of food must have a beginning. An anthropologist would choose to begin the story in the dim, distant ages of human prehistory some two million years ago with the earliest fossils human paleontologists place in the genus *Homo*. From the two-million-year point, our evolving ancestors and their food quests and habits will be traced to the present time. Important aspects of their lifestyles that bear on the evolution of their diets and general subsistence economies will also be briefly sketched. This approach takes us first, into the realms of physical anthropology and archaeology (paleoanthropology); second, into ancient history; and third, into history. We move from the earliest foraging and scavanging subsistence economies through gathering and hunting to the invention of agriculture and beyond. Finally, some modern cuisines will be briefly addressed in order to demonstrate the varied world origins of ingredients in given dishes and meals.

The earliest known writing appeared in the Near East about 5500 yr ago. All human experience before that, there and in the rest of the world, falls under prehistory. Prehistory is the province of the prehistorian. Most prehistorians are archaeologists. All research is archaeological and based on the few imperishable remains left to us. For example, for early times we must rely on stone, bone and antler tools, bones from animals that were consumed, charcoal and ash from campfires, traces of camps, and so on. Much of what people did and ate was not recorded or preserved (see below). Because our data are limited, there are massive holes in our story. Consequently, caution is necessary when presenting it.

For roughly two million years, humans foraged, gathered, and hunted for wild food. As time passed, we grew more efficient in exploiting our planet's wild resources. This was accomplished through the gradual development of more efficient and sophisticated tool technology based on stone, bone, wood, antler, and such and through developments in social and political organization and probably ideology, which allowed greater cooperation within social groups. These developments were accompanied by increasing brain size and intelligence. This long episode in human history is called the Paleolithic or Old Stone Age. The chronological subdivisions of the Paleolithic are used below for ease of presentation of the history of food (Fig. 1).

About 10,000 or 11,000 years ago in various areas of the Old World, strides were taken that resulted in the domestication of some plants and animals. The "agricultural revolution" had begun. By 9000 yr ago, the revolution was in full swing. There was still no writing. Neither were there cities and states, only small farming villages. Some groups in many world areas still hunted and gathered for their livelihood. The so-called invention of agriculture is prehistoric. In the Near East (sometimes called the Fertile Crescent and including southern Turkey, western Iran, Iraq, Syria, Jordan, Lebanon, and Israel), this earliest agricultural period is called the Neolithic (Fig. 1), and was truly revolutionary for much of humanity. While the invention of agriculture was not simultaneous in all Old World areas, developments in India and the Far East

Period		Years before Present	Subsistence Economies
Neolithic			Agricultural subsistence economy in the Near East
		—9,000	
Mesolithic			Incipient agriculture in some world areas; beginning domestication of some plants and animals
		—12,000	
P a l e o l i t h i c	U p p e r		Homo sapiens gathers wild plants and hunts small and large game
		—40,000	
	M i d d l e		
		—100,000	
	L o w e r	—200,000	Homo erectus gathers wild plants, hunts small and large game; and scavanges carnivore kills
		—1.5 mya	
		—2.5 mya	Homo habilis gathers wild plants, forages for small animals, and scavanges carnivore kills
		?	?

Figure 1. Simplified chronological chart not to scale.

(Southeast Asia and China) were roughly concordant. For none of these areas do we have a complete picture of the development of agriculture.

In the New World, agriculture began independently a few thousand years later than it did in the Near East. There were dramatic developments in Mexico, tropical Central and South America, and in the South American Andes Mountain region. Like those of the Old World, New World domesticates were to have worldwide consequences. Also like that of the Old World, the picture is frustratingly incomplete.

The development of agriculture was revolutionary in several senses: (1) it eliminated dependence on wild resources; (2) in its areas of origin and in areas that ultimately adopted it, it permitted population growth well beyond what wild resources could support; and (3) it laid the economic foundation for the eventual development of civilization. Without agriculture, there would be no cities and nation-states with their centralized government bureaucracies, monumental public works, social class systems and inequities, exact and predictive sciences, and all other civilizational characteristics such as taxes, welfare systems, and terrorists. Humanity could not afford them.

As indicated earlier, the major part of this presentation is based on archaeological research. It is essentially an archaeological odyssey through time and around the globe; thus, and because the reader is most likely not an archaeologist, it is necessary to present a brief discussion of the problems and methods involved in reconstructing the prehistory and history of food. A glossary of technical terms may be found at the end of the article.

BACKGROUND

Reconstructing the history of food is fraught with dangers and difficulties. Much of what is written, especially that for prehistoric times, is based on a number of assumptions, some warranted, and some not. Those based on human biological necessities probably have some validity. Those based on archaeological research are subject to two major problems. First, not everything preserves. Second, archaeological interpretation of what does remain is difficult and has itself been based on assumptions. We are all familiar with the image of "man the hunter" and the portrayal of Stone Age humans as mighty, intrepid hunters of massive, dangerous, and cunning game like mammoths and cave bears. This image comes to us through the magnificent Late Stone Age paintings of western Europe, from food bones found in caves and other living sites and from kill and butchery sites found throughout the Stone Age world. Because plant foods rarely preserve and are even more rarely portrayed, incautious interpretation by some has left the impression that prehistoric humans ate mostly meat. Except for environments like the arctic and the subarctic, where edible plants are rare, this is unlikely to be the case for a variety of reasons.

Humans are primates. Most primates are herbivorous and only accidentally omnivorous, and some are omnivorous (1,2). Higher primates tend to be plant eaters, although a few, such as chimpanzees, will deliberately hunt and eat meat (2). Humans are omnivorous today, and it is likely that our hominid ancestors, as primates, were also omnivorous. Early hominid teeth, like those of modern humans, are those of omnivores (2). The greater majority of modern humans cannot extract sufficient vitamins and minerals from meat alone to survive. Eskimos are an exception, but even they eat the stomach contents of the herbivorous animals they kill and take advantage of plant foods during the short growing season (1). Nor can we live on plants alone without bringing together the right combination of plants to supply us with complete proteins. Nutritionists tell us that humans need a diet balanced so that we receive all the vitamins and minerals the body demands for efficient functioning. This means a combination of plant and animal foods with heavy emphasis on plants as opposed to animal protein.

Anthropological study of historic and modern gatherer-hunters shows that, with the exception of the Siberian and North American Eskimos of the subarctic, the ratio of wild plant food gathered is almost always greater than that of hunted or fished animal protein. The actual ratio depends on both available environmental resources and cultural factors and differs between groups. An additional factor affecting the ratio is the seasonality of resources. In some seasons plant foods are more available than in others, such as winter in the temperate northern hemisphere, when, in the absence of plant preservation techniques or even with them, more meat will be eaten—or perhaps, less food will be consumed in general.

In a given environment various small animals are foraged, such as rodents, birds, reptiles, and insects. While larger game may be preferred, it is not always present or present in sufficient quantity to appease human appetites for animal protein.

In short, gatherer-hunters adapt to the environment in which they live and the ratio of plant to animal foods will reflect that adaptation. It will also reflect cultural factors such as religious taboos and the cultural definition of what is and is not food. We also recognize that hunting is high risk and not always successful. Gathering provides an ensured food supply.

Archaeologically, little trace of the plants consumed by recent gatherer-hunters will remain. Without the ethnographic documentation of these living people, we would never have complete knowledge of their subsistence economies. This is because plant material is notoriously perishable and subject to selective preservation. In some dry desert areas, plant remains do preserve. So do human coprolites. Ancient coprolites contain seeds, pollens, and husks or shell parts that aid in reconstructing paleodiet. But human coprolites are rarely found. In wet areas, such as bogs, plants often preserve, as do human bodies. Perhaps one of the most famous and scientifically valuable examples of preservation is Tollund Man, in whose preserved stomach were found the remains of his last meal, a porridge prepared from a wide variety of seeds of domesticated and wild plants (3). A very recent technique of extracting collagen from human bone and conducting trace-element and isotopic analysis enables us to determine whether and what kinds of cereal grasses were being consumed, for example, maize. But, of course, the bone itself must be preserved and unfossilized (4). Fossilized bone is completely mineralized and contains no organic material.

Animal bone is also subject to selective preservation. Large bones survive better than do small bones, and not all bone fossilizes. Acid soil destroys bone. Insect parts rarely preserve. In sum, little of what was originally deposited at a site by human subsistence activity preserves. There is no real way of reconstructing the ratio of plant to animal food consumed by our remote ancestors or even those closest to us in time.

We are left on shaky ground. Our assumptions are: (1) both presapiens and sapiens hominids were omnivorous; (2) they adapted to the varied environments in which they lived and exploited their edible resources using the then current technology and sociopolitical organization; (3) gathering plants and foraging small animals provided them with a more reliable food supply than did hunting large game; and (4) by the Late Stone Age, fully human cultural factors surfaced that led to food selection among the edibles when people could afford it.

Figure 1 is provided as a convenient guide through the time periods discussed below. The chart is highly simplified and applies only to Europe, northern Africa, and the Near East. Although data from other world areas are discussed within the time divisions, it is fully acknowledged that the terms Paleolithic (Old Stone Age), Mesolithic (Middle Stone Age), and Neolithic (New Stone Age) do not apply well to Subsaharan Africa, the Far East and North and South America. With this in mind, the periods used and their finer subdivisions are briefly defined below.

The Paleolithic or Old Stone Age began about 2,500,000 years ago (2.5 mya) and is subdivided into three major chronological stages on the basis of advances in stone tool technology: Lower or Early Paleolithic (2.5 mya–100,000 yr ago), Middle Paleolithic (100,000–40,000 yr ago), and Upper or Late Paleolithic (40,000–12,000 yr ago) (Fig. 1). The Paleolithic ended with the end of the last ice age, called the Wurm or Weichsel in the Old World and the Wisconsin in the New World.

The Mesolithic or Middle Stone Age was a short period that began with the final retreat of the last glacier about 12,000 years ago (Fig. 1). No definitive end date that applies to the planet can be given because the proper end of the period is with the beginning of agriculture. Agriculture began at different times in different parts of the globe. In the Near East, the period fades into what is sometimes called the Proto-Neolithic. The Proto-Neolithic (about 11,000–9000 yr ago) is called such because it was the era of incipient agriculture. Various animals and plants were brought under domestication, but wild animals and plants still constituted an important part of subsistence. There appear to have been some similar developments in Thailand dating to this period.

The Neolithic began in the Near East about 9000 years ago, when human subsistence was based fully on agriculture (Fig. 1). Fully agricultural economies were later in Europe, the Far East, and the New World.

PALEOLITHIC

Lower Paleolithic (Early Stone Age): 2.5 mya–100,000 Years Ago

East Africa; Olduvai Gorge, Tanzania between 1.85 and 1.5 mya; Hominid Form: *Homo habilis*. A fully bipedal form with a manipulative hand and a complex brain, the upper range of which was more than half the size of the present human brain, *H. habilis* was probably an omnivore. Although no primate is a carnivore, *H. habilis* looks like one because of the nature of what has been found on its habitation sites (living floors). Isolated and fragmentary remains of almost every conceivable available animal from mice and turtles to gazelles and saber-toothed tigers have been found. Although some paleoanthropologists argue about just how these bones got there, the consensus is that *H. habilis* was a scavenger of carnivore kills who brought pieces back to eat raw. In some cases, as at Olduvai Gorge, Tanzania, and Koobi Fora, Kenya, it camped at large animal carcasses such as elephant and hippopotamus and ate them on the spot, bringing fragments of other scavenged animals to the same place (5). Although there is no way of knowing, it is possible that the meat was not always fresh. The smaller animals could have been hunted and/or foraged, but given the hominid's size (about 4 ft) and rudimentary stone tool technology, it is doubtful the large and more dangerous animals were hunted (2,5). Like historic Australian aborigines and Amazonian Tukanoan Indians, they may also have consumed grubs and insects (6,7). One assumes they would have eaten any eggs they found.

It seems obvious that *H. habilis* sought meat. Did it also eat plants? There are no remains, but the likelihood is that it did and probably in a higher proportion than meat. Its primate physiology would have demanded this. Seeds, nuts, edible leaves and twigs, and fruits were available.

Europe, Africa, China, and the General Far East; 1.5 mya–300,000–200,000 years ago. *Homo erectus*, the presumed descendant of *Homo habilis*, took the stage around 1.5 million years ago. Although the size of the *H. erectus* brain was not yet equal to that of the *H. sapiens sapiens* brain, *H. erectus* was otherwise fully human physically. Taller, heavier, more muscular, and with a brain the upper range of which was within the lower range of ours, *H. erectus* was more intrepid than *H. habilis* and accomplished more, such as controlling fire. *H. erectus* preferred warm to temperate cold climates. The latter would be characterized by seasonality of resources and in some cases, seasonal low availability of plant foods. It is likely that in the colder climates a high degree of nomadism was required in pursuit of plant foods. Of course, in the colder environments, more meat could and would have been eaten.

True to form, *H. erectus* ate everything that was available and edible, including it seems, itself (8). *Homo erectus* more than likely continued the practice of scavanging carnivore kills and foraging for small animals. However, we know that large game was hunted. Abundant evidence from Spain (elephant and horse) (9), France (10), and the Zhoukoudien Caves in northern China indicate this (5,8). No stone spear points are yet known, but wooden spears with fire-hardened tips would have served well. Evidence from China indicates that at least some meat was cooked, including *H. erectus* (8).

Evidence from France (10), East Africa (11), and China (8), although scarce, indicates that plants were eaten.

Middle Paleolithic (Middle Stone Age): 100,000–40,000 Years Ago

When this period began depends on what is being emphasized. For present purposes the discussion here begins at 100,000 yr ago and ends 40,000 yr ago. This period is characterized by the emergence of *Homo sapiens*. The form most commonly known is Neanderthal, but there were other varieties who are given other taxonomic names.

Homo sapiens lived in a variety of environments from the Ice Age northern European subarctic tundra to tropical forests and therefore evolved a variety of environmental adaptations specific to each. The ratio of plant to animal consumption probably reflected differential environmental adaptations. In cold and temperate climates mobility was the main pattern owing to the seasonality of resources. Hunting emerged as an important pattern around the then inhabited world and stone spear points are known. Massive game such as woolly mammoth and woolly rhinoceros were taken in cold climates, but smaller game including such large cold-climate animals as reindeer and warm-climate camel and gazelle and small game such as birds and rabbits were also hunted. Harpoon heads and fish bones indicate that some fishing was done on the tundra of northern Europe (9). Opportunistic scavenging probably continued. The diet appears to have been broad-spectrum and unspecialized. Meat was probably roasted.

While no plant food remains have been found except for Kalambo Falls, Zambia (11), digging stick tips from northern Europe (9) and tooth wear patterns indicate plant food consumption. How plant foods were prepared and whether they were cooked remains a mystery, as does the existence of methods of storing them for winter. The proportion of plant to animal foods would have depended on the environment.

Upper Paleolithic (Late Stone Age): 40,000–12,000 Years Ago

The modern human form called *Homo sapiens sapiens* characterizes this period. During these 30,000 years of the latter part of the last Ice Age, humans expanded into all inhabitable environments from the tropical forests of Africa to the subarctic tundra of Siberia and into Australia and North and South America. A wide variety of macroenvironmental and microenvironmental adaptations were necessary. Humanity—through its flexible mode of adaptation anthropology refers to as culture and society—adapted. The result was the emergence of a great deal of cultural diversity. If we were to emphasize the diversity, a discussion of the history of food would become immediately unmanageable. Therefore, only the broadest picture will be drawn.

Northern peoples made cold climate adaptations that led them to specialize in hunting certain large game. That is not to say they did not take smaller game. It seems, however, that the habits of the animals on which they depended dictated their basic lifestyle. The degree of nomadism a people pursued depended on the migratory patterns of the main animal. Large game ranged from woolly mammoth and woolly rhinoceros to bison, horse, and reindeer (9). The decimation and extinction of horse and bison herds in Europe suggest that meat was stored for the winter. Fishing was carried out, especially of salmon. In southern coastal areas people not only fished for salmon but also gathered large quantities of mollusks and shellfish (12). There are no plant food remains or indisputable evidence for the storage of plants for winter (13). It is possible that more meat was consumed during the ferocious winters. The ratio of plant to animal food is impossible to determine, but at least in some areas, a wide variety of berries and other fruits, roots, tubers, and nuts would have been available in summer and autumn. Given the intelligence, knowledge, and experience of these people, one would assume they had a variety of preservation and storage techniques for at least some plant foods.

Of course, one can never say anything about the consumption of eggs and honey, or even insects, since these have no way of preserving, but their consumption is likely throughout these time periods.

During the last Ice Age, northern Africa and the Middle East enjoyed more temperate climates. The massive cold climate game did not exist. Although temperate cold, temperatures were warmer and seasons more varied than in northern Europe. As a consequence, the ratios of plant to animal food probably reflected that difference. But little can be stated with certainty. Further south, one would expect a high ratio of plant to animal food and in desert regions, the highest. Today, large game is almost absent in most desert regions and, when present, is sparse. Small game such as rodents, birds, and reptiles and a variety of insects are characteristic fauna. The ratio of plant to animal consumption is very heavily in favor of plants.

For late Ice Age North and South America, depending on latitude and altitude, the same picture may be painted. There is a dearth of information on plant food consumption and a great deal bearing on the hunting of massive, large and small game.

Mesolithic: 12,000 Yr Ago to a Regionally Variable End Date

The Mesolithic or Middle Stone Age (sometimes called the Epipaleolithic) began with the final retreat of the last glacier about 12,000 yr ago. The period's end date is variable depending on when agriculture came into being in a given area.

With the end of the last glaciation, the climatic and vegetational picture characteristic of the modern world took over. Consequently, and especially in northern latitudes, there was considerable environmental change. Humanity had to respond and did. The emerging cultural adaptations were more complex than before and an even greater amount of cultural diversity based on, among other things, microenvironmental adaptations came into being. Again, only the broadest picture will be given.

Throughout the world, the Ice Age megafauna disappeared. Mammoths, cave bears, giant sloths, gigantic deer, and their like became extinct. Smaller game, fish and other seafood became the focus of humanity's search for protein. People adapted to forest, grassland, desert, lake, riverine, and seacoast niches (9). Most people exploited several adjacent niches, such as forest and lake,

Table 1. Origins of Major Domesticated Food Animals

Animal	Region of Original Domestication
Cattle	Near East
Chicken	India
Goat	Near East
Pig	Near East
Sheep	Near East
Turkey	Mexico

and in most areas, because of the seasonality of resources, were nomadic in their lifestyles. The ratio of plant to animal food would have varied considerably depending on the adaptive strategy in a given set of niches.

One of the consequences of environmental change was the establishment of vast stands of wild wheat and barley extending in appropriate niches from east of the Caspian Sea west and south into the Near East (14). Concurrent with this was the spread of herds of wild sheep and goats in the same region (15).

During what is sometimes called the Proto-Neolithic of the Near East, gatherer-hunters took full advantage of these abundant wild resources. Some even settled down into small, permanent communities and built permanent houses. They were able to gather sufficient wild wheat to do so. Jack Harlan conducted an experiment in the same area (the ancestral wild wheat still grows there) and estimates that in 3 weeks a quantity sufficient to last a full year could be gathered (16). It is likely that it is this that permitted sedentary living. People supplemented these plant resources with hunting sheep, goats, deer, pigs, cattle, and so forth. At some point, perhaps 11000 yr ago, they began managing herds of wild sheep and/or goats and animal husbandry was born (17,18). Somewhere around 9000 yr ago, wheat and barley were domesticated. Shortly thereafter, peas and lentils, the wild varieties of which grow in the same area, were added to the list of domesticates. So were pigs. Agriculture was born in the Near East around 9000 yr ago (7000 BC) (see Tables 1 and 2).

Neolithic (New Stone Age): 7000 BC to an End-Date Variable Depending on Other Cultural Developments

In the Near East, the plant/animal food ratio appears to have been greatly in favor of plants. Sickles, grinding stones, storage pits, and ovens are highly characteristic artifacts of the period. Impressions of wheat and barley kernels appear in mud bricks, hearth materials, and pottery. Bread and porridge probably formed the main part of meals. Sheep and goat herds seem to have been hus-

Table 2. Origins of Selected Domesticated Food Plants

Plant	Region of Original Domestication
Grains	
Amaranth	Mexico
Barley	Near East
Maize (corn)	Mexico
Millet	Asia
Oats	Europe
Quinoa	South American Andes
Rice	China and Southeast Asia
Rye	Turkey
Sorghum	India
Wheat	Near East
Legumes	
Common beans	Mexico and South America depending on species
Lentils	Near East
Lima Beans	Peru
Peas	Near East
Peanuts	Bolivia
Soybeans	Far East
Sugar	
Sugarcane	Indonesia or New Guinea
Sugarbeet	Europe
Root crops	
Beet	Europe
Potato	South America
Sweet Potato	South America
Vegetables[a]	
Avocado	Mexico
Cabbage (cole) family	Mediterranean
Chives	Near East
Eggplant	India
Garlic	Near East
Leeks	Near East
Olive	Mediterranean
Onion	Near East
Peppers	Mexico
Squash	Mexico
Tomato	Mexico
Fruits[a]	
Apple	Western Europe and Asia depending on variety
Banana	Southeast Asia
Grape	Near East
Kiwi	China
Lemon	Southeast Asia
Lime	Southeast Asia
Mango	India
Orange	Southeast Asia
Pear	Western Europe and Asia depending on species
Pineapple	Brazil
Nuts and edible seeds	
Almond	Mediterranean
Brazil nut	Brazil
Cashew	Brazil
Hazelnut	Europe
Pecan	United States
Pistachio	Mediterranean
Sunflower	United States
Walnut	Iran
Spices	
Black pepper	India
Chili pepper	Mexico
Mustard seed	Mediterranean
Nutmeg	Malay Peninsula
Vanilla	Tropical America
Beverages	
Chocolate	Mexico
Coffee	Ethiopia
Tea	India and China

[a] Common, not botanical classification.

banded mainly for wool, hides, and milk. Cheese may have been invented and added to the diet. Animal slaughter was selective by age and sex. Meat, therefore, was not consumed as a daily ration. As in ancient historic times, domesticated sheep and perhaps goats may have been religious feast foods.

True to form, humans remained omnivorous, but in the Near East, at least, dairy foods in the form of milk and perhaps, cheese took the place of meat as a daily food. For a while, some hunting was done that would have placed meat on at least some tables. People may also have traded for meat with nonagriculturalists such as hunters and pastoralists, but it is likely, as mentioned above, that it was primarily a feast-day item.

Since meat would have been scarce, salt would have become necessary. Without the salt naturally present in meat, humans cannot survive. A substitution is necessary. Salt production and trade became important.

Ultimately, certain fruits such as figs, dates, and grapes came out of the Near East (see Table 2).

Let us look at the development of agriculture elsewhere. Many Near Eastern crops and animals made their ways into Europe and together with native plants (eg, perhaps cherry and plums) ended gathering and hunting there as the major subsistence mode about 6000 BC (19). Northern China domesticated millet, which eventually was adopted by eastern Africa. Rice was domesticated in Southeast Asia and has made its way all over the world. South America gave the world the potato, peanuts, manioc for tapioca, and the sweet potato, among other crops. Regarding animals, southern Europe and/or Turkey gave us cattle, the jungles of India the chicken, and Mexico the turkey (20) (see Tables 1 and 2). Charles Heiser's book, *Seed to Civilization: The Story of Food* (20), is a valuable source for the geographic origins of domesticated plants and animals.

Mexico has given much to the world. It was one of the planet's primary centers of domestication. It took approximately four thousand years for Mexico to domesticate the wealth of foods it has given (21). Some 21 domesticated plants have their origins in Mexico (22). Maize—corn, as we call it—is the result of 4000 yr of painfully slow selection and hybridization that began around 6000 BC. Corn is a staple food in parts of Africa and South America as well as in Middle America. Mexico has also given us a large variety of beans, chili peppers, and squashes, to say nothing of a variety of tree and cactus fruits. It also gave us the tomato, avocado, and chocolate. With the limitations of space here, it is impossible to list its valuable and varied contributions to the world's diet (see Table 2).

The development of agriculture in Mexico lasted from at least 6000 to 2000 BC, and there were several centers of domestication, some humid tropical, and some semiarid desert. Throughout most of its prehistory and history, plant foods have superseded animal foods in dietary importance. In the desert areas of the southern part of the state of Tamaulipas and in the Tehuacan Valley we have the plant preservation to prove this (21). We also have the historical records from conquest times. Some people rarely, if ever, saw meat prior to the arrival of the Spaniards, and this is true even now in the most poverty-stricken rural areas. Traditionally in Mexico and over much of Middle America, meat is a fiesta food, a special-occasion food eaten only a few times a year. (This has changed among the more affluent populace.) How, then, eating plant foods almost exclusively, did these people survive?

The fortuitous congruence of maize and beans, provided they are eaten together, provides complete protein. Therefore, a meal of tortillas and beans, in terms of protein, is healthful. The lime water in which the dried maize is soaked prior to grinding into flour for tortillas provides calcium. A few chilis and other vegetables or fruits added to the meal provides additional vitamins and minerals. The use of salt provides sodium. Adults do well on such a diet. Small children, however, are often malnourished because they need more animal protein. Infants who are weaned too soon die. Infant mortality is still very high in rural areas of Mexico and Guatemala.

Of course, before the arrival of the Spaniards, the population was not completely without sources of animal protein. Turkey eggs, ant larvae, other insects, and grubs were eaten. Eggs of other birds were probably gathered. In the coastal areas marine resources, including iguana, turtle and turtle eggs, fish, crustaceans, and mollusks were consumed, although perhaps some were not allowed to everyone in these socially stratified societies. In the tropical rainforests reptiles, monkeys, deer, wild pig, and other animals existed and were hunted. But again, the consumption of some of these was probably strictly controlled. Deer, wild pig, and small game such as badger, squirrels of various kinds, rabbit, and hare abounded in the central Mexican highlands, and it is probable that the consumption of the small game was not regulated. Throughout the Middle American region, the common people probably made do with beans and corn as their main source of protein and supplemented these with small game, insects, and eggs and in the region around modern Mexico City, with fish and larvae from lakes now drained.

It is of interest to note that the beginnings of agriculture in the Near East, Far East, and Middle and South America appear to have been roughly concurrent. This is not likely to have been due to any communication of ideas. Why agriculture began is subject to much debate, and there are no clear answers. Environmental change, population pressure, religion, and the human propensity to experiment have all been invoked (1,16, 17,20). It is beyond the scope of this article to discuss these debates.

Another interesting fact is that in the probably unintentional search for complete protein in the virtual absence of meat consumption, humanity in various parts of the globe domesticated plants that, when consumed together, provide complete protein. The case for Middle America has been briefly discussed. Rice and beans and other combinations also provide complete proteins (1).

A final fact that should be added is that in all world areas, the development of agriculture was a long, slow process. In a given area, not all plants were domesticated and brought under cultivation at the same time. Nor were all animals domesticated simultaneously. It took centuries and, in some cases, millennia to bring together the agricultural complexes that are prehistorically and historically known.

CONCLUSION

It would appear that for most of humanity's history, we were omnivores like many other primates and that, like other primates, we consumed more plant food than animal. Under special circumstances the ratio might swing in favor of animal over plant, but this is rare and due to special environmental circumstances. Even then, plant food is consumed, even if this means consuming the stomach contents of dead herbivores. Also rare is the society that does not seek some form of animal protein even if it means eating insects, grubs, or raw bird eggs.

The ratio of plant to animal food consumption is probably still in favor of plants around the planet, even for those of us who like our meat so much that we jokingly call ourselves carnivores. If we look at the modern Western diet, we find that even fast foods like hamburgers and pizza are heavy on the plant end. The hamburger comes on a bun made from flour and is accompanied by french fried potatoes, potato chips, or potato salad, is relished by plant products such as tomato catsup, mustard (seed, vinegar), pickles or pickle relish, onions, tomatoes, lettuce, and sometimes mushrooms. The basic pizza is, of course, made from flour and tomato sauce, regardless of the topping on the pizza. If we look at the Far Eastern diets of China, Southeast Asia, and Japan, we find that while meat or seafood may form part of the meal, the major constituent is vegetables and other plant foods like rice or noodles.

We have reached a point in the world where the history of any given dish or meal is utterly fascinating. The ancient northern Chinese somehow received wheat and the idea of flour from the Near East and ultimately invented the noodle. In the thirteenth century, Marco Polo went to China and allegedly, inter alia, encountered the noodle and found it to taste good. He took it back to Italy, where, ultimately, a variety of pastas were developed, including spaghetti. A moderately well-off Mexican sits down to a several-course meal. The first dish is *arroz à la mexicana* (Mexican-style rice). It consists of rice (Southeast Asia) toasted raw in lard (a pork product originating from the Near East) or vegetable oil (an eastern Mediterranean idea) seasoned with a sauce made from tomatoes (Mexico), onion, and garlic (both Near East) and cooked in chicken broth (ultimately India). It is served with corn tortillas (Mexico) or perhaps, flour tortillas (made from wheat flour which has its ultimate origins in the Near East). The next course is *carne de cerdo con salsa verde* (pork in green sauce). The pork is ultimately Near Eastern. The green sauce is made from *tomates* (*tomatillos* or green husk tomatoes) and chilis (Mexico), onion (Near East), and *cilantro* (Near East). The final course is *frijoles*, which, if done right, are cooked with lard or oil and also, perhaps, a little onion. Without the fats and onion, the *frijoles* are thoroughly Mexican. All may be accompanied by beer (a thoroughly European idea incorporating many Near Eastern ingredients) or carbonated soft drinks (a U.S. invention).

Many of the "foreign" ingredients in the Mexican cuisine can, of course, be attributed to the sixteenth-century Spanish conquest, one of the results of which was a revolution in cooking. A glance at the history of Spain would illuminate how some of these ingredients got into its cuisine!

The cross-fertilization in our various cuisines is enormous and universal. A simple plate of bacon and hen's eggs served with orange juice, coffee, toast, and jam has its ultimate origins in the Near East, India, Southeast Asia, Ethiopia, and Europe or North America (if the jam is raspberry). If you add hashbrowns or grits, you must acknowledge South America or Mexico. A bowl of vegetable beef soup or chicken and vegetable soup contains ingredients from around the world. A vegan's vegetarian plate is a veritable travelog of time and space.

The history of food and cuisine is, among other needs and things, very much a product of trial and error experimentation and the human propensity to travel, trade, and even conquer and to bring home new and delightful products and ideas and to modify and invent new products, new combinations, and new dishes. Humans are indeed omnivores in every sense of the word.

BIBLIOGRAPHY

1. P. Farb and G. Armelagos, *Consuming Passions: The Anthropology of Eating*, Houghton-Mifflin, Boston, 1980.
2. R. Jurmain, H. Nelson, and W. A. Turnbaugh, *Understanding Physical Anthropology and Archaeology*, 4th ed., West Publishing Co., St. Paul, Minn., 1990.
3. P. V. Glob, *The Bog People: Iron Age Man Preserved*, Ballentine Books, New York, 1969.
4. A. Sillen, J. C. Sealy, and N. J. Van der Merwe, "Chemistry and Paleodietary Research: No More Easy Answers," *American Antiquity* **54**, 504–512 (1989).
5. B. G. Campbell, *Humankind Emerging*, Scott, Foresman and Co., Boston, 1988.
6. N. Tindale, *Aboriginal Tribes of Australia*, University of California Press, Berkeley, 1974.
7. D. L. Dufour, "Insects as Food: A Case Study from the Northwest Amazon," *American Anthropologist* **89**, 383–387 (1987).
8. K. C. Chang, *The Archaeology of Ancient China*, 4th ed., Yale University Press, New Haven, Conn., 1986.
9. K. W. Butzer, *Environment and Archaeology: An Ecological Approach to Prehistory*, 2nd ed., Aldine-Atherton, Chicago, 1971.
10. H. de Lumley, "A Paleolithic Camp at Nice," *Scientific American* **220**, 42–50 (1969).
11. D. J. Clark, *The Prehistory of Africa*, Praeger, New York, 1970.
12. L. G. Straus, G. A. Clark, J. Altuna, and J. A. Ortega, "Ice Age Subsistence in Northern Spain," *Scientific American* **242**, 142–152 (1980).
13. R. Dennell, *European Economic Prehistory: A New Approach*, Academic Press, Orlando, Fla., 1985.
14. J. R. Harlan and D. Zohary, "Distribution of Wild Wheats and Barley," *Science* **153**, 1074–1080 (1966).
15. J. R. Harlan, "Plant and Animal Distribution in Relation to Domestication," *Philosophical Transactions of the Royal Society of London* **275**, 13–25 (1976).

16. J. R. Harlan, "A Wild Wheat Harvest in Turkey," *Archaeology* **20**, 197–201 (1967).
17. C. Redman, *The Rise of Civilization: From Early Farmers to Urban Society in the Ancient Near East,* Freeman, San Francisco, 1978.
18. H. J. Nissen, *The Early History of the Ancient Near East 9000–2000 B.C.,* University of Chicago Press, Chicago, 1988.
19. A. Whittle, *Neolithic Europe: A Survey,* Cambridge University Press, Cambridge, 1985.
20. C. B. Heiser, Jr., *Seed to Civilization: The Story of Food,* Harvard University Press, Cambridge, 1990.
21. R. S. MacNeish and D. S. Byers, *The Prehistory of the Tehuacan Valley,* Vol. 1, *Environment and Subsistence,* University of Texas Press, Austin, 1967.
22. R. C. West and J. P. Augelli, *Middle America: Its Lands and Peoples,* Prentice Hall, Englewood Cliffs, N.J., 1989.

GLOSSARY

Anthropology. The field that studies humans as biological and cultural beings. Such study includes human culture, biology, language, and history. See also *Culture.*

Archaeology. The subfield of anthropology that studies and reconstructs past cultures and societies.

Coprolites. Naturally dried fecal matter.

Culture. The patterned thought and behavior—such as social, political, economic, and religious—that individuals learn and are taught as members of social groups and that is transmitted from one generation to the next.

Ethnography. The anthropological study and description of the cultures of living groups: their social, political, economic, and ideological systems and all the behavior that relates to these.

Food bones. The bones of animals consumed by humans.

Foraging. The collection of edible wild plants and small animals, such as birds, rodents, reptiles, and insects.

Hominid. The common name for those primates referred to in the taxonomic family Hominidae: modern humans and their nearest evolutionary ancestors.

Homo. The genus to which modern humans and their nearest evolutionary ancestors belong: *Homo habilis* (handy human), *Homo erectus* (upright human), *Homo sapiens* (knowing human), and *Homo sapiens sapiens* (modern human).

Kill and butchery sites. Locations where an animal or animals were killed by humans and butchered on the same spot.

Living floors. Locations where concentrations of living debris are found. Such debris may be tools, debris from tool manufacture, food bones, and other evidence for human occupation.

Lower Paleolithic. Dating from 2.5 mya to about 100,000 yr ago. The beginning of the period is marked by the presence of the earliest known stone tools and probably, the first appearance of the *Homo* genus. The period is characterized by increasing human brain size, and capacity, intelligence, and the evolution toward greater complexity of human technology and culture.

Middle America. Mexico and Central America.

Mesolithic. A term that designates immediately preagricultural societies in the Old World. A diagnostic technological characteristic is the presence of microliths, small stone blades set into bone or wood. In the Near East, sickels used for the harvesting of wild grains were made using this technique.

Middle Paleolithic. Dating from about 100,000 to 40,000 yr ago. The period is marked by increased sophistication in stone tool technology, such as the making of tools from prepared cores, and by the presence of *Homo sapiens*. Human culture became more complex, particularly with regard to the ideological system. Religious ritual, as manifested archaeologically in the burial of the dead with grave goods, came into being.

mya. Million years ago.

Neolithic. A stage in cultural evolution generally marked by the appearance of ground stone tool technology (bowls, adzes, axes, etc) and frequently by domesticated plants and animals and permanent villages.

Paleoanthropology. The multidisciplinary approach to the study of human biological and cultural evolution. It includes physical anthropology, archaeology, geology, ecology, and many other fields.

Paleolithic. Dating from about 2.5 mya to 10,000 BC. The period during which stone tools were produced by percussion flaking (chipping). This period is characterized by the origin and evolution of modern humans and culture.

Physical anthropology. The subfield of anthropology that studies human biology and evolution.

Presapiens. Members of the genus *Homo* who lived before the appearance of the *sapiens* species: *Homo habilis* and *Homo erectus.*

Primates. The order of mammals that includes prosimians, Old and New World monkeys, apes, and humans.

sapiens. The species to which Middle and Upper Paleolithic and modern humans belong.

Scavanging. The procurement of meat from animals killed by carnivores.

Site. A confined geographic area or location of interest to archaeologists in which the remains of earlier human activity are concentrated.

Subsistence economy. A term referring to food resources and their modes of procurement: foraging, gathering, and hunting for wild plants and animals, agriculture, marketing, etc.

Upper Paleolithic. Dating from about 40,000 to 12,000 yr ago. The final stage of the Paleolithic. It is characterized by the prevalence of modern humans with more sophisticated culture. Characteristic remains in western Europe are polychrome cave paintings, sculpture, engraving, and stone tools made from blades.

DARLENA K. BLUCHER
Humboldt State University
Arcata, California

HOME CANNING. See FOODBORNE DISEASES.

HOMOGENIZERS

The homogenizer, which is used today in many food and dairy applications, was invented in the 1890s. Even though many changes and modifications have been made to the machine over the years, the basic components of the homogenizer are the same as those early machines. Before describing the homogenizer in detail, it is important to distinguish the type of homogenizer used in food and dairy processing from the more generic term homogenizer. Today, the term homogenizer is often applied to any piece of equipment that disperses or emulsifies. This equipment may include a turbine blade mixer, an ultrasonic probe, a high shear mixer, a colloid mill, a blender, or even a mortar and pestle. The more precise definition of homogenizer is a machine consisting of a positive displacement pump and a homogenizing valve that forms a restricted orifice through which product flows.

Figure 1. Modern day dairy homogenizer.

EARLY HISTORY

The first homogenizers were invented at the turn of the century and were used for making artificial butter (1). Gaulin invented and patented his homogenizer for the processing of milk and first showed his machine to the public at the 1900 World's Fair in Paris (2, 3). The early literature attributes the term homogenizing, or homogenization, to Gaulin (1). The homogenization of milk, ie, reducing the milk fat globules in size to retard separation and the resulting cream layer, was at least 25 years ahead of its time, because the pasteurization of milk had not been perfected and public acceptance of this product was yet to be realized. Even though very little homogenized milk was produced in the early years, homogenizers were sold for ice cream and evaporated milk.

Few changes were made to the homogenizer design from 1900 to 1930, but after that, improvements were made to make machines more cleanable and sanitary (1). Homogenized milk became more popular in the 1940s. Some of the benefits of homogenized milk that helped to sell it were reduction of curd tension (which made milk more digestible, especially for infants), uniformity of fat throughout the product, and improvement in the appearance and palatability of the milk (4,5). Today, of course, homogenized milk is universally accepted. The dairy industry is one of the largest users of homogenizers, but homogenizers, are also used for other food products.

DESCRIPTION

Figure 1 shows a modern dairy homogenizer. As previously mentioned, the homogenizer consists of a pump and homogenizing valve. The pump is usually a reciprocating positive-displacement pump, which delivers a relatively constant flow rate despite the pressure or restriction to flow made by the homogenizing valve. The positive displacement pump has a power end and a liquid end. The power, or drive, end converts rotating motion to reciprocating motion. In most machines, an electric motor is connected to a drive shaft by means of V-belts and sheaves. The drive shaft turns an eccentric shaft by means of gears. In other cases, the motor is connected directly to the eccentric shaft by V-belts and sheaves.

The eccentric shaft has cams that drive the plungers or pistons by means of connecting rods and crossheads. The crosshead couples the connecting rod to the plungers or pistons. A piston and a plunger are both solid cylinders that displace fluid; however, for a piston the sealing elements are moving with the piston, and for a plunger the sealing elements are stationary. The sealing elements are called plunger packing (6).

The liquid end includes the pump chamber, or block, and all its components. Figure 2 shows a typical pumping chamber. The plungers move back and forth in the chamber. The eccentric shaft cams that drive the plungers are offset to provide a steady flow. If the pump has three plungers, they would be offset by 120 degrees. This means that the flow profile is represented by the summation of three overlapping sine curves producing some peaks and valleys but, in general, a continuous flow would be produced. Some homogenizers have five, six, or seven plungers depending on the model.

In addition to the reciprocating plungers, the pumping chamber has suction and discharge valves. On the rearward motion of the plunger, the suction valve opens and liquid is drawn in while the discharge valve closes. On the forward discharge stroke the suction valve closes and the discharge valve opens. The plunger then displaces or pushes the liquid out of the chamber and through the discharge valve. These pump valves can be simply a ball sitting on a seat or a guided valve with a pilot aligning the valve to the seat. The piloted valve is called a poppet valve and the other a ball valve. The liquid end of the pump, especially components in contact with the liquid, is usually made of ceramics, stainless steel, or special alloys. These materials include 17-4, 15-5, 304, 316, or Rexalloy.

Homogenizers can cover a wide range of flow rates and pressures. The range of flow rates for different machine

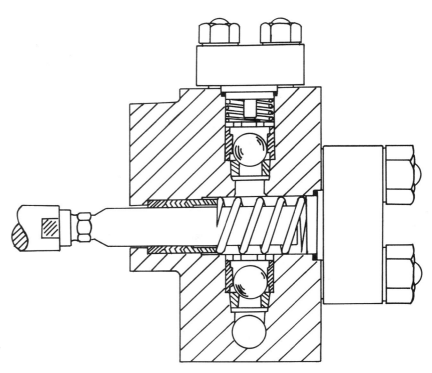

Figure 2. Pumping chamber of a dairy homogenizer.

sizes can go from 57 L (15 gal) per hour for a laboratory machine up to 52,996 L (14,000 gal) per hour for a large production machine. The maximum operating pressure can be 6.90 MPa (1,000 psi) up to 103.4 MPa (15,000 psi), but the higher the pressure rating, the lower the capacity for a given size machine. This is due to the limitation of force allowed on the drive end of the machine. For a given size machine, increasing capacity, in most cases, requires larger diameter plungers or pistons with a corresponding decrease in operating pressure due to the limiting thrust loading. Because the operating pressure is exerting force against the face of the plunger or piston, the larger diameter means that the power stroke requires greater force. Therefore, the size of the plunger or piston is limited by the maximum tolerable force on the drive end. Of course, as the pressure and capacity increase, the horsepower needed also increases because the required horsepower is directly proportional to the product of the flow rate times the operating pressure.

HOMOGENIZING VALVE

The positive-displacement pump delivers fluid to the homogenizing valve, which is usually mounted by means of a valve body onto the pumping chamber of the machine. Figure 3 shows a simple homogenizing valve assembly and its designated parts. The fluid to be processed flows into the homogenizing valve seat and pushes against the face of the homogenizing valve. After flowing out through the restricted orifice between the valve and seat, the liquid impinges on the impact or wear ring. Figure 4 shows a two-stage manually actuated homogenizing valve assembly. The actuation of the valve requires applying force on the valve to counteract the force pushing the valve open due to the pressure of the liquid. This pressure is caused by the reduction of flow area when the valve is pushed toward the seat, while the pump is delivering a constant flow rate. The force exerted by the liquid is equal to the area of the valve in contact with the liquid times the pressure generated. For example, a valve with a contact diameter of 0.25 in. at 68.97 MPa (10,000 psi) needs a counteracting force of 2184 N (491 lb). A valve with a contact diameter of 1 in. at 10,000 psi requires 34,936 N (7,854 lb) of counteracting force.

If the counteracting forces are large, then a spring-loaded handwheel will not deliver enough mechanical advantage to achieve high pressures and some other means of actuation such as a hydraulic system must be used. For some products, a two-stage homogenizing valve consisting of two valves in series is used, and the valves can be manual or hydraulic depending on the valve size and pressure.

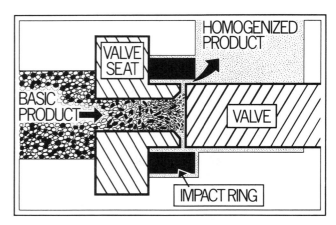

Figure 3. Simple homogenizing valve assembly.

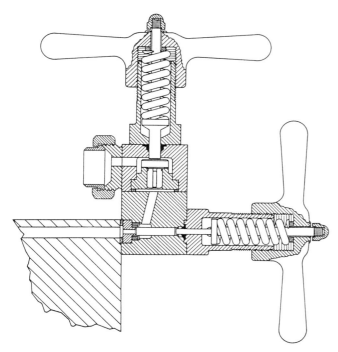

Figure 4. Two-stage manually actuated homogenizing valve assembly.

The homogenizing valve, seat, and impact ring are usually made of special wear-resistant materials because of the high fluid velocities in the valve assembly and because some products contain suspended solids.

As previously described, Figure 3 shows the flow through the homogenizing valve. Even though this flow profile may look simple, the fluid dynamics occurring are quite profound. Intense energy changes occur in the homogenizing valve as the liquid goes from high pressure and low velocity to low pressure and high velocity. The best way to understand these changes is to consider an example. A homogenizer operating at 13.8 MPa (2,000 psi) and 11,356 L (3,000 gal) per hour with a conventional homogenizing valve would, typically have a calculated gap of about 0.178 mm (0.007 in.) between the valve and seat. The pressure before the homogenizing valve seat is 13.8 MPa (2,000 psi) and the velocity is about 6.1 m (20 ft) per second. As the fluid enters the gap, its pressure drops to the vapor pressure of the liquid, and the velocity increases to about 122 m (400 ft) per second. These changes occur over distances less than 0.254 mm (0.1 in.) in less than 0.000003 s (7–9). Homogenization is completed before the liquid leaves the gap area between the valve and seat. Therefore, the potential energy of the liquid stored in the pump chamber is rapidly converted to kinetic energy producing the homogenization effect.

THEORIES OF HOMOGENIZATION

Because of the dynamic forces occurring in the homogenizing valve, there has been some interest in determining the mechanism of homogenization. Many researchers have suggested theories to account for homogenization by relating the flow condition in the homogenizing valve to known concepts of fluid dynamics. Homogenization includes the formation of emulsions (one immiscible liquid uniformly distributed into another) or dispersions (solid particles dispersed throughout a liquid) but most theories of homogenization only consider the formation of emulsions. Therefore, these theories usually relate to the mechanism by which dispersed oil droplets are disrupted into smaller droplets and distributed throughout a continuous water phase. This approach is a consequence of the fact that most researchers were investigating the homogenization of milk when they developed their theories.

Some of the theories of homogenization include shear, impact, wire drawing, acceleration and deceleration, homogenizing valve vibration, turbulence, and cavitation (1,10–12). Many of these theories have been discounted over the years. For example, impact of the dispersed droplets on the impact ring is not the cause of homogenization because the required velocities at impact are not great enough to disrupt the droplet, and increasing the distance from the valve seat to the impact ring does not significantly affect homogenization. Wire drawing is the elongation and subsequent disruption of the thinned droplet, but this is unlikely to occur due to the flow conditions in the valve. Shear is commonly mentioned as a mechanism, but the velocity gradients in the homogenizing valve do not fit the classic shear flow profile. Also, the successful homogenization of oils with viscosities greater than that allowed for shearing action to occur indicate a mechanism other than shear (13).

The two most likely mechanisms of homogenization involve turbulence and cavitation (14,15). The intense turbulent eddies generated in the liquid at the instant of energy conversion (potential to kinetic energy) produce significant local velocity gradients that disrupt the droplets. Cavitation theory suggests that the extreme pressure drop in the homogenizing valve generates cavitation bubbles, and when these bubbles collapse, the shock waves in the fluid cause the droplets to break apart. Actual measurements of cavitation noise in the homogenizing valve have demonstrated that the greater the intensity of cavitation, the greater is the homogenizing effect (15). However, there are fluid dynamic forces that would be common to both theories; therefore, it is not a simple matter to distinguish one from the other.

HOMOGENIZING VALVE DESIGN

One consequence of understanding the mechanism of homogenization is the ability to improve the efficiency of homogenization. The geometry of the homogenizing valve is important in regard to the quality of the finished product. Gaulin realized this, and he experimented with different valve designs at the turn of the century (1). The objective of changing valve designs is to find one that gives the best product possible at the lowest pressure (16). Some of the basic designs include the plug valve (Fig. 3), the piloted valve, and a grooved valve. The piloted valve has a guide on the valve that fits into the seat to align the two. The grooved valve has concentric grooves on the valve and seat. These grooves are machined in such a way that the peaks on the valve fit into the grooves on the seat. There are different configurations of these grooves depending on the design of the manufacturer. Variations of the plug and piloted valve include a knife edge on the seat having a short land or travel distance, replaceable screens on the faces of the valve and seat, and a cone-shaped valve fitting into a hollowed seat. There is even a valve consisting of a tightly compressed wire bundle through which the product flows. For large flow rates, the Micro-Gap valve is used, and this valve consists of stacked valve plates that split the flow into equal parts for optimization of homogenization (16,17). All these designs seek to make efficient use of the available homogenization energy. Along with the geometry of the valves it is important to consider the size of the valves, because as the flow rate increases, valve size also increases to maintain efficient homogenization. A laboratory homogenizer operating at 114 L (30 gal per hour may have a valve diameter of 9.53 mm (0.375 in.), while a production-size homogenizer operating at 15,142 L (4,000 gal) per hour might have a valve diameter of 76.2 mm (3 in.). A homogenizer at 37,854 L (10,000 gal) per hour with a Micro-Gap valve would have the equivalent diameter (by summing the stacked valves) of a valve 787.4 mm (31 in.) across. Of course, as the size of the homogenizing valve increases, the required actuating force becomes larger and this must be considered when homogenizing valves are designed.

APPLICATIONS

The homogenizer is used in the processing of many food and dairy products. Table 1 lists some common applications of the homogenizer. The pressure ranges given are for conventional homogenizing valves, but the actual pressures can vary depending on the product formulation, the required shelf life and the product specifications of each processor. Also, some of these applications might require more than one pass through the homogenizer or the use of a two-stage homogenizing valve. When using the two-stage valve, it has been determined that the second-stage pressure should be in the range of 10–15% of the total homogenizing pressure (11,16).

OPERATION

Certain requirements must be met for successful operation of the homogenizer. The first of these is adequate in-feed pressure. A positive-displacement pump must have a positive in-feed pressure so that the pump is not starved. If it is starved, transient high—pressure shock loading can occur, eventually resulting in severe damage to the pump (6). The required in-feed pressure will depend on the characteristics of the product and the size of the pump. Excessive amounts of entrained air in a product will cause a similar type of shock loading, and some products require deaeration to eliminate large amounts of entrained air.

The correct type of pump valves should be used depending on the nature of the product. Viscous products may require ball valves. Abrasive products may require ball valves and special wear-resistant materials for the pump-valve seats and homogenizing valve. The homogenizer should be sanitary. Parts should be easily removable for

Table 1. Common Applications of the Homogenizer

Product	Pressure Range, psi
Dairy	
Evaporated milk	2,400–3,000
Half-and-half	1,800–2,000
Ice cream mix	2,000–2,500
Ice milk	1,800–2,200
Light cream	1,500–2,000
Pasteurized milk	1,800–2,200
Soft cheese	1,000–3,000
Sour cream	2,000–2,500
UHT milk	3,000–4,000
Yogurt base	2,500–3,000
Food	
Baby foods	2,000–5,000
Coffee whiteners	2,000–5,000
Cream-base cocktails	3,000–4,000
Cream soup base	2,000–3,000
Flavor emulsions	3,000–5,000
Frozen whipped toppings	5,000–8,000
Fruit nectar	3,000–5,000
Infant formula	3,000–5,000
Liquid egg	1,000–3,000
Orange juice concentrate	3,000–5,000
Peanut butter	5,000–8,000
Puddings	2,000–5,000
Salad dressings	1,000–3,000
Soy beverages	3,000–5,000
Tomato ketchup	3,000–4,000
Tomato juice	500–1,000
Tomato sauce	3,000–4,000
Toppings (hot fudge, etc)	3,000–5,000

cleaning or replacement, and the machine should have no flow areas where product can be trapped. Of course, the homogenizer must be rugged and reliable, for example, some dairies may require the homogenizer to operate more than 16 h a day, six days a week.

TESTING

There are many methods used to test a product for homogenization quality. These methods vary with the type of product and quality control requirements. A homogenized substance might be checked for viscosity change, either an increase or a decrease. The appearance of a product may be important; for example, texture, color, smoothness, graininess or pulpiness. A dispersion might be checked microscopically for changes in the size of dispersed solids. An emulsion, such as milk, can be examined microscopically for the size of the milk fat globules, or more sophisticated spectroturbidimetric methods may be used to determine the average diameter and size distribution of the fat globules (18).

SUMMARY

The homogenizer is a unique piece of equipment, which was invented at the turn of the century for the processing of artificial butter and for milk but now is used for many applications in the food and dairy industries. The homogenizer consists of a pump and homogenizing valve. The pump is a positive displacement pump, and the homogenizing valve can be made in different sizes and configurations. Even after more than 80 y, the mechanism of homogenization is not completely understood, but homogenization has improved many consumer products.

BIBLIOGRAPHY

1. G. Malcolm Trout, *Homogenized Milk,* Michigan State College Press, East Lansing, 1950.
2. W. Clayton, *The Theory of Emulsions and Their Technical Treatment,* 2nd ed., J. & A. Churchill, London, 1928.
3. U. S. Pat. 756,953 (Apr. 12, 1904), Auguste Gaulin.
4. F. J. Doan, "Changes in the Physico-Chemical Characteristics of Milk as a Result of Homogenization," *American Milk Review,* 54 (June 1954).
5. F. J. Doan and C. H. Minster, "The Homogenization of Milk and Cream," School of Agriculture and Experiment Station, Pennsylvania State College Bulletin **287,** 1933.
6. T. L. Henshaw, "Reciprocating Pumps," *Chemical Engineering,* 104 (Sept. 21, 1981).
7. C. C. Loo and W. M. Carleton, "Further Studies of Cavitation in the Homogenization of Milk Products," *Journal of Dairy Science* **36,** 64 (1953).
8. A. A. McKillop, W. L. Dunkley, R. L. Brockmeyer, and R. L. Perry, "The Cavitation Theory of Homogenization," *Journal of Dairy Science* **38,** 273 (1955).
9. L. W. Phipps, "Action of the High Pressure Homogenizer," *Bienn. Rev. Natn. Inst. Res. Dairy,* 61 (1976).
10. H. Mulder and P. Walstra, *The Milk Fat Globule,* PUDOC (Centre for Agricultural Publishing and Documentation), The Netherlands, 1974.
11. H. G. Kessler, *Food Engineering and Dairy Technology,* Verlag A. Kessler, Freising, FRG, 1981.
12. L. H. Rees, "The Theory and Practical Application of Homogenization," *Journal of the Society of Dairy Technology* **21**(4), 172 (1968).
13. W. D. Pandolfe, "Effect of Dispersed and Continuous Phase Viscosity on Droplet Size of Emulsions Generated by Homogenization," *Journal of Disp. Sci. and Technol.* **2**(4), 459 (1981).
14. P. Walstra, "Preliminary Note on the Mechanism of Homogenization," *Neth. Milk Dairy J.* **23,** 290 (1969).
15. H. A. Kurzhals, Ph.D. dissertation, Technical University Hannover, FRG, 1977.
16. W. D. Pandolfe, "Development of the New Gaulin Micro-Gap Homogenizing Valve," *Journal of Dairy Sci.* **65,** 2035 (1982).
17. U. S. Pat. 4,383,769 (May 17, 1983), W. D. Pandolfe and H. Graglia (to APV Gaulin, Inc.).
18. W. D. Pandolfe and S. F. Masucci, "Rapid Spectroturbidimetric Analysis," *American Laboratory,* 40 (Aug. 1984).

WILLIAM D. PANDOLFE
APV Gaulin, Inc.
Wilmington, Massachusetts

HONEY ANALYSIS

Honey-storing social bees existed long before humans. Therefore, honey has probably served as an important food from the time that humans first walked the earth. Refined sugar (sucrose) from sugarcane developed slowly as a commodity and did not become competitive with honey in price until the industrial revolution, when sugar was more easily transported to the increasingly urban populations. About this time, honey and sugar changed roles, with sugar becoming cheaper and easier to obtain and honey developing as a more expensive, less available luxury. With the price differential came the opportunity for unscrupulous merchants to sell sugar solutions at honey prices, so the necessity for some type of honey analysis developed.

Today, honey has an enviable image among food commodities as a natural sweetener. However, with sugar still considerably less expensive than honey and general environmental concerns, the consumer demands assurance that honey is of the highest quality and is not adulterated. These demands and the significant international trade in honey have contributed to the development of honey analysis to the point where honey is probably one of the most analyzed food commodities.

The most thorough description of honey is given in the book edited by E. Crane (1).

CARBOHYDRATES

The most obvious and still most useful analysis to carry out on honey is to measure the major component, the carbohydrates. The most comprehensive honey analysis reported (Table 1) indicated that with surprisingly limited variability, the major three components in honey were the monosaccharides fructose and glucose and water. Fructose levels are almost always higher than glucose levels. In fact, although there are definitely some honey samples that contain more glucose than fructose, some countries consider a honey adulterated if fructose is not the major carbohydrate component. Other than the monosaccharides and water, the remaining portion of honey contains a great variety of materials, but by far the greatest of these are a complex mixture oligosaccharides, which are mainly disaccharides and trisaccharides of glucose and fructose (unreported in Table 1).

It is now generally acknowledged that although floral nectars can vary in carbohydrate composition, they probably contribute to honey only the sugars glucose, fructose, and sucrose (3). With only a very few known exceptions, melezitose in honeydew honey for example (4), it is now thought that the vast majority of the oligosaccharides found in honey originate from enzymes added by the bees (5,6).

Several review articles are available on the carbohydrate composition of honey (6–8).

Simple Analysis Systems

The traditional method of honey carbohydrate analysis involves direct analysis of a diluted honey sample for reducing sugars before and after acid hydrolysis using either the Munson-Walker or Lane-Eynon methods (9). The reducing sugars are usually reported as invert sugar (half fructose and half glucose), and nonreducing sugars (the difference between the results before and after acid hydrolysis) are reported as sucrose. This analysis compromise does not reflect the complexity of the honey carbohydrate composition, but it does serve to define reducing and nonreducing limits for honey. For example, the joint FAO/

Table 1. Average Composition of U.S. Honey[a]

Component	Average Percentage[b]	Range
Moisture	17.2 ± 1.5	13.4–22.9
Fructose	38.2 ± 2.1	27.2–44.3
Glucose	31.3 ± 3.0	22.0–40.7
Free acid (as gluconic)	0.43 ± 0.16	0.13–0.92
Lactone (as δ-gluconolactone)	0.14 ± 0.07	0.00–0.37
Ash	0.169 ± 0.15	0.020–1.028

[a] Ref. 2.
[b] ± standard deviation.

WHO Codex Alimentarius Commission recommended that, except for a few noted exceptions, blossom honey should contain not less than 65% reducing sugar calculated as invert sugar and not more than 5% apparent sucrose (nonreducing sugar as sucrose) (10).

A more accurate carbohydrate analysis of honey can be performed by initially using a charcoal-celite column separation of the sugars in honey to monosaccharides, disaccharides, and higher sugars (9). Even with all the modern packing materials available, the charcoal-celite separation of carbohydrates is of such utility that it is still the method of choice for preliminary oligosaccharide separation. Once separated into fractions, further selective chemical analyses or enzymatic methods such as the use of glucose oxidase for glucose determination can be performed (9).

Glucose is of particular interest in the supersaturated sugar solution that is honey, because it is glucose that crystallizes from honey. The crystallization of glucose from honey is desirable in creamed honey, but undesirable in liquid honey. A fairly rapid enzymatic method for glucose detection in honey has been developed to aid in selection of the honey samples with the required glucose content for processing (11).

High Performance Liquid Chromatography (HPLC)

The development of HPLC methods has been a great boon to honey carbohydrate analysis because both separation and detection of the carbohydrates are possible with one fairly rapid analysis. Three main separation systems have been used for HPLC of honey carbohydrates. The stationary phases of these three systems are aminopropyl-bonded silica, cation exchange resins, and anion exchange resins. Comparison of a single honey sample separated on different versions of these systems is given in Figures 1 through 4.

The most common HPLC method used for carbohydrate honey analysis uses the aminopropyl-bonded silica stationary phase (Fig. 1) (12). This system has the advantage of some separation of the oligosaccharides. For example, sucrose and maltose can be separated. There are several disadvantages with this system also, however. The HPLC columns have considerable variability from manufacturer to manufacturer (13) and have limited lifetimes due to reducing sugar-amine reactions (14). High percentages of expensive, toxic acetonitrile solvent are required for separations. Eluting carbohydrates are determined with refractive index detectors, which are generally less sensitive than other detectors. The low sensitivity is particularly annoying for the late-eluting oligosaccharides that are present in small amounts and not well resolved when it is considered that up to 20 di- and trisaccharides can be present in significant amounts in a honey sample (5,15,16).

Carbohydrate separation systems based on cation exchange resins can use resins in a variety of forms. Silver, lead, calcium, potassium, sodium, and hydrogen forms of these resins are available. These columns have the advantages of being more resilient than aminopropyl-bonded

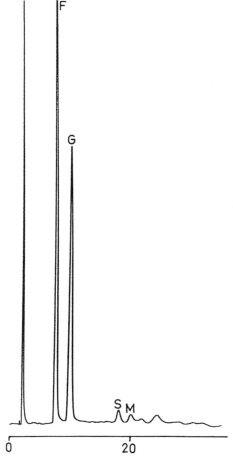

Figure 1. HPLC honey analysis on μ-Bondapak 10 NH$_2$, 300 × 3.9 mm column; refractive index detection; solvent, acetonitrile:water (83:17); room temperature; 20 μL injection of 10% filtered honey solution; flow rate 1.5 mL/min; F = fructose, G = glucose, S = sucrose, N = maltose; bottom axis, time in min.

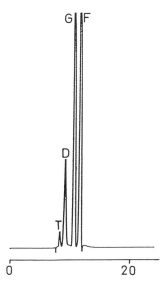

Figure 2. HPLC honey analysis on Rezex RHM-Monosaccharide, 300 × 7.8 mm column; refractive index detection; solvent, 0.01 N sulfuric acid; room temperature; 20 µL injection of filtered 10% honey solution (same honey as in Fig. 1); flow rate 0.6 mL/min; T = trisaccharides, D = disaccharides, G = glucose, F = fructose; bottom axis, time in min.

silica columns and using water as the elution solvent. The disadvantages are that honey oligosaccharides are poorly separated, generally eluting in a trisaccharide peak, then a disaccharide peak, and finally separated monosaccharides (Figs. 2 and 3). Somewhat better separations can be obtained with some of the resin forms, especially at elevated temperatures, but care should be taken. For example, the lead form of these resins causes decomposition of ketoses (mainly fructose in honey, see Fig. 3) (17). Also, the acid form can only be used at room temperature, with somewhat higher back pressures, to prevent hydrolyzing the oligosaccharides (see Fig. 2). Sucrose is the oligosaccharide most susceptible to acid hydrolysis. Another minor inconvenience of the metal-complexed columns if only water is used as the solvent is that the columns must be periodically regenerated with metal salt solutions.

The newest separation system is one based on the weak acidity of the sugar hydroxyl groups and uses an anion exchange system (18). Coupled with a more sensitive pulsed amperometric detector (19), this system offers the advantage of increased resolution (mainly because gradient elution is possible with this system) and greatly increased sensitivity (Fig. 4, note that monosaccharides have largely been removed with a charcoal-celite column before HPLC injection). Because the oligosaccharides in honey can be so informative of the honey composition, this system will probably become increasingly popular for honey analysis.

Adulteration of Honey with Syrups

The amount and identity of oligosaccharides in honey is of importance because of the possibility of adulteration with a variety of less expensive sweeteners. With larger markets and declining costs of high fructose corn syrups,

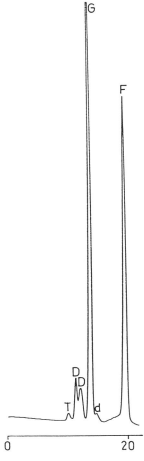

Figure 3. HPLC honey analysis on Aminex HPX-87P, 300 × 7.8 mm column; refractive index detection; solvent, water; 85°C; 20 µL injection of filtered 10% honey solution (same honey as in Fig. 1); flow rate 0.6 mL/min; T = trisaccharides, D = disaccharides, G = glucose, d = decomposition peak, F = fructose; bottom axis, time in min.

which have sugar compositions very similar to honey, the potential for adulteration of honey with these syrups has grown. One of the simplest methods of detecting such adulteration is by using thin layer chromatographic (TLC) examination of the higher sugar, oligosaccharide fraction from charcoal-celite chromatography of honey (20). An even more difficult to detect adulterant sweetener is enzymatically hydrolyzed beet sugar. Analysis for this adulteration again involves examination of the higher sugar, oligosaccharide fraction, but this time for galactose oligosaccharides using galactose oxidase (21).

An entirely different approach to detection of carbohydrate adulteration is the use of isotope ratio methods. Most honey-producing plants (although not all of them) fix carbon dioxide through the Calvin photosynthetic cycle, whereas sugarcane and corn fix carbon dioxide through the more efficient Hatch-Slack cycle. What this means is that the $^{13}C/^{12}C$ ratio is different for the sugars in honey compared to the sucrose from sugarcane or to any corn syrup products. Using combustion of the carbohydrate to carbon dioxide followed by mass spectrometry, the plant source of the carbohydrate can be differentiated (22).

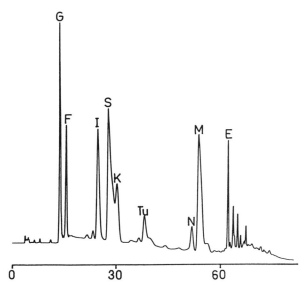

Figure 4. HPLC honey analysis on series of two 10 μm Carbo Pac PA1 pellicular anion exchange columns, 250 × 4 mm columns; pulsed amperometric detection; solvent, 100 mM sodium hydroxide for 4 min, then linear gradient to 100 mM sodium hydroxide and 3 mM sodium acetate at 20 min, then linear gradient to 100 mM sodium hydroxide and 100 mM sodium acetate at 50 min and hold; also 300 mM sodium hydroxide post-column injection at 0.8 mL/min; 85°C; 100 μL injection of 10% honey solution (same honey as in Fig. 1) which was first deionized and treated on a charcoal-celite column to remove most of the monosaccharides; flow rate 0.7 mL/min; G = glucose, F = fructose, I = isomaltose, S = sucrose, K = kojibiose, Tu = turanose, N = nigerose, M = maltose, E = erlose; bottom axis, time in min. Chromatogram courtesy K. W. Swallow and N. H. Low, Department of Applied Microbiology and Food Science, University of Saskatchewan, Saskatoon, Saskatchewan, Canada

Since beets are also Calvin cycle plants, this method will not work for beet sucrose or products of beet sucrose. There is the potential in the future for isotope ratio methods using other isotopes such as $^2H/^1H$ and $^{18}O/^{16}O$ to further differentiate plant sources, including beet sugar, as has been observed with orange juice (23). In general, however, these isotope ratio methods require expensive equipment and considerable expertise and can only detect larger amounts of adulteration because of necessity to allow for the natural variation found in different honeys.

BEE-ADDED MATERIALS

Bees add a number of materials to nectar. Some of these materials, such as some enzymes, have a definite role to play in the development and storage of honey. As it would be very difficult to mimic all these added materials with artificial additives, analysis of bee-added materials is often used as an indicator of the purity and quality of honey.

Enzymes

A number of enzymes, added by bees, are found in all honey samples. The most tested is diastase enzyme. Diastase in honey is a mixture of α- and β-amylase activity. The reason that bees add this enzyme system to nectar is obscure, as there is no starch in nectar. The activity of this enzyme is most often measured by noting the decrease in iodine-starch color with time (24). Because different starches can have different amounts of amylose and amylopectin, which in turn react to give different absorbances with iodine, it is important for valid comparison of results that a standard starch be used for all diastatic honey tests. Abusive heating of honey or very long storage will lead to low enzyme levels. Because diastase is the most frequently measured honey enzyme, acceptable levels of activity are considered to be greater than 8 mL 1% starch solution hydrolyzed/g honey/h at 40°C for most honeys (10).

Honey invertase is actually an α-glucosidase (maltase) added to honey by bees to hydrolyze sucrose in nectar to its monosaccharides. Because sucrose crystallizes much easier than the hydrolyzed product, the hydrolysis facilitates the conversion of nectar to honey by allowing for the preparation of the required supersaturated sugar solution. Crystallization of honey in the comb can be disastrous for a beehive because the bees must redissolve the crystals before ingesting the sugar, requiring water, which is unavailable in cold winter months. A number of methods have been developed for the analysis of invertase in honey, but the easiest method involves use of an enzyme substrate (p-nitrophenyl-α-D-glucopyranoside) that develops color after enzymatic hydrolysis (25).

Glucose oxidase is also a very important enzyme added to honey by bees. The products of this enzyme are responsible for a great deal of the antimicrobial activity of honey. Hydrogen peroxide, one product of this enzyme, has been identified as the inhibine compound in honey (26). The most common honey analysis involving this enzyme is a determination of the other enzymatic product, δ-D-gluconolactone, and its chemical hydrolysis product, gluconic acid (Table 1) (27). The acidity and resulting low pH developed by the hydrolysis product also contributes to the well-established antimicrobial properties of honey.

Other enzymes have occasionally been reported in honey, but they have not received the same amount of analytical attention as the above enzymes. Recently, β-glucosidase, which is almost certainly added by the bees, has been reported, explaining the origin of β-linked oligosaccharides in honey (28).

Proline

Although honey contains a variety of free amino acids, proline is by far present in the greatest amount. It is believed that proline is important as an osmoregulator when the bees add enzymes to nectar, counterbalancing the high osmotic pressure of nectar (29). Honey analysis for proline has been developed into a relatively simple test using ninhydrin to form color (30).

OTHER ANALYSES

Moisture

Moisture analysis of honey is one of the most important measurements, as honey is hygroscopic and once it contains much more than 20–21% moisture it is likely to be fermented by ubiquitous yeast. Excessive moisture in honey is a great problem in humid tropical countries. The analysis of moisture in honey is accurately carried out using a simple refractometer (31).

Color

Color can vary a great deal in honeys from different floral sources, but many major commercial honeys are quite light in color. In fact, as any heating of honey leads to nonenzymatic browning, color is often a very important quality indicator for honey. Premium honeys are often the lightest in color. A special color classification has been developed for honey, and it is widely used (32). The honey colors from lightest to darkest are: water white, extra white, white, extra light amber, light amber, amber.

Hydroxymethylfurfural (HMF)

Honey contains mainly monosaccharides, and because it is acidic, the stable compound HMF builds up on heating. High levels (usually considered to be 40 mg/kg or higher) (10) of HMF indicate abusive heating of honey and/or addition of an acid hydrolyzed sucrose or cornstarch product. Older HMF tests required barbituric acid which is difficult to obtain, but newer methods can be carried out using a simple ultraviolet procedure (33).

Ash

Honey is generally very low in ash (mostly honeys are required to be below 0.6%, see ref. 10 and Table 1). Honey higher in ash can be detrimental to overwintering bees because bees eliminate the ash by excretion, which must be performed outside the hive. Honeydew honey, which is honey produced from carbohydrate fluid excretions of sucking insects such as aphids, is high in ash. The percentage ash is one of the major factors identifying honeydew honeys from floral honeys (34). High ash in honey can also be an indication of honey adulteration with acid hydrolyzed carbohydrates, which are later neutralized. Ash analysis of honey is carried out using a standard dry ash methodology (35).

Pollen

Although there have been numerous attempts to develop alternative methods, pollen analysis, or mellissopalynology, is still the only general method to determine the floral origin of honey. Correct pollen analysis requires considerable expertise and time. The basic methods for pollen analysis of honey have been reviewed (36).

Chemical Residues

Honey, like any food commodity, has been analyzed for environmental contaminants. In fact, because bees collect nectar over large areas, bees and bee products have been used as environmental monitors. There are, however, some chemical residues that are unique to honey because of beekeepers' hive management techniques. These chemicals include: bee repellents (37,38), certain antibiotics (39–41), and acaricides (42), and there has been considerable analytic effort expended to ensure that honey is free of these residues.

BIBLIOGRAPHY

1. E. Crane, *Honey: A Comprehensive Survey,* Heinemann, London, 1979.
2. J. W. White, Jr. and co-workers, "Composition of American Honeys," *Technical Bulletin, U.S. Department of Agriculture* No. 1261, 1962.
3. H. G. Baker and I. Baker, "A Brief Historical Review of the Chemistry of Floral Nectar," in *The Biology of Nectaries,* Columbia University Press, New York, 1983.
4. J. S. D. Bacon and B. Dickinson, "The Origin of Melezitose: A Biochemical Relation Between the Lime Tree (*Tilia* spp.) and an Aphis (*Eucalliptus tiliae*)," *Biochemistry Journal* **66,** 289–297 (1957).
5. N. H. Low and co-workers, "Carbohydrate Analysis of Western Canadian Honeys and Their Nectar Sources to Determine the Origin of Honey Oligosaccharides," *Journal of Apicultural Research* **27,** 245–251 (1988).
6. I. R. Siddiqui, "The Sugars of Honey," *Advances in Carbohydrate Chemistry and Biochemistry* **25,** 285–309 (1970).
7. J. W. White, Jr., "Honey," *Advances in Food Research* **24,** 287–375 (1978).
8. L. W. Doner, "The Sugars of Honey," *Journal of the Science of Food and Agriculture* **28,** 443–456 (1977).
9. Association of Official Analytical Chemists, *Official Methods of Analysis,* 15th ed., AOAC, Arlington, Va., 1990, pp 1025–1033.
10. FAO/WHO Codex Alimentarious Commission, *Recommended European Regional Standard for Honey,* FAO and WHO, Italy, 1969. Also *Bee World* **51,** 79–91 (1970).
11. J. W. White, Jr., "Dextrose Determination in Honey: A Rapid Photometric Determination," *Journal of the Association of Official Analytical Chemists* **47,** 488–491 (1964).
12. Ref. 9, Method 977.20.
13. Z. L. Nikolov and co-workers, "High-Performance Liquid Chromatography of Disaccharides on Amine-Bonded Silica Columns," *Journal of Chromatography* **319,** 51–57 (1985).
14. C. Brons and C. Olieman, "Study of the High-Performance Liquid Chromatographic Separation of Reducing Sugars, Applied to the Determination of Lactose in Milk," *Journal of Chromatography* **259,** 79–86 (1983).
15. I. R. Siddiqui and B. Furgala, "Isolation and Characterization of Oligosaccharides from Honey. Part I Disaccharides," *Journal of Apicultural Research* **6,** 139–145 (1967).
16. I. R. Siddiqui and B. Furgala, "Isolation and Characterization of Oligosaccharides from Honey. Part II Trisaccharides," *Journal of Apicultural Research* **7,** 51–59 (1968).
17. J. O. Baker and co-workers, "Degradation of Ketoses During Aqueous High-Performance Liquid Chromatography on Lead-form Cation-exchange Resins," *Journal of Chromatography* **437,** 387–397 (1988).
18. D. P. Lee and M. T. Bunker, "Carbohydrate Analysis by Ion Chromatography," *Journal of Chromatographic Science* **27,** 496–503 (1989).
19. D. C. Johnson and T. Z. Polta, "Amperometric Detection in Liquid Chromatography With Pulsed Cleaning and Reaction of Noble Metal Electrodes," *Chromatographic Forum* 37–43, (November–December 1986).
20. Ref. 9, Method 979.22.

21. J. W. White, Jr., and co-workers, "Detection of Beet Sugar Adulteration of Honey," *Journal of the Association of Official Analytical Chemists* **69**, 652–654 (1986).
22. Ref. 9, Method 978.17.
23. L. W. Doner and co-workers, "Detecting Sugar Beet Syrups in Orange Juice by D/H and $^{18}O/^{16}O$ Analysis of Sucrose," *Journal of Agricultural and Food Chemistry* **35**, 610–612 (1987).
24. Ref. 9, Method 958.09.
25. Siegenthaler, U, "Eine Einfache und Rasche Methode zur Bestimmung der α-Glucosidase (Saccharase) in Honig," *Mitt. Geb. Lebensmittelunters Hyg.* **68**, 251–258 (1977).
26. J. W. White, Jr., and co-workers "The Identification of Inhibine, the Antibacterial Factor in Honey, as Hydrogen Peroxide and Its Origin in a Honey Glucose-Oxidase System," *Biochimica et Biophysica Acta* **73**, 57–70 (1963).
27. Ref. 9, Method 962.19.
28. N. H. Low and co-workers, "A New Enzyme, β-Glucosidase in Honey," *Journal of Apicultural Research* **25**, 178–181 (1986).
29. A. M. C. Davies, "Proline in Honey: An Osmoregulatory Hypothesis," *Journal of Apicultural Research* **17**, 227–233 (1978).
30. Ref. 9, Method 979.20.
31. Ref. 9, Method 969.38.
32. Ref. 9, Method 960.44.
33. Ref. 9, Method 980.23.
34. K. C. Kirkwood and co-workers, "An Examination of the Occurence of Honeydew in Honey," *Analyst* **86**, 164–165 (1961).
35. Ref. 9, Method 920.181.
36. J. Louveaux and co-workers, "Methods of Melissopalynology," *Bee World* **59**, 139–157 (1979).
37. P. A. Daharu and P. Sporns, "Evaluation of Analytical Methods for the Determination of Residues of the Bee Repellent, Phenol, in Honey and Beeswax," *Journal of Agricultural and Food Chemistry* **32**, 108–111 (1984).
38. S. Kwan and P. Sporns, "Analysis of Bee Repellents in Honey," *Journal of Apicultural Research* **27**, 162–168 (1988).
39. P. Sporns and co-workers, "HPLC Analysis of Oxytetracycline Residues in Honey," *Journal of Food Protection* **49**, 383–388 (1986).
40. L. Roth and co-workers, "The Use of a Disc-Assay System to Detect Oxytetracycline Residues in Honey," *Journal of Food Protection* **49**, 436–441 (1986).
41. E. Neidert and co-workers "Rapid Quantitative Thin Layer Chromatographic Screening Procedure for Sulfathiazole Residues in Honey," *Journal of the Association of Official Analytical Chemists* **69**, 641–643 (1986).
42. G. Formica "Gas Chromatographic Determination of Residues of Bromopropylate and Two of Its Degradation Products in Honey," *Journal of the Association of Official Analytical Chemists* **67**, 896–898 (1984).

PETER SPORNS
University of Alberta
Edmonton, Alberta

HORMONES. See APPETITE.
HORTICULTURE. See APRICOTS AND APRICOT PROCESSING; APPLES AND APPLE PROCESSING; VEGETABLE PRODUCTION; and entries under FOOD CROPS.

HYDROGEN-ION ACTIVITY (pH)

The effective concentration of hydrogen ion in solution is expressed in terms of pH, which is the negative logarithm of the hydrogen-ion activity:

$$\mathrm{pH} = -\log_{10} a_{\mathrm{H}+} \quad (1)$$

The relationship between activity and concentration is

$$a = \gamma c \quad (2)$$

where the activity coefficient γ is a function of the ionic strength of the solution and approaches unity as the ionic strength decreases; ie, the difference between the activity and the concentration of hydrogen ion diminishes as the solution becomes more dilute. The pH of a solution may have little relationship to the titratable acidity of a solution that contains weak acids or buffering substances; the pH of a solution indicates only the free hydrogen-ion activity. If total acid concentration is to be determined, an acid-base titration must be performed.

Thermodynamically, the activity of a single ionic species (and, eg, the pH) is an inexact quantity, and a conventional pH scale has been adopted that is defined by reference solutions with assigned pH values. These reference solutions, in conjunction with equation 3, define the pH.

$$\mathrm{pH}(X) = \mathrm{pH}(S) - \frac{(E_x - E_s)\mathrm{F}}{2.303\ \mathrm{RT}} \quad (3)$$

E_s is the electromotive force (emf) of the cell:

reference electrode |KCl($\geq 3.5\ M$)||solution S|$H_2(g)$, Pt,

and E_x is the emf of the same cell when the reference buffer solution S is replaced by the sample solution X. The quantities R, T, and F are the gas constant, the thermodynamic temperature, and the Faraday constant, respectively. For routine pH measurements, the hydrogen gas electrode $[H_2(g), Pt]$ usually is replaced by a glass membrane electrode.

The availability of multiple pH reference solutions makes possible an alternative definition of pH:

$$\mathrm{pH}(X) = \mathrm{pH}(S_1) + [\mathrm{pH}(S_2) - \mathrm{pH}(S_1)] \frac{(E_{x2} - E_{s1})}{(E_{s2} - E_{s1})} \quad (4)$$

where E_{s1} and E_{s2} are the measured cell potentials when the sample solution X is replaced in the cell by the two reference solutions S_1 and S_2 such that the values E_{s1} and E_{s2} are on either side of, and as near as possible to, E_x. Equation 4 assumes linearity of the pH versus E response between the two reference solutions, whereas equation 3 assumes both linearity and ideal Nernstian response of the pH electrode. The two-point calibration procedure is recommended if a glass pH electrode is used for the measurements.

pH DETERMINATION

Two methods are used to measure pH (1–6). The most common uses the commercial pH meter with a glass electrode. This procedure is based on the measurement of the difference between the pH of an unknown or test solution and that of a standard solution. The instrument measures the emf developed between the glass electrode and a reference electrode of constant potential. The difference in emf when the electrodes are removed from the standard solution and placed in the test solution is converted to a difference in pH. The second method, which has more limited applications, is the indicator method. The success of this procedure depends on matching the color that is produced by the addition of a suitable indicator dye to a portion of the unknown solution with the color produced by adding the same quantity of the same dye to a series of standard solutions of known pH. The results obtained by the indicator method are less accurate relative to those obtained using a pH meter. The indicator method, however, is simple to apply and inexpensive.

Reference Buffer Solutions

The uncertainties introduced by the reference electrode liquid junction that exist in conventional electrochemical cells can be avoided by using a cell without transference, eg:

$$\text{Ag,AgCl} | \text{KCl}(m), \text{ solution } S | \text{H}_2(\text{g}), \text{Pt}$$

Potassium chloride, KCl, of molality m is added to each reference solution. If the standard potential, $E°$, of the cell and the molality of the chloride ion, m_{Cl}, are known, emf measurements yield values of the acidity function $p(a_H \gamma_{Cl})$, as shown by the following equation:

$$p(a_H \gamma_{Cl}) \equiv -\log(m_H \gamma_H \gamma_{Cl}) = \frac{(E° - E)F}{2.303\ RT} + \log m_{Cl} \quad (5)$$

To eliminate the effect of the added KCl on the acidity of the buffer solution, $p(a_H \gamma_{Cl})$ is determined for three or more portions of the buffer solution that contain different amounts of added chloride. The limiting value of the acidity function is obtained by extrapolation to zero molality of chloride ion. If the activity coefficient of chloride ion in the buffer solution could be obtained, the activity of hydrogen ion would be readily accessible.

$$pa_H \equiv -\log(m_H \gamma_H) = p(a_H \gamma_{Cl}) + \log \gamma_{Cl} \quad (6)$$

To establish a conventional scale of hydrogen-ion activity, it has been suggested (7) that the activity coefficient of chloride ion in selected reference buffer solutions having ionic strengths of ≤ 0.1 be defined as:

$$-\log \gamma_{Cl} = \frac{A\sqrt{I}}{1 + 1.5\sqrt{I}} \quad (7)$$

where A is a constant of the Debye-Hückel theory and I is the ionic strength. The pa_H for these selected reference

Table 1. pH Standards, Molality Scale[a]

Solution Composition	pH(S) at 25°C
Primary Standards	
Potassium hydrogen tartrate (saturated at 25°C)	3.557
0.05 m Potassium dihydrogen citrate	3.776
0.05 m Potassium hydrogen phthalate	4.006
0.025 m KH$_2$PO$_4$ + 0.025 m Na$_2$HPO$_4$	6.863
0.008695 m KH$_2$PO$_4$ + 0.03043 m Na$_2$HPO$_4$	7.410
0.01 m Na$_2$B$_4$O$_7 \cdot$ 10H$_2$O	9.180
0.025 m NaHCO$_3$ + 0.025 m Na$_2$CO$_3$	10.011
Secondary Standards	
0.05 m Potassium tetroxalate $\cdot$ 2H$_2$O	1.681
0.05 m HEPES[b] + 0.05 m NaHEPESate	7.504
Ca(OH)$_2$ (saturated at 25°C)	12.454

[a] Ref. 2.
[b] HEPES $\equiv$ N-2-hydroxyethylpiperazine-N'-2-ethanesulfonic acid.

solutions are identified with pH(S) in the operational definition:

$$pa_H \equiv \text{pH}(S) \quad (8)$$

The pH(S) values at 25°C of the NIST primary and secondary reference buffer solutions are listed in Table 1 (2). The International Union of Pure and Applied Chemistry recommends these primary standards plus a series of operational standards, measured versus the phthalate reference value standard in a cell with liquid junction, for the definition of the pH scale (8).

Accuracy and Interpretation of Measured pH Values

The acidity function $p(a_H \gamma_{Cl})$, which is the experimental basis for the assignment of pH(S), is reproducible within about 0.003 pH unit from 10 to 40°C. If the ionic strength is known, the assignment of numerical values to the activity coefficient of chloride ion does not add to the uncertainty. However, errors in the standard potential of the cell, in the composition of the buffer materials, and in the preparation of the solutions may raise the uncertainty to 0.005 pH unit.

The reproducibility of the practical scale that has been defined using the seven primary standards includes the possible inconsistencies introduced in the standardization of the instrument with seven different standards of different composition and concentration. These inconsistencies are the result of variations in the liquid-junction potential when one solution is replaced by another and are unavoidable. The accuracy of the practical scale from 10 to 40°C therefore appears to be from 0.008 to 0.01 pH unit.

Variations in the liquid-junction potential may be increased when the standard solutions are replaced by test solutions that do not closely match the standards with respect to the types and concentrations of solutes, or to the composition of the solvent. Under these circumstances, the pH remains a reproducible number, but it may have little or no meaning in terms of the conventional hydrogen-ion activity of the medium. The use of experimental pH numbers as a measure of the extent of acid-base reactions or to obtain thermodynamic equilibrium constants is justified only when the pH of the medium is between 2.5 and 11.5 and when the mixture is an aqueous solution of simple solutes in total concentration of ca ≤ 0.2 M.

Sources of Error

Although subject to fewer interferences and other types of error than most potentiometric ionic-activity sensors, ie, ion-selective electrodes (qv), pH electrodes must be used with an awareness of their particular response characteristics as well as the potential sources of error that may affect the other components of the measurement system, especially the reference electrode (see also pH Measurement System Electrodes). Several common causes of measurement problems are electrode interferences and/or fouling of the pH sensor, sample matrix effects, reference electrode instability, and improper calibration of the measurement system (9).

In general, the potential of an electrochemical cell, E_{cell}, is the sum of three potential terms:

$$E_{cell} = E_{pH} - E_{ref} + E_{lj} \tag{9}$$

where E_{pH} and E_{ref} are the potentials of the pH and reference electrodes, respectively, and E_{lj} is the ubiquitous liquid-junction potential. After substitution of the Nernst equation for the pH electrode potential term in equation 9,

$$E_{cell} = E_{pH}^\circ - \frac{RT}{F} \ln a_H - E_{ref} + E_{lj} \tag{10}$$

it can be calculated that a 1 mV error in any of the potential terms corresponds to an error of ca 4% in the hydrogen-ion activity. Under carefully controlled experimental conditions, the potential of a pH cell can be measured with an uncertainty as small as 0.3 mV, which corresponds to a ± 0.005 pH unit uncertainty.

pH MEASUREMENT SYSTEM ELECTRODES

Glass Electrodes

The glass electrode is the hydrogen-ion sensor in most pH-measurement systems. The pH-responsive surface of the glass electrode consists of a thin membrane formed from a special glass that, after suitable conditioning, develops a surface potential that is an accurate index of the acidity of the solution in which the electrode is immersed. To permit changes in the potential of the active surface of the glass membrane to be measured, an inner reference electrode of constant potential is placed in the internal compartment of the glass membrane. The inner reference compartment contains a solution that has a stable hydrogen-ion concentration and counterions to which the inner electrode is reversible. The choice of the inner cell components has a bearing on the temperature coefficient of the emf of the pH assembly. The inner cell commonly consists of a silver–silver chloride electrode or calomel electrode in a buffered chloride solution.

Immersion electrodes are the most common glass electrodes. These are roughly cylindrical and consist of a barrel or stem of inert glass that is sealed at the lower end to a tip, which is often hemispherical, of special pH-responsive glass. The tip is completely immersed in the solution during measurements. Miniature and microelectrodes are used widely, particularly in physiological studies. Capillary electrodes permit the use of small samples and protect them from exposure to air during the measurements, eg, for the determination of blood pH. This type of electrode may be provided with a water jacket for temperature control.

The membrane of pH-responsive glass usually is made as thin as is consistent with adequate mechanical strength; nevertheless, its electrical resistance is high, eg, from 10–250 MΩ. Therefore, electronic amplifiers must be used to obtain adequate accuracy in the measurement of the surface potentials of glass electrodes. The versatility of the glass electrode results from its mechanism of operation, which is one of proton exchange rather than electron transfer; hence, oxidizing and reducing agents in the solution do not affect the pH response.

Most modern electrode glasses contain silicon dioxide, either sodium or lithium oxide, and either calcium, barium, cesium, or lanthanum oxide. The latter oxides are added to reduce spurious response to alkali metal ions in high pH solutions. The composition of the glass has a profound effect on the electrical resistance, the chemical durability of the pH-sensitive surface, and the accuracy of the pH response in alkaline solutions. Both the electrical and the chemical resistance of the electrode glasses decrease rapidly with a rise in temperature. Therefore, it is difficult to design an electrode that is sufficiently durable for extended use at high temperatures, and yet, when used at room temperature, free from the sluggish response often characteristic of pH cells of excessively high resistance. Most manufacturers use different glass compositions for electrodes, depending on their intended use.

The mechanism of the glass electrode response is not entirely clear. It is clear, however, that when a freshly blown membrane of pH-responsive glass is first conditioned in water, the sodium or lithium ions that occupy the interstices of the silicon-oxygen network in the glass surface are exchanged for protons from the water. The protons find stable sites in the conditioned gel layer of the glass surface. Exchange of the labile protons between these sites and the solution phase appears to be the mechanism by which the surface potential reflects changes in the acidity of the external solution. When the glass electrode and the hydrogen gas electrode are immersed in the same solution, their potentials usually differ by a constant amount, even though the pH of the medium is raised from 1 to 10 or greater. In this range, the potential E_g of a glass electrode may be written

$$E_g = E_g^\circ + \frac{RT}{F} \ln a_{H^+} \tag{11}$$

where E_g° is the "standard" potential of that particular glass electrode on the hydrogen scale.

Departures from the ideal behavior expressed by equation 11 usually are found in alkaline solutions containing alkali metal ions in appreciable concentration, and often in solutions of strong acids. The supposition that the alkaline error is associated with the development of an imperfect response to alkali metal ions is substantiated by the successful design of cation-sensitive electrodes that are used to determine sodium, silver, and other monovalent cations (3).

The advantage of the lithium glasses over the sodium glasses in the reduction of alkaline error is attributed to the smaller size of the proton sites remaining after elution of the lithium ions from the glass surface. This view is consistent with the relative magnitudes of the alkaline errors for various cations. These errors decrease rapidly as the diameter of the cation becomes larger. The error observed in concentrated solutions of the strong acids is characterized by a marked drift of potential with time, which is thought to result from the penetration of acid anions, as well as protons, into the glass surface (10).

The immersion of glass electrodes in strongly dehydrating media should be avoided. If the electrode is used in solvents of low water activity, frequent conditioning in water is advisable, as dehydration of the gel layer of the surface causes a progressive alteration in the electrode potential with a consequent drift of the measured pH. Slow dissolution of the pH-sensitive membrane is unavoidable, and it leads to mechanical failure. Standardization of the electrode with two buffer solutions is the best means of early detection of electrode failure.

Fouling of the pH sensor may occur in solutions containing surface-active constituents that coat the electrode surface and result in sluggish response and drift of the pH reading. Prolonged measurements in blood, sludges, and various industrial process materials and wastes can cause such drift; therefore, it is necessary to clean the membrane mechanically or chemically at intervals that are consistent with the magnitude of the effect and the precision of the results required.

Reference Electrodes and Liquid Junctions

The electrical circuit of the pH cell is completed through a salt bridge that usually consists of a concentrated solution of potassium chloride. The solution makes contact at one end with the test solution and at the other with a reference electrode of constant potential. The liquid junction is formed at the area of contact between the salt bridge and the test solution. The mercury–mercurous chloride electrode—the calomel electrode—provides a highly reproducible potential in the potassium chloride bridge solution and is the most widely used reference electrode. However, mercurous chloride is converted readily into mercuric ion and mercury when in contact with concentrated potassium chloride solutions above 80°C. This disproportionation reaction causes an unstable potential with calomel electrodes. Therefore, the silver–silver chloride electrode and the thallium amalgam–thallous chloride electrode often are used for measurements above 80°C. However, because silver chloride is relatively soluble in concentrated solutions of potassium chloride, the solution in the electrode chamber must be saturated with silver chloride.

The commercially used reference electrode–salt bridge combination usually is of the immersion type. The salt bridge chamber may surround the electrode element. Some provision is made to allow a slow leakage of the bridge solution out of the tip of the electrode to establish the liquid junction with the standard solution or test solution in the pH cell. An opening is usually provided through which the chamber may be refilled. Various devices are used to impede the outflow of bridge solution, eg, fibers, porous ceramics, capillaries, ground-glass joints, and controlled cracks. Such commercial electrodes normally give very satisfactory results, but there is some evidence that the type and structure of the junction may affect the reference potential when measurements are made at very low pH and, possibly, at high alkalinities.

Combination electrodes have increased in use and are a consolidation of the glass and reference electrodes in a single probe, usually in a concentric arrangement, with the reference electrode compartment surrounding the pH sensor. The advantages of combination electrodes include the convenience of using a single probe and the ability to measure small volumes of sample solution or in restricted-access containers, eg, test tubes and narrow-neck flasks. In addition, the surrounding electrolyte solution in the reference electrode compartment provides excellent electrical shielding of the pH sensor, which reduces noise and susceptibility to polarization.

Theoretical considerations favor junctions by which cylindrical symmetry and a steady state of diffusion are achieved. Special cells in which a stable junction can be achieved are not difficult to construct and are available commercially.

A solution of potassium chloride that is saturated at room temperature usually is used for the salt bridge. It has been shown that the higher the concentration of the solution of potassium chloride, the more effective the bridge solution is in reducing the liquid-junction potential (11). Also, the saturated calomel reference electrode is stable, reproducible, and easy to prepare. However, the saturated electrode is not without its disadvantages. For example, it shows a marked hysteresis with changes of temperature. After long periods and on temperature-lowering, the salt-bridge chamber may become filled with large crystals of potassium chloride that block the flow of bridge solution and thereby impair the reproducibility of the junction potential and raise the resistance of the cell. A slightly undersaturated (eg, 3.5 M) solution of potassium chloride is preferred.

Samples that contain suspended matter are among the most difficult types from which to obtain accurate pH readings because of the suspension effect, ie, the suspended particles produce abnormal liquid-junction potentials at the reference electrode (12). This effect is especially noticeable with soil slurries, pastes, and other types of colloidal suspensions. In the case of a slurry that separates into two layers, pH differences of several units may result, depending on the placement of the electrodes in the layers. Internal consistency is achieved by pH measurement using carefully prescribed measurement protocols, as has been used in the determination of soil pH (13).

Another effect that may result in spurious pH readings is caused by streaming potentials. Presumably, these are attributable to changes in the reference electrode liquid junction that are caused by variations in the flow rate of the sample solution. Factors that affect the observed pH include the magnitude of the flow-rate changes, the geometry of the electrode system, and the concentration of the salt-bridge electrolyte; therefore, this problem may be avoided by maintaining constant flow and geometry characteristics and calibrating the system under operating conditions that are identical to those of the sample measurement.

pH INSTRUMENTATION

The pH meter is an electronic voltmeter that provides a direct conversion of voltage differences to differences of pH at the measurement temperature. One class of instruments is the direct-reading analogue, which has a deflection meter with a large scale calibrated in mV and pH units. Most modern direct-reading meters have digital displays of the emf or pH. The types range from very inexpensive meters that read to the nearest 0.1 pH unit to the research models capable of measuring pH with a precision of 0.001 pH unit and drifting less than 0.003 pH unit over 24 h; however, the fundamental meaning of these measured values is considerably less certain than the precision of the measurement.

Because of the very large resistance of the glass membrane in a conventional pH electrode, an input amplifier of high impedance (usually 10^{12}–10^{14} Ω) is required to avoid errors in the pH (or mV) readings. Most pH meters have field-effect transistor (FET) amplifiers that typically exhibit bias currents of only a picoampere (10^{-12} ampere), which, for an electrode resistance of 100 MΩ, results in an emf error of only 0.1 mV (0.002 pH unit).

In addition, most of these devices provide operator control of settings for temperature and/or response slope, isopotential point, zero or standardization, and function (pH, mV, or monovalent/bivalent cation/anion). Microprocessors are incorporated in advanced-design meters to facilitate calibration, calculation of measurement parameters, and automatic temperature compensation. Furthermore, pH meters are provided with output connectors for continuous readout via a strip-chart recorder and often with BCD (binary-coded decimal) output for computer interconnections or connection to a printer. Although the accuracy of the measurement is not increased by the use of a recorder, the readability of the displayed pH (on analogue models) can be expanded, and recording provides a permanent record and information on response and equilibration times during measurement (5).

TEMPERATURE EFFECTS

The emf E of a pH cell may be written

$$E_g = E_g^\circ - k\text{pH} \quad (12)$$

where k is (2.303 RT)/F, and E_g° includes the liquid-junction potential and the half-cell emf on the reference side of the glass membrane. Changes of temperature alter the scale slope because k is proportional to T. The scale position also is changed because the "standard" potential is temperature dependent: E_g° is usually a quadratic function of the temperature.

The objective of temperature compensation in a pH meter is to nullify changes in emf from any source except changes in the true pH of the test solution. Nearly all pH meters provide automatic or manual adjustment for the change of k with T. If correction is not made for the change of standard potential, however, the instrument must always be standardized at the temperature at which the pH is to be determined. In industrial pH control, standardization of the assembly at the temperature of the measurements is not always possible, and compensation for shift of the scale position, though imperfect, is useful. If the value of E_g° were a linear function of T, it would be easy to show that the straight lines representing the variation of E_g and pH at different temperatures would intersect at a point, the isopotential point or pH_i. Even though E_g° does not usually vary linearly with T, these plots intersect at about pH_i when the range of temperatures is narrow. By providing a temperature-dependent bias potential of $k\text{pH}_i$, which is measured in volts, an approximate correction for the change of the standard potential with temperature can be applied automatically (1,5).

NONAQUEOUS SOLVENTS

The activity of the hydrogen ion is affected by the properties of the solvent in which it is measured. Scales of pH only apply to the medium (single solvent or mixed solvents, eg, water–alcohol) for which they are developed. The comparison of the pH values of a buffer in aqueous solution to one in a nonaqueous solvent has neither direct quantitative nor thermodynamic significance. Consequently, operational pH scales must be developed for the individual solvent systems. In certain cases, correlation to the aqueous pH scale can be made but, in others, pH values are used only as relative indicators of the hydrogen-ion activity.

Other difficulties of measuring pH in nonaqueous solvents are the complications that result from dehydration of the glass pH membrane, increased sample resistance, and large liquid-junction potentials. These effects are complex and are highly dependent on the type of solvent or mixture used (1,5).

INDICATOR pH MEASUREMENTS

The indicator method is especially convenient when the pH of a well-buffered colorless solution must be measured at room temperature with an accuracy no greater than 0.5 pH unit. Under optimum conditions an accuracy of 0.2 pH unit is obtainable. A list of representative acid–base indicators is given in Table 2 with their corresponding transformation ranges. A more complete listing, including the theory of the indicator color change and of the salt effect, is given in Reference 1.

Because they are weak acids or bases, the indicators

Table 2. Acid-Base Indicators

Indicator	pH Range	Color Change
Crystal violet	0.0–1.8	yellow-blue
Acid cresol red	0.2–1.8	red-yellow
Acid thymol blue	1.2–2.8	red-yellow
Bromophenol blue	3.0–4.6	yellow-blue
Methyl orange	3.2–4.4	red-yellow
Bromocresol green	3.8–5.4	yellow-blue
Methyl red	4.4–6.0	red-yellow
Bromocresol purple	5.2–6.8	yellow-purple
Bromothymol blue	6.0–7.6	yellow-blue
Phenol red	6.8–8.4	yellow-red
Cresol red	7.2–8.8	yellow-red
Thymol blue	8.0–9.6	yellow-blue
Phenolphthalein	8.2–10.0	colorless-pink
Tolyl red	10.0–11.6	red-yellow
2,4,6-Trinitrotoluene	11.5–13.0	colorless-orange
1,3,5-Trinitrobenzene	12.0–14.0	colorless-orange

may affect the pH of the sample, especially in the case of a poorly buffered solution. Variations in the ionic strength or solvent composition, or both, also can produce large uncertainties in pH measurements, presumably caused by changes in the equilibria of the indicator species. Specific chemical reactions also may occur between solutes in the sample and the indicator to produce appreciable pH errors. Examples of such interferences include binding of the indicator forms by proteins and colloidal substances and direct reaction with sample components, eg, bleaches and heavy metal ions.

INDUSTRIAL PROCESS CONTROL

Specialized equipment for industrial measurements and automatic control have been developed (14) (see Instrumentation and Control). In general, the pH of an industrial process need not be controlled with great accuracy. Consequently, frequent standardization of the cell assembly may be unnecessary. On the other hand, the temperature and humidity conditions under which the industrial control measurements are made may be such that the pH meter must be much more rugged than those intended for laboratory use. To avoid costly downtime for repairs, pH instruments may be constructed of modular units, permitting rapid removal and replacement of a defective subassembly.

The pH meter usually is coupled to a recorder and often to a pneumatic or electric controller. The controller governs the addition of reagent so that the pH of the process stream is maintained at the desired level.

Immersion-cell assemblies are designed for continuous pH measurement in tanks, troughs, or other vessels containing process solutions at different levels under various conditions of agitation and pressure. The electrodes are protected from mechanical damage and sometimes are provided with devices to remove surface deposits as they accumulate. Process flow chambers are designed to introduce the pH electrodes directly into piped sample streams or bypass sample loops that may be pressurized. Electrode chambers of both types usually contain a temperature-sensing element that controls the temperature-compensating circuits of the measuring instrument.

Glass electrodes for process control do not differ materially from those used for pH measurements in the laboratory, but the emphasis in industrial application is on rugged construction to withstand both mechanical stresses and high pressures. Pressurized salt bridges, which ensure slow leakage of bridge solution into the process stream even under very high pressures, have been developed. For less-severe process-monitoring conditions, reference electrodes are available with no-flow polymeric or gel-filled junctions that can be used without external pressurization.

BIBLIOGRAPHY

R. A. Durst and R. G. Bates, "Hydrogen-Ion Activity," in M. Grayson, ed., *Kirk-Othmer Encyclopedia of Chemical Technology*, 3rd ed., Vol. 13, 1981, pp. 1–11.

1. R. G. Bates, *Determination of pH*, 2nd ed., Wiley-Interscience, New York, 1973.
2. Y. C. Wu, W. F. Koch, and R. A. Durst, *Standardization of pH Measurements*, National Bureau of Standards Special Publication 260-53, U.S. Government Printing Office, Washington, D.C., 1988.
3. G. Eisenman, ed., *Glass Electrodes for Hydrogen and Other Cations*, Marcel Dekker, New York, 1967.
4. G. Mattock, *pH Measurement and Titration*, Macmillan, New York, 1961.
5. C. C. Westcott, *pH Measurements*, Academic Press, New York, 1978.
6. "pH of Aqueous Solutions with the Glass Electrode," *ASTM Method E 70-77*, American Society for Testing and Materials, Philadelphia, Pa., 1977.
7. R. G. Bates and E. A. Guggenheim, *Pure and Applied Chemistry* **1**, 163 (1960).
8. A. K. Covington, R. G. Bates, and R. A. Durst, *Pure and Applied Chemistry* **57**, 531 (1985).

9. R. A. Durst, "Sources of Error in Ion-Selective Electrode Potentiometry" in H. Freiser, ed., *Ion-Selective Electrodes in Analytical Chemistry*, Plenum Publishing, New York, 1978, Chapt. 5.
10. K. Schwabe, "pH Measurements and Their Applications" in H. W. Nürnberg, ed., *Electroanalytical Chemistry*, John Wiley & Sons, New York, 1974, Chapt. 7.
11. E. A. Guggenheim, *Journal of the American Chemical Society* **52**, 1315 (1930).
12. H. Jenny, T. R. Nielson, N. T. Coleman, and D. E. Williams, *Science* **112**, 164 (1950).
13. A. M. Pommer, "Glass Electrodes for Soil Waters and Soil Suspensions" in G. Eisenman, ed., *Glass Electrodes for Hydrogen and Other Cations*, Marcel Dekker, New York, 1967, Chapt. 14.
14. F. G. Shinskey, *pH and pIon: Control in Process and Waste Streams*, John Wiley & Sons, New York, 1973.

General References

Refs. 1–6 are also general references.

RICHARD A. DURST
Cornell University
Geneva, New York

ROGER G. BATES
University of Florida
Gainesville, Florida

HYDROGENATION

Hydrogenation is an important process utilized to alter the chemical and physical properties of vegetable oils. Hydrogen is reacted with oils in the presence of a catalyst to combine with unsaturated fatty acids in the triglyceride molecule. The reaction results in a physical hardening of the oil and can produce a range of semisolid or solid fat products dependent on the quantity of hydrogen added. These hydrogenated oils can be used for shortenings, margarines, and frying fats. The oxidative stability of hydrogenated oils is improved as highly reactive sites are occupied by hydrogen thus leaving fewer, less reactive sites for oxidation.

The hydrogenation of oils is complex, involving not only the chemical reaction with hydrogen but also isomerization, which also effects the physical properties of the processed oil. Oil processors have learned to use reaction conditions to achieve the desired product physical properties.

Hydrogenation is a heterogeneous reaction where gaseous hydrogen is reacted with oil in the presence of a solid nickel catalyst. Catalysts have been developed specifically for vegetable oil hydrogenation and typically contain 25% nickel, 15% substrate, and fully hardened vegetable oil as a binder. Other metal catalysts, eg, copper chromite, and palladium, have been investigated but are not widely used industrially.

Catalyst is generally specified on the basis of activity and selectivity ratio. The activity is the rate of hydrogenation normally defined by the drop in iodine value per minute when the reaction is carried out under controlled conditions. The iodine value (IV) is a measure of the degree of saturation of an oil. The selectivity ratio measures the relative rates of hydrogenation of different fatty acids in the oil. For soybean oil the selectivity ratio is defined as the rate of hydrogenation of linoleic acid double bonds compared to the rate of hydrogenation of oleic acid. The calculation of selectivity ratio requires analysis of fatty acid composition before and after hydrogenation. As all fatty acids within the triglyceride molecule are reacting with hydrogen simultaneously, the solution of rate equations to determine selectivity ratio is complex. A simplified equation is available to determine selectivity ratio assuming first-order kinetics (ref. 1). A linolenic selectivity ratio is of importance when hydrogenation conditions are set to produce a maximum stability oil by reducing linolenic acid preferentially. This situation occurs for high-stability liquid frying oils where minimum linolenic acid is desired while maintaining a liquid hydrogenated oil. The selectivity ratio and linoleic selectivity ratio are measured under a standard set of conditions to measure catalyst performance.

The hydrogenation of oils is usually carried out as a batch reaction. Catalyst performance and the hydrogenated product physical properties are strongly dependent on reaction conditions of temperature, pressure, and agitation as well as catalyst concentration. The hydrogenation reaction will not proceed below 250°F. The reaction is exothermic, and as the reaction proceeds, the reacting oil will heat. The temperature of starting the reaction and the temperature at which the reaction is allowed to proceed are important variables that have significant effects on the product obtained. As with most chemical reactions, increased temperature will increase rate, ie, activity. This will also increase selectivity ratio and increase the formation of trans acids. The latter change is important as trans acids have significantly higher melting points than do the corresponding cis acids that occur naturally. Thus, as the reaction proceeds at higher temperature, the increased trans acid production will yield a product that is physically harder than a product saturated to the same IV at a lower temperature. In practice, hydrogenation reactions may be run at temperatures of 250–300°F with exothermic heat removal by cooling coils or the temperature may be allowed to rise to 450°F and then controlled at this temperature by cooling coils. Reaction conditions above 450°F are unusual as mechanical restrictions on the equipment become limiting. Pressure of hydrogen will affect product properties and the normal range of pressure is 5–50 psig. Increasing pressure will increase activity, decrease selectivity ratio and decrease isomerization, ie, trans acid formation. The changes are not as large as temperature effects but nevertheless are significant.

Increasing the rate of agitation in the reactor will decrease selectivity ratio, increase rate and decrease trans acid formation. In large batch hydrogenation reactors, the agitator speed of rotation is frequently fixed and the design of the agitator is optimized to minimize power consumption while maintaining a desired rate of reaction.

The addition of catalyst is normally tightly controlled as catalyst is a significant cost associated with the hydrogenation process. Increasing catalyst concentration will increase catalyst activity and the selectivity ratio with little direct effect on trans acid formation. Catalyst is typically used at a level of 100–250-ppm nickel concentration.

The operation of a hydrogenation plant involves selection of a temperature and pressure for the reaction at an economically viable catalyst level that will produce the desired product physical properties. The reaction may be monitored by sampling the reactor and measuring the refractive index of the hydrogenated oil. The IV, or degree of saturation, is proportional to the refractive index, and thus an end point for the reaction may be determined rapidly. The correlation of refractive index and iodine value is dependent on the type of oil being utilized.

The hydrogenation reaction can be severely impeded by catalyst poisons that may be present in the oil. The presence of soap or phosphorus in refined, bleached oil will decrease catalyst activity and alter the selectivity ratio during the hydrogenation reaction. Soap levels above 10 ppm and phosphorus levels above 1 ppm will effect hydrogenation to produce higher trans acid and harder products. Sulfur may be present in corn or canola oil, and this poison has a severe effect at low levels. It may be necessary to treat the oil to absorb sulfur and have predictable reactions.

Hydrogen gas purity above 99.9% is necessary and above 99.99% is preferred. Most large hydrogenation facilities have their own hydrogen gas manufacturing plant. The purification of hydrogen gas produced from natural gas is desirable and pressure swing adsorption is frequently used. The presence of carbon monoxide in hydrogen will poison catalyst at low temperature (300–350°F) but is desorbed at higher temperature. The important consideration with inert gas impurities is the dilution effect on hydrogen gas. As hydrogen is consumed by the reaction, inert gases become increasingly concentrated in the reactor. Venting is necessary to remove inert gas with consequent loss of hydrogen.

To reduce overall catalyst costs, reuse of catalyst may be practiced. Theoretically, the catalyst should be usable almost indefinitely as it is unaffected by the reaction itself. In actuality, partial poisoning of the catalyst almost always occurs and reused catalyst will exhibit reduced activity and reduced selectivity ratio.

After a reaction is completed and the batch cooled to a suitable filtration temperature, the nickel catalyst is removed on a precoated filter. As it is essential that all nickel be removed from the oil, it is common practice to employ a second polish filter to ensure complete nickel removal.

Batch reactors normally used for hydrogenation will cope with a variety of different oils hydrogenated to a variety of different products. A typical reactor batch of oil weighs 20,000–60,000 lb and the reactor will contain heating coils, cooling coils, a hydrogen gas sparge ring at the bottom of the reactor, and an agitator with top and bottom blades designed for vortex formation at the oil surface. There will also be provision for vacuum and nitrogen to protect the oil and remove hydrogen when necessary. Automatic control of the reaction is achieved by programming temperature and pressure for a specific product, and this will control the cooling coils and hydrogen addition rate.

Continuous vegetable oil hydrogenation reactors have been constructed and used in situations where one product is made and the necessity for changing conditions to accommodate different product requirements does not exist.

Reactors have been designed in a variety of forms, but the most common in United States is an enclosed vessel with a hydrogen headspace maintained at reaction pressure.

Recirculation systems with external heat exchange are available. Drop tanks to accommodate a batch of oil from the reactor after hydrogenation permit holding a batch before and during filtration. This generally increases capacity by allowing the reactor to be refilled immediately. A wide variety of commercial filters may be used for catalyst removal and polish filtration.

The products of hydrogenation are characterized by melting point, IV, and solid fat index. These three analytical measurements are sufficient in most cases to define the basic physical and chemical properties for a particular application. Several other analytical tests may be appropriate for products required to meet special requirements. For example, a fatty acid composition may be required for a high-stability liquid soybean frying oil to determine the level of the undesirable linolenic acid.

BIBLIOGRAPHY

1. R. R. Allen, *Journal of American Oil Chemists' Society*, **59**, 204, (1982)

General References

D. Swern, ed., *Bailey's Industrial Oil and Fat Products,* 4th ed., John Wiley & Sons, New York, 1982.

D. R. Erickson, E. H. Pryde, O. L. Brekke, T. L. Mounts, and R. A. Falb, eds., *Handbook of Soy Oil Processing and Utilization,* American Soybean Association–American Oil Chemists' Society, 1980.

ALLAN S. HODGSON
Bunge Foods
Bradley, Illinois

HYDROPHOBICITY IN FOOD PROTEIN SYSTEMS

BACKGROUND

Definition

The term lyophobic is used to describe a solute that has little or no affinity for the solvent medium in which it is placed. When the medium is composed of water or an aqueous solution, the more specific term hydrophobic is used as the descriptor. Similarly, lyophilic or "hydrophilic" are used to indicate affinity of a solute for solvent or water.

Earlier literature referred to the formation of hydrophobic bonds between nonpolar solutes or moieties of a macromolecule, but in fact no such bonds are actually formed. The association of nonpolar groups arises from enthalpically favorable interactions (mainly London forces between polarizable groups) in conjunction with the entropically driven tendency to minimize contact of the nonpolar groups with water. This self-association is more correctly referred to as hydrophobic effect, or hydrophobic interaction, rather than hydrophobic bonding. Furthermore, it is difficult to assign an absolute value of hydrophobicity to a solute or functional group. Commonly, hydrophobicity is described by an empirically measured or calculated parameter, which is related to the strength of hydrophobic interaction compared to a standard or with respect to an arbitrary scale.

Importance of Hydrophobic Interactions in Food Protein Systems

Almost all biological systems are composed of molecules that involve at least some nonpolar groups. Because water is a common constituent in these systems, it has been suggested that hydrophobic interactions must be ubiquitous in nature, playing a key role in both equilibrium and kinetic phenomena. Originally, interest in the hydrophobic effect was focused on its contribution to the stabilization of the structure of native protein molecules and on the mechanisms of protein folding. In the older literature, it was presumed that nonpolar or hydrophobic groups should be restricted to the interior of folded protein molecules, where they would be buried ie, not exposed to the solvent water molecules. In fact, crystallographic analysis of the three-dimensional structures of many proteins has indicated that many hydrophobic groups are at least partly exposed on the surface of the protein molecules. The resulting hydrophobic patches play a key role in intermolecular interactions, such as the binding of small ligands or the association with other macromolecules, including protein–protein or protein–lipid interactions in biological membranes and micellar systems. A model has been proposed to correlate biological data with physicochemical parameters describing native protein structure (1). Three types of parameter were proposed: hydrophobic or lipophilic, electronic, and steric parameters. Using this type of approach, biological properties may be explained by a quantitative structure–activity relationship.

In food systems, the role of the hydrophobic effect in stabilizing native protein molecules and structural networks is exemplified by the existence of stable casein micelles in milk. However, processing of foods leads to a variety of changes in both intramolecular and intermolecular interactions. The transformation of milk into cheese by enzymatic treatment and heating is one example of the dramatic effect of processing. Hydrophobic, electronic, and steric effects must all be taken into consideration to explain not only the biological properties of the native protein molecules in foods, such as enzymatic activity, but also the functional properties that give the food its characteristic texture and form. These properties include solubility, viscosity, gellability or coagulability, emulsifying and foaming properties. For example, food products such as meringue and angel food cake rely on the unique properties of egg white proteins, including their excellent foaming properties and ability to form thermostable foams, as well as their capability to yield strong yet elastic gels on heating. The understanding of the basic molecular mechanisms responsible for these properties is essential to control the various aspects of food processing and ingredient selection.

THEORY AND MEASUREMENT OF PROTEIN HYDROPHOBICITY

Theory

The existence of hydrophobic interactions between solutes in an aqueous medium may be explained by two hypotheses (2,3). One approach is to explain the hydrophobic effect as simply arising from the attractive forces between nonpolar moieties and the phobia experienced by those groups when placed in a medium of water molecules. The other approach emphasizes the disruption of the structure of water molecules by the presence of the nonpolar solutes. It is generally well recognized that water molecules exhibit strong attractive interactions with each other through hydrogen bonding in a roughly tetrahedral arrangement. These intermolecular hydrogen bonds produce a continuous yet dynamic network. The second hypothesis, based on entropic forces, proposes that when nonpolar solutes are introduced into water the hydrophobic effect results from disruption of the dynamic water structure, rather than an actual repulsion or phobia of the nonpolar groups by the water molecules. The ordering of water molecules around the nonpolar solutes is reflected in a large negative partial molar entropy change, ie, a decrease in entropy or randomness. To minimize this unfavorable change, nonpolar solutes or moieties associate to reduce their total surface area of contact with the water molecules. However, this hypothesis emphasizes only the disruption of water structure and does not explain the observation that the enthalpy and free energy of transfer of solutes such as phenol from an organic solvent to water differs considerably depending on the solvent, eg, octane versus toluene. Thus the contribution of favorable attractive forces between nonpolar groups to the net changes in enthalpy cannot be neglected. It is probably safe to conclude that the hydrophobic effect culminates from both the tendency to reduce the entropically unfavorable contact between nonpolar groups with water, as well as the need to form enthalpically favorable noncovalent associations, including those interactions broadly classified as van der Waals forces, which are the net effect of attractive London interactions, repulsive electron cloud overlap, and inducible dipole orientation and induction effects.

The attractive forces between nonpolar solutes can be calculated if the solutes are assumed to be approximately spherical bodies (4). Repulsion arising from the interaction of the diffuse double layers surrounding these molecules can also be explained by the DLVO theory (5). The overall attraction or repulsion potential resulting from the net effect of these forces can thus be assessed using these theories. The underlying assumptions, principles, and mathematics for this calculation have been clearly explained in standard textbooks and will not be elaborated here. However, it must be noted that in real food protein systems, much of the theory no longer holds true because ideal dilute solutions of spherical and uniform molecules are rarely encountered. Thus while the theory enables us to understand the phenomenon of attraction or repulsion between solute and solvent molecules, it is not useful for quantitative assessment of the strength of interaction of protein molecules in food.

Empirical Assessment of Hydrophobicity

Hydrophobicity Scales. Proteins found in nature are composed of twenty different amino acids. Numerous scales have been proposed to assess the relative hydrophobicities of these amino acids. Broadly speaking, these scales can be classified into (1) those that are based on the solubility behavior in solvents of different polarity and (2) those that are calculated using crystallographic or other data of location of amino acid residues in the molecular structure, assuming that hydrophobic residues will locate in the interior of the molecule. The former scales include those based on the free energy of transfer of the amino acid residues or their derivatives from water to an organic solvent or vapor, or the partition coefficients measured as a solubility ratio between water and a nonpolar immiscible organic solvent (6–8), as well as those based on relative retention times of amino acids and peptides during reverse-phase liquid chromatography (9). The latter scales include those based on accessible or buried area of the residues (10), or on the location of amino acid residues in proteins assessed either in terms of the distance from the protein center of mass and average orientation of the side chain (11) or in terms of the average surroundings of residue types (12). A hydropathy scale has been reported based on both the average extent of buriedness and the energy of transfer from water to vapor (13). Calculation of hydrophobicity values based on the hydrophobic fragmental constant of constituent fragments has also been proposed (14). Table 1 shows some of the values of hydrophobicity for the twenty amino acids that have been reported in the literature; the different scales have been compared (15,16).

Using the hydrophobicity scales of the constituent amino acids, various approaches have been taken to calculate values for proteins. An equation has been formulated to calculate average hydrophobicity ($H\emptyset_{avg}$) values of proteins based on Tanford's scale for free energies of transfer of amino acid side chains from an organic to an aqueous environment (17). Bigelow's average hydrophobicity values may be calculated using only information on the amino acid composition of the protein. In contrast, most other values require knowledge of the amino acid sequence or primary structure of the proteins as well. Computer programs are now commonly used to calculate histograms or plots showing hydrophobicity profiles or hydropathy profiles (15,18) or amphipathic helix patterns (19).

A significant drawback to the use of hydrophobicity scales of amino acid residues to calculate corresponding

Table 1. Scales of Hydrophobicity for Amino Acid Residues in Proteins

	$\Delta G°_{transfer}$, Kcal/mol		Distribution Coefficient[c], $\log\left[\frac{C_{octanol}}{C_{water}}\right]$	Σf from RP-HPLC[d]	Molar Fraction, Percent Buried[e]	Average Surrounding Hydrophobicity[f], Kcal	Hydrophobicity Index[f], Kcal	Hydropathy Index[g]	Rekker's Fragmental Constant[h]	Average Area Buried[i], Å²
	Organic Solvent→Water[a]	Vapor→Water[b]								
ala	0.5	1.94	−2.74	−0.3	11.2	12.97	0.87	1.8	0.53	86.6
arg	—	−19.92	−4.08	−1.1	0.5	11.72	0.85	−4.5	−0.82	162.2
asn	—	−9.68	—	−0.2	2.9	11.42	0.09	−3.5	−1.05	103.3
asp	—	−10.95	—	−1.4	2.9	10.85	0.66	−3.5	−0.02	97.8
cys/2	—	−1.24	—	—	—	—	—	2.5	0.93	—
cys	—	—	—	6.3	4.1	14.63	1.52	—	1.11	132.3
gln	—	−9.38	—	−0.2	1.6	11.76	0.00	−3.5	−1.09	119.2
glu	—	−10.20	−3.69	0.0	1.8	11.89	0.67	−3.5	−0.07	113.9
gly	—	2.39	−3.11	1.2	11.8	12.43	0.10	−0.4	0.00	62.9
his	0.5	−10.27	−1.95	−1.3	2.0	12.16	0.87	−3.2	−0.23	155.8
ile	2.6	2.15	−1.69	4.3	8.6	15.67	3.15	4.5	1.99	158.0
leu	1.8	2.28	−1.79	6.6	11.7	14.90	2.17	3.8	1.99	164.1
lys	—	−9.52	−3.05	−3.6	0.5	11.36	1.64	−3.9	0.52	115.5
met	1.3	−1.48	−1.87	2.5	1.9	14.39	1.67	1.9	1.08	172.9
phe	2.5	−0.76	−1.43	7.5	5.1	14.00	2.87	2.8	2.24	194.1
pro	—	—	−2.54	2.2	2.7	11.37	2.77	−1.6	1.01	92.9
ser	−0.3	−5.06	−3.07	−0.6	8.0	11.23	0.07	−0.8	−0.56	85.6
thr	0.4	−4.88	−2.94	−2.2	4.9	11.69	0.07	−0.7	−0.26	106.5
trp	3.4	−5.88	−1.11	7.9	2.2	13.93	3.77	−0.9	2.31	224.6
tyr	2.3	−6.11	−2.05	7.1	2.6	13.42	2.67	−1.3	1.70	177.7
val	1.5	1.99	−2.26	5.9	12.9	15.71	1.87	4.2	1.46	141.0

[a] Ref. 6.
[b] Ref. 7.
[c] Ref. 8.
[d] Ref. 9.
[e] Ref. 10.
[f] Ref. 12.
[g] Ref. 13.
[h] Ref. 14.
[i] Ref. 15.

hydrophobicity values or profiles for proteins is the lack of consideration of the effect of three-dimensional structure of each individual protein on the extent of exposure of the residues. The inadequacy of this approach is especially true for calculations using scales developed based on the behavior of amino acids or small peptides. Scales that were formulated on the basis of location or buriedness of residues measured in different proteins attempt to address this problem. However, this approach is still limited in universality of application, due to the need to extrapolate the behavior of residues to proteins other than those for which data are available. In the case of food systems, the problem is even more complex, due to the simultaneous occurrence of many different types of protein in the system. The calculation of a net or average value of hydrophobicity for all the proteins present requires calculation of a hydrophobicity value for each protein as well as knowledge of how each value might change through possible interactions between the proteins. Furthermore, the calculated values fail to take into account the effects of processing on buriedness or surface exposure of residues. For these reasons, various methods of measuring parameters that may relate to the hydrophobicity of complex food proteins are usually favored over calculation of average values or profiles based on the constituent amino acids.

Partition Methods: Partition in Aqueous Two-Phase System. An indication of the relative hydrophobicity of proteins may be given by their partition coefficients measured as the solubility ratio between a polar and an immiscible nonpolar solvent. Due to the virtual insolubility of most proteins in organic solvents, a two-phase aqueous system containing dextran and poly(ethylene glycol) or PEG was devised (20). Esterification of PEG with a fatty acid such as palmitate is used to alter nonpolarity of the system. The difference in partition of proteins in phase systems with and without the fatty acyl hydrocarbon group bound to PEG is taken as a measure of hydrophobic interaction. Length of the hydrocarbon chain may be varied to investigate the effects on the partition behavior of different proteins. The method has been shown to give useful data (20,21), but suffers from the tedious nature of the procedure and difficulties in solubilizing certain proteins.

Partition Methods: Reverse-Phase or Hydrophobic-Interaction Chromatography. The measurement of relative retention times of solutes during chromatography has been used, in analogy with partition coefficients, to indicate the relative solubility between a nonpolar stationary phase and mobile phase of differing polarity. According to the solvophobic effect theory (22), the retention time of peptides depends mainly on their nonpolar and polar surfaces and thus may be a good measure of hydrophobicity. In fact, this hypothesis has proved true for the behavior of small peptides (9). However, the chromatographic behavior of proteins is not as straightforward. Possible denaturation of proteins under the harsh solvent conditions often required in reverse-phase chromatography has led to recommendations to use the milder conditions of hydrophobic-interaction chromatography, but the relevance of these retention data to hydrophobicity of native protein molecules has still been questioned. Nevertheless, differences in chromatographic retention between proteins have been demonstrated, and it has been suggested that differences observed during the binding to aliphatic versus aromatic types of adsorbent (23) may confirm the need to differentiate between these types of interaction in the hydrophobic effect (24–26).

Binding Methods. Various methods have been proposed for quantitating the binding of a nonpolar or hydrophobic ligand to proteins as a measure of the protein hydrophobicity. The ligands used have included aliphatic and aromatic hydrocarbons (27), sodium dodecylsulfate (28), simple triglycerides (29), and corn oil (30). Commonly, the mixture of protein solution and ligand are incubated for a specified time to allow interaction, followed by removal of any unbound or free ligand by dialysis, extraction, microfiltration or other similar techniques. The protein-bound ligand may then quantitated by gas chromatographic or radioactive count analysis, or by a fluorescence probe method. Although the procedures are time-consuming, these binding methods can be useful in determining empirical parameters that may be relevant in assessing interaction of proteins with various nonpolar components in foods, such as flavor compounds, fatty acids or triglycerides.

Contact Angle Measurement. The quantitative determination of the Lifshitz-van der Waals (LW) and electron donor–acceptor or Lewis acid–base (AB) interactions contributing to surface tension has been extended to the case of proteins. The method is based on Young's equation describing the relationship of the advancing contact angle of drops of a liquid on a flat solid surface (31). The solid surface in this case consists of protein sample prepared as flat layers. The advancing contact angles of droplets of three different well-characterized liquids (eg, water, glycerol, and alpha-bromonaphthalene) on the protein surface are measured. Based on Young's equation and the known surface tensions of the three liquids, the contact angles are then used to calculate the contributions of LW and AB interactions to the surface tension of the liquids on the layered protein. This approach has been used to obtain values for native hydrated proteins using one liquid only, namely drops of saline water, which correlated with the hydrophobicity of proteins indicated by relative retention on hydrophobic chromatography (32). However, a much higher surface tension value is obtained by contact angle measurement of a drop of protein solution on a solid polymer surface, compared to the value obtained by measuring the angle for the air-dried protein layer, or by other measurements for surface tension at an air–liquid interface such as platinum ring tensiometry, Wilhelmy plate, pendent drop shape, and weight. This suggests that when protein is exposed at the interface, it may reorient to a much more hydrophobic configuration than its native hydrated state.

Spectroscopic Methods: Intrinsic Fluorescence. The intrinsic fluorescence spectra of proteins is primarily attributed to the aromatic amino acid residues of tryptophan, tyrosine, and phenylalanine. In practice, tryptophan fluorescence is the most commonly studied aspect of the spectrum, because phenylalanine has a low quantum yield and tyrosine fluorescence is frequently weakened due to quenching by ionization or by interactions with amino, carboxyl, or tryptophan residues (33). The fluorescence of both tryptophan and tyrosine depend significantly on their environment, both the magnitude of fluorescence intensity as well as the wavelength of maximum fluorescence emission intensity being sensitive to the polarity of the environment. Three spectral classes of tryptophan residues in proteins have been reported. These are the residues that are completely buried in nonpolar regions of the molecule, those that are completely exposed to the surrounding water, and those that have limited contact with water and are probably immobilized at the protein surface. The wavelengths of maximum emission for these three groups of tryptophan residues are 330–332 nm, 350–353 nm and 340–342 nm, respectively (34). Measurements of intrinsic fluorescence can thus give information on buriedness of aromatic residues and on the effect of interactions with other molecules through the effect on the spectrum, including quenching (35). However, it is often difficult to directly relate this information to hydrophobic interactions as the fluorescence characteristics may also be altered by general changes in conformation.

Spectroscopic Methods: Derivative Spectroscopy. The ultraviolet absorption spectrum of proteins depends on the chromophoric properties of the constituent aromatic amino acids. Although the spectra of the individual chromophores are different, it is difficult to resolve their contributions in the resulting broad absorption spectrum. Derivative spectroscopy, particularly of the second order ($d^2A/d\lambda^2$) and fourth order ($d^4A/d\lambda^4$), has the ability to resolve overlapping bands in the original spectrum. The commercial availability over the last decade of spectrophotometers with built-in derivative mode has spurred on application of derivative spectroscopy for analysis of protein absorption spectra. It has been used to quantitatively determine contents of phenylalanine, tyrosine, and tryptophan residues and to probe for changes in polarity of the microenvironment around the chromophores and thus evaluate extent of exposure or buriedness of these aromatic amino acid residues (36,37). Changes in the state of aromatic amino acid residues in soybean globulin proteins as a function of heating or sulfhydryl modification were monitored by using difference-second derivative absorption as well as spectrofluorimetric spectra of native and heated or modified proteins (38).

Spectroscopic Methods: Fluorescence Probes. Compounds whose quantum yields of fluorescence and wavelength of maximum emission depend on the polarity of their environment have been used to probe the hydrophobic or nonpolar nature of proteins. The most popular types used include the anionic probes of the aromatic sulfonic acid class, such as the amphiphilic 1-anilinonaphthalene-8-sulfonate (ANS), or its dimeric form bis-ANS. These probes have high quantum yields of fluorescence in organic solvents but not in water; they thus fluoresce when bound to membranes or relatively hydrophobic cavities in many proteins (39,40). However, the direct interpretation of ANS fluorescence with hydrophobicity should be made with great caution, as enhancement of fluorescence and blue shift of the emission maximum have also been observed even in strong aqueous $MgCl_2$ solutions (41). It was suggested that solvents or environments that are not necessarily nonpolar but that favor the rigid, planar configuration of the ANS molecule may influence fluorescence (42).

Another group of anionic fluorescence probes is of the fatty acid analogue type, including cis-parinaric acid (CPA), which has been used as a probe for proteins and biological membranes in particular (43). The parinaric acids are among the few nonaromatic fluorophores known; their similarity to natural fatty acids, nonfluorescence in water and good Stokes shift characteristics are among the advantages of their use to probe for hydrophobic regions that may be important in protein–lipid interactions in food systems. Good correlations were obtained between the relative hydrophobicity values of proteins determined by CPA fluorescence, and properties related to protein–lipid interactions such as interfacial tension and emulsifying activity (44).

Limitations in using anionic probes such as ANS and CPA to determine protein hydrophobicity include the possibility that electrostatic as well as hydrophobic interactions may contribute to the binding of the probes. The use of neutral or uncharged probes may circumvent this problem. Diphenylhexatriene (DPH), which clearly labels the interior of membranes, is the most commonly used neutral lipophilic fluorescent probe. The primary application of DPH has been to estimate membrane fluidity and the effects of various drugs or treatments on the membrane structure. Fluorescence of DPH outside of membranes or nonpolar solvents is low. Unfortunately, the nonpolar nature of DPH restricts its solubility in aqueous systems and thus limits its use as a probe for protein solutions. A novel approach has been proposed to use DPH by first dissolving it in a corn oil–heptane mixture; after removal of the heptane by evaporation, the DPH in corn oil is allowed to interact with aqueous protein solution. The resulting fluorescence may be measured as an indication of binding of DPH-in-oil by the protein molecules and used as an index of relative hydrophobicity (30).

Other Methods. Most of the spectroscopic methods outlined above provide information on the contribution to hydrophobic interactions from aromatic amino acids; there is a great dearth of methods for aliphatic amino acids. A number of other spectroscopic methods provide useful information on protein structure and environment of constituent amino acid residues in general. These include circular dichroism (cd), optical rotatory dispersion (ord), nuclear magnetic resonance (nmr), infrared spectroscopy (ir), and laser Raman spectroscopy (lrs). Each of these techniques has its particular advantages and limitations; more specialized volumes on the theory and applications of the individual techniques are available. Unfortunately, with the present state of knowledge, there are no general empirical rules to explain observed spectra in terms of the specific secondary or tertiary structure, or electrostatic or hydrophobic interactions. The interpretation of spectra must be made individually for each protein,

Table 2. Relative Hydrophobicity Values of Some Proteins Measured by Different Methods.[a,b]

Proteins	$H\emptyset_{ave}$	Fluorescence Probes			DPH Probe, TG Binding	Chromatography		Binding			Partition $\Delta \log K$
		S_OANS	S_OCPA	S_eCPA		Rpc	Hic	Heptane	SDS	TG	
Albumin, bovine	1,120	1,000	100	100	100	100	100	100	100	100	100
Albumin, chicken	1,110	7	2	96	15	0	9–86	2	14	3	1–25
Casein, α-	1,200	57[c]	6–30	—	150	400	—	—	—	160	—
Casein, β-	1,320	60[c]	50	107	—	315	—	—	—	—	—
Casein, κ-	1,210	83	13–93	89	—	340	—	—	30	—	—
Chymotrypsin, α-	1,030	0	3	—	41	—	57–139	9	—	3	9–33
Globulin, soy 7S	1,090	47	9–77	—	50	165	—	—	21	—	—
Globulin, soy 11S	950	27	2–17	—	25	150	—	—	—	—	—
Lactalbumin, α-	1,150	33	9–54	—	98	225	10	—	—	—	—
Lactoglobulin, β-	1,230	13	54–146	80	49	0	67–102	100	62	130	41–78
Lysozyme	970	0.7	7	—	—	—	42–113	0	19	1	0
Ovomucoid, chicken	920	—	0	22	—	—	—	24	18	108	—
Pepsin	1,063	0.7	2	57	32	—	92	—	—	95	—
Ribonuclease A	870	—	1	24	—	—	4–71	17	18	2	—
Transferrin, chicken	1,080	—	4	108	—	—	16–31	29	20	118	9–20
Trypsin	940	8	3	34	—	—	—	—	13	—	9–27
Trypsin inhibitor	1,040	0	0	—	1	0	—	—	—	—	—

[a] Data were compiled from references 16, 17, 45, references cited therein, and unpublished data. The majority of the data are expressed relative to 100 for bovine albumin to facilitate comparison of values. Where varying values were reported by investigators using essentially the same method, a range of values is presented.

[b] Abbreviations: $H\emptyset_{ave}$ = average hydrophobicity calculated by Bigelow's method (17); S_O = initial slope of relative fluorescence intensity versus protein concentration plot, using native proteins; S_e = S_O measured for protein solutions after treatment with 1.5% SDS at 100°C for 10 min; ANS = 1-anilino-8-naphthalene sulfonic acid; CPA = cis-parinaric acid; DPH = 1, 6-diphenyl-1,3,5-hexatriene; TG = triglyceride; Rpc = reverse-phase chromatography; Hic = hydrophobic-interaction chromatography; SDS = sodium dodecyl sulfate.

[c] At the isoelectric pH, 530 and 800 for α- and β-casein, respectively.

based usually on data obtained from several methods coupled with known sequence and whenever possible the crystallographic data of the protein.

HYDROPHOBICITY OF FOOD PROTEINS

Table 2 is a compilation of relative hydrophobicity values of some proteins, either calculated from their amino acid composition or measured by some of the empirical methods outlined above. Unfortunately at present there is no consensus on a standard, absolute method for hydrophobicity measurement. In comparing these values, differences between average, total, and surface-exposed hydrophobicity should be kept in mind. In addition, variability in published data may arise from different sources and purity of protein, and different conditions during measurement (eg, pH, ionic strength, buffer salts). Until a more theoretically acceptable, universal method is established, simple methods that give a measure of the hydrophobicity that is actually available for participation in functionality may be the methods of choice. An example is given in Table 3, which lists typical values obtained by measurement with aliphatic and aromatic probes for some food proteins, including complex systems composed of several molecular species.

IMPORTANCE OF HYDROPHOBICITY IN FUNCTIONALITY OF FOOD PROTEINS

Functional properties of food proteins have been defined as those characteristics, other than nutritional ones, that affect their utilization. This includes a diverse range of functionality, such as solubility, viscosity, water- and fat-binding properties, emulsification, foaming, film formation, gellability or coagulability, elasticity, and flavor. Functionality depends on the interaction of the protein molecules with other components in the food system, in-

Table 3. Hydrophobicity Values of Some Food Proteins Measured by Fluorescence Probe Method[a]

Protein[b]	ANSS$_O$	CPAS$_O$
Albumin, bovine serum	1,600	2,750
Canola isolate	110	370
Casein, bovine	150	400
Egg albumen	40	135
Gelatin	0	0
Lactalbumen	145	1,700
Lactalbumin, α-	90	280
Lactoglobulin, β-	40	7,000
Lysozyme, hen	1	15
Muscle, chicken breast	75	350
Myosin, chicken breast	100	—
Ovalbumin, hen	10	40
Pea protein isolate	280	825
Soy protein isolate	250	1,000
Sunflower protein isolate	155	910
Whey protein concentrate	70	3,600
Zein	410	390

[a] See Table 2 for abbreviations.

[b] Measurements for S_O determination were carried out for protein solutions in 0.01 M sodium phosphate buffer, pH 7.0–7.4; 0.6 M NaCl and 0.3 M NaCl were included in the buffer for salt extracts of chicken breast muscle and for isolated chicken myosin, respectively.

cluding water; macromolecules such as other proteins, complex carbohydrates, and lipids; and small molecules such as salts, simple sugars, and flavor compounds. These interactions may take place in the bulk phase, or at a surface or interface. The interplay of electrostatic, hydrophobic and steric parameters, important in predicting biological activity of molecules (1), also holds true for the elucidation of the relationship between structure and functionality of food proteins. Quantitative analysis of the structure–activity relationship (QSAR) of small molecules has been progressing rapidly with the aid of computer technology (46), but similar advances in QSAR for functionality of food protein analyses have not been evident (47). Although the qualitative importance of the physicochemical properties on functionality is well recognized, quantitative prediction based on QSAR analysis of food proteins is still not a facile task, being limited by the need to measure relevant parameters, including hydrophobic interactions, which can take into account the effects of processing for incorporation into structure–functionality models. The following highlight current concepts of the role of hydrophobic interactions in explaining some of the important functional properties of food proteins.

Solubility

Solubility has long been considered a critical property because of its effect on many other functional properties, eg, emulsifying, foaming, and gelation. It was proposed that charge frequency and hydrophobicity are the two major factors affecting protein solubility (17). Generally speaking, proteins are soluble in water when electrostatic or hydration repulsion between protein molecules is greater than the driving forces for hydrophobic interactions. The repulsive forces are at a minimum near the isoelectric pH of the protein. However, proteins having relatively few exposed hydrophobic regions on the molecular surface may remain soluble even at pH near their isoelectric point. For example, undenatured proteins in whey remain soluble during acidification of milk to precipitate the casein fraction. On the other hand, proteins such as casein with large areas of hydrophobic amino acids on their surface tend to be insoluble near their isolectric point.

It has been suggested that aromatic amino acids play a greater role than aliphatic amino acids in their contribution to hydrophobic interactions affecting insolubilization (24). At pH values where charge effects are minimized (zero zeta potential), the insolubility of some food proteins could be explained by hydrophobicity parameters measured using the aromatic fluorescence probe ANS or by reverse phase chromatographic behavior on phenyl-Sepharose (PSC), but not by hydrophobicity measured using the aliphatic probe CPA. At other pH values, charge frequency measured as zeta potential (ZP) as well as hydrophobicity measured by ANS or PSC are significant parameters to explain the insolubility of proteins. The following models were obtained by multiple regression analyses of the data for some milk and soy proteins:

Percent insoluble protein =
$4.44 + 3.43\text{PSC} - 0.0416\text{ZP}^2 - 0.0573\text{PSCxZP}$
$(n = 189; R^2 = 0.612; P < 0.001)$.

Percent insoluble protein $= 13.4 +$
$0.781\text{ANS} - 0.0014\text{ANS}^2 - 0.042\text{ZP}^2 - 0.010\text{ANSxZP}$
$(n = 183; R^2 = 0.480; P < 0.001)$.

Although calculations of hydrophobicity scale values based on free energy of transfer from organic solvent to water indicate that aromatic amino acids are more hydrophobic than aliphatic amino acids, statistical scales based on frequency of location of amino acid residues at the surface versus interior of protein molecules indicate that aromatic amino acids are often considered less hydrophobic than aliphatic ones. It has been suggested that while inherently more hydrophobic, the bulky nature of the aromatic residues may discourage their effective burial in the protein interior (25). These surface-exposed residues determine the strength of hydrophobic interactions that affect properties such as solubility.

The effects of various compounds on protein solubility may be explained on the basis of their opposite effects on electrostatic and hydrophobic interactions of the proteins. Various salts were ranked in terms of a lyotropic series describing their water structure-making or -breaking effects based on molal surface tension values (22). At high ionic strength, ammonium sulfate and sodium chloride increase the surface tension of water, whereas tetraethylammonium chloride and guanidinium chloride reduce it. The salting-in or salting-out of proteins by different salts can thus be explained theoretically by the lyotropic series. It was also shown that the salting-out constant of a protein is a function of its surface hydrophobicity, calculated as the hydrophobic contribution of each amino acid (6) and the fraction of exposed hydrophobic residues (10,48). Similarly, the stabilizing effects of sugars and polyols such as glycerol may be related to their water structure-enhancing effects, intensifying intramolecular hydrophobic interactions that stabilize protein structure. The effect of urea in solubilizing proteins has been explained by solvation of the hydrophobic moieties in urea, as well as by water structure-breaking effects. The precipitating action of trichloracetic or sulfosalicylic acid has also been postulated to be due to an increase of accessible hydrophobicity of peptide moieties (49).

Emulsifying and Foaming Properties

Emulsion and foam-related properties depend on interactions of protein molecules at the oil–water interface and air–water interface, respectively. Although protein solubility is a key determinant in these functionalities, additional important considerations are molecular flexibility and hydrophobic interactions of the protein at the surface or interface, which result in formation and stabilization of the emulsion or foam. Decrease in interfacial tension and improvement in emulsifying activity were not related to hydrophobicity values calculated from the total content of hydrophobic amino acids, but were correlated with effective or surface hydrophobicity of proteins measured by hydrophobic partition, hydrophobic interaction chromatography, or fluorescence probes (21,44). The amphiphilic nature of proteins is vital, in analogy to the hydrophile–lipophile balance (HLB) concept originally developed for nonprotein, synthetic emulsifiers. A balance in accessible hydrophobic and hydrophilic areas is required for optimal functionality. Solubility of the protein molecules facilitates diffusion to the surface or interface; however, once there, the ability of the protein to interact with the oil or with other protein molecules to form an interfacial layer depends on the flexibility and accessibility of surface groups, especially through hydrophobic interactions. The importance of hydrophobic interactions has been demonstrated by the observation that functionality can often be improved by mild denaturing treatments that increase surface hydrophobicity of proteins without impairing their solubility (50). When solubility is a constant parameter, emulsifying properties are improved by increasing surface hydrophobicity, up to some critical point. However, in analogy to the HLB concept, excessive hydrophobicity is detrimental to functionality, and emulsifying properties are poor when proteins have too high values of hydrophobicity (51,52). The extent of unfolding to expose areas for hydrophobic interactions is thought to be greater for foaming than emulsifying properties, which may be related to the higher tension at an air–water surface than an oil–water interface. Hydrophobicity parameters measured by fluorescence probe assay after denaturation of protein molecules by heating in sodium dodecyl sulfate solution were found to be related to the total or average hydrophobicity values calculated according to Bigelow; these exposed hydrophobicity values were correlated with foaming capacity (45).

Thermal Functional Properties

The importance of intramolecular hydrophobic interactions in the stability of proteins to thermally induced denaturation has been assumed, due to the increasing strength of these interactions with increasing temperature up to 60–70°C. However, extensive comparison of hydrophobic indexes in thermophilic versus mesophilic proteins has not demonstrated any definitive relationship between hydrophobicity and thermal stability. Stabilization may be thought of as an increase in the internal and a decrease of external or surface hydrophobicity. Because aliphatic residues have a greater tendency to be located in the protein interior than bulky aromatic side chains, it has been suggested that an aliphatic index may be more relevant than a general hydrophobicity index (25).

Thermally induced gelation or coagulation are important properties affecting textural characteristics of foods. Coagulation-type (concentration dependent) and gelation-type (concentration-independent) proteins have been differentiated depending on molecular weight and relative content of hydrophobic amino acid residues (53). Proteins with a high molar percent of hydrophobic residues tend to be of the coagulation type. Hydrophobic interactions may be involved in the initial intramolecular stage of unfolding as well as the later stage of network or aggregate formation through intermolecular interactions. Both surface and exposed hydrophobicity may be important in the nature of the resulting product. It has been suggested that a low value of surface hydrophobicity of the native protein, coupled with a high value of exposed hydrophobicity after thermal denaturation, may lead to the formation of strong gels (54). The ability to unfold, or molecular flexibility, may be hindered by intramolecular stabilization through noncovalent interactions as well as covalent bonds, particularly of the disulfide type. The involvement of intermolecular sulfhydryl–disulfide reactions usually appears in later stages of these thermally induced phenomena and may be concentration dependent (55). At temperatures below 70°C, sulfhydryl reactivity is not typically involved and hydrophobic interactions may be the major driving force in network formation. A striking example of this is the setting phenomonon observed in particular fish species, wherein an elastic gel forms at mild heating temperatures, eg, 40°C. The importance of hydrophobic interactions in this phenomenon was demonstrated (56). Introducing aromatic and aliphatic hydrophobic groups through chemical modification induced nonsetting species to behave like easily setting species, exhibiting characteristics of increased viscosity and gelling on low temperature heating.

Flavor

While most proteins do not have a strong intrinsic flavor, they do influence flavor due to their ability to bind flavor compounds. The nonpolar nature of many of the flavor components suggests that hydrophobic interactions may be critical in their binding to proteins. The literature in this area of research is scarce, but preliminary investigations indicate that indeed, protein binding of flavor components such as aldehydes and ketones appears to be hydrophobic in nature, the strength of binding increasing with chain length of the ligand and being affected by neutral salts according to the Hofmeister or lyotropic series (57).

Hydrolysis of proteins to form peptides also has a great influence on flavor. Formation of bitter peptides has been observed, especially from casein and soy proteins. Bitterness appears to be related to the average hydrophobicity (HQ) of the peptides, with those peptides having HQ values above 1,400 cal/mol residue being bitter (58). However, the extent of exposure of the hydrophobic residues and their incorporation into peptides rather than as the free amino acids also affects bitterness. Thus bitterness depends not only on the content of hydrophobic amino acids in the original protein, but also on the degree of hydrolysis (59).

FUTURE TRENDS

Long-term, systematic research is required to establish quantitative, empirical rules to describe the structure–function relationship. This basic research is required to explain at the molecular basis well-known but little-understood functional properties such as the thermostability of egg white foams, elasticity of egg white gels, coagulation of casein to form cheese, binding and texture formation of muscle proteins, elasticity and extensibility of wheat gluten dough, etc. Although only one of the physicochemical parameters involved in elucidation of structure–function relationships, hydrophobic interactions often play a key role. Improvement of functionality including whipping and emulsifying properties by modifying the hydrophobic–hydrophilic balance of proteins is already emerging, as exemplified by the preparation of proteinaceous surfactants with different HLB values through attachment of hydrophobic amino acid alkyl esters to various food proteins (60). Development of novel functionality such as antifreeze or cryoprotectant properties has also been described by this approach. Another rapidly developing area is the enhancement of hydrophobic nature of enzymes through attachment of amphiphilic groups such as poly(ethylene glycol), coupled with selection of organic solvent media to alter the hydrophobic environment. This area has interesting applications such as the possible resolution of alcohols and acids by lipase-catalyzed esterification, protease-catalyzed synthesis of proteins, and lipase-catalyzed interesterification of fats and oils (61). By quantitatively establishing the relationship between food protein structure and function, systematic and predictable enzyme and protein engineering for tailoring of specific biological and functional properties should become a reality.

BIBLIOGRAPHY

1. C. Hansch, A. Leo, S. H. Unger, and co-workers, "Aromatic Substituent Constants for Structure-Activity Correlations," *Journal of Medicinal Chemistry* **16,** 1207 (1973); C. Hansch and A. J. Leo, *Substitute Constants for Correlation Analysis in Chemistry and Biology,* John Wiley & Sons, Inc., New York, 1979.
2. C. Tanford, *The Hydrophobic Effect: Formation of Micelles and Biological Membranes,* 2nd ed., John Wiley & Sons, Inc., New York, 1980.
3. C. J. van Oss, R. J. Good, and M. K. Chaudbury, "The Role of van der Waals Forces and Hydrogen Bonds in 'Hydrophobic Interactions' between Biopolymers and Low Energy Surfaces," *Journal of Colloid Interface Science* **111,** 378 (1976).
4. H. C. Hamaker, "The London-van der Waals Attraction between Spherical Particles," *Physica* **4,** 1058 (1937).
5. E. Dickinson and G. Stainsby, eds., *Colloids in Food,* Elsevier Applied Science Publishers, Ltd., Barking, UK, 1982, pp. 33–66.
6. Y. Nozaki and C. Tanford, "The Solubility of Amino Acids and Two Glycine Peptides in Aqueous Ethanol and Dioxane Solutions. Establishment of a Hydrophobicity Scale," *Journal of Biological Chemistry* **246,** 2211 (1971).
7. R. Wolfenden, L. Andersson, P. M. Cullis, and C. C. B. Southgate," Affinities of Amino Acid Side Chains for Solvent Water," *Biochemistry* **20,** 849 (1981).
8. L. M. Yunger and R. D. Cramer III, "Measurement and Correlation of Partition Coefficients of Polar Amino Acids," *Molecular Pharmacology* **20,** 602 (1981).
9. K. J. Wilson, A. Honneger, R. P. Stotzel, and G. J. Hughes, "The Behavior of Peptides on Reverse-Phase Supports During High Pressure Liquid Chromatography," *Biochemical Journal* **199,** 31 (1981).
10. C. Chothia, "Principles That Determine the Structure of Proteins," *Annual Review of Biochemistry* **54,** 537 (1984).
11. H. Meirovitch, S. Rackovsky, and H. A. Scheraga, "Empirical Studies of Hydrophobicity. I. Effect of Protein Size on the Hydrophobic Behavior of Amino Acids," *Macromolecules* **13,** 1398 (1980).
12. P. Manavalan and P. K. Ponnuswamy, "Hydrophobic Character of Amino Acid Residues in Globular Proteins," *Nature* **275,** 673 (1978).
13. J. Kyte and R. F. Doolittle, "A Simple Method for Displaying the Hydropathic Character of a Protein," *Journal of Molecular Biology* **157,** 105 (1982).
14. R. F. Rekker, *The Hydrophobic Fragmental Constant,* Elsevier Science Publishing Co., Inc., New York, 1977.
15. G. D. Rose, L. M. Gierasch, and J. A. Smith, "Turns in Peptides and Proteins," *Advances in Protein Chemistry* **37,** 1 (1985).
16. S. Nakai and E. Li-Chan, *Hydrophobic Interactions in Food Systems,* CRC Press, Inc., Boca Raton, Fla., 1988.
17. C. C. Bigelow, "On the Average Hydrophobicity of Proteins and the Relation between It and Protein Structure, *Journal of Theoretical Biology* **16,** 187 (1967).
18. S. R. Krystek, Jr., and T. T. Anderson, "A Program for Hydropathy and Antigenicity Analysis of Protein Sequences," *Endocrinology* **117,** 1118 (1985).
19. J. P. Segrest and R. J. Feldman, "Amphipathic Helixes and Plasma Lipoproteins: A Computer Study," *Biopolymers* **16,** 2053 (1977).
20. V. P. Shanbhag and C. -G. Axelsson, "Hydrophobic Interaction Determined by Partition in Aqueous Two-Phase Systems. Partition of Proteins in Systems Containing Fatty-Acid Esters of Poly(ethylene glycol)," *European Journal of Biochemistry* **60,** 17 (1975).
21. E. Keshavarz and S. Nakai, "The Relationship between Hydrophobicity and Interfacial Tension of Proteins," *Biochimica et Biophysica Acta* **576,** 269 (1979).
22. W. Melander and C. Horvath, "Salt Effects on Hydrophobic Interactions in Precipitation and Chromatography of Proteins: An Interpretation of the Lyotropic Series," *Archives of Biochemistry and Biophysics* **183,** 200 (1977).
23. B. H. Hofstee and N. F. Otillio, "Modifying Factors in Hydrophobic Protein Binding by Substituted Agaroses," *Journal of Chromatography* **161,** 153 (1978).
24. S. Hayakawa and S. Nakai, "Relationships of Hydrophobicity and Net Charge to the Solubility of Milk and Soy Proteins," *Journal of Food Science* **50,** 486 (1985).
25. B. B. Mozhaev and K. Martinek, "Structure-Stability Relationships in Proteins: New Approaches to Stabilizing Enzymes," *Enzyme Microb. Technol.* **6,** 50 (1984).
26. S. K. Burley and G. A. Petsko, "Aromatic-Aromatic Interaction: A Mechanism of Protein Structure Stabilization, *Science* **229,** 23 (1985); "Weakly Polar Interactions in Proteins," *Advances in Protein Chemistry* **39,** 125 (1988).
27. A. Mohammadzadeh-K., R. E. Feeney, and L. M. Smith, "Hydrophobic Binding of Hydrocarbons by Proteins. I. Relationship of Hydrocarbon Structure," *Biochimica et Biophysica Acta* **194,** 246 (1969); "Hydrophobic Binding of Hydrocarbons by Proteins. II. Relationship of Protein Structure," *ibid.,* p. 256.
28. A. Kato, T. Matsuda, N. Matsudomi, and K. Kobayashi, "Determination of Protein Hydrophobicity Using a Sodium Dodecyl Sulfate Binding Method," *Journal of Agricultural and Food Chemistry* **32,** 284 (1984).
29. L. M. Smith, P. Fantozzi, and R. K. Creveling, "Study of Triglyceride-Protein Interaction Using a Microemulsion-Filtration Method," *Journal of American Oil Chemists' Society* **60,** 960 (1983).

30. T. Tsutsui, E. Li-Chan, and S. Nakai, "A Simple Fluorometric Method for Fat-Binding Capacity as an Index of Hydrophobicity of Proteins, *Journal of Food Science* 51, 1268 (1986).
31. C. J. van Oss, "Energetics of Cell-Cell and Cell-Biopolymer Interactions," *Cell Biophysics* 14, 1 (1989).
32. C. J. van Oss, D. R. Absolom, A. W. Neumann, and W. Zingg, "Determination of the Surface Tension of Proteins. I. Surface Tension of Native Serum Proteins in Aqueous Media," *Biochimica et Biophysica Acta* 670, 64 (1981); "Determination of the Surface Tension of Proteins. II. Surface Tension of Serum Albumin, Altered at the Protein-Air Interface," *ibid.*, p. 74.
33. D. Freifelder, *Physical Biochemistry. Application to Biochemistry and Molecular Biology*, W. H. Freeman & Co., San Francisco, 1976, pp. 410–421.
34. E. A. Burstein, N. S. Vedenkina, and M. N. Ivkova, "Fluorescence and the Location of Tryptophan Residues in Protein Molecules," *Photochemistry and Photobiology* 18, 263 (1973).
35. M. R. Eftink, J. L. Zajicek, and C. A. Ghiron, "A Hydrophobic Quencher of Protein Fluorescence: 2,2,2-Trichloroethanol," *Biochimica et Biophysica Acta* 491, 473 (1977).
36. T. Ichikawa and H. Terada, "Estimation of State and Amount of Phenylalanine Residues in Proteins by Second Derivative Spectrophotometry," *Biochimica et Biophysica Acta* 580, 120 (1979); "Determination of Phenylalanine, Tryptophan and Tyrosine in a Mixture of Amino Acids by Second Derivative Spectrophotometry," *Chemical and Pharmaceutical Bulletin* 29, 438 (1981).
37. E. Padros, M. Dunach, and co-workers, "Fourth-Derivative Spectrophotometry of Proteins," *TIBS* 508 (Dec. 1984).
38. Y. Yamagishi, F. Ebina, and F. Yamauchi, "Analysis of the State of Aromatic Amino Acid Residues in Heated Soybean 7S Globulin by Absorption Derivative Spectrophotometry and Spectrofluorimetry," *Agricultural and Biological Chemistry*, 46, 2441 (1982); T. Yamagishi, F. Yamauchi, and K. Shibasaki, "State of Aromatic Amino Acid Residues in Soybean 11S Globulin Heated in the Presence of N-Ethylmaleimide by Derivative Spectrophotometry," *Agricultural and Biological Chemistry* 45, 459 (1981).
39. L. Stryer, "The Interaction of a Naphthalene Dye with Apomyoglobin and Apohemoglobin. A Fluorescent Probe of Nonpolar Binding Sites," *Journal of Molecular Biology* 13, 482 (1965).
40. G. Weber and L. B. Young, "Fragmentation of Bovine Serum Albumin by Pepsin. 1. The Origin of the Acid Expansion of the Albumin Molecules," *Journal of Biological Chemistry* 239, 1415 (1964).
41. G. Penzer, "1-Anilinonaphthalene-8-Sulfonate. The Dependence of Emission Spectra on Molecular Formation Studied by Fluorescence and Proton Magnetic Resonance," *European Journal of Biochemistry* 25, 218 (1972).
42. S. Ainsworth and M. T. Flanagan, "The Effects That the Environment Exerts on the Spectroscopic Properties of Certain Dyes That Are Bound by Bovine Serum Albumin," *Biochimica et Biophysica Acta* 194, 213 (1969).
43. L. A. Sklar, B. S. Hudson, and R. D. Simoni, "Conjugated Polyene Fatty Acids as Membrane Probes: Preliminary Characterization, *Proceedings of the National Academy of Sciences of the United States of America* 72, 1649 (1975); "Model System Studies," *Journal of Supramolecular Structure* 4, 449 (1976); "Binding to Bovine Serum Albumin," *Biochemistry* 16, 5100 (1977).
44. A. Kato and S. Nakai, "Hydrophobicity Determined by a Fluorescence Probe Method and Its Correlation with Surface Properties of Proteins," *Biochimica et Biophysica Acta* 624, 13 (1980).
45. A. -A. Townsend and S. Nakai, "Relationships between Hydrophobicity and Foaming Characteristics of Food Proteins," *Journal of Food Science* 48, 588 (1983).
46. A. J. Stuper, W. E. Brugger, and P. C. Jurs, *Computer Assisted Studies of Chemical Structure and Biological Function*, John Wiley & Sons, Inc., New York, 1977.
47. S. Nakai, "Structure-Function Relationships of Food Proteins with an emphasis on the Importance of Protein Hydrophobicity," *Journal of Agricultural and Food Chemistry* 31, 676 (1983).
48. A. Salahuddin, A. Waseem, M. Yahiya Khan, and co-workers, "A Possible Relation Between the Salting-Out Behavior of Proteins and Their Surface Hydrophobicity," *Indian Journal of Biochemistry and Biophysics*. 20, 127 (1983).
49. M. Yvon, C. Chabanet, and J. -P. Pelissier, "Solubility of Peptides in Trichloroacetic Acid (TCA) Solutions. Hypothesis on the Precipitation Mechanism," *International Journal of Peptide and Protein Research* 34, 166 (1989).
50. A. Kato, N. Tsutsui, N. Matsudomi, and co-workers, "Effects of Partial Denaturation on Surface Properties of Ovalbumin and Lysozyme," *Agricultural and Biological Chemistry* 45, 2755 (1981).
51. L. P. Voutsinas, E. Cheung, and S. Nakai, "Relationships of Hydrophobicity to Emulsifying Properties of Heat Denatured Proteins," *Journal of Food Science* 48, 26 (1983).
52. E. Li-Chan, S. Nakai, and D. F. Wood, "Hydrophobicity and Solubility of Meat Proteins and Their Relationship to Emulsifying Properties," *Journal of Food Science* 49, 345 (1984).
53. K. Shimada and S. Matsushita, "Effects of Salts and Denaturants on Thermocoagulation of Proteins," *Journal of Agricultural and Food Chemistry* 29, 15 (1981); "Relationship between Thermocoagulation of Proteins and Amino Acid Compositions," *Journal of Agricultural and Food Chemistry* 28, 413 (1980).
54. E. Li-Chan, S. Nakai, and D. F. Wood, "Muscle Protein Structure-Function Relationships and Discrimination of Functionality by Multivariate Analysis," *Journal of Food Science* 52, 37 (1987).
55. S. Hayakawa and S. Nakai, "Contribution of Hydrophobicity, Net Charge and Sulfhydryl Groups to Thermal Properties of Ovalbumin," *Can. Inst. Food Sci. Technol. J.* 18, 290 (1985).
56. E. Niwa, "Role of Hydrophobic Bonding in Gelation of Fish Flesh Paste," *Bull. Jpn. Soc. Sci. Fish.* 41, 907 (1975); E. Niwa and co-workers, "Arylsulfonyl Chloride Induced Setting of Dolphinfish Flesh Sol," 47, 179 (1981); "Setting of Flesh Sol Induced by Ethylsulfonation," 47, 915 (1981).
57. S. Damodaran and J. E. Kinsella, "Interaction of Carbonyls with Soy Protein: Thermodynamic Effects," *Journal of Agricultural and Food Chemistry* 29, 1249 (1981); "Flavor Protein Interactions," *Journal of Agricultural and Food Chemistry* 28, 567 (1980); "Stabilization of Proteins By Solvents," *Journal of Biological Chemistry* 255, 8503 (1980); "The Effects of Neutral Salts on the Stability of Macromolecules," *Journal of Biological Chemistry* 256, 3394 (1981).
58. K. H. Ney, "Aminosaurezusammensetzung von Proteinen und die Bitterkeit ihrer Peptide," *Zeitschrift fuer Lebensmittel-Untersuchung und-Forschung* 149, 321 (1972); *Zeitschrift fuer Lebensmittel-Untersuchung und-Forshung* 147, 64 (1971).
59. J. Adler-Nissen and H. S. Olsen, "The Influence of Peptide Chain Length on Taste and Functional Properties of Enzymatically Modified Soy Protein," in A. Pour-El, ed., *Function-*

ality and Protein Structure, ACS Symposium Series **92**, American Chemical Society, Washington D.C., 1979, Chapt. 7.

60. S. Arai, M. Watanabe, and N. Hirao, "Modification to Change Physical and Functional Properties of Food Proteins," in R. E. Feeney and J. R. Whitaker, eds., *Protein Tailoring for Food and Medical Uses*, Marcel Dekker Inc., New York, 1986, Chapt. 3.

61. H. W. Blanch and A. M. Klibanov, eds., *Enzyme Engineering 9*, Vol. 542, Annals of the New York Academy of Sciences, 1988.

General References

References 2, 5, 14 and 16 are good general references.

J. D. Andrade, ed., *Surface and Interfacial Aspects of Biomedical Polymers, Vol. 1, Surface Chemistry and Physics; Vol. 2, Protein Adsorption,* Plenum Press, New York, 1985.

A. Ben-Naim, *Hydrophobic Interactions,* Plenum Press, New York, 1980.

A. Ben-Naim, "Hydrophobic Interactions in Biological Systems," in A. S. V. Burgen and G. C. K. Roberts, eds., *Topics in Molecular Pharmacology*, Vol. 2, Elsevier Science Publishing Co., New York, 1983, pp. 1–52.

C. C. Bigelow and M. Channon, "Hydrophobicities of Amino Acids and Proteins," in G. D. Fasman, ed., *Handbook of Biochemistry and Molecular Biology,* CRC Press, Vol. 1, 3rd ed., Boca Raton, Fla., 1976, pp. 209–243.

K. S. Birdi, *Lipid and Biopolymer Monolayers at Liquid Interfaces,* Plenum Press, New York, 1989.

J. P. Cherry, ed., *Protein Functionality in Foods,* American Chemical Society, Washington, D.C., 1981.

J. E. Kinsella, "Relationships between Structure and Functional Properties of Food Proteins," in P. F. Fox and J. J. Condon, eds., *Food Proteins,* Elsevier Applied Science Publishers, Ltd., Barking, UK, 1982, Chapt. 3.

C. A. Miller and P. Neogi, *Interfacial Phenomena, Equilibrium and Dynamic effects,* Marcel Dekker, Inc., New York, 1985.

J. R. Mitchell and D. A. Ledward, *Functional Properties of Food Macromolecules,* Elsevier Applied Science Publishers, Ltd., Barking, UK, 1986.

A. Pour-El, ed., *Functionality and Protein Structure, ACS Symposium Series* **92**, American Chemical Society, Washington, D.C., 1979.

E. Li-Chan
University of British Columbia
Vancouver, Canada

HYGIENE. See Cleaning-in-place; disinfectants; Food surface sanitation; Freezing. systems for the food industry; meat slaughtering and processing equipment.